CHILTON®

FORD

DIAGNOSTIC SERVICE
2005 Edition

THOMSON

DELMAR LEARNING™

Australia • Canada • Mexico • Singapore • Spain • United Kingdon • United States

THOMSON
™
DELMAR LEARNING

Chilton®

Ford Diagnostic Service

2005 Edition

Vice President, Technology and Trades SBU:

Alar Elken

Executive Director, Professional Business Unit:

Gregory L. Clayton

Publisher, Professional Business Unit:

David Koontz

Marketing Director:

Beth A. Lutz

Production Director:
Mary Ellen Black

Marketing Specialist:

Brian McGrath

Marketing Coordinator:
Marissa Mariella

Production Editor:

Elizabeth Hough

Editorial Assistant:
Christine Wade

Editor:
Timothy A. Crain

Publishing Assistant:
Paula Baillie

Cover Design:
Melinda Possinger

ISBN: 1-4180-0551-7

NOTICE TO THE READER

TABLE OF CONTENTS

1 - INTRODUCTION TO FORD SYSTEMS

UNDERSTANDING ON-BOARD DIAGNOSTICS . . . . i - xxv
FORD ENGINE PERFORMANCE 1-2
ELECTRONIC ENGINE CONTROLS 1-3
VEHICLE IDENTIFICATION 1-4
TECHNICIAN REQUIREMENTS 1-6
SERIAL DATA COMMUNICATION 1-7
FLASH REPROGRAMMING 1-9
POWERTRAIN DIAGNOSTICS 1-10
ON-BOARD DIAGNOSTICS 1-38
REFERENCE INFORMATION: PCM LOCATION TABLES 1-49
MAF SENSOR CALIBRATION TABLES 1-54
OPERATING RANGE CHARTS 1-56
FORD OBD GLOSSARY 1-57

2 - FORD CARS

2001 FORD FOCUS POWERTRAIN: VEHICLE IDENTIFICATION 2-5
ELECTRONIC ENGINE CONTROL SYSTEM 2-6
INTEGRATED ELECTRONIC IGNITION SYSTEM 2-7
EVAPORATIVE EMISSION SYSTEM 2-12
EXHAUST GAS RECIRCULATION SYSTEM 2-19
FUEL DELIVERY SYSTEM 2-25
OXYGEN SENSOR 2-33
MODULE COMMUNICATION NETWORK 2-37
REFERENCE INFORMATION: PCM PID TABLES - INPUTS 2-38
PCM PID TABLES - OUTPUTS 2-39
PCM PIN VOLTAGE TABLES 2-40
PCM WIRING DIAGRAMS 2-42
2001 FORD FOCUS ANTI-THEFT SYSTEM: ACTIVE SYSTEM 2-48
PASSIVE SYSTEM (SECURILOCK) 2-53
1996 TAURUS POWERTRAIN: VEHICLE IDENTIFICATION 2-55
ELECTRONIC ENGINE CONTROL SYSTEM 2-56
INTEGRATED ELECTRONIC IGNITION SYSTEM 2-57
EVAPORATIVE EMISSION SYSTEM 2-67
EXHAUST GAS RECIRCULATION SYSTEM 2-79
FUEL DELIVERY SYSTEM 2-86
IDLE AIR CONTROL SOLENOID 2-97
MASS AIRFLOW SENSOR 2-99
OXYGEN SENSOR 2-101
TRANSMISSION CONTROLS 2-105
REFERENCE INFORMATION: DIAGNOSTIC MONITORING TEST RESULTS . 2-107
PCM PID TABLES - INPUTS 2-108
PCM PID TABLES - OUPUTS 2-109
PCM PIN VOLTAGE TABLES 2-110
PCM WIRING DIAGRAMS 2-112

3 - TRUCK APPLICATIONS

1998 F150 PICKUP POWERTRAIN: INTRODUCTION 3-3
VEHICLE IDENTIFICATION 3-4
ELECTRONIC ENGINE CONTROL SYSTEM 3-5
INTEGRATED ELECTRONIC IGNITION SYSTEM 3-6
EVAPORATIVE EMISSION SYSTEM 3-16
EXHAUST GAS RECIRCULATION SYSTEM 3-29
FUEL DELIVERY SYSTEM 3-38
IDLE AIR CONTROL VALVE 3-49
INPUT SENSORS 3-51
MASS AIRFLOW SENSOR 3-54
OXYGEN SENSOR 3-57
TRANSMISSION CONTROLS 3-60
REFERENCE INFORMATION: PCM PID TABLES - INPUTS 3-62
PCM PID TABLES - OUTPUTS 3-63
PCM PIN VOLTAGE TABLES 3-64
PCM WIRING DIAGRAMS 3-66

4 - MINIVAN APPLICATIONS

2001 WINDSTAR POWERTRAIN: INTRODUCTION 4-3
VEHICLE IDENTIFICATION 4-4
ELECTRONIC ENGINE CONTROL SYSTEM 4-5
COMPUTER CONTROLLED CHARGING SYSTEM 4-6
INTEGRATED ELECTRONIC IGNITION SYSTEM 4-8
EVAPORATIVE EMISSION SYSTEM 4-14
EXHAUST GAS RECIRCULATION SYSTEM 4-23
FRONT ELECTRONIC MODULE 4-28
FUEL DELIVERY SYSTEM 4-33
IDLE AIR CONTROL SOLENOID 4-43
MASS AIRFLOW SENSOR 4-45
OXYGEN SENSOR 4-48
TRANSMISSION CONTROLS 4-52
REFERENCE INFORMATION: PCM PID TABLE - INPUTS 4-55
PCM PID TABLE OUTPUTS 4-56
PCM PIN VOLTAGE TABLES 4-57
PCM WIRING DIAGRAMS 4-59

5 - SUV APPLICATIONS

2000 EXPLORER POWERTRAIN: INTRODUCTION 5-3
VEHICLE IDENTIFICATION 5-4
ELECTRONIC ENGINE CONTROL SYSTEM 5-5
INTEGRATED ELECTRONIC IGNITION SYSTEM 5-6
ENHANCED EVAPORATIVE EMISSION SYSTEM 5-16
EXHAUST GAS RECIRCULATION SYSTEM 5-29
FUEL DELIVERY SYSTEM 5-38
COMMUNICATIONS NETWORK 5-49
GENERIC ELECTRONIC MODULE 5-50
IDLE AIR CONTROL VALVE 5-55
INPUT SENSORS 5-57
MASS AIRFLOW SENSOR 5-58
OXYGEN SENSOR 5-60
TRANSMISSION CONTROLS 5-63
REFERENCE INFORMATION: PCM PID TABLES - INPUTS 5-65
PCM PID TABLES - OUTPUTS 5-66
PCM PIN VOLTAGE TABLES 5-67
PCM WIRING DIAGRAMS 5-69

6 - OBD II SYSTEMS

ABOUT THIS SECTION 6-2
OBD II SYSTEMS 6-3
OBD II SYSTEM TERMINOLOGY 6-6
OBD II SYSTEM MONITORS 6-13
REFERENCE INFORMATION: PID DATA EXPLANATION 6-30
OBD II PID MODE 6-31
PIN VOLTAGE TABLE EXPLANATION 6-32
USING A BREAKOUT BOX 6-33
PCM ACTUATOR & SENSOR TABLES 6-34
PID DATA & PIN CHART GLOSSARY 6-39
GAS ENGINE OBD II TROUBLE CODES 6-41
DIESEL ENGINE OBD II TROUBLE CODES 6-133

7 - FORD CAR PID TABLES

ABOUT THIS SECTION 7-4
FORD & MERCURY INFORMATION: ASPIRE PID DATA 7-5
CONTOUR & MYSTIQUE PID DATA 7-6
COUGAR PID DATA 7-10
CROWN VICTORIA & GRAND MARQUIS PID DATA 7-14
ESCORT, TRACER & ZX2 PID DATA 7-20
MUSTANG PID DATA 7-33
PROBE PID DATA 7-43
TAURUS PID DATA 7-54
TEMPO & TOPAZ PID DATA 7-65
THUNDERBIRD PID DATA 7-66
THUNDERBIRD & COUGAR PID DATA 7-69
LINCOLN PID INFORMATION: CONTINENTAL PID DATA 7-75
LS6 PID DATA 7-80
LS8 PID DATA 7-82
MARK VIII PID DATA 7-84
TOWN CAR PID DATA 7-87

8 - FORD TRUCKS, VANS & SUVS PID TABLES

ABOUT THIS SECTION. 8-4
F-SERIES TRUCK PID DATA 8-5
RANGER PICKUP PID DATA. 8-33
AEROSTAR VAN PID DATA. 8-50
E-SERIES VAN PID DATA 8-56
VILLAGER VAN PID DATA. 8-88
WINDSTAR VAN PID DATA 8-90
BRONCO PID DATA 8-9
ESCAPE PID DATA 8-113
EXCURSION PID DATA 8-117
EXPEDITION PID DATA 8-94
EXPLORER & MOUNTAINEER PID DATA 8-104
BLACKWOOD PID DATA. 8-97
NAVIGATOR ID DATA 8-98

9 - FORD CAR PIN TABLES

ABOUT THIS SECTION. 9-4
ASPIRE PIN TABLES 9-5
CONTOUR PIN TABLES 9-9
CROWN VICTORIA PIN TABLES. 9-27
ESCORT & ZX2 PIN TABLES 9-49
FOCUS PIN TABLES. 9-81
MUSTANG PIN TABLES 9-91
PROBE PIN TABLES. 9-128
TAURUS PIN TABLES 9-148
TEMPO PIN TABLES 9-196
THUNDERBIRD PIN TABLES 9-202

10 - LINCOLN CAR PIN TABLES

ABOUT THIS SECTION. 10-2
CONTINENTAL PIN TABLES 10-3
LS 3.0L PIN TABLES 10-15
LS 3.9L PIN TABLES 10-21
MARK VII PIN TABLES 10-27
MARK VIII PIN TABLES. 10-29
TOWN CAR PIN TABLES 10-35

11 - MERCURY CAR PIN TABLES

ABOUT THIS SECTION. 11-3
COLONY PARK PIN TABLES. 11-4
COUGAR PIN TABLES 11-6
GRAND MARQUIS PIN TABLES 11-28
MARAUDER PIN TABLES 11-42
MYSTIQUE PIN TABLES 11-44
SABLE PIN TABLES 11-58
TOPAZ PIN TABLES 11-82
TRACER PIN TABLES 11-90

12 - FORD & LINCOLN TRUCK PIN TABLES

ABOUT THIS SECTION. 12-3
F-SERIES TRUCKS. 12-4
LIGHTNING PICKUPS. 12-97
RANGER PICKUPS. 12-99
BLACKWOOD PICKUP 12-133

13 - FORD SUV PIN TABLES

ABOUT THIS SECTION. 13-3
AVIATOR PIN TABLES 13-4
BRONCO PIN TABLES 13-7
ESCAPE PIN TABLES. 13-38
EXCURSION PIN TABLES. 13-42
EXPEDITION PIN TABLES 13-58
EXPLORER PIN TABLES 13-76
MOUNTAINEER PIN TABLES 13-101
NAVIGATOR PIN TABLES 13-115

14 - FORD VAN PIN TABLES

ABOUT THIS SECTION. 14-3
AEROSTAR PIN TABLES 14-4
E-SERIES VAN PIN TABLES 14-16
VILLAGER PIN TABLES 14-95
WINDSTAR PIN TABLES 14-103

USING THIS MANUAL

Manufacturer and Model Coverage

This manual does not cover every Ford Motor Company model that is currently available on the market. Rather, the Chilton editorial staff makes judicious decisions as to which models warrant coverage, based on which vehicles are serviced by most technicians.

Model Year Information

This manual is published toward the end of the year prior to the edition year. Every effort is made to gather current data from the Original Vehicle Manufacturers (OEMs) when they publish it. Different OEMs choose to release their new model information at different times of the year. Indeed, the same OEM can publish information early one season and late the next season. As a result, not all models are equally current when each edition of this manual is published.

Although information in this manual is based on industry sources and is as complete as possible at the time of publication, some vehicle manufacturers may make changes which cannot be included here. Information on late models may not be available in some circumstances. While striving for total accuracy, the publisher cannot assume responsibility for any errors, changes, or omissions that may occur in the compilation of this data.

Safety Notice

Proper service and repair procedures are vital to the safe, reliable operation of all motor vehicles, as well as the personal safety of those performing the repairs. This manual outlines procedures for diagnosing and serving vehicles using safe, effective methods. The procedures may contain many NOTES and CAUTIONS which should be followed along with standard safety procedures to reduce the possibility of personal injury or improper service which could damage the vehicle or compromise its safety.

Diagnostic procedures, tools, parts, and technician skill and experience vary widely. It is not possible to anticipate all conceivable ways or conditions under which vehicles may be serviced, or to provide cautions for all possible hazards that may result. Standard and accepted safety precautions and equipment should be used when handling toxic or flammable substances, and safety goggles or other protection should be used during any process that may cause sparking, material removal or projectiles.

Some procedures require the use of tools specially designed for a specific purpose. Before substituting another tool or procedure, you must be completely satisfied that neither your personal safety, nor the performance of the vehicle will be endangered.

Special Tools

Special tools are recommended by the vehicle manufacturer to perform specific jobs. When necessary, special tools may be referred to in the text by part number. These tools may be purchased, under the appropriate part number, from your local dealer or regional distributor, or an equivalent tool can be purchased locally from a tool supplier or parts outlet. Before substituting any tool for the one recommended, read the previous Safety Notice.

ACKNOWLEDGEMENT

This publication contains material that is reproduced and distributed under a license from Ford Motor Company. No further reproduction or distribution of the Ford Motor Company material is allowed without the express written permission of Ford Motor Company.

The publisher would like to express appreciation to Ford Motor Company for its assistance in producing this publication.

Understanding On-Board Diagnostics

Introduction

OBD II OVERVIEW

The OBD II system was developed as a step toward compliance with California and Federal regulations that set standards for vehicle emission control monitoring for all automotive manufacturers. The primary goal of this system is to detect when the degradation or failure of a component or system will cause emissions to rise by 50%. Every manufacturer must meet OBD II standards by the 1996 model year. Some manufacturers began programs that were OBD II mandated as early as 1992, but most manufacturers began an OBD II phase-in period starting in 1994.

The changes to On-Board Diagnostics influenced by this new program include:

- Common Diagnostic Connector
- Expanded Malfunction Indicator Light Operation
- Common Trouble Code and Diagnostic Language
- Common Diagnostic Procedures
- New Emissions-Related Procedures, Logic and Sensors
- Expanded Emissions-Related Monitoring

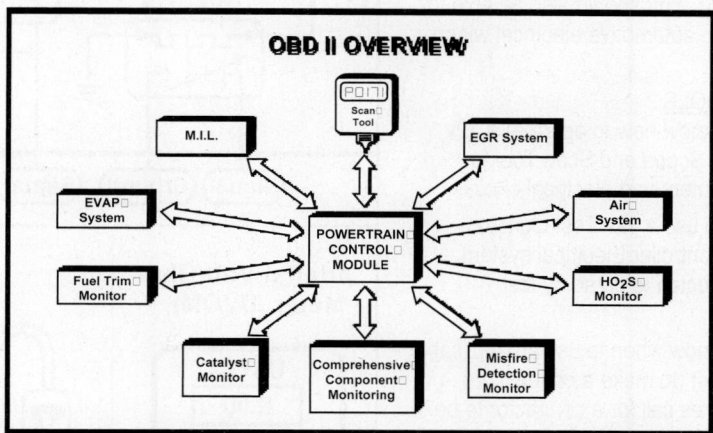

COMMON TERMINOLOGY

OBD II introduces common terms, connectors, diagnostic language and new emissions-related monitoring procedures. The most important benefit of OBD II is that all vehicles will have a common data output system with a common connector. This allows equipment Scan Tool manufacturers to read data from every vehicle and pull codes with common names and similar descriptions of fault conditions. In the future, emissions testing will require the use of an OBD II certifiable Scan Tool.

TECHNICIAN REQUIREMENTS

As an automotive repair technician, you should have a basic understanding of how to use the hand tools and meters necessary to effectively use the information in this OBD II manual.

■ **NOTE:** *Lack of basic knowledge of the Powertrain when performing test procedures could cause incorrect diagnosis or damage to Powertrain components. Do not attempt to diagnose a Powertrain problem without having this basic knowledge.*

ELECTRICITY AND ELECTRICAL CIRCUITS

You should understand basic electricity and know the meaning of voltage (volts), current (amps), and resistance (ohms). You should be able to identify a *Series* circuit as well as a *Parallel* circuit in an automotive wiring diagram. Refer to the examples in the Graphic to the right.

You should understand what happens in an electrical circuit with an open circuit or a shorted wire, and you should be able to identify an open or shorted circuit condition using a DVOM. You should also be able to read and understand an automotive electrical wiring diagram.

CIRCUIT TESTING TOOLS

You should have (and know how to operate) a 12v Test Light, DVOM, Lab Scope and Scan Tool to diagnose vehicle computers and electrical circuits.

You should know not to use a 12v Test Light to diagnose the Engine Controller Electrical system unless specifically instructed to do so by test procedures.

You should have and know when to use an applicable aftermarket connector kit (to make a connection) whenever test procedures call for a connector to be probed in order to make a measurement.

ELECTRICAL CIRCUITS

When you encounter a wiring problem during testing, and need to refer to electrical circuit information, you should be comfortable with this type of information:

- Wiring schematics (including circuit numbers and colors)
- Electrical component connector, splice and ground locations
- Wiring repair procedures and wiring repair parts information

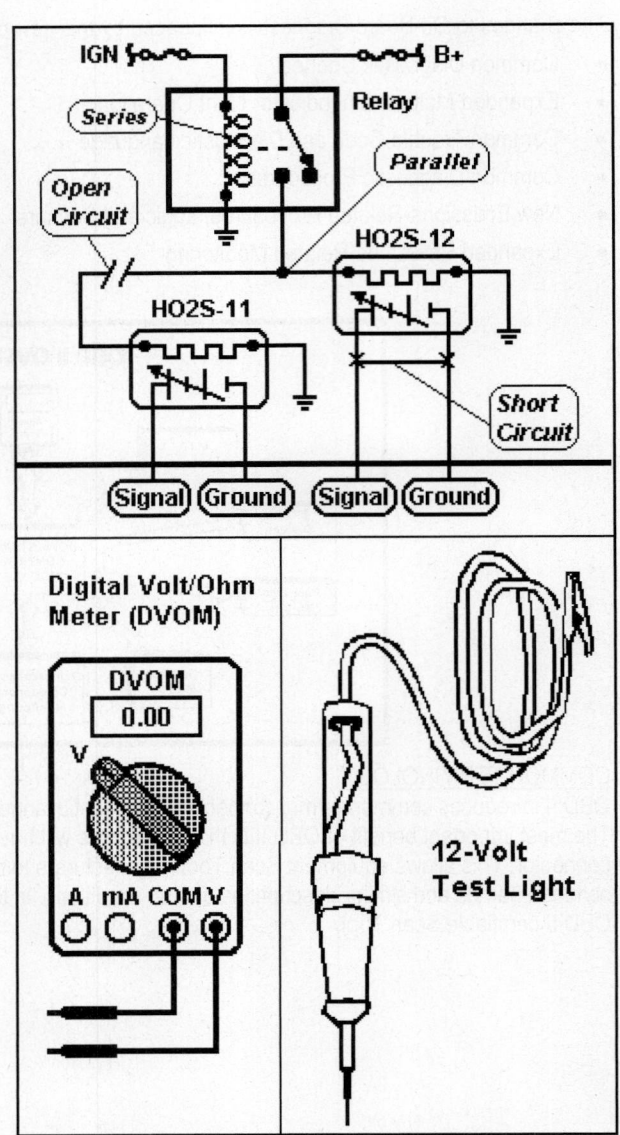

OBD II System

History of OBD Systems

INTRODUCTION

Starting in 1978, several vehicle manufacturers introduced a new type of control for several vehicle systems and computer control of engine management systems. These computer-controlled systems included programs to test for problems in the engine mechanical area, electrical fault identification and tests to help diagnose the computer control system. Early attempts at diagnosis involved expensive and specialized diagnostic testers that hooked up externally to the computer in series with the wiring connector and monitored the input/output operations of the computer.

By early 1980, vehicle manufacturers had designed systems in which the onboard computer incorporated programs to monitor selected components, and to store a trouble code in its memory that could be retrieved at a later time. These trouble codes identified failure conditions that could be used to refer a technician to diagnostic repair charts or test procedures to help pinpoint the problem area.

EVOLUTION OF DAIMLER CHRYSLER COMPUTERIZED ENGINE CONTROLS

The evolution of Computerized Engine Controls on Chrysler vehicles equipped with fuel injection is highlighted in the Graphic below.

Computerized Engine Controls Evolution Graphic - Chrysler & Jeep

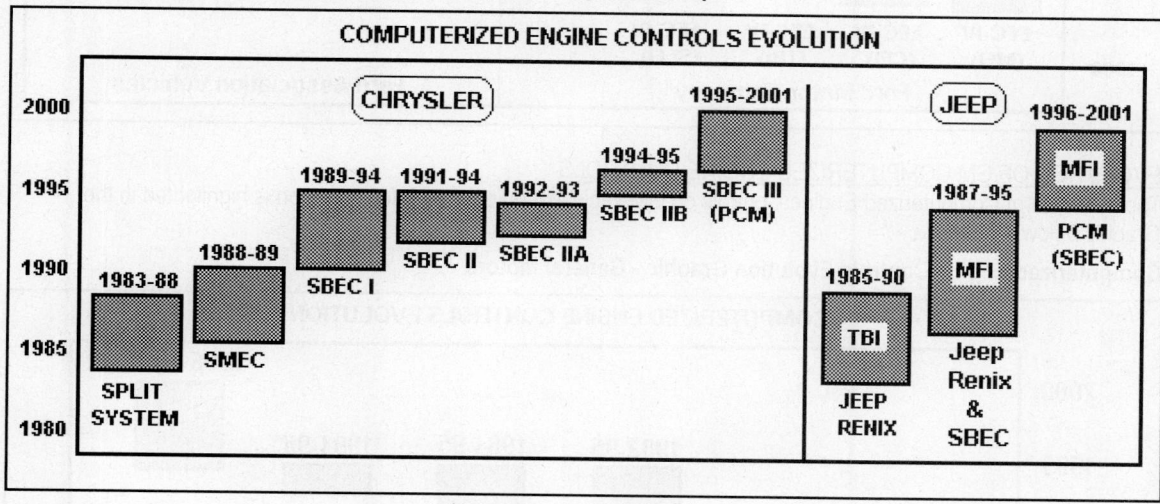

EVOLUTION OF FORD MOTOR COMPANY COMPUTERIZED ENGINE CONTROLS

The evolution of Computerized Engine Controls on Ford vehicles equipped with fuel injection is highlighted in the Graphic below.

Computerized Engine Controls Evolution Graphic - Ford Motor Company

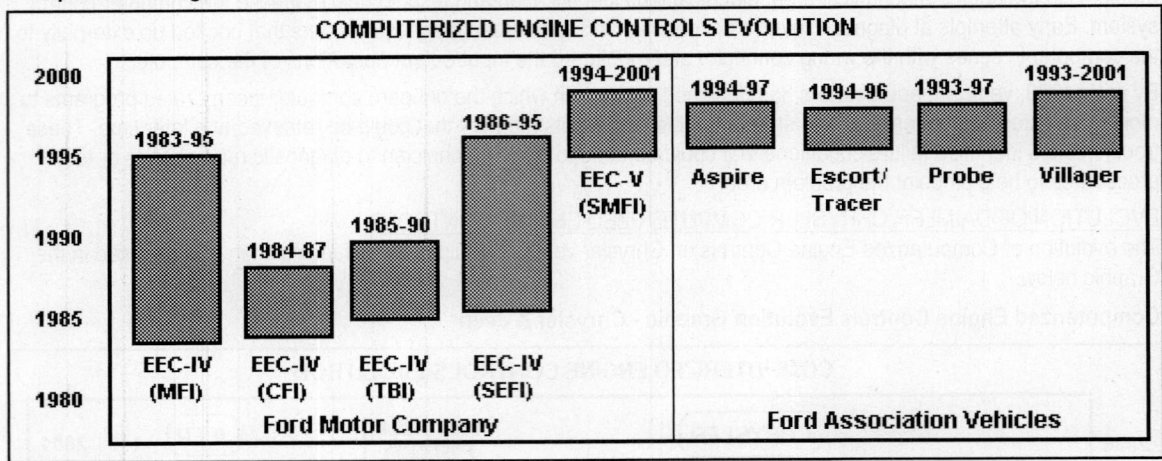

EVOLUTION OF GM COMPUTERIZED ENGINE CONTROLS

The evolution of Computerized Engine Controls on GM vehicles equipped with fuel injection is highlighted in the Graphic below.

Computerized Engine Controls Evolution Graphic - General Motors

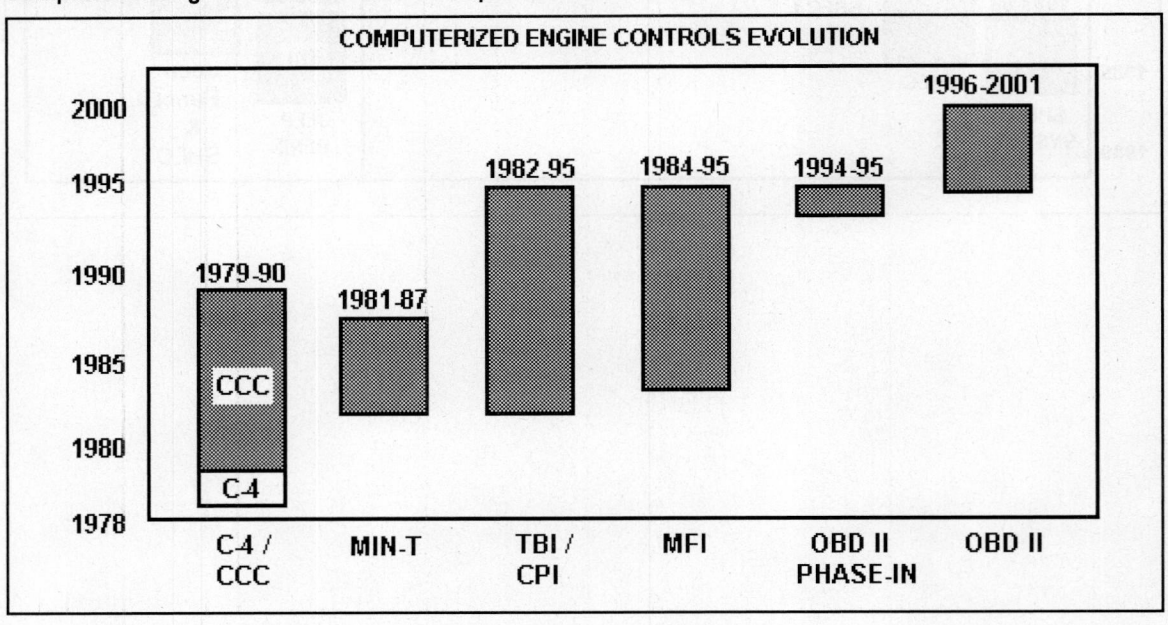

Computer Diagnostics

INTRODUCTION

General diagnostics of computers fall into these two categories:

- External Onboard Diagnostics
- Internal Onboard Diagnostics

The first level of diagnostics uses an external tool that taps into the computer and runs a series of diagnostic tests. This method of diagnostics was popular in the 1970's and was used in the 1980's on many European vehicles. The second level incorporates diagnostics into the circuit board of the computer and is known in the industry as "On-Board" diagnostics because the diagnostics are on the computer circuit board.

In 1980 General Motors incorporated an On-Board Computer Program where the "check engine" light came on to inform the vehicle owner that there was a fault in the computer system. The light was turned on when a diagnostic code was set to alert the driver that service was needed on the vehicle.

California formed a government agency, the California Air Resources Board (CARB) to monitor the air quality and establish regulations to reduce air pollution. The California Health and Safety Code authorized the Air Resources Board to adopt motor vehicle emissions standards and in-use performance standards that it finds necessary, cost effective and technologically feasible. In 1988, CARB required that all vehicles sold in California incorporate a system with an On-Board Diagnostic program where a "check engine" light would come on to notify the vehicle owner of a potential failure of computer sensors and/or their systems. This system is known as On-Board Diagnostics First Generation and is now referred to as OBD I.

PROBLEMS WITH OBD I SYSTEMS

One of the problems with OBD I was that the code retrieval methods varied from manufacturer to manufacturer and there was no consistency between systems. Most manufacturers looked at similar computer sensors and circuits, but codes were inconsistent and difficult to identify and define. Some manufacturers require special tools to retrieve trouble codes or required special test procedures for these tools which self-tested circuits and systems or energized the components for testing in the field.

SCAN TOOL INTRODUCTION

Domestic vehicle manufacturers (Chrysler, Ford and GM) designed their computers to have an accessible data line where a diagnostic tester could retrieve data on sensors and the status of operation for components.

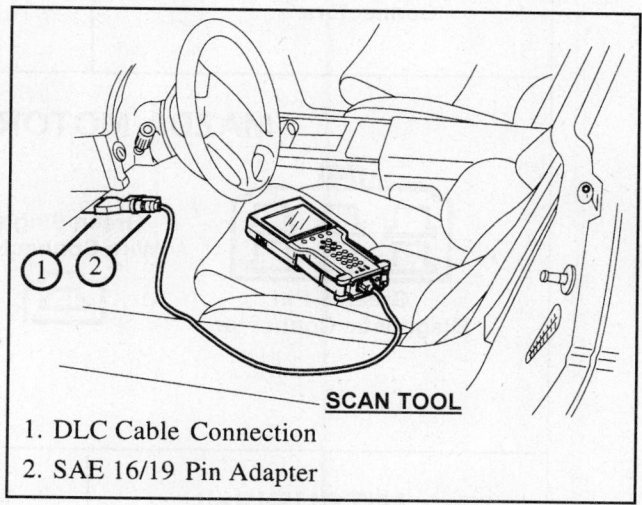

SCAN TOOL

1. DLC Cable Connection
2. SAE 16/19 Pin Adapter

These testers became known in the automotive repair industry as "Scan Tools" because they scanned the data on the computers and provided information for the technician.

Ford Motor Company developed a tester that would access codes, activate sensors and perform limited tests and adjustments, however they did not incorporate data stream features until 1988.

<u>OBD I SYSTEM CONNECTORS</u>

ACURA/HONDA

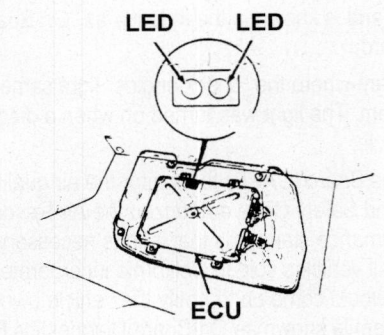

LED   LED

ECU

CHRYSLER

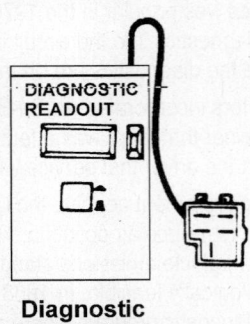

**Diagnostic
Read-Out Box**

FORD

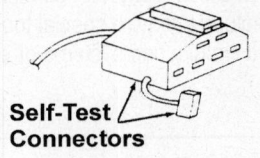

**Self-Test
Connectors**

GENERAL MOTORS

**Diagnostic Test
Terminal**   Ground

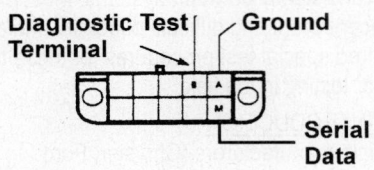

**Serial
Data**

MAZDA MOTORS CORP.

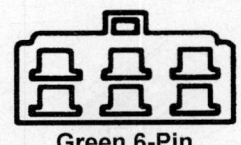

**Green 6-Pin
Diagnostic Connector**

**Green Single
Wire Connector**

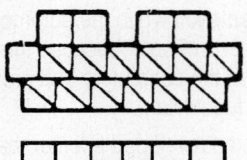

**25-Pin
Diagnostic Connector**

MITSUBISHI

Diagnostic Terminal

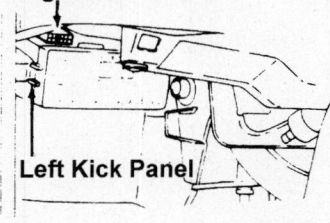

Left Kick Panel

NISSAN MOTOR CO.

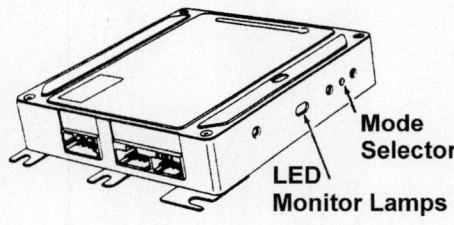

**Mode
Selector**

**LED
Monitor Lamps**

Government Regulations

INTRODUCTION

The California Air Resources Board (CARB) conducted research on OBD I vehicle emissions and the study resulted in the following conclusions:

- The research found a significant number of pre-1988 vehicles with degraded emissions components. These components were not failing outright, but deterioration increased emissions levels. This problem did not usually set codes alerting the vehicle owner or technician that there was a problem, therefore the condition was not perceived as a problem in the field. However, CARB viewed this as a problem due to the increased emission levels.

- Vehicle testing programs found failures in Canister Purge systems and Secondary Air Management systems. Many of these failures occurred under road load conditions and were not quickly or easily detectable in the service bay. These failures resulted in increased emissions.

- Catalytic Converters were failing and vehicles were being driven with deteriorated catalysts. A leading cause of this failure was engine misfire.

- The On-Board Monitoring Systems did not detect fuel system faults that were responsible for increasing emissions even though fuel systems were deteriorated enough to have excessive emissions.

- The monitoring systems did not detect oxygen sensors that were "lazy" or slow in response. This condition was found to result in an increase in emissions levels.

- EGR monitoring did not verify if the system was operating within a range that could result in an increase of emissions. There was a need to monitor the flow of EGR gases through the system in order to verify the EGR passages were not clogged.

- Codes were different for each manufacturer and this was confusing for a technician working on different vehicles.

DEVELOPMENT OF OBD I STANDARDS

CARB reviewed the system of monitoring Engine Control sensors and systems developed by Chrysler Motors and General Motors. They incorporated this concept into their regulations.

The result was that the California regulations required that all vehicle manufacturers develop a set of diagnostics that would incorporate a system where codes and data are made available through a Scan Tool accessible to every technician.

These California standards, originally published in October of 1988, generally apply to 1994 and later passenger cars, light duty trucks and medium duty vehicles. Similar diesel and alternative fuel vehicle regulations took affect in 1996. After 1988, California made the decision to accept the Federal (EPA) OBD II regulations.

FEDERAL TEST PROCEDURE

OBD II requires that the on-board computer monitors vehicle emissions and in some cases perform an "Active" diagnostic test of those systems. These tests were developed by the EPA and are a reflection of the Federal Test Procedure (FTP).

The FTP is a series of programmed tests where a vehicle is driven through specific drive cycles while emissions are being monitored. These tests are conducted at various mileage levels and test emissions under very specific conditions. The amount of fuel in the gas tank is monitored, the type of fuel and octane level are all controlled. These tests are conducted on a dynamometer and are performed under hot and cold vehicle conditions. They are conducted under EPA supervision and are required to certify a vehicle for sale in the USA.

The OBD II system was designed to monitor these same systems and a Malfunction Indicator Lamp (MIL) must illuminate if a system or component either fails or deteriorates to a point where the vehicle emissions could rise beyond 1.5 times the FTP standard.

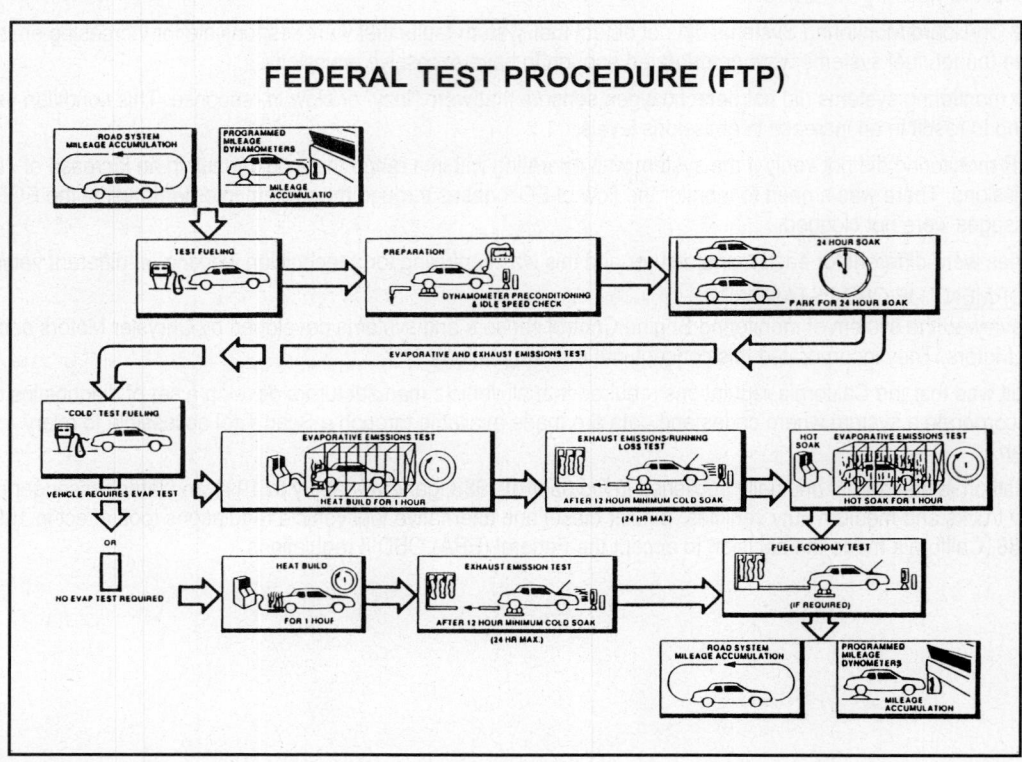

CARB Regulations

SUMMARY OF CARB REGULATIONS

- All vehicles are equipped with a Malfunction Indicator Light (MIL) that will remain "on" when certain faults are detected by the controller. If the MIL is "on", the program should be able to turn it "off" if the fault does not reappear on two consecutive trips. Codes can be erased from memory if the fault does not occur for 40 warmup periods.

- The catalytic converter closest to the engine (if more than one converter is on the vehicle) must be monitored to see if it has deteriorated.

- The oxygen sensor must be monitored for the output voltage and the response rate. The monitor should include the rich/lean transitions (cross counts or oxygen sensor switch rate) and should check the oxygen sensor transition from rich to lean and lean to rich. The oxygen sensor heater should be monitored for proper operation.

- The diagnostic system must monitor the engine for misfire and identify the specific cylinders misfiring. This system must include identification informing the technician that the catalyst is deteriorated, damaged, or has failed a Federal Test Procedure type drive cycle. If the engine is misfiring at a level that could cause damage to the converter catalyst, the MIL should flash when these conditions are present and the controller should switch to a backup program designed to reduce the level of misfire.

- The fuel delivery system should be monitored for its ability to control fuel. It should determine if the system has a condition that is over the range for optimal fuel control. This system is referred to as fuel trim monitor.

- The EGR system should be monitored for both low and high flow rates as well as verify that solenoids controlling the EGR are working properly.

- The EVAP System should be monitored for HC flow changes based on fuel tank fill level and the monitor should calculate for these differences.

- The Secondary Air Management system should be monitored for the control valve operation and for the airflow throughout the system.

- Any emissions-related component or system not otherwise described that provides input directly or indirectly to the processor should be monitored. This includes all input and output components.

EPA Regulations

The Federal Environmental Protection Act (EPA) also established diagnostic regulations, but in an effort to simplify the development process for vehicle manufacturers, the EPA decided to comply with the California OBD II regulations from 1994 to 1998.

SAE Forms Committees to Develop Standards

The Society of Automotive Engineers formed committees to help its member engineers coordinate efforts and develop a second generation of diagnostics. The new diagnostics, known as OBD II, would standardize the diagnostic connector, diagnostic data retrieval and code identification.

EPA Expands the SAE Standards

The EPA reviewed the SAE standards and added regulations requiring all manufacturers to meet Federal OBD II standards by 1996. To assist in development of expanded diagnostics, the EPA decided to use the California regulations as federal standards through 1998. Federal and California regulations established a two-year phase-in to take effect from 1994-1996 to allow for design and phase-in of expanded diagnostics.

Government (CARB & EPA) and SAE Regulations

OVERVIEW

The Society of Automobile Engineers (SAE), CARB and the EPA set the standards that relate to changes in the industry (terminology, common Scan Tool interface, etc.) while others set the diagnostic standard for how information is handled by vehicle controllers.

Industry Regulations

Government Mandated Regulations	SAE "Recommended" Compliance
J1930 - Industry Terminology Standardization	J2201 - Additional Guidelines for Generic Scan Tool Interface
J1978 - Standards for Generic OBD II Scan Tool Interface Protocol	J2190 - Enhanced Test Mode Standards
J2205 - Standards for Expanded Scan Tool Interface Protocol	J2008 - Guidelines for Repair Service Information (CARB Standards)
J2008 - Standards for Repair and Service Information (EPA Guidelines)	

On-Board Computer Regulations

Government Mandated Regulations	SAE "Recommended" Compliance
J2012 - Standards for a Diagnostic Trouble Code (DTC)	J2186 - CARB approved standards for anti-tamper procedures
J1962 - Standard 16-Pin diagnostic connector	J2178 - Scan Tool message strategy guidelines
J1979 - Standards for Diagnostic Test Modes	J2190 - Enhanced Test Mode Standards
J1850 - Scan Tool Communication guidelines for Class 'B' Data Interface	J1724 - Vehicle Electronic Identification Standards
J2186 - EPA mandated Anti-Tamper procedures	

CARB AND EPA REGULATIONS

The government agencies that set OBD regulations are CARB and the EPA. The tables below compare differences between the EPA and CARB requirements for OBD II.

Part One - Industry Regulations

CARB	EPA	Government Requirements
	X	1994 model year - all service information must conform to J1930.
X	X	Service manuals must publish a normal range for calculated load values and Mass Air Flow Rate at idle and at 2,500 RPM.
	X	The vehicle manufacturer is responsible for ensuring information is available even if the information is provided by an intermediary.
	X	The cost of repair information to the independent technician shall not exceed the lowest price that is available to a dealership.
	X	All other information available at a fair and reasonable price, otherwise it is considered not available (a fine of $25K per day could be applied).
	X	Electronic service information must be available by the 1998 model year.
X		Repair procedures must be available which allow effective diagnosis and repairs using a J1978 Generic Scan Tool and readily available repair tools.
X	X	J1978 Scan Tool compatibility - communication protocol. All serial data and enhanced tests must be available to a Scan Tool. Scan Tool must inform user which emissions systems are monitored. EPA added requirements that the VIN be accessible off the DLC.
	X	Requires Bi-directional diagnostic control of the computer be available on the Scan Tool meeting J2205 and J1979 standards.
	X	1996 model year - all service information must be in the J2008 format.
	X	Labeling requirements must meet J1877 and J1892 standards.

Part Two - On-Board Computer Regulations

CARB	EPA	Government Requirements
X	X	J2012 Diagnostic Trouble Codes - If the PCM detects a fault, it must set a code to identify a fault (uniform identification), include conditions that describe how the PCM reverts to default mode and erase the code after 40 warm up cycles with the MIL off.
X	X	J1962 Diagnostic Connector mounted on Instrument Panel driver's side of vehicle with standard pins for serial data, power and ground.
X	X	J1979 Diagnostic Test Mode Messages - defines standard messages for access to trouble codes, vehicle data stream and Freeze Frame data.
X	X	Scan Tool Interface for DLC - SAE J1850 serial data link required to access all emission-related data. Requires 3 byte headers (does not allow IBS or checksum).
X	X	Vehicle manufacturers must provide tampering deterrence for a PCM that is programmable (where the PROM is rewritten to change operating parameters). This must include write-protect standards for programmable memory and references J2186 - Data Link Security for write protect.
	X	Access to vehicle calibration data, odometer and keyless entry codes can be limited, but OEMs must provide "the best means available for providing non-dealer technicians with calibration data necessary to perform repairs".
X	X	Freeze Frame Data Stored with the first fault of any component or system and replace data if there is a subsequent fuel system or misfire fault. EPA added "airflow rate" to required Freeze Frame data.
X	X	Signal access to the required diagnostic data must be made available through the diagnostic connector using standard messages. Actual values should be identified separately from default or limp home values.
X	X	The vehicle must have only one Malfunction Indicator Light (MIL) for emission related problems. The MIL must remain "on" if a malfunction is detected and stays on until three trips indicate the fault is gone. The MIL must blink if a catalyst-threatening misfire condition is present. Note: A few manufacturers received exemptions from this standard for specific 1994-96 models.

Part Three - On-Board Computer Monitor Regulations

CARB	EPA	Government Requirements
X	X	Oxygen Sensor Monitor - It must check the output voltage and response rate for all oxygen sensors once per trip. The results of most recent oxygen sensor evaluation test must be available over the data link connector as serial data. The EPA added a requirement that the results of the most recent on-board monitoring data and test limits for all systems with specific Monitor evaluation tests must be made available.
X	X	Catalyst Monitor - it must verify the catalyst is functioning at steady state efficiency and that it does not deteriorate over 1.5X the standard. This test is done once per trip.
X	X	Misfire Monitor - It must run continuously under all conditions in order to identify a misfiring cylinder. On vehicles with SFI, it must cutoff fuel to a misfiring cylinder.
X		EGR System - It must monitor for both low and high EGR flow rate once per trip.
X		EVAP Purge Monitor - It must check the system function and for leaks once per trip.
X		Secondary AIR Management Monitor - It must perform a functional test of the AIR system and switching valves (a test for proper function and airflow once per trip).
X		Fuel System Monitor - It must check the ability of the controller to control fuel delivery.
X	X	Comprehensive Component Monitor - It must monitor all components or systems that send data to or receives data from the PCM. CARB requires a check for out-of-range signals and a functional response test of related outputs. CARB requires continuous monitoring while the EPA requires evaluation periodically once per "Drive Cycle".
	X	The system must monitor any deterioration or malfunction which occurs which can cause exhaust or evaporative emissions to increase 1.5X the Federal Test Standard.
X		Air Conditioning system must be monitored for loss of reactive refrigerant once per trip. Non-reactive refrigerant does not have to be monitored.

Explanation of SAE Standards

J1930 - Common Names for Components

J1930 established common nomenclature for emissions and computer-related components and systems. This includes common definitions, abbreviations and acronyms.

This standard is designed to provide the technician with a recognizable name for components that apply to all vehicles. This nomenclature has been determined to be beneficial for technicians who work on multiple lines of vehicles as well as vehicles from different manufacturers.

J2008 - Service Information Availability

J2008 requires that "all information" must be made available to "any person engaged" in the repair of the vehicle. The legislation is very specific and requires that "no such information on vehicle repair" may be withheld from any technician. J2008 is still being finalized and will continue to be interpreted by the vehicle manufacturer. It also sets standards for the organization of vehicle service information. This includes the data model, data type definition, graphics standards and electronic transmission of data.

EPA Guidelines on Repair Information

The EPA published guidelines that state that information availability requirements include 1994 model year vehicles, and that vehicle manufacturer must furnish to "any person" engaged in the repair or service of a motor vehicle with "all information" required to make emission related diagnosis and repairs. Includes, but is not limited to service manuals, technical service bulletins, vehicle recalls, engine control emissions system information, bi-directional control and training information.

None of this information may be withheld if provided directly or indirectly to the dealers. Information cost to independent technicians shall not exceed the lowest price of the same information to the dealerships. Other repair information must be made available at a "fair and reasonable" cost, otherwise it is considered unavailable.

J2205 - Expanded Diagnostic Protocol for Scan Tools

Some Scan Tools incorporate a protocol that allows the technician to access information not specifically required by OBD II standards. The information in the messages on this tool will be specified in factory service information and provided to the technician. Refer to the examples under Diagnostic Function in the Graphic to the right.
Source: OTC Scan Tool with a 1999 Pathfinder cartridge.

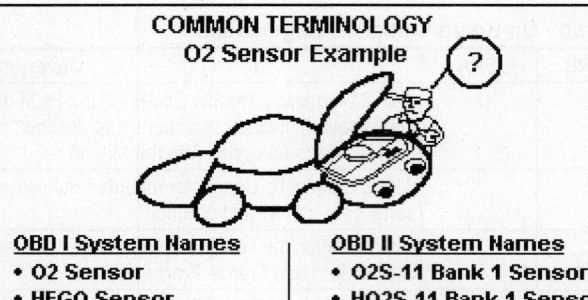

COMMON TERMINOLOGY
O2 Sensor Example

OBD I System Names	OBD II System Names
• O2 Sensor	• O2S-11 Bank 1 Sensor 1
• HEGO Sensor	• HO2S-11 Bank 1 Sensor 1
• EGO Sensor	• HO2S-12 Bank 1 Sensor 2
• LAMBDA Sensor	• HO2S-13 Bank 1 Sensor 3
• Feedback Sensor	• HO2S-21 Bank 2 Sensor 1
	• HO2S-22 Bank 2 Sensor 2

SCAN TOOL MENUS

Press:
> 1-OEM Tests
  2-OBD II

DIAGNOSTIC FUNCTION

Press:
> 1- Datastream
  2-Diagnostic Codes
  4-Record/Playback
  5-Special Test
  7-Monitor Setup

1-DATASTREAM

BARO	29.4" HG
BATT TEMP	49°F
ENGINE RPM	750
IAC DESIRED	37
IAC MOTOR	37

2-DIAGNOSTIC CODES

Press:
> 1-Read Codes
  2-Clear Codes
  3-Code History

EXPLANATION OF ONBOARD COMPUTER REGULATIONS

J1978 - Generic Scan Tool Usage

J1978 requires that all vehicle manufacturers make readily available to the automotive repair industry all data, codes and emissions-related information that can be accessed by a generic Scan Tool. The values for all trouble codes, sensors and components along with Freeze Frame data stored in the computer must be accessible for download to a Generic Scan Tool.

Once a Generic Scan Tool is connected to the 16-pin OBD II connector, it can retrieve certain data from the computer data stream, retrieve Freeze Frame data, read any 5-digit codes and clear these codes from memory.

The EPA expanded this regulation to include the ability to perform bi-directional diagnostic control. The EPA did not define "bi-directional", and the vehicle manufacturers requested that more specific standards be written and incorporated into J2205 after review by SAE committees.

J2178 sets the standards for how vehicle interface messages are displayed on the Generic Scan Tool.

J2201 - Generic Scan Tool Terminal Designation

J2201 sets additional guidelines for Generic Scan Tool interface and assigns the designation of terminals for voltage feed, ground and data transmission.

Generic Scan Tool Menu Example

The Scan Tool Parameter Identification (PID) Mode allows access to certain data values, analog and digital input and output signals, calculated values and system status information.

Generic Scan Tool Navigation

An example of how to navigate through the Scan Tool menus (Snap On example) to locate the Generic PID information is shown in the Graphic to the right.

There are 16 engine related parameters for this vehicle on a Generic OBD II Scan Tool. The parameters in the last frame of this example represent known good values.

Parameter ID (PID) Information

The proper sequence to follow to obtain a complete Generic PID list for this vehicle is shown in the Graphic.

1) Scroll through the main menu and line up the tilde (~) with the desired choice (in this case, GENERIC).
2) Scroll to CODES & DATA. Select it with the tilde (~).
3) Connect an OBD II K2 Adapter to the test connector to allow the tool to read OBD II Generic information.
4) Scroll through the menu and then line up the tilde (~) with the desired choice (DATA - NO CODES).

Source: Snap On Scan Tool with a 1999 cartridge.

SCAN TOOL MENUS

> **GM/SATURN (1980-1999)**
> **CHRYSLER (1983-1999)**
> **JEEP (1984-1999)**
> **FORD (1981-1999)**
> **~GENERIC OBD II**

(**OBD II GENERIC SCREENS**)

> **MAIN MENU-EMISSIONS**
> **[PRESS N FOR HELP]**
> **~CODES & DATA MENU**
> **CUSTOM SETUP**

> **CONNECT OBD-II**
> **K2 ADAPTER TO**
> **16-PIN OBD II TEST**
> **CONNECTOR.**
> **NO REPAIR TIPS**
> **AVAILABLE IN**
> **GENERIC MODE.**
> **PRESS Y TO CONTINUE**

> **CODES & DATA MENU**
> **CODES ONLY**
> **~DATA (NO CODES)**
> **O2 MONITORS**
> **FREEZE FRAME**
> **PENDING CODES**

> **OBD II DATA**
> **(CODES NOT AVAILABLE)**
>
> | | |
> |---|---|
> | ENGINE RPM | 720 |
> | THROTTLE(%) | 16.4 |
> | FUEL SYS1 | OL |
> | FUEL SYS2 | N/A |
> | COOLANT(°F) | 117 |
> | MAP("Hg) | 29.00 |
> | IGN ADVANCE(°) | 0.0 |
> | ST TRIM B1(%) | 0.00 |
> | LT TRIM B1(%) | 00.0 |
> | O2 B1-S1(V) | 0.470 |
> | TRIM B1-S1(%) | 0.00 |
> | O2 B1-S2(V) | 0.510 |
> | TRIM B1-S2(%) | N/A |
> | VEH SPEED(MPH) | 0 |
> | ENG LOAD(%) | 17.0 |
> | MIL STATUS | OFF |

J1724 - Vehicle Electronic Identification

SAE has developed a recommended practice to provide electronic access to vehicle content information necessary to diagnose, service, test and repair passenger cars and light duty trucks. The SAE committee in charge of this area continues to look at a wide range of interpretations for this standard.

J1850 - Scan Tool Access to Emission Related Data
Access to emission related data must be made available on a standard Diagnostic Data Link (DDL) defined in the J1850 standard (Class 'A' data). There are also other systems on the vehicle that use PCM data. For example, the Climate Control Automatic Air Conditioning System uses signals from the ECT sensor to help determine when to operate the Air Conditioning and Electric Cooling Fan. SAE also developed standards for Vehicle Network and Multiplexing Data Communications (referred to as Class 'B' data). Class 'B' data communications use a system where data is transferred between one or more controllers (or Modules) to eliminate redundant sensors and other system duplication. The modules in this type of system form a multiplex of interactive systems.

Class 'B' Data Communication
Class 'B' data communications have to be able to perform all Class 'A' data functions. However, these two types of communication protocols differ from each other and usually do not communicate in the same format. Scan Tools that communicate with both formats will be available as this standard is defined further and may be made available from the vehicle manufacturer. This means that the Generic Scan Tool (GST) may not be able to access information from computers that control ABS, Air Conditioning, Steering and Suspension, Electronic Transmissions and other related systems.

J1962 - Common Diagnostic Connector
J1962 establishes a set of standards for the OBD II 16-Pin diagnostic connector. The 8 pins assigned by SAE include two pins for a Serial Data Link, two pins for an ISO 9141 Serial Data Link (European) and pins for battery power and ground.

J1979 - Diagnostic Test Mode Messages
Defines standard messages for access to trouble codes, vehicle data stream information and Freeze Frame data.

J2012 - Diagnostic Trouble Codes (DTC) Standardization
Diagnostic Trouble Code (DTC) is a term used to describe the method used when a vehicle computer detects a problem in a component or system that it is monitoring.

OBD I system trouble codes were one (1), two (2) or three (3) digit numbers. SAE J2012 set standards for trouble codes and definitions for emission-related systems.

OBD II codes use a five-digit code. OBD II codes begin with a letter followed by four numbers. Refer to the example in the Graphic to the right.

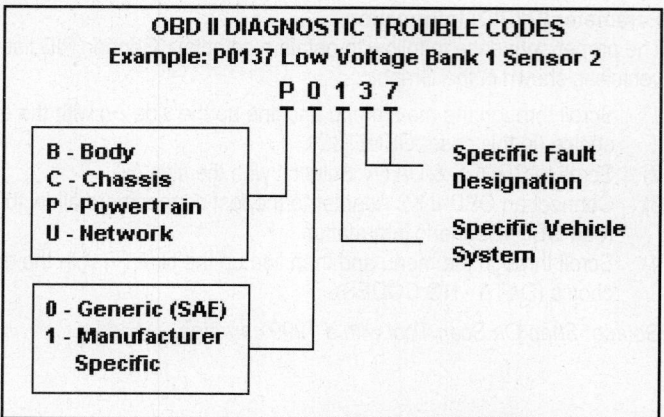

J2186 - Diagnostic Data Link Security Standards

Procedures used in tamper protection must discourage tampering yet allow for the service industry to reprogram or service the PCM as deemed necessary by a vehicle manufacturer. Legitimate service of the PROM will be referenced in later EPA regulations. This service can only be performed if the proper security codes are transmitted from the Scan Tool. However, normal Generic Scan Tool (GST) communications are not affected by this standard.

SAE J2186 defines EE Data Link Security standards for computers with electronically erasable and re-programmable PROMS. There are established standards intended to eliminate "hot rod" PROMS that could disable/defeat emissions-related control systems.

Engine operating parameters should not be changed without the use of manufacturer specialized tools, codes and procedures. CARB specified that any re-programmable computer coded system should include proven write-protection procedures and hardware. The Federal EPA requires that any re-programmable computer codes or operating parameters must be tamper resistant and must conform to SAE J2186 EE Data Link Security standards.

Tamper Protection Explanation

The possibility that On-Board programs may be tampered with from manufacturer specifications introduces the possibility of additional vehicle problems that would need to be diagnosed. Even before vehicles used computers, it was difficult to diagnose a driveability problem in a modified or performance vehicle. The potential of these changes could make it difficult for committees to balance between allowing individuals to install performance changes and protecting the repair industry from the changes.

There is a need for standards that would allow a PCM to identify a vehicle ID number and current calibration, whether the change is a factory update, or an aftermarket modification. This information could also be made accessible to a repair technician.

This would allow the technician to know at the start of a repair procedure that the diagnostic situation might not follow published diagnostic information. To avoid this situation, EPA proposed regulations that would force vehicle manufacturers to utilize complicated methods to deter unauthorized reprogramming. They would include executable routines that could have copyright protection with encrypted data and mandated electronic access by service facilities to an off-site computer maintained by the vehicle manufacturer. Access to an executable routine would be controlled by the vehicle manufacturer and made available only at one of their authorized dealers.

OBD II Warmup Cycle Definition

It is important to understand the meaning of the expression *warmup cycle*. Once the MIL is turned off and the fault that caused the code does not reappear, the OBD II code that was stored in PCM memory will be erased after 40 warmup cycles. The exceptions to this rule are codes related to a Fuel system or Misfire problem. These trouble codes require that 80 warmup cycles occur without the reoccurrence of the fault before the code will be erased from the PCM memory.

A warmup cycle is defined as engine operation (after a key off and engine cool-down period) in which engine temperature increases at least 40°F and reaches at least 160°F.

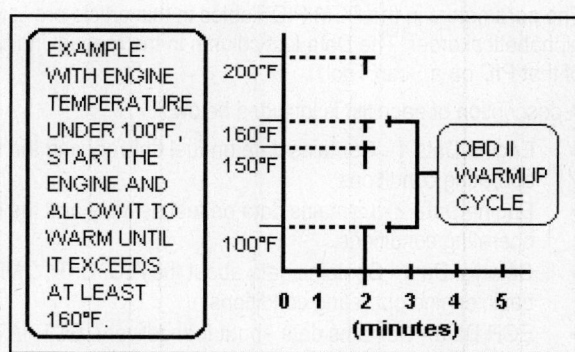

EXPLANATION OF SAE STANDARDS

J2190 - Enhanced Diagnostic Test Modes

This standard identifies test modes and diagnostics for issues not covered by the EPA and CARB regulations. They include Enhanced Test Modes (including an expanded diagnostic routine), and include the protocols required to establish the screens on the Scan Tool for these items:

- Request a diagnostic session
- Request trouble code related Freeze Frame data
- Request all diagnostic trouble codes
- Request status of Main Monitors, clear all test data
- Request diagnostic data, security access
- Disable or enable normal message transmission
- Request diagnostic data packets, test routine results
- Enter or exit diagnostic routines
- Substitute sensor values, substitute output controls
- Read or write to the PROM
- Messages from Enhanced Diagnostic Test Modes are available through an Enhanced Scan Tool, but may not be available through a Generic Scan Tool.

Scan Tool Enhanced Menu Example

An example of how to navigate through the Vetronix Scan Tool menus to locate the OEM PID information is shown in the Graphic. An example of first seven steps to follow is shown. Step (8) contains examples of PID data.

Parameter ID (PID) Information

The PID information for this vehicle is organized into various Data Lists (Engine Data 1, etc). Each PID is categorized into a particular list.

The parameters in the PCM PID Tables in this article are listed in alphabetical order. The Data List column in the manual indicates the location of that PID on a Scan Tool.

A description of each list is included below:

- Engine Data 1 - Contains data on fuel delivery and the basic engine operating conditions.
- Engine Data 2 - Contains data on fuel delivery and the basic engine operating conditions.
- Catalyst Data - Contains data about the A/C, CKP, CMP, KS, and the basic engine operating conditions.
- EGR Data - Contains data about fuel delivery, ECT, IAT, VTD, and the basic engine operating conditions.
- EVAP Data - Contains data that allows it to display parameters needed to verify EVAP system operation.
- HO2S Data - Contains data on the Oxygen sensor.
- Misfire Data - Contains data for Misfire diagnostics.
- Output Device Driver Data - Contains data specific to the ODD operation.

Source: Vetronix Scan Tool & 2000 Mass Storage Unit.

SCAN TOOL MENUS

(1)
```
SELECT APPLICATION

GLOBAL OBDII (MT)
GLOBAL OBDII (T1)
GM P/T
GM CHASSIS
GM BODY SYSTEMS
FORD P/T
FORD CHASSIS
CHRYSLER P/T
ACURA
CHRYSLER IMPORTS
```

(2)
```
2000 SELECT:
F0:VIN     F4:CAD
F1:MFI     F5:CSFI
F2:TBI     F6:DIESL
```

(3)
```
F0: C-CAR
F1: F-CAR
F2: H-CAR
F3: W-CAR
```

(4)
```
SELECT MFI ENG.
3800 SFI (VIN=K)
  C,F,H,W-CAR?
2000 (YES/NO)
```

(5)
```
SELECT TRANS.
F0:  3 SPD AUTO
F1:  4 SPD AUTO
F2:  5 SPD MAN
```

(6)
```
SELECT MODE:↑↓
F0:Data List
F1:Capture Info.
F2:DTC
F3:Snapshot
F4:OBD Controls
F8:Information
```

(7)
```
SELECT DATA:↑↓
F0:Engine 1
F1:Engine 2
F4:Specific Eng.
F5:A/T
F9:Specific A/T
```

(8)
```
Inj. Pulse Width
    2.9 ms
Air/Fuel Ratio
   14.7:1
```

Malfunction Indicator Lamp

INTRODUCTION

The CARB and Federal EPA regulations require that a Malfunction Indicator Lamp (MIL) be illuminated when an emissions related fault is detected and that a Diagnostic Trouble Code be stored in the vehicle controller (PCM) memory.

Most vehicle manufacturers provided the "Check Engine" light diagnostics required by the 1988 California regulations in time to meet this deadline.

OBD II regulations established changes in the "Check Engine" light operation. A new universal term identified this "light" as a Malfunction Indicator Light (MIL). However, the light on the dash may still be identified with the term "Check Engine" or "Service Engine Soon" for ease of customer understanding.

OBD II guidelines set tight conditions for activating and de-activating the MIL (lamp). This strict set of guidelines has resulted in multiple "levels" of diagnostics with different criteria and conditions for *when* an emissions-related fault will cause the MIL to activate and set a code. Also, there are other codes available that will not cause the PCM to activate the MIL. The guidelines established how quickly the onboard diagnostics must be able to identify a fault, set the trouble code in memory and activate the MIL (lamp).

REGULATIONS FOR CLEARING CODES AND CONTROLLING THE MIL

There are strict regulations for conditions to turn off the light and to clear trouble codes. In the past, some vehicle manufacturers had the technician remove battery voltage from the computer to clear the codes. These new regulations contain significant changes in how and when the controller turns off the MIL. The vehicle must be driven under specific conditions while the emission systems are monitored. Once a fault is detected, the system or component that failed must pass three consecutive tests (three trips) without failing before the MIL will be turned off. OBD II regulations include:

- A standard to regulate how quickly a computer must identify a fault, activate the MIL and set a trouble code.
- A standard to regulate criteria that can turn off the MIL when a fault is not present.
- A standard that establishes how long a trouble code remains in the computer memory once the problem has been repaired and the code is cleared.
- A standard to regulate what information must be available from the vehicle manufacturer that the repair technician can use to assist them in identifying the cause of a fault (i.e., Scan Tool and Freeze Frame data).

FAULTS NOT RELATED TO EMISSION CONTROL SYSTEMS

On an OBD II system, the MIL is not activated unless the computer determines that a failure in a component or system will affect the emissions levels of the vehicle. In effect, this means that **only emissions-related faults (codes)** will cause the PCM to activate the MIL. Be aware that some driveability-related problems not related to emission control components or systems can cause a code to set without the PCM activating the MIL. However, any trouble codes associated with the fault will still be set in memory for a technician to access with an OBD II certified Scan Tool.

KEY POINTS

Just like with OBD I systems, there can be trouble codes without activating the MIL, and there can be failure conditions on some systems not related to emission controls that do not set a trouble code. However, on OBD II systems, when diagnosing any driveability or emissions-related problems, all codes are considered "hard" codes. You should first read and record the codes and related data, then make the repairs. Once these steps are done, you can clear the trouble codes and related Freeze Frame data.

UNDERSTANDING MIL CONDITIONS
The three (3) possible MIL conditions are explained next.

Condition 1: MIL Off
This condition indicates that the PCM has not detected any faults in an emission-related component or system, or that the MIL power or control circuit is not working properly.

Condition 2: MIL On Steady
This condition indicates a fault in an emission-related component or system that could increase tailpipe emissions.

Condition 3: MIL Flashing
This condition indicates either a misfire or fuel system related fault that could cause damage to a catalytic converter.

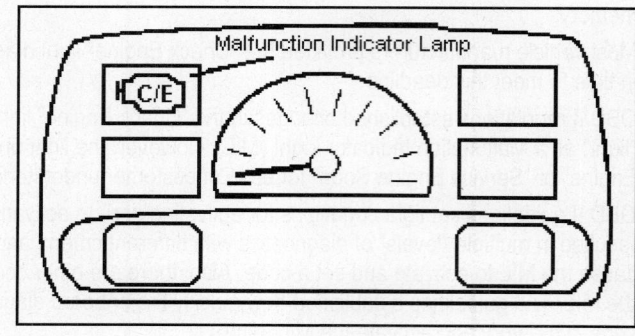

Note: *If a misfire condition exists with the MIL "on" steady, and the driver reaches a vehicle speed and load condition where the engine misfires at a level that could cause catalyst damage, the MIL will begin to flash. It will continue to flash until the engine speed and load conditions that caused that level of misfire subside. Then the MIL will return to the MIL "on" steady condition. This situation may result in a customer complaint as described next: "The MIL in my instrument cluster comes on and then flashes intermittently".*

ACTIONS OR CONDITIONS TO TURN OFF THE MIL
The PCM will turn off the MIL if any of the following actions or conditions occurs:

- The codes are cleared with a Generic or Proprietary Scan Tool
- Power to the PCM is removed (at the battery or with the PCM power fuse)
- A vehicle is driven on three consecutive trips **(including three warmup cycles)** and meets all of the particular code set conditions without the PCM detecting any faults

The PCM will set a code if a fault is detected that could cause tailpipe emissions to exceed 1.5 times the FTP Standard. However, the PCM will not de-activate the MIL until the vehicle has been driven on three consecutive trips with vehicle conditions similar to actual conditions present when the fault was detected. *This is not just three (3) vehicle startups and trips. It means three trips where certain engine operating conditions are met so that the OBD II Monitor that found the fault can "rerun" and pass that diagnostic test.*

Once the MIL is de-activated, the original code will remain in memory until forty warmup cycles are completed without the fault reappearing. A warmup cycle is defined as a trip where with an engine temperature change of at least 40°F, and where the engine temperature reaches at least 160°F.

SIMILAR CONDITIONS (FUEL TRIM AND MISFIRE CODES)
If a Fuel Control system (fuel trim) or misfire-related code is set, the vehicle must be driven under conditions similar to conditions present when the fault was detected before the PCM will de-activate the MIL (lamp). These "similar conditions: are described next:

- The vehicle must be driven with engine speed within 375 RPM of the engine speed stored in the Freeze Frame data when the code set
- The vehicle must be driven within engine load ± 10% of the engine load value stored in the Freeze Frame data when the code set
- The vehicle must be driven with engine temperature conditions similar to the temperature value stored in Freeze Frame data when the code set

Diagnostic Trouble Codes

INTRODUCTION

One of the key features in the OBD II system was an attempt to standardize the wording that describes a diagnostic trouble code or DTC (a term used to describe the method applied when the onboard controller recognizes and identifies a problem in one or more of the circuits or components that it monitors). As a point of review, keep in mind that the trouble codes used with OBD I systems consisted of codes identified with one (1), two (2), or three (3) digits. In effect, trouble codes were only identified with numbers.

DIAGNOSTIC TROUBLE CODES (5-DIGIT)

As previously discussed, SAE J2012 set standards for trouble codes and definitions for emission-related systems. OBD II trouble codes use a five-digit code, and these codes begin with a letter and are followed by four numbers. Since a letter is involved in the sequence, the correct way to read this type of code is with an OBD II certified Scan Tool.

The range of the code designations was designed to allow for future expansion and to allow for manufacturer specific usage on some systems. The illustration in the Graphic includes an explanation of OBD II Code Standardization for a DTC P0137.

UNIVERSAL CODE DESIGNATION EXPLANATION

The number in the thousandths position indicates that the trouble code is common to all manufacturers (a "P0" code).

Most vehicle manufacturers use this designation and then assign a common number and fault message to the problem. The code repair chart is not universal and service procedures will vary between the different vehicle manufacturers. However, the fault described in the code title is common to all systems on the vehicles (e.g., it was assigned a universal code designation). The first letter in the code identifies the system that controlled the device that failed (refer to the table below).

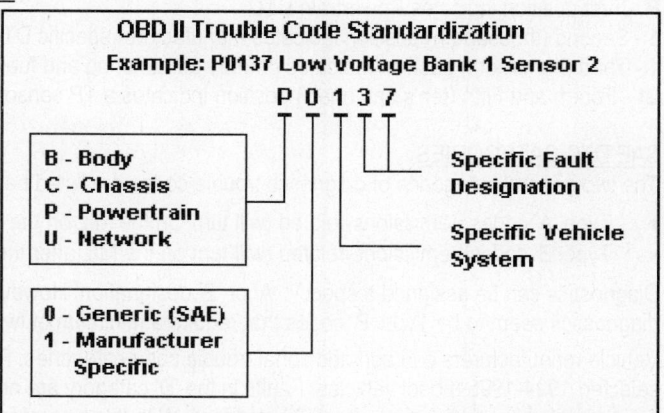

Code Description Table

System ID	System Description
B	Body Control System
C	Chassis Control System
P	Powertrain System
U	UART Data Link, Network Code

MANUFACTURER SPECIFIC DESIGNATION EXPLANATION

Vehicle manufacturers had some code conditions that are specific to the design of their individual system. Not all vehicle manufacturers have chosen to use P1xxx series codes due to differences in their basic systems, diagnostic strategy and their implementation.

These codes are designated as manufacturer specific codes (e.g., a "P1xxx" code), and each manufacturer can define the code and fault description for this designation. Although it was expected that each vehicle manufacturer would remain consistent across their product line, there has been considerable variation on P1xxx designations.

DTC NUMBERING EXPLANATION

The Number in the hundredth position indicates the specific vehicle system or subgroup that failed. This position should be consistent for P0xxx and P1xxx type trouble codes. An SAE committee established the numbers and systems listed below:

- **P0100** - Air Metering and Fuel System fault
- **P0200** - Fuel System (fuel injector only) fault
- **P0300** - Ignition System or Misfire fault
- **P0400** - Emissions Control System fault
- **P0500** - Idle Speed Control, Vehicle Speed Sensor fault
- **P0600** - Computer Output Circuit (relay, solenoid, etc.) fault
- **P0700** - Transaxle, Transmission faults

Note: The "ten's" and "one's" in the numbers indicate the part of the system at fault.

DTC NUMBERING EXAMPLE

DTC P1121 - GM Throttle Position Sensor Circuit Intermittent High Voltage:

P - First position indicates Powertrain DTC
1 - Second (thousandth) position indicates manufacturer specific DTC
1 - Third (hundredth) position indicates primary air metering and fuel system
21 - Fourth and Fifth (ten's and one's) position indicates a TP sensor fault

SAE DTC CATEGORIES

The two general categories of diagnostic trouble codes are listed below:

- Type 'A' codes - emissions-related (will turn On the MIL on the first failure)
- Type 'B' codes - emissions-related (will turn on the MIL after the second consecutive trip with a failure)

Diagnostics can be assigned a specific 'A' or 'B' designation. However, most vehicle manufacturer expanded diagnostics seem to be Type 'B' codes that require a minimum of two consecutive trips with a fault to activate the MIL.

Vehicle manufacturers can add additional trouble code categories. For example, GM has a 'D' category for a few selected 1994-1995 model vehicles. Faults in the 'D' category are non-emissions faults that will not cause tailpipe emissions to exceed 1.5 times the FTP standard. With this type of fault, the PCM does not activate any lamps or store any fault data in the Freeze Frame buffer (used for Type 'A' and Type 'B' faults).

COMMON CODE NAMES AND DESCRIPTIONS

OBD II guidelines set standards to universalize the Code Name and Description. These standards only apply to P0xxx codes. You need to be careful because there are several OBD II trouble codes where the same code number can have a different code title.

CODE VARIATION EXAMPLE

Note the use of the same code number for 2 different electrical faults in the table below.

DTC Number	Code Description & Conditions
P1641 (N/MIL) 1996-98 A, L, N & W Body: VIN M, X	**MIL Control Circuit Conditions** Key on, then the PCM received an improper voltage level on the MIL driver circuit (ODM 'A' output 1), condition met for 30 seconds. • Refer to the correct code repair chart.
P1641 (N/MIL) 1996 B Body: VIN P & W 1996 Y Body: VIN 5 & P	**Fan Control Relay 1 Control Circuit Conditions** No A/C or ECT codes set, engine speed over 600 rpm, then the PCM detected that the commanded state of the FC Relay 1 driver and Actual state did not match for 5 seconds. • Refer to the correct code repair chart.

DTC DESCRIPTOR DEFINITIONS

The SAE J2012 document further defines most circuit, component or system codes into the four basic categories explained next.

Circuit Malfunction - Indicates a fixed value or no response from the system. This descriptor can be used instead of a dual High/Low Voltage Code or used to indicate another failure mode.

Range/Performance - Indicates that the circuit is functional, but not operating normally. This descriptor may also indicate a stuck, erratic, intermittent or skewed value that could cause poor performance of an emission control circuit, component or system.

Low Input - Indicates that a signal circuit voltage, frequency or other measurement at a PCM input terminal is at or near zero. The test is made with the external circuit, component or system connected. The signal type is used in place of the word "input."

High Input - Indicates that a signal circuit voltage, frequency or other test measurement at a PCM input terminal is at or near full scale. This test is made with the external circuit, component, or system connected. Signal type is used in place of the word "input."

CONDITIONS TO CLEAR TROUBLE CODES

Diagnostic trouble codes are cleared from the PCM memory using several different methods (the actual method varies between vehicle manufacturers). An example of the Scan Tool navigation screens that appear on a 1999 Ford Windstar during a code clearing procedure is shown in the Graphic (Source: Snap On).

The list below contains a summary of a few of the methods that can be used to clear OBD II trouble codes. The actual conditions for each vehicle manufacturer must be determined and followed exactly.

- Regulations adopted with OBD II allow codes to be cleared by the PCM once 40 warmup cycles occur after the "last test failed" message clears and after 40 "last test passed" messages occur. Refer to Page 1-17 for the definition of a "warmup cycle".
- The Scan Tool can be used to clear any stored codes (and Freeze Frame data).
- On some vehicles, if battery voltage to the PCM is removed, the trouble codes, Freeze Frame data, "trip" or "drive cycle" status and I/M Readiness status will be lost. The battery voltage must be removed for 5 minutes or longer for this action to occur.

■ NOTE: *Do not clear the trouble codes unless the code repair chart diagnostic procedure instructs you to do so. Most manufacturers will clear Freeze Frame data (that can be used to diagnose the cause of the fault) at the same time a code is cleared. In effect, this step will result in the loss of the Freeze Frame data on most systems.*

Source: Snap On Scan Tool with a 1999 cartridge.

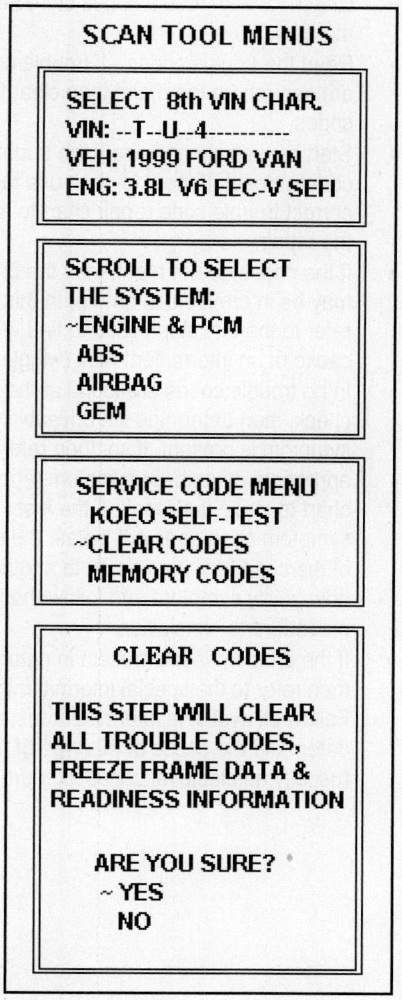

SCAN TOOL MENUS

SELECT 8th VIN CHAR.
VIN: --T--U--4----------
VEH: 1999 FORD VAN
ENG: 3.8L V6 EEC-V SEFI

SCROLL TO SELECT
THE SYSTEM:
~ENGINE & PCM
ABS
AIRBAG
GEM

SERVICE CODE MENU
KOEO SELF-TEST
~CLEAR CODES
MEMORY CODES

CLEAR CODES

THIS STEP WILL CLEAR
ALL TROUBLE CODES,
FREEZE FRAME DATA &
READINESS INFORMATION

ARE YOU SURE?
~ YES
NO

Diagnostic Routines

OBD I SYSTEM DIAGNOSTICS

One of the most important things to understand about the automotive repair industry is the fact that you have to continually learn new systems and new diagnostic routines (the test procedures designed to isolate a problem on a vehicle system). For OBD I and II systems, a diagnostic routine can be defined as a procedure (a series of steps) that you follow to find the cause of a problem, make a repair and then verify the problem is fixed.

CHANGES IN DIAGNOSTIC ROUTINES

In some cases, a new Engine Control system may be similar to an earlier system, but it can have more indepth control of vehicle emissions, input and output devices and it may include a diagnostic "monitor" embedded in the engine controller designed to run a thorough set of emission control system tests.

OBD I Diagnostic Flowchart

The OBD I Diagnostic Flowchart on this page can be used to find the cause of problems related to Engine Control system trouble codes or driveability symptoms detected on OBD I systems. It includes a step-by-step procedure to use to repair these systems. To compare this flowchart with the one used on OBD II systems, refer to the next page.

The steps in this flow chart should be followed as described below (from top to bottom).

- Do the Pre-Computer Checks.
- Check for any trouble codes stored in memory.
- Read the trouble codes - If trouble codes are set, record them and then clear the codes.
- Start the vehicle and see if the trouble code(s) reset. If they do, then use the correct trouble code repair chart to make the repair.
- If the codes do not reset, than the problem may be intermittent in nature. In this case, refer to the test steps used to find the cause of an intermittent fault (wiggle test).
- In no trouble codes are found at the initial check, then determine if a driveability symptom is present. If so, then refer to the approriate driveability symptom repair chart to make the repair. If the first symptom chart does not isolate the cause of the condition, then go on to another driveability symptom and follow that procedure to conclusion.
- If the problem is intermittent in nature, then refer to the special intermittent tests. Follow all available intermittent tests to determine the cause of this type of fault (usually an electrical connection problem).

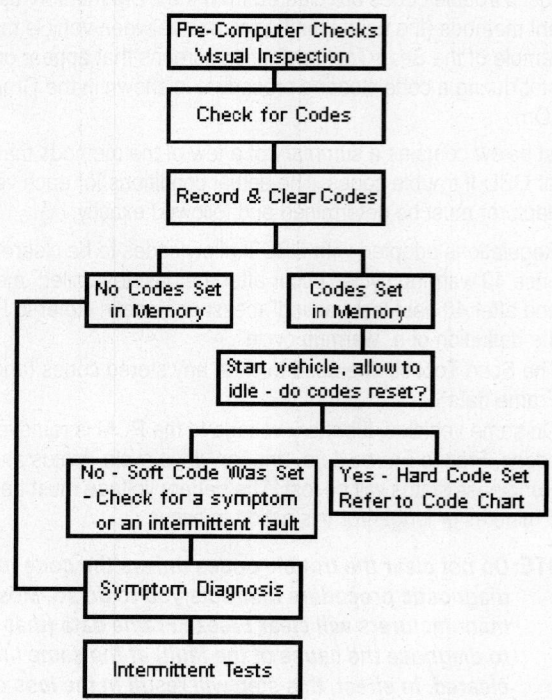

OBD I DIAGNOSTIC FLOWCHART

OBD II SYSTEM DIAGNOSTICS

The diagnostic approach used in OBD II systems is more complex than that of the one for OBD I systems. This complexity will effect how you approach diagnosing the vehicle. On an OBD II system, the onboard diagnostics will identify sensor faults (i.e., open, shorted or grounded circuits) as well as those that lose calibration. Another new test that arrived with OBD II is the rationality test (a test that checks whether the value for one input makes rational sense when compared against other sensor input values). The changes plus the use of OBD II Monitors have dramatically changed OBD II diagnostics.

The use of a repeatable test routine can help you quickly get to the root cause of a customer complaint, save diagnostic time and result in a higher percentage of properly repaired vehicles. You can use this Diagnostic Flow Chart to keep on track as you diagnose an Engine Control problem or a base engine fault on vehicles with OBD II.

OBD II Diagnostic Flowchart

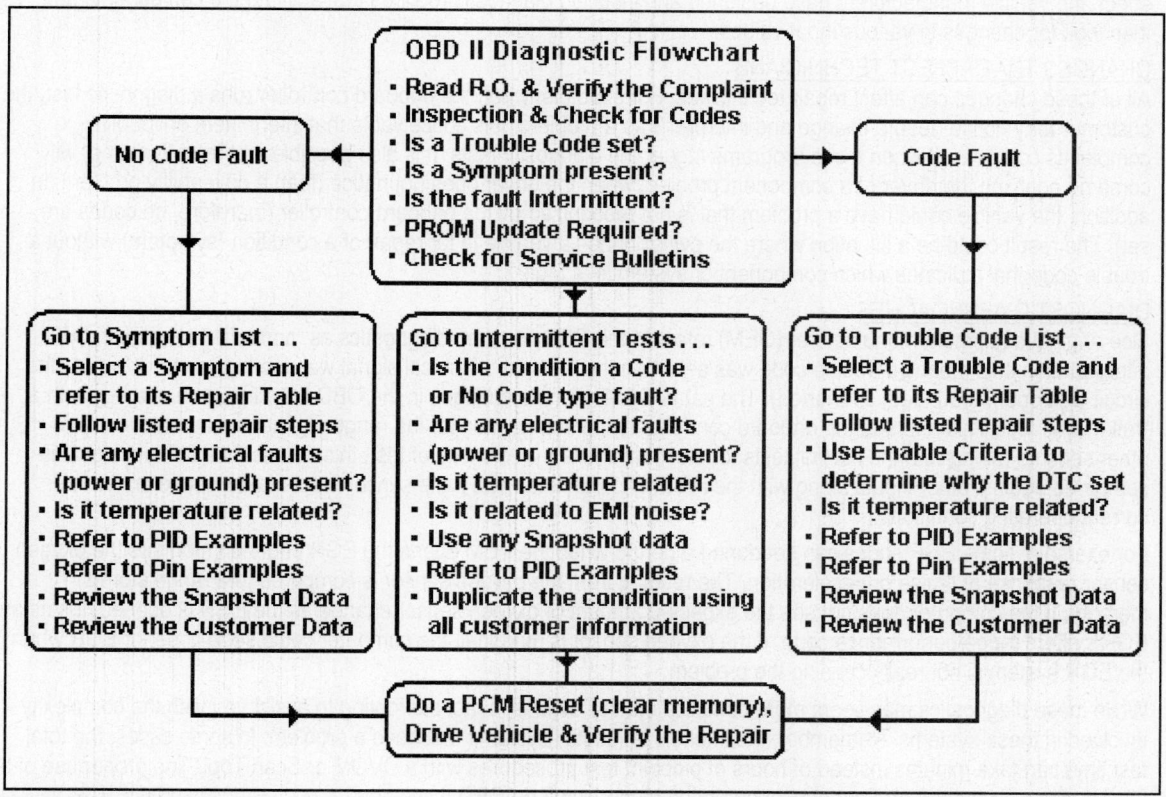

FLOW CHART STEPS

Here are some of the steps included in the Diagnostic Routine:

- Review the repair order and verify the customer complaint as described
- Perform a Visual Inspection of underhood or engine related items
- If the engine will not start, refer to No Start Tests
- If codes are set, refer to the trouble code list, select a code and use the repair chart
- If no codes are set, and a symptom is present, refer to the Symptom List
- Check for any related technical service bulletins (for both Code and No Code Faults)
- If the problem is intermittent in nature, refer to the special Intermittent Tests

Expanded Diagnostics

INTRODUCTION

The primary focus of OBD Expanded Diagnostics is to verify that all Emission Control systems continue to operate efficiently and that they do not deteriorate to a point where tailpipe emissions would increase to a point more than 1.5 times the FTP standard.

On a vehicle equipped with an OBD II system, all emissions faults must set diagnostic codes. However, instead of just identifying a circuit or component failure, this system will identify some problems where the component is deteriorating. These types of problems often occur before the vehicle driver notices a problem. If the fault is driveability related, the driver may not notice it while driving the vehicle. Some faults do not impact vehicle operation (and may not be noticed by the driver - they are not symptom-related faults).

All of these faults can be identified by an OBD II system when it monitors an emissions-related system and determines that it has deteriorated to a point where it would increase emissions beyond the FTP standard. Evaluation of these systems requires expanded programs that monitor all emissions-related components and systems for deterioration. In effect, the vehicle manufacturers have designed and installed diagnostic routines that activate the components and then look for changes in various input values.

CHANGES THAT AFFECT TECHNICIANS

All of these changes can affect repair technicians. For example, when the onboard controller runs a diagnostic test, the customer may notice (feel) a change and interpret it as a problem. It is conceivable that intermittent driveability complaints could result when these programs trigger the diagnostic tests. It is also possible that the MIL (lamp) will come on notifying the driver of a component problem when the driver does not notice (feel) a driveability problem. In addition, the vehicle could have a problem that is not recognized by the onboard controller (therefore, no codes are set). The result could be a situation where the owner brings a vehicle in for repair of a condition (symptom) without a trouble code that indicates which component or system is at fault.

DIAGNOSTIC APPROACHES

One original equipment manufacturer (OEM) refers to the OBD I system diagnostics as "normality" checks. In this interpretation, this means a trouble code was set when the "normal" electrical signal was too high or too low (i.e., the circuit was open, grounded, or shorted). The same OEM refers to changes in the OBD II system as a strategy with a "rationality" test. This refers to the onboard controller being able to monitor the range of a sensor in relationship to other sensors (along with the normal tests for electrical faults). In this type of test, this means that an input signal is compared against other inputs along with the information to determine if the sensor input makes sense under the current operating conditions.

For example, some EGR codes can set during a period when the PCM opens the EGR and then monitors the oxygen sensor response at cruise or deceleration. The reading from the oxygen sensor is compared to a range stored in memory. If the computer does not see the expected amount of oxygen sensor change as the EGR is opened, it sets an EGR system range/performance code. If the oxygen sensor is marginal, the computer could set a code for EGR when the EGR system is not really causing the problem.

While these diagnostics may seem more complicated, the OEM is in fact attempting to assist you with the complexity involved in these systems. Remember, the intent is to identify systems that have a problem. In some cases, the total test time can take minutes instead of hours of pinpoint test procedures with a DVOM or Scan Tool. The proper use of a DVOM, Scan Tool and Lab Scope must be understood to work on OBD II systems.

Expanded Diagnostics

SCAN TOOL INFORMATION

CARB regulations require that vehicle manufacturers make available to repair technicians procedures which allow effective emission related diagnostic and repair using a Generic Scan Tool. This regulation was developed into SAE J1978 that sets guidelines for a common Scan Tool to access On-Board information.

The actual information shown on a Scan Tool can vary between different vehicle systems. Each manufacturer emphasized certain programs and then displayed the information in their own format. SAE J2205 sets the standard for how the Scan Tool will interface with the computer and access the computer information. This standard was necessary because computers are interfacing in bi-directional formats on these vehicles.

There is a difference in the amount of information available on each brand of Scan Tool. Review how this information is accessed on your OBD II certified Scan Tool. The Scan Tool gets its power from the vehicle being tested and talks or "interfaces" with the vehicle diagnostic system or the diagnostic executive program.

TROUBLE CODE INFORMATION

The trouble code information on a Scan Tool includes:

- Current and History trouble codes
- MIL Requested Information ("MIL ON" data)
- Diagnostic test status (test run/test pass or fail)
- Last test pass or fail message
- Freeze Frame data for the 1st emission fault
- Some Scan Tools can display Failure Records

FREEZE FRAME INFORMATION

CARB and EPA regulations require that the controller store specific Freeze Frame (engine related) data when the first emission related fault is detected. The data stored in Freeze Frame can only be replaced by data from a trouble code with a higher priority (i.e., a trouble related to a Fuel system or Misfire Monitor fault).

The Freeze Frame has to contain data values that occurred at the time the code was set (these values are provided in standard units of measurements). As a result, OBD II systems record the data present at the time an emission related code is recorded and the MIL activated. This data can be accessed and displayed on a Scan Tool. Freeze Frame data is one frame or one instant in time. It records the data that set the code.

REQUIRED FREEZE FRAME DATA ITEMS

- Calculated load value, Engine Speed (rpm), Short and Long Term fuel trim values
- Fuel system pressure value (where applicable)
- Vehicle speed (MPH) & Closed / Open Loop status
- Engine coolant temperature and Intake manifold pressure
- Trouble Code that triggered the Freeze Frame
- If misfire code set - identify which cylinder is misfiring

```
┌──────────────────────────┐
│ (MAIN MENU (CHRYSLER))    │
│ ┌──────────────────────┐ │
│ │ Press:               │ │
│ │  OTHER SYSTEMS       │ │
│ │  FUNCTIONAL TESTS    │ │
│ │  CODES & DATA MENU   │ │
│ │  CUSTOM SETUP        │ │
│ │ ~SYSTEM TESTS        │ │
│ └──────────────────────┘ │
└──────────────────────────┘

┌──────────────────────────┐
│    ( SYSTEM TESTS )       │
│ ┌──────────────────────┐ │
│ │ Press:               │ │
│ │ SYSTEM TESTS:        │ │
│ │  EGR SYSTEMS TEST    │ │
│ │  GENERATOR FIELD TEST│ │
│ │  INJ. KILL TEST      │ │
│ │  MISFIRE COUNTERS    │ │
│ │ ~PURGE VAPORS TEST   │ │
│ │  READ VIN            │ │
│ └──────────────────────┘ │
└──────────────────────────┘

┌──────────────────────────┐
│  ( PURGE VAPORS TEST )    │
│ ┌──────────────────────┐ │
│ │ Press:               │ │
│ │ * Y TO SWITCH BETWEEN│ │
│ │ NORM, FLOW & BLOCK   │ │
│ │ PURGE STATUS__NORM   │ │
│ │ ENGINE RPM_____736  │ │
│ │ NO CODES IN THIS MODE│ │
│ │ UPSTRM O2S(V)___0.63 │ │
│ │ DWNSTRM O2S(V)__0.14 │ │
│ │ PURGE(mA)_____120  │ │
│ │ ST ADAP(%)_____1.4  │ │
│ └──────────────────────┘ │
└──────────────────────────┘
```

Expanded Diagnostics

TROUBLE CODE "TEST CONDITIONS"

Some vehicle emission control components and systems are "continuously" monitored by the Comprehensive Component, Fuel System and Misfire Monitors while some of the OBD II Main Monitors only run their diagnostic tests after certain test conditions or enable criteria have been met (e.g., the EGR system and EVAP system Monitors).

Key Point - Certain code "test conditions" must be met to "run" certain Monitors, and the conditions vary by vehicle and engine configuration. Also, the information related to each trouble code contains the actual conditions present when that particular code set.

INTRODUCTION CONTENTS

Ford Motors Systems

FORD ENGINE PERFORMANCE
    About This Manual ....Page 1-2
ELECTRONIC ENGINE CONTROLS
    Introduction ....Page 1-3
VEHICLE IDENTIFICATION
    Vehicle Identification Number ....Page 1-4
    Vehicle Certification Label ....Page 1-5
TECHNICIAN REQUIREMENTS
    Introduction ....Page 1-6
    Where to Start the Diagnosis ....Page 1-6
SERIAL DATA COMMUNICATION
    Standard Corporate Protocol ....Page 1-7
    Scan Tool will not Operate ....Page 1-7
    Scan Tool Communication ....Page 1-8
FLASH REPROGRAMMING
    Introduction ....Page 1-9
    Aftermarket Reprogramming ....Page 1-9
POWERTRAIN DIAGNOSTICS
    Where To Begin ....Page 1-10
    Using Test Procedures ....Page 1-11
    Repair Verification ....Page 1-11
    Base Engine Tests ....Page 1-12
    Engine Vacuum Tests ....Page 1-13
    Ignition System Tests ....Page 1-15
    Symptom Diagnosis ....Page 1-17
    Intermittent Test Procedures ....Page 1-27
    Pinpoint Text Z (Intermittent Faults) ....Page 1-30
    Ford Quick Test ....Page 1-35
    Read Trouble Codes ....Page 1-36
    Pinpoint Test QT (Diagnostic Starting Point) ....Page 1-37
ON BOARD DIAGNOSTICS
    Introduction ....Page 1-38
    PCM Reset (Clear Codes Step) ....Page 1-39
    Diagnostic Trouble Codes ....Page 1-40
    Cylinder Bank Identification ....Page 1-41
    Diagnostic Executive ....Page 1-42
    Comprehensive Component Monitor ....Page 1-43
    Main Monitors ....Page 1-44
    OBD III Diagnostics ....Page 1-45
    How to Access Generic PID Information ....Page 1-47
    How to Access OEM PID Information ....Page 1-48
REFERENCE INFORMATION
    PCM Location Tables (1992-2002) ....Page 1-49
    MAF Sensor Calibration Tables ....Page 1-54
    Operating Range Charts ....Page 1-56
    Ford OBD Glossary ....Page 1-57

Ford Engine Performance

ABOUT THIS MANUAL
This manual was developed to provide you with information that explains the theory of operation, diagnosis and repair of the Electronic Engine Controls on several late model Ford Motor Company vehicles.

ASE Test Help
The information in this manual can be a valuable asset as you prepare to take the ASE A-8 or L-1 tests. It can also expand on your previous repair knowledge if you have already passed both of these certification tests.

Main Features
The main features of this product are divided into individual sections that explain how to use Ford vehicle diagnostics along with a DVOM, Lab Scope and Scan Tool to test the Powertrain and Body Control Module input and output signals used on Ford applications.

SECTION OVERVIEWS
Introduction Section - This section contains separate articles on these subjects: Engine Controls, Vehicle Identification, Problem Solving, Serial Data Communication, Flash Reprogramming, Powertrain Diagnostics, Ford Quick Test and OBD Systems.

Car Section - This section contains information specific to a 1996 Taurus and a 2001 Focus including these subjects: Anti-Theft System, Electronic Ignition, EGR & EVAP systems, Fuel Delivery system, Idle Air Control Valve, MAF, Oxygen Sensor, Transmission Controls, PID Data, Pin Voltage Tables and Wiring Diagram examples.

F150 Section - This section contains information specific to a 1998 F150 Series Pickup including these subjects: Electronic Ignition, EGR & EVAP systems, Fuel Delivery system, Idle Air Control Valve, MAF, Oxygen Sensor, Transmission Controls, PID Data, Pin Voltage Tables and Wiring Diagram examples.

Minivan Section - This section contains information specific to a 2001 Windstar including these subjects: Electronic Ignition, EGR & EVAP systems, Front Electronic Module, Fuel Delivery system, Idle Air Control Valve, MAF, Oxygen Sensor, Transmission Controls, PID Data, Pin Voltage Tables and Wiring Diagram examples.

SUV Section - This section contains information specific to a 2000 Explorer including these subjects: Electronic Ignition, EGR & EVAP systems, Fuel Delivery system, Generic Electronic Module, Idle Air Control Valve, MAF, Oxygen Sensor, Transmission Controls, PID Data, Pin Voltage Tables and Wiring Diagram examples.

Diagnostic Help
Sections 2 through 5 contain articles that include "real world" test examples and results that can be used with a DVOM, Breakout Box, Lab Scope or Scan Tool.

Electronic Engine Controls

INTRODUCTION

The evolution of the various Electronic Engine Control (EEC) systems used on Ford vehicles from 1983-2002 is shown in the Graphic below. Refer to the information in the vehicle sections to read more about the fuel injection system for a particular model. Some of the information required to diagnose problems a particular vehicle is included in each section as well as information on the OBD II system and its related diagnostics.

EEC Evolution Graphic

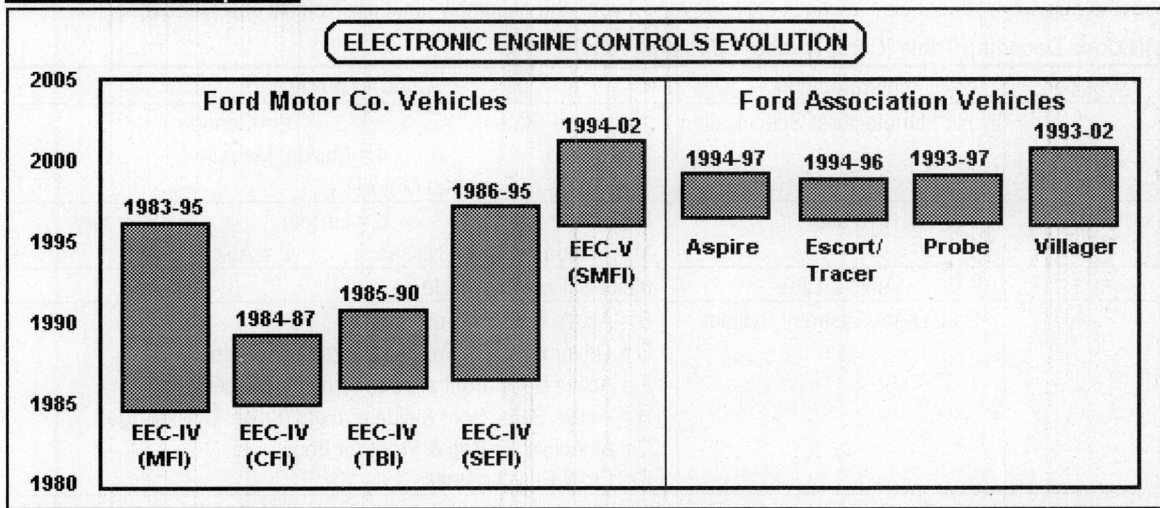

ELECTRONIC FUEL INJECTION

Ford Motor Co. Vehicle Applications

CFI System - The EEC-IV system with central fuel injection (CFI - dual injectors) was introduced in 1984 on models with 3.8L V6 (VIN 3) and 5.0L V8 (VIN F & M) engines.

MFI System - The first EEC-IV system with multiport fuel injection (MFI) was introduced in 1983 on several models with a 1.6L I4 engine (VIN 5). A 2.3L I4 engine (VIN W) with a turbocharger was introduced in mid-1983 on a few models with a 40-Pin PCM connector.

TBI System - The EEC-IV system with throttle body injection (TBI - single injector) was introduced in 1985 on models with 2.3L I4 (VIN S & X) and 2.5L I4 (VIN D) engines.

SFI or SMFI System - The EEC-IV system with SFI (4, 6 or 8 injectors) was introduced in 1986. The EEC-V version of SFI was introduced in 1994 and was in use though 2002.

Association Vehicle Applications

Aspire (MFI) - 1994-97 Aspire models (1.3L I4 VIN H engine) were equipped with an EEC system and electronic fuel injection.

Escort/Tracer 1.8L I4 (MFI) - 1994-96 Escort/Tracer models (1.8L I4 VIN 8 engine) were equipped with an EEC system and electronic fuel injection.

Probe 2.5L V6 (MFI) - 1993-97 Ford Probe (2.5L V6 VIN B engine) were equipped with an EEC system and electronic fuel injection.

Villager (MFI) - 1993-2002 Mercury Villager (3.0L and 3.3L V6 engines) were equipped with an EEC system and electronic fuel injection.

Vehicle Identification

VEHICLE IDENTIFICATION NUMBER

The vehicle identification number (VIN) is a seventeen digit legal identifier of the vehicle. It is fastened to a plate attached to the upper left corner of the instrument panel where it can be seen through the windshield from outside the vehicle. It is also stamped on the right-hand floor panel and is included on the vehicle certification label on the LH B-pillar.

This VIN information includes the country of origin, the make, the vehicle type, the passenger safety equipment, the car line, the body style, the engine, a check digit, the model year, the assembly plant and vehicle build sequence.

VIN Code Decoding Table (Cars, Trucks, SUVs and Vans)

Position	Interpretation	Code = Description
1	World Manufacturer Specification	1 = United States 2 = Canada 3 = Mexico 4 = Mazda, Mercury 5 = Nissan or Ford (Kia Motors)
2	Make	F = Ford L = Lincoln M = Mercury M = Mazda N = Nissan Z = Auto Alliance
3	Vehicle Type	Make & Type of Vehicle
4	Passenger Restraint System	B = Active Belts (all positions) C = Driver Airbag & Front Car Restraint System F = Active Belts, front airbags (driver & passenger) H = Active Belts, front & side airbags (driver & passenger) L = All Active Air Bag & Front Air Bags P = Front Passive Belts
5	Car or Vehicle Line	M = Passenger Car (Lincoln, Mercury, N. American Vehicles P = Passenger Car (Ford North American Vehicle) T = Passenger Vehicle (All passenger vehicles imported from outside N. America or Non-Ford built passenger vehicles
6	Body Chassis Type	Indicates the Body and Chassis Type
7	Body Chassis Type	Indicates the Body and Chassis Type
8	Engine Type & Make Note: The examples to the right are from various Truck, Utility Vehicle and Van applications.	A = 2.3L I4 MPI - OHC M = 3.0L V6 MPI E = 4.0L V6 MPI (OHC) U = 3.0L V6 MPI E = 4.0L V6 MPI (OHC) X = 4.0L V6 MPI F = 7.3L V8 Turbo Diesel 2 = 4.2L V6 MPI G = 7.5L V8 6 = 4.6L V8 MPI
9	Check Digit	Varies
10	Model Year	Code Year 'R' 1994 'S' 1995 'T' 1996 'V' 1997 'W' 1998 'X' 1999 'Y' 2000 '1' 2001 '2' 2002
11	Assembly Plant	Plant Location (e.g., 'A' = Atlanta: Hapeville, Georgia)
12 to 17	Plant Sequential Number	Ford Division Vehicles =100,001-500,999 600,001-999,999 = Lincoln/Mercury Division Vehicles

VEHICLE CERTIFICATION LABEL

The vehicle certification label is positioned on the left hand 'B' pillar above the latch. This label also contains the 17-digit vehicle identification number (VIN Code).

Vehicle Certification Label Table (Cars, Trucks, SUVs and Vans)

Item	Part Number	Description
1	Ford Motor Co. in USA	Name and location of Vehicle Manufacturer
2	5100LB / 2313KG	Gross vehicle weight ratings in pounds (LB) and Kilograms (KG)
3	F6006, T10156	Accessory reserve load codes
4	-	District/special order codes
5	RG 71	Sales region codes
6	OUSIA, 405	Suspension identification
7	D	Transmission/Transaxle codes D = 5-Speed Automatic Overdrive (AS LDF) M = 5-Speed Manual (Mazda) U = 4-Speed Automatic Overdrive (4R70W)
8	45	Axle ratio code (e.g., 41 = 3.27 [non-limited slip], 45 = 3.55 [non-limited slip], D = 3.73 [limited slip])
9	R, K	Radio codes (e.g., K= Luxury AM/FM Stereo Cassette, CD Player, DSP
10	TP, PS	Tape stripe/paint stripe codes
11	IN TR: F2	Interior trim codes (F = leather sport bucket seats, 2 = medium graphite)
12	2	Brake type code (e.g., 2 = Antilock Brakes or ABS)
13	UA	Exterior paint codes (e.g., UA = Ebony)
14	1FMZU63E0YZC11515	Vehicle Identification (VIN Code) Number
15	2900 LB / 1315 KG	Rear gross axle weight ratings in pounds (LB) and Kilograms (KG)
16	2510LB / 1138 KG	Front gross axle weight ratings in pounds (LB) and Kilograms (KG)
17	Date: 5/00	Build date (i.e., May of 2000)

Vehicle Certification Label Graphic

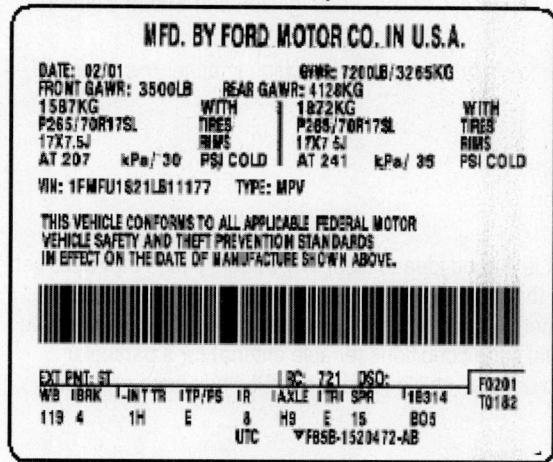

Technician Requirements

INTRODUCTION

As an automotive technician, there are certain fundamentals that you should understand:

Hand Tools and Meter Operation

You should have a good basic understanding of how to operate required hand tools and test meters in order to effectively use the information in this manual or in any other service repair manual or electronic repair media.

Electronic Engine Controls

You should have a basic knowledge of Electronic Engine Controls when performing test procedures to keep from making an incorrect diagnosis or from damaging the PCM or its devices. Do not attempt to diagnose a Powertrain problem without this basic knowledge.

Electricity and Electrical Circuits

You should understand basic electricity and know the meaning of voltage (volts), current (amps), and resistance (ohms). You should understand what happens in an electrical circuit with a circuit has an open or shorted condition, and you should be able to identify an open circuit or shorted circuit using a DVOM. You should also be able to read and understand automotive electrical wiring diagrams.

Circuit Testing Tools

You should know when to use and <u>when not to use a 12-volt test light</u> during diagnosis of a Powertrain electrical system (do not use this tester unless specifically instructed to do so by a test procedure). Instead of using a 12-volt test light, you should use a DVOM or Lab Scope with the whenever a diagnostic procedure calls for a measurement at a PCM connector or component wiring harness.

Where to Start the Diagnosis

If you are reasonably certain that the problem is related to the Chrysler Powertrain, start by checking the system by performing a DTC Test to check for any stored codes.

On vehicles with more than one vehicle computer (i.e., BCM or TCM), if you are unsure whether the problem is Powertrain related, start by checking for codes in the other controllers to determine if the problem is related to another vehicle system.

If there are no codes set, and you are certain which Powertrain subsystem has a problem, you can start with the system check for that subsystem. The subsystems include the Charging, Cooling, Fuel, Ignition and Speed Control systems.

If a wiring problem is found during testing, you will need to refer to electrical wiring information in either this manual, other manuals or electronic media that includes:

- Wiring schematics (including circuit numbers and colors)
- Electrical component connector, splice and ground locations
- Wiring repair procedures and wiring repair parts information

After a Repair is Complete

Once you decide on a repair, in addition to repairing the fault, it is a good idea to clear any trouble codes that were set and to verify they will not reset. To clear trouble codes, refer to the information in the PCM Reset step in this section. To verify a repair, you must duplicate the conditions present when the customer complaint occurred or when a trouble code was set. For OBD II systems, you must duplicate the actual code conditions (enable criteria) for a particular trouble code. This information can be used to determine how to drive a vehicle to determine if it has been repaired properly.

Serial Data Communication

STANDARD CORPORATE PROTOCOL

Standard Corporate Protocol is a communication language used by Ford Motor Co. for exchanging bi-directional messages between stand-alone modules and devices. With this system, two or more signals can be sent over the same circuit. Messages may include diagnostic data that is output over Bus (+) and Bus (-) lines to the Data Link Connector (DLC). This information is then accessible with a Generic or OEM Scan Tool.

DATA COMMUNICATION LINK

In 1988, Continental models added a Data Communication Link (DCL) to connect the PCM, Instrument Cluster and Electronic Message Center together so they could share serial data information with one another. The DCL also allowed connection to the Ford Service Bay Diagnostic System (SBDS) to assist technicians during vehicle diagnosis.

SERIAL DATA

Serial data refers to data transferred in a linear fashion over a single line, one bit at a time. The serial data on the Bus circuits during communication is similar to this example.

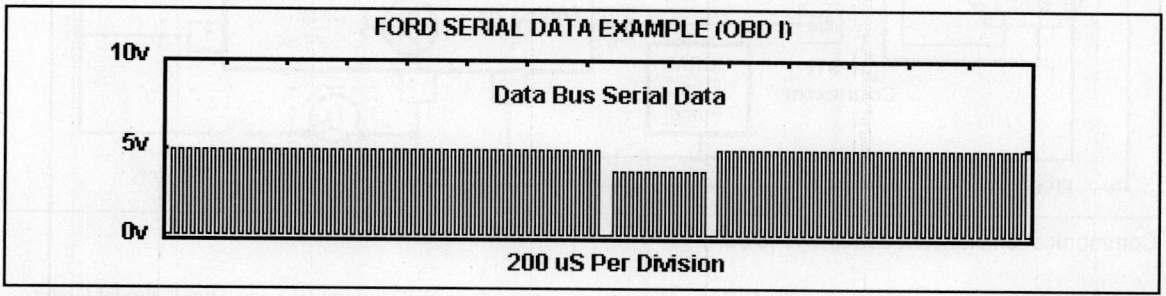

Serial Data Example

CLASS 2 SERIAL DATA

Ford vehicles utilize the "Class 2" communications system where each bit of data can have a long or short length. This allows multiple signals to be transmitted on one wire.

Messages carried on Class 2 data streams are also prioritized. If two messages attempt to establish communications on the data line at the same time, only the message with higher priority will continue. The device with the lower priority message must wait.

Class 2 Serial Data Graphic

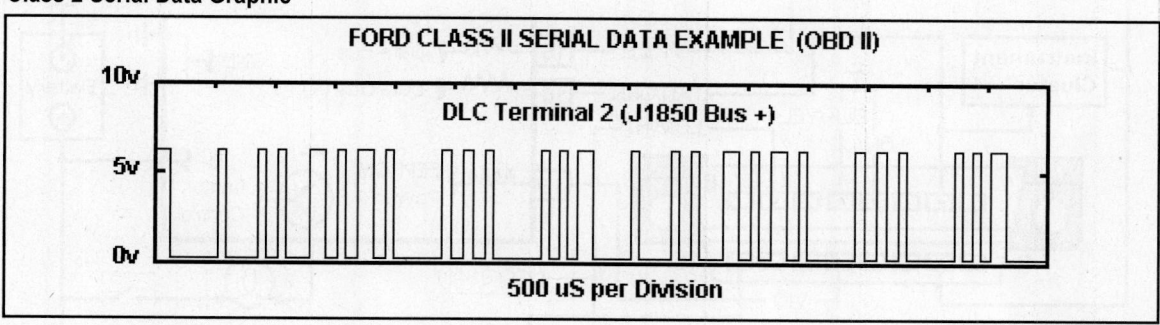

Scan Tool Will Not Operate

If the Scan Tool display is blank, check the STI, STO, Data Bus (+) and (-) circuits for an open or grounded circuit condition. The SCP Data Bus (+) and (-) circuits as well as the system power and ground circuits can be checked at the DLC on an OBD II system. Try the tool on another vehicle before condemning the tool or the test connector circuits.

SCAN TOOL COMMUNICATION

For the Scan Tool to communicate with various vehicle stand-alone modules, it must be connected into the system. On OBD I systems, connect the Scan Tool to the underhood test connector. On OBD II systems, connect the Scan Tool to the DLC under the dash.

OBD I Systems - To access data in the PCM on this system, connect an OBD I Scan Tool to the DLC and enter the Quick Test. This test allows you to run the KOEO, KOER and Continuous Memory Self-Tests, read trouble codes and access serial data (on some 1988-95 models). These Scan Tool functions are performed via the data bus circuits.

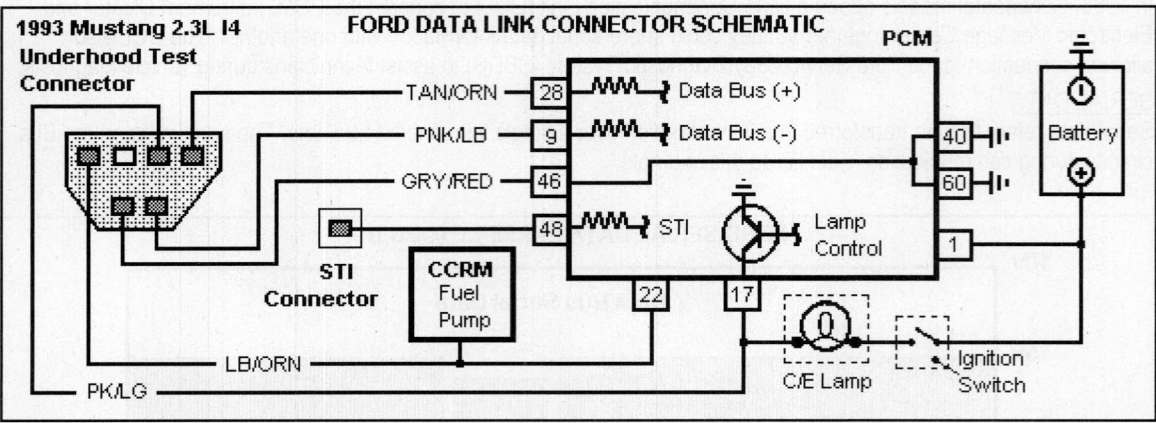

Communication Network Schematic (OBD I)

OBD II Systems - To access data in the PCM on this system, connect an OBD II certified Scan Tool to the DLC and select from the menu. The menu allows you to run the KOEO, KOER and Continuous Memory Self-Tests, read trouble codes and access serial data.

The Scan Tool communicates with the vehicle controllers on separate networks as shown in the Graphic. It communicates with the PCM and I/P Cluster on Pin 2 and 10 (SCP data bus). The ABS Module, Restraint Control Module and GEM communicate with the tool on the ISO 9141 line. If one of the SCP bus circuits fails, the module will continue to communicate. If the ISO line fails, communication with the Scan Tool stops.

Communication Network Schematic (OBD II)

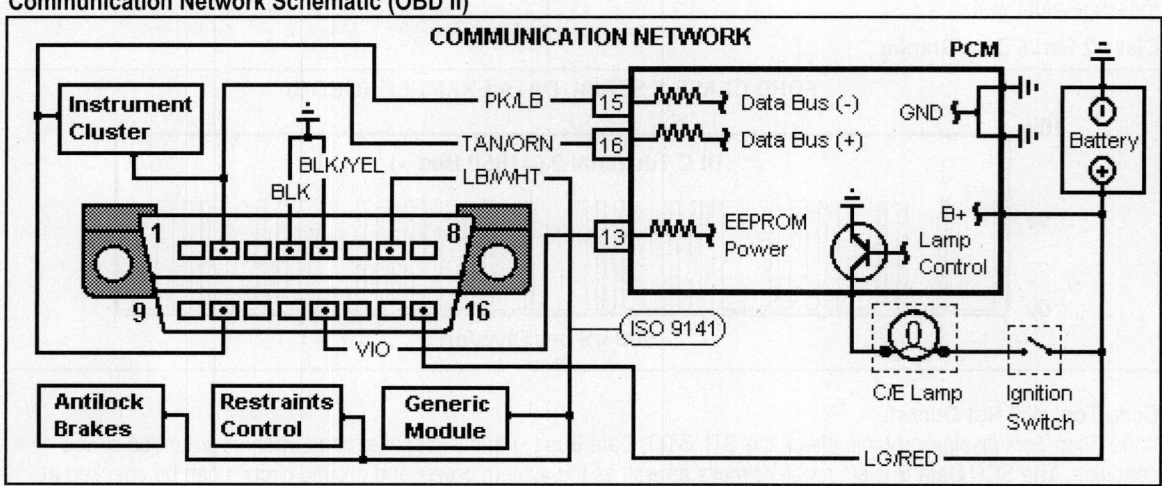

Flash Reprogramming

INTRODUCTION

The Flash Electrically Erasable Programmable Read Only Memory (EEPROM) portion of the integrated circuit (IC) contains the vehicle strategy as well as calibration information for a particular vehicle. The IC is reprogrammable and at times it may be necessary to reprogram or "flash" its contents. This step is usually done as an after production strategy change or if the Vehicle Identification (VID) area has been previously reprogrammed and has reached its limit. The VID is a section of memory inside the IC. It includes the octane adjust, fuel type, tire size and axle ratio items. There are 3 different methods used during an update: PMI, calibration update and programmable parameters.

Module Reprogramming

The act of "flashing" the EEPROM requires the use of the NGS Scan Tool, the Worldwide Diagnostic System (WDS) or the Ford Dealership Service Bay Diagnostic System (SBDS). These are the only tools that can reprogram the PCM. An example of the decal placed under the hood after the PCM is "flashed" is shown here.

Flash Reprogramming

Flashing the EEPROM can be done in Connected or Disconnected mode. To flash the EEPROM in Connected Mode, the vehicle is connected to the SBDS and NGS and then the VIN is entered into the NGS or SBDS to identify the vehicle. Next, the SBDS updates the EEPROM. The NGS can also be used by itself (Disconnected Mode). It must contain the latest software calibrations prior to this step (it retains the calibrations for 3-4 hours).

If the NGS tool is used, the Ford Service Function card is used along with the NGS Flash Cable. Plug the cable into the underdash OBD II connector. From the main menu select "Service Bay Functions", "PCM" and then "Programmable Module Installation". The screen will now display two selections - the first one is for old information to be retrieved and stored in the NGS. The second display is for restoring the new PCM with information that has been retrieved from the old PCM.

Follow the screen instructions to either transfer data from the old module to the new module or to do a manual update to the VID block. If the VID block has been reprogrammed previously, the NGS will display a message. This message will indicate the need to "flash" the entire IC. Follow the Flash EEPROM procedure.

The NGS Flash Kit is used to reprogram the IC contents. The Service Bay Diagnostic System is used to download the new strategy and calibration into the NGS static RAM Card (data transfer card). Plug the NGS into the DLC and follow the PCM Programming steps. The VID block data can also be viewed and changed using this procedure.

Flash Reprogramming Graphic

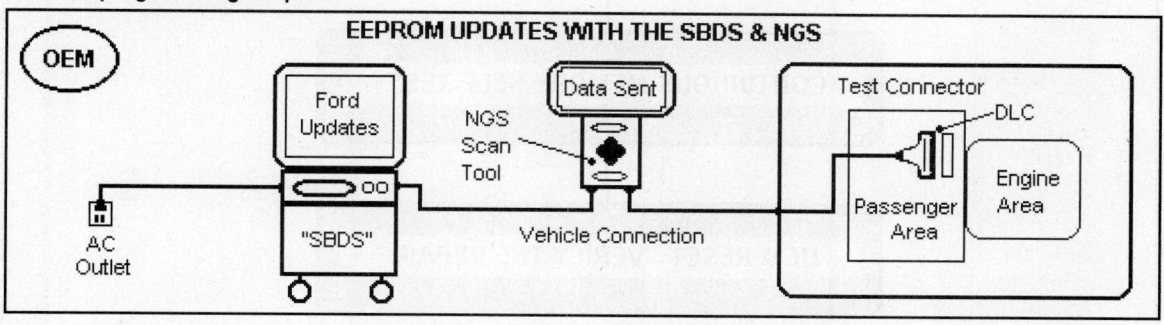

Powertrain Diagnostics

WHERE TO BEGIN

Ford recommends that you start with the Quick Test Procedure to diagnose the vehicle. This test procedure provides you with a plan of action for each specific repair situation, and checks the integrity and function of the Electronic Engine Controls. Test results are output (when requested) to a Scan Tool. The Quick Test provides a quick check of the PCM - it is performed at the start of each diagnosis and at the end any vehicle repairs.

Ford Test Procedure Graphic

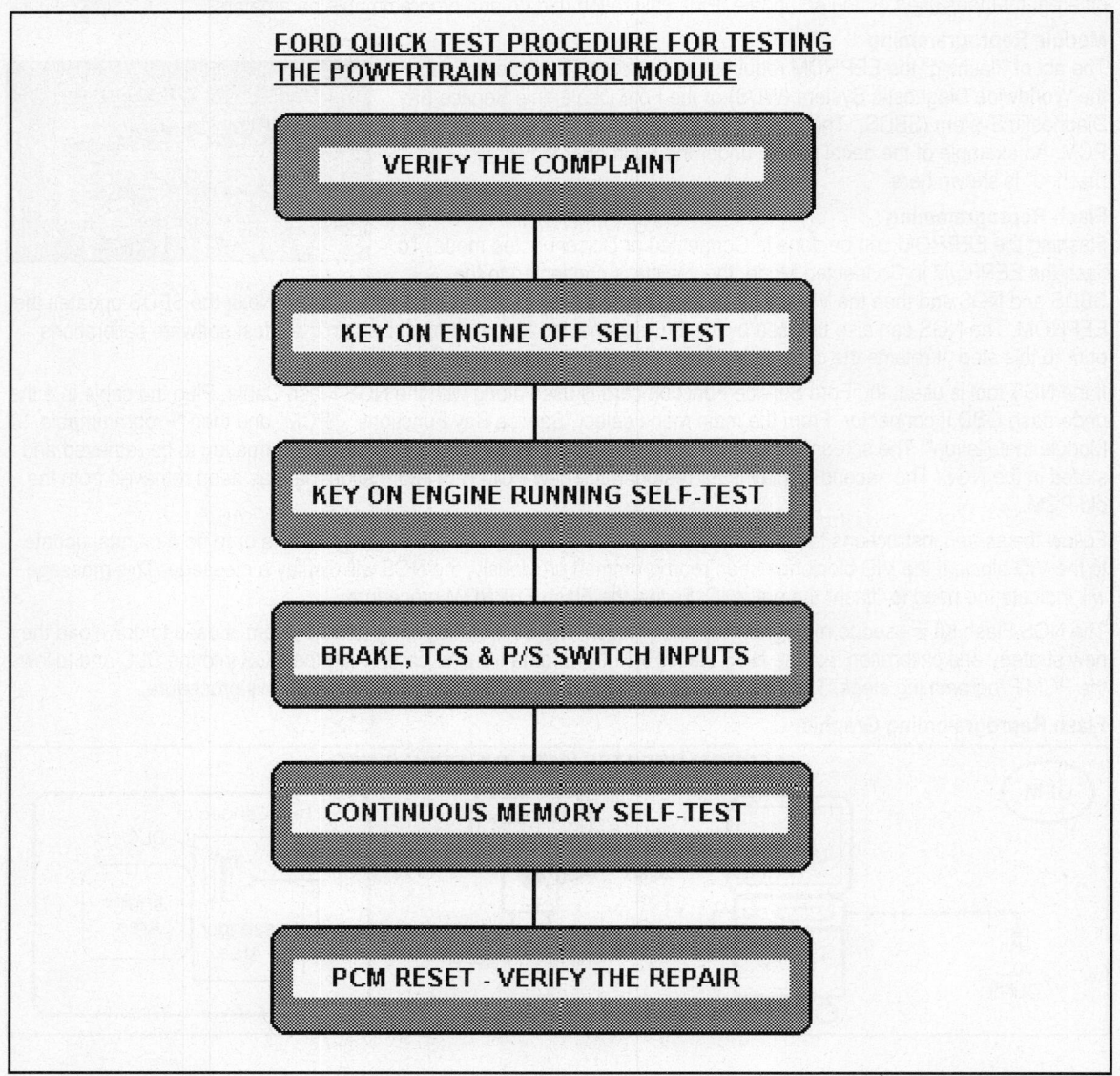

FORD QUICK TEST PROCEDURE FOR TESTING
THE POWERTRAIN CONTROL MODULE

- VERIFY THE COMPLAINT
- KEY ON ENGINE OFF SELF-TEST
- KEY ON ENGINE RUNNING SELF-TEST
- BRAKE, TCS & P/S SWITCH INPUTS
- CONTINUOUS MEMORY SELF-TEST
- PCM RESET - VERIFY THE REPAIR

■ NOTE: If an emission related component has been replaced, the battery disconnected or a PCM Reset is performed during a repair step, it is necessary to drive the vehicle in order to allow the PCM to "relearn" several critical operating parameters.

<u>USING TEST PROCEDURES</u>

As discussed, if a fault is suspected in the PCM or one of its related circuits, start by performing a Quick Test, This step is the same for both OBD I and OBD II systems. On vehicles with more than one type of computer, start by checking for related faults in other vehicle controllers. If you are absolutely certain that the problem is in the PCM or one of its related systems, then start by checking a particular Engine Control system.

During a typical diagnosis, you will need to access certain vehicle components. Whenever you perform this type of testing, make sure to read and understand any of the "special" recommended procedures before accessing and testing a component.

If a wiring problem is found or suspected, look up the electrical information for the vehicle you are testing. This includes wiring schematics with circuit numbers and colors, component connectors, splices and ground locations and wiring repair procedures.

Repair Verification

After diagnosing a problem, in addition to repairing the discovered fault, you should clear any codes that were set and then verify proper operation. To clear codes on 1996-2002 vehicles, use the Scan Tool and do a PCM Reset. To verify the results of a No Code or Symptom repair, try and repeat the conditions present when the condition occurred.

To verify a trouble code is repaired, drive the vehicle under the test conditions for that particular trouble code (called enable criteria for OBD II). If the Quick Test completes without setting any new trouble codes, there is a good chance that the fault has been repaired and is no longer present. In effect, the Quick Test can be used to verify a repair.

Diagnostic Tips:

Step 1 - Try to understand the customer's complaint. Failure to understand the complaint can lead to a misdiagnosis. Among other things, you must know whether the condition is present at all times, only under certain circumstances, or random (intermittent). This information can assist you when you try to duplicate and diagnose the problem. Knowing what type of problem occurs can help you determine whether the complaint requires service or if it is normal vehicle operation. You will only waste time if you try to diagnose a complaint that is really normal vehicle operation!

Step 2 - Do the Vehicle Diagnostics work properly? Use the Quick Test to perform the KOEO, KOER and Continuous Memory Self-Tests. This is the starting point for all PCM diagnostic procedures, so you should always start with a Quick Test!

Step 3 - Are trouble codes displayed? If a code is identified during testing, a trouble code index will direct you to the correct Pinpoint test in Ford information (look in other repair manuals or in electronic media for Pinpoint tests). An explanation of how to use the Quick Test to test for faults and verify a fault is repaired is included in this section.

Step 4 - Is the customer complaint related to a specific PCM subsystem? If no related codes are set, the logical way to locate the problem is to narrow it down to a specific Engine Control system. If a specific subsystem can be pinpointed as the cause of a problem, it is easier to diagnose. Refer to the correct Symptom Diagnosis section in other manual or media to find the Symptom repair charts you need to make the repair.

Step 5 - Is the problem PCM related? Some customer complaints may appear to be PCM related but are actually caused by other vehicle systems. A list of Driveability Symptoms is included in this section. This information can be used to help determine if the problem is in the Base Engine, the PCM or located in a non-Powertrain system.

BASE ENGINE TESTS

To determine that an engine is mechanically sound, certain tests need to be performed to determine that the correct A/F mixture enters the engine, is compressed, ignited, burnt, and then discharged out of the Exhaust system. The tests in this article can be used to help determine the mechanical condition of the engine.

To diagnose an engine-related complaint, compare the results of the Compression, Cylinder Balance, Engine Cylinder Leakage (not included) and Engine Vacuum Tests.

Engine Compression Test

The Engine Compression Test is used to determine if each cylinder is contributing its equal share of power. The compression readings of all the cylinders are recorded and then compared to each other and to the manufacturer's specification (if available).

Cylinders that have low compression readings have lost their ability to seal. It this type of problem exists, the location of the compression leak must be identified. The leak can be in any of these areas: piston, head gasket, spark plugs, exhaust valve or intake valve.

The results of this test can be used to determine the overall condition of the engine and to identify any problem cylinders as well as the most likely cause of the problem.

✸✸CAUTION: **Prior to starting this procedure, set the parking brake, place the gear selector in P/N and block the drive wheels for safety. The battery must be fully charged.**

Test Procedure

- Allow the engine to run until it is fully warmed up.
- Remove the spark plugs and disable the Ignition system. It may be necessary to disable the Fuel system to prevent fuel from entering the cylinders during cranking.
- Carefully block the throttle to the wide-open position.
- Insert the compression gauge into the cylinder and tighten it firmly by hand.
- Use a remote starter switch or ignition key and crank the engine for 3-5 complete engine cycles. If the test is interrupted for any reason, release the gauge pressure and retest. Repeat this test procedure on all cylinders and record the readings.

The lowest cylinder compression reading should not be less than 70% of the highest cylinder compression reading and no cylinder should read less than 100 psi.

Evaluating the Test Results

To determine why an individual cylinder has a low compression reading, insert a small amount of engine oil (three squirts) into the suspect cylinder. Reinstall the compression gauge and retest the cylinder and record the reading. Review the explanations below.

Reading is higher - If the reading is higher at this point, oil inserted into the cylinder helped to seal the piston rings against the cylinder walls. Look for worn piston rings.

Reading did not change - If the reading didn't change, the most likely cause of the low cylinder compression reading is the head gasket or valves.

Low readings on companion cylinders - If low compression readings were recorded from cylinders located next to each other, the most likely cause is a blown head gasket.

A readings that is higher than normal - If the compression readings are higher than normal, excessive carbon may have collected on the piston and in the exhaust areas. One way to remove the carbon is with an approved brand of Top Engine Cleaner.

◼**Note: Always clean spark plug threads and seat with a spark plug thread chaser and seat cleaning tool prior to reinstallation. Use anti-seize compound on Aluminum heads.**

CYLINDER BALANCE TEST

The Cylinder Balance Test is used to test the contribution of each cylinder to the total output of the engine. The results of this test can help identify any potential Engine Mechanical, Fuel and Ignition system problems that are not common to all cylinders. Although the results of the Cylinder Balance Test do not always point to the exact fault area, the results can help to determine if further testing is needed.

✳✳**CAUTION:** **Vehicle manufacturers do not recommend defeating each cylinder by removing or opening a spark plug cable circuit. This type of action can cause a shock to the technician. It could also damage the coil or module. Their recommendation is to do this test using the cylinder defeat function of the engine analyzer.**

During the test, voltage to one or more spark plugs is defeated so that no power is produced in the selected cylinder(s). This test is run in manual or automatic mode. The engine speed (rpm) drop or vacuum change results are recorded for evaluation.

Most automotive engine analyzers are equipped with a button that allows the technician to defeat each cylinder. Depressing the button for Cylinder 1 cancels the firing of cylinder number one in the firing order and so on as each button is depressed.

■ **Note: For best results, perform this test while maintaining as steady an engine speed as possible (IAC & O2S disconnected). Do not use this test on 1994 and later vehicles.**

As each cylinder is defeated, the amount of rpm drop can be observed and compared against the drop that occurs in other cylinders. Cylinders with the least amount of drop are not contributing their share of power. They become the "suspect" cylinders!

Evaluating the Test Results

All cylinders should show an even rpm drop within 50-75 rpm of each other. If the rpm drop between cylinders is more than this amount, run a compression or cylinder leakage test on the "suspect" cylinder to determine the cause of the difference.

ENGINE VACUUM TESTS

An engine vacuum test is used to determine if each cylinder is contributing an equal share of power. Engine vacuum, defined as any pressure lower than atmospheric pressure, is produced in each cylinder during the intake stroke. If each cylinder produces an equal amount of vacuum, the measured vacuum in the intake manifold will be even during engine cranking, idle and at off-idle speeds.

Engine vacuum is measured with a vacuum gauge calibrated to show the difference between the vacuum and the lack of pressure in the intake manifold and atmospheric pressure. Vacuum gauge measurements are usually shown in inches of Mercury (" Hg).

■ **Note: In the tests described in this article, connect the vacuum gauge to an intake manifold vacuum source at a point below the throttle plate on the throttle body.**

The Engine Cranking Vacuum Test can be used to verify that low engine vacuum is not the cause of a No Start, Hard Start, Long Cranking Times, Starts and Dies or Rough Idle symptom.

The vacuum gauge needle fluctuations during the engine cranking test are strong indications of individual cylinder problems. If a cylinder produces less than normal engine vacuum, the needle will respond by fluctuating between a steady high reading (from normal cylinders) and a lower reading (from the faulty cylinder). If more than one cylinder has a low vacuum reading, the needle will fluctuate very rapidly.

ENGINE CRANKING VACUUM TEST

Prior to starting this test, set the parking brake, place gear selector in P/N and block drive wheels for safety. Then block the PCV valve and disable the idle air control device.

1) If possible, allow the engine to fully warmup. Disable the Fuel and/or Ignition system to prevent the vehicle from starting and injecting fuel into the cylinders during cranking. Close the throttle plate and connect a vacuum gauge to an intake manifold vacuum source. Verify that the hose connection point is not restricted with carbon.

2) Crank the engine for at least three seconds. Repeat the test at least once. Test results will vary due to engine design characteristics, the type of PCV valve and the use of different idle speed control devices that allow air to bypass the throttle plate.

Evaluating the Test Results

Steady vacuum reading between 1.0" - 4.0" Hg - This is a normal reading and indicates the engine mechanical condition is okay during engine cranking mode.

Steady vacuum reading below 1.0" Hg - This reading indicates that there is low compression on all cylinders, that there are air leaks in devices connected to the intake manifold (manifold vacuum), wrong valve timing or a severely restricted exhaust.

Uneven pulsating vacuum reading between 1.0" - 4.0" Hg - This reading indicates possible faults in the head gasket, pistons, rings, valves or valve train.

ENGINE RUNNING VACUUM TEST

This test can be used to verify that the engine is not the cause of a Hard Start, Starts and Dies, Rough Idle, Loss of Power and Cuts-Out condition at idle or cruise speeds.

Test Procedure

1) Allow the engine to run until fully warmed up. Connect a vacuum gauge to a clean intake manifold source. Connect a tachometer or Scan Tool to read engine speed.

2) Start the engine and let the idle speed stabilize. Note the vacuum reading at idle. Raise the engine speed rapidly to just over 2000-rpm and again note the vacuum reading. Repeat the test three times. Compare the idle and vehicle cruise readings.

Evaluating the Test Results

If the engine wear is even, the gauge reading will be steady, but well below the normal expected reading of at least 16" Hg. The test results can vary due to various engine characteristics and the altitude above or below sea level.

At idle speed, a typical hot idle vacuum gauge reading should be at least 16" Hg and steady. The vacuum gauge reading at cruise should exceed the idle reading. If it does not, there is a problem somewhere in the Engine Mechanical, Fuel or Ignition systems.

Engine Running Vacuum Test Graphic

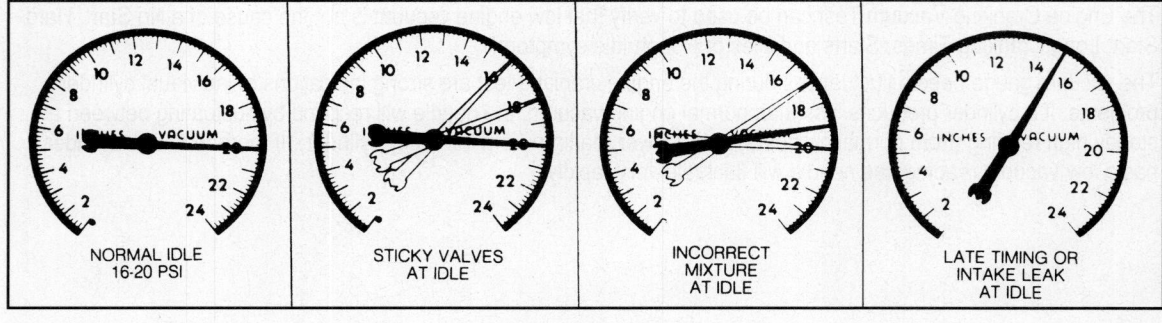

IGNITION SYSTEM TESTS – DISTRIBUTOR

This article provides an overview of ignition tests with examples of Engine Analyzer patterns for the Distributor Ignition system used on various Ford vehicles.

Preliminary Inspection - Perform these checks prior to connecting the Engine Analyzer:
- Check the battery condition (verify that it can sustain a cranking voltage of 9.6v).
- Inspect the ignition coil for signs of damage or carbon tracking at the coil tower.
- Remove the coil wire and check for signs of corrosion on the wire or tower.
- Test the coil wire resistance with a DVOM (it should be less than 7 Kohms per foot).

Connect a high output spark tester to the coil wire and engine ground. Verify that the ignition coil can sustain adequate spark output while cranking for 3-6 seconds.

Ignition System Scope Patterns

Connect the Engine Analyzer to the Ignition system. Turn the scope selector to view the Parade Display of the ignition secondary. Start the engine in P/N and slowly increase the engine speed from idle to 2000 rpm. Compare actual display to the examples below.

Ignition System Scope Patterns (4-Cylinder Engine)

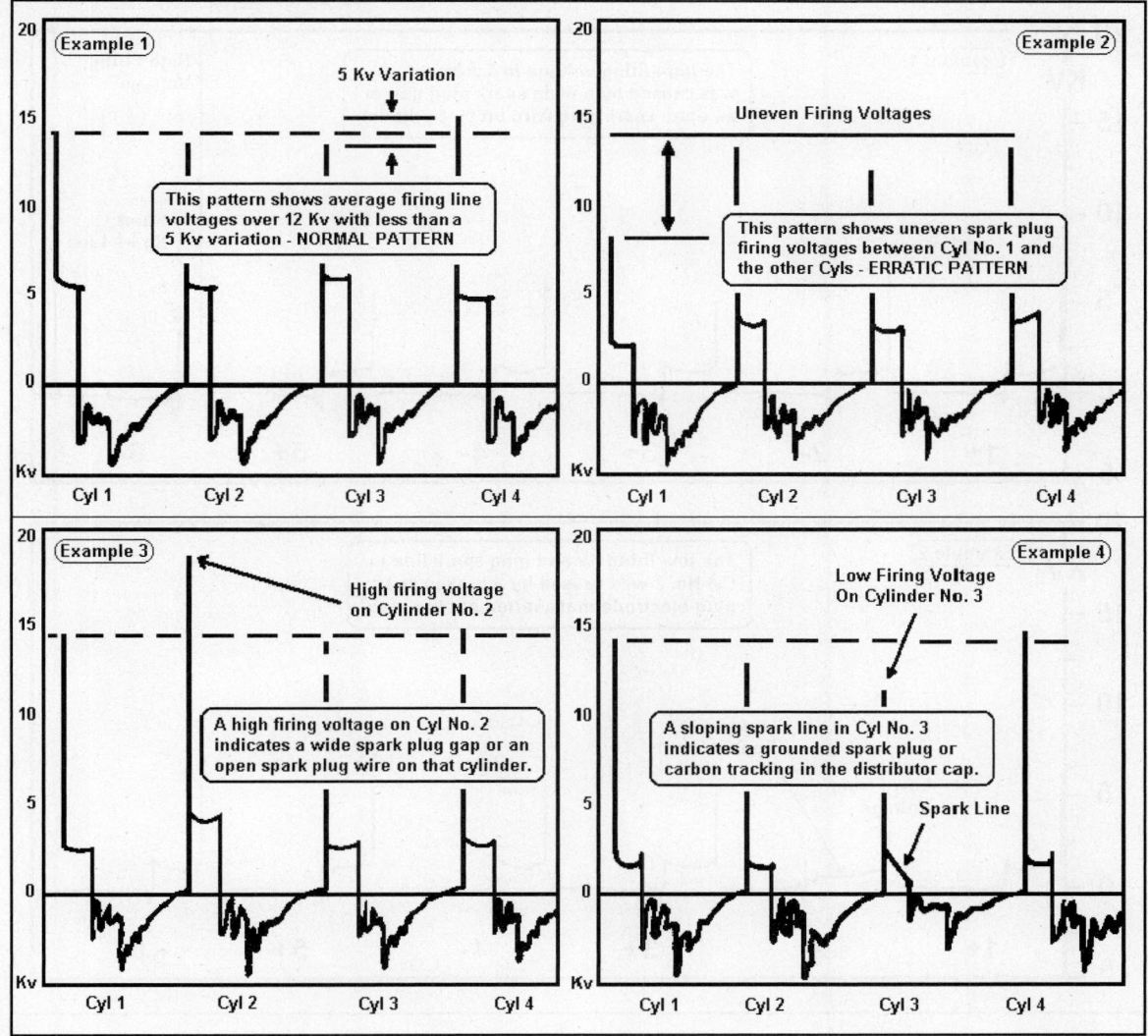

IGNITION SYSTEM TESTS - DISTRIBUTORLESS

This article provides an overview of ignition tests with examples of Engine Analyzer patterns for the Electronic Ignition (EI) system used on various Ford vehicles.

Preliminary Inspection - Perform these checks prior to connecting the Engine Analyzer:

*   Check the battery condition (verify that it can sustain a cranking voltage of 9.6v).
*   Inspect the ignition coils for signs of damage or carbon tracking at the coil towers.
*   Remove the ignition coil wires and check for signs of corrosion on the wires or tower.
*   Test the plug wire resistance with a DVOM (specification varies from 15-30 Kohms).

Connect a high output spark tester to a plug wire and to engine ground. Verify that the ignition coil can sustain adequate spark output for 5-10 seconds.

Secondary Ignition System Scope Patterns (V6 Engine)

Connect the Engine Analyzer to the Ignition system. Turn the scope selector to view the Parade Display of the ignition secondary. Start the engine in P/N and slowly increase the engine speed from idle to 2000 rpm. Compare actual display to the examples below.

Secondary Ignition System Graphic (V6 Engine)

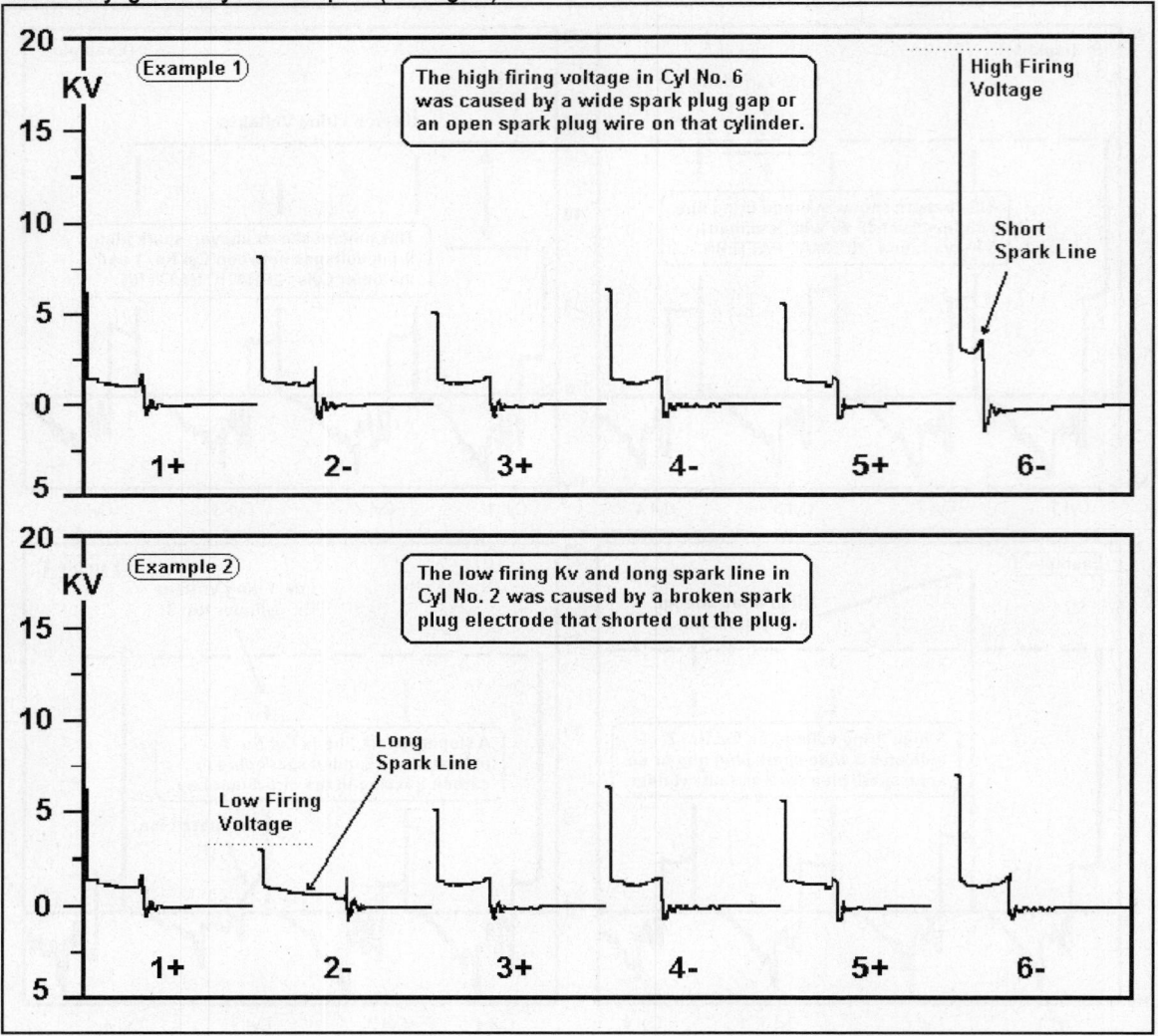

SYMPTOM DIAGNOSIS

Do not attempt to diagnose a Driveability Symptoms without having a plan of attack to use to determine which Engine Control system is the cause of the symptom - this plan should include a way to determine which systems do not have a problem! Driveability symptom diagnosis is a part of an organized approach to problem solving and repair.

Driveability Symptom List

To use this list, locate the symptom that matches the problem and refer to the areas to test. The items listed under each symptom may not apply to all models, engines or vehicle systems. The repair steps indicate what vehicle component or system to test.

Some of the symptoms in the table below are linked to examples on the next few pages.

Symptom Description	Table Number in the Manual
Starting Problems	
• No Crank	1
• Hard Start, Long Crank, Erratic Crank or Start	1
• Stalls or Quits Running	7
• No Start, Normal Crank	1
Unique Idle Problems	---
• Slow Return to Idle	---
• Rolling Idle	7
• Fast Idle	---
• Low or Slow Idle	---
Stalls, Quits Running	---
• At idle	7
• During acceleration	7
• At cruise	7
• At deceleration	---
Runs Rough	---
• At idle	7
• During acceleration	7
• At cruise	7
Cuts-out, Misses	---
• At idle	7
• During acceleration	7
• At cruise	7
Bucks, Jerks	---
• During acceleration, Cruise speed or deceleration	7
Hesitation, Stumble	---
• During acceleration	7
Surge	---
• During acceleration	7
• At cruise	7
Backfires	---
• At idle	8
• During acceleration or deceleration	8
Lack of Power, Loss of Power	---
• During acceleration	9
• At cruise	9
Spark Knock	---
• During acceleration or at Cruise speed	10
DTC P1000 set in Emission Test (I/M Readiness Tests not done)	---
Poor Fuel Economy	---
Emissions Compliance	---

DRIVEABILITY SYMPTOMS

This repair table is for 1996-2002 vehicles. It should be used to test for the cause of a No Start condition (including checking the operation of the Powertrain Control Module).

⁂ **CAUTION:**   **Stop this test sequence at the first sign of fuel leaking and service as required.**

Repair Table 1: No Start

Step	Action	Yes	No
1	Step description: Check Anti-Theft System The vehicle may be equipped with a Passive Anti-Theft system that may be activated and cause the No Start condition. This can be determined by viewing the Anti-Theft System Indicator Light on the instrument panel or by using the Scan Tool to check for DTC P1260. Verify the Anti-Theft system. Is it activated?	Refer to Electrical Anti-Theft information in the workshop manual or electronic media for instructions on how to diagnose and test this system.	Go to Step 2.
2	Step description: Attempt to crank the engine Verify that the inertia fuel shutoff (IFS) switch is set (push the button in). Refer to the Owner's Guide for location information. Does the engine crank?	Go to Step 3.	Refer to Starting system information in other repair manuals or electronic media as needed.
3	Step description: Identify the type of No Start The purpose of this test step is to identify an intermittent No Start in order to guide the technician to the proper repair procedure. Does the vehicle start now?	The No Start problem is intermittent. Refer to the Intermittent Test Index in this or other repair manuals.	Natural Gas Vehicles: Turn the key off. Go to Step HA47 in other repair information. All Others - go to Step 4.
4	Step description: Check VREF to TP Sensor Disconnect the TP sensor. Turn the key on. Measure voltage the between VREF and SIG RTN circuits at TP sensor harness connector. Is the voltage 4.9-5.1v?	Reconnect TP sensor. Go to Step 5. 	Find and repair the cause of loss of VREF to TP sensor. Check the PCM power and ground connections.
5	Step description: Check FEPS for short Turn the key on. Measure the voltage between Pin 13 at the DLC and chassis ground or battery negative. Is the voltage more than 9v?	Repair the short to power in the Flash EEPROM supply circuit. Retest for the No Start condition.	For COP engines - go to Step 17. For All Other engines - go to Step 6.
6	Step description: Check PIP Signal to PCM A known good Scan Tool and fully charged battery is required for this next test step. Turn the key to on and Access the PIP PID. Crank the engine and watch for changes in the PIP PID. Does it change from On to Off?	Go to pinpoint test JB1 to check ignition coils, wires and spark plugs. If okay, for natural gas vehicles, go to 15. All Others - go to Step 7.	If the PIP PID does not switch from On to Off during cranking, refer to pinpoint test JD1 in other repair manuals or electronic media.
7	Step description: Check fuel pressure Release the fuel pressure following all related safety, warning and handling directions. Install fuel pressure tester and the Scan Tool. Turn to key on and engine off. Access Output Test Mode and run fuel pump to obtain maximum fuel pressure. Is the pressure from 35-40 psi?	Go to Step 8.	Turn the key off. Go to pinpoint test HC in other repair manuals or electronic media for instructions on how to test the Fuel Delivery system.

Repair Table 1: No Start (Continued)

Step	Action	Yes	No
8	Step description: Check pressure leak down Follow all related safety, warning and handling directions. Fuel pressure tester installed. Turn to key on and engine off. Access Output Test Mode and run fuel pump to obtain maximum fuel pressure. Then exit Output Test Mode. Does pressure remain with 5 psi for 1 minute?	Go to Step 9.	Turn the key off. Go to pinpoint test HC in other repair manuals or electronic media for instructions on how to test the Fuel Delivery system.
9	Step description: Check power to injectors Check at least two fuel injectors (one in each bank). Disconnect two fuel injectors. Turn the key on and measure VPWR circuit voltage at each of the injector harness connectors. Is the voltage more than 10.5v?	Go to Step 10.	Repair the open in the VPWR circuit. Then retest the No Start condition.
10	Step description: Check Injector triggering Connect a fuel pressure gauge to Schrader valve. Cycle the key on to off several times. Monitor the gauge while cranking the engine for at least five seconds. Does the pressure drop by over 5 psi during engine cranking?	The PCM and its related systems are not the cause of the No Start problem. Check other Base Engine related items.	Replace the PCM and "flash" the EEPROM. Retest the No Start condition.
15	Step description: Check fuel pressure (NGV) Connect the Scan Tool. Access the FRP PID and record the fuel pressure. Connect fuel pressure gauge to Schrader valve. Turn to key on and record the pressure. Is fuel pressure at 80-120 psi on the tool and pressure gauge?	Go to Step 16.	Turn the key off. Go to pinpoint test HB1 for NG vehicles to check the Fuel Delivery system.
16	Step description: Check Injector Signal This test requires a known good 12v-test light. A properly operating system will show a dim test light bulb. Remove a fuel injector connector. Connect test light between the injector signal and the VPWR pins at the injector harness connector. Crank the engine. Does the test lamp have a dim glow during cranking?	Replace the PCM (NGV Module) and "flash" the EEPROM. Retest the No Start condition.	If there is no light or a continuously bright light, go to HA47 to test the Fuel Delivery system on Natural Gas Vehicles.
17	Step description: Check PCM driver coils Connect an incandescent test lamp between B+ and each ignition coil driver circuit at the harness connector. Crank the engine. Does the test light blink continuously and brightly (there should be one blink per revolution)?	Turn the key off. Go to Step 7.	Turn the key off. Go to pinpoint test JD1 in other repair manuals or electronic media.

Prior to starting this symptom repair table, inspect these underhood items:

- All related vacuum lines for proper routing and integrity
- All related electrical connectors and wiring harnesses for faults
- Air Intake system for restrictions (air inlet tubes, dirty air filter, etc.)
- Intake manifold and components for leaks (EGR valve, IAC assembly)
- Ignition secondary components (coil, coil wire, spark plugs and wires)
- Check Fuel quality (contaminated fuel, wrong octane level or blend)
- Also, the PCM has an Overspeed function to protect against high rpm

Repair Table 7: Stalls or Quits Running

Step	Action	Yes	No
1	Step description: Verify the engine stalls • Start the engine and allow it to idle in Park or Neutral. • Does the engine stall or almost stall at idle speed in Park or Neutral?	Perform the KOEO Self-Test to output codes. Any codes present? If yes, repair the code. If no codes exist, go to Step 4.	If the engine does not stall or almost stall, go to Step 2.
2	Step description: Check for rough idle • Does the engine have a rough idle condition in Park or Neutral?	If the engine does have a rough idle condition, go to Step 4.	Do the KOEO Self-Test (record codes). If codes are set, go to DTC Index. If no codes are set, go to Step 3.
3	Step description: Compare PID values • Connect a Scan Tool. • Turn off all of the accessories. • Start engine and allow it to warm to normal operating temperature. • Use Scan Tool to access each PID (verify all PIDs are in the normal range). • Are all of the PIDs in their normal range?	If all of the PIDs are okay), go to Step 6.	If one or more PIDs are out of range, go to the Pinpoint Test, Symptom Chart or repair manual section to repair the component related to that PID.
4	Step description: Check for stall or near stall with the IAC solenoid disconnected • Start engine at part throttle in P/N. • Does the engine start and run smoothly at part throttle in Park or Neutral?	Refer to Pinpoint Test KE2 (in this or other manuals) to check the operation of the IAC system.	Go to Step 5.
5	Step description: Check IAC Solenoid • Start the engine and let it idle. • Disconnect the IAC Solenoid connector. • Check for rpm drop or engine stall. • After testing, turn engine off and reconnect the IAC solenoid connector. • Did the rpm drop or engine stall with IAC connector off?	Return to Step 3 to read and record the appropriate PID values. Refer to the Reference Information in this section for examples of PID Charts.	Refer to Pinpoint Test KE1 in this or other manuals for instructions on how to test the complete IAC system.
6	Step description: Check Ignition secondary • Inspect the Ignition secondary for damaged or defective components. • Check spark output with spark tester, connect engine analyzer to check ignition secondary (refer to Pinpoint Test JB1). • Are any Ignition system faults suspected?	Go to Pinpoint Test JB1 in other manuals to test the Ignition system. Make repairs as needed. Then do a PCM Reset and Retest the system.	Go to Step 7.

Repair Table 7: Stalls or Quits Running (Continued)

Step	Action	Yes	No
7	Step description: Check Fuel system • Inspect the Fuel system for leaks. • For NG Crown Victoria - Go to Test HB1 • For Probe - Go to Pinpoint Test HC • All Others - Go to Pinpoint Test HC1 • Are any Fuel system faults suspected?	Go to correct Pinpoint Test in other manuals (select from list in column 2). Make needed repairs, then Reset PCM and retest.	Go to Step 8.
8	Step description: Check Exhaust system • Inspect for leaking or damaged components. • Test the Exhaust system for leaks, damage or a restriction (Go to Pinpoint Test HF1). • Is an exhaust restriction suspected?	Go to Pinpoint Test HF1 in this or other manuals. Locate the exhaust restriction and make any needed repairs. Then do a PCM Reset and retest.	Go to Step 9.
9	Step description: Check the PCV system • Inspect the PCV system components for any loose, disconnected or broken parts. • Test the PCV valve operation (Go to HG1). • Are any PCV system faults suspected?	Go to Pinpoint Test HG1. Locate the PCV system fault and make needed repairs. Then do a PCM Reset and Retest the system.	Go to Step 10.
10	Step description: Check Secondary AIR system • Inspect Secondary AIR system components for any loose, disconnected or broken parts. • Test Secondary AIR system (Go to HM80). • Are any Secondary Air faults suspected?	Go to Pinpoint Test HM80 and make any needed repairs. Then do a PCM Reset and Retest the system.	Go to Step 11.
11	Step description: Check EVAP system • Inspect EVAP system components for any loose, disconnected or broken parts. • For Taurus/Sable, Crown Victoria/Grand Marquis, Town Car, F-Series 4.2L and Explorer 4.0L, 4.6L - Go to Pinpoint Test STEP HX40. • For Probe - Go to Pinpoint Test STEP HV1. • All Others - Go to Pinpoint Test STEP HW10. • Are any EVAP system faults suspected?	Go to the correct Pinpoint Test. Make any needed repairs. Then do a PCM Reset and Retest the system.	Test the Base Engine items and Air Intake system (Go to Test HU1). Test transaxle or transmission, Traction Control (Continental), and A/C Pressure sensor with Pinpoint Test DS22.

Prior to starting this symptom repair table, inspect these underhood items:

- All related vacuum lines for proper routing and integrity
- All related electrical connectors and wiring harnesses for faults
- Ignition secondary components (coil, coil wire, spark plugs and wires)
- Spark plug wires for proper routing and for correct firing order
- Verify that engine backfires at idle or during acceleration or deceleration

Repair Table 8: Backfires

Step	Action	Yes	No
1	Step description: Perform Quick Check • Perform a complete visual inspection and then make any needed repairs. • Connect the Scan Tool to DLC. • Read and record all trouble codes. • Were any codes recorded?	Go to the correct Pinpoint Test in other repair manuals to make the repair.	Go to Step 2.
2	Step description: Check Ignition secondary • Inspect the ignition secondary for damaged or defective components. • Check the spark output with spark tester, connect engine analyzer to check ignition secondary (Pinpoint Test JB1). • Are Ignition system faults suspected?	Go to Pinpoint Test JB1 in other repair manuals to test the Ignition system. Make any needed repairs. Then do a PCM Reset and Retest the system.	Go to Step 3.
3	Step description: Check Fuel system • Inspect the Fuel system for leaks. • For NG Crown Victoria - Go to Test HB1. • For Probe - Go to Pinpoint Test HC. • All Others - Go to Pinpoint Test HC1. • Are any Fuel system faults suspected?	Go to the correct Pinpoint Test in other repair manuals (select from list in column 2). Make the needed repairs. Then do a PCM Reset and Retest the system.	Go to Step 4
4	Step description: Inspect Base Engine • Test the engine compression. • Check the valve timing and timing chain. • Check for worn camshaft or valve train. • Check for any manifold gasket leaks. • Check for any intake vacuum leaks. • Are any Base Engine faults present?	Make repairs as needed. Then do a PCM Reset and Retest the system.	Go to Step 5.
5	Step description: Check Exhaust system • Inspect for any leaking or damaged Exhaust system components. • Check the Exhaust system for leaks, damage or a restriction (Go to Pinpoint Test HF1). • Is an exhaust restriction suspected?	Locate the cause of the exhaust restriction. Make the needed repairs. Then do a PCM Reset and Retest the system.	Confirm the test results from the steps in this chart. If okay, go to Pinpoint Test Z1 in this or other repair manuals or return to Driveability Symptom List to identify and test other related symptoms.

Prior to starting this symptom repair table, inspect these underhood items:

- All related vacuum lines for proper routing and integrity
- All related electrical connectors and wiring harnesses for faults
- Verify that engine has a lack of power or loss of power condition during acceleration or at cruise speeds

Repair Table 9: Lack of Power

Step	Action	Yes	No
1	Step description: Check these items • Inspect for a severely restricted air filter. • Inspect Air Intake system for restriction. • Inspect for binding throttle linkage. • Inspect for plugged/restricted radiator. • Check the transmission fluid level. • Check the Overspeed Limiting Function. • Were any problems found?	Make the needed repairs. Then do a PCM Reset and Retest the system.	Go to Step 2.
2	Step description: Perform Quick Test • Perform a complete visual inspection and make any needed repairs. • Connect Scan Tool to DLC. • Read and record all trouble codes. • Were any codes recorded?	Go to correct Pinpoint Test in other repair manuals and make the repairs. Then do a PCM Reset and Retest the system.	Go to Step 3
3	Step description: Check PID values • Connect a Scan Tool and turn off all accessories. • Start the engine and allow it to warm to normal operating temperature. • Access each PID and verify that all of the PIDs are in their normal range. • Are all PIDs in their normal range?	If all of the PIDs are in normal range, pick from the list below: F-Series Truck (4.6L) - Go to Step 4. Probe - Go to Step 5. All Others - go to Step 6.	If one or more of the PIDs are out of normal range, Go to the Pinpoint Test or Symptom Chart in other repair manuals to make the needed repairs.
4	Step description: Check Intake Manifold Tuning Valve (F-Series Trucks with 4.6L) • Connect a Scan Tool to DLC. • Access the IMTVF PID value and then record the IMTVF PID at KOEO and in P/N with the engine speed at over 2800 rpm. • Does the IMTVF PID read "YES" (meaning on) in KOEO or at 2800 rpm?	A fault is present in the IMT system. Go to Pinpoint Test HU65 in other repair manuals to make the repair. Then do a PCM Reset and Retest the system.	Go to Step 6.
5	Step description: Check Fuel Pressure Regulator Control solenoid (Probe 2.0L only) • Inspect and check the Fuel Pressure Regulator Control solenoid for damage (Go to Pinpoint Test KN1). • Is the FPR solenoid faulty or damaged?	Repair or replace the damaged Fuel Pressure Regulator Control solenoid. Then do a PCM Reset and Retest the system.	Go to Step 6
6	Step description: Check Fuel system • Inspect the Fuel system for leaks. • For NG Crown Victoria, go to Test HB1. • For Probe - Go to Pinpoint Test HC. • All Others - Go to Pinpoint Test HC1. • Are any Fuel system faults suspected?	Go to Pinpoint Test in other repair manuals. Make repairs as needed. Then do a PCM Reset and Retest the system.	Go to Step 7

Repair Table 9: Lack of Power (Continued)

Step	Action	Yes	No
7	Step description: Check Ignition secondary • Inspect the ignition secondary components. • Check spark output with spark tester and test ignition secondary with analyzer (Test JB1). • Are any Ignition system faults suspected?	Go to Test JB1 in other manuals to test the Ignition system. Make repairs as needed. Do a PCM Reset & retest.	Go to Step 8.
8	Step description: Check Exhaust system • Inspect for leaking or damaged components. • Test the Exhaust system for leaks, damage or a restriction (Go to Pinpoint Test HF1). • Is an exhaust restriction suspected?	Go to Pinpoint Test HF1 in this or other repair manuals. Repair restriction. Do a PCM Reset and retest.	For Continental, go to Step 9. For All Others, Go to Step 10.
9	Step description: Check operation of the Series Throttle Assembly (Continental Only) • Turn the key off and then remove the air tube to the Series Throttle Assembly. • Inspect the throttle body for binding/sticking, loose stepper motor or TP Sensor 'B' faults. • Check the Series Throttle Plate for sticking (it should be wide open). Push it closed and release it a few times (it should spring open). • Turn to key on engine off and watch for T/C initialization as the plate is closed briefly (the PCM reads the TP Sensor 'B' voltage). • Did the Series Throttle pass all of its checks?	If the symptom was "No Traction" or for "Poor/Erratic Traction", go to Pinpoint Test HT23 in other repair manuals. If the symptom was Lack of Power or Loss of Power, return to the next step (10) in this repair table and continue testing.	Go to Pinpoint Test HT21 in other repair manuals (or the Ford Continental repair manual) and isolate the fault.
10	Step description: Test Base Engine • Test the base engine items (valve train, etc.). • Check for intake manifold vacuum leaks. • Are any Base Engine faults present?	Make repairs as needed. Then do a PCM Reset and Retest the system.	If equipped with an A/T, go to Step 11. If equipped with a M/T, go to Step 12.
11	Step description: Check A/T operation • Inspect for low fluid levels or burnt fluid. • Refer to Transmission Repair information. • Are any A/T faults present?	Make repairs as needed. Then do a PCM Reset and Retest the system.	Go to Step 12.
12	Step description: Check Brake System • Inspect for binding or dragging brakes. • Refer to the Brake Repair information. • Are any Brake system faults present?	Make repairs as needed. Then do a PCM Reset and Retest the system.	Go to Step 13
13	Step description: Check for no A/C Cut Out under WOT (with WAC Relay or CCRM) • Does the A/C Cut Out function operate (does it turn off) under WOT conditions?	For Mustang 4.6L 4V - Go to Step 14. For All Others - Go to Step 15.	Go to A/C Repair Chart in other manuals. Then do a PCM Reset and Retest the system.
14	Step description: Check high speed fuel pump • Inspect and check the operation of the High Speed Fuel Pump (Go to Pinpoint Test X210) • Is a fault indicated in the fuel pump?	Go to Pinpoint Test X210 in other manuals. Make the repair. Do a PCM Reset & retest.	Go to Step 15.
15	Step description: Other inspections and tests • Check for excessive loads/towing conditions. • Check operation of IMRC linkage. • Check for faulty or slipping clutch (M/T). • Check Optional Traction Control system operation on Contour/Mystique models. • Were any faults located in these steps?	Make repairs as needed. Then do a PCM Reset and Retest the system.	Confirm the test results from previous steps. If okay, go to Intermittent Tests or Driveability Symptom List to test for other symptoms.

Prior to starting this symptom repair table, inspect these underhood items:

- All related vacuum lines for proper routing and integrity
- Check the fuel quality and grade
- All related electrical connectors and wiring harnesses for faults
- Check for signs of an overheating condition
- Inspect the Air Intake system for restrictions (e.g., very dirty air filter element)
- Verify that engine has a spark knock condition during acceleration or cruise speeds

Repair Table 10: Spark Knock

Step	Action	Yes	No
1	Step description: Perform an inspection • Inspect Air Intake system for restriction. • Inspect for signs of engine overheating. • Inspect for a plugged or restricted radiator. • Check for fuel contamination and for the proper grade of fuel (octane level). • Were any problems found?	Make the needed repairs. Then do a PCM Reset and Retest the system.	Go to Step 2.
2	Step description: Check for overheating • Access the ECT PID with the Scan Tool. • Read and record the value for the ECT PID. • Does the engine appear to be overheating? Note: With the engine running and temperature stabilized, the ECT PID should read >0.30v indicating a temperature of less than 240°F.	Go to cooling system repair information in other manuals and make the repairs. Then do a PCM Reset and Retest the system.	Go to Step 3
3	Step description: Perform Quick Test • Perform a complete visual inspection and make any needed repairs. • Connect Scan Tool to DLC. • Read and record all trouble codes. • Were any codes recorded?	Go to correct Pinpoint Test in other repair manuals and make the repairs. Then do a PCM Reset and Retest the system.	Go to Step 4
4	Step description: Test Base Engine • Test the engine compression. • Test valve timing, timing chain. • Check for worn camshaft or valve train. • Check for manifold gasket leaks. • Check for intake vacuum leaks. • Are any Base Engine faults present?	Make repairs as needed. Then do a PCM Reset and Retest the system.	Go to Step 5.
5	Step description: Check Fuel system • Inspect the Fuel system for leaks. • For NG Crown Victoria, go to Test HB1. • For Probe - Go to Pinpoint Test HC. • All Others - Go to Pinpoint Test HC1. • Are any Fuel system faults suspected?	Go to correct Pinpoint Test in other repair manuals (select from list in column 2). Make repairs as needed. Then do a PCM Reset and Retest the system.	Go to Step 6.
6	Step description: Check Ignition secondary • Inspect the ignition secondary for damaged or defective components. • Check spark output with spark tester, connect engine analyzer to check ignition secondary (Pinpoint Test JB1). • Are any Ignition system faults suspected?	Go to Pinpoint Test JB1 in other manuals to test the Ignition system. Make repairs as needed. Then do a PCM Reset and Retest the system.	Go to Step 7.

Repair Table 10: Spark Knock (Continued)

Step	Action	Yes	No
7	Step description: Check PCV system • Inspect the PCV system for damaged, leaking or defective components. • Are any PCV system faults suspected?	Make repairs as needed. Then do a PCM Reset and Retest the system.	Go to Step 8.
8	Step description: Check Octane Adjust input • For applications with an Octane Adjust circuit (PCM Pin 30), follow this procedure. • For all other models, Go to Step 10. • Start the engine and allow it to warm to normal operating temperature. • Turn the key off and connect the Scan Tool. • Enter the KOEO Self-Test. Read and record the trouble codes. • Is DTC P1390 present?	Refer to the P1390 repair chart in other manuals and make repairs as needed. Then do a PCM Reset and Retest the system.	Go to Step 9.
9	Step description: Shorting Bar Installed? • Visually inspect the Octane Adjust in-line connector. • Is the shorting bar installed in the connector?	If it is installed, check for a TSB that okays its removal.	If it is not installed, check for modification decal that authorizes removal of the bar. If the decal is not there, replace the shorting bar and continue.
10	Step description: Check Knock Sensor • Verify that the vehicle has a Knock sensor. • Inspect the Knock sensor for a loose connection and for the proper torque. • Check the operation of the Knock sensor (refer to Pinpoint Test DG1 as needed). • Were any faults located?	Make repairs as needed. To run a complete check of the Knock Sensor system, refer to Pinpoint Test DG1 in other repair manuals. Then do a PCM Reset and Retest the system.	Go to Step 11.
11	Step description: Additional Checks • Check the ignition base timing (if not already checked). • Inspect the Air Intake system for restrictions. • Check the engine oil quality and level. • Check for an Octane Adjust related TSB. • Were any faults located?	Make repairs as needed. Then do a PCM Reset and Retest the system.	Confirm the test results from previous steps in this chart. If okay, go to Pinpoint Test Z1 in this or other repair manuals or return to Driveability Symptom List to identify and test other related symptoms.

INTERMITTENT TEST PROCEDURES

One way to find an intermittent problem is to gather the information that was present when the problem occurred. In the case of a Code Fault, this can be done in two ways: by capturing the data in Snapshot or Movie mode or by driver observations.

The PCM has to see the fault for a specific period of time before a fault is detected and a code set. While intermittent problems may appear to be occasional in nature, they usually occur under specific conditions. Therefore, technicians must identify and duplicate these conditions. Since intermittent faults are difficult to duplicate, a logical and systematic routine (checklist) must be followed when attempting to find the faulty component, system or circuit. The list on the next page can be used to help find the cause of an intermittent fault.

Some intermittent faults occur due to a loose connection, wiring problem or warped circuit board. Also, incorrect test methods that cause damage to the male or female ends of a connector and cause an intermittent problem to occur.

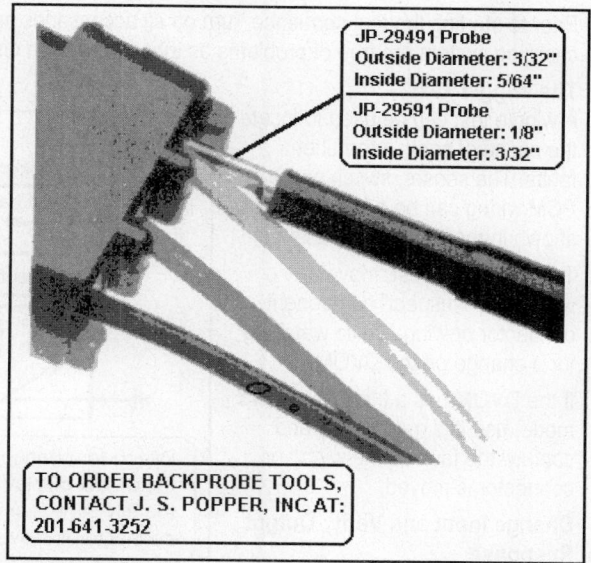

JP-29491 Probe
Outside Diameter: 3/32"
Inside Diameter: 5/64"

JP-29591 Probe
Outside Diameter: 1/8"
Inside Diameter: 3/32"

TO ORDER BACKPROBE TOOLS, CONTACT J. S. POPPER, INC AT: 201-641-3252

Fault Not Present at this Time!

Many code repair charts end with a result that reads "Fault Not Present at this Time." This expression means that the conditions that were present when a code set or driveability symptom occurred are not happening now or that the conditions required that caused the intermittent fault were not met. In effect, the problem was present at least once, but is not present at this time. However, the fault is likely to return in the future, so it should be diagnosed and repaired if at all possible.

One way to find an intermittent problem is to gather the information that was present when the problem occurred. In the case of a No Code Fault, this can be done in two ways: by capturing the data in Snapshot or Movie mode or by driver observations.

Test for Loose Connections

To test for a loose or damaged connection, take the male end of a connector from another wiring harness and carefully push it into the "suspect" female terminal to verify that the opening is tight. You should feel some resistance as the male connector is inserted in the terminal connector.

12-Volt Test Lights

The 12-volt test light, whether it is a self-powered or stand-alone light, remains a useful tool for certain automotive electrical circuit tests. Most of us have used a 12-volt test light to check for source voltage and ground continuity in automotive electrical circuits for many years.

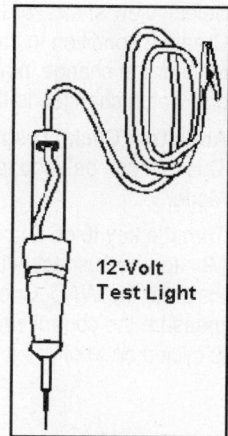

12-Volt
Test Light

However, with the advent of computerized engine controls in 1978, the type of tool to use to test a vehicle controller (ABS, Engine or TCM) circuit has changed. In place of the 12-volt test light we now have the DVOM, Lab Scope and Scan Tool.

All of the preliminary checks listed below should be done prior to testing for the cause of an intermittent fault. These checks include an inspection or check of the following items:

- All related electrical connections
- For vacuum leaks in related components or mounting hardware
- Fuel level - both quantity and quality
- Ignition wiring connections
- Air intake system filters, tubes and gaskets
- Any aftermarket add-on devices

How to Repair an Intermittent Fault

Intermittent faults are generally due to circuit problems. To pinpoint the fault to a particular component, its wiring or connectors requires thorough inspection and testing.

Prior to starting the test sequence, turn off all accessories and vehicle lighting. Also, verify that the battery and vehicle charging system are free of problems as these areas can disguise or mask a problem.

The Wiggle Test

A wiggle test can be used to locate the cause of some intermittent faults. The sensor, switch or the PCM wiring can be back probed as shown in the Wiggle Test Example.

To perform the test, move or wiggle the "suspect" component, connector or wiring while watching for a change on the DVOM.

If the DVOM has a Min/Max record mode, use this mode to try and capture the fault as the wiring or connector is moved.

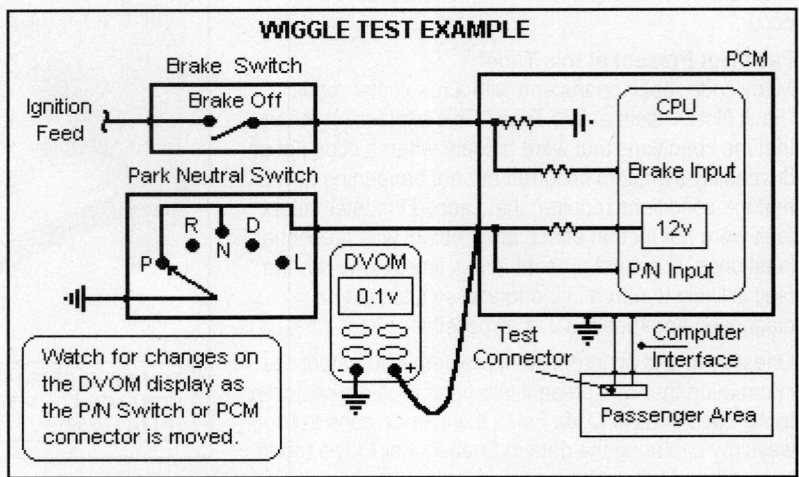

Change Input and Verify Output Response

This test can be used to monitor how the PCM and its output devices respond to changes in sensor or switch inputs.

Select, view and record any PIDs related to a code, a symptom or suspect circuit.
Create a condition to cause the selected input condition to change.
Monitor the change in a particular PID signal on the Scan Tool (i.e., change the TP sensor angle under engine load and watch changes to the IAC and TP sensor PIDs).

Actuator "Click" Testing

This test can be used to activate a particular relay or solenoid through the PCM using the Scan Tool and Output Test Mode.

Turn the key to on and enter the Output Test Mode to cycle an actuator on to off
Monitor the suspect actuator as it is cycled from on and then to off.
Listen to the WAC, Cooling Fan and Power relays click on and then off. Connect a breakout box to the PCM and measure the control circuit while turning the devices on and then off. The circuit should cycle from 0-12v as the device is cycled on to off.

This test can be used to check a suspect wiring harness for an open circuit condition.

- Turn the key off and disconnect the Scan Tool from the test connector.
- Disconnect the suspect component from the vehicle wiring harness.
- Install the breakout box (BOB).
- Use a DVOM (select the ohmmeter low-range) to measure from one end of the suspect circuit at the BOB to the other end of the circuit at the particular component connector pin (this is a continuity test of the suspect circuit).

The continuity test result should read under 5 ohms (actual test results will read near 0.1 ohms if the test leads are okay and the meter is properly zeroed).

Test for Shorted Conditions in a Harness

This test can be used to check a suspect wiring harness for a short-to-power or short-to-ground condition.

- Turn the key off and disconnect the PCM.
- Disconnect the suspect component from its wiring harness connector.
- Use a DVOM (select the ohmmeter high-range) at the sensor harness connector to measure between the suspect circuit and signal return (ground) and to vehicle power or the voltage reference circuit (this is a short to VPWR or VREF test).

The continuity test result read more than 10,000 ohms (Infinity, >> or OL on some meters). If the actual reading is less than this amount, the two circuits are making contact somewhere in the wiring harness.

Road Test Repair Verification

In many cases you will need to drive the vehicle through a specific drive pattern as part of a Road Test to verify a repair. In some cases you may want to write down or make available a list of known good PIDs from the Reference tables in one of the vehicle sections or in the Ford OBD System Handbooks.

Once you have the known good data collected, drive the vehicle and have an assistant record the actual PID values or use the Scan Tool to record the actual values. At the end of the Road Test, compare the values you collect to the "known good" values.

Specific Intermittent Tests

The test index below includes a list of the examples of Ford Intermittent Tests contained on the next few pages.

To use this information to repair an actual intermittent problem, start with Test Step Z1 on the next page to navigate to the most appropriate test step for your particular vehicle problem. If you know the type of intermittent test that you want to run, refer to the index directly below. This step can save you time if you are sure of the type of test to perform!

Specific Intermittent Tests

Pinpoint Test Z (All Codes and Symptoms) ....Page 1-30
Pinpoint Test Z12 (Input Devices Test) ....Page 1-32
Pinpoint Test Z14 (Water Soak Tests for Input Devices) ....Page 1-33
Pinpoint Test Z16 (Output Devices Test) ....Page 1-33
Pinpoint Test Z18 (Water Soak Tests for Output Devices) ....Page 1-34
Pinpoint Test Z50 (Intermittent Ignition Test) ....Page 1-34

PINPOINT TEST Z (INTERMITTENT FAULTS)

The first step to perform is to look at the PIDs collected earlier in the diagnosis. Then look at the list of components or circuits suggested at the end of the test steps in Pinpoint Test Z, or refer to a Symptom Chart at the end of a Pinpoint Test as needed.

Pinpoint Test Z - Directions for how to use this Intermittent Test Sequence

Step	Action	Yes	No
Z1	Step description: Determine the type of fault Note: There are two main procedures used in this section to isolate and repair an intermittent concern. One will utilize the Rotunda Distributorless Ignition System Tester (DIST) and the other, a Scan Tool and a DVOM. The DIST is only available for use on vehicles with coil pack ignition systems. If a DIST is not available, Go to Step Z2. • Is this a predetermined Ignition system fault?	Go to Step Z50 for a coil pack application except for the 2.0L Cougar/Focus. On these vehicles go to Step Z2.	Go to Step Z2.
Z2	Step description: Do PCM Reset clear FMEM Note: Proceed with this step only if a PCM Reset was not done earlier; otherwise, go to Step Z3. By eliminating FMEM, you will insure reproduction of any PCM related symptom. • Connect a Scan Tool. • Turn the key on and do a PCM Reset step. • Was the PCM Reset procedure completed? Note: Be sure all Freeze Frame data has been recorded before resetting the PCM.	Go to Step Z3.	Complete the PCM Reset procedure. Then go to Step Z3.
Z3	Step description: Select PIDs related to fault Note: A list of PIDs and/or signals is needed for use with the Scan Tool to indicate the area of fault. Obtain the description of the problem from the customer. Use the Reference Value Symptom Chart and proceed to the Reference Value PID/Measurement tables in handbook. • Highlight each PID and/or Signal recommended by the charts under the PID/Signal selection menu on the Scan Tool. • Have all PIDs and/or Signals been chosen?	Go to Step Z4.	Complete the PCM Reset procedure. Then go to Step Z3.
Z4	Step description: Decide to verify a Symptom Note: The path to symptom verification is optional, but may need to be done because the vehicle is in for repeat repair, there are not trouble codes set or the customer is having difficulty describing the condition or symptom. • Does a Symptom need to be verified?	Go to Step Z5.	Go to Step Z11.
Z5	Step description: Collect all required data Note: Only MIL codes trigger Freeze Frame data. Gather the previously recorded data used with the Symptom Tables. Trouble codes should already be recorded from an earlier pinpoint test. Access information from the customer worksheet or any other information. • Has all available data been recorded?	Go to Step Z6.	Gather as much data as possible to aid in isolation of the intermittent fault area. Repeat this step (Z5).

Pinpoint Test Z (Continued)

Step	Action	Yes	No
Z6	Step description: Recreate the Symptom Note: This step may require that the vehicle be driven as part of the test. The concern must be verified by recreating the conditions that originally set the code or caused the symptom. • Select and monitor PIDs displayed in Freeze Frame along with any previous highlighted PIDs and/or Signals from Step Z3. Use the previously recorded Freeze Frame data to recreate the conditions described by each of these PIDs. Pay special attention to the ECT, LOAD, RPM and VSS. Also, use any available customer data to aid in producing the correct conditions for recreating the symptom. • When the symptom occurs, press trigger or capture to begin recording relevant PID data. • Were you able to recreate the symptom?	Go to Step Z11.	Go to Step Z7.
Z7	Step description: Do Symptom Test (KOEO) This test is the last attempt to locate the fault or condition before disturbing vehicle circuits. Note: PIDs for outputs in the Reference Value Tables represent commanded values only. Circuit measurements with a DVOM indicate the actual output status. So, in the case of a fault, the PID and circuit reading on the vehicle may not correspond with each other. PIDs for PCM inputs with a mismatch to the circuit test may indicate a possible PCM concern. The Intermittent Road Test Procedure is a set of instructions for monitoring PIDs and signals with a Scan Tool and circuits with a DVOM. This step is done under (4) different conditions - KOEO, Hot Idle, 30 and 55 mph. Use the typical reference values in the JI Handbook to compare with the actual vehicle values. If you plan to use the 30 and 55 mph procedures, drive over the planned route with a helper. • Set up the vehicle to measure circuits with a DVOM, Lab Scope or a Scan Tool. • Connect the Scan Tool. Turn to key on. • Select and monitor the desired PIDs and/or measure the desired circuits with a DVOM. • Compare the PIDs and DVOM measurements to the JI Handbook tables. • Were any values out of range during testing?	Go to Step Z11.	Go to Step Z8.
Z8	Step description: Do the Road Test (KOER) • Engine running (ECT PID at least 195ºF). • Continue to monitor the list of PIDs and circuits as in the previous step at hot idle. • Were any of the monitored PID out of range?	Go to Step Z11.	Go to Step Z9.

Pinpoint Test Z (Continued)

Step	Action	Yes	No
Z9	Step description: Repeat symptom (35 mph) • Drive the vehicle over the planned route. • Continue to monitor the list of PIDs and circuits as in the previous step at 35 mph. • Were any of the monitored PID out of range?	Go to Step Z11.	Go to Step Z10.
Z10	Step description: Repeat symptom (55 mph) • Drive the vehicle over the planned route. • Continue to monitor the list of PIDs and circuits as in the previous step at 55 mph. • Were any of the monitored PID out of range?	Go to Step Z11.	The next few steps require that you disturb the vehicle circuits in an attempt to recreate the intermittent fault. Go to Step Z11.
Z11	Step description: Select a Circuit to Test • Leave Scan Tool in the PID or Signal menu. • If the Intermittent Road Test was used to verify the symptom, highlight PIDs or Signals that displayed a mismatch to the Reference Values in the Handbook. Otherwise, highlight only the PIDs and/or Signals from Step Z3. • Proceed to the Intermittent Test Chart located at the beginning of this pinpoint test. • Match selected PIDs and/or Signals to the corresponding circuit in the chart. If the PIDs were recorded, replay the snapshot data now. • From the same chart, select and proceed to the correct Input Test or Output Device Test. • Has a specific test been chosen?	For Input Tests, go to Step Z12. For Output Tests, go to Step Z16.	To diagnose other Driveability Symptoms go to Symptom Diagnosis in this section.
Z12	Step description: KOEO Input Test Steps Caution: When performing any of the test steps, always be aware of hands, clothing or tools near cooling fans, or hot surfaces. • Using circuits chosen from the Test Chart, select the recommended PIDs and/or Signals to check with a Scan Tool or with the DVOM. • Locate the suspect wiring or the component. • Turn to key on, engine off. • If the device is a switch, turn on manually. • Tap the component and/or wiggle the sensor harness while watching the signal value. • Look for any abrupt change in the value. • Compare actual values to the KOEO value. • Did any values fluctuate in or out of range?	Make any repairs as necessary. Restore the vehicle and then verify the repair is complete.	Go to Step Z13.
Z13	Step description: KOER Input Test Steps • Start the engine and allow it to idle. • If the device is a switch, turn on manually. • Tap the component and/or wiggle the sensor harness while watching the signal value. • Look for any abrupt change in the value. • Compare actual values to the KOER value. • Did any values fluctuate in or out of range?	Make any repairs as necessary. Restore the vehicle and then verify the repair is complete.	Go to Step Z14.

Pinpoint Test Z (Continued)

Step	Action	Yes	No
Z14	Step description: Water Soak Test Steps • Turn to key on, engine off. • Continue to monitor the list of selected PIDs. • Locate the suspect wiring or component. • If the device is a switch, turn on manually. • Monitor the PID or DVOM reading while lightly spraying water mist on the component. • Look for any abrupt change in the value. • Compare actual values to the KOEO value. • Did any values fluctuate in or out of range?	Make any repairs as necessary. Restore the vehicle and then verify the repair is complete.	Go to Step Z15.
Z15	Step description: Water Soak Test (KOEO) • Turn to key on, engine off. • Continue to monitor the list of selected PIDs. • Locate the suspect wiring or component. • If the device is a switch, turn on manually. • Monitor the PID or DVOM reading while lightly spraying water mist on the component. • Look for any abrupt change in the value. • Compare actual values to the KOEO value. • Did any values fluctuate in or out of range?	Make any repairs as necessary. Restore the vehicle and then verify the repair is complete.	Go to Step Z16.
Z16	Step description: KOEO Output Test Steps Caution: When performing any of the test steps, always be aware of hands, clothing or tools near cooling fans, or hot surfaces. Note: Output Test Mode may not control some outputs, such as injectors and ignition coils. To test these output types, Go to Step Z17. • Using circuits chosen from the Test Chart, select the recommended PIDs and/or Signals to check with a Scan Tool or with the DVOM. • Turn to key on, engine off. • Turn all outputs on using Output Test Mode. • Locate the suspect wiring or the component. • Tap the component and/or wiggle the sensor harness while watching the signal value. • Look for any abrupt change in the value. • Compare actual values to the KOEO value. • Did any values fluctuate in or out of range?	Make any repairs as necessary. Restore the vehicle and then verify the repair is complete.	Go to Step Z17.
Z17	Step description: Output Test Mode (KOER) Note: If you suspect a problem with an ignition coil, check for an intermittent fault with key off. • Start the engine and allow it to idle. • Tap the component and/or wiggle the sensor harness while watching the signal value. • Look for any abrupt change in the value. • Compare actual values to the KOER value. • Did the idle fluctuate or a reading fluctuate?	Make any repairs as necessary. Restore the vehicle and then verify the repair is complete.	Go to Step Z18.

Pinpoint Test Z (Continued)

Step	Action	Yes	No
Z18	Step description: Water Soak Test (KOEO) Caution: When performing any of the test steps, always be aware of hands, clothing or tools near cooling fans, or hot surfaces. Note: Output Test Mode may not control some outputs, such as injectors and ignition coils. To test these output types, Go to Step Z19. • Turn to key on, engine off. • Turn all outputs on using Output Test Mode. • Locate the suspect wiring or the component. • Tap the component and/or wiggle the sensor harness while spraying with a light water mist. • Look for any abrupt change in the value. • Did any values fluctuate in or out of range?	Make any repairs as necessary. Restore the vehicle and then verify the repair is complete.	Go to Step Z19.
Z19	Step description: Water Soak Test (KOER) • Start the engine and allow it to idle. • Continue to monitor the list of selected PIDs. • Locate the suspect wiring or component. • Monitor the PID or DVOM value while lightly spraying water mist on device wiring to PCM. • Look for any abrupt change in the value. • Compare actual values to the KOER value. • Did the idle fluctuate or a reading fluctuate?	Make any repairs as necessary. Restore the vehicle and then verify the repair is complete.	Go to Step Z20.
Z20	Step description: Check for Mechanical Fault Note: An intermittent mechanical concern can cause a good PCM system to react incorrectly. • The mechanical systems that could cause a trouble code or symptom should have already been done previously. If not, do them now. • Look for wires, vacuum lines or hoses that may short or kink during these conditions: - Engine rock during acceleration. - Components moving during conditions of hash vibration (high speed or rough road). - Accelerator or transmission linkage contact. • Was a mechanical concern detected?	Make any repairs as necessary. Restore the vehicle and then verify the repair is complete.	Look in additional sources for help. Refer to a complete list of technical service bulletins for this particular vehicle application or to a Hotline. A customer flight recorder may also be useful. It is very likely that this fault has occurred on a similar vehicle!
Z50	Step description: Intermittent Ignition Test Note: This pinpoint test must be used with the Rotunda DIST Tester 418-F024 (007-00075) or equivalent for non-Coil on Plug applications. The DIST cannot be used on a Cougar/Focus 2.0L coilpack application. The Quick Test must be performed and all test procedures in any Pinpoint Test steps done before using this test. • Check the CKP sensor shield connector. • Verify the generator output and battery state. • Turn all accessories off during Ignition tests. • Is the vehicle prepared for equipment setup?	Refer to the manufacturer's instructions on how to install and use the DIST (Ignition Tester). Follow all of the related instructions to completion to isolate an intermittent fault in the Ignition system. Verify the repair when all tests are complete.	Repeat Step Z50 until the vehicle is prepared for the equipment to be installed.

FORD QUICK TEST

The Ford Quick Test is divided into three specialized tests:

- The KOEO Self-Test
- The KOER Self-Test
- The Continuous Memory Diagnostic Trouble Code Access

The Quick Test is used to check the integrity and function of the Powertrain Diagnostics. This test is performed at the start of each diagnostic procedure, and at the end of a Pinpoint Test for repair verification purposes.

■ **Note: The KOEO and KOER tests in an EEC-V system are traditional Ford "on-demand" tests. They do not cause the MIL to illuminate.**

Quick Test System Pass

If no codes are set at the end of a Quick Test, and no communication faults with the Scan Tool exist, a System Pass indication is shown on the Scan Tool. System Pass means that inputs and outputs monitored by the PCM are operating within normal limits.

■ **Note: During the "read codes" step, no codes set means that there are no Continuous Memory Codes for emissions related faults (it is possible that a fault exists in other non-emission related inputs or outputs circuits). Also, on EEC-V systems, Continuous Memory Codes should be read separately from the Quick Test.**

Key On, Engine Off Self-Test

The KOEO Self-Test is a functional test of the PCM System performed ON DEMAND with the key on and the engine off. A fault has to be present at the time of testing for this test to detect a fault. If a fault is detected during this test, a DTC will be output on the data link and can be read by the Scan Tool.

Key On, Engine Running Self-Test

The KOER Self-Test is a functional test of the PCM System performed on demand with the key on, engine running and the vehicle stopped. A check of inputs and outputs is made during operating conditions at normal temperature. A fault has to be present at the time of testing for this test to detect a fault. If a fault is detected during this test, a DTC will be output on the data link and can be read by the Scan Tool. Certain vehicle tests require a particular action be performed (push brake pedal, turn the steering wheel).

Continuous Memory DTC Access

Continuous Memory DTC access is also a Self-Test of the PCM system. However, unlike the KOEO and KOER Self-Tests that are initiated only On Demand, the Continuous Memory Test is always active. On an OBD II system, this test consists of running the Monitors (including the CCM). These monitors are designed to detect faults that could cause the vehicle tailpipe emissions to exceed 1.5 times the FTP Standard. When this Self-Test detects a fault, a code is stored in PCM memory. A code can be retrieved at a later time even if the fault no longer exists (useful for intermittent faults).

Expanded Diagnostic Protocol

Most diagnostic procedures require that all trouble codes be retrieved. The Ford NGS DDL Menu or the Expanded Diagnostics Protocol (EDP) mode can be used with the Generic Scan Tool function to accomplish this task.

The Expanded Diagnostic Protocol (EDP) mode on the Generic Scan Tool accesses enhanced diagnostic information to aid in the repair of emission related faults. This information may be required by the manufacturer to diagnose sensors or components that affect vehicle emissions.

READ TROUBLE CODES

During the Quick Test, trouble codes that are detected can be read and repaired. At the end of a Pinpoint Test, certain drive cycles can be "run" before repeating the Quick Test. This repeated action is necessary to detect faults not found in the KOEO or KOER Self-Tests. You can select the command "Retrieve Continuous Codes" on a Scan Tool in both the NGS Diagnostic Data Link (DDL) and Generic menu.

If you only want to read trouble codes that activate the MIL, use the Generic OBD II functions menu. If you want to read all of the Continuous Memory Codes (Non-MIL and MIL trouble codes), you should select the NGS DDL menu on from the Scan Tool menu.

■ **Note:** **MIL codes activate the Check Engine Lamp in the instrument panel and are only set for emission related faults. Generally, non-MIL related codes are used to identify faults in certain controls like A/C or Transmission Range switches.**

A Generic Scan Tool that meets OBD II requirements can access the Continuous Memory Test to retrieve MIL codes. However, when performing a KOEO or KOER Self-Test or retrieving Continuous Memory codes (both non-MIL and MIL) with a Generic Scan Tool, an additional set of commands may need to be entered manually.

Continuous Tests

These tests are done during normal driving and are classified into two types: "TRIP and BGLOOP." TRIP refers to a test that executes completely only once per trip or drive cycle during normal driving conditions. TRIP tests include the Catalyst, EVAP, HO2S and Secondary AIR Monitors. BGLOOP is a term chosen to refer to tests that execute every background loop during normal driving.

Brake On/Off Test

On vehicles with a Brake switch input, after the KOER test begins and the Identification (ID) code appears, depress and release the brake pedal to allow the PCM to detect a change (from on to off) in the Stop Lamp switch.

Power Steering Pressure Switch Test

On vehicles equipped with a PSP switch, turn the steering wheel 1/2 turn within 1-2 seconds after the KOER test starts so the PCM can detect a change in the PSP switch.

Overdrive Cancel Switch Test

On vehicles with E4OD A/T, cycle the Overdrive Cancel switch within 1-2 seconds after the KOER test starts so the PCM can detect a change in the Overdrive Cancel switch.

Also, remember to cycle the Transmission Control switch right after the KOER test starts to allow the PCM to detect a change of state in the Transmission Control switch.

4x4 Low Switch Test

On vehicles with a 4x4 Low Switch, cycle the 4x4 switch within 1-2 seconds after the KOER test starts so the PCM can detect a change in status of the 4x4 Low Switch.

PINPOINT TEST QT

The Output State Check (or test) can be used to check the operation of PCM controlled actuators and relays by commanding them on and off. Refer to the Graphic to the right for instructions on how to run this test.

Quick Test Preparation

- Apply the parking brake and block the drive wheels for safety.
- Place the gearshift selector in Park (Neutral position on M/T applications)
- Turn off all electrical accessory loads

```
┌─────────────────────────────────────────────┐
│          OUTPUT STATE CHECK                   │
│ 1. Key Off (10 seconds), connect Scan Tool,   │
│    select OEM tests, then Output State Check   │
│ 2. Key On, start KOEO Self Test               │
│ 3. After KOEO and CMC codes are displayed,    │
│    do a brief wide open throttle movement      │
│ 4. The PCM should immediately energize all     │
│    related output devices (actuators, relays, etc.) │
│ 5. Repeat the brief wide open throttle movement │
│    and the PCM should de-energize all devices  │
└─────────────────────────────────────────────┘
```

PCM Quick Test QT (Diagnostic Starting Point)

Step	Action	Yes	No
QT1	Description: Perform the Quick Test Note: If the vehicle was brought in with an Emission Compliance Failure, go directly to Pinpoint Test EM1in other manuals. • Complete all preliminary checks while looking for obvious concerns that relate to this symptom. - Check related electrical connectors & fuses. - Check related vacuum lines and hose routing. - Check intake air system for leaks/restrictions. - Check the fuel octane for the correct rating. - Check the fuel supply for any contamination. - Check the operation of the Cooling system. • Look up any related service bulletins (use OASIS if it is available, otherwise use your normal source for Technical Service Bulletins). • Connect the Scan Tool to the DLC (connector). • Perform the Quick Test to read any codes. Record any KOEO, KOER and Continuous Memory Codes. Note: If the Scan Tool does not communicate or you cannot read any trouble codes, go to Pinpoint Test QA1. • Were any trouble codes present?	If the engine runs "rough" at idle and any KOER or CMC codes are present, go to Step QT2 to check Injector PIDs. All Other Conditions: Go to correct DTC repair table while noting the following: Service the codes in this order (start with 1st code output). 1. KOEO codes 2. KOER codes 3. CMC (codes)	Go to Step QT2 (no codes set). Refer to the Ford Symptom Index in other repair manuals or electronic media. Note: If symptom is not present, go to Intermittent Test Index in other repair manuals or electronic media. Start with Pinpoint Test Z1.
QT2	Description: Engine runs rough at idle, or Key On Engine Running or Continuous Memory Codes set (check the Fuel Injector PIDs on the Scan Tool). Note: An injector circuit fault can cause the PCM to detect incorrect or unrelated trouble codes. • Turn to key on, engine off. • Access the Injector PIDs (INJxF with the 'x' standing for the injector number). There will be (1) INJxF PID for each cylinder number. • Did any of the INJxF PIDs indicate "YES"?	Turn the key off. An injector fault exists so disregard these trouble codes. For Natural Gas applications: Go to Pinpoint Test HA60. For All Others: Go to Pinpoint Test H56 Refer to these tests in other manuals.	There are no Injector related faults. Turn the key off. Return to Step QT1 and follow all of the "Yes" steps under the title "All Others".

Onboard Diagnostics

INTRODUCTION

In order to diagnose Ford vehicles equipped with an OBD II System, it is important that you understand the terms related to these test procedures. Some of these terms and their definitions are discussed in the next few articles.

Two-Trip Detection

In most cases, an emission related system or component must fail a Monitor test more than once before it activates the MIL. The first time an OBD II Monitor detects a fault during a related trip, it sets a "pending code" in PCM memory. These codes appear when the Memory or Continuous codes are read. For a "pending code" to mature into a hard code (and illuminate the MIL), the original fault must occur for two consecutive trips (two-trip detection). However, a "pending code" can remain in the PCM for a long time before the conditions that caused the code to set <u>reappear</u>.

Fuel Trim and Misfire Detection trouble codes can cause the PCM to flash the MIL after <u>one</u> trip because faults in these systems can cause damage to the catalytic converter.

Pending Code

The term "pending code" is used to describe a fault that has been detected once and is stored in memory. This type of fault has not been detected on two consecutive trips (i.e., it has not matured into a hard code).

It is possible to access a "pending code" with a Generic Scan Tool (GST) except for some Continental, Ranger and Windstar 3.0L models. Be aware that you may not be able to read a pending code with a Generic Scan Tool on some Ford 1994-95 phase-in models.

Similar Conditions

If a "pending code" is set because of a Fuel System or Misfire Monitor detected fault, the vehicle must meet similar conditions for two consecutive trips before the code matures and the PCM activates the MIL and stores the code in memory. The meaning of similar conditions is important when you attempt to diagnose a fault detected by the Fuel System Monitor or Misfire Detection Monitor.

To achieve similar conditions, the vehicle must reach the following engine running conditions simultaneously (for the first failure recorded that set the code):

- Engine speed must be within 375 rpm
- Engine load must be within 10%
- Engine warmup state must match the previous state (cold or warm)

Summary - Similar conditions are defined as conditions that match those recorded when the fault was detected and a code was set.

Warmup Cycle

A warmup cycle is defined as vehicle operation (after a cool-down period) when the engine temperature increases by at least 40°F and reaches at least 160°F.

Most trouble codes are cleared from the PCM memory after 40 "warmup cycles" if the fault does not reappear.

```
SCAN TOOL DISPLAY

   PENDING DTCs
ECU: $10 (Engine)
Number of Codes:1
 P0113  Intake Air
        Temperature
        Circuit High Input
```

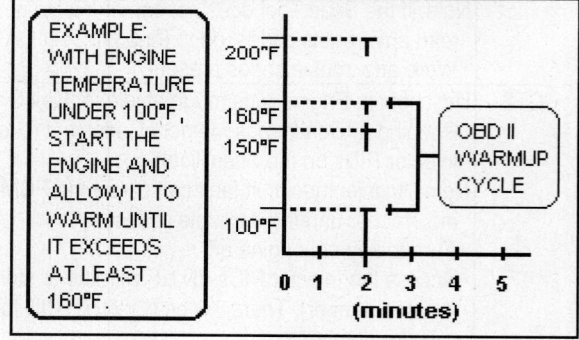

EXAMPLE: WITH ENGINE TEMPERATURE UNDER 100°F, START THE ENGINE AND ALLOW IT TO WARM UNTIL IT EXCEEDS AT LEAST 160°F.

PCM RESET (CLEAR CODES STEP)

The PCM Reset step allows the Scan Tool to command the PCM to clear all emission-related diagnostic information. Each time this step is done, a trouble code (DTC P1000) is stored in the PCM until all of the OBD II system Monitors or components have been tested to satisfy their particular "trip" (without any faults occurring). To do a PCM Reset on an OTC Scan Tool - select the Special Test Menu, on a Snap-On Scan Tool - select the Functional Test Menu and on a Vetronix Scan Tool - select the Misc Test Menu.

```
CLEAR INFO

THIS OPERATION
WILL CLEAR ALL DTC,
FREEZE FRAME, AND
READINESS TEST
DATA.

DO YOU WISH TO
CONTINUE?

PRESS [YES] OR [NO]
```

The following events occur when a PCM Reset is done:

- All emission-related trouble codes are cleared
- Clears the Freeze Frame data
- Clears the Diagnostic Monitoring results
- Resets the status of the OBD II system Monitors
- Sets DTC P1000 in the PCM memory

An example of the warning displayed on a Vetronix Scan Tool prior to completing a PCM Reset step is shown in the example in the Graphic at the top of this page.

Freeze Frame Data

The term Freeze Frame is used to describe the engine conditions recorded in the PCM at the time an emission related fault is detected. These conditions include the fuel control state, spark timing, engine speed and load. Frame data is recorded during the first trip on a two-trip fault. It can be overwritten by codes with a higher priority.

Standard Corporate Protocol

A standard corporate protocol (SCP) communication language is used to exchange bi-directional messages between stand-alone modules and devices. Two or more messages can be sent over one circuit with this system.

Fuel Trim & Misfire Detection Codes

The PCM clears most emission codes from memory after 40 warmup periods if the same fault does not repeat. The exceptions to this rule are codes related to the Fuel Trim and Misfire Detection Monitors. Trouble codes related to these Monitors require that 80 "warmup cycles" occur without the same fault being detected before the PCM will clear these trouble codes from memory.

Malfunction Indicator Lamp

If the PCM detects an emission related fault for two consecutive "trips", the MIL is activated and a code is set. The MIL can be turned off after (3) consecutive "trips" occur without the same fault being detected. To turn the MIL "off" after a repair, use the Scan Tool PCM Reset function.

```
SCAN TOOL MENUS

DTC MENU
F1: READ DTCs
F2: FREEZE DATA
F3: PENDING DTCs
F4: CLEAR INFO
```

```
DIAG. TROUBLE CODES
ECU: $40 (Engine)
Number of DTCs: 1
*P0455 EVAP Control
       System Leak (gross
       leak) Detected

ENTER=FREEZE FRAME
DTC....................P0455
ENGINE SPD..........704RPM
ECT (°)...............135°F
VEHICLE SPD...........0MPH
ENGINE LOAD..........6.6%
MAP (P)..........15.1inHg
FUEL STAT 1..............CL
FUEL STAT 2.........UNUSED
ST FT 1................-3.0%
LT FT 1................-1.5%
```

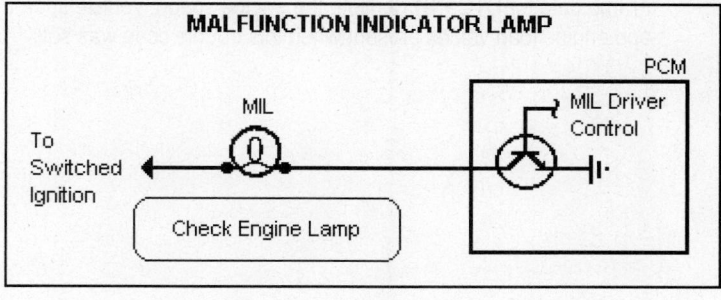

MALFUNCTION INDICATOR LAMP

DIAGNOSTIC TROUBLE CODES

The OBD II system uses a Diagnostic Trouble Code (DTC) identification system that was established by SAE and the EPA. The first letter of a DTC is used to identify the type of vehicle computer system that failed. The types of systems are shown below:

*   The letter 'P' indicates a Powertrain related device
*   The letter 'C' indicates a Chassis related device
*   The letter 'B' indicates a Body Control related device
*   The letter 'U' indicates a Data Link or Network code.

The first number of a diagnostic trouble code (DTC) indicates either a generic (P0xxx) or a manufacturer specific (P1xxx) type of trouble code.

The number in the hundredth position indicates the specific vehicle system or subgroup that failed (i.e., P0300 for a Misfire code, P0400 for an emission system code, etc.).

Trouble Code Example

An example of how to navigate through the Vetronix Scan Tool menus to read a trouble code is shown step by step in the Graphic to the right. The Generic Scan Tool function was used in this example.

The vehicle application in this example was a 2001 Ford Taurus V6 VIN U (2v per cylinder) engine. In this example, the cause of the problem in the Intake Air Temperature (IAT) sensor signal circuit was an open circuit condition. The Comprehensive Component Monitor (CCM) identified the fault on the first failure and stored the code conditions (engine operating conditions) in Freeze Frame. When the fault occurred on the second consecutive trip, the MIL was illuminated and a hard trouble code was set (two-trip fault detection). Note the Freeze Frame code conditions shown in the Graphic.

1) Select from the three (3) choices on the screen. In this case, F1: SCAN TEST was selected.
2) Select the application from the choices on the screen. In this case, Global OBD II was selected.
3) Select the type of Test from the choices on the screen. In this case, F1: OBD II Functions.
4) Select the type of function from the choices. In this case, F1: READ DTCs was selected.
5) This screen shows an example of the DTC Menu and the Freeze Frame data for DTC P0113. Note the engine speed, vehicle speed and engine load values present when this trouble code was set.

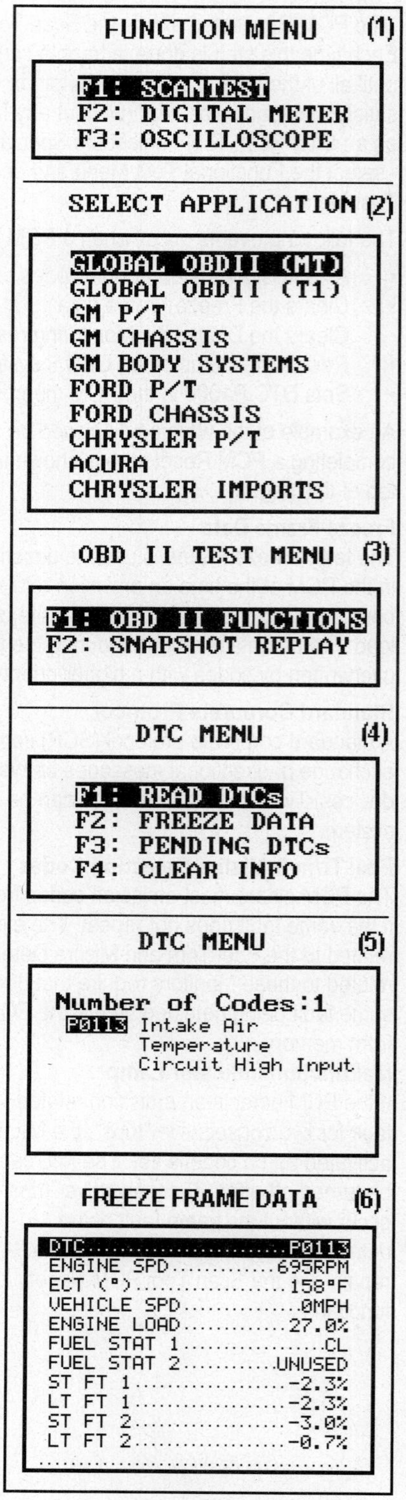

```
FUNCTION MENU         (1)

F1: SCANTEST
F2: DIGITAL METER
F3: OSCILLOSCOPE

SELECT APPLICATION (2)

GLOBAL OBDII (MT)
GLOBAL OBDII (T1)
GM P/T
GM CHASSIS
GM BODY SYSTEMS
FORD P/T
FORD CHASSIS
CHRYSLER P/T
ACURA
CHRYSLER IMPORTS

OBD II TEST MENU     (3)

F1: OBD II FUNCTIONS
F2: SNAPSHOT REPLAY

DTC MENU             (4)

F1: READ DTCs
F2: FREEZE DATA
F3: PENDING DTCs
F4: CLEAR INFO

DTC MENU             (5)

Number of Codes:1
P0113  Intake Air
       Temperature
       Circuit High Input

FREEZE FRAME DATA    (6)

DTC.................P0113
ENGINE SPD...........695RPM
ECT (°)..............158°F
VEHICLE SPD..........0MPH
ENGINE LOAD..........27.0%
FUEL STAT 1...............CL
FUEL STAT 2..........UNUSED
ST FT 1..............-2.3%
LT FT 1..............-2.3%
ST FT 2..............-3.0%
LT FT 2..............-0.7%
.........................
```

CYLINDER BANK IDENTIFICATION

Cylinder banks and related sensors are defined as follows:

Cylinder Bank - A specific group of engine cylinders that share a common control sensor (e.g., Bank 1 identifies the location of cylinder 1 while Bank 2 identifies cylinders on the opposite bank). Refer to the Ford example in the Graphic.

Sensor - If sensors are numbered (Bank 1 Sensor 1), they follow the convention described above. If they are identified with letters ('A', 'B', 'C'), they are manufacturer defined. If only (1) sensor is used, the letter or number may be omitted.

Data Link Connector

Ford vehicles equipped with OBD II Systems use a standardized Data Link Connector (DLC). It is typically located between the left end of the instrument panel and 12 inches past vehicle centerline. The connector is mounted out of sight from the passengers, but should be easy to see from outside by a technician in a kneeling position (door open).

DLC Features

The DLC is rectangular in design. It can accept up to 16 terminals.

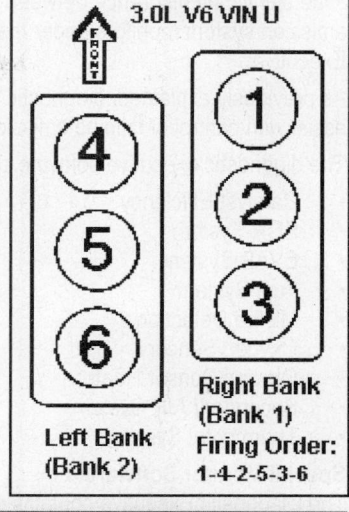

Both the DLC and Scan Tool have latching features that ensure that the Scan Tool will remain connected to the vehicle during operation.

Common uses of the Scan Tool while connected to the DLC include:

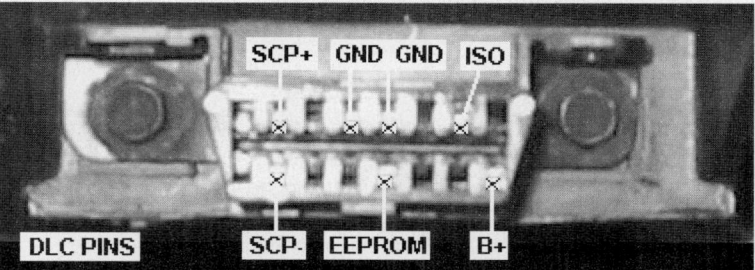

- Display the results of the most current I/M Readiness Tests
- Read and clear any diagnostic trouble codes
- Read the Serial Data from the PCM
- Perform Enhanced Diagnostic Tests (manufacturer specific)

DLC Pin Assignments Table

Cavity	Pin Assignment	Cavity	Pin Assignment
1	Ignition Control	9	Ignition Switch Power
2	Bus (+) SCP	10	Bus (-) SCP
3	Not Assigned	11	Not Assigned
4	Chassis Ground	12	Not Assigned
5	Signal Ground (Return)	13	Flash EEPROM
6	Class C Link Bus (+)	14	Class C Link Bus (-)
7	K Line of ISO 9141	15	L Line of ISO 9141
8	Trigger Signal Input	16	Fused Battery Power

DIAGNOSTIC EXECUTIVE

One significant difference between early Ford EEC Systems and the EEC-V version is the use of several dedicated emission system monitors under the control of a special piece of software inside the PCM referred to as the Diagnostic Executive.

As previously explained, diagnostic monitors are required to comply with regulations mandated by the EPA designed to assist with control of tailpipe emissions.

The diagnostic executive software controls the operation of these OBD II Monitors:

- Catalyst Efficiency
- EGR System
- EVAP System
- Fuel System
- Misfire Detection
- Oxygen Sensor
- Oxygen Sensor Heater
- Secondary AIR System
- Thermostat System

Special Monitor Software

The Diagnostic Executive contains special software designed to allow the PCM to organize and prioritize all of the Main Monitor tests and procedures, to record and display the test results (in Freeze Frame) and any related diagnostic trouble codes.

The functions controlled by this software include:

- To control and arrange changes between the "states" of the diagnostic system so that the vehicle will continue to operate in a normal manner during testing.
- To verify that all OBD II Monitors run during the first two (2) sample periods of the Federal Test Procedure (FTP).
- To verify that all OBD II Monitors and their related tests are sequenced so that the required inputs (enable criteria) for a particular Monitor are present prior to running that Monitor.
- To sequence the running of the OBD II Monitors to eliminate the chance that one of the Monitor tests might interfere with another or upset the normal vehicle operation.
- To provide a Scan Tool interface by coordinating the operation of special tests or data requests.

OBD II Monitor Test Results

Generally, when a particular OBD II Monitor is run and fails a test during a "trip", a "pending code" is set. If that Monitor detects the same fault for two consecutive trips, the MIL is activated and a hard code set in memory. The results of a particular Monitor test indicate the emission related system (and sometimes the component) that failed, but they do not always indicate the cause of the failure.

OBD II Problem Diagnosis

To find the cause of a problem, select the correct trouble code repair chart, a symptom from the Symptom List or select an appropriate Intermittent Test. Although it may not be necessary to do all of the test steps in a trouble code repair chart, these charts remain the backbone of any OBD II System diagnosis and repair.

■ **NOTE: Two important pieces of information that can help speed up the diagnosis are the DTC code conditions (including all enable criteria), and the parameter information (PIDs) stored in the Freeze Frame related to a stored code.**

COMPREHENSIVE COMPONENT MONITOR

OBD II regulations require that all emission related circuits and components controlled by the PCM that could affect emissions be monitored for circuit continuity and out-of-range faults. The comprehensive component monitor (CCM) consists of four different monitoring strategies: two for inputs and two for outputs. Some tests run continuously, some only after actuation. The CCM Monitor is a 2- trip monitor for emission devices.

Input Strategies

One input strategy is used to check devices with analog inputs for an open or shorted condition, or an input value that is out-of-range (i.e., IAT, ECT, MAF, TP and TR sensor).

Input Rationality Tests

The input signals to the PCM are constantly monitored for electrical circuit faults. As discussed, some input devices are also tested for rationality. In effect, the signal is compared against other inputs and information to see if it makes sense under current engine operating conditions.

Rationality Definition

Rationality is defined as a type of CCM test in which component input signals are compared against other component inputs to verify that the conditions match.

Output Strategies

An Output State Monitor in the PCM checks outputs for opens or shorts by watching the control voltage level of the related device. The control voltage should be low with it "on" and high with the device "off" (i.e., EPC, SS1-SS3, TCC, HFC, VMV and HO2S Heater).

Output Functionality Tests

The output signals to the PCM are constantly monitored for electrical circuit faults. Some of the PCM outputs are tested for functionality in addition to testing for electrical circuit faults. The PCM can send a command to an output device and then monitor certain input signals related to that device for expected changes to verify the command was carried out (i.e., the PCM commands the IAC valve to a specific opening position, it expects to see a target idle speed). If it does not detect a change, the CCM fails and a code is set.

Functionality Definition

Functionality is defined as a type of comprehensive component test in which the PCM output commands are verified by monitoring specific input signals from other PCM components for an expected change.

CCM Repair Verification "Trip" Graphic

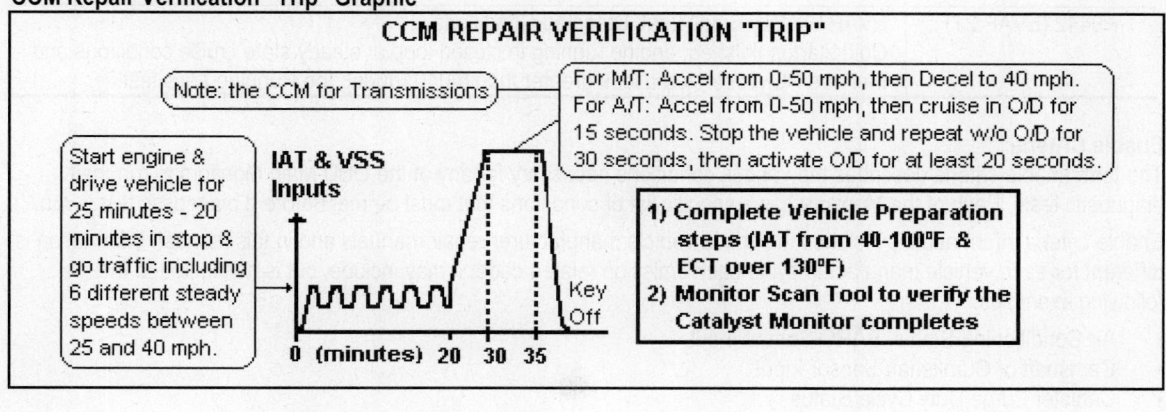

CCM REPAIR VERIFICATION "TRIP"

Note: the CCM for Transmissions

For M/T: Accel from 0-50 mph, then Decel to 40 mph.
For A/T: Accel from 0-50 mph, then cruise in O/D for 15 seconds. Stop the vehicle and repeat w/o O/D for 30 seconds, then activate O/D for at least 20 seconds.

Start engine & drive vehicle for 25 minutes - 20 minutes in stop & go traffic including 6 different steady speeds between 25 and 40 mph.

IAT & VSS Inputs

Key Off

1) Complete Vehicle Preparation steps (IAT from 40-100°F & ECT over 130°F)
2) Monitor Scan Tool to verify the Catalyst Monitor completes

0 (minutes) 20 30 35

MAIN MONITORS

One of the key features of OBD Diagnostics is the use of several PCM controlled diagnostic monitors contained within the PCM software structure. These monitors are needed to meet CARB and U.S. EPA OBD II regulations. These monitors (Main Monitors) are advanced beyond OBD I tests. They include the following tests:

- Air Injection System (requires specific driving conditions)
- Catalytic Converter Efficiency (requires specific driving conditions)
- EGR System Monitor (requires idle, acceleration periods to complete)
- EVAP system function and flow monitoring.
- Fuel System Monitor (completes anytime during a trip)
- Misfire Detection (includes identification of the misfiring cylinder if possible)
- Oxygen Sensor Heater
- Oxygen Sensor Monitor (requires steady cruise speed to complete)

OBD II Trip Definition

The term OBD II "Trip" describes a method of driving the vehicle during which the following Main Monitors complete their tests: The actual "trip" requirements are different for each CCM and Main Monitor. Examples of these requirements are included below.

CCM "Trip" Requirements

The CCM "Trip" requirements for DTC P0116 are shown in the table below.

DTC	Trouble Code Title & Conditions
P0116 (2T CCM)	ECT Sensor Range Conditions Engine running for a calibrated period of time after an engine cold soak period of over 6 hours, then the PCM detected the ECT sensor signal exceeded the IAT sensor signal by more than a calibrated value (e.g., 30°F), the ECT sensor signal exceeded a calibrated value (e.g., 225°F), the Catalyst, Fuel System, HO2S and Misfire Monitors did not complete, or the Calibrated Timer has expired.

Main Monitor "Trip" Requirements

The EVAP "Trip" requirements for DTC P0442 are shown in the table below.

DTC	Trouble Code Title & Conditions
P0442 (EVAP 2T)	EVAP System Small Leak Detected Conditions Cold startup finished, engine running in closed loop at steady state cruise conditions and then the PCM detected a leak greater than 0.040" during the Running Loss test.

Enable Criteria

The term enable criteria describes the various conditions necessary for any of the OBD Main Monitors to run their diagnostic tests. Each of the Monitors has a specific list of conditions that must be met before a diagnostic test is run.

Enable criteria information is included in various vehicle manufacturer repair manuals and in this manual. Information is different for each vehicle manufacturer and each emission related code. It may include, but is not limited to the following examples:

- Air Conditioning Status, BARO Sensor Input
- Camshaft or Crankshaft Sensor Input
- Canister Purge Duty Cycle Status
- ECT, IAT and TFT Sensor Inputs

OBD III DIAGNOSTICS

 A third version of On Board Diagnostics (OBD III) may be introduced in the next decade. This article includes comments about what this version of OBD might include and the role of the California Air Resources Board (CARB) in the design of this new system.

How will it work? While no final documents have been released about OBD III at this time, it now appears likely that several new technologies are being considered to detect and relay emissions faults as part of OBD III. The highlights of one program under study by Sierra Research (Sacramento, CA) for the Air Resources Board are presented next.

OBD III Highlights

Allows wireless communication of the OBD system status to a manufacturer (e.g., GM "On Star" system) or to a state contractor responsible for emissions-related data. This could be similar to the role MCI has under the current Smog Check program.

Vehicle owner decides whether the system is active or not active.

Eliminates need for Smog Check (with system activated).

Eliminates need for re-inspection of faulty vehicle after repair (system activated).

Life-cycle cost would be lower than the cost of the Smog Check program test fees.

OBD III System Graphic

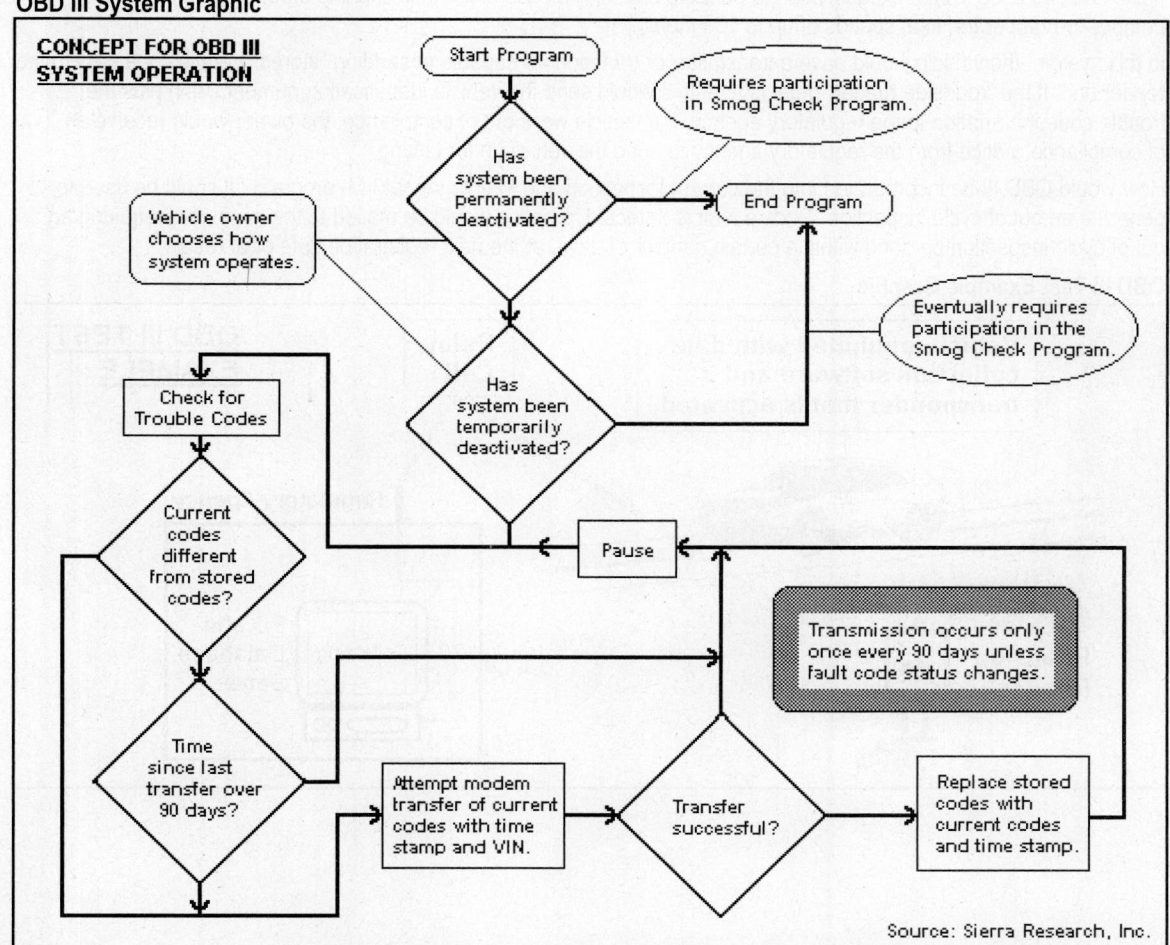

OBD III DIAGNOSTICS

It also appears that the OBD II System diagnostics will continue to be the diagnostic "backbone" of the third version of On Board Diagnostics. However, with OBD III, the vehicle owner may have the option of having a vehicle equipped with this system either activated or deactivated. Vehicles in this program may be equipped with a transmitter or transponder that would allow the vehicle to communicate with a monitoring service by wireless communication (i.e., a cell phone or a roadside reader).

As stated, the vehicle owner would have the choice of having the system "activated" or "deactivated". If the system were "on", they would not have to participate in the Smog Check program. Owners who decide not to activate the system would be required to participate in the Smog Check program.

Cell Phone

A method of wireless communication that uses the vehicle cell phone has been studied. In this case, the phone would communicate with the regulatory agency when the vehicle has a trouble code set, and the agency would alert the owner of the emissions problem.

Roadside Reader

Another method of wireless communication that uses a roadside reader has also been studied (since 1994). This device can be used from a fixed location, a portable unit, or a mobile unit and is capable of reading eight lanes of bumper-to-bumper traffic at speeds of up to 100 mph.

In this system, the vehicle would have a transmitter or transponder capable of sending "stored" trouble code data to the reader unit. If the "roadside reader" detects a fault, it would send the vehicle identification number (VIN) plus the trouble code information to the regulatory agency. If a vehicle were out of compliance, the owner would receive an "out of compliance" notice from the regulatory agency to bring the vehicle in for testing.

How would OBD III be incorporated into the current Inspection and Maintenance (I/M) program? It could be used to generate an out of cycle inspection. Once a fault is detected, a notice would be mailed to the car owner requiring an out of cycle inspection be done within a certain number of days, at the next registration date or at resale.

OBD III Test Example Graphic

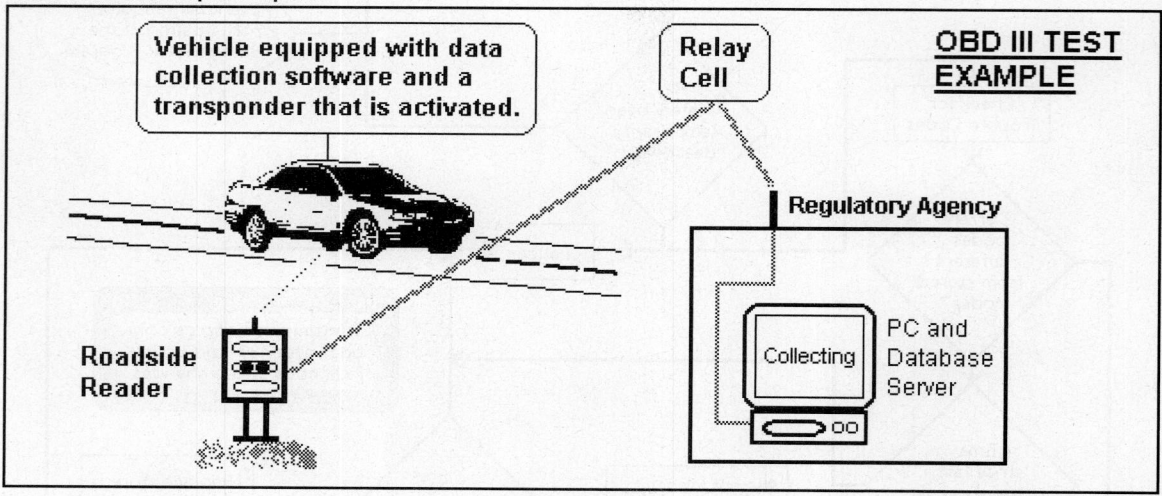

HOW TO ACCESS GENERIC PID INFORMATION

The Scan Tool Parameter Identification (PID) Mode allows access to certain data values, analog and digital input and output signals, calculated values and system status information.

The Generic OBD II PID list is an example of the PIDs that all OBD II compatible Scan Tools must be able to access and display.

Scan Tool PID Menus
An example of how to navigate through the Vetronix Mastertech Scan Tool menus to locate the Generic PID information is shown in the Graphic to the right.

The Graphic also contains a total of twenty-five (25) engine related parameters that are available for this vehicle on an OBD II compatible Scan Tool.

Parameter ID (PID) Information
The proper sequence to follow to obtain a complete Generic PID list for this vehicle is shown in the Graphic.

(1) Connect a Scan Tool with OBD adapter to the DLC under the drivers side of instrument panel, to allow the tool to read the OBD II Generic information

(2) Select GLOBAL OBDII (M/T) from the APPLICATION menu. (This is the title for Vetronix Generic software).

(3) The Scan Tool will attempt to communicate with the PCM using the Generic protocol (or language).

(4) Select F1: DATA LIST from the FUNCTIONS menu.

(5) Select F1: ALL DATA from the PARAMETER SELECTION menu to view all available PIDs. Select F2: USER LIST to view specific PIDs. Selecting only a few PIDs will significantly increase scan tool speed.

(6) The last frame contains all available Generic PIDs for this application.

Diagnostic Tip
The Typical Values in the example on this page should be used after the following conditions are met:

• The Ford Quick Test has been completed.
• The engine is at hot engine idle speed.
• The throttle is closed unless checking a value at a specified speed (rpm).
• The vehicle is in Park or Neutral.
• The vehicle is operating in Closed Loop mode.
• The vehicle accessories are turned off.

■ NOTE: A Scan Tool that displays faulty data should not be used. The use of a faulty Scan Tool can result in misdiagnosis and unnecessary parts replacement.

GENERIC SCAN TOOL MENUS

SELECT APPLICATION
GLOBAL OBDII (MT)
GM P/T
FORD P/T
CHRYSLER P/T
CHRYSLER CHASSIS

INITIALIZING THE OBD II INTERFACE
PLEASE WAIT.

OBD II FUNCTIONS
F1: DATA LIST
F2: DTCs
F3: SNAPSHOT
F4: OBD CONTROLS
F5: SYSTEM TESTS
F6: READINESS TESTS
F8: INFORMATION
F9: UNIT CONVERSION

PARAMETER SELECTION
F1: ALL DATA
F2: USER LIST

```
ENGINE SPD      697RPM
ECT (°)          181°F
VEHICLE SPD      0MPH
IGN. TIMING      19.0°
ENGINE LOAD      23.5%
MAF (R)        5.37gm/s
TPS (%)          19.6%
IAT (°)           70°F
FUEL STAT 1         CL
FUEL STAT 2     UNUSED
ST FT 1           2.3%
LT FT 1          -0.7%
ST FT 2           0.0%
LT FT 2           3.1%
O2S B1 S1       0.535V
FT O2S B1 S1     -1.5%
O2S B1 S2       0.000V
FT O2S B1 S2    UNUSED
O2S B2 S1       0.735V
FT O2S B2 S1      0.0%
O2S B2 S2       0.000V
FT O2S B2 S2    UNUSED
MIL STATUS         OFF
STORED DTCs          0
OBD CERT.      OBD II
```

HOW TO ACCESS OEM PID INFORMATION

The OEM PID List contains all of the engine related parameters that are available on the Scan Tool.

If all of the PID values are within normal range, refer to the Symptoms Diagnosis information found in manuals.

Scan Tool PID Menus

An example of how to navigate through the Vetronix Mastertech scan tool menus to locate the PID information is shown in the Graphic to the right.

Many vehicles have PIDs that displaying all of them at the same time would slow down the refresh rate substantially. For this application, the OEM data is arranged in four (4) separate lists. There is overlap between the lists, but each is a selection of the most relevant data for the subject.

Parameter ID (PID) Information

The proper sequence to follow to obtain a complete OEM PID list for this vehicle is as follows:

Steps (1)-(4) are not shown in the graphic to the right.

(1) Select Ford P/T software from the application list.
(2) Select the year (2001).
(3) At this point, there are a few options for selecting the proper vehicle. Select the VIN method to avoid choosing from a long list of models.
(4) Select VIN 4.
(5) The scan tool display will read: 3.8L SFI MAF WINDSTAR (VIN 4). Press the YES button.
(6) F0: DATA LIST – Press the F0 button to obtain the Data List selection menu.
(7) F2: FUEL – Press the F2 button to obtain the list of PIDs related to fuel control.
(8) The list of PID items inside the last box of the Graphic does not represent the complete PID list, only those PIDs in the list selected for this example.

Diagnostic Tip

The Typical Values for the PCM Inputs and Outputs on the next two (2) pages should be used after the following conditions are met:

- The Ford Quick Test has been completed.
- The engine is at hot engine idle speed.
- The throttle is closed unless checking a value at a specified speed (rpm).
- The vehicle is in Park or Neutral.
- The vehicle is operating in Closed Loop mode.
- The vehicle accessories are turned off.

■ **Note:** A Scan Tool that displays faulty data should not be used. The use of a faulty Scan Tool can result in misdiagnosis and unnecessary parts replacement.

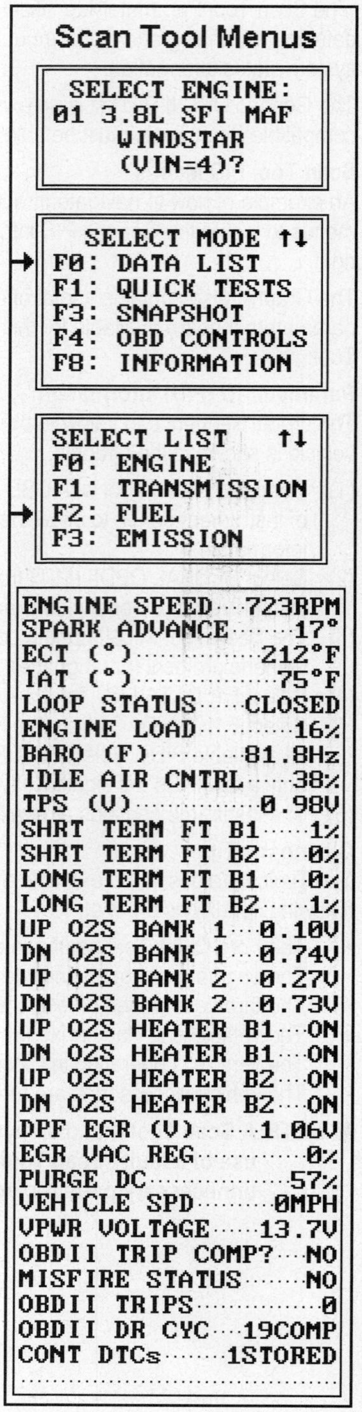

Scan Tool Menus

REFERENCE INFORMATION

PCM Location Tables - Cars (Ford, Lincoln & Mercury)

Year	Ford	Lincoln	Mercury
1992-94	Aspire • Located behind LH side of the IP Crown Victoria • Located behind LH side of the I/P Escort • Located below the center of dash Mustang, Thunderbird • Located at the RH side of firewall Probe • Located forward of center console Taurus • Located at RH side of safety wall Tempo • Located behind LH side of dash	Continental • Located at top RH safety wall under the splash cover Mark VII (1992-93) • Located behind the right kick panel Mark VIII (1994) • Located behind the LH side of instrument panel Town Car • Located at LH side of cowl below wiper motor	Cougar • Located behind the RH cowl panel inside the vehicle Grand Marquis • Located at LH side of the I/P Sable • Located at RH of safety wall Topaz • Located under the left side of instrument panel Tracer • Located below center of I/P
1995-96	Aspire • Located behind LH side of the IP Contour • Located at RH corner of engine area Crown Victoria • Located behind LH side of firewall Escort • Located below the center of dash Mustang, Thunderbird • Located behind the RH cowl panel Probe • Located forward of center console Taurus • Located at top RH of safety wall	Continental • Located at top RH safety wall under the splash cover Mark VIII • Located behind the LH side of instrument panel (on LH side of firewall) Town Car • Located at the LR of the engine area on the safety wall	Cougar • Located behind the RH cowl panel inside the vehicle Grand Marquis • Located behind left side of instrument panel on firewall Mystique • Located at right rear corner of engine by P/S reservoir Sable • Located on top RH right of the safety wall Tracer • Located below center of I/P
1997	Aspire • Located below dash by steering column Contour • Located on the RH engine bulkhead (access is from panel in engine area) Crown Victoria • Located behind kick panel cover near I/P on the driver's side Escort • Located below the center of I/Panel Mustang, Thunderbird • Located behind kick panel cover near I/P on passenger's side Probe • Located under console by shifter Taurus • Located behind glove box (access is from engine side of dash panel)	Continental • Located at top RH safety wall under the splash cover Mark VIII • Located near steering column on driver side Town Car • Located at the LR of the engine area on the safety wall	Cougar • Located behind kick panel cover near dash panel (passenger side) Mystique • Located on RH engine bulkhead (access is from a panel in engine area) Grand Marquis • Located behind kick panel cover near dash panel on driver side Sable • Located behind the glove box (access is from dash panel in engine area) Tracer • Located under dash on passenger side

PCM Location Tables - Cars (Ford, Lincoln & Mercury)

Year	Ford	Lincoln	Mercury
1998	Contour • Located on the RH engine bulkhead (access is from a panel in engine area) Crown Victoria • Located behind kick panel cover near instrument panel on driver side Escort • Located below the center of the I/Panel Mustang • Located behind kick panel cover near instrument panel on passenger side Taurus • Located behind the glove box (access is from the dash panel in engine area)	Continental • Located at top RH safety wall under the splash cover Mark VIII • Located near steering column on driver side Town Car • Located at the LR of the engine area on the safety wall	Grand Marquis • Located behind LH kick panel cover by the dash Mystique • Located on RH engine bulkhead (access is from a panel in engine area) Sable • Located behind the glove box (access is from dash panel in engine area) Tracer • Located under dash on passenger side
1999	Contour • Located on the RH engine bulkhead (access is from a panel in engine area) Crown Victoria • Located behind kick panel cover near instrument panel on driver side Escort • Located below the center of the I/Panel Mustang • Located behind kick panel cover near instrument panel on passenger side Taurus • Located behind the glove box (access is from the dash panel in engine area)	Continental • Located at top RH safety wall under the splash cover LS6 • Located in engine area on RH side cowl LS8 • Located in engine area on RH side cowl Town Car • Located at the LR of the engine area on the safety wall	Cougar • Located behind RH kick panel cover near the dash Grand Marquis • Located behind LH kick panel cover by the dash Mystique • Located on RH bulkhead on cowl (access panel) Sable • Located behind the glove box (from access panel) Tracer • Located under dash on passenger side
2000-02	Contour (2000) • Located on the RH engine bulkhead (access is from a panel in engine area) Crown Victoria • Located behind kick panel cover near instrument panel on driver side Escort • Located below the center of the I/Panel Focus • Located near glove box over kick panel Mustang • Located above kick panel cover near instrument panel on passenger side Taurus • Located behind the glove box (access is from the dash panel in engine area) Thunderbird (2002) • Located in the engine compartment in the RH cowl area under the vent screen	Continental • Located at top RH safety wall under the splash cover LS6 • Located in engine area on RH side cowl LS8 • Located in engine area on RH side cowl Town Car • Located at the LR of the engine area on the safety wall	Cougar • Located behind RH kick panel cover near the dash Grand Marquis • Located behind LH side of I/Panel on the left hand safety wall Mystique (2000) • Located on RH bulkhead on cowl (access panel) Sable • Located behind the glove box (from access panel)

PCM Location Tables - Sports Utility Vehicles (Ford, Lincoln & Mercury)

Year	Ford	Lincoln	Mercury
1992-93	Bronco • <u>LOCATED BELOW THE BRAKE FLUID RESERVOIR</u> Explorer • Located at right side of dash near kick panel	---	---
1993-94	Bronco • Located at left side of firewall near harness Explorer • Located under the dash at the RH kick panel	---	---
1995	Bronco • Located at left side of firewall near harness Explorer • Located at RH (engine) side of cowl panel	---	---
1996	Bronco • Located at left side of firewall near harness Explorer • Located at RH (engine) side of cowl panel	---	---
1997	Expedition • Located at RH side of engine near bulkhead Explorer • Located at RH (engine) side of cowl panel	Navigator • Located at RH (engine) side of cowl panel	Mountaineer • Located at RH (engine) side of cowl panel
1998	Expedition • Located at RH side of engine near bulkhead Explorer • Located at RH (engine) side of cowl panel	Navigator • Located at RH (engine) side of cowl panel	Mountaineer • Located at RH (engine) side of cowl panel
1999	Expedition • Located at RH side of engine near bulkhead Explorer • Located at RH (engine) side of cowl panel	Navigator • Located at RH (engine) side of cowl panel	Mountaineer • Located at RH (engine) side of cowl panel
2000	Expedition • Located at RH side of engine near bulkhead Explorer • Located at RH (engine) side of cowl panel	Navigator • Located at RH (engine) side of cowl panel	Mountaineer • Located at RH (engine) side of cowl panel
2001	Escape • Located at rear center of the engine area Expedition • Located at RH side of engine near bulkhead Explorer • Located at RH (engine) side of cowl panel	Navigator • Located at RH (engine) side of cowl panel	Mountaineer • Located at RH (engine) side of cowl panel
2002	Escape • Located at rear center of the engine area Expedition • Located at RH side of engine near bulkhead Explorer • Located at RH (engine) side of cowl panel	Navigator • Located at RH (engine) side of cowl panel	Mountaineer • Located at RH (engine) side of cowl panel

PCM Location Tables - Trucks (Ford)

Year	Ford	Lincoln	Mercury
1992	F-Series Pickup • Located in engine area below brake cylinder Ranger Pickup • Located behind right side of dash near kick panel	---	---
1993	F-Series Pickup • Located at left side of engine near harness connectors Ranger Pickup • Located at left rear corner of the engine compartment	---	---
1994	F-Series Pickup • Located at left side of engine near harness connectors Ranger Pickup • Located at left rear corner of the engine compartment	---	---
1995	F-Series Pickup • Located at left side of engine near harness connectors Ranger Pickup • Located at right rear of the engine through the firewall	---	---
1996	F-Series Pickup • Located at left side of engine near harness connectors Ranger Pickup • Located at right rear of the engine through the firewall	---	---
1997	F-Series Pickup • Located at lower RH side of dash (visible at engine area) Ranger Pickup • Located at top of the engine cowl (visible at engine area)	---	---
1998	F-Series Pickup • Located at lower RH side of dash (visible at engine area) Ranger Pickup • Located at top of the engine cowl (visible at engine area)	---	---
1999	F-Series Pickup • Located at lower RH side of dash (visible at engine area) Ranger Pickup • Located at top of the engine cowl (visible at engine area)	---	---
2000	F-Series Pickup • Located at lower RH side of dash (visible at engine area) Ranger Pickup • Located at top of the engine cowl (visible at engine area)	---	---
2001	F-Series Pickup • Located at lower RH side of dash (visible at engine area) Ranger Pickup • Located at top of the engine cowl (visible at engine area)	---	---
2002	F-Series Pickup • Located at lower RH side of dash (visible at engine area) Ranger Pickup • Located at top of the engine cowl (visible at engine area)	Blackwood Pickup • Located behind the RH side passenger cowl trim panel	---

PCM Location Tables - Vans (Ford, Lincoln & Mercury)

Year	Ford	Lincoln	Mercury
1992-95	Aerostar • Located at the left side of firewall near the master cylinder E-Series Van • Located on left side of firewall near the master cylinder	---	Villager (1993-95) • Located behind top RH side of the Instrument Panel behind glove box
1996	Aerostar • Located on driver side of the firewall near master cylinder E-Series Van • Located at the left rear of engine compartment by master cylinder Windstar • Located at the RH side of the cowl area (access is from the engine side)	---	Villager • Located behind top RH side of the Instrument Panel behind glove box
1997	Aerostar • Located on driver side behind the dash panel (inside vehicle) E-Series Van • Located at the left rear of engine compartment by master cylinder Windstar • Located at the RH side of the cowl area (access is from the engine side)	---	Villager • Located behind top RH side of the Instrument Panel behind glove box
1998	E-Series Van • Located at the left rear of engine compartment by master cylinder Windstar • Located at the RH side of the cowl area (access is from the engine side)	---	Villager • Located behind top RH side of the Instrument Panel behind glove box
1999	E-Series Van • Located on driver side of lower dash panel Windstar • Located at the RH side of the cowl area (access is from the engine side)	---	Villager • Located behind top RH side of the Instrument Panel behind glove box
2000	E-Series Van • Located on driver side of lower dash panel Windstar • Located at the RH side of the cowl area (access is from the engine side)	---	Villager • Located behind top RH side of the Instrument Panel behind glove box
2001-02	E-Series Van • Located on driver side of lower dash panel Windstar • Located at the RH side of the cowl area (access is from the engine side)	---	Villager • Located behind top RH side of the Instrument Panel behind glove box

REFERENCE INFORMATION

MAF Sensor Calibration Table

This calibration table can be used to determine if the MAF sensor is out of calibration. The PID that shows gm/sec can be found in the Global OBD II PID List (the Emissions List for Snap-On). The Ford OEM PID List does not show a MAF PID with the gm/sec.

Voltage to Mass Airflow Conversion Table (1996-98 Models)

Vehicle Application	Voltage		Signal		Level	
Voltage Values	0.34v	0.39v	0.60v	1.00v	1.96v	3.90v
Contour/Mystique with 2.0L	0.39 gm/sec	0.45 gm/sec	0.72 gm/sec	2.10 gm/sec	7.64 gm/sec	40.59 gm/sec
Contour/Mystique with 2.5L	0.33 gm/sec	0.95 gm/sec	1.69 gm/sec	5.53 gm/sec	22.92 gm/sec	137.98 gm/sec
Cougar 2.0L & 2.5L	1.44 gm/sec	1.66 gm/sec	2.73 gm/sec	8.06 gm/sec	20.33 gm/sec	153.31 gm/sec
Escort/Tracer 1.9L & 2.0L, Aerostar 3.0L	0.94 gm/sec	1.07 gm/sec	1.68 gm/sec	4.77 gm/sec	16.87 gm/sec	88.91 gm/sec
Probe 2.0L and Windstar with 3.0L	0.82 gm/sec	0.88 gm/sec	1.25 gm/sec	3.44 gm/sec	14.48 gm/sec	135.95 gm/sec
Ranger Pickup with 2.3L & 3.0L	0.93 gm/sec	1.07 gm/sec	1.71 gm/sec	4.58 gm/sec	16.23 gm/sec	135.17 gm/sec
Sable/Taurus 3.0L 2v and Taurus FF 3.0L	1.31 gm/sec	1.11 gm/sec	1.72 gm/sec	4.82 gm/sec	17.00 gm/sec	146.84 gm/sec
Sable/Taurus 3.0L 4v, 3.8L Windstar, 4.2L E & F-Series	1.43 gm/sec	1.64 gm/sec	2.58 gm/sec	7.53 gm/sec	26.57 gm/sec	154.40 gm/sec
Taurus 3.4L, Continental, Mark VIII & Mustang with 4.6L	1.39 gm/sec	1.60 gm/sec	2.69 gm/sec	8.36 gm/sec	32.99 gm/sec	193.69 gm/sec
Mustang 3.8L, Aerostar, Explorer & Ranger with 4.0L, Explorer & Mountaineer 5.0L	1.46 gm/sec	1.68 gm/sec	2.52 gm/sec	7.64 gm/sec	27.09 gm/sec	151.79 gm/sec
Cougar/Thunderbird with 3.8L & 4.6L	1.44 gm/sec	1.66 gm/sec	2.73 gm/sec	8.06 gm/sec	29.33 gm/sec	153.31 gm/sec
Crown Victoria/Grand Marquis 4.6L, Crown Victoria & Town Car 4.6L NGV, Town Car 4.6L	1.28 gm/sec	1.48 gm/sec	2.37 gm/sec	6.97 gm/sec	26.16 gm/sec	148.18 gm/sec
E-Series 4.9L, 5.0L & 5.8L	1.24 gm/sec	1.42 gm/sec	2.98 gm/sec	8.38 gm/sec	29.88 gm/sec	151.25 gm/sec
F-Series Pickup 4.9L and Bronco with 5.0L & 5.8L	1.68 gm/sec	1.93 gm/sec	3.32 gm/sec	8.43 gm/sec	28.97 gm/sec	151.44 gm/sec
Sable/Taurus 3.0L 4v, 3.8L Windstar & F-Series 4.2L	1.43 gm/sec	1.64 gm/sec	2.58 gm/sec	7.53 gm/sec	26.57 gm/sec	154.40 gm/sec
E & F-Series 4.6L & 5.4L, E-Series 6.8L, F-Series 7.5L	1.35 gm/sec	1.56 gm/sec	2.66 gm/sec	8.31 gm/sec	32.99 gm/sec	190.18 gm/sec
E-Series 5.8L	1.24 gm/sec	1.42 gm/sec	2.98 gm/sec	8.38 gm/sec	29.88 gm/sec	151.25 gm/sec
Expedition 4.6L & 5.4L, F-Series 5.8L, Navigator 5.4L	1.68 gm/sec	1.93 gm/sec	3.32 gm/sec	8.43 gm/sec	28.97 gm/sec	151.44 gm/sec
Windstar 3.0L	0.82 gm/sec	0.88 gm/sec	1.25 gm/sec	3.44 gm/sec	14.48 gm/sec	135.95 gm/sec

■ **Note: To use this chart, locate the MAF sensor voltage and then compare it against the known good grams per second (gm/sec) reading for the applicable vehicle/engine.**

Voltage to Mass Airflow Conversion Table (1999-2002 Models)

Vehicle Application	Voltage		Signal	Level		
Voltage Values	0.23v	0.27v	0.46v	2.44v	4.60v	4.79v
Continental & Mustang 4.6L	1.07 gm/sec	1.22 gm/sec	2.08 gm/sec	59.50 gm/sec	266.89 gm/sec	296.52 gm/sec
Contour, Mystique & Cougar 2.0L	0.70 gm/sec	0.80 gm/sec	1.36 gm/sec	30.86 gm/sec	125.00 gm/sec	138.89 gm/sec
Contour, Mystique & Cougar 2.5L	1.00 gm/sec	1.14 gm/sec	1.95 gm/sec	52.50 gm/sec	213.33 gm/sec	236.11 gm/sec
Crown Victoria, Grand Marquis & Town Car 4.6L	1.01 gm/sec	1.16 gm/sec	1.98 gm/sec	50.26 gm/sec	206.94 gm/sec	223.62 gm/sec
Escort 2.0L 2v & Focus	0.67 gm/sec	0.77 gm/sec	1.32 gm/sec	30.55 gm/sec	124.44 gm/sec	138.89 gm/sec
Escort 2.0L 4v	0.62 gm/sec	0.71 gm/sec	1.22 gm/sec	30.66 gm/sec	124.36 gm/sec	138.28 gm/sec
Excursion, Expedition & Navigator	1.06 gm/sec	1.21 gm/sec	2.07 gm/sec	58.33 gm/sec	263.89 gm/sec	295.18 gm/sec
Explorer & Mountaineer 4.0L & 5.0L	0.97 gm/sec	1.11 gm/sec	1.90 gm/sec	49.97 gm/sec	206.70 gm/sec	233.91 gm/sec
E-Series 4.2L	0.97 gm/sec	1.11 gm/sec	1.90 gm/sec	49.97 gm/sec	206.70 gm/sec	233.91 gm/sec
E-Series 4.6L & 5.4L	1.15 gm/sec	1.31 gm/sec	2.24 gm/sec	63.37 gm/sec	259.88 gm/sec	289.02 gm/sec
F-150 Series 4.2L, 4.6L, 5.4L	1.06 gm/sec	1.21 gm/sec	2.07 gm/sec	58.33 gm/sec	263.89 gm/sec	295.18 gm/sec
F-250 & F-350 Series 4.2L, 5.4L & 6.8L	1.06 gm/sec	1.21 gm/sec	2.07 gm/sec	58.33 gm/sec	263.89 gm/sec	295.18 gm/sec
Lighting	1.17 gm/sec	1.34 gm/sec	2.28 gm/sec	62.50 gm/sec	397.22 gm/sec	375.00 gm/sec
LS6 & LS8	0.90 gm/sec	1.03 gm/sec	1.76 gm/sec	52.27 gm/sec	222.22 gm/sec	240.48 gm/sec
Mustang 3.8L & Ranger 4.0L	1.00 gm/sec	1.14 gm/sec	1.94 gm/sec	52.44 gm/sec	211.20 gm/sec	236.85 gm/sec
Ranger 2.5L & 3.0L	0.67 gm/sec	0.77 gm/sec	1.32 gm/sec	30.55 gm/sec	124.44 gm/sec	138.89 gm/sec
Sable & Taurus 3.0L 2v and Taurus FF 3.0L	0.67 gm/sec	0.77 gm/sec	1.32 gm/sec	30.55 gm/sec	124.44 gm/sec	138.89 gm/sec
Taurus 3.0L 4v	1.00 gm/sec	1.14 gm/sec	1.95 gm/sec	52.50 gm/sec	213.33 gm/sec	236.11 gm/sec
Windstar 3.0L	0.62 gm/sec	0.71 gm/sec	1.22 gm/sec	30.66 gm/sec	124.36 gm/sec	138.28 gm/sec
Windstar 3.8L	0.97 gm/sec	1.11 gm/sec	1.90 gm/sec	49.97 gm/sec	206.70 gm/sec	233.91 gm/sec

■ **Note:** **To use this chart, locate the MAF sensor voltage and then compare it against the known good grams per second (gm/sec) reading for the applicable vehicle/engine.**

OPERATING RANGE CHARTS
ECT, IAT and TFT Sensor Range Charts

Temperature		Voltage Resistance	
Degrees F	Degrees C	Volts	Kohms
267	131	0.2	0.8
248	120	0.28	1.18
230	110	0.36	1.55
212	100	0.47	2.07
194	90	0.61	2.8
176	80	0.8	3.84
158	70	1.04	5.37
140	60	1.35	7.6
104	40	2.16	16.15
86	30	2.62	24.27
68	20	3.06	37.3
50	10	3.52	58.75
32	0	3.97	65.85
14	-10	4.42	78.19
-4	-20	4.87	90.54
-22	-30	4.89	102.88

DPFE EGR Sensor Chart

Pressure			Voltage
PSI	Inches HG	kPa	Volts
4.34	8.83	29.81	4.56
3.25	6.62	22.36	3.54
2.17	4.41	14.90	2.51
1.08	2.21	7.46	1.48
0	0	0	0.45

Transmission Range Sensor Chart

TR Sensor Position	Voltage Value	Resistance in Ohms
P	4.41v	4.16K ohms
R	3.60v	1.44K ohms
N	2.83v	733 ohms
D	2.09v	401 ohms
2	1.37v	211 ohms
1	0.68v	81 ohms

TP Sensor Chart / Output Device Resistance Chart

Throttle Angle	Voltage	Component	Ohm Value	Component	Ohm Value
0 Degrees	0.50v	CANP Solenoid	50-100 ohms	IAC Motor	7-13 ohms
10 Degrees	0.97v	CCRM	1K Minimum	IMRC	50-100 ohms
20 Degrees	1.44v	EPC Solenoid	3.0-5.1 ohms	Injectors	7-16 ohms
30 Degrees	1.90v	EVR Solenoid	25-70 ohms	SS1 Solenoid	15-25 ohms
40 Degrees	2.37v	EDF Relay	65-110 ohms	SS2 Solenoid	15-25 ohms
50 Degrees	2.84v	FC/LFC Relay	65-110 ohms	SS3 (AX4S)	15-25 ohms
60 Degrees	3.31v	FP Relay	40-90 ohms	TCC Solenoid	0.9-1.9 ohms
70 Degrees	3.78v	HFC Relay	50-100 ohms	WAC Relay	1K Minimum

Ford OBD Glossary

Glossary of Terms & Acronyms	
(<) Indicates less than the listed value	(>) Indicates more than the listed value
A/C: Air Conditioning	ACC: Air Conditioning Clutch
ACCS: Air Conditioning Cycling Switch	ACD: Air Conditioning Demand Switch
ACON: Air Conditioning On Signal	ACP: Air Conditioning Pressure Signal
ACPSW: Air Conditioning Pressure Switch	ACT: Air Charge Temperature Sensor
A/D: Analog to Digital Converter	AIR: Secondary Air Injection Reactor
AIRB: AIR Bypass Solenoid	AIRD: AIR Diverter Solenoid
AM1: Air Management 1, AIR Bypass	AM2: Air Management 2, AIR Diverter
AODE: Automatic Overdrive Electronic Transmission	A4R70W: Automatic Overdrive Electronic Wide Ratio 4-Speed Transmission
ARC: Automatic Ride Control	A/T: Automatic Transmission
ATDC: After Top Dead Center	AX4S: Automatic 4-Speed Transmission
BANK 1: Fuel Bank 1 (bank fired engines)	BANK 2: Fuel Bank 2 (bank fired engines)
B/MAP: Barometric/Manifold Absolute Press.	BOO: Brake On/Off Switch
BPA: Mechanical Bypass Air	BTDC: Before Top Dead Center
Bus (+): Standard Corporate Protocol Communication Link Positive Terminal	Bus (-): Standard Corporate Protocol Communication Link Negative Terminal
CANP: Canister Purge Solenoid	CCD: Computer Controlled Dwell
CCNT: Count Code or DTC CCNT	CCRM: Constant Control Relay Module
CCS: Coast Clutch Solenoid	CCSP: Carbon Canister Storage/Purge
CID: Cylinder Identification Signal	CKP: Crankshaft Position Sensor
CMP: Camshaft Position Sensor	CPP: Clutch Pedal Position
CSM: Central Security Module	CTVS: Closed Throttle Vacuum Switch
Data (+): Data Positive Circuit	Data (-): Data Negative Circuit
DCL: Data Communication Link	DDL: Diagnostic Data Link
DI: Distributor Type Ignition System	DLC: Data Link Connector (OBD II)
DOHC: Dual Overhead Cam	DOL: Data Output Line to IPC
DPFE: Delta Pressure Feedback EGR	DRL: Daylight Running Lights
DSM: Driver Seat Module	DTC: Diagnostic Trouble Code
DTM: Diagnostic Test Mode	DVOM: Digital Volt/Ohm Meter
EATC: Electronic Automatic Temperature Control Module	ECA: Electronic Control Assembly
ECT: Engine Coolant Temperature	EEC: Electronic Engine Control System
EEC-IV: Fourth Generation EEC	EEC-V: Fifth Generation EEC
HEGO: Heated Exhaust Gas Oxygen Sensor	EGRC: Exhaust Gas Recirculation Control
EGR Monitor: EGR System Diagnostic Test	EGRV: Exhaust Gas Recirculation Vent
EI: Electronic Ignition Type System	EPC: Electronic Pressure Control
EPT: EGR Pressure Transducer	EVP: EGR Valve Position Sensor
EVR: EGR Vacuum Regulator	FC: Fan Control Cooling Fan
FEM: Front Electronic Module	FPDM: Fuel Pump Driver Module
FPM: Fuel Pump Monitor	FTP: Federal Test Procedure

Ford OBD Glossary

Glossary of Terms & Acronyms	
FTP: Federal Test Procedure	GEM: Generic Electronic Module
HFC: High Fan Control (high speed fan)	HLOS: Hardware Limited Operating Strategy
HO2S-11: Rear Heated Oxygen Sensor (B1)	HO2S-12: Front Heated Oxygen Sensor (B1)
IAC: Idle Air Control (solenoid)	IAT: Intake Air Temperature
IC: Integrated Circuit	ICM: Ignition Control Module
IDM: Ignition Diagnostic Monitor	IFS: Inertia Fuel Switch
IGN GND: Ignition Ground	IMRC: Intake Manifold Runner Control
ISO 9141: International Standards Organization Network	ITS: Idle Tracking Switch
KAM: Keep Alive Memory	KAPWR: Direct Battery Power
KOEC: Key On, Engine Cranking	KOEO: Key On, Engine Off
KOER: Key On, Engine Running	KS: Knock Sensor
LAMBSE: Short Term Fuel Trim	LCD: Liquid Crystal Display
LFC: Low Fan Control for low speed fan	LFP: Low Speed Fuel Pump
LONGFT: Long Term Fuel Trim	LOS: Limited Operating Strategy
LPG: Liquid Petroleum Gas	LTS: Low Coolant Switch
M/T: Manual Transmission	MAF: Mass Airflow sensor
MAF RTN: Mass Airflow Sensor Ground	MAP: Manifold Air Pressure
MCU: Microprocessor Control Unit	MFI: Multiport Fuel Injection
MIL: Malfunction Indicator Lamp	MLP: Manual Lever Position Sensor
OBD I: On Board Diagnostics (Version One)	OBD II -On Board Diagnostics (Version Two)
OCT ADJ: Octane Adjust Fuel	OSM: Output State Monitor
OSS: Output Shaft Speed (sensor)	O2S: Oxygen Sensor
PATS: Passive Anti-Theft System	PCM: Powertrain Control Module
PFE: Pressure Feedback EGR	PIP: Profile Ignition Pickup Signal
PMI: Programmable Module Installation	PSP: Power Steering Pressure Switch
PSOM: Programmable Speed/Odometer Unit	RAM: Random Access Memory
RCM: Restraint Control Module	REM: Rear Electronic Module
RKE: Remote Keyless Entry Module	ROM: Read Only Memory
SCP: Standard Communications Protocol	SHRTFT: Short Term Fuel Trim
SPOUT: Spark Output signal	SRS: Supplemental Inflatable Restraint Sys.
STI: Self-Test Input	STO: Self-Test Output
TCIL: Trans. Control Indicator Lamp	TCS: Transmission Control Switch
TFT: Transmission Fluid Temperature	TOT: Transmission Oil Temperature
TR: Transmission Range sensor	TSS: Transmission Speed Sensor
UBP: UART Based Protocol Network	VAF: Vane Airflow Meter or Sensor
VAT: Vane Air Temperature sensor	VLCM: Variable Load Control Module
VID: Vehicle Identification Block (backup)	VIM: Vehicle Interface Module
VPWR: Ignition switched Power	VREF: Voltage Reference
VRS: Variable Reluctance Sensor	WAC: Wide Open Throttle A/C Cutout Relay

SECTION 2 CONTENTS

Introduction

How To Use This Section ....Page 2-4

2001 Focus - Powertrain

VEHICLE IDENTIFICATION

VECI Decal ....Page 2-5

ELECTRONIC ENGINE CONTROL SYSTEM

Operating Strategies ....Page 2-6

PCM Location ....Page 2-6

INTEGRATED ELECTRONIC IGNITION SYSTEM

Introduction ....Page 2-7

Ignition Coil Operation ....Page 2-8

Lab Scope Test (Coil Primary) ....Page 2-9

CKP Sensor (Magnetic Design) ....Page 2-10

DVOM Test & Lab Scope Test (CKP Sensor) ....Page 2-11

EVAPORATIVE EMISSION SYSTEM

Introduction ....Page 2-12

Evaporative Charcoal Canister ....Page 2-13

Canister Vent Solenoid ....Page 2-13

Fuel Vapor Control Valves ....Page 2-13

Canister Purge Valve & Fuel Tank Pressure Sensor ....Page 2-14

Scan Tool Test (CV Solenoid & Purge Solenoid) ....Page 2-15

DVOM Test (Purge Solenoid) ....Page 2-16

Lab Scope Test (CV Solenoid) ....Page 2-17

Lab Scope Test (Purge Solenoid) ....Page 2-18

EXHAUST GAS RECIRCULATION SYSTEM

Introduction ....Page 2-19

Exhaust Gas Recirculation Valve ....Page 2-20

Orifice Tube Assembly ....Page 2-20

DPFE Sensor Information ....Page 2-20

Lab Scope Test (DPFE Sensor) ....Page 2-21

Vacuum Regulator Solenoid ....Page 2-22

Scan Tool Test (VR Solenoid) ....Page 2-22

Lab Scope Test (VR Solenoid) ....Page 2-23

EGR System Monitor ....Page 2-24

FUEL DELIVERY SYSTEM

Returnless Fuel System ....Page 2-25

Fuel System Operation ....Page 2-26

Fuel Pump Driver Module ....Page 2-27

Lab Scope Test (FPDM) ....Page 2-28

Scan Tool Test (FDPM) ....Page 2-29

Fuel Pump Module ....Page 2-30

CMP Sensor (Variable Reluctance) ....Page 2-31

DVOM Test (CMP Sensor) ....Page 2-31

Lab Scope Test (CMP Sensor) ....Page 2-32

OXYGEN SENSOR
General Description ....Page 2-33
Scan Tool Test (HO2S) ....Page 2-34
Lab Scope Test (HO2S) ....Page 2-35
HO2S Monitor ....Page 2-36
MODULE COMMUNICATION NETWORK
System Overview ....Page 2-37
REFERENCE INFORMATION
PCM PID Tables - Inputs ....Page 2-38
PCM PID Tables - Outputs ....Page 2-39
PCM Pin Voltage Tables ....Page 2-40
PCM Wiring Diagrams ....Page 2-42

2001 Focus - Anti-theft system

ACTIVE SYSTEM
Introduction ....Page 2-48
Perimeter Anti-Theft ....Page 2-49
Pinpoint Test 'A' & 'B' ....Page 2-50
Wiring Diagrams ....Page 2-51
CSM DTC & PID Lists ....Page 2-52
PASSIVE SYSTEM (SECURILOCK)
Introduction ....Page 2-53
PCM DTC & PID Lists ....Page 2-54

1996 TAURUS - Powertrain

VEHICLE IDENTIFICATION
VECI Decal ....Page 2-55
ELECTRONIC ENGINE CONTROL SYSTEM
Operating Strategies..Page 2-56
INTEGRATED ELECTRONIC IGNITION SYSTEM
Introduction ....Page 2-57
Ignition Coil Operation ....Page 2-58
Lab Scope Test (Coil Primary) ....Page 2-59
CKP Sensor (Magnetic Design) ....Page 2-60
CKP Sensor (Signal Explanation) ....Page 2-61
DVOM & Lab Scope Test (CKP Sensor) ....Page 2-62
JI Diagnostic System - DTC P0300 ....Page 2-63
EVAPORATIVE EMISSION SYSTEM
Introduction ....Page 2-67
System Components ....Page 2-68
Evaporative Charcoal Canister ....Page 2-69
Canister Vent Solenoid ....Page 2-70
EVAP Emission Valves ....Page 2-71
EVAP System Operation ....Page 2-72
Scan Tool Test (Purge Solenoid) ....Page 2-73
DVOM Test (FTP Sensor) ....Page 2-73

EVAPORATIVE EMISSION SYSTEM (CONTINUED)
 Lab Scope Test (VMV Solenoid) ....Page 2-74
 EVAP System Monitor ....Page 2-75
 JI Diagnostic System - DTC P0442 ....Page 2-76
EXHAUST GAS RECIRCULATION SYSTEM
 Introduction ....Page 2-79
 DPFE Sensor Information ....Page 2-80
 DVOM Test & Scan Tool Test (DPFE Sensor) ....Page 2-81
 Vacuum Regulator Solenoid ....Page 2-82
 Lab Scope Test (VR Solenoid) ....Page 2-83
 JI Diagnostic System - DTC P0401 ....Page 2-84
FUEL DELIVERY SYSTEM
 Introduction ....Page 2-86
 Returnable Fuel System ....Page 2-86
 Constant Control Relay Module ....Page 2-87
 Lab Scope Test (Fuel Pump) ....Page 2-88
 Fuel Pressure Regulator ....Page 2-89
 Inertia Switch ....Page 2-90
 Fuel Injectors ....Page 2-91
 Scan Tool Test (Fuel Injector & Fuel Trim) ....Page 2-92
 Lab Scope Test (Fuel Injector) ....Page 2-93
 CMP Sensor (Hall-Effect Design) ....Page 2-94
 CMP Synchronizer Replacement ....Page 2-95
 DVOM & Lab Scope Test (CMP Sensor) ....Page 2-96
IDLE AIR CONTROL SOLENOID
 General Description ....Page 2-97
 Scan Tool Test (IAC Solenoid) ....Page 2-97
 Lab Scope Test (IAC Solenoid) ....Page 2-98
MASS AIRFLOW SENSOR
 General Description ....Page 2-99
 Lab Scope Test (MAF Sensor) ....Page 2-100
OXYGEN SENSOR
 General Description ....Page 2-101
 Scan Tool Test (HO2S) ....Page 2-102
 Lab Scope Test (HO2S) ....Page 2-103
TRANSMISSION CONTROLS
 Component Tests ....Page 2-105
 Scan Tool Test (EPC, TCC & TSS) ....Page 2-106
 Scan Tool Test (TR & TFT Sensors) ....Page 2-106
REFERENCE INFORMATION
 Diagnostic Monitoring Test Results ....Page 2-107
 PCM PID Tables - Inputs ....Page 2-108
 PCM PID Tables - Outputs ....Page 2-109
 PCM Pin Voltage Tables ....Page 2-110
 PCM Wiring Diagrams ....Page 2-112

HOW TO USE THIS SECTION

This section of the training manual includes diagnostic and repair information for Ford Car applications. This information can be used to help you understand the Theory of Operation and Diagnostics of the electronic controls and devices on this vehicle.

The articles in this section are separated into three sub categories:

- Anti-Theft System (Active & Passive)
- Powertrain Control Module
- Transmission Controls

2001 Focus 2.0L VIN P (FN Automatic Transaxle)

The articles listed for this vehicle include a wide variety of subjects from the Base Engine to the PCM to how to use the vehicle diagnostics. Several of the Powertrain sensor inputs and output devices are featured along with detailed descriptions of how they operate, what can go wrong with them, and most importantly, how to test them.

The Malfunction Indicator Lamp (MIL) can help you during diagnosis. You should note if the customer complaint included whether the MIL remained on during engine operation. And be sure to determine if the MIL comes on at key on, and if it goes out after startup.

1996 Taurus 3.0L 2v VIN U (AX4N Automatic Transaxle)

The articles listed for this vehicle include a wide variety of subjects from the Base Engine to the PCM to how to use the vehicle diagnostics. Several of the Powertrain sensor inputs and output devices are featured along with detailed descriptions of how they operate, what can go wrong with them, and most importantly, how to test them.

Choose The Right Diagnostic Path

In most cases, the first step in any vehicle diagnosis is to follow the manufacturers recommended diagnostic path. In the case of Ford vehicle applications, the first step is to connect a Scan Tool (OBD II certified for this vehicle) and attempt to communicate with the engine controller or any other vehicle onboard controllers that apply. From this point on, the recommended procedure is to use the Ford Quick Test to check for faults.

Diagnostic Points to Consider

If a problem is detected in a vehicle similar to this one (whether it is a trouble code or no code condition), you need to be able to determine which vehicle system (or systems) are involved before you proceed with any testing. While this may sound like a fairly routine way to approach problem solving, it is a key point to consider before you start testing.

Due to the complexity of these vehicles, a trouble code does not always point to a failed component - instead, it may point to a system that needs to be tested for some kind of problem. For example, if an Oxygen sensor trouble code is set, you need to consider what other systems or devices could cause this component to set a trouble code.

Component Diagrams

You will find numerous "customized" component diagrams used throughout this section. These diagrams are really a mini-schematic of how a particular device (where an input or output device) is connected to the engine or transmission control module.

Each diagram contains the power, ground, signal and control circuits that relate to a particular device. Of particular note is the use of "code check" identifiers in the diagrams that indicate circuits monitored by the module in order to set a particular trouble code(s).

Vehicle Identification

VECI DECAL

The Vehicle Emission Control Information (VECI) Decal is located on the underside of the hood. Notice that the idle speed and timing are not adjustable.

TWC/ 2HO2S/EGR/SFI

These designators indicate this vehicle is equipped with a three-way catalyst (TWC), 2 Heated Oxygen Sensors (HO2S), an EGR system and sequential fuel injection (SFI).

OBD II Certified

This designator indicates that this vehicle has been certified as OBD II compliant.

ULEV

If this designator is used, the vehicle conforms to U.S. EPA & State of California Ultra-Low Emissions Vehicle standards.

U.S and California

If these designators are used, the vehicle conforms to either U.S. EPA 50-state regulations or California regulations.

Underhood View Graphic

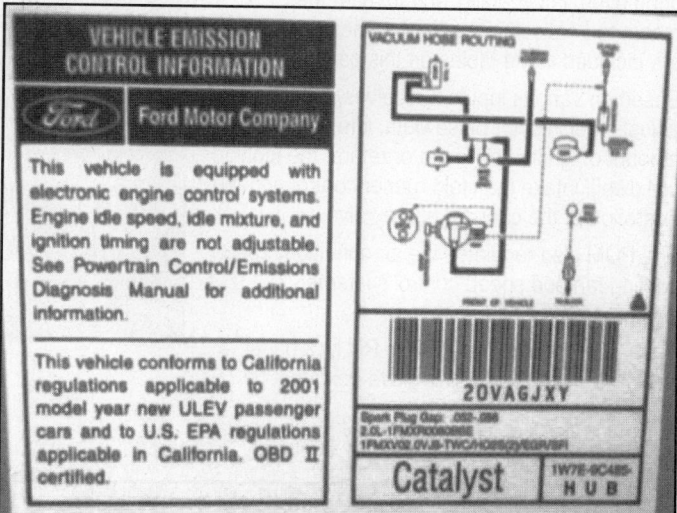

Electronic Engine Control System

OPERATING STRATEGIES

The Electronic Engine Control (EEC) system uses several operating strategies to maintain good overall driveability and to meet the EPA mandated emission standards. These strategies are included in the tables on this page.

Based on various inputs it receives, the PCM adjusts fuel injector pulsewidth, idle speed, the amount of ignition advance or retard, the ignition coil dwell, intake manifold runner control and the operation of the canister purge valve.

The PCM also regulates the air conditioning, cooling fan and speed control systems.

PCM Location

The PCM is located above the RH kick panel behind the glove box. The glove box must be removed.

Operating Strategies

A/C Compressor Clutch Cycling Control	Closed Loop Fuel Control
Electric Fuel Pump Control (FPDM)	Exhaust Gas Recirculation Control
Fuel Metering for Sequential Injection	Fuel Pump Monitor
Idle Speed Control	Ignition Timing Control for A/F Change
Vapor Canister Purge Control	

General Operating Strategies

Adaptive Fuel Trim Strategy	Adaptive Idle Speed Control Strategy
Base Engine/Transmission Strategy	Failure Mode Effects Mgmt. Strategy

■ NOTE: Hardware Limited Operating Strategy or HLOS is a part of the PCM hardware.

Operational Control of Components

A/C Compressor Clutch	EGR System (VR Solenoid)
Engine Cooling System Electric Fan	Fuel Delivery (Injector Pulsewidth)
Fuel Pump Driver Module	EVAP System Solenoids
Fuel Pump Relay	Intake Man. Runner Control (Split Port)
Idle Speed	Malfunction Indicator Lamp
Shift Points for the A/T	Integrated Speed Control
Torque Converter Clutch Lockup (TCC)	Transmission Line Pressure (EPC)

Items Provided (or Stored) By the EEC System

Data to the Operational Control Centers	DTC Data for Non-Emission Faults
DTC Data for the FN Transaxle	DTC Data for Emission Related Faults

Integrated Electronic Ignition System

INTRODUCTION

The PCM controls the operation of the Integrated Electronic Ignition (EI) system on this vehicle application. Battery voltage is supplied to the coilpack (2 coils in one assembly) via the Ignition Relay and #1 fuse (30A) in the battery junction box.

The PCM controls the individual ground circuits of the two (2) ignition coils. By switching the ground path for each coil "on" and "off" at the appropriate time, the PCM controls the ignition timing for each cylinder pair correctly to meet all engine operating conditions. The coil pairs are identified as Coilpacks 1 and 2. Coil '1' fires cylinders 1 & 4, and Coil '2' fires cylinders 2 & 3. The firing order on this engine is 1-3-4-2.

Integrated EI System Graphic

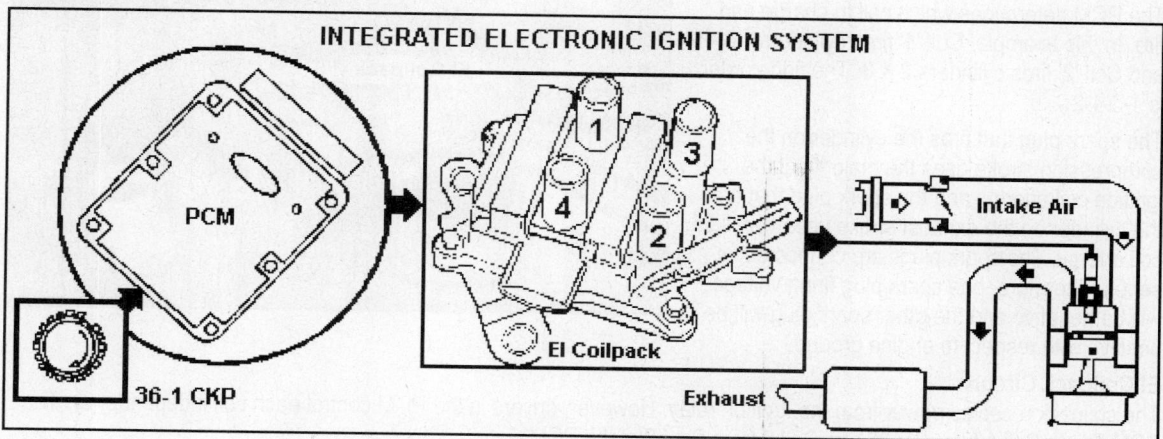

Multi-Strike Ignition Strategy

The multi-strike strategy is used to achieve longer burn times to completely burn the lean idle mixture without overheating the coils. At engine speeds over 1500 RPM, the PCM changes to a single-strike strategy.

System Components

The main components of this Integrated EI system are the coilpack, crankshaft position sensor, the PCM and EI system related wiring.

Ignition Spark Timing

The ignition timing on this EI system is entirely controlled by the PCM - base timing is not adjustable. Do not attempt to check the base timing - you will receive false readings.

The PCM uses the CKP sensor signal to indicate crankshaft position and speed by sensing teeth on a pulse wheel mounted on the transaxle. It also uses this signal to calculate a spark target and then fire the coils to that spark target.

Once the engine starts, the PCM calculates the spark advance with these signals:

- BARO Sensor (a signal calculated from the MAF Sensor reading)
- ECT and IAT Sensors
- Engine Speed Signal (rpm)
- MAF Sensor
- Throttle Position

IGNITION COIL OPERATION

The ignition coilpack consists of two (2) independent coils molded together. The coil assembly is mounted to the intake manifold. Spark plug cables rout to each cylinder from the coil assembly.

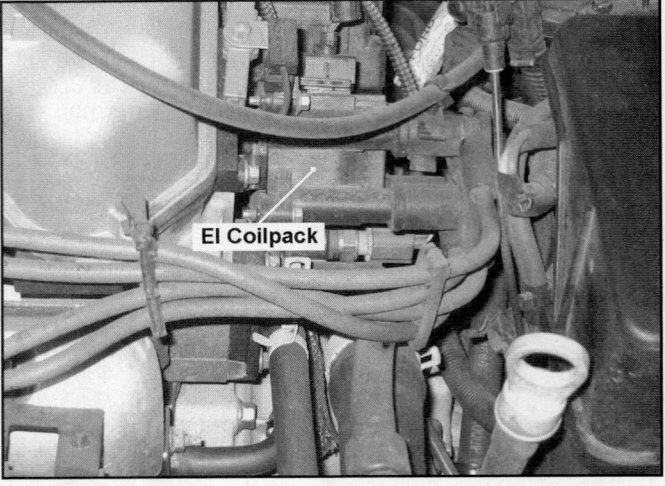

El Coilpack

The coil fires two spark plugs before each power stroke. One spark plug fires a cylinder under compression, while the other plug fires the cylinder on the exhaust stroke (a true waste spark method).

The PCM determines which coil to charge and fire. In this example, Coil '1' fires cylinders 1 & 4, and Coil '2' fires cylinders 2 & 3. The firing order is 1-3-4-2.

The spark plug that fires the cylinder on the compression stroke uses the majority of the ignition coil energy, and the spark plug that fires the cylinder on the exhaust stroke uses very little coil energy. The spark plugs are connected in series. Therefore, one spark plug firing voltage will be negative and the other spark plug will be positive with respect to engine ground.

El Coilpack Circuits

The coilpack receives power from the Ignition relay. However, drivers in the PCM control each coil independently. The PCM drives Coil 1 from Pin 26 and Coil 2 from Pin 52 of the PCM.

Trouble Code Help

Note the two driver circuits that are shown with dotted lines in the schematic. The PCM monitors these circuits as it switches the coil drivers "on" and then "off". If the PCM does not receive a valid IDM signal due to a coil or circuit problem, it will set DTC P0351 for Coil 1, or it will set DTC P0352 for Coil 2.

Ignition Coil Primary Schematic

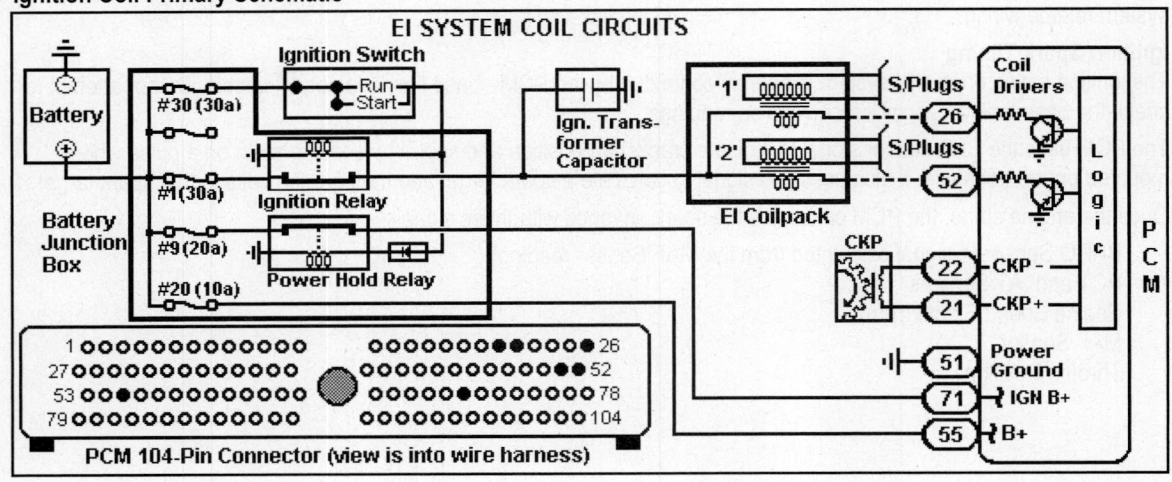

DVOM TEST (CKP SENSOR)

A DVOM can be used to test the CKP when sensor values are desired, such as voltage output (AC) or frequency (Hz). Place the shift selector in Park, block drive wheels for safety.

DVOM Connections

Turn the key off and connect the breakout box (BOB) if available. Select AC volts on the DVOM. Connect the DVOM positive lead to the CKP (+) circuit (White/Red wire) at Pin 21 and the negative lead to CKP (-) circuit (Brown/Red wire) at Pin 22.

Test Results

This capture was taken at 2500 RPM. The DVOM display shows a CKP signal of 5.85v AC at a rate of 1.6 kHz. For a no-start, the DVOM should show a minimum of 400mv AC during cranking. At idle, the DVOM should show approximately 425 Hz.

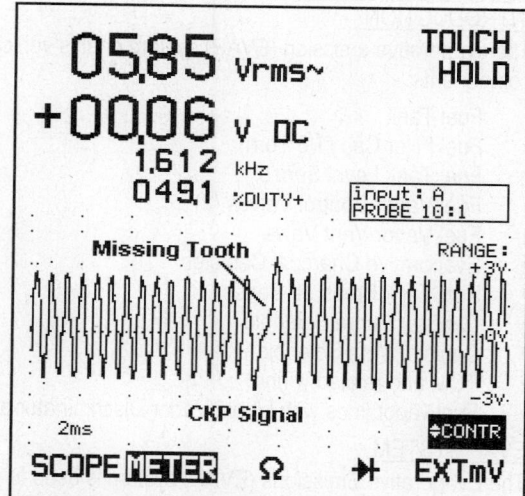

LAB SCOPE TEST (CKP SENSOR)

The Lab Scope can be used to test the CKP sensor as it provides a very accurate view of sensor waveform and of any glitches. Place the shift selector in Park and block the drive wheels for safety.

Scope Connections

Install a BOB if available. Connect the Channel 'B' positive probe to the CKP (+) circuit at Pin 21 (White/Red wire) and the negative probe to the CKP (-) circuit at Pin 22 (Brown/Red wire).

Scope Settings

To make the waveforms as clear as possible, set the scope settings to match the examples.

Lab Scope Example

In this example, the Channel 'A' trace shows the Coil 1 Primary Signal and the CKP sensor signal. This waveform was captured at idle, and represents around 200 degrees of crank rotation (each pulse represents 10 degrees).

Trigger Help

If a waveform is erratic, or the scope used has trouble locking a signal on the screen, use the extra channel as a trigger. In this example, the CKP signal amplitude varied, (probably crank pulley run-out). This variance is not severe enough to cause a driveability problem, but could cause a scope to lose the intended trigger point, (the higher amplitude at the missing tooth in the CKP waveform). Choose another circuit that has a direct relationship to help stabilize the waveform on the screen. In this example, coil primary was used as the trigger, and the CKP signal remained stable.

Evaporative Emission System

INTRODUCTION

The evaporative emission (EVAP) system on this vehicle is a Running Loss/Monitor system, and includes the following components:

- Fuel Tank
- Fuel Filler Cap (1/8 Turn)
- Fuel Tank Level Sensor
- Fuel Vapor Control Valves (2)
- Fuel Vapor Vent Valve
- Evaporative Charcoal Canister
- Fuel Tank Pressure Sensor
- Canister Purge Solenoid
- Canister Vent Solenoid
- PCM and related wiring
- Fuel vapor lines with Liquid/Vapor Discriminator and Test Port

EVAP SYSTEM

The Evaporative Emissions (EVAP) system is used to prevent the escape of fuel vapors to the atmosphere under hot soak, refueling and engine off conditions. Any fuel vapor pressure trapped in the fuel tank is vented through the fuel vapor vent and control valves on top of the fuel tank, and routed to the EVAP Canister for storage.

The most efficient way to dispose of fuel vapors without causing pollution is to burn them in the normal combustion process. The PCM commands the purge valve 'On' (to open) during normal operation. With the valve open, manifold vacuum is applied to the canister and this allows it to draw in fresh air and fuel vapors from the canister into the intake manifold. These vapors are drawn into each cylinder to be burned.

The PCM uses various sensor inputs to calculate the desired amount of EVAP purge flow. The PCM meters the purge flow by varying the duty cycle of the EVAP Canister Purge Valve control signal.

EVAP System Graphic

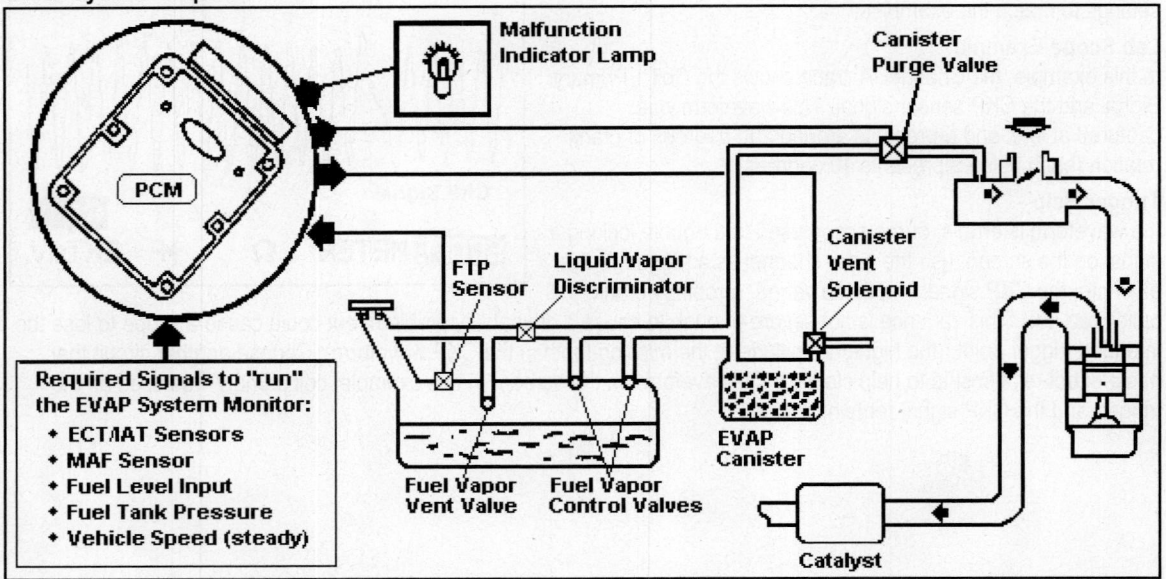

EVAPORATIVE CHARCOAL CANISTER

This vehicle uses a charcoal canister to store fuel vapors. The canister is located at the forward edge of the fuel tank. Fuel vapors from the fuel tank are stored in the canisters.

Once the engine is started, fuel vapors stored in charcoal canister are purged into the engine where they are burned during normal combustion.

The activated charcoal has the ability to purge or release any stored fuel vapors when it is exposed to fresh air.

CANISTER VENT SOLENOID

The canister vent (CV) solenoid is a normally open (N.O.) solenoid with a resistance of 48-65 ohms. Under normal conditions, the CV solenoid is open to allow fuel system vapors to escape after the hydrocarbons have been absorbed by the canisters. While the system is purging, the open solenoid allows fresh air to enter the canisters, which frees up the trapped hydrocarbons to be drawn into the intake manifold to be burned.

The main function of the CV solenoid is to isolate the charcoal canister from outside air for testing purposes.

During OBD II EVAP Monitor operation, the PCM activates (closes) the CV solenoid. At the same time, the purge solenoid is activated, (open). This allows manifold vacuum to draw fuel tank vacuum down to a specified level for leak test purposes.

Canister Vent Solenoid Location

The CV solenoid is located on the EVAP canister, forward of the fuel tank. The CV solenoid can be serviced separately.

FUEL VAPOR VENT VALVE & FUEL VAPOR CONTROL VALVES

The vapor valves are located on top of the fuel tank. The valves allow fuel vapors to flow from the

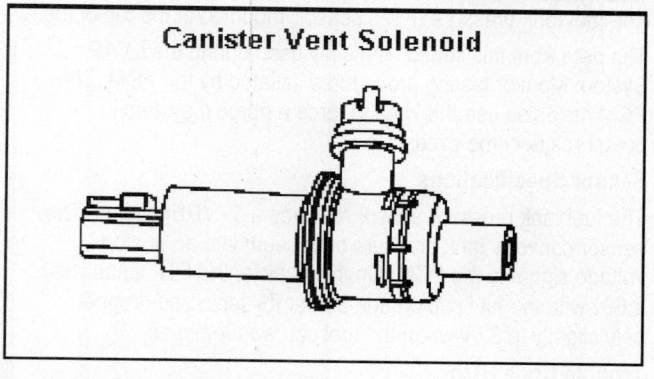

tank to the canister, while preventing liquid fuel from entering the lines during over-filling or aggressive driving.

This EVAP system is designed to recover fuel vapors displaced during refueling. The fuel vapor control valves were designed to handle these large quantities of fuel vapors.

CANISTER PURGE VALVE

The EVAP canister purge valve is a normally closed (N.C.) solenoid controlled by the PCM. A single purge valve,

located on the firewall, allows vapors from the charcoal canister to be drawn into the intake manifold when energized.

The PCM controls purging by commanding a driver to ground the purge valve solenoid. By varying the duty cycle, (% of time driver is grounded and solenoid is operating), the PCM can precisely control the volume of fuel vapors being drawn into the manifold.

Leak Testing Tip

All OBD II vehicles must have an EVAP test port. This application incorporates the test port into the purge valve. The test port is for use with Ford Evaporative Emissions Leak Tester, part No. 310-F007 (134-00056), or equivalent. However, smoke machines have proven effective in isolating EVAP system leaks using the test port.

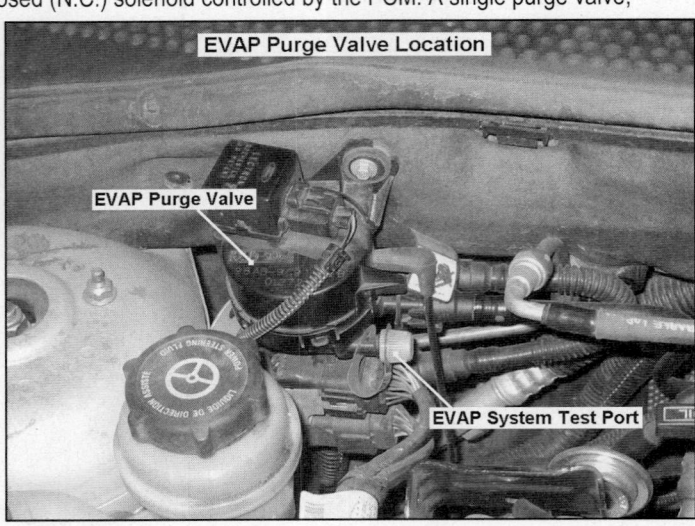

FUEL TANK PRESSURE SENSOR

The fuel tank pressure (FTP) sensor, mounted in the top of the fuel tank, monitors system vapor pressure levels.

The data from this sensor is mainly used during the EVAP System Monitor testing procedures initiated by the PCM. The PCM may also use this data to force a purge if system pressures become excessive.

Sensor Specifications

The fuel tank pressure sensor receives a 5v VREF signal. The sensor converts tank pressure or vacuum into an analog voltage signal to the PCM. On this vehicle, the FTP value was 2.69v with the fuel cap on and a near full tank, and dropped only slightly to 2.6v when the fuel cap was removed.

Trouble Code Help

Whenever a trouble code is retrieved, it is important to read the code description in the Ford Handbook so that you understand the conditions present when it set. For instance, DTC P0452 will set whenever the FTP sensor signal is below 0.22v. This trouble code only sets when the signal voltage falls outside of its parameters.

In this case, you don't need to check for leaks in the EVAP system (broken components or disconnects) because they cannot an electrical circuit fault. In effect, you should limit your testing to the actual sensor and its related circuit for this type of trouble code.

SCAN TOOL TEST (CV SOLENOID)

Connect Scan Tool to DLC under driver's side of dash.

EVAP CV DC (Duty Cycle) PID

The canister vent solenoid is activated by the PCM during the EVAP Running Loss Test (part of OBD II). To obtain more information on how to run the EVAP System Monitor, refer to this heading in Section 1 of the Ford Handbook. During normal engine operation, the PID value should be 0%. The PID value will change to a high duty cycle when the EVAP system monitor begins testing.

Forcing the CV Solenoid Closed

The CV solenoid can be difficult to test, since it operates only once per drive cycle. To force CV solenoid operation with a scan tool, access OBD Controls option from Generic scan tool menus.

Follow the sequence at the right. The Generic EVAP Leak Test is used with EVAP system leak test equipment. However, the procedure involves the PCM commanding the CV solenoid closed to seal off the EVAP system.

Once test conditions have been enabled, the CV solenoid can be tested using a DVOM or Lab Scope. For a Lab Scope example of a CV solenoid during an EVAP Leak Test, refer to the Lab Scope Test (CV solenoid).

Generic OBD Control Limitations

Because these controls are not manufacturer specific, they are not consistent and will not work on all applications. This example from a 2001 Focus can be duplicated on many other, but not all, OBD II vehicles.

SCAN TOOL TEST (PURGE SOLENOID)

Start the engine and monitor EVAP PURGE DC PID. The PID value will be a varying percentage between 0% and 100%. A higher percentage indicates more purge commanded. The PID value will be high at the start of the leak check portion of the EVAP system monitor.

The PID may be monitored to observe the PCM strategy as the vehicle is operated under different driving conditions. The Purge solenoid may also be monitored with a DVOM or Lab Scope to compare the PCM strategy to the driver command received by the component.

The EVAP duty cycle command varies over time, generally in patterns. This means that the EVAP PURGE DC PID from the example above, (46%), may read anywhere from 0% to 100%. It is important to verify that the PCM purge command varies over time and varying conditions.

It should be noted that these PIDs are not included in the Generic OBD II PID List.

```
GENERIC OBD CONTROLS

┌──────────────────────────┐
│    OBD CONTROL MENU       │
│ ▮F1: EVAP LEAK TEST▮      │
└──────────────────────────┘

┌──────────────────────────┐
│    EVAP LEAK TEST         │
│                          │
│ This test mode           │
│ enables conditions       │
│ required to conduct      │
│ an evaporative           │
│ system leak test,        │
│ but does not run the     │
│ test.                    │
│                          │
│    Press [ENTER]         │
│     to begin.            │
└──────────────────────────┘

┌──────────────────────────┐
│    EVAP LEAK TEST         │
│                          │
│ Test conditions have     │
│ been enabled.            │
│                          │
│ Turn ignition off        │
│ to terminate test.       │
│                          │
│    PRESS [ENTER]         │
└──────────────────────────┘
```

```
Feb 9, 02        2:18:17 pm
┌──────────────────────────┐
│ EVAP PURGE DC            │
│    46 % DC               │
│ UP O2S BANK1             │
│    0.69 VOLTS            │
└──────────────────────────┘

     0.00 ᵥ

[RCV] FOR ENHANCED
[*EXIT] TO EXIT
```

DVOM TEST (PURGE SOLENOID)

Place the gear selector in Park and block the drive wheels for safety.

DVOM Connections

Connect the DVOM positive probe to the Purge solenoid control circuit (Black/Blue wire) at the component, the BOB or PCM Pin 56.

Connect the negative probe to the battery negative post. Start the engine and change the engine speed while monitoring the DVOM.

The reading should vary as the engine speed, load and temperature conditions are changed. The frequency (Hz) will remain fixed, (except when a 0% or 100% purge is commanded). In effect, the voltage reading varies in proportion to the duty cycle of the device. Compare these readings to the Scan Tool duty cycle readings.

DVOM Example (1) & (2)

These examples show the pattern of driver commands to the solenoids. The readings (9.38v and 7.09v) shown are averages of the voltage fluctuations. The higher the duty cycle command is, the lower the average voltage.

The frequency is fixed at 100 Hz by the PCM (only the duty cycle varies). The duty cycle in example (1) was 52.3% at idle and 71.1% at cruise in example (2).

The duty cycle is constantly varying according to temperature, driving conditions and test status. This test can be used to verify that the PCM is in control of the purge process.

EVAP System Test Schematic

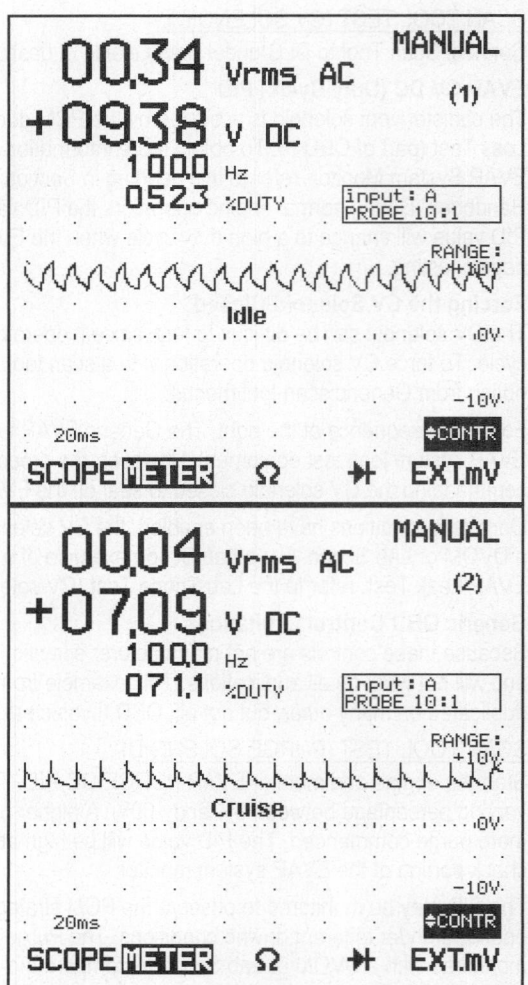

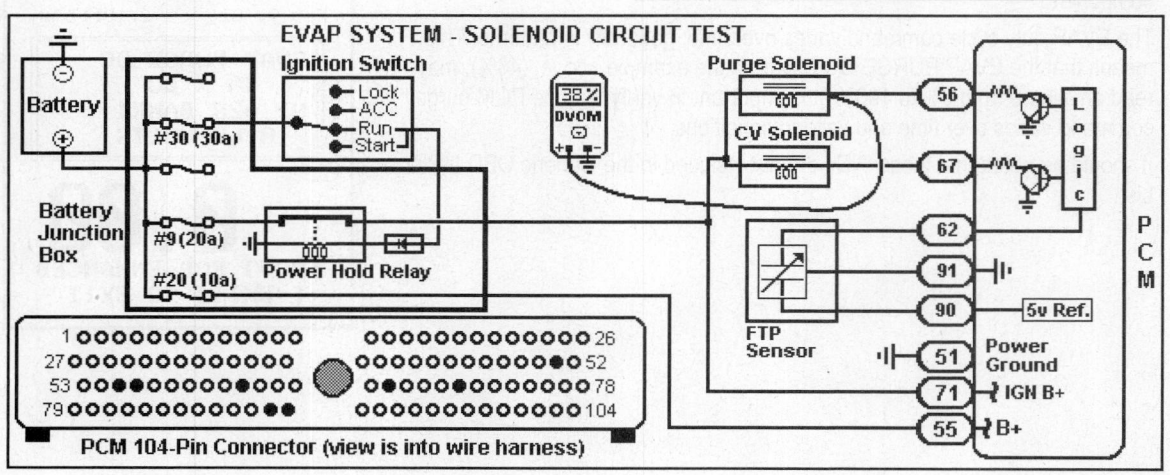

LAB SCOPE TEST (CV SOLENOID)

The EVAP Running Loss Monitor is activated once per drive cycle. If you want to activate the CV solenoid without an 8-hour cold soak, you can command the CV solenoid on using a Generic Scan Tool. Refer to the Scan Tool Test (CV Solenoid) heading on Page 2-15.

Example (1)

In example (1), the trace shows the CV solenoid with system voltage present (the circuit is not complete to ground through the driver in the PCM (the valve is not operating).

This waveform indicates the circuit has proper voltage and there are no problems in the CV solenoid circuit (i.e., it is not open or shorted).

Example (2)

In example (2), the trace shows the CV solenoid commanded on using the EVAP Leak Test from the Generic scan tool software. The CV solenoid is energized (closed) when the PCM grounds the circuit, and the signal is pulled low. This can be seen on the Lab Scope as a drop from system voltage to ground.

The circuit will remain grounded until the test is terminated when the key is turned off. The signal will remain low because battery voltage has been removed from the circuit. When the key is cycled back on, system voltage is applied to the circuit. The PCM is no longer applying ground, so the trace shows system voltage again.

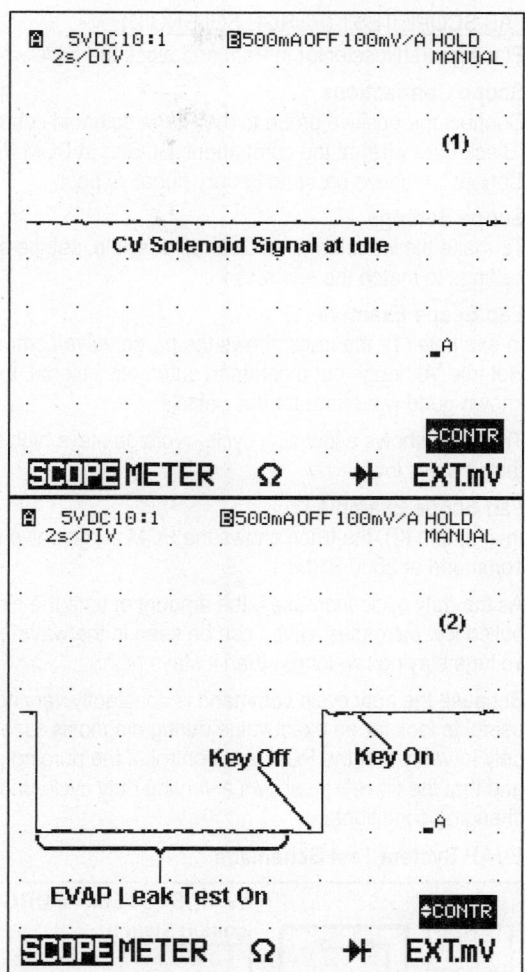

EVAP System Test Schematic

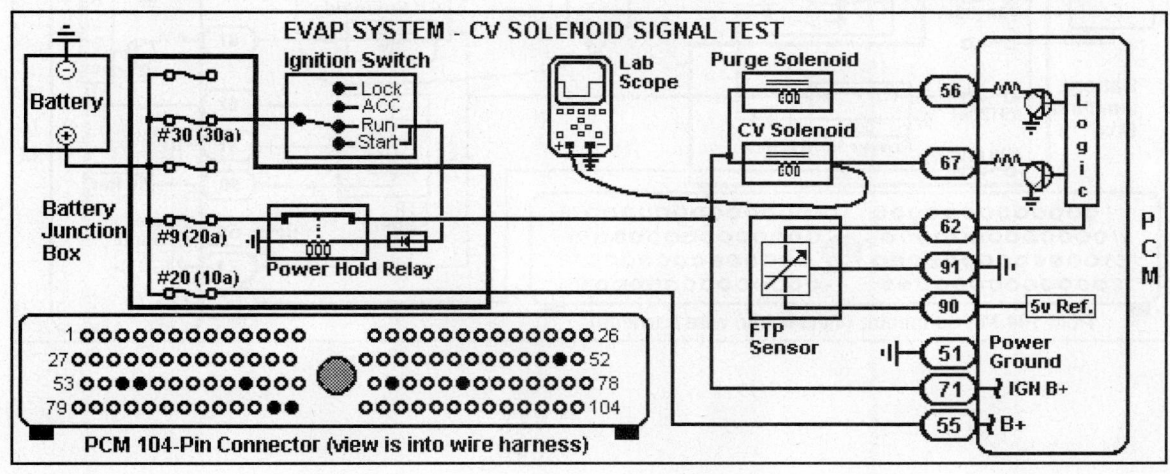

LAB SCOPE TEST (PURGE SOLENOID)
Place gearshift selector in Park and block the drive wheels for safety.

Scope Connections
Connect the positive probe to the Purge solenoid control circuit (Black/Blue wire) at the component, BOB or at PCM Pin 56. Connect negative probe to battery negative post.

Scope Settings
To make the waveforms as clear as possible, set the scope settings to match the examples.

Lab Scope Example (1)
In example (1), the trace shows the purge valve command at Hot Idle. Although not a common automotive signal, this is a known good waveform for this vehicle.

The trace shows a low duty cycle, (voltage stays high longer than it stays low).

Lab Scope Example (2)
In example (2), the trace shows the PCM purge valve solenoid command at 2500 RPM.

As the duty cycle increases, the amount of time the circuit is pulled low increases, which can be seen in the waveform as voltage staying low longer than it stays high.

Because the duty cycle command is constantly varying it is not useful to look for an exact value during diagnosis. Use this test only to verify that the PCM is in control of the purging process, and that the waveform shows a varying duty cycle under changing conditions.

EVAP System Test Schematic

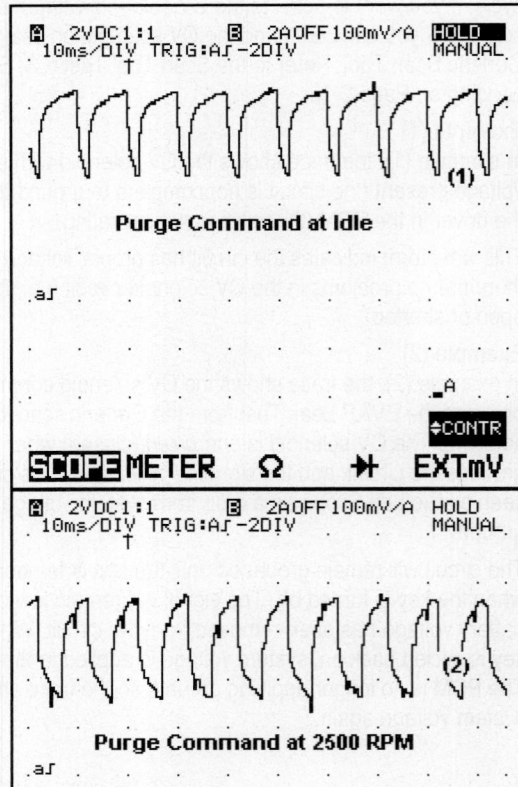

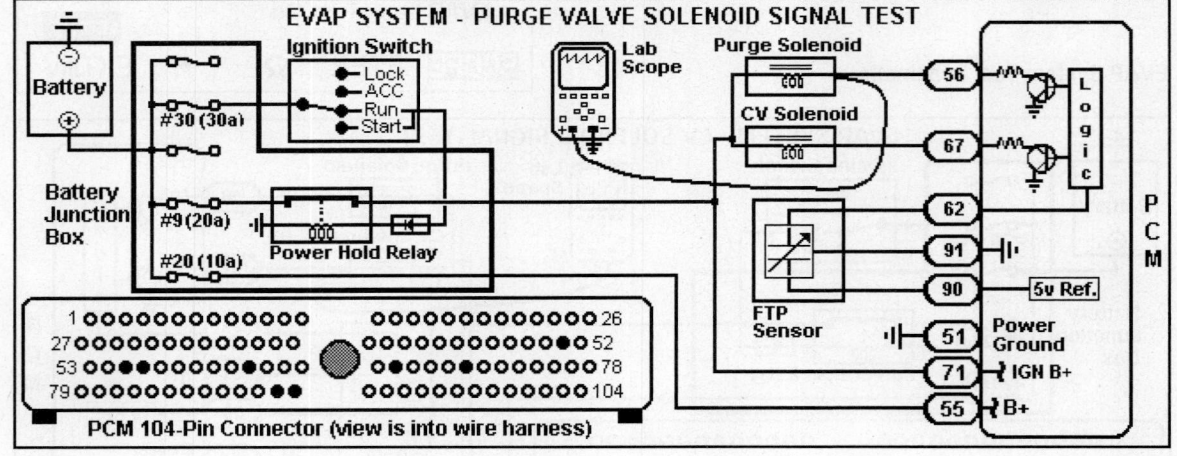

Exhaust Gas Recirculation System

INTRODUCTION

The EGR system is designed to control the formation of NOx during combustion by controlling combustion temperatures. Exhaust gas has few oxygen or hydrocarbon molecules, the two main ingredients of combustion. The EGR system allows small amounts of exhaust gases to be recirculated back into the combustion chamber to mix with the A/F mixture. This reduces combustion temperatures and less NOx is formed.

The DPFE system includes a DPFE sensor, EGR valve and orifice tube assembly, EGR Vacuum Regulator (VR) solenoid valve assembly and PCM and related vacuum hoses.

Differential Pressure Concepts

The DPFE system is based on the principle that airflow through an orifice will create a pressure differential between the two sides of the orifice. This pressure drop can be used for precise EGR flow monitoring and control.

The DPFE sensor is connected to both sides of this orifice where it is exposed to pressure in vacuum lines. One side of the orifice is exposed to exhaust backpressure and the other side to manifold vacuum (EGR valve open). This design creates a pressure drop across the orifice whenever there is EGR flow. When the EGR valve closes, there is no longer flow across the metering orifice and the pressure equalizes.

DPFE System Operation

The operation of the DPFE System is explained next:

1. The PCM monitors the CKP, ECT, IAT, MAF and TP sensor signals to determine the optimum operating conditions for this system. The engine must be warm, running at a stable engine speed under moderate load before this system is activated. It will not operate at idle, extended WOT periods or if a fault is detected in the EGR system.

2. The PCM calculates a desired amount of EGR flow for each engine condition. Then it determines the desired pressure drop across the metering orifice that it needs to achieve the correct amount of flow, and issues a command to the VR solenoid.

3. The VR solenoid receives a variable duty cycle signal (0-100%). The higher the duty cycle rate, the more vacuum the solenoid diverts to the EGR valve.

4. The increase in vacuum acting upon the EGR valve diaphragm overcomes the valve spring pressure and lifts the valve pintle off its seat. This allows exhaust gas to flow.

5. As exhaust gases are flowing through the orifice, the DPFE sensor measures the pressure drop and relays a proportional voltage signal (from 0-5 volts) to the PCM. The PCM uses this signal to verify correct EGR flow or correct for any errors.

EGR System Graphic

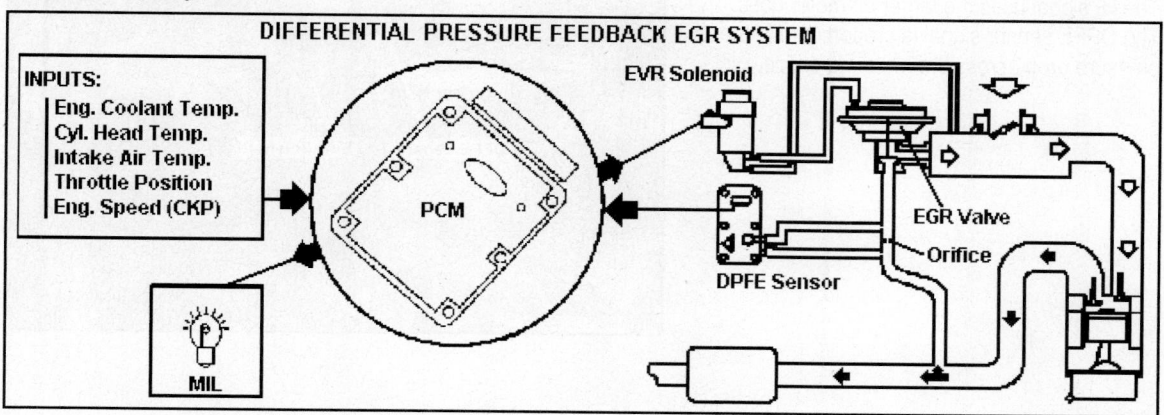

DIFFERENTIAL PRESSURE FEEDBACK EGR SYSTEM

INPUTS:
Eng. Coolant Temp.
Cyl. Head Temp.
Intake Air Temp.
Throttle Position
Eng. Speed (CKP)

EVR Solenoid

PCM

EGR Valve

Orifice

DPFE Sensor

MIL

EXHAUST GAS RECIRCULATION VALVE

The EGR valve used with the DPFE EGR system is a conventional vacuum actuated EGR valve. The valve increases or decreases the flow of exhaust gas. A spring forces the valve closed at rest. As vacuum applied to the valve diaphragm, the spring pressure is overcome and the valve begins to open. As the vacuum signal weakens, EGR spring pressure closes the valve.

The valve starts to open at about 1.6" Hg of vacuum, and is fully open over about 5" Hg. During operation, the EGR valve will be somewhere between fully open and fully closed. In effect, the PCM uses the VR solenoid to maintain precise control over EGR valve position.

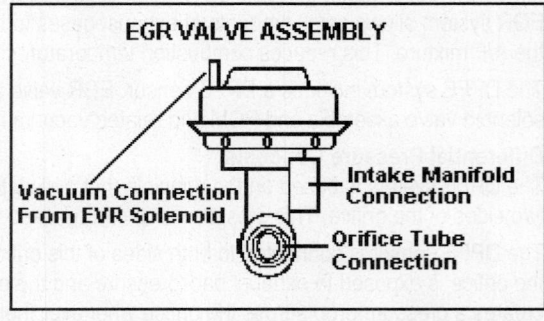

ORIFICE TUBE ASSEMBLY

The orifice tube assembly is a section of tubing placed between the exhaust system and intake manifold. The tube provides a flow path for EGR gases to the intake manifold. It has a metering orifice and two pressure pickup tubes.

The internal metering orifice creates a measurable pressure drop as the EGR gases flow. This pressure differential across the orifice is picked up by the DPFE sensor and converted to feedback data on system operation by the PCM.

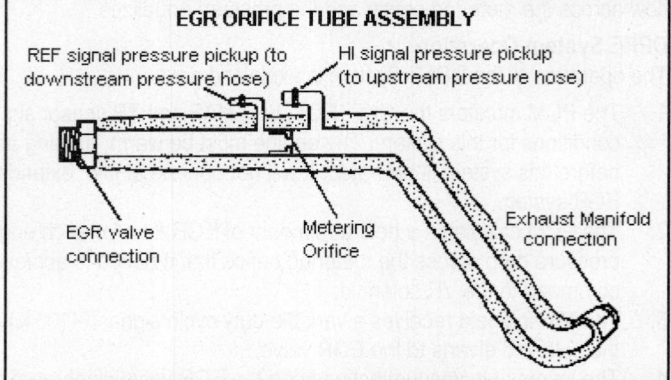

DPFE SENSOR INFORMATION

The DPFE sensor is a capacitive-type pressure transducer that monitors the differential pressure across the metering orifice located in the orifice tube assembly. The sensor receives vacuum or pressure signals through two hoses referred to as the downstream pressure (REF signal) and upstream pressure hose (HI signal).

The HI signal uses the larger diameter hose. The DPFE sensor signal is proportional to the pressure drop across the metering orifice.

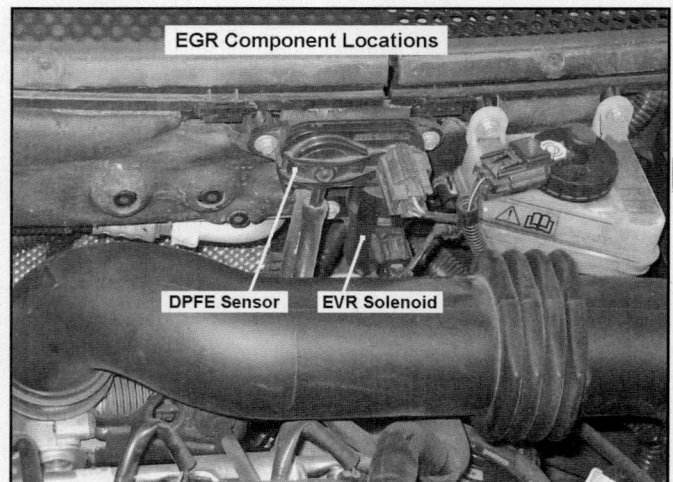

LAB SCOPE TEST (DPFE SENSOR)

The DPFE sensor signal can be checked for exact voltage values with a DVOM when tests for proper calibration and range are needed. However, the Lab Scope is the tool of choice for testing both the DPFE sensor and EGR system *performance.*

DPFE Sensor Functional Test

This test can also be performed with a hand vacuum pump. Disconnect vacuum hoses from DPFE sensor. By applying vacuum to the small port of the DPFE sensor and then releasing, the sensor performance can be observed in the service bay. The voltage value at rest should be 0.2-1.3v. Apply 8-9" Hg vacuum to the small port. Voltage should exceed 4v. Voltage should drop under 1.5v within 3 seconds when vacuum is released.

Lab Scope Connections

Locate the DPFE sensor circuits (underhood, at the BOB or at the PCM). Connect the Lab Scope positive lead to the DPFE signal (White/Blue wire) and negative lead to chassis ground. Start the engine and record sensor signal under various conditions.

Lab Scope Test Description

The Lab Scope and hand vacuum pump can be used to check the condition of the EGR system. Reinstall the DPFE sensor as needed. Disconnect the EGR vacuum hose (from VR solenoid). Connect the vacuum pump and apply 8-9" Hg vacuum to the EGR valve. Do this step at 2000 RPM to avoid an engine stall. Observe the DPFE signal to verify that exhaust gases are flowing when vacuum is applied, and stop flowing when vacuum is released.

This test can help verify the EGR valve is functional and has no diaphragm leaks, the system has no blockages, the orifice tube is clear and the DPFE sensor with normal range. It is an effective test because several devices can be checked with a single procedure.

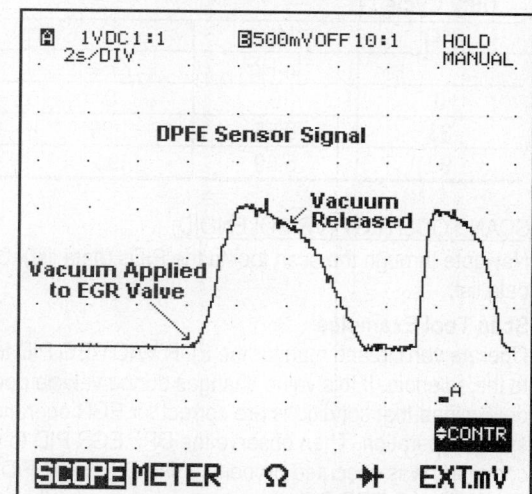

DPFE Sensor Test Schematic

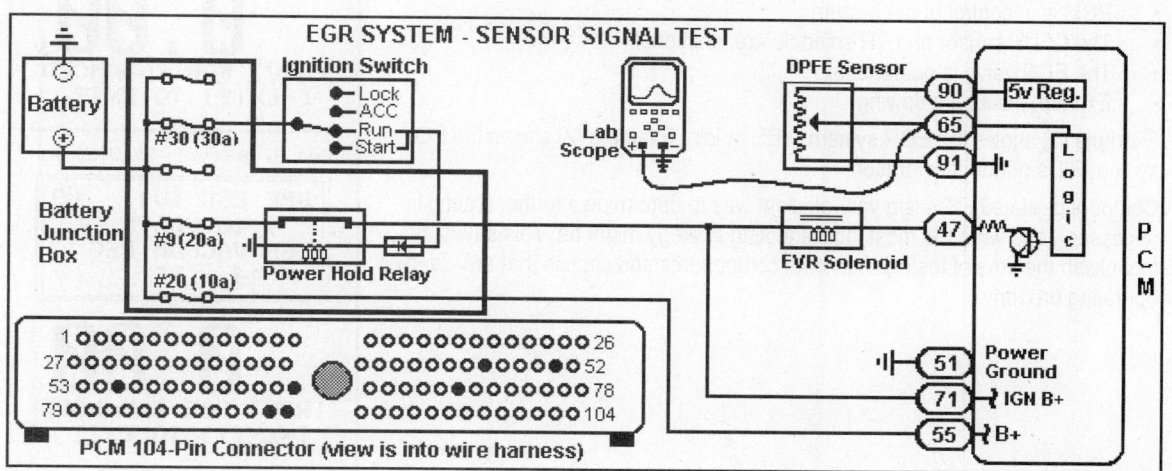

VACUUM REGULATOR SOLENOID

The EGR vacuum regulator (VR) solenoid is used to regulate the vacuum supply to the EGR valve. As the duty cycle to the coil is increased, the vacuum signal passed through the solenoid to the EGR valve increases.

Any source vacuum that is not directed to the EGR valve is released through a vent valve into the atmosphere. At 0% duty cycle (no solenoid command), the EGR vacuum regulator solenoid allows some vacuum to pass, but not enough to open the valve.

EGR VACUUM REGULATOR SOLENOID

Vacuum Bleed

Electrical connector

To EGR Valve

Source Vacuum

Resistance Specification at room temperature: 26-40 ohms

EGR Vacuum Regulator Solenoid Table

Duty Cycle	Vacuum Output					
(%)	Min		Nominal		Maximum	
	In-Hg	KPa	In-Hg	KPa	In-Hg	KPa
0	0	0	0.38	1.28	0.75	2.53
33	0.55	1.86	1.3	4.39	2.05	6.90
90	5.69	19.2	6.32	21.3	6.95	23.47

SCAN TOOL TEST (VR SOLENOID)

Navigate through the scan tool to the PIDs (data list). Select DPF EGR and EGR VAC REG PIDs from the scan tool data list.

Scan Tool Examples

Operate vehicle and monitor the EGR VAC REG PID to see the PCM command to the solenoid. If this value changes during vehicle operation, the PCM has determined that conditions are correct for EGR operation, and is commanding system operation. Then observe the DPF EGR PID to verify that the VR command was executed properly. If the DPF EGR PID reacts proportionally to the EGR VAC REG PID, then it has been proven that:

- PCM is in control of the system
- The DFPE sensor and VR solenoid are functional.
- The EGR valve is operational.
- Exhaust gases are flowing.

Example (1) shows the EGR system PIDS at idle. Example (2) shows the EGR system PIDs under acceleration.

Comparing related PIDs is a very efficient way to determine if further testing is necessary, and what the most logical testing strategy might be. This saves the technician the time of testing individual components and circuits that are clearly operating properly.

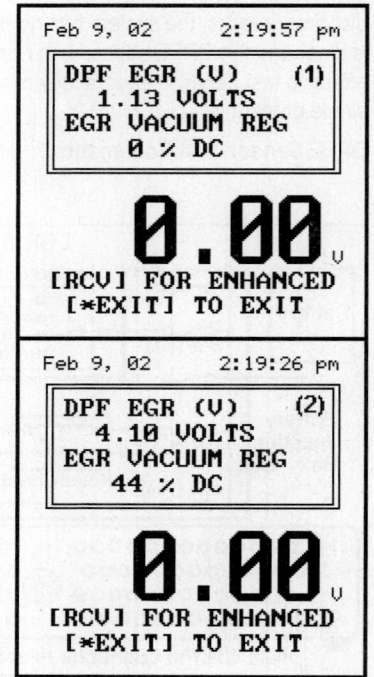

Feb 9, 02   2:19:57 pm

DPF EGR (V)   (1)
   1.13 VOLTS
EGR VACUUM REG
   0 % DC

0.00 v

[RCV] FOR ENHANCED
[*EXIT] TO EXIT

Feb 9, 02   2:19:26 pm

DPF EGR (V)   (2)
   4.10 VOLTS
EGR VACUUM REG
   44 % DC

0.00 v

[RCV] FOR ENHANCED
[*EXIT] TO EXIT

LAB SCOPE TEST (VR SOLENOID)

The Lab Scope is the "tool of choice" to test the VR solenoid as it provides an accurate view of the solenoid response and frequency rate. Place the gearshift selector in Park and block the drive wheels for safety.

Scope Connections

Connect the Channel 'A' positive probe to the VR control circuit (BRN/GRN wire) at PCM Pin 47 and the negative probe to battery ground.

Scope Settings

To make the waveforms as clear as possible, set the scope settings to match the examples.

Lab Scope Tests

The engine was allowed to run long enough to enter closed loop operation prior to testing. Note that the two examples were captured under different operating conditions.

Lab Scope Example (1)

In example (1), the trace shows the VR solenoid in the off position. This is a good test to use to verify the correct system voltage, and verify that there are no problems present in the circuit (i.e., it is not open or grounded).

Lab Scope Example (2)

In example (2), the trace shows the VR solenoid command. This waveform was captured during initial acceleration from a stop. The VR PID would read close to 25% on a Scan Tool in this example.

VR Solenoid Signal Test Schematic

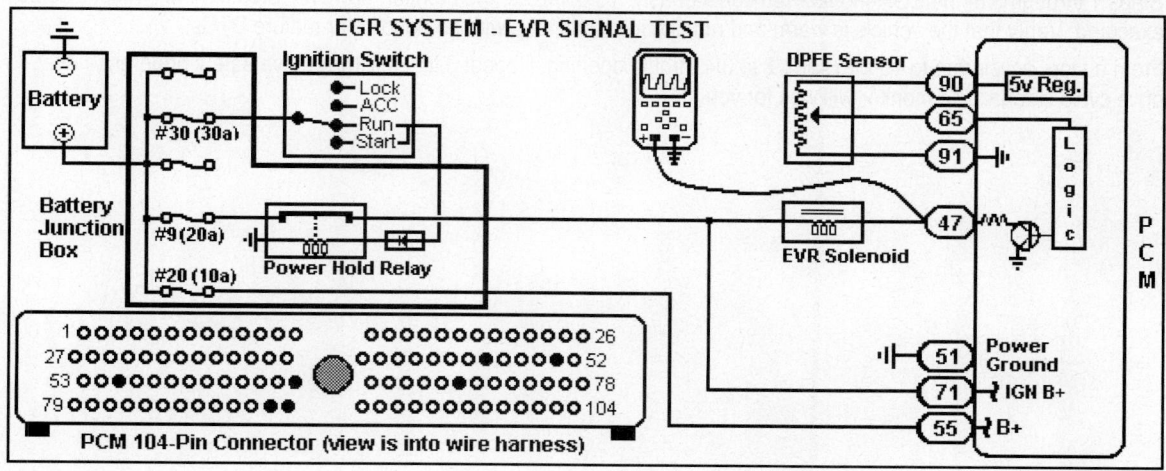

EGR SYSTEM MONITOR

The Differential Pressure Feedback EGR System Monitor is a PCM strategy designed to test the integrity and flow characteristics of the EGR system. The monitor is activated during EGR system operation and after certain base engine conditions are satisfied. Input from the ECT, CHT (Cylinder Head Temperature sensor), IAT, TP and CKP sensors is required to activate the EGR System Monitor. Once activated, the EGR System Monitor will perform each of the tests described below during the operating conditions listed.

1. The DPFE sensor and circuit are continually tested for opens and shorts. The monitor looks for the DPFE sensor circuit voltage to exceed the maximum or minimum allowable limits.

   The DTCs associated with Step 1 are P1400 and P1401.

2. The EGR Vacuum Regulator (VR) solenoid is continually tested for open and shorts. The monitor looks for an EGR vacuum regulator circuit voltage that is inconsistent with the VR circuit commanded output state.
   The DTC associated with Step 2 is P1409.

3. The test for stuck-open EGR valve or EGR flow at idle is continuously performed whenever at idle, (TP sensor indicating closed throttle). The monitor compares the DPFE circuit voltage at idle to the DPFE circuit voltage stored KOEO. The monitor compares these two voltages to determine if EGR flow is present at idle.
   The DTC associated with Step 3 is P0402.

4. The DPFE sensor upstream hose, (the larger hose, on the exhaust side of the orifice), is tested once per drive cycle for disconnect and plugging. The test is performed with EGR valve closed and during a period of acceleration. The PCM will momentarily command the EGR valve closed, (VR commanded 0%). The monitor looks for the DPFE sensor voltage to be inconsistent with expected voltage for a no-flow condition. A voltage increase or decrease during acceleration while the EGR valve is closed may indicate a fault with the signal hose during this test.
   The DTC associated with Step 4 is P1405.

5. The EGR flow rate test is performed once per drive cycle during a steady state when engine speed and load are moderate and VR duty cycle is high. The monitor compares the actual DPFE circuit voltage to a predetermined voltage corresponding to the desired EGR flow. If the voltage is higher or lower than expected, EGR flow is greater or less than desired. This is a system test and may trigger a DTC for any fault causing the EGR system to fail.
   The DTC associated with Step 5 is P1401. DTC P1408 is similar to P0401, but is performed during KOER self-test conditions.

6. The MIL is activated after one of the above tests fails on two consecutive trips.

Running the EGR Monitor

Steps 1 through 3 of the EGR monitor run continuously. To complete the monitor, however, steps 4 and 5 must also be executed. Verify that the vehicle is warm and running in closed loop with no pending or mature DTCs.

From a stop, accelerate to 45 MPH at 1/2 to 3/4 throttle opening. Repeat 3 times. Always give safety preference over drive cycle criteria. The monitor will wait for you.

RETURNLESS FUEL SYSTEM

The Returnless Fuel system delivers fuel at a controlled pressure and correct volume for efficient combustion. The PCM controls the fuel pump relay operation, the fuel pump driver module (FPDM), injector timing and duration.

Fuel Delivery

The vehicle is equipped with a sequential fuel injection (SFI) system that delivers the correct A/F mixture to the engine at the precise time throughout its entire speed range and under all operating conditions.

A high-pressure (in-tank mounted) fuel pump delivers fuel to the fuel injection supply manifold. The fuel injection supply manifold incorporates electrically actuated fuel injectors, (Bosch), mounted directly above each of the intake ports of the engine.

Air Induction

Air enters the system through the fresh air duct and flows through the air cleaner where it is monitored by the MAF sensor. The metered air passes through the air duct and enters the throttle body. The incoming air passes through the throttle body and into the intake plenum to the intake manifold where it is mixed with fuel for combustion.

System Operation

The Electronic Returnless Fuel System does not return fuel to the fuel tank by means of a fuel return line. The fuel supply manifold does not include a fuel pressure regulator.

The PCM grounds the Fuel Pump relay to provide power to the fuel pump, (through the inertia switch and FPDM). The PCM also sends a duty cycle command to the FPDM to control fuel pump operation according to operating conditions. The FPDM controls the fuel pressure by varying the duty cycle of a high frequency ground signal. In this way, fuel pressure can be precisely controlled between 35-65 psi.

The FPDM sends a 1 Hz signal back to the PCM as a way of communicating circuit condition. The duty cycle of the signal is varied to communicate different messages.

Returnless Fuel System Graphic

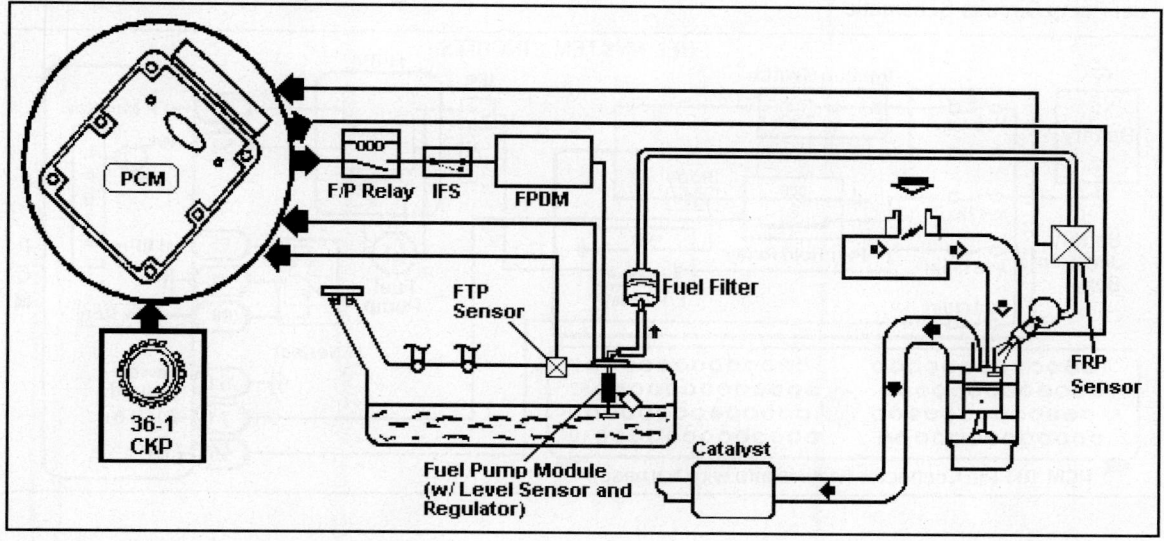

Fuel Delivery System

FUEL SYSTEM OPERATION

The Fuel Pump Driver Module (FPDM) varies the fuel pump output according to a PCM command. The fuel pressure is varied from 35-65 psi, according to operating conditions. The PCM controls the fuel pump through a fuel pump relay in the Battery Junction Box, and a Fuel Pump Driver Module (FPDM) located under the rear seat. The relays provide power to the system under the following operation conditions:

1) With the ignition turned "off", the Power Hold relay, PCM and fuel pump relay are not energized.
2) When the ignition switch is first turned to the "run" position, these events occur:
   a. The Power Hold (PH) relay is energized, which supplies power to the PCM and the fuel pump (FP) relay.
   b. The FP relay supplies power to the FPDM through the inertia switch (IFS).
   c. The PCM commands the FPDM to provide the fuel pump a timed ground.
   d. The FPDM responds by controlling voltage to the fuel pump.
3) If the ignition switch is not turned to the "start" position, the timing device in the PCM will cease commanding the FPDM (PCM Pin 54). This circuitry provides for pre-pressurization of the fuel system.
4) When the ignition switch is turned to the "start" position, the PCM operates the fuel pump by re-commanding the FPDM. This provides pressure fuel for starting the engine while cranking.
5) After the engine starts and the ignition switch is returned to the "run" position:
   a. The PCM continues to issue duty cycle commands to the FPDM from Pin 54.
   b. The PCM monitors the CKP sensor to determine if engine continues to run.
   c. The Fuel Rail Pressure (FRP) sensor reports pressure to the PCM. The PCM controls desired fuel pressure by varying the duty cycle command to the FPDM.
   d. The PCM shuts off the fuel pump by commanding 75% duty cycle to the FPDM if engine stops, or if the engine speed is below 120 rpm.
   e. The FPDM sends a duty cycle signal to the fuel pump monitor on PCM Pin 40.

Fuel Pump Circuits Schematic

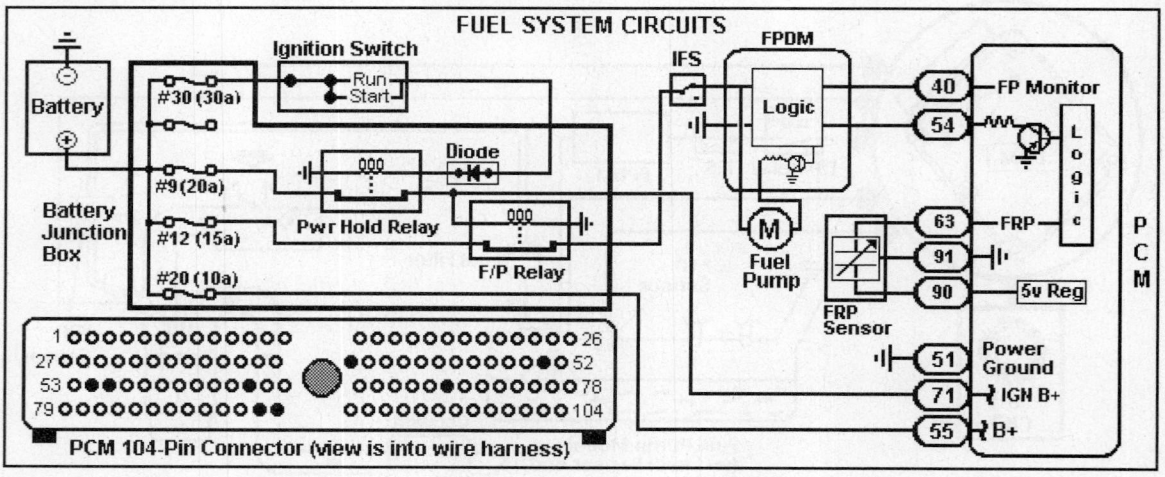

FUEL PUMP DRIVER MODULE

The fuel pump driver module (FPDM), located under the rear seat, regulates voltage to the in-tank fuel pump module to precisely control fuel pressure under all operating conditions. The PCM sends a duty cycle command representing desired fuel pressure.

The FPDM outputs a frequency command to the fuel pump module according to the PCM command. There is a 2:1 ratio between the command and fuel pump operation. If the PCM command is 40%, the FPDM will run the fuel pump at 80% of its capacity.

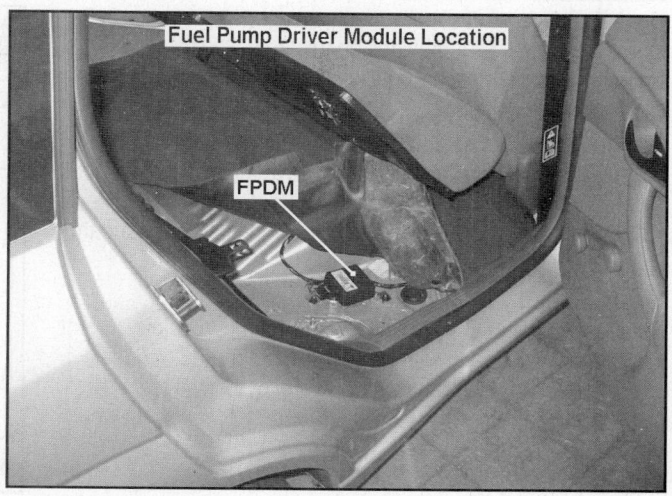
Fuel Pump Driver Module Location
FPDM

Fuel Pump Duty Cycle Command Table

Duty Cycle (%)	PCM Status	FPDM Actions
0-5%	PCM will not command this %	Invalid %. See Fuel Pump Monitor Table
5-51%	Normal Range	Fuel pump operated at requested speed
51-67.5%	PCM will not command this %	Invalid %. See Fuel Pump Monitor Table
67.5-82.5%	Requesting fuel pump off	Valid signal. Fuel pump off
82.5-100%	PCM will not command this %	Invalid %. See Fuel Pump Monitor Table

Fuel Pump Monitor

The FPDM also monitors the FP command for faults. The FPDM sends a 1 Hz feedback signal to the PCM at Pin 40. If the FPDM receives an invalid signal or detects a circuit fault, it will change the duty cycle to the PCM. Refer to the table below for specifications.

Fuel Pump Monitor Signal Table

FPM Signal Duty Cycle	FPDM Actions
50%	All OK. PCM can verify the FPDM powered and communicating
25%	FPDM determined PCM command missing or invalid
75%	FPDM detected circuit faults between PCM and FPDM

Trouble Code Help

The description for both DTC P1233 and P1234 is "Fuel system is disabled or offline". However, DTC P1233 will illuminate the MIL, and P1234 will not. It is difficult to repair this type of code without understanding why it set. However, the conditions to set this code can easily viewed on a Scan Tool (you can view the FPM and FPDM on the tool).

The PCM monitors the FPDM for a 50% duty cycle signal (the feedback signal). If at any time, the PCM receives a signal with a different duty cycle, or stops receiving a signal at all, the PCM will set a code. Also, a tripped inertia switch can set DTC P1233.

Measure the signal at the FPDM, PCM or the BOB with the engine running. Compare the value to the Fuel Pump Monitor Signal Table to determine the next step in diagnosis.

LAB SCOPE TEST (FPDM)

The Lab Scope is an effective tool to diagnose problems in the fuel pump circuit, as it provides an accurate way to view the command and feedback signals in this circuit. Place gear selector in Park; block drive wheels for safety.

Scope Connections, (Examples 1 & 2)

Connect the Channel 'A' positive probe to the FPDM connector Pin 3 (Black wire), and the negative probe to ground. Connect a low amp probe to Channel 'B' and clamp the probe around the same Black wire as Channel 'A'.

Scope Setting Tip

Note that Channel 'B' was set to AC coupling. Absolute current values cannot be seen with this setting, but the AC voltage setting allows for high resolution viewing of the waveform itself. If you use DC coupling to view this signal, the waveform would be off the top of the screen.

Example (1) & (2)

In example (1), the Lab Scope was used to monitor the FPDM command to the fuel pump. The frequency of this command is fixed at about 9.5 kHz. The duty cycle varies according to desired fuel pump command, about 60% in this example (1) and about 75% in example (2). A slight rise in the current waveform can also be seen with the higher duty cycle.

Example (3)

In example (3), the trace shows the feedback signal from the FPDM to the PCM on the fuel pump monitor circuit. Connect Channel 'A' positive probe to Green/Black wire at FPDM connector, PCM Pin 40 or BOB.

The frequency of this signal is fixed at exactly 1Hz (the duty cycle is varied to communicate messages). In this example the duty cycle is 50%, indicating no circuit problems.

Summary

Use the DVOM and Lab Scope to monitor the PCM to FPDM command, the FPDM to fuel pump command and the FPDM feedback signals to help isolate the cause of driveability problems, no starts or DTCs.

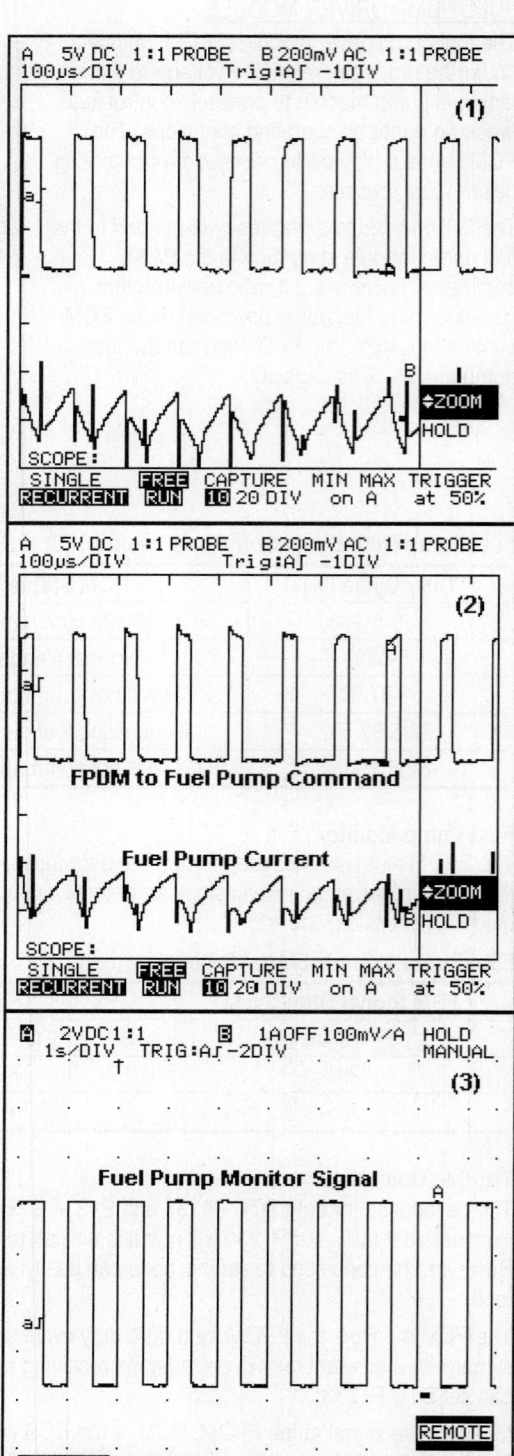

<u>SCAN TOOL TEST (FPDM)</u>

Although a Scan Tool cannot determine the condition of actual circuits, it is an excellent tool for determining the necessity and direction of further testing.

Connect Scan Tool to DLC and navigate through the Scan Tool menus until you get to the Ford (OEM) Data List (a list of PID items). View the FUEL INJ (P) and FUEL PUMP DC PIDs under various operating conditions.

This PID selection provides a way to see the command to the fuel system, (FUEL PUMP DC), and the result of that command, (FUEL INJ (P)), on the same screen.

This example shows that the PCM sent a 31% duty cycle command to the FPDM. This command results in a fuel rail pressure of 54.4 psi (as measured by the FRP sensor).

Diagnostic Help

You can compare these two PIDs to verify the operation of the fuel pump relay, IFS, FPDM, fuel pump module, FRP sensor and PCM. The relationship between these two PID values can help you diagnose problem in the Fuel system or its related trouble codes.

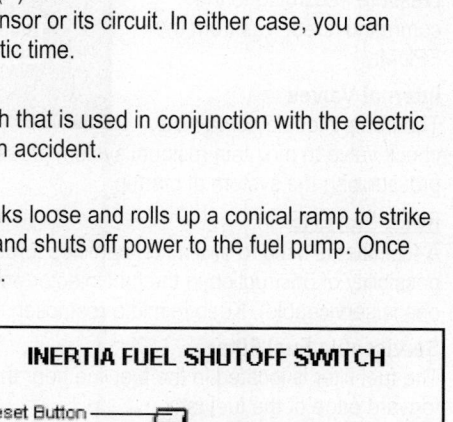

For example, a "high" FUEL PUMP DC PID value with a "low" FUEL INJ (P) PID value would most likely be caused by a weak fuel pump or a fault FRP sensor or its circuit. In either case, you can compare the PIDs to narrow the number of possibilities and save diagnostic time.

<u>INERTIA FUEL SHUTOFF SWITCH</u>

This vehicle application is equipped with an inertia fuel shutoff (IFS) switch that is used in conjunction with the electric fuel pump. The IFS is designed to shut off the fuel pump in the event of an accident.

Inertia Switch Operation

When a sharp impact occurs, a steel ball, held in place by a magnet, breaks loose and rolls up a conical ramp to strike a target plate. This action opens the electrical contacts of the IFS switch and shuts off power to the fuel pump. Once the switch is open, it must be manually reset before the engine will restart.

Inertia Switch Location

On the Focus, the IFS switch is located behind a small access door in the right kick panel.

Inertia Switch Reset Instructions

1) With key off, check for any signs of leaking fuel near the fuel tank or in the underhood area. If any leaks are found, repair the source of the leak before proceeding to the next step.

2) If there are no fuel leaks present, push the reset button to reset the switch (it can be closed an extra 0.16" in the closed position).

3) Cycle the key to "on" for a few seconds, and then to off (in order to recheck for signs of any fuel leaks anywhere in the system).

FUEL PUMP MODULE

The fuel pump module (FPM) is mounted in the fuel tank. Fuel initially enters the module through the bottom of the module, and primes the fuel pump.

A portion of the high-pressure flow from the pump is diverted to operate a Venturi jet pump. The jet pump draws fuel from the fuel tank into the fuel pump module. This design permits optimum fuel pump operation during vehicle maneuvers with low fuel levels.

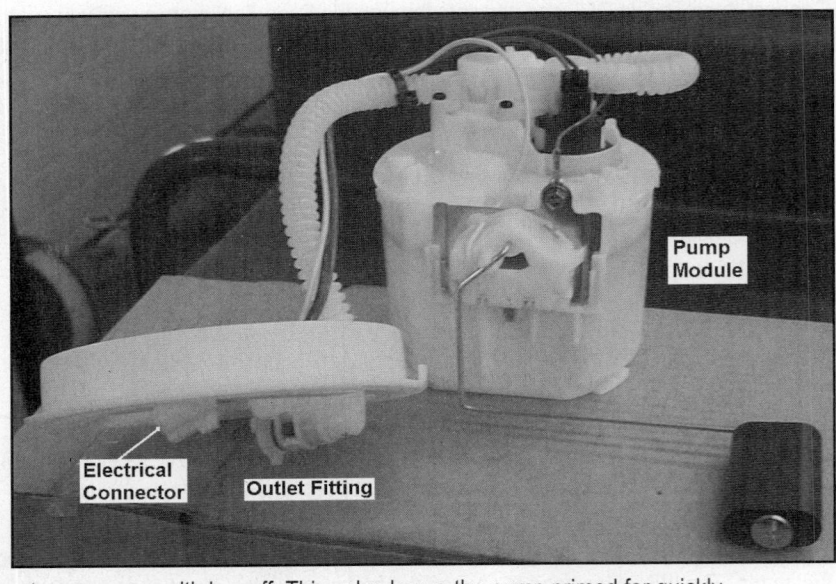

Pressure Regulation

The electric returnless system does not use a fuel pressure regulator. The pump module accurately controls fuel pressure according to the command it receives from the FPDM.

Internal Valves

The pump module contains a check valve to maintain residual system pressure with key off. This valve keeps the pump primed for quickly pressurizing the system at startup.

FUEL FILTERS

A fuel filter is used to strain fuel particles through a screen or paper element. This filtration process reduces the possibility of obstruction in the fuel injector orifices. The Focus has four (4) types of filtration elements (although only one is serviceable). Keep in mid a restriction can occur in any of the filter (especially if the fuel pump has failed).

Serviceable Fuel Filter

The fuel filter is located in the fuel line near the forward edge of the fuel tank.

Fuel Inlet Screen

The inlet of the pump includes a non-replaceable nylon filter to prevent dirt and other particulate matter from entering the system. This protects the entire fuel system, but is the only filter that protects the pump as well. Water accumulation in the tank can pass through the filter without restriction.

Other Fuel Screens

The system utilizes fuel debris screens at each fuel injector on the fuel supply manifold. These are part of the injector assemblies and cannot be serviced separately.

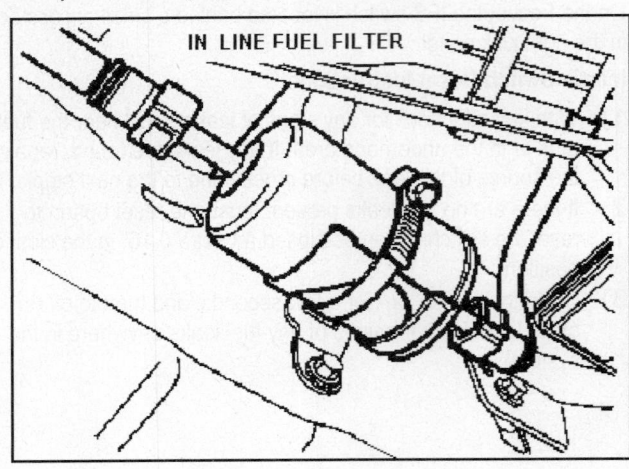

CMP SENSOR (VARIABLE RELUCTANCE)

The camshaft position (CMP) sensor detects the position of the camshaft when engine piston No. 1 is at a specific point of the compression stroke.

The PCM uses the CMP signal to synchronize the fuel injector timing. Since the ignition system is a waste-spark type, the cam position is not needed for ignition timing.

The CMP sensor is a 2-pin variable reluctance (VR) sensor. The CMP sensor is mounted on the intake side of the cylinder head adjacent to the camshaft (it senses a target on the camshaft).

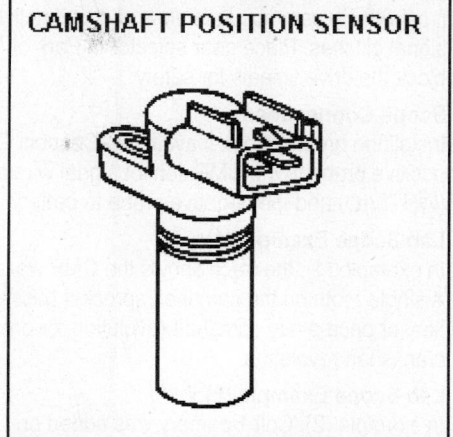

CAMSHAFT POSITION SENSOR

Trouble Code Help

The CMP sensor failure cannot cause a no-start condition.

If the PCM does not receive a valid CMP sensor signal due to a problem in the sensor, the PCM or circuits, it will set a related trouble code (DTC P0340 - CMP Sensor Circuit Malfunction). A failure in the CMP sensor circuit may also set a DTC P1309 – Misfire Monitor Disabled, due to the PCMs inability to keep track of cylinders for misfire detection.

DVOM TEST (CMP SENSOR)

Place the shift selector in Park and block the drive wheels prior to starting this test.

DVOM Connections

Turn the key off and connect the breakout box, if available. Connect the DVOM positive lead to the CMP signal (WHT/VIO wire) at PCM Pin 85 and the negative lead to the battery negative post. Refer to the Cam Sensor schematic on next page. Start or crank the engine and record the readings.

DVOM Test Results

The CMP sensor AC voltage should never be below 100 mv. If it is below this threshold, a DTC P0340 will set.

The CMP sensor signal frequency and voltage should vary with engine speed. In this example, with the engine at Hot Idle, the CMP signal cycled at about 6 Hz.

The PCM counts a CMP pulse when the voltage crosses -300 mv on the downward slope, and watches the rate at which pulses (Hz) occur.

Although the AC voltage changes greatly according to engine speed, the PCM does not use this data to determine cam position, as long as the signal is above 100 mv.

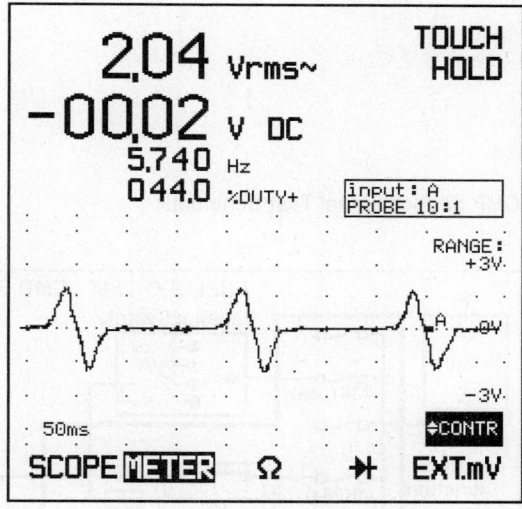

LAB SCOPE TEST (CMP SENSOR)

The Lab Scope can be used to test the CMP sensor circuit as it provides a very accurate view of circuit activity and any signal glitches. Place gear selector in Park or Neutral, and block the drive wheels for safety.

Scope Connections

Install the breakout box if available. Connect Channel 'A' positive probe to the CMP sensor signal wire at PCM Pin 85 (WHT/VIO) and the negative probe to battery ground.

Lab Scope Example (1)

In example (1), the trace shows the CMP waveform at cruise. A single tooth on the camshaft sprocket passes the CMP sensor once every camshaft revolution, or once every other crankshaft revolution.

Lab Scope Example (2)

In example (2), Coil 1 primary was added on Channel B. This can help to stabilize the pattern to check for glitches in the CMP sensor signal. Coil 1 is an excellent trigger to view an intermittent in the CMP signal.

Because the EI system does not use the CMP as an input, a loss of the CMP signal would not affect the waveform stability. It can also be concluded that the CMP signal should not be tested during ignition system diagnosis. Testing of the CMP signal should only be done for related DTCs or sequential fuel injection system concerns.

CMP Sensor Signal Test Schematic

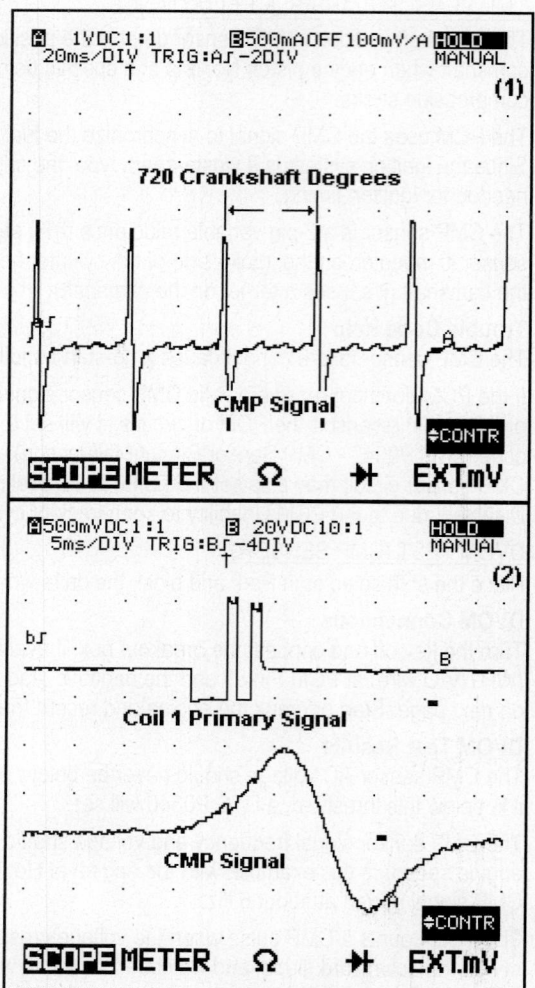

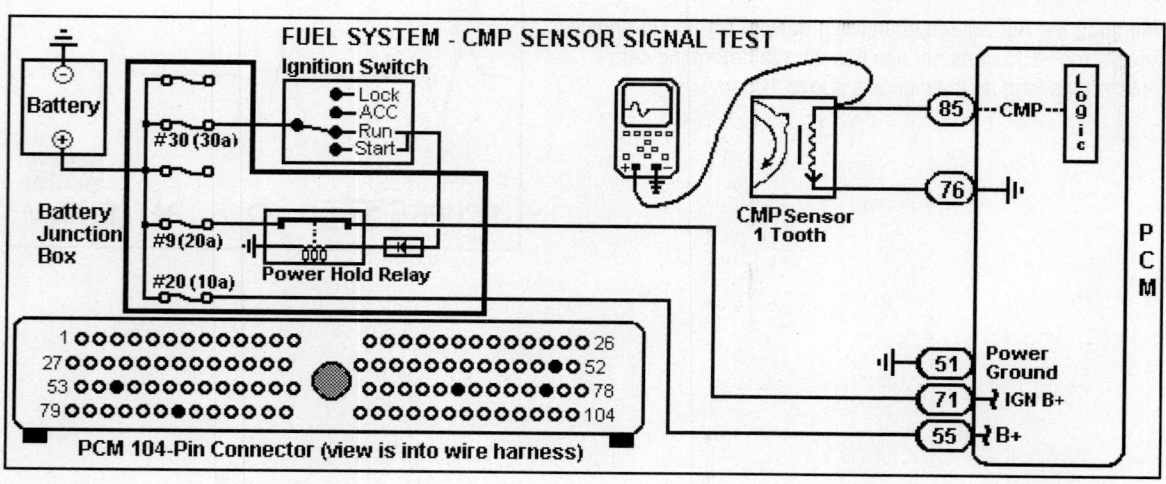

Oxygen Sensor

GENERAL DESCRIPTION

The heated oxygen sensor (HO2S) is mounted in the Exhaust system where it monitors the oxygen content of the exhaust stream. During engine operation, oxygen present in the exhaust reacts with the HO2S to produce a voltage output.

Simply stated, if the A/F mixture has a high concentration of oxygen in the exhaust, the signal to the PCM from the HO2S will be less than 0.4v. If the A/F mixture has a low concentration of oxygen in the exhaust, the signal to the PCM will be over 0.6v.

Heated Oxygen Sensors

This vehicle application is equipped with two oxygen sensors that have internal heaters. The heaters allow for quicker sensor operation in cold weather, and to maintain sensor efficiency during vehicle operation.

On this vehicle (OBD II diagnostics), the front (or pre-catalyst) oxygen sensor is mounted in front of the catalytic converter. The rear (or post-catalyst) oxygen sensor is mounted after the catalytic converter.

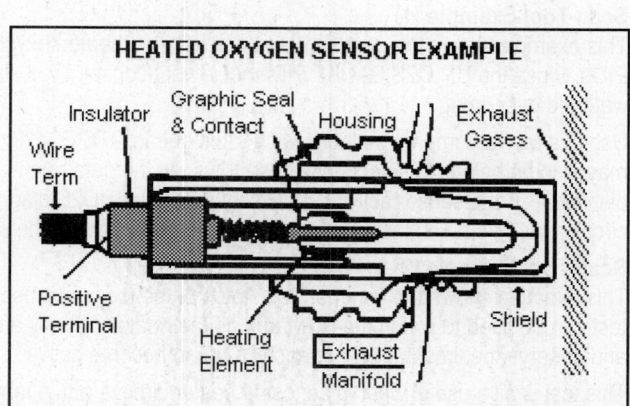

Heated Oxygen Sensor Identification and Location

On this application the front heated oxygen sensor (HO2S-11) is mounted in the exhaust manifold. The rear heated oxygen sensor (HO2S-12) is mounted just after the converter.

Circuit Description

The heater inside the oxygen sensor is connected to power by the PCM power relay and to ground through the heater control circuit at the PCM as shown in the Graphic below. This type of electrical circuit configuration allows the PCM to control when the heaters are activated. In effect, it can determine when to delay the heater activation to protect the sensor ceramic element, and when to activate the HO2S Monitor heater tests.

Each heated oxygen sensor generates a voltage that switches rapidly between 0.0v and 1.0v. The sensor signal is sent to the PCM on the appropriate pin (e.g., Pin 35 or 60).

Heated Oxygen Sensor Schematic

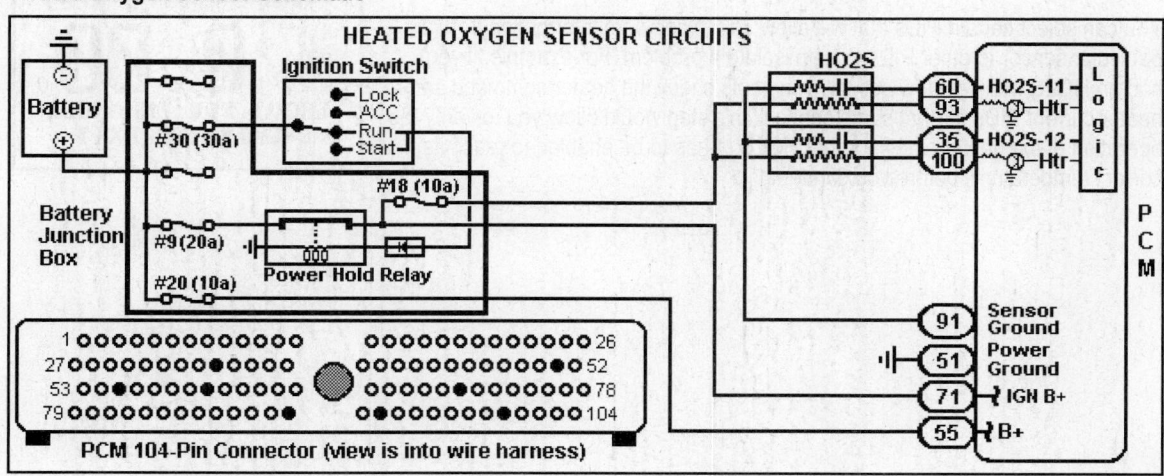

SCAN TOOL TEST (HO2S)

Connect the Scan Tool to the DLC underdash connector. Navigate through the Scan Tool menus until you get to the Ford (OEM) Data List (a list of PID items). Raise the engine speed to 2000 rpm for 3 minutes to set up the test. The examples shown here were captured in Park at Hot Idle.

Then select the type of oxygen sensor (O2S) information that you want to view from the various options in the menu.

Scan Tool Examples

These scan tool examples were used to show the various combinations of PIDs and what they can be used for. The scan tool menus allow you to select certain PIDs that are relevant to your diagnosis.

Scan Tool Example (1)

This example shows the upstream and downstream heated oxygen sensor PIDs. Since the UP O2S BANK1 PID should vary constantly, it should be watched over time.

Use this test to compare sensor switch rates during DTC P0420 diagnosis. It may also be helpful to select only these PIDs, so the data rate is fast enough to display accurate sensor activity. This is important for HO2S diagnosis, since the circuit activity is much faster than the PCM normally talks to the scan tool.

Scan Tool Example (2)

This example shows the commanded status of the oxygen sensor heaters. This test can be used to watch the PCM turn on the heaters shortly after a cold start, and observe the time it takes for each HO2S to become active.

This test is also useful when diagnosing heater circuits with a lab scope or DVOM, because it is difficult to determine the condition of the heater circuit without first knowing if the PCM is attempting to turn it on.

Scan Tool Example (3)

This example shows the PIDs for the current consumption in the heater circuits. The PCM monitors these circuits to verify heater operation. Use these PIDs to check for suspected opens or shorts in the heater circuits. Shorting of the heater circuits is a common failure in this type of sensor.

Summary

There are several screens available on the Scan Tool that can help you diagnose the condition of the heated oxygen sensors and their respective heaters.

You can select certain PIDs that will allow you to compare information from a particular sensor to other PIDs to help isolate a problem. For example, if you had an HO2S with very low activity, you could check the heater command and heater current PIDs for that same sensor. This step would allow you to verify the operation of the heater circuit for that HO2S (it has to be enabled to reach its correct temperature) before you condemn it.

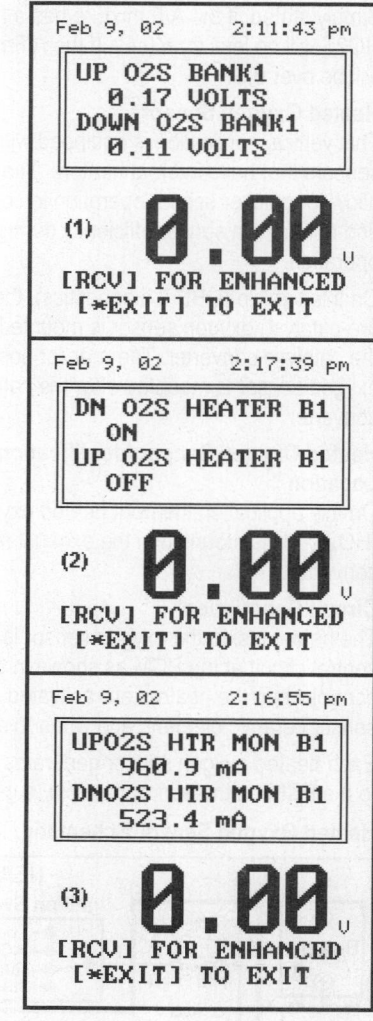

LAB SCOPE TEST (HO2S)

The Lab Scope is the "tool of choice" to test the oxygen sensor as it provides an accurate view of the sensor response and switch rate. Place the gearshift selector in Park and block the drive wheels for safety.

Lab Scope Connections

Connect the Channel 'A' positive probe to the HO2S-11 signal (WHT wire) at Pin 60 of the BOB or PCM connector. Connect the negative probe to the battery negative post.

Connect the Channel 'B' positive probe to the HO2S-12 signal (WHT/RED wire) at Pin 35 of the BOB or PCM connector.

Scope Settings

To make the waveforms as clear as possible, set the scope settings to match the examples.

Depending upon the scope capabilities, the sensor waveforms will have slight differences.

Lab Scope Example (1)

In example (1), the Channel 'A' trace shows the HO2S-11, and the Channel 'B' trace shows HO2S-12. These waveforms were captured at Hot Idle.

With a catalytic converter in perfect condition, the downstream sensor trace (Channel 'B') should remain steady. Use this test to compare the switch rates between the front and rear sensors at steady cruise to diagnose converter efficiency concerns or trouble codes.

In this example, the flat rear sensor waveform shows that the converter is efficient. Note the front sensor range is from just less than 200 mv to greater than 900 mv.

Lab Scope Example (2)

In example (2), the trace shows the front sensor, (HO2S-11). The cursor and trigger settings were used to evaluate the performance of the sensor. Note that the time/div. has been changed to 50 ms to show proper resolution for measuring the lean-to-rich switch rate.

To setup this test, the A/F mixture was driven rich at cruise speed, then driven lean on decel, followed by a snap throttle event. This test method caused the HO2S to switch at a very rapid rate, and it provides a clear view of switch rates on screen.

Note that the lean-to-rich switch rate was 28.0 milliseconds. A switch rate of 100 ms or higher will cause the PCM to set DTC P0133. It can also cause a rolling idle condition or several other driveability concerns.

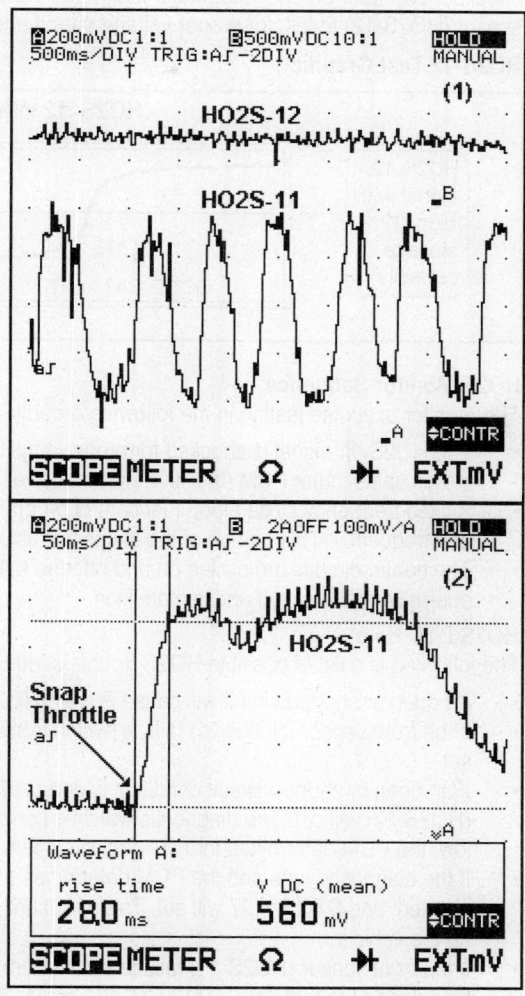

HO2S MONITOR

The Heated Oxygen Sensor Monitor is an OBD II diagnostic designed to monitor the HO2S sensors for circuit malfunctions or deterioration that can effect tailpipe emissions. Under specific conditions, the front sensor (HO2S-11) is checked for proper output voltage and lean-to-rich switch rates. The Catalyst Monitor output signals from the rear sensor (HO2S-12) to test for proper voltage output and switch ratios.

HO2S-12 Test Graphic

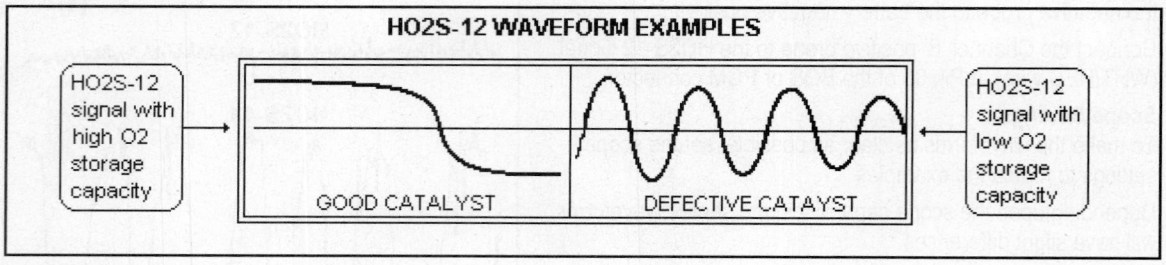

HO2S Monitor Sequence

The monitor executes testing in the following order:

- The HO2S-11 signal is checked for proper amplitude and frequency. Peak voltage is compared to a maximum stored value in the PCM (this is a "non-intrusive" test).
- A fixed frequency closed loop fuel control program is executed, and the HO2S-11 is signal is tested for amplitude and frequency. This portion of the test is "intrusive".
- The heater circuits are cycled on and off, and the heater circuit current is monitored by the PCM to verify the operation of the circuit and its condition.

HO2S DTC Help

The following is a list of possible HO2S trouble codes that are applicable to this vehicle:

- An open or shorted circuit will cause P0131 (HO2S-11) or P0136 (HO2S-12) to set.
- If the front sensor (HO2S-11) fails to switch from lean to rich during testing in less than 100 ms, DTC P0133 will set.
- If an open or shorted circuit condition is detected in the heater circuits, DTC P0135 (HO2S-11) or DTC P0141 (HO2S-12) will set. The diagnostic watches current draw when the heaters are cycled on and off. If the draw is too low, the PCM determines that the circuit is open. If the draw is high, it determines the circuit is shorted.
- If the diagnostic runs and the PCM determines the exhaust is not hot enough, the downstream sensor test will be aborted, and DTC P1127 will set. This will usually occur if a technician initiated the KOER self-test before warming up the vehicle.
- If the front sensor (HO2S-11) fails to switch from 200-900 mv across a 450 mv threshold for a calibrated period of time during the test, DTC P1130, P1131 or P1132 will set. The actual code depends on the severity and direction of the failure.

These tests and codes can help you simplify testing, since the trouble code definitions and conditions are very specific. Remember that it is important to check for external causes for any failure due to this monitor, (exhaust leaks, mechanical problems, etc.).

In cases where several codes are set for multiple sensors, it is highly unlikely that all of the sensors are defective. Look for common causes, like shared heater circuit power, or an exhaust leak that could affect both sensors. Low fuel pressure or a contaminated MAF sensor could cause A/F ratio problems that may be "seen" as a HO2S circuit fault.

Module Communication Network

SYSTEM OVERVIEW

This vehicle is equipped with an antilock brake and electronic stability program module. Module communication is established via two different bus systems; the data link connector (DLC) forms the interface between the modules and the diagnostic tester. The DLC is a standard 16-pin connector that connects the communication wires (SCP and ISO 9141), the programming line (EEPROM) and the Scan Tool power supply.

The data bus Standard Communication Protocol (SCP) consists of two twisted wires. Communication takes place along these wires both between the modules themselves and between the modules and the Scan Tool. Even if one of the wires is open, shorted to ground or shorted to power, communication between the modules and tool continues.

The ISO 9141 data bus consists of a single line. If this wire is open, shorted to ground or to power, communication between the module and the Scan Tool is not possible.

ESP Module

The steering position sensor passes on the angle of turn of the steering wheel to the electronic stability program (ESP) module and communicates with the ESP module and the data link connector via the data bus (SCP). The ESP module controls the antilock braking system. It is connected to the PCM, the steering position sensor and the DLC via the SCP data bus. A connection to the DLC is also possible via the ISO 9141 bus.

The ABS control module is connected to the DLC via the ISO 9141 data bus.

PCM Controlled Functions

The PCM controls all functions related to engine management and emission controls. It communicates with the ESP module, the ABS module, the instrument cluster and the DLC via the data bus (SCP). The instrument cluster is connected to the DLC and the PCM via the data bus (SCP), and consists of the following components:

• Engine Temperature Gauge	• Fuel Gauge
• Odometer	• Tachometer
• Trip Recorder	• Various Indicator & Warming Lamps

CTM Functions

The central timer module (CTM) is connected to the DLC via the ISO 9141 data bus. It controls time-dependent functions for the following systems:

• Acoustic Warning Signal	• Front & Rear Intermittent Wipers
• Front & Rear Washers	• Heated Rear Window & Windscreen
• Ignition Key Removal Block	• Interior Lighting
• Power Saving Relay	• Seat Belt Warning

Auxiliary Warning System

The auxiliary warning system module is connected to the DLC via the data bus (ISO 9141) and controls the following functions:

• Ice Warning	• Fault Code Display
• Outside Air Temperature Display	• Washer Fluid Level Low-Warning Lamp

Airbag Module Functions

The airbag module controls the driver and passenger airbags, the side airbags and the pyrotechnic seat belt pretensioners. It is connected to the DLC via the ISO 9141 bus.

Diagnosis - The networks are diagnosed using a visual inspection and Pinpoint Test 'A'.

Reference Information
PCM PID TABLES - INPUTS

■ **NOTE:** *The following readings were obtained with the engine at idle speed, radiator hose hot, throttle closed, gear selector in Park and all accessories turned off.*

Acronym	Scan Tool Parameter	Range	Typical Value
ACCS	A/C Cycling Clutch Switch Signal	ON / OFF	OFF (ON with A/C on)
ACP	A/C Pressure Switch Signal	0-12v	12v (Open)
BPP	Brake Pedal Position Switch	ON / OFF	OFF (ON with brake on)
CMP (CID)	CMP Sensor Signal	0-1000 Hz	5-7
CKP	CKP Sensor Signal	0-10,000 Hz	390-450
CPP	Clutch Pedal Position Signal (M/T)	0-5.1v	5v (Clutch out)
DPFE	DPFE Sensor Signal	0-5.1v	0.95-1.05
ECT ºF	Engine Coolant Temp. Sensor	-40ºF to 304ºF	160-200
EPC SW	EPC Switch Signal	0-12v	12v
FEPS	Flash EEPROM	0-5.1v	0.5-0.6
FPM	Fuel Pump Monitor	0-100%	50%
FRP	Fuel Rail Pressure Signal	0-5.1v	2.8v
FTP	Fuel Tank Pressure Sensor Signal	0-5.1v (0-10" H2O)	Cap Off: 2.6v (0" H2O)
GEAR	Transmission Gear Position	1-2-3-4	1
GFS	Generator Field Sense	0-100%	30%
HO2S-11	HO2S-11 (Bank 1 Sensor 1)	0-1100 mv	100-900 mv
HO2S-12	HO2S-12 (Bank 1 Sensor 2)	0-1100 mv	100-900 mv
HO2S-21	HO2S-21 (Bank 2 Sensor 1)	0-1100 mv	100-900 mv
HO2S-22	HO2S-22 (Bank 2 Sensor 2)	0-1100 mv	100-900 mv
IAT ºF	Intake Air Temperature Sensor	-40ºF to 304ºF	50-120
IMRC-M	Intake Runner Manifold Control	0-5.1v	5v (or 2.5v)
KS	Knock Sensor Signal	AC Voltage	0v AC
LOAD	Calculated Engine Load	0-100%	10-20
MAF V	Mass Airflow Sensor Signal	0-5.1v	0.6-0.9
MISF	Misfire Detection	ON / OFF	OFF
OCT ADJ	Octane Adjust Signal	Closed / Open	Closed
OSS	Output Shaft Speed Sensor Signal	0-9999 RPM	0 RPM
PNP	Park/Neutral Position Signal (A/T)	0-5.1v	5v (Clutch out)
PSP V/PSP	Power Steering Pressure Switch	0-5.1v / HIGH-LOW	300 mv / LOW
RPM	Engine Speed	0-9999	730-790
TCS	Traction Control Switch Signal	ON / OFF	OFF
TFT ºF	Transmission Fluid Temp. Sensor	-40ºF to 304ºF	110-210
TP V	Throttle Position	0-5.1v	0.53-1.27
TRR/TR	Gear Selector Reverse Signal	0v or 12v	0v
TRL/TR	Gear Selector Low (1st) Signal	0v or 12v	0v
TRD/TR	Gear Position Drive Signal	0v or 12v	0v
TROD/TR	Gear Position Overdrive Signal	0v or 12v	0v
TSS	TSS Sensor Signal	0-10,000 Hz/RPM	340-380 / 680-720
VPWR	Vehicle Power	0-25.5v	14.1v
VSO	Vehicle Speed (M/T)	0-159 MPH	0
VSS	Vehicle Speed (A/T)	0-159 MPH	0

PCM PID TABLES - OUTPUTS

■ NOTE: *The following readings were obtained with the engine at idle speed, radiator hose hot, throttle closed, gear selector in Park and all accessories turned off.*

Acronym	Scan Tool Parameter	Range	Typical Value
CDA	Coil 'A' Driver	0-12v	12v
CDB	Coil 'B' Driver	0-12v	12v
DPC1	Shift Solenoid 'C' Control	0-12v	0v
DPC2	Shift Solenoid 'D' Control	0-12v	0v
DPC3	Shift Solenoid 'E' Control	0-12v	0v
EGR VR	EGR Vacuum Regulator Valve	0-100%	0%
EVAPPDC	EVAP Canister Purge Valve	0 Or 10 Hz (0-100%)	0 Or 10 (0-100)
EVAP VT	EVAP Canister Vent Valve	0 Or 10 Hz (0-100%)	0 Or 10 (0-100)
EPC	EPC Solenoid Control	0-12v / 0-100%	2.5v / 73%
FP	Fuel Pump Control	0.0-5.0v / 0-100%	1.5v / 33%
FUELPW1	Fuel Injector Pulsewidth Bank 1	0-99.9 ms	3.3-3.7
HFC	High Speed Fan Control	ON / OFF	OFF (ON with fan on)
HTR-11	HO2S-11 Heater Control	ON / OFF	ON
HTR-12	HO2S-12 Heater Control	ON / OFF	ON
IAC	Idle Air Control Motor	0-100%	20-40
IMRC	Intake Manifold Runner Control	0-12v	12v (off)
INJ1	Fuel Injector 1 Control	0-99.9 ms	3.3-3.7
INJ2	Fuel Injector 2 Control	0-99.9 ms	3.3-3.7
INJ3	Fuel Injector 3 Control	0-99.9 ms	3.3-3.7
INJ4	Fuel Injector 4 Control	0-99.9 ms	3.3-3.7
LFC	Low Speed Fan Control	ON / OFF	OFF (On with fan on)
LONGFT1	Long Term Fuel Trim Bank 1	-20% to 20%	0-1
MIL	Malfunction Indicator Lamp Control	ON / OFF	OFF (ON with DTC)
PATSIL	Passive Anti-Theft Indicator Lamp	ON / OFF	OFF (ON with DTC)
PATSTRT	Vehicle Start Signal	YES / NO	YES
SHRTFT1	Short Term Fuel Trim Bank 1	-10% to 10%	-1 to 1
SPARKADV	Spark Advance	-90° to 90°	15-22
SS1	Transmission Shift Solenoid 1	ON / OFF	ON
SS2	Transmission Shift Solenoid 2	ON / OFF	OFF
TCC	Torque Converter Clutch Solenoid	0-100%	0
TCIL	Traction Control Indicator Lamp	ON / OFF	OFF (ON with TCIL on)
VREF	Voltage Reference	0-5.1v	4.9-5.1v
WAC	Wide Open Throttle A/C Relay	ON / OFF	OFF (ON at WOT)

PCM PIN VOLTAGE TABLES

2001 Focus 2.0L I4 MFI VIN P (All) 104-Pin Connector

Pin Number	Wire Color	Application	Value at Hot Idle
1	GN/BL	Shift Solenoid 'B' Control	1v, Cruise: 1v
2, 5-6, 9-10	---	Not Used	---
3	W/VT	IMRC Solenoid Control	IMRC Off: 5v, On: 2.5v
4	GN/BL	TR Reverse Position Signal	In 'R': 12v, others: 0v
7	GN/W	TR 1st Position Signal	MAN 1: 12v, others: 1v
8	GN/O	TR Drive Position Signal	In Drive: 12v, others: 1v
12, 14, 18, 23	---	Not Used	---
11	GN/BK	TR Overdrive Position Signal	In O/D: 12v, others: 1v
13	O/BK	Flash EEPROM Power	0.5v
15	BU/BK	SCP Data Bus (-) Signal	Digital Signals
16	GY/O	SCP Data Bus (+) Signal	Digital Signals
17	BK/W	High Speed Fan Control	Fan On: 1v, Off: 12v
19	GY/O	Passive Anti-Theft System	Digital Signals
20	BK/BL	Injector 3 Control	3.3-3.7 ms
21	W/R	CKP Sensor (+) Signal	400-425 Hz
22	BN/R	CKP Sensor (-) Signal	400-425 Hz
24, 51	BK/Y	Power Ground	<0.1v
25	BK	Shield Ground	<0.050v
26	BK/O	Coil 'A' Driver Control	5° dwell
27	BK/R	Vehicle Start Signal	KOEC: 9-11v
28	W/VT	M/T: VSS (+) Signal	0 Hz, 55 mph: 125 Hz
29	GN/BK	TR Overdrive OFF Switch	O/D Off: 0v
30, 32	---	Not Used	---
31	BK/Y	PSP Switch Signal	Straight: 0.6v, turned: 2v
33	BN/BL	OSS Sensor (-) Signal	0 Hz, 55 mph: 120 Hz
34	W/VT	TSS Sensor (+) Signal	360 Hz, 55 mph: 990 Hz
35	W/R	HO2S-12 (B1 S2) Signal	0-1.1v (use Lab Scope)
36	BN/BL	MAF Sensor Return	<0.050v
37	W/GN	TFT Sensor Signal	2.10-2.40v
38	W	ECT Sensor Signal	0.5-0.6v
39	W/VT	IAT Sensor Signal	1.5-2.5v
40	GN/BK	Fuel Pump Driver Module	3.5v (100% on time)
41	GN/Y	A/C Cycling Clutch Switch	Closed: 12v, Open: 0v
42	BK/O	Passive Anti-Theft System	Digital Signals
43, 45-46, 48-50	---	Not Used	---
44	GN/O	EPC Switch Signal	12v
47	BK/GN	EGR VR Solenoid Control	0%, 55 mph: 40%
52	BK/GN	Coil 'B' Driver Control	5° dwell
53	W/GN	Passive Anti-Theft System	Digital Signals
54	BK/W	Fuel Pump Driver Module	1.5v (33% on time)
55	R	Keep Alive Power	12-14v
56	BK/BL	EVAP Purge Valve Control	0-10 Hz (0-100%)
57	W/BK	Knock Sensor (-) 1 Signal	0v (knocking: ACV)
58	W/BL	OSS Sensor (+) Signal	0 Hz, 55 mph: 120 Hz
59	GY	Generator Field Sense	130 Hz (30% d/cycle)

2001 Focus 2.0L I4 MFI VIN P (All) 104-Pin Connector

Pin Number	Wire Color	Application	Value at Hot Idle
60	W	HO2S-11 (B1 S1) Signal	0-1.1v (use Lab Scope)
61, 66, 74-75	---	Not Used	---
62	WH/VT	FTP Sensor Signal	2.6v (0" H20 - cap off)
63	WH/GN	FRP Sensor Signal	2.8v (39 psi)
64	GN/Y	A/T: Neutral Start Position	In 'N': 0v, others: 5v
64	WH	M/T: Clutch Pedal Switch	Clutch In: 0v, Out: 5v
65	WH/BL	DPFE Sensor Signal	0.95-1.05v
67	BK/OG	EVAP CV Solenoid Control	0-10 Hz (0-100%)
68	BK/BL	Low Speed Fan Control	Fan On: 1v, Off: 12v
69	BK/YE	A/C WOT Relay Control	Relay On: 1v, Off: 12v
70	BK/WH	Injector 1 Control	Idle: 3.3-3.7 ms
71, 97	GN/YE	Vehicle Power	12-14v
72	BL	Generator Field Control	0 Hz (0% d/cycle)
73	GN/YE	Shift Solenoid 'A' Control	0v, 55 mph: 12v
76	BN/WH	Sensor Reference Ground	<0.050v
77, 103	BK/Y	Power Ground	<1v
80	BK/W	IMRC Solenoid Control	12v (off)
81	BK/R	EPC Solenoid Control	2.5v (73% d/cycle)
82	GN/BK	Shift Solenoid 'C' Control	0v, 55 mph: 4.10v
83	BK/Y	IAC Motor Control	10v (30% on time)
78-79, 84, 98, 94	---	Not Used	---
85	WH/VT	CMP Sensor Signal (+)	5-9 Hz
86	WH/RD	A/C Pressure Switch	A/C On: 12v (Open)
87	BN/Y	Knock Sensor 1 Signal	0v (knocking: ACV)
88	W/BL	MAF Sensor Signal	0.6-0.9v
89	W	TP Sensor Signal	0.53-1.27v
90	Y	Reference Voltage	4.9-5.1v
91	BN	Analog Signal Return	<0.050v
92	GN/R	Brake Pedal Position Switch	Brake On: 12v, Off: 0v
93	BK/Y	HO2S-11 (B1 S1) Heater	Heater On: 1v, Off: 12v
95	BK/O	Injector 4 Control	3.3-3.7 ms
96	BK/Y	Injector 2 Control	3.3-3.7 ms
99	GN/O	Shift Solenoid 'D' Control	0v, 55 mph: 0v
100	BK/BL	HO2S-12 (B1 S2) Heater	Heater On: 1v, Off: 12v
101, 104	---	Not Used	---
102	GN/W	Shift Solenoid 'E' Control	0v, 55 mph: 0v

PCM 104-Pin Connector

TERMINAL VIEW OF WIRING HARNESS CONNECTOR FOR PCM (104 PIN)

Note: back-probing the 104-pin connector with a DVOM probe can damage the PCM - use a breakout box!

Fuel & Ignition System Diagrams (1 of 7)

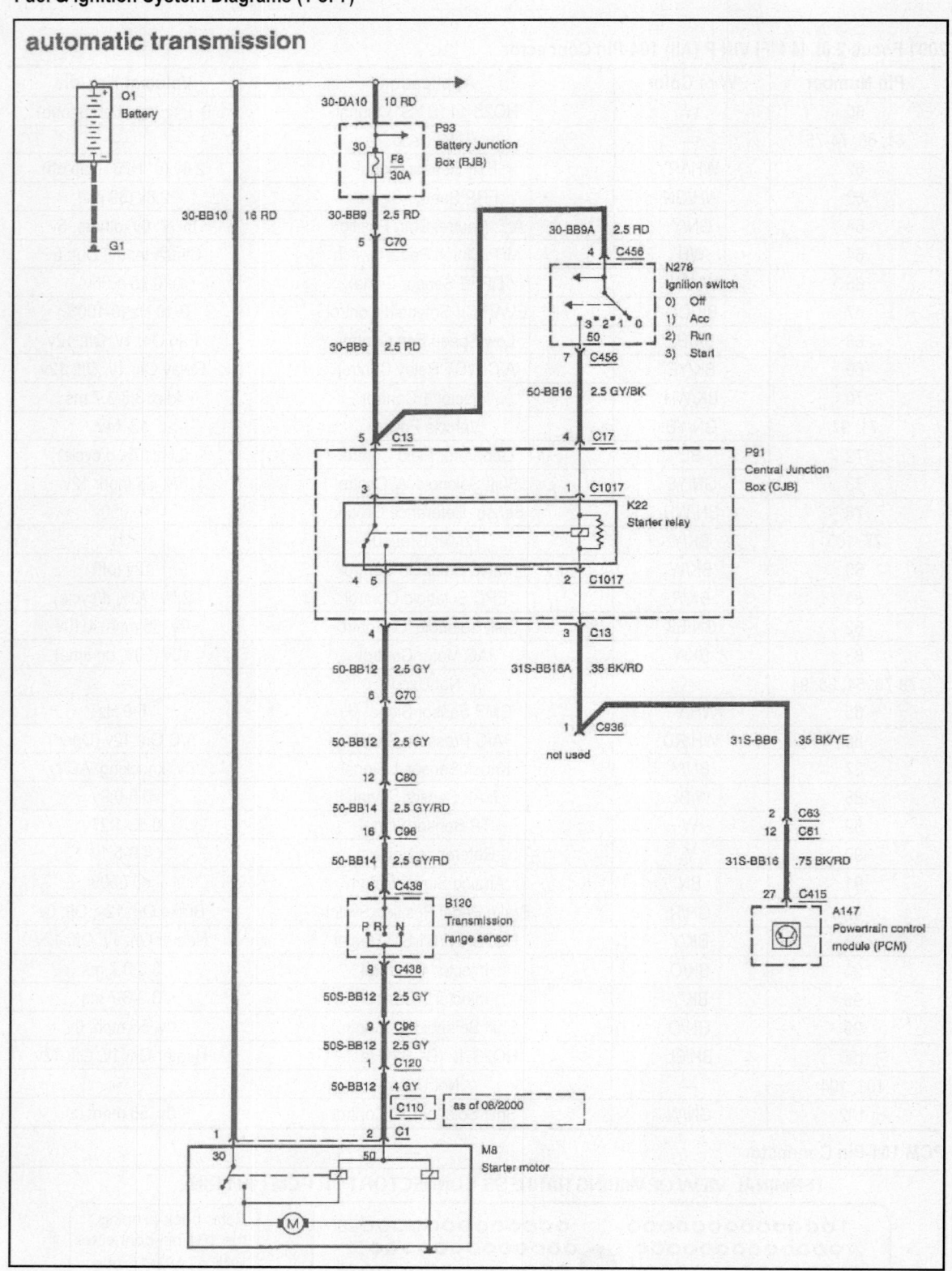

Fuel & Ignition System Diagrams (2 of 7)

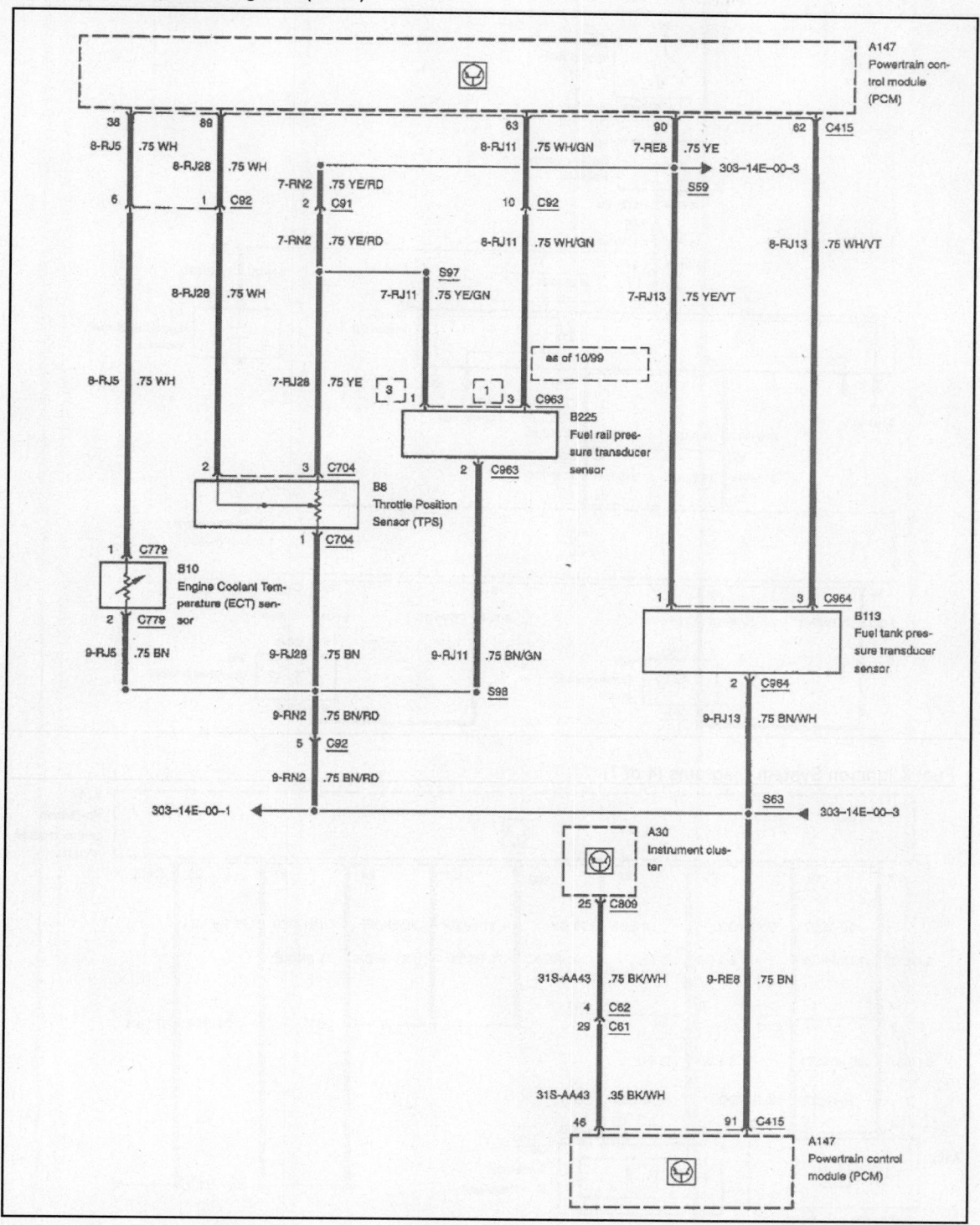

Fuel & Ignition System Diagrams (3 of 7)

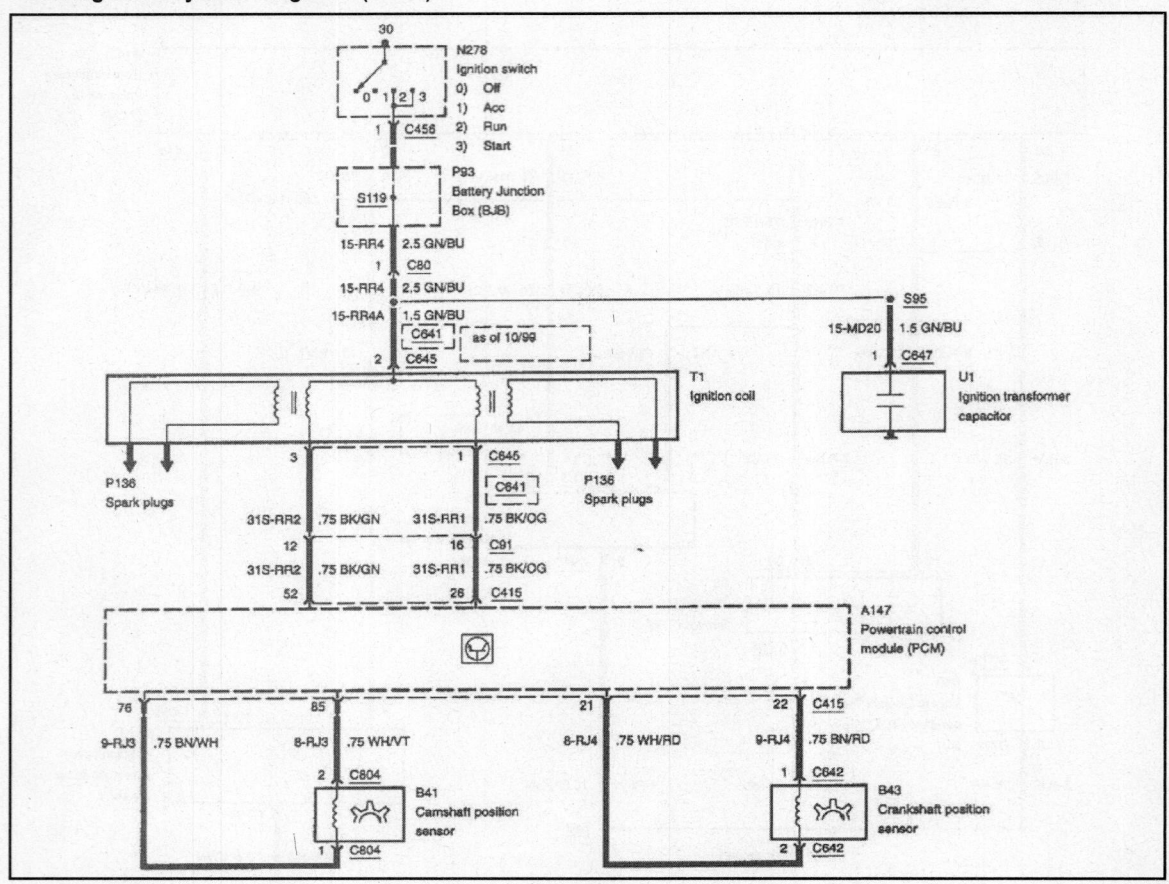

Fuel & Ignition System Diagrams (4 of 7)

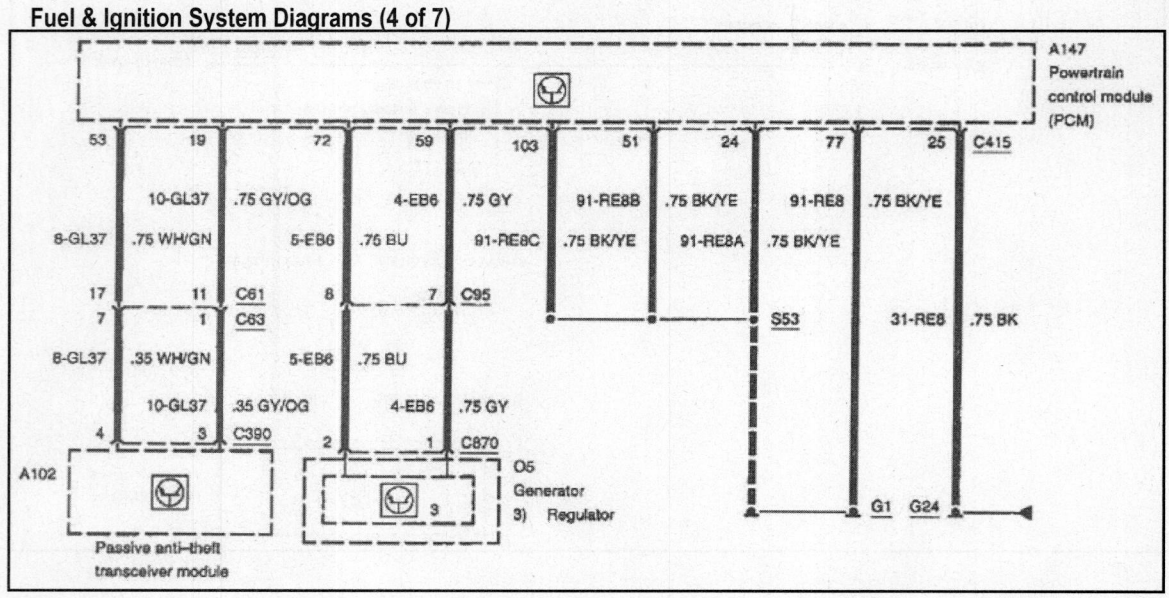

Fuel & Ignition System Diagrams (5 of 7)

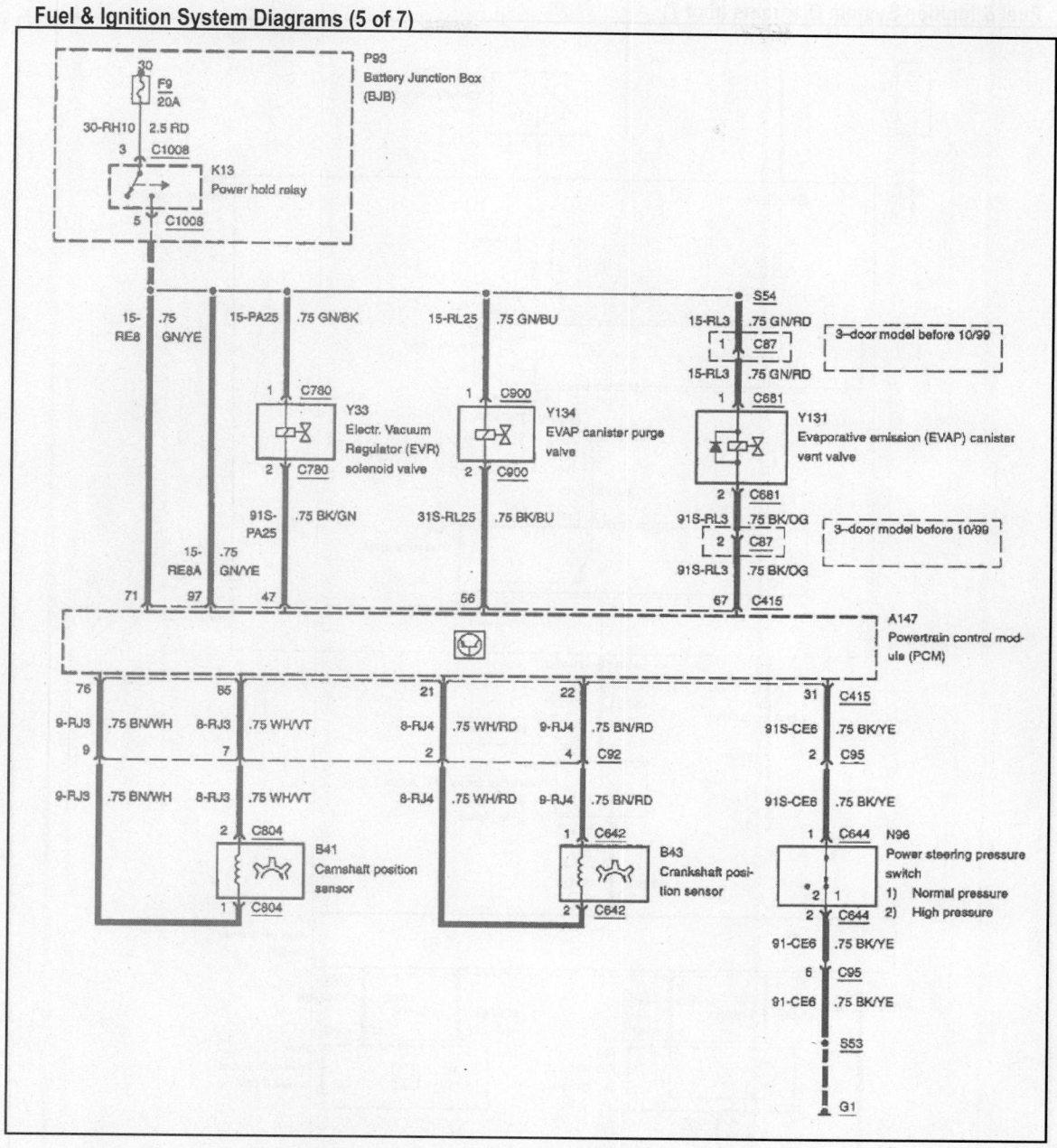

Fuel & Ignition System Diagrams (6 of 7)

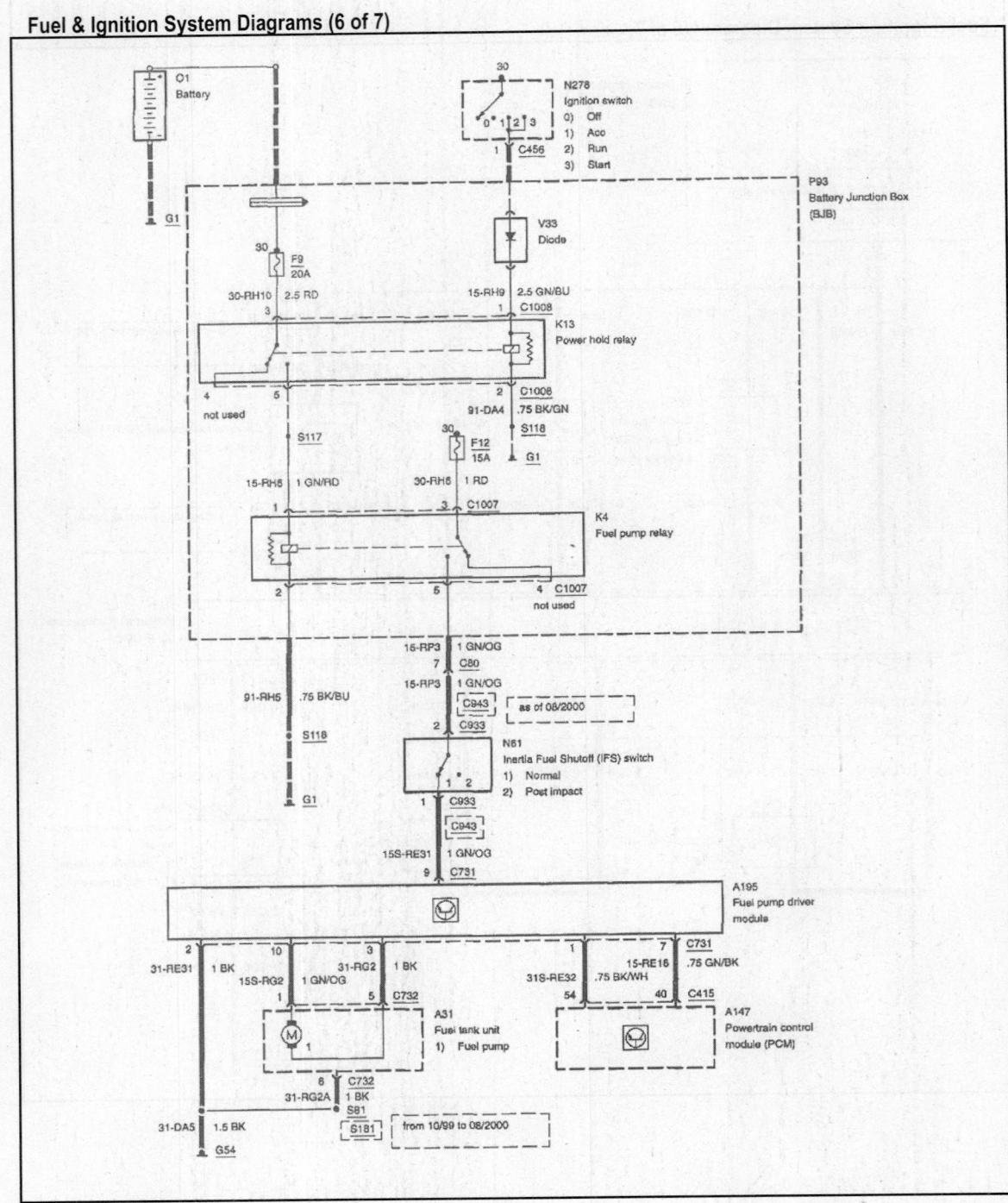

Fuel & Ignition System Diagrams (7 of 7)

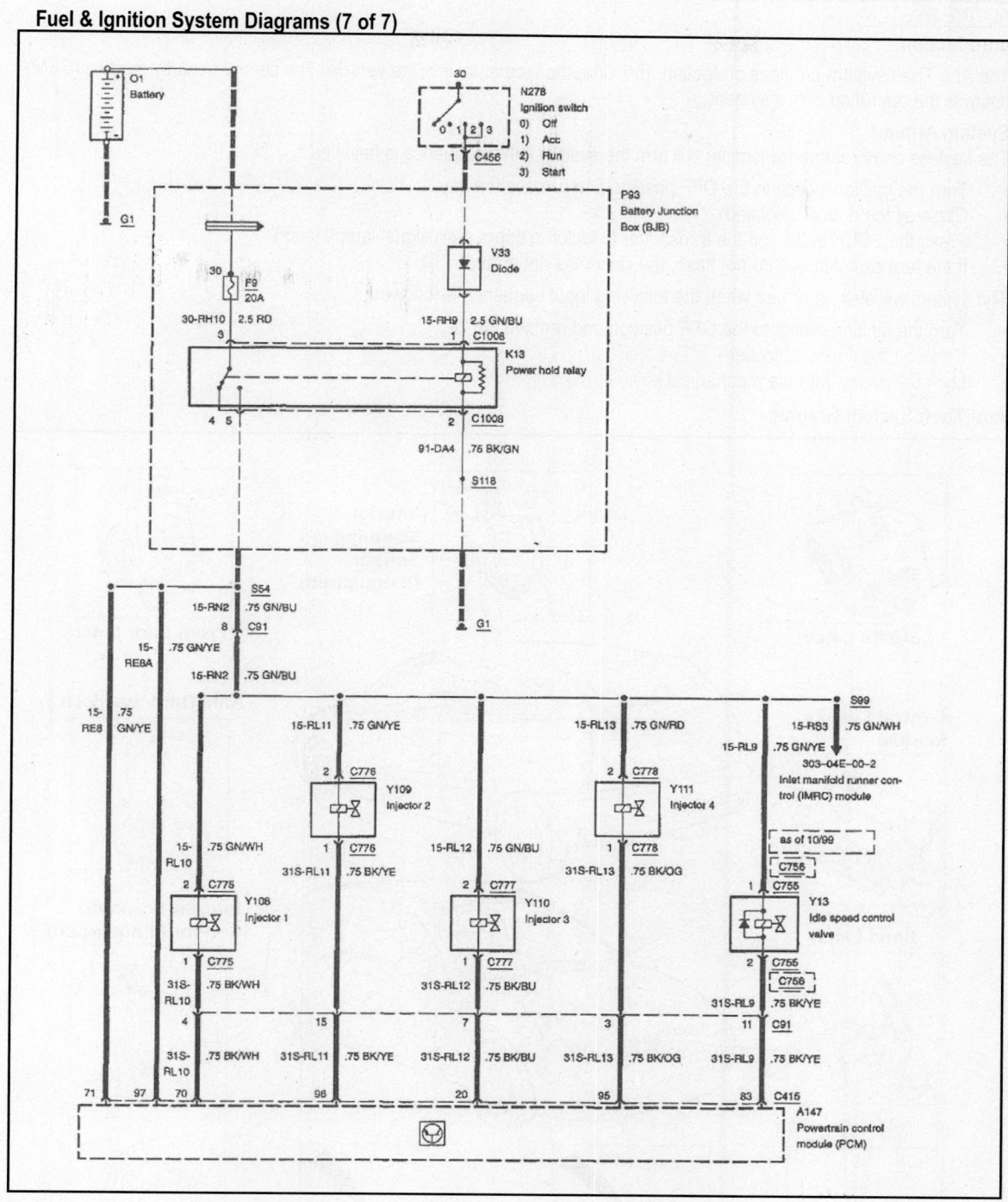

2001 Focus - Anti-Theft system

ACTIVE SYSTEM

Introduction

The Anti-Theft system provides protection from unauthorized entry into the vehicle. The central security module (CSM) controls the operation of this system.

System Arming

The keyless entry remote transmitter will arm the system if this sequence is followed:

- Turn the ignition switch to the OFF position and remove the key.
- Close all the doors (unlocked).
- Press the LOCK button on the transmitter to lock the doors (turn signal lamps flash).
- If the turn signal lamps do not flash, the system is not armed.

The system will also be armed when the following input sequence is followed:

- Turn the ignition switch to the OFF position and remove the key.
- Close all the doors (unlocked).
- Lock the doors, with the mechanical key.

Anti-Theft System Graphic

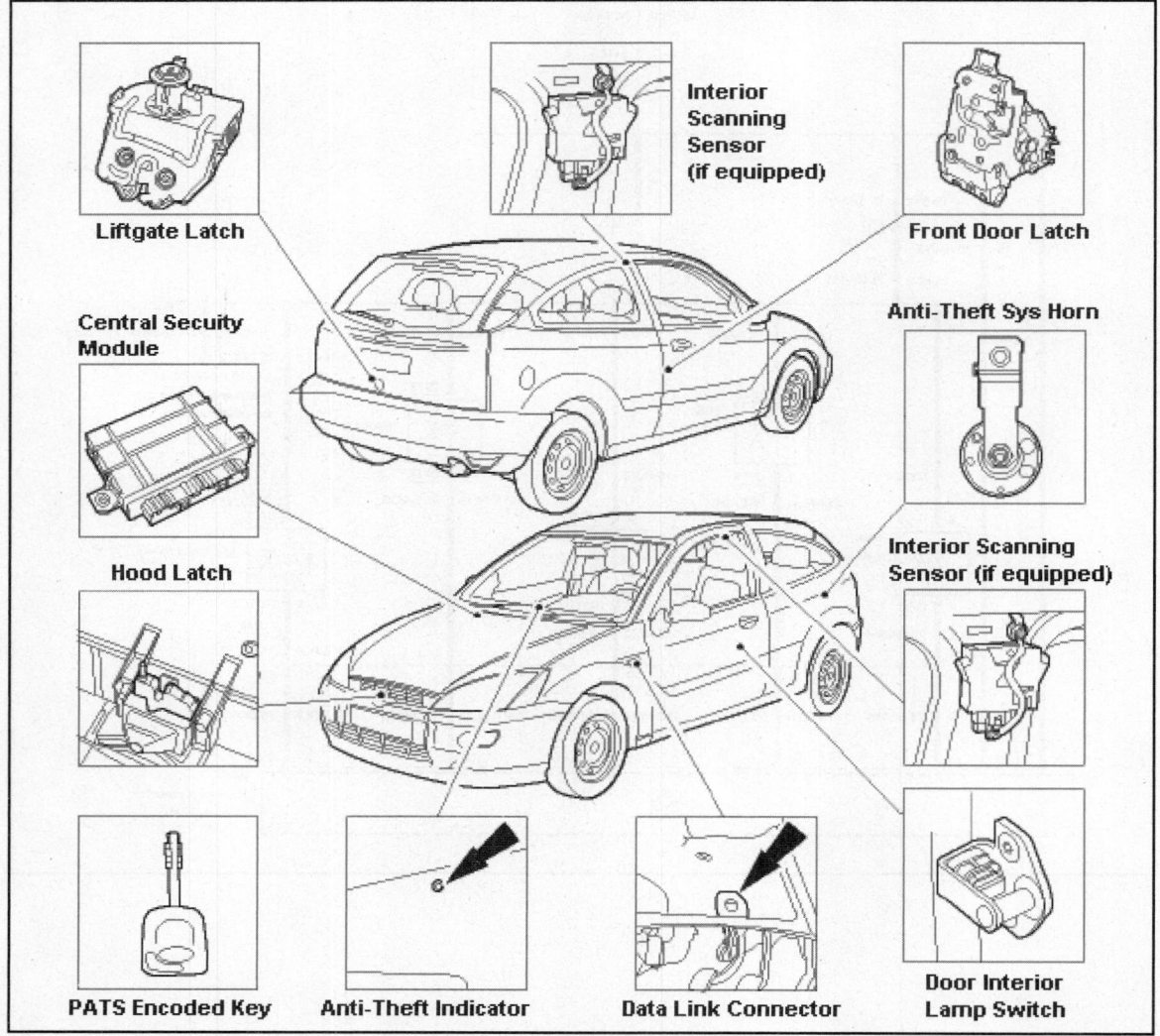

Liftgate Latch

Interior Scanning Sensor (if equipped)

Front Door Latch

Central Secuity Module

Anti-Theft Sys Horn

Hood Latch

Interior Scanning Sensor (if equipped)

PATS Encoded Key

Anti-Theft Indicator

Data Link Connector

Door Interior Lamp Switch

PERIMETER ANTI-THEFT

If the central security module (CSM) receives a signal from a monitored switch that indicates an intrusion into the vehicle, it supplies a ground to the horn control and hazard warning, turn signal lamp relays. The relays close, supplying power to the horn and the turn signal lamp flasher. The CSM cycles the output to the relays on and off whenever the PAT system is activated.

The perimeter anti-theft feature does not arm until 20 seconds elapses after the vehicle is locked. The perimeter anti-theft device will disarm if any driver door, passenger door, luggage compartment or liftgate key cylinders is moved to the "unlock" position, or if the ignition switch is turned to key Position II (run position) or key Position III (start position). The perimeter anti-theft device will also disarm if an unlock request is made through the remote transmitter.

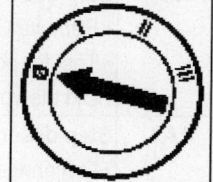

Inspection and Verification

Verify the concern by operating the system. Inspect for mechanical or electrical damage.

Visual Inspection Chart

Mechanical Checks:		Electrical Checks:
Key Cylinder		Central J/B Fuse #63 (20a)
Linkage		Wiring Harness
CSM (module)		Loose connections, switch problems

System Diagnosis - Step One

If an obvious cause of an observed concern is found, correct the cause (if possible) before proceeding. Connect a Scan Tool. If the tool does not power up:

- Check that the program card is correctly installed.
- Check the connections to the vehicle.
- Check that the ignition switch is in Position II (run position)
- If the Scan tool does not power up after these checks refer to the Scan Tool manual.

System Diagnosis - Step Two

Carry out the DATA LINK DIAGNOSTICS test. If the Scan Tool responds with:

- CKT914, CKT915 or CKT70 = ALL ECUS NO RESP/NOT EQUIP. Refer to the Module Communications Network Inspection and Tests article in other information.
- NO RESP/NOT EQUIP for CSM, go to Pinpoint Test 'A'.
- SYSTEM PASSED, read, record and clear the CMC (DTCs) and then perform the self-test diagnostics for the central security module (CSM).
- If any codes are retrieved that related to the concern, refer to the CSM code index.
- If no codes are retrieved that relate to the concern, refer to the Symptom Chart.

Symptom Repair Chart

Problem	Items to Check/Inspect	Instructions
No Communication with CSM	DLC circuit, the CSM (module)	Go to Pinpoint Test 'A'
Alarm System does not operate	Alarm system horn, hood switch, remote transmitter, door set/reset switch, courtesy lamp switch, turn signal lamps, luggage compartment set/reset switch, luggage compartment courtesy lamp switch, door ajar switch, the CSM, the Alarm circuit	Go to Pinpoint Test 'B' Note: These Pinpoint Tests are not included in this manual. Look in other repair manuals.

PINPOINT TEST 'A'

If the Scan Tool cannot establish communication with the CSM, use Pinpoint Test 'A'.

Pinpoint Test 'A' Repair Table

Step	Action	Yes	No
A1	**Step description:** *Check CJB Fuse 63 (20A)* • Remove CJB Fuse (20A). Turn the key to on. • Test for power between Pin 1 and ground. • Did the voltage read greater than 10v?	Go to Step A2.	Repair the power circuit that feeds the Pin 1 connection of the CJB Fuse and retest.
A2	**Step description:** *Check power from junction* • Disconnect the C451 harness connector (key off). • Measure the voltage between Pin 1 of the CSM C451 harness connector and chassis ground. • Did the voltage read greater than 10v?	Go to Step A2.	Repair the power circuit after the Pin 1 to the C451 harness connector for an open condition. Then retest for normal operation.
A3	**Step description:** *Check power from junction* • Measure the resistance between the harness side of CSM C451 Pin 1 and ground (with the key off). • Did the resistance read less than 5 ohms?	Refer to article on the Module Communication Network.	Repair the open circuit as needed. Then retest for normal operation.

PINPOINT TEST 'B'

If the Alarm System does not operate, use Pinpoint Test 'B' to repair the problem.

Pinpoint Test 'B' Repair Table

Step	Action	Yes	No
B1	**Step description:** *Check the self-test inputs/outputs* • Verify that the doors, luggage compartment and hood are fully closed properly. • Connect a Scan Tool to the DLC. • Use the Scan Tool to do the CSM self-test. • Did all of the tested inputs respond correctly? (Alarm system, turn signal lamps, door key operation, luggage compartment operation and luggage compartment courtesy lamp switch operation).	Go to Step B2 to test the hood switch.	If the alarm system does not sound when the hood is opened, go to Step B3. If the alarm system horn does not sound, go to Step B5. If the turn signal lamps do not flash, go to B6. For door key operation (seat reset switch), go to Step B8. For the door courtesy lamp switch, go to Step B13. For luggage compartment operation (set/reset/switch), go to B14. For luggage compartment courtesy lamp switch, go to Step B16.
B2	**Step description:** *Check the hood switch* • Disconnect the hood switch C897 connector. • Measure resistance between Pins 1 and 2 of C897 hood switch (at the component side). • Did DVOM read less than 5 ohms (at rest) and over 10K ohms (switch open)?	Go to Step B3.	Install a new hood switch. Then retest for normal operation.
B3	**Step description:** *Check hood switch signal* • Measure the resistance between Pin 3 of the CSM C451 main connector (at the harness side) and ground. • Did the DVOM read less than 5 ohms (hood closed) and over 10K ohms (hood open)?	Replace the CSM. Then retest the system for normal operation.	Repair CKT 31S-GL7 (BLK/YEL wire). Retest system for normal operation. Refer to Handles, Locks, Latches and Entry Systems for additional information about this component and circuit.

■ NOTE: *Refer to other repair manuals for the rest of the steps in Pinpoint Test 'B'.*

WIRING DIAGRAMS (PARTIAL)

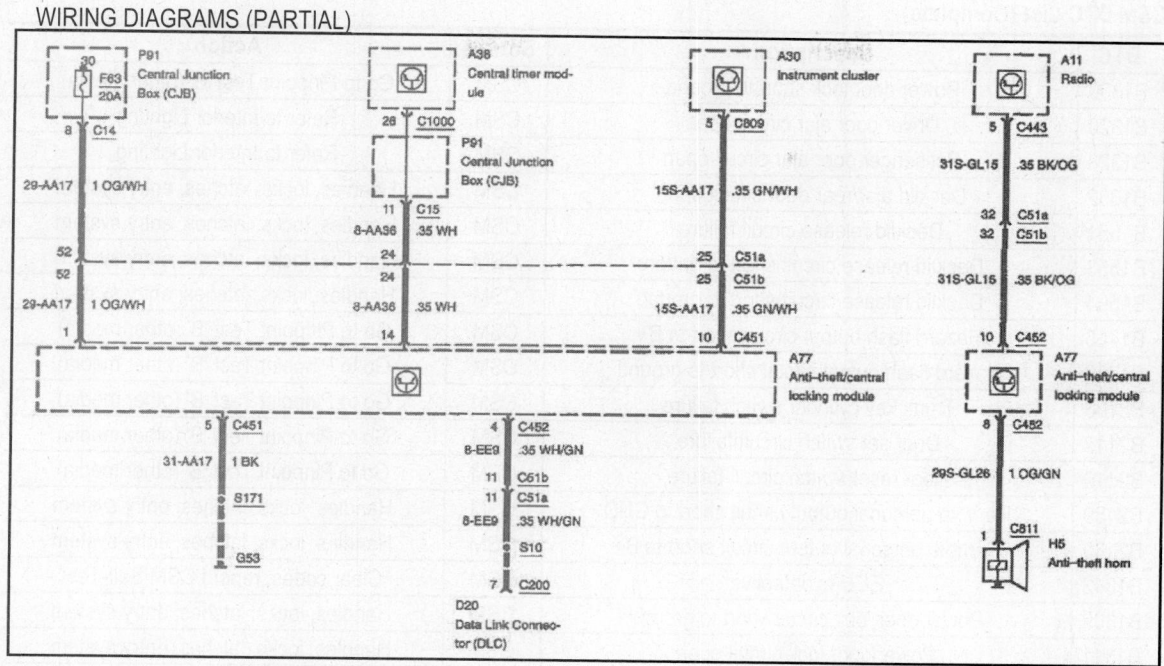

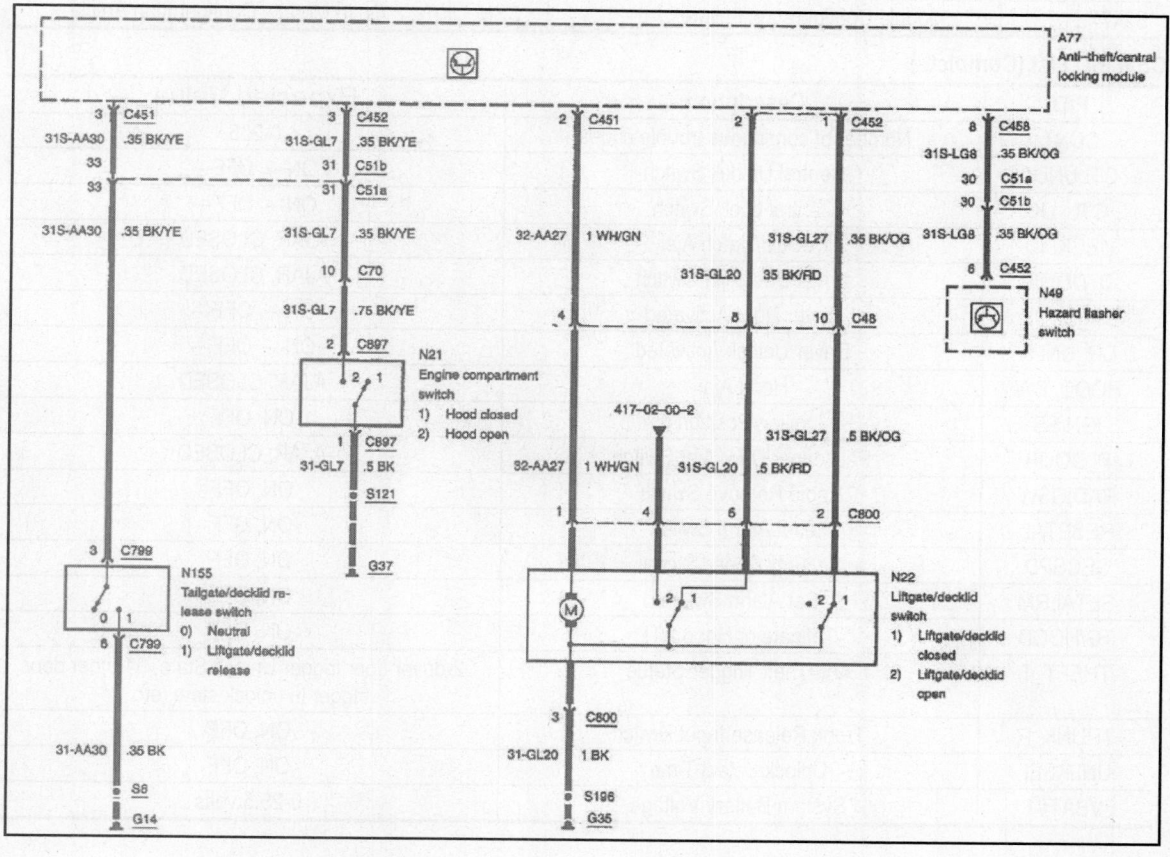

CSM DTC & PID Lists

CSM DTC List (Complete)

DTC	Description	Source	Action
B1300	Power door lock short to ground	CSM	Go to Pinpoint Test 'B' (other media)
B1320	Driver door ajar circuit open	CSM	Refer to Interior Lighting
B1328	Passenger door ajar circuit open	CSM	Refer to Interior Lighting
B1332	Decklid ajar/rear door circuit open	CSM	Handles, locks, latches, entry system
B1551	Decklid release circuit failure	CSM	Handles, locks, latches, entry system
B1553	Decklid release circuit short to battery	CSM	Handles, locks, latches, entry system
B1554	Decklid release circuit short to ground	CSM	Handles, locks, latches, entry system
B1755	Hazard flash output circuit short to B+	CSM	Go to Pinpoint Test 'B' (other media)
B1756	Hazard flash output circuit short to ground	CSM	Go to Pinpoint Test 'B' (other media)
B2108	Trunk key cylinder switch failure	CSM	Go to Pinpoint Test 'B' (other media)
B2112	Door set switch circuit failure	CSM	Go to Pinpoint Test 'B' (other media)
B2116	Door reset switch circuit failure	CSM	Go to Pinpoint Test 'B' (other media)
B2159	Remote personal output circuit short to GND	CSM	Handles, locks, latches, entry system
B2160	Remote personal output circuit short to B+	CSM	Handles, locks, latches, entry system
B1342	ECU is defective	CSM	Clear codes, repeat CSM Self-Test
B1309	Power door lock circuit short to ground	CSM	Handles, locks, latches, entry system
B1311	Power door lock circuit open	CSM	Handles, locks, latches, entry system
B1341	Power door unlock circuit short to ground	CSM	Handles, locks, latches, entry system
B2494	Horn/Panic output circuit short to B+	CSM	Handles, locks, latches, entry system
B2496	Horn/Panic output driver short to ground	CSM	Go to Pinpoint Test 'B' (other media)
B2477	Module Configuration Failure	CSM	Go to Module Configuration Article

CSM PID List (Complete)

PID	Description	Expected Value
CCNT	Number of continuous trouble codes	0-255
CTLUNLK	Central Unlock Switch	ON---, OFF---
CTL_LK	Central Lock Switch	ON---, OFF---
DECKLID	Decklid/Hatch Ajar	AJAR, CLOSED
D_DOOR	Driver Door Ajar Switch	AJAR, CLOSED
DR_LOCK	Driver Door Activated	ON---, OFF---
DR_UNLK	Driver Unlock Activated	ON---, OFF---
HOOD_SW	Hood Ajar	AJAR, CLOSED
IGN/SS	Ignition Position II	ON, OFF
P_DOOR	Passenger Door Ajar Switch	AJAR, CLOSED
RADIOSW	Radio Remove Switch	ON, OFF
RESETAL	Reset Alarm Switch	ON, OFF
SECSPD	Security Speed Signal	ON, OFF
SETALRM	Set Alarm Switch	ON, OFF
TG/HOOD	Tailgate or Hood ATI	ON, OFF
THEFT_T	Anti-Theft Trigger Status	02-driver door trigger Unlock State, 04-other door trigger in unlock state, etc.
TRUNK_R	Trunk Release Input Switch	ON, OFF
UNLKSEL	Unlock Select Time	ON, OFF
VBAT#1	System Battery Voltage	0-25.5 volts

PASSIVE SYSTEM (SECURILOCK)
Introduction

The passive anti-theft system (PATS) uses radio frequency identification technology to deter a driveway theft. Passive means that it does not require any activity from the user.

This system uses a specially encoded ignition key. Each encoded ignition key contains a permanently installed electronic device called a transponder. Each transponder contains a unique electronic identification code that is one of many combinations.

Each encoded ignition key must be programmed into the vehicle PCM before it can be used to start the engine. There are special diagnostic repair procedures outlined in other repair information that must be carried out if a new encoded ignition key is necessary.

The transceiver module communicates with the encoded ignition key. This module is located behind the steering column shroud and contains an antenna connected to a small electronics module. During each vehicle start sequence, the transceiver module reads the encoded ignition key identification code and sends data to the PCM.

The control functions are contained in the PCM. This module carries out all of the PATS functions (e.g., verifying the ID code from an encoded ignition key and controlling starter enable). The PCM initiates the key interrogation sequence with the key in Start or Run. All elements of the PATS must be functional before the engine is allowed to start. If any of the components are not working correctly, the vehicle will not start.

The PATS uses a visual theft indicator. This indicator will prove out for three seconds when the ignition switch is turned to Start or Run position. If a PATS problem exists, this indicator will either flash rapidly or glow steadily when the ignition switch in Start or Run. The PATS system also flashes the theft indicator every two seconds with the key "off".

The PATS will disable the vehicle from starting if any of these conditions are present:

Damaged encoded ignition key	A key that is not programmed
Non-Encoded Key (a key with no electronics)	Damaged Wiring or Connectors
Damaged Transceiver	Damaged PCM or Red Marked PATS Key

Inspection and Verification
Verify the concern by operating the system. Inspect for mechanical or electrical damage.

Visual Inspection Chart	
Mechanical Checks:	**Electrical Checks:**
Ignition Lock Cylinder	PCM or the PATS Transceiver
PATS key or use of a Non-PATS key	Ignition Switch Functions
More than one PATS key on a chain	Loose or corroded connections

System Diagnosis
Step One - If an obvious cause of an observed concern is found, correct the cause (if possible) before proceeding. Connect a Scan Tool. If the tool does not power up:
- Check that the program card is correctly installed.
- Check the connections to the vehicle.
- Check that the ignition switch is in Position II (run position)
- If the Scan tool does not power up after these checks refer to the Scan Tool manual.

Step Two - CKT914, CKT915 or CKT 70 = ALL ECUS NO RESP/NOT EQUIP, refer to the Module Communications Network Inspection and Tests article in other information.

- NO RESP/NOT EQUIP for CSM, refer to the Powertrain Control Emissions Manual.
- SYSTEM PASSED, read, record and clear any codes. Perform a Quick Test - If any codes are set, refer to the correct code repair chart or use the Symptom Tests.

PCM DTC & PID LISTS

PCM DTC List (Complete)

DTC	Description	Source	Action
B1213	Number of encoded ignition keys below minimum	PCM	Program additional keys
B1342	ECU is defective	PCM	Install a new PCM
B1600	No key code received	PCM	Go to Pinpoint Test 'B'
B1601	Incorrect key code received	PCM	Go to Pinpoint Test 'C'
B1602	Invalid key code format from encoded key transponder	PCM	Go to Pinpoint Test 'D'
B1681	Anti-theft transceiver module signal not received	PCM	Go to Pinpoint Test 'E'
B2103	Antenna not connected	PCM	Go to Pinpoint Test 'A'
B2431	Transponder programming failure	PCM	Program an additional key

PCM PID List (Complete)

PID	Description	Expected Value
ACCESS	Security Access Status	TIMED, CODED
ANTISCM	Anti-Scan Status	ACTIVE, NOT ACTIVE
NUMKEYS	Number of ignition key codes programmed	0 to 8
ENABL_S	Vehicle starting status	ENABLED, DISABLED
M_KEY	Master Key Present	YES, NO
MIN#KEYS	Number of key codes required to be initially programmed to start the vehicle	2 to 8
SPARE_KY	Spare key programming allowed	YES, NO
SERV_MOD	Service module	YES, NO

ANTI-THEFT SECURITY ACCESS

This procedure is utilized to obtain passive anti-theft system (PATS) security access. This security access must be granted to erase ignition keys or enable/disable the spare key programming switch (PID SPARE KEY). This procedure invokes a time delay prior to granting security access (the Scan Tool must remain connected during this period). Once security access is granted, a security access command menu is displayed which offers command options (refer to the Configuration Command Index in the tool menu).

1) Connect a Scan Tool and select SECURITY ACCESS. This procedure will take about 10 minutes to carry out with the ignition switch in Position II (run position).

2) Once this procedure is finished a new menu will appear with command options. Select only the functions you require to finish before exiting out of this menu.

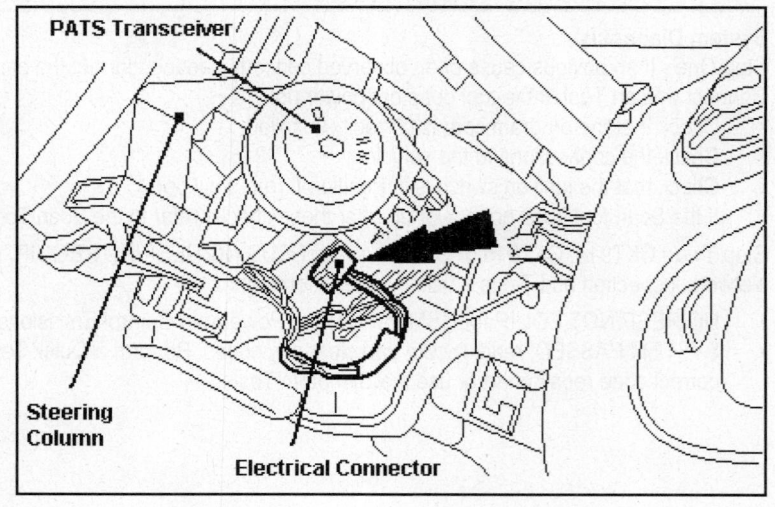

1996 TAURUS - Powertrain

VEHICLE IDENTIFICATION
VECI Decal

The Vehicle Emission Control Information (VECI) decal is located on the underside of the hood. *This example is from a 1996 Ford Taurus station wagon.*

SFI, TWC (2), HO2S, EGR, 3.0 LITER

These designators indicate this vehicle is equipped with 3.0L V6 with SFI, three-way catalysts (TWC), Heated Oxygen Sensors (HO2S) and an EGR system.

OBD II Certified

This designator (located on the VECI label) indicates that this vehicle has been certified as OBD II compliant.

50ST (50 States)

If this designator is used, the vehicle conforms to U.S. EPA & State of California regulations for 1996 model year new motor vehicles.

CAL (California)

If this designator is used, the vehicle conforms to U.S. EPA regulations applicable to 1996 model year new motor vehicles sold in California.

Underhood View Graphic

ELECTRONIC ENGINE CONTROL SYSTEM

Operating Strategies

The Electronic Engine Control (EEC) system uses several operating strategies to maintain good overall driveability and to meet the EPA mandated emission standards. These strategies are included in the tables on this page.

Based on various inputs it receives, the PCM adjusts the fuel injector pulsewidth, the idle speed, the amount of ignition advance or retard, the ignition coil dwell and the operation of the canister purge valve.

It also regulates the air conditioning, cooling fan and speed control systems.

PCM Location

The PCM is located behind the right hand (RH) cowl panel as shown in the Graphic to the right.

Operating Strategies

A/C Compressor Clutch Cycling Control	Closed Loop Fuel Control
Electric Fuel Pump Control	Exhaust Gas Recirculation Control
Fuel Metering for Multiport Fuel Injectors	Fuel Pump Monitor
Idle Speed Control	Ignition Timing Control for A/F Change
Vapor Canister Purge Control	

General Operating Strategies

Adaptive Fuel Trim Strategy	Adaptive Idle Speed Control Strategy
Base Engine/Transmission Strategy	Failure Mode Effects Mgmt. Strategy

■ **NOTE:** *Hardware Limited Operating Strategy or HLOS is a part of the PCM hardware.*

Operational Control of Components

A/C Compressor Clutch	EGR System Control
Engine Cooling System Electric Fan	Fuel Delivery (Injector Pulsewidth)
Fuel Pump Operation	Fuel Vapor Recovery System
Idle Speed	Malfunction Indicator Lamp
Shift Points for the A/T	Speed for Integrated Speed Control
Torque Converter Clutch Lockup (TCC)	Transmission Line Pressure (EPC)

Items Provided (or Stored) By the EEC System

Data to the Operational Control Centers	Data to the Suspension Control Centers
DTC Data for the AX4N	DTC Data for Emission Related Faults
DTC Data for Non-Emission Faults	

INTEGRATED ELECTRONIC IGNITION SYSTEM
Introduction

The engine controller (PCM) controls the operation of the Integrated Electronic Ignition (EI) system on this vehicle. Battery voltage is supplied to the coilpack (3 coils in one assembly) through the ignition switch and the #10 (20 amp) fuse in the I/P fuse panel.

The PCM controls the individual ground circuits of the three (3) ignition coils by switching the ground path for each coil "on" and "off" at the appropriate time. The PCM adjusts the ignition timing for each cylinder pair correctly to meet changes in the engine operating conditions. The ignition coil pairs are as follows: 1/5, 2/6 and 3/4. The firing order of this engine application is as follows: 1-4-2-5-3-6.

System Components

The main components of this Integrated EI system are the coilpack, crankshaft position sensor, the PCM and the related wiring. Note that the camshaft position sensor is not an integral part of the EI system as it is used only as an input for sequential fuel injection.

EI System Graphic

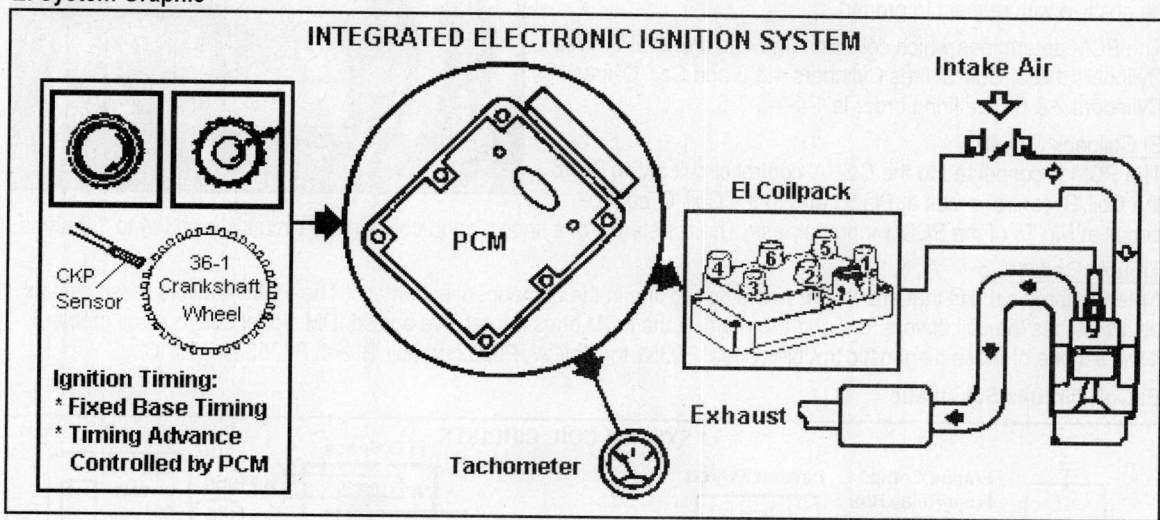

Ignition Spark Timing

The ignition timing on this EI system is entirely controlled by the PCM - base timing is not adjustable. Do not attempt to check the base timing, as you will get false readings.

The PCM uses the CKP sensor signal to indicate crankshaft position and speed by sensing a missing tooth on a pulse wheel mounted behind the harmonic balancer. It also uses this signal to calculate a spark target and then fire the coils to that spark target.

After engine startup, the PCM calculates the ignition spark advance from these signals:

- BARO Sensor (this is a signal calculated from the MAF sensor)
- ECT Sensor
- Engine Speed Signal (rpm)
- IAT Sensor
- MAF Sensor
- Throttle Position

Ignition Coil Operation

The PCM controls the operation of the EI system coilpack in order to provide ignition secondary spark at the correct time in the combustion process.

The coilpack consists of three (3) independent coils molded together. The coil assembly is mounted to the intake manifold. Spark plug cables rout to each of the six (6) cylinders from the coilpack assembly.

Each ignition coil fires two (2) spark plugs. This is a form of waste spark ignition where one spark plug fires the cylinder under compression and the other spark plug (same coil) fires the cylinder on the exhaust stroke.

The spark plug that fires the cylinder on the compression stroke uses the majority of the ignition coil energy and the spark plug that fires the cylinder on the exhaust stroke uses very little coil energy. The spark plugs are connected in series, so one spark plug firing voltage will be negative and the other spark plug will be positive with respect to ground.

The PCM determines which coil to charge and fire. Coil 'A' fires Cylinders 1 & 5, Coil 'B' fires Cylinders 4 & 3 and Coil 'C' fires Cylinders 2 & 6. The firing order is 1-4-2-5-3-6.

EI Coilpack Circuits
The PCM is connected to the Coil 'A' control circuit at Pin 26, to the Coil 'B' control circuit at Pin 52 and to the Coil 'C' control circuit at Pin 78 of the PCM wiring harness. The resistance of a known good coil primary circuit is from 0.4 to 1.0 ohms.

Trouble Code Help
Note the three circuits that are shown with a dotted line in the component schematic. The PCM monitors these circuits as it switches the coil drivers "on" and then "off". If the PCM does not receive a valid IDM signal due to a coil problem, it will set one of these diagnostic trouble codes: P0351 for Coil 'A', P0352 or Coil 'B' and P0353 for Coil 'C'.

EI Coil Circuits Schematic

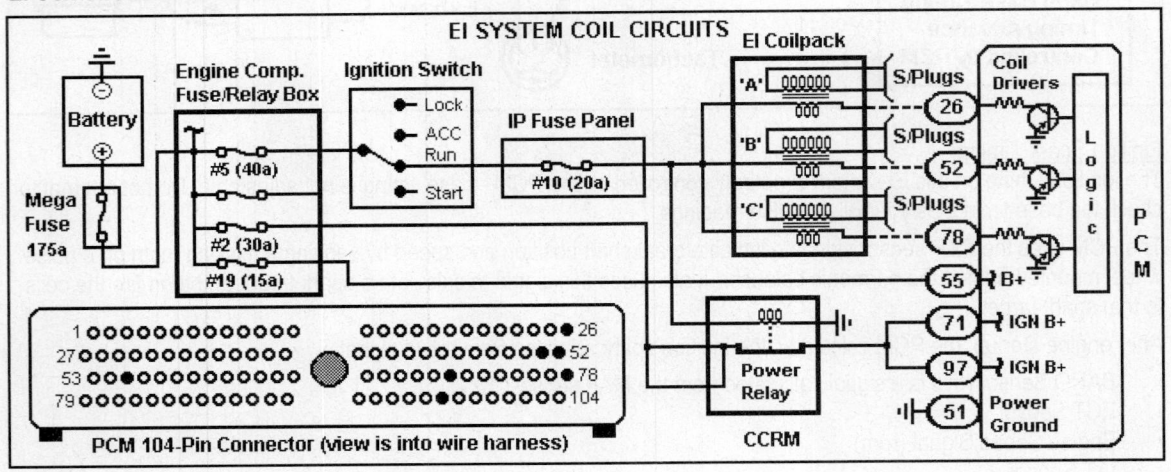

INTEGRATED ELECTRONIC IGNITION SYSTEM
Lab Scope Test (Coil Primary)

The ignition coil primary circuit includes the switched ignition feed circuit, coil primary circuit and the coil driver control circuits at the PCM. Coil 'A' is connected to Pin 26, Coil 'B' connects to Pin 52 and Coil 'C' connects to Pin 78 at the PCM (refer to previous page).

Coil Saturation
Once maximum current flows through the ignition coil primary windings, a maximum magnetic field is present in the coil windings, and the windings are considered saturated.

Coil Dwell Period
The coil dwell period is the length of time that current flows through the coil primary windings (it is measured in milliseconds and represents the degrees of engine rotation). A complete turn of the camshaft equals 360 degrees and represents one firing of all engine cylinders.

Lab Scope Connections
Connect the Channel 'A' positive probe to one of the three (3) coil control circuits and the negative probe to the battery negative post.

Scope Settings
To make the waveforms as clear as possible, set the scope settings to match the examples.

Lab Scope Example (1)
In example (1), the trace shows the coil primary circuit. The firing line reached 325v.

Lab Scope Example (2)
In example (2), the volts per division settings were changed from 50v to 10v per division. The trace shows a spark line "intersect point" (the range is 30v to 40v). If it over 45v, the A/F mixture is too lean, the cylinder compression is too high or the spark timing is retarded.

If it is below 30v, the A/F mixture is too rich, the cylinder compression is too low or the spark timing is over advanced.

Lab Scope Example (3)
The low current amp probe was set to 100 mv (100 mv equals 1 amp on the scope) and adjusted to zero. Note that each vertical grid equals 1 amp. The probe was clamped around a coil primary control wire. The current peaked at 5.0 amps on this known good ignition coil.

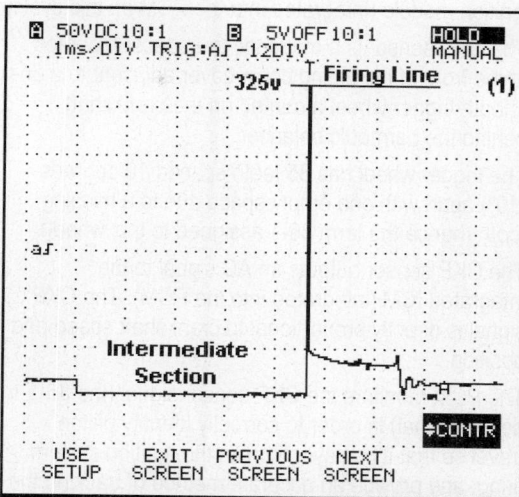

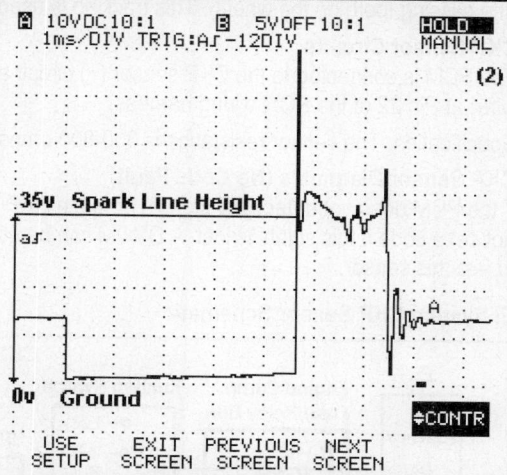

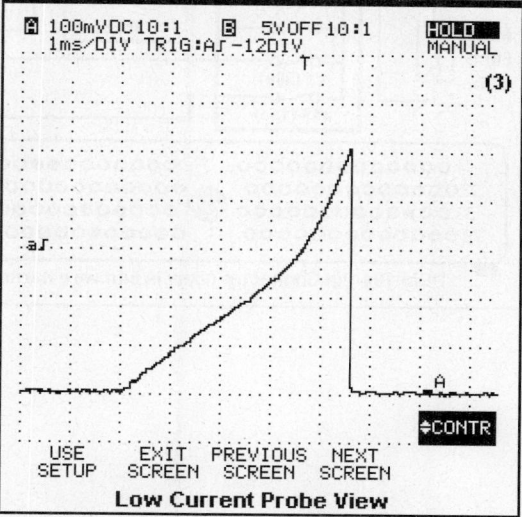

Low Current Probe View

Integrated Electronic Ignition System

CKP SENSOR (MAGNETIC DESIGN)

The crankshaft position (CKP) sensor used with this EI system is the primary sensor for crankshaft position data to the ignition module (integrated into the PCM on this system).

The CKP sensor is a magnetic transducer mounted at the front of the timing chain cover adjacent to a 36-1 tooth trigger wheel mounted on the crankshaft behind the harmonic balancer.

The trigger wheel has 35 teeth spaced 10 degrees (10º) apart with one empty space due to a missing tooth (hence the term 36-1 assigned to this wheel).

The CKP sensor outputs an AC signal to the integrated ICM (integrated into the PCM). The CKP signal is directly proportional to crankshaft speed and position.

The PCM monitors the CKP sensor signal (the 36-1 design signal) in order to correctly identify piston travel so that it can synchronize the Ignition system firing, and provide an accurate method of tracking the angular position of the crankshaft, relative to a fixed reference (the missing tooth on the wheel). This tracking is used for misfire diagnosis.

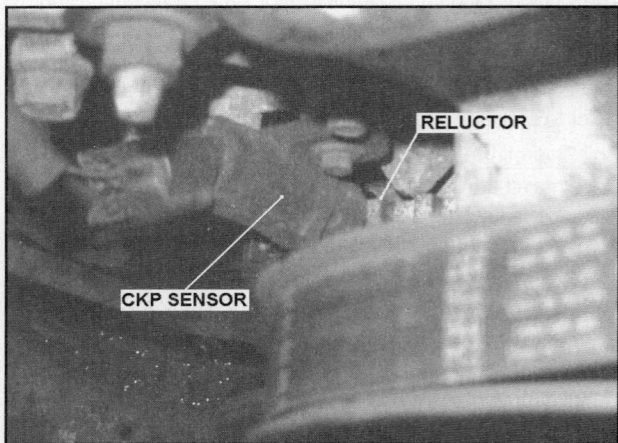

CKP Sensor Circuits

The PCM is connected to the CKP sensor (+) circuit at Pin 21 (Gray wire) and to the CKP sensor (-) circuit (Dark BLU wire) at Pin 22 of the PCM wiring harness.

Specification: The sensor resistance is 300-800 ohms (actual was 460 ohms at 76ºF).

CKP Sensor Diagnosis (No Code Fault)

If the PCM does not detect any signals from the CKP sensor with the engine cranking, it will not start. The PCM does not set a code if this signal is lost. A DVOM can be used to check for "no output", but a Lab Scope is the tool of choice to test this sensor.

EI System CKP Sensor Schematic

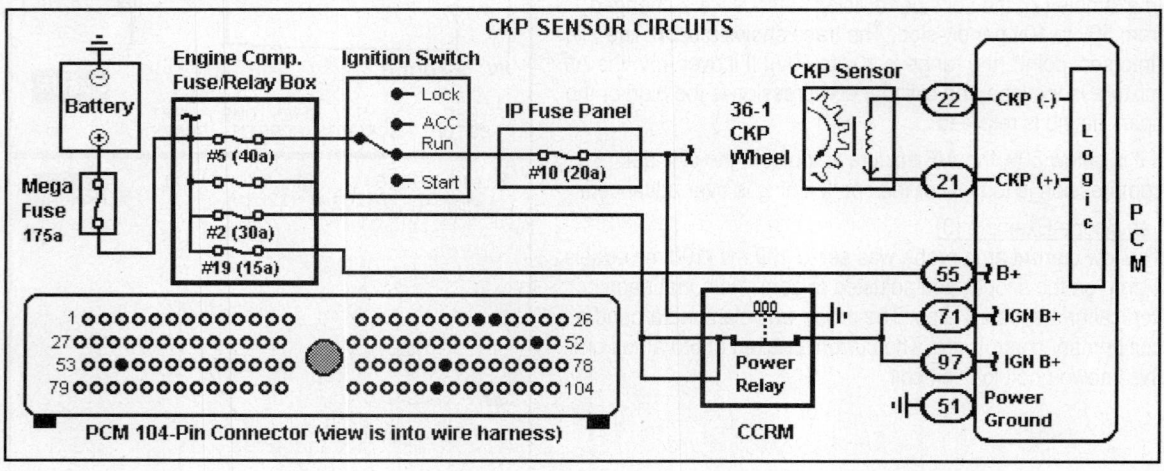

CKP SENSOR (SIGNAL EXPLANATION)

The CKP sensor (also referred to as a variable reluctance sensor or VRS) signal is derived by sensing the passage of teeth from a 36 minus one (36-1) tooth wheel mounted on the crankshaft.

The ferromagnetic-toothed timing wheel is rigidly mounted to the crankshaft and designed to have the center of each tooth spaced at 10-degree increments. Since there are 360 degrees in one revolution, there should be 36 teeth. However, in this system, there is one tooth missing. The CKP sensor is located directly opposite the missing tooth at a point where Cylinder No. 1 is located at 90° BTDC on a 4-cylinder engine, 60° BTDC on a 6-cylinder engine and 50° BTDC on an 8-cylinder engine. It should be noted that on a 10-cylinder engine that a 40-1 tooth wheel is used.

In addition to crankshaft-mounted 36-1 tooth wheel located at the front of the engine, some applications utilize holes in the flex plate at the rear of the engine to trigger the VRS sensor.

A VRS sensor is a magnetic transducer with a pole piece wrapped with fine wire. When the VRS encounters the rotating ferromagnetic teeth, the varying reluctance induces a voltage proportional to the rate of change in magnetic flux and the number of coil windings. As the tooth approaches the pole piece, a positive differential voltage is induced. As the tooth and the pole piece align, the signal rapidly transitions form maximum positive to maximum negative. As the tooth moves away, the voltage returns to a zero differential voltage. The negative zero crossing of the VRS signal corresponds to the alignment of the center of each tooth with the center of each pole piece. The ICM input circuitry (in the PCM) triggers on this negative zero crossing at around -0.3 volts and uses this to establish crankshaft position.

At normal engine speeds, the VRS signal waveform approximates a sine wave with increased amplitude adjacent to the missing tooth region. At 30 rpm, the VRS output should be 150 mv peak-to-peak, at 300 rpm, the VRS output should be 1.5 volts peak-to-peak and at 6000 rpm, the VRS output should be 24.0 volts peak-to-peak. At 8000 rpm, the VRS output should not exceed 300 volts peak-to-peak. The normal sensor air gap is 1.0 mm nominal; 2.0 mm max. The resistance of the sensor should be 300-800 ohms. The inductance should be 170 to 930 Millihenries.

CMP SENSOR (SIGNAL EXPLANATION)

The CMP sensor input is not needed for the EDIS system since the spark plugs are fired in pairs. The CMP sensor is, however, used by the PCM to determine Cylinder 1 for sequential fuel control. This vehicle application utilizes a Hall-effect sensor (not a VRS).

The Hall-effect device and shutter are designed to produce a 50% duty cycle with the rising edge located at 26° ATDC of Cylinder 1 and the falling edge 180° of camshaft rotation later. On models with a VRS design Camshaft sensor, the VRS produces a very narrow sinusoidal signal when a tooth on the camshaft passes the pole piece. The negative-going edge is typically located at 22° ATDC of cylinder No. 1.

For systems that have Camshaft Position Control (also known as Variable Cam Timing), there are 3 + 1 teeth (V6 engine) or 4 + 1 teeth (I4 or V8) located on each camshaft. The PCM uses the extra (+1) tooth on Bank 1 to identify Cylinder 1 and the extra tooth on Bank 2 to identify the corresponding paired cylinder.

DVOM TEST (CKP SENSOR)

Place the shift selector in Park and block the drive wheels for safety.

DVOM Connections

Turn the key off and connect the breakout box. Select AC volts on the DVOM. Connect the DVOM positive lead to the CKP sensor (+) circuit (GRY wire) at Pin 21 and the negative lead the CKP sensor (-) circuit (DK BLU wire).

Crank the engine for 3-5 seconds or start the engine and record the readings.

Test Results

The DVOM display shows a CKP signal of 1.6v AC at a rate of 514 Hz. Note that the true amplitude of the AC voltage reading is about 3.3v in this example.

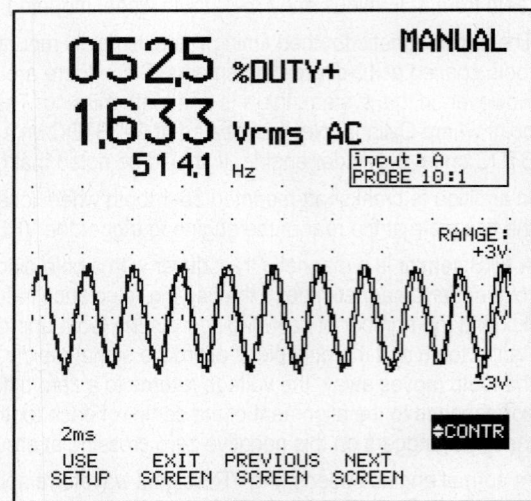

LAB SCOPE TEST (CKP SENSOR)

The Lab Scope can be used to test the CKP sensor as it provides a very accurate view of sensor waveform and of any glitches. Place the shift selector in Park (A/T) for safety.

Scope Connections

Install a breakout box (BOB) if available. Connect the Channel 'A' positive probe to the CKP sensor positive circuit at PCM Pin 21 (GRY wire) and the negative probe to the CKP sensor negative circuit at PCM Pin 22 (DK BLU wire). Connect the Channel 'B' positive probe to the CMP sensor signal at PCM Pin 85 (DB/ORN wire).

Scope Settings

To make the waveforms as clear as possible, set the scope settings to match the examples.

Lab Scope Tests

The CMP and CKP sensor signals can be checked during cranking, idle and at off-idle speeds with the engine cold or fully warm.

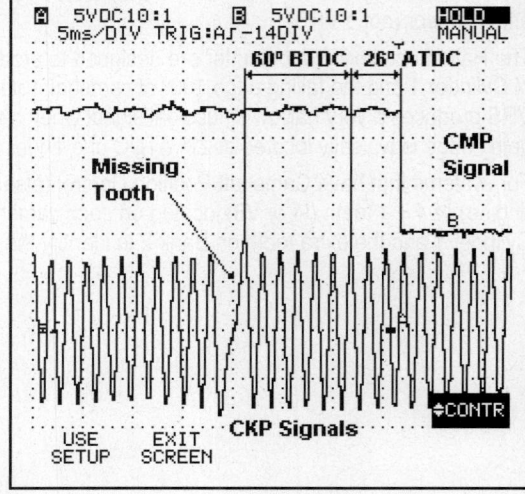

Lab Scope Example

In this example, the top trace shows the CMP sensor signal. The bottom trace shows the CKP sensor signal. These waveforms were captured with the gear selector in Park, the engine running at idle speed in closed loop.

The top trace represents 135° of camshaft rotation from the CMP sensor signal. Note the voltage level of the CMP sensor signal (13.5v).

The bottom trace represents around 270 degrees of crankshaft revolution from the CKP sensor (a 6v positive peak signal and a 4v negative peak signal). This signal is 10v AC peak-to-peak and is occurring at 520 hertz. Note the point at which the missing tooth occurs (60° BTDC) and where the CMP signal occurs (26° ATDC).

CHILTON DIAGNOSTIC SYSTEM - DTC P0300

The purpose of the Chilton Diagnostic System is to provide you with one or more tests that can be done with a DVOM, Lab Scope or Scan Tool *prior* to entering the complete trouble code repair procedure. The quick checks listed in the DVOM, Lab Scope or Scan Tool tests on the previous pages may help you quickly find the cause of the problem. If you cannot resolve the problem with these tests, a code repair chart is also included.

Code Description

DTC P0300 - The Misfire Monitor detected a random misfire condition in multiple cylinders (in effect, the PCM cannot determine which cylinder(s) are misfiring.

Code Conditions (Failure)

This code is set with the engine running under a positive torque condition, the PCM detected a random misfire condition in one or more cylinders.

Quick Check Items

Inspect or physically check the following items as described below:

- Check for a faulty CMP sensor and for intermittent loss of the CMP sensor signal
- Check the EGR valve to determine if it is stuck open
- Determine if the vehicle was driven with less than 1/8 of a tank of fuel (low on fuel!)

Drive Cycle Explanation

Start the vehicle and drive it under any positive load condition.

Drive Cycle Preparation

There are no drive cycle preparation steps required to run the Misfire Monitor.

Install the scan tool. Turn to key on engine off. Cycle the key off, then on. Select the appropriate vehicle and engine qualifier. Clear the codes and do a PCM reset.

Scan Tool Help

Use the Scan Tool menu to view and then select the desired PIDs (see the PIDs below). Start the engine without turning the key off. Refer to the article titled "How to Access and Use Generic PID Information" at the end of Section 1 for information on this subject.

■ **NOTE: An example of how to read a trouble code, view its Freeze Frame Data and then clear the Diagnostic information from P0300 is shown in the Graphic below (Vetronix).**

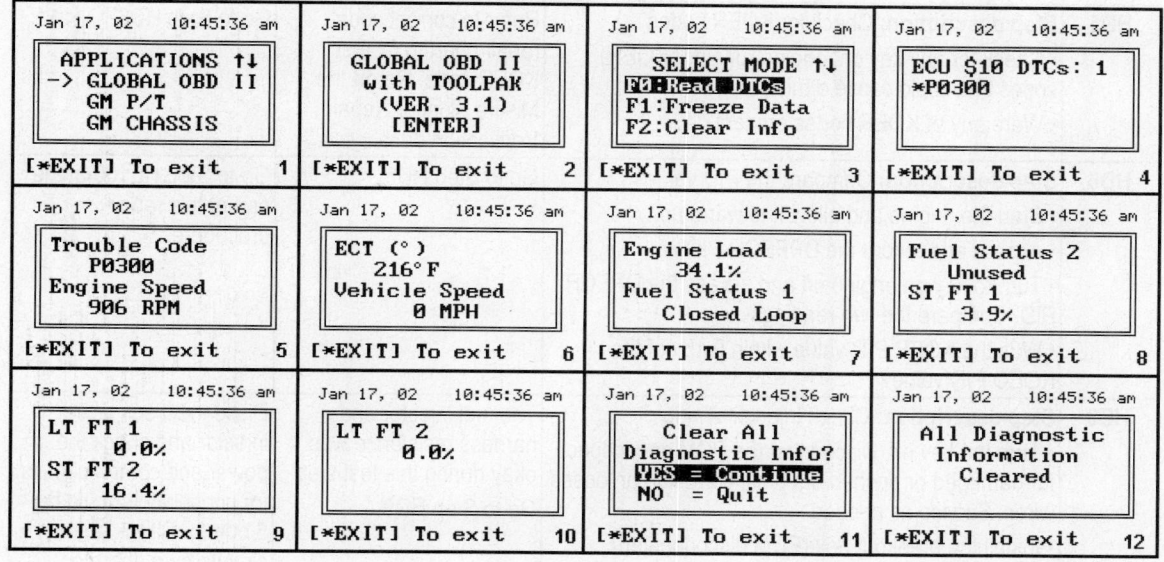

If DTC P0300 was set, Pinpoint Test HD can be used to diagnose the following devices and/or circuits:

- Camshaft Position Sensor Signal
- CKP Sensor (Physical Check)
- EGR Position Sensor Signal
- Power Control Module (PCM)

Pinpoint Test HD Repair Table

Step	Action	Yes	No
HD1	**Step description:** Did vehicle run out of fuel? • Running out of fuel can cause this code to set. Was the vehicle run out of fuel?	OBD II System is okay. Do a PCM Reset to clear the codes.	Go to Step HD2.
HD2	**Step description:** Check damper assembly? • Turn the key to off. Observe the crank pulley for wobble. • Examine the EI pulse ring that is fastened to the harmonic fastener. • Does the crank pulley wobble or is the pulse ring loose or damaged?	Replace the pulley or damper assembly. Disconnect the battery for 5 minutes to clear the crank data. Then run the full Crank Relearn procedure.	Go to Step HD3.
HD3	**Step description:** Check for any Fuel Trim and HO2S Continuous Memory codes • Check for non-misfire Continuous Memory codes that could cause this code. Look for P0136, P0156, P0171, P0172, P0175, P1130 and P1150. • Were any of the listed codes present?	Refer to appropriate code repair chart to fix the first Non-Misfire related trouble code. Rerun the Misfire Monitor Repair Verification drive cycle.	Go to Step HD4.
HD4	**Step description:** Check for KOEO codes • Check for any key on, engine off (KOEO) On Demand codes that could cause a misfire. • Were any of KOEO codes present?	Refer to correct code repair chart to fix first KOEO code. Rerun the Misfire Monitor Repair Verification drive cycle.	Go to Pinpoint Test JB1 to test spark plugs and secondary wires. If these components are okay, go to Step HD6.
HD5	**Step description:** Check for KOER codes • Check for any key on, engine running (KOER) codes that could cause a misfire. • Were any of KOER codes present?	Refer to correct code repair chart to fix first KOER code. Rerun the Misfire Monitor Repair Verification drive cycle.	Go to Step HD6
HD6	**Step description:** Compare the PID values • Start the engine and allow it to warm up. • Access and record the DPFEGR PID. • Turn to key on engine off and access the DPEGR PID. Compare the two readings. • Was the KOER PID value within 0.15v of the KOEO PID value?	Go to Step HD8.	Go to Step HE100 at the end of this repair procedure.
HD8	**Step description:** Check Injector Wiring • Turn the key off. Disconnect the PCM and inspect for damaged or pushed out pins, corrosion or looses wires. Service as needed. • Install the breakout box (PCM disconnected). • Measure the resistance between the injector test pin (one at a time) and Test Pin 71 or 91. • Was the resistance value from 11-18 ohms?	The fuel injector and harness resistance was okay during this test step. Go to Step HD9.	Disconnect suspect injector and check the power and control circuits for continuity (should be < 5 ohms). Check both circuits for a short to ground or short to power fault. This is repair step H56.

Pinpoint Test HD Repair Table (Continued)

Step	Action	Yes	No
HD9	**Step description:** Check the Fuel Injectors • Turn the key off. • Connect the PCM to the breakout box. • Connect a 12v test light between the injector test pin (one at a time) and Test Pin 71 or 91. • Start the engine and allow it to idle. • Did the test light have a dim glow (with the engine running)?	Go to Step HD10.	If the "known good" 12v test light did not glow, the PCM is the problem. Replace the PCM. Rerun the Misfire Monitor Repair Verification drive cycle.
HD10	**Step description:** Check the Fuel Pressure • Turn the key off. Install fuel pressure gauge. • Start the engine and allow it to idle. Record the fuel pressure reading on the gauge. • Raise the engine speed to 2500 rpm in Park and hold it for one minute. Record the reading. • Was the fuel pressure 30-45 psi in the test?	Go to Step HD11.	Refer to the fuel pressure tests in other repair manuals to test and repair the Fuel system. Rerun the Misfire Monitor Repair Verification drive cycle.
HD11	**Step description:** Check the Hold Pressure • Start the engine and allow it to idle. Record the fuel pressure reading on the gauge. • Raise the engine speed to 2500 rpm in Park and hold it for one minute. Record the reading. • Look for signs of fuel leaking at the injector O-rings or the fuel pressure regulator. • Turn the key off and wait 60 seconds. • Did the pressure hold at 16 psi (60 sec's)?	Go to Step HD12.	Refer to the fuel pressure tests in other repair manuals to test and repair the Fuel system. Rerun the Misfire Monitor Repair Verification drive cycle.
HD12	**Step description:** Check Injector Flow Rate • Use a Flow Test to verify the flow rate for each fuel injector (follow all related Safety Precautions and Test Procedures). • Was injector flow rate within specifications?	Go to Step HD20.	Replace or clean the inoperative injectors as needed. Rerun the Misfire Monitor Repair Verification drive cycle.
HD20	**Step description:** Check the Vacuum System • Listen for signs of an engine vacuum leak. • Inspect all vacuum lines for signs of pinched, cracked or misrouting and/or poor assembly. • Refer to the underhood vacuum label. • Was the Vacuum system okay?	Go to Step HD21.	Service the Vacuum system components as needed. Rerun the Misfire Monitor Repair Verification drive cycle.
HD21	**Step description:** Check EVAP Canister • Inspect the canister for signs of excess fuel. • Was there excess fuel in the EVAP canister?	Replace the fuel vapor canister. Rerun the Misfire Monitor Repair Verification drive cycle.	Go to Step HD22.
HD22	**Step description:** EVAP Pressure Test • Turn the key off. Install the appropriate EVAP tester at EVAP service port or fuel filler cap. • Follow test instructions to check the system. • Did the EVAP system hold pressure?	Go to Step HD23.	Service the EVAP system components as needed. Rerun the Misfire Monitor Repair Verification drive cycle.
HD23	**Step description:** Inspect EVAP System • Inspect the system for blockage/restriction. • Was any blockage located in the system?	Make repairs to the system as needed. Run the Misfire Monitor Verification Test.	Go to Step HD26.

Pinpoint Test HD Repair Table (Continued)

Step	Action	Yes	No
HD25	**Step description:** Check Base Engine • Perform an Engine Compression test. • Perform a dynamic Valve Train test. • Check the PCV system for problems. • Was a problem found during this step?	Service or repair the engine components as needed. Rerun the Misfire Monitor Repair Verification drive cycle.	The cause of the Misfire is intermittent. Use an Ignition system analyzer to diagnose the Ignition system.
HD26	**Step description:** Check VMV for Leaks • Turn the key off (VMV unit still connected). • Install a hand vacuum pump to fuel vapor port to the fuel vapor canister on the VMV line. • Apply 16" Hg of vacuum (using the pump). • Did VMV hold vacuum at room temperature?	Go to Step HD27.	Remove the vacuum pump and then replace the damaged VMV. Rerun the Misfire Monitor Repair Verification drive cycle.
HD27	**Step description:** Check VMV Filter • Turn the key off. Remove vacuum line from the input vacuum port to intake manifold port on VMV (control solenoid portion of the valve). • Connect hand vacuum pump to the open input port of the VMV. • Apply 10-15" Hg of vacuum to the VMV. • Did the VMV hold vacuum, or did the valve very slowly release vacuum to atmosphere?	Service the VMV filter. If unable to clean the filter, replace the VMV. Rerun the Misfire Monitor Repair Verification drive cycle.	Remove the vacuum pump and reconnect all of the components. Go to Step HD25.

Pinpoint Test HE Repair Table (Step HE 100 only)

Step	Action	Yes	No
HE100	**Step description:** Check for EGR Flow (this test will check for possible EGR flow at idle speed with no related trouble codes stored) • Disconnect and plug EGR hose at the valve. • Turn to key on, engine off. • Access the DPFEGR PID on the Scan Tool. • The PID voltage should be from 0.35-0.80v. • Start the engine and allow it to return to idle. • Monitor the DPFEGR PID value. Compare it the engine off reading. An increase in the PID voltage reading indicates that the differential pressure feedback EGR sensor is sensing EGR flow at idle speed. • Was the DPFEGER PID voltage reading at idle speed at least 0.15v more than the key on, engine off PID voltage reading?	The DPFEGR PID voltage is indicating EGR flow at idle. Since the EGR vacuum hose is disconnected and plugged, the fault is most likely in the EGR valve. Remove and check the EGR valve for signs of contamination, unusual wear, carbon deposits, binding or damage. Service the EGR valve as needed. Then rerun the test to verify the repair is completed.	This result indicates a fault in the EGR valve vacuum supply. Inspect the EGR vacuum regulator (VR) solenoid vent and vent filter for restrictions. Service the component as needed. If they are okay, replace the EGR vacuum regulator solenoid. Then rerun the test to verify the repair is completed and the original condition not longer is present.

Evaporative Emission System

INTRODUCTION

This vehicle application uses a Vapor Management Flow (VMV) design of Evaporative Emission (EVAP) system to prevent fuel vapor buildup in the sealed fuel tank.

System Overview

Fuel vapors trapped in the sealed fuel tank are vented through a vapor valve assembly at the top of the tank through a single vapor line to the charcoal canister for storage. The fuel vapors are then purged into the intake manifold where they are burned during normal engine operation. The purge valve is called a vapor management valve (VMV).

EVAP System Graphic

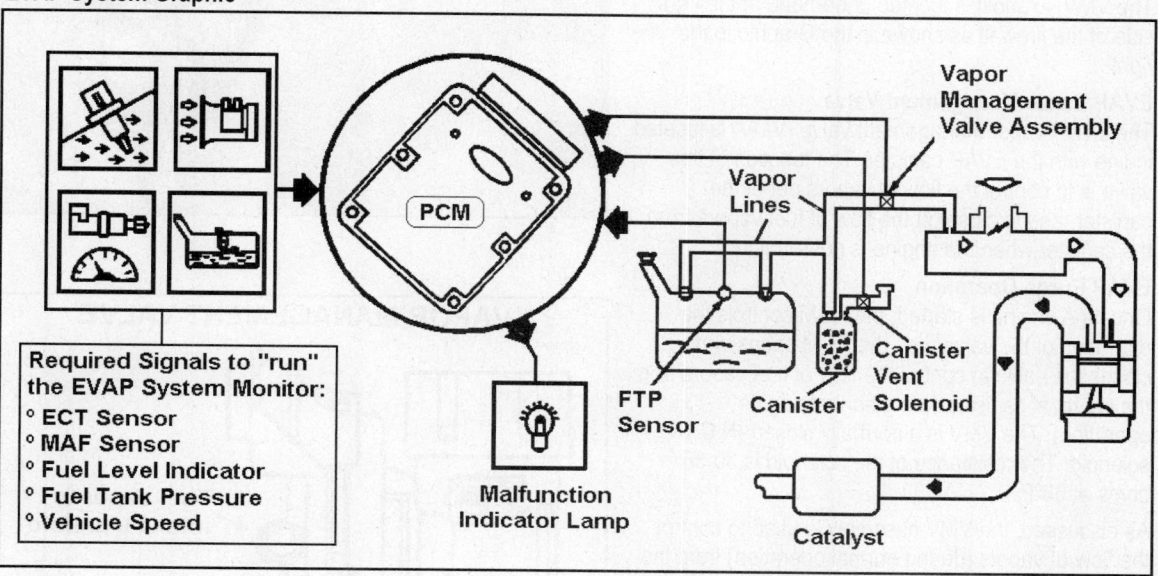

Fuel Tank Pressure Sensor

On this vehicle application, a fuel tank pressure (FTP) sensor is used to monitor the pressure or vacuum value in the fuel tank during the Running Loss Monitor Test portion of the OBD II Diagnostics (this is the EVAP Purge Test).

The EVAP Monitor performs a Fuel System integrity check to determine if a leak is present in the EVAP system.

The FTP sensor converts variations in pressure into an analog voltage signal that is sent to the PCM.

Trouble Code Help

If the PCM detects an open or short circuit condition in the FTP sensor or its circuit during the OBD II Test (Comprehensive Component Test), it will set DTC P0452. If the FTP sensor fails on the next consecutive trip, the code will mature and the PCM will turn on the MIL (this is two-trip fault detection).

Fuel Tank Pressure Sensor

SYSTEM COMPONENTS

The Vapor Management Flow system includes the following components:

- Crankcase ventilation tube assembly
- Fuel tank and fuel filler cap
- Fuel vapor storage canister
- Fuel vapor control valve
- Related fuel vapor hoses
- PCM and related connecting wires
- Vapor management valve (VMV)

VMV Solenoid Location

The VMV solenoid is located underhood at the right side of the firewall as shown in the Graphic to the right.

EVAP Vapor Management Valve

The EVAP vapor management valve (VMV) is located in-line with the EVAP canister. The function of this valve is to control the flow of vapors out of the canister, and to close off the flow of fuel vapors from the canister when the engine is not running.

EVAP Purge Operation

Once the engine is started, the PCM controls the operation of the valve (i.e., the PCM opens and closes the valve to control the flow of fuel vapors from the charcoal canister during normal engine operation). The VMV is a normally closed (N.C.) solenoid. The resistance of the solenoid is 30-36 ohms at 68°F.

As discussed, the VMV assembly is used to control the flow of vapors (during normal operation) from the fuel vapor storage canister into the intake manifold.

The actual conditions under which the VMV solenoid is opened are determined by the PCM from various sensor inputs.

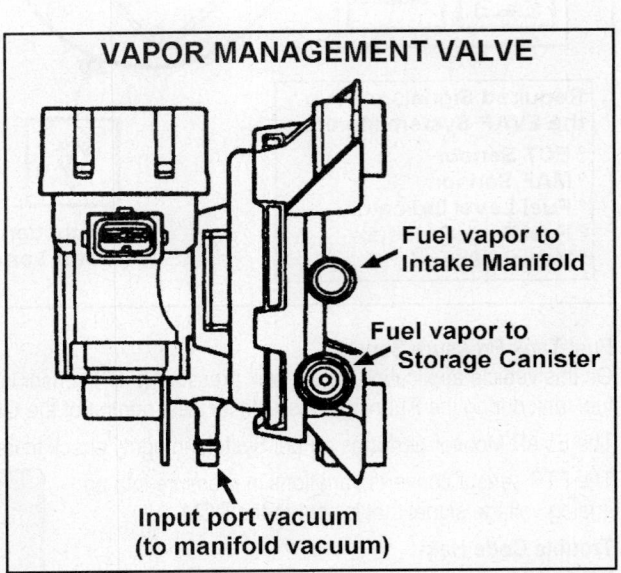

VAPOR MANAGEMENT VALVE

Fuel vapor to Intake Manifold

Fuel vapor to Storage Canister

Input port vacuum (to manifold vacuum)

Trouble Code Help

The first time that the PCM detects a problem in the VMV solenoid operation or one of its circuits, it will set a pending trouble code (DTC P0443).

If the same fault is detected on two consecutive trips, the trouble code matures and the PCM will turn on the MIL to indicate that an emission related fault has been detected.

The repair steps for this trouble code (DTC P0443) include a check of the VMV solenoid and its related circuits.

EVAPORATIVE CHARCOAL CANISTER

The charcoal canister is located at the rear of the vehicle below the spare tire.

Fuel vapors from the fuel tank and the air cleaner are stored in the charcoal canister.

Once the engine is started, fuel vapors stored in charcoal canister are purged into the engine where they are burned during normal combustion.

The vehicle application is equipped with a single canister that contains 2.0L of activated carbon in the carbon bed.

The activated charcoal has the ability to purge or release any stored fuel vapors when it is exposed to fresh air.

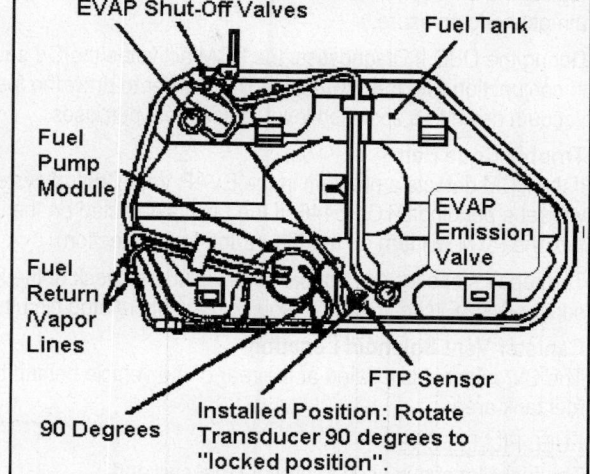

Fuel Tank Pressure Sensor
The fuel tank pressure (FTP) sensor is located on top of the fuel tank as shown in the Graphic to the right.

EVAP Canister Location Graphic

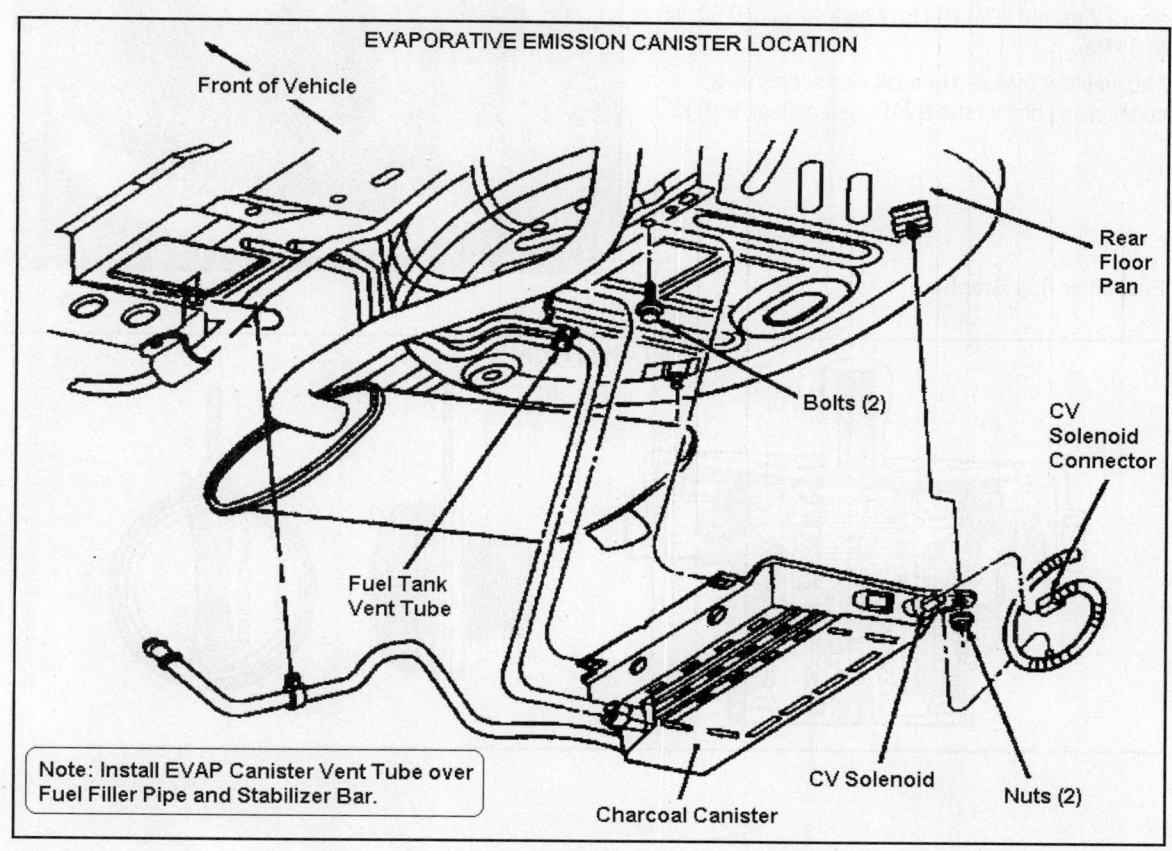

CANISTER VENT SOLENOID

The canister vent (CV) solenoid is a normally open (N.O.) solenoid with a resistance of 40-50 ohms at 68°F. The CV solenoid is used to seal the atmosphere port of the charcoal canister from atmospheric pressure.

During the OBD II Diagnostics, the PCM activates the CV solenoid, in conjunction with the VMV solenoid, in order to draw the fuel tank vacuum down to a specified level for leak test purposes.

Trouble Code Help

If the PCM detects a problem in the EVAP Vent Control system, it will set a pending DTC P0446. If the fault is detected on the next trip, the PCM will turn on the MIL (2-trip fault detection)

The repair steps for this trouble code include a check of the FTP sensor (VREF) circuit, the CV solenoid and its related circuits.

Canister Vent Solenoid Location

The CV solenoid is located at the rear of the vehicle behind the fuel tank area.

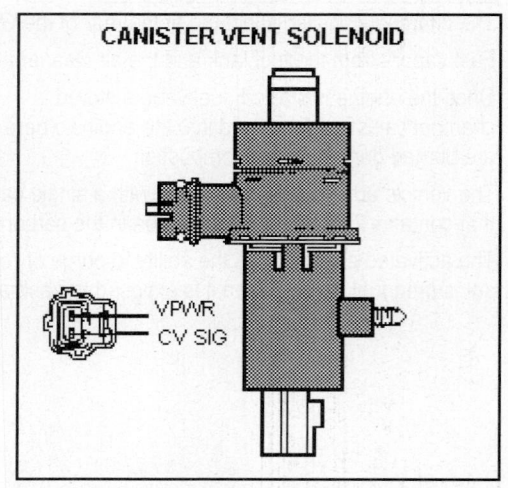

FUEL FILLER CAP

The fuel filler cap is used to prevent fuel spill and close the EVAP and Fuel systems to atmosphere.

The fuel tank cap is designed to relieve pressure above 2 psi (14 kPa) and to relieve vacuum 0.53 psi (3.8 kPa).

The fuel filler tank cap or neck also serves as a connection point for the EVAP system (leak test) kit.

Fuel Filler Cap Graphic

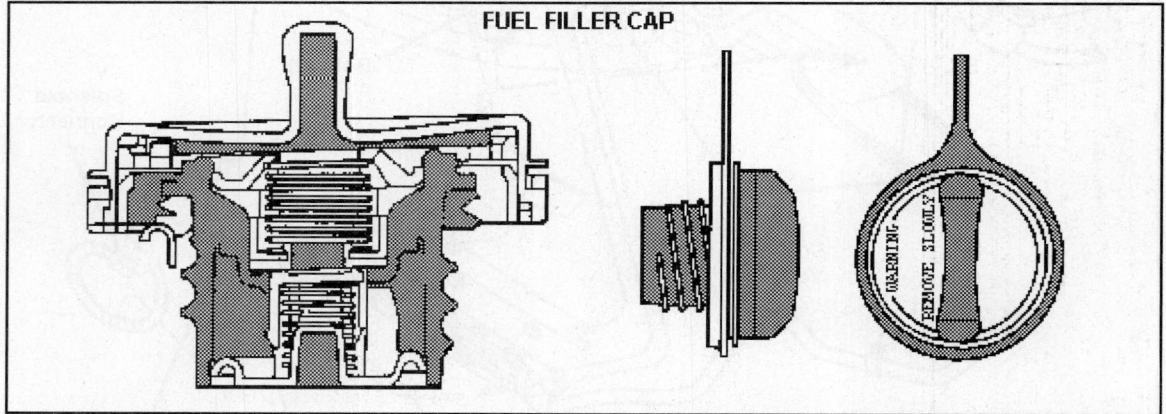

EVAP EMISSION VALVES

The fuel vapor control valve is located in series between the fuel vapor valve and the fuel vapor storage canister. It is mounted on a fuel tank bracket and connected to the fuel filler pipe. This valve is used to close the path of flow from the fuel vapor valve to the fuel vapor storage canister during refueling to prevent overfilling the fuel tank.

The shut-off valve detects fuel filler cap removal by sensing a change in pressure (through the fuel filler pipe sensing tube) that occurs during the removal of the fuel filler cap. The valve closes when the cap is removed and opens when the cap is installed.

EVAP Component Location Graphic

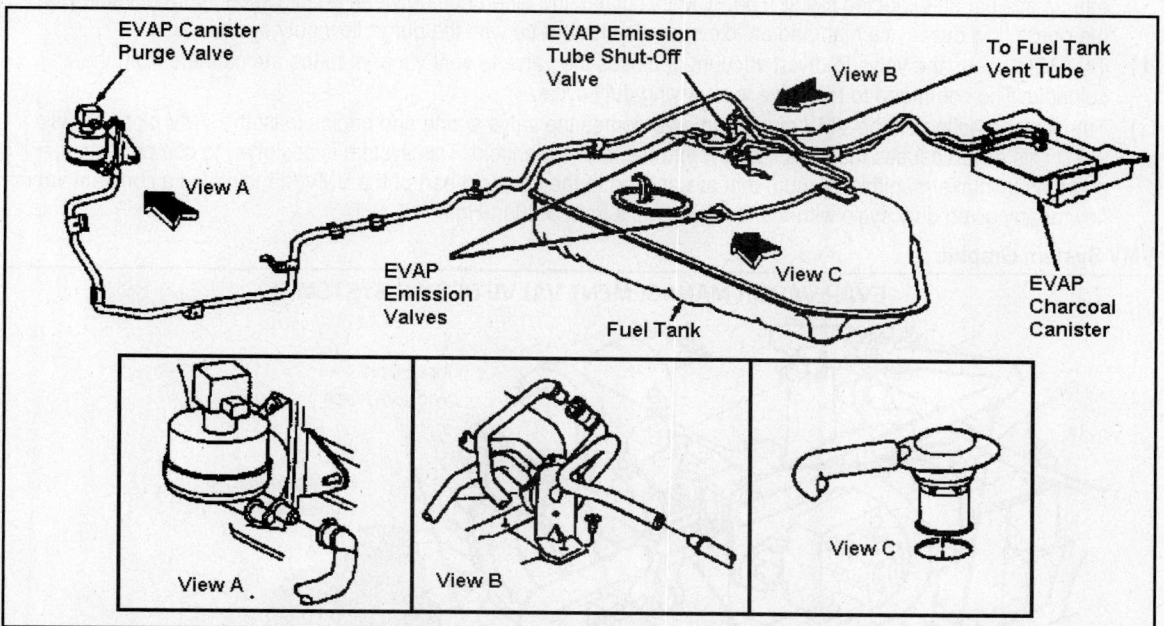

Fuel Vapor Control Valve Operation

Fuel vapor in the fuel tank is vented to the fuel vapor storage charcoal canister through the fuel vapor assembly.

The valve is mounted in a rubber grommet at a central location in the upper surface of the fuel tank.

A vapor space, between the fuel level and upper surface of the tank, combines with a small orifice and float shutoff (rollover) valve in the vapor valve assembly to prevent liquid fuel from passing to the fuel vapor storage canister. The vapor space allows for thermal expansion of fuel in the tank.

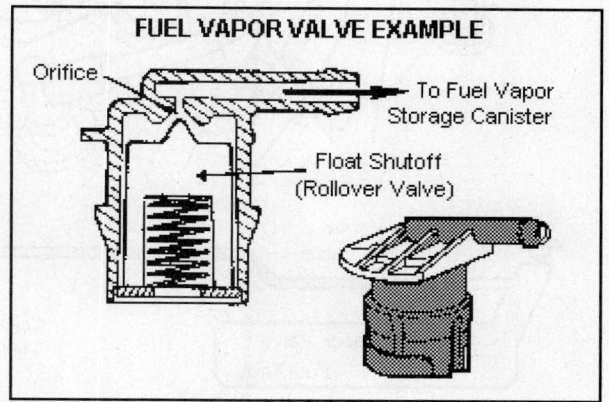

FUEL VAPOR VALVE EXAMPLE

EVAP SYSTEM OPERATION

The operation of the EVAP system used on this vehicle application is discussed next.

1) The PCM receives signals from the CKP, ECT, IAT, MAF, TP and VSS to determine when to operate the VMV system (ECT and IAT inputs must indicate a warm engine, at stable idle, running in closed loop at moderate load and open or part throttle position with the vehicle moving at an elevated steady speed for a period of time).

2) The PCM deactivates fuel vapor management flow during idle periods, or whenever a fault is detected in the VMV valve, or any fuel vapor management flow inputs.

3) The PCM monitors the effects of Canister Purge Compensation strategy on the IAC strategy and compares the effects against an expected result. The PCM calculates the difference between an idle speed control value with the purge flow duty cycle high and an idle speed control value with the purge flow duty cycle low.

4) The PCM opens the valve to divert vacuum or closes the valve to vent vacuum to the atmosphere from the solenoid. The command to the valve is a varying duty cycle.

5) The vacuum acting on the VMV diaphragm overcomes the valve spring and begins to lift the VMV pintle off its seat. This action causes fuel vapor to flow into the intake manifold. This system is designed to compensate for changes in intake manifold vacuum that are applied to the solenoid part of the VMV. It maintains a constant vapor flow at any given duty cycle with manifold vacuums from 5-20 In-Hg.

VMV System Graphic

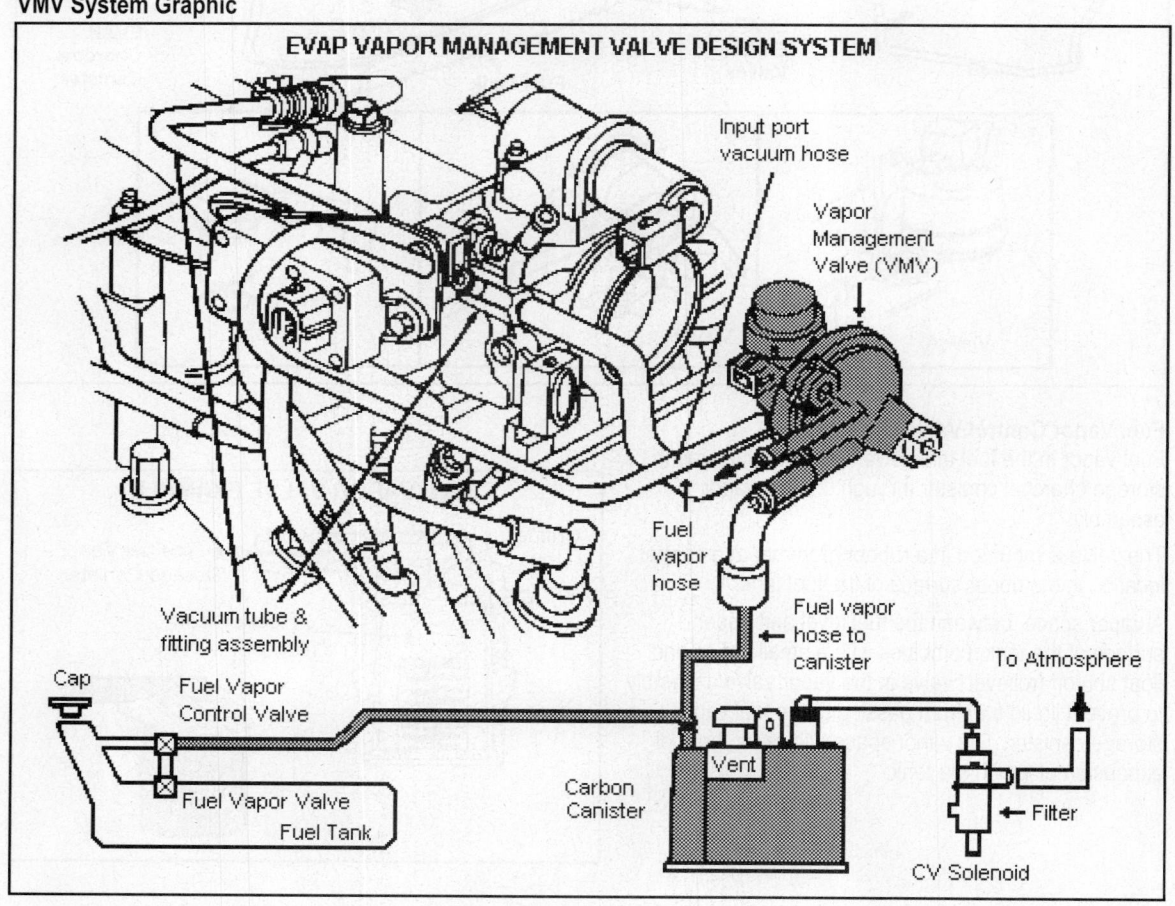

EVAP VAPOR MANAGEMENT VALVE DESIGN SYSTEM

SCAN TOOL TEST (EVAP PURGE

Connect the Scan Tool to the DLC underdash connector. Navigate through the Scan Tool menus until you get to the Ford (OEM) Data List (a list of PID items). Select the EVAP PURGE DC from the menu items along with the voltage signal from the Upstream O2S for Cylinder Bank 1.

EVAP PURGE DC PID Example

Start with the engine cold or warm and then monitor the EVAP PURGE PID. The PID (%) can read anywhere from 0-100% DC (it will read 100% after a cold engine startup).

SCAN TOOL TEST (FTP SENSOR)

Connect the Scan Tool to the DLC underdash connector. Navigate through the Scan Tool menus until you get to the Ford (OEM) Data List (a list of PID items). Select the Fuel Tank Pressure sensor from the menu items (in this case, both the sensor (P) and (V) were selected for the capture).

FTP PID Example

The FTP Sensor PID is shown in Example (2). If the fuel cap is removed, the FTP PID will read very close to 2.5v.

DVOM TEST (VMV SOLENOID)

To test the VMV solenoid with a DVOM, place the gear selector in Park (A/T) and block the drive wheels for safety.

DVOM Connections

Connect the DVOM positive probe to the VMV solenoid control circuit (GRY/YEL wire) and the negative probe to the battery negative ground post.

Start the engine and change the engine speed while monitoring the VMV control signal on the DVOM for signs of a steady hertz rate and proper voltage level. A sudden change in engine speed will cause a corresponding change in the VMV control signal.

Jan 15, 02   4:51:22 pm **(1)**

```
EVAP PURGE DC
    25 % DC
UP O2S BANK1
    0.89 VOLTS
```

Always use [HELP] to
   list keys & modes
[RCV] For enhanced
[*EXIT] To exit

Jan 15, 02   4:53:55 pm **(2)**

```
FUEL TNK SNSR(P)
  - 1.8 "H2O
FUEL TNK SNSR(V)
    2.88 VOLTS
```

Always use [HELP] to
   list keys & modes
[RCV] For enhanced
[*EXIT] To exit

VMV Solenoid Test Schematic

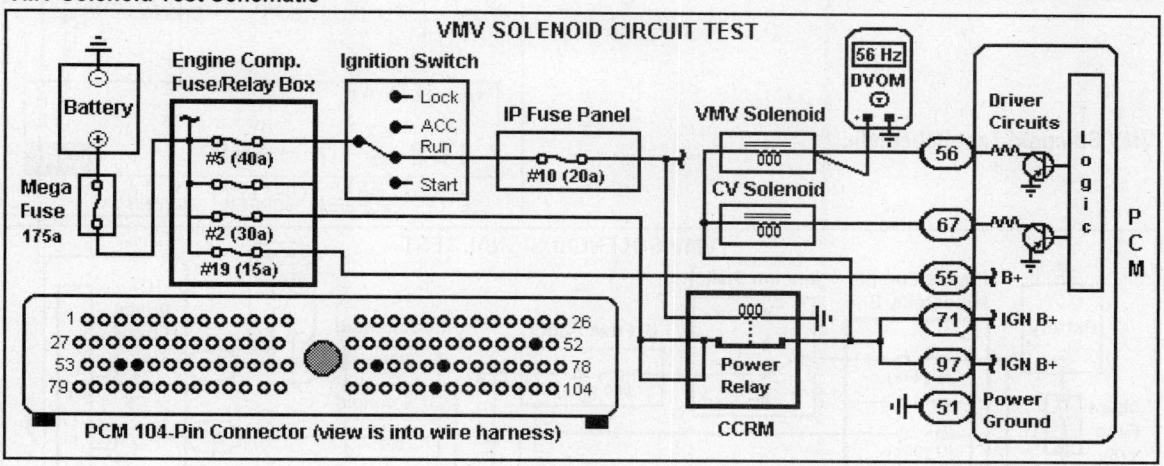

VMV SOLENOID CIRCUIT TEST

LAB SCOPE TEST (CV & VMV SOLENOIDS)

The Lab Scope is the "tool of choice" to test the CV and VMV solenoids as it provides an accurate view of the solenoid response and frequency rate. Place the gearshift selector in Park and block the drive wheels for safety.

Scope Connections

Connect the Channel 'A' positive probe to the CV solenoid control circuit (PPL/WHT wire) at PCM Pin 67 and the negative probe to the battery ground post. Connect the Channel 'A' positive probe to the VMV solenoid control circuit (GRY/YEL wire) at PCM Pin 56.

Scope Settings

To make the waveforms as clear as possible, set the scope settings to match the examples.

Lab Scope Tests

The examples on this page were taken from a vehicle with 144,000 miles on the odometer. The waveforms in these examples are from the original solenoids. The engine was allowed to run long enough for the engine to enter closed loop operation during testing.

Lab Scope Example (1)

In example (1), the trace shows the CV solenoid. Note that this waveform was captured after a hot engine startup (with the EVAP Leak Test). The CV PID will read 0% under these conditions (Purge Test disabled).

Lab Scope Example (2)

In example (2), the trace shows the VMV solenoid with the engine running at idle speed. Note the frequency value (102 Hz) in the box.

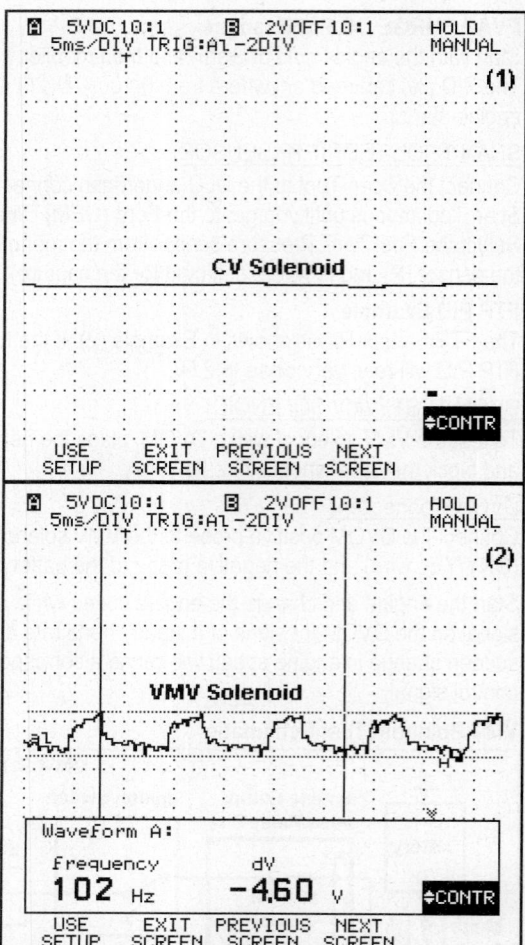

VMV Solenoid Test Schematic

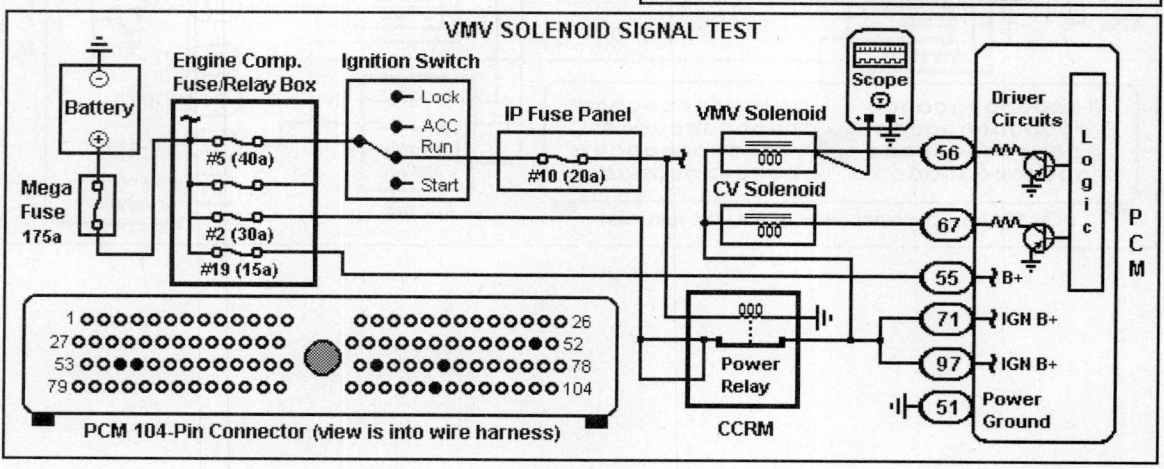

EVAP SYSTEM MONITOR

The EVAP System Monitor is an onboard strategy designed to test for proper operation of the EVAP system by checking the function of its components and its ability to flow fuel vapor (hydrocarbons) to the engine. In addition, the EVAP Monitor detects leaks equal to or greater than 0.040 inch by performing a vacuum check of the entire system.

The EVAP Monitor relies upon the EVAP canister vent (CV) solenoid to seal the entire EVAP system from atmosphere and the vapor management valve (VMV) to pull engine vacuum on the fuel tank. Then, with the system sealed and vacuum maintained, the Monitor uses the signal from the fuel tank pressure (FTP) sensor to observe the rate at which vacuum is lost (the rate of decay) during a period of system bleed-up. If the rate of decay is not within specifications for two consecutive trips, the MIL is illuminated.

The PCM receives signals from the ECT, FLI, FTP, IAT, MAF and VSS signals. The FLI signal is used to determine when to activate the EVAP Monitor (fuel sloshing, etc.).

1) Monitor requirements: Excessive pressure or vacuum not present in the fuel tank, ECT and IAT inputs indicate the engine is warm, engine running in closed loop at moderate load, and the vehicle moving at an elevated steady speed for a period of time. The PCM stops fuel vapor purging at idle speed, or if a fault is detected in the FTP sensor, VMV valve, canister solenoid, or one of the fuel vapor purge signals.

2) The PCM uses the FTP sensor input and the required amount of purge vapor flow to the intake manifold to calculate the purge command to the VMV. This action occurs after the PCM Purge Compensation strategy has been activated for a period of time (usually when fuel vapors are being purged from the canister). Once the Monitor startup criteria are satisfied, the PCM determines when to run the EVAP Monitor.

3) The PCM uses engine vacuum to evacuate the system using the VMV solenoid, and seals the system from atmosphere using the canister vent solenoid. It observes the rate at which vacuum is lost using the FTP sensor. The amount of vacuum lost or bleed-up is a function of the integrity of the system. A system with no leak has a small vacuum bleed-up over time while a system with a leak has a large vacuum bleed-up over time. If a leak is detected, the PCM will set a related EVAP code.

4) The PCM provides the FTP sensor with VREF and SIG RTN circuits. It controls the ground circuits to the canister vent (CV) and VMV solenoids through driver signals.

5) The CV solenoid is used to either seal or open the EVAP system to atmosphere during appropriate phases of the EVAP Monitor test. The canister vent solenoid is a normally open (0% duty cycle). When commanded closed (from 0-100% duty cycle), the fuel tank can be drawn down to a specific level of vacuum during the test period.

6) The PCM sends a duty cycle command to the VMV to divert vacuum to the VMV valve port (to open it) or to vent vacuum to atmosphere at the valve port (to close it). The increase in vacuum acting on the VMV diaphragm overcomes valve spring and begins to lift the VMV pintle off its seat causing vapor flow into the intake manifold. The valve compensates for changes in vacuum applied to the solenoid. It maintains a constant vapor flow at any given duty cycle for manifold vacuum (from 5-20" Hg).

7) The FTP sensor monitors the pressure or vacuum level in the fuel tank at key on, engine off, and with the engine running. The PCM continuously monitors the FTP signal. During testing, it monitors the EVAP system fuel pressure or vacuum buildup.

8) The EVAP Running Loss/Monitor system includes all of the fuel vapor hoses. These hoses are checked for any possible leaks by pressurizing the EVAP system using an EVAP Test Kit (Rotunda Kit 134-00056 or equivalent) and a frequency leak detector.

CHILTON DIAGNOSTIC SYSTEM - DTC P0442

The purpose of the Chilton Diagnostic System is to provide you with one or more tests that can be done with a DVOM, Lab Scope or Scan Tool *prior* to entering the complete trouble code repair procedure. The quick checks listed in the DVOM, Lab Scope or Scan Tool tests on the previous pages may help you quickly find the cause of the problem.

If you cannot resolve the problem with these tests, a code repair chart is also included.

Code Description

DTC P0442 - The EVAP Purge Monitor detected a leak as small as 0.040 inch in the system. The MIL is activated if this fault is detected on two consecutive trips (2T DTC).

Code Conditions (Failure)

This code is set if, following a cold engine startup, the vehicle is driven as described in the Drive Cycle Explanation, and the PCM detects the following conditions: The EVAP system bleed-up was less than 0.625 kPa (2.5 inches H2O) during a 15 second period with the fuel tank less than 75% full or the EVAP system vapor generation was more than 0.625 kPa (2.5 inches H2O) during a 120 second period.

Quick Check Items

Inspect or physically check the following items as described below:

*   A loose fill cap (tighten the fuel cap 1/8 of a turn until it clicks)
*   Small vapor leaks at the plastic vapor line connection to VMV
*   Small vapor hose leaks at connections to VMV, FTP sensor or canister
*   Small cuts or holes in vacuum lines or vapor hoses in the EVAP system

Drive Cycle Explanation

Start the vehicle (cold) and allow it to idle for at least 15 seconds. Following the cold engine startup (ECT sensor input less than 100°F and the IAT sensor input more than 40°F), drive the vehicle at a speed of at least 40 mph for 60 seconds until the ECT sensor input reaches at least 170°F.

Then with the EVAP DC PID more than 75%, the FLI PID from 15-85% (1998-2002) and the TP MODE PID at PT, drive at steady cruise speed (at over 40 mph) for 10 minutes.

EVAP "Cold Soak" Bypass Procedure

Install the Scan Tool. Turn the key on with the engine off. Cycle the key off, then on. Select the correct vehicle and engine qualifier. Clear the codes and do a PCM reset.

■ **NOTE:** *This step bypasses the engine soak timer and resets the OBD II monitor status.*

Use the Scan Tool menu to view and then select the following PIDs: ECT, EVAPDC, FLI (if available) and TP MODE. Start the engine without turning the key off. Refer to the article titled "How to Access and Use OEM PID Information" in Section 1 of this manual.

1996 Taurus 3.0L V6 MFI VIN U (A/T) DTC P0442 Parameters

PCM PID Acronym	Parameter Identification	PID Range	PID Value at Hot Idle	PID Value at 30 mph	PID Value at 55 mph
ECT (°F)	ECT Sensor (°F)	-40-304°F	160-200°F	160-200°F	160-200°F
FLI	Fuel Level Indicator (1998-02)	0-100%	50% (1.7v)	25-75%	25-75%
TP MODE	TP Sensor MODE	CT or PT	CT	PT	PT
EVAP CV	EVAP Vent Valve	0-100%	0%	0%	0%
EVAP DC	EVAP Purge Valve	0-100%	0-100%	0-100%	0-100%

DTC P0442 or P1442 - The EVAP Purge Monitor detected a leak as small as 0.040 inch in the system. The only difference between these two trouble codes is that DTC P0442 is a MIL trouble code. The MIL is not activated when the PCM sets a DTC P1442.

Pinpoint Test HX

Pinpoint Test HX is used to diagnose the following devices and/or circuits:

- Canister Vent Solenoid
- Carbon Canister, Fuel Vapor and Vacuum Hoses
- Fuel Tank, Fuel Fill Cap, Fuel Fill Cap Valve
- Fuel Tank Pressure (FTP) Sensor
- Fuel Vapor Control Valve, Fuel Vapor Valve
- Power Control Module (PCM)
- Vapor Management Valve (VMV)
- These PCM related circuits: CV CONT, VMV CONT, SIG RTN, VPWR and VREF

Possible Causes for this Trouble Code

- A loose fill cap (tighten the fuel cap 1/8 of a turn until it clicks)
- Small vapor leaks at the plastic vapor line connection to VMV
- Small vapor hose leaks at connections to VMV, FTP sensor or canister
- Small cuts or holes in vacuum lines or vapor hoses in the EVAP system

EVAP System Schematic

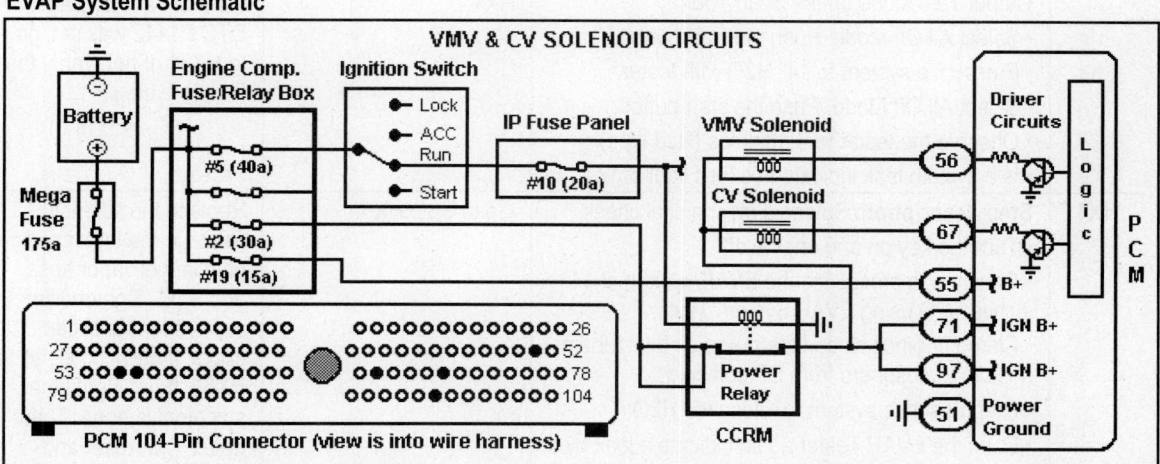

Pinpoint Test HX Repair Table

Step	Action	Yes	No
HX1	**Step description:** Check for a loose fuel cap • Verify that gas cap is tightened properly, and that the threads are not worn out. • Is the fill cap loose or worn out? Note: If the fuel filler cap is not tightened properly (1/8 turn until it clicks), the EVAP Monitor may fail its diagnostic test.	Tighten the cap 1/8 of a turn until it clicks. If the code is a memory code, go to HX61. If it is not a memory code, do a PCM Reset and retest the system.	Go to Step HX2.

PINPOINT TEST HX REPAIR TABLE (CONTINUED)

Step	Action	Yes	No
HX2	**Step description:** Inspect the connections • Pull slightly on the vacuum hose to VMV, FTP sensor, and carbon canister. • Check again for loose connections • Visually inspect for cuts or small hoses in the EVAP fuel vapor tubes or hoses. • Is a fault detected?	Make the needed repairs. If the code is a memory code, go to HX61. If not, do a PCM Reset and retest the system.	Go to Step HX3.
HX3	**Step description:** Perform EVAP Leak Test • Turn the key off and remove the fuel fill cap. • Install an approved EVAP System Tester. • Connect a compressed gas (nitrogen or argon) with tester pressure regulator closed. • Disconnect and plug (cap) the fuel vapor hose from the VMV at the intake manifold port. • To close the Canister Vent solenoid, access the Output Test Mode on the Scan Tool. • Select All Off Mode. Push the start button. • Pressurize system to 14" H2O with tester. • Select All Off Mode. Push the start button. • Observe the tester for 2 minutes (Red light). • Is a system leak indicated (is Red light on)?	Use the 2-position control valve on the tester to provide a continuing flow of gas to the EVAP system. Maintain 14 inches H2O pressure on the system (watch the gauges while performing this task). Go to Step HX4.	The pressure will discharge. A Green Light On indicates the test passed. Remove the tester, gas source, BOB and the plug in the fuel vapor hose. Tighten filler cap only 1/8 turn so cap initially clicks. Reconnect the PCM and all parts. If DTC P1442 was set, go to HX61. If not, verify the fault is fixed.
HX4	**Step description:** Solenoid mechanical check • Turn the key on and engine off. • Continue to pressurize the EVAP system at 14 inches H2O using EVAP System Tester. • Clamp or pinch closed the purge air inlet tube or hose to atmosphere from CP solenoid. • Re-pressurize system to 14 inches H2O. • Does the EVAP Tester again indicate a leak by a Red light On sequence?	Go to Step HX5.	Replace the solenoid. Remove the tester and plug in fuel vapor hose and BOB. Tighten filler cap 1/8 turn until cap clicks. If MEM code, go to HX61. If not, verify that symptom is gone. Then do a PCM Reset and retest the system.
HX5	**Step description:** *Check for leak with Ultra-Sonic Leak Detector or approved leak tester* • Turn the key on and engine off. • Verify that the fuel vapor hose from the VMV at the engine intake manifold is still plugged on all vehicles except Explorer/Sable/Taurus. • Verify the CV Solenoid is still closed. • Verify the tester still pressurizing the system. • Obtain the Ultra-Sonic detector. • Place the earphones from the detector over your ears and adjust the detector audio dial. • Closely pass the probe of the detector over vacuum lines, vapor hoses and connections from the fuel tank to the canister and VMV, and from the VMV to the intake manifold • Listen for audible sound changes (from noticeably quiet to a louder pitch) • Is a sudden audible change indicated?	Isolate the pressure leak in EVAP system and replace any damaged tubes/hoses. If the fault is in the fuel tank or fuel vapor control valve, use the Fuel system test. Remove the tester, gas source, BOB and the plug in fuel hose. Tighten gas filler cap only 1/8 turn so cap clicks. If it is a memory code, go to HX61. If it is not, verify that the fault no longer exists.	The EVAP system is okay or a small leak was not detected at this time. Remove the tester, gas source, BOB and the plug in the fuel vapor hose. Tighten gas filler cap only 1/8 turn so cap initially clicks. If it is a memory code, go to HX61. If not, verify that the symptom no longer exists.

Exhaust Gas Recirculation System

INTRODUCTION

The EGR system is designed to control NOx emissions by allowing small amounts of exhaust gases to be recirculated back into the combustion chamber to mix with the A/F mixture. This action causes a reduction in the amount of combustion, and less NOx. This vehicle is equipped with a differential pressure feedback EGR (DPFE) sensor.

Differential Pressure Feedback EGR System

The DPFE system includes the following components:

*   DPFE sensor
*   EGR valve and orifice tube assembly
*   Vacuum regulator valve assembly
*   PCM and related vacuum hoses.

The DPFE system operates as described next:

1) The PCM receives signals from the CKP, ECT, IAT, MAF and TP sensor in order to determine the optimum operating conditions for the DPFE EGR system. The engine must be warm, running at a stable engine speed under moderate load before the EGR system can be activated. The PCM deactivates the EGR system at idle speed, at extended wide-open throttle period or if a fault is detected in the EGR system.
2) The PCM calculates a desired amount of EGR flow for each engine condition. Then it determines the desired pressure drop across the metering orifice required to achieve the correct amount of flow and outputs the signal to the EGR VR solenoid.
3) The VR solenoid receives a variable duty cycle signal (0-100%). The higher the duty cycle rate, the more vacuum the solenoid diverts to the EGR valve.
4) The increase in vacuum acting upon the EGR valve diaphragm overcomes the valve spring pressure and lifts the valve pintle off its seat. This allows exhaust gas to flow.
5) The exhaust gas that flows through the EGR valve must first pass through the EGR metering orifice. One side of the orifice us exposed to exhaust backpressure and the other side is exposed to the intake manifold. This design creates a pressure drop across the orifice whenever there is EGR flow. When the EGR valve closes, there is no longer flow across the metering orifice and the pressure is equal on both sides.
6) The DPFE sensor measures the "actual" pressure drop across the metering orifice and relays a proportional voltage signal (from 0-5 volts) to the PCM. The PCM uses this feedback signal to correct for any errors in achieving the "desired" EGR flow.

EGR System Schematic

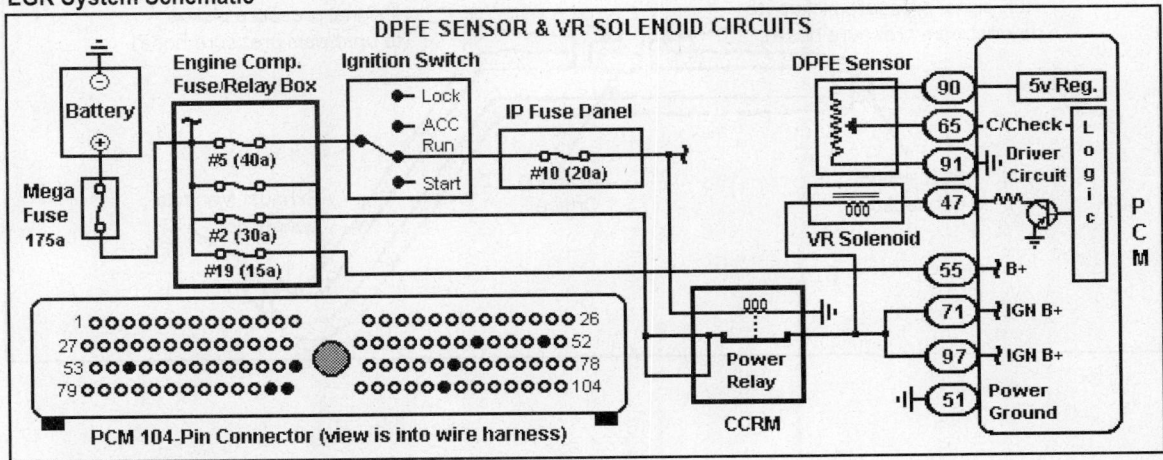

DPFE SENSOR INFORMATION

The DPFE sensor is a ceramic, capacitive-type pressure transducer that monitors the differential pressure across the metering orifice located in the orifice tube assembly. The sensor receives vacuum signals through two hoses referred to as the downstream pressure (REF signal) and upstream pressure hose (HI signal).

The REF and HI signals are marked on the aluminum sensor housing. The HI signal uses the larger diameter hose. The DPFE sensor signal is proportional to the pressure drop across the metering orifice. This input is referred to as the EGR flow rate feedback signal.

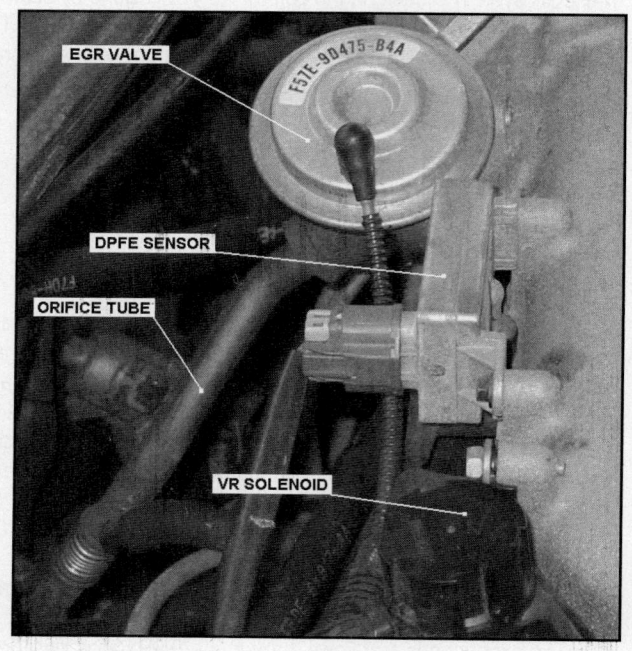

Aluminum Cast DPFE Sensor

This vehicle application uses an aluminum cast-housing design of DPFE sensor. The part number from Ford for this sensor is F48E-9J460-BB: The KOEO reading is 0.55v on this vehicle.

Note: *Always identify the part number of the DPFE sensor to verify the correct sensor is used.*

Orifice Tube Assembly

The orifice tube assembly is a section of tubing connecting the Exhaust system to the intake manifold. This assembly provides a flow path for EGR gases to the intake manifold. It also contains the metering orifice and two pressure pickup tubes.

The internal metering orifice creates a measurable pressure drop across it as the EGR valve open and closes. This pressure differential across the orifice is picked up by the DPFE sensor and converted to feedback data on EGR system operation to the PCM.

Orifice Tube Assembly Graphic

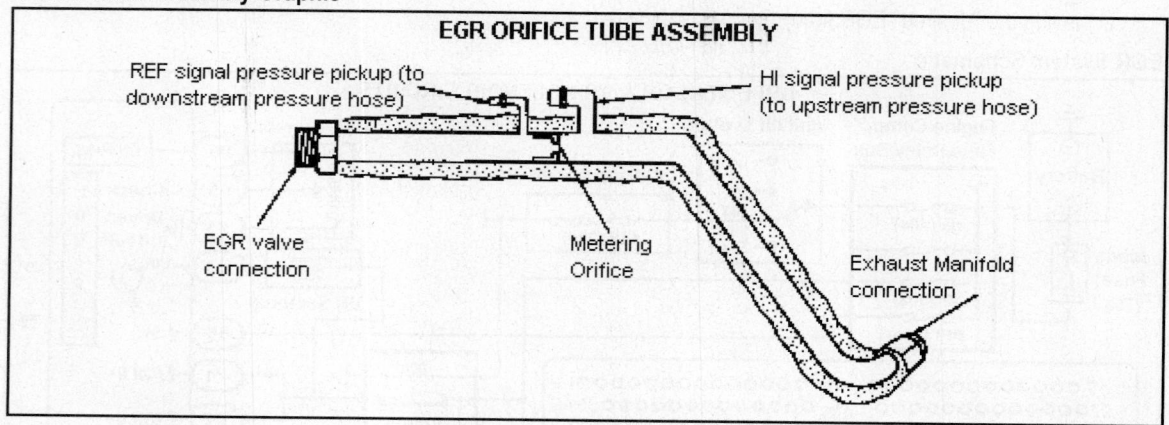

LAB SCOPE TEST (VR SOLENOID)

The Lab Scope is the "tool of choice" to test the VR solenoid as it provides an accurate view of the solenoid response and frequency rate. Place the gearshift selector in Park and block the drive wheels for safety.

Scope Connections

Connect the Channel 'A' positive probe to the VR control circuit (BRN/PNK wire) at PCM Pin 47 and the negative probe to battery ground.

Scope Settings

To make the waveforms as clear as possible, set the scope settings to match the examples.

Lab Scope Tests

The examples on this page were taken from a vehicle with 144,000 miles on the odometer. The waveforms in these examples are from the original solenoids. The engine was allowed to run long enough until it entered closed loop operation prior to testing. Note that the examples were captured at different speeds.

Lab Scope Example (1)

In example (1), the trace shows the VR solenoid command signal with the cursors set as indicated. This waveform was captured at idle speed. The VR PID will read 0% with the VR solenoid not activated during the capture.

Lab Scope Example (2)

In example (2), the trace shows the VR solenoid command with the cursors set as indicated. This waveform was captured at Cruise speed. The VR PID will read 45% with the VR solenoid activated during this test.

VR Solenoid Test Schematic

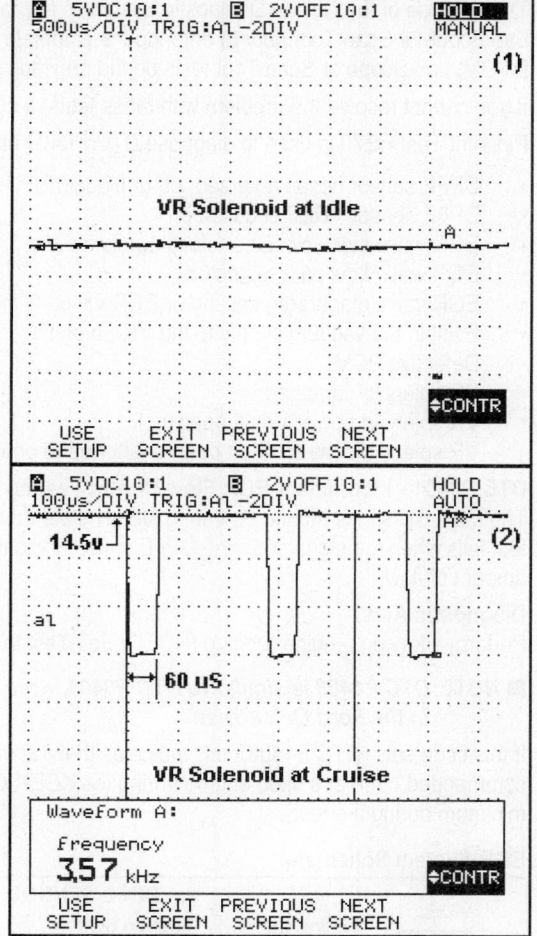

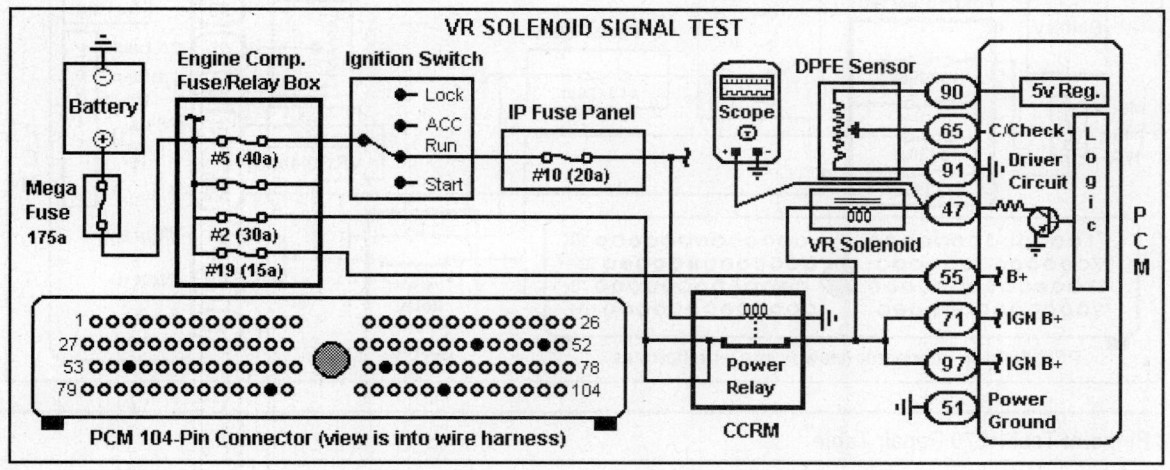

CHILTON DIAGNOSTIC SYSTEM - DTC P0401

The purpose of the Chilton Diagnostic System is to provide you with one or more tests that can be done with a DVOM, Lab Scope or Scan Tool *prior* to entering the complete trouble code repair procedure. The quick checks listed in the DVOM, Lab Scope or Scan Tool tests on the previous pages may help you quickly find the cause of the problem.

If you cannot resolve the problem with these tests, a code repair chart is also included.

Pinpoint Test HE70 is used to diagnose DTC P0401 and the following items:

- DPFE sensor hoses reversed, off or plugged
- DPFE sensor VREF circuit open
- EGR valve stuck closed or stuck due to ice
- EGR valve flow path restricted
- EGR valve diaphragm leaking or EGR vacuum line off, plugged or leaking
- Fault in the vacuum supply to the VR Solenoid
- Defective PCM
- VR solenoid damaged
- VR solenoid power circuit open
- VR solenoid control circuit open or shorted to power

DTC P0401 - Insufficient EGR Flow During Testing (CMC) Code Conditions

The EGR System is monitored during steady state driving conditions while the EGR valve is commanded open. The test fails when the signal from the DPFE sensor indicates that the EGR flow rate was less than the desired minimum amount of flow.

Diagnostic Aids

Perform a key on, engine running (KOER) Self-Test to determine if DTC P1408 sets.

■ **NOTE:** *DTC P1408 is similar to DTC P0401, except that DTC P1408 is an engine running test that is included in the Ford Quick Test.*

If this code sets (DTC P1408), this indicates that the PCM detected an incorrect flow rate with the EGR valve commanded open at a fixed engine during the KOER Quick Test (i.e., the PCM detected a flow rate below a calibrated minimum during the test).

EGR System Schematic

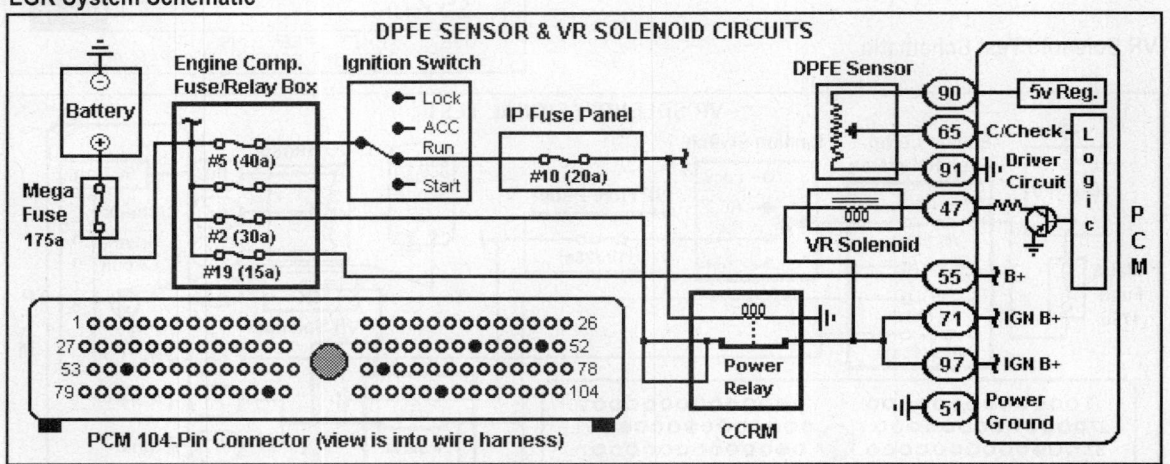

DPFE SENSOR & VR SOLENOID CIRCUITS

Pinpoint Test HE70 Repair Table

Step	Action	Yes	No
HE70	**Step description:** Was KOER self-test done? • Was the KOER Self-Test performed?	Go to Step HE90.	Perform the KOER Self-Test.
HE90	**Step description:** Inspect for Intermittent fault • Turn the key to off. • Inspect EGR System for signs of intermittent failure (loose connections or hoses off).	Make repairs as needed. Do a PCM Reset and then recheck for the condition.	Go to Step HE91.

	• Were any faults found?		
HE91	**Step description:** EGR functional check • Turn the key to off. • Disconnect the vacuum hose at EGR valve. • Connect a hand vacuum pump to the valve. • Connect a Scan Tool and start the engine. • Read the DPEGR and RPM PID values. • Slowly apply 5-10" Hg of vacuum to the valve (hold it for at least 10 seconds). If the engine starts to stall, increase the engine speed slightly to allow a minimum speed of 800 rpm. • Look for these events: The EGR valve starts to open at about 1.6" Hg of vacuum; DPFEGR PID value increases until the valve opens (it should read at least 2.5v at full vacuum); the DPFEGR PID value is steady while vacuum is held (if the vacuum drops off, there is a leak). • Does the DPFEGR PID indicate that the EGR valve operated as described in this text?	Go to Step HE92.	Remove and inspect the EGR valve for signs of contamination, unusual wear, carbon deposits, binding, a leaking diaphragm or other damage. Make repairs as needed. Then do a PCM Reset and retest for the condition.
HE92	**Step description:** Check source vacuum • Turn the key to off and disconnect the PCM. • Check for loose or damaged pins at PCM. • Install a BOB and connect the PCM to BOB. • Remove the vacuum hose at the EGR valve and connect a vacuum gauge to the hose. • Start the engine and allow it to idle. • Jumper test pin 47 (VR) to chassis ground (this will activate the VR to full on position). • The vacuum gauge should read a minimum of 4.0" Hg at idle under these conditions. • Observe vacuum gauge during these events: Tap lightly on VR solenoid; wiggle the VR solenoid harness connector, vacuum lines and wiring harness to the VR solenoid. A fault is indicated if the vacuum gauge reading drops. • Was a fault detected during this step?	Locate and repair the problem that caused the vacuum reading to drop during testing. Restore the vehicle to normal condition. Do a PCM Reset and then perform a Quick Test.	The fault did not appear or was intermittent at this time. Refer to the Intermittent tests in this manual or in other repair information.

Fuel Delivery System

INTRODUCTION

The Fuel system delivers fuel at a controlled pressure and correct volume for efficient combustion. The PCM controls the Fuel system and provides the correct injector timing.

The vehicle is equipped with a sequential fuel injection (SFI) system that delivers the correct A/F mixture to the engine at the precise time throughout its entire speed range and under all operating conditions.

Returnable Fuel System

This vehicle application is equipped with a "returnable" fuel system. In this type of Fuel system, excess fuel delivered by the fuel pump to the fuel rail is returned to the fuel tank through a fuel return line.

System Operation

A high-pressure (in-tank mounted) fuel pump delivers fuel to the fuel injection supply manifold. The fuel injection supply manifold incorporates electrically actuated fuel injectors mounted directly above each of the intake ports of the engine.

A constant amount of fuel pressure is maintained at the injectors by a fuel pressure regulator. The fuel pressure is a constant while fuel demand is not, so the system includes a fuel return line to allow excess fuel to flow through the regulator to the tank.

The PCM determines the amount of fuel the injectors spray under all engine-operating conditions. The PCM receives electrical signals from engine control sensors that monitor various factors such as airflow pressure, temperature of the air entering the engine, engine coolant temperature, throttle position and vehicle speed. The PCM evaluates the sensor information it receives and then signals the fuel injectors in order to control the fuel injector pulsewidth.

Fuel Delivery System Graphic

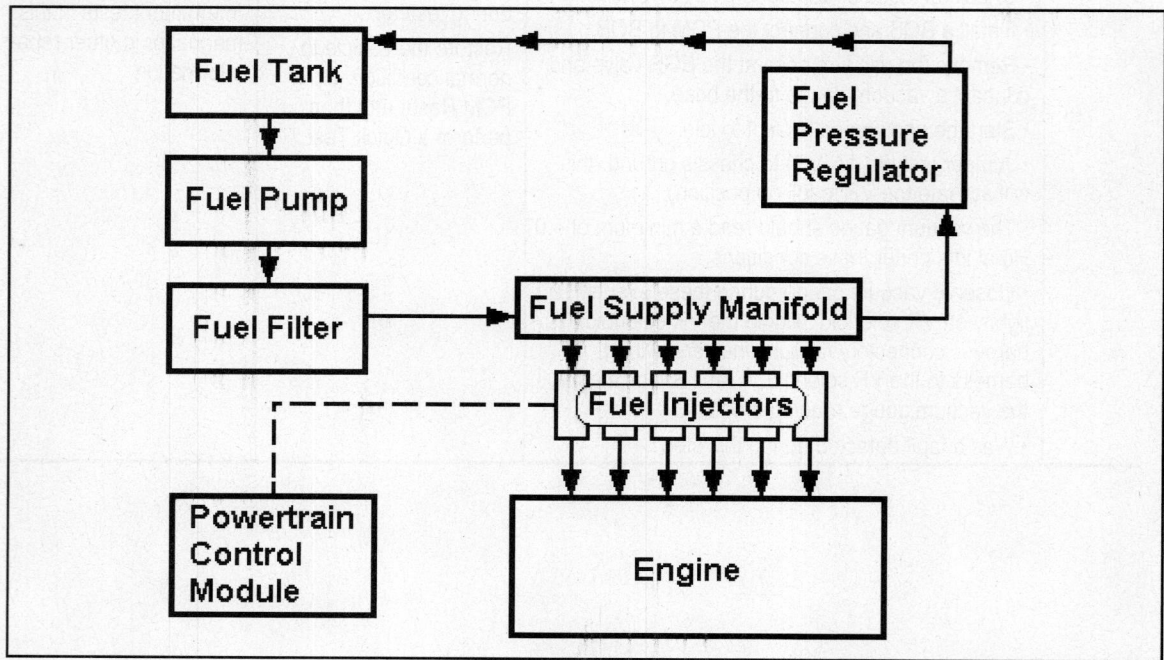

CONSTANT CONTROL RELAY MODULE

The PCM controls the fuel pump through a fuel pump relay that is integral to the constant control relay module (CCRM).

Component Location

The CCRM is located to the left of the battery near the headlamp assembly.

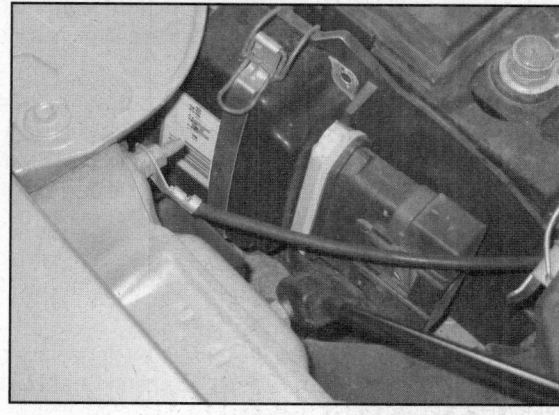

The fuel pump relay, which is controlled by the PCM, provides power to the fuel pump under the following operation conditions:

1) With the ignition turned "off", the PCM and fuel pump relay are not energized.

2) When the ignition switch is first turned to the "run" position, these events occur:
   a. The power relay is energized.
   b. Power is provided to the fuel pump relay and a timing device in the PCM.
   c. The fuel pump is supplied power by the fuel pump relay.

3) If the ignition switch is not turned to the "start" position, the timing device in the PCM will open the fuel pump relay control circuit at Pin 80 (LB/ORN wire) after one second, and this action de-energizes the fuel pump relay, which in turn de-energizes the fuel pump. This circuitry provides for pre-pressurization of the fuel system.

4) When the ignition switch is turned to the "start" position, the PCM operates the fuel pump relay to provide fuel for starting the engine while cranking.

5) After the engine starts and the ignition switch is returned to the "run" position:
   a. Power to the fuel pump is again supplied through the fuel pump relay.
   b. The PCM shuts off the fuel pump by opening the ground circuit to the fuel pump relay if the engine stops, or if the engine speed is below 120 rpm.

Fuel Pump Monitor

The PCM determines that the fuel pump circuit has power through the FPM circuit. This circuit is used to determine the fuel pump circuit status for DTC P0231 and P0232.

Fuel Pump Schematic

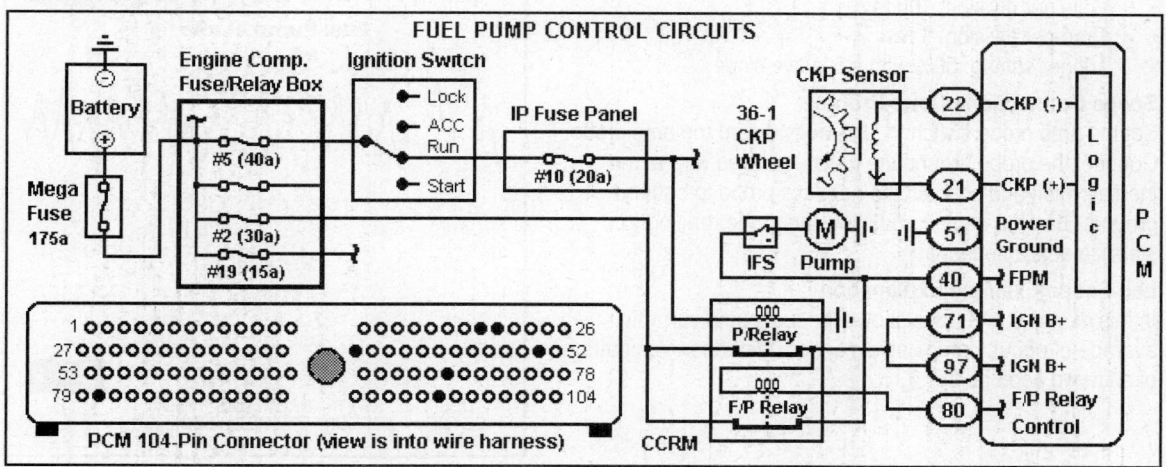

FUEL PUMP ASSEMBLY

The fuel pump module is mounted in the module retainer (integral to fuel tank). Fuel initially enters the module through a flapper valve, located at the bottom of the module, and primes the fuel pump.

A portion of the high-pressure flow from the pump is diverted to operate a Venturi jet pump. The jet pump draws fuel from the fuel tank into the fuel pump module. This design permits optimum fuel pump operation during vehicle maneuvers and driving on steep hills with low-tank full levels.

3.0L 2V: The fuel pump in the 3.0L 2V engine is capable of supplying 80 liters (21.1 gallons) of fuel per hour at 269 kPa (39 psi) at 13.2 volts.

3.0L 4v: The fuel pump in the 3.0L 4V engine is capable of supplying 120 liters (31.7 gallons) of fuel per hour at 269 kPa (39 psi) at 13.2 volts

3.0L 2V FF: The pump in the 3.0L 2V FF engine is capable of supplying 125 liters (33 gallons) of fuel per hour at 269 kPa (39 psi) at 13.2 volts.

Internal Relief Valve

The pump has an internal pressure relief valve that restricts fuel pressure to 124 psi if the fuel flow becomes restricted due to a clogged fuel filter.

LAB SCOPE TEST (FUEL PUMP)

If you suspect the fuel pump is faulty, you can use a low amp probe with an inductive pickup to check the operation of the fuel pump and its circuits.

Lab Scope Settings

Set the Lab Scope to these initial settings:

- Volts per division: 100 mv
- Time per division: 1 ms
- Trigger setting: 50% with a positive slope

Scope Connections (Amp Probe)

Set the amp probe switch to 100 mv and zero the amp probe. Connect the probe around the fuel pump feed wire at the inertia switch and connect the negative probe to battery ground. Start the engine and allow the probe reading to stabilize at idle speed.

Lab Scope Example Explanation

In this example, the trace shows the fuel pump current average of about 5 amps at idle speed. Note the even pattern of a known good pump.

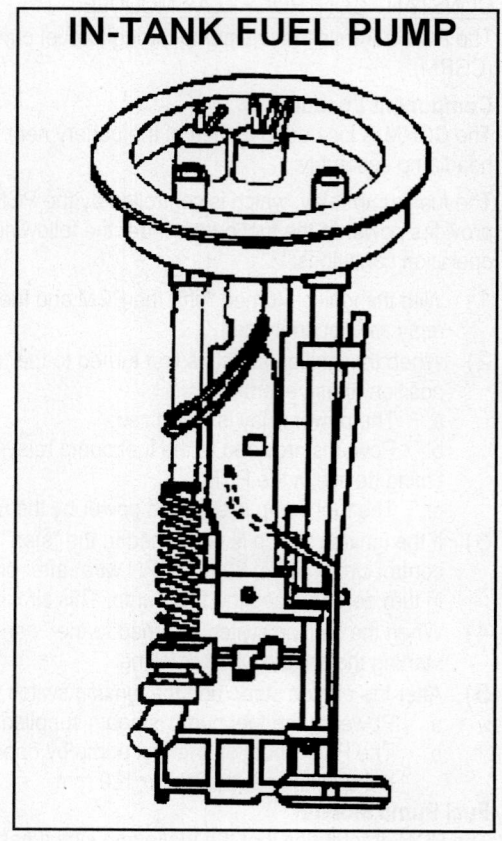

IN-TANK FUEL PUMP

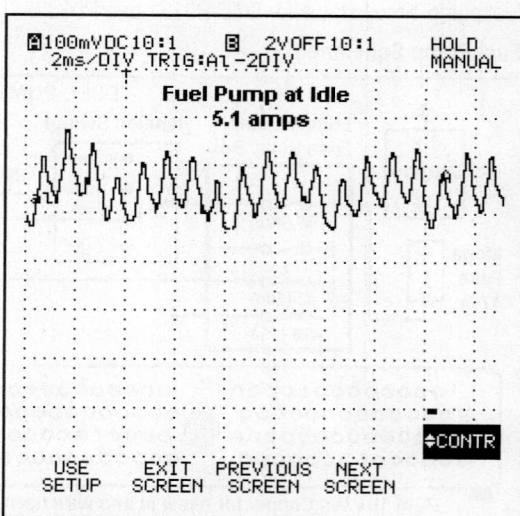

FUEL PRESSURE REGULATOR

The Fuel Pressure Regulator is attached to the return side of the fuel rail downstream of the fuel injectors. The fuel pressure regulator is a diaphragm-operated relief valve. It function is to regulate the fuel pressure supplied to the injectors. It also traps fuel during engine shutdown. This eliminates the possibility of vapor formation in the fuel line, and provides instant restarts at initial idle speed.

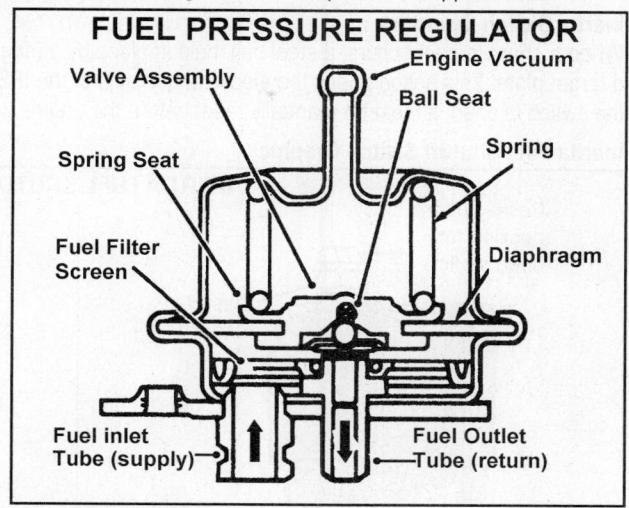

One side of the diaphragm senses fuel pressure. The other side is connected to intake manifold vacuum. A spring pre-load applied to the diaphragm establishes the nominal fuel pressure.

A constant fuel pressure drop across the injectors is maintained due to the balancing affect of applying manifold vacuum to one side of the diaphragm. Any excess fuel in the fuel delivery system is bypassed through the regulator and returned to the fuel tank.

Fuel Pressure Regulator Location (Station Wagon)

The fuel pressure regulator is located on the fuel rail in the engine compartment as shown in the Graphic to the right.

Ford Recall Notice 96V151000

Ford Motor Co. has a recall notice for this vehicle application. The notice refers to the fact the fuel pressure regulator diaphragm may have been damaged during production.

If the diaphragm is torn or ruptured, liquid fuel could enter the intake manifold plenum through the regulator vacuum line and result in either the release of fuel into the air cleaner assembly or into the Exhaust system.

Remember to inspect the fuel pressure regulator for damage and/or leakage. Any fuel pressure regulators produced during the suspect time period should be replaced.

■ NOTE: *Owners who take their vehicles to an authorized dealer on an agreed upon service date and do not receive the free remedy within a reasonable time should contact Ford at 1.800.392.3673. They can also contact National Highway Traffic Safety Administration Auto Safety Hotline at 1.800.424.9393.*

Fuel Filters

A fuel filter is used to strain fuel particles through a paper element to remove dirt particles. The fuel filter is located on the right side frame rail near the fuel tank.

The inlet of the pump includes a nylon filter to prevent dirt and other particulates from entering the system, but water in the tank will pass through the filter without restriction.

INERTIA SWITCH

This vehicle application is equipped with an inertia fuel shutoff (IFS) switch that is used in conjunction with the electric fuel pump. The inertia switch is designed to shut off the fuel pump in the event of an accident.

Inertia Switch Operation

When a sharp impact occurs, a steel ball, held in place by a magnet, breaks loose and rolls up a conical ramp to strike a target plate. This action opens the electrical contacts of the IFS switch and shuts off power to the fuel pump. Once the switch is open, it must be manually reset before the engine will restart.

Inertia Fuel Shutoff Switch Graphic

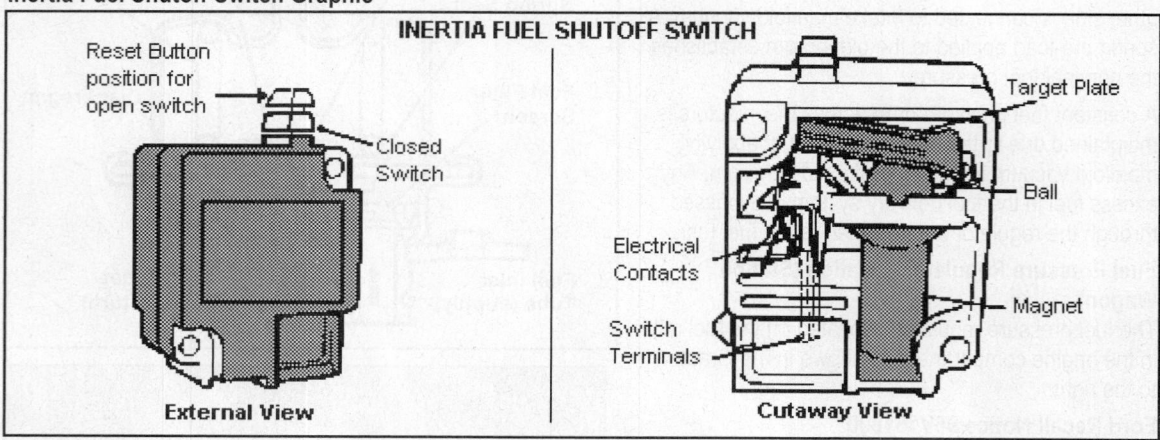

Inertia Switch Location (Station Wagon)

One of the challenges in dealing with resetting or checking this switch is locating the switch.

On the 1996 TAURUS station wagon application used in this article, the IFS switch is located behind the removable cover at the right side tailgate area (looking into the tailgate area from behind the vehicle). Refer to the example in the Graphic to the right.

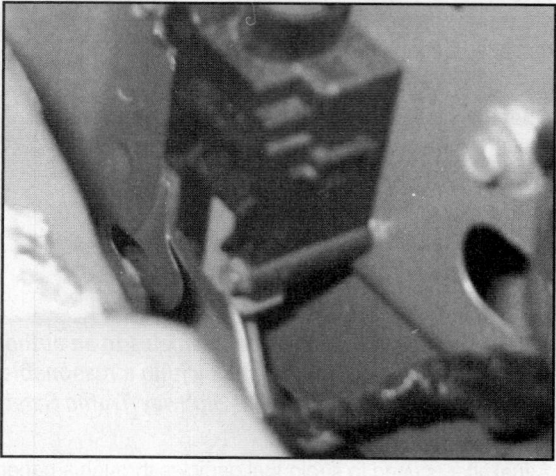

Inertia Switch Reset Instructions

4) Turn the key off and check for any signs of leaking fuel near the fuel tank or in the underhood area. If any leaks are found, repair the source of the leak before proceeding to the next step.

5) If there are no fuel leaks present, push the reset button to reset the switch (it can be closed an extra 0.16" in the closed position).

6) Cycle the key to "on" for a few seconds, and then to off (in order to recheck for signs of any fuel leaks anywhere in the system).

FUEL INJECTORS

The fuel injector is a solenoid-operated valve designed to meter the fuel flow to each combustion chamber. The injector is opened and closed a constant number of times per crankshaft revolution. The fuel injectors are a deposit resistant injection (DRI) design and do not have to be cleaned.

The amount of fuel delivered is controlled by the length of time the injector is held open (injector pulsewidth). The injector is supplied power through the battery feed circuit that connects to power relay inside the constant control relay module (CCRM).

Fuel Injection Timing

The PCM determines the fuel flow rate needed to maintain the correct A/F ratio from the ECT, IAT and MAF sensor inputs.

The PCM computes the desired injector pulsewidth and then it sends a command to the fuel injectors to deliver the correct quantity of fuel.

The injectors are energized in this sequence: 1-4-2-5-3-6.

Injectors 4 - 5 - 6

Spark Plugs 4 - 5 - 6

Circuit Description

The EEC power relay connects the PCM and fuel injectors to direct battery power (B+) as shown in the schematic. With the key in the "start or run" position, the power relay coil is connected to ground and this action supplies power to the PCM and fuel injectors.

The PCM begins to control (turn them on and off) the fuel injectors once it begins to receive signals from the CKP and CMP sensors.

Fuel Injection Timing Modes

The PCM operates the fuel injectors in two modes: Simultaneous and Sequential. In Simultaneous mode, fuel is supplied to all six cylinders at the same time by sending the same injector pulsewidth signal to all 6 injectors, twice for each engine cycle. This mode is used during startup and in the event of a fault in the CKP sensor or the PCM.

Sequential mode is used during all other engine operating conditions. In this mode, fuel is injected into each cylinder once per engine cycle in the firing order (1-4-2-5-3-6).

Sequential Fuel Injection Graphic

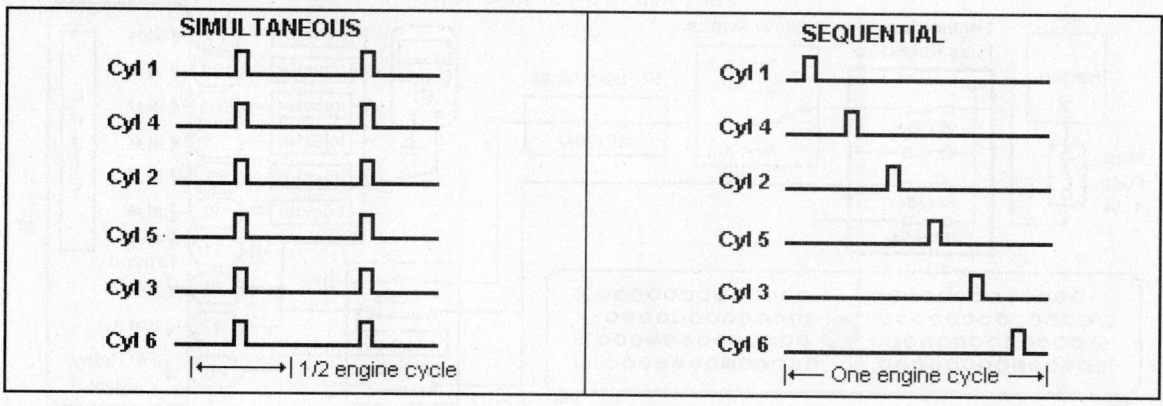

SCAN TOOL TEST (FUEL INJECTOR)

The Scan Tool can be used to quickly determine the fuel injector pulsewidth or duration. However, it is not the tool of choice for certain types of injector problems.

Connect the Scan Tool to the DLC underdash connector. Navigate through the Scan Tool menus until you get to the Ford (OEM) PID Data List.

Then select Injector Pulsewidth from the menu. The example to the right was captured with the vehicle in Park and running at Hot Idle speed.

SCAN TOOL TEST (FUEL TRIM)

Short Term fuel trim (SHRTFT) is an operating parameter that indicates the amount of short term fuel adjustment made by the PCM to compensate for operating conditions that vary from an ideal A/F ratio condition.

A negative SHRTFT number (-15%) means the HO2S is indicating to the PCM that the A/F ratio is rich, and that the PCM is trying to lean the A/F mixture. If A/F ratio conditions are ideal, the SHRTFT number will be close to 0%.

Long Term Fuel Trim

Long Term fuel trim (LONGFT) is an engine operating parameter that indicates the amount of long term fuel adjustment made by the PCM to compensate for operating conditions that vary from an ideal A/F ratio.

A positive LONGFT number (+15%) means the HO2S is indicating to the PCM that the A/F ratio is lean, and that the PCM is trying to add more fuel to the A/F mixture. If A/F ratio conditions are ideal the LONGFT number will be 0%.

The LONGFT number is displayed on the Scan Tool in percentage. Refer to the examples in the Graphic.

```
Jan 15, 02        4:56:16 pm

 INJECTOR PW B1
    3.56 MSEC
 INJECTOR PW B2
    3.67 MSEC


Always use [HELP] to
 list keys & modes
```

```
Jan 15, 02        4:55:34 pm

 SHORT TERM FT B1
      0 %
 LONG TERM FT B1
      4 %


Always use [HELP] to
 list keys & modes
[RCV] For enhanced
[*EXIT] To exit
```

```
Jan 15, 02        4:54:53 pm

 SHORT TERM FT B2
  -   4 %
 LONG TERM FT B2
      5 %


Always use [HELP] to
 list keys & modes
[RCV] For enhanced
[*EXIT] To exit
```

Fuel Injector Test Schematic

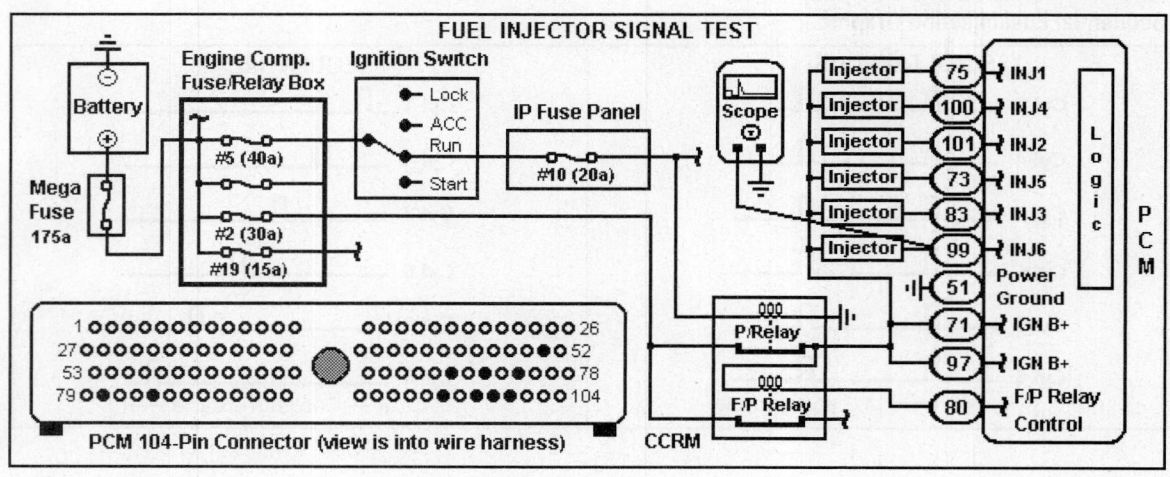

LAB SCOPE TEST (FUEL INJECTOR)

The Lab Scope is a useful tool to test the fuel injectors as it provides an accurate view of the injector operation. Place the gearshift selector in Park. Block the drive wheels for safety.

Scope Connections

Connect the Channel 'A' positive probe to the injector control wire and negative probe to the battery ground (refer to the previous page).

Scope Settings

To make the waveforms as clear as possible, set the scope settings to match the examples.

Set the amp probe to 100 mv (this setting equals 0.2 amps with the scope set to 20 mv).

Lab Scope Example (1)

In example (1), the trace shows the injector waveform with a sweep rate setting of 1 ms. Note how this setting allows you to accurately determine the injector on time (4.4 ms in this example). The inductive kick was 60 volts (viewed at a 10v per division setting).

Lab Scope Example (2)

In example (2), the trace shows the injector current signal (current ramping) at idle speed. Note the point in the injector current signal where the injector opened and closed. The current flow in the circuit was near 720 mA.

Lab Scope Example (3)

In example (3), the voltage setting was changed from 10v to 2v per division. This setting allows for an accurate calculation of the injector feed voltage and a way to monitor injector driver performance.

Injector Driver Test

In example (3), the voltage drop through the injector driver control circuit in the PCM rises to 400 mv above the ground level point. This is caused by current flow through the injector winding, the injector driver and PCM ground circuit that connects to battery ground.

If the voltage drop amount exceeds 750 mv, look for a problem in the injector driver control circuit or the PCM main ground circuit. If the voltage drop across the main ground circuit is less than 50 mv, the problem is in the injector driver circuit inside the PCM.

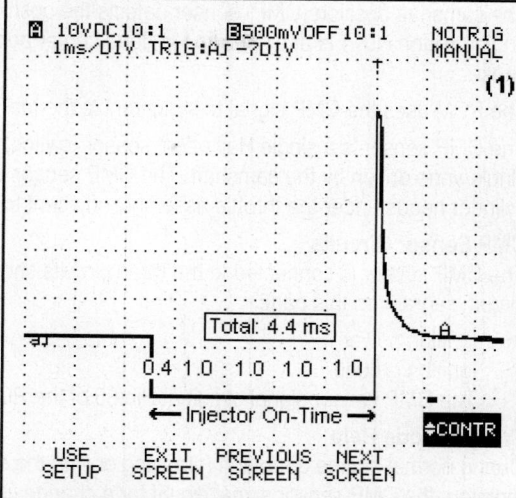

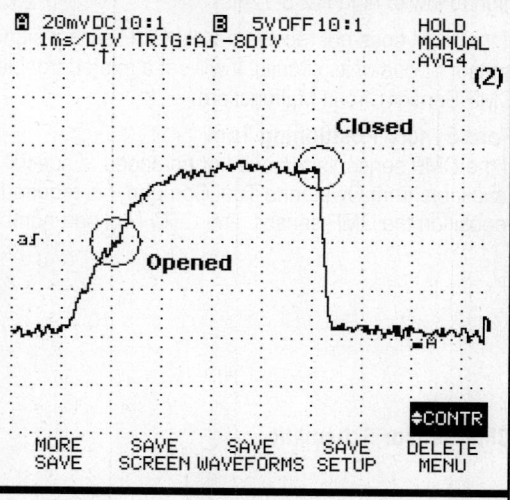

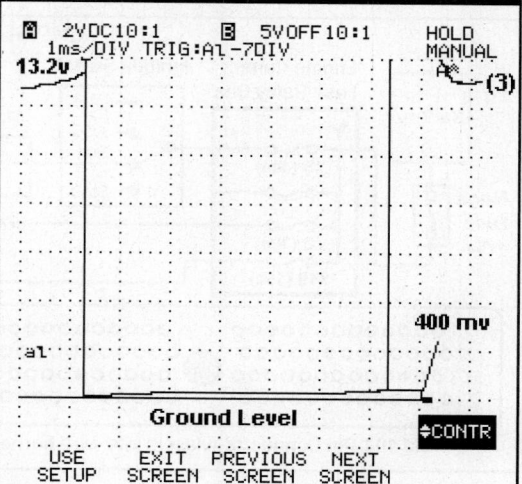

CMP SENSOR (HALL-EFFECT)

The camshaft position (CMP) sensor detects the position of the camshaft when engine piston No. 1 is at a specific top dead center point of the compression stroke.

The PCM uses the CMP signal to synchronize the fuel injector timing.

The CMP sensor is a single Hall effect sensor (switch) that is activated by a single vane driven by the camshaft. The CMP sensor is mounted between the cylinder heads under the throttle assembly adjacent to the camshaft.

CMP Sensor Circuits

The CMP sensor is connected to the three circuits shown in the CMP sensor circuit diagram on this page.

*   Ignition power
*   Ignition ground
*   The CMP sensor signal circuit at Pin 85 of the PCM.

Trouble Code Help

During normal engine operation (cranking or with the engine running), the PCM monitors the CMP sensor signal circuit for a change in voltage as it switches high to low to high (12-0-12v).

If the PCM does not receive a valid CMP sensor signal due to a problem in the sensor or one of its circuits, it will set a related trouble code (DTC P0340 - CMP Sensor Circuit Malfunction).

Ford Syncro Positioning Tool

If the CMP sensor is removed or damaged, a special Syncro Positioning Tool (available from Owatonna Tool Company) is required in order to correctly reposition the CMP sensor. The OTC Tool part number is #303-529.

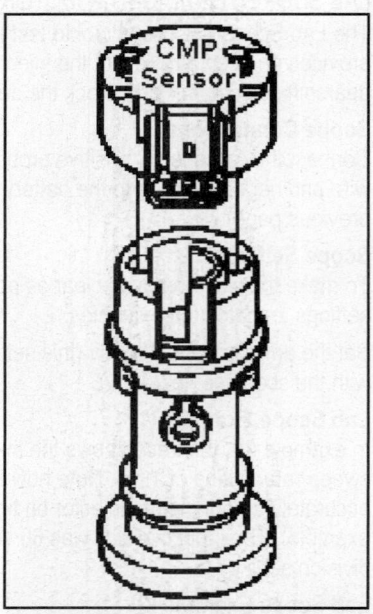

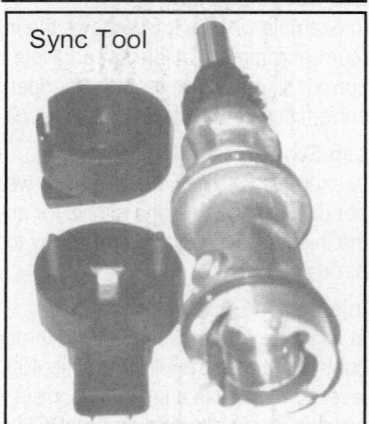

Sync Tool

CMP Sensor Schematic

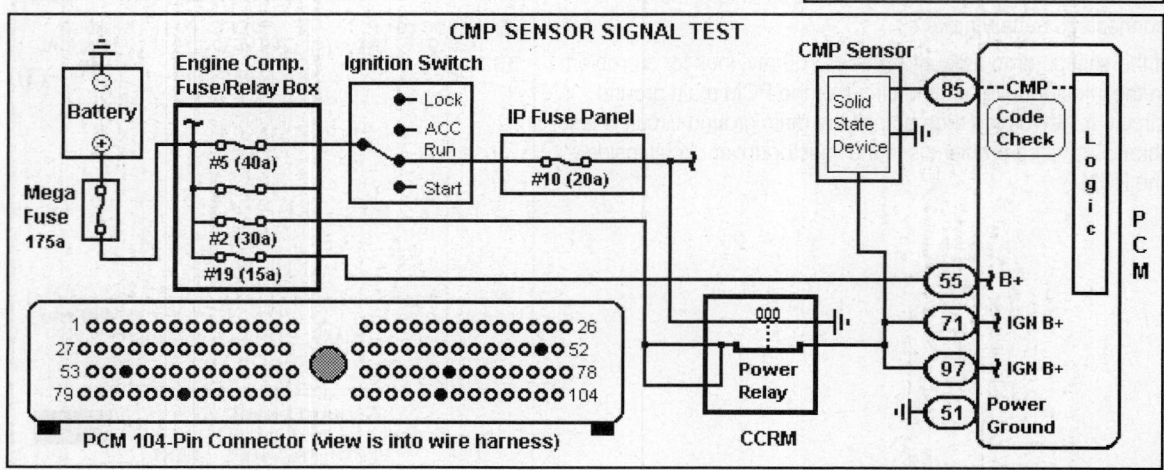

CMP SYNCHRONIZER REPLACEMENT

If the CMP sensor (synchronizer unit) is damaged or defective, and needs to be replaced, follow the replacement procedures carefully. A Syncro Positioning Tool from Ford is required. The OTC Tool part number is #303-529.

Removal

1) Disconnect the battery ground cable.
2) Set Cylinder No. 1 to zero degrees TDC of the compression stroke.
3) Disconnect the ECT sensor wiring at the ECT sensor connector.
4) Remove the retaining screws and the CMP sensor from the CMP sensor housing.
5) If the housing needs to be removed, remove the hold-down clamp and washer.
6) Remove CMP sensor housing from the block.

Installation:

1) Attach the Syncro Tool by engaging the CMP sensor housing into the radial slot of tool. Rotate tool on the CMP sensor housing until tool boss engages notch in the housing.
2) Pre-lubricate the synchronizer gear with engine oil. Install the sensor housing so that drive gear engagement occurs when the arrow on locator tool is pointed 75 degrees counterclockwise from the rear face of the engine block. This step will locate the sensor connector in the pre-removal position.
3) Rotate the tool clockwise slightly to engage oil pump intermediate shaft. Push down while adjusting the tool until the Syncro Gear engages the camshaft gear.
4) When the synchronizer flange is fully seated against the block, the installation tool should be facing counterclockwise of engine centerline (33 to 45 degrees).
5) Install the hold-down washer and bolt (tighten to 14-22 lb-ft. and the remove the Syncro Tool).
6) Install CMP sensor and retaining screws, reconnect ECT sensor and battery cable.

■ NOTE:*If the CMP sensor connector is not in the correct position, do not rotate the sensor housing to reposition it, as this action will put the fuel system out of time.*

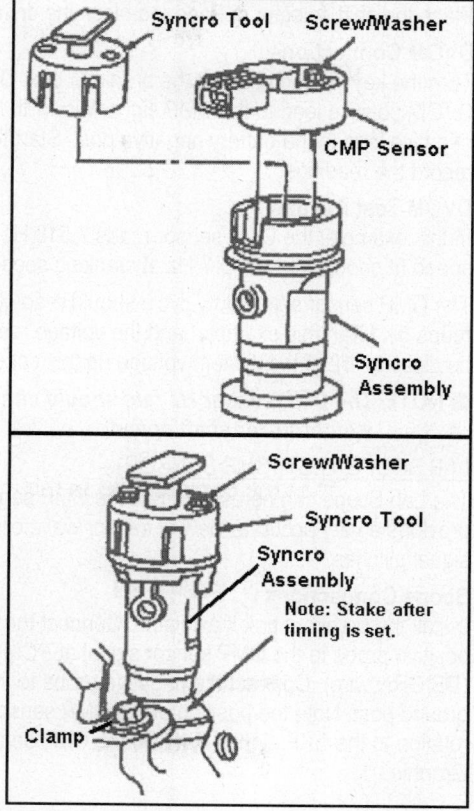

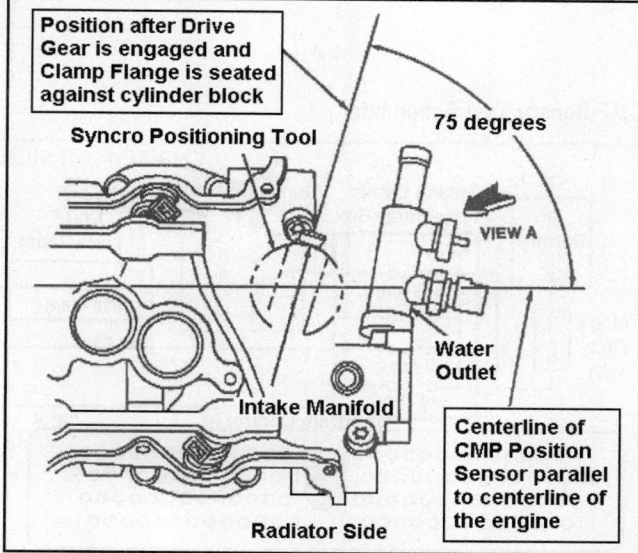

DVOM TEST (CMP SENSOR)

Place the shift selector in Park and block the drive wheels prior to starting this test.

DVOM Connections

Turn the key off and connect the breakout box. Connect the DVOM positive lead to the CMP signal wire (Pin 85) and the negative lead to the battery negative post. Start the engine and record the readings.

DVOM Test Results

In this example, the CMP sensor reads 7.518 Hz at hot idle speed (it should be from 5-7 Hz at cranking speed).

The CMP sensor signal duty cycle should read close to 50% (it reads 51.1% in this example) and the voltage reading should be close to 1/2 of the system voltage (in this case 6.77v).

■ **NOTE:** *The CMP sensor Hz rate should change with synchronizer shaft speed.*

LAB SCOPE TEST (CMP SENSOR)

The Lab Scope can be used to test the CMP sensor as it provides a very accurate view of sensor waveform and of any signal glitches.

Scope Connections

Install the breakout box if available. Connect the Channel 'B' positive probe to the CMP sensor signal at PCM Pin 85 (DB/ORN wire). Connect the negative probe to the battery ground post. Note the position of the CMP sensor signal in relation to the CKP signal. Refer to the CKP Sensor Test Graphic.

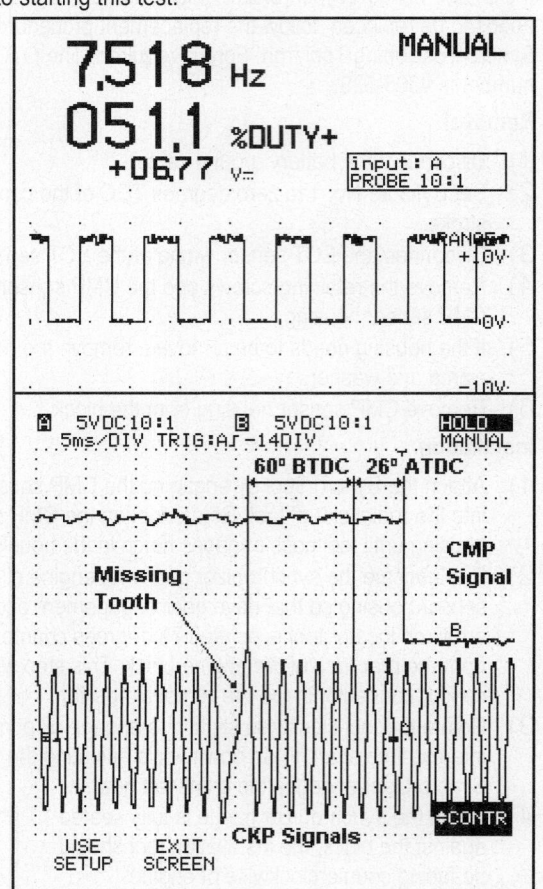

CMP Sensor Test Schematic

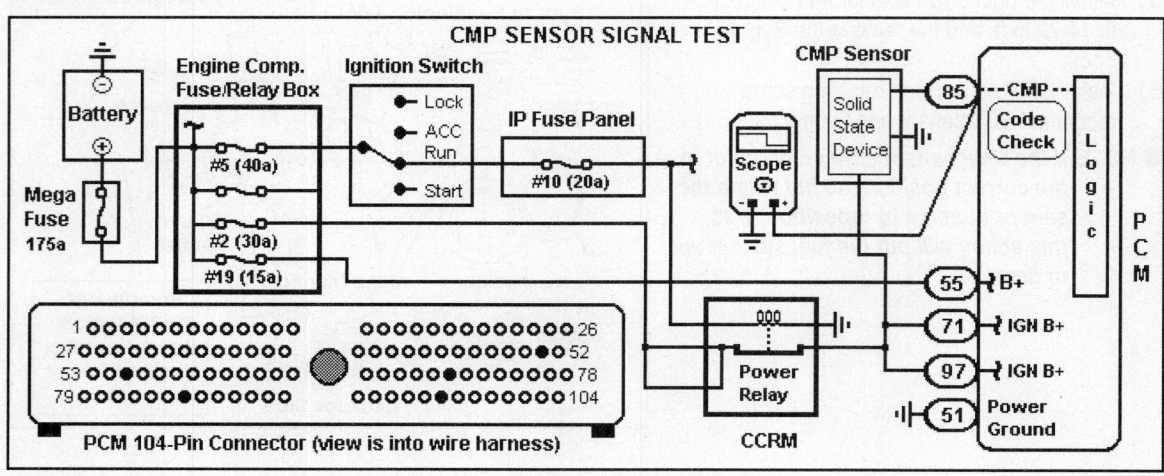

Idle Air Control Solenoid

GENERAL DESCRIPTION

The Idle Air Control (IAC) solenoid is mounted to the side of the throttle body. This device is designed to control engine idle speed (rpm) under all engine-operating conditions and to provide a dashpot function.

The IAC solenoid meters the intake air that flows past the throttle plate through a bypass area within the IAC assembly and throttle body.

IAC Solenoid Location
Refer to the Graphic to the right for the IAC solenoid location on this application.

The PCM determines the desired idle speed (amount of bypass air) and controls the IAC solenoid through a duty cycle command under all engine conditions.

The IAC solenoid responds by changing the position of the valve to control the amount of bypass air.

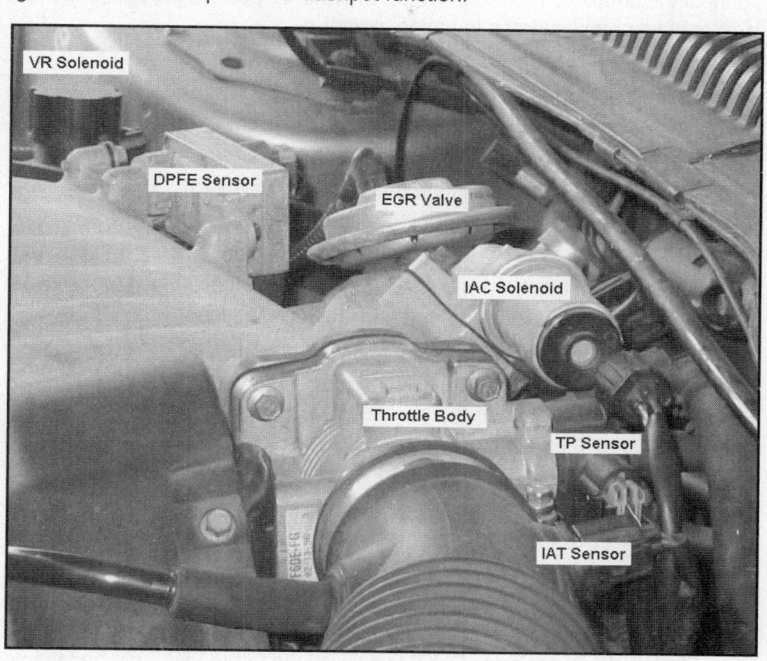

SCAN TOOL TEST (IAC SOLENOID)
Connect the Scan Tool to the DLC underdash connector. Navigate through the Scan Tool menus until you get to the Ford (OEM) Engine Data List (a list of PID items).

Select the RPM and IAC PID items from the menu so that you can view them separately on the Scan Tool display.

Scan Tool Example (1)
An example of the ENGINE SPEED (RPM) and IDLE AIR CONTROL (%) PID for this vehicle at Hot Idle speed is shown in the Graphic. The first example was captured at Hot Idle speed in Park with all accessories turned off.

Scan Tool Example (2)
In Example (2), the IAC PID was captured at Hot Idle speed with the A/C on in Drive (brake on). Note the change in the IAC reading (37% to 46%) with the A/C turned on in Drive.

Diagnostic Tip
To test the operation of the IAC solenoid, record the Hot Idle reading in Park with all accessories off. Then drive the vehicle. The IAC reading should return to the initial reading each time you stop and place the gear selector in Park.

```
Jan 11, 90      1:41:46 pm (1)

┌──────────────────────────┐
│ ENGINE SPEED             │
│   901 RPM                │
│ IDLE AIR CNTRL           │
│   37 %                   │
└──────────────────────────┘

Always use [HELP] to
 list keys & modes
[RCV] For enhanced
[*EXIT] To exit
```

```
Jan 15, 02      4:31:58 pm (2)

┌──────────────────────────┐
│ ENGINE SPEED             │
│   706 RPM                │
│ IDLE AIR CNTRL           │
│   46 %                   │
└──────────────────────────┘

Always use [HELP] to
 list keys & modes
[RCV] For enhanced
[*EXIT] To exit
```

LAB SCOPE TEST (IAC SOLENOID)

The Lab Scope can be used to monitor the IAC solenoid control signal under various engine load, speed and temperature conditions. Place the gearshift selector in Park or Neutral (A/T). Block the drive wheels prior to starting this test.

Scope Connections

Connect the breakout box. Connect the Channel 'A' positive probe to the IAC control circuit at PCM Pin 74 (BRN/YEL wire) and the negative probe to battery negative.

Scope Settings

To make the waveforms as clear as possible, set the scope settings to match the examples.

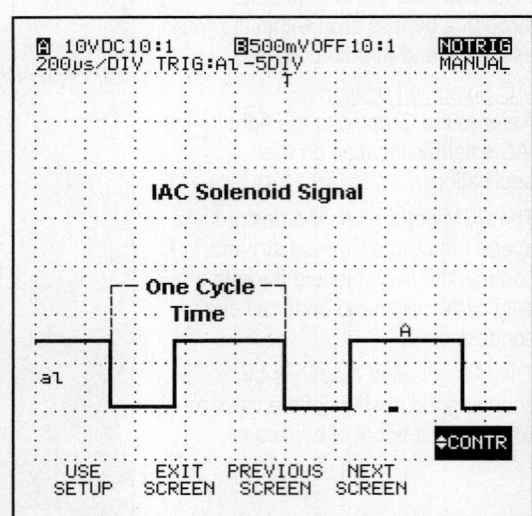

Lab Scope Example

In this example, the trace shows the IAC solenoid waveform. Note the time period to complete one cycle of the IAC command.

These signals were captured with the vehicle in Park, the engine at idle, all accessories off and running in closed loop.

The IAC solenoid trace should show a change in the digital on/off signal as the engine speed (rpm) is changed up and down. If it does not change or the square wave pattern is broken or jagged, a problem may exist in the IAC solenoid or its related control or feed circuit.

Circuit Description

The IAC solenoid connects to battery power through engine compartment fuse box and power relay. The solenoid includes a diode to protect the PCM against voltage spikes.

Internal Diode Test

To test the diode, connect the DVOM positive lead to the VBAT pin and the negative lead to the control pin (use the diode test). Look carefully at the harness connectors and mating component to view the proper orientation when testing the internal diode. If the diode is open, the DVOM will read 0.8v in both directions.

IAC Solenoid Test Schematic

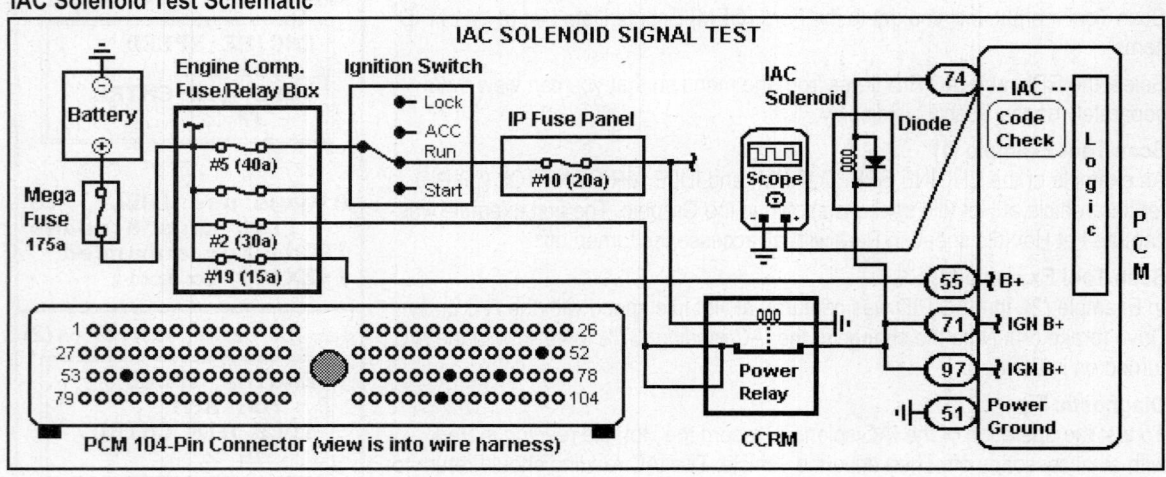

Mass Airflow Sensor

GENERAL DESCRIPTION

The mass airflow (MAF) sensor is located between the air cleaner and the throttle body. This MAF sensor uses a Hot-wire sensing element to measure the amount of air that enters the engine. Air passing through the device flows over the hot wire element and causes it to cool. The hot wire element is maintained at 392°F (200°C) above the current ambient temperature as measured by the constant cold wire part of the sensor.

The current level required to maintain the temperature of the hot wire element is proportional to amount of air flowing that passes the sensor. The MAF input to the PCM is an analog signal proportional to the intake air mass.

The PCM calculates the desired fuel injector on time in order to deliver the desired A/F ratio. The PCM also uses the MAF sensor to determine transmission EPC solenoid shifting and the TCC solenoid scheduling.

Circuit Description

The MAF sensor signal and ground circuits connect directly to the PCM.

With the ignition key in the "start or run" position, the power relay provides battery power to the MAF sensor. The MAF signal goes directly to the PCM.

MAF Sensor

MAF Sensor Schematic

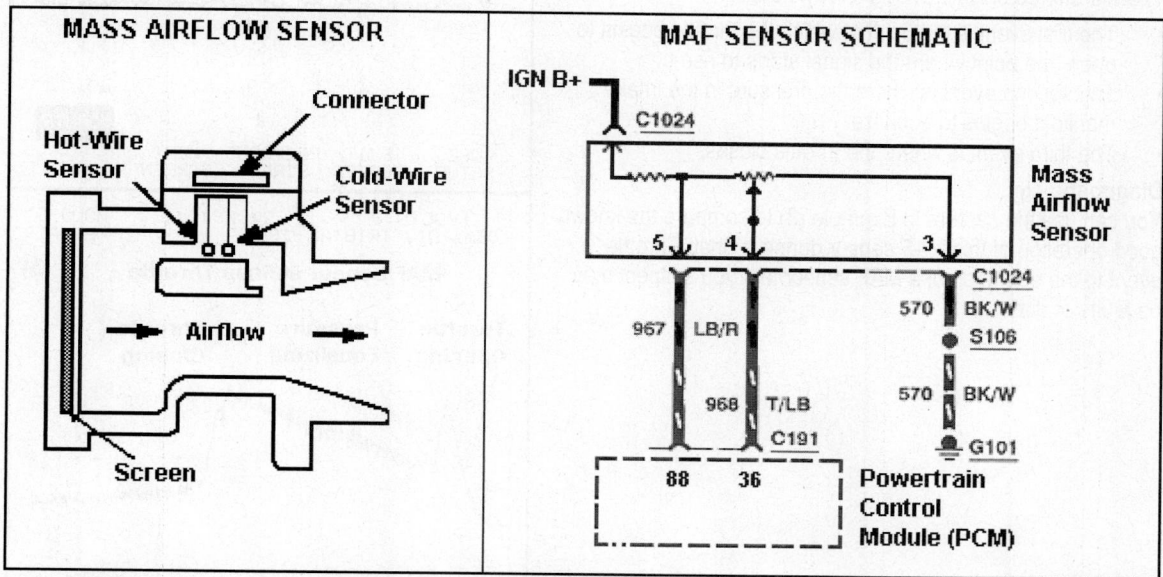

MASS AIRFLOW SENSOR

MAF SENSOR SCHEMATIC

Ford TSB No. 94-19-4 (High Resistance at MAF Sensor Terminals)

Vehicles with a 3.0L V6 engine may exhibit a rough idle, bucking on deceleration and/or hesitation with the engine temperature cold or hot. Check for signs of high resistance in the MAF connector. A new connector (Part No. F4DZ-14A411-A) is available from Ford.

LAB SCOPE TEST (MAF SENSOR)

The Lab Scope can be used to monitor the MAF sensor output signal under various engine load and speed conditions. Place the gearshift selector in Park prior to testing.

Scope Connections

Connect the Channel 'A' positive probe to the MAF signal at the PCM Pin 88 (LB/RD wire) and connect the Channel 'A' negative probe to the battery negative ground post.

Scope Settings

To make the waveforms as clear as possible, set the scope settings to match the examples.

The MAF sensor waveforms can have slight differences from one scope to another.

Lab Scope Example (1)

In Example (1) the trace shows the MAF sensor signal at idle speed. This waveform was captured with the vehicle in Park, the engine at idle and running in closed loop.

The MAF sensor trace should show an even change in its analog signal as the engine speed is changed up and down. It if drops or rises suddenly without a corresponding change in engine speed, there may be a problem in the MAF sensor.

Lab Scope Example (2)

In Example (2) the trace shows the MAF sensor signal at cruise speed (2000 rpm). Note the change in voltage from the Hot Idle value (1.1v) to the Cruise value (1.9v).

Lab Scope Example (3)

In Example (3) the trace shows the MAF sensor signal during a Snap Throttle event.

There are three distinct events shown here:

- The first event is the point at which the throttle begins to open (the point where the signal starts to rise).
- The second event is where the pressure in the intake manifold begins to equalize.
- The third event is where the throttle closes.

Diagnostic Tip

You can use the capture in Example (3) to compare the known good operation of this MAF sensor during a Snap Throttle event to the operation of a MAF sensor that you suspect may be faulty or dirty.

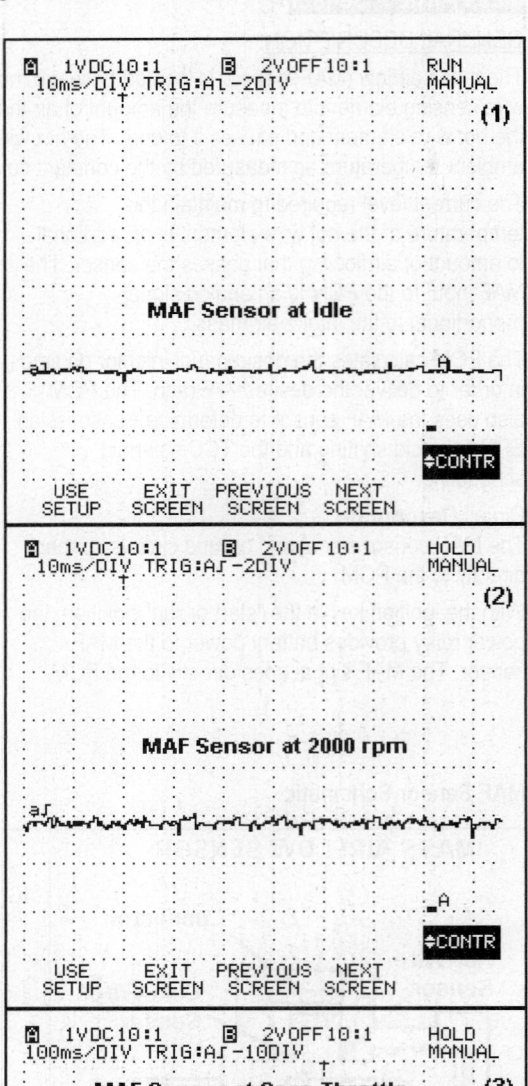

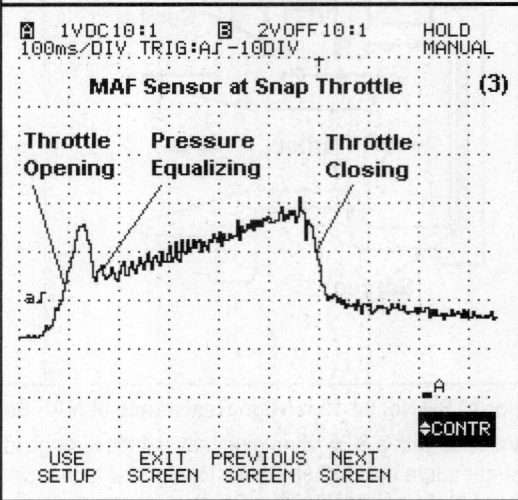

Oxygen Sensor

GENERAL DESCRIPTION

The heated oxygen sensor (HO2S) is mounted in the Exhaust system where it monitors the oxygen content of the exhaust. The oxygen present in the exhaust reacts with the HO2S to produce a signal during engine operation.

If the A/F mixture has a high concentration of oxygen in the exhaust, the HO2S signal to the PCM will be less than 0.4v. If the A/F mixture has a low concentration of oxygen in the exhaust, the HO2S signal will be over 0.6v.

HO2S-21 (Front Bank 2)

HO2S-22 (Rear Bank 2)

Heated Oxygen Sensors

This vehicle application is equipped with four oxygen sensors with internal heaters that allow the sensors to warmup quickly in cold weather.

On this vehicle, the front (pre-catalyst) oxygen sensor is mounted in front of the catalytic converter. The rear (post-catalyst) oxygen sensor is mounted after the catalytic converter.

Oxygen Sensor Identification

The front heated oxygen sensor is identified as either a Bank 1 HO2S-11 or a Bank 2 HO2S-21. The Bank 1 O2 sensor is always located in the engine bank cylinder No. 1.

Note that the "Code Check" circuit for the HO2S-11 signal connects to the PCM at Pin 60. The rest of the HO2S monitored circuits are shown in the Oxygen Sensor Graphic.

Circuit Description

The EEC power relay is used to supply battery voltage to the internal heater circuits to the front and rear oxygen sensors. The heaters are also connected to the heater "control" circuits in the PCM. Refer to the Graphic directly below to view the circuits.

In this case, the PCM can decide when to enable the HO2S heaters to protect the ceramic elements of the sensors and when to run the HO2S main monitor heater tests.

Heated Oxygen Sensor Schematic

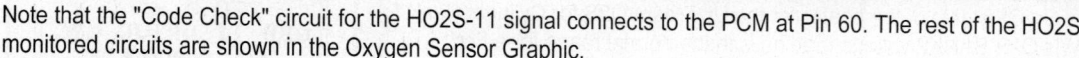

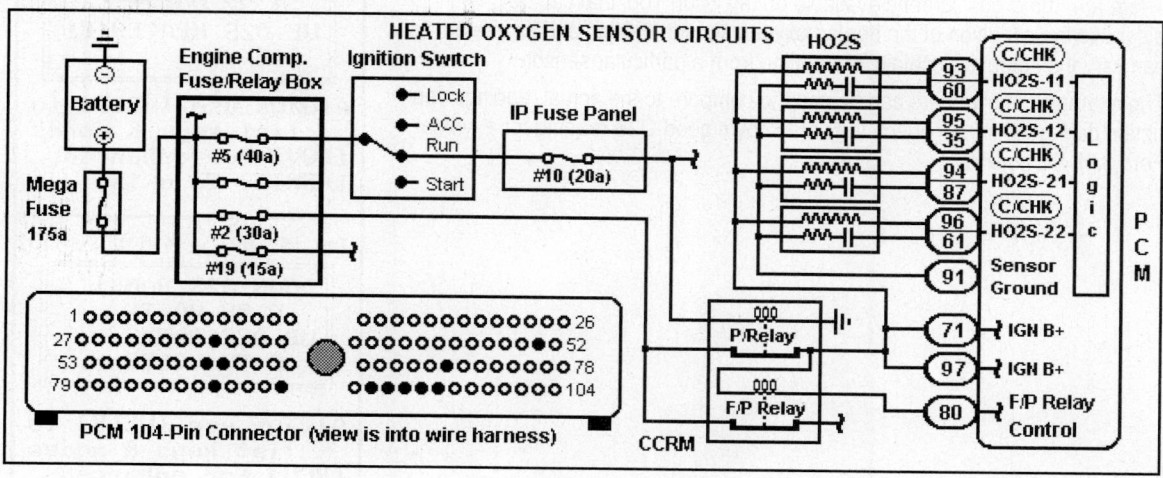

SCAN TOOL TEST (HO2S)

Connect the Scan Tool to the DLC underdash connector. Navigate through the Scan Tool menus until you get to the Ford (OEM) Data List (a list of PID items). Raise the engine speed to 2000 rpm for 3 minutes to set up the test. The examples shown here were captured at Hot Idle in 'P'.

Then select the type of oxygen sensor (O2S) information that you want to view from the various options in the menu.

Scan Tool Example (1)

This example shows the On/Off status of the heaters for the Downstream O2S for Cylinder Bank 1 and Upstream O2S for Cylinder Bank 2. The operation of these heaters is controlled by the PCM (shortly after engine startup).

Scan Tool Example (2)

This example shows the On/Off status of the heater for the Downstream O2S for Cylinder Bank 2 and the current level of the heater for the Upstream O2S for Cylinder Bank 1. The current level shown here (937.5 mA) is the correct amount of current for a Ford HO2S (a known good value).

Scan Tool Example (3)

This example shows the current level of the heater for the Downstream O2S for Cylinder Bank 2 along with the status of the OBD II Drive Cycle ("321 Complete" indicates that the OBD II HO2S Monitor has finished testing this heater).

Scan Tool Example (4)

This example shows the voltage level of the Downstream O2S for Cylinder Bank 2 and the On/Off status of the heater for the Upstream O2S for Cylinder Bank 1. The DOWN O2S BANK2 voltage reading is in the normal range for a Ford heated oxygen sensor (0.100 to 900 Millivolts).

Scan Tool Example (5)

This example shows the voltage level of the Downstream O2S for Cylinder Bank 1 and the voltage level of the Upstream O2S for Cylinder Bank 2. Both of the voltage readings in this example are within the normal range for a Ford heated oxygen sensor (0.100 to 900 Millivolts).

Summary

There are numerous screens available on the Scan Tool that can help you diagnose the condition of the heated oxygen sensor (HO2S) and its heater. You can vary the screens to obtain information from a particular sensor.

The examples shown here can be used to compare to the actual readings you obtain during testing. In effect, they are known good O2S reading for a typical Ford 3.0L V6 engine.

```
Jan 15, 02      4:48:16 pm (1)
  ┌──────────────────────────┐
  │ DN O2S HEATER B1          │
  │ ON                        │
  │ UP O2S HEATER B2          │
  │ ON                        │
  └──────────────────────────┘
  Always use [HELP] to
    list keys & modes
  [RCV] For enhanced
  [*EXIT] To exit
```

```
Jan 15, 02      4:46:59 pm (2)
  ┌──────────────────────────┐
  │ DN O2S HEATER B2          │
  │ ON                        │
  │ UPO2S HTR MON B1          │
  │ 937.5 mA                  │
  └──────────────────────────┘
  Always use [HELP] to
    list keys & modes
  [RCV] For enhanced
  [*EXIT] To exit
```

```
Jan 15, 02      4:43:52 pm (3)
  ┌──────────────────────────┐
  │ DNO2S HTR MON B2          │
  │ 750.0 mA                  │
  │ OBD II DRIVE CYC          │
  │ 321 COMPLETE              │
  └──────────────────────────┘
  Always use [HELP] to
    list keys & modes
  [RCV] For enhanced
  [*EXIT] To exit
```

```
Jan 15, 02      4:49:05 pm (4)
  ┌──────────────────────────┐
  │ DOWN O2S BANK2            │
  │ 0.32 VOLTS                │
  │ UP O2S HEATER B1          │
  │ ON                        │
  └──────────────────────────┘
  Always use [HELP] to
    list keys & modes
  [RCV] For enhanced
  [*EXIT] To exit
```

```
Jan 15, 02      4:50:13 pm (5)
  ┌──────────────────────────┐
  │ DOWN O2S BANK1            │
  │ 0.08 VOLTS                │
  │ UP O2S BANK2              │
  │ 0.73 VOLTS                │
  └──────────────────────────┘
  Always use [HELP] to
    list keys & modes
  [RCV] For enhanced
  [*EXIT] To exit
```

LAB SCOPE TEST (HO2S)

The Lab Scope is the "tool of choice" to test the oxygen sensor as it provides an accurate view of the sensor response and switch rate. Place the shift selector in Park and block the drive wheels prior to testing.

Scope Settings

To make the waveforms as clear as possible, set the scope settings to match the examples.

Lab Scope Connections (1)

Connect the Channel 'A' probe to the HO2S-11 and the Channel 'B' probe to the HO2S-12 circuit. Connect the negative probe to ground.

Lab Scope Connections (2)

Connect the Channel 'A' probe to HO2S-22 and the Channel 'B' probe to the HO2S-21 circuit. Connect the negative probe to ground.

Lab Scope Tests

The examples to the right are from the original oxygen sensor (144,000 miles). The captures were made with the vehicle in closed loop.

Lab Scope Example (1)

In example (1), the two traces were captured with the vehicle at Cruise speed (55 mph) with the ECT at 199°F and the IAT at 78°F.

Lab Scope Example (2)

In example (2), the two traces were captured with the cruise control set at 55 mph. This trace shows what the waveform should look like with the vehicle at steady cruise.

Lab Scope Example (3)

In example (3), the two traces include the use of the vertical cursors in order to view the true half-cycle time phase of two Zirconia design oxygen sensors tested at Cruise mode.

There are 4.25 subdivisions between the vertical cursors. If you multiply 4.25 x 0.040 (40 ms per sub-division) you get a total time of 170 ms per half-cycle.

The HO2S signal should switch L-R and R-L in less than 100 ms, and should switch at least 3-5 times per second. This oxygen sensor was defective (the normal half cycle time phase for a good Zirconia sensor is less than 100 ms).

The sensor was replaced, the adaptive fuel trim was cleared and a Fuel Trim relearn was performed during a Road Test.

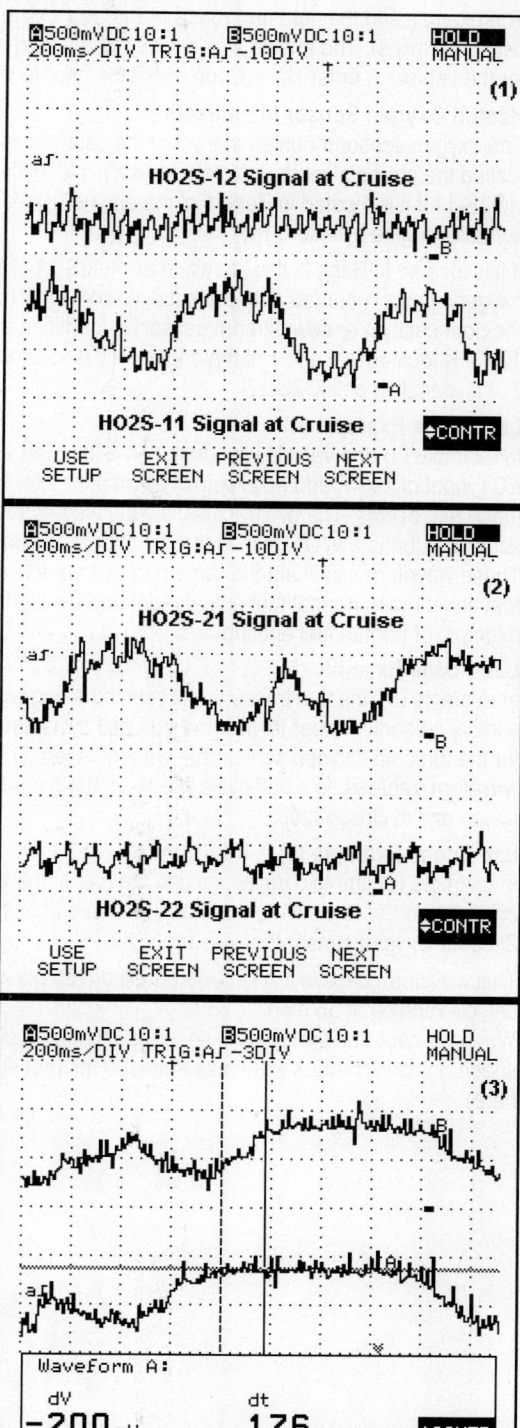

This page contains the waveforms captured after the faulty oxygen sensor was replaced.

Lab Scope Tests

The examples to the right are from a brand new oxygen sensor (0 miles). The engine was allowed to run long enough for the vehicle to enter closed loop operation prior to testing.

Heated Oxygen Sensor Identification

The oxygen sensor mounted in front of the catalytic converter (called the pre-catalyst or upstream sensor) is identified as the HO2S-11 if it is located in Bank 1 of the engine (Bank 1 always contains engine cylinder 1).

If it is located in Bank 2, it is identified as HO2S-21. The oxygen sensor mounted after the catalytic converter (called the post-catalyst or downstream sensor) is identified as HO2S-12 if it is located in Bank 1 of the engine. If it is located in Bank 2, it is identified as HO2S-22.

Lab Scope Example (4)

In example (4), the vehicle was driven on a flat road with an ECT input of 199°F and all accessories off prior to entering these test events. The two traces show the waveforms achieved during a WOT, closed throttle and WOT conditions. These waveforms indicate the pattern from a vehicle with good fuel enrichment, normal fuel enleanment and a second example of normal fuel enrichment.

Lab Scope Example (5)

In example (5), the two traces are presented along with the vertical cursors in order to view the true half cycle time phase for the Zirconia Oxygen sensor that was replaced. This waveform represents a half cycle phase of 104 ms at idle speed and in closed loop.

Lab Scope Example (6)

In example (6), the two traces include the use of the vertical cursors in order to view the true half cycle time phase for the Zirconia Oxygen sensor that was replaced.

This waveform depicts a half cycle phase of 120 ms with the vehicle running at 55 mph.

When we captured the Oxygen sensor switch rates at cruise speed, the switch rates were less than 100 ms (not shown here).

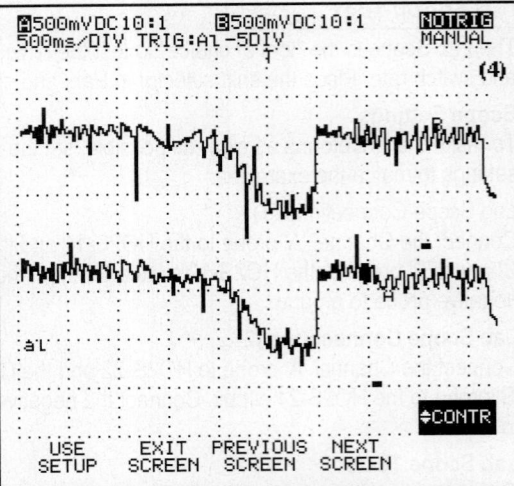

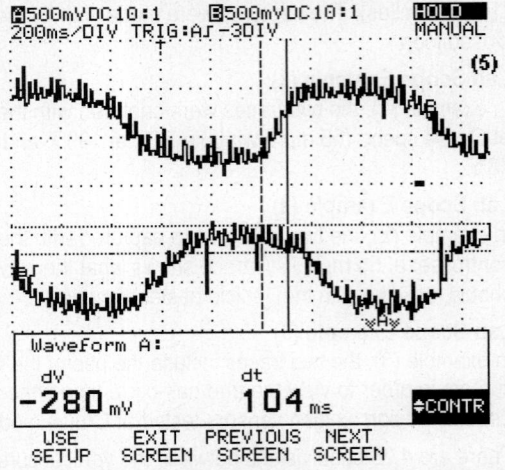

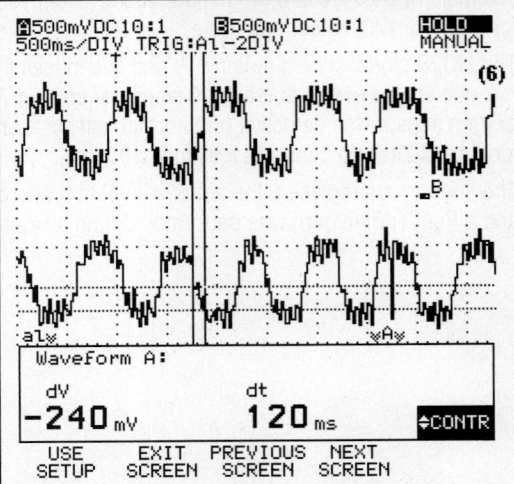

Transmission Controls

COMPONENT TESTS

This vehicle is equipped with a 4-speed AX4N automatic transaxle with automatic shift control. The PCM controls all upshift, downshift, line pressure and TCC operations. The captures in the examples on this page were done with the vehicle in Park at Hot Idle.

Turbine Shaft Speed Sensor

The turbine shaft speed (TSS) sensor is a magnetic pickup that outputs an AC signal that indicates the turbine shaft speed. The TSS positive circuit is connected to the PCM at Pin 84 (DG/WHT wire) and the TSS negative circuit is connected at the analog signal ground at Pin 91(GRY/RED wire). The PCM uses the information from this sensor to determine the TCC lockup strategy and to determine the static EPC solenoid pressure.

Electronic Pressure Control

The electronic pressure control (EPC) solenoid is a variable force style (VFS) solenoid (it combines a solenoid and a regulator valve). This device is used to regulate transaxle line and line modulator pressure by producing resisting forces to the main regulator and line modulator circuits. The pressures are used to control clutch application pressures.

Torque Converter Clutch Solenoid

The torque converter clutch (TCC) solenoid is used in this automatic transaxle (AX4N) to control the application, modulation and release of the torque converter clutch.

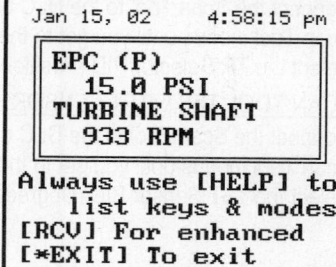

SCAN TOOL TESTS (EPC & TSS)

Connect the Scan Tool to the DLC underdash connector. Navigate through the Scan Tool menus until you get to the Ford (OEM) Data List (a list of PID items). Select the EPC (P) PID and the TSS (RPM) PID from the menu selections.

SCAN TOOL TESTS (TCC & EPC)

Connect the Scan Tool to the DLC underdash connector. Navigate through the Scan Tool menus until you get to the Ford (OEM) Data List (a list of PID items). Select the TCC (%) PID and EPC (V) PID selections from the tool menu.

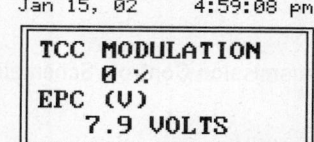

Transmission Control Schematic

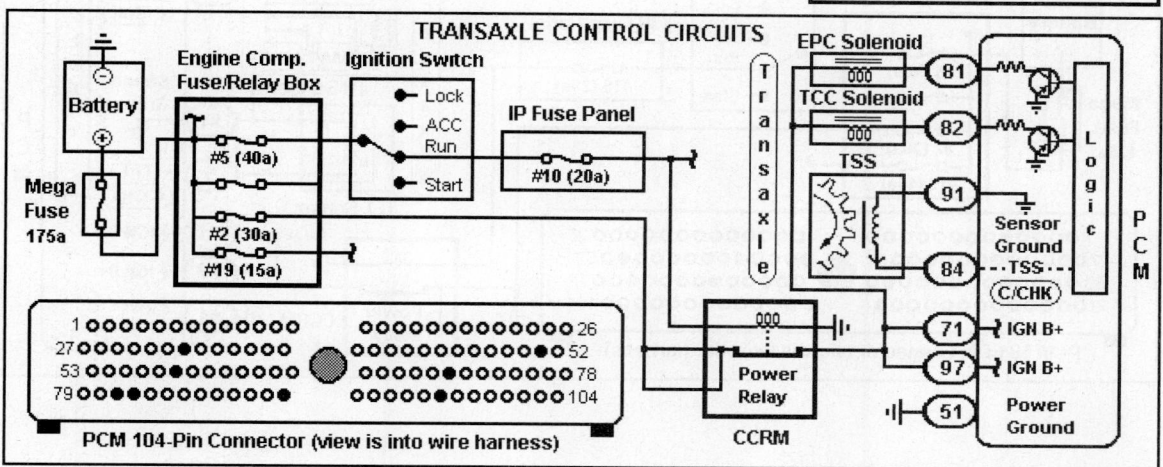

This vehicle is equipped with a 4-Speed AX4N automatic transaxle with automatic shift control. The PCM controls all upshift, downshift, line pressure and TCC operations. The captures in the examples on this page were done with the vehicle in Park at Hot Idle.

Transmission Range Sensor

The transmission range (TR) sensor is a radiometric sensor with six discrete resistors in series. The PCM decodes the signals by monitoring differential voltages produced in each shift lever position. The PCM uses this input to determine the manual shift lever position, determine the desired gear and the pressure of the EPC solenoid. The TR sensor is mounted on the manual lever. It also contains the Backup and Start circuits. Refer to the Operating Range Charts in Section 1 for examples of the TR sensor values.

Transmission Fluid Temperature Sensor

The transmission fluid temperature (TFT) sensor is located on the transaxle main control body. The PCM monitors the voltage drop across the sensor (the sensor value changes) to determine the transmission fluid temperature. The PCM uses the initial TFT sensor signal to determine whether a cold start shift schedule is necessary. The cold start shift schedule allows for quicker shifts, allows the PCM to inhibit TCC lockup and to correct the EPC pressures when the transaxle fluid is cold.

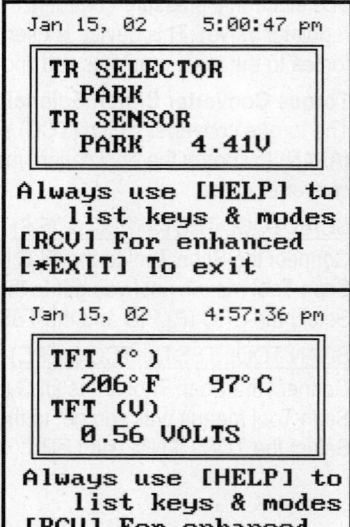

SCAN TOOL TEST (TR SENSOR)

Connect the Scan Tool to the DLC underdash connector. Navigate through the Scan Tool menus until you get to the Ford (OEM) Data List (a list of PID items). Select the TR Selector PID (Position and Volts) from the menu selections.

SCAN TOOL TEST (TFT SENSOR)

Connect the Scan Tool to the DLC underdash connector. Navigate through the Scan Tool menus until you get to the Ford (OEM) Data List (a list of PID items). Select the TFT Sensor PID (Degrees and Volts) from the menu selections.

Transmission Controls Schematic

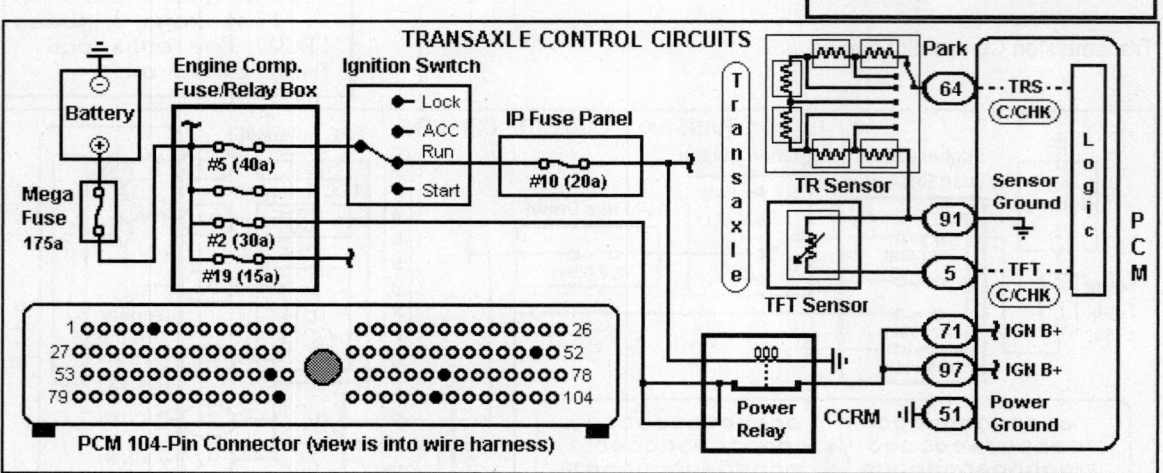

Reference Information

DIAGNOSTIC MONITORING TEST RESULTS

The purpose of this test mode (Mode 6) is to allow access to the results of certain OBD II Monitor diagnostic tests. The test values stored at the time that a particular monitor completes a test are displayed if requested. *These test results do not indicate a Monitor passed or failed (only that a part of the test passed or failed as noted in the test results).*

Module ID h*	Test ID h*	Component ID h*	Test Description (Note: h* is hexadecimal)
---	---	---	**Oxygen Sensor Monitor** (01-0F) - **Examples 4 & 5**
10	01	11, 21	HO2S-11 voltage amplitude (Bank 1 & Bank 2, Sensor 1)
10	02	11	HO2S-11 upstream static shift, lean & rich shift (B1 S1)
10	02	21	HO2S-11 upstream static shift, lean & rich shift (B2 S1)
10	03	01	Upstream Switch Point
10	03	02	Downstream Switch Point
---	---	---	**Catalyst Monitor** (10-1F) - **Example 6**
10	10	11, 21	Rear to Front Switch Ratio Test, Bank 1 & Bank 1 Test
---	---	---	**EVAP Monitor** (21-2F) - **Examples 6 & 7**
10	21	00	Fuel Tank Pressure Low & High Pressure Test
10	22	00	EVAP Phase 2 - change in Pressure Test
10	23	00	EVAP Phase 4 - change in Pressure Test
10	24	00	EVAP Phase 2 - change in Pressure Test too small
---	---	---	**EGR System Monitor** (41-4F) - **Example 8 & 9**
10	41	11	Upstream Hose Disconnected Test
10	41	12	Downstream Hose Disconnected Test
10	45	20	Stuck Open Valve Test
10	4A	30	EGR Flow Test
10	4B	30	Flow Test
---	---	---	**Misfire Monitor** (51-5F) - **Examples 10, 11 & 12**
10	50	00	Total Misfires that exceeded threshold
10	51	01, 02, 03, 04, 05, 06	Cylinder 1 misfires that exceeded threshold
10	51	09	Cylinder 9 misfires that exceeded threshold
10	51	0A	Cylinder 10 misfires that exceeded threshold
10	52	00	Consecutive cylinder events during Active Test

CID's & TID's Graphic

```
Jan 17, 02   10:52:36 am
  SELECT TEST
 F0:O2S Results
 F1:Other Results

[*EXIT] To exit    1
```

```
Jan 17, 02   10:52:36 am
 O2 Sensor Tests
 NOT Completed
 YES = Continue
 NO  = Quit

[*EXIT] To exit    2
```

```
Jan 17, 02   10:52:36 am
   SELECT BANK
 F0: O2S BANK 1
 F1: O2S BANK 2

[*EXIT] To exit    3
```

```
Jan 17, 02   10:52:36 am
 ↓TID/CID Result
 * 01/11 ....Pass
   01/21 ....Pass
   02/11 ....Fail

[*EXIT] To exit    4
```

```
Jan 17, 02   10:52:36 am
 ↑↓TID/CID Result
   02/21 ....Fail
   03/01 ....Pass
 * 03/02 ....Pass

[*EXIT] To exit    5
```

```
Jan 17, 02   10:52:36 am
 ↑↓TID/CID Result
 * 10/11 ....Pass
   10/21 ....Pass
   21/00 ....Fail

[*EXIT] To exit    6
```

```
Jan 17, 02   10:52:36 am
 ↑↓TID/CID Result
 * 22/00 ....Pass
   23/00 ....Fail
   24/00 ....Pass

[*EXIT] To exit    7
```

```
Jan 17, 02   10:52:36 am
 ↑↓TID/CID Result
 * 41/11 ....Pass
   41/12 ....Fail
   45/20 ....Pass

[*EXIT] To exit    8
```

```
Jan 17, 02   10:52:36 am
 ↑↓TID/CID Result
 * 4A/30 ....Pass
   4B/30 ....Pass
   50/00 ....Pass

[*EXIT] To exit    9
```

```
Jan 17, 02   10:52:36 am
 ↑↓TID/CID Result
 * 51/01 ....Pass
   51/02 ....Pass
   51/03 ....Pass

[*EXIT] To exit   10
```

```
Jan 17, 02   10:52:36 am
 ↑↓TID/CID Result
 * 51/04 ....Pass
   51/05 ....Pass
   51/06 ....Pass

[*EXIT] To exit   11
```

```
Jan 17, 02   10:52:36 am
 ↑ TID/CID Result
 * 51/07 ....Fail
   51/08 ....Fail
   52/00 ....Fail

[*EXIT] To exit   12
```

PCM PID TABLES - INPUTS

■ **NOTE:** *The following readings were obtained with the engine at idle speed, radiator hose hot, throttle closed, gear selector in Park and all accessories turned off.*

Acronym	Scan Tool Parameter	Range	Typical Value
ACCS	A/C Cycling Clutch Switch Signal	ON / OFF	OFF (ON with A/C on)
ACP	A/C Pressure Switch Signal	Open / Closed	Open
AIR-M	Electronic Air Monitor (California)	ON / OFF	ON
BPP	Brake Pedal Position Switch	ON / OFF	OFF (ON with brake on)
CMP (CID)	CMP Sensor Signal	0-1000 Hz	13-16
CKP	CKP Sensor Signal	0-10,000 Hz	510-540
DPFE	DPF EGR Sensor Signal	0-5.1v	0.4-0.6
ECT ºF	Engine Coolant Temp. Sensor	-40ºF to 304ºF	160-200
FEPS	Flash EEPROM	0-5.1v	0.5-0.6
FPM	Fuel Pump Monitor	ON / OFF	ON
Gear	Transmission Gear Position	1-2-3-4	1
HO2S-11	HO2S-11 (Bank 1 Sensor 1)	0-1100 mv	100-900 mv
HO2S-12	HO2S-12 (Bank 1 Sensor 2)	0-1100 mv	100-900 mv
HO2S-21	HO2S-21 (Bank 2 Sensor 1)	0-1100 mv	100-900 mv
HO2S-22	HO2S-22 (Bank 2 Sensor 2)	0-1100 mv	100-900 mv
IAT ºF	Intake Air Temperature Sensor	-40ºF to 304ºF	50-120
LOAD	Calculated Engine Load	0-100%	10-20
MAF V	Mass Airflow Sensor Signal	0-5.1v	0.6-0.9
MISF	Misfire Detection	ON / OFF	OFF
OCT ADJ	Octane Adjust Signal	Closed / Open	Closed
PNP	Park Neutral Position Switch	Neutral / Drive	Neutral
PSP	Power Steering Position Switch	HI / LOW	HI (while turning)
RPM	Engine Speed	0-9999	760-850
TFT ºF	Transmission Fluid Temp. Sensor	-40ºF to 304ºF	110-210
TP V	Throttle Position	0-5.1v	0.5-1.27v
TR1	Gear Position TR1 Signal	0v or 12v	0v
TR2	Gear Position TR2 Signal	0v or 12v	0v
TR4	Gear Position TR4 Signal	0v or 12v	0v
TRV/TR	Gear Position TR Sensor Volts	0-5.1v	P/N: 4.4v
TSS	TSS Sensor Signal	0-10,000 Hz	790-820
Vacuum	Engine Vacuum	0-30" Hg	19-20" Hg
VPWR	Vehicle Power	0-25.5v	14.1v
VSS	Vehicle Speed	0-159 mph / 0-1000 Hz	55 mph = 125 Hz

PCM PID TABLES - OUTPUTS

■ **NOTE:** *The following readings were obtained with the engine at idle speed, radiator hose hot, throttle closed, gear selector in Park and all accessories turned off.*

Acronym	Scan Tool Parameter	Range	Typical Value
AIR	Electronic Air System (California)	ON / OFF	OFF
CD1	Coil Driver 1	0-60° dwell	7°
CD2	Coil Driver 2	0-60° dwell	7°
CD3	Coil Driver 3	0-60° dwell	7°
CTO	Clean Tachometer Output Signal	0-1,000 Hz	42-50
EVAP CV	EVAP Canister Vent Valve	0-100%	0-10%
EGR VR	EGR Vacuum Regulator Valve	0-100%	0%
EPC	EPC Solenoid Control	0-100 psi	40
FP	Fuel Pump Control	ON / OFF	ON
Fuel B1	Fuel Injector Pulsewidth Bank 1	0-99.9 ms	3.8-4.7
Fuel B2	Fuel Injector Pulsewidth Bank 2	0-99.9 ms	3.8-4.7
HFC	High Speed Fan Control	ON / OFF	OFF (ON with fan on)
HTR-11	HO2S-11 Heater Control	ON / OFF	ON
HTR-12	HO2S-12 Heater Control	ON / OFF	ON
HTR-21	HO2S-21 Heater Control	ON / OFF	ON
HTR-22	HO2S-22 Heater Control	ON / OFF	ON
IAC	Idle Air Control Motor	0-100%	34
INJ1	Fuel Injector 1 Control	0-99.9 ms	3.8-4.7
INJ2	Fuel Injector 2 Control	0-99.9 ms	3.8-4.7
INJ3	Fuel Injector 3 Control	0-99.9 ms	3.8-4.7
INJ4	Fuel Injector 4 Control	0-99.9 ms	3.8-4.7
INJ5	Fuel Injector 5 Control	0-99.9 ms	3.8-4.7
INJ6	Fuel Injector 6 Control	0-99.9 ms	3.8-4.7
LFC	Low Speed Fan Control	ON / OFF	OFF (On with fan on)
LONGFT1	Long Term Fuel Trim Bank 1	-20% to 20%	0-1
LONGFT2	Long Term Fuel Trim Bank 2	-20% to 20%	0-1
MIL	Malfunction Indicator Lamp Control	ON / OFF	OFF (ON with DTC)
SHRTFT1	Short Term Fuel Trim Bank 1	-10% to 10%	-1 to 1
SHRTFT2	Short Term Fuel Trim Bank 2	-10% to 10%	-1 to 1
SPARK	Spark Advance	-90° to 90°	24-30
SSA	Transmission Shift Solenoid 'A'	ON / OFF	OFF
SSB	Transmission Shift Solenoid 'B'	ON / OFF	ON
SSC	Transmission Shift Solenoid 'C'	ON / OFF	OFF
TCC	Torque Converter Clutch Solenoid	0-100%	0%
VMV	EVAP Vapor Management Valve	0-100%	0-100%
VREF	Voltage Reference	0-5.1v	4.9-5.1v
WAC	Wide Open Throttle A/C Relay	ON / OFF	OFF (ON at WOT)

PCM PIN VOLTAGE TABLES

PCM Pin Voltage Table: 104-Pin Connector

Pin Number	Wire Color	Application	Value at Hot Idle
1	P/O	Transmission Shift Solenoid 'B' Control	1v
2	PK/LG	MIL (lamp) Control	Lamp off: 12v, On: 1v
3-12, 14	---	Not Used	---
13	P	Flash EEPROM Power	0.1v
15	PK/LB	SCP Data Bus (+)	Digital Signals
16	T/O	SCP Data Bus (-)	Digital Signals
17-20, 23	---	Not Used	---
21	DB	CKP+ Sensor Signal	510-540 Hz
22	GY	CKP- Sensor Signal	510-540 Hz
24	BK/W	Power Ground	0.050v
25	BK	Chassis Ground	0.050v
26	W/BK	Coil Driver 1 Control	5° dwell
27	O/Y	Transmission Shift Solenoid 'A' Control	12v
28	T/O	Low Speed Fan Control	Fan Off: 12v, On: 1v
29, 32, 34	---	Not Used	---
30	DG	Octane Adjust Signal	Closed: 0v
31	Y/LG	PSP Switch Signal	Turning: 12v, straight: 0v
33	PK/O	VSS- Signal	55 mph: 125 Hz
35	R/BK	HO2S-12 (Bank 1 Sensor 2) Signal	100-900 mv
36	T/LB	MAF Sensor Return	0.050v
37	O/BK	Transmission Fluid Temperature Sensor	2.20v
38	LG/R	Engine Coolant Temperature Sensor	0.6v
39	GY	Intake Air Temperature Sensor	1.6v
40	DG/Y	Fuel Pump Monitor	Pump On: 12v, Off: 0v
41	PK/LB	A/C Cycling Clutch Switch Signal	A/C Switch On: 12v
42-45, 49-50	---	Not Used	---
46	LG/P	High Speed Fan Control	Fan On: 1v, Off: 12v
47	BR/PK	EGR Vacuum Regulator Solenoid	0% (55 mph: 0-45%)
48	T/Y	Clean Tachometer Output	42-50 Hz
51	BK/W	Power Ground	0.050v
52	Y/R	Coil Driver 2 Control	5° dwell
53	PK/BK	Transmission Shift Solenoid 'C' Control	12v
54, 57, 59, 63	---	Not Used	---
55	P	Keep Alive Power	12-14v
56	GY/Y	EVAP Vapor Management Valve	0-10 Hz
58	GY/BK	VSS+ Signal	At 55 mph: 125 Hz

PCM 104-Pin Connector

TERMINAL VIEW OF WIRING HARNESS CONNECTOR FOR PCM (104 PIN)

Note: back-probing the 104-pin connector with a DVOM probe can damage the PCM - use a breakout box!

PCM PIN VOLTAGE TABLE: 104-PIN CONNECTOR

Pin Number	Wire Color	Application	Value at Hot Idle
60	GY/LB	HO2S-11 (Bank 1 Sensor 1) Signal	100-900 mv
61	P/LG	HO2S-22 (Bank 2 Sensor 2) Signal	100-900 mv
62	R/PK	Fuel Tank Pressure Sensor Signal	Cap Off: 2.6v
64	LB/Y	Transmission TR Sensor Signal	In 'P': 0v, at 55 mph: 1.7v
65	BR/LG	EGR DPFE Sensor Signal	0.5v
66, 68, 70, 72	---	Not Used	---
67	P/W	EVAP Vent Valve Control	0-10 Hz
69	PK/Y	A/C WOT Relay Control	12v
71, 97	R	Vehicle Power	12-14v
72, 79	---	Not Used	---
73	T/BK	Fuel Injector 5 Control	3.8-4.7 ms
74	BR/Y	Idle Air Control Motor	10v (32% duty cycle)
75	T	Fuel Injector 1 Control	3.8-4.7 ms
76-77, 103	BK/W	Power Ground	0.050 mv
78	Y/W	Coil Driver 3 Control	5° dwell
80	LB/O	Fuel Pump Relay Control	Relay On: 1v, Off: 12v
81	W/Y	Transmission EPC Solenoid Control	10.6v
82	P/Y	Torque Converter Clutch Solenoid	12v (off)
83	W/LB	Fuel Injector 3 Control	3.8-4.7 ms
84	DG/W	Transmission Shaft Speed Sensor	50-65 Hz
85	DB/O	Camshaft Position Sensor Signal	6-8 Hz
86	BK/Y	A/C Pressure Switch	12v (open - A/C off)
87	R/LG	HO2S-21 (Bank 2 Sensor 1) Signal	100-900 mv
88	LB/R	Mass Airflow Sensor Signal	0.9v
89	GY/W	Throttle Position Sensor	0.53-1.27v
90	BR/W	Voltage Reference	4.9-5.1v
91	GY/R	Analog Sensor Return	0.050v
92	R/LG	Brake Pedal Position Switch	Brake On: 12v, Off: 0v
93	R/W	HO2S-11 (Bank 1 Sensor 1) Heater	Heater On: 1v, off: 12v
94	W/BK	HO2S-21 (Bank 2 Sensor 1) Heater	Heater On: 1v, off: 12v
95	Y/LB	HO2S-12 (Bank 1 Sensor 2) Heater	Heater On: 1v, off: 12v
96	T/Y	HO2S-22 (Bank 2 Sensor 2) Heater	Heater On: 1v, off: 12v
98, 102, 104	---	Not Used	---
99	LG/O	Fuel Injector 6 Control	3.8-4.7 ms
100	BR/LB	Fuel Injector 4 Control	3.8-4.7 ms
101	W	Fuel Injector 2 Control	3.8-4.7 ms

PCM 104-Pin Connector Graphic

TERMINAL VIEW OF WIRING HARNESS CONNECTOR FOR PCM (104 PIN)

Note: back-probing the 104-pin connector with a DVOM probe can damage the PCM - use a breakout box!

PCM Wiring Diagrams
FUEL & IGNITION SYSTEM DIAGRAMS (1 OF 7)

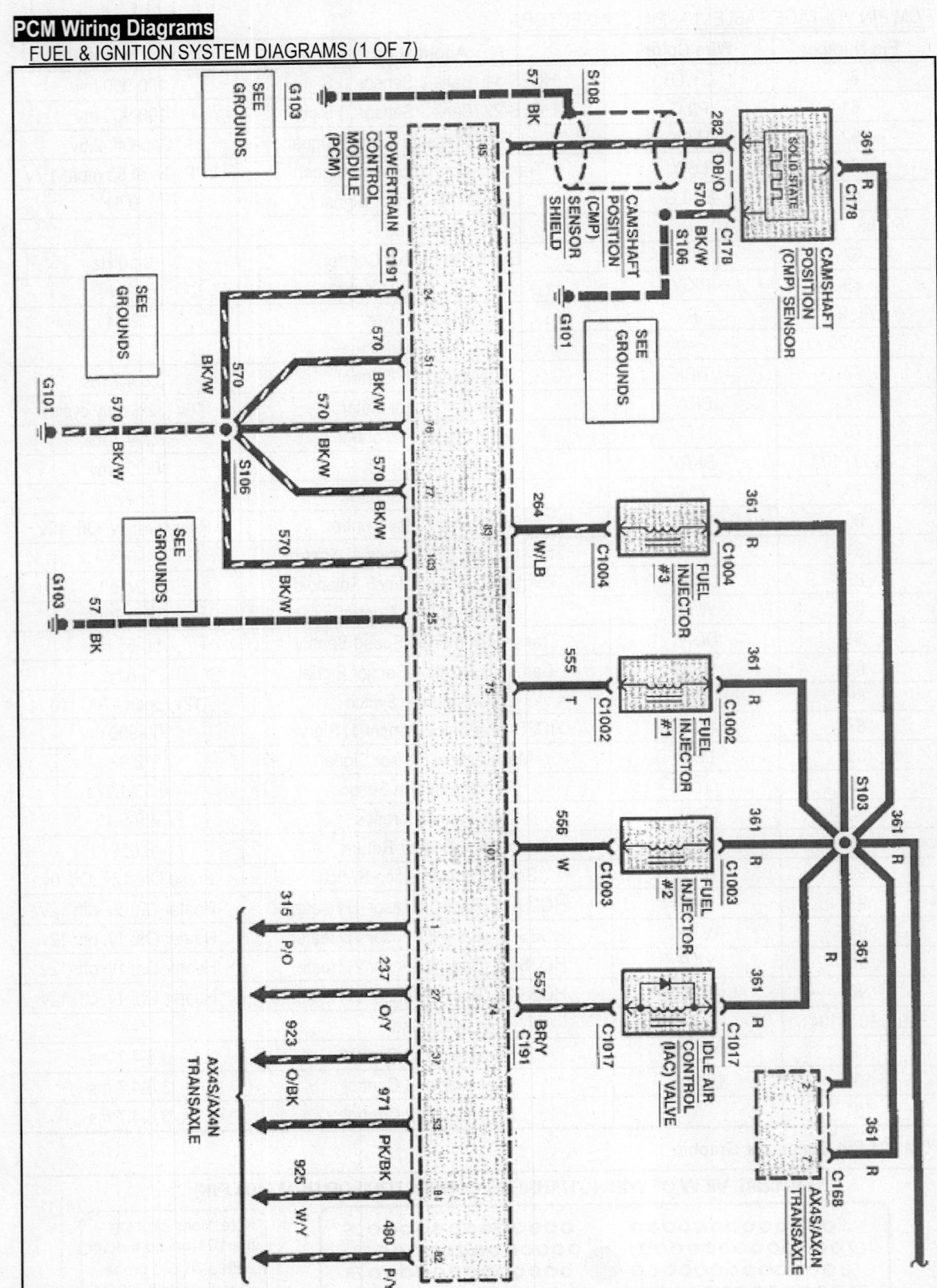

FUEL & IGNITION SYSTEM DIAGRAMS (2 OF 7)

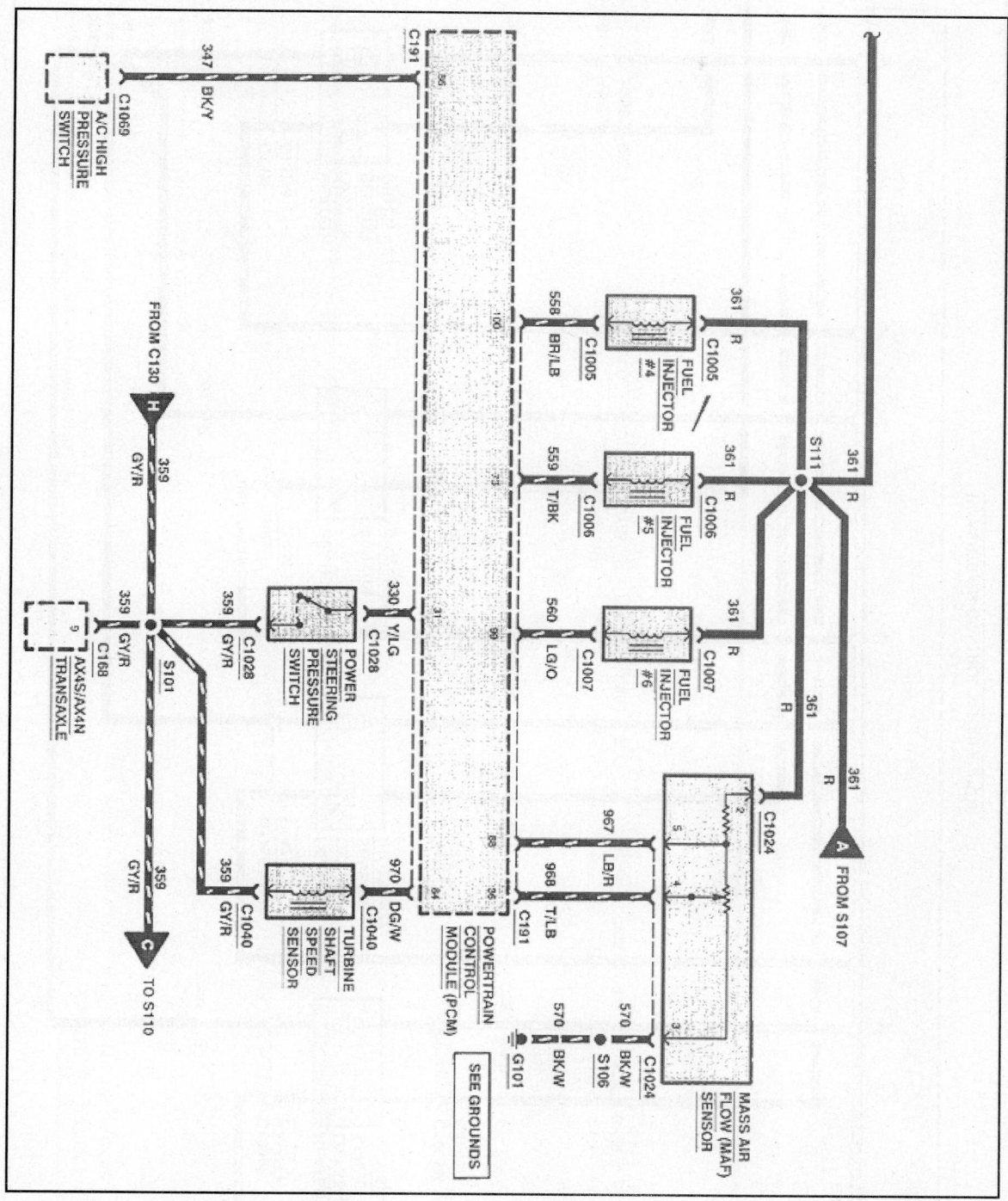

FUEL & IGNITION SYSTEM DIAGRAMS (3 OF 7)

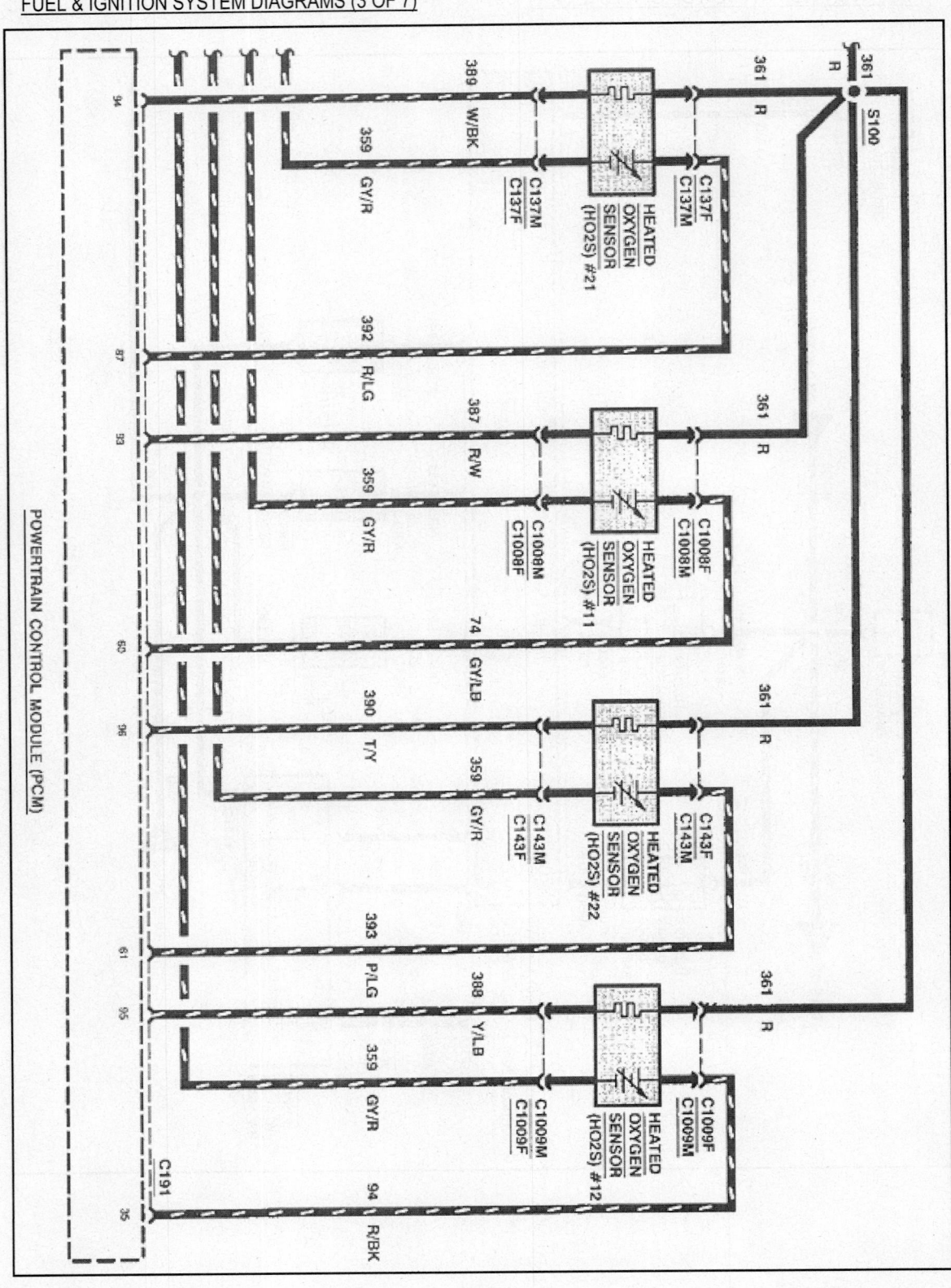

FUEL & IGNITION SYSTEM DIAGRAMS (4 OF 7)

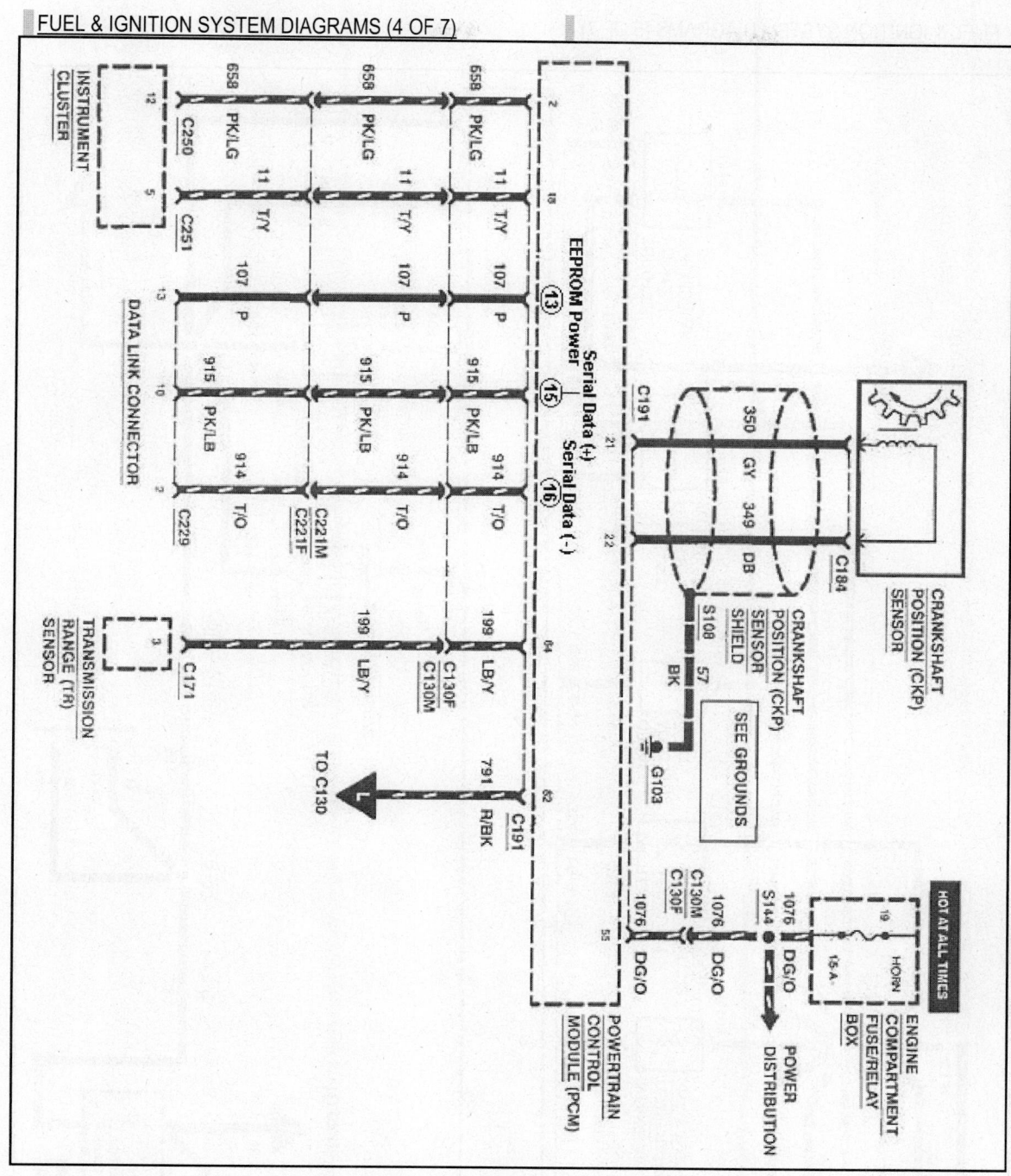

FUEL & IGNITION SYSTEM DIAGRAMS (5 OF 7)

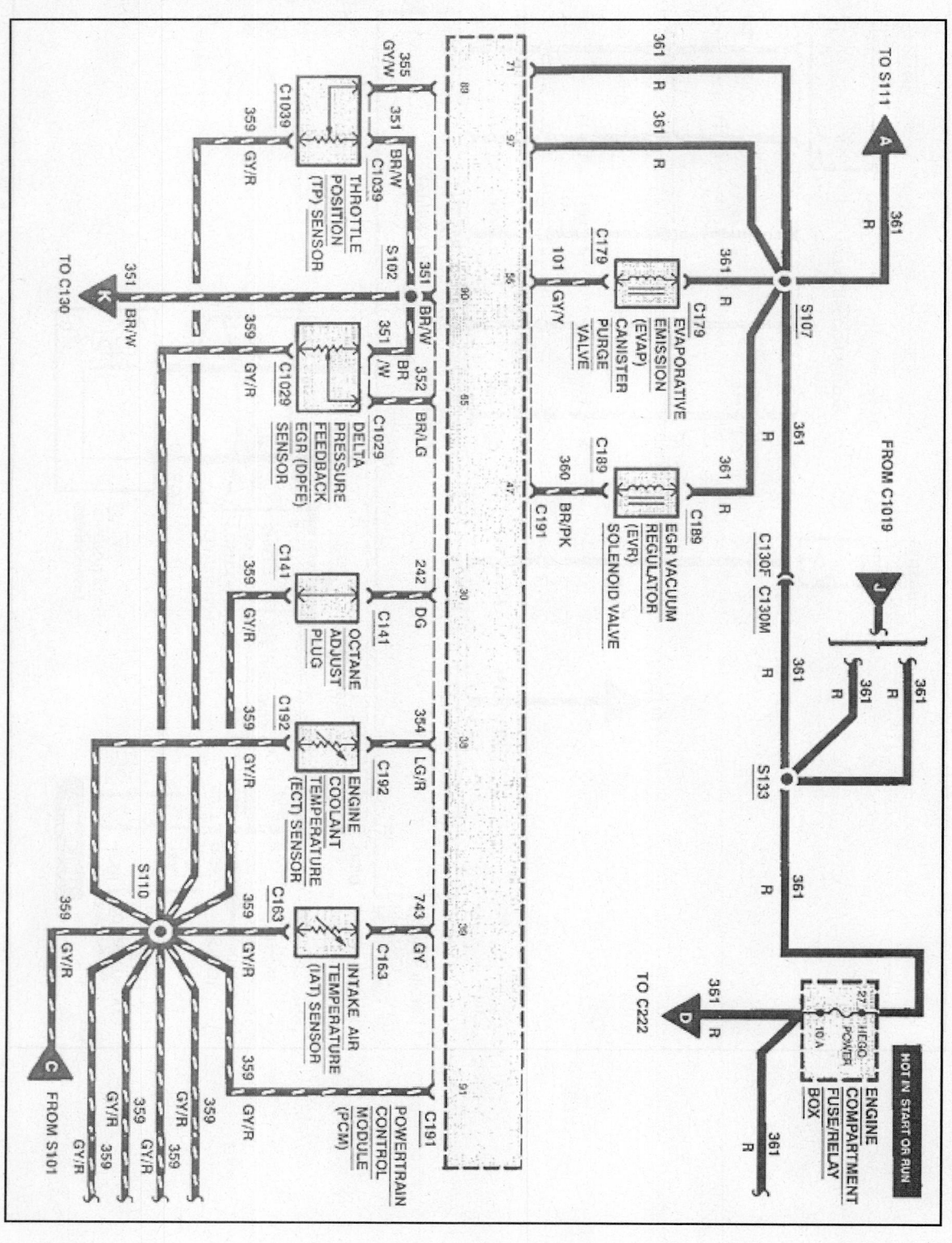

FUEL & IGNITION SYSTEM DIAGRAMS (6 OF 7)

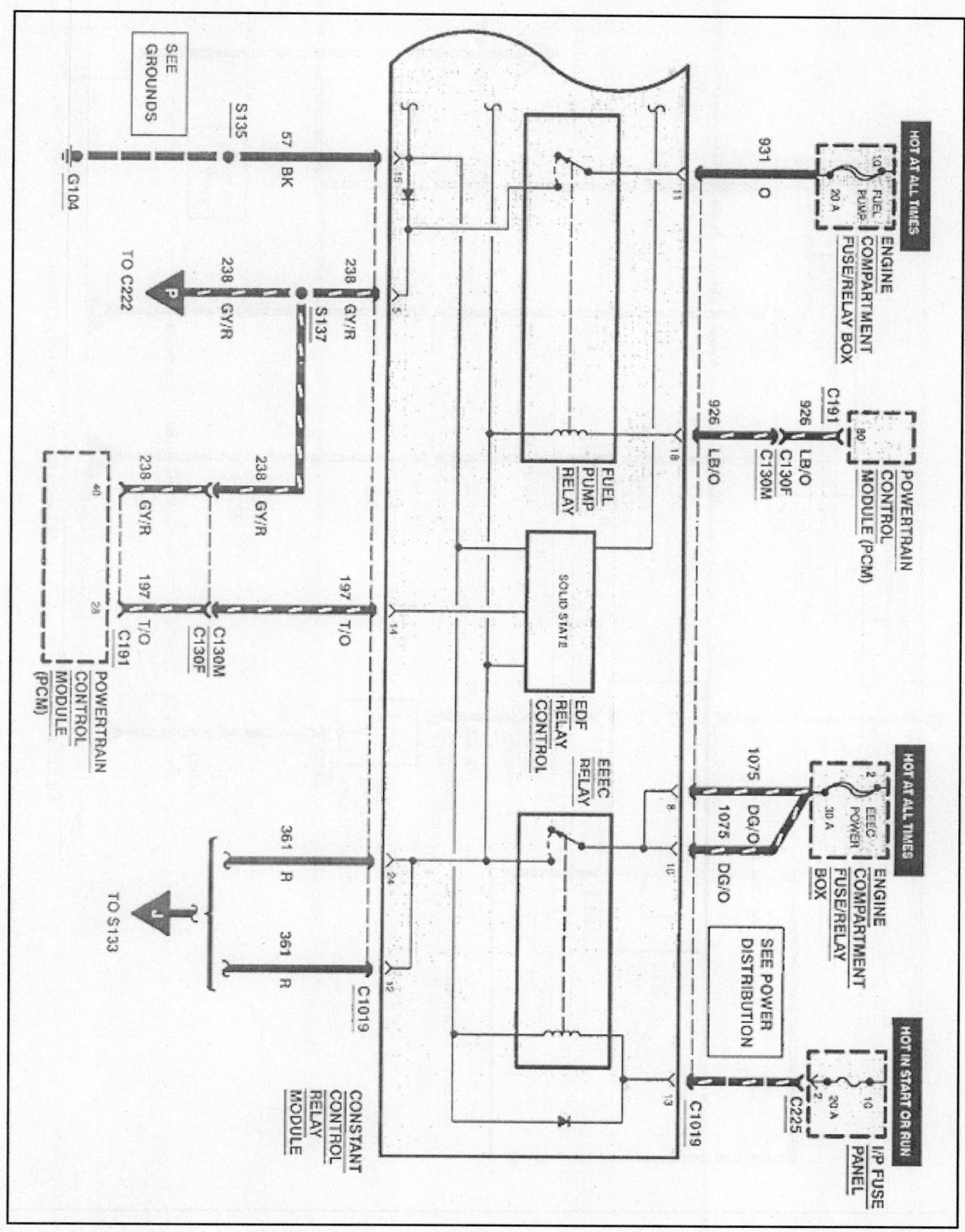

Fuel & Ignition System Diagrams (7 of 7)

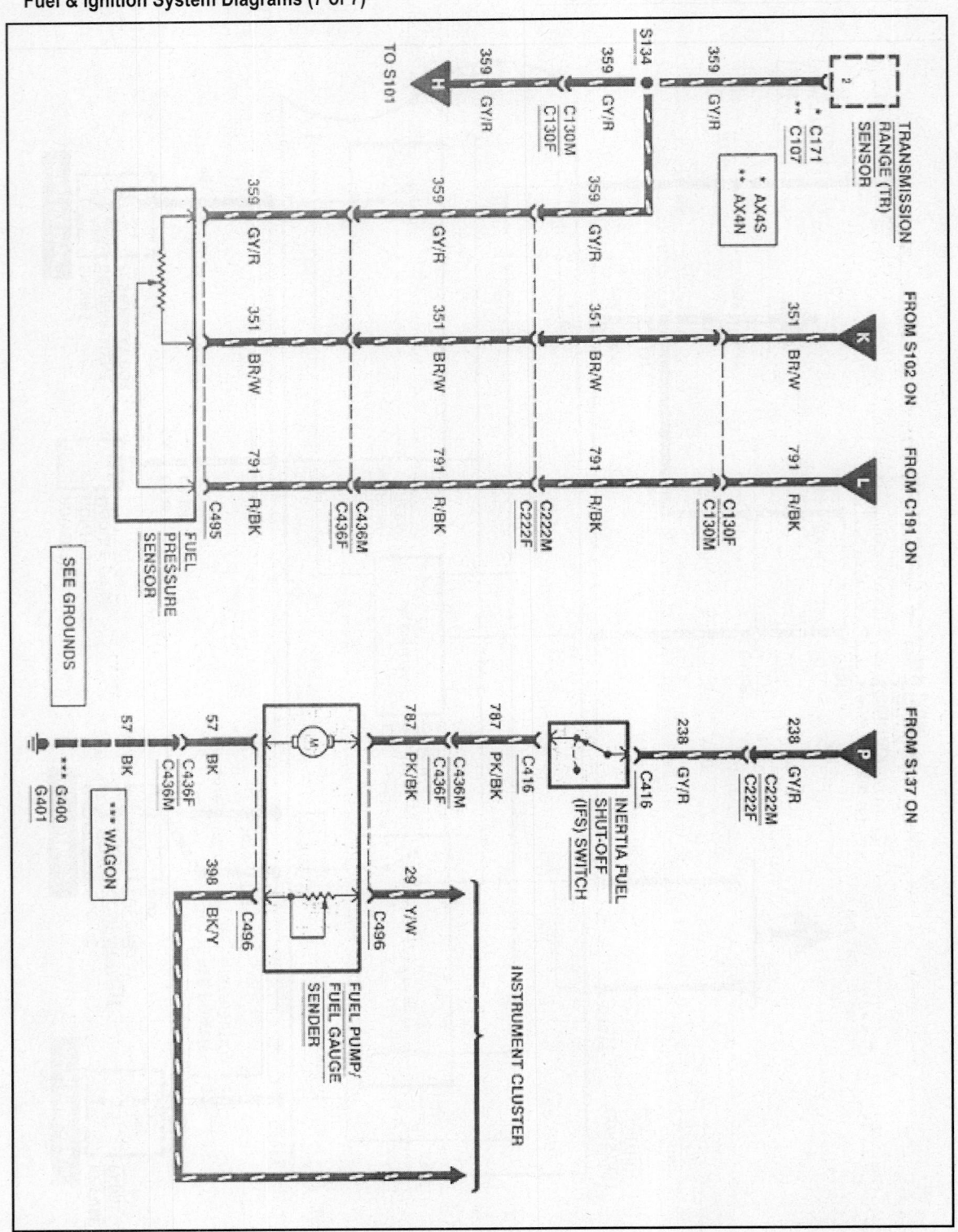

SECTION 3 CONTENTS

1998 F150 Pickup - Powertrain

Introduction
How To Use This Section ....Page 3-3

VEHICLE IDENTIFICATION
VECI Decal ....Page 3-4

ELECTRONIC ENGINE CONTROL SYSTEM
Operating Strategies ....Page 3-5
PCM Location ....Page 3-5

INTEGRATED ELECTRONIC IGNITION SYSTEM
Introduction ....Page 3-6
Ignition Coil Operation ....Page 3-7
Lab Scope Test (Coil Primary) ....Page 3-8
CKP Sensor (Magnetic) ....Page 3-9
CKP Sensor Diagnosis (No Code Fault) ....Page 3-9
DVOM Test (CKP Sensor) ....Page 3-11
Lab Scope Test (CKP Sensor) ....Page 3-11
JI Diagnostic System - DTC P0351 ....Page 3-12

EVAPORATIVE EMISSION SYSTEM
Vapor Management Valve System..Page 3-16
Fuel Tank Pressure Sensor ....Page 3-16
VMV System Components ....Page 3-17
Evaporative Charcoal Canister ....Page 3-18
Canister Vent Solenoid ....Page 3-19
Fuel Tank Filler Cap ....Page 3-19
Fuel Tank Vapor Valve ....Page 3-20
VMV System Operation ....Page 3-21
Scan Tool Test (Purge Solenoid) ....Page 3-22
Scan Tool Test (FTP Sensor) ....Page 3-22
Lab Scope Test (VMV Solenoid) ....Page 3-23
EVAP System Monitor ....Page 3-24
JI Diagnostic System - DTC P0442 ....Page 3-25

EXHAUST GAS RECIRCULATION SYSTEM
General Information ....Page 3-29
Differential Pressure Feedback EGR System ....Page 3-29
DPFE Sensor Information ....Page 3-30
Orifice Tube Assembly ....Page 3-30
Scan Tool Test (DPFE Sensor) ....Page 3-31
DVOM Test (DPFE Sensor) ....Page 3-31
Vacuum Regulator Solenoid ....Page 3-32
Lab Scope Test (VR Solenoid) ....Page 3-33
JI Diagnostic System - DTC P0401 ....Page 3-34

FUEL DELIVERY SYSTEM

    Introduction .... Page 3-38

    Fuel Pump Relay & Fuel Pump Monitor .... Page 3-39

    Fuel Pump Module .... Page 3-40

    Lab Scope Test (Fuel Pump) .... Page 3-40

    Fuel Pump Components .... Page 3-41

    External Fuel Filter .... Page 3-41

    Fuel Pressure Regulator .... Page 3-42

    Inertia Switch .... Page 3-43

    Fuel Injectors .... Page 3-44

    Scan Tool Test (Fuel Injector & Fuel Trim) .... Page 3-45

    Lab Scope Test (Fuel Injector) .... Page 3-46

    Technical Service Bulletins: .... Page 3-46

    CMP Sensor (Magnetic) .... Page 3-47

    CMP Sensor (Signal Explanation) .... Page 3-48

    Lab Scope Test (CMP Sensor) .... Page 3-48

IDLE AIR CONTROL VALVE

    General Description .... Page 3-49

    Scan Tool Test (IAC Valve) .... Page 3-49

    Lab Scope Test (IAC Valve) .... Page 3-50

INPUT SENSORS

    Cylinder Head Temperature Sensor .... Page 3-51

    Engine Coolant Temperature Sensor .... Page 3-52

    Throttle Position Sensor .... Page 3-53

MASS AIRFLOW SENSOR

    General Description .... Page 3-54

    BARO Sensor Calibration .... Page 3-55

    Scan Tool Test (MAF Sensor) .... Page 3-55

    Lab Scope Test (MAF Sensor) .... Page 3-56

OXYGEN SENSOR

    General Description .... Page 3-57

    Scan Tool Test (HO2S) .... Page 3-58

    Lab Scope Test (HO2S) .... Page 3-59

TRANSMISSION CONTROLS

    Component Tests .... Page 3-60

    Scan Tool Tests (EPC, TCC & OSS) .... Page 3-60

    Scan Tool Tests (TR & TFT Sensors) .... Page 3-61

REFERENCE INFORMATION

    PCM PID Tables - Inputs .... Page 3-62

    PCM PID Tables - Outputs .... Page 3-63

    PCM Pin Voltage Tables .... Page 3-64

    PCM Wiring Diagrams .... Page 3-66

1998 F150 PICKUP - POWERTRAIN

Introduction

HOW TO USE THIS SECTION

This section of the training manual includes diagnostic and repair information for Ford F150 applications. This information can be used to help you understand the Theory of Operation and Diagnostics of the electronic controls and devices on this vehicle.

The articles in this section are separated into two sub categories:

- Powertrain Control Module
- Transmission Controls

1998 F150 Pickup 4.6L V8 VIN W (4R70W Automatic Transmission)

The articles listed for this vehicle include a wide variety of subjects that cover the key engine controls and vehicle diagnostics. Several of the Powertrain sensor inputs and output devices are featured along with detailed descriptions of how they operate, what can go wrong with them, and most importantly, how to use the PCM onboard diagnostics and common shop tools to determine if one or more devices has failed.

This vehicle uses an Integrated Electronic Ignition (distributorless) system along with a camshaft position sensor (Magnetic) to control the fuel injection timing. These devices and others (i.e., the MAF and Oxygen sensors) are covered along with how to test them.

Choose The Right Diagnostic Path

In most cases, the first step in any vehicle diagnosis is to follow the manufacturers recommended diagnostic path. In the case of Ford vehicle applications, the first step is to connect a Scan Tool (OBD II certified for this vehicle) and attempt to communicate with the engine controller or any other vehicle onboard controllers that apply. From this point on, the recommended procedure is to use the Ford Quick Test to check for faults.

The Malfunction Indicator Lamp (MIL) can be a great help to you during diagnosis. You should note if the customer complaint included the fact that the MIL remained on during engine operation. And be sure to determine if the MIL comes on at key on, and if it goes out once the engine starts. There is more information on using the MIL in Section 1.

Diagnostic Points to Consider

If a problem is detected in a vehicle similar to this one (whether it is a trouble code or no code condition), you need to be able to determine which vehicle system (or systems) are involved before you proceed with any testing. While this may sound like a fairly routine way to approach problem solving, it is a key point to consider before you start testing.

Due to the complexity of these vehicles, a trouble code does not always point to a failed component - instead, it may point to a system that needs to be tested for some kind of problem. For example, if an Oxygen sensor trouble code is set, you need to consider what other systems or devices could cause this component to set a trouble code.

Component Diagrams

You will find numerous "customized" component diagrams used throughout this section. These diagrams are really a mini-schematic of how a particular device (where an input or output device) is connected to the engine or transmission control module.

Each diagram contains the power, ground, signal and control circuits that relate to a particular device. Of particular note is the use of "code check" identifiers in the diagrams that indicate circuits monitored by the module in order to set a particular trouble code(s).

Vehicle Identification

The Vehicle Emission Control Information (VECI) decal is located on the underside of the hood. *This example is from a 1998 Ford F150 4.6L V8 (VIN W) with a 4R70W A/T.*

Ford Motor Company IMPORTANT VEHICLE INFORMATION

This vehicle is equipped with electronic engine control systems. Engine idle speed, idle mixture, and ignition timing are not adjustable. Refer to the Powertrain Control Emissions Diagnosis Manual as necessary.

Oil Recommendations

Use SAE 5W-30 American Petroleum Institute Certified Oils for Gasoline Engines.

Vehicle Certification

This vehicle conforms to U.S. EPA and California regulations applicable to 1998 model year new light-duty trucks certified for sale in California, and it is OBD II Certified.

Emissions: F6AE-9C455, Catalyst, **Spark Plug Gap:** 0.52-0.56
Engine: 4.6L - WFMXT04.6AAA-2TWC(2)/2HO2S (2)/EGR/SFI

Underhood View Graphic

Electronic Engine Control System

OPERATING STRATEGIES

The Electronic Engine Control (EEC) system uses several operating strategies to maintain good overall driveability and to meet the EPA mandated emission standards. These strategies are included in the tables on this page.

Based on various inputs it receives, the PCM adjusts the fuel injector pulsewidth, the idle speed, the amount of ignition advance or retard, the ignition coil dwell and the operation of the canister purge valve.

It also regulates the air conditioning and speed control systems.

PCM Location
The PCM is located under the right hand (RH) dash panel behind the RH kick panel as shown in the Graphic.

Operating Strategies

A/C Compressor Clutch Cycling Control	Closed Loop Fuel Control
Electric Fuel Pump Control	Exhaust Gas Recirculation Control
Fuel Metering of Sequential Fuel Injectors	Fuel Pump Monitor
Idle Speed Control	Ignition Timing Control for A/F Change
Vapor Canister Purge Control	

General Operating Strategies

Adaptive Fuel Trim Strategy	Adaptive Idle Speed Control Strategy
Base Engine/Transmission Strategy	Failure Mode Effects Mgmt. Strategy

■ **NOTE:** *Hardware Limited Operating Strategy (HLOS) is a part of the PCM hardware.*

Operational Control of Components

A/C Compressor Clutch	EGR System Control
Engine Cooling System Fan	Fuel Delivery (Injector Pulsewidth)
Fuel Pump Operation	Fuel Vapor Recovery System
Idle Speed	Malfunction Indicator Lamp
Shift Points for the A/T	Speed for Integrated Speed Control
Torque Converter Clutch Lockup (TCC)	Transmission Line Pressure (EPC)

Items Provided (or Stored) By the EEC System

Data for the Operational Control Centers	Data for the Suspension Control Centers
DTC Data for the A/T (4R70W)	DTC Data for Emission Related Faults
DTC Data for Non-Emission Faults	DTC Data for the GEM (module)

Integrated Electronic Ignition System

INTRODUCTION

The engine controller (PCM) controls the operation of the Integrated Electronic Ignition (EI) system on this vehicle. Battery voltage is supplied to the right (RH) and left hand (LH) coilpacks from the ignition through fuse No. 30 (30 amp) in the J/B fuse panel.

The PCM controls the ground circuits of RH coils (1/2) and LH coils (3/4) by switching the ground path for each coil "on" and "off" at the appropriate time. The PCM adjusts the ignition timing for each cylinder as needed to meet changes in the engine operating conditions. The LH coilpack connects to cylinders 2, 4, 7 and 8 while the RH coilpack connects to cylinders 1, 3, 5 and 6. The firing order of this engine is 1-3-7-2-6-5-4-8.

System Components

The main components of this Integrated EI system are the coilpack, crankshaft position sensor, the PCM and the related wiring. Note that the camshaft position sensor is not an integral part of the EI system as it is used only as an input for sequential fuel injection.

EI System Graphic

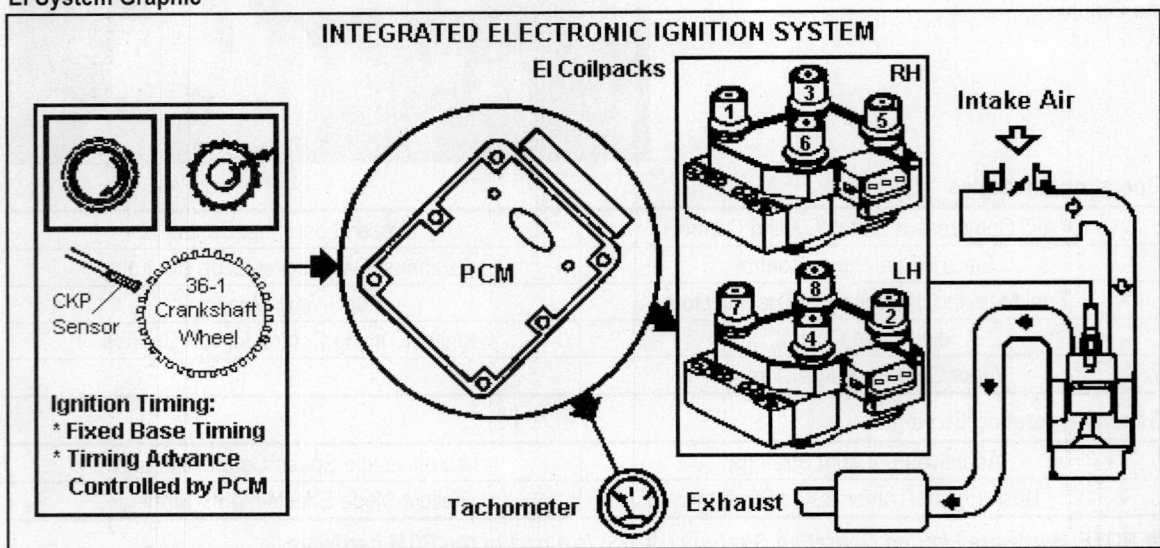

INTEGRATED ELECTRONIC IGNITION SYSTEM

Ignition Spark Timing

The ignition timing on this EI system is entirely controlled by the PCM (base timing is not adjustable). Do not attempt to check the base timing, as you will get false readings.

The PCM uses the CKP sensor signal to indicate crankshaft position and speed by sensing a missing tooth on a 36-1 pulse wheel mounted behind the harmonic balancer. It uses this signal to calculate a spark target and then fire the coils to that spark target.

After engine startup, the PCM calculates the ignition spark advance with these signals:

- BARO Sensor (the PCM calculates this signal at key on, engine off)
- ECT Sensor
- Engine Speed Signal (rpm)
- IAT Sensor
- MAF Sensor
- Throttle Position

IGNITION COIL OPERATION

The PCM controls the operation of the EI system coilpack in order to provide ignition secondary spark at the correct time in the combustion process.

The coilpack consists of two coilpacks mounted at the front of each side of the engine. Spark plug cables rout to each of the eight (8) cylinders from the coilpacks.

Each ignition coil fires two (2) spark plugs. This is a form of waste spark ignition where one spark plug fires the cylinder under compression and the other spark plug (same coil) fires the cylinder on the exhaust stroke.

The spark plug that fires the cylinder on the compression stroke uses the majority of the coil energy and the spark plug that fires the cylinder on the exhaust stroke uses very little coil energy. The spark plugs are connected in series, so one spark plug firing voltage is negative and the other spark plug is positive with respect to ground.

The PCM determines which coil to charge and fire. The coil marked RH fires cylinders 1, 3, 6 and 5. The coil marked LH fires cylinders 2, 4, 7 and 8 (firing order is 1-3-7-2-6-5-4-8).

RH Coilpack

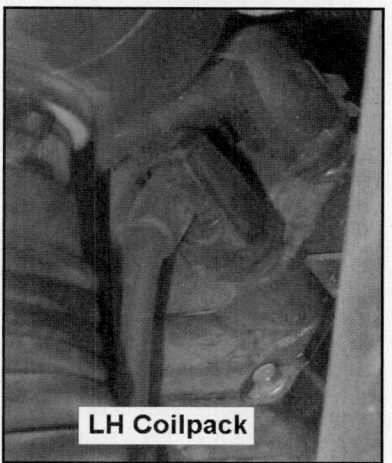
LH Coilpack

EI Coilpack Circuits
The PCM is connects to the Coil "1" & "2" control circuits at Pins 26 and 52. The PCM is connects to the Coil "3" & "4" control circuits at Pins 78 and 104 of the PCM harness. The resistance of this coil is 0.4 to 1.0 ohms.

Trouble Code Help
Note the four circuits with a dotted line in the schematic. The PCM monitors these circuits as it switches the coil drivers "on" and then "off". If the PCM does not receive a valid IDM signal due to a coil problem it will set one of these trouble codes: P0351 for CD1, P0352 for CD2, P0353 for CD3 and P0354 for CD4.

EI Coil Circuits Schematic

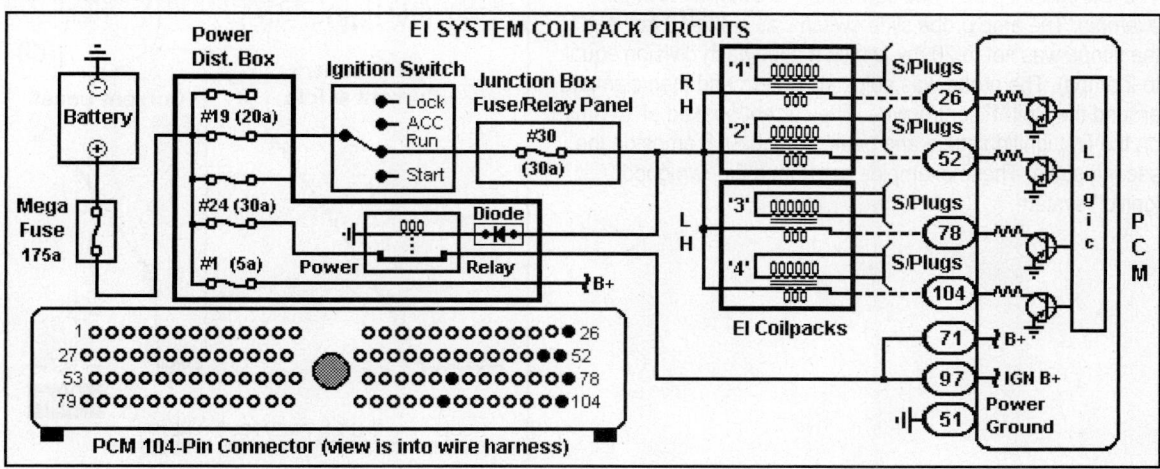

LAB SCOPE TEST (COIL PRIMARY)

The ignition coil primary circuit includes the switched ignition feed circuit, coil primary circuit and coil driver control circuit in the PCM. Coil '1' connects to Pin 26; Coil '2' connects to Pin 52; Coil '3' connects to Pin 78 and Coil '4' connects to Pin 104 at the PCM.

Dual Strike Method

The EI system on this engine is equipped with a Dual Strike Method of coil firing. The ignition coils are fired twice at idle speed and just once at engine speeds over 1000 rpm.

Coil Dwell Period

The coil dwell period is the length of time that current flows through the coil primary windings (it is measured in milliseconds and represents the degrees of engine rotation). A complete turn of the camshaft equals 360 degrees and represents one firing of all engine cylinders.

Lab Scope Connections

The Channel 'A' positive probe was connected to Coil 1 at Pin 26 (CD1 control circuit) and the negative probe to the battery negative post.

Scope Settings

To make the waveforms as clear as possible, set the scope settings to match the examples.

Lab Scope Example (1)

In example (1), the trace shows the coil primary circuit. The firing line reached 300v.

Lab Scope Example (2)

In example (2), the Min/Max setting was selected on the scope to allow an accurate view of the firing line. Note that the firing line reached 325v on the first and second strike (Dual Strike firing method).

Lab Scope Example (3)

The low current probe was connected to Channel 'A in this example. The amp probe slide switch was set to 10 mv, and the scope was set to 20 mv (which makes each division equal to 2 amps). The probe was adjusted to zero and then clamped around the Coil '1' control wire. The current peaked at 13 amps on the first ignition strike and peaked at about 8 amps on the second strike. These examples are from a "known good" ignition system.

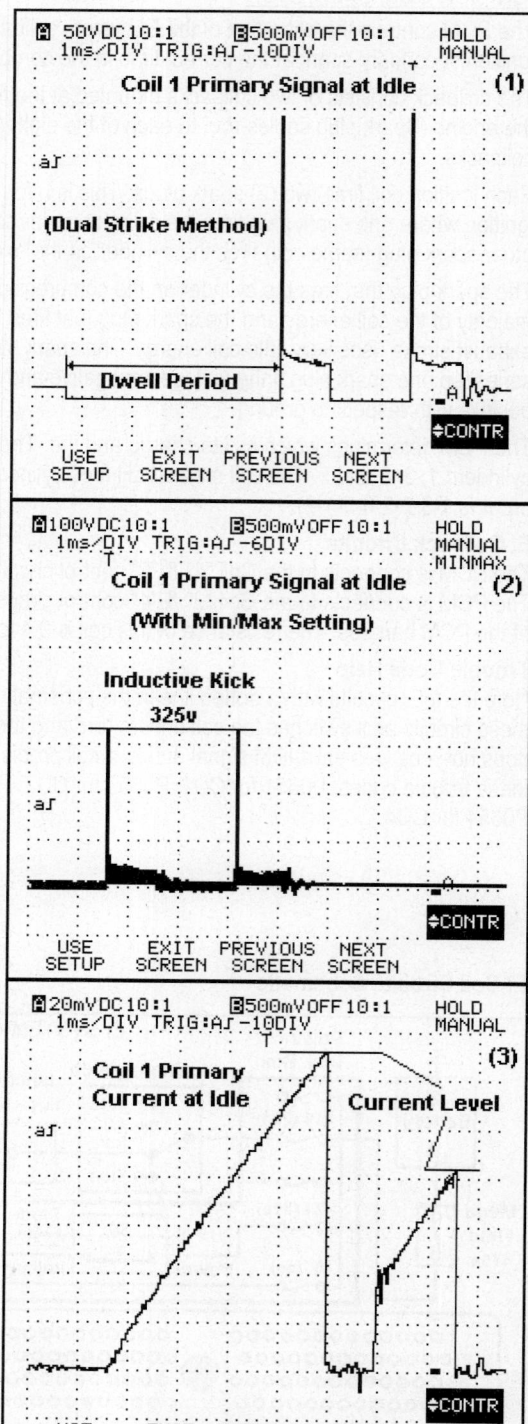

CKP SENSOR (MAGNETIC)

The crankshaft position (CKP) sensor used with this EI system is the primary sensor for crankshaft position data to the ignition module (integrated into the PCM on this system).

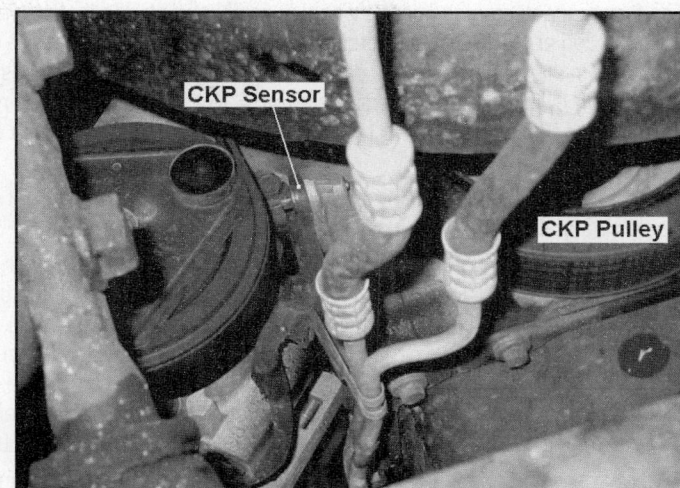

The CKP sensor is a magnetic transducer mounted at the front of the timing chain cover adjacent to a 36-1 tooth trigger wheel mounted on the crankshaft behind the harmonic balancer.

The trigger wheel has 35 teeth spaced 10 degrees (10º) apart with one empty space due to a missing tooth (hence the term 36-1 assigned to this wheel).

The CKP sensor outputs an AC signal to the integrated ICM (integrated into the PCM). The CKP signal directly proportional to crankshaft speed and position.

The PCM monitors the CKP sensor signal (the 36-1 design signal) in order to correctly identify piston travel so that it can synchronize the Ignition system firing, and provide an accurate method of tracking the angular position of the crankshaft relative to a fixed reference (the missing tooth of the CKP sensor).

CKP Sensor Circuits

The PCM is connected to the CKP sensor (+) circuit at Pin 21 (GRY wire) and to the CKP sensor (-) circuit (DK BLU wire) at Pin 22 of the PCM wiring harness. The sensor resistance is 300-800 ohms.

CKP Sensor Diagnosis (No Code Fault)

If the PCM does not detect any signals from the CKP sensor with the engine cranking, it will not start. This sensor input is not monitored for a loss of signals. A DVOM can be used to check for "no output", but a Lab Scope is the tool of choice to test this sensor.

EI System CKP Sensor Schematic

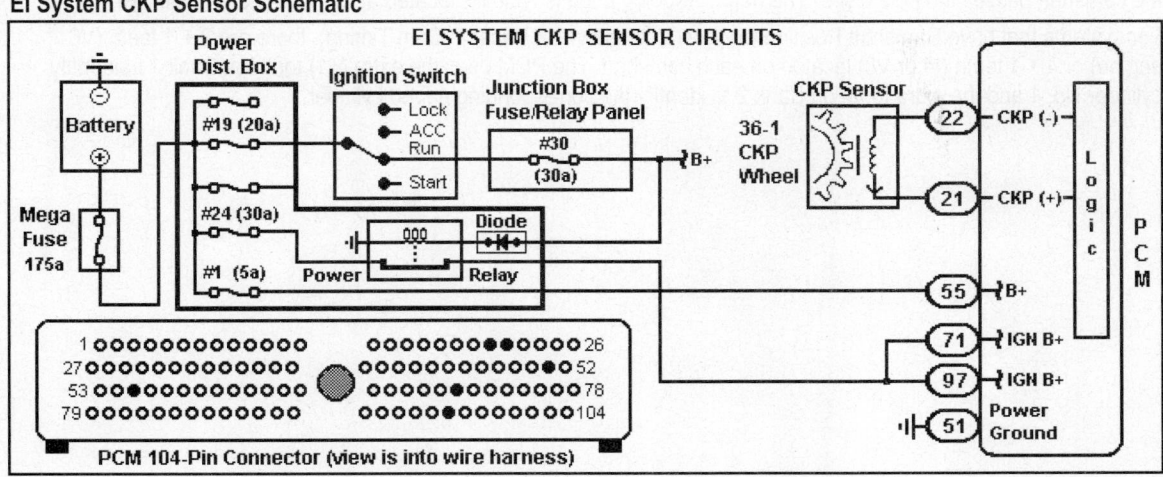

CKP SENSOR (SIGNAL EXPLANATION)

The CKP sensor (also referred to as a variable reluctance sensor or VRS) signal is derived by sensing the passage of teeth from a 36 minus one (36-1) tooth wheel mounted on the crankshaft.

The ferromagnetic-toothed timing wheel is rigidly mounted to the crankshaft and designed to have the center of each tooth spaced at 10-degree increments. Since there are 360 degrees in one revolution, there should be 36 teeth. However, in this system, there is one tooth missing. The CKP sensor is located directly opposite the missing tooth at a point where cylinder No. 1 is located at 90° BTDC for a 4-cylinder engine, 60° BTDC for a 6-cylinder engine and 50° BTDC for an 8-cylinder engine. It should be noted that on a 40-1 tooth wheel is used on engines with a 10-cylinder engine.

In addition to crankshaft-mounted 36-1 tooth wheel located at the front of the engine, some applications utilize holes in the flex plate at the rear of the engine to trigger the VRS sensor.

A VRS sensor is a magnetic transducer with a pole piece wrapped with fine wire. When the VRS encounters the rotating ferromagnetic teeth, the varying reluctance induces a voltage proportional to the rate of change in magnetic flux and the number of coil windings. As the tooth approaches the pole piece, a positive differential voltage is induced. As the tooth and the pole piece align, the signal rapidly transitions form maximum positive to maximum negative. As the tooth moves away, the voltage returns to a zero differential voltage. The negative zero crossing of the VRS signal corresponds to the alignment of the center of each tooth with the center of each pole piece. The ICM input circuitry triggers on this negative zero crossing at approximately -0.3 volts and uses this to establish crankshaft position.

At normal engine speeds, the VRS signal waveform approximates a sine wave with increased amplitude adjacent to the mission tooth region. At 30 rpm, the VRS output should be 150 mv peak-to-peak, at 300 rpm, the VRS output should be 1.5 volts peak-to-peak and at 6000 rpm, the VRS output should be 24.0 volts peak-to-peak. At 8000 rpm, the VRS output should not exceed 300 volts peak-to-peak. The normal sensor air gap is 1.0 mm nominal; 2.0 mm max. The resistance of the sensor should be 290 to 790 ohms. The inductance should be 170 to 930 Millihenries.

CMP SENSOR (SIGNAL EXPLANATION)

The CMP sensor input is not needed for the EDIS system since the spark plugs are fired in pairs. The CMP sensor is, however, used by the PCM to determine cylinder No. 1 for sequential fuel control. This vehicle utilizes a variable reluctance design sensor.

On models with a VRS design Camshaft sensor, the VRS produces a very narrow sinusoidal signal when a tooth on the camshaft passes the pole piece. The negative-going edge is typically located at 22° ATDC of cylinder No. 1.

For systems that have Camshaft Position Control (also known as Variable Cam Timing), there are 3 + 1 teeth (V6 engine) or 4 + 1 teeth (I4 or V8) located on each camshaft. The PCM uses the extra (+1) tooth on Bank 1 to identify cylinder No. 1 and the extra tooth on Bank 2 to identify the corresponding paired cylinder.

DVOM TEST (CKP SENSOR)

Place the shift selector in Park and block the drive wheels for safety.

DVOM Connections

Turn the key off and connect the breakout box. Select AC volts on the DVOM. Connect the DVOM positive lead to the CKP sensor (+) circuit (GRY wire) at Pin 21 and the negative lead the CKP sensor (-) circuit (DK BLU wire).

Crank the engine for 3-5 seconds or start the engine and record the readings.

Test Results

The DVOM display shows a CKP signal of 1.6v AC at a rate of 514 Hz. Note that the true amplitude of the AC voltage reading is about 3.3v in this example.

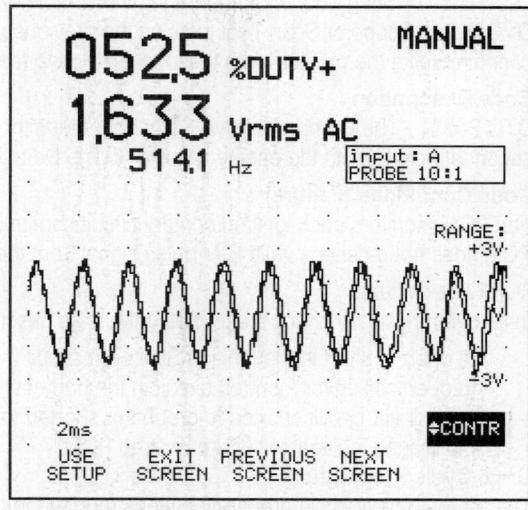

LAB SCOPE TEST (CKP SENSOR)

The Lab Scope can be used to test the CKP sensor as it provides a very accurate view of sensor waveform and of any glitches. Place the shift selector in Park (A/T) for safety.

Scope Connections

Install a breakout box if available. Connect the Channel 'B' positive probe to the CKP sensor positive circuit at PCM Pin 21 (GRY wire) and the negative probe to the CKP sensor negative circuit at PCM Pin 22 (DK BLU wire). Connect the Channel 'A' positive probe to the CMP sensor signal at PCM Pin 85 (DG wire).

Scope Settings

To make the waveforms as clear as possible, set the scope settings to match the examples.

Lab Scope Tests

The CMP and CKP sensor signals can be checked during cranking, idle and at off-idle speeds with the engine cold or fully warm.

Lab Scope Example

In this example, the top trace shows the CKP sensor signal. The bottom trace shows the CMP sensor signal. These waveforms were captured with the gear selector in Park, the engine running at idle speed in closed loop.

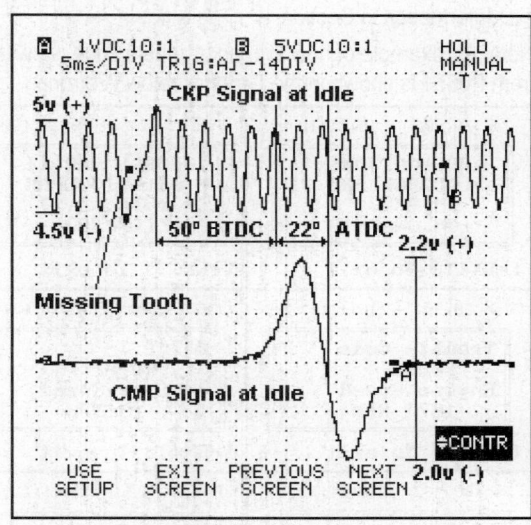

The top trace represents around 200 degrees of crankshaft revolution from the CKP sensor. This is a 9.5v AC peak-to-peak signal (+5v and -4.5v peak to peak) that is occurring at 410 hertz. Note the point at which the missing tooth occurs (50° BTDC) and where the CMP signal occurs (22° ATDC).

The bottom trace represents 100° of camshaft rotation from the CMP sensor signal. Note the voltage level of the CMP sensor signal (4.2v AC peak-to-peak).

CHILTON DIAGNOSTIC SYSTEM - DTC P0351

The purpose of the Chilton Diagnostic System is to provide you with one or more tests that can be done with a DVOM, Lab Scope or Scan Tool *prior* to entering the complete trouble code repair procedure. The quick checks listed in the DVOM, Lab Scope or Scan Tool tests on the previous pages may help you quickly find the cause of the problem. If you cannot resolve the problem with these tests, a code repair chart is also included.

Code Description

DTC P0351 - The Comprehensive Component Monitor (CCM) detected a fault in the Coil 1 (or 'A') primary or secondary circuit with the engine running (2-Trip Detection).

Code Conditions (Failure)

The PCM monitors each ignition primary circuit continuously whenever the engine is running. The test fails when the PCM does not receive a valid IDM pulse signal from the ignition module (integrated in PCM).

Quick Check Items

Inspect or physically check the following items as described below:

- An open or short in the Ignition Start/Run circuit
- An open coil driver (control) circuit in the harness
- A coil driver circuit shorted to ground or shorted to power
- A damaged ignition coil or a damaged PCM

Drive Cycle Preparation

There are no drive cycle preparation steps required to run the Component Monitor.

Install the Scan Tool. Turn to key on, engine off. Cycle the key off, than back to on. Select the correct vehicle and engine qualifier. Clear the codes and do a PCM reset.

Scan Tool Help

A Scan Tool can be used to view the trouble code ID and Freeze Frame Data (PIDs) for DTC P0351. It can also be used to clear stored trouble codes and Freeze Frame Data. Refer to the article titled "How to Access and Use Generic PID Information in Section 1.

Note: An example of how to read a trouble code, view its Freeze Frame Data and then clear the Diagnostic information from P0351 is shown in the Graphic below (Vetronix).

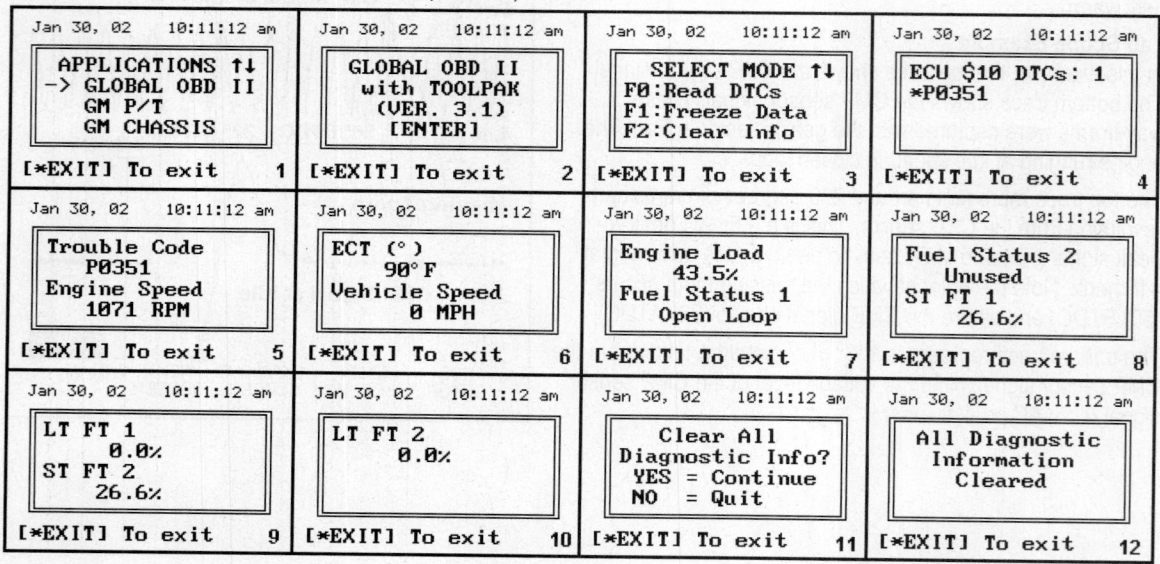

CHILTON DIAGNOSTIC SYSTEM - DTC P0351

If DTC P0351 is set, Pinpoint Test JE is used to diagnose these devices and circuits:

- The Ignition Start/Run circuit for a short or open condition
- The Coil 1 driver circuit for an open, short to power or short to ground condition
- A damaged ignition coil (Coil 1) or a damaged PCM

■ NOTE: *This code repair chart that follows does not contain all of the possible repair steps.*

Coil Identification Tables

Cylinder Number	Coil Driver Number	PCM Pin Number	Related DTC
4-Cylinder Applications (2.5L Ranger)			
Cyl 1	CD1, CD3	Pin 26, Pin 78	P0351, P0353
Cyl 3	CD2, CD4	Pin 52, Pin 104	P0352, P0354
Cyl 4	CD1, CD3	Pin 26, Pin 78	P0351, P0353
Cyl 2	CD2, CD4	Pin 52, Pin 104	P0352, P0354
4-Cylinder Applications (Except 2.5L Ranger)			
Cyl 1	CD1	Pin 26	P0351
Cyl 3	CD2	Pin 52	P0352
Cyl 4	CD1	Pin 26	P0351
Cyl 2	CD2	Pin 52	P0352
6-Cylinder Applications			
Cyl 1	CD1	Pin 26	P0351
Cyl 4	CD2	Pin 52	P0352
Cyl 2	CD3	Pin 78	P0353
Cyl 5	CD1	Pin 26	P0351
Cyl 3	CD2	Pin 52	P0352
Cyl 6	CD3	Pin 78	P0353
8-Cylinder Applications			
Cyl 1	CD1	Pin 26	P0351
Cyl 3	CD2	Pin 52	P0352
Cyl 7	CD3	Pin 78	P0353
Cyl 2	CD4	Pin 104	P0354
Cyl 6	CD1	Pin 26	P0351
Cyl 5	CD2	Pin 52	P0352
Cyl 4	CD3	Pin 78	P0353
Cyl 8	CD4	Pin 104	P0354

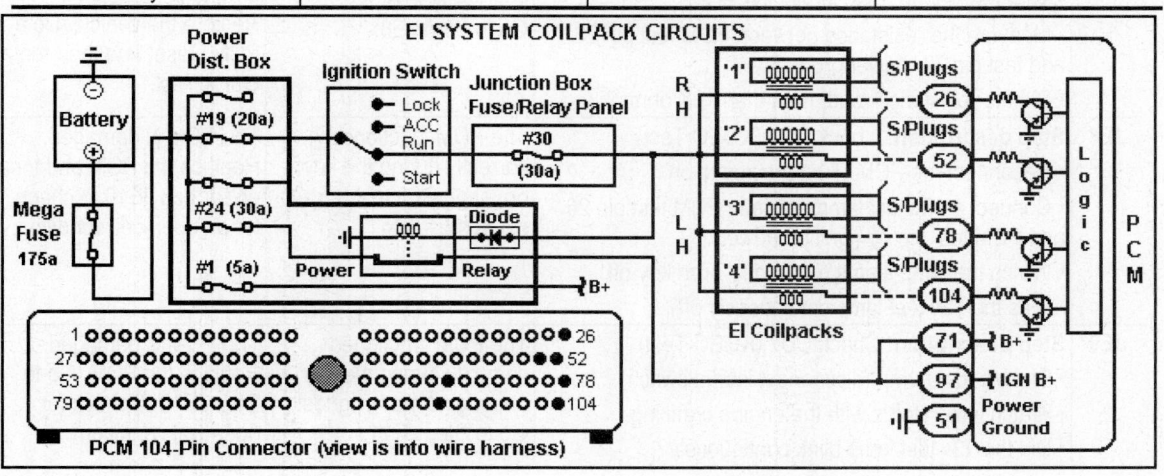

Pinpoint Test JE1 Repair Table

Step	Action	Yes	No
JE1	**Step description:** Which coil is not firing? • Determine the coil that is not firing using the table on the previous page. • Is the vehicle a 2.5L Ranger or does the vehicle have a V8 engine?	For engines with a damaged CD1 or CD2 circuit, go to Step JE2. For engines with a damaged CD3 or CD4 circuit, go to Step JE3.	For 4-cylinder and 6-cylinder engines, go to Step JE4.
JE2	**Step description:** Test for power to coilpack • Disconnect the Coil 1 or 2. Turn the key on. • Measure the voltage between the Start / Run circuit at the harness connector and ground. • Did the voltage read greater than 10.0v?	The Start/Run circuit is okay. For DTC P0351, go to Step JE5. For DTC P0352, go to Step JE15	A fault exists in Start / Run circuit. Look for an open fuse. If the fuses are okay, look for an open circuit. Make the repair. Do Quick Test.
JE3	**Step description:** Test for power to coilpack • Disconnect CD1, CD2, CD3. Turn key on. • Measure the voltage between the Start / Run circuit at the harness connector and ground. • Did the voltage read greater than 10.0v?	The Start/Run circuit is okay. For DTC P0353, go to Step JE5. For DTC P0354, go to Step JE35	A fault exists in Start / Run circuit. Look for an open fuse. If the fuses are okay, look for an open circuit. Make the repair. Do Quick Test.
JE4	**Step description:** Test for power to coilpack • Disconnect the Coil 1 or 2. Turn the key on. • Measure the voltage between the Start / Run circuit at the harness connector and ground. • Did the voltage read greater than 10.0v?	The Start/Run circuit is okay. For DTC P0351, go to Step JE5. For DTC P0352, go to Step JE15. For DTP P0353, go to Step JE25	A fault exists in Start / Run circuit. Look for an open fuse. If the fuses are okay, look for an open circuit. Make the repair. Do Quick Test.
JE5	**Step description:** Check CD1 (open harness) • Install a breakout box (if available). • Measure the resistance of the CD1 circuit between Pin 26 and the harness connector. • Did the resistance read less than 5.0 ohms?	The CD1 circuit is not open. Go to Step JE6.	Make repairs to the open circuit in CD1. Restore the vehicle. Do a PCM Reset and then run a Quick Test.
JE6	**Step description:** Check CD1 (short to power) • Measure the voltage between the PCM test pin 26 and chassis ground with the key turned on. • Did the voltage read less than 1.0 volt?	The CD1 circuit is not shorted to power. Go to Step JE7.	Make repairs to the short circuit to power. Restore the vehicle. Do a PCM Reset and then run a Quick Test.
JE7	**Step description:** Check CD1 (short to ground) • Turn the key off. Disconnect the PCM. • Measure the resistance between PCM test pin 26 and test pin 77 (power ground). • Did the resistance read more than 10K ohms?	The CD1 circuit is not shorted to ground. Go to Step JE8.	Make repairs to short to ground in harness. Restore the vehicle. Do a PCM Reset and then run a Quick Test.
JE8	**Step description:** Check CD1 (KOEO Test) • Reconnect the PCM. Turn the key to on. • Connect a 12v test lamp between PCM test pin 26 and PCM test pin 71 (ignition power). • Watch the lamp status (on or off). Turn key off. • Was the 12v test lamp continuously off?	The PCM functioned correctly during the key on, engine off test. Go to Step JE9.	The PCM is damaged. Replace the PCM and then go to Step JE10 to check for an open coil condition.
JE9	**Step description:** Check CD1 (KOEC Test) • Disable fuel pump (disconnect inertia switch). • Watch lamp status with the engine cranking. • Did the 12v test lamp blink continuously?	The PCM functioned correctly during the key on, engine cranking test. Go to Step JE11.	The PCM is damaged. Replace the PCM. Go to Step JE10 to check for open coil condition.

Pinpoint Test JE1 Repair Table

Step	Action	Yes	No
JE10	**Step description:** Confirm Coil is not Open • Measure Coil 1 primary resistance between CD1 and the Ignition Start / Run circuit at the coilpack harness connector (coil connector). • Did the resistance read less than 5 ohms.	Coil is not damaged. Restore the vehicle. Do the Complete Misfire Monitor Repair Verification Drive Cycle and rerun Quick Test.	The Coil 1 is damaged. Replace the coilpack. Restore the vehicle. Do a PCM Reset and Misfire Monitor Repair Verification Drive Cycle and rerun test.
JE11	**Step description:** Confirm Coil 1 Primary • Measure Coil 1 primary resistance between CD1 and the Ignition Start / Run circuit at the coilpack harness connector (at the coil pins). • Did the resistance read less than 5 ohms (actual was 0.4 to 1.0 ohms)?	The coil primary is not damaged. Go to Step JE12.	The Coil 1 is damaged. Replace the coilpack. Restore the vehicle. Do a PCM Reset and Misfire Monitor Repair Verification Drive Cycle and rerun Quick Test.
JE12	**Step description:** Confirm Coil 1 Secondary • Remove both plug wires from the Coil 1 (CD1) secondary circuits at the coilpack. • Measure the resistance of Coil 1 secondary between the coilpack secondary towers. • Did resistance read from 12K to 14.5K ohms?	The coil secondary is not damaged. If DTC P0350 was set, go to Step JE15. If no other codes were set, go to Z1 in Section 1.	The Coil 1 is damaged. Replace the coilpack. Restore the vehicle. Do a PCM Reset and Misfire Monitor Repair Verification Drive Cycle and rerun Quick Test.
JE15	**Step description:** Check CD2 (open harness) • Install a breakout box (if available). • Measure the resistance of the CD2 circuit between Pin 52 and the harness connector. • Did the resistance read less than 5.0 ohms?	The CD2 circuit is not open. Go to Step JE16.	Make repairs to the open circuit in CD2. Restore the vehicle. Do a PCM Reset and then run a Quick Test.
JE16	**Step description:** Check CD2 (short to power) • Measure the voltage between the PCM test pin 52 and chassis ground with the key turned on. • Did the voltage read less than 1.0 volt?	The CD2 circuit is not shorted to power. Go to Step JE17.	Make repairs to the short circuit to power. Do a PCM Reset and then run a Quick Test.
JE17	**Step description:** Check CD2 (short to ground) • Turn the key off. Disconnect the PCM. • Measure the resistance between PCM test pin 26 and test pin 77 (chassis ground). • Did the resistance read more than 10K ohms?	The CD2 circuit is not shorted to ground. Go to Step JE18.	Make repairs to short to ground in harness. Restore the vehicle. Do a PCM Reset and then run a Quick Test.
JE18	**Step description:** Check CD2 (KOEO Test) • Reconnect the PCM. Turn the key to on. • Connect a 12v test lamp between PCM test pin 52 and PCM test pin 71 (ignition power). • Watch the lamp status (on or off). Turn key off. • Was the 12v test lamp continuously off?	The PCM functioned correctly during the key on, engine off test. Go to Step JE19.	The PCM is damaged. Replace the PCM and then go to Step JE20 to check for an open coil condition.
JE19	**Step description:** Check CD2 (KOEC Test) • Disable fuel pump (disconnect inertia switch). • Watch lamp status with the engine cranking. • Did the 12v test lamp blink continuously?	The PCM functioned correctly during the key on, engine cranking test. Go to Step JE21.	The PCM is damaged. Replace the PCM. Go to Step JE20 to check for open coil condition.
JE20	**Step description:** Confirm Coil is not Open • Measure Coil 2 resistance between CD1 and Start / Run circuit at the coilpack connector. • Did the resistance read less than 5 ohms.	Coil 2 is not damaged. Do Complete Misfire Monitor Repair Drive Cycle and Quick Test.	The Coil 2 is damaged. Replace the coilpack. Do a PCM Reset and rerun the Quick Test.

Evaporative Emission System

VAPOR MANAGEMENT VALVE SYSTEM

This vehicle application uses a Vapor Management Flow (VMV) design of Evaporative Emission (EVAP) system to prevent fuel vapor buildup in the sealed fuel tank.

Any fuel vapors trapped in the sealed fuel tank are vented through a vapor valve assembly at the top of the tank through a single vapor line to the charcoal canister for storage. Fuel vapors are purged into the intake manifold where they are burned during normal engine operation.

EVAP System Graphic

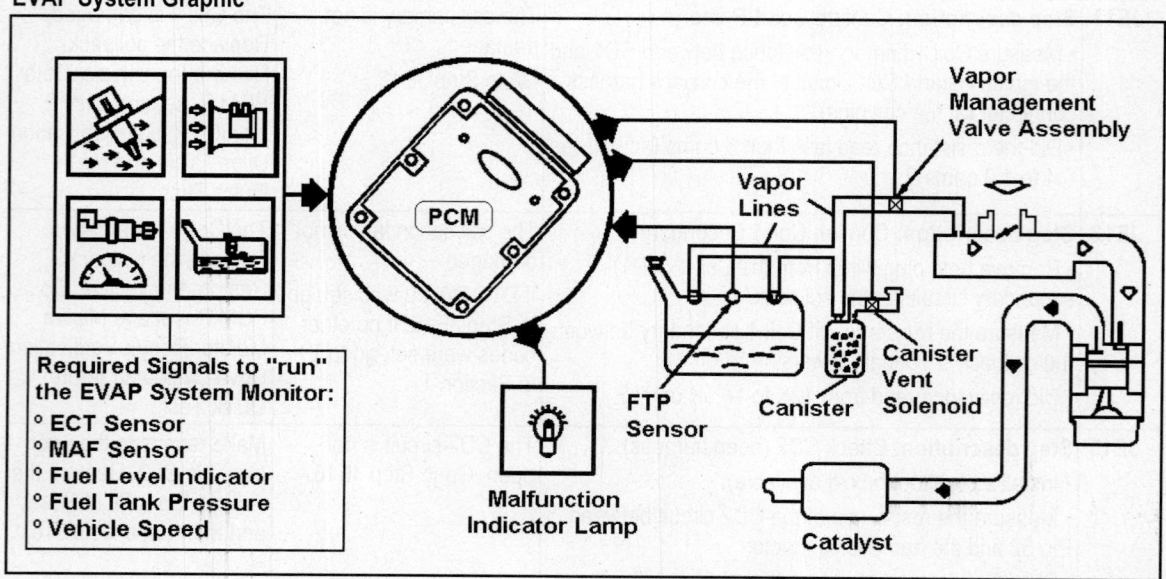

Fuel Tank Pressure Sensor

On this vehicle application, a fuel tank pressure (FTP) sensor is used to monitor the pressure or vacuum value in the fuel tank during the Running Loss Monitor Test portion of the EVAP Purge Test (part of the OBD II Diagnostics). The FTP sensor is built into the fuel tank assembly, but is removable.

The EVAP Monitor performs a Fuel System integrity check to determine if a leak is present in the EVAP system.

The FTP sensor converts variations in pressure into an analog voltage signal that is sent to the PCM.

Trouble Code Help

If the PCM detects an open or short circuit condition in the FTP sensor or its circuit during the OBD II Test (Comprehensive Component Test), it will set DTC P0452. If the FTP sensor fails on the next consecutive trip, the code will mature and the PCM will turn on the MIL (this is two-trip fault detection).

VMV SYSTEM COMPONENTS

The Vapor Management Flow system includes the following components:

- Crankcase ventilation tube
- Fuel tank and fuel filler cap
- Fuel vapor storage canister
- Fuel vapor control valve
- Related fuel vapor hoses
- PCM and related connecting wires
- Vapor management valve (VMV)

VMV Solenoid Location

The VMV solenoid is located on driver's side of the cowl as shown in the Graphic.

EVAP Vapor Management Valve

The EVAP vapor management valve (VMV) is located in-line with the EVAP canister. The function of this valve is to control the flow of vapors out of the canister, and to close off the flow of fuel vapors from the canister when the engine is not running.

EVAP Purge Operation

Once the engine is started, the PCM controls the operation of the valve (i.e., the PCM opens and closes the valve to control the flow of fuel vapors from the charcoal canister during normal engine operation). The VMV is a normally closed (N.C.) solenoid. The resistance of the solenoid is 30-36 ohms at 68°F.

As discussed, the VMV assembly is used to control the flow of vapors (during normal operation) from the fuel vapor storage canister into the intake manifold.

The actual conditions under which the VMV solenoid is opened are determined by the PCM from various sensor inputs.

Trouble Code Help

The first time that the PCM detects a problem in the VMV solenoid operation or one of its circuits, it will set a pending trouble code (DTC P0443).

If the same fault is detected on two consecutive trips, the trouble code matures and the PCM will turn on the MIL to indicate that an emission related fault has been detected.

The repair steps for this trouble code (DTC P0443) include a check of the VMV solenoid and its related circuits.

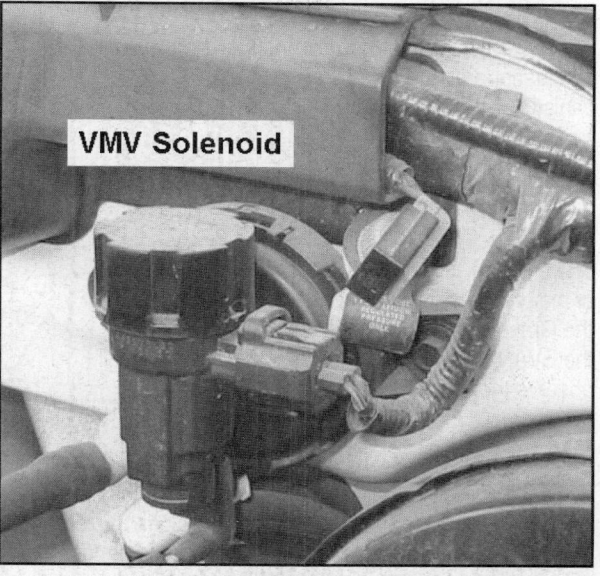

VMV Solenoid

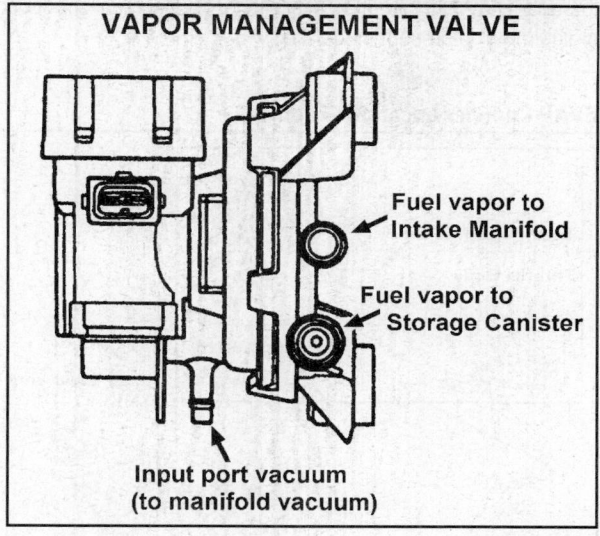

VAPOR MANAGEMENT VALVE

Fuel vapor to Intake Manifold

Fuel vapor to Storage Canister

Input port vacuum (to manifold vacuum)

EVAPORATIVE CHARCOAL CANISTER

The EVAP canister assembly contains 2.8L of activated carbon in the carbon bed. The activated charcoal has the ability to purge or release any stored fuel vapors when it is exposed to fresh air.

Fuel vapors from the fuel tank and the air cleaner are stored in the charcoal canister.

Once the engine is started, fuel vapors stored in charcoal canister are purged into the engine where they are burned during normal combustion.

The charcoal canister is located in front of the rear axle assembly as shown in the Graphic at the bottom of this page. If the EVAP canister has to be removed, observe the *Caution* note below.

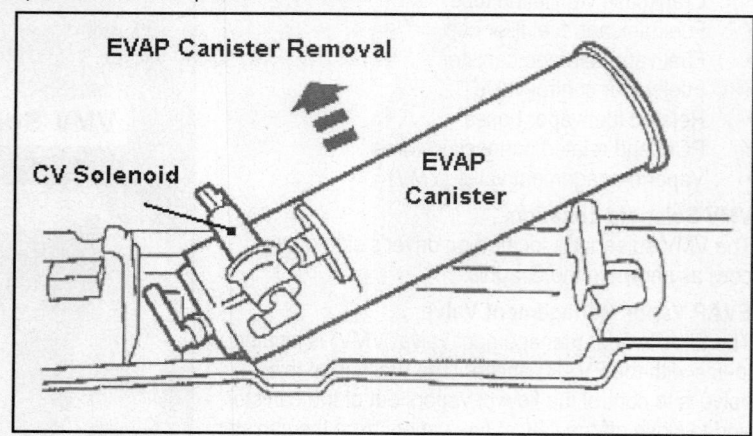

✳✳ **CAUTION:** The Evaporative Emissions System contains fuel vapor and condensed fuel vapor. Although not present in large quantities, it still represents the danger of explosion of fire. Disconnect the battery ground cable from the battery to minimize the possibility of an electrical spark occurring, possibly causing a fire or explosion if fuel vapor or liquid fuel is present in the area.

EVAP Canister Location Graphic

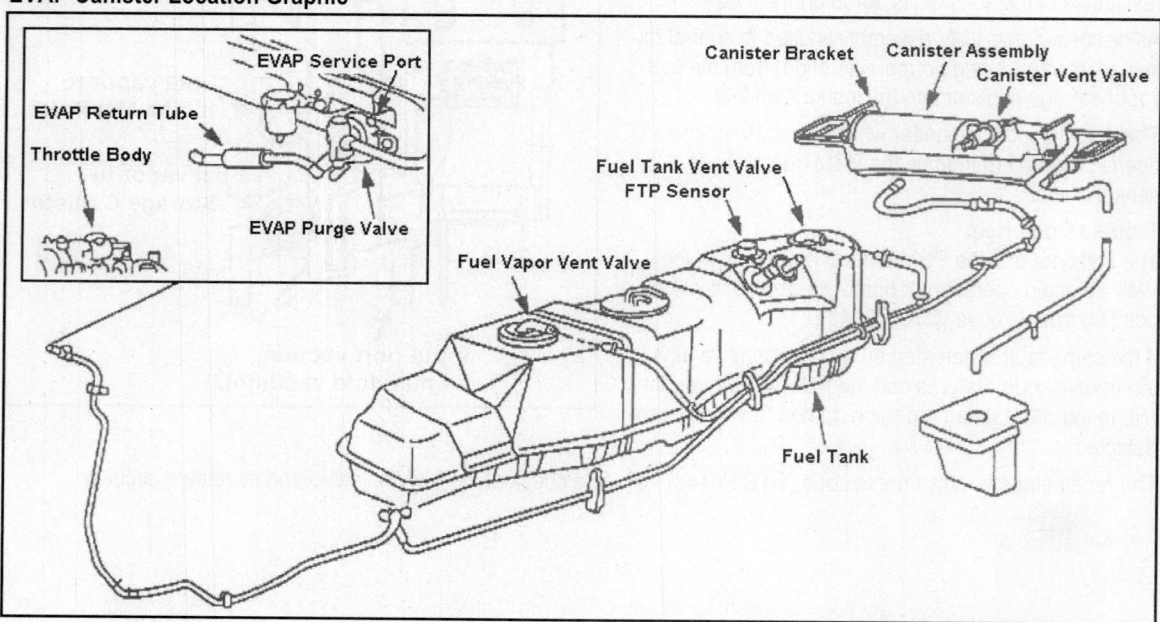

CANISTER VENT SOLENOID

The canister vent (CV) solenoid is a normally open (N.O.) solenoid with a resistance of 40-50 ohms at 68ºF. The CV solenoid is used to seal the atmosphere port of the charcoal canister from atmospheric pressure.

During the OBD II Diagnostics, the PCM activates the CV solenoid, in conjunction with the VMV solenoid, in order to draw fuel tank vacuum down to a specified level for the leak test.

Trouble Code Help

If the PCM detects a problem in the EVAP Vent Control system, it will set a pending DTC P0446. If the fault is detected on the next trip, the PCM will turn on the MIL (2-trip fault detection)

The repair steps for this trouble code include a check of the FTP sensor (VREF) circuit, CV solenoid and circuits.

Canister Vent Solenoid

Canister Vent Solenoid Location

The CV solenoid is located at the rear of the vehicle on the EVAP canister.

FUEL TANK FILLER CAP

The fuel filler cap is used to prevent fuel spill and close the EVAP and Fuel systems to atmosphere.

The fuel tank cap is designed to relieve pressure above 2 psi (14 kPa) and to relieve vacuum 0.53 psi (3.8 kPa).

The fuel filler tank cap also serves as a connection point for the EVAP Leak Test kit.

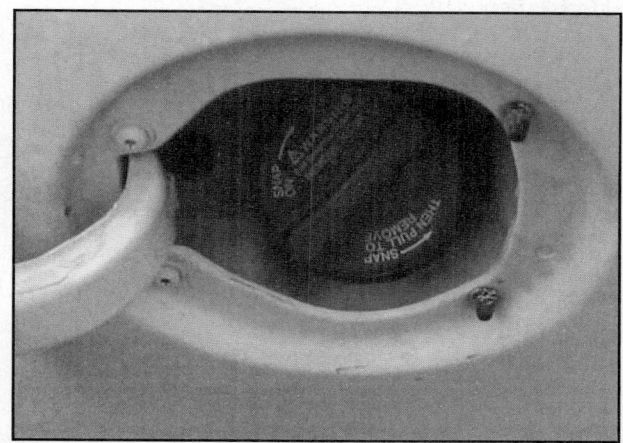

Fuel Filler Cap Graphic

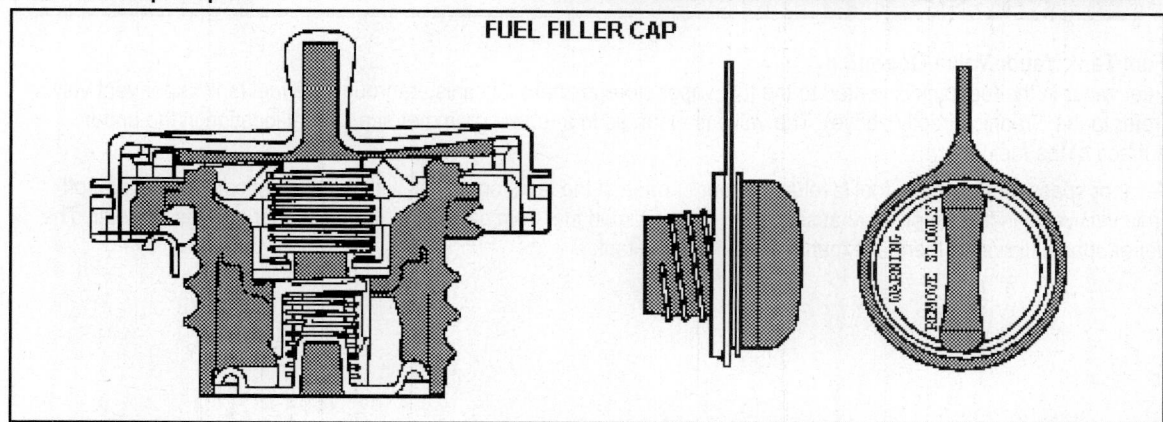

FUEL FILLER CAP

FUEL TANK VAPOR VALVE

The Fuel Tank Vapor valve (or liquid separator valve) is located in series between the fuel tank and the fuel vapor storage canister. It is mounted on top of the fuel tank as shown in the Graphic to the right.

This valve is closes the path of flow from the fuel vapor valve to the fuel vapor storage canister during refueling to prevent overfilling the fuel tank.

The fuel tank vapor valve detects fuel filler cap removal by sensing a change in pressure (through the fuel filler pipe sensing tube) that occurs during the removal of the fuel filler cap.

In effect, the valve closes when the cap is removed and opens when the fuel filler tank cap is installed.

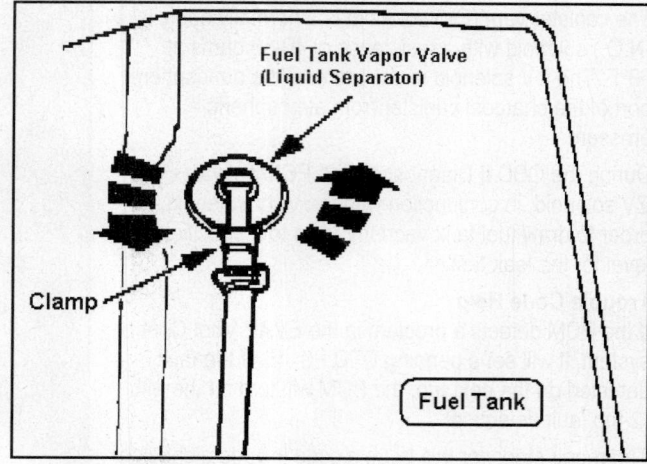

EVAP Component Location Graphic

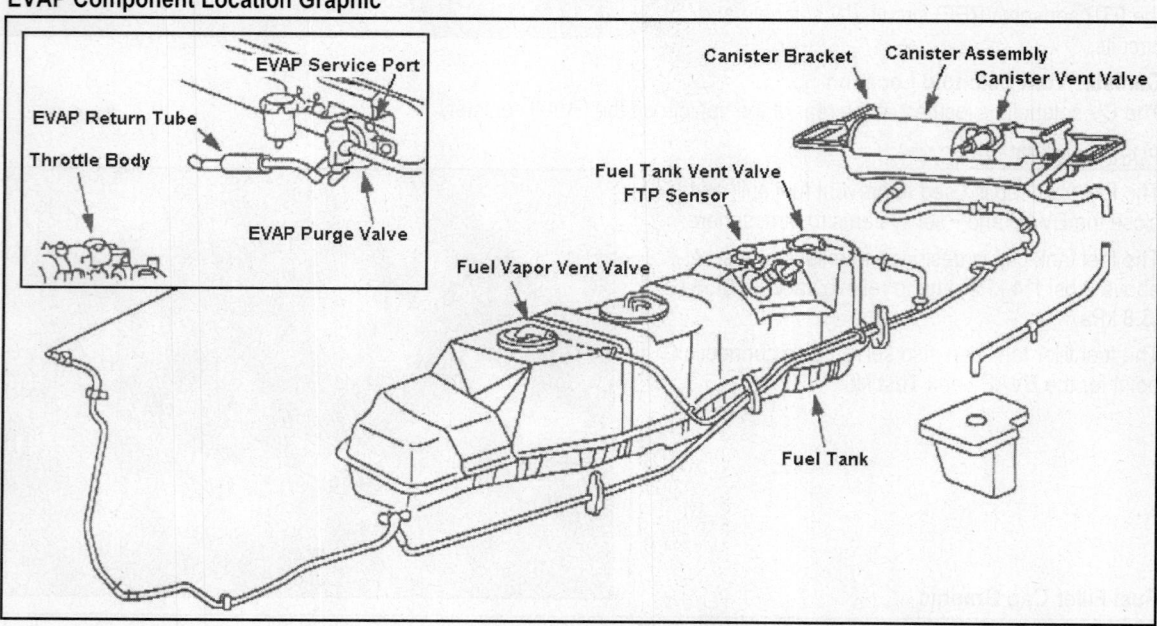

Fuel Tank Vapor Valve Operation

Fuel vapor in the fuel tank is vented to the fuel vapor storage charcoal canister through the fuel tank vapor vent valve (refer to the Graphic directly above). The valve is mounted in a rubber grommet at a central location in the upper surface of the fuel tank.

A vapor space, between the fuel level and upper surface of the tank, combines with a small orifice and float shutoff (rollover) valve in the vapor valve assembly to prevent liquid fuel from passing to the fuel vapor storage canister. The vapor space allows for thermal expansion of fuel in the tank.

VMV SYSTEM OPERATION

The operation of the VMV system used on this vehicle application is discussed next.

1) The PCM receives signals from the CKP, ECT, IAT, MAF, TP and VSS to determine when to operate the VMV system (ECT and IAT inputs must indicate a warm engine, at stable idle, running in closed loop at moderate load and open or part throttle position with the vehicle moving at an elevated steady speed for a period of time).

2) The PCM deactivates fuel vapor management flow during idle periods, or whenever a fault is detected in the VMV valve, or any fuel vapor management flow inputs.

3) The PCM monitors the effects of Canister Purge Compensation strategy on the IAC strategy and compares the effects against an expected result. The PCM calculates the difference between an idle speed control value with the purge flow duty cycle high and an idle speed control value with the purge flow duty cycle low.

4) The PCM opens the valve to divert vacuum or closes the valve to vent vacuum to the atmosphere from the solenoid. The command to the valve is a varying duty cycle.

5) The vacuum acting on the VMV diaphragm overcomes the valve spring and begins to lift the VMV pintle off its seat. This action causes fuel vapor to flow into the intake manifold. This system is designed to compensate for changes in intake manifold vacuum that are applied to the solenoid part of the VMV. It maintains a constant vapor flow at any given duty cycle with manifold vacuums from 5-20 In-Hg.

VMV System Example Graphic

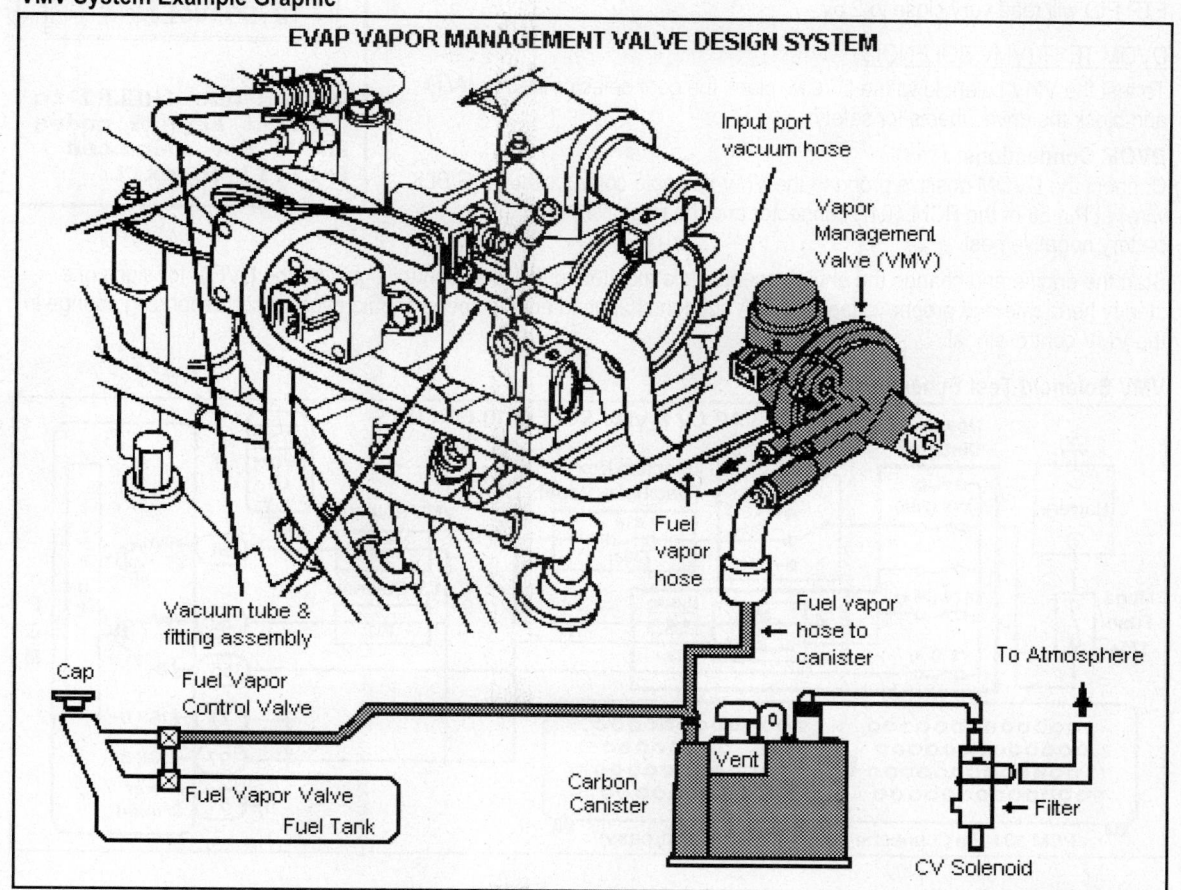

SCAN TOOL TEST (PURGE SOLENOID)

Connect the Scan Tool to the DLC underdash connector. Navigate through the Scan Tool menus until you get to the Ford (OEM) Data List (a list of PID items). Select the EVAP PURGE DC from the menu items along with the voltage signal from the Upstream HO2S for Cylinder Bank 1.

EVAP PURGE DC PID Example

Start with the engine cold or warm and then monitor the EVAP PURGE PID. The PID (%) can read anywhere from 0-100% DC (it will read 100% after a cold engine startup).

SCAN TOOL TEST (FTP SENSOR)

Connect the Scan Tool to the DLC underdash connector. Navigate through the Scan Tool menus until you get to the Ford (OEM) Data List (a list of PID items). Select the Fuel Tank Pressure sensor from the menu items (in this case, both the sensor (P) and (V) were selected for the capture).

FTP PID Example

The FTP Sensor PID is shown in Example (2). If the fuel cap is removed, the FTP PID will read very close to 2.5v.

DVOM TEST (VMV SOLENOID)

To test the VMV solenoid with a DVOM, place the gear selector in Park (A/T) and block the drive wheels for safety.

DVOM Connections

Connect the DVOM positive probe to the VMV solenoid control circuit (LG/BLK wire) at Pin 56 of the PCM 104P connector and the negative probe to the battery negative post.

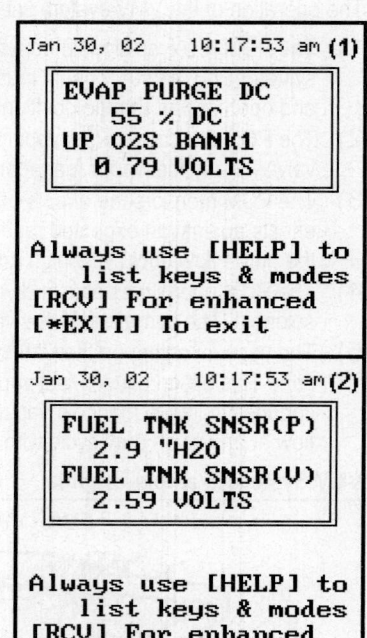

Start the engine and change the engine speed while monitoring the VMV control signal on the DVOM for signs of a steady hertz rate and proper voltage level. A sudden change in engine speed should cause a corresponding change in the VMV control signal.

VMV Solenoid Test Schematic

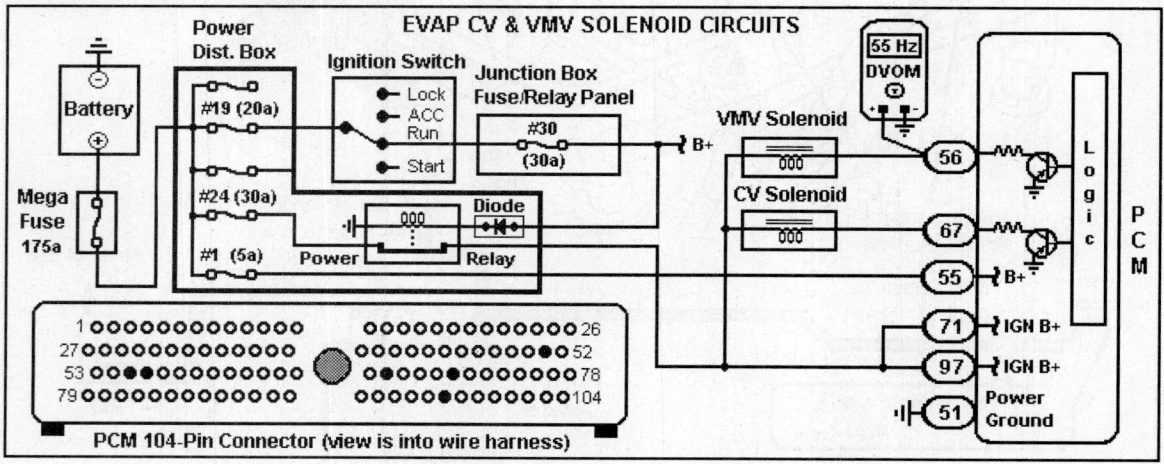

LAB SCOPE TEST (VMV SOLENOID)

The Lab Scope is the tool of choice to test the VMV solenoid as it provides an accurate view of the solenoid response and frequency rate.

Place the gearshift selector in Park and block the drive wheels for safety.

Scope Connections

Connect the Channel 'A' positive probe to the VMV control circuit (LG/BLK wire) at Pin 56 and the negative probe to the battery ground.

Scope Settings

To make the waveforms as clear as possible, set the scope settings to match the examples.

Lab Scope Tests

The examples on this page were taken from a vehicle with 54,400 miles on the odometer. The waveforms in these examples are from the original solenoid.

The engine was allowed to run long enough for the engine to enter closed loop operation before testing in Park.

Lab Scope Example (1)

In example (1), the trace shows the VMV solenoid waveform at idle speed.

Lab Scope Example (2)

In example (2), the trace shows the VMV solenoid waveform at cruise speed with the valve opened and closed at 100 Hz.

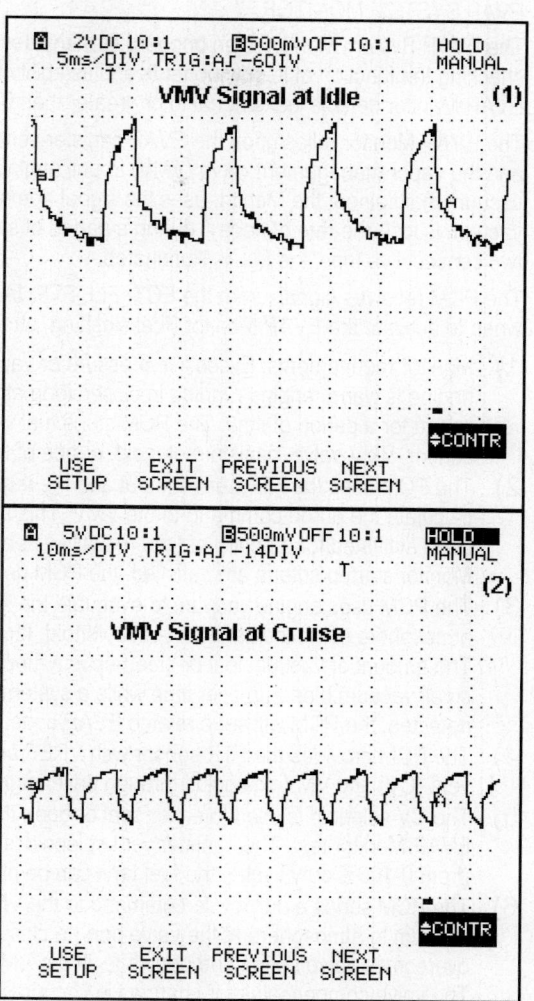

VMV Solenoid Test Schematic

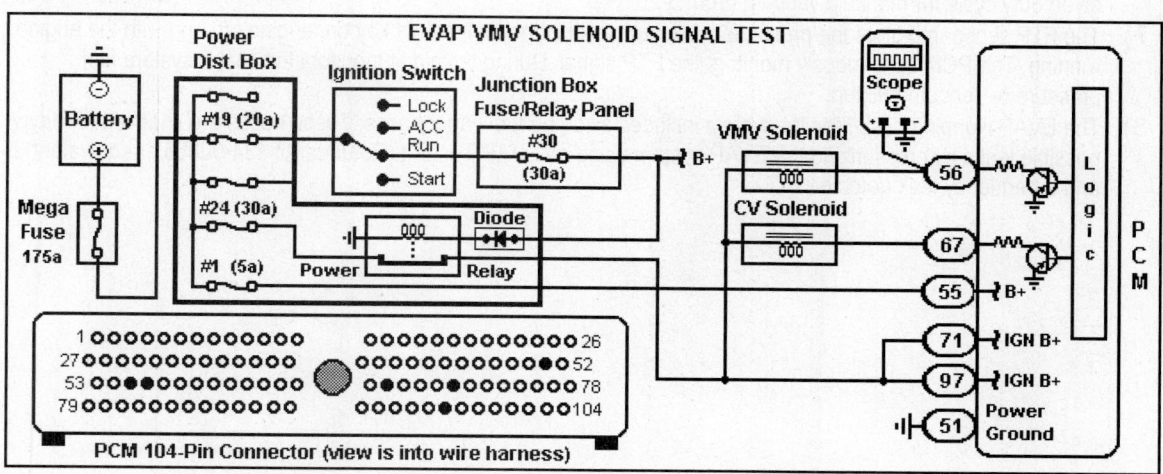

EVAP SYSTEM MONITOR

The EVAP System Monitor is an onboard strategy designed to test for proper operation of the EVAP system by checking the function of its components and its ability to flow fuel vapor (hydrocarbons) to the engine. In addition, the EVAP Monitor detects leaks equal to or greater than 0.040 inch by performing a vacuum check of the entire system.

The EVAP Monitor relies upon the EVAP canister vent (CV) solenoid to seal the entire EVAP system from atmosphere and the vapor management valve (VMV) to pull engine vacuum on the fuel tank. Then, with the system sealed and vacuum maintained, the Monitor uses the signal from the fuel tank pressure (FTP) sensor to observe the rate at which vacuum is lost (the rate of decay) during a period of system bleed-up. If the rate of decay is not within specifications for two consecutive trips, the MIL is illuminated.

The PCM receives signals from the ECT, FLI, FTP, IAT, MAF and VSS signals. The FLI signal is used to determine when to activate the EVAP Monitor (fuel sloshing, etc.).

1) Monitor requirements: Excessive pressure or vacuum not present in the fuel tank, ECT and IAT inputs indicate the engine is warm, engine running in closed loop at moderate load, and the vehicle moving at an elevated steady speed for a period of time. The PCM stops fuel vapor purging at idle speed, or if a fault is detected in the FTP sensor, VMV valve, canister solenoid, or one of the fuel vapor purge signals.

2) The PCM uses the FTP sensor input and the required amount of purge vapor flow to the intake manifold to calculate the purge command to the VMV. This action occurs after the PCM Purge Compensation strategy has been activated for a period of time (usually when fuel vapors are being purged from the canister). Once the Monitor startup criteria are satisfied, the PCM determines when to run the EVAP Monitor.

3) The PCM uses engine vacuum to evacuate the system using the VMV solenoid, and seals the system from atmosphere using the canister vent solenoid. It observes the rate at which vacuum is lost using the FTP sensor. The amount of vacuum lost or bleed-up is a function of the integrity of the system. A system with no leak has a small vacuum bleed-up over time while a system with a leak has a large vacuum bleed-up over time. If a leak is detected, the PCM will set a related EVAP code.

4) The PCM provides the FTP sensor with VREF and SIG RTN circuits. It controls the ground circuits to the canister vent (CV) and VMV solenoids through driver signals.

5) The CV solenoid is used to either seal or open the EVAP system to atmosphere during appropriate phases of the EVAP Monitor test. The canister vent solenoid is a normally open (0% duty cycle). When commanded closed (from 0-100% duty cycle), the fuel tank can be drawn down to a specific level of vacuum during the test period.

6) The PCM sends a duty cycle command to the VMV to divert vacuum to the VMV valve port (to open it) or to vent vacuum to atmosphere at the valve port (to close it). The increase in vacuum acting on the VMV diaphragm overcomes valve spring and begins to lift the VMV pintle off its seat causing vapor flow into the intake manifold. The valve compensates for changes in vacuum applied to the solenoid. It maintains a constant vapor flow at any given duty cycle for manifold vacuum (from 5-20" Hg).

7) The FTP sensor monitors the pressure or vacuum level in the fuel tank at key on, engine off, and with the engine running. The PCM continuously monitors the FTP signal. During testing, it monitors the EVAP system fuel pressure or vacuum buildup.

8) The EVAP Running Loss/Monitor system includes all of the fuel vapor hoses. These hoses are checked for any possible leaks by pressurizing the EVAP system using an EVAP Test Kit (Rotunda Kit 134-00056 or equivalent) and a frequency leak detector.

CHILTON DIAGNOSTIC SYSTEM - DTC P0442

The purpose of the Chilton Diagnostic System is to provide you with one or more tests that can be done with a DVOM, Lab Scope or Scan Tool *prior* to entering the complete trouble code repair procedure. The quick checks listed in the DVOM, Lab Scope or Scan Tool tests on the previous pages may help you quickly find the cause of the problem. If you cannot resolve the problem with these tests, a code repair chart is also included.

Code Description

DTC P0442 - The EVAP Purge Monitor detected a leak as small as 0.040 inch in the system. The MIL is activated if this fault is detected on two consecutive trips (2T DTC).

Code Conditions (Failure)

This code is set if, following a cold engine startup, the vehicle is driven as described in the Drive Cycle Explanation, and the PCM detects the following conditions: The EVAP system bleed-up was less 0.625 kPa (2.5 inches H2O) during a 15 second period with the fuel tank at least 75% full or the EVAP system vapor generation was more than 0.625 kPa (2.5 inches H2O) during a 120 second period.

Quick Check Items

Inspect or physically check the following items as described below:

- A loose fill cap (tighten the fuel cap 1/8 of a turn until it clicks)
- Small vapor leaks at the plastic vapor line connection to VMV
- Small vapor hose leaks at connections to VMV, FTP sensor or canister
- Small cuts or holes in vacuum lines or vapor hoses in the EVAP system

Drive Cycle Explanation

Start with a cold engine (ECT PID less than 100°F and IAT PID more than 40°F) and allow the engine to idle for 15 seconds. Then drive the vehicle at speeds over 40 mph for 60 seconds until the ECT PID reaches 170°F. Then with the EVAP PID over 75%, the FLI PID at 15-85% and TP PID at PT, drive at steady cruise for 10 minutes.

Technical Service Bulletin

Ford issued TSB #99-1-9 on 11-15-99 that provides help with situations where certain trouble codes (DTC P0442 and P0455) are set in memory without driveability concerns. This TSB explains how to diagnose these codes using the EVAP Running Loss Monitor Procedure, Service Bay Static Leak Test and EVAP Running Loss Monitor Drive Cycle.

EVAP "Cold Soak" Bypass Procedure

Install the Scan Tool. Turn the key on with the engine off. Cycle the key off, then on. Select the correct vehicle and engine qualifier. Clear the codes and do a PCM reset.

■ **NOTE:** *This step bypasses the engine soak timer and resets the OBD II monitor status.*

Use the Scan Tool menu to view and then select the following PIDs: ECT, EVAPDC, FLI (if available) and TP MODE. Start the engine without turning the key off. Refer to the article titled "How to Access and Use OEM PID Information" at the end of this section.

1998 F150 Pickup 4.6L V8 SFI VIN W (A/T) DTC P0442 Parameters

PID Acronym	Parameter Identification	PID Range	At Hot Idle	At 30 mph	At 55 mph
ECT (°F)	ECT Sensor (°F)	-40-304°F	160-200°F	160-200°F	160-200°F
FLI	Fuel Level Indicator (1998-02)	0-100%	50% (1.7v)	25-75%	25-75%
TP MODE	TP Sensor MODE	CT or PT	CT	PT	PT
EVAP CV	EVAP Vent Valve	0-100%	0%	0%	0%
EVAP DC	EVAP Purge Valve	0-100%	0-100%	0-100%	0-100%

Pinpoint Test HX

Pinpoint Test HX is used to diagnose the following devices and/or circuits:

- Canister Vent Solenoid
- Carbon Canister, Fuel Vapor and Vacuum Hoses
- Fuel Tank, Fuel Fill Cap, Fuel Fill Cap Valve
- Fuel Tank Pressure (FTP) Sensor
- Fuel Vapor Control Valve, Fuel Vapor Valve
- Power Control Module (PCM)
- Vapor Management Valve (VMV)
- These PCM related circuits: CV CONT, VMV CONT, SIG RTN, VPWR and VREF

Possible Causes for this Trouble Code

- A loose fill cap (tighten the fuel cap 1/8 of a turn until it clicks)
- Small vapor leaks at the plastic vapor line connection to VMV
- Small vapor hose leaks at connections to VMV, FTP sensor or canister
- Small cuts or holes in the vacuum lines or vapor hoses in the EVAP system

EVAP System Schematic

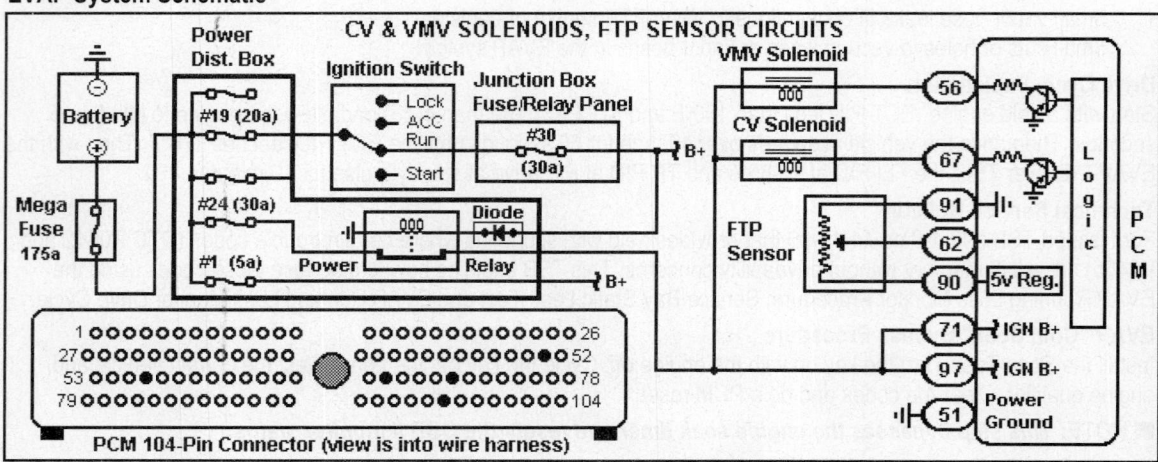

Pinpoint Test HX Repair Table

Step	Action	Yes	No
HX1	**Step description:** Check for a loose fuel cap • Verify that gas cap is tightened properly, and that the threads are not worn out. • Check the vacuum hose connections to the CV and VMV solenoid, and carbon canister. • Visually inspect for cuts or small hoses in the EVAP fuel vapor tubes or hoses. • Look for aftermarket parts that don't conform. • Was a problem detected in this test step?	Tighten the cap 1/8 of a turn until it clicks. If the code is a memory code, go to HX61. If it is not a memory code, do a PCM Reset and retest the system.	Go to Step HX2. Note: If the fuel filler cap is not tightened properly (1/8 turn until it clicks), the EVAP Monitor may fail its diagnostic test.

Pinpoint Test HX Repair Table (Continued)

Step	Action	Yes	No
HX2	**Step description:** Perform EVAP Leak Test • Disconnect and plug (cap) the EVAP return tube (VMV to intake manifold) at the source. • Connect a Scan Tool and turn the key on. • Access the VPWR PID. If the voltage is not 12 volts or greater, go to Step HX61. • Locate the EVAP test port (do not use unregulated pressure over 1 psi in this test). • If there is no EVAP test port, go to Step HX3. • Install the appropriate EVAP leak tester. • Close the CV solenoid with the Scan Tool. • Regulate argon or nitrogen gas pressure to 14" in-H2O to pressurize the EVAP system. • Perform the EVAP system leak test. • Did the leak test pass (and pressure hold)?	If the pressure stayed above 6 in-H2O (1.49 kPa) check for signs of physical damage to the fuel filler cap. Remove the EVAP leak tester from the EVAP test port. Go to Step HX3.	Remove the EVAP leak tester from the EVAP test port. Go to Step HX3.
HX3	**Step description:** Check for leak with Ultra-Sonic Leak Detector or approved leak tester • Install EVAP leak tester at the fuel filler pipe. • Use the tester to Test at the Fuel Filler Neck. • Access the ultra-sonic detector from the kit. • Close the CV solenoid with the Scan Tool. • Pressure system to 14 in-H2O or 3.48 kPa. • Perform the EVAP system leak test. • Slowly pass the detector probe around the fuel filler cap and EVAP test port. • Was there an audible change around the fuel filler cap or near the EVAP test port, and did the EVAP test fail?	Remove the EVAP leak tester from the fuel filler cap or EVAP test port. Repeat the Step HX3. If the EVAP leak test passed, prepare the vehicle to run an EVAP Monitor Verification drive cycle.	Prepare the vehicle to run the EVAP Monitor verification drive cycle. Then run the EVAP Running Loss Monitor Repair Verification drive cycle procedure.
HX4	**Step description:** Check for Small Leaks • Install the leak tester at the EVAP test port. • Close the CV solenoid with the Scan Tool. • Select All Off Mode and push Start Button. • Pressure system to 14" H2O or 3.48 kPa. • Perform the EVAP system leak test. • Turn the selector on the tester to Fill position. • Pressure system to 14" H2O or 3.48 kPa • Did the pressure hold at 13.80-14.20" H2O?	Go to Step HX5.	Continue to pressurize the EVAP system. Go to Step HX6.
HX5	**Step description:** Check for Small Leaks • Turn to key on, engine off. • Do not energize the solenoid over 9 times. • Pressurize the EVAP system to 14" H2O. • Access the ultra-sonic detector from the kit. • Close the CV solenoid with the Scan Tool. • Slowly pass the ultra-sonic detector from the Purge valve and CV solenoid, from canister to the CV solenoid and around the pipe and cap. • Was a sudden audible change indicated?	Reconnect the loose or damaged fuel vapor hose or tubes, canister purge outlet tube and canister inlet tube. Go to Step HX6.	Stop pressurizing the EVAP system. Go to Step HX6.

Pinpoint Test HX Repair Table (Continued)

Step	Action	Yes	No
HX6	**Step description:** Check for Small Leaks • Disconnect the EVAP Canister tube from the fuel tank at the fuel vapor tee between the Canister Purge valve and the canister (or at the 'F' fitting on the EVAP canister). • Plug (cap) the fuel vapor tee (or 'F' fitting). • Access the Output Test Mode with the Scan Tool and close the CV solenoid. • Pressurize the EVAP system to 14 in-H2O. • Access the ultra-sonic detector from the kit. • Slowly pass the ultra-sonic detector from the EVAP return tube (intake manifold to EVAP canister purge valve), EVAP canister purge outlet tube (canister purge valve to canister-CV solenoid) and the canister vent hose. • Was a sudden audible change indicated?	Reconnect loose or damaged EVAP return tube, EVAP canister purge outlet tube or canister vent hose assembly. Repeat Step HX6 to verify that the fuel vapor leak no longer exists. Go to Step HX7.	Remove plug from fuel vapor tee (or 'F' fitting) on the EVAP canister. Go to Step HX7.
HX7	**Step description:** *Check for Small Leaks* • Remove EVAP tester from the test port. • Remove the fuel filler cap. • Install EVAP leak tester at the fuel filler pipe. • Plug the open end of the EVAP canister tube (from the fuel tank) at the fuel vapor tee (or at the 'F' fitting on the EVAP canister). • Turn to key on, engine off. • Pressurize the EVAP system to 26-28" H2O (6.47-6.97 kPa). • Access the ultra-sonic detector from the kit. • Slowly pass the ultra-sonic detector from the EVAP canister tube to the fuel tank while checking the FTP sensor, fuel vapor vent valve(s) and fuel filler pipe. • Was a sudden audible change indicated?	Reconnect the loose or replace the EVAP canister vent tube. Repeat Step HX7 to verify that the fuel vapor leak no longer exists.	Go to Step HX8.
HX8	**Step description:** Check for Small Leaks • Reconnect the EVAP canister tube to the fuel vapor tee (or 'F' fitting on the canister). • Turn to key on, engine off. • Close the CV solenoid with the Scan Tool. • Regulate the argon or nitrogen gas pressure to 14" H2O (3.48 kPa). • Pressurize the EVAP system to 14" H2O. • Follow the instructions that come with the EVAP Leak Tester. • Perform an EVAP Leak Test. • Did the EVAP system pass the leak test?	Remove the EVAP leak tester from the fuel filler pipe. Reinstall the fuel filler cap and tighten it only 1/8 turn so that the cap initially clicks by sound. Reconnect the EVAP return tube to intake manifold vacuum. Perform the EVAP Monitor Running Loss Verification drive cycle.	Return to Step HX4 and try to locate the small leak that still exists in the system. Then return to this step (HX8) to verify that a small leak is not present in the system. Reinstall the fuel filler cap and tighten it only 1/8 turn so that the cap initially clicks by sound. Reconnect the EVAP return tube to intake manifold vacuum. Do the EVAP Monitor Running Loss Verification drive cycle

Exhaust Gas Recirculation System

GENERAL INFORMATION

The EGR system is designed to control NOx emissions by allowing small amounts of exhaust gases to be recirculated back into the combustion chamber to mix with the A/F mixture. This action causes a reduction in the amount of combustion, and less NOx. This vehicle is equipped with a differential pressure feedback EGR (DPFE) sensor.

Differential Pressure Feedback EGR System

The DPFE system includes the following components:

- DPFE sensor
- EGR valve and orifice tube assembly
- Vacuum regulator valve assembly
- PCM and related vacuum hoses.

The DPFE system operates as described next:

1) The PCM receives signals from the CKP, ECT, IAT, MAF and TP sensor in order to determine the optimum operating conditions for the DPFE EGR system. The engine must be warm, running at a stable engine speed under moderate load before the EGR system can be activated. The PCM deactivates the EGR system at idle speed, at extended wide-open throttle period or if a fault is detected in the EGR system.

2) The PCM calculates a desired amount of EGR flow for each engine condition. Then it determines the desired pressure drop across the metering orifice required to achieve the correct amount of flow and outputs the signal to the EGR VR solenoid.

3) The VR solenoid receives a variable duty cycle signal (0-100%). The higher the duty cycle rate, the more vacuum the solenoid diverts to the EGR valve.

4) The increase in vacuum acting upon the EGR valve diaphragm overcomes the valve spring pressure and lifts the valve pintle off its seat. This allows exhaust gas to flow.

5) The exhaust gas that flows through the EGR valve must first pass through the EGR metering orifice. One side of the orifice is exposed to exhaust backpressure and the other side is exposed to the intake manifold. This design creates a pressure drop across the orifice whenever there is EGR flow. When the EGR valve closes, there is no longer flow across the metering orifice and the pressure is equal on both sides.

6) The DPFE sensor measures the "actual" pressure drop across the metering orifice and relays a proportional voltage signal (from 0-5 volts) to the PCM. The PCM uses this feedback signal to correct for any errors in achieving the "desired" EGR flow.

EGR System Schematic

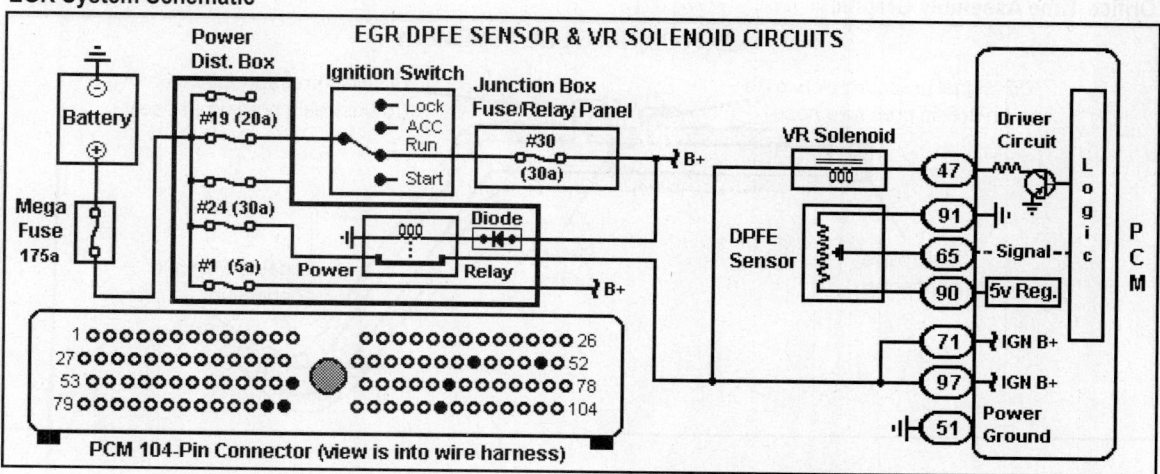

EGR DPFE SENSOR & VR SOLENOID CIRCUITS

DPFE SENSOR INFORMATION

The DPFE sensor is a ceramic, capacitive-type pressure transducer that monitors the differential pressure across the metering orifice located in the orifice tube assembly. The sensor receives vacuum signals through two hoses referred to as the downstream pressure (REF signal) and upstream pressure hose (HI signal). The HI signal uses the larger diameter hose. The DPFE sensor signal is proportional to a pressure drop across the metering orifice. This input is referred to as the EGR flow rate feedback signal.

EGR System Components Graphic

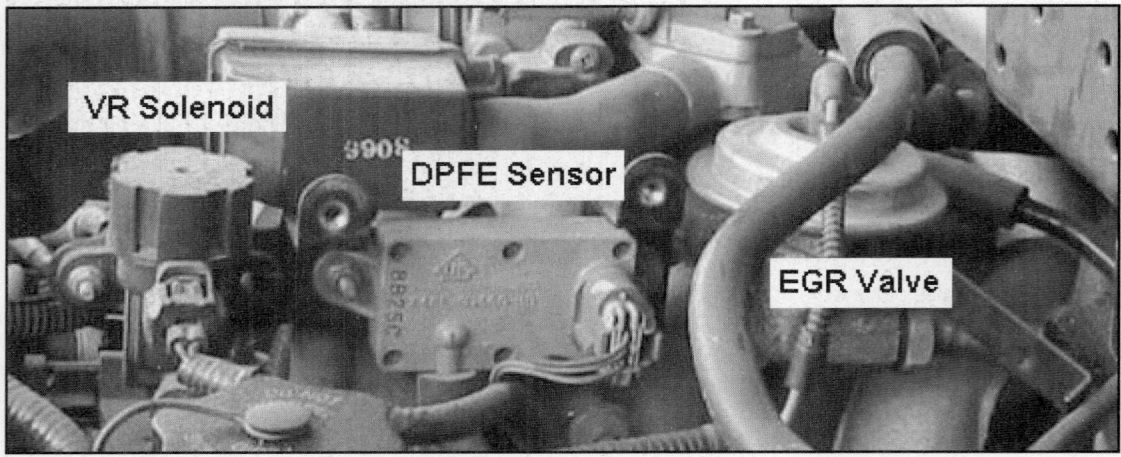

Orifice Tube Assembly

The orifice tube assembly is a section of tubing connecting the Exhaust system to the intake manifold. This assembly provides a flow path for EGR gases to the intake manifold. It also contains the metering orifice and two pressure pickup tubes.

The internal metering orifice creates a measurable pressure drop across it as the EGR valve open and closes. This pressure differential across the orifice is picked up by the DPFE sensor and converted to feedback data on EGR system operation to the PCM.

Orifice Tube Assembly Graphic

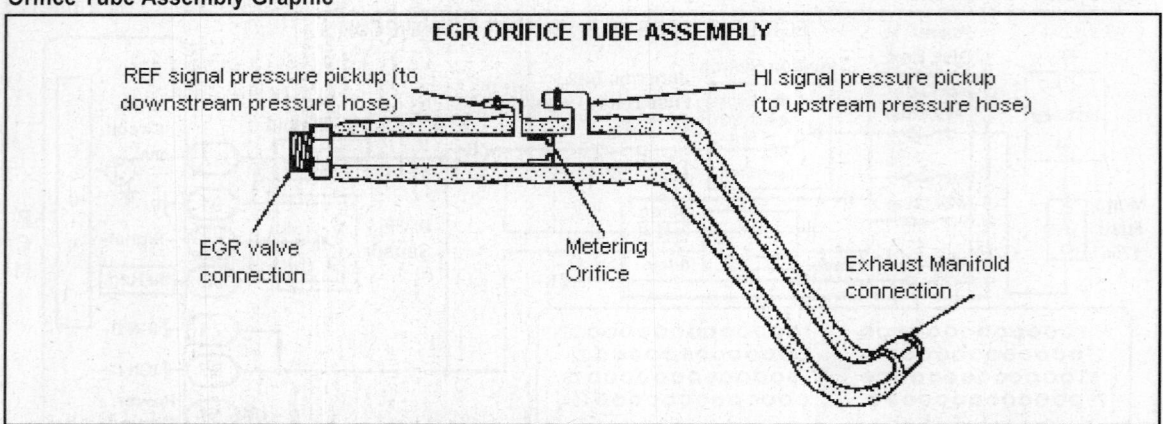

DPFE SENSOR TABLES

The tables below contain examples of "known good" specifications for use during a DPFE sensor static test.

Null 0.55v DPFE Sensor (Part #F48E-9J460-BB)

Differential Pressure			
In - H20	In - Hg	kPa	Volts
120	8.83	29.81	4.66v
90	6.62	22.36	3.64v
60	4.41	14.90	2.61v
30	2.21	7.46	1.58v
0	0	0	0.55v

Null 1.00v DPFE Sensor (Part #F7UE-9J460-AB)

Differential Pressure Volts			
In - H20	In - Hg	kPa	Volts
116	8.56	28.9	4.95v
58	4.3	14.4	2.97v
0	0	0	1.0v

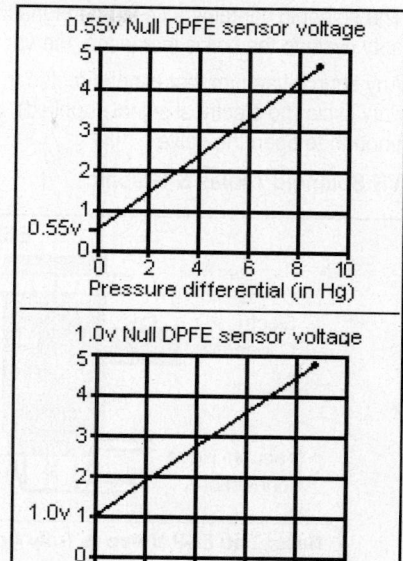

SCAN TOOL TEST (DPFE SENSOR)

Connect the Scan Tool to the DLC underdash connector. Select the DPFE and EGRVR PID from the Scan Tool menu. Record the values at Hot Idle, 30 and 55 mph. An example of the DPFE sensor and VR PID readings in Park (2000 rpm) appear in the Graphic.

DVOM TEST (DPFE SENSOR)

Locate the DPFE sensor circuits (underhood, at the BOB or at the PCM). Connect the DVOM positive lead to the DPFE signal (BRN/LG wire) and negative lead to chassis ground. Start the engine and record the desired DPFE readings. Compare the actual readings to the Pin Voltage Tables.

DPFE Sensor Test Schematic

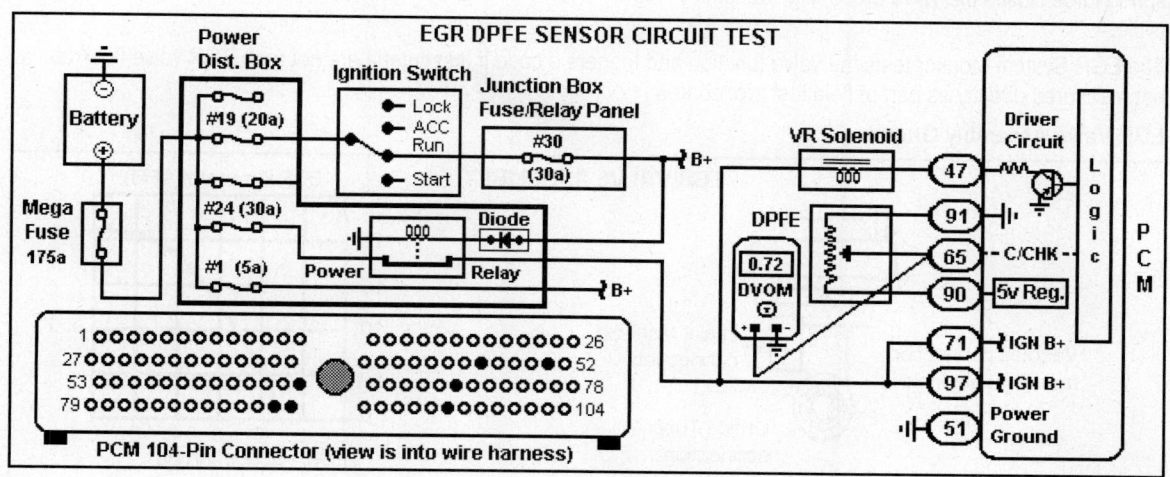

EGR DPFE SENSOR CIRCUIT TEST

PCM 104-Pin Connector (view is into wire harness)

VACUUM REGULATOR SOLENOID

The EGR vacuum regulator (VR) is an electromagnetic solenoid used to regulate the vacuum supply to the EGR valve. The solenoid contains a coil that magnetically controls the position of an internal disc to regulate the vacuum. As the duty cycle to the coil is increased, the vacuum signal passed through the solenoid to the EGR valve increases.

Any source vacuum that is not directed to the EGR valve is vented through a vent valve into the atmosphere. At 0% duty cycle (no electrical signal supplied), the EGR vacuum regulator solenoid allows some vacuum to pass, but not enough to open the valve.

VR Solenoid Tables & Graphic

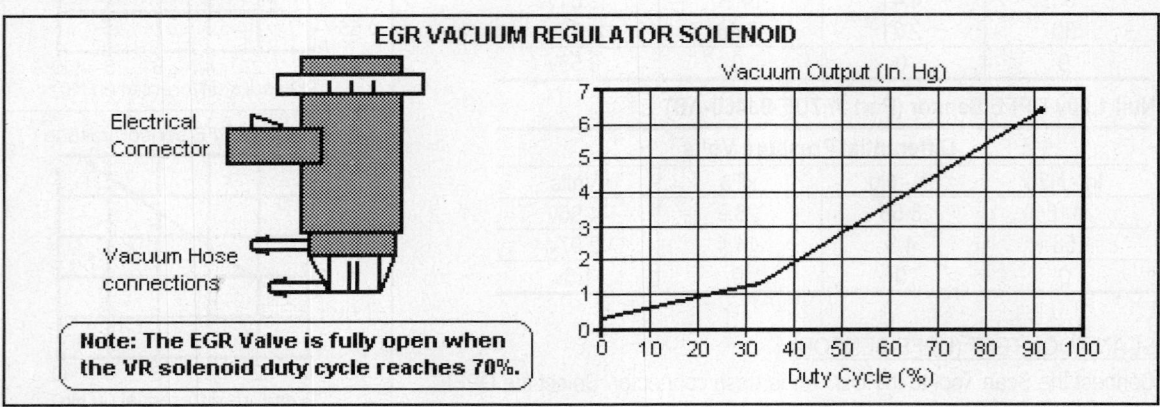

EGR VACUUM REGULATOR SOLENOID

Note: The EGR Valve is fully open when the VR solenoid duty cycle reaches 70%.

EGR Vacuum Regulator Solenoid Table

Duty Cycle	Vacuum Output					
(%)	Min		Nominal		Maximum	
	In-Hg	KPa	In-Hg	KPa	In-Hg	KPa
0	0	0	0.38	1.28	0.75	2.53
33	0.55	1.86	1.3	4.39	2.05	6.90
90	5.69	19.2	6.32	21.3	6.95	23.47

EGR Valve Assembly

The EGR valve is a conventional vacuum actuated valve that increases or decreases the flow of exhaust gas recirculation. The vacuum applied to the diaphragm overcomes the spring force and the valve begins to open. The spring force closes the valve at 1.6" Hg vacuum or less. The valve is wide open at 4.5" Hg of vacuum with a 70% duty cycle.

The EGR System Monitor tests the valve function and triggers a code if test criteria are not met. EGR valve flow rate is not measured directly as part of field test procedures.

EGR Valve Assembly Graphic

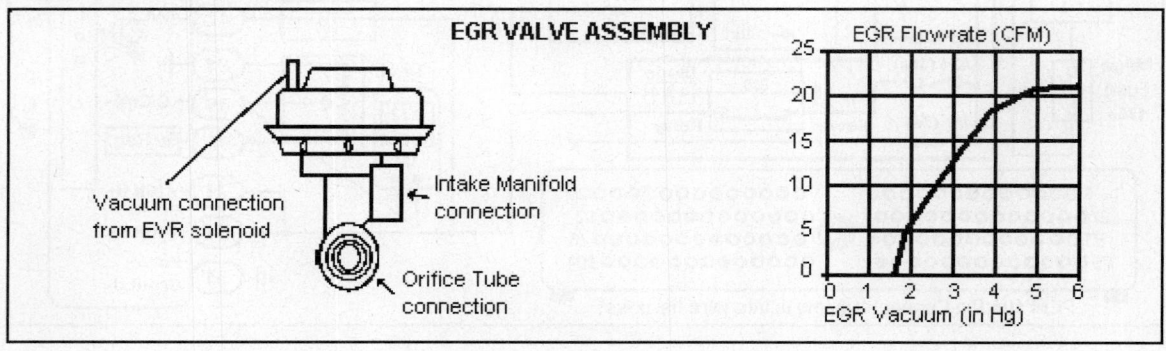

EGR VALVE ASSEMBLY

LAB SCOPE TEST (VR SOLENOID)

The Lab Scope is the "tool of choice" to test the VR solenoid as it provides an accurate view of the solenoid response and frequency rate. Place the gearshift selector in Park and block the drive wheels for safety.

Scope Connections

Connect the Channel 'A' positive probe to the VR control circuit (BRN/PNK wire) at PCM Pin 47 and the negative probe to battery ground.

Scope Settings

To make the waveforms as clear as possible, set the scope settings to match the examples.

Lab Scope Tests

The examples on this page were taken from a vehicle with 54,400 miles on the odometer. The waveforms in these examples are from the original solenoids. The engine was allowed to run long enough until it entered closed loop operation prior to testing. Note that the examples were captured at different speeds.

Lab Scope Example (1)

In example (1), the trace shows the solenoid command signal at idle speed (the PCM is not controlling the solenoid). The VR PID will read 0% with the VR solenoid in this state.

Lab Scope Example (2)

In example (2), the trace shows the VR solenoid command signal captured during an acceleration event. The VR PID will read 45% with the VR solenoid activated during this test.

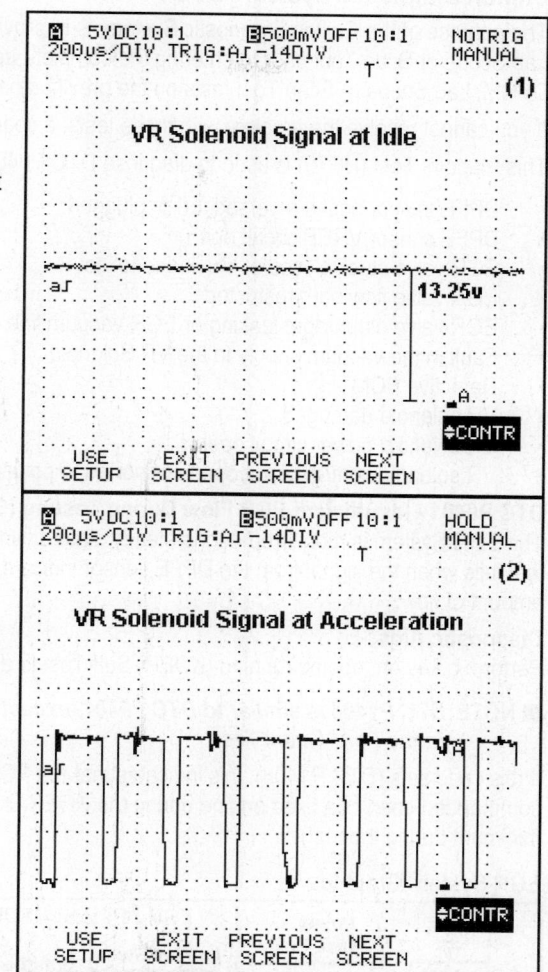

VR Solenoid Test Schematic

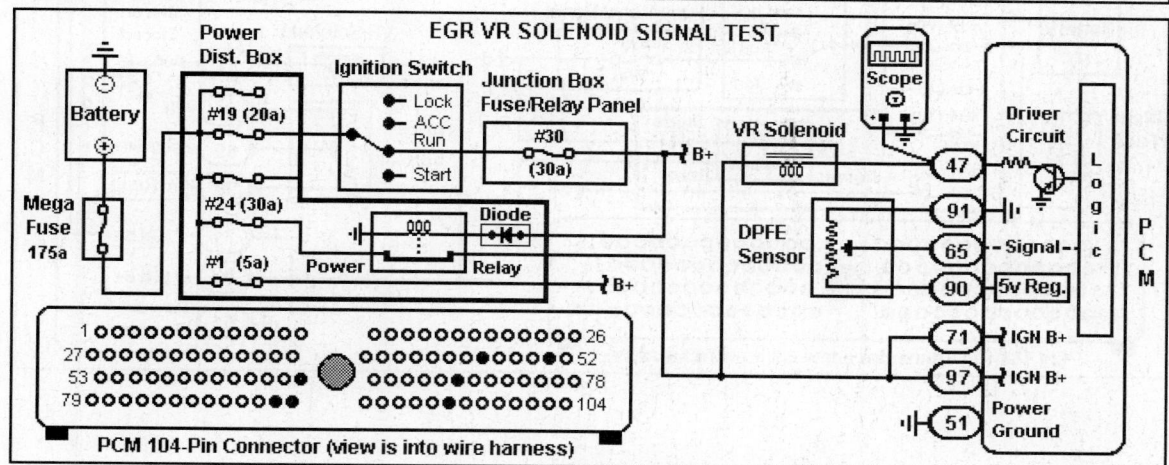

Chilton Diagnostic System - DTC P0401

The purpose of the Chilton Diagnostic System is to provide you with one or more tests that can be done with a DVOM, Lab Scope or Scan Tool *prior* to entering the complete trouble code repair procedure. The quick checks listed in the DVOM, Lab Scope or Scan Tool tests on the previous pages may help you quickly find the cause of the problem.

If you cannot resolve the problem with these tests, a code repair chart is also included.

This Pinpoint Test (HE70) is used to diagnose DTC P0401 and the following items:

- DPFE sensor hoses reversed, off or plugged
- DPFE sensor VREF circuit open
- EGR valve stuck closed or stuck due to ice
- EGR valve flow path restricted
- EGR valve diaphragm leaking or EGR vacuum line off, plugged or leaking
- Fault in the vacuum supply to the VR Solenoid
- Defective PCM
- VR solenoid damaged
- VR solenoid power circuit open
- VR solenoid control circuit open or shorted to power

DTC P0401 - Insufficient EGR Flow During Testing (CMC) Code Conditions

The EGR System is monitored during steady state driving conditions while the EGR valve is commanded open. The test fails when the signal from the DPFE sensor indicates that the EGR flow rate was less than the desired minimum amount of flow.

Diagnostic Aids

Perform a key on, engine running (KOER) Self-Test to determine if DTC P1408 sets.

■ **NOTE:** *DTC P1408 is similar to DTC P0401, except that DTC P1408 is an engine running test that is included in the Ford Quick Test.*

If this code sets (DTC P1408), this indicates that the PCM detected an incorrect flow rate with the EGR valve commanded open at a fixed engine during the KOER Quick Test (i.e., the PCM detected a flow rate below a calibrated minimum during the test).

EGR System Graphic

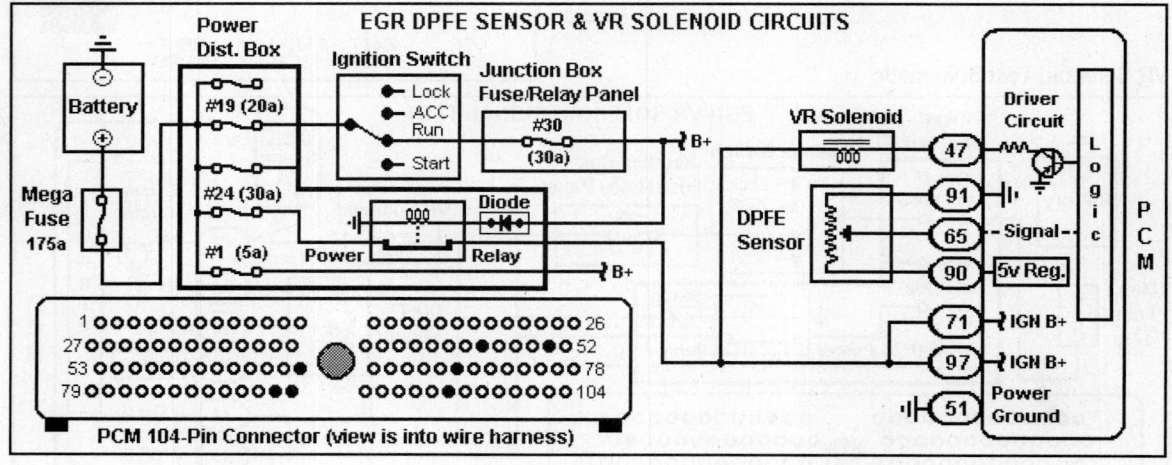

Repair Table for DTC P0401 (Pinpoint Test HE70)

Step	Action	Yes	No
HE70	**Step description:** Run the KOER self-test • Run the KOER Self-Test. • Was KOER DTC P1408 set during the test?	DTC P1408 is a current code. Go to Step HE71.	Remove and inspect the EGR valve and intake manifold port for a restriction. If okay, the fault is intermittent. Go to Step HE91.
HE71	**Step description:** Run KOER self-test while monitoring the EGR vacuum signal • Read the continuous memory trouble codes. • Was CMC DTC P1406 set during the test?	Go to Step HE60.	Go to HE72.
HE72	**Step description:** Read Continuous Codes • Disconnect vacuum hose at the EGR valve and connect a vacuum gauge to the hose. • Ignore any codes that set with the hose off. • Run the KOER self-test while monitoring the vacuum level on the gauge. The PCM will enable the EGR solenoid about 30 seconds into the test. The vacuum level on the gauge should increase to 1.6" Hg at this time. • Turn the key off. • Did the vacuum increase to 3.0" Hg or more at anytime during the KOER self-test?	This result indicates the vacuum level was sufficient to open the EGR valve. The fault is probably not in the EGR vacuum control system. Go to Step HE73.	This result indicates the vacuum level was insufficient to open the EGR valve. Go to Step HE80.
HE73	**Step description:** Check DPFE sensor hoses • Check both DPFE sensor hoses to detect if they are reversed at the sensor or orifice tube. • Inspect the hoses for signs of being pinched or dips where water could collect and freeze. • Check the DPFE sensor and orifice tube for blockage or damage at the pickup tubes. • Were any faults detected during this step?	Make repairs to the pressure hoses or devices as needed. Restore the vehicle to its normal operation. Do a PCM Reset to clear any codes and then do a Quick Test.	Go to Step HE74.
HE74	**Step description:** Check DPFE sensor output • Disconnect pressure hoses at DPFE sensor. • Connect hand vacuum pump to downstream connection at the DPFE sensor (at the intake manifold side of the sensor - the smaller tube). • Turn to key on, engine off. • Access the DPFEGR PID on the Scan Tool. • Apply 8-9" Hg of vacuum to sensor and hold it for 3 seconds - then release the vacuum. • For applications with a 1.0v offset, the DPFE sensor PID must be between 0.75-1.25v at KOEO with no vacuum applied. The voltage should increase to 4.0v with vacuum applied. The vacuum should drop to less than 3.0v within 3 seconds after vacuum is released. • Did DPFE PID indicate a fault in this step?	Turn the key off. Go to Step HE75.	Turn the key off and reconnect the pressure hoses to the sensor. Go to Step HE76.

Repair Table for DTC P0401 (Pinpoint Test HE70)

Step	Action	Yes	No
HE75	**Step description:** Test DPFE sensor VREF • Disconnect the DPFE sensor connector. • Turn to key on, engine off. • Measure the voltage between the DPFE sensor VREF circuit and the sensor ground. • Did the DVOM read from 4.9-5.1 in this step?	Replace the damaged DPFE sensor. Restore the vehicle to normal operation. Do a PCM Reset to clear any trouble codes. Do a Quick Test.	The DPFE sensor VREF voltage is out of normal range. Go to Step C1 in other repair manuals to determine the cause of the low VREF signal condition.
HE76	**Step description:** Check EGR Valve Function • Disconnect and plug EGR vacuum hose. • Connect a hand vacuum pump to the valve. • Start the engine and allow it to idle. • Access the EGR & RPM PID's on Scan Tool. • Slowly apply 8-10" Hg vacuum to valve and hold that vacuum level for 10 seconds. If the engine starts to stall, increase to 1000 RPM. • The EGR valve should start to open at about 1.6" Hg vacuum as indicated by an increase in the DPFE PID value. The voltage will continue to increase to at least 2.5v under full vacuum. The voltage should hold steady with vacuum held steady. If it drops, the valve is leaking. • Did the PID value indicate as described?	Restore the vehicle to normal operation. Go to Step HE85.	Remove and inspect EGR valve for signs of contamination, unusual wear, carbon deposits, binding or a leaking diaphragm. If the valve is okay, look for a restricted EGR port to the intake manifold or a plugged orifice tube assembly. Repair as needed. Restore the vehicle to normal operation. Do a PCM Reset to clear codes. Do a Quick Test.
HE80	**Step description:** Check VR Vacuum Source • Inspect the vacuum lines between vacuum source and VR solenoid, and between the VR solenoid and EGR valve for disconnects, leaks, kinks, blockage, routing or damage. • Disconnect vacuum hoses to VR solenoid and connect a vacuum gauge to the source vacuum hose to the VR solenoid. • Start the engine, allow it to idle and note the reading on the vacuum gauge at idle speed. • Did the gauge read at least 15" Hg vacuum?	Restore the vehicle to normal operation. Go to Step HE81.	Isolate the source vacuum fault and make repairs as needed to correct the low vacuum condition. Restore the vehicle to normal operation. Do a PCM Reset to clear any trouble codes and then do a Quick Test.
HE81	**Step description:** Check VPWR to Solenoid • Disconnect the VR Solenoid connector. • Turn to key on, engine off. • Measure the VPWR voltage to the VR solenoid connector (the RED wire connection). • Disconnect the VR solenoid hoses. Connect a vacuum gauge to the source vacuum hose. • Did the solenoid VPWR read at least 10.5v?	Turn the key to off. Go to Step HE82.	Locate and repair the open circuit condition in the VR solenoid VPWR circuit. Restore the vehicle to normal operation. Do a PCM Reset to clear any trouble codes and then do a Quick Test.
HE82	**Step description:** Test VR Sol. Resistance • With the VR Solenoid connector removed, measure the resistance of the VR solenoid. • Did the resistance read from 26-40 ohms?	Go to Step HE83.	Replace the solenoid. Restore the vehicle and do a PCM Reset. Do a Quick Test.
HE83	**Step description:** Test for Short to Power • Install a breakout box. Measure the voltage between PCM Test Pin 47 and ground. • Did the voltage read from 1 volt?	Repair short circuit condition in VR solenoid control circuit. Restore the vehicle and reset PCM	Go to Step HE84.

Repair Table for DTC P0401 (Pinpoint Test HE70)

Step	Action	Yes	No
HE84	**Step description:** Check for an Open Circuit • Measure the resistance of the EGRVR circuit from Pin 47 and the VR harness connector. • Did the resistance read less than 5 ohms?	Reconnect the VR solenoid connector. Go to Step HE85.	Repair the open circuit condition in the VR solenoid control circuit. Restore the vehicle to normal operation. Do a PCM Reset and then run a Quick Test.
HE85	**Step description:** Test VR Solenoid Function • Reconnect the PCM and the VR solenoid. • Disconnect the vacuum hose at EGR valve and connect a vacuum gauge to the hose. • Start the engine and allow it to idle. • Jumper PCM Pin 47 (EGRVR pin) to chassis ground at the breakout box. • Did vacuum gauge read 4.0" Hg or greater?	Replace the damaged PCM. Restore the vehicle to normal operation. Do a PCM Reset and then run a Quick Test.	Replace the damaged VR solenoid. Restore the vehicle to normal operation. Do a PCM Reset and then run a Quick Test.
HE90	**Step description:** Test for Intermittent Fault • Visually check the EGR system components for signs of an intermittent fault. • Were any problems found in this step?	Make repairs as needed. Restore the vehicle. Do a PCM Reset and Quick Test.	Go to Step HE91.
HE91	**Step description:** EGR functional check • Disconnect the vacuum hose at EGR valve. Connect a hand vacuum pump to the valve. • Connect a Scan Tool and start the engine. • Access the DPEGR and RPM PID values. • Slowly apply 5-10" Hg of vacuum to the valve (hold it for at least 10 seconds). If the engine starts to stall, increase the engine speed to a minimum engine speed of 800 rpm. • Look for these events: The EGR valve starts to open at about 1.6" Hg of vacuum; DPFEGR PID value increases until the valve opens (it should read at least 2.5v at full vacuum); the DPFEGR PID value is steady while vacuum is held (if the vacuum drops off, there is a leak). • Does the DPFEGR PID indicate that the EGR valve operated as described in this text?	Turn the key to off. Go to Step HE92.	Turn the key to off. Remove and inspect the EGR valve for signs of contamination, unusual wear, carbon deposits, binding, a leaking diaphragm or other damage. Make repairs as needed. Restore the vehicle to normal operation. Do a PCM Reset and then do a Quick Test.
HE92	**Step description:** Check the Source Vacuum • Install a breakout box and connect the PCM. • Disconnect the plugged hose at EGR valve and connect a vacuum gauge to the hose. • Start the engine and allow it to idle. • Jumper test pin 47 (EVR) to chassis ground (the idle vacuum should read at least 4" Hg). • Observe vacuum gauge during these events: Tap lightly on EVR solenoid; wiggle the EVR solenoid harness connector, vacuum lines and wiring harness to the EVR solenoid. A fault is indicated if the vacuum gauge reading drops. • Was a fault indicated during this test step?	Turn the key off. Locate and repair the problem that caused the vacuum reading to drop during testing. Restore the vehicle to normal operation. Do a PCM Reset and then do a Quick Test.	The fault did not appear or was intermittent at this time. Refer to the Intermittent tests in this manual or in other repair information.

Fuel Delivery System

<u>INTRODUCTION</u>

The Fuel system delivers fuel at a controlled pressure and correct volume for efficient combustion. The PCM controls the Fuel system and provides the correct injector timing.

Fuel Delivery

This vehicle is equipped with a sequential fuel injection (SFI) system that delivers the correct A/F mixture to the engine at the precise time under all operating conditions.

Air Induction

Air enters the system through the fresh air duct and flows through the air cleaner where it is monitored by the MAF sensor. The metered air passes through the air duct and enters the throttle body. The incoming air passes through the throttle body and into the intake plenum to the intake manifold where it is mixed with fuel for combustion.

System Operation

A high-pressure (in-tank mounted) fuel pump delivers fuel to the fuel injection supply manifold. The fuel injection supply manifold incorporates electrically actuated fuel injectors mounted directly above each of the intake ports of the engine.

A constant amount of fuel pressure is maintained at the injectors by a fuel pressure regulator. The fuel pressure is a constant while fuel demand is not, so the system includes a fuel return line to allow excess fuel to flow through the regulator to the tank.

The PCM determines the amount of fuel the injectors spray under all engine-operating conditions. The PCM receives electrical signals from engine control sensors that monitor various factors such as airflow pressure, temperature of the air entering the engine, engine coolant temperature, throttle position and vehicle speed. The PCM evaluates the sensor information it receives and then signals the fuel injectors in order to control the fuel injector pulsewidth.

Fuel Delivery System Graphic

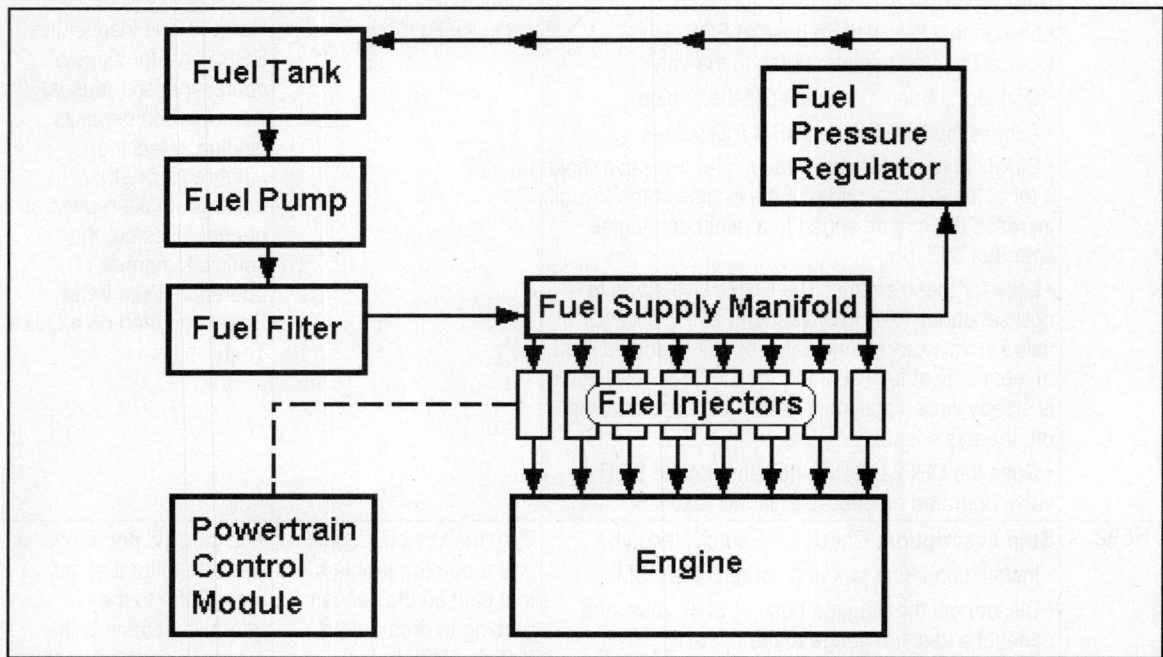

Fuel Pump Relay

The PCM controls power to the fuel pump by controlling the fuel pump relay.

<u>Component Location</u>

The fuel pump relay is located in the Power Distribution Box located on the driver side underhood area. Refer to the Graphic to the right.

The PCM energizes the fuel pump relay under the following operation conditions:

1) With the ignition turned "off", the PCM and fuel pump relay are not energized.

2) When the ignition switch is first turned to the "run" position, these events occur:
   a. The power relay is energized.
   b. Power is provided to the fuel pump relay and a timing device in the PCM.
   c. The fuel pump is supplied power through the fuel pump relay.

3) If the ignition switch is not turned to the "start" position, the timing device in the PCM will open the fuel pump control circuit at Pin 80 (LB/ORN wire) after one second, and this action de-energizes the fuel pump relay, which in turn de-energizes the fuel pump. This circuitry also provides for pre-pressurization of the fuel system.

4) When the ignition switch is turned to the "start" position, the PCM operates the fuel pump relay to provide fuel for starting the engine while cranking.

5) After the engine starts and the ignition switch is returned to the "run" position:
   a. Power to the fuel pump is again supplied through the fuel pump relay.
   b. The PCM shuts off the fuel pump by opening the ground circuit to the fuel pump relay (Pin 80) if the engine stops, or if the engine speed is below 120 rpm.

Fuel Pump Monitor

The PCM determines if the fuel pump circuit has power through the FPM circuit (Pin 40). This circuit is used to determine the fuel pump circuit status for DTC P0231 and P0232.

Fuel Pump Circuits Schematic

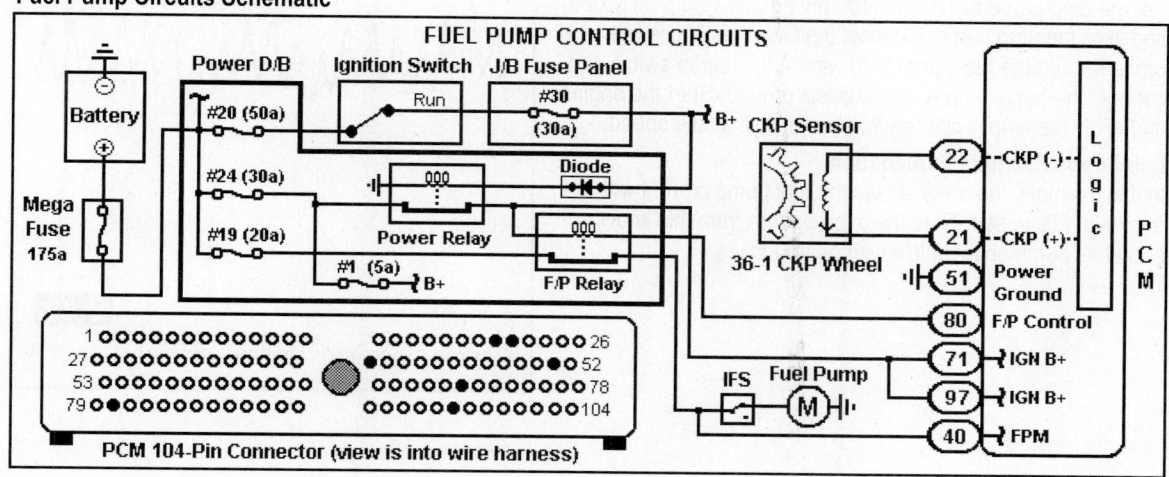

FUEL PUMP MODULE

The fuel pump module in a "returnable" fuel system is mounted in a reservoir inside the fuel tank. It has a discharge check ball that maintains system pressure after engine shutdown to help a hot restart.

The reservoir prevents fuel low interruptions during hard vehicle maneuvers during periods when the fuel tank level is low.

A portion of the high-pressure flow from the pump is diverted to operate a Venturi jet pump. The jet pump draws fuel from the fuel tank into the fuel pump module.

This design permits optimum fuel pump operation during vehicle maneuvers and driving on steep hills with low-tank full levels.

Internal Relief Valve

The fuel pump has an internal pressure relief valve that restricts fuel pressure to 124 psi if the fuel flow becomes restricted due to a clogged fuel filter or damaged fuel line.

LAB SCOPE TEST (FUEL PUMP)

If you suspect that the fuel pump is faulty, you can use a low amp probe with an inductive pickup to monitor the operation of the fuel pump and its related circuits.

Lab Scope Settings

Set the Lab Scope to these initial settings:

- Volts per division: 100 mv
- Time per division: 1 ms
- Trigger setting: 50% with a positive slope

Scope Connections (Amp Probe)

Set the amp probe to 100 mv (100 mv equals 1 amp on scope) and zero the amp probe. Connect the Channel 'A' positive amp probe around the fuel pump feed wire at the inertia switch and connect the negative probe to chassis ground. Start the engine and allow the amp probe reading to stabilize at idle speed.

Lab Scope Example Explanation

In this example, the trace shows the fuel pump current with the engine at idle speed. Note the even pattern from this known good fuel pump on a vehicle with 54,400 miles.

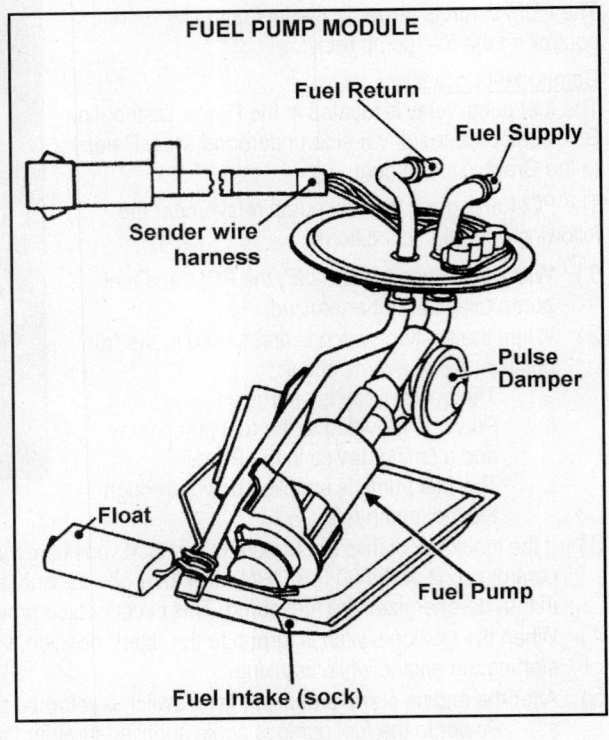

FUEL PUMP MODULE

Fuel Return
Fuel Supply
Sender wire harness
Pulse Damper
Float
Fuel Pump
Fuel Intake (sock)

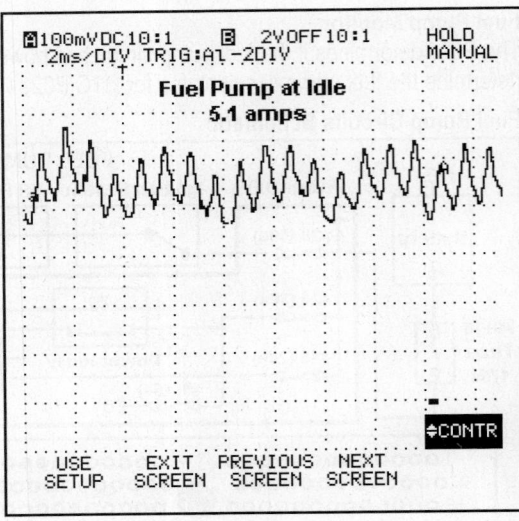

100mVDC10:1   2VOFF10:1   HOLD
2ms/DIV TRIG:A1 -2DIV   MANUAL

**Fuel Pump at Idle
5.1 amps**

USE SETUP   EXIT SCREEN   PREVIOUS SCREEN   NEXT SCREEN

FUEL PUMP COMPONENTS

The internal portions of the fuel pump module are described next.

Check Valve

The fuel pump includes a normally closed (N.C.) check valve in the manifold outlet of the module.

The valve opens when outlet pressure from the pump exceeds the opposing check valve spring force.

When the pump is off, the check valve closes to maintain pump prime and fuel line pressure.

Fuel Reservoir

A fuel reservoir is used to prevent fuel flow interruptions during extreme vehicle maneuvers with low tank fill levels.

The reservoir is welded into the tank or included in the pump sender housing.

Fuel Inlet Screen

The inlet of the pump includes a nylon filter to prevent dirt and other particulate matter from entering the system.

Water accumulation in the tank can pass through the filter without restriction.

External Fuel Filter

A frame mounted external fuel filter is used to strain fuel particles through a paper element.

This filtration process reduces the possibility of obstruction in the fuel injector orifices.

The fuel filter is located at the rear at the right side frame rail near the fuel tank. Refer to the location Graphic to the right.

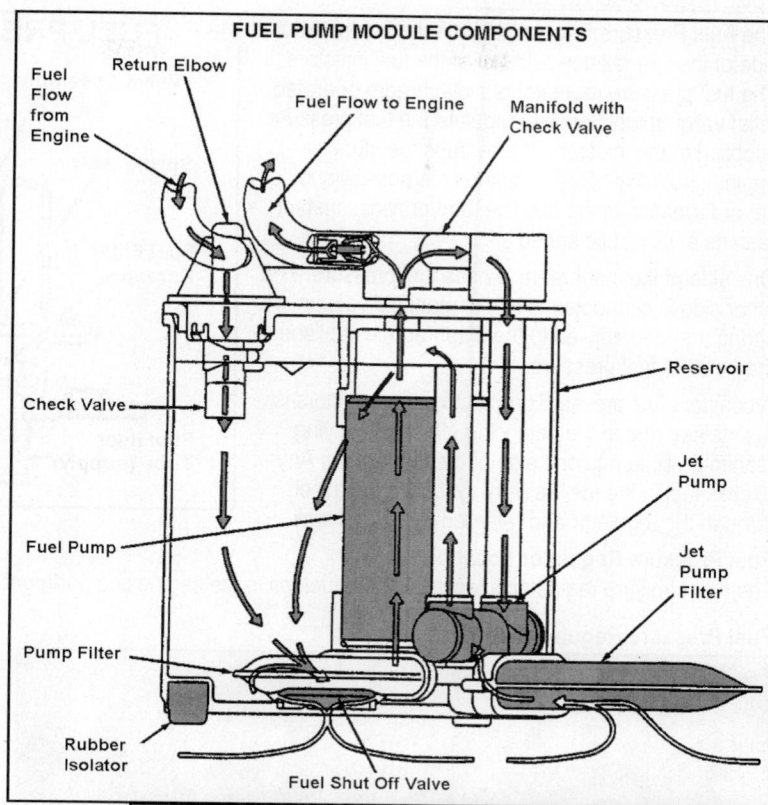

FUEL PUMP MODULE COMPONENTS

FUEL PRESSURE REGULATOR

The Fuel Pressure Regulator is attached to the return side of the fuel rail downstream of the fuel injectors. The fuel pressure regulator is a diaphragm-operated relief valve. It function is to regulate the fuel pressure supplied to the injectors. It also traps fuel during engine shutdown. This eliminates the possibility of vapor formation in the fuel line, and provides instant restarts at initial idle speed.

One side of the diaphragm senses fuel pressure. The other side is connected to intake manifold vacuum. A spring pre-load applied to the diaphragm establishes the nominal fuel pressure.

A constant fuel pressure drop across the injectors is maintained due to the balancing affect of applying manifold vacuum to one side of the diaphragm. Any excess fuel in the fuel delivery system is bypassed through the regulator and returned to the fuel tank.

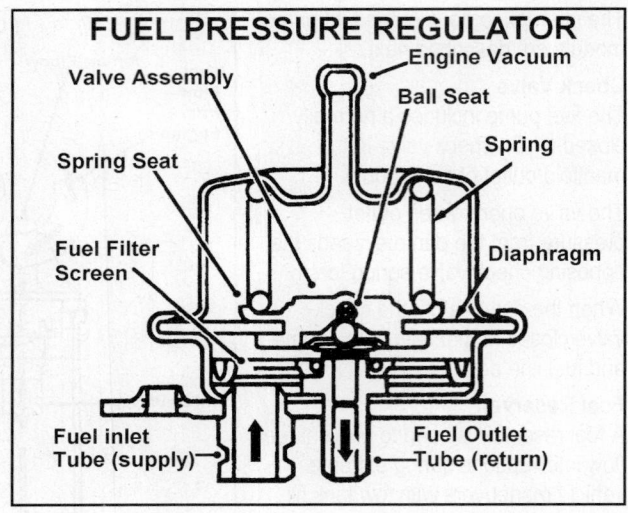

Fuel Pressure Regulator Location

The fuel pressure regulator is located on the fuel rail in the engine compartment as shown in the Graphic below.

Fuel Pressure Regulator Location Graphic

INERTIA SWITCH

This vehicle application is equipped with an inertia fuel shutoff (IFS) switch that is used in conjunction with the electric fuel pump. The inertia switch is designed to shut off the fuel pump in the event of an accident.

Inertia Switch Operation

When a sharp impact occurs, a steel ball, held in place by a magnet, breaks loose and rolls up a conical ramp to strike a target plate. This action opens the electrical contacts of the IFS switch and shuts off power to the fuel pump. Once the switch is open, it must be manually reset before the engine will restart.

Inertia Fuel Shutoff Switch Graphic

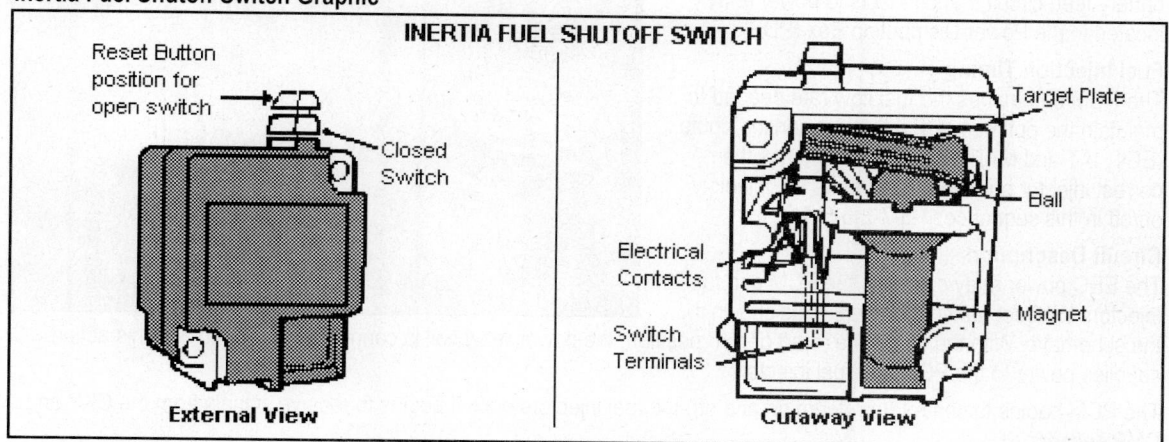

INERTIA FUEL SHUTOFF SWITCH

Reset Button position for open switch — Closed Switch

External View

Target Plate — Ball — Magnet — Switch Terminals — Electrical Contacts

Cutaway View

Inertia Switch Location

One of the challenges in dealing with resetting or checking this switch is locating the switch.

On this vehicle application (a 1998 F150 pickup), the IFS switch is located behind the right side kick panel (look for the IFS label on the panel).

Refer to the Graphic to the right as needed.

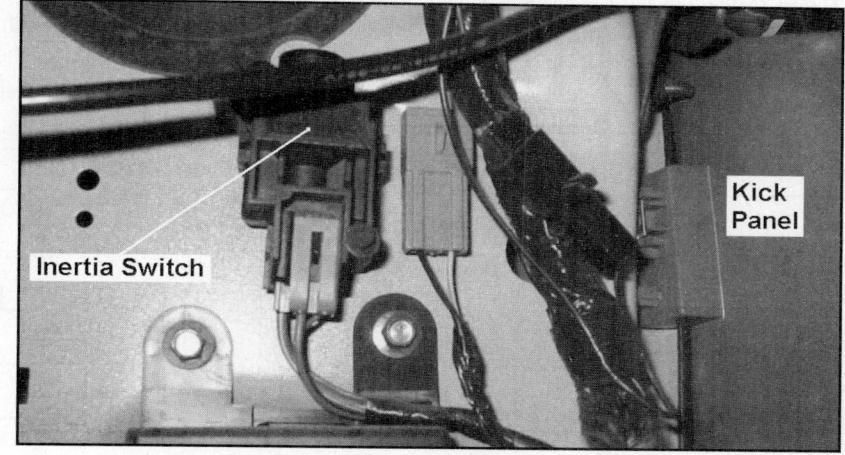

Inertia Switch

Kick Panel

IFS Reset Instructions

1) Turn the key off and check for any signs of leaking fuel near the fuel tank or in the underhood area. If any leaks are found, repair the source of the leak before proceeding to the next step.

2) If there are no fuel leaks present, push the reset button to reset the switch (it can be closed an extra 0.16" in the closed position).

3) Cycle the key to "on" for a few seconds, and then to off (in order to recheck for signs of any fuel leaks anywhere in the system).

FUEL INJECTORS

The fuel injector is a solenoid-operated valve designed to meter the fuel flow to each combustion chamber (SFI). The fuel injector is opened and closed a constant number of times per crankshaft revolution. The injectors are a deposit resistant injection (DRI) design and do not have to be cleaned.

The amount of fuel delivered is controlled by the length of time the injector is open (pulsewidth).

The fuel injectors are supplied power through the battery feed circuit that connects to power relay located in the Power Distribution Box (PDB).

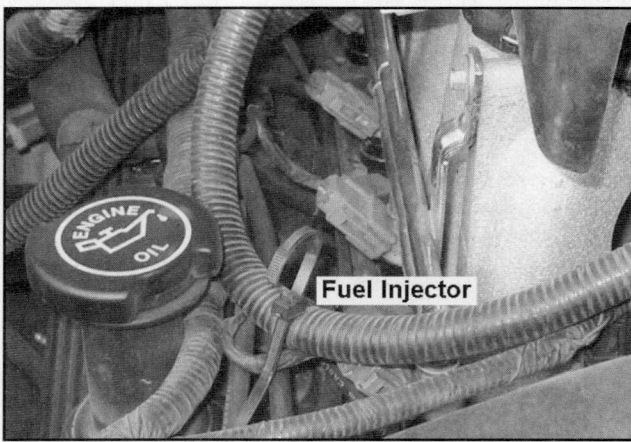

Fuel Injector

Fuel Injection Timing

The PCM determines the fuel flow rate needed to maintain the optimum A/F ratio from sensor inputs (ECT, IAT and MAF). The PCM computes the correct injector pulsewidth and turns the injectors on/off in this sequence: 1-3-7-2-6-5-4-8.

Circuit Description

The EEC power relay connects the PCM and fuel injectors to direct battery power (B+) as shown in the schematic. With the key in the "start or run" position, the power relay coil is connected to ground and this action supplies power to the PCM and fuel injectors.

The PCM begins to control (turn them on and off) the fuel injectors once it begins to receive signals from the CKP and CMP sensors.

Fuel Injection Timing Modes

The PCM operates the fuel injectors in two modes: Simultaneous and Sequential. In Simultaneous mode, fuel is supplied to all eight cylinders at the same time by sending the same injector pulsewidth signal to all 8 injectors, twice for each engine cycle. This mode is used during startup and in the event of a fault in the CKP sensor or the PCM.

Sequential mode is used during all other engine operating conditions. In this mode, fuel is injected into each cylinder once per engine cycle in the firing order (1-3-7-2-6-5-4-8).

Sequential Fuel Injection Graphic (V6 Example Shown)

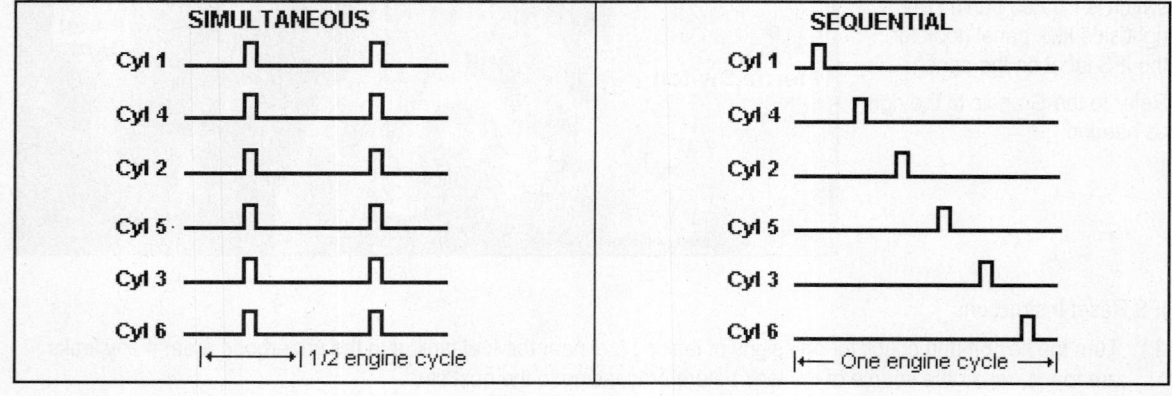

SCAN TOOL TEST (FUEL INJECTOR)

The Scan Tool can be used to quickly determine the fuel injector pulsewidth or duration. However, it is not the tool of choice for certain types of injector problems.

Connect the Scan Tool to the DLC underdash connector. Navigate through the Scan Tool menus until you get to the Ford (OEM) PID Data List.

Then select Injector Pulsewidth from the menu. The example to the right was captured with the vehicle in Park and running at Hot Idle speed.

SCAN TOOL TEST (FUEL TRIM)

Short Term fuel trim (SHRTFT) is an operating parameter that indicates the amount of short term fuel adjustment made by the PCM to compensate for operating conditions that vary from an ideal A/F ratio condition.

A negative SHRTFT number (-15%) means the HO2S is indicating to the PCM that the A/F ratio is rich, and that the PCM is trying to lean the A/F mixture. If A/F ratio conditions are ideal, the SHRTFT number will be close to 0%.

Long Term Fuel Trim

Long Term fuel trim (LONGFT) is an engine operating parameter that indicates the amount of long term fuel adjustment made by the PCM to compensate for operating conditions that vary from an ideal A/F ratio.

A positive LONGFT number (+15%) means the HO2S is indicating to the PCM that the A/F ratio is lean, and that the PCM is trying to add more fuel to the A/F mixture. If A/F ratio conditions are ideal the LONGFT number will be 0%.

The LONGFT number is displayed on the Scan Tool in percentage. Refer to the examples in the Graphic.

```
Jan 30, 02      10:17:53 am

  INJECTOR PW B1
   2.54 MSEC
  INJECTOR PW B2
   2.51 MSEC


Always use [HELP] to
  list keys & modes
[RCV] For enhanced
[*EXIT] To exit
```

```
Jan 30, 02      10:17:53 am

  SHORT TERM FT B1
   0 %
  LONG TERM FT B1
   2 %


Always use [HELP] to
  list keys & modes
[RCV] For enhanced
[*EXIT] To exit
```

```
Jan 30, 02      10:17:53 am

  SHORT TERM FT B2
   0 %
  LONG TERM FT B2
   0 %


Always use [HELP] to
  list keys & modes
[RCV] For enhanced
[*EXIT] To exit
```

Fuel Injector Test Schematic

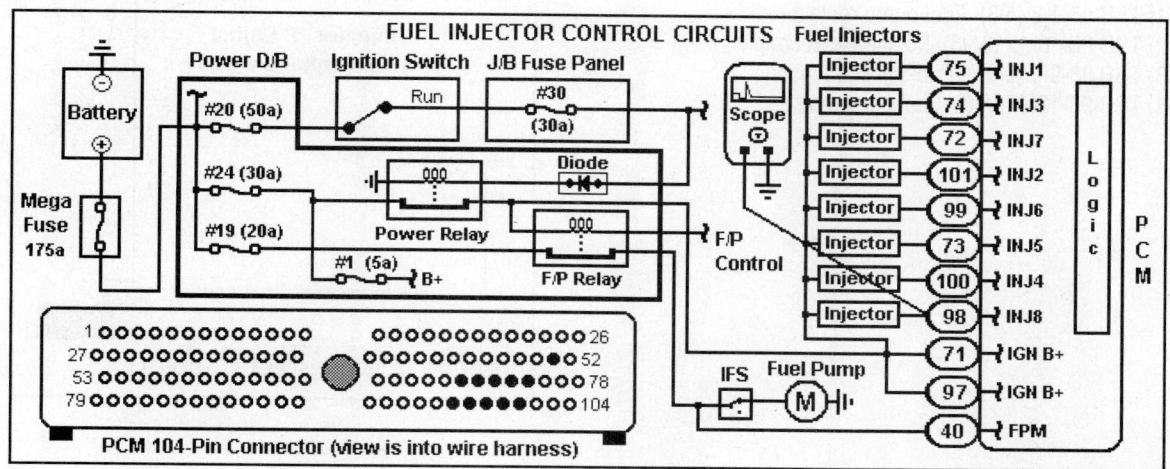

LAB SCOPE TEST (FUEL INJECTOR)

The Lab Scope is the tool of choice to test the injectors as it provides an accurate view of the injector operation. Place the gearshift selector in Park. Block the drive wheels for safety.

Scope Connections

Connect the Channel 'A' positive probe to the injector '1' control wire (Pin 75) and the negative probe to the battery negative post.

Scope Settings

To make the waveforms as clear as possible, set the scope settings to match the examples.

Lab Scope Example (1)

In example (1), the trace shows the injector waveform with the scope sweep rate set to 500 us. This setting allows a good view of the injector pulsewidth (3.5 ms in this example).

The inductive kick was 60 volts (when viewed at a 10v per division setting).

Lab Scope Example (2)

In example (2), a current probe was connected to Channel 'A' and the low amp probe was set to 100 mv (each division equals 0.2a).

The trace shows the injector current ramping signal at idle speed. Note the point at which the fuel injector opened.

The fuel injector current level in this example was less than 1 amp on this good injector.

Lab Scope Example (3)

In example (3), the voltage setting was changed from 10v to 5v per division.

This setting allows for an accurate calculation of the injector feed voltage (13.9v) and the ability to monitor the injector driver performance.

Technical Service Bulletins

The following Ford TSB's relate to problems in the Fuel system or its components:

1) TSB 98-1-8 (/98): Fuel pump replacement
2) TSB 98-16-12 (8/98): Fuel leaks at coupler
3) TSB 98-9-9 (5/98): Fuel injector testing
4) TSB 99-12-9 (9/98): Fuel pump noise

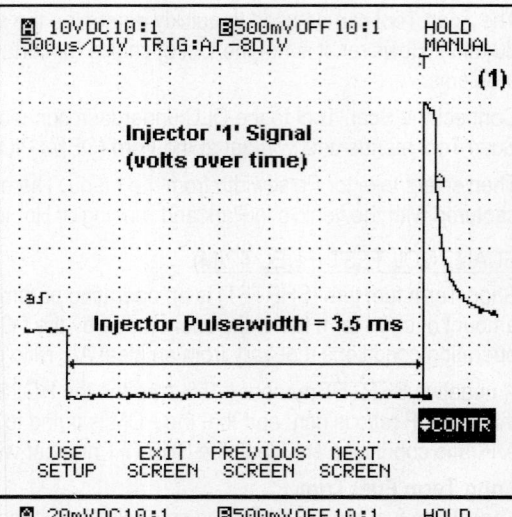

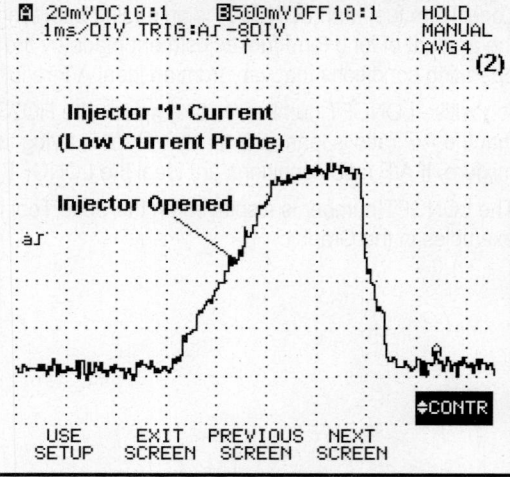

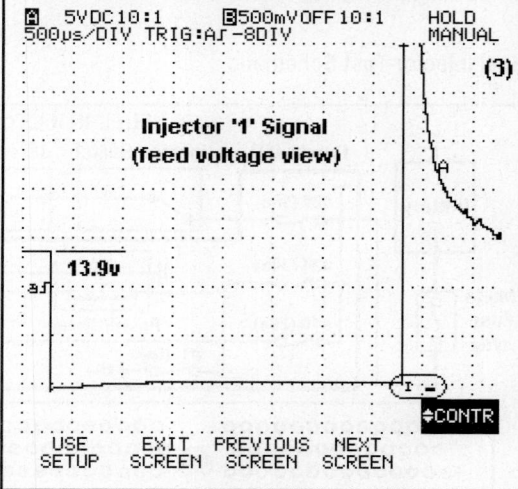

CMP SENSOR (MAGNETIC)

The camshaft position (CMP) sensor used with this application sends a signal to the PCM that indicates the camshaft position. This signal is used for fuel synchronization.

The CMP sensor is a magnetic transducer mounted on the engine underneath the driver's side coilpack adjacent to a reluctor wheel mounted on the front of the camshaft.

The reluctor on this system has one (1) tooth (there is one CMP signal for each 360º of camshaft rotation).

The CMP sensor outputs an AC signal to the PCM that indicates the top dead center (TDC) position of engine cylinder No. 1

The PCM monitors the CMP sensor signal in order to correctly synchronize the firing of the eight (8) injectors on this engine application.

CMP Sensor Circuits

The PCM is connected to the CMP sensor positive (+) circuit (DG wire) at Pin 85. The CMP sensor ground circuit is connected to Pin 91 of the PCM 104P connector.

Specification: The CMP sensor resistance is 290-790 ohms at 68ºF (actual: 488 ohms).

CMP Sensor Diagnosis (Code Help)

If the PCM does not detect any signals from the CMP sensor with the engine running it will set DTC P0340 for Camshaft Position Sensor Circuit Fault. A DVOM can be used to check for "no output", but a Lab Scope is the tool of choice to test this sensor.

CMP Sensor Schematic

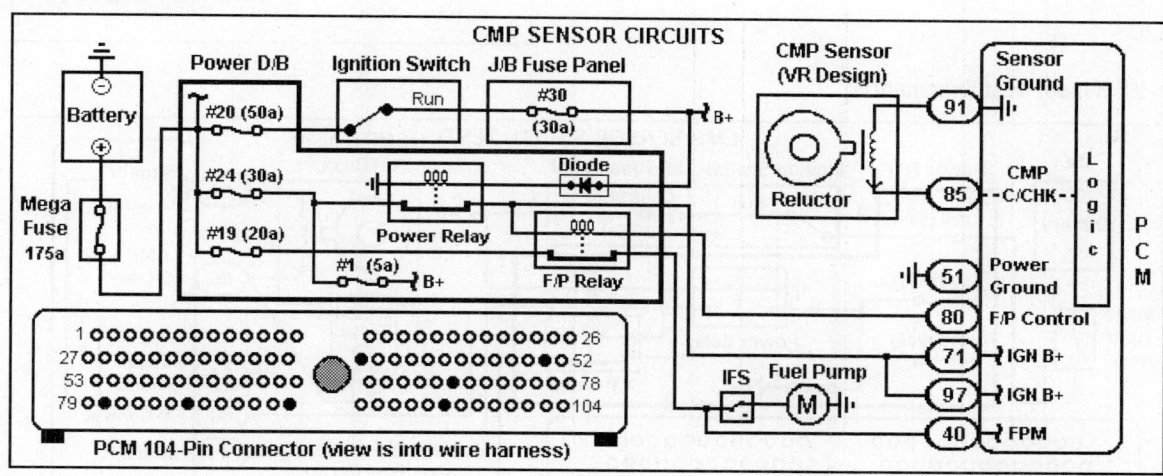

CMP SENSOR (SIGNAL EXPLANATION)

The CMP sensor input is not needed by the EDIS system since the spark plugs are fired in pairs. The CMP sensor is used by the PCM to determine cylinder No. 1 for sequential fuel control. The CMP sensor on this vehicle is a variable reluctance sensor (VRS).

The VRS produces a narrow sinusoidal signal when a tooth on the camshaft passes the pole piece. The negative-going edge is typically located at 22° ATDC of cylinder No. 1.

A VRS sensor is a magnetic transducer with a pole piece wrapped with fine wire. When the VRS encounters the rotating ferromagnetic teeth, the varying reluctance induces a voltage proportional to the rate of change in magnetic flux and the number of coil windings. As the tooth approaches the pole piece, a positive differential voltage is induced. As the tooth and the pole piece align, the signal rapidly transitions form maximum positive to maximum negative. As the tooth moves away, the voltage returns to a zero differential voltage.

The VRS signal waveform approximates a sine wave with increased amplitude adjacent to the mission tooth region at normal engine speeds. The VRS output should be 150 Millivolts peak-to-peak at 30 rpm, about 1.5 volts peak-to-peak at 300 rpm, and about 24 volts peak-to-peak at 6000 rpm.

The VRS output should not exceed 300 volts peak-to-peak at speeds above 8000 rpm.

The VRS resistance is 290 to 790 ohms.

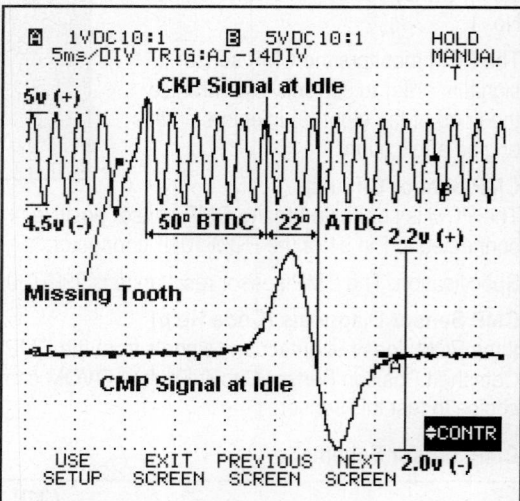

LAB SCOPE TEST (CMP SENSOR)

Place the shift selector in Park and block the drive wheels prior to starting this test.

Scope Connections

Install a breakout box if available. Connect the Channel 'A' positive probe to the CMP sensor signal at PCM Pin 85 (DG wire). Connect the negative probe to the battery ground post.

CMP Sensor Test Schematic

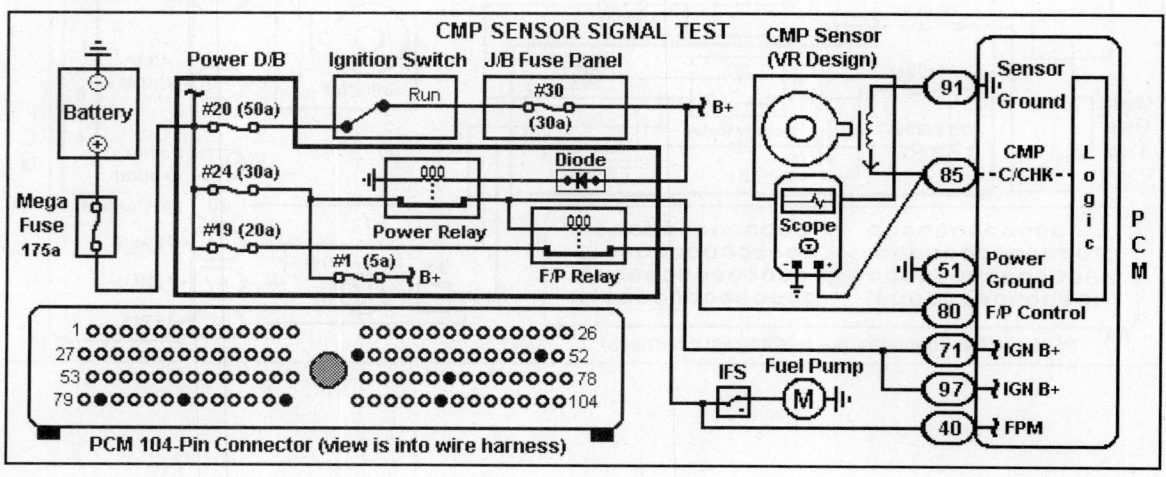

Idle Air Control Valve

GENERAL DESCRIPTION

The Idle Air Control (IAC) valve assembly is mounted to the top of the throttle body. It is designed to control engine idle speed (rpm) under all operating conditions and to provide a dashpot function.

The IAC assembly meters the intake air that flows past the throttle plate through a bypass area within the IAC assembly and throttle body.

IAC Valve

IAC Valve Location

The IAC valve is located on top of the throttle body (refer to Graphic).

The PCM determines the desired idle speed and controls the IAC motor through a duty cycle signal that changes the position of the valve to control the amount of bypass air.

Scan Tool Test (IAC Valve)

Connect the Scan Tool to the DLC underdash connector. Navigate through the Scan Tool menus until you get to the Ford (OEM) Engine Data List (a list of PID items).

Select the RPM AND IAC PID items from the menu so that you can view them separately on the Scan Tool display.

Scan Tool Example (1)

An example of the ENGINE SPEED (RPM) and IDLE AIR CONTROL (%) PID for this vehicle at Hot Idle speed is shown in the Graphic. The first example was captured at Hot Idle speed in Park with all accessories turned off. The second example was captured with the A/C on in Drive.

Scan Tool Example (2)

In Example (2), the IAC PID was captured at Hot Idle speed with the A/C on in Drive (brake on). Note the change in the IAC reading (38% to 44%) with the A/C turned on in Drive.

```
Jan 30, 02    10:17:53 am (1)

ENGINE SPEED
  750 RPM
IDLE AIR CNTRL
  38 %


Always use [HELP] to
  list keys & modes
[RCV] For enhanced
[*EXIT] To exit
```

```
Jan 30, 02    10:17:53 am (2)

ENGINE SPEED
  642 RPM
IDLE AIR CNTRL
  44 %


Always use [HELP] to
  list keys & modes
[RCV] For enhanced
[*EXIT] To exit
```

Diagnostic Tip

To test the IAC valve, record the reading (accessories off) at Idle in Park. Drive the vehicle and then return to the shop. The IAC PID readings should be similar each time you place the gearshift selector in Park.

Technical Service Bulletins

There are TSB's that relate to problems on this application.

1) TSB # 98-21-10 from 10-26-98 (a no-code idle concern).
2) TSB # 98-21-21 from 10-26-98. This TSB deals with a fast idle concern that occurs with DTC P1506. The fix is to clean any hose fragments out of IAC valve.

LAB SCOPE TEST (IAC VALVE)

The Lab Scope can be used to monitor the IAC Valve control signal under various engine load, speed and temperature conditions. Place the gearshift selector in Park or Neutral (A/T) and block the drive wheels.

Scope Connections

Connect the breakout box (BOB). Connect the Channel 'A' positive probe to the IAC Valve circuit at PCM Pin 83 (WHT/LB wire) and negative probe to battery negative.

Scope Settings

To make the waveforms as clear as possible, set the scope settings to match the examples.

Lab Scope Example (1)

In this example, the trace shows the IAC valve waveform in Park with an accessory load.

Lab Scope Example (2)

In this example, the trace shows the IAC valve waveform in Park without any load.

The IAC Valve trace should show a change in the digital on/off signal as the engine speed (rpm) is changed up and down.

If it does not change or the square wave pattern is broken or jagged, a problem may exist in the IAC Valve or its related control or feed circuit.

Circuit Description

The IAC valve is connected to battery power through the Power D/B and Power Relay.

The IAC assembly includes an internal diode to protect the PCM against voltage spikes.

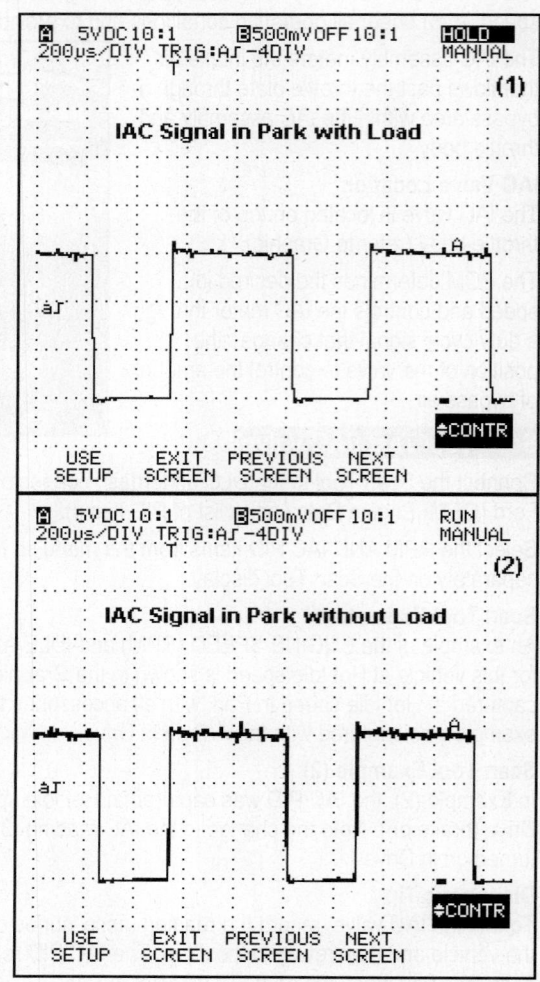

IAC Valve Test Schematic

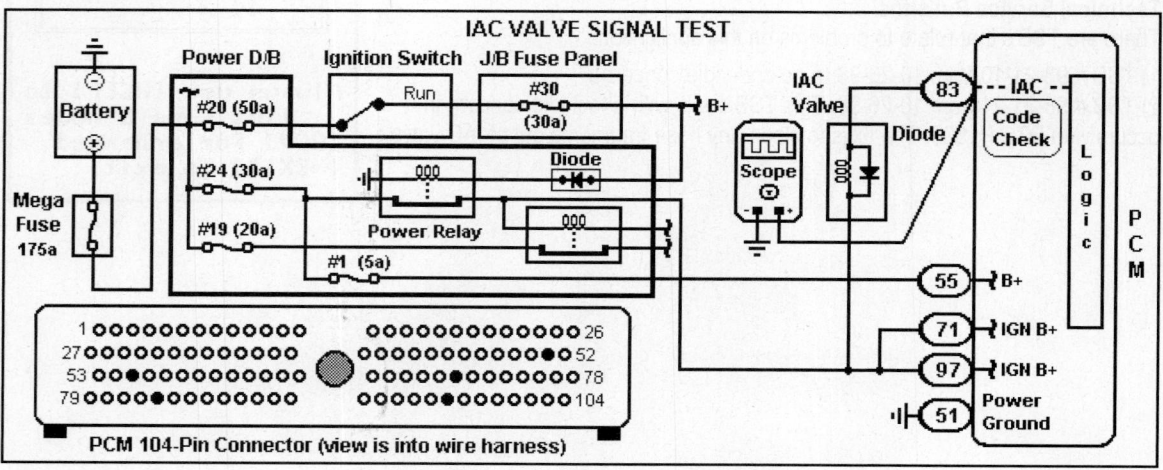

Input Sensors

CYLINDER HEAD TEMPERATURE SENSOR

The cylinder head temperature (CHT) sensor is mounted in the aluminum cylinder head where it measures the metal temperature (it is not connected to any coolant passages).

The CHT sensor communicates an overheating condition to the PCM. If this occurs, the PCM will initiate a cooling strategy based on signals from the ECT sensor. The Cooling system strategy is used to prevent damage by allowing the air-cooling of the engine.

Thermistor Operation

The resistance of the thermistor decreases as the temperature of the cylinder head increases, and its resistance increases as the temperature of the cylinder head decreases (an inverse ratio).

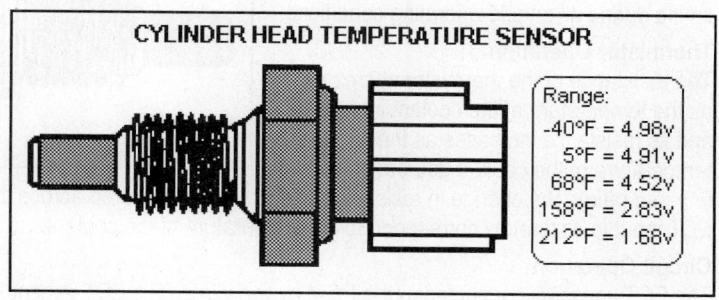

CYLINDER HEAD TEMPERATURE SENSOR

Range:
-40°F = 4.98v
5°F = 4.91v
68°F = 4.52v
158°F = 2.83v
212°F = 1.68v

The change in resistance affects the voltage drop across the sensor circuit and this action causes the CHT input to accurately correspond to the cylinder head temperature.

Circuit Operation

The CHT sensor is connected to the PCM by two wires. The CHT sensor is connected to the PCM by a 5v signal and a sensor ground circuit. Refer to the Schematic below.

The PCM supplies the 5v signal circuit to the CHT sensor through a resistor in series with it. As the resistance value of the sensor changes in accordance with changes in cylinder head temperature, the voltage potential of the CHT sensor signal circuit changes. Note the "Code Check" circuit for this sensor connects to the PCM at Pin 66.

DVOM TEST (CHT SENSOR)

Connect the DVOM positive probe to the CHT sensor signal circuit (YEL/LG wire) at PCM Pin 66. Connect the negative probe to the sensor ground wire (GRY/RED) at PCM Pin 91 of the 104P connector or chassis ground. The readings taken with a DVOM can be used to determine if a circuit is operational, or if the CHT sensor had drifted out of its normal operating range. Refer to the range chart shown above.

CHT Sensor Test Schematic

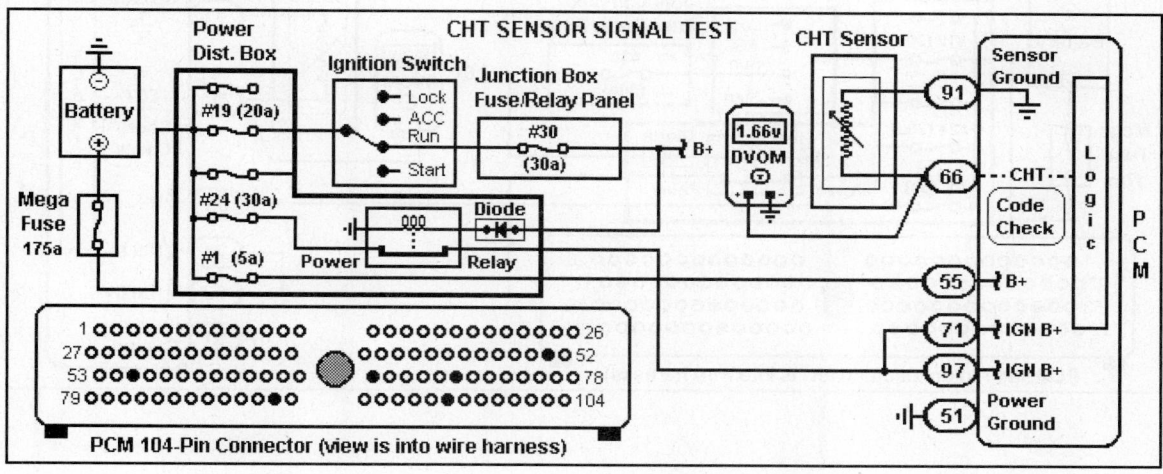

ENGINE COOLANT TEMPERATURE SENSOR

The engine coolant temperature (ECT) sensor is located at the top of the engine in a water jacket. This device is used to sense the temperature of the engine coolant.

The ECT sensor includes a thermistor designed to change its internal resistance value according to changes in the temperature of engine coolant.

The PCM uses the information from the ECT sensor in order to determine the amount of fuel injector pulsewidth, idle speed and spark timing during all engine-operating conditions.

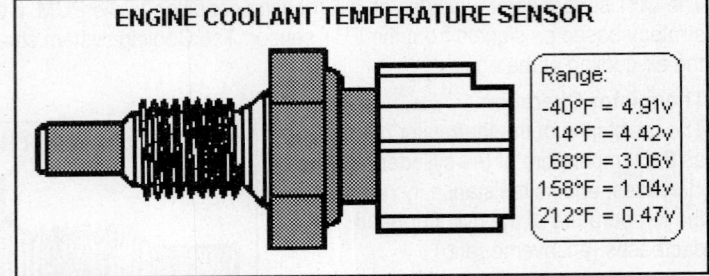

Thermistor Operation

The resistance of the thermistor decreases as the temperature of the coolant increases, and its resistance increases as the temperature of the coolant decreases (inverse ratio). The change in resistance affects the voltage drop across the sensor circuit and this action causes the ECT input to accurately correspond to the temperature of the coolant.

Circuit Operation

The ECT sensor is connected to the PCM by two wires. The ECT sensor is connected to the PCM by a 5v signal and a sensor ground circuit. Refer to the Schematic below.

The PCM supplies the 5v signal circuit to the ECT sensor through a resistor in series with it. As the resistance value of the sensor changes in accordance with changes in coolant temperature, the voltage potential of the ECT sensor signal circuit also changes. Note that the "Code Check" circuit for the ECT sensor connects to the PCM at Pin 38.

DVOM TEST (ECT SENSOR)

Connect the DVOM positive probe to the ECT sensor signal circuit (LG/RED wire) at PCM Pin 38. Connect the negative probe to the sensor ground wire (GRY/RED) at PCM Pin 91 of the 104P connector or to the battery negative ground post. The readings taken with a DVOM can be used to determine if a circuit is operational, or if the ECT sensor had drifted out of its normal operating range. Refer to the range chart shown above.

ECT Sensor Test Schematic

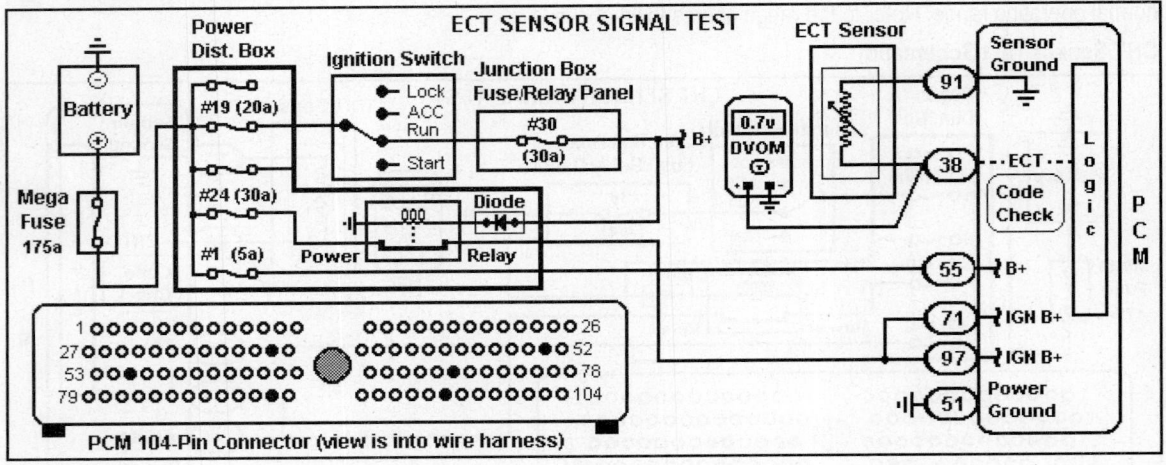

THROTTLE POSITION SENSORS

The Throttle Position (TP) sensor is mounted to the throttle body where it detects changes in the throttle valve angle.

The PCM uses the TP sensor analog voltage DC signal to detect the following vehicle driving conditions:

- Air Fuel Ratio Correction
- Fuel Cut Control
- Power Increase Correction

TP Sensor Circuits

The three circuits that connect the TP sensor to the PCM are listed below:

- The VREF 5v circuit
- The TP sensor signal circuit
- The TP sensor ground return circuit

Circuit Operation

With the throttle fully closed, the TP sensor input to the PCM is from 0.5-1.27v at Pin 89.

The TP sensor voltage increases in proportion to the throttle valve-opening angle. The TP sensor signal is from 3.6-4.8v with the throttle valve fully open.

Note that the "Code Check" circuit for the TP sensor connects to the PCM at Pin 89. In effect, this is the critical circuit that the PCM monitors in order to detect a circuit fault or a fault in the relationship between the TP sensor input and the MAF sensor input.

TP Sensor Signal Test Schematic

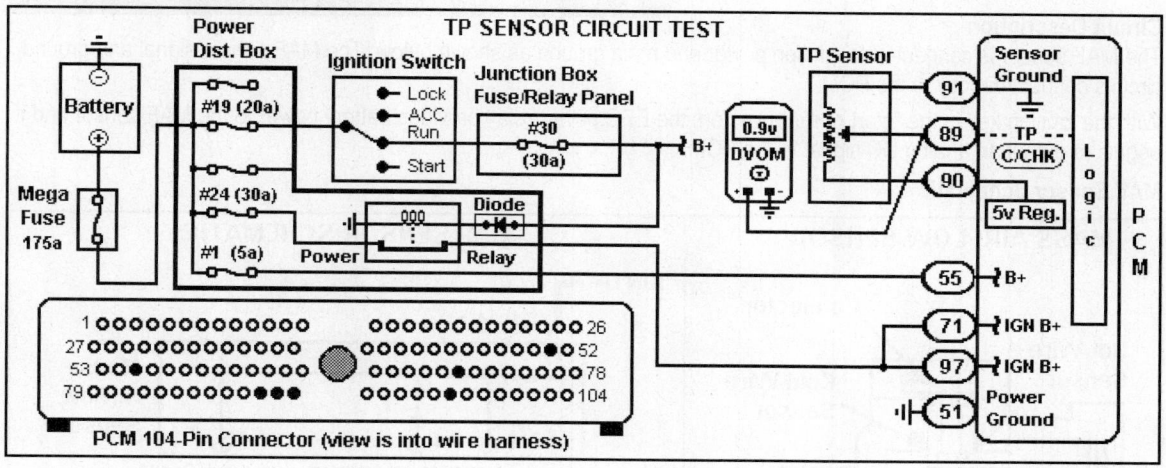

Mass Airflow Sensor

GENERAL DESCRIPTION

he mass airflow (MAF) sensor is located in the air cleaner before the throttle body. This MAF sensor uses a Hot-wire sensing element to measure the amount of air that enters the engine. Air passing through the device flows over the hot wire element and causes it to cool. The hot wire element is maintained at 392ºF (200ºC) above the current ambient temperature as measured by the constant cold wire part of the sensor.

The current level required to maintain the temperature of the hot wire element is proportional to amount of air flowing passing the sensor. The MAF input to the PCM is an analog signal proportional to the intake air mass.

The PCM calculates the desired fuel injector on time in order to deliver the desired A/F ratio. The PCM also uses the MAF sensor to determine transmission EPC solenoid shifting and the TCC solenoid scheduling.

■ NOTE: *When this sensor fails, it may set a trouble code or cause a No Code Fault (engine cold or hot).*

Circuit Description

The MAF sensor is connected to ignition power and main ground as shown below. The MAF sensor signal and ground circuits connect directly to the PCM.

With the ignition key in the "start or run" position, the EEC power relay provides battery power to the MAF sensor and it begins to output an analog DC signal to the PCM.

MAF Sensor Schematic

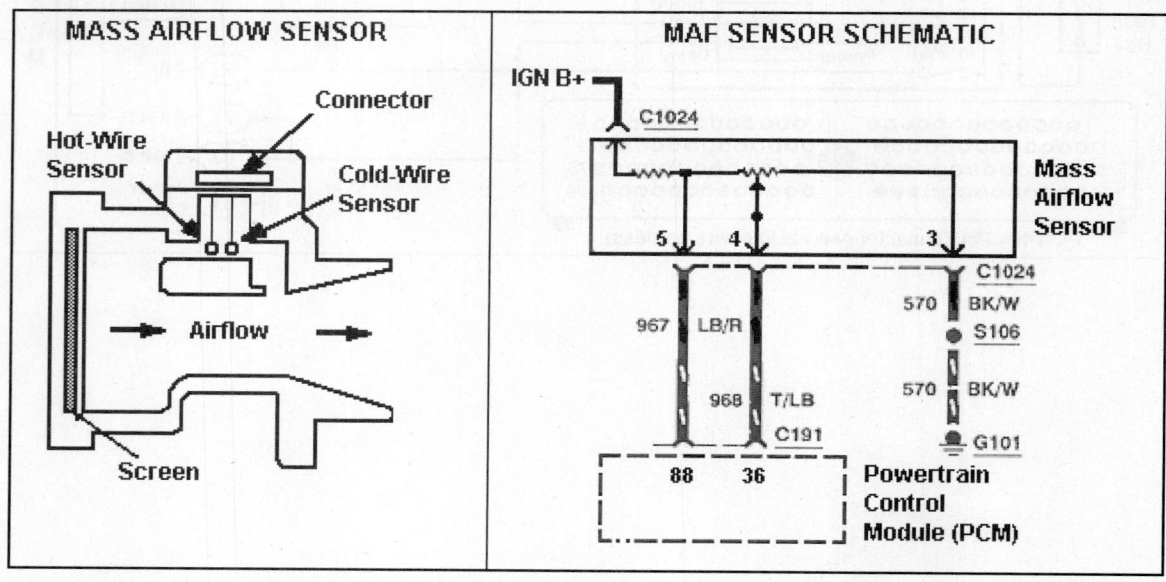

BARO SENSOR CALIBRATION

The barometric pressure (BARO) sensor signal, which is an inferred value by the PCM, can be used to diagnose a "suspect" MAF sensor. The PCM updates the BARO reading during conditions of a high throttle opening. Drive the vehicle using 3-4 sustained heavy throttle operations and then record the updated BARO PID. Compare the reading to the values in the table. If the BARO reading indicates a higher altitude than actual, the MAF sensor is contaminated. The updated reading should be ± 4 Hz of the values below.

The MAF sensor signal should also be checked for loss of calibration. This step can be done using the MAF PID from the Scan Tool at cold or hot engine startup at various engine speeds. Refer to the article on the next page for more information on this test.

BARO Sensor Calibration Table

BARO Pressure (" Hg)	BARO Pressure (kPa)	BARO/MAP PID (Hz)	Altitude over Sea Level
30 In. Hg	101.3 kPa	158.9 Hz	0 feet (sea level)
29 In. Hg	97.9 kPa	155.8 Hz	1,000 feet.
28 In. Hg	94.6 kPa	152.8 Hz	2,000 feet
27 In. Hg	91.2 kPa	149.8 Hz	3,000 feet
26 In. Hg	87.8 kPa	146.8 Hz	4,000 feet
25 In. Hg	84.4 kPa	144.0 Hz	5,000 feet
24 In. Hg	81.1 kPa	141.1 Hz	6,000 feet
23 In. Hg	77.7 kPa	138.3 Hz	7,000 feet
22 In. Hg	74.3 kPa	135.4 Hz	8,000 feet

SCAN TOOL TEST (MAF SENSOR)

Connect the Scan Tool to the underdash connector. Navigate through the Scan Tool menus until you get to the Ford OEM Data List (a list of PID items). Select the MAF and BARO PID items from the menu so that you can view them on the Scan Tool display.

Scan Tool Examples
An example of the MAF (V) PID and BARO (F) PID for this vehicle at Hot Idle speed is shown in the Graphic.

MAP Sensor PID Explanation
Start with the engine cold or warm and monitor the MAF PID. The MAF PID voltage is 0.75v at hot idle. The voltage should change voltage smoothly as the engine speed is increased (the second reading was captured at 2000 rpm)

Compare the "actual" MAF PID reading to the values shown in the Voltage to Mass Airflow Conversion Table. The "actual" MAF PID reading (the voltage reading) should be very close to the MAF PID reading in the calibration table in the Reference Information in Section 1.

BARO Sensor PID Explanation
Turn to key on, engine off and monitor the BARO PID on the Scan Tool. Record the reading for comparison.

Compare the "actual" BARO PID reading to the values shown in the BARO Sensor Calibration Table on this page. The "actual" BARO PID hertz reading should be within ± 4 Hz of the reading in the BARO calibration table above.

```
Jan 30, 02    10:17:53 am

 MAF (V)
  0.75 VOLTS
 BARO (F)
  158.9 HZ


Always use [HELP] to
  list keys & modes
[RCV] For enhanced
[*EXIT] To exit
```

```
Jan 30, 02    10:17:53 am

 MAF (V)
  1.43 VOLTS
 BARO (F)
  158.9 HZ


Always use [HELP] to
  list keys & modes
[RCV] For enhanced
[*EXIT] To exit
```

LAB SCOPE TEST (MAF SENSOR)

The Lab Scope can be used to monitor the MAF sensor output signal under various engine load and speed conditions. Place the gearshift selector in Park prior to testing.

Scope Connections

Connect the Channel 'A' positive probe to the MAF signal at the PCM Pin 88 (LB/RED wire) and connect the Channel 'A' negative probe to the battery negative ground post.

Scope Settings

To make the waveforms as clear as possible, set the scope settings to match the examples.

The MAF sensor waveforms can have slight differences from one scope to another.

Lab Scope Example (1)

In Example (1) the trace shows the MAF sensor signal at idle speed. This waveform was captured with the vehicle in Park, the engine at idle and running in closed loop.

The MAF sensor trace should show an even change in its analog signal as the engine speed is changed up and down. It if drops or rises suddenly without a corresponding change in engine speed, there may be a problem in the MAF sensor.

Lab Scope Example (2)

In Example (2) the trace shows the MAF sensor signal at cruise speed (2500 rpm). Note the change in voltage from the Hot Idle value (0.7v) to the Cruise value (1.5v).

Lab Scope Example (3)

In Example (3) the trace shows the MAF sensor signal during a Snap Throttle event.
There are three distinct events shown here:

- The first event is the point at which the throttle begins to open (the point where the signal starts to rise).
- The second event is where the pressure in the intake manifold begins to equalize.
- The third event is where the throttle closes.

Diagnostic Tip

You can use the capture in Example (3) to compare the known good operation of this MAF sensor during a Snap Throttle event to the operation of a MAF sensor that you suspect may be faulty or dirty.

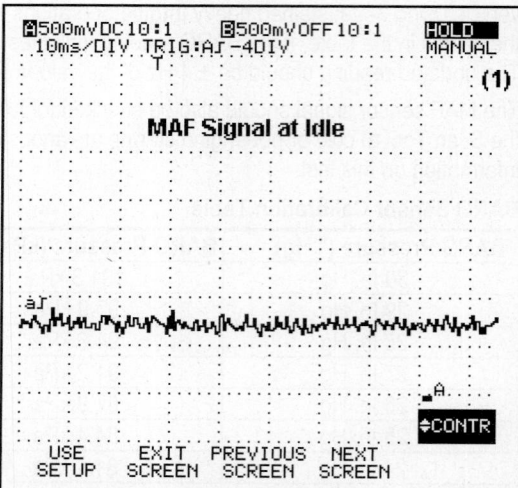

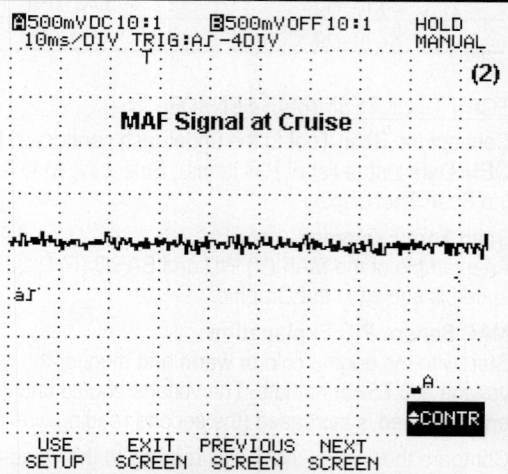

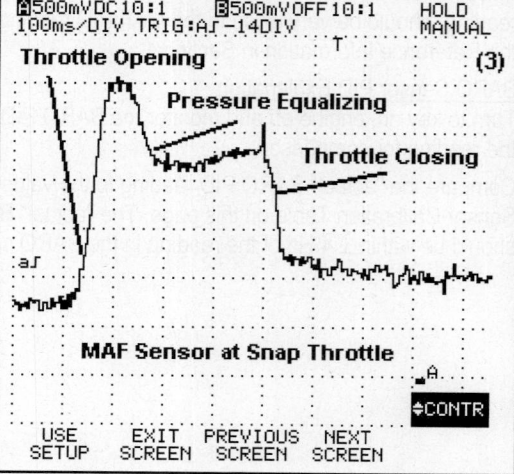

Oxygen Sensor

GENERAL DESCRIPTION

The heated oxygen sensor (HO2S) is mounted in the Exhaust system where it monitors the oxygen content of the exhaust stream. During engine operation, oxygen present in the exhaust reacts with the HO2S to produce a voltage output.

Simply stated, if the A/F mixture has a high concentration of oxygen in the exhaust, the signal to the PCM from the HO2S will be less than 0.4v. If the A/F mixture has a low concentration of oxygen in the exhaust, the signal to the PCM will be over 0.6v.

Heated Oxygen Sensors

This vehicle application is equipped with four oxygen sensors that have internal heaters allow the sensors to warmup very quickly in cold weather.

The front (or pre-catalyst) oxygen sensor is mounted in front of the catalytic converter. The rear oxygen sensor is mounted after the catalytic converter.

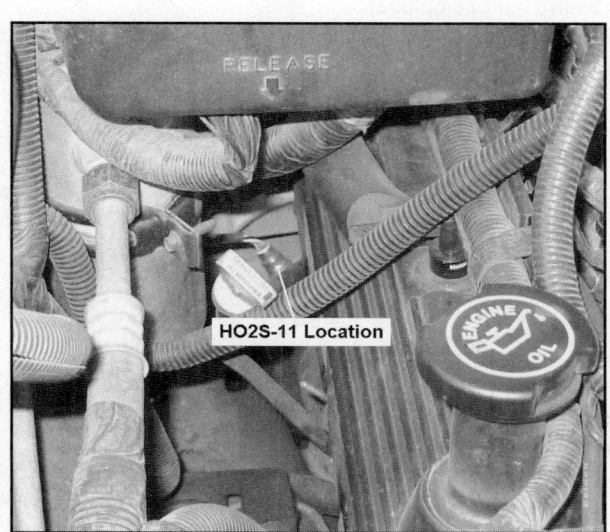

Oxygen Sensor Identification

The front heated oxygen sensor is identified as either a Bank 1 HO2S-11 or a Bank 2 HO2S-21. The heated oxygen sensor identified as Bank 1 is always located in the engine bank that contains cylinder No. 1.

Note that the "Code Check" circuit for the HO2S-11 signal connects to the PCM at Pin 60. All of the HO2S circuits for this engine are shown in the Graphic.

Circuit Description

The power relay in the Power D/B supplies battery voltage to the internal heater circuits to the front and rear oxygen sensors. The heaters are also connected to the heater "control" circuits at the PCM. Refer to the Graphic directly below to view the circuits.

The PCM decides when to enable the HO2S heaters to protect their ceramic elements and when to run the HO2S Heater Tests. Note the Code Check circuits in the Graphic.

Heated Oxygen Sensor Schematic

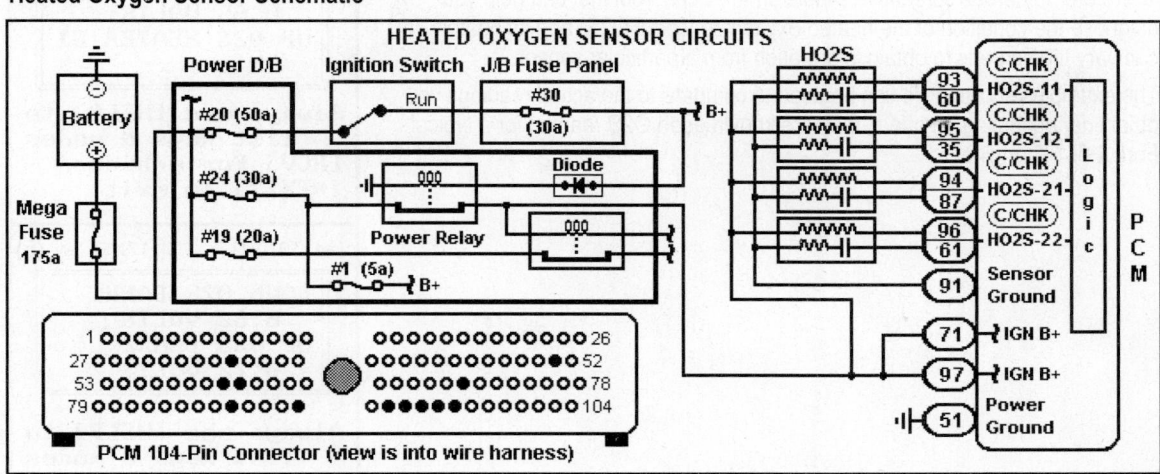

SCAN TOOL TEST (HO2S)

Connect the Scan Tool to the DLC underdash connector. Navigate through the Scan Tool menus until you get to the Ford (OEM) Data List (a list of PID items). Raise the engine speed to 2000 rpm for 3 minutes to set up the test. The examples shown here were captured at Hot Idle in 'P'.

Then select the type of oxygen sensor (O2S) information that you want to view from the various options in the menu.

Scan Tool Example (1)

This example shows the On/Off status of the heaters for the Downstream O2S for Cylinder Bank 1 and Upstream O2S for Cylinder Bank 2. The operation of these heaters is controlled by the PCM (shortly after engine startup).

Scan Tool Example (2)

This example shows the On/Off status of the heater for the Downstream O2S for Cylinder Bank 2 and the current level of the heater for the Upstream O2S for Cylinder Bank 1. The current level shown here (812.5 mA) is the correct amount of current for a Ford HO2S (a known good value).

Scan Tool Example (3)

This example shows the current level of the heater for the Downstream O2S for (914.1 mA) Cylinder Bank 2 and the current level (878.9 mA) for the Downstream O2S for Cylinder Bank 1.

Scan Tool Example (4)

This example shows the voltage level of the Downstream O2S for Cylinder Bank 2 and the On/Off status of the heater for the Upstream O2S for Cylinder Bank 1. The DOWN O2S BANK2 voltage reading is in the normal range for a Ford heated oxygen sensor (0.100 to 900 Millivolts).

Scan Tool Example (5)

This example shows the voltage level of the Downstream O2S for Cylinder Bank 1 and the voltage level of the Upstream O2S for Cylinder Bank 2. Both of the voltage readings in this example are within the normal range for a Ford heated oxygen sensor (0.100 to 900 Millivolts).

Summary

There are numerous screens available on the Scan Tool that can help you diagnose the condition of the heated oxygen sensor (HO2S) and its heater. You can vary the screens to obtain information from a particular sensor.

The examples shown here can be used to compare to the actual readings you obtain during testing. In effect, they are known good O2S reading for a typical Ford 4.6L V8 engine.

```
Jan 30, 02    10:17:53 am (1)

  DN O2S HEATER B1
    ON
  UP O2S HEATER B2
    ON

Always use [HELP] to
 list keys & modes
[RCV] For enhanced
[*EXIT] To exit
```

```
Jan 30, 02    10:17:53 am (2)

  DN O2S HEATER B2
    ON
  UPO2S HTR MON B1
    812.5 mA

Always use [HELP] to
 list keys & modes
[RCV] For enhanced
[*EXIT] To exit
```

```
Jan 30, 02    10:17:53 am (3)

  DNO2S HTR MON B2
    914.1 mA
  DNO2S HTR MON B1
    878.9 mA

Always use [HELP] to
 list keys & modes
[RCV] For enhanced
[*EXIT] To exit
```

```
Jan 30, 02    10:17:53 am (4)

  DOWN O2S BANK2
    0.65 VOLTS
  UP O2S HEATER B1
    ON

Always use [HELP] to
 list keys & modes
[RCV] For enhanced
[*EXIT] To exit
```

```
Jan 30, 02    10:17:53 am (5)

  DOWN O2S BANK1
    0.62 VOLTS
  UP O2S BANK2
    0.72 VOLTS

Always use [HELP] to
 list keys & modes
[RCV] For enhanced
[*EXIT] To exit
```

LAB SCOPE TEST (HO2S)

The Lab Scope is the tool of choice to test the oxygen sensor as it provides an accurate view of the sensor response and switch rate. Place the gearshift selector in Park and block the drive wheels prior to starting the test.

Lab Scope Connections for Example (1)
Connect the Channel 'A' positive probe to the HO2S-11 signal and the Channel 'B' probe to the HO2S-12 signal. Connect the Channel 'A' negative probe to the battery negative post.

Lab Scope Connections for Example (2)
Connect the Channel 'A' positive probe to the HO2S-21 signal and the Channel 'B' probe to the HO2S-22 signal. Connect the Channel 'A' negative probe to the battery negative post.

Scope Settings
To make the waveforms as clear as possible, set the scope settings to match the examples.

Depending upon the scope capabilities, the sensor waveforms will have slight differences.

Lab Scope Tests
The examples to the right are from the original oxygen sensors (54,400 miles). The vehicle was driven long enough for the engine to enter closed loop operation and tested in Park.

Lab Scope Example (1)
In example (1), the two traces show a "known good" set of heated oxygen sensor waveforms for engine bank No. 1 (the right bank). The top trace is from the oxygen sensor labeled as HO2S-12 (the post catalyst oxygen sensor). The bottom trace is from the oxygen sensor labeled as HO2S-11 (the front or pre-catalyst oxygen sensor).

Lab Scope Example (2)
In example (1), the two traces show a "known good" set of heated oxygen sensor waveforms for engine bank No. 2 (the left bank). The top trace is from the oxygen sensor labeled as HO2S-22 (the post-catalyst oxygen sensor). The bottom trace is from the oxygen sensor labeled as HO2S-21 (the pre-catalyst oxygen sensor).

Oxygen Sensor Test Summary
The HO2S signal should switch L-R and R-L in less than 100 ms, and should switch at least 3-5 times per second.

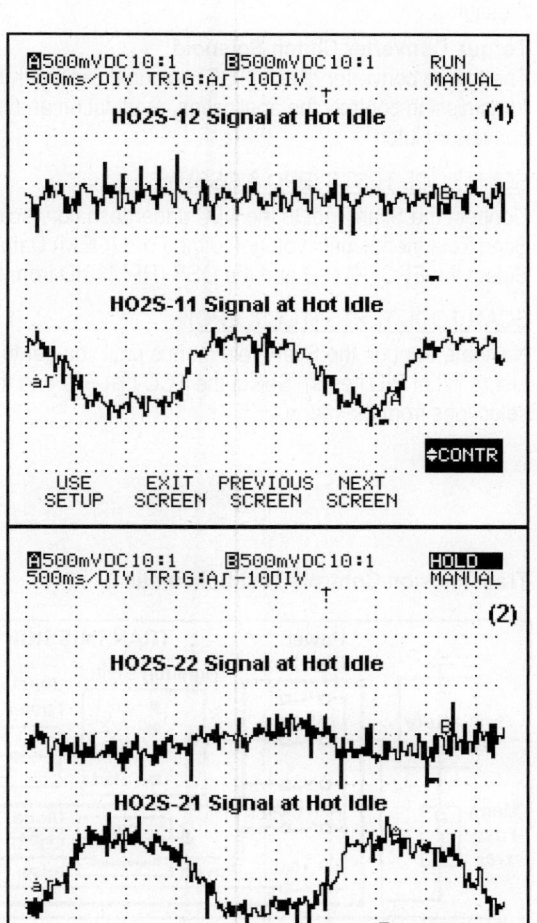

Transmission Controls

COMPONENT TESTS

This vehicle is equipped with a 4-speed 4R70W automatic transmission with a Ravigneaux-style double-pinion gear set with 2 bands, 2 one-way roller clutches and friction clutches that produce 4 Forward gears and a Reverse gear. The PCM controls upshift, downshift and TCC functions.

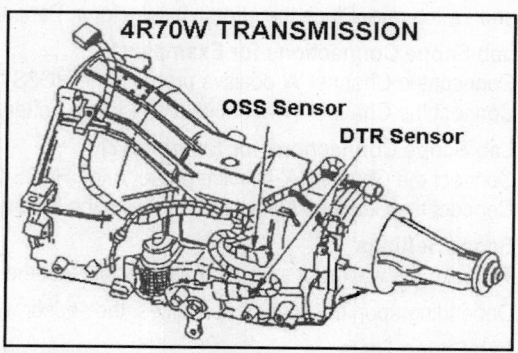

Output Shaft Speed Sensor

The output shaft speed (OSS) sensor is a magnetic pickup located at the output shaft ring gear. Its signal is used for TCC lockup, shift scheduling and EPC solenoid pressure.

Electronic Pressure Control

The EPC solenoid is a device used to regulate transmission line and line modulator pressure by producing resisting forces to the main regulator and line modulator circuits. The pressures control the clutch application pressures.

Torque Converter Clutch Solenoid

The torque converter clutch (TCC) solenoid used with the 4R70W automatic transmission controls the application, modulation and release of the torque converter clutch.

SCAN TOOL TESTS (EPC & OSS)

Connect the Scan Tool to the DLC underdash connector. Navigate through the Scan Tool menus until you get to the Ford (OEM) Data List (a list of PID items). Select the EPC (P) PID and the OSS (RPM) PID from the menu selections.

SCAN TOOL TESTS (TCC & EPC)

Navigate through the Scan Tool menus until you get to the Ford (OEM) Data List (a list of PID items). Select the TCC (%) PID and the EPC (V) PID selections from the menu.

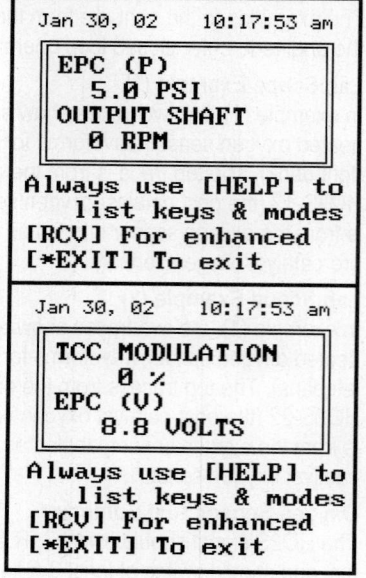

Transmission Controls Test Schematic

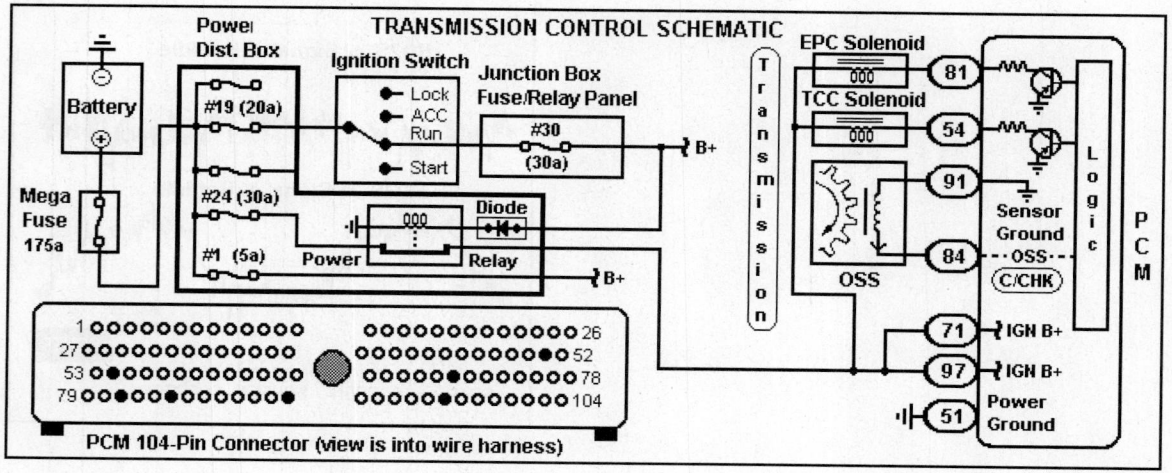

COMPONENT TESTS

This vehicle is equipped with a 4-speed 4R70W automatic transmission with automatic shift control. The PCM controls all upshift, downshift, line pressure and TCC operations. The captures on this page were done in Park at Hot Idle.

Digital Transmission Range Sensor

The digital transmission range (TR) sensor is located on the outside of the transmission at the manual lever. The TR sensor completes the start circuit in Park and Neutral, the backup lamp circuit in Reverse and the Neutral sense circuit in Neutral (4X4 only). The TR sensor also opens and closes a set of switches that are monitored by the PCM. Refer to the Operating Range Charts in Section 1 for examples of the TR sensor values.

Transmission Fluid Temperature Sensor

The transmission fluid temperature (TFT) sensor is located on the transmission main control body. The PCM monitors the voltage drop across the sensor (the sensor value changes) to determine the transmission fluid temperature. The PCM uses the initial TFT sensor signal to determine whether a cold start shift schedule is necessary. The cold start shift schedule allows for quicker shifts, allows the PCM to inhibit TCC lockup and to correct the EPC pressures when the transmission fluid is cold.

SCAN TOOL TESTS (TR SENSOR)

Connect the Scan Tool to the DLC underdash connector. Navigate through the Scan Tool menus until you get to the Ford (OEM) Data List (a list of PID items). Select the TR Selector PID (Position and Volts) from the menu selections.

SCAN TOOL TESTS (TFT SENSOR)

Connect the Scan Tool to the DLC underdash connector. Navigate through the Scan Tool menus until you get to the Ford (OEM) Data List (a list of PID items). Select the TFT Sensor PID (Degrees and Volts) from the menu selections.

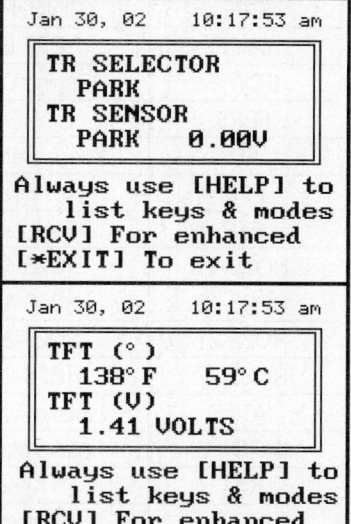

```
Jan 30, 02      10:17:53 am

TR SELECTOR
    PARK
TR SENSOR
    PARK    0.00V

Always use [HELP] to
  list keys & modes
[RCV] For enhanced
[*EXIT] To exit
```

```
Jan 30, 02      10:17:53 am

TFT (°)
   138° F    59° C
TFT (V)
   1.41 VOLTS

Always use [HELP] to
  list keys & modes
[RCV] For enhanced
[*EXIT] To exit
```

Transmission Controls Test Schematic

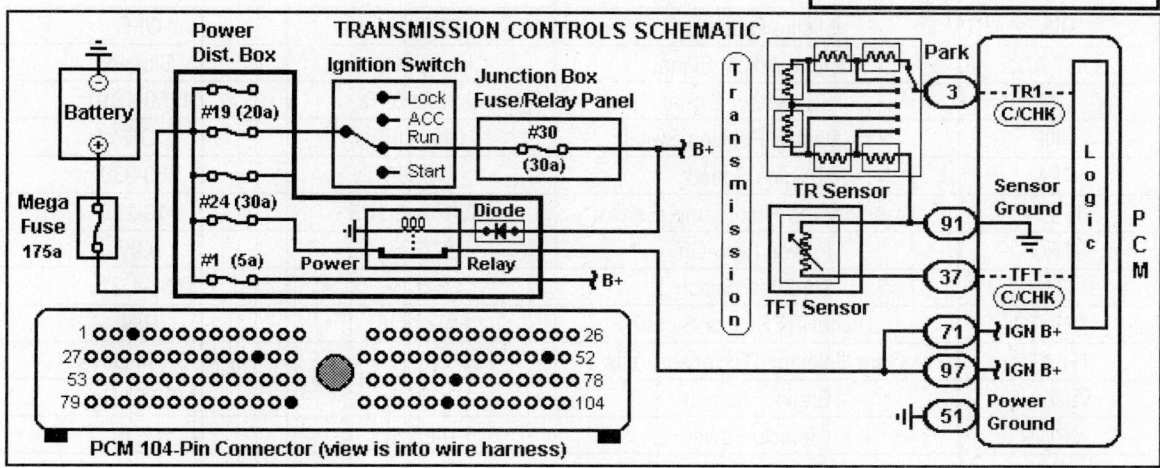

TRANSMISSION CONTROLS SCHEMATIC

PCM 104-Pin Connector (view is into wire harness)

Reference Information

PCM PID TABLES - INPUTS

■ NOTE: *The following readings were obtained with the engine at idle speed, radiator hose hot, throttle closed, gear selector in Park and all accessories turned off.*

Acronym	Scan Tool Parameter	Range	Typical Value
4x4L	4x4 Signal	ON / OFF	OFF (ON if on)
ACCS	A/C Cycling Clutch Switch Signal	ON / OFF	OFF (ON with A/C on)
BPP	Brake Pedal Position Switch	ON / OFF	OFF (ON with brake on)
CMP (CID)	CMP Sensor Signal	0-1000 Hz	13-16
CKP	CKP Sensor Signal	0-10,000 Hz	430-475
CPP	Clutch Pedal Position Switch	ON / OFF	OFF (ON with clutch in)
DPFE	DPF EGR Sensor Signal	0-5.1v	0.4-0.6
ECT °F	Engine Coolant Temp. Sensor	-40°F to 304°F	160-200
FEPS	Flash EEPROM	0-5.1v	0.5-0.6
FLI	Fuel Level Indicator Signal (volts)	0-5.1v	1.7v (1/2 full)
FPM	Fuel Pump Monitor	ON / OFF	ON
FTP	Fuel Tank Pressure Sensor	0-5.1v or 0-10" H2O	2.6v (0 in. Hg - cap off)
HO2S-11	HO2S-11 (Bank 1 Sensor 1)	0-1100 mv	100-900 mv
HO2S-12	HO2S-12 (Bank 1 Sensor 2)	0-1100 mv	100-900 mv
HO2S-21	HO2S-21 (Bank 2 Sensor 1)	0-1100 mv	100-900 mv
HO2S-22	HO2S-22 (Bank 2 Sensor 2)	0-1100 mv	100-900 mv
IAT °F	Intake Air Temperature Sensor	-40°F to 304°F	50-120
IMRC1-M	Intake Manifold Runner Control	0-5.1v	5v
KS1	Knock Sensor Signal	0-5.1v	0v (no knocking)
LOAD	Calculated Engine Load	0-100%	10-20
MAF V	Mass Airflow Sensor Signal	0-5.1v	0.6-0.9
MISF	Misfire Detection	ON / OFF	OFF
OCT ADJ	Octane Adjust Signal	Closed / Open	Closed
OSS	TSS Sensor Signal	0-10,000 Hz	1250-1310
PNP	Park Neutral Position Switch	ON / OFF	ON
RPM	Engine Speed	0-9999	680-830
TFT °F	Transmission Fluid Temp. Sensor	-40°F to 304°F	110-210
TPO	Power Take-Off	0-25.5v	0.9v
TP V	Throttle Position	0-5.1v	0.53-1.27v
TRV/TR	Digital TR Sensor Signal	PN or O/D	P/N
TRV/TR	Gear Selector TR Sensor Volts	0-5.1v	In 'P': 0v, In Drive: 12v
Vacuum	Engine Vacuum	0-30" Hg	19-20" Hg
VPWR	Vehicle Power	0-25.5v	14.1v
VSS	Vehicle Speed	0-159 mph / 0-1000 Hz	55 mph = 125 Hz

PCM PID TABLES - OUTPUTS

■ NOTE: *The following readings were obtained with the engine at idle speed, radiator hose hot, throttle closed, gear selector in Park and all accessories turned off.*

Acronym	Scan Tool Parameter	Range	Typical Value
CD1	Coil 1 Driver (Dwell)	0-60° dwell	6°
CD2	Coil 2 Driver (Dwell)	0-60° dwell	6°
CD3	Coil 3 Driver (Dwell)	0-60° dwell	6°
CD4	Coil 4 Driver (Dwell)	0-60° dwell	6°
CHTIL	Cylinder Head Temp. Lamp Control	ON / OFF	OFF (ON if on)
CTO	Clean Tachometer Output Signal	0-1,000 Hz	35-49
EGR VR	EGR Vacuum Regulator Valve	0-100%	0%
EVAP CV	EVAP Canister Vent Valve	0-100%	0%
EVAP CP	EVAP Canister Purge Valve	0-100%	0-100%
EPC	EPC Solenoid Control	0-100 psi	4
FP	Fuel Pump Control	ON / OFF	ON
Fuel B1	Fuel Injector Pulsewidth Bank 1	0-99.9 ms	2.7-3.5
Fuel B2	Fuel Injector Pulsewidth Bank 2	0-99.9 ms	2.7-3.5
HTR-11	HO2S-11 Heater Control	ON / OFF	ON
HTR-12	HO2S-12 Heater Control	ON / OFF	ON
HTR-21	HO2S-21 Heater Control	ON / OFF	ON
HTR-22	HO2S-22 Heater Control	ON / OFF	ON
IAC	Idle Air Control Valve	0-100%	25-32
INJ1	Fuel Injector 1 Control	0-99.9 ms	2.7-4.1
INJ2	Fuel Injector 2 Control	0-99.9 ms	2.7-4.1
INJ3	Fuel Injector 3 Control	0-99.9 ms	2.7-4.1
INJ4	Fuel Injector 4 Control	0-99.9 ms	2.7-4.1
INJ5	Fuel Injector 5 Control	0-99.9 ms	2.7-4.1
INJ6	Fuel Injector 6 Control	0-99.9 ms	2.7-4.1
INJ7	Fuel Injector 5 Control	0-99.9 ms	2.7-4.1
INJ8	Fuel Injector 6 Control	0-99.9 ms	2.7-4.1
LONGFT1	Long Term Fuel Trim Bank 1	-20% to 20%	0
LONGFT2	Long Term Fuel Trim Bank 2	-20% to 20%	0
MIL	Malfunction Indicator Lamp Control	ON / OFF	OFF (ON with DTC)
SHRTFT1	Short Term Fuel Trim Bank 1	-10% to 10%	-1 to 1
SHRTFT2	Short Term Fuel Trim Bank 2	-10% to 10%	-1 to 1
SPARK	Spark Advance	-90° to 90°	11-20
SS1	Transmission Shift Solenoid '1'	ON / OFF	ON
SS2	Transmission Shift Solenoid '2'	ON / OFF	OFF
TCC	Torque Converter Clutch Solenoid	0-100%	0, at 55 mph: 95%
TCIL	Trans. Control Indicator Lamp	ON / OFF	OFF (ON if on)
VREF	Voltage Reference	0-5.1v	4.9-5.1

PCM PIN VOLTAGE TABLES

PCM Pin Voltage Table: 104-Pin Connector

Pin Number	Wire Color	Application	Value at Hot Idle
1	P/O	Transmission Shift Solenoid 'B' Control	12v
2	PK/LG	MIL (lamp) Control	Lamp off: 12v, On: 1v
3	Y/BK	Digital TR1 Sensor Signal	Idle: 0v, 55 mph: 12v
4-11	---	Not Used	---
12	Y/W	Fuel Level Indicator Signal	1.7v (1/2 full)
13	P	Flash EEPROM Power	0.1v
14	LB/BK	4x4 Indicator Lamp Control	Lamp On: 1v, Off: 12v
15	PK/LB	SCP Data Bus (-)	Digital Signals
16	T/O	SCP Data Bus (+)	Digital Signals
17-20	---	Not Used	---
21	DB	CKP+ Sensor Signal	510-540 Hz
22	GY	CKP- Sensor Signal	<0.050v
24	BK/W	Power Ground	0.050v
25	BK/LB	Chassis Ground	0.050v
26	DB/LG	Coil 1 Driver Control	6° dwell
27	O/Y	Transmission Shift Solenoid 'A' Control	1v, 55 mph: 12v
29	T/W	TCS (Switch) Signal	TCS & O/D On: 12v
30	DG	Octane Adjustment	Closed: 1v, Open: 12v
31-32, 34	---	Not Used	---
33	PK/O	VSS- Signal	55 mph: 125 Hz
35	R/LG	HO2S-12 (Bank 1 Sensor 2) Signal	100-900 mv
36	T/LB	MAF Sensor Return	0.050v
37	O/BK	Transmission Fluid Temperature Sensor	2.20-2.40v
38	LG/R	Engine Coolant Temperature Sensor	0.6v
39	GY	Intake Air Temperature Sensor	1.6v
40	DG/Y	Fuel Pump Monitor	Pump On: 12v, Off: 0v
41	BK/Y	A/C Cycling Clutch Switch Signal	A/C Switch On: 12v
42-44, 53, 59	---	Not Used	---
45	R/W	Cylinder Head Temp. Lamp Control	Lamp On: 1v, Off: 12v
46	BN	Intake Manifold Tuning Valve	12v, 55 mph: 12v
47	BR/PK	EGR Vacuum Regulator Solenoid	0%, 55 mph: 47%
48	W/PK	Clean Tachometer Output	39-49 Hz
49	LB/BK	Digital TR2 Sensor Signal	0v, 55 mph: 12v
50	LG/R	Digital TR4 Sensor Signal	0v, 55 mph: 12v
51	BK/W	Power Ground	<0.1v
52	R/LB	Coil 2 Driver Control	6° dwell
54	P/Y	Torque Converter Clutch Control	0%, 55 mph: 95%

PCM 104-Pin Connector

TERMINAL VIEW OF WIRING HARNESS CONNECTOR FOR PCM (104 PIN)

Note: back-probing the 104-pin connector with a DVOM probe can damage the PCM - use a breakout box!

PCM Pin Voltage Table: 104-Pin Connector

Pin Number	Wire Color	Application	Value at Hot Idle
55	BK/LG	Keep Alive Power	12-14v
56	LG/BK	EVAP Purge Control Valve	0-10 Hz
57	Y/R	Knock Sensor 1 Signal	0v (use Lab Scope)
58	GY/BK	VSS+ Signal	At 55 mph: 125 Hz
59, 63	---	Not Used	---
60	GY/LB	HO2S-11 (Bank 1 Sensor 1) Signal	100-900 mv
61	PK/LG	HO2S-22 (Bank 2 Sensor 2) Signal	100-900 mv
62	R/PK	Fuel Tank Pressure Sensor Signal	Cap Off: 2.6v
64	LB/Y	Transmission TR Sensor Signal	In 'P': 0v, at 55 mph: 1.7v
65	BN/LG	DPFE Sensor Signal	0.5v
66	Y/LG	Cylinder Head Temp. Sensor Signal	0.6v (94°F)
67	P/W	EVAP Canister Vent Valve Control	0 Hz
68-70	---	Not Used	---
71, 97	R	Vehicle Power	12-14v
72	T/R	Fuel Injector 7 Control	2.7-4.1 ms
73	T/BK	Fuel Injector 5 Control	2.7-4.1 ms
74	BN/Y	Injector 3 Control	2.7-4.1 ms
75	T	Fuel Injector 1 Control	2.7-4.1 ms
76-77, 103	BK/W	Power Ground	0.1v
78	PK/W	Coil 3 Driver Control	6° dwell
79	W/LG	Transmission Control Indicator Lamp	Lamp On: 1v, Off: 12v
80	LB/O	Fuel Pump Relay Control	Relay On: 1v, Off: 12v
81	W/Y	Transmission EPC Solenoid Control	9.1v (6 psi)
82	---	Not Used	---
83	W/LB	Idle Air Control Valve	10.7v (33% duty cycle)
84	DB/Y	Output Shaft Speed Sensor Signal	0 Hz, 55 mph: 250 Hz
85	DG	Camshaft Position Sensor Signal	6-8 Hz
86	---	Not Used	---
87	R/BK	HO2S-21 (Bank 2 Sensor 1) Signal	100-900 mv
88	LB/R	Mass Airflow Sensor Signal	0.6v
89	GY/W	Throttle Position Sensor	0.53-1.27v
90	BR/W	Voltage Reference	4.9-5.1v
91	GY/R	Analog Sensor Return	0.050v
92	LG	Brake Pedal Position Switch	Brake Off: 0v, On: 12v
93	R/W	HO2S-11 (Bank 1 Sensor 1) Heater	Heater On: 1v, Off: 12v
94	Y/LB	HO2S-21 (Bank 2 Sensor 1) Heater	Heater On: 1v, Off: 12v
95	W/BK	HO2S-12 (Bank 1 Sensor 2) Heater	Heater On: 1v, Off: 12v
96	T/Y	HO2S-22 (Bank 2 Sensor 2) Heater	Heater On: 1v, Off: 12v
98	LB	Fuel Injector 8 Control	2.7-4.1 ms
99	LG/O	Fuel Injector 6 Control	2.7-4.1 ms
100	BN/LB	Fuel Injector 4 Control	2.7-4.1 ms
101	W	Fuel Injector 2 Control	2.7-4.1 ms
102-103	---	Not Used	---
104	R/Y	Coil 4 Driver Control	6° dwell

PCM Wiring Diagrams

FUEL & IGNITION SYSTEM DIAGRAMS (1 OF 11)

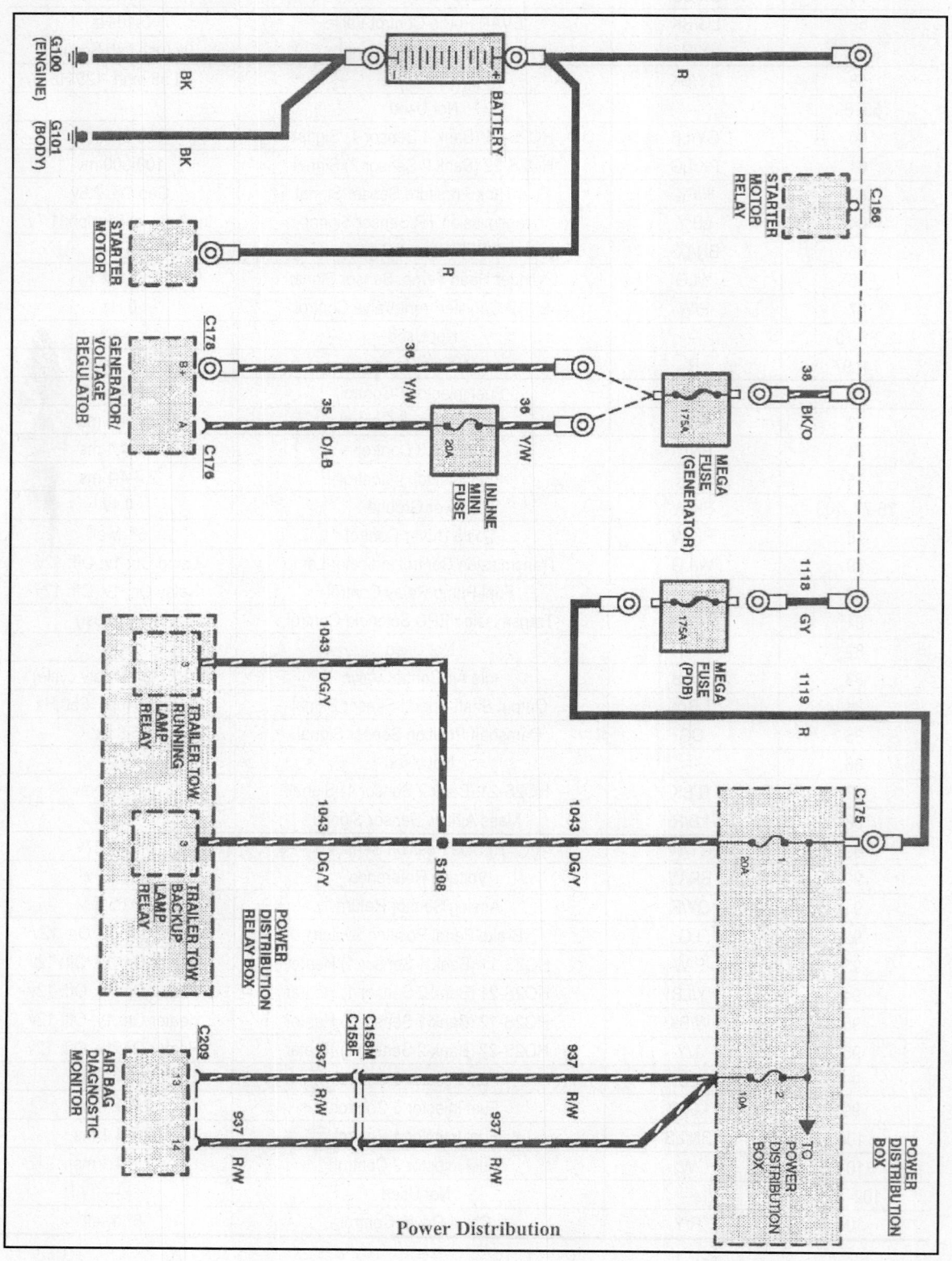

Power Distribution

FUEL & IGNITION SYSTEM DIAGRAMS (2 OF 11)

FUEL & IGNITION SYSTEM DIAGRAMS (3 OF 11)

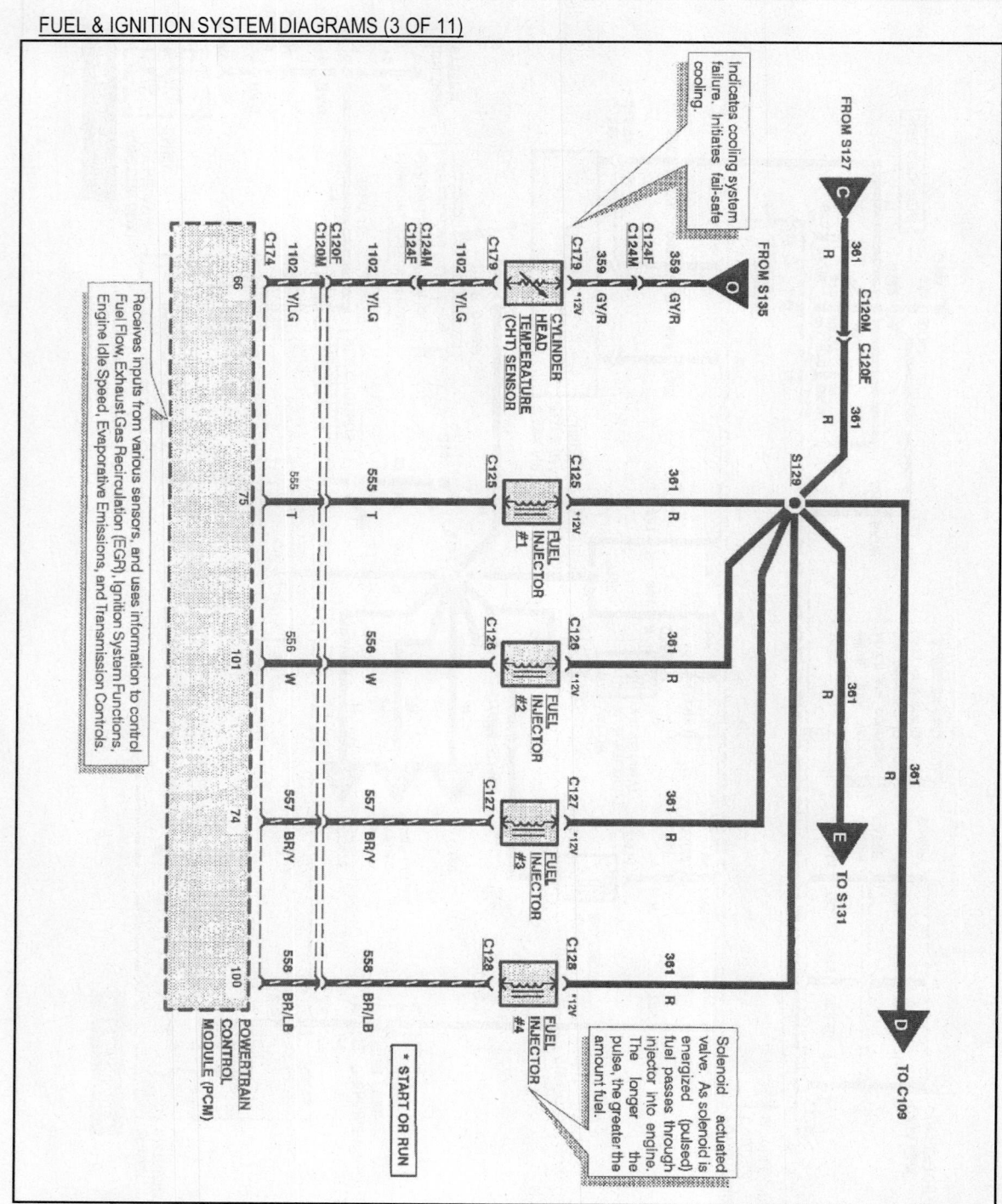

FUEL & IGNITION SYSTEM DIAGRAMS (4 OF 11)

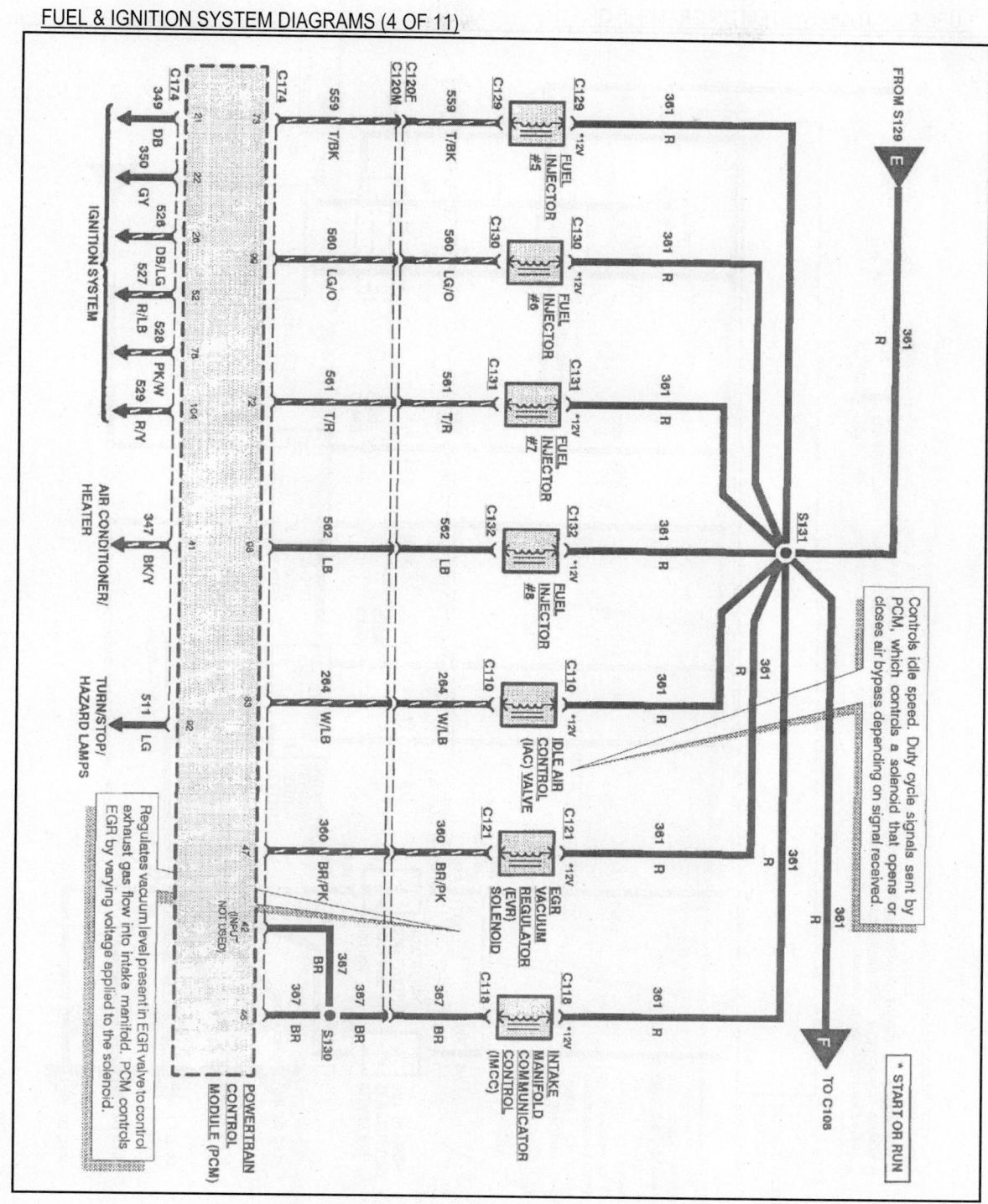

FUEL & IGNITION SYSTEM DIAGRAMS (5 OF 11)

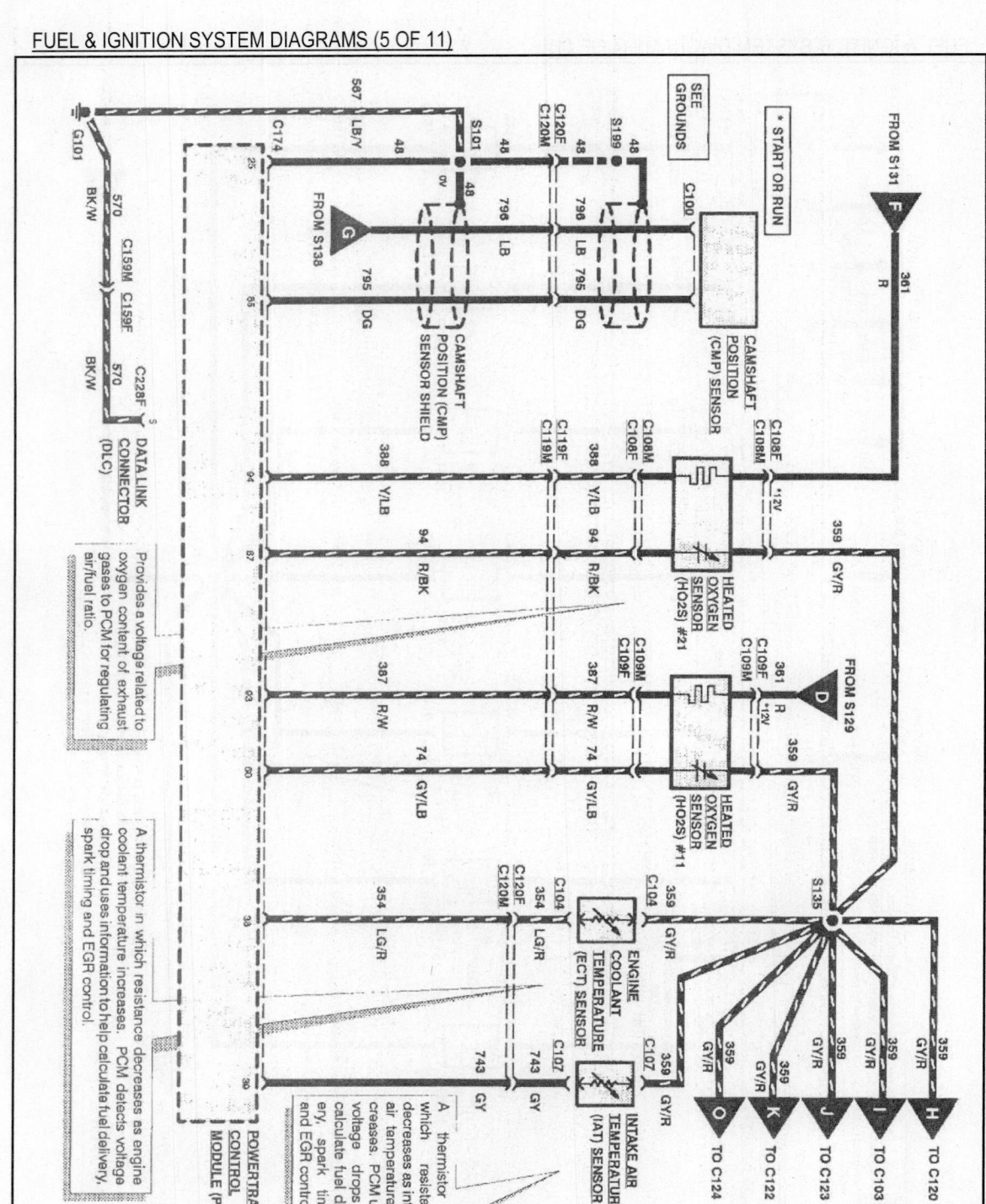

FUEL & IGNITION SYSTEM DIAGRAMS (6 OF 11)

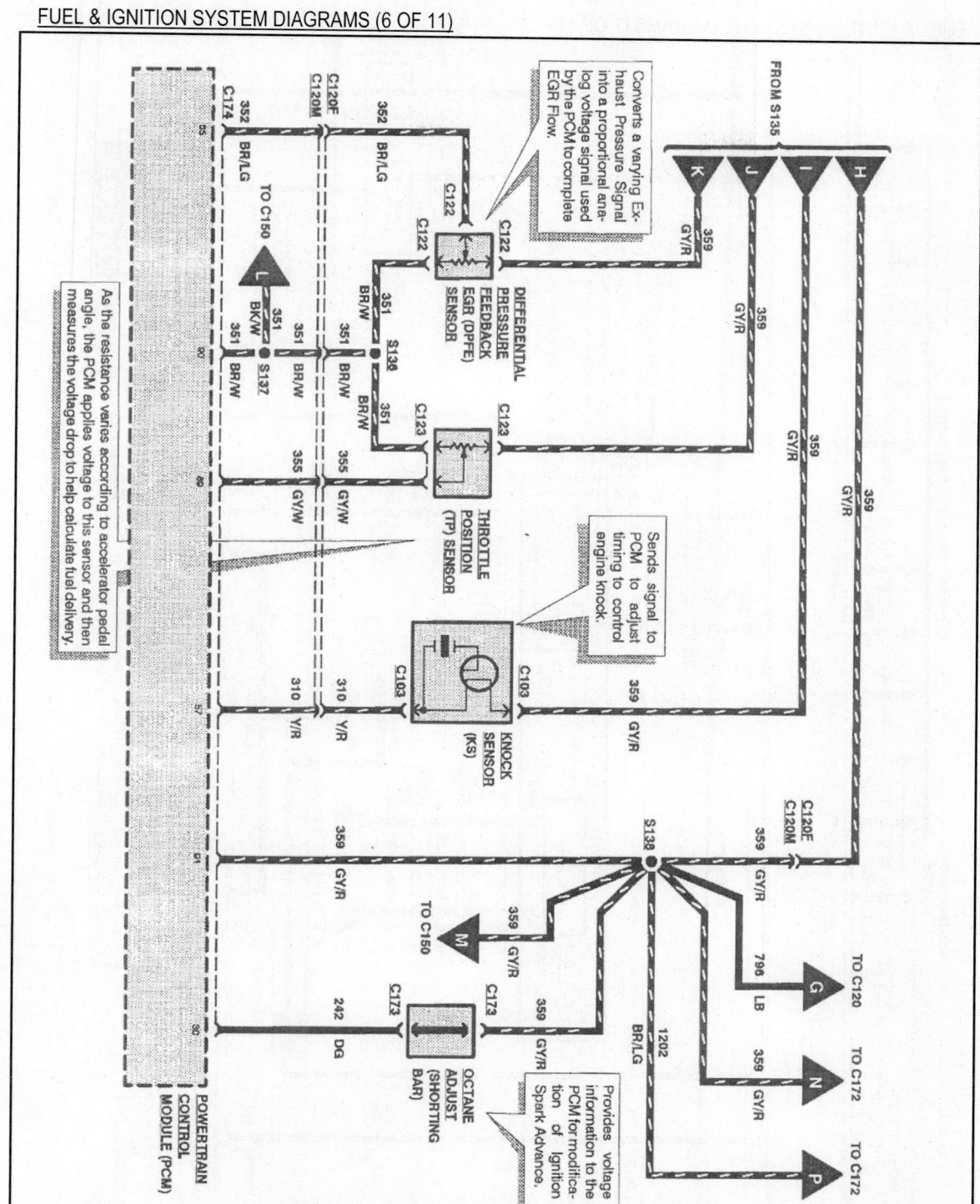

FUEL & IGNITION SYSTEM DIAGRAMS (7 OF 11)

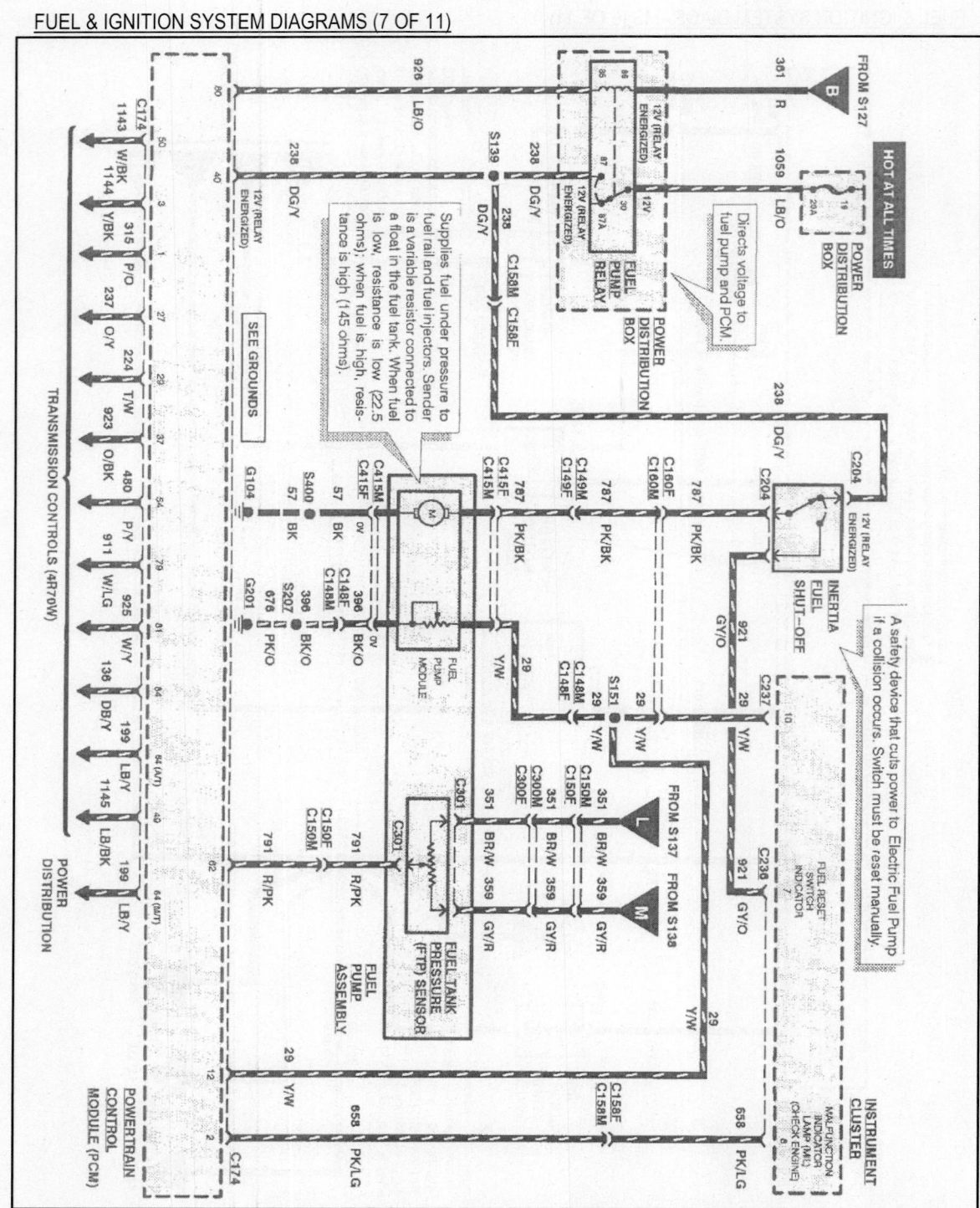

FUEL & IGNITION SYSTEM DIAGRAMS (8 OF 11)

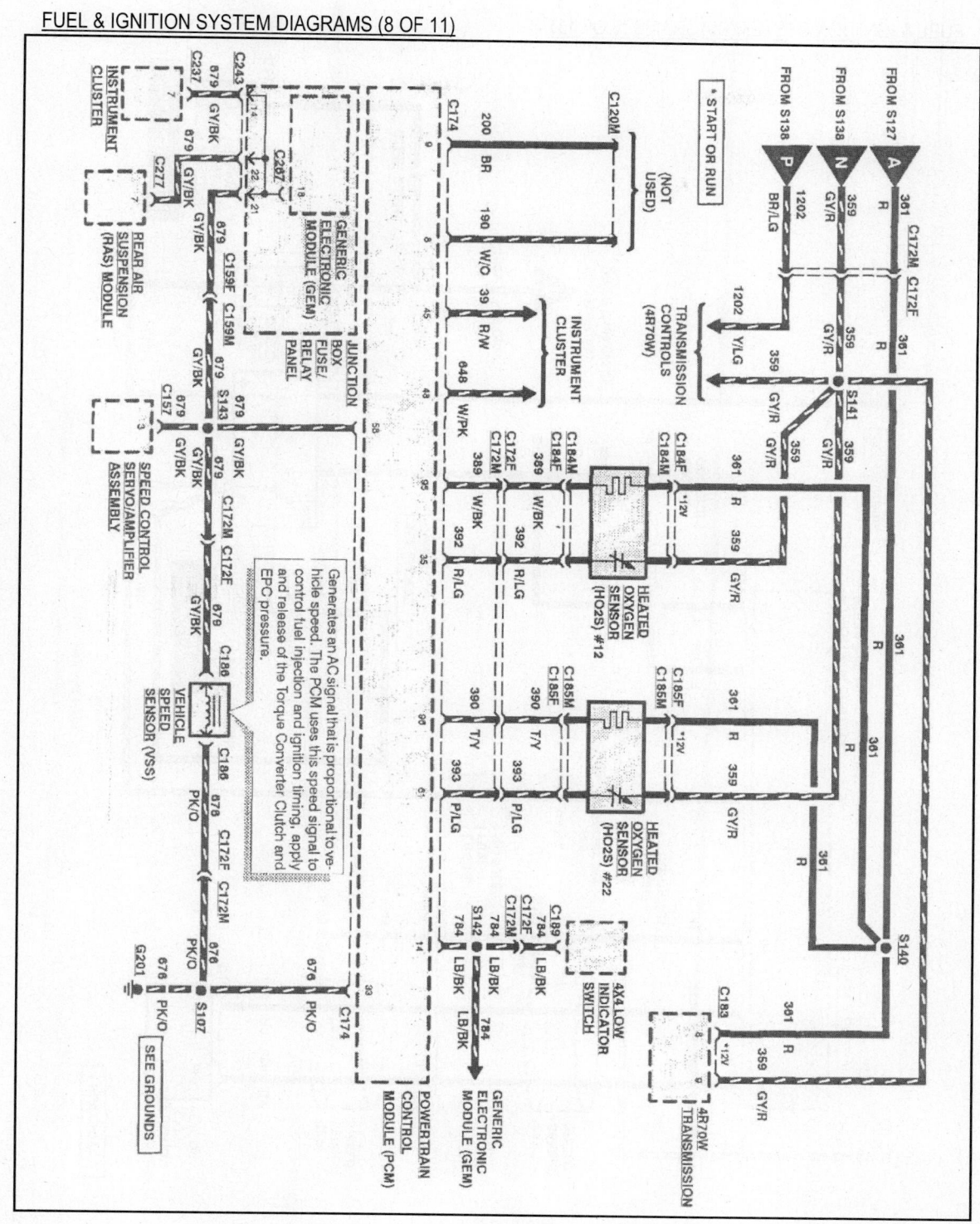

FUEL & IGNITION SYSTEM DIAGRAMS (9 OF 11)

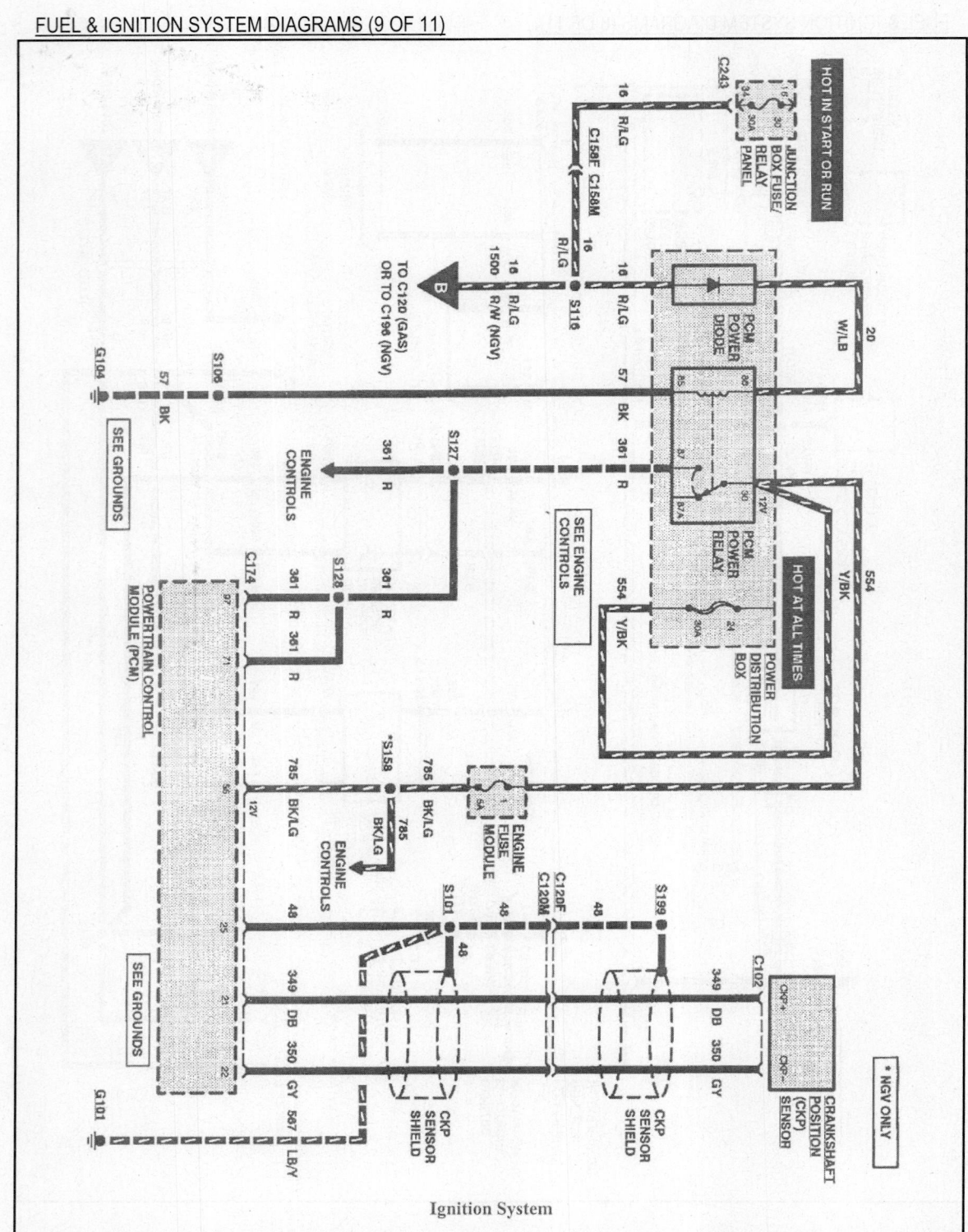

Ignition System

FUEL & IGNITION SYSTEM DIAGRAMS (10 OF 11)

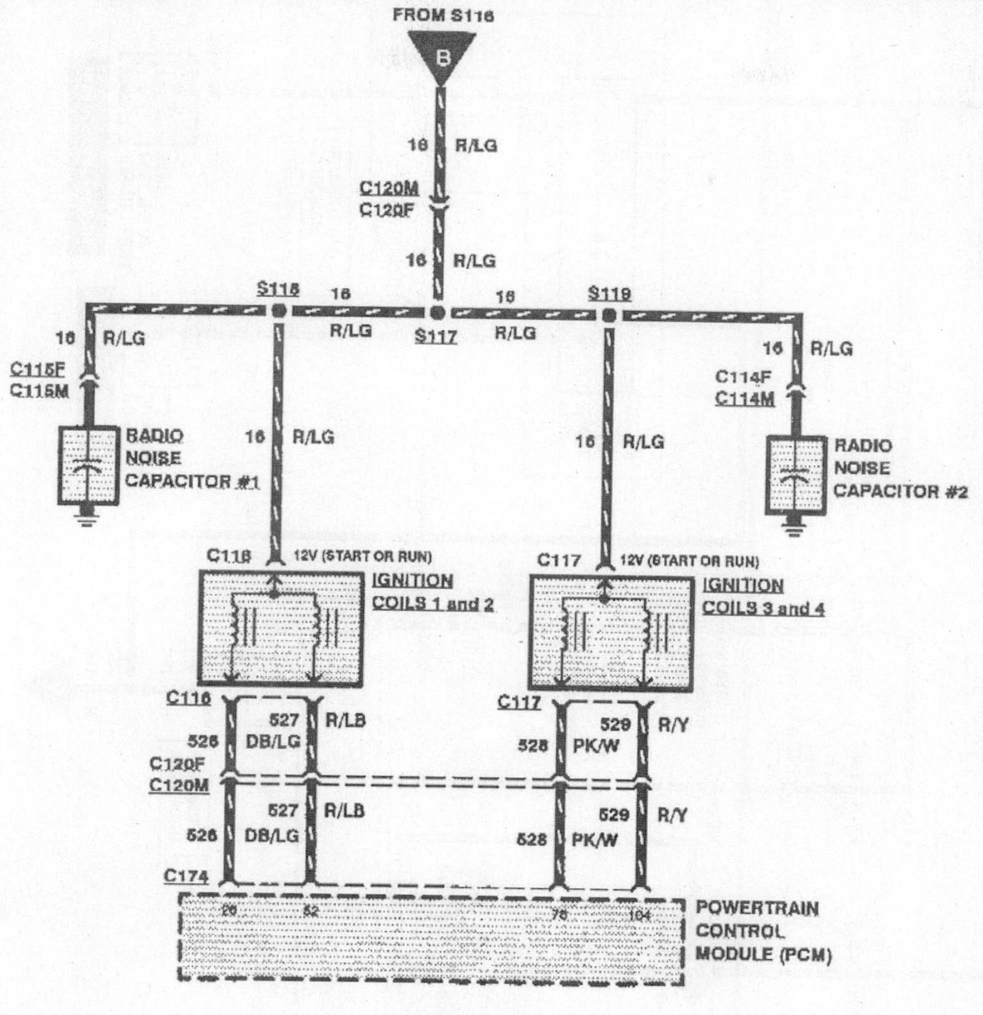

FROM S116

B

16   R/LG

C120M
C120F

16   R/LG

S118   16   16   S119
      R/LG   S117   R/LG

16   R/LG   16   R/LG

C115F
C115M   C114F
                                 C114M

RADIO   16   R/LG   16   R/LG   RADIO
NOISE   NOISE
CAPACITOR #1   CAPACITOR #2

C118   12V (START OR RUN)   C117   12V (START OR RUN)

             IGNITION   IGNITION
             COILS 1 and 2   COILS 3 and 4

C116   C117

       527   R/LB   529   R/Y
526   DB/LG   528   PK/W

C120F
C120M

       527   R/LB   529   R/Y
526   DB/LG   528   PK/W

C174

      20   52   76   104   **POWERTRAIN**
                                     CONTROL
                                     MODULE (PCM)

Ignition System

FUEL & IGNITION SYSTEM DIAGRAMS (11 OF 11)

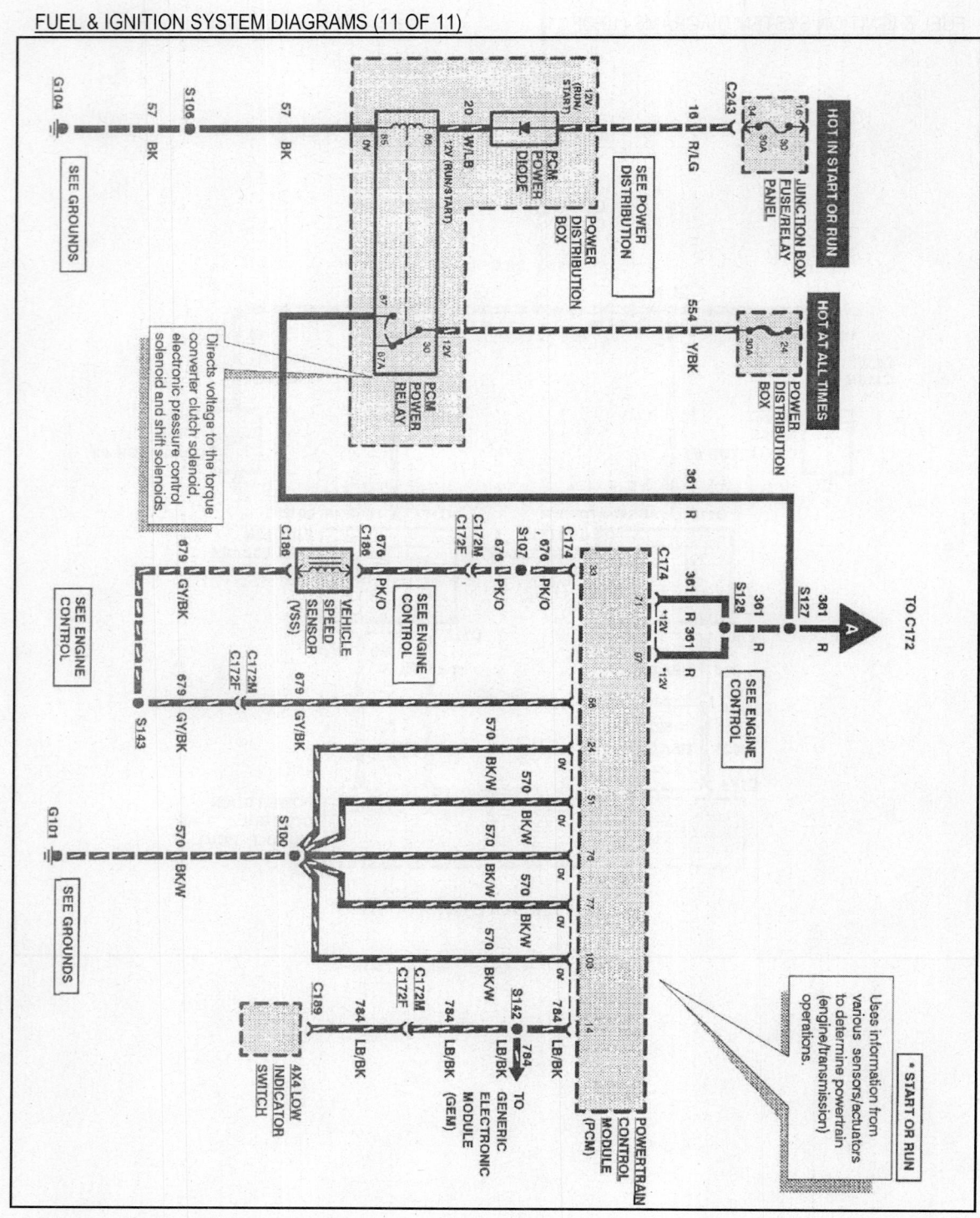

SECTION 4 CONTENTS

2001 Windstar - Powertrain

INTRODUCTION
    How To Use This Section ....Page 4-3
VEHICLE IDENTIFICATION
    VECI Decal ....Page 4-4
ELECTRONIC ENGINE CONTROL SYSTEM
    Operating Strategies ....Page 4-5
COMPUTER CONTROLLED CHARGING SYSTEM
    Introduction ....Page 4-6
    Lab Scope Test (Generator) ....Page 4-7
INTEGRATED ELECTRONIC IGNITION SYSTEM
    Introduction ....Page 4-8
    Ignition Coil Operation ....Page 4-9
    Lab Scope Test (Coil Primary) ....Page 4-10
    Lab Scope Test (Coil Secondary) ....Page 4-11
    CKP Sensor Overview (Magnetic) ....Page 4-12
    DVOM Test & Lab Scope Test (CKP Sensor) ....Page 4-13
EVAPORATIVE EMISSION SYSTEM
    Introduction ....Page 4-14
    Evaporative Charcoal Canister ....Page 4-15
    Canister Vent Solenoid ....Page 4-15
    Fuel Vapor Vent Valves ....Page 4-16
    Canister Purge Valve With Test Port ....Page 4-16
    Fuel Tank Pressure Sensor ....Page 4-16
    Scan Tool Test (Canister Vent Solenoid) ....Page 4-17
    Scan Tool Test (Purge Solenoid) ....Page 4-17
    DVOM Test (Purge Solenoid) ....Page 4-18
    Lab Scope test (Canister Vent Solenoid) ....Page 4-19
    Lab Scope Test (Purge Solenoid) ....Page 4-20
    EVAP Running Loss Monitor ....Page 4-21
    EVAP Monitor Cold Soak Bypass ....Page 4-22
    JI Diagnostic System - P0451 ....Page 4-22
EXHAUST GAS RECIRCULATION SYSTEM
    Introduction ....Page 4-23
    Exhaust Gas Recirculation Valve ....Page 4-24
    Orifice Tube Assembly ....Page 4-24
    DPFE Sensor Information ....Page 4-25
    DVOM Test (DPFE Sensor) ....Page 4-25
    EGR Vacuum Regulator Solenoid ....Page 4-26
    Scan Tool Test (VR Solenoid) ....Page 4-26
    Lab Scope Test (VR Solenoid) ....Page 4-27

SECTION 4 CONTENTS

FRONT ELECTRONIC MODULE
    Introduction ....Page 4-28
    Scan Tool Test (FEM) ....Page 4-29
    FEM PID, DTC & Active Control Lists (Partial) ....Page 4-30
    Lighting & Horns - Pinpoint Test 'I' (DTC B1319) ....Page 4-31
    FEM Wiring Diagram ....Page 4-32

FUEL DELIVERY SYSTEM
    Introduction ....Page 4-33
    Fuel System Operation ....Page 4-34
    Fuel Pump Module & Fuel Filters ....Page 4-35
    Fuel Pressure Regulator & Inertia Switch ....Page 4-36
    Fuel Injectors ....Page 4-37
    Scan Tool Test (Fuel Injectors & Fuel Trim) ....Page 4-38
    Lab Scope Test (Fuel Injector) ....Page 4-39
    CMP Sensor (Variable Reluctance) ....Page 4-41
    DVOM Test (CMP Sensor) ....Page 4-41
    Lab Scope Test (CMP Sensor) ....Page 4-42

IDLE AIR CONTROL SOLENOID
    General Description ....Page 4-43
    Scan Tool Test (IAC Solenoid) ....Page 4-43
    Lab Scope Test (IAC Solenoid) ....Page 4-44

MASS AIRFLOW SENSOR
    General Description ....Page 4-45
    BARO Sensor Calibration ....Page 4-46
    Scan Tool Test (MAF Sensor) ....Page 4-46
    Lab Scope Test (MAF Sensor) ....Page 4-47

OXYGEN SENSOR
    General Description ....Page 4-48
    Scan Tool Test (HO2S) ....Page 4-49
    Lab Scope Test (HO2S) ....Page 4-50

TRANSMISSION CONTROLS
    Electronic Pressure Control ....Page 4-52
    Output Shaft Speed Sensor ....Page 4-52
    Turbine Shaft Speed Sensor ....Page 4-52
    DVOM Test (OSS Sensor) ....Page 4-53
    DVOM Test (TSS Sensor) ....Page 4-53
    Lab Scope Test (EPC Solenoid) ....Page 4-54
    Lab Scope Test (OSS & TSS Sensor) ....Page 4-54

REFERENCE INFORMATION
    PCM PID Table - Inputs ....Page 4-55
    PCM PID Table - Outputs ....Page 4-56
    PCM Pin Voltage Tables ....Page 4-57
    PCM Wiring Diagrams ....Page 4-59

2001 WINDSTAR - POWERTRAIN

Introduction

HOW TO USE THIS SECTION

This section of the training manual includes diagnostic and repair information for Ford Minivan applications. This information can be used to help you understand the Theory of Operation and Diagnostics of the electronic controls and devices on this vehicle.

The articles in this section are separated into three sub categories:

- Front Electronic Module
- Powertrain Control Module
- Transmission Controls

2001 Windstar 3.8L V6 VIN 4 (AX4N Automatic Transaxle)

The articles listed for this vehicle include a wide variety of subjects that cover the key engine controls and vehicle diagnostics. Several of the Powertrain sensor inputs and output devices are featured along with detailed descriptions of how they operate, what can go wrong with them, and most importantly, how to use the PCM onboard diagnostics and common shop tools to determine if one or more devices has failed.

This vehicle uses an Integrated Electronic Ignition (distributorless) system along with a camshaft position sensor (magnetic) to control the fuel injection timing. These devices and others (i.e., the MAF and Oxygen sensors) are covered along with how to test them.

Choose The Right Diagnostic Path

In most cases, the first step in any vehicle diagnosis is to follow the manufacturers recommended diagnostic path. In the case of Ford vehicle applications, the first step is to connect a Scan Tool (OBD II certified for this vehicle) and attempt to communicate with the engine controller or any other vehicle onboard controllers that apply. From this point on, the recommended procedure is to use the Ford Quick Test to check for faults.

The Malfunction Indicator Lamp (MIL) can be a great help to you during diagnosis. You should note if the customer complaint included the fact that the MIL remained on during engine operation. And be sure to determine if the MIL comes on at key on, and if it goes out once the engine starts. There is more information on using the MIL in Section 1.

Diagnostic Points to Consider

If a problem is detected in a vehicle similar to this one (whether it is a trouble code or no code condition), you need to be able to determine which vehicle system (or systems) are involved before you proceed with any testing. While this may sound like a fairly routine way to approach problem solving, it is a key point to consider before you start testing.

Due to the complexity of these vehicles, a trouble code does not always point to a failed component - instead, it may point to a system that needs to be tested for some kind of problem. For example, if an Oxygen sensor trouble code is set, you need to consider what other systems or devices could cause this component to set a trouble code.

Component Diagrams

You will find numerous "customized" component diagrams used throughout this section. These diagrams are really a mini-schematic of how a particular device (where an input or output device) is connected to the engine or transmission control module.

Each diagram contains the power, ground, signal and control circuits that relate to a particular device. Of particular note is the use of "code check" identifiers in the diagrams that indicate circuits monitored by the module in order to set a particular trouble code(s).

Vehicle Identification

VECI DECAL

The Vehicle Emission Control Information (VECI) decal is located in the engine compartment. The information in this decal relates specifically to this vehicle and engine application. The specifications on this decal are critical to emissions system service. Note that the idle speed, fuel mixture and ignition timing are not adjustable.

2 TWC/ 2HO2S(2)/EGR/SFI

These designators indicate this vehicle is equipped with 2 three-way catalysts (TWC), 4 Heated Oxygen Sensors (HO2S), an EGR system and sequential fuel injection (SFI).

OBD II Certified

This designator indicates that this vehicle has been certified as OBD II compliant.

ULEV

If this designator is used, the vehicle conforms to U.S. EPA and State of California Ultra-Low Emissions Vehicle standards.

Emission Decal

This decal includes a black and white schematic of the engine vacuum system.

Underhood View Graphic

Electronic Engine Control System

OPERATING STRATEGIES

The Electronic Engine Control (EEC) system uses several operating strategies to maintain good overall driveability and to meet the EPA mandated emission standards. These strategies are included in the tables on this page.

Based on various inputs it receives, the PCM adjusts the fuel injector pulsewidth, the idle speed, the amount of ignition advance or retard, the ignition coil dwell and the operation of the canister purge valve.

It also regulates the air conditioning, cooling fan and speed control systems.

PCM Location

The PCM is located on the firewall behind the right hand (RH) strut tower as shown in the Graphic to the right. Access requires removal of wiper linkage and some trim.

Operating Strategies

A/C Compressor Clutch Cycling Control	Closed Loop Fuel Control
Electric Fuel Pump Control	Exhaust Gas Recirculation Control
Fuel Metering for Multiport Fuel Injectors	Fuel Pump Monitor
Idle Speed Control	Ignition Timing Control for A/F Change
Vapor Canister Purge Control	

General Operating Strategies

Adaptive Fuel Trim Strategy	Adaptive Idle Speed Control Strategy
Base Engine/Transmission Strategy	Failure Mode Effects Mgmt. Strategy

■ NOTE: *Hardware Limited Operating Strategy or HLOS is a part of the PCM hardware.*

Operational Control of Components

A/C Compressor Clutch	EGR System Control
Engine Cooling System Electric Fan	Fuel Delivery (Injector Pulsewidth)
Fuel Pump Operation	Fuel Vapor Recovery System
Idle Speed	Malfunction Indicator Lamp
Shift Points for the A/T	Speed for Integrated Speed Control
Torque Converter Clutch Lockup (TCC)	Transmission Line Pressure (EPC)

Items Provided (or Stored) By the EEC System

Data to the Operational Control Centers	Data to the Suspension Control Centers
DTC Data for the AX4N Transaxle	DTC Data for Emission Related Faults
DTC Data for Non-Emission Faults	

Computer Controlled Charging System

INTRODUCTION

The PCM Controlled Charging system provides additional benefits over the Integral Generator Regulator system. The first benefit is improved battery life. The regulator set point in an integral generator regulator system is established by a temperature sensor in the regulator that estimates battery temperature. Field data indicated that this approach lacked accuracy. In this new system, the regulator voltage set point is determined by the PCM and communicated to the regulator through the generator communication line. The PCM uses a calibratable algorithm to estimate battery temperature. Improving battery temperature estimates reduces the chance of over or undercharging the battery.

The second benefit is improved engine performance. Whenever the PCM senses a WOT condition, the PCM momentarily lowers the regulator voltage set point. This action reduces the torque load of the generator on the engine and improves acceleration. The PCM has a calibratable time limit on this reduced voltage feature. This prevents the generator output from being cut back too long which could cause battery discharge.

The third benefit is improved idle stability. In response to the PCM GENCOM (generator communication) signal, the regulator uses a generator monitor (GENMON) signal to provide feedback to the PCM. The generator monitor signal provides the PCM with charging system information. Specifically, it lets the PCM know when the charging system receives a transient electrical load that would normally affect idle stability.

Because the PCM can anticipate additional loads, actions can be taken to minimize idle sag. The PCM can choose to either reduce the regulator set point or increase engine idle speed, both of which are calibratable features. The regulator uses a charging system voltage line (battery sense) to sense battery voltage at the power distribution box in order to determine whether the regulator is accurately maintaining the desired voltage set point.

The fourth benefit is reduced cranking efforts. The PCM can reduce the mechanical load on the starter by initially commanding a low voltage set point to improve startup.

Charging System Faults

If the PCM detects a charging system error (DTC P1246), it will broadcast a low voltage telltale "on" command that tells the cluster to light the charge indicator. The charge indicator is illuminated if the PCM fails to see a signal on the generator monitor line for more than 500 ms. This telltale command is also be used if an over voltage condition is detected by the PCM controlled generator.

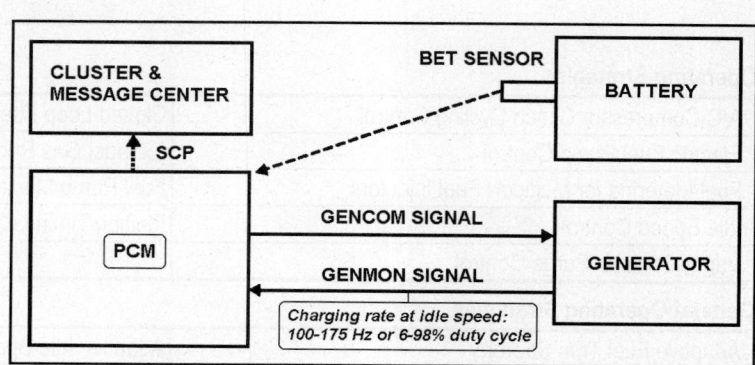

Charge Indicator Lamp Control

Each time the ignition switch is cycled to the run, the cluster will initiate a bulb check by illuminating the charge indicator. The PCM has the responsibility to issue a low voltage telltale "off" command if the charging system is functioning properly. This message should be sent during Network Initialization in the voluntary phase (250-450 ms after the ignition switch is cycled to the run position). If the cluster does not receive a low voltage telltale 'off' command, it will continue to light the charge light indefinitely.

DVOM TEST (GENERATOR)

A DVOM can be used to test the Computer Controlled Generator Monitor circuit voltage output (AC) or frequency (Hz). Place the shift selector in Park, block drive wheels for safety.

DVOM Connections

Turn the key off and connect the breakout box (BOB) if available. Select DC volts on the DVOM. Connect the DVOM positive lead to the GENMON circuit (VIO wire) at Pin 20 and the negative lead the battery ground post.

Test Results

This capture was taken at idle speed. The DVOM display shows a GENMON signal of 04.49v DC at a rate of 129.7 Hz and a duty cycle rate of 57.3%.

To test the signal change, turn on the Air Conditioning and set the blower motor to high. The voltage and duty cycle should change.

LAB SCOPE TEST (GENERATOR)

The Lab Scope can be used to test the Computer Controlled Generator as it provides an accurate view of generator input and output signals. Place the shift selector in Park and block the drive wheels for safety.

Scope Connections (GENMON Signal)

Install a BOB if available. Connect the Channel 'A' positive probe to the GENMON signal at Pin 20 (VIO wire) and the negative probe to the battery negative post.

Scope Settings (GENCOM Signal)

Install a BOB if available. Connect the Channel 'A' positive probe to the GENCOM signal at Pin 45 (RD/PK wire) and the negative probe to the battery negative post.

Lab Scope Tests

The Generator Monitor and Communication signals can be checked at idle speed or cruise speed with the engine cold or warm.

Lab Scope Examples

In these examples, the traces show the GENMON and GENCOM signals at idle speed. Note that the GENMON signal is 10v digital signal while the GENCOM signal is a battery voltage digital pulse train signal.

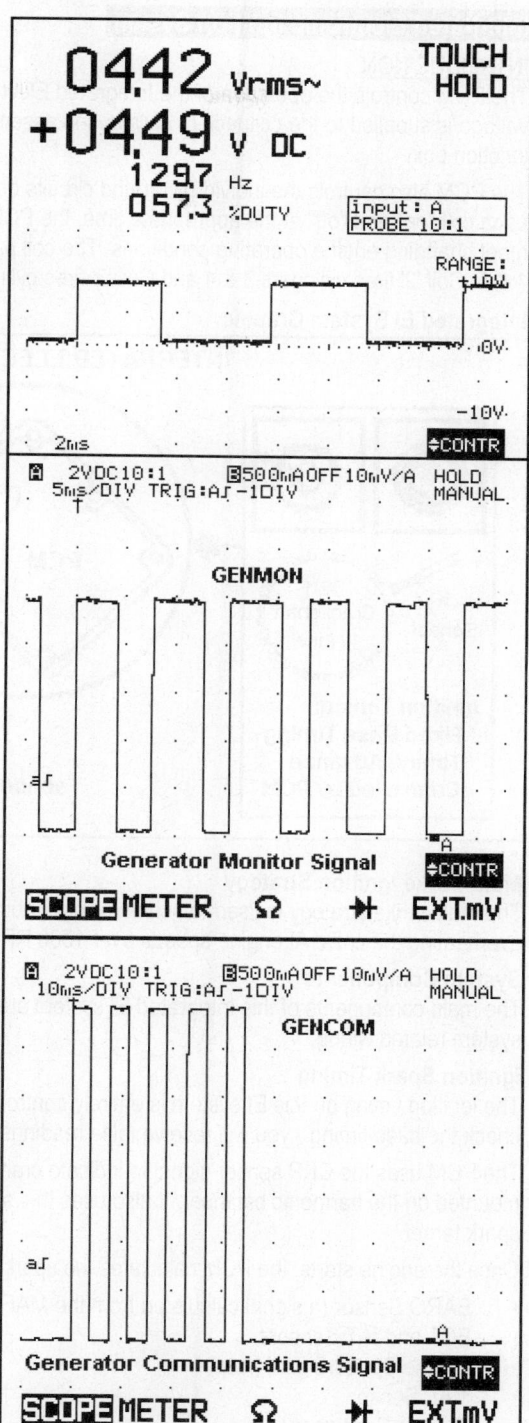

Generator Monitor Signal

Generator Communications Signal

Integrated Electronic Ignition System

INTRODUCTION

The PCM controls the operation of the Integrated Electronic Ignition (EI) system on this vehicle application. Battery voltage is supplied to the coilpack (3 coils in one assembly) via the PCM power relay and #20 fuse (15A) in the battery junction box.

The PCM also controls the individual ground circuits of the three (3) ignition coils. By switching the ground path for each coil "on" and "off" at the appropriate time, the PCM adjusts the ignition timing for each cylinder pair correctly to meet changing engine operating conditions. The coil pairs are identified as Coilpacks 1, 2 and 3. Coil '1' fires cylinders 1 & 5; Coil '2' fires cylinders 3 & 4 and Coil '3' fires cylinders 2 & 6. The firing order on this engine is 1-4-2-5-3-6.

Integrated EI System Graphic

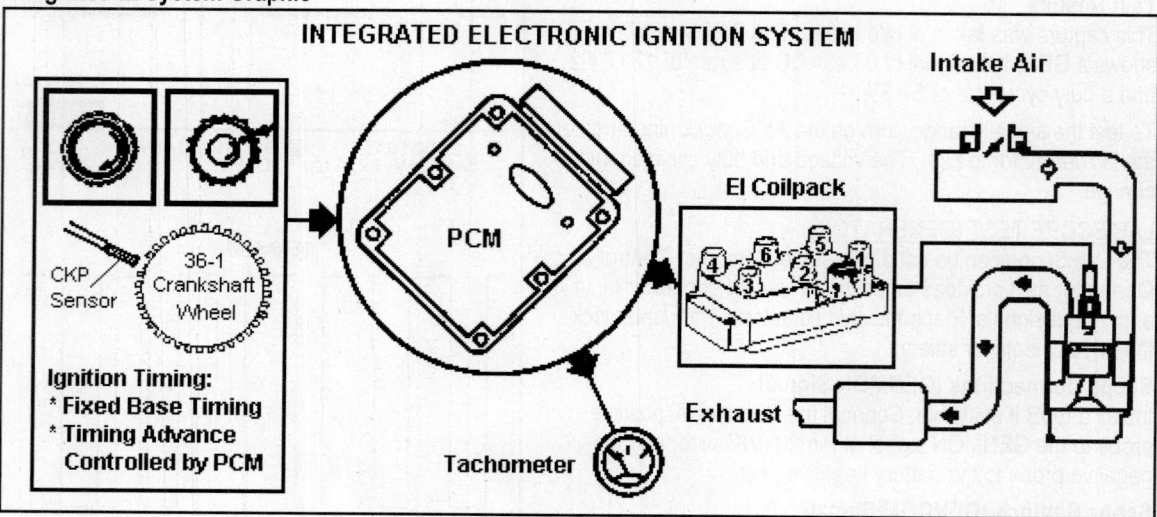

Multi-Strike Ignition Strategy

The multi-strike strategy is used to achieve longer burn times to completely burn the lean idle mixture without overheating the coils. At engine speeds over 1500 RPM, the driver changes to a single-strike strategy.

System Components

The main components of this Integrated EI system are the coilpack, crankshaft position sensor, the PCM and EI system related wiring.

Ignition Spark Timing

The ignition timing on this EI system is entirely controlled by the PCM - base timing is not adjustable. Do not attempt to check the base timing - you will receive false readings.

The PCM uses the CKP sensor signal to indicate crankshaft position and speed by sensing teeth on a pulse wheel mounted on the harmonic balancer. It also uses this signal to calculate a spark target and then fire the coils to that spark target.

Once the engine starts, the PCM calculates the spark advance with these signals:

- BARO Sensor (a signal calculated from the MAF Sensor reading)
- ECT and IAT Sensors
- Engine Speed Signal (rpm)
- MAF Sensor
- Throttle Position

IGNITION COIL OPERATION

The ignition coilpack consists of three (3) independent coils molded together. The coil assembly is mounted to the intake manifold. Spark plug cables rout to each cylinder from the coil assembly.

The coil fires two spark plugs before each power stroke. One plug is the cylinder under compression, the other cylinder fires on the exhaust stroke (waste spark method).

The PCM determines which coil to charge and fire. In this example, Coil '1' fires cylinders 1 & 5; Coil '2' fires cylinders 3 & 4 and Coil '3' fires cylinders 2 & 6. The firing order is 1-4-2-5-3-6.

The spark plug that fires the cylinder on the compression stroke uses the majority of the ignition coil energy and the spark plug that fires the cylinder on the exhaust stroke uses very little coil energy. The spark plugs are connected in series, so one spark plug firing voltage will be negative and the other spark plug will be positive with respect to ground.

El Coilpack Circuits

The entire coil pack is powered by the PCM power relay. However, drivers in the PCM control each coil independently. The PCM drives Coil 1 from Pin 26, Coil 2 from Pin 52 and Coil 3 from Pin 78 of the PCM harness connector.

Trouble Code Help

Note the three driver circuits that are shown with a dotted line in the schematic. The PCM monitors these circuits as it switches the coil drivers "on" and then "off". If the PCM does not receive a valid IDM signal due to a coil problem it will set any one of three trouble codes (DTC P0351 for Coil 1, P0352 for Coil 2 or P0353 for Coil 3).

Ignition Coil Primary Schematic

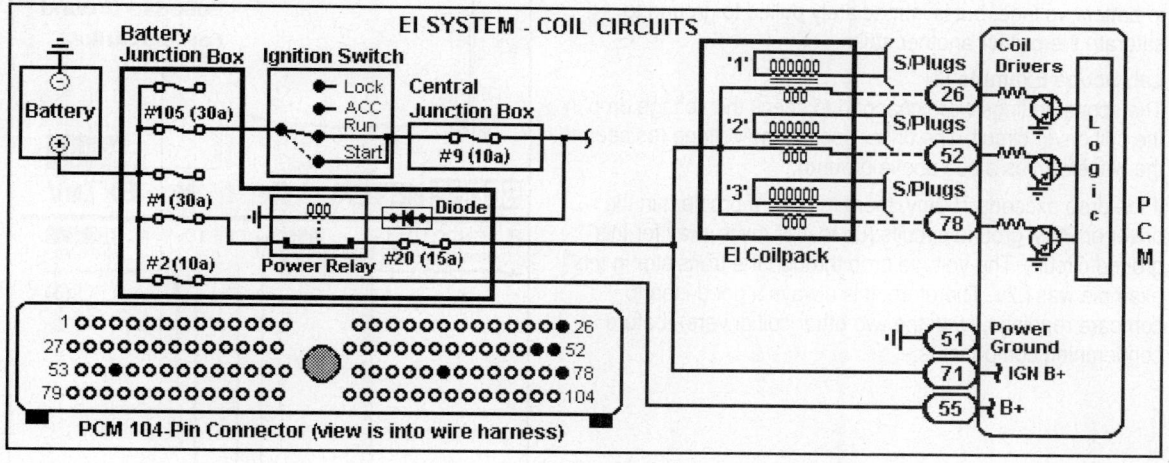

LAB SCOPE TEST (COIL PRIMARY)

The ignition coil primary includes the switched ignition feed circuit, coil primary circuit and the coil driver control circuits at the PCM. Coil 1 connects to Pin 26, Coil 2 connects to Pin 56 and Coil 3 connects to Pin 78 of the 104P connector (refer to Graphic on previous page).

Coil Dwell Period

The coil dwell period is the length of time that current flows through the coil primary windings (it is measured in milliseconds and represents the degrees of engine rotation). A complete turn of the camshaft equals 360 degrees and represents one firing of all engine cylinders on the compression stroke.

Lab Scope Connections

Connect the Channel 'A' positive probe to one of the three (3) coil control circuits and the negative probe to the battery negative post.

Scope Settings

To make the waveforms as clear as possible, set scope settings to match the examples.

Lab Scope Example (1)

In example (1), the trace shows the coil driver (primary) circuit at idle. Note that this vehicle uses a multiple-strike strategy. This strategy is used to achieve longer burn times at idle without overheating the coils.

Lab Scope Example (2)

In example (2), the time and voltage per division were changed. At 50v/div. the peak firing voltage can be seen (350v). Note that after the spark line, the circuit does not return to battery voltage, but is immediately pulled to ground to saturate the coil for another strike.

Lab Scope Example (3)

The scope settings were changed to check the voltage drop in the coil driver circuit. The driver uses some voltage (as seen in the waveform as a rise above ground).

If the drop exceeds 750mv, there may be a problem in the driver or PCM ground circuits (up to 100 mv is okay for this ground circuit). The voltage drop through the transistor in this example was1.2v. Therefore, it is always a good idea to compare readings, (with the two other coil drivers), before condemning components.

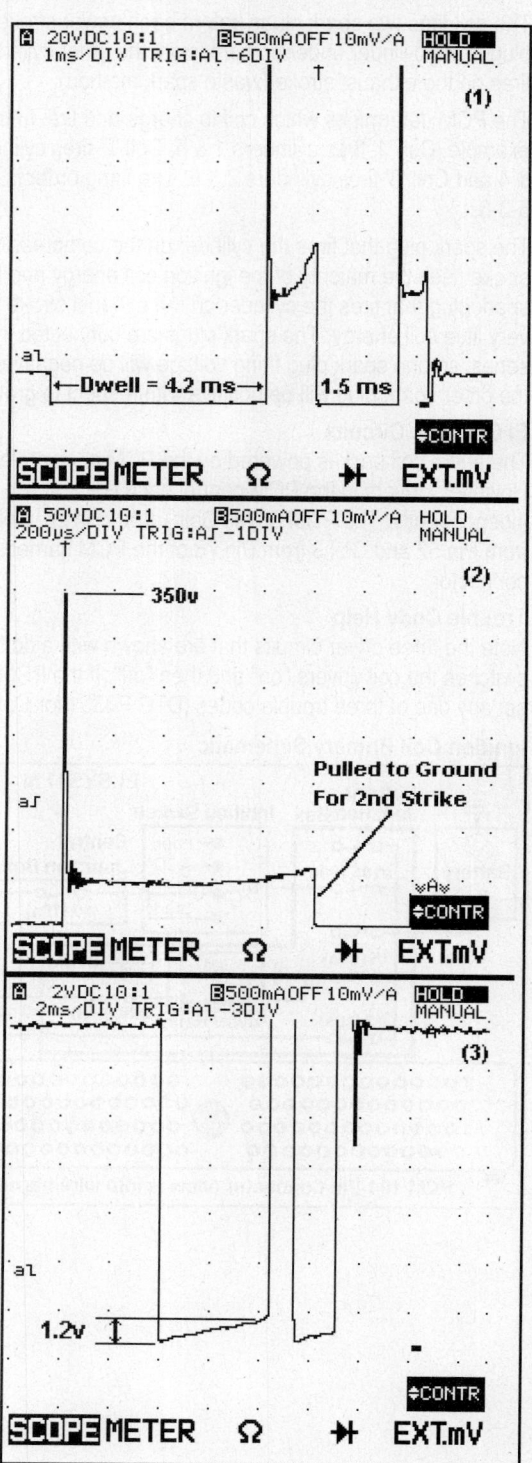

LAB SCOPE TEST (COIL SECONDARY)

Each of the three (3) coils in the coil pack fires a pair of cylinders simultaneously. Each coil has a positive (+) and a negative (-) tower. In this example, two cylinders are paired, and as one cylinder fires on the compression stroke, while the other cylinder fires on the exhaust stroke. Because each pair of cylinders is fired twice during each engine cycle, this form of distributorless ignition (DIS) is called a Waste Spark system.

Lab Scope Connections

Select the proper vehicle or configuration from the scope menus. Connect the individual DIS leads and trigger pickup as prompted by scope menus. Pay close attention to the polarity of the DIS leads and the coil towers.

Scope Settings

To make the waveforms as clear as possible, set scope settings to match the examples.

Lab Scope Example

In this example, the trace shows the secondary multi-strike ignition pattern. This strategy is used to achieve longer burn times at idle without overheating the coils or the drivers.

Waste Spark ignition system secondary ignition patterns vary much more than those of conventional ignition systems.

This makes it difficult to diagnose mixture, plug gap, engine mechanical or other problems easily seen in a conventional ignition pattern. However, modern vehicles experience these problems less frequently, and can be diagnosed with other tools, (i.e. mixture can be seen as a fuel trim PID on a scan tool).

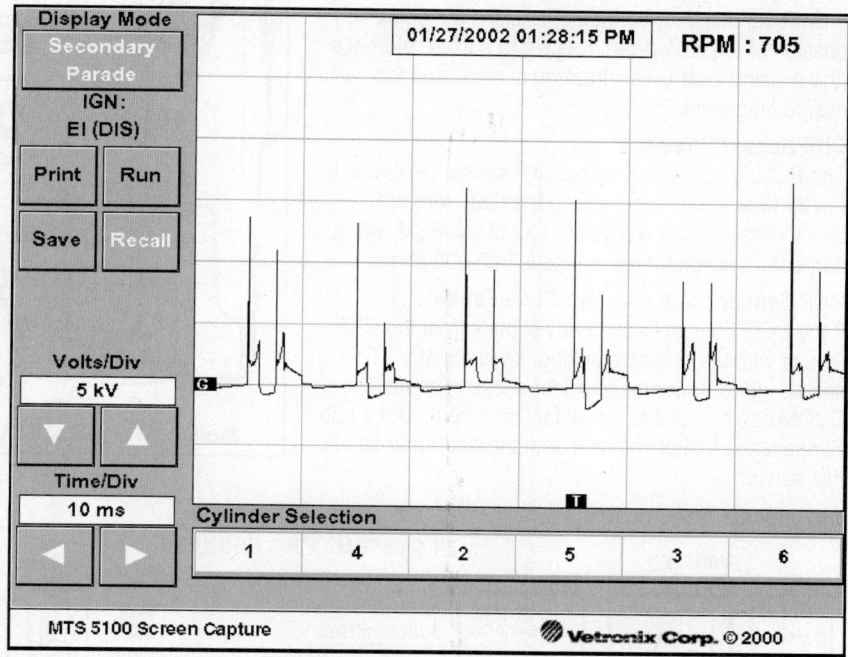

Problems with modern Waste Spark systems are more often caused by open or shorted secondary wires, failed components and driver problems. These failures can be easily seen on an ignition analyzer.

Use this test to check for extremely high or low firing voltage. Also check to see if problems occurring in one cylinder are present in the companion cylinder as well. (Remember, these coils are paired as follows: 1 & 5, 4 & 3 and 2 & 6).

The ignition analyzer can also be used to verify proper operation of the multi-strike strategy. Raise engine RPM slowly and verify PCM switches to single-strike at 1500 RPM. (Some Ford vehicle applications switch at 1800 RPM).

CKP SENSOR OVERVIEW (MAGNETIC)

The crankshaft position (CKP) sensor used with this EI system is the primary sensor for crankshaft position data to the powertrain control module (PCM).

The CKP sensor is a magnetic transducer mounted at the front of the timing chain cover adjacent to a 36-1 tooth trigger wheel mounted on the crankshaft on the harmonic balancer. The trigger wheel has 35 teeth spaced 10 degrees (10°) apart with one empty space due to a missing tooth (hence the term 36-1 assigned to this wheel).

The CKP sensor sends an AC voltage signal to the ignition control module (ICM) integrated into the PCM. The frequency (Hz) of the signal is directly proportional to crankshaft speed.

The PCM monitors the CKP sensor signal (the 36-1 design signal) in order to correctly identify piston travel so that it can synchronize the Ignition system firing, and provide an accurate method of tracking the angular position of the crankshaft, relative to a fixed reference (the missing tooth). This tracking is also used for misfire diagnosis.

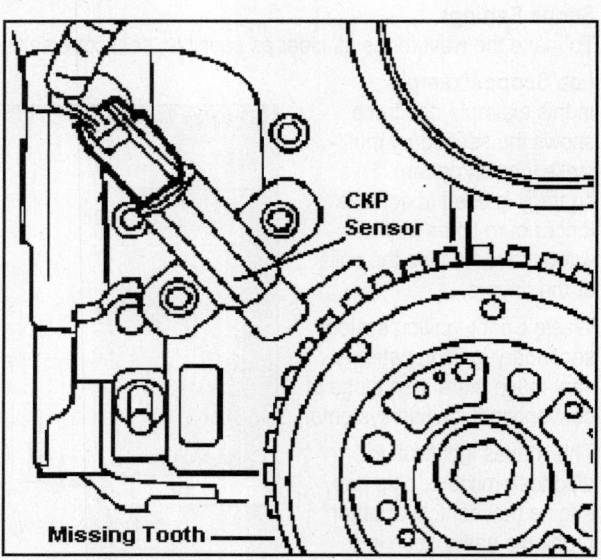

CKP Sensor Circuits

The PCM is connected to the CKP sensor (+) circuit at Pin 21 (Black/Pink wire) and to the CKP sensor (-) circuit (Gray/Yellow wire) at Pin 22 of the PCM wiring harness. The sensor resistance is 300-800 ohms.

CKP Sensor Diagnosis (No Code Fault)

If the PCM does not detect any signals from the CKP sensor with the engine cranking, it will not start. This sensor input is not monitored for a loss of signals. A DVOM can be used to check for "no output", but a Lab Scope is the tool of choice to test the performance of this sensor.

CKP Sensor Schematic

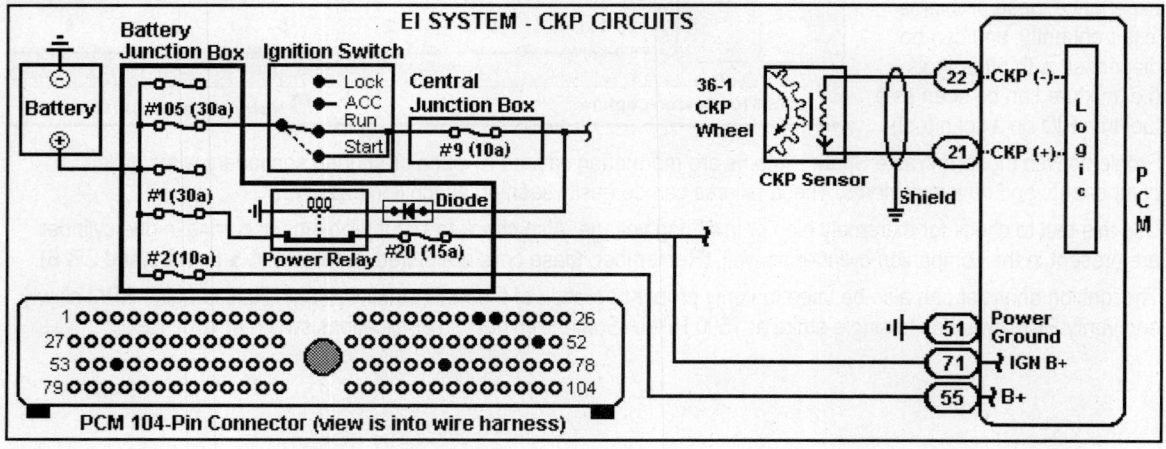

DVOM TEST (CKP SENSOR)

A DVOM can be used to test the CKP sensor signals (i.e., the sensor AC voltage output and/or frequency). Place the shift selector in Park and block the drive wheels for safety before performing this test procedure.

DVOM Connections

Connect the breakout box (BOB) if available. Select AC volts on the DVOM. Connect the DVOM positive lead to the CKP (+) circuit (BK/PK wire) at Pin 21 and the negative lead to CKP (-) circuit (GRY/YEL wire) at Pin 22.

Test Results

This capture was taken at 2500 RPM. The DVOM display shows a CKP signal of 15.8v AC at a rate of 1.3 kHz. Note that the true amplitude of the AC voltage reading is about 32v in this example.

For a no-start, the DVOM should show a minimum of 400mv AC during cranking. At idle, the DVOM should show at least 3v AC at 350-450 Hz.

LAB SCOPE TEST (CKP SENSOR)

The Lab Scope can be used to test the CKP sensor as it provides a very accurate view of sensor waveform and of any glitches. Place the shift selector in Park and block the drive wheels for safety.

Scope Connections

Install a BOB if available. Connect the Channel 'A' positive probe to the CKP (+) circuit at Pin 21 (BK/PK wire) and the negative probe to the CKP (-) circuit at Pin 22 (GRY/YEL wire).

Scope Settings

To make the waveforms as clear as possible, set the scope settings to match the examples.

Lab Scope Tests

The CKP sensor signal can be checked during cranking, idle and at off-idle speeds with the engine cold or fully warm.

Lab Scope Example

In this example, the trace shows the CKP sensor signal. This waveform was captured at idle, and represents around 80° of crankshaft rotation when counted using the negative pulse, (each pulse represents 10°).

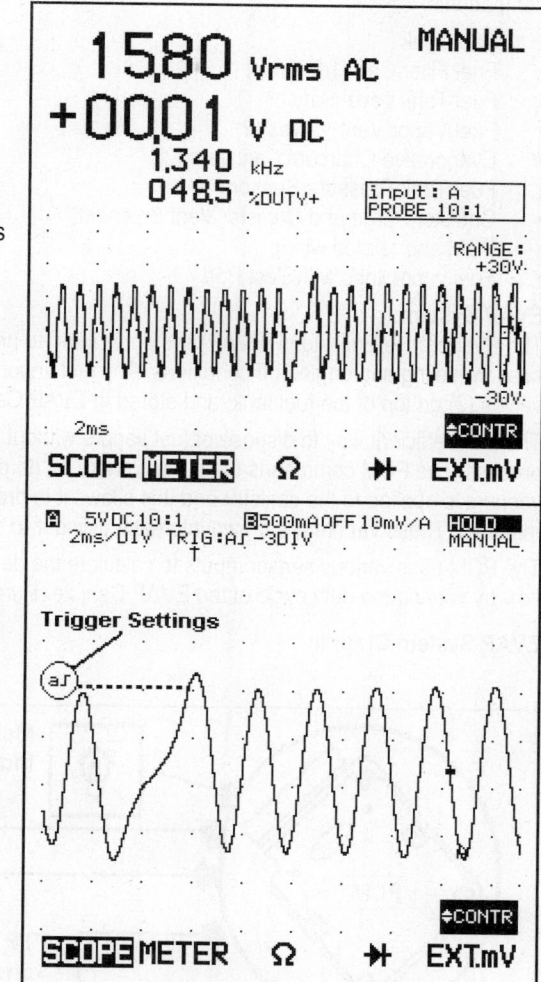

Trigger Help

Note the trigger level and slope in the example. The scope was set to trigger on channel A on the rising edge of the waveform. The waveform amplitude is higher when the missing tooth of the trigger wheel passes the CKP sensor. This higher voltage can be used to trigger the scope, and helps to stabilize the waveform on the screen.

Evaporative Emission System

INTRODUCTION

The evaporative emission (EVAP) system on this vehicle is a Running Loss/Monitor system, and includes the following components:

- Fuel Tank
- Fuel Filler Cap (1/8 Turn)
- Fuel Tank Level Sensor
- Fuel Vapor Vent Valves (2)
- Evaporative Charcoal Canisters (2)
- Fuel Tank Pressure Sensor
- Canister Purge and Canister Vent Solenoid
- PCM and related wiring
- Fuel vapor lines with Test Port

EVAP System

The Evaporative Emissions (EVAP) system is used to prevent the escape of fuel vapors to the atmosphere under hot soak, refueling and engine off conditions. Any fuel vapor pressure trapped in the fuel tank is vented through fuel vapor valves (2) on top of the fuel tank, and stored in EVAP Canisters (2).

The most efficient way to dispose of fuel vapors without causing pollution is to burn them in the normal combustion process. The PCM commands the purge valve 'On' (to open) during normal operation. With the valve open, manifold vacuum is applied to the canister and this allows it to draw in fresh air and fuel vapors from the canister into the intake manifold. These vapors are drawn into each cylinder to be burned.

The PCM uses various sensor inputs to calculate the desired amount of EVAP purge flow. The PCM meters the purge flow by varying the duty cycle of the EVAP Canister Purge Valve control signal.

EVAP System Graphic

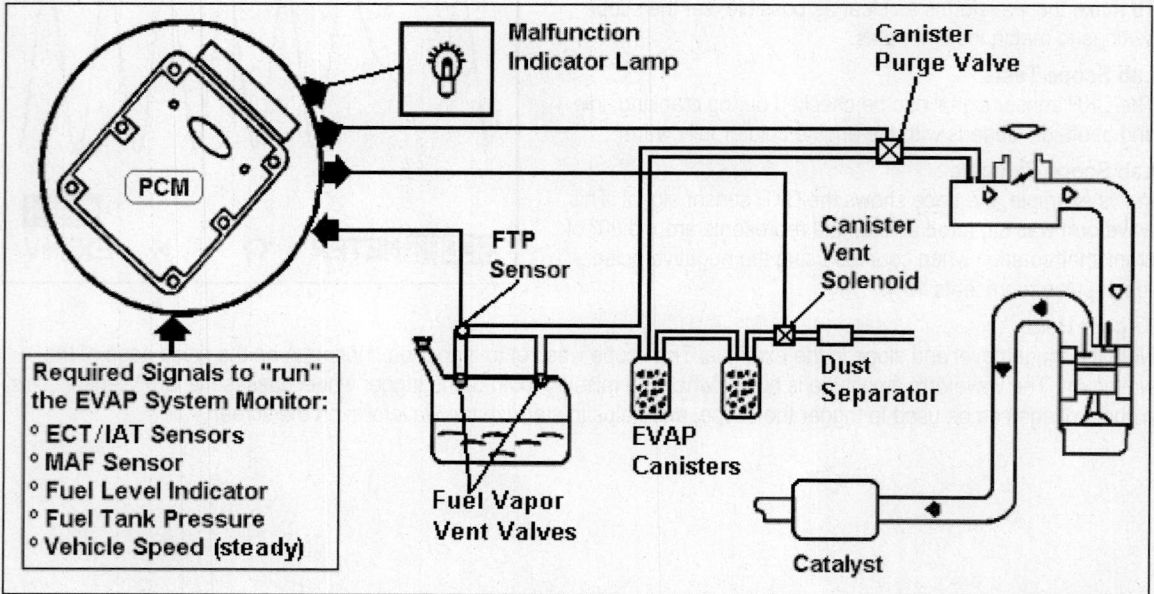

EVAPORATIVE CHARCOAL CANISTER

This vehicle uses two (2) charcoal canisters in series. The canisters are located to the left of the fuel tank. Fuel vapors from the fuel tank are stored in the canisters.

Once the engine is started, fuel vapors stored in charcoal canister are purged into the engine where they are burned during normal combustion.

The activated charcoal has the ability to purge or release any stored fuel vapors when it is exposed to fresh air.

The upper and lower canisters are connected in series by fuel vapor lines.

CANISTER VENT SOLENOID

The canister vent (CV) solenoid is a normally open (N.O.) solenoid with a resistance of 48-65 ohms. Under normal conditions, the CV solenoid is open to allow fuel system vapors to escape after the hydrocarbons have been absorbed by the canisters. While the system is purging, the open solenoid allows fresh air to enter the canisters, which frees up the trapped hydrocarbons to be drawn into the intake manifold to be burned.

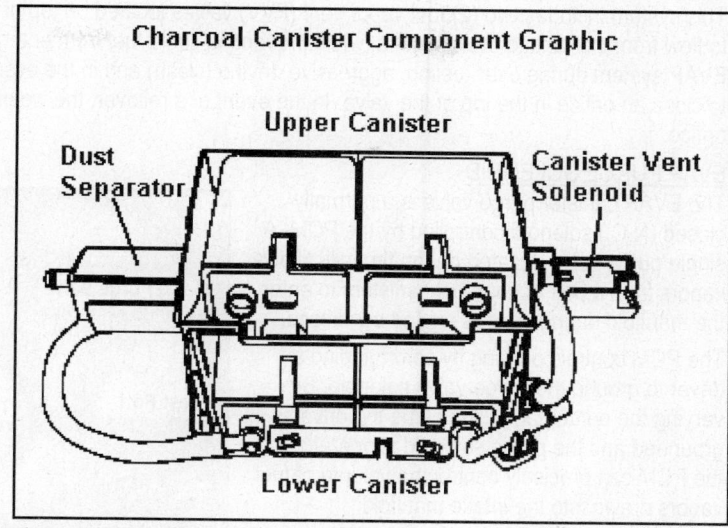

Charcoal Canister Component Graphic

Upper Canister

Dust Separator

Canister Vent Solenoid

Lower Canister

The main function of the CV solenoid is to isolate the charcoal canister from outside air for testing purposes.

Canister Vent Solenoid Location

The CV solenoid is located on the upper EVAP canister, to the left of the fuel tank. The solenoid can be serviced separately.

During OBD II EVAP Monitor operation, the PCM activates (closes) the CV solenoid. At the same time, the purge solenoid is activated, (open). This allows manifold vacuum to draw fuel tank vacuum down to a specified level for leak test purposes.

Trouble Code Help

If the PCM detects a large leak or no flow in the EVAP

Canister Vent Solenoid

system, it will set a pending DTC P0455. If the fault is detected on the next trip, the PCM will turn on the MIL (2-trip fault detection).

Although a loose or missing fuel cap is often the cause of this code, it is important to remember that if anything prevents the PCM from drawing a vacuum on the EVAP system during a leak check, this code will set. A CV solenoid that is mechanically stuck open or electrically faulty will cause a leak check to fail.

EVAP FUEL VAPOR VENT VALVES

This system includes two (2) fuel vapor vent (FVV) valves located on top of the fuel tank. The valves allow fuel vapors to flow from the tank to the canisters, while preventing liquid fuel from entering the lines. Liquid fuel may enter the EVAP system during over-fueling, aggressive driving (slosh) and in the event of a rollover. An open-bottom float rises to close an orifice in the top of the valve. In the event of a rollover, the weight of the open-bottom float closes the orifice.

EVAP PURGE SOLENOID

The EVAP canister purge valve is a normally closed (N.C.) solenoid controlled by the PCM. A single purge valve, located on the firewall, allows vapors from a pair of charcoal canisters to enter the manifold during various engine conditions.

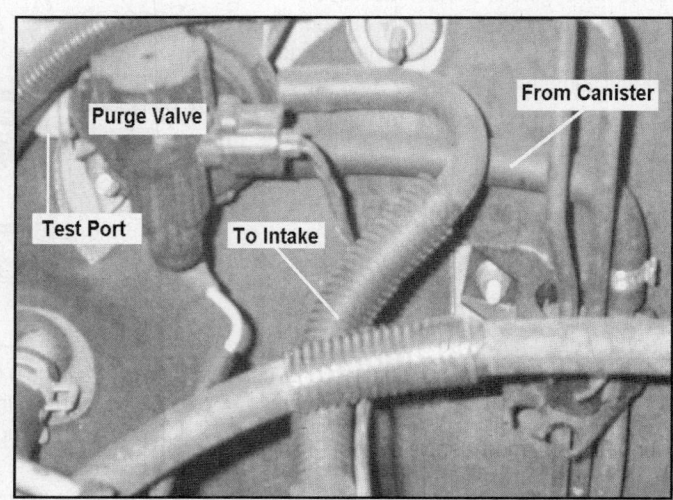

The PCM controls purging by commanding a driver to ground the purge valve solenoid. By varying the percentage (%) of time the driver is grounded and the purge solenoid is operating, the PCM can precisely control the volume of fuel vapors drawn into the intake manifold.

All OBDII vehicles must have an EVAP test port. This application incorporates the test port into the purge valve. The test port is for use with Ford Evaporative Emissions Leak Tester, part No. 310-F007 (134-00056), or equivalent. A green cap identifies the EVAP test port.

FUEL TANK PRESSURE SENSOR

The fuel tank pressure (FTP) sensor monitors the pressure levels inside the fuel tank. The sensor is mounted in the main vapor line of the EVAP system, near the fuel tank.

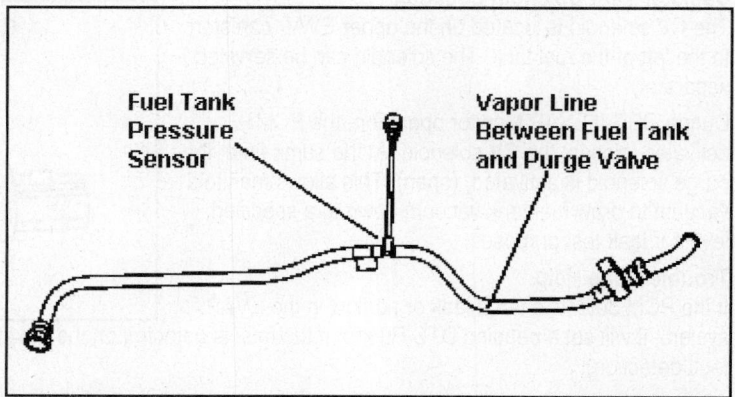

The data from this sensor is mainly used during the OBDII testing procedures initiated by the PCM during the EVAP system monitor. The PCM may also use this data to force a purge if tank pressures become excessive.

FTP Sensor Specifications

The fuel tank pressure sensor receives a 5v VREF signal. The sensor converts tank pressure or vacuum into an analog voltage signal to the PCM. This value is not an available PID on most aftermarket scan tools, and can only be seen with a DVOM or lab scope. On this vehicle, the FTP value was 2.0v with the fuel cap on and a near full tank, but rose quickly to 2.6v when the fuel cap was removed.

SCAN TOOL TEST (CV SOLENOID)

Connect Scan Tool to DLC under driver's side of I/P. Place the gearshift selector in Park and block the drive wheels for safety.

EVAP CV DC (Duty Cycle) PID

The canister vent solenoid is activated by the PCM during the EVAP System Monitor. Refer to EVAP Running Loss Monitor heading in this article for the complete criteria to run this test. During normal engine operation, PID value should be 0%. The PID value will change to a high duty cycle when the EVAP system monitor begins testing.

Forcing the CV Solenoid Closed

The CV solenoid can be difficult to test, since it operates only once per drive cycle, during the EVAP System Monitor test. To force CV solenoid operation with a scan tool, access Canister Vent Closing procedure from OEM Scan Tool menus, or OBD Controls option from Generic scan tool menus.

Follow the sequence at the right. The Generic EVAP Leak Test is for use with EVAP system leak test equipment. However, the procedure involves the PCM commanding the CV solenoid closed to seal off the EVAP system.

Once test conditions have been enabled, the CV solenoid can be tested using a DVOM or Lab Scope. To view an example of how to use a scope to test this solenoid during an EVAP Test, refer to the Lab Scope Test (CV Solenoid).

Generic OBD Control Limitations

Because the Generic controls are not manufacturer specific, they do not consistently function on all applications. This example from a 2001 Windstar can be duplicated on many other, but not all, OBDII vehicles.

SCAN TOOL TEST (PURGE SOLENOID)

Start the engine and monitor EVAP PURGE DC PID. The PID value will be a varying percentage between 0% and 100%. A higher percentage indicates more purge commanded. The PID value will be 100% at the start of the leak check portion of the EVAP system monitor.

The PID may be monitored to observe the PCM strategy as the vehicle is operated under different driving conditions. The Purge solenoid may also be monitored with a DVOM or Lab Scope to compare the PCM strategy to the driver command received by the component.

■ NOTE: *The EVAP PURGE DC PID is not included in the Generic OBDII data list by this vehicle manufacturer.*

GENERIC OBD CONTROLS

```
OBD CONTROL MENU
F1: EVAP LEAK TEST
```

```
EVAP LEAK TEST

This test mode
enables conditions
required to conduct
an evaporative
system leak test,
but does not run the
test.

     Press [ENTER]
      to begin.
```

```
EVAP LEAK TEST

Test conditions have
been enabled.

Turn ignition off
to terminate test.




   PRESS [ENTER]
```

```
Jan 27, 02     5:26:16 pm

EVAP PURGE DC
   40 % DC
UP O2S BANK1
  0.73 VOLTS


0.00 v

[RCV] FOR ENHANCED
  [*EXIT] TO EXIT
```

DVOM TEST (PURGE SOLENOID)

Place the gear selector in Park and block the drive wheels for safety.

DVOM Connections

Connect the DVOM positive probe to the Purge solenoid control circuit (LT GRN/BLK wire) and the negative probe to the battery negative post. Start engine and change the engine speed while monitoring the DVOM.

The reading should vary as the engine speed, load and temperature conditions are changed. The frequency (Hz) will remain fixed, (except when a 0% or 100% purge is commanded). In effect, the voltage reading varies in proportion to the duty cycle of the device. Compare these readings to the Scan Tool duty cycle readings.

DVOM Example (1) & (2)

These examples show the pattern of driver commands to the solenoids. The readings (9.50v and 7.41v) shown are averages of the voltage fluctuations. The frequency is fixed at 100 Hz by the PCM (the duty cycle varies). The duty cycle in example (1) was 51.9% at idle and 71.7% at cruise in example (2).

The duty cycle is constantly varying according to temperature, driving conditions and test status. The duty cycle also varies according to a larger cycle, which can be seen on a Lab Scope. Refer to Lab Scope Test (Canister Purge Valve Solenoid) heading. This test can be used to verify that the PCM is in control of the purge process.

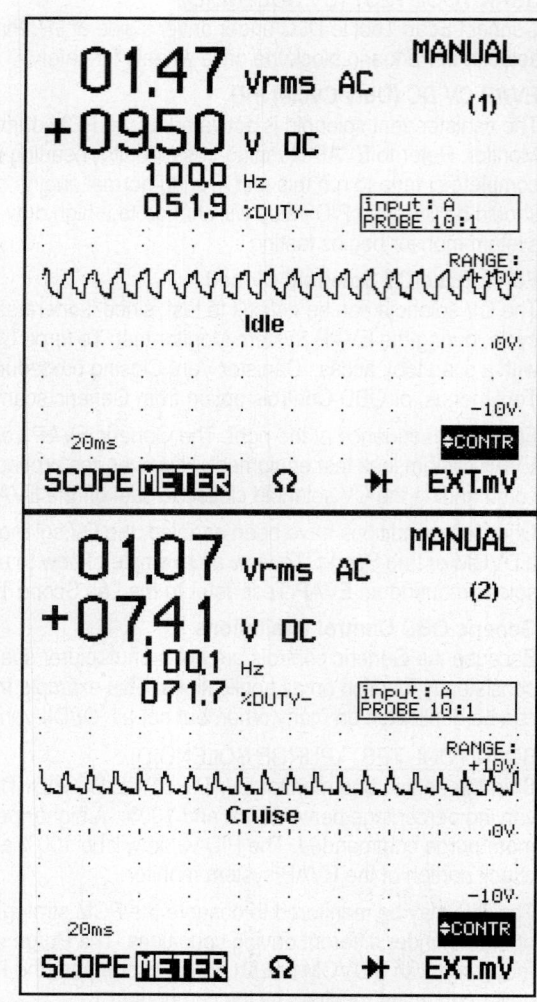

EVAP System Test Schematic

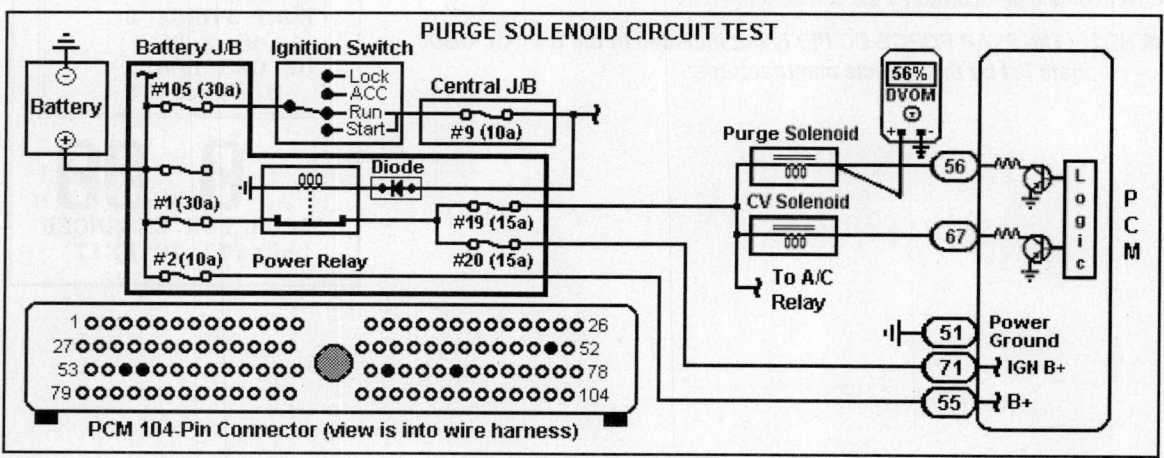

LAB SCOPE TEST (CV SOLENOID)

The CV solenoid operates only once per drive cycle during EVAP System Monitor testing. This makes the CV solenoid difficult to test. An Alternative to meeting the monitor criteria is to command the CV solenoid on with a Scan Tool using either Generic or OEM software. Refer to the Scan Tool Test (CV Solenoid) heading to view how to force the solenoid to operate.

Example (1)

In example (1), the trace shows the CV solenoid with system voltage present the circuit is not complete to ground through the driver in the PCM (the valve is not operating).

This waveform indicates the entire CV solenoid circuit had proper voltage but was not enabled.

Example (2)

In example (2), the trace shows the CV solenoid commanded on during the EVAP Leak Test or Canister Vent Closing procedure from Generic or OEM scan tool software.

The CV solenoid is energized (closed) to seal the system for manual leak testing. The closing of the solenoid can be seen on the Lab Scope as a drop from system voltage to ground.

This is a good test to verify proper circuit voltage, solenoid operation and solenoid driver condition.

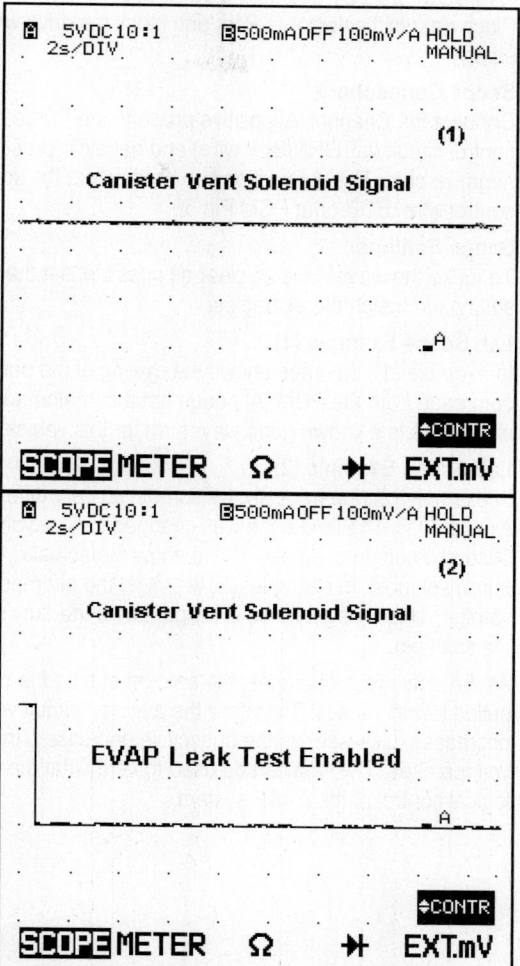

EVAP System Test Schematic

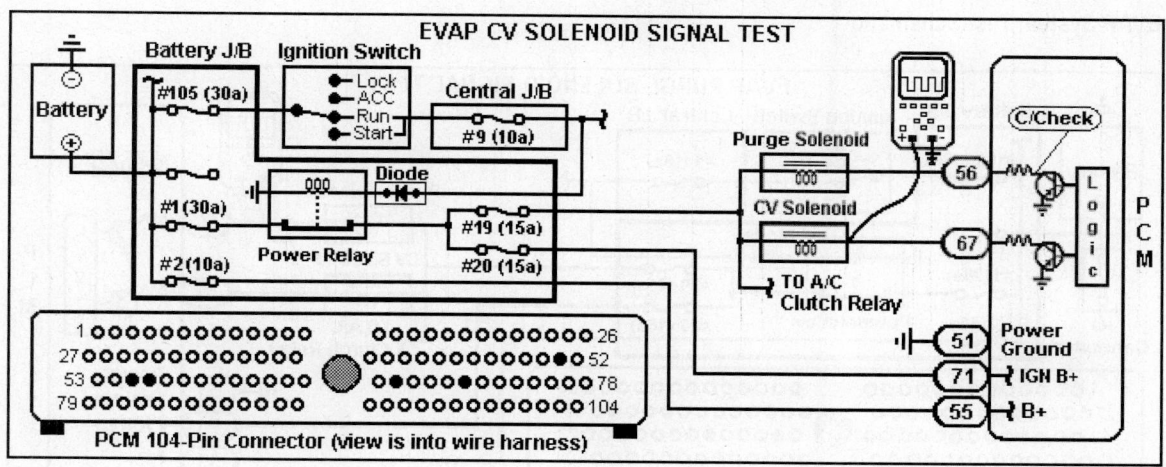

LAB SCOPE TEST (PURGE SOLENOID)

Place gearshift selector in Park and block the drive wheels for safety.

Scope Connections

Connect the Channel 'A' positive probe to the Purge solenoid control circuit (LT GRN/BLK wire) and negative probe to battery negative post. The connection can be made at the solenoid control wire, BOB or at PCM Pin 56.

Scope Settings

To make the waveforms as clear as possible, set the scope settings to match the examples.

Lab Scope Example (1)

In example (1), the trace shows the cycling of the purge valve command from the PCM. Although not a common automotive signal, this is a known good waveform for this vehicle.

Lab Scope Example (2)

In example (2), the trace shows something very different from example (1). The time base was changed to 5 sec/div, or 50sec/screen. In example (1), we show the actual cycling of the purge solenoid. In example (2), we show the average circuit voltage. This voltage varies in proportion to the duty cycle of the solenoid.

As the duty cycle increases, the amount of time the circuit is pulled low increases. Therefore the average circuit voltage decreases. Likewise, as the duty cycle decreases, the average voltage rises. This test can be used to verify that the PCM has logical control of the EVAP system.

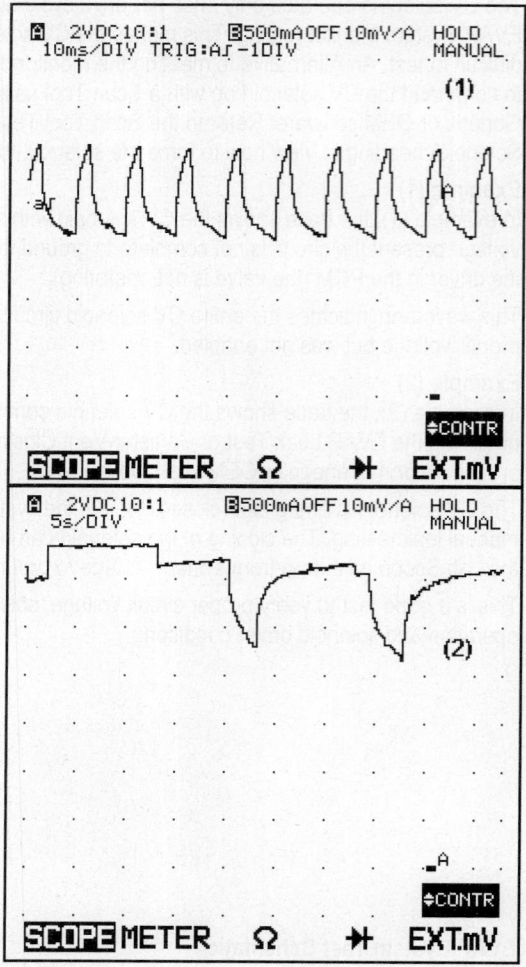

EVAP System Test Schematic

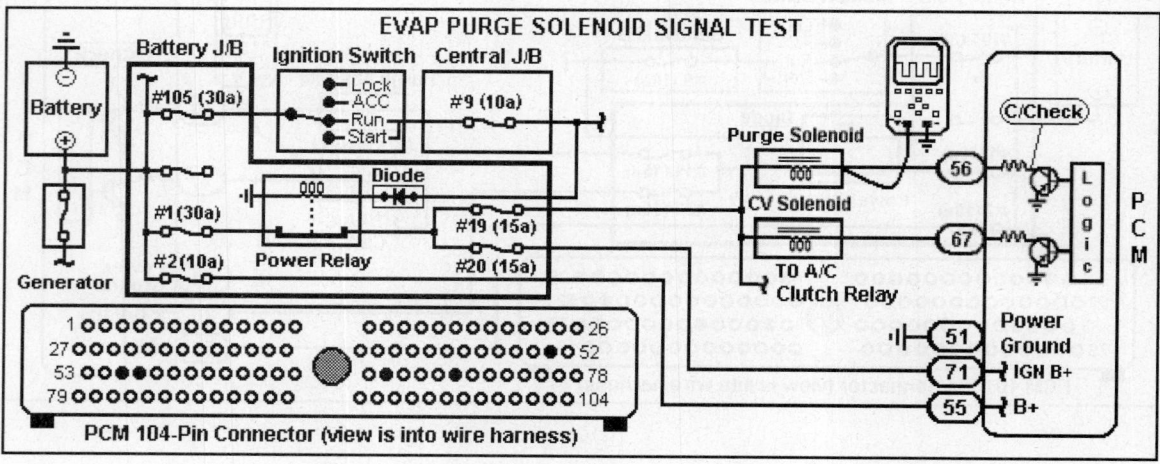

EVAP RUNNING LOSS MONITOR

The EVAP System Running Loss Monitor, referred to simply as EVAP Monitor, is an onboard strategy designed to test for proper operation of the EVAP system by checking the function of its components and its ability to flow fuel vapor (hydrocarbons) to the engine. In addition, the EVAP Monitor detects leaks equal to or greater than 0.020 inch by performing a two or three stage performance check of the entire system.

The enable criteria that must be met to allow the EVAP Monitor to run is shown below:

- Engine cold soak of at least 8 hours, unless bypassed).
- Fuel tank level PID FLI between 15% and 85%.
- Closed loop engine operation, but run time of less than 30 minutes.
- IAT between 40º and 80ºF.
- BARO calculation less than 8000 ft above sea level.
- PID TP MODE value should read PT (part throttle).
- PID EVAPDC (purge solenoid duty cycle) value greater than 75%.
- Cruise between 45 and 65 MPH for 10 minutes.

Here are a few tips that can help you run the EVAP Monitor more quickly:

- Although the PID FLI should be 15-85%, the Monitor runs best close to 3/4 full. Drive the vehicle as steady as possible. Any movement of the fuel (fuel sloshing) causes vapor generation, (loss of vacuum), and may lengthen or abort the testing process.
- Many PCM inputs are indirectly used to enable this monitor. If there are problems in the VSS, ECT, FLI, MAF or TP sensor, this Monitor will not meet its enable criteria.

The testing strategy of the EVAP Monitor is described next:

First Stage

The engine must be under stable purge conditions and a specific vehicle speed must be reached before this test will start. Then the PCM closes the canister vent (CV) solenoid and allows the canister purge valve to remain open. This action connects the fuel tank and EVAP system to the intake manifold. The fuel tank pressure (FTP) sensor monitors the vacuum being created in the fuel tank and vapor lines. If the FTP sensor reports a negative pressure (vacuum) in the system, no gross leaks are present.

Second Stage

During the second stage of the test, the canister purge valve closes to seal the system once a preprogrammed vacuum level is reached during the first stage. The target vacuum is 3.48 KPa (14 in. H_2O). Then the FTP sensor monitors the rate of decay of vacuum. If the vacuum remains greater than 2.0 KPa (8 in. H_2O) for 2 minutes, the test passes and the PCM terminates the test. A leak in the EVAP system as small as 0.020" will cause this test to fail.

Third Stage

If a failure is detected during the second stage (if a small leak is detected), the PCM will try to determine if the failure was due to excessive vapor generation. If the vehicle is being driven roughly, fuel slosh can cause a large increase in vapor generation and this could be interpreted as a small leak during the second stage. The CV solenoid is opened with the purge solenoid closed to establish atmospheric pressure in the system. Then the PCM closes the CV solenoid to seal the system. The FTP sensor monitors the rate of vapor generation by reporting the rise in system pressure over time. If the rate of vapor generation is too high, the results of the second stage of testing are discarded and the test is stopped. If the rate is normal, a small leak is reported as a pending code.

EVAP MONITOR COLD SOAK BYPASS

The conditions to run the EVAP Monitor include an 8-hour cold soak period, and if the Monitor starts to run and aborts for any reason, you must allow for the cold soak period.

It is possible to bypass the cold soak timer to enable the EVAP Monitor (all criteria and driving patterns must still be met). The bypass step can be initiated after you perform a PCM reset. If you do a PCM Reset without following the bypass procedure, the cold soak timer must be allowed to count up to 8 hours. In effect, if the EVAP Monitor status reads "Incomplete" before you do a PCM reset, it is too late to bypass the timer.

1.   Install a Scan Tool. Turn to key on, engine off. Cycle the key off, and back on again. Follow scan tool menus to select appropriate vehicle and engine.
2.   Do a PCM reset with a Scan Tool (clear codes). Do not turn key off when prompted.
3.   Without turning off or disconnecting the Scan Tool, select the appropriate PIDs so you can ensure that the enable criteria are met and maintained.
4.   Start and drive vehicle through the EVAP Monitor drive pattern. When the Monitor status changes to 'Complete', check for pending codes and repair them as needed.

CHILTON DIAGNOSTIC SYSTEM - DTC P0451

The purpose of the Chilton Diagnostic System is to provide you with one or more tests that can be done with a DVOM, Lab Scope or Scan Tool *prior* to entering the complete trouble code repair procedure. These explanations, along with the DVOM, Lab Scope and Scan Tool tests on the previous pages may help you quickly find the problem. If you cannot resolve the problem with these tests, we have included an example repair chart.

Code Description

DTC P0451 - Trouble codes P0452 and P0453 are set if the PCM detects that the FTP sensor signal was too low or too high. However, the PCM can also set a code P0451 if it detects an intermittent fluctuation in the FTP signal (not enough to set P0452 or P0453).

Code Conditions (Failure)

The PCM monitors for any fuel tank pressure fluctuation of over 14 in H_2O in less than 100 ms. Pressure changes of that amount in either direction in such a short period of time are interpreted by the PCM as an intermittent circuit fault and a pending code P0451 (FTP "noisy" circuit) is set. (MIL illuminates on second consecutive failure).

The PCM assumes an intermittent must be the cause, because under normal conditions, an EVAP system would not experience such rapid fluctuations in pressure.

Test Tips

A Scan Tool can be used to observe the FTP PID for rapid fluctuations, indicating an intermittent open or short. Keep in mind though that a Scan Tool is limited because the condition you observe may change faster than the update rate of the PCM to the tool.

A DVOM or Lab Scope can be used with the Scan Tool to help isolate the problem. By using these different tools at the same time, you can compare the signal the PCM is receiving with the processed value, or what the PCM 'thinks' it sees. For example, this will help you to determine if the signal is missing, or if it is being processed incorrectly.

It is generally more efficient to diagnose circuits and components using the concepts just described. For P0451, it is even more so, since the code repair chart is essentially a general strategy for getting an intermittent problem to occur while you are watching. To read more about how to "chase down" an intermittent fault (similar to an intermittent P0451), refer to Intermittent Diagnosis under Powertrain Diagnostics in Section 1.

Exhaust Gas Recirculation System

INTRODUCTION

The EGR system is designed to control the formation of NOx during combustion by controlling combustion temperatures. Exhaust gas has few oxygen or hydrocarbon molecules, the two main ingredients of combustion. The EGR system allows small amounts of exhaust gases to be recirculated back into the combustion chamber to mix with the A/F mixture. This reduces combustion temperatures and less NOx is formed.

The DPFE system includes a DPFE sensor, EGR valve and orifice tube assembly, EGR Vacuum Regulator (VR) solenoid valve assembly and PCM and related vacuum hoses.

Differential Pressure Concepts

The DPFE system is based on the principle that airflow through an orifice will create a pressure differential between the two sides of the orifice. This pressure drop can be used for precise EGR flow monitoring and control.

The DPFE sensor is connected to both sides of this orifice where it is exposed to pressure in vacuum lines. One side of the orifice is exposed to exhaust backpressure and the other side to manifold vacuum (EGR valve open). This design creates a pressure drop across the orifice whenever there is EGR flow. When the EGR valve closes, there is no longer flow across the metering orifice and the pressure equalizes.

DPFE System Operation

The operation of the DPFE System is explained next:

1. The PCM monitors the CKP, ECT, IAT, MAF and TP sensor signals to determine the optimum operating conditions for this system. The engine must be warm, running at a stable engine speed under moderate load before this system is activated. It will not operate at idle, extended WOT periods or if a fault is detected in the EGR system.
2. The PCM calculates a desired amount of EGR flow for each engine condition. Then it determines the desired pressure drop across the metering orifice that corresponds to the correct amount of EGR flow. A command is issued to the VR solenoid.
3. The VR solenoid receives a variable duty cycle signal (0-100%). The higher the duty cycle rate, the more vacuum the solenoid diverts to the EGR valve.
4. The increase in vacuum acting upon the EGR valve diaphragm overcomes the valve spring pressure and lifts the valve pintle off its seat. This allows exhaust gas to flow.
5. As exhaust gases are flowing through the orifice, the DPFE sensor measures the pressure drop and relays a proportional voltage signal (from 0-5 volts) to the PCM. The PCM uses this signal to verify correct EGR flow or correct for any errors.

EGR System Graphic

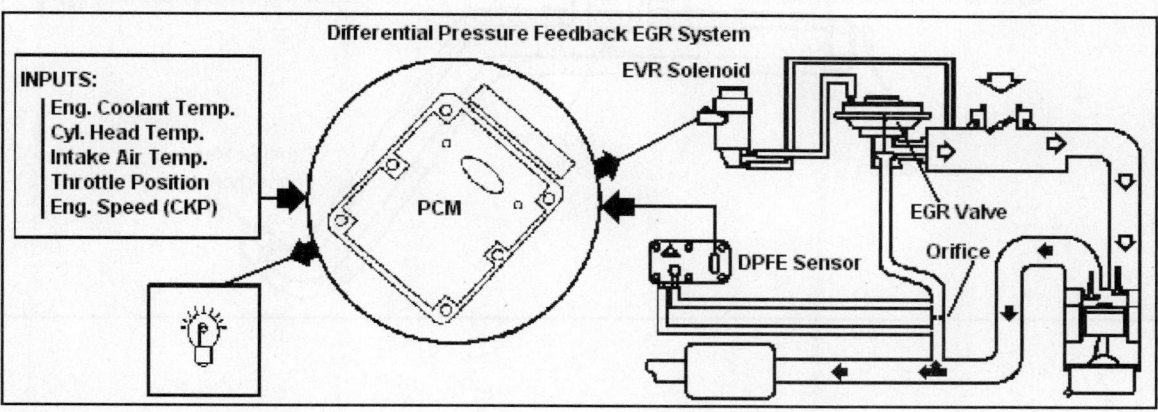

Differential Pressure Feedback EGR System

INPUTS:
Eng. Coolant Temp.
Cyl. Head Temp.
Intake Air Temp.
Throttle Position
Eng. Speed (CKP)

PCM

EVR Solenoid

DPFE Sensor

EGR Valve

Orifice

EXHAUST GAS RECIRCULATION VALVE

The EGR valve used with the DPFE EGR system is a conventional vacuum actuated EGR valve. The valve increases or decreases the flow of exhaust gas. A spring forces the valve closed at rest. As vacuum applied to the valve diaphragm, the spring pressure is overcome and the valve begins to open. As the vacuum signal weakens, EGR spring force closes the valve.

The valve will begin to open at about 1.6" Hg of vacuum, and is fully open over about 5" Hg. Most EGR operating positions are somewhere between fully open and fully closed. Using the VR solenoid, the PCM has very precise control over EGR valve position.

EGR Valve Assembly Graphic

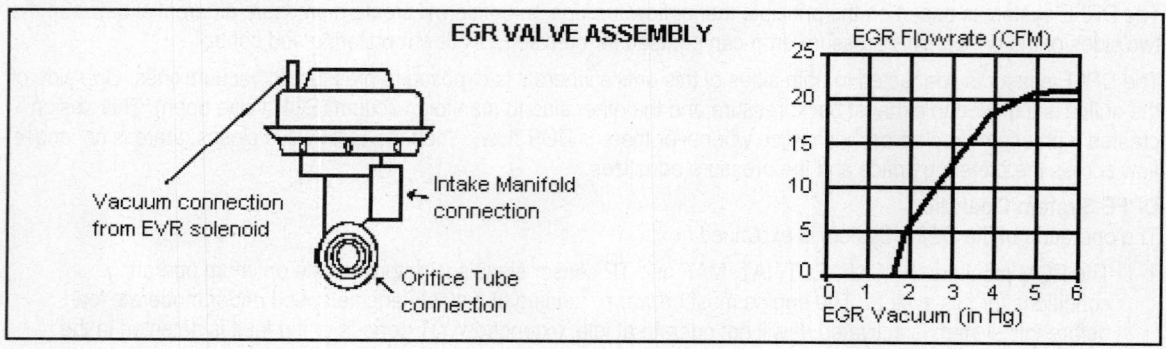

ORIFICE TUBE ASSEMBLY

The orifice tube assembly is a section of tubing placed between the exhaust system and intake manifold. The tube provides a flow path for EGR gases to the intake manifold. It has a metering orifice and two pressure pickup tubes.

The internal metering orifice creates a measurable pressure drop as the EGR gases flow. The DPFE sensor monitors the pressure differential across the orifice and sends the PCM a feedback signal. The PCM uses the DPFE sensor signal to determine the optimum amount of EGR flow under all engine operating conditions.

Orifice Tube Assembly Graphic

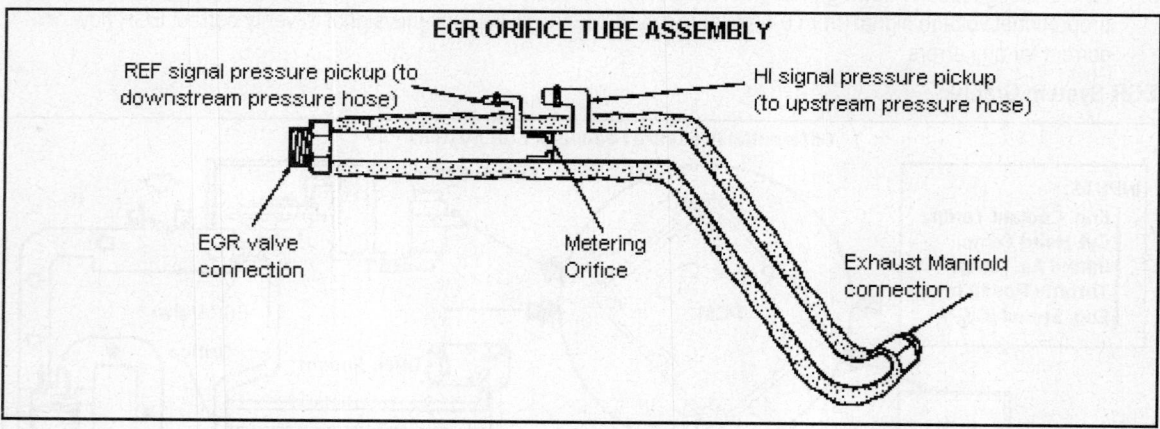

DPFE SENSOR INFORMATION

The DPFE sensor is a capacitive-type pressure transducer that monitors the differential pressure across the metering orifice located in the orifice tube assembly. The sensor receives vacuum signals through two hoses referred to as the downstream pressure (REF signal) and upstream pressure hose (HI signal).

The REF and HI signals are marked on the sensor housing. The HI signal uses the larger diameter hose. The DPFE sensor signal is proportional to the pressure drop across the metering orifice. This input is referred to as the EGR flow rate feedback signal.

Note: *Always identify the part number of the DPFE sensor to verify the correct sensor is used.*

The DPFE signal is used in place of an EGR valve position sensor, and is also used by the PCM for EGR system diagnostics.

DVOM TEST (DPFE SENSOR)

Locate the DPFE sensor circuits (underhood, at the BOB or at the PCM). Connect the DVOM positive lead to the DPFE signal (BRN/LG wire) and negative lead to the battery ground post. Start the engine and record the readings under various engine conditions. Compare the actual readings to the Pin Voltage Tables located in this section.

The DVOM test can also be carried out with a hand vacuum pump. Remove DPFE sensor from orifice tube. By applying vacuum to the small port of the DPFE sensor and then releasing, the sensor can be observed for glitches, erratic operation or performance problems. The voltage value with vacuum hose disconnected should be 0.2-1.3v. Apply 8-9" Hg vacuum to the small port. Voltage should exceed 4v. Release vacuum and verify that voltage drops to under 1.5v within 3 seconds.

DPFE Sensor Circuit Test Schematic

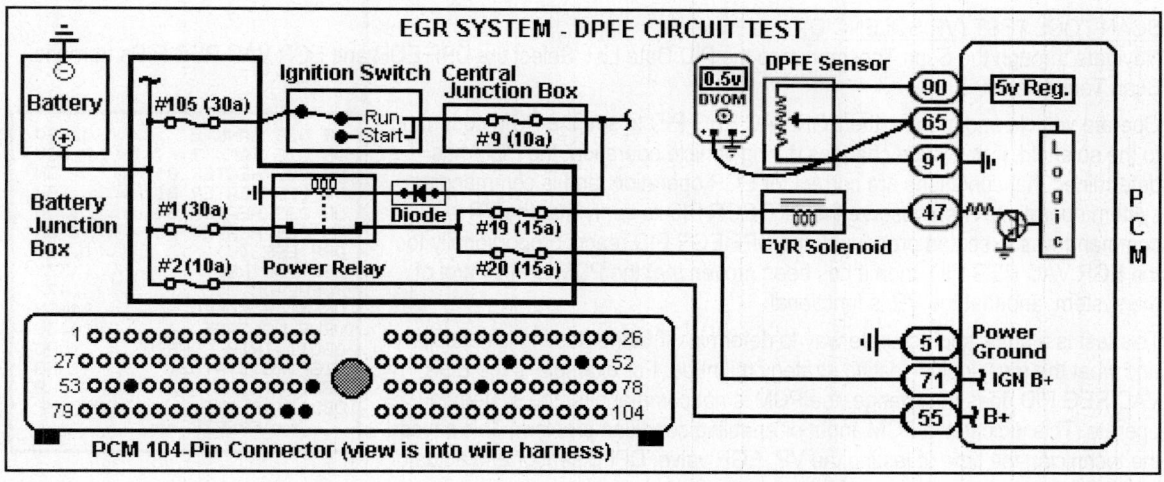

VACUUM REGULATOR SOLENOID

The EGR vacuum regulator (VR) is an electromagnetic solenoid used to regulate the vacuum supply to the EGR valve. The solenoid contains a coil that magnetically controls the position of an internal disc to regulate the vacuum. As the duty cycle to the coil is increased, the vacuum signal passed through the solenoid to the EGR valve increases.

Any source vacuum that is not directed to the EGR valve is vented through a vent valve into the atmosphere. At 0% duty cycle (no electrical signal supplied), the EGR vacuum regulator solenoid allows some vacuum to pass, but not enough to open the valve.

Vacuum Regulator Solenoid Graphic

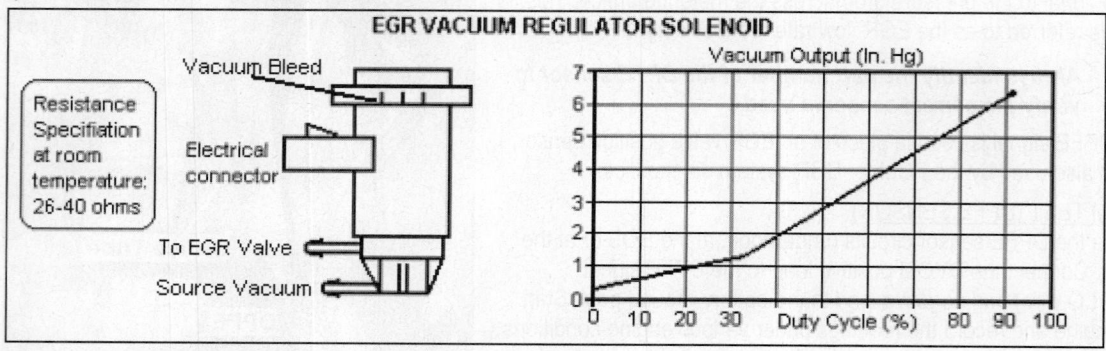

Vacuum Regulator Solenoid Table

Duty Cycle	Vacuum Output					
(%)	Min		Nominal		Maximum	
	In-Hg	KPa	In-Hg	KPa	In-Hg	KPa
0	0	0	0.38	1.28	0.75	2.53
33	0.55	1.86	1.3	4.39	2.05	6.90
90	5.69	19.2	6.32	21.3	6.95	23.47

SCAN TOOL TEST (VR SOLENOID)

Navigate through the Scan Tool menus to the PID Data List. Select the DPF EGR and EGR VAC REG PIDs from the Scan Tool data list.

Operate vehicle and monitor the EGR VAC REG PID to see the PCM command to the solenoid. If this value changes during vehicle operation, the PCM has determined that conditions are correct for EGR operation, and is commanding system operation. Then observe the DPF EGR PID to verify that the VR command was executed properly. If the DPF EGR PID reacts proportionally to the EGR VAC REG PID, then it has been proven that the PCM is in control of the system, and that the VR is functional.

This test is a quick and accurate way to determine if further testing is necessary, and what the most logical testing strategy might be. For example, if the EGR VAC REG PID does not change, the PCM is not commanding the system to operate. This indicates a PCM, input or enabling condition problem. This saves the technician the time of testing the VR, EGR valve, DPFE sensor and actuator wiring that are clearly not the problem.

```
UP O2S BANK 2...........0.69V
DN O2S BANK 2...........0.66V
UP O2S HEATER B1............ON
DN O2S HEATER B1............ON
UP O2S HEATER B2............ON
DN O2S HEATER B2............ON
DPF EGR (V).............1.52V
EGR VAC REG..............37%
PURGE DC.................47%
VEHICLE SPEED.........24MPH
VPWR VOLTAGE.........14.3V
OBDII TRIP COMP?..........NO
MISFIRE STATUS............NO
OBDII TRIPS................0
OBDII DR CYC.........25COMP
   ** FREEZE **
```

LAB SCOPE TEST (VR SOLENOID)

The Lab Scope is the "tool of choice" to test the VR solenoid as it provides an accurate view of the solenoid response and frequency rate. Place the gearshift selector in Park and block the drive wheels for safety.

Scope Connections

Connect the Channel 'A' positive probe to the VR control circuit (BRN/PNK wire) at PCM Pin 47 and the negative probe to battery ground.

Scope Settings

To make the waveforms as clear as possible, set the scope settings to match the examples.

Lab Scope Tests

The engine was allowed to run long enough until it entered closed loop operation prior to testing. Note that the examples were captured at different speeds.

Lab Scope Example (1)

In example (1), the trace shows the VR solenoid in the off position. This is a good test to use to verify the correct system voltage, and verify that there are no problems present in the circuit (i.e., it is not open or grounded).

Lab Scope Example (2)

In example (2), the trace shows the VR solenoid command signal. This waveform was captured at Cruise speed.

The VR PID would read approximately 37% in this example on a Scan Tool.

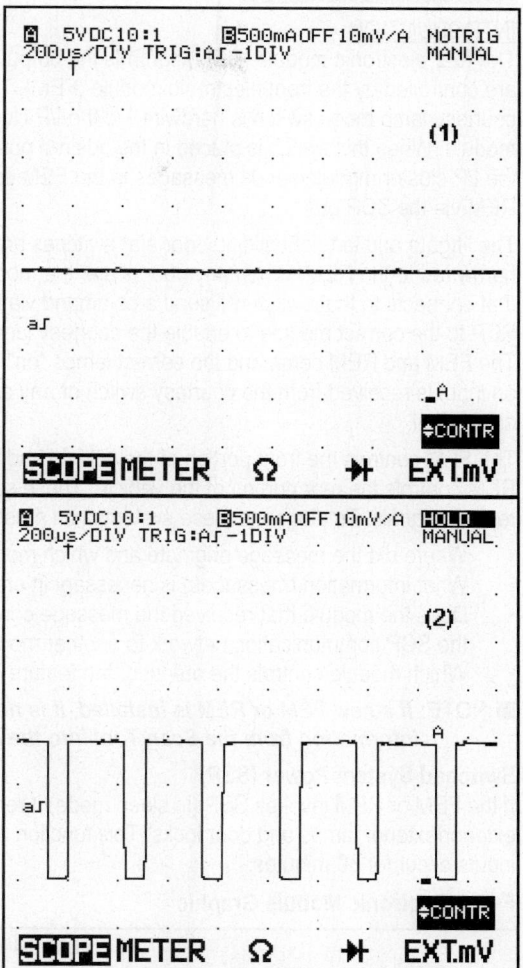

VR Solenoid Test Schematic

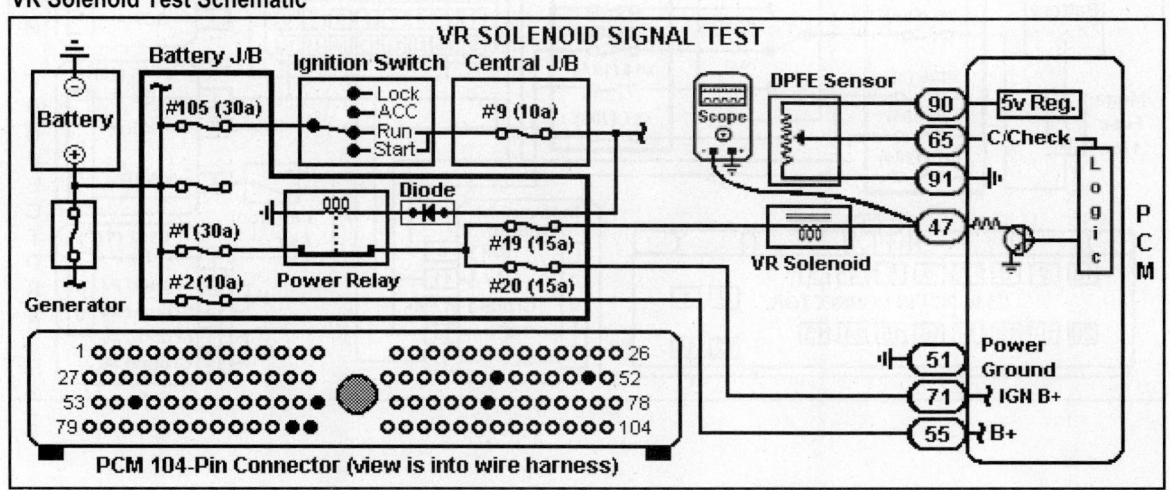

Front Electronic Module

INTRODUCTION

The rear electronic module (REM) controls the outputs to all interior lighting except for the stepwell/puddle lamps that
are controlled by the front electronic module (FEM). The
courtesy lamp mode switch is hardwired to the I/P cluster
module. When this switch is placed in the desired position,
the I/P cluster module sends messages to the FEM and the
REM via the SCP bus.

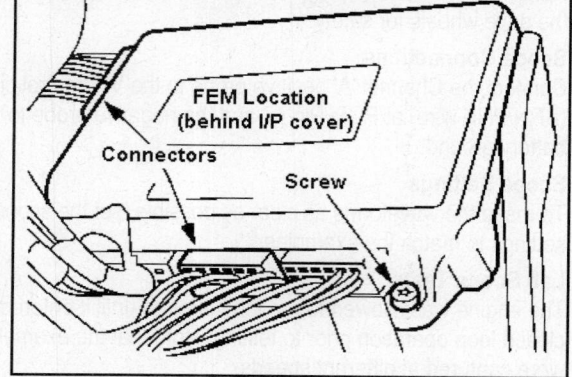

The liftgate and left/right sliding door ajar switches are
hardwired to the REM. When any door is ajar the module
that connects to that switch will send a command via the
SCP to the correct module to enable the courtesy lamps.
The FEM and REM command the correct lamps "on" when
an input is received from the courtesy switch or any door
ajar switch.

The FEM controls the front portion of the vehicle and the
REM controls the rear portion of the vehicle. These systems rely heavily on the SCP network in order to transmit and
receive signals. To diagnose these systems you must understand these key points:

- Where did the message originate and which modules received the command?
- What information (messages) is necessary in order for a feature to operate?
- Does the module that received the message control the output of the feature, or does it output a message over
  the SCP communication network to another module?
- Which module controls the output of the feature?

■ **NOTE:** *If a new FEM or REM is installed, it is necessary to download the updated module configuration
information from the Scan Tool into the new module.*

Switched System Power (SSP)

If the FEM or REM invokes SSP (in sleep mode) it removes power to the relays (SS1-4) and provides power to the
exterior/exterior lamps and door locks. This function places the modules in sleep mode with the key off if no wake up
inputs occur for 30 minutes.

Front Electronic Module Graphic

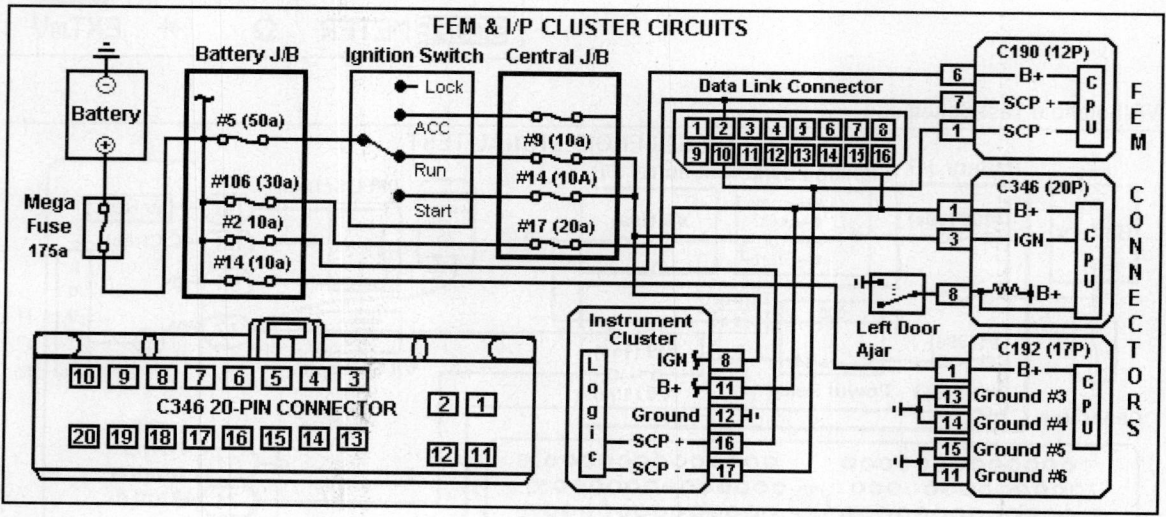

SCAN TOOL TEST (FEM)
If there is a problem in one of the modules, follow the inspection steps described here. A Scan Tool can be used to access Message Center, FEM and REM data.

To diagnose a problem in one of the electronic modules (or a particular trouble code), follow these steps:

1) Verify the customer concern by operating the modules and instrument cluster units.

2) Inspect for signs of mechanical damage to the door locks and check these items:

- Battery Junction Box (BJB) fuses:
  - Fuse 2 (10A), Fuse 13 (30A), Fuse 23 (15A)
- Central Junction Box (CJB) fuses:
  - Fuse 28 (10A), Fuse 14 (10A), Fuse 10 (10A)
  - Fuse 16 (10A), Fuse 9 (10A)
- Circuitry and Connectors
- Door Lock Actuators & Door Lock Switches
- FEM, REM, RKE (DDM)
- SCP Communication Network
- Remote Transmitter and its batteries

3) If an obvious cause of a reported concern is found, make the repair before proceeding to the next step.

4) If you did not find the cause of the concern during this step, connect the Scan Tool to the data link connector (DLC) located beneath the instrument panel and select the vehicle and system from the tool menu. If it does not communicate with the vehicle:

- Verify that the program card is properly installed.
- Check the connections to the vehicle.
- Check the ignition switch position.

5) If the tool still does not communicate with vehicle, refer to the Scan Tool manual and NO DATA tests.

6) Perform the Data Link Diagnostic Test. If the tool displays any of these messages, perform the test:

- CKT 914, CKT 915 or CKT 70 = ALL ECUS NO RESP/NOT EQUIP, refer to Information Bus in other repair manuals or electronic media.
- NO RESP/NOT EQUIP for IC - Pinpoint Test A.
- SYSTEM PASSED - retrieve, record and then erase all the continuous codes. Perform the appropriate self-test for the FEM, REM or RKE.

7) If codes retrieved are related to the Message Center Module or Instrument Cluster DTC Index - refer to the Message Center DTC Index for repair information.

8) If a DTC is set (B1319), refer to DTC list (next page).

Vetronix

SCAN TOOL MAIN MENU

SELECT APPLICATION

GLOBAL OBDII (MT)
GLOBAL OBDII (T1)
GM P/T
GM CHASSIS
GM BODY SYSTEMS
FORD P/T
FORD CHASSIS
FORD BODY
CHRYSLER P/T
CHRYSLER CHASSIS

SCREEN 1

SELECT
MODEL YEAR
2001 (1) ↑↓
[ENTER]

SCREEN 2

SELECT ENGINE:
01 3.8L SFI MAF
WINDSTAR
(VIN=4)?

SCREENS 3 & 4

SELECT SYSTEM:
F1: FEM

SELECT TEST ↑↓
F1: CONTINUOUS
F2: ON-DEMAND
F3: FRONT WIPER

SCREENS 5 & 6

ON-DEM. DTCs: 1
>B1319

LF DOOR AJAR
NO
IGN SW RUN POS
YES

SCREENS 6 & 7

BODY MODULE
ON DEMAND
QUICK TEST
START? [YES/NO]

SYSTEM PASS.
NO DTCs
AVAILABLE
[EXIT]

FEM DTC, PID & ACTIVE CONTROL LISTS

The tables on this page contain a "partial" list of the DTC, PID data Active command lists for this vehicle application. If a FEM trouble code or particular symptom is present, look up the trouble code (or symptom) and refer to the corresponding repair table.

FEM DTC List (Partial)

DTC	Description	Source	Action
B1309	Power Door Lock Circuit Short to Ground	FEM	Go to Pinpoint Test 'D'
B1319	Driver Door Ajar Circuit Fault (Refer to the example on the next page)	FEM	Refer to Lighting & Horns
B1327	Passenger Door Ajar Circuit Fault	FEM	Refer to Lighting & Horns
B1332	Decklid Door Ajar Circuit Fault	REM	Refer to Lighting & Horns
B1338	Door Ajar RR Door Circuit Open	REM	Refer to Lighting & Horns
B1341	Power Door Unlock Circuit short to ground	FEM	Go to Pinpoint Test 'D'
B1342	ECU is Defective	FEM	Install a new REM & retest
B1676	Battery Pack Voltage Out-of-Range	FEM	Refer to Charging System
B2477	Module Configuration failure	FEM	Refer to Comm. Network
B2595	Antitheft Input Signal Input circuit fault	FEM	Refer to Section 419-01a
U1041	SCP Invalid / Missing Data for Vehicle Speed	ABS/TC	Perform ABS Self-Test
U1059	SCP Invalid / Missing Data for Transmission	PCM	Perform PCM Self-Test
U1178	SCP Invalid / Missing Data for Climate/HVAC	IC	Perform I/C Self-Test
U1262	SCP (J1850) Communication Bus Fault	---	Refer to Comm. Network

FEM PID List (Partial)

PID	Description	Expected Value
CCNT	Number of Continuous DTC in module	DROPEN, HOODTR, IGNTAM, PANIC
D_DOOR	Left Front Door Ajar Switch	CLOSED, AJAR
D_DSRM	Driver door Unlock Disarm Switch	NO, YES
D_PWRK	Driver's Power Peak Current	AMPS (amperage value)
D_SBELT	Driver's Seat Belt	OUT, IN
DD_UNLK	Driver's Door Unlock Output	OFF, OFF-G, OFF-B, OFF-BG, OFFO-
DR_LOCK	Drivers Door Lock Output State	NO, YES
DR_UNLK	All Doors Unlock Output State	NO, YES
HOOD_SW	Hood Ajar Switch	CLOSED, AJAR
IGN_R	Ignition Switch - Run Position	NO, YES
P_DOOR	Passenger Door Ajar Switch	CLOSED, AJAR
P_DSRM	Passenger Door Unlock disarm switch	NO, YES
PRK_BRK	Parking Brake Switch Input	OFF, ON
VBAT	Battery Voltage	VOLTS (voltage value)

FEM Active Command List (Partial)

Display	Active Command	Action
HORN	Antitheft Indicator Command	OFF, ON
COURTESYL	Battery Saver & Courtesy Entry	OFF, ON
DDLOCK	Door Lock Control	OFF, ON

LIGHTING & HORNS - PINPOINT TEST 'I' (DTC B1319)
If DTC B1319 is set (Driver Door Ajar Circuit Fault), use Pinpoint Test 'I' to test the fault.

Pinpoint Test 'I' Repair Table

Step	Action	Yes	No
'I'1	**Step description:** *Read the I/P Cluster codes* • Were any Instrument Cluster codes set?	Refer to the Instrument Cluster code index.	Go to Step 'I'2.
'I'2	**Step description:** *Check Dimmer Switch* • Disconnect the dimmer switch and measure the resistance between switch Pin 1 and Pin 3. • Was the resistance greater than 10K ohms?	Go to Step 'I'3.	Replace the dimmer switch. Restore the vehicle and repeat the self-test for verification.
'I'3	**Step description:** *I/P Circuit Test to Ground* • Measure the resistance between I/C C241 Pin 1 and CKT 404 (harness side) to ground. • Was the resistance greater than 10K ohms?	Go to Step 'I'4.	Replace the dimmer switch. Restore the vehicle and repeat the self-test for verification.
'I'4	**Step description:** *Check Instrument Cluster* • Close the doors - check the courtesy lamps. • Are the courtesy lamps on continuously?	Reconnect the dimmer switch and instrument cluster. Go to Step 'I'5.	Replace the instrument cluster. Clear codes and repeat the self-test
'I'5	**Step description:** *Use Recorded REM codes* • Were any FEM trouble codes recorded?	B1319 set - go to 'I'12. B1327 set - go to 'I'17.	Go to Step 'I'6 (not included in this chart).
'I'12	**Step description:** *Monitor FEM PID D-DOOR* • Disconnect the LF door ajar switch. Connect a jumper wire from C500 CKT 1312 to ground. • Did the lights turn off and the PID agree?	Replace the LF door ajar switch. Clear the codes and repeat the self-test to verification.	Go to Step 'I'13. 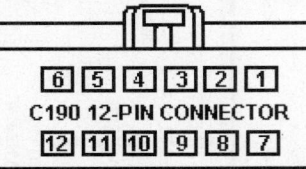
'I'13	**Step description:** *Test for an open circuit* • Were any FEM trouble codes recorded?	B1319 set - go to 'I'12. B1327 set - go to 'I'17.	Go to Step 'I'6.

```
 ┌─┐
┌┴─┴┐
│6 5 4 3 2 1│
C190 12-PIN CONNECTOR
│12 11 10 9 8 7│
```

PIN	CIRCUIT	CIRCUIT FUNCTION
1	PK/LB	DATA BUS (-)
2	DG	LEFT FRONT PARK LAMP
3	VT/OG	HOOD AJAR SWITCH
4	DG/VT	LIFTGATE WIPER MOTOR OUTPUT
5	---	NOT USED
6	RD	V/BATT BATTERY FEED
7	TN/OG	DATA BUS (+)
8	OG/RD	HORN RELAY OUTPUT
9	BN/OG	ENGINE OIL PRESSURE SWITCH
10	BK/PK	RIGHT FRONT PARK LAMP OUTPUT
11	DB	BRAKE FLUID LEVEL SWITCH
12	BK	GROUND

Note: Only 2 of the 4 available connectors are shown.

```
C346 20-PIN CONNECTOR
10 9 8 7 6 5 4 3        2 1
20 19 18 17 16 15 14 13   12 11
```

PIN	CIRCUIT	CIRCUIT FUNCTION
1	LB/RED	V/BATT4 SWITCHED POWER RELAY FEED
2	YE	WIPER MODE SELECT
3	RD/YE	POWER FEED (ST/RUN)
4	LG/YE	DRIVER UNLOCK SW.
5, 7	---	NOT USED
6	LB/OG	DRIVER LOCK SWITCH
8	LG/BK	DRIVER AJAR SWITCH
9-11	---	NOT USED
12	OG/WH	COURTESY LAMPS OUT
13	VT	WIPER DELAY WASHER
14	WT/LG	WIPER SWITCH GROUND
15-18	---	NOT USED
19	RD/YE	PARK BRAKE SWITCH
20	BN/WH	FRONT BLOWER MOTOR RELAY

FEM Wiring Diagrams

FEM WIRING DIAGRAM (PARTIAL)

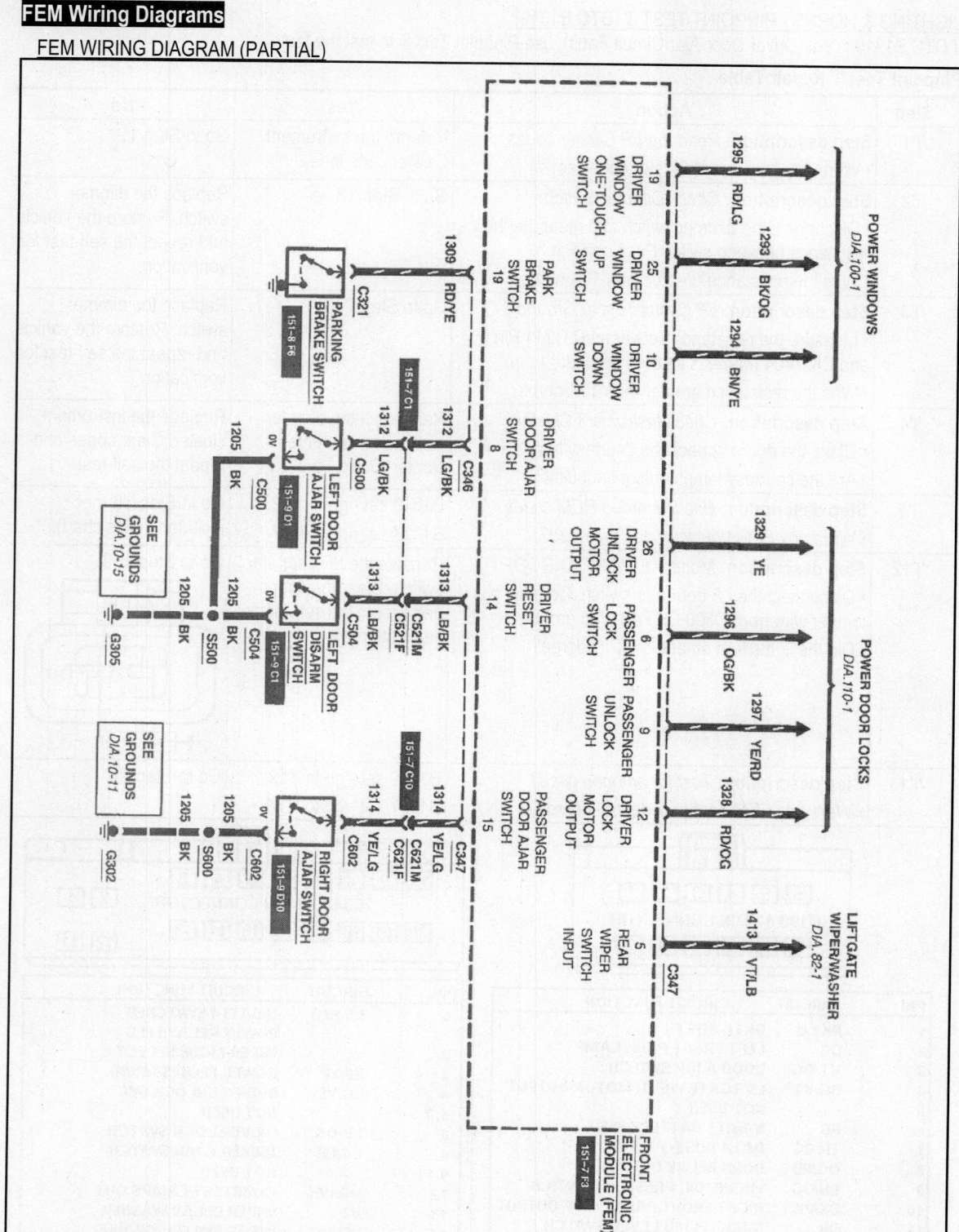

Fuel Delivery System

INTRODUCTION

The Fuel system delivers fuel at a controlled pressure and correct volume for efficient combustion. The PCM controls the Fuel system and injector timing and duration.

Fuel Delivery

The vehicle is equipped with a sequential fuel injection (SFI) system that delivers the correct A/F mixture to the engine at the precise time throughout its entire speed range and under all operating conditions.

Air Induction

Air enters the system through the fresh air duct and flows through the air cleaner where it is monitored by the MAF sensor. The metered air passes through the air duct and enters the throttle body. The incoming air passes through the throttle body and into the intake plenum to the intake manifold where it is mixed with fuel for combustion.

System Overview

A high-pressure (in-tank mounted) fuel pump delivers fuel to the fuel injection supply manifold. The fuel injection supply manifold incorporates electrically actuated fuel injectors mounted directly above each of the intake ports of the engine.

A constant amount of fuel pressure is maintained at the injectors by a fuel pressure regulator. The fuel pressure is a constant while fuel demand is not, so the system includes a fuel return line to allow excess fuel to flow through the regulator to the tank.

The PCM determines the amount of fuel the injectors spray under all engine-operating conditions. The PCM receives electrical signals from engine control sensors that monitor various factors such as airflow, temperature of the air entering the engine, engine coolant temperature, throttle position and vehicle speed. The PCM evaluates the sensor information it receives and then signals the fuel injectors in order to control the fuel injector pulsewidth.

Fuel Delivery System Graphic

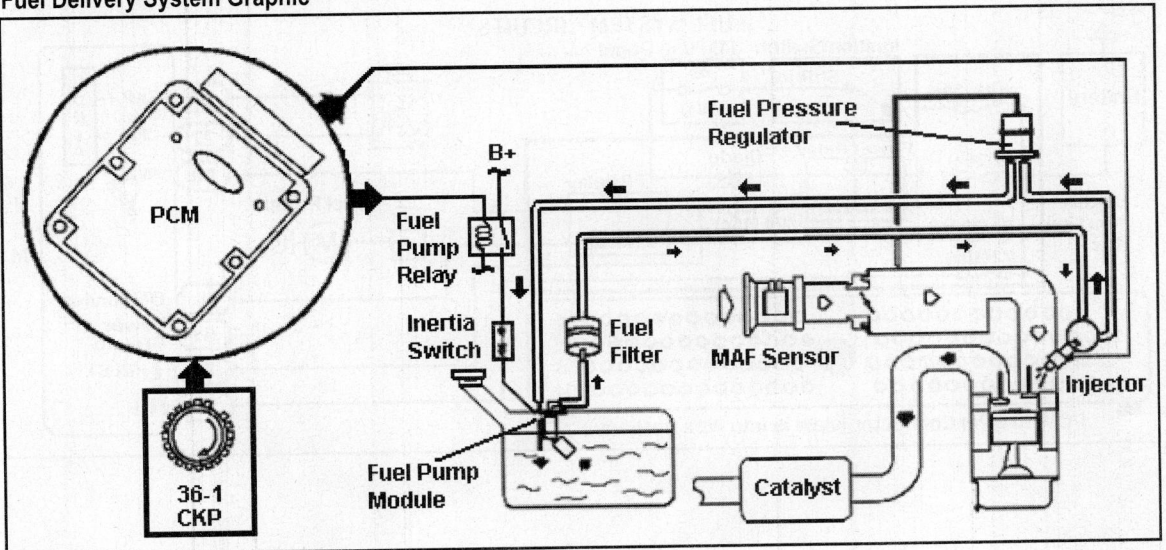

FUEL SYSTEM OPERATION

The PCM controls the fuel pump through a fuel pump relay in the Battery Junction Box. The fuel pump relay provides power to the fuel pump under the following operation conditions:

1) With the ignition turned "off", the PCM and fuel pump relay are not energized.
2) When the ignition switch is first turned to the "run" position, these events occur:
   a. The PCM power relay is energized.
   b. The power relay provides power to the fuel pump relay.
   c. The PCM supplies the fuel pump relay a timed ground.
   d. The fuel pump is supplied power by the fuel pump relay.
3) If the ignition switch is not turned to the "start" position, the timing device in the PCM will open ground (pin 80) after one second, and this action de-energizes the fuel pump relay, which in turn de-energizes the fuel pump. This circuitry provides for pre-pressurization of the fuel system.
4) When the ignition switch is turned to the "start" position, the PCM operates the fuel pump relay by re-grounding the fuel pump relay (pin 80). This provides pressure fuel for starting the engine while cranking.
5) After the engine starts and the ignition switch is returned to the "run" position:
   a. The PCM continues to ground Pin 80 to energize the fuel pump relay and supply power to the fuel pump.
   b. The PCM monitors the CKP sensor to determine if engine continues to run. The PCM shuts off the fuel pump by opening the ground circuit to the fuel pump relay if the engine stops, or if the engine speed is below 120 rpm.

Fuel Pump Monitor

The PCM determines that the fuel pump circuit has power through the FPM circuit (Pin 40). This circuit is used to determine the status of the fuel pump circuit as it relates to DTC P0231 and DTC P0232.

Fuel Pump Circuits Schematic

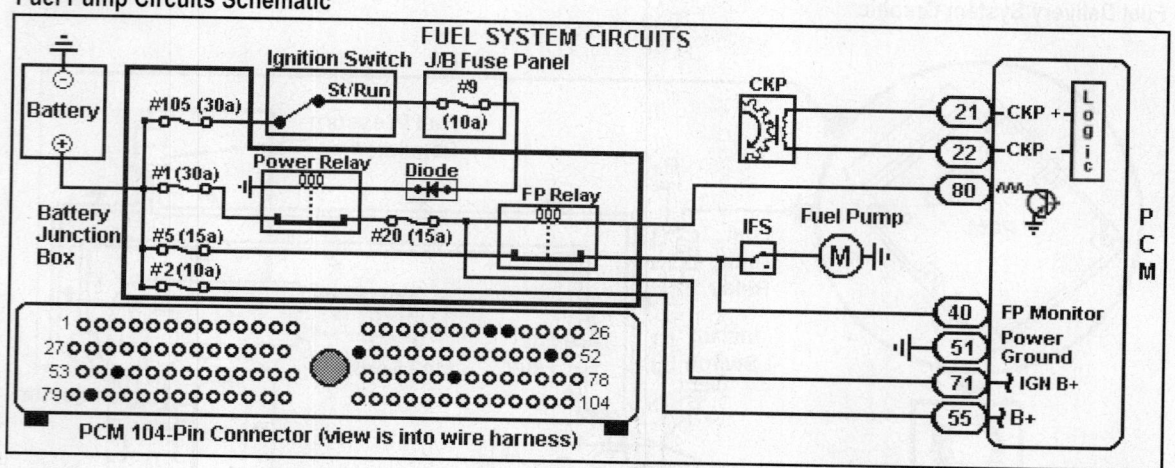

FUEL PUMP MODULE

The fuel pump module is mounted in the module retainer (integral to fuel tank). Fuel initially enters the module through the bottom of the module, and primes the fuel pump.

A portion of the high-pressure flow from the pump is diverted to operate a Venturi jet pump. The jet pump draws fuel from the fuel tank into the fuel pump module. This design permits optimum fuel pump operation during vehicle maneuvers with low fuel levels.

Internal Valves

The pump module has an internal pressure relief valve that restricts fuel pressure if the fuel flow becomes restricted due to a clogged fuel filter or damaged line.

The module also contains a check valve to maintain residual system pressure with key off. This valve keeps the pump primed for quickly pressurizing the system at startup.

FUEL FILTERS

A fuel filter, located along the left frame rail, is used to strain fuel particles through a screen or paper element. This filtration process reduces the possibility of obstruction in the fuel injector orifices. The Windstar has four types of filtration components, though only one is serviceable. It is important to remember that fuel systems can become clogged at any of these filters, especially if a fuel pump has failed.

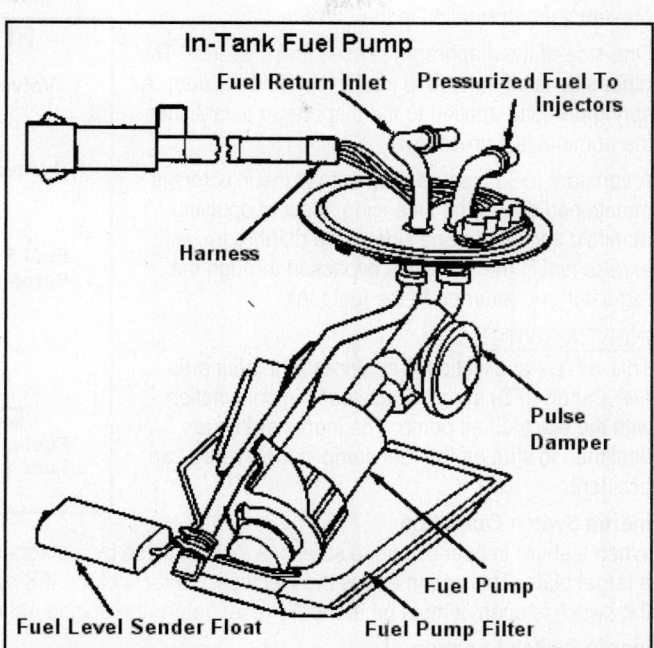

In-Tank Fuel Pump

Serviceable Fuel Filter

A conventional style fuel filter is located in the high-pressure fuel line on the left hand frame rail, near the fuel tank.

Fuel Inlet Screen

The inlet of the pump includes a non-replaceable nylon filter to prevent dirt and other particulate matter from entering the system. This protects the entire fuel system, but is the only filter that protects the pump as well. Water accumulation in the tank can pass through the filter without restriction.

Other Fuel Screens

The Windstar utilizes fuel debris screens in the fuel line at the inlet of the fuel pressure regulator, and at each fuel injector line on the fuel manifold.

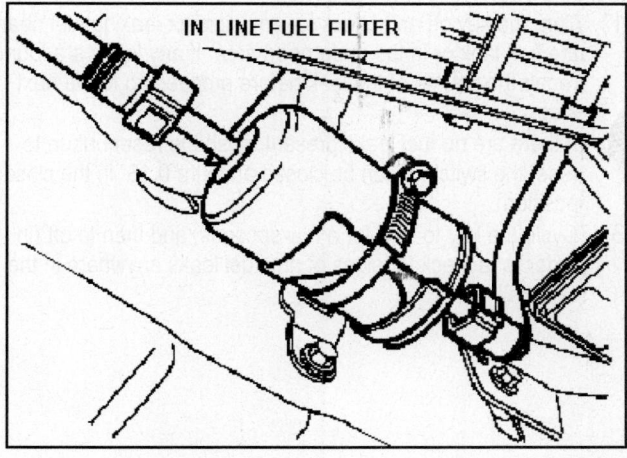

IN LINE FUEL FILTER

FUEL PRESSURE REGULATOR

The Fuel Pressure Regulator is a diaphragm-operated relief valve attached to the return side of the fuel rail. Its function is to regulate the fuel pressure supplied to the injectors. With engine off, it also traps fuel for quick restarts, and to prevent vapor formation in the fuel line.

One side of the diaphragm senses fuel pressure. The other side is connected to intake manifold vacuum. A spring pre-load applied to the diaphragm establishes the nominal fuel pressure.

A constant fuel pressure drop across the injectors is maintained due to the balancing affect of applying manifold vacuum to one side of the diaphragm. Any excess fuel in the system is bypassed through the regulator and returned to the fuel tank.

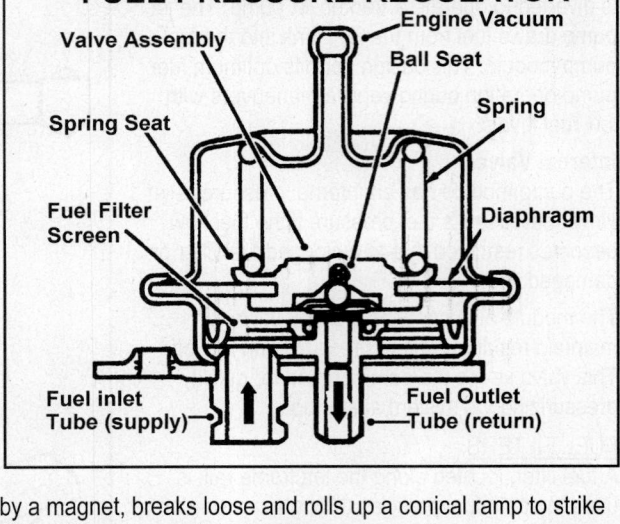

INERTIA SWITCH

This vehicle application is equipped with an inertia fuel shutoff (IFS) switch that is used in conjunction with the electric fuel pump. The inertia switch is designed to shut off the fuel pump in the event of an accident.

Inertia Switch Operation

When a sharp impact occurs, a steel ball, held in place by a magnet, breaks loose and rolls up a conical ramp to strike a target plate. This action opens the electrical contacts of the IFS switch and shuts off power to the fuel pump. Once the switch is open, it must be manually reset before the engine will restart.

Inertia Switch Location

One of the challenges in dealing with resetting or checking this switch is locating it.

On the Windstar, the IFS switch is located in the jack access panel on the right rear interior trim panel.

Inertia Switch Reset Instructions

1) Turn the key off and check for any signs of leaking fuel near the fuel tank or in the underhood area. If any leaks are found, repair the source of the leak before proceeding to the next step.
2) If there are no fuel leaks present, push the reset button to reset the switch (it can be closed an extra 0.16" in the closed position).
3) Cycle the key to "on" for a few seconds, and then to off (in order to recheck for signs of any fuel leaks anywhere in the system).

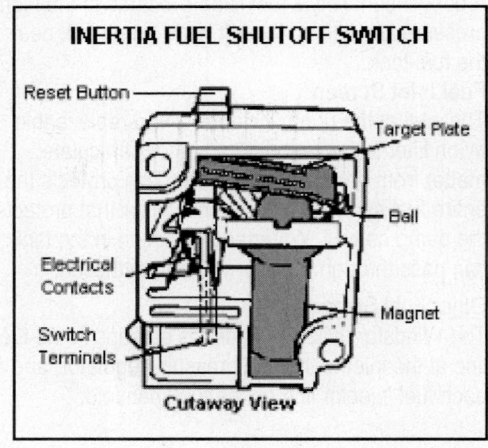

FUEL INJECTORS

The Windstar uses Visteon fuel injectors, with a resistance of 8.5-15.5 Ohms. Under normal operation, each injector is opened and closed at a specific time, once every other crankshaft revolution, (once every engine cycle).

The amount of fuel delivered is controlled by the length of time the injector is grounded (injector pulsewidth). The injectors are supplied voltage through the PCM power relay.

Fuel Injection Timing

The PCM determines the fuel flow rate, (injector pulsewidth), from the ECT, IAT and MAF sensors, but needs to see CKP and CMP signals to determine injector timing.

The PCM computes the desired injector pulsewidth and then grounds each fuel injector in sequence. Whenever an injector is grounded, the circuit through the injector is complete, and the current flow creates a magnetic field. This magnetic filed pulls a pintle back against spring pressure and pressurized fuel escapes from the injector nozzle.

The injectors are energized in this sequence: 1-4-2-5-3-6.

Circuit Description

The PCM power relay connects the PCM and fuel injectors to direct battery power (B+) as shown in the schematic. With the key in the "start or run" position, the power relay coil is connected to B+, which creates a magnetic field that closes relay contacts and supplies power to the PCM and fuel injectors.

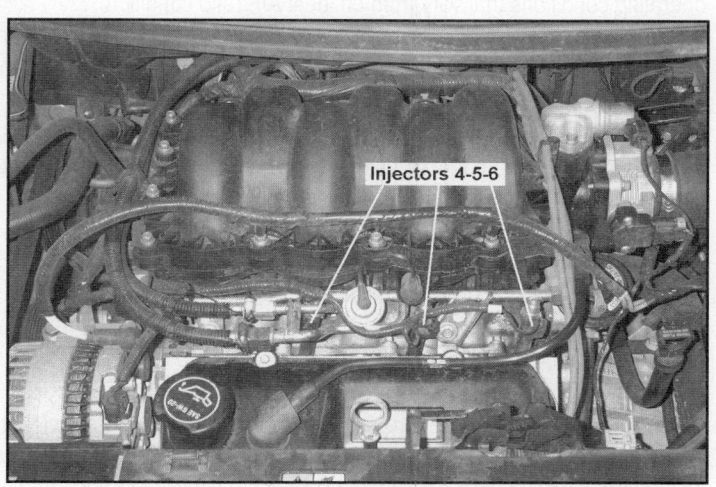

Injectors 4-5-6

Fuel Injection Timing Modes

In Simultaneous mode, fuel is supplied to all six cylinders at the same time by sending the same injector pulsewidth signal to all 6 injectors, twice for each engine cycle. This mode is used during startup and in the event of a fault in the CKP sensor or the PCM.

Sequential mode is used during all other engine operating conditions. In this mode, fuel is injected into each cylinder once per engine cycle in the firing order (1-4-2-5-3-6).

Sequential Fuel Injection Graphic

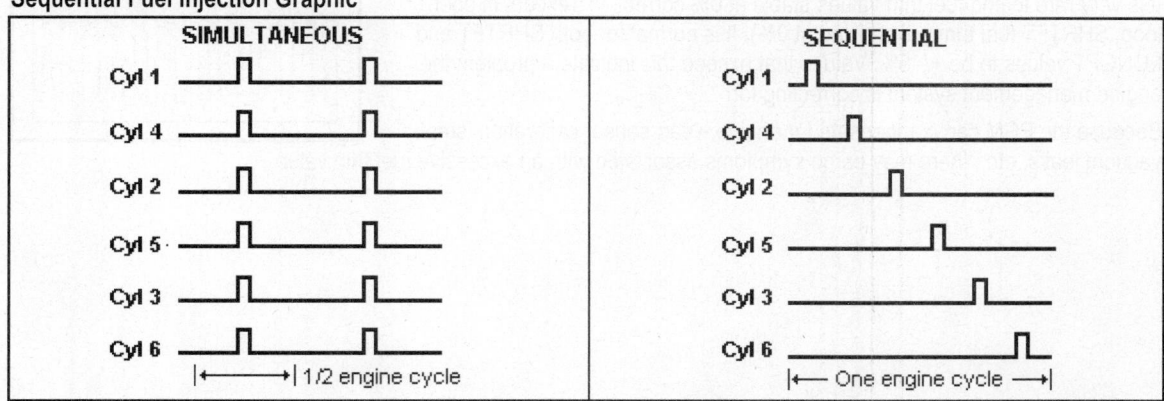

SCAN TOOL TEST (FUEL INJECTOR)

The Scan Tool can be used to quickly determine the fuel injector pulsewidth. However, it is not the tool of choice for certain types of fuel injector problems.

Connect the Scan Tool to DLC. Navigate through the menus until you get to the Ford (OEM) PID Data List.

Select Injector Pulsewidth from the menu. The example to the right was captured at 55 MPH at steady throttle in closed loop. The injector pulsewidth at idle speed was 8.17 ms for Bank 2, and 8.90 ms for Bank 1. The injector status can be viewed on a Scan Tool as "FAULT: OK / NOT OK".

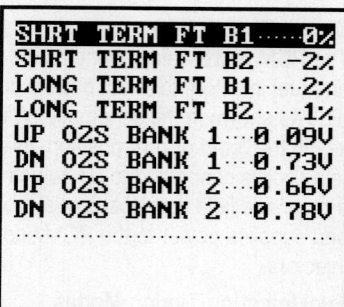

SCAN TOOL TEST (FUEL TRIM)

Short Term fuel trim (SHRTFT) is an operating parameter that indicates the amount of short term fuel adjustment made by the PCM to compensate for operating conditions that vary from an ideal A/F ratio condition.

A negative SHRTFT number (-15%) means the HO2S is indicating to the PCM that the A/F ratio is rich, and that the PCM is trying to lean the A/F mixture. If A/F ratio conditions are ideal, the SHRTFT number will be close to 0%.

Long Term fuel trim (LONGFT) is an engine operating parameter that indicates the amount of long term fuel adjustment made by the PCM to compensate for operating conditions that vary from an ideal A/F ratio.

A positive LONGFT number (+15%) means the HO2S is indicating to the PCM that the A/F ratio is lean, and that the PCM is trying to add more fuel to the A/F mixture. If A/F ratio conditions are ideal the LONGFT number will be 0%.

The LONGFT number is displayed on the Scan Tool in percentage. Refer to the examples in the Graphic.

Short Term and Long Term as a Team

SHRTFT and LONGFT work together by altering the injector pulsewidth to achieve better fuel control. Just as each cylinder bank pulsewidth is calculated independently, each bank also has it's own SHRTFT and LONGFT.

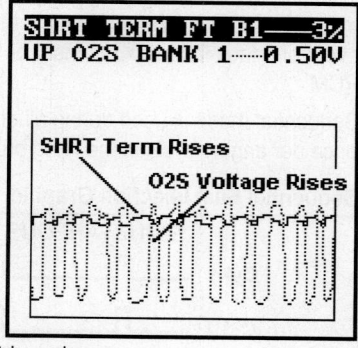

In the Scan Tool example to the right, you can see that SHRTFT fuel trim is responsible for very quick adjustments that are followed by the HO2S. LONGFT is responsible to keep SHRTFT cycling near 0% correction.

It is very rare to find fuel trim values stable at 0% correction, (except in open loop, SHRTFT fuel trim will be fixed at 0%). It is normal for both SHRTFT and LONGFT values to be +/- 5%. Values that exceed this indicate a problem the engine management system is correcting for.

Because the PCM can compensate for engine wear, sensor calibration, small vacuum leaks, etc., there may be no symptoms associated with an excessive fuel trim value.

LAB SCOPE TEST (FUEL INJECTOR)

With a Lab Scope, a variety of tests can be performed on injectors. Place the gear selector in Park. Block the drive wheels for safety.

Scope Connections

Connect the Channel 'A' positive probe to the injector control wire and negative probe to the battery negative post. See schematic below.

Scope Settings

To make the waveforms as clear as possible, set the scope settings to match the examples.

Lab Scope Example (1)

In example (1), the trace shows the injector voltage signal at idle speed. The cursors were used to determine the injector pulsewidth and inductive kick voltage. The window at the bottom of the screen shows the time and voltage difference between the cursor positions. Note that the inductive kick was 'clipped' at 52.8v, (flat top on spike).

Lab Scope Example (2)

In example (2), the time and voltage settings were changed to view the injector pintle closing.

A normal fuel injector will have a single hump. However, if it is physically worn out, it will bounce and produce an oscillation in the waveform. If this condition exists, the vehicle will jerk or buck on deceleration. If there is no hump at all, the pintle is stuck. An injector *circuit* may be fine, but have a physical failure that can only be seen using this test, unless the injector is removed and bench tested.

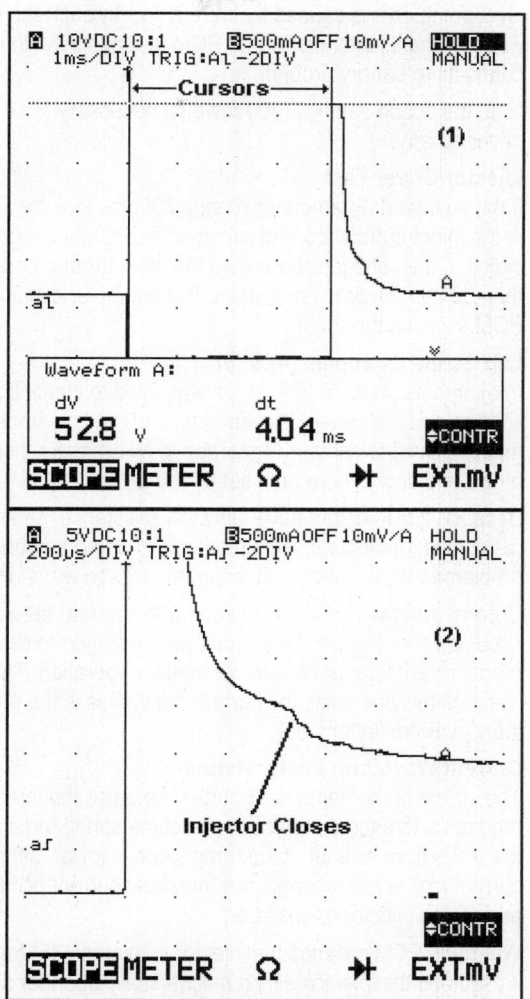

Fuel Injector Test Schematic

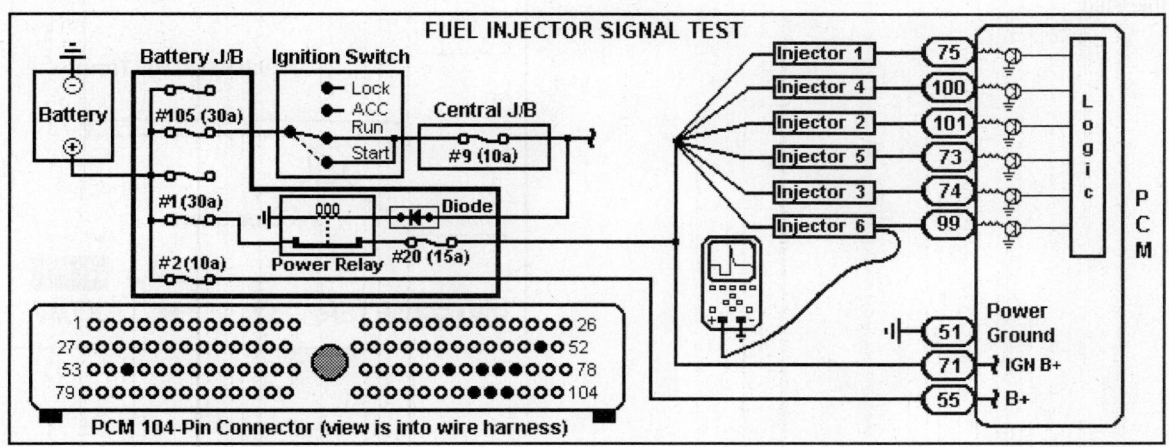

LAB SCOPE TEST (FUEL INJECTOR)

In example (3), the voltage drop through the injector driver control circuit in the PCM rises to 330 mv above the ground level point. This is caused by current flow through the injector winding, the injector driver and PCM ground circuit that connects to battery ground.

Note the scope settings (200 mv/div.) necessary to see the driver activity.

Injector Driver Test

If the voltage drop amount exceeds 750 mv, look for a problem in the injector driver control circuit or the PCM main ground circuit. If the voltage drop across the main ground circuit is less than 50 mv, the problem is in the injector driver circuit in the PCM (replace the PCM).

Lab Scope Examples (4) & (5)

In examples (4) & (5), the scope was used to check both the voltage and the current draw of the circuit. A low-amps probe must be used to convert Amperage to Millivolts for display. The probe and scope were both set to 100 mv/A.

Because fuel injectors have fairly low resistance, small resistance problems in the circuit may cause driveability problems without greatly affecting the voltage waveforms.

Current "ramping" is a way of seeing the current draw in a circuit as a waveform. This circuit was designed to draw much more current than necessary for injector operation. The PCM watches this and limits the current draw. This is the point where the waveform flattens out.

Current Waveform Interpretation

The shape of the 'ramp' is important because the injector requires a threshold current to overcome spring force and lift the pintle from its seat. If the ramp slope is lower, although the current limit is still reached, the threshold current occurs later, and so the injector opens later.

When the PCM grounds the injector in example (5) for 6.4 ms, it assumes the injector will be *mechanically* open for about 5.6 ms. If the current draw in the circuit is incorrect, the amount of fuel actually delivered will be different from what the PCM intended.

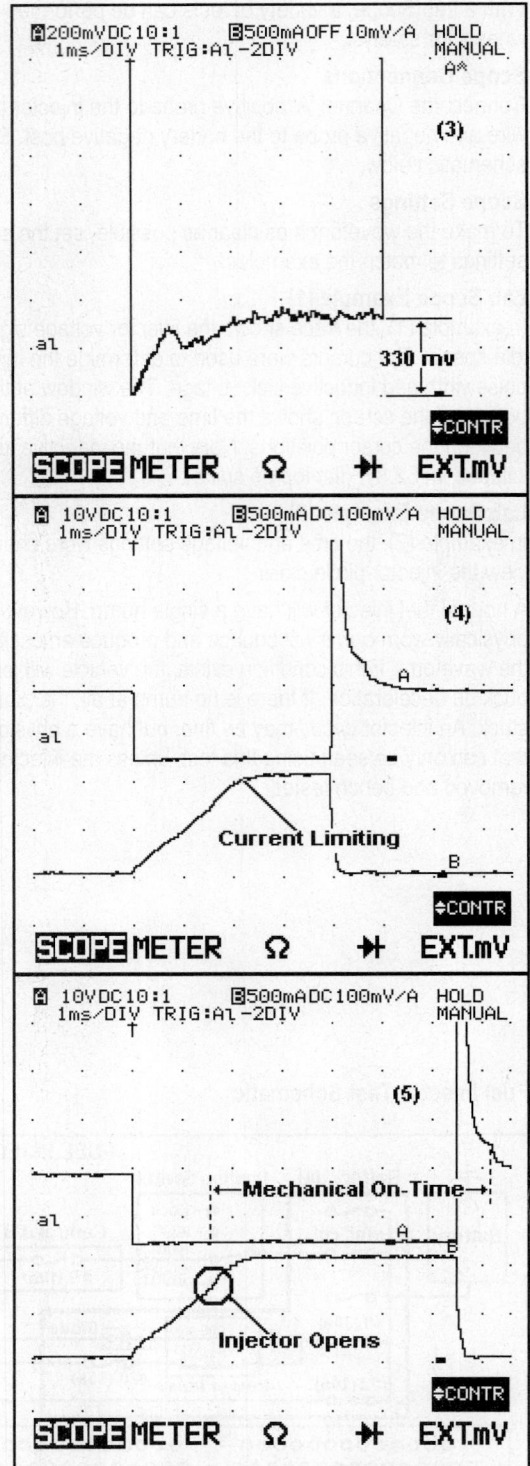

CMP SENSOR (VARIABLE RELUCTANCE)

The camshaft position (CMP) sensor detects the position of the camshaft when engine piston No. 1 is at a specific point of the compression stroke.

The PCM uses the CMP signal to synchronize the fuel injector timing. Since the ignition system is a waste-spark type, the cam position is not needed for ignition timing.

The CMP sensor is a 2-pin variable reluctance (VR) sensor. The CMP sensor is mounted in the front engine cover adjacent to the camshaft, and senses a target on the camshaft sprocket.

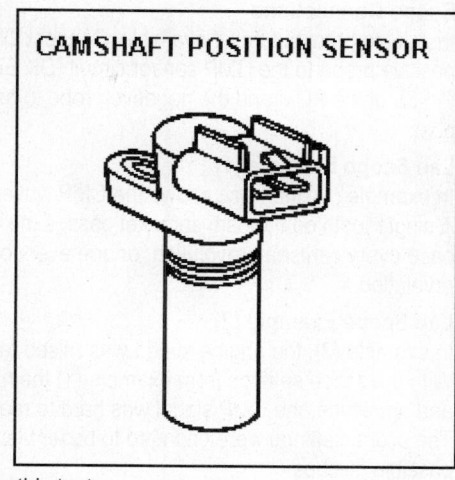

CAMSHAFT POSITION SENSOR

Trouble Code Help

The CMP sensor failure cannot cause a no-start condition. If the PCM does not receive a valid CMP sensor signal due to a problem in the sensor, the PCM or circuits, it will set a related trouble code (DTC P0340 - CMP Sensor Circuit Malfunction).

A failure in the CMP sensor circuit may also set a DTC P1309 - Misfire Monitor Disabled, due to the PCMs inability to keep track of cylinders for misfire detection.

DVOM TEST (CMP SENSOR)

Place the shift selector in Park and block the drive wheels prior to starting this test.

DVOM Connections

Connect the breakout box, if available. Connect the DVOM positive lead to the CMP signal (DK BLU/ORN wire) at Pin 85 of the PCM 104P connector. Connect the negative lead to the battery negative ground post. Refer to the Cam Sensor schematic on the next page. Start or crank the engine and record the readings.

DVOM Test Results

The CMP sensor AC voltage should never be below 100 mv. If it is below this threshold, a DTC P0340 will set.

The CMP sensor signal duty cycle should read close to 50% (it reads 46.0% in this example). The voltage reading should vary with engine speed.

The PCM counts a CMP pulse when the voltage crosses the (-) 300 mv point on the downward slope, and watches the rate at which pulses (Hz) occur.

Although the AC voltage changes greatly according to engine speed, the PCM does not use this data to determine cam position, as long as the signal is above 100 mv.

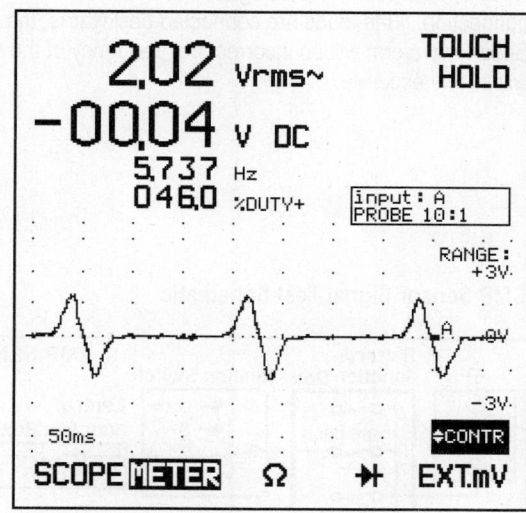

LAB SCOPE TEST (CMP SENSOR)

The Lab Scope can be used to test the CMP sensor circuit as it provides a very accurate view of circuit activity and any signal glitches.

Scope Connections

Install the breakout box if available. Connect Channel 'A' positive probe to the CMP sensor circuit (DK BL/ORN wire) at Pin 85 of the PCM and the negative probe to battery ground post.

Lab Scope Example (1)

In example (1), the trace shows the CMP waveform at hot idle. A single tooth on the cam sprocket passes the CMP sensor once every camshaft revolution, or one every other crankshaft revolution.

Lab Scope Example (2)

In example (2), the engine speed was raised to 2000 RPM. With the scope settings from example (1) the higher frequency and amplitude, the CMP signal was hard to read on screen. The scope settings were changed to better view the signal for possible glitches.

How the PCM Monitors VR Sensors

Note that the Lab Scope was connected to ground rather than the sensor (-) circuit, as in the case of CKP. For the CKP sensor test, scope leads were connected to CKP (+) and CKP (-). Be sure to review a wiring diagram to determine which wires are the positive and negative circuits before making a connection. If the leads are connected backwards, the Lab Scope waveform will be incorrect but frequency of the signal will still be accurate.

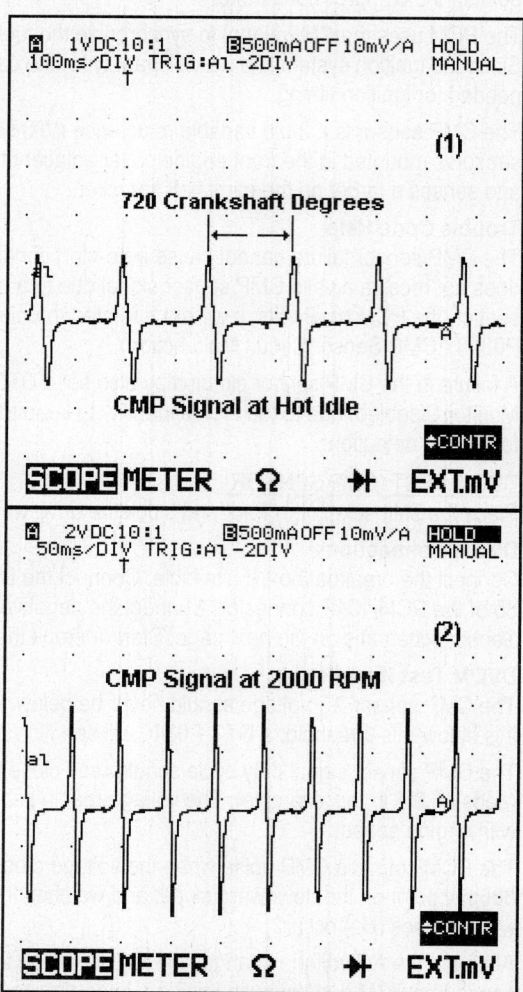

CMP Sensor Signal Test Schematic

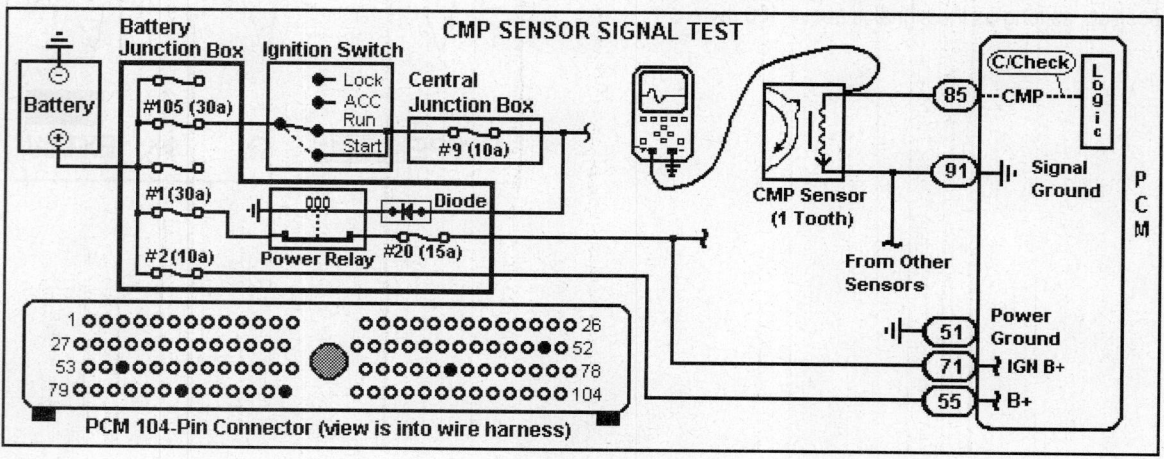

Idle Air Control Solenoid

GENERAL DESCRIPTION

The Idle Air Control (IAC) valve assembly is mounted to the top of the throttle body.

This device is designed to control engine idle speed (rpm) under all engine-operating conditions and to provide a dashpot function, (controls rate of deceleration for emissions and idle quality).

The IAC valve meters the intake air that flows past the throttle plate through a bypass area within the IAC assembly and the throttle body.

IAC Valve

The PCM determines the desired idle speed (amount of bypass air) and controls the IAC solenoid through a duty cycle command. The duty cycle (on time) creates a magnetic field that pulls a pintle back against spring pressure. The longer the on time, the more the pintle overcomes the force of the spring, and the more air bypasses the throttle plate. By balancing these magnetic and mechanical forces, the IAC solenoid can precisely control bypass air under all operating conditions.

SCAN TOOL TEST (IAC SOLENOID

Connect the Scan Tool to the DLC underdash connector. Navigate through the Scan Tool menus until you get to the Ford (OEM) Engine Data List (a list of PID items).

Select the RPM AND IAC PID items from the menu so that you can view them separately on the Scan Tool display.

Scan Tool Example (1)

Example (1) shows the ENGINE SPEED and IDLE AIR CNTRL (%) PID values at idle speed. The first example was captured with all accessories turned off.

Scan Tool Example (2)

In Example (2), the PIDs were captured at Hot Idle speed with the headlights and A/C on. Note the change in the IAC reading (38% to 43%). The PCM increased IAC command to compensate for engine load.

This is an excellent test to determine if the PCM reacts to changing engine conditions, and the IAC reacts to changing commands.

To check for a sticking IAC solenoid, verify that the PCM returns to the same IAC command each time the vehicle is operated and returned to idle, (all other conditions, like cooling fan mode and gear position, must be the same).

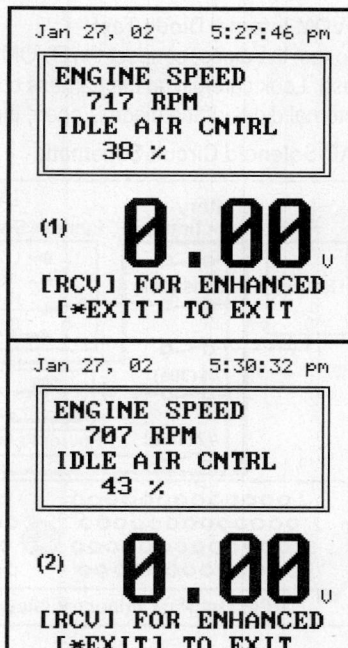

LAB SCOPE TEST (IAC SOLENOID)

The Lab Scope can be used to monitor the IAC solenoid control signal under various engine load, speed and temperature conditions. Place the gearshift selector in Park and block drive wheels for safety.

Scope Connections

Connect the breakout box (BOB), if available. Connect the Channel 'A' positive probe to the IAC solenoid circuit (WHT/LB wire) at the solenoid connector or at PCM Pin 83. Connect the negative probe to the battery negative ground post.

Scope Settings

To make the waveforms as clear as possible, set the scope settings to match the examples.

Lab Scope Example

This IAC solenoid signal was captured with the vehicle in Park and the engine at 2500 RPM. As engine RPM or load are increased, the trace should show an increase in duty cycle (on-time). In this example, the duty cycle is near 50%. On this application the frequency of the IAC command also fluctuates, (between 1.4 kHz and 2.3 kHz).

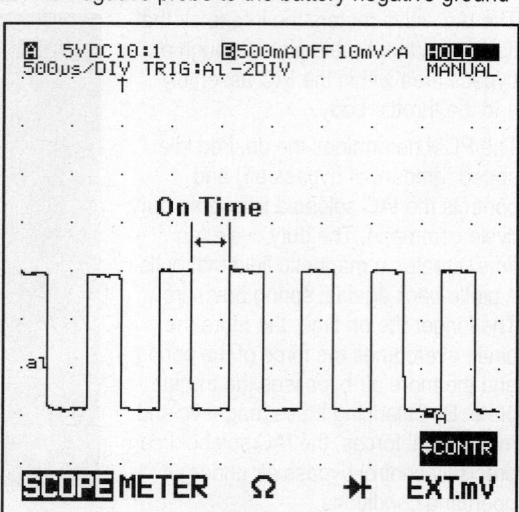

Circuit Description

The IAC solenoid connects to battery power through the power relay. The IAC solenoid operates when the circuit is completed by the PCM. The PCM grounds the circuit at Pin 83 at a high frequency to rapidly open and close the solenoid. The IAC assembly includes an internal diode to suppress voltage spikes. If the diode has failed these spikes will be visible on rising edge of the trace.

DVOM Internal Diode Test

To test the diode, connect the DVOM positive lead to the VBAT pin and the negative lead to the control pin (the Diode test). Look carefully at the harness connectors and mating component to view the proper orientation when testing the internal diode. If the diode is open, the DVOM reading will be 0.8v with the leads connected either way.

IAC Solenoid Circuit Schematic

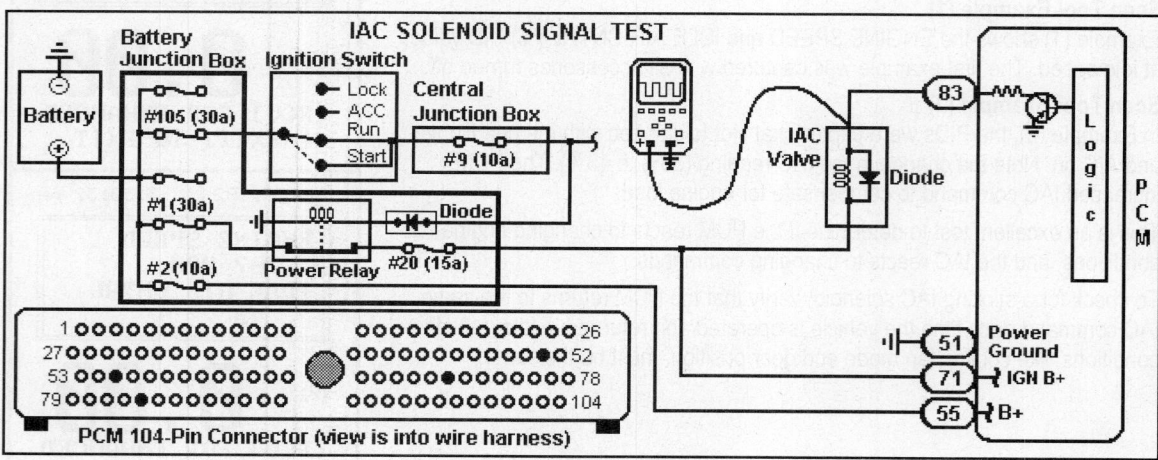

Mass Airflow Sensor

GENERAL DESCRIPTION

The mass airflow (MAF) sensor is located inside the air cleaner housing and contains an integrated IAT sensor. The MAF sensor uses a Hot-wire sensing element to measure the amount of air that enters the engine.

Air passing through the device flows over the hot wire element and causes it to cool. The hot wire element is maintained at 392°F (200°C) above the current ambient temperature as measured by the constant cold wire part of the sensor.

The current level required to maintain the temperature of the hot wire element is proportional to amount of air flowing pass the sensor. The MAF input to the PCM is an analog signal proportional to the intake air mass.

The PCM calculates the desired fuel injector on time in order to deliver the desired A/F ratio. The PCM also uses the MAF sensor to determine transmission EPC solenoid shifting and the TCC solenoid scheduling.

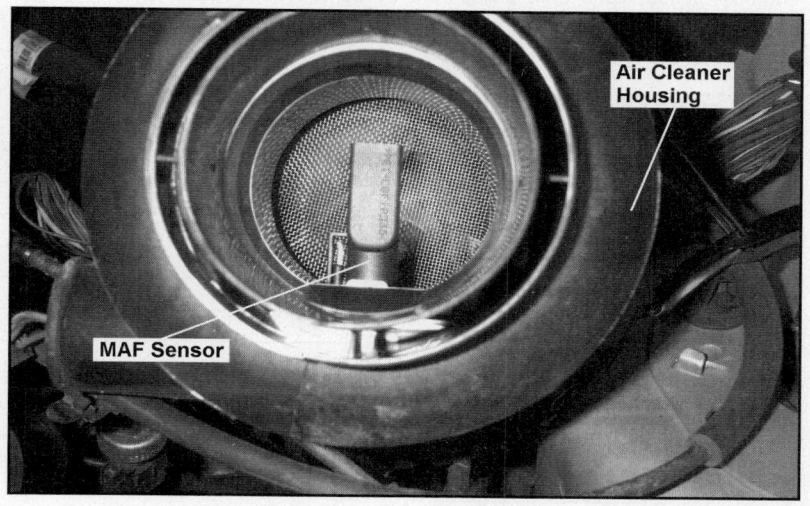

■ **NOTE:** *When this sensor fails, it may set a trouble code or cause a No Code Fault (engine cold or hot).*

Circuit Description

The MAF sensor is powered by the PCM power relay. The MAF sensor signal and MAF ground circuits connect directly to the PCM at Pins 88 and 36. The integrated IAT signal connects to the PCM on Pin 39, and to sensor ground on Pin 91.

MAF Sensor Schematic

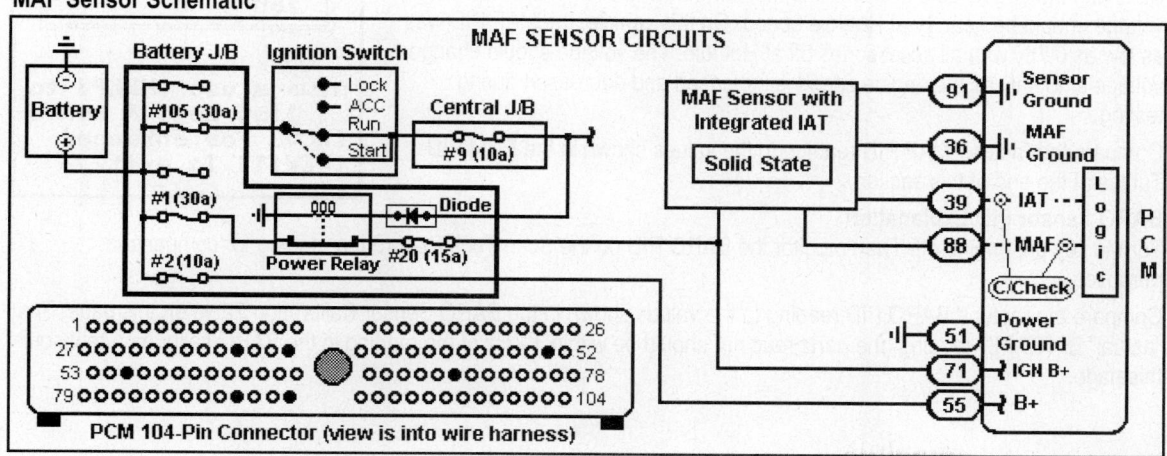

BARO SENSOR CALIBRATION

The barometric pressure (BARO) sensor signal, which is an inferred value by the PCM, can be used to diagnose a "suspect" MAF sensor. The PCM updates the BARO reading during conditions of a high throttle opening. Drive the vehicle using 3-4 sustained heavy throttle operations and then record the updated BARO PID. Compare the reading to the values in the table. If the BARO reading indicates a higher altitude than actual, the MAF sensor is contaminated. The updated reading should be ±4 Hz of the values below.

The MAF sensor signal should also be checked for loss of calibration. This step can be done using the MAF PID from the Scan Tool at cold or hot engine startup at various engine speeds. Refer to the MAF Sensor Calibration Table in Section 1 of the manual.

BARO Sensor Calibration Table

BARO Pressure (" Hg)	BARO Pressure (kPa)	BARO/MAP PID (Hz)	Altitude over Sea Level
30 In. Hg	101.3 kPa	158.9 Hz	0 feet (sea level)
29 In. Hg	97.9 kPa	155.8 Hz	1,000 feet.
28 In. Hg	94.6 kPa	152.8 Hz	2,000 feet
27 In. Hg	91.2 kPa	149.8 Hz	3,000 feet
26 In. Hg	87.8 kPa	146.8 Hz	4,000 feet
25 In. Hg	84.4 kPa	144.0 Hz	5,000 feet
24 In. Hg	81.1 kPa	141.1 Hz	6,000 feet
23 In. Hg	77.7 kPa	138.3 Hz	7,000 feet
22 In. Hg	74.3 kPa	135.4 Hz	8,000 feet

SCAN TOOL TEST (MAF SENSOR)

Connect the Scan Tool to the underdash connector. Navigate through the Scan Tool menus until you get to the OBD II Data List (a list of PID items). Select the MAF and BARO PID items from the menu so that you can view them on the Scan Tool display.

Scan Tool Examples

An example of the MAF (V) PID and BARO (F) PID for this vehicle at Hot Idle speed is shown in the Graphic.

MAP Sensor PID Explanation

Start with the engine cold or warm and monitor the MAF PID. The MAF PID voltage should be near 1v at Hot Idle speed. On this vehicle, the MAF PID was as low as 0.76v with all accessories off at Hot Idle. The voltage should change voltage smoothly as the engine speed is increased and decreased during testing.

Compare the "actual" MAF PID reading to the values shown in the PCM PID Tables at the end of this section.

BARO Sensor PID Explanation

Turn to key on, engine off. Then monitor the BARO PID on the Scan Tool. Record the reading for comparison purposes.

Compare the "actual" BARO PID reading to the values shown in the BARO Sensor Calibration Table on this page. The "actual" BARO PID reading (the hertz reading) should be within ±4 Hz of the reading in the BARO calibration table on this page.

```
Jan 15, 02      1:57:21 pm

MAF (V)
   1.05 VOLTS
BARO (F)
   160.0 HZ


Always use [HELP] to
   list keys & modes
[RCV] For enhanced
[*EXIT] To exit
```

LAB SCOPE TEST (MAF SENSOR)

The Lab Scope can be used to monitor the MAF sensor output signal under various engine load and speed conditions. Place the gearshift selector in Park and block drive wheels for safety.

Scope Connections

Connect the Channel 'A' positive probe to the MAF signal at the PCM Pin 88 (LB/RED wire) and connect the Channel 'A' negative probe to the battery negative ground post.

Scope Settings

To make the waveforms as clear as possible, set the scope settings to match the examples.

The MAF sensor waveforms can have slight differences from one scope to another.

Lab Scope Example

In the example, the trace shows the MAF sensor signal during a Snap Throttle event. The MAF signal peaked at about 3.6v, a change of 2.6v from the idle signal.

It is important to note that this newer generation MAF sensor, integrated into the air cleaner housing, responds much slower than the previous generation of Ford MAF sensors. In this trace, the time from snap throttle to peak MAF signal voltage

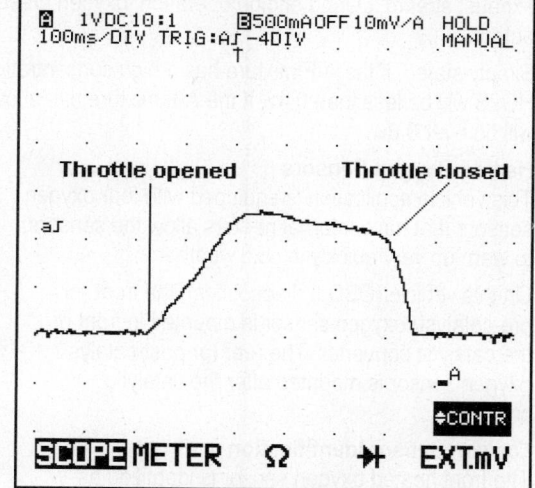

(200 ms) is almost 4 times longer than for the previous generation. The trace also shows much less signal detail due to the design of the sensor. However, this is a known good waveform for this type of sensor.

MAF Signal Explanation

The MAF sensor trace should show an even change in its analog signal as the engine speed is changed up and down. It if drops or rises suddenly without a corresponding change in engine speed, there may be a problem in the MAF sensor or its circuit.

Look for changes in MAF voltage under various driving conditions. On this application, the MAF sensor signal was 1.06v at idle, 1.92v at 25 MPH and 2.02v at 55 MPH.

MAF Sensor Schematic

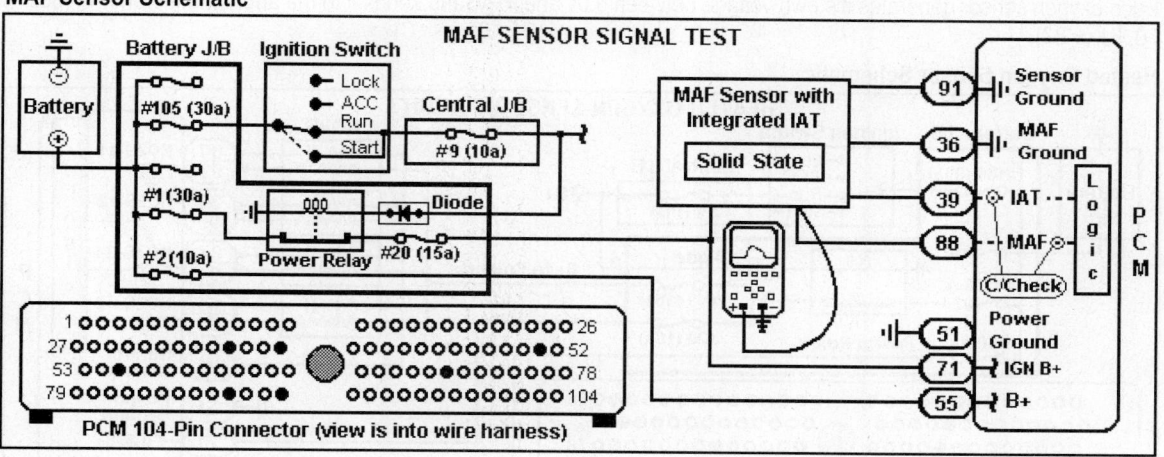

Oxygen Sensor

GENERAL DESCRIPTION

The heated oxygen sensor (HO2S) is mounted in the Exhaust system where it monitors the oxygen content of the exhaust stream. During engine operation, oxygen present in the exhaust reacts with the HO2S to produce a voltage output.

Simply stated, if the A/F mixture has a high concentration of oxygen in the exhaust, the signal to the PCM from the HO2S will be less than 0.4v. If the A/F mixture has a low concentration of oxygen in the exhaust, the signal to the PCM will be over 0.6v.

Heated Oxygen Sensors

This vehicle application is equipped with four oxygen sensors that have internal heaters allow the sensors to warmup very quickly in cold weather.

On this vehicle (OBD II diagnostics), the front (or pre-catalyst) oxygen sensor is mounted in front of the catalytic converter. The rear (or post-catalyst) oxygen sensor is mounted after the catalytic converter.

Oxygen Sensor Identification

The front heated oxygen sensor is identified as either a Bank 1 HO2S-11 or a Bank 2 HO2S-21. The heated oxygen sensor identified as Bank 1 is always located in the engine bank that contains cylinder No. 1.

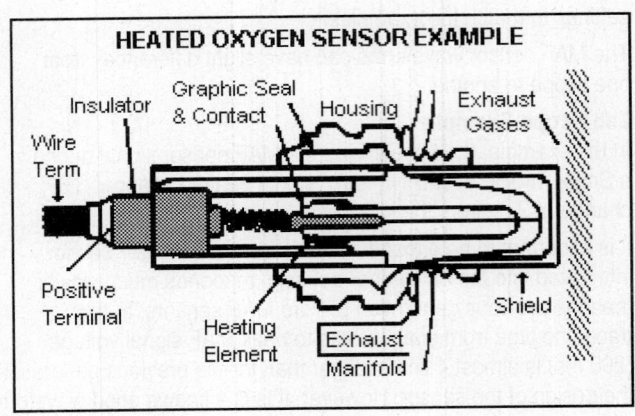

Circuit Description

The PCM power relay supplies battery voltage to the HO2S internal heater circuits.

The heaters are also connected to the Oxygen sensor "control" circuits in the PCM as shown in the Graphic below. In this case, the PCM can determine when to enable the HO2S heaters to protect the ceramic elements of the sensors and to be able to run the HO2S main monitor heater tests.

Each oxygen sensor generates it's own voltage between 0.0v and 1.0v, and sends it to the appropriate PCM Pin (35, 60, 61 or 87).

Heated Oxygen Sensor Schematic

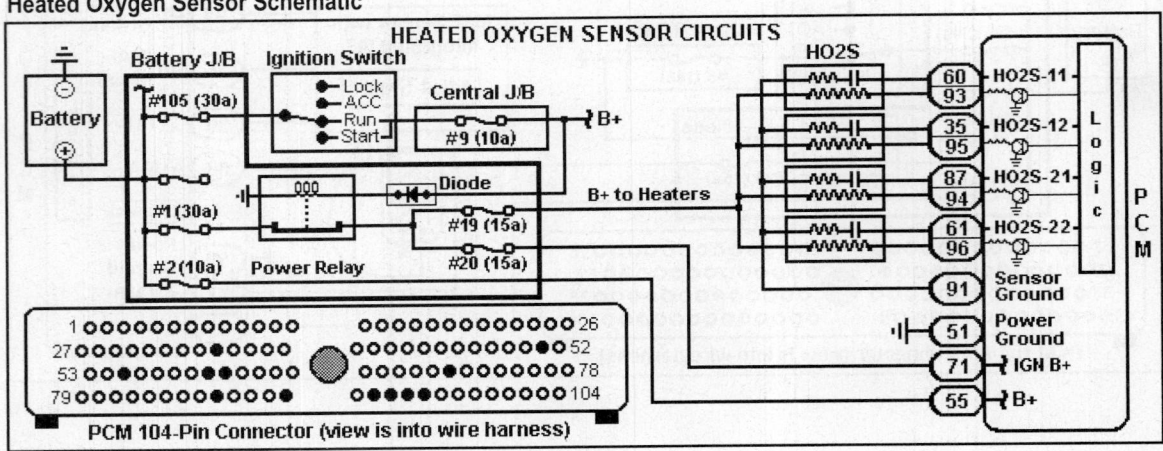

SCAN TOOL TEST (HO2S)

Connect the Scan Tool to the DLC underdash connector. Navigate through the Scan Tool menus until you get to the Ford (OEM) Data List (a list of PID items). Raise the engine speed to 2000 rpm for 3 minutes to set up the test. The examples shown here were captured at Hot Idle in 'P'.

Then select the type of oxygen sensor (O2S) information that you want to view from the various options in the menu.

Scan Tool Examples

These scan tool examples were used to show the various combinations of PIDs and what they can be used for. The scan tool menus allow you to select certain PIDs that are relevant to your diagnosis.

Scan Tool Example (1)

This example shows the upstream and downstream heated oxygen sensor signals, as well as the On/Off status for the heaters.

This test can be used to watch the PCM turn on the heaters shortly after a cold start, and observe the time it takes for each of the HO2S signals to become active.

Scan Tool Example (2)

This example shows a selection of PIDs that allows the user to compare each HO2S signal with the Fuel Trim value calculated by the PCM. These Fuel Trim values are short-term values associated with each HO2S.

This test can help to determine if the PCM was compensating for a problem, and can also be used to watch the PCM react to created conditions, (such as propane enrichment or vacuum leaks).

Note that on this vehicle, the downstream sensors (O2S B1 S2 and O2S B2 S2) are not used for Fuel Trim.

Scan Tool Example (3)

This example shows only the four (4) HO2S signals. Every PCM can process data faster than it can transmit that data. By trimming down the list of Generic or OEM PIDs, the scan tool refreshes the data much faster. This is important for HO2S diagnosis, since the circuit activity is much faster than the PCM talks to the scan tool.

Summary

There are numerous screens available on the Scan Tool that can help you diagnose the condition of the heated oxygen sensor (HO2S) and its heater.

You can select PIDs that will allow you to see only what is necessary, or compare a particular sensor to other PIDs to isolate a problem. For example, an unmetered air leak may cause HO2S voltage to drop, and could be diagnosed by viewing MAF, HO2S and Fuel Trim PIDs.

```
UP O2S BANK 1····0.10V
DN O2S BANK 1····0.74V
UP O2S BANK 2····0.27V
DN O2S BANK 2····0.73V
UP O2S HEATER B1···ON
DN O2S HEATER B1···ON
UP O2S HEATER B2···ON
DN O2S HEATER B2···ON
DPF EGR (V)·······1.06V
EGR VAC REG········0%
```
(1)

```
LT FT 2···········3.1%
O2S B1 S1·······0.535V
FT O2S B1 S1·····-1.5%
O2S B1 S2·······0.173V
FT O2S B1 S2···UNUSED
O2S B2 S1·······0.735V
FT O2S B2 S1······0.0%
O2S B2 S2·······0.310V
FT O2S B2 S2···UNUSED
```
(2)

```
UP O2S BANK 1····0.17V
DN O2S BANK 1····0.76V
UP O2S BANK 2····0.11V
DN O2S BANK 2····0.79V
```
(3)

LAB SCOPE TEST (HO2S)

The Lab Scope is the "tool of choice" to test the oxygen sensor as it provides an accurate view of the sensor response **Lab Scope Test (HO2S)** and switch rate. Place the gearshift selector in Park and block the drive wheels for safety.

Lab Scope Connections

Connect the Channel 'A' positive probe to the sensor signal wire at the BOB, PCM or sensor connector. Connect the negative probe to the battery negative post.

Scope Settings

To make the waveforms as clear as possible, set the scope settings to match the examples.

Depending upon the scope capabilities, the sensor waveforms will have slight differences.

Lab Scope Example (1)

In example (1), the trace shows HO2S 11, and was captured at cold idle soon after startup. Because of the sensor design, and efficient heaters, these oxygen sensors are active within a few seconds of startup.

Note the sensor range is from about 100 mv to about 800 mv.

Lab Scope Example (2)

In example (2), the trace shows the same sensor, but cursors and trigger settings were used to evaluate the performance of the sensor.

The vehicle was driven rich at cruise, then lean on decel, followed by a snap throttle. This method causes the HO2S to switch its fastest, and provides a clear view of switch rates on screen.

Note that the lean-to-rich switch rate was 40.0 ms, whereas a switch rate of 100 ms or more will cause DTCs and driveability concerns.

Lab Scope Example (3)

In example (3), the trace shows HO2S 12 during a snap throttle. Note that the time/div. has been changed to 2 seconds to show sensor activity over a longer period of time.

With a catalytic converter in perfect condition, the downstream sensor trace should remain fairly steady at steady cruise. However, this test shows that the downstream sensor reacts very quickly to sudden throttle movements.

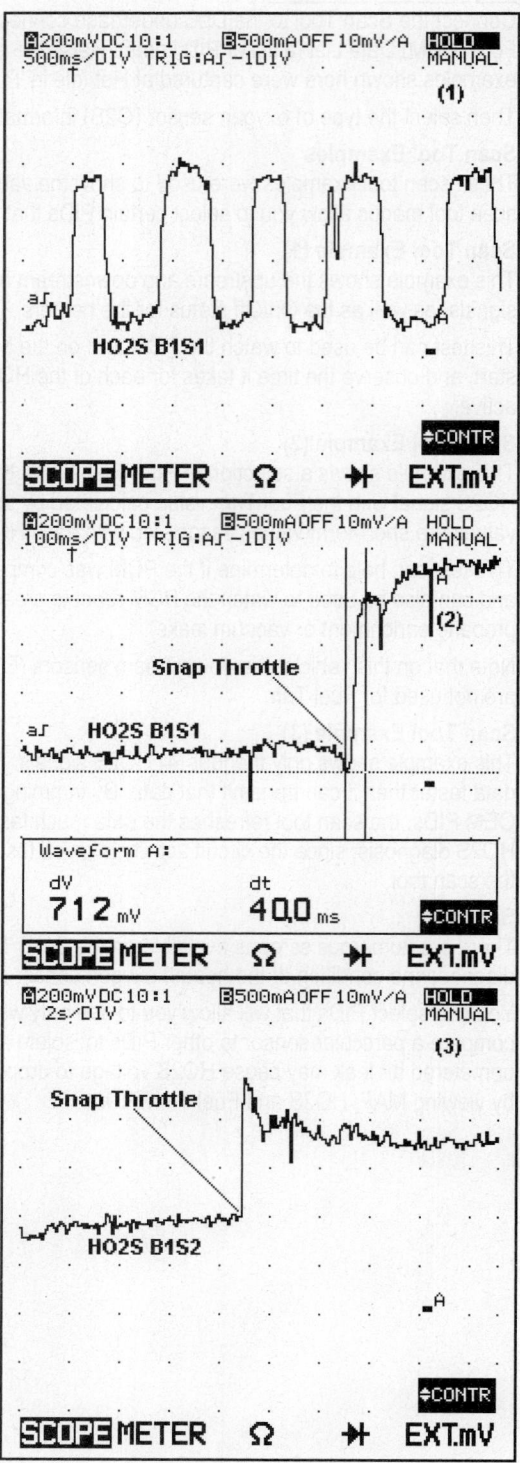

LAB SCOPE TEST (HO2S)

The PCM must detect circuit activity outside of test parameters over long periods of time to set some Oxygen sensor trouble codes. In order to diagnose the cause of an Oxygen sensor or Catalyst related trouble code, you need to be able to monitor the same inputs (HO2S activity and switch rates) the PCM uses to determine if a problem is present.

Scope Settings

To make the traces as clear as possible, set the scope settings to match the examples.

Lab Scope Test

This example shows the signals from all four (4) heated oxygen sensors on one screen. This allows comparison of switch rates between banks, and between upstream and downstream sensor to determine catalytic converter efficiency and sensor condition.

The PCM uses the Catalyst Monitor to test for converter efficiency. During the test, the Monitor compares the Switch rates for upstream and downstream sensors. The Oxygen Sensor Monitor tests for switch rates, time-to-activity, voltage thresholds and frequency. All of these parameters can be clearly seen in the example.

This capture was taken with the vehicle cold, shortly after startup. Note that switch rate for HO2S - 12 is diminishing as the Bank 1 catalytic converter warms up. The Bank 2 catalytic converter has already warmed up, as seen in the HO2S – 22 trace.

Four (4) Channel HO2S Test

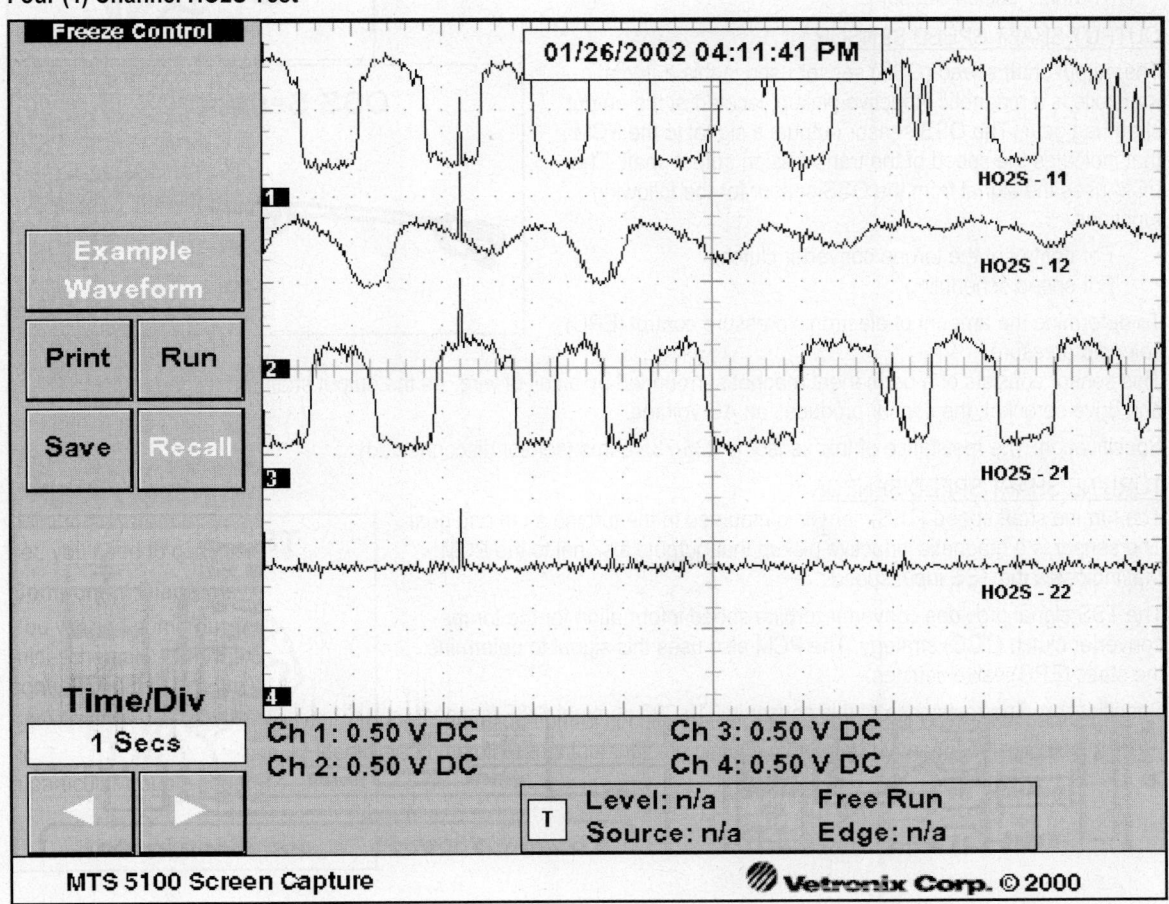

Transmission Controls

ELECTRONIC PRESSURE CONTROL

This vehicle is equipped with a 4-speed AX4N automatic transaxle with automatic shift control. The PCM controls all upshift, downshift, line pressure and TCC operations. The captures in the examples on this page were done with the vehicle in Park at Hot Idle.

The electronic pressure control (EPC) solenoid used on this automatic transaxle is a variable force style (VFS) solenoid. The VFS-type solenoid is an electro-hydraulic actuator that combines a solenoid and a regulator valve.

The PCM uses the EPC solenoid to regulate the transmission line pressure and line modulator pressure. This task is accomplished by producing resisting forces to the main regulator and line modulator circuits. These two pressures are used to control clutch application pressures.

EPC Solenoid

The PCM controls the EPC and thereby the transmission line pressure and modulator based on these input signals:

*   Engine Coolant Temp. Sensor
*   Intake Air Temperature Sensor
*   Transmission Fluid Temp. Sensor
*   Throttle Position Sensor

OUTPUT SHAFT SPEED SENSOR

The output shaft speed (OSS) sensor used in this automatic transaxle is a magnetic inductive pickup, located at the output shaft ring gear. The OSS sensor outputs a signal to the PCM that indicates the speed of the transmission output shaft. The PCM uses the signal from the OSS sensor for the following functions:

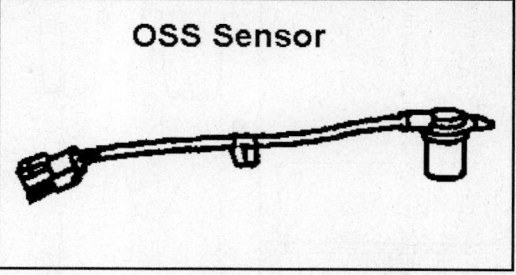
OSS Sensor

*   For control of the torque converter clutch
*   For speed scheduling

To determine the amount of electronic pressure control (EPC) solenoid pressure

This sensor consists of a permanent magnet surrounded by a coil of wire. As the output shaft rotates the reluctor wheel and drive sprocket, the sensor produces an AC voltage.

Specification: The resistance of this sensor is 235-735 ohms (sensor disconnected).

TURBINE SHAFT SPEED SENSOR

The turbine shaft speed (TSS) sensor is mounted in the turbine shaft ring gear. The sensor is a magnetic inductive pickup that outputs a signal to the PCM that indicates the TSS input speed.

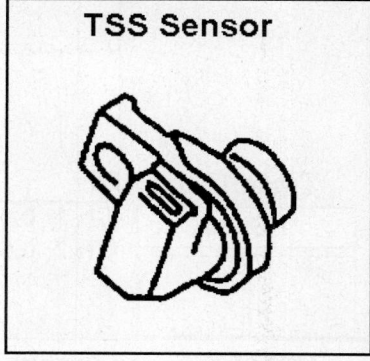
TSS Sensor

The TSS signal provides converter turbine speed information for the torque converter clutch (TCC) strategy. The PCM also uses this signal to determine the static (EPC) valve settings.

Specification: The resistance of this sensor is 310-390 ohms at 68°F (sensor disconnected).

DVOM TEST (OSS SENSOR)

The DVOM can be used to test the OSS sensor as it provides an alternate method to use to test the sensor output. Prior to starting the test, place the gearshift selector in Park (A/T) and block the drive wheels for safety. Install the breakout box (BOB) as required.

DVOM Settings

To make the readings as clear as possible, set the DVOM settings to match the examples.

DVOM Tests

The OSS sensor signals can be checked with the vehicle moving at low and high speeds with the engine temperature cold or warm.

DVOM Example

In this example, the DVOM shows the OSS sensor output voltage is 0.673 volts AC at a rotation speed of 60.29 hertz. In this case, the Lab Scope feature of the DVOM was also used to capture the OSS signals. The vehicle speed was 10 mph during the capture. Note the voltage range was +3v to -3v.

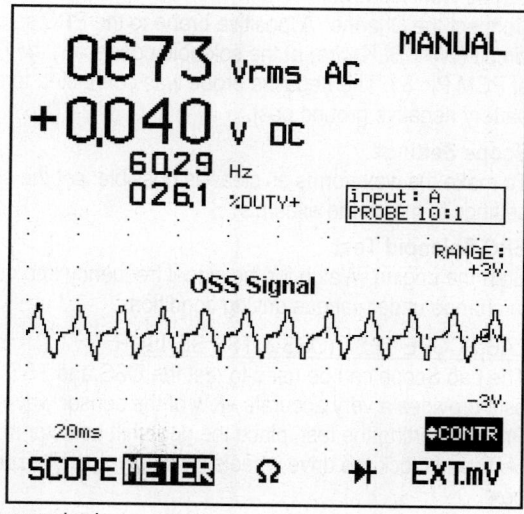

DVOM TEST (TSS SENSOR)

The DVOM can be used to test the TSS sensor as it provides an alternate method to use to test the sensor output. Prior to starting the test, place the gearshift selector in Park (A/T) and block the drive wheels for safety. Install the breakout box (BOB) as required.

DVOM Settings

To make the readings as clear as possible, set the DVOM settings to match the examples.

DVOM Tests

The TSS sensor signals can be checked with the vehicle moving at low and high speeds with the engine temperature cold or warm.

DVOM Example

In this example, the DVOM shows the TSS sensor output voltage was 0.386 volts AC at a rotation speed of 51.28 hertz.

In this case, the Lab Scope feature of the DVOM was also used to capture the TSS waveform. The vehicle speed was 10 mph during the capture.

Specification: The resistance of this sensor is 310-390 ohms at 68°F (sensor disconnected).

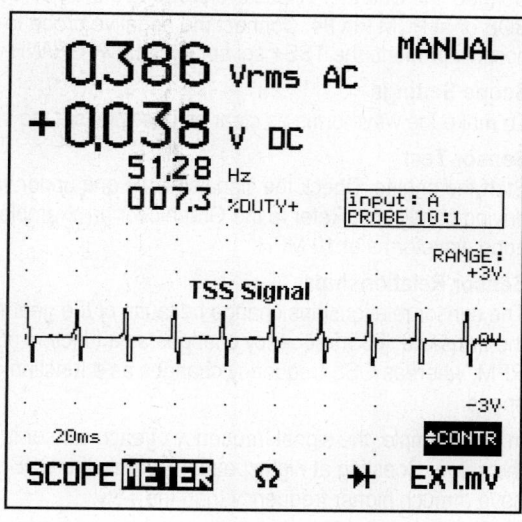

LAB SCOPE TEST (EPC SOLENOID)

The Lab Scope can be used to test the EPC solenoid as it provides an accurate view of the solenoid waveform. Install a breakout box if available. Prior to starting the test, place the gearshift selector in Park (A/T) and block the drive wheels for safety.

Scope Connections

Connect the Channel 'A' positive probe to the EPC solenoid circuit (WHT/BLK wire) at the solenoid connector, the BOB or at PCM Pin 81. The negative probe was connected to the battery negative ground post.

Scope Settings

To make the waveforms as clear as possible, set the scope settings to match the example.

EPC Solenoid Test

Start the engine. Watch for the signal frequency and duty cycle to change under various driving conditions.

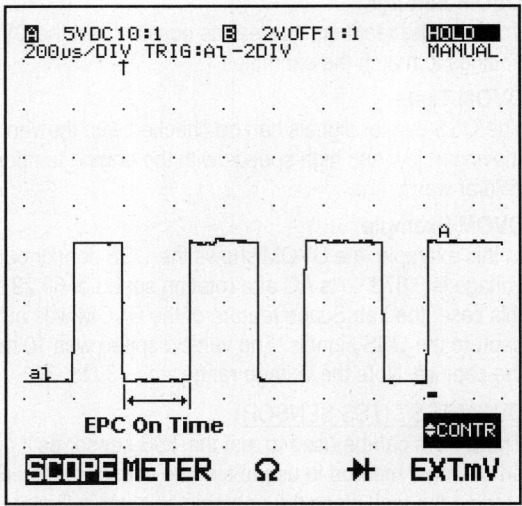

LAB SCOPE TEST (OSS & TSS SENSOR

The Lab Scope can be used to test the OSS and TSS sensors as it provides a very accurate view of the sensor waveforms. Prior to starting the test, place the gearshift selector in Park (A/T) and block the drive wheels for safety. Install a breakout box.

Scope Connections

Connect the Channel 'A' positive probe to the OSS+ sensor circuit (DK GRY/BLK wire) at the solenoid connector, the BOB or at PCM Pin 84. Connect the negative probe to the battery negative ground post. Connect the Channel 'B' positive probe to the TSS+ sensor circuit (DK GR/WH wire) at the solenoid connector, the BOB or at PCM Pin 6.

Scope Settings

To make the waveforms as clear as possible, set the scope settings to match the example.

Sensor Test

Start the engine. Check the signals at idle and under various driving conditions. Refer to the Graphic for an example known good waveforms at 10 MPH.

Sensor Relationships

The sensor relationships change because of the gear ratios in the transaxle. TSS frequency changes as a function of engine RPM, whereas OSS frequency changes as a function of vehicle speed.

In the example, the signal frequency of each sensor is almost the same. Coasting at high speed, however, the OSS signal will have a much higher frequency than the TSS.

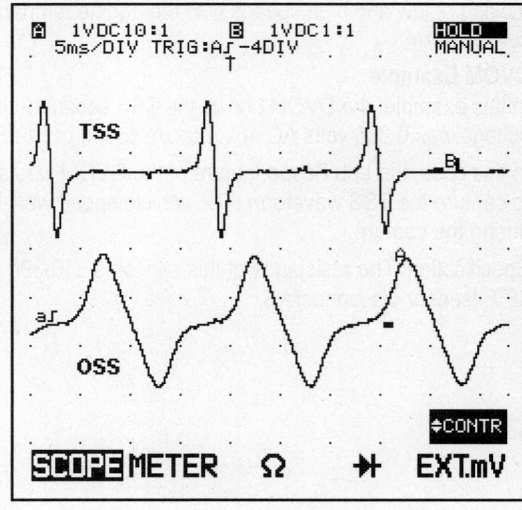

Reference Information

PCM PID TABLES - INPUTS

■ **NOTE:** *The following readings were obtained with the engine at idle speed, radiator hose hot, throttle closed, gear selector in Park and all accessories turned off.*

Acronym	Scan Tool Parameter	Range	Typical Value
ACCS	A/C Cycling Clutch Switch Signal	ON / OFF	OFF (ON with A/C on)
ACP	A/C Pressure Sensor Signal	0-5.1v	1v (54 psi)
BPP	Brake Pedal Position Switch	ON / OFF	OFF (ON with brake on)
CMP (CID)	CMP Sensor Signal	0-1000 Hz	5-7
CKP	CKP Sensor Signal	0-10,000 Hz	390-450
DPFE	DPF EGR Sensor Signal	0-5.1v	0.95-1.05
ECT °F	Engine Coolant Temp. Sensor	-40°F to 304°F	160-200
FEPS	Flash EEPROM	0-5.1v	0.1
FLI	Fuel Level Input Signal	0-100%	50
FPM	Fuel Pump Monitor	ON / OFF	ON
FTP	Fuel Tank Pressure Sensor	0.5.1v (0-10" H$_2$O)	2.6v (0" H20 - cap off)
Gear	Transmission Gear Position	1-2-3-4	1
GFS	Generator Field Signal	0-1,000 Hz	130 (37% duty cycle)
HO2S-11	HO2S-11 (Bank 1 Sensor 1)	0-1100 mv	100-900
HO2S-12	HO2S-12 (Bank 1 Sensor 2)	0-1100 mv	100-900
HO2S-21	HO2S-21 (Bank 2 Sensor 1)	0-1100 mv	100-900
HO2S-22	HO2S-22 (Bank 2 Sensor 2)	0-1100 mv	100-900
IAT °F	Intake Air Temperature Sensor	-40°F to 304°F	50-120
IMRC-M	Intake Runner Manifold Control	0-5.1v	5v (On: 2.5v)
LOAD	Calculated Engine Load	0-100%	15-20
MAF V	Mass Airflow Sensor Signal	0-5.1v	0.6-0.8
MISF	Misfire Detection	ON / OFF	OFF
OCT ADJ	Octane Adjust Signal	Closed / Open	Closed
OSS	Output Shaft Speed Sensor	0-10,000 Hz	0, at 55 mph: 475
PATSIN	PAT Signal Input	0-25.5v	12-14v
PNP	Park Neutral Position Switch	ON or OFF	ON
PSP	Power Steering Pressure Switch	LOW or HIGH	LOW
RPM	Engine Speed	0-10K RPM	700-730
TFT °F	Transaxle Fluid Temp. Sensor	-40°F to 304°F	210-240
TP V	Throttle Position	0-5.1v	0.8-1.1
TR1	Gear Selector TR1 Signal	0v or 11.5v	0v
TR2	Gear Selector TR2 Signal	0v or 11.5v	0v
TR4	Gear Position TR4 Signal	0v or 11.5v	0v
TRV/TR	Transmission TR Sensor Signal	0-5.1v	In 'P': 0, in 'D': 1.7
TSS	TSS Sensor Signal	0-10,000 Hz	43, at 55 mph: 101
VPWR	Vehicle Power	0-25.5v	14.1v
VSS	Vehicle Speed	0-159 mph / 0-1000 Hz	0 (55 mph = 125 Hz)

PCM PID TABLES - OUTPUTS

■ NOTE: *The following readings were obtained with the engine at idle speed, radiator hose hot, throttle closed, gear selector in Park and all accessories turned off.*

Acronym	Scan Tool Parameter	Range	Typical Value
CDA	Coil 'A' Driver (Dwell)	0-60° dwell	7°
CDB	Coil 'B' Driver (Dwell)	0-60° dwell	7°
CDC	Coil 'C' Driver (Dwell)	0-60° dwell	7°
EGR VR	EGR Vacuum Regulator Valve	0-100%	0-40
EVAP CP	EVAP Canister Purge Valve	0-1000 Hz (0-100%)	0-10 (0-100)
EVAP CV	EVAP Canister Vent Valve	0-1000 Hz (0-100%)	0 (0)
EPC	EPC Solenoid Control	0-100 psi	9.2 (15 psi)
FP	Fuel Pump Control	0-100%	100
Fuel B1	Fuel Injector Pulsewidth Bank 1	0-99.9 ms	3.0-4.0
Fuel B2	Fuel Injector Pulsewidth Bank 2	0-99.9 ms	3.0-4.0
GENFDC	Generator Field Control	0-1,000 Hz	130
HFC	High Speed Fan Control	ON / OFF	OFF (ON with fan on)
HTR-11	HO2S-11 Heater Control	ON / OFF	ON
HTR-12	HO2S-12 Heater Control	ON / OFF	ON
HTR-21	HO2S-21 Heater Control	ON / OFF	ON
HTR-22	HO2S-22 Heater Control	ON / OFF	ON
IAC	Idle Air Control Motor	0-100%	25-35
IMRC	Intake Manifold Runner Control	ON / OFF	OFF
INJ1	Fuel Injector 1 Control	0-99.9 ms	3.0-4.0
INJ2	Fuel Injector 2 Control	0-99.9 ms	3.0-4.0
INJ3	Fuel Injector 3 Control	0-99.9 ms	3.0-4.0
INJ4	Fuel Injector 4 Control	0-99.9 ms	3.0-4.0
INJ5	Fuel Injector 5 Control	0-99.9 ms	3.0-4.0
INJ6	Fuel Injector 6 Control	0-99.9 ms	3.0-4.0
LFC	Low Speed Fan Control	ON / OFF	OFF (On with fan on)
LONGFT1	Long Term Fuel Trim Bank 1	-20% to 20%	0-1
LONGFT2	Long Term Fuel Trim Bank 2	-20% to 20%	0-1
MIL	Malfunction Indicator Lamp Control	ON / OFF	OFF (ON with DTC)
PATSOUT	PAT Output Signal	0-25.5v	12-14v
SHRTFT1	Short Term Fuel Trim Bank 1	-10% to 10%	-1 to 1
SHRTFT2	Short Term Fuel Trim Bank 2	-10% to 10%	-1 to 1
SPARK	Spark Advance	-90° to 90°	15-20
SS1	Transmission Shift Solenoid 1	ON / OFF	ON
SS2	Transmission Shift Solenoid 2	ON / OFF	OFF
SS3	Transmission Shift Solenoid 3	ON / OFF	OFF
TCC	Torque Converter Clutch Solenoid	0-100%	0, at 55 mph: 95
VREF	Voltage Reference	0-5.1v	4.9-5.1v
WAC	Wide Open Throttle A/C Relay	ON / OFF	OFF (ON at WOT)

PCM PIN VOLTAGE TABLES

PCM Pin Voltage Table: 104-Pin Connector

Pin Number	Wire Color	Application	Value at Hot Idle
1	VIO/O	Transaxle Shift Solenoid ' B' Control	1v
2	---	Not Used	---
3	O/BK	Transaxle Digital TR1 Sensor	In 'P', 0v, In Drive: 11.5v
4-5, 7, 9-12	---	Not Used	---
6	DG/WH	Turbine Shaft Speed Sensor	43 Hz
8	O/W	Intake Manifold Runner Control Monitor	5v
13	VIO	EEPROM Reflashing Power	0.1v
14, 19	---	Not Used	---
15	PK/LB	SCP Data Bus (-)	Digital Signals
16	T/O	SCP Data Bus (+)	Digital Signals
17	GY/O	RX Signal (Passive Anti-Theft System)	Battery Voltage
18	WH/LG	TX Signal (Passive Anti-Theft System)	Battery Voltage
20	VIO	Generator Monitor Input	130 Hz (37% duty cycle)
21	BK/PK	CKP+ Sensor Signal	390-450 Hz
22	GY/Y	CKP- Sensor Signal	390-450 Hz
23, 25	---	Not Used	---
24	BK/W	Power Ground	0.050v
26	DB/LG	Coil 1 Driver Control	5° dwell
27	O/Y	Transaxle Shift Solenoid 'A' Control	12v
28	DB	Low Speed Fan Control	Fan Off: 12v, On: 1v
29	---	Not Used	---
30	DB/LG	Anti-Theft Indicator Input	Battery Voltage
31	Y/LG	PSP Switch Signal	Turning: 12v
32-34	---	Not Used	---
35	R/LG	HO2S-12 (Bank 1 Sensor 2) Signal	100-900 mv
36	T/LB	MAF Sensor Return	0.050v
37	O/BK	Transaxle Fluid Temperature Sensor	2.10v
38	LG/R	Engine Coolant Temperature Sensor	0.6v
39	GY	Intake Air Temperature Sensor	1.6v
40	DG/Y	Fuel Pump Monitor	Pump On: 12v, Off: 0v
41	BK/Y	A/C Clutch Cycling Switch Signal	A/C Switch On: 12v
42	BR	Intake Manifold Runner Control Solenoid	12v
43, 48	---	Not Used	---
44	LG/Y	Starter Relay	0v
45	R/PK	Generator Communication Input	Digital Signals
46	LG/VIO	High Speed Fan Control	Fan On: 1v, Off: 12v
47	BR/PK	Vacuum Regulator Solenoid	0% (55 mph: 0-45%)
49	LG/BK	Transaxle Digital TR2 Sensor	In 'P': 0v, 55 mph: 11.5v
50	WH/BK	Transaxle Digital TR4 Sensor	In 'P': 0v, 55 mph: 11.5v
51	BK/W	Power Ground	0.050v
52	R/LB	Coil 2 Driver Control	5° dwell
53	PK/BK	Transaxle Shift Solenoid 'C' Control	12v
54	R/LB	Torque Converter Clutch Solenoid	0%, at 55 mph: 95%
55	R	Keep Alive Power	12-14v
56	LG/BK	EVAP Canister Purge Solenoid	0-10 Hz (0-100% DC)

PCM Pin Voltage Table: 104-Pin Connector

Pin Number	Wire Color	Application	Value at Hot Idle
57-59, 63	---	Not Used	---
60	GY/LB	HO2S-11 (Bank 1 Sensor 1) Signal	100-900 mv
61	VIO/LG	HO2S-22 (Bank 2 Sensor 2) Signal	100-900 mv
62	R/PK	Fuel Tank Pressure Sensor Signal	2.6v (0 in. HG - cap off)
66, 68, 70	---	Not Used	---
64	LB/Y	Transaxle Range Sensor TR3	In 'P': 0v, 55 mph: 1.7v
65	BR/LG	DPFE Sensor Signal	0.95-1.05v
67	VIO/W	EVAP Vent Valve Control	0-10 Hz (0-100%)
69	PK/Y	A/C Clutch Relay Control	12v
71	R	Vehicle Power	12-14v
72, 79, 82	---	Not Used	---
73	T/BK	Fuel Injector 5 Control	3.5-3.8 ms
74	BR/Y	Fuel Injector 3 Control	3.5-3.8 ms
75	T	Fuel Injector 1 Control	3.5-3.8 ms
76-77, 103	BK/W	Power Ground	<0.050 mv
78	P/W	Coil 3 Driver Control	5° dwell
80	LB/O	Fuel Pump Relay Control	Relay On: 1v, Off: 12v
81	W/Y	Transaxle EPC Solenoid Control	9.9v (15 psi)
83	W/LB	Idle Air Control Motor	10.7v (33% duty cycle)
84	GY/BK	Output Shaft Speed (OSS) Sensor	0 Hz, at 55 mph: 350 Hz
85	DB/O	Camshaft Position Sensor Signal	6 Hz
86	T/LG	A/C Pressure Switch Signal	12v (open - A/C off)
87	R/BK	HO2S-21 (Bank 2 Sensor 1) Signal	100-900 mv
88	LB/R	Mass Airflow Sensor Signal	0.8v, at 55 mph: 1.8v
89	GY/W	Throttle Position Sensor Signal	0.9v, at 55 mph: 1.2v
90	BR/W	Voltage Reference	4.9-5.1v
91	GY/R	Sensor Return Signal	<0.050v
92, 97-98	---	Not Used	---
93	R/W	HO2S-11 (Bank 1 Sensor 1) Heater	1v
94	Y/LB	HO2S-21 (Bank 2 Sensor 1) Heater	1v
95	W/BK	HO2S-12 (Bank 1 Sensor 2) Heater	1v
96	T/Y	HO2S-22 (Bank 2 Sensor 2) Heater	1v
99	LG/O	Fuel Injector 6 Control	3.5-3.8 ms
100	BR/LB	Fuel Injector 4 Control	3.5-3.8 ms
101	W	Fuel Injector 2 Control	3.5-3.8 ms
102, 104	---	Not Used	---

PCM 104-Pin Connector

TERMINAL VIEW OF WIRING HARNESS CONNECTOR FOR PCM (104 PIN)

Note: back-probing the 104-pin connector with a DVOM probe can damage the PCM - use a breakout box!

PCM Wiring Diagrams

FUEL & IGNITION SYSTEM DIAGRAMS (1 OF 10)

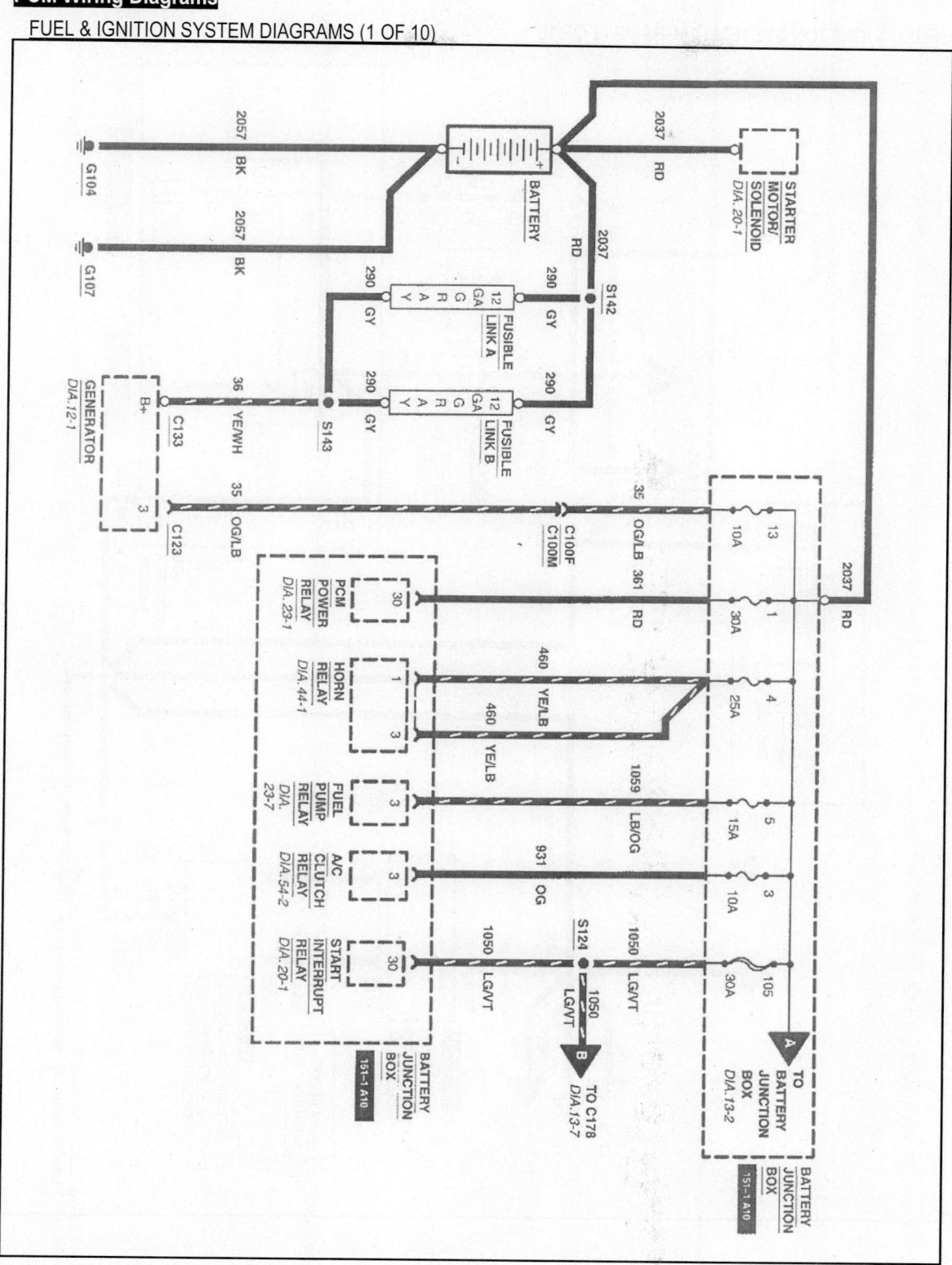

FUEL & IGNITION SYSTEM DIAGRAMS (2 OF 10)

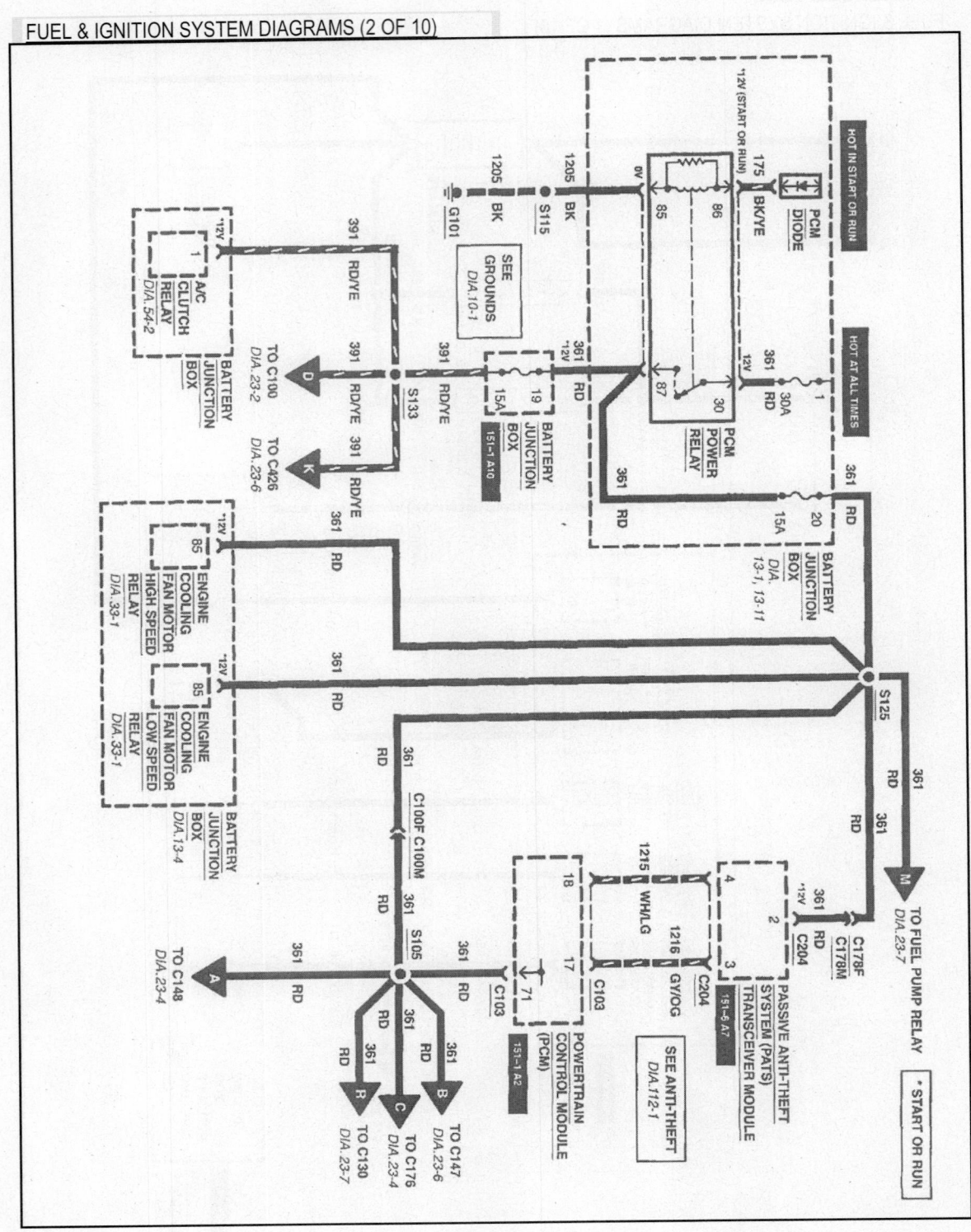

FUEL & IGNITION SYSTEM DIAGRAMS (3 OF 10)

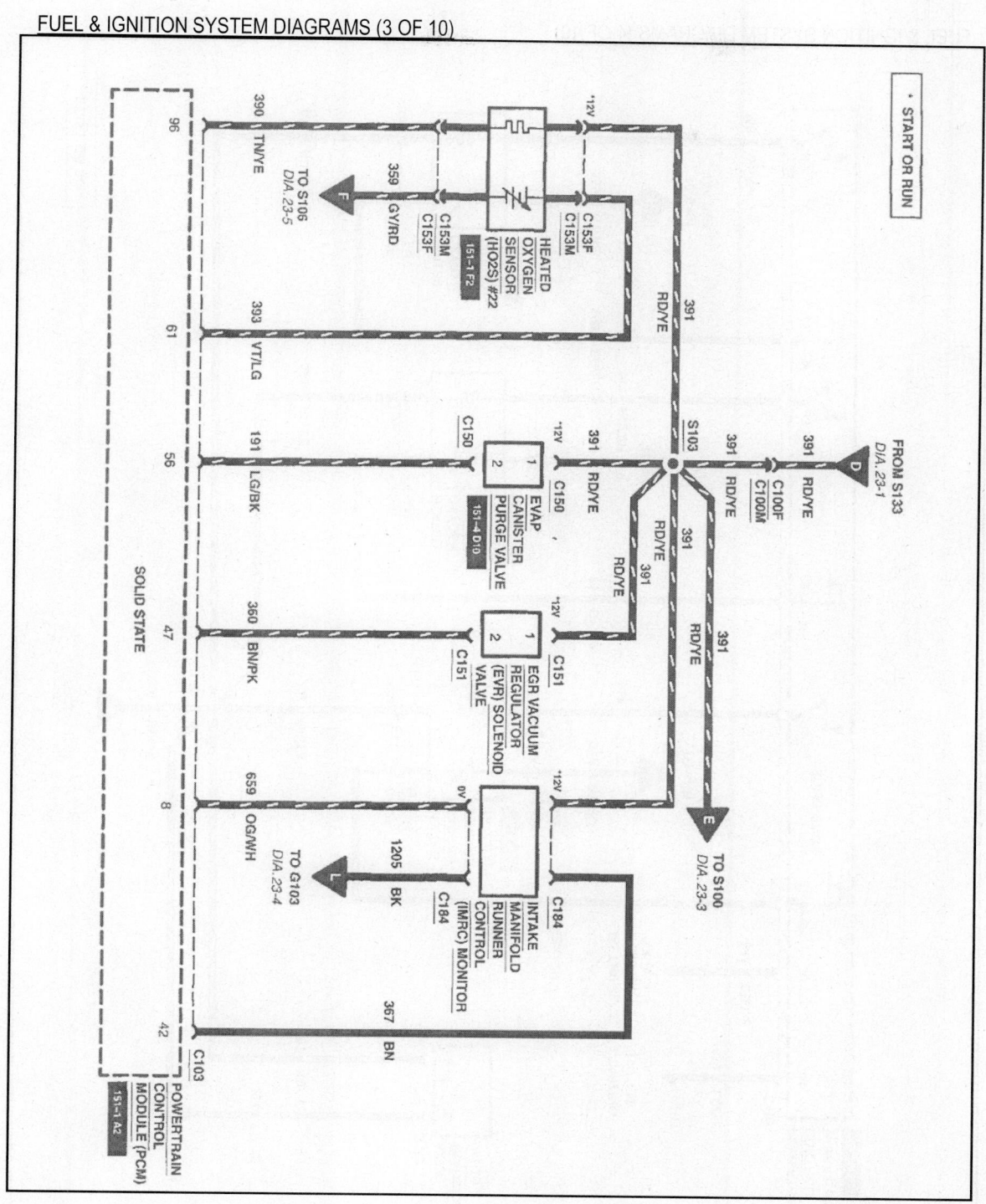

FUEL & IGNITION SYSTEM DIAGRAMS (4 OF 10)

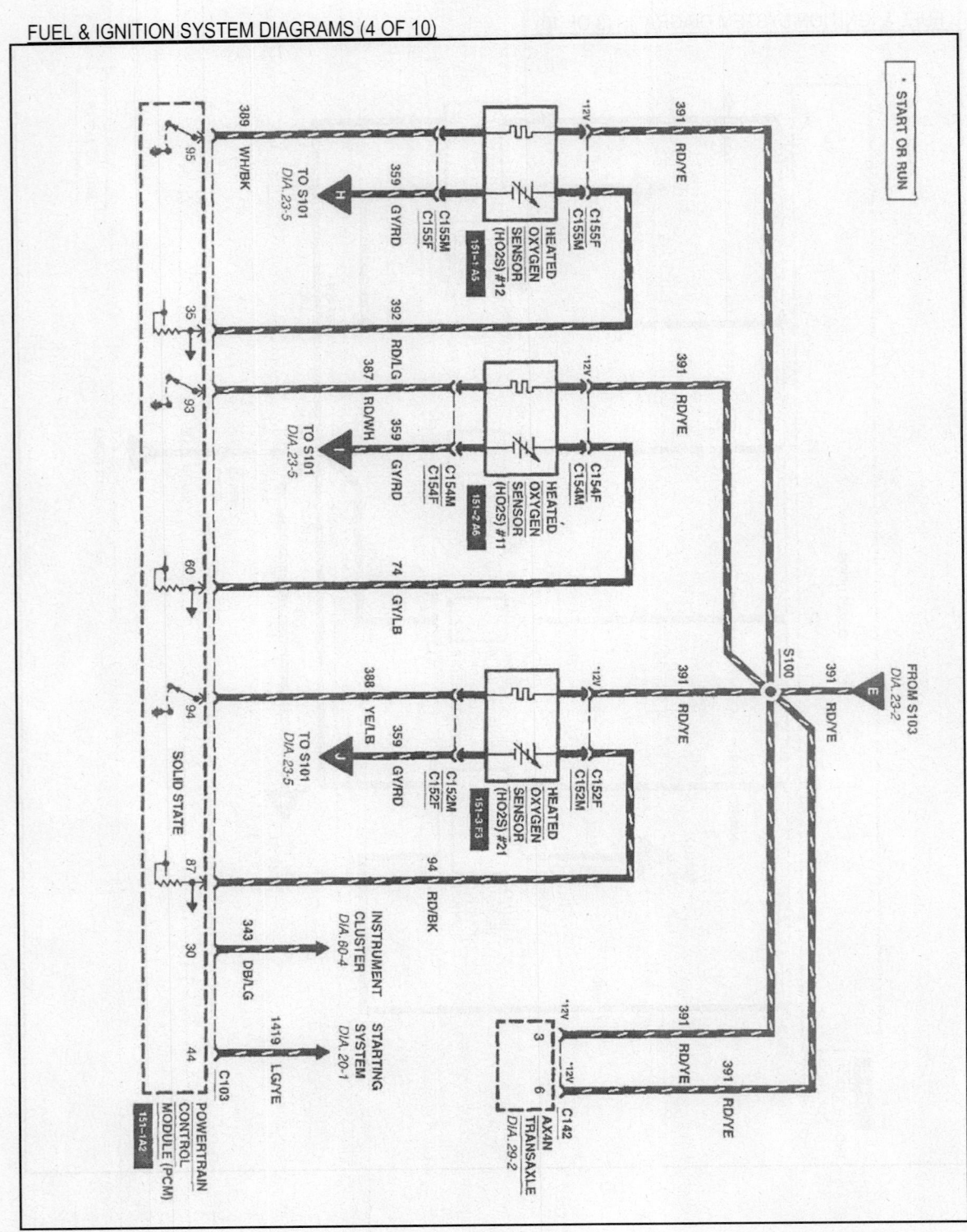

FUEL & IGNITION SYSTEM DIAGRAMS (5 OF 10)

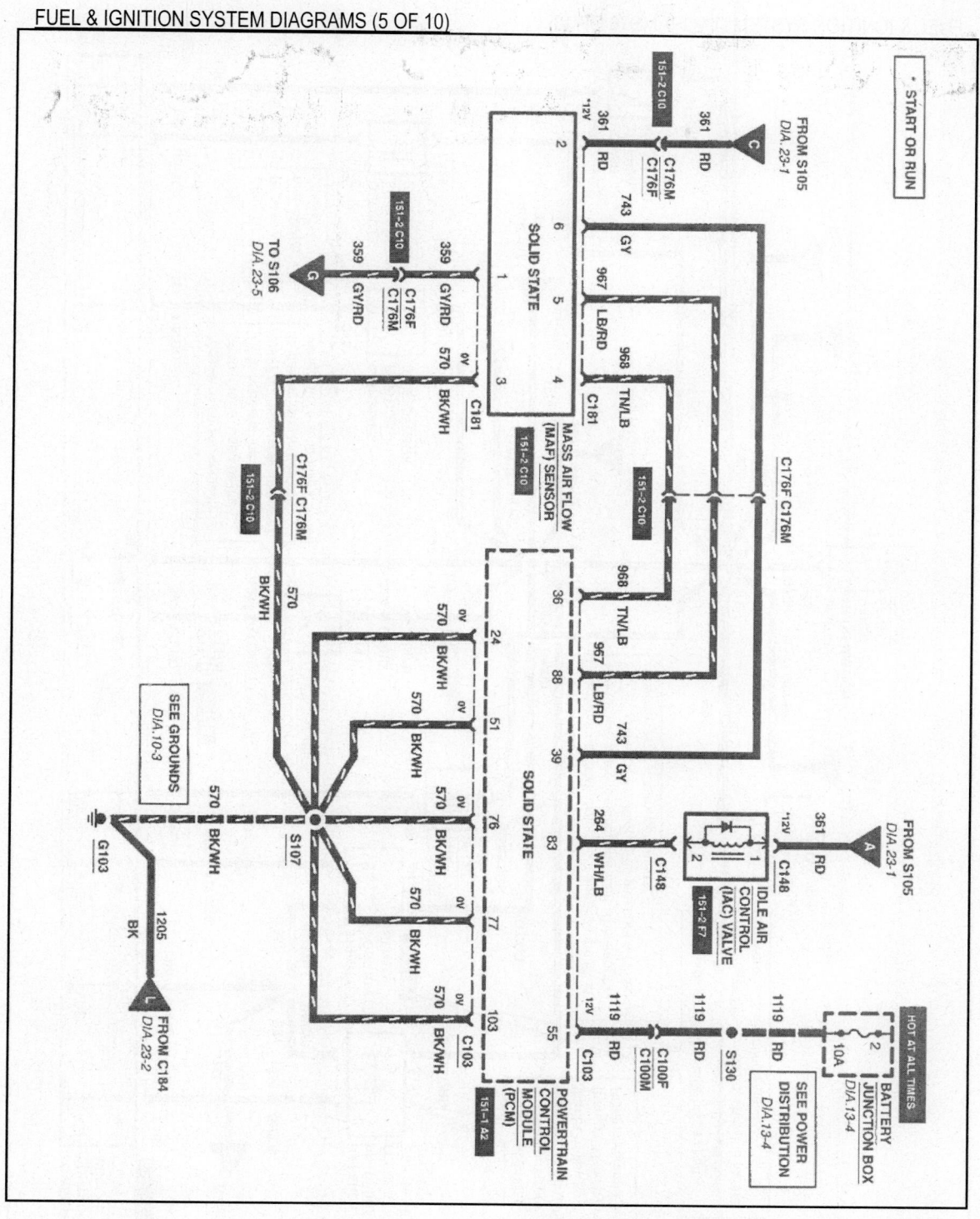

FUEL & IGNITION SYSTEM DIAGRAMS (6 OF 10)

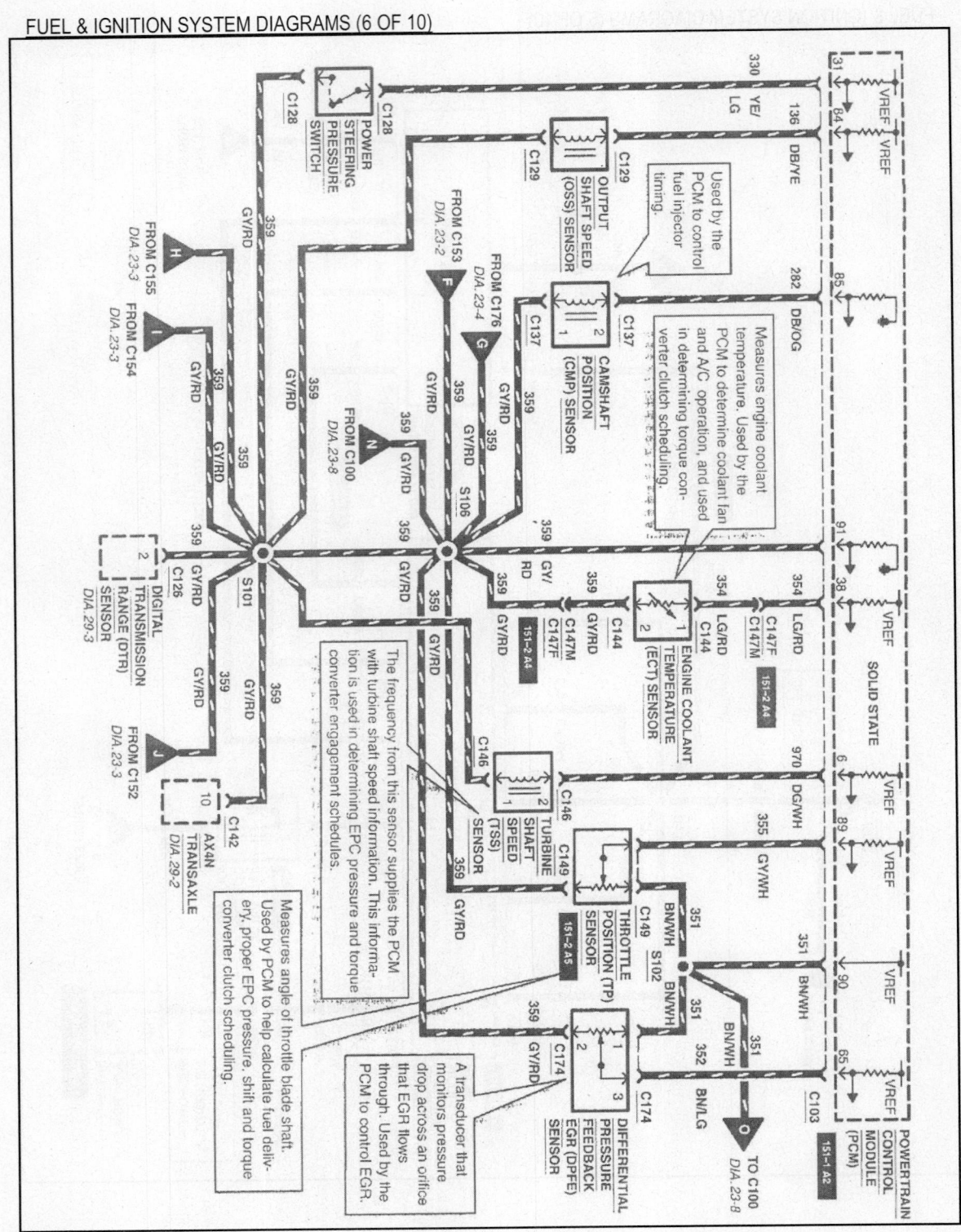

FUEL & IGNITION SYSTEM DIAGRAMS (7 OF 10)

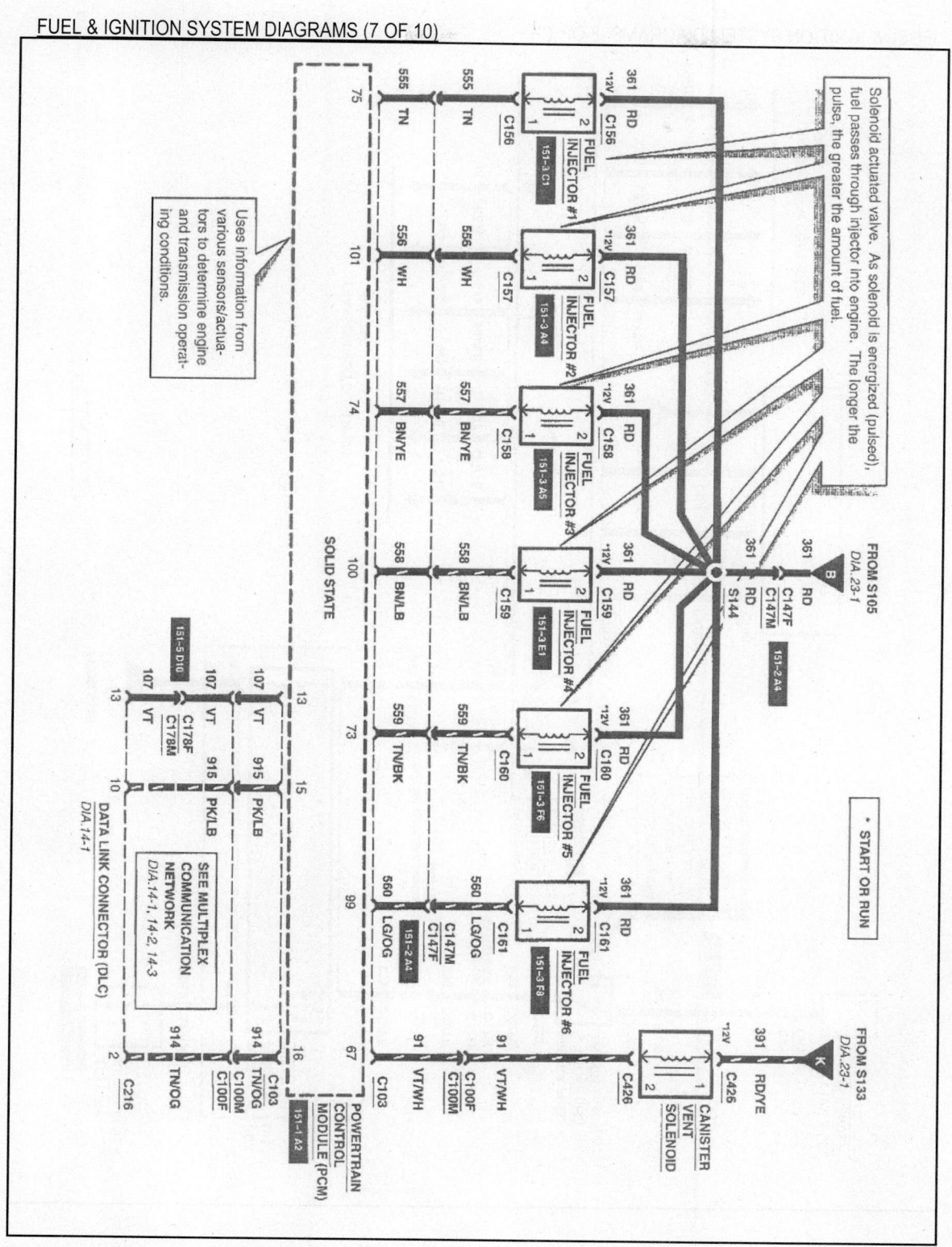

FUEL & IGNITION SYSTEM DIAGRAMS (8 OF 10)

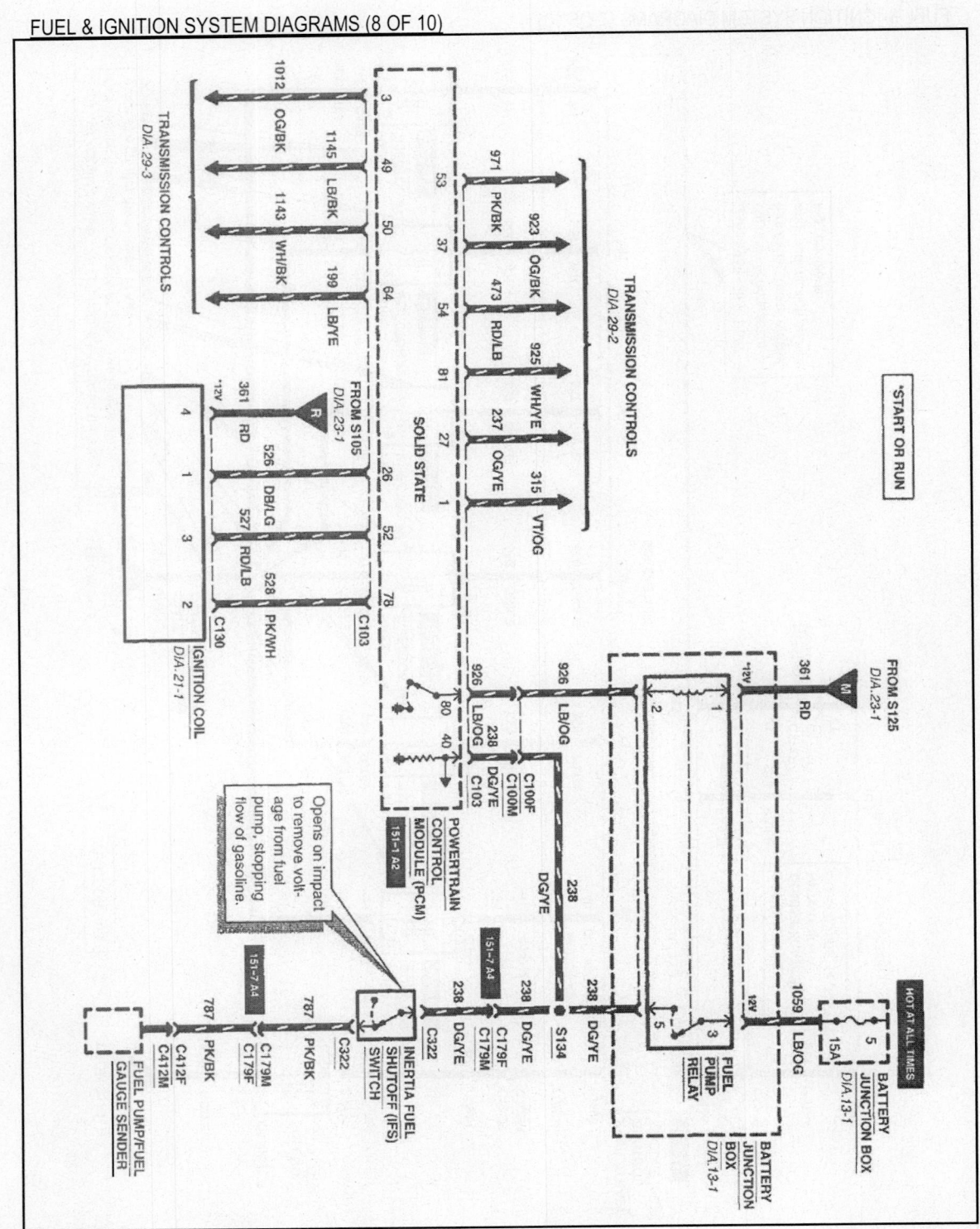

Fuel & Ignition System Diagrams (9 of 10)

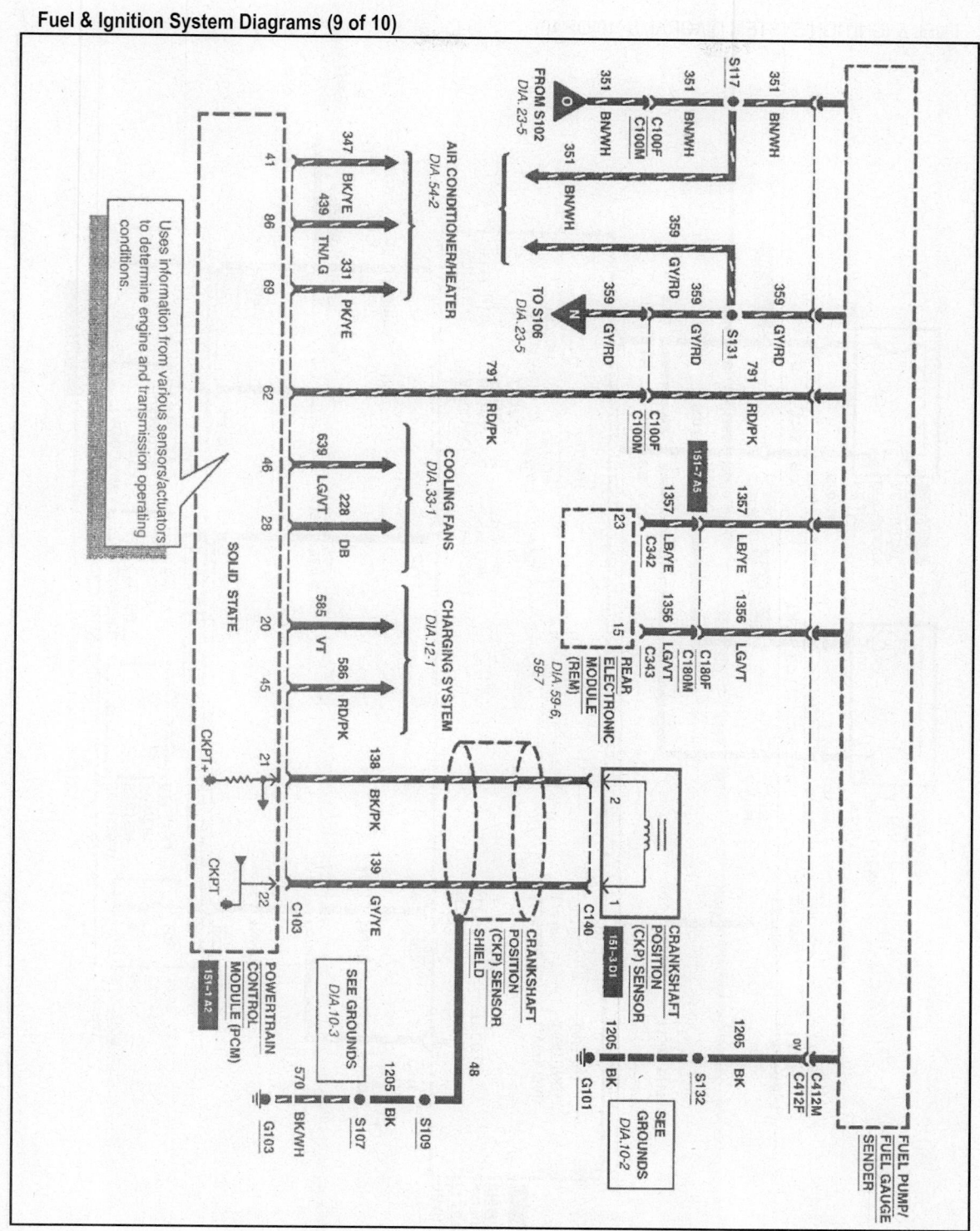

FUEL & IGNITION SYSTEM DIAGRAMS (10 OF 10)

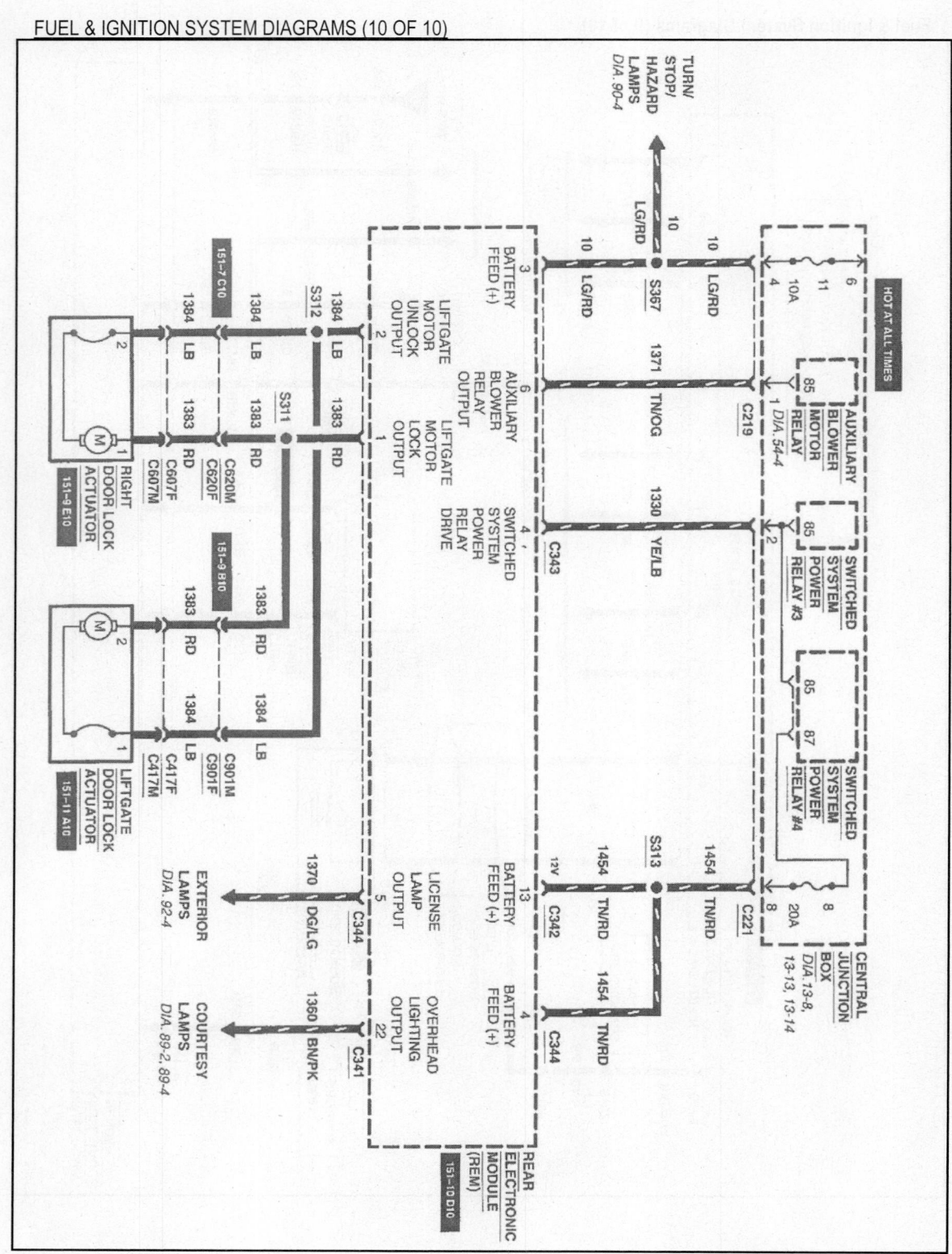

SPORT UTILITY VEHICLE CONTENTS

2000 Explorer - Powertrain

INTRODUCTION
How To Use This Section ....Page 5-3
VEHICLE IDENTIFICATION
VECI Decal ....Page 5-4
ELECTRONIC ENGINE CONTROL SYSTEM
Operating Strategies ....Page 5-5
PCM Location ....Page 5-5
INTEGRATED ELECTRONIC IGNITION SYSTEM
Introduction ....Page 5-6
Ignition Coil Operation ....Page 5-7
Lab Scope Test (Coil Primary) ....Page 5-8
CKP Sensor (Magnetic) ....Page 5-9
CKP Sensor Diagnosis (No Code Fault) ....Page 5-9
DVOM Test (CKP Sensor) ....Page 5-11
Lab Scope Test (CKP Sensor) ....Page 5-11
JI Diagnostic System - DTC P0300 ....Page 5-12
ENHANCED EVAPORATIVE EMISSION SYSTEM
Introduction ....Page 5-16
Canister Purge Solenoid ....Page 5-17
Charcoal Canister ....Page 5-18
Fuel Vapor Vent Valve ....Page 5-18
Canister Vent Solenoid ....Page 5-19
Fuel Tank Filler Cap ....Page 5-19
Fuel Tank Pressure Sensor ....Page 5-19
Fuel Vapor Control Valve ....Page 5-20
EVAP Leak Tester ....Page 5-20
ORVR System ....Page 5-21
Scan Tool Test (Purge Solenoid) ....Page 5-22
Scan Tool Test (FTP Sensor) ....Page 5-22
DVOM Test (Purge Solenoid) ....Page 5-22
Lab Scope Test (CV Solenoid) ....Page 5-23
EVAP System Monitor ....Page 5-24
JI Diagnostic System - DTC P0442 ....Page 5-25
EXHAUST GAS RECIRCULATION SYSTEM
General Information ....Page 5-29
Differential Pressure Feedback EGR System ....Page 5-29
DPFE Sensor Information ....Page 5-30
Orifice Tube Assembly ....Page 5-30
Scan Tool & DVOM Tests (DPFE Sensor) ....Page 5-31
Vacuum Regulator Solenoid ....Page 5-32
Lab Scope Test (VR Solenoid) ....Page 5-33
JI Diagnostic System - DTC P0401 ....Page 5-34

FUEL DELIVERY SYSTEM
     Introduction ....Page 5-38
     Fuel Pump Relay & Fuel Pump Monitor..Page 5-39
     Fuel Pump Module & Lab Scope Test (Fuel Pump) ....Page 5-40
     Fuel Pump Components ....Page 5-41
     Fuel Pressure Regulator ....Page 5-42
     Inertia Switch ....Page 5-43
     Fuel Injectors ....Page 5-44
     Scan Tool Test (Fuel Injector & Fuel Trim) ....Page 5-45
     Lab Scope Test (Fuel Injector) ....Page 5-46
     CMP Sensor (Magnetic) ....Page 5-47
     CMP Sensor (Signal Explanation) ....Page 5-48
     Lab Scope Test (CMP Sensor) ....Page 5-48

COMMUNICATIONS NETWORK
     Principles of Operation ....Page 5-49

GENERIC ELECTRONIC MODULE
     Introduction ....Page 5-50
     Scan Tool Test (GEM & CTM) ....Page 5-51
     GEM & CTM DTC & Symptom List ....Page 5-52
     GEM & CTM PID List ....Page 5-52
     Pinpoint Test 'G' (No Communication) ....Page 5-53
     GEM & CTM Wiring Diagram ....Page 5-54

IDLE AIR CONTROL VALVE
     General Description ....Page 5-55
     Scan Tool Test (IAC Solenoid) ....Page 5-55
     Lab Scope Test (IAC Solenoid) ....Page 5-56

INPUT SENSORS
     Throttle Position Sensor ....Page 5-57

MASS AIRFLOW SENSOR
     General Description ....Page 5-58
     Lab Scope Test (MAF Sensor) ....Page 5-59

OXYGEN SENSOR
     General Description ....Page 5-60
     Scan Tool Test (HO2S) ....Page 5-61
     Lab Scope Test (HO2S) ....Page 5-62

TRANSMISSION CONTROLS
     Component Tests ....Page 5-63
     Scan Tool Tests (EPC, TCC & TSS) ....Page 5-63
     Scan Tool Tests (DTR & TFT Sensor) ....Page 5-64

REFERENCE INFORMATION
     PCM PID Tables - Inputs ....Page 5-65
     PCM PID Tables - Outputs ....Page 5-66
     PCM Pin Voltage Tables ....Page 5-67
     PCM Wiring Diagrams ....Page 5-69

2000 EXPLORER - POWERTRAIN

Introduction

HOW TO USE THIS SECTION

This section of the training manual includes diagnostic and repair information for Ford SUV applications. This information can be used to help you understand the Theory of Operation and Diagnostics of the electronic controls and devices on this vehicle.

The articles in this section are separated into three sub categories:

- Generic Electronic Module
- Powertrain Control Module
- Transmission Controls

2000 Explorer 4.0L V6 VIN E (5R55E Automatic Transmission)

The articles listed for this vehicle include a wide variety of subjects that cover the key engine controls and vehicle diagnostics. Several of the Powertrain sensor inputs and output devices are featured along with detailed descriptions of how they operate, what can go wrong with them, and most importantly, how to use the PCM onboard diagnostics and common shop tools to determine if one or more devices has failed.

This vehicle uses an Integrated Electronic Ignition (distributorless) system along with a camshaft position sensor (Magnetic) to control the fuel injection timing. These devices and others (i.e., the MAF and Oxygen sensors) are covered along with how to test them.

Choose The Right Diagnostic Path

In most cases, the first step in any vehicle diagnosis is to follow the manufacturers recommended diagnostic path. In the case of Ford vehicle applications, the first step is to connect a Scan Tool (OBD II certified for this vehicle) and attempt to communicate with the engine controller or any other vehicle onboard controllers that apply. From this point on, the recommended procedure is to use the Ford Quick Test to check for faults.

The Malfunction Indicator Lamp (MIL) can be a great help to you during diagnosis. You should note if the customer complaint included the fact that the MIL remained on during engine operation. And be sure to determine if the MIL comes on at key on, and if it goes out once the engine starts. There is more information on using the MIL in Section 1.

Diagnostic Points to Consider

If a problem is detected in a vehicle similar to this one (whether it is a trouble code or no code condition), you need to be able to determine which vehicle system (or systems) are involved before you proceed with any testing. While this may sound like a fairly routine way to approach problem solving, it is a key point to consider before you start testing.

Due to the complexity of these vehicles, a trouble code does not always point to a failed component - instead, it may point to a system that needs to be tested for some kind of problem. For example, if an Oxygen sensor trouble code is set, you need to consider what other systems or devices could cause this component to set a trouble code.

Component Diagrams

You will find numerous "customized" component diagrams used throughout this section. These diagrams are really a mini-schematic of how a particular device (where an input or output device) is connected to the engine or transmission control module.

Each diagram contains the power, ground, signal and control circuits that relate to a particular device. Of particular note is the use of "code check" identifiers in the diagrams that indicate circuits monitored by the module in order to set a particular trouble code(s).

Vehicle Identification

<u>VECI DECAL</u>

The Vehicle Emission Control Information (VECI) decal is located on the underside of the hood. *This example is from a 2000 Ford Explorer 4.0L V6 SOHC VIN E.*

Ford Motor Company IMPORTANT VEHICLE INFORMATION

This vehicle is equipped with electronic engine control systems. Engine idle speed, idle mixture, and ignition timing are not adjustable. See Powertrain Control/Emissions Diagnosis Manual for additional information.

Oil Type Recommendations

Use SAE 5W-30 American Petroleum Institute Certified Oils for Gasoline Engines.

Vehicle Certification

This vehicle conforms to U.S. E.P.A NLEV (non-low emissions vehicle) regulations applicable to gasoline fueled 2000 model year new LEV light-duty trucks and California regulations applicable to 2000 model year LEV light-duty trucks. It is OBD II Certified.

Other Identification Items

Emissions: YW7E-9C485 **G C Y, Catalyst, Spark Plug Gap:** 0.52-0.56
Engine: 4.0L - YFMXE04.02F4-2TWC, TWC (2) /2HO2S, HO2S/EGR/SFI

Underhood View Graphic

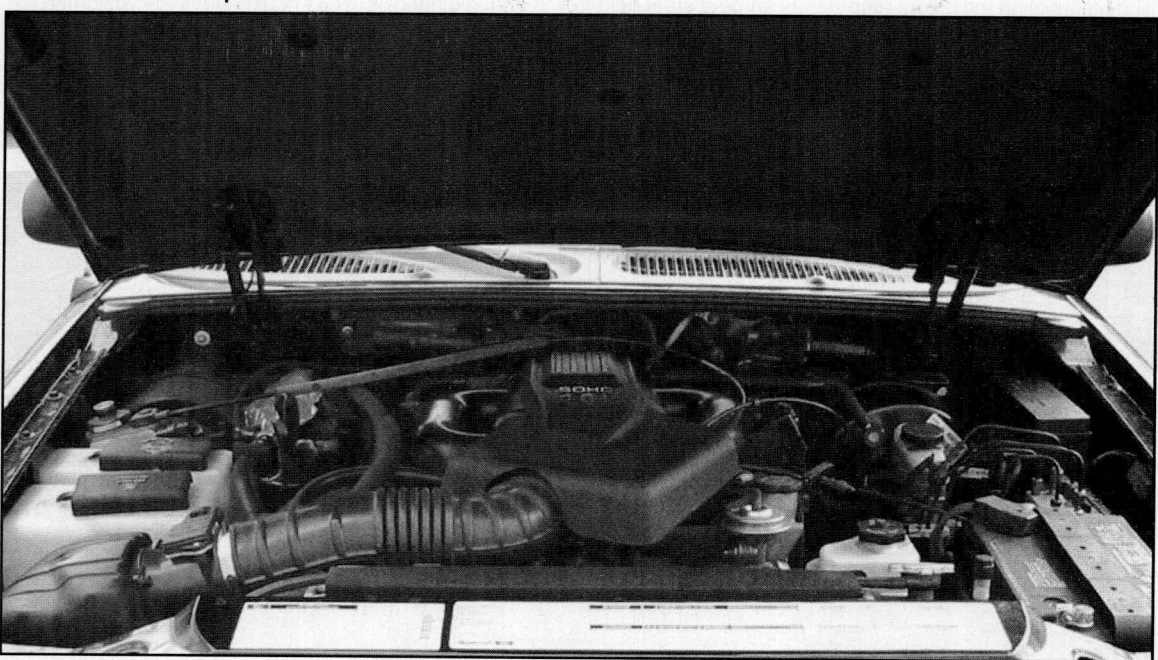

Electronic Engine Control System

OPERATING STRATEGIES

The Electronic Engine Control (EEC) system provides optimum control of the engine and transmission through the enhanced capability of the PCM. The PCM uses several operating strategies to maintain good overall driveability and to meet the EPA mandated emission standards. These strategies are included in the tables on this page.

Based on various inputs it receives, the PCM adjusts the fuel injector pulsewidth, the idle speed, the amount of ignition advance or retard, the ignition coil dwell and operation of the canister purge valve.

The PCM controls the speed control and air conditioning systems.

PCM Location

The PCM is located at the left side of the engine cowling area as shown in the Graphic.

PCM Location at Cowling

Operating Strategies

A/C Compressor Clutch Cycling Control	Closed Loop Fuel Control
Electric Fuel Pump Control	Exhaust Gas Recirculation Control
Fuel Metering of Sequential Fuel Injectors	Fuel Pump Monitor
Idle Speed Control	Ignition Timing Control for A/F Change
Vapor Canister Purge Control	

General Operating Strategies

Adaptive Fuel Trim Strategy	Adaptive Idle Speed Control Strategy
Base Engine/Transmission Strategy	Failure Mode Effects Mgmt. Strategy

■ NOTE: *Hardware Limited Operating Strategy (HLOS) is a part of the PCM hardware.*

Operational Control of Components

A/C Compressor Clutch	EGR System Control
Fuel Delivery (Injector Pulsewidth)	Fuel Pump Operation
Fuel Vapor Recovery System	Idle Speed
Malfunction Indicator Lamp	Shift Points for the A/T
Speed for Integrated Speed Control	Torque Converter Clutch Lockup (TCC)
Transmission Line Pressure (EPC)	

Items Provided (or Stored) By the EEC System

Data for the Operational Control Centers	Data for the Suspension Control Centers
DTC Data for the A/T (5R55E)	DTC Data for Emission Related Faults
DTC Data for Non-Emission Faults	

Integrated Electronic Ignition System

INTRODUCTION

The engine controller (PCM) controls the operation of the Integrated Electronic Ignition (EI) system on this vehicle. Battery voltage is supplied to the coilpack (three coils in one assembly) through the ignition switch and the #19 (25 amp) fuse in the I/P fuse panel.

The PCM controls the individual ground circuits of the three (3) ignition coils by switching the ground path for each coil "on" and "off" at the appropriate time. The PCM adjusts the ignition timing for each cylinder pair correctly to meet changes in the engine operating conditions. The ignition coil pairs are as follows: 1/5, 2/6 and 3/4. The firing order of this engine application is as follows: 1-4-2-5-3-6.

System Components

The main components of this Integrated EI system are the coilpack, crankshaft position sensor, the PCM and the related wiring. Note that the camshaft position sensor is not an integral part of the EI system as it is used only as an input for sequential fuel injection.

EI System Graphic

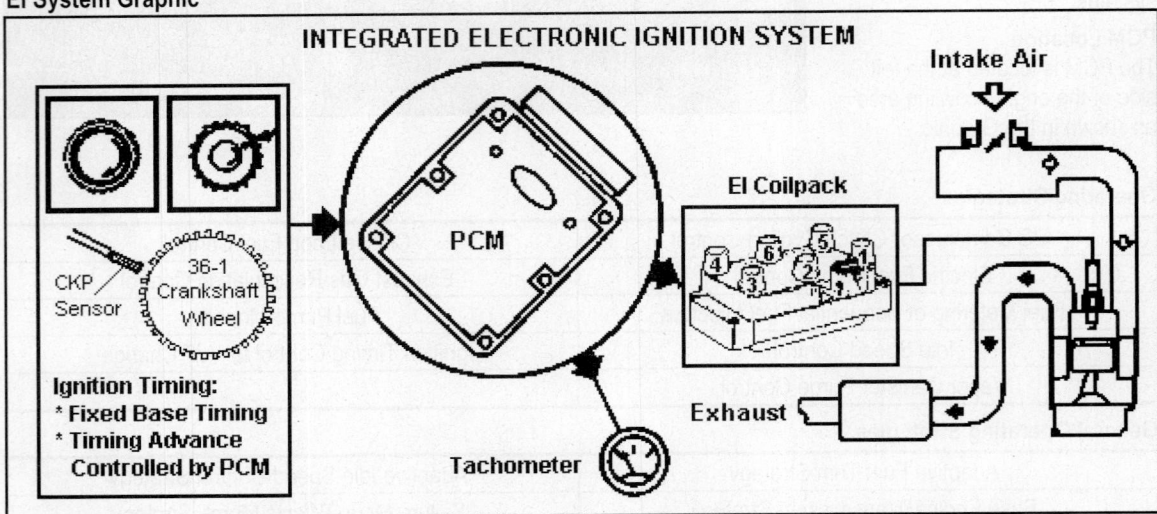

Ignition Spark Timing

The ignition timing on this EI system is entirely controlled by the PCM - base timing is not adjustable. Do not attempt to check the base timing, as you will get false readings.

The PCM uses the CKP sensor signal to indicate crankshaft position and speed by sensing a missing tooth on a pulse wheel mounted behind the harmonic balancer. It also uses this signal to calculate a spark target and then fire the coils to that spark target.

After engine startup, the PCM calculates the ignition spark advance with these signals:

- BARO Sensor (this is a signal calculated from the MAF sensor)
- ECT Sensor
- Engine Speed Signal (rpm)
- IAT Sensor
- MAF Sensor
- Throttle Position

IGNITION COIL OPERATION

The PCM controls the operation of the EI system coilpack in order to provide ignition secondary spark at the correct time in the combustion process.

As discussed, the coilpack consists of three (3) independent coils molded together. The coil assembly is mounted to the intake manifold. Spark plug cables rout to each of the six (6) cylinders from the coilpack assembly.

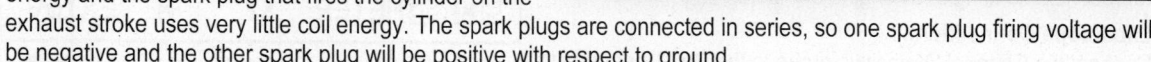

Each ignition coil fires two (2) spark plugs. This is a form of waste spark ignition where one spark plug fires the cylinder under compression and the other spark plug (same coil) fires the cylinder on the exhaust stroke.

The spark plug that fires the cylinder on the compression stroke uses the majority of the ignition coil energy and the spark plug that fires the cylinder on the exhaust stroke uses very little coil energy. The spark plugs are connected in series, so one spark plug firing voltage will be negative and the other spark plug will be positive with respect to ground.

The PCM determines which coil to charge and fire. Coil '1' fires Cylinders 1 & 5; Coil '2' fires Cylinders 4 & 3; and Coil '3' fires Cylinders 2 & 6. The firing order is 1-4-2-5-3-6.

EI Coilpack Control Circuits

The PCM is connected to the Coil '1' control circuit at Pin 26, to the Coil '2' control circuit at Pin 52 and to the Coil '3' control circuit at Pin 78 of the PCM wiring harness. The resistance of a known good coil is from 0.4 to 1.0 ohms.

Trouble Code Help

Note the three circuits that are shown with a dotted line in the component schematic. The PCM monitors these circuits as it switches the coil drivers "on" and then "off". If the PCM does not receive a valid IDM signal due to a coil problem, it will set one of these diagnostic trouble codes: P0351 for Coil '1', P0352 for Coil '2' or P0353 for Coil '3'.

EI Coil Circuits Schematic

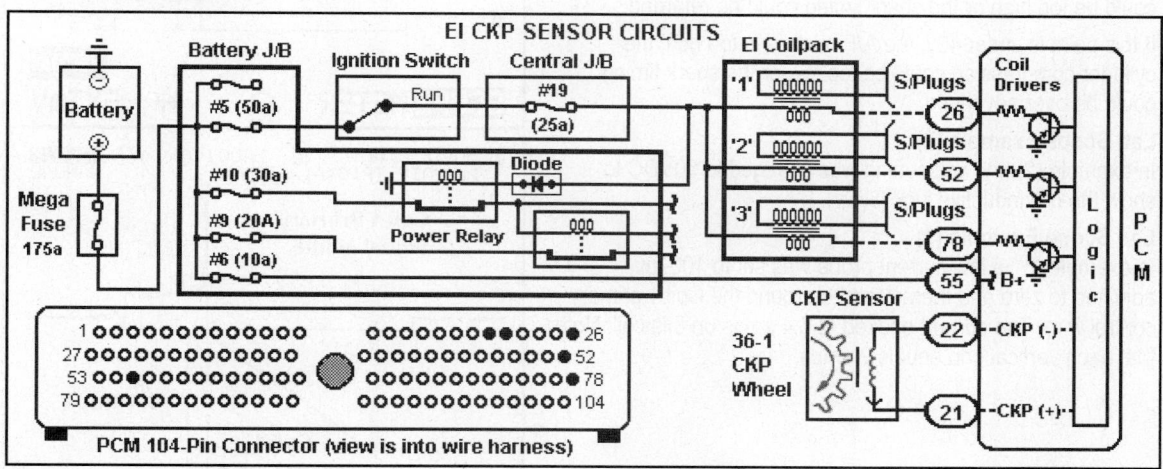

LAB SCOPE TEST (COIL PRIMARY)

The ignition coil primary circuit includes the battery junction box, switched ignition feed circuit, coil primary circuits and coil driver control circuits in the PCM.

In this example, Coil '1' connects to Pin 26; Coil '2' connects to Pin 52 and Coil '3' connects to Pin 78 of the 104P connector.

Coil Saturation

Once maximum current flows through the ignition coil primary windings, a maximum magnetic field is present in the coil windings, and the windings are considered saturated.

Coil Dwell Period

The coil dwell period is the length of time that current flows through the coil primary windings (it is measured in milliseconds and represents the degrees of engine rotation). A complete turn of the camshaft equals 360 degrees and represents one firing of all engine cylinders.

Lab Scope Connections

Connect the Channel 'A' positive probe to the Coil 1 primary at Pin 26 (TN/WHT wire) and negative probe to the battery negative post.

Scope Settings

To make the waveforms as clear as possible, set the scope settings to match the examples.

Lab Scope Example (1)

In example (1), the trace shows the coil primary circuit. The dwell period was close to 4.2 ms. The trace shows a spark line "intersect point" (the range is 40v to 50v). If this point is over 50v, the A/F mixture could be lean, the cylinder compression could be too high or the spark timing could be retarded.

If this point is under 40v, the A/F mixture is too rich, the cylinder compression could be too low or the spark timing could be over advanced.

Lab Scope Example (2)

In example (2), the volt setting was changed to 50VDC to show the coil inductive kick (330v).

Lab Scope Example (3)

In example (3), a low current probe was set to 100 mv, adjusted to zero and then clamped around the Coil 1 primary control wire. The current peaked at 6.4 amps on this coil. Note that each vertical grid equals 2 amps.

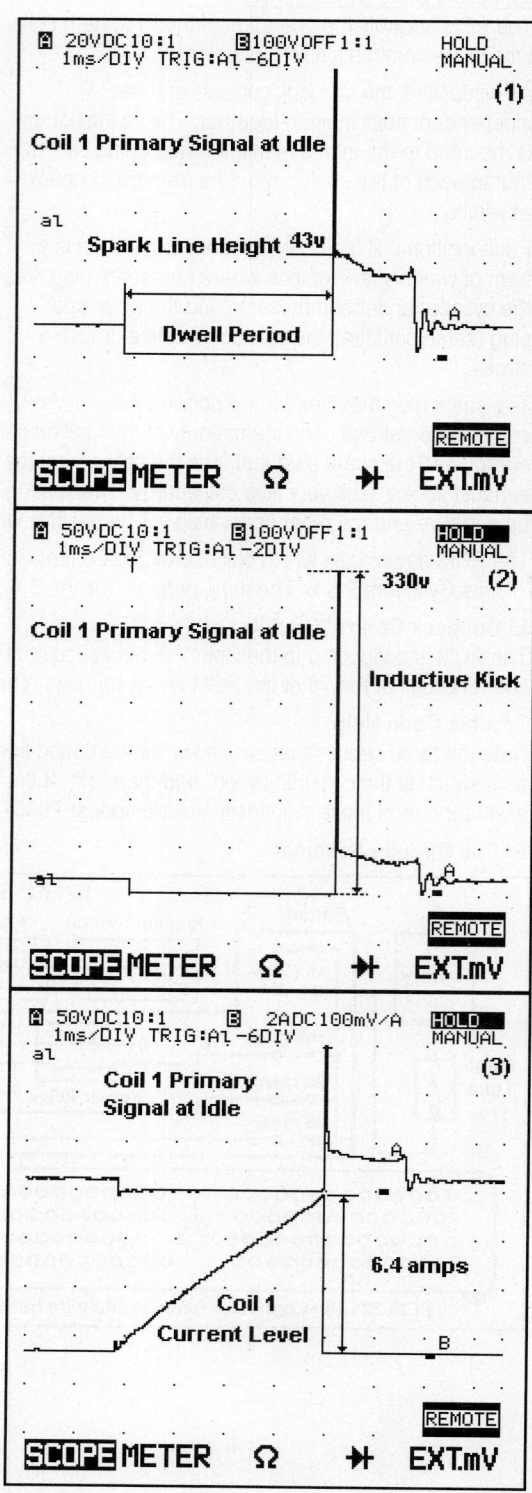

CKP SENSOR (MAGNETIC)

The crankshaft position (CKP) sensor used with this EI system is the primary sensor for crankshaft position data to the ignition control module that is integrated into the PCM.

The CKP sensor is a magnetic transducer mounted at the front of the timing chain cover adjacent to a 36-1 tooth trigger wheel mounted on the crankshaft behind the harmonic balancer.

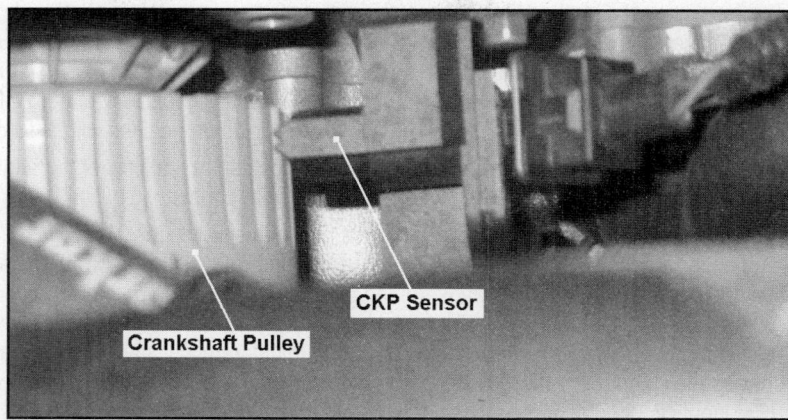

The trigger wheel has 35 teeth spaced 10 degrees (10°) apart with one empty space due to a missing tooth (hence the term 36-1 assigned to the trigger wheel).

The CKP sensor outputs an AC signal to the integrated ICM (integrated into the PCM). The CKP signal directly proportional to crankshaft speed and position.

The PCM monitors the CKP sensor signal (the 36-1 design signal) in order to correctly identify piston travel so that it can synchronize the firing of the coils, and provide an accurate method of tracking the angular position of the crankshaft, relative to a fixed reference (the missing tooth on the wheel). This tracking is used for misfire diagnosis.

CKP Sensor Circuits

The PCM connects to the CKP sensor (+) circuit at Pin 21 (DB wire) and to the CKP sensor (-) circuit (GRY wire) at Pin 22 of the 104P harness connector. The CKP sensor resistance is 290-790 ohms at 68°F.

CKP Sensor Diagnosis (No Code Fault)

If the PCM does not detect any signals from the CKP sensor with the engine cranking, it will not start. This sensor input is not monitored for a loss of signals. A DVOM can be used to check for "no output", but a Lab Scope is the tool of choice to test this sensor.

EI System CKP Sensor Schematic

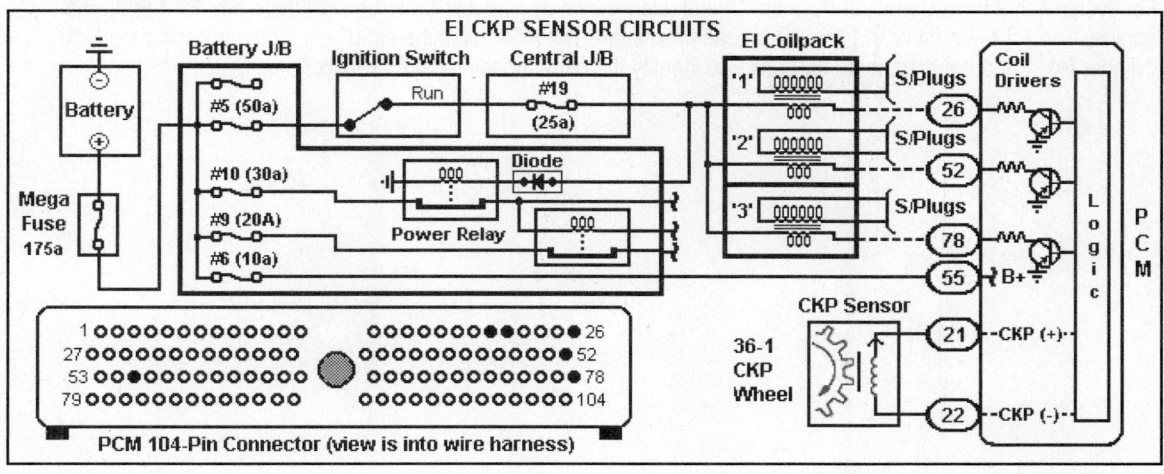

CKP SENSOR (SIGNAL EXPLANATION)

The CKP sensor (also referred to as a variable reluctance sensor or VRS) signal is derived by sensing the passage of teeth from a 36 minus one (36-1) tooth wheel mounted on the crankshaft.

The ferromagnetic-toothed timing wheel is rigidly mounted to the crankshaft and designed to have the center of each tooth spaced at 10-degree increments. Since there are 360 degrees in one revolution, there should be 36 teeth. However, in this system, there is one tooth missing. The CKP sensor is located directly opposite the missing tooth at a point where cylinder No. 1 is located at 90º BTDC for a 4-cylinder engine, 60º BTDC for a 6-cylinder engine and 50º BTDC for an 8-cylinder engine. A 40-1 tooth wheel is used on engines with a 10-cylinder engine.

In addition to crankshaft-mounted 36-1 tooth wheel located at the front of the engine, some applications utilize holes in the transaxle at the rear of the engine to trigger the VRS sensor.

A VRS sensor is a magnetic transducer with a pole piece wrapped with fine wire. When the VRS encounters the rotating ferromagnetic teeth, the varying reluctance induces a voltage proportional to the rate of change in magnetic flux and the number of coil windings. As the tooth approaches the pole piece, a positive differential voltage is induced. As the tooth and the pole piece align, the signal rapidly transitions form maximum positive to maximum negative. As the tooth moves away, the voltage returns to a zero differential voltage. The negative zero crossing of the VRS signal corresponds to the alignment of the center of each tooth with the center of each pole piece. The ICM portion of the PCM input circuitry triggers on this negative zero crossing at approximately -0.3 volts and it uses this point to establish crankshaft position.

At normal engine speeds, the VRS signal waveform approximates a sine wave with increased amplitude adjacent to the mission tooth region. At 30 rpm, the VRS output should be 150 mv peak-to-peak, at 300 rpm, the VRS output should be 1.5 volts peak-to-peak and at 6000 rpm, the VRS output should be 24.0 volts peak-to-peak. At 8000 rpm, the VRS output should not exceed 300 volts peak-to-peak. The normal sensor air gap is 1.0 mm nominal; 2.0 mm max. The resistance of the sensor should be 290 to 790 ohms. The inductance should be 170 to 930 Millihenries.

CMP SENSOR (SIGNAL EXPLANATION)

The CMP sensor input is not needed for the EDIS system since the spark plugs are fired in pairs. The CMP sensor is, however, used by the PCM to determine cylinder No. 1 for sequential fuel control. This vehicle application utilizes a VRS design camshaft sensor.

On models with a VRS design Camshaft sensor, the VRS produces a very narrow sinusoidal signal when a tooth on the camshaft passes the pole piece. The negative-going edge is usually located at 22º ATDC of Cylinder 1.

For systems that have Camshaft Position Control (also known as Variable Cam Timing), there are 3 + 1 teeth (V6 engine) or 4 + 1 teeth (I4 or V8) located on each camshaft. The PCM uses the extra (+1) tooth on Bank 1 to identify cylinder No. 1 and the extra tooth on Bank 2 to identify the corresponding paired cylinder.

DVOM TEST (CKP SENSOR)

Place the shift selector in Park and block the drive wheels for safety.

DVOM Connections

Turn the key off and connect the breakout box. Select AC volts on the DVOM. Connect the DVOM positive lead to the CKP sensor (+) circuit (DB wire) at Pin 21 and the negative lead the CKP sensor (-) circuit (GRY wire).

Crank the engine for 3-5 seconds or start the engine and then record the readings.

Test Results

The DVOM display shows a CKP signal of 1.633v AC at a rate of 389.3 Hz. Note that the true amplitude of the AC voltage reading is about 3.00v in this example.

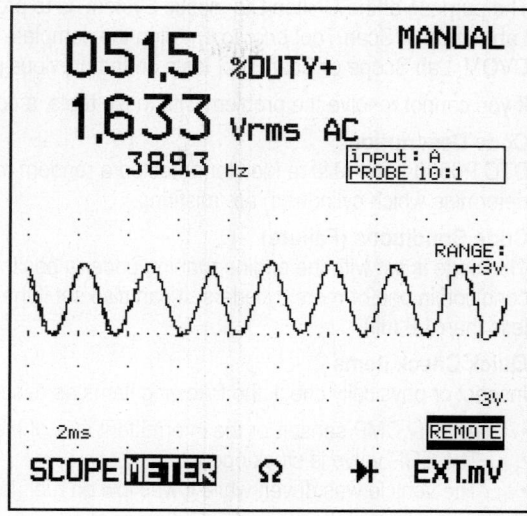

LAB SCOPE TEST (CKP SENSOR)

The Lab Scope can be used to test the CKP sensor as it provides a very accurate view of sensor waveform and of any glitches. Place the shift selector in Park (A/T) for safety.

Scope Connections

Install a breakout box (BOB) if available. Connect the Channel 'A' positive probe to the CMP signal at Pin 85 (DG wire) of the 104P connector. Connect the Channel 'B' positive probe to the CKP sensor positive circuit at Pin 21 (DB wire) and the negative probe to the CKP sensor negative circuit at Pin 22 (GRY wire) of the PCM 104P connector.

Scope Settings

To make the waveforms as clear as possible, set the scope settings to match the examples.

Lab Scope Tests

The CMP and CKP sensor signals can be checked during cranking, idle and at off-idle speeds with the engine cold or fully warm.

Lab Scope Example

In this example, the top trace shows the CMP sensor signal. The bottom trace shows the CKP sensor signal. These waveforms were captured with the gear selector in Park, the engine running at idle speed in closed loop.

The top trace represents 95° of camshaft rotation from the CMP sensor signal. Note the position of the missing tooth in this example.

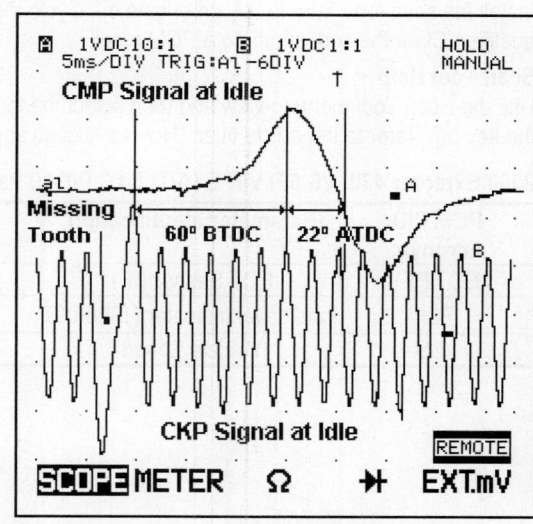

The bottom trace represents around 190 degrees of crankshaft revolution from the CKP sensor (a 1.5v positive peak signal and a 1.5v negative peak signal). This signal is 3v AC peak-to-peak and is occurring at 385 hertz. Note the point at which the missing tooth occurs (60° BTDC) and where the CMP signal crosses its mid-point (22° ATDC).

<u>CHILTON DIAGNOSTIC SYSTEM - DTC P0300</u>

The purpose of the Chilton Diagnostic System is to provide you with one or more tests that can be done with a DVOM, Lab Scope or Scan Tool *prior* to entering the complete trouble code repair procedure. The quick checks listed in the DVOM, Lab Scope or Scan Tool tests on the previous pages may help you quickly find the cause of the problem.

If you cannot resolve the problem with these tests, a code repair chart is also included.

Code Description

DTC P0300 - The Misfire Monitor detected a random misfire condition in multiple cylinders (in effect, the PCM cannot determine which cylinder(s) are misfiring.

Code Conditions (Failure)

This code is set with the engine running under a positive torque condition, the PCM detected a random misfire condition in one or more cylinders. It can also set if the vehicle is driven with a "low fuel condition" (with the fuel tank less than 1/8 full).

Quick Check Items

Inspect or physically check the following items as described below:

- A faulty CMP sensor, or the intermittent loss of the CMP sensor signal
- The EGR valve is stuck open
- The vehicle was driven while it was low on fuel (less than 1/8 of a tank of fuel)

Drive Cycle Explanation

Start the vehicle (with the engine cold or warm) and then drive it under a positive load condition (any condition except deceleration).

Drive Cycle Preparation

There are no drive cycle preparation steps required to run the Misfire Monitor.

Install the scan tool. Turn to key on engine off. Cycle the key off, then on. Select the appropriate vehicle and engine qualifier. Clear the codes and do a PCM reset.

Scan Tool Help

Use the Scan Tool menu to view and then select the following PIDs: FLI (if available). Start the engine without turning the key off. Refer to the article titled "How to Access and Use OEM PID Information at the end of this section.

2000 Explorer 4.0L V6 SFI VIN E (A/T) DTC P0300 Parameters

PCM PID Acronym	Parameter Identification	PID Range	PID Value at Hot Idle	PID Value at 30 mph	PID Value at 55 mph
ECT (°F)	ECT Sensor (°F)	-40-304°F	160-200°F	160-200°F	160-200°F
FLI	Fuel Level Indicator (1998-02)	0-100%	50% (1.7v)	25-75%	25-75%
VSS	Vehicle Speed	0-255 mph	0 mph	30 mph	55 mph

CHILTON DIAGNOSTIC SYSTEM - DTC P0300
If DTC P0300 is set, Pinpoint Test HD can be used to diagnose the following devices and/or circuits:

- Camshaft Position Sensor Signal
- EGR Position Sensor Signal
- Power Control Module (PCM)

Pinpoint Test HD Repair Table

Step	Action	Yes	No
HD1	**Step description:** Check for any Fuel Trim and HO2S Continuous Memory codes • Check for non-misfire Continuous Memory codes that could cause this trouble code. Look for P0136, P0156, P0171, P0172, P0175, P1130 and P1150. • Were any of the listed codes present?	Go to Step HD4	Go to Step HD2.
HD2	**Step description:** Check for other Non-Misfire Continuous Memory codes • Check for other non-misfire Continuous Memory codes that could cause this code. • Were any other Non-Misfire codes present?	Address the next Continuous Memory code (disregard the Misfire code at this time). Refer to the correct Powertrain code repair chart.	Go to Step HD3.
HD3	**Step description:** Check for KOEO codes • Check for any KOEO On Demand codes that could cause this misfire code to set. • Were any of KOEO codes present?	Refer to correct code repair chart to fix first KOEO code. Rerun the Misfire Monitor Repair Verification drive cycle.	Go to Pinpoint Test JB1 to test spark plugs and secondary wires. If these components are okay, go to Step HD4.
HD4	**Step description:** Check for KOER codes • Check for any KOER codes that could cause a misfire. If P1131, P1137, P1151 or P1157 are set, go to Step HD8. • Were any of KOER codes present?	Refer to correct code repair chart to fix first KOER code. Rerun the Misfire Monitor Repair Verification drive cycle.	If a misfire exists, go to Step HD6 (DPEGR). For others go to Step HD7. If intermittent, go to Step Z1 (Section 1).
HD6	**Step description:** Compare the PID values • Start the engine and allow it to warm up. • Access and record the DPFEGR PID. • Turn to key on engine off and access the DPEGR PID. Compare the two readings. • Was the KOER PID value within 0.15v of the KOEO PID value?	Go to Step HD7.	Go to Step HE100 (at the end of this repair procedure).
HD7	**Step description:** EGR Restriction/Flow Test • Record and clear any codes. Other codes may set during this test step. If so, clear them. • Disconnect / plug vacuum line to EGR valve. • Drive the vehicle through the complete Misfire Detection Drive Cycle. • Did the original Misfire code set?	Go to Step HD8.	Inspect the EGR valve and intake port for blockage or restriction. Make repairs as needed and then retest for the original fault.
HD8	**Step description:** Check Injector Driver PID • Turn to key on, engine off. • Access the appropriate Injector PID(s). • Did the INJ PID value indicate YES?	Go to Step HD9.	Go to Step HD10.

Pinpoint Test HD Repair Table (Continued)

Step	Action	Yes	No
HD9	**Step description:** Check Injector Wiring • Turn the key off. Disconnect the PCM and inspect for the PCM pins (service as needed). • Install the breakout box (PCM disconnected). • Measure the resistance between the injector test pin (one at a time) and Test Pin 71 or 91. • Was the resistance value from 11-18 ohms?	Replace the PCM. Refer to Section 1 for information on how to Flash the Electrically Program Read Only Memory in PCM to complete this test step.	Disconnect suspect injector. Check power and control circuits for continuity (<5 ohms). Check both circuits for a short to ground or short to power fault.
HD10	**Step description:** Check the Fuel Pressure • Turn the key off. Install fuel pressure gauge. • Start the engine and allow it to idle. Record the fuel pressure reading on the gauge. • Raise the engine speed to 2500 rpm in Park and hold it for one minute. Record the reading. • Did the pressure hold at 30-45 psi in the test?	Go to Step HD11.	Refer to fuel pressure tests in other repair manuals to test and repair the Fuel system. Rerun the Misfire Monitor Repair Verification drive cycle.
HD11	**Step description:** Check the Fuel Volume • Start the engine and then raise the engine speed to 2500 rpm in Park. Hold that engine speed for one minute. Record the reading. • Look for signs of fuel leaking at the injector O-rings or the fuel pressure regulator. • Did the fuel pressure drop more than 5 psi at any time during the 60-second period?	Refer to the fuel pressure tests in other repair manuals to test and repair the Fuel system. Rerun the complete Misfire Monitor Repair Verification drive cycle.	The Fuel pressure held during this test step (it was okay). Go to Step HD12.
HD12	**Step description:** Check Injector Flow Rate • Use a Flow Test to verify the flow rate for each fuel injector (follow all related Safety Precautions and Test Procedures). • Was injector flow rate within specifications?	The Fuel system is not the cause of the Misfire trouble code(s). Go to Step HD20.	Replace or clean the inoperative injectors as needed. Rerun Misfire Monitor Repair Verification drive cycle.
HD20	**Step description:** Check the Vacuum System • Listen for signs of an engine vacuum leak. • Inspect all vacuum lines for signs of pinched, cracked or misrouting and/or poor assembly. (Refer to the underhood vacuum label) • Was the Vacuum system okay?	Go to Step HD21.	Service the Vacuum system components as needed. Rerun the Misfire Monitor Repair Verification drive cycle.
HD21	**Step description:** Check damper assembly? • Observe the crank pulley for wobble. • Inspect pulse ring on the harmonic fastener. • Does the crank pulley wobble or is the pulse ring loose or damaged?	Replace the damper or pulley assembly. Clear the crank relearn data by disconnecting the batter for 5 minutes. Do a Crank Relearn.	Go to Step HD22.
HD22	**Step description:** Check the EVAP Canister • Inspect the canister for signs of excess fuel. • Was there excess fuel in the EVAP canister?	Replace the fuel vapor canister. Run Misfire Monitor Drive Cycle.	Check the fuel tank vent system. Go to Step HD23.
HD23	**Step description:** EVAP Pressure Test • Turn the key off. Install the appropriate EVAP tester at EVAP service port or fuel filler cap. • Follow test instructions to check the system. • Did the EVAP system hold pressure?	Go to Step HD24.	Service the EVAP system components as needed. Rerun the Misfire Monitor Repair Verification drive cycle.

Pinpoint Test HD Repair Table (Continued)

Step	Action	Yes	No
HD24	**Step description:** Inspect EVAP for Blockage • Inspect the system for blockage/restriction between the engine vacuum port and canister. • Was any blockage located in the system?	Make repairs to the system as needed. Rerun Misfire Monitor Repair verification test.	Go to Step HD25.
HD26	**Step description:** Test the solenoid for leaks • Turn the key off (purge solenoid connected). • Install a hand vacuum pump to fuel vapor port to fuel vapor canister on the valve line. • Apply 16" Hg of vacuum to valve with pump. • Did it hold vacuum at room temperature?	Go to Step HD27.	Remove the vacuum pump and then replace the damaged valve. Rerun the Misfire Monitor Repair Verification drive cycle.
HD27	**Step description:** Check Filter to the solenoid • Turn the key off. Remove vacuum line from the input vacuum port to intake manifold port on valve (control solenoid portion of the valve). • Connect vacuum pump to valve open port. • Apply 10-15" Hg of vacuum to the valve. • Did the valve hold vacuum, or did the valve very slowly release vacuum to atmosphere?	Service the valve filter. If unable to clean the filter, replace the valve. Rerun the Misfire Monitor Repair Verification Drive Cycle.	Remove the vacuum pump and reconnect all of the components. Go to Step HD29.
HD29	**Step description:** Check the Base Engine • Perform an Engine Compression test. • Perform a dynamic Valve Train test. • Check the PCV system for problems. • Was a problem found during this step?	Service or repair the engine components as needed. Rerun the Quick Test to recheck for the condition.	The cause of the Misfire is intermittent. Test Ignition system with an analyzer. If okay, go to Step HD30.
HD30	**Step description:** Check for Misfire DTCS • DTC P0300 indicates multiple cylinders were misfiring or PCM could not identify a cylinder. • Were any other Misfire trouble codes set?	Go to Step HD1.	Go to Step HD31.
HD31	**Step description:** Check for Additional DTCS • Were any Continuous Memory codes set?	Refer to correct code repair chart to fix code.	Compare PID values; check the CMP output.

Pinpoint Test HE Repair Table (HE100 only)

Step	Action	Yes	No
HE100	**Step description:** This EGR Flow test checks for EGR flow at idle speed with no codes set). • Disconnect and plug EGR hose at the valve. • Access the DPFEGR PID on the Scan Tool. • The PID voltage should be from 0.35-0.80v. • Start the engine and allow it to return to idle. • Monitor the DPFEGR PID value. Compare it the engine off reading. An increase in the PID voltage reading indicates that the differential pressure feedback EGR sensor is sensing EGR flow at idle speed. • Was the DPFEGER PID voltage reading at idle speed at least 0.15v more than the key on, engine off PID voltage reading?	The DPFEGR PID indicates EGR flow at idle. Since the EGR vacuum hose is removed and plugged, the fault is most likely in the EGR valve. Remove and check the EGR valve for signs of contamination, unusual wear, carbon deposits, binding or damage. Service the EGR valve as needed. Then rerun the test to verify the repair is completed.	This result indicates a fault in the vacuum supply to the valve. Inspect the vacuum regulator (VR) solenoid vent and vent filter for restrictions. Service the component as needed. If they are okay, replace the EGR VR solenoid. Then rerun the test to verify the repair is completed and the original condition is not longer present.

Enhanced Evaporative Emission System

INTRODUCTION

This vehicle application uses an Enhanced Evaporative Emission system to prevent fuel vapor buildup in the sealed fuel tank. Fuel vapors trapped in the sealed fuel tank are vented through a vapor valve on top of the tank via a single vapor line to the canister for storage. The vapors are purged to the intake manifold for burning in the engine.

The Enhanced EVAP system consists of a fuel tank, fuel filler cap, fuel tank mounted or in-line fuel vapor control valve, fuel vapor vent valve, EVAP canister, fuel tank mounted fuel tank pressure (FTP) sensor, canister purge valve, intake manifold hose assembly, canister vent solenoid, PCM and connecting wire and the fuel vapor hoses.

EVAP System Graphic

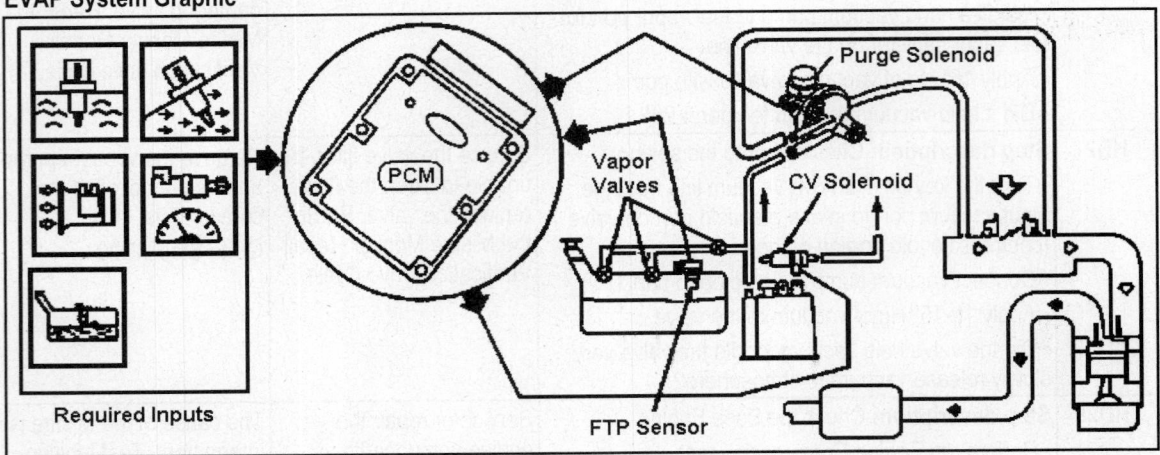

System Operation

The operation of the Enhanced EVAP system is described next.

- The PCM uses the ECT, IAT, FLI, FTP, MAF and VSS inputs to detect the engine conditions. The FLI and FTP sensors are used to determine when to activate the EVAP Monitor based on the presence of vapor generation or fuel sloshing.
- The PCM calculates the desired amount of purge vapor flow to the intake manifold for a given engine condition and then outputs a variable duty cycle command to the canister purge valve. During the evacuation period, the PCM uses the canister purge valve to purge the system, the canister vent solenoid valve to seal the system and the FTP sensor to determine the total vacuum lost over a period of time.
- The canister vent solenoid seals the system to atmosphere during the Leak Check.
- The PCM outputs a variable duty cycle signal to the EVAP canister purge solenoid.
- The fuel tank pressure (FTP) sensor monitors the fuel tank pressure during engine operation and continuously sends a signal to the PCM. During the EVAP Monitor test period, the FTP sensor monitors the fuel tank pressure or vacuum bleed-up.
- The fuel vapor vent valve and fuel tank vapor control valves (on the fuel tank) are used to control the flow of fuel vapor to the canister. They also prevent fuel tank overfilling during refueling and prevent liquid fuel from entering the canister and canister purge valve under any handling situation or rollover condition.
- This system (including all of the fuel vapor hoses) can be checked when a leak is suspected by the PCM. This step is accomplished using an EVAP Test Kit and a leak (frequency) detector. This kit is available from Ford and Aftermarket sources.

CANISTER PURGE SOLENOID

The EVAP canister purge solenoid is located underhood at the left front of the engine area. Refer to the Graphic at the bottom of this page for additional location information.

Canister Purge Valve Operation

The EVAP canister purge valve is located in-line with the canister. The function of this valve is to control the flow of vapors out of the canister, and to close off the flow of fuel vapors from the canister with the engine off.

Once the engine is started, the PCM controls the operation of the valve (i.e., the PCM opens and closes the valve to control the flow of fuel vapors from the charcoal canister during normal engine operation).

This device is a normally closed (N.C.) solenoid. The resistance of the solenoid is 30-36 ohms at 68°F.

As previously discussed, the canister purge valve is used to control the flow of vapors (during normal engine operation) from the fuel vapor storage canister into the intake manifold during certain conditions.

Canister Purge Valve Assembly

Underhood Component Locations

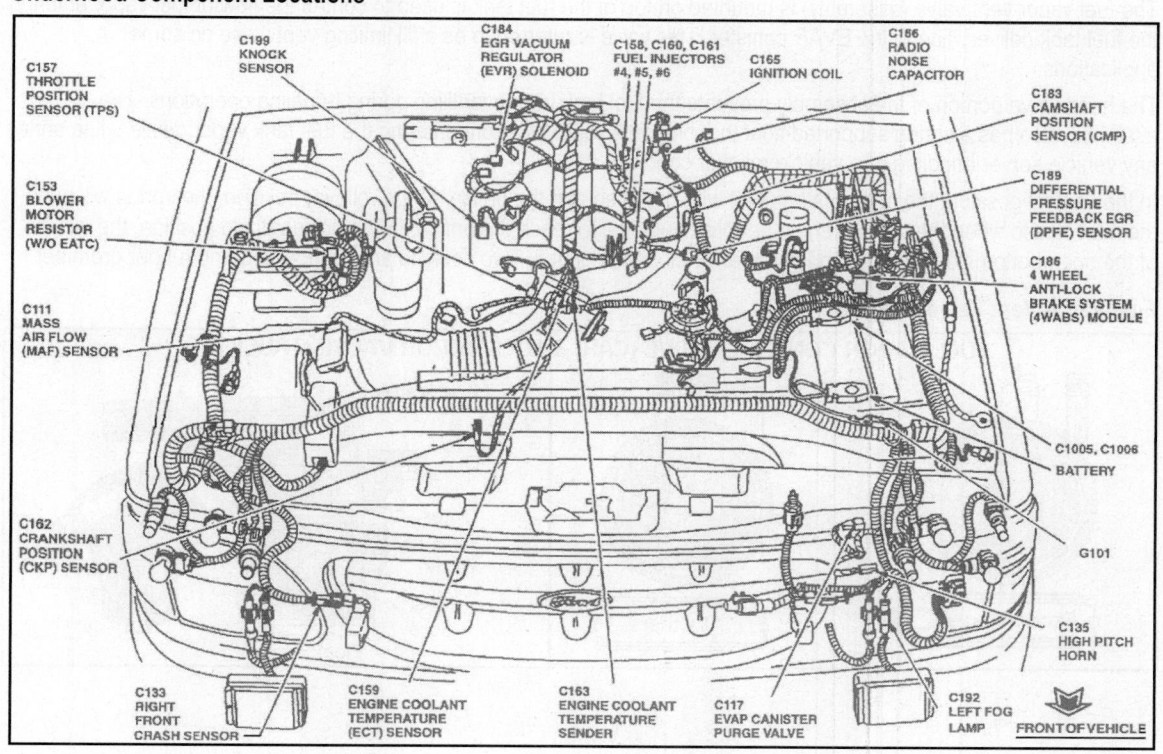

CHARCOAL CANISTER

The EVAP canisters contain 1.0L (2.113 pints) of activated carbon in the carbon bed. The activated charcoal has the ability to purge or release any stored fuel vapors when it is exposed to fresh air.

Fuel vapors from the fuel tank and the air cleaner are stored in the canister. This system has an emission dust separator.

Once the engine starts, fuel vapors stored in canister are purged into the engine to be burned during normal combustion.

The dual canisters are located at the rear of the vehicle above the spare tire. If the EVAP canister has to be removed, observe the *Caution* note below.

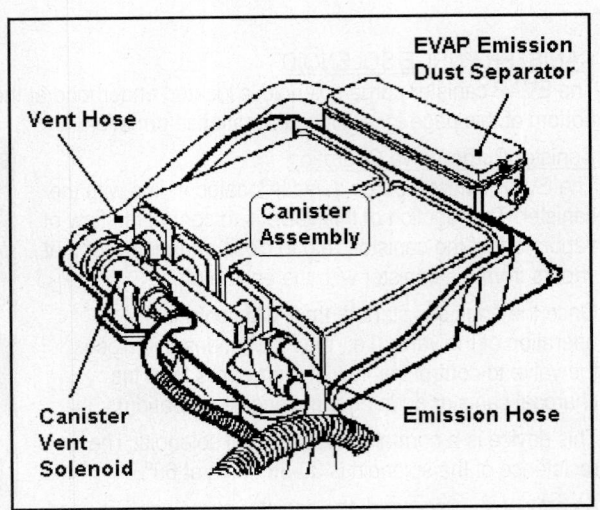

✳✳ Caution: ***The EVAP system contains fuel vapor and condensed fuel vapor. Although not present in large quantities, it still represents the danger of explosion of fire. Disconnect the battery ground cable at the battery to minimize the possibility of an electrical spark that could cause a fire or explosion if fuel vapor or liquid fuel is present in the area.***

FUEL VAPOR VENT VALVE

The fuel vapor vent valve (assembly) is mounted on top of the fuel tank is used to control the flow of fuel vapor entering the fuel tank delivery line to the EVAP canister. This valve is referred to as a fill limiting vent valve on some applications.

The head valve portion of the assembly prevents the fuel tank from overfilling during refueling operations. The assembly also has a spring supported float that prevents liquid fuel from entering the fuel tank vapor delivery line under any vehicle server handling or a vehicle rollover condition.

In the vertical position, the open bottom float will lift and shutoff the orifice. In the rollover position, the spring will push the float closed when the rollover angle permits liquid fuel to reach the orifice. In the upside down position, the weight of the open bottom float and the spring force will close the orifice. Two designs are used: o-ring and rubber grommet.

Fuel Vapor Vent Valve Graphic

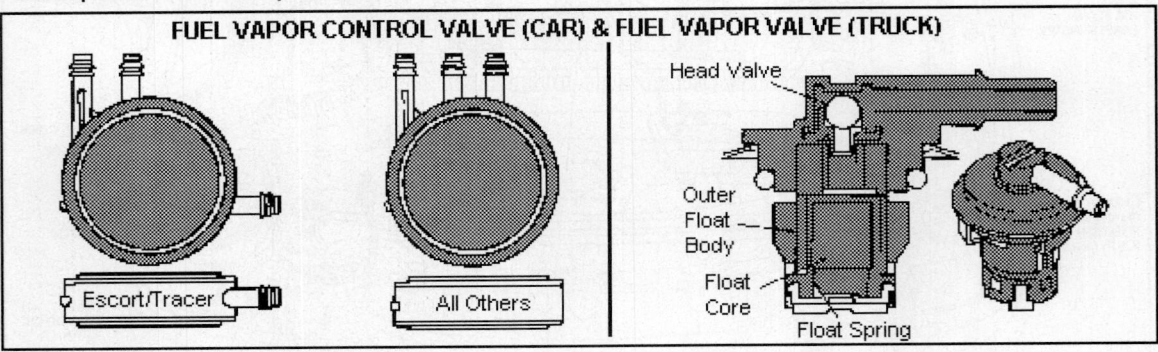

CANISTER VENT SOLENOID

The canister vent (CV) solenoid is a normally open (N.O.) solenoid with a resistance of 40-50 ohms at 68°F. The CV solenoid is used to seal the atmosphere port of the charcoal canister from atmospheric pressure.

During the OBD II Diagnostics, the PCM activates the CV solenoid, in conjunction with the purge solenoid, in order to draw fuel tank vacuum down to a specified level for the leak test.

Trouble Code Help

If the PCM detects a problem in the EVAP Vent Control system, it will set a pending DTC P0446. If the fault is detected on the next trip, the PCM will turn on the MIL (2-trip fault detection)

The repair steps for this trouble code include a check of the FTP sensor (VREF) circuit, CV solenoid and circuits.

Canister Vent Solenoid Location

The CV solenoid is located at the rear of the vehicle near the EVAP canister.

FUEL TANK FILLER CAP

The fuel filler cap is used as an entry point for refueling, to close the EVAP and Fuel systems to atmosphere and to prevent fuel spillage during operation.

The filler cap relieves pressure above 56.21" H2O (14 kPa) and relieves vacuum below 15.26" H2O (3.8 kPa). The fuel cap also serves as a connection point for the EVAP Leak Test kit.

FUEL TANK PRESSURE SENSOR

On this vehicle application, a fuel tank pressure (FTP) sensor is used to monitor the pressure or vacuum value in the fuel tank during the Running Loss Monitor Test portion of the OBD II Diagnostics (the EVAP Purge Test).

The EVAP Monitor performs a Fuel System integrity check to determine if a leak is present in the system. The FTP sensor converts variations in pressure into an analog voltage signal sent to the PCM

Canister Vent Solenoid

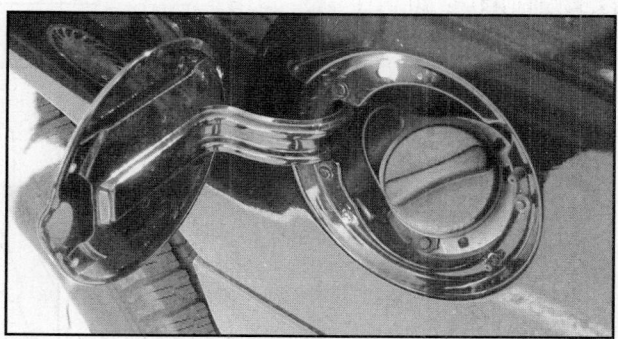

Fuel Tank Pressure Sensor

FUEL VAPOR CONTROL VALVE

The Fuel Vapor Control valve (or liquid separator valve) is located in series between the fuel tank and the fuel vapor storage canister.

Component Location

The vapor control valve is mounted on top of the fuel tank as shown in the Graphic to the right.

This valve closes the path of flow from the fuel vapor valve to the fuel vapor storage canister during refueling to prevent overfilling the fuel tank.

The fuel tank vapor valve detects fuel filler cap removal by sensing a change in pressure (through the fuel filler pipe sensing tube) that occurs during the removal of the fuel filler cap.

In effect, the valve closes when the cap is removed and opens when the fuel filler tank cap is installed.

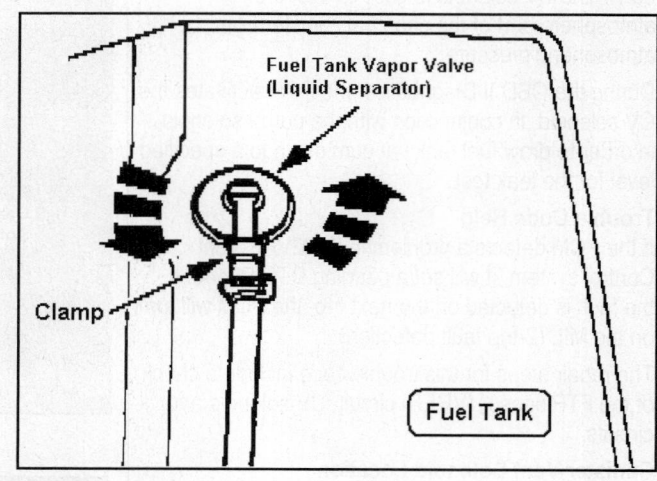

FUEL VAPOR CONTROL VALVE

The EVAP Emission Leak Tester is used to perform various leak check of the complete EVAP system.

It is an integral part of any repair verification procedure (in effect, it is used to verify that there are no leaks in the system during diagnosis and after a repair is complete.

EVAP System Leak Tester

An EVAP System Leak Tester is available from Ford Motor Co. as well as from various Aftermarket sources. Refer to the example in the Graphic.

The EVAP Emission Test Port is located on the EVAP purge outlet tube near the valve.

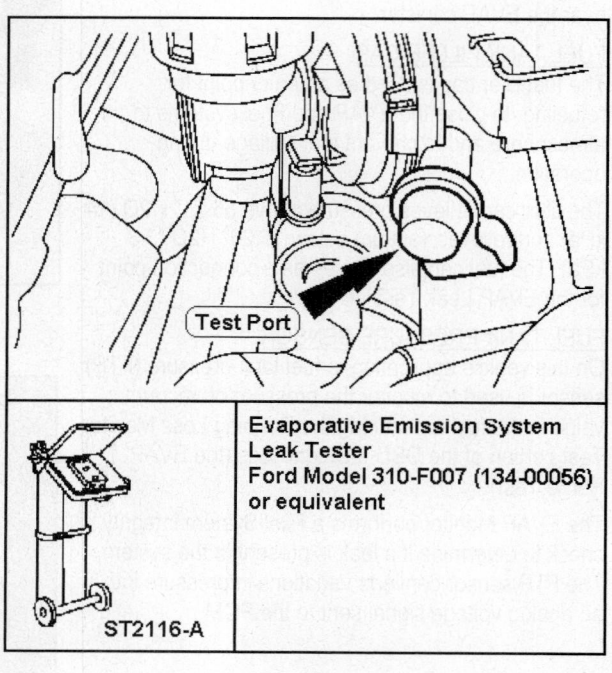

ORVR SYSTEM

The On-Board Refueling Vapor Recovery (ORVR) system is an on-board vehicle system designed to recover fuel vapors during the vehicle refueling operation. The main components of this system are shown in the Graphic below.

ORVR System Graphic

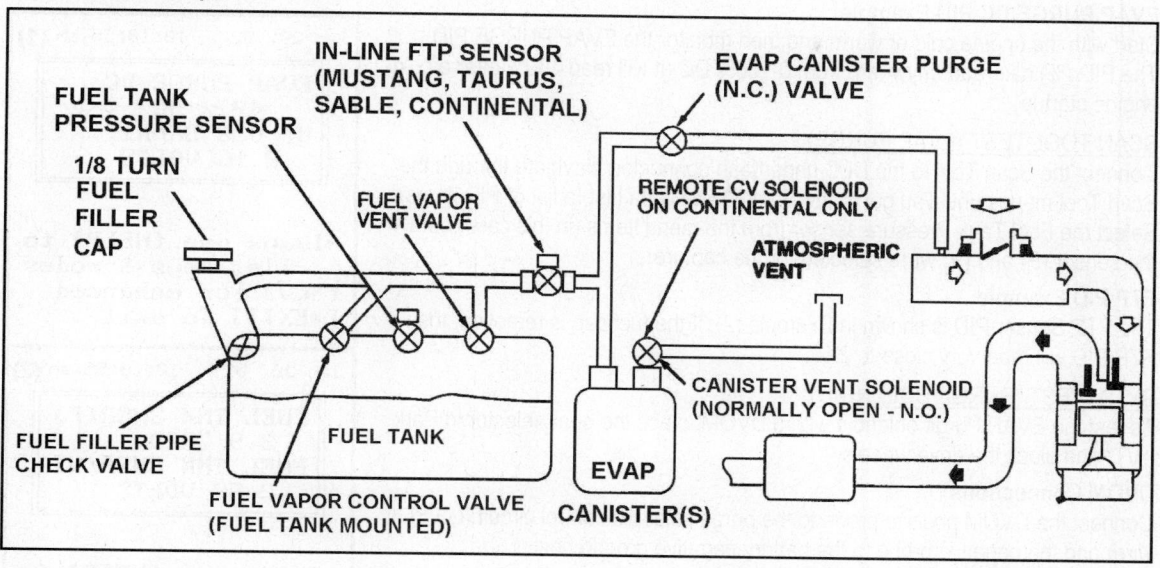

System Operation

The operation of the ORVR system when fuel is dispensed is described next:

- The fuel filler pipe forms a seal to prevent vapors from escaping the fuel tank, while liquid is entering the fuel tank (liquid in the 1" diameter tube blocks vapors from rushing back up the fuel filler pipe).
- A fuel vapor control valve controls the flow of vapors out of the fuel tank (the valve closes when the liquid level reaches a height associated with the fuel tank usable capacity). This valve accomplishes the following:
- Limits the total amount of fuel that can be dispensed into the fuel tank.
- Prevents liquid gasoline from exiting the fuel tank when submerged (and also when tipped well beyond a horizontal plane as part of the vehicle rollover protection in a road accident).
- Minimizes vapor flow resistance during anticipated refueling conditions.
- Fuel vapor tubing connects the fuel vapor control valve to the EVAP canister. This routes the fuel tank vapors (that are displaced by the incoming fuel) to the canister.
- A check valve in the bottom of the pipe prevents any liquid from rushing back up the fuel filler pipe during liquid flow variations associated with the filler nozzle shut-off.
- Between refueling events, the canister is purged with fresh air so that it may be used again to store vapors accumulated during engine soak periods or subsequent refueling events. The vapors drawn from the canister are consumed in the engine.

Fuel Filler Check Valve

The ORVR system includes a check valve in the bottom of the fuel filler pipe (located internal to the pipe where the tube joins the tank). The valve prevents liquid from rushing back up the fuel filler pipe during liquid flow variations related to the filler nozzle shut-off.

SCAN TOOL TEST (EVAP PURGE)

Connect the Scan Tool to the DLC underdash connector. Navigate through the Scan Tool menus until you get to the Ford (OEM) Data List (a list of PID items). Select the EVAP PURGE DC from the menu items along with the voltage signal from the Upstream HO2S for Cylinder Bank 1.

EVAP PURGE DC PID Example

Start with the engine cold or warm and then monitor the EVAP PURGE PID. The PID (%) can read anywhere from 0-100% DC (it will read 100% after a cold engine startup).

SCAN TOOL TEST (EVAP PURGE)

Connect the Scan Tool to the DLC underdash connector. Navigate through the Scan Tool menus until you get to the Ford (OEM) Data List (a list of PID items). Select the Fuel Tank Pressure sensor from the menu items (in this case, both the sensor (P) and (V) were selected for the capture).

FTP PID Example

The FTP Sensor PID is shown in Example (2). If the fuel cap is removed, the FTP PID will read very close to 2.5v.

DVOM TEST (PURGE SOLENOID)

To test the EVAP Purge solenoid with a DVOM, place the gear selector in Park (A/T) and block the drive wheels.

DVOM Connections

Connect the DVOM positive probe to the purge solenoid control circuit (LG/BLK wire) and the negative probe to the battery negative ground post.

Start the engine and change the engine speed. Monitor the solenoid control signal on the DVOM for signs of a steady hertz rate and proper voltage level. A sudden change in engine speed should cause a corresponding change in the purge solenoid signal.

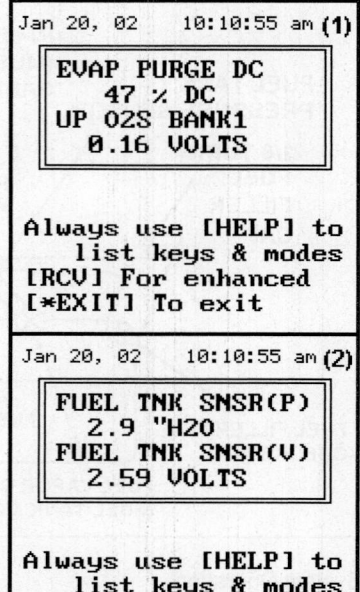

Purge Solenoid Test Schematic

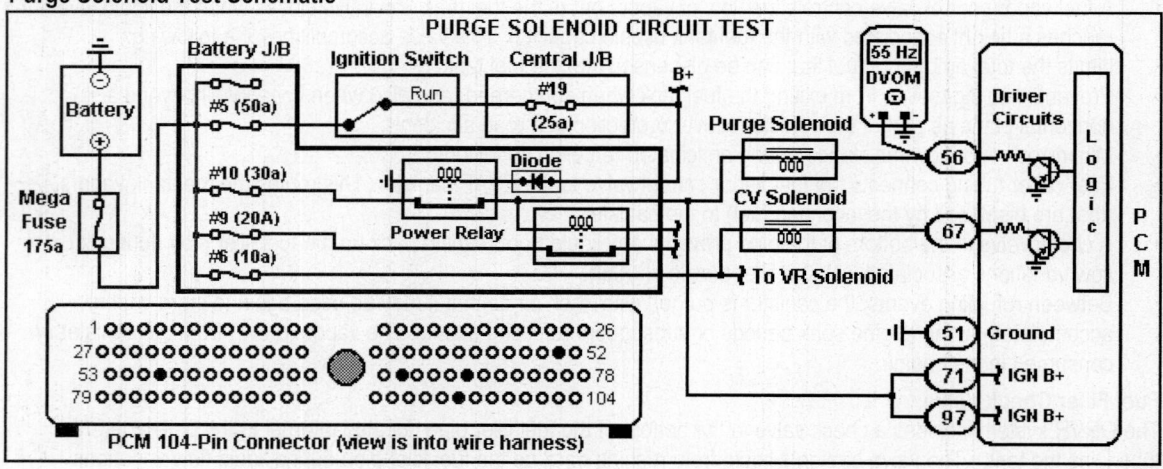

LAB SCOPE TEST (CV SOLENOID)

The Lab Scope is the tool of choice to test the CV solenoid as it provides an accurate view of the solenoid response and frequency rate. Place the gearshift selector in Park and block the drive wheels for safety.

Scope Connections (1)

Connect the Channel 'A' positive probe to the CV control circuit (PPL/WHT wire) at Pin 67 and the negative probe to the battery ground.

Scope Settings

To make the waveforms as clear as possible, set the scope settings to match the examples.

Lab Scope Example (1)

In example (1), the trace shows the CV solenoid with the solenoid off (not energized). This waveform was captured at idle speed. The CV PID will read 0% duty cycle under these conditions (with the CV solenoid off).

Lab Scope Example (2)

In example (2), the trace shows the CV solenoid with the solenoid energized. This waveform was also captured at idle speed. The CV PID will read 100% duty cycle under these conditions (with the solenoid enabled).

Key Point

One easy way to energize the CV solenoid is to use the EVAP Test function in the Scan Tool. Depending upon the Scan Tool manufacturer, this test can be found in the OBD II Generic or OEM selection. This example shows a known good solenoid driver.

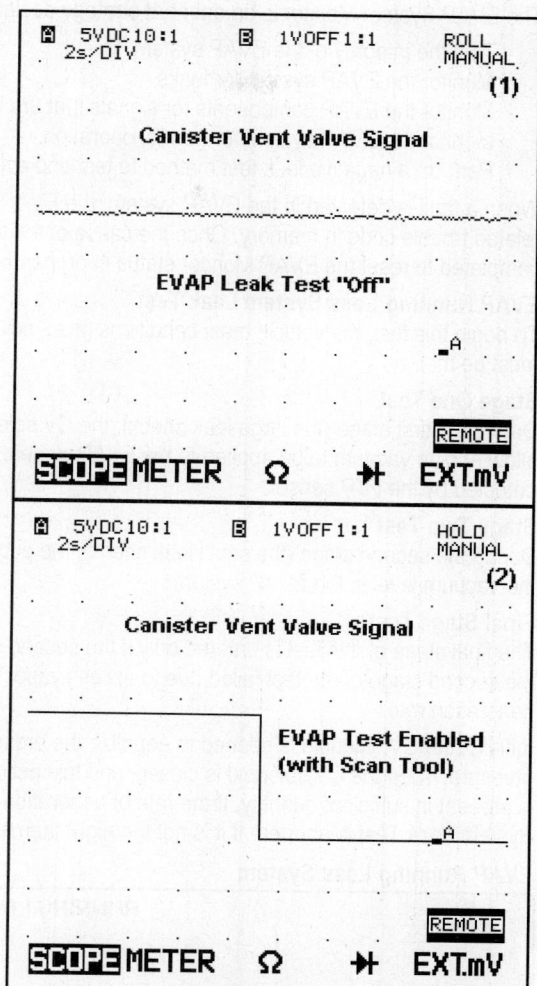

CV Solenoid Test Schematic

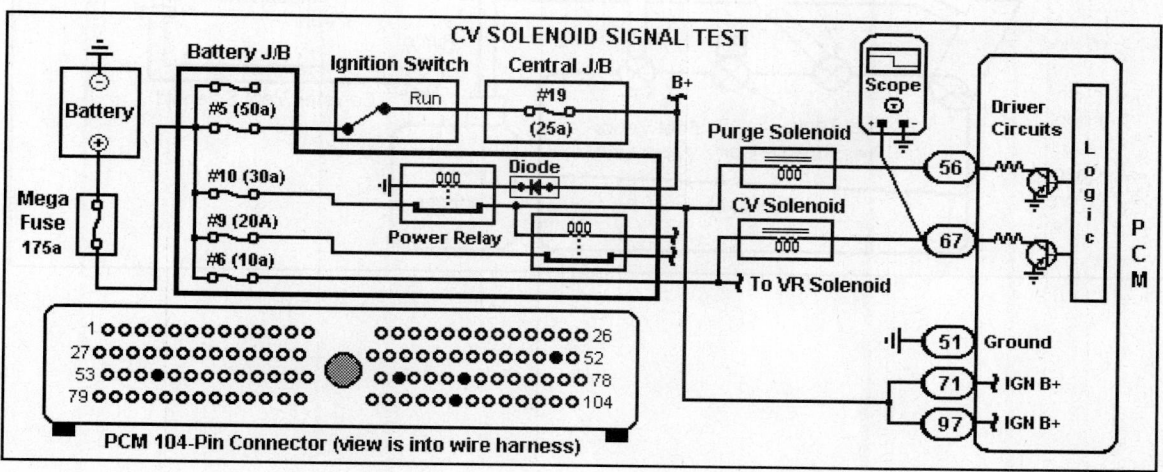

EVAP SYSTEM MONITOR

The EVAP System Monitor is an onboard strategy designed to perform these functions:

- Test the integrity of the EVAP system
- Monitor the EVAP system for leaks
- Monitor the EVAP components for signals that are irrationally high or low
- Monitor the EVAP system for normal operation
- Perform a negative leak test method to test and activate the EVAP system

When a fault is detected in the EVAP system, the EVAP Monitor status in the PCM is reset to NO and the PCM sets a related trouble code in memory. Once the cause of the trouble code is repaired, the EVAP Monitor Drive Cycle must be completed to reset the EVAP Monitor status in preparation for any Inspection and Maintenance testing.

EVAP Running Loss System Leak Test

To begin this test, the vehicle must conditions must include stable EVAP purging and the vehicle speed requirement must be met.

Stage One Test

During the first stage (the large leak check), the CV solenoid is closed, while the purge or purge valve remains open to allow engine vacuum to be applied to the EVAP system. This action allows the vacuum to build up to a level that is detected by the FTP sensor.

Stage Two Test

During the second stage (the small leak check), the purge valve closes, and the PCM looks for a minimal decay rate in the vacuum level in the EVAP system.

Final Stage Test

The final stage of this test is entered only if the second stage of the leak test fails. The last stage test check whether the second stage of the test failed due to excess vapor generation in the fuel tank. This test monitors the fuel vapor generation rate.

Initially, the CV solenoid is opened to equalize the pressure in the EVAP system to be the same as the atmospheric pressure. Then the CV solenoid is closed, and this action allows the pressure in the system to build if vapor generation is present in sufficient quantity. If the rate of generation is found to be too high, the EVAP Running Loss Monitor System Leak Test is aborted. If it is not too high, then a small leak is diagnosed.

EVAP Running Loss System

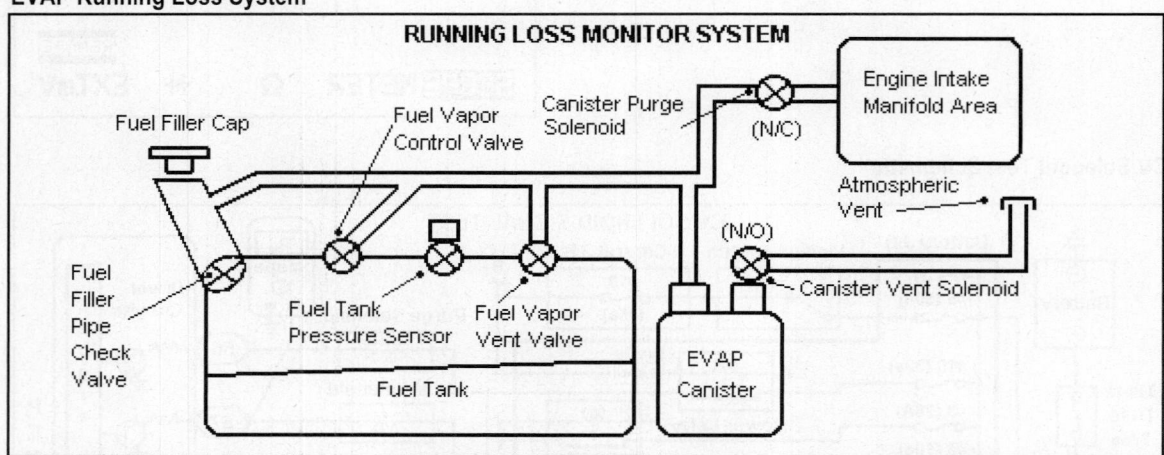

CHILTON DIAGNOSTIC SYSTEM - DTC P0442

The purpose of the Chilton Diagnostic System is to provide you with one or more tests that can be done with a DVOM, Lab Scope or Scan Tool *prior* to entering the complete trouble code repair procedure. The quick checks listed in the DVOM, Lab Scope or Scan Tool tests on the previous pages may help you quickly find the cause of the problem.

If you cannot resolve the problem with these tests, a code repair chart is also included.

Code Description

DTC P0442 - The EVAP Purge Monitor detected a leak as small as 0.040 inch in the system. The MIL is activated if this fault is detected on two consecutive trips (2T DTC).

Code Conditions (Failure)

This code is set if, following a cold engine startup, the vehicle is driven as described in the EVAP Monitor Drive Cycle Explanation, and the PCM detected the following conditions: The EVAP system bleed-up was less than 0.625 kPa (2.5 inches H2O) during a 15 second period with the fuel tank at least 75% full or the EVAP system vapor generation was more than 0.625 kPa (2.5 inches H2O) during a 120 second period.

Quick Check Items

Inspect or physically check the following items as described below:

* A loose fill cap (tighten the fuel cap 1/8 of a turn until it clicks)
* Small vapor leaks at the plastic vapor line connection to purge solenoid
* Small vapor hose leaks at connections to the purge solenoid, FTP sensor or canister
* Small cuts or holes in vacuum lines or vapor hoses in the EVAP system

Drive Cycle Explanation

Start the vehicle (cold) and allow it to idle for at least 15 seconds. Following the cold engine startup (ECT sensor input less than 100°F and the IAT sensor input more than 40°F), drive the vehicle at a speed of at least 40 mph for 60 seconds until the ECT sensor input reaches at least 170°F.

Then with the EVAP DC PID more than 75%, the FLI PID from 15-85% (1998-2002) and the TP MODE PID at PT, drive at steady cruise speed (at over 40 mph) for 10 minutes.

EVAP "Cold Soak" Bypass Procedure

Install the Scan Tool. Turn the key on with the engine off. Cycle the key off, then on. Select the correct vehicle and engine qualifier. Clear the codes and do a PCM reset.

■ **NOTE: *This step bypasses the engine soak timer and resets the OBD II monitor status.***

Use the Scan Tool menu to view and then select the following PIDs: ECT, EVAPDC, FLI (if available) and TP MODE. Start the engine without turning the key off. Refer to the article titled "How to Access and Use OEM PID Information" at the end of this section.

2000 Explorer 4.0L V6 SFI VIN E (A/T) DTC P0442 Parameters

PCM PID Acronym	Parameter Identification	PID Range	PID Value at Hot Idle	PID Value at 30 mph	PID Value at 55 mph
ECT (°F)	ECT Sensor (°F)	-40-304°F	160-200°F	160-200°F	160-200°F
FLI	Fuel Level Indicator (1998-02)	0-100%	50% (1.7v)	25-75%	25-75%
TP MODE	TP Sensor MODE	CT or PT	CT	PT	PT
EVAP CV	EVAP Vent Valve	0-100%	0%	0%	0%
EVAP DC	EVAP Purge Valve	0-100%	0-100%	0-100%	0-100%

DTC P0442 - This trouble code indicates that the EVAP Purge Monitor detected a leak as small as 0.040 inch in the system during the EVAP leak test part of the OBD II test.

Pinpoint Test HX

Pinpoint Test HX is used to diagnose the following devices and/or circuits:

- Canister Vent Solenoid
- Carbon Canister, Fuel Vapor and Vacuum Hoses
- Fuel Cap, Fuel Filler Check Valve, Fuel Tank
- Fuel Tank Pressure (FTP) Sensor
- Fuel Vapor Control Valve, Fuel Vapor Valve
- Power Control Module (PCM)
- Canister Purge Solenoid Valve
- These PCM related circuits: EVAPCP, EVAPCV, SIG RTN, VPWR and VREF

Possible Causes for this Trouble Code

- A loose fill cap (tighten the fuel cap 1/8 of a turn until it clicks)
- Small vapor leaks at the plastic vapor line connection to the Purge solenoid
- Small vapor hose leaks at connections to the Purge solenoid, FTP sensor or canister
- Small cuts or holes in the vacuum lines or vapor hoses in the EVAP system

EVAP System Schematic

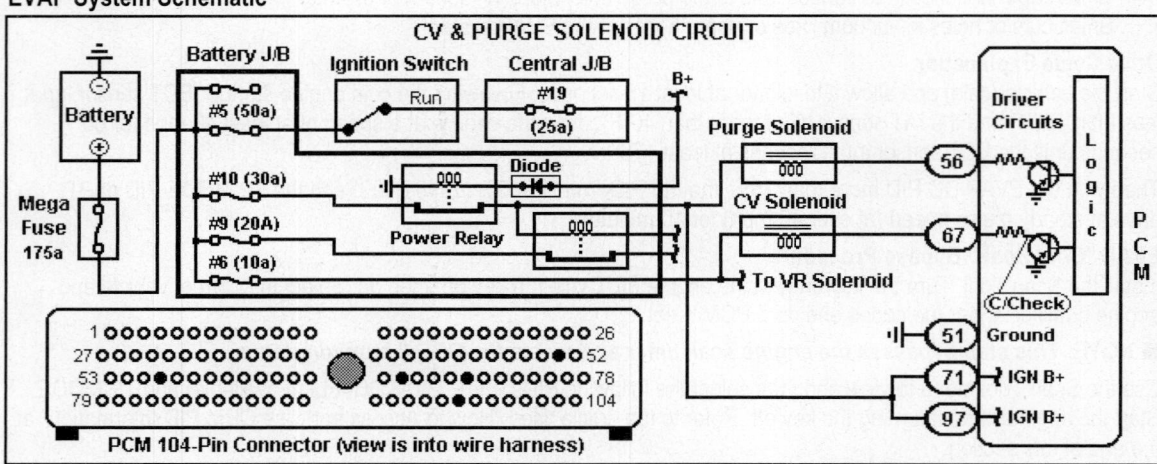

Technical Service Bulletins

There are two (2) technical service bullitens that relate to problems in the EVAP system.

1) TSB # 99-23-4 dated 11-15-99. This TSB deals with trouble codes P0442 and P0455 that set without any noticeable driveability concerns. The fix requires extensive testing.

2) TSB # 1-7-8 dated 4-16-01. This TSB deals with with trouble codes P0442, P0455, P0456, P0457, P1442, P1443, P1450 or P1455. The fix requires the use of the Vacutec #522 Leak Deteciton Smoke Machine following very specific test and repair procedures.

Pinpoint Test 'HF' Repair Table

Step	Action	Yes	No
HF1	**Step description:** Check for a loose fuel cap • Check for the presence of a fuel cap. Do not check for correct installation at this time. • Verify the CV solenoid is properly seated. • Check for cut or loose connections to the fuel vapor hoses, tubes and connections at the canister, purge and fuel vapor vent valves. • Check the fuel filler pipe for damage. • Was a problem detected in this test step?	Make repairs as needed and install any new components. Go to Step HF2.	Go to Step HF2.
HF2	**Step description:** Perform EVAP Leak Test • Disconnect and plug (cap) the evaporative emission return tube at the intake manifold. • Perform the EVAP system leak test. • Did the EVAP system pass the Leak Test?	Go to Step HF3.	Go to Step HF4.
HF3	**Step description:** Inspect Fuel Filler Cap • Visually check the fuel filler cap for damage. • Was there any damage indicated to the cap?	Replace the fuel filler cap. Go to Step HF4.	Go to Step HF4.
HF4	**Step description:** Check for leak in fuel cap • Install EVAP leak tester at the fuel filler pipe. • Close the CV solenoid with the Scan Tool. • Pressure system to 14" H2O (3.48 kPa). • Use the ultrasonic leak detector to check for leaks at the fuel filler cap and EVAP test port. • Was a leak detected (e.g., was there an audible change around the fuel filler cap or near the EVAP test port)?	Make repairs or install new components as needed to make repair. Go to Step HF5.	Install the fuel filler cap and perform the EVAP leak test. If the system passes this leak test, perform an EVAP Repair Verification Drive Cycle.
HF5	**Step description:** Check for other leaks • Did the system pass the EVAP leak test carried out in Step HF2?	Perform an EVAP leak test. If it passes do the Verification drive cycle.	Install the fuel filler cap. Go to Step HF6.
HF6	**Step description:** Check for leaks (Fill mode) • Connect the EVAP System Leak Tester at the EVAP leak test port. • Close the CV solenoid with the Scan Tool. • Turn the selector on the tester to Fill position. • Pressurize the system to 14" H2O (3.48 kPa) • Does the system hold pressure (between 13.80-14.20" H2O (3.43-3.53 kPa)?	Go to Step HF7.	Stop pressurizing the EVAP system with the tester. Go to Step HF8.
HF7	**Step description:** Check for leaks (Complete) • Connect the EVAP System Leak Tester at the EVAP leak test port. • Close the CV solenoid with the Scan Tool. • Pressurize the system to 14" H2O (3.48 kPa) • Use the leak detector to check for leaks at the return tube to purge valve, purge valve to canister and canister vent assembly, EVAP canister to canister vent assembly to the fuel tank, and fuel filler cap to the fuel filler tube. • Was a leak detected at any of these areas?	Make repairs or install new components as needed to make repair. If the system passes this leak test, perform an EVAP Repair Verification Drive Cycle.	Stop pressurizing the EVAP system with the tester. Go to Step HF8.

Step	Action	Yes	No
HF8	**Step description:** Check for leak (return tube) • Disconnect the fuel vapor tube at the fuel vapor tee. Plug the opening in the line. • Connect the EVAP System Leak Tester at the EVAP leak test port. • Close the CV solenoid with the Scan Tool. • Pressurize the EVAP system to 14" H2O. • Access the ultra-sonic detector from the kit. • Use the ultrasonic leak detector to check for leaks from the intake manifold to CV solenoid. • Was a leak detected at any of these areas?	Make repairs or install new components as needed to make repair. Then repeat the steps in Step A6 to verify the repair is complete. Go to Step HF9.	Open the CV solenoid with the Scan Tool. Go to Step HF9.
HF9	**Step description:** *Check for leak (fuel tank)* • Connect the EVAP System Leak Tester at the fuel filler pipe and transfer the plug from the fuel vapor tee to the fuel tank vapor tube. • Turn the leak tester selector to Fill position. • Pressurize the EVAP system to 14" H2O. • Use the leak detector to check for leaks from the fuel tank vapor tube to the tank, the FTP sensor, fuel tank vapor tube and the filler pipe. • Was a leak detected at any of these areas?	Make repairs or install new components as needed to make repair. Go to Step HF10.	Go to Step HF10.
HF10	**Step description:** Check for leak (fill pipe) • Reconnect the fuel tank vapor tube to the fuel vapor tee. • Complete the EVAP System leak test. • Did the EVAP System pass the leak test?	Restore the system to normal. Perform the EVAP system leak test. If it passes the test, perform an EVAP Repair Verification Drive Cycle.	Go to Step HF6.

ORVR System Graphic

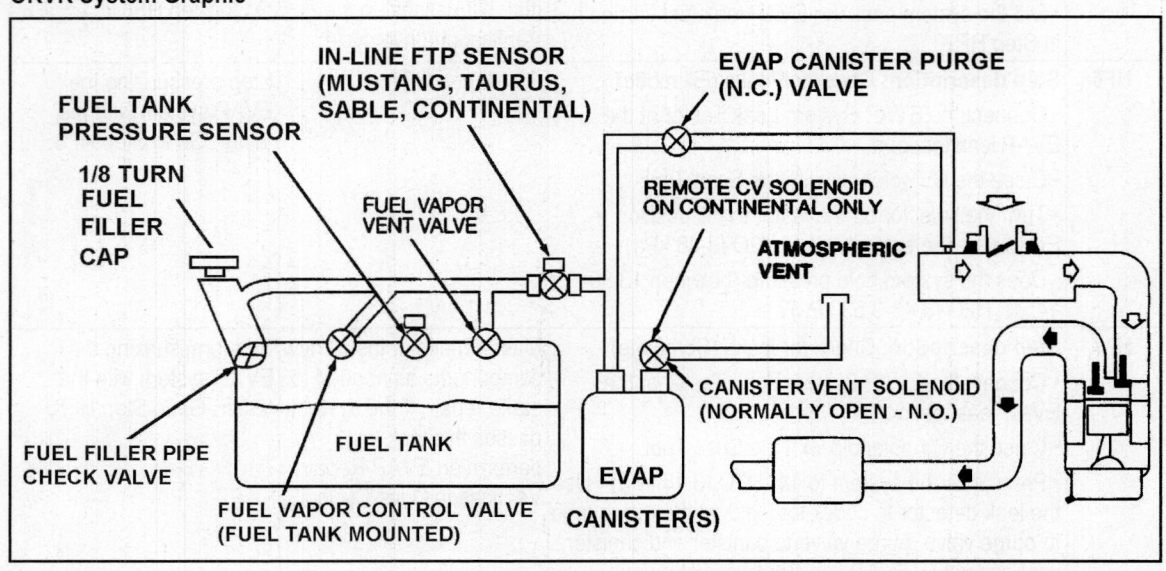

Exhaust Gas Recirculation System

GENERAL INFORMATION

The EGR system is designed to control NOx emissions by allowing small amounts of exhaust gases to be recirculated back into the combustion chamber to mix with the A/F mixture. This action causes a reduction in the amount of combustion, and less NOx. This vehicle is equipped with a differential pressure feedback EGR (DPFE) sensor.

Differential Pressure Feedback EGR System

The DPFE system includes the following components:

- DPFE sensor
- EGR valve and orifice tube assembly
- Vacuum regulator valve assembly
- PCM and related vacuum hoses.

The DPFE system operates as described next:

1) The PCM receives signals from the CKP, ECT, IAT, MAF and TP sensor in order to determine the optimum operating conditions for the DPFE EGR system. The engine must be warm, running at a stable engine speed under moderate load before the EGR system can be activated. The PCM deactivates the EGR system at idle speed, at extended wide-open throttle period or if a fault is detected in the EGR system.

2) The PCM calculates a desired amount of EGR flow for each engine condition. Then it determines the desired pressure drop across the metering orifice required to achieve the correct amount of flow and outputs the signal to the EGR VR solenoid.

3) The VR solenoid receives a variable duty cycle signal (0-100%). The higher the duty cycle rate, the more vacuum the solenoid diverts to the EGR valve.

4) The increase in vacuum acting upon the EGR valve diaphragm overcomes the valve spring pressure and lifts the valve pintle off its seat. This allows exhaust gas to flow.

5) The exhaust gas that flows through the EGR valve must first pass through the EGR metering orifice. One side of the orifice us exposed to exhaust backpressure and the other side is exposed to the intake manifold. This design creates a pressure drop across the orifice whenever there is EGR flow. When the EGR valve closes, there is no longer flow across the metering orifice and the pressure is equal on both sides.

6) The DPFE sensor measures the "actual" pressure drop across the metering orifice and relays a proportional voltage signal (from 0-5 volts) to the PCM. The PCM uses this feedback signal to correct for any errors in achieving the "desired" EGR flow.

EGR System Circuits Schematic

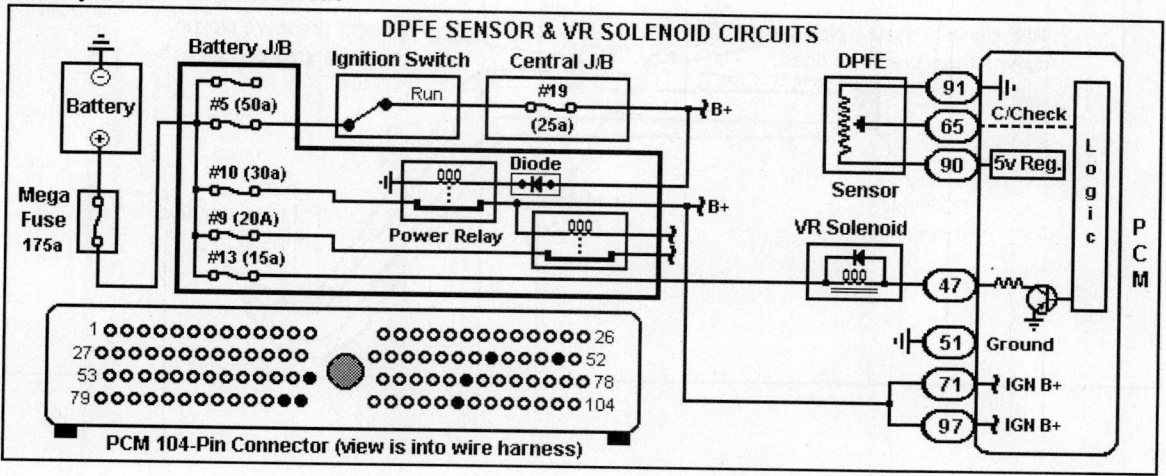

DPFE SENSOR INFORMATION

The DPFE sensor is a ceramic, capacitive-type pressure transducer that monitors the differential pressure across the metering orifice located in the orifice tube assembly. The sensor receives vacuum signals through two hoses referred to as the downstream pressure (REF signal) and upstream pressure hose (HI signal). The HI signal uses the larger diameter hose. The DPFE sensor signal is proportional to a pressure drop across the metering orifice. This input is referred to as the EGR flow rate feedback signal.

EGR System Components Graphic

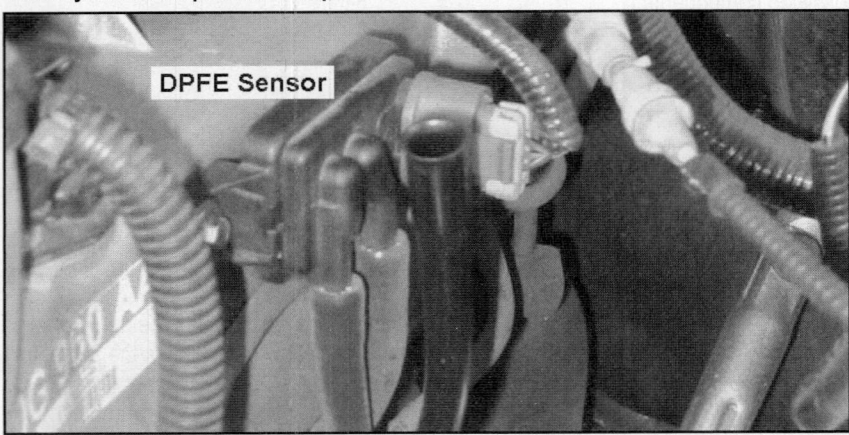

Orifice Tube Assembly

The orifice tube assembly is a section of tubing connecting the Exhaust system to the intake manifold. This assembly provides a flow path for EGR gases to the intake manifold. It also contains the metering orifice and two pressure pickup tubes.

The internal metering orifice creates a measurable pressure drop across it as the EGR valve open and closes. This pressure differential across the orifice is picked up by the DPFE sensor and converted to feedback data on EGR system operation to the PCM.

Orifice Tube Assembly Graphic

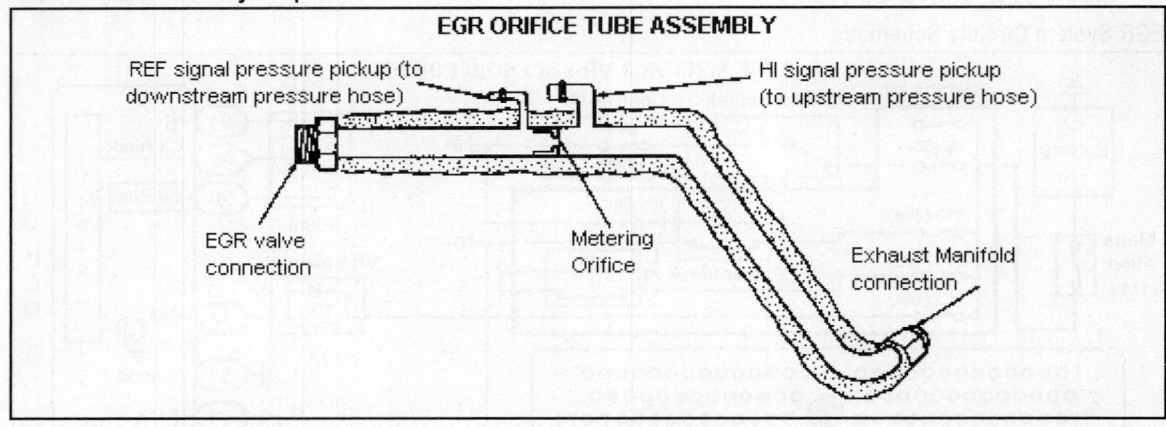

DPFE SENSOR TABLES

The tables below contain examples of "known good" specifications for use during a DPFE sensor static test.

Null 0.55v DPFE Sensor (Part #F48E-9J460-BB)

Differential Pressure			
In - H20	In - Hg	kPa	Volts
120	8.83	29.81	4.66v
90	6.62	22.36	3.64v
60	4.41	14.90	2.61v
30	2.21	7.46	1.58v
0	0	0	0.55v

Null 1.00v DPFE Sensor (Part #F7UE-9J460-AB)

Differential Pressure Volts			
In - H20	In - Hg	kPa	Volts
116	8.56	28.9	4.95v
58	4.3	14.4	2.97v
0	0	0	1.0v

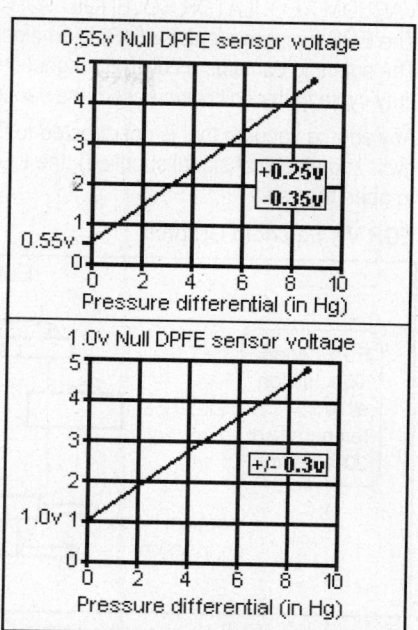

SCAN TOOL TEST (DPFE SENSOR)

Connect the Scan Tool to the DLC underdash connector. Select the DPFE and EGRVR PID from the Scan Tool menu. Record the values at Hot Idle, 30 and 55 mph. An example of the DPFE sensor and VR PID readings in Park (700 rpm) appear in the Graphic to the right.

DVOM TEST (DPFE SENSOR)

Locate the DPFE sensor circuits (underhood, at the BOB or at the PCM). Connect the DVOM positive lead to the DPFE signal (BRN/LG wire) and negative lead to chassis ground. Start the engine and record the desired DPFE readings. Compare the actual readings to the Pin Voltage Tables.

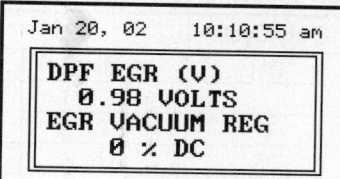

```
Jan 20, 02    10:10:55 am

DPF EGR (V)
    0.98 VOLTS
EGR VACUUM REG
    0 % DC


Always use [HELP] to
  list keys & modes
[RCV] For enhanced
[*EXIT] To exit
```

DPFE Sensor Test Schematic

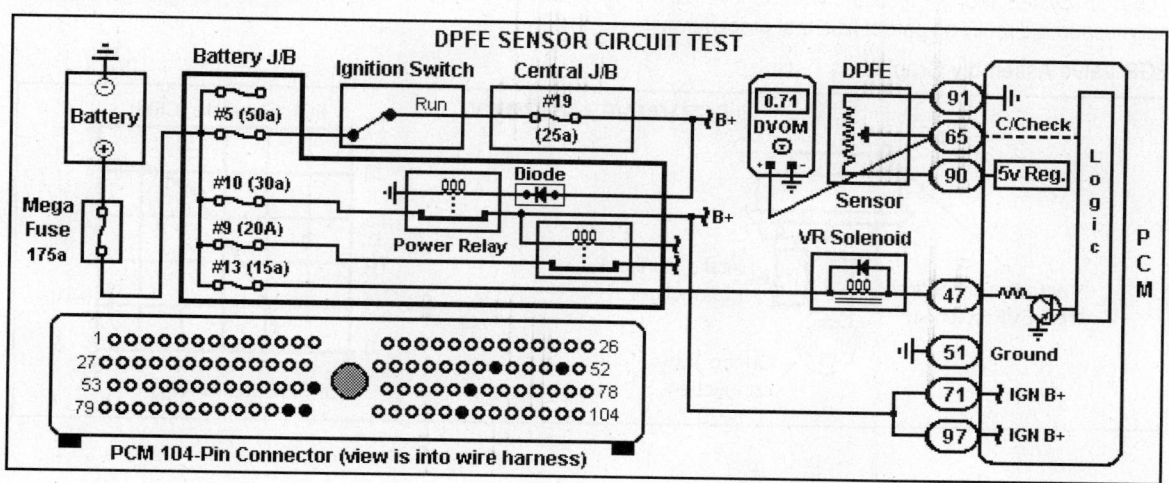

DPFE SENSOR CIRCUIT TEST

VACUUM REGULATOR SOLENOID

The EGR vacuum regulator (VR) is an electromagnetic solenoid used to regulate the vacuum supply to the EGR valve. The solenoid contains a coil that magnetically controls the position of an internal disc to regulate the vacuum. As the duty cycle to the coil is increased, the vacuum signal passed through the solenoid to the EGR valve increases.

Any source vacuum that is not directed to the EGR valve is vented through a vent filter to the atmosphere. At 0% duty cycle (no electrical signal supplied), the EGR vacuum regulator solenoid allows some vacuum to pass, but not enough to open the valve.

EGR VR Solenoid Graphic

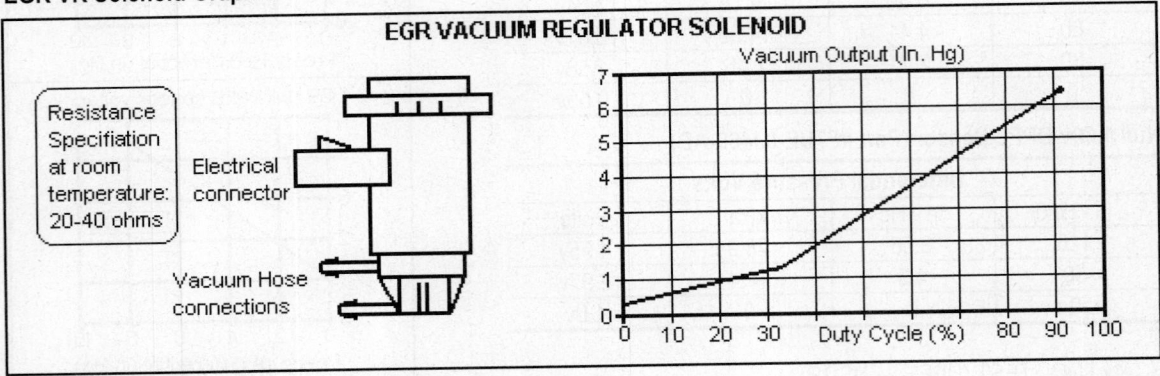

EGR Vacuum Regulator Solenoid Table

Duty Cycle	Vacuum Output					
(%)	Min		Nominal		Maximum	
	In-Hg	KPa	In-Hg	KPa	In-Hg	KPa
0	0	0	0.38	1.28	0.75	2.53
33	0.55	1.86	1.3	4.39	2.05	6.90
90	5.69	19.2	6.32	21.3	6.95	23.47

EGR Valve Assembly

The EGR valve used with the DPFE EGR system is a conventional vacuum actuated EGR valve. The valve increases or decreases the flow of exhaust gas recirculation. As vacuum applied to the valve diaphragm overcomes the spring force, the valve begins to open. As the vacuum signal weakens, EGR spring force closes the valve at 1.6" Hg or less. The valve is fully open at about 4.5" Hg of vacuum.

The EGR System Monitor tests the valve function and triggers a code if test criteria are not met. EGR valve flow rate is not measured directly as part of field test procedures.

EGR Valve Assembly Graphic

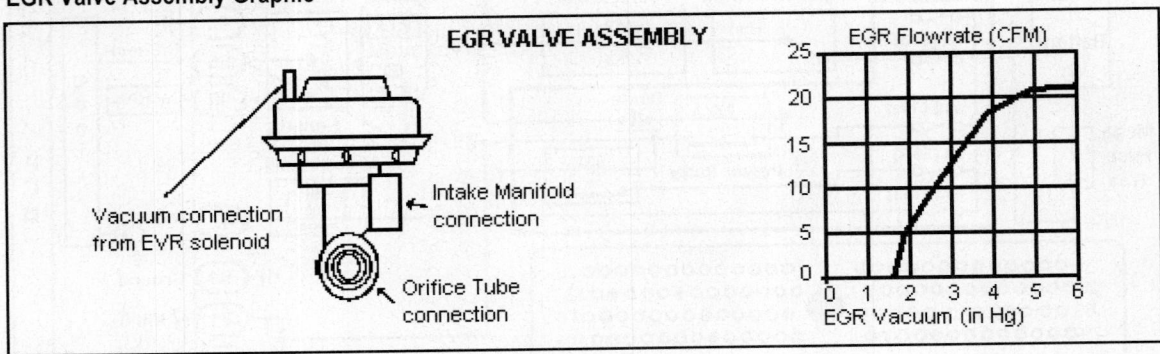

LAB SCOPE TEST (VR SOLENOID)

The Lab Scope is the tool of choice to test the VR solenoid as it provides an accurate view of the solenoid response and frequency rate. Place the gearshift selector in Park and block the drive wheels for safety.

Scope Connections

Connect the Channel 'A' positive probe to the VR control circuit (BRN/PNK wire) at Pin 47 and the negative probe to the battery ground.

Scope Settings

To make the waveforms as clear as possible, set the scope settings to match the examples.

Lab Scope Tests

The waveforms in these captures are from the original solenoid. The engine was allowed to run long enough until it entered closed loop operation prior to testing. Note that the examples were captured at different speeds.

Lab Scope Example (1)

In example (1), the trace shows the VR solenoid command signal at hot idle speed. Note the waveform voltage indicates system voltage. The VR PID will read 0% under these conditions (with the solenoid not activated).

Lab Scope Example (2)

In example (2), the trace shows the VR solenoid command signal at Cruise speed. The Scan Tool VR PID will read 60% with the VR solenoid activated as indicated in this test.

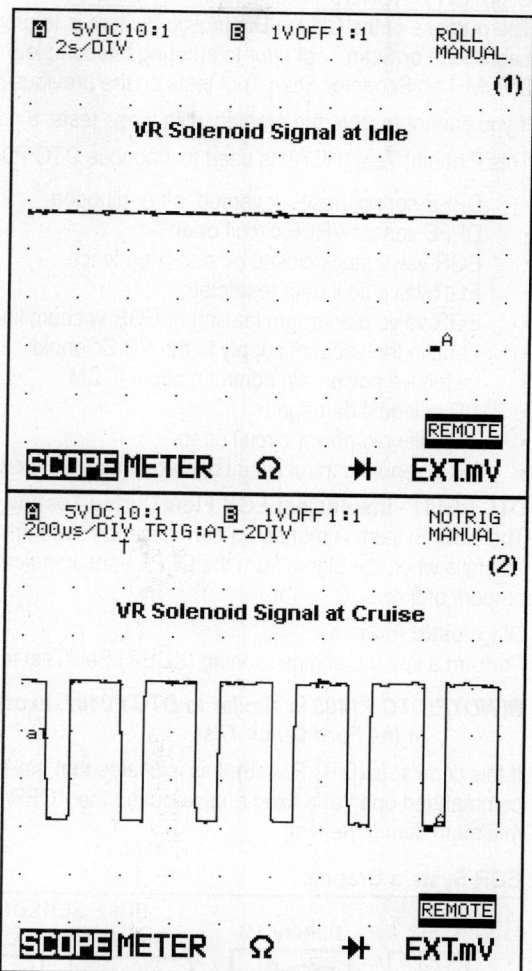

VR Solenoid Test Schematic

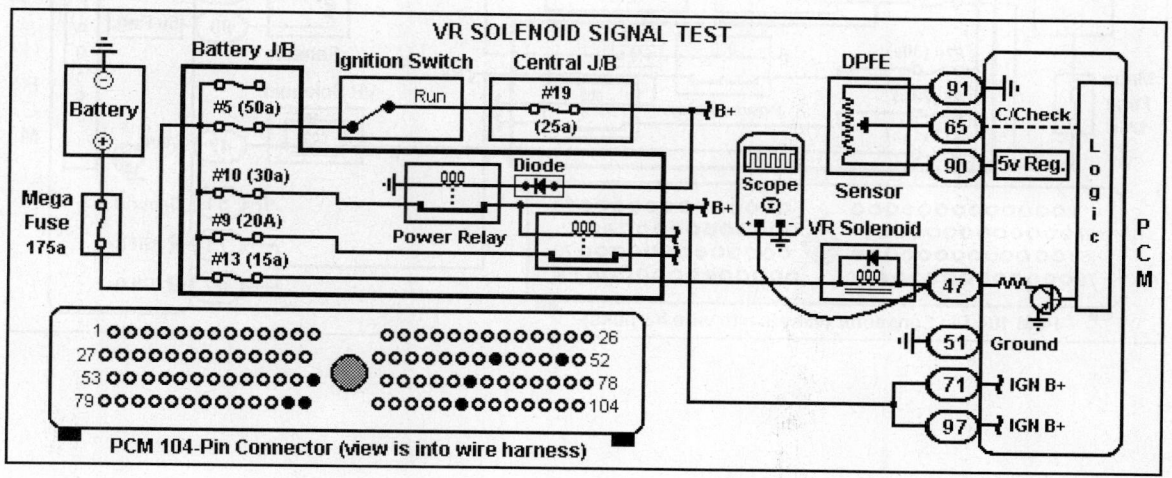

CHILTON DIAGNOSTIC SYSTEM - DTC P0401

The purpose of the Chilton Diagnostic System is to provide you with one or more tests that can be done with a DVOM, Lab Scope or Scan Tool *prior* to entering the complete trouble code repair procedure. The quick checks listed in the DVOM, Lab Scope or Scan Tool tests on the previous pages may help you quickly find the cause of the problem.

If you cannot resolve the problem with these tests, a code repair chart is also included.

This Pinpoint Test (HE70) is used to diagnose DTC P0401 and the following items:

- DPFE sensor hoses reversed, off or plugged
- DPFE sensor VREF circuit open
- EGR valve stuck closed or stuck due to ice
- EGR valve flow path restricted
- EGR valve diaphragm leaking or EGR vacuum line off, plugged or leaking
- Fault in the vacuum supply to the VR Solenoid
- Defective powertrain control module (PCM)
- VR solenoid damaged
- VR solenoid power circuit open
- VR solenoid control circuit open or shorted to power

DTC P0401 - Insufficient EGR Flow During Testing (CMC) Code Conditions

The EGR System is monitored during steady state driving conditions while the EGR valve is commanded open. The test fails when the signal from the DPFE sensor indicates that the EGR flow rate was less than the desired minimum amount of flow.

Diagnostic Aids

Perform a key on, engine running (KOER) Self-Test to determine if DTC P1408 sets.

■ *NOTE:DTC P1408 is similar to DTC P0401, except that DTC P1408 is an engine running test that is included in the Ford Quick Test.*

If this code sets (DTC P1408), this indicates that the PCM detected an incorrect flow rate with the EGR valve commanded open at a fixed engine during the KOER Quick Test (i.e., the PCM detected a flow rate below a calibrated minimum during the test).

EGR System Graphic

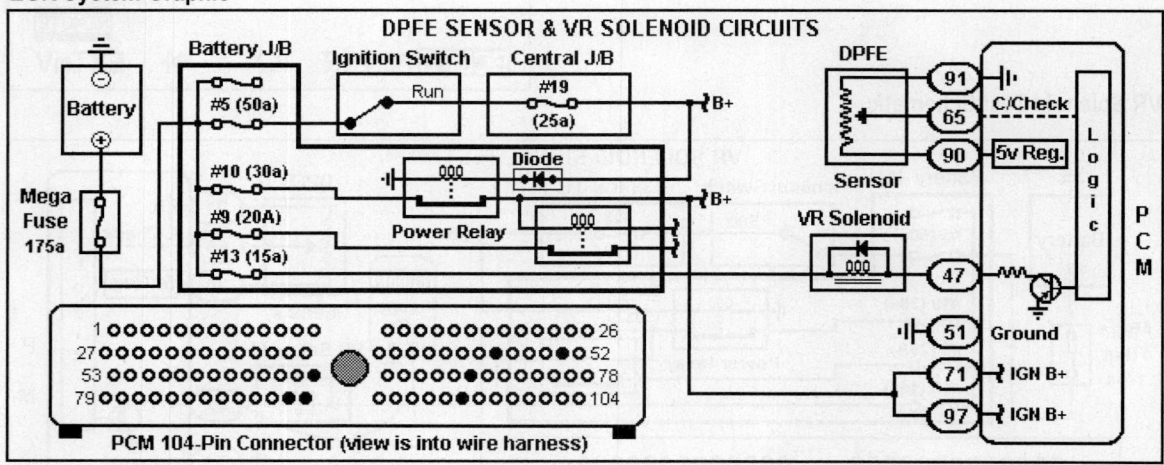

Repair Table for DTC P0401 (Pinpoint Test HE70)

Step	Action	Yes	No
HE70	**Step description:** Run the KOER self-test • Run the KOER Self-Test. • Was KOER DTC P1408 set during the test?	DTC P1408 is a current code. Go to Step HE71.	Go to Step HE91.
HE71	**Step description:** Read Continuous Codes • Read the continuous memory trouble codes. • Was CMC DTC P1406 set during the test?	Go to Step HE60.	Go to HE72.
HE72	**Step description:** KOER Test - test vacuum • Disconnect vacuum hose at the EGR valve and connect a vacuum gauge to the hose. • Ignore any codes that set with the hose off. • Run the KOER self-test while monitoring the vacuum level on the gauge. The PCM will enable the EGR solenoid about 30 seconds into the test. The vacuum level on the gauge should increase to 1.6" Hg at this time. • Did the vacuum increase to 3.0" Hg or more at anytime during the KOER self-test?	Turn the key to off. This result indicates the vacuum level was sufficient to open the EGR valve. The fault is probably not in the EGR vacuum control system. Go to Step HE73.	Turn the key to off. This result indicates the vacuum level was insufficient to open the EGR valve. Go to Step HE80.
HE73	**Step description:** Check DPFE sensor hoses • Check both DPFE sensor hoses to detect if they are reversed at the sensor or orifice tube. • Inspect the hoses for signs of being pinched or dips where water could collect and freeze. • Check the DPFE sensor and orifice tube for blockage or damage at the pickup tubes. • Were any faults detected during this step?	Make repairs to the pressure hoses or devices as needed. Restore the vehicle to its normal operation. Do a PCM Reset to clear any codes and then do a Quick Test.	Go to Step HE74.
HE74	**Step description:** Check DPFE sensor output • Disconnect pressure hoses at DPFE sensor. • Connect hand vacuum pump to downstream connection at the DPFE sensor (at the intake manifold side of the sensor - the smaller tube). • Turn to key on, engine off. • Access the DPFEGR PID on the Scan Tool. • Apply 8-9" Hg of vacuum to sensor and hold it for 3 seconds - then release the vacuum. • Did DPFE PID indicate a fault in this step?	Turn the key off. Note: The DPFE PID should be between 0.2-1.3v at KOEO with no vacuum applied. The reading should increase to 4.0v with vacuum applied and should drop to under 3.0v within 3 seconds after release. Go to Step HE75.	Turn the key off and reconnect the pressure hoses to the sensor. Go to Step HE76.
HE76	**Step description:** Check EGR Valve Function • Disconnect and plug EGR vacuum hose. • Connect a hand vacuum pump to the valve. • Start the engine and allow it to idle. • Access the EGR & RPM PID's on Scan Tool. • Slowly apply 8-10" Hg vacuum to valve and hold that vacuum level for 10 seconds. If the engine starts to stall, increase to 1000 RPM. • Did the PID value indicate as described?	Note: The EGR valve should start to open at 1.6" Hg as indicated by an increase in the DPFE PID value. The voltage will continue to increase to at least 2.5v under full vacuum. It should hold steady with a steady vacuum. If it drops, the valve is leaking. Restore the vehicle back to normal. Go to Step HE85.	Remove and inspect the valve for signs of contamination, carbon deposits, binding or a leaking diaphragm. If the valve is okay, look for a restricted EGR port to intake manifold or a plugged orifice tube assembly. Repair as needed. Restore the vehicle back to normal. Do a PCM Reset and Quick Test.

Repair Table for DTC P0401 (Pinpoint Test HE70)

Step	Action	Yes	No
HE80	**Step description:** Check VR Vacuum Source • Inspect the vacuum lines between vacuum source and VR solenoid, and between the VR solenoid and EGR valve for disconnects, leaks, kinks, blockage, routing or damage. • Disconnect vacuum hoses to VR solenoid and connect a vacuum gauge to the source vacuum hose to the VR solenoid. • Start the engine, allow it to idle and note the reading on the vacuum gauge at idle speed. • Did the gauge read at least 15" Hg vacuum?	Restore the vehicle to normal operation. Go to Step HE81.	Isolate the source vacuum fault and make repairs as needed to correct the low vacuum condition. Restore the vehicle to normal operation. Do a PCM Reset to clear any trouble codes and then do a Quick Test.
HE81	**Step description:** Check VPWR to Solenoid • Disconnect the VR Solenoid connector. • Turn to key on, engine off. • Measure the VPWR voltage to the VR solenoid connector (the RED wire connection). • Disconnect the VR solenoid hoses. Connect a vacuum gauge to the source vacuum hose. • Did the solenoid VPWR read at least 10.5v?	Turn the key to off. Go to Step HE82.	Locate and repair the open circuit condition in the VR solenoid VPWR circuit. Restore the vehicle to normal operation. Do a PCM Reset to clear any trouble codes and then do a Quick Test.
HE82	**Step description:** Test VR Sol. Resistance • With the VR Solenoid connector removed, measure the resistance of the VR solenoid. • Did the resistance read from 20-40 ohms?	Go to Step HE83.	Replace the solenoid. Restore the vehicle and do a PCM Reset. Do a Quick Test.
HE83	**Step description:** Test for Short to Power • Install a breakout box. Measure the voltage between PCM Test Pin 47 and ground. • Did the voltage read from 1 volt?	Repair the short circuit condition in the VR solenoid control circuit. Restore the vehicle and do a PCM Reset.	Go to Step HE84.
HE84	**Step description:** Check for an Open Circuit • Measure the resistance of the EGRVR circuit from Pin 47 and the VR harness connector. • Did the resistance read less than 5 ohms?	Reconnect the VR solenoid connector. Go to Step HE85.	Repair the open circuit condition. Restore the vehicle to normal operation. Do a PCM Reset and then run a Quick Test.
HE85	**Step description:** Test VR Solenoid Function • Reconnect the PCM and the VR solenoid. • Disconnect the vacuum hose at EGR valve and connect a vacuum gauge to the hose. • Start the engine and allow it to idle. • Jumper PCM Pin 47 (EGRVR pin) to chassis ground at the breakout box. • Did vacuum gauge read 4.0" Hg or greater?	Replace the damaged PCM. Restore the vehicle to normal operation. Refer to the PCM replacement instructions in Section 1 (flash requirements). Do a PCM Reset and then run a Quick Test.	Replace the damaged VR solenoid. Restore the vehicle to normal operation. Do a PCM Reset and then run a Quick Test.

■ NOTE: *If the problem is intermittent in nature, go to the next page and use Step HE90.*

Repair Table for DTC P0401 (Intermittent EGR Failure)

Step	Action	Yes	No
HE90	**Step description:** Test for Intermittent Fault • Visually check the EGR system components for signs of an intermittent fault. • Were any problems found in this step?	Make repairs as needed. Restore the vehicle. Do a PCM Reset and Quick Test.	Go to Step HE91.
HE91	**Step description:** EGR functional check • Disconnect the vacuum hose at EGR valve. Connect a hand vacuum pump to the valve. • Connect a Scan Tool and start the engine. • Access the DPEGR and RPM PID values. • Slowly apply 5-10" Hg of vacuum to the valve (hold it for at least 10 seconds). If the engine starts to stall, increase the engine speed to a minimum engine speed of 800 rpm. • Look for these events: The EGR valve starts to open at about 1.6" Hg of vacuum; DPFEGR PID value increases until the valve opens (it should read at least 2.5v at full vacuum); the DPFEGR PID value is steady while vacuum is held (if the vacuum drops off, there is a leak). • Does the DPFEGR PID indicate that the EGR valve operated as described above?	Turn the key to off. Go to Step HE92. 	Turn the key to off. Remove and inspect the EGR valve for signs of contamination, unusual wear, carbon deposits, binding, a leaking diaphragm or other damage. Make repairs as needed. Restore the vehicle to normal operation. Do a PCM Reset and then do a Quick Test.
HE92	**Step description:** Check the Source Vacuum • Install a breakout box and connect PCM. • Disconnect the plugged hose at EGR valve and connect a vacuum gauge to the hose. • Start the engine and allow it to idle. • Jumper test pin 47 (EVR) to chassis ground (the idle vacuum should read at least 4" Hg). • Observe vacuum gauge during these events: Tap lightly on EVR solenoid; wiggle the EVR solenoid harness connector, vacuum lines and wiring harness to the EVR solenoid. A fault is indicated if the vacuum gauge reading drops. • Was a fault indicated during this test step?	Turn the key off. Locate and repair the problem that caused the vacuum reading to drop during testing. Restore the vehicle to normal operation. Do a PCM Reset and then do a Quick Test.	The fault did not appear or was intermittent at this time. Refer to the Intermittent tests in this manual or in other repair information.

Fuel Delivery System

INTRODUCTION

The Fuel system delivers fuel at a controlled pressure and correct volume for efficient combustion. The PCM controls the Fuel system and provides the correct injector timing.

Fuel Delivery

This vehicle is equipped with a sequential fuel injection (SFI) system that delivers the correct A/F mixture to the engine at the precise time under all operating conditions.

Air Induction

Air enters the system through the fresh air duct and flows through the air cleaner where it is monitored by the MAF sensor. The metered air passes through the air duct and enters the throttle body. The incoming air passes through the throttle body and into the intake plenum to the intake manifold where it is mixed with fuel for combustion.

System Operation

A high-pressure (in-tank mounted) fuel pump delivers fuel to the fuel injection supply manifold. The fuel injection supply manifold incorporates electrically actuated fuel injectors mounted directly above each of the intake ports of the engine.

A constant amount of fuel pressure is maintained at the injectors by a fuel pressure regulator. The fuel pressure is a constant while fuel demand is not, so the system includes a fuel return line to allow excess fuel to flow through the regulator to the tank.

The PCM determines the amount of fuel the injectors spray under all engine-operating conditions. The PCM receives electrical signals from engine control sensors that monitor various factors such as airflow pressure, temperature of the air entering the engine, engine coolant temperature, throttle position and vehicle speed. The PCM evaluates the sensor information it receives and then signals the fuel injectors in order to control the fuel injector pulsewidth.

Fuel Delivery System Graphic

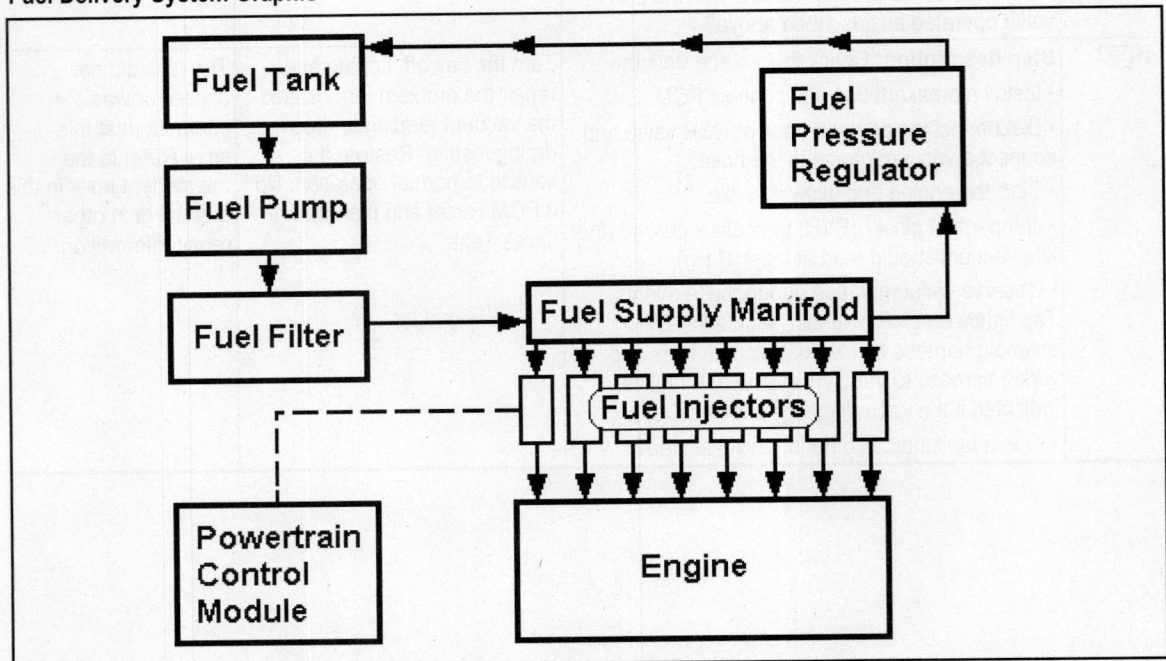

<u>FUEL PUMP RELAY</u>

The PCM controls the fuel pump through a fuel pump relay that is located under the hood in the Power Distribution Box (PDB).

Component Location

The fuel pump relay is located in the Power Distribution Box located at the driver's side underhood area. Refer to the Graphic to the right.

The fuel pump relay, which is controlled by the PCM, provides power to the fuel pump under the following operation conditions:

1) With the ignition off, the PCM and fuel pump relay are both "off".

2) When the ignition is first turned to "run", these events occur:
   a. The power relay is energized.
   b. Power is provided to the fuel pump relay and a timing device in the PCM.
   c. The fuel pump is supplied power through the fuel pump relay.

3) If the ignition switch is not turned to the "start" position, the timing device in the PCM will open the fuel pump control circuit at Pin 80 (LB/ORN wire) after one second, and this action de-energizes the fuel pump relay, which in turn de-energizes the fuel pump. This circuitry also provides for pre-pressurization of the fuel system.

4) When the ignition switch is turned to the "start" position, the PCM operates the fuel pump relay to provide fuel for starting the engine while cranking.

5) After the engine starts and the ignition switch is returned to the "run" position:
   a. Power to the fuel pump is again supplied through the fuel pump relay.
   b. The PCM shuts off the fuel pump by opening the F/P relay control circuit (Pin 80) if the engine stops (no CKP signals), or if engine speed falls below 120 rpm.

Fuel Pump Monitor

The PCM determines if the fuel pump circuit has power through the FPM circuit (Pin 40). This circuit is used to determine the fuel pump circuit status for DTC P0231 and P0232.

Fuel Pump Circuits Schematic

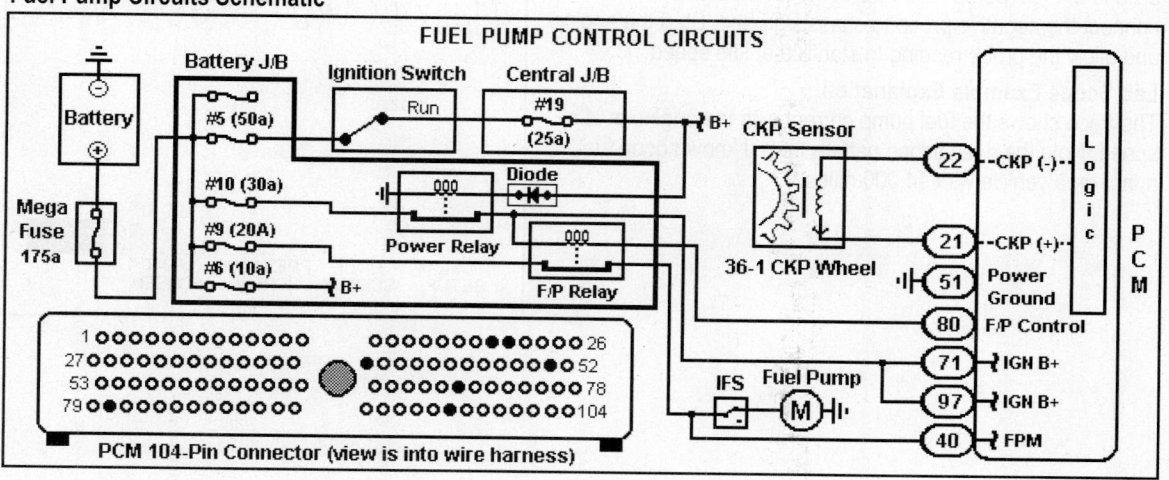

FUEL PUMP CONTROL CIRCUITS

PCM 104-Pin Connector (view is into wire harness)

FUEL PUMP MODULE

The fuel pump module in this "returnable" fuel system is mounted inside the fuel tank in a reservoir. It has a discharge check that maintains system pressure after the key has been turned off to minimize starting concerns.

The reservoir prevents fuel low interruptions during hard vehicle maneuvers during periods when the fuel tank level is low.

A portion of the high-pressure flow from the pump is diverted to operate a Venturi jet pump. The jet pump draws fuel from the fuel tank into the fuel pump module.

This design permits optimum fuel pump operation during vehicle maneuvers and driving on steep hills with low-tank full levels.

Internal Relief Valve

The fuel pump has an internal pressure relief valve that restricts fuel pressure to 124 psi if the fuel flow becomes restricted due to a clogged fuel filter or damaged fuel line.

LAB SCOPE TEST (FUEL PUMP)

If you suspect that the fuel pump is faulty, you can use a low amp probe with an inductive pickup to monitor the operation of the fuel pump and its related circuits.

Lab Scope Settings

Set the Lab Scope to these initial settings:

*   Volts per division: 100 mv
*   Time per division: 1 ms
*   Trigger setting: 50% with a positive slope

Scope Connections (Amp Probe)

Set the amp probe to 100 mv (a 100 mv setting equals 1 amp on the lab scope) and zero the amp probe. Connect the probe around the fuel pump feed wire at the inertia switch and connect the negative probe to chassis ground. Start the engine and allow the probe reading to stabilize at idle speed.

Lab Scope Example Explanation

The trace shows the fuel pump current with the engine at idle speed. Note the even scope pattern from a known good fuel pump on a vehicle with 14,000 miles.

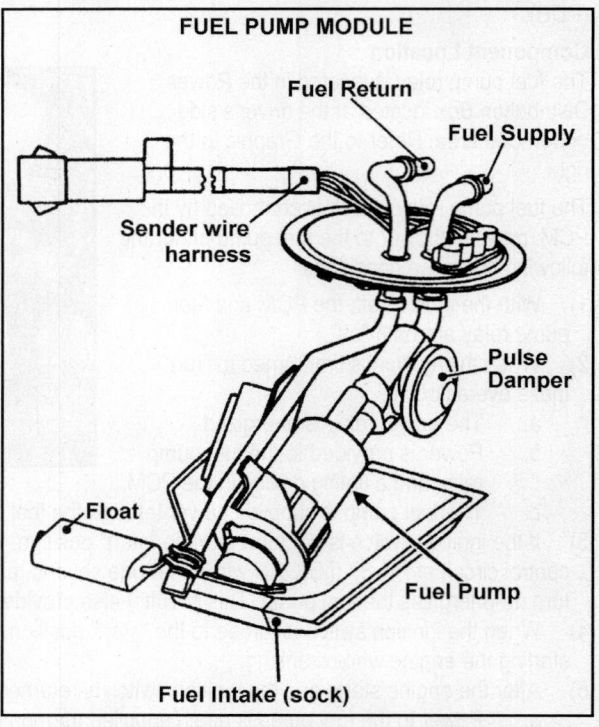

FUEL PUMP MODULE

Fuel Return

Fuel Supply

Sender wire harness

Pulse Damper

Float

Fuel Pump

Fuel Intake (sock)

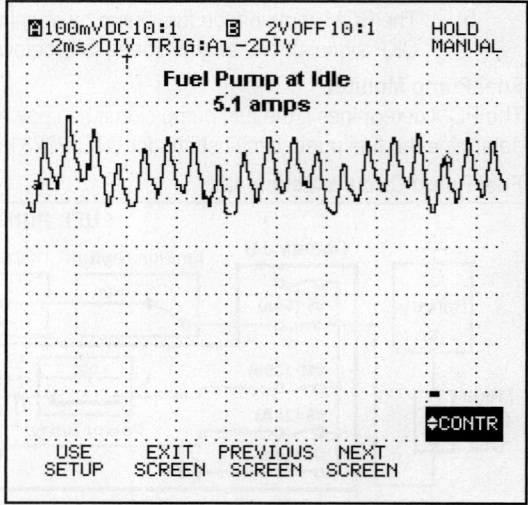

🅐100mVDC10:1   🅑 2VOFF10:1   HOLD
2ms/DIV TRIG:A1 -2DIV   MANUAL

Fuel Pump at Idle
5.1 amps

◆CONTR

USE   EXIT   PREVIOUS   NEXT
SETUP   SCREEN   SCREEN   SCREEN

FUEL PUMP COMPONENTS

The internal portions of the fuel pump module are described next.

Check Valve

The fuel pump includes a normally closed (N.C.) check valve in the manifold outlet of the module.

The valve opens when outlet pressure from the pump exceeds the opposing check valve spring force.

When the pump is off, the check valve closes to maintain pump prime and fuel line pressure.

Fuel Reservoir

A fuel reservoir is used to prevent fuel flow interruptions during extreme vehicle maneuvers with low tank fill levels.

The reservoir is welded into the tank or included in the pump sender housing.

Fuel Inlet Screen

The inlet of the pump includes a nylon filter to prevent dirt and other particulate matter from entering the system. Water accumulation in the tank can pass through the filter without restriction.

EXTERNAL FUEL FILTER

A frame mounted external fuel filter is used to strain fuel particles through a paper element.

This filtration process reduces the possibility of obstruction in the fuel injector orifices.

The fuel filter is located at the rear of the vehicle at the right side frame rail near the fuel tank.

Refer to the location example in the Graphic to the right.

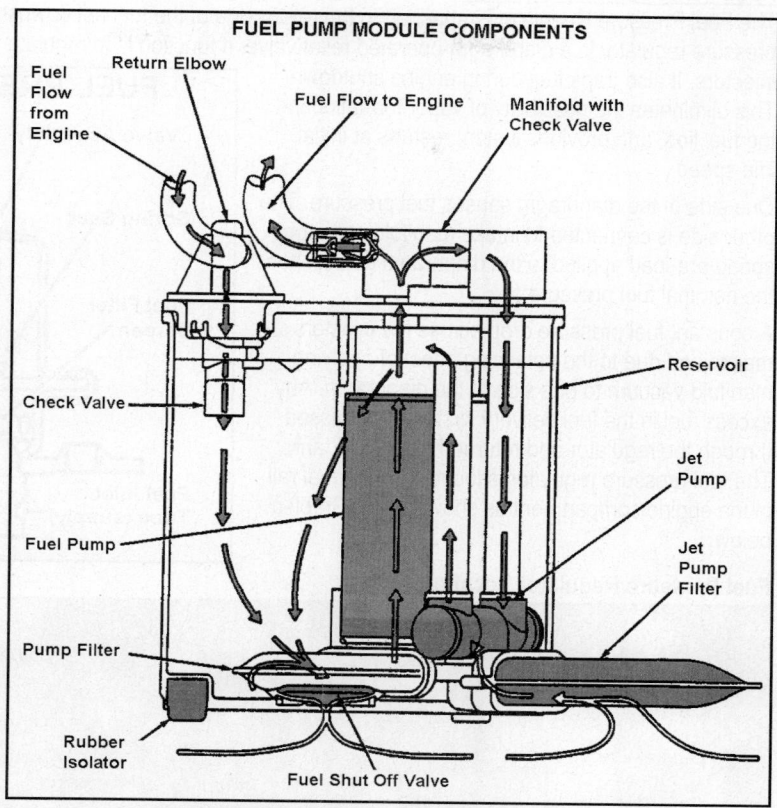

FUEL PUMP MODULE COMPONENTS

Fuel Flow from Engine

Return Elbow

Fuel Flow to Engine

Manifold with Check Valve

Reservoir

Check Valve

Jet Pump

Fuel Pump

Jet Pump Filter

Pump Filter

Rubber Isolator

Fuel Shut Off Valve

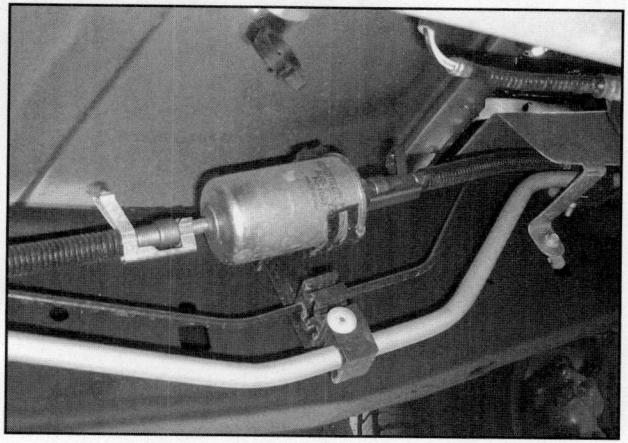

FUEL PRESSURE REGULATOR

The Fuel Pressure Regulator is attached to the return side of the fuel rail downstream of the fuel injectors. The fuel pressure regulator is a diaphragm-operated relief valve. It function is to regulate the fuel pressure supplied to the injectors. It also traps fuel during engine shutdown. This eliminates the possibility of vapor formation in the fuel line, and provides instant restarts at initial idle speed.

One side of the diaphragm senses fuel pressure. The other side is connected to intake manifold vacuum. A spring pre-load applied to the diaphragm establishes the nominal fuel pressure.

A constant fuel pressure drop across the injectors is maintained due to the balancing affect of applying manifold vacuum to one side of the diaphragm. Any excess fuel in the fuel delivery system is bypassed through the regulator and returned to the fuel tank. The fuel pressure regulator is located on the fuel rail in the engine compartment as shown in the Graphic below.

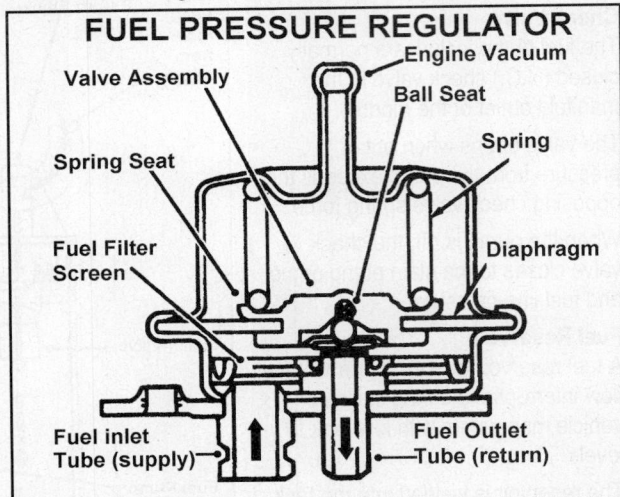

Fuel Pressure Regulator Location Graphic

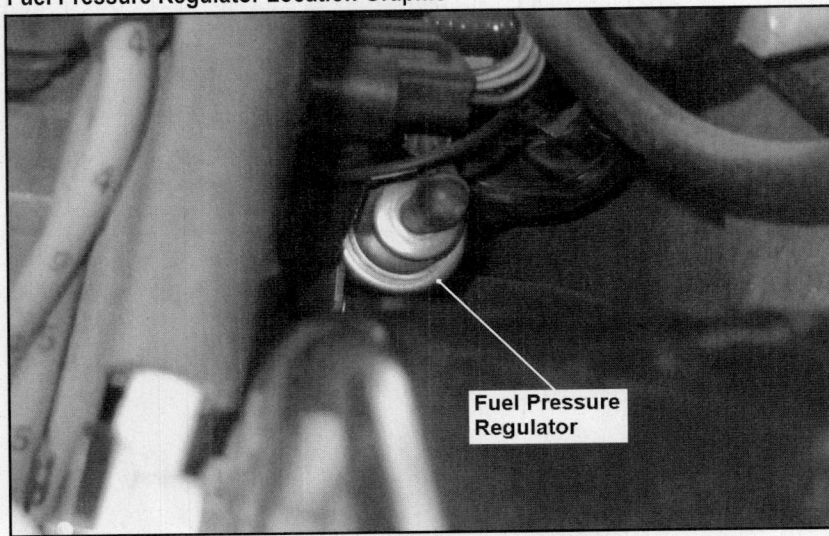

INERTIA SWITCH

This vehicle application is equipped with an inertia fuel shutoff (IFS) switch that is used in conjunction with the electric fuel pump. The inertia switch is designed to shut off the fuel pump in the event of an accident.

Inertia Switch Operation

When a sharp impact occurs, a steel ball, held in place by a magnet, breaks loose and rolls up a conical ramp to strike a target plate. This action opens the electrical contacts of the IFS switch and shuts off power to the fuel pump. Once the switch is open, it must be manually reset before the engine will restart.

Inertia Fuel Shutoff Switch Graphic

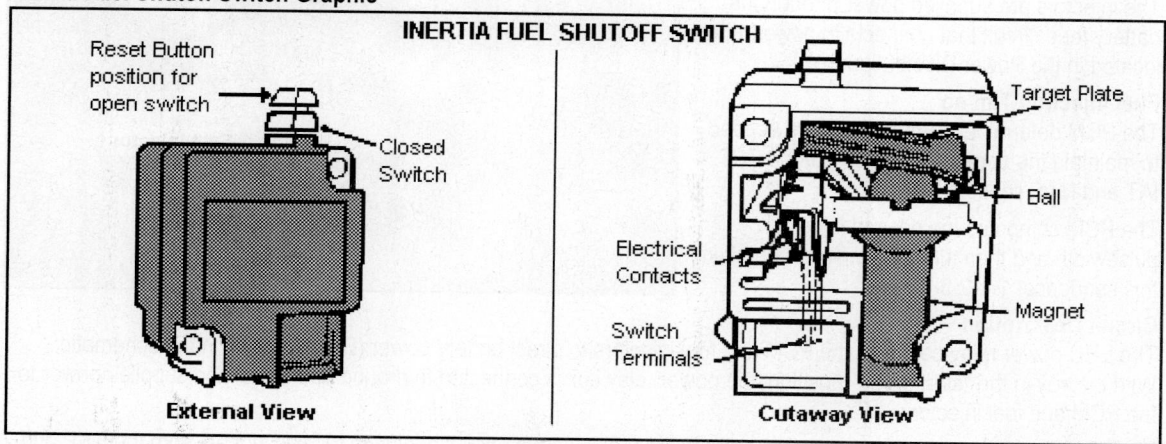

Inertia Switch Location

One of the challenges in dealing with resetting or checking this switch is locating the switch.

On the 2000 Explorer application used in this article, the IFS switch is located under the dash behind the radio (refer to the example view from under the I/P cluster in the Graphic to the right).

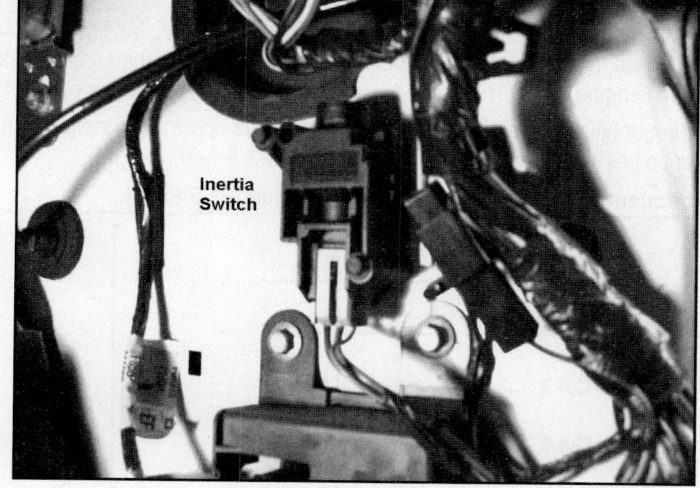

Inertia Switch Reset Instructions

1) Turn the key off and check for any signs of leaking fuel near the fuel tank or in the underhood area. If any leaks are found, repair the source of the leak before proceeding to the next step.

2) If there are no fuel leaks present, push the reset button to reset the switch (it can be closed an extra 0.16" in the closed position).

3) Cycle the key to "on" for a few seconds, and then to off (in order to recheck for signs of any fuel leaks anywhere in the system).

FUEL INJECTORS

The fuel injector is a solenoid-operated valve designed to meter the fuel flow to each combustion chamber (SFI). The fuel injector is opened and closed a constant number of times per crankshaft revolution.

The injectors are deposit resistant injection (DRI) design (they do not require cleaning).

The amount of fuel delivered is controlled by the length of time the injector is held open (injector pulsewidth).

The injectors are supplied power through the battery feed circuit that connects to power relay located in the Power Distribution Box.

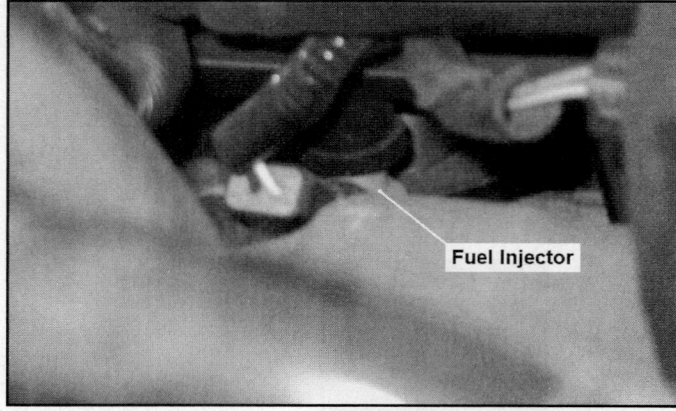

Fuel Injector

Fuel Injection Timing

The PCM determines the fuel flow rate needed to maintain the correct A/F ratio from the ECT, IAT and MAF sensors.

The PCM computes the desired injector pulsewidth and then turns the injectors on/off in this sequence: 1-4-2-5-3-6.

Circuit Description

The EEC power relay connects the PCM and fuel injectors to direct battery power (B+) as shown in the schematic. With the key in the "start or run" position, the power relay coil is connected to ground and this action supplies power to the PCM and fuel injectors.

The PCM begins to control (turn them on and off) the fuel injectors once it begins to receive signals from the CKP and CMP sensors.

Fuel Injection Timing Modes

The PCM operates the fuel injectors in two modes: Simultaneous and Sequential. In Simultaneous mode, fuel is supplied to all six cylinders at the same time by sending the same injector pulsewidth signal to all 6 injectors, twice for each engine cycle. This mode is used during startup and in the event of a fault in the CMP sensor or the PCM.

Sequential mode is used during all other engine operating conditions. In this mode, fuel is injected into each cylinder once per engine cycle in the firing order (1-4-2-5-3-6).

Sequential Fuel Injection Graphic (V6 Example Shown)

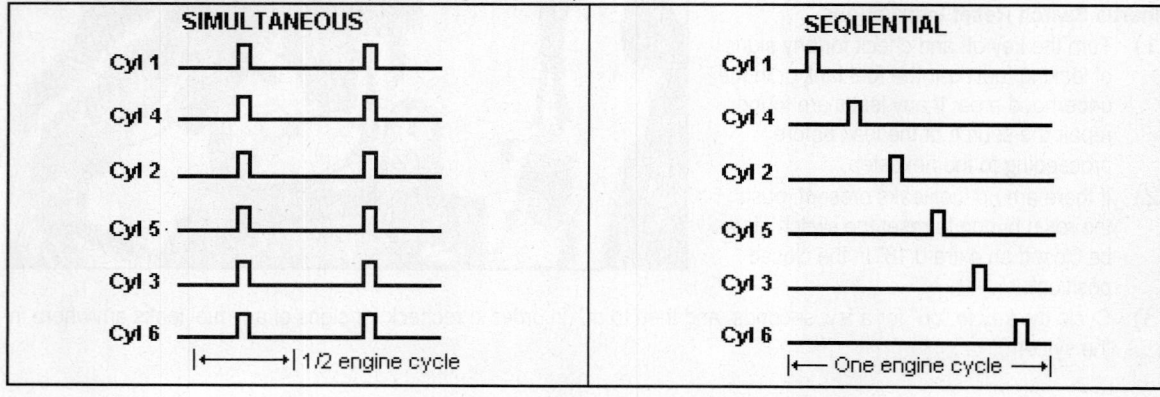

SCAN TOOL TEST (FUEL INJECTOR)

The Scan Tool can be used to quickly determine the fuel injector pulsewidth or duration. However, it is not the tool of choice for certain types of injector problems.

Connect the Scan Tool to the DLC underdash connector. Navigate through the Scan Tool menus until you get to the Ford (OEM) PID Data List.

Then select Injector Pulsewidth from the menu. The example to the right was captured with the vehicle in Park and running at Hot Idle speed.

SCAN TOOL TEST (FUEL TRIM)

Short Term fuel trim (SHRTFT) is an operating parameter that indicates the amount of short term fuel adjustment made by the PCM to compensate for operating conditions that vary from an ideal A/F ratio condition.

A negative SHRTFT number (-15%) means the HO2S is indicating to the PCM that the A/F ratio is rich, and that the PCM is trying to lean the A/F mixture. If A/F ratio conditions are ideal, the SHRTFT number will be close to 0%.

Long Term Fuel Trim

Long Term fuel trim (LONGFT) is an engine operating parameter that indicates the amount of long term fuel adjustment made by the PCM to compensate for operating conditions that vary from an ideal A/F ratio.

A positive LONGFT number (+15%) means the HO2S is indicating to the PCM that the A/F ratio is lean, and that the PCM is trying to add more fuel to the A/F mixture. If A/F ratio conditions are ideal the LONGFT number will be 0%.

The LONGFT number is displayed on the Scan Tool in percentage. Refer to the examples in the Graphic.

```
Jan 20, 02      10:10:55 am

  INJECTOR PW B1
    3.38 MSEC
  INJECTOR PW B2
    3.41 MSEC


Always use [HELP] to
  list keys & modes
[RCV] For enhanced
[*EXIT] To exit
```

```
Jan 20, 02      10:10:55 am

  SHORT TERM FT B1
       2 %
  LONG TERM FT B1
       4 %


Always use [HELP] to
  list keys & modes
[RCV] For enhanced
[*EXIT] To exit
```

```
Jan 20, 02      10:10:55 am

  SHORT TERM FT B2
       1 %
  LONG TERM FT B2
       5 %


Always use [HELP] to
  list keys & modes
[RCV] For enhanced
[*EXIT] To exit
```

Fuel Injector Test Schematic

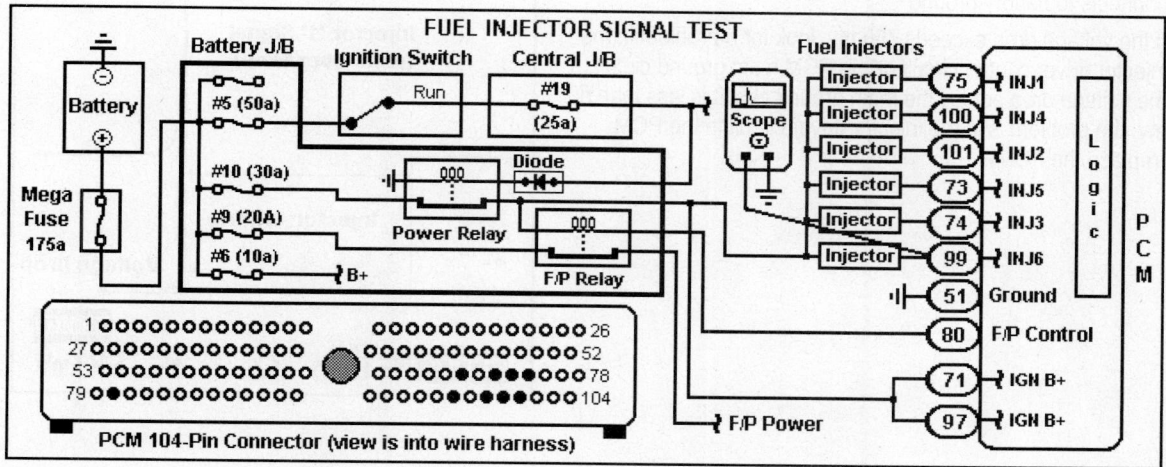

LAB SCOPE TEST (FUEL INJECTOR)

The Lab Scope is a useful tool to test the fuel injectors as it provides an accurate view of the injector operation. Place the gearshift selector in Park and block the drive wheels.

Scope Connections

Connect the Channel 'A' positive probe to the injector control wire and negative probe to the battery negative post.

Scope Settings

To make the waveforms as clear as possible, set the scope settings to match the examples.

Set the amp probe to 100 mv (this probe setting equals 500 mA per scope division).

Lab Scope Example (1)

In example (1), the trace shows the injector waveform with the sweep rate set to 1 ms. Note how this setting allows you to determine the injector on time at idle speed (3.2 ms in this example). The inductive kick was 55 volts (viewed with a 10v per division setting).

Note the point at which the injector closed (referred to as the pintle bounce point here).

Lab Scope Example (2)

In example (2), the trace shows the injector current signal (current ramping) along with the injector volts over time signal at idle speed. Note the point in the injector current signal where the injector opens. The current flow in the circuit was near 2 amps.

Lab Scope Example (3)

In example (3), the voltage setting was changed to 5v per division to allow for a more accurate calculation of the feed voltage and to calculate the injector on time in a snap test.

Injector Driver Test

In example (3), the voltage drop through the injector driver control circuit in the PCM rises to 450 mv above the ground level point. This is caused by current flow through the injector winding, the injector driver and PCM ground circuit that connects to battery ground.

If the voltage drop exceeds 750 mv, look for a problem in the injector driver control circuit or the PCM main ground circuit. If the voltage drop across the main ground circuit is less than 50 mv, the problem is in the injector driver circuit in the PCM (replace the PCM).

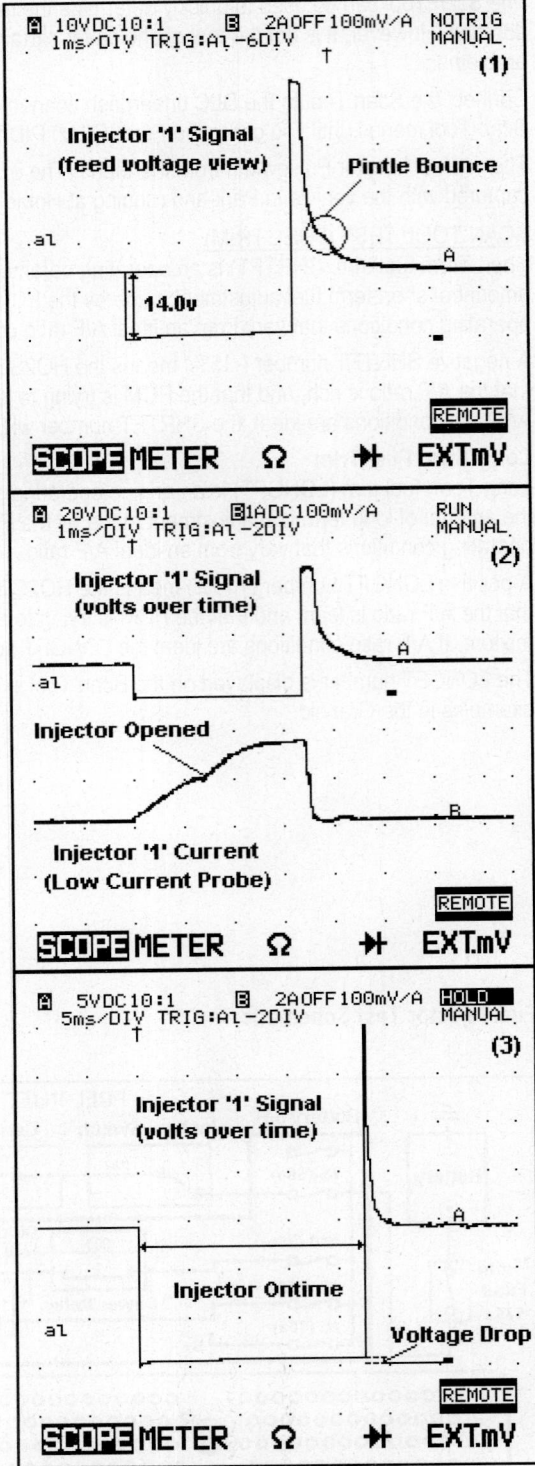

CMP SENSOR (MAGNETIC)

The camshaft position (CMP) sensor used with this application sends a signal to the PCM that indicates the camshaft position. This signal is used for fuel synchronization.

The CMP sensor is a magnetic transducer mounted at the top of the valve cover as shown in the Graphic.

This sensor monitors a single tooth on the CMP reluctor.

The CMP sensor outputs an AC signal to the PCM that indicates the top dead center (TDC) position of cylinder No. 1

The PCM monitors the CMP sensor signal in order to correctly synchronize the firing of the six (6) fuel injectors on this engine application.

CMP Sensor Circuits

The PCM is connected to the CMP sensor positive (+) circuit (DB/OG wire) at Pin 85 of the PCM 104P connector. Connect the negative probe to the battery negative ground post or to Pin 91 of the PCM 104P connector.

Specification: The resistance of the CMP sensor is 290-790 ohms at 68°F.

CMP Sensor Diagnosis (Code Help)

If the PCM does not detect any signals from the CMP sensor with the engine running it will set DTC P0340 for Camshaft Position Sensor Circuit Fault. A DVOM can be used to check for "no output", but a Lab Scope is the tool of choice to test this sensor.

EI System CMP Sensor Schematic

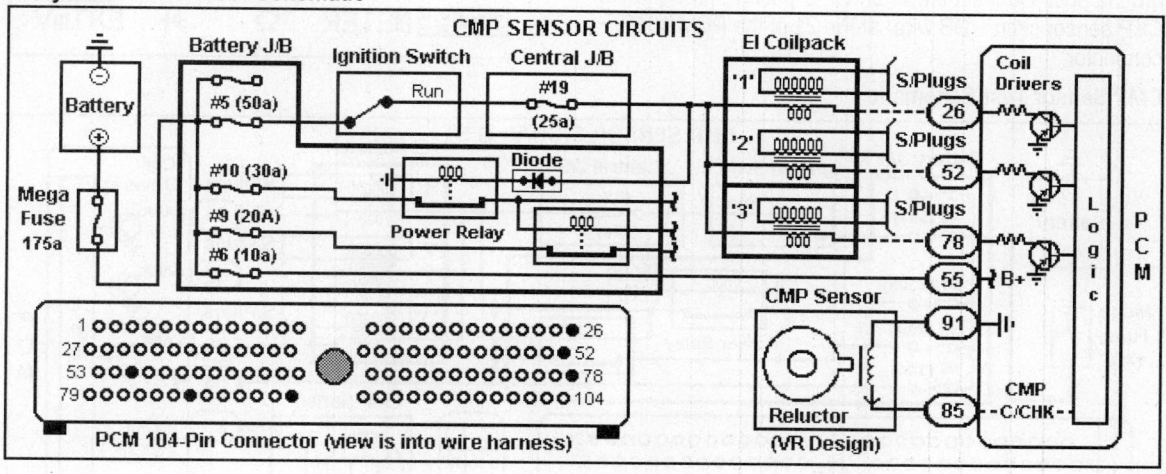

Electronic Ignition System

CMP SENSOR (SIGNAL EXPLANATION)

The CMP sensor input is not needed by the EDIS system since the spark plugs are fired in pairs. The CMP sensor is used by the PCM to determine cylinder No. 1 for sequential fuel control. The CMP sensor on this vehicle is a variable reluctance sensor (VRS).

The VRS produces a narrow sinusoidal signal when a tooth on the camshaft passes the pole piece. The negative-going edge is typically located at 22° ATDC of Cylinder 1.

As the VRS encounters the rotating ferromagnetic teeth, the varying reluctance induces a voltage proportional to the rate of change in magnetic flux and the number of coil windings. As the tooth approaches the pole piece, a positive differential voltage is induced. As the tooth and the pole piece align, the signal rapidly transitions form maximum positive to maximum negative. As the tooth moves away, the voltage returns to a zero differential voltage.

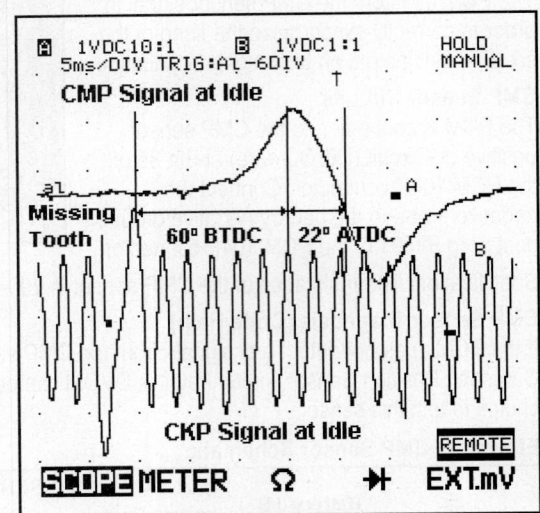

At normal engine speeds, the VRS signal waveform approximates a sine wave with increased amplitude adjacent to the mission tooth region. At 30 rpm, the VRS output should be 150 mv peak-to-peak and at 300 rpm, the VRS output should be 1.5 volts peak-to-peak. At 6000 rpm, the VRS output should be 24.0 volts peak-to-peak.

The sensor resistance is 290 to 790 ohms.

LAB SCOPE TEST (CMP SENSOR)

Place the shift selector in Park and block the drive wheels prior to starting this test.

Scope Connections

Install a breakout box, if available. Connect the Channel 'A' positive probe to the CMP sensor circuit (DG wire) at Pin 85 and the Channel 'A' negative probe to the battery negative ground post. Connect the Channel 'B' positive probe to the CKP sensor circuit (DB wire) at Pin 21 of the PCM 104P connector.

CMP Sensor Test Schematic

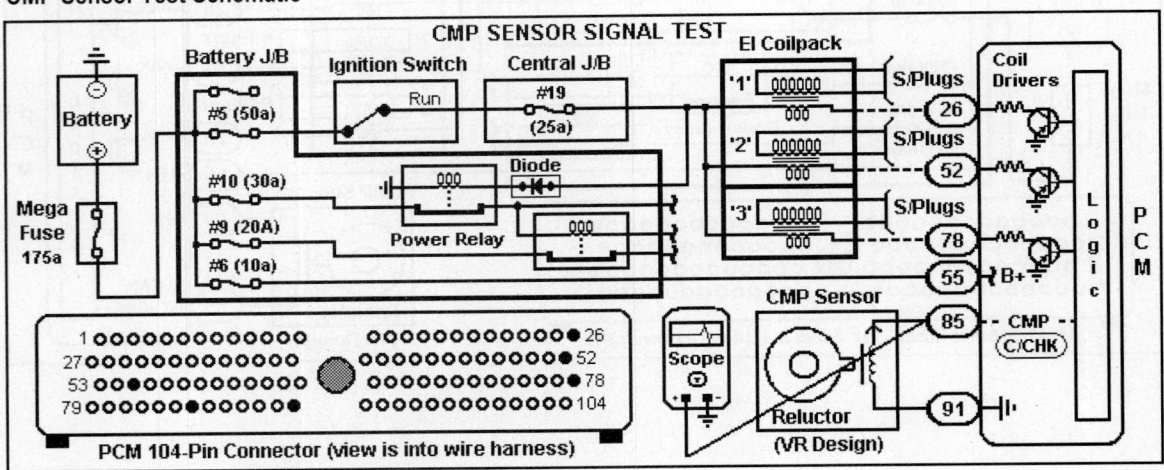

Communications Network

PRINCIPLES OF OPERATION

The vehicle has three module communication networks:

- The standard corporate protocol (SCP) communications network
- The international standards organization (ISO) 9141 communications network
- The UART based protocol (UBP) communications network

The Scan Tool connects to all three networks through the DLC. This makes diagnosis and testing of these systems easier by allowing one smart tester to be able to diagnose and control any module on the three networks from one connector. The DLC can be found under the instrument panel between the steering column and the radio.

Communications Network Descriptions

ISO 9141 Network - The ISO network does not allow inter-module communication. When the tool communicates to modules on the ISO 9141 network, it must request all information; or initiate module commands. The ISO 9141 network will not function if the wire is shorted to ground or to battery voltage. If one of the modules on the ISO 9141 network loses power or shorts internally, communications to that module will stop.

SCP Network - The SCP network will remain operational even with the severing of one of the bus wires. Communications will also continue if one of the bus wires is shorted to ground or voltage, or if some but not all termination resistors are lost. The SCP network does allow inter-module communication.

UBP Network - The UBP network will not function if the wire is shorted to ground or to voltage. The UBP communication network also allows inter-module communication.

Vehicle Module Descriptions

Antilock Brakes - The 4-wheel (4ABS) control module is connected to the ISO 9141 communication network. The module controls the brake pressure to the four wheels to keep the vehicle under control while braking. It also controls the traction control system.

Driver Seat Module - The DSM is connected to the ISO 9141 network where it controls the driver power seat. The DSM is programmable to store seat and pedal position.

Generic Electronic Module - The GEM connects to the ISO 9141 bus on this vehicle.

Instrument Cluster - The IC is connected to the SCP communication network where it displays driver information, including tachometer reading, fuel level, engine coolant level and speed. The message center is included in the high option level cluster and displays door/hatch ajar information, lamp outage and low washer fluid level.

Powertrain Control Module - The PCM is connected on the SCP network where it controls engine performance, electronic ignition, emission controls and on-board diagnostics. The passive anti-theft system (PATS) is integrated into the PCM. The PCM controls the PATS functions as well as illumination of the anti-theft indicator. The illumination command is sent from the PCM through the SCP network to the instrument cluster, which in turn sends power to the anti-theft indicator. The PCM stores the ignition key codes and controls engine disable. The vehicle speed control is integrated into the PCM. The PCM controls the system functions as well as illumination of the speed control indicator located in the instrument cluster. The illumination command is sent from the PCM through the SCP network to the instrument cluster, which in turn sends power to the speed control indicator.

Restraint Control Module - The RCM is connected to the ISO 9141 communication network. The RCM controls the deployment of the air bags based on sensor inputs.

Generic Electronic Module

INTRODUCTION

The generic electronic module (GEM) and central timer module (CTM) on this vehicle control the following features:

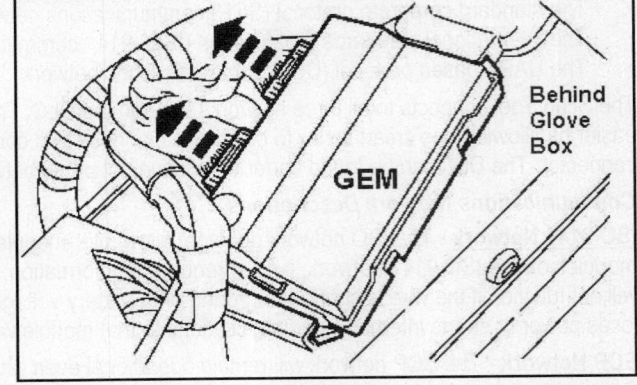

- Accessory Delay Relay
- Auto Lamps
- Battery Saver
- Warning Chime
- Driver One-Touch Down Power Window
- Illuminated Entry and Courtesy Lamps
- Wipers and Washers (front & rear)

The GEM and CTM constantly monitor their subsystems for faults. If a fault is found in one of the subsystems, the module will record the concern in the form of a diagnostic trouble code (DTC). The GEM and CTM modules are located behind the glove box.

The ignition switch input is important to the GEM functions. An erratic or unexpected GEM function can often be traced to a problem with in one of the ignition switch inputs.

■ NOTE: *If the GEM is replaced, the new module must be reconfigured.*

Circuit Operation

In this example, the GEM & CTM Module connect to power through the four (4) fuses shown in the Schematic below. The main grounds for the modules are at Pin 14 and 26. The Sensor Return connects to Pin 21. Note the connections for the "door ajar" inputs.

The "door ajar control" circuit to the Instrument Panel connects at Pin 9 of the module. When a door is opened, the switch closes and the GEM senses the signal change and closes the "door ajar control" switch to send a signal to the Instrument Cluster which turns on the "door ajar lamp". The GEM and CTM controllers communicate with other stand-alone modules and the Scan Tool at Pin 25 using the ISO 9141 protocol.

Generic Electronic Module Schematic

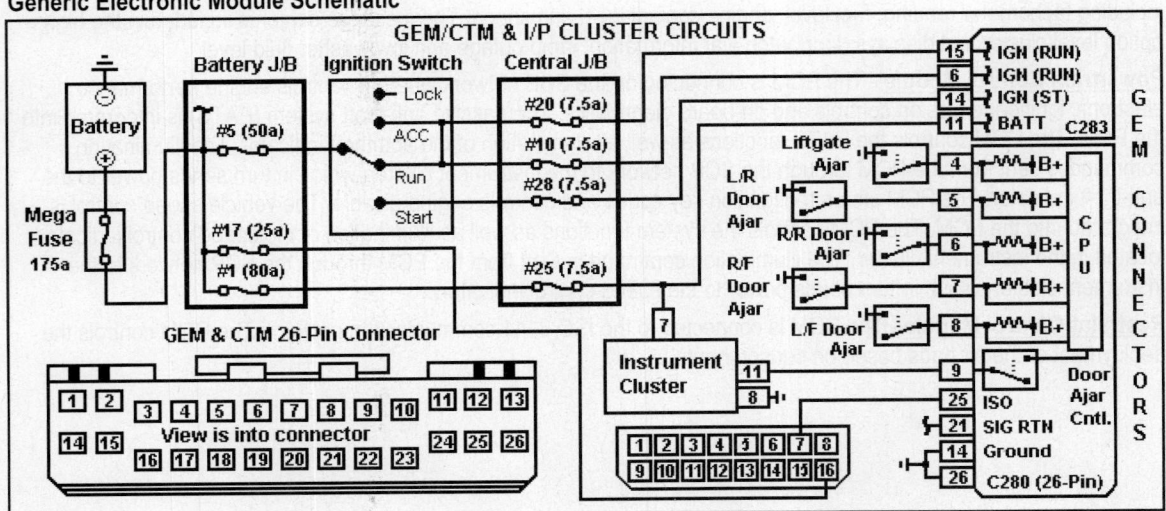

SCAN TOOL TEST (GEM & CTM)

If there is a problem in one of the modules, follow the inspection steps described here. A Scan Tool can be used to access the Message Center, GEM and CTM information.

To diagnose a condition of "No Data" from the GEM, PCM or Message Center, follow these steps:

1) Verify the customer concern by operating the system.

2) Visually inspect for signs of mechanical damage to the door lock mechanisms and check these items:

- Battery Junction Box (BJB) fuses
  - Fuse 1 (80A), Fuse 5 (50A)
  - Fuse 17 (25A)
- Central Junction Box (CJB) fuses:
  - Fuse 10 (7.5A), Fuse 20 (7.5A)
  - Fuse 25 (7.5A), Fuse 28 (7.5A)
- Circuitry and Connectors
- Door Lock Actuators
- Door Lock Switches
- FEM & CTM Units
- ISO Communication Network

3) If an obvious cause of a reported concern is found, repair the fault before proceeding to the next step.

4) If the cause of a concern is not found during this step, connect the Scan Tool to the DLC located under the dash and select the vehicle and system from the tool menu. If it does not communicate with the vehicle:

- Verify that the program card is properly installed
- Check the connections to the vehicle
- Check the ignition switch position

5) If the Scan Tool does not communicate with vehicle, refer to the instruction manual under NO DATA tests.

6) Do the Data Link Diagnostic Test: If the response is:

- CKT930 = ALL ECUS NO RESP/NOT EQUIP: refer to Pinpoint Test A.
- NO RESP/NOT EQUIP for GEM: do pinpoint test A
- SYSTEM PASSED: read, record and clear any memory codes and do the Self-Test for the GEM.

7) If the codes are retrieved related to the concern, look up the code and do the repair.

8) If no codes related to the concern are retrieved, proceed to the FEM Symptom chart.

Symptom Diagnosis: The Symptom Index for the GEM includes Pinpoint Test A-G. A list of available symptoms for the electronic modules is included on the next page.

■ NOTE: *A wiring diagram of the FEM and CTM circuits is included at the end of this article.*

Vetronix
SCAN TOOL MAIN MENU

SELECT APPLICATION

GLOBAL OBDII (MT)
GLOBAL OBDII (T1)
GM P/T
GM CHASSIS
GM BODY SYSTEMS
FORD P/T
FORD CHASSIS
FORD BODY
CHRYSLER P/T
CHRYSLER CHASSIS

SCREEN 1

SELECT
MODEL YEAR
2000 (Y) ↑↓
[ENTER]

SCREEN 2

SELECTION TYPE:
F0: VIN
F1: CAR
F2: TRUCK

SCREEN 3

SELECT ENGINE:
00 4.0L SFI MAF
EXPLORER SPORT
A/T 2WD(VIN=E)?

SCREEN 4

SELECT MODE ↑↓
F0: DATA LIST
F1: QUICK TESTS
F3: SNAPSHOT

SCREEN 5

NO DATA OR GEM
RESPONSE. CHECK:
IGN. KEY ON, DLC
CONN. OR PPT G

Generic Electronic Module

GEM/CTM SYMPTOM, DTC & PID TABLES

The repair tables on this page contain a "partial" list of the trouble codes and PID data for this vehicle application. If a GEM/CTM trouble code or particular symptom is present, look up the trouble code (or symptom) and refer to the corresponding repair information.

Electronic Module Symptom List (Complete)

Condition	Possible Source	Action
No Communication with RAP	Check Fuse #20 and Fuse #9	Go to Pinpoint Test A
Doors do not lock/unlock w/pad	Keyless entry pad & RAP module	Go to Pinpoint Test B
Doors do not lock/unlock w/remote	Remote, antenna, ECU program	Go to Pinpoint Test C
No illuminated entry w/remote	GEM, lighting circuits, RAP unit	Go to Lighting/Horns
Auto-Lock does not operate	Circuit, door open switch, BPP	Go to Pinpoint Test D
Memory seat does not operate	Circuitry, RAP and DSM units	Go to Pinpoint Test E
All door locks inoperative	Circuitry, switches, relay, GEM	Go to Pinpoint Test F
No communication with GEM unit	Circuitry, Fuse 1 & 7, GEM unit	Go to Pinpoint Test G
Panic Button does not operate	Keyless entry remote, RAP unit	Check both remotes

GEM & CTM DTC List (Partial)

DTC	Trouble Code Description	Source	Action
B1317	Battery Voltage High	GEM/TCM	Refer to Charging System
B1318	Battery Voltage Low	GEM/TCM	Refer to Charging System
B1322	Door Ajar LF Circuit Short to Ground	GEM/TCM	Refer to Warning Indicator
B1323	Door Ajar Lamp Circuit Failure	GEM/TCM	Refer to Warning Indicator
B1325	Door Ajar Lamp Circuit Short to Battery	GEM/TCM	Refer to Warning Indicator
B1330	Door Ajar RF Circuit Short to Ground	GEM/TCM	Refer to Warning Indicator
B1334	Decklid Ajar Rear Door Circuit Short to Ground	GEM/TCM	Refer to Warning Indicator
B1338	RR Door Ajar Circuit Short to Ground	GEM/TCM	Refer to Warning Indicator
B1342	GEM/TCM is Defective	GEM/TCM	Clear codes, read codes - if B1342 sets GEM/TCM is bad
B1352	Ignition Key - In Circuit Fault	GEM/TCM	Refer to Ignition Switch Table
B1354	Ignition RUN Circuit Fault	GEM/TCM	Refer to Ignition Switch Table
B1359	Ignition RUN/ACC Circuit Fault	GEM/TCM	Refer to Ignition Switch Table
B1574	LR Door Ajar Circuit Short to Ground	GEM/TCM	Refer to Warning Indicator

GEM & CTM PID List (Partial)

PID	PID Description	Expected Value
D_DR_SW	Driver Door Ajar Switch Status	CLOSED, AJAR
DRAJR_L	Door Ajar Warning Lamp Circuit	OFF, ON
DR_DSRM	Door Disarm Switch Status	L_DOOR, R_DOOR, LIFT_G, OFF
IGN_KEY	Key-In-Ignition Status	IN, OUT
IGN_GEM	Ignition Switch Status	START, RUN, OFF, ACCY
LGATESW	Liftgate Ajar Switch Status	CLOSED, AJAR
LRDR_SW	Left Rear Door Ajar Switch Status	CLOSED, AJAR
P_DR_SW	Passenger Door Ajar Switch Status	CLOSED, AJAR
PARK_SW	External Access Ajar Switch Status	OFF, ON
RRDR_SW	Right Rear Door Ajar Switch Status	CLOSED, AJAR

Pinpoint Test 'G' (No Communication)

The GEM/CTM are diagnosed by first connecting a Scan Tool (it reads Pin 7 ISO 9141). To diagnose the cause of "No Data" from the GEM/CTM, use this test procedure.

Pinpoint Test 'G' Repair Table

Step	Action	Yes	No
G1	**Step description:** *Check the Ignition Fuse* • Inspect the #5 Fuse in Battery Junction Box. • Was the #5 Fuse (50a) okay?	Go to Step G2.	Locate and repair short to the circuit. Replace the fuse and retest the system operation.
G2	**Step description:** *Check Battery Backup fuse* • Inspect the Fuse 25 in Central Junction Box. • Was the #25 Fuse (7.5a) okay?	Go to Step G3.	Locate and repair short to the circuit. Replace the fuse and retest the system operation.
G3	**Step description:** *Test backup circuit at Fuse Box* • Measure the voltage between Fuse #25 in the Central J/B (after the fuse) and ground. • Did the voltage read greater than 10 volts?	Go to Step G4.	Repair circuit that read low. Clear the codes and retest the system operation when done.
G4	**Step description:** *Test backup circuit at GEM* • Measure the voltage between Pin 11 (W/Y wire) at GEM connector C283 (harness side) and ground. • Did the voltage read greater than 10 volts?	Go to Step G5.	Repair the open circuit. Clear the codes and then retest the system operation when done.
G5	**Step description:** *Test for open ground circuit* • Measure the resistance from Pin 14 (BK/W wire) at GEM connector C280 (harness side) to ground. • Did the resistance read less than 5 ohms?	The ground and power circuits are okay. Go to Data Link DIAG Test.	Repair the open circuit. Clear the codes and then retest the system operation when done.

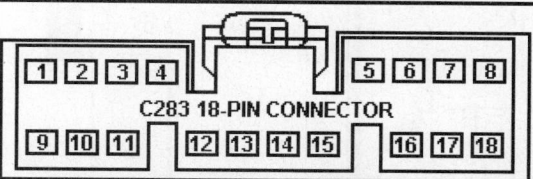

C283 18-PIN CONNECTOR

PIN	CIRCUIT	CIRCUIT FUNCTION
1, 4-5	---	NOT USED
2	OG/WH	CURRENT SENSE LOW
3	TN/BK	DOWN INPUT
6	GY/YE	IGNITION (HOT IN RUN)
7	YE/RED	ONE TOUCH DOWN RELAY
8	VT/OG	BATTERY SAVER RELAY
9	GY/BK	VSS (+)
10	GY	CURRENT SENSE HIGH
11	WH/YE	POWER (HOT AT ALL TIMES)
12	LG/OG	INTERIOR LAMP RELAY
13	YE/WH	THROTTLE POSITION INPUT
14	RD/BK	IGNITION (HOT IN START)
15	BK/PK	IGNITION (HOT IN ACC / RUN
16	DB/YE	DEFROST RELAY OUTPUT
17	LB/RD	ACCESSORY DELAY RELAY
18	---	NOT USED

C280 26-PIN CONNECTOR
View is into connector

PIN	CIRCUIT	CIRCUIT FUNCTION
1	DB/OG	REAR WINDOW DEFROST INPUT
2-3	---	NOT USED
4	WH/VT	LIFTGATE DISABLE AND AJAR
5	LG/YE	LH REAR DOOR AJAR INPUT
6	PK/LB	RH REAR DOOR AJAR INPUT
7	GY/RD	RH FRONT DOOR AJAR INPUT
8	YEL/BK	LH FRONT DOOR AJAR INPUT
9	BK/OR	DOOR AJAR INDICATOR OUTPUT
10	BK/PK	KEY WARNING SWITCH
11	WH/PK	HEADLAMPS ON INPUT
12	YE	FASTEN SEATBELTS INDICATOR
13	LB/OG	WINSHIELD WIPER INTERVAL DELAY
14, 26	BK/WH	GROUND
15	BK	PARK SENSE INPUT
16-17	---	NOT USED
18	GY/LB	FRONT WIPER SPEED RELAY OUTPUT
19	YE/WH	BRAKE/RUN RELAY OUTPUT
20	TN/YE	TONE REQUEST INPUT
21	GY/RD	SIGNAL RETURN
22	PK/YE	WINDSHIELD WIPER MODE SELECT
23	BN/LB	SAFETY BELT WARNING SWITCH
24	TN/RD	WASHER PUMP RELAY OUTPUT
25	LB/WH	ISO 9141 SIGNAL

■ NOTE: Only two (2) of the four (4) connectors used with the GEM/CTM assembly for this vehicle are shown in this repair example.

GEM & CTM Wiring Diagram

GEM & CTM DIAGRAM (PARTIAL)

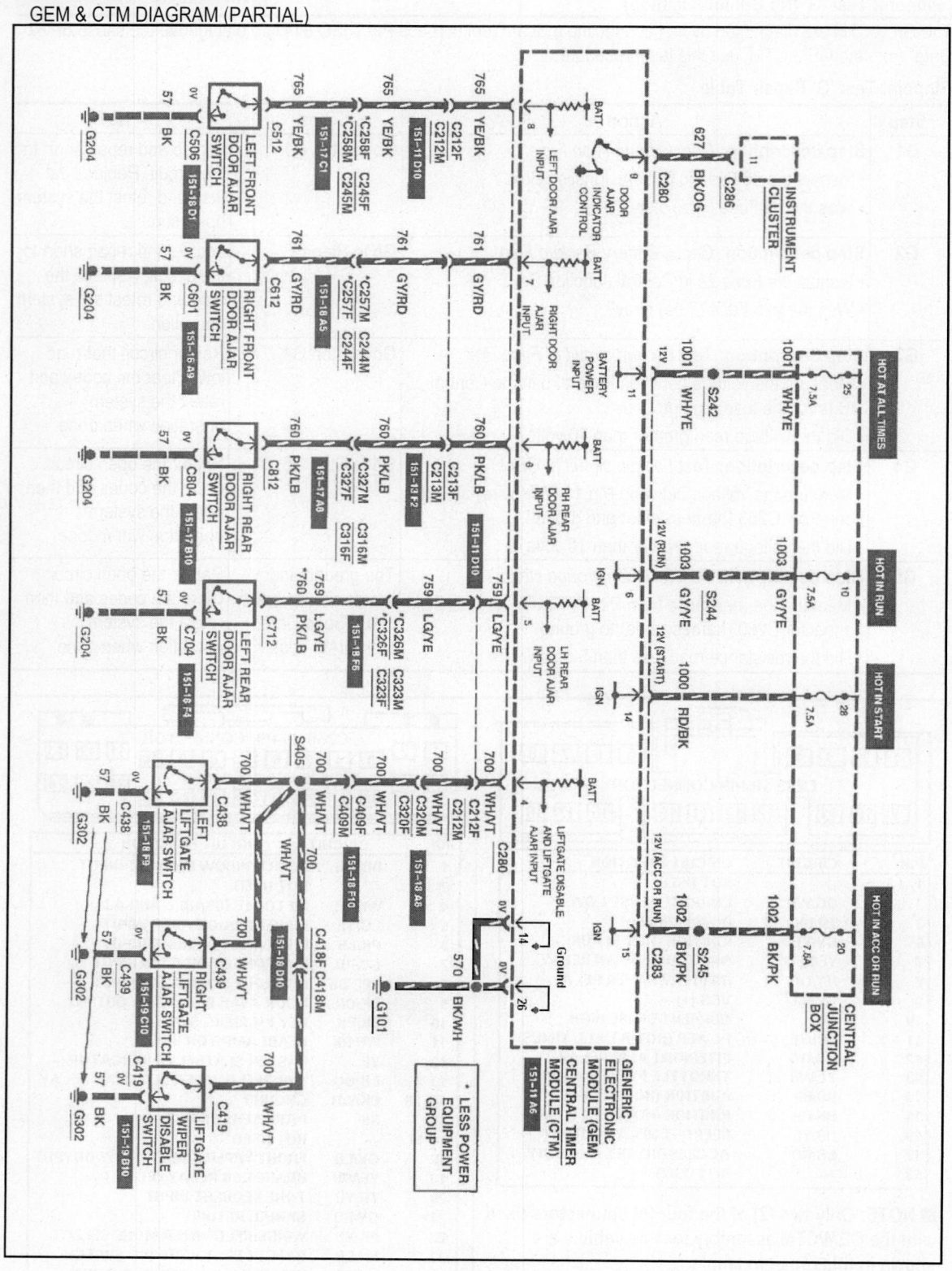

Idle Air Control Valve

GENERAL DESCRIPTION

The Idle Air Control (IAC) solenoid is mounted to the side of the throttle body. It is designed to control engine idle speed (rpm) under all operating conditions and to provide a dashpot function.

The IAC solenoid meters the intake air that flows past the throttle plate through a bypass area within the IAC assembly and throttle body.

IAC Valve Location

Refer to the Graphic to the right for the IAC solenoid location on this application.

The PCM calculates a desired idle speed and controls the IAC solenoid through a duty cycle signal that changes the valve position to control the amount of bypass air.

IAC Valve Assembly

SCAN TOOL TEST (IAC SOLENOID)

Connect the Scan Tool to the DLC underdash connector. Navigate through the Scan Tool menus until you get to the Ford (OEM) Engine Data List (a list of PID items).

Select the RPM AND IAC PID items from the menu so that you can view them separately on the Scan Tool display.

Scan Tool Example (1)

An example of the ENGINE SPEED (RPM) and IDLE AIR CONTROL (%) PID for this vehicle at Hot Idle speed is shown in the Graphic. The first example was captured at Hot Idle speed in Park with all accessories turned off. The second example was captured with the A/C on in Drive.

Scan Tool Example (2)

In Example (2), the IAC PID was captured at Hot Idle speed with the A/C on in Drive (brake on). Note the change in the IAC reading (38% to 41%) after shifting into Drive position.

Diagnostic Tip

To test the IAC solenoid operation, record the idle speed PID reading with the accessories off in Park. Then drive the vehicle. The IAC reading should return to the initial reading each time you stop and place the gear selector in Park.

```
Jan 20, 02    10:10:55 am (1)

ENGINE SPEED
  674 RPM
IDLE AIR CNTRL
  38 %


Always use [HELP] to
  list keys & modes
[RCV] For enhanced
[*EXIT] To exit
```

```
Jan 20, 02    10:10:55 am (2)

ENGINE SPEED
  644 RPM
IDLE AIR CNTRL
  41 %


Always use [HELP] to
  list keys & modes
[RCV] For enhanced
[*EXIT] To exit
```

Idle Air Control Solenoid

LAB SCOPE TEST (IAC SOLENOID)

The Lab Scope can be used to monitor the IAC solenoid control signal under engine load, speed and temperature conditions. Place the gearshift selector in Park and block the drive wheels prior to starting the test.

Scope Connections

Connect the breakout box (BOB). Connect the Channel 'A' positive probe to the IAC solenoid circuit at PCM Pin 83 (WHT/LB wire) and the negative probe to the battery negative post.

Scope Settings

To make the waveforms as clear as possible, set the scope settings to match the examples.

Lab Scope Example (1)

In this example, the trace shows the IAC solenoid signal in Park without an engine load.

Lab Scope Example (2)

In this example, the trace shows the IAC solenoid signal in Park with the A/C enabled.

The IAC signal should show a change in the digital on/off signal as a load is placed on the engine (note the change in these 2 examples).

If it does not change or the square wave pattern is broken or jagged, a problem may exist in the IAC solenoid or its related control or ignition feed circuit.

Circuit Description

The IAC solenoid connects to power at Fuse #10 (30a) in the Battery J/B and power relay. The IAC solenoid includes an internal diode to protect the PCM against voltage spikes.

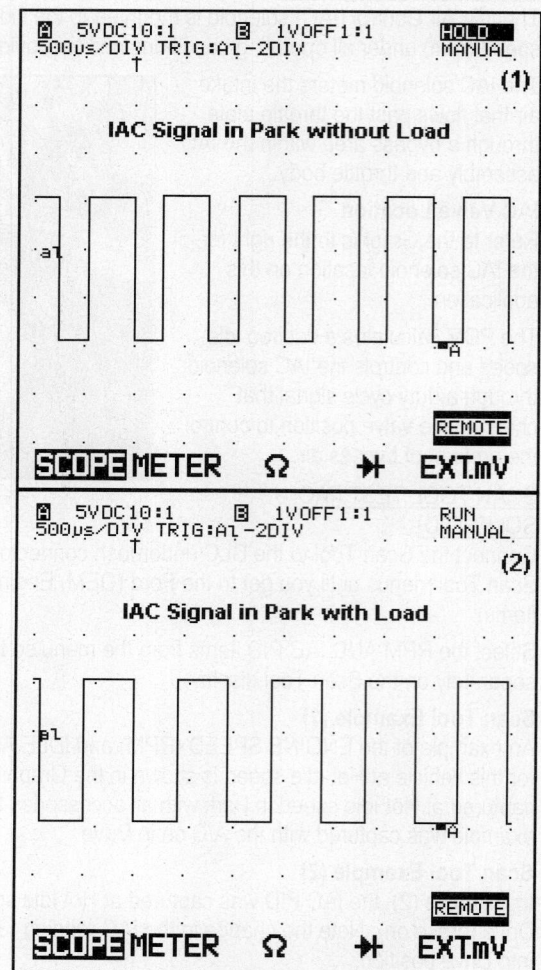

Idle Air Control Solenoid Test Schematic

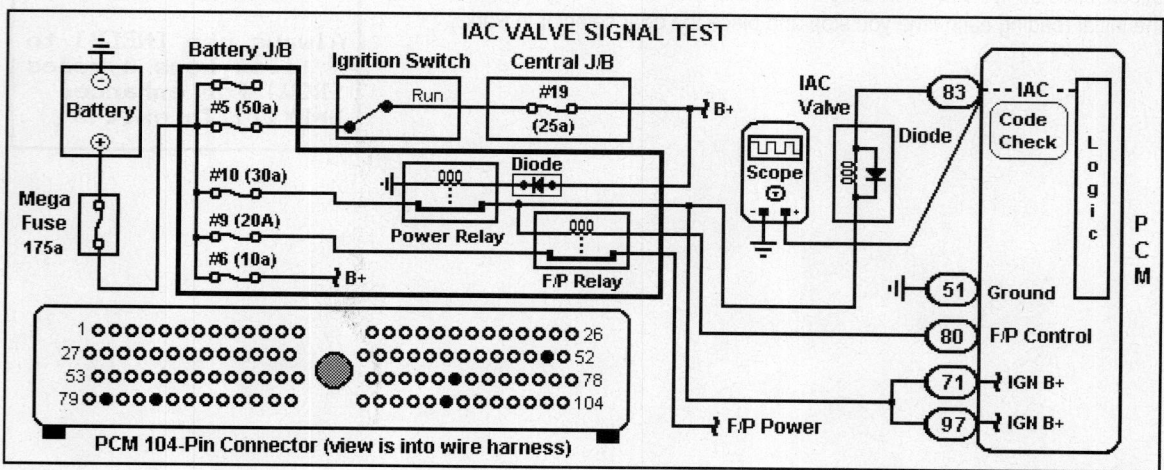

Input Sensors

THROTTLE POSITION SENSOR

The Throttle Position (TP) sensor is mounted to the throttle body where it detects changes in the throttle valve angle.

The PCM uses the TP sensor analog voltage DC signal to detect the following vehicle driving conditions:

- Air Fuel Ratio Correction
- Fuel Cut Control
- Power Increase Correction

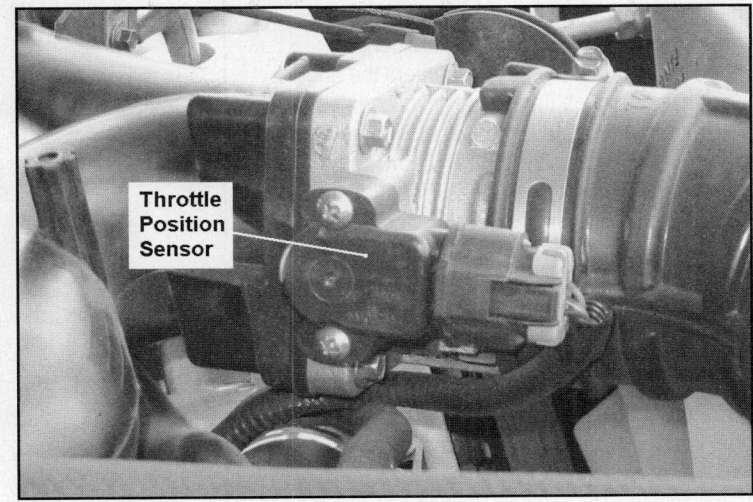

Throttle Position Sensor

TP Sensor Circuits

The three circuits that connect the TP sensor to the PCM are listed below:

- VREF 5v circuit
- TP sensor signal circuit
- TP sensor ground return circuit

Circuit Operation

With the throttle fully closed, the TP sensor input to the PCM is from 0.5-1.27v at Pin 89.

The TP sensor voltage increases in proportion to the throttle valve-opening angle. The TP sensor signal is from 3.6-4.8v with the throttle valve fully open.

Note that the "Code Check" circuit for the TP sensor connects to the PCM at Pin 89. In effect, this is the critical circuit that the PCM monitors in order to detect a circuit fault or a fault in the relationship between the TP sensor input and the MAF sensor input.

TP Sensor Test Schematic

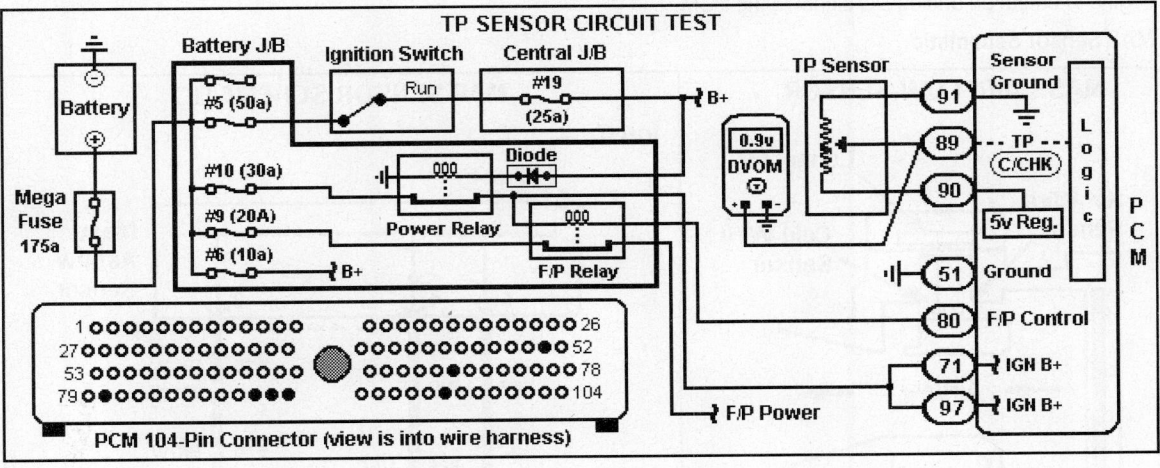

Mass Airflow Sensor

GENERAL DESCRIPTION

The mass airflow (MAF) sensor is located between the air cleaner and the throttle body. This MAF sensor uses a Hot-wire sensing element to measure the amount of air that enters the engine. Air passing through the device flows over the hot wire element and causes it to cool. The hot wire element is maintained at 392ºF (200ºC) above the current ambient temperature as measured by the constant cold wire part of the sensor,

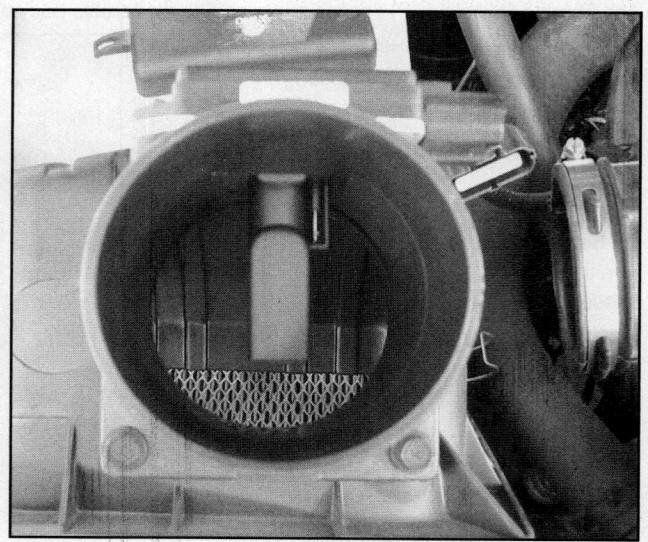

The current level required to maintain the temperature of the hot wire element is proportional to amount of air flowing pass the sensor. The MAF input to the PCM is an analog signal proportional to the intake air mass.

The PCM calculates the desired fuel injector on time in order to deliver the desired A/F ratio. The PCM also uses the MAF sensor to determine transmission EPC solenoid shifting and the TCC solenoid scheduling.

■ **NOTE:** *When this sensor fails, it may set a trouble code or cause a No Code Fault (engine cold or hot).*

Circuit Description

The MAF sensor is connected to ignition power and main ground as shown below. The MAF sensor signal and ground circuits connect directly to the PCM.

With the ignition key in the "start or run" position, the EEC power relay provides battery power to the MAF sensor and it begins to output an analog DC signal to the PCM.

MAF Sensor Schematic

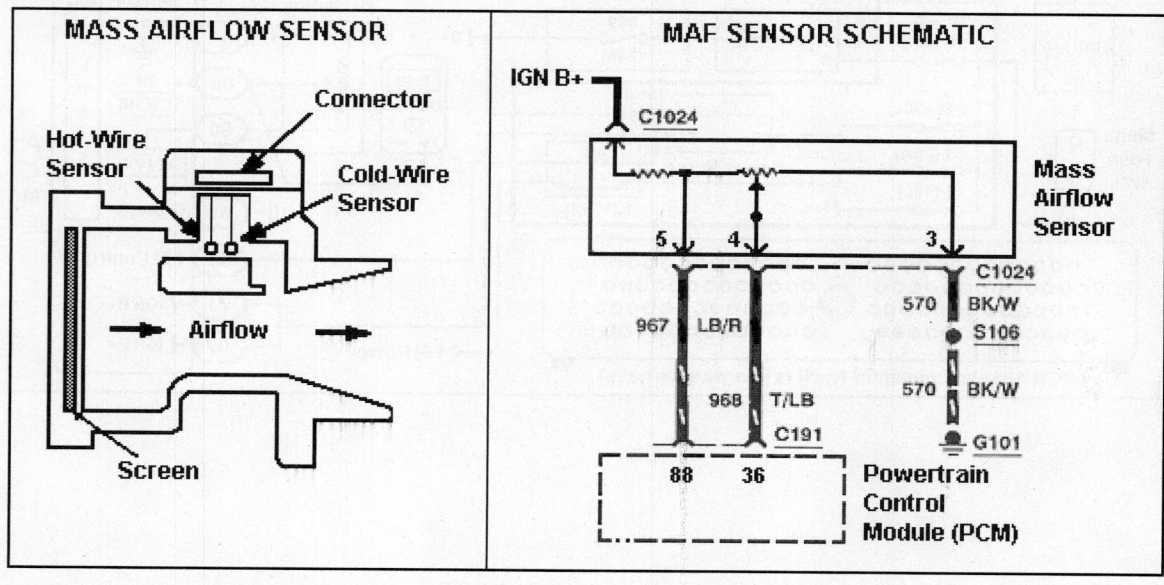

LAB SCOPE TEST (MAF SENSOR)

The Lab Scope can be used to monitor the MAF sensor output signal under various engine load and speed conditions. Place the gearshift selector in Park prior to testing.

Scope Connections

Connect the Channel 'A' positive probe to the MAF signal at the PCM Pin 88 (LB/RED wire) and connect the Channel 'A' negative probe to the battery negative ground post.

Scope Settings

To make the waveforms as clear as possible, set the scope settings to match the examples.

The MAF sensor waveforms can have slight differences from one scope to another.

Lab Scope Example (1)

In Example (1) the trace shows the MAF sensor signal at idle speed. This waveform was captured with the vehicle in Park, the engine at idle and running in closed loop.

The MAF sensor trace should show an even change in its analog signal as the engine speed is changed up and down. It if drops or rises suddenly without a corresponding change in engine speed, there may be a problem in the MAF sensor.

Lab Scope Example (2)

In Example (2) the trace shows the MAF sensor signal at cruise speed (2000 rpm). Note the change in voltage from the Hot Idle value (0.9v) to the Cruise value (1.8v).

Lab Scope Example (3)

In Example (3) the trace shows the MAF sensor signal during a Snap Throttle event.

There are three distinct events shown here:

- The first event is the point at which the throttle begins to open (the point where the signal starts to rise).
- The second event is where the pressure in the intake manifold begins to equalize.
- The third event is where the throttle closes.

Diagnostic Tip

You can use the capture in Example (3) to compare the known good operation of this MAF sensor during a Snap Throttle event to the operation of a MAF sensor that you suspect may be faulty or dirty.

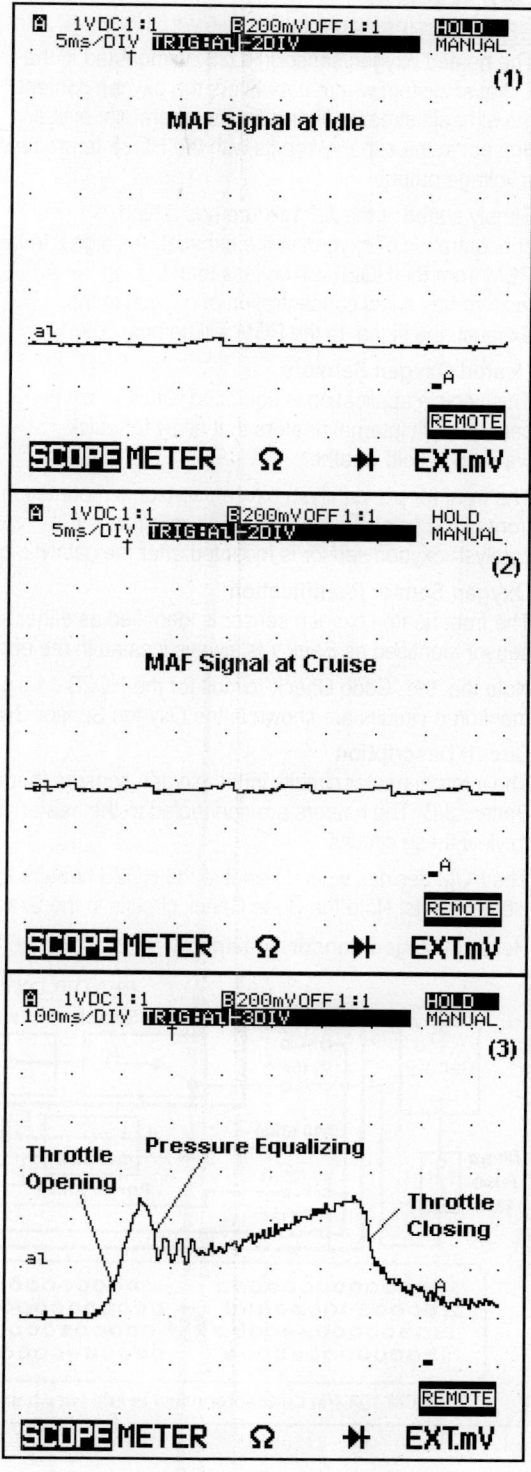

Oxygen Sensor

GENERAL DESCRIPTION

The heated oxygen sensor (HO2S) is mounted in the Exhaust system where it monitors the oxygen content of the exhaust stream. During engine operation, oxygen present in the exhaust reacts with the HO2S to produce a voltage output.

Simply stated, if the A/F mixture has a high concentration of oxygen in the exhaust, the signal to the PCM from the HO2S will be less than 0.4v. If the A/F mixture has a low concentration of oxygen in the exhaust, the signal to the PCM will be over 0.6v.

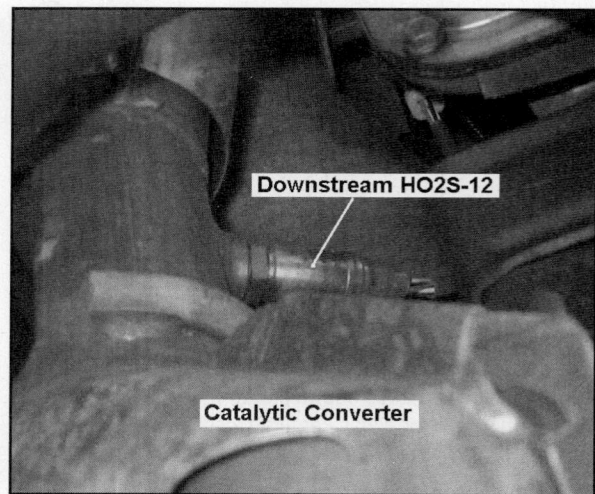

Heated Oxygen Sensors

This vehicle application is equipped with four oxygen sensors with internal heaters that allow for quick warmup in cold weather.

The front (or pre-catalyst) oxygen sensor is mounted in front of the catalytic converter. The rear (or post-catalyst) oxygen sensor is mounted after the catalytic converter.

Oxygen Sensor Identification

The front heated oxygen sensor is identified as either a Bank 1 HO2S-11 or a Bank 2 HO2S-21. The heated oxygen sensor identified as Bank 1 is always located in the engine bank that contains cylinder No. 1.

Note that the "Code Check" circuit for the HO2S-11 signal connects to the PCM at Pin 60. The rest of the HO2S monitored circuits are shown in the Oxygen Sensor Graphic.

Circuit Description

The internal heater circuits in the oxygen sensors (both front and rear pairs) receive power through Fuse #13 in the Battery J/B. The heaters are connected to the heater "control" circuits at the PCM. Refer to the Graphic directly below to view these circuits.

The PCM decides when to enable the HO2S heaters to protect their ceramic elements and when to run the HO2S Heater Tests. Note the Code Check circuits in the Graphic.

Heated Oxygen Sensor Schematic

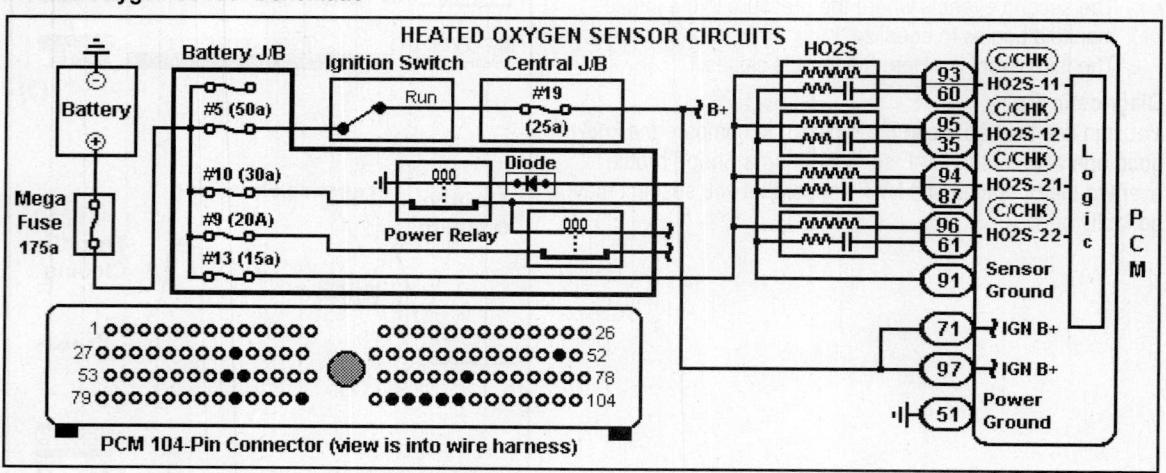

SCAN TOOL TEST (HO2S)

Connect the Scan Tool to the DLC underdash connector. Navigate through the Scan Tool menus until you get to the Ford (OEM) Data List (a list of PID items). Raise the engine speed to 2000 rpm for 3 minutes to set up the test. The examples shown here were captured at Hot Idle in 'P'.

Then select the type of oxygen sensor (O2S) information that you want to view from the various options in the menu.

Scan Tool Example (1)

This example shows the On/Off status of the heaters for the Downstream O2S for Cylinder Bank 1 and Upstream O2S for Cylinder Bank 2. The operation of these heaters is controlled by the PCM (shortly after engine startup).

Scan Tool Example (2)

This example shows the On/Off status of the heater for the Downstream O2S for Cylinder Bank 2 and the current level of the heater for the Upstream O2S for Cylinder Bank 1. The current level shown here (812.2 mA) is the correct amount of current for a Ford HO2S (a known good value).

Scan Tool Example (3)

This example shows the current level of the heater for the Downstream O2S for Cylinder Bank 2 along with the current level for the Downstream O2S for Cylinder Bank 2. These values are within the normal range of this sensor.

Scan Tool Example (4)

This example shows the voltage level of the Downstream O2S for Cylinder Bank 2 and the On/Off status of the heater for the Upstream O2S for Cylinder Bank 1. The DOWN O2S BANK2 voltage reading is in the normal range for a Ford heated oxygen sensor (0.100 to 900 Millivolts).

Scan Tool Example (5)

This example shows the voltage level of the Downstream O2S for Cylinder Bank 1 and the voltage level of the Upstream O2S for Cylinder Bank 2. Both of the voltage readings in this example are within the normal range for a Ford heated oxygen sensor (0.100 to 900 Millivolts).

Summary

There are numerous screens available on the Scan Tool that can help you diagnose the condition of the heated oxygen sensor (HO2S) and its heater. You can vary the screens to obtain information from a particular sensor.

The examples shown here can be used to compare to the actual readings you obtain during testing. In effect, they are known good O2S reading for a typical Ford 4.0L V6 engine.

```
Jan 20, 02    10:10:20 am (1)

  DN O2S HEATER B1
    ON
  UP O2S HEATER B2
    ON

Always use [HELP] to
  list keys & modes
[RCV] For enhanced
[*EXIT] To exit
```

```
Jan 20, 02    10:10:20 am (2)

  DN O2S HEATER B2
    ON
  UPO2S HTR MON B1
    812.2 mA

Always use [HELP] to
  list keys & modes
[RCV] For enhanced
[*EXIT] To exit
```

```
Jan 20, 02    10:10:20 am (3)

  DNO2S HTR MON B2
    911.4 mA
  DNO2S HTR MON B1
    878.1 mA

Always use [HELP] to
  list keys & modes
[RCV] For enhanced
[*EXIT] To exit
```

```
Jan 20, 02    10:10:21 am (4)

  DOWN O2S BANK2
    0.32 VOLTS
  UP O2S HEATER B1
    ON

Always use [HELP] to
  list keys & modes
[RCV] For enhanced
[*EXIT] To exit
```

```
Jan 20, 02    10:10:21 am (5)

  DOWN O2S BANK1
    0.31 VOLTS
  UP O2S BANK2
    0.71 VOLTS

Always use [HELP] to
  list keys & modes
[RCV] For enhanced
[*EXIT] To exit
```

LAB SCOPE TEST (HO2S)

The Lab Scope is the "tool of choice" to test the oxygen sensor as it provides an accurate view of the sensor response and switch rate. Place the gearshift selector in Park and block the drive wheels prior to starting the test.

Lab Scope Connections

Connect the Channel 'A' positive probe to the sensor signal wire and the negative probe to the battery negative post.

Scope Settings

To make the waveforms as clear as possible, set the scope settings to match the examples.

Depending upon the scope capabilities, the sensor waveforms will have slight differences.

Lab Scope Tests

The examples to the right are from the original oxygen sensor (18,000 miles). The vehicle was driven long enough for the engine to enter closed loop operation prior to testing.

Lab Scope Example (1)

In example (1), the two traces from Bank 1 were captured as the vehicle at Hot Idle speed (with the engine warmed up and operating in closed loop). The HO2S signal should switch L-R and R-L in less than 100 ms, and should switch at least 3-5 times per second.

Lab Scope Example (2)

In example (1), the two traces from the Bank 2 sensors were captured as the vehicle was operated at Hot Idle speed (with the engine warmed up and operating in closed loop.

Lab Scope Example (3)

In example (3), the traces was captured during a snap throttle event in order to view the true half-cycle time phase of a Zirconia design HO2S in a vehicle with 18,000 miles on it.

Note the rapid rise and fall of the HO2S-11 signal during the Snap Throttle test in this example. You can see that the HO2S signal rose and fell well within one subdivision of time (40 ms) in this known good example.

This oxygen sensor passed because the half cycle time phase for this Zirconia oxygen sensor was less than 100 ms. This waveform shows the HO2S-11 signal from a known good oxygen sensor during Snap Throttle.

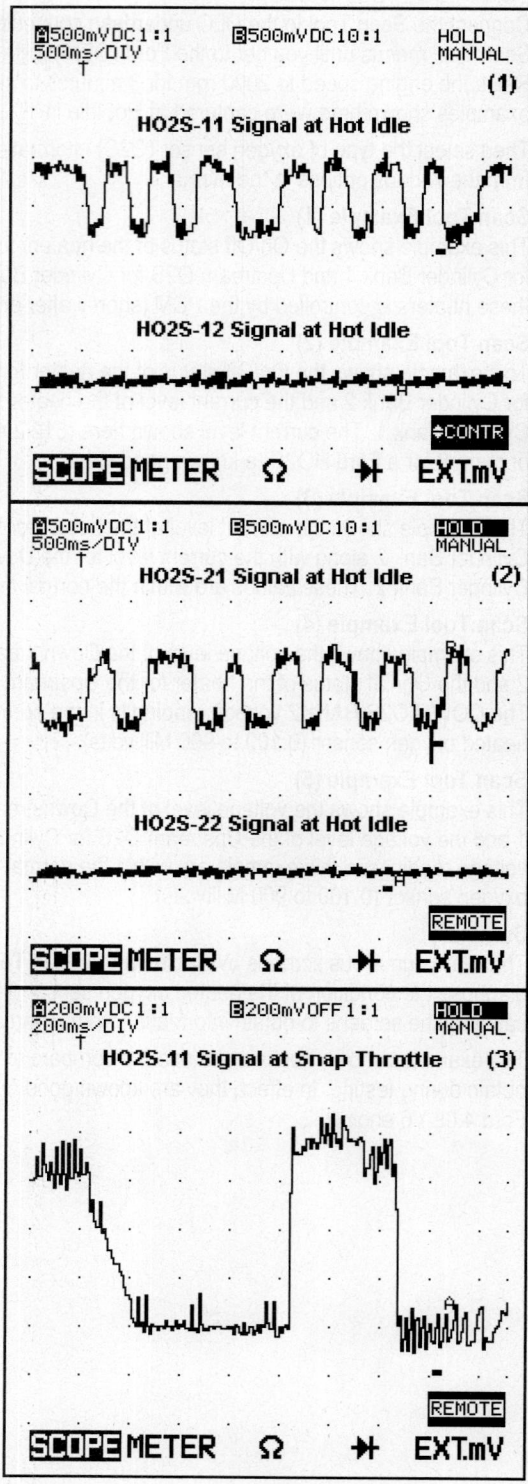

Transmission Controls

COMPONENT TESTS

This vehicle is equipped with a 5-speed 5R55E automatic transmission. The PCM controls all upshift, downshift, line pressure and TCC operations. It has three compound planetary gear sets, three bands, three multi-plate clutches and two one-way clutches.

Turbine Shaft Speed Sensor

The turbine shaft speed (TSS) sensor is a magnetic pickup that is mounted externally on the case. The PCM uses the TSS signal to decide the correct operating pressures and TCC operation.

Electronic Pressure Control

The electronic pressure control (EPC) solenoid is used to regulate transaxle line and line modulator pressure by producing resisting forces to the main regulator and line modulator circuits.

Torque Converter Clutch Solenoid

The torque converter clutch (TCC) solenoid used with this application controls the application, modulation and release of the torque converter clutch. The control signal is PWM.

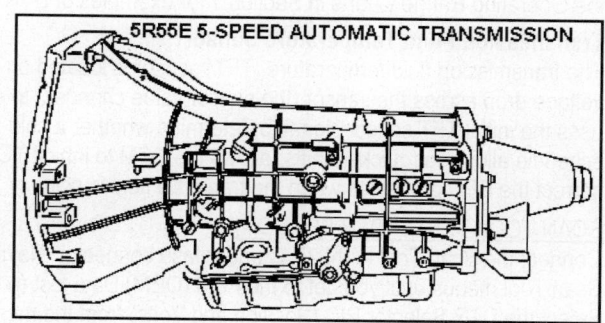

SCAN TOOL TESTS (EPC & TSS)

Connect the Scan Tool to the DLC underdash connector. Navigate through the Scan Tool menus until you get to the Ford (OEM) Data List (a list of PID items). Select the EPC (P) PID and the TSS (RPM) PID from the menu selections. The captures on this page were done in Park at Hot Idle.

SCAN TOOL TESTS (TCC & EPC)

Navigate through the Scan Tool menus until you get to the Ford (OEM) Data List (a list of PID items). Select the TCC (%) PID and the EPC (V) PID selections from the menu.

```
Jan 20, 02,   10:10:22 am

EPC (P)
  15.5 PSI
TURBINE SHAFT
  750 RPM

Always use [HELP] to
list keys & modes
[RCV] For enhanced
[*EXIT] To exit
```

```
Jan 20, 02   10:10:22 am

TCC MODULATION
  0 %
EPC (V)
  7.0 VOLTS

Always use [HELP] to
list keys & modes
[RCV] For enhanced
[*EXIT] To exit
```

Transmission Control Schematic

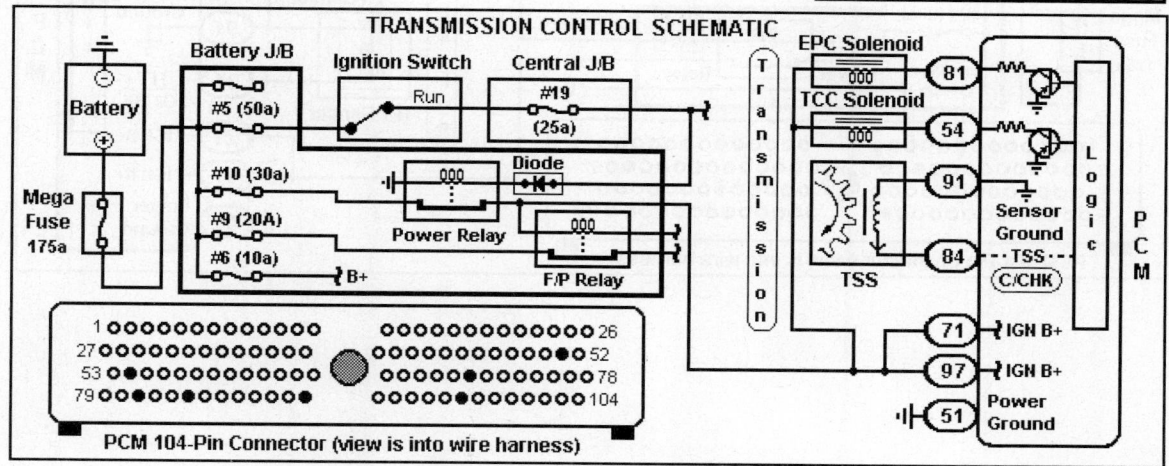

Digital Transmission Range Sensor

The digital transmission range (DTR) sensor is located outside the transmission on the manual lever. The DTR sensor completes the start circuit in Park and Neutral, the backup lamp circuit in Reverse and the Neutral sense circuit in Neutral (4X4 only). The DTR sensor also opens and closes a set of four (4) switches monitored by the PCM. Refer to the Operating Range Charts in Section 1 for examples of DTR sensor values.

Transmission Fluid Temperature Sensor

The transmission fluid temperature (TFT) sensor is located on the transmission control body. The PCM monitors the voltage drop across the sensor (the sensor value changes) to determine the transmission fluid temperature. The PCM uses the initial TFT sensor signal to determine whether a cold start shift schedule is necessary. The cold start shift schedule allows for quicker shifts, allows the PCM to inhibit TCC lockup and to correct the EPC pressures when the transaxle fluid is cold.

SCAN TOOL TESTS (DTR SENSOR)

Connect the Scan Tool to the DLC underdash connector. Navigate through the Scan Tool menus until you get to the Ford (OEM) Data List (a list of PID items). Select the DTR Selector PID (Position and Volts) from the menu selections.

SCAN TOOL TESTS (TFT SENSOR)

Connect the Scan Tool to the DLC underdash connector. Navigate through the Scan Tool menus until you get to the Ford (OEM) Data List (a list of PID items). Select the TFT Sensor PID (Degrees and Volts) from the menu selections.

Transmission Control Schematic

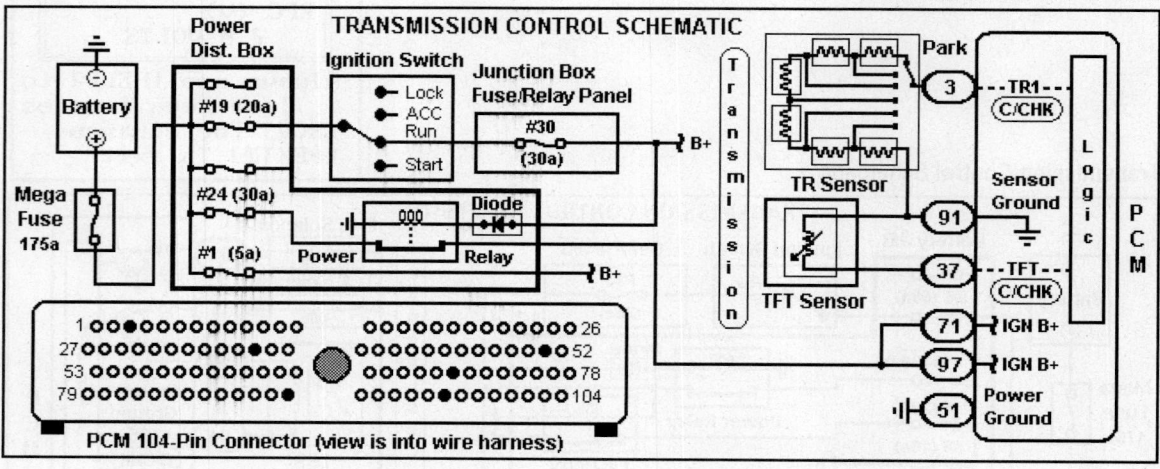

Reference Information

PCM PID TABLES - INPUTS

■ NOTE: *The following readings were obtained with the engine at idle speed, radiator hose hot, throttle closed, gear selector in Park and all accessories turned off.*

Acronym	Scan Tool Parameter	Range	Typical Value
4x4L	4x4 Signal	ON / OFF	OFF (ON in 4x4)
ACCS	A/C Cycling Clutch Switch Signal	ON / OFF	OFF (ON with A/C on)
ACP	A/C Pressure Switch Signal	OPEN/CL	OPEN (CL with A/C on)
BPP	Brake Pedal Position Switch	ON / OFF	OFF (ON with brake on)
CMP (CID)	CMP Sensor Signal	0-1000 Hz	13-16
CKP	CKP Sensor Signal	0-10,000 Hz	430-475
DPFE	EGR DPFE Sensor Signal	0-5.1v	0.20-1.30
ECT °F	Engine Coolant Temp. Sensor	-40°F to 304°F	160-200
FEPS	Flash EEPROM	0-5.1v	0.1
FLI	Fuel Level Indicator Signal (volts)	0-5.1v	1.7v at 50% full
FPM	Fuel Pump Monitor	ON / OFF	ON
FTP	Fuel Tank Pressure Sensor	0-5.1v or 0-10" H2O	2.6v (0 in. H2O - cap off)
Gear	Gear Position	1-2-3-4-5	1
HO2S-11	HO2S-11 (Bank 1 Sensor 1)	0-1100 mv	100-900 mv
HO2S-12	HO2S-12 (Bank 1 Sensor 2)	0-1100 mv	100-900 mv
HO2S-21	HO2S-21 (Bank 2 Sensor 1)	0-1100 mv	100-900 mv
HO2S-22	HO2S-22 (Bank 2 Sensor 2)	0-1100 mv	100-900 mv
IAT °F	Intake Air Temperature Sensor	-40°F to 304°F	50-120
KS1	Knock Sensor Signal	0-5.1v	0 (use a Lab Scope)
LOAD	Calculated Engine Load	0-100%	17-28
MAF V	Mass Airflow Sensor Signal	0-5.1v	0.6-0.9
MISF	Misfire Detection	ON / OFF	OFF
OCT ADJ	Octane Adjust Signal	Closed / Open	Closed
OSS	Output Shaft Speed Sensor	0-10,000 rpm	0
PNP	Park Neutral Position Switch	ON / OFF	ON
RPM	Engine Speed	0-9999	750
TFT °F	Transmission Fluid Temp. Sensor	-40°F to 304°F	110-210
TP V	Throttle Position	0-5.1v	0.53-1.27v
TR1	Digital TR1 Sensor Signal	0v or 12v	0, in Drive: 11.5
TR2	Digital TR2 Sensor Signal	0v or 12v	0, in Drive: 11.5
TR4	Digital TR4 Sensor Signal	0v or 12v	0, in Drive: 11.5
TR V/TR	Digital TR Sensor Signal	PN or O/D	P/N
TSS	Turbine Shaft Speed Sensor	0-10,000 rpm	750
Vacuum	Engine Vacuum	0-30" Hg	19-20" Hg
VPWR	Vehicle Power	0-25.5v	14.1v
VSS	Vehicle Speed	0-159 mph / 0-1000 Hz	55 mph = 125 Hz

PCM PID TABLES - OUTPUTS

■ NOTE: *The following readings were obtained with the engine at idle speed, radiator hose hot, throttle closed, gear selector in Park and all accessories turned off.*

Acronym	Scan Tool Parameter	Range	Typical Value
CD1	Coil 1 Driver (Dwell)	0-60° dwell	6
CD2	Coil 2 Driver (Dwell)	0-60° dwell	6
CD3	Coil 3 Driver (Dwell)	0-60° dwell	6
CTO	Clean Tachometer Output Signal	0-1,000 Hz	35-49
EGR VR	EGR Vacuum Regulator Valve	0-100%	0
EVAP CV	EVAP Canister Vent Valve	0-100%	0
EVAP CP	EVAP Canister Purge Valve	0-100%	0-100
EPC	EPC Solenoid Control	0-25.5v	7
FP	Fuel Pump Control	ON / OFF	ON
Fuel B1	Fuel Injector Pulsewidth Bank 1	0-99.9 ms	2.9-3.3
Fuel B2	Fuel Injector Pulsewidth Bank 2	0-99.9 ms	2.9-3.3
HTR-11	HO2S-11 Heater Control	ON / OFF	ON
HTR-12	HO2S-12 Heater Control	ON / OFF	ON
HTR-21	HO2S-21 Heater Control	ON / OFF	ON
HTR-22	HO2S-22 Heater Control	ON / OFF	ON
IAC	Idle Air Control Motor	0-100%	33 (10.7v)
INJ1	Fuel Injector 1 Control	0-99.9 ms	2.9-3.3
INJ2	Fuel Injector 2 Control	0-99.9 ms	2.9-3.3
INJ3	Fuel Injector 3 Control	0-99.9 ms	2.9-3.3
INJ4	Fuel Injector 4 Control	0-99.9 ms	2.9-3.3
INJ5	Fuel Injector 5 Control	0-99.9 ms	2.9-3.3
INJ6	Fuel Injector 6 Control	0-99.9 ms	2.9-3.3
LONGFT1	Long Term Fuel Trim Bank 1	-20% to 20%	0
LONGFT2	Long Term Fuel Trim Bank 2	-20% to 20%	0
MIL	Malfunction Indicator Lamp Control	ON / OFF	OFF (ON with DTC)
PATSOUT	PATS Output Signal	0v or 12v	1
SHRTFT1	Short Term Fuel Trim Bank 1	-10% to 10%	-1 to 1
SHRTFT2	Short Term Fuel Trim Bank 2	-10% to 10%	-1 to 1
SPARKADV	Spark Advance	-90° to 90°	20-25
SS1	Transmission Shift Solenoid 'A'	ON / OFF	ON
SS2	Transmission Shift Solenoid 'B'	ON / OFF	OFF
SS3	Transmission Shift Solenoid 'C'	ON / OFF	OFF
SS4	Transmission Shift Solenoid 'D'	ON / OFF	ON
TCC	Torque Converter Clutch Solenoid	0-100%	0, at 55 mph: 95
TCIL	Trans. Control Indicator Lamp	ON / OFF	OFF (ON if on)
TPO	Transmission Power Takeoff	0-100%	15
VREF	Voltage Reference	0-5.1v	4.9-5.1v

PCM Pin Voltage Tables

PCM PIN VOLTAGE TABLE: 104-PIN CONNECTOR

Pin Number	Wire Color	Application	Value at Hot Idle
1	VT/OG	Transmission Shift Solenoid 'B' Control	12v
2	PK/LG	MIL (lamp) Control	Lamp off: 12v, On: 1v
3	YE/BK	Digital TR1 Sensor Signal	0v, 55 mph: 12v
4-5	---	Not Used	---
6	DB/YE	Output Shaft Speed Sensor Signal	0 Hz
7-11	---	Not Used	---
12	YE/WH	Fuel Gauge Signal	1.7v (1/2 full)
13	VT	Flash EEPROM Power	0.1v
14	LB/BK	4x4 Indicator Lamp Control	Lamp On: 1v, Off: 12v
15	PK/LB	SCP Data Bus (-)	Digital Signals
16	TN/OG	SCP Data Bus (+)	Digital Signals
17-18, 20, 23	---	Not Used	---
19	OG/BK	Load Leveling Acceleration Signal	0.1v (Off)
21	DB	CKP+ Sensor Signal	430-460 Hz
22	GY	CKP- Sensor Signal	<0.050v
24, 51	BK/WH	Power Ground	<1v
25	BK	Chassis Ground	<0.050v
26	TN/WH	Coil Driver 1 Control	6° dwell
27	OG/YE	Transmission Shift Solenoid 'A' Control	1v, 55 mph: 12v
28	BN/OG	Transmission Shift Solenoid 'D' Control	1v, 55 mph: 12v
29	TN/WH	TCS (Switch) Signal	TCS & O/D On: 12v
30-31, 33-34	---	Not Used	---
32	DG/VT	Knock Sensor Return	<0.050v
35	RD/LG	HO2S-12 (Bank 1 Sensor 2) Signal	100-900 mv
36	TN/LB	MAF Sensor Return	<0.050v
37	OG/BK	Transmission Fluid Temperature Sensor	1.90v
38	LG/RD	Engine Coolant Temperature Sensor	0.7v
39	GY	Intake Air Temperature Sensor	1.2v
40	DG/Y	Fuel Pump Relay Monitor	Pump On: 12v, Off: 0v
41	VT	A/C Cycling Clutch Switch Signal	A/C Switch On: 12v
42, 44	---	Not Used	---
43	LB/PK	Message Center (Fuel Flow Data)	Digital Signals
45	YE/WH	PCM to Generic Electronic Module	Digital Signals
46	BKWH	Fuel Cap Indicator	Lamp On: 1v, Off: 12v
47	BN/PK	Vacuum Regulator Solenoid	0% (55 mph: 45%)
48	TN/YE	Clean Tachometer Output	35-49 Hz

PCM 104-Pin Connector

TERMINAL VIEW OF WIRING HARNESS CONNECTOR FOR PCM (104 PIN)

1 oooooooooooooo   oooooooooooooo 26
27 oooooooooooooo   oooooooooooooo 52
53 oooooooooooooo ● oooooooooooooo 78
79 oooooooooooooo   oooooooooooooo 104

Note: back-probing the 104-pin connector with a DVOM probe can damage the PCM - use a breakout box!

PCM PIN VOLTAGE TABLE: 104-PIN CONNECTOR

Pin Number	Wire Color	Application	Value at Hot Idle
49	LB/BK	Digital TR2 Sensor Signal	0v, 55 mph: 12v
50	WH/BK	Digital TR4 Sensor Signal	0v, 55 mph: 12v
52	TN/OG	Coil Driver 2 Control	6° dwell
53	PK/BK	Transmission Shift Solenoid 'C' Control	12v, 55 mph: 0v
54	VT/YE	Torque Converter Clutch Control	0%, 55 mph: 0-90%
55	YE	Keep Alive Power	12-14v
56	LG/BK	EVAP Canister Purge Solenoid Control	0-10 Hz
57	YE/RD	Knock Sensor 1 Signal	2.5v (use Lab Scope)
58	GY/BK	VSS+ Signal	At 55 mph: 125 Hz
59, 63	---	Not Used	---
60	GY/LB	HO2S-11 (Bank 1 Sensor 1) Signal	100-900 mv
61	VT/LG	HO2S-22 (Bank 2 Sensor 2) Signal	100-900 mv
62	RD/PK	Fuel Tank Pressure Sensor Signal	2.6v (0 in. H2O - cap off)
64	LB/YE	Digital TR Sensor Signal	In P: 0v, at 55 mph: 1.7v
65	BN/LG	DPFE Sensor Signal	0.20-1.30v
66, 68	---	Not Used	---
67	VT/WH	EVAP Canister Vent Solenoid	0 Hz
70, 72, 82	---	Not Used	---
71, 97	RD	Vehicle Power	12-14v
73	TN/BK	Fuel Injector 5 Control	3.4-3.8 ms
74	BN/YE	Fuel Injector 3 Control	3.4-3.8 ms
75	TN	Fuel Injector 1 Control	3.4-3.8 ms
76-77, 103	BK/WH	Power Ground	<0.1v
78	TN/LG	Coil Driver 3 Control	6° dwell
79	WH/LG	Transmission Control Indicator Lamp	Lamp On: 1v, Off: 12v
80	LB/OG	Fuel Pump Relay Control	Relay On: 1v, Off: 12v
81	WH/YE	Transmission EPC Solenoid Control	9.1v (6 psi)
83	WH/LB	Idle Air Control Motor Signal	10v (25-32% duty cycle)
84	DG/WH	Turbine Shaft Speed Sensor Signal	85-100 Hz
85	DB/OG	Camshaft Position Sensor Signal	6-8 Hz
86	BK/YE	PCM to A/C Cycling Pressure Switch	Digital Signals
87	RD/BK	HO2S-21 (Bank 2 Sensor 1) Signal	100-900 mv
88	LB/RD	Mass Airflow Sensor Signal	0.6v
89	GY/WH	Throttle Position Sensor	0.53-1.27v
90	BN/WH	Voltage Reference	4.9-5.1v
91	GY/RD	Sensor Signal Return	<0.050v
92	RD/LG	Brake Pedal Position Switch Signal	Brake Off: 0v, On: 12v
93	RD/WH	HO2S-11 (Bank 1 Sensor 1) Heater	Heater On: 1v, Off: 12v
94	YE/LB	HO2S-21 (Bank 2 Sensor 1) Heater	Heater On: 1v, Off: 12v
95	WH/BK	HO2S-12 (Bank 1 Sensor 2) Heater	Heater On: 1v, Off: 12v
96	TN/YE	HO2S-22 (Bank 2 Sensor 2) Heater	Heater On: 1v, Off: 12v
98, 102, 104	---	Not Used	---
99	LG/OG	Fuel Injector 6 Control	3.4-3.8 ms
100	BN/LB	Fuel Injector 4 Control	3.4-3.8 ms
101	WH	Fuel Injector 2 Control	3.4-3.8 ms

PCM Wiring Diagrams

FUEL & IGNITION SYSTEM DIAGRAMS (1 OF 9)

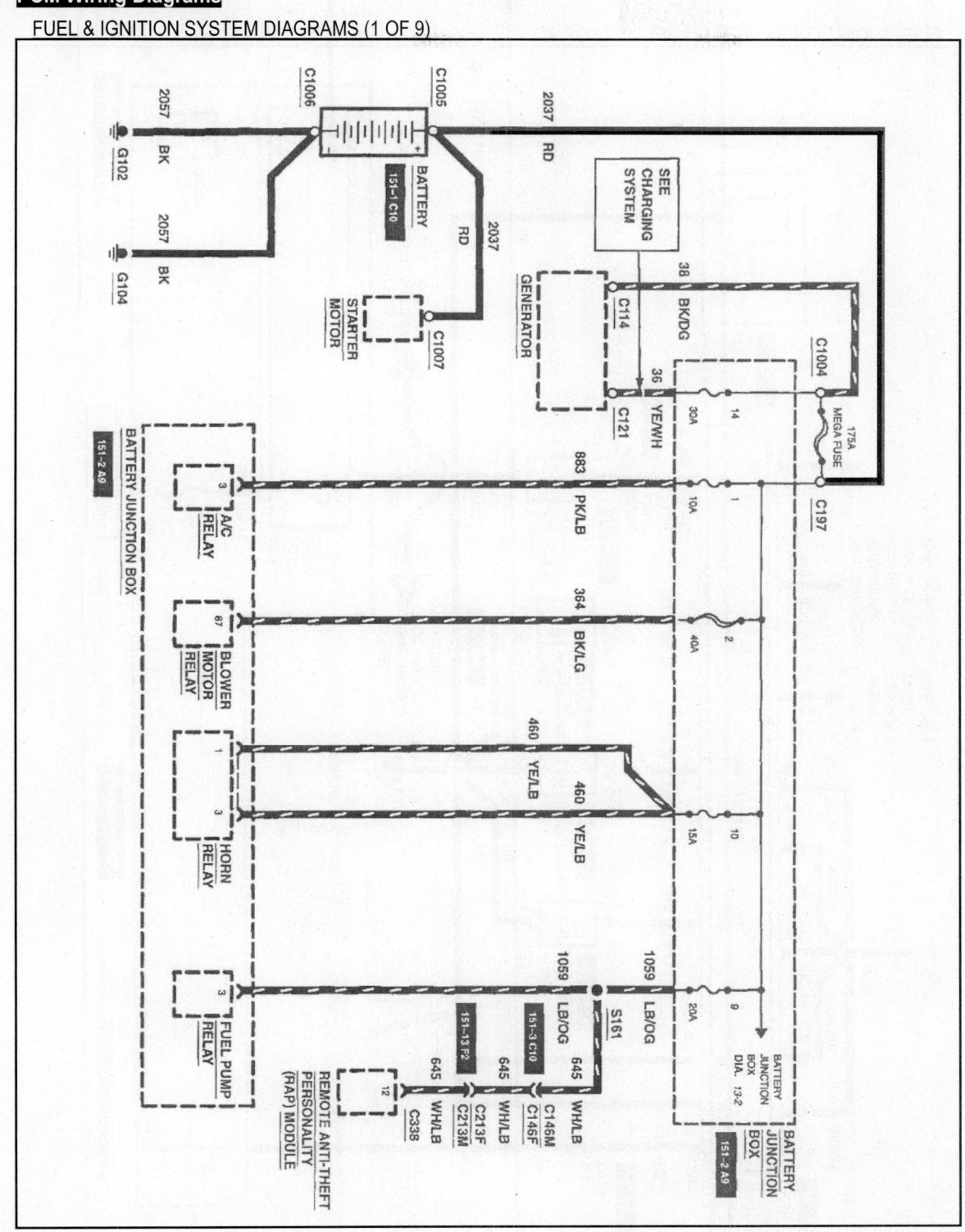

FUEL & IGNITION SYSTEM DIAGRAMS (2 OF 9)

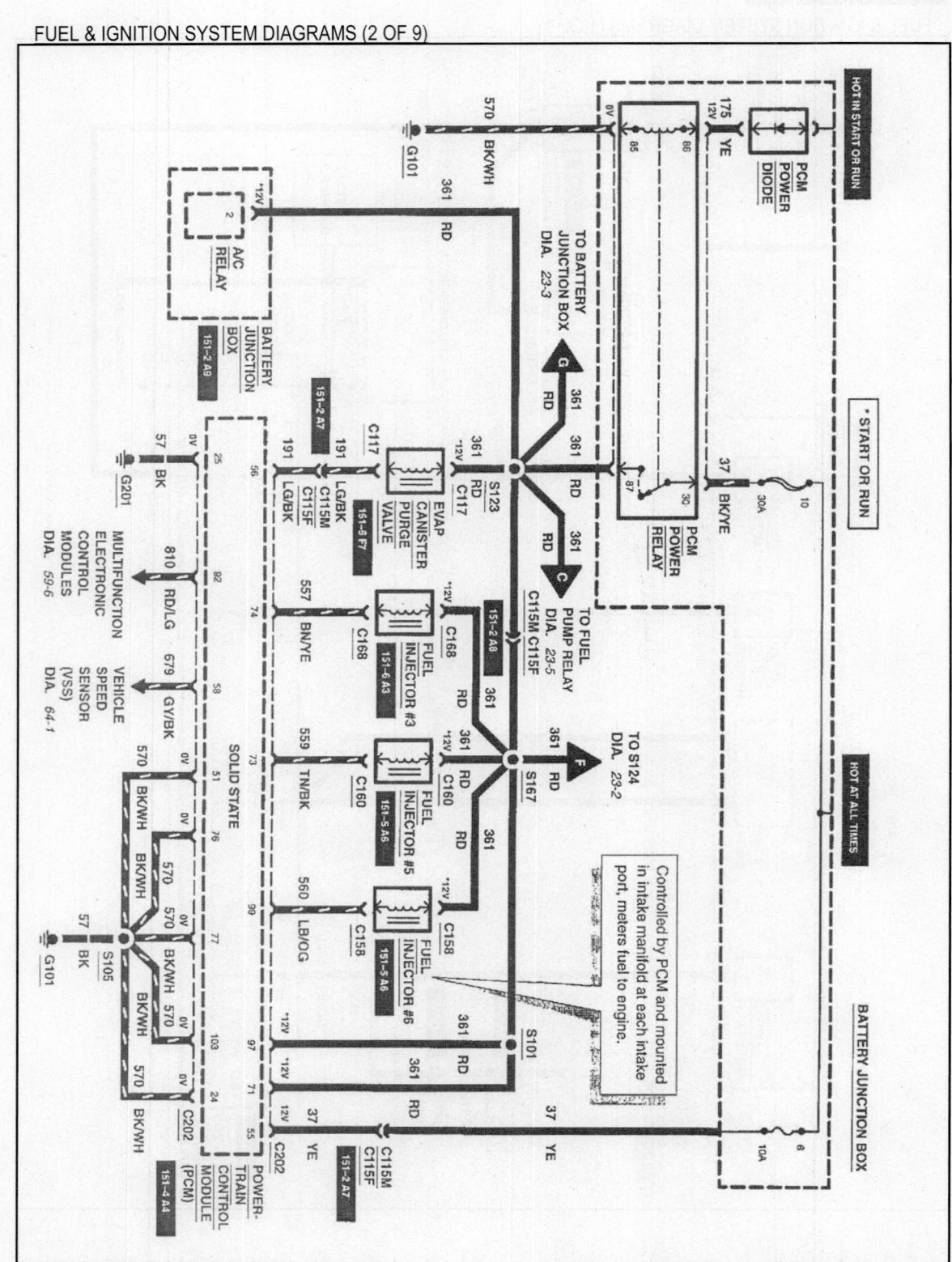

FUEL & IGNITION SYSTEM DIAGRAMS (3 OF 9)

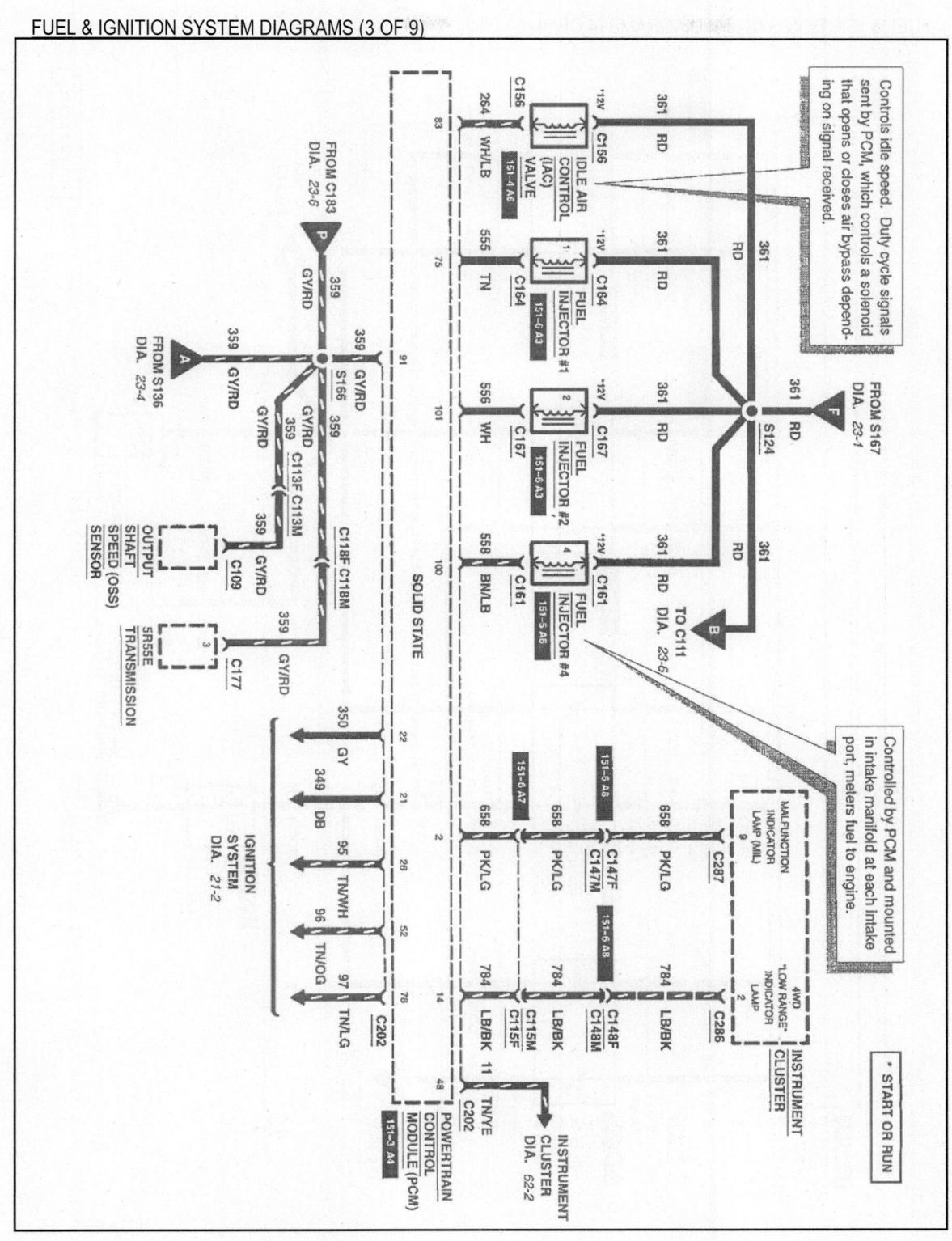

FUEL & IGNITION SYSTEM DIAGRAMS (4 OF 9)

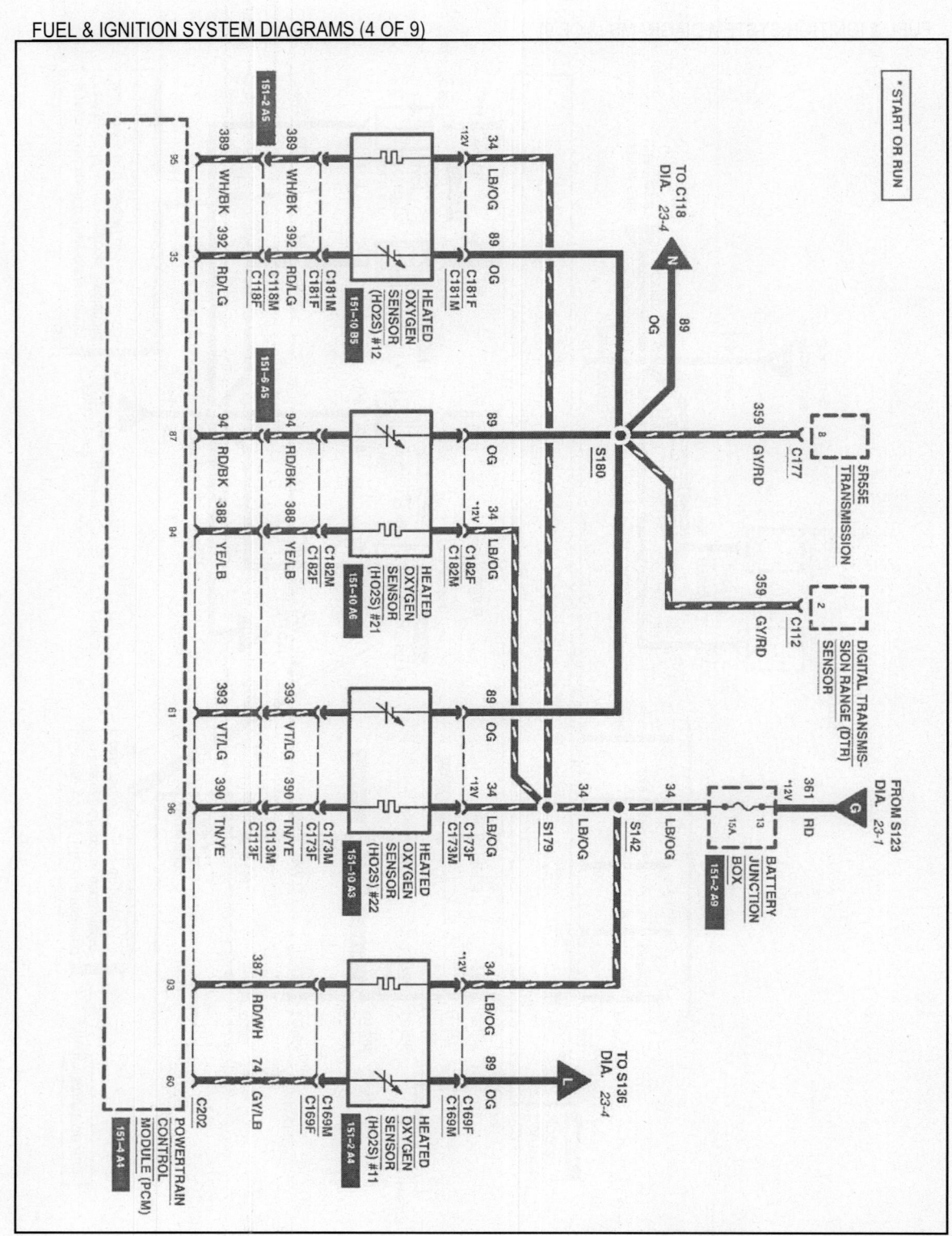

FUEL & IGNITION SYSTEM DIAGRAMS (5 OF 9)

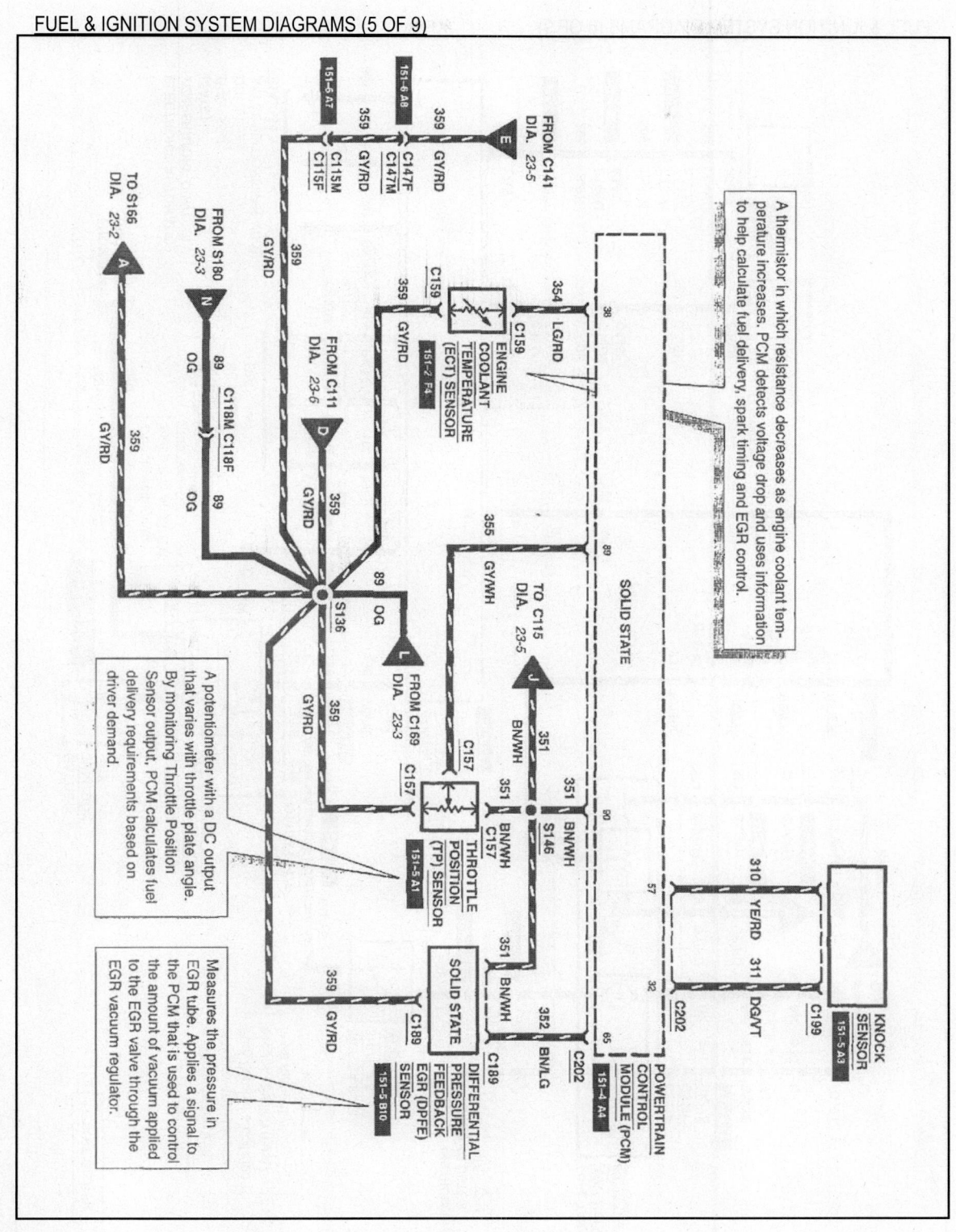

FUEL & IGNITION SYSTEM DIAGRAMS (6 OF 9)

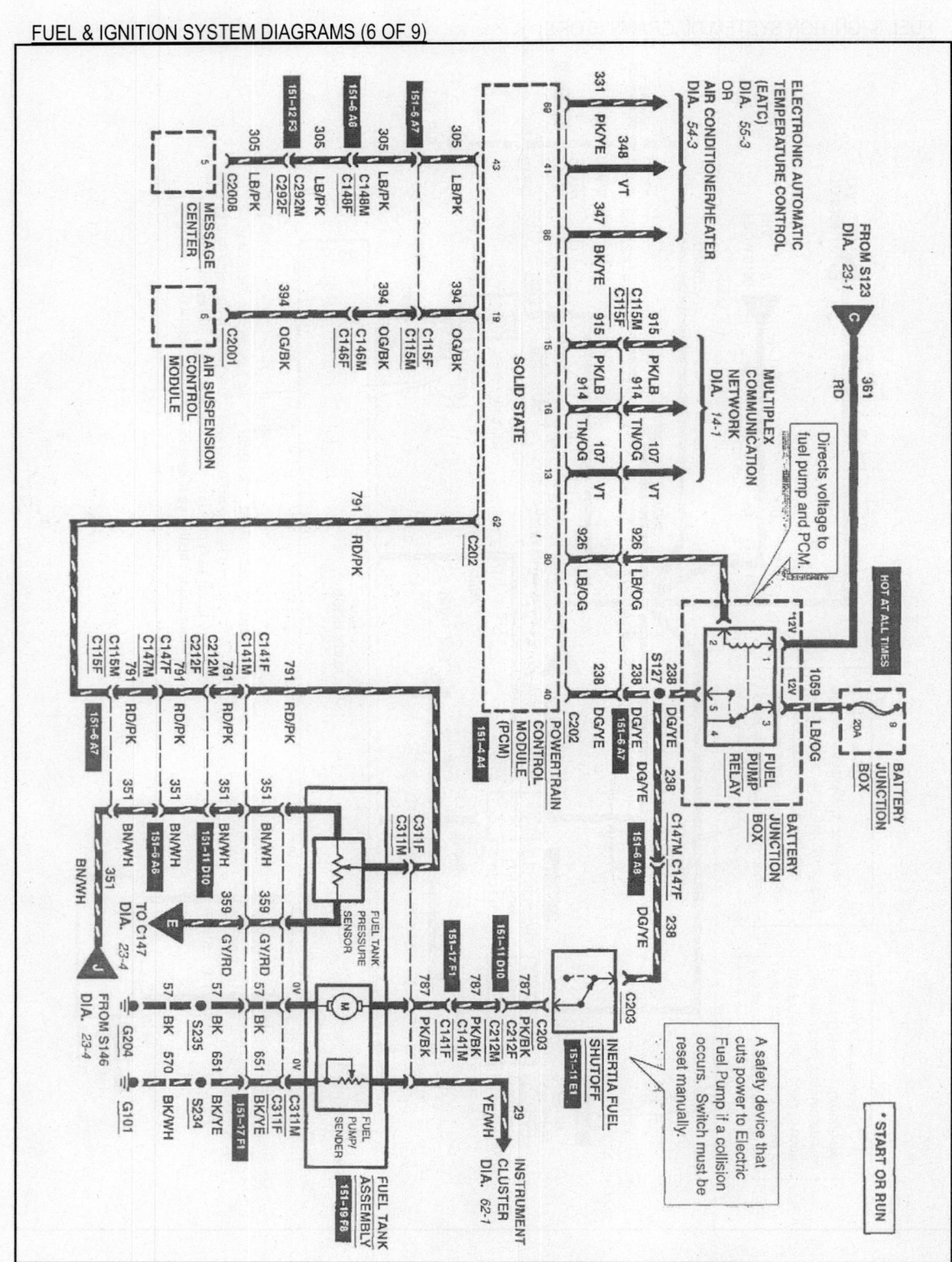

FUEL & IGNITION SYSTEM DIAGRAMS (7 OF 9)

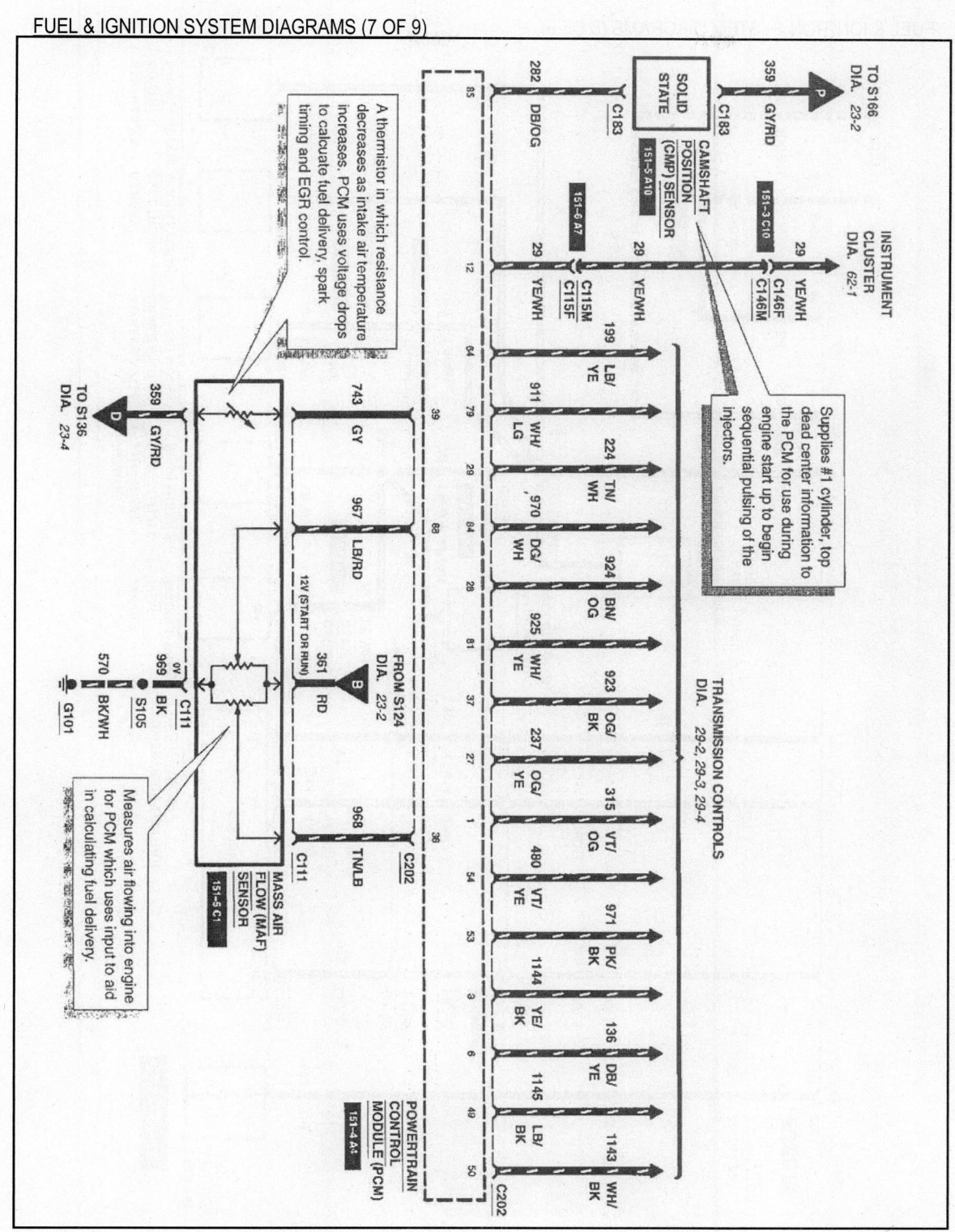

FUEL & IGNITION SYSTEM DIAGRAMS (8 OF 9)

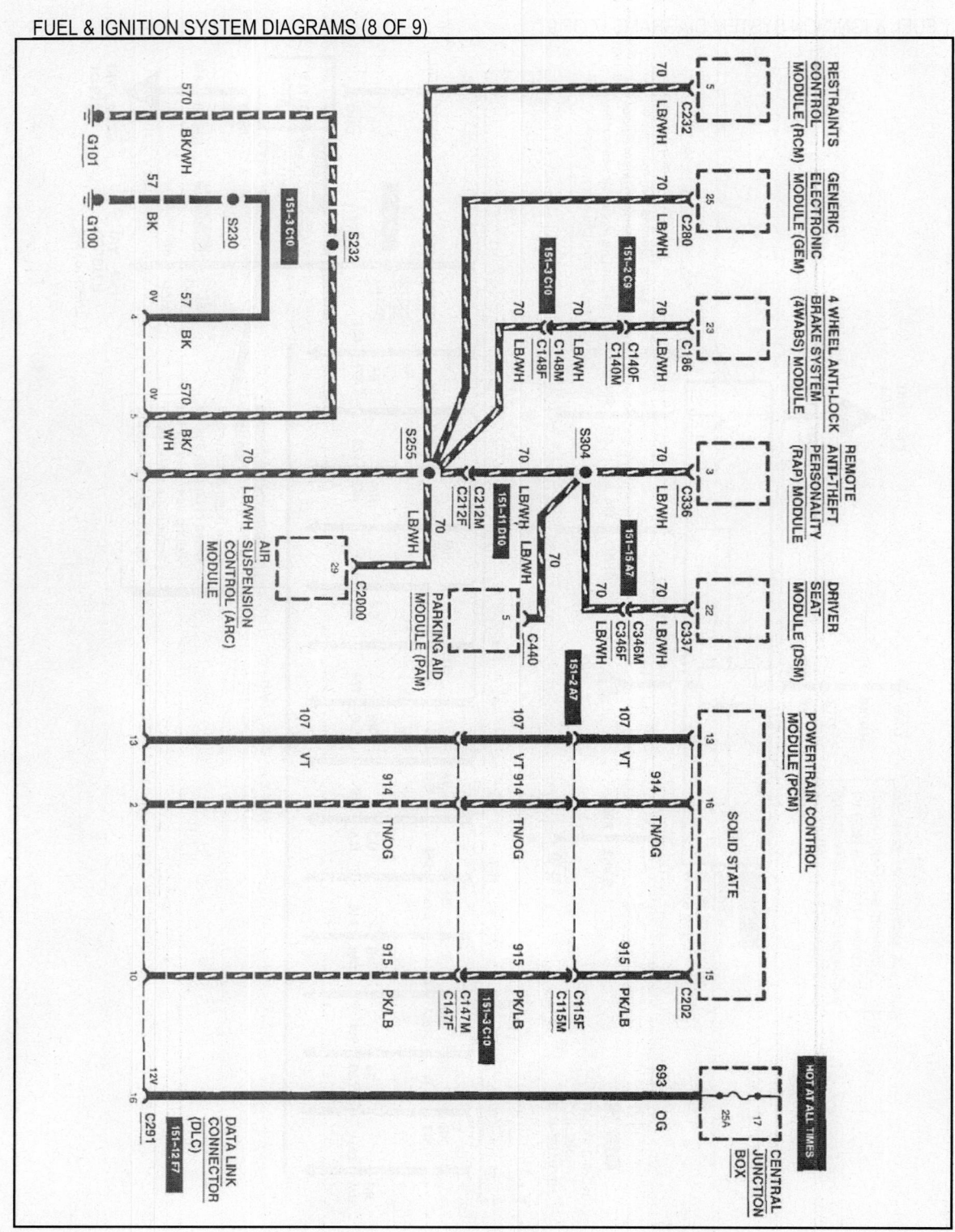

FUEL & IGNITION SYSTEM DIAGRAMS (9 OF 9)

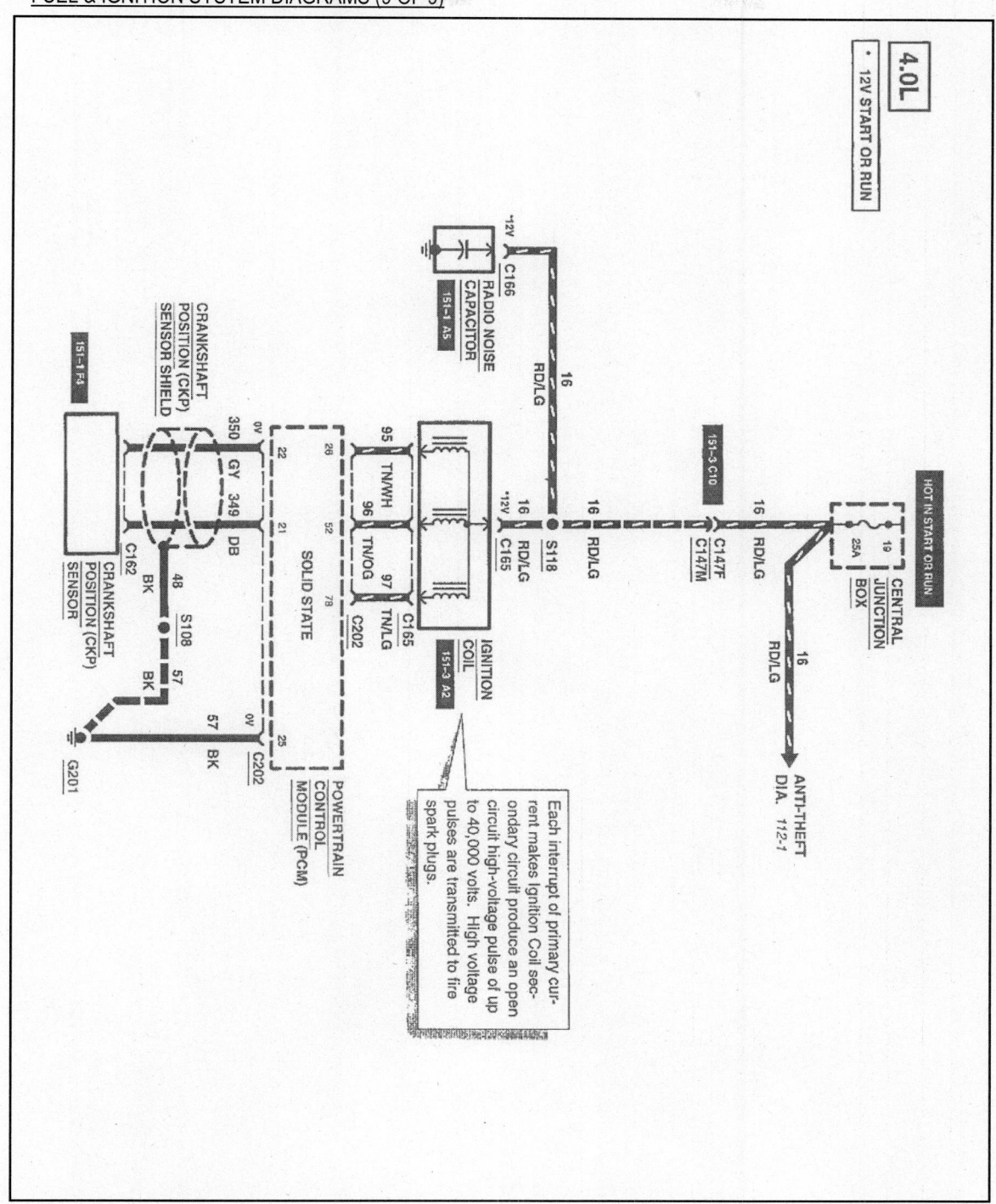

Ford OBD II Diagnostics Contents

ABOUT THIS SECTION
    How to Use Each Section ....Page 6-2

OBD II SYSTEMS
    Powertrain Control Module ....Page 6-3
    Flash EEPROM ....Page 6-3
    OBD II Phase In ....Page 6-4
    Diagnostic Test Modes ....Page 6-4
    On-board Diagnostics ....Page 6-5

OBD SYSTEM TERMINOLOGY
    Two-Trip Detection & Similar Conditions ....Page 6-6
    OBD II Warmup Cycle ....Page 6-7
    Freeze Frame Data ....Page 6-7
    Cylinder Bank & Oxygen Sensor Identification ....Page 6-9
    Adaptive Fuel Control Strategy ....Page 6-10
    Enable Criteria ....Page 6-11
    OBD II Drive Cycle Procedure ....Page 6-12

OBD II SYSTEM MONITORS
    Comprehensive Component Monitor ....Page 6-13
    Catalyst Efficiency Monitor ....Page 6-14
    EGR System Monitor ....Page 6-16
    EVAP System Monitor ....Page 6-19
    Fuel System Monitor ....Page 6-24
    Misfire Detection Monitor ....Page 6-25
    Oxygen Sensor Monitor ....Page 6-27
    Oxygen Sensor Heater Monitor ....Page 6-28
    Air Injection System Monitor ....Page 6-29

REFERENCE INFORMATION
    PID Data Explanation ....Page 6-30
    OBD II PID Mode ....Page 6-31
    Pin Voltage Table Explanation ....Page 6-32
    Using a Breakout Box ....Page 6-33
    PCM Actuator & Sensor Tables...Page 6-34
    PID Data & Pin Chart Glossary ....Page 6-39

GAS ENGINE OBD II TROUBLE CODES
    P0xxx Trouble Code List ....Page 6-41
    P1xxx Trouble Code List ....Page 6-90
    P2xxx Trouble Code List...Page 6-121
    U1xxx Trouble Code List ....Page 6-129
    U2xxx Trouble Code List ....Page 6-132

DIESEL ENGINE OBD II TROUBLE CODES
    B1xxx Trouble Code List ....Page 6-133
    P0xxx Trouble Code List ....Page 6-135
    P1xxx Trouble Code List ....Page 6-160
    P2xxx Trouble Code List ....Page 6-176
    U1xxx Trouble Code List ....Page 6-180

How to Use Each Section

Section 6: OBD II Diagnostics
Refer to this section to learn more about OBD II system operation on Ford applications. If you want to clear codes, learn how to drive a particular vehicle to run the OBD II Main Monitors, look in this section. If you want to know why a code set, or how to drive the vehicle to verify the repair of a trouble code, look in this section.

Section 7: PID Data for Ford, Lincoln & Mercury Applications
• 1991-97 Association Vehicles (Aspire, Escort/Tracer & Probe V6)
• 1990-2003 Domestic Car applications
Refer to this section to look up PID Data information for Ford, Lincoln, Mercury, and Association applications from 1990-2003.

Section 8: PID Data for Ford Truck, Van & SUV applications
• 1990-2003 Ford Trucks, Vans & SUVs
Refer to this section to look up PID Data information for Ford applications from 1990-2003.

Section 9: Pin Voltage Tables for Ford Car Applications
• 1990-2003 Ford Cars
Refer to this section to identify PCM circuit descriptions, terminals, wire colors & pin values for Ford Cars from 1990-2003.

Section 10: Pin Voltage Tables for Lincoln Car Applications
• 1990-2003 Lincoln Cars
Refer to this section to identify PCM circuit descriptions, terminals, wire colors & pin values for Lincoln Cars from 1990-2003.

Section 11: Pin Voltage Tables for Mercury Car Applications
• 1990-2003 Mercury Cars
Refer to this section to identify PCM circuit descriptions, terminals, wire colors & pin values for Mercury Cars from 1990-2003.

Section 12: Pin Voltage Tables for Ford and Lincoln Truck Applications
• 1990-2003 Ford & Lincoln Trucks
Refer to this section to identify PCM circuit descriptions, terminals, wire colors & pin values for Ford & Lincoln Trucks from 1990-2003.

Section 13: Pin Voltage Tables for Ford SUV Applications
• 1990-2003 Ford SUV
Refer to this section to identify PCM circuit descriptions, terminals, wire colors & pin values for Ford SUVs from 1990-2003.

Section 14: Pin Voltage Tables for Ford Van Applications
• 1990-2003 Ford Vans
Refer to this section to identify PCM circuit descriptions, terminals, wire colors & pin values for Ford Vans from 1990-2003.

Diagnostic Help

The OBD II Diagnostics, PID information and Pin Voltage Tables in this handbook contain *diagnostic help* in the form of "known good" values and examples of what caused a trouble code to set. These examples were obtained with a BOB, DVOM and a Scan Tool.

OBD II SYSTEMS

The California Air Resources Board (CARB) began regulating On-Board Diagnostic (OBD) systems for vehicles sold in California beginning with the 1988 model year. The initial requirements, known as OBD I, required the identification of the likely area of a fault with regard to the fuel metering system, EGR system, emission-related components and the PCM. Implementation of this new vehicle emission control monitoring regulation was done in several phases.

OBD I Systems

A malfunction indicator lamp (MIL) labeled *Check Engine Lamp* or *Service Engine Soon* was required to illuminate and alert the driver of a fault, and the need to service the emission controls. A diagnostic trouble code (DTC) was required to assist in identifying the system or component associated with the fault. If the fault that caused the MIL goes away, the MIL will go out and the code associated with the fault will disappear after a predetermined number of ignition cycles.

Following extensive research, CARB determined that by the time an Emission System component failed and caused the MIL to illuminate, that the vehicle could have emitted excess emissions over a long period of time. CARB also concluded that semi-annual or annual tailpipe tests were not catching enough of the vehicles with Emission Control systems operating at less than normal efficiency.

To take advantage of improvements in vehicle manufacturer adaptive and failsafe strategies, CARB developed new requirements designed to monitor the performance of Emission Control components, as well as to detect circuit and component hard faults. The new diagnostics were designed to operate under normal driving conditions, and the results of its tests would be viewable without any special equipment.

Enhanced OBD Systems

Beginning in the 1994 model year, both CARB and the EPA mandated Enhanced OBD systems, commonly known as OBD II. The objectives of OBD II were to improve air quality by reducing high in-use emissions caused by emission-related faults, reduce the time between the occurrence of a fault and its detection and repair, and assist in the diagnosis and repair of an emissions-related fault.

Differences Between OBD I & OBD II

As with OBD I, if an emission related problem is detected on a vehicle with OBD II, the MIL is activated and a code is set. However, that is the only real similarity between these systems. OBD II procedures that define emissions component and system tests, code clearing and drive cycles are more comprehensive than tests in the OBD I system.

Powertrain Control Module

The PCM in the OBD II system monitors almost all Emission Control systems that affect tailpipe or evaporative emissions. In most cases, the fault must be detected before tailpipe emissions exceed 1.5 times applicable 50K or 100K-mile FTP standards. If a component exceeds emission levels or fails to operate within the design specifications, the MIL is illuminated and a code is stored within two OBD II drive cycles.

The OBD II test runs continuously or once per trip (it depends on the driving mode requirement). Tests are run once per drive cycle during specific drive patterns called trips. Codes are stored in the PCM memory when a fault is first detected. In most cases, the MIL is turned on after two trips with a fault present. If the MIL is "on", it will go off after three consecutive trips if the same fault does not reappear. If the same fault is not detected after 40 engine warmup periods, the code will be erased (Fuel and Misfire faults require 80 warmup cycles).

OBD II Standardization

OBD II diagnostics require the use of a standardized diagnostic link connector (DLC), standard communication protocol and messages, and standardized trouble codes and terminology. Examples of this standardization are Freeze Frame Data and I/M Readiness Monitors.

Changes in MIL Operation

An important change for OBD II involves when to activate the MIL. The MIL must be activated by at least the second trip if vehicle emissions could exceed 1.5 times the FTP standard. If any single component or system failure would allow the emissions to exceed this level, the MIL is activated and a related code is stored in the PCM.

1994 OBD II Phase-In Systems

Starting in 1994, Ford began to "phase-in" the OBD II system on certain Mustang models (equipped with 3.8L V6 engines) and certain Thunderbird and Cougar models (equipped with 4.6L V8 engines). The OBD II "phase-in" system on these vehicles included the use of a Misfire Monitor that operated with a "lower threshold" Misfire Detection system designed to monitor misfires without setting any codes. In addition, the EVAP Monitor was not operational on these vehicles.

1996 & Later OBD II Systems

By the 1996 model year, all California passenger cars and trucks up to 14,000 lb. GVWR, and all Federal passenger cars and trucks up to 8,600 lb. GWVR were required to comply with the CARB-OBD II or EPA OBD requirements. The requirements applied to diesel and gasoline vehicles, and were phased in on alternative-fuel vehicles.

Flash EEPROM

The PCM on an EEC-V system includes a flash electrically erasable and programmable read-only memory (EEPROM) module. This software, in the form of an integrated circuit, contains the program used by the PCM to control the vehicle Powertrain. The EEPROM can be updated (reprogrammed) at a Ford dealership through the DLC and SBDS without removing the PCM. It can also be updated if it is removed and then taken to the parts counter. Changes to vehicle calibration are performed as directed by Recall Letters and Technical Service Bulletins. Refer to TSB 98-26-3 for an explanation of the complete procedure.

EEC-V Powertrain Controller

The purpose of the EEC-V system is to provide optimum control of the engine and transmission while meeting the objectives of the OBD II regulations. The PCM connects to various input and output devices through a wiring harness via a 104-pin connector (88 pins on the Villager and 150 pins on LS models). The PCM receives inputs from various sensors and switches, performs calculations based on data stored in an integrated circuit called Keep Alive Memory (KAM), and controls various output devices (i.e., actuators, relays, and solenoids).

EEC-V Hardware & Software

The EEC-V system hardware components include:

- All related actuators, relays, solenoids, sensors and switches
- The CCRM, PCM and VLCM (modules) and connecting wiring

The EEC-V system software components include:

- Programs that make up the strategies used by the PCM to control operation of the engine, electronic transmission, Failure Mode Effects Management, idle speed and fuel delivery systems.
- The EEC-V system includes backup or fail-safe circuitry should the Central Processing Unit (CPU) or EEPROM in the PCM fail.

Diagnostic Test Modes

The "test mode" messages available on a Scan Tool are listed below:

- Mode $01: Used to display Powertrain Data (PID data)
- Mode $02: Used to display any stored Freeze Frame data
- Mode $03: Used to request any trouble codes stored in memory
- Mode $04: Used to request that any trouble codes be cleared
- Mode $05: Used to monitor the Oxygen sensor test results
- Mode $06: Used to monitor Non-Continuous Monitor test results
- Mode $07: Used to monitor the Continuous Monitor test results
- Mode $08: Used to request control of a special test (EVAP Leak)
- Mode $09: Used to request vehicle information (INFO MENU)

Onboard Diagnostics

OBD II Systems incorporate the dedicated Ford test procedures built into the system. In effect, the key on, engine off (KOEO) and key on, engine running (KOER) Self-Tests are still an important functional part of the diagnostics as with earlier Ford diagnostics for OBD I systems.

Trouble codes associated with OBD II System are linked to the Ford code repair charts (Pinpoint Tests) using the customary CONT or MEM, KOEO, and KOER designators. In addition, the OBD II Main Monitors frequently run as part of the dedicated Ford Self-Tests.

Diagnostic Procedure

The Diagnostic Repair Chart on this page should be used as follows:

- Trouble Code Diagnosis - Refer to the Code List (in this section) or electronic media for a repair chart for a particular trouble code.
- Driveability Symptoms - Refer to the Driveability Symptom List in other manuals or in electronic media.
- Intermittent Faults - Refer to the Intermittent Test Procedures.
- OBD II Drive Cycles - Refer to the Comprehensive Component Monitor or a Main Monitor drive cycle articles in this section.

OBD II Repair Chart Graphic

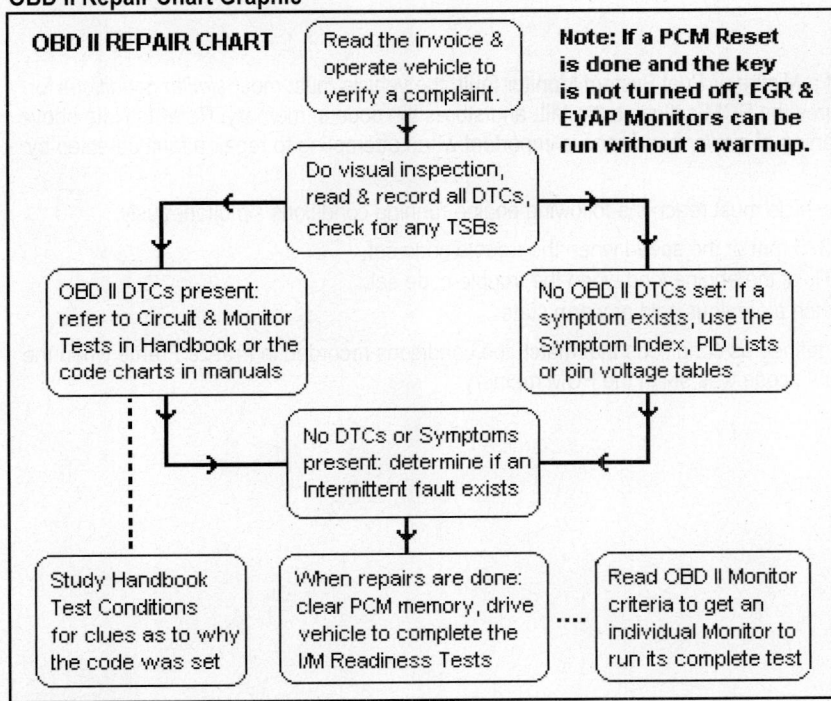

OBD SYSTEM TERMINOLOGY

It is very important that service technicians understand terminology related to OBD II test procedures. Several of the essential OBD II terms and definitions are discussed on the next few pages.

Two-Trip Detection

Frequently, an emission system or component must fail a Monitor test more than once before the MIL is activated. In these cases, the first time an OBD II Monitor detects a fault during any drive cycle it sets a pending code in the PCM memory.

A pending code, which is read by selecting DDL from the Scan Tool menu, appears when Memory or Continuous codes are read. In order for a pending code to cause the MIL to activate, the original fault must be repeated under *similar conditions*.

This is a critical issue to understand as a pending code could remain in the PCM for a long time before the conditions that caused the code to set reappear. This type of OBD II trouble code logic is frequently referred to as the "Two-Trip Detection Logic".

■ **NOTE:** *Codes related to a Misfire fault and Fuel Trim can cause the PCM to activate the MIL after <u>one</u> trip because these codes are related to critical emission systems that could cause emissions to exceed the federally mandated limits.*

Similar Conditions

If a pending code is set because of a Misfire or Fuel System Monitor fault, the vehicle must meet *similar conditions* for a second trip before the code matures the PCM activates the MIL and stores the code in memory. Refer to Note above for exceptions to this rule. The meaning of *similar conditions* is important when attempting to repair a fault detected by a Misfire or Fuel System Monitor.

To achieve *similar conditions*, the vehicle must reach the following engine running conditions simultaneously:

- Engine speed must be within 375 rpm of the speed when the trouble code set.
- Engine load must be within 10% of the engine load when the trouble code set.
- Engine warmup state must match a previous cold or warm state.

Summary - Similar conditions are defined as conditions that match the conditions recorded in Freeze Frame when the fault was first detected and the trouble code was set in the PCM memory.

OBD II Warmup Cycle

The meaning of the expression *warmup cycle* is important. Once the fault that caused an OBD II trouble code to set is gone and the MIL is turned off, the PCM will not erase that code until after 40 warmup cycles. *This is the purpose of the warmup cycle - to help clear stored codes.*

OBD II Warmup Cycle Graphic

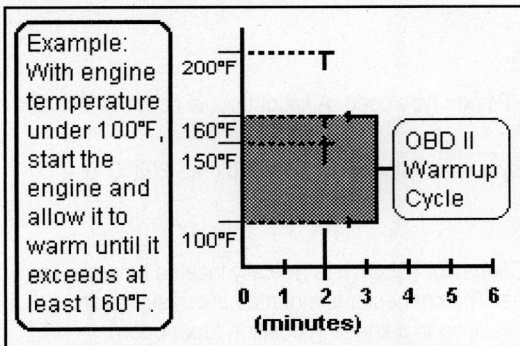

However, trouble codes related to a Fuel system or Misfire fault require that 80 warmup cycles occur without the fault reappearing before codes related to these monitors will be erased from the PCM memory.

■ **NOTE:** *A warmup cycle is defined as vehicle operation (after an engine off and cool-down period) when the engine temperature rises to at least 40ºF and reaches at least 160ºF.*

Malfunction Indicator Lamp

If the PCM detects an emission related component or system fault for two consecutive drive cycles on OBD II systems, the MIL is turned on and a trouble code is stored. The MIL is turned off if three consecutive drive cycles occur without the same fault being detected.

Most trouble codes related to a MIL are erased from KAM after 40 warmup periods if the same fault is not repeated. The MIL can be turned off after a repair by using the Scan Tool PCM Reset function.

Freeze Frame Data

The term *Freeze Frame* is used to describe the engine conditions that are recorded in PCM memory at the time a Monitor detects an emissions related fault. These conditions include fuel control state, spark timing, engine speed and load.

Freeze Frame data is recorded when a system fails the first time for two-trip type faults. The Freeze Frame Data will only be overwritten by a different fault with a "higher emission priority."

Scan Tool Freeze Frame Graphic

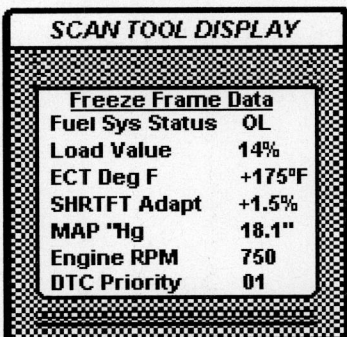

Diagnostic Trouble Codes

The OBD II system uses a *Diagnostic Trouble Code (DTC)* identification system established by the Society of Automotive Engineers (SAE) and the EPA. The first letter of a DTC is used to identify the type of computer system that has failed as shown below:

- The letter 'P' indicates a Powertrain related device
- The letter 'C' indicates a Chassis related device
- The letter 'B' indicates a Body related device
- The letter 'U' indicates a Data Link or Network device code.

The first DTC number indicates a generic (P0xxx) or manufacturer (P1xxx) type code. A list of trouble codes is included in this section.

The number in the hundreds position indicates the specific vehicle system or subgroup that failed (i.e., P0300 for a Misfire code, P0400 for an emission system code, etc.).

Data Link Connector

Ford vehicles equipped with OBD II use a standardized Data Link Connector (DLC). It is typically located between the left end of the instrument panel and 12 inches past vehicle centerline. The connector is mounted out of sight from vehicle passengers, but should be easy to see from outside by a technician in a kneeling position (door open). *However, not all of the connectors are located in this exact area.*

The DLC is rectangular in design and capable of accommodating up to 16 terminals. It has keying features to allow easy connection to the Scan Tool. Both the DLC and Scan Tool have latching features used to ensure that the Scan Tool will remain connected to the vehicle during testing.

Once the Scan Tool is connected to the DLC, it can be used to:

- Display the results of the most current I/M Readiness Tests
- Read and clear any diagnostic trouble codes
- Read the Parameter ID (PID) data from the PCM
- Perform Enhanced Diagnostic Tests (manufacturer specific)

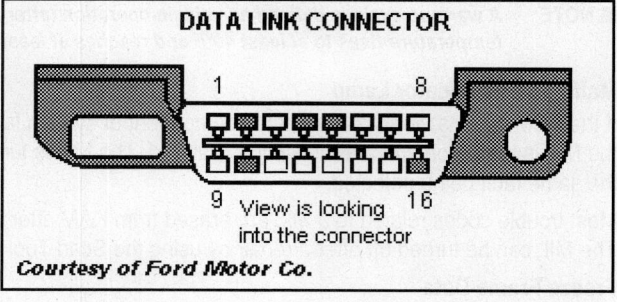

DATA LINK CONNECTOR

1   8
9   16
View is looking into the connector

Courtesy of Ford Motor Co.

Standard Corporate Protocol

On vehicles equipped with OBD II, a Standard Corporate Protocol (SCP) communication language is used to exchange bi-directional messages between stand-alone modules and devices. With this type of system, two or more messages can be sent over one circuit.

OBD II Monitor Software

The Diagnostic Executive contains software designed to allow the PCM to organize and prioritize the Main Monitor tests and procedures, and to record and display test results and diagnostic trouble codes.

The functions controlled by this software include:

- To control the diagnostic system so the vehicle continues to operate in a normal manner during testing.
- To ensure the OBD II Monitors run during the first two sample periods of the Federal Test Procedure.
- To ensure that all OBD II Monitors and their related tests are sequenced so that required inputs (enable criteria) for a particular Monitor are present prior to running that particular Monitor.
- To sequence the running of the Monitors to eliminate the possibility of different Monitor tests interfering with each other or upsetting normal vehicle operation.
- To provide a Scan Tool interface by coordinating the operation of special tests or data requests.

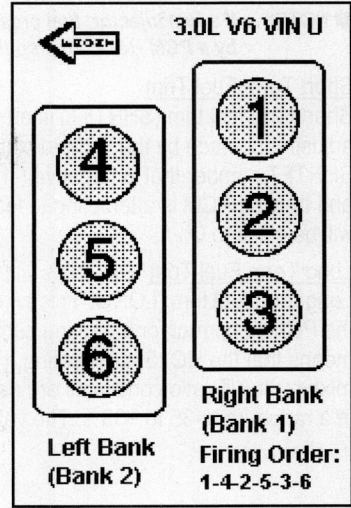

Cylinder Bank Identification

Engine sensors are identified on each engine cylinder bank as explained next.

Bank - A specific group of engine cylinders that share a common control sensor (e.g., Bank 1 identifies the location of Cyl 1 while Bank 2 identifies the cylinders on the opposite bank).

An example of the cylinder bank configuration for a Ford Taurus with FWD and a 3.0L V6 (VIN U) engine is shown in the Graphic on this page.

Oxygen Sensor Identification

Oxygen sensors are identified in each cylinder bank as the front O2S (pre-catalyst) or rear O2S (post-catalyst). The acronym HO2S-11 identifies the front oxygen sensor located (Bank 1) while the HO2S-21 identifies the front oxygen sensor in Bank 2 of the engine, and so on.

OBD II Monitor Test Results

Generally, when an OBD II Monitor runs and fails a particular test during a trip, a pending code is set. If the same Monitor detects a fault for two consecutive trips, the MIL is activated and a code is set in PCM memory. The results of a particular Monitor test indicate that an emission system or component failed - not the circuit that failed!

To determine where the fault is located, follow the correct code repair chart, symptom diagnosis or intermittent test. The code and symptom repair charts are the most efficient way to repair an OBD II system.

■ **NOTE::** *Two important pieces of information that can help speed up a diagnosis are code conditions (including all enable criteria), and the parameter information (PID) stored in the Freeze Frame at the time a trouble code is set and stored in memory.*

Adaptive Fuel Control Strategy

The PCM incorporates an Adaptive Fuel Control Strategy that includes an adaptive fuel control table stored in KAM to compensate for normal changes in fuel system devices due to age or engine wear.

During closed loop operation, the Fuel System Monitor has two methods of attempting to maintain an ideal A/F ratio of 14:7 to 1 (they are referred to as short term fuel trim and long term fuel trim).

■ **NOTE:** *If a fuel injector, fuel pressure regulator or oxygen sensor is replaced the KAM in the PCM should be cleared by a PCM Reset step so that the PCM will not use a previously learned strategy.*

Short Term Fuel Trim

Short term fuel trim (SHRTFT) is an engine operating parameter that indicates the amount of short term fuel adjustment made by the PCM to compensate for operating conditions that vary from the ideal A/F ratio condition. A SHRTFT number that is negative (-15%) means that the HO2S is indicating a richer than normal condition to the PCM, and that the PCM is attempting to lean the A/F mixture. If the A/F ratio conditions are near ideal, the SHRTFT number will be close to 0%.

Long Term Fuel Trim

Long term fuel trim (LONGFT) is an engine parameter that indicates the amount of long term fuel adjustment made by the PCM to correct for operating conditions that vary from ideal A/F ratios. A LONGFT number that is positive (+15%) means that the HO2S is indicating a leaner than normal condition, and that it is attempting to add more fuel to the A/F mixture. If A/F ratio conditions are near ideal, the LONGFT number will be close to 0%. The PCM adjusts the LONGFT in a range from -35 to +35%. The values are in percentage on a Scan Tool.

Enable Criteria

The term *enable criteria* describe the conditions necessary for any of the OBD II Monitors to run their diagnostic tests. Each Monitor has specific conditions that must be met before it will run its test.

Enable criteria information can be found in the vehicle manufacturer repair manuals, in the JENDHAM Ford Manual and in this Handbook. Look under Diagnostics and then Trouble Code Conditions. This type of data can be different for each vehicle and engine type. Examples of trouble code conditions for DTC P0460 and P1168 are shown below:

DTC	Trouble Code Title & Conditions
	EVAP System Small Leak Conditions: Cold startup, engine running at off-idle conditions, then the PCM detected a small leak (a leak of more than 0.040") in the EVAP system.
	FRP Sensor in Range but Low Conditions: Engine running, then the PCM detected that the FRP sensor signal was out-of-range low. *Scan Tool Tip: Monitor the FRP PID for a value below 80 psi (551 kPa).*

Code information includes any of the following examples:

- Air Conditioning Status
- BARO, ECT, IAT, TFT, TP and Vehicle Speed sensors
- Camshaft (CMP) and Crankshaft (CKP) sensors
- Canister Purge (duty cycle) and Ignition Control Module Signals
- Short (SHRTFT) and Long Term (LONGFT) Fuel Trim Values
- Transmission Shift Solenoid On/Off Status

Drive Cycle

The term *drive cycle* has been used to describe a drive pattern used to verify that a trouble code, driveability symptom or intermittent fault had been fixed. With OBD II systems, this term is used to describe a vehicle drive pattern that would allow all the OBD II Monitors to initiate and run their diagnostic tests. For OBD II purposes, a minimum *drive cycle* includes an engine startup with continued vehicle operation that exceeds the amount of time required to enter closed loop fuel control.

OBD II Trip

The term *OBD II Trip* describes a method of driving the vehicle so that one or more of the following OBD II Monitors complete their tests:

- Comprehensive Component Monitor (completes anytime in a trip)
- Fuel System Monitor (completes anytime during a trip)
- EGR System Monitor (completes after accomplishing a specific idle and acceleration period)
- Oxygen Sensor Monitor (completes after accomplishing a specific steady state cruise speed for a certain amount of time)

OBD II Drive Cycle

The ambient or inlet air temperature must be from 40-100°F to initiate the OBD II drive cycle. Allow the engine to warm to 130°F prior to starting the test (except for Escort and Tracer models that require the engine be less than 100°F to run the EVAP Monitor).

Connect the Scan Tool prior to beginning the drive cycle. Some tools are designed to emit a three-pulse beep when all of the OBD II Monitors complete their tests and DTC P1000 has been erased.

■ **NOTE::** *The IAT PID must be from 50-100°F to start the drive cycle. If it is less than 50°F at any time during the highway part of the drive cycle, the EVAP Monitor may not complete. The engine should reach 130°F before starting the trip except on Escort &Tracer models where a cold engine startup of less than 100°F is used before attempting to verify an EVAP system fault. Disengage the PTO before proceeding (PTO PID will show OFF) if applicable. For the EVAP Running Loss system, verify FLI PID is at 15-85%. Some Monitors require very specific idle and acceleration steps.*

Drive Cycle Procedure

The primary intention of the Ford OBD II drive cycle is to clear a DTC P1000. The drive cycle can also be used to assist in identifying any OBD II concerns present through total Monitor testing. Perform all of the Vehicle Preparation steps. Then refer to the Drive Cycle Table and Graphic below for details on how to run a Ford OBD II Drive Cycle.

Connect a Scan Tool and have an assistant watch the Scan Tool I/M Readiness Status to determine when the Catalyst, EGR, EVAP, Fuel System, O2 Sensor, Secondary AIR and Misfire Monitors complete.

OBD II Drive Cycle Graphic

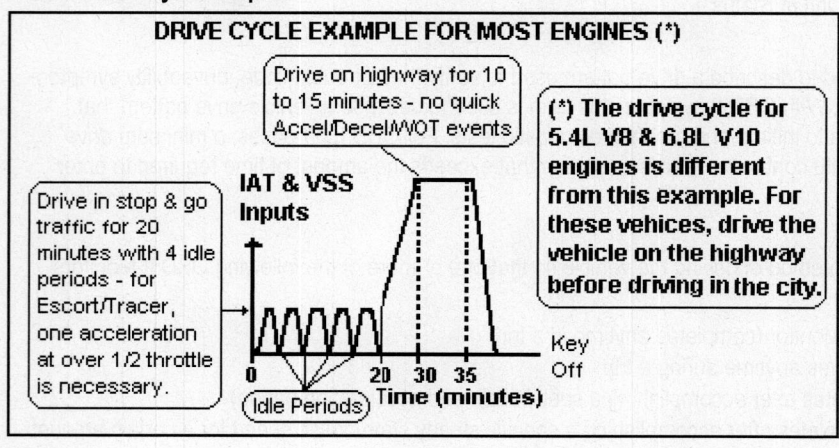

OBD II System Monitors

Comprehensive Component Monitor

OBD II regulations require that all emission related circuits and components controlled by the PCM that could affect emissions are monitored for circuit continuity and out-of-range faults. The Comprehensive Component Monitor (CCM) consists of four different monitoring strategies: two for inputs and two for output signals. *The CCM is a two trip Monitor for emission faults on Ford vehicles.*

<u>Input Strategies</u>
One input strategy is used to check devices with analog inputs for opens, shorts, or out-of-range values. The CCM accomplishes this task by monitoring A/D converter input voltages. The analog inputs monitored include the ECT, IAT, MAF, TP and Transmission Range Sensors signals.

A second input strategy is used to check devices with digital and frequency inputs by performing rationality checks. The PCM uses other sensor readings and calculations to determine if a sensor or switch reading is correct under existing conditions. Some tests run continuously, some only after actuation. The inputs monitored by the PCM include the CMP, IDM, PIP, OSS, TSS and VSS signals.

<u>Output Strategies</u>
An Output State Monitor in the PCM checks outputs for opens or shorts by observing the control voltage level of the related device. The control voltage is low with it on, and high with the device off. Monitored outputs include the EPC, SS1, SS2, SS3, TCC, HFC, VMV, WOT A/C Cutout and the HO2S Heater.

Component Monitor Graphic

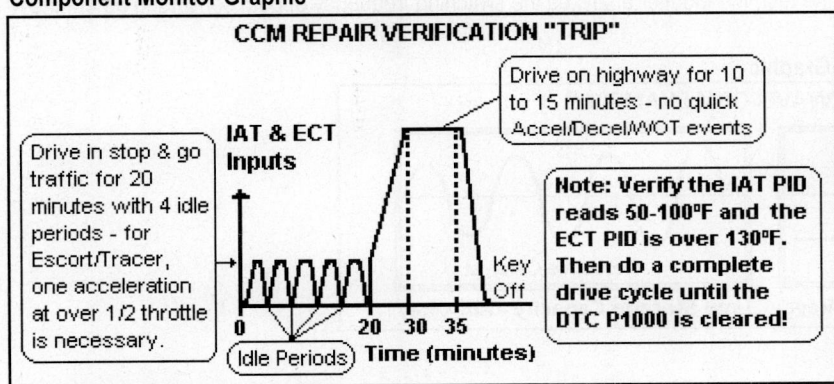

IAC Motor Test

The PCM monitors the IAC system in order to "learn" the closed loop correlation it needs to reposition the IAC solenoid (a rationality check).

Catalyst Efficiency Monitor

The Catalyst Monitor is a PCM diagnostic run once per drive cycle that uses the downstream heated Oxygen Sensor (HO2S-12) to determine if a catalyst falls below a minimum level of effectiveness in its ability to control exhaust emissions. The PCM uses a program to determine the catalyst efficiency based on the oxygen storage capacity of the catalytic converter.

<u>Catalyst Monitor Operation</u>

The Catalyst Monitor is a diagnostic that tests the oxygen storage capacity of the catalyst. The PCM determines the capacity by comparing the switching frequency of the rear oxygen sensor to the switching frequency of the front oxygen sensor. If the catalyst is okay, the switching frequency of the rear oxygen sensor will be much slower than the frequency of the front oxygen sensor.

However, as the catalyst efficiency deteriorates its ability to store oxygen declines. This deterioration causes the rear oxygen sensor to switch more rapidly. If the PCM detects the switching frequency of the rear oxygen sensor is approaching the frequency of the front oxygen sensor, the test fails and a pending code is set. If the PCM detects a fault on consecutive trips *(from two to six consecutive trips)* the MIL is activated, and a trouble code is stored in the PCM memory.

The Catalyst Monitor runs after startup once a specified time has elapsed and the vehicle is in closed loop. The amount of time is subject to each PCM calibration. Certain inputs (enable criteria) from various engine sensors (i.e., CKP, ECT, IAT, TPS and VSS) are required before the Catalyst Monitor can run.

Once the Catalyst Monitor is activated, closed loop fuel control is temporarily transferred from the front oxygen sensor to the rear oxygen sensor. During the test, the Monitor analyzes the switching frequency of both sensors to determine if a catalyst has degraded.

Rear Oxygen Sensor Waveform Graphic

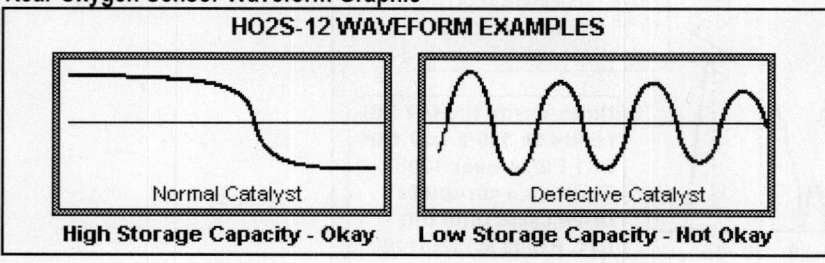

Catalyst Efficiency Monitor

Catalyst Test - Steady State Catalyst Efficiency Test

The PCM transfers the input for closed loop fuel control from the front HO2S-11 to the rear HO2S-21 during this test. The PCM measures the output frequency of the rear HO2S. This "test frequency" indicates the current oxygen storage capacity of the converter. The slower the frequency of the test result, the higher the efficiency of the converter.

Catalyst Test - Calibrated Frequency Test

In Part 2 of the test a second frequency is calculated based on engine speed and load. This frequency serves as a high limit threshold for the test frequency. If the PCM detects the test frequency is less than the calibrated frequency the catalyst passes the test. If the frequency is too high, the converter or system has failed (a pending code is set).

The sequence of counting the front and rear O2S switches continues until the drive cycle completes. The ratio of total HO2S-21 switches to the total of the HO2S-11 switches is calculated. If the switch ratio is over the stored threshold, the catalyst has failed and a code is set.

Trouble Codes associated with this OBD II Monitor are listed below:

- DTC P0420, P0421 - The catalyst in Bank 1 has failed the test
- DTC P0430, P0431 - The catalyst in Bank 2 has failed the test

Catalytic Monitor Repair Verification Trip

Start the engine, and drive in stop and go traffic for over 20 minutes. (Ambient air temperature must be over 50°F to run this test). Drive at speeds from 25-40 mph (6 times) and then at cruise for five minutes.

Catalyst Monitor Graphic

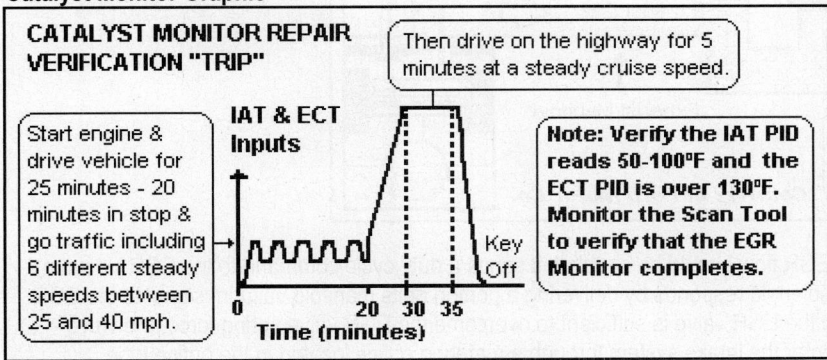

Possible Causes of a Catalyst Efficiency Fault

- Base Engine faults (engine mechanical)
- Exhaust leaks or contaminated fuel

EGR System Monitor

The EGR System Monitor is a PCM diagnostic run once per trip that monitors EGR system component functionality and components for faults that could cause vehicle tailpipe levels to exceed 1.5 times the FTP Standard. A series of sequenced tests is used to test the system.

Differential Pressure Feedback EGR System

This system includes a DPFE sensor, vacuum regulator solenoid, EGR valve, orifice tube assembly, the PCM and related wiring/hoses.

EGR DPFE System Graphic

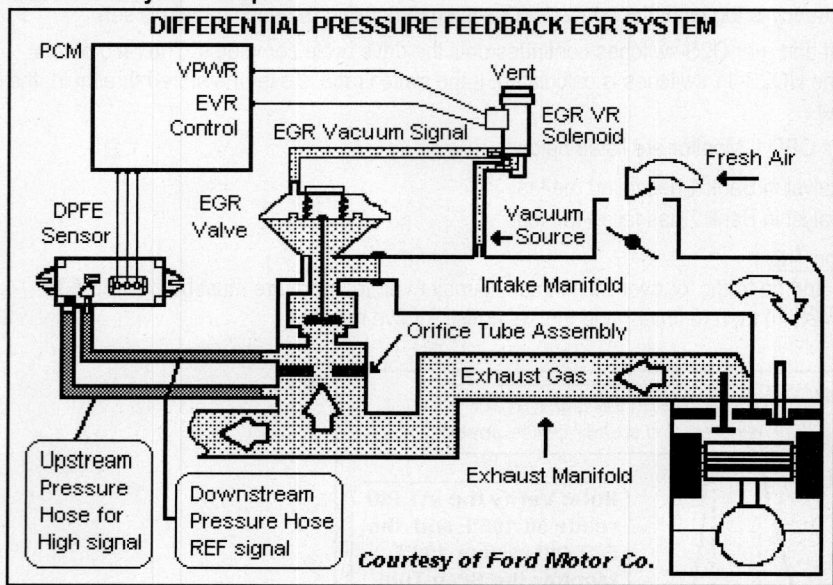

DIFFERENTIAL PRESSURE FEEDBACK EGR SYSTEM

Courtesy of Ford Motor Co.

When PCM strategy dictates that EGR flow should be enabled, it sends a duty cycle command to the EGR Vacuum Regulator (VR) solenoid. The VR solenoid responds by delivering a portion of its manifold vacuum signal to the EGR valve. Once the vacuum applied to the EGR valve is sufficient to overcome the EGR valve spring force, the valve opens allowing exhaust gases to enter the intake system through a metering orifice located in the orifice tube assembly. A pressure drop is created across this orifice that is proportional to the rate of EGR flow entering the intake manifold.

The DPFE sensor senses the differential pressure signal across the orifice and sends an analog signal to the PCM that indicates the rate of EGR flow. This feedback signal is used to adjust the EVR duty cycle to achieve the desired amount of EGR flow.

EGR System Monitor (Continued)

DPFE EGR Valve - Test One

First, the Monitor tests the DPFE sensor signal to determine if it is out of normal operating range. Once this test is passed, and all enable criteria are met, the Monitor checks for a pressure differential with the engine at idle speed and with the EGR valve closed. At this point, both the upstream and downstream ports of the DPFE sensor should be reading exhaust pressure. Any differential pressure reading at this point indicates that the valve is open (and it should not be open).

Next, with the EGR valve commanded open, the differential pressure is checked. With the valve open, the sensor should signal a positive change. If there is no positive change, the downstream hose is off or plugged at the DPFE sensor. If the DPFE sensor does not indicate an upstream pressure change anytime, this indicates that it is plugged.

DPFE EGR Valve - Test Two

Next, the EGR DPFE sensor is tested with the EGR valve commanded open. If a negative DPFE sensor reading is detected, the indication is that the downstream and upstream hoses are reversed on the EGR valve. Then, the EGR valve flow is checked.

Once the engine reaches cruise speed with the ECT, MAP and TP sensor signals constant, the PCM compares the actual and expected DPFE sensor values. If the actual reading is lower than the expected value due to a restriction, the PCM fails the test and sets a code.

Possible Causes of an EGR System Failure

- Leaks or disconnects in upstream or downstream vacuum hoses
- Damaged DPFE or EGR EVP sensor
- Plugged or restricted DPFE or EGR VP sensor or orifice assembly

EGR Monitor Graphic

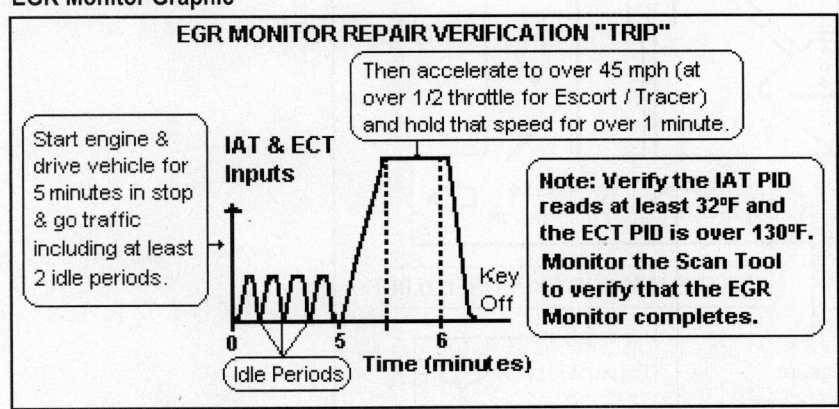

EGR System Monitor (Continued)

<u>Sonic EGR System (Explorer 5.0L V8)</u>

The Sonic EGR system used on these models includes an EGR valve position (VP) sensor, EGR vacuum regulator (VR) solenoid, EGR valve, system wiring and vacuum hoses, and the PCM.

<u>System Operation</u>

When the PCM strategy dictates the need for EGR flow, it outputs a duty cycle signal to the EGR VR solenoid. The solenoid responds by delivering a calibrated amount of manifold vacuum to the EGR valve. The remainder of the source vacuum is vented into the atmosphere.

At some point, engine vacuum applied to the EGR valve is sufficient to overcome the spring force of the valve, and the valve begins to open. This action allows exhaust gases to enter the intake manifold. As the EGR valve pintle lifts upward, it causes the EGR VP sensor to also lift upward in direct proportion to the amount of EGR opening.

In this Sonic EGR system, the PCM uses the VP sensor signal as an indication of the amount of EGR flow into the engine. It adjusts the VR solenoid duty cycle signal to achieve the desired amount of EGR flow.

If a fault is detected during the EGR Monitor test that could cause the tailpipe emissions to exceed 1.5 times the FTP Standard, the test fails and a pending code is set. If the EGR test fails on two consecutive trips, the MIL is activated and a hard code is set.

Purge Flow System Graphic

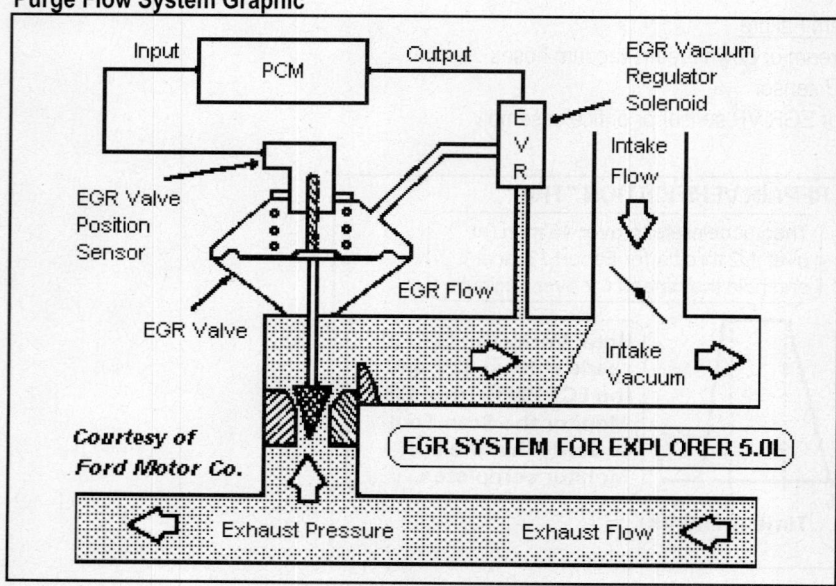

EVAP System Monitor

The EVAP System Monitor is a PCM diagnostic run once per trip that monitors the EVAP system in order to detect a loss of system integrity or leaks in the system (anywhere from 0.020" to 0.040" in diameter).

The Ford vehicles included in this article are equipped with three different EVAP systems: Purge Flow, Vapor Management Valve and On Board Refueling Vapor Recovery.

Purge Flow System (Explorer 5.0L V8 Engine)

The Purge Flow system consists of a purge flow (PF) sensor, purge solenoid, fuel vapor valve, a gas tank, charcoal canister (with internal atmospheric vent) and related wiring and fuel vapor hoses.

In this EVAP system, the PCM commands a normally closed (N.C.) purge solenoid "on" and "off" to purge fuel vapors from the charcoal canister into the intake manifold during various engine conditions. The PCM monitors changes in the fuel pressure (FP) sensor analog signal to detect fuel vapor flow through the purge solenoid during the EVAP diagnostic test. Changes are based on "flow" or "no flow" conditions.

Purge Flow System Graphic (Explorer)

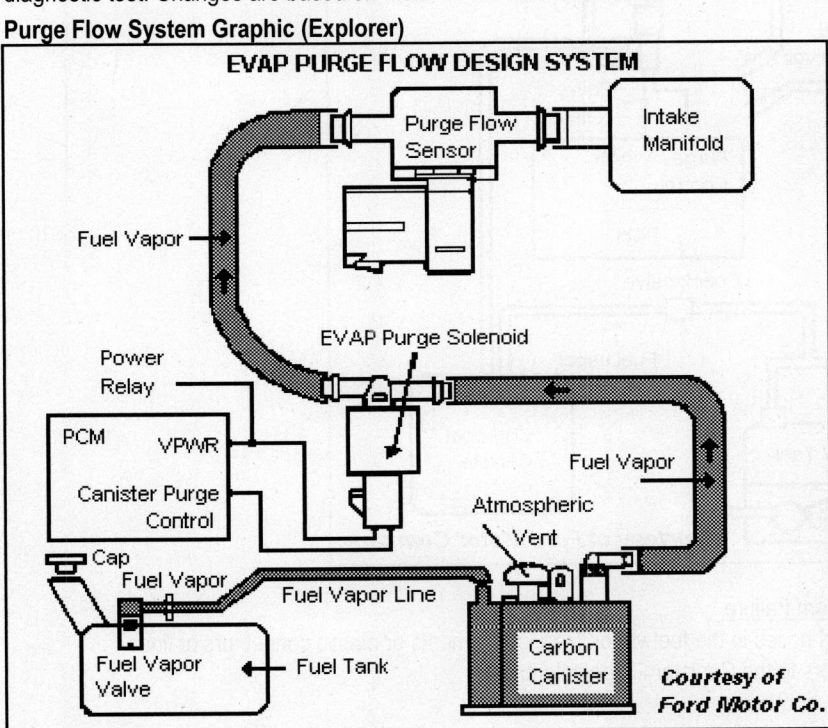

EVAP PURGE FLOW DESIGN SYSTEM

Courtesy of Ford Motor Co.

EVAP System Monitor (Continued)

Purge Flow System (Probe 2.0L I4)

The Purge Flow system used on these models includes a canister purge solenoid, fuel line check valve, charcoal canister, fuel tank, related wiring and vacuum hoses, and the PCM.

System Operation

On this system, the PCM commands a normally closed (N.C.) canister purge solenoid "on" and "off" to control purging of fuel vapors from the charcoal canister into the intake manifold under various engine conditions.

Purge Flow System Graphic (Probe)

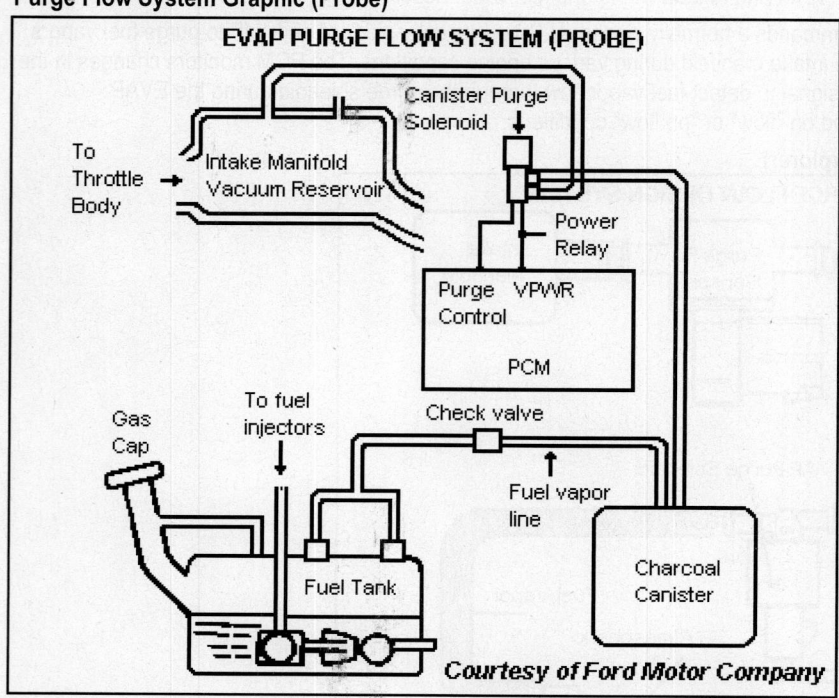

EVAP PURGE FLOW SYSTEM (PROBE)

Courtesy of Ford Motor Company

Possible Causes of an EVAP System Failure

- Cracks, leaks or disconnected hoses in the fuel vapor lines, components or plastic connectors or lines
- Backed-out or loose connectors to the Canister Purge solenoid
- Fuel filler cap (gas cap) loose or missing
- PCM has failed

EVAP System Monitor (Continued)

EVAP VMV Design System

This system consists of a Vapor Management Valve (VMV), fuel vapor valve, gas tank, the charcoal canister with internal atmospheric vent, related wiring and fuel vapor hoses, and the PCM. The PCM commands a normally closed (N.C.) valve "on" and "off" to control when to purge fuel vapors from the canister to the intake manifold.

Fuel vapors trapped in the sealed fuel tank are vented through a vapor valve assembly on top of fuel tank. The vapors leave the valve assembly through a single vapor line and continue on to the carbon canister for storage until they are purged to the engine for recycling.

The VMV and vent solenoid control signals are cycled "on" and "off" at a frequency of 10 hertz with a variable duty cycle. The duty cycle is ramped-up to slowly draw the canister vapors into the intake manifold.

VMV System Graphic

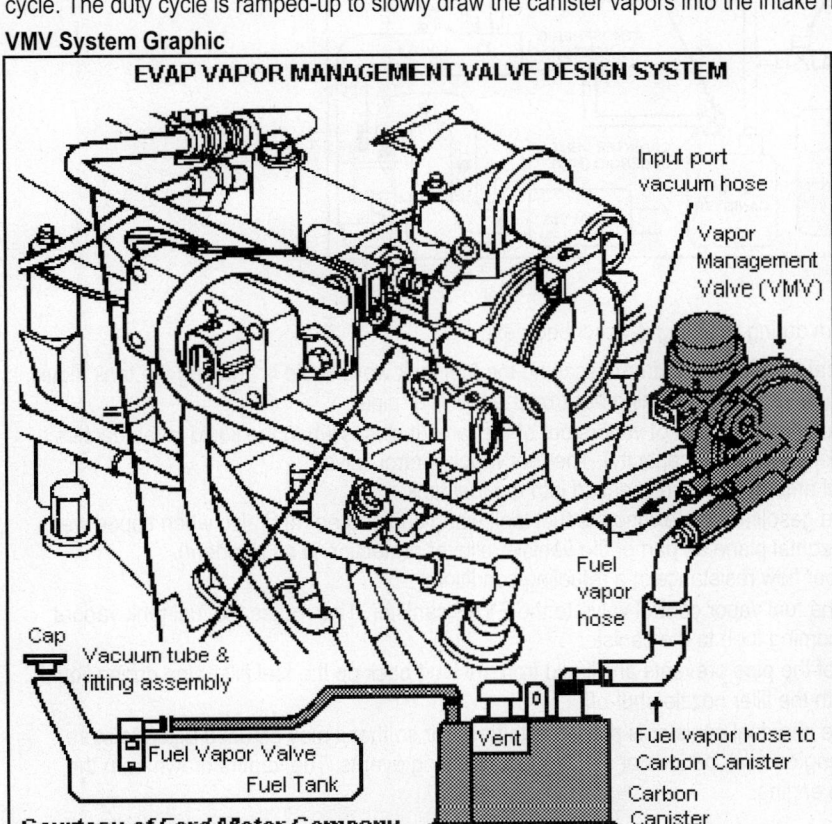

EVAP VAPOR MANAGEMENT VALVE DESIGN SYSTEM

Courtesy of Ford Motor Company

EVAP System Monitor (Continued)

On-Board Refueling Vapor Recovery System

An On-Board Refueling Vapor Recovery (ORVR) system is used on late model vehicles to recover fuel vapors during vehicle refueling.

ORVR System Graphic

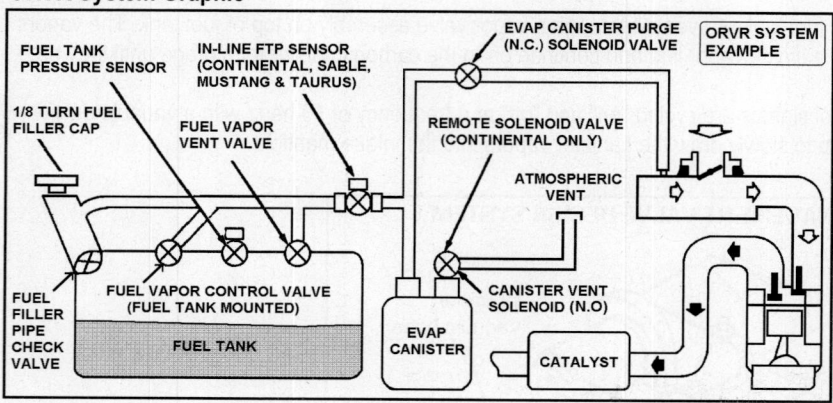

System Operation

The operation of the ORVR system during refueling is described next:

- The fuel filler pipe forms a seal to stop vapors from escaping the fuel tank while liquid is entering the tank (liquid in the 1" diameter tube blocks fuel vapor from rushing back up the fuel filler pipe).
- The fuel vapor control valve controls the flow of vapors out of the tank (it closes when the liquid level reaches a height associated with the fuel tank usable capacity). The fuel vapor control valve:
    - Limits the total amount of fuel dispensed into the fuel tank.
    - Prevents liquid gasoline from exiting the fuel tank when submerged (and also when tipped well beyond a horizontal plane as part of the vehicle rollover protection in an accident).
    - Minimizes vapor flow resistance in a refueling condition.
- Fuel vapor tubing connects the fuel vapor control valve to the EVAP canister. This routes the fuel tank vapors (that are displaced by the incoming fuel) to the canister.
- A check valve in the bottom of the pipe prevents any liquid from rushing back up the fuel filler pipe during liquid flow variations associated with the filler nozzle shut-off.
- Between refueling events, the charcoal canister is purged with fresh air so that it may be used again to store vapors accumulated during engine soak periods or subsequent refueling events. The vapors drawn from the canister are consumed in the engine.

EVAP System Monitor (Continued)

EVAP Monitor Test Conditions

The PCM allows canister purge to occur when the engine is warm, at wide open or part throttle (as long as the engine is not overheated). The engine can be in open or closed loop fuel control during purging.

MIL Operation, How to Clear History Trouble Codes

If the EVAP Monitor detects a fault during a drive cycle, it will set a pending code. If it detects the fault for two consecutive trips, the MIL is activated and a code is set. The MIL will remain on for more than one trip, but will go out if conditions that caused the Monitor to fail do not reappear on three consecutive trips. After the MIL is off, the code will be erased after 40 consecutive trips if the fault does not reappear.

EVAP Running Loss Monitor Graphic

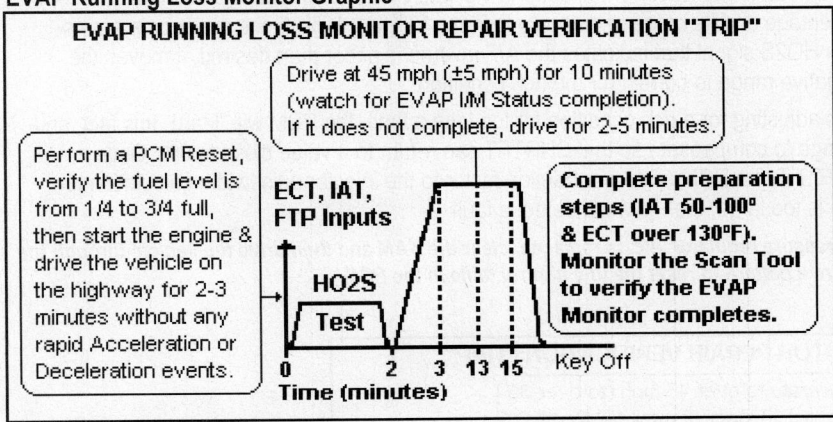

EVAP Leak Check Monitor Graphic

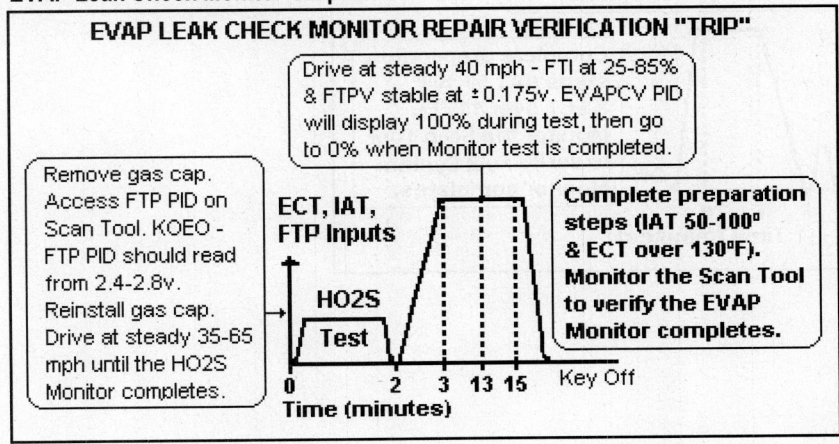

Fuel System Monitor

The Fuel System Monitor is a PCM diagnostic that monitors the Adaptive Fuel Control system. The PCM uses adaptive fuel tables that are updated constantly and stored in long term memory (KAM) to compensate for wear and aging in the Fuel system components.

Fuel System Monitor Operation

Once the PCM determines all the enable criteria has been are met (ECT, IAT and MAF PIDs in range and closed loop enabled), the PCM uses its adaptive strategy to "learn" changes needed to correct a Fuel system that is biased either rich or lean. The PCM accomplishes this task by monitoring Short Term and Long Term fuel trim in closed loop mode.

Long and Short Term Fuel Trim

Short Term fuel trim is a PCM parameter identification (PID) used to indicate Short Term fuel adjustments. This parameter is expressed as a percentage and its range of authority is from -10% to +10%. Once the engine enters closed loop, if the PCM receives a HO2S signal that indicates the A/F mixture is richer than desired, it moves the SHRTFT command to a more negative range to correct for the rich condition.

If the PCM detects the SHRTFT is adjusting for a rich condition for too long a time, the PCM will "learn" this fact, and move LONGFT into a negative range to compensate so that SHRTFT can return to a value close to 0%. Once a change occurs to LONGFT or SHRTFT, the PCM adds a correction factor to the injector pulsewidth calculation to adjust for variations. If the change is too large, the PCM will detect a fault.

■ **NOTE:** *If a fuel injector, fuel pressure regulator, etc. is replaced, clear the KAM and then drive the vehicle through the Fuel System Monitor drive pattern to reset the fuel control table in the PCM.*

Fuel System Monitor Graphic

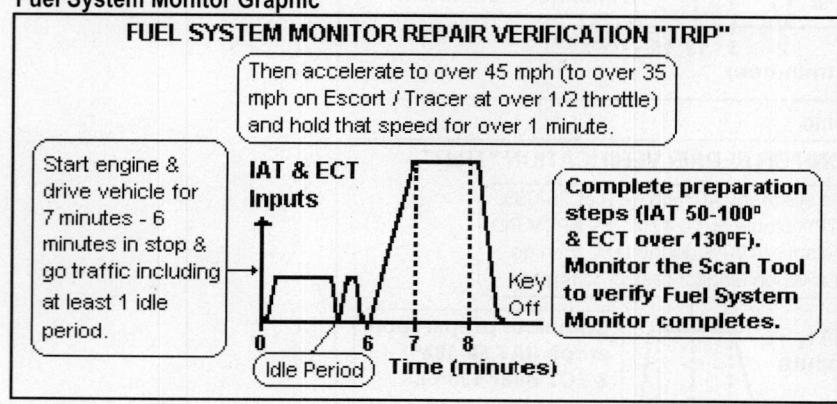

Misfire Detection Monitor

The Misfire Monitor is a PCM diagnostic that continuously monitors for engine misfires under all engine positive load and speed conditions (accelerating, cruising and idling). The Misfire Monitor detects misfires caused by fuel, ignition or mechanical misfire conditions. If a misfire is detected, engine conditions present at the time of the fault are written to the Freeze Frame Data. These conditions overwrite existing data.

Misfire Monitor Operation

The Misfire Monitor is designed to measure the amount of power that each cylinder contributes to the engine. The amount of contribution is calculated based upon measurements determined by crankshaft acceleration (TDC of compression stroke to BDC of the power stroke) for each cylinder. This calculation requires accurate measurement of the crankshaft angle. Crankshaft angle measurement is determined using a low data rate system on 4-Cyl engines. The high data rate system is used to determine crankshaft angle on all other engines.

Crankshaft Position Sensor Graphic

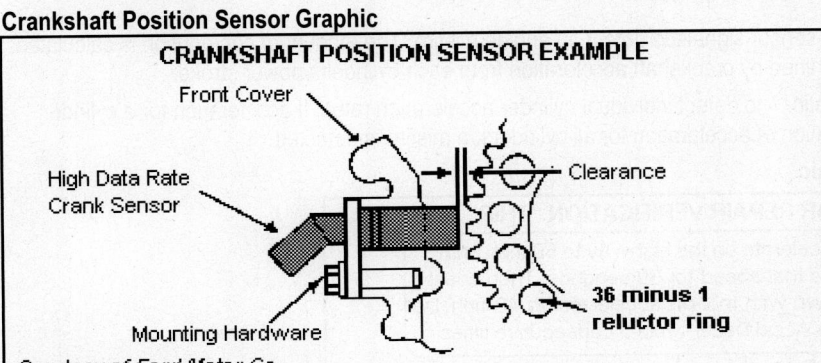

Catalyst Damaging Misfire (One-Trip Detection)

If the PCM detects a Catalyst Damaging Misfire, the MIL will flash once per second within 200 engine revolutions from the point where misfire is detected. The MIL will stop flashing and remain on if the engine stops misfiring in a manner that could damage the catalyst.

High Emissions Misfire (Two-Trip Detection)

A High Emissions Misfire is set if a misfire condition is present that could cause the tailpipe emissions to exceed the FTP emissions standard by 1.5 times. If this fault is detected for two consecutive trips under similar engine speed, load and temperature conditions, the MIL is activated. It is also activated if a misfire is detected under *similar conditions* for two non-consecutive trips that are not 80 trips apart.

Misfire Detection Monitor (Continued)

<u>State Emissions Failure Misfire (Two-Trip Detection)</u>

A State Emissions Failure Misfire is set if the misfire is sufficient to cause the vehicle to fail a State Inspection or Maintenance (I/M) Test. This fault is determined by identifying misfire percentages that would cause a "durability demonstration vehicle" to fail an Inspection Maintenance (I/M) Test. If the Misfire Monitor detects the fault for two consecutive trips with the engine at similar engine speed, load and temperature conditions, the MIL is activated and a code is set. The MIL is also activated if this type of misfire is detected under similar conditions for two non-consecutive trips of not more than 80 trips apart.

■ **NOTE:** *Some vehicles set Misfire codes because of an early version of OBD II hardware and software. If a misfire code is set and the cause of the fault is not found, clear the code and retest. Search the TSB list for possible answers or contact the dealer.*

<u>Misfire Detection</u>

The Misfire Monitor uses the CKP sensor signals to detect an engine misfire. The amount of contribution is calculated based upon measurements determined by crankshaft acceleration from each cylinder's power stroke.

The PCM performs various calculations to detect individual cylinder acceleration rates. If acceleration for a cylinder deviates beyond the average variation of acceleration for all cylinders, a misfire is detected.

Misfire Detection Monitor Graphic

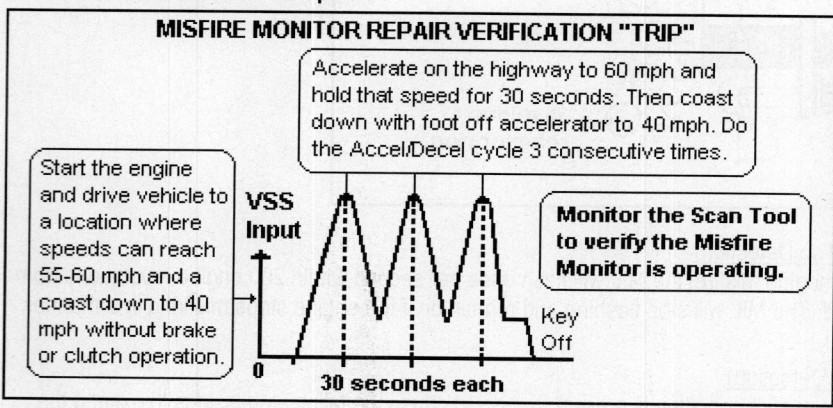

Faults detected by the Misfire Monitor

- Engine mechanical faults, restricted intake or exhaust system
- Dirty or faulty fuel injectors, loose or damaged injector connectors
- The vehicle has been run low on fuel or run until it ran out of fuel

Oxygen Sensor Monitor

The Oxygen Sensor Monitor is a PCM diagnostic designed to monitor the front and rear oxygen sensor for faults or deterioration that could cause tailpipe emissions to exceed 1.5 times the FTP standard. The front oxygen sensor voltage and response time are also monitored.

HO2S Monitor Operation

Fuel System and Misfire Monitors must be run and complete before the PCM will start the HO2S Monitor. Additionally, parts of the HO2S Sensor Monitor are enabled during the KOER Self-Test. The HO2S Monitor is run during each drive cycle after the CKP, ECT, IAT and MAF sensor signals are within a predetermined range.

Fixed Frequency Closed Loop Test

The HO2S Monitor constantly monitors the sensor voltage and frequency. The PCM detects a high voltage condition by comparing the HO2S signal to a preset level.

A Fixed Frequency Closed Loop Test is used to check the HO2S voltage and frequency. A sample of the HO2S signal is checked to determine if the sensor is capable of switching properly or has a slow response time (referred to as a lazy sensor).

Fixed Frequency Test Graphic

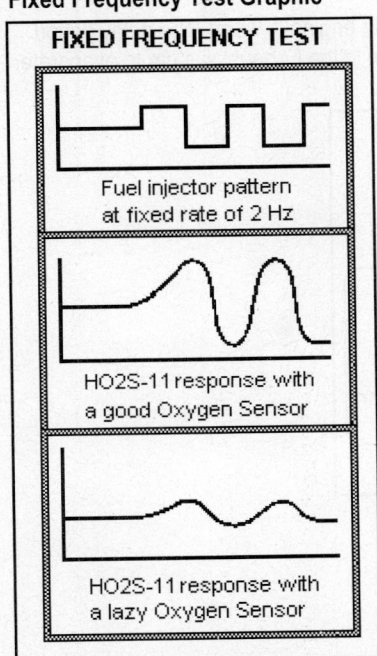

FIXED FREQUENCY TEST

Fuel injector pattern at fixed rate of 2 Hz

HO2S-11 response with a good Oxygen Sensor

HO2S-11 response with a lazy Oxygen Sensor

Oxygen Sensor Monitor Graphic

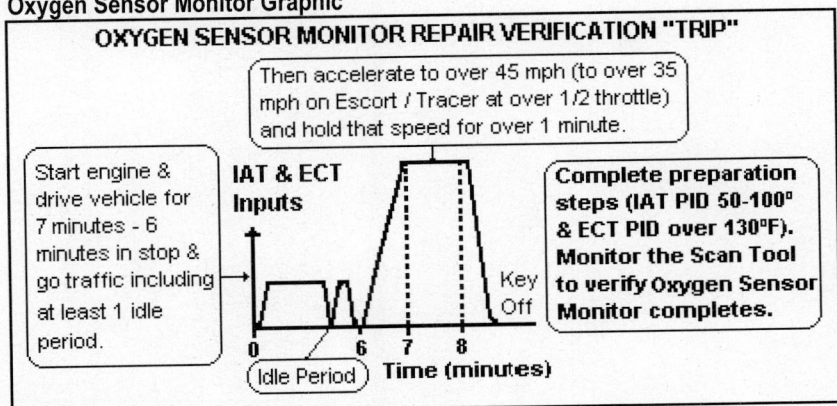

OXYGEN SENSOR MONITOR REPAIR VERIFICATION "TRIP"

Then accelerate to over 45 mph (to over 35 mph on Escort / Tracer at over 1/2 throttle) and hold that speed for over 1 minute.

Start engine & drive vehicle for 7 minutes - 6 minutes in stop & go traffic including at least 1 idle period.

IAT & ECT Inputs

Key Off

Complete preparation steps (IAT PID 50-100° & ECT PID over 130°F). Monitor the Scan Tool to verify Oxygen Sensor Monitor completes.

Idle Period   Time (minutes)

0   6 7   8

Oxygen Sensor Heater Monitor

The Oxygen Sensor Heater Monitor is a PCM diagnostic designed to monitor the Oxygen Sensor Heater and its related circuits for faults.

Oxygen Sensor Heater Monitor Operation

The Oxygen Sensor Heater Monitor performs its task by detecting whether the proper amount of O2 sensor voltage change occurred as the HO2S Heater is turned from "on" to "off" with the engine in closed loop. The time it takes for the HO2S-11 and HO2S-12 signal to switch (the response time) is constantly monitored by the Oxygen Sensor Monitor. Once the Oxygen Sensor Heater Monitor is enabled, if the switch time for the HO2S-11 or HO2S-12 signal is too long, the PCM fails the test, the MIL is activated and a trouble code is set.

Note: *Response time is defined as the amount of time it takes for a HO2S signal to switch from Rich to Lean, and then Lean to Rich.*

Front and Rear Oxygen Sensor Heater Operation

Both upstream and downstream Oxygen sensors are used on the OBD II system. These sensors are designed with additional protection around the ceramic core to protect them from condensation that could crack them if the heater is turned on with condensation present.

The HO2S heaters are not turned on until the ECT sensor signal indicates that the engine is warm. The delay period can last for as long as 5 minutes from startup. The delay allows any condensation in the Exhaust system to evaporate.

Oxygen Sensor Monitor Graphic

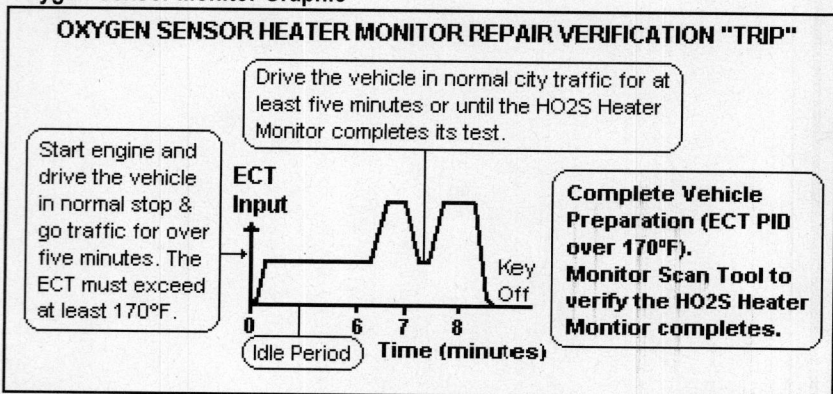

Faults detected by the HO2S or HO2S Heater Monitor

- A fault in the HO2S, the HO2S heater or its related circuits
- A fault in the HO2S connectors (look for moisture tracking)
- A defective Power Control Module

Air Injection System Monitor

The Air Injection System Monitor is an OBD diagnostic controlled by the PCM that monitors the Air Injection (AIR) system. The Oxygen Sensor Monitor must run and complete before the PCM will run this test. The PCM enables this test during AIR system operation after certain engine conditions are met and these enable criteria are met:

- Crankshaft Position sensor signal must be present
- ECT and IAT sensor input signals must be within limits

AIR Monitor - Electric Pump Design

The AIR Monitor consists of these Solid State Monitor tests:

- A check of the Solid State relay for electrical faults.
- A check of the secondary side of the relay for electrical faults.
- A test to determine if the AIR system can inject additional air.

AIR Monitor - Mechanical Pump Design

The AIR Monitor for the mechanical (belt-driven air pump) design uses two Output State Monitor configurations to perform two different circuit tests. One test is used to check for faults in the Secondary Air Bypass (AIRB) solenoid circuit. The normal function of the AIRB solenoid and valve assembly is to dump air into the atmosphere.

A second test is used to check for electrical faults in the Secondary Air Divert (AIRD) solenoid. The normal function of the AIRD solenoid and valve assembly is to direct the air either upstream or downstream.

Functional Check

An AIR system functional check is done at startup with the AIR pump on or during a hot idle period if the startup part of the test was not performed. A flow test is included that uses the HO2S signal to indicate the presence of extra air injected into the exhaust stream.

Secondary AIR Monitor Graphic

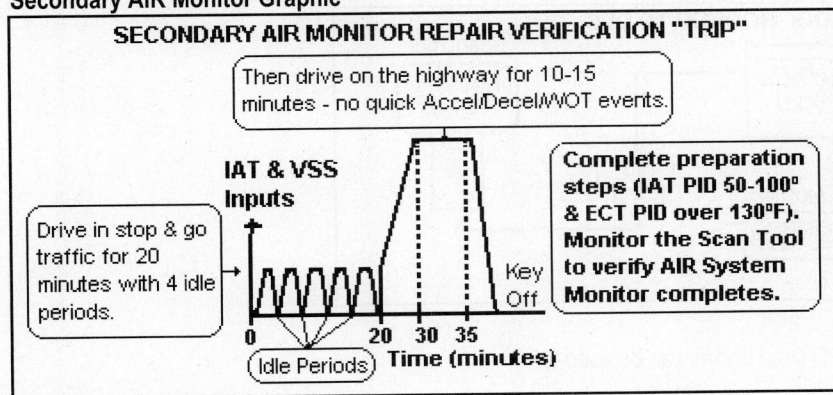

SECONDARY AIR MONITOR REPAIR VERIFICATION "TRIP"

Then drive on the highway for 10-15 minutes - no quick Accel/Decel/WOT events.

IAT & VSS Inputs

Drive in stop & go traffic for 20 minutes with 4 idle periods.

Complete preparation steps (IAT PID 50-100° & ECT PID over 130°F). Monitor the Scan Tool to verify AIR System Monitor completes.

Key Off

0   20 30 35

Idle Periods   Time (minutes)

Reference Information

PID Data Explanation

PID is an acronym for Parameter Identification Data. It is used to identify PCM related items on both OEM and aftermarket Scan Tools. The PID Data or data stream items available for display include:

- PCM analog & digital input signals (ACT, ECT, IAT, MAP, TP)
- PCM analog & digital output signals (EVR, FUELPW1, MIL)
- PCM calculated values (VACUUM, LOAD, MISF, SPARK)
- PCM system status information (OBD II I/M Readiness Status)

OBD I & OBD II Data Stream Information

PID Data is separated by OBD System type as listed below:

- OBD I System PID Data for Ford Cars from 1990-95
- OBD II System PID Data for Ford Cars from 1994-2002

How to View PID Data

To view PID Data on a Scan Tool, connect the Scan Tool to the vehicle connector; then identify the vehicle by following the setup instructions, and select Data Stream or Parameter from the menu.

■ **NOTE:** *If a Scan Tool does not power up or read PID Data, verify that the power, ground and tool cable connections are all okay. To test if the tool is working properly, try it on another vehicle.*

Example of Scan Tool Connection

In this example, the Scan Tool is connected to the DLC in order to view "live" HO2S PID Data once it has been converted in the PCM.

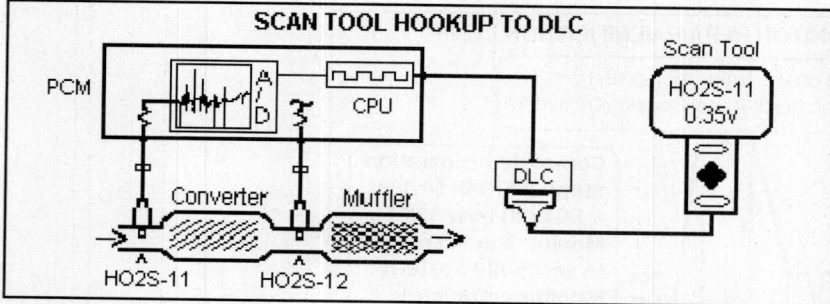

How to use PID Data

Information contained within the PID Data Charts can be used to:

- Validate a repair procedure
- Check the operation of a component before or after a repair
- Check the operation of a component or system by viewing "live" data from the vehicle computer data stream

OBD II PID Mode

Vehicles equipped with OBD II diagnostics have a unique PID Mode that can be selected to allow access to vehicle Parameter Data information. Part of the PID Data is generic in nature and can be accessed with an OBD II compatible Scan Tool in Generic Mode.

However, some of the PID Data can only be accessed using an OEM New Generation Star (NGS) Scan Tool or an Aftermarket Scan Tool with the capability to read Ford Enhanced Data. To learn how to view Generic or Enhanced Data, refer to the Scan Tool operating manual.

PID Data Display

An example of Enhanced Data captured with a Scan Tool is shown in the table below. This data is from a 1996 Ford 3.0L V6 Taurus VIN U.

PID #	Acronym	Description	Measurement Units
000F	IAT	Intake Air Temperature	110°F
0001	DTC CNTM	Continuous DTC Counter	Numbers
0003	FUEL SYS1	Fuel System Feedback (BK1)	CL (Closed Loop)
0005	ECT	Engine Coolant Temperature	188°F
0014	O2S-11	Bank 1 Front Oxygen Sensor	0.635v (varies at cruise)
0015	O2S-12	Bank 1 Rear Oxygen Sensor	750 mv (varies at cruise)

Scan Tool Graphing Mode

Most Scan Tools include a function called Scan Tool Graphing Mode that can be used to capture signals from various devices (TP sensor, IAC Motor, etc.). Note how this feature was used to capture changes in the TP sensor signal and IAC Motor control signals in the Graphic below. This is a good test to use to detect an intermittent fault.

Scan Tool Graphing Mode Examples

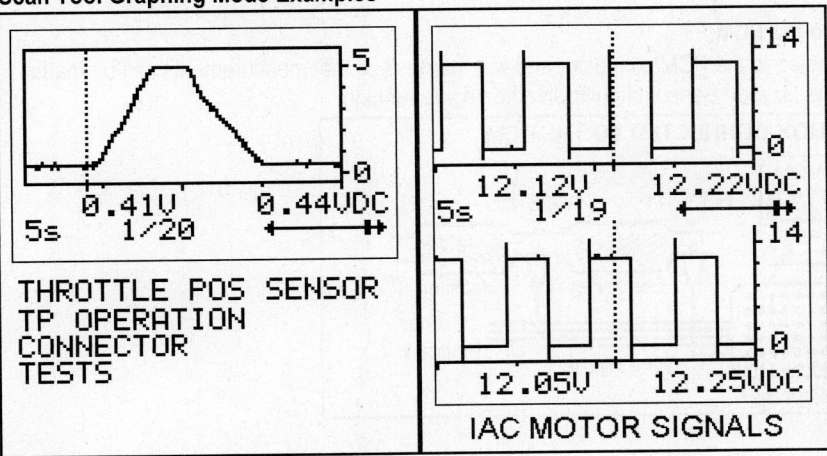

Pin Voltage Table Explanation

In this handbook, a Pin Voltage Table is a term used to describe a table that identifies PCM and Breakout Box (BOB) pins, wire colors of the PCM circuits and "known good" values for devices that connect to the PCM. These tables include the following information:

- Signals from various sensors (ECT, IAT, MAP, TOT, TP, VAF)
- Signals from various switches (BOO, CES, IDL, NDS, PSP)
- Signals from oxygen sensors (O2S-11, HO2S-11, HO2S-12)
- Signals from output devices (CANP, EVR, IAC, INJ, TCC)
- Power & ground signals (KAPWR, PWR GND, VPWR, VREF)

■ NOTE *Acronyms shown above are listed in the Glossary in this section.*

OBD I & OBD II Pin Voltage Tables

Pin Voltages are separated by OBD System type as listed below:

- OBD I System pin voltage values for Ford Cars from 1990-95
- OBD II System pin voltage values for Ford Cars from 1994-2002

How to Connect a Breakout Box

To use pin voltage information with a DVOM, a Breakout Box (BOB) should be installed. To connect a BOB, first turn the ignition off and remove the wire harness at the engine controller (PCM). Next, connect the correct wiring adapter to the PCM and BOB connectors. This places the BOB between the PCM and wiring so that circuit measurements can be made at the pin connections on the BOB.

■ NOTE   *Be sure to read and record all trouble code and freeze frame data in the PCM before connecting the BOB as all codes and data are lost when the PCM connector is removed.*

Example of a BOB Connected to the PCM

In this example, the BOB is connected to the PCM connector and wire harness so that measurements can be made with a DVOM (or Lab Scope) of the Oxygen Sensor circuits with the engine running.

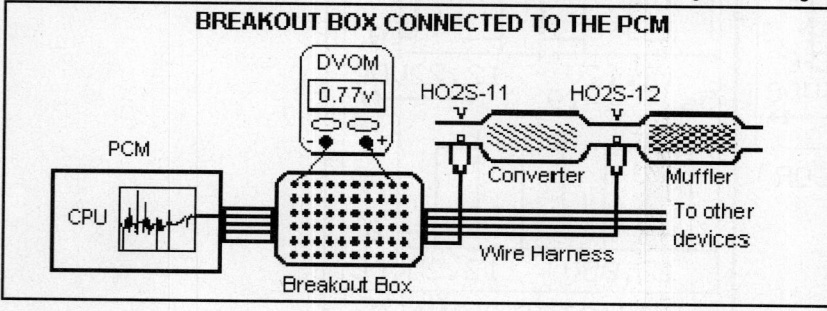

Handbook Pin Voltage Tables

Information contained within the Pin Voltage Tables can be used to:

Test circuits for open, short to power or short to ground faults

Check the operation of a component before or after a repair

Check the operation of a component or system by viewing signals on PCM input/output circuits with
a DVOM or Lab Scope

Using a Breakout Box

There are several Breakout Box designs available for use to test the PCM and its input and output circuits. However, all of them require removal of the wire harness to the PCM so that the BOB can be installed between the PCM and wire harness connector. Several breakout boxes require the use of overlays in order to allow the tool to be used on more than one year or engine type. Always verify that the correct adapter and overlays are used to prevent connection to the wrong circuits and a misdiagnosis.

Power and Ground Circuit Checks

Measurements made at the BOB are accomplished via test leads and probes from the DVOM or a Lab Scope. If any of the terminals on the PCM or BOB are damaged or loose, test measurements made at the Breakout Box will be inaccurate. To verify the PCM battery power (KAPWR and VPWR) and ground circuits (PWR GND) are normal (correct) at the BOB, test the condition of the circuit between the battery negative (-) post and these circuits prior to starting a test sequence.

■ **NOTE**   *With the key on, the voltage drop from the battery positive (+) terminal to KAPWR or VPWR at the BOB should be less than 0.1v. The voltage drop from the battery negative (-) post to PWR GND at the BOB should be less than 0.1v (key on).*

Diagnosis with Pin Voltage Tables

Once an actual PCM pin voltage reading is recorded, it can be compared to an example from a vehicle with "known good" values. In the example below (from a 1990 Crown Victoria 5.0L V8), the Value at Hot Idle for the EVP sensor signal (0.4v) is the "known good" value.

PCM Pin #	W/Color	Circuit Description (60-Pin)	Value at Hot Idle
27	BN/LG	EVP Sensor Signal	0.4v

Wire Color Changes

Every effort has been made to obtain and list the correct circuit wire colors for vehicles in this handbook. However, running changes from the vehicle manufacturers can cause the wrong colors to be listed.

PCM Actuator & Sensor Tables

Application: Association Vehicles

- Aspire 1.3L I4
- Escort/Tracer 1.8L I4
- Probe 2.5L V6

ACT, ECT, IAT & TFT Sensor Conversion Chart

Temperature		Voltage Resistance	
Degrees F	Degrees C	Volts	Kohms
194°F	90°C	0.90v	0.2 Kohms
122°F	50°C	2.20v	0.8 Kohms
68°F	20°C	3.50v	2.5 Kohms
14°F	-10°C	4.40v	9.2 Kohms

Application: Vehicles with EEC-IV & EEC-V Systems

ACT, ECT, IAT, TFT & TOT Sensor Conversion Chart

Temperature		Voltage Resistance	
Degrees F	Degrees C	Volts	Kohms
302	160	0.12	0.54
267	131	0.20	0.80
248	120	0.28	1.18
230	110	0.36	1.55
212	100	0.47	2.07
194	90	0.61	2.8
176	80	0.8	3.84
158	70	1.04	5.37
140	60	1.35	7.6
104	40	2.16	16.15
86	30	2.62	24.27
68	20	3.06	37.3
50	10	3.52	58.75
32	0	3.97	65.85
14	-10	4.42	78.19
-4	-20	4.87	90.54
-22	-30	4.89	102.88
-40	-40	4.91	115.23

Differential Pressure Feedback EGR Sensor

Pressure			Voltage
PSI	Inches Hg	KPa	Volts
4.34	8.83	29.81	4.56
3.25	6.62	22.36	3.54
2.17	4.41	14.90	2.51
1.08	2.21	7.46	1.48
0	0	0	0.45

Application: All vehicles with a BARO/MAP Sensor
Barometric Pressure Sensor Chart

Approximate Altitude in Feet	BARO Hertz Rate
0 feet (sea level)	159 Hz
1000 feet	156 Hz
2000 feet	153 Hz
3000 feet	150 Hz
4000 feet	147 Hz
5000 feet	144 Hz
6000 feet	141 Hz
7000 feet	139 Hz
8000 feet	136 Hz

Manifold Absolute Pressure Sensor Chart

Manifold Vacuum (" Hg)	MAP Hertz Rate
KOEO at Sea Level	159 Hz
5" Hg	146 Hz
8" Hg	134 Hz
10" Hg	129 Hz
11" Hg	126 Hz
12" Hg	124 Hz
13" Hg	121 Hz
14" Hg	119 Hz
15" Hg	116 Hz
16" Hg	114 Hz
17" Hg	111 Hz
18" Hg	109 Hz
19" Hg	107 Hz
20" Hg	104 Hz
21" Hg	102 Hz
22" Hg	99 Hz
23" Hg	97 Hz
24" Hg	95 Hz

Application: Association & Domestic Vehicles
- Aspire 1.3L I4, Probe 2.5L V6 Engine
- Vehicles with EEC-V and 5.0L V8 (VIN P) Engine

EGR Valve Position (EVP) Sensor

Percent of EGR	Voltage	Code Set Values
0% (valve closed)	0.40v	Hot Idle: A code sets if the EVP
10% (valve open)	0.75v	Sensor voltage is less than 0.24v or
30% (valve open)	1.45v	If the EVP signal is more than 0.67v
50% (valve open)	2.15v	
70% (valve open)	2.85v	
90% (valve open)	3.55v	

Application: Domestic Vehicles

- All applicable EEC-IV & EEC-V Vehicles

Pressure Feedback EGR (PFE) Sensor

PSI	Inches HG	Volts	Code Set Values
1.82	3.70	4.75	Hot idle: A code sets if the PFE
1.36	2.79	4.38	Sensor signal is under 2.60v or if
0.91	1.85	4.00	the PFE signal is more than 4.20v
0.46	0.94	3.63	
0	0	3.25	
-2.47	-5.03	1.22	
-3.63	-7.40	0.25	

EGR Vacuum Regulator (EVR) Solenoid Vacuum Examples

Duty	Vacuum				Output	
Cycle (%)	Min		Nom		Max	
	In-Hg	KPa	In-Hg	KPa	In-Hg	KPa
0	0	0	0.38	1.28	0.75	2.53
33	0.55	1.86	1.3	4.39	2.05	6.90
90	5.69	19.2	6.32	21.3	6.95	23.47

■ **NOTE** *EVR resistance values - for 4.9L and 5.8L engines: 20-45 ohms, for 7.5L engines: 100 ohms. For all other engines: 30-70 ohms.*

Output Device Resistance Chart for OBD I and OBD II Vehicles

Component	Ohm Value	Component	Ohm Value
AM1 Valve	40-80 ohms	AM2 Valve	40-80 ohms
Canister Purge Valve	40-80 ohms	Idle Air Control Motor	7-13 ohms
CCRM (module)	1K minimum	IRCM (module) - FP	65-110 ohms
EPC Solenoid	36-43 ohms	High Pressure Injector	12-16 ohms
EGR VR Solenoid	26-40 ohms	Shift Solenoid 1 (SS1)	15-25 ohms
Electro Drive Fan (EDF)	65-110 ohms	Shift Solenoid 2 (SS2)	15-25 ohms
Fan or Low Fan Control	65-110 ohms	Shift Solenoid 3 (SS3)	15-25 ohms
Fuel Pump (FP) Relay	40-90 ohms	TCC Solenoid	0.9-1.9 ohms
High Electro Drive Fan	50-100 ohms	AC WOT Cutout Relay	1K minimum

Output Device Resistance Chart for Association Vehicles

Component	Ohm Value	Component	Ohm Value
Canister Purge Valve	30-34 ohms	A/C WOT Cutout Relay	1K minimum
EPC Solenoid	36-43 ohms	Shift Solenoid 1, 2, 3	11-27 ohms
EGR Control, Vent	30-45 ohms	VRIS Solenoid 1, 2	35-40 ohms
EGR Boost Solenoid	90-110 ohms	Shift Solenoid 1, 2, 3	11-27 ohms
Idle Air Control Valve	10.7-12 ohms	TCC Solenoid	11-27 ohms

■ **NOTE** *An Amp Draw Test should not be done on output devices with a resistance of less than 25 ohms (or duty cycle controlled).*

Voltage to Mass Airflow Conversation Table (1996-98 Models)

Vehicle & Engine Application	MAF Volts 0.34v =	MAF Volts 0.39v =	MAF Volts 0.60v =	MAF Volts 1.00v =
Aerostar with 3.0L Engines	0.94 gm/sec	1.07 gm/sec	1.68 gm/sec	4.77 gm/sec
Aerostar with 4.0L Engines	1.46 gm/sec	1.68 gm/sec	2.52 gm/sec	7.64 gm/sec
Continental, Mark VII & Mustang with 4.6L Engine	1.39 gm/sec	1.60 gm/sec	2.69 gm/sec	8.36 gm/sec
Contour, Mystique with 2.0L	0.39 gm/sec	0.45 gm/sec	0.72 gm/sec	2.10 gm/sec
Contour & Mystique with 2.5L	0.33 gm/sec	0.95 gm/sec	1.69 gm/sec	5.53 gm/sec
Cougar, Thunderbird with 3.8L & 4.6L Engines	1.44 gm/sec	1.66 gm/sec	2.73 gm/sec	8.06 gm/sec
Crown Victoria 4.6L, Crown Victoria CNG 4.6L	1.28 gm/sec	1.48 gm/sec	2.37 gm/sec	6.97 gm/sec
Escort and Tracer with 1.9L Engines	0.94 gm/sec	1.07 gm/sec	1.68 gm/sec	4.77 gm/sec
Explorer and Ranger with 4.0L Engines	1.46 gm/sec	1.68 gm/sec	2.52 gm/sec	7.64 gm/sec
E-Series Van with 4.9L, 5.0L & 5.8L	1.24 gm/sec	1.42 gm/sec	2.98 gm/sec	8.38 gm/sec
F-Series Truck 4.6L & 7.5L Engines	1.35 gm/sec	1.56 gm/sec	2.66 gm/sec	8.31 gm/sec
F-Series Truck 4.9L and Bronco with 5.0L & 5.8L	1.68 gm/sec	1.93 gm/sec	3.32 gm/sec	8.43 gm/sec
Grand Marquis, Town Car with 4.6L Engines	1.28 gm/sec	1.48 gm/sec	2.37 gm/sec	6.97 gm/sec
Mustang 3.8L and Aerostar, Explorer and Ranger with 4.0L	1.46 gm/sec	1.68 gm/sec	2.52 gm/sec	7.64 gm/sec
Probe 2.0L and Windstar with 3.0L Engines	0.82 gm/sec	0.88 gm/sec	1.25 gm/sec	3.44 gm/sec
Ranger Pickup with 2.3L & 3.0L Engines	0.93 gm/sec	1.07 gm/sec	1.71 gm/sec	4.58 gm/sec
Sable, Taurus 3.0L 2v and Taurus FF 3.0L Engines	1.31 gm/sec	1.11 gm/sec	1.72 gm/sec	4.82 gm/sec
Sable/Taurus 3.0L 4v, 3.8L Windstar and F-Series Pickup with 4.2L	1.43 gm/sec	1.64 gm/sec	2.58 gm/sec	7.53 gm/sec
Taurus SHO 3.4L Engines	1.39 gm/sec	1.60 gm/sec	2.69 gm/sec	8.36 gm/sec

■ **NOTE:** *To use this table, connect a Scan Tool and select the MAF PID. Read the value and then compare it against the known good grams per second (gm/sec) reading for the vehicle and engine in this table. A faulty MAF sensor will read out of normal range.*

Voltage to Mass Airflow Conversation Table (1999-2002 Models)

Vehicle & Engine Application	MAF Volts 0.23v =	MAF Volts 0.27v =	MAF Volts 0.46v =	MAF Volts 2.44v =
Continental & Mustang with 4.6L Engine	1.07 gm/sec	1.22 gm/sec	2.08 gm/sec	59.52 gm/sec
Contour, Mystique & Cougar with 2.0L Engine	0.70 gm/sec	0.80 gm/sec	1.36 gm/sec	30.86 gm/sec
Contour, Mystique & Cougar 2.5L and Taurus 3.0L 4v Engines	1.00 gm/sec	1.14 gm/sec	1.95 gm/sec	52.50 gm/sec
Crown Victoria, Grand Marquis, & Town Car with 4.6L Engines	1.01 gm/sec	1.16 gm/sec	1.98 gm/sec	50.26 gm/sec
Escort & Tracer 2.0L 2v Engines	0.67 gm/sec	0.77 gm/sec	1.32 gm/sec	30.55 gm/sec
Escort & Tracer 2.0L 42v Engines	0.62 gm/sec	0.71 gm/sec	1.22 gm/sec	30.66 gm/sec
Expedition & Navigator 4.6L & 5.4L Engines	1.68 gm/sec	1.93 gm/sec	3.32 gm/sec	58.33 gm/sec
Explorer & Mountaineer 4.0L & 5.0L Engines	0.97 gm/sec	1.11 gm/sec	1.90 gm/sec	49.97 gm/sec
E-Series Van with 4.2L Engines	0.97 gm/sec	1.11 gm/sec	1.90 gm/sec	49.97 gm/sec
E-Series Van with 4.6L & 5.4L Engines	1.15 gm/sec	1.31 gm/sec	2.24 gm/sec	63.37 gm/sec
F-150 Truck with 4.2L, 4.6L & 5.4L Engines	1.68 gm/sec	1.93 gm/sec	3.32 gm/sec	58.43 gm/sec
F-250, F-350 Truck 4.2L, 5.4L & 6.8L Engines	1.06 gm/sec	1.21 gm/sec	2.07 gm/sec	58.33 gm/sec
Lighting 5.4L Engine	1.17 gm/sec	1.34 gm/sec	1.28 gm/sec	62.52 gm/sec
LS6, LS8 3.0L & 3.9L Engines	0.90 gm/sec	1.03 gm/sec	1.76 gm/sec	52.27 gm/sec
Mustang 3.8L, Ranger 4.0L Engines	1.00 gm/sec	1.14 gm/sec	1.94 gm/sec	52.44 gm/sec
Ranger 2.5L and 3.0L Engines	0.67 gm/sec	0.77 gm/sec	1.32 gm/sec	30.55 gm/sec
Sable & Taurus 3.0L 2v, Taurus FFV 3.0L Engines	0.67 gm/sec	0.77 gm/sec	1.32 gm/sec	30.55 gm/sec
Taurus SHO 3.4L Engine	1.06 gm/sec	1.21 gm/sec	2.07 gm/sec	58.33 gm/sec
Windstar 3.0L Engine	0.62 gm/sec	0.71 gm/sec	1.90 gm/sec	30.66 gm/sec
Windstar 3.8L Engine	0.62 gm/sec	0.71 gm/sec	1.90 gm/sec	30.66 gm/sec

■ **NOTE::** *Connect a Scan Tool and select the MAF PID. Read the value and compare it against the known good grams per second reading of that vehicle in the table. A faulty MAF sensor will be out of range.*

PID Data & Pin Chart Glossary

Glossary of Terms and Acronyms	
(<): Indicates less than the value	(>): Indicates more than the value
ACCS: A/C Cycling Clutch Switch	ACP: Air Conditioning Pressure Sensor
AODE: Automatic Overdrive Electronic	A4R70W: Automatic Overdrive Electronic
BOO: Brake On/Off Switch	CANP: EVAP Canister Purge Solenoid
CES: Clutch Engage Switch	CCS: Coast Clutch Solenoid
CKP: Crankshaft Position	CMP: Camshaft Position
COP: Coil On Plug Electronic Ignition	CTM: Central Timer Module
CTO: Clean Tachometer Output	DI: Distributor Ignition
DIS: Direct Ignition (Waste Spark)	DLC: Data Link Connector
DPFE: Differential Pressure Feedback	DRL: Daytime Running Lights
EBCM: Electronic Brake Control Module	EBTCM: Electronic Brake T/C Module
EFI: Electronic Fuel Injection	EGR: Exhaust Gas Recirculation
EGRC: EGR Control Solenoid	EGR Monitor: OBD II EGR Test
EGRV: EGR Vent Solenoid	EI: Electronic Ignition System
EPC: Electronic Pressure Control	EPT: EGR Pressure Transducer
EVAP: Evaporative Emission System	EVAP CP: Evaporative Canister Purge
EVAP CV: Evaporative Canister Vent	EVP: EGR Valve Position Sensor
EVR: EGR Vacuum Regulator	FAN: Cooling Fan (Low or High Speed)
FF: Flexible Fuel Vehicle	FIC: Fast Idle Control
FEM: Front Electronic Module	FMEM: Failure Mode Effect Management
FDPM: Fuel Pump Driver Module	FTO: Filtered Tachometer Output
FPM: Fuel Pump Monitor (in the PCM)	FTP: Fuel Tank Pressure
GEM: Generic Electronic Module	GOOSE: Brief throttle open/close
GND: Electrical ground connection	GVW: Gross Vehicle Weight
HDR-CKP: High Data Rate Crankshaft Position Sensor	HFC: High Fan Control, controls the high speed cooling fan
HFP: High Fuel Pump (Relay) Control	HLOS: Hardware Limited Operating Strategy
HO: High Output	HO2S: Heated Oxygen Sensor
HO2S-11 (Bank 1 Sensor 1) Signal	HO2S-12 (Bank 1 Sensor 2) Signal
HO2S-21 (Bank 2 Sensor 1) Signal	HO2S-22 (Bank 2 Sensor 2) Signal
HSC: High Swirl Combustion	IAC: Idle Air Control (solenoid)
IAT: Intake Air Temperature	IC: Integrated Circuit
ICM: Ignition Control Module	IDL: Idle Position Switch
IDM: Ignition Diagnostic Monitor	IFS: Inertia Fuel Switch
IGN GND: Ignition Ground	IMRC: Intake Manifold Runner Control
IMRC: Intake Manifold Runner Control	INJ 1 to INJ 10: Fuel Injectors 1 to 10
EI: Integrated-Electronic Ignition	ISO: International Standards Org.
ITS: Idle Tracking Switch	KAM: Keep Alive Memory
KAPWR: Direct Battery Power	Kg/cm^2: Kilograms/Cubic Centimeters
KOEC: Key On, Engine Cranking	KOEO: Key On, Engine Off
KOER: Key On, Engine Running	KS: Knock Sensor (Bank 1 or Bank 2)
LAMBSE: Short Term Fuel Trim	LCD: Liquid Crystal Display
LFC: Low Fan Control	LFP: Low Speed Fuel Pump (control)
LONGFT: Long Term Fuel Trim	LOS: Limited Operating Strategy
LOOP: Engine Operating Loop Status	LPG: Liquid Petroleum Gas

PID Data & Pin Chart Glossary

Glossary of Terms and Acronyms	
M/T: Manual Transmission	MAF: Mass Airflow
MAF RTN: Mass Airflow Sensor Return	MAP: Manifold Absolute Pressure
MCU: Microprocessor Control Unit	MFI: Multiport Fuel Injection
MIL: Malfunction Indicator Lamp	MLP: Manual Lever Position (Sensor)
MPH: Miles Per Hour	N.C.: Normally Closed
NDS: Neutral Drive (Switch)	N/V: Input shaft speed to vehicle speed
NGV: Natural Gas Vehicles	N.O.: Normally Open
OASIS: Online Automotive Service Information System	OBD I: On Board Diagnostics Version I
	OBD II: On Board Diagnostics Version II
OCIL: Overdrive Cancel Indicator Lamp	OCS: Overdrive Cancel Switch
OCT ADJ: Octane Adjust Fuel (Switch)	OHC: Overhead Cam Engine
OSM: Output State Monitor	OSS: Output Speed Shaft
O2S-11 (Bank 1 Sensor 1) Signal	O2S-21 (Bank 2 Sensor 1) Signal
PAS: Passive Anti-Theft System	PCM: Powertrain Control Module
PCV: Positive Crankcase Ventilation	PFE: Pressure Feedback EGR System
PFI: Port Fuel Injection	PID: Parameter Identification Location
PIP: Profile Ignition Pickup Signal	PMI: Programmable Module Installation
PNP: Park Neutral Position	PSP: Power Steering Pressure (switch)
PSOM: Programmable Speedometer Odometer Module	RAM: Random Access Memory
RCM: Restraint Control Module	REM: Rear Electronic Module
ROM: Read Only Memory	RPM: Revolutions Per Minute
RTN: Dedicated sensor ground circuit	RWD: Rear Wheel Drive
SC: Supercharged Engine	SCP: Standard Corporate Protocol
SFI: Sequential Fuel Injection	SHRTFT1: Short Term Fuel Trim Bank 1
SHO: Super High Output Engine	SIG RTN: Signal Return
SIL: Shift Indicator Lamp	SPOUT: Spark Output signal
SS: Shift Solenoid	STAR: Self Test Automatic Readout
STI: Self Test Input	STO: Self Test Output
TAB: Thermactor Air Bypass	TAD: Thermactor Air Diverter
TC: Turbo Charged	TCC: Torque Converter Clutch
TCC: Torque Converter Clutch Control	TCS: Traction Control Switch
TCIL: Transmission Control Indicator Lamp	TFT: Transmission Fluid Temperature
TKS: Throttle Kicker Solenoid	TOT: Transmission Oil Temperature
TP: Throttle Position (Sensor)	TP Mode: Throttle Position Mode
TR: Transmission Range sensor	TSB: Technical Service Bulletin
TSS: Transmission Shaft Speed Sensor	TWC: Three Way Catalyst
TWC + OC: 3-Way Catalyst	UBP: UART Based Protocol Network
VAF: Vane Airflow Meter or sensor	VAT: Vane Air Temperature sensor
VID: Vehicle Information Block (memory)	VIM: Vehicle Interface Module
VBAT: Vehicle Battery Voltage	VLCM: Variable Load Control Module
VPWR: Ignition switched Power	VREF: Reference Voltage (from PCM)
VRS: Variable Reluctance Sensor	VMV: Vapor Management Valve (EVAP)
VCT: Variable Camshaft Timing	WAC: WOT A/C Cutout Relay
WACA- A/C WOT Cutout Relay Monitor	WOT: Wide Open Throttle

Gas Engine Trouble Code List

Introduction

To use this information, first read and record all codes in memory along with any Freeze Frame data. *If the PCM reset function is done prior to recording any data, all codes and freeze frame data will be lost!* Look up the desired code by DTC number, Code Title and Conditions (enable criteria) that indicate why a code set, and how to drive the vehicle. **1T and 2T** indicate a 1-trip or 2-trip fault and the Monitor type.

Gas Engine OBD II Trouble Code List (P0xxx Codes)

DTC	Trouble Code Title, Conditions & Possible Causes
P0010 **2T CCM, MIL: Yes** 2002, 2003 Cougar, Focus, LS, Thunderbird & ZX2 Transmissions: All	**Variable Cam Timing Solenoid 'A' Circuit Malfunction Conditions:** Key on or engine running; and the PCM detected an unexpected high voltage or low voltage condition on the Variable Cam Timing (VCT) Solenoid 'A' control circuit during testing. **Possible Causes** • VCT 'A' solenoid connector is damaged, loose or shorted • VCT 'A' solenoid control circuit is open, shorted to ground or shorted to power • VCT 'A' solenoid is damaged or the PCM has failed
P0011 **2T CCM, MIL: Yes** 2002, 2003 Cougar, Focus, LS, Thunderbird & ZX2 Transmissions: All	**Variable Cam Timing Over Advanced (Bank 1) Conditions:** Engine started; and the PCM detected the camshaft timing exceeded the maximum calibrated advance value, or the camshaft remained in an advanced position during the CCM test. **Possible Causes** • Camshaft timing improperly set, or continuous oil flow to the VCT piston chamber • Camshaft advance mechanism (the VCT unit) is sticking or binding mechanically • VCT solenoid valve is stuck in open position
P0012 **2T CCM, MIL: Yes** 2002, 2003 Cougar, Focus, LS, Thunderbird & ZX2 Transmissions: All	**Variable Cam Timing Over Retarded (Bank 1) Conditions:** Engine started; and the PCM detected the camshaft timing exceeded the maximum calibrated retard value, or the camshaft remained in a retarded position during the CCM test. **Possible Causes** • Camshaft timing improperly set, or continuous oil flow to the VCT piston chamber • Camshaft advance mechanism (the VCT unit) is sticking or binding mechanically • VCT solenoid valve is stuck in open position
P0020 **2T CCM, MIL: Yes** 2002, 2003 Cougar, Focus, LS, Thunderbird & ZX2 Transmissions: All	**Variable Cam Timing Solenoid 'B' Circuit Malfunction Conditions:** Key on or engine running; and the PCM detected an unexpected high or low voltage condition on the Variable Cam Timing (VCT) Solenoid 'B' control circuit during testing. **Possible Causes** • VCT 'B' solenoid connector is damaged, loose or shorted • VCT 'B' solenoid control circuit is open, shorted to ground or shorted to power • VCT 'B' solenoid is damaged or the PCM has failed
P0020 **2T CCM, MIL: Yes** 2002, 2003 Cougar, Focus, LS, Thunderbird & ZX2 Transmissions: All	**Variable Cam Timing Over Advanced (Bank 2) Conditions:** Engine started; and the PCM detected the camshaft timing exceeded the maximum calibrated advance value, or the camshaft remained in an advanced position during the CCM test. **Possible Causes** • Camshaft timing improperly set, or continuous oil flow to the VCT piston chamber • Camshaft advance mechanism (the VCT unit) is sticking or binding mechanically • VCT solenoid valve is stuck in open position
P0022 **2T CCM, MIL: Yes** 2002, 2003 Cougar, Focus, LS, Thunderbird & ZX2 Transmissions: All	**Variable Cam Timing Over Retarded (Bank 2) Conditions:** Engine started; and the PCM detected the camshaft timing exceeded the maximum calibrated retard value, or the camshaft remained in a retarded position during the CCM test. **Possible Causes** • Camshaft timing improperly set, or continuous oil flow to the VCT piston chamber • Camshaft advance mechanism (the VCT unit) is sticking or binding mechanically • VCT solenoid valve is stuck in open position
P0040 **2T CCM, MIL: Yes** 2003 Crown Victoria, Marauder & Grand Marquis Transmissions: All	**Upstream Oxygen Sensors Swapped From Bank To Bank Conditions:** Engine started; and after the PCM performed a brief fuel shift in engine Bank 1, it did not detect the correct response in the HO2S-11 and/or HO2S-21 (sensors) during testing. **Possible Causes** • HO2S-11 and HO2S-21 harness connectors are swapped • HO2S-11 and HO2S-21 wiring is crossed inside the harness • HO2S-11 and HO2S-21 wires are crossed at 104-pin connector

Gas Engine OBD II Trouble Code List (P0xxx Codes)

DTC	Trouble Code Title, Conditions & Possible Causes
P0041 **2T CCM, MIL: Yes** 2003 Crown Victoria, Marauder & Grand Marquis Transmissions: All	**Downstream Oxygen Sensors Swapped Bank To Bank Conditions:** Engine started; and after the PCM performed a brief fuel shift in engine Bank 1, it did not detect the correct response in the HO2S-12 and/or HO2S-22 (sensors) during testing. **Possible Causes** • HO2S-12 and HO2S-22 harness connectors are swapped • HO2S-12 and HO2S-22 wiring is crossed inside the harness • HO2S-12 and HO2S-22 wires are crossed at 104-pin connector
P0041 **2T CCM, MIL: Yes** 2003 Crown Victoria, Marauder & Grand Marquis Transmissions: All	**TP Sensor, ETC TP Sensor Inconsistent With MAF Sensor Conditions:** Engine started, KOER Self Test enabled, and the PCM detected the MAF and ETC TP sensor signals were not consistent with the calibrated values expected for these sensors in the test. **TP Sensor:** Drive the vehicle and then monitor the TPNS PID in all gears. A TP PID under 0.24v (4.82%) with a LOAD PID over 55%, or a TP V PID over 2.44v (49.05%) with a LOAD PID under 30% will set this code. **ETC TP:** Drive the vehicle and then monitor the TPNS PID in all gears. A TPNS PID under 0.24v (4.82%) with a LOAD PID over 55% or TPNS PID over 2.44v (49.05%) with LOAD PID under 30% will set a code. **Possible Causes** • Air leak exists between MAF sensor and the throttle body • MAF sensor is out-of-calibration or it has failed • TP sensor is not seated properly or the TP sensor is damaged
P0102 **2T CCM, MIL: Yes** 1996-1997-1998-1999-2000-2001-2002-2003 All Models Transmissions: All	**MAF Sensor Circuit Low Input Conditions:** DTC P0505 not set, engine started, and the PCM detected the MAF sensor signal was less than 0.23v during the CCM test period. **Possible Causes** • Check for leaks at air outlet tube • Sensor power circuit open, sensor ground circuit open • Sensor signal circuit open (may be disconnected) • Check for loose tube clamps near the MAF sensor
P0103 **2T CCM, MIL: Yes** 1996-1997-1998-1999-2000-2001-2002-2003 All Models Transmissions: All	**MAF Sensor Circuit High Input Conditions:** DTC P0505 not set, engine started, and the PCM detected the MAF sensor signal was more than 4.60v during the CCM test period. **Possible Causes** • Check for a restricted inlet screen on the MAF sensor. • MAF sensor signal circuit is shorted to system power (B+) • MAF sensor is damaged or the PCM has failed
P0106 **2T CCM, MIL: Yes** 1996-1997-1998-1999-2000-2001-2002-2003 All Models Transmissions: All	**Barometric Pressure Sensor Circuit Performance Conditions:** Engine started, and the PCM detected the BARO sensor was out of range during the CCM test. The BARO sensor signal should be in a range of 4.0-6.0v. The Scan Tool displays the sensor reading as a frequency. *Note: The sensor VREF should be less than 6.0v at all times.* **Possible Causes** • Sensor has deteriorated (response time too slow) or has failed • PCM has failed
P0107 **2T CCM, MIL: Yes** 1996-1997-1998-1999-2000-2001-2002-2003 All Models Transmissions: All	**Barometric Pressure Sensor Circuit Low Input Conditions:** Engine started, and the PCM detected the BARO sensor indicated less than the minimum calibrated parameter. The BARO sensor is a variable capacitance unit used to detect altitude. **Possible Causes** • BARO sensor signal circuit is shorted to ground • BARO sensor VREF circuit (5v) is open • BARO sensor is damaged or it has failed • PCM has failed
P0107 **2T CCM, MIL: Yes** 1999-2000 2.0L I4 VIN Z CNG engine Contour Transmissions: All	**MAP/BARO Sensor (Compuvalve) Circuit Low Input Conditions:** Engine started, vehicle driven under heavy load conditions, and the PCM detected the Compuvalve MAP/BARO sensor signal indicated too low a value. Note: The MAP/BARO sensor is mounted (internal) in the Compuvalve assembly. The PCM monitors this signal. **Possible Causes** • Vacuum hoses from the intake manifold and the Compuvalve may be open or restricted • Compuvalve MAP/BARO sensor has failed

Gas Engine OBD II Trouble Code List (P0xxx Codes)

DTC	Trouble Code Title, Conditions & Possible Causes
P0108 **2T CCM, MIL: Yes** 1996-1997-1998-1999-2000-2001-2002-2003 All Models Transmissions: All	**Barometric Sensor Circuit High Input Conditions:** Engine started, and the PCM detected the BARO sensor signal was more than the maximum calibrated parameter of 5.0v. Note: The sensor VREF should be 4.0v to 6.0v at all times. **Possible Causes** • BARO sensor connector is damaged, open or shorted • BARO sensor signal circuit is open or shorted to VREF (5v) • BARO sensor is damaged or it has failed • PCM has failed
P0108 **2T CCM, MIL: Yes** 1999-2000 2.0L I4 VIN Z CNG engine Contour Transmissions: All	**MAP/BARO Sensor (Compuvalve) Circuit High Input Conditions:** Engine started, vehicle driven under heavy load conditions, and the PCM detected the Compuvalve MAP/BARO sensor signal indicated too high a value during the CCM test. The MAP/BARO sensor is mounted (internal) in the Compuvalve assembly on this application. **Possible Causes** • Vacuum hoses from the intake manifold and Compuvalve may be open and/or restricted • Compuvalve MAP/BARO sensor has failed
P0109 **2T CCM, MIL: Yes** 1996-1997-1998-1999-2000-2001-2002-2003 All Models Transmissions: All	**Barometric Sensor Circuit Intermittent Input Conditions:** Engine started, and the PCM detected the BARO sensor signal had an intermittent failure during normal engine operation. **Possible Causes** • BARO sensor signal circuit is open or shorted to ground (intermittent) • BARO sensor VREF circuit (5v) is open • BARO sensor signal circuit is shorted to ground (Intermittent) • BARO sensor is damaged or it has failed
P0112 **2T CCM, MIL: Yes** 1996-1997-1998-1999-2000-2001-2002-2003 All Models Transmissions: All	**Intake Air Temperature Sensor Circuit Low Input Conditions:** Key on or engine running; and the PCM detected the IAT sensor signal was less than the self-test minimum of 0.20v (Scan Tool reads over 250ºF). This is a thermistor-type sensor with a variable resistance that changes when exposed to different temperatures. **Possible Causes** • IAT sensor signal circuit is grounded (check wiring & connector) • IAT sensor is damaged or it has failed • PCM has failed
P0113 **2T CCM, MIL: Yes** 1996-1997-1998-1999-2000-2001-2002-2003 All Models Transmissions: All	**Intake Air Temperature Sensor Circuit High Input Conditions:** Key on or engine running; and the PCM detected the IAT sensor signal was more than the self-test maximum 4.60v (Scan Tool reads under -46ºF). This is a thermistor-type sensor with a variable resistance that changes when exposed to different temperatures. **Possible Causes** • IAT sensor signal circuit is open (inspect wiring & connector) • IAT sensor signal circuit is shorted to VREF (5v) • IAT sensor is damaged or it has failed • PCM has failed
P0116 **2T ECT, MIL: Yes** 2001-2002-2003 All Models Transmissions: All	**ECT Sensor / CHT Sensor Signal Range/Performance Conditions:** Engine off for a calibrated period of time after an engine cold soak period over 6 hours, engine started, and the PCM detected the ECT sensor exceeded the IAT sensor by more than a calibrated value (e.g., 30ºF), or the ECT sensor was more than a calibrated value of 225ºF; the Catalyst, Fuel System, HO2S and Misfire Monitor did not complete, or the timer expired. **Possible Causes** • Check for low coolant level or incorrect coolant mixture • CHT sensor is out-of-calibration or it has failed • ECT sensor is out-of-calibration or it has failed
P0117 **2T CCM, MIL: Yes** 1996-1997-1998-1999-2000-2001-2002-2003 All Models Transmissions: All	**ECT Sensor Circuit Low Input Conditions:** Key on or engine running; and the PCM detected the ECT sensor signal was less than the self-test minimum of 0.20v (Scan Tool reads over 250ºF). This is a thermistor-type sensor with a variable resistance that changes when exposed to different temperatures **Possible Causes** • ECT sensor signal circuit is grounded in the wiring harness • ECT sensor is damaged or the PCM has failed

Gas Engine OBD II Trouble Code List (P0xxx Codes)

DTC	Trouble Code Title, Conditions & Possible Causes
P0118 **2T CCM, MIL: Yes** 1996-1997-1998-1999-2000-2001-2002-2003 All Models Transmissions: All	**ECT Sensor Circuit High Input Conditions:** Key on or engine running; and the PCM detected the ECT sensor signal was more than the self-test maximum of 4.60v (Scan Tool reads under -46ºF). This is a thermistor-type sensor with a variable resistance that changes when exposed to different temperatures **Possible Causes** • ECT sensor signal circuit is open (inspect wiring & connector) • ECT sensor signal circuit is shorted to VREF (5v) • ECT sensor is damaged or it has failed • PCM has failed
P0121 **2T CCM, MIL: Yes** 1996-1997-1998-1999-2000-2001-2002-2003 All Models Transmissions: All	**TP Sensor Signal Range/Performance Conditions:** Engine started; then immediately following a condition where the engine was running under at off-idle, the PCM detected the TP sensor signal indicated the throttle did not return to its previous closed position during the CCM Rationality test. **Possible Causes** • Throttle plate is binding, dirty or sticking • TP sensor signal circuit open (inspect wiring & connector) • TP sensor ground circuit open (inspect wiring & connector) • TP Sensor is damaged or has failed
P0121 **2T CCM, MIL: Yes** 2003 Crown Victoria, Marauder & Grand Marquis Transmissions: All	**ETC Throttle Position Sensor Signal Range/Performance Conditions:** Key on or engine running; then immediately after the throttle is open or following a closed throttle (deceleration) event, the PCM detected the TP2 PID (TP Sensor 2) signal indicated more than 93% (4.65v) from its previous closed position during the CCM Rationality test. **Possible Causes** • ECT TP sensor connector is damaged, loose or shorted • ETC TP sensor signal circuit is open • ETC TP sensor ground circuit is open • ETC TP sensor circuit VREF (5v) is open • ECT TP sensor is not seated properly, or it is sticking • ETC TP Sensor is damaged or it has failed
P0122 **2T CCM, MIL: Yes** 1996-1997-1998-1999-2000-2001-2002-2003 All Models Transmissions: All	**TP Sensor Circuit Low Input Conditions:** Key on or engine running; and the PCM detected the TP sensor was less than 0.17v (Scan Tool TP PID reads under 3.42%) in the test. **Possible Causes** • TP sensor signal circuit open (inspect wiring & connector) • TP sensor signal shorted to ground (inspect wiring & connector) • TP sensor VREF circuit is open (between the sensor and PCM) • TP sensor is damaged or has failed • PCM has failed
P0122 **2T CCM, MIL: Yes** 2003 Crown Victoria, Marauder & Grand Marquis Transmissions: All	**ETC Throttle Position Sensor Circuit Low Input Conditions:** Key on or engine running; and the PCM detected the ETC TP sensor was below the self-test minimum of 0.17v (Scan Tool below 3.42%). **Possible Causes** • TP sensor signal circuit open (inspect wiring & connector) • TP sensor signal shorted to ground (inspect wiring & connector) • TP sensor VREF circuit is open (between the sensor and PCM) • TP sensor is damaged or has failed • PCM has failed
P0123 **2T CCM, MIL: Yes** 1996-1997-1998-1999-2000-2001-2002-2003 All Models Transmissions: All	**TP Sensor Circuit High Input Conditions:** Engine started, and the PCM detected the TP sensor signal was more than the self-test maximum of 4.65v (equivalent to a Scan Tool TP PID of more than 93%) during testing. **Possible Causes** • TP sensor not seated correctly in housing (may be damaged) • TP sensor signal is circuit shorted to VREF or system voltage • TP sensor ground circuit is open (check the wiring harness) • Perform a "sensor sweep test" and monitor for any glitches • PCM has failed

Gas Engine OBD II Trouble Code List (P0xxx Codes)

DTC	Trouble Code Title, Conditions & Possible Causes
P0122 **2T CCM, MIL: Yes** 2003 Crown Victoria, Marauder & Grand Marquis Transmissions: All	**ETC Throttle Position Sensor Circuit High Input Conditions:** Key on or engine running; and the PCM detected the ETC TP sensor was less than the self-test maximum of 4.65v (Scan Tool TP PID reads more than 93%) during the CCM test. **Possible Causes** • TP sensor connector is damaged, loose or shorted • TP sensor signal circuit is open or shorted to ground • TP sensor VREF circuit is open (between the sensor and PCM) • TP sensor is damaged or has failed • PCM has failed
P0125 **2T ECT, MIL: Yes** 1996-1997-1998-1999- 2000-2001-2002-2003 All Models Transmissions: All	**Insufficient Coolant Temperature For Closed Loop Conditions:** Engine runtime at road load more than 6 minutes, and the PCM detected that the ECT sensor (or CHT sensor) signal did not indicate the required engine temperature value to enter closed loop within a specified amount of time. The amount of time is calculated from the point at which the engine is started, and depends upon the ECT or CHT sensor signal value at startup. **Possible Causes** • Check the coolant mixture for an incorrect mixture • Check the operation of the thermostat (it may be stuck open) • ECT sensor (or the CHT sensor) has failed • Inspect for low coolant level
P0127 **2T CCM, MIL: Yes** 1999-2000-2001-2002- 2003 F150 Series Truck, Mustang with a 4.6L VIN Y & 5.4L VIN 3 engines Transmissions: All	**IAT Sensor 2 Circuit High Input Conditions:** Engine started, engine running for a calibrated period of time, and the PCM detected the IAT Sensor 2 (PID IAT2) signal was too high. This code indicates a potential fault present in the Intercooler system (the Supercharger Boost is bypassed when this code is set). **Possible Causes** • Blockage present in the Heat Exchangers • Low fluid level, or a fluid leak is present • Intercooler pump or relay has failed • Intercooler coolant lines may be crossed
P0128 **2T CCM, MIL: Yes** 2003 All Models Transmissions: All	**Intake Air Temperature Sensor 2 Circuit High Input Conditions:** Engine started, vehicle driven for over 10 minutes, and the PCM detected the engine did not reach an engine operating temperature of 160°F after an additional runtime of 2 minutes. **Possible Causes** • Check the operation of the thermostat (it may be stuck open) • ECT sensor or CHT sensor is out-of-calibration, or has failed • Inspect for low coolant level or an incorrect coolant mixture
P0131 **2T CCM, MIL: Yes** 1996-1997-1998-1999- 2000-2001-2002-2003 All Models Transmissions: All	**HO2S-11 (Bank 1 Sensor 1) Circuit Low Input Conditions:** Engine running for more than 5 minutes, and the PCM detected the HO2S signal was in a negative voltage range referred to as "character shift downward". This code sets when the HO2S signal remains in a low state (usually less than 156 mv). In effect, it does not switch properly between 0.1v and 1.1v in closed loop operation. **Possible Causes** • HO2S is contaminated (due to presence of silicone in fuel) • HO2S signal and ground circuit wires crossed in wiring harness • HO2S signal circuit is shorted to sensor or chassis ground • HO2S element has failed (internal short condition) • PCM has failed
P0132 **2T CCM, MIL: Yes** 1996-1997-1998-1999- 2000-2001-2002-2003 All Models Transmissions: All	**HO2S-11 (Bank 1 Sensor 1) Circuit High Input Conditions:** Engine running for more than 5 minutes, and the PCM detected the HO2S signal remained in a high state (i.e., more than 1.5v). Note: The HO2S signal circuit may be shorted to the heater power circuit due to "tracking inside of the HO2S connector. Remove the connector and visually inspect the connector for signs of oil or water. **Possible Causes** • HO2S signal shorted to heater power circuit inside connector • HO2S signal circuit shorted to VREF or to system voltage • PCM has failed

Gas Engine OBD II Trouble Code List (P0xxx Codes)

DTC	Trouble Code Title, Conditions & Possible Causes
P0133 **2T O2S, MIL: Yes** 1996-1997-1998-1999-2000-2001-2002-2003 All Models Transmissions: All	**HO2S-11 (Bank 1 Sensor 1) Circuit Slow Response Conditions:** Engine started, engine running in closed loop for over 5 minutes, and the PCM detected the HO2S amplitude and frequency were out of the normal range (e.g., the HO2S rich to lean switch time was more than 100 ms) during the HO2S Monitor test. **Possible Causes** • HO2S is contaminated (due to presence of silicone in fuel) • HO2S signal circuit open • Leaks present in the exhaust manifold or exhaust pipes • HO2S is damaged or has failed • PCM has failed
P0135 **2T O2HTR, MIL: Yes** 1996-1997-1998-1999-2000-2001-2002-2003 All Models Transmissions: All	**HO2S-11 (Bank 1 Sensor 1) Heater Circuit Malfunction Conditions:** Engine started, engine running for 5 minutes, and the PCM detected an unexpected voltage condition, or it detected excessive current draw in the heater circuit during the CCM test. **Possible Causes** • HO2S heater power circuit is open or heater ground circuit open • HO2S signal tracking (due to oil or moisture in the connector) • HO2S is damaged or has failed • PCM has failed
P0136 **2T O2S, MIL: Yes** 1996-1997-1998-1999-2000-2001-2002-2003 All Models Transmissions: All	**HO2S-12 (Bank 1 Sensor 1) Circuit No Activity Conditions:** Engine started, engine running in closed loop for over 5 minutes, and the PCM detected the HO2S signal failed to meet the maximum or minimum voltage levels (i.e., it failed the voltage range check). **Possible Causes** • Leaks present in the exhaust manifold or exhaust pipes • HO2S signal wire and ground wire crossed in connector • HO2S element is fuel contaminated or has failed • PCM has failed
P0138 **2T CCM, MIL: Yes** 1996-1997-1998-1999-2000-2001-2002-2003 All Models Transmissions: All	**HO2S-12 (Bank 1 Sensor 2) Circuit High Input Conditions:** Engine running for more than 5 minutes, and the PCM detected the HO2S signal remained in a high state (i.e., more than 1.5v). *Note: The HO2S signal circuit may be shorted to the heater power circuit due to "tracking inside of the HO2S connector. Remove the connector and visually inspect the connector for signs of oil or water.* **Possible Causes** • HO2S signal shorted to heater power circuit in the connector • HO2S signal circuit shorted to VREF or to system voltage • PCM has failed
P0141 **2T O2HTR, MIL: Yes** 1996-1997-1998-1999-2000-2001-2002-2003 All Models Transmissions: All	**HO2S-12 (Bank 1 Sensor 2) Heater Circuit Malfunction Conditions:** Engine running for 5 minutes, and the PCM detected an open or shorted condition, or excessive current draw in the heater circuit. **Possible Causes** • HO2S heater power circuit is open or heater ground circuit open • HO2S signal tracking (due to oil or moisture in the connector) • HO2S is damaged or has failed • PCM has failed
P0148 **2T CCM, MIL: Yes** 2001-2002-2003 All Models Transmissions: All	**Fuel Delivery Error Conditions:** Engine started, engine running for a specified period of time in closed loop, then after at least one WOT event was recorded, the PCM detected a lean air/fuel condition in at least one engine bank during the wide-open throttle event. **Possible Causes** • Severely restricted fuel filter • Severely pinched or restricted fuel delivery line

Gas Engine OBD II Trouble Code List (P0xxx Codes)

DTC	Trouble Code Title, Conditions & Possible Causes
P0151 **2T CCM, MIL: Yes** 1996-1997-1998-1999-2000-2001-2002-2003 Models with a 4-Cyl, V6 or V8 engine Transmissions: All	**HO2S-21 (Bank 2 Sensor 1) Circuit Low Input Conditions:** Engine started, engine running for over 5 minutes, and the PCM detected the HO2S signal was in a negative voltage range referred to as "character shift downward". This code sets when the HO2S signal remains in a low state (less than 156 mv). In effect, the sensor did not switch properly between 0.1v and 1.1v in closed loop. **Possible Causes** • HO2S connector is damaged or shorted • HO2S signal and ground circuit wires crossed in wiring harness • HO2S signal circuit is shorted to sensor or chassis ground • HO2S element has failed (internal short condition) • PCM has failed
P0152 **2T CCM, MIL: Yes** 1996-1997-1998-1999-2000-2001-2002-2003 Models with a 4-Cyl, V6 or V8 engine Transmissions: All	**HO2S-21 (Bank 2 Sensor 1) Circuit High Input Conditions:** Engine started, engine runtime over 5 minutes, and the PCM detected the HO2S signal remained in a high state (more than 1.5v). Note: The HO2S signal circuit may be shorted to the heater power circuit due to "tracking inside of the HO2S connector. Remove the connector and visually inspect the connector for signs of oil or water. **Possible Causes** • HO2S is contaminated (due to presence of silicone in fuel) • HO2S signal tracking (due to oil or moisture in the connector) • HO2S signal circuit is open or shorted to VREF • PCM has failed
P0153 **2T O2S, MIL: Yes** 1996-1997-1998-1999-2000-2001-2002-2003 Models with a 4-Cyl, V6 or V8 engine Transmissions: All	**HO2S-21 (Bank 2 Sensor 1) Circuit Slow Response Conditions:** Engine started, engine running in closed loop for over 5 minutes, and the PCM detected the HO2S amplitude and frequency were out of the normal range (e.g., the HO2S rich to lean switch time was more than 100 ms) during the HO2S Monitor test. **Possible Causes** • HO2S is contaminated (due to presence of silicone in fuel) • Leaks present in the exhaust manifold or exhaust pipes • HO2S is damaged or has failed • PCM has failed
P0155 **2T O2HTR, MIL: Yes** 1996-1997-1998-1999-2000-2001-2002-2003 Models with a 4-Cyl, V6 or V8 engine Transmissions: All	**HO2S-21 (Bank 2 Sensor 1) Heater Circuit Malfunction Conditions:** Engine running for 5 minutes, and the PCM detected an open or shorted condition, or excessive current draw in the heater circuit. **Possible Causes** • HO2S heater power circuit is open • HO2S heater ground circuit is open • HO2S signal tracking (due to oil or moisture in the connector) • HO2S is damaged or has failed • PCM has failed
P0156 **2T O2S, MIL: Yes** 1996-1997-1998-1999-2000-2001-2002-2003 Models with a 4-Cyl, V6 or V8 engine Transmissions: All	**HO2S-22 (Bank 2 Sensor 2) Circuit No Activity Conditions:** Engine running in closed loop for more than 5 minutes, and the PCM detected the HO2S signal failed to meet the maximum or minimum voltage (i.e., it failed the voltage check). **Possible Causes** • Leaks present in the exhaust manifold or exhaust pipes • HO2S signal wire and ground wire crossed in connector • HO2S element is fuel contaminated or has failed • PCM has failed
P0158 **2T CCM, MIL: Yes** 1996-1997-1998-1999-2000-2001-2002-2003 Models with a 4-Cyl, V6 or V8 engine Transmissions: All	**HO2S-22 (Bank 2 Sensor 2) Circuit High Input Conditions:** Engine running for more than 5 minutes, and the PCM detected the HO2S signal remained in a high state (i.e., more than 1.5v). Note: The HO2S signal circuit may be shorted to the heater power circuit due to "tracking inside of the HO2S connector. Remove the connector and visually inspect the connector for signs of oil or water. **Possible Causes** • HO2S signal shorted to the heater power circuit (due to oil or moisture in the connector) • HO2S signal circuit shorted to VREF or to system voltage • PCM has failed

Gas Engine OBD II Trouble Code List (P0xxx Codes)

DTC	Trouble Code Title, Conditions & Possible Causes
P0161 **2T O2HTR, MIL: Yes** 1996-1997-1998-1999-2000-2001-2002-2003 Models with a 4-Cyl, V6 or V8 engine Transmissions: All	**HO2S-22 (Bank 2 Sensor 2) Heater Circuit Malfunction Conditions:** Engine running for 5 minutes, and the PCM detected an open or shorted condition, or excessive current draw in the heater circuit. **Possible Causes** • HO2S heater power circuit or the heater ground circuit is open • HO2S signal tracking (due to oil or moisture in the connector) • HO2S has failed, or the PCM has failed
P0171 **2T FUEL, MIL: Yes** 1996-1997-1998-1999-2000-2001-2002-2003 Some Models Transmissions: All	**Fuel System Too Lean (Cylinder Bank 1) Conditions:** Engine started, engine running at cruise speed for 3 to 4 minutes, and the PCM detected the Bank 1 Adaptive Fuel Control System reached its rich correction limit (a lean A/F condition). **Possible Causes** • Air leaks after the MAF sensor, or leaks in the PCV system • Exhaust leaks before or near where the HO2S is mounted • Fuel injector(s) restricted or not supplying enough fuel • Fuel pump not supplying enough fuel during high fuel demand conditions • Leaking EGR gasket, or leaking EGR valve diaphragm • MAF sensor dirty (causes PCM to underestimate airflow) • Vehicle running out of fuel or engine oil dip stick not seated • TSB 01-20-5 contains a repair procedure for this trouble code
P0171 **2T FUEL, MIL: Yes** 1996-1997-1998-1999-2000-2001-2002-2003 Windstar Models Transmissions: All	**Fuel System Too Lean (Cylinder Bank 1) Conditions:** Engine started, engine running at cruise speed for 3 to 4 minutes, and the PCM detected the Bank 1 Adaptive Fuel Control System reached its rich correction limit (a lean A/F condition). **Possible Causes** • Air leaks after the MAF sensor, or leaks in the PCV system • Exhaust leaks before or near where the HO2S is mounted • Fuel injector(s) restricted or not supplying enough fuel • Fuel system not supplying enough fuel during high fuel demand conditions (e.g., the fuel pump may not supply enough fuel) • Leaking EGR gasket, or leaking EGR valve diaphragm • MAF sensor dirty (causes PCM to underestimate airflow) • Vehicle running out of fuel or engine oil dip stick not seated • TSB 03-16-1 contains a repair procedure for this trouble code
P0172 **2T FUEL, MIL: Yes** 1996-1997-1998-1999-2000-2001-2002-2003 All Models Transmissions: All	**Fuel System Too Rich (Cylinder Bank 1) Conditions:** Engine started, engine running at cruise speed for 3 to 4 minutes, and the PCM detected the Bank 1 Adaptive Fuel Control System reached its rich correction limit (a rich A/F condition). **Possible Causes** • Camshaft timing is incorrect, or the engine has an oil overfill condition • EVAP vapor recovery system failure (may be pulling vacuum) • Fuel pressure regulator is damaged or leaking • HO2S element is contaminated with alcohol or water • MAF or MAP sensor values are incorrect or out-of-range • One of more fuel injectors is leaking
P0174 **2T FUEL, MIL: Yes** 1996-1997-1998-1999-2000-2001-2002-2003 Some Models (V6, V8) Transmissions: All	**Fuel System Too Lean (Cylinder Bank 2) Conditions:** Engine started, engine running at cruise speed for 3 to 4 minutes, and the PCM detected the Bank 2 Adaptive Fuel Control System reached its rich correction limit (a lean A/F condition). **Possible Causes** • Air leaks after the MAF sensor, or leaks in the PCV system • Exhaust leaks before or near where the HO2S is mounted • Fuel injector(s) restricted or not supplying enough fuel • Fuel system not supplying enough fuel during high fuel demand conditions (e.g., the fuel pump may not supply enough fuel) • Leaking EGR gasket, or leaking EGR valve diaphragm • MAF sensor dirty (causes PCM to underestimate airflow) • Vehicle running out of fuel or engine oil dip stick not seated • TSB 1-20-5 contains a repair procedure for this trouble code

Gas Engine OBD II Trouble Code List (P0xxx Codes)

DTC	Trouble Code Title, Conditions & Possible Causes
P0174 **2T FUEL, MIL: Yes** 1996-1997-1998-1999-2000-2001-2002-2003 Windstar Models Transmissions: All	**Fuel System Too Lean (Cylinder Bank 2) Conditions:** Engine started, engine running at cruise speed for 3 to 4 minutes, and the PCM detected the Bank 2 Adaptive Fuel Control System reached its rich correction limit (a lean A/F condition). **Possible Causes** • Air leaks after the MAF sensor, or leaks in the PCV system • Exhaust leaks before or near where the HO2S is mounted • Fuel injector(s) restricted or not supplying enough fuel • Fuel system not supplying enough fuel during high fuel demand conditions (e.g., the fuel pump may not supply enough fuel) • Leaking EGR gasket, or leaking EGR valve diaphragm • MAF sensor dirty (causes PCM to underestimate airflow) • Vehicle running out of fuel or engine oil dip stick not seated • TSB 03-16-1 contains a repair procedure for this trouble code
P0175 **2T FUEL, MIL: Yes** 1996-1997-1998-1999-2000-2001-2002-2003 Some Models (V6, V8) Transmissions: All	**Fuel System Too Rich (Cylinder Bank 2) Conditions:** Engine started, engine running at cruise speed for 3 to 4 minutes, and the PCM detected the Bank 2 Adaptive Fuel Control System reached its rich correction limit (a rich A/F condition). **Possible Causes** • Camshaft timing is incorrect • Engine oil overfill condition • EVAP vapor recovery system failure (may be pulling vacuum) • Fuel pressure regulator is damaged or leaking • HO2S element is contaminated with alcohol or water • MAF or MAP sensor values are incorrect or out-of-range • One of more fuel injectors is leaking
P0176 **2T CCM, MIL: Yes** 1996-1997-1998-1999-2000-2001-2002-2003 Taurus with a 3.0L VIN 1 or VIN 2 engine Transmissions: All	**Flexible Fuel Sensor Circuit Malfunction Conditions:** Engine started, and the PCM detected the Flexible Fuel sensor (FFV) signal was not within its maximum or minimum calibrated range (the normal calibrated range is from 40-160 Hz). Note: Prefix 610C indicates an ethanol vehicle and prefix 610G indicates a methanol vehicle. **Possible Causes** • FFV sensor VPWR circuit is open or shorted to ground • FFV sensor ground circuit open or circuit shorted to ground • FFV sensor has a fuel contamination condition • FFV sensor has a fuel separation condition • PCM has failed
P0180 **2T CCM, MIL: Yes** 1998-1999-2000-2001-2002-2003 Crown Victoria, Grand Marquis, E & F Series Vans & Trucks with a 4.6L VIN 9, 5.4L VIN M, 5.4L VIN Z engine Transmissions: All	**Engine Fuel Temperature Sensor 'A' Circuit Malfunction Conditions:** Engine runtime over 2 minutes, and the PCM detected the Engine Fuel Temperature (EFT) sensor 'A' signal was out-of-range (i.e., it was more than 4.54v [-46ºF] or less than 0.21v [275ºF]. Note: Monitor the EFT PID value to identify an open or short circuit. **Possible Causes** • Engine operating under "low" ambient temperature conditions • EFT sensor signal circuit open or shorted in the wiring harness • EFT sensor is damaged or it has failed • PCM has failed
P0181 **2T CCM, MIL: Yes** 1998-1999-2000-2001-2002-2003 Crown Victoria, Grand Marquis, E & F Series Vans & Trucks with a 4.6L VIN 9, 5.4L VIN M, 5.4L VIN Z engine Transmissions: All	**Fuel Temperature Sensor 'A' Signal Performance Conditions:** Engine runtime over 2 minutes, and the PCM detected the Engine Fuel Temperature (EFT) sensor 'A' signal was more than 4.50v [-46ºF] or less than 0.21v [275ºF] the calibrated limit in the self-test. **Possible Causes** • Engine operating under "low" ambient temperature conditions • EFT sensor signal circuit open or shorted in the wiring harness • EFT sensor is damaged or it has failed • PCM has failed

Gas Engine OBD II Trouble Code List (P0xxx Codes)

DTC	Trouble Code Title, Conditions & Possible Causes
P0182 **2T CCM, MIL: Yes** 1998-1999-2000-2001-2002-2003 Crown Victoria, Grand Marquis, E & F Series Vans & Trucks with a 4.6L VIN 9, 5.4L VIN M, 5.4L VIN Z engine Transmissions: All	**Engine Fuel Temperature Sensor 'A' Circuit Low Input Conditions:** Key on or engine running; and the PCM detected the Engine Fuel Temperature (EFT) Sensor 'A' signal was over 260ºF in the self-test. **Possible Causes** • EFT sensor connector is damaged or shorted • EFT sensor VREF circuit is open or shorted to ground • EFT sensor circuit is shorted to chassis or to sensor ground • EFT sensor is damaged or it has failed • PCM has failed
P0182 **2T CCM, MIL: Yes** 1999-2000 2.0L I4 VIN Z CNG engine Contour Transmissions: All	**Engine Fuel Temperature Sensor 'A' Circuit Low Input Conditions:** Key on or engine running; and the PCM detected the EFT sensor 'A' signal from the Compuvalve was less than -38ºF during the self-test. **Possible Causes** • EFT sensor connector is damaged or shorted • EFT sensor VREF circuit is open or shorted to ground • EFT sensor circuit is shorted to chassis or to sensor ground • EFT sensor is damaged or it has failed • PCM has failed
P0183 **2T CCM, MIL: Yes** 1998-1999-2000-2001-2002-2003 Crown Victoria, Grand Marquis, E & F Series Vans & Trucks with a 4.6L VIN 9, 5.4L VIN M, 5.4L VIN Z engine Transmissions: All	**Engine Fuel Temperature Sensor 'A' Circuit High Input Conditions:** Key on or engine running; and the PCM detected the EFT sensor 'A' signal was more than the limit of 4.50v [-46ºF] during the self-test. **Possible Causes** • EFT sensor connector is damaged, loose or open • EFT sensor signal circuit is shorted to VREF (5v) • EFT sensor signal circuit is open • EFT sensor is damaged or it has failed • PCM has failed
P0183 **2T CCM, MIL: Yes** 1999-2000 2.0L I4 VIN Z CNG engine Contour Transmissions: All	**Engine Fuel Temperature Sensor 'A' Circuit High Input Conditions:** Key on or engine running; and the PCM detected the EFT sensor 'A' signal from the Compuvalve was more than 260ºF in the self-test. **Possible Causes** • EFT sensor connector is damaged, loose or open • EFT sensor signal circuit is open or it is shorted to VREF (5v) • EFT sensor is damaged or it has failed • PCM has failed
P0186 **2T CCM, MIL: Yes** 1998-1999-2000-2001-2002-2003 Crown Victoria, Grand Marquis, E & F Series Vans & Trucks with a 4.6L VIN 9, 5.4L VIN M, 5.4L VIN Z engine Transmissions: All	**Engine Fuel Temperature Sensor 'B' Signal Performance Conditions:** Engine started, engine runtime over 2 minutes, and the PCM detected the EFT sensor 'B' signal was more than 4.50v (Scan Tool reads less than -46ºF) or less than 0.21v (Scan Tool reads more than 275ºF) the calibrated limit during the self-test. **Possible Causes** • Engine operating under "low" ambient temperature conditions • EFT sensor signal circuit open or shorted in the wiring harness • EFT sensor is damaged or it has failed • PCM has failed
P0187 **2T CCM, MIL: Yes** 1998-1999-2000-2001-2002-2003 Crown Victoria, Grand Marquis, E & F Series Vans & Trucks with a 4.6L VIN 9, 5.4L VIN M, 5.4L VIN Z engine Transmissions: All	**Engine Fuel Temperature Sensor 'B' Circuit Low Input Conditions:** Engine started, engine runtime over 2 minutes, and the PCM detected the EFT Sensor 'B' signal was less than the calibrated limit of 0.21v (Scan Tool reads less than 275ºF) during the self-test. **Possible Causes** • Engine operating under "low" ambient temperature conditions • EFT sensor signal circuit shorted to ground in the harness • EFT sensor is damaged or it has failed • PCM has failed

Gas Engine OBD II Trouble Code List (P0xxx Codes)

DTC	Trouble Code Title, Conditions & Possible Causes
P0187 **2T CCM, MIL: Yes** 1999-2000 2.0L I4 VIN Z CNG engine Contour Transmissions: All	**Fuel Tank Temperature Sensor Circuit Low Input Conditions:** Engine started, and the PCM detected the Fuel Tank Temperature (FTT) sensor voltage was too low (Scan Tool reads less than 140ºF) during the self-test. Note that when this fault is detected, the PCM sets the FFT sensor default value to a reading of 32ºF. **Possible Causes** • FTT sensor signal circuit is shorted to ground in the harness • FFT sensor is damaged or it has failed • PCM has failed
P0188 **2T CCM, MIL: Yes** 1996, 1997 Crown Victoria, Grand Marquis, E & F Series Vans & Trucks with a 4.6L VIN 9, 5.4L VIN M, 5.4L VIN Z engine Transmissions: All	**Engine Fuel Temperature Sensor 'B' Circuit High Conditions:** Engine running for more than 2 minutes, and the PCM detected the EFT sensor 'A' signal was more than the calibrated limit of 4.50v [-46ºF] during the self-test. **Possible Causes** • EFT sensor signal circuit open in the wiring harness • EFT sensor signal circuit shorted to VREF in the wiring harness • EFT sensor is damaged or it has failed • PCM has failed
P0188 **2T CCM, MIL: Yes** 1999-2000 2.0L I4 VIN Z CNG engine Contour Transmissions: All	**Fuel Tank Temperature Sensor Circuit High Input Conditions:** Engine started, and the PCM detected the Fuel Tank Temperature (FTT) sensor voltage was too high (Scan Tool read less than -38ºF) during the self-test. Note that when this fault is detected, the PCM sets the FFT sensor default value to a reading of 32ºF. **Possible Causes** • FTT sensor signal circuit is open in the wiring harness • FTT sensor signal circuit is shorted to VREF or to power • FFT sensor is damaged or it has failed • PCM has failed
P0190 **2T CCM, MIL: Yes** 1998-1999-2000-2001-2002-2003 Crown Victoria, Grand Marquis, E & F Series Vans & Trucks with a 4.6L VIN 9, 5.4L VIN M, 5.4L VIN Z engine Transmissions: All	**Fuel Rail Pressure Sensor Circuit Malfunction Conditions:** Key on, and the PCM detected that the FRP sensor VREF was less than its acceptable range (minimum value of 4.0v) during the CCM test). The FRP sensor signal should be in a range of 4.0-6.0v during testing. *Note: The sensor VREF should be between 4.0 to 6.0v at all times.* **Possible Causes** • Sensor VREF circuit open in the wiring harness • Sensor VREF circuit open at the sensor • Sensor VREF circuit open between the sensor and the PCM • PCM has failed
P0191 **2T CCM, MIL: Yes** 1996-1997-1998-1999-2000-2001-2002-2003 Crown Victoria, Grand Marquis, E & F Series Vans & Trucks with a 4.6L VIN 9, 5.4L VIN M, 5.4L VIN Z engine Transmissions: All	**Fuel Rail Pressure Sensor Range/Performance Conditions:** Engine started, and the PCM detected the FRP sensor signal was less than the minimum acceptable range or was more than the maximum acceptable range. With the engine running, the FRP PID should read between 20 psi [138 kPa] and 60 psi [413 kPa] for gasoline powered vehicles, or between 85 psi [586 kPa] and 105 psi [725 kPa] for natural gas (NG) powered vehicles. **Possible Causes** • Fuel pressure too high or too low during engine operation • FRP sensor signal circuit has high resistance • FRP sensor is damaged or has failed
P0192 **2T CCM, MIL: Yes** 1996-1997-1998-1999-2000-2001-2002-2003 Crown Victoria, Grand Marquis, E & F Series Vans & Trucks with a 4.6L VIN 9, 5.4L VIN M, 5.4L VIN Z engine Transmissions: All	**Fuel Rail Pressure Sensor Circuit Low Conditions:** Engine started, and the PCM detected the FRP sensor signal was less than 0.3v for gasoline vehicles or less than 0.5v for NG vehicles during the self-test. **Possible Causes** • FRP sensor signal shorted to chassis ground or sensor ground • FRP sensor signal circuit open (NG usage only) • FRP sensor is damaged or has failed • PCM has failed

Gas Engine OBD II Trouble Code List (P0xxx Codes

DTC	Trouble Code Title, Conditions & Possible Causes
P0192 **2T CCM, MIL: Yes** 1999-2000 2.0L I4 VIN Z CNG engine Contour Transmissions: All	**Fuel Absolute Pressure Sensor Circuit Low Input Conditions:** Engine started, and the PCM detected the Fuel Absolute Pressure (FAP) sensor inside the Compuvalve signal was too low (less than 4.9 psi). The FRPPREAB PID displays the actual PSI value. Bleed the system (to 0 psi), the FRPPREAB and MAP PID should agree. **Possible Causes** • FAP sensor signal circuit is shorted to ground in the harness • FAP sensor is damaged or it has failed • PCM has failed
P0193 **2T CCM, MIL: Yes** 1996-1997-1998-1999- 2000-2001-2002-2003 Crown Victoria, Grand Marquis, E & F Series Vans & Trucks with a 4.6L VIN 9, 5.4L VIN M, 5.4L VIN Z engine Transmissions: All	**Fuel Absolute Pressure Sensor Circuit High Input Conditions:** Engine started, and the PCM detected the FRP sensor was more than 4.5v for gasoline vehicles or more than 4.8v for NG vehicles. **Possible Causes** • FRP sensor signal shorted to VREF or system voltage • FRP sensor signal circuit open (gasoline usage only) • Low fuel pressure (NG vehicle only) • FRP sensor is damaged or has failed • PCM has failed
P0193 **2T CCM, MIL: Yes** 1999-2000 2.0L I4 VIN Z CNG engine Contour Transmissions: All	**Fuel Rail Pressure Sensor Circuit High Input Conditions:** Engine started, and the PCM detected the Fuel Absolute Pressure (FAP) sensor in the Compuvalve unit was too high (over 167 psi). Note: The FRPPREAB PID displays the actual PSI value. Bleed the system to 0 psi, and the FRPPREAB and MAP PID should agree). **Possible Causes** • FAP sensor signal circuit is open in the wiring harness • FAP sensor is damaged or it has failed • PCM has failed
P0196 **2T CCM, MIL: Yes** 2003 Crown Victoria, Marauder & Grand Marquis Transmissions: All	**Engine Oil Temperature Sensor Signal Performance Conditions:** Engine started, KOER Self Test enabled, and the PCM detected the Engine Oil Temperature (EOT) sensor signal was not within a calibrated amount of the ECT sensor signal during the test. The EOT sensor value should be close to the engine oil temperature. **Possible Causes** • Cooling system malfunction, or the thermostat is stuck • Engine not operating at normal operating temperature • EOT sensor is damaged or it has failed • PCM has failed
P0197 **2T CCM, MIL: Yes** 2003 Crown Victoria, Marauder & Grand Marquis Transmissions: All	**Engine Oil Temperature Sensor Circuit Low Input Conditions:** Key on or engine running; and the PCM detected that the Engine Oil Temperature (EOT) sensor was less than 0.20v during the test. The EOT sensor value should be close to the engine oil temperature. **Possible Causes** • EOT sensor connector is damaged or shorted • EOT sensor signal circuit shorted to chassis or sensor ground • EOT sensor is damaged or it has failed • PCM has failed
P0198 **2T CCM, MIL: Yes** 2003 Crown Victoria, Marauder & Grand Marquis Transmissions: All	**Engine Oil Temperature Sensor Circuit High Input Conditions:** Key on or engine running; and the PCM detected that the Engine Oil Temperature (EOT) sensor was less than 4.50v during the test. The EOT sensor value should be close to the engine oil temperature. **Possible Causes** • EOT sensor connector is damaged or open • EOT sensor signal circuit is shorted to VREF (5v) • EOT sensor is damaged or it has failed • PCM has failed

Gas Engine OBD II Trouble Code List (P0xxx Codes)

DTC	Trouble Code Title, Conditions & Possible Causes
P0201 **2T CCM, MIL: Yes** 1996-1997-1998-1999- 2000-2001-2002-2003 All Models Transmissions: All	**Cylinder 1 Injector Circuit Malfunction Conditions:** Engine started, and the PCM detected the fuel injector "1" control circuit was in a high state when it should have been low, or in a low state when it should have been high (wiring harness & injector okay). Note: Monitor the INJIF PID Fault "flags" with the Scan Tool. The appropriate INJF PID "flag" will read Yes when this code is set. **Possible Causes** • Injector 1 connector is damaged, open or shorted • Injector 1 control circuit is open, shorted to ground or to power • PCM has failed (the injector driver circuit may be damaged)
P0202 **2T CCM, MIL: Yes** 1996-1997-1998-1999- 2000-2001-2002-2003 All Models Transmissions: All	**Cylinder 2 Injector Circuit Malfunction Conditions:** Engine started, and the PCM detected the fuel injector "2" control circuit was in a high state when it should have been low, or in a low state when it should have been high (wiring harness & injector okay). Note: Monitor the INJIF PID Fault "flags" with the Scan Tool. The appropriate INJF PID "flag" will read Yes when this code is set. **Possible Causes** • Injector 2 connector is damaged, open or shorted • Injector 2 control circuit is open, shorted to ground or to power • PCM has failed (the injector driver circuit may be damaged)
P0203 **2T CCM, MIL: Yes** 1996-1997-1998-1999- 2000-2001-2002-2003 All Models Transmissions: All	**Cylinder 3 Injector Circuit Malfunction Conditions:** Engine started, and the PCM detected the fuel injector "3" control circuit was in a high state when it should have been low, or in a low state when it should have been high (wiring harness & injector okay). Note: Monitor the INJIF PID Fault "flags" with the Scan Tool. The appropriate INJF PID "flag" will read Yes when this code is set. **Possible Causes** • Injector 3 connector is damaged, open or shorted • Injector 3 control circuit is open, shorted to ground or to power • PCM has failed (the injector driver circuit may be damaged)
P0204 **2T CCM, MIL: Yes** 1996-1997-1998-1999- 2000-2001-2002-2003 All Models Transmissions: All	**Cylinder 4 Injector Circuit Malfunction Conditions:** Engine started, and the PCM detected the fuel injector "4" control circuit was in a high state when it should have been low, or in a low state when it should have been high (wiring harness & injector okay). Note: Monitor the INJIF PID Fault "flags" with the Scan Tool. The appropriate INJF PID "flag" will read Yes when this code is set. **Possible Causes** • Injector 4 connector is damaged, open or shorted • Injector 4 control circuit is open, shorted to ground or to power • PCM has failed (the injector driver circuit may be damaged)
P0205 **2T CCM, MIL: Yes** 1996-1997-1998-1999- 2000-2001-2002-2003 Models equipped with V6, V8 or V10 engine Transmissions: All	**Cylinder 5 Injector Circuit Malfunction Conditions:** Engine started, and the PCM detected the fuel injector "5" control circuit was in a high state when it should have been low, or in a low state when it should have been high (wiring harness & injector okay). Note: Monitor the INJIF PID Fault "flags" with the Scan Tool. The appropriate INJF PID "flag" will read Yes when this code is set. **Possible Causes** • Injector 5 connector is damaged, open or shorted • Injector 5 control circuit is open, shorted to ground or to power • PCM has failed (the injector driver circuit may be damaged)
P0206 **2T CCM, MIL: Yes** 1996-1997-1998-1999- 2000-2001-2002-2003 Models equipped with V6, V8 or V10 engine Transmissions: All	**Cylinder 6 Injector Circuit Malfunction Conditions:** Engine started, and the PCM detected the fuel injector control circuit was in a high state when it should have been low, or in a low state when it should have been high (wiring harness & injector okay). Note: Monitor the INJIF PID Fault "flags" with the Scan Tool. The appropriate INJF PID "flag" will read Yes when this code is set. **Possible Causes** • Injector 6 connector is damaged, open or shorted • Injector 6 control circuit is open, shorted to ground or to power • PCM has failed (the injector driver circuit may be damaged)

Gas Engine OBD II Trouble Code List (P0xxx Codes)

DTC	Trouble Code Title, Conditions & Possible Causes
P0207 **2T CCM, MIL: Yes** 1996-1997-1998-1999-2000-2001-2002-2003 Models equipped with a V8 or V10 engine Transmissions: All	**Cylinder 7 Injector Circuit Malfunction Conditions:** Engine started, and the PCM detected the fuel injector "7" control circuit was in a high state when it should have been low, or in a low state when it should have been high (wiring harness & injector okay). Note: Monitor the INJIF PID Fault "flags" with the Scan Tool. The appropriate INJF PID "flag" will read Yes when this code is set. **Possible Causes** • Injector 7 connector is damaged, open or shorted • Injector 7 control circuit is open, shorted to ground or to power • PCM has failed (the injector driver circuit may be damaged)
P0208 **2T CCM, MIL: Yes** 1996-1997-1998-1999-2000-2001-2002-2003 a V8 or V10 engine Transmissions: All	**Cylinder 8 Injector Circuit Malfunction Conditions:** Engine started, and the PCM detected the fuel injector "8" control circuit was in a high state when it should have been low, or in a low state when it should have been high (wiring harness & injector okay). Note: Monitor the INJIF PID Fault "flags" with the Scan Tool. The appropriate INJF PID "flag" will read Yes when this code is set. **Possible Causes** • Injector 8 connector is damaged, open or shorted • Injector 8 control circuit is open, shorted to ground or to power • PCM has failed (the injector driver circuit may be damaged)
P0209 **2T CCM, MIL: Yes** 1997-1998-1999-2000-2001-2002-2003 E-Van, F-Series Truck & Excursion with 6.8L VIN S or VIN Z engine Transmissions: All	**Cylinder 9 Injector Circuit Malfunction Conditions:** Engine started, and the PCM detected the fuel injector "9" control circuit was in a high state when it should have been low, or in a low state when it should have been high (wiring harness & injector okay). Note: Monitor the INJIF PID Fault "flags" with the Scan Tool. The appropriate INJF PID "flag" will read Yes when this code is set. **Possible Causes** • Injector 9 connector is damaged, open or shorted • Injector 9 control circuit is open, shorted to ground or to power • PCM has failed (the injector driver circuit may be damaged)
P0210 **2T CCM, MIL: Yes** 1997-1998-1999-2000-2001-2002-2003 E-Van, F-Series Truck & Excursion with 6.8L VIN S or VIN Z engine Transmissions: All	**Cylinder 10 Injector Circuit Malfunction Conditions:** Engine started, and the PCM detected the fuel injector "10" control circuit was in a high state when it should have been low, or in a low state when it should have been high (wiring harness & injector okay). Note: Monitor the INJIF PID Fault "flags" with the Scan Tool. The appropriate INJF PID "flag" will read Yes when this code is set. **Possible Causes** • Injector 10 connector is damaged, open or shorted • Injector 10 control circuit is open, shorted to ground or to power • PCM has failed (the injector driver circuit may be damaged)
P0217 **2T CCM, MIL: Yes** 2003 F-Series Truck with a 5.4L VIN 3 SC engine Transmissions: All	**Engine Coolant Over-Temperature Condition Conditions:** Engine started, vehicle driven, and the PCM detected a signal from the ECT or CHT sensor (depends on how the vehicle is equipped) that indicated an engine overheat condition. The PCM will cause the boost from the Supercharger to be bypassed when this code is set. **Possible Causes** • Base engine problems that could cause the engine to overheat • Cooling system faults that could cause the engine to overheat • Low engine coolant or incorrect mixture of engine coolant
P0219 **2T CCM, MIL: Yes** 2003 All Models Transmissions: All	**Engine Over-Speed Condition Conditions:** Engine started, and the PCM determined the vehicle had been driven in a manner that caused the engine to over-speed, and to exceed the engine speed calibration limit stored in memory. **Possible Causes** • Engine operated in the wrong transmission gear position • Excessive engine speed with gear selector in Neutral position • Wheel slippage due to wet, muddy or snowing conditions

Gas Engine OBD II Trouble Code List (P0xxx Codes)

DTC	Trouble Code Title, Conditions & Possible Causes
P0221 **2T CCM, MIL: Yes** 2003 Crown Victoria, Grand Marquis, LS & Thunderbird equipped with a 3.0L VIN S, 3.9L VIN A, 4.6L VIN 9, 4.6L VIN W engine Transmissions: All	**Throttle Position Sensor 'B' Signal Performance Conditions:** Engine started; and the PCM detected the TP Sensor 'B' circuit was out of its normal operating range during a condition with the throttle wide open, or with it completely closed. **Possible Causes** • Throttle body is damaged • Throttle linkage is binding or sticking • ETC TP Sensor 'B' signal circuit to the PCM is open • ETC TP Sensor 'B' ground circuit is open • ETC TP Sensor 'B' is damaged or it has failed
P0222 **2T CCM, MIL: Yes** 2003 Crown Victoria, Grand Marquis, LS & Thunderbird equipped with a 3.0L VIN S, 3.9L VIN A, 4.6L VIN 9, 4.6L VIN W engine Transmissions: All	**Throttle Position Sensor 'B' Circuit Low Input Conditions:** Key on or engine running; and the PCM detected the TP Sensor 'B' indicated less than 0.17v (Scan Tool reads less than 3.42%). **Possible Causes** • ETC TP Sensor 'B' connector is damaged or shorted • ETC TP Sensor 'B' signal circuit is shorted to ground • ETC TP Sensor 'B' is damaged or it has failed • PCM has failed
P0223 **2T CCM, MIL: Yes** 2003 Crown Victoria, Grand Marquis, LS & Thunderbird equipped with a 3.0L VIN S, 3.9L VIN A, 4.6L VIN 9, 4.6L VIN W engine Transmissions: All	**Throttle Position Sensor 'B' Circuit High Input Conditions:** Key on or engine running; and the PCM detected the TP Sensor 'B' indicated more than 4.65v (Scan Tool reads more than 93%) during the CCM test period. **Possible Causes** • ETC TP Sensor 'B' connector is damaged or open • ETC TP Sensor 'B' signal circuit is open • ETC TP Sensor 'B' signal circuit is shorted to VREF (5v) • ETC TP Sensor 'B' is damaged or it has failed
P0230 **2T CCM, MIL: Yes** 1996-1997-1998-1999-2000-2001-2002-2003 All Models Transmissions: All	**Fuel Pump Primary Circuit Malfunction Conditions:** Key on, and the PCM detected high current in fuel pump or fuel shutoff valve (FSV) circuit (NG only), or it detected voltage with the valve off, or it did not detect voltage on the circuit. The circuit is used to energize the fuel pump relay for 20 seconds at key on or while running. **Possible Causes** • FP or FSV circuit is open or shorted • Fuel pump relay VPWR circuit open • Fuel pump relay is damaged or has failed • PCM has failed
P0231 **2T CCM, MIL: Yes** 1996-1997-1998-1999-2000-2001-2002-2003 All Models Transmissions: All	**Fuel Pump Primary Circuit Low Input Conditions:** Key on, and the PCM detected a lack of voltage on the FP Monitor circuit with the fuel pump commanded on. The fuel pump control circuit is used by the PCM to energize the fuel pump relay. At key on, the relay is energized for 20 seconds, and all the time the engine is running. **Possible Causes** • FP or FSV circuit is open or shorted to ground • Fuel pump relay VPWR circuit open or fuel pump relay failed • PCM has failed
P0232 **2T CCM, MIL: Yes** 1996-1997-1998-1999-2000-2001-2002-2003 All Models Transmissions: All	**Fuel Pump Secondary Circuit High Input Conditions:** Key on, and the PCM detected voltage on the FP Monitor circuit with fuel pump "off". The PCM uses the fuel pump control circuit to energize the fuel pump relay. At key on, the relay is "on" for 20 seconds or while running. This circuit is used to check voltage to the pump. **Possible Causes** • Fuel pump relay contacts always closed • Fuel pump ground circuit has high resistance • Fuel pump secondary circuit is shorted to power • Low speed fuel pump relay damaged or related circuit problem

Gas Engine OBD II Trouble Code List (P0xxx Codes)

DTC	Trouble Code Title, Conditions & Possible Causes
P0234 **2T CCM, MIL: Yes** 1999-2000-2001-2002-2003 F-Series Truck with a 5.4L VIN 3 SC engine Transmissions: All	**Supercharger Overboost Condition Conditions:** Engine started, vehicle driven, and the PCM detected an operating condition that could harm the engine or automatic transmission. **Possible Causes** • Brake "on" with throttle under wide open throttle condition • Ignition misfire condition exceeds the calibrated threshold • Knock sensor circuit has failed, or excessive knock detected • Low speed fuel pump relay not switching properly • Transmission oil temperature beyond the calibrated threshold
P0243 **2T CCM, MIL: Yes** 1999-2000-2001-2002-2003 F-Series Truck with a 5.4L VIN 3 SC engine Transmissions: All	**Supercharger Boost Bypass Solenoid Circuit Malfunction Conditions:** Key on or engine running; and the PCM detected an unexpected low or high voltage condition on the S/C Bypass Solenoid control circuit **Possible Causes** • SCB power supply circuit (VPWR) is open • SCB control circuit is open, shorted to ground or system power • SCB assembly is damaged or has failed • PCM has failed
P0297 **2T CCM, MIL: Yes** 2003 All Models Transmissions: All	**Vehicle Over-Speed Condition Conditions:** Engine started, vehicle driven at a very high engine speed, and the PCM detected the vehicle speed exceeded the calibration limit, and then enabled the High Vehicle Speed Strategy to control the speed. **Possible Causes** • The code indicates the vehicle was driven at very high engine speed (rpm) for too long. The PCM temporarily prohibits high engine speed by disabling the fuel injectors with this code set.
P0298 **2T CCM, MIL: Yes** 2003 LS Models all engines Transmissions: All	**Engine Oil Over-Temperature Condition Conditions:** Engine started, engine running for several minutes, and the PCM detected an engine overheating condition, and then enabled the Engine Oil Temperature Protection Strategy (injectors off). **Possible Causes** • Check for signs of base engine concern or engine overheating • EOP sensor or related circuit fault • Very high engine speed (rpm) for an extended period of time
P0300 **2T MISFIRE** **MIL: Yes** 1996-1997-1998-1999-2000-2001-2002-2003 All Models Transmissions: All	**Random Misfire Detected Conditions:** DTC P0136, P0156, P0171, P0172, P0175, P1130 and P1150 not set, engine running under positive torque conditions, and the PCM detected a misfire in 1000 revolution (High Emissions) or the 200 revolution (Catalyst Damaging 1T) range in two or more cylinders. Note: If the misfire is severe, the MIL will flash on/off on the 1st trip! **Possible Causes** • Base engine mechanical fault that affects two or more cylinders • Fuel metering fault that affects two or more cylinders • Fuel pressure too low or too high, fuel supply contaminated • EVAP system problem or the EVAP canister is fuel saturated • EGR valve is stuck open or the PCV system has a vacuum leak • Ignition system fault (coil, plug) affecting two or more cylinders • MAF sensor contamination (it can cause a very lean condition) • Vehicle driven while very low on fuel (less than 1/8 of a tank) • TSB 03-14-4 contains repair help for this code for COP ignition

Gas Engine OBD II Trouble Code List (P0xxx Codes)

DTC	Trouble Code Title, Conditions & Possible Causes
P0301 **2T MISFIRE** **MIL: Yes** 1996-1997-1998-1999-2000-2001-2002-2003 All Models Transmissions: All	**Cylinder Number 1 Misfire Detected Conditions:** DTC P0136, P0156, P0171, P0172, P0175, P1130 and P1150 not set, engine started, engine running under positive torque conditions, and the PCM detected a misfire a misfire in Cylinder 1 during the 200 revolution (Catalyst) or 1000 revolution (High Emissions) period. Note: If the misfire is severe, the MIL will flash on/off on the 1st trip! **Possible Causes** • Air leak in the intake manifold, or in the EGR or PCM system • Base engine mechanical problem that affects only Cylinder 1 • Fuel delivery component problem that affects only Cylinder 1 (i.e., a contaminated, dirty or sticking fuel injector) • Ignition system problem (coil, plug) that affects only Cylinder 1 • TSB 02-16-2 contains repair help for this code (LS & T-Bird) • TSB 03-14-4 contains repair help for this code for COP ignition
P0302 **2T MISFIRE** **MIL: Yes** 1996-1997-1998-1999-2000-2001-2002-2003 All Models Transmissions: All	**Cylinder Number 2 Misfire Detected Conditions:** DTC P0136, P0156, P0171, P0172, P0175, P1130 and P1150 not set, engine started, engine running under positive torque conditions, and the PCM detected a misfire a misfire in Cylinder 2 during the 200 revolution (Catalyst) or 1000 revolution (High Emissions) period. Note: If the misfire is severe, the MIL will flash on/off on the 1st trip! **Possible Causes** • Air leak in the intake manifold, or in the EGR or PCM system • Base engine mechanical problem that affects only Cylinder 2 • Fuel delivery component problem that affects only Cylinder 2 (i.e., a contaminated, dirty or sticking fuel injector) • Ignition system problem (coil, plug) that affects only Cylinder 2 • TSB 02-16-2 contains repair help for this code (LS & T-Bird) • TSB 03-14-4 contains repair help for this code for COP ignition
P0303 **2T MISFIRE** **MIL: Yes** 1996-1997-1998-1999-2000-2001-2002-2003 All Models Transmissions: All	**Cylinder Number 3 Misfire Detected Conditions:** DTC P0136, P0156, P0171, P0172, P0175, P1130 and P1150 not set, engine started, engine running under positive torque conditions, and the PCM detected a misfire a misfire in Cylinder 3 during the 200 revolution (Catalyst) or 1000 revolution (High Emissions) period. Note: If the misfire is severe, the MIL will flash on/off on the 1st trip! **Possible Causes** • Air leak in the intake manifold, or in the EGR or PCM system • Base engine mechanical problem that affects only Cylinder 3 • Fuel delivery component problem that affects only Cylinder 3 (i.e., a contaminated, dirty or sticking fuel injector) • Ignition system problem (coil, plug) that affects only Cylinder 3 • TSB 02-16-2 contains repair help for this code (LS & T-Bird) • TSB 03-14-4 contains repair help for this code for COP ignition
P0304 **2T MISFIRE** **MIL: Yes** 1996-1997-1998-1999-2000-2001-2002-2003 All Models Transmissions: All	**Cylinder Number 4 Misfire Detected Conditions:** DTC P0136, P0156, P0171, P0172, P0175, P1130 and P1150 not set, engine started, engine running under positive torque conditions, and the PCM detected a misfire a misfire in Cylinder 4 during the 200 revolution (Catalyst) or 1000 revolution (High Emissions) period. Note: If the misfire is severe, the MIL will flash on/off on the 1st trip! **Possible Causes** • Air leak in the intake manifold, or in the EGR or PCM system • Base engine mechanical problem that affects only Cylinder 4 • Fuel delivery component problem that affects only Cylinder 4 (i.e., a contaminated, dirty or sticking fuel injector) • Ignition system problem (coil, plug) that affects only Cylinder 4 • TSB 02-16-2 contains repair help for this code (LS & T-Bird) • TSB 03-14-4 contains repair help for this code for COP ignition

Gas Engine OBD II Trouble Code List (P0xxx Codes)

DTC	Trouble Code Title, Conditions & Possible Causes
P0305 **2T MISFIRE** **MIL: Yes** 1996-1997-1998-1999-2000-2001-2002-2003 Models equipped with V6, V8 or V10 engine Transmissions: All	**Cylinder Number 5 Misfire Detected Conditions:** DTC P0136, P0156, P0171, P0172, P0175, P1130 and P1150 not set, engine started, engine running under positive torque conditions, and the PCM detected a misfire a misfire in Cylinder 5 during the 200 revolution (Catalyst) or 1000 revolution (High Emissions) period. Note: If the misfire is severe, the MIL will flash on/off on the 1st trip! **Possible Causes** • Air leak in the intake manifold, or in the EGR or PCM system • Base engine mechanical problem that affects only Cylinder 5 • Fuel delivery component problem that affects only Cylinder 5 (i.e., a contaminated, dirty or sticking fuel injector) • Ignition system problem (coil, plug) that affects only Cylinder 5 • TSB 02-16-2 contains repair help for this code (LS & T-Bird) • TSB 03-14-4 contains repair help for this code for COP ignition
P0306 **2T MISFIRE** **MIL: Yes** 1996-1997-1998-1999-2000-2001-2002-2003 Models equipped with V6, V8 or V10 engine Transmissions: All	**Cylinder Number 6 Misfire Detected Conditions:** DTC P0136, P0156, P0171, P0172, P0175, P1130 and P1150 not set, engine started, engine running under positive torque conditions, and the PCM detected a misfire a misfire in Cylinder 6 during the 200 revolution (Catalyst) or 1000 revolution (High Emissions) period. Note: If the misfire is severe, the MIL will flash on/off on the 1st trip! **Possible Causes** • Air leak in the intake manifold, or in the EGR or PCM system • Base engine mechanical problem that affects only Cylinder 6 • Fuel delivery component problem that affects only Cylinder 6 (i.e., a contaminated, dirty or sticking fuel injector) • Ignition system problem (coil, plug) that affects only Cylinder 6 • TSB 02-16-2 contains repair help for this code (LS & T-Bird) • TSB 03-14-4 contains repair help for this code for COP ignition
P0307 **2T MISFIRE** **MIL: Yes** 1996-1997-1998-1999-2000-2001-2002-2003 Models equipped with V8 or V10 engine Transmissions: All	**Cylinder Number 7 Misfire Detected Conditions:** DTC P0136, P0156, P0171, P0172, P0175, P1130 and P1150 not set, engine started, engine running under positive torque conditions, and the PCM detected a misfire a misfire in Cylinder 7 during the 200 revolution (Catalyst) or 1000 revolution (High Emissions) period. Note: If the misfire is severe, the MIL will flash on/off on the 1st trip! **Possible Causes** • Air leak in the intake manifold, or in the EGR or PCM system • Base engine mechanical problem that affects only Cylinder 7 • Fuel delivery component problem that affects only Cylinder 7 (i.e., a contaminated, dirty or sticking fuel injector) • Ignition system problem (coil, plug) that affects only Cylinder 7 • TSB 02-16-2 contains repair help for this code (LS & T-Bird) • TSB 03-14-4 contains repair help for this code for COP ignition
P0308 **2T MISFIRE** **MIL: Yes** 1996-1997-1998-1999-2000-2001-2002-2003 Models equipped with V8 or V10 engine Transmissions: All	**Cylinder Number 8 Misfire Detected Conditions:** DTC P0136, P0156, P0171, P0172, P0175, P1130 and P1150 not set, engine started, engine running under positive torque conditions, and the PCM detected a misfire a misfire in Cylinder 8 during the 200 revolution (Catalyst) or 1000 revolution (High Emissions) period. Note: If the misfire is severe, the MIL will flash on/off on the 1st trip! **Possible Causes** • Air leak in the intake manifold, or in the EGR or PCM system • Base engine mechanical problem that affects only Cylinder 8 • Fuel delivery component problem that affects only Cylinder 8 (i.e., a contaminated, dirty or sticking fuel injector) • Ignition system problem (coil, plug) that affects only Cylinder 8 • TSB 02-16-2 contains repair help for this code (LS & T-Bird) • TSB 03-14-4 contains repair help for this code for COP ignition

Gas Engine OBD II Trouble Code List (P0xxx Codes)

DTC	Trouble Code Title, Conditions & Possible Causes
P0309 **2T MISFIRE** **MIL: Yes** 1997-1998-1999-2000-2001-2002-2003 Models equipped with a V10 engine Transmissions: All	**Cylinder Number 9 Misfire Detected Conditions:** DTC P0136, P0156, P0171, P0172, P0175, P1130 and P1150 not set, engine started, engine running under positive torque conditions, and the PCM detected a misfire a misfire in Cylinder 9 during the 200 revolution (Catalyst) or 1000 revolution (High Emissions) period. Note: If the misfire is severe, the MIL will flash on/off on the 1st trip! **Possible Causes** • Air leak in the intake manifold, or in the EGR or PCM system • Base engine mechanical problem that affects only Cylinder 9 • Fuel delivery component problem that affects only Cylinder 9 (i.e., a contaminated, dirty or sticking fuel injector) • Ignition system problem (coil, plug) that affects only Cylinder 9 • TSB 03-14-4 contains repair help for this code for COP ignition
P0310 **2T MISFIRE** **MIL: Yes** 1997-1998-1999-2000-2001-2002-2003 Models equipped with a V10 engine Transmissions: All	**Cylinder Number 10 Misfire Detected Conditions:** DTC P0136, P0156, P0171, P0172, P0175, P1130 and P1150 not set, engine started, engine running under positive torque conditions, and the PCM detected a misfire a misfire in Cylinder 10 during the 200 revolution (Catalyst) or 1000 revolution (High Emissions) period. Note: If the misfire is severe, the MIL will flash on/off on the 1st trip! **Possible Causes** • Air leak in the intake manifold, or in the EGR or PCM system • Base engine mechanical problem that affects only Cylinder 10 • Fuel delivery component problem that affects only Cylinder 10 i.e., a contaminated, dirty or sticking fuel injector) • Ignition system problem (coil, plug) that affects only Cylinder 10 • TSB 03-14-4 contains repair help for this code for COP ignition
P0315 **2T CCM, MIL: Yes** 1996-1997-1998-1999-2000-2001-2002-2003 All Models Transmissions: All	**Unable to Learn Crankshaft Variation Conditions:** Engine started, and the PCM determined that it was unable to correct for mechanical inaccuracies in the CKP wheel tooth spacing. Note: The Misfire Monitor will be disabled. **Possible Causes** • Inspect the CKP sensor for damage • Inspect the CKP sensor for debris on the rotor • Inspect the crankshaft pulse wheel for damaged teeth • Inspect the crankshaft pulse wheel for wobble (loose condition)
P0316 **2T MISFIRE** **MIL: Yes** 2002-2003 All Models Transmissions: All	**Misfire in the First 1000 Revolutions Conditions:** Engine started, and the PCM detected a severe misfire within the first 1000 engine revolutions. **Possible Causes** • Check for CMC DTC P0136, P0156, P0171, P0172, P0175, P1130 and P1150. Repair these adaptive fuel and HO2S codes • Check for any other CMC in memory. Repair these codes first! • Ignore P1000 codes that set during KOEO and KOER Self-Test
P0320 **2T CCM, MIL: Yes** 1996 E Van, F Series Truck equipped with a 4.9L VIN Y, 5.0L VIN N, 5.8L VIN H, 7.5L VIN G engine Transmissions: All	**Ignition Engine Speed Input Circuit Malfunction Conditions:** Engine started, and the PCM detected two or more successive erratic PIP signals during the self-test. **Possible Causes** • Verify that the vehicle Antitheft system is operational • Inspect for any Aftermarket 2-way radio problems • Inspect for signs of "arcing" at one or more of the ignition coils • Verify that the Inertia Fuel Switch (IFS) is set properly (reset)
P0320 **2T CCM, MIL: Yes** 1997 F Series Truck equipped with a 7.5L VIN G engine Transmissions: All	**Ignition Engine Speed Input Circuit Malfunction Conditions:** Engine started, and the PCM detected 2 or more successive erratic PIP signals while testing. **Possible Causes** • Verify that the vehicle Antitheft system is operational • Inspect for any Aftermarket 2-way radio problems • Inspect for signs of "arcing" at one or more of the ignition coils • Verify that the Inertia Fuel Switch (IFS) is set properly (reset)

Gas Engine OBD II Trouble Code List (P0xxx Codes)

DTC	Trouble Code Title, Conditions & Possible Causes
P0320 **2T CCM, MIL: Yes** 2002-03 All Models Transmissions: All	**Ignition Engine Speed Input Circuit Malfunction Conditions:** Engine started, and the PCM detected 2 or more successive erratic PIP signals during testing. **Possible Causes** • Inspect for problems with an Aftermarket 2-way radio • Inspect for signs of "arcing" at one or more of the ignition coils • Inspect the Profile Ignition Pickup (PIP) unit inside distributor (check for damage or corrosion at the PIP sensor connector) • PIP sensor is damaged or it has failed (distributor models) • Ignition control module (ICM) has failed (Distributorless models)
P0325 **2T CCM, MIL: Yes** 1996-1997-1998-1999- 2000-2001-2002-2003 All Models Transmissions: All	**Knock Sensor 1 Circuit Malfunction Conditions:** Key on or engine running; and the PCM detected the knock Sensor 1 (KS1) signal was more than 0.5v at key on, engine off, or the KS1 signal was out of normal range (engine running). **Possible Causes** • Knock sensor circuit is open • Knock sensor circuit is shorted to ground, or shorted to power • Knock sensor is damaged or it has failed • PCM has failed
P0326 **2T CCM, MIL: Yes** 1996-1997-1998-1999- 2000-2001-2002-2003 All Models Transmissions: All	**Knock Sensor 1 Signal Range/Performance Conditions:** Engine started, vehicle driven, and the PCM detected the Knock Sensor 1 (KS1) signal was more than the calibrated value. This code can set at key on, engine off, if the KS 1 signal is more than 0.5v. **Possible Causes** • Knock sensor circuit is open • Knock sensor circuit is shorted to ground, or shorted to power • Knock sensor is damaged or it has failed • PCM has failed
P0330 **2T CCM, MIL: Yes** 1996-1997-1998-1999- 2000-2001-2002-2003 All Models Transmissions: All	**Knock Sensor 2 Circuit Malfunction Conditions:** Key on or engine running; and the PCM detected the knock Sensor 2 (KS2) signal was more than 0.5v at key on, engine off, or that the KS2 signal was out of the normal range with the engine running. **Possible Causes** • Knock sensor circuit is open • Knock sensor circuit is shorted to ground, or shorted to power • Knock sensor is damaged or it has failed • PCM has failed
P0331 **2T CCM, MIL: Yes** 1996-1997-1998-1999- 2000-2001-2002-2003 All Models Transmissions: All	**Knock Sensor 2 Signal Range/Performance Conditions:** Engine started, vehicle driven, and the PCM detected that the Knock Sensor 2 (KS2) signal was more than the calibrated value. This code can set at key on, engine off, if the KS2 signal is more than 0.5v. **Possible Causes** • Knock sensor circuit is open • Knock sensor circuit is shorted to ground, or shorted to power • Knock sensor is damaged or it has failed • PCM has failed
P0340 **2T CCM, MIL: Yes** 1996-1997-1998-1999- 2000-2001-2002-2003 All Models Transmissions: All	**Camshaft Position Sensor Circuit Malfunction Conditions:** Engine started, and the PCM detected the CMP sensor signal was missing or it was erratic. **Possible Causes** • CMP sensor circuit is open or shorted to ground • CMP sensor circuit is shorted to power • CMP sensor ground (return) circuit is open • CMP sensor installation incorrect (Hall-effect type) • CMP sensor is damaged or CMP sensor shielding damaged • PCM has failed • TSB 02-22-1 contains repair information for this trouble code

Gas Engine OBD II Trouble Code List (P0xxx Codes)

DTC	Trouble Code Title, Conditions & Possible Causes
P0350 **2T CCM, MIL: Yes** 1996-1997-1998-1999-2000-2001-2002-2003 Some Models Transmissions: All	**Ignition Coil (Undetermined) Primary/Secondary Circuit Malfunction Conditions:** Engine started, and the PCM did not receive valid IDM pulses from the ignition module. The PCM did not identify the coil with a problem. **Possible Causes** • Ignition START/RUN circuit is open or shorted to ground • Ignition coil driver circuit is open or shorted to ground • Ignition coil circuit is shorted to power • Ignition coil damaged or it has failed • PCM has failed
P0350 **2T CCM, MIL: Yes** 2000-2001-2002-2003 Crown Victoria, Grand Marquis, Excursion, LS, E & F Series Vans & Trucks & Town Car equipped with a 3.0L VIN S, 3.9L VIN A, 4.6L VIN 9, 4.6L VIN W, 6.8L VIN S engine Transmissions: All	**Ignition Coil Primary/Secondary Circuit Malfunction Conditions:** Engine started, and the PCM did not receive valid IDM pulses from the ignition module. The PCM did not identify the coil with a problem. **Possible Causes** • Ignition START/RUN circuit is open or shorted to ground • Ignition coil driver circuit is open • Ignition coil driver circuit is shorted to ground • Ignition coil driver circuit is shorted to system power (B+) • Ignition coil damaged or it has failed • PCM has failed • TSB 01-1-6 contains a repair procedure for this trouble code
P0351-P0354 **2T CCM, MIL: Yes** 1998-1999-2000-2001-2002-2003 Escort, ZX2 & Focus equipped with 2.0L VIN 3, 2.0L VIN P engine (w/Coilpack) Transmissions: All	**Ignition Coilpack 1-4 Primary/Secondary Circuit Malfunction Conditions:** Engine started, and the PCM did not receive any valid IDM pulses from the ignition module for the Ignition Coilpack 1-4 primary circuit. **Possible Causes** • Ignition START/RUN circuit is open or shorted to ground • Ignition coilpack 1-4 control circuit is open or shorted to ground • Ignition coilpack 1-4 control circuit is shorted to power • Ignition coilpack 1-4 is damaged or it has failed • PCM has failed
P0351-P0356 **2T CCM, MIL: Yes** 2003 Focus equipped with a 3.0L VIN Z engine (Coil on Plug design) Transmissions: All	**Ignition COP 1-6 Primary/Secondary Circuit Malfunction Conditions:** Engine started, and the PCM did not detect a valid IDM pulse from the ignition module for the Ignition Coil on Plug 1-6 primary circuit. **Possible Causes** • Ignition START/RUN circuit is open or shorted to ground • Ignition COP driver 1-6 control circuit open or shorted to ground • Ignition COP driver 1-6 control circuit is shorted to power • Ignition COP 1-6 is damaged or it has failed • PCM has failed
P0351-P0310 **2T CCM, MIL: Yes** 1998-1999-2000-2001-2002-2003 Aviator, Blackwood, Crown Victoria, Grand Marquis, Explorer, E & F Series Van & Truck, Explorer, Town Car, Mark VII and Ranger equipped with 2.5L VIN C, 4.6L VIN 9, 4.6L VIN B, 4.6L VIN H, 4.6L VIN W, 5.4L VIN A, 5.4L VIN L, 5.4L VIN R, 6.8L VIN S engine (COP) Transmissions: All	**Ignition Coil 1-10 Primary/Secondary Circuit Malfunction Conditions:** Engine started, and the PCM did not receive any valid IDM pulses from the ignition module for the Ignition Coil 1-10 primary circuit. **Possible Causes** • Ignition START/RUN circuit is open or shorted to ground • Ignition coil driver 1-10 circuit is open or shorted to ground • Ignition coil 1-10 circuit is shorted to power • Ignition coil -101 damaged or it has failed • PCM has failed

Gas Engine OBD II Trouble Code List (P0xxx Codes)

DTC	Trouble Code Title, Conditions & Possible Causes
P0351-P0358 **2T CCM, MIL: Yes** 2000-2001-2002-2003 Escape, LS, Ranger & Thunderbird equipped with a 2.0L VIN B, 3.0L (VIN 1, VIN 2, VIN S, VIN U), 3.9L VIN A, 4.0L VIN E, 4.0L VIN K, 4.6L VIN W, 5.4L VIN S engine (Coil on Plug) Transmissions: All	**Ignition Coil 1-8 Primary/Secondary Circuit Malfunction Conditions:** Engine started, and the PCM did not receive any valid IDM pulses from the ignition module for the Ignition Coil 1-8 primary circuit. **Possible Causes** • Ignition START/RUN circuit is open or shorted to ground • Ignition coil driver 1-8 circuit is open or shorted to ground • Ignition coil 1-8 circuit is shorted to power • Ignition coil 1-8 damaged or it has failed • PCM has failed
P0351-P0310 **2T CCM, MIL: Yes** 1997-1998-1999-2000- 2001-2002-2003 Aviator, Explorer, Mountaineer, E & F Series Vans & Trucks equipped with a 4.0L VIN E, 4.0L VIN K, 4.2L VIN 2, 4.2L VIN E, 4.6L VIN 6, 4.6L VIN W, 5.4L VIN 3, 5.4L VIN E, 5.4L VIN L, 5.4L VIN M, 5.4L VIN Z, 6.8L VIN S, 6.8L VIN Z engine Transmissions: All	**Ignition Coil 1-10 Primary/Secondary Circuit Malfunction Conditions:** Engine started, and the PCM did not receive any valid IDM pulses from the ignition module for the Ignition Coil 1-10 primary circuit. **Possible Causes** • Ignition START/RUN circuit is open or shorted to ground • Ignition coilpack or COP 1-10 circuit is open or shorted to ground • Ignition coilpack or COP 1-10 circuit is shorted to power • Ignition coilpack or COP 1-10 damaged or it has failed • PCM has failed
P0351-P0358 **2T CCM, MIL: Yes** 1998-1999 Expedition, Navigator equipped with 4.6L VIN 6, 4.6L VIN W engine (w/Coilpack) Transmissions: All	**Ignition Coilpack 1-8 Primary/Secondary Circuit Malfunction Conditions:** Engine started, and the PCM did not receive any valid IDM pulses from the ignition module for the Ignition Coilpack 1-8 primary circuit. **Possible Causes** • Ignition START/RUN circuit is open or shorted to ground • Ignition coilpack 1-8 control circuit is open or shorted to ground • Ignition coilpack 1-8 control circuit is shorted to power • Ignition coilpack 1-8 is damaged or it has failed • PCM has failed
P0351-P0358 **2T CCM, MIL: Yes** 2000-2001-2002-2003 Expedition, Navigator equipped with 4.6L VIN 6, 4.6L VIN W engine (Coil on Plug) Transmissions: All	**Ignition COP 1-8 Primary/Secondary Circuit Malfunction Conditions:** Engine started, and the PCM did not detect a valid IDM pulse from the Ignition Module on the Ignition Coil on Plug 1-8 primary circuit. **Possible Causes** • Ignition START/RUN circuit is open or shorted to ground • Ignition COP driver 1-8 circuit is open or shorted to ground • Ignition COP driver 1-8 circuit is shorted to power • Ignition COP 1-8 is damaged or it has failed • PCM has failed
P0351-P0358 **2T CCM, MIL: Yes** 1998-1999-2000-2001- 2002-2003 Expedition, Navigator with a 5.4L VIN A, 5.4L VIN L & VIN R engine (Coil on Plug) Transmissions: All	**Ignition COP 1-8 Primary/Secondary Circuit Malfunction Conditions:** Engine started, and the PCM did not detect a valid IDM pulse from the Ignition Module on the Ignition Coil on Plug 1 primary circuit. **Possible Causes** • Ignition START/RUN circuit is open or shorted to ground • Ignition COP driver 1-8 circuit is open or shorted to ground • Ignition COP driver 1-8 circuit is shorted to power • Ignition COP 1-8 is damaged or it has failed • PCM has failed

Gas Engine OBD II Trouble Code List (P0xxx Codes)

DTC	Trouble Code Title, Conditions & Possible Causes
P0351-P0358 **2T CCM, MIL: Yes** 2000 Crown Victoria equipped with a 4.6L VIN 9, 4.6L VIN W engine (Coilpack) Transmissions: All	**Ignition Coil 1-8 Primary/Secondary Circuit Malfunction Conditions:** Engine started, and the PCM did not receive any valid IDM pulses from the ignition module for the Ignition Coil 1-8 primary circuit. **Possible Causes** • Ignition START/RUN circuit is open or shorted to ground • Ignition coil driver 1-8 circuit open, shorted to ground or to power • Ignition coil 1-8 is damaged or it has failed • PCM has failed • TSB 01-1-6 contains a repair procedure for this trouble code
P0351-P0356 **2T CCM, MIL: Yes** 1996-1997-1998-1999 E Series Vans with a 3.4L VIN N engine Transmissions: All	**Ignition Coil 1-6 Primary/Secondary Circuit Malfunction Conditions:** Engine started, and the PCM did not receive any valid IDM pulses from the ignition module for the Ignition Coil 1-6 primary circuit. **Possible Causes** • Ignition START/RUN circuit is open or shorted to ground • Ignition coil driver 1-6 circuit is open or shorted to ground • Ignition coil 1-6 circuit is shorted to power • Ignition coil 1-6 damaged or it has failed • PCM has failed
P0351-P0354 **2T CCM, MIL: Yes** 1999-2000 Contour & Mystique equipped with a 2.0L engine (Coil on Plug) Engines: All Transmissions: All	**Ignition Coil 1-4 Primary/Secondary Circuit Malfunction Conditions:** Engine started, and the PCM did not receive any valid IDM pulses from the ignition module for the Ignition Coil 1-4 primary circuit. **Possible Causes** • Ignition START/RUN circuit is open or shorted to ground • Ignition coil driver 1-4 circuit open, shorted to ground or to power • Ignition coil 1-4 is damaged or it has failed • PCM has failed • TSB 99-21-7 contains a repair procedure for this trouble code
P0400 **2T EGR, MIL: Yes** 1996-1997 Probe equipped with a 2.0L I4 VIN A or 2.5L V6 VIN B engine Transmissions: All	**EGR Flow Malfunction Conditions:** DTC P0106, P0107, P0108, P1400, P1401 and P1408 not set, engine started, vehicle driven at a steady speed in closed loop for over 1 minute, and the PCM detected the EGR flow rate was less than or more than the calibrated flow rate limits during the test. **Possible Causes** • EGR atmospheric solenoid connector is damaged or loose • EGR atmospheric solenoid is damaged or it has failed • EGR check solenoid connector is damaged, loose or shorted • EGR check solenoid is damaged or it has failed • EGR vacuum regulator connector is damaged, loose or shorted • EGR vacuum regulator solenoid is damaged or it has failed • EGR valve position sensor connector is damaged or loose • EGR valve position sensor is damaged or it has failed • Exhaust system is restricted • PCM has failed
P0400 **2T EGR, MIL: Yes** 2003 Aviator, Escort, Focus, Explorer, Mountaineer, E Van, F Series Truck, Mustang, Sable, Taurus Navigator, Escape, Explorer, Expedition, Navigator, ZX2 Transmissions: All	**Exhaust EGR System Malfunction Conditions:** DTC P0102, P0103, P0107, P0108, P1100 and P1101 not set, vehicle driven at over 48 mph at a steady speed in closed loop for 1 minute, and the PCM detected the EGR flow rate was less than or more than the calibrated flow rate limits during the EGR test period. **Possible Causes** • DPFE EGR sensor connector is damaged, loose or shorted • EGR valve is sticking, damaged or it has failed • MAP sensor is damaged or out of calibration • PCM has failed • TSB 03-31-3 contains repair information for this trouble code

Gas Engine OBD II Trouble Code List (P0xxx Codes)

DTC	Trouble Code Title, Conditions & Possible Causes
P0400 **2T EGR, MIL: Yes** 2001-2002-2003 Focus & Ranger equipped with a 2.3L VIN D or 2.3L VIN Z engine Transmissions: All	**Electronic EGR System Malfunction Conditions:** DTC P0102, P0103, P0107, P0108, P1100 and P1101 not set, engine started, vehicle driven at over 48 mph at a steady speed in closed loop for 1 minute, and the PCM detected the EGR flow rate was less than or more than the calibrated limits during the EGR test. **Possible Causes** • EEGR valve connector is damaged, loose or shorted • EEGR valve is sticking, damaged or it has failed • MAP or TMAP sensor is damaged or not properly seated • PCM has failed
P0400 **2T EGR, MIL: Yes** 2003 Crown Victoria, Grand Marquis, Aviator, LS, Marauder, Town Car & Thunderbird Models Transmissions: All	**Exhaust EGR System Malfunction (ESM System) Conditions:** DTC P0102, P0103, P0107, P0108, P1100 and P1101 not set, vehicle driven at over 48 mph at a steady speed in closed loop for 1 minute, and the PCM detected the EGR flow rate was less than or more than the calibrated flow rate limits during the EGR test period. **Possible Causes** • DPFE EGR sensor connector is damaged, loose or shorted • EGR valve is sticking, damaged or it has failed • MAP sensor is damaged or out of calibration • PCM has failed
P0401 **2T EGR, MIL: Yes** 1996-1997 Explorer, Mountaineer equipped with a 5.0L VIN P engine Transmissions: All	**Insufficient EGR Flow Detected Conditions:** Engine running at a steady cruise speed in closed loop, and the PCM detected insufficient EGR gas flow. The EGR valve actuator and EGR valve position sensor are in one unit. **Possible Causes** • EVP sensor is damaged or has failed • EGR valve source vacuum supply problem (hose off or leaking) • EGR valve is stuck partially open or closed (check for carbon) • EGR valve may be leaking vacuum (the diaphragm is broken) • VR solenoid is damaged or the VR control circuit is open • PCM has failed
P0401 **2T EGR, MIL: Yes** 1996-1997-1998-1999-2000-2001-2002-2003 All Models Transmissions: All	**Insufficient EGR Flow Detected Conditions:** Engine started, engine running in closed loop under steady cruise conditions, and the PCM detected the DPFE sensor input indicated insufficient EGR gas flow. Run the KOER Self-Test, and if DTC P1408 is present, the fault is currently present. **Possible Causes** • DPFE sensor signal circuit is shorted to ground • DPFE sensor VREF circuit is open between sensor and PCM • DPFE sensor downstream hose off or plugged • DPFE sensor hoses both off, loose or damaged • DPFE sensor hoses connected wrong (reversed) • EGR orifice tube is damaged or restricted • TSB 03-31-3 contains repair information for this trouble code
P0401 **2T EGR, MIL: Yes** 2003 Escort, Escape, Focus, E Van, F Series Truck, Excursion, Expedition, Explorer, Mustang, Ranger, Navigator, Sable, Mountaineer Transmissions: All	**Exhaust Gas Recirculation Malfunction Conditions:** Engine started, engine running under at cruise speed in closed loop, and the PCM detected a problem in the EGR system. Run the KOER self-test. If DTC P1406 is set, test the EGR valve operation. **Possible Causes** • DPFE EGR valve hoses are damaged, leaking or restricted • DPFE EGR valve hoses may be reversed at the sensor • EGR valve connector is damaged, loose or shorted • EGR valve is damaged or it has failed • PCM has failed • TSB 4-3-1 contains repair information for this trouble code

Gas Engine OBD II Trouble Code List (P0xxx Codes)

DTC	Trouble Code Title, Conditions & Possible Causes
P0401 **2T EGR, MIL: Yes** 2003 Crown Victoria, Grand Marquis, Aviator, LS, Marauder, Town Car & Thunderbird Models Transmissions: All	**Exhaust Gas Recirculation Malfunction (ESM System) Conditions:** Engine running under at cruise speed in closed loop, and the PCM detected a problem in the EGR ESM system. Run the KOER self-test. If DTC P1408 is present, inspect the EGR valve. **Possible Causes** • DPFE EGR valve hoses are damaged, leaking or restricted • EGR valve connector is damaged, loose or shorted • EGR valve is damaged or it has failed • PCM has failed
P0402 **2T EGR, MIL: Yes** 1996-1997-1998-1999-2000-2001-2002 All Models Transmissions: All	**Excessive EGR Flow Detected Conditions:** Engine started, engine running in hot idle speed, and the PCM detected the Actual DPFE sensor value indicated more than the KOEO DPFE sensor value stored in the PCM memory. **Possible Causes** • DPFE EGR valve source hoses loose or connected wrong • DPFE sensor slow to respond or sluggish (it may have failed) • DPFE sensor signal circuit is open or shorted to ground • DPFE EGR sensor is damaged or the PCM has failed
P0402 **2T EGR, MIL: Yes** 1996-1997 Explorer, Mountaineer equipped with a 5.0L VIN P engine Transmissions: All	**Excessive EGR Flow Detected Conditions:** Engine running under at a steady cruise speed in closed loop, and the PCM detected the EGR valve position did not match the Desired EGR position based on engine speed and load. **Possible Causes** • EGR valve actuator is damaged or sticking (perform a KOEO or KOER Self Test to test the actuator for an On Demand code) • EGR position sensor signal circuit shorted to bias voltage • EP sensor signal circuit shorted to bias voltage • PCM has failed
P0402 **2T EGR, MIL: Yes** 2003 Crown Victoria, Grand Marquis, Aviator, LS, Marauder, Town Car & Thunderbird Models Transmissions: All	**EGR Flow At Idle Speed Detected (ESM System) Conditions:** Engine started, engine running in hot idle speed, and the PCM detected the Actual DPFE sensor value indicated more than the KOEO DPFE sensor value stored in the memory. If DTC P1405 is set, repair the cause of that trouble code prior to repairing P0402. **Possible Causes** • DPFE EGR sensor is damaged • DPFE EGR valve source hoses loose or connected wrong • DPFE sensor slow to respond or sluggish (it may have failed) • DPFE sensor signal circuit is open or shorted to ground • PCM has failed • TSB 03-31-3 contains repair information for this trouble code
P0402 **2T EGR, MIL: Yes** 2003 Escort, Escape, Focus, E Van, F Series Truck, Excursion, Expedition, Explorer, Mustang, Ranger, Navigator, Sable, Mountaineer Transmissions: All	**EGR Flow At Idle Speed Detected (ESM System) Conditions:** Engine started, engine running in hot idle speed, and the PCM detected the Actual DPFE sensor value indicated more than the KOEO DPFE sensor value stored in the memory. If DTC P1405 is set, repair the cause of that trouble code prior to repairing P0402. **Possible Causes** • DPFE EGR sensor is damaged • DPFE EGR valve source hoses loose or connected wrong • DPFE sensor slow to respond or sluggish (it may have failed) • DPFE sensor signal circuit is open or shorted to ground • PCM has failed • TSB 03-31-3 contains repair information for this trouble code
P0403 **2T CCM, MIL: Yes** 2001-2002-2003 Ranger equipped with a 2.3L VIN D engine Transmissions: All	**EGR Solenoid Circuit Malfunction Conditions:** Engine started, and the PCM detected an unexpected "high" or "low" voltage condition on the EEGR solenoid control circuit at idle speed. **Possible Causes** • EEGR solenoid control circuit is open, or shorted to ground • EEGR solenoid power circuit is open (check power to relay) • EEGR motor winding is open or shorted to power • EEGR solenoid connector not seated correctly or the EGR solenoid has failed • PCM has failed

Gas Engine OBD II Trouble Code List (P0xxx Codes)

DTC	Trouble Code Title, Conditions & Possible Causes
P0403 **2T CCM, MIL: Yes** 2003 Crown Victoria, Grand Marquis, Aviator, LS, Marauder, Town Car & Thunderbird Models Transmissions: All	**EGR Solenoid Circuit Malfunction (ESM System) Conditions:** Engine started, and the PCM detected an unexpected high or low voltage condition on the ESM EGR solenoid control circuit at idle. **Possible Causes** • EGR solenoid connector is damaged, loose or shorted • EGR solenoid control circuit is open, or shorted to ground • EGR solenoid power circuit is open (check power to relay) • EGR solenoid is damaged or the PCM has failed
P0403 **2T CCM, MIL: Yes** 2003 Aviator, Escort, Focus, Explorer, Mountaineer, E Van, F Series Truck, Mustang, Sable, Taurus Navigator, Escape, Explorer, Expedition, Navigator, ZX2 Transmissions: All	**EGR Solenoid Circuit Malfunction Conditions:** Engine started, and the PCM detected an unexpected high or low voltage condition on the EGR solenoid control circuit at idle speed. **Possible Causes** • EGR solenoid connector is damaged, loose or shorted • EGR solenoid control circuit is open, or shorted to ground • EGR solenoid power circuit is open (check power to relay) • EGR motor winding is open or shorted to power • EGR solenoid is damaged or has failed • PCM has failed
P0405 **2T CCM, MIL: Yes** 2003 Crown Victoria, Grand Marquis, Aviator, LS, Marauder, Town Car & Thunderbird Models Transmissions: All	**DPFE Sensor Circuit Low Input (ESM System) Conditions:** Engine started, and the PCM detected an unexpected low voltage condition (less than 0.20v) on the ESM DPFE sensor circuit. **Possible Causes** • DPFE sensor connector is damaged or shorted • DPFE sensor power supply circuit is open or shorted to ground • DPFE sensor signal circuit is shorted to ground • DPFE sensor is damaged or the PCM has failed
P0405 **2T CCM, MIL: Yes** 2003 Aviator, Escort, Focus, Explorer, Mountaineer, E Van, F Series Truck, Mustang, Sable, Taurus Navigator, Escape, Explorer, Expedition, Navigator, ZX2 Transmissions: All	**DPFE Sensor Circuit Low Input Conditions:** Engine started, and the PCM detected an unexpected low voltage condition (less than 0.20v) on the ESM DPFE sensor circuit. **Possible Causes** • DPFE sensor connector is damaged or shorted • DPFE sensor power supply circuit is open or shorted to ground • DPFE sensor signal circuit is shorted to ground • DPFE sensor is damaged or has failed • PCM has failed
P0406 **2T CCM, MIL: Yes** 2003 Crown Victoria, Grand Marquis, Aviator, LS, Marauder, Town Car & Thunderbird Models Transmissions: All	**DPFE Sensor Circuit High Input (ESM System) Conditions:** Key on or engine running at idle speed, and the PCM detected an unexpected high voltage condition (more than 4.00v) on the ESM DPFE sensor circuit during the CCM test period. **Possible Causes** • DPFE sensor connector is damaged or open • DPFE sensor signal circuit is open or it is shorted to VREF (5v) • DPFE sensor is damaged or has failed • PCM has failed
P0406 **2T CCM, MIL: Yes** 2003 Aviator, Escort, Focus, Explorer, Mountaineer, E Van, F Series Truck, Mustang, Sable, Taurus Navigator, Escape, Explorer, Expedition, Navigator, ZX2 Transmissions: All	**DPFE Sensor Circuit High Input Conditions:** Key on or engine running at idle speed, and the PCM detected an unexpected high voltage condition (more than 4.00v) on the ESM DPFE sensor circuit during the CCM test period. **Possible Causes** • DPFE sensor connector is damaged or open • DPFE sensor signal circuit is shorted to VREF (5v) • DPFE sensor signal circuit is open • DPFE sensor is damaged or has failed • PCM has failed

Gas Engine OBD II Trouble Code List (P0xxx Codes)

DTC	Trouble Code Title, Conditions & Possible Causes
P0411 **2T AIR, MIL: Yes** 1996-1997-1998-1999 Some Models Transmissions: All	**Secondary AIR System Incorrect Upstream Flow Detected Conditions:** Engine started, engine running at idle speed in closed loop, and the PCM detected the secondary AIR pump airflow was not diverted correctly when requested during the self-test. **Possible Causes** • Air pump output is blocked • AIR bypass solenoid leaking or blocked • AIR bypass solenoid is stuck open or stuck closed • Air injection pump hose(s) leaking • PCM has failed
P0411 **2T AIR, MIL: Yes** 2000-2001-2002-2003 LS, Mustang, Taurus equipped with a 3.0L VIN 2, 3.0L VIN S or 3.8L VIN 4 engine Transmissions: All	**Secondary Air Injection System Upstream Flow Detected Conditions:** Engine started, engine runtime from 20-120 seconds at any speed, and the PCM detected the Secondary AIR pump airflow was not diverted correctly when requested during the self-test. **Possible Causes** • Air pump output is blocked or restricted • AIR bypass solenoid is leaking or it is restricted • AIR bypass solenoid is stuck open or stuck closed • Check valve (one or more) is damaged or leaking • Electric air injection pump hose(s) leaking • PCM has failed
P0412 **2T CCM, MIL: Yes** 1996-1997-1998-1999 Some Models Transmissions: All	**Secondary Air Injection Solenoid Circuit Malfunction Conditions:** Engine started, and the PCM detected an unexpected low or high voltage condition on the AIR solenoid control circuit during testing. **Possible Causes** • AIR solenoid power circuit (B+) is open (check dedicated fuse) • AIR bypass solenoid control circuit is open or shorted to ground • AIR diverter solenoid control circuit open or shorted to ground • AIR pump control circuit is open or shorted to ground • Check valve (one or more) is damaged or leaking • Solid State relay is damaged or it has failed • PCM has failed
P0412 **2T CCM, MIL: Yes** 2000-2001-2002-2003 LS, Mustang, Taurus equipped with a 3.0L VIN 2, 3.0L VIN S or 3.8L VIN 4 engine Transmissions: All	**Secondary AIR Solenoid Control Circuit Malfunction Conditions:** Engine started, and the PCM detected an unexpected low or high voltage condition on the AIR solenoid control circuit during testing. **Possible Causes** • AIR solenoid power circuit (B+) is open (check dedicated fuse) • AIR bypass solenoid control circuit is open or shorted to ground • AIR diverter solenoid control circuit open or shorted to ground • Electric AIR pump control circuit is open or shorted to ground • PCM has failed
P0413 **2T CCM, MIL: Yes** 1996-1997-1998-1999 Some Models Transmissions: All	**Secondary AIR System Switching Valve 'A' Circuit Malfunction Conditions:** Engine started; Air Injection solenoid commanded "on", and the PCM detected an unexpected "high" voltage condition on the Secondary AIR control circuit during the CCM test. **Possible Causes** • AIR solenoid control circuit is shorted to system power • AIR pump solenoid is damaged or has failed • PCM has failed
P0414 **2T CCM, MIL: Yes** 1996-1997-1998-1999 Some Models Transmissions: All	**Secondary AIR System Switching Valve 'A' Circuit Malfunction Conditions:** Engine started; AIR solenoid disabled, and the PCM detected an unexpected low voltage condition on the AIR Solenoid control circuit **Possible Causes** • AIR solenoid control circuit is shorted to ground • AIR solenoid power circuit is open (no power to the solenoid) • AIR pump solenoid is damaged or has failed • Solid State relay is damaged or the PCM has failed

Gas Engine OBD II Trouble Code List (P0xxx Codes)

DTC	Trouble Code Title, Conditions & Possible Causes
P0416 **2T CCM, MIL: Yes** 1996-1997-1998-1999 Some Models Transmissions: All	**Secondary AIR System Switching Valve 'B' Circuit Fault Conditions:** Engine started, AIR solenoid enabled, and the PCM detected an unexpected high voltage condition on AIR solenoid control circuit. **Possible Causes** • AIR solenoid control circuit is open or it is shorted to system power • AIR pump is damaged or the Solid State relay is damaged or has failed • PCM has failed
P0417 **2T CCM, MIL: Yes** 1996-1997-1998-1999 Some Models Transmissions: All	**Secondary AIR System Switching Valve 'B' Circuit Fault Conditions:** Engine started; Air Injection solenoid commanded "on", and the PCM detected an unexpected "low" voltage condition on the AIR Solenoid control circuit during the CCM test. **Possible Causes** • AIR solenoid control circuit is shorted to ground or there is no power to the circuit • AIR pump is damaged or the Solid State relay is damaged or has failed • PCM has failed
P0420 **2T CAT, MIL: Yes** 1996-1997-1998-1999-2000-2001-2002-2003 All Models Transmissions: All	**Catalyst System Efficiency Bank 1 Below Threshold Conditions:** Vehicle driven at steady cruise speed for 5 minutes, and the PCM detected the switch rate of the rear HO2S-12 was close to the switch rate of front HO2S (it should be much slower). **Possible Causes** • Air leaks at the exhaust manifold or in the exhaust pipes • Catalytic converter is damaged, contaminated or it has failed • ECT/CHT sensor has lost its calibration (the signal is incorrect) • Engine cylinders misfiring, or the ignition timing is over retarded • Engine oil is contaminated • Front HO2S or rear HO2S is contaminated with fuel or moisture • Front HO2S and/or the rear HO2S is loose in the mounting hole • Front HO2S much older than the rear HO2S (HO2S-11 is lazy) • Fuel system pressure is too high (check the pressure regulator) • Rear HO2S wires improperly connected or the HO2S has failed
P0421 **2T CAT, MIL: Yes** 1996-1997 Probe equipped with a 2.5L V6 VIN B engine Transmissions: All	**Catalyst System Efficiency Bank 1 Below Threshold Conditions:** Vehicle driven at steady cruise speed for 5 minutes, and the PCM detected the switch rate of the rear HO2S-12 was close to the switch rate of front HO2S (it should be much slower). **Possible Causes** • Air leaks at the exhaust manifold or in the exhaust pipes • Catalytic converter is damaged, contaminated or it has failed • ECT/CHT sensor has lost its calibration (the signal is incorrect) • Engine cylinders misfiring, or the ignition timing is over retarded • Front HO2S or rear HO2S is contaminated with fuel or moisture • Front HO2S and/or the rear HO2S is loose in the mounting hole • Front HO2S much older than the rear HO2S (HO2S-11 is lazy) • Fuel system pressure is too high (check the pressure regulator) • Rear HO2S wires improperly connected or the HO2S has failed
P0430 **2T CAT, MIL: Yes** 1996-1997-1998-1999-2000-2001-2002-2003 All Models Transmissions: All	**Catalyst System Efficiency Bank 2 Below Threshold Conditions:** Vehicle driven at steady cruise speed for 5 minutes, and the PCM detected the switch rate of the rear HO2S-12 was close to the switch rate of front HO2S (it should be much slower). **Possible Causes** • Air leaks at the exhaust manifold or in the exhaust pipes • Catalytic converter is damaged, contaminated or it has failed • ECT/CHT sensor has lost its calibration (the signal is incorrect) • Engine cylinders misfiring, or the ignition timing is over retarded • Engine oil is contaminated • Front HO2S or rear HO2S is contaminated with fuel or moisture • Front HO2S and/or the rear HO2S is loose in the mounting hole • Front HO2S much older than the rear HO2S (HO2S-11 is lazy) • Fuel system pressure is too high (check the pressure regulator) • Rear HO2S wires improperly connected or the HO2S has failed

Gas Engine OBD II Trouble Code List (P0xxx Codes)

DTC	Trouble Code Title, Conditions & Possible Causes
P0431 **2T CAT, MIL: Yes** 1996-1997 Probe equipped with a 2.5L V6 VIN B engine Transmissions: All	**Catalyst System Efficiency Bank 2 Below Threshold Conditions:** Vehicle driven at steady cruise speed for 5 minutes, and the PCM detected the switch rate of the rear HO2S-12 was close to the switch rate of front HO2S (it should be much slower). **Possible Causes** • Air leaks at the exhaust manifold or in the exhaust pipes • Catalytic converter is damaged, contaminated or it has failed • ECT/CHT sensor has lost its calibration (the signal is incorrect) • Engine cylinders misfiring, or the ignition timing is over retarded • Front HO2S or rear HO2S is contaminated with fuel or moisture • Front HO2S and/or the rear HO2S is loose in the mounting hole • Front HO2S much older than the rear HO2S (HO2S-11 is lazy) • Fuel system pressure is too high (check the pressure regulator) • Rear HO2S wires improperly connected or the HO2S has failed
P0440 **2T EVAP, MIL: Yes** 1996-1997 Probe equipped with a 2.0L VIN A or 2.5L V6 VIN B engine Transmissions: All	**EVAP System Malfunction Conditions:** ECT sensor less than 90ºF at startup (cold engine), engine running in closed loop at a steady cruise speed, and the PCM detected a problem in the EVAP system during the EVAP System Monitor test. Note: If DTC P0443 is set, repair that diagnostic trouble code first. **Possible Causes** • Canister Purge valve is damaged (DTC P0443 not set) • Vapor line between Purge solenoid and intake manifold vacuum reservoir is damaged, or vapor line between EVAP Canister Purge solenoid and charcoal canister is damaged • Vapor line between charcoal canister and check valve, or vapor line between check valve and fuel vapor valves is damaged • PCM has failed
P0442 **2T EVAP, MIL: Yes** 1996-1997 All Models Transmissions: All	**EVAP Control System Small Leak Detected Conditions:** ECT sensor less than 90ºF at startup (cold engine), engine running in closed loop at a steady cruise speed, and the PCM detected a leak in the EVAP system as small as 0.040" in the test. **Possible Causes** • Aftermarket EVAP parts that do not conform to specifications • CV solenoid stays partially open when commanded to close • EVAP component seals leaking (i.e., leaks in the Purge valve, fuel vapor control valve tube assembly or fuel vapor vent valve) • Fuel filler cap damaged, cross-threaded or loosely installed • Loose fuel vapor hose/tube connections to EVAP components • Small holes or cuts in fuel vapor hoses or EVAP canister tubes
P0442 **2T EVAP, MIL: Yes** 1998-1999-2000-2001-2002-2003 All Models Transmissions: All	**EVAP Control System Small Leak (0.040") Detected Conditions:** ECT sensor less than 90ºF and within 10ºF of the IAT sensor at startup (cold engine), engine started, the with the engine running in closed loop at a steady cruise speed, the PCM detected a leak in the EVAP system as small as 0.040" during the EVAP Monitor Test. **Possible Causes** • Aftermarket EVAP parts that do not conform to specifications • CV solenoid remains partially open when commanded to close • EVAP component seals leaking (i.e., leaks in the Purge valve, fuel tank pressure sensor, canister vent solenoid, fuel vapor control valve tube assembly or fuel vapor vent valve). • Fuel filler cap damaged, cross-threaded or loosely installed • Loose fuel vapor hose/tube connections to EVAP components • Small holes or cuts in fuel vapor hoses or EVAP canister tubes • TSB 99-23-4, TSB 3-9-8 & TSB 3-20-3 contain a repair procedure for this trouble code
P0443 **2T CCM, MIL: Yes** 1996-1997 All Models Transmissions: All	**EVAP Canister Purge Solenoid Circuit Malfunction Conditions:** Engine started, and the PCM detected an unexpected high or low voltage condition on the Purge solenoid control circuit when the device was cycled "on" and "off" during testing. **Possible Causes** • EVAP purge solenoid supply circuit is open • EVAP purge solenoid control circuit open, shorted to ground • EVAP purge solenoid control circuit is shorted to power (B+) • EVAP canister purge solenoid valve is damaged or the PCM has failed

Gas Engine OBD II Trouble Code List (P0xxx Codes)

DTC	Trouble Code Title, Conditions & Possible Causes
P0443 **2T CCM, MIL: Yes** 1996-1997 Contour & Mystique Models Transmissions: All	**EVAP Vapor Management Valve Circuit Malfunction Conditions:** Engine started, and the PCM detected an unexpected high or low voltage condition on the Vapor Management Valve (VMV) circuit when the device was cycled On/Off during testing. **Possible Causes** • EVAP VMV power supply circuit is open • EVAP VMV solenoid control circuit is open or shorted to ground • EVAP VMV solenoid control circuit is shorted to power (B+) • EVAP VMV solenoid valve is damaged or it has failed • PCM has failed
P0443 **2T CCM, MIL: Yes** 1996-1997 E Van, F Series Truck Models equipped with a 5.8L VIN H or 7.5L VIN G engine Transmissions: All	**EVAP Vapor Management Valve Circuit Malfunction Conditions:** Engine started, and the PCM detected an unexpected high or low voltage condition on the Vapor Management Valve (VMV) circuit after the device was cycled On/Off during testing. **Possible Causes** • EVAP VMV power supply circuit is open • EVAP VMV solenoid control circuit is open or shorted to ground • EVAP VMV solenoid control circuit is shorted to power (B+) • EVAP VMV solenoid valve is damaged or it has failed • PCM has failed
P0443 **2T CCM, MIL: Yes** 1998-1999-2000-2001- 2002-2003 All Models Transmissions: All	**EVAP Canister Purge Solenoid Circuit Malfunction Conditions:** Engine started, and the PCM detected an unexpected high or low voltage condition on the EVAP Purge solenoid control circuit when the device was cycled On/Off during testing. **Possible Causes** • EVAP purge solenoid supply circuit is open • EVAP purge solenoid control circuit open, shorted to ground • EVAP purge solenoid control circuit is shorted to power (B+) • EVAP canister purge solenoid valve is damaged or it has failed • PCM has failed
P0446 **2T EVAP, MIL: Yes** 1996-1997-1998 All Models Transmissions: All	**EVAP Canister Vent System Performance Conditions:** ECT sensor less than 90°F, engine started, engine running at a steady cruise speed, and the PCM detected excessive vacuum was present in the EVAP system during the test period. **Possible Causes** • Canister vent (CV) solenoid is stuck closed (partially or fully) • EVAP canister purge outlet tube blocked or kinked between the canister purge valve and the EVAP canister, or EVAP canister tube blocked between the fuel tank and canister • EVAP canister restricted, or plugged CV solenoid filter unit • Plugged or contaminated CV solenoid filter • EVAP canister purge valve stuck open • Fuel filler cap stuck closed (no vacuum relief) • FTP sensor VREF circuit open, or FTP sensor is damaged • Fuel vapor elbow at the EVAP canister contaminated
P0446 **2T EVAP, MIL: Yes** 2003 All Models Transmissions: All	**EVAP Canister Vent Solenoid Circuit Malfunction Conditions:** Engine started, and the PCM detected an unexpected high or low voltage condition on the EVAP Canister Vent solenoid control circuit after the device was cycled On/Off in the test. **Possible Causes** • Canister vent solenoid supply circuit is open • Canister vent solenoid control circuit open, shorted to ground • Canister vent solenoid control circuit is shorted to power (B+) • Canister vent solenoid valve is damaged or the PCM has failed
P0451 **2T CCM, MIL: Yes** 1999-2000-2001-2002- 2003 All Models Transmissions: All	**Fuel Tank Pressure Sensor Intermittent Signal Conditions:** Engine started, and the PCM detected the FTP sensor signal changed from over +15" H2O to under -15" H2O within 100 ms. **Possible Causes** • FTP sensor signal circuit has an intermittent open condition • FTP sensor signal circuit has an intermittent shorted condition • FTP sensor is damaged or it has failed

Gas Engine OBD II Trouble Code List (P0xxx Codes)

DTC	Trouble Code Title, Conditions & Possible Causes
P045 **2T CCM, MIL: Yes** 1999-2000-2001-2002-2003 All Models Transmissions: All	**FTP Sensor Circuit Low Input Conditions:** Key on or engine running; and the PCM detected the FTP sensor indicated less than the minimum calibrated limit of 0.22v in the test. **Possible Causes** • FTP sensor connector has internal damage or contamination • FTP sensor signal circuit is shorted to chassis or signal ground • FTP sensor is damaged • PCM has failed
P0453 **2T CCM, MIL: Yes** 1999-2000-2001-2002-2003 All Models Transmissions: All	**FTP Sensor Circuit High Input Conditions:** Key on or engine running; and the PCM detected the FTP sensor indicated more than the maximum calibrated limit (4.50v) in the test. **Possible Causes** • FTP sensor signal circuit is open or the ground circuit is open • FTP sensor signal circuit is shorted to VREF (5v) • FTP sensor is damaged or the PCM has failed
P0455 **2T EVAP, MIL: Yes** 1996-1997 All Models Transmissions: All	**EVAP Control System Large Leak Detected Conditions:** ECT sensor less than 90ºF at startup, engine running, and the PCM detected several small fuel vapor leaks or a large leak in the system. **Possible Causes** • Aftermarket EVAP hardware non-conforming to specifications • Canister vent (CV) solenoid stuck open • EVAP canister purge valve stuck closed, or canister damaged • EVAP canister tube, EVAP canister purge outlet tube or EVAP return tube disconnected or cracked, or canister is damaged • Fuel filler cap missing, loose (not tightened) or the wrong part • Loose fuel vapor hose/tube connections to EVAP components • Purge sensor or FTP sensor is out of calibration or has failed
P0455 **2T EVAP, MIL: Yes** 1998-1999-2000-2001-2002-2003 All Models Transmissions: All	**EVAP Control System Large Leak (0.080") Detected Conditions:** ECT sensor less than 90ºF, engine running at a steady cruise speed, and the PCM detected multiple small fuel vapor leaks; or it detected a large leak in the system during the leak test. **Possible Causes** • Aftermarket EVAP hardware non-conforming to specifications • EVAP canister tube, EVAP canister purge outlet tube or EVAP return tube disconnected or cracked, or canister is damaged • EVAP canister purge valve stuck closed, or canister damaged • Fuel filler cap missing, loose (not tightened) or the wrong part • Loose fuel vapor hose/tube connections to EVAP components • Canister vent (CV) solenoid stuck open • Fuel tank pressure (FTP) sensor has failed mechanically • TSB 99-23-4 contains a repair procedure for this trouble code • TSB 03-9-8 contains a repair procedure for this trouble code • TSB 3-20-3 contains a repair procedure for this trouble code
P0456 **2T EVAP, MIL: Yes** 2000-2001-2002-2003 All Models Transmissions: All	**EVAP Control System Very Small Leak (0.020") Detected Conditions:** ECT sensor less than 90ºF (cold engine), engine started, engine running at a steady cruise speed, and the PCM detected a very small fuel vapor leak (0.020") during the leak test. **Possible Causes** • Canister tube, EVAP canister purge outlet tube or return tube disconnected or cracked • EVAP canister purge valve stuck closed, or canister damaged • Fuel vapor hoses/tubes that have very small holes and/or cuts • Fuel vapor hose/tube connections are loose or damaged • EVAP component seals are leaking (i.e., Purge valve, fuel tank pressure sensor, canister vent solenoid, fuel vapor control valve tube assembly or fuel vapor vent valve assembly) • TSB 03-9-8 contains a repair procedure for this trouble code • TSB 3-20-3 contains a repair procedure for this trouble code

Gas Engine OBD II Trouble Code List (P0xxx Codes)

DTC	Trouble Code Title, Conditions & Possible Causes
P0457 **2T EVAP, MIL: Yes** 1999-2000-2001-2002-2003 All Models Transmissions: All	**EVAP Control System Leak Detected (Fuel Cap Missing) Conditions:** ECT sensor less than 90ºF at startup, engine running at a steady state cruise speed, and the PCM detected the fuel tank pressure changed more than minus (-) 7" H2O in 30 seconds, or excessive purge flow (over 0.06 pounds per minute) occurred in the EVAP Running Loss Monitor Test ("Check Fuel Cap" Lamp may be "on"). **Possible Causes** • Fuel filler cap not installed after refueling (CMC P0457 is set) • Fuel filler cap missing, loose or cross-threaded • TSB 03-9-8 contains a repair procedure for this trouble code • TSB 3-20-3 contains a repair procedure for this trouble code
P0460 **2T CCM, MIL: Yes** 1997-1998-1999-2000-2001-2002-2003 All Models Transmissions: All	**Fuel Level Sensor Signal Range/Performance Conditions:** Engine started, and the PCM detected the FLI sensor did not match the fuel level (e.g., FLI V PID below 0.90v with FLI PID at 25%, or FLIV PID more than 2.45v with FLI PID at 75%). **Possible Causes** • Fuel tank is empty • FP module is stuck open • Fuel gauge is incorrectly installed • Instrument cluster damaged • PCM Case ground circuit open • Fuel level indicator (FLI) circuit is shorted to power, or is open • Fuel tank has been overfilled, or fuel gauge is damaged • Fuel pump (FP) module is stuck closed, or is stuck open • Fuel level indicator circuit shorted to Case or to power ground • PCM Case ground shorted to VPWR (shorted to system power) • TSB 03-1-7 contains repair help for this code (LS & T-Bird)
P0462 **2T CCM, MIL: Yes** 2003 All Models Transmissions: All	**Fuel Level Sensor Circuit Low Input Conditions:** Key on or engine running; and the PCM detected the FLI sensor indicated less than 0.20v at any time during the CCM test period. **Possible Causes** • Fuel tank is empty • FLI signal circuit is open • FLI signal circuit is shorted to case or chassis ground • PCM has failed
P0463 **2T CCM, MIL: Yes** 2003 All Models Transmissions: All	**Fuel Level Sensor Circuit High Input Conditions:** Key on or engine running; and the PCM detected that the FLI sensor indicated more than 4.50v at any time during the CCM test period. **Possible Causes** • Fuel level sensor connector is damaged or shorted • Fuel tank has been over-filled • FLI signal circuit is shorted to VREF (5v or 12v) • PCM has failed
P0480 **2T CCM, MIL: Yes** 2002-2003 Crown Victoria, Grand Marquis, Town Car, LS Thunderbird, Mustang, Expedition, Navigator, ZX2 Models Transmissions: All	**Visctronic Drive Fan Primary Circuit Malfunction Conditions:** Key on or engine running; and the PCM detected an unexpected high or low voltage condition on the Visctronic Drive Fan (VDF) primary circuit during the CCM test period. **Possible Causes** • VDF variable control circuit is open • VDF variable control circuit is shorted to chassis ground • VDF variable control circuit shorted to Fan Speed Sensor circuit • VDF clutch power supply (VPWR) circuit is open • VDF clutch is damaged or it has failed • PCM has failed

Gas Engine OBD II Trouble Code List (P0xxx Codes)

DTC	Trouble Code Title, Conditions & Possible Causes
P0480 **2T CCM, MIL: Yes** 2003 Focus, E Van, Escape, Taurus, Sable, Ranger Models equipped with a 2.0L VIN B, 2.0L VIN 3, 2.0L VIN 5, 2.0L VIN P, 2.3L VIN D, 2.3L VIN Z, 3.0L VIN 1, 3.0L VIN 2, 3.0L VIN S, 3.0L VIN U, 3.0L VIN V, 4.0L VIN E engine Transmissions: All	**Fan Control Relay Circuit Malfunction Conditions:** Key on or engine running; and the PCM detected an unexpected high or low voltage condition on the Fan Control relay control circuit. **Possible Causes** • High/Low/Medium FC relay control circuit is open • High/Low/Medium FC relay control circuit is shorted to ground • High/Low/Medium FC relay VPWR circuit is open • High/Low/Medium FC relay direct battery (B+) circuit is open • High/Low/Medium FC relay is damaged or it has failed • PCM has failed
P0481 **2T CCM, MIL: Yes** 2003 Escort, ZX2, Mustang equipped with a 2.0L VIN 3, 3.8L VIN 4, 4.6L VIN X, 4.6L VIN X, 4.6L VIN Y Transmissions: All	**Constant Control Relay Module Circuit Malfunction Conditions:** Key on or engine running; and the PCM detected an unexpected high or low voltage condition on the High Fan Control (HFC) relay control circuit (located inside the CCRM). **Possible Causes** • HFC relay control circuit is open • HFC relay control circuit is shorted to chassis ground • HFC relay power supply (VPWR) circuit is open • HFC relay direct battery (B+) circuit is open • HFC relay (inside the CCRM) is damaged or it has failed • PCM has failed
P0481 **2T CCM, MIL: Yes** 2003 Escape, Focus, Taurus, Sable, Ranger equipped with a 2.0L (VIN B, VIN 3, VIN 5, VIN P), 2.3L VIN D, 2.3L VIN Z, 3.0L VIN 1, 3.0L VIN 2, 3.0L VIN S, 3.0L VIN U, 3.0L VIN V, 4.0L VIN E engine Transmissions: All	**Fan Control Relay Circuit Malfunction Conditions:** Key on or engine running; and the PCM detected an unexpected high or low voltage condition on the Fan Control relay control circuit. **Possible Causes** • FC relay control circuit is open • FC relay control circuit is shorted to chassis ground • FC relay power supply (VPWR) circuit is open • FC relay direct battery (B+) circuit is open • FC relay is damaged or it has failed • PCM has failed
P0482 **2T CCM, MIL: Yes** 2003 Aviator, Escape, Focus, Taurus, Sable, Ranger, Explorer, Mountaineer, Expedition, Navigator equipped with a 2.0L VIN B, 2.0L VIN 3, 2.0L VIN 5, 2.0L VIN P, 2.3L VIN D, 2.3L VIN Z, 3.0L VIN 1, 3.0L VIN 2, 3.0L VIN S, 3.0L VIN U, 3.0L VIN V, 3.8L VIN 4, 4.0L VIN E, 4.6L VIN H engine Transmissions: All	**Constant Control Relay Module Circuit Malfunction Conditions:** Key on or engine running; and the PCM detected an unexpected high or low voltage condition on the High Fan Control (HFC) relay control circuit (located inside the CCRM) during the CCM test period. **Possible Causes** • FC relay control circuit is open • FC relay control circuit is shorted to chassis ground • FC relay power supply (VPWR) circuit is open • FC relay direct battery (B+) circuit is open • PCM has failed

Gas Engine OBD II Trouble Code List (P0xxx Codes)

DTC	Trouble Code Title, Conditions & Possible Causes
P0500 **2T CCM, MIL: Yes** 1996-1997-1998-1999 All Models Transmissions: All	**Vehicle Speed Sensor Malfunction Conditions:** Engine running, then with the engine speed more than the TCC stall speed, the PCM detected a lack of vehicle speed data occurred. Note: The PCM receives vehicle speed data from the VSS, TCSS, ABS module, CTM or GEM controller, depending up the application. **Possible Causes** • Modules connected to VSC/VSS harness circuits are damaged • Mechanical drive mechanism for the VSS or TCSS is damaged • VSS+ or VSS- harness circuit is open • TCSS signal or TCSS signal return harness circuit is open • VSS harness circuit, TCSS harness circuit is shorted to ground • VSS harness circuit, CSS harness circuit is shorted to power • VSS circuit open between the PCM and related control module • VSS or TCSS, or wheel speed sensors circuits are damaged • TSB 01-21-13 contains a repair procedure for this trouble code
P0500 **2T CCM, MIL: Yes** 2000-2001-2002-2003 All Models Transmissions: All	**Vehicle Speed Sensor Circuit Malfunction Conditions:** Engine running, then with the engine speed more than the TCC stall speed, the PCM detected a lack of vehicle speed data occurred. Note: The PCM receives vehicle speed data from the Vehicle Speed Sensor on these vehicle applications. **Possible Causes** • VSS signal circuit is open or shorted to ground • VSS ground circuit is open or VSS power circuit is open • VSS is damaged or it has failed • PSOM is damaged or it has failed (some models) • PCM has failed
P0500 **2T CCM, MIL: Yes** 2000-2001-2002-2003 Continental, Town Car, LS, Windstar Models Transmissions: All	**Vehicle Speed Sensor Circuit Malfunction Conditions:** Engine running, then with the engine speed more than the TCC stall speed, the PCM detected a lack of vehicle speed data occurred. Note: The PCM receives vehicle speed data from the ABS module. **Possible Causes** • The vehicle speed information on this vehicle application is provided to the PCM by the Antilock Brake System module. • Refer to the ABS diagnostics and trouble codes to diagnose this particular trouble code.
P0500 **2T CCM, MIL: Yes** 2000-2001-2002 Contour, Cougar & Mystique Models Transmissions: All	**Vehicle Speed Sensor Circuit Malfunction Conditions:** Engine running, then with the engine speed more than the TCC stall speed, the PCM detected a lack of vehicle speed data occurred. Note: The PCM receives vehicle speed data from the VSS sensor. **Possible Causes** • VSS+ signal circuit is open or shorted to ground • VSS- signal circuit is open • VSS power supply (VPWR) circuit is open • VSS is damaged or it has failed • PCM has failed
P0500 **2T CCM, MIL: Yes** 2000-2001-2002-2003 F Series Trucks equipped a 4.6L VIN W, 5.4L VIN 3, 5.4L VIN L, 5.4L VIN M, 5.4L VIN Z engine Transmissions: All	**Vehicle Speed Sensor Circuit Malfunction Conditions:** Engine running, then with the engine speed more than the TCC stall speed, the PCM detected a lack of vehicle speed data for a period of time. Note: The PCM receives vehicle speed data from the Transfer Case Speed Sensor on these vehicle applications. **Possible Causes** • TCSS signal circuit is open or shorted to ground • TCSS ground circuit is open • TCSS is damaged or it has failed • PCM has failed

Gas Engine OBD II Trouble Code List (P0xxx Codes)

DTC	Trouble Code Title, Conditions & Possible Causes
P0500 **2T CCM, MIL: Yes** 2000-2001-2002-2003 Aviator, Explorer, Mountaineer, Mustang, F Series Truck, Ranger equipped with a 2.5L VIN C, 3.0L (VIN U, VIN V), 4.0L (VIN E, VIN K, VIN X), 5.0L VIN P, 5.4L (VIN 3, VIN L, VIN M, VIN Z) 6.8L VIN S engines Transmissions: All	**Vehicle Speed Sensor Circuit Malfunction Conditions:** Engine started; then with the engine speed more than the TCC stall speed, the PCM detected a lack of vehicle speed data occurred. Note: The PCM receives vehicle speed data from the Rear Wheel ABS (RABS) or 4-Wheel ABS (4WABS) on these applications. **Possible Causes** • VSC positive signal circuit is open or shorted to ground • VSC negative signal circuit is open • RABS or 4WABS control unit is damaged or has failed • One of the other modules (CTM or GEM) may be the cause of this trouble code. Diagnose other codes from these modules.
P0501 **1T CCM, MIL: Yes** 1996-1997-1998-1999 All Models Transmissions: All	**Vehicle Speed Sensor or PSOM Range/Performance Conditions:** Engine started; engine speed above the TCC stall speed, and the PCM detected a loss of the VSS signal over a period of time. Note: The PCM receives vehicle speed data from the VSS, TCSS, ABS module, CTM or GEM controller, depending up the application. **Possible Causes** • VSS+ or VSS- signal circuit is open or shorted to ground • TCSS signal or TCSS signal return harness circuit is open • VSS harness circuit, TCSS harness circuit is shorted to ground • VSS harness circuit, CSS harness circuit is shorted to power • VSS circuit open between the PCM and related control module • VSS or TCSS, or wheel speed sensors circuits are damaged • Modules connected to VSC/VSS harness circuits are damaged • Mechanical drive mechanism for the VSS or TCSS is damaged
P0501 **1T CCM, MIL: Yes** 2000-2001-2002-2003 All Models Transmissions: All	**Vehicle Speed Sensor Range/Performance Conditions:** Engine started; then with the engine speed more than the TCC stall speed, the PCM detected a problem with the vehicle speed data. Note: The PCM receives vehicle speed data from the Vehicle Speed Sensor on these vehicle applications. **Possible Causes** • VSS signal circuit is open or shorted to ground • VSS ground circuit is open • VSS power circuit (VPWR) is open • VSS is damaged or it has failed • PCM has failed
P0500 **2T CCM, MIL: Yes** 2000-2001-2002-2003 Continental, Town Car, LS, Windstar Models Transmissions: All	**Vehicle Speed Sensor Signal Range/Performance Conditions:** Engine started; then with the engine speed more than the TCC stall speed, the PCM detected a problem with the vehicle speed data. Note: The PCM receives vehicle speed data from the ABS module. **Possible Causes** • The vehicle speed information on this vehicle application is provided to the PCM by the Antilock Brake System module. • Refer to the ABS diagnostics and trouble codes to diagnose this particular trouble code.
P0501 **2T CCM, MIL: Yes** 2000-2001-2002 Contour, Cougar & Mystique Models Transmissions: All	**Vehicle Speed Sensor Signal Range/Performance Conditions:** Engine started; then with the engine speed more than the TCC stall speed, the PCM detected a problem with the vehicle speed data. Note: The PCM receives vehicle speed data from the VSS sensor. **Possible Causes** • VSS+ signal circuit is open or shorted to ground • VSS- signal circuit is open • VSS power supply (VPWR) circuit is open • VSS is damaged or it has failed • PCM has failed

Gas Engine OBD II Trouble Code List (P0xxx Codes)

DTC	Trouble Code Title, Conditions & Possible Causes
P0501 **2T CCM, MIL: Yes** 2000-2001-2002-2003 F Series Trucks equipped a 4.6L VIN W, 5.4L VIN 3, 5.4L VIN L, 5.4L VIN M, 5.4L VIN Z engine Transmissions: All	**Vehicle Speed Sensor Signal Range/Performance Conditions:** Engine running, then with the engine speed more than the TCC stall speed, the PCM detected a problem with the vehicle speed data. Note: The PCM receives vehicle speed data from the Transfer Case Speed Sensor on these vehicle applications. **Possible Causes** • TCSS signal circuit is open or shorted to ground • TCSS ground circuit is open • TCSS is damaged or it has failed • PCM has failed
P0501 **2T CCM, MIL: Yes** 2000-2001-2002-2003 Aviator, Explorer, Mountaineer, Mustang, F Series Truck, Ranger equipped with a 2.5L VIN C, 3.0L (VIN U, VIN V), 4.0L (VIN E, VIN K, VIN X), 5.0L VIN P, 5.4L (VIN 3, VIN L, VIN M, VIN Z) 6.8L VIN S engines Transmissions: All	**Vehicle Speed Sensor Signal Range/Performance Conditions:** Engine running, then with the engine speed more than the TCC stall speed, the PCM detected a problem with the vehicle speed data. Note: The PCM receives vehicle speed data from the Rear Wheel ABS (RABS) or 4-Wheel ABS (4WABS) on these applications. **Possible Causes** • VSC positive signal circuit is open or shorted to ground • VSC negative signal circuit is open • RABS or 4WABS control unit is damaged or has failed • One of the other modules (CTM or GEM) may be the cause of this trouble code. Diagnose other codes from these modules.
P0503 **2T CCM, MIL: Yes** 1997-1998-1999-2000- 2001-2002-2003 All Models Transmissions: A/T	**Vehicle Speed Sensor Signal Intermittent Conditions:** Engine started, engine speed above the TCC stall speed, and the PCM detected the vehicle speed data was "noisy" or intermittent. Note: The PCM receives vehicle speed data from the VSS, TCSS, ABS module, CTM or GEM controller, depending up the application. **Possible Causes** • Module or circuits connected to VSS/TCSS circuit are damaged • VSS/TCSS wiring harness or connector is damaged or loose • VSS/TCSS signal is "noisy" due to RFI or EMI interference from sources such as ignition components or charging system • VSS/TCSS gears are damaged or there is debris on the sensor
P0503 **2T CCM, MIL: Yes** 1997-1998-1999-2000- 2001-2002-2003 Aviator, Contour, Mystique, Explorer, Mountaineer, Mustang, F Series Truck, Ranger equipped with a 2.0L (VIN 3, VIN Z, VIN B) 2.5L VIN L, 2.5L VIN G, 4.2L VIN 2, 4.6L VIN 6, 4.6L VIN W, 5.4L VIN L engine Transmissions: M/T	**Vehicle Speed Sensor Signal Intermittent Conditions:** Engine started, engine speed above the TCC stall speed, and the PCM detected the vehicle speed data was "noisy" or intermittent. Note: The PCM receives vehicle speed data from the VSS or TCSS. **Possible Causes** • TCSS or VSS signal circuit is open or shorted to ground • TCSS or VSS ground circuit is open (an intermittent problem) • TCSS or VSS power supply (VREF) circuit is open (intermittent) • TCSS or VSS is damaged or it has failed (intermittent problem) • PCM has failed

Gas Engine OBD II Trouble Code List (P0xxx Codes)

DTC	Trouble Code Title, Conditions & Possible Causes
P0505 **2T CCM, MIL: Yes** 1996-1997-1998-1999-2000-2001-2002-2003 All Models Transmissions: All	**Idle Air Control System Malfunction Conditions:** Engine started, engine running at hot idle speed, and the PCM detected the Actual Idle Speed was too low or too high when compared to the Target Idle Speed during the KOER self-test. Specification: The IAC valve resistance is 6-13 ohms at 68°F. **Possible Causes** • Air inlet dirty, restricted or the air cleaner is severely restricted • IAC solenoid control circuit is open, shorted to ground or to B+ • IAC solenoid power circuit (VPWR) is open from the relay • IAC valve is damaged or has failed • PCM has failed • TSB 03-3-5 contains repair information for this trouble code
P0506 **2T CCM, MIL: Yes** 2003 All Models Transmissions: All	**Idle Air Control System RPM Lower Than Expected Conditions:** DTC P0402 not set, engine started, engine running in closed loop, and the PCM detected it could not control the idle speed correctly. **Possible Causes** • Air inlet is plugged or the air filter element is severely clogged • IAC circuit is open or shorted to the VPWR circuit • IAC circuit VPWR circuit is open • IAC solenoid is damaged or has failed • PCM has failed • TSB 03-3-5 contains repair information for this trouble code
P0507 **2T CCM, MIL: Yes** 2003 All Models Transmissions: All	**Idle Air Control System RPM Higher Than Expected Conditions:** DTC P0402 not set, engine started, engine running in closed loop, and the PCM detected it could not control the idle speed correctly. **Possible Causes** • Air intake leak located somewhere after the throttle body • IAC control circuit is shorted to chassis ground • IAC solenoid is damaged or has failed • PCM has failed • TSB 03-3-5 contains repair information for this trouble code
P0511 **2T CCM, MIL: Yes** 2003 All Models Transmissions: All	**Idle Air Control Valve Circuit Malfunction Conditions:** DTC P0402 not set, engine started, engine running in closed loop, and the PCM detected it could not control the idle speed correctly. **Possible Causes** • IAC control circuit is open • IAC control circuit is shorted to power (B+) • IAC power supply circuit (VPWR) is open • IAC solenoid is damaged or the PCM has failed
P0528 **2T CCM, MIL: Yes** 2002-2003 Crown Victoria, Grand Marquis, Town Car, Mustang, Thunderbird, Escort, Expedition, LS, Navigator & ZX2 Models Transmissions: All	**Visctronic Drive Fan Speed Sensor Circuit Malfunction Conditions:** Engine started, Visctronic Drive Fan (VDF) commanded to a 100% duty cycle position, and the PCM detected the VDF Speed Sensor signal was less than a calibrated value in the test. **Possible Causes** • VDF fan motor has a mechanical interference fault or is binding • VDF speed sensor circuit is open or shorted to ground • Vehicle Buffered Power (VBPWR) circuit is open or shorted • VDF speed sensor power ground circuit is open • VDF speed sensor is damaged or the PCM has failed
P0534 **2T CCM, MIL: Yes** 2003 All Models Transmissions: All	**Low Air Conditioning Cycle Period Conditions:** Engine started; A/C enabled, and the PCM detected frequent A/C compressor clutch cycling during the CCM test period. Note that this trouble code and test was designed to protect the transmission. In some cases, the PCM will unlock TCC operation. **Possible Causes** • A/C cycling pressure switch signal to PCM open (intermittent) • A/C cycling pressure switch IGN (B+) circuit open (intermittent) • A/C mechanical problem (low A/C refrigerant charge or a damaged A/C cycling switch)

Gas Engine OBD II Trouble Code List (P0xxx Codes)

DTC	Trouble Code Title, Conditions & Possible Causes
P0537 **2T CCM, MIL: Yes** 2003 All Models Transmissions: All	**A/C Evaporator Temperature Circuit Sensor Low Input Conditions:** Engine started; A/C enabled, and the PCM detected an unexpected low voltage condition on the A/C Evaporator Temperature (ACET) sensor circuit during the CCM test period. **Possible Causes** • ACET sensor signal circuit shorted to sensor or chassis ground • ACET sensor is damaged or it has failed • PCM has failed
P0538 **2T CCM, MIL: Yes** 2003 All Models Transmissions: All	**A/C Evaporator Temperature Sensor Circuit High Input Conditions:** Engine started; A/C enabled, and the PCM detected an unexpected high voltage condition on the A/C Evaporator Temperature (ACET) sensor circuit during the CCM test period. **Possible Causes** • ACET sensor signal circuit is open • ACET sensor signal circuit is shorted to VREF (5v) • ACET sensor ground circuit is open • ACET sensor is damaged or it has failed • PCM has failed
P0552 **2T CCM, MIL: Yes** 1996-1997-1998-1999-2000-2001-2002-2003 All Models Transmissions: All	**Power Steering Pressure Sensor Circuit Low Input Conditions:** Engine started, and the PCM detected an unexpected low voltage condition on the Power Steering Pressure (PSP) sensor circuit. **Possible Causes** • PSP sensor signal circuit is shorted to sensor ground • PSP sensor signal circuit is shorted to chassis ground • PSP sensor VREF (5v) circuit is open • PSP sensor is damaged or it has failed • PCM has failed
P0553 **2T CCM, MIL: Yes** 1996-1997-1998-1999-2000-2001-2002-2003 All Models Transmissions: All	**Power Steering Pressure Sensor Circuit High Input Conditions:** Engine started, and the PCM detected an unexpected high voltage condition on the Power Steering Pressure (PSP) sensor circuit. **Possible Causes** • PSP sensor ground circuit is open • PSP sensor ground circuit is shorted to VREF (5v) • PSP sensor signal circuit is shorted to VREF (5v) • PSP sensor is damaged or the PCM has failed
P0597 **2T CCM, MIL: Yes** 2003 Ranger equipped with a 2.3L I4 VIN D engine Transmissions: All	**Thermostat Heater Control Circuit Malfunction Conditions:** Engine started, and the PCM detected an unexpected low or high voltage condition on the Thermostat Heater Control (THTRC) circuit. **Possible Causes** • THTRC circuit is open or shorted to ground • THTRC power (VPWR) circuit is open • Thermostat assembly is damaged or the PCM has failed
P0602 **1T PCM, MIL: Yes** 1998-1999-2000-2001-2002-2003 All Models Transmissions: All	**Control Module Programming Error Conditions:** Key on, and the PCM detected a programming error in the VID block. This fault requires that the VID Block be reprogrammed, or that the EEPROM be re-flashed. **Possible Causes** • During the VID reprogramming function, the Vehicle ID (VID) data block failed during reprogramming wit the Scan Tool.
P0603 **1T PCM, MIL: Yes** 1996-1997-1998-1999-2000-2001-2002-2003 All Models Transmissions: All	**PCM Keep Alive Memory Test Error Conditions:** Key on, and the PCM detected an internal memory fault. This code will set if KAPWR to the PCM is interrupted (at the initial key on). **Possible Causes** • Battery terminal corrosion, or loose battery connection • KAPWR to PCM interrupted, or the circuit has been opened • Reprogramming error has occurred • PCM has failed and needs replacement. Remember to check for Aftermarket Performance Products before replacing a PCM.

Gas Engine OBD II Trouble Code List (P0xxx Codes)

DTC	Trouble Code Title, Conditions & Possible Causes
P0605 **1T PCM, MIL: Yes** 1996-1997-1998-1999- 2000-2001-2002-2003 All Models Transmissions: All	**PCM Read Only Memory Test Error Conditions:** Key on, and the PCM detected a ROM test error (ROM inside PCM is corrupted). The PCM is normally replaced if this code has set. **Possible Causes** • An attempt was made to change the module calibration, or a Module programming error may have occurred • Clear the trouble codes and then check for this trouble code. If it resets, the PCM has failed and needs replacement. • Remember to check for signs of Aftermarket Performance Products installation before replacing the PCM.
P0606 **1T PCM, MIL: Yes** 2002-2003 All Models Transmissions: All	**PCM Internal Communication Error Conditions:** Key on, and the PCM detected an internal communications register read back error during the initial key on check period. **Possible Causes** • Clear the trouble codes and then check for this trouble code. If it resets, the PCM has failed and needs replacement. • Remember to check for signs of Aftermarket Performance Products installation before replacing the PCM.
P0622 **1T CCM, MIL: Yes** 2003 All Models Transmissions: All	**Generator Regulator System Malfunction Conditions:** Engine started; and the PCM detected an unexpected voltage condition on the Generator control circuit. **Possible Causes** • Generator belt is loose or worn out • Generator or regulator is damaged or has failed • PCM has failed
P0645 **1T CCM, MIL: Yes** 2003 All Models Transmissions: All	**Wide Open Throttle A/C Output Primary Circuit Malfunction Conditions:** Key on or engine running; and the PCM detected an unexpected low or high voltage condition WAC output primary circuit during the test. **Possible Causes** • WAC relay control circuit is open or shorted to ground • WAC relay power circuit (VPWR) is open • WAC relay is damaged or it has failed • PCM has failed
P0645 **1T CCM, MIL: Yes** 2003 Escort, Mustang, ZX2 equipped with a 2.0L VIN 3, 3.8L VIN 4, 4.6L VIN X, 4.6L VIN X, 4.6L VIN Y engine Transmissions: All	**Wide Open Throttle A/C Output Primary Circuit Malfunction Conditions:** Key on or engine running; and the PCM detected an unexpected low or high voltage condition on the WAC output circuit (in the CCRM). **Possible Causes** • WAC relay control circuit is open or shorted to ground • WAC relay power circuit (VPWR) is open • WAC relay is damaged or it has failed • PCM has failed
P0660 **1T CCM, MIL: Yes** 2002-2003 All Models Transmissions: All Escort, Focus, E Van & F Series Truck, ZX2 & Windstar Models equipped with a 2.0L VIN P, 3.8L VIN 4, 4.2L VIN 2 engine Transmissions: All	**Intake Manifold Runner Control Valve Circuit Malfunction Conditions:** Key on or engine running; and the PCM detected an unexpected low or high voltage condition on the Intake Manifold Runner Control (IMRC) signal circuit during the CCM test. **Possible Causes** • IMRC signal circuit is open • IMRC signal circuit is shorted to chassis ground • IMRC actuator assembly is damaged or failed • PCM has failed

Gas Engine OBD II Trouble Code List (P0xxx Codes)

DTC	Trouble Code Title, Conditions & Possible Causes
P0660 **1T CCM, MIL: Yes** 2002-2003 Blackwood, Expedition Navigator, LS, Taurus, Sable, Mustang Models equipped with a 3.0L VIN S, 4.6L VIN W, 5.4L VIN A, 5.4L VIN L, 5.4L VIN R engine Transmissions: All	**Intake Manifold Tuning Valve Circuit Malfunction Conditions:** Key on or engine running; and the PCM detected an unexpected low or high voltage condition on the Intake Manifold Tuning Valve (ITMV) signal circuit during the CCM test. **Possible Causes** • ITMV signal circuit is open • ITMV signal circuit is shorted to chassis ground • ITMV electric actuator assembly is damaged or failed • PCM has failed
P0660 **1T CCM, MIL: Yes** 2002-2003 Ranger Models equipped with a 2.3L VIN D engine Transmissions: All	**Intake Manifold Swirl Control Actuator Circuit Malfunction Conditions:** Key on or engine running; and the PCM detected an unexpected low or high voltage condition on the Intake Manifold Swirl Control (IMSC) signal circuit during the CCM test. **Possible Causes** • IMSC signal circuit is open • IMSC signal circuit is shorted to chassis ground • IMSC actuator assembly is damaged or failed • PCM has failed
P0660 **1T CCM, MIL: Yes** 2003 Aviator equipped with a 4.6L VIN H engine Transmissions: All	**Intake Manifold Communication Control Circuit Malfunction Conditions:** Key on or engine running; and the PCM detected an unexpected low or high voltage condition on the Intake Manifold Communication Control (IMCC) signal circuit in the test. **Possible Causes** • IMCC signal circuit is open • IMCC signal circuit is shorted to chassis ground • Long / Short actuator assembly is damaged or failed • PCM has failed
P0703 **2T CCM, MIL: Yes** 1996-1997-1998-1999- 2000-2001-2002-2003 Continental, Town Car & Windstar Models Transmissions: A/T	**Brake Switch Circuit Malfunction Conditions:** Engine started, and the PCM did not detect any change in the Brake Pedal Position switch status, or with the vehicle running at Cruise speed, followed by a short deceleration periods, the PCM did not detect any change in the Brake Pedal Position switch status. **Possible Causes** • BPP switch circuit is open • BPP switch is damaged or it is out of adjustment • BPP switch power circuit is open (test switch inline fuse) • Module(s) connected the BPP switch circuit have a problem (e.g., Rear Electronic Module on Windstar or LS6/LS8, or the Lighting Control Module (LCM) Continental and Town Car)
P0703 **2T CCM, MIL: Yes** 2000-2001-2002-2003 LS Models Transmissions: A/T	**Brake Switch Circuit Malfunction Conditions:** Engine started, and the PCM did not detect any change in the Brake Pedal Position switch status, or with the vehicle running at Cruise speed, followed by a short deceleration periods, the PCM did not detect any change in the Brake Pedal Position switch status. **Possible Causes** • BPA/ BPP switch circuit is open • BPA/ BPP switch is damaged or it is out of adjustment • BPA/ BPP switch power circuit is open (test switch inline fuse) • One or more of the Module(s) that connect to the BPA or the BPP switch circuits have a problem (e.g., Rear Electronic Module or the Vehicle Speed Control)
P0703 **2T CCM, MIL: Yes** 1996-1997-1998-1999- 2000-2001-2002-2003 Some Models Transmissions: A/T	**Brake Switch Circuit Malfunction Conditions:** Engine started, and the PCM did not detect any change in the Brake Pedal Position (BPP) switch status, or with the vehicle at Cruise speed, followed by one or more short deceleration periods, the PCM did not detect any change in the Brake Pedal Position switch status. **Possible Causes** • BPP switch circuit is open • BPP switch is damaged or it is out of adjustment • BPP switch power circuit is open (check the switch inline fuse)

Gas Engine OBD II Trouble Code List (P0xxx Codes)

DTC	Trouble Code Title, Conditions & Possible Causes
P0703 **2T CCM, MIL: Yes** 1996-1997-2002-2003 Thunderbird Models Transmissions: A/T	**Brake Switch Circuit Malfunction Conditions:** Engine started, and the PCM did not detect any change in the Brake Pedal Position switch status, or with the vehicle running at Cruise speed, followed by a short deceleration periods, the PCM did not detect any change in the Brake Pedal Position switch status. **Possible Causes** • BPP switch circuit open, stop lamp switch circuit open/shorted • BPP/SLW switch is damaged or it is out of adjustment • BPP/SLW switch power circuit is open (test switch inline fuse) • One or more of the Module(s) that connect to the BPA or Stop Lamp switch circuits have a problem (e.g., Generic Electronic Module, ABS or Shift Lock Actuator or Vehicle Speed Control)
P0704 **1T CCM, MIL: No** 1996-1997-1998-1999-2000-2001-2002-2003 Contour, Mystique, LS, Escort, Focus, E Van & F Series Truck, ZX2 & Windstar Models Transmissions: M/T	**Clutch Pedal Position Switch Circuit Malfunction Conditions:** Engine running in gear, followed by several gearshift changes, and the PCM did not detect any change in the clutch switch status. *Note: The CCP PID should change (5v to 0v) with clutch depressed.* **Possible Causes** • CPP switch signal circuit shorted to power • CPP switch ground (return) circuit is open • CPP switch is damaged or out of adjustment • PCM has failed
P0705 **2T CCM, MIL: Yes** 1996-1997-1998-1999-2000-2001-2002-2003 All Models Transmissions: A/T	**DTR Sensor / TR Sensor Circuit Malfunction Conditions:** Key on or engine running; and the PCM detected that one or more of the Digital Transmission Range (DTR) or Transmission Range sensor (TR) signals (TR4, TR3, TR2 and TR1) were invalid (e.g., two TR or DR sensor signals received at the same time). **Possible Causes** • DTR or TR sensor connector is damaged or shorted • DTR or TR sensor signal circuit is open or shorted to ground • DTR or TR sensor signal circuit is shorted to VREF (5v) • DTR or TR sensor damaged • PCM has failed
P0707 **2T CCM, MIL: Yes** 1996-1997-1998-1999-2000-2001-2002-2003 All Models Transmissions: A/T	**DTR Sensor / TR Sensor Circuit Low Input Conditions:** Key on or engine running; and the PCM detected the Digital Transmission Range (DTR) or Transmission Range sensor (TR) signal was less than the self-test minimum value in the test. **Possible Causes** • DTR or TR sensor connector is damaged or it is shorted • DTR or TR sensor signal circuit is shorted to sensor ground • DTR or TR sensor damaged • PCM has failed
P0708 **2T CCM, MIL: Yes** 1996-1997-1998-1999-2000-2001-2002-2003 All Models Transmissions: A/T	**DTR Sensor or TR Sensor Circuit High Input Conditions:** Key on or engine running; and the PCM detected the Digital Transmission Range (DTR) or Transmission Range sensor (TR) input was more than the self-test maximum range in the test. **Possible Causes** • DTR or TR sensor connector is damaged or open • DTR or TR sensor signal circuit is open • DTR or TR sensor is shorted to VREF (5v) • DTR or TR sensor is damaged or the PCM has failed
P0711 **2T CCM, MIL: No** 1996-1997-1998-1999-2000-2001-2002-2003 All Models Transmissions: A/T	**TFT Sensor Signal Range/Performance Conditions:** Engine started, KOER Self-Test enabled, engine running for over 10 minutes, and the PCM detected the Transmission Fluid Temperature (TFT) sensor value was not close its normal operating temperature. **Possible Causes** • ATF is low, contaminated, dirty or burnt • TFT sensor signal circuit has a high resistance condition • TFT sensor is out-of-calibration ("skewed") or it has failed • PCM has failed

Gas Engine OBD II Trouble Code List (P0xxx Codes)

DTC	Trouble Code Title, Conditions & Possible Causes
P0712 **2T CCM, MIL: No** 1996-1997-1998-1999-2000-2001-2002-2003 All Models Transmissions: A/T	**TFT Sensor Circuit Low Input Conditions:** Key on or engine running; and the PCM detected the Transmission Fluid Temperature (TFT) sensor was less than its minimum self-test range (Scan Tool reads below -40°F) in the test. **Possible Causes** • TFT sensor signal circuit is shorted to chassis ground • TFT sensor signal circuit is shorted to sensor ground • TFT sensor is damaged, or out-of-calibration, or has failed • PCM has failed
P0713 **2T CCM, MIL: No** 1996-1997-1998-1999-2000-2001-2002-2003 All Models Transmissions: A/T	**TFT Sensor Circuit High Input Conditions:** Key on or engine running; and the PCM detected the Transmission Fluid Temperature (TFT) sensor was more than its maximum self-test range (Scan Tool reads over 315°F) in the test. **Possible Causes** • TFT sensor signal circuit is open between the sensor and PCM • TFT sensor ground circuit is open between sensor and PCM • TFT sensor is damaged or has failed • PCM has failed
P0715 **2T CCM, MIL: No** 1996-1997-1998-1999-2000-2001-2002-2003 All Models Transmissions: A/T	**Transmission Speed Shaft Sensor Circuit Malfunction Conditions:** Engine started, vehicle driven with the vehicle speed sensor indicating more than 1 mph, and the PCM detected the TSS signals were erratic, or that they were missing for a period of time. **Possible Causes** • TSS signal circuit is open • TSS signal is shorted to chassis ground • TSS signal is shorted to sensor ground • TSS assembly is damaged or it has failed • PCM has failed
P0717 **2T CCM, MIL: No** 1996-1997-1998-1999-2000-2001-2002-2003 All Models Transmissions: A/T	**Transmission Speed Shaft Sensor Signal Intermittent Conditions:** Engine started, vehicle speed sensor indicating over 1 mph, and the PCM detected an intermittent loss of TSS signals (i.e., the TSS signals were erratic, irregular or missing). **Possible Causes** • TSS connector is damaged, loose or shorted • TSS signal circuit has an intermittent open condition • TSS signal circuit has an intermittent short to ground condition • TSS assembly is damaged or is has failed • PCM has failed
P0718 **2T CCM, MIL: No** 1996-1997-1998-1999-2000-2001-2002-2003 All Models Transmissions: A/T	**Transmission Speed Shaft Sensor Signal Noisy Conditions:** Engine started, vehicle speed sensor signal over 1 mph, and the PCM detected the "noise" interference on the TSS signal circuit. **Possible Causes** • TSS signal is "noisy" due to RFI or EMI interference from sources such as ignition components or charging system • TSS signal wiring is damaged or contacting other signal wiring • PCM has failed
P0718 **2T CCM, MIL: Yes** 1996-1997-1998-1999-2000-2001-2002-2003 All Models Transmissions: A/T	**A/T Output Shaft Speed Sensor Insufficient Input Conditions:** Engine started, VSS signal more than 1 mph, and the PCM detected the Output Shaft Speed signal did not correlate to the incoming signals received from the VSS or TCSS devices or related modules. **Possible Causes** • OSS sensor signal circuit is shorted to ground or • OSS sensor signal circuit is open • OSS sensor circuit is shorted to power • OSS sensor is damaged or it has failed • PCM has failed

Gas Engine OBD II Trouble Code List (P0xxx Codes)

DTC	Trouble Code Title, Conditions & Possible Causes
P0721 **2T CCM, MIL: No** 1997-1998-1999-2000-2001-2002-2003 All Models Transmissions: A/T	**A/T Output Shaft Speed Sensor Noise Interference Conditions:** Engine started, VSS signal more than 1 mph, and the PCM detected "noise" interference on the Output Shaft Speed (OSS) sensor circuit. **Possible Causes** • After market add-on devices interfering with the OSS signal • OSS connector is damaged, loose or shorted, or the wiring is misrouted or it is damaged • OSS assembly is damaged or it has failed • PCM has failed
P072 **2T CCM, MIL: No** 1999-2000-2001-2002-2003 All Models Transmissions: A/T	**A/T Output Speed Sensor No Signal Conditions:** Engine started, and the PCM did not detect any Output Shaft Speed (OSS) sensor signals upon initial vehicle movement. **Possible Causes** • After market add-on devices interfering with the OSS signal • OSS sensor wiring is misrouted or damaged, or the OSS sensor is damaged • PCM has failed
P0723 **2T CCM, MIL: No** 1999-2000-2001-2002-2003 All Models Transmissions: A/T	**A/T Output Speed Sensor Signal Intermittent Conditions:** Engine started, and the PCM detected the Output Shaft Speed (OSS) sensor signal was interrupted or irregular during testing. **Possible Causes** • OSS harness connector is damaged, loose or shorted, or the connector is not seated • OSS signal is open or it is shorted to ground (intermittent fault) • OSS assembly is damaged or it has failed
P0731 **2T CCM, MIL: No** 1996-1997-1998-1999-2000-2001-2002-2003 Some Models Transmissions: A/T	**Incorrect First Gear Ratio Conditions:** Engine started, vehicle operating with 1st Gear commanded "on", and the PCM detected an incorrect 1st gear ratio during the test. **Possible Causes** • 1st Gear solenoid harness connector not properly seated • 1st Gear solenoid signal shorted to ground, or open • 1st Gear solenoid wiring harness connector is damaged • 1st Gear solenoid is damaged or not properly installed
P0731 **2T CCM, MIL: No** 1996-1997-1998-1999-2000-2001-2002-2003 Escape, Contour, Probe Cougar, Mystique Transmissions: A/T	**Incorrect First Gear Ratio Conditions:** Engine started, vehicle operating with 1st Gear commanded "on", and the PCM detected an incorrect 1st gear ratio during the test. **Possible Causes** • 1st Gear solenoid harness connector not properly seated • 1st Gear solenoid signal shorted to ground, or open • 1st Gear solenoid wiring harness connector is damaged • 1st Gear solenoid is damaged or not properly installed • TSB 02-2-4 contains a repair procedure for this trouble code
P0732 **2T CCM, MIL: No** 1996-1997-1998-1999-2000-2001-2002-2003 Some Models Transmissions: A/T	**Incorrect Second Gear Ratio Conditions:** Engine started, vehicle operating with 2nd Gear commanded "on", and the PCM detected an incorrect 2nd gear ratio during the test. **Possible Causes** • 2nd Gear solenoid harness connector not properly seated • 2nd Gear solenoid signal shorted to ground, or open • 2nd Gear solenoid wring harness connector is damaged • 2nd Gear solenoid is damaged or not properly installed
P0732 **2T CCM, MIL: No** 1996-1997-1998-1999-2000-2001-2002-2003 Escape, Contour, Probe Cougar, Mystique Transmissions: A/T	**Incorrect Second Gear Ratio Conditions:** Engine started, vehicle operating with 2nd Gear commanded "on", and the PCM detected an incorrect 2nd gear ratio during the test. **Possible Causes** • 2nd Gear solenoid harness connector not properly seated • 2nd Gear solenoid signal shorted to ground, or open • 2nd Gear solenoid wring harness connector is damaged • 2nd Gear solenoid is damaged or not properly installed • TSB 02-2-4 contains a repair procedure for this trouble code

Gas Engine OBD II Trouble Code List (P0xxx Codes)

DTC	Trouble Code Title, Conditions & Possible Causes
P0733 **2T CCM, MIL: No** 1996-1997-1998-1999-2000-2001-2002-2003 All Models Transmissions: A/T	**Incorrect Third Gear Ratio Conditions:** Engine started, vehicle operating with 3rd Gear commanded "on", and the PCM detected an incorrect 3rd gear ratio during the test. **Possible Causes** • 3rd Gear solenoid harness connector not properly seated • 3rd Gear solenoid signal shorted to ground, or open • 3rd Gear solenoid wiring harness connector is damaged • 3rd Gear solenoid is damaged or not properly installed
P0734 **2T CCM, MIL: No** 1996-1997-1998-1999-2000-2001-2002-2003 Some Models Transmissions: A/T	**Incorrect Fourth Gear Ratio Conditions:** Engine started, vehicle operating with 4th Gear commanded "on", and the PCM detected an incorrect 4th gear ratio during the test. **Possible Causes** • 4th Gear solenoid harness connector not properly seated • 4th Gear solenoid signal shorted to ground, or open • 4th Gear solenoid wiring harness connector is damaged • 4th Gear solenoid is damaged or not properly installed
P0734 **2T CCM, MIL: No** 1996-1997-1998-1999-2000-2001-2002-2003 Escape, Contour, Probe Cougar, Mystique Transmissions: A/T	**Incorrect Fourth Gear Ratio Conditions:** Engine started, vehicle operating with 4th Gear commanded "on", and the PCM detected an incorrect 4th gear ratio during the test. **Possible Causes** • 4th Gear solenoid harness connector not properly seated • 4th Gear solenoid signal shorted to ground, or open • 4th Gear solenoid wiring harness connector is damaged • 4th Gear solenoid is damaged or not properly installed • TSB 02-2-4 contains a repair procedure for this trouble code
P0735 **2T CCM, MIL: No** 1996-1997-1998-1999-2000-2001-2002-2003 All Models Transmissions: A/T	**Incorrect Fifth Gear Ratio Conditions:** Engine started, vehicle operating with 5th Gear commanded "on", and the PCM detected an incorrect 5th gear ratio during the test. **Possible Causes** • 5th Gear solenoid harness connector not properly seated • 5th Gear solenoid signal shorted to ground, or open • 5th Gear solenoid wiring harness connector is damaged • 5th Gear solenoid is damaged or not properly installed
P0736 **2T CCM, MIL: No** 1996-1997-1998-1999-2000-2001-2002-2003 All Models Transmissions: A/T	**Incorrect Reverse Gear Ratio Conditions:** Engine started, vehicle operating with Reverse Gear commanded "on", and the PCM detected an incorrect reverse gear ratio occurred. **Possible Causes** • Reverse Gear solenoid harness connector not properly seated • Reverse Gear solenoid signal shorted to ground, or open • Reverse Gear solenoid wiring harness connector is damaged • Reverse Gear solenoid is damaged or not properly installed
P0740 **2T CCM, MIL: No** 2000-2001-2002-2003 All Models Transmissions: A/T	**TCC Solenoid Circuit Malfunction Conditions:** Engine started, KOER Self-Test enabled, vehicle driven at cruise speed, and the PCM did not detect any voltage drop across the TCC solenoid circuit during the test period. **Possible Causes** • TCC solenoid control circuit is open or shorted to ground • TCC solenoid wiring harness connector is damaged • TCC solenoid is damaged or has failed • PCM has failed
P0741 **2T CCM, MIL: No** 1996-1997-1998-1999-2000-2001-2002-2003 All Models Transmissions: A/T	**TCC Mechanical System Range/Performance Conditions:** Engine started, vehicle driven in gear with VSS signals received, and the PCM detected excessive slippage while in normal operation. **Possible Causes** • TCC solenoid has a mechanical failure • TCC solenoid has a hydraulic failure • PCM has failed

Gas Engine OBD II Trouble Code List (P0xxx Codes)

DTC	Trouble Code Title, Conditions & Possible Causes
P0741 **2T CCM, MIL: No** 1996-1997-1998-1999- 2000-2001-2002-2003 Continental, Sable, Taurus, Windstar Transmissions: A/T	**TCC Mechanical System Range/Performance Conditions:** Engine started, vehicle driven in gear with VSS signals received, and the PCM detected excessive slippage while in normal operation. **Possible Causes** • TCC solenoid has a mechanical failure • TCC solenoid has a hydraulic failure • PCM has failed • TSB 3-12-3 contains a repair procedure for this trouble code
P0741 **2T CCM, MIL: No** 1996-1997-1998-1999- 2000-2001-2002-2003 Escape, Contour, Probe Cougar, Mystique Transmissions: A/T	**TCC Mechanical System Range/Performance Conditions:** Engine started, vehicle driven in gear with VSS signals received, and the PCM detected excessive slippage while in normal operation. **Possible Causes** • TCC solenoid has a mechanical failure • TCC solenoid has a hydraulic failure • PCM has failed • TSB 02-2-4 contains a repair procedure for this trouble code
P0743 **2T CCM, MIL: Yes** 1996-1997-1998-1999- 2000-2001-2002-2003 All Models Transmissions: A/T	**TCC Solenoid Circuit Malfunction Conditions:** Key on, KOEO Self-Test enabled and the PCM did not detect any voltage drop across the TCC solenoid circuit during the test period. **Possible Causes** • TCC solenoid control circuit is open • TCC solenoid control circuit is shorted to ground • TCC solenoid wiring harness connector is damaged • TCC solenoid is damaged or it has failed • PCM has failed
P0746 **2T CCM, MIL: No** 1996-1997-1998-1999- 2000-2001-2002-2003 All Models Transmissions: A/T	**A/T EPC Solenoid Circuit Malfunction Conditions:** Key on, KOEO Self-Test enabled and the PCM did not detect any voltage drop across the EPC solenoid circuit during the test period. **Possible Causes** • EPC solenoid control circuit is open • EPC solenoid control circuit is shorted to ground • EPC solenoid wiring harness connector is damaged • EPC solenoid is damaged or it has failed • PCM has failed
P0750 **2T CCM, MIL: Yes** 1996-1997-1998-1999- 2000-2001-2002-2003 All Models Transmissions: A/T	**A/T Shift Solenoid 1/A Circuit Malfunction Conditions:** Engine started, vehicle driven with the solenoid applied, and the PCM detected an unexpected voltage condition on the SS1/A solenoid circuit was incorrect during the test. **Possible Causes** • SS1/A solenoid control circuit is open • SS1/A solenoid control circuit is shorted to ground • SS1/A solenoid wiring harness connector is damaged • SS1/A solenoid is damaged or has failed • PCM has failed
P0751 **2T CCM, MIL: No** 1996-1997-1998-1999- 2000-2001-2002-2003 All Models Transmissions: A/T	**A/T Shift Solenoid 1/A Function Range/Performance Conditions:** Engine started, vehicle driven with the solenoid applied, and the PCM detected a mechanical failure while operating the Shift Solenoid 1/A during the CCM test period. **Possible Causes** • SS1/A solenoid is stuck in the "off" position • SS1/A solenoid has a mechanical failure • SS1/A solenoid has a hydraulic failure • PCM has failed

Gas Engine OBD II Trouble Code List (P0xxx Codes)

DTC	Trouble Code Title, Conditions & Possible Causes
P0752 **1T CCM, MIL: No** 2000-2001-2002-2003 All Models Transmissions: A/T	**A/T Shift Solenoid 1/A Function Range/Performance Conditions:** Engine started, vehicle driven with the solenoid applied, and the PCM detected a mechanical failure while operating the Shift Solenoid 1/A during the CCM test period. **Possible Causes** • SS1/A solenoid is stuck in the "on" position • SS1/A solenoid has a mechanical failure • SS1/A solenoid has a hydraulic failure • PCM has failed
P0753 **1T CCM, MIL: Yes** 1996-1997-1998-1999- 2000-2001-2002-2003 All Models Transmissions: A/T	**A/T Shift Solenoid 1/A Circuit Malfunction Conditions:** Engine started, vehicle driven with the solenoid applied, and the PCM detected an unexpected voltage condition on the SS1/A solenoid circuit was incorrect during the test. **Possible Causes** • SS1/A solenoid control circuit is open • SS1/A solenoid control circuit is shorted to ground • SS1/A solenoid wiring harness connector is damaged • SS1/A solenoid is damaged or has failed • PCM has failed
P0755 **1T CCM, MIL: Yes** 1996-1997-1998-1999- 2000-2001-2002-2003 All Models Transmissions: A/T	**A/T Shift Solenoid 2/B Circuit Malfunction Conditions:** Engine started, vehicle driven with the solenoid applied, and the PCM detected an unexpected voltage condition on the SS2/B solenoid circuit was incorrect during the test. **Possible Causes** • SS2/B solenoid control circuit is open • SS2/B solenoid control circuit is shorted to ground • SS2/B solenoid wiring harness connector is damaged • SS2/B solenoid is damaged or has failed • PCM has failed
P0756 **1T CCM, MIL: Yes** 1996-1997-1998-1999- 2000-2001-2002-2003 All Models Transmissions: A/T	**A/T Shift Solenoid 2/B Function Range/Performance Conditions:** Engine started, vehicle driven with the solenoid applied, and the PCM detected a mechanical failure while operating the Shift Solenoid 2/B during the CCM test period. **Possible Causes** • SS2/B solenoid is stuck in the "on" position • SS2/B solenoid has a mechanical failure • SS2/B solenoid has a hydraulic failure • PCM has failed
P0757 **1T CCM, MIL: Yes** 1996-1997-1998-1999- 2000-2001-2002-2003 All Models Transmissions: A/T	**A/T Shift Solenoid 2/B Function Range/Performance Conditions:** Engine started, vehicle driven with the solenoid applied, and the PCM detected a mechanical failure while operating the Shift Solenoid 2/B during the CCM test period. **Possible Causes** • SS2/B solenoid is stuck in the "on" position • SS2/B solenoid has a mechanical failure • SS2/B solenoid has a hydraulic failure • PCM has failed
P0758 **1T CCM, MIL: Yes** 2000-2001-2002-2003 All Models Transmissions: A/T	**A/T Shift Solenoid 2/B Circuit Malfunction Conditions:** Key on, KOEO Self-Test enabled, Shift Solenoid 2/B applied, and the PCM detected an unexpected voltage condition on the Shift Solenoid 2/B circuit during the CCM test period. **Possible Causes** • Shift Solenoid 2/B connector is damaged, open or shorted • Shift Solenoid 2/B control circuit is open • Shift Solenoid 2/B control circuit is shorted to ground • Shift Solenoid 2/B is damaged or it has failed • PCM has failed

Gas Engine OBD II Trouble Code List (P0xxx Codes)

DTC	Trouble Code Title, Conditions & Possible Causes
P0760 **1T CCM, MIL: Yes** 1996-1997-1998-1999-2000-2001-2002-2003 All Models Transmissions: A/T	**A/T Shift Solenoid 3/C Circuit Malfunction Conditions:** Engine started, vehicle driven with Shift Solenoid 3/C applied, and the PCM detected an unexpected voltage condition on the Shift Solenoid 3/C circuit during the CCM test period. **Possible Causes** • Shift Solenoid 3/C connector is damaged, open or shorted • Shift Solenoid 3/C control circuit is open • Shift Solenoid 3/C control circuit is shorted to ground • Shift Solenoid 3/C is damaged or it has failed • PCM has failed
P0761 **1T CCM, MIL: No** 1996-1997-1998-1999-2000-2001-2002-2003 All Models Transmissions: A/T	**A/T Shift Solenoid 3/C Function Range/Performance Conditions:** Engine started, vehicle driven with Shift Solenoid 3/C applied, and the PCM detected a mechanical failure occurred (stuck "off") while operating Shift Solenoid 3/C during the test. **Possible Causes** • SS3/C solenoid may be stuck "off" • SS3/C solenoid has a mechanical failure • SS3/C solenoid has a hydraulic failure • PCM has failed
P0762 **1T CCM, MIL: No** 1996-1997-1998-1999-2000-2001-2002-2003 All Models Transmissions: A/T	**A/T Shift Solenoid 3/C Function Range/Performance Conditions:** Engine started, vehicle driven with Shift Solenoid 3/C applied, and the PCM detected a mechanical failure occurred (stuck "on") while operating Shift Solenoid 3/C during the test. **Possible Causes** • SS3/C solenoid may be stuck "on" • SS3/C solenoid has a mechanical failure • SS3/C solenoid has a hydraulic failure • PCM has failed
P0765 **1T CCM, MIL: Yes** 1996-1997-1998-1999-2000-2001-2002-2003 All Models Transmissions: A/T	**A/T Shift Solenoid 4/D Circuit Malfunction Conditions:** Engine started, vehicle driven with Shift Solenoid 4/D applied, and the PCM detected an unexpected voltage condition on Shift Solenoid 4/D circuit during the CCM continuous test. **Possible Causes** • Shift Solenoid 4/D wiring harness or connector is damaged • Shift Solenoid 4/D control circuit is open or shorted to ground • Shift Solenoid 4/D is damaged or it has failed • PCM has failed
P0781 **1T CCM, MIL: No** 1996-1997-1998-1999-2000-2001-2002-2003 All Models Transmissions: A/T	**A/T 1 to 2 Shift Error Conditions:** Engine started, vehicle driven in gear with VSS signals received, and the PCM detected the engine speed (rpm) did not decrease properly (i.e., an incorrect 1-2 gear ratio was detected during a shift event). **Possible Causes** • SS1/A solenoid may be stuck • SS1/A solenoid has a hydraulic problem • SS2/B solenoid may be stuck • SS2/B has a hydraulic problem • Transmission may have damaged friction material • Transmission has internal damage and needs replacement
P0782 **1T CCM, MIL: No** 1996-1997-1998-1999-2000-2001-2002-2003 All Models Transmissions: A/T	**A/T 2 to 3 Shift Error Conditions:** Engine started, vehicle driven in gear with VSS signals received, and the PCM detected the engine speed (rpm) did not decrease properly (i.e., an incorrect 2-3 gear ratio was detected during a shift event). **Possible Causes** • SS1/A solenoid may be stuck • SS1/A solenoid has a hydraulic problem • SS2/B solenoid may be stuck • SS2/B has a hydraulic problem • Transmission may have damaged friction material • Transmission has internal damage and needs replacement

Gas Engine OBD II Trouble Code List (P0xxx Codes)

DTC	Trouble Code Title, Conditions & Possible Causes
P0783 **1T CCM, MIL: No** 1996-1997-1998-1999-2000-2001-2002-2003 All Models Transmissions: A/T	**A/T 3 to 4 Shift Error Conditions:** Engine started, vehicle driven in gear with VSS signals received, and the PCM detected the engine speed (rpm) did not change properly (i.e., an incorrect 3-4 gear ratio was detected during the shift event). **Possible Causes** • SS1/A solenoid may be stuck, or a hydraulic failure exists • SS2/B solenoid may be stuck, or a hydraulic failure exists • Transmission may have damaged friction material
P0784 **1T CCM, MIL: No** 1996-1997-1998-1999-2000-2001-2002-2003 All Models Transmissions: A/T	**A/T 4 to 5 Shift Error Conditions:** Engine started, vehicle driven in gear with VSS signals received, and the PCM detected the engine speed (rpm) did not change properly (i.e., an incorrect 4-5 gear ratio was detected during a shift event). **Possible Causes** • SS2/B solenoid may be stuck, or a hydraulic failure exists • SS3/C solenoid may be stuck, or a hydraulic failure exists • Transmission may have damaged friction material
P0812 **1T CCM, MIL: No** 1999-2000-2001-2002-2003 All Models Transmissions: A/T	**A/T Reverse Switch Circuit Malfunction Conditions:** Key on, engine off, KOEO Self Test enabled, and the PCM detected the reverse switch signal did not change as the selector was shifted in or out of reverse gear. Note: The RS PID should change from ON to OFF while shifting. **Possible Causes** • Transmission shift not indicating neutral during the self-test • RS switch circuit shorted to VREF or VPWR • RS switch circuit is open or shorted to ground (signal return) • Reverse switch is damaged • PCM has failed
P0813 **1T CCM, MIL: Yes** 1999-2000-2001-2002-2003 All Models Transmissions: A/T	**Transmission Control System Malfunction Conditions:** Engine started, vehicle speed more than 1 in gear, and the PCM detected a problem in the Transmission Control System operation. **Possible Causes** • Refer to the information in the Transmission Section of the appropriate Workshop Repair manual (i.e., the information for the particular vehicle that set this trouble code).
P0815 **1T CCM, MIL: Yes** 1999-2000-2001-2002-2003 All Models Transmissions: A/T	**Transmission Control System Malfunction Conditions:** Key on, engine off, KOEO Self Test enabled, and the PCM detected the reverse switch input did not change as the selector was shifted in or out of reverse (i.e., it was high when it should have been low). Note: The RS PID should change from ON to OFF while shifting. **Possible Causes** • Refer to the information in the Transmission Section of the appropriate Workshop Repair manual (i.e., the information for the particular vehicle that set this trouble code).

Gas Engine OBD II Trouble Code List (P1xxx Codes)

DTC	Trouble Code Title, Conditions & Possible Causes
P1000 **1T PCM, MIL: No** 1996-1997-1998-1999- 2000-2001-2002-2003 All Models Transmissions: All	**OBD II Monitor Testing Not Complete Conditions:** Key on or engine running; and the PCM detected one the conditions shown under Possible Causes (i.e., this code cannot be cleared manually - it must clear itself after all of the OBD II Monitors complete). Note: This code must be cleared to pass an Inspection/Maintenance Test required to register a vehicle in certain states. **Possible Causes** • Battery keep alive power (KAPWR) was removed to the PCM • One or more OBD II Monitors did not complete during an official OBD II Drive Cycle • PCM Reset step was performed with an OBD II Scan Tool
P1001 **1T CCM, MIL: No** 1996-1997-1998-1999- 2000-2001-2002-2003 All Models Transmissions: All	**KOER Self-Test Not Completed, KOER Test Aborted Conditions:** Key on, engine running self-test not completed during the normal allowable time period. **Possible Causes** • Engine speed (rpm) out of specification during the KOER test • Incorrect Self-Test Procedure • Scan Tool has a communication problem • Unexpected response from Self-Test monitors
P1100 **2T CCM, MIL: Yes** 1996-1997-1998-1999- 2000-2001-2002-2003 All Models Transmissions: All	**MAF Sensor Signal Intermittent Conditions:** Engine started, engine running at idle or cruise speed, and the PCM detected the MAF sensor signal above or below the calibrated limit. **Possible Causes** • MAF sensor continuity problems at the connector • MAF sensor continuity through the wiring harness • MAF sensor circuit intermittent open inside the sensor • PCM has failed
P1101 **2T CCM, MIL: Yes** 1996-1997-1998-1999- 2000-2001-2002-2003 All Models Transmissions: All	**MAF Sensor Out Of Self-Test Range Conditions:** Key on and engine off, and the PCM detected the MAF sensor was more than 0.27v, or with the engine running, the MAF sensor voltage was not within a normal range of 0.46v to 2.44v. **Possible Causes** • Low battery charge • MAF sensor partially connected, or the sensor is contaminated • MAF sensor power ground circuit or sensor signal (return) open • MAF sensor is damaged or it has failed • PCM has failed
P1112 **2T CCM, MIL: Yes** 1996-1997-1998-1999- 2000-2001-2002-2003 All Models Transmissions: All	**IAT Sensor Circuit Intermittent Conditions:** Engine started, and the PCM detected an intermittent condition in the IAT sensor signal during the self-test. Note: Select the IAT PID and monitor the signal for sudden changes. **Possible Causes** • IAT sensor wiring harness is damaged (wire may be open) • IAT sensor harness connector is damaged • IAT sensor is damaged or the PCM has failed
P1114 **2T CCM, MIL: Yes** 1996-1997-1998-1999- 2000-2001-2002-2003 All Models Transmissions: All	**IAT Sensor Circuit Low Input Conditions:** Engine started, and the PCM detected the IAT sensor signal was less than the self-test minimum of 0.20v (equivalent to 250ºF). Monitor the IAT PID for very low signal. **Possible Causes** • IAT sensor wiring harness is damaged (wire may be grounded) • IAT sensor harness connector is damaged (may be grounded) • IAT sensor is damaged or the PCM has failed
P1115 **2T CCM, MIL: Yes** 1999-2000-2001-2002- 2003 All Models Transmissions: All	**IAT Sensor 2 Circuit High Input Conditions:** Engine started, and the PCM detected the IAT Sensor 2 signal was more than the self-test maximum of 4.60v (equivalent to 250ºF). Monitor the IAT PID for very high signal. **Possible Causes** • IAT sensor wiring harness or harness connector is damaged (wire may be open) • IAT sensor signal circuit is open, or the ground circuit is open • IAT sensor is damaged or has failed • PCM has failed

Gas Engine OBD II Trouble Code List (P1xxx Codes)

DTC	Trouble Code Title, Conditions & Possible Causes
P1116 **1T CCM, MIL: Yes** 1996-1997-1998-1999-2000-2001-2002-2003 All Models Transmissions: All	**CHT or ECT Sensor Out Of Self-Test Range Conditions:** Key on, KOEO Self-Test enabled, and the PCM detected the ECT sensor was more than the expected range (50ºF), or engine running, KOER Self-Test enabled, and the PCM detected the ECT senor signal was less than 180ºF during the self test period. The ECT PID must be above 50ºF in the KOEO test or above 180ºF in the KOER self-test to pass these parameters. **Possible Causes** • ECT sensor harness connector is damaged, loose or shorted • ECT sensor is damaged • KOER or KOER Self-Test performed with the engine "too cold"
P1117 **2T CCM, MIL: Yes** 1996-1997-1998-1999-2000-2001-2002-2003 All Models Transmissions: All	**CHT or ECT Sensor Signal Intermittent Conditions:** Engine started, and the PCM detected an intermittent loss of the CHT or ECT sensor signal (it may have an open circuit condition). Note: Select the CHT or IAT PID and monitor the signal for sudden changes while wiggling the CHT or IAT sensor connector. On the 5.4L V8, if the temperature exceeds 258ºF, the PCM disables four fuel injectors at a time. It alternates which four fuel injectors are disabled every 32-engine cycles. The cylinders that are disabled do not inject fuel, so they act as air pumps to aid in cooling the engine. If the temperature exceeds 310ºF, the PCM disables all of the fuel injectors until the engine temperature drops below 310ºF. **Possible Causes** • ECT sensor harness connector is damaged, loose or shorted • ECT sensor is damaged or it has failed • Engine overheating condition present • Thermostat is faulty, or engine coolant level is low
P1120 **2T CCM, MIL: Yes** 1996-1997-1998-1999-2000-2001-2002-2003 All Models Transmissions: All	**TP Sensor Signal Out-of-Range Low Conditions:** Key on or engine running; and the PCM detected the TP sensor signal was between 0.17-0.49v (3.42-9.85%) with the signal within the calibrated self-test range. **Possible Causes** • ECT sensor harness connector is damaged • ECT sensor is damaged • Engine coolant level is low • PCM has failed
P1121 **2T CCM, MIL: Yes** 1996-1997-1998-1999-2000-2001-2002-2003 All Models Transmissions: All	**TP Sensor Inconsistent With MAF Sensor Conditions:** Engine started; and the PCM detected the MAF and TP sensor signals were not consistent the calibrated values expected for these two sensors during the self-test. Note: Drive the vehicle and monitor the TP PID in all gears. A TP PID of less than 0.24v (4.82%) with a LOAD PID over 55%, or a TP PID over 2.44v (49.05%) with a LOAD PID under 30% will set this code. **Possible Causes** • Air leak exists between MAF sensor and the throttle body • MAF sensor is damaged or it has failed • TP sensor is not seated properly • TP sensor is damaged
P1124 **1T CCM, MIL: Yes** 1996-1997-1998-1999-2000-2001-2002-2003 All Models Transmissions: All	**TP Sensor Out of Self-Test Range Conditions:** Key on, KOEO Self-Test enabled, and the PCM detected the TP sensor signal was less than 0.66v (13.27%), or with the engine running, KOER Self-Test enabled, the PCM detected the TP sensor signal was approximately 1.17v (23.52%). Note: A TP V PID less than 4.82 % (0.24 volt) with a LOAD PID more than 55%; or the TP V PID more than 49.05% (2.44 volts) with a LOAD PID less than 30% indicates a hard fault is present. **Possible Causes** • Throttle linkage is binding, or TP sensor is not seated properly • Throttle plate below closed throttle position • Throttle plate screw is misadjusted • TP sensor is damaged or it has failed • PCM has failed

Gas Engine OBD II Trouble Code List (P1xxx Codes)

DTC	Trouble Code Title, Conditions & Possible Causes
P1125 **2T CCM, MIL: Yes** 1996-1997-1998-1999- 2000-2001-2002-2003 All Models Transmissions: All	**TP Sensor Circuit Malfunction (Intermittent) Conditions:** Engine started, and the PCM detected the TP sensor rotational angle changed beyond the minimum or maximum calibrated limit. Note: Monitor the TP V PID, and tap lightly on the TP sensor housing and wiggle the wiring harness. Watch for the value to suddenly go below 0.49v or over 4.65v. **Possible Causes** • TP sensor wiring harness or connector has an intermittent open • TP sensor has an intermittent open or shorted condition
P1127 **2T CCM, MIL: Yes** 1996-1997-1998-1999- 2000-2001-2002-2003 All Models Transmissions: All	**Exhaust Not Warm, Downstream Sensor Not Tested Conditions:** Engine started, KOER Self-Test enabled, and the PCM detected the inferred exhaust temperature was less than a minimum value. Note: Monitor the HO2S Heater PID to determine their ON/OFF status (the heaters must work properly in order to pass this test). **Possible Causes** • Engine not operating long enough prior to the KOER Self-Test • Exhaust system temperature too cold to run the self-test
P1128 **2T CCM, MIL: Yes** 1996-1997-1998-1999- 2000-2001-2002-2003 All Models Transmissions: All	**Upstream Oxygen Sensors Swapped From Bank-to-Bank Conditions:** Engine started, KOER Self-Test enabled, and the PCM detected the HO2S signal response to a related fuel shift did not correspond to the correct engine cylinder bank (e.g., the HO2S-11 and the HO2S-21 wires were crossed) during the test period. **Possible Causes** • Upstream HO2S-11, HO2S-21 wiring crossed at the connector • Upstream HO2S-11, HO2S-21 crossed in the wiring harness • Upstream HO2S-11, HO2S-21 crossed at PCM pin connector
P1129 **2T CCM, MIL: Yes** 1996-1997-1998-1999- 2000-2001-2002-2003 All Models Transmissions: All	**Downstream Oxygen Sensors Swapped From Bank-to-Bank Conditions:** Engine started, KOER Self-Test enabled, and the PCM detected the HO2S signal response to a related fuel shift did not correspond to the correct engine cylinder bank (e.g., the HO2S-12 and HO2S-22 wires were crossed) during the test period. **Possible Causes** • Upstream HO2S-12, HO2S-21 wiring crossed at the connector • Upstream HO2S-12, HO2S-21 crossed in the wiring harness • Upstream HO2S-12, HO2S-21 crossed at PCM pin connector
P1130 **2T O2S, MIL: Yes** 1996-1997-1998-1999- 2000-2001-2002-2003 All Models Transmissions: All	**Lack of HO2S-11 Switching, Fuel Trim at Rich/Lean Limit Conditions:** DTC P0300-P0310 not set, engine running in closed loop, and the PCM detected the HO2S circuit was too lean or too rich, or that it could no longer change Fuel Trim because it was at its rich limit or its lean limit. **Possible Causes** • Air intake system leaking, vacuum hoses leaking or damaged • Air leaks located after the MAF sensor mounting location • EGR valve sticking, EGR diaphragm leaking, or gasket leaking • EVAP vapor recovery system has failed • Excessive fuel pressure, leaking or contaminated fuel injectors • Exhaust leaks before or near the HO2S(s) mounting location • Fuel pressure regulator is leaking or damaged • HO2S circuits wet or oily, corroded, or poor terminal contact • HO2S is damaged or it has failed • HO2S signal circuit open, shorted to ground, shorted to power • Low fuel pressure or vehicle driven until it was out of fuel • Oil dipstick not seated or engine oil level too high (overfilled)

Gas Engine OBD II Trouble Code List (P1xxx Codes)

DTC	Trouble Code Title, Conditions & Possible Causes
P1131 **2T O2S, MIL: Yes** 1996-1997-1998-1999-2000-2001-2002-2003 All Models Transmissions: All	**Lack of HO2S-11 Switching, HO2S Signal Low Input Conditions:** DTC P0300-P0310 not set, engine started, engine running in closed loop, and the PCM detected the HO2S-11 was not switching (i.e., the HO2S-11 indicated a lean A/F mixture). **Possible Causes** • Air intake system leaking, vacuum hoses leaking or damaged • Air leaks located after the MAF sensor mounting location • Base engine mechanical fault (i.e., compression, valve timing) • HO2S circuits wet or oily, corroded, or poor terminal contact • HO2S signal circuit open, shorted to ground, shorted to power, or the sensor has failed • Low fuel pressure or vehicle driven until it was out of fuel • Possible air leaks at the PCV valve or at the related hoses
P1132 **2T O2S, MIL: Yes** 1996-1997-1998-1999-2000-2001-2002-2003 All Models Transmissions: All	**Lack of HO2S-11 Switching, HO2S Signal High Input Conditions:** DTC P0300-P0310 not set, engine started, engine running in closed loop, and the PCM detected the HO2S-11 was not switching (i.e., the HO2S-11 indicated a rich A/F mixture). **Possible Causes** • Check air cleaner element and air cleaner housing for blockage • EVAP vapor recovery system has failed (canister full of fuel) • Fuel pressure too high, contaminated or leaking fuel injectors • HO2S is fuel contaminated, or coated with silicone or moisture
P1137 **2T O2S, MIL: Yes** 1996-1997-1998-1999-2000-2001-2002-2003 All Models Transmissions: All	**Lack of HO2S-12 Switching, HO2S Signal Low Input Conditions:** DTC P0300-P0310 not set, engine started, engine running in closed loop, and the PCM detected the HO2S-12 was not switching (i.e., the HO2S-12 indicated a lean A/F mixture). **Possible Causes** • Air intake system leaking, vacuum hoses leaking or damaged • Air leaks located after the MAF sensor mounting location • Base engine mechanical fault (i.e., compression, valve timing) • HO2S circuits wet or oily, corroded, or poor terminal contact • HO2S is damaged or it has failed • HO2S signal circuit open, shorted to ground, shorted to power • Low fuel pressure or vehicle driven until it was out of fuel • Possible air leaks at the PCV valve or at the related hoses
P1138 **2T O2S, MIL: Yes** 1996-1997-1998-1999-2000-2001-2002-2003 All Models Transmissions: All	**Lack of HO2S-12 Switching, HO2S Signal High Input Conditions:** DTC P0300-P0310 not set, engine started, engine running in closed loop, and the PCM detected the HO2S-12 was not switching (i.e., the HO2S-12 indicated a rich A/F mixture). **Possible Causes** • Check air cleaner element and air cleaner housing for blockage • EVAP vapor recovery system has failed (canister full of fuel) • Fuel pressure too high, contaminated or leaking fuel injectors • HO2S is fuel contaminated, or coated with silicone or moisture
P1150 **2T O2S, MIL: Yes** 1996-1997-1998-1999-2000-2001-2002-2003 All Models Transmissions: All	**Lack of HO2S-21 Switching, Fuel Trim At Rich/Lean Limit Conditions:** DTC P0300-P0310 not set, engine running in closed loop, and the PCM detected the HO2S circuit was too lean or too rich, or that it could no longer correct Fuel Trim (i.e., the Fuel Trim was at its calibrated rich limit or its calibrated lean limit). **Possible Causes** • Air intake system leaking, vacuum hoses leaking or damaged • Air leaks located after the MAF sensor mounting location • EGR valve sticking, EGR diaphragm leaking, or gasket leaking • EVAP vapor recovery system has failed • Excessive fuel pressure, leaking or contaminated fuel injectors • Exhaust leaks before or near the HO2S(s) mounting location • Fuel pressure regulator is leaking or damaged • HO2S circuits wet or oily, corroded, or poor terminal contact • HO2S signal circuit open, shorted to ground, shorted to power, or the sensor has failed • Low fuel pressure or vehicle driven until it was out of fuel • Oil dipstick not seated or engine oil level too high (overfilled)

Gas Engine OBD II Trouble Code List (P1xxx Codes)

DTC	Trouble Code Title, Conditions & Possible Causes
P1151 **2T O2S, MIL: Yes** 1996-1997-1998-1999-2000-2001-2002-2003 All Models Transmissions: All	**Lack of HO2S-21 Switching, HO2S Signal Low Input Conditions:** DTC P0300-P0310 not set, engine started, engine running in closed loop, and the PCM detected the HO2S-21 was not switching (i.e., the HO2S-21 indicated a lean A/F mixture). **Possible Causes** • Air intake system leaking, vacuum hoses leaking or damaged • Air leaks located after the MAF sensor mounting location or in the PCV system • Base engine mechanical fault (i.e., compression, valve timing) • HO2S circuits wet or oily, corroded, or poor terminal contact • HO2S signal circuit open, shorted to ground, shorted to power, or the sensor has failed • Low fuel pressure or vehicle driven until it was out of fuel
P1152 **2T O2S, MIL: Yes** 1996-1997-1998-1999-2000-2001-2002-2003 All Models Transmissions: All	**Lack of HO2S-21 Switching, HO2S Signal High Input Conditions:** DTC P0300-P0310 not set, engine started, engine running in closed loop, and the PCM detected the HO2S-21 was not switching (i.e., the HO2S-21 indicated a rich A/F mixture). **Possible Causes** • Check air cleaner element and air cleaner housing for blockage • EVAP vapor recovery system has failed (canister full of fuel) • Fuel pressure too high, contaminated or leaking fuel injectors • HO2S is fuel contaminated, or coated with silicone or moisture
P1157 **2T O2S, MIL: Yes** 1996-1997-1998-1999-2000-2001-2002-2003 All Models Transmissions: All	**Lack of HO2S-22 Switching, HO2S Signal Low Input Conditions:** DTC P0300-P0310 not set, engine started, engine running in closed loop, and the PCM detected the HO2S-22 was not switching (i.e., the HO2S-22 indicated a lean A/F mixture). **Possible Causes** • Air intake system leaking, vacuum hoses leaking or damaged • Air leaks located after the MAF sensor mounting location • Base engine mechanical fault (i.e., compression, valve timing) • HO2S circuits wet or oily, corroded, or poor terminal contact • HO2S is damaged or it has failed • HO2S signal circuit open, shorted to ground, shorted to power • Low fuel pressure or vehicle driven until it was out of fuel • Possible air leaks at the PCV valve or at the related hoses
P1158 **2T O2S, MIL: Yes** 1996-1997-1998-1999-2000-2001-2002-2003 All Models Transmissions: All	**Lack of HO2S-22 Switching, HO2S Signal High Input Conditions:** DTC P0300-P0310 not set, engine started, engine running in closed loop, and the PCM detected the HO2S-22 was not switching (i.e., the HO2S-22 indicated a rich A/F mixture). **Possible Causes** • Check air cleaner element and air cleaner housing for blockage • EVAP vapor recovery system has failed (canister full of fuel) • Fuel pressure too high, contaminated or leaking fuel injectors • HO2S is fuel contaminated, or coated with silicone or moisture
P1168 **2T CCM, MIL: Yes** 1998-1999-2000-2001-2002-2003 Crown Victoria, E Van & F Series Truck with a 4.6L VIN 9 or 5.4L VIN M CNG engine Transmissions: All	**Fuel Rail Pressure Sensor In Range But Low Conditions:** DTC P0230, P0231 and P0232 not set, Engine started; and the PCM detected the FRP sensor signal was less than the normal self-test range for these operating conditions. Note: A FRP PID value below 551 kPa (80 psi) indicates a failure. **Possible Causes** • Low fuel level or the vehicle is out of fuel • Low fuel pressure • FRP sensor is damaged • FRP sensor signal circuit has a high resistance condition
P1169 **2T CCM, MIL: Yes** 1998-1999-2000-2001-2002-2003 Crown Victoria, E Van & F Series Truck with a 4.6L VIN 9 or 5.4L VIN M CNG engine Transmissions: All	**Fuel Rail Pressure Sensor In Range But High Conditions:** DTC P0230, P0231 and P0232 not set, Engine started; and the PCM detected the FRP sensor signal was more than the normal self-test range for these operating conditions. Note: A FRP PID value above 896 kPa (130 psi) indicates a failure. **Possible Causes** • Low fuel level, no fuel or low fuel pressure • FRP sensor is damaged • FRP sensor signal circuit has a high resistance condition

Gas Engine OBD II Trouble Code List (P1xxx Codes)

DTC	Trouble Code Title, Conditions & Possible Causes
P1180 **2T CCM, MIL: Yes** 1998-1999-2000-2001-2002-2003 Crown Victoria, E Van & F Series Truck with a 4.6L VIN 9 or 5.4L VIN M CNG engine Transmissions: All	**Fuel Delivery System Low Conditions:** Engine started, and the PCM detected the FTP sensor signal from the NG module data indicated an inferred fuel deliver pressure that was less than a minimum calibrated value. **Possible Causes** • Fuel line is restricted • Fuel filter is plugged or restricted
P1181 **2T CCM, MIL: Yes** 1998-1999-2000-2001-2002-2003 Crown Victoria, E Van & F Series Truck with a 4.6L VIN 9 or 5.4L VIN M CNG engine Transmissions: All	**Fuel Delivery System High Conditions:** Engine started, and the PCM detected the FTP sensor signal from the NG module data indicated an inferred fuel deliver pressure that was higher than a maximum calibrated value. **Possible Causes** • Fuel pressure regulator is damaged or has failed
P1183 **2T CCM, MIL: Yes** 1999-2000-2001-2002-2003 All Models Transmissions: All	**Engine Oil Temperature Sensor Circuit Malfunction Conditions:** Engine started, and the PCM detected the engine oil temperature (EOT) sensor circuit was open or shorted to ground (i.e., this fault is usually caused by an interruption of the signal - intermittent fault). **Possible Causes** • EOT sensor circuit is open or shorted to ground • EOT sensor has failed • PCM has failed
P1184 **2T CCM, MIL: Yes** 1999-2000-2001-2002-2003 All Models Transmissions: All	**Engine Oil Temperature Sensor Out Of Self-Test Range Conditions:** Engine started, and the PCM detected the engine oil temperature (EOT) sensor circuit was open or shorted to ground (i.e., this fault can be caused by an intermittent loss of this signal). **Possible Causes** • EOT sensor circuit is open or shorted to ground (intermittent) • EOT sensor is corroded, damaged or it has failed • PCM has failed
P1220 **1T CCM, MIL: Yes** 1996-1997 Continental with 4.6L DOHC VIN V engine Transmissions: All	**Series Throttle Control System Malfunction Conditions:** Engine started and the PCM detected a malfunction in the Series Throttle Control system. **Possible Causes** • Series Throttle Stepper Motor circuit open, ground or shorted • STC module VPWR circuit open (no power) • STC module ground or power circuit open • TP-B sensor ground (return) circuit is open • TAPW circuit open, shorted to ground, or shorted to VPWR • Series Throttle (ST), Stepper Motor or TP-B sensor damaged • STC Module has failed
P1224 **2T CCM, MIL: Yes** 1996-1997 Continental with 4.6L DOHC VIN V engine Transmissions: All	**TP Sensor 'B' Out Of Self-Test Range Conditions:** Key on or engine running; no other Traction Control codes set, and the PCM detected the TP-B signal was out-of-range during the Continuous self test. **Possible Causes** • TP-B sensor binding/sticking, throttle stop screws misadjusted • TP-B sensor VREF circuit out of range • TP-B sensor is damaged • Series Throttle is damaged

Gas Engine OBD II Trouble Code List (P1xxx Codes)

DTC	Trouble Code Title, Conditions & Possible Causes
P1229 **2T CCM, MIL: Yes** 1999-2000-2001-2002, 2003 F Series Truck with a 5.4L VIN 3 SC engine Transmissions: All	**Supercharger Intercooler Pump Not Operating Conditions:** Engine running at Cruise speed, then after the PCM commanded the Intercooler pump (ICP) to operate it did not detect any current flow in the ICP circuit (with the intercooler pump energized). Note: The Scan Tool ICP PID should change from ON to OFF. **Possible Causes** • Pump motor windings are open, or the pump relay coil is open • The circuit between the relay and the pump is open, or the circuit between the relay and the PCM is open • Pump motor is shorted, or the motor ground connection is open • PCM has failed
P1230 **1 CCM, MIL: Yes** 1996 Mark VIII Transmissions: All	**Low Speed Fuel Pump Malfunction Conditions:** Key on, Low Speed fuel pump energized; and the PCM detected an open condition in the Power-to-Pump circuit between the VLCM and the FPM splice to this fuel pump circuit. **Possible Causes** • An open condition exists in the Power-To-Pump circuit between the VLCM and the FPM splice to the circuit.
P1231 **1T CCM, MIL: No** 1996 All Models Transmissions: All	**Fuel Pump Secondary Low, High Speed Pump On Conditions:** Key on, KOEO Self-Test enabled; High Speed Fuel Pump (HFP) relay energized, fuel pump driver in VLCM off (to VLCM Pin 7) off, the PCM detected voltage on the FPM circuit. **Possible Causes** • HFP relay circuit to battery power (B+) is open • HFP relay is damaged or it has failed • Power-To-Pump circuit between HFP relay and splice is open
P1232 **2T CCM, MIL: Yes** 1996-1997-1998-1999-2000-2001-2002-2003 All Models Transmissions: All	**Low Speed Fuel Pump Primary Circuit Malfunction Conditions:** Engine started, Low Speed Fuel Pump (LFP) relay energized, the PCM detected excessive current on the LFP circuit; or with LFP commanded off it detected power on the LFP circuit. **Possible Causes** • Low fuel pump (LFP) circuit open or shorted • Low speed fuel pump relay VPWR circuit open • Low speed fuel pump relay is damaged • PCM has failed
P1233 **2T CCM, MIL: Yes** 1996-1997-1998-1999-2000-2001-2002-2003 All Models Transmissions: All	**Fuel System Disabled Or Offline Conditions:** Key on or engine running; and the PCM did not receive any diagnostic information (via duty cycle signals) from the FPDM. **Possible Causes** • Inertia fuel shutoff (IFS) switch needs to be reset • FPDM ground circuit is open, or FPM circuit is open or shorted • Mark VIII: FPDM PWR circuit is open, the FPDM power supply relay VPWR circuit is opened or grounded, or power relay failed • Escort/Tracer: FPDM PWR circuit is open, or the CCRM pin 11 is open to power (B+), or the CCRM (relay) is damaged • Continental: FPDM circuit to VPWR is open, or the FPDM or the IFS is damaged • Refer to the GEM or REM controllers for related trouble codes
P1234 **1T CCM, MIL: Yes** 1996-1997-1998-1999-2000-2001-2002-2003 All Models Transmissions: All	**Fuel System Disabled Or Offline Conditions:** Key on, and the PCM did not receive any diagnostic information from the FPDM. **Possible Causes** • Inertia fuel shutoff (IFS) switch needs to be reset or has failed • FPDM ground circuit is open, or FPM circuit is open or shorted • Mark VIII: FPDM PWR circuit is open, the FPDM power supply relay VPWR circuit is opened or grounded, or power relay failed • Escort/Tracer: FPDM PWR circuit is open, or the CCRM pin 11 is open to power (B+), or the CCRM (relay) is damaged • Continental: FPDM circuit to VPWR is open, or the FPDM or the IFS is damaged • LS6, LS8 Models: This code indicates the PCM is not receiving data about the fuel level on the SCP data line from the Rear Electronics Module (REM). Test the REM first! • PCM has failed

Gas Engine OBD II Trouble Code List (P1xxx Codes)

DTC	Trouble Code Title, Conditions & Possible Causes
P1235 **2T CCM, MIL: Yes** 1996-1997-1998-1999-2000-2001-2002-2003 Some Models Transmissions: All	**Fuel Pump Control Out Of Range Conditions:** Key on or engine running; and the PCM received a signal from the FPM over the SCP bus that the FPDM had received an invalid or missing fuel pump command from the PCM. **Possible Causes** • FP circuit is open or shorted • FPDM is damaged • PCM has failed
P1235 **2T CCM, MIL: Yes** 2000-2001-2002-2003 LS Models Transmissions: All	**Fuel Pump Control Out Of Range Conditions:** Key on or engine running; and the PCM received a signal from the FPM over the SCP bus that the FPDM had received an invalid or missing fuel pump command from the PCM. Note that the FPDM commands to the Rear Electronics Module (REM) are sent over the SCP bus communication circuits. **Possible Causes** • FP circuit is open or it is shorted • FPDM is damaged • REM is damaged or it has failed • PCM has failed
P1236 **2T CCM, MIL: No** 1996-1997-1998-1999-2000-2001-2002-2003 Some Models Transmissions: All	**Fuel Pump Control Out Of Range Conditions:** Key on or engine running; and the PCM received a signal (from the FPM over the SCP bus) that the FPDM had received an invalid or missing fuel pump command from the PCM. **Possible Causes** • FP circuit is open or it is shorted • FPDM is damaged or the PCM has failed
P1236 **1T CCM, MIL: No** 2000-2001-2002-2003 LS Models Transmissions: All	**Fuel Pump Control Out Of Range Conditions:** Key on or engine running; and the PCM received a signal (from the FPM over the SCP bus) that the FPDM had received an invalid or missing fuel pump command from the PCM. Note that the FPDM commands to the Rear Electronics Module (REM) are sent over the SCP bus. **Possible Causes** • FP circuit is open or it is shorted • FPDM is damaged or the PCM has failed
P1237 **2T CCM, MIL: Yes** 1996-1997-1998-1999-2000-2001-2002-2003 Some Models Transmissions: All	**Fuel Pump Secondary Circuit Malfunction Conditions:** Key on or engine running; and the PCM received a signal from the FPDM that it had detected a fault in the fuel pump secondary circuit. **Possible Causes** • FP PWR circuit is open or shorted • FPDM fuel pump return circuit is open • Fuel pump windings are open or shorted, or the rotor is locked • FPDM is damaged
P1237 **2T CCM, MIL: Yes** 2000-2001-2002-2003 LS Models Transmissions: All	**Fuel Pump Secondary Circuit Malfunction Conditions:** Key on or engine running; and the PCM received a signal from the FPDM that it had detected a fault in the fuel pump secondary circuit. Note that the FPDM commands to the Rear Electronics Module (REM) are sent over the SCP bus communication circuits. **Possible Causes** • FP PWR circuit is open or shorted • FPDM fuel pump return circuit is open • Fuel pump windings are open or shorted, or the rotor is locked • FPDM is damaged
P1238 **2T CCM, MIL: Yes** 1996-1997-1998-1999-2000-2001-2002-2003 Some Models Transmissions: All	**Fuel Pump Secondary Circuit Malfunction Conditions:** Key on or engine running; and the PCM received a signal from the FPDM that it had detected a fault in the fuel pump secondary circuit. **Possible Causes** • FP PWR circuit is open or shorted • FPDM fuel pump return circuit is open • Fuel pump windings are open or shorted • Fuel pump rotor is locked • FPDM is damaged

Gas Engine OBD II Trouble Code List (P1xxx Codes)

DTC	Trouble Code Title, Conditions & Possible Causes
P1238 **1T CCM, MIL: Yes** 2000-2001-2002-2003 LS Models Transmissions: All	**Fuel Pump Secondary Circuit Malfunction Conditions:** Key on or engine running; and the PCM received a signal from the FPDM that it had detected a fault in the fuel pump secondary circuit. Note that the FPDM commands to the Rear Electronics Module (REM) are sent over the SCP bus communication circuits. **Possible Causes** • FP PWR circuit is open or shorted • FPDM fuel pump return circuit is open • Fuel pump windings are open or shorted • Fuel pump rotor is locked • FPDM is damaged
P1244 **2T CCM, MIL: Yes** 1999-2000-2001-2002-2003 Some Models Transmissions: All	**Generator Load Circuit Low Input Conditions:** Engine started, and the PCM detected the GLI signal was less than the calibrated limit for a calibrated amount of time. **Possible Causes** • GLI circuit is open or shorted • Voltage regulator/generator is damaged • PCM has failed
P1245 **2T CCM, MIL: Yes** 1999-2000-2001-2002-2003 Some Models Transmissions: All	**Generator Load circuit High Input Conditions:** Engine started, and the PCM detected the GLI signal was more than the calibrated limit for a calibrated amount of time. **Possible Causes** • GLI circuit is open or shorted • Voltage regulator/generator is damaged • PCM has failed
P1246 **2T CCM, MIL: Yes** 1999-2000-2001-2002-2003 Some Models Transmissions: All	**Generator Load Circuit Malfunction Conditions:** Engine started, and the PCM detected the GLI was more than or less than a calibrated amount for too long a period of time. **Possible Causes** • Generator circuit is open, shorted to ground or shorted to power • Generator drive mechanism has failed • Generator/regulator assembly is damaged or the PCM has failed
P1246 **2T CCM, MIL: Yes** 2003 Crown Victoria, Grand Marquis & Town Car Transmissions: All	**Generator Load Circuit Malfunction Conditions:** Engine started, and the PCM detected the GLI was more than or less than a calibrated amount for too long a period of time. **Possible Causes** • Generator circuit is open, shorted to ground or shorted to power • Generator drive mechanism has failed • Generator/regulator assembly is damaged or the PCM has failed • TSB 03-14-2 contains repair information for this trouble code
P1250 **2T CCM, MIL: Yes** 1996, 1997 Probe Models with a 2.0L I4 VIN A, 2.5L V6 VIN B engine Transmissions: All	**Fuel Pressure Regulator Control Circuit Malfunction Conditions:** KOEO or KOER Self-Test enabled, and the PCM detected a lack of power (VPWR) to the Fuel Pressure Regulator Control (FPRC) solenoid circuit. **Possible Causes** • FPRC solenoid valve harness circuits are open or shorted • FPRC input port or output port vacuum lines are damaged • FRPC solenoid is damaged • PCM has failed
P1260 **2T PCM, MIL: Yes** 1996-1997-1998-1999-2000-2001-2002-2003 All Models Transmissions: All	**Theft Detected, Vehicle Immobilized Conditions:** Key on, and the PCM received a signal from the Anti-Theft System that a theft condition had occurred. The theft indicator on the dash will flash rapidly or remain on "solid" with the ignition switch in the "on" position. The engine may "start and stall", or may not crank if the vehicle is equipped with the PATS starter disable feature. **Possible Causes** • A Previous theft condition has occurred • Anti-Theft System is damaged or has failed • TSB 01-6-2 (superseded from 76-65-4) contains an updated repair procedure

Gas Engine OBD II Trouble Code List (P1xxx Codes)

DTC	Trouble Code Title, Conditions & Possible Causes
P1270 **1T CCM, MIL: Yes** 1996-1997-1998-1999-2000-2001-2002-2003 All Models Transmissions: All	**Engine Speed/Vehicle Speed Limiter Fault Conditions:** Engine started, and after the PCM monitored the engine speed and VSS signals), it detected the vehicle was operated in a manner where the engine or vehicle speed to exceeded its limit. **Possible Causes** • Excessive wheel slippage due to water, ice, mud and snow • Excessive engine speed (rpm) with the gearshift in Neutral • Vehicle driven at a high rate of speed
P1285 **2T CCM, MIL: Yes** 1996-1997-1998-1999-2000-2001-2002-2003 All Models with CHT Transmissions: All	**Cylinder Head Over-Temperature Sensed Conditions:** Key on or engine running; and the PCM detected an engine overheat condition through inputs from the cylinder head temperature sensor. Engine started, and the PCM detected the CHT or ECT sensor signal was intermittent (it may have an intermittent open condition). Note: Select the CHT or IAT PID and monitor the signal for sudden changes while wiggling the CHT or IAT sensor connector. On the 5.4L V8, if the temperature exceeds 258°F, the PCM disables four fuel injectors at a time. It alternates which four fuel injectors are disabled every 32-engine cycles. The cylinders that are disabled do not inject fuel, so they act as air pumps to aid in cooling the engine. If the temperature exceeds 310°F, the PCM disables all of the fuel injectors until the engine temperature drops below 310°F. **Possible Causes** • Base engine problems or related concerns • CHT sensor has deteriorated or it has failed • Engine coolant level is too low • Engine cooling system has a problem • TSB 10-29-1 contains repair help for this code (LS & T-Bird)
P1288 **2T CCM, MIL: Yes** 1996-1997-1998-1999-2000-2001-2002-2003 All Models with CHT Transmissions: All	**Cylinder Head Temperature Sensor Out of Self-Test Range Conditions:** Key on and KOEO Self-Test enabled, or engine running with the KOER Self-Test enabled, and the PCM detected the CHT sensor was out of its self-test range (i.e., the engine was too hot or it did not warm to its normal operating temperature) during the test period. **Possible Causes** • CHT sensor harness connector is damaged • CHT sensor is damaged • Engine coolant level is too low • Engine is cold, or the engine is overheated
P1289 **2T CCM, MIL: Yes** 1996-1997-1998-1999-2000-2001-2002-2003 All Models with CHT Transmissions: All	**Cylinder Head Temperature Sensor Circuit High Input Conditions:** Key on or engine running; and the PCM detected a Cylinder Head Temperature (CHT) sensor signal that was more than 4.60v. This code may be due to an intermittent fault. Wiggle the CHT sensor wiring and connector while monitoring the CHT PID for a sudden change in voltage. DTC P0118 may also be reported when this code is set, and either code will cause the PCM to activate the MIL. **Possible Causes** • CHT sensor circuit is open in the wiring harness, or an open circuit exists in the CHT sensor circuit at the harness connector • CHT sensor is damaged or has failed • Engine coolant level is too low or the thermostat has failed • PCM has failed
P1290 **2T CCM, MIL: Yes** 1996-1997-1998-1999-2000-2001-2002-2003 All Models with CHT Transmissions: All	**Cylinder Head Temperature Sensor Circuit Low Input Conditions:** Key on or engine running; and the PCM detected a Cylinder Head Temperature (CHT) sensor signal that was less than 0.2v. Note that this trouble code may be due to an intermittent type of fault. Wiggle the CHT sensor wiring and connector while monitoring the CHT V PID for signs of a sudden change in the voltage. DTC P0118 may also set along with this code (both codes will cause a MIL to be on). **Possible Causes** • CHT sensor connector is damaged or a short circuit exists • CHT sensor signal circuit is shorted to sensor ground • CHT sensor is damaged or the PCM has failed

Gas Engine OBD II Trouble Code List (P1xxx Codes)

DTC	Trouble Code Title, Conditions & Possible Causes
P1299 **2T CCM, MIL: Yes** 1996-1997-1998-1999-2000-2001-2002-2003 All Models Transmissions: All	**Cylinder Head Over-Temperature Protection Active Conditions:** Engine started, and after a period of time with the engine running, the PCM detected the engine was in an overheated condition. Note: The PCM enables the Fail-Safe Cooling whenever this code is set to cool the engine (a Failure Mode Effects Strategy or FMEM). **Possible Causes** • Cooling system has a problem • Engine coolant level is too low • A Base Engine problem may be present • TSB 10-29-1 contains repair help for this code (LS & T-Bird)
P1309 **1T MISFIRE** **MIL: Yes** 1997-1998-1999-2000-2001-2002-2003 All Models Transmissions: All	**Misfire Monitor Disabled Conditions:** DTC P0136, P0156, P0171, P0172, P0174, P0175, P1130 and P1150 not set, engine started, and the PCM disabled the Misfire Monitor in order to verify that the CMP sensor is synchronized. Note that this code can be caused by an incorrect input from the CMP sensor (i.e., it senses the passage of teeth from the CMP wheel). **Possible Causes** • Camshaft position sensor is damaged or it has failed • CKP, ECT or MAF sensors may be out-of-calibration or failed • PCM has failed • TSB 02-22-1 contains repair information for this trouble code
P1336 **2T CCM, MIL: Yes** 2002-2003 All Models Transmissions: All	**CKP or CMP Signal Malfunction Conditions:** Engine started, and the PCM detected an erratic signal from CKP sensor or the CMP sensor. It is possible for EMI/RFI interference to cause this code when they occur on these circuits. **Possible Causes** • Base Engine problem or concern exists • CKP sensor or CMP signal circuit is open or shorted to ground • CKP sensor or CMP sensor is damaged or failed (check for EMI/RFI on this circuit). • PCM has failed • TSB 02-22-1 contains repair information for this trouble code
P1351 **2T CCM, MIL: Yes** 1996-1997 All Models Transmissions: All	**Ignition Diagnostic Monitor Circuit Malfunction Conditions:** Engine started, and the PCM detected a loss of the IDM circuit from the ignition module in the distributor (the fault may be intermittent). Note: If DTC P0350, P0351, P0352, P0353 or P0354 is set, repair these trouble codes and then recheck to see if DTC P1351 resets. **Possible Causes** • Camshaft position sensor may have failed • IDM signal circuit may be open or grounded • CKP, ECT or MAF sensors may be damaged or have failed • PCM has failed
P1356 **1T CCM, MIL: No** 1996 All Models Transmissions: All	**PIP Signals Present With Engine Off Conditions:** Key on, and the PCM detected the presence of PIP signals, yet the Ignition Diagnostic Monitor (IDM) signals indicated the engine was not turning. **Possible Causes** • Ignition Module has failed • PCM has failed
P1358 **2T CCM, MIL: Yes** 1996 All Models Transmissions: All	**IDM Signals Out Of Self-Test Range Conditions:** Engine started; and the PCM detected PIP signals that indicated that the Ignition Diagnostic Monitor signals were out of the self-test range under these operating conditions. **Possible Causes** • Ignition Module has failed • PCM has failed
P1359 **2T CCM, MIL: Yes** 1996-1997 All Models Transmissions: All	**Spark Output Circuit Malfunction Conditions:** Engine started, and the PCM did not detect any change in the Spark Output (SPOUT) signals during the test period. **Possible Causes** • SPOUT signal circuit may be open (check the connector) • SPOUT signal circuit may be grounded • Ignition Module is damaged or has failed

Gas Engine OBD II Trouble Code List (P1xxx Codes)

DTC	Trouble Code Title, Conditions & Possible Causes
P1380 **2T CCM, MIL: Yes** 1998-1999-2000-2001-2002-2003 Cougar, Focus, Escort, LS, Thunderbird, ZX2 Equipped with a 2.0L (VIN 3, VIN 5, VIN P) 2.3L VIN Z, 3.0L VIN S, 3.9L VIN A engine Transmissions: All	**Variable Cam Timing Solenoid 'A' Circuit Malfunction (Bank 1) Conditions:** Key on or engine running; and the PCM detected an unexpected voltage condition on the Variable Cam Timing signal circuit. Note that this code is due to an electrical fault (not a mechanical fault). **Possible Causes** • VCT solenoid circuit is open or shorted • VCT solenoid is damaged or has failed • PCM has failed
P1381 **2T CCM, MIL: Yes** 1998-1999-2000-2001-2002-2003 Cougar, Focus, Escort, LS, Thunderbird, ZX2 Equipped with a 2.0L (VIN 3, VIN 5, VIN P) 2.3L VIN Z, 3.0L VIN S, 3.9L VIN A engine Transmissions: All	**Variable Cam Timing Over-Advanced (Cylinder Bank 1) Conditions:** Engine started, and the PCM detected the Variable Cam Timing position indicated the camshaft timing was over-advanced when compared to a maximum calibrated limit in an advanced position. Note: This code is a mechanical problem - not an electrical problem. The engine may be hard to start or idle rough when this code is set. **Possible Causes** • Cam timing improperly set • No oil flow to VCT piston chamber, or low engine oil pressure • VCT solenoid valve stuck in closed position • Camshaft advance mechanism binding (inside the VCT unit)
P1383 **2T CCM, MIL: Yes** 1998-1999-2000-2001-2002-2003 Cougar, Focus, Escort, LS, Thunderbird, ZX2 Equipped with a 2.0L (VIN 3, VIN 5, VIN P) 2.3L VIN Z, 3.0L VIN S, 3.9L VIN A engine Transmissions: All	**Variable Cam Timing Over-Retarded (Cylinder Bank 1) Conditions:** Engine started, and the PCM detected the Variable Cam Timing position indicated the camshaft timing was over-retarded when compared to a maximum calibrated limit in a retarded position. Note that this code is a mechanical problem - not an electrical problem. The engine may be hard to start or idle rough when this code is set. **Possible Causes** • Cam timing improperly set • Low oil pressure, VCT piston chamber not receiving any oil flow • VCT solenoid valve stuck in closed position • Camshaft advance mechanism binding (inside the VCT unit)
P1385 **2T CCM, MIL: Yes** 2002-2003 LS & Thunderbird Models with 3.0L VIN S, 3.9L VIN A engine Transmissions: All	**Variable Cam Timing Solenoid 'A' Circuit Malfunction (Cylinder Bank 2) Conditions:** Key on or engine running; and the PCM detected an unexpected voltage condition on the Variable Cam Timing signal circuit. Note that this code is due to an electrical fault (not a mechanical fault). **Possible Causes** • VCT solenoid circuit is open or shorted • VCT solenoid is damaged or has failed • PCM has failed
P1386 **2T CCM, MIL: Yes** 2002-2003 LS & Thunderbird Models with 3.0L VIN S, 3.9L VIN A engine Transmissions: All	**Variable Cam Timing Over-Advanced (Cylinder Bank 2) Conditions:** Engine started, and the PCM detected the Variable Cam Timing position indicated the camshaft timing was over-advanced when compared to a maximum calibrated limit in an advanced position. Note: This code indicates the presence of a mechanical problem - not an electrical problem. The engine may be hard to start or idle rough when this code is set. **Possible Causes** • Cam timing improperly set • No oil flow to VCT piston chamber, or low engine oil pressure • VCT solenoid valve stuck in closed position • Camshaft advance mechanism binding (inside the VCT unit)

Gas Engine OBD II Trouble Code List (P1xxx Codes)

DTC	Trouble Code Title, Conditions & Possible Causes
P1388 **2T CCM, MIL: Yes** 2002-2003 LS & Thunderbird Models with 3.0L VIN S, 3.9L VIN A engine Transmissions: All	**Variable Cam Timing Over-Retarded (Cylinder Bank 2) Conditions:** Engine started, and the PCM detected the Variable Cam Timing position indicated the camshaft timing was over-retarded when compared to a maximum calibrated limit in a retarded position. This code indicates a mechanical problem - not an electrical problem. The engine may be hard to start or idle rough when this code is set. **Possible Causes** • Cam timing improperly set • Low oil pressure, VCT piston chamber not receiving any oil flow • VCT solenoid valve stuck in closed position • Camshaft advance mechanism binding (inside the VCT unit)
P1390 **1T CCM, MIL: No** 1996-1997-1998 All Models Transmissions: All	**Octane Adjust Circuit Malfunction Conditions:** Key on, KOEO Self-Test enabled, and with the octane adjust software activated, the PCM detected a malfunction in the OCT circuit. **Possible Causes** • OCT shorting bar removed • OCT circuit open • PCM has failed
P1400 **2T CCM, MIL: Yes** 1996-1997-1998-1999-2000-2001-2002-2003 Some Models Transmissions: All	**DPFE Sensor Circuit Low Input Conditions:** Key on, and the PCM detected the DPF EGR sensor signal was less than the minimum calibrated value of 0.2v. Note: The DPF EGR PID will read less than 0.2v with this code set. **Possible Causes** • DPF EGR signal circuit shorted to ground • DPF EGR signal VREF circuit open • DPF EGR sensor is damaged or has failed • PCM has failed • TSB 4-3-1 contains repair information for this trouble code
P1400 **2T CCM, MIL: Yes** 2003 Crown Victoria, Grand Marquis, Aviator, LS, Marauder, Town Car & Thunderbird Models Transmissions: All	**DPFE Sensor Circuit Low Input Conditions:** Engine started, and the PCM detected the DPF EGR sensor signal was less than the minimum calibrated value of 0.2v. The DPFE, EGR valve and EVR solenoid are integrated into the ESM assembly. **Possible Causes** • DPFE sensor signal circuit is shorted to ground • DPFE sensor VREF circuit (5v) is open • DPFE sensor is damaged or it has failed • PCM has failed
P1400 **2T CCM, MIL: Yes** 1996-1997 Explorer, Mountaineer equipped with a 5.0L VIN P engine Transmissions: All	**EVP EGR Sensor Circuit Low Input Conditions:** Key on or engine running; and the PCM detected the EVP sensor signal was below the self-test minimum value of 0.2v. The EVP EGR PID will read less than 0.2v with this code set. **Possible Causes** • EVP sensor circuit is open or shorted to ground • EVP sensor VREF circuit is open • EGR valve position sensor is damaged or has failed • PCM has failed
P1401 **2T CCM, MIL: Yes** 1996-1997-1998-1999-2000-2001-2002-2003 Some Models Transmissions: All	**DPFE Sensor Circuit High Input Conditions:** Key on; and the PCM detected the DPF EGR sensor signal was more than the maximum calibrated value of 4.5v. The DPF EGR PID will read more than 4.5v with this code set. **Possible Causes** • DPF EGR signal circuit open, or sensor ground circuit open • DPF EGR signal shorted to VREF or to power • DPF EGR sensor is damaged or has failed • PCM has failed • TSB 4-3-1 contains repair information for this trouble code

Gas Engine OBD II Trouble Code List (P1xxx Codes)

DTC	Trouble Code Title, Conditions & Possible Causes
P1401 **2T CCM, MIL: Yes** 2003 Crown Victoria, Grand Marquis, Aviator, LS, Marauder, Town Car & Thunderbird Models Transmissions: All	**DPFE Sensor Circuit High Input Conditions:** Key on or engine running; and the PCM detected the DPF EGR sensor signal was more than the maximum calibrated value of 4.5v. On this vehicle application, the DPFE, EGR valve and EVR solenoid are integrated into the ESM assembly. **Possible Causes** • DPFE sensor signal circuit is open • DPFE sensor ground circuit is open • DPFE sensor signal is shorted to VREF (5v) • DPFE sensor is damaged or it has failed • PCM has failed
P1405 **2T CCM, MIL: Yes** 1996-1997-1998-1999-2000-2001-2002-2003 Some Models Transmissions: All	**DPFE Sensor Upstream Hose Off Or Plugged Conditions:** Engine started; and the PCM detected the DPF EGR sensor indicated EGR flow in a negative direction (a closed EGR valve). Check for signs of icing in the hose, or wrong hose routing. **Possible Causes** • DPF EGR sensor upstream hose is disconnected • DPF EGR sensor upstream hose is plugged (ice) • EGR tube is plugged or damaged
P1406 **2T CCM, MIL: Yes** 1996-1997-1998-1999-2000-2001-2002-2003 Some Models Transmissions: All	**DPFE Sensor Downstream Hose Off Or Plugged Conditions:** Engine started; and the PCM detected the DPF EGR sensor signal indicated EGR flow existed with the EGR valve commanded closed. **Possible Causes** • Check for signs of icing in the hose, or for a restricted tube • DPF EGR sensor downstream hose is disconnected • DPF EGR sensor downstream hose is plugged (ice) • EGR tube is plugged or damaged
P1406 **2T CCM, MIL: Yes** 2003 Crown Victoria, Grand Marquis, Aviator, LS, Marauder, Town Car & Thunderbird Models Transmissions: All	**DPFE Sensor Downstream Hose Off Or Plugged Conditions:** Engine started, and the PCM detected the DPFE sensor indicated that EGR flow was present with the EGR valve commanded closed. On this vehicle application, the DPFE, EGR valve and EVR solenoid are integrated into the ESM assembly. **Possible Causes** • Check for signs of icing in the hose, or for a restricted tube • DPF EGR sensor downstream hose is disconnected • DPF EGR sensor downstream hose is plugged (ice) • EGR tube is plugged or damaged
P1407 **2T EGR, MIL: Yes** 1996-1997 Probe Models with a 2.0L I4 VIN A, 2.5L V6 VIN B engine Transmissions: All	**EGR Flow Out Of Self-Test Range Conditions:** DTC P0107, P0108, P1400, P1401 and P1408 not set, engine running, and the PCM detected the EGR valve did not move. Note: The EGR valve may be stuck in the closed position. **Possible Causes** • EGR VR solenoid circuit open or VR solenoid is damaged • EGR Atmospheric solenoid circuit open or solenoid is damaged • EGR Valve Position sensor circuit open or sensor is damaged • EGR Check solenoid circuit open or solenoid is damaged • Exhaust system is restricted • PCM has failed
P1408 **1T EGR, MIL: Yes** 1996-1997-1998-1999-2000-2001-2002-2003 Some Models Transmissions: All	**EGR Flow Out Of Self-Test Range Conditions:** KOER Self-Test enabled, and the PCM detected the EGR flow was out of the self-test range during the self-test with the engine running. **Possible Causes** • EGR vacuum regulator solenoid vacuum supply problem • EVR valve stuck closed or iced up, or the flow path is restricted • EGR valve diaphragm leaking, hose is off, plugged or leaking • EGR VR solenoid open or the VPWR circuit is open • DPF EGR sensor pressure hoses connected wrong (reversed) • DPF EGR downstream hose connection leaking or plugged • EGR Orifice tube assembly is damaged • DPF EGR sensor or the EGR VR solenoid is damaged, or the PCM has failed

Gas Engine OBD II Trouble Code List (P1xxx Codes)

DTC	Trouble Code Title, Conditions & Possible Causes
P1408 **1T EGR, MIL: No** 1996-1997 Probe Models with a 2.0L I4 VIN A, 2.5L V6 VIN B engine Transmissions: All	**EGR Flow Out Of Self-Test Range Conditions:** DTC P1400, P1401 and P1407 not set, engine started, KOER Self-Test enabled, and the PCM detected a lack of EGR valve movement while monitoring the EGR EVP sensor signal. **Possible Causes** • EGR valve vacuum lines are loose or damaged • EGR valve is damaged or has failed • EGR VR solenoid circuit open or the solenoid is damaged • EGR Atmospheric solenoid circuit open or solenoid is damaged • PCM has failed
P1408 **1T EGR, MIL: No** 2001-2002-2003 Focus, Ranger Models with a 2.3L VIN D, 2.3L VIN Z engine Transmissions: All	**EGR Flow Out Of Self-Test Range Conditions:** Engine started, KOER Self-Test enabled, and the PCM detected the EGR flow was out of the self-test range. There is no DPFE sensor or orifice tube assembly on this vehicle application. **Possible Causes** • E-EGR valve connector is damaged, loose or shorted • E-EGR valve is sticking, damaged or it has failed • MAP or TMAP sensor is damaged or out of calibration • PCM has failed
P1408 **1T EGR, MIL: No** 2003 Crown Victoria, Grand Marquis, Aviator, LS, Marauder, Town Car & Thunderbird Models Transmissions: All	**EGR Flow Out Of Self-Test Range Conditions:** Engine started, KOER Self-Test enabled, and the PCM detected the EGR flow was out of the self-test range during the self-test. On this vehicle application, the DPFE, EGR valve and EVR solenoid are integrated into the ESM assembly. **Possible Causes** • EGR vacuum regulator solenoid vacuum supply problem • EVR valve stuck closed or iced up, or the flow path is restricted • EGR valve diaphragm leaking, hose is off, plugged or leaking • EGR VR solenoid open or the VPWR circuit is open • DPF EGR sensor pressure hoses connected wrong (reversed) • DPF EGR sensor VREF circuit is open • DPF EGR downstream hose connection leaking or plugged • EGR Orifice tube assembly is damaged • DPF EGR sensor or the EGR VR solenoid is damaged • PCM has failed
P1409 **2T CCM, MIL: Yes** 1996-1997-1998-1999-2000-2001-2002-2003 Some Models Transmissions: All	**EGR Vacuum Regulator Solenoid Circuit Malfunction Conditions:** Engine started, and the PCM detected a fault in the EGR VR solenoid circuit (i.e., the VR circuit was too high or low when compared to the expected range with the solenoid enabled). **Possible Causes** • EGR VR solenoid circuit is open, or shorted to ground • EGR VR circuit is shorted to power or the VPWR circuit is open • EGR vacuum regulator solenoid is damaged or the PCM has failed
P1409 **1T CCM, MIL: Yes** 2003 Crown Victoria, Grand Marquis, Aviator, LS, Marauder, Town Car & Thunderbird Models Transmissions: All	**EGR Vacuum Regulator Solenoid Circuit Malfunction Conditions:** Engine started, and the PCM detected a fault in the EGR VR solenoid circuit (i.e., the VR circuit was too high or low when compared to its expected range with the solenoid enabled). **Possible Causes** • EGR VR solenoid circuit is open, or shorted to ground • EGR VR circuit is shorted to power or the VPWR circuit is open • EGR vacuum regulator solenoid is damaged or the PCM has failed
P1410 **2T CCM, MIL: Yes** 1996-1997 Probe Models with a 2.0L I4 VIN A, 2.5L V6 VIN B engine Transmissions: All	**EGR Check Or Fuel Pressure Solenoid Circuit Malfunction Conditions:** Key on or engine running; and the PCM detected a fault in the EGR Check solenoid or the EGR Fuel Pressure Control Solenoid circuits. **Possible Causes** • EGR check solenoid control circuit is open or shorted to ground • FPC solenoid control circuit is open or shorted to ground • Inspect the EGR Check Solenoid and Fuel Pressure Control solenoid harness connectors to determine if they are swapped • PCM has failed

Gas Engine OBD II Trouble Code List (P1xxx Codes)

DTC	Trouble Code Title, Conditions & Possible Causes
P1411 **2T AIR, MIL: Yes** 1996-1997-1998 All Models Transmissions: All	**Secondary Air Injection System Downstream Flow Conditions:** Engine started, engine running with AIR system "on", and the PCM detected the HO2S signal did not go lean with the AIR system "on". **Possible Causes** • Secondary AIR System Electric Pump is damaged • Secondary AIR System Mechanical Pump is damaged • Secondary AIR pump hose is leaking • Secondary AIR pump hose is blocked • Secondary AIR Bypass solenoid passage leaking or blocked • Secondary AIR Bypass solenoid stuck open or stuck closed
P1413 **2T CCM, MIL: Yes** 1996-1997-1998-1999- 2000-2001-2002-2003 All Models Transmissions: All	**Secondary AIR System Monitor Circuit Low Input Conditions:** Engine started, engine running with AIR system "off", and the PCM detected an unexpected "low" voltage condition on the Secondary AIR monitor during the CCM test. **Possible Causes** • AIR solenoid control circuit is open or it is shorted to ground • AIR pump is damaged or it has failed • Solid State relay is damaged or it has failed • Solid State relay battery power circuit (B+) is open • PCM has failed
P1414 **2T CCM, MIL: Yes** 1996-1997-1998-1999- 2000-2001-2002-2003 All Models Transmissions: All	**Secondary AIR System Monitor Circuit High Input Conditions:** Engine started, AIR system not active, and the PCM detected a high voltage signal present on the Secondary AIR monitor signal circuit. **Possible Causes** • AIR Monitor circuit from the pump to the PCM is open • AIR solenoid control circuit is shorted to power • Solid State relay is damaged or has failed • AIR pump ground circuit is open • AIR pump is damaged or has failed • PCM has failed
P1432 **1T CCM, MIL: Yes** 2001-2002-2003 All Models Transmissions: All	**Thermostat Heater Control Circuit Malfunction Conditions:** Engine started; and the PCM detected the Thermostat Heater Control circuit was less than or more than a calibrated limit for too long a period of time during the CCM self-test. **Possible Causes** • Thermostat Heater Control (THTRC) circuit open or shorted • Thermostat Heater Control (THTRC) VPWR circuit open • Thermostat Heater assembly is damaged or has failed • PCM has failed
P1436 **2T CCM, MIL: Yes** 2002-2003 All Models Transmissions: All	**A/C Evaporator Temperature (ACET) Circuit Low Input Conditions:** Key on or engine running; and the PCM detected the ACET signal was less than the self-test minimum amount of 0.13v in the self-test. **Possible Causes** • ACET signal circuit shorted to sensor ground (return) • ACET signal circuit shorted to chassis ground • ACET sensor is damaged or has failed • PCM has failed
P1437 **2T CCM, MIL: Yes** 2002-2003 All Models Transmissions: All	**A/C Evaporator Temperature (ACET) Circuit High Input Conditions:** Key on or engine running; and the PCM detected the ACET signal was more than the self-test maximum amount of 4.5v in the self-test. **Possible Causes** • ACET signal circuit is open, or the ground circuit is open • ACET signal is shorted to VREF • ACET sensor is damaged or has failed • PCM has failed

Gas Engine OBD II Trouble Code List (P1xxx Codes)

DTC	Trouble Code Title, Conditions & Possible Causes
P1442 **2T EVAP, MIL: Yes** 1996-1997-1998-1999- 2000-2001-2002-2003 All Models Transmissions: All	**EVAP System Small Leak (0.040") Detected Conditions:** Cold startup requirement met, engine running at Cruise speed in closed loop for 2-3 minutes, and the PCM detected a leak (as small as 0.040") in the EVAP system. Note: Inspect the CV solenoid for contamination (as contamination can hold the CV open set DTC P0442 and also plugs the port to atmosphere enough to keep system from being vented quickly). **Possible Causes** • Fuel filler cap damaged, cross-threaded or loosely installed • Aftermarket EVAP parts that do not conform to specifications • Small holes or cuts in fuel vapor hoses or EVAP canister tubes • CV solenoid stays partially open when commanded closed • Loose fuel vapor hose/tube connections to EVAP components • EVAP component seals leaking (i.e., leaks in the Purge valve, fuel tank pressure sensor, canister vent solenoid, fuel vapor control valve tube assembly or fuel vapor vent valve) • TSB 03-9-8 contains a repair procedure for this trouble code • TSB 3-20-3 contains a repair procedure for this trouble code
P1443 **2T EVAP, MIL: Yes** 1996-1997-1998-1999 All Models Transmissions: All	**EVAP Canister Purge System Malfunction Conditions:** Engine started, engine warmup completed engine running at a steady cruise speed, and the PCM detected a leak or blockage was present somewhere in the vapor line between the intake manifold, EVAP purge valve and the charcoal canister during the Continuous self test. **Possible Causes** • EVAP canister purge valve is damaged or has failed • PF sensor is out-of-calibration or it is "skewed" • PCM has failed
P1443 **2T EVAP, MIL: Yes** 1996-1997 Contour & Mystique Models Transmissions: All	**EVAP VMV System Malfunction Conditions:** Engine started, engine warmup completed engine running at a steady cruise speed, and the PCM detected a leak or blockage was present somewhere in the vapor line between the intake manifold, VMV valve and the charcoal canister during the CCM self-test. **Possible Causes** • VMV (valve) is sticking or has failed mechanically • PF sensor is out-of-calibration or it is "skewed" • PCM has failed
P1443 **2T EVAP, MIL: Yes** 2000-2001-2002-2003 All Models Transmissions: All	**Low Purge Flow Or No Purge Flow Condition Detected Conditions:** ECT sensor less than 90°Fat startup (cold engine), engine running at a steady cruise speed, and the PCM detected a fuel tank pressure change occurred of more than -7" H2O within 30 seconds with the purge flow less than 0.02 pounds per minute during testing. **Possible Causes** • EVAP canister purge valve stuck closed (mechanically) • Fuel vapor hose blocked between EVAP purge valve and FTP sensor, or blocked between purge valve and intake manifold, or vacuum hose blocked between purge valve and intake manifold
P1444 **2T CCM, MIL: Yes** 1996-1997-1998 All Models Transmissions: All	**Purge Flow Sensor Circuit Low Input Conditions:** Key on or engine running; and the PCM detected the Purge Flow (PF) sensor signal was less than the minimum calibrated limit of 0.40v during the Continuous self test. **Possible Causes** • PF sensor signal circuit is shorted to sensor or chassis ground • PF sensor is damaged or has failed • PCM has failed
P1445 **2T CCM, MIL: Yes** 1996-1997-1998 All Models Transmissions: All	**Purge Flow Sensor Circuit High Input Conditions:** Key on or engine running; and the PCM detected the Purge Flow (PF) sensor was more than the maximum calibrated limit of 4.80v. **Possible Causes** • PF sensor signal circuit shorted to VREF or power (VPWR) • PF sensor signal circuit open or sensor ground circuit open • PF sensor is damaged or has failed • PCM has failed

Gas Engine OBD II Trouble Code List (P1xxx Codes)

DTC	Trouble Code Title, Conditions & Possible Causes
P1450 **2T EVAP, MIL: Yes** 1996-1997 All Models Transmissions: All	**Unable to Bleed Up Fuel Tank Vacuum Conditions:** ECT sensor less than 90ºF at startup (cold engine), engine running at a steady cruise speed, and the PCM detected a high fuel tank vacuum condition was present during the EVAP test. **Possible Causes** • CV solenoid is stuck partially or fully open or filter is plugged • EVAP canister tube or EVAP canister purge outlet tube blocked or kinked between fuel tank, purge valve and EVAP canister • Fuel filler cap stuck closed (vacuum relief cannot occur) • Contaminated fuel vapor elbow at the EVAP canister, or the EVAP canister is restricted or canister purge valve stuck open
P1450 **2T EVAP, MIL: Yes** 1998-1999-2000-2001-2002-2003 All Models Transmissions: All	**Unable to Bleed Up Fuel Tank Vacuum Conditions:** ECT sensor less than 90ºF at startup (cold engine), engine running at a steady cruise speed, and the PCM detected a high fuel tank vacuum condition was present during the EVAP test. **Possible Causes** • CV solenoid is stuck partially or fully open or filter is plugged • EVAP canister tube or EVAP canister purge outlet tube blocked or kinked between fuel tank, purge valve and EVAP canister • Fuel filler cap stuck closed (vacuum relief cannot occur) • Contaminated fuel vapor elbow at the EVAP canister, or the EVAP canister is restricted or canister purge valve stuck open • FTP sensor is damaged
P1451 **2T CCM, MIL: Yes** 1996-1997-998-1999-2000-2001-2002-2003 All Models Transmissions: All	**EVAP System Canister Vent Solenoid Circuit Malfunction Conditions:** Engine started, engine running at a steady cruise speed, canister vent solenoid enabled, and the PCM detected an unexpected voltage condition on the Canister Vent solenoid circuit. **Possible Causes** • CV solenoid circuit is open, shorted to ground or system power • CV solenoid is damaged or has failed • PCM has failed
P1452 **2T CCM, MIL: Yes** 1996-1997 All Models Transmissions: All	**Fuel Tank Pressure Sensor Circuit Malfunction Conditions:** Key on or engine running; and the PCM detected that the Fuel Tank Pressure (FTP) sensor signal was less than or more than the calibrated amount during the self-test. Note that the FTP V PID should read from 2.40-2.80 with the cap off. **Possible Causes** • FTP sensor signal circuit is open or shorted to ground • FTP sensor ground return circuit is open • FTP sensor is damaged or has failed • PCM has failed
P1455 **2T EVAP, MIL: Yes** 1996-1997 All Models Transmissions: All	**EVAP System Gross Leak Detected Conditions:** ECT sensor less than 90ºF at startup (cold engine), engine running at a steady cruise speed for 2-3 minutes, and the PCM detected a gross leak in the EVAP system during the test. **Possible Causes** • Fuel filler cap missing, loose (not tightened) or the wrong part • FTP sensor signal circuit open or sensor ground circuit open • FTP sensor ground circuit open • FTP sensor is damaged or has failed • PCM has failed
P1460 **1T CCM, MIL: Yes** 1996-1997-998-1999-2000-2001-2002-2003 All Models Transmissions: All	**Wide Open Throttle A/C Cutout Relay Circuit Malfunction Conditions:** Key on, and the PCM detected a malfunction in the A/C wide-open throttle (WOT) circuit during the test. Note: If this code sets on vehicles without an A/C system, ignore this code. **Possible Causes** • WOT A/C Relay control circuit is open or shorted to ground • WOT A/C Relay VREF circuit is open • WOT A/C Relay is damaged or has failed • PCM has failed

Gas Engine OBD II Trouble Code List (P1xxx Codes)

DTC	Trouble Code Title, Conditions & Possible Causes
P1461 **2T CCM, MIL: Yes** 1996-1997-998-1999-2000-2001-2002-2003 All Models Transmissions: All	**A/C Pressure Sensor Circuit High Input Conditions:** Engine started, and the PCM detected the A/C Pressure sensor signal was over the test limit. **Possible Causes** • ACP sensor circuit shorted to VREF or to power (VPWR) • ACP sensor circuit is open, or the ground circuit is open • ACP sensor is damaged or has failed • PCM has failed
P1462 **2T CCM, MIL: Yes** 1996-1997-998-1999-2000-2001-2002-2003 All Models Transmissions: All	**A/C Pressure Sensor Circuit Low Input Conditions:** Engine started, and the PCM detected the A/C Pressure sensor signal was under the test limit. **Possible Causes** • ACP sensor circuit shorted to VREF or to power (VPWR) • ACP sensor circuit is open, or the ground circuit is open • ACP sensor is damaged or has failed • PCM has failed
P1463 **2T CCM, MIL: Yes** 1996-1997-998-1999-2000-2001-2002-2003 All Models Transmissions: All	**A/C Pressure Sensor Insufficient Pressure Change Conditions:** Engine started, and with the A/C compressor operating, the PCM detected the A/C refrigerant pressure did not change as the compressor cycled during the self-test period. **Possible Causes** • A/C system mechanical failure, or A/C clutch always engaged • ACP sensor signal open, or sensor ground circuit open • A/C sensor is damaged or the PCM has failed
P1464 **1T CCM, MIL: No** 1996-1997-998-1999-2000-2001-2002-2003 All Models Transmissions: All	**A/C Demand Out of Self-Test Range Conditions:** Key on, KOEO Self-Test enabled, or with the engine running, KOER Self-Test enabled, and the PCM detected the A/C demand switch signal was high during the self-test period. **Possible Causes** • A/C switch was left "on" during the KOER self-test • A/C PWR circuit is shorted to power (N/C WAC relay contacts) • ACCS circuit is shorted to power • A/C Demand Switch, WAC relay or CCRM is damaged
P1469 **2T CCM, MIL: Yes** 1996-1997-998-1999-2000-2001-2002-2003 All Models Transmissions: All	**Low A/C Cycling Period Conditions:** Engine started, and with the A/C selected, PCM detected frequent cycling of the A/C compressor clutch. This test was designed to protect the transmission. In some strategies, the PCM will unlock the torque converter during A/C clutch engagement. If a concern is present that results in frequent A/C clutch cycling, damage could occur if the torque converter was cycled at these intervals. This test will detect this condition, set the code and prevent the torque converter from excessive cycling. **Possible Causes** • Cycling pressure switch circuit open between pin 41 (ACCS) and the PCM, or the IGN RUN circuit is open to the cycling pressure switch circuit (if applicable) • Mechanical A/C system concern (i.e., low refrigerant charge, damaged A/C switch)
P1473 **2T CCM, MIL: Yes** 1996-1997-998 All Models Transmissions: All	**Fan Secondary High with Fan(s) Off Conditions:** Key on, KOEO Self-Test enabled, and the PCM detected an unexpected voltage condition on the Power-To-Cooling fan circuit **Possible Causes** • Power-to-Cooling fan circuit open in the wiring harness • Power-to-Cooling fan circuit shorted to power in wiring harness • Cooling fan motor windings open, or fan ground circuit is open • VLCM is damaged or has failed
P1474 **2T CCM, MIL: Yes** 1996-1997-998-1999-2000-2001-2002-2003 All Models Transmissions: All	**Hydraulic Cooling Fan Primary Circuit Malfunction Conditions:** Key on or engine running; low cooling fan enabled, and the PCM detected the voltage to the Hydraulic Cooling Fan (HFC) motor was higher or lower than the expected range for the fan primary circuit. **Possible Causes** • FC circuit is open or shorted to power • LFC circuit is open or shorted to power (CCRM models) • FC or LFC relay VPWR circuit is open (Start/Run on Probe)

Gas Engine OBD II Trouble Code List (P1xxx Codes)

DTC	Trouble Code Title, Conditions & Possible Causes
P1474 **2T CCM, MIL: Yes** 2000-2001-2002-2003 LS Models with a 3.0L VIN S, 3.9L VIN A Transmissions: All	**Hydraulic Cooling Fan Primary Circuit Malfunction Conditions:** Key on or engine running; hydraulic cooling fan "on", and the PCM detected an unexpected voltage (too high or too low) condition on the Hydraulic Cooling Fan (HFC) motor circuit. **Possible Causes** • HCF control circuit is open or shorted to ground • HCF control circuit is shorted to power (VPWR) • HCF motor is damaged or has failed • PCM has failed
P1477 **2T CCM, MIL: Yes** 2001-2002-2003 All Models Transmissions: All	**Medium Fan Control Primary Circuit Malfunction Conditions:** Key on, medium cooling fan (MFC) enabled; and the PCM detected excessive current draw in the circuit; or with the MFC disabled (off), it detected voltage present on the MFC circuit. **Possible Causes** • MFC circuit is open • MFC relay circuit to IGN START/RUN is open • MFC relay is damaged or has failed • PCM has failed
P1479 **2T CCM, MIL: Yes** 1996-1997-998-1999-2000-2001-2002-2003 All Models Transmissions: All	**High Fan Control Primary Circuit Malfunction Conditions:** Key on, high cooling fan (HFC) enabled, and the PCM detected excessive current draw in the circuit; or with the HFC commanded off, it detected voltage present on the HFC circuit. **Possible Causes** • HFC circuit is open • HFC circuit is shorted to ground • HFC relay power circuit (VPWR) is open • High speed FC relay is damaged or it has failed • PCM has failed
P1481 **2T CCM, MIL: Yes** 1996 All Models Transmissions: All	**Fan Secondary Low With High Fan On Conditions:** Key on or engine running; high speed cooling fan enabled, and the PCM detected the fan secondary circuit was low with the High Speed cooling fan commanded "on" during testing. **Possible Causes** • High speed cooling fan circuit is open • High speed cooling fan circuit is shorted to ground • High speed cooling fan relay power circuit (VPWR) is open • High speed FC relay is damaged or has failed • PCM has failed
P1483 **2T CCM, MIL: Yes** 1996-1997-1998 All Models Transmissions: All	**Power To Fan Circuit Over-Current Detected Conditions:** Key on or engine running; cooling fan enabled, and the PCM detected the current in the Fan PWR circuit exceeded the limit. **Possible Causes** • Power-to-Cooling Fan circuit shorted to ground • Cooling fan motor is damaged or has failed • VLCM is damaged or has failed • PCM has failed
P1487 **2T CCM, MIL: Yes** 1996-1997-1998 Mark VIII Models Transmissions: All	**VLCM Power Ground Circuit Malfunction Conditions:** Key on or engine running; and the PCM detected an unexpected voltage condition on the Variable Load Control Module (VLCM) power ground circuit during the CCM test period. **Possible Causes** • VLCM power ground circuit is open • VLCM is damaged or it has failed
P1487 **2T CCM, MIL: Yes** 1996-1997 Probe Models with a 2.0L I4 VIN A, 2.5L V6 VIN B engines Transmissions: All	**EGR Check Solenoid Circuit Malfunction Conditions:** Key on or engine running; and the PCM detected an unexpected voltage condition on the EGR Check solenoid circuit during testing. **Possible Causes** • EGR Check solenoid circuit is open or shorted in the harness • EGR Check solenoid circuit is open or shorted in the connector • EGR Check solenoid power circuit (VREF) is open • EGR Check solenoid is damaged or the PCM has failed

Gas Engine OBD II Trouble Code List (P1xxx Codes)

DTC	Trouble Code Title, Conditions & Possible Causes
P1500 **2T CCM, MIL: Yes** 1996-1997-998-1999-2000-2001-2002-2003 All Models Transmissions: All	**VSS Signal Or POSM Signal Intermittent Conditions:** Engine running in gear with a VSS signal present, and the PCM detected that the VSS signal was intermittent Note: The VSS signal is received from the VSS, transfer case speed sensor, ABS Control module, the GEM or the Central Timer module (CTM), depending upon the vehicle application. **Possible Causes** • VSS pins damaged, loose or pushed in at the connector • VSS circuit open or shorted in the wiring harness (insulation) • VSS wiring harness routing incorrect or VSS mounting incorrect • TSB 01-21-13 contains a repair procedure for this trouble code
P1501 **1T CCM, MIL: No** 1996-1997 All Models Transmissions: All	**VSS Signal Out Of Self-Test Range Conditions:** Engine started, KOER Self-Test enabled, and the PCM detected a VSS signal during the self-test (i.e., with the vehicle not moving). **Possible Causes** • VSS signal is noisy due to Radio Frequency Interference/ Electro-Magnetic Interference (RFI/EMI) from outside devices (ignition wires, charging circuit or aftermarket devices)
P1501 **1T CCM, MIL: Yes** 1996-1997-1998-1999-2000-2001-2002-2003 All Models Transmissions: All	**VSS Signal Intermittent Conditions:** Engine started, and the PCM detected the VSS signal dropped out. The TCIL will flash on the first trip this code sets. The speed signal is received from the VSS, transfer case speed sensor, ABS Control module, GEM or the Central Timer module (depends upon the vehicle). **Possible Causes** • VSS signal is noisy due to Radio Frequency Interference/ Electro-Magnetic Interference (RFI/EMI) from outside devices (ignition wires, charging circuit or aftermarket devices)
P1502 **1T CCM, MIL: Yes** 1996-1997-1998-1999-2000-2001-2002-2003 All Models Transmissions: All	**VSS Signal Intermittent Conditions:** Engine started, and the PCM detected an intermittent VSS signal. The TCIL will flash on the first trip that this code is set. The VSS signal is received from the VSS, transfer case speed sensor, ABS Control module, GEM or the Central Timer module (depends upon the vehicle). **Possible Causes** • VSS+ or VSS- harness circuit is open • TCSS signal or TCSS signal return harness circuit is open • VSS harness circuit, TCSS harness circuit is shorted to ground • VSS harness circuit, CSS harness circuit is shorted to power • VSS circuit open between the PCM and related control module • VSS or TCSS, or wheel speed sensors circuits are damaged • Modules connected to VSC/VSS harness circuits are damaged • Mechanical drive mechanism for the VSS or TCSS is damaged
P1504 **2T CCM, MIL: Yes** 1996-1997-1998-1999-2000-2001-2002-2003 All Models Transmissions: All	**Idle Air Control Circuit Malfunction Conditions:** Engine started, engine running for 1 minute, and the PCM detected an electrical load failure on the IAC motor circuit during the self-test. **Possible Causes** • IAC circuit is open, shorted to ground or to the VPWR circuit • IAC solenoid VPWR circuit is open • IAC valve is damaged or has failed • PCM has failed
P1505 **2T CCM, MIL: Yes** 1996-1997-1998 Models Transmissions: All	**Idle Air Control System At Adaptive Clip Conditions:** Engine running for over one minute, and the PCM detected the idle speed control had reached its "idle air trim limit" during the Continuous self test. **Possible Causes** • Base engine air leaks are present • Air cleaner element is dirty, plugged or restricted • Throttle body/linkage is binding • IAC valve body is damaged or contaminated • Throttle body is damaged

Gas Engine OBD II Trouble Code List (P1xxx Codes)

DTC	Trouble Code Title, Conditions & Possible Causes
P1506 **2T CCM, MIL: Yes** 1996-1997-1998 Crown Victoria, Grand Marquis Models Transmissions: All	**Idle Air Control Overspeed Error Conditions:** Engine started, engine running for 1 minute, and the PCM detected the idle speed was more than the desired engine Target Idle Speed. **Possible Causes** • Base engine vacuum leaks present • EVAP system has a problem • IAC circuit shorted to ground • IAC valve is stuck open, or it is damaged • Throttle body or throttle plate is contaminated or very dirty • TSB 98-25-19 contains a repair procedure for this trouble code
P1506 **2T CCM, MIL: Yes** 1999-2000-2001-2002-2003 Some Models Transmissions: All	**Idle Air Control Overspeed Error Conditions:** Engine started, engine running for 1 minute, and the PCM detected the idle speed was more than the desired engine Target Idle Speed. **Possible Causes** • Base engine vacuum leaks present • EVAP system has a problem • IAC circuit shorted to ground • IAC valve is stuck open, or it is damaged • Throttle body or throttle plate is contaminated or very dirty
P1507 **2T CCM, MIL: Yes** 1996-1997-1998-1999-2000-2001-2002-2003 Some Models Transmissions: All	**Idle Air Control Underspeed Error Conditions:** Engine started, engine running for 1 minute, and the PCM detected the idle speed was less than the desired engine Target Idle Speed. **Possible Causes** • Air inlet is plugged or the air filter element is severely clogged • IAC circuit is open, or shorted to the VPWR circuit • IAC circuit VPWR circuit is open • IAC solenoid is damaged or has failed • Throttle body or throttle plate is contaminated or very dirty
P1512 **2T CCM, MIL: Yes** 1996-1997-1998-1999-2000 Some Models Transmissions: All	**Intake Manifold Runner Control System Malfunction Conditions:** Engine started, and the PCM detected the IMRC Monitor indicated that the IMRC was stuck closed during the Continuous self test. **Possible Causes** • Leaky vacuum reservoir, vacuum lines loose or damaged • Vacuum solenoid or vacuum actuator is damaged • IMRC actuator cable/gears are seized, or the cables are improperly routed or seized • IMRC housing return springs are damaged or disconnected • Lever/shaft return stop may be obstructed or bent, or the lever/shaft wide open stop may be obstructed or bent, or the IMRC lever/shaft may be sticking, binding or disconnected • IMRC control circuit open, shorted or the VPWR circuit is open • PCM has failed
P1513 **2T CCM, MIL: Yes** 1996-1997-1998-1999-2000 Some Models Transmissions: All	**Intake Manifold Runner Control Malfunction (Bank 1) Conditions:** Engine started, and the PCM detected the IMRC Monitor indicated the IMRC was not functioning correctly during the self-test period. **Possible Causes** • IMRC actuator cable/gears are seized, or the cables are improperly routed or seized • IMRC control circuit open, shorted or the VPWR circuit is open • IMRC housing return springs are damaged or disconnected • Leaky vacuum reservoir, vacuum lines loose or damaged • Lever/shaft return stop may be obstructed or bent, or the lever/shaft wide open stop may be obstructed or bent, or the IMRC lever/shaft may be sticking, binding or disconnected • Vacuum solenoid or vacuum actuator is damaged • PCM has failed

Gas Engine OBD II Trouble Code List (P1xxx Codes)

DTC	Trouble Code Title, Conditions & Possible Causes
P1516 **2T CCM, MIL: Yes** 1996-1997-1998-1999- 2000-2001-2002-2003 Some Models Transmissions: All	**Intake Manifold Runner Control Input Error (Bank 1) Conditions:** Key on or engine running; and the PCM detected the IMRC Monitor signal for Bank 1 was outside of its expected calibrated range during the Continuous self test. **Possible Causes** • IMRC mechanical fault - the linkage may be bound or seized • Inspect for binding or improper routing. The cable core wire at the IMRC/IMSC housing attachment must have slack and lever must contact close plate stop screw
P1517 **2T CCM, MIL: Yes** 1996-1997-1998-1999- 2000-2001-2002-2003 Some Models Transmissions: All	**Intake Manifold Runner Control Input Error (Bank 2) Conditions:** Key on or engine running; and the PCM detected the IMRC Monitor signal for Bank 2 was outside of its expected calibrated range during the Continuous self test. **Possible Causes** • IMRC mechanical fault - the linkage may be bound or seized • Visually inspect for binding or improper routing. The cable core wire at the IMRC or IMSC housing attachment must have slack and lever must contact close plate stop screw
P1518 **2T CCM, MIL: Yes** 1996-1997-1998-1999- 2000-2001-2002-2003 Some Models Transmissions: All	**Intake Manifold Runner Control Malfunction (Stuck Open) Conditions:** Engine started, and the PCM detected the IMRC Monitor signal was less than its expected calibrated range at closed throttle. An IMRCM PID of 1v at closed throttle indicates a fault. **Possible Causes** • IMRC monitor signal circuit shorted to power ground • IMRC Monitor signal circuit shorted to signal ground (return) • IMRC actuator is damaged or has failed • PCM has failed
P1519 **2T CCM, MIL: Yes** 1996-1997-1998-1999- 2000-2001-2002-2003 Some Models Transmissions: All	**Intake Manifold Runner Control Stuck Closed Conditions:** Key on, and the PCM detected the IMRC Monitor was more than the expected calibrated range at closed throttle. Note: An IMRCM PID of VREF at 3000 rpm may indicate a fault. **Possible Causes** • IMRC monitor signal circuit shorted to power ground • IMRC Monitor signal circuit shorted to signal ground (return) • IMRC actuator is damaged or has failed (e.g., there may be a small leak in the vacuum diaphragm of the actuator) • PCM has failed
P1520 **2T CCM, MIL: Yes** 1996-1997-1998-1999- 2000-2001-2002-2003 Some Models Transmissions: All	**Intake Manifold Runner Control Input Error Conditions:** Key on or engine running; and the PCM detected the IMRC Monitor signal for was outside of its expected calibrated range. Use the Active Command or Output State Control on a Generic Scan Tool to help determine if an electrical fault is present. **Possible Causes** • IMRC control circuit is open or shorted to ground • IMRC Monitor VREF circuit is open • IMRC is damaged or the PCM has failed
P1530 **2T CCM, MIL: Yes** 1996-1997-1998 All Models Transmissions: All	**Air Conditioning Clutch Circuit Malfunction Conditions:** Key on or engine running; and the PCM detected a circuit fault in the A/C Clutch power (Power To Clutch) circuit. **Possible Causes** • A/C Clutch power circuit open or shorted to VPWR in harness • A/C clutch ground circuit is open • A/C clutch is open • VLCM is damaged or has failed
P1537 **2T CCM, MIL: Yes** 2002-2003 Some Models Transmissions: All	**Intake Manifold Runner Control Malfunction (Bank 1 Stuck Open) Conditions:** Key on or engine running; and the PCM detected the Bank 1 IMRC Monitor signal was less than its expected calibrated range at closed throttle (it may be stuck in open position). An IMRCM PID of 1v at closed throttle may indicate a fault is present. **Possible Causes** • IMRC monitor signal circuit shorted to power ground • IMRC Monitor signal circuit shorted to signal ground (return) • IMRC actuator is damaged or the PCM has failed

Gas Engine OBD II Trouble Code List (P1xxx Codes)

DTC	Trouble Code Title, Conditions & Possible Causes
P1538 **2T CCM, MIL: Yes** 1996-1997-1998-1999-2000-2001-2002-2003 Some Models Transmissions: All	**Intake Manifold Runner Control Stuck Open (Bank 2) Conditions:** Key on or engine running; and the PCM detected the Bank 2 IMRC Monitor signal was more than its expected calibrated range at closed throttle (it may be stuck in open position). An IMRCM PID of VREF at 3000 rpm may indicate a fault is present. **Possible Causes** • IMRC monitor signal circuit shorted to power ground • IMRC Monitor signal circuit shorted to signal ground (return) • IMRC actuator is damaged or has failed • PCM has failed
P1539 **2T CCM, MIL: No** 1996-1997-1998 All Models Transmissions: All	**Power To A/C Clutch Circuit Over-Current Conditions:** Key on or engine running; and with the A/C switch "on", the PCM detected the current in the A/C Clutch power (PWR) circuit exceeded the normal current level during the self-test. **Possible Causes** • A/C Clutch power circuit open or shorted to VPWR in harness • A/C clutch ground circuit is open • A/C clutch is open • VLCM is damaged or has failed
P1549 **1T CCM, MIL: No** 1997-1998-1999-2000-2001-2002-2003 Escort, Focus, E Van & F Series Truck, Ranger, LS, Windstar, ZX2 Models equipped with a 2.0L (VIN 3, VIN 5, VIN P) 2.3L VIN D, 4.2L VIN 2 engine Transmissions: All	**Intake Manifold Tuning Valve Circuit Malfunction Conditions:** Key on or engine running and the PCM detected an unexpected voltage on the Intake Manifold Tuning Valve (IMTV) during testing. An IMT valve (IMTVF) PID of YES status indicates a fault is present. **Possible Causes** • IMTV circuit is open or shorted to ground • IMTV power circuit (VPWR) is open • IMTV assembly is damaged or it has failed • PCM has failed
P1549 **1T CCM, MIL: No** 1997-1998-1999-2000-2001-2002-2003 Blackwood, LS, Sable, Taurus, LS, Expedition & Navigator Models with a 3.0L VIN S, 4.6L VIN W, 5.4L VIN A, 5.4L VIN L, 5.4L VIN R engine Transmissions: All	**Intake Manifold Runner Control Circuit Malfunction Conditions:** Key on or engine running and the PCM detected an unexpected voltage on the Intake Manifold Runner Control (IMRC) during testing. An IMRC valve PID of YES status indicates a fault is present. **Possible Causes** • IMRC circuit is open or shorted to ground • IMRC power circuit (VPWR) is open • IMRC assembly is damaged or it has failed • PCM has failed
P1549 **1T CCM, MIL: No** 2003 Aviator, Explorer & Mountaineer with a 4.6L VIN H engine Transmissions: All	**Long / Short Runner Control Circuit Malfunction Conditions:** Key on or engine running and the PCM detected an unexpected voltage on the Long / Short Runner Control (LSRC) circuit. An LSRC valve PID of YES status indicates a fault exists. **Possible Causes** • LSRC circuit is open or shorted to ground • LSRC power circuit (VPWR) is open • LSRC assembly is damaged or it has failed • PCM has failed
P1549 **1T CCM, MIL: No** 2001-2002-2003 Ranger Models with a 2.3L VIN D engine Transmissions: All	**Intake Manifold Swirl Control Circuit Malfunction Conditions:** Key on or engine running and the PCM detected an unexpected voltage on the Intake Manifold Swirl Control (IMSC) circuit. An IMSC valve PID of YES status indicates a fault. **Possible Causes** • IMSC circuit is open or shorted to ground • IMSC power circuit (VPWR) is open • IMSC assembly is damaged or the PCM has failed

Gas Engine OBD II Trouble Code List (P1xxx Codes)

DTC	Trouble Code Title, Conditions & Possible Causes
P1550 **1T CCM, MIL: No** 1997-1998-1999-2000-2001-2002-2003 All Models Transmissions: All	**Power Steering Pressure Switch Circuit Malfunction Conditions:** KOER Self-Test enabled, and the PCM detected the PSP switch signal did not change during the self-test. This code indicates the PSP input is out of its self-test range. **Possible Causes** • PSP switch circuit open or shorted, or the ground circuit open • Steering wheel was not rotated during the KOER Self-Test • PCM has failed
P1572 **2T CCM, MIL: Yes** 2001-2002-2003 All Models Transmissions: All	**Brake Pedal Switch Circuit Malfunction Conditions:** Engine started, and the PCM detected the Brake Pedal switch and Brake Pressure switch inputs failed the Rationality test (i.e., one or both of these inputs did not change as expected). DTC P1572 is set when the PCM does not see the proper sequence of the brake pedal input signal from both the BPP and BPA when the brake pedal is pressed and released. **Possible Causes** • BPP or BPA switches are out of adjustment (one or both) • Blown fuse to switch power circuit • BPP switch or BPA switch is damaged (one or both) • BPP or BPA switch circuit is open or shorted • PCM has failed
P1605 **1T PCM, MIL: Yes** 1996-1997-1998-1999-2000-2001-2002-2003 All Models Transmissions: All	**PCM Keep Alive Memory Test Error Conditions:** Key on, and the PCM detected an internal memory fault. This code can be set if KAPWR to the PCM is interrupted. This trouble code will set at first key on if a battery circuit is opened. **Possible Causes** • Battery terminals loose or corroded (high resistance in circuit) • Keep Alive Memory circuit to PCM interrupted or open • Reprogramming function not performed • PCM has failed
P1625 **2T CCM, MIL: Yes** 1996-1997-1998 Mark VIII Transmissions: All	**Battery Supply to VLCM Fan Circuit Malfunction Conditions:** Key on or engine running; and the PCM detected no battery power at the VLCM fan circuit. **Possible Causes** • B+ circuit to VLCM (pins 4 and 5) has been open (check for an intermittent condition)
P1626 **2T CCM, MIL: Yes** 1996-1997-1998 Mark VIII Transmissions: All	**Battery Supply To VLCM A/C Circuit Malfunction Conditions:** Key on or engine running; and the PCM detected no battery power at the VLCM fan circuit. **Possible Causes** • B+ circuit to VLCM (pins 4 and 5) is open (check for an intermittent condition)
P1633 **1T PCM, MIL: Yes** 1999-2000-2001-2002-2003 All Models Transmissions: All	**PCM Keep Alive Memory Voltage Too Low Conditions:** Key on, and the PCM detected that the KAM power circuit to the battery was interrupted. **Possible Causes** • KAPWR circuit has been interrupted (this problem may be an intermittent condition) • PCM has failed
P1635 **1T PCM, MIL: Yes** 1999-2000-2001-2002-2003 All Models Transmissions: All	**Tire Axle/Ratio Out Of Acceptable Range Conditions:** Key on, and the PCM detected the tire and axle information in the VID Block does not match the vehicle hardware. Note: This code indicates that the PCM needs to be reprogrammed. **Possible Causes** • Incorrect tire size or Incorrect axle ratio • Incorrect VID configuration parameters • PCM need to be reprogrammed • TSB 02-23-4 contains repair information for this trouble code
P1636 **1T PCM, MIL: Yes** 1999-2000-2001-2002-2003 All Models Transmissions: All	**Inductive Signature Chip Communication Error Conditions:** Key on, and the PCM determined it had lost communication with the Inductive Signature Chip. The PCM has internal damage when this trouble code is present. **Possible Causes** • PCM has failed and needs to be replaced

Gas Engine OBD II Trouble Code List (P1xxx Codes)

DTC	Trouble Code Title, Conditions & Possible Causes
P1639 **1T PCM, MIL: Yes** 2000-2001-2002-2003 All Models Transmissions: All	**Vehicle ID Block Not Programmed Or Is Corrupt Conditions:** Key on, and the PCM determined the Vehicle ID Block information was incorrect. **Possible Causes** • PCM may not be the correct application • PCM may need to be reprogrammed • VID configuration may not be correct • TSB 02-23-4 contains repair information for this trouble code
P1640 **1T PCM, MIL: Yes** 1999-2000-2001-2002-2003 All Models Transmissions: All	**PCM Trouble Codes Available In Another Module Conditions:** Engine started, and the PCM received a request from another module to turn on the MIL due to a fault that could affect emissions. Note: Vehicles using a secondary Engine Control Module can request that the PCM turn on the Check Engine Light when a failure occurs that could affect emissions. Request PID 0946 to determine which module made the request. Then select that module to read the related trouble code(s). **Possible Causes** • Trouble codes are stored in a secondary module, which in turn, requested that the PCM turn on the MIL when this code is set.
P1650 **2T CCM, MIL: Yes** 1996-1997-1998-1999-2000-2001-2002-2003 All Models Transmissions: All	**Power Steering Pressure Switch Circuit Malfunction Conditions:** Engine started, and the PCM detected the PSP switch signal did not change after a certain number of vehicle speed transitions. The PCM counts the number of times that the vehicle speed transitions from 0 mph to a calibrated speed. The PCM expects the PSP switch input to change after a certain number of transitions. **Possible Causes** • Steering wheel must be turned during the KOER Self-Test • PSP switch/shorting bar is damaged • PSP signal circuit is open or shorted to ground • PSP switch ground (return) circuit is open • PCM has failed
P1651 **2T CCM, MIL: Yes** 1996-1997-1998-1999-2000-2001-2002-2003 All Models Transmissions: All	**Power Steering Pressure Switch Circuit Malfunction Conditions:** Engine started, and the PCM detected the PSP switch signal did not change after a certain number of vehicle speed transitions. Note: The PCM counts the number of times that the vehicle speed transitions from 0 mph to a calibrated speed. The PCM expects the PSP switch input to change after a certain number of transitions. **Possible Causes** • Steering wheel must be turned during the KOER Self-Test • PSP switch/shorting bar is damaged • PSP signal circuit is open or shorted to ground • PSP switch ground (return) circuit is open • PCM has failed
P1700 **1T CCM, MIL: No** 1996-1997-1998-1999-2000-2001-2002-2003 All Models Transmissions: A/T	**Transaxle Mechanical Malfunction Conditions:** Engine started, vehicle driven in gear, and the PCM detected a transmission mechanical fault. **Possible Causes** • This code can set due to low transmission fluid level • Refer to the appropriate Transmission Repair Manual or information in electronic media to perform a complete diagnosis of the automatic transmission when this code is set
P1701 **1T CCM, MIL: No** 1996-1997-1998-1999-2000-2001-2002-2003 All Models Transmissions: A/T	**Reverse Engagement Error Conditions:** Engine started, and the PCM detected a Transmission Range (TR) sensor signal that indicated a reverse engagement error. **Possible Causes** • Refer to the appropriate Transmission Repair Manual or information in electronic media to perform a complete diagnosis of the automatic transmission when this code is set
P1702 **1T CCM, MIL: No** 1998-1999-2000-2001-2002-2003 All Models Transmissions: A/T	**TR Sensor Signal Intermittent Conditions:** Key on or engine running; and the PCM detected the failure Trouble Code Conditions for DTC P0705 or P0708 were met intermittently. **Possible Causes** • Refer to the appropriate Transmission Repair Manual or information in electronic media to perform a complete diagnosis of the automatic transmission when this code is set

Gas Engine OBD II Trouble Code List (P1xxx Codes)

DTC	Trouble Code Title, Conditions & Possible Causes
P1703 **1T CCM, MIL: No** 1996-1997-1998-1999- 2000-2001-2002-2003 All Models Transmissions: A/T	**Brake Switch Circuit Out of Self-Test Range Conditions:** Key on, KOEO Self-Test enabled; and the PCM detected the brake switch signal was high, or with the KOER Self-Test enabled, the PCM detected the switch signal did not cycle On / Off. **Possible Causes** • BPP switch circuit open or shorted • Brake Switch is misadjusted, damaged or has failed • Stop lamp circuits open or shorted • Malfunction in the module(s) connected to BPP circuit (i.e., the Rear Electronic Module on Windstar and LS, or the Lighting Control Module on the Continental and Town Car) • PCM has failed
P1704 **1T CCM, MIL: No** 1996-1997-1998-1999- 2000-2001-2002-2003 All Models Transmissions: A/T	**Transmission Range Sensor Circuit Out Of Self-Test Range Conditions:** Key on, KOEO Self Test enabled, and the PCM detected a Transmission Range (TR) sensor signal occurred in between gear positions. **Possible Causes** • Digital TR sensor or shift cable misadjusted • Digital TR sensor circuit is open or shorted to ground • Digital TR sensor has failed
P1705 **1T CCM, MIL: No** 1996-1997-1998-1999- 2000-2001-2002-2003 All Models Transmissions: A/T	**Transmission Range Sensor Out of Self-Test Range Conditions:** Key on, KOEO Self Test enabled, and the PCM detected it did not receive a Transmission Range (TR) sensor signal in Park or Neutral position. **Possible Causes** • Gear selector not in Park or Neutral during the self-test • Digital TR sensor circuit is open or shorted to ground • Digital TR sensor has failed • PCM has failed
P1705 **1T CCM, MIL: Yes** 1996-1997-1998-1999- 2000-2001-2002 Villager Models Transmissions: A/T	**Throttle Position Sensor Circuit Malfunction Conditions:** Key on, KOEO Self-Test enabled, or engine running, KOER Self-Test enabled, and the PCM detected the TP sensor was too high or low. **Possible Causes** • TP sensor signal circuit is open • TP sensor signal circuit is shorted to ground • TP sensor circuit is shorted to VREF or power • TP switch is open, shorted to ground or power • TP sensor or the TP switch has failed, or the PCM has failed
P1708 **1T CCM, MIL: Yes** 1996-1997-1998-1999- 2000-2001-2002-2003 All Models Transmissions: A/T	**Digital Transmission Range Sensor Circuit Malfunction Conditions:** Engine started, and the PCM detected it did not receive a change in the Digital Transmission Range (TR) sensor signal after the vehicle was driven in gear. **Possible Causes** • Digital TR sensor circuit open • Digital TR sensor ground circuit open • Digital TR sensor is damaged or has failed • PCM has failed
P1709 **1T CCM, MIL: No** 1996-1997-1998-1999- 2000-2001-2002-2003 All Models Transmissions: A/T	**PNP Switch Out Of Self-Test Range Conditions:** Key on, KOEO Self-Test enabled, and the PCM detected the PNP switch was high when is should have been low (wrong gearshift position). **Possible Causes** • PNP switch ground circuit is open • PNP switch circuit short to power (VPWR) • PNP switch is damaged or has failed • PCM has failed
P1710 **2T CCM, MIL: Yes** 1998-1999-2000-2001- 2002-2003 All Models Transmissions: A/T	**TFT Sensor In-Range Circuit Malfunction Conditions:** Engine started, vehicle driven to a speed over 1 mph, TFT sensor signal in-range, and the PCM did not detect any change in the TFT signal in the self-test. **Possible Causes** • Refer to the appropriate Transmission Repair Manual or information in electronic media to perform a complete diagnosis of the automatic transmission when this code is set

Gas Engine OBD II Trouble Code List (P1xxx Codes)

DTC	Trouble Code Title, Conditions & Possible Causes
P1711 **1T CCM, MIL: No** 1998-1999-2000-2001-2002-2003 All Models Transmissions: A/T	**TFT Sensor Out of Self-Test Range Conditions:** Key on, KOER Self Test enabled; or engine running with the KOER Self Test enabled, and the PCM detected the Transmission Fluid Temperature (TFT) sensor was more than or less than the calibrated range (25ºF to 240ºF) during the self-test. **Possible Causes** • Refer to the appropriate Transmission Repair Manual or information in electronic media to perform a complete diagnosis of the automatic transmission when this code is set
P1712 **1T CCM, MIL: No** 2000-2001-2002-2003 All Models Transmissions: A/T	**TFT Sensor Circuit Low Input Conditions:** **Engine started, and the PCM detected the TFT sensor signal was less than 0.2v (equivalent to a temperature of more than 357ºF).** **Possible Causes** • Refer to the appropriate Transmission Repair Manual or information in electronic media to perform a complete diagnosis of the automatic transmission when this code is set
P1713 **1T CCM, MIL: No** 1998-1999-2000-2001-2002-2003 All Models Transmissions: A/T	**TFT Sensor No Activity or TFT Sensor Circuit Low Input Conditions:** **Engine started, VSS over 1 mph, and the PCM did not detect any change in the TFT low range circuit during the self-test.** **Possible Causes** • Refer to the appropriate Transmission Repair Manual or information in electronic media to perform a complete diagnosis of the automatic transmission when this code is set
P1714 **1T CCM, MIL: Yes** 1998-1999-2000-2001-2002-2003 All Models Transmissions: A/T	**Transmission Control System Malfunction Conditions:** **Engine started, VSS over 1 mph, and the PCM did not detect any change in the TFT low range circuit during the self-test.** **Possible Causes** • Refer to the appropriate Transmission Repair Manual or information in electronic media to perform a complete diagnosis of the automatic transmission when this code is set
P1715 **1T CCM, MIL: Yes** 1998-1999-2000-2001-2002-2003 All Models Transmissions: A/T	**Transmission Control System Malfunction Conditions:** **Engine started, VSS over 1 mph, and the PCM detected a mechanical problem in the Shift Solenoid 'B' (SSB) during the test.** **Possible Causes** • Refer to the appropriate Transmission Repair Manual or information in electronic media to perform a complete diagnosis of the automatic transmission when this code is set
P1716 **2T CCM, MIL: No** 1998-1999-2000-2001-2002-2003 All Models Transmissions: A/T	**Transmission Control System Malfunction Conditions:** **Engine started, VSS over 1 mph, and the PCM detected a problem in the Transmission Control system during the self-test.** **Possible Causes** • Refer to the appropriate Transmission Repair Manual or information in electronic media to perform a complete diagnosis of the automatic transmission when this code is set
P1717 **1T CCM, MIL: No** 1998-1999-2000-2001-2002-2003 All Models Transmissions: A/T	**Transmission Control System Malfunction Conditions:** **Engine started, VSS over 1 mph, and the PCM detected a problem in the Transmission Control system during the self-test.** **Possible Causes** • Refer to the appropriate Transmission Repair Manual or information in electronic media to perform a complete diagnosis of the automatic transmission when this code is set
P1718 **1T CCM, MIL: No** 1998-1999-2000-2001-2002-2003 All Models Transmissions: A/T	**TFT Sensor No Activity Or TFT Sensor Circuit High Input Conditions:** **Engine started, VSS over 1 mph, and the PCM did not detect any change in the TFT high range circuit during the self-test.** **Possible Causes** • Refer to the appropriate Transmission Repair Manual or information in electronic media to perform a complete diagnosis of the automatic transmission when this code is set
P1719 **1T CCM, MIL: No** 1998-1999-2000-2001-2002-2003 All Models Transmissions: A/T	**Transmission Control System Malfunction Conditions:** **Engine started, VSS over 1 mph, and the PCM detected a problem in the Transmission Control system during the self-test.** **Possible Causes** • Refer to the appropriate Transmission Repair Manual or information in electronic media to perform a complete diagnosis of the automatic transmission when this code is set

Gas Engine OBD II Trouble Code List (P1xxx Codes)

DTC	Trouble Code Title, Conditions & Possible Causes
P1727 **1T CCM, MIL: No** 2000-2001-2002-2003 All Models Transmissions: A/T	**Transmission Coast Clutch Solenoid Slip Malfunction Conditions:** Engine started, VSS over 1 mph in gear, and the PCM detected a signal that indicated the coast clutch solenoid had a slippage fault. **Possible Causes** • Refer to the appropriate Transmission Repair Manual or information in electronic media to perform a complete diagnosis of the automatic transmission when this code is set
P1728 **1T CCM, MIL: No** 1996-1997-1998-1999- 2000-2001-2002-2003 All Models Transmissions: A/T	**Transmission Slip Malfunction Conditions:** Engine started, VSS over 1 mph in gear, and the PCM detected a signal that indicated the transmission was slipping while in gear. **Possible Causes** • Refer to the appropriate Transmission Repair Manual or information in electronic media to perform a complete diagnosis of the automatic transmission when this code is set
P1729 **1T CCM, MIL: No** 1996-1997-1998-1999- 2000-2001-2002-2003 All Models Transmissions: A/T	**4 x 4 Low Switch Circuit Malfunction Conditions:** Engine started and the PCM detected the 4x4 switch did not go low after the switch was on. **Possible Causes** • Speedometer out of calibration • 4x4L wiring harness is open or shorted, 4x4L switch is damaged or has failed • Electronic Shift Control Module is damaged or has failed, or the PCM has failed
P1740 **1T CCM, MIL: Yes** 1998-1999-2000-2001- 2002-2003 All Models Transmissions: A/T	**TCC Solenoid Mechanical Malfunction Conditions:** Engine started, vehicle speed more than 20 mph, and the PCM detected that TCC lockup did not occur (the lockup event is inferred from other inputs). **Possible Causes** • Refer to the appropriate Transmission Repair Manual or information in electronic media to perform a complete diagnosis of the automatic transmission when this code is set
P1741 **1T CCM, MIL: No** 1996-1997-1998-1999- 2000-2001-2002-2003 Some Models Transmissions: A/T	**TCC Engagement Error Conditions:** Engine started, vehicle in gear at Cruise speed, and the PCM detected an error due to excessive TCC engagement. Note: This problem can cause speed changes or vehicle surges. **Possible Causes** • Refer to the appropriate Transmission Repair Manual or information in electronic media to perform a complete diagnosis of the automatic transmission when this code is set
P1741 **1T CCM, MIL: No** 1996-1997-1998-1999- 2000-2001-2002-2003 Escape, MF, Contour, Mystique, Probe Transmissions: A/T	**TCC Engagement Error Conditions:** Engine started, vehicle in gear at Cruise speed, and the PCM detected an error due to excessive TCC engagement. Note: This problem can cause speed changes or vehicle surges. **Possible Causes** • Refer to the appropriate Transmission Repair Manual for more information on this code • TSB 02-2-4 contains a repair procedure for this trouble code
P1742 **1T CCM, MIL: Yes** 1996-1997-1998-1999- 2000-2001-2002-2003 Some Models Transmissions: A/T	**TCC Solenoid Failed On (Electrical Or Mechanical Fault) Conditions:** Engine started, vehicle in gear at Cruise speed, and the PCM detected that the Torque Converter Clutch system had failed "on". **Possible Causes** • Refer to the appropriate Transmission Repair Manual or information in electronic media to perform a complete diagnosis of the automatic transmission when this code is set.
P1744 **1T CCM, MIL: Yes** 1996-1997-1998-1999- 2000-2001-2002-2003 Some Models Transmissions: A/T	**TCC System Mechanically Stuck In Off Position Conditions:** Engine started, vehicle in gear at Cruise speed, and the PCM detected the Torque Converter Clutch system had failed with the TCC in the mechanically "off" position. **Possible Causes** • Refer to the appropriate Transmission Repair Manual or information in electronic media to perform a complete diagnosis of the automatic transmission when this code is set.
P1744 **2T CCM, MIL: No** 1996-1997-1998-1999- 2000-2001-2002-2003 Continental, Taurus, Sable, Windstar models Transmissions: A/T	**EPC Solenoid Circuit Malfunction Conditions:** Engine started, vehicle in gear, and the PCM detected the Electronic Pressure Control (EPC) solenoid control circuit was "open". This fault can cause maximum can cause harsh shifts. **Possible Causes** • Refer to the appropriate Transmission Repair Manual for more information on this code • PCM has failed • TSB 3-12-3 contains a repair procedure for this trouble code

GAS ENGINE OBD II TROUBLE CODE LIST (P1XXX CODES)

DTC	Trouble Code Title, Conditions & Possible Causes
P1746 **1T CCM, MIL: No** 1996-1997-1998-1999-2000-2001-2002-2003 Some Models Transmissions: A/T	**EPC Solenoid Circuit Malfunction Conditions:** Engine started, vehicle in gear, and the PCM detected the Electronic Pressure Control (EPC) solenoid circuit indicated "open". This fault can cause harsh engagements and shifts. **Possible Causes** • Refer to the appropriate Transmission Repair Manual or information in electronic media to perform a complete diagnosis of the automatic transmission when this code is set
P1747 **1T CCM, MIL: No** 1996-1997-1998-1999-2000-2001-2002-2003 Some Models Transmissions: A/T	**A/T EPC Solenoid Circuit Malfunction Conditions:** Engine started, vehicle in gear at Cruise speed, and the PCM detected a shorted output driver or the TCC solenoid was shorted. **Possible Causes** • Refer to the appropriate Transmission Repair Manual or information in electronic media to perform a complete diagnosis of the automatic transmission when this code is set.
P1749 **1T CCM, MIL: No** 1996-1997-1998-1999-2000-2001-2002-2003 Some Models Transmissions: A/T	**A/T EPC Solenoid Failed Low Conditions:** Engine started, vehicle in gear at Cruise speed, and the PCM detected the Torque Converter Clutch solenoid had failed "low". **Possible Causes** • Refer to the appropriate Transmission Repair Manual or information in electronic media to perform a complete diagnosis of the automatic transmission when this code is set.
P1751 **1T CCM, MIL: No** 1996-1997-1998-1999-2000-2001-2002-2003 Some Models Transmissions: A/T	**A/T Shift Solenoid 1 Performance Conditions:** Engine started, vehicle in gear at Cruise speed, and the PCM detected a mechanical fault in the Shift Solenoid 1 (SS1) operation. **Possible Causes** • Refer to the appropriate Transmission Repair Manual or information in electronic media to perform a complete diagnosis of the automatic transmission when this code is set.
P1754 **1T CCM, MIL: No** 1996-1997-1998-1999-2000-2001-2002-2003 Some Models Transmissions: A/T	**A/T Coast Clutch Solenoid Circuit Malfunction Conditions:** Engine started, vehicle in gear at Cruise speed, and the PCM detected an unexpected voltage condition on the Coast Clutch Solenoid (CCS) circuit during the CCM test period. **Possible Causes** • Refer to the appropriate Transmission Repair Manual or information in electronic media to perform a complete diagnosis of the automatic transmission when this code is set.
P1756 **1T CCM, MIL: No** 1996-1997-1998-1999-2000-2001-2002-2003 Some Models Transmissions: A/T	**A/T Shift Solenoid 2 Performance Conditions:** Engine started, vehicle in gear at Cruise speed, and the PCM detected a mechanical fault in the Shift Solenoid 2 (SS2) operation. **Possible Causes** • Refer to the appropriate Transmission Repair Manual or information in electronic media to perform a complete diagnosis of the automatic transmission when this code is set
P1760 **1T CCM, MIL: No** 1996-1997-1998-1999-2000-2001-2002-2003 Some Models Transmissions: A/T	**A/T EPC Solenoid Circuit Malfunction Conditions:** Engine started, vehicle in gear at Cruise speed, and the PCM detected a shorted output driver or the TCC solenoid was shorted. **Possible Causes** • Refer to the appropriate Transmission Repair Manual or information in electronic media to perform a complete diagnosis of the automatic transmission when this code is set.
P1761 **1T CCM, MIL: No** 1996-1997-1998-1999-2000-2001-2002-2003 Some Models Transmissions: A/T	**A/T Shift Solenoid 3 Performance Conditions:** Engine started, vehicle in gear at Cruise speed, and the PCM detected a malfunction in the Shift Solenoid 3 (SS3) operation. **Possible Causes** • Refer to the appropriate Transmission Repair Manual or information in electronic media to perform a complete diagnosis of the automatic transmission when this code is set.
P1762 **1T CCM, MIL: No** 1996-1997-1998-1999-2000-2001-2002-2003 Some Models Transmissions: A/T	**Transmission System Malfunction Conditions:** Engine started, vehicle in gear at Cruise speed, and the PCM detected a malfunction in the Transmission System operation. **Possible Causes** • Refer to the appropriate Transmission Repair Manual or information in electronic media to perform a complete diagnosis of the automatic transmission when this code is set.

GAS ENGINE OBD II TROUBLE CODE LIST (P1XXX CODES)

DTC	Trouble Code Title, Conditions & Possible Causes
P1767 **1T CCM, MIL: No** 1998-1999-2000-2001-2002-2003 Some Models Transmissions: A/T	**A/T Shift Solenoid Performance Conditions:** Engine started, vehicle in gear at Cruise speed, and the PCM detected a malfunction in the Shift Solenoid operation. **Possible Causes** • Refer to the appropriate Transmission Repair Manual or information in electronic media to perform a complete diagnosis of the automatic transmission when this code is set.
P1780 **1T CCM, MIL: No** 1996-1997-1998-1999-2000-2001-2002-2003 Some Models Transmissions: A/T	**Transmission Control Switch Out of Self-Test Range Conditions:** Engine started, KOER Self-Test enabled, and the PCM detected the Transmission Control Switch (TCS) was out of range during the test. **Possible Causes** • TCS circuit open or shorted in the wiring harness • TCS not cycled during the self-test • TCS is damaged, or the PCM has failed
P1781 **1T CCM, MIL: No** 1996-1997-1998-1999-2000-2001-2002-2003 Some Models Transmissions: All	**4 x 4 Low Switch Out Of Self-Test Range Conditions:** Key on, KOEO Self-Test enabled, and the PCM detected the 4x4 switch input was not low with the switch engaged or "on". **Possible Causes** • 4x4L switch circuit is open or shorted in the wiring harness • Electronic Shift Module is damaged or has failed • PCM has failed
P1783 **1T CCM, MIL: No** 1996-1997-1998-1999-2000-2001-2002-2003 Some Models Transmissions: A/T	**Transmission Over-Temperature Malfunction Conditions:** Engine started, engine runtime more than 5 minutes, vehicle in gear at Cruise speed, and the PCM detected the TFT sensor signal was more than 300ºF during the CCM test period. **Possible Causes** • Refer to the appropriate Transmission Repair Manual or information in electronic media to perform a complete diagnosis of the automatic transmission when this code is set.
P1784 **1T CCM, MIL: No** 1996-1997-1998-1999-2000-2001-2002-2003 Some Models Transmissions: A/T	**Transmission System First Or Reverse Gear Malfunction Conditions:** Engine started, vehicle speed over 1 mph in gear, shift command received for First or Reverse gear, and the PCM detected a problem in the Transmission Control system. **Possible Causes** • Refer to the appropriate Transmission Repair Manual or information in electronic media to perform a complete diagnosis of the automatic transmission when this code is set.
P1785 **1T CCM, MIL: No** 1996-1997-1998-1999-2000-2001-2002-2003 Some Models Transmissions: A/T	**Transmission System First Or Second Gear Malfunction Conditions:** Engine started, vehicle speed over 1 mph in gear, shift command received for First or Second gear, and the PCM detected a problem in the Transmission Control system during the test. **Possible Causes** • Refer to the appropriate Transmission Repair Manual or information in electronic media to perform a complete diagnosis of the automatic transmission when this code is set
P1786 **1T CCM, MIL: No** 1996-1997-1998-1999-2000-2001-2002-2003 Some Models Transmissions: A/T	**Transmission System Second Or Third Gear Malfunction Conditions:** Engine started, vehicle speed over 1 mph in gear, shift command received for Second or Third gear, and the PCM detected a problem in the Transmission Control system. **Possible Causes** • Refer to the appropriate Transmission Repair Manual or information in electronic media to perform a complete diagnosis of the automatic transmission when this code is set.
P1787 **1T CCM, MIL: No** 1996-1997-1998-1999-2000-2001-2002-2003 Some Models Transmissions: A/T	**Transmission System Third Or Fourth Gear Malfunction Conditions:** Engine started, vehicle speed over 1 mph in gear, shift command received for Third or Fourth gear, and the PCM detected a problem in the Transmission Control system during the test. **Possible Causes** • Refer to the appropriate Transmission Repair Manual or information in electronic media to perform a complete diagnosis of the automatic transmission when this code is set.
P1788 **1T CCM, MIL: No** 1996-1997-1998-1999-2000-2001-2002-2003 Some Models Transmissions: A/T	**3-2 Timing/Coast Clutch Solenoid Signal High Input Conditions:** Engine started, vehicle in gear at Cruise speed, and the PCM detected the malfunction 3-2 Timing or Coast Clutch solenoid circuit. **Possible Causes** • 3-2 Timing or Coast Clutch solenoid circuit open or grounded, or the solenoid has failed • Coast Clutch solenoid is damaged or has failed

Gas Engine OBD II Trouble Code List (P1xxx Codes) - (P2xxx Codes)

DTC	Trouble Code Title, Conditions & Possible Causes
P1789 **1T CCM, MIL: No** 1996-1997-1998-1999-2000-2001-2002-2003 Some Models Transmissions: A/T	**3-2 Timing/Coast Clutch Solenoid Signal Low Input Conditions:** Engine started, vehicle in gear at Cruise speed, and the PCM detected the malfunction 3-2 Timing or Coast Clutch solenoid circuit. **Possible Causes** • 3-2 Timing or Coast Clutch solenoid circuit is shorted • 3-2 Timing solenoid is damaged or has failed • Coast Clutch solenoid is damaged or has failed
P1900 **1T CCM, MIL: No** 1998-1999-2000-2001-2002-2003 Some Models Transmissions: A/T	**Transmission System Malfunction Conditions:** Engine started, vehicle in gear at Cruise speed, and the PCM detected a malfunction in the Transmission System operation. **Possible Causes** • Refer to the appropriate Transmission Repair Manual or information in electronic media to perform a complete diagnosis of the automatic transmission when this code is set
P1901 **1T CCM, MIL: No** 1998-1999-2000-2001-2002-2003 Some Models Transmissions: A/T	**Transmission System Malfunction Conditions:** Engine started, vehicle in gear at Cruise speed, and the PCM detected a malfunction in the Transmission System operation. **Possible Causes** • Refer to the appropriate Transmission Repair Manual or information in electronic media to perform a complete diagnosis of the automatic transmission when this code is set.
P2004 **1T CCM, MIL: No** 2003 Escort, Focus, Aviator, Mountaineer, Explorer, LS, Ranger, Blackwood E Van, F Series Truck, Expedition, Navigator, Windstar, ZX2 with a 2.0L (VIN 3, VIN 5, 2.0L VIN P), 2.3L VIN D, 3.0L VIN S, 4.2L VIN 2, 4.6L VIN H, 4.6L VIN W, 5.4L VIN A, 5.4L VIN L, 5.4L VIN R engine Transmissions: All	**Intake Air System Malfunction Conditions:** Engine started, engine running at hot idle speed for one minute, and the PCM detected a problem in the Intake Air System operation. It should be noted that the throttle bore cannot be cleaned as any attempt to clean it will damage the throttle bore and plate. **Possible Causes** • Test for a sticking Accelerator or speed control cable condition: Turn the key off and disconnect accelerator and speed control cable from the throttle body. Rotate the throttle body linkage to determine if it rotates freely (the throttle body may have failed). • Check the air cleaner and air inlet assembly for restrictions • Check the IAC motor response (it may be damaged or sticking) • Check the PCV system (valve and hoses) for leaks or plugging • Check for signs of vacuum leaks in the engine or components • Test TP sensor signal (due a sweep test at key on, engine off)
P2005 **1T CCM, MIL: No** 2003 Escort, Focus, Aviator, Mountaineer, Explorer, LS, Ranger, Blackwood E Van, F Series Truck, Expedition, Navigator, Windstar, ZX2 with a 2.0L (VIN 3, VIN 5, 2.0L VIN P), 2.3L VIN D, 3.0L VIN S, 4.2L VIN 2, 4.6L VIN H, 4.6L VIN W, 5.4L VIN A, 5.4L VIN L, 5.4L VIN R engine Transmissions: All	**Intake Air System Malfunction Conditions:** Engine started, engine running at hot idle speed for one minute, and the PCM detected a problem in the Intake Air System operation. It should be noted that the throttle bore cannot be cleaned as any attempt to clean it will damage the throttle bore and plate. **Possible Causes** • Test for a sticking Accelerator or speed control cable condition: Turn the key off and disconnect accelerator and speed control cable from the throttle body. Rotate the throttle body linkage to determine if it rotates freely (the throttle body may have failed). • Check the air cleaner and air inlet assembly for restrictions • Check the IAC motor response (it may be damaged or sticking) • Check the PCV system (valve and hoses) for leaks or plugging • Check for signs of vacuum leaks in the engine or components • Test TP sensor signal (due a sweep test at key on, engine off)

Gas Engine OBD II Trouble Code List (P2xxx Codes)

DTC	Trouble Code Title, Conditions & Possible Causes
P2006 **1T CCM, MIL: No** 2003 Escort, Focus, Aviator, Mountaineer, Explorer, LS, Ranger, Blackwood E Van, F Series Truck, Expedition, Navigator, Windstar, ZX2 with a 2.0L (VIN 3, VIN 5, 2.0L VIN P), 2.3L VIN D, 3.0L VIN S, 4.2L VIN 2, 4.6L VIN H, 4.6L VIN W, 5.4L VIN A, 5.4L VIN L, 5.4L VIN R engine Transmissions: All	**Intake Air System Malfunction Conditions:** Engine started, engine running at hot idle speed for one minute, and the PCM detected a problem in the Intake Air System operation. It should be noted that the throttle bore cannot be cleaned as any attempt to clean it will damage the throttle bore and plate. **Possible Causes** • Test for a sticking Accelerator or speed control cable condition: Turn the key off and disconnect accelerator and speed control cable from the throttle body. Rotate the throttle body linkage to determine if it rotates freely (the throttle body may have failed). • Check the air cleaner and air inlet assembly for restrictions • Check the IAC motor response (it may be damaged or sticking) • Check the PCV system (valve and hoses) for leaks or plugging • Check for signs of vacuum leaks in the engine or components • Test TP sensor signal (due a sweep test at key on, engine off)
P2008 **1T CCM, MIL: No** 2003 Escort, Focus, Aviator, Mountaineer, Explorer, LS, Ranger, Blackwood E Van, F Series Truck, Expedition, Navigator, Windstar, ZX2 with a 2.0L (VIN 3, VIN 5, 2.0L VIN P), 2.3L VIN D, 3.0L VIN S, 4.2L VIN 2, 4.6L VIN H, 4.6L VIN W, 5.4L VIN A, 5.4L VIN L, 5.4L VIN R engine Transmissions: All	**Intake Air System Malfunction Conditions:** Engine started, engine running at hot idle speed for one minute, and the PCM detected a problem in the Intake Air System operation. It should be noted that the throttle bore cannot be cleaned as any attempt to clean it will damage the throttle bore and plate. **Possible Causes** • Accelerator or speed control cable sticking or binding. To test for this condition, turn the key off. Then disconnect the accelerator and speed control cable from the throttle body. Then rotate the throttle body linkage to determine if it rotates freely. If it is sticking, the throttle body may need replacement. • Check the air cleaner and air inlet assembly for restrictions • Check the IAC motor response (it may be damaged or sticking) • Check the PCV system (valve and hoses) for leaks or plugging • Check for signs of vacuum leaks in the engine or components • Test TP sensor signal (due a sweep test at key on, engine off)
P2014 **1T CCM, MIL: No** 2003 Escort, Focus, Aviator, Mountaineer, Explorer, LS, Ranger, Blackwood E Van, F Series Truck, Expedition, Navigator, Windstar, ZX2 with a 2.0L (VIN 3, VIN 5, 2.0L VIN P), 2.3L VIN D, 3.0L VIN S, 4.2L VIN 2, 4.6L VIN H, 4.6L VIN W, 5.4L VIN A, 5.4L VIN L, 5.4L VIN R engine Transmissions: All	**Intake Air System Malfunction Conditions:** Engine started, engine running at hot idle speed for one minute, and the PCM detected a problem in the Intake Air System operation. It should be noted that the throttle bore cannot be cleaned as any attempt to clean it will damage the throttle bore and plate. **Possible Causes** • Accelerator or speed control cable sticking or binding. To test for this condition, turn the key off. Then disconnect the accelerator and speed control cable from the throttle body. Then rotate the throttle body linkage to determine if it rotates freely. If it is sticking, the throttle body may need replacement. • Check the air cleaner and air inlet assembly for restrictions • Check the IAC motor response (it may be damaged or sticking) • Check the PCV system (valve and hoses) for leaks or plugging • Check for signs of vacuum leaks in the engine or components • Test TP sensor signal (due a sweep test at key on, engine off)

Gas Engine OBD II Trouble Code List (P2xxx Codes)

DTC	Trouble Code Title, Conditions & Possible Causes
P2019 **1T CCM, MIL: No** 2003 Escort, Focus, Aviator, Mountaineer, Explorer, LS, Ranger, Blackwood E Van, F Series Truck, Expedition, Navigator, Windstar, ZX2 with a 2.0L (VIN 3, VIN 5, 2.0L VIN P), 2.3L VIN D, 3.0L VIN S, 4.2L VIN 2, 4.6L VIN H, 4.6L VIN W, 5.4L VIN A, 5.4L VIN L, 5.4L VIN R engine Transmissions: All	**Intake Air System Malfunction Conditions:** Engine started, engine running at hot idle speed for one minute, and the PCM detected a problem in the Intake Air System operation. It should be noted that the throttle bore cannot be cleaned as any attempt to clean it will damage the throttle bore and plate. **Possible Causes** • Accelerator or speed control cable sticking or binding. To test these devices, turn the key off and disconnect the accelerator and speed control cable from the throttle body. Then rotate the throttle body linkage to determine if it rotates freely. • Check the air cleaner and air inlet assembly for restrictions • Check the IAC motor response (it may be damaged or sticking) • Check the PCV system (valve and hoses) for leaks or plugging • Check for signs of vacuum leaks in the engine or components • Test TP sensor signal (due a sweep test at key on, engine off)
P2070 **1T CCM, MIL: No** 2003 Escort, Focus, E Van, F Series Truck, Windstar, ZX2 engines with a 2.0L (VIN 3, VIN 5, VIN P), 2.3L VIN D, 4.2L VIN 2 engine Transmissions: All	**Intake Manifold Tuning Valve Malfunction (Stuck Open) Conditions:** Key on or engine running; and the PCM detected an unexpected low voltage condition on the Intake Manifold Tuning Valve circuit during the CCM test period (i.e., the valve may be stuck open). **Possible Causes** • IMTV signal circuit shorted to chassis ground • IMTV signal circuit shorted to sensor ground • IMTV actuator is damaged or has failed • PCM has failed
P2070 **2T CCM, MIL: No** 2003 E Van, F Series Truck, Blackwood, Expedition Navigator & LS with a 3.0L VIN S, 4.6L VIN W, 5.4L VIN A, 5.4L VIN L, 5.4L VIN R Transmissions: All	**Intake Manifold Runner Control Malfunction (Stuck Open) Conditions:** Key on or engine running; and the PCM detected an unexpected low voltage condition on the Intake Manifold Runner Control circuit during the CCM test period (i.e., the valve may be stuck open). **Possible Causes** • IMRC signal circuit shorted to chassis ground • IMRC signal circuit shorted to sensor ground • IMRC actuator is damaged or has failed • PCM has failed
P2070 **2T CCM, MIL: No** 2003 Aviator, Mountaineer, Explorer equipped with a 4.6L VIN H engine Transmissions: All	**Long / Short Runner Control Circuit Malfunction (Open) Conditions:** Key on or engine running; and the PCM detected an unexpected low voltage condition on the Long / Short Runner Control (LSRC) signal circuit during the test (i.e., the valve may be stuck open). The PCM uses the Intake Manifold Communication Control (IMCC) signal to monitor the status of the Long / Short Runner Control position. **Possible Causes** • LSRC signal circuit is shorted to chassis ground • LSRC signal circuit is shorted to sensor ground • LSRC actuator assembly is damaged or failed • PCM has failed
P2070 **2T CCM, MIL: No** 2003 Ranger equipped with a 2.3L VIN D engine Transmissions: All	**Intake Manifold Swirl Control Actuator Circuit Malfunction Conditions:** Key on or engine running; and the PCM detected an unexpected low voltage condition on the Intake Manifold Swirl Control (IMSC) circuit (i.e., the valve may be stuck open). **Possible Causes** • IMSC signal circuit is shorted to chassis ground • IMSC signal circuit is shorted to sensor ground • IMSC actuator assembly is damaged or failed • PCM has failed

Gas Engine OBD II Trouble Code List (P2xxx Codes)

DTC	Trouble Code Title, Conditions & Possible Causes
P2071 **2T CCM, MIL: No** 2003 Escort, Focus, E Van, F Series Truck, Windstar, ZX2 engines with a 2.0L (VIN 3, VIN 5, VIN P), 2.3L VIN D, 4.2L VIN 2 engine Transmissions: All	**Intake Manifold Tuning Valve Circuit Malfunction (Stuck Closed) Conditions:** Key on or engine running; and the PCM detected an unexpected high voltage condition on the Intake Manifold Tuning Valve circuit during the CCM test period (i.e., the valve may be stuck closed). **Possible Causes** • IMTV signal circuit is open • IMTV power circuit (VPWR) is open • IMTV actuator is damaged or has failed • PCM has failed
P2071 **2T CCM, MIL: No** 2003 E Van, F Series Truck, Blackwood, Expedition Navigator & LS with a 3.0L VIN S, 4.6L VIN W, 5.4L VIN A, 5.4L VIN L, 5.4L VIN R Transmissions: All	**Intake Manifold Runner Control Circuit Malfunction (Stuck Closed) Conditions:** Key on or engine running; and the PCM detected an unexpected high voltage condition on the Intake Manifold Runner Control (IMRC) circuit during the CCM test (i.e., the valve may be stuck closed). **Possible Causes** • IMRC monitor signal circuit is open • IMRC power circuit (VPWR) is open • IMRC actuator is damaged or has failed • PCM has failed
P2071 **2T CCM, MIL: No** 2003 Aviator, Mountaineer, Explorer equipped with a 4.6L VIN H engine Transmissions: All	**Long / Short Runner Control Circuit Malfunction (Stuck Closed) Conditions:** Key on or engine running; and the PCM detected an unexpected high voltage condition on the Long / Short Runner Control (LSRC) circuit (i.e., the valve may be stuck closed). The PCM uses the IMCC signal to monitor the status of the Long / Short Runner Control valve. **Possible Causes** • LSRC control (signal) circuit is open • LSRC power circuit (VPWR) is open • LSRC actuator assembly is damaged or failed • PCM has failed
P2071 **2T CCM, MIL: No** 2003 Ranger equipped with a 2.3L VIN D engine Transmissions: All	**Intake Manifold Swirl Control Circuit Malfunction (Stuck Closed) Conditions:** Key on or engine running; and the PCM detected an unexpected high voltage condition on the Intake Manifold Swirl Control (IMSC) Monitor circuit (i.e., the valve is stuck closed). **Possible Causes** • IMSC control (signal) circuit is open • IMSC power circuit (VPWR) is open • IMSC actuator assembly is damaged or failed • PCM has failed
P2075 **12T CCM, MIL: No** 2003 Escort, Focus, E Van, F Series Truck, Windstar, ZX2 engines with a 2.0L (VIN 3, VIN 5, VIN P), 2.3L VIN D, 4.2L VIN 2 engine Transmissions: All	**Intake Manifold Tuning Valve Monitor Circuit Malfunction Conditions:** Key on or engine running; and the PCM detected an unexpected low or high voltage condition on the Intake Manifold Tuning Valve Monitor circuit during the CCM test period. **Possible Causes** • IMTV monitor signal circuit is open • IMTV monitor signal circuit shorted to chassis ground • IMTV actuator is damaged or has failed • PCM has failed
P2075 **2T CCM, MIL: No** 2003 E Van, F Series Truck, Blackwood, Expedition Navigator & LS with a 3.0L VIN S, 4.6L VIN W, 5.4L VIN A, 5.4L VIN L, 5.4L VIN R Transmissions: All	**Intake Manifold Runner Control Monitor Circuit Malfunction Conditions:** Key on or engine running; and the PCM detected an unexpected high voltage condition on the Intake Manifold Runner Control (IMRC) Monitor circuit during the CCM test period. **Possible Causes** • IMRC monitor signal circuit is open • IMRC monitor signal circuit shorted to chassis ground • IMRC actuator is damaged or has failed • PCM has failed

Gas Engine OBD II Trouble Code List (P2xxx Codes)

DTC	Trouble Code Title, Conditions & Possible Causes
P2075 **2T CCM, MIL: No** 2003 Aviator, Mountaineer, Explorer equipped with a 4.6L VIN H engine Transmissions: All	**Long / Short Runner Control Monitor Circuit Malfunction Conditions:** Key on or engine running; and the PCM detected an unexpected low or high voltage condition on the Intake Manifold Communication Control (IMCC) Monitor circuit during the CCM test period. **Possible Causes** • LSRC control (signal) circuit is open • LSRC control (signal) circuit is shorted to chassis ground • LSRC actuator assembly is damaged or failed • PCM has failed
P2075 **2T CCM, MIL: No** 2003 Ranger equipped with a 2.3L VIN D engine Transmissions: All	**Intake Manifold Swirl Control Monitor Circuit Malfunction Conditions:** Key on or engine running; and the PCM detected an unexpected high voltage condition on the Intake Manifold Swirl Control (IMSC) Actuator monitor circuit during the CCM test period. **Possible Causes** • IMSC monitor circuit is open • IMSC monitor circuit is shorted to chassis ground • IMSC actuator assembly is damaged or failed • PCM has failed
P2075 **2T PCM, MIL: No** 2003 LS & Thunderbird Models Transmissions: All	**Throttle Actuator Control Motor Circuit Malfunction (Open) Conditions:** Key on or engine running; and the PCM detected an unexpected voltage condition on the Throttle Actuator Control Motor (TACM) circuit during the CCM test period. **Possible Causes** • TACM wiring harness connector is damaged or open • TACM (motor) circuit is open • TACM assembly is damaged or it has failed (an open circuit) • PCM has failed
P2101 **2T PCM, MIL: No** 2003 LS & Thunderbird Models Transmissions: All	**Throttle Actuator Control Motor Range/Performance Conditions:** Key on or engine running; and the PCM detected an unexpected low or high voltage condition on the Throttle Actuator Control Motor (TACM) circuit during the CCM test. **Possible Causes** • TACM wiring harness connector is damaged or open • TACM wiring may be crossed in the wire harness assembly • TACM (motor) circuit is open, or TACM assembly is damaged (possible open circuit) • PCM has failed
P2104 **2T PCM, MIL: No** 2003 LS & Thunderbird Models Transmissions: All	**Throttle Actuator Control System - Forced Idle Mode Conditions:** Key on, and the PCM detected the Throttle Actuator Control Motor (TACM) system was in Forced Idle mode while operating in Failure Mode Effect Management (FMEM). **Possible Causes** • PCM is damaged. Clear the codes (do a PCM reset), and if the same trouble code resets, the PCM will have to be replaced.
P2105 **2T PCM, MIL: No** 2003 LS & Thunderbird Models Transmissions: All	**Throttle Actuator Control System - Forced Engine Shutdown Conditions:** Key on, and the PCM detected the Throttle Actuator Control Motor (TACM) system was in Forced Engine Shutdown mode while in Failure Mode Effect Management (FMEM). **Possible Causes** • A total system failure has occurred • PCM is damaged. Clear the codes (do a PCM reset), and if the same trouble code resets, the PCM will have to be replaced.
P2106 **2T PCM, MIL: No** 2003 LS & Thunderbird Models Transmissions: All	**Title: Throttle Actuator Control System - Forced Limited Power** **Trouble Code Conditions** Key on, and the PCM detected the Throttle Actuator Control Motor (TACM) system was in Forced Limited Power mode while operating in Failure Mode Effect Management (FMEM). **Possible Causes** • TACM (motor) wiring harness is disconnected or loose • TACM (motor) circuits are shorted to power • PCM may have failed. Clear the codes (do a PCM reset), and if the same code resets, the PCM will have to be replaced.

Gas Engine OBD II Trouble Code List (P2xxx Codes)

DTC	Trouble Code Title, Conditions & Possible Causes
P2107 **2T PCM, MIL: No** 2003 LS & Thunderbird Models Transmissions: All	**Throttle Actuator Control Motor Processor Malfunction Conditions:** Key on or engine running; and the PCM detected the Throttle Actuator Control Motor processor received an invalid command or the TACM processor did not execute a command. **Possible Causes** • TACM (motor) wiring harness is shorted, or TACM signal wires are shorted together • TACM (motor) circuits are shorted to power • PCM is damaged. Clear the codes. If the same code resets, the PCM needs replacement.
P2110 **2T PCM, MIL: No** 2003 LS & Thunderbird Models Transmissions: All	**Throttle Actuator Control System - Forced Limited RPM Conditions:** Key on or engine running; and the PCM detected the Throttle Actuator Control System was operating in Forced Limited RPM mode while in Failure Mode Effect Management (FMEM). **Possible Causes** • PCM is damaged. Clear the codes. If the same code resets, the PCM needs replacement.
P2111 **2T PCM, MIL: No** 2003 LS & Thunderbird Models Transmissions: All	**Throttle Actuator Control System - Stuck Open Conditions:** Key on or engine running; and the PCM detected the throttle plate angle (opening) was more the commanded amount during testing. **Possible Causes** • PCM is damaged. Clear the codes. If the same code resets, the PCM needs replacement.
P2112 **2T PCM, MIL: No** 2003 LS & Thunderbird Models Transmissions: All	**Throttle Actuator Control System - Stuck Closed Conditions:** Key on or engine running; and the PCM detected the throttle plate angle (opening) was less the commanded amount during testing. **Possible Causes** • TACM wiring may be crossed in the wire harness assembly • Throttle body is binding - the throttle is stuck closed when these conditions are present • PCM has failed
P2119 **2T PCM, MIL: No** 2003 LS & Thunderbird Models Transmissions: All	**Throttle Actuator Control Throttle Body Range/Performance Conditions:** Key on or engine running; and the PCM detected a signal that indicated the throttle return spring was damaged or it had failed. **Possible Causes** • Throttle body is binding or sticking • Throttle return spring is damaged or it is broken • PCM has failed
P2121 **2T PCM, MIL: No** 2003 LS & Thunderbird Models Transmissions: All	**Accelerator Pedal Position Sensor 'D' Signal Range/Performance Conditions:** Key on or engine running; and the PCM detected the Accelerator Pedal Position Sensor 'D' signal circuit was out of the normal operating range during the CCM test. **Possible Causes** • APP sensor signal circuits are shorted together • APP sensor is damaged or the PCM has failed
P2122 **2T PCM, MIL: No** 2003 LS & Thunderbird Models Transmissions: All	**Accelerator Pedal Position Sensor 'D' Circuit Low Input Conditions:** Key on or engine running; and the PCM detected the Accelerator Pedal Position Sensor 'D' signal circuit was less than the normal range during the test period. **Possible Causes** • APP sensor signal circuit is open • APP sensor signal circuit is shorted to ground • APP sensor is damaged or it has failed • PCM has failed
P2123 **2T PCM, MIL: No** 2003 LS & Thunderbird Models Transmissions: All	**Accelerator Pedal Position Sensor 'D' Circuit High Input Conditions:** Key on or engine running; and the PCM detected the Accelerator Pedal Position Sensor 'D' signal circuit was more than the normal range during the test period. **Possible Causes** • APP sensor connector is damaged or shorted • APP sensor signal circuit is shorted to VREF (5v) • APP sensor is damaged or it has failed • PCM has failed

Gas Engine OBD II Trouble Code List (P2xxx Codes)

DTC	Trouble Code Title, Conditions & Possible Causes
P2126 **2T PCM, MIL: No** 2003 LS & Thunderbird Models Transmissions: All	**Accelerator Pedal Position Sensor 'E' Signal Range/Performance Conditions:** Key on or engine running; and the PCM detected the Accelerator Pedal Position Sensor 'E' signal circuit was more than the normal range during the test period. **Possible Causes** • APP sensor connector is damaged or shorted • APP sensor signal circuit is shorted to VREF (5v) • APP sensor is damaged or the PCM has failed
P212 **2T PCM, MIL: No** 2003 LS & Thunderbird Models Transmissions: All	**Accelerator Pedal Position Sensor 'E' Circuit Low Input Conditions:** Key on or engine running; and the PCM detected the Accelerator Pedal Position Sensor 'D' signal circuit was less than the normal range during the test period. **Possible Causes** • APP sensor signal circuit is open • APP sensor signal circuit is shorted to ground • APP sensor is damaged or the PCM has failed
P2128 **2T PCM, MIL: No** 2003 LS & Thunderbird Models Transmissions: All	**Accelerator Pedal Position Sensor 'E' Circuit High Input Conditions:** Key on or engine running; and the PCM detected the Accelerator Pedal Position Sensor 'E' signal circuit was more than the normal range during the test period. **Possible Causes** • APP sensor connector is damaged or shorted • APP sensor signal circuit is shorted to VREF (5v) • APP sensor is damaged or it has failed • PCM has failed
P2131 **2T PCM, MIL: No** 2003 LS & Thunderbird Models Transmissions: All	**Accelerator Pedal Position Sensor 'F' Signal Range/Performance Conditions:** Key on or engine running; and the PCM detected the Accelerator Pedal Position Sensor 'F' signal circuit was more than the normal range during the test period. **Possible Causes** • APP sensor connector is damaged or shorted • APP sensor signal circuit is shorted to VREF (5v) • APP sensor is damaged or it has failed • PCM has failed
P2132 **2T PCM, MIL: No** 2003 LS & Thunderbird Models Transmissions: All	**Accelerator Pedal Position Sensor 'F' Circuit Low Input Conditions:** Key on or engine running; and the PCM detected the Accelerator Pedal Position Sensor 'F' signal circuit was less than the normal range during the test period. **Possible Causes** • APP sensor signal circuit is open • APP sensor signal circuit is shorted to ground • APP sensor is damaged or it has failed • PCM has failed
P2133 **2T PCM, MIL: No** 2003 LS & Thunderbird Models Transmissions: All	**Accelerator Pedal Position Sensor 'F' Circuit High Input Conditions:** Key on or engine running; and the PCM detected the Accelerator Pedal Position Sensor 'F' signal circuit was more than the normal range during the test period. **Possible Causes** • APP sensor connector is damaged or shorted • APP sensor signal circuit is shorted to VREF (5v) • APP sensor is damaged or it has failed • PCM has failed
P2135 **2T PCM, MIL: No** 2003 LS & Thunderbird Models Transmissions: All	**ETC Throttle Position Sensor A/B Voltage Correlation Conditions:** Key on or engine running; and the PCM detected the Throttle Position 'A' (TPA) and Throttle Position 'B' (TPB) sensors disagreed, or that the TPA sensor should not be in its detected position, or that the TPB sensor should not be in its detected position during testing. **Possible Causes** • ETC TP sensor connector is damaged or shorted • ETC TP sensor circuits shorted together in the wire harness • ETC TP sensor signal circuit is shorted to VREF (5v) • ETC TP sensor is damaged or the PCM has failed

Gas Engine OBD II Trouble Code List (P2xxx Codes)

DTC	Trouble Code Title, Conditions & Possible Causes
P2195 **2T CCM, MIL: No** 2003 All Models Transmissions: All	**Lack of HO2S-11 Switching, Sensor Indicates Lean Conditions:** DTC P0300-P0310 not set, engine running in closed loop, and the PCM detected the HO2S indicated a lean signal, or it could no longer control Fuel Trim because it was at lean limit. **Possible Causes** • Base engine problems: engine oil level high, camshaft timing error, cylinder compression low, exhaust leaks in front of HO2S • EGR System problem: EGR valve is stuck open, the gasket is leaking, or the EVR diaphragm is leaking • Fuel System problem: damaged fuel pressure regulator or extremely low fuel pressure • HO2S problems: HO2S circuit is open or shorted in the wiring harness or the HO2S is damaged or it has failed • Induction System problems: air leaks after the MAF sensor, PCV system leaks, engine vacuum leaks or dip stick not seated
P2196 **2T CCM, MIL: No** 2003 All Models Transmissions: All	**Lack of HO2S-21 Switching, Sensor Indicates Rich Conditions:** DTC P0300-P0310 not set, engine running in closed loop, and the PCM detected the HO2S indicated a rich signal, or it could no longer control Fuel Trim because it was at its rich limit. **Possible Causes** • Base engine problems: engine oil level high, camshaft timing error, cylinder compression low, exhaust leaks in front of HO2S • Fuel System problem: excessive fuel pressure, leaking fuel injectors, fuel pressure regulator leaking • HO2S problems: HO2S circuit is open or shorted in the wiring harness, the HO2S signal circuit is contacting moisture in harness connector, or the HO2S is damaged or it has failed
P2197 **2T CCM, MIL: No** 2003 All Models Transmissions: All	**Lack of HO2S-21 Switching, Sensor Indicates Lean Conditions:** DTC P0300-P0310 not set, engine running in closed loop, and the PCM detected the HO2S indicated a lean signal, or it could no longer control Fuel Trim because it was at lean limit. **Possible Causes** • Base engine problems: engine oil level high, camshaft timing error, cylinder compression low, exhaust leaks in front of HO2S • EGR System problem: EGR valve is stuck open, the gasket is leaking, or the EVR diaphragm is leaking • Fuel System problem: damaged fuel pressure regulator or extremely low fuel pressure • HO2S problems: HO2S circuit is open or shorted in the wiring harness or the HO2S is damaged or it has failed • Induction System problems: air leaks after the MAF sensor, PCV system leaks, engine vacuum leaks or dip stick not seated
P2198 **2T CCM, MIL: No** 2003 All Models Transmissions: All	**Lack of HO2S-21 Switching, Sensor Indicates Rich Conditions:** DTC P0300-P0310 not set, engine running in closed loop, and the PCM detected the HO2S indicated a rich signal, or it could no longer control Fuel Trim because it was at its rich limit. **Possible Causes** • Base engine problems: engine oil level high, camshaft timing error, cylinder compression low, exhaust leaks in front of HO2S • Fuel System problem: excessive fuel pressure, leaking fuel injectors, fuel pressure regulator leaking • HO2S problems: HO2S circuit is open or shorted in the wiring harness, the HO2S signal circuit is contacting moisture in harness connector, or the HO2S is damaged or it failed

Gas Engine OBD II Trouble Code List (P2xxx Codes) – (U1xxx)

DTC	Trouble Code Title, Conditions & Possible Causes
P2270 **2T CCM, MIL: No** 2003 All Models Transmissions: All	**Lack of HO2S-12 Switching, Sensor Indicates Lean Conditions:** DTC P0300-P0310 not set, engine running in closed loop, and the PCM detected the HO2S indicated a lean signal, or it could no longer control Fuel Trim because it was at lean limit. **Possible Causes** • Base engine problems: engine oil level high, camshaft timing error, cylinder compression low, exhaust leaks in front of HO2S • EGR System problem: EGR valve is stuck open, the gasket is leaking, or the EVR diaphragm is leaking • Fuel System problem: damaged fuel pressure regulator or extremely low fuel pressure • HO2S problems: HO2S circuit is open or shorted in the wiring harness or the HO2S is damaged or it has failed • Induction System problems: air leaks after the MAF sensor, PCV system leaks, engine vacuum leaks or dip stick not seated
P2271 **2T CCM, MIL: No** 2003 All Models Transmissions: All	**Lack of HO2S-12 Switching, Sensor Indicates Rich Conditions:** DTC P0300-P0310 not set, engine running in closed loop, and the PCM detected the HO2S indicated a rich signal, or it could no longer control Fuel Trim because it was at its rich limit. **Possible Causes** • Base engine problems: engine oil level high, camshaft timing error, cylinder compression low, exhaust leaks in front of HO2S • Fuel System problem: excessive fuel pressure, leaking fuel injectors, fuel pressure regulator leaking • HO2S problems: HO2S circuit is open or shorted in the wiring harness, the HO2S signal circuit is contacting moisture in harness connector, or the HO2S is damaged or it failed
P2272 **2T CCM, MIL: No** 2003 All Models Transmissions: All	**Lack of HO2S-22 Switching, Sensor Indicates Lean Conditions:** DTC P0300-P0310 not set, engine running in closed loop, and the PCM detected the HO2S indicated a lean signal, or it could no longer control Fuel Trim because it was at lean limit. **Possible Causes** • Base engine problems: engine oil level high, camshaft timing error, cylinder compression low, exhaust leaks in front of HO2S • EGR System problem: EGR valve is stuck open, the gasket is leaking, or the EVR diaphragm is leaking • Fuel System problem: damaged fuel pressure regulator or extremely low fuel pressure • HO2S problems: HO2S circuit is open or shorted in the wiring harness or the HO2S is damaged or it has failed • Induction System problems: air leaks after the MAF sensor, PCV system leaks, engine vacuum leaks or dip stick not seated
P2273 **2T CCM, MIL: No** 2003 All Models Transmissions: All	**Title: Lack of HO2S-22 Switching, Sensor Indicates Rich** **Trouble Code Conditions** DTC P0300-P0310 not set, engine running in closed loop, and the PCM detected the HO2S indicated a rich signal, or it could no longer control Fuel Trim because it was at its rich limit. **Possible Causes** • Base engine problems: engine oil level high, camshaft timing error, cylinder compression low, exhaust leaks in front of HO2S • Fuel System problem: excessive fuel pressure, leaking fuel injectors, fuel pressure regulator leaking • HO2S problems: HO2S circuit is open or shorted in the wiring harness, the HO2S signal circuit is contacting moisture in harness connector, or the HO2S is damaged or it failed
U1011 **1T PCM, MIL: No** 1996-1997-1998-1999-2000-2001-2002-2003 All Models Transmissions: All	**Data Circuit Message Conditions:** Key on, and the PCM detected that invalid or Missing Data from the Engine Air Intake system was received on the SCP data bus. Note: Network codes occur during module-to-module communication failures. Invalid and Missing data network faults are outlined below. **Possible Causes** • Invalid Data: Data transferred in normal inter-module messages with known invalid data. Transmitting module will set the code. • Missing Network Data: Missing message fault logged by a module upon failure to receive a message from another module within a defined retry period.

Gas Engine OBD II Trouble Code List (U1xxx Codes)

DTC	Trouble Code Title, Conditions & Possible Causes
U1020 **1T PCM, MIL: No** 1996-1997-1998-1999-2000-2001-2002-2003 All Models Transmissions: All	**Data Circuit Message Conditions:** Key on, and the PCM detected that invalid or Missing Data from the Air Conditioning system was received on the SCP data bus. Note: Network codes occur during module-to-module communication failures. Invalid and Missing data network faults are outlined below. **Possible Causes** • Invalid Data: Data transferred in normal inter-module messages with known invalid data. Transmitting module will set the code. • Missing Network Data: Missing message fault logged by a module upon failure to receive a message from another module within a defined retry period.
U1021 **1T PCM, MIL: No** 1996-1997-1998-1999-2000-2001-2002-2003 All Models Transmissions: All	**Data Circuit Message Conditions:** Key on, and the PCM detected that invalid or Missing Data from the Air Conditioning Clutch status was received on the SCP data bus. Note: Network codes occur during module-to-module communication failures. Invalid and Missing data network faults are outlined below. **Possible Causes** • Invalid Data: Data transferred in normal inter-module messages with known invalid data. Transmitting module will set the code. • Missing Network Data: Missing message fault logged by a module upon failure to receive a message from another module within a defined retry period.
U1037 **1T PCM, MIL: No** 1996-1997-1998-1999-2000-2001-2002-2003 All Models Transmissions: All	**Data Circuit Message Conditions:** Key on, and the PCM detected that invalid or Missing Data from the Telltale Lamp Module was received on the SCP data bus. Note: Network codes occur during module-to-module communication failures. Invalid and Missing data network faults are outlined below. **Possible Causes** • Invalid Data: Data transferred in normal inter-module messages with known invalid data. Transmitting module will set the code. • Missing Network Data: Missing message fault logged by a module upon failure to receive a message from another module within a defined retry period.
U1039 **1T PCM, MIL: No** 1996-1997-1998-1999-2000-2001-2002-2003 All Models Transmissions: All	**Data Circuit Message Conditions:** Key on, and the PCM detected that invalid or Missing Data from the Vehicle Speed Sensor was received on the SCP data bus. Note: Network codes occur during module-to-module communication failures. Invalid and Missing data network faults are outlined below. **Possible Causes** • Invalid Data: Data transferred in normal inter-module messages with known invalid data. Transmitting module will set the code. • Missing Network Data: Missing message fault logged by a module upon failure to receive a message from another module within a defined retry period. • TSB 01-21-13 contains a repair procedure for this trouble code
U1041 **1T PCM, MIL: No** 1996-1997-1998-1999-2000-2001-2002-2003 All Models Transmissions: All	**Data Circuit Message Conditions:** Key on, and the PCM detected that invalid or Missing Data from the Vehicle Speed Sensor was received on the SCP data bus. Note: Network codes occur during module-to-module communication failures. Invalid and Missing data network faults are outlined below. **Possible Causes** • Invalid Data: Data transferred in normal inter-module messages with known invalid data. Transmitting module will set the code. • Missing Network Data: Missing message fault logged by a module upon failure to receive a message from another module within a defined retry period.
U1051 **1T PCM, MIL: No** 1996-1997-1998-1999-2000-2001-2002-2003 All Models Transmissions: All	**Data Circuit Message Conditions:** Key on, and the PCM detected that invalid or Missing Data from the Antilock Brake System was received on the SCP data bus. Note: Network codes occur during module-to-module communication failures. Invalid and Missing data network faults are outlined below. **Possible Causes** • Invalid Data: Data transferred in normal inter-module messages with known invalid data. Transmitting module will set the code. • Missing Network Data: Missing message fault logged by a module upon failure to receive a message from another module within a defined retry period.

Gas Engine OBD II Trouble Code List (U1xxx Codes)

DTC	Trouble Code Title, Conditions & Possible Causes
U1071 **1T PCM, MIL: No** 1996-1997-1998-1999-2000-2001-2002-2003 All Models Transmissions: All	**Data Circuit Message Conditions:** Key on, and the PCM detected that invalid or Missing Data from the Engine Sensor was received on the SCP data bus. Note: Network codes occur during module-to-module communication failures. Invalid and Missing data network faults are outlined below. **Possible Causes** • Invalid Data: Data transferred in normal inter-module messages with known invalid data. Transmitting module will set the code. • Missing Network Data: Missing message fault logged by a module upon failure to receive a message from another module within a defined retry period.
U1073 **1T PCM, MIL: No** 1996-1997-1998-1999-2000-2001-2002-2003 All Models Transmissions: All	**Data Circuit Message Conditions:** Key on, and the PCM detected that invalid or Missing Data from the Engine Coolant Fan Status was received on the SCP data bus. Note: Network codes occur during module-to-module communication failures. Invalid and Missing data network faults are outlined below. **Possible Causes** • Invalid Data: Data transferred in normal inter-module messages with known invalid data. Transmitting module will set the code. • Missing Network Data: Missing message fault logged by a module upon failure to receive a message from another module within a defined retry period.
U1089 **1T PCM, MIL: No** 1996-1997-1998-1999-2000-2001-2002-2003 All Models Transmissions: All	**Data Circuit Message Conditions:** Key on, and the PCM detected that invalid or Missing Data from the Suspension System Module was received on the SCP data bus. Note: Network codes occur during module-to-module communication failures. Invalid and Missing data network faults are outlined below. **Possible Causes** • Invalid Data: Data transferred in normal inter-module messages with known invalid data. Transmitting module will set the code. • Missing Network Data: Missing message fault logged by a module upon failure to receive a message from another module within a defined retry period.
U1098 **1T PCM, MIL: No** 1996-1997-1998-1999-2000-2001-2002-2003 All Models Transmissions: All	**Data Circuit Message Conditions:** Key on, and the PCM detected that invalid or Missing Data from the Vehicle Speed Control Module was received on the SCP data bus. Note: Network codes occur during module-to-module communication failures. Invalid and Missing data network faults are outlined below. **Possible Causes** • Invalid Data: Data transferred in normal inter-module messages with known invalid data. Transmitting module will set the code. • Missing Network Data: Missing message fault logged by a module upon failure to receive a message from another module within a defined retry period.
U1130 **1T PCM, MIL: No** 1996-1997-1998-1999-2000-2001-2002-2003 All Models Transmissions: All	**Data Circuit Message Conditions:** Key on, and the PCM detected that invalid or Missing Data from the Fuel System was received on the SCP data bus. Note: Network codes occur during module-to-module communication failures. Invalid and Missing data network faults are outlined below. **Possible Causes** • Invalid Data: Data transferred in normal inter-module messages with known invalid data. Transmitting module will set the code. • Missing Network Data: Missing message fault logged by a module upon failure to receive a message from another module within a defined retry period.
U1131 **1T PCM, MIL: No** 1996-1997-1998-1999-2000-2001-2002-2003 All Models Transmissions: All	**Data Circuit Message Conditions:** Key on, and the PCM detected that invalid or Missing Data from the Fuel System was received on the SCP data bus. Note: Network codes occur during module-to-module communication failures. Invalid and Missing data network faults are outlined below. **Possible Causes** • Invalid Data: Data transferred in normal inter-module messages with known invalid data. Transmitting module will set the code. • Missing Network Data: Missing message fault logged by a module upon failure to receive a message from another module within a defined retry period.

Gas Engine OBD II Trouble Code List (U1xxx Codes)

DTC	Trouble Code Title, Conditions & Possible Causes
U1135 **1T PCM, MIL: No** 1996-1997-1998-1999- 2000-2001-2002-2003 All Models Transmissions: All	**Data Circuit Message Conditions:** Key on, and the PCM detected that invalid or Missing Data from the Ignition Switch Signal was received on the SCP data bus. Note: Network codes occur during module-to-module communication failures. Invalid and Missing data network faults are outlined below. **Possible Causes** • Invalid Data: Data transferred in normal inter-module messages with known invalid data. Transmitting module will set the code. • Missing Network Data: Missing message fault logged by a module upon failure to receive a message from another module within a defined retry period.
U1147 **1T PCM, MIL: No** 1996-1997-1998-1999- 2000-2001-2002-2003 All Models Transmissions: All	**Data Circuit Message Conditions:** Key on, and the PCM detected that invalid or Missing Data from the Vehicle Security System was received on the SCP data bus. Note: Network codes occur during module-to-module communication failures. Invalid and Missing data network faults are outlined below. **Possible Causes** • Invalid Data: Data transferred in normal inter-module messages with known invalid data. Transmitting module will set the code. • Missing Network Data: Missing message fault logged by a module upon failure to receive a message from another module within a defined retry period.
U1243 **1T PCM, MIL: No** 1996-1997-1998-1999- 2000-2001-2002-2003 All Models Transmissions: All	**Data Circuit Message Conditions:** Key on, and the PCM detected that invalid or Missing Data from the Exterior Environment System was received on the SCP data bus. Note: Network codes occur during module-to-module communication failures. Invalid and Missing data network faults are outlined below. **Possible Causes** • Invalid Data: Data transferred in normal inter-module messages with known invalid data. Transmitting module will set the code. • Missing Network Data: Missing message fault logged by a module upon failure to receive a message from another module within a defined retry period.
U1256 **1T PCM, MIL: No** 1996-1997-1998-1999- 2000-2001-2002-2003 All Models Transmissions: All	**Data Circuit Message Conditions:** Key on, and the PCM detected a signal indicating a communication error had occurred with another module over the SCP data bus. Note: Network codes occur during module-to-module communication failures. Invalid and Missing data network faults are outlined below. **Possible Causes** • Invalid Data: Data transferred in normal inter-module messages with known invalid data. Transmitting module will set the code. • Missing Network Data: Missing message fault logged by a module upon failure to receive a message from another module within a defined retry period.
U1260 **1T PCM, MIL: No** 1996-1997-1998-1999- 2000-2001-2002-2003 All Models Transmissions: All	**Data Circuit Message Conditions:** Key on, and the PCM detected a signal that indicated an open or shorted condition was present in the SCP (+) bus circuit. Note: Network codes occur during module-to-module communication failures. Invalid and Missing data network faults are outlined below. **Possible Causes** • Invalid Data: Data transferred in normal inter-module messages with known invalid data. Transmitting module will set the code. • Missing Network Data: Missing message fault logged by a module upon failure to receive a message from another module within a defined retry period.
U1261 **1T PCM, MIL: No** 1996-1997-1998-1999- 2000-2001-2002-2003 All Models Transmissions: All	**Data Circuit Message Conditions:** Key on, and the PCM detected a signal that indicated an open or shorted condition was present in the SCP (-) bus circuit. Note: Network codes occur during module-to-module communication failures. Invalid and Missing data network faults are outlined below. **Possible Causes** • Invalid Data: Data transferred in normal inter-module messages with known invalid data. Transmitting module will set the code. • Missing Network Data: Missing message fault logged by a module upon failure to receive a message from another module within a defined retry period.

Gas Engine OBD II Trouble Code List (U1xxx Codes)

DTC	Trouble Code Title, Conditions & Possible Causes
U1262 **1T PCM, MIL: No** 1996-1997-1998-1999-2000-2001-2002-2003 All Models Transmissions: All	**Data Circuit Message Conditions:** Key on, and the PCM detected a signal that indicated a fault was present in the SCP bus (perform the network communication tests). Note: Network codes occur during module-to-module communication failures. Invalid and Missing data network faults are outlined below. **Possible Causes** • Invalid Data: Data transferred in normal inter-module messages with known invalid data. Transmitting module will set the code. • Missing Network Data: Missing message fault logged by a module upon failure to receive a message from another module within a defined retry period.
U1341 **1T PCM, MIL: No** 1996-1997-1998-1999-2000-2001-2002-2003 All Models Transmissions: All	**Data Circuit Message Conditions:** Key on, and the PCM detected that invalid or Missing Data from the Function Read Vehicle Speed was received on the SCP data bus. Note: Network codes occur during module-to-module communication failures. Invalid and Missing data network faults are outlined below. **Possible Causes** • Invalid Data: Data transferred in normal inter-module messages with known invalid data. Transmitting module will set the code. • Missing Network Data: Missing message fault logged by a module upon failure to receive a message from another module within a defined retry period.
U1451 **1T PCM, MIL: No** 1996-1997-1998-1999-2000-2001-2002-2003 All Models Transmissions: All	**Data Circuit Message Conditions:** Key on, and the PCM detected that invalid or Missing Data from the Vehicle Antitheft Module was received on the SCP data bus. Note: Network codes occur during module-to-module communication failures. Invalid and Missing data network faults are outlined below. **Possible Causes** • Invalid Data: Data transferred in normal inter-module messages with known invalid data. Transmitting module will set the code. • Missing Network Data: Missing message fault logged by a module upon failure to receive a message from another module within a defined retry period.
U2015 **1T PCM, MIL: No** 1996-1997-1998-1999-2000-2001-2002-2003 All Models Transmissions: All	**Data Circuit Message Conditions:** Key on, and the PCM detected that invalid or Missing Data from the Function Read Vehicle Speed was received on the SCP data bus. Note: Network codes occur during module-to-module communication failures. Invalid and Missing data network faults are outlined below. **Possible Causes** • Invalid Data: Data transferred in normal inter-module messages with known invalid data. Transmitting module will set the code. • Missing Network Data: Missing message fault logged by a module upon failure to receive a message from another module within a defined retry period.
U2195 **1T PCM, MIL: No** 1996-1997-1998-1999-2000-2001-2002-2003 All Models Transmissions: All	**Data Circuit Message Conditions:** Key on, and the PCM detected an open or shorted condition in the Signal Link circuit (not on the SCP data bus circuits). Note: Network codes occur during module-to-module communication failures. Invalid and Missing data network faults are outlined below. **Possible Causes** • Invalid Data: Data transferred in normal inter-module messages with known invalid data. Transmitting module will set the code. • Missing Network Data: Missing message fault logged by a module upon failure to receive a message from another module within a defined retry period.
U2243 **1T PCM, MIL: No** 1996-1997-1998-1999-2000-2001-2002-2003 All Models Transmissions: All	**Data Circuit Message Conditions:** Key on, and the PCM detected that invalid or Missing Data from the SCLM Status was received on the SCP data bus. Note: Network codes occur during module-to-module communication failures. Invalid and Missing data network faults are outlined below. **Possible Causes** • Invalid Data: Data transferred in normal inter-module messages with known invalid data. Transmitting module will set the code. • Missing Network Data: Missing message fault logged by a module upon failure to receive a message from another module within a defined retry period.

Diesel Engine OBD II Trouble Code List (B1xxx Codes)

DTC	Trouble Code Title, Conditions & Possible Causes
B1213 **1T PCM, MIL: No** 2003 F-Series Truck & Excursion with a 6.0L VIN P Diesel engine Transmissions: All	**Less Than Two Keys Programmed To The System Conditions:** Key on, and the PCM received a signal that indicated there were less than two (2) keys programmed to the system. **Possible Causes** • Refer to test procedures Section 419 of the Workshop Manual. Then reprogram the correct amount of keys.
B1342 **1T PCM, MIL: No** 2003 F-Series Truck & Excursion with a 6.0L VIN P Diesel engine Transmissions: All	**ECU Damaged (EEPROM Inside PCM Not Working) Conditions:** Key on, and the PCM received a signal that indicated there were less than two (2) keys programmed to the system. **Possible Causes** • Refer to test procedures Section 419 of the Workshop Manual. • Replace the PCM. Follow the correct procedures to "flash" new PCM. Then recheck the system for trouble codes.
B1600 **1T PCM, MIL: No** 2003 F-Series Truck & Excursion with a 6.0L VIN P Diesel engine Transmissions: All	**PATS Ignition Key Transponder Signal Not Received Conditions:** Key on, and the PATS did not receive Ignition Key Transponder Signal was not received. **Possible Causes** • Refer to test procedures Section 419 of the Workshop Manual to determine why the Passive Antitheft System (PATS) did not receive the Ignition Key Transponder signal correctly.
B1601 **1T PCM, MIL: No** 2003 F-Series Truck & Excursion with a 6.0L VIN P Diesel engine Transmissions: All	**PATS Received Incorrect Key Code From Ignition Key Transponder Conditions:** Key on, and the PATS received an incorrect Key Code from the Ignition Key Transponder. **Possible Causes** • Refer to test procedures Section 419 of the Workshop Manual to determine why the Passive Antitheft System (PATS) received an incorrect Key Code from the Ignition Key Transponder.
B1602 **1T PCM, MIL: No** 2003 F-Series Truck & Excursion with a 6.0L VIN P Diesel engine Transmissions: All	**PATS Received Invalid Format of Key Code From Ignition Key Transponder Conditions:** Key on, and the PATS received an invalid format of Key Code from the Ignition Key Transponder. **Possible Causes** • Refer to test procedures Section 419 of the Workshop Manual to determine why the Passive Antitheft System (PATS) received an invalid format of Key Code from the Ignition Key Transponder.
B1681 **1T PCM, MIL: No** 2003 F-Series Truck & Excursion with a 6.0L VIN P Diesel engine Transmissions: All	**PATS Receiver Module Signal Not Received Conditions:** Key on, and the PATS Receiver Module Signal was not received. **Possible Causes** • Refer to test procedures Section 419 of the Workshop Manual to determine why the Passive Antitheft System (PATS) Receiver Module Signal was not received.
B2103 **1T PCM, MIL: No** 2003 F-Series Truck & Excursion with a 6.0L VIN P Diesel engine Transmissions: All	**Antenna Not Connected Conditions:** Key on, and the PATS detected that the Antenna was not connected. **Possible Causes** • Refer to test procedures Section 419 of the Workshop Manual to determine why the Passive Antitheft System (PATS) detected the Antenna was not connected.

Diesel Engine OBD II Trouble Code List (B2xxx Codes)

DTC	Trouble Code Title, Conditions & Possible Causes
B2103 **1T PCM, MIL: No** 2003 F-Series Truck & Excursion with a 6.0L VIN P Diesel engine Transmissions: All	**Transponder Programming Failed Conditions:** Key on, and the PATS detected that the Transponder Programming failed. **Possible Causes** • Refer to test procedures Section 419 of the Workshop Manual to determine why the Passive Antitheft System (PATS) detected the Transponder Programming failed.
P0046 **1T CCM, MIL: No** 2003 F-Series Truck & Excursion with a 6.0L VIN P Diesel engine Transmissions: All	**Turbo/Supercharger Boost Solenoid Signal Performance Conditions:** Key on or engine running; and the PCM detected a high or low voltage on the Variable Geometry Turbo (VGT) control circuit. The VGT actuator is a variable position valve that controls the vane position in the turbine housing. The VGT is located on top of the turbocharger. The valve position signal is controlled inside the PCM. **Possible Causes** • VGT solenoid connector is damaged, open or shorted • VGT solenoid control circuit is open or shorted • VGT solenoid is damaged, or it has failed • PCM has failed
P0069 **1T CCM, MIL: Yes 2003** F-Series Truck & Excursion with a 6.0L VIN P Diesel engine Transmissions: All	**MAP/BARO Sensor Signal Correlation Conditions:** Engine started, engine running at idle speed, and the PCM detected the difference between the BARO sensor and MAP sensor signal was more than a specified value during the CCM continuous test. **Possible Causes** • MAP/BARO sensor connector is damaged, open or shorted • MAP/BARO sensor circuit is open or shorted • MAP/BARO sensor is damaged, or it has failed • PCM has failed
P0096 **1T CCM, MIL: Yes 2003** F-Series Truck & Excursion with a 6.0L VIN P Diesel engine Transmissions: All	**Intake Air Temperature Sensor 2 Signal Performance Conditions:** Engine started, and the PCM detected an unexpected voltage condition on the IAT Sensor 2 signal during the CCM test period. **Possible Causes** • IAT2 assembly connector is damaged, open or shorted • IAT2 assembly signal circuit is open or shorted • IAT2 assembly is damaged, or it has failed • PCM has failed
P0097 **1T CCM, MIL: Yes 2003** F-Series Truck & Excursion with a 6.0L VIN P Diesel engine Transmissions: All	**Intake Air Temperature Sensor 2 Circuit Low Input Conditions:** Key on or engine running; and the PCM detected an unexpected low voltage condition on the IAT Sensor 2 signal during the CCM test. **Possible Causes** • IAT2 assembly connector is damaged or shorted • IAT2 assembly signal circuit is shorted to ground • IAT2 assembly is damaged, or it has failed • PCM has failed
P0098 **1T CCM, MIL: Yes 2003** F-Series Truck & Excursion with a 6.0L VIN P Diesel engine Transmissions: All	**Intake Air Temperature Sensor 2 Circuit High Input Conditions:** Key on or engine running; and the PCM detected an unexpected high voltage condition on the IAT Sensor 2 signal during the test. **Possible Causes** • IAT2 sensor connector is damaged, open or shorted • IAT sensor signal circuit is open or shorted to 5v VREF • IAT2 sensor ground circuit is open • IAT2 assembly is damaged, or it has failed • PCM has failed

Diesel Engine OBD II Trouble Code List (P0xxx Codes)

DTC	Trouble Code Title, Conditions & Possible Causes
P0101 **1T CCM, MIL: No** 2003 F-Series Truck & Excursion with a 6.0L VIN P Diesel engine Transmissions: All	**MAF Sensor Or VAF Sensor Signal Range/Performance Conditions:** DTC P0102 and DTC P0103 not set, engine started, and the PCM detected the Actual MAF sensor value was not within a preset range of the Calculated MAF sensor value while testing. **Possible Causes** • Base engine vacuum leak, PCV valve leaking or stuck open • Check for leaks at air outlet tube, or loose tube clamps • Engine oil dipstick missing or not fully seated • MAF sensor element (wire) is contaminated or dirty • MAF sensor signal or ground circuit fault or sensor has failed
P0102 **2T CCM, MIL: Yes** 2003 F-Series Truck & Excursion with a 6.0L VIN P Diesel engine Transmissions: All	**MAF Sensor Circuit Low Input Conditions:** DTC P0505 not set, engine started, and the PCM detected the MAF sensor signal was less than 0.23v during the CCM test period. **Possible Causes** • Check for leaks at air outlet tube • Sensor power circuit open, sensor ground circuit open • Sensor signal circuit open (may be disconnected) • Check for loose tube clamps near the MAF sensor
P0103 **2T CCM, MIL: Yes** 2003 F-Series Truck & Excursion with a 6.0L VIN P Diesel engine Transmissions: All	**MAF Sensor Circuit High Input Conditions:** DTC P0505 not set, engine started, and the PCM detected the MAF sensor signal was more than 4.60v during the CCM test period. **Possible Causes** • Check for a restricted inlet screen on the MAF sensor. • MAF sensor signal circuit is shorted to system power (B+) • MAF sensor is damaged or has failed • PCM has failed
P0107 **2T CCM, MIL: Yes** 1996-1997-1998-1999-2000-2001-2002-2003 E-Van, F-Series Truck & Excursion with 6.0L or 7.3L Diesel engine Transmissions: All	**Barometric Pressure Sensor Circuit Low Input Conditions:** Key on, KOEO Self-Test enabled, or engine running, and the PCM detected the BARO sensor signal was less than the minimum calibrated parameter. The BARO sensor is a variable capacitance sensor used to determine altitude. Prior to the 1999 1/2 model year, this sensor was mounted under the Instrument Cluster. This sensor is internal to the PCM on later models, and it is serviced separately. **Possible Causes** • BARO sensor signal circuit is shorted to ground • BARO sensor VREF circuit is open • BARO sensor is damaged or it has failed • PCM has failed
P0108 **2T CCM, MIL: Yes** 1996-1997-1998-1999-2000-2001-2002-2003 E-Van, F-Series Truck & Excursion with 6.0L or 7.3L Diesel engine Transmissions: All	**Barometric Pressure Sensor Circuit High Input Conditions:** Key on, KOEO Self-Test enabled, or engine running, and the PCM detected the BARO sensor signal was more than the minimum calibrated parameter. The BARO sensor is a variable capacitance sensor used to determine altitude. Prior to the 1999 1/2 model year, this sensor was mounted under the Instrument Cluster. This sensor is internal to the PCM on later models, and it is serviced separately. **Possible Causes** • BARO ground signal circuit is open • BARO sensor signal circuit is shorted to VREF (5v) • BARO sensor is damaged or it has failed • PCM has failed
P0112 **2T CCM, MIL: No** 1996-1997-1998-1999-2000-2001-2002-2003 E-Van, F-Series Truck & Excursion with 6.0L or 7.3L Diesel engine Transmissions: All	**IAT Sensor Circuit Low Input Conditions:** Key on or engine running; and the PCM detected the IAT sensor signal was less than the self-test minimum of 0.20v (Scan Tool reads over 250°F). This is a thermistor-type sensor with a variable resistance that changes when exposed to different temperatures. **Possible Causes** • IAT sensor signal circuit is grounded (check wiring & connector) • IAT sensor is damaged or it has failed • PCM has failed

Diesel Engine OBD II Trouble Code List (P0xxx Codes)

DTC	Trouble Code Title, Conditions & Possible Causes
P0113 **2T CCM, MIL: No** 1996-1997-1998-1999-2000-2001-2002-2003 E-Van, F-Series Truck & Excursion with 6.0L or 7.3L Diesel engine Transmissions: All	**IAT Sensor Circuit High Input Conditions:** Key on or engine running; and the PCM detected the IAT sensor was more than the self-test maximum of 4.60v (Scan Tool reads -46°F) during the self-test. This is a thermistor-type sensor with a variable resistance that changes as the temperature changes. **Possible Causes** • Sensor signal circuit open (check wiring harness/connector) • Sensor signal circuit shorted to VREF or system voltage • Sensor ground circuit is open • Sensor is damaged or it has failed • PCM has failed
P0117 **2T CCM, MIL: No** 1996-1997-1998-1999-2000-2001-2002-2003 E-Van, F-Series Truck & Excursion with 6.0L or 7.3L Diesel engine Transmissions: All	**ECT Sensor Circuit Low Input Conditions:** Key on or engine running; and the PCM detected the ECT sensor signal was less than the self-test minimum of 0.20v (Scan Tool reads over 250°F). This is a thermistor-type sensor with a variable resistance that changes when exposed to different temperatures **Possible Causes** • ECT sensor signal circuit is grounded in the wiring harness • ECT sensor is damaged or it has failed • PCM has failed
P0118 **2T CCM, MIL: No** 1996-1997-1998-1999-2000-2001-2002-2003 E-Van, F-Series Truck & Excursion with 6.0L or 7.3L Diesel engine Transmissions: All	**ECT Sensor Circuit High Input Conditions:** Key on or engine running; and the PCM detected the ECT sensor signal was more than the self-test maximum of 4.60v (Scan Tool reads under -46°F). This is a thermistor-type sensor with a variable resistance that changes when exposed to different temperatures **Possible Causes** • ECT sensor signal circuit is open (inspect wiring & connector) • ECT sensor signal circuit is shorted to VREF (5v) • ECT sensor is damaged or it has failed • PCM has failed
P0122 **2T CCM, MIL: Yes** 1996-1997-1998-1999-2000-2001-2002-2003 E-Van, F-Series Truck & Excursion with 6.0L or 7.3L Diesel engine Transmissions: All	**Accelerator Position Sensor Circuit Low Input Conditions:** Key on or engine running; and the PCM detected the APP sensor was less than the minimum calibrated limit during the CCM test. **Possible Causes** • APP sensor signal circuit is open (inspect wiring & connector) • APP sensor signal is shorted to ground (check the connector) • APP sensor VREF circuit (5v) is open (from sensor to PCM) • APP sensor is damaged or it has failed • PCM has failed
P0123 **2T CCM, MIL: Yes** 1996-1997-1998-1999-2000-2001-2002-2003 E-Van, F-Series Truck & Excursion with 6.0L or 7.3L Diesel engine Transmissions: All	**Accelerator Position Sensor Circuit High Input Conditions:** Key on or engine running; and the PCM detected the APP sensor was more than the minimum calibrated limit during the CCM test. **Possible Causes** • APP sensor ground circuit is open (inspect wiring & connector) • APP sensor signal is shorted to the VREF circuit (5v) • APP sensor is damaged or it has failed • PCM has failed
P0148 **2T CCM, MIL: Yes** 2003 F-Series Truck & Excursion with a 6.0L VIN P Diesel engine Transmissions: All	**FICM Fuel Delivery Circuit Error Conditions:** Engine started, and the PCM detected a Fuel Delivery Error related to the FICM circuit. The FICM monitor input line informs the Diesel Engine Power Monitor (DEPM) when the injectors are turned "on" and "off". If the FICM line is open or shorted, the monitor strategy assumes that the fuel injectors are always turned "on", and it sets DTC P2552. In some cases, the DEPM may also set DTC P0148. **Possible Causes** • Severely restricted fuel filter • Severely pinched or restricted fuel delivery line

Diesel Engine OBD II Trouble Code List (P0xxx Codes)

DTC	Trouble Code Title, Conditions & Possible Causes
P00196 **2T CCM, MIL: Yes** 2003 F-Series Truck & Excursion with a 6.0L VIN P Diesel engine Transmissions: All	**Engine Oil Temperature Sensor Signal Performance Conditions:** Engine started, KOER Self Test enabled, and the PCM detected that the Engine Oil Temperature (EOT) sensor signal was not within a calibrated amount of the ECT sensor signal. The EOT sensor signal is used to determine the timing and quality of the fuel required to optimize engine startup over all temperature conditions. **Possible Causes** • Cooling system malfunction, or the thermostat is stuck • Engine not operating at normal operating temperature • EOT sensor is damaged or it has failed • PCM has failed
P0197 **2T CCM, MIL: Yes** 1996-1997-1998-1999-2000-2001-2002-2003 E-Van, F-Series Truck & Excursion with 6.0L or 7.3L Diesel engine Transmissions: All	**Engine Oil Temperature Sensor Circuit Low Input Conditions:** Key on or engine running; and the PCM detected that the Engine Oil Temperature (EOT) sensor was less than 0.20v. The EOT sensor signal is used to determine the timing and quality of the fuel required to optimize engine startup over all temperature conditions. **Possible Causes** • EOT sensor connector is damaged or shorted to ground • EOT sensor signal circuit shorted to sensor ground • EOT sensor is damaged or it has failed • PCM has failed
P0198 **2T CCM, MIL: Yes** 1996-1997-1998-1999-2000-2001-2002-2003 E-Van, F-Series Truck & Excursion with 6.0L or 7.3L Diesel engine Transmissions: All	**Engine Oil Temperature Sensor Circuit High Input Conditions:** Key on or engine running; and the PCM detected that the Engine Oil Temperature (EOT) sensor was less than 4.50v. The EOT sensor signal is used to determine the timing and quality of the fuel required to optimize engine startup over all temperature conditions. **Possible Causes** • EOT sensor connector is damaged or open • EOT sensor signal circuit is shorted to VREF (5v) • EOT sensor is damaged or it has failed • PCM has failed
P0219 **1T CCM, MIL: Yes** 2003 F-Series Truck & Excursion with a 6.0L VIN P Diesel engine Transmissions: All	**Engine Overspeed Condition Conditions:** Engine started, and the PCM detected an Engine Overspeed condition had occurred. **Possible Causes** • Repeat the KOEO and KOER Self Test • Refer to the Symptom Charts for further information
P0220 **2T CCM, MIL: Yes** 1996-1997-1998-1999-2000-2001-2002-2003 E-Van, F-Series Truck & Excursion with 6.0L or 7.3L Diesel engine Transmissions: All	**Throttle Position Sensor 'B' Circuit Malfunction Conditions:** Key on or engine running; and the PCM detected an unexpected voltage condition on the TP Sensor 'B' circuit during the CCM test. The Idle Validation Switch (IVS) provides the PCM with a signal to verify when the accelerator is in the idle position. **Possible Causes** • ETC TP Sensor 'B' connector is damaged or shorted • ETC TP Sensor 'B' signal circuit is shorted to ground • ETC TP Sensor 'B' is damaged or it has failed
P0221 **2T CCM, MIL: Yes** 1999-2000-2001-2002 E-Vans & F-Series Trucks & Excursion with a 7.3L Diesel engine Transmissions: All	**Idle Validation/Throttle Switch 'B' Performance Conditions:** Engine started, KOER Self Test enabled, and the PC'M detected the TP-B sensor was less than or more than its calibrated minimum. Note: Wait 5 seconds before starting the KOER switch test after first pressing the trigger to start running the driver-operated controls. **Possible Causes** • Blown fuse, or open in power circuit to IVS switch • IVS signal circuit open • IVS has failed, or the IVS transition is out of range

Diesel Engine OBD II Trouble Code List (P0xxx Codes)

DTC	Trouble Code Title, Conditions & Possible Causes
P0222 **2T CCM, MIL: Yes** 1996-1997-1998-1999-2000-2001-2002-2003 E-Van, F-Series Truck & Excursion with 6.0L or 7.3L Diesel engine Transmissions: All	**Throttle Position Sensor 'B' Circuit Low Input Conditions:** Key on or engine running; and the PCM detected the TP Sensor 'B' circuit was out of its normal operating range during the CCM test. The Idle Validation Switch (IVS) provides the PCM with a signal to verify when the accelerator is in the idle position. **Possible Causes** • Throttle body is damaged • Throttle linkage is binding or sticking • ETC TP Sensor 'B' signal circuit to the PCM is open • ETC TP Sensor 'B' ground circuit is open • ETC TP Sensor 'B' is damaged or it has failed
P0223 **2T CCM, MIL: Yes** 2003 E Van, F-Series Truck & Excursion with 7.3L VIN F Diesel engine Transmissions: All	**Throttle Position Sensor 'B' Circuit High Input Conditions:** Key on or engine running; and the PCM detected the TP Sensor 'B' signal indicated more than 4.65v (Scan Tool reads more than 93%). **Possible Causes** • ETC TP Sensor 'B' connector is damaged or open • ETC TP Sensor 'B' signal circuit is open • ETC TP Sensor 'B' signal circuit is shorted to VREF (5v) • ETC TP Sensor 'B' is damaged or it has failed
P0230 **2T CCM, MIL: Yes** 1996-1997-1998-1999-2000-2001-2002-2003 E-Van, F-Series Truck & Excursion with 6.0L or 7.3L Diesel engine Transmissions: All	**Fuel Pump Relay Driver Circuit Malfunction Conditions:** Key on or engine running; and the PCM detected an unexpected voltage condition on the Fuel Pump Relay Driver control circuit. The fuel pump control circuit is used by the PCM to energize the fuel pump relay. At key on, the relay is energized for 20 seconds, and all the time the engine is running. The Fuel Pump Monitor (FPM) circuit is after the inertia switch. It is used to monitor voltage to the pump. **Possible Causes** • Fuel pump relay driver control circuit is open • Fuel pump relay driver control circuit is shorted to ground • Fuel pump relay is damaged or it has failed • PCM has failed
P0231 **2T CCM, MIL: Yes** 1996-1997-1998-1999-2000-2001-2002-2003 E-Van, F-Series Truck & Excursion with 6.0L or 7.3L Diesel engine Transmissions: All	**Fuel Pump Relay Driver Failed On Conditions:** Key on or engine running; and the PCM did not detect any voltage on the FP Monitor circuit with the fuel pump commanded on. The fuel pump control circuit is used by the PCM to energize the fuel pump relay. At key on, the relay is energized for 20 seconds, and all the time the engine is running. The Fuel Pump Monitor (FPM) circuit is after the inertia switch. **Possible Causes** • FPM circuit is open or it is shorted to ground • Fuel pump relay output circuit (VPWR) is open • Fuel pump relay is damaged or it has failed (no output voltage) • PCM has failed
P0232 **2T CCM, MIL: Yes** 1996-1997-1998-1999-2000-2001-2002-2003 E-Van, F-Series Truck & Excursion with 6.0L or 7.3L Diesel engine Transmissions: All	**Fuel Pump Relay Driver Failed Off Conditions:** Key on or engine running; and the PCM detected a voltage on the FP Monitor circuit with fuel pump commanded off. The fuel pump control circuit is used by the PCM to energize the fuel pump relay. At key on, the relay is energized for 20 seconds, and all the time the engine is running. The Fuel Pump Monitor (FPM) circuit is after the inertia switch. **Possible Causes** • Fuel pump relay contacts always closed • Fuel pump ground circuit has high resistance • Fuel pump secondary circuit shorted to power • Low speed fuel pump relay damaged or related circuit problem

Diesel Engine OBD II Trouble Code List (P0xxx Codes)

DTC	Trouble Code Title, Conditions & Possible Causes
P0236 **2T CCM, MIL: Yes** 1996-1997-1998-1999-2000-2001-2002-2003 E-Van, F-Series Truck & Excursion with 6.0L or 7.3L Diesel engine Transmissions: All	**Turbo Boost Sensor 'A' Signal Range/Performance Conditions:** Key on or engine running; and the PCM detected an unexpected low or high voltage condition for the specific operating conditions on the MAP sensor circuit. The MAP sensor is a variable capacitance sensor that, when supplied with a 5v reference signal, produces an analog signal that indicates pressure. The MAP sensor is used to control (diesel) smoke by limiting fuel quality during acceleration until a specified boost pressure is obtained. **Possible Causes** • MAP sensor wiring harness or connector is damaged or open • MAP sensor signal circuit is open or shorted to ground • MAP sensor VREF circuit (5v) is open or shorted to ground • MAP sensor is damaged or it has failed • PCM has failed
P0237 **2T CCM, MIL: Yes** 1996-1997-1998-1999-2000-2001-2002-2003 E-Van, F-Series Truck & Excursion with 6.0L or 7.3L Diesel engine Transmissions: All	**Turbo Boost Sensor 'A' Circuit Low Input Conditions:** Key on or engine running; and the PCM detected an unexpected low voltage on the MAP sensor circuit. The MAP sensor is a variable capacitance sensor supplied with a 5v reference signal. It produces an analog signal that indicates pressure. The MAP sensor is used to control (diesel) smoke by limiting fuel quality during acceleration until a specified boost pressure is obtained. **Possible Causes** • MAP sensor wiring harness or connector is shorted • MAP sensor signal circuit is shorted to ground • MAP sensor VREF circuit (5v) is open • MAP sensor is damaged or it has failed • PCM has failed
P0238 **2T CCM, MIL: Yes** 1996-1997-1998-1999-2000-2001-2002-2003 E-Van, F-Series Truck & Excursion with 6.0L or 7.3L Diesel engine Transmissions: All	**Turbo Boost Sensor 'A' Circuit High Input Conditions:** Key on or engine running; and the PCM detected an unexpected high voltage on the MAP sensor circuit. The MAP sensor is a variable capacitance sensor produces an analog signal that indicates pressure. The signal from this sensor is used to control (diesel) smoke by limiting fuel quality during acceleration until a specified boost pressure is obtained. **Possible Causes** • MAP sensor wiring harness or connector is open • MAP sensor signal circuit is open • MAP sensor signal circuit is shorted to VREF (5v) • MAP sensor is damaged or it has failed • PCM has failed
P0261 **2T CCM, MIL: Yes** 1996-1997-1998-1999-2000-2001-2002-2003 E-Van, F-Series Truck & Excursion with 6.0L or 7.3L Diesel engine Transmissions: All	**Fuel Injector Circuit Low Input - Cylinder 1 Conditions:** Key on or engine running; and the PCM detected an unexpected low voltage condition on the Injector 1 control circuit. The High Side driver output function is to distribute energy to the correct bank based on cylinder identification and to provide regulated current to the unit injectors, based on fuel delivery command signal from the injector driver module internal 115v supply. Injector timing and duration is commanded by the PCM in the FDCS module. **Possible Causes** • Injector 1 control circuit is open • Injector 1 power circuit (B+) is open • Injector 1 control circuit is shorted to chassis ground • Injector 1 is damaged or has failed • PCM has failed
P0262 **2T CCM, MIL: No** 1996-1997-1998-1999-2000-2001-2002-2003 E-Van, F-Series Truck & Excursion with 6.0L or 7.3L Diesel engine Transmissions: All	**Fuel Injector Circuit High Input - Cylinder 1 Conditions:** Key on or engine running; and the PCM detected an unexpected high voltage condition on the Injector 1 control circuit. The High Side driver output function is to distribute energy to the correct bank based on cylinder identification and to provide regulated current to the unit injectors, based on fuel delivery command signal from the injector driver module internal 115v supply. Injector timing and duration are commanded by the PCM in the FDCS module. **Possible Causes** • Injector 1 control circuit is shorted to system power (B+) • PCM has failed

Diesel Engine OBD II Trouble Code List (P0xxx Codes)

DTC	Trouble Code Title, Conditions & Possible Causes
P0263 **2T CCM, MIL: No** 1996-1997-1998-1999- 2000-2001-2002-2003 E-Van, F-Series Truck & Excursion with 6.0L or 7.3L Diesel engine Transmissions: All	**Fuel Injector Cylinder 1 Contribution/Balance Malfunction Conditions:** Engine started, and the PCM detected an engine condition with low cylinder contribution or "no" cylinder contribution condition on Engine Cylinder 1 during the CCM test period. **Possible Causes** • Base engine problem affecting only Cylinder 1 • Fuel injection delivery problem affecting only Cylinder 1 • PCM has failed
P0264 **2T CCM, MIL: Yes** 1996-1997-1998-1999- 2000-2001-2002-2003 E-Van, F-Series Truck & Excursion with 6.0L or 7.3L Diesel engine Transmissions: All	**Fuel Injector Circuit Low Input - Cylinder 2 Conditions:** Key on or engine running; and the PCM detected an unexpected low voltage condition on the Injector 2 control circuit. The High Side driver output function is to distribute energy to the correct bank based on cylinder identification and to provide regulated current to the unit injectors, based on fuel delivery command signal from the injector driver module internal 115v supply. Injector timing and duration is commanded by the PCM in the FDCS module. **Possible Causes** • Injector 2 power circuit (B+) is open • Injector 2 control circuit is open or shorted to chassis ground • Injector 2 is damaged or has failed • PCM has failed
P0265 **2T CCM, MIL: No** 1996-1997-1998-1999- 2000-2001-2002-2003 E-Van, F-Series Truck & Excursion with 6.0L or 7.3L Diesel engine Transmissions: All	**Fuel Injector Circuit High Input - Cylinder 2 Conditions:** Key on or engine running; and the PCM detected an unexpected high voltage condition on the Injector 2 control circuit. The High Side driver output function is to distribute energy to the correct bank based on cylinder identification and to provide regulated current to the unit injectors, based on fuel delivery command signal from the injector driver module internal 115v supply. Injector timing and duration are commanded by the PCM in the FDCS module. **Possible Causes** • Injector 2 control circuit is shorted to system power (B+) • PCM has failed
P0266 **, T CCM** **MIL: No** 1996-1997-1998-1999- 2000-2001-2002-2003 E-Van, F-Series Truck & Excursion with 6.0L or 7.3L Diesel engine Transmissions: All	**Fuel Injector 2 Cylinder Contribution/Balance Malfunction Conditions:** Engine started, and the PCM detected an engine condition with low cylinder contribution or "no" cylinder contribution condition on Engine Cylinder 2 during the CCM test period. **Possible Causes** • Base engine problem affecting only Cylinder 2 • Fuel injection delivery problem affecting only Cylinder 2 • PCM has failed
P0267 **2T CCM, MIL: Yes** 1996-1997-1998-1999- 2000-2001-2002-2003 E-Van, F-Series Truck & Excursion with 6.0L or 7.3L Diesel engine Transmissions: All	**Fuel Injector Circuit Low Input - Cylinder 3 Conditions:** Key on or engine running; and the PCM detected an unexpected low voltage condition on the Injector 3 control circuit. The High Side driver output function is to distribute energy to the correct bank based on cylinder identification and to provide regulated current to the unit injectors, based on fuel delivery command signal from the injector driver module internal 115v supply. Injector timing and duration is commanded by the PCM in the FDCS module. **Possible Causes** • Injector 3 control circuit is open • Injector 3 power circuit (B+) is open • Injector 3 control circuit is shorted to chassis ground
P0268 **2T CCM, MIL: No** 1996-1997-1998-1999- 2000-2001-2002-2003 E-Van, F-Series Truck & Excursion with 6.0L or 7.3L Diesel engine Transmissions: All	**Fuel Injector Circuit High Input - Cylinder 3 Conditions:** Key on or engine running; and the PCM detected an unexpected high voltage condition on the Injector 3 control circuit. The High Side driver output function is to distribute energy to the correct bank based on cylinder identification and to provide regulated current to the unit injectors, based on fuel delivery command signal from the injector driver module internal 115v supply. Injector timing and duration is commanded by the PCM in the FDCS module. **Possible Causes** • Injector 3 control circuit is shorted to system power (B+) • PCM has failed

Diesel Engine OBD II Trouble Code List (P0xxx Codes)

DTC	Trouble Code Title, Conditions & Possible Causes
P0269 **2T CCM, MIL: No** 1996-1997-1998-1999-2000-2001-2002-2003 E-Van, F-Series Truck & Excursion with 6.0L or 7.3L Diesel engine Transmissions: All	**Fuel Injector Cylinder 3 Contribution/Balance Malfunction Conditions:** Engine started, and the PCM detected an engine condition with low cylinder contribution or "no" cylinder contribution condition on Engine Cylinder 3 during the CCM test period. **Possible Causes** • Base engine problem affecting only Cylinder 3 • Fuel injection delivery problem affecting only Cylinder 3 • PCM has failed
P0270 **2T CCM, MIL: Yes** 1996-1997-1998-1999-2000-2001-2002-2003 E-Van, F-Series Truck & Excursion with 6.0L or 7.3L Diesel engine Transmissions: All	**Fuel Injector Circuit Low Input - Cylinder 4 Conditions:** Key on or engine running; and the PCM detected an unexpected low voltage condition on the Injector 4 control circuit. The High Side driver output function is to distribute energy to the correct bank based on cylinder identification and to provide regulated current to the unit injectors, based on fuel delivery command signal from the injector driver module internal 115v supply. Injector timing and duration is commanded by the PCM in the FDCS module. **Possible Causes** • Injector 4 control circuit is open • Injector 4 power circuit (B+) is open • Injector 4 control circuit is shorted to chassis ground • PCM has failed
P0271 **2T CCM, MIL: No** 1996-1997-1998-1999-2000-2001-2002-2003 E-Van, F-Series Truck & Excursion with 6.0L or 7.3L Diesel engine Transmissions: All	**Fuel Injector Circuit High Input - Cylinder 4 Conditions:** Key on or engine running; and the PCM detected an unexpected high voltage condition on the Injector 4 control circuit. The High Side driver output function is to distribute energy to the correct bank based on cylinder identification and to provide regulated current to the unit injectors, based on fuel delivery command signal from the injector driver module internal 115v supply. Injector timing and duration are commanded by the PCM in the FDCS module. **Possible Causes** • Injector 4 control circuit is shorted to system power (B+) • PCM has failed
P0272 **2T CCM, MIL: No** 1996-1997-1998-1999-2000-2001-2002-2003 E-Van, F-Series Truck & Excursion with 6.0L or 7.3L Diesel engine Transmissions: All	**Fuel Injector Cylinder 4 Contribution/Balance Malfunction Conditions:** Engine started, and the PCM detected an engine condition with low cylinder contribution or "no" cylinder contribution condition on Engine Cylinder 4 during the CCM test period. **Possible Causes** • Base engine problem affecting only Cylinder 4 • Fuel injection delivery problem affecting only Cylinder 4.
P0273 **2T CCM, MIL: Yes** 1996-1997-1998-1999-2000-2001-2002-2003 E-Van, F-Series Truck & Excursion with 6.0L or 7.3L Diesel engine Transmissions: All	**Fuel Injector Circuit Low Input - Cylinder 5 Conditions:** Key on or engine running; and the PCM detected an unexpected low voltage condition on the Injector 5 control circuit. The High Side driver output function is to distribute energy to the correct bank based on cylinder identification and to provide regulated current to the unit injectors, based on fuel delivery command signal from the injector driver module internal 115v supply. Injector timing and duration are commanded by the PCM in the FDCS module. **Possible Causes** • Injector 5 control circuit is open • Injector 5 power circuit (B+) is open • Injector 5 control circuit is shorted to chassis ground
P0274 **2T CCM, MIL: No** 1996-1997-1998-1999-2000-2001-2002-2003 E-Van, F-Series Truck & Excursion with 6.0L or 7.3L Diesel engine Transmissions: All	**Fuel Injector Circuit High Input - Cylinder 5 Conditions:** Key on or engine running; and the PCM detected an unexpected high voltage condition on the Injector 5 control circuit. The High Side driver output function is to distribute energy to the correct bank based on cylinder identification and to provide regulated current to the unit injectors, based on fuel delivery command signal from the injector driver module internal 115v supply. Injector timing and duration are commanded by the PCM in the FDCS module. **Possible Causes** • Injector 5 control circuit is shorted to system power (B+) • PCM has failed

Diesel Engine OBD II Trouble Code List (P0xxx Codes)

DTC	Trouble Code Title, Conditions & Possible Causes
P0275 **2T CCM, MIL: No** 1996-1997-1998-1999- 2000-2001-2002-2003 E-Van, F-Series Truck & Excursion with 6.0L or 7.3L Diesel engine Transmissions: All	**Fuel Injector Cylinder 5 Contribution/Balance Malfunction Conditions:** Engine started, and the PCM detected an engine condition with low cylinder contribution or "no" cylinder contribution condition on Engine Cylinder 5 during the CCM test period. **Possible Causes** • Base engine problem affecting only Cylinder 5 • Fuel injection delivery problem affecting only Cylinder 5 • PCM has failed
P0276 **2T CCM, MIL: Yes** 1996-1997-1998-1999- 2000-2001-2002-2003 E-Van, F-Series Truck & Excursion with 6.0L or 7.3L Diesel engine Transmissions: All	**Fuel Injector Circuit Low Input - Cylinder 6 Conditions:** Key on or engine running; and the PCM detected an unexpected low voltage condition on the Injector 6 control circuit. The High Side driver output function is to distribute energy to the correct bank based on cylinder identification and to provide regulated current to the unit injectors, based on fuel delivery command signal from the injector driver module internal 115v supply. Injector timing and duration are commanded by the PCM in the FDCS module. **Possible Causes** • Injector 6 control circuit is open • Injector 6 power circuit (B+) is open • Injector 6 control circuit is shorted to chassis ground • PCM has failed
P0277 **2T CCM, MIL: No** 1996-1997-1998-1999- 2000-2001-2002-2003 E-Van, F-Series Truck & Excursion with 6.0L or 7.3L Diesel engine Transmissions: All	**Fuel Injector Circuit High Input - Cylinder 6 Conditions:** Key on or engine running; and the PCM detected an unexpected high voltage condition on the Injector 6 control circuit. The High Side driver output function is to distribute energy to the correct bank based on cylinder identification and to provide regulated current to the unit injectors, based on fuel delivery command signal from the injector driver module internal 115v supply. Injector timing and duration are commanded by the PCM in the FDCS module. **Possible Causes** • Injector 6 control circuit is shorted to system power (B+) • PCM has failed
P0278 **2T CCM, MIL: No** 1996-1997-1998-1999- 2000-2001-2002-2003 E-Van, F-Series Truck & Excursion with 6.0L or 7.3L Diesel engine Transmissions: All	**Fuel Injector Cylinder 6 Contribution/Balance Malfunction Conditions:** Engine started, and the PCM detected an engine condition with low cylinder contribution or "no" cylinder contribution condition on Engine Cylinder 6 during the CCM test period. **Possible Causes** • Base engine problem affecting only Cylinder 6 • Fuel injection delivery problem affecting only Cylinder 6
P0279 **2T CCM, MIL: Yes** 1996-1997-1998-1999- 2000-2001-2002-2003 E-Van, F-Series Truck & Excursion with 6.0L or 7.3L Diesel engine Transmissions: All	**Fuel Injector Circuit Low Input - Cylinder 7 Conditions:** Key on or engine running; and the PCM detected an unexpected low voltage condition on the Injector 7 control circuit. The High Side driver output function is to distribute energy to the correct bank based on cylinder identification and to provide regulated current to the unit injectors, based on fuel delivery command signal from the injector driver module internal 115v supply. Injector timing and duration are commanded by the PCM in the FDCS module. **Possible Causes** • Injector 7 control circuit is open • Injector 7 power circuit (B+) is open • Injector 7 control circuit is shorted to chassis ground
P0280 **2T CCM, MIL: No** 1996-1997-1998-1999- 2000-2001-2002-2003 E-Van, F-Series Truck & Excursion with 6.0L or 7.3L Diesel engine Transmissions: All	**Fuel Injector Circuit High Input - Cylinder 7 Conditions:** Key on or engine running; and the PCM detected an unexpected high voltage condition on the Injector 7 control circuit. The High Side driver output function is to distribute energy to the correct bank based on cylinder identification and to provide regulated current to the unit injectors, based on fuel delivery command signal from the injector driver module internal 115v supply. Injector timing and duration are commanded by the PCM in the FDCS module. **Possible Causes** • Injector 7 control circuit is shorted to system power (B+) • PCM has failed

Diesel Engine OBD II Trouble Code List (P0xxx Codes)

DTC	Trouble Code Title, Conditions & Possible Causes
P0281 **2T CCM, MIL: Yes** 1996-1997-1998-1999- 2000-2001-2002-2003 E-Van, F-Series Truck & Excursion with 6.0L or 7.3L Diesel engine Transmissions: All	**Fuel Injector Cylinder 7 Contribution/Balance Malfunction Conditions:** Engine started, and the PCM detected an engine condition with low cylinder contribution or "no" cylinder contribution condition on Engine Cylinder 7 during the CCM test period. **Possible Causes** • Base engine problem affecting only Cylinder 7 • Fuel injection delivery problem affecting only Cylinder 7
P0282 **2T CCM, MIL: Yes** 1996-1997-1998-1999- 2000-2001-2002-2003 E-Van, F-Series Truck & Excursion with 6.0L or 7.3L Diesel engine Transmissions: All	**Fuel Injector Circuit Low Input - Cylinder 8 Conditions:** Key on or engine running; and the PCM detected an unexpected low voltage condition on the Injector 8 control circuit. The High Side driver output function is to distribute energy to the correct bank based on cylinder identification and to provide regulated current to the unit injectors, based on fuel delivery command signal from the injector driver module internal 115v supply. Injector timing and duration is commanded by the PCM in the FDCS module. **Possible Causes** • Injector 8 control circuit is open • Injector 8 power circuit (B+) is open • Injector 8 control circuit is shorted to chassis ground
P0283 **2T CCM, MIL: Yes** 1996-1997-1998-1999- 2000-2001-2002-2003 E-Van, F-Series Truck & Excursion with 6.0L or 7.3L Diesel engine Transmissions: All	**Fuel Injector Circuit High Input - Cylinder 8 Conditions:** Key on or engine running; and the PCM detected an unexpected high voltage condition on the Injector 8 control circuit. The High Side driver output function is to distribute energy to the correct bank based on cylinder identification and to provide regulated current to the unit injectors, based on fuel delivery command signal from the injector driver module internal 115v supply. Injector timing and duration is commanded by the PCM in the FDCS module. **Possible Causes** • Injector 8 control circuit is shorted to system power (B+) • PCM has failed
P0284 **2T CCM, MIL: Yes** 1996-1997-1998-1999- 2000-2001-2002-2003 E-Van, F-Series Truck & Excursion with 6.0L or 7.3L Diesel engine Transmissions: All	**Fuel Injector Cylinder 8 Contribution/Balance Malfunction Conditions:** Engine started, and the PCM detected an engine condition with low cylinder contribution or "no" cylinder contribution condition on Engine Cylinder 8 during the CCM test period. **Possible Causes** • Base engine problem affecting only Cylinder 8 • Fuel injection delivery problem affecting only Cylinder 8 • PCM has failed
P0301 **2T Misfire, MIL: Yes** 1998-1999-2000-2001- 2002-2003 E-Van, F-Series Truck & Excursion with 6.0L or 7.3L Diesel engine Transmissions: All	**Fault Cylinder 'A' - Misfire Detected Conditions:** DTP P0263 not set, engine started, engine running under positive torque conditions, and the PCM detected a low cylinder contribution (misfire) condition related to Engine Cylinder 1 or Engine Cylinder 'A'. **Possible Causes** • Base engine problems: broken compression rings, leaking or bent valves, bent push rod, broken rocker arm bolts or bent connecting rod • Fuel metering problem on Cylinder 'A' (a fault with the injector)
P0302 **2T Misfire, MIL: Yes** 1998-1999-2000-2001- 2002-2003 E-Van, F-Series Truck & Excursion with 6.0L or 7.3L Diesel engine Transmissions: All	**Fault Cylinder 'B' - Misfire Detected Conditions:** DTP P0266 not set, engine started, engine running under positive torque conditions, and the PCM detected a low cylinder contribution (misfire) condition related to Engine Cylinder 2 or Engine Cylinder 'B'. **Possible Causes** • Base engine problems: broken compression rings, leaking or bent valves, bent push rod, broken rocker arm bolts or bent connecting rod • Fuel metering problem on Cylinder 'B' (a fault with the injector)

Diesel Engine OBD II Trouble Code List (P0xxx Codes)

DTC	Trouble Code Title, Conditions & Possible Causes
P0303 **2T Misfire, MIL: Yes** 1998-1999-2000-2001-2002-2003 E-Van, F-Series Truck & Excursion with 6.0L or 7.3L Diesel engine Transmissions: All	**Fault Cylinder 'C' - Misfire Detected Conditions:** DTP P0269 not set, engine started, engine running under positive torque conditions, and the PCM detected a low cylinder contribution (misfire) condition related to Engine Cylinder 3 or Engine Cylinder 'C'. **Possible Causes** • Base engine problems: broken compression rings, leaking or bent valves, bent push rod, broken rocker arm bolts or bent connecting rod • Fuel metering problem on Cylinder 'C' (a fault with the injector)
P0304 **2T Misfire** **MIL: Yes** 1998-1999-2000-2001-2002-2003 E-Van, F-Series Truck & Excursion with 6.0L or 7.3L Diesel engine Transmissions: All	**Fault Cylinder 'D' - Misfire Detected Conditions:** DTP P0272 not set, engine started, engine running under positive torque conditions, and the PCM detected a low cylinder contribution (misfire) condition related to Engine Cylinder 4 or 'D'. **Possible Causes** • Base engine problems: broken compression rings, leaking or bent valves, bent push rod, broken rocker arm bolts or bent connecting rod • Fuel metering problem on Cylinder 'D' (a fault with the injector)
P0305 **2T Misfire, MIL: Yes** 1998-1999-2000-2001-2002-2003 E-Van, F-Series Truck & Excursion with 6.0L or 7.3L Diesel engine Transmissions: All	**Fault Cylinder 'E' - Misfire Detected Conditions:** DTP P0275 not set, engine started, engine running under positive torque conditions, and the PCM detected a low cylinder contribution (misfire) condition related to Engine Cylinder 5 or 'E'. **Possible Causes** • Base engine problems: broken compression rings, leaking or bent valves, bent push rod, broken rocker arm bolts or bent connecting rod • Fuel metering problem on Cylinder 'E' (a fault with the injector)
P0306 **2T Misfire, MIL: Yes** 1998-1999-2000-2001-2002-2003 E-Van, F-Series Truck & Excursion with 6.0L or 7.3L Diesel engine Transmissions: All	**Fault Cylinder 'F' - Misfire Detected Conditions:** DTP P0278 not set, engine started, engine running under positive torque conditions, and the PCM detected a low cylinder contribution (misfire) condition related to Engine Cylinder 6 or 'F'. **Possible Causes** • Base engine problems: broken compression rings, leaking or bent valves, bent push rod, broken rocker arm bolts or bent connecting rod • Fuel metering problem on Cylinder 'F' (a fault with the injector)
P0307 **2T Misfire, MIL: Yes** 1998-1999-2000-2001-2002-2003 E-Van, F-Series Truck & Excursion with 6.0L or 7.3L Diesel engine Transmissions: All	**Fault Cylinder 'G' - Misfire Detected Conditions:** DTP P0281 not set, engine started, engine running under positive torque conditions, and the PCM detected a low cylinder contribution (misfire) condition related to Engine Cylinder 7 or 'G'. **Possible Causes** • Base engine problems: broken compression rings, leaking or bent valves, bent push rod, broken rocker arm bolts or bent connecting rod • Fuel metering problem on Cylinder 'G' (a fault with the injector)
P0308 **2T Misfire, MIL: Yes** 1998-1999-2000-2001-2002-2003 E-Van, F-Series Truck & Excursion with 6.0L or 7.3L Diesel engine Transmissions: All	**Fault Cylinder 'H' - Misfire Detected Conditions:** DTP P0284 not set, engine started, engine running under positive torque conditions, and the PCM detected a low cylinder contribution (misfire) condition related to Engine Cylinder 8 or 'H'. **Possible Causes** • Base engine problems: broken compression rings, leaking or bent valves, bent push rod, broken rocker arm bolts or bent connecting rod • Fuel metering problem on Cylinder 'H' (a fault with the injector)

Diesel Engine OBD II Trouble Code List (P0xxx Codes)

DTC	Trouble Code Title, Conditions & Possible Causes
P0335 **1T CCM, MIL: Yes** 2003 F-Series Truck & Excursion with a 6.0L VIN P Diesel engine Transmissions: All	**Crankshaft Position Sensor 'A' Circuit Malfunction Conditions:** Engine cranking or running; and the PCM did not detect any Crankshaft Position Sensor 'A' circuit signals during the test. **Possible Causes** • CKP sensor signal circuit is open or shorted to ground • CKP sensor ground (return) circuit is open • CKP sensor is damaged or CKP sensor shielding is damaged • PCM has failed
P0336 **1T CCM, MIL: Yes** 2003 F-Series Truck & Excursion with a 6.0L VIN P Diesel engine Transmissions: All	**Crankshaft Position Sensor 'A' Circuit Performance Conditions:** Engine started, and the PCM detected an erratic Crankshaft Position Sensor 'A' signal or an intermittent loss of the CKP Sensor 'A' signal. **Possible Causes** • CKP sensor signal circuit is open or shorted to ground • CKP sensor ground (return) circuit is open • CKP sensor is damaged or CKP sensor shielding is damaged • PCM has failed
P0340 **2T CCM, MIL: No** 1996-1997-1998-1999-2000-2001-2002-2003 E-Van, F-Series Truck & Excursion with 6.0L or 7.3L Diesel engine Transmissions: All	**Camshaft Position Sensor Circuit Malfunction Conditions:** Engine started, and the PCM did not detect any CMP sensor signals. This device is a Hall Effect design sensor that generates a digital frequency, as windows in a target wheel pass through its magnetic field. The frequency of the windows passing by the sensor as well as the width of the selected windows allows the PCM to detect the engine speed and position. The engine speed is determined by counting the 12 the windows on the cam gear in each camshaft revolution. The position of cylinders 1 and 4 is determined by distinguishing a narrow or wide window on the camshaft gear. **Possible Causes** • CMP (Hall effect) sensor circuit is open or shorted to ground • CMP (Hall effect) sensor power (B+) circuit is open • CMP (Hall effect) sensor is incorrectly installed or damaged • PCM is damaged
P0341 **2T CCM, MIL: Yes** 1996-1997-1998-1999-2000-2001-2002-2003 E-Van, F-Series Truck & Excursion with 6.0L or 7.3L Diesel engine Transmissions: All	**Camshaft Position Sensor Signal Range/Performance Conditions:** Engine started, and the PCM detected an irregular or out-of-phase CMP sensor signal during the CCM test. The target wheel spokes are each 15 degrees apart except for a narrow spoke that identifies Cylinder 1 and a wide spoke that identifies Cylinder 4. As the camshaft rotates, this sensor generates a digital frequency. **Possible Causes** • CMP sensor connector is damaged, open or shorted • CMP sensor signal circuit is open or shorted to ground • CMP sensor signal circuit is erratic (check for EMI or RFI) • CMP sensor is damaged or it has failed • PCM has failed
P0340 **2T CCM, MIL: Yes** 1996-1997-1998-1999-2000-2001-2002-2003 E-Van, F-Series Truck & Excursion with 6.0L or 7.3L Diesel engine Transmissions: All	**Camshaft Position Sensor Circuit Malfunction (Intermittent) Conditions:** Engine started, and the PCM detected that the CMP sensor signal was interrupted during testing. The target wheel spokes are each 15 degrees apart except for a narrow spoke that identifies Cylinder 1 and a wide spoke that identifies Cylinder 4. As the camshaft rotates, this sensor generates a digital frequency. **Possible Causes** • CMP (Hall effect) sensor circuit is open (an intermittent fault) • CMP (Hall effect) sensor circuit shorted to ground (intermittent) • CMP (Hall effect) sensor is damaged or it has failed • PCM has failed

Diesel Engine OBD II Trouble Code List (P0xxx Codes)

DTC	Trouble Code Title, Conditions & Possible Causes
P0380 **2T CCM, MIL: Yes** 1996-1997-1998-1999-2000-2001-2002-2003 E-Van, F-Series Truck & Excursion with 6.0L or 7.3L Diesel engine Transmissions: All	**Glow Plug Relay Circuit Malfunction Conditions:** Key on, and the PCM detected an unexpected voltage condition on the Glow Plug Relay control circuit during the CCM test. The Glow Plug Lamp remains "on" for 1-12 seconds (depending on the Glow Plug relay on-time which can vary from 1 and 120 seconds). **Possible Causes** • Glow plug relay control circuit is open or shorted to ground • Glow plug relay power circuit is open (test the 12GA fuse link) • Glow plug relay is damaged or it has failed
P0381 **2T CCM, MIL: Yes** 1996-1997-1998-1999-2000-2001-2002-2003 E-Van, F-Series Truck & Excursion with 6.0L or 7.3L Diesel engine Transmissions: All	**Glow Plug Indicator Circuit Malfunction Conditions:** Key on, and the PCM detected an unexpected voltage condition on the Glow Plug Lamp circuit during the CCM test. The Glow Plug Lamp remains "on" for 1-12 seconds (depending on the Glow Plug relay on-time which can vary from 1 and 120 seconds). **Possible Causes** • Glow plug lamp circuit is open or shorted to ground • Glow plug relay control circuit is open or shorted to ground • Glow plug relay power circuit is open (test the 12GA fuse link) • Glow plug relay is damaged or it has failed
P0401 **2T EGR, MIL: Yes** 2003 F-Series Truck & Excursion with a 6.0L VIN P Diesel engine Transmissions: All	**Insufficient EGR Flow Detected Conditions:** Engine started, engine running at steady cruise speed, and the PCM detected the EGR valve position did not match the Desired EGR position based on engine speed and load. **Possible Causes** • EGR valve actuator is damaged or sticking (perform a KOEO or KOER Self Test to test the actuator for an On Demand code) • EGR position sensor signal circuit shorted to bias voltage • EP sensor signal circuit shorted to bias voltage • PCM has failed
P0402 **2T EGR, MIL: Yes** 2003 F-Series Truck & Excursion with a 6.0L VIN P Diesel engine Transmissions: All	**Excessive EGR Flow Detected Conditions:** Engine started, engine running at steady cruise speed, and the PCM detected excessive EGR gas flow. The EGR valve actuator and EGR valve position sensor are in one unit. **Possible Causes** • EVP sensor is damaged or has failed • EGR valve source vacuum supply problem (hose off or leaking) • EGR valve is stuck partially open or closed (check for carbon) • EGR valve may be leaking vacuum (the diaphragm is broken) • VR solenoid is damaged or the VR control circuit is open • PCM has failed
P0403 **2T CCM, MIL: Yes** 2003 F-Series Truck & Excursion with a 6.0L VIN P Diesel engine Transmissions: All	**EGR Solenoid Control Circuit Malfunction Conditions:** Engine started, and the PCM detected an unexpected high or low voltage condition on the EGR solenoid control circuit at idle speed. **Possible Causes** • EGR solenoid connector is damaged, loose or shorted • EGR solenoid control circuit is open, shorted to ground or B+ • EGR solenoid is damaged or has failed • PCM has failed
P0404 **2T CCM, MIL: Yes** 2003 F-Series Truck & Excursion with a 6.0L VIN P Diesel engine Transmissions: All	**EGR Solenoid Control Signal Range/Performance Conditions:** Engine started, and the PCM detected and EGRP error (± 0.10v) from the Actual to the Commanded EGR valve position at cruise. **Possible Causes** • EGR position sensor circuit is open or shorted (intermittent) • EGR position sensor is damaged or it has failed • EGR valve is damaged or it has failed • PCM has failed

Diesel Engine OBD II Trouble Code List (P0xxx Codes)

DTC	Trouble Code Title, Conditions & Possible Causes
P0405 **2T CCM, MIL: Yes** 2003 F-Series Truck & Excursion with a 6.0L VIN P Diesel engine Transmissions: All	**EGR Position Sensor 'A' Circuit Low Input Conditions:** Engine started, and the PCM detected an unexpected low voltage condition (less than 0.30v) on the EGR valve position sensor circuit. **Possible Causes** • EGR position sensor connector is damaged, open or shorted • EGR position sensor circuit is open or shorted to ground • EGR position sensor is damaged or it has failed • PCM has failed
P0406 **2T CCM, MIL: Yes** 2003 F-Series Truck & Excursion with a 6.0L VIN P Diesel engine Transmissions: All	**EGR Position Sensor 'A' Circuit High Input Conditions:** Engine started, and the PCM detected an unexpected high voltage condition (more than 4.9v) on the EGR valve position sensor circuit. **Possible Causes** • EGR position sensor connector is damaged or shorted • EGR position sensor circuit is shorted to VREF or power (B+) • EGR position sensor is damaged or it has failed • PCM has failed
P0460 **2T CCM, MIL: Yes** 1998-1999-2000-2001-2002-2003 E-Van, F-Series Truck & Excursion with 6.0L or 7.3L Diesel engine Transmissions: All	**Fuel Level Sensor Circuit Malfunction Conditions:** Engine started, and the PCM detected the FLI signal did not match the fuel level during the CCM test. For example, a FLI V PID below 0.90v (Scan Tool reads FLI PID = 25%), or a FLI over 2.45v (Scan Tool reads FLI PID = 75). **Possible Causes** • Fuel tank is empty, or fuel pump (FP) module is stuck open • Fuel gauge incorrectly installed or Instrument cluster damaged • PCM Case ground circuit is open • Fuel level indicator (FLI) circuit shorted to VPWR, or is open • Fuel tank has been overfilled, or fuel gauge is damaged • Fuel pump (FP) module is stuck closed, or is stuck open • Fuel level indicator circuit shorted to Case or to power ground • PCM Case ground shorted to VPWR (shorted to system power)
P0470 **2T CCM, MIL: Yes** 1996-1997-1998-1999-2000-2001-2002-2003 E-Van, F-Series Truck & Excursion with 6.0L or 7.3L Diesel engine Transmissions: All	**Exhaust Backpressure Sensor Circuit Malfunction Conditions:** Key on or engine running; and the PCM detected an unexpected low or high voltage condition on the Exhaust Backpressure Sensor (EBP) signal circuit. The EBP sensor is a variable capacitance sensor that, when supplied with a 5v reference signal, produces a linear analog signal that indicates pressure. The primary function of this sensor is to measure exhaust backpressure so that the PCM can control the Exhaust Back Pressure Regulator (EBP) operation. **Possible Causes** • EBP sensor signal circuit is open or shorted to ground • EBP sensor signal circuit is shorted to VREF (5v) • EBP sensor power circuit (VREF) is open from the PCM • EBP sensor is damaged or it has failed • PCM is damaged
P0471 **2T CCM, MIL: Yes** 1996-1997-1998-1999-2000-2001-2002-2003 E-Van, F-Series Truck & Excursion with 6.0L or 7.3L Diesel engine Transmissions: All	**Exhaust Backpressure Sensor Signal Range/Performance Conditions:** Key on or engine running; and the PCM detected an unexpected voltage condition on the Exhaust Backpressure Sensor signal circuit. The EBP sensor is a variable capacitance sensor that, when supplied with a 5v reference signal, produces a linear analog signal that indicates pressure. The primary function of this sensor is to measure exhaust backpressure so that the PCM can control the Exhaust Back Pressure Regulator (EBP) operation. **Possible Causes** • EBP sensor is damaged • EBP sensor pressure supply tube is damaged or clogged • EPR linkage is damaged or sticking, or the butterfly is damaged • PCM is damaged

Diesel Engine OBD II Trouble Code List (P0xxx Codes)

DTC	Trouble Code Title, Conditions & Possible Causes
P0472 **2T CCM, MIL: Yes** 1996-1997-1998-1999-2000-2001-2002-2003 E-Van, F-Series Truck & Excursion with 6.0L or 7.3L Diesel engine Transmissions: All	**Exhaust Backpressure Sensor Circuit Low Input Conditions:** Key on or engine running; and the PCM detected an unexpected low voltage condition on the Exhaust Backpressure Sensor signal circuit. This sensor is used to measure exhaust backpressure so the PCM can control the Exhaust Back Pressure Regulator (EBP) operation. **Possible Causes** • EBP sensor signal circuit is shorted to ground • EBP sensor power circuit (VREF) is open • EBP sensor is damaged or it has failed • PCM is damaged
P0473 **2T CCM, MIL: Yes** 1996-1997-1998-1999-2000-2001-2002-2003 E-Van, F-Series Truck & Excursion with 6.0L or 7.3L Diesel engine Transmissions: All	**Exhaust Backpressure Sensor Circuit High Input Conditions:** Key on or engine running; and the PCM detected an unexpected high voltage condition on the Exhaust Backpressure Sensor signal circuit. This sensor is used to measure exhaust backpressure so the PCM can control the Exhaust Back Pressure Regulator operation. **Possible Causes** • EBP sensor is shorted to VREF (5v) • EBP sensor ground circuit is open • EBP sensor is damaged or it has failed • PCM is damaged
P0475 **2T CCM, MIL: Yes** 1996-1997-1998-1999-2000-2001-2002-2003 E-Van, F-Series Truck & Excursion with 6.0L or 7.3L Diesel engine Transmissions: All	**Exhaust Backpressure Control Valve Malfunction Conditions:** Key on, KOEO Self-Test enabled and the PCM detected an unexpected voltage condition on the Exhaust Backpressure Control Valve circuit. The EPR is a variable position valve that is used (along with signals from the IAT sensor and engine load) to control exhaust backpressure during cold ambient temperature to increase cab heat and decrease the amount of time needed to defrost the windshield. **Possible Causes** • EPR (regulator) valve circuit is open or shorted to ground • EPR (regulator) valve circuit is shorted to system power (B+) • EPR (regulator) valve is damaged or it has failed • PCM is damaged
P0476 **2T CCM, MIL: Yes** 1996-1997-1998-1999-2000-2001-2002-2003 E-Van, F-Series Truck & Excursion with 6.0L or 7.3L Diesel engine Transmissions: All	**Exhaust Backpressure Regulator Malfunction Conditions:** Engine started, KOER Self-Test enabled, and the PCM detected a problem in the operation of the Exhaust Backpressure Regulator. The EPR is a variable position valve that is used (along with signals from the IAT sensor and engine load) to control exhaust backpressure during cold ambient temperature to increase cab heat and decrease the amount of time needed to defrost the windshield. **Possible Causes** • EPR (regulator) control valve is binding or sticking • EPR (regulator) control valve is damaged or it has failed • EPR (regulator) resistance is 2.5-12.0 ohms at 68°F • PCM is damaged
P0478 **2T CCM, MIL: Yes** 1996-1997-1998-1999-2000-2001-2002-2003 E-Van, F-Series Truck & Excursion with 6.0L or 7.3L Diesel engine Transmissions: All	**Exhaust Backpressure Control Valve Excessive Conditions:** Engine started, vehicle driven under normal conditions, and the PCM detected an excessive backpressure condition. The EPR is a variable position valve that is used (along with signals from the IAT sensor and engine load) to control exhaust backpressure during cold ambient temperature to increase cab heat and decrease the amount of time needed to defrost the windshield. **Possible Causes** • EBP butterfly valve is stuck • EBP sensor line (or exhaust system) is clogged or restricted • EPR linkage is adjusted incorrectly or it is binding • Wastegate turbo may be in Overboost mode • PCM is damaged

Diesel Engine OBD II Trouble Code List (P0xxx Codes)

DTC	Trouble Code Title, Conditions & Possible Causes
P0480 **1T CCM, MIL: Yes** 2003 F-Series Truck & Excursion with a 6.0L VIN P Diesel engine Transmissions: All	**VDF Fan Control Circuit Malfunction Conditions:** Engine started, and the PCM detected an unexpected voltage condition on the Visctronic Drive Fan (VDF) control circuit. The VDF is a viscous coupling. The viscous drag should be smooth during fan rotation. The amount of resistance is dependant upon the final VDF operational state before engine shutdown. Check the PCM to determine if it has the latest calibration. **Possible Causes** • VDF assembly is damaged or it has failed (the VDF resistance is 6-10 ohms at 68ºF) • PCM has failed
P0500 **2T CCM, MIL: Yes** 1996-1997-1998-1999-2000-2001-2002-2003 E-Van, F-Series Truck & Excursion with 6.0L or 7.3L Diesel engine Transmissions: All	**Vehicle Speed Sensor Malfunction Conditions:** Engine running, then with the engine speed more than the TCC stall speed, the PCM detected a lack of vehicle speed data for a period of time. The PCM receives vehicle speed data from the ABS, VSS, TCSS, GEM, or CTM controller, depending up the application. **Possible Causes** • Mechanical drive mechanism for the VSS is damaged • VSS (+) or (-) signal circuit is open, shorted to ground or power • VSS is damaged or it has failed • PCM has failed • TSB 01-21-13 contains a repair procedure for this code
P0501 **2T CCM, MIL: Yes** 1998-1999 E-Vans & F-Series Trucks with a 7.3L Diesel engine Transmissions: All	**Vehicle Speed Sensor Signal Range/Performance Conditions:** Engine running, then with the engine speed more than the TCC stall speed, the PCM detected a problem with the VSS signal data during testing. The PCM receives vehicle speed data from the VSS (sensor) on this vehicle application. **Possible Causes** • Mechanical drive mechanism for the VSS is damaged • VSS (+) or (-) signal circuit is open, shorted to ground or power • VSS is damaged or it has failed • PCM has failed
P0502 **2T CCM, MIL: Yes** 1998-1999 E-Vans & F-Series Trucks with a 7.3L Diesel engine Transmissions: All	**Vehicle Speed Sensor Intermittent Signal Conditions:** Engine running, then with the engine speed more than the TCC stall speed, the PCM detected an intermittent loss of the VSS signal. The PCM receives vehicle speed data from a vehicle speed sensor on this vehicle application. **Possible Causes** • Mechanical drive mechanism for the VSS is damaged • VSS (+) or (-) signal circuit is open, shorted to ground or power • VSS is damaged or it has failed • PCM has failed
P0503 **2T CCM, MIL: Yes** 1999-2000-2001-2002-2003 E-Van, F-Series Truck & Excursion with 6.0L or 7.3L Diesel engine Transmissions: All	**Vehicle Speed Sensor Intermittent Signal Conditions:** Engine running, then with the engine speed more than the TCC stall speed, the PCM detected an intermittent or "noisy" signal on the VSS signal circuit. The PCM receives vehicle speed data from the ABS, VSS, TCSS, GEM, or CTM controller, depending up the vehicle. **Possible Causes** • An Aftermarket add-on device causing "noise" on the circuit • Module or circuits connected to VSS/TCSS circuit are damaged • VSS/TCSS signal "noise" due to RFI or EMI interference from sources such as ignition components or charging system • VSS/TCSS gears are damaged • VSS/TCSS wiring harness or connectors are damaged
P0528 **1T CCM, MIL: Yes** 2003 F-Series Truck & Excursion with a 6.0L VIN P Diesel engine Transmissions: All	**VDF Fan Control Circuit Malfunction Conditions:** Engine started, and the PCM detected an unexpected voltage condition on the Visctronic Drive Fan (VDF) control circuit. The VDF is a viscous coupling. The viscous drag should be smooth during fan rotation. The amount of resistance is dependant upon the final VDF operational state before engine shutdown. Check PCM calibration. **Possible Causes** • VDF assembly is damaged or it has failed (the VDF resistance is 6-10 ohms at 68ºF) • PCM has failed

Diesel Engine OBD II Trouble Code List (P0xxx Codes)

DTC	Trouble Code Title, Conditions & Possible Causes
P0541 **2T CCM, MIL: Yes** 1999-2000-2001-2002-2003 E-Van, F-Series Truck & Excursion with a 7.3L Diesel engine Transmissions: All	**Manifold Intake Air Heater Circuit Low Input Conditions:** Engine started, IAT sensor under 32°F, EOT sensor less than 131°F, system voltage at 11.8-15.0v, and the PCM detected an unexpected low voltage condition on the Manifold Intake Air Heater circuit. **Possible Causes** • MIAH circuit is open or shorted to ground from relay to the PCM • MIAH heater relay power (B+) circuit is open to the Alternator • MIAH is damaged or it has failed • PCM has failed
P0542 **2T CCM, MIL: No** 1999-2000-2001-2002-2003 E-Van, F-Series Truck & Excursion with a 7.3L Diesel engine Transmissions: All	**Manifold Intake Air Heater Circuit High Input Conditions:** Engine started, IAT sensor less than 32°F, system voltage from 11.8-15.0v, EOT sensor less than 131°F, and the PCM detected an unexpected high voltage condition on the Manifold Intake Air Heater circuit during the CCM test period. **Possible Causes** • MIAH circuit shorted to power between MIAH relay and heater • MIAH is damaged or it has failed • PCM has failed
P0560 **2T CCM, MIL: No** 1996-1997-1998-1999-2000-2001-2002-2003 E-Van, F-Series Truck & Excursion with 6.0L or 7.3L Diesel engine Transmissions: All	**System Voltage Malfunction Conditions:** Engine started, engine running at idle or cruise speed, and the PCM detected the system voltage was too high or too low during the test. **Possible Causes** • Battery connections corroded (high resistance) or loose • Generator is damaged or it has failed (output is too high or low) • Ignition system voltage circuit is open at the PCM terminals • PCM has failed
P0562 **2T CCM, MIL: Yes** 1996-1997-1998-1999-2000-2001-2002-2003 E-Van, F-Series Truck & Excursion with 6.0L or 7.3L Diesel engine Transmissions: All	**System Voltage Low Input Conditions:** Engine started, engine running at idle or cruise speed, and the PCM detected the system voltage was too low during the CCM test. This code can set due to a low battery voltage condition during cranking. **Possible Causes** • Battery connections corroded (high resistance) or loose • Generator is damaged or it has failed (output is too low) • Ignition system voltage circuit is open at the PCM terminals • PCM has failed
P0563 **2T CCM, MIL: No** 1996-1997-1998-1999-2000-2001-2002-2003 E-Van, F-Series Truck & Excursion with 6.0L or 7.3L Diesel engine Transmissions: All	**System Voltage High Input Conditions:** Engine started, engine running at idle or cruise speed, and the PCM detected the system voltage was too high during the CCM test. This code can set due to a 24v battery jump-start event. **Possible Causes** • Battery connections corroded (high resistance) or loose • Generator is damaged or it has failed (output is too high) • Ignition system voltage circuit is open at the PCM terminals • PCM has failed
P0565 **2T CCM, MIL: No** 1996-1997-1998-1999-2000-2001-2002-2003 E-Van, F-Series Truck & Excursion with 6.0L or 7.3L Diesel engine Transmissions: All	**Speed Control ON, Not Pressed - KOER Switch Test Conditions:** Engine started, KOER Self-Test enabled, and the PCM detected the Cruise Control "On" switch was not pressed during the KOER test. The Speed Control function is integrated into the PCM. The speed command switches are "momentary" devices located on the face of the steering wheel. These are On/Off switches and one 3-position SET/ACCEL-COAST-RESUME switch. When these switches are pressed, they select one of several resistance values to the PCM. **Possible Causes** • Cruise Control "On" switch not pressed during the Self-Test • Cruise Control "On" switch is damaged, or the circuit has failed • PCM has failed

Diesel Engine OBD II Trouble Code List (P0xxx Codes)

DTC	Trouble Code Title, Conditions & Possible Causes
P0566 **2T CCM, MIL: No** 1996-1997-1998-1999- 2000-2001-2002-2003 E-Van, F-Series Truck & Excursion with 6.0L or 7.3L Diesel engine Transmissions: All	**Speed Control OFF, Not Pressed - KOER Switch Test Conditions:** Engine started, KOER Self-Test enabled, and the PCM detected the Cruise Control "Off" switch was not pressed. The speed command switches are "momentary" devices located on the face of the steering wheel. These are On/Off switches and one 3-position SET/ACCEL-COAST-RESUME switch. When these switches are pressed, they select one of several resistance values to the PCM. **Possible Causes** • Cruise Control "Off" switch not pressed during the Self-Test • Cruise Control "Off" switch is damaged, or the circuit has failed • PCM has failed
P0567 **2T CCM, MIL: No** 1996-1997-1998-1999- 2000-2001-2002-2003 E-Van, F-Series Truck & Excursion with 6.0L or 7.3L Diesel engine Transmissions: All	**Speed Control RESUME, Not Pressed - KOER Switch Test Conditions:** Engine started, KOER Self-Test enabled, and the PCM detected the Cruise Control "RESUME" switch was not pressed. The speed command switches are "momentary" devices located on the face of the steering wheel. These are On/Off switches and one 3-position SET/ACCEL-COAST-RESUME switch. When these switches are pressed, they select one of several resistance values to the PCM. **Possible Causes** • Cruise Control "Resume" switch not pressed in the Self-Test • Cruise Control "Resume" switch is damaged has a circuit fault • PCM has failed
P0568 **2T CCM, MIL: No** 1996-1997-1998-1999- 2000-2001-2002-2003 E-Van, F-Series Truck & Excursion with 6.0L or 7.3L Diesel engine Transmissions: All	**Speed Control "Set", Not Pressed - KOER Switch Test Conditions:** Engine started, KOER Self-Test enabled, and the PCM detected the Cruise Control "Set" switch was not pressed. The speed command switches are "momentary" devices located on the face of the steering wheel. These are On/Off switches and one 3-position SET/ACCEL-COAST-RESUME switch. As a switch is pressed, a resistance value is selected in the PCM. **Possible Causes** • Cruise Control "Set" switch not pressed in the Self-Test • Cruise Control "Set" switch is damaged has a circuit fault • PCM has failed
P0569 **2T CCM, MIL: No** 1996-1997-1998-1999- 2000-2001-2002-2003 E-Van, F-Series Truck & Excursion with 6.0L or 7.3L Diesel engine Transmissions: All	**Cruise Control "Coast", Not Pressed - KOER Switch Test Conditions:** Engine started, KOER Self-Test enabled, and the PCM detected the Cruise Control "Coast" switch was not pressed. The speed command switches are "momentary" devices located on the face of the steering wheel. These are On/Off switches and one 3-position SET/ACCEL-COAST-RESUME switch. As a switch is pressed, a resistance value is selected in the PCM. **Possible Causes** • Cruise Control "Coast" switch not pressed in the Self-Test • Cruise Control "Coast" switch is damaged has a circuit fault • PCM has failed
P0571 **2T CCM, MIL: No** 1996-1997-1998-1999- 2000-2001-2002-2003 E-Van, F-Series Truck & Excursion with 6.0L or 7.3L Diesel engine Transmissions: All	**Brake Pressure Applied Switch Circuit Malfunction Conditions:** Engine started, KOER Self-Test enabled, and the PCM detected the Brake Pressure Applied (BPA) switch did not change status after the brake was pressed. This is a pressure switch that senses brake pressure, and is redundant to the Brake On/Off (BOO) switch to provide a backup signal to deactivate the Speed Control system. **Possible Causes** • Brake Pedal switch is open between switch and the PCM • Brake Pedal switch shorted to ground between switch and PCM • Brake Pedal switch power circuit is open from J/Box to switch • Brake Pedal switch is damaged or it has failed
P0602 **1T PCM, MIL: Yes** 1998-1999-2000-2001- 2002-2003 E-Van, F-Series Truck & Excursion with 6.0L or 7.3L Diesel engine Transmissions: All	**Control Module Programming Error Conditions:** Key on, and the PCM detected a programming error in the VID block. This fault requires that the VID Block be reprogrammed, or that the EEPROM be re-flashed. **Possible Causes** • During the VID reprogramming function, the Vehicle ID (VID) data block failed during reprogramming wit the Scan Tool.

Diesel Engine OBD II Trouble Code List (P0xxx Codes)

DTC	Trouble Code Title, Conditions & Possible Causes
P0603 **1T PCM, MIL: Yes** 1996-1997-1998-1999-2000-2001-2002-2003 E-Van, F-Series Truck & Excursion with 6.0L or 7.3L Diesel engine Transmissions: All	**PCM Keep Alive Memory Test Error Conditions:** Key on or engine running; and the PCM detected an internal memory fault. This code can be set if KAPWR to the PCM is interrupted. The code will set the first time during the initial key "on". **Possible Causes** • Battery terminal corrosion, or loose battery connection • KAPWR to PCM interrupted, or circuit opened • Reprogramming error occurred • PCM has failed
P0603 **1T PCM, MIL: Yes** 1996-1997-1998-1999-2000-2001-2002-2003 E-Van, F-Series Truck & Excursion with 6.0L or 7.3L Diesel engine Transmissions: All	**PCM Read Only Memory Test Error Conditions:** Key on, and the PCM detected a ROM test error (the ROM is corrupted). The PCM is normally replaced if this code has set. **Possible Causes** • An attempt was made to change the control module calibration • Module programming error has occurred • PCM has failed
P0606 **1T PCM, MIL: Yes** 1996-1997-1998-1999-2000-2001-2002-2003 E-Van, F-Series Truck & Excursion with 6.0L or 7.3L Diesel engine Transmissions: All	**PCM Internal Communication Error Conditions:** Key on, and the PCM detected an internal communications register read back error at initial key on. **Possible Causes** • PCM has failed
P0611 **1T CCM, MIL: Yes** 2003 F-Series Truck & Excursion with a 6.0L VIN P Diesel engine Transmissions: All	**Fuel Injector Control Module Performance Conditions:** Key on or engine cranking; and the PCM detected an invalid signal from the Fuel Injector Control Module (FICM) during the test period. **Possible Causes** • FICM is damaged or it has failed. Replace the FICM and then retest the system for trouble codes. • PCM has failed
P0615 **2T CCM, MIL: Yes 2003** F-Series Truck & Excursion with a 6.0L VIN P Diesel engine Transmissions: All	**Starter Relay Signal Circuit Malfunction Conditions:** Key on or engine cranking; and the PCM detected an invalid signal from the Starter Relay signal under these conditions. **Possible Causes** • Starter relay connector is damaged, loose or shorted • Starter relay signal circuit is open, shorted to ground or power • Starter relay is damaged or it has failed • PCM has failed
P0620 **2T CCM, MIL: Yes** 2003 F-Series Truck & Excursion with a 6.0L VIN P Diesel engine Transmissions: All	**Generator 1 Control Circuit Malfunction Conditions:** Engine started; and the PCM detected an unexpected voltage condition on the Generator 1 control circuit during the CCM test. **Possible Causes** • Generator 1 connector is damaged, loose or shorted • Generator 1 control circuit is open or shorted to ground • Generator or regulator function is damaged or has failed • PCM has failed

Diesel Engine OBD II Trouble Code List (P0xxx Codes)

DTC	Trouble Code Title, Conditions & Possible Causes
P0623 **2T CCM, MIL: Yes** 2003 F-Series Truck & Excursion with a 6.0L VIN P Diesel engine Transmissions: All	**Generator Lamp Control Circuit Malfunction Conditions:** Engine started; and the PCM detected an unexpected voltage condition on the Generator lamp control circuit during the CCM test. **Possible Causes** • Generator lamp control circuit is open or shorted to ground • Generator or regulator function is damaged or has failed • PCM has failed
P0640 **2T PCM, MIL: Yes** 1999-2000-2001-2002-2003 E-Van, F-Series Truck & Excursion with a 7.3L Diesel engine Transmissions: All	**Manifold Intake Air Heater Relay Circuit Malfunction Conditions:** Engine started, IAT sensor less than 32ºF, Engine Oil Temperature (EOT) sensor less than 131ºF, system voltage from 11.8-15.0v, and the PCM detected an unexpected voltage condition on the Manifold Intake Air Heater (MIAH) Relay control circuit. The PCM activates the MIAH relay under the conditions listed to reduce the amount of white smoke during long idle periods at low ambient temperatures. **Possible Causes** • MIAH relay circuit is open, shorted to ground or to power (B+) • MIAH heater relay power (B+) circuit is open to the Alternator • MIAH heater relay is damaged or it has failed • PCM has failed
P0645 **1T CCM, MIL: Yes** 2003 F-Series Truck & Excursion with a 6.0L VIN P Diesel engine Transmissions: All	**Air Conditioning Clutch Relay Control Circuit Malfunction Conditions:** Key on or engine running; and the PCM detected an unexpected low or high voltage condition A/C clutch relay control circuit. **Possible Causes** • A/C clutch relay control circuit is open or shorted to ground • A/C clutch relay power circuit (VPWR) is open • A/C clutch relay is damaged or it has failed • PCM has failed
P0649 **2T CCM, MIL: Yes** 2003 F-Series Truck & Excursion with a 6.0L VIN P Diesel engine Transmissions: All	**Cruise Control Lamp Control Circuit Malfunction Conditions:** Engine started, vehicle driven at cruise with the Cruise switch set, and the PCM detected a fault in the C/C lamp control circuit. The Cruise Control function is integrated into the PCM. The Speed Control (S/C) command switches are momentary switches located in the face of the steering wheel. They include a 3-position SET/ACCEL/COAST/RESUME switch and On/Off toggle switch. When a switch is depressed, a resistance values is selected in the PCM. **Possible Causes** • Cruise Control lamp control circuit is open or shorted • Cruise Control lamp is damaged or it has failed • PCM has failed
P0670 **2T CCM, MIL: Yes** 2000-2001-2002-2003 E-Van, F-Series Truck & Excursion with 6.0L or 7.3L Diesel engine Transmissions: All	**Glow Plug Control Circuit Malfunction Conditions:** Engine started, IAT sensor less than 32ºF, Engine Oil Temperature (EOT) sensor less than 131ºF, system voltage from 11.8-15.0v, and the PCM detected an unexpected voltage condition on the Glow Plug control circuit. Glow Plug "on" time, controlled by the PCM, is a function of the inputs listed here. The Glow Plug "on" time varies from 1-120 seconds. **Possible Causes** • Glow Plug control circuit is open or shorted to ground • PCM has failed • TSB 99-25-10 contains a repair procedure for this code
P0671 **2T CCM, MIL: Yes** 2000-2001-2002-2003 E-Van, F-Series Truck & Excursion with 6.0L or 7.3L Diesel engine Transmissions: All	**Glow Plug No. 1 Circuit Malfunction Conditions:** Engine started, system voltage from 11-15v, and the PCM detected an unexpected voltage condition on the Glow Plug 1 circuit. The Glow Plug "on" time varies from 1-120 seconds. **Possible Causes** • Glow Plug No. 1 power circuit is open or shorted to ground • Glow Plug No. 1 is damaged or it has failed • Glow Plug Green and Black connectors could be mismatched • PCM has failed • TSB 99-25-10 contains a repair procedure for this code

Diesel Engine OBD II Trouble Code List (P0xxx Codes)

DTC	Trouble Code Title, Conditions & Possible Causes
P0672 **2T CCM, MIL: Yes** 2000-2001-2002-2003 E-Van, F-Series Truck & Excursion with 6.0L or 7.3L Diesel engine Transmissions: All	**Glow Plug No. 2 Circuit Malfunction Conditions:** Engine started, system voltage from 11-15v, and the PCM detected an unexpected voltage condition on the Glow Plug 2 circuit. The Glow Plug "on" time varies from 1-120 seconds. **Possible Causes** • Glow Plug No. 2 power circuit is open or shorted to ground • Glow Plug No. 2 is damaged or it has failed • Glow Plug Green and Black connectors could be mismatched • PCM has failed • TSB 99-25-10 contains a repair procedure for this code
P0673 **, T CCM** **MIL: Yes** 2000-2001-2002-2003 E-Van, F-Series Truck & Excursion with 6.0L or 7.3L Diesel engine Transmissions: All	**Glow Plug No. 3 Circuit Malfunction Conditions:** Engine started, system voltage from 11-15v, and the PCM detected an unexpected voltage condition on the Glow Plug 3 circuit. The Glow Plug "on" time varies from 1-120 seconds. **Possible Causes** • Glow Plug No. 3 power circuit is open or shorted to ground • Glow Plug No. 3 is damaged or it has failed • Glow Plug Green and Black connectors could be mismatched • PCM has failed • TSB 99-25-10 contains a repair procedure for this code
P0674 **2T CCM, MIL: Yes** 2000-2001-2002-2003 E-Van, F-Series Truck & Excursion with 6.0L or 7.3L Diesel engine Transmissions: All	**Glow Plug No. 4 Circuit Malfunction Conditions:** Engine started, system voltage from 11-15v, and the PCM detected an unexpected voltage condition on the Glow Plug 4 circuit. The Glow Plug "on" time varies from 1-120 seconds. **Possible Causes** • Glow Plug No. 4 power circuit is open or shorted to ground • Glow Plug No. 4 is damaged or it has failed • Glow Plug Green and Black connectors could be mismatched • PCM has failed • TSB 99-25-10 contains a repair procedure for this code
P0675 **2T CCM, MIL: Yes** 2000-2001-2002-2003 E-Van, F-Series Truck & Excursion with 6.0L or 7.3L Diesel engine Transmissions: All	**Glow Plug No. 5 Circuit Malfunction Conditions:** Engine started, system voltage from 11-15v, and the PCM detected an unexpected voltage condition on the Glow Plug 5 circuit. The Glow Plug "on" time varies from 1-120 seconds. **Possible Causes** • Glow Plug No. 5 power circuit is open or shorted to ground • Glow Plug No. 5 is damaged or it has failed • Glow Plug Green and Black connectors could be mismatched • PCM has failed • TSB 99-25-10 contains a repair procedure for this code
P0676 **2T CCM, MIL: Yes** 2000-2001-2002-2003 E-Van, F-Series Truck & Excursion with 6.0L or 7.3L Diesel engine Transmissions: All	**Glow Plug No. 6 Circuit Malfunction Conditions:** Engine started, system voltage from 11-15v, and the PCM detected an unexpected voltage condition on the Glow Plug 6 circuit. The Glow Plug "on" time varies from 1-120 seconds. **Possible Causes** • Glow Plug No. 6 power circuit is open or shorted to ground • Glow Plug No. 6 is damaged or it has failed • Glow Plug Green and Black connectors could be mismatched • PCM has failed • TSB 99-25-10 contains a repair procedure for this code
P0677 **2T CCM, MIL: Yes** 2000-2001-2002-2003 E-Van, F-Series Truck & Excursion with 6.0L or 7.3L Diesel engine Transmissions: All	**Glow Plug No. 7 Circuit Malfunction Conditions:** Engine started, system voltage from 11-15v, and the PCM detected an unexpected voltage condition on the Glow Plug 7 circuit. The Glow Plug "on" time varies from 1-120 seconds. **Possible Causes** • Glow Plug No. 7 power circuit is open or shorted to ground • Glow Plug No. 7 is damaged or it has failed • Glow Plug Green and Black connectors could be mismatched • PCM has failed • TSB 99-25-10 contains a repair procedure for this code

Diesel Engine OBD II Trouble Code List (P0xxx Codes)

DTC	Trouble Code Title, Conditions & Possible Causes
P0678 **2T CCM, MIL: Yes** 2000-2001-2002-2003 E-Van, F-Series Truck & Excursion with 6.0L or 7.3L Diesel engine Transmissions: All	**Glow Plug No. 8 Circuit Malfunction Conditions:** Engine started, system voltage from 11-15v, and the PCM detected an unexpected voltage condition on the Glow Plug 8 circuit. The Glow Plug "on" time varies from 1-120 seconds). **Possible Causes** • Glow Plug No. 8 power circuit is open or shorted to ground • Glow Plug No. 8 is damaged or it has failed • Glow Plug Green and Black connectors could be mismatched • PCM has failed • TSB 99-25-10 contains a repair procedure for this code
P0683 **2T CCM, MIL: Yes** 2000-2001-2002-2003 E-Van, F-Series Truck & Excursion with 6.0L or 7.3L Diesel engine Transmissions: All	**Glow Plug Diagnostic Communication Circuit Malfunction Conditions:** Engine started, system voltage at 11-15.0v, and the PCM detected an unexpected voltage condition on the Glow Plug data line circuit. **Possible Causes** • Glow Plug diagnostic circuit is open or shorted to ground • Glow Plug diagnostic circuit is shorted to system power (B+) • Glow Plug control module is damaged or it has failed • PCM has failed • TSB 99-25-10 contains a repair procedure for this code
P0684 **2T CCM, MIL: Yes** 2000-2001-2002-2003 E-Van, F-Series Truck & Excursion with 6.0L or 7.3L Diesel engine Transmissions: All	**Glow Plug Module Communication Signal Performance Conditions:** Engine started, system voltage at 11-15.0v, and the PCM detected an unexpected voltage condition on the Glow Plug Communication Module (GPCM) to PCM signal circuit. **Possible Causes** • Glow Plug Module communication circuit to the PCM is open or shorted to ground • GPCM circuit is picking up EMI or RFI "noise"
P0645 **1T CCM, MIL: Yes** 2003 F-Series Truck & Excursion with a 6.0L VIN P Diesel engine Transmissions: A/T	**Transmission Control System MIL Request Detected Conditions:** Engine started; and the PCM received a message from the TCM to turn "on" the MIL after it detected a fault in a monitored function. **Possible Causes** • Read and record any trouble codes stored in the TCM as this is only an indicator code. There will be other TCM related codes stored along with DTC P0700.
P0703 **1T CCM, MIL: No** 1996-1997-1998-1999- 2000-2001-2002-2003 E-Van, F-Series Truck & Excursion with 6.0L or 7.3L Diesel engine Transmissions: A/T	**Brake Switch Circuit Malfunction Conditions:** Engine started, KOER Test enabled, and the PCM did not detect a change in the Brake Switch signal status after the brake pedal was pressed. This BPP switch is wired to the stoplamp switch. The PCM uses the BBP switch signal to detect when the brake is applied to determine when to disengage Speed Control and Auxiliary devices. **Possible Causes** • Brake pedal not depressed correctly during the KOER Self-Test • Brake switch circuit is open or shorted to ground • Brake switch is damaged or it has failed • PCM has failed
P0704 **1T CCM, MIL: No** 1996-1997-1998-1999- 2000-2001-2002-2003 E-Van, F-Series Truck & Excursion with 6.0L or 7.3L Diesel engine Transmissions: A/T	**Clutch Pedal Position Switch Circuit Malfunction Conditions:** Engine started, KOER Self-Test enabled, and the PCM did not detect any change in the CPP switch status during the Self-Test. Note: The CCP PID should change (5v to 0v) with clutch depressed. **Possible Causes** • Clutch pedal was not depressed during the KOER Self-Test • Clutch pedal switch circuit is open or shorted to ground • Clutch pedal switch is damaged or it has failed • PCM has failed

Diesel Engine OBD II Trouble Code List (P0xxx Codes)

DTC	Trouble Code Title, Conditions & Possible Causes
P0705 **2T CCM, MIL: Yes** 1996-1997-1998-1999-2000-2001-2002-2003 E-Van, F-Series Truck & Excursion with 6.0L or 7.3L Diesel engine Transmissions: A/T	**DTR Sensor / TR Sensor Circuit Malfunction Conditions:** Engine started, and the PCM detected the Transmission Range (TR) sensor signal was out of its normal operating range during the test. **Possible Causes** • TR sensor signal circuit open or shorted to ground or to VREF • TR sensor is damaged or it has failed • PCM has failed
P0707 **2T CCM, MIL: Yes** 1996-1997-1998-1999-2000-2001-2002-2003 E-Van, F-Series Truck & Excursion with 6.0L or 7.3L Diesel engine Transmissions: A/T	**DTR Sensor / TR Sensor Circuit Low Input Conditions:** Engine started, and the PCM detected the Transmission Range (TR) sensor signal was less than its self-test minimum voltage. **Possible Causes** • TR sensor signal circuit is shorted to ground • TR sensor is damaged or it has failed • PCM has failed
P0708 **2T CCM, MIL: Yes** 1996-1997-1998-1999-2000-2001-2002-2003 E-Van, F-Series Truck & Excursion with 6.0L or 7.3L Diesel engine Transmissions: A/T	**DTR Sensor / TR Sensor Circuit High Input Conditions:** Engine running, and the PCM detected the Digital Transmission Range (DTR) or Transmission Range sensor (TR) signal was more than its self-test maximum range. **Possible Causes** • DTR or TR sensor ground circuit is open • DTR or TR sensor signal circuit is open or shorted to power • DTR or TR sensor damaged • PCM has failed
P0712 **2T CCM, MIL: No** 1996-1997-1998-1999-2000-2001-2002-2003 E-Van, F-Series Truck & Excursion with 6.0L or 7.3L Diesel engine Transmissions: A/T	**TFT Sensor Circuit Low Input Conditions:** Key on or engine running; and the PCM detected the Transmission Fluid Temperature (TFT) sensor signal voltage was less than its self-test minimum range. Note: The TFT PID will read -40ºF when this problem is present. **Possible Causes** • TFT sensor signal circuit is shorted to sensor or chassis ground • TFT sensor is damaged, or out-of-calibration, or it has failed • PCM has failed
P0713 **2T CCM, MIL: No** 1996-1997-1998-1999-2000-2001-2002-2003 E-Van, F-Series Truck & Excursion with 6.0L or 7.3L Diesel engine Transmissions: A/T	**TFT Sensor Circuit High Input Conditions:** Key on or engine running; and the PCM detected the Transmission Fluid Temperature (TFT) sensor signal voltage was more than the maximum self-test value. Note: The TFT PID will read 315ºF when this fault is present. **Possible Causes** • TFT sensor signal circuit is open between the sensor and PCM • TFT sensor ground circuit is open between sensor and PCM • TFT sensor is damaged or it has failed • PCM has failed
P0715 **2T CCM, MIL: No** 1998-1999-2000-2001-2002-2003 E-Van, F-Series Truck & Excursion with 6.0L or 7.3L Diesel engine Transmissions: A/T	**Transmission Speed Shaft Sensor Circuit Malfunction Conditions:** Engine started, vehicle driven to a speed of over 1 mph, and the PCM detected the TSS signals were missing or erratic. **Possible Causes** • TSS signal circuit open or shorted to ground • TSS is damaged • PCM has failed

Diesel Engine OBD II Trouble Code List (P0xxx Codes)

DTC	Trouble Code Title, Conditions & Possible Causes
P0717 **2T CCM, MIL: No** 1998-1999-2000-2001-2002-2003 E-Van, F-Series Truck & Excursion with 6.0L or 7.3L Diesel engine Transmissions: A/T	**Transmission Speed Shaft Sensor Circuit Intermittent Conditions:** Engine running, vehicle speed sensor signal more than 1 mph, and the PCM detected the TSS signals were erratic, irregular or missing. **Possible Causes** • TSS signal circuit has an intermittent open or short to ground • TSS is damaged • PCM has failed
P0718 **2T CCM, MIL: No** 1998-1999-2000-2001-2002-2003 E-Van, F-Series Truck & Excursion with 6.0L or 7.3L Diesel engine Transmissions: A/T	**Transmission Speed Shaft Sensor Circuit Noisy Conditions:** Engine started, vehicle driven to a speed of over 1 mph, and the PCM detected "noise" interference on the TSS signal circuit. **Possible Causes** • TSS signal is "noisy" due to RFI or EMI interference from sources such as ignition components or charging system • TSS signal wiring is damaged or contacting other signal wiring • PCM has failed
P0720 **2T CCM, MIL: Yes** 1998-1999-2000-2001-2002-2003 E-Van, F-Series Truck & Excursion with 6.0L or 7.3L Diesel engine Transmissions: A/T	**Output Shaft Speed Sensor Circuit Low Input Conditions:** Engine started, VSS signal over 1 mph, and the PCM detected the Output Shaft Speed signal did not correlate to the incoming signals received from the VSS or TCSS devices/modules. **Possible Causes** • OSS sensor circuit shorted to ground or open • OSS sensor circuit shorted to VPWR • OSS sensor is damaged • PCM has failed
P0721 **2T CCM, MIL: No** 1998-1999-2000-2001-2002-2003 E-Van, F-Series Truck & Excursion with 6.0L or 7.3L Diesel engine Transmissions: A/T	**Output Shaft Speed Sensor Noise Interference Conditions:** Engine started, VSS input more than 1 mph, and the PCM detected "noise" interference on the Output Shaft Speed (OSS) sensor circuit. **Possible Causes** • OSS sensor wiring misrouted or damaged • After market add-on devices interfering with the OSS signal • OSS sensor damaged • PCM has failed
P0721 **2T CCM, MIL: No** 1998-1999-2000-2001-2002-2003 E-Van, F-Series Truck & Excursion with 6.0L or 7.3L Diesel engine Transmissions: A/T	**Output Speed Sensor No Signal Conditions:** Engine started, and the PCM did not detect any OSS sensor signals upon initial vehicle movement during the test. **Possible Causes** • OSS sensor wiring misrouted or damaged • After market add-on devices interfering with the OSS signal • OSS sensor damaged • PCM has failed
P0732 **2T CCM, MIL: No** 1998-1999-2000-2001-2002-2003 E-Van, F-Series Truck & Excursion with 6.0L or 7.3L Diesel engine Transmissions: A/T	**Incorrect Second Gear Ratio Conditions:** Engine started, vehicle operating with 2nd Gear commanded "on", and the PCM detected an incorrect 2nd gear ratio. **Possible Causes** • 2nd Gear solenoid harness connector not properly seated • 2nd Gear solenoid wring harness connector damaged • 2nd Gear solenoid signal shorted to ground, or open • 2nd Gear solenoid is damaged or not properly installed
P0733 **2T CCM, MIL: No** 1998-1999-2000-2001-2002-2003 E-Van, F-Series Truck & Excursion with 6.0L or 7.3L Diesel engine Transmissions: A/T	**Incorrect Third Gear Ratio Conditions:** Engine started, vehicle operating with 3rd Gear commanded "on", and the PCM detected an incorrect 3rd gear ratio. **Possible Causes** • 3rd Gear solenoid harness connector not properly seated • 3rd Gear solenoid wiring harness connector damaged • 3rd Gear solenoid signal shorted to ground, or open • 3rd Gear solenoid is damaged or not properly installed

Diesel Engine OBD II Trouble Code List (P0xxx Codes)

DTC	Trouble Code Title, Conditions & Possible Causes
P0741 **2T CCM, MIL: No** 1996-1997-1998-1999-2000-2001-2002-2003 E-Van, F-Series Truck & Excursion with 6.0L or 7.3L Diesel engine Transmissions: A/T	**A/T TCC Mechanical System Range/Performance Conditions:** Engine started, vehicle driven in gear with VSS signals received, and the PCM detected excessive TCC slippage while in normal vehicle operation. **Possible Causes** • TCC solenoid has a mechanical failure • TCC solenoid has a hydraulic failure • PCM has failed
P0743 **1T CCM, MIL: Yes** 1996-1997-1998-1999-2000-2001-2002-2003 E-Van, F-Series Truck & Excursion with 6.0L or 7.3L Diesel engine Transmissions: A/T	**A/T TCC Solenoid Circuit Malfunction Conditions:** KOEO Self-Test enabled, and the PCM did not detect any voltage drop from the TCC solenoid circuit during testing. **Possible Causes** • TCC solenoid control circuit open or shorted to ground • TCC solenoid wiring harness connector damaged • TCC solenoid is damaged or it has failed • PCM has failed
P0750 **1T CCM, MIL: Yes** 1996-1997-1998-1999-2000-2001-2002-2003 E-Van, F-Series Truck & Excursion with 6.0L or 7.3L Diesel engine Transmissions: A/T	**A/T Shift Solenoid 1/A Circuit Malfunction Conditions:** Engine started, vehicle driven in 1st gear, and the PCM detected an unexpected voltage condition on the SS1/A solenoid circuit with the SS1/A solenoid applied during testing. **Possible Causes** • SS1/A solenoid control circuit is open or shorted to ground • SS1/A solenoid wiring harness connector damaged • SS1/A solenoid is damaged or it has failed • PCM has failed
P0755 **1T CCM, MIL: Yes** 1996-1997-1998-1999-2000-2001-2002-2003 E-Van, F-Series Truck & Excursion with 6.0L or 7.3L Diesel engine Transmissions: A/T	**Shift Solenoid 2/B Circuit Malfunction Conditions:** Engine started, vehicle driven in 1st gear, and the PCM detected an unexpected voltage condition on the SS1/A solenoid circuit with the SS2/B solenoid applied during testing. **Possible Causes** • SS2/B solenoid control circuit open or shorted to ground • SS2/B solenoid wiring harness connector damaged • SS2/B solenoid is damaged or it has failed • PCM has failed
P0781 **1T CCM, MIL: No** 1996-1997-1998-1999-2000-2001-2002-2003 E-Van, F-Series Truck & Excursion with 6.0L or 7.3L Diesel engine Transmissions: A/T	**A/T 1 to 2 Shift Error Conditions:** Engine started, vehicle driven with VSS signals received, and the PCM detected the engine speed (rpm) did not change properly (i.e., an incorrect 1-2 gear ratio was detected during shifting). **Possible Causes** • SS1/A solenoid may be stuck, or a hydraulic failure exists • SS2/B solenoid may be stuck, or a hydraulic failure exists • Transmission may have damaged friction material
P0782 **1T CCM, MIL: No** 1996-1997-1998-1999-2000-2001-2002-2003 E-Van, F-Series Truck & Excursion with 6.0L or 7.3L Diesel engine Transmissions: A/T	**A/T 2 to 3 Shift Error Conditions:** Engine started, vehicle driven with VSS signals received, and the PCM detected the engine speed (rpm) did not change properly (an incorrect 2-3 gear ratio was detected while shifting). **Possible Causes** • SS1/A solenoid may be stuck, or a hydraulic failure exists • SS2/B solenoid may be stuck, or a hydraulic failure exists • Transmission may have damaged friction material
P0783 **1T CCM, MIL: No** 1996-1997-1998-1999-2000-2001-2002-2003 E-Van, F-Series Truck & Excursion with 6.0L or 7.3L Diesel engine Transmissions: A/T	**A/T 3 to 4 Shift Error Conditions:** Engine started, vehicle driven with VSS signals received, and the PCM detected the engine speed (rpm) did not change properly (an incorrect 3-4 gear ratio was detected while shifting). **Possible Causes** • SS1/A solenoid may be stuck, or a hydraulic failure exists • SS2/B solenoid may be stuck, or a hydraulic failure exists • Transmission may have damaged friction material

Diesel Engine OBD II Trouble Code List (P1xxx Codes)

DTC	Trouble Code Title, Conditions & Possible Causes
P1000 **1T PCM, MIL: No** 2000-2001-2002-2003 E-Van, F-Series Truck & Excursion with 6.0L or 7.3L Diesel engine Transmissions: All	**OBD II Monitor Testing Not Complete Conditions:** Key on or engine running; and the PCM detected one the conditions shown under Possible Causes (i.e., this code cannot be cleared manually - it must clear itself after all the OBD II Monitors complete). Note: This code must be cleared to pass an I/M Test for registration. **Possible Causes** • Battery keep alive power (KAPWR) was removed to the PCM • One or more OBD II Monitors did not complete during an official OBD II Drive Cycle • PCM Reset step was performed with an OBD II Scan Tool
P1001 **1T CCM, MIL: No** 1997-1998-1999-2000- 2001-2002-2003 E-Van, F-Series Truck & Excursion with 6.0L or 7.3L Diesel engine Transmissions: All	**KOER Self-Test Not Completed, KOER Test Aborted Conditions:** Key on, engine running self-test not completed during the normal allowable time period. **Possible Causes** • Engine speed (rpm) out of specification during the KOER test • Incorrect Self-Test Procedure • Scan Tool has a communication problem • Unexpected response from Self-Test monitors
P1105 **1T CCM, MIL: No** 2000-2001-2002-2003 E-Van, F-Series Truck & Excursion with 6.0L or 7.3L Diesel engine Transmissions: All	**Dual Alternator Upper Fault (Monitor) Conditions:** Key on or engine running; and the PCM detected an unexpected voltage condition on the Dual Alternator Upper circuit during the test. **Possible Causes** • Dual Alternator Upper circuit connection problem (corrosion) • Dual Alternator Upper monitor circuit open or shorted to ground • Dual Alternator is damaged or it has failed • PCM has failed
P1106 **1T CCM, MIL: No** 2000-2001-2002-2003 E-Van, F-Series Truck & Excursion with 6.0L or 7.3L Diesel engine Transmissions: All	**Dual Alternator Lower Fault (Control) Conditions:** Key on or engine running; and the PCM detected an unexpected voltage condition on the Dual Alternator Lower circuit during the test. **Possible Causes** • D Dual Alternator Lower circuit connection problem (corrosion) • Dual Alternator Lower control circuit open or shorted to ground • Dual Alternator is damaged or it has failed • PCM has failed
P1107 **1T CCM, MIL: Yes** 2000-2001-2002-2003 E-Van, F-Series Truck & Excursion with 6.0L or 7.3L Diesel engine Transmissions: All	**Dual Alternator Lower Circuit Malfunction Conditions:** Key on or engine running; and the PCM detected an unexpected voltage condition on the Dual Alternator Lower circuit during the test. **Possible Causes** • Dual Alternator Lower circuit (IGN) is open or shorted to ground • Dual Alternator is damaged or it has failed • PCM has failed
P1108 **1T CCM, MIL: No** 2000-2001-2002-2003 E-Van, F-Series Truck & Excursion with 6.0L or 7.3L Diesel engine Transmissions: All	**Dual Alternator BATT Lamp Circuit Malfunction Conditions:** Key on or engine running; and the PCM detected an unexpected voltage condition on the Dual Alternator Lower circuit during the test. **Possible Causes** • Dual Alternator BATT Lamp circuit is open or shorted to ground • Instrument Panel fuse is open (check for cause of the short) Instrument Panel is damaged or it has failed • PCM has failed
P1118 **1T CCM, MIL: Yes** 2000-2001-2002-2003 E-Van, F-Series Truck & Excursion with 6.0L or 7.3L Diesel engine Transmissions: All	**Manifold Air Temperature Sensor Low Input Conditions:** Key on or engine running; and the PCM detected an unexpected "low" voltage condition on the MAT sensor circuit during the test. This is a thermistor-type sensor with a variable resistance that changes when exposed to changes in the air temperature. **Possible Causes** • MAT sensor signal circuit is shorted to sensor ground • MAT sensor signal circuit is shorted to chassis ground • MAT sensor is damaged or it has failed • PCM has failed

Diesel Engine OBD II Trouble Code List (P1xxx Codes)

DTC	Trouble Code Title, Conditions & Possible Causes
P1119 **1T CCM, MIL: Yes** 2000-2001-2002-2003 E-Van, F-Series Truck & Excursion with 6.0L or 7.3L Diesel engine Transmissions: All	**Manifold Air Temperature Sensor High Input Conditions:** Key on or engine running; and the PCM detected an unexpected "high" voltage condition on the MAT sensor circuit during the test. This is a thermistor-type sensor with a variable resistance that changes when exposed to changes in the air temperature. **Possible Causes** • MAT sensor signal circuit is open between sensor and the PCM • MAT sensor ground circuit is open between sensor and PCM • MAT sensor signal circuit is shorted to VREF or system power • MAT sensor is damaged or it has failed • PCM has failed
P1139 **1T CCM, MIL: No** 2000-2001-2002-2003 E-Van, F-Series Truck & Excursion with 6.0L or 7.3L Diesel engine Transmissions: All	**Water-In-Fuel Indicator Circuit Malfunction Conditions:** Key on or engine running; and the PCM detected an unexpected voltage condition on the MAT sensor circuit during the test. **Possible Causes** • Water-In-Fuel circuit is open between Indicator and the PCM • Water-In-Fuel circuit is shorted to ground • Water-In-Fuel power circuit is open (check fuse in I/P panel) • Water is present in the fuel filter housing (drain it and retest) • PCM has failed
P1140 **1T CCM, MIL: No** 2000-2001-2002-2003 E-Van, F-Series Truck & Excursion with 6.0L or 7.3L Diesel engine Transmissions: All	**Water-In-Fuel Condition Conditions:** Key on or engine running; and the PCM detected an unexpected voltage condition on the MAT sensor circuit during the test. **Possible Causes** • Drain the water from the fuel filter into a 1-quart clear container • Inspect the fuel that was previously drained. If there is no water or contaminants in the container, check for an intermittent fault • PCM has failed
P1148 **2T CCM, MIL: No** 2003 F-Series Truck & Excursion with a 6.0L VIN P Diesel engine Transmissions: All	**Generator 2 Control Circuit Malfunction Conditions:** Engine started, and the PCM detected an unexpected voltage condition on the Generator 2 control circuit during the CCM test. **Possible Causes** • Generator 2 control circuit is open or shorted, or the connector is damaged • Generator has failed • PCM has failed
P1149 **2T CCM, MIL: No** 2003 F-Series Truck & Excursion with a 6.0L VIN P Diesel engine Transmissions: All	**Generator 2 Control Circuit High Input Conditions:** Engine started, and the PCM detected an unexpected high voltage condition on the Generator 2 control circuit during the CCM test. **Possible Causes** • Generator 2 control circuit is shorted to system power, or the connector is damaged • Generator has failed • PCM has failed
P1184 **2T CCM, MIL: No** 2003 F-Series Truck & Excursion with a 6.0L VIN P Diesel engine Transmissions: All	**Engine Oil Temperature Sensor Out Of Self Test Range Conditions:** Key on, KOEO Self Test enabled, and the PCM detected that the engine temperature was below the self test minimum - test aborted. **Possible Causes** • Engine temperature to cold to run the KOEO Self Test • Thermostat is damaged or leaking (engine will not warm up) • EOT sensor signal circuit is open or shorted, or the PCM has failed
P1209 **2T CCM, MIL: Yes** 2000-2001-2002-2003 E-Van, F-Series Truck & Excursion with 6.0L or 7.3L Diesel engine Transmissions: All	**Injection Control System Pressure Peak Malfunction Conditions:** Engine cranking and the PCM detected an unexpected voltage condition on the Injection Control Pressure circuit during the test. Note: The engine may not start if this code is set. **Possible Causes** • ICP sensor signal circuit is open, shorted to ground or shorted to system power • ICP sensor is damaged or it has failed • PCM has failed

Diesel Engine OBD II Trouble Code List (P1xxx Codes)

DTC	Trouble Code Title, Conditions & Possible Causes
P1210 **2T CCM, MIL: Yes** 2000-2001-2002-2003 E-Van, F-Series Truck & Excursion with 6.0L or 7.3L Diesel engine Transmissions: All	**Injection Control System Pressure Above Expected Level Conditions:** Key on, and the PCM detected an unexpected "high" pressure condition on the Injection Control Pressure circuit during the test. **Possible Causes** • ICP sensor signal circuit is open or shorted to VREF or system power • ICP sensor ground circuit is open • ICP sensor is damaged or it has failed • PCM has failed
P1211 **1T CCM, MIL: Yes** 1996-1997-1998-1999- 2000-2001-2002-2003 E-Van, F-Series Truck & Excursion with 6.0L or 7.3L Diesel engine Transmissions: All	**Injection Control System Pressure Not Controllable - Pressure Above/Below Desired** **Code Conditions:** Engine started, and the PCM detected the engine was operating in open loop with the Injection Control Pressure too "high" or too "low". **Possible Causes** • ICP sensor circuit is open or shorted to ground (intermittent) • ICP sensor is damaged or it has failed • ICP system has failed, or the PCM has failed
P1212 **1T CCM, MIL: No** 1996-1997-1998-1999- 2000-2001-2002-2003 E-Van, F-Series Truck & Excursion with 6.0L or 7.3L Diesel engine Transmissions: All	**Injection Control System Pressure Not At Expected Level Conditions:** Engine started, and the PCM detected the engine was operating in open loop with the Injection Control Pressure too "high" or too "low". **Possible Causes** • ICP sensor circuit is open or shorted to ground (intermittent) • ICP sensor is damaged or it has failed • ICP system is damaged or it has failed • PCM has failed
P1218 **2T CCM, MIL: Yes** 1996-1997-1998-1999- 2000-2001-2002-2003 E-Van, F-Series Truck & Excursion with 6.0L or 7.3L Diesel engine Transmissions: All	**Cylinder ID Signal Stuck High Conditions:** Engine started, and the PCM detected the engine was operating in open loop with the Injection Control Pressure too "high" or too "low". **Possible Causes** • CID signal circuit to the IDM is open or shorted to system power • CID signal return circuit is open between the IDM and the PCM • IDM is damaged or it has failed • PCM has failed
P1219 **2T CCM, MIL: Yes** 1996-1997-1998-1999- 2000-2001-2002-2003 E-Van, F-Series Truck & Excursion with 6.0L or 7.3L Diesel engine Transmissions: All	**Cylinder ID Signal Stuck Low Conditions:** Engine started, and the PCM detected the engine was operating in open loop with the Injection Control Pressure too "high" or too "low". **Possible Causes** • CID signal circuit to the IDM is shorted to ground • IDM is damaged or it has failed • PCM has failed
P1247 **2T CCM, MIL: Yes** 1998-1999-2000-2001- 2002-2003 E-Van, F-Series Truck & Excursion with 6.0L or 7.3L Diesel engine Transmissions: All	**Turbo Boost Pressure Circuit Low Input Conditions:** Engine started, and the PCM detected a signal from the Injection Pressure sensor indicating the Boost Pressure was too "low". **Possible Causes** • ICP sensor signal circuit is shorted to ground (intermittent fault) • Injection Pressure system is damaged or it has failed • PCM has failed
P1248 **2T CCM, MIL: Yes** 1998-1999-2000-2001- 2002-2003 E-Van, F-Series Truck & Excursion with 6.0L or 7.3L Diesel engine Transmissions: All	**Turbo Boost Pressure Not Detected Conditions:** Engine started, and the PCM detected a signal from the Injection Pressure sensor indicating a lack of Boost Pressure was present. **Possible Causes** • Wastegate Control hose going to the actuator is leaking, loose or damaged • Wastegate solenoid is damaged, sticking or it has failed • PCM has failed

Diesel Engine OBD II Trouble Code List (P1xxx Codes)

DTC	Trouble Code Title, Conditions & Possible Causes
P1249 **2T CCM, MIL: Yes** 1998-1999-2000-2001-2002-2003 E-Van, F-Series Truck & Excursion with 6.0L or 7.3L Diesel engine Transmissions: All	**Wastegate Fail Steady State Test Conditions:** Engine started, and the PCM detected an Injection Pressure sensor signal indicating the Wastegate failed the Steady state test. **Possible Causes** • WGC circuit is shorted to ground • WGC hose going to the actuator is leaking, loose or damaged • WGC actuator or valve is damaged or it has failed • Wastegate solenoid is damaged or the PCM has failed
P1250 **1T PCM, MIL: No** 1998-1999-2000-2001-2002-2003 E-Van, F-Series Truck & Excursion with 6.0L or 7.3L Diesel engine Transmissions: All	**Theft Detected, Vehicle Immobilized Conditions:** Key on, and the PCM received a signal from the Anti-Theft System that a theft condition had occurred. The theft indicator on the dash will flash rapidly or remain on "solid" with the ignition switch in the "on" position. The engine may "start and stall", or may not crank if the vehicle is equipped with the PATS starter disable feature. **Possible Causes** • Anti-Theft System is damaged or it has failed • Previous theft condition may have occurred and set this code • TSB 01-6-02 contains a repair procedure for this trouble code
P1260 **1T PCM, MIL: No** 2003 F-Series Truck & Excursion with a 6.0L VIN P Diesel engine Transmissions: All	**Theft Detected, Vehicle Immobilized Conditions:** Key on or engine cranking; and the PCM received a message that indicated an invalid key had been inserted during a start sequence. **Possible Causes** • Ignition key is invalid or not coded properly • Theft detection module or circuit is damaged or it has failed • PCM has failed
P1261 **1T CCM, MIL: No** 1996-1997-1998-1999-2000-2001-2002-2003 E-Van, F-Series Truck & Excursion with a 7.3L Diesel engine Transmissions: All	**Cylinder 1 - High To Low Side Short Detected Conditions:** Engine started, and the PCM detected a shorted condition on the High to Low side of the Injector 1 control circuit. The High side driver output function provides power and regulated current to the correct injector bank based on the CID signal and Fuel Delivery Command Signal from the Injector Driver Module that controls the 115v supply. **Possible Causes** • Injector 1 "high" side to "low" side shorted circuit detected • IDM or the PCM is damaged or it has failed
P1262 **1T CCM, MIL: No** 1996-1997-1998-1999-2000-2001-2002-2003 E-Van, F-Series Truck & Excursion with a 7.3L Diesel engine Transmissions: All	**Cylinder 2 - High To Low Side Short Detected Conditions:** Engine started, and the PCM detected a shorted condition on the High to Low side of the Injector 2 control circuit. The High side driver output function provides power and regulated current to the correct injector bank based on the CID signal and Fuel Delivery Command Signal from the Injector Driver Module that controls the 115v supply. **Possible Causes** • Injector 2 "high" side to "low" side shorted circuit detected • IDM or the PCM is damaged or it has failed
P1263 **1T CCM, MIL: No** 1996-1997-1998-1999-2000-2001-2002-2003 E-Van, F-Series Truck & Excursion with a 7.3L Diesel engine Transmissions: All	**Cylinder 3 - High To Low Side Short Detected Conditions:** Engine started, and the PCM detected a shorted condition on the High to Low side of the Injector 3 control circuit. The High side driver output function provides power and regulated current to the correct injector bank based on the CID signal and Fuel Delivery Command Signal from the Injector Driver Module that controls the 115v supply. **Possible Causes** • Injector 3 "high" side to "low" side shorted circuit detected • IDM or the PCM is damaged or it has failed
P1264 **1T CCM, MIL: No** 1996-1997-1998-1999-2000-2001-2002-2003 E-Van, F-Series Truck & Excursion with a 7.3L Diesel engine Transmissions: All	**Cylinder 4 - High To Low Side Short Detected Conditions:** Engine started, and the PCM detected a shorted condition on the High to Low side of the Injector 4 control circuit. The High side driver output function provides power and regulated current to the correct injector bank based on the CID signal and Fuel Delivery Command Signal from the Injector Driver Module that controls the 115v supply. **Possible Causes** • Injector 4 "high" side to "low" side shorted circuit detected • IDM or the PCM is damaged or it has failed

Diesel Engine OBD II Trouble Code List (P1xxx Codes)

DTC	Trouble Code Title, Conditions & Possible Causes
P1265 **1T CCM, MIL: No** 1996-1997-1998-1999-2000-2001-2002-2003 E-Van, F-Series Truck & Excursion with a 7.3L Diesel engine Transmissions: All	**Cylinder 5 - High To Low Side Short Detected Conditions:** Engine started, and the PCM detected a shorted condition on the High to Low side of the Injector 5 control circuit. The High side driver output function provides power and regulated current to the correct injector bank based on the CID signal and Fuel Delivery Command Signal from the Injector Driver Module that controls the 115v supply. **Possible Causes** • Injector 5 "high" side to "low" side shorted circuit detected • IDM or the PCM is damaged or it has failed
P1266 **1T CCM, MIL: No** 1996-1997-1998-1999-2000-2001-2002-2003 E-Van, F-Series Truck & Excursion with a 7.3L Diesel engine Transmissions: All	**Cylinder 6 - High To Low Side Short Detected Conditions:** Engine started, and the PCM detected a shorted condition on the High to Low side of the Injector 6 control circuit. The High side driver output function provides power and regulated current to the correct injector bank based on the CID signal and Fuel Delivery Command Signal from the Injector Driver Module that controls the 115v supply. **Possible Causes** • Injector 6 "high" side to "low" side shorted circuit detected • IDM or the PCM is damaged or it has failed
P1267 **1T CCM, MIL: No** 1996-1997-1998-1999-2000-2001-2002-2003 E-Van, F-Series Truck & Excursion with a 7.3L Diesel engine Transmissions: All	**Cylinder 7 - High To Low Side Short Detected Conditions:** Engine started, and the PCM detected a shorted condition on the High to Low side of the Injector 7 control circuit. The High side driver output function provides power and regulated current to the correct injector bank based on the CID signal and Fuel Delivery Command Signal from the Injector Driver Module that controls the 115v supply. **Possible Causes** • Injector 7 "high" side to "low" side shorted circuit detected • IDM or the PCM is damaged or it has failed
P1268 **1T CCM, MIL: No** 1996-1997-1998-1999-2000-2001-2002-2003 E-Van, F-Series Truck & Excursion with a 7.3L Diesel engine Transmissions: All	**Cylinder 8 - High To Low Side Short Detected Conditions:** Engine started, and the PCM detected a shorted condition on the High to Low side of the Injector 8 control circuit. The High side driver output function provides power and regulated current to the correct injector bank based on the CID signal and Fuel Delivery Command Signal from the Injector Driver Module that controls the 115v supply. **Possible Causes** • Injector8 "high" side to "low" side shorted circuit detected • IDM or the PCM is damaged or it has failed
P1271 **1T CCM, MIL: No** 1996-1997-1998-1999-2000-2001-2002-2003 E-Van, F-Series Truck & Excursion with a 7.3L Diesel engine Transmissions: All	**Cylinder 1 - High To Low Side Open Detected Conditions:** Engine started, and the PCM detected an open condition on the High to Low side of the Injector 1 control circuit. The High side driver output function provides power and regulated current to the correct injector bank based on the CID signal and Fuel Delivery Command Signal from the Injector Driver Module that controls the 115v supply. **Possible Causes** • Injector 1 high side circuit is open between the feed and injector • IDM or the PCM is damaged or it has failed
P1272 **1T CCM, MIL: No** 1996-1997-1998-1999-2000-2001-2002-2003 E-Van, F-Series Truck & Excursion with a 7.3L Diesel engine Transmissions: All	**Cylinder 2 - High To Low Side Open Detected Conditions:** Engine started, and the PCM detected an open condition on the High to Low side of the Injector 2 control circuit. The High side driver output function provides power and regulated current to the correct injector bank based on the CID signal and Fuel Delivery Command Signal from the Injector Driver Module that controls the 115v supply. **Possible Causes** • Injector 2 high side circuit is open between the feed and injector • IDM or the PCM is damaged or it has failed

Diesel Engine OBD II Trouble Code List (P1xxx Codes)

DTC	Trouble Code Title, Conditions & Possible Causes
P1273 **1T CCM, MIL: No** 1996-1997-1998-1999-2000-2001-2002-2003 E-Van, F-Series Truck & Excursion with a 7.3L Diesel engine Transmissions: All	**Cylinder 3 - High To Low Side Open Detected Conditions:** Engine started, and the PCM detected an open condition on the High to Low side of the Injector 3 control circuit. The High side driver output function provides power and regulated current to the correct injector bank based on the CID signal and Fuel Delivery Command Signal from the Injector Driver Module that controls the 115v supply. **Possible Causes** • Injector 3 high side circuit is open between the feed and injector • IDM or the PCM is damaged or it has failed
P1274 **1T CCM, MIL: No** 1996-1997-1998-1999-2000-2001-2002-2003 E-Van, F-Series Truck & Excursion with a 7.3L Diesel engine Transmissions: All	**Cylinder 4 - High To Low Side Open Detected Conditions:** Engine started, and the PCM detected an open condition on the High to Low side of the Injector 4 control circuit. The High side driver output function provides power and regulated current to the correct injector bank based on the CID signal and Fuel Delivery Command Signal from the Injector Driver Module that controls the 115v supply. **Possible Causes** • Injector 4 high side circuit is open between the feed and injector • IDM or the PCM is damaged or it has failed
P1275 **1T CCM, MIL: No** 1996-1997-1998-1999-2000-2001-2002-2003 E-Van, F-Series Truck & Excursion with a 7.3L Diesel engine Transmissions: All	**Cylinder 5 - High To Low Side Open Detected Conditions:** Engine started, and the PCM detected an open condition on the High to Low side of the Injector 5 control circuit. The High side driver output function provides power and regulated current to the correct injector bank based on the CID signal and Fuel Delivery Command Signal from the Injector Driver Module that controls the 115v supply. **Possible Causes** • Injector 5 high side circuit is open between the feed and injector • IDM or the PCM is damaged or it has failed
P1276 **1T CCM, MIL: No** 1996-1997-1998-1999-2000-2001-2002-2003 E-Van, F-Series Truck & Excursion with a 7.3L Diesel engine Transmissions: All	**Cylinder 6 - High To Low Side Open Detected Conditions:** Engine started, and the PCM detected an open condition on the High to Low side of the Injector 6 control circuit. The High side driver output function provides power and regulated current to the correct injector bank based on the CID signal and Fuel Delivery Command Signal from the Injector Driver Module that controls the 115v supply. **Possible Causes** • Injector 6 high side circuit is open between the feed and injector • IDM or the PCM is damaged or it has failed
P1277 **1T CCM, MIL: No** 1996-1997-1998-1999-2000-2001-2002-2003 E-Van, F-Series Truck & Excursion with a 7.3L Diesel engine Transmissions: All	**Cylinder 7 - High To Low Side Open Detected Conditions:** Engine started, and the PCM detected an open condition on the High to Low side of the Injector 7 control circuit. The High side driver output function provides power and regulated current to the correct injector bank based on the CID signal and Fuel Delivery Command Signal from the Injector Driver Module that controls the 115v supply. **Possible Causes** • Injector 7 high side circuit is open between the feed and injector • IDM or the PCM is damaged or it has failed
P1278 **1T CCM, MIL: No** 1996-1997-1998-1999-2000-2001-2002-2003 E-Van, F-Series Truck & Excursion with a 7.3L Diesel engine Transmissions: All	**Cylinder 8 - High To Low Side Open Detected Conditions:** Engine started, and the PCM detected an open condition on the High to Low side of the Injector 8 control circuit. The High side driver output function provides power and regulated current to the correct injector bank based on the CID signal and Fuel Delivery Command Signal from the Injector Driver Module that controls the 115v supply. **Possible Causes** • Injector 8 high side circuit is open between the feed and injector • IDM or the PCM is damaged or it has failed

Diesel Engine OBD II Trouble Code List (P1xxx Codes)

DTC	Trouble Code Title, Conditions & Possible Causes
P1280 **1T CCM, MIL: Yes** 1996-1997-1998-1999-2000-2001-2002-2003 E-Van, F-Series Truck & Excursion with a 7.3L Diesel engine Transmissions: All	**Injection Control System Pressure Out-Of-Range Low Conditions:** Engine started, and the PCM detected the Injector Pressure Sensor (ICP) indicated an out-of-range "low" condition. The ICP sensor is a variable capacitance sensor that, when supplied with a 5v reference signal, produces an analog signal that indicates pressure. It is designed to provide a feedback signal that indicates the fuel rail pressure so the PCM can command injector timing and pressure. **Possible Causes** • ICP sensor circuit is shorted to ground • ICP sensor is "skewed" to the "low" side of its range, or it has failed • PCM has failed
P1281 **1T CCM, MIL: Yes** 1996-1997-1998-1999-2000-2001-2002-2003 E-Van, F-Series Truck & Excursion with a 7.3L Diesel engine Transmissions: All	**Injection Control Pressure Circuit Out-Of-Range High Conditions:** Engine started, and the PCM detected the Injector Pressure Sensor (ICP) indicated an out-of-range "high" condition. The ICP sensor is a variable capacitance sensor that, when supplied with a 5v reference signal, produces an analog signal that indicates pressure. It is designed to provide a feedback signal that indicates the fuel rail pressure so the PCM can command injector timing and pressure. **Possible Causes** • ICP sensor circuit is open or shorted to VREF (5v) • ICP sensor is "skewed" to the "high" side of its range, or it has failed • PCM has failed
P1282 **1T CCM, MIL: Yes** 1996-1997-1998-1999-2000-2001-2002-2003 E-Van, F-Series Truck & Excursion with a 7.3L Diesel engine Transmissions: All	**Injection Pressure Regulator Excessive Pressure Conditions:** Engine started, and the PCM detected the Injector Pressure Sensor signal was more than 3,675 psi for 1.5 seconds during the CCM test. **Possible Causes** • IPR control circuit is shorted to ground • ICP sensor is damaged or it has failed • PCM has failed
P1283 **1T CCM, MIL: No** 1996-1997-1998-1999-2000-2001-2002-2003 E-Van, F-Series Truck & Excursion with a 7.3L Diesel engine Transmissions: All	**Injection Pressure Regulator Circuit Malfunction Conditions:** Key on, KOEO Self-Test enabled and the PCM detected an unexpected voltage condition on the Injector Pressure Regulator circuit during the test period. **Possible Causes** • IPR control circuit is open or shorted to ground • IPR control circuit is shorted to system power (B+) • IPR assembly is damaged or it has failed • PCM has failed
P1284 **1T CCM, MIL: No** 1996-1997-1998-1999-2000-2001-2002-2003 E-Van, F-Series Truck & Excursion with a 7.3L Diesel engine Transmissions: All	**Injection Pressure Regulator Failure - KOER Test Aborted Conditions:** Engine started, KOER Self-Test enabled, and the PCM detected an unexpected voltage condition on the Injector Pressure Regulator and "aborted" the rest of the KOER test. **Possible Causes** • IPR control circuit is open or shorted to ground • IPR control circuit is shorted to system power • IPR is damaged or it has failed • PCM has failed
P1291 **1T CCM, MIL: No** 1996-1997-1998-1999-2000-2001-2002-2003 E-Van, F-Series Truck & Excursion with a 7.3L Diesel engine Transmissions: All	**High Side No. 1 (Right) Short To B+ or Ground Detected Conditions:** Engine started, and the PCM detected the Injector Bank 1 "high" side circuit was shorted to ground or shorted to system power (B+). Note: The "high" side driver output function is to provide power to the proper bank based on a Fuel Delivery Command Signal (FDCS) and Injector Driver Module (IDM) which controls the 115v supply. **Possible Causes** • Injector 1 "high" side circuit is shorted to ground or to power • IDM or the PCM is damaged or it has failed

Diesel Engine OBD II Trouble Code List (P1xxx Codes)

DTC	Trouble Code Title, Conditions & Possible Causes
P1292 **1T CCM, MIL: No** 1996-1997-1998-1999-2000-2001-2002-2003 E-Van, F-Series Truck & Excursion with a 7.3L Diesel engine Transmissions: All	**High Side No. 2 (Left) Short To Ground Or B+ Detected Conditions:** Engine started, and the PCM detected the Injector Bank 2 "high" side circuit was shorted to ground or shorted to system power (B+). Note: The "high" side driver output function is to provide power to the proper bank based on a Fuel Delivery Command Signal (FDCS) and Injector Driver Module (IDM) which controls the 115v supply. **Possible Causes** • Injector 2 "high" side circuit is shorted to ground or to power • IDM or the PCM is damaged or it has failed
P1293 **1T CCM, MIL: No** 1996-1997-1998-1999-2000-2001-2002-2003 E-Van, F-Series Truck & Excursion with a 7.3L Diesel engine Transmissions: All	**High Side Bank No. 1 (Right) Open Conditions:** Engine started, and the PCM detected an open circuit condition on Injector Bank 1 (right bank). Note: The "high" side driver output function is to provide power to the proper bank based on a Fuel Delivery Command Signal (FDCS) and Injector Driver Module (IDM) which controls the 115v supply. **Possible Causes** • Injector Bank 1 "high" side circuit is open (the right side bank) • IDM or the PCM is damaged or it has failed
P1294 **1T CCM, MIL: No** 1996-1997-1998-1999-2000-2001-2002-2003 E-Van, F-Series Truck & Excursion with a 7.3L Diesel engine Transmissions: All	**High Side Bank No. 2 (Left) Open Conditions:** Engine started, and the PCM detected an open circuit condition on Injector Bank 2 (left bank). Note: The "high" side driver output function is to provide power to the proper bank based on a Fuel Delivery Command Signal (FDCS) and Injector Driver Module (IDM) which controls the 115v supply. **Possible Causes** • Injector "high" side circuit is open (the left side bank) • IDM or the PCM is damaged or it has failed
P1295 **1T CCM, MIL: Yes** 1996-1997-1998-1999-2000-2001-2002-2003 E-Van, F-Series Truck & Excursion with a 7.3L Diesel engine Transmissions: All	**Multiple Faults On Bank No. 1 (Right) Conditions:** Engine started, and the PCM detected multiple circuit faults on Cylinder Bank 1 (right side bank). Note: The "high" side driver output function is to provide power to the proper bank based on a Fuel Delivery Command Signal (FDCS) and Injector Driver Module (IDM) which controls the 115v supply. **Possible Causes** • Multiple circuit faults exist on Cylinder Bank 1 (the right side) • IDM or the PCM is damaged or it has failed
P1296 **1T CCM, MIL: Yes** 1996-1997-1998-1999-2000-2001-2002-2003 E-Van, F-Series Truck & Excursion with a 7.3L Diesel engine Transmissions: All	**Multiple Faults On Bank No. 2 (Left) Conditions:** Engine started, and the PCM detected multiple circuit faults on Cylinder Bank 2 (left side bank). Note: The "high" side driver output function is to provide power to the proper bank based on a Fuel Delivery Command Signal (FDCS) and Injector Driver Module (IDM) which controls the 115v supply. **Possible Causes** • Multiple circuit faults exist on Cylinder Bank 2 (the left side) • IDM or the PCM is damaged or it has failed
P1297 **1T CCM, MIL: No** 1996-1997-1998-1999-2000-2001-2002-2003 E-Van, F-Series Truck & Excursion with a 7.3L Diesel engine Transmissions: All	**High Sides Shorted Together Conditions:** Engine started, and the PCM detected the "high" side circuit were shorted together. Note: The "high" side driver output function is to provide power to the proper bank based on a Fuel Delivery Command Signal (FDCS) and Injector Driver Module (IDM) which controls the 115v supply. **Possible Causes** • Injector "high" side circuits are shorted together • IDM or the PCM is damaged or it has failed

Diesel Engine OBD II Trouble Code List (P1xxx Codes)

DTC	Trouble Code Title, Conditions & Possible Causes
P1298 **1T CCM, MIL: No** 1996-1997-1998-1999-2000-2001-2002-2003 E-Van, F-Series Truck & Excursion with a 7.3L Diesel engine Transmissions: All	**Injector Driver Module Malfunction Conditions:** Engine started, and the PCM detected a signal from the Injector Driver Module (IDM) indicating that it had failed during the CCM test. The "high" side driver output function is to provide power to the proper bank based on a Fuel Delivery Command Signal (FDCS) and Injector Driver Module (IDM) which controls the 115v supply. **Possible Causes** • Injector Driver Module is damaged or it has failed
P1316 **1T CCM, MIL: Yes** 1998-1999-2000-2001-2002-2003 E-Van, F-Series Truck & Excursion with a 7.3L Diesel engine Transmissions: All	**Injector Circuit/IDM Codes Detected Conditions:** Engine started, and the PCM detected a signal from the Injector Driver Module (IDM) indicating that IDM codes were in its memory. The "high" side driver output function is to provide power to the proper bank based on a Fuel Delivery Command Signal (FDCS) and Injector Driver Module (IDM) which controls the 115v supply. **Possible Causes** • IDM contains one or more IDM related trouble codes
P1378 **1T CCM, MIL: Yes** 2003 F-Series Truck & Excursion with a 6.0L VIN P Diesel engine Transmissions: All	**FICM Supply Voltage Low Input Conditions:** Key on or engine started, and the PCM detected an unexpected low voltage condition on the FICM supply voltage circuit during the test. **Possible Causes** • FICM supply voltage connector is damaged or open • FICM supply voltage circuit is open or shorted to ground • FICM is damaged or it has failed • PCM has failed
P1379 **1T CCM, MIL: Yes** 2003 F-Series Truck & Excursion with a 6.0L VIN P Diesel engine Transmissions: All	**FICM Supply Voltage High Input Conditions:** Key on or engine started, and the PCM detected an unexpected high voltage condition on the FICM supply voltage circuit during the test. **Possible Causes** • FICM supply voltage connector is damaged or shorted • FICM supply voltage circuit is shorted to system power • FICM is damaged or it has failed • PCM has failed
P1391 **1T CCM, MIL: Yes** 1998-1999-2000 E-Van, F-Series Truck & Excursion with a 7.3L Diesel engine Transmissions: All	**Glow Plug Circuit Low Input On Bank 1 (Right Side) Conditions:** Engine started, engine running and the PCM detected all four Glow Plug circuits were open on Cylinder Bank 1 (the right side bank). The "high" side driver output function is to provide power to the proper bank based on a Fuel Delivery Command Signal (FDCS) and Injector Driver Module (IDM) which controls the 115v supply. **Possible Causes** • Glow Plugs circuits or fusible links open on Bank 1 (right side) • Glow plugs are open on Cylinder Bank 1 (right side) • Glow plug relay circuit is open • IDM or PCM is damaged or it has failed
P1393 **1T CCM, MIL: Yes** 1998-1999-2000 E-Van, F-Series Truck & Excursion with a 7.3L Diesel engine Transmissions: All	**Glow Plug Circuit Low Input On Bank 2 (Left Side) Conditions:** Engine started, and the PCM detected that all four Glow Plug circuits were open on Cylinder Bank 2 (the left side bank). The "high" side driver output function is to provide power to the proper bank based on a Fuel Delivery Command Signal (FDCS) and Injector Driver Module (IDM) which controls the 115v supply. **Possible Causes** • Glow Plugs circuits or fusible links open on Bank 2 (left side) • Glow plugs are open on Cylinder Bank 2 (left side) • Glow plug relay circuit is open • PCM is damaged or it has failed

Diesel Engine OBD II Trouble Code List (P1xxx Codes)

DTC	Trouble Code Title, Conditions & Possible Causes
P1395 **1T CCM, MIL: Yes** 1998-1999-2000 E-Van, F-Series Truck & Excursion with a 7.3L Diesel engine Transmissions: All	**Glow Plug Monitor Fault On Bank 1 (Right Side) Conditions:** Key on, glow plugs enabled, system voltage over 11.4v, and the PCM detected the Bank 1 Glow Plug current flow was less than 39 amps. The "high" side driver output function is to provide power to the proper bank based on a Fuel Delivery Command Signal (FDCS) and Injector Driver Module (IDM) which controls the 115v supply. **Possible Causes** • Glow plug circuit or one or two Glow Plugs with high resistance • One or more Glow Plugs shorted (low resistance) on Bank 2
P1396 **1T CCM, MIL: Yes** 1998-1999-2000 E-Van, F-Series Truck & Excursion with a 7.3L Diesel engine Transmissions: All	**Glow Plug Monitor Fault On Bank 2 (Left Side) Conditions:** Key on, glow plugs enabled, system voltage over 11.4v, and the PCM detected the Bank 1 Glow Plug current flow was less than 39 amps. The "high" side driver output function is to provide power to the proper bank based on a Fuel Delivery Command Signal (FDCS) and Injector Driver Module (IDM) which controls the 115v supply. **Possible Causes** • Glow plug circuit or one or two Glow Plugs with high resistance • One or more Glow Plugs shorted (low resistance) on Bank 1
P1397 **1T CCM, MIL: Yes** 1998-1999-2000-2001-2002-2003 E-Van, F-Series Truck & Excursion with 6.0L or 7.3L Diesel engine Transmissions: All	**Glow Plug Voltage Out Of Self-Test Range Conditions:** Engine started, and the PCM detected the system voltage (to the Glow Plugs) was out of its normal operating range. The "high" side driver output function is to provide power to the proper bank based on a Fuel Delivery Command Signal (FDCS) and Injector Driver Module (IDM) which controls the 115v supply. **Possible Causes** • Glow Plug system voltage is out of its self-test range
P1408 **1T EGR, MIL: No** 2003 F-Series Truck & Excursion with a 6.0L VIN P Diesel engine Transmissions: All	**EGR Flow Out Of Self-Test Range Conditions:** Engine started, KOER Self-Test enabled, and the PCM detected the EGR flow was out of the self-test range during the self-test. The EGR valve and EGR sensor are included in one unit. **Possible Causes** • EVR valve stuck closed or iced up, or the flow path is restricted • EGR valve diaphragm leaking, hose is off, plugged or leaking • EGR sensor or the EGR solenoid is damaged or has failed • PCM has failed
P1464 **1T CCM, MIL: No** 1996-1997-1998-1999-2000-2001-2002-2003 E-Van, F-Series Truck & Excursion with 6.0L or 7.3L Diesel engine Transmissions: All	**A/C Demand Out of Self-Test Range Conditions:** Engine started, KOER Self-Test enabled, and the PCM detected the ACCS input was high during the self-test. This code can set if the A/C is turned "on" during the KOER Self-Test. **Possible Causes** • A/C switch was "on" during self-test • A/C PWR circuit shorted to VPWR (N/C WAC relay contacts) • ACCS circuit shorted to power • A/C Demand Switch, WAC relay or CCRM is damaged
P1501 **1T CCM, MIL: No** 1996-1997-1998-1999-2000-2001-2002-2003 E-Van, F-Series Truck & Excursion with 6.0L or 7.3L Diesel engine Transmissions: All	**VSS Signal Out of Self-Test Range Conditions:** Engine started, KOER Self-Test enabled, and the PCM detected a VSS signal during the KOER test (i.e., with the vehicle not moving). **Possible Causes** • VSS signal is noisy due to Radio Frequency Interference/ Electro-Magnetic Interference (RFI/EMI) • Check for RFI or EMI "noise" from external sources such as ignition wires, charging circuit or aftermarket devices
P1502 **1T CCM, MIL: No** 1999-2000-2001-2002-2003 E-Van, F-Series Truck & Excursion with 6.0L or 7.3L Diesel engine Transmissions: All	**Invalid Self-Test - APCM Functioning Conditions:** Engine started, KOER Self-Test enabled, and the PCM detected the APCM was "on" during the test period. **Possible Causes** • APCM is "on" during the KOER Self-Test • Repeat the KOER Self-Test with the APCM turned "off"

Diesel Engine OBD II Trouble Code List (P1xxx Codes)

DTC	Trouble Code Title, Conditions & Possible Causes
P1531 **1T CCM, MIL: No** 1996-1997-1998-1999-2000-2001-2002-2003 E-Van, F-Series Truck & Excursion with 6.0L or 7.3L Diesel engine Transmissions: All	**Invalid Self-Test - Accelerator Pedal Movement Conditions:** Engine started, KOER Self-Test enabled, and the PCM detected the accelerator pedal was moved during the test period. **Possible Causes** • Accelerator pedal movement during the KOER Self-Test • Repeat KOER Self-Test without moving the accelerator pedal
P1536 **1T CCM, MIL: No** 1996-1997-1998-1999-2000-2001-2002-2003 E-Van, F-Series Truck & Excursion with 6.0L or 7.3L Diesel engine Transmissions: All	**Parking Brake Applied Failure Conditions:** Engine started, KOER Self-Test enabled, and the PCM detected the Parking Brake application failed during the PTO / Raised Idle Mode portion of the KOER test. **Possible Causes** • Parking brake switch signal is open or shorted to ground • Parking brake switch is damaged or it has failed • PCM has failed
P1610 **1T PCM, MIL: No** 2003 F-Series Truck & Excursion with a 6.0L VIN P Diesel engine Transmissions: All	**Interactive Reprogramming Code - Replace The PCM Conditions:** Key on, and the PCM detected an error that indicated the PCM had failed during the initial test phase. **Possible Causes** • Perform the correct procedures to replace the PCM. Once a new PCM is installed, perform the correct procedures to "flash" the new PCM. Then retest the system for any trouble codes.
P1611 **1T PCM, MIL: No** 2003 F-Series Truck & Excursion with a 6.0L VIN P Diesel engine Transmissions: All	**Interactive Reprogramming Code - Diagnose Further Conditions:** Key on, and the PCM detected an error that indicated further diagnosis of the PCM is required to complete the test phase. **Possible Causes** • Perform the correct procedures to "flash" the control module. Then retest the system for any trouble codes.
P1615 **1T PCM, MIL: No** 2003 F-Series Truck & Excursion with a 6.0L VIN P Diesel engine Transmissions: All	**Interactive Reprogramming Code - Flash Erase Error Conditions:** Key on, and the PCM detected a Flash Error during its initial startup. **Possible Causes** • Perform the correct procedures to "flash" the control module. Then retest the system for any trouble codes.
P1616 **1T PCM, MIL: No** 2003 F-Series Truck & Excursion with a 6.0L VIN P Diesel engine Transmissions: All	**Interactive Reprogramming Code - Flash Erase Error, Low Voltage Conditions:** Key on, and the PCM detected a Flash Erase Error due to a low voltage condition during a previous "flash" event. **Possible Causes** • Perform the correct procedures to "flash" the control module. Then retest the system for any trouble codes.
P1617 **1T PCM, MIL: No** 2003 F-Series Truck & Excursion with a 6.0L VIN P Diesel engine Transmissions: All	**Interactive Reprogramming Code - Block Programming Error Conditions:** Key on, and the PCM detected a Block Programming Error existed. **Possible Causes** • Perform the correct procedures to "flash" the control module. Then retest the system for any trouble codes.

Diesel Engine OBD II Trouble Code List (P1xxx Codes)

DTC	Trouble Code Title, Conditions & Possible Causes
P1618 **1T PCM, MIL: No** 2003 F-Series Truck & Excursion with a 6.0L VIN P Diesel engine Transmissions: All	**Interactive Reprogramming Code - Block Programming Error, Low Voltage Conditions:** Key on, and the PCM detected a Block Programming Error existed due to a low voltage condition during a previous reflash event. **Possible Causes** • Perform the correct procedures to "flash" the control module. Then retest the system for any trouble codes.
P1660 **1T CCM, MIL: No** 1996-1997-1998 E-Van, F-Series Truck & Excursion with 7.3L VIN F Diesel engine Transmissions: All	**OCC Signal High Conditions:** Key on, KOEO Self-Test enabled, and the PCM detected high system voltage during the test due to an internal PCM failure. **Possible Causes** • PCM is damaged or it has failed. Clear the codes, and then rerun the KOEO Self-Test. If the same code resets, the PCM has failed and must be replaced.
P1660 **1T CCM, MIL: No** 1996-1997-1998 E-Van, F-Series Truck & Excursion with 7.3L VIN F Diesel engine Transmissions: All	**OCC Signal Low Conditions:** Key on, KOEO Self-Test enabled, and the PCM detected low system voltage during the test due to an internal PCM failure. **Possible Causes** • PCM is damaged or it has failed. Clear the codes, and then rerun the KOEO Self-Test. If the same code resets, the PCM has failed and must be replaced.
P1662 **1T CCM, MIL: No** 1996-1997-1998-1999-2000-2001-2002-2003 E-Van, F-Series Truck & Excursion with 6.0L or 7.3L Diesel engine Transmissions: All	**EF Feedback Signal Not Detected Conditions:** Engine started, and the PCM did not detect an Electronic Feedback signal from the IDM during the CCM test. **Possible Causes** • EF signal circuit is open or shorted to ground • EF ground circuit is open • IDM is damaged or it has failed • PCM has failed
P1663 **1T CCM, MIL: No** 1996-1997-1998-1999-2000-2001-2002-2003 E-Van, F-Series Truck & Excursion with 6.0L or 7.3L Diesel engine Transmissions: All	**Fuel Delivery Command Signal Circuit Malfunction Conditions:** Key on, KOEO Self-Test enabled and the PCM detected an unexpected low or high voltage condition on the Fuel Delivery Command Signal (FDCS) circuit during the initial test period. **Possible Causes** • FDCS signal circuit is open • FDCS signal circuit is shorted to ground • FDCS ground circuit is open • PCM has failed
P1667 **1T CCM, MIL: No** 1996-1997-1998-1999-2000-2001-2002-2003 E-Van, F-Series Truck & Excursion with 6.0L or 7.3L Diesel engine Transmissions: All	**CID Circuit Malfunction Conditions:** Engine started, KOEO Self-Test enabled, and the PCM detected an unexpected low or condition on the Cylinder Identification (CID) circuit during the test period. **Possible Causes** • CID signal circuit is open • CID signal circuit is shorted to ground • CID ground circuit is open • PCM has failed
P1668 **1T CCM, MIL: No** 1996-1997-1998-1999-2000-2001-2002-2003 E-Van, F-Series Truck & Excursion with 6.0L or 7.3L Diesel engine Transmissions: All	**PCM TO IDM Diagnostic Communication Error Conditions:** Engine started, and the PCM detected an unexpected low or high voltage condition on the Electronic Feedback signal circuit during the test period. **Possible Causes** • EF signal circuit is open or shorted to ground • EF ground circuit is open • PCM has failed

Diesel Engine OBD II Trouble Code List (P1xxx Codes)

DTC	Trouble Code Title, Conditions & Possible Causes
P1670 **2T CCM, MIL: Yes** 1998-1999-2000-2001-2002-2003 E-Van, F-Series Truck & Excursion with 6.0L or 7.3L Diesel engine Transmissions: All	**EF Feedback Signal Not Detected Conditions:** Engine started, and the PCM did not detect any signals on the EF feedback signal circuit during the test. **Possible Causes** • EF signal circuit is open or shorted to ground • EF ground circuit is open • PCM has failed
P1690 **2T CCM, MIL: Yes** 1998-1999-2000-2001-2002-2003 E-Van, F-Series Truck & Excursion with 6.0L or 7.3L Diesel engine Transmissions: All	**Wastegate Control Valve Malfunction Conditions:** Engine started, and the PCM did not detect any signals on the EF feedback signal circuit during the test. A Wastegate type of turbo is designed to reach maximum boost sooner than a conventional turbo. However, over-boosting will cause damage to the turbo assembly. The PCM controls the boost pressure with a duty cycle signal to the solenoid to maximize boost performance (no more than 16.5 psi). When pressure is supplied on the Red hose to the actuator (with the solenoid not energized), the valve will open, dumping boost. When low or no pressure is supplied to the Red hose to the actuator (solenoid is being energized), the actuator valve will stay closed. **Possible Causes** • EF signal circuit is open or shorted to ground • EF ground circuit is open • PCM has failed
P1702 **1T CCM, MIL: No** 1998-1999-2000-2001-2002-2003 E-Van, F-Series Truck & Excursion with a 7.3L Diesel engines Transmissions: A/T	**Transmission Range Sensor Intermittent Signal Conditions:** Key on or engine running; and the PCM detected the failure conditions were met intermittently for DTC P0705 or P0708 due to a malfunction of the Transmission Range (TR) Sensor or its circuit. **Possible Causes** • Refer to the appropriate Transmission Repair Manual or information in electronic media to perform a complete diagnosis of the automatic transmission when this code is set
P1703 **1T CCM, MIL: No** 2003 F-Series Truck & Excursion with a 6.0L VIN P Diesel engine Transmissions: A/T	**Brake Switch Out Of Self Test Range Conditions:** Key on with KOEO Self Test enabled, or engine started with KOER Self Test enabled, and the PCM detected an invalid Brake Switch signal during the CCM test period. Rerun the appropriate self test. **Possible Causes** • Brake switch not cycled properly during the self test • Brake switch connector is damaged, loose or shorted • Brake switch is damaged or it has failed
P1704 **1T CCM, MIL: No** 1998-1999-2000-2001-2002-2003 E-Van, F-Series Truck & Excursion with a 7.3L Diesel engines Transmissions: A/T	**Transmission Range Sensor Out Of Self-Test Range Conditions:** Key on or engine running; and the PCM detected a Transmission Range (TR) sensor signal occurred in between gear positions. **Possible Causes** • Digital TR sensor or shift cable misadjusted • Digital TR sensor circuit open or shorted to ground • Digital TR sensor is damaged or it has failed
P1705 **1T CCM, MIL: No** 1996-1997-1998-1999-2000-2001-2002-2003 E-Van, F-Series Truck & Excursion with 6.0L or 7.3L Diesel engine Transmissions: A/T	**Transmission Range Sensor Out Of Self-Test Range Conditions:** Key on, KOEO Self-Test enabled, or engine running, KOER Self-Test enabled, and the PCM detected it did not receive a Transmission Range (TR) sensor signal while in Park or Neutral. **Possible Causes** • Gear selector not in Park or Neutral during the self-test • Digital TR sensor circuit open or shorted to ground • Digital TR sensor is damaged or it has failed

Diesel Engine OBD II Trouble Code List (P1xxx Codes)

DTC	Trouble Code Title, Conditions & Possible Causes
P1711 **1T CCM, MIL: No** 1996-1997-1998-1999-2000-2001-2002-2003 E-Van, F-Series Truck & Excursion with a 7.3L Diesel engines Transmissions: A/T	**TFT Sensor Out of Self-Test Range Conditions:** **KOEO or KOER Self-Test Enabled** Key on, KOEO Self-Test enabled, or engine running, KOER Self-Test enabled, and the PCM detected the Transmission Fluid Temperature (TFT) sensor was more than or less than the calibrated range (25ºF to 240ºF) during the self-test period. **Possible Causes** • Refer to the appropriate Transmission Repair Manual or information in electronic media to perform a complete diagnosis of the automatic transmission when this code is set.
P1713 **2T CCM, MIL: No** 1998-1999-2000-2001-2002-2003 E-Van, F-Series Truck & Excursion with a 7.3L Diesel engines Transmissions: A/T	**TFT Sensor Stuck In Low Range Conditions:** Engine started, vehicle driven to over 1 mph for 2-3 minutes, and the PCM detected the TFT sensor signal was stuck in low range (less than 50°F) during the CCM test. **Possible Causes** • Refer to the appropriate Transmission Repair Manual or information in electronic media to perform a complete diagnosis of the automatic transmission when this code is set.
P1714 **2T CCM, MIL: Yes** 1998-1999-2000-2001-2002-2003 E-Van, F-Series Truck & Excursion with 6.0L or 7.3L Diesel engine Transmissions: A/T	**A/T Shift Solenoid 'A' Inductive Signature Malfunction Conditions:** Engine started, VSS over 1 mph, and the PCM detected a problem with the Shift Solenoid 'A' Inductive signature during the CCM test. **Possible Causes** • Refer to the appropriate Transmission Repair Manual or information in electronic media to perform a complete diagnosis of the automatic transmission when this code is set.
P1715 **2T CCM, MIL: Yes** 1998-1999-2000-2001-2002-2003 E-Van, F-Series Truck & Excursion with 6.0L or 7.3L Diesel engine Transmissions: A/T	**A/T Shift Solenoid 'B' Inductive Signature Malfunction Conditions:** Engine running, VSS over 1 mph, and the PCM detected a problem with the Shift Solenoid 'B' Inductive signature during the CCM test. **Possible Causes** • Refer to the appropriate Transmission Repair Manual or information in electronic media to perform a complete diagnosis of the automatic transmission when this code is set
P1718 **1T CCM, MIL: No** 1998-1999-2000-2001-2002-2003 E-Van, F-Series Truck & Excursion with a 7.3L Diesel engines Transmissions: A/T	**TFT Sensor Stuck In High Range Conditions:** Engine started, vehicle driven to over 1 mph for 2-3 minutes, and the PCM detected the TFT sensor signal was stuck in high range (more than 250°F) during the CCM test. **Possible Causes** • Refer to the appropriate Transmission Repair Manual or information in electronic media to perform a complete diagnosis of the automatic transmission when this code is set.
P1725 **1T CCM, MIL: No** 2003 F-Series Truck & Excursion with a 6.0L VIN P Diesel engine Transmissions: A/T	**Insufficient Engine Speed Increase During The Self Test Conditions:** Engine started, KOER Self Test enabled, and the PCM detected an insufficient amount of engine speed increase during the self test. **Possible Causes** • Check for other trouble codes in the PCM • Rerun the Self Test as directed
P1726 **1T CCM, MIL: No** 2003 F-Series Truck & Excursion with a 6.0L VIN P Diesel engine Transmissions: A/T	**Insufficient Engine Speed Decrease During The Self Test Conditions:** Engine started, KOER Self Test enabled, and the PCM detected an insufficient amount of engine speed decrease during the self test. **Possible Causes** • Check for other trouble codes in the PCM • Rerun the self test as directed

Diesel Engine OBD II Trouble Code List (P1xxx Codes)

DTC	Trouble Code Title, Conditions & Possible Causes
P1727 **1T CCM, MIL: No** 1996-1997-1998 E-Van, F-Series Truck & Excursion with a 7.3L Diesel engines Transmissions: A/T	**Coast Clutch Solenoid Inductive Signature Malfunction Conditions:** Engine started, VSS over 1 mph in gear, and the PCM detected a signal that indicated a problem had been detected in the Coast Clutch Solenoid Inductive Signature value during the test period. **Possible Causes** • Refer to the appropriate Transmission Repair Manual or information in electronic media to perform a complete diagnosis of the automatic transmission when this code is set.
P1728 **1T CCM, MIL: No** 1996-1997-1998-1999-2000-2001-2002-2003 E-Van, F-Series Truck & Excursion with 6.0L or 7.3L Diesel engine Transmissions: A/T	**Transmission Slip Malfunction Conditions:** Engine running, VSS over 1 mph in gear, and the PCM detected a signal that indicated the transmission was slipping while in gear. **Possible Causes** • Refer to the appropriate Transmission Repair Manual or information in electronic media to perform a complete diagnosis of the automatic transmission when this code is set.
P1729 **1T CCM, MIL: No** 1996-1997-1998-1999-2000-2001-2002-2003 E-Van, F-Series Truck & Excursion with 6.0L or 7.3L Diesel engine Transmissions: A/T	**4 X 4 Low Switch Circuit Malfunction Conditions:** Key on or engine running; and the PCM detected the 4x4 switch input did not go low with the switch cycled "on" during the self-test. **Possible Causes** • Speedometer out of calibration • 4x4L wiring harness is open or shorted • 4x4L switch is damaged or it has failed • Electronic Shift Control Module is damaged or it has failed • PCM has failed
P1744 **1T CCM, MIL: Yes** 1998-1999-2000-2001-2002-2003 E-Van, F-Series Truck & Excursion with 6.0L or 7.3L Diesel engine Transmissions: A/T	**TCC System Mechanical Malfunction (Stuck Off) Conditions:** Engine started, vehicle in gear at Cruise speed, and the PCM detected the Torque Converter Clutch system had failed with the TCC in the mechanically "off" position. **Possible Causes** • Refer to the appropriate Transmission Repair Manual or information in electronic media to perform a complete diagnosis of the automatic transmission when this code is set.
P1746 **1T CCM, MIL: Yes** 1996-1997-1998-1999-2000-2001-2002-2003 E-Van, F-Series Truck & Excursion with 6.0L or 7.3L Diesel engine Transmissions: A/T	**A/T EPC Solenoid Circuit Malfunction (Open) Conditions:** Engine started, vehicle in gear, and the PCM detected an unexpected high voltage condition on the Electronic Pressure Control (EPC) solenoid circuit. This fault causes maximum EPC pressure and results in harsh engagements and shifts. **Possible Causes** • EPC solenoid control circuit open between solenoid and PCM • Refer to the appropriate Transmission Repair Manual or information in electronic media to perform a complete diagnosis of the automatic transmission when this code is set
P1747 **1T CCM, MIL: Yes** 1996-1997-1998-1999-2000-2001-2002-2003 E-Van, F-Series Truck & Excursion with 6.0L or 7.3L Diesel engine Transmissions: A/T	**A/T EPC Solenoid Circuit Malfunction (Shorted) Conditions:** Engine started, vehicle driven at a steady cruise speed, and the PCM detected an unexpected low voltage condition on the TCC solenoid driver circuit during the CCM test. **Possible Causes** • EPC solenoid control circuit is shorted to ground or power (B+) • EPC solenoid power circuit is open • Refer to the appropriate Transmission Repair Manual or information in electronic media to perform a complete diagnosis of the automatic transmission when this code is set
P1748 **1T CCM, MIL: Yes** 1996-1997-1998-1999-2000-2001-2002-2003 E-Van, F-Series Truck & Excursion with 6.0L or 7.3L Diesel engine Transmissions: A/T	**A/T EPC Solenoid Circuit Malfunction Conditions:** Engine started, vehicle driven at a steady cruise speed, and the PCM detected an unexpected voltage condition on the TCC solenoid driver circuit during the CCM test. **Possible Causes** • EPC solenoid control circuit is open • EPC solenoid power circuit is open • Refer to the appropriate Transmission Repair Manual or information in electronic media to perform a complete diagnosis of the automatic transmission when this code is set

Diesel Engine OBD II Trouble Code List (P1xxx Codes)

DTC	Trouble Code Title, Conditions & Possible Causes
P1754 **1T CCM, MIL: No** 1996-1997-1998-1999-2000-2001-2002-2003 E-Van, F-Series Truck & Excursion with 6.0L or 7.3L Diesel engine Transmissions: A/T	**A/T Coast Clutch Solenoid Circuit Malfunction Conditions:** Key on, KOEO Self-Test enabled and the PCM detected an unexpected low or high voltage condition on the Coast Clutch Solenoid (CCS) circuit during the test period. **Possible Causes** • CCS control circuit is open or shorted to ground • Refer to the appropriate Transmission Repair Manual or information in electronic media to perform a complete diagnosis of the automatic transmission when this code is set
P1760 **1T CCM, MIL: No** 1998-1999-2000-2001-2002-2003 E-Van, F-Series Truck & Excursion with 6.0L or 7.3L Diesel engine Transmissions: A/T	**A/T EPC Solenoid Circuit Malfunction (Intermittent) Conditions:** Engine started, vehicle driven at a steady cruise speed, and the PCM detected an interruption of the TCC solenoid driver signal (the TCC control circuit) during the CCM test. **Possible Causes** • EPC solenoid control circuit is open or shorted (intermittent) • Refer to the appropriate Transmission Repair Manual or information in electronic media to perform a complete diagnosis of the automatic transmission when this code is set
P1779 **1T CCM, MIL: No** 1996-1997-1998 E-Van, F-Series Truck & Excursion with 7.3L Diesel engine Transmissions: A/T	**Transmission Control Switch Out Of Self-Test Range Conditions:** Key on, KOEO Self-Test enabled and the PCM detected an unexpected voltage condition on the Transmission Control Switch (TCS) circuit during the test period. **Possible Causes** • TCS not cycled during the self-test • TCS connector is damaged, open or shorted • TCS circuit is open • TCS circuit is shorted to ground • TCS is damaged • PCM has failed
P1780 **1T CCM, MIL: No** 1996-1997-1998-1999-2000-2001-2002-2003 E-Van, F-Series Truck & Excursion with 6.0L or 7.3L Diesel engine Transmissions: A/T	**Transmission Control Switch Out Of Self-Test Range Conditions:** Engine started, KOER Self-Test enabled, and the PCM detected the Transmission Control Switch (TCS) was out of range in the self-test. **Possible Causes** • TCS not cycled during the self-test • TCS connector is damaged, open or shorted • TCS circuit is open • TCS circuit is shorted to ground • TCS is damaged • PCM has failed
P1781 **1T CCM, MIL: No** 1996-1997-1998-1999-2000-2001-2002-2003 E-Van, F-Series Truck & Excursion with 6.0L or 7.3L Diesel engine Transmissions: All	**4 X 4 Low Switch Out Of Self-Test Range Conditions:** Key on, KOEO Self-Test enabled, and the PCM detected the 4x4 switch input was not low with the switch engaged ("on"). **Possible Causes** • 4x4L switch circuit is open or shorted in the wiring harness • Electronic Shift Module is damaged or it has failed • PCM has failed
P1783 **1T CCM, MIL: No** 1996-1997-1998-1999-2000-2001-2002-2003 E-Van, F-Series Truck & Excursion with 6.0L or 7.3L Diesel engine Transmissions: A/T	**Transmission Over-Temperature Conditions:** Engine runtime over 5 minutes, vehicle in gear at Cruise speed, and the PCM detected the TFT sensor signal was more than 300°F. **Possible Causes** • Refer to the appropriate Transmission Repair Manual or information in electronic media to perform a complete diagnosis of the automatic transmission when this code is set

Diesel Engine OBD II Trouble Code List (P2xxx Codes)

DTC	Trouble Code Title, Conditions & Possible Causes
P2121 **1T CCM, MIL: No** 2003 F-Series Truck & Excursion with a 6.0L VIN P Diesel engine Transmissions: All	**Accelerator Pedal Position Sensor 'D' Signal Range/Performance Conditions:** Key on or engine running; and the PCM detected the Accelerator Pedal Position Sensor 'D' signal circuit was out of the normal operating range during the CCM test. **Possible Causes** • APP sensor signal circuits are shorted together • APP sensor is damaged or it has failed • PCM has failed
P2122 **1T CCM, MIL: No** 2003 F-Series Truck & Excursion with a 6.0L VIN P Diesel engine Transmissions: All	**Accelerator Pedal Position Sensor 'D' Circuit Low Input Conditions:** Key on or engine running; and the PCM detected the Accelerator Pedal Position Sensor 'D' signal circuit was less than the normal range during the test period. **Possible Causes** • APP sensor signal circuit is open • APP sensor signal circuit is shorted to ground • APP sensor is damaged or it has failed • PCM has failed
P2123 **1T CCM, MIL: No** 2003 F-Series Truck & Excursion with a 6.0L VIN P Diesel engine Transmissions: All	**Accelerator Pedal Position Sensor 'D' Circuit High Input Conditions:** Key on or engine running; and the PCM detected the Accelerator Pedal Position Sensor 'D' signal circuit was more than the normal range during the test period. **Possible Causes** • APP sensor connector is damaged or shorted • APP sensor signal circuit is shorted to VREF (5v) • APP sensor is damaged or it has failed • PCM has failed
P2124 **1T CCM, MIL: No** 2003 F-Series Truck & Excursion with a 6.0L VIN P Diesel engine Transmissions: All	**Accelerator Pedal Position Sensor 'D' Signal Intermittent Conditions:** Key on or engine running; and the PCM detected an intermittent signal from the Accelerator Pedal Position Sensor 'D' during the test. **Possible Causes** • APP sensor connector is damaged or shorted (intermittent) • APP sensor signal circuit is open or shorted (intermittent fault) • APP sensor is damaged or it has failed (intermittent fault) • PCM has failed
P2126 **1T CCM, MIL: No** 2003 F-Series Truck & Excursion with a 6.0L VIN P Diesel engine Transmissions: All	**Accelerator Pedal Position Sensor 'E' Signal Range/Performance Conditions:** Key on or engine running; and the PCM detected the Accelerator Pedal Position Sensor 'E' signal circuit was more than the normal range during the test period. **Possible Causes** • APP sensor connector is damaged or shorted • APP sensor signal circuit is shorted to VREF (5v) • APP sensor is damaged or it has failed • PCM has failed
P2127 **1T CCM, MIL: No** 2003 F-Series Truck & Excursion with a 6.0L VIN P Diesel engine Transmissions: All	**Accelerator Pedal Position Sensor 'E' Circuit Low Input Conditions:** Key on or engine running; and the PCM detected the Accelerator Pedal Position Sensor 'D' signal circuit was less than the normal range during the test period. **Possible Causes** • APP sensor signal circuit is open • APP sensor signal circuit is shorted to ground • APP sensor is damaged or it has failed • PCM has failed
P2128 **1T CCM, MIL: No** 2003 F-Series Truck & Excursion with a 6.0L VIN P Diesel engine Transmissions: All	**Accelerator Pedal Position Sensor 'E' Circuit High Input Conditions:** Key on or engine running; and the PCM detected the Accelerator Pedal Position Sensor 'E' signal circuit was more than the normal range during the test period. **Possible Causes** • APP sensor connector is damaged or shorted • APP sensor signal circuit is shorted to VREF (5v) • APP sensor is damaged or it has failed • PCM has failed

Diesel Engine OBD II Trouble Code List (P2xxx Codes)

DTC	Trouble Code Title, Conditions & Possible Causes
P2129 **1T CCM, MIL: No** 2003 F-Series Truck & Excursion with a 6.0L VIN P Diesel engine Transmissions: All	**Accelerator Pedal Position Sensor 'E' Signal Intermittent Conditions:** Key on or engine running; and the PCM detected an intermittent signal from the Accelerator Pedal Position Sensor 'E' during the test. **Possible Causes** • APP sensor connector is damaged or shorted (intermittent) • APP sensor signal circuit is open or shorted (intermittent fault) • APP sensor is damaged or it has failed (intermittent fault) • PCM has failed
P2131 **1T CCM, MIL: No** 2003 F-Series Truck & Excursion with a 6.0L VIN P Diesel engine Transmissions: All	**Accelerator Pedal Position Sensor 'F' Signal Range/Performance Conditions:** Key on or engine running; and the PCM detected the Accelerator Pedal Position Sensor 'F' signal circuit was more than the normal range during the test period. **Possible Causes** • APP sensor connector is damaged or shorted • APP sensor signal circuit is shorted to VREF (5v) • APP sensor is damaged or it has failed • PCM has failed
P2132 **1T CCM, MIL: No** 2003 F-Series Truck & Excursion with a 6.0L VIN P Diesel engine Transmissions: All	**Accelerator Pedal Position Sensor 'F' Circuit Low Input Conditions:** Key on or engine running; and the PCM detected the Accelerator Pedal Position Sensor 'F' signal circuit was less than the normal range during the test period. **Possible Causes** • APP sensor signal circuit is open • APP sensor signal circuit is shorted to ground • APP sensor is damaged or it has failed • PCM has failed
P2133 **1T CCM, MIL: No** 2003 F-Series Truck & Excursion with a 6.0L VIN P Diesel engine Transmissions: All	**Accelerator Pedal Position Sensor 'F' Circuit High Input Conditions:** Key on or engine running; and the PCM detected the Accelerator Pedal Position Sensor 'F' signal circuit was more than the normal range during the test period. **Possible Causes** • APP sensor connector is damaged or shorted • APP sensor signal circuit is shorted to VREF (5v) • APP sensor is damaged or it has failed • PCM has failed
P2134 **1T CCM, MIL: No** 2003 F-Series Truck & Excursion with a 6.0L VIN P Diesel engine Transmissions: All	**Accelerator Pedal Position Sensor 'F' Signal Intermittent Conditions:** Key on or engine running; and the PCM detected an intermittent signal from the Accelerator Pedal Position Sensor 'F' during the test. **Possible Causes** • APP sensor connector is damaged or shorted (intermittent) • APP sensor signal circuit is open or shorted (intermittent fault) • APP sensor is damaged or it has failed (intermittent fault) • PCM has failed
P2138 **1T CCM, MIL: No** 2003 F-Series Truck & Excursion with a 6.0L VIN P Diesel engine Transmissions: All	**ETC Throttle Position Sensor D/E Voltage Correlation Conditions:** Key on or engine running; and the PCM detected the Throttle Position 'D' (TPD) and Throttle Position 'E' (TPE) sensors disagreed, or that the TPD sensor should not be in its detected position, or that the TPE sensor should not be in its detected position during testing. **Possible Causes** • ETC TP sensor connector is damaged or shorted • ETC TP sensor circuits shorted together in the wire harness • ETC TP sensor signal circuit is shorted to VREF (5v) • ETC TP sensor is damaged or it has failed • PCM has failed

Diesel Engine OBD II Trouble Code List (P2xxx Codes)

DTC	Trouble Code Title, Conditions & Possible Causes
P2139 **1T CCM, MIL: No** 2003 F-Series Truck & Excursion with a 6.0L VIN P Diesel engine Transmissions: All	**ETC Throttle Position Sensor D/F Voltage Correlation Conditions:** Key on or engine running; and the PCM detected the Throttle Position 'D' (TPD) and Throttle Position 'F' (TPF) sensors disagreed, or that the TPD sensor should not be in its detected position, or that the TPF sensor should not be in its detected position during testing. **Possible Causes** • ETC TP sensor connector is damaged or shorted • ETC TP sensor circuits shorted together in the wire harness • ETC TP sensor signal circuit is shorted to VREF (5v) • ETC TP sensor is damaged or it has failed • PCM has failed
P2140 **1T CCM, MIL: No** 2003 F-Series Truck & Excursion with a 6.0L VIN P Diesel engine Transmissions: All	**ETC Throttle Position Sensor E/F Voltage Correlation Conditions:** Key on or engine running; and the PCM detected the Throttle Position 'E' (TPE) and Throttle Position 'F' (TPF) sensors disagreed, or that the TPE sensor should not be in its detected position, or that the TPF sensor should not be in its detected position during testing. **Possible Causes** • ETC TP sensor connector is damaged or shorted • ETC TP sensor circuits shorted together in the wire harness • ETC TP sensor signal circuit is shorted to VREF (5v) • ETC TP sensor is damaged or it has failed • PCM has failed
P2199 **1T CCM, MIL: No** 2003 F-Series Truck & Excursion with a 6.0L VIN P Diesel engine Transmissions: All	**IAT Sensor 1-2 Voltage Correlation Conditions:** Key on or engine running; and the PCM detected the Intake Air Temperature Sensor 1-2 signals did not agree during the CCM test. **Possible Causes** • IAT Sensor 1 connector is damaged or has high resistance • IAT Sensor 1 signal circuit has a high resistance condition • IAT Sensor 1 is damaged or it has failed • IAT Sensor 2 connector is damaged or has high resistance • IAT Sensor 2 signal circuit has a high resistance condition • IAT Sensor 2 is damaged or it has failed
P2262 **1T CCM, MIL: No** 2003 F-Series Truck & Excursion with a 6.0L VIN P Diesel engine Transmissions: All	**Turbo/Super Charger Boost Pressure Not Detected Conditions:** Engine started, vehicle driven under boost requirement conditions, and the PCM did not detect sufficient boost pressure during the test. **Possible Causes** • CAC system is leaking (cold air cooler assembly) • EP sensor is damaged or it has failed • Exhaust system is leaking or severely restricted • Intake system has air leaks • MAP sensor vacuum hose is disconnected, restricted or the MAP sensor has failed
P2263 **1T CCM, MIL: No** 2003 F-Series Truck & Excursion with a 6.0L VIN P Diesel engine Transmissions: All	**Turbo/Super Charger System Performance Conditions:** Engine started, vehicle driven under boost requirement conditions, and the PCM did not detect sufficient boost pressure during the test. **Possible Causes** • CAC system is leaking (cold air cooler assembly) • EP sensor is damaged or it has failed • Exhaust system is leaking or severely restricted • Intake system has air leaks • MAP sensor vacuum hose is disconnected, restricted or the MAP sensor has failed
P2269 **1T CCM, MIL: No** 2003 F-Series Truck & Excursion with a 6.0L VIN P Diesel engine Transmissions: All	**Water In Fuel Condition Conditions:** Key on or engine running; and the PCM detected a signal from the Water In Fuel (WIF) sensor that there was water in the Fuel system. **Possible Causes** • Drain the water from the fuel separator and retest the system • WIF sensor circuit is damaged, open or shorted • WIF sensor is damaged or it has failed

Diesel Engine OBD II Trouble Code List (P2xxx Codes)

DTC	Trouble Code Title, Conditions & Possible Causes
P2284 **1T CCM, MIL: No** 2003 F-Series Truck & Excursion with a 6.0L VIN P Diesel engine Transmissions: All	**Injector Control Pressure Sensor Signal Performance Conditions:** Engine started; and the PCM detected an unexpected voltage condition on the Injector Control Pressure sensor circuit. **Possible Causes** • ICP sensor connector is damaged, loose or shorted • ICP sensor signal circuit is open or shorted • ICP sensor is damaged or it has failed • PCM has failed
P2285 **1T CCM, MIL: No** 2003 F-Series Truck & Excursion with a 6.0L VIN P Diesel engine Transmissions: All	**Injector Control Pressure Sensor Circuit Low Input Conditions:** Engine started; and the PCM detected an unexpected low voltage condition (less than 0.04v) on the Injector Control Pressure sensor circuit during the CCM test. **Possible Causes** • ICP sensor connector is damaged or shorted • ICP sensor signal circuit is shorted to ground • ICP sensor is damaged or it has failed • PCM has failed
P2286 **1T CCM, MIL: No** 2003 F-Series Truck & Excursion with a 6.0L VIN P Diesel engine Transmissions: All	**Injector Control Pressure Sensor Circuit High Input Conditions:** Engine started; and the PCM detected an unexpected high voltage condition (more than 4.91v) on the Injector Control Pressure sensor circuit during the CCM test. **Possible Causes** • ICP sensor signal circuit is open or shorted to VREF (5v) • ICP sensor ground circuit is open • ICP sensor is damaged or it has failed • PCM has failed
P2288 **1T CCM, MIL: No** 2003 F-Series Truck & Excursion with a 6.0L VIN P Diesel engine Transmissions: All	**Injector Control Pressure Too High Conditions:** Engine started; and the PCM detected the Injector Control Pressure was too high (more than 3675 psi) during the CCM test period. **Possible Causes** • Refer to ICP system diagnosis information • ICP sensor is damaged or it has failed • PCM has failed
P2289 **1T CCM, MIL: No** 2003 F-Series Truck & Excursion with a 6.0L VIN P Diesel engine Transmissions: All	**Injector Control Pressure Too High - Engine Off Conditions:** Key on, KOEO Self Test enabled; and the PCM detected the Injector Control Pressure was too high (more than 3675 psi) during the test. **Possible Causes** • ICP sensor ground circuit is open • ICP sensor signal circuit is open • ICP sensor is damaged or it has failed • PCM has failed
P2290 **1T CCM, MIL: No** 2003 F-Series Truck & Excursion with a 6.0L VIN P Diesel engine Transmissions: All	**Injector Control Pressure Too Low Conditions:** Engine started; and the PCM detected the Injector Control Pressure was less than the Desired pressure during the CCM test period. **Possible Causes** • Refer to ICP system diagnosis information • ICP sensor is damaged or it has failed • PCM has failed
P2291 **1T CCM, MIL: No** 2003 F-Series Truck & Excursion with a 6.0L VIN P Diesel engine Transmissions: All	**Injector Control Pressure Too Low - Engine Cranking Conditions:** Engine cranking; and the PCM detected the Injector Control Pressure was too low (less than 725 psi) during the CCM test period. **Possible Causes** • Refer to ICP system diagnosis information • ICP sensor is damaged or it has failed • PCM has failed

Diesel Engine OBD II Trouble Code List (P2xxx Codes)

DTC	Trouble Code Title, Conditions & Possible Causes
P2552 **1T CCM, MIL: No** 2003 F-Series Truck & Excursion with a 6.0L VIN P Diesel engine Transmissions: All	**FICMM Circuit - Throttle/Fuel Inhibit Circuit Malfunction Conditions:** Key on or engine running; and the PCM did not detect a signal from the FICM during the CCM test. **Possible Causes** • FICM connector is damaged, open or shorted • FICMM circuit is damaged, open or shorted • FICM is damaged or it has failed • PCM has failed
P2614 **1T CCM, MIL: No** 2003 F-Series Truck & Excursion with a 6.0L VIN P Diesel engine Transmissions: All	**Camshaft Position Output Circuit Intermittent Conditions:** Key on or engine running; and the PCM detected an intermittent loss of the Camshaft Position (CMP) sensor circuit during the CCM test. **Possible Causes** • CMP sensor connector is damaged, open or shorted • CMP sensor circuit is damaged, open or shorted • CMP is damaged or it has failed • PCM has failed
P2617 **1T CCM, MIL: No** 2003 F-Series Truck & Excursion with a 6.0L VIN P Diesel engine Transmissions: All	**Crankshaft Position Output Circuit Intermittent Conditions:** Key on or engine running; and the PCM detected an intermittent loss of the Crankshaft Position (CKP) sensor circuit during the CCM test. **Possible Causes** • CKP sensor connector is damaged, open or shorted • CKP sensor circuit is damaged, open or shorted • CKP is damaged or it has failed • PCM has failed
P2623 **1T CCM, MIL: No** 2003 F-Series Truck & Excursion with a 6.0L VIN P Diesel engine Transmissions: All	**Injector Control Pressure Regulator Circuit Malfunction Conditions:** Key on or engine running; and the PCM detected an unexpected voltage condition on the Injector Pressure regulator circuit. **Possible Causes** • IPR (regulator) connector is damaged, open or shorted • IPR (regulator) circuit is damaged, open or shorted • IPR (regulator) is damaged, sticking or it has failed • PCM has failed

Diesel Engine OBD II Trouble Code List (U1xxx Codes)

DTC	Trouble Code Title, Conditions & Possible Causes
U0101 **1T PCM, MIL: No** 2003 F-Series Truck & Excursion with a 6.0L VIN P Diesel engine Transmissions: All	**Lost Communication With TCM Conditions:** Key on, and the PCM detected that it has lost communication with the Transmission Control Module (TCM) during its initial startup. **Possible Causes** • Replace the PCM and retest the system for codes.
U0105 **1T PCM, MIL: No** 2003 F-Series Truck & Excursion with a 6.0L VIN P Diesel engine Transmissions: All	**Lost Communication With FICM Conditions:** Key on, and the PCM detected that it has lost communication with the Fuel Injection Control Module (FICM) during its initial startup. **Possible Causes** • Replace the PCM and retest the system for codes.
U0155 **1T PCM, MIL: No** 2003 F-Series Truck & Excursion with a 6.0L VIN P Diesel engine Transmissions: All	**Lost Communication With Instrument Cluster Conditions:** Key on, and the PCM detected that it has lost communication with the Instrument Cluster Panel (I/P) during its initial startup. **Possible Causes** • Refer to the test procedures found in the Workshop Manual in Section 418.
U0306 **1T PCM, MIL: No** 2003 F-Series Truck & Excursion with a 6.0L VIN P Diesel engine Transmissions: All	**Software Incompatibility With Fuel Injector Control Module Conditions:** Key on, and the PCM detected a software incompatibility condition with the Fuel Injector Control Module (FICM) during its initial startup. **Possible Causes** • Refer to the test procedures found in the Workshop Manual in Section 418.

FORD PARAMETER ID (PID) CONTENTS

About This Section
Introduction ....Page 7-4
How to Use This Section ....Page 7-4

Ford & Mercury Information

ASPIRE PID DATA
1.3L I4 MFI VIN H (All) **(1996-97)** ....Page 7-5

CONTOUR & MYSTIQUE PID DATA
2.0L I4 MFI VIN 3 (All) **(1996-2000)** ....Page 7-6
2.5L V6 MFI VIN L (All) **(1996-2000)** ....Page 7-8

COUGAR PID DATA
2.0L I4 4v MFI VIN 3 (All) **(1999-2002)** ....Page 7-10
2.5L V6 MFI VIN L (All) **(1999-2002)** ....Page 7-12

CROWN VICTORIA & GRAND MARQUIS PID DATA
4.6L V8 MFI VIN W (A/T - AOD) **(1992)** ....Page 7-14
4.6L V8 MFI VIN W (A/T - AODE) **(1992-94)** ....Page 7-15
4.6L V8 MFI VIN W (A/T) **(1995-2003)** ....Page 7-16
4.6L V8 NGV VIN 9 (A/T) **(1996-2003)** ....Page 7-18

ESCORT, TRACER & ZX2 PID DATA
1.8L I4 MFI VIN 8 (All) **(1996)** ....Page 7-20
1.9L I4 MFI VIN J (All) **(1991-95)** ....Page 7-21
1.9L I4 MFI VIN J (A/T) **(1996)** ....Page 7-22
1.9L I4 MFI VIN J (M/T) **(1996)** ....Page 7-24
2.0L I4 MFI VIN P (A/T) **(1997-2002)** ....Page 7-25
2.0L I4 MFI VIN P (M/T) **(1997-2002)** ....Page 7-27
2.0L I4 MFI VIN 3 (A/T) **(1998-2003)** ....Page 7-29
2.0L I4 MFI VIN 3 (M/T) **(1998-2003)** ....Page 7-31

MUSTANG PID DATA
2.3L I4 MFI VIN M (All) **(1991-93)** ....Page 7-33
3.8L V6 MFI VIN 4 (A/T) **(1994-2003)** ....Page 7-34
3.8L V6 MFI VIN 4 (M/T) **(1994-2003)** ....Page 7-36
4.6L V8 MFI VIN V, Y (All) **(1996-2003)** ....Page 7-38
4.6L V8 MFI VIN W, X (All) **(1996-2003)** ....Page 7-40
5.0L V8 MFI VIN T (All) **(1994-95)** ....Page 7-42

Ford & Mercury PID Information

PROBE PID DATA

2.0L I4 MFI VIN A (CD4E) **(1994-95)** .... Page 7-43
2.0L I4 MFI VIN A (All) **(1996-97)** .... Page 7-44
2.5L V6 MFI VIN B (All) **(1996-97)** .... Page 7-46
3.0L V6 MFI VIN U (All) **(1990-92)** .... Page 7-47

TAURUS PID DATA

3.0L V6 FFV VIN 1 (A/T) **(1993-95)** .... Page 7-48
3.0L V6 FFV VIN 1, 2 (A/T) **(1996-2003)** .... Page 7-49
3.2L V6 MFI SHO VIN P (A/T) **(1993-95)** .... Page 7-51
3.4L V6 MFI SHO VIN N (A/T) **(1996-99)** .... Page 7-52

TAURUS & SABLE PID DATA

3.0L V6 MFI VIN S (A/T) **(1996-2003)** .... Page 7-54
3.0L V6 MFI VIN U (A/T) California **(1990)** .... Page 7-56
3.0L V6 MFI VIN U (A/T) **(1991-95)** .... Page 7-57
3.0L V6 MFI VIN U (A/T) California **(1995)** .... Page 7-58
3.0L V6 MFI VIN U (A/T) **(1996-2003)** .... Page 7-59
3.8L V6 MFI VIN 4 (A/T) **(1990)** .... Page 7-61
3.8L V6 MFI VIN 4 (A/T) **(1991)** .... Page 7-62
3.8L V6 MFI VIN 4 (A/T) **(1992-95)** .... Page 7-63
3.8L V6 MFI VIN 4 (A/T) California **(1994-95)** .... Page 7-64

TEMPO & TOPAZ PID DATA

2.3L I4 MFI VIN X (All) **(1992-94)** .... Page 7-65
3.0L V6 MFI VIN U (All) **(1992-94)** .... Page 7-65

THUNDERBIRD PID DATA

3.8L V6 MFI SC VIN R (All) **(1990-95)** .... Page 7-66
3.9L V8 MFI VIN A (A/T) **(2002-2003)** .... Page 7-67

THUNDERBIRD & COUGAR PID DATA

3.8L V6 MFI VIN 4 (A/T) **(1990-95)** .... Page 7-69
3.8L V6 MFI VIN 4 (A/T) **(1996-97)** .... Page 7-70
4.6L V8 MFI VIN W (A/T) **(1994-97)** .... Page 7-72
5.0L V8 MFI VIN T (A/T) **(1991-93)** .... Page 7-74

Lincoln PID Information

CONTINENTAL PID DATA
3.8L V6 MFI VIN 4 (A/T) **(1990-94)** ....Page 7-75
4.6L V8 MFI VIN V (A/T) **(1995-2000)** ....Page 7-76
4.6L V8 MFI VIN V (A/T) **(2001-02)** ....Page 7-78

LS6 PID DATA
3.0L V6 MFI VIN S (All) **(2000-2003)** ....Page 7-80

LS8 PID DATA
3.9L V8 MFI VIN A (A/T) **(2000-2003)** ....Page 7-82

MARK VIII PID DATA
4.6L V8 MFI VIN V (A/T) **(1993-95)** ....Page 7-84
4.6L V8 MFI VIN V (A/T) **(1996-98)** ....Page 7-85

TOWN CAR PID DATA
4.6L V8 MFI VIN W (A/T) **(1991-94)** ....Page 7-87
4.6L V8 MFI VIN W (A/T) **(1995-2003)** ....Page 7-88
5.0L V8 MFI VIN F (A/T) **(1990)** ....Page 7-90

About This Section

Introduction

This section of the Domestic Car Handbook contains Parameter ID (PID) tables for Ford, Lincoln and Mercury vehicles from 1990-2003. It can be used to assist in the repair of both Code & No Code problems.

VEHICLE COVERAGE

- Aspire Applications (1996-97)
- Continental Applications (1990-2002)
- Contour & Mystique Applications (1996-2000)
- Cougar & Mustang Applications (1991-2003)
- Crown Victoria & Grand Marquis Applications (1992-2003)
- Escort & Tracer Applications (1991-2002)
- Escort ZX2 Applications (2003)
- LS6 & LS8 Applications (1999-2003)
- Mark VIII Applications (1993-98)
- Probe Applications (1990-97)
- Sable & Taurus Applications (1990-2003)
- Tempo Applications (1992-94)
- Thunderbird Applications (1990-97)
- Town Car Applications (1990-2003)

How to Use This Section

This Section of the Handbook can be used to look up PID Data so that you can compare the "known good" values to the actual values you see on the Scan Tool display. To locate the PID Data, find the model, correct engine size (with VIN Code) and finally the year of the vehicle.

For example, to look up the PID Data for a 1999 Taurus 3.0L VIN U, go to Contents Page 2 and find the text string shown below:

3.0L V6 MFI VIN U (A/T) - PIDs **(1996-2002)** ....Page 7-59

Then turn to Page 7-59 to find the following PCM PID information.

1996-2002 TAURUS/SABLE 3.0L V6 MFI VIN U (A/T) - INPUTS

PCM PID Acronym	Parameter Identification	PID Range	PID Value at Hot Idle	PID Value at 30	PID Value at 55
FTP	FTP Sensor (1998-02)	0-5.1v or 0-10" H20	2.6v at 0" of H20	2.6v at 0" of H20	2.6v at 0" of H20

In this example, the Fuel Tank Pressure Sensor should read near 2.6v with the cap removed. The actual PID value from the vehicle can be used to check the sensor calibration by comparing it against the known good PID value in pin voltage table.

Aspire PID Data - OBD II

1996-97 ASPIRE 1.3L I4 MFI VIN H (ALL) - INPUTS / OUTPUTS

PCM PID Acronym	Parameter Identification	PID Value Range	PID Value at Hot Idle	PID Value at 30	PID Value at 55
ACR	AC Relay	ON / OFF	ON (if on)	OFF	OFF
ACCS	AC Relay	ON / OFF	ON (if on)	OFF	OFF
ATP	Atmospheric Press.	0-5.1v	3.9	3.9	3.9
B+	Battery Power	0-25.5v	12-14	12-14	12-14
BARO	BARO Pressure	0-30 in. Hg	29.4	29.4	29.4
BLT SW	Blower Motor SW	ON / OFF	ON (if on)	OFF	OFF
DEF SW	Defrost Switch	ON / OFF	ON (if on)	OFF	OFF
DRL SW	Daytime Running	ON / OFF	ON (if on)	OFF	OFF
ECT (°F)	ECT Sensor	-40-304°F	177	177	177
ECT (V)	ECT Sensor	0-5.1v	0.8	0.8	0.8
EGRS (V)	EGR Sensor Volts	0-5.1v	0.7	0.7-1.6	2.5-3.2
EGR-A	EGR 'A' Solenoid	0-100 ms	6400	0	2.7
EGR-V	EGR Vent Solenoid	0-25.5v	14.3	0.1	0.1
HDP SW	Headlamp Switch	ON / OFF	ON (if on)	OFF	OFF
HOS2-11	HO2S-11 (Bank 1)	0.1-1.1v	0.1-1.1	0.1-1.1	0.1-1.1
HOS2-12	HO2S-12 (Bank 1)	0.1-1.1v	0.1-1.1	0.1-1.1	0.1-1.1
IAT (°F)	IAT Sensor	-40-304°F	77	86	68
IAT (V)	IAT Sensor	0-5.1v	2.4	2.4	2.7
IDL SW	Throttle Idle Switch	ON / OFF	ON	OFF	OFF
IGN SW	Ignition Switch	ON / OFF	ON-KOEC	OFF	OFF
INJ1	Fuel Injector 1	0-99.9 ms	4.7	4.8	6.4
INJ2	Fuel Injector 2	0-99.9 ms	4.7	4.8	6.4
ISC	Idle Speed Control	0-100 ms	2.2	2.4	2.1
LOAD	Engine Load	0-100%	26	27	56
LONGFT	Long Term F/T B1	-20 to 20%	-5 to +5	-5 to +5	-5 to +5
MAF	MAF Sensor LB/M	0-10 LBM	0.0	0.85	1.43
MAF (V)	MAF Sensor	0-5.1v	2.0v	2.4	3.2
MIL	MIL Control	ON / OFF	ON (if on)	OFF	OFF
PSP	PSP Switch	ON / OFF	ON turning	OFF	OFF
PUMP	Fuel Pump	ON / OFF	ON	ON	ON
Purge	EVAP Purge	0-100%	0	40-80	84-90
RPM	Engine Speed	0-10K rpm	750	1690	2713
DES RPM	Desired Engine Speed	0-10K rpm	750	750	750
SHRTFT	Short Term F/T B1	-5 to +5	-5 to +5	-5 to +5	-5 to +5
SPARK	Spark Advance (°)	-90° to 90°	10	12	20
STP SW	Stop Lamp Switch	ON / OFF	ON (if on)	OFF	OFF
TPV	TP Sensor	0-5.1v	0.5	0.6	0.4v
VSS	Vehicle Speed	0-159 mph	0	30	55

Contour & Mystique PID Data - OBD II

1996-2000 CONTOUR/MYSTIQUE 2.0L I4 MFI VIN 3 (ALL) INPUTS

PCM PID Acronym	Parameter Identification	PID Range	PID Value at Hot Idle	PID Value at 30	PID Value at 55
ACCS	A/C Clutch Switch	ON / OFF	ON (if on)	OFF	OFF
ACP	A/C Press. Switch	CL/OPEN	Open	OPEN	OPEN
BPP	Brake Pedal Switch	ON / OFF	ON (if on)	OFF	OFF
CID	CID Sensor	0-1 KHz	5-7	11-15	17-21
CKP	CKP Sensor	0-10 KHz	470-530	990-1200	1350-1460
CPP	Clutch Pedal Switch	ON / OFF	ON (if in)	OFF	OFF
DPFE	DPFE Sensor	0-5.1v	0.4-0.6	0.7-0.9	0.7-1.2v
ECT (°F)	ECT Sensor	-40-304°F	160-200	160-200	160-200
FEPS	Flash EEPROM	0-5.1v	0.1	0.1	0.1
FLI	Fuel Level Indicator (1998-2002)	0-100%	50 (1.7v)	50 (1.7v)	50 (1.7v)
FPM	Fuel Pump Monitor	ON / OFF	ON	ON	ON
FRP	Fuel Rail Pressure (1999-2002)	0-100 psi	39	39	39
FTP	Fuel Tank Pressure (1998-2002)	0-5.1v or 0-10" H2O	2.6v / 0" H2O (with cap off)	2.6v / 0" H2O (with cap off)	2.6v / 0" H2O (with cap off)
HO2S-11	HO2S-11 (Bank 1)	0.1-1.1v	0.1-1.1	0.1-1.1	0.1-1.1
HO2S-12	HO2S-12 (Bank 1)	0.1-1.1v	0.1-1.1	0.1-1.1	0.1-1.1
IAT (°F)	IAT Sensor	-40-304°F	50-120	50-120	50-120
KS1	Knock Sensor 1	2.5v	0	0	0
GEAR	Gear Position	1-2-3-4	1	3	4
LOAD	Engine Load	0-100%	10-20	20-28	34-42
MAFV	MAF Sensor	0-5.1v	0.6-0.9	0.8-1.5	1.2-2.5
MISF	Misfire Detection	ON / OFF	OFF	OFF	OFF
OCTADJ	Octane Adjustment	CL/OPEN	CLOSED	CLOSED	CLOSED
PNP	PNP Switch	ON / OFF	ON (in 'P')	OFF	OFF
PSP	PSP Switch	HI / LOW	HI: turning	LOW	LOW
RPM	Engine Speed	0-10K rpm	790-900	1785-1835	2390-2430
TCS	TCS Switch	ON / OFF	OFF	ON	ON
TFT (°F)	TFT Sensor	-40-304°F	110-210	110-210	110-210
TPV	TP Sensor	0-5.1v	0.5-0.9	1.0-1.3	1.1-1.4
TR/V	TR Sensor	0-5.1v	4.4	2.1	2.1
TSS	TSS Sensor	0-10K rpm	680-710	1100-1220	2000-2800
Vacuum	Engine Vacuum	0-30" Hg	19-20	18-20	8-15
VPWR	Vehicle Power	0-25.5v	12.8-14.0	12.8-14.0	12.8-14.0
VSS	Vehicle Speed (MPH & Hertz Rate)	0-159 mph	0 = 0 Hz	30 = 65 Hz	55 = 125 Hz

1996-2000 CONTOUR/MYSTIQUE 2.0L I4 MFI VIN 3 (ALL) OUTPUTS

PCM PID Acronym	Parameter Identification	PID Range	PID Value at Hot Idle	PID Value at 30	PID Value at 55
CD1	Coil 1 Driver	0-90° dwell	7	10	12
CD2	Coil 2 Driver	0-90° dwell	7	10	12
CTO	Clean Tach Output	0-10 KHz	23-31	54-68	73-90
EGRVR	Vacuum Regulator	0-100%	0	0-40	0-40
EPC	EPC Solenoid	0-500 psi	15-18	23-28	30-35
EVAP DC	EVAP Purge Valve	0-100%	0-100	80	90
EVAPCV	EVAP CV Valve	0-100%	0	0	0
FP	Fuel Pump Relay	0-100%	100	100	100
FUELPW	INJ 1 Pulsewidth	0-99.9 ms	2.3-2.9	2.9-4.7	4.6-6.5
HFC	High Fan Control	ON / OFF	ON (if on)	OFF	OFF
HTR11	HO2S-11 Heater	ON / OFF	ON	ON	ON
HTR12	HO2S-12 Heater	ON / OFF	ON	ON	ON
IAC	Idle Air Control	0-100%	33-35	40-48	40-55
INJ1	INJ 1 Pulsewidth	0-999 ms	2.3-2.9	2.9-4.7	4.6-6.5
INJ2	INJ 2 Pulsewidth	0-999 ms	2.3-2.9	2.9-4.7	4.6-6.5
INJ3	INJ 3 Pulsewidth	0-999 ms	2.3-2.9	2.9-4.7	4.6-6.5
INJ4	INJ 4 Pulsewidth	0-999 ms	2.3-2.9	2.9-4.7	4.6-6.5
LFC	Low Fan Control	ON / OFF	ON (if on)	OFF	OFF
LONGFT	Long Term F/T B1	-20 to 20%	-5 to +5	-5 to +5	-5 to +5
MIL	MIL Control	ON / OFF	ON (if on)	OFF	OFF
SHRTFT	Short Term F/T B1	-20 to 20%	-5 to +5	-5 to +5	-5 to +5
SPARK	Spark Advance	-90° to 90°	15-25	19-30	25-36
SS1	Shift Solenoid 1	ON / OFF	OFF	ON	ON
SS2	Shift Solenoid 2	ON / OFF	ON	OFF	OFF
SS3	Shift Solenoid 3	0-25.5v	7-9.5v	8.3-9.5	8.3-9.5
TCC	TCC Solenoid	0-100%	0	50-100	95-100
TCIL	TCIL (lamp)	ON / OFF	ON (if on)	OFF	OFF
VCT	VCT Solenoid (1998-99)	0-25.5v	12-14	10.5-14	10.5-14
VREF	Voltage Reference	0-5.1v	4.9-5.1	4.9-5.1	4.9-5.1
WAC	A/C WOT Cutout Relay	ON / OFF	ON (if on)	OFF	OFF

1996-2000 CONTOUR/MYSTIQUE 2.5L V6 MFI VIN L (ALL) INPUTS

PCM PID Acronym	Parameter Identification	PID Range	PID Value at Hot Idle	PID Value at 30	PID Value at 55
ACCS	A/C Clutch Switch	ON / OFF	ON (if on)	OFF	OFF
ACPSW	A/C Press. Switch	CL/OPEN	OPEN	OPEN	OPEN
BPP	Brake Pedal Switch	ON / OFF	ON (if on)	OFF	OFF
CID	CID Sensor	0-1 KHz	5-7	7-9	11-13
CKP	CKP Sensor	0-10 KHz	410-440	770-830	1190-1120
CPP	Clutch Pedal Switch	ON / OFF	ON (if in)	OFF	OFF
DPFE	DPFE Sensor	0-5.1v	0.4-0.6	0.7-1.1v	1.1-1.7
ECT (°F)	ECT Sensor	-40-304°F	160-200	160-200	160-200
FEPS	Flash EEPROM	0-5.1v	0.1	0.1	0.1
FLI	Fuel Level Indicator (1998-2002)	0-100%	50 (1.7v)	50 (1.7v)	50 (1.7v)
FPM	Fuel Pump Monitor	0-100%	50	50	50
FRP	Fuel Rail Pressure (1999-2000)	0-100 psi	39	39	39
FTP	Fuel Tank Pressure (1998-2002)	0-5.1v or 0-10" H2O	2.6v / 0" H2O (with cap off)	2.6v / 0" H2O (with cap off)	2.6v / 0" H2O (with cap off)
Gear	Transmission Gear	1-2-3-4	1	3	4
HO2S-11	HO2S-11 (Bank 1)	0.1-1.1v	0.1-1.1	0.1-1.1	0.1-1.1
HO2S-12	HO2S-12 (Bank 1)	0.1-1.1v	0.1-1.1	0.1-1.1	0.1-1.1
HO2S-21	HO2S-21 (Bank 2)	0.1-1.1v	0.1-1.1	0.1-1.1	0.1-1.1
HO2S-22	HO2S-22 (Bank 2)	0.1-1.1v	0.1-1.1	0.1-1.1	0.1-1.1
IAT (°F)	IAT Sensor	-40-304°F	50-120	50-120	50-120
IMRCM	IMRC Monitor	ON / OFF	OFF	OFF	OFF
KS1	Knock Sensor 1	0-5.1v	2.5v	2.5v	2.5v
Load	Engine Load	0-100%	10-20	20-28	34-46
MAF (V)	MAF Sensor	0-5.1v	0.6-0.9	0.9-1.1	1.2-2.6v
MISF	Misfire Detection	ON / OFF	OFF	OFF	OFF
OCTADJ	Octane Adjustment	CL/OPEN	CLOSED	CLOSED	CLOSED
PNP	PNP Switch	ON / OFF	ON (in 'P')	OFF	OFF
PSP	PSP Switch	HI / LOW	HI: turning	LOW	LOW
RPM	Engine Speed	0-10K rpm	700-760	1340-1380	1950-2100
TCS	TCS (switch)	ON / OFF	ON (if on)	OFF	OFF
TFT (°F)	TFT Sensor	-40-304°F	110-210	110-210	110-210
TPV	TP Sensor	0-5.1v	0.5-0.9	1.0-1.3	1.1-1.4
TRS	TR Sensor	0-5.1v	4.4	2.1	2.1
TSS	TSS Sensor	0-10 KHz	680-710	1150-1240	2030-3020
Vacuum	Engine Vacuum	0-30" Hg	19-20	18-20	8-15
VPWR	Vehicle Power	0-25.5v	12-14	12-14	12-14
VSS	Vehicle Speed (MPH & Hertz Rate)	0-159 mph	0 = 0 Hz	30 = 65 Hz	55 = 125 Hz

1996-99 CONTOUR/MYSTIQUE 2.5L V6 MFI VIN L (ALL) OUTPUTS

PCM PID Acronym	Parameter Identification	PID Range	PID Value at Hot Idle	PID Value at 30	PID Value at 55
CD1-CD4	Coil Driver 1-4	0-60° dwell	7	10	11
CTO	Clean Tach Output	0-10 KHz	25-42	62-73	88-115
EGRVR	VR Solenoid	0-100%	0	0-40	35-45
EPC	EPC Solenoid	0-500 psi	11-18	23-28	30-35
EVAPCV	EVAP CV Valve	0-100%	0	0	0
EVAPDC	EVAP Purge Valve	0-100%	0-100	0-100	0-100
FP	Fuel Pump Relay	0-100%	100	100	100
FuelPW1	INJ Pulsewidth B1	0-99.9 ms	2.1-2.5	3.5-5.5	3.3-5.9
FuelPW2	INJ Pulsewidth B2	0-99.9 ms	2.1-2.5	3.5-5.5	3.3-5.9
HFC	High Fan Control	ON / OFF	ON (if on)	OFF	OFF
HTR11	HO2S-11 Heater	ON / OFF	ON	ON	ON
HTR12	HO2S-12 Heater	ON / OFF	ON	ON	ON
HTR21	HO2S-21 Heater	ON / OFF	ON	ON	ON
HTR22	HO2S-22 Heater	ON / OFF	ON	ON	ON
IAC	Idle Air Control	0-100%	32-40	30-55	40-60
IMRC	IMRC Solenoid	ON / OFF	OFF	OFF	OFF
INJ1	INJ 1 Pulsewidth	0-99.9 ms	2.1-2.5	3.5-5.5	3.3-5.9
INJ2	INJ 2 Pulsewidth	0-99.9 ms	2.1-2.5	3.5-5.5	3.3-5.9
INJ3	INJ 3 Pulsewidth	0-99.9 ms	2.1-2.5	3.5-5.5	3.3-5.9
INJ4	INJ 4 Pulsewidth	0-99.9 ms	2.1-2.5	3.5-5.5	3.3-5.9
INJ5	INJ 5 Pulsewidth	0-99.9 ms	2.1-2.5	3.5-5.5	3.3-5.9
INJ6	INJ 6 Pulsewidth	0-99.9 ms	2.1-2.5	3.5-5.5	3.3-5.9
LFC	Low Fan Control	ON / OFF	ON (if on)	OFF	OFF
LongFT1	Long Term F/T B1	-20 to 20%	-5 to +5	-5 to +5	-5 to +5
LongFT2	Long Term F/T B2	-20 to 20%	-5 to +5	-5 to +5	-5 to +5
MIL	MIL Control	ON / OFF	ON (if on)	OFF	OFF
SHTFT1	Short Term F/T B1	-20 to 20%	-5 to +5	-5 to +5	-5 to +5
SHTFT2	Short Term F/T B2	-20 to 20%	-5 to +5	-5 to +5	-5 to +5
SPARK	Spark Advance (°)	-90° to 90°	4-7	19-30	25-36
SS1	Shift Solenoid 1	ON / OFF	OFF	ON	ON
SS2	Shift Solenoid 2	ON / OFF	ON	OFF	OFF
SS3	Shift Solenoid 3	7.0-8.1v	7.0-9.5	8.3-9.5	8.3-9.5
TCC	TCC Solenoid	0-100%	0	50-100	95-100
TCIL	TCIL (lamp)	ON (if on)	OFF	OFF	OFF
VREF	Voltage Reference	0-5.1v	4.9-5.1	4.9-5.1	4.9-5.1
WAC	A/C WOT Cutout Relay	ON / OFF	ON (if on)	OFF	OFF

Cougar PID Data - OBD II

1999-2002 COUGAR 2.0L I4 4V MFI VIN 3 (ALL) INPUTS

PCM PID Acronym	Parameter Identification	PID Range	PID Value at Hot Idle	PID Value at 30	PID Value at 55
ACCS	Cycling Clutch SW	ON / OFF	ON (if on)	OFF	OFF
ACP	A/C Press. Switch	CL/OPEN	OPEN	OPEN	OPEN
BOO	BPP Switch	ON / OFF	ON (if on)	OFF	OFF
CID	Cylinder ID Signal	0-1 KHz	5-7	11-15	17-21
CKP	CKP Sensor	0-10 KHz	470-650	995-1210	1345-1455
CPP	Clutch Switch	ON / OFF	ON (if in)	OFF	OFF
ECT (°F)	ECT Sensor	-40-304°F	160-200	160-200	160-200
FEPS	Flash EEPROM	0-5.1v	0.1	0.1	0.1
FLI	Fuel Level Indicator	0-100%	50 (1.7v)	25-75	25-75
FPM	Fuel Pump Monitor	0-100%	87-100	87-100	87-100
FRP	Fuel Rail Pressure	0-100 psi	39	39	39
FTP	Fuel Tank Pressure	0-5.1v or 0-10" H2O	2.6v / 0" H2O (with cap off)	2.6v / 0" H2O (with cap off)	2.6v / 0" H2O (with cap off)
IAT (°F)	IAT Sensor	-40-304°F	50-120	50-120	50-120
GEAR	Gear Position	1-2-3-4	1	3	4
GLI	Generator Load	0-25.5v	8.2 (high)	7.6	7.9
HO2S-11	HO2S-11 (Bank 1)	0.1-1.1v	0.1-1.1	0.1-1.1	0.1-1.1
HO2S-12	HO2S-12 (Bank 1)	0.1-1.1v	0.1-1.1	0.1-1.1	0.1-1.1
KS1	Knock Sensor 1	0-5.1v	0	0	0
LOAD	Engine Load	0-100%	10-20	19-30	40-47
MAFV	MAF Sensor	0-5.1v	0.6-0.9	0.8-1.5	1.2-2.5
MISF	Misfire Detection	ON / OFF	OFF	OFF	OFF
OCTADJ	Octane Adjustment	CL/OPEN	CLOSED	CLOSED	CLOSED
PATSIN	PAT System	See PATS	See PATS	See PATS	See PATS
PNP	PNP Switch	ON / OFF	ON (in 'P')	OFF	OFF
PSP	PSP Switch	HI / LOW	HI: turning	LOW	LOW
RPM	Engine Speed	0-10K rpm	790-900	1680-1840	2250-2385
TFT (°F)	TFT Sensor	-40-304°F	110-210	110-210	110-210
TPV	TP Sensor	0-5.1v	0.53-1.27	1.0-1.3	1.1-1.4
Vacuum	Engine Vacuum	0-30" Hg	19-20	18-20	8-15
VPWR	Vehicle Power	0-25.5v	12.8-14.0	12.8-14.0	12.8-14.0
VSS	Vehicle Speed (MPH & Hertz Rate)	0-159 mph	0 = 0 Hz	30 = 65 Hz	55 = 125 Hz

1999-2002 COUGAR 2.0L I4 4V MFI VIN 3 (ALL) OUTPUTS

PCM PID Acronym	Parameter Identification	PID Range	PID Value at Hot Idle	PID Value at 30	PID Value at 55
CDA	Coil Driver 'A'	0-90° dwell	7	10	11
CD2	Coil Driver 'B'	0-90° dwell	7	10	11
CTO	Clean Tach Output	0-10K KHz	23-31	54-68	73-90
EPC	EPC Solenoid	0-500 psi	0	23-28	35-48
EVAPCV	EVAP CV Valve	0-100%	0	0	0
EVAPDC	EVAP Purge Valve	0-100%	0-100	80	90
FP	Fuel Pump Relay	0-100%	100	100	100
FuelPW1	INJ Pulsewidth B1	0-999 ms	2.9-2.4.0	2.9-5.6	6.5-9.3
HFC	High Fan Control	ON / OFF	ON (if on)	OFF	OFF
HTR11	HO2S-11 Heater	ON / OFF	ON	ON	ON
HTR12	HO2S-12 Heater	ON / OFF	ON	ON	ON
IAC	Idle Air Control	0-100%	32	35-48	40-55
INJ1	INJ 1 Pulsewidth	0-999 ms	2.9-2.4.0	2.9-5.6	6.5-9.3
INJ2	INJ 2 Pulsewidth	0-999 ms	2.9-2.4.0	2.9-5.6	6.5-9.3
INJ3	INJ 3 Pulsewidth	0-999 ms	2.9-2.4.0	2.9-5.6	6.5-9.3
INJ4	INJ 4 Pulsewidth	0-999 ms	2.9-2.4.0	2.9-5.6	6.5-9.3
LFC	Low Fan Control	ON / OFF	ON (if on)	OFF	OFF
LONGFT	Long Term F/T B1	-20 to 20%	-5 to +5	-5 to +5	-5 to +5
MIL	MIL Control	ON / OFF	ON (if on)	OFF	OFF
PATSIL	PAT System	See PATS	See PATS	See PATS	See PATS
PATSOUT	PAT System	See PATS	See PATS	See PATS	See PATS
PATSTRT	PAT System	See PATS	See PATS	See PATS	See PATS
SHRTFT	Short Term F/T B1	-20 to 20%	-5 to +5	-5 to +5	-5 to +5
SPARK	Spark Advance	-90° to 90°	15-25	19-33	25-36
SS1	Shift Solenoid 1	ON / OFF	OFF	ON	ON
SS2	Shift Solenoid 2	ON / OFF	OFF	ON	ON
SS3	Shift Solenoid 3	0-25.5v	7.0-9.5	8.3-9.5	8.3-9.5
TCC	TCC Solenoid	0-100%	0	0	95-100
TCIL	TCIL Lamp	ON / OFF	OFF	OFF	OFF
VCT	VCT Solenoid	0-25.5v	14	10.5-14	10.5-14
VSO	VSO Signal	0-1K Hz	0	65	125
VREF	Voltage Reference	0-5.1v	4.9-5.1	4.9-5.1	4.9-5.1
WAC	A/C WOT Cutout Relay	ON / OFF	ON (if on)	OFF	OFF

1999-2002 COUGAR 2.5L V6 MFI VIN L (ALL) INPUTS

PCM PID Acronym	Parameter Identification	PID Range	PID Value at Hot Idle	PID Value at 30	PID Value at 55
ACCS	A/C Clutch Switch	ON / OFF	ON (if on)	OFF	OFF
ACP	A/C Press. Switch	CL/OPEN	OPEN	OPEN	OPEN
BPP	Brake Pedal Switch	ON / OFF	ON (if on)	OFF	OFF
CID	CID Sensor	0-1 KHz	5-7	10-14	15-21
CKP	CKP Sensor	0-10 KHz	410-480	830-1000	1220-1400
CPP	Clutch Pedal Switch	ON / OFF	ON (if in)	OFF	OFF
DPFE	DPFE Input (99-'00)	0-5.1v	0.2-1.3	0.2-4.5	0.2-4.5
DPFE	DPFE Input (01-'02)	0-5.1v	0.95-1.05	0.95-4.65	0.95-4.65
ECT (°F)	ECT Sensor	-40-304°F	160-200	160-200	160-200
FEPS	Flash EEPROM	0-5.1v	0.1	0.1	0.1
FLI	Fuel Level Indicator	0-5.1v	50 (1.7v)	50 (1.7v)	50 (1.7v)
FPM	Fuel Pump Monitor	0-100%	100	100	100
FRP	Fuel Rail Pressure	0-100 psi	39	39	39
FTP	Fuel Tank Pressure	0-5.1v or 0-10" H2O	2.6v / 0" H2O (with cap off)	2.6v / 0" H2O (with cap off)	2.6v / 0" H2O (with cap off)
Gear	Transmission Gear	1-2-3-4	1	3	4
HO2S-11	HO2S-11 (Bank 1)	0.1-1.1v	0.1-1.1	0.1-1.1	0.1-1.1
HO2S-12	HO2S-12 (Bank 1)	0.1-1.1v	0.1-1.1	0.1-1.1	0.1-1.1
HO2S-21	HO2S-21 (Bank 2)	0.1-1.1v	0.1-1.1	0.1-1.1	0.1-1.1
HO2S-22	HO2S-22 (Bank 2)	0.1-1.1v	0.1-1.1	0.1-1.1	0.1-1.1
IMRCM	IMRC Monitor	ON / OFF	OFF	OFF	OFF
IAT (°F)	IAT Sensor	-40-304°F	50-120	50-120	50-120
KS1	Knock Sensor 1	0-5.1v	0	0	0
Load	Engine Load	0-100%	10-20	20-28	30-42
MAF (V)	MAF Sensor	0-5.1v	0.6-0.9	1.1-1.3	1.2-1.6
MISF	Misfire Detection	ON / OFF	OFF	OFF	OFF
OCTADJ	Octane Adjustment	CL/OPEN	CLOSED	CLOSED	CLOSED
OSS	OSS Sensor Hertz	0-10 KHz	0	590	1050
PATSIN	PAT System	See PATS	See PATS	See PATS	See PATS
PNP	PNP Switch	ON / OFF	ON (in 'P')	OFF	OFF
PSP	PSP Switch	HI / LOW	HI: turning	LOW	LOW
RPM	Engine Speed	0-10K rpm	700-760	1700-1800	2100-2390
TCS	TCS (switch)	ON / OFF	ON (if on)	OFF	OFF
TFT (°F)	TFT Sensor	-40-304°F	110-210	110-210	110-210
TPV	TP Sensor	0-5.1v	0.53-1.27	1.0-1.3	1.1-1.4
TRV	TR Sensor	0-5.1v	4.4 (Park)	2.1 (O/D)	2.1 (O/D)
TSS	TSS Sensor	0-10 KHz	46-50	70-95	130-145
VPWR	Vehicle Power	0-25.5v	12-14	12-14	12-14
VSS (+)	Vehicle Speed (MPH & Hertz Rate)	0-159 mph	0 = 0 Hz	30 = 65 Hz	55 = 125 Hz

1999-2002 COUGAR 2.5L V6 MFI VIN L (ALL) OUTPUTS

PCM PID Acronym	Parameter Identification	PID Range	PID Value at Hot Idle	PID Value at 30	PID Value at 55
CDA	Coil Driver 'A'	0-60° dwell	7	10	11
CDB	Coil Driver 'B'	0-60° dwell	7	10	11
CDC	Coil Driver 'C'	0-60° dwell	7	10	11
CTO	Clean Tach Output	0-10 KHz	35-42	73-86	88-120
EGRVR	VR Solenoid	0-100%	0	0-40	0-40
EPC	EPC Solenoid	0-500 psi	0	23-28	30-35
EVAPCV	EVAP CV Valve	0-100%	0	0	0
EVAPDC	EVAP Purge Valve	0-100%	0-100	0-100	0-100
FP	Fuel Pump Relay	0-100%	30	30	30
FuelPW1	Injector on time B1	0-99.9 ms	3.5-4.0	3.0-6.0	3.3-7.3
FuelPW2	Injector on time B2	0-99.9 ms	3.5-4.0	3.0-6.0	3.3-7.3
HFC	High Fan Control	ON / OFF	ON (if on)	OFF	OFF
HTR11	HO2S-11 Heater	ON / OFF	ON	ON	ON
HTR12	HO2S-12 Heater	ON / OFF	ON	ON	ON
HTR21	HO2S-21 Heater	ON / OFF	ON	ON	ON
HTR22	HO2S-22 Heater	ON / OFF	ON	ON	ON
IAC	Idle Air Control	0-100%	33	40-48	40-55
IMRC	IMRC Solenoid	ON / OFF	OFF	OFF	OFF
INJ1	INJ 1 Pulsewidth	0-99.9 ms	3.5-4.0	3.0-6.0	3.3-7.3
INJ2	INJ 2 Pulsewidth	0-99.9 ms	3.5-4.0	3.0-6.0	3.3-7.3
INJ3	INJ 3 Pulsewidth	0-99.9 ms	3.5-4.0	3.0-6.0	3.3-7.3
INJ4	INJ 4 Pulsewidth	0-99.9 ms	3.5-4.0	3.0-6.0	3.3-7.3
INJ5	INJ 5 Pulsewidth	0-99.9 ms	3.5-4.0	3.0-6.0	3.3-7.3
INJ6	INJ 6 Pulsewidth	0-99.9 ms	3.5-4.0	3.0-6.0	3.3-7.3
LFC	Low Fan Control	ON / OFF	ON (if on)	OFF	OFF
LongFT1	Long Term F/T B1	-20 to 20%	-5 to +5	-5 to +5	-5 to +5
LongFT2	Long Term F/T B2	-20 to 20%	-5 to +5	-5 to +5	-5 to +5
MIL	MIL Control	ON / OFF	ON (if on)	OFF	OFF
PATSIL	PAT System	See PATS	See PATS	See PATS	See PATS
PATSOUT	PAT System	See PATS	See PATS	See PATS	See PATS
PATSTRT	PAT System	See PATS	See PATS	See PATS	See PATS
SHTFT1	Short Term F/T B1	-20 to 20%	-5 to +5	-5 to +5	-5 to +5
SHTFT2	Short Term F/T B2	-20 to 20%	-5 to +5	-5 to +5	-5 to +5
SPARK	Spark Advance (°)	-90° to 90°	4-7	19-30	25-36
SS1	Shift Solenoid 1	ON / OFF	OFF	ON	ON
SS2	Shift Solenoid 2	ON / OFF	ON	OFF	OFF
SS3	Shift Solenoid 3	0-25.5v	7.0-9.5	8.3-9.5	8.3-9.5
TCC	TCC Solenoid	0-100%	0	0	95-100
TCIL	TCIL (lamp)	ON (if on)	OFF	OFF	OFF
VREF	Voltage Reference	0-5.1v	4.9-5.1	4.9-5.1	4.9-5.1
VSO	VSO Signal	0-1K Hz	0	65	125
WAC	A/C WOT Cutout Relay	ON / OFF	ON (if on)	OFF	OFF

Crown Victoria & Grand Marquis PID Data - OBD I

1992 CROWN VICTORIA/GRAND MARQUIS 4.6L V8 VIN W (A/T - AOD)

PCM PID Acronym	Parameter Identification	PID Range	PID Value at Hot Idle	PID Value at 30	PID Value at 55
CANP	EVAP Purge Valve	ON / OFF	OFF	ON	ON
ECT (°F)	ECT Sensor	-40-304°F	160-200	160-200	160-200
ECT (V)	ECT Sensor	0-5.1v	0.6	0.6	0.6
EVR	VR Solenoid	0-100%	0	0-40	0-60
FuelPW1	INJ Pulsewidth B1	0-99.9 ms	6.4-6.8	8.2-8.8	8.4-9.0
FuelPW2	INJ Pulsewidth B2	0-99.9 ms	6.4-6.8	8.2-8.8	8.4-9.0
HO2S-11	HO2S-11 (Bank 1)	0.1-1.1v	0.1-1.1	0.1-1.1	0.1-1.1
HO2S-21	HO2S-21 (Bank 2)	0.1-1.1v	0.1-1.1	0.1-1.1	0.1-1.1
IAC	Idle Air Control	0-100%	20-40	34-40	45-55
IAT (°F)	IAT Sensor	-40-304°F	50-120	50-120	50-120
IAT (V)	IAT Sensor	-40-304°F	1.7-3.5	1.7-3.5	1.7-3.5
LOOP	Loop Status	CL or OL	CL	CL	CL
LongFT1	Long Term F/T B1	-20 to 20%	-5 to +5	-5 to +5	-5 to +5
LongFT2	Long Term F/T B2	-20 to 20%	-5 to +5	-5 to +5	-5 to +5
MAF (V)	MAF Sensor	0-5.1v	0.6	0.9-1.1	1.1-1.3
MLP	MLP Sensor	Park/Drive	PARK	DRIVE	DRIVE
MLP (V)	MLP Sensor	0-5.1v	0v	5v	5v
PFE	PFE Sensor	0-5.1v	3.2	3.3	3.4
PNP	PNP Switch	Neutral/DR	NEUTRAL	DRIVE	DRIVE
RPM	Engine Speed	0-10K rpm	750-820	1450-1630	1750-2100
SHTFT1	Short Term F/T B1	-20 to 20%	-5 to +5	-5 to +5	-5 to +5
SHTFT2	Short Term F/T B2	-20 to 20%	-5 to +5	-5 to +5	-5 to +5
SPARK	Spark Advance (°)	-90° to 90°	15-22	28-35	25-35
TP	TP Sensor	0-5.1v	0.7	0.8-1.0	1.1-1.3
TP Mode	TP Mode	C/T or P/T	C/T	P/T	P/T
VPWR	Vehicle Power	0-25.5v	12-14	12-14	12-14
VREF	Voltage Reference	0-5.1v	4.9-5.1	4.9-5.1	4.9-5.1
VSS	Vehicle Speed	0-159 mph	0	30	55
WAC	A/C WOT Cutout Relay	ON / OFF	ON (if on)	OFF	OFF

1992-94 CROWN VICTORIA/GRAND MARQUIS 4.6L VIN W (A/T AODE)

PCM PID Acronym	Parameter Identification	PID Range	PID Value at Hot Idle	PID Value at 30	PID Value at 55
ACCS	A/C Clutch Switch	ON / OFF	ON (if on)	OFF	ON
BOO	Brake Switch	ON / OFF	ON (if on)	OFF	OFF
DPFE	DPFE Sensor	0-5.1v	0.4-0.6	0.6-1.8	1.9-4.8
ECT (°F)	ECT Sensor	-40-304°F	160-200	160-200	160-200
ECT (V)	ECT Sensor	0-5.1v	0.6	0.6	0.6
EPC	EPC Solenoid	0-500 psi	20	12	30
EVR	VR Solenoid	0-100%	0	0-40	0-60
FP	Fuel Pump Relay	ON / OFF	ON	ON	ON
GEAR	Gear Position	P-R-N-D	P-R-N-D	DRIVE 3	DRIVE 4
FuelPW1	INJ Pulsewidth B1	0-99.9 ms	5.5-5.8	6.4-8.4	9.4-12
FuelPW2	INJ Pulsewidth B2	0-99.9 ms	5.5-5.8	6.4-8.4	9.4-12
HO2S-11	HO2S-11 (Bank 1)	0.1-1.1v	0.1-1.1	0.1-1.1	0.1-1.1
HO2S-21	HO2S-21 (Bank 1)	0.1-1.1v	0.1-1.1	0.1-1.1	0.1-1.1
IAC	Idle Air Control	0-100%	20-40	34-40	45-55
IAT (°F)	IAT Sensor	-40-304°F	50-120	50-120	50-120
IAT (V)	IAT Sensor	0-5.1v	1.7-3.5	1.7-3.5	1.7-3.5
LOOP	Loop Status	CL or OL	CL	CL	CL
LongFT1	Long Term F/T B1	-20 to 20%	-5 to +5	-5 to +5	-5 to +5
LongFT2	Long Term F/T B2	-20 to 20%	-5 to +5	-5 to +5	-5 to +5
MAF	MAF Sensor	0-5.1v	1.0v	1.3-1.5	1.6-2.0
MLP	MLP Switch	DRIVE or PARK	PARK	DRIVE	DRIVE
MLP (V)	MLP Switch	0-5.1v	0	5	5
PNP	PNP Switch	Neutral/DR	NEUTRAL	DRIVE	DRIVE
RPM	Engine Speed	0-10K rpm	750-820	1450-1630	1750-2100
SHTFT1	Short Term F/T B1	-20 to 20%	-5 to +5	-5 to +5	-5 to +5
SHTFT2	Short Term F/T B2	-20 to 20%	-5 to +5	-5 to +5	-5 to +5
SPARK	Spark Advance	-90° to 90°	15-22	28-35	25-35
TCC	TCC Solenoid	0-100%	0	42-44	92-100
TOT (°F)	TOT Sensor	-40-304°F	110-210	110-210	110-210
TOT (V)	TOT Sensor	0-5.1v	1.7-3.5	1.7-3.5	1.7-3.5
TPV	TP Sensor	0-5.1v	0.7	0.8-1.0	1.1-1.3
TP MIN	TP Minimum	0-5.1v	0.7	0.8-1.0	1.1-1.3
TP Mode	TP Mode	C/T or P/T	C/T	P/T	P/T
VPWR	Vehicle Power	0-25.5v	12-14	12-14	12-14
VREF	Voltage Reference	0-5.1v	4.9-5.1	4.9-5.1	4.9-5.1
VSS	Vehicle Speed MPH	0-159 mph	0	30	55
WAC	A/C WOT Cutout Relay	OFF	ON (if on)	OFF	OFF

1995-2003 CROWN VICTORIA/GRAND MARQUIS 4.6L V8 VIN W (A/T)

PCM PID Acronym	Parameter Identification	PID Range	PID Value at Hot Idle	PID Value at 30	PID Value at 55
ACCS	A/C Clutch Switch	ON / OFF	ON (if on)	OFF	OFF
ACP	A/C Pressure	OPEN or CLOSED	CLOSED (if low)	OPEN	OPEN
BOO	BPP Switch	ON / OFF	ON (if on)	OFF	OFF
CHT (°F)	CHT Sensor (°F)	-40-304°F	190-194	190-194	190-194
CID	CID Sensor Signal	0-1 KHz	6-7	10-11	13-14
CKP	CKP Sensor Signal	0-10 KHz	440-490	550-780	900-1100
DPFE	DPFE Sensor	0-5.1v	0.25-0.6	0.25-4.65	0.25-4.65
ECT (°F)	ECT Sensor	-40-304°F	160-200	160-200	160-200
FEPS	Flash EEPROM	0-5.1v	0.1	0.1	0.1
FLI	Fuel Level Indicator	0-100%	50 (1.7v)	25-75	25-75
FPM	Fuel Pump Monitor	ON / OFF	ON	ON	ON
FTP	Fuel Tank Pressure	0-5.1v or 0-10" H2O	2.6v / 0" H2O (with cap off)	2.6v / 0" H2O (with cap off)	2.6v / 0" H2O (with cap off)
GEAR	Transmission Gear	1-2-3-4	1	3	4
HO2S-11	HO2S-11 (Bank 1)	0.1-1.1v	0.1-1.1	0.1-1.1	0.1-1.1
HO2S-12	HO2S-12 (Bank 1)	0.1-1.1v	0.1-1.1	0.1-1.1	0.1-1.1
HO2S-21	HO2S-21 (Bank 2)	0.1-1.1v	0.1-1.1	0.1-1.1	0.1-1.1
HO2S-22	HO2S-22 (Bank 2)	0.1	0.1-1.1	0.1-1.1	0.1-1.1
IAT (°F)	IAT Sensor	-40-304°F	50-120	50-120	50-120
LOAD	Engine Load	0-100%	15-19	20-26	28-38
MAFV	MAF Sensor	0-5.1v	0.6-0.9	0.9-1.5	1.4-2.1
MISF	Misfire Detection	ON / OFF	OFF	OFF	OFF
OCTADJ	Octane Adjustment	CL/OPEN	CLOSED	CLOSED	CLOSED
OSS	OSS Sensor	0-10 KHz	0 rpm	1260-1330	2265-2400
PNP	PNP Switch	ON / OFF	ON (in 'P')	OFF	OFF
RPM	Engine Speed	0-10K rpm	790-815	1250-1400	1540-1620
TCS	TCS (Momentary ON Switch)	ON when depressed	OFF	OFF	OFF
TFT (°F)	TFT Sensor	110-210	110-210	110-210	110-210
TPV	TP Sensor	0.5-0.9	0.5-0.9	1.0-1.2	1.0-1.3
TR1	TR1 Sensor (2003)	0-11.5v	0	11.5	11.5
TR2 (V)	TR2 Sensor	0-25.5v	0	11.5	11.5
TR4 (V)	TR4 Sensor	0-25.5v	0	11.5	11.5
TR/V	TR Sensor	0-5v	4.4	2.1	2.1
Vacuum	Engine Vacuum	0-30" Hg	19-20	18-20	8-15
VPWR	Vehicle Power	12-14v	12-14	12-14	12-14
VSS/VSO	Vehicle Speed (MPH & Hertz Rate)	0-159 mph	0 = 0 Hz	30 = 65 Hz	55 = 125 Hz
WACA	A/C WOT Cutout Relay Monitor	ON / OFF	ON (if on)	OFF	OFF

<u>1995-2003 CROWN VICTORIA/GRAND MARQUIS 4.6L V8 VIN W (A/T)</u>

PCM PID Acronym	Parameter Identification	PID Range	PID Value at Hot Idle	PID Value at 30	PID Value at 55
CD1-CD8	Coil Driver 1-8	0-45° dwell	8	10	12
CHTIL	Cylinder Head Temperature Lamp	ON / OFF	OFF	OFF	OFF
CTO	Clean Tach Output	0-10 KHz	50-57	82-90	100-110
EGRVR	VR Solenoid	0-100%	0	0-40	40-50
EPC	EPC Solenoid	0-500 psi	20	22	22
EVAPCV	EVAP CV Valve	0-100%	0	0	0
EVAPDC	EVAP Purge Valve	0-100%	0-100	80	90
FP	Fuel Pump Monitor	0-100%	100	100	100
FuelPW1	INJ Pulsewidth B1	0-99.9 ms	3.4-3.7	3.7-6.0	5.5-9.0
FuelPW2	INJ Pulsewidth B2	0-99.9 ms	3.4-3.7	3.7-6.0	5.5-9.0
HTR-11	HO2S-11 Heater	ON / OFF	Switch	Switch	Switch
HTR-12	HO2S-12 Heater	ON / OFF	ON	ON	ON
HTR-21	HO2S-21 Heater	ON / OFF	Switch	Switch	Switch
HTR-22	HO2S-22 Heater	ON / OFF	ON	ON	ON
IAC	Idle Air Control	0-100%	32-40	30-55	40-60
INJ1	INJ 1 Pulsewidth	0-99.9 ms	3.4-3.7	3.7-6.0	5.5-9.0
INJ2	INJ 2 Pulsewidth	0-99.9 ms	3.4-3.7	3.7-6.0	5.5-9.0
INJ3	INJ 3 Pulsewidth	0-99.9 ms	3.4-3.7	3.7-6.0	5.5-9.0
INJ4	INJ 4 Pulsewidth	0-99.9 ms	3.4-3.7	3.7-6.0	5.5-9.0
INJ5	INJ 5 Pulsewidth	0-99.9 ms	3.4-3.7	3.7-6.0	5.5-9.0
INJ6	INJ 6 Pulsewidth	0-99.9 ms	3.4-3.7	3.7-6.0	5.5-9.0
INJ7	INJ 7 Pulsewidth	0-99.9 ms	3.4-3.7	3.7-6.0	5.5-9.0
INJ8	INJ 8 Pulsewidth	0-99.9 ms	3.4-3.7	3.7-6.0	5.5-9.0
LFC	Low Fan Control	ON / OFF	ON (if on)	OFF	OFF
LongFT1	Long Term F/T B1	-20 to 20%	-5 to +5	-5 to +5	-5 to +5
LongFT2	Long Term F/T B2	-20 to 20%	-5 to +5	-5 to +5	-5 to +5
MIL	MIL Control	ON / OFF	OFF	OFF	OFF
SHTFT1	Short Term F/T B1	-20 to 20%	-5 to +5	-5 to +5	-5 to +5
SHTFT2	Short Term F/T B2	-20 to 20%	-5 to +5	-5 to +5	-5 to +5
SPARK	Spark Advance	-90° to 90°	15-18	33-36	32-38
SS1	Shift Solenoid 1	ON / OFF	ON	OFF	ON
SS2	Shift Solenoid 2	ON / OFF	OFF	ON	ON
TCC	TCC Solenoid	0-100%	0	42-44	90-100
TCIL	TCIL (lamp)	ON (if on)	OFF	OFF	OFF
VREF	Voltage Reference	0-5.1v	4.9-5.1	4.9-5.1	4.9-5.1
WAC	A/C WOT Cutout Relay	ON / OFF	ON (if on)	OFF	OFF

1996-2003 CROWN VICTORIA/GRAND MARQUIS 4.6L V8 NGV VIN 9

PCM PID Acronym	Parameter Identification	PID Range	PID Value at Hot Idle	PID Value at 30	PID Value at 55
ACCS	A/C Clutch Switch	ON / OFF	ON (if on)	OFF	OFF
BPP	Brake Pedal Switch	ON / OFF	ON (if on)	OFF	OFF
CID	CID Sensor	0-1 KHz	6-7	8-9	16-17
CKP	CKP Sensor	0-10 KHz	440-490	580-770	850-1100
DPFE	DPFE Sensor	0-5.1v	0.25-1.30	0.25-4.65	0.25-4.65
ECT (°F)	ECT Sensor	-40-304°F	160-200	160-200	160-200
EFTA (°F)	EFT Sensor B1	-40-304°F	50-120	50-120	50-120
EFTB (°F)	EFT Sensor B2	-40-304°F	50-120	50-120	50-120
FEPS	Flash EEPROM	0-5.1v	0.5-0.6	0.5-0.6	0.5-0.6
FSV M	Fuel Shutoff Valve Monitor	ON / OFF	ON	ON	ON
FRP (psi)	Fuel Pressure	0-500 psi	105-130	105-130	105-130
FRP	Fuel Rail Pressure	0-5v	2.7-3.7	2.7-3.7	2.7-3.7
GEAR	Transmission Gear	1-2-3-4	1	3	4
HO2S-11	HO2S-11 (Bank 1)	0.1-1.1v	0.1-1.1	0.1-1.1	0.1-1.1
HO2S-12	HO2S-12 (Bank 1)	0.1-1.1v	0.1-1.1	0.1-1.1	0.1-1.1
HO2S-21	HO2S-21 (Bank 2)	0.1-1.1v	0.1-1.1	0.1-1.1	0.1-1.1
HO2S-22	HO2S-22 (Bank 2)	0.1-1.1v	0.1-1.1	0.1-1.1	0.1-1.1
IAT (°F)	IAT Sensor	-40-304°F	50-120	50-120	50-120
IP ('96)	Injector Pressure	0-5v	2.7-3.7	2.7-3.7	2.7-3.7
LOAD	Engine Load	0-100%	15-19	22-30	31-46
MAF (V)	MAF Sensor	0-5.1v	0.6-0.9	0.9-1.5	1.2-2.1
MISF	Misfire Detection	ON / OFF	OFF	OFF	OFF
OCTADJ	Octane Adjustment	CL/OPEN	CLOSED	CLOSED	CLOSED
OSS	Output Speed Shaft	0-10K rpm	0	950-1100	1750-1900
PNP	PNP Switch	ON / OFF	ON (in 'P')	OFF	OFF
RPM	Engine Speed	0-10K rpm	790-815	925-1125	1320-1350
TCS	TCS (Switch)	ON / OFF	ON (if OD & TCS on)	OFF	OFF
TFT (°F)	TFT Sensor	-40-304°F	110-210	110-210	110-210
TPV	TP Sensor	0-5.1v	0.5-0.9	0.8-1.1	0.9-1.3
TRV/TR	TR Sensor	0-5.1v	0-PARK	1.7-OD	1.7-OD
VPWR	Vehicle Power	12-14v	12-14	12-14	12-14
VSS	Vehicle Speed (MPH & Hertz Rate)	0-159 mph	0 / 0	30 = 65 Hz	55 = 125 Hz
WACA	A/C WOT Cutout Relay Monitor	ON / OFF	ON (if AC is on)	OFF	OFF

1996-2003 CROWN VICTORIA/GRAND MARQUIS 4.6L V8 NGV VIN 9

PCM PID Acronym	Parameter Identification	PID Range	PID Value at Hot Idle	PID Value at 30	PID Value at 55
CD1-CD8	Coil Driver 1-8	0-45° dwell	8	10	12
CHTIL	Cylinder Head Temperature Lamp	ON / OFF	ON (if on)	OFF	OFF
CTO	Clean Tach Output	0-10 KHz	50-57	60-65	85-95
EGRVR	VR Solenoid	0-100%	0	0-40	0-50
EPC	EPC Solenoid	0-500 psi	20	12	30
FSV	Fuel Shutoff Valve	ON / OFF	ON	ON	ON
FuelPW1	INJ Pulsewidth B1	0-99.9 ms	3.9-4.6	4.7-12	4.7-12.2
FuelPW2	INJ Pulsewidth B2	0-99.9 ms	3.9-4.6	4.7-12	4.7-12.2
HTR-11	HO2S-11 Heater	ON / OFF	ON	ON	ON
HTR-12	HO2S-12 Heater	ON / OFF	ON	ON	ON
HTR-21	HO2S-21 Heater	ON / OFF	ON	ON	ON
HTR-22	HO2S-22 Heater	ON / OFF	ON	ON	ON
IAC	Idle Air Control	0-100%	32-36	30-55	40-60
INJ1	INJ 1 Pulsewidth	0-99.9 ms	3.9-4.6	4.7-12.0	4.7-12.2
INJ2	INJ 2 Pulsewidth	0-99.9 ms	3.9-4.6	4.7-12.0	4.7-12.2
INJ3	INJ 3 Pulsewidth	0-99.9 ms	3.9-4.6	4.7-12.0	4.7-12.2
INJ4	INJ 4 Pulsewidth	0-99.9 ms	3.9-4.6	4.7-12.0	4.7-12.2
INJ5	INJ 5 Pulsewidth	0-99.9 ms	3.9-4.6	4.7-12.0	4.7-12.2
INJ6	INJ 6 Pulsewidth	0-99.9 ms	3.9-4.6	4.7-12.0	4.7-12.2
INJ7	INJ 7 Pulsewidth	0-99.9 ms	3.9-4.6	4.7-12.0	4.7-12.2
INJ8	INJ 8 Pulsewidth	0-99.9 ms	3.9-4.6	4.7-12.0	4.7-12.2
LFC	Low Fan Control	ON / OFF	ON (if on)	OFF	OFF
LongFT1	Long Term F/T B1	-20 to 20%	-5 to 5%	-5 to 5%	-5 to 5%
LongFT2	Long Term F/T B2	-20 to 20%	-5 to 5%	-5 to 5%	-5 to 5%
MIL	MIL Control	ON / OFF	ON (if on)	OFF	OFF
SHTFT1	Short Term F/T B1	-20 to 20%	-5 to 5%	-5 to 5%	-5 to 5%
SHTFT2	Short Term F/T B2	-20 to 20%	-5 to 5%	-5 to 5%	-5 to 5%
SPARK	Spark Advance	-90° to 90°	5-10	20-31	20-28
SS1	Shift Solenoid 1	ON / OFF	ON	OFF	ON
SS2	Shift Solenoid 2	ON / OFF	OFF	ON	ON
TCC	TCC Solenoid	0-100%	0	0-50	80-100
TCIL	TCIL (lamp)	ON / OFF	ON (if on)	OFF	OFF
VREF	Voltage Reference	0-5.1v	4.9-5.1	4.9-5.1	4.9-5.1
WAC	A/C WOT Cutout Relay	ON / OFF	ON (if on)	OFF	OFF

Escort & Tracer PID Data - OBD II

1996 ESCORT/TRACER 1.8L I4 MFI VIN 8 (ALL) INPUTS / OUTPUTS

PCM PID Acronym	Parameter Identification	PID Range	PID Value at Hot Idle	PID Value at 30	PID Value at 55
ACR	A/C Relay (On with A/C blower at 3-4)	ON / OFF	ON (if on)	OFF	OFF
ACS	A/C Relay (On with A/C blower at 3-4)	ON / OFF	ON (if on)	OFF	OFF
ATP	Atmospheric Press.	0-5.1v	3.9	3.9	3.9
BARO	BARO Sensor " Hg	0-30" Hg	29.4	29.4	29.4
BLMTSW	Blower Motor SW	ON / OFF	ON (if on)	OFF	OFF
DEF SW	Defrost Switch	ON / OFF	ON (if on)	OFF	OFF
DRL SW	Daytime Run Lamp	ON / OFF	ON (if on)	OFF	OFF
ECT (°F)	ECT Sensor	-40-304°F	185	180	192
ECT (V)	ECT Sensor	0-5.1v	0.7	0.7	0.6
HO2S-11	HO2S-11 (Bank 1)	0.1-1.1v	0.1-1.1	0.1-1.1	0.1-1.1
HO2S-12	HO2S-12 (Bank 1)	0.1-1.1v	0.1-1.1	0.1-1.1	0.1-1.1
HTR1	HO2S-11 Heater	ON / OFF	ON	ON	ON
IAT (°F)	IAT Sensor	-40-304°F	69	69	71
IAT (V)	IAT Sensor	0-5.1v	2.2	2.2	2.2
IDL SW	Throttle Idle Switch	ON / OFF	ON	OFF	OFF
IGN SW	Ignition Switch	ON-KOEC	OFF	OFF	OFF
INJ1 & 2	Fuel Injector B1, B2	0-99.9 ms	2.3	3.4	4.2
ISC	Idle Speed Control	0-99.9 ms	2.8	3.2	3.3
LOAD	Engine Load	0-100%	18	21	38
LongFT1	Long Term F/T B1	-20 to 20%	-5 to 5%	-5 to 5%	-5 to 5%
LUP	Lockup Solenoid	ON / OFF	OFF	OFF	OFF
MAF LB	MAF Sensor LB/M	0-25 LB/M	0.8	2.35	3.0
MAF (V)	MAF Sensor	0-5.1v	1.3	1.9	1.9
MIL	MIL Control	ON / OFF	ON (if on)	OFF	OFF
PRC	Pressure Regulator	ON / OFF	ON (if on)	OFF	OFF
PSP	PSP Switch	ON / OFF	ON turning	OFF	OFF
PUMP	Fuel Pump	ON / OFF	ON	ON	ON
Purge	EVAP Purge	0-100%	0	0-50	34-67
RPM	Engine Speed	0-10K rpm	765	1690	1850
RPMDES	Desired RPM	0-10K rpm	750	750	750
SOL1-2-3	Shift Solenoid 1-2-3	ON / OFF	OFF	ON	ON
SHTFT1	Short Term F/T B1	-20 to 20%	-5 to +5	-5 to +5	-5 to +5
SPARK	Spark Advance	-90° to 90°	10	11	20
STP SW	Stop Lamp Switch	ON / OFF	ON (if on)	OFF	OFF
TFT	TFT Sensor Volts	0-5.1v	4.0	3.9	3.1
TPV	TP Sensor	0-5.1v	0.5	0.7	0.8
NS SW	Neutral Start Switch	ON / OFF	ON (in 'P')	OFF	OFF
TSS	TSS Sensor	0-10 KHz	712	1575	1712
VICS	VICS Solenoid	ON / OFF	ON	ON	ON
VSS	Vehicle Speed	0-159 mph	0	30	55

1991-95 ESCORT/TRACER 1.9L I4 MFI VIN J (ALL) INPUTS / OUTPUTS

PCM PID Acronym	Parameter Identification	PID Range	PID Value at Hot Idle	PID Value at 30	PID Value at 55
BOO	Brake Switch (1993-95)	ON / OFF	ON (if on)	OFF	OFF
CANP	EVAP Purge Valve (1992-95)	0-100%	0	0-40	40-50
ECT (°F)	ECT Sensor	-40-304°F	160-200	160-200	160-200
ECT (V)	ECT Sensor	0-5.1v	0.6	0.6	0.6
EVR	VR Solenoid	0-100%	0%	0-40	39-50%
HFC	High Fan Control	ON / OFF	ON (if on)	ON	ON
INJPW1	INJ Pulsewidth B1	0-99.9 ms	4.8-5.0	7.8-8.4	10-11
IAC	Idle Air Control	0-100%	20-40	34-40	45-55
IAT (°F)	IAT Sensor	-40-304°F	50-120	50-120	50-120
IAT (V)	IAT Sensor	0-5.1v	1.7-3.5	1.7-3.5	1.7-3.5
LFC	Low Fan Control	ON / OFF	ON (if on)	OFF	OFF
LOOP	Loop Status	CL or OL	CL	CL	CL
MAF (V)	MAF Sensor	0-5.1v	0.7	1.3-1.5	1.8-2.2
HO2S-11	HO2S-11 (Bank 1)	0.1-1.1v	0.1-1.1	0.1-1.1	0.1-1.1
LongFT1	Long Term F/T B1	-20 to 20%	-5 to +5	-5 to +5	-5 to +5
PFE (V)	PFE Sensor	0-5.1v	3.2	3.3	3.4
PNP	PNP Switch	DRIVE or NEUTRAL	NEUTRAL	DRIVE	DRIVE
RPM	Engine Speed	0-10K rpm	750-820	1450-1630	1750-2100
SHTFT1	Short Term F/T B1	-20 to +20	-5 to +5	-5 to +5	-5 to +5
SPARK	Spark Advance	-90° to 90°	15-22	28-35	25-35
TP	TP Sensor	0-5.1v	0.7	0.8-1.0	1.1-1.3
TP Mode	TP Mode	C/T or P/T	C/T	P/T	P/T
VPWR	Vehicle Power	0-25.5v	12-14	12-14	12-14
VREF	Voltage Reference	0-5.1v	4.9-5.1	4.9-5.1	4.9-5.1
VSS	Vehicle Speed	0-159 mph	0	30	55
WAC	A/C WOT Cutout Relay	ON / OFF	ON (if on)	OFF	OFF

1996 ESCORT/TRACER 1.9L I4 MFI VIN J (A/T) - INPUTS

PCM PID Acronym	Parameter Identification	PID Range	PID Value at Hot Idle	PID Value at 30	PID Value at 55
1ST	First Position	OFF / 1ST	1ST	1ST	1ST
2ND	Second Position	OFF / 2ND	2ND	2ND	2ND
ACCS	A/C Clutch Switch	ON / OFF	ON (if on)	OFF	OFF
ACPSW	AC Pressure Switch	0-5.1v	1.5-1.9v	0.6-0.8	0.6-0.8
BPP	Brake Pedal Switch	ON / OFF	ON (if on)	OFF	OFF
CID	CID Sensor	0-1 KHz	5-7	11-15	17-21
CKP	CKP Sensor	0-10 KHz	435-475	770-900	1200-1400
DRV	Drive Position	OFF / DR	OFF	DR	DR
DPFE	DPFE Sensor	0-5.1v	0.4-0.6	0.6-1.0v	1.1-3.8v
ECT (°F)	ECT Sensor	-40-304°F	160-200	160-200	160-200
FEPS	Flash EEPROM	0.1	0.1	0.1	0.1
FRP PSI	Fuel Pressure	29-32 psi	30-32	30-32	30-35
FPM	Fuel Pump Monitor	ON / OFF	ON	ON	ON
GEAR	Transmission Gear	P-R-N-D	P-R-N-D	DRIVE 3	DRIVE 4
HO2S-11	HO2S-11 (Bank 1)	0.1-1.1v	0.1-1.1	0.1-1.1	0.1-1.1
HO2S-12	HO2S-12 (Bank 1)	0.1-1.1v	0.1-1.1	0.1-1.1	0.1-1.1
IAT (°F)	IAT Sensor	-40-304°F	50-120	50-120	50-120
LOAD	Engine Load	0-100%	10-20	20-31%	36-52%
MAF (V)	MAF Sensor	0-5.1v	0.6-0.9	1.0-1.6v	1.3-1.9v
MISF	Misfire Detection	ON / OFF	OFF	OFF	OFF
OCTADJ	Octane Adjustment	CL/OPEN	CLOSED	CLOSED	CLOSED
PF (V)	Purge Flow Sensor	0-5.1v	1.55	2.10	2.30
PNP	PNP Switch	ON / OFF	ON (in 'P')	OFF	OFF
RPM	Engine Speed	0-10K rpm	760-820	1450-1630	1750-2100
REV	Reverse Position	OFF / REV	REV	OFF	OFF
TFT (°F)	TFT Sensor	-14-304°F	110-210	110-210	110-210
TPV	TP Sensor	0-5.1v	0.5-0.9	1.0-1.2	1.0-1.3
TSS	TSS Sensor	0-10 KHz	340-380	620-680	1090-1150
Vacuum	Engine Vacuum	0-30" Hg	19-20	18-20	8-15
VPWR	Vehicle Power	0-25.5v	12-14	12-14	12-14
VSS	Vehicle Speed (MPH & Hertz Rate)	0-159 mph	0 = 0 Hz	30 = 65 Hz	55 = 125 Hz

1996 ESCORT/TRACER 1.9L I4 MFI VIN J (A/T) - OUTPUTS

PCM PID Acronym	Parameter Identification	PID Range	PID Value at Hot Idle	PID Value at 30	PID Value at 55
CD1	Coil 1 Driver	0-90° dwell	7	10	12
CD2	Coil 2 Driver	0-90° dwell	7	10	12
CTO	Clean Tach Output	0-10 KHz	25-38	40-48	55-85 Hz
EGRVR	VR Solenoid	0-100%	0	0-40	0-60
EVAPDC	EVAP Purge Valve	0-100%	0-100	0-100	0-100
FP	Fuel Pump Relay	ON / OFF	ON	ON	ON
FuelPW1	INJ Pulsewidth B1	0-99.9 ms	4.0-4.5	4.1-8.0	4.4-10
HFC	High Fan Control	ON / OFF	ON (if on)	OFF	OFF
HTR-11	HO2S-11 Heater	ON / OFF	ON	ON	ON
HTR-12	HO2S-12 Heater	ON / OFF	ON	ON	ON
IAC	Idle Air Control	0-100%	33	40-52	66
INJ1	INJ 1 Pulsewidth	0-99.9 ms	4.0-4.5	4.1-8.0	4.4-10
INJ2	INJ 2 Pulsewidth	0-99.9 ms	4.0-4.5	4.1-8.0	4.4-10
INJ3	INJ 3 Pulsewidth	0-99.9 ms	4.0-4.5	4.1-8.0	4.4-10
INJ4	INJ 4 Pulsewidth	0-99.9 ms	4.0-4.5	4.1-8.0	4.4-10
LFC	Low Fan Control	ON / OFF	ON (if on)	OFF	OFF
LongFT1	Long Term F/T B1	-20 to 20%	-5 to +5	-5 to +5	-5 to +5
MIL	MIL Control	ON / OFF	ON (if on)	OFF	OFF
SHTFT1	Short Term F/T B1	-20 to 20%	-5 to +5	-5 to +5	-5 to +5
SPARK	Spark Advance	-90° to 90°	15-22	28-35	25-35
SS1	Shift Solenoid 1	ON / OFF	OFF	ON	ON
SS2	Shift Solenoid 2	ON / OFF	ON	OFF	OFF
SS3	Shift Solenoid 3	ON / OFF	ON	ON	ON
TCC	TCC Solenoid	0-100%	0	55-100	95-100
VREF	Voltage Reference	0-5.1v	4.9-5.1	4.9-5.1	4.9-5.1
WAC	A/C WOT Cutout Relay	ON / OFF	ON (if on)	OFF	OFF

1996 ESCORT/TRACER 1.9L I4 MFI VIN J (M/T) - INPUTS / OUTPUTS

PCM PID Acronym	Parameter Identification	PID Range	PID Value at Hot Idle	PID Value at 30	PID Value at 55
ACCS	A/C Clutch Switch	ON / OFF	ON (if on)	OFF	OFF
ACPSW	AC Pressure Switch	0-5.1v	1.5-1.9v	0.6-0.8	0.6-0.8
BPP	Brake Switch	ON / OFF	ON (if on)	OFF	OFF
CID	CID Sensor	0-1 KHz	5-7	11-15	17-21
CKP	CKP Sensor	0-10 KHz	435-475	770-900	1200-1400
CPP	CPP Switch	ON / OFF	ON (if in)	OFF	OFF
DPFE	DPFE Sensor	0-5.1v	0.4-0.6	0.6-1.0v	1.1-3.8v
ECT (°F)	ECT Sensor	-40-304°F	160-200	160-200	160-200
FEPS	Flash EEPROM	0-5.1v	0.1	0.1	0.1
FPM	Fuel Pump Monitor	ON / OFF	ON	ON	ON
FRP	Fuel Rail Pressure	0-100 psi	30-32	30-32	30-35
HO2S-11	HO2S-11 (Bank 1)	0.1-1.1v	0.1-1.1	0.1-1.1	0.1-1.1
HO2S-12	HO2S-12 (Bank 1)	0.1-1.1v	0.1-1.1	0.1-1.1	0.1-1.1
IAT (°F)	IAT Sensor	-40-304°F	50-120	50-120	50-120
LOAD	Engine Load	0-100%	10-20	20-31%	36-52%
MAF (V)	MAF Sensor	0-5.1v	0.6-0.9	1.0-1.6v	1.3-1.9v
MISF	Misfire Detection	OFF	OFF	OFF	OFF
OCTADJ	Octane Adjustment	CL or OP	CLOSED	CLOSED	CLOSED
PF (V)	Purge Flow Sensor	0-5.1v	1.55	2.10	2.30
RPM	Engine Speed	0-10K rpm	760-820	1450-1630	1750-2100
TPV	TP Sensor	0-5.1v	0.5-0.9	1.0-1.2	1.0-1.3
Vacuum	Engine Vacuum	0-30" Hg	19-20	18-20	8-15
VPWR	Vehicle Power	0-25.5v	12-14	12-14	12-14

1996 ESCORT/TRACER 1.9L I4 MFI VIN J (M/T) - OUTPUTS

PCM PID Acronym	Parameter Identification	PID Range	PID Value at Hot Idle	PID Value at 30	PID Value at 55
CD1-CD2	Coil Driver 1-2	0-90° dwell	7	10	12
CTO	Clean Tach Output	0-10 KHz	25-38	40-48	55-85 Hz
EGRVR	VR Solenoid	0-100%	0%	0-40	0-60
EVAPDC	EVAP Purge Valve	0-100%	0-100%	0-100%	0-100%
FP	Fuel Pump Relay	ON / OFF	ON	ON	ON
FUELPW	INJ 1 Pulsewidth	0-99.9 ms	4.0-4.5	4.1-8.0	4.4-10
HFC	High Fan Control	ON / OFF	ON (if on)	OFF	OFF
HTR-11	HO2S-11 Heater	ON / OFF	ON	ON	ON
HTR-12	HO2S-12 Heater	ON / OFF	ON	ON	ON
IAC	Idle Air Control	0-100%	30%	40-52%	60-66%
INJ1-4	INJ 1-4 On-time	0-99.9 ms	4.0-4.5	4.1-8.0	4.4-10
LFC	Low Fan Control	ON / OFF	ON (if on)	OFF	OFF
LongFT1	LONGFT (Bank 1)	-20 to 20%	-5 to +5	-5 to +5	-5 to +5
MIL	MIL Control	ON / OFF	ON (if on)	OFF	OFF
SIL	Shift Indicator Lamp	ON / OFF	OFF	ON (if on)	OFF
SHTFT1	SHRTFT (Bank 1)	-20 to 20%	-5 to +5	-5 to +5	-5 to +5
SPARK	Spark Advance	-90° to 90°	15-22	28-35	25-35
WAC	A/C WOT Relay	ON / OFF	ON (if on)	OFF	OFF

1997-2002 ESCORT/TRACER 2.0L 2V I4 MFI VIN P (A/T) - INPUTS

PCM PID Acronym	Parameter Identification	PID Range	PID Value at Hot Idle	PID Value at 30	PID Value at 55
ACCS	A/C Clutch Switch	ON / OFF	ON (if on)	OFF	OFF
ACP	AC Press. Sensor	0-5.1v	1.5-2.6	0.0-0.8	0.0-0.8
BPP	Brake Pedal Switch	ON / OFF	ON (if on)	OFF	OFF
CID	CID Sensor	0-1 KHz	5-7	11-15	17-21
CKP	CKP Sensor	0-10 KHz	400-425	770-900	1200-1400
DPFE	DPFE Sensor	0-5.1v	0.4-0.6	0.6-1.5	0.5-3.8
ECT (°F)	ECT Sensor	-14-304°F	160-200	160-200	160-200
EFTA (°F)	EFT 'A' Sensor	-14-304°F	50-120	50-120	50-120
FEPS	Flash EEPROM	0-5.1v	0.1	0.1	0.1
FLI	Fuel Level Indicator (1998-99)	0-100%	50 (1.7v)	25-75	25-75
FPM	Fuel Pump Monitor	ON / OFF	ON	ON	ON
FRP	Fuel Rail Pressure (1998-99)	0-100 psi	39	39	39
FTPV	Fuel Tank Pressure (1998-99)	0-5.1v or 0-10" H2O	2.6v / 0" H2O (with cap off)	2.6v / 0" H2O (with cap off)	2.6v / 0" H2O (with cap off)
GEAR	Gear Position	1-2-3-4	1	3	4
HO2S-11	HO2S-11 (Bank 1)	0.1-1.1v	0.1-1.1	0.1-1.1	0.1-1.1
HO2S-12	HO2S-12 (Bank 1)	0.1-1.1v	0.1-1.1	0.1-1.1	0.1-1.1
IAT (°F)	IAT Sensor	-14-304°F	50-120	50-120	50-120
IMRCM	IMRC Monitor	0-5.1v	5	5	5
KS1	Knock Sensor	0-5.1v	0	0	0
LOAD	Engine Load	0-100%	10-20	20-31	25-52
MAF (V)	MAF Sensor	0-5.1v	0.6-0.9	1.0-1.6	1.3-2.3
MISF	Misfire Detection	OFF	OFF	OFF	OFF
OCTADJ	Octane Adjustment	CL/OPEN	CLOSED	CLOSED	CLOSED
PF (V)	Purge Flow Sensor	0-5.1v	1.55	2.10	2.30
PNP	PNP Switch	ON / OFF	ON (in 'P')	OFF	OFF
PSP (V)	PSP Sensor	0-5.1v	1.5-1.8	0.5-0.8	0.5-0.8
RPM	Engine Speed	0-10K rpm	730-790	1450-1630	1750-2100
TFT (°F)	TFT Sensor	-14-304°F	110-210	110-210	110-210
TPV	TP Sensor	0-5.1v	0.5-0.9	1.0-1.3	1.1-1.9
TRD/TR	TR Sensor	N / OD	N	OD	OD
TRL/TR	TR Sensor Low	MAN1 /OD	MAN1	OD	OD
TROD/TR	TR Overdrive	N / OD	OD	OD	OD
TRR/TR	TR Reverse	R / OD	R	OD	OD
TSS	TSS Sensor	0-10 KHz	340-380	620-680	1090-1250
Vacuum	Engine Vacuum	0-30" Hg	19-20	18-20	8-15
VPWR	Vehicle Power	0-25.5v	12-14	12-14	12-14
VSS	Vehicle Speed (MPH & Hertz Rate)	0-159 mph	0 = 0 Hz	30 = 65 Hz	55 = 125 Hz

<u>1997-2002 ESCORT/TRACER 2.0L 2V I4 MFI VIN P (A/T) - OUTPUTS</u>

PCM PID Acronym	Parameter Identification	PID Range	PID Value at Hot Idle	PID Value at 30	PID Value at 55
CD1-CD2	Coil Driver 1-2	0-90° dwell	7	10	11
CTO	Clean Tach Output	0-10 KHz	25-38	40-48	72-85
EGRVR	VR Solenoid	0-100%	0	0-40	0-60
EVAPCV	EVAP CV Valve	0-100%	0-100	0-100	0-100
EVAPDC	EVAP Purge Valve	0-100%	0-100	0-100	0-100
FP	Fuel Pump Relay	0-100%	100	100	100
FuelPW1	INJ Pulsewidth B1	0-99.9 ms	3.3-3.7	4.1-8.0	4.4-10
HFC	High Fan Relay	ON / OFF	ON (if on)	OFF	OFF
HTR-11	HO2S-11 Heater	ON / OFF	ON	ON	ON
HTR-12	HO2S-12 Heater	ON / OFF	ON	ON	ON
IAC	Idle Air Control	0-100%	20-40	34-40	45-55
IMRC	IMRC Solenoid	ON / OFF	OFF	OFF	OFF
INJ1	INJ 1 Pulsewidth	0-99.9 ms	3.3-3.7	4.1-8.0	4.4-10
INJ2	INJ 2 Pulsewidth	0-99.9 ms	3.3-3.7	4.1-8.0	4.4-10
INJ3	INJ 3 Pulsewidth	0-99.9 ms	3.3-3.7	4.1-8.0	4.4-10
INJ4	INJ 4 Pulsewidth	0-99.9 ms	3.3-3.7	4.1-8.0	4.4-10
LFC	Low Fan Relay	ON / OFF	ON (if on)	OFF	OFF
LongFT1	Long Term F/T B1	-20 to 20%	-5 to +5%	-5 to +5%	-5 to +5%
MIL	MIL Control	ON / OFF	ON (if on)	OFF	OFF
SS1	Shift Solenoid 1	ON / OFF	ON	OFF	OFF
SS2	Shift Solenoid 2	ON / OFF	OFF	ON	ON
SS3	Shift Solenoid 3	0-25.5v	7-9.5v	8.3-9.5	8.3-9.5
SHTFT1	Short Term F/T B1	-20 to 20%	-5 to +5	-5 to +5	-5 to +5
SPARK	Spark Advance	-90° to 90°	15-22	28-35	25-35
TCC	TCC Solenoid	0-100%	0	55-100	95-100
VREF	Voltage Reference	0-5.1v	4.9-5.1	4.9-5.1	4.9-5.1
WAC	A/C WOT Cutout Relay	ON / OFF	ON (if on)	OFF	OFF

1997-2002 ESCORT/TRACER 2.0L 2V I4 MFI VIN P (M/T) - INPUTS

PCM PID Acronym	Parameter Identification	PID Range	PID Value at Hot Idle	PID Value at 30	PID Value at 55
ACCS	A/C Clutch Switch	ON / OFF	ON (if on)	OFF	OFF
ACP	A/C Press. Switch	0-5.1v	1.5-2.6	0.0-0.8	0.0-0.8
CID	CID Sensor	0-1 KHz	5-7	11-15	17-21
CKP	CKP Sensor	0-10 KHz	400-425	770-900	1200-1400
CPP	Clutch Pedal Switch	ON / OFF	ON (if in)	OFF	OFF
DPFE	DPFE Sensor	0-5.1v	0.4-0.6	0.6-1.5	0.5-3.8
ECT (°F)	ECT Sensor	-40-304°F	160-200	160-200	160-200
FEPS	Flash EEPROM	0-5.1v	0.1	0.1	0.1
FLI	Fuel Level Indicator (1998-99)	0-100%	50 (1.7v)	25-75	25-75
FPM	Fuel Pump Monitor	ON / OFF	ON	ON	ON
FRP (V)	Fuel Rail Pressure (1998-99)	0-5.1v or 0-100 psi	2.8v / 39	2.8v / 39	2.8v / 39
FTPV	Fuel Tank Pressure (1998-99)	0-5.1v or 0-10" H2O	2.6v / 0" H2O (with cap off)	2.6v / 0" H2O (with cap off)	2.6v / 0" H2O (with cap off)
HO2S-11	HO2S-11 (Bank 1)	0.1-1.1v	0.1-1.1	0.1-1.1	0.1-1.1
HO2S-12	HO2S-12 (Bank 1)	0.1-1.1v	0.1-1.1	0.1-1.1	0.1-1.1
IAT (°F)	IAT Sensor	-40-304°F	50-120	50-120	50-120
LOAD	Engine Load	0-100%	10-20	20-31	25-52
IMRCM	IMRC Monitor	0-5.1v	5	5	5
MAF (V)	MAF Sensor	0-5.1v	0.6-0.9	1.0-1.6	1.3-2.3
MISF	Misfire Detection	OFF	OFF	OFF	OFF
OCT ADJ	Octane Adjustment	CL/OPEN	CLOSED	CLOSED	CLOSED
PF (V)	Purge Flow Sensor (1997 only)	0-5.1v	1.55	2.10	2.30
PSPV	PSP Switch	0-5.1v	1.5-1.8	0.5-0.8	0.5-0.8
RPM	Engine Speed	0-10K rpm	750-820	1450-1630	1750-2100
TPV	TP Sensor	0-5.1v	0.5-0.9	1.0-1.3	1.1-1.9
Vacuum	Engine Vacuum	0-30" Hg	19-20	18-20	8-15
VPWR	Vehicle Power	0-25.5v	12-14	12-14	12-14
VSS	Vehicle Speed (MPH & Hertz Rate)	0-159 mph	0 = 0 Hz	30 = 65 Hz	55 = 125 Hz

1997-2002 ESCORT/TRACER 2.0L 2V I4 MFI VIN P (M/T) - OUTPUTS

PCM PID Acronym	Parameter Identification	PID Range	PID Value at Hot Idle	PID Value at 30	PID Value at 55
CD1	Coil 1 Driver	0-90° dwell	7	10	11
CD2	Coil 2 Driver	0-90° dwell	7	10	11
CTO	Clean Tach Output	0-10 KHz	25-38	40-48	72-85
EGRVR	VR Solenoid	0-100%	0	0-40	0-60
EVAPCV	EVAP CV Valve	0-100%	0-100	0-100	0-100
EVAPDC	EVAP Purge Valve	0-100%	0-100	0-100	0-100
FP	Fuel Pump Relay	0-100%	100	100	100
FuelPW1	INJ Pulsewidth B1	0-99.9 ms	3.3-3.7	4.1-8.0	4.4-10
HFC	High Fan Relay	ON / OFF	ON (if on)	OFF	OFF
HTR-11	HO2S-11 Heater	ON / OFF	ON	ON	ON
HTR-12	HO2S-12 Heater	ON / OFF	ON	ON	ON
IAC	Idle Air Control	0-100%	20-40	34-40	45-55
INJ1	INJ 1 Pulsewidth	0-99.9 ms	3.3-3.7	4.1-8.0	4.4-10
INJ2	INJ 2 Pulsewidth	0-99.9 ms	3.3-3.7	4.1-8.0	4.4-10
INJ3	INJ 3 Pulsewidth	0-99.9 ms	3.3-3.7	4.1-8.0	4.4-10
INJ4	INJ 4 Pulsewidth	0-99.9 ms	3.3-3.7	4.1-8.0	4.4-10
IMRC	IMRC Solenoid	ON / OFF	OFF	OFF	OFF
LFC	Low Fan Relay	ON / OFF	ON (if on)	OFF	OFF
LongFT1	Long Term F/T B1	-20 to 20%	-5 to +5%	-5 to +5%	-5 to +5%
MIL	MIL Control	ON / OFF	ON (if on)	OFF	OFF
SHTFT1	Short Term F/T B1	-20 to 20%	-5 to +5	-5 to +5	-5 to +5
SPARK	Spark Advance	-90° to 90°	15-22	28-35	25-35
VREF	Voltage Reference	0-5.1v	4.9-5.1	4.9-5.1	4.9-5.1
WAC	A/C WOT Cutout Relay	ON / OFF	ON (if on)	OFF	OFF

1998-2003 ESCORT/TRACER/ZX2 2.0L 4V I4 MFI VIN 3 (A/T) - INPUTS

PCM PID Acronym	Parameter Identification	PID Range	PID Value at Hot Idle	PID Value at 30	PID Value at 55
ACCS	A/C Clutch Switch	ON / OFF	ON (if on)	OFF	OFF
ACP	A/C Press. Sensor	CLOSED or OPEN	OPEN	OPEN	OPEN
BPP	Brake Pedal Switch	ON / OFF	ON (if on)	OFF	OFF
CID	CID Sensor	0-1 KHz	5-7	11-15	17-21
CKP	CKP Sensor	0-10 KHz	400-425	770-900	1200-1400
DPFE	DPFE Sensor	0-5.1v	0.4-0.6	0.6-1.5	0.5-3.8
ECT (ºF)	ECT Sensor	-40-304ºF	160-200	160-200	160-200
EFTA (ºF)	EFT 'A' Sensor	-40-304ºF	50-120	50-120	50-120
FEPS	Flash EEPROM	0-5.1v	0.5-0.6	0.5-0.6	0.5-0.6
FLI	Fuel Level Indicator (1998-99)	0-100%	50 (1.7v)	50 (1.7v)	50 (1.7v)
FPM	Fuel Pump Monitor	ON / OFF	ON	ON	ON
FRP (V)	Fuel Rail Pressure (1998-99)	0-100 psi	39	39	39
FTPV	Fuel Tank Pressure (1998-99)	0-5.1v or 0-10" H2O	2.6v / 0" H2O (with cap off)	2.6v / 0" H2O (with cap off)	2.6v / 0" H2O (with cap off)
GEAR	Gear Position	1-2-3-4	1	3	4
HO2S-11	HO2S-11 (Bank 1)	0.1-1.1v	0.1-1.1	0.1-1.1	0.1-1.1
HO2S-12	HO2S-12 (Bank 1)	0.1-1.1v	0.1-1.1	0.1-1.1	0.1-1.1
IAT (ºF)	IAT Sensor	-40-304ºF	50-120	50-120	50-120
KS1	Knock Sensor	0-5.1v	0v	0v	0v
LOAD	Engine Load	0-100%	10-20	20-31	25-52
MAF (V)	MAF Sensor	0-5.1v	0.6-0.9	1.0-1.6	1.3-2.3
MISF	Misfire Detection	ON / OFF	OFF	OFF	OFF
OCTADJ	Octane Adjustment	CL/OPEN	CLOSED	CLOSED	CLOSED
PF (V)	Purge Flow Sensor (1997 only)	0-5.1v	1.55	2.10	2.30
PNP	PNP Switch	ON / OFF	ON	OFF	OFF
PSP (V)	PSP Sensor	0-5.1v	1.5-1.8	0.5-0.8	0.5-0.8
RPM	Engine Speed	0-10K rpm	750-820	1450-1630	1750-2100
TFT (ºF)	TFT Sensor	-40-304ºF	110-210	110-210	110-210
TPV	TP Sensor	0-5.1v	0.5-0.9	1.0-1.3	1.1-1.9
TRD/TR	TR Sensor	N / OD	N	OD	OD
TRL/TR	TR Sensor Low	MAN1 /OD	MAN1	OD	OD
TROD/TR	TR Overdrive	N / OD	OD	OD	OD
TRR/TR	TR Reverse	R / OD	R	OD	OD
TSS	TSS Sensor	0-10 KHz	340-380	620-680	1090-1250
VPWR	Vehicle Power	0-25.5V	12-14	12-14	12-14
VSS	Vehicle Speed (MPH & Hertz Rate)	0-159 mph	0 = 0 Hz	30 = 65 Hz	55 = 125 Hz

1998-2003 ESCORT/TRACER/ZX2 2.0L 4V I4 MFI VIN 3 (A/T) - OUTPUTS

PCM PID Acronym	Parameter Identification	PID Range	PID Value at Hot Idle	PID Value at 30	PID Value at 55
CD1-CD2	Coil Driver 1-2	0-90° dwell	7	10	11
CTO	Clean Tach Output	0-10 KHz	25-38	40-48	72-85
EGRVR	VR Solenoid	0-100%	0	0-40	0-60
EVAPCV	EVAP CV Valve	0-100%	0-100	0-100	0-100
EVAPDC	EVAP Purge Valve	0-100%	0-100	0-100	0-100
FP	Fuel Pump Relay	0-100%	100	100	100
FuelPW1	INJ Pulsewidth B1	0-99.9 ms	3.3-3.7	4.1-8.0	4.4-10
HFC	High Fan Relay	ON / OFF	ON (if on)	OFF	OFF
HTR-11	HO2S-11 Heater	ON / OFF	ON	ON	ON
HTR-12	HO2S-12 Heater	ON / OFF	ON	ON	ON
IAC	Idle Air Control	0-100%	20-40	34-40	45-55
INJ1	INJ 1 Pulsewidth	0-99.9 ms	3.3-3.7	4.1-8.0	4.4-10
INJ2	INJ 2 Pulsewidth	0-99.9 ms	3.3-3.7	4.1-8.0	4.4-10
INJ3	INJ 3 Pulsewidth	0-99.9 ms	3.3-3.7	4.1-8.0	4.4-10
INJ4	INJ 4 Pulsewidth	0-99.9 ms	3.3-3.7	4.1-8.0	4.4-10
LFC	Low Fan Relay	ON / OFF	ON (if on)	OFF	OFF
LongFT1	Long Term F/T B1	-20 to 20%	-5 to +5%	-5 to +5%	-5 to +5%
MIL	MIL Control	ON / OFF	OFF	OFF	OFF
SHTFT1	Short Term F/T B1	-20 to 20%	-5 to +5	-5 to +5	-5 to +5
SPARK	Spark Advance	-90° to 90°	15-22	28-35	25-35
SS1	Shift Solenoid 1	ON / OFF	ON	OFF	OFF
SS2	Shift Solenoid 2	ON / OFF	OFF	ON	ON
SS3	Shift Solenoid 3	0-25.5v	7.0-9.5	8.3-9.5	8.3-9.5
TCC	TCC Solenoid	0-100%	0	55-100	95-100
VCT	Variable Cam Timing Solenoid	0-25.5v	12-14	10.5-14	10.5-14
VREF	Voltage Reference	0-5.1v	4.9-5.1	4.9-5.1	4.9-5.1
WAC	A/C WOT Cutout Relay	ON / OFF	ON (if on)	OFF	OFF

1998-2003 ESCORT/TRACER/ZX2 2.0L 4V I4 MFI VIN 3 (M/T) - INPUTS

PCM PID Acronym	Parameter Identification	PID Range	PID Value at Hot Idle	PID Value at 30	PID Value at 55
ACCS	A/C Clutch Switch	ON / OFF	ON (if on)	OFF	OFF
ACP	A/C Press. Switch	CL/OPEN	OPEN	OPEN	OPEN
CID	CID Sensor	0-1 KHz	5-7	11-15	17-21
CKP	CKP Sensor	0-10 KHz	400-425	770-900	1200-1400
CPP	Clutch Pedal Switch	ON / OFF	ON (if in)	OFF	OFF
DPFE	DPFE Sensor	0-5.1v	0.4-0.6	0.6-1.5	0.5-3.8
ECT (°F)	ECT Sensor	-40-304°F	160-200	160-200	160-200
FEPS	Flash EEPROM	0-5.1v	0.5-0.6	0.5-0.6	0.5-0.6
FLI	Fuel Level Indicator (1998-99)	0-100%	50 (1.7v)	50 (1.7v)	50 (1.7v)
FPM	Fuel Pump Monitor	ON / OFF	ON	ON	ON
FRP (V)	Fuel Rail Pressure (1998-99)	0-100 psi	39	39	39
FTPV	Fuel Tank Pressure (1998-99)	0-5.1v or 0-10" H2O	2.6v / 0" H2O (with cap off)	2.6v / 0" H2O (with cap off)	2.6v / 0" H2O (with cap off)
HO2S-11	HO2S-11 (Bank 1)	0.1-1.1v	0.1-1.1	0.1-1.1	0.1-1.1
HO2S-12	HO2S-12 (Bank 1)	0.1-1.1v	0.1-1.1	0.1-1.1	0.1-1.1
IAT (°F)	IAT Sensor	-40-304°F	50-120	50-120	50-120
LOAD	Engine Load	0-100%	10-20	20-31	25-52
MAF (V)	MAF Sensor	0-5.1v	0.6-0.9	1.0-1.6	1.3-2.3
MISF	Misfire Detection	ON / OFF	OFF	OFF	OFF
OCT ADJ	Octane Adjustment	CL/OPEN	CLOSED	CLOSED	CLOSED
PF (V)	Purge Flow Sensor (1997 only)	0-5.1v	1.55	2.10	2.30
PSP	PSP Switch	0-5.1v	1.5-1.8	0.5-0.8	0.5-0.8
RPM	Engine Speed	0-10K rpm	750-820	1450-1630	1750-2100
TPV	TP Sensor	0-5.1v	0.5-0.9	1.0-1.3	1.1-1.9
VPWR	Vehicle Power	0-25.5V	12-14	12-14	12-14
VSS	Vehicle Speed (MPH & Hertz Rate)	0-159 mph	0 = 0 Hz	30 = 65 Hz	55 = 125 Hz

1998-2003 ESCORT/TRACER/ZX2 2.0L 4V I4 MFI VIN 3 (M/T) - OUTPUTS

PCM PID Acronym	Parameter Identification	PID Range	PID Value at Hot Idle	PID Value at 30	PID Value at 55
CD1	Coil 1 Driver	0-90º dwell	7	10	11
CD2	Coil 2 Driver	0-90º dwell	7	10	11
CTO	Clean Tach Output	0-10 KHz	25-38	40-48	72-85
EGRVR	VR Solenoid	0-100%	0	0-40	0-60
EVAPCV	EVAP CV Valve	0-100%	0	0	0
EVAPDC	EVAP Purge Valve	0-100%	0-100	0-100	0-100
FP	Fuel Pump Relay	0-100%	100	100	100
FuelPW1	INJ Pulsewidth B1	0-99.9 ms	3.3-3.7	4.1-8.0	4.4-10
HFC	High Fan Relay	ON / OFF	ON (if on)	OFF	OFF
HTR-11	HO2S-11 Heater	ON / OFF	ON	ON	ON
HTR-12	HO2S-12 Heater	ON / OFF	ON	ON	ON
IAC	Idle Air Control	0-100%	20-40	34-40	45-55
INJ1	INJ 1 Pulsewidth	0-99.9 ms	3.3-3.7	4.1-8.0	4.4-10
INJ2	INJ 2 Pulsewidth	0-99.9 ms	3.3-3.7	4.1-8.0	4.4-10
INJ3	INJ 3 Pulsewidth	0-99.9 ms	3.3-3.7	4.1-8.0	4.4-10
INJ4	INJ 4 Pulsewidth	0-99.9 ms	3.3-3.7	4.1-8.0	4.4-10
LFC	Low Fan Relay	ON / OFF	ON (if on)	OFF	OFF
LongFT1	Long Term F/T B1	-20 to 20%	-5 to +5	-5 to +5	-5 to +5
MIL	MIL Control	ON / OFF	ON (if on)	OFF	OFF
SHTFT1	Short Term F/T B1	-20 to 20%	-5 to +5	-5 to +5	-5 to +5
SPARK	Spark Advance	-90º to 90º	15-22	28-35	25-35
VCT	Variable Cam Timing Solenoid	0-25.5v	12-14	10.5-14	10.5-14
VREF	Voltage Reference	0-5.1v	4.9-5.1	4.9-5.1	4.9-5.1
WAC	A/C WOT Cutout Relay	ON / OFF	ON (if on)	OFF	OFF

Mustang PID Data

1991-93 MUSTANG 2.3L I4 MFI VIN M (ALL) INPUTS / OUTPUTS

PCM PID Acronym	Parameter Identification	PID Range	PID Value at Hot Idle	PID Value at 30	PID Value at 55
BOO	Brake Switch	ON / OFF	ON (if on)	OFF	OFF
BARO	Barometric Pressure Sensor	0-1 KHz	159 Hz at Sea Level	159 Hz at Sea Level	159 Hz at Sea Level
CANP	EVAP Purge Valve	0-100%	0%	0-40	40-50
ECT (ºF)	ECT Sensor	-40-304ºF	160-200	160-200	160-200
ECT (V)	ECT Sensor	0-5.1v	0.6	0.6	0.6
EVP (V)	EGR EVP Sensor	0-5.1v	0.3	0.4-1.0	0.9-2.0
EVR	VR Solenoid	0-100%	0%	0-40	30-50
HO2S-11	HO2S-11 (Bank 1)	0.1-1.1v	0.1-1.1	0.1-1.1	0.1-1.1
IAC	Idle Air Control	0-100%	20-40	34-40	45-55
IAT (ºF)	IAT Sensor	-40-304ºF	50-120	50-120	50-120
IAT (V)	IAT Sensor	0-5.1v	1.7-3.5	1.7-3.5	1.7-3.5
INJPW1	INJ Pulsewidth B1	0-99.9 ms	3.0-3.8 ms	4.5-5.5 ms	6.0-6.8 mv
LFAN	Low Speed Fan	ON / OFF	ON (if on)	OFF	OFF
LOOP	Loop Status	CL or OL	CL	CL	CL
MAF (V)	Mass Airflow	0-5.1v	0.5	1.3-1.4v	1.5-2.0v
PNP	PNP Switch	Neutral/DR	NEUTRAL	DRIVE	DRIVE
RPM	Engine Speed	0-10K rpm	750-820	1450-1630	1750-2100
SHTFT1	Short Term F/T B1	-20 to 20%	-5 to +5	-5 to +5	-5 to +5
SPARK	Spark Advance	-90º to 90º	15-22	28-35	25-35
TPV	TP Sensor	0-5.1v	0.7	0.8-1.0	1.1-1.3
TP Mode	TP Mode	C/T or P/T	C/T	P/T	P/T
VPWR	Vehicle Power	0-25.5v	12-14	12-14	12-14
VREF	Voltage Reference	0-5.1v	4.9-5.1	4.9-5.1	4.9-5.1
VSS	Vehicle Speed	0-159 mph	0	30	55
WAC	A/C WOT Cutout Relay	ON / OFF	ON (if on)	OFF	OFF

1994-2003 MUSTANG 3.8L 2V V6 MFI VIN 4 (A/T) - INPUTS

PCM PID Acronym	Parameter Identification	PID Range	PID Value at Hot Idle	PID Value at 30	PID Value at 55
ACCS	A/C Clutch Switch	ON / OFF	ON (if on)	OFF	OFF
AIR-M	AIR System Monitor	ON / OFF	OFF	OFF	OFF
BPP	Brake Pedal Switch	ON / OFF	ON (if on)	OFF	OFF
CID	CID Sensor	0-1 KHz	5-7	9-11	10-15
CKP	CKP Sensor	0-10 KHz	390-450	650-700	875-1000
DPFE	DPFE Sensor	0-5.1v	0.25-1.30	0.25-4.65	0.25-4.65
ECT (°F)	ECT Sensor	-40-304°F	160-200	160-200	160-200
EFT (°F)	EFT Sensor (°F)	-40-304°F	50-120	50-120	50-120
FEPS	Flash EEPROM	0-5.1v	0.5-0.6	0.5-0.6	0.5-0.6
FLI	Fuel Level Indicator	0-100%	50 (1.7v)	25-75	25-75
FPM	Fuel Pump Monitor	0-100%	100	100	100
FRP	Fuel Rail Pressure (1998-2001)	0-100 psi	39	39	39
FTPV	Fuel Tank Pressure (1998-2001)	0-5.1v or 0-10" H2O	2.6v / 0" H2O (with cap off)	2.6v / 0" H2O (with cap off)	2.6v / 0" H2O (with cap off)
GEAR	Transmission Gear	1-2-3-4	1	3	4
HO2S-11	HO2S-11 (Bank 1)	0.1-1.1v	0.1-1.1	0.1-1.1	0.1-1.1
HO2S-12	HO2S-12 (Bank 1)	0.1-1.1v	0.1-1.1	0.1-1.1	0.1-1.1
HO2S-21	HO2S-21 (Bank 2)	0.1-1.1v	0.1-1.1	0.1-1.1	0.1-1.1
HO2S-22	HO2S-22 (Bank 2)	0.1-1.1v	0.1-1.1	0.1-1.1	0.1-1.1
IAT (°F)	IAT Sensor	-40-304°F	50-120	50-120	50-120
LOAD	Engine Load	0-100%	10-20	16-36	25-35
MAF (V)	MAF Sensor	0-5.1v	0.6-0.9	0.8-1.6	1.1-2.3
MISF	Misfire Detection	ON / OFF	OFF	OFF	OFF
OCTADJ	Octane Adjustment	CL/OPEN	CLOSED	CLOSED	CLOSED
OSS	Output Speed Shaft Sensor RPM	0-10 KHz rpm	0	1150-1300	2400
PF (V)	Purge Flow Sensor Volts (1996-97)	0-5.1v	1.55	2.10	2.30
PNP	PNP Switch	ON/OFF	ON in Park	OFF	OFF
RPM	Engine Speed	0-10K rpm	700-730	1000-1200	1500-1700
TCS	TCS (Switch)	ON / OFF	ON (if on)	OFF	OFF
TFT (°F)	TFT Sensor	-40-304°F	110-210	110-210	110-210
TPV	TP Sensor	0-5.1v	0.5-1.27	0.8-1.1	0.8-1.2
TR1	TR1 Sensor	0-25.5v	0	11.5	11.5
TRV/TR	TR Sensor	0-5.1v	0v (PN)	1.7v (OD)	1.7v (OD)
VPWR	Vehicle Power	0-25.5V	12-14	12-14	12-14
VSS	Vehicle Speed (MPH & Hertz Rate)	0-159 mph	0 = 0 Hz	30 = 65 Hz	55 = 125 Hz

1994-2003 MUSTANG 3.8L 2V V6 MFI VIN 4 (A/T) - OUTPUTS

PCM PID Acronym	Parameter Identification	PID Range	PID Value at Hot Idle	PID Value at 30	PID Value at 55
AIR-GS	AIR Solenoid	ON / OFF	OFF	OFF	OFF
CD1-CD3	Coil Driver 1-3	0-60° dwell	7	11	12
CTO	Clean Tach Output	0-10 KHz	33-37	57-63	68-74
EGRVR	VR Solenoid	0-100%	0	0-40	35-45
EPC	EPC Solenoid	0-500 psi	8	12-18	18-22
EVAPCV	EVAP CV Valve	0-100%	0-100	0-100	0-100
EVAPDC	EVAP Purge Valve	0-100%	0-100	0-100	0-100
FLI	Fuel Level Indicator (1998-99)	0-100%	50 (1.7v)	25-75	25-75
FP	Fuel Pump Control	0-100%	50	50	50
FuelPW1	INJ Pulsewidth B1	0-99.9 ms	4.9-5.1	5.3-10	6.5-12
FuelPW2	INJ Pulsewidth B2	0-99.9 ms	4.9-5.1	5.3-10	6.5-12
HTR-11	HO2S-11 Heater	ON / OFF	ON	ON	ON
HTR-12	HO2S-12 Heater	ON / OFF	ON	ON	ON
HTR-21	HO2S-21 Heater	ON / OFF	ON	ON	ON
HTR-22	HO2S-22 Heater	ON / OFF	ON	ON	ON
IAC	Idle Air Control	0-100%	34-39	44-73	50-75
INJ1	INJ 1 Pulsewidth	0-99.9 ms	4.9-5.1	5.3-10	6.5-12
INJ2	INJ 2 Pulsewidth	0-99.9 ms	4.9-5.1	5.3-10	6.5-12
INJ3	INJ 3 Pulsewidth	0-99.9 ms	4.9-5.1	5.3-10	6.5-12
INJ4	INJ 4 Pulsewidth	0-99.9 ms	4.9-5.1	5.3-10	6.5-12
INJ5	INJ 5 Pulsewidth	0-99.9 ms	4.9-5.1	5.3-10	6.5-12
INJ6	INJ 6 Pulsewidth	0-99.9 ms	4.9-5.1	5.3-10	6.5-12
LFC	Low Fan Relay	ON / OFF	ON (if on)	OFF	OFF
LongFT1	Long Term F/T B1	-20 to 20%	-5 to +5	-5 to +5	-5 to +5
LongFT2	Long Term F/T B2	-20 to 20%	-5 to +5	-5 to +5	-5 to +5
MIL	MIL Control	ON / OFF	ON (if on)	OFF	OFF
SHTFT1	Short Term F/T B1	-20 to 20%	-5 to +5	-5 to +5	-5 to +5
SHTFT2	Short Term F/T B2	-20 to 20%	-5 to +5	-5 to +5	-5 to +5
SPARK	Spark Advance	-90° to 90°	15-20	25-35	31-40
SS1	Shift Solenoid 1	ON / OFF	ON	OFF	ON
SS2	Shift Solenoid 2	ON / OFF	OFF	ON	ON
TCC	TCC Solenoid	0-100%	0	0-45	95-100
TCIL	TCIL (lamp)	ON / OFF	ON (if on)	OFF	OFF
VREF	Voltage Reference	0-5.1v	4.9-5.1	4.9-5.1	4.9-5.1
VSO	VSS Output Signal	0-1 KHz	0	55	125
WAC	A/C WOT Cutout Relay	ON / OFF	ON (if on)	OFF	OFF

1994-2003 MUSTANG 3.8L 2V V6 MFI VIN 4 (M/T) - INPUTS

PCM PID Acronym	Parameter Identification	PID Range	PID Value at Hot Idle	PID Value at 30	PID Value at 55
ACCS	A/C Clutch Switch	ON / OFF	ON (if on)	OFF	OFF
CID	CID Sensor	0-1 KHz	5-7	9-11	10-15
CKP	CKP Sensor	0-10 KHz	390-450	700-740	825-875
CPP/NDS	Clutch Pedal Position Switch	ON/OFF	ON when clutch depressed	OFF	OFF
DPFE	DPFE Sensor	0-5.1v	0.25-1.30	0.25-4.65	0.25-4.65
ECT (°F)	ECT Sensor	-40-304°F	160-200	160-200	160-200
FEPS	Flash EEPROM	0-5.1v	0.5-0.6	0.5-0.6	0.5-0.6
FLI	Fuel Level Indicator	0-100%	50 (1.7v)	25-75	25-75
FPM	Fuel Pump Monitor	0-100%	100	100	100
FRP	Fuel Rail Pressure (1998-2001)	0-100 psi	39	39	39
FTPV	Fuel Tank Pressure (1998-2001)	0-5.1v or 0-10" H2O	2.6v / 0" H2O (with cap off)	2.6v / 0" H2O (with cap off)	2.6v / 0" H2O (with cap off)
HO2S-11	HO2S-11 (Bank 1)	0.1-1.1v	0.1-1.1	0.1-1.1	0.1-1.1
HO2S-12	HO2S-12 (Bank 1)	0.1-1.1v	0.1-1.1	0.1-1.1	0.1-1.1
HO2S-21	HO2S-21 (Bank 2)	0.1-1.1v	0.1-1.1	0.1-1.1	0.1-1.1
HO2S-22	HO2S-22 (Bank 2)	0.1-1.1v	0.1-1.1	0.1-1.1	0.1-1.1
IAT (°F)	IAT Sensor	-40-304°F	50-120	50-120	50-120
MAF (V)	MAF Sensor	0-5.1v	0.6-0.9	0.8-1.6	1.1-2.3v
MISF	Misfire Detection	ON / OFF	OFF	OFF	OFF
OCTADJ	Octane Adjustment	CL/OPEN	CLOSED	CLOSED	CLOSED
LOAD	Engine Load	0-100%	10-20	16-36	25-35
PF (V)	Purge Flow Sensor Volts (1996-97)	0-5.1v	1.55	2.10	2.30
RPM	Engine Speed	0-10K rpm	700-780	1000-1200	1500-1700
TPV	TP Sensor	0-5.1v	0.53-1.27	0.8-1.1	0.8-1.2
VPWR	Vehicle Power	0-25.5v	12-14	12-14	12-14
VSS	Vehicle Speed (MPH & Hertz Rate)	0-159 mph	0 = 0 Hz	30 = 65 Hz	55 = 125 Hz

1994-2003 MUSTANG 3.8L 2V V6 MFI VIN 4 (M/T) - OUTPUTS

PCM PID Acronym	Parameter Identification	PID Range	PID Value at Hot Idle	PID Value at 30	PID Value at 55
AIR-GS	AIR Solenoid	ON / OFF	OFF	OFF	OFF
CD1-CD3	Coil Driver 1-3	0-60° dwell	7	11	12
CTO	Clean Tach Output	0-10 KHz	33-37	57-63	68-74
EGRVR	VR Solenoid	0-100%	0	0-40	35-45
EVAPCV	EVAP CV Valve	0-100%	0-100	0-100	0-100
EVAPDC	EVAP Purge Valve	0-100%	0-100	0-100	0-100
FP	Fuel Pump Control	0-100%	26	26	26
FuelPW1	INJ Pulsewidth B1	0-99.9 ms	4.9-5.1	5.3-10	6.5-12
FuelPW2	INJ Pulsewidth B2	0-99.9 ms	4.9-5.1	5.3-10	6.5-12
HTR-11	HO2S-11 Heater	ON / OFF	ON	ON	ON
HTR-12	HO2S-12 Heater	ON / OFF	ON	ON	ON
HTR-21	HO2S-21 Heater	ON / OFF	ON	ON	ON
HTR-22	HO2S-22 Heater	ON / OFF	ON	ON	ON
IAC	Idle Air Control	0-100%	34-39	44-73	50-75
INJ1	INJ 1 Pulsewidth	0-99.9 ms	4.9-5.1	5.3-10	6.5-12
INJ2	INJ 2 Pulsewidth	0-99.9 ms	4.9-5.1	5.3-10	6.5-12
INJ3	INJ 3 Pulsewidth	0-99.9 ms	4.9-5.1	5.3-10	6.5-12
INJ4	INJ 4 Pulsewidth	0-99.9 ms	4.9-5.1	5.3-10	6.5-12
INJ5	INJ 5 Pulsewidth	0-99.9 ms	4.9-5.1	5.3-10	6.5-12
INJ6	INJ 6 Pulsewidth	0-99.9 ms	4.9-5.1	5.3-10	6.5-12
LFC	Low Fan Control	ON / OFF	ON (if on)	OFF	OFF
LongFT1	Long Term F/T B1	-20 to 20%	-5 to +5	-5 to +5	-5 to +5
LongFT2	Long Term F/T B2	-20 to 20%	-5 to +5	-5 to +5	-5 to +5
MIL	MIL Control	ON / OFF	ON (if on)	OFF	OFF
SHTFT1	Short Term F/T B1	-20 to 20%	-5 to +5	-5 to +5	-5 to +5
SHTFT2	Short Term F/T B2	-20 to 20%	-5 to +5	-5 to +5	-5 to +5
SPARK	Spark Advance	-90° to 90°	15-20	25-35	31-40
VREF	Voltage Reference	0-5.1v	4.9-5.1	4.9-5.1	4.9-5.1
VSO	VSS Output Signal	0-1 KHz	0	65	125
WAC	A/C WOT Cutout Relay	ON / OFF	ON (if on)	OFF	OFF

1996-2003 MUSTANG 4.6L 4V V8 MFI VIN V, Y (ALL) - INPUTS

PCM PID Acronym	Parameter Identification	PID Range	PID Value at Hot Idle	PID Value at 30	PID Value at 55
ACCS	A/C Clutch Switch	ON / OFF	ON (if on)	OFF	OFF
ACP	A/C Press. Sensor	CL/OPEN	CLOSED (if low)	OPEN	OPEN
AIR-M	AIR System Monitor	ON / OFF	OFF	OFF	OFF
BPP	Brake Pedal Switch	ON / OFF	ON (if on)	OFF	OFF
CID	CID Sensor	0-1 KHz	5-7	9-11	11-14
CKP	CKP Sensor	0-10 KHz	360-420	680-800	860-910
DPFE	DPFE Sensor	0-5.1v	0.25-1.30	0.25-4.65	0.25-4.65
ECT (°F)	ECT Sensor	-40-304°F	160-200	160-200	160-200
EFTA (°F)	EFTA Sensor (°F)	-40-304°F	50-120	50-120	50-120
FEPS	Flash EEPROM	0-5.1v	0.5-0.6	0.5-0.6	0.5-0.6
FLI	Fuel Level Indicator (1998-99)	0-100%	50 (1.7v)	25-75	25-75
FPM	Fuel Pump Monitor	0-100%	87-100	87-100	87-100
FTPV	Fuel Tank Pressure (1998-99)	0-5.1v or 0-10" H2O	2.6v / 0" H2O (with cap off)	2.6v / 0" H2O (with cap off)	2.6v / 0" H2O (with cap off)
HO2S-11	HO2S-11 (Bank 1)	0.1-1.1v	0.1-1.1	0.1-1.1	0.1-1.1
HO2S-12	HO2S-12 (Bank 1)	0.1-1.1v	0.1-1.1	0.1-1.1	0.1-1.1
HO2S-21	HO2S-21 (Bank 2)	0.1-1.1v	0.1-1.1	0.1-1.1	0.1-1.1
HO2S-22	HO2S-22 (Bank 2)	0.1-1.1v	0.1-1.1	0.1-1.1	0.1-1.1
IAT (°F)	IAT Sensor	-40-304°F	50-120	50-120	50-120
IMRCM	IMRC Monitor	0-5.1v	5	5	5
KS1	Knock Sensor B1	0-5.1v	0	0	0
KS2	Knock Sensor B2	0-5.1v	0	0	0
LOAD	Engine Load	0-100%	10-20	16-30	20-30
MAF (V)	MAF Sensor	0-5.1v	0.5-0.8	0.8-1.3	1.3-1.7
MISF	Misfire Detection	ON / OFF	OFF	OFF	OFF
OCTADJ	Octane Adjustment	CL/OPEN	CLOSED	CLOSED	CLOSED
OSS	Output Speed Shaft Sensor RPM	0-10K rpm	0 rpm	1365 rpm	2440 rpm
PF (V)	Purge Flow Sensor Volts (1996-97)	0-5.1v	1.55	2.10	2.30
PNP	PNP Switch	Neutral/DR	NEUTRAL	DRIVE	DRIVE
RPM	Engine Speed	0-10K rpm	630-750	1180-1360	1530-1750
TPV	TP Sensor	0-5.1v	0.5-0.9	0.9-1.0	1.0-1.1
VPWR	Vehicle Power	0-25.5V	12-14	12-14	12-14
VSS	Vehicle Speed (MPH & Hertz Rate)	0-159 mph	0 = 0 Hz	30 = 65 Hz	55 = 125 Hz

1996-2003 MUSTANG 4.6L 4V V8 MFI VIN V, Y (ALL) - OUTPUTS

PCM PID Acronym	Parameter Identification	PID Range	PID Value at Hot Idle	PID Value at 30	PID Value at 55
AIR	AIR Solenoid	ON / OFF	OFF	OFF	OFF
CD1-CD8	Coil Driver 1-8	0-90° dwell	7	11	12
CTO	Clean Tach Output	0-10 KHz	37-44	86-92	87-115
EGRVR	VR Solenoid	0-100%	0	0-40	0-40
EVAP CP	EVAP CV Valve	0-100%	0-100	0-100	0-100
EVAPPC	EVAP Purge Valve	0-100%	0-100	0-100	0-100
FP	Fuel Pump Control	0-100%	33	28	100
FPL	Fuel Pump Low	0-25.5v	0.1	0.1	0.1
FuelPW1	INJ Pulsewidth B1	0-99.9 ms	2.4-2.8	1.6-5.0	3.3-5.0
FuelPW2	INJ Pulsewidth B2	0-99.9 ms	2.4-2.8	1.6-5.0	3.3-5.0
HFC	High Fan Control	ON / OFF	ON (if on)	OFF	OFF
HTR-11	HO2S-11 Heater	ON / OFF	ON	ON	ON
HTR-12	HO2S-12 Heater	ON / OFF	ON	ON	ON
HTR-21	HO2S-21 Heater	ON / OFF	ON	ON	ON
HTR-22	HO2S-22 Heater	ON / OFF	ON	ON	ON
IAC	Idle Air Control	0-100%	32	34-46	34-46
IMRC	IMRC Solenoid	ON / OFF	OFF	OFF	OFF
INJ1	INJ 1 Pulsewidth	0-99.9 ms	2.4-2.8	1.6-5.0	3.3-5.0
INJ2	INJ 2 Pulsewidth	0-99.9 ms	2.4-2.8	1.6-5.0	3.3-5.0
INJ3	INJ 3 Pulsewidth	0-99.9 ms	2.4-2.8	1.6-5.0	3.3-5.0
INJ4	INJ 4 Pulsewidth	0-99.9 ms	2.4-2.8	1.6-5.0	3.3-5.0
INJ5	INJ 5 Pulsewidth	0-99.9 ms	2.4-2.8	1.6-5.0	3.3-5.0
INJ6	INJ 6 Pulsewidth	0-99.9 ms	2.4-2.8	1.6-5.0	3.3-5.0
INJ7	INJ 7 Pulsewidth	0-99.9 ms	2.4-2.8	1.6-5.0	3.3-5.0
INJ8	INJ 8 Pulsewidth	0-99.9 ms	2.4-2.8	1.6-5.0	3.3-5.0
LFC	Low Fan Control	ON / OFF	ON (if on)	OFF	OFF
LongFT1	Long Term F/T B1	-20 to 20%	-5 to +5	-5 to +5	-5 to +5
LongFT2	Long Term F/T B2	-20 to 20%	-5 to +5	-5 to +5	-5 to +5
MIL	MIL Control	ON / OFF	ON (if on)	OFF	OFF
SHTFT1	Short Term F/T B1	-20 to 20%	-5 to +5	-5 to +5	-5 to +5
SHTFT2	Short Term F/T B2	-20 to 20%	-5 to +5	-5 to +5	-5 to +5
SPARK	Spark Advance	-90° to 90°	11-15	17-20	19-24
SS1	Shift Solenoid 1	ON / OFF	ON	OFF	ON
SS2	Shift Solenoid 2	ON / OFF	OFF	ON	ON
VREF	Voltage Reference	0-5.1v	4.9-5.1	4.9-5.1	4.9-5.1
VSO	VSS Output	0-1 KHz	0	65	125
WAC	A/C WOT Cutout Relay	ON / OFF	ON (if on)	OFF	OFF

1996-2003 MUSTANG 4.6L 2V V8 MFI VIN W, X (ALL) - INPUTS

PCM PID Acronym	Parameter Identification	PID Range	PID Value at Hot Idle	PID Value at 30	PID Value at 55
ACCS	A/C Clutch Switch	ON / OFF	ON (if on)	OFF	OFF
ACP	A/C Press. Sensor	CL/OPEN	OPEN	OPEN	OPEN
AIR-M	AIR System Monitor	ON / OFF	OFF	OFF	OFF
BPP	Brake Switch	ON / OFF	ON (if on)	OFF	OFF
CID	CID Sensor	0-1 KHz	5-7	10-12	12-16
CKP	CKP Sensor	0-10 KHz	390-450	650-760	980-1020
CPP/NDS	Clutch Pedal Switch	ON/OFF	ON with clutch depressed	OFF	OFF
DPFE	DPFE Sensor	0-5.1v	0.25-1.30	0.25-4.65	0.25-4.65
ECT (°F)	ECT Sensor	-40-304°F	160-200	160-200	160-200
EFTA (°F)	EFTA Sensor (°F)	-40-304°F	50-120	50-120	50-120
FEPS	Flash EEPROM	0-5.1v	0.5-0.6	0.5-0.6	0.5-0.6
FLI	Fuel Level Indicator (1998-99)	0-100%	50 (1.7v)	50 (1.7v)	50 (1.7v)
FPM	Fuel Pump Monitor	0-100%	87-100	87-100	87-100
FTP	Fuel Tank Pressure (1998-99)	0-5.1v or 0-10" H2O	2.6v / 0" H2O (with cap off)	2.6v / 0" H2O (with cap off)	2.6v / 0" H2O (with cap off)
FRP	Fuel Rail Pressure	0-100 psi	39	39	39
GEAR	Transmission Gear	1-2-3-4	1	3	4
HO2S-11	HO2S-11 (Bank 1)	0.1-1.1v	0.1-1.1	0.1-1.1	0.1-1.1
HO2S-12	HO2S-12 (Bank 1)	0.1-1.1v	0.1-1.1	0.1-1.1	0.1-1.1
HO2S-21	HO2S-21 (Bank 2)	0.1-1.1v	0.1-1.1	0.1-1.1	0.1-1.1
HO2S-22	HO2S-22 (Bank 2)	0.1-1.1v	0.1-1.1	0.1-1.1	0.1-1.1
IAT (°F)	IAT Sensor	-40-304°F	50-120	50-120	50-120
LOAD	Engine Load	0-100%	10-20	16-30	20-30
MAF (V)	MAF Sensor	0-5.1v	0.6-0.9	0.8-1.2	1.4-1.9
MISF	Misfire Detection	ON / OFF	OFF	OFF	OFF
OCTADJ	Octane Adjustment	CL/OPEN	CLOSED	CLOSED	CLOSED
OSS	Output Speed Shaft	0-10 KHz	0	1385-1420	2475-2500
PF (V)	Purge Flow Sensor (1996-97)	0-5.1v	1.55	2.10	2.30
PNP	PNP Switch	Neutral/DR	NEUTRAL	DRIVE	DRIVE
RPM	Engine Speed	0-10K rpm	660-700	1380-1420	1700-1740
TCS	TCS (switch)	ON / OFF	ON (if on)	OFF	OFF
TFT (°F)	TFT Sensor	-40-304°F	110-210	110-210	110-210
TPV	TP Sensor	0-5.1v	0.5-0.9	1.0-1.2	1.2-1.5
TRV/TR	TR Sensor	0-5.1v	4.4 / PN	2.1 / OD	2.1 / OD
VPWR	Vehicle Power	0-25.5V	12-14	12-14	12-14
VSS	Vehicle Speed (MPH & Hertz Rate)	0-159 mph	0 = 0 Hz	30 = 65 Hz	55 = 125 Hz

1996-2003 MUSTANG 4.6L 2V V8 MFI VIN W, X (ALL) - OUTPUTS

PCM PID Acronym	Parameter Identification	PID Range	PID Value at Hot Idle	PID Value at 30	PID Value at 55
AIR-GS	AIR Solenoid	ON / OFF	OFF	OFF	OFF
CD1-CD8	Coil Driver 1-8	0-90° dwell	7	11	12
CTO	Clean Tach Output	0-10 KHz	37-44	86-92	87-115
EGRVR	VR Solenoid	0-100%	0	0-40	0-40
EPC	EPC Solenoid	0-100 psi	15-18	12-16	40
EVAP CP	EVAP CV Valve	0-100%	0-100%	0-100%	0-100%
EVAPDC	EVAP Purge Valve	0-100%	0-100%	0-100%	0-100%
FP	Fuel Pump Control	0-100%	33	28	100
FPL	Fuel Pump Low	0-25.5v	0.1	0.1	0.1
FuelPW1	INJ Pulsewidth B1	0-99.9 ms	3.5-3.7	3.8-5.5	4.5-9.6
FuelPW2	INJ Pulsewidth B2	0-99.9 ms	3.5-3.7	3.8-5.5	4.5-9.6
HFC	High Fan Control	ON / OFF	ON (if on)	OFF	OFF
HTR-11	HO2S-11 Heater	ON / OFF	ON	ON	ON
HTR-12	HO2S-12 Heater	ON / OFF	ON	ON	ON
HTR-21	HO2S-21 Heater	ON / OFF	ON	ON	ON
HTR-22	HO2S-22 Heater	ON / OFF	ON	ON	ON
IAC	Idle Air Control	0-100%	37	50-57	60-65
INJ1	INJ 1 Pulsewidth	0-99.9 ms	3.5-3.7	3.8-5.5	4.5-9.6
INJ2	INJ 2 Pulsewidth	0-99.9 ms	3.5-3.7	3.8-5.5	4.5-9.6
INJ3	INJ 3 Pulsewidth	0-99.9 ms	3.5-3.7	3.8-5.5	4.5-9.6
INJ4	INJ 4 Pulsewidth	0-99.9 ms	3.5-3.7	3.8-5.5	4.5-9.6
INJ5	INJ 5 Pulsewidth	0-99.9 ms	3.5-3.7	3.8-5.5	4.5-9.6
INJ6	INJ 6 Pulsewidth	0-99.9 ms	3.5-3.7	3.8-5.5	4.5-9.6
INJ7	INJ 7 Pulsewidth	0-99.9 ms	3.5-3.7	3.8-5.5	4.5-9.6
INJ8	INJ 8 Pulsewidth	0-99.9 ms	3.5-3.7	3.8-5.5	4.5-9.6
LFC	Low Fan Control	ON / OFF	ON (if on)	OFF	OFF
LongFT1	Long Term F/T B1	-20 to 20%	-5 to +5	-5 to +5	-5 to +5
LongFT2	Long Term F/T B2	-20 to 20%	-5 to +5	-5 to +5	-5 to +5
MIL	MIL Control	ON / OFF	ON (if on)	OFF	OFF
SHTFT1	Short Term F/T B1	-20 to 20%	-5 to +5	-5 to +5	-5 to +5
SHTFT2	Short Term F/T B2	-20 to 20%	-5 to +5	-5 to +5	-5 to +5
SPARK	Spark Advance	-90° to 90°	15-20	29-38	25-34
SS1	Shift Solenoid 1	ON / OFF	ON	OFF	ON
SS2	Shift Solenoid 2	ON / OFF	OFF	ON	ON
TCC	TCC Solenoid	0-100%	0	37-43	95-100
TCIL	TCIL (lamp)	ON / OFF	ON (if on)	OFF	OFF
VREF	Voltage Reference	0-5.1v	4.9-5.1	4.9-5.1	4.9-5.1
VSO	VSS Output	0-1 KHz	0	65	125 Hz
WAC	A/C WOT Cutout Relay	ON / OFF	ON (if on)	OFF	OFF

1994-95 Mustang 5.0L V8 MFI VIN T (All) Inputs / Outputs

PCM PID Acronym	Parameter Identification	PID Range	PID Value at Hot Idle	PID Value at 30	PID Value at 55
ACCS	A/C Clutch Switch	ON / OFF	ON (if on)	OFF	OFF
BOO	Brake Switch	ON / OFF	ON (if on)	OFF	OFF
CANP	EVAP Purge	0-100%	0	0-40	40-50
ECT (°F)	ECT Sensor	-40-304°F	160-200	160-200	160-200
ECT (V)	ECT Sensor	0-5.1v	0.6	0.6	0.6
EPC	EPC Solenoid	0-100 psi	20	12	30
EVP (V)	EGR EVP Sensor	0-5.1v	0.3	0.9-1.0	2.0-3.0v
EVR	VR Solenoid	0-100%	0%	0-40	0-60
GEAR	Gear Position	P-R-N-D	P-R-N-D	DRIVE 3	DRIVE 4
HO2S-11	HO2S-11 (Bank 1)	0.1-1.1v	0.1-1.1	0.1-1.1	0.1-1.1
HO2S-12	HO2S-12 (Bank 1)	0.1-1.1v	0.1-1.1	0.1-1.1	0.1-1.1
IAC	Idle Air Control	0-100%	20-40	34-40	45-55
IAT (°F)	IAT Sensor	-40-314F	50-120	50-120	50-120
IAT (V)	IAT Sensor	0-5.1v	1.7-3.5	1.7-3.5	1.7-3.5
INJPW1	INJ Pulsewidth B1	0-99.9 ms	5.5-5.8	6.4-8.4	9.4-12
INJPW2	INJ Pulsewidth B2	0-99.9 ms	5.5-5.8	6.4-8.4	9.4-12
LOOP	Loop Status	CL or OL	CL	CL	CL
LongFT1	Long Term F/T B1	-20 to 20%	-5 to +5	-5 to +5	-5 to +5
LongFT2	Long Term F/T B2	-20 to 20%	-5 to +5	-5 to +5	-5 to +5
MAF	MAF Sensor	0-5.1v	1.0v	1.3-1.5	1.6-2.0
MLP	MLP Sensor	Park / OD	PARK	OD	OD
MLP (V)	MLP Sensor	0-5.1v	0v	5v	5v
PNP	PNP Switch	Neutral/DR	NEUTRAL	DRIVE	DRIVE
RPM	Engine Speed	0-10K rpm	750-820	1450-1630	1750-2100
SHTFT1	Short Term F/T B1	-20 to 20%	-5 to +5	-5 to +5	-5 to +5
SHTFT2	Short Term F/T B2	-20 to 20%	-5 to +5	-5 to +5	-5 to +5
SPARK	Spark Advance	-90° to 90°	15-22	28-35	25-35
TCC	TCC Solenoid	0-100%	0%	42-44	92-100
TOT (°F)	TOT Sensor	-40-304°F	110-210	110-210	110-210
TOT (V)	TOT Sensor	0-5.1v	1.7-3.5	1.7-3.5	1.7-3.5
TP	TP Sensor	0-5.1v	0.7	0.8-1.0	1.1-1.3
TP C/T	TP Closed Throttle	0-5.1v	0.7	0.8-1.0	1.1-1.3
TP Mode	TP Mode	C/T or P/T	C/T	P/T	P/T
VPWR	Vehicle Power	0-25.5v	12-14	12-14	12-14
VREF	Voltage Reference	0-5.1v	4.9-5.1	4.9-5.1	4.9-5.1
VSS	Vehicle Speed	0-159 mph	0	30	55
WAC	A/C WOT Cutout Relay	ON / OFF	ON (if on)	OFF	OFF

Probe PID Data

1994-95 Probe 2.0L I4 MFI VIN A (CD4E) Inputs / Outputs

PCM PID Acronym	Parameter Identification	PID Range	PID Value at Hot Idle	PID Value at 30	PID Value at 55
ACCS	A/C Clutch Switch	ON / OFF	ON (if on)	OFF	OFF
BOO	Brake Switch	ON / OFF	ON (if on)	OFF	ON
CANP	EVAP Purge Valve	0-100%	0	0-40	40-50
ECT (°F)	ECT Sensor	-40-304°F	160-200	160-200	160-200
ECT (V)	ECT Sensor	0-5.1v	0.6	0.6	0.6
EPC	EPC Solenoid	0-100 psi	8-9 psi	24 psi	20-30
EVP (V)	EGR EVP Sensor	0-5.1v	0.4	0.5-1.1	0.9-2.1
EVR	VR Solenoid	0-100%	0	0	0-40
GEAR	Gear Position	P-R-N-DR	P-R-N-D	DRIVE 3	DRIVE 4
HFAN	High Speed Fan	ON / OFF	ON (if on)	OFF	OFF
FuelPW1	INJ Pulsewidth B1	0-99.9 ms	1.7-2.3	2.4-4.6	3.5-8.4
HO2S-11	HO2S-11 (Bank 1)	0.1-1.1v	0.1-1.1	0.1-1.1	0.1-1.1
IAC	Idle Air Control	0-100%	41	20-35	50-75
IAT (°F)	IAT Sensor	-40-304°F	50-120	50-120	50-120
IAT (V)	IAT Sensor	0-5.1v	1.7-3.5	1.7-3.5	1.7-3.5
LFAN	Low Speed Fan	ON / OFF	ON (if on)	OFF	OFF
LongFT1	Long Term F/T B1	-20 to 20%	-5 to +5	-5 to +5	-5 to +5
LOOP	Loop Status	CL or OL	CL	CL	CL
MAF (V)	MAF Sensor	0-5.1v	0.4-0.7	0.7-1.6	1.2-2.2
MLP (V)	MLP Sensor	0-5.1v	0	5	5
PNP	PNP Switch	Neutral/DR	NEUTRAL	DRIVE	DRIVE
RPM	Engine Speed	0-10K rpm	680-720	1575-1635	2200-2450
SHTFT1	Short Term F/T B1	-20 to 20%	-5 to +5	-5 to +5	-5 to +5
SPARK	Spark Advance	-90° to 90°	15-20	25-35	28-33
TCC	TCC Solenoid	0-100%	0	0-45	95-100
TOT (°F)	TOT Sensor	-40-304°F	110-210	110-210	110-210
TOT (V)	TOT Sensor	0-5.1v	0.5-2.0	0.5-2.0	0.5-2.0
TSS	TSS Sensor	0-10 KHz	680-720	1575-1635	2200-2450
TP	TP Sensor	0-5.1v	0.7	0.8-1.0	1.1-1.3
TP MIN	TP Minimum	0-5.1v	0.5	0.8-1.0	1.1-1.3
TP Mode	TP Mode	C/T or P/T	C/T	P/T	P/T
VPWR	Vehicle Power	0-25.5v	12-14	12-14	12-14
VREF	Voltage Reference	0-5.1v	4.9-5.1	4.9-5.1	4.9-5.1
VSS	Vehicle Speed	0-159 mph	0	30	55
WAC	A/C WOT Cutout Relay	ON / OFF	ON (if on)	OFF	OFF

1996-97 Probe 2.0L I4 MFI VIN A (All) - Inputs

PCM PID Acronym	Parameter Identification	PID Range	PID Value at Hot Idle	PID Value at 30	PID Value at 55
ACCS	A/C Clutch Switch	ON / OFF	ON (if on)	OFF	OFF
ACP	A/C Press. Sensor	0-5.1v	0.8-1.9v	0.8-1.9v	0.8-1.9v
BPP	Brake Pedal Switch	ON / OFF	ON (if on)	OFF	OFF
CID	CID Sensor	0-1 KHz	5-7	11-15	17-21
CKP	CKP Sensor	0-10 KHz	390-450	925-950	1320-1410
CPP	M/T: CPP Switch	ON / OFF	ON (if in)	OFF	OFF
DRL	Daytime Running Lamps	0-5.1v	0.1	0.1	0.1
ECT (°F)	ECT Sensor	-40-304°F	160-200	160-200	160-200
EGRB	EGR BARO Sensor	0-5.1v	0.4-4.8	0.4-4.8	0.4-4.8
EGR VP	EGR EVP Sensor	0-5.1v	0.5-0.8	0.7-1.6	2.5-3.2
FEPS	Flash EEPROM	0-5.1v	0.1	0.1	0.1
FPM	Fuel Pump Monitor	ON / OFF	ON	ON	ON
FRP	Fuel Rail Pressure	0-100 psi	30-32	30-32	32-35
IAT	IAT Sensor	-40-304°F	50-120	50-120	50-120
GEAR	Gear Position	P-N-R-D or DRIVE	P-R-N-D	DRIVE 3	DRIVE 4
HDLMP	Headlamp Switch	ON / OFF	ON (if on)	OFF	OFF
HO2S-11	HO2S-11	0.1-1.1v	0.1-1.1	0.1-1.1	0.1-1.1
HO2S-12	HO2S-12	0.1-1.1v	0.1-1.1	0.1-1.1	0.1-1.1
LOAD	Engine Load	0-100%	10-20	16-26	24-36
MAF (V)	MAF Sensor	0-5.1v	0.4-0.7	0.7-1.6	1.2-2.2
MISF	Misfire Detection	OFF	OFF	OFF	OFF
OCTADJ	Octane Adjustment	CL/OPEN	CLOSED	CLOSED	CLOSED
PNP	A/T: PNP Switch	Neutral/DR	NEUTRAL	DRIVE	DRIVE
PSP	PSP Switch	0-25.5v	0v turning	12-14	12-14
RDEF	Rear Defrost Switch	ON / OFF	ON (if on)	OFF	OFF
RPM	Engine Speed	0-10K rpm	680-720	1575-1635	2200-2450
TCS	TCS (Switch)	ON / OFF	ON (if on)	OFF	OFF
TFT (°F)	TFT Sensor	-40-304°F	110-210	110-210	110-210
TPV	TP Sensor	0-5.1v	0.5-0.9	0.6-0.8	0.7-1.0
TSS	TSS Sensor	0-10 KHz	680-720	1575-1635	2200-2450
VACUUM	Engine Vacuum	0-30" Hg	19-20	18-20	8-15
VPWR	Vehicle Power	0-25.5v	12-14	12-14	12-14
VSS	Vehicle Speed in MPH & Hz	0-159 mph	0 = 0 Hz	30 = 65 Hz	55 = 125 Hz

1996-97 Probe 2.0L I4 MFI VIN A (All) - Outputs

PCM PID Acronym	Parameter Identification	PID Range	PID Value at Hot Idle	PID Value at 30	PID Value at 55
BLWMTR	Blower Motor	ON / OFF	ON (if on)	OFF	OFF
CD1	Coil Driver 1	0-90° dwell	7	10	12
EGRC	EGR Control Solenoid	ON / OFF	OFF	ON	ON
EGRCS	EGR Check Solenoid	ON / OFF	OFF	OFF	OFF
EGRV	EGR Vent Solenoid	ON / OFF	ON	ON	ON
EVAP CP	EVAP Purge Valve	ON / OFF	OFF	ON	ON
FP	Fuel Pump Control	ON / OFF	ON	ON	ON
FPRC	Fuel Pump Regulator Control	ON / OFF	ON (if on)	OFF	OFF
FuelPW1	INJ Pulsewidth B1	0-99.9 ms	1.0-2.3	2.4-4.6	3.5-8.4
HFC	High Fan Control	ON / OFF	ON (if on)	OFF	OFF
HTR-11	HO2S-11 Heater	ON / OFF	ON	ON	ON
HTR-12	HO2S-12 Heater	ON / OFF	ON	ON	ON
IAC	Idle Air Control	0-100%	41	20-35	50-75
INJ1	INJ 1 Pulsewidth	0-99.9 ms	1.0-2.3	2.4-4.6	3.5-8.4
INJ2	INJ 2 Pulsewidth	0-99.9 ms	1.0-2.3	2.4-4.6	3.5-8.4
INJ3	INJ 3 Pulsewidth	0-99.9 ms	1.0-2.3	2.4-4.6	3.5-8.4
INJ4	INJ 4 Pulsewidth	0-99.9 ms	1.0-2.3	2.4-4.6	3.5-8.4
LFC	Low Fan Control	ON / OFF	ON (if on)	OFF	OFF
LongFT1	Long Term F/T B1	-20 to 20%	-5 to +5	-5 to +5	-5 to +5
MIL	MIL Control	ON / OFF	ON (if on)	OFF	OFF
SS1	Shift Solenoid 1	ON / OFF	OFF	ON	ON
SS2	Shift Solenoid 2	ON / OFF	ON	OFF	OFF
SS3	Shift Solenoid 3	0-25.5v	6.7-7.7	6.7-7.7	6.7-7.7
SHTFT1	Short Term F/T B1	-20 to 20%	-5 to +5	-5 to +5	-5 to +5
SPARK	Spark Advance	-90° to 90°	15-20	25-35	28-33
TCC	TCC Solenoid	0-100%	0	0-45	95-100
TCIL	TCIL (lamp)	ON / OFF	ON (if on)	OFF	OFF
VREF	Voltage Reference	0-5.1v	4.9-5.1	4.9-5.1	4.9-5.1
WAC	A/C WOT Cutout Relay	ON / OFF	ON (if on)	OFF	OFF

1996-97 Probe 2.5L V6 MFI VIN B (All) - Inputs / Outputs

PCM PID Acronym	Parameter Identification	PID Range	PID Value at Hot Idle	PID Value at 30	PID Value at 55
ACR	A/C Relay (on with A/C & BLMT at 3-4)	ON / OFF	ON (if on)	OFF	OFF
ACCS	A/C Relay (on with A/C & BLMT at 3-4)	ON / OFF	ON (if on)	OFF	OFF
ATP	Atmospheric Pressure Sensor	0-5.1v	3.9	3.9	3.9
B+	Vehicle Power	0-25.5v	12-14	12-14	12-14
BARO	BARO Sensor " Hg (varies with altitude)	0-30" Hg	29.4	29.4	29.4
BLMT SW	Blower Motor Switch	ON / OFF	ON (if on)	OFF	OFF
DEF SW	Defroster Switch	ON / OFF	ON (if on)	OFF	OFF
DRL SW	Daytime Running Lamp	ON / OFF	ON (if on)	ON	ON
ECT (°F)	ECT Sensor	-40-304°F	177	185°F	186
ECT (V)	ECT (°F) Sensor	0-5.1v	0.6	0.4	0.4v
EGRS	EGR Sensor	0-5.1v	0.8	0.9	1.4v
EGR-A	EGR 'A' Solenoid	0-99.9 ms	64-99.9	64-99.9	19-99.9
HDLPSW	Headlamp Switch	ON / OFF	ON (if on)	OFF	OFF
HOS2-11	HO2S-11 (Bank 1)	0.1-1.1v	0.1-1.1	0.1-1.1	0.1-1.1
HOS2-12	HO2S-12 (Bank 1)	0.1-1.1v	0.1-1.1	0.1-1.1	0.1-1.1
IAT (°F)	IAT Sensor	-40-304°F	77	86	69°F
IAT (V)	IAT Sensor	0-5.1v	2.4	2.6v	2.7
IDL SW	Throttle Idle Switch	ON / OFF	ON	OFF	OFF
IGN SW	Ignition Switch	ON / OFF	ON-KOEC	OFF	OFF
INJ1	Fuel Injector 1	0-99.9 ms	4.0-5.0	5.0-5.8	5.0-6.0
INJ2	Fuel Injector 2	0-99.9 ms	4.0-5.0	5.0-5.8	5.0-6.0
KS	Knock Sensor	0-5.1v	2.5	2.5	2.5
MC-VAF	Measuring Core Vane Airflow sensor	0-5.1v	3.06	1.50	1.62
MIL	MIL Control	ON / OFF	ON (if on)	OFF	OFF
PNP SW	PNP Switch	0-25.5v	0v (in 'P')	14	14
PSP	PSP Switch	ON / OFF	ON turning	OFF	OFF
PUMP	Fuel Pump	ON / OFF	ON	ON	ON
PURGE	EVAP Purge Valve	0-100%	0	40-50	84-90
SPARK	Spark Advance	-90° to 90°	11	20	20
STP SW	Stop Lamp Switch	ON / OFF	ON (if on)	OFF	OFF
TPV	TP Sensor	0-5.1v	0.5	0.6	0.7
VSS	Vehicle Speed	0-159 mph	0	30	55

1996-2003 Taurus 3.0L V6 FFV VIN 1, 2 (A/T) - Inputs

PCM PID Acronym	Parameter Identification	PID Range	PID Value at Hot Idle	PID Value at 30	PID Value at 55
ACCS	A/C Clutch Switch	ON / OFF	ON (if on)	OFF	OFF
ACP	A/C Press. Switch	OPEN or CLOSED	OPEN	OPEN	OPEN
AIR-M	AIR System Monitor	ON / OFF	ON	ON	ON
BPP	Brake Pedal Switch	ON / OFF	ON (if on)	OFF	OFF
CID	CID Sensor	0-1 KHz	6-8	11-13	14-17
CKP	CKP Sensor	0-10 KHz	410-510	810-950	1050-1820
DPFE	DPFE Sensor	0-5.1v	0.2-1.3v	0.2-4.5v	0.2-4.5v
ECT (°F)	ECT Sensor	-40-304°F	160-200	160-200	160-200
FEPS	Flash EEPROM	0-5.1v	0.5-0.6	0.5-0.6	0.5-0.6
FFS	Flexible Fuel sensor (100% gas mixture)	0-1 KHz	40-60	40-60	40-60
FP-M	Fuel Pump Monitor	ON / OFF	ON	ON	ON
FRP	Fuel Rail Pressure	0-100 psi	54	42	40
FTP	FTP Sensor	0-5.1v or 0-10" H20	2.6v at 0" of H20	2.6v at 0" of H20	2.6v at 0" of H20
GEAR	Transmission Gear	P-R-N-D	P-R-N-D	DRIVE 3	DRIVE 4
HO2S-11	HO2S-11 (Bank 1)	0.1-1.1v	0.1-1.1	0.1-1.1	0.1-1.1
HO2S-12	HO2S-12 (Bank 1)	0.1-1.1v	0.1-1.1	0.1-1.1	0.1-1.1
HO2S-21	HO2S-21 (Bank 2)	0.1-1.1v	0.1-1.1	0.1-1.1	0.1-1.1
HO2S-22	HO2S-22 (Bank 2)	0.1-1.1v	0.1-1.1	0.1-1.1	0.1-1.1
IAT (°F)	IAT Sensor	-40-304°F	50-120	50-120	50-120
LOAD	Engine Load	0-100%	10-20	16-30	13-50
MAF (V)	MAF Sensor	0-5.1v	0.6-0.9	0.9-1.9	1.5-2.5
MISF	Misfire Detection	OFF	OFF	OFF	OFF
OCTADJ	Octane Adjustment	CL/OPEN	CLOSED	CLOSED	CLOSED
TSS	TSS Sensor	0-10 KHz	790-820	1190-1450	1950-2200
PNP	PNP Switch	ON / OFF	ON	OFF	OFF
PSP	PSP Switch	HI / LOW	HI: turning	LOW	LOW
RPM	Engine Speed	0-10K rpm	660-800	1440-1625	1830-1970
TFT (°F)	TFT Sensor	-40-304°F	110-210	110-210	110-210
TPV	TP Sensor	0-5.1v	0.53-1.27	0.8-1.2	0.8-1.2
TRV/TR	TR Sensor	0-5.1v	4.4 / PN	2.1 / OD	2.1 / OD
Vacuum	Engine Vacuum	0-30" Hg	19-20	18-20	8-15
VPWR	Vehicle Power	0-25.5v	12-14	12-14	12-14
VSS	Vehicle Speed	0-159 mph 0-1 KHz	0 = 0 Hz	30 = 65 Hz	55 = 125 Hz

1996-2003 Taurus 3.0L V6 FFV VIN 1, 2 (A/T) - Outputs

PCM PID Acronym	Parameter Identification	PID Range	PID Value at Hot Idle	PID Value at 30	PID Value at 55
AIR	AIR Solenoid	ON / OFF	OFF	OFF	OFF
CD1	Coil Driver 1	0-60º dwell	7	11	12
CD2	Coil Driver 2	0-60º dwell	7	11	12
CD3	Coil Driver 3	0-60º dwell	7	11	12
CTO	Clean Tach Output	0-10 KHz	42-50	65-78	91-105
EGRVR	VR Solenoid	0-100%	0	0-40	40-50
EPC	EPC Solenoid	0-100 psi	38-42	26-30	28-30
EVAPCV	EVAP CV Valve	0-100%	0-100	0-100	0-100
EVAPDC	EVAP Purge Valve	0-100%	0-100	0-100	0-100
FP	Fuel Pump Control	ON / OFF	ON	ON	ON
FuelPW1	INJ Pulsewidth B1	0-99.9 ms	2.3-2.8	2.5-6.0	3.0-7.0
FuelPW2	INJ Pulsewidth B2	0-99.9 ms	2.3-2.8	2.5-6.0	3.0-7.0
HFC	High Fan Control	ON / OFF	ON (if on)	OFF	OFF
HTR-11	HO2S-11 Heater	ON / OFF	ON	ON	ON
HTR-12	HO2S-12 Heater	ON / OFF	ON	ON	ON
HTR-21	HO2S-21 Heater	ON / OFF	ON	ON	ON
HTR-22	HO2S-22 Heater	ON / OFF	ON	ON	ON
IAC	Idle Air Control	0-100%	40-45	40-60	40-60
INJ1	INJ 1 Pulsewidth	0-99.9 ms	2.3-2.8	2.5-6.0	3.0-7.0
INJ2	INJ 2 Pulsewidth	0-99.9 ms	2.3-2.8	2.5-6.0	3.0-7.0
INJ3	INJ 3 Pulsewidth	0-99.9 ms	2.3-2.8	2.5-6.0	3.0-7.0
INJ4	INJ 4 Pulsewidth	0-99.9 ms	2.3-2.8	2.5-6.0	3.0-7.0
INJ5	INJ 5 Pulsewidth	0-99.9 ms	2.3-2.8	2.5-6.0	3.0-7.0
INJ6	INJ 6 Pulsewidth	0-99.9 ms	2.3-2.8	2.5-6.0	3.0-7.0
LFC	Low Fan Control	ON / OFF	ON (if on)	OFF	OFF
LongFT1	Long Term F/T B1	-20 to 20%	-5 to +5	-5 to +5	-5 to +5
LongFT2	Long Term F/T B2	-20 to 20%	-5 to +5	-5 to +5	-5 to +5
MIL	MIL Control	ON / OFF	ON (if on)	OFF	OFF
SHTFT1	Short Term F/T B1	-20 to 20%	-5 to +5	-5 to +5	-5 to +5
SHTFT2	Short Term F/T B2	-20 to 20%	-5 to +5	-5 to +5	-5 to +5
SPARK	Spark Advance	-90º to 90º	24-30	30-45	33-43
SS1	Shift Solenoid 1	ON / OFF	OFF	OFF	ON
SS2	Shift Solenoid 2	ON / OFF	ON	OFF	OFF
SS3	Shift Solenoid 3	ON / OFF	OFF	ON	ON
TCC	TCC Solenoid	0-100%	0	90-100	95-100
VMV	EVAP VMV Valve (1996 only)	0-100%	0-100	0-100	0-100
VREF	Voltage Reference	0-5.1v	4.9-5.1	4.9-5.1	4.9-5.1
WAC	A/C WOT Cutout Relay	ON / OFF	ON (if on)	OFF	OFF

1993-95 Taurus 3.2L V6 MFI SHO VIN P (A/T) Inputs / Outputs

PCM PID Acronym	Parameter Identification	PID Range	PID Value at Hot Idle	PID Value at 30	PID Value at 55
ACCS	A/C Clutch Switch	ON / OFF	ON (if on)	OFF	OFF
BOO	Brake Switch	ON / OFF	ON (if on)	OFF	OFF
CANP	EVAP Purge Valve	0-100%	0	0-40	85-95
ECT (°F)	ECT Sensor	-40-304°F	160-200	160-200	160-200
ECT (V)	ECT Sensor	0-5.1v	0.6	0.6	0.6
EPC	EPC Solenoid	0-100 psi	40	15	42
EVR	VR Solenoid	0-100%	0	0-40	40-50
FPM	Fuel Pump Monitor	ON / OFF	ON	ON	ON
GEAR	Gear Position	P-R-N-D	P-R-N-D	DRIVE 3	DRIVE 4
HFC	High Fan Control	ON / OFF	ON (if on)	OFF	OFF
FLPW1	INJ On-time (B1)	0-99.9 ms	3.6-3.8	4.5-4.7	5.0-10
FLPW2	INJ On-time (B2)	0-99.9 ms	3.6-3.8	4.5-4.7	5.0-10
IAC	Idle Air Control	0-100%	20-40	34-40	45-55
IAT (°F)	IAT Sensor	-40-304°F	50-120	50-120	50-120
IAT (V)	IAT Sensor	0-5.1v	1.7-3.5	1.7-3.5	1.7-3.5
LFC	Low Fan Control	ON / OFF	ON (if on)	OFF	ON
LOOP	Loop Status	CL or OL	CL	CL	CL
LongFT1	Long Term F/T B1	-20 to 20%	-5 to +5	-5 to +5	-5 to +5
LongFT2	Long Term F/T B2	-20 to 20%	-5 to +5	-5 to +5	-5 to +5
MAF (V)	MAF Sensor	0-5.1v	0.6	1.0-1.8v	1.4-1.8v
MLP (V)	MLP Sensor	0-5.1v	0	5	5
HO2S-11	HO2S-11 (Bank 1)	0.1-1.1v	0.1-1.1	0.1-1.1	0.1-1.1
HO2S-21	HO2S-21 (Bank 2)	0.1-1.1v	0.1-1.1	0.1-1.1	0.1-1.1
PFE	PFE Sensor	0-5.1v	3.2	3.3	3.4
PNP	PNP Switch	NEUTRAL	NEUTRAL	DRIVE	DRIVE
PSP	PSP Switch	ON / OFF	ON turning	OFF	OFF
RPM	Engine Speed	0-10K rpm	700-750	1410-1510	1760-1860
SHTFT1	Short Term F/T B1	-20 to 20%	-5 to +5	-5 to +5	-5 to +5
SHTFT2	Short Term F/T B2	-20 to 20%	-5 to +5	-5 to +5	-5 to +5
SPARK	Spark Advance	-90° to 90°	16-21	32-36	30-36°
TCC	TCC Solenoid	0-100%	0	90-100	90-100
TOT °F	TOT Sensor	-40-304°F	110-210	110-210	110-210
TOT (V)	TOT Sensor	0-5.1v	0.5-2.0	0.5-2.0	0.5-2.0
TP	TP Sensor	0-5.1v	0.8	0.9-1.0	1.1-1.2v
TP MIN	TP Minimum	0-5.1v	0.5	0.9-1.0	1.1-1.2v
TP Mode	TP Mode	C/T or P/T	C/T	P/T	P/T
TSS	TSS Sensor	0-10K rpm	550-600	1350-1440	1640-1795
VPWR	Vehicle Power	0-25.5v	12-14	12-14	12-14
VREF	Voltage Reference	0-5.1v	4.9-5.1	4.9-5.1	4.9-5.1
VSS	Vehicle Speed	0-159 mph	0	30	55
WAC	A/C WOT Cutout Relay	ON / OFF	ON (if on)	OFF	OFF

1996-99 Taurus 3.4L V6 MFI SHO VIN N (A/T) - Inputs

PCM PID Acronym	Parameter Identification	PID Range	PID Value at Hot Idle	PID Value at 30	PID Value at 55
ACCS	A/C Clutch Switch	ON / OFF	ON (if on)	OFF	OFF
ACP	A/C Press. Switch	0-5.1v	0.8-1.9v	0.8-1.9v	0.8-1.9v
AIR-M	AIR System Monitor	ON / OFF	ON	ON	ON
BPP	Brake Pedal Switch	ON / OFF	ON (if on)	OFF	OFF
CID	CID Sensor	0-1 KHz	5-7	11-13	12-15
CKP	CKP Sensor	0-10 KHz	380-420	850-900	950-1130
DPFE	DPFE Sensor	0-5.1v	0.4-0.6	0.4-0.7	0.8-2.7
ECT (°F)	ECT Sensor	-40-304°F	160-200	160-200	160-200
FEPS	Flash EEPROM	0-5.1v	0.1	0.1	0.1
FLI	Fuel Level Indicator (1998-99)	0-100%	50 (1.7v)	25-75	25-75
FTP	FTP Sensor (1998-99)	0-5.1v or 0-10" H20	2.6v at 0" of H20	2.6v at 0" of H20	2.6v at 0" of H20
FP-M	Fuel Pump Monitor	ON / OFF	ON	ON	ON
FRP	Fuel Rail Pressure	0-100 psi	30-32	30-32	32-35
FTP	FTP Sensor	0-5.1 or 0-10" H20	2.6v at 0" of H20	2.6v at 0" of H20	2.6v at 0" of H20
GEAR	Transmission Gear	1-2-3-4	1	3	4
HO2S-11	HO2S-11 (Bank 1)	0.1-1.1v	0.1-1.1	0.1-1.1	0.1-1.1
HO2S-12	HO2S-12 (Bank 1)	0.1-1.1v	0.1-1.1	0.1-1.1	0.1-1.1
HO2S-21	HO2S-21 (Bank 2)	0.1-1.1v	0.1-1.1	0.1-1.1	0.1-1.1
HO2S-22	HO2S-22 (Bank 2)	0.1-1.1v	0.1-1.1	0.1-1.1	0.1-1.1
IAT (°F)	IAT Sensor	-40-304°F	50-120	50-120	50-120
IMRCM	IMRC Monitor	0-5.1v	5v	5v	5v
KS1	Knock Sensor B1	0-5.1v	0	0	0
KS2	Knock Sensor B2	0-5.1v	0	0	0
LOAD	Engine Load	0-100%	10-20	16-30	13-32
MAF (V)	MAF Sensor	0-5.1v	0.6-0.9	0.8-1.4	1.1-1.9
MISF	Misfire Detection	OFF	OFF	OFF	OFF
OCTADJ	Octane Adjustment	CL/OPEN	CLOSED	CLOSED	CLOSED
PNP	PNP Switch	NEUTRAL	NEUTRAL	DRIVE	DRIVE
PSP	PSP Switch	ON / OFF	ON turning	OFF	OFF
RPM	Engine Speed	0-10K rpm	650-730	1360-1495	1780-1900
TCS	TCS (switch)	ON / OFF	ON (if on)	OFF	OFF
TFT (°F)	TFT Sensor	-40-304°F	110-210	110-210	110-210
TPV	TP Sensor	0-5.1v	0.5-0.9	0.8-1.2	0.0-1.3v
TRV/TR	TR Sensor	0-5.1v	4.4 / PN	2.1 / OD	2.1 / OD
TSS	TSS Sensor	0-10 KHz	550-600	1350-1440	1640-1795
Vacuum	Engine Vacuum	0-30" Hg	19-20	18-20	8-15
VPWR	Vehicle Power	0-25.5v	12-14	12-14	12-14
VSS	Vehicle Speed	0-159 mph 0-1 KHz	0 = 0 Hz	30 = 65 Hz	55 = 125 Hz

1996-99 Taurus 3.4L V6 SHO MFI VIN N (A/T) - Outputs

PCM PID Acronym	Parameter Identification	PID Range	PID Value at Hot Idle	PID Value at 30	PID Value at 55
AIR	AIR Solenoid	OFF	OFF	OFF	OFF
CD1-CD6	COP Driver 1-6	0-60° dwell	7	10	11
CTO	Clean Tach Output	0-10 KHz	40-50	85-110	100-140
EGRVR	VR Solenoid	0-100%	0%	0-50	0-50
EPC	EPC Solenoid	0-100 psi	40	15	42
EVAPCV	EVAP CV Valve	0-100%	0%	0%	0%
EVAPDC	EVAP Purge Valve	0-100%	0-100%	0-100%	0-100%
FP	Fuel Pump Control	0-100%	100%	100%	100%
FuelPW1	INJ Pulsewidth B1	0-99.9 ms	2.3-2.8	2.8-5.5	2.9-6.0
FuelPW2	INJ Pulsewidth B2	0-99.9 ms	2.3-2.8	2.8-5.5	2.9-6.0
HFC	High Fan Control	ON / OFF	ON	OFF	OFF
HTR-11	HO2S-11 Heater	ON / OFF	ON	ON	ON
HTR-12	HO2S-12 Heater	ON / OFF	ON	ON	ON
HTR-21	HO2S-21 Heater	ON / OFF	ON	ON	ON
HTR-22	HO2S-22 Heater	ON / OFF	ON	ON	ON
IAC	Idle Air Control	0-100%	34	40-60	40-50
INJ1	INJ 1 Pulsewidth	0-99.9 ms	2.3-2.8	2.8-5.5	2.9-6.0
INJ2	INJ 2 Pulsewidth	0-99.9 ms	2.3-2.8	2.8-5.5	2.9-6.0
INJ3	INJ 3 Pulsewidth	0-99.9 ms	2.3-2.8	2.8-5.5	2.9-6.0
INJ4	INJ 4 Pulsewidth	0-99.9 ms	2.3-2.8	2.8-5.5	2.9-6.0
INJ5	INJ 5 Pulsewidth	0-99.9 ms	2.3-2.8	2.8-5.5	2.9-6.0
INJ6	INJ 6 Pulsewidth	0-99.9 ms	2.3-2.8	2.8-5.5	2.9-6.0
IMRC	IMRC Solenoid	ON / OFF	OFF	OFF	OFF
LFC	Low Fan Control	ON / OFF	ON (if on)	OFF	OFF
LongFT1	Long Term F/T B1	-20 to 20%	-5 to +5	-5 to +5	-5 to +5
LongFT2	Long Term F/T B2	-20 to 20%	-5 to +5	-5 to +5	-5 to +5
MIL	MIL Control	ON / OFF	ON (if on)	OFF	OFF
SHTFT1	Short Term F/T B1	-20 to 20%	-5 to +5	-5 to +5	-5 to +5
SHTFT2	Short Term F/T B2	-20 to 20%	-5 to +5	-5 to +5	-5 to +5
SPARK	Spark Advance	-90° to 90°	5-10	31-39	29-41
SS1	Shift Solenoid 1	ON / OFF	OFF	ON	ON
SS2	Shift Solenoid 2	ON / OFF	ON	OFF	OFF
SS3	Shift Solenoid 3	ON / OFF	OFF	ON	ON
TCC	TCC Solenoid	0-100%	0	50-95	90-95
TCIL	TCIL (lamp)	ON / OFF	On (if on)	OFF	OFF
VMV	EVAP VMV Valve (1996-97)	0-100%	0-100	0-100	0-100
VREF	Voltage Reference	0-5.1v	4.9-5.1	4.9-5.1	4.9-5.1
WAC	A/C WOT Cutout Relay	ON / OFF	ON (if on)	OFF	OFF

Taurus & Sable PID Data - OBD II

1996-2002 Taurus/Sable 3.0L V6 MFI VIN S (A/T) - Inputs

PCM PID Acronym	Parameter Identification	PID Range	PID Value at Hot Idle	PID Value at 30	PID Value at 55
ACCS	A/C Clutch Switch	ON / OFF	ON (if on)	OFF	OFF
ACP	A/C Press. Sensor	CL/OPEN	OPEN	OPEN	OPEN
AIR-M	AIR System Monitor	ON / OFF	OFF	OFF	OFF
BPP	Brake Pedal Switch	ON / OFF	ON (if on)	OFF	OFF
CID	CID Sensor	0-1 KHz	5-7	10-13	14-17
CKP	CKP Sensor	0-10 KHz	390-520	850-1120	1140-1220
DPFE	DPFE Sensor	0-5.1v	0.4-0.6	0.4-1.5v	2.0-2.3v
ECT (ºF)	ECT Sensor	-40-304ºF	160-200	160-200	160-200
FRP	Fuel Rail Pressure	0-100 psi	39	41	40
FEPS	Flash EEPROM	0-5.1v	0.5-0.6	0.5-0.6	0.5-0.6
FLI	Fuel Level Indicator (1998-99)	0-100%	50 (1.7v)	50 (1.7v)	50 (1.7v)
FTP	FTP Sensor (1998-99)	0-5.1v or 0-10" H20	2.6v at 0" of H20	2.6v at 0" of H20	2.6v at 0" of H20
FP-M	Fuel Pump Monitor	ON / OFF	ON	ON	ON
GEAR	Transmission Gear	1-2-3-4	1	3	4
HO2S-11	HO2S-11 (Bank 1)	0.1-1.1v	0.1-1.1	0.1-1.1	0.1-1.1
HO2S-12	HO2S-12 (Bank 1)	0.1-1.1v	0.1-1.1	0.1-1.1	0.1-1.1
HO2S-21	HO2S-21 (Bank 2)	0.1-1.1v	0.1-1.1	0.1-1.1	0.1-1.1
HO2S-22	HO2S-22 (Bank 2)	0.1-1.1v	0.1-1.1	0.1-1.1	0.1-1.1
IAT (ºF)	IAT Sensor	-40-304ºF	50-120	50-120	50-120
IMRCM	IMRC Monitor	0-5.1v	5v	5v	5v
KS1	Knock Sensor B1	0-5.1v	0v	0v	0v
LOAD	Engine Load	0-100%	15-20%	20-35	15-35
MAF (V)	MAF Sensor	0-5.1v	0.6-0.9	0.7-1.5v	1.3-2.0
MISF	Misfire Detection	OFF	OFF	OFF	OFF
OCTADJ	Octane Adjustment	CL/OPEN	CLOSED	CLOSED	CLOSED
PNP	PNP	NEUTRAL	NEUTRAL	DRIVE	DRIVE
PSPV	PSP Switch	0-25.5v	14 (wheel turning)	0.1	0.1
RPM	Engine Speed	0-10K rpm	700-730	1350-1650	1800-2060
TCS	TCS (switch)	ON / OFF	ON (if on)	OFF	OFF
TFT (ºF)	TFT Sensor	-40-304ºF	110-210	110-210	110-210
TPV	TP Sensor	0-5.1v	0.5-0.9	0.8-1.1	1.0-1.5v
TRV/TR	TR Sensor	0-5.1v	4.4 / PN	2.1 / OD	2.1 / OD
TSS	TSS Sensor	0-10 KHz	650	1480-1570	1690-1855
Vacuum	Engine Vacuum	0-30" Hg	19-20	18-20	8-15
VPWR	Vehicle Power	0-25.5v	12-14	12-14	12-14
VSS	Vehicle Speed	0-159 mph 0-1 KHz	0 = 0 Hz	30 = 65 Hz	55 = 125 Hz

1996-2002 Taurus/Sable 3.0L V6 MFI VIN S (A/T) - Outputs

PCM PID Acronym	Parameter Identification	PID Range	PID Value at Hot Idle	PID Value at 30	PID Value at 55
AIR	AIR Solenoid	ON / OFF	OFF	OFF	OFF
CD1-CD3	Coil Driver 1-3	0-90° dwell	7	11	12
CTO	Clean Tach Output	0-10 KHz	33-37	75-85	92-120
EGRVR	VR Solenoid	0-100%	0	0-50	0-60
EPC	EPC Solenoid	0-100 psi	9-15	16-25	27-29
EVAPCV	EVAP CV Valve	0-100%	0-100	0-100	0-100
EVAPDC	EVAP Purge Valve	0-100%	0-100	0-100	0-100
FP	Fuel Pump Control	0-100%	100	100	100
FuelPW1	INJ Pulsewidth B1	0-99.9 ms	2.2-2.7	2.3-5.5	2.0-7.0
FuelPW2	INJ Pulsewidth B2	0-99.9 ms	2.2-2.7	2.3-5.5	2.0-7.0
HFC	High Fan Control	OFF	ON (if on)	OFF	OFF
HTR-11	HO2S-11 Heater	ON / OFF	ON	ON	ON
HTR-12	HO2S-12 Heater	ON / OFF	ON	ON	ON
HTR-21	HO2S-21 Heater	ON / OFF	ON	ON	ON
HTR-22	HO2S-22 Heater	ON / OFF	ON	ON	ON
IAC	Idle Air Control	0-100%	25-35	30-35	50-59
INJ1	INJ 1 Pulsewidth	0-99.9 ms	2.2-2.7	2.3-5.5	2.0-7.0
INJ2	INJ 2 Pulsewidth	0-99.9 ms	2.2-2.7	2.3-5.5	2.0-7.0
INJ3	INJ 3 Pulsewidth	0-99.9 ms	2.2-2.7	2.3-5.5	2.0-7.0
INJ4	INJ 4 Pulsewidth	0-99.9 ms	2.2-2.7	2.3-5.5	2.0-7.0
INJ5	INJ 5 Pulsewidth	0-99.9 ms	2.2-2.7	2.3-5.5	2.0-7.0
INJ6	INJ 6 Pulsewidth	0-99.9 ms	2.2-2.7	2.3-5.5	2.0-7.0
LFC	Low Fan Control	ON / OFF	ON (if on)	OFF	OFF
LongFT1	Long Term F/T B1	-20 to 20%	-5 to +5	-5 to +5	-5 to +5
LongFT2	Long Term F/T B2	-20 to 20%	-5 to +5	-5 to +5	-5 to +5
MIL	MIL Control	ON / OFF	ON (if on)	OFF	OFF
SHTFT1	Short Term F/T B1	-20 to 20%	-5 to +5	-5 to +5	-5 to +5
SHTFT2	Short Term F/T B2	-20 to 20%	-5 to +5	-5 to +5	-5 to +5
SPARK	Spark Advance	-90° to 90°	12-27	25-42	20-40
SS1	Shift Solenoid 1	ON / OFF	OFF	OFF	ON
SS2	Shift Solenoid 2	ON / OFF	ON	OFF	OFF
SS3	Shift Solenoid 3	ON / OFF	OFF	ON	ON
TCC	TCC Solenoid	0-100%	0	0-70	90-100
VMV	EVAP VMV Valve (1996-97)	0-100%	0	0-100	0-100
VREF	Voltage Reference	0-5.1v	4.9-5.1	4.9-5.1	4.9-5.1
WAC	A/C WOT Cutout Relay	ON / OFF	ON (if on)	OFF	OFF

1990 Taurus/Sable 3.0L V6 MFI VIN U (A/T) - California

PCM PID Acronym	Parameter Identification	PID Range	PID Value at Hot Idle	PID Value at 30	PID Value at 55
ARC	Auto Ride Control	ON / OFF	ON	ON	ON
BOO	Brake Switch	ON / OFF	ON (if on)	OFF	OFF
CANP	EVAP Purge Valve	ON / OFF	OFF	ON	ON
ECT (°F)	ECT Sensor	-40-304°F	160-200	160-200	160-200
ECT (V)	ECT Sensor	0-5.1v	0.6	0.6	0.6
EVR	VR Solenoid	0-100%	0	0-40	40-50
GEAR	Gear Position	P-R-N-D	P-R-N-D	DRIVE 3	DRIVE 4
HFC	High Fan Control	ON / OFF	ON (if on)	OFF	OFF
FuelPW1	INJ Pulsewidth B1	0-99.9 ms	5.4-5.6	4.8-5.4	12-13
FuelPW2	INJ Pulsewidth B2	0-99.9 ms	5.4-5.6	4.8-5.4	12-13
HO2S-11	HO2S-11 (Bank 1)	0.1-1.1v	0-1v	0-1v	0-1v
HO2S-21	HO2S-21 (Bank 2)	0.1-1.1v	0-1v	0-1v	0-1v
IAC	Idle Air Control	0-100%	20-40	34-40	45-55
IAT (°F)	IAT Sensor	-40-304°F	50-120	50-120	50-120
IAT (V)	IAT Sensor	0-5.1v	1.7-3.5	1.7-3.5	1.7-3.5
LFC	Low Fan Control	ON / OFF	ON (if on)	OFF	OFF
LOOP	Loop Status	CL or OL	CL	CL	CL
MAP	MAP Sensor Hertz	0-1 KHz	107 Hz	112	128 Hz
PFE (V)	PFE Sensor	0-5.1v	3.2	3.3-3.4	3.1-3.5v
PNP	PNP Switch	NEUTRAL	NEUTRAL	DRIVE	DRIVE
PSP	PSP Switch	ON / OFF	ON turning	OFF	OFF
RPM	Engine Speed	0-10K rpm	700-750	1410-1510	1760-1860
SCCS	S/C Command Switch	0-25.5v	6.7	6.7	6.7
SCVAC	S/C Vacuum Solenoid	ON / OFF	OFF	OFF	OFF
SCVENT	S/C Vent Solenoid	ON / OFF	OFF	ON	ON
SHTFT1	Short Term F/T B1	-20 to +20	-5 to +5	-5 to +5	-5 to +5
SHTFT2	Short Term F/T B2	-20 to +20	-5 to +5	-5 to +5	-5 to +5
SPARK	Spark Advance	-90° to 90°	15-22	28-35	25-35
TCC	TCC Solenoid	ON / OFF	OFF	ON	ON
THS 3-2	Trans. 3-2 Switch	0-25.5v	11v	0.1	0.1
THS 4-3	Trans. 4-3 Switch	0-25.5v	11v	0.1	0.1
TPV	TP Sensor	0-5.1v	0.7	0.9	1.2-1.4
TP Mode	TP Mode	C/T or P/T	C/T	P/T	P/T
VPWR	Vehicle Power	0-25.5v	12-14	12-14	12-14
VREF	Voltage Reference	0-5.1v	4.9-5.1	4.9-5.1	4.9-5.1
VSS	Vehicle Speed	0-159 mph	0	30	55
WAC	A/C WOT Cutout Relay	ON / OFF	ON (if on)	OFF	OFF

1991-95 Taurus/Sable 3.0L V6 MFI VIN U (A/T) Inputs / Outputs

PCM PID Acronym	Parameter Identification	PID Range	PID Value at Hot Idle	PID Value at 30	PID Value at 55
ACCS	A/C Clutch Switch	ON / OFF	ON (if on)	OFF	OFF
BOO	Brake Switch	ON / OFF	ON (if on)	OFF	OFF
CANP	EVAP Purge Valve	0-100%	0	0-40	85-95
ECT (°F)	ECT Sensor	-40-304°F	160-200	160-200	160-200
ECT (V)	ECT Sensor	0-5.1v	0.6	0.6	0.6
EPC	EPC Solenoid	0-100 psi	40	15	42
EVR	VR Solenoid	0-100%	0	0-40	40-50
FPM	Fuel Pump Monitor	ON / OFF	ON	ON	ON
GEAR	Gear Position	P-R-N-D	P-R-N-D	DRIVE 3	DRIVE 4
HFC	High Fan Control	ON / OFF	ON (if on)	OFF	OFF
FUELB1	INJ On-time B1	0-99.9 ms	5.4-5.6	4.8-5.4	12-13
FUELB2	INJ On-time B2	0-99.9 ms	5.4-5.6	4.8-5.4	12-13
HO2S-11	HO2S-11 (Bank 1)	0.1-1.1v	0.1-1.1	0.1-1.1	0.1-1.1
HO2S-21	HO2S-21 (Bank 2)	0.1-1.1v	0.1-1.1	0.1-1.1	0.1-1.1
IAC	Idle Air Control	0-100%	20-40	34-40	45-55
IAT (°F)	IAT Sensor	-40-304°F	50-120	50-120	50-120
IAT (V)	IAT Sensor	0-5.1v	1.7-3.5	1.7-3.5	1.7-3.5
LFC	Low Fan Control	ON / OFF	ON (if on)	OFF	ON
LOOP	Loop Status	CL or OL	CL	CL	CL
LongFT1	Long Term F/T B1	-20 to 20%	-5 to +5	-5 to +5	-5 to +5
LongFT2	Long Term F/T B2	-20 to 20%	-5 to +5	-5 to +5	-5 to +5
MAF (V)	MAF Sensor	0-5.1v	0.7	1.3-1.6	1.9-2.2
MLP (V)	MLP Sensor	0-5.1v	0v	5v	5v
PFE (V)	PFE Sensor volts	0-5.1v	3.2	3.3	3.4
PNP	PNP Switch	NEUTRAL	NEUTRAL	DRIVE	DRIVE
PSP	PSP Switch	ON / OFF	ON turning	OFF	OFF
RPM	Engine Speed	0-10K rpm	700-750	1410-1510	1760-1860
SHTFT1	Short Term F/T B1	-20 to 20%	-5 to +5	-5 to +5	-5 to +5
SHTFT2	Short Term F/T B2	-20 to 20%	-5 to +5	-5 to +5	-5 to +5
SPARK	Spark Advance	-90° to 90°	26-30	44-48	46-52
TCC	TCC Solenoid	0-100%	0	90-100	90-100
TOT °F	TOT Sensor	-40-304°F	110-210	110-210	110-210
TOT (V)	TOT Sensor	0-5.1v	0.5-2.0	0.5-2.0	0.5-2.0
TPV	TP Sensor	0-5.1v	0.7	0.9	1.2-1.4
TP MIN	TP Minimum	0-5.1v	0.5	0.9	1.2-1.4
TP Mode	TP Mode	C/T or P/T	C/T	P/T	P/T
TSS	TSS Sensor	0-10 KHz	790-820	1400-1450	1740-1800
VPWR	Vehicle Power	0-25.5v	12-14	12-14	12-14
VREF	Voltage Reference	0-5.1v	4.9-5.1	4.9-5.1	4.9-5.1
VSS	Vehicle Speed	0-159 mph	0	30	55
WAC	A/C WOT Cutout	ON / OFF	ON (if on)	OFF	OFF

1995 Taurus/Sable 3.0L V6 MFI VIN U (A/T) - California

PCM PID Acronym	Parameter Identification	PID Range	PID Value at Hot Idle	PID Value at 30	PID Value at 55
ACCS	A/C Clutch Switch	ON / OFF	ON (if on)	OFF	OFF
BOO	Brake Switch	ON / OFF	ON (if on)	OFF	OFF
CANP	EVAP Purge Valve	0-100%	0	0-40	85-95
DPFE	DPFE Sensor	0-5.1v	0.4-0.6	0.4-0.9	0.4-2.3
ECT (°F)	ECT Sensor	-40-304°F	160-200	160-200	160-200
ECT (V)	ECT Sensor	0-5.1v	0.6	0.6	0.6
EPC	EPC Solenoid	0-100 psi	40	15	42
EVR	VR Solenoid	0-100%	0%	0-40	40-50
FPM	Fuel Pump Monitor	ON / OFF	ON	ON	ON
GEAR	Gear Position	P-R-N-D	P-R-N-D	DRIVE 3	DRIVE 4
HFC	High Fan Control	ON / OFF	ON (if on)	OFF	OFF
FUELB1	INJ On-time B1	0-99.9 ms	5.4-5.6	4.8-5.4	12-13
FUELB2	INJ On-time B2	0-99.9 ms	5.4-5.6	4.8-5.4	12-13
HO2S-11	HO2S-11 (Bank 1)	0.1-1.1v	0-1v	0-1v	0-1v
HO2S-21	HO2S-21 (Bank 2)	0.1-1.1v	0-1v	0-1v	0-1v
IAC	Idle Air Control	0-100%	20-40	34-40	45-55
IAT (°F)	IAT Sensor	-40-304°F	50-120	50-120	50-120
IAT (V)	IAT Sensor	0-5.1v	1.7-3.5	1.7-3.5	1.7-3.5
LFC	Low Fan Control	ON / OFF	ON (if on)	OFF	ON
LOOP	Loop Status	CL or OL	CL	CL	CL
LongFT1	Long Term F/T B1	-20 to 20%	-5 to +5	-5 to +5	-5 to +5
LongFT2	Long Term F/T B2	-20 to 20%	-5 to +5	-5 to +5	-5 to +5
MAF (V)	MAF Sensor	0-5.1v	0.7	1.3-1.6	1.9-2.2
MLP (V)	MLP Sensor	0-5.1v	0v	5v	5v
PNP	PNP Switch	NEUTRAL	NEUTRAL	DRIVE	DRIVE
PSP	PSP Switch	ON / OFF	ON turning	OFF	OFF
RPM	Engine Speed	0-10K rpm	700-750	1410-1510	1760-1860
SHTFT1	Short Term F/T B1	-20 to 20%	-5 to +5	-5 to +5	-5 to +5
SHTFT2	Short Term F/T B2	-20 to 20%	-5 to +5	-5 to +5	-5 to +5
SPARK	Spark Advance	-90° to 90°	26-30	44-48	46-52
TCC	TCC Solenoid	0-100%	0%	90-100	90-100
TOT °F	TOT Sensor	-40-304°F	110-210	110-210	110-210
TOT (V)	TOT Sensor	0-5v	0.5-2.0	0.5-2.0	0.5-2.0
TPV	TP Sensor	0-5.1v	0.7	0.9	1.2-1.4
TP MIN	TP Minimum	0-5.1v	0.5	0.9	1.2-1.4
TP Mode	TP Mode	C/T or P/T	C/T	P/T	P/T
TSS	TSS Sensor	0-10 KHz	790-820	1400-1450	1740-1800
VPWR	Vehicle Power	0-25.5v	12-14	12-14	12-14
VREF	Voltage Reference	0-5.1v	4.9-5.1	4.9-5.1	4.9-5.1
VSS	Vehicle Speed	0-159 mph	0	30	55
WAC	A/C WOT Cutout	ON / OFF	ON (if on)	OFF	OFF

1996-2003 Taurus/Sable 3.0L V6 MFI VIN U (A/T) - Inputs

PCM PID Acronym	Parameter Identification	PID Range	PID Value at Hot Idle	PID Value at 30	PID Value at 55
ACCS	A/C Clutch Switch	ON / OFF	ON (if on)	OFF	OFF
ACET	A/C Temp. Sensor	0-12v	4.95-5.15	4.95-5.15	4.95-5.15
ACP	A/C Press. Sensor	CL/OPEN	OPEN	OPEN	OPEN
AIR-M	EAM Monitor (CAL)	ON / OFF	ON	ON	ON
BPP	Brake Pedal Switch	ON / OFF	ON (if on)	OFF	OFF
CID	CID Sensor	0-1 KHz	6-8	12-14	13-16
CKP	CKP Sensor	0-10 KHz	410-510	810-950	1050-1820
DPFE	DPFE Input (96-'00)	0-5.1v	0.2-1.3	0.2-4.5	0.2-4.5
DPFE	DPFE Sensor	0-5.1v	0.95-1.05	0.9-4.5	0.9-4.5
ECT (°F)	ECT Sensor	-40-304°F	160-200	160-200	160-200
FRP	Fuel Rail Pressure	0-100 psi	54	42	40
FEPS	Flash EEPROM	0-5.1v	0.5-0.6	0.5-0.6	0.5-0.6
FLI	Fuel Level Indicator (1998-02)	0-100%	50 (1.7v)	25-75	25-75
FPM	FP Monitor ('96-'99)	ON / OFF	ON	ON	ON
FPM	FP Monitor ('00-'02)	0-100%	90	90	90
FTP	FTP Sensor (1998-02)	0-5.1v or 0-10" H20	2.6v at 0" of H20	2.6v at 0" of H20	2.6v at 0" of H20
GEAR	Transmission Gear	1-2-3-4	1	3	4
GFS	Generator Field	0-1 KHz	130 (30%)	130 (27%)	130 (33%)
HO2S-11	HO2S-11 (Bank 1)	0.1-1.1v	0.1-1.1	0.1-1.1	0.1-1.1
HO2S-12	HO2S-12 (Bank 1)	0.1-1.1v	0.1-1.1	0.1-1.1	0.1-1.1
HO2S-21	HO2S-21 (Bank 2)	0.1-1.1v	0.1-1.1	0.1-1.1	0.1-1.1
HO2S-22	HO2S-22 (Bank 2)	0.1-1.1v	0.1-1.1	0.1-1.1	0.1-1.1
IAT (°F)	IAT Sensor	-40-304°F	50-120	50-120	50-120
KS1	Knock Sensor 1	0-25v	0	0	0
LOAD	Engine Load	0-100%	10-20	16-30	13-50
MAF (V)	MAF Sensor	0-5.1v	0.6-0.9	1.0-1.5	1.1-2.0
MISF	Misfire Detection	OFF	OFF	OFF	OFF
OCTADJ	Octane Adjust	Retard/No	No Retard	No Retard	No Retard
OSS	Output Shaft Speed	0-10 KHz	0	300	500
PATSIN	PAT System	See PATS	See PATS	See PATS	See PATS
PNP	PNP Switch	ON / OFF	ON (in 'P')	OFF	OFF
PSP	PSP Switch	HI / LOW	HI: turning	LOW	LOW
RPM	Engine Speed	0-10K rpm	660-800	1440-1625	1830-1970
TFT (°F)	TFT Sensor	-40-304°F	110-210	110-210	110-210
TPV	TP Sensor	0-5.1v	0.53-1.27	0.8-1.2	0.8-1.2
TR1	TR Sensor 1	0-12v	0	10.5	10.5
TR2	TR Sensor 2	0-12v	0	10.5	10.5
TR4	TR Sensor 4	0-12v	0	10.5	10.5
TRV/TR	TR Sensor	0-5.1v	4.4 / PN	1.7 / OD	1.7 / OD
TSS	Turbine shaft speed	0-10 KHz	50-65	82-99	88-120
Vacuum	Engine Vacuum	0-30" Hg	19-20	18-20	8-15
VPWR	Vehicle Power	0-25.5v	12-14	12-14	12-14
VSS	Vehicle Speed	0-159 mph 0-1 KHz	0 = 0 Hz	30 = 65 Hz	55 = 125 Hz

1996-2003 Taurus/Sable 3.0L V6 MFI VIN U (A/T) - Outputs

PCM PID Acronym	Parameter Identification	PID Range	PID Value at Hot Idle	PID Value at 30	PID Value at 55
AIR	EAM Signal (CAL)	ON / OFF	OFF	OFF	OFF
CD1-CD3	Coil Driver 1-3	0-60° dwell	7	10	12
CTO	Clean Tach Output	0-10 KHz	35-50	65-78	91-105
EGRVR	VR Solenoid	0-100%	0	0-40	36-50
EPC AX4N	EPC Solenoid	0-100 psi	40	15	42
EPC AX4S	EPC Solenoid	0-100 psi	15	17	40
EVAPCV	EVAP CV Valve	0-100%	0-100	0-100	0-100
EVAPDC	EVAP Purge Valve	0-100%	0-100	0-100	0-100
FP	Fuel Pump (96-00)	ON / OFF	ON	ON	ON
FP	Fuel Pump (01-02)	0-12v	1v (100%)	1v (100%)	1v (100%)
FUELB1	INJ Pulsewidth B1	0-99.9 ms	3.8-4.7	3.9-8.0	3.0-9.0
FUELB2	INJ Pulsewidth B2	0-99.9 ms	3.8-4.7	3.9-8.0	3.0-9.0
GENFDC	Generator Field	0-1 KHz	0	0	0
HFC	High Fan Control	ON / OFF	ON (if on)	OFF	OFF
HTR-11	HO2S-11 Heater	ON / OFF	ON	ON	ON
HTR-12	HO2S-12 Heater	ON / OFF	ON	ON	ON
HTR-21	HO2S-21 Heater	ON / OFF	ON	ON	ON
HTR-22	HO2S-22 Heater	ON / OFF	ON	ON	ON
IAC	Idle Air Control	0-100%	40	40-60	40-47
INJ1	INJ 1 Pulsewidth	0-99.9 ms	3.8-4.7	3.9-8.0	3.0-9.0
INJ2	INJ 2 Pulsewidth	0-99.9 ms	3.8-4.7	3.9-8.0	3.0-9.0
INJ3	INJ 3 Pulsewidth	0-99.9 ms	3.8-4.7	3.9-8.0	3.0-9.0
INJ4	INJ 4 Pulsewidth	0-99.9 ms	3.8-4.7	3.9-8.0	3.0-9.0
INJ5	INJ 5 Pulsewidth	0-99.9 ms	3.8-4.7	3.9-8.0	3.0-9.0
INJ6	INJ 6 Pulsewidth	0-99.9 ms	3.8-4.7	3.9-8.0	3.0-9.0
LFC	Low Fan Control	ON / OFF	ON (if on)	OFF	OFF
LongFT1	Long Term F/T B1	-20 to 20%	-5 to +5	-5 to +5	-5 to +5
LongFT2	Long Term F/T B2	-20 to 20%	-5 to +5	-5 to +5	-5 to +5
MIL	MIL Control	ON / OFF	ON (if on)	OFF	OFF
PATSOUT	PAT System	See PATS	See PATS	See PATS	See PATS
PATSIL	PAT System	See PATS	See PATS	See PATS	See PATS
PATSTRT	PAT System	See PATS	See PATS	See PATS	See PATS
SHTFT1	Short Term F/T B1	-20 to 20%	-5 to +5	-5 to +5	-5 to +5
SHTFT2	Short Term F/T B2	-20 to 20%	-5 to +5	-5 to +5	-5 to +5
SPARK	Spark Advance	-90° to 90°	24-30	34-42	33-46
SS1	Shift Solenoid 1	ON / OFF	OFF	ON	ON
SS2	Shift Solenoid 2	ON / OFF	ON	OFF	ON
SS3	Shift Solenoid 3	ON / OFF	OFF	ON	ON
TCC	TCC Solenoid	0-100%	0	42	95-100
VMV	EVAP VMV '96-'97	0-100	0-100	0-100	0-100
VREF	Voltage Reference	0-5.1v	4.9-5.1	4.9-5.1	4.9-5.1
VSO	VSS Output Signal	0-1 KHz	0	65	125
WAC	A/C WOT Cutout Relay	ON / OFF	ON (if on)	OFF	OFF

1990 Taurus/Sable 3.8L V6 MFI VIN 4 (A/T) Inputs / Outputs

PCM PID Acronym	Parameter Identification	PID Range	PID Value at Hot Idle	PID Value at 30	PID Value at 55
ARC	Auto Ride Control	ON / OFF	OFF	OFF	OFF
BOO	Brake Switch	ON / OFF	ON (if on)	OFF	OFF
CANP	EVAP Purge Valve	ON / OFF	OFF	ON	ON
ECT (°F)	ECT Sensor	-40-304°F	160-200	160-200	160-200
ECT (V)	ECT Sensor	0-5.1v	0.6	0.6	0.6
EVR	VR Solenoid	0-100%	0	0-40	0-60
GEAR	Gear Position	P-R-N-D	P-R-N-D	DRIVE 3	DRIVE 4
FLPW1	INJ On-time (B1)	0-99.9 ms	5.9-6.1	7.4-8.4	8.0-9.0
FLPW2	INJ On-time (B2)	0-99.9 ms	5.9-6.1	7.4-8.4	8.0-9.0
HFC	High Fan Control	ON / OFF	ON (if on)	OFF	OFF
HO2S-11	HO2S-11 (Bank 1)	0.1-1.1v	0.1-1.1	0.1-1.1	0.1-1.1
HO2S-21	HO2S-21 (Bank 2)	0.1-1.1v	0.1-1.1	0.1-1.1	0.1-1.1
IAC	Idle Air Control	0-100%	20-40	34-40	45-55
IAT (°F)	IAT Sensor	-40-304°F	50-120	50-120	50-120
IAT (V)	IAT Sensor	0-5.1v	1.7-3.5	1.7-3.5	1.7-3.5
LFC	Low Fan Control	ON / OFF	ON (if on)	OFF	OFF
LongFT1	Long Term F/T B1	-20 to 20%	-5 to +5	-5 to +5	-5 to +5
LongFT2	Long Term F/T B2	-20 to 20%	-5 to +5	-5 to +5	-5 to +5
LOOP	Loop Status	CL or OL	CL	CL	CL
MAP	MAP Sensor Hertz	0-1 KHz	107	112	128
PFE (V)	PFE Sensor	0-5.1v	3.2	3.3	3.4
PNP	PNP Switch	NEUTRAL	NEUTRAL	DRIVE	DRIVE
RPM	Engine Speed	0-10K rpm	650-750	1340-1440	2410-2510
SCCS	S/C Command Switch	0-25.5v	6.7	6.7	6.7
SCVAC	S/C Vacuum Solenoid	0-100%	0	0	0
SCVENT	S/C Vent Solenoid	0-100%	0	98	98
SHTFT1	Short Term F/T B1	-20 to 20%	-5 to +5	-5 to +5	-5 to +5
SHTFT2	Short Term F/T B2	-20 to 20%	-5 to +5	-5 to +5	-5 to +5
SPARK	Spark Advance	-90° to 90°	18-22	34-38	41-45
TCC	TCC Solenoid	ON / OFF	OFF	ON	ON
THS 3-2	Trans. 3-2 Switch	0-25.5v	11	0.1	0.1
THS 4-3	Trans. 4-3 Switch	0-25.5v	11	0.1	0.1
TPV	TP Sensor	0-5.1v	0.8	0.9-1.0	1.0-1.1
TP Mode	TP Mode	C/T or P/T	C/T	P/T	P/T
VPWR	Vehicle Power	0-25.5v	12-14	12-14	12-14
VREF	Voltage Reference	0-5.1v	4.9-5.1	4.9-5.1	4.9-5.1
VSS	Vehicle Speed	0-159 mph	0	30	55
WAC	A/C WOT Cutout	ON / OFF	ON (if on)	Off	Off

1991 Taurus/Sable 3.8L V6 MFI VIN 4 (A/T) Inputs / Outputs

PCM PID Acronym	Parameter Identification	PID Range	PID Value at Hot Idle	PID Value at 30	PID Value at 55
BOO	Brake Switch	ON / OFF	ON (if on)	OFF	OFF
CANP	EVAP Purge Valve	ON / OFF	OFF	ON	ON
ECT (°F)	ECT Sensor	-40-304°F	160-200	160-200	160-200
ECT (V)	ECT Sensor	0-5.1v	0.6	0.6	0.6
EVR	VR Solenoid	0-100%	0	0-40	0-60
GEAR	Gear Position	P-R-N-D	P-R-N-D	DRIVE 3	DRIVE 4
FuelPW1	INJ Pulsewidth B1	0-99.9 ms	5.9-6.1	7.4-8.4	8.0-9.0
FuelPW2	INJ Pulsewidth B2	0-99.9 ms	5.9-6.1	7.4-8.4	8.0-9.0
HFC	High Fan Control	ON / OFF	ON (if on)	OFF	ON
HO2S-11	HO2S-11 (Bank 1)	0.1-1.1v	0.1-1.1	0.1-1.1	0.1-1.1
HO2S-21	HO2S-21 (Bank 2)	0.1-1.1v	0.1-1.1	0.1-1.1	0.1-1.1
IAC	Idle Air Control	0-100%	20-40	34-40	45-55
IAT (°F)	IAT Sensor	-40-304°F	50-120	50-120	50-120
IAT (V)	IAT Sensor	0-5.1v	1.7-3.5	1.7-3.5	1.7-3.5
LFC	Low Fan Control	ON / OFF	ON (if on)	OFF	OFF
LOOP	Loop Status	CL or OL	CL	CL	CL
LongFT1	Long Term F/T B1	-20 to 20%	-5 to +5	-5 to +5	-5 to +5
LongFT2	Long Term F/T B2	-20 to 20%	-5 to +5	-5 to +5	-5 to +5
MAF (V)	MAF Sensor	0-5.1v	0.1	0.6-0.8	1.9-2.2
PFE (V)	PFE Sensor	0-5.1v	3.2	3.3	3.4
PNP	PNP Switch	NEUTRAL	NEUTRAL	DRIVE	DRIVE
RPM	Engine Speed	0-10K rpm	650-750	1340-1440	2410-2510
SHTFT1	Short Term F/T B1	-20 to 20%	-5 to +5	-5 to +5	-5 to +5
SHTFT1	Short Term F/T B2	-20 to 20%	-5 to +5	-5 to +5	-5 to +5
SPARK	Spark Advance	-90° to 90°	18-22	34-38	41-45
TCC	TCC Solenoid	ON / OFF	OFF	ON	ON
TPV	TP Sensor	0-5.1v	0.8	0.9-1.0	1.0-1.1
TP Mode	TP Mode	C/T or P/T	C/T	P/T	P/T
VPWR	Vehicle Power	0-25.5v	12-14	12-14	12-14
VREF	Voltage Reference	0-5.1v	4.9-5.1	4.9-5.1	4.9-5.1
VSS	Vehicle Speed	0-159 mph	0	30	55
WAC	A/C WOT Cutout Relay	ON / OFF	ON (if on)	OFF	OFF

1992-95 Taurus/Sable 3.8L V6 MFI VIN 4 (A/T) Inputs / Outputs

PCM PID Acronym	Parameter Identification	PID Range	PID Value at Hot Idle	PID Value at 30	PID Value at 55
ACCS	A/C Clutch Switch	ON / OFF	ON (if on)	OFF	OFF
BOO	Brake Switch	ON / OFF	ON (if on)	OFF	OFF
CANP	EVAP Purge Valve	0-100%	0	0-40	85-95
ECT (°F)	ECT Sensor	-40-304°F	160-200	160-200	160-200
ECT (V)	ECT Sensor	0-5.1v	0.6	0.6	0.6
EPC	EPC Solenoid	0-100 psi	16-18	12-16	18-22
EVR	VR Solenoid	0-100%	0%	0-40	35-45
FPM	Fuel Pump Monitor	ON / OFF	ON	ON	ON
GEAR	Gear Position	P-R-N-D	P-R-N-D	DRIVE 3	DRIVE 4
HFC	High Fan Control	ON / OFF	ON (if on)	OFF	OFF
FUEL B1	INJ On-time B1	0-99.9 ms	5.9-6.1	7.4-8.4	8.0-9.0
FUEL B2	INJ On-time B2	0-99.9 ms	5.9-6.1	7.4-8.4	8.0-9.0
HO2S-11	HO2S-11 (Bank 1)	0.1-1.1v	0-1v	0-1v	0-1v
HO2S-21	HO2S-21 (Bank 2)	0.1-1.1v	0-1v	0-1v	0-1v
IAC	Idle Air Control	0-100%	34	44-74	50-75
IAT (°F)	IAT Sensor	-40-304°F	50-120	50-120	50-120
IAT (V)	IAT Sensor	0-5.1v	1.7-3.5	1.7-3.5	1.7-3.5
LFC	Low Fan Control	ON / OFF	ON (if on)	OFF	ON
LOOP	Loop Status	CL or OL	CL	CL	CL
LongFT1	Long Term F/T B1	-20 to 20%	-5 to +5	-5 to +5	-5 to +5
LongFT2	Long Term F/T B2	-20 to 20%	-5 to +5	-5 to +5	-5 to +5
MAF (V)	MAF Sensor	0-5.1v	0.7	1.3-1.6	1.9-2.2
MLP (V)	MLP Sensor	0-5.1v	0v	5	5
PFE (V)	PFE Sensor Volts	0-5.1v	3.2	3.3	3.4
PNP	PNP Switch	NEUTRAL	NEUTRAL	DRIVE	DRIVE
PSP	PSP Switch	ON / OFF	ON turning	OFF	OFF
RPM	Engine Speed	0-10K rpm	650-750	1340-1440	2410-2510
SHTFT1	Short Term F/T B1	-20 to 20%	-5 to +5	-5 to +5	-5 to +5
SHTFT2	Short Term F/T B2	-20 to 20%	-5 to +5	-5 to +5	-5 to +5
SPARK	Spark Advance	-90° to 90°	26-30	44-48	46-52
TCC	TCC Solenoid	0-100%	0	90-100	90-100
TOT (°F)	TOT Sensor	-40-304°F	110-210	110-210	110-210
TOT (V)	TOT Sensor	0-5.1v	.5-2.0	0.5-2.0	0.5-2.0
TPV	TP Sensor	0-5.1v	0.7	0.9	1.2-1.4
TP MIN	TP Minimum	0-5.1v	0.5	0.9	1.2-1.4
TP Mode	TP Mode	C/T or P/T	C/T	P/T	P/T
TSS	TSS Sensor	0-10 KHz	790-820	1400-1450	1740-1800
VPWR	Vehicle Power	0-25.5v	12-14	12-14	12-14
VREF	Voltage Reference	0-5.1v	4.9-5.1	4.9-5.1	4.9-5.1
VSS	Vehicle Speed	0-159 mph	0	30	55
WAC	A/C WOT Cutout	ON / OFF	ON (if on)	OFF	OFF

1994-95 Taurus/Sable 3.8L V6 MFI VIN 4 (A/T) - California

PCM PID Acronym	Parameter Identification	PID Range	PID Value at Hot Idle	PID Value at 30	PID Value at 55
ACCS	A/C Clutch Switch	ON / OFF	ON (if on)	OFF	OFF
BOO	Brake Switch	ON / OFF	ON (if on)	OFF	OFF
CANP	EVAP Purge Valve	0-100%	0	0-40	85-95
DPFE	DPFE Sensor	0-5.1v	0.4-0.6	0.4-0.7	0.4-1.6v
ECT (°F)	ECT Sensor	-40-304°F	160-200	160-200	160-200
ECT (V)	ECT Sensor	0-5.1v	0.6	0.6	0.6
EPC	EPC Solenoid	0-100 psi	16-18	12-16	18-22
EVR	VR Solenoid	0-100%	0	0-40	35-45
FPM	Fuel Pump Monitor	ON / OFF	ON	ON	ON
GEAR	Gear Position	P-R-N-D	P-R-N-D	DRIVE 3	DRIVE 4
HFC	High Fan Control	ON / OFF	ON (if on)	OFF	OFF
FUEL B1	INJ On-time B1	0-99.9 ms	5.9-6.1	7.4-8.4	8.0-9.0
FUEL B2	INJ On-time B2	0-99.9 ms	5.9-6.1	7.4-8.4	8.0-9.0
HO2S-11	HO2S-11 (Bank 1)	0.1-1.1v	0.1-1.1	0.1-1.1	0.1-1.1
HO2S-21	HO2S-21 (Bank 2)	0.1-1.1v	0.1-1.1	0.1-1.1	0.1-1.1
IAC	Idle Air Control	0-100%	34	44-74	50-75
IAT (°F)	IAT Sensor	-40-304°F	50-120	50-120	50-120
IAT (V)	IAT Sensor	0-5.1v	1.7-3.5	1.7-3.5	1.7-3.5
LFC	Low Fan Control	ON / OFF	ON (if on)	OFF	ON
LOOP	Loop Status	CL or OL	CL	CL	CL
LongFT1	Long Term F/T B1	-20 to 20%	-5 to +5	-5 to +5	-5 to +5
LongFT2	Long Term F/T B2	-20 to 20%	-5 to +5	-5 to +5	-5 to +5
MAF (V)	MAF Sensor	0-5.1v	0.7	1.3-1.6	1.9-2.2
MLP (V)	MLP Sensor	0-5.1v	0	5	5
PNP	PNP Switch	NEUTRAL	NEUTRAL	DRIVE	DRIVE
PSP	PSP Switch	ON / OFF	ON turning	OFF	OFF
RPM	Engine Speed	0-10K rpm	650-750	1340-1440	2410-2510
SHTFT1	Short Term F/T B1	-20 to 20%	-5 to +5	-5 to +5	-5 to +5
SHTFT2	Short Term F/T B2	-20 to 20%	-5 to +5	-5 to +5	-5 to +5
SPARK	Spark Advance	-90° to 90°	26-30	44-48	46-52
TCC	TCC Solenoid	0-100%	0	90-100	90-100
TOT (°F)	TOT Sensor	-40-304°F	110-210	110-210	110-210
TOT (V)	TOT Sensor	0-5.1v	0.5-2.0	0.5-2.0	0.5-2.0
TPV	TP Sensor	0-5.1v	0.7	0.9	1.2-1.4
TP MIN	TP Minimum	0-5.1v	0.5	0.9	1.2-1.4
TP Mode	TP Mode	C/T or P/T	C/T	P/T	P/T
TSS	TSS Sensor	0-10 KHz	790-820	1400-1450	1740-1800
VPWR	Vehicle Power	0-25.5v	12-14	12-14	12-14
VREF	Voltage Reference	0-5.1v	4.9-5.1	4.9-5.1	4.9-5.1
VSS	Vehicle Speed	0-159 mph	0	30	55
WAC	A/C WOT Cutout	ON / OFF	ON (if on)	OFF	OFF

Tempo & Topaz PID Data

1992-94 Tempo/Topaz 2.3L VIN X, 3.0L VIN X (All) Inputs/Outputs

PCM PID Acronym	Parameter Identification	PID Range	PID Value at Hot Idle	PID Value at 30	PID Value at 55
ACCS	A/C Clutch Switch	ON / OFF	ON (if on)	OFF	OFF
CANP	EVAP Purge Valve	ON / OFF	OFF	ON	ON
ECT (°F)	ECT Sensor	-40-304°F	160-200	160-200	160-200
ECT (V)	ECT Sensor	0-5.1v	0.6	0.6	0.6
EVR	VR Solenoid	0-100%	0%	0%	30-40
FPM	Fuel Pump Monitor	ON / OFF	ON	ON	ON
FuelPW1	INJ Pulsewidth B1	0-99.9 ms	3.6-6.0	4.0-7.6	4.0-10
HFAN	High Fan Control	ON / OFF	ON (if on)	OFF	OFF
HO2S-11	HO2S-11 (Bank 1)	0.1-1.1v	0.1-1.1	0.1-1.1	0.1-1.1
HO2S-21	HO2S-21 (Bank 2)	0.1-1.1v	0.1-1.1	0.1-1.1	0.1-1.1
IAC	Idle Air Control	0-100%	20-40	34-40	45-55
IAT (°F)	IAT Sensor	-40-304°F	50-120	50-120	50-120
IAT (V)	IAT Sensor	0-5.1v	1.7-3.5	1.7-3.5	1.7-3.5
LFAN	Low Fan Control	ON / OFF	ON (if on)	OFF	OFF
LongFT1	Long Term F/T B1	-20 to 20%	-5 to +5	-5 to +5	-5 to +5
LongFT2	Long Term F/T B2	-20 to 20%	-5 to +5	-5 to +5	-5 to +5
LOOP	Loop Status	CL or OL	CL	CL	CL
MAF (V)	Mass Airflow	0-5.1v	0.5	1.3-1.4	1.5-2.0
PFE (V)	PFE Sensor	0-5.1v	3.3	3.4	2.5-3.5
PNP	PNP Switch	NEUTRAL	NEUTRAL	DRIVE	DRIVE
PSP	PSP Switch	HI / LOW	HI: turning	LOW	LOW
RPM	Engine Speed	0-10K rpm	820-860	1500-1600	2550-2650
SHTFT1	Short Term F/T B1	-20 to 20%	-5 to +5	-5 to +5	-5 to +5
SHTFT2	Short Term F/T B2	-20 to 20%	-5 to +5	-5 to +5	-5 to +5
SPARK	Spark Advance	-90° to 90°	16-30	30-34	34-40
TPV	TP Sensor	0-5.1v	0.8	1.0	1.3-1.4
TP MIN	TP Minimum	0-5.1v	0.8	1.0	1.3-1.4
TP Mode	TP Mode	C/T or P/T	C/T	P/T	P/T
VPWR	Vehicle Power	0-25.5v	12-14	12-14	12-14
VREF	Voltage Reference	0-5.1v	4.9-5.1	4.9-5.1	4.9-5.1
VSS	Vehicle Speed	0-159 mph	0	30	55
WAC	A/C WOT Cutout Relay	ON / OFF	ON (if on)	OFF	OFF

Thunderbird PID Data

1990-95 Thunderbird 3.8L V6 MFI SC VIN R (A/T) Inputs / Outputs

PCM PID Acronym	Parameter Identification	PID Range	PID Value at Hot Idle	PID Value at 30	PID Value at 55
ACCS	A/C Clutch Switch	ON / OFF	ON (if on)	OFF	OFF
BARO	Barometric Pressure Sensor	0-1 KHz	159 Hz (sea level)	159 Hz (sea level)	159 Hz (sea level)
BOO	Brake Switch	ON / OFF	ON (if on)	OFF	OFF
CANP	EVAP Purge Valve	0-100%	0	0-40	85-95
DPFE	DPFE Sensor	0.5.1v	0.4-0.6	0.4-0.9	0.4-2.3
ECT (°F)	ECT Sensor	-40-304°F	160-200	160-200	160-200
ECT (V)	ECT Sensor	0-5.1v	0.6	0.6	0.6
EPC	EPC Solenoid	0-100 psi	40	15	42
EVR	VR Solenoid	0-100%	0%	0-40	40-50
FPM	Fuel Pump Monitor	ON / OFF	ON	ON	ON
GEAR	Gear Position	P-R-N-D	P-R-N-D	DRIVE 3	DRIVE 4
HFC	High Fan Control	ON / OFF	ON (if on)	OFF	OFF
HO2S-11	HO2S-11 (Bank 1)	0.1-1.1v	0-1v	0-1v	0-1v
FUEL B1	INJ On-time B1	0-99.9 ms	3.6-3.8	4.5-6.2	5.8-7.0
FUEL B2	INJ On-time B2	0-99.9 ms	3.6-3.8	4.5-6.2	5.8-7.0
IAC	Idle Air Control	0-100%	20-40	34-40	45-55
IAT (°F)	IAT Sensor	-40-304°F	50-120	50-120	50-120
IAT (V)	IAT Sensor	0-5.1v	1.7-3.5	1.7-3.5	1.7-3.5
LFC	Low Fan Control	ON / OFF	ON (if on)	OFF	ON
LOOP	Loop Status	OL or CL	CL	CL	CL
LNGFT1	LONGFT (Bank 1)	-20 to 20%	-5 to +5	-5 to +5	-5 to +5
LNGFT2	LONGFT (Bank 2)	-20 to 20%	-5 to +5	-5 to +5	-5 to +5
MAF (V)	MAF Sensor	0-5.1v	0.7	1.0-1.4	1.5-2.0v
MLP (V)	MLP Sensor	0-5.1v	0v	5v	5v
PNP	PNP Switch	NEUTRAL	NEUTRAL	DRIVE	DRIVE
PSP	PSP Switch	ON / OFF	ON turning	OFF	OFF
RPM	Engine Speed	0-10K rpm	770-830	1150-1250	1460-1560
SHTFT1	Short Term F/T B1	-20 to 20%	-5 to +5	-5 to +5	-5 to +5
SHTFT2	Short Term F/T B2	-20 to 20%	-5 to +5	-5 to +5	-5 to +5
SPARK	Spark Advance	-90° to 90°	17-22	28-32	28-32
TCC	TCC Solenoid	0-100%	0%	90-100	90-100
TOT °F	TOT Sensor	-40-304°F	110-210	110-210	110-210
TOT (V)	TOT Sensor	0-5.1v	0.5-2.0	0.5-2.0	0.5-2.0
TPV	TP Sensor	0-5.1v	0.7	0.9	1.2-1.4
TP MIN	TP Minimum	0-5.1v	0.5	0.9	1.2-1.4
TP Mode	TP Mode	C/T or P/T	C/T	P/T	P/T
TSS	TSS Sensor	0-10 KHz	790-820	1400-1450	1740-1800
VPWR	Vehicle Power	0-25.5v	12-14	12-14	12-14
VREF	Voltage Reference	0-5.1v	4.9-5.1	4.9-5.1	4.9-5.1
VSS	Vehicle Speed	0-159 mph	0	30	55
WAC	A/C WOT Cutout	ON / OFF	ON (if on)	OFF	OFF

2002-03 Thunderbird 3.9L 4v V8 MFI VIN A (A/T) - Inputs

PCM PID Acronym	Parameter Identification	PID Range	PID Value at Hot Idle	PID Value at 30	PID Value at 55
ACP	A/C Press. Switch	CL/OPEN	OPEN	OPEN	OPEN
BPP	BPP Switch	ON / OFF	ON (if on)	OFF	OFF
BPS	Brake Pedal Switch	0v or 12v	0v (12 on)	0v	0v
CHT (ºF)	CHT Sensor (ºF)	-40-304ºF	160-200	160-200	160-200
CID	CID Sensor	0-1 KHz	6.6	10	17
CKP	CKP Sensor	0-10 KHz	420	665-800	1160-1200
DPFE	DPFE Sensor	0-5.1v	0.25-1.30	0.25-4.65	0.25-4.65
ECT (ºF)	ECT Sensor	-40-304ºF	160-200	160-200	160-200
EFTA (ºF)	EFT Sensor (ºF)	-40-304ºF	50-120	50-120	50-120
FEPS	Flash EEPROM	0-5.1v	0.1	0.1	0.1
FP-M	Fuel Pump Monitor	0-100%	50	50	50
FRP	Fuel Rail Pressure	0-100 psi	44	44	44
FTP	FTP Sensor	0-5.1v or 0-10" H20	2.6v at 0" of H20	2.6v at 0" of H20	2.6v at 0" of H20
FuelPW1	Fuel On-time B1	0-99.9 ms	2.9-3.6	5.1	6.5-7.5
FuelPW2	Fuel On-time B2	0-99.9 ms	2.9-3.6	5.1	6.5-7.5
GEAR	Transmission Gear	1-2-3-4-5	1	4	5
GFS	Generator Field	0-1K Hz	130 (45%)	130 (25%)	130 (20%)
HO2S-11	HO2S-11 (Bank 1)	0.1-1.1v	0.1-1.1	0.1-1.1	0.1-1.1
HO2S-12	HO2S-12 (Bank 1)	0.1-1.1v	0.1-1.1	0.1-1.1	0.1-1.1
HO2S-21	HO2S-21 (Bank 2)	0.1-1.1v	0.1-1.1	0.1-1.1	0.1-1.1
HO2S-22	HO2S-22 (Bank 2)	0.1-1.1v	0.1-1.1	0.1-1.1	0.1-1.1
IAT (ºF)	IAT Sensor	-40-304ºF	50-120	50-120	50-120
ISS	Input Shaft Speed	0-10 KHz	235-380	725	1370
KS1	Knock Sensor B1	0-5.1v	0	0	0
KS2	Knock Sensor B2	0-5.1v	0	0	0
LOAD	Engine Load	0-100%	17-18.6	26-35.7	30-50
MAF (V)	MAF Sensor	0-5.1v	0.7	1.6-1.8	2.1-2.3
MISF	Misfire Detection	ON / OFF	OFF	OFF	OFF
OCTADJ	Octane Adj. Switch	Retard/NO	No Retard	No Retard	No Retard
OSS	OSS Sensor Hertz	0-10 KHz	0	536-595	956-1100
PNP	PNP Switch	ON / OFF	ON	OFF	OFF
PS1	PS1 Switch	0-25.5v	12	11.7	11.7
RPM	Engine Speed	0-10K rpm	660-700	1425	1950
TCS	TCS (switch)	ON / OFF	ON (if on)	OFF	OFF
TSS	TSS Sensor	0-10K rpm	340	539	1025
TFT (ºF)	TFT Sensor	-40-304ºF	110-210	110-210	110-210
TPV	TP Sensor	0-5.1v	0.53-1.05	1.1-1.3	1.3-1.5
TR 1	TR Sensor 1 Volts	0-12v	0.1	11.5	11.5
TR 2	TR2 Sensor	0-12v	0.1	11.5	11.5
TR 4	TR4 Sensor	0-12v	0.1	11.5	11.5
TRV/TR	TR Sensor	0-5.1v	0.1 / PN	1.7 / OD	1.7 / OD
VPWR	Vehicle Power	0-25.5v	12-14	12-14	12-14
VSS	Vehicle Speed	0-159 mph 0-1 KHz	0 = 0 Hz	30 = 65 Hz	55 = 125 Hz

2002-03 T-Bird 3.9L 4v V8 MFI VIN A (A/T) - Outputs

PCM PID Acronym	Parameter Identification	PID Range	PID Value at Hot Idle	PID Value at 30	PID Value at 55
ALDFDC	Alternator Control	0-1 KHz	0-130	0	0
CD1-8	Coil Driver 1-8	0-45° dwell	8	11	12
CHTIL	CHTIL (lamp)	ON / OFF	ON (if on)	OFF	OFF
EGRVR	VR Solenoid	0-100%	0	40	40
EPC	EPC Solenoid (V)	0-25.5v	8.8	9.3	9.9
EPC2	EPC2 Solenoid (V)	0-25.5v	10.9	9.3	9.9
EPC3	EPC3 Solenoid (V)	0-25.5v	8.5	12-14	12-14
EVAPCV	EVAP CV Valve	0-100%	0-100%	0-100%	0-100%
EVAPDC	EVAP Purge Valve	0-100%	0-100%	0-100%	0-100%
FP	Fuel Pump Control	0-100%	26	29	27
HCFD	Hydraulic Cooling Fan Drive	ON / OFF	ON (if on)	OFF	OFF
HTR-11	HO2S-11 Heater	ON / OFF	ON	ON	ON
HTR-12	HO2S-12 Heater	ON / OFF	ON	ON	ON
HTR-21	HO2S-21 Heater	ON / OFF	ON	ON	ON
HTR-22	HO2S-22 Heater	ON / OFF	ON	ON	ON
IAC	Idle Air Control	0-100%	25	54	72
IMRC	IMRC Solenoid	ON / OFF	OFF	OFF	OFF
INJ1	INJ 1 Pulsewidth	0-99.9 ms	2.9-3.6	5.1	6.5-7.5
INJ2	INJ 2 Pulsewidth	0-99.9 ms	2.9-3.6	5.1	6.5-7.5
INJ3	INJ 3 Pulsewidth	0-99.9 ms	2.9-3.6	5.1	6.5-7.5
INJ4	INJ 4 Pulsewidth	0-99.9 ms	2.9-3.6	5.1	6.5-7.5
INJ5	INJ 5 Pulsewidth	0-99.9 ms	2.9-3.6	5.1	6.5-7.5
INJ6	INJ 6 Pulsewidth	0-99.9 ms	2.9-3.6	5.1	6.5-7.5
INJ7	INJ 7 Pulsewidth	0-99.9 ms	2.9-3.6	5.1	6.5-7.5
INJ8	INJ 8 Pulsewidth	0-99.9 ms	2.9-3.6	5.1	6.5-7.5
LNGFT1	L/T Fuel Trim (B1)	-20 to 20%	-5 to +5	-5 to +5	-5 to +5
LNGFT2	L/T Fuel Trim (B2)	-20 to 20%	-5 to +5	-5 to +5	-5 to +5
MIL	MIL Control	ON / OFF	ON (if on)	OFF	OFF
SCC	S/C Signal	0-25.5v	12-14	12-14	12-14
SCCS	S/C Switch Signal	0-5.1v	0.1	4.6	4.6
SCMA	S/C MA Signal	0-25.5v	12-14	9-12	9-12
SCMB	S/C MB Signal	0-25.5v	12-14	9-12	9-12
SCMC	S/C MC Signal	0-25.5v	12-14	9-12	9-12
SHTFT1	Short Term F/T B1	-20 to 20%	-5 to +5	-5 to +5	-5 to +5
SHTFT2	Short Term F/T B2	-20 to 20%	-5 to +5	-5 to +5	-5 to +5
SPARK	Spark Advance	-90° to 90°	10-17	37	36
SS1	Shift Solenoid 1	ON / OFF	ON	OFF	OFF
SS2	Shift Solenoid 2	ON / OFF	OFF	OFF	OFF
SS3	Shift Solenoid 3	ON / OFF	OFF	ON	ON
SS4	Shift Solenoid 4	ON / OFF	ON	ON	ON
TCC	TCC Solenoid	0-100%	0	0	100
VREF	Voltage Reference	0-5.1v	4.9-5.1	4.9-5.1	4.9-5.1

Thunderbird & Cougar PID Data

1990-95 Thunderbird/Cougar 3.8L V6 MFI VIN 4 (A/T)

PCM PID Acronym	Parameter Identification	PID Range	PID Value at Hot Idle	PID Value at 30	PID Value at 55
ACCS	A/C Clutch Switch	ON / OFF	ON (if on)	OFF	OFF
ARC	Auto Ride Control (1990)	ON / OFF	ON	ON	ON
CANP	EVAP Purge Valve	0-100%	0	0-40	85-95
ECT (°F)	ECT Sensor	-40-304°F	160-200	160-200	160-200
ECT (V)	ECT Sensor	0-5.1v	0.6	0.6	0.6
EVR	VR Solenoid	0-100%	0%	0-40	35-45
FPM	Fuel Pump Monitor	ON / OFF	ON	ON	ON
FLPW1	INJ On-time (B1)	0-99.9 ms	5.9-6.1	7.4-8.4	8.0-9.0
FLPW2	INJ On-time (B2)	0-99.9 ms	5.9-6.1	7.4-8.4	8.0-9.0
HFC	High Fan Control	ON / OFF	ON (if on)	OFF	OFF
HO2S-11	HO2S-11 (Bank 1)	0.1-1.1v	0.1-1.1	0.1-1.1	0.1-1.1
HO2S-21	HO2S-21 (Bank 2)	0.1-1.1v	0.1-1.1	0.1-1.1	0.1-1.1
IAC	Idle Air Control	0-100%	34	44-74	50-75
IAT (°F)	IAT Sensor	-40-304°F	50-120	50-120	50-120
IAT (V)	IAT Sensor	0-5.1v	1.7-3.5	1.7-3.5	1.7-3.5
LFC	Low Fan Control	ON / OFF	ON (if on)	OFF	OFF
LOOP	Loop Status	OL or CL	CL	CL	CL
LongFT1	Long Term F/T B1	-20 to 20%	-5 to +5	-5 to +5	-5 to +5
LongFT2	Long Term F/T B2	-20 to 20%	-5 to +5	-5 to +5	-5 to +5
MAF (V)	MAF Sensor	0-5.1v	0.7	1.3-1.6	1.9-2.2
PFE (V)	PFE Sensor volts	0-5.1v	3.2	3.3	3.4
PNP	PNP Switch	NEUTRAL	NEUTRAL	DRIVE	DRIVE
PSP	PSP Switch	ON / OFF	ON turning	OFF	OFF
RPM	Engine Speed	0-10K rpm	650-750	1340-1440	2410-2510
SCCS	S/C Command Switch	0-25.5v	6.7	6.7	6.7
SCVAC	S/C Vacuum Solenoid	0-100%	0	0	0
SCVENT	S/C Vent Solenoid	0-100%	0	98	98
SHTFT1	Short Term F/T B1	-20 to 20%	-5 to +5	-5 to +5	-5 to +5
SHTFT2	Short Term F/T B2	-20 to 20%	-5 to +5	-5 to +5	-5 to +5
SPARK	Spark Advance	-90° to 90°	26-30	44-48	46-52
TPV	TP Sensor	0-5.1v	0.7	0.9	1.2-1.4
TP Mode	TP Mode	C/T or P/T	C/T	P/T	P/T
TSS	TSS Sensor	0-10 KHz	790-820	1400-1450	1740-1800
VPWR	Vehicle Power	0-25.5v	12-14	12-14	12-14
VREF	Voltage Reference	0-5.1v	4.9-5.1	4.9-5.1	4.9-5.1
VSS	Vehicle Speed	0-159 mph	0	30	55
WAC	A/C WOT Cutout Relay	ON / OFF	ON (if on)	OFF	OFF

1996-97 Thunderbird/Cougar 3.8L V6 VIN 4 (A/T) Inputs

PCM PID Acronym	Parameter Identification	PID Range	PID Value at Hot Idle	PID Value at 30	PID Value at 55
ACCS	A/C Clutch Switch	ON / OFF	ON (if on)	OFF	OFF
ACP	A/C Press. Sensor	CL/OPEN	OPEN	OPEN	OPEN
BPP	Brake Pedal Switch	ON / OFF	ON (if on)	OFF	OFF
CID	CID Sensor	0-1 KHz	5-7	9-11	11-15
CKP	CKP Sensor	0-10 KHz	390-450	710-760	875-940
DPFE	DPFE Sensor	0-5.1v	0.4-0.6	0.4-0.9	1.0-1.3
ECT (°F)	ECT Sensor	-40-304°F	160-200	160-200	160-200
FRP	Fuel Rail Pressure	0-100 psi	30-32	30-32	32-35
FEPS	Flash EEPROM	0-5.1v	0.1	0.1	0.1
FPM	Fuel Pump Monitor	ON / OFF	ON	ON	ON
GEAR	Transmission Gear	1-2-3-4	1	3	4
HO2S-11	HO2S-11 (Bank 1)	0.1-1.1v	0.1-1.1	0.1-1.1	0.1-1.1
HO2S-12	HO2S-12 (Bank 1)	0.1-1.1v	0.1-1.1	0.1-1.1	0.1-1.1
HO2S-21	HO2S-21 (Bank 2)	0.1-1.1v	0.1-1.1	0.1-1.1	0.1-1.1
HO2S-22	HO2S-22 (Bank 2)	0.1-1.1v	0.1-1.1	0.1-1.1	0.1-1.1
IAT	IAT Sensor	-40-304°F	50-120	50-120	50-120
LOAD	Engine Load	0-100%	10-20	16-30	25-38
MAF (V)	MAF Sensor	0-5.1v	0.6-0.9	1.0-1.2	1.3-2.0
MISF	Misfire Detection	ON / OFF	OFF	OFF	OFF
OCTADJ	Octane Adjustment	CL/OPEN	CLOSED	CLOSED	CLOSED
OSS	OSS Sensor	0-10K rpm	0	1150-1300	2260-2295
PF (V)	Purge Flow Sensor	0-5.1v	1.55	2.10	2.30
PNP	PNP Switch	ON / OFF	ON (in 'P')	OFF	OFF
RPM	Engine Speed	0-10K rpm	740-760	1200-1325	1580-1640
TCS	TCS (switch)	ON / OFF	ON (if on)	OFF	OFF
TFT (°F)	TFT Sensor	-40-304°F	110-210	110-210	110-210
TPV	TP Sensor	0-5.1v	0.5-0.9	0.8-0.9	0.9-1.2
TRV/TR	TR Sensor	0-5.1v	4.4 / PN	2.1 / OD	2.1 / OD
Vacuum	Engine Vacuum	0-30" Hg	19-20	18-20	8-15
VPWR	Vehicle Power	0-25.5v	12-14	12-14	12-14
VSS	Vehicle Speed	0-159 mph 0-1 KHz	0 = 0 Hz	30 = 65 Hz	55 = 125 Hz

1996-97 Thunderbird/Cougar 3.8L V6 VIN 4 (A/T) Outputs

PCM PID Acronym	Parameter Identification	PID Range	PID Value at Hot Idle	PID Value at 30	PID Value at 55
CD1-CD3	Coil Driver 1-3	0-60° dwell	7	10	12
EGRVR	VR Solenoid	0-100%	0	0-40	40-50
EPC	EPC Solenoid	0-100 psi	20	24-30	25-30
EVAPDC	EVAP Purge Valve	0-100%	0-100	0-100	0-100
FP	Fuel Pump Control	0-100%	100	100	100
FuelPW1	INJ Pulsewidth B1	0-99.9 ms	4.5-4.8	5.0-8.0	6.0-11
FuelPW2	INJ Pulsewidth B2	0-99.9 ms	4.5-4.8	5.0-8.0	6.0-11
HFC	High Fan Control	ON / OFF	ON (if on)	OFF	OFF
HTR-11	HO2S-11 Heater	ON / OFF	ON	ON	ON
HTR-12	HO2S-12 Heater	ON / OFF	ON	ON	ON
HTR-21	HO2S-21 Heater	ON / OFF	ON	ON	ON
HTR-22	HO2S-22 Heater	ON / OFF	ON	ON	ON
IAC	Idle Air Control	0-100%	40	45-55	65-75
INJ1	INJ 1 Pulsewidth	0-99.9 ms	4.5-4.8	5.0-8.0	6.0-11
INJ2	INJ 2 Pulsewidth	0-99.9 ms	4.5-4.8	5.0-8.0	6.0-11
INJ3	INJ 3 Pulsewidth	0-99.9 ms	4.5-4.8	5.0-8.0	6.0-11
INJ4	INJ 4 Pulsewidth	0-99.9 ms	4.5-4.8	5.0-8.0	6.0-11
INJ5	INJ 5 Pulsewidth	0-99.9 ms	4.5-4.8	5.0-8.0	6.0-11
INJ6	INJ 6 Pulsewidth	0-99.9 ms	4.5-4.8	5.0-8.0	6.0-11
LFC	Low Fan Control	ON / OFF	ON (if on)	OFF	OFF
LongFT1	Long Term F/T B1	-20 to 20%	-5 to +5	-5 to +5	-5 to +5
LongFT2	Long Term F/T B2	-20 to 20%	-5 to +5	-5 to +5	-5 to +5
MIL	MIL Control	ON / OFF	ON (if on)	OFF	OFF
SHTFT1	Short Term F/T B1	-20 to 20%	-5 to +5	-5 to +5	-5 to +5
SHTFT2	Short Term F/T B2	-20 to 20%	-5 to +5	-5 to +5	-5 to +5
SPARK	Spark Advance	-90° to 90°	15-20	25-35	31-40
SS1	Shift Solenoid 1	ON / OFF	ON	OFF	ON
SS2	Shift Solenoid 2	OFF	OFF	ON	ON
TCC	TCC Solenoid	0-100%	0	35-45	95-100
TCIL	TCIL (lamp)	ON / OFF	ON (if on)	OFF	OFF
VREF	Voltage Reference	0-5.1v	4.9-5.1	4.9-5.1	4.9-5.1
WAC	A/C WOT Cutout Relay	ON / OFF	ON (if on)	OFF	OFF

1994-97 Thunderbird/Cougar 4.6L V8 VIN W (A/T) Inputs

PCM PID Acronym	Parameter Identification	PID Range	PID Value at Hot Idle	PID Value at 30	PID Value at 55
ACCS	A/C Clutch Switch	ON / OFF	ON (if on)	OFF	OFF
ACP	A/C Press. Sensor	CL/OPEN	OPEN	OPEN	OPEN
BPP	Brake Pedal Switch	ON / OFF	ON (if on)	OFF	OFF
CID	CID Sensor	0-1 KHz	5-9	10-12	11-15
CKP	CKP Sensor	0-10 KHz	450-480	720-750	830-870
DPFE	DPFE Sensor	0-5.1v	0.4-0.6	0.6-1.5	1.1-1.7
ECT (°F)	ECT Sensor	-40-304°F	160-200	160-200	160-200
FEPS	Flash EEPROM	0-5.1v	0.1	0.1	0.1
FRP	Fuel Rail Pressure	0-100 psi	30-32	30-32	32-35
FP-M	Fuel Pump Monitor	ON / OFF	ON	ON	ON
GEAR	Transmission Gear	1-2-3-4	1	3	4
HO2S-11	HO2S-11 (Bank 1)	0.1-1.1v	0.1-1.1	0.1-1.1	0.1-1.1
HO2S-12	HO2S-12 (Bank 1)	0.1-1.1v	0.1-1.1	0.1-1.1	0.1-1.1
HO2S-21	HO2S-21 (Bank 2)	0.1-1.1v	0.1-1.1	0.1-1.1	0.1-1.1
HO2S-22	HO2S-22 (Bank 2)	0.1-1.1v	0.1-1.1	0.1-1.1	0.1-1.1
IAT (°F)	IAT Sensor	-40-304°F	50-120	50-120	50-120
LOAD	Engine Load	0-100%	10-20	15-25	14-31
MAF (V)	MAF Sensor	0-5.1v	0.6-0.9	0.8-1.6	0.6-3.1
MISF	Misfire Detection	ON / OFF	OFF	OFF	OFF
OCTADJ	Octane Adjustment	CL/OPEN	CLOSED	CLOSED	CLOSED
OSS	OSS Sensor	0-10 KHz	0 rpm	1210-1235	2050-2200
PF (V)	Purge Flow Sensor	0-5.1v	1.55	2.10	2.30
PNP	PNP Switch	ON / OFF	ON (in 'P')	OFF	OFF
RPM	Engine Speed	0-10K rpm	660-700	1170-1200	1460-1500
TCS	TCS (switch)	ON / OFF	ON (if on)	OFF	OFF
TFT (°F)	TFT Sensor	-40-304°F	110-210	110-210	110-210
TPV	TP Sensor	0-5.1v	0.5-0.9	0.9-1.1	0.9-1.2
TRV/TR	TR Sensor	0-5.1v	4.4 / PN	2.1 / OD	2.1 / OD
Vacuum	Engine Vacuum	0-30" Hg	19-20	18-20	8-15
VPWR	Vehicle Power	0-25.5v	12-14	12-14	12-14
VSS	Vehicle Speed	0-159 mph 0-1 KHz	0 = 0 Hz	30 = 65 Hz	55 = 125 Hz

1994-97 Thunderbird/Cougar 4.6L V8 VIN W (A/T) Outputs

PCM PID Acronym	Parameter Identification	PID Range	PID Value at Hot Idle	PID Value at 30	PID Value at 55
CD1-CD4	Coil Driver 1-4	0-45° dwell	8	11	12
EGRVR	VR Solenoid	0-100%	0	0-40	40-50
EVAPDC	EVAP Purge Valve	0-100%	0-100	0-100	0-100
FP	Fuel Pump Control	ON / OFF	ON	ON	ON
FuelPW1	INJ Pulsewidth B1	0-99.9 ms	3.1-3.5	3.1-6.6	4.7-7.2
FuelPW2	INJ Pulsewidth B2	0-99.9 ms	3.1-3.5	3.1-6.6	4.7-7.2
HFC	High Fan Control	ON / OFF	ON (if on)	OFF	OFF
HTR-11	HO2S-11 Heater	ON / OFF	ON	ON	ON
HTR-12	HO2S-12 Heater	ON / OFF	ON	ON	ON
HTR-21	HO2S-21 Heater	ON / OFF	ON	ON	ON
HTR-22	HO2S-22 Heater	ON / OFF	ON	ON	ON
IAC	Idle Air Control	0-100%	37	50-57	60-65
INJ1	INJ 1 Pulsewidth	0-99.9 ms	3.1-3.5	3.1-6.6	4.7-7.2
INJ2	INJ 2 Pulsewidth	0-99.9 ms	3.1-3.5	3.1-6.6	4.7-7.2
INJ3	INJ 3 Pulsewidth	0-99.9 ms	3.1-3.5	3.1-6.6	4.7-7.2
INJ4	INJ 4 Pulsewidth	0-99.9 ms	3.1-3.5	3.1-6.6	4.7-7.2
INJ5	INJ 5 Pulsewidth	0-99.9 ms	3.1-3.5	3.1-6.6	4.7-7.2
INJ6	INJ 6 Pulsewidth	0-99.9 ms	3.1-3.5	3.1-6.6	4.7-7.2
INJ7	INJ 7 Pulsewidth	0-99.9 ms	3.1-3.5	3.1-6.6	4.7-7.2
INJ8	INJ 8 Pulsewidth	0-99.9 ms	3.1-3.5	3.1-6.6	4.7-7.2
LFC	Low Fan Control	ON / OFF	ON (if on)	OFF	OFF
LongFT1	Long Term F/T B1	-20 to 20%	-5 to +5	-5 to +5	-5 to +5
LongFT2	Long Term F/T B2	-20 to 20%	-5 to +5	-5 to +5	-5 to +5
MIL	MIL Control	ON / OFF	ON (if on)	OFF	OFF
SHTFT1	Short Term F/T B1	-20 to 20%	-5 to +5	-5 to +5	-5 to +5
SHTFT2	Short Term F/T B2	-20 to 20%	-5 to +5	-5 to +5	-5 to +5
SPARK	Spark Advance	-90° to 90°	15-20	30-41	29-40
SS1	Shift Solenoid 1	ON / OFF	ON	OFF	ON
SS2	Shift Solenoid 2	ON / OFF	OFF	ON	ON
TCC	TCC Solenoid	0-100%	0	37-43	95-100
TCIL	TCIL (lamp)	ON / OFF	ON (if on)	OFF	OFF
VREF	Voltage Reference	0-5.1v	4.9-5.1	4.9-5.1	4.9-5.1
WAC	A/C WOT Cutout Relay	ON / OFF	ON (if on)	OFF	OFF

1991-93 Thunderbird/Cougar 5.0L V8 MFI VIN T (A/T)

PCM PID Acronym	Parameter Identification	PID Range	PID Value at Hot Idle	PID Value at 30	PID Value at 55
ACCS	A/C Clutch Switch	ON / OFF	ON (if on)	OFF	OFF
ARC	Auto Ride Control	ON / OFF	ON	ON	ON
BARO	Barometric Pressure Sensor	0-1 KHz	159 Hz (sea level)	159 Hz (sea level)	159 Hz (sea level)
CANP	EVAP Purge Valve	0-100%	0	0-40	85-95
ECT (°F)	ECT Sensor	-40-304°F	160-200	160-200	160-200
ECT (V)	ECT Sensor	0-5.1v	0.6	0.6	0.6
EVP	EGR EVP Sensor	0-5.1v	0.4v	0.4v	1.3-1.6
EVR	VR Solenoid	0-100%	0%	0-40	35-45
FPM	Fuel Pump Monitor	ON / OFF	ON	ON	ON
FuelPW1	INJ Pulsewidth B1	0-99.9 ms	4.9-5.2	4.8-6.0	7.4-8.6
FuelPW2	INJ Pulsewidth B2	0-99.9 ms	4.9-5.2	4.8-6.0	7.4-8.6
HFC	High Fan Control	ON / OFF	ON (if on)	OFF	OFF
HO2S-11	HO2S-11 (Bank 1)	0.1-1.1v	0.1-1.1	0.1-1.1	0.1-1.1
HO2S-21	HO2S-21 (Bank 2)	0.1-1.1v	0.1-1.1	0.1-1.1	0.1-1.1
IAC	Idle Air Control	0-100%	34	44-74	50-75
IAT (°F)	IAT Sensor	-40-304°F	50-120	50-120	50-120
IAT (V)	IAT Sensor	0-5.1v	1.7-3.5	1.7-3.5	1.7-3.5
LFC	Low Fan Control	ON / OFF	ON (if on)	OFF	OFF
LOOP	Loop Status	OL or CL	CL	CL	CL
LongFT1	Long Term F/T B1	-20 to 20%	-5 to +5	-5 to +5	-5 to +5
LongFT2	Long Term F/T B2	-20 to 20%	-5 to +5	-5 to +5	-5 to +5
MAF (V)	MAF Sensor	0-5.1v	0.7	1.3-1.6	1.9-2.2
PNP	PNP Switch	ON / OFF	ON (in 'P')	OFF	OFF
PSP	PSP Switch	ON / OFF	ON turning	OFF	OFF
RPM	Engine Speed	0-10K rpm	715-755	1140-1180	1390-1420
SHTFT1	Short Term F/T B1	-20 to 20%	-5 to +5	-5 to +5	-5 to +5
SHTFT2	Short Term F/T B2	-20 to 20%	-5 to +5	-5 to +5	-5 to +5
SPARK	Spark Advance	-90° to 90°	18-22	32-36	40-44
TP	TP Sensor	0-5.1v	1.0	1.1	1.2-1.3
TP Mode	TP Mode	C/T or P/T	C/T	P/T	P/T
VPWR	Vehicle Power	0-25.5v	12-14	12-14	12-14
VREF	Voltage Reference	0-5.1v	4.9-5.1	4.9-5.1	4.9-5.1
VSS	Vehicle Speed	0-159 mph 0-1 KHz	0 = 0 Hz	30 = 65 Hz	55 = 125 Hz
WAC	A/C WOT Cutout Relay	ON / OFF	ON (if on)	OFF	OFF

Continental PID Data

1990-94 Continental 3.8L V6 MFI VIN 4 (A/T) Inputs / Outputs

PCM PID Acronym	Parameter Identification	PID Range	PID Value at Hot Idle	PID Value at 30	PID Value at 55
ACCS	A/C Clutch Switch	ON / OFF	ON (if on)	OFF	OFF
ARC	Auto Ride Control	ON / OFF	ON	ON	ON
BOO	Brake Switch	ON / OFF	ON (if on)	OFF	OFF
CANP	EVAP Purge Valve	0-100%	0	0-40	85-95
ECT (°F)	ECT Sensor	-40-304°F	160-200	160-200	160-200
ECT (V)	ECT Sensor	0-5.1v	0.6	0.6	0.6
EPC	EPC Solenoid	0-100 psi	40	15	42
EVR	VR Solenoid	0-100%	0%	0-40	40-50
FPM	Fuel Pump Monitor	ON / OFF	ON	ON	ON
GEAR	Gear Position	P-R-N-D	P-R-N-D	DRIVE 3	DRIVE 4
HFC	High Fan Control	ON / OFF	ON (if on)	OFF	OFF
FUEL B1	INJ On-time B1	0-99.9 ms	3.6-3.8	4.5-6.2	5.8-7.0
FUEL B2	INJ On-time B2	0-99.9 ms	3.6-3.8	4.5-6.2	5.8-7.0
HO2S-11	HO2S-11 (Bank 1)	0.1-1.1v	0.1-1.1	0.1-1.1	0.1-1.1
HO2S-21	HO2S-21 (Bank 2)	0.1-1.1v	0.1-1.1	0.1-1.1	0.1-1.1
IAC	Idle Air Control	0-100%	20-40	34-40	45-55
IAT (°F)	IAT Sensor	-40-304°F	50-120	50-120	50-120
IAT (V)	IAT Sensor	0-5.1v	1.7-3.5	1.7-3.5	1.7-3.5
LFC	Low Fan Control	ON / OFF	ON (if on)	OFF	ON
LOOP	Loop Status	OL or CL	CL	CL	CL
LGFT1	L/T Fuel Trim (B1)	-20 to 20%	-5 to +5	-5 to +5	-5 to +5
LGFT2	L/T Fuel Trim (B2)	-20 to 20%	-5 to +5	-5 to +5	-5 to +5
MAF (V)	MAF Sensor	0-5.1v	0.7	1.0-1.4	1.5-2.0v
MLP (V)	MLP Sensor	0-5.1v	0v	5v	5v
PFE (V)	PFE Sensor	0-5.1v	3.3	3.2	2.8-3.2
PNP	PNP Switch	NEUTRAL	NEUTRAL	DRIVE	DRIVE
PSP	PSP Switch	ON / OFF	ON turning	OFF	OFF
RPM	Engine Speed	0-10K rpm	700-750	1300-1400	1650-1750
SHTFT1	Short Term F/T B1	-20 to 20%	-5 to +5	-5 to +5	-5 to +5
SHTFT2	Short Term F/T B2	-20 to 20%	-5 to +5	-5 to +5	-5 to +5
SPARK	Spark Advance	-90° to 90°	20-22	42-48	42-48
TCC	TCC Solenoid	0-100%	0	90-100	90-100
TOT °F	TOT Sensor	-40-304°F	110-210	110-210	110-210
TOT (V)	TOT Sensor	0-5.1v	0.5-2.0	0.5-2.0	0.5-2.0
TPV	TP Sensor	0-5.1v	0.7	0.9	1.2-1.4
TP MIN	TP Minimum	0-5.1v	0.5	0.9	1.2-1.4
TP Mode	TP Mode	C/T or P/T	C/T	P/T	P/T
TSS	TSS Sensor	0-10 KHz	790-820	1400-1450	1740-1800
VPWR	Vehicle Power	0-25.5v	12-14	12-14	12-14
VREF	Voltage Reference	0-5.1v	4.9-5.1	4.9-5.1	4.9-5.1
VSS	Vehicle Speed	0-159 mph 0-1 KHz	0 = 0 Hz	30 = 65 Hz	55 = 125 Hz
WAC	A/C WOT Cutout	OFF	ON (if on)	OFF	OFF

1995-2000 Continental 4.6L V8 MFI VIN V (A/T) - Inputs

PCM PID Acronym	Parameter Identification	PID Range	PID Value at Hot Idle	PID Value at 30	PID Value at 55
ACCS	A/C Clutch Switch	ON / OFF	ON (if on)	OFF	OFF
ACP	A/C Press. Switch	OP / CL	OPEN	OPEN	OPEN
AIR-M	AIR System Monitor	ON / OFF	OFF	OFF	OFF
BPP	Brake Switch	ON / OFF	ON (if on)	OFF	OFF
CID	CID Sensor	0-1 KHz	4-6	10-13	12-16
CKP	CKP Sensor	0-10 KHz	330-420	800-850	990-1100
DPFE	DPFE Sensor	0-5.1v	0.2-1.3	0.2-4.5	0.2-4.5
ECT (°F)	ECT Sensor	-40-304°F	160-200	160-200	160-200
FEPS	Flash EEPROM	0-5.1v	0.1	0.1	0.1
FLI	Fuel Level Indicator	0-100%	50 (1.7v)	25-75	25-75
FP M	Fuel Pump Monitor	0-100%	50-100	50-100	50-100
FRP	Fuel Rail Pressure	0-100 psi	39 (2.8v)	39 (2.8v)	39 (2.8v)
FTP	FTP Sensor (1998-2002)	0-5.1v or 0-10" H20	2.6v at 0" of H20	2.6v at 0" of H20	2.6v at 0" of H20
Gear	Gear Position	1 - 5	1	3	4
HO2S-11	HO2S-11 (Bank 1)	0.1-1.1v	0.1-1.1	0.1-1.1	0.1-1.1
HO2S-12	HO2S-12 (Bank 1)	0.1-1.1v	0.1-1.1	0.1-1.1	0.1-1.1
HO2S-21	HO2S-21 (Bank 2)	0.1-1.1v	0.1-1.1	0.1-1.1	0.1-1.1
HO2S-22	HO2S-22 (Bank 2)	0.1-1.1v	0.1-1.1	0.1-1.1	0.1-1.1
IAT (°F)	IAT Sensor	-40-304°F	50-120	50-120	50-120
IMRCM	IMRC Monitor	0-5.1v	5	5	5
KS1	Knock Sensor B1	0-5.1v	0	0	0
KS2	Knock Sensor B2	0-5.1v	0	0	0
LOAD	Engine Load	0-100%	10-20	16-36	23-33
MAF (V)	MAF Sensor	0-5.1v	0.5-0.8	0.8-1.6	1.1-1.9
MISF	Misfire Detection	ON / OFF	OFF	OFF	OFF
OCTADJ	Octane Adjust	Retard/NO	No Retard	No Retard	No Retard
PNP	PNP Switch	ON / OFF	ON	OFF	OFF
PSP	PSP Switch	HI / LOW	HI: turning	LOW	LOW
RPM	Engine Speed	0-10K rpm	695-760	1350-1440	1700-1820
TCS	TCS (switch)	ON / OFF	ON (if on)	OFF	OFF
TFT (°F)	TFT Sensor	-40-304°F	110-210	110-210	110-210
TP-B	SEC TP Sensor	0-5.1v	0.5-0.7	0.5-0.7	0.5-0.7
TPV	TP Sensor	0-5.1v	0.53-1.27	0.8-1.1	0.8-1.2
TRAC	Traction Control	ON / OFF	ON	ON	ON
TR1	TR Sensor 1 Volts	0-12v	0	10.7	10.7
TR2	TR2 Sensor	0-12v	0	10.7	10.7
TR4	TR4 Sensor	0-12v	0	10.7	10.7
TRV/TR	TR Sensor	0-5.1v	4.4 / PN	2.1 / OD	2.1 / OD
TSS	Turbine shaft speed	0-1 KHz	40-45	85-105	110-118
VPWR	Vehicle Power	0-25.5v	12-14	12-14	12-14
VSS	Vehicle Speed	0-159 mph 0-1 KHz	0 = 0 Hz	30 = 65 Hz	55 = 125 Hz
WACA	A/C WOT Relay	ON / OFF	ON (if on)	OFF	OFF

1995-2000 Continental 4.6L V8 MFI VIN V (A/T) - Outputs

PCM PID Acronym	Parameter Identification	PID Range	PID Value at Hot Idle	PID Value at 30	PID Value at 55
AIR	AIR Solenoid	ON / OFF	OFF	OFF	OFF
CD1-8	Coil Driver 1-8	0-45° dwell	8	11	12
CTO	Clean Tach Output	0-10 KHz	33-37	57-63	68-74
EGRVR	VR Solenoid	0-100%	0	0-40	35-55
EPC	EPC Solenoid	0-100 psi	16-18	12-16	18-27
EVAPCV	EVAP CV Valve	0-100%	0-100	0-100	0-100
EVAPDC	EVAP Purge Valve	0-100%	0-100	0-100	0-100
FP	Fuel Pump Control	0-100%	33	33	33
FLPW1	INJ On-time (B1)	0-99.9 ms	2.8-2.9	2.9-5.5	4.5-8.0
FLPW2	INJ On-time (B2)	0-99.9 ms	2.8-2.9	2.9-5.5	4.5-8.0
HFC	High Fan Control	ON / OFF	ON (if on)	OFF	OFF
HTR-11	HO2S-11 Heater	ON / OFF	ON	ON	ON
HTR-12	HO2S-12 Heater	ON / OFF	ON	ON	ON
HTR-21	HO2S-21 Heater	ON / OFF	ON	ON	ON
HTR-22	HO2S-22 Heater	ON / OFF	ON	ON	ON
IAC	Idle Air Control	0-100%	35	44-55%	51-59%
IMRC	IMRC Solenoid	ON / OFF	OFF	OFF	OFF
INJ1	INJ 1 Pulsewidth	0-99.9 ms	2.8-2.9	2.9-5.5	4.5-8.0
INJ2	INJ 2 Pulsewidth	0-99.9 ms	2.8-2.9	2.9-5.5	4.5-8.0
INJ3	INJ 3 Pulsewidth	0-99.9 ms	2.8-2.9	2.9-5.5	4.5-8.0
INJ4	INJ 4 Pulsewidth	0-99.9 ms	2.8-2.9	2.9-5.5	4.5-8.0
INJ5	INJ 5 Pulsewidth	0-99.9 ms	2.8-2.9	2.9-5.5	4.5-8.0
INJ6	INJ 6 Pulsewidth	0-99.9 ms	2.8-2.9	2.9-5.5	4.5-8.0
INJ7	INJ 7 Pulsewidth	0-99.9 ms	2.8-2.9	2.9-5.5	4.5-8.0
INJ8	INJ 8 Pulsewidth	0-99.9 ms	2.8-2.9	2.9-5.5	4.5-8.0
LFC	Low Fan Control	ON / OFF	ON (if on)	OFF	OFF
LongFT1	Long Term F/T B1	-20 to 20%	-5 to +5	-5 to +5	-5 to +5
LongFT2	Long Term F/T B2	-20 to 20%	-5 to +5	-5 to +5	-5 to +5
MIL	MIL Control	ON / OFF	ON (if on)	OFF	OFF
SHTFT1	Short Term F/T B1	-20 to 20%	-5 to +5	-5 to +5	-5 to +5
SHTFT2	Short Term F/T B2	-20 to 20%	-5 to +5	-5 to +5	-5 to +5
SPARK	Spark Advance	-90° to 90°	8-10	35-40	30-40
SS1	Shift Solenoid 1	ON / OFF	ON	OFF	ON
SS2	Shift Solenoid 2	ON / OFF	OFF	ON	ON
SS3	Shift Solenoid 3	ON / OFF	OFF	ON	ON
TCC	TCC Solenoid	0-100%	0	90	90-95
TC SEC	T/C Secondary Throttle	0-100%	0	0	0
TCIL ('96)	TCIL (lamp)	ON / OFF	ON (if on)	OFF	OFF
VREF	Voltage Reference	0-5.1v	4.9-5.1	4.9-5.1	4.9-5.1
WAC	A/C WOT Cutout	ON / OFF	ON (if on)	OFF	OFF

2001-02 Continental 4.6L V8 MFI VIN V (A/T) - Inputs

PCM PID Acronym	Parameter Identification	PID Range	PID Value at Hot Idle	PID Value at 30	PID Value at 55
ACCS	A/C Clutch Switch	ON / OFF	ON (if on)	OFF	OFF
ACP	A/C Press. Switch	OP / CL	OPEN	OPEN	OPEN
BPP	Brake Switch	ON / OFF	ON (if on)	OFF	OFF
CID	CID Sensor	0-1 KHz	4-6	10-13	12-16
CKP	CKP Sensor	0-10 KHz	330-420	800-850	990-1100
DPFE	DPFE Sensor	0-5.1v	0.95-1.05	0.95-4.65	0.95-4.65
ECT (°F)	ECT Sensor	-40-304°F	160-200	160-200	160-200
FEPS	Flash EEPROM	0-5.1v	0.1	0.1	0.1
FLI	Fuel Level Indicator	0-100%	50 (1.7v)	25-75	25-75
FP M	Fuel Pump Monitor	0-100%	50-100	50-100	50-100
FRP	Fuel Rail Pressure	0-100 psi	39 (2.8v)	39 (2.8v)	39 (2.8v)
FTP	FTP Sensor (1998-2002)	0-5.1v or 0-10" H20	2.6v at 0" of H20	2.6v at 0" of H20	2.6v at 0" of H20
Gear	Gear Position	1 - 5	1	3	4
HO2S-11	HO2S-11 (Bank 1)	0.1-1.1v	0.1-1.1	0.1-1.1	0.1-1.1
HO2S-12	HO2S-12 (Bank 1)	0.1-1.1v	0.1-1.1	0.1-1.1	0.1-1.1
HO2S-21	HO2S-21 (Bank 2)	0.1-1.1v	0.1-1.1	0.1-1.1	0.1-1.1
HO2S-22	HO2S-22 (Bank 2)	0.1-1.1v	0.1-1.1	0.1-1.1	0.1-1.1
IAT (°F)	IAT Sensor	-40-304°F	50-120	50-120	50-120
IMRCM	IMRC Monitor	0-5.1v	5	5	5
KS1	Knock Sensor B1	0-5.1v	0	0	0
KS2	Knock Sensor B2	0-5.1v	0	0	0
LOAD	Engine Load	0-100%	10-20	16-36	23-33
MAF (V)	MAF Sensor	0-5.1v	0.5-0.8	0.8-1.6	1.1-1.9
MISF	Misfire Detection	ON / OFF	OFF	OFF	OFF
OCTADJ	Octane Adjust	Retard/NO	No Retard	No Retard	No Retard
PNP	PNP Switch	ON / OFF	ON	OFF	OFF
PSP	PSP Switch	HI / LOW	HI: turning	LOW	LOW
RPM	Engine Speed	0-10K rpm	695-760	1350-1440	1700-1820
TCS	TCS (switch)	ON / OFF	ON (if on)	OFF	OFF
TFT (°F)	TFT Sensor	-40-304°F	110-210	110-210	110-210
TP-B	SEC TP Sensor	0-5.1v	0.5-0.7	0.5-0.7	0.5-0.7
TPV	TP Sensor	0-5.1v	0.53-1.27	0.8-1.1	0.8-1.2
TRAC	Traction Control	ON / OFF	ON	ON	ON
TR1	TR Sensor 1 Volts	0-12v	0	10.7	10.7
TR2	TR2 Sensor	0-12v	0	10.7	10.7
TR4	TR4 Sensor	0-12v	0	10.7	10.7
TRV/TR	TR Sensor	0-5.1v	4.4 / PN	2.1 / OD	2.1 / OD
TSS	Turbine shaft speed	0-1 KHz	40-45	85-105	110-118
VPWR	Vehicle Power	0-25.5v	12-14	12-14	12-14
VSS	Vehicle Speed	0-159 mph 0-1 KHz	0 = 0 Hz	30 = 65 Hz	55 = 125 Hz
WACA	A/C WOT Relay	ON / OFF	ON (if on)	OFF	OFF

2001-02 Continental 4.6L V8 MFI VIN V (A/T) - Outputs

PCM PID Acronym	Parameter Identification	PID Range	PID Value at Hot Idle	PID Value at 30	PID Value at 55
CD1-8	Coil Driver 1-8	0-45º dwell	8	11	12
CTO	Clean Tach Output	0-10 KHz	33-37	57-63	68-74
EGRVR	VR Solenoid	0-100%	0	0-40	35-55
EPC	EPC Solenoid	0-100 psi	16-18	12-16	18-27
EVAPCV	EVAP CV Valve	0-100%	0-100	0-100	0-100
EVAPDC	EVAP Purge Valve	0-100%	0-100	0-100	0-100
FP	Fuel Pump Control	0-100%	33	31	33
FLPW1	INJ On-time (B1)	0-99.9 ms	2.8-2.9	2.9-5.5	4.5-8.0
FLPW2	INJ On-time (B2)	0-99.9 ms	2.8-2.9	2.9-5.5	4.5-8.0
HFC	High Fan Control	ON / OFF	ON (if on)	OFF	OFF
HTR-11	HO2S-11 Heater	ON / OFF	ON	ON	ON
HTR-12	HO2S-12 Heater	ON / OFF	ON	ON	ON
HTR-21	HO2S-21 Heater	ON / OFF	ON	ON	ON
HTR-22	HO2S-22 Heater	ON / OFF	ON	ON	ON
IAC	Idle Air Control	0-100%	28-30	44-55	51-75
IMRC	IMRC Solenoid	ON / OFF	OFF	OFF	OFF
INJ1	INJ 1 Pulsewidth	0-99.9 ms	2.8-2.9	2.9-5.5	4.5-8.0
INJ2	INJ 2 Pulsewidth	0-99.9 ms	2.8-2.9	2.9-5.5	4.5-8.0
INJ3	INJ 3 Pulsewidth	0-99.9 ms	2.8-2.9	2.9-5.5	4.5-8.0
INJ4	INJ 4 Pulsewidth	0-99.9 ms	2.8-2.9	2.9-5.5	4.5-8.0
INJ5	INJ 5 Pulsewidth	0-99.9 ms	2.8-2.9	2.9-5.5	4.5-8.0
INJ6	INJ 6 Pulsewidth	0-99.9 ms	2.8-2.9	2.9-5.5	4.5-8.0
INJ7	INJ 7 Pulsewidth	0-99.9 ms	2.8-2.9	2.9-5.5	4.5-8.0
INJ8	INJ 8 Pulsewidth	0-99.9 ms	2.8-2.9	2.9-5.5	4.5-8.0
LFC	Low Fan Control	ON / OFF	ON (if on)	OFF	OFF
LongFT1	Long Term F/T B1	-20 to 20%	-5 to +5	-5 to +5	-5 to +5
LongFT2	Long Term F/T B2	-20 to 20%	-5 to +5	-5 to +5	-5 to +5
MIL	MIL Control	ON / OFF	ON (if on)	OFF	OFF
SHTFT1	Short Term F/T B1	-20 to 20%	-5 to +5	-5 to +5	-5 to +5
SHTFT2	Short Term F/T B2	-20 to 20%	-5 to +5	-5 to +5	-5 to +5
SPARK	Spark Advance	-90º to 90º	8-10	35-45	30-45
SS1	Shift Solenoid 1	ON / OFF	OFF	ON	ON
SS2	Shift Solenoid 2	ON / OFF	ON	OFF	ON
SS3	Shift Solenoid 3	ON / OFF	OFF	ON	ON
TCC	TCC Solenoid	0-100%	0	0	100
VREF	Voltage Reference	0-5.1v	4.9-5.1	4.9-5.1	4.9-5.1
WAC	A/C WOT Cutout	ON / OFF	ON (if on)	OFF	OFF

LS6 PID Data

2000-2003 LS6 3.0L 4v V6 MFI VIN S (All) - Inputs

PCM PID Acronym	Parameter Identification	PID Range	PID Value at Hot Idle	PID Value at 30	PID Value at 55
AIR-M	AIR Monitor	ON / OFF	OFF	OFF	OFF
ACP	A/C Press. Switch	CL/OPEN	OPEN	OPEN	OPEN
AFS	AFS Signal Hertz	0-1 KHz	130	130	130
BPP	BPP Switch	ON / OFF	ON (if on)	OFF	OFF
BPS	Brake Pedal Switch	0-25.5v	0.1 (if on)	12-14	12-14
CHT (°F)	CHT Sensor (°F)	-40-304°F	160-200	160-200	160-200
CID	CID Sensor	0-1 KHz	6.6	10	17
CKP	CKP Sensor	0-10 KHz	435	780-710	1170-1180
CSTT	Clutch Switch	0-25.5v	12v (if in)	0.1	0.1
DPFE	DPFE input (99-'00)	0-5.1v	0.2-1.3	0.2-4.5	0.2-4.5
DPFE	DPFE input (01-'02)	0-5.1v	0.95-1.05	0.95-4.65	0.95-4.65
ECT (°F)	ECT Sensor	-40-304°F	160-200	160-200	160-200
EFTA (°F)	EFT Sensor (°F)	-40-304°F	50-120	50-120	50-120
FEPS	Flash EEPROM	0-5.1v	0.1	0.1	0.1
FLI	Fuel Level Indicator	0-100%	50 (1.7v)	25-75	25-75
FP-M	Fuel Pump Monitor	0-100%	50	50	50
FRP	Fuel Rail Pressure	0-100 psi	39	39	39
FTP	FTP Sensor	0-5.1v or 0-10" H20	2.6v at 0" of H20	2.6v at 0" of H20	2.6v at 0" of H20
GEAR	Transmission Gear	1-2-3-4-5	1	4	5
HO2S-11	HO2S-11 (Bank 1)	0.1-1.1v	0.1-1.1	0.1-1.1	0.1-1.1
HO2S-12	HO2S-12 (Bank 1)	0.1-1.1v	0.1-1.1	0.1-1.1	0.1-1.1
HO2S-21	HO2S-21 (Bank 2)	0.1-1.1v	0.1-1.1	0.1-1.1	0.1-1.1
HO2S-22	HO2S-22 (Bank 2)	0.1-1.1v	0.1-1.1	0.1-1.1	0.1-1.1
IAT (°F)	IAT Sensor	-40-304°F	50-120	50-120	50-120
KS1	Knock Sensor B1	0-5.1v	0	0	0
KS2	Knock Sensor B2	0-5.1v	0	0	0
LOAD	Engine Load	0-100%	18.6	35	40
MAF (V)	MAF Sensor	0-5.1v	0.7	1.6-1.8	2.1-2.3
OCTADJ	Octane Adjustment	CLOSED	CLOSED	CLOSED	CLOSED
OSS	OSS Sensor	0-10 KHz	0 rpm	1500	2660
PSP	PSP Switch	HI / LOW	HI: turning	LOW	LOW
PS1	PS1 Switch	0-25.5v	11.7	11.7	11.7
REV	Reverse Signal	0-25.5v	0.1 (in 'R')	12-14	12-14
SCCS	S/C Switch	0-5.1v	0.1 (if on)	4.6	4.6
TCS	TCS (switch)	ON / OFF	ON (if on)	OFF	OFF
TSS	TSS Sensor	0-10 KHz	680	1080	2060
TFT (°F)	TFT Sensor	-40-304°F	110-210	110-210	110-210
TPV	TP Sensor	0-5.1v	0.5-0.9	1.1v	1.3v
TR1	TR Sensor 1 Volts	0-12v	0	10.7	10.7
TR2	TR2 Sensor	0-12v	0	10.7	10.7
TR4	TR4 Sensor	0-12v	0	10.7	10.7
TRV/TR	TR Sensor	0-5.1v	0v / PN	1.7 / OD	1.7 / OD
WACA	A/C WOT Relay	ON / OFF	ON (if on)	OFF	OFF

2000-2003 LS6 3.0L 4v V6 MFI VIN S (All) - Outputs

PCM PID Acronym	Parameter Identification	PID Range	PID Value at Hot Idle	PID Value at 30	PID Value at 55
AIR	AIR Solenoid	ON / OFF	OFF	OFF	OFF
ALDFDC	Alternator Control	0-1 KHz	0-130	0 -10	0-10
CD1-CD6	Coil Driver 1-6	0-60° dwell	7	10	11
CHTIL	CHTIL (lamp)	ON / OFF	ON (if on)	OFF	OFF
CTO	Clean Tach Output	0-10 KHz	39	89	100
EGRVR	VR Solenoid	0-100%	0	0-40	55
EPC	EPC Solenoid	0-25.5v	9.3	10.4	10.5
EPC2	EPC2 Solenoid PSI	0-25.5v	11.1	10.4	10.5
EPC3	EPC3 Solenoid PSI	0-25.5v	7.5	12-14	12-14
EVAPCV	EVAP CV Valve	0-100%	0-100	0-100	0-100
EVAPDC	EVAP Purge Valve	0-100%	0-100	0-100	0-100
FP	Fuel Pump Control	0-100%	25	25	25
HTR-11	HO2S-11 Heater	ON / OFF	ON	ON	ON
HTR-12	HO2S-12 Heater	ON / OFF	ON	ON	ON
HTR-21	HO2S-21 Heater	ON / OFF	ON	ON	ON
HTR-22	HO2S-22 Heater	ON / OFF	ON	ON	ON
IAC	Idle Air Control	0-100%	34	53	67
INJ1	INJ 1 Pulsewidth	0-99.9 ms	3.2	4.9	6.7-7.1
INJ2	INJ 2 Pulsewidth	0-99.9 ms	3.2	4.9	6.7-7.1
INJ3	INJ 3 Pulsewidth	0-99.9 ms	3.2	4.9	6.7-7.1
INJ4	INJ 4 Pulsewidth	0-99.9 ms	3.2	4.9	6.7-7.1
INJ5	INJ 5 Pulsewidth	0-99.9 ms	3.2	4.9	6.7-7.1
INJ6	INJ 6 Pulsewidth	0-99.9 ms	3.2	4.9	6.7-7.1
LongFT1	Long Term F/T B1	-20 to 20%	-5 to +5	-5 to +5	-5 to +5
LongFT2	Long Term F/T B2	-20 to 20%	-5 to +5	-5 to +5	-5 to +5
MIL	MIL Control	ON / OFF	ON (if on)	OFF	OFF
SCC	S/C C Signal	0-25.5v	12-14	12-14	12-14
SCMA	S/C MA Signal	0-25.5v	12-14	12-14	12-14
SCMB	S/C MB Signal	0-25.5v	12-14	12-14	12-14
SCMC	S/C MC Signal	0-25.5v	12-14	12-14	12-14
SHTFT1	Short Term F/T B1	-20 to 20%	-5 to +5	-5 to +5	-5 to +5
SHTFT2	Short Term F/T B2	-20 to 20%	-5 to +5	-5 to +5	-5 to +5
SPARK	Spark Advance	-90° to 90°	12-17	34	40
S1	Shift Solenoid 1	ON / OFF	ON	OFF	ON
S2	Shift Solenoid 2	ON / OFF	OFF	OFF	OFF
S3	Shift Solenoid 3	ON / OFF	OFF	ON	ON
SS4	Shift Solenoid 4	ON / OFF	ON	ON	ON
TCC	TCC Solenoid	0-100%	0	0	100
VREF	Voltage Reference	0-5.1v	4.9-5.1	4.9-5.1	4.9-5.1
WAC	WOT Cutout Relay	ON / OFF	ON (if on)	OFF	OFF

LS8 PID Data

2000-2003 LS8 3.9L 4v V8 MFI VIN A (A/T) - Inputs

PCM PID Acronym	Parameter Identification	PID Range	PID Value at Hot Idle	PID Value at 30	PID Value at 55
AIR-M	AIR Monitor	ON / OFF	OFF	OFF	OFF
ACP	A/C Press. Switch	CL/OPEN	OPEN	OPEN	OPEN
AFS	AFS Signal Hertz	0-1 KHz	130	130	130
BPP	BPP Switch	ON / OFF	ON (if on)	OFF	OFF
BPS	Brake Pedal Switch	0-25.5v	12v (if on)	0v	0v
CHT (°F)	CHT Sensor (°F)	-40-304°F	160-200	160-200	160-200
CID	CID Sensor	0-1 KHz	6	11	18
CKP	CKP Sensor	0-10 KHz	420	665-800	1160-1200
CTO	Clean Tach Output	0-10 KHz	46	106	123
DPFE	DPFE input	0-5.1v	0.20-1.3	0.2-4.5	0.2-4.5
ECT (°F)	ECT Sensor	-40-304°F	160-200	160-200	160-200
EFTA (°F)	EFT Sensor (°F)	-40-304°F	50-120	50-120	50-120
EOT	Engine Oil Temp.	0-5.1v	1.0	0.9	1.2
FEPS	Flash EEPROM	0-5.1v	0.1	0.1	0.1
FLI	Fuel Level Indicator	0-100%	50 (1.7v)	25-75	25-75
FP-M	Fuel Pump Monitor	0-100%	50	50	50
FRP	Fuel Rail Pressure	0-100 psi	44	44	44
FTP	FTP Sensor	0-5.1v	2.6 (0")	2.6 (0")	2.6 (0")
GEAR	Transmission Gear	1-2-3-4-5	1	4	5
GFS	Generator Field	0-1 KHz	130 (35%)	130 (25%)	130 (20%)
HO2S-11	HO2S-11 (Bank 1)	0.1-1.1v	0.1-1.1	0.1-1.1	0.1-1.1
HO2S-12	HO2S-12 (Bank 1)	0.1-1.1v	0.1-1.1	0.1-1.1	0.1-1.1
HO2S-21	HO2S-21 (Bank 2)	0.1-1.1v	0.1-1.1	0.1-1.1	0.1-1.1
HO2S-22	HO2S-22 (Bank 2)	0.1-1.1v	0.1-1.1	0.1-1.1	0.1-1.1
IAT (°F)	IAT Sensor	-40-304°F	50-120	50-120	50-120
ISS	Input Shaft Speed	0-10 KHz	235-380	725	1370
KS1	Knock Sensor B1	0-5.1v	0	0	0
KS2	Knock Sensor B2	0-5.1v	0	0	0
LOAD	Engine Load	0-100%	17-18.6	26-35.7	30-50
MAF (V)	MAF Sensor	0-5.1v	0.78	1.2-1.4	1.5-1.9
MISF	Misfire Detection	ON / OFF	OFF	OFF	OFF
OSS	Output Shaft Speed	0-10 KHz	0	595	1070
PS1	PS1 Switch	0-25.5v	12	11.7	11.7
PNP	PNP Switch	ON / OFF	ON	OFF	OFF
PSP	PSP Switch	HI / LOW	HI: turning	LOW	LOW
RPM	Engine Speed	0-10K rpm	660-700	1425	1950
TCS	TCS (switch)	ON / OFF	ON (if on)	OFF	OFF
TSS	Turbine shaft speed	0-10 KHz	340	590	1070
TFT (°F)	TFT Sensor	-40-304°F	110-210	110-210	110-210
TPV	TP Sensor	0-5.1v	0.53-1.37	1.1-1.3	1.3-1.5
TR1	TR Sensor 1 Volts	0-12v	0.1	12	12
TR2	TR2 Sensor	0-12v	0.1	12	12
TR4	TR4 Sensor	0-12v	0.1	12	12
TRV/TR	TR Sensor	0-5.1v	0 / PN	1.7 / OD	1.7 / OD

2000-2002 LS8 3.9L 4v V8 MFI VIN A (A/T) - Outputs

PCM PID Acronym	Parameter Identification	PID Range	PID Value at Hot Idle	PID Value at 30	PID Value at 55
AIR	AIR Solenoid	ON / OFF	OFF	OFF	OFF
CD1-8	Coil Driver 1-8	0-45° dwell	8	11	12
CHTIL	CHTIL (lamp)	ON / OFF	ON (if on)	OFF	OFF
EGRVR	VR Solenoid	0-100%	0	40	40
EPC	EPC Solenoid	0-25.5v	7.4	9.3	9.9
EPC2	EPC2 Solenoid PSI	0-25.5v	12-14	9.3	9.9
EPC3	EPC3 Solenoid PSI	0-25.5v	7.5	12-14	12-14
EVAPCV	EVAP CV Valve	0-100%	0-100	0-100	0-100
EVAPDC	EVAP Purge Valve	0-100%	0-100	0-100	0-100
FP	Fuel Pump Control	0-100%	26 (3.5v)	29 (4v)	27 (3.8v)
GENFDC	Alternator Control	0-1 KHz	0-130	0	0
HFC	High Fan Control	ON / OFF	ON (if on)	OFF	OFF
HTR-11	HO2S-11 Heater	ON / OFF	ON	ON	ON
HTR-12	HO2S-12 Heater	ON / OFF	ON	ON	ON
HTR-21	HO2S-21 Heater	ON / OFF	ON	ON	ON
HTR-22	HO2S-22 Heater	ON / OFF	ON	ON	ON
IAC	Idle Air Control	0-100%	32	54	72
IMRC	IMRC Solenoid	ON / OFF	OFF	OFF	OFF
INJ1	INJ 1 Pulsewidth	0-99.9 ms	2.9-3.6	5.0-5.1	6.5-7.5
INJ2	INJ 2 Pulsewidth	0-99.9 ms	2.9-3.6	5.0-5.1	6.5-7.5
INJ3	INJ 3 Pulsewidth	0-99.9 ms	2.9-3.6	5.0-5.1	6.5-7.5
INJ4	INJ 4 Pulsewidth	0-99.9 ms	2.9-3.6	5.0-5.1	6.5-7.5
INJ5	INJ 5 Pulsewidth	0-99.9 ms	2.9-3.6	5.0-5.1	6.5-7.5
INJ6	INJ 6 Pulsewidth	0-99.9 ms	2.9-3.6	5.0-5.1	6.5-7.5
INJ7	INJ 7 Pulsewidth	0-99.9 ms	2.9-3.6	5.0-5.1	6.5-7.5
INJ8	INJ 8 Pulsewidth	0-99.9 ms	2.9-3.6	5.0-5.1	6.5-7.5
LGFT1	L/T Fuel Trim (B1)	-20 to 20%	-5 to +5	-5 to +5	-5 to +5
LGFT2	L/T Fuel Trim (B2)	-20 to 20%	-5 to +5	-5 to +5	-5 to +5
MIL	MIL Control	ON / OFF	ON (if on)	OFF	OFF
SCCS	S/C Switch Signal	0-5.1v	0.1	4.6	4.6
SCMA	S/C MA Signal	0-25.5v	12-14	12-14	12-14
SCMB	S/C MB Signal	0-25.5v	12-14	12-14	12-14
SCMC	S/C MC Signal	0-25.5v	12-14	12-14	12-14
SHTFT1	Short Term F/T B1	-20 to 20%	-5 to +5	-5 to +5	-5 to +5
SHTFT2	Short Term F/T B2	-20 to 20%	-5 to +5	-5 to +5	-5 to +5
SPARK	Spark Advance	-90° to 90°	10-20	36	33
S1	Shift Solenoid 1	ON / OFF	ON	OFF	OFF
S2	Shift Solenoid 2	ON / OFF	OFF	OFF	OFF
S3	Shift Solenoid 3	ON / OFF	OFF	ON	ON
SS4	Shift Solenoid 4	ON / OFF	ON	ON	ON
TCC	TCC Solenoid	0-100%	0	0	100
TCIL	Trans. Control lamp	ON / OFF	ON (if on)	OFF	OFF
WAC	WOT Cutout Relay	ON / OFF	ON (if on)	OFF	OFF

Mark VIII PID Data

1993-95 Mark VIII 4.6L V8 MFI VIN V (A/T) Inputs / Outputs

PCM PID Acronym	Parameter Identification	PID Range	PID Value at Hot Idle	PID Value at 30	PID Value at 55
ACCS	A/C Clutch Switch	ON / OFF	ON (if on)	OFF	OFF
BOO	Brake Switch	On / OFF	ON (if on)	OFF	OFF
DPFE	DPFE Sensor	0-5.1v	0.4-0.6	0.6-1.8	1.9-4.8
ECT (°F)	ECT Sensor	-40-304°F	160-200	160-200	160-200
ECT (V)	ECT Sensor	0-5.1v	0.6	0.6	0.6
EPC	EPC Solenoid	0-100 psi	20	12	30
EVR	VR Solenoid	0-100%	0	0-40	0-60
FP	Fuel Pump Control	ON / OFF	ON	ON	ON
GEAR	Gear Position	P-R-N-D	P-R-N-D	DRIVE 3	DRIVE 4
FuelPW1	INJ Pulsewidth B1	0-99.9 ms	5.5-5.8	6.4-8.4	9.4-12
FuelPW2	INJ Pulsewidth B2	0-99.9 ms	5.5-5.8	6.4-8.4	9.4-12
HO2S-11	HO2S-11 (Bank 1)	0.1-1.1v	0.1-1.1	0.1-1.1	0.1-1.1
HO2S-21	HO2S-21 (Bank 2)	0.1-1.1v	0.1-1.1	0.1-1.1	0.1-1.1
IAC	Idle Air Control	0-100%	20-40	34-40	45-55
IAT (°F)	IAT Sensor	-40-304°F	50-120	50-120	50-120
IAT (V)	IAT Sensor	1.7-3.5	1.7-3.5	1.7-3.5	1.7-3.5
LOOP	Loop Status	OL or CL	CL	CL	CL
LongFT1	Long Term F/T B1	-20 to 20%	-5 to +5	-5 to +5	-5 to +5
LongFT2	Long Term F/T B2	-20 to 20%	-5 to +5	-5 to +5	-5 to +5
MAF	MAF Sensor	0-5.1v	1.0v	1.3-1.5	1.6-2.0
MLP	MLP Sensor	PARK	PARK	DRIVE	O/D
MLP (V)	MLP Sensor	0-5.1v	0v	5v	5v
PNP	PNP Switch	NEUTRAL	NEUTRAL	DRIVE	DRIVE
RPM	Engine Speed	0-10K rpm	750-820	1450-1630	1750-2100
SHTFT1	Short Term F/T B1	-20 to 20%	-5 to +5	-5 to +5	-5 to +5
SHTFT2	Short Term F/T B2	-20 to 20%	-5 to +5	-5 to +5	-5 to +5
SPARK	Spark Advance	-90° to 90°	15-22	28-35	25-35
TCC	TCC Solenoid	0-100%	0	42-44	92-100
TOT (V)	TFT Sensor Volts	0-5.1v	1.7-3.5	1.7-3.5	1.7-3.5
TOT (°F)	TFT Sensor	-40-304°F	110-210	110-210	110-210
TPV	TP Sensor	0-5.1v	0.7	0.8-1.0	1.1-1.3
TP MIN	TP Minimum	0-5.1v	0.5	0.8-1.0	1.1-1.3
TP Mode	TP Mode	C/T or P/T	C/T	P/T	P/T
VPWR	Vehicle Power	0-25.5v	12-14	12-14	12-14
VREF	Voltage Reference	0-5.1v	4.9-5.1	4.9-5.1	4.9-5.1
VSS	Vehicle Speed	0-159 mph	0	30	55
WAC	A/C WOT Cutout Relay	ON / OFF	ON (if on)	OFF	OFF

1996-98 Mark VIII 4.6L V8 MFI VIN V (A/T) - Inputs

PCM PID Acronym	Parameter Identification	PID Range	PID Value at Hot Idle	PID Value at 30	PID Value at 55
ACCS	A/C Clutch Switch	ON / OFF	ON (if on)	OFF	OFF
ACP	ACP Sensor Volts	0-5.1v	0.8-1.4	0.8-1.4	0.8-1.4
BPP	Brake Pedal Switch	ON / OFF	ON (if on)	OFF	OFF
AIR-M	AIR System Monitor	ON / OFF	ON	ON	ON
CID	CID Sensor	0-1 KHz	5-7	7-9	11-14
CKP	CKP Sensor	0-10 KHz	365-395	540-550	850-895
DPFE	DPFE Sensor	0-5.1v	0.2-1.3v	0.2-4.5v	0.2-4.5v
ECT (°F)	ECT Sensor	-40-304°F	160-200	160-200	160-200
FEPS	Flash EEPROM	0-5.1v	0.1	0.1	0.1
FLI ('98)	Fuel Level Indicator	0-100%	50 (1.7v)	50 (1.7v)	50 (1.7v)
FP-M	Fuel Pump Monitor	0-100%	50%	50%	50%
FTP ('98)	FTP Sensor	0-5.1v or 0-10" H20	2.6v at 0" of H20	2.6v at 0" of H20	2.6v at 0" of H20
GEAR	Gear Position	1-2-3-4	1	3	4
HO2S-11	HO2S-11 (Bank 1)	0.1-1.1v	0.1-1.1	0.1-1.1	0.1-1.1
HO2S-12	HO2S-12 (Bank 1)	0.1-1.1v	0.1-1.1	0.1-1.1	0.1-1.1
HO2S-21	HO2S-21 (Bank 2)	0.1-1.1v	0.1-1.1	0.1-1.1	0.1-1.1
HO2S-22	HO2S-22 (Bank 2)	0.1-1.1v	0.1-1.1	0.1-1.1	0.1-1.1
IAT (°F)	IAT Sensor	-40-304°F	50-120	50-120	50-120
IMRCM	IMRC Monitor	0-5.1v	5	5	5
KS1	Knock Sensor B1	0-5.1v	0	0	0
KS2	Knock Sensor B2	0-5.1v	0	0	0
LOAD	Engine Load	0-100%	10-20	23-35	25-35
MAF (V)	MAF Sensor	0-5.1v	0.6-0.9	0.9-1.1	1.2-1.6v
MISF	Misfire Detection	ON / OFF	OFF	OFF	OFF
OCTADJ	Octane Adjustment	CL/OPEN	CLOSED	CLOSED	CLOSED
OSS	OSS Sensor	0-10K rpm	0	1345-1355	2170-2210
PNP	PNP Switch	ON / OFF	ON	OFF	OFF
RPM	Engine Speed	0-10K rpm	660-700	1380-1420	1700-1740
TCS	TCS (switch)	ON / OFF	ON (if on)	OFF	OFF
TFT (°F)	TFT Sensor	-40-304°F	110-210	110-210	110-210
TP-B	SEC TP Sensor	0-5.1v	0.5-0.7	0.5-0.7	0.5-0.7
TPV	TP Sensor	0-5.1v	0.5-0.9	0.8-1.1	0.8-1.2
TR1	TR Sensor 1 Volts	0-12v	0	11.5	11.5
TR2	TR2 Sensor	0-12v	0	11.5	11.5
TRV/TR	TR Sensor	0-5.1v	0	1.7 / OD	1.7 / OD
TR4	TR4 Sensor	0-12v	0	11.5	11.5
Vacuum	Engine Vacuum	0-30" Hg	19-20	18-20	8-15
VPWR	Vehicle Power	0-25.5v	12-14	12-14	12-14
VSS	Vehicle Speed	0-159 mph 0-1 KHz	0 = 0 Hz	30 = 65 Hz	55 = 125 Hz

1996-98 Mark VIII 4.6L V8 MFI VIN V (A/T) - Outputs

PCM PID Acronym	Parameter Identification	PID Range	PID Value at Hot Idle	PID Value at 30	PID Value at 55
AIR	AIR Solenoid	ON / OFF	OFF	OFF	OFF
CD1-CD8	COP Driver 1-8	0-45° dwell	8	11	12
EGRVR	VR Solenoid	0-100%	0	0-45	45-90
EPC	EPC Solenoid	0-100 psi	15	40	40
EVAPCV	EVAP CV Valve	0-100%	0	0	0
EVAPDC	EVAP Purge Valve	0-100%	0-100	0-100	0-100
FP	Fuel Pump Control	0-100%	35	35	35
FuelPW1	INJ Pulsewidth B1	0-99.9 ms	2.5-3.0	3.5-8.0	3.8-9.0
FuelPW2	INJ Pulsewidth B2	0-99.9 ms	2.5-3.0	3.5-8.0	3.8-9.0
HFC	High Fan Control	ON / OFF	ON (if on)	OFF	OFF
HTR-11	HO2S-11 Heater	ON / OFF	ON	ON	ON
HTR-12	HO2S-12 Heater	ON / OFF	ON	ON	ON
HTR-21	HO2S-21 Heater	ON / OFF	ON	ON	ON
HTR-22	HO2S-22 Heater	ON / OFF	ON	ON	ON
IAC	Idle Air Control	0-100%	32	46-53	60-75
IMRC	IMRC Solenoid	ON / OFF	OFF	OFF	OFF
INJ1	INJ 1 Pulsewidth	0-99.9 ms	2.5-3.0	3.5-8.0	3.8-9.0
INJ2	INJ 2 Pulsewidth	0-99.9 ms	2.5-3.0	3.5-8.0	3.8-9.0
INJ3	INJ 3 Pulsewidth	0-99.9 ms	2.5-3.0	3.5-8.0	3.8-9.0
INJ4	INJ 4 Pulsewidth	0-99.9 ms	2.5-3.0	3.5-8.0	3.8-9.0
INJ5	INJ 5 Pulsewidth	0-99.9 ms	2.5-3.0	3.5-8.0	3.8-9.0
IN6	INJ 6 Pulsewidth	0-99.9 ms	2.5-3.0	3.5-8.0	3.8-9.0
INJ7	INJ 7 Pulsewidth	0-99.9 ms	2.5-3.0	3.5-8.0	3.8-9.0
INJ8	INJ 8 Pulsewidth	0-99.9 ms	2.5-3.0	3.5-8.0	3.8-9.0
LFC	Low Fan Control	ON / OFF	ON (if on)	OFF	OFF
LongFT1	Long Term F/T B1	-20 to 20%	-5 to +5	-5 to +5	-5 to +5
LongFT2	Long Term F/T B2	-20 to 20%	-5 to +5	-5 to +5	-5 to +5
MIL	MIL Control	ON / OFF	ON (if on)	OFF	OFF
SHTFT1	Short Term F/T B1	-20 to 20%	-5 to +5	-5 to +5	-5 to +5
SHTFT2	Short Term F/T B2	-20 to 20%	-5 to +5	-5 to +5	-5 to +5
SPARK	Spark Advance	-90° to 90°	13-19	19-32	25-36
SS1	Shift Solenoid 1	ON / OFF	ON	OFF	ON
SS2	Shift Solenoid 2	ON / OFF	OFF	ON	ON
SS3	Shift Solenoid 3	ON / OFF	OFF	ON	ON
TCC	TCC Solenoid	0-100%	0	90-95	90-95
VREF	Voltage Reference	0-5.1v	4.9-5.1	4.9-5.1	4.9-5.1

Town Car PID Data

1991-94 Town Car 4.6L V8 MFI VIN W (A/T) Inputs / Outputs

PCM PID Acronym	Parameter Identification	PID Range	PID Value at Hot Idle	PID Value at 30	PID Value at 55
ACCS	A/C Clutch Switch	ON / OFF	ON (if on)	OFF	OFF
BOO	Brake Switch	ON / OFF	ON (if on)	OFF	OFF
CANP	EVAP Purge Valve	ON / OFF	OFF	ON	ON
DPFE	DPFE Sensor	0-5.1v	0.4-0.6	0.6-1.8	1.9-4.8
ECT (ºF)	ECT Sensor	-40-304ºF	160-200	160-200	160-200
ECT (V)	ECT Sensor	0-5.1v	0.6	0.6	0.6
EPC	EPC Solenoid	0-100 psi	20	12	30
EVR	VR Solenoid	0-100%	0%	0-40	38-57
FP	Fuel Pump Control	ON / OFF	ON	ON	ON
GEAR	Gear Position	NEUTRAL	NEUTRAL	DRIVE	DRIVE
FuelPW1	INJ Pulsewidth B1	0-99.9 ms	4.2-4.4	4.6-5.6	6.4-7.0
FuelPW2	INJ Pulsewidth B2	0-99.9 ms	4.2-4.4	4.6-5.6	6.4-7.0
HO2S-11	HO2S-11 (Bank 1)	0.1-1.1v	0.1-1.1	0.1-1.1	0.1-1.1
HO2S-21	HO2S-21 (Bank 2)	0.1-1.1v	0.1-1.1	0.1-1.1	0.1-1.1
IAC	Idle Air Control	0-100%	32-36	42-47	58-65
IAT (ºF)	IAT Sensor	-40-304ºF	50-120	50-120	50-120
IAT (V)	IAT Sensor	0-5.1v	1.7-3.5	1.7-3.5	1.7-3.5
LongFT1	Long Term F/T B1	-20 to 20%	-5 to +5	-5 to +5	-5 to +5
LongFT2	Long Term F/T B2	-20 to 20%	-5 to +5	-5 to +5	-5 to +5
LOOP	Loop Status	OL or CL	CL	CL	CL
MAF (V)	MAF Sensor	0-5.1v	0.7	1.1-1.3	1.4-1.6
MLP (V)	MLP Sensor	0-5.1v	0	5	5
PNP	PNP Switch	PARK	PARK	DRIVE	DRIVE
RPM	Engine Speed	0-10K rpm	750-820	1450-1630	1750-2100
SHTFT1	Short Term F/T B1	-20 to 20%	-5 to +5	-5 to +5	-5 to +5
SHTFT2	Short Term F/T B2	-20 to 20%	-5 to +5	-5 to +5	-5 to +5
SPARK	Spark Advance	-90º to 90º	15-22	28-35	25-35
TCC	TCC Solenoid	ON / OFF	OFF	ON	ON
TOT (V)	TOT Sensor	0-5.1v	0.5-2.0v	0.5-2.0	0.5-2.0
TPV	TP Sensor	0-5.1v	0.9	0.9-1.0	1.1-1.3
TP C/T	TP Mode	C/T or P/T	C/T	P/T	P/T
VPWR	Vehicle Power	0-25.5v	12-14	12-14	12-14
VREF	Voltage Reference	0-5.1v	4.9-5.1	4.9-5.1	4.9-5.1
VSS	Vehicle Speed	0-159 mph	0	30	55
WAC	A/C WOT Cutout Relay	ON / OFF	ON (if on)	OFF	OFF

1995-2003 Town Car 4.6L V8 MFI VIN W (A/T) - Inputs

PCM PID Acronym	Parameter Identification	PID Range	PID Value at Hot Idle	PID Value at 30	PID Value at 55
ACCS	A/C Clutch Switch	ON / OFF	ON (if on)	OFF	Off
BPP	Brake Pedal Switch	ON / OFF	ON (if on)	OFF	OFF
CID	CID Sensor	0-1 KHz	6-7	9-10	12-14
CKP	CKP Sensor	0-10 KHz	440-490	680-700	870-885
DPFE	DPFE Sensor	0-5.1v	0.4-0.6	0.5-0.9	1.6-4.4
ECT (°F)	ECT Sensor	-40-304°F	160-200	160-200	160-200
FEPS	Flash EEPROM	0-5.1v	0.1	0.1	0.1
FLI	Fuel Level Indicator (1998-99)	0-100%	50 (1.7v)	50 (1.7v)	50 (1.7v)
FP-M	Fuel Pump Monitor	0-100%	100	100	100
FTPV	FTP Sensor (1998-99)	0-5.1v or 0-10" H20	2.6v at 0" of H20	2.6v at 0" of H20	2.6v at 0" of H20
GEAR	Gear Position	1-2-3-4	1	3	4
IAT (°F)	IAT Sensor	-40-304°F	50-120	50-120	50-120
LOAD	Engine Load	0-100%	12-18	17-23	24-28
MAF (V)	MAF Sensor	0-5.1v	0.6-0.9	0.9-1.3	1.3-2.0
MISF	Misfire Detection	ON / OFF	OFF	OFF	OFF
OCTADJ	Octane Adjustment	CL/OPEN	CLOSED	CLOSED	CLOSED
OSS	OSS Sensor	0-10 KHz	0	1200	2130
HO2S-11	HO2S-11 (Bank 1)	0.1-1.1v	0.1-1.1	0.1-1.1	0.1-1.1
HO2S-12	HO2S-12 (Bank 1)	0.1-1.1v	0.1-1.1	0.1-1.1	0.1-1.1
HO2S-21	HO2S-21 (Bank 2)	0.1-1.1v	0.1-1.1	0.1-1.1	0.1-1.1
HO2S-22	HO2S-22 (Bank 2)	0.1-1.1v	0.1-1.1	0.1-1.1	0.1-1.1
PNP	PNP Switch	ON / OFF	ON (in 'P')	OFF	OFF
RPM	Engine Speed	0-10K rpm	790-815	1150-1180	1480-1530
TCS	TCS (switch)	ON / OFF	ON (if on)	OFF	OFF
TFT (°F)	TFT Sensor	-40-304°F	110-210	110-210	110-210
TPV	TP Sensor	0-5.1v	0.5-0.9	1.0-1.2	1.0-1.4
TRV/TR	TR Sensor	0-5.1v	4.4 / PN	2.1 / OD	2.1 / OD
Vacuum	Engine Vacuum	0-30" Hg	19-20	18-20	8-15
VPWR	Vehicle Power	0-25.5v	12-14	12-14	12-14
VSS	Vehicle Speed	0-159 mph 0-1 KHz	0 = 0 Hz	30 = 65 Hz	55 = 125 Hz
WACA	A/C WOT Cutout Relay Monitor	ON / OFF	ON (if on)	OFF	OFF

1995-2003 Town Car 4.6L V8 MFI VIN W (A/T) - Outputs

PCM PID Acronym	Parameter Identification	PID Range	PID Value at Hot Idle	PID Value at 30	PID Value at 55
CD1-CD8	Coil Driver 1-8	0-45° dwell	8	11	12
EGRVR	VR Solenoid	0-100%	0	0-40	38-57
EPC	EPC Solenoid	0-100 psi	20	12	30
EVAPCV	EVAP CV Valve	0-100%	0-100	0-100	0-100
EVAPDC	EVAP Purge Valve	0-100%	0-100	0-100	0-100
FP	Fuel Pump Control	0-100%	100	100	100
FuelPW1	INJ Pulsewidth B1	0-99.9 ms	3.3-3.5	3.7-4.4	5.2-5.6
FuelPW2	INJ Pulsewidth B2	0-99.9 ms	3.3-3.5	3.7-4.4	5.2-5.6
HTR-11	HO2S-11 Heater	ON / OFF	ON	ON	ON
HTR-12	HO2S-12 Heater	ON / OFF	ON	ON	ON
HTR-21	HO2S-21 Heater	ON / OFF	ON	ON	ON
HTR-22	HO2S-22 Heater	ON / OFF	ON	ON	ON
IAC	Idle Air Control	0-100%	32-36	42-47	58-61
INJ1	INJ 1 Pulsewidth	0-99.9 ms	3.3-3.5	3.7-4.4	5.2-5.6
INJ2	INJ 2 Pulsewidth	0-99.9 ms	3.3-3.5	3.7-4.4	5.2-5.6
INJ3	INJ 3 Pulsewidth	0-99.9 ms	3.3-3.5	3.7-4.4	5.2-5.6
INJ4	INJ 4 Pulsewidth	0-99.9 ms	3.3-3.5	3.7-4.4	5.2-5.6
INJ5	INJ 5 Pulsewidth	0-99.9 ms	3.3-3.5	3.7-4.4	5.2-5.6
INJ6	INJ 6 Pulsewidth	0-99.9 ms	3.3-3.5	3.7-4.4	5.2-5.6
INJ7	INJ 7 Pulsewidth	0-99.9 ms	3.3-3.5	3.7-4.4	5.2-5.6
INJ8	INJ 8 Pulsewidth	0-99.9 ms	3.3-3.5	3.7-4.4	5.2-5.6
LFC	Low Fan Control	ON / OFF	ON (if on)	OFF	OFF
LongFT1	Long Term F/T B1	-20 to 20%	-5 to +5	-5 to +5	-5 to +5
LongFT2	Long Term F/T B2	-20 to 20%	-5 to +5	-5 to +5	-5 to +5
MIL	MIL Control	ON / OFF	ON (if on)	OFF	OFF
SHTFT1	Short Term F/T B1	-20 to 20%	-5 to +5	-5 to +5	-5 to +5
SHTFT2	Short Term F/T B2	-20 to 20%	-5 to +5	-5 to +5	-5 to +5
SPARK	Spark Advance	-90° to 90°	18	33-36	32-36
SS1	Shift Solenoid 1	ON / OFF	ON	OFF	OFF
SS2	Shift Solenoid 2	ON / OFF	OFF	ON	ON
TCC	TCC Solenoid	0-100%	0	40-47	85-93
TCIL	TCIL (lamp)	ON / OFF	ON (if on)	OFF	OFF
VREF	Voltage Reference	0-5.1v	4.9-5.1	4.9-5.1	4.9-5.1
WAC	A/C WOT Cutout Relay	ON / OFF	ON (if on)	OFF	OFF

1990 Town Car 5.0L V8 MFI VIN F (A/T) Inputs / Outputs

PCM PID Acronym	Parameter Identification	PID Range	PID Value at Hot Idle	PID Value at 30	PID Value at 55
ACCS	A/C Clutch Switch	ON / OFF	ON (if on)	OFF	OFF
BARO	Barometric Pressure Sensor	0-1 KHz	159 Hz (sea level)	159 Hz (sea level)	159 Hz (sea level)
BOO	Brake Switch	ON / OFF	ON (if on)	OFF	OFF
CANP	EVAP Purge Valve	ON / OFF	OFF	ON	ON
ECT (°F)	ECT Sensor	-40-304°F	160-200	160-200	160-200
ECT (V)	ECT Sensor	0-5.1v	0.6	0.6	0.6
EVP	EGR EVP Sensor	0-5.1v	0.3	0.3	0.8-1.7
EVR	VR Solenoid	0-100%	0	0-40	0-42
FP	Fuel Pump Control	ON / OFF	ON	ON	ON
FuelPW1	INJ Pulsewidth B1	0-99.9 ms	4.7-5.2	5.5-6.5	7.8-9.5
FuelPW2	INJ Pulsewidth B2	0-99.9 ms	4.7-5.2	5.5-6.5	7.8-9.5
HO2S-11	HO2S-11 (Bank 1)	0.1-1.1v	0.1-1.1	0.1-1.1	0.1-1.1
HO2S-21	HO2S-21 (Bank 2)	0.1-1.1v	0.1-1.1	0.1-1.1	0.1-1.1
IAC	Idle Air Control	0-100%	20-40	34-40	45-55
IAT (°F)	IAT Sensor	-40-304°F	50-120	50-120	50-120
IAT (V)	IAT Sensor	0-5.1v	1.7-3.5	1.7-3.5	1.7-3.5
LOOP	Loop Status	OL or CL	CL	CL	CL
MAP	MAP Sensor Hertz	0-1 KHz	110	112	126
MLP (V)	MLP Sensor	0-5.1v	0	5	5
PNP	PNP Switch	NEUTRAL	NEUTRAL	DRIVE	DRIVE
RPM	Engine Speed	0-10K rpm	650-750	1120-1270	1410-1510
SHTFT1	Short Term F/T B1	-20 to +20	-5 to +5	-5 to +5	-5 to +5
SHTFT2	Short Term F/T B2	-20 to +20	-5 to +5	-5 to +5	-5 to +5
SCCS	S/C Command Switch	0-25.5v	6.7	6.7	6.7
SCVAC	S/C Vacuum Solenoid	0-100%	0	0	0
SCVENT	S/C Vent Solenoid	0-100%	0	98	98
SPARK	Spark Advance	-90° to 90°	15-22	28-35	25-35
TCC	TCC Solenoid	ON / OFF	OFF	ON	ON
TPV	TP Sensor	0-5.1v	0.7	0.8-1.0	1.1-1.3
TP Mode	TP Mode	C/T or P/T	C/T	P/T	P/T
VPWR	Vehicle Power	0-25.5v	12-14	12-14	12-14
VREF	Voltage Reference	0-5.1v	4.9-5.1	4.9-5.1	4.9-5.1
VSS	Vehicle Speed	0-159 mph	0	30	55
WAC	A/C WOT Cutout Relay	ON / OFF	ON (if on)	OFF	OFF

FORD TRUCKS, VANS AND SUVS PARAMETER ID (PID) CONTENTS

About This Section
    Introduction ....Page 8-4
    How to Use This Section ....Page 8-4
F-SERIES TRUCK PID DATA
    4.2L V6 MFI VIN 2 (All) **(1997-2003)** ....Page 8-5
    4.6L V8 MFI VIN W, 6 (All) **(1997-2003)** ....Page 8-7
    4.9L I6 MFI VIN Y (All) **(1990-95)** ....Page 8-9
    4.9L I6 MFI VIN Y (E4OD) **(1990-95)** ....Page 8-9
    4.9L I6 MFI VIN Y (All) **(1996)** ....Page 8-10
    4.9L I6 MFI VIN Y (E4OD) **(1996)** ....Page 8-11
    5.0L V8 MFI VIN N (All) **(1990-95)** ....Page 8-13
    5.0L V8 MFI VIN N (E4OD) **(1990-95)** ....Page 8-13
    5.0L V8 MFI VIN N (All) **(1996)** ....Page 8-14
    5.0L V8 MFI VIN N (E4OD) **(1996)** ....Page 8-16
    5.4L V8 MFI VIN L (E4OD) **(1997-2003)** ....Page 8-18
    5.4L V8 Lightning SC MFI VIN 3 (E4OD) **(1999-03)** ....Page 8-20
    5.4L V8 CNG VIN M (E4OD) **(1997-2003)** ....Page 8-22
    5.4L V8 NGV VIN Z (All) **(1998-99)** ....Page 8-24
    5.4L V8 NGV VIN Z (E4OD) **(1998-99)** ....Page 8-24
    5.8L V8 MFI VIN H (All) **(1990-95)** ....Page 8-26
    5.8L V8 MFI VIN H (E4OD) **(1990-95)** ....Page 8-26
    5.8L V8 MFI VIN R (A/T) **(1993-95)** ....Page 8-27
    5.8L V8 MFI VIN H (E4OD) **(1996-97)** ....Page 8-28
    7.3L V8 Diesel VIN F (E4OD) **(1996-2003)** ....Page 8-30
    7.5L V8 MFI VIN G (All) **(1992-97)** ....Page 8-31
    7.5L V8 MFI VIN G (E4OD) **(1992-97)** ....Page 8-32
RANGER PICKUP PID DATA
    2.3L I4 MFI VIN A (All) **(1992-94)** ....Page 8-33
    2.3L I4 MFI VIN A (All) **(1995-97)** ....Page 8-34
    2.3L I4 SOHC MFI VIN D (All) **(2001-03)** ....Page 8-36
    2.5L I4 SOHC MFI VIN C (All) **(1998-2001)** ....Page 8-38
    3.0L V6 MFI VIN U (All) **(1993-94)** ....Page 8-40
    3.0L V6 MFI VIN U (All) **(1995-2003)** ....Page 8-41
    3.0L V6 OHV FFV VIN V (All) **(1999-2001)** ....Page 8-43
    4.0L V6 MFI VIN E (A/T) **(2002-03)** ....Page 8-45
    4.0L V6 MFI VIN X (All) **(1990-94)** ....Page 8-47
    4.0L V6 MFI VIN X (All) **(1995-2001)** ....Page 8-48
AEROSTAR VAN PID DATA
    3.0L V6 MFI VIN U (A/T) **(1993-95)** ....Page 8-50
    3.0L V6 MFI VIN U (A/T) **(1996-97)** ....Page 8-51
    4.0L V6 MFI VIN X (A/T) **(1990-95)** ....Page 8-53
    4.0L V6 MFI VIN X (A/T) **(1996-97)** ....Page 8-54

E-SERIES VAN PID DATA

4.2L V6 MFI VIN 2 (A/T) **(1997-2003)** .... Page 8-56
4.6L V8 MFI VIN 6 (A/T) **(1997-2001)** .... Page 8-58
4.6L V8 MFI VIN W (A/T) **(1999-2003)** .... Page 8-60
4.9L I6 MFI VIN Y (All) **(1990-95)** .... Page 8-62
4.9L I6 MFI VIN Y (E4OD) **(1990-95)** .... Page 8-62
4.9L I6 MFI VIN Y (All) **(1996)** .... Page 8-63
4.9L I6 MFI VIN Y (E4OD) **(1996)** .... Page 8-65
5.0L V8 MFI VIN N (All) **(1990-95)** .... Page 8-67
5.0L V8 MFI VIN N (E4OD) **(1990-95)** .... Page 8-67
5.0L V8 MFI VIN N (All) **(1996)** .... Page 8-68
5.0L V8 MFI VIN N (E4OD) **(1996)** .... Page 8-70
5.4L V8 MFI VIN L (E4OD) **(1997-2003)** .... Page 8-72
5.4L V8 CNG VIN M (E4OD) **(1997-2003)** .... Page 8-74
5.4L V8 NGV VIN Z (All) **(1998-99)** .... Page 8-76
5.4L V8 NGV VIN Z (E4OD) **(1998-99)** .... Page 8-76
5.8L V8 MFI VIN H, R (All) **(1990-95)** .... Page 8-78
5.8L V8 MFI VIN H (E4OD) **(1990-95)** .... Page 8-78
5.8L V8 MFI VIN H (E4OD) **(1996)** .... Page 8-79
6.8L V10 MFI VIN S (E4OD) **(1997-2001)** .... Page 8-81
7.3L V8 Turbo Diesel VIN F (E4OD) **(1996-2003)** .... Page 8-83
7.5L V8 MFI VIN G (All) **(1992-95)** .... Page 8-85
7.5L V8 MFI VIN G (E4OD) **(1992-95)** .... Page 8-85
7.5L V8 MFI VIN G (E4OD) **(1996)** .... Page 8-86

VILLAGER VAN PID DATA

3.0L V6 SOHC MFI VIN W, 1 (A/T) **(1996-98)** .... Page 8-88
3.3L V6 SOHC MFI VIN T (A/T) **(1999-2003)** .... Page 8-89

WINDSTAR VAN PID DATA

3.0L V6 MFI VIN U (A/T) **(1996-2000)** .... Page 8-90
3.8L V6 MFI VIN R (A/T) **(1996-2003)** .... Page 8-92

BRONCO PID DATA
    4.9L I6 MFI VIN Y (All) **(1990-95)** ....Page 8-9
    4.9L I6 MFI VIN Y (E4OD) **(1990-95)** ....Page 8-9
    4.9L I6 MFI VIN Y (All) **(1996)** ....Page 8-10
    4.9L I6 MFI VIN Y (E4OD) **(1996)** ....Page 8-11
    5.0L V8 MFI VIN N (All) **(1990-95)** ....Page 8-13
    5.0L V8 MFI VIN N (E4OD) **(1990-95)** ....Page 8-13
    5.0L V8 MFI VIN N (All) **(1996)** ....Page 8-14
    5.0L V8 MFI VIN N (E4OD) **(1996)** ....Page 8-16
    5.8L V8 MFI VIN H (E4OD) **(1996)** ....Page 8-28
ESCAPE PID DATA
    2.0L I4 4v SFI VIN B (All) **(2001-03)** ....Page 8-113
    3.0L V6 4v SFI VIN 1 (A/T) **(2001-03)** ....Page 8-115
EXCURSION PID DATA
    5.4L V8 MFI VIN L (E4OD) **(2000-03)** ....Page 8-117
    6.8L V10 MFI VIN S (E4OD) **(2000-03)** ....Page 8-119
    7.3L V8 Turbo Diesel VIN F (E4OD) **(2000-03)** ....Page 8-121
EXPEDITION PID DATA
    4.6L V8 SOHC MFI VIN 6 (A/T) **(1997-2001)** ....Page 8-94
    4.6L V8 SOHC MFI VIN W (A/T) **(1997-2003)** ....Page 8-94
    5.4L V8 SOHC MFI VIN L (E4OD) **(1997-2003)** ....Page 8-98
EXPLORER & MOUNTAINEER PID DATA
    4.0L V6 MFI VIN X (All) **(1991-95)** ....Page 8-104
    4.0L V6 MFI VIN X (All) **(1996-2001)** ....Page 8-105
    4.0L V6 MFI VIN E (A/T) **(1997-2003)** ....Page 8-107
    4.0L V6 MFI VIN K (A/T) **(1997-2001)** ....Page 8-107
    4.6L V8 MFI SOHC VIN W (A/T) **(2002-03)** ....Page 8-109
    5.0L V8 MFI VIN P (A/T) **(1996-2001)** ....Page 8-111
BLACKWOOD PID DATA
    5.4L V8 MFI DOHC 4v VIN A (A/T) **(2002-03)** ....Page 8-97
NAVIGATOR PID DATA
    5.4L V8 MFI SOHC 2v VIN L (E4OD) **(1998-99)** ....Page 8-98
    5.4L V8 MFI DOHC 4v VIN A (E4OD) **(1999-2001)** ....Page 8-100
    5.4L V8 MFI DOHC 4v VIN R (E4OD) **(2001-03)** ....Page 8-102

About This Section

Introduction

This section contains Parameter ID (PID) tables for Ford vehicles from 1990-2003. It can be used to assist in repair of Trouble Code and No Code problems related to the PCM.

VEHICLE COVERAGE

- Aerostar Van Applications (1993-97)
- Aviator SUV Applications (2003)
- Blackwood Pickup Applications (2002-03)
- Bronco Applications (1990-96)
- E-Series Van Applications (1990-2003)
- Escape SUV Applications (2001-03)
- Excursion Applications (2000-03)
- Expedition Applications (1997-2003)
- Explorer Applications (1991-2003)
- Mountaineer Applications (1997-2003)
- Navigator Applications (1997-2003)
- Ranger Pickup Applications (1990-2003)
- Villager Applications (1996-2003)
- Windstar Applications (1989-2003)

How to Use This Section

This section can be used to look up PID Data so that you can compare the "known good" values to actual PID Data values you see on the Scan Tool display. To locate the PID Data, find the model, correct engine size (with VIN Code) and finally the year of the vehicle.

For example, to look up the PID Data for a 1999 Ranger 4.0L VIN X, go to Contents Page 1 and find the text string shown below:

4.0L V6 MFI VIN X (All) PIDs **(1995-2001)** ....Page 8-49

Then turn to Page 8-49 to find the following PCM PID information.

1995-2001 Ranger 4.0L V6 MFI VIN X (All) Inputs

PCM PID Acronym	Parameter Identification	PID Range	PID Value at Hot Idle	PID Value at 30	PID Value at 55
CKP	CKP Sensor	0-9999 Hz	430-475	810-870	1180-1230
DPFE	DPFE Sensor	0-5.1v	0.4-0.6	0.4-1.0	0.4-1.1
FLI	Fuel Level Indicator	0-100%	50% (1.7)	50% (1.7)	50% (1.7)
FTP	Fuel Tank Pressure	0-5.1v or 0-10" H2O	2.6v / 0" H2O (with cap off)	2.6v / 0" H2O (with cap off)	2.6v / 0" H2O (with cap off)

F-Series Truck PID DATA

1997-2003 F-SERIES 4.2L V6 VIN 2 (ALL) INPUTS

PCM PID Acronym	Parameter Identification	PID Range	PID Value at Hot Idle	PID Value at 30	PID Value at 55
4x4L	4x4 Switch Signal	ON / OFF	ON (if on)	OFF	OFF
ACCS	A/C Switch Signal	ON / OFF	ON (if on)	OFF	OFF
BPP	Brake Position	ON / OFF	ON (if on)	OFF	OFF
CID	CMP Sensor	0-999 Hz	5-7	10-12	13-17
CKP	CKP Sensor	0-9999 Hz	430-475	900-1000	1140-1300
CPP	CPP Switch Signal	ON / OFF	ON (if in)	OFF	OFF
DPFE	DPFE Sensor	0-5.1v	0.4-0.6	0.4-1.0	0.6-1.1
ECT (°F)	ECT Sensor	-40-304°F	160-200	160-200	160-200
FEPS	Flash EEPROM	0-5.1v	0.1	0.1	0.1
FLI	Fuel Level Indicator	0-100%	50% (1.7)	50% (1.7)	50% (1.7)
FPM	Fuel Pump Monitor	ON / OFF	ON	ON	ON
FTP	Fuel Tank Pressure	0-5.1v or 0-10" H2O	2.6v / 0" H2O (with cap off)	2.6v / 0" H2O (with cap off)	2.6v / 0" H2O (with cap off)
IAT (°F)	IAT Sensor	-40-304°F	50-120	50-120	50-120
IMRC-M	IMRC 1 Monitor	5v / 0	5v	5v	5v
KS1	Knock Sensor 1	0-5.1v	0	0	0
LOAD	Engine Load	0-100%	10-20	20-27	30-45
MAFV	MAF Sensor Signal	0-5.1v	0.6-0.9	1.3-1.7	1.2-2.3
MISF	Misfire Detection	ON / OFF	OFF	OFF	OFF
OCTADJ	Octane Adjustment	CL/OPEN	CLOSED	CLOSED	CLOSED
HO2S-11	HO2S-11 (Bank 1)	0.1-1.1v	0.1-1.1	0.1-1.1	0.1-1.1
HO2S-12	HO2S-12 (Bank 1)	0.1-1.1v	0.1-1.1	0.1-1.1	0.1-1.1
HO2S-21	HO2S-21 (Bank 2)	0.1-1.1v	0.1-1.1	0.1-1.1	0.1-1.1
HO2S-22	HO2S-22 (Bank 2)	0.1-1.1v	0.1-1.1	0.1-1.1	0.1-1.1
PNP	PNP Switch Signal	ON / OFF	ON	OFF	OFF
RPM	Engine Speed	0-10K rpm	680-830	1200-1300	1600-1800
TFT	TFT Sensor Signal	-40-304°F	110-210	110-210	110-210
TPV	TP Sensor	0-5.1v	0.53-1.27	1.0-1.3	1.1-1.6
TRV/TR	TR Sensor	P/N or O/D	P/N	O/D	O/D
OSS	OSS Sensor	0-10K rpm	0	1250-1310	2450-2550
Vacuum	Engine Vacuum	0-30" Hg	19-20	18-20	8-15
VPWR	Vehicle Power	0-25.5v	12-14	12-14	12-14
VSS	Vehicle Speed (MPH & Hertz Rate)	0-159 mph	0 = 0 Hz	30 = 65 Hz	55 = 125 Hz

1997-2003 F-SERIES 4.2L V6 VIN 2 (ALL) OUTPUTS

PCM PID Acronym	Parameter Identification	PID Range	PID Value at Hot Idle	PID Value at 30	PID Value at 55
CD1	Coil 1 Driver	0-60º	6	8	12
CD2	Coil 2 Driver	0-60º	6	8	12
CD3	Coil 3 Driver	0-60º	6	8	12
CTO	Clean Tachometer signal	0-9999 Hz	35-49	65-90	90-120
EGR VR	EGR VR Solenoid	0-100%	0-100	0-40	0-50
EPC	EPC Solenoid	0-300 psi	4	20	20
EVAPCV	EVAP CV Valve	0-100%	0-100	0-100	0-100
EVAPPC	EVAP Purge Valve	0-100%	0-100	0-100	0-100
FP	Fuel Pump Control	ON / OFF	ON	ON	ON
FuelPW1	INJ Pulsewidth - Bank 1	0-999 ms	2.7-4.1	4.5-8.0	5.5-11.0
FuelPW2	INJ Pulsewidth - Bank 2	0-999 ms	2.7-4.1	4.5-8.0	5.5-11.0
HTR-11	HO2S-11 Heater	ON / OFF	ON	ON	ON
HTR-12	HO2S-12 Heater	ON / OFF	ON	ON	ON
HTR-21	HO2S-21 Heater	ON / OFF	ON	ON	ON
HTR-22	HO2S-22 Heater	ON / OFF	ON	ON	ON
IAC	Idle Air Control	0-100%	25-32	30-55	60-70
IMRC	IMRC Solenoid	ON / OFF	OFF	OFF	OFF
INJ1	INJ 1 Pulsewidth	0-999 ms	2.7-4.1	4.5-8.0	5.5-11.0
INJ2	INJ 2 Pulsewidth	0-999 ms	2.7-4.1	4.5-8.0	5.5-11.0
INJ3	INJ 3 Pulsewidth	0-999 ms	2.7-4.1	4.5-8.0	5.5-11.0
INJ4	INJ 4 Pulsewidth	0-999 ms	2.7-4.1	4.5-8.0	5.5-11.0
INJ5	INJ 5 Pulsewidth	0-999 ms	2.7-4.1	4.5-8.0	5.5-11.0
INJ6	INJ 6 Pulsewidth	0-999 ms	2.7-4.1	4.5-8.0	5.5-11.0
LongFT1	Long Term FT - Bank 1	-20 to 20%	-5 to +5	-5 to +5	-5 to +5
LongFT2	Long Term FT - Bank 2	-20 to 20%	-5 to +5	-5 to +5	-5 to +5
MIL	MIL (lamp) Control	ON / OFF	OFF	OFF	OFF
SHTFT1	Short Term F/T - Bank 1	-10 to 10%	-5 to +5	-5 to +5	-5 to +5
SHTFT2	Short Term FT - Bank 2	-10 to 10%	-5 to +5	-5 to +5	-5 to +5
SPARK	Spark Advance	-90º to 90º	11-20	15-35	20-39
SS1	Shift Solenoid 1	ON / OFF	ON	OFF	ON
SS2	Shift Solenoid 2	ON / OFF	OFF	ON	ON
TCC	TCC Solenoid	0-100%	0	0-45	90-95
TCIL	TCIL (lamp) Control	ON / OFF	ON (if on)	OFF	OFF
VREF	Vehicle Reference	0-5.1v	4.9-5.1	4.9-5.1	4.9-5.1

1997-2003 F-SERIES 4.6L V8 VIN W, VIN 6 (ALL) INPUTS

PCM PID Acronym	Parameter Identification	PID Range	PID Value at Hot Idle	PID Value at 30	PID Value at 55
4x4L	4x4 Switch Signal	ON / OFF	ON (if on)	OFF	OFF
ACCS	A/C Switch Signal	ON / OFF	ON (if on)	OFF	OFF
BPP	Brake Position	ON / OFF	ON (if on)	OFF	OFF
CHT	CHT Sensor	-40-304°F	194	194	194
CID	CMP Sensor	0-999 Hz	5-7	10-12	13-17
CKP	CKP Sensor	0-9999 Hz	430-475	900-1100	1140-1220
CPP	CPP Switch Signal	ON / OFF	ON (if in)	OFF	OFF
DPFE	DPFE Sensor	0-5.1v	0.4-0.6	0.4-1.0	0.6-1.1
ECT	ECT Sensor	-40-304°F	160-200	160-200	160-200
FEPS	Flash EEPROM	0-5.1v	0.1	0.1	0.1
FLI	Fuel Level Indicator	0-100%	50% (1.7)	0-100%	0-100%
FPM	Fuel Pump Monitor	ON / OFF	ON	ON	ON
FTP	Fuel Tank Pressure	0-5.1v or 0-10" H2O	2.6v / 0" H2O (with cap off)	2.6v / 0" H2O (with cap off)	2.6v / 0" H2O (with cap off)
IAT	IAT Sensor	-40-304°F	50-120	50-120	50-120
IMRC1-M	IMRC 1 Monitor	0-5.1v	5v	5v	5v
KS1	Knock Sensor 1	0-5.1v	0	0	0
LOAD	Engine Load	0-100%	10-20	20-27	30-45
MAFV	MAF Sensor	0-5.1v	0.6-0.9	0.7-1.0	1.2-2.3
MISF	Misfire Detection	ON / OFF	OFF	OFF	OFF
OCTADJ	Octane Adjustment	CL/OPEN	CLOSED	CLOSED	CLOSED
HO2S-11	HO2S-11 (Bank 1)	0.1-1.1v	0.1-1.1	0.1-1.1	0.1-1.1
HO2S-12	HO2S-12 (Bank 1)	0.1-1.1v	0.1-1.1	0.1-1.1	0.1-1.1
HO2S-21	HO2S-21 (Bank 2)	0.1-1.1v	0.1-1.1	0.1-1.1	0.1-1.1
HO2S-22	HO2S-22 (Bank 2)	0.1-1.1v	0.1-1.1	0.1-1.1	0.1-1.1
PNP	PNP Switch Signal	ON / OFF	ON	OFF	OFF
RPM	Engine Speed	0-10K rpm	680-830	1200-1500	1600-1800
TFT	TFT Sensor Signal	-40-304°F	110-210	110-210	110-210
TPO	Power Take-Off	0-25.5v	0.9	1.4	1.8
TPV	TP Sensor	0-5.1v	0.53-1.27	1.0-1.3	1.1-1.6
TRV/TR	TR Sensor	P/N or O/D	P/N or O/D	O/D	O/D
OSS	OSS Sensor	0-10K rpm	0	1250-1310	2450-2550
Vacuum	Engine Vacuum	0-30" Hg	19-20	18-20	8-15
VPWR	Vehicle Power	0-25.5v	12-14	12-14	12-14
VSS	Vehicle Speed (Hertz or MPH)	0-159 mph 0-999 Hz	0 = 0 Hz	30 = 65 Hz	55 = 125 Hz

1997-2003 F-SERIES 4.6L V8 VIN W, VIN 6 (ALL) OUTPUTS

PCM PID Acronym	Parameter Identification	PID Range	PID Value at Hot Idle	PID Value at 30	PID Value at 55
CD1	Coil 1 Driver	0-45	5	6	8
CD2	Coil 2 Driver	0-45	5	6	8
CD3	Coil 3 Driver	0-45	5	6	8
CD4	Coil 4 Driver	0-45	5	6	8
CHTIL	CHT (lamp) Control	ON / OFF	OFF	OFF	OFF
CTO	Clean Tachometer signal	0-9999 Hz	35-49	65-90	90-120
EGR VR	EGR VR Solenoid	0-100%	0-100	0-100	0-100
EVAP CP	EVAP Purge Valve	0-100%	0-100	0-100	0-100
EPC	EPC Solenoid	0-300 psi	4	20	20
FP	Fuel Pump Control	ON / OFF	ON	ON	ON
FuelPW1	INJ Pulsewidth - Bank 1	0-999 ms	2.7-3.5	4.5-8.0	5.5-9.0
FuelPW2	INJ Pulsewidth - Bank 2	0-999 ms	2.7-4.1	4.5-8.0	5.5-9.0
HTR-11	HO2S-11 Heater	ON / OFF	ON	ON	ON
HTR-12	HO2S-12 Heater	ON / OFF	ON	ON	ON
HTR-21	HO2S-21 Heater	ON / OFF	ON	ON	ON
HTR-22	HO2S-22 Heater	ON / OFF	ON	ON	ON
IAC	Idle Air Control	0-100%	25-32	30-55	60-70
IMRC	IMRC Signal	0-5.1v	0.1	0.1	0.1
INJ1	INJ 1 Pulsewidth	0-999 ms	2.7-4.1	4.5-8.0	5.5-9.0
INJ2	INJ 2 Pulsewidth	0-999 ms	2.7-4.1	4.5-8.0	5.5-9.0
INJ3	INJ 3 Pulsewidth	0-999 ms	2.7-4.1	4.5-8.0	5.5-9.0
INJ4	INJ 4 Pulsewidth	0-999 ms	2.7-4.1	4.5-8.0	5.5-9.0
INJ5	INJ 5 Pulsewidth	0-999 ms	2.7-4.1	4.5-8.0	5.5-9.0
INJ6	INJ 6 Pulsewidth	0-999 ms	2.7-4.1	4.5-8.0	5.5-9.0
INJ7	INJ 7 Pulsewidth	0-999 ms	2.7-4.1	4.5-8.0	5.5-9.0
INJ8	INJ 8 Pulsewidth	0-999 ms	2.7-4.1	4.5-8.0	5.5-9.0
LongFT1	Long Term FT - Bank 1	-20 to 20%	-5 to +5	-5 to +5	-5 to +5
LongFT2	Long Term FT - Bank 2	-20 to 20%	-5 to +5	-5 to +5	-5 to +5
MIL	MIL (lamp) Control	ON / OFF	ON (if on)	OFF	OFF
SS1	Shift Solenoid 1	ON / OFF	ON	ON	ON
SS2	Shift Solenoid 2	ON / OFF	OFF	ON	ON
SHTFT1	Short Term F/T - Bank 1	-10 to 10%	-5 to +5	-5 to +5	-5 to +5
SHTFT2	Short Term F/T - Bank 2	-10 to 10%	-5 to +5	-5 to +5	-5 to +5
SPARK	Spark Advance	-90-90°	11-20	15-35	20-39
TCC	TCC Solenoid	0-100%	0	0-45	90-95
TCIL	TCIL (lamp) Control	ON / OFF	ON (if on)	OFF	OFF
VREF	Vehicle Reference	0-5.1v	4.9-5.1	4.9-5.1	4.9-5.1

F-Series & Bronco PID DATA -

1990-95 F-SERIES & BRONCO 4.9L I6 VIN Y (A/T, M/T, E4OD)

PCM PID Acronym	Parameter Identification	PID Range	PID Value at Hot Idle	PID Value at 30	PID Value at 55
ACCS	A/C Switch Signal	ON / OFF	ON (if on)	OFF	OFF
CANP	EVAP Purge Valve	ON / OFF	OFF	ON	ON
CPP	Clutch Pedal Switch	ON / OFF	ON (if in)	OFF	OFF
ECT (°F)	ECT Sensor	-40-304°F	160-200	160-200	160-200
ECTV	ECT Sensor	0-5.1v	0.6	0.6	0.6
EPC	EPC Solenoid	0-300 psi	5	5	15
EVP	EGR Valve Position	0-5.1v	0.3	1.2-2.0	2.5-3.5
EVR	EGR VR Solenoid	0-100%	0	0-40	0-40
FuelPW1	INJ Pulsewidth - Bank 1	0-999 ms	6.8-7.0	9.5-10	12-13
GEAR	Gear Position	P-R-N-D	P-R-N-D	DRIVE 3	DRIVE 4
HO2S-11	HO2S-11 (Bank 1)	0.1-1.1v	0.1-1.1	0.1-1.1	0.1-1.1
IAC	Idle Air Control	0-100%	35	44-50	59-65
IAT (°F)	IAT Sensor	-40-304°F	50-120	50-120	50-120
IATV	IAT Sensor	0-5.1v	1.5-3.5	1.5-3.5	1.5-3.5
LOOP	Loop Status	CL or OL	CL	CL	CL
MLPV	MLP Switch (E4OD)	0-5.1v	0v	5	5
MAP	MAP Sensor	0-999 Hz	107	114-120	120-130
PNP	PNP Switch Signal	Neutral/DR	NEUTRAL	DRIVE	DRIVE
RPM	Engine Speed	0-10K rpm	600-700	1050-1150	1840-1940
SHTFT1	Short Term F/T - Bank 1	-10 to 10%	-5 to +5	-5 to +5	-5 to +5
SPARK	Spark Advance	-90° to 90°	17-20	24-28	24-30
TP	TP Sensor	0-5.1v	1.0	1.2-1.3	1.5-1.6
TCC	TCC Sol. (E4OD)	ON / OFF	OFF	ON	ON
TOT	TOT sensor (E4OD)	0-5.1v	2.10-2.40	2.10-2.40	2.10-2.40
TP	TP Sensor	0-5.1v	1.0	1.2-1.3	1.5-1.6
TP Mode	TP Sensor Mode	C/T or P/T	C/T	P/T	P/T
VPWR	Vehicle Power	0-25.5v	12-14	12-14	12-14
VREF	Vehicle Reference	0-5.1v	4.9-5.1	4.9-5.1	4.9-5.1
VSS	Vehicle Speed	0-159 mph	0	30	55

1996 F-SERIES & BRONCO 4.9L I6 MFI VIN Y INPUTS

PCM PID Acronym	Parameter Identification	PID Range	PID Value at Hot Idle	PID Value at 30	PID Value at 55
ACCS	A/C Cycling Clutch	ON / OFF	ON	OFF	OFF
CPP	CPP Switch Signal	ON / OFF	ON (if in)	OFF	OFF
DPFE	DPFE Sensor	0-5.1v	0.4-0.6	0.4-0.9	0.6-1.0
ECT	ECT Sensor	-40-304°F	160-200	160-200	160-200
FEPS	Flash EEPROM	0-5.1v	0.1	0.1	0.1
FPM	Fuel Pump Monitor	ON / OFF	ON	ON	ON
IAT	IAT Sensor	-40-304°F	50-120	50-120	50-120
IDM	IDM Signal	0-9999 Hz	32-38	59-65	88-95
KS1	Knock Sensor 1	0-5.1v	0	0	0
LOAD	Engine Load	0-100%	12-14	16-25	30-40
MAFV	MAF Sensor	0-5.1v	0.5-0.7	1.1-1.5	1.7-2.2
MISF	Misfire Detection	ON / OFF	OFF	OFF	OFF
OCTADJ	Octane Adjustment	OPEN/CL	CLOSED	CLOSED	CLOSED
HO2S-11	HO2S-11 (Bank 1)	0.1-1.1v	0.1-1.1	0.1-1.1	0.1-1.1
HO2S-12	HO2S-12 (Bank 1)	0.1-1.1v	0.1-1.1	0.1-1.1	0.1-1.1
HO2S-21	HO2S-21 (Bank 2)	0.1-1.1v	0.1-1.1	0.1-1.1	0.1-1.1
PIP	PIP Sensor	0-9999 Hz	32-38	59-65	88-95
PNP	PNP Switch Signal	NEUT/DR	NEUTRAL	DRIVE	DRIVE
PTO	Power Takeoff Sig.	ON / OFF	ON (if on)	OFF	OFF
RPM	Engine Speed	0-10K rpm	680-730	1200-1300	1810-2000
TPV	TP Sensor	0-5.1v	0.53-1.27	0.8-1.1	0.9-1.2
Vacuum	Engine Vacuum	0-30" Hg	19-20	18-20	8-15
VPWR	Vehicle Power	0-25.5v	12-14	12-14	12-14
WACA	A/C WOT Monitor	ON / OFF	OFF (if on)	ON	ON

1996 F-SERIES & BRONCO 4.9L I6 MFI VIN Y OUTPUTS

PCM PID Acronym	Parameter Identification	PID Range	PID Value at Hot Idle	PID Value at 30	PID Value at 55
AIRB	AIR System Bypass	0v / VBAT	0v	0v	0v
AIRD	AIR System Divert	0v / VBAT	VBAT	VBAT	VBAT
EGR VR	EGR VR Solenoid	0-100%	0	0-40	0-40
FP	Fuel Pump Control	ON / OFF	ON	ON	ON
PW - Bank 1, 2	INJ Pulsewidth	0-999 ms	4.9-5.1	5.3-10.0	9.4-13.0
HTR-11	HO2S-11 Heater	ON / OFF	ON	ON	ON
HTR-12	HO2S-12 Heater	ON / OFF	ON	ON	ON
HTR-21	HO2S-21 Heater	ON / OFF	ON	ON	ON
IAC	Idle Air Control	0-100%	35	44-50	59-65
INJ1-6	INJ 1 Pulsewidth	0-999 ms	4.9-5.1	5.3-10.0	9.4-13.0
LTB1, - Bank 2	Long Term F/T - Bank 1	-20 to 20%	-5 to +5	-5 to +5	-5 to +5
MIL	MIL (lamp) Control	ON / OFF	OFF	OFF	OFF
STB1, - Bank 2	Short Term F/T - Bank 1	-10 to 10%	-5 to +5	-5 to +5	-5 to +5
SPARK	Spark Advance	-90° to 90°	13-16	17-20	17-22
SPOUT	Spark Output Signal	0-999 Hz	32-38	59-65	88-95
VMV	EVAP Solenoid	0-999 Hz	0-10	0-10	0-10
VREF	Vehicle Reference	0-5.1v	4.9-5.1	4.9-5.1	4.9-5.1
WAC	A/C WOT Relay	ON / OFF	ON (if on)	OFF	OFF

1996 F-SERIES & BRONCO 4.9L I6 MFI VIN Y (E4OD) INPUTS

PCM PID Acronym	Parameter Identification	PID Range	PID Value at Hot Idle	PID Value at 30	PID Value at 55
4x4L	4x4 Switch Signal	ON / OFF	ON (if on)	OFF	OFF
ACCS	A/C Switch Signal	ON / OFF	ON (if on)	OFF	OFF
BPP	Brake Position	ON / OFF	ON (if on)	OFF	OFF
DPFE	DPFE Sensor	0-5.1v	0.4-0.6	0.4-0.9	0.4-0.9
ECT	ECT Sensor	-40-304°F	160-200	160-200	160-200
FEPS	Flash EEPROM	0-5.1v	0.1	0.1	0.1
FPM	Fuel Pump Monitor	ON / OFF	ON	ON	ON
IAT	IAT Sensor	-40-304°F	50-120	50-120	50-120
GEAR	Gear Position	1-2-3-4	1	3	4
IDM	IDM Signal	0-9999 Hz	35-42	58-69	71-82
KS1	Knock Sensor 1	0-5.1v	0	0	0
LOAD	Engine Load	0-100%	14-16	21-32	33-42
MAFV	MAF Sensor	0-5.1v	0.7-0.9	1.2-1.8	1.7-2.0
MISF	Misfire Detection	ON / OFF	OFF	OFF	OFF
OCTADJ	Octane Adjustment	CL/OPEN	CLOSED	CLOSED	CLOSED
HO2S-11	HO2S-11 (Bank 1)	0.1-1.1v	0.1-1.1	0.1-1.1	0.1-1.1
HO2S-12	HO2S-12 (Bank 1)	0.1-1.1v	0.1-1.1	0.1-1.1	0.1-1.1
HO2S-21	HO2S-21 (Bank 2)	0.1-1.1v	0.1-1.1	0.1-1.1	0.1-1.1
PIP	PIP Sensor	0-9999 Hz	35-42	58-69	71-82
PNP	PNP Switch Signal	Neutral/DR	NEUTRAL	DRIVE	DRIVE
PTO	PTO Signal	ON / OFF	OFF	OFF	OFF
RPM	Engine Speed	0-10K rpm	760-830	1200-1270	1510-1570
TCS	TCS Switch	ON / OFF	OFF	OFF	OFF
TFT	TFT Sensor Signal	-40-304°F	110-210	110-210	110-210
TPV	TP Sensor	0-5.1v	0.53-1.27	1.0-1.3	1.2-1.7
TRV/TR	TR Sensor	0v / 1.7v	0	1.7	1.7
Vacuum	Engine Vacuum	0-30" Hg	19-20	18-20	8-15
VPWR	Vehicle Power	0-25.5v	12-14	12-14	12-14
VSS	Vehicle Speed (Hertz or MPH)	0-999 Hz (0-159 mph)	0 Hz = 0 mph	65 Hz = 30 mph	125 Hz = 55 mph
WACA	A/C WOT Relay Monitor	ON / OFF	OFF	ON	ON

1996 F-SERIES & BRONCO 4.9L I6 VIN Y (E4OD) OUTPUTS

PCM PID Acronym	Parameter Identification	PID Range	PID Value at Hot Idle	PID Value at 30	PID Value at 55
AIRB	AIR System Bypass	0v / VBAT	0v	0v	0v
AIRD	AIR System Divert	0v / VBAT	VBAT	VBAT	VBAT
CCS	Coast Clutch Solenoid Control	ON / OFF	OFF	OFF	OFF
EGR VR	EGR VR Solenoid	0-100%	0-100	0-40	0-40
EPC	EPC Solenoid	0-300 psi	5	6	7
FP	Fuel Pump Control	ON / OFF	ON	ON	ON
FuelPW1	INJ Pulsewidth - Bank 1	0-999 ms	4.9-5.3	5.5-10.0	9.5-13.0
FuelPW2	INJ Pulsewidth - Bank 2	0-999 ms	4.9-5.3	5.5-10.0	9.5-13.0
HTR-11	HO2S-11 Heater	ON / OFF	ON	ON	ON
HTR-12	HO2S-12 Heater	ON / OFF	ON	ON	ON
HTR-21	HO2S-21 Heater	ON / OFF	ON	ON	ON
IAC	Idle Air Control	0-100%	30-34	30-55	58-61
INJ1	INJ 1 Pulsewidth	0-999 ms	4.9-5.3	5.5-10.0	9.5-13.0
INJ2	INJ 2 Pulsewidth	0-999 ms	4.9-5.3	5.5-10.0	9.5-13.0
INJ3	INJ 3 Pulsewidth	0-999 ms	4.9-5.3	5.5-10.0	9.5-13.0
INJ4	INJ 4 Pulsewidth	0-999 ms	4.9-5.3	5.5-10.0	9.5-13.0
INJ5	INJ 5 Pulsewidth	0-999 ms	4.9-5.3	5.5-10.0	9.5-13.0
INJ6	INJ 6 Pulsewidth	0-999 ms	4.9-5.3	5.5-10.0	9.5-13.0
LongFT1	Long Term FT - Bank 1	-20 to 20%	-5 to +5	-5 to +5	-5 to +5
LongFT2	Long Term FT - Bank 2	-20 to 20%	-5 to +5	-5 to +5	-5 to +5
MIL	MIL (lamp) Control	ON / OFF	OFF	OFF	OFF
SS1	Shift Solenoid 1	ON / OFF	ON	OFF	OFF
SS2	Shift Solenoid 2	ON / OFF	OFF	ON	OFF
SHTFT1	Short Term F/T - Bank 1	-10 to 10%	-5 to +5	-5 to +5	-5 to +5
SHTFT2	Short Term FT - Bank 2	-10 to 10%	-5 to +5	-5 to +5	-5 to +5
SPARK	Spark Advance	-90° to 90°	13-16	17-20	17-22
SPOUT	Spark Output Signal	0-9999 Hz	47-52	82-88	105-110
TCC	TCC Solenoid	ON / OFF	OFF	ON	ON
TCIL	TCIL (lamp) Control	ON / OFF	OFF	OFF	OFF
VMV	EVAP Solenoid	0-10	0-10	0-10	0-10
VREF	Vehicle Reference	0-5.1v	4.9-5.1	4.9-5.1	4.9-5.1
WAC	A/C WOT Relay Control	ON / OFF	ON (AC on at WOT)	OFF	OFF

1990-95 F-SERIES & BRONCO 5.0L V8 VIN N (A/T, M/T, E4OD)

PCM PID Acronym	Parameter Identification	PID Range	PID Value at Hot Idle	PID Value at 30	PID Value at 55
ACCS	A/C Switch Signal	ON / OFF	ON (if on)	OFF	OFF
CANP	EVAP Purge Valve	ON / OFF	OFF	ON	ON
CPP	Clutch Pedal Switch	ON / OFF	ON (if in)	OFF	OFF
DPFE	DPFE Sensor ('95)	0-5.1v	0.4-0.6	0.4-0.9	0.4-0.9
ECT (°F)	ECT Sensor	-40-304°F	160-200	160-200	160-200
ECTV	ECT Sensor	0-5.1v	0.6	0.6	0.6
EPC	EPC Solenoid	0-300 psi	4	4	14
EVP	EGR Valve Position	0-5.1v	0.4	0.4-0	3.5-4.5
EVR	EGR VR Solenoid	0-100%	0-100	0-40	0-40
FuelPW1	INJ Pulsewidth - Bank 1	0-999 ms	3.8-4.8	4.4-7.8	7.5-12.0
GEAR (E4OD)	Gear Selector Position Signal	P-R-N-D	P-R-N-D	DRIVE 3	DRIVE 4
IAC	Idle Air Control	0-100%	30-34	43-48	58-61
IAT (°F)	IAT Sensor	-40-304°F	50-120	50-120	50-120
IATV	IAT Sensor	0-5.1v	1.5-3.5	1.5-3.5	1.5-3.5
LOOP	Loop Status	CL or OL	CL	CL	CL
MLPV	MLP Switch (E4OD)	0-5.1v	0	5	5
MAP	MAP Sensor	0-999 Hz	103-105	112-120	122-140
HO2S-11	HO2S-11 (Bank 1)	0.1-1.1v	0.1-1.1	0.1-1.1	0.1-1.1
PNP	PNP Switch Signal	Neutral/DR	NEUTRAL	DRIVE	DRIVE
RPM	Engine Speed	0-10K rpm	680-780	1240-1340	1650-1750
SHTFT1	Short Term F/T - Bank 1	-10 to 10%	-5 to +5	-5 to +5	-5 to +5
SPARK	Spark Advance	-90° to 90°	14-20	28-36	30-40
TCC	TCC Sol. (E4OD)	ON / OFF	OFF	ON	ON
TOTV	TOT sensor (E4OD)	0-5.1v	2.10-2.40	2.10-2.40	2.10-2.40
TP	TP Sensor	0-5.1v	1.0	1.2-1.3	1.5-1.6
TP Mode	TP Sensor Mode	C/T or P/T	C/T	P/T	P/T
VPWR	Vehicle Power	0-25.5v	12-14	12-14	12-14
VREF	Vehicle Reference	0-5.1v	4.9-5.1	4.9-5.1	4.9-5.1
VSS	Vehicle Speed	0-159 mph	0	30	55

F-Series & Bronco PID DATA - OBD II

1996 F-SERIES & BRONCO 5.0L V8 VIN N (ALL) INPUTS

PCM PID Acronym	Parameter Identification	PID Range	PID Value at Hot Idle	PID Value at 30	PID Value at 55
4x4L	4x4 Switch Signal	ON / OFF	ON (if on)	OFF	OFF
ACCS	A/C Switch Signal	ON / OFF	ON (if on)	OFF	OFF
BPP	Brake Switch Signal	ON / OFF	ON (if on)	OFF	OFF
CPP	CPP Switch Signal	ON / OFF	ON (if in)	OFF	OFF
DPFE	DPFE Sensor	0-5.1v	0.4-0.6	0.4-0.9	0.4-0.9
ECT	ECT Sensor	-40-304°F	160-200	160-200	160-200
FEPS	Flash EEPROM	0-5.1v	0.1	0.1	0.1
FPM	Fuel Pump Monitor	ON / OFF	ON	ON	ON
IAT	IAT Sensor	-40-304°F	50-120	50-120	50-120
GEAR	Gear Position	1-2-3-4	1	3	4
IDM	IDM Signal	0-9999 Hz	47-52	82-88	105-110
KS1	Knock Sensor 1	0	0	0	0
LOAD	Engine Load	0-100%	14-16	19-25	26-35
MAFV	MAF Sensor	0-5.1v	0.7-0.9	1.1-1.6	1.7-2.4
MISF	Misfire Detection	ON / OFF	OFF	OFF	OFF
OCTADJ	Octane Adjustment	CL/OPEN	CLOSED	CLOSED	CLOSED
OSS	OSS Sensor	0-9999 Hz	0	122-133	235-250
HO2S-11	HO2S-11 (Bank 1)	0.1-1.1v	0.1-1.1	0.1-1.1	0.1-1.1
HO2S-12	HO2S-12 (Bank 1)	0.1-1.1v	0.1-1.1	0.1-1.1	0.1-1.1
HO2S-21	HO2S-21 (Bank 2)	0.1-1.1v	0.1-1.1	0.1-1.1	0.1-1.1
PIP	PIP Sensor	0-9999 Hz	47-52	82-88	105-110
PNP	PNP Switch Signal	Neutral/DR	NEUTRAL	DRIVE	DRIVE
PTO	Power Takeoff Switch	ON / OFF	ON (if on)	OFF	OFF
RPM	Engine Speed	0-10K rpm	760-830	1200-1270	1590-1675
TCS	TCS Switch	ON / OFF	ON	ON	ON
TFT	TFT Sensor Signal	-40-304°F	110-210	110-210	110-210
TPV	TP Sensor	0-5.1v	0.53-1.27	0.8-1.0	1.0-1.3
TRV/TR	TR Sensor	P/N or O/D	P/N	O/D	O/D
Vacuum	Engine Vacuum	0-30" Hg	19-20	18-20	8-15
VPWR	Vehicle Power	0-25.5v	12-14	12-14	12-14
VSS	Vehicle Speed (Hertz or MPH)	0-999 or 0-159 mph	0 Hz = 0 mph	65 Hz = 30 mph	125 Hz = 55 mph

1996 F-SERIES & BRONCO 5.0L V8 VIN N (ALL) OUTPUTS

PCM PID Acronym	Parameter Identification	PID Range	PID Value at Hot Idle	PID Value at 30	PID Value at 55
AIRB	AIR System Bypass	0-25.5v	0.1	0.1	0.1
AIRD	AIR System Divert	0-25.5v	VBAT	VBAT	VBAT
EGR VR	EGR VR Solenoid	0-100%	0	0-40	35-40
EPC	EPC Solenoid	0-300 psi	4	5	5
FP	Fuel Pump Control	ON / OFF	ON	ON	ON
FuelPW1	INJ Pulsewidth - Bank 1	0-999 ms	3.2-3.8	4.1-6.9	6.5-12
FuelPW2	INJ Pulsewidth - Bank 2	0-999 ms	3.2-3.8	4.1-6.9	6.5-12
HTR-11	HO2S-11 Heater	ON / OFF	ON	ON	ON
HTR-12	HO2S-12 Heater	ON / OFF	ON	ON	ON
HTR-21	HO2S-21 Heater	ON / OFF	ON	ON	ON
IAC	Idle Air Control	0-100%	30-34	43-48	58-61
INJ1	INJ 1 Pulsewidth	0-999 ms	3.2-3.8	4.1-6.9	6.5-12
INJ2	INJ 2 Pulsewidth	0-999 ms	3.2-3.8	4.1-6.9	6.5-12
INJ3	INJ 3 Pulsewidth	0-999 ms	3.2-3.8	4.1-6.9	6.5-12
INJ4	INJ 4 Pulsewidth	0-999 ms	3.2-3.8	4.1-6.9	6.5-12
INJ5	INJ 5 Pulsewidth	0-999 ms	3.2-3.8	4.1-6.9	6.5-12
INJ6	INJ 6 Pulsewidth	0-999 ms	3.2-3.8	4.1-6.9	6.5-12
INJ7	INJ 7 Pulsewidth	0-999 ms	3.2-3.8	4.1-6.9	6.5-12
INJ8	INJ 8 Pulsewidth	0-999 ms	3.2-3.8	4.1-6.9	6.5-12
LongFT1	Long Term FT - Bank 1	-20 to 20%	-5 to +5	-5 to +5	-5 to +5
LongFT2	Long Term FT - Bank 2	-20 to 20%	-5 to +5	-5 to +5	-5 to +5
MIL	MIL (lamp) Control	ON / OFF	OFF	OFF	OFF
SS1	Shift Solenoid 1	ON / OFF	ON	OFF	OFF
SS2	Shift Solenoid 2	ON / OFF	OFF	ON	OFF
SHTFT1	Short Term F/T - Bank 1	-10 to 10%	-5 to +5	-5 to +5	-5 to +5
SHTFT2	Short Term FT - Bank 2	-10 to 10%	-5 to +5	-5 to +5	-5 to +5
SPARK	Spark Advance	-90° to 90°	12-17	35-40	28-37
SPOUT	Spark Output Signal	0-9999 Hz	47-52	82-88	105-110
TCC	TCC Solenoid	0-100%	0	0-40	90-100
TCIL	TCIL (lamp) Control	ON / OFF	OFF	OFF	OFF
VMV	EVAP Solenoid	0-999 Hz	0-10	0-10	0-10
VREF	Voltage Reference	0-5.1v	4.9-5.1	4.9-5.1	4.9-5.1

1996 F-SERIES & BRONCO 5.0L V8 MFI VIN N (E4OD) INPUTS

PCM PID Acronym	Parameter Identification	PID Range	PID Value at Hot Idle	PID Value at 30	PID Value at 55
4x4L	4x4 Switch Signal	ON / OFF	ON (if on)	OFF	OFF
ACCS	A/C Switch Signal	ON / OFF	ON (if on)	OFF	OFF
BPP	Brake Switch Signal	ON / OFF	ON (if on)	OFF	OFF
DPFE	DPFE Sensor	0-5.1v	0.4-0.6	0.4-0.9	0.4-0.9
ECT	ECT Sensor	-40-304°F	160-200	160-200	160-200
FEPS	Flash EEPROM	0-5.1v	0.1	0.1	0.1
FPM	Fuel Pump Monitor	ON / OFF	ON	ON	ON
IAT	IAT Sensor	-40-304°F	50-120	50-120	50-120
GEAR	Gear Position	1-2-3-4	1	3	4
IDM	IDM Signal	0-9999 Hz	47-52	82-88	105-110
KS1	Knock Sensor 1	0-5.1v	0	0	0
LOAD	Engine Load	0-100%	14-16	19-25	26-35
MAFV	MAF Sensor	0-5.1v	0.7-0.9	1.1-1.6	1.7-2.4
MISF	Misfire Detection	ON / OFF	OFF	OFF	OFF
OCTADJ	Octane Adjustment	CL/OPEN	CLOSED	CLOSED	CLOSED
HO2S-11	HO2S-11 (Bank 1)	0.1-1.1v	0.1-1.1	0.1-1.1	0.1-1.1
HO2S-12	HO2S-12 (Bank 1)	0.1-1.1v	0.1-1.1	0.1-1.1	0.1-1.1
HO2S-21	HO2S-21 (Bank 2)	0.1-1.1v	0.1-1.1	0.1-1.1	0.1-1.1
HO2S-22	HO2S-22 (Bank 2)	0.1-1.1v	0.1-1.1	0.1-1.1	0.1-1.1
PIP	PIP Sensor	0-9999 Hz	47-52	82-88	105-110
PNP	PNP Switch Signal	Neutral/DR	NEUTRAL	DRIVE	DRIVE
PTO	Power Takeoff Switch	ON / OFF	ON (if on)	OFF	OFF
RPM	Engine Speed	0-10K rpm	760-830	1200-1270	1600-1650
TCS	TCS Switch	ON / OFF	OFF	OFF	OFF
TFT	TFT Sensor Signal	-40-304°F	110-210	110-210	110-210
TPV	TP Sensor	0-5.1v	0.53-1.27	0.8-1.0	1.0-1.3
TRV/TR	TR Sensor	P/N or O/D	P/N	O/D	O/D
Vacuum	Engine Vacuum	0-30" Hg	19-20	18-20	8-15
VPWR	Vehicle Power	0-25.5v	12-14	12-14	12-14
VSS	Vehicle Speed (Hertz or MPH)	0-999 Hz (0-159 mph)	0 Hz = 0 mph	65 Hz = 30 mph	125 Hz = 55 mph

1996 F-SERIES & BRONCO 5.0L V8 VIN N (E4OD) OUTPUTS

PCM PID Acronym	Parameter Identification	PID Range	PID Value at Hot Idle	PID Value at 30	PID Value at 55
AIRB	AIR System Bypass	0-25.5v	0.1	0.1	0.1
AIRD	AIR System Divert	0-25.5v	VBAT	VBAT	VBAT
CCS	Coast Clutch Solenoid Control	0-25.5v	VBAT	VBAT	VBAT
EGR VR	EGR VR Solenoid	0-100%	0	0-40	35-40
EPC	EPC Solenoid	0-300 psi	4	5	5
FP	Fuel Pump Control	ON / OFF	ON	ON	ON
FuelPW1	INJ Pulsewidth - Bank 1	0-999 ms	3.2-3.8	4.1-6.9	6.5-12
FuelPW2	INJ Pulsewidth - Bank 2	0-999 ms	3.2-3.8	4.1-6.9	6.5-12
HTR-11	HO2S-11 Heater	ON / OFF	ON	ON	ON
HTR-12	HO2S-12 Heater	ON / OFF	ON	ON	ON
HTR-21	HO2S-21 Heater	ON / OFF	ON	ON	ON
IAC	Idle Air Control	0-100%	30-34	43-48	58-61
INJ1	INJ 1 Pulsewidth	0-999 ms	3.2-3.8	4.1-6.9	6.5-12
INJ2	INJ 2 Pulsewidth	0-999 ms	3.2-3.8	4.1-6.9	6.5-12
INJ3	INJ 3 Pulsewidth	0-999 ms	3.2-3.8	4.1-6.9	6.5-12
INJ4	INJ 4 Pulsewidth	0-999 ms	3.2-3.8	4.1-6.9	6.5-12
INJ5	INJ 5 Pulsewidth	0-999 ms	3.2-3.8	4.1-6.9	6.5-12
INJ6	INJ 6 Pulsewidth	0-999 ms	3.2-3.8	4.1-6.9	6.5-12
INJ7	INJ 7 Pulsewidth	0-999 ms	3.2-3.8	4.1-6.9	6.5-12
INJ8	INJ 8 Pulsewidth	0-999 ms	3.2-3.8	4.1-6.9	6.5-12
LongFT1	Long Term FT - Bank 1	-20 to 20%	-5 to +5	-5 to +5	-5 to +5
LongFT2	Long Term FT - Bank 2	-20 to 20%	-5 to +5	-5 to +5	-5 to +5
MIL	MIL (lamp) Control	ON	OFF	OFF	OFF
SS1	Shift Solenoid 1	ON	ON	OFF	OFF
SS2	Shift Solenoid 2	OFF	OFF	ON	OFF
SHTFT1	Short Term F/T - Bank 1	-10 to 10%	-5 to +5	-5 to +5	-5 to +5
SHTFT2	Short Term FT - Bank 2	-10 to 10%	-5 to +5	-5 to +5	-5 to +5
SPARK	Spark Advance	-90° to 90°	12-17	35-40	28-37
SPOUT	Spark Output Signal	0-9999 Hz	47-52	82-88	105-110
TCC	TCC Solenoid	ON / OFF	OFF	ON	ON
TCIL	TCIL (lamp) Control	ON / OFF	OFF	OFF	OFF
VMV	EVAP Solenoid	0-999 Hz	0-10	0-10	0-10
VREF	Vehicle Reference	0-5.1v	4.9-5.1	4.9-5.1	4.9-5.1

1997-2003 F-SERIES 5.4L V8 MFI VIN L (E4OD) INPUTS

PCM PID Acronym	Parameter Identification	PID Range	PID Value at Hot Idle	PID Value at 30	PID Value at 55
4x4L	4x4 Switch Signal	ON / OFF	ON (if on)	OFF	OFF
ACCS	A/C Switch Signal	ON / OFF	ON (if on)	OFF	OFF
BPP	Brake Switch Signal	ON / OFF	ON (if on)	OFF	OFF
CHT	CHT Sensor	-40-304°F	194	194	194
CID	CMP Sensor	0-999 Hz	6	11	11
CKP	CKP Sensor	0-9999 Hz	41	800-840	900-1125
DPFE	DPFE Sensor	0-5.1v	0.20-1.30	0.9-1.36	0.9-1.4
ECT	ECT Sensor	-40-304°F	160-200	160-200	160-200
FEPS	Flash EEPROM	0-5.1v	0.1	0.1	0.1
FLI	Fuel Level Input	1.7 / 50%	1.7 / 50%	1.7 / 50%	1.7 / 50%
FPM	Fuel Pump Monitor	ON / OFF	ON	ON	ON
FTP	Fuel Tank Pressure	2.6v / 0 psi	2.6v / 0 psi	2.6v / 0 psi	2.6v / 0 psi
IAT	IAT Sensor	-40-304°F	50-120	50-120	50-120
GEAR	Gear Position	1-2-3-4	1	3	4
KS1	Knock Sensor 1	0-5.1v	0	0	0
LOAD	Engine Load	0-100%	14-16	19-25	26-35
MAFV	MAF Sensor	0-5.1v	0.7-0.9	1.1-1.6	1.7-2.4
MISF	Misfire Detection	ON / OFF	OFF	OFF	OFF
OCTADJ	Octane Adjustment	CL/OPEN	CLOSED	CLOSED	CLOSED
HO2S-11	HO2S-11 (Bank 1)	0.1-1.1v	0.1-1.1	0.1-1.1	0.1-1.1
HO2S-12	HO2S-12 (Bank 1)	0.1-1.1v	0.1-1.1	0.1-1.1	0.1-1.1
HO2S-21	HO2S-21 (Bank 2)	0.1-1.1v	0.1-1.1	0.1-1.1	0.1-1.1
HO2S-22	HO2S-22 (Bank 2)	0.1-1.1v	0.1-1.1	0.1-1.1	0.1-1.1
PNP	PNP Switch Signal	Neutral/DR	NEUTRAL	DRIVE	DRIVE
PTO	Power Takeoff Switch	OFF	ON (if on)	OFF	OFF
RPM	Engine Speed	0-10K rpm	760-830	1200-1270	1590-1675
TCS	TCS Switch	ON / OFF	OFF	OFF	OFF
TFT	TFT Sensor Signal	-40-304°F	110-210	110-210	110-210
TPV	TP Sensor	0-5.1v	0.53-1.27	0.8-1.0	1.0-1.3
TR1	TR1 Sensor	0v / 11.5v	0v	11.5	11.5
TR2	TR2 Sensor	0v / 11.5v	0v	11.5	11.5
TR3/TR	TR3A Sensor	0v / 1.7v	0v / PN	O/D	O/D
TR4	TR4 Sensor	0v / 11.5v	0	11.5	11.5
Vacuum	Engine Vacuum	0-30" Hg	19-20	18-20	8-15
VPWR	Vehicle Power	0-25.5v	12-14	12-14	12-14
VSS	Vehicle Speed (Hertz or MPH)	0-999 Hz (0-159 mph)	0 Hz = 0 mph	65 Hz = 30 mph	125 Hz = 55 mph

1997-2003 F-SERIES 5.4L V8 MFI VIN L (E4OD) OUTPUTS

PCM PID Acronym	Parameter Identification	PID Range	PID Value at Hot Idle	PID Value at 30	PID Value at 55
CD1-CD8	COP Driver 1-8	0-45° dwell	5	6	8
CHTIL	CHT (lamp) Control	ON / OFF	OFF	OFF	OFF
CTO	Clean Tachometer signal	0-9999 Hz	46	90	115
EGR VR	EGR VR Solenoid	0-100%	0	0-40	35-45
EVAP CP	EVAP Purge Valve	0-100%	0-100	0-100	0-100
EVAP CV	EVAP CV Valve	0-100%	0-100	0-100	0-100
EPC	EPC Solenoid	0-300 psi	4	5	5
FP	Fuel Pump Control	ON / OFF	ON	ON	ON
FuelPW1	INJ Pulsewidth - Bank 1	0-999 ms	3.2-3.8	4.2-6.9	6.5-12
FuelPW2	INJ Pulsewidth - Bank 2	0-999 ms	3.2-3.8	4.2-6.9	6.5-12
HTR-11	HO2S-11 Heater	ON / OFF	ON	ON	ON
HTR-12	HO2S-12 Heater	ON / OFF	ON	ON	ON
HTR-21	HO2S-21 Heater	ON / OFF	ON	ON	ON
HTR-22	HO2S-22 Heater	ON / OFF	ON	ON	ON
IAC	Idle Air Control	0-100%	30-34	43-48	58-61
INJ1	INJ 1 Pulsewidth	0-999 ms	3.2-3.8	4.2-6.9	6.5-12
INJ2	INJ 2 Pulsewidth	0-999 ms	3.2-3.8	4.2-6.9	6.5-12
INJ3	INJ 3 Pulsewidth	0-999 ms	3.2-3.8	4.2-6.9	6.5-12
INJ4	INJ 4 Pulsewidth	0-999 ms	3.2-3.8	4.2-6.9	6.5-12
INJ5	INJ 5 Pulsewidth	0-999 ms	3.2-3.8	4.2-6.9	6.5-12
INJ6	INJ 6 Pulsewidth	0-999 ms	3.2-3.8	4.2-6.9	6.5-12
INJ7	INJ 7 Pulsewidth	0-999 ms	3.2-3.8	4.2-6.9	6.5-12
INJ8	INJ 8 Pulsewidth	0-999 ms	3.2-3.8	4.2-6.9	6.5-12
LongFT1	Long Term FT - Bank 1	-20 to 20%	-5 to +5	-5 to +5	-5 to +5
LongFT2	Long Term FT - Bank 2	-20 to 20%	-5 to +5	-5 to +5	-5 to +5
MIL	MIL (lamp) Control	ON / OFF	OFF	OFF	OFF
SS1	Shift Solenoid 1	ON / OFF	ON	OFF	OFF
SS2	Shift Solenoid 2	ON / OFF	OFF	ON	OFF
SHTFT1	Short Term F/T - Bank 1	-10 to 10%	-5 to +5	-5 to +5	-5 to +5
SHTFT2	Short Term FT - Bank 2	-10 to 10%	-5 to +5	-5 to +5	-5 to +5
SPARK	Spark Advance	-90° to 90°	12-17	35-40	28-37
TCC	TCC Solenoid	0-100%	0	0-40	90-95
TCIL	TCIL (lamp) Control	ON / OFF	OFF	OFF	OFF
VREF	Vehicle Reference	0-5.1v	4.9-5.1	4.9-5.1	4.9-5.1

Ford Lightning

1999-03 LIGHTNING 5.4L V8 SC MFI VIN 3 (E4OD) INPUTS

PCM PID Acronym	Parameter Identification	PID Range	PID Value at Hot Idle	PID Value at 30	PID Value at 55
ACCS	A/C Switch Signal	ON / OFF	ON (if on)	OFF	OFF
BARO	BARO Sensor	0-999 Hz	159	159	159
BPP	Brake Switch Signal	ON / OFF	ON (if on)	OFF	OFF
CHT	CHT Sensor	-40-304°F	194	194	194
CID	CMP Sensor	0-999 Hz	6-8	10-12	14-17
CKP	CKP Sensor	0-9999 Hz	410	650	1060
DPFE	DPFE Sensor	0-5.1v	0.75-1.25	0.9-1.36	0.9-1.4
FEPS	Flash EEPROM	0-5.1v	0.1	0.1	0.1
FLI	Fuel Level Input	1.7v / 50%	1.7 / 50%	1.7 / 50%	1.7 / 50%
FPM	Fuel Pump Monitor	ON / OFF	ON	ON	ON
FTP	Fuel Tank Pressure	2.6v / 0 psi	2.6v / 0 psi	2.6v / 0 psi	2.6v / 0 psi
IAT1	IAT Sensor 1 (°F)	-40-304°F	50-120	50-120	50-120
IAT2	IAT Sensor 2 (°F)	-40-304°F	50-120	50-120	50-120
GEAR	Gear Position	1-2-3-4	1	3	4
KS1	Knock Sensor 1	0-5.1v	0	0	0
LOAD	Engine Load	0-100%	18-21	27-32	35-45
MAFV	MAF Sensor	0-5.1v	0.7-0.9	1.1-1.6	1.7-2.4
MISF	Misfire Detection	ON / OFF	OFF	OFF	OFF
HO2S-11	HO2S-11 (Bank 1)	0.1-1.1v	0.1-1.1	0.1-1.1	0.1-1.1
HO2S-12	HO2S-12 (Bank 1)	0.1-1.1v	0.1-1.1	0.1-1.1	0.1-1.1
HO2S-21	HO2S-21 (Bank 2)	0.1-1.1v	0.1-1.1	0.1-1.1	0.1-1.1
HO2S-22	HO2S-22 (Bank 2)	0.1-1.1v	0.1-1.1	0.1-1.1	0.1-1.1
OCTADJ	Octane Adjustment	CL/OPEN	CLOSED	CLOSED	CLOSED
PNP	PNP Switch Signal	ON / OFF	ON	OFF	OFF
RPM	Engine Speed	0-10K rpm	700	1270-1330	1575-1670
TCS	TCS Switch	ON / OFF	OFF	OFF	ON
TFT	TFT Sensor Signal	-40-304°F	110-210	110-210	110-210
TPV	TP Sensor	0-5.1v	0.53-1.27	0.8-1.0	1.0-1.3
TR1	TR1 Sensor	0v / 11.5v	0v	11.5	11.5
TR2	TR2 Sensor	0v / 11.5v	0v	11.5	11.5
TR4	TR4 Sensor	0v / 11.5v	0v	11.5	11.5
TRV/TR	TR Sensor	0v / 1.7v	0v / PN	O/D	O/D
TSS	TSS Sensor	0-9999 Hz	110	710	1700
Vacuum	Engine Vacuum	0-30" Hg	19-20	18-20	8-15
VPWR	Vehicle Power	0-25.5v	12-14	12-14	12-14
VSS	Vehicle Speed (Hertz or MPH)	0-999 Hz (0-159 mph)	0 Hz = 0 mph	65 Hz = 30 mph	125 Hz = 55 mph

<u>1999-03 LIGHTNING 5.4L V8 SC MFI VIN 3 (E4OD) OUTPUTS</u>

PCM PID Acronym	Parameter Identification	PID Range	PID Value at Hot Idle	PID Value at 30	PID Value at 55
CD1-CD8	COP Driver 1-8	0-45° dwell	5	6	8
CHTIL	CHT (lamp) Control	ON / OFF	OFF	OFF	OFF
CTO	Clean Tachometer signal	0-9999 Hz	46	90	115
EGR VR	EGR VR Solenoid	0-100%	0	0-40	35-45
EVAP CP	EVAP Purge Valve	0-100%	0-100	0-100	0-100
EVAP CV	EVAP CV Valve	0-100%	0-100	0-100	0-100
EPC	EPC Solenoid	0-300 psi	4	5	5
FP	Fuel Pump Control	ON / OFF	ON	ON	ON
FuelPW1	INJ Pulsewidth - Bank 1	0-999 ms	3.2-3.8	4.2-6.9	6.5-12
FuelPW2	INJ Pulsewidth - Bank 2	0-999 ms	3.2-3.8	4.2-6.9	6.5-12
HTR-11	HO2S-11 Heater	ON / OFF	ON	ON	ON
HTR-12	HO2S-12 Heater	ON / OFF	ON	ON	ON
HTR-21	HO2S-21 Heater	ON / OFF	ON	ON	ON
HTR-22	HO2S-22 Heater	ON / OFF	ON	ON	ON
IAC	Idle Air Control	0-100%	30-34	43-48	58-61
ICP	Injector Control Pressure Solenoid	ON / OFF	ON	ON	ON
INJ1	INJ 1 Pulsewidth	0-999 ms	3.2-3.8	4.2-6.9	6.5-12
INJ2	INJ 2 Pulsewidth	0-999 ms	3.2-3.8	4.2-6.9	6.5-12
INJ3	INJ 3 Pulsewidth	0-999 ms	3.2-3.8	4.2-6.9	6.5-12
INJ4	INJ 4 Pulsewidth	0-999 ms	3.2-3.8	4.2-6.9	6.5-12
INJ5	INJ 5 Pulsewidth	0-999 ms	3.2-3.8	4.2-6.9	6.5-12
INJ6	INJ 6 Pulsewidth	0-999 ms	3.2-3.8	4.2-6.9	6.5-12
INJ7	INJ 7 Pulsewidth	0-999 ms	3.2-3.8	4.2-6.9	6.5-12
INJ8	INJ 8 Pulsewidth	0-999 ms	3.2-3.8	4.2-6.9	6.5-12
LongFT1	Long Term FT - Bank 1	-20 to 20%	-5 to +5	-5 to +5	-5 to +5
LongFT2	Long Term FT - Bank 2	-20 to 20%	-5 to +5	-5 to +5	-5 to +5
MIL	MIL (lamp) Control	ON / OFF	OFF	OFF	OFF
SCB	S/C Bypass Control	ON / OFF	OFF	OFF	ON
SHTFT1	Short Term F/T - Bank 1	-10 to 10%	-5 to +5	-5 to +5	-5 to +5
SHTFT2	Short Term FT - Bank 2	-10 to 10%	-5 to +5	-5 to +5	-5 to +5
SPARK	Spark Advance	-90° to 90°	12-17	35-40	28-37
SS1	Shift Solenoid 1	ON / OFF	ON	OFF	OFF
SS2	Shift Solenoid 2	ON / OFF	OFF	OFF	OFF
TCC	TCC Solenoid	0-100%	0	0-40	90-95
TCIL	TCIL (lamp) Control	ON / OFF	OFF	OFF	OFF
VREF	Vehicle Reference	0-5.1v	4.9-5.1	4.9-5.1	4.9-5.1
VSO	VSO Control	ON / OFF	ON	ON	ON

Ford F-Series Truck

1997-2003 F-SERIES 5.4L V8 CNG VIN M (E4OD) INPUTS

PCM PID Acronym	Parameter Identification	PID Range	PID Value at Hot Idle	PID Value at 30	PID Value at 55
ACCS	A/C Switch Signal	ON / OFF	ON (if on)	OFF	OFF
BPP	Brake Switch Signal	ON / OFF	ON (if on)	OFF	OFF
CHT	CHT Sensor	-40-304°F	194	194	194
CID	CMP Sensor	0-999 Hz	6-8	9-12	15-17
CKP	CKP Sensor	0-9999 Hz	400-490	790-870	1000-1089
EFTA	EFT Sensor	-40-275F	50-120	50-120	50-120
FEPS	Flash EEPROM	0-5.1v	0.1	0.1	0.1
FPM	Fuel Pump Monitor	ON / OFF	ON	ON	ON
FSV-M	Fuel Solenoid Valve	ON / OFF	ON	ON	ON
IAT	IAT Sensor	-40-304°F	50-120	50-120	50-120
FRP	FRP Sensor	0-150 psi	90-100	90-100	90-100
GEAR	Gear Position	1-2-3-4	1	3	4
LOAD	Engine Load	0-100%	13-19	26-33	38-46
MAFV	MAF Sensor	0-5.1v	0.7-0.9	0.9-1.7	1.2-2.4
MISF	Misfire Detection	ON / OFF	OFF	OFF	OFF
OCTADJ	Octane Adjustment	CL/OPEN	CLOSED	CLOSED	CLOSED
HO2S-11	HO2S-11 (Bank 1)	0.1-1.1v	0.1-1.1	0.1-1.1	0.1-1.1
HO2S-21	HO2S-21 (Bank 2)	0.1-1.1v	0.1-1.1	0.1-1.1	0.1-1.1
PNP	PNP Switch Signal	Neutral/DR	NEUTRAL	DRIVE	DRIVE
PTO	Power Takeoff Switch	ON / OFF	ON (if on)	OFF	OFF
RPM	Engine Speed	0-10K rpm	900-930	1470-1490	1840-1860
TCS	TCS Switch	ON / OFF	OFF	OFF	OFF
TPV	TP Sensor	0-5.1v	0.53-1.27	0.8-1.2	0.9-1.6v
TR1	TR1 Sensor	0v / 11.5v	0	11.5	11.5
TR2	TR2 Sensor	0v / 11.5v	0	11.5	11.5
TR3/TR	TR3A Sensor	0v / 1.7v	0	1.7	1.7
TR4	TR4 Sensor	0v / 11.5v	0	11.5	11.5
Vacuum	Engine Vacuum	0-30" Hg	19-20	18-20	8-15
VPWR	Vehicle Power	0-25.5v	12-14	12-14	12-14
VSS	Vehicle Speed (Hertz or MPH)	0-999 Hz (0-159 mph)	0 Hz = 0 mph	65 Hz = 30 mph	125 Hz = 55 mph

1997-2003 F-SERIES 5.4L V8 CNG VIN M (E4OD) OUTPUTS

PCM PID Acronym	Parameter Identification	PID Range	PID Value at Hot Idle	PID Value at 30	PID Value at 55
CCS	Coast Clutch Solenoid Control	ON / OFF	OFF	OFF	OFF
CD1-CD8	COP Driver 1-8	0-45° dwell	5	6	8
CHTIL	CHT (lamp) Control	ON / OFF	OFF	OFF	OFF
CTO	Clean Tachometer signal	0-9999 Hz	50-58	90-100	115-125
EPC	EPC Solenoid	0-300 psi	5	6	10 psi
FP	Fuel Pump Control	ON / OFF	ON	ON	ON
FSV	Fuel Solenoid Valve	ON / OFF	ON	ON	ON
FuelPW1	INJ Pulsewidth - Bank 1	0-999 ms	3.9-4.6	4.7-12.0	4.7-12.0
FuelPW2	INJ Pulsewidth - Bank 2	0-999 ms	3.9-4.6	4.7-12.0	4.7-12.0
HTR-11	HO2S-11 Heater	ON / OFF	ON	ON	ON
HTR-21	HO2S-21 Heater	ON / OFF	ON	ON	ON
IAC	Idle Air Control	0-100%	30-34	43-48	58-61
IMTV	Intake Manifold Tuning Valve	0-100%	0	0	0
INJ1	INJ 1 Pulsewidth	0-999 ms	3.9-4.6	4.7-12.0	4.7-12.0
INJ2	INJ 2 Pulsewidth	0-999 ms	3.9-4.6	4.7-12.0	4.7-12.0
INJ3	INJ 3 Pulsewidth	0-999 ms	3.9-4.6	4.7-12.0	4.7-12.0
INJ4	INJ 4 Pulsewidth	0-999 ms	3.9-4.6	4.7-12.0	4.7-12.0
INJ5	INJ 5 Pulsewidth	0-999 ms	3.9-4.6	4.7-12.0	4.7-12.0
INJ6	INJ 6 Pulsewidth	0-999 ms	3.9-4.6	4.7-12.0	4.7-12.0
INJ7	INJ 7 Pulsewidth	0-999 ms	3.9-4.6	4.7-12.0	4.7-12.0
INJ8	INJ 8 Pulsewidth	0-999 ms	3.9-4.6	4.7-12.0	4.7-12.0
LongFT1	Long Term FT - Bank 1	-20 to 20%	-5 to 5%	-5 to 5%	-5 to 5%
LongFT2	Long Term FT - Bank 2	-20 to 20%	-5 to 5%	-5 to 5%	-5 to 5%
MIL	MIL (lamp) Control	ON / OFF	OFF	OFF	OFF
SHTFT1	Short Term F/T - Bank 1	-10 to 10%	-5 to 5%	-5 to 5%	-5 to 5%
SHTFT2	Short Term FT - Bank 2	-10 to 10%	-5 to 5%	-5 to 5%	-5 to 5%
SPARK	Spark Advance	-90° to 90°	15-25	20-35	20-30
SS1	Shift Solenoid 1	ON / OFF	ON	OFF	OFF
SS2	Shift Solenoid 2	ON / OFF	OFF	ON	OFF
TCC	TCC Solenoid	0-100%	0%	0-40	80-100
TCIL	TCIL (lamp) Control	ON / OFF	OFF	OFF	OFF
VREF	Vehicle Reference	0-5.1v	4.9-5.1	4.9-5.1	4.9-5.1

1998-99 F-SERIES 5.4L V8 NGV VIN Z (A/T, M/T, E4OD)

PCM PID Acronym	Parameter Identification	PID Range	PID Value at Hot Idle	PID Value at 30	PID Value at 55
ACCS	A/C Switch Signal	ON / OFF	ON (if on)	OFF	OFF
BPP	Brake Position	ON / OFF	ON (if on)	OFF	OFF
CHT	CHT Sensor	-40-500ºF	194	194	194
CID	CMP Sensor	0-999 Hz	6-7	9-12	15-17.5
CKP	CKP Sensor	0-9999 Hz	440-490	810-855	1088-1089
CPP	Clutch Pedal Switch	ON / OFF	ON (if in)	ON	ON
EFTA	EFT Sensor	-40-275F	50-120	50-120	50-120
FEPS	Flash EEPROM	0-5.1v	0.1	0.1	0.1
FPM	Fuel Pump Monitor	ON / OFF	ON	ON	ON
FRP (psi)	FRP Sensor	0-300 psi	90-100	90-100	90-100
FRP (V)	FRP Sensor	0-5.1v	2.7-3.7	2.7-3.7	2.7-3.7
FSV-M	Fuel Solenoid Valve	ON / OFF	ON	ON	ON
IAT	IAT Sensor	-40-304ºF	50-120	50-120	50-120
GEAR	Gear Position	1-2-3-4	1	3	4
LOAD	Engine Load	0-100%	13-19	26-33	38-46
MAFV	MAF Sensor	0-5.1v	0.7-0.9	0.9-1.7	1.2-2.4
MISF	Misfire Detection	ON / OFF	OFF	OFF	OFF
O2S-11	O2S-11 (Bank 1)	0.1-1.1v	0.1-1.1	0.1-1.1	0.1-1.1
PNP	PNP Switch Signal	Neutral/DR	NEUTRAL	DRIVE	DRIVE
PTO	Power Takeoff Switch	OFF	ON (if on)	OFF	OFF
RPM	Engine Speed	0-10K rpm	900-930	1470-1490	1840-1860
TCS	TCS Switch	ON / OFF	OFF	OFF	OFF
TFT	TFT Sensor Signal	-40-304ºF	110-210	110-210	110-210
TPV	TP Sensor	0-5.1v	0.53-1.27	0.8-1.0	0.9-1.2
TR1	TR1 Sensor	0v / 11.5v	0	11.5	11.5
TR2	TR2 Sensor	0v / 11.5v	0	11.5	11.5
TR4	TR4 Sensor	0v / 11.5v	0	11.5	11.5
TRV/TR	TR Sensor	0v / 1.7v	0	1.7	1.7
Vacuum	Engine Vacuum	0-30" Hg	19-20	18-20	8-15
VPWR	Vehicle Power	0-25.5v	12-14	12-14	12-14
VSS	Vehicle Speed (Hertz or MPH)	0-999 Hz (0-159 mph)	0 Hz = 0 mph	65 Hz = 30 mph	125 Hz = 55 mph

<u>1998-99 F-SERIES 5.4L V8 NGV VIN Z (A/T, M/T, E4OD)</u>

PCM PID Acronym	Parameter Identification	PID Range	PID Value at Hot Idle	PID Value at 30	PID Value at 55
CCS	Coast Clutch Solenoid Control	ON / OFF	OFF	OFF	OFF
CD1-CD8	COP Driver 1-8	0-45° dwell	5	6	8
CHTIL	CHT (lamp) Control	ON / OFF	OFF	OFF	OFF
CTO	Clean Tachometer signal	0-9999 Hz	50-57	90-100	115-125
EPC	EPC Solenoid (E4OD)	0-300 psi	5	6	11
FP	Fuel Pump Control	ON / OFF	ON	ON	ON
FSV	Fuel Solenoid Valve	ON / OFF	ON	ON	ON
FuelPW1	INJ Pulsewidth - Bank 1	0-999 ms	3.9-4.6	4.7-12.0	4.7-12.0
FuelPW2	INJ Pulsewidth - Bank 2	0-999 ms	3.9-4.6	4.7-12.0	4.7-12.0
HTR-11	HO2S-11 Heater	ON / OFF	ON	ON	ON
HTR-21	HO2S-21 Heater	ON / OFF	ON	ON	ON
IAC	Idle Air Control	0-100%	30-34	43-48	58-61
IMTV	Intake Manifold Tuning Valve	0-100%	0	0	0
INJ1	INJ 1 Pulsewidth	0-999 ms	3.9-4.6	4.7-12.0	4.7-12.0
INJ2	INJ 2 Pulsewidth	0-999 ms	3.9-4.6	4.7-12.0	4.7-12.0
INJ3	INJ 3 Pulsewidth	0-999 ms	3.9-4.6	4.7-12.0	4.7-12.0
INJ4	INJ 4 Pulsewidth	0-999 ms	3.9-4.6	4.7-12.0	4.7-12.0
INJ5	INJ 5 Pulsewidth	0-999 ms	3.9-4.6	4.7-12.0	4.7-12.0
INJ6	INJ 6 Pulsewidth	0-999 ms	3.9-4.6	4.7-12.0	4.7-12.0
INJ7	INJ 7 Pulsewidth	0-999 ms	3.9-4.6	4.7-12.0	4.7-12.0
INJ8	INJ 8 Pulsewidth	0-999 ms	3.9-4.6	4.7-12.0	4.7-12.0
LongFT1	Long Term FT - Bank 1	-20 to 20%	-5 to 5%	-5 to 5%	-5 to 5%
LongFT2	Long Term FT - Bank 2	-20 to 20%	-5 to 5%	-5 to 5%	-5 to 5%
MIL	MIL (lamp) Control	ON / OFF	OFF	OFF	OFF
SHTFT1	Short Term F/T - Bank 1	-10 to 10%	-5 to 5%	-5 to 5%	-5 to 5%
SHTFT2	Short Term FT - Bank 2	-10 to 10%	-5 to 5%	-5 to 5%	-5 to 5%
SPARK	Spark Advance	-90° to 90°	15-25	20-35	20-30
SS1	Shift Solenoid 1	ON / OFF	ON	OFF	ON
SS2	Shift Solenoid 2	ON / OFF	OFF	ON	OFF
TCC	TCC Solenoid	0-100%	0	50-100	90-100
TCIL	TCIL (lamp) Control	ON / OFF	OFF	OFF	OFF
VPWR	Vehicle Power	0-25.5v	12-14	12-14	12-14
VREF	Vehicle Reference	5v	4.9-5.1	4.9-5.1	4.9-5.1

1990-95 F-Series 5.8L V8 MFI VIN H (A/T, M/T E4OD)

PCM PID Acronym	Parameter Identification	PID Range	PID Value at Hot Idle	PID Value at 30	PID Value at 55
ACCS	A/C Switch Signal	ON / OFF	ON (if on)	OFF	OFF
CANP	EVAP Purge Valve	ON / OFF	OFF	ON	ON
CPP	Clutch Pedal Switch	ON / OFF	ON (if in)	OFF	OFF
ECT (°F)	ECT Sensor	-40-304°F	160-200	160-200	160-200
ECTV	ECT Sensor	0-5.1v	0.6	0.6	0.6
EPC	EPC Solenoid (E4OD)	0-300 psi	4	4	14
EVP	EGR Valve Position	0-5.1v	0.3	0.4	0.6-3.0
EVR	EGR VR Solenoid	0-100%	0	0-40	0-40
FuelPW1	INJ Pulsewidth - Bank 1	0-999 ms	5.0-5.8	6.0-6.8	6.4-7.0
GEAR	Gear Position	P-R-N-D	P-R-N-D	DRIVE 3	DRIVE 4
IAC	Idle Air Control	0-100%	30-34	28-32	63-70
IAT (°F)	IAT Sensor	-40-304°F	50-120	50-120	50-120
IATV	IAT Sensor	0-5.1v	1.5-3.5	1.5-3.5	1.5-3.5
LOOP	Loop Status	CL or OL	CL	CL	CL
MLPV	MLP Switch (E4OD)	0-5.1v	0	5	5
MAF	MAF Sensor (Calif.)	0-5.1v	0.7-0.9	1.2-1.7	1.6-2.4
MAP	MAP Sensor	0-999 Hz	108-112	110-120	130-140
HO2S-11	HO2S-11 (Bank 1)	0.1-1.1v	0.1-1.1	0.1-1.1	0.1-1.1
PNP	PNP Switch Signal	Neutral/DR	NEUTRAL	DRIVE	DRIVE
RPM	Engine Speed	0-10K rpm	750-850	1250-1350	1600-1700
SHTFT1	Short Term F/T - Bank 1	-10 to 10%	-5 to +5	-5 to +5	-5 to +5
SPARK	Spark Advance	-90° to 90°	14-18	30-36	38-44
TP	TP Sensor	0-5.1v	0.9	1.0-1.1	1.1-1.2
TCC	TCC Sol. (E4OD)	ON / OFF	OFF	ON	ON
TOTV	TOT sensor (E4OD)	0-5.1v	2.10-2.40	2.10-2.40	2.10-2.40
TP	TP Sensor	0-5.1v	1.0	1.2-1.3	1.5-1.6
TP Mode	TP Sensor Mode	C/T or P/T	C/T	P/T	P/T
VPWR	Vehicle Power	0-25.5v	12-14	12-14	12-14
VREF	Vehicle Reference	0-5.1v	4.9-5.1	4.9-5.1	4.9-5.1
VSS	Vehicle Speed	0-159 mph	0	30	55

1993-95 F-Series 5.8L V8 MFI VIN R (A/T) Inputs & Outputs

PCM PID Acronym	Parameter Identification	PID Range	PID Value at Hot Idle	PID Value at 30	PID Value at 55
CANP	EVAP Purge Valve	ON / OFF	OFF	ON	ON
ECT (°F)	ECT Sensor	-40-304°F	160-200	160-200	160-200
ECTV	ECT Sensor	0-5.1v	0.6	0.6	0.6
EVP	EGR Valve Position	0-5.1v	0.3	0.4	0.6-3.0
EVR	EGR VR Solenoid	0-100%	0	0-40	0-40
FuelPW1	INJ Pulsewidth - Bank 1	0-999 ms	5.0-5.8	6.0-6.8	6.4-7.0
IAC	Idle Air Control	0-100%	30-34	43-48	58-61
IAT (°F)	IAT Sensor	-40-304°F	50-120	50-120	50-120
IATV	IAT Sensor	0-5.1v	1.5-3.5	1.5-3.5	1.5-3.5
LOOP	Loop Status	CL or OL	CL	CL	CL
MAP	MAP Sensor	0-999 Hz	108-112	110-120	130-140
HO2S-11	HO2S-11 (Bank 1)	0.1-1.1v	0.1-1.1	0.1-1.1	0.1-1.1
PNP	PNP Switch Signal	Neutral/DR	NEUTRAL	DRIVE	DRIVE
RPM	Engine Speed	0-10K rpm	750-850	1250-1350	1600-1700
SHTFT1	Short Term F/T - Bank 1	-10 to 10%	-5 to +5	-5 to +5	-5 to +5
SPARK	Spark Advance	-90° to 90°	14-18	30-36	36-44
TP	TP Sensor	0-5.1v	0.9	1.0-1.1	1.1-1.2
TP Mode	TP Sensor Mode	C/T or P/T	C/T	P/T	P/T
VPWR	Vehicle Power	0-25.5v	12-14	12-14	12-14
VREF	Vehicle Reference	0-5.1v	4.9-5.1	4.9-5.1	4.9-5.1
VSS	Vehicle Speed	0-159 mph	0	30	55

F-Series & Bronco PID Data

1996-97 F-SERIES & BRONCO 5.8L V8 MFI VIN H (E4OD)

PCM PID Acronym	Parameter Identification	PID Range	PID Value at Hot Idle	PID Value at 30	PID Value at 55
4x4L	4x4 Switch Signal	ON / OFF	ON (if on)	OFF	OFF
ACCS	A/C Switch Signal	ON / OFF	ON (if on)	OFF	OFF
BPP	Brake Switch Signal	ON / OFF	ON (if on)	OFF	OFF
DPFE	DPFE Sensor	0-5.1v	0.4-0.6	0.4-0.9	0.8-1.1
ECTV	ECT Sensor	0-5.1v	0.6	0.6	0.6
ECT (°F)	ECT Sensor	-40-304°F	160-200	160-200	160-200
FEPS	Flash EEPROM	0-5.1v	0.1	0.1	0.1
FLI	Fuel Level Input	0-100%	1.7 / 50%	1.7 / 50%	1.7 / 50%
FPM	Fuel Pump Monitor	ON / OFF	ON	ON	ON
FuelPW1	INJ Pulsewidth - Bank 1	0-999 ms	4.2-4.6	4.4-7.0	7.4-12.0
GEAR	Gear Selector	P-R-N-D	P-R-N-D	DRIVE 3	DRIVE 4
IAC	Idle Air Control	0-100%	30-34	28-32	63-70
IAT (°F)	IAT Sensor	-40-304°F	50-120	50-120	50-120
IDM	IDM Signal	0-9999 Hz	42-50	82-88	105-120
LOAD	Engine Load	0-100%	14-16	20-25	24-35
MAFV	MAF Sensor	0-5.1v	0.7-0.9	1.2-1.7	1.6-2.4
MISF	Misfire Detection	ON / OFF	OFF	OFF	OFF
MLPV	MLP Switch (V)	0-5.1v	0	5	5
OCTADJ	Octane Adjustment	OPEN/CL	CLOSED	CLOSED	CLOSED
HO2S-11	HO2S-11 (Bank 1)	0.1-1.1v	0.1-1.1	0.1-1.1	0.1-1.1
HO2S-12	HO2S-12 (Bank 1)	0.1-1.1v	0.1-1.1	0.1-1.1	0.1-1.1
HO2S-21	HO2S-21 (Bank 2)	0.1-1.1v	0.1-1.1	0.1-1.1	0.1-1.1
PIP	PIP Sensor	0-9999 Hz	42-50	82-88	105-120
PNP	PNP Switch Signal	ON / OFF	ON in P/N	OFF	OFF
PTO	Power Takeoff	ON / OFF	ON (if on)	OFF	OFF
RPM	Engine Speed	0-10K rpm	675-715	1250-1390	1590-1700
TCS	TCS Switch	ON / OFF	OFF	OFF	OFF
TFTV	TFT Sensor Signal	-40-304°F	110-210	110-210	110-210
TPV	TP Sensor	0-5.1v	0.53-1.27	0.8-1.0	0.9-1.2
TP Mode	TP Sensor Mode	C/T or P/T	C/T	P/T	P/T
TRV/TR	TR Sensor	P/N or O/D	P/N	O/D	O/D
Vacuum	Engine Vacuum	0-30" Hg	19-20	18-20	8-15
VPWR	Vehicle Power	0-25.5v	12-14	12-14	12-14
VSS	Vehicle Speed (Hertz or MPH)	0-999 Hz (0-159 mph)	0 Hz = 0 mph	65 Hz = 30 mph	125 Hz = 55 mph

1996-97 F-SERIES & BRONCO 5.8L V8 MFI VIN H (E4OD)

PCM PID Acronym	Parameter Identification	PID Range	PID Value at Hot Idle	PID Value at 30	PID Value at 55
CCS	Coast Clutch Solenoid Control	ON / OFF	OFF	OFF	OFF
EGR VR	EGR VR Solenoid	0-100%	0	0-40	35-45
EVAP CP	EVAP Purge Valve	0-100%	0-100	0-100	0-100
EPC	EPC Solenoid	0-300 psi	5	5	5
FP	Fuel Pump Control	ON / OFF	ON	ON	ON
FuelPW1	INJ Pulsewidth - Bank 1	0-999 ms	4.2-4.7	4.4-7.0	7.4-12.0
FuelPW2	INJ Pulsewidth - Bank 2	0-999 ms	4.2-4.7	4.4-7.0	7.4-12.0
HTR-11	HO2S-11 Heater	ON / OFF	ON	ON	ON
HTR-12	HO2S-12 Heater	ON / OFF	ON	ON	ON
HTR-21	HO2S-21 Heater	ON / OFF	ON	ON	ON
HTR-22	HO2S-22 Heater	ON / OFF	ON	ON	ON
IAC	Idle Air Control	0-100%	30-34	28-32	63-70
INJ1	INJ 1 Pulsewidth	0-999 ms	4.2-4.7	4.4-7.0	7.4-12.0
INJ2	INJ 2 Pulsewidth	0-999 ms	4.2-4.7	4.4-7.0	7.4-12.0
INJ3	INJ 3 Pulsewidth	0-999 ms	4.2-4.7	4.4-7.0	7.4-12.0
INJ4	INJ 4 Pulsewidth	0-999 ms	4.2-4.7	4.4-7.0	7.4-12.0
INJ5	INJ 5 Pulsewidth	0-999 ms	4.2-4.7	4.4-7.0	7.4-12.0
INJ6	INJ 6 Pulsewidth	0-999 ms	4.2-4.7	4.4-7.0	7.4-12.0
INJ7	INJ 7 Pulsewidth	0-999 ms	4.2-4.7	4.4-7.0	7.4-12.0
INJ8	INJ 8 Pulsewidth	0-999 ms	4.2-4.7	4.4-7.0	7.4-12.0
LongFT1	Long Term FT - Bank 1	-20 to 20%	-5 to +5	-5 to +5	-5 to +5
LongFT2	Long Term FT - Bank 2	-20 to 20%	-5 to +5	-5 to +5	-5 to +5
MIL	MIL (lamp) Control	ON / OFF	OFF	OFF	OFF
SS1	Shift Solenoid 1	ON / OFF	ON	OFF	OFF
SS2	Shift Solenoid 2	ON / OFF	OFF	ON	OFF
SHTFT1	Short Term F/T - Bank 1	-10 to 10%	-5 to +5	-5 to +5	-5 to +5
SHTFT2	Short Term FT - Bank 2	-10 to 10%	-5 to +5	-5 to +5	-5 to +5
SPARK	Spark Advance	-90° to 90°	15-18	23-29	26-32
SPOUT	Spark Output Signal	0-9999 Hz	42-50	82-88	105-120
TCC	TCC Solenoid	0-100%	0	90-100	90-100
TCIL	TCIL (lamp) Control	ON / OFF	OFF	OFF	OFF
VREF	Vehicle Reference	0-5.1v	4.9-5.1	4.9-5.1	4.9-5.1

1996-2003 F-SERIES 7.3L V8 DIESEL VIN F (E4OD)

PCM PID Acronym	Parameter Identification	PID Range	PID Value at low idle	PID Value at high idle
4x4L	4x4 Low Switch Input	ON / OFF	ON	OFF
ACCS	A/C Switch Signal	ON / OFF	ON	OFF
AP	Accelerator Pedal Sensor	0-5.1v	0.5-1.6	0.5-4.5
ASMM	A/T Shift Modulator (M)	0v / 12	12	12
BAROV	BARO Sensor (sea level)	0-5.1v	4.75	4.75
BPP	Brake Switch Signal	ON / OFF	OFF	OFF
BPA	Brake Pressure Applied	0v / 12	12	0v
CCS	Coast Clutch Solenoid Control	ON / OFF	OFF	ON
C/S (TCS)	Cancel Switch & TCS On/Off Status (depressed is on)	0v/ VBAT	Off: 0v	Off: 0v
CPP	CPP Switch Signal	0v / 5v	0v (In)	5v
CRUISE	Cruise Control Module	---	---	---
DTC CNT	DTC Count	0-256	0	0
EBP V	Exhaust Back Pressure Actual	0-5.1v	0.8-0.95	0.9-3.0
EOT	Engine Oil Temperature	0-5.1v	0.35-4.7	0.35-4.7
EPC	Electronic Pressure Control	---	---	---
EPC V	EPC Solenoid - Actual	0-25.5v	7.5	12
EPR	Exhaust Back Pressure Regulator	ON / OFF	OFF	ON
FDCS	Fuel Delivery Control Signal	0-999 Hz	49	40-240
GEAR	Gear Position	1-2-3-4	1	4
GPC	Glow Plug Control Duty Cycle	---	---	---
GPMH	Glow Plug Monitor (high side)	ON / OFF	OFF	OFF
GPML	Glow Plug Monitor (left side)	ON / OFF	OFF	OFF
GPMR	Glow Plug Monitor (right side)	ON / OFF	OFF	OFF
GPL	Glow Plug Lamp	ON / OFF	OFF	OFF
IAT	IAT Sensor	0-5.1v	1.5-3.5	1.5-3.5
IAT V	IAT Sensor	-40-304°F	50-120	50-120
ICP	ICP Sensor (startup is 0.83v)	0-5.1v	0.25-0.40	0.25-0.40
ICP	Injector Control Pressure Actual	0-5.1v	0.25-0.40	0.25-0.40
IPR	Injector Control Pressure Regulator	0-100%	35	40-100
IVS	Idle Validation Switch	ON / OFF	ON	OFF
MAP	MAP Sensor Expected (4.6-4.8v)	0-5.1v	1-2	1-3
MAP H	MAP Sensor - Actual	0-5.1v	1-2	1-3
MIAH	Manifold Intake Air Heater	0v / 12	12	12
MIAHM	Manifold Intake Air Heater (M)	ON / OFF	OFF	OFF
MGP	Manifold Gauge Pressure	---	---	---
PBA	Parking Brake Applied	Brake On: 0v	Brake Off: 12	Brake Off: 12
RPM	Engine Speed	0-10K rpm	---	---
SCCS	Speed Control Command Switch	0-25.5v	S/C On: 12	S/C On: 12
SCCS-M	Speed Control Command Switch Mode	---	---	---
TCC	Transmission Converter Clutch	ON / OFF	OFF	ON / OFF
SS1	Shift Solenoid 1 Control	ON / OFF	ON	OFF
SS2	Shift Solenoid 2 Control	ON / OFF	OFF	ON
TCIL	Trans. Control Indicator Lamp	ON / OFF	OFF	OFF
TCS	TCS Switch	TCS On: 0.1	TCS On: 0.1	TCS On: 0.1
TFT V	TFT Sensor	0-5.1v	2.10-2.40	2.10-2.40
TORQUE	Engine Torque	---	---	---
TPREL	Low Idle TP Sensor	---	---	---
TR2	TR2 Sensor	0 / 10.7	0v	10.7
TR3	TR4 Sensor	0 / 5v	4.5	2.2
TR4	TR4 Sensor	0 / 10.7	0v	10.7
TR V	Transmission Range Sensor Actual	0-5.1v	P/N: 4.45v	OD: 2.87v
VFDES	Volume Flow Desired	---	---	---
VPWR	Vehicle Power Supply	0-25.5v	12-14	12-14
VREF	Vehicle Reference	0-5.1v	4.9-5.1	4.9-5.1
VS SET	Vehicle Speed Setting	---	---	---
VSS	Vehicle Speed (MPH)	0-159 mph	0	Actual Speed

<u>1992-97 F-SERIES 7.5L V8 MFI VIN G (A/T, M/T, E4OD)</u>

PCM PID Acronym	Parameter Identification	PID Range	PID Value at Hot Idle	PID Value at 30	PID Value at 55
ACCS	A/C Cycling Clutch Switch	ON / OFF	ON (A/C on)	OFF	OFF
CPP	Clutch Pedal Switch	ON / OFF	ON (if in)	OFF	OFF
DPFE	DPFE Sensor	0-5.1v	0.4-0.6	0.4-0.7	0.8-1.1
ECT	ECT Sensor	-40-304°F	160-200	160-200	160-200
FEPS	Flash EEPROM	0-5.1v	0.1	0.1	0.1
FPM	Fuel Pump Monitor	ON / OFF	ON	ON	ON
HO2S-11	HO2S-11 (Bank 1)	0.1-1.1v	0.1-1.1	0.1-1.1	0.1-1.1
HO2S-12	HO2S-12 (Bank 1)	0.1-1.1v	0.1-1.1	0.1-1.1	0.1-1.1
HO2S-21	HO2S-21 (Bank 2)	0.1-1.1v	0.1-1.1	0.1-1.1	0.1-1.1
HO2S-22	HO2S-22 (Bank 2)	0.1-1.1v	0.1-1.1	0.1-1.1	0.1-1.1
IAT	IAT Sensor	-40-304°F	50-120	50-120	50-120
IDM	IDM Signal	0-9999 Hz	45-55	88-120	125-160
LOAD	Engine Load	0-100%	14-16	20-25	25-45
MAFV	MAF Sensor	0-5.1v	0.9-1.1	1.1-1.6	2.0-2.6
MISF	Misfire Detection	ON / OFF	OFF	OFF	OFF
OCTADJ	Octane Adjust Switch	CLOSED or OPEN	CLOSED	CLOSED	CLOSED
PIP	PIP Sensor	0-9999 Hz	45-55	88-120	110-220
PNP	PNP Switch Signal	ON / OFF	ON in P/N	OFF	OFF
PTO	Power Takeoff	ON / OFF	ON (if on)	OFF	OFF
RPM	Engine Speed	0-10K rpm	780-810	1280-1360	2150-2400
TPV	TP Sensor	0-5.1v	0.53-1.27	1.1-1.3v	1.3-1.7
Vacuum	Engine Vacuum	0-30" Hg	19-20	18-20	8-15
VPWR	Vehicle Power	0-25.5v	12-14	12-14	12-14
VSS	Vehicle Speed (Hertz or MPH)	0-999 Hz (0-159 mph)	0 Hz = 0 mph	65 Hz = 30 mph	125 Hz = 55 mph

1992-97 F-SERIES 7.5L V8 MFI VIN G (A/T, M/T, E4OD)

PCM PID Acronym	Parameter Identification	PID Range	PID Value at Hot Idle	PID Value at 30	PID Value at 55
AIRB	AIR System Bypass	0-25.5v	0.1	0.1	0.1
AIRD	AIR System Divert	0-25.5v	VBAT	VBAT	VBAT
EGR VR	EGR VR Solenoid	0-100%	0-100	40-45	50-60
EVAP CP	EVAP Purge Valve	0-100%	0-100	0-100	0-100
EPC	EPC Solenoid	0-300 psi	5	5	10-15
FP	Fuel Pump Control	ON / OFF	ON	ON	ON
FuelPW1	INJ Pulsewidth - Bank 1	0-999 ms	4.3-4.6	5.2-6.5	6.6-9.0
FuelPW2	INJ Pulsewidth - Bank 2	0-999 ms	4.3-4.6	5.2-6.5	6.6-9.0
HTR-11	HO2S-11 Heater	ON / OFF	ON	ON	ON
HTR-12	HO2S-12 Heater	ON / OFF	ON	ON	ON
HTR-21	HO2S-21 Heater	ON / OFF	ON	ON	ON
HTR-22	HO2S-22 Heater	ON / OFF	ON	ON	ON
IAC	Idle Air Control	0-100%	30-34	60-65	63-70
INJ1	INJ 1 Pulsewidth	0-999 ms	4.3-4.6	5.2-6.5	6.6-9.0
INJ2	INJ 2 Pulsewidth	0-999 ms	4.3-4.6	5.2-6.5	6.6-9.0
INJ3	INJ 3 Pulsewidth	0-999 ms	4.3-4.6	5.2-6.5	6.6-9.0
INJ4	INJ 4 Pulsewidth	0-999 ms	4.3-4.6	5.2-6.5	6.6-9.0
INJ5	INJ 5 Pulsewidth	0-999 ms	4.3-4.6	5.2-6.5	6.6-9.0
INJ6	INJ 6 Pulsewidth	0-999 ms	4.3-4.6	5.2-6.5	6.6-9.0
INJ7	INJ 7 Pulsewidth	0-999 ms	4.3-4.6	5.2-6.5	6.6-9.0
INJ8	INJ 8 Pulsewidth	0-999 ms	4.3-4.6	5.2-6.5	6.6-9.0
LongFT1	Long Term FT - Bank 1	-20 to 20%	-5 to +5	-5 to +5	-5 to +5
LongFT2	Long Term FT - Bank 2	-20 to 20%	-5 to +5	-5 to +5	-5 to +5
MIL	MIL (lamp) Control	ON / OFF	OFF	OFF	OFF
SS1	Shift Solenoid 1	ON / OFF	ON	OFF	OFF
SS2	Shift Solenoid 2	ON / OFF	OFF	ON	OFF
SHTFT1	Short Term F/T - Bank 1	-10 to 10%	-5 to +5	-5 to +5	-5 to +5
SHTFT2	Short Term FT - Bank 2	-10 to 10%	-5 to +5	-5 to +5	-5 to +5
SPARK	Spark Advance	-90° to 90°	22-26	25-28	27-32
SPOUT	Spark Output Signal	0-9999 Hz	45-55	88-120	110-220
TCC	TCC Solenoid	0-100%	0	0-100	90-95
TCIL	TCIL (lamp) Control	ON / OFF	OFF	OFF	OFF
VREF	Vehicle Reference	0-5.1v	4.9-5.1	4.9-5.1	4.9-5.1

Ranger PID DATA

1992-94 RANGER 2.3L I4 VIN A (ALL) INPUTS & OUTPUTS

PCM PID Acronym	Parameter Identification	PID Range	PID Value at Hot Idle	PID Value at 30	PID Value at 55
ACCS	A/C Switch Signal	ON / OFF	ON (if on)	OFF	OFF
BOO	Brake Switch Signal	ON / OFF	ON (if on)	OFF	OFF
ECT (ºF)	ECT Sensor	-40-304ºF	160-200	160-200	160-200
ECTV	ECT Sensor	0-5.1v	0.7	0.7	0.7
EVP	EVP Sensor	0-5.1v	0.4	0.5-0.7	1.2-1.6
EVR	EGR VR Solenoid	0-100%	0%	0-40	30-50
FPM	Fuel Pump Monitor	ON / OFF	ON	ON	ON
FuelPW1	INJ Pulsewidth - Bank 1	0-999 ms	3.3-3.5	4.7-5.4	6.2-7.2
HO2S-11	HO2S-11 (Bank 1)	0.1-1.1v	0.1-1.1	0.1-1.1	0.1-1.1
IAC	Idle Air Control	0-100%	26	42-48	45-52
IAT (ºF)	IAT Sensor	-40-304ºF	50-120	50-120	50-120
IATV	IAT Sensor	0-5.1v	1.5-3.5	1.5-3.5	1.5-3.5
MAFV	MAF Sensor	0-5.1v	0.1	1.3-1.4	1.8-2.2
PNP	PNP Switch Signal	Neutral/DR	NEUTRAL	DRIVE	DRIVE
PSP	PSP Switch Signal	ON / OFF	ON: turned	OFF	OFF
RPM	Engine Speed	0-10K rpm	760-820	1500-1630	1930-2100
SHTFT1	Short Term F/T - Bank 1	-10 to 10%	-5 to +5	-5 to +5	-5 to +5
SPARK	Spark Advance	-90º to 90º	18-22	28-32	30-34
TP	TP Sensor	0-5.1v	0.9	1.0-1.1	1.3-1.6v
TPCT	TP sensor minimum	0-5.1v	0.1	1.0-1.1	1.3-1.6v
TP Mode	TP Sensor Mode	C/T or P/T	C/T	P/T	P/T
VPWR	Vehicle Power	0-25.5v	12-14	12-14	12-14
VREF	Vehicle Reference	0-5.1v	4.9-5.1	4.9-5.1	4.9-5.1
VSS	Vehicle Speed	0-159 mph	0	30	55
WAC	A/C WOT Relay Control	ON / OFF	ON (AC on at WOT)	OFF	OFF

1995-97 RANGER 2.3L I4 VIN A (ALL) INPUTS

PCM PID Acronym	Parameter Identification	PID Range	PID Value at Hot Idle	PID Value at 30	PID Value at 55
4x4L	4x4 Low Signal	ON / OFF	ON (if on)	OFF	OFF
ACCS	A/C Switch Signal	ON / OFF	ON (if on)	OFF	OFF
BPP	Brake Switch Signal	ON / OFF	ON (if on)	OFF	OFF
CD1A	Coil 1 Driver	0v / VBAT	VBAT	VBAT	VBAT
CD2A	Coil 2 Driver	0v / VBAT	VBAT	VBAT	VBAT
CID	CMP Sensor	0-999 Hz	6-8	13-15	17-19
CKP	CKP Sensor	0-9999 Hz	518-540	840-850	1180-1250
CPP	Clutch Pedal Switch	ON / OFF	ON (if in)	OFF	OFF
DPFE	DPFE Sensor	0-5.1v	0.2-1.3	0.2-4.5	0.2-4.5
ECT	ECT Sensor	-40-304°F	160-200	160-200	160-200
FEPS	Flash EEPROM	0-5.1v	0.1	0.1	0.1
FPM	Fuel Pump Monitor	ON / OFF	ON	ON	ON
HO2S-11	HO2S-11 (Bank 1)	0.1-1.1v	0.1-1.1	0.1-1.1	0.1-1.1
HO2S-12	HO2S-12 (Bank 1)	0.1-1.1v	0.1-1.1	0.1-1.1	0.1-1.1
IAT	IAT Sensor	-40-304°F	50-120	50-120	50-120
GEAR	Gear Position	1-2-3-4	1	3	4
LOAD	Engine Load	0-100%	10-20	16-36	30-45
MAFV	MAF Sensor	0-5.1v	0.6-0.9	0.8-1.1	1.5-2.8
MISF	Misfire Detection	ON / OFF	OFF	OFF	OFF
OCTADJ	Octane Adjustment	CL/OPEN	CLOSED	CLOSED	CLOSED
PF	Purge Flow Sensor	0-5.1v	1.55	2.10	2.30
PNP	PNP Switch Signal	Neutral/DR	NEUTRAL	DRIVE	DRIVE
PSP	PSP Switch Signal	LOW or HI	HI: turned	LOW	LOW
RPM	Engine Speed	0-10K rpm	760-820	1500-1630	1930-2100
TCS	TCS Switch	ON / OFF	OFF	OFF	OFF
TFT	TFT Sensor Signal	-40-304°F	110-210	110-210	110-210
TPV	TP Sensor	0-5.1v	0.53-1.27	1.0-1.3	1.1-1.9
TRV/TR	TR Sensor	P/N or O/D	P/N	O/D	O/D
TR1	TR1 Sensor	0v / 11.5v	0	11.5	11.5
TR2	TR2 Sensor	0v / 11.5v	0	11.5	11.5
TR4	TR4 Sensor	0v / 11.5v	0	11.5	11.5
TSS	TSS Sensor	0-9999 Hz	115-120	192-196	268-275
Vacuum	Engine Vacuum	0-30" Hg	19-20	18-20	8-15
VPWR	Vehicle Power	0-25.5v	12-14	12-14	12-14
VSS	Vehicle Speed (Hertz or MPH)	0-999 Hz (0-159 mph)	0 Hz = 0 mph	65 Hz = 30 mph	125 Hz = 55 mph

1995-97 RANGER 2.3L I4 VIN A (ALL) OUTPUTS

PCM PID Acronym	Parameter Identification	PID Range	PID Value at Hot Idle	PID Value at 30	PID Value at 55
CD1	Coil 1 Driver	0-90° dwell	5	6	8
CD2	Coil 2 Driver	0-90° dwell	5	6	8
CCS	Coast Clutch Solenoid Control	0-25.5v	VBAT	VBAT	VBAT
CTO	Clean Tachometer signal	0 Hz	25-38	50-60	60-74
EGR VR	EGR VR Solenoid	0-100%	0	0-40	0-40
EVAP CP	EVAP Purge Valve	0-100%	0-100	0-100	0-100
EPC	EPC Solenoid	0-300 psi	24	27	26-30
FP	Fuel Pump Control	ON / OFF	ON	ON	ON
FuelPW1	INJ Pulsewidth - Bank 1	0-999 ms	4.0-4.5	5.3-10.0	10-18.0
HTR-11	HO2S-11 Heater	ON / OFF	ON	ON	ON
HTR-12	HO2S-12 Heater	ON / OFF	ON	ON	ON
IAC	Idle Air Control	0-100%	33	40-48	40-55
INJ1	INJ 1 Pulsewidth	0-999 ms	4.0-4.5	5.3-10.0	10-18.0
INJ2	INJ 2 Pulsewidth	0-999 ms	4.0-4.5	5.3-10.0	10-18.0
INJ3	INJ 3 Pulsewidth	0-999 ms	4.0-4.5	5.3-10.0	10-18.0
INJ4	INJ 4 Pulsewidth	0-999 ms	4.0-4.5	5.3-10.0	10-18.0
LongFT1	Long Term FT - Bank 1	-20 to 20%	-5 to 5%	-5 to 5%	-5 to 5%
MIL	MIL (lamp) Control	ON / OFF	OFF	OFF	OFF
SS1	Shift Solenoid 1	ON / OFF	ON	OFF	OFF
SS2	Shift Solenoid 2	ON / OFF	OFF	ON	ON
SHTFT1	Short Term F/T - Bank 1	-10 to 10%	-5 to 5%	-5 to 5%	-5 to 5%
SPARK	Spark Advance	-90° to 90°	15-22	25-35	15-30
TCC	TCC Solenoid	0-100%	0	0-100	95-100
TCIL	TCIL (lamp) Control	ON / OFF	OFF	OFF	OFF
VREF	Vehicle Reference	0-5.1v	4.9-5.1	4.9-5.1	4.9-5.1
WAC	A/C WOT Relay Control	ON / OFF	ON (AC on at WOT)	OFF	OFF

2001-03 RANGER 2.3L I4 MFI VIN D (ALL) INPUTS

PCM PID Acronym	Parameter Identification	PID Range	PID Value at Hot Idle	PID Value at 30	PID Value at 55
ACCS	A/C Switch Signal	ON / OFF	ON (if on)	OFF	OFF
ACP	A/C Press. Switch	CL/OPEN	OPEN	OPEN	OPEN
BPP	Brake Switch Signal	ON / OFF	ON (if on)	OFF	OFF
CID	CMP Sensor	0-999 Hz	6-8	12-15	12-18
CD1A	Coil 1 Driver	0-25.5v	VBAT	VBAT	VBAT
CD2A	Coil 2 Driver	0-25.5v	VBAT	VBAT	VBAT
CHT	CHT Sensor	-40-304°F	194	194	194
CKP	CKP Sensor	0-9999 Hz	390-450	1000-1220	1220-1500
CPP	Clutch Pedal Switch	ON / OFF	ON (if in)	OFF	OFF
ECT	ECT Sensor	-40-304°F	160-200	160-200	160-200
EGR MS	EGR Motor Speed	0-60 Steps	3	3-31	3-31
FEPS	Flash EEPROM	0-5.1v	0.1	0.1	0.1
FLI	Fuel Level Indicator	0-100%	1.7 (1/2)	0-100	0-100
FPM	Fuel Pump Monitor	ON / OFF	ON	ON	ON
FTP	Fuel Tank Pressure	0-5.1v or 0-10" H2O	2.6v / 0" H2O (with cap off)	2.6v / 0" H2O (with cap off)	2.6v / 0" H2O (with cap off)
IAT	IAT Sensor	-40-304°F	50-120	50-120	50-120
LOAD	Engine Load	0-100%	10-20	16-36	30-45
MAFV	MAF Sensor	0-5.1v	0.6-0.9	0.8-1.1	1.5-2.8
MAP	MAP Sensor	0-5.1v	1.7	3.0	3.5
O2S-11	HO2S-11 (Bank 1)	0.1-1.1v	0.1-1.1	0.1-1.1	0.1-1.1
O2S-12	HO2S-12 (Bank 1)	0.1-1.1v	0.1-1.1	0.1-1.1	0.1-1.1
PF	Purge Flow Sensor	0-5.1v	1.55	2.10	2.30
PNP	PNP Switch Signal	NEUT/DR	NEUTRAL	DRIVE	DRIVE
RPM	Engine Speed	0-10K rpm	800-950	1400-1760	1930-2150
TFT	TFT Sensor Signal	-40-304°F	110-210	110-210	110-210
TPV	TP Sensor	0-5.1v	0.53-1.27	1.0-1.3	1.1-1.9
OSS	OSS Sensor	0-9999 Hz	0	213	385
Vacuum	Engine Vacuum	0-30" Hg	19-20	18-20	8-15
VPWR	Vehicle Power	0-25.5v	12-14	12-14	12-14
VSS	Vehicle Speed (Hertz or MPH)	0-999 Hz (0-159 mph)	0 Hz = 0 mph	65 Hz = 30 mph	125 Hz = 55 mph

2001-03 RANGER 2.3L I4 MFI VIN D (ALL) OUTPUTS

PCM PID Acronym	Parameter Identification	PID Range	PID Value at Hot Idle	PID Value at 30	PID Value at 55
CD1	Coil 1 Driver	0-90° dwell	5	6	8
CD2	Coil 2 Driver	0-90° dwell	5	6	8
CHTIL	CHT (lamp) Control	ON / OFF	OFF	OFF	OFF
CTO	Clean Tachometer signal	0 Hz	25-38	50-60	60-74
EGRMC1	EGR Motor Cont. 1	0-25.5v	12-14	12-14	12-14
EGRMC2	EGR Motor Cont. 2	0-25.5v	0.2	0.2	0.2
EGRMC3	EGR Motor Cont. 3	0-25.5v	12-14	12-14	12-14
EGRMC4	EGR Motor Cont. 4	0-25.5v	0.2	0.2	0.2
EVAP CP	EVAP Purge Valve	0-100%	0-100	0-100	0-100
EVAP CV	EVAP CV Valve	0-100%	0-100	0-100	0-100
FP	Fuel Pump Control	ON / OFF	ON	ON	ON
FuelPW1	INJ Pulsewidth - Bank 1	0-999 ms	3.0-4.5	5.3-10.0	10-18.0
HTR-11	HO2S-11 Heater	ON / OFF	ON	ON	ON
HTR-12	HO2S-12 Heater	ON / OFF	ON	ON	ON
IAC	Idle Air Control	0-100%	38	42-48	45-52
INJ1	INJ 1 Pulsewidth	0-999 ms	3.0-4.5	5.3-10.0	10-18.0
INJ2	INJ 2 Pulsewidth	0-999 ms	3.0-4.5	5.3-10.0	10-18.0
INJ3	INJ 3 Pulsewidth	0-999 ms	3.0-4.5	5.3-10.0	10-18.0
INJ4	INJ 4 Pulsewidth	0-999 ms	3.0-4.5	5.3-10.0	10-18.0
LongFT1	Long Term FT - Bank 1	-20 to 20%	-5 to +5	-5 to +5	-5 to +5
MIL	MIL (lamp) Control	ON / OFF	OFF	OFF	OFF
MISF	Misfire Monitor	ON / OFF	ON	ON	ON
SCVM	S/C Vacuum Motor	HI/LOW	LOW	LOW	LOW
SHTFT1	Short Term F/T - Bank 1	-10 to 10%	-5 to +5	-5 to +5	-5 to +5
SPARK	Spark Advance	-90° to 90°	15-22	25-35	15-30
TCIL	TCIL (lamp) Control	ON / OFF	OFF	OFF	OFF
VREF	Vehicle Reference	0-5.1v	4.9-5.1	4.9-5.1	4.9-5.1
WAC	A/C WOT Relay Control	ON / OFF	ON (AC on at WOT)	OFF	OFF

1998-2001 RANGER 2.5L I4 MFI VIN C (ALL) INPUTS

PCM PID Acronym	Parameter Identification	PID Range	PID Value at Hot Idle	PID Value at 30	PID Value at 55
ACCS	A/C Switch Signal	ON / OFF	ON (if on)	OFF	OFF
ACP	A/C Press. Switch	CL/OPEN	OPEN	OPEN	OPEN
BPP	Brake Switch Signal	ON / OFF	ON (if on)	OFF	OFF
CID	CMP Sensor	0-999 Hz	5-7	9-12	10-16
CD1A	Coil 1 Driver	0-25.5v	VBAT	VBAT	VBAT
CD2A	Coil 2 Driver	0-25.5v	VBAT	VBAT	VBAT
CKP	CKP Sensor	0-9999 Hz	390-450	850-950	1160-1220
CPP	Clutch Pedal Switch	ON / OFF	ON (if in)	OFF	OFF
DPFE	DPFE Sensor	0-5.1v	0.2-1.3	0.2-4.5	0.2-4.5
ECT	ECT Sensor	-40-304°F	160-200	160-200	160-200
FEPS	Flash EEPROM	0-5.1v	0.1	0.1	0.1
FLI	Fuel Level Indicator	0-100%	1.7 (1/2)	0-100%	0-100%
FPM	Fuel Pump Monitor	ON / OFF	ON	ON	ON
FTP	Fuel Tank Pressure	0-5.1v or 0-10" H2O	2.6v / 0" H2O (with cap off)	2.6v / 0" H2O (with cap off)	2.6v / 0" H2O (with cap off)
IAT	IAT Sensor	-40-304°F	50-120	50-120	50-120
GEAR	Gear Position	1-2-3-4	1	3	4
LOAD	Engine Load	0-100%	10-20	16-36	30-45
MAFV	MAF Sensor	0-5.1v	0.6-0.9	0.8-1.1	1.5-2.8
MISF	Misfire Detection	ON / OFF	OFF	OFF	OFF
OCTADJ	Octane Adjustment	CL or OP	CLOSED	CLOSED	CLOSED
O2S-11	HO2S-11 (Bank 1)	0.1-1.1v	0.1-1.1	0.1-1.1	0.1-1.1
O2S-12	HO2S-12 (Bank 1)	0.1-1.1v	0.1-1.1	0.1-1.1	0.1-1.1
PF	Purge Flow Sensor	0-5.1v	1.55	2.10	2.30
PNP	PNP Switch Signal	NEUT/DR	NEUTRAL	DRIVE	DRIVE
PSP	PSP Switch Signal	LOW or HI	HI: turning	LOW	LOW
RPM	Engine Speed	0-10K rpm	760-820	1400-1630	1930-2100
TCS	TCS Switch	ON / OFF	OFF	OFF	OFF
TFT	TFT Sensor Signal	-40-304°F	110-210	110-210	110-210
TPV	TP Sensor	0-5.1v	0.53-1.27	1.0-1.3	1.1-1.9
TR1	TR1 Sensor	0v / 11.5v	0v	11.5	11.5
TR2	TR2 Sensor	0v / 11.5v	0v	11.5	11.5
TR3/TR	TR Sensor	P/N or O/D	P/N	O/D	O/D
TR4	TR4 Sensor	0v / 11.5v	0	11.5	11.5
TSS	TSS Sensor	0-9999 Hz	95-120	160-180	260-280
Vacuum	Engine Vacuum	0-30" Hg	19-20	18-20	8-15
VPWR	Vehicle Power	0-25.5v	12-14	12-14	12-14
VSS	Vehicle Speed (Hertz or MPH)	0-999 Hz (0-159 mph)	0 Hz = 0 mph	65 Hz = 30 mph	125 Hz = 55 mph

1998-2001 RANGER 2.5L I4 MFI VIN C (ALL) OUTPUTS

PCM PID Acronym	Parameter Identification	PID Range	PID Value at Hot Idle	PID Value at 30	PID Value at 55
CD1	Coil 1 Driver	0-90° dwell	5	6	8
CD2	Coil 2 Driver	0-90° dwell	5	6	8
CCS	Coast Clutch Solenoid Control	0-25.5v	VBAT	VBAT	VBAT
CTO	Clean Tachometer signal	0 Hz	25-38	50-60	60-74
EGR VR	EGR VR Solenoid	0-100%	0	0-100	0-40
EVAP CP	EVAP Purge Valve	0-100%	0-100	0-100	0-100
EVAP CV	EVAP CV Valve	0-100%	0-100	0-100	0-100
EPC	EPC Solenoid	0-300 psi	25	27	26-30
FP	Fuel Pump Control	ON / OFF	ON	ON	ON
FuelPW1	INJ Pulsewidth - Bank 1	0-999 ms	4.0-4.5	5.3-10.0	10-18.0
HTR-11	HO2S-11 Heater	ON / OFF	ON	ON	ON
HTR-12	HO2S-12 Heater	ON / OFF	ON	ON	ON
IAC	Idle Air Control	0-100%	38	42-48	45-52
INJ1	INJ 1 Pulsewidth	0-999 ms	4.0-4.5	5.3-10.0	10-18.0
INJ2	INJ 2 Pulsewidth	0-999 ms	4.0-4.5	5.3-10.0	10-18.0
INJ3	INJ 3 Pulsewidth	0-999 ms	4.0-4.5	5.3-10.0	10-18.0
INJ4	INJ 4 Pulsewidth	0-999 ms	4.0-4.5	5.3-10.0	10-18.0
LongFT1	Long Term FT - Bank 1	-20 to 20%	-5 to +5	-5 to +5	-5 to +5
MIL	MIL (lamp) Control	ON / OFF	OFF	OFF	OFF
SHTFT1	Short Term F/T - Bank 1	-10 to 10%	-5 to +5	-5 to +5	-5 to +5
SPARK	Spark Advance	-90° to 90°	15-22	25-35	15-30
SS1	Shift Solenoid 1	ON / OFF	ON	OFF	OFF
SS2	Shift Solenoid 2	ON / OFF	OFF	ON	OFF
SS3	Shift Solenoid 3	ON / OFF	OFF	ON	ON
TCC	TCC Solenoid	0-100%	0%	0-100	95-100
TCIL	TCIL (lamp) Control	ON / OFF	OFF	OFF	OFF
VREF	Vehicle Reference	0-5.1v	4.9-5.1	4.9-5.1	4.9-5.1
WAC	A/C WOT Relay Control	ON / OFF	ON (AC on at WOT)	OFF	OFF

1993-94 RANGER 3.0L V6 VIN U (ALL) INPUTS & OUTPUTS

PCM PID Acronym	Parameter Identification	PID Range	PID Value at Hot Idle	PID Value at 30	PID Value at 55
ACCS	A/C Switch Signal	ON / OFF	ON (if on)	OFF	OFF
BOO	Brake Switch Signal	ON / OFF	ON (if on)	OFF	OFF
CANP	EVAP Purge Valve	ON / OFF	OFF	ON	ON
CPP	Clutch Pedal Switch	ON / OFF	ON (if in)	OFF	OFF
DPFE	DPFE Sensor	0-5.1v	0.4-0.6	0.4-0.8	0.4-0.9
ECT (°F)	ECT Sensor	-40-304°F	160-200	160-200	160-200
ECTV	ECT Sensor	0-5.1v	0.7	0.7	0.7
EVR	EGR VR Solenoid	0-100%	0	0-40	0-40
FPM	Fuel Pump Monitor	ON / OFF	ON	ON	ON
FuelPW1	INJ Pulsewidth - Bank 1	0-999 ms	3.7-3.9	4.4-4.6	5.0-6.0
FuelPW2	INJ Pulsewidth - Bank 2	0-999 ms	3.7-3.9	4.4-4.6	5.0-6.0
HO2S-11	HO2S-11 (Bank 1)	0.1-1.1v	0.1-1.1	0.1-1.1	0.1-1.1
HO2S-21	HO2S-21 (Bank 2)	0.1-1.1v	0.1-1.1	0.1-1.1	0.1-1.1
IAC	Idle Air Control	0-100%	33	35-41	57-68
IAT (°F)	IAT Sensor	-40-304°F	50-120	50-120	50-120
IATV	IAT Sensor	0-5.1v	1.5-3.5	1.5-3.5	1.5-3.5
LongFT1	Long Term F/T - Bank 1	-20 to 20%	-5 to +5	-5 to +5	-5 to +5
LongFT2	Long Term F/T - Bank 1	-20 to 20%	-5 to +5	-5 to +5	-5 to +5
MAF	MAF Sensor	0-5.1v	0.9	1.4-1.6	1.9-2.4
PNP	PNP Switch Signal	NEUT/DR	NEUTRAL	DRIVE	DRIVE
RPM	Engine Speed	0-10K rpm	880-920	1430-1470	1550-1750
SHTFT1	Short Term F/T - Bank 1	-10 to 10%	-5 to +5	-5 to +5	-5 to +5
SHTFT2	Short Term FT - Bank 2	-10 to 10%	-5 to +5	-5 to +5	-5 to +5
SPARK	Spark Advance	-90° to 90°	15-22	28-35	25-35
TP	TP Sensor	0-5.1v	0.7	0.9-1.0	1.0-1.2
TPCT	TP sensor Minimum	0-5.1v	0.1	0.9-1.0	1.0-1.2
TP Mode	TP Sensor Mode	C/T or P/T	C/T	P/T	P/T
VPWR	Vehicle Power	0-25.5v	12-14	12-14	12-14
VREF	Vehicle Reference	0-5.1v	4.9-5.1	4.9-5.1	4.9-5.1
VSS	Vehicle Speed	0-159 mph	0	30	55
WAC	A/C WOT Relay Control	ON / OFF	ON (AC on at WOT)	OFF	OFF

1995-2003 RANGER 3.0L V6 MFI VIN U (ALL) INPUTS

PCM PID Acronym	Parameter Identification	PID Range	PID Value at Hot Idle	PID Value at 30	PID Value at 55
4x4L	4x4 Low Signal	ON / OFF	ON (if on)	OFF	OFF
ACCS	A/C Switch Signal	ON / OFF	ON (if on)	OFF	OFF
BOO	Brake Switch Signal	ON / OFF	ON (if on)	OFF	OFF
CID	CMP Sensor	0-999 Hz	6-8	13-15	17-19
CKP	CKP Sensor	0-9999 Hz	518-540	840-860	1180-1210
CPP	Clutch Pedal Switch	ON / OFF	ON (if in)	OFF	OFF
DPFE	DPFE Sensor	0-5.1v	0.4-0.6	0.4-0.8	0.4-0.9
ECT	ECT Sensor	-40-304°F	160-200	160-200	160-200
FEPS	Flash EEPROM	0-5.1v	0.1	0.1	0.1
FLI	Fuel Level Indicator	0-100%	50% (1.7)	0-100%	0-100%
FPM	Fuel Pump Monitor	ON / OFF	ON	ON	ON
FTP	Fuel Tank Pressure	0-5.1v or 0-10" H2O	2.6v / 0" H2O (with cap off)	2.6v / 0" H2O (with cap off)	2.6v / 0" H2O (with cap off)
IAT	IAT Sensor	-40-304°F	50-120	50-120	50-120
GEAR	Gear Position	1-2-3-4	1	3	4
LOAD	Engine Load	0-100%	17-19	21-27	27-34
MAFV	MAF Sensor	0-5.1v	0.6-0.9	1.4-1.7	2.0-0
MISF	Misfire Detection	ON / OFF	OFF	OFF	OFF
OCTADJ	Octane Adjustment	CL/OPEN	CLOSED	CLOSED	CLOSED
HO2S-11	HO2S-11 (Bank 1)	0.1-1.1v	0.1-1.1	0.1-1.1	0.1-1.1
HO2S-12	HO2S-12 (Bank 1)	0.1-1.1v	0.1-1.1	0.1-1.1	0.1-1.1
HO2S-21	HO2S-21 (Bank 2)	0.1-1.1v	0.1-1.1	0.1-1.1	0.1-1.1
PF	Purge Flow Sensor	0-5.1v	1.55	2.10	2.30
PNP	PNP Switch Signal	Neutral/DR	NEUTRAL	DRIVE	DRIVE
PSP	PSP Switch Signal	LOW or HI	HI: turned	LOW	LOW
RPM	Engine Speed	0-10K rpm	880-920	1430-1470	1550-1750
TCS	TCS Switch	ON / OFF	OFF	OFF	OFF
TFT	TFT Sensor Signal	-40-304°F	110-210	110-210	110-210
TPV	TP Sensor	0-5.1v	0.53-1.27	0.8-1.7	1.2-1.7
TRV/TR	TR Sensor	P/N or O/D	P/N	O/D	O/D
TSS	TSS Sensor	0-9999 Hz	115-120	192-196	268-275
Vacuum	Engine Vacuum	0-30" Hg	19-20	18-20	8-15
VPWR	Vehicle Power	0-25.5v	12-14	12-14	12-14
VSS	Vehicle Speed (Hertz or MPH)	0-999 Hz (0-159 mph)	0 Hz = 0 mph	65 Hz = 30 mph	125 Hz = 55 mph

1995-2003 RANGER 3.0L V6 MFI VIN U (ALL) OUTPUTS

PCM PID Acronym	Parameter Identification	PID Range	PID Value at Hot Idle	PID Value at 30	PID Value at 55
CD1	Coil 1 Driver	0-60°	6	8	12
CD2	Coil 2 Driver	0-60°	6	8	12
CD3	Coil 3 Driver	0-60°	6	8	12
CTO	Clean Tachometer signal	0-9999 Hz	42-48	73-79	99-115
CCS	Coast Clutch Solenoid Control	0-25.5v	VBAT	VBAT	VBAT
EGR VR	EGR VR Solenoid	0-100%	0	0-40	0-45
EVAP CV	EVAP CV Valve	0-100%	0-100	0-100	0-100
EVAP CP	EVAP Purge Valve	0-100%	0-100	0-100	0-100
EPC	EPC Solenoid	0-300 psi	24	23-28	34-38
FP	Fuel Pump Control	ON / OFF	ON	ON	ON
FuelPW1	INJ Pulsewidth - Bank 1	0-999 ms	4.5-4.8	6.3-8.0	7.0-13.0
FuelPW2	INJ Pulsewidth - Bank 2	0-999 ms	4.5-4.8	6.3-8.0	7.0-13.0
HTR-11	HO2S-11 Heater	ON / OFF	ON	ON	ON
HTR-12	HO2S-12 Heater	ON / OFF	ON	ON	ON
HTR-21	HO2S-21 Heater	ON / OFF	ON	ON	ON
IAC	Idle Air Control	0-100%	33	35-41	57-68
INJ1	INJ 1 Pulsewidth	0-999 ms	4.5-4.8	6.3-8.0	7.0-13.0
INJ2	INJ 2 Pulsewidth	0-999 ms	4.5-4.8	6.3-8.0	7.0-13.0
INJ3	INJ 3 Pulsewidth	0-999 ms	4.5-4.8	6.3-8.0	7.0-13.0
INJ4	INJ 4 Pulsewidth	0-999 ms	4.5-4.8	6.3-8.0	7.0-13.0
INJ5	INJ 5 Pulsewidth	0-999 ms	4.5-4.8	6.3-8.0	7.0-13.0
INJ6	INJ 6 Pulsewidth	0-999 ms	4.5-4.8	6.3-8.0	7.0-13.0
LongFT1	Long Term FT - Bank 1	-20 to 20%	-5 to +5	-5 to +5	-5 to +5
LongFT2	Long Term FT - Bank 2	-20 to 20%	-5 to +5	-5 to +5	-5 to +5
MIL	MIL (lamp) Control	ON / OFF	OFF	OFF	OFF
SHTFT1	Short Term F/T - Bank 1	-10 to 10%	-5 to +5	-5 to +5	-5 to +5
SHTFT2	Short Term FT - Bank 2	-10 to 10%	-5 to +5	-5 to +5	-5 to +5
SPARK	Spark Advance	-90° to 90°	11-15	26-31	25-32
SS1	Shift Solenoid 1	ON / OFF	ON	OFF	OFF
SS2	Shift Solenoid 2	ON / OFF	OFF	ON	ON
SS3	Shift Solenoid 3	ON / OFF	OFF	OFF	ON
TCC	TCC Solenoid	0-100%	0	62-67	85-95
TCIL	TCIL (lamp) Control	ON / OFF	OFF	OFF	OFF
VREF	Vehicle Reference	0-5.1v	4.9-5.1	4.9-5.1	4.9-5.1
WAC	A/C WOT Relay Control	ON / OFF	ON (AC on at WOT)	OFF	OFF

1999-2001 RANGER 3.0L V6 FFV VIN V (ALL) INPUTS

PCM PID Acronym	Parameter Identification	PID Range	PID Value at Hot Idle	PID Value at 30	PID Value at 55
4x4L	4x4 Low Signal	ON / OFF	ON (if on)	OFF	OFF
ACCS	A/C Switch Signal	ON / OFF	ON (if on)	OFF	OFF
BOO	Brake Switch Signal	ON / OFF	ON (if on)	OFF	OFF
CID	CMP Sensor	0-999 Hz	6-8	13-15	17-19
CKP	CKP Sensor	0-9999 Hz	518-540	860-1000	1210-1250
CPP	Clutch Pedal Switch	ON / OFF	ON (if in)	OFF	OFF
DPFE	DPFE Sensor	0-5.1v	0.4-0.6	0.4-0.8	0.4-0.9
ECT	ECT Sensor	-40-304°F	160-200	160-200	160-200
FEPS	Flash EEPROM	0-5.1v	0.1	0.1	0.1
FFS	FF Sensor (value at 100% gas mixture)	0-999 Hz	40-60	40-60	40-60
FLI	Fuel Level Indicator	0-100%	50% (1.7)	0-100%	0-100%
FPM	Fuel Pump Monitor	ON / OFF	ON	ON	ON
FTP	Fuel Tank Pressure	0-5.1v or 0-10" H2O	2.6v / 0" H2O (with cap off)	2.6v / 0" H2O (with cap off)	2.6v / 0" H2O (with cap off)
HO2S-11	HO2S-11 (Bank 1)	0.1-1.1v	0.1-1.1	0.1-1.1	0.1-1.1
HO2S-12	HO2S-12 (Bank 1)	0.1-1.1v	0.1-1.1	0.1-1.1	0.1-1.1
HO2S-21	HO2S-21 (Bank 2)	0.1-1.1v	0.1-1.1	0.1-1.1	0.1-1.1
IAT	IAT Sensor	-40-304°F	50-120	50-120	50-120
GEAR	Gear Position	1-2-3-4	1	3	4
LOAD	Engine Load	0-100%	17-19	21-27	27-34
MAFV	MAF Sensor	0-5.1v	0.6-0.9	1.4-1.7	2.0-0
MISF	Misfire Detection	ON / OFF	OFF	OFF	OFF
OCTADJ	Octane Adjustment	CL/OPEN	CLOSED	CLOSED	CLOSED
PF	Purge Flow Sensor	0-5.1v	1.55	2.10	2.30
PNP	PNP Switch Signal	Neutral/DR	NEUTRAL	DRIVE	DRIVE
PSP	PSP Switch Signal	LOW or HI	HI: turned	LOW	LOW
RPM	Engine Speed	0-10K rpm	880-920	1430-1470	1550-1750
TCS	TCS Switch	ON / OFF	OFF	OFF	OFF
TFT	TFT Sensor Signal	-40-304°F	110-210	110-210	110-210
TPV	TP Sensor	0-5.1v	0.53-1.27	0.8-1.7	1.2-1.7
TRV/TR	TR Sensor	P/N or O/D	P/N	O/D	O/D
TSS	TSS Sensor	0-9999 Hz	115-120	192-196	268-275
Vacuum	Engine Vacuum	0-30" Hg	19-20	18-20	8-15
VPWR	Vehicle Power	0-25.5v	12-14	12-14	12-14
VSS	Vehicle Speed (Hertz or MPH)	0-999 Hz (0-159 mph)	0 Hz = 0 mph	65 Hz = 30 mph	125 Hz = 55 mph

1995-2001 RANGER 3.0L V6 FFV VIN V (ALL) OUTPUTS

PCM PID Acronym	Parameter Identification	PID Range	PID Value at Hot Idle	PID Value at 30	PID Value at 55
CD1	Coil 1 Driver	0-60°	6	8	12
CD2	Coil 2 Driver	0-60°	6	8	12
CD3	Coil 3 Driver	0-60°	6	8	12
CTO	Clean Tachometer signal	0-9999 Hz	42-48	73-79	99-115
CCS	Coast Clutch Solenoid Control	0-25.5v	VBAT	VBAT	VBAT
EGR VR	EGR VR Solenoid	0-100%	0	0-40	0-45
EVAP CP	EVAP Purge Valve	0-100%	0-100	0-100	0-100
EVAP CV	EVAP CV Valve	0-100%	0-100	0-100	0-100
EPC	EPC Solenoid	0-300 psi	39	23-38	34-38
FP	Fuel Pump Control	ON / OFF	ON	ON	ON
FuelPW1	INJ Pulsewidth - Bank 1	0-999 ms	1.9-2.8	3.6-4.8	4.6-6.0
FuelPW2	INJ Pulsewidth - Bank 2	0-999 ms	1.9-2.8	3.6-4.8	4.6-6.0
HTR-11	HO2S-11 Heater	ON / OFF	ON	ON	ON
HTR-12	HO2S-12 Heater	ON / OFF	ON	ON	ON
HTR-21	HO2S-21 Heater	ON / OFF	ON	ON	ON
IAC	Idle Air Control	0-100%	33	35-50	57-68
INJ1	INJ 1 Pulsewidth	0-999 ms	1.9-2.8	3.6-4.8	4.6-6.0
INJ2	INJ 2 Pulsewidth	0-999 ms	1.9-2.8	3.6-4.8	4.6-6.0
INJ3	INJ 3 Pulsewidth	0-999 ms	1.9-2.8	3.6-4.8	4.6-6.0
INJ4	INJ 4 Pulsewidth	0-999 ms	1.9-2.8	3.6-4.8	4.6-6.0
INJ5	INJ 5 Pulsewidth	0-999 ms	1.9-2.8	3.6-4.8	4.6-6.0
INJ6	INJ 6 Pulsewidth	0-999 ms	1.9-2.8	3.6-4.8	4.6-6.0
LongFT1	Long Term FT - Bank 1	-20 to 20%	-5 to +5	-5 to +5	-5 to +5
LongFT2	Long Term FT - Bank 2	-20 to 20%	-5 to +5	-5 to +5	-5 to +5
MIL	MIL (lamp) Control	ON / OFF	OFF	OFF	OFF
SHTFT1	Short Term F/T - Bank 1	-10 to 10%	-5 to +5	-5 to +5	-5 to +5
SHTFT2	Short Term FT - Bank 2	-10 to 10%	-5 to +5	-5 to +5	-5 to +5
SPARK	Spark Advance	-90° to 90°	11-15	26-31	25-32
SS1	Shift Solenoid 1	ON / OFF	ON	OFF	OFF
SS2	Shift Solenoid 2	ON / OFF	OFF	ON	ON
SS3	Shift Solenoid 3	ON / OFF	OFF	OFF	ON
TCC	TCC Solenoid	0-100%	0	62-67	85-95
TCIL	TCIL (lamp) Control	ON / OFF	OFF	OFF	OFF
VREF	Vehicle Reference	0-5.1v	4.9-5.1	4.9-5.1	4.9-5.1
WAC	A/C WOT Relay Control	ON / OFF	ON (AC on at WOT)	OFF	OFF

2002-03 RANGER 4.0L V6 MFI VIN E (A/T) INPUTS

PCM PID Acronym	Parameter Identification	PID Range	PID Value at Hot Idle	PID Value at 30	PID Value at 55
4x4L	4x4 Low Signal	ON / OFF	ON (if on)	OFF	OFF
ACCS	A/C Switch Signal	ON / OFF	ON (if on)	OFF	OFF
ACP	A/C Pressure	OPEN/CL	OPEN	OPEN	OPEN
ARC	Auto Ride Control	ON / OFF	OFF	OFF	OFF
BPP	Brake Position	ON / OFF	ON (if on)	OFF	OFF
CID	CMP Sensor	0-999 Hz	6-8	11-15	16-19
CKP	CKP Sensor	0-9999 Hz	400-475	800-1100	1140-1220
DPFE	DPFE Sensor	0-5.1v	0.4-0.6	0.4-1.0	0.6-1.1
ECT	ECT Sensor	-40-304°F	160-200	160-200	160-200
FEPS	Flash EEPROM	0-5.1v	0.1	0.1	0.1
FLI	Fuel Level Input	1.7 / 50%	1.7 / 50%	1.7 / 50%	1.7 / 50%
FPM	Fuel Pump Monitor	ON / OFF	ON	ON	ON
FTP	Fuel Tank Pressure	0-5.1v at 0-10" H2O	2.6v at 0" H2O (gas cap "off")	2.6v at 0" H2O (gas cap "off")	2.6v at 0" H2O (gas cap "off")
IAT	IAT Sensor	-40-304°F	50-120	50-120	50-120
GEAR	Gear Position	1-2-3-4-5	1	4	5
HO2S-11	HO2S-11 (Bank 1)	0.1-1.1v	0.1-1.1	0.1-1.1	0.1-1.1
HO2S-12	HO2S-12 (Bank 1)	0.1-1.1v	0.1-1.1	0.1-1.1	0.1-1.1
HO2S-21	HO2S-21 (Bank 2)	0.1-1.1v	0.1-1.1	0.1-1.1	0.1-1.1
LOAD	Engine Load	0-100%	14-20	21-27	30-45
MAFV	MAF Sensor	0-5.1v	0.6-0.9	1.3-1.7	1.5-2.3
MISF	Misfire Detection	ON / OFF	OFF	OFF	OFF
OCTADJ	Octane Adjustment	CL/OPEN	CLOSED	CLOSED	CLOSED
PNP	PNP Switch Signal	ON / OFF	ON	OFF	OFF
PSP	PSP Switch Signal	LOW	HI: turning	LOW	LOW
RPM	Engine Speed	0-10K rpm	670-750	1400-1600	1800-2100
TCS	TCS Switch	ON / OFF	OFF	OFF	OFF
TFT	TFT Sensor Signal	-40-304°F	110-210	110-210	110-210
TPV	TP Sensor	0-5.1v	0.53-1.27	0.8-1.7	1.2-1.7
TR1	TR1 Sensor	0v / 11.5v	0v	11.5	11.5
TR2	TR2 Sensor	0v / 11.5v	0v	11.5	11.5
TR4	TR4 Sensor	0v / 11.5v	0v	11.5	11.5
TRV/TR	TR Sensor	0v / 1.7v	0v	1.7	1.7
TSS	TSS Sensor	0-9999 Hz	100-125	185-205	260-280
Vacuum	Engine Vacuum	0-30" Hg	19-20	18-20	8-15
VPWR	Vehicle Power	0-25.5v	12-14	12-14	12-14
VSS	Vehicle Speed (Hertz or MPH)	0-999 Hz (0-159 mph)	0 Hz = 0 mph	65 Hz = 30 mph	125 Hz = 55 mph

2002-03 RANGER 4.0L V6 MFI VIN E (A/T) OUTPUTS

PCM PID Acronym	Parameter Identification	PID Range	PID Value at Hot Idle	PID Value at 30	PID Value at 55
CCS	Coast Clutch Solenoid Control	VBAT	VBAT	VBAT	VBAT
CD1	Coil 1 Driver	0-60°	5	6	8
CD2	Coil 2 Driver	0-60°	5	6	8
CD3	Coil 3 Driver	0-60°	5	6	8
CTO	Clean Tachometer signal	0-10K Hz	35-49	65-90	90-120
EGR VR	EGR VR Solenoid	0-100%	0-100	0-40	0-40
EVAP CP	EVAP Purge Valve	0-100%	0-100	0-100	0-100
EVAP CV	EVAP CV Valve	0-100%	0-100	0-100	0-100
EPC	EPC Solenoid	0-300 psi	26	23-38	34-38
FP	Fuel Pump Control	ON / OFF	ON	ON	ON
FuelPW1	INJ Pulsewidth - Bank 1	0-999 ms	3.4-3.8	3.6-7.5	6.0-9.8
FuelPW2	INJ Pulsewidth - Bank 2	0-999 ms	3.4-3.8	3.6-7.5	6.0-9.8
HTR-11	HO2S-11 Heater	ON / OFF	ON	ON	ON
HTR-12	HO2S-12 Heater	ON / OFF	ON	ON	ON
HTR-21	HO2S-21 Heater	ON / OFF	ON	ON	ON
IAC	Idle Air Control	0-100%	25-32	35-49	30-68
INJ1	INJ 1 Pulsewidth	0-999 ms	3.4-3.8	3.6-7.5	6.0-9.8
INJ2	INJ 2 Pulsewidth	0-999 ms	3.4-3.8	3.6-7.5	6.0-9.8
INJ3	INJ 3 Pulsewidth	0-999 ms	3.4-3.8	3.6-7.5	6.0-9.8
INJ4	INJ 4 Pulsewidth	0-999 ms	3.4-3.8	3.6-7.5	6.0-9.8
INJ5	INJ 5 Pulsewidth	0-999 ms	3.4-3.8	3.6-7.5	6.0-9.8
INJ6	INJ 6 Pulsewidth	0-999 ms	3.4-3.8	3.6-7.5	6.0-9.8
LongFT1	Long Term FT - Bank 1	-20 to 20%	-5 to +5	-5 to +5	-5 to +5
LongFT2	Long Term FT - Bank 2	-20 to 20%	-5 to +5	-5 to +5	-5 to +5
MIL	MIL (lamp) Control	ON / OFF	OFF	OFF	OFF
SHTFT1	Short Term F/T - Bank 1	-10 to 10%	-5 to +5	-5 to +5	-5 to +5
SHTFT2	Short Term FT - Bank 2	-10 to 10%	-5 to +5	-5 to +5	-5 to +5
SPARK	Spark Advance	-90° to 90°	11-20	31-36	32-40
SS1	Shift Solenoid 1	ON / OFF	ON	OFF	OFF
SS2	Shift Solenoid 2	ON / OFF	OFF	OFF	OFF
SS3	Shift Solenoid 3	ON / OFF	OFF	OFF	ON
SS4	Shift Solenoid 4	ON / OFF	OFF	OFF	ON
TCC	TCC Solenoid	0-100%	0	0-100	90-95
TCIL	TCIL (lamp) Control	ON / OFF	OFF	OFF	OFF
VREF	Vehicle Reference	0-5.1v	4.9-5.1	4.9-5.1	4.9-5.1
WAC	A/C WOT Relay Control	ON / OFF	ON (AC on at WOT)	ON / OFF	ON / OFF

1990-94 RANGER 4.0L V6 VIN X (ALL) INPUTS & OUTPUTS

PCM PID Acronym	Parameter Identification	PID Range	PID Value at Hot Idle	PID Value at 30	PID Value at 55
ACCS	A/C Switch Signal	ON / OFF	ON (if on)	OFF	OFF
BARO	BARO Sensor	0-999 Hz	159	159	159
BOO	Brake Switch Signal	ON / OFF	ON (if on)	OFF	OFF
CANP	Purge Solenoid	0-100%	0	0-40	85-95
DPFE	DPFE Sensor	0-5.1v	0.4-0.6	0.5-1.0	0.7-1.1
ECT (°F)	ECT Sensor	-40-304°F	160-200	160-200	160-200
ECTV	ECT Sensor	0-5.1v	0.7	0.7	0.7
EVR	EGR VR Solenoid	0-100%	0	0-40	0-40
FPM	Fuel Pump Monitor	ON / OFF	ON	ON	ON
FuelPW1	INJ Pulsewidth - Bank 1	0-999 ms	3.3-3.5	4.0-4.6	5.0-6.0
FuelPW2	INJ Pulsewidth - Bank 2	0-999 ms	3.3-3.5	4.0-4.6	5.0-6.0
HO2S-11	HO2S-11 (Bank 1)	0.1-1.1v	0.1-1.1	0.1-1.1	0.1-1.1
HO2S-21	HO2S-21 (Bank 2)	0.1-1.1v	0.1-1.1	0.1-1.1	0.1-1.1
IAC	Idle Air Control	0-100%	20-40	34-40	45-55
IAT (°F)	IAT Sensor	-40-304°F	50-120	50-120	50-120
IATV	IAT Sensor	0-5.1v	1.5-3.5	1.5-3.5	1.5-3.5
LongFT1	Long Term F/T - Bank 1	-20 to 20%	-5 to +5	-5 to +5	-5 to +5
LongFT2	Long Term F/T - Bank 2	-20 to 20%	-5 to +5	-5 to +5	-5 to +5
LOOP	Loop Status	CL or OL	CLOSED	CLOSED	CLOSED
MAF	MAF Sensor	0-5.1v	0.7	1.3-1.4	1.7-2.0v
PNP	PNP Switch Signal	NEUT/DR	NEUTRAL	DRIVE	DRIVE
RPM	Engine Speed	0-10K rpm	750-830	1500-1650	1800-2100
SHTFT1	Short Term F/T - Bank 1	-10 to 10%	-5 to +5	-5 to +5	-5 to +5
SHTFT2	Short Term FT - Bank 2	-10 to 10%	-5 to +5	-5 to +5	-5 to +5
SPARK	Spark Advance	-90° to 90°	11-20	26-31	20-32
TP	TP Sensor	0-5.1v	0.9	1.2-1.3	1.4-1.6
TPCT	TP sensor Minimum	0-5.1v	0.1	1.2-1.3	1.4-1.6
TP Mode	TP Sensor	C/T or P/T	C/T	P/T	P/T
VPWR	Vehicle Power	0-25.5v	12-14	12-14	12-14
VREF	Vehicle Reference	0-5.1v	4.9-5.1	4.9-5.1	4.9-5.1
VSS	Vehicle Speed	0-159 mph	0	30	55
WAC	A/C WOT Relay Control	OFF	ON (AC on at WOT)	OFF	OFF

1995-2001 RANGER 4.0L V6 MFI VIN X (ALL) INPUTS

PCM PID Acronym	Parameter Identification	PID Range	PID Value at Hot Idle	PID Value at 30	PID Value at 55
4x4L	4x4 Low Signal	ON / OFF	ON (if on)	OFF	OFF
ACCS	A/C Switch Signal	ON / OFF	ON (if on)	OFF	OFF
BPP	Brake Switch Signal	ON / OFF	ON (if on)	OFF	OFF
CID	CMP Sensor	0-999 Hz	6-8	13-15	17-19
CKP	CKP Sensor	0-10K Hz	430-475	810-870	1180-1230
CPP	Clutch Pedal Switch	ON / OFF	ON (if in)	OFF	OFF
DPFE	DPFE Sensor	0-5.1v	0.4-0.6	0.4-1.0	0.4-1.1
ECT	ECT Sensor	-40-304°F	160-200	160-200	160-200
FEPS	Flash EEPROM	0-5.1v	0.1	0.1	0.1
FLI	Fuel Level Indicator	0-100%	50% (1.7)	50% (1.7)	50% (1.7)
FTP	Fuel Tank Pressure	0-5.1v or 0-10" H2O	2.6v / 0" H2O (with cap off)	2.6v / 0" H2O (with cap off)	2.6v / 0" H2O (with cap off)
FPM	Fuel Pump Monitor	ON / OFF	ON	ON	ON
IAT	IAT Sensor	-40-304°F	50-120	50-120	50-120
GEAR	Gear Position	1-2-3-4	1	3	4
HO2S-11	HO2S-11 (Bank 1)	0.1-1.1v	0.1-1.1	0.1-1.1	0.1-1.1
HO2S-12	HO2S-12 (Bank 1)	0.1-1.1v	0.1-1.1	0.1-1.1	0.1-1.1
HO2S-21	HO2S-21 (Bank 2)	0.1-1.1v	0.1-1.1	0.1-1.1	0.1-1.1
LOAD	Engine Load	0-100%	16-20	21-27	27-34
MAFV	MAF Sensor	0-5.1v	0.8-1.0	1.4-1.8	2.0-2.7
MISF	Misfire Detection	ON / OFF	OFF	OFF	OFF
OCTADJ	Octane Adjustment	CL/OPEN	CLOSED	CLOSED	CLOSED
PF	Purge Flow Sensor	0-5.1v	1.55	2.10	2.30
PNP	PNP Switch Signal	Neutral/DR	NEUTRAL	DRIVE	DRIVE
PSP	PSP Switch Signal	LOW or HI	HI: turned	LOW	LOW
RPM	Engine Speed	0-10K rpm	760-830	1400-1475	1525-1760
TCS	TCS Switch	ON / OFF	OFF	OFF	OFF
TFT	TFT Sensor Signal	-40-304°F	110-210	110-210	110-210
TPV	TP Sensor	0-5.1v	0.53-1.27	0.8-1.0	1.2-1.7
TRV/TR	TR Sensor	P/N or O/D	P/N	O/D	O/D
TSS	TSS Sensor	0-9999 Hz	100-125	185-205	260-280
Vacuum	Engine Vacuum	0-30" Hg	19-20	18-20	8-15
VPWR	Vehicle Power	0-25.5v	12-14	12-14	12-14
VSS	Vehicle Speed (Hertz or MPH)	0-999 Hz (0-159 mph)	0 Hz = 0 mph	65 Hz = 30 mph	125 Hz = 55 mph

1995-2001 RANGER 4.0L V6 MFI VIN X (ALL) OUTPUTS

PCM PID Acronym	Parameter Identification	PID Range	PID Value at Hot Idle	PID Value at 30	PID Value at 55
CD1	Coil 1 Driver	VBAT	VBAT	VBAT	VBAT
CD2	Coil 2 Driver	VBAT	VBAT	VBAT	VBAT
CD3	Coil 3 Driver	VBAT	VBAT	VBAT	VBAT
CTO	Clean Tachometer signal	0-9999 Hz	35-49	70-82 Hz	90-120
CCS	Coast Clutch Solenoid Control	0-14v	VBAT	VBAT	VBAT
EGR VR	EGR VR Solenoid	0-100%	0	0-40	0-40
EVAP CP	EVAP Purge Valve	0-100%	0-100	0-100	0-100
EVAP CV	EVAP CV Valve	0-100%	0-100	0-100	0-100
EPC	EPC Solenoid	0-300 psi	24	23-28	34-38
FP	Fuel Pump Control	ON / OFF	ON	ON	ON
FuelPW1	INJ Pulsewidth - Bank 1	0-999 ms	4.5-4.8	6.3-8.0	7.0-13.0
FuelPW2	INJ Pulsewidth - Bank 2	0-999 ms	4.5-4.8	6.3-8.0	7.0-13.0
HTR-11	HO2S-11 Heater	ON / OFF	ON	ON	ON
HTR-12	HO2S-12 Heater	ON / OFF	ON	ON	ON
HTR-21	HO2S-21 Heater	ON / OFF	ON	ON	ON
IAC	Idle Air Control	0-100%	33	35-41	57-68
INJ1	INJ 1 Pulsewidth	0-999 ms	4.5-4.8	6.3-8.0	7.0-13.0
INJ2	INJ 2 Pulsewidth	0-999 ms	4.5-4.8	6.3-8.0	7.0-13.0
INJ3	INJ 3 Pulsewidth	0-999 ms	4.5-4.8	6.3-8.0	7.0-13.0
INJ4	INJ 4 Pulsewidth	0-999 ms	4.5-4.8	6.3-8.0	7.0-13.0
INJ5	INJ 5 Pulsewidth	0-999 ms	4.5-4.8	6.3-8.0	7.0-13.0
INJ6	INJ 6 Pulsewidth	0-999 ms	4.5-4.8	6.3-8.0	7.0-13.0
LongFT1	Long Term FT - Bank 1	-20 to 20%	-5 to +5	-5 to +5	-5 to +5
LongFT2	Long Term FT - Bank 2	-20 to 20%	-5 to +5	-5 to +5	-5 to +5
MIL	MIL (lamp) Control	ON / OFF	OFF	OFF	OFF
SHTFT1	Short Term F/T - Bank 1	-10 to 10%	-5 to +5	-5 to +5	-5 to +5
SHTFT2	Short Term FT - Bank 2	-10 to 10%	-5 to +5	-5 to +5	-5 to +5
SPARK	Spark Advance	-90° to 90°	11-15	26-31	25-32
SS1	Shift Solenoid 1	ON / OFF	ON	OFF	OFF
SS2	Shift Solenoid 2	ON / OFF	OFF	OFF	OFF
SS3	Shift Solenoid 3	ON / OFF	OFF	OFF	ON
TCC	TCC Solenoid	0-100%	0	55-72	80-95
TCIL	TCIL (lamp) Control	ON / OFF	OFF	OFF	OFF
VREF	Vehicle Reference	0-5.1v	4.9-5.1	4.9-5.1	4.9-5.1
WAC	A/C WOT Relay Control	ON / OFF	ON (AC on at WOT)	OFF	OFF

Aerostar PID Data

1993-95 AEROSTAR 3.0L V6 MFI VIN U (A/T) INPUTS & OUTPUTS

PCM PID Acronym	Parameter Identification	PID Range	PID Value at Hot Idle	PID Value at 30	PID Value at 55
ACCS	A/C Switch Signal	ON / OFF	ON (if on)	OFF	OFF
BOO	Brake Switch Signal	ON / OFF	ON (if on)	OFF	OFF
CANP	EVAP Purge Valve	OFF	OFF	ON	ON
ECT (°F)	ECT Sensor	-40-304°F	160-200	160-200	160-200
ECTV	ECT Sensor	0-5.1v	0.7	0.7	0.7
FPM	Fuel Pump Monitor	ON / OFF	ON	ON	ON
FuelPW1	INJ Pulsewidth - Bank 1	0-999 ms	3.7-3.9	4.4-4.6	5.0-6.0
HO2S-11	HO2S-11 (Bank 1)	0.1-1.1v	0.1-1.1	0.1-1.1	0.1-1.1
IAC	Idle Air Control	0-100%	33	35-41	57-68
IAT (°F)	IAT Sensor	-40-304°F	50-120	50-120	50-120
IATV	IAT Sensor	0-5.1v	1.5-3.5	1.5-3.5	1.5-3.5
LongFT1	Long Term F/T - Bank 1	-20 to 20%	-5 to +5	-5 to +5	-5 to +5
MAF	MAF Sensor	0-5.1v	0.9	1.4-1.6	1.9-2.4
PNP	PNP Switch Signal	Neutral/DR	NEUTRAL	DRIVE	DRIVE
RPM	Engine Speed	0-10K rpm	880-920	1430-1470	1550-1750
SHTFT1	Short Term F/T - Bank 1	-10 to 10%	-5 to +5	-5 to +5	-5 to +5
SPARK	Spark Advance	-90° to 90°	15-22	28-35	25-35
TP	TP Sensor	0-5.1v	0.7	0.9-1.0	1.0-1.2
TPCT	TP sensor Minimum	0-5.1v	0.1	0.9-1.0	1.0-1.2
TP Mode	TP Sensor Mode	C/T or P/T	C/T	P/T	P/T
VPWR	Vehicle Power	0-25.5v	12-14	12-14	12-14
VREF	Vehicle Reference	0-5.1v	4.9-5.1	4.9-5.1	4.9-5.1
VSS	Vehicle Speed	0-159 mph	0	30	55
WAC	A/C WOT Relay Control	OFF / ON	ON (AC on at WOT)	OFF	OFF

1996-97 AEROSTAR 3.0L V6 MFI VIN U (A/T) INPUTS

PCM PID Acronym	Parameter Identification	PID Range	PID Value at Hot Idle	PID Value at 30	PID Value at 55
ACCS	A/C Switch Signal	ON / OFF	ON (if on)	OFF	OFF
BPP	Brake Position	ON / OFF	ON (if on)	OFF	OFF
CID	CMP Sensor	0-999 Hz	6-8	13-15	17-19
CKP	CKP Sensor	0-9999 Hz	518-540	917-1020	1290-1330
DPFE	DPFE Sensor	0-5.1v	0.4-0.6	0.4-0.9	0.4-0.9
ECT	ECT Sensor	-40-304°F	160-200	160-200	160-200
FEPS	Flash EEPROM	0-5.1v	0.1	0.1	0.1
FPM	Fuel Pump Monitor	ON / OFF	ON	ON	ON
GEAR	Gear Position	1-2-3-4	1	3	4
HO2S-11	HO2S-11 (Bank 1)	0.1-1.1v	0.1-1.1	0.1-1.1	0.1-1.1
HO2S-12	HO2S-12 (Bank 1)	0.1-1.1v	0.1-1.1	0.1-1.1	0.1-1.1
IAT	IAT Sensor	-40-304°F	50-120	50-120	50-120
LOAD	Engine Load	0-100%	13-18	25-35	39-46
MAFV	MAF Sensor	0-5.1v	0.8-1.0	1.5-1.9	2.1-2.9
MISF	Misfire Detection	ON / OFF	OFF	OFF	OFF
OCTADJ	Octane Adjustment	CL/OPEN	CLOSED	CLOSED	CLOSED
PF	Purge Flow Sensor	0-5.1v	1.55	2.10	2.30
PNP	PNP Switch Signal	Neutral/DR	NEUTRAL	DRIVE	DRIVE
PSP	PSP Switch Signal	HI or LOW	HI: turning	LOW	LOW
RPM	Engine Speed	0-10K rpm	880-910	1675-1730	2200-2275
TCS	TCS Switch	ON / OFF	OFF	OFF	OFF
TFT	TFT Sensor Signal	-40-304°F	110-210	110-210	110-210
TPV	TP Sensor	0-5.1v	0.53-1.27	0.8-1.7	1.2-1.7
TRV/TR	TR Sensor	P/N or O/D	P/N or O/D	O/D	O/D
TSS	TSS Sensor	0-9999 Hz	118-122	218-230	2200-2275
Vacuum	Engine Vacuum	0-30" Hg	19-20	18-20	8-15
VPWR	Vehicle Power	0-25.5v	12-14	12-14	12-14
VSS	Vehicle Speed (Hertz or MPH)	0-999 Hz (0-159 mph)	0 Hz = 0 mph	65 Hz = 30 mph	125 Hz = 55 mph

1996-97 AEROSTAR 3.0L V6 MFI VIN U (A/T) OUTPUTS

PCM PID Acronym	Parameter Identification	PID Range	PID Value at Hot Idle	PID Value at 30	PID Value at 55
CD1	Coil 1 Driver	0-60° dwell	6	8	12
CD2	Coil 2 Driver	0-60° dwell	6	8	12
CD3	Coil 3 Driver	0-60° dwell	6	8	12
CTO	Clean Tachometer signal	0-9999 Hz	42-48	79-87	99-115
CCS	Coast Clutch Solenoid Control	0-25.5v	VBAT	VBAT	VBAT
EGR VR	EGR VR Solenoid	0-100%	0	0-40	0-55
EVAP CP	EVAP Purge Valve	0-100%	0-100	0-100	0-100
EPC	EPC Solenoid	0-300 psi	27	27	27
FP	Fuel Pump Control	ON / OFF	ON	ON	ON
FuelPW1	INJ Pulsewidth - Bank 1	0-999 ms	4.5-4.8	5.7-7.0	8.0-14.0
HTR-11	HO2S-11 Heater	ON / OFF	ON	ON	ON
HTR-12	HO2S-12 Heater	ON / OFF	ON	ON	ON
IAC	Idle Air Control	0-100%	33	35-41	57-68
INJ1	INJ 1 Pulsewidth	0-999 ms	4.5-4.8	5.7-7.0	8.0-14.0
INJ2	INJ 2 Pulsewidth	0-999 ms	4.5-4.8	5.7-7.0	8.0-14.0
INJ3	INJ 3 Pulsewidth	0-999 ms	4.5-4.8	5.7-7.0	8.0-14.0
INJ4	INJ 4 Pulsewidth	0-999 ms	4.5-4.8	5.7-7.0	8.0-14.0
INJ5	INJ 5 Pulsewidth	0-999 ms	4.5-4.8	5.7-7.0	8.0-14.0
INJ6	INJ 6 Pulsewidth	0-999 ms	4.5-4.8	5.7-7.0	8.0-14.0
LongFT1	Long Term FT - Bank 1	-20 to 20%	-5 to +5	-5 to +5	-5 to +5
MIL	MIL (lamp) Control	ON / OFF	OFF	OFF	OFF
SHTFT1	Short Term F/T - Bank 1	-10 to 10%	-5 to +5	-5 to +5	-5 to +5
SPARK	Spark Advance	-90° to 90°	15-22	29-39	25-32
SS1	Shift Solenoid 1	ON / OFF	ON	OFF	OFF
SS2	Shift Solenoid 2	ON / OFF	OFF	ON	ON
SS3	Shift Solenoid 3	ON / OFF	OFF	OFF	ON
TCC	TCC Solenoid	0-100%	0	59-66	72-81
TCIL	TCIL (lamp) Control	ON / OFF	OFF	OFF	OFF
VREF	Vehicle Reference	0-5.1v	4.9-5.1	4.9-5.1	4.9-5.1
WAC	A/C WOT Relay Control	ON / OFF	ON (AC on at WOT)	OFF	OFF

1990-95 AEROSTAR 4.0L V6 MFI VIN X (A/T)

PCM PID Acronym	Parameter Identification	PID Range	PID Value at Hot Idle	PID Value at 30	PID Value at 55
ACCS	A/C Switch Signal	ON / OFF	ON (if on)	OFF	OFF
BARO	BARO Sensor	0-999 Hz	159	159	159
BOO	Brake Switch Signal	ON / OFF	ON (if on)	OFF	OFF
CANP	EVAP Purge Valve	0-100%	0-100	0-100	0-100
ECT (°F)	ECT Sensor	-40-304°F	160-200	160-200	160-200
ECTV	ECT Sensor	0-5.1v	0.7	0.7	0.7
FPM	Fuel Pump Monitor	ON / OFF	ON	ON	ON
FuelPW1	INJ Pulsewidth - Bank 1	0-999 ms	3.3-3.5	4.0-4.6	5.0-6.0
FuelPW2	INJ Pulsewidth - Bank 2	0-999 ms	3.3-3.5	4.0-4.6	5.0-6.0
HO2S-11	HO2S-11 (Bank 1)	0.1-1.1v	0.1-1.1	0.1-1.1	0.1-1.1
HO2S-21	HO2S-21 (Bank 2)	0.1-1.1v	0.1-1.1	0.1-1.1	0.1-1.1
IAC	Idle Air Control	0-100%	20-40	34-40	45-55
IAT (°F)	IAT Sensor	-40-304°F	50-120	50-120	50-120
IATV	IAT Sensor	0-5.1v	1.5-3.5	1.5-3.5	1.5-3.5
LongFT1	Long Term F/T - Bank 1	-20 to 20%	-5 to +5	-5 to +5	-5 to +5
LongFT2	Long Term F/T - Bank 2	-20 to 20%	-5 to +5	-5 to +5	-5 to +5
LOOP	Loop Status	CL or OL	CLOSED	CLOSED	CLOSED
MAF	MAF Sensor	0-5.1v	0.7	1.3-1.4	1.7-2.0v
PNP	PNP Switch Signal	Neutral/DR	NEUTRAL	DRIVE	DRIVE
RPM	Engine Speed	0-10K rpm	790-820	1600-1660	1990-2100
SHTFT1	Short Term F/T - Bank 1	-20 to 20%	-5 to +5	-5 to +5	-5 to +5
SHTFT2	Short Term FT - Bank 2	-20 to 20%	-5 to +5	-5 to +5	-5 to +5
SPARK	Spark Advance	-90° to 90°	15-22	25-29	32-35
TP	TP Sensor	0-5.1v	0.9	1.2-1.3	1.4-1.6
TPCT	TP sensor Minimum	0-5.1v	0.1	1.2-1.3	1.4-1.6
TP Mode	TP Sensor	C/T or P/T	C/T	P/T	P/T
VPWR	Vehicle Power	0-25.5v	12-14	12-14	12-14
VREF	Vehicle Reference	0-5.1v	4.9-5.1	4.9-5.1	4.9-5.1
VSS	Vehicle Speed	0-159 mph	0	30	55
WAC	A/C WOT Relay Control	OFF	ON (AC on at WOT)	OFF	OFF

1996-97 AEROSTAR 4.0L V6 MFI VIN X (A/T) INPUTS

PCM PID Acronym	Parameter Identification	PID Range	PID Value at Hot Idle	PID Value at 30	PID Value at 55
ACCS	A/C Switch Signal	ON / OFF	ON (if on)	OFF	OFF
BPP	Brake Position	ON / OFF	ON (if on)	OFF	OFF
CID	CMP Sensor	0-999 Hz	5-7	13-15	15-17
CKP	CKP Sensor	0-9999 Hz	450-480	1000-1100	1140-1220
DPFE	DPFE Sensor	0-5.1v	0.4-0.6	0.5-1.0	0.7-1.1
ECT	ECT Sensor	-40-304°F	160-200	160-200	160-200
FEPS	Flash EEPROM	0-5.1v	0.1	0.1	0.1
FPM	Fuel Pump Monitor	ON / OFF	ON	ON	ON
GEAR	Gear Position	1-2-3-4	1	3	4
HO2S-11	HO2S-11 (Bank 1)	0.1-1.1v	0.1-1.1	0.1-1.1	0.1-1.1
HO2S-12	HO2S-12 (Bank 1)	0.1-1.1v	0.1-1.1	0.1-1.1	0.1-1.1
IAT	IAT Sensor	-40-304°F	50-120	50-120	50-120
LOAD	Engine Load	0-100%	13-18	18-33	35-50
MAFV	MAF Sensor	0-5.1v	0.7-0.9	1.1-1.6	1.5-2.3
MISF	Misfire Detection	ON / OFF	OFF	OFF	OFF
OCTADJ	Octane Adjustment	CL/OPEN	CLOSED	CLOSED	CLOSED
PF	Purge Flow Sensor	0-5.1v	1.55	2.10	2.30
PNP	PNP Switch Signal	Neutral/DR	NEUTRAL	DRIVE	DRIVE
PSP	PSP Switch Signal	LOW	HI: turning	LOW	LOW
RPM	Engine Speed	0-10K rpm	790-820	1610-1860	1990-2100
TCS	TCS Switch	ON / OFF	OFF	OFF	OFF
TFT	TFT Sensor Signal	-40-304°F	110-210	110-210	110-210
TPV	TP Sensor	0-5.1v	0.53-1.27	0.8-1.7	1.2-1.7
TRV/TR	TR Sensor	P/N or O/D	P/N or O/D	O/D	O/D
TSS	TSS Sensor	0-9999 Hz	95-110	190-200	260-280
Vacuum	Engine Vacuum	0-30" Hg	19-20	18-20	8-15
VPWR	Vehicle Power	0-25.5v	12-14	12-14	12-14
VSS	Vehicle Speed (Hertz or MPH)	0-999 Hz (0-159 mph)	0 Hz = 0 mph	65 Hz = 30 mph	125 Hz = 55 mph

1996-97 AEROSTAR 4.0L V6 MFI VIN X (A/T) OUTPUTS

PCM PID Acronym	Parameter Identification	PID Range	PID Value at Hot Idle	PID Value at 30	PID Value at 55
CD1	Coil 1 Driver	0-90° dwell	6	8	12
CD2	Coil 2 Driver	0-90° dwell	6	8	12
CD3	Coil 3 Driver	0-90° dwell	6	8	12
CTO	Clean Tachometer signal	0-999	38-42	79-87	96-105
CCS	Coast Clutch Solenoid Control	0-25.5v	VBAT	VBAT	VBAT
EGR VR	EGR VR Solenoid	0-100%	0	0-40	0-45
EVAP CP	EVAP Purge Valve	0-100%	0-100	0-100	0-100
EPC	EPC Solenoid	0-300 psi	26	30	38
FP	Fuel Pump Control	ON / OFF	ON	ON	ON
FuelPW1	INJ Pulsewidth - Bank 1	0-999 ms	3.9-4.1	4.3-6.9	6.0-12.0
HTR-11	HO2S-11 Heater	ON / OFF	ON	ON	ON
HTR-12	HO2S-12 Heater	ON / OFF	ON	ON	ON
IAC	Idle Air Control	0-100%	39	46-52	49-53
INJ1	INJ 1 Pulsewidth	0-999 ms	3.9-4.1	4.3-6.9	6.0-12.0
INJ2	INJ 2 Pulsewidth	0-999 ms	3.9-4.1	4.3-6.9	6.0-12.0
INJ3	INJ 3 Pulsewidth	0-999 ms	3.9-4.1	4.3-6.9	6.0-12.0
INJ4	INJ 4 Pulsewidth	0-999 ms	3.9-4.1	4.3-6.9	6.0-12.0
INJ5	INJ 5 Pulsewidth	0-999 ms	3.9-4.1	4.3-6.9	6.0-12.0
INJ6	INJ 6 Pulsewidth	0-999 ms	3.9-4.1	4.3-6.9	6.0-12.0
LongFT1	Long Term FT - Bank 1	-20 to 20%	-5 to +5	-5 to +5	-5 to +5
MIL	MIL (lamp) Control	ON / OFF	OFF	OFF	OFF
SHTFT1	Short Term F/T - Bank 1	-10 to 10%	-5 to +5	-5 to +5	-5 to +5
SPARK	Spark Advance	-90° to 90°	15-22	25-29	19-25
SS1	Shift Solenoid 1	ON / OFF	ON	OFF	OFF
SS2	Shift Solenoid 2	ON / OFF	OFF	ON	OFF
SS3	Shift Solenoid 3	ON / OFF	OFF	OFF	ON
TCC	TCC Solenoid	0-100%	0	0-100	75-81
TCIL	TCIL (lamp) Control	ON / OFF	OFF	OFF	OFF
VREF	Vehicle Reference	0-5.1v	4.9-5.1	4.9-5.1	4.9-5.1
WAC	A/C WOT Relay Control	ON / OFF	ON (AC on at WOT)	OFF	OFF

E-Series Van PID Data

1997-2003 E-SERIES 4.2L V6 MFI VIN 2 (A/T) INPUTS

PCM PID Acronym	Parameter Identification	PID Range	PID Value at Hot Idle	PID Value at 30	PID Value at 55
ACCS	A/C Switch Signal	OFF	ON	OFF	OFF
BPP	Brake Position	ON / OFF	ON (if on)	OFF	OFF
CID	CMP Sensor	0-999 Hz	5-7	10-12	13-17
CHT	CHT Sensor	-40-500°F	194	194	194
CKP	CKP Sensor	0 Hz	430-4500	700-900	1000-1200
DPFE	DPFE Sensor	0-5.1v	0.4-0.6	0.4-1.0	0.6-1.1
FEPS	Flash EEPROM	0-5.1v	0.1	0.1	0.1
FLI	Fuel Level Indicator	0-100%	50% (1.7)	50% (1.7)	50% (1.7)
FPM	Fuel Pump Monitor	ON / OFF	ON	ON	ON
FTP	Fuel Tank Pressure	0-5.1v or 0-10" H2O	2.6v / 0" H2O (with cap off)	2.6v / 0" H2O (with cap off)	2.6v / 0" H2O (with cap off)
GEAR	Gear Position	1-2-3-4	1	3	4
HO2S-11	HO2S-11 (Bank 1)	0.1-1.1v	0.1-1.1	0.1-1.1	0.1-1.1
HO2S-12	HO2S-12 (Bank 1)	0.1-1.1v	0.1-1.1	0.1-1.1	0.1-1.1
HO2S-21	HO2S-21 (Bank 2)	0.1-1.1v	0.1-1.1	0.1-1.1	0.1-1.1
HO2S-22	HO2S-22 (Bank 2)	0.1-1.1v	0.1-1.1	0.1-1.1	0.1-1.1
IAT	IAT Sensor	-40-304°F	50-120	50-120	50-120
IMRC-M	IMRC Monitor	5v / 0	5	5	5
KS1	Knock Sensor 1	0-5.1v	0	0	0
LOAD	Engine Load	0-100%	10-20	20-27	30-45
MAFV	MAF Sensor	0-5.1v	0.6-0.9	0.7-1.0	1.2-2.3
MISF	Misfire Detection	ON / OFF	OFF	OFF	OFF
OCTADJ	Octane Adjustment	CL/OPEN	CLOSED	CLOSED	CLOSED
OSS	OSS Sensor	0-9999 Hz	0	125-131	245-255
PNP	PNP Switch Signal	Neutral/DR	NEUTRAL	DRIVE	DRIVE
RPM	Engine Speed	0-10K rpm	680-830	1200-1500	1600-1800
TFT	TFT Sensor Signal	-40-304°F	110-210	110-210	110-210
TPV	TP Sensor	0-5.1v	0.53-1.27	1.0-1.3	1.1-1.6
TR1	TR1 Sensor	0v / 11.5v	0v	11.5	11.5
TR2	TR2 Sensor	0v / 11.5v	0v	11.5	11.5
TR3/TR	TR Sensor	0v / 1.7v	0v	1.7	1.7
TR4	TR4 Sensor	0v / 11.5v	0v	11.5	11.5
Vacuum	Engine Vacuum	0-30" Hg	19-20	18-20	8-15
VPWR	Vehicle Power	0-25.5v	12-14	12-14	12-14
VSS	Vehicle Speed (Hertz or MPH)	0-999 Hz (0-159 mph)	0 Hz = 0 mph	65 Hz = 30 mph	125 Hz = 55 mph
4x4L	4x4 Switch Signal	ON / OFF	ON (if on)	OFF	OFF

1997-2003 E-SERIES 4.2L V6 MFI VIN 2 (A/T) OUTPUTS

PCM PID Acronym	Parameter Identification	PID Range	PID Value at Hot Idle	PID Value at 30	PID Value at 55
CD1	Coil 1 Driver	0-60° dwell	6	8	12
CD2	Coil 2 Driver	0-60° dwell	6	8	12
CD3	Coil 3 Driver	0-60° dwell	6	8	12
CHIL	CHIL (lamp) Control	ON / OFF	OFF	OFF	OFF
CTO	Clean Tachometer signal	0-9999 Hz	35-49	65-90	90-120
EGR VR	EGR VR Solenoid	0-100%	0	0-40	0-40
EVAP CP	EVAP Purge Valve	0-100%	0-100	0-100	0-100
EVAP CV	EVAP Purge Valve	0-100%	0-100	0-100	0-100
EPC	EPC Solenoid	0-300 psi	15-20	35-40	40
FP	Fuel Pump Control	ON / OFF	ON	ON	ON
FuelPW1	INJ Pulsewidth - Bank 1	0-999 ms	2.7-3.5	4.5-8.0	5.5-9.0
FuelPW2	INJ Pulsewidth - Bank 2	0-999 ms	2.7-3.5	4.5-8.0	5.5-9.0
HTR-11	HO2S-11 Heater	ON / OFF	ON	ON	ON
HTR-12	HO2S-12 Heater	ON / OFF	ON	ON	ON
HTR-21	HO2S-21 Heater	ON / OFF	ON	ON	ON
HTR-22	HO2S-22 Heater	ON / OFF	ON	ON	ON
IAC	Idle Air Control	0-100%	25-32	30-55	60-70
IMRC	IMRC Solenoid	ON / OFF	OFF	OFF	OFF
INJ1	INJ 1 Pulsewidth	0-999 ms	2.7-3.5	4.5-8.0	5.5-9.0
INJ2	INJ 2 Pulsewidth	0-999 ms	2.7-3.5	4.5-8.0	5.5-9.0
INJ3	INJ 3 Pulsewidth	0-999 ms	2.7-3.5	4.5-8.0	5.5-9.0
INJ4	INJ 4 Pulsewidth	0-999 ms	2.7-3.5	4.5-8.0	5.5-9.0
INJ5	INJ 5 Pulsewidth	0-999 ms	2.7-3.5	4.5-8.0	5.5-9.0
INJ6	INJ 6 Pulsewidth	0-999 ms	2.7-3.5	4.5-8.0	5.5-9.0
LongFT1	Long Term FT - Bank 1	-20 to 20%	-5 to +5	-5 to +5	-5 to +5
LongFT2	Long Term FT - Bank 2	-20 to 20%	-5 to +5	-5 to +5	-5 to +5
MIL	MIL (lamp) Control	ON / OFF	OFF	OFF	OFF
SHTFT1	Short Term F/T - Bank 1	-10 to 10%	-5 to +5	-5 to +5	-5 to +5
SHTFT2	Short Term FT - Bank 2	-10 to 10%	-5 to +5	-5 to +5	-5 to +5
SPARK	Spark Advance	-90° to 90°	11-20	15-35	20-39
SS1	Shift Solenoid 1	ON / OFF	ON	OFF	ON
SS2	Shift Solenoid 2	ON / OFF	OFF	ON	ON
TCC	TCC Solenoid	0-100%	0	0-45	90-95
TCIL	TCIL (lamp) Control	ON / OFF	OFF	OFF	OFF
VREF	Vehicle Reference	0-5.1v	4.9-5.1	4.9-5.1	4.9-5.1
WAC	A/C WOT Relay Control	ON / OFF	ON (AC on at WOT)	OFF	OFF

1997-2001 E-SERIES 4.6L V8 MFI VIN 6 (A/T) INPUTS

PCM PID Acronym	Parameter Identification	PID Range	PID Value at Hot Idle	PID Value at 30	PID Value at 55
4x4L	4x4 Switch Signal	ON / OFF	ON (if on)	OFF	OFF
ACCS	A/C Switch Signal	ON / OFF	ON (if on)	OFF	OFF
BPP	Brake Position	ON / OFF	ON (if on)	OFF	OFF
CHT	CHT Sensor	-40-500ºF	194	194	194
CID	CMP Sensor	0-999 Hz	5-7	10-12	13-17
CKP	CKP Sensor	0-9999 Hz	430-475	900-1000	1140-1220
DPFE	DPFE Sensor	0-5.1v	0.4-0.6	0.4-1.0	0.6-1.1
ECT	ECT Sensor	-40-304ºF	160-200	160-200	160-200
FEPS	Flash EEPROM	0-5.1v	0.1	0.1	0.1
FLI	Fuel Level Input	1.7 / 50%	1.7 / 50%	1.7 / 50%	1.7 / 50%
FPM	Fuel Pump Monitor	ON / OFF	ON	ON	ON
FTP	Fuel Tank Pressure	2.6v / 0 psi	2.6v / 0 psi	2.6v / 0 psi	2.6v / 0 psi
GEAR	Gear Position	1-2-3-4	1	3	4
HO2S-11	HO2S-11 (Bank 1)	0.1-1.1v	0.1-1.1	0.1-1.1	0.1-1.1
HO2S-12	HO2S-12 (Bank 1)	0.1-1.1v	0.1-1.1	0.1-1.1	0.1-1.1
HO2S-21	HO2S-21 (Bank 2)	0.1-1.1v	0.1-1.1	0.1-1.1	0.1-1.1
HO2S-22	HO2S-22 (Bank 2)	0.1-1.1v	0.1-1.1	0.1-1.1	0.1-1.1
IAT	IAT Sensor	-40-304ºF	50-120	50-120	50-120
KS1	Knock Sensor 1	0	0	0	0
LOAD	Engine Load	0-100%	10-20	20-27	30-45
MAFV	MAF Sensor	0-5.1v	0.6-0.9	0.7-1.0	1.2-2.3
MISF	Misfire Detection	OFF	OFF	OFF	OFF
OCTADJ	Octane Adjustment	CLOSED	CLOSED	CLOSED	CLOSED
OSS	OSS Sensor	0-9999 Hz	0	125-131	245-255
PNP	PNP Switch Signal	Neutral/DR	NEUTRAL	DRIVE	DRIVE
RPM	Engine Speed	0-10K rpm	680-830	1200-1500	1600-1800
TFT	TFT Sensor Signal	-40-304ºF	110-210	110-210	110-210
TPV	TP Sensor	0-5.1v	0.53-1.27	1.0-1.3	1.1-1.6
TR1	TR1 Sensor	0v / 11.5v	0v	11.5	11.5
TR2	TR2 Sensor	0v / 11.5v	0v	11.5	11.5
TR3/TR	TR Sensor	0v / 1.7v	0v	1.7	1.7
TR4	TR4 Sensor	0v / 11.5v	0v	11.5	11.5
Vacuum	Engine Vacuum	0-30" Hg	19-20	18-20	8-15
VPWR	Vehicle Power	0-25.5v	12-14	12-14	12-14
VSS	Vehicle Speed (Hertz or MPH)	0-999 Hz (0-159 mph)	0 Hz = 0 mph	65 Hz = 30 mph	125 Hz = 55 mph

1997-2001 E-SERIES 4.6L V8 MFI VIN 6 (A/T) OUTPUTS

PCM PID Acronym	Parameter Identification	PID Range	PID Value at Hot Idle	PID Value at 30	PID Value at 55
CD1	Coil 1 Driver	0-45° dwell	5	6	8
CD2	Coil 2 Driver	0-45° dwell	5	6	8
CD3	Coil 3 Driver	0-45° dwell	5	6	8
CD4	Coil 4 Driver	0-45° dwell	5	6	8
CHTIL	CHT (lamp) Control	ON / OFF	OFF	OFF	OFF
CTO	Clean Tachometer signal	0-9999 Hz	35-49	65-90	90-120
EGR VR	EGR VR Solenoid	0-100%	0	0-40	0-40
EVAP CP	EVAP Purge Valve	0-100%	0-100	0-100	0-100
EVAP CV	EVAP CV Valve	0-100%	0-100	0-100	0-100
EPC	EPC Solenoid	0-300 psi	6	40	40
FP	Fuel Pump Control	ON / OFF	ON	ON	ON
FuelPW1	INJ Pulsewidth - Bank 1	0-999 ms	2.7-4.1	4.5-8.0	5.5-9.0
FuelPW2	INJ Pulsewidth - Bank 2	0-999 ms	2.7-4.1	4.5-8.0	5.5-9.0
FRP	Fuel Rail Pressure	29-32 psi	30-32	30-32	32-35
HTR-11	HO2S-11 Heater	ON / OFF	ON	ON	ON
HTR-12	HO2S-12 Heater	ON / OFF	ON	ON	ON
HTR-21	HO2S-21 Heater	ON / OFF	ON	ON	ON
HTR-22	HO2S-22 Heater	ON / OFF	ON	ON	ON
IAC	Idle Air Control	0-100%	25-32	30-55	60-70
IMTV	Intake Manifold Tuning Valve	0v / VBAT	VBAT	VBAT	VBAT
INJ1	INJ 1 Pulsewidth	0-999 ms	2.7-4.1	4.5-8.0	5.5-9.0
INJ2	INJ 2 Pulsewidth	0-999 ms	2.7-4.1	4.5-8.0	5.5-9.0
INJ3	INJ 3 Pulsewidth	0-999 ms	2.7-4.1	4.5-8.0	5.5-9.0
INJ4	INJ 4 Pulsewidth	0-999 ms	2.7-4.1	4.5-8.0	5.5-9.0
INJ5	INJ 5 Pulsewidth	0-999 ms	2.7-4.1	4.5-8.0	5.5-9.0
INJ6	INJ 6 Pulsewidth	0-999 ms	2.7-4.1	4.5-8.0	5.5-9.0
INJ7	INJ 7 Pulsewidth	0-999 ms	2.7-4.1	4.5-8.0	5.5-9.0
INJ8	INJ 8 Pulsewidth	0-999 ms	2.7-4.1	4.5-8.0	5.5-9.0
LongFT1	Long Term FT (B1)	-20 to 20%	-5 to +5	-5 to +5	-5 to +5
LongFT2	Long Term FT (B2)	-20 to 20%	-5 to +5	-5 to +5	-5 to +5
MIL	MIL (lamp) Control	ON / OFF	OFF	OFF	OFF
SHTFT1	Short Term F/T - Bank 1	-10 to 10%	-5 to +5	-5 to +5	-5 to +5
SHTFT2	Short Term FT - Bank 2	-10 to 10%	-5 to +5	-5 to +5	-5 to +5
SPARK	Spark Advance	-90° to 90°	11-20	15-35	20-39
SS1	Shift Solenoid 1	ON / OFF	ON	ON	ON
SS2	Shift Solenoid 2	ON / OFF	OFF	ON	ON
TCC	TCC Solenoid	0-100%	0%	0-45	90-100
TCIL	TCIL (lamp) Control	ON / OFF	OFF	OFF	OFF
VREF	Vehicle Reference	0-5.1v	4.9-5.1	4.9-5.1	4.9-5.1
WAC	A/C WOT Relay	ON / OFF	ON (if on)	OFF	OFF

1999-2003 E-SERIES 4.6L V8 MFI VIN W (A/T) INPUTS

PCM PID Acronym	Parameter Identification	PID Range	PID Value at Hot Idle	PID Value at 30	PID Value at 55
ACCS	A/C Switch Signal	ON / OFF	ON (if on)	OFF	OFF
BPP	Brake Position	ON / OFF	ON (if on)	OFF	OFF
CHT	CHT Sensor	-40-500°F	194	194	194
CID	CMP Sensor	0-999 Hz	5-7	10-12	13-17
CKP	CKP Sensor	0-9999 Hz	430-475	900-1000	1140-1220
DPFE	DPFE Sensor	0-5.1v	0.4-0.6	0.4-1.0	0.6-1.1
ECT	ECT Sensor	-40-304°F	160-200	160-200	160-200
FEPS	Flash EEPROM	0-5.1v	0.1	0.1	0.1
FLI	Fuel Level Input	1.7 / 50%	1.7 / 50%	1.7 / 50%	1.7 / 50%
FPM	Fuel Pump Monitor	ON / OFF	ON	ON	ON
FTP	Fuel Tank Pressure	2.6v / 0 psi	2.6v / 0 psi	2.6v / 0 psi	2.6v / 0 psi
GEAR	Gear Position	1-2-3-4	1	3	4
HO2S-11	HO2S-11 (Bank 1)	0.1-1.1v	0.1-1.1	0.1-1.1	0.1-1.1
HO2S-12	HO2S-12 (Bank 1)	0.1-1.1v	0.1-1.1	0.1-1.1	0.1-1.1
HO2S-21	HO2S-21 (Bank 2)	0.1-1.1v	0.1-1.1	0.1-1.1	0.1-1.1
HO2S-22	HO2S-22 (Bank 2)	0.1-1.1v	0.1-1.1	0.1-1.1	0.1-1.1
IAT	IAT Sensor	-40-304°F	50-120	50-120	50-120
KS1	Knock Sensor 1	0	0	0	0
LOAD	Engine Load	0-100%	10-20	20-27	30-45
MAFV	MAF Sensor	0-5.1v	0.6-0.9	0.7-1.0	1.2-2.3
MISF	Misfire Detection	OFF	OFF	OFF	OFF
OCTADJ	Octane Adjustment	CLOSED	CLOSED	CLOSED	CLOSED
OSS	OSS Sensor	0-9999 Hz	0	125-131	245-255
PNP	PNP Switch Signal	Neutral/DR	NEUTRAL	DRIVE	DRIVE
RPM	Engine Speed	0-10K rpm	680-830	1200-1500	1600-1800
TFT	TFT Sensor Signal	-40-304°F	110-210	110-210	110-210
TPV	TP Sensor	0-5.1v	0.53-1.27	1.0-1.3	1.1-1.6
TR1	TR1 Sensor	0v / 11.5v	0v	11.5	11.5
TR2	TR2 Sensor	0v / 11.5v	0v	11.5	11.5
TR3/TR	TR Sensor	0v / 1.7v	0v	1.7	1.7
TR4	TR4 Sensor	0v / 11.5v	0v	11.5	11.5
Vacuum	Engine Vacuum	0-30" Hg	19-20	18-20	8-15
VPWR	Vehicle Power	0-25.5v	12-14	12-14	12-14
VSS	Vehicle Speed (Hertz or MPH)	0-999 Hz (0-159 mph)	0 Hz = 0 mph	65 Hz = 30 mph	125 Hz = 55 mph
4x4L	4x4 Switch Signal	ON / OFF	ON (if on)	OFF	OFF

1999-2003 E-SERIES 4.6L V8 MFI VIN W (A/T) OUTPUTS

PCM PID Acronym	Parameter Identification	PID Range	PID Value at Hot Idle	PID Value at 30	PID Value at 55
CD1	Coil 1 Driver	0-45° dwell	5	6	8
CD2	Coil 2 Driver	0-45° dwell	5	6	8
CD3	Coil 3 Driver	0-45° dwell	5	6	8
CD4	Coil 4 Driver	0-45° dwell	5	6	8
CHTIL	CHT (lamp) Control	ON / OFF	OFF	OFF	OFF
CTO	Clean Tachometer signal	0-9999 Hz	35-49	65-90	90-120
EGR VR	EGR VR Solenoid	0-100%	0	0-40	0-40
EVAP CP	EVAP Purge Valve	0-100%	0-100	0-100	0-100
EVAP CV	EVAP CV Valve	0-100%	0-100	0-100	0-100
EPC	EPC Solenoid	0-300 psi	6	40	40
FP	Fuel Pump Control	ON / OFF	ON	ON	ON
FuelPW1	INJ Pulsewidth - Bank 1	0-999 ms	2.7-4.1	4.5-8.0	5.5-9.0
FuelPW2	INJ Pulsewidth - Bank 2	0-999 ms	2.7-4.1	4.5-8.0	5.5-9.0
FRP	Fuel Rail Pressure	29-32 psi	30-32	30-32	32-35
HTR-11	HO2S-11 Heater	ON / OFF	ON	ON	ON
HTR-12	HO2S-12 Heater	ON / OFF	ON	ON	ON
HTR-21	HO2S-21 Heater	ON / OFF	ON	ON	ON
HTR-22	HO2S-22 Heater	ON / OFF	ON	ON	ON
IAC	Idle Air Control	0-100%	25-32	30-55	60-70
IMTV	Intake Manifold Tuning Valve	0v / VBAT	VBAT	VBAT	VBAT
INJ1	INJ 1 Pulsewidth	0-999 ms	2.7-4.1	4.5-8.0	5.5-9.0
INJ2	INJ 2 Pulsewidth	0-999 ms	2.7-4.1	4.5-8.0	5.5-9.0
INJ3	INJ 3 Pulsewidth	0-999 ms	2.7-4.1	4.5-8.0	5.5-9.0
INJ4	INJ 4 Pulsewidth	0-999 ms	2.7-4.1	4.5-8.0	5.5-9.0
INJ5	INJ 5 Pulsewidth	0-999 ms	2.7-4.1	4.5-8.0	5.5-9.0
INJ6	INJ 6 Pulsewidth	0-999 ms	2.7-4.1	4.5-8.0	5.5-9.0
INJ7	INJ 7 Pulsewidth	0-999 ms	2.7-4.1	4.5-8.0	5.5-9.0
INJ8	INJ 8 Pulsewidth	0-999 ms	2.7-4.1	4.5-8.0	5.5-9.0
LongFT1	Long Term FT (B1)	-20 to 20%	-5 to +5	-5 to +5	-5 to +5
LongFT2	Long Term FT (B2)	-20 to 20%	-5 to +5	-5 to +5	-5 to +5
MIL	MIL (lamp) Control	ON / OFF	OFF	OFF	OFF
SHTFT1	Short Term F/T - Bank 1	-10 to 10%	-5 to +5	-5 to +5	-5 to +5
SHTFT2	Short Term FT - Bank 2	-10 to 10%	-5 to +5	-5 to +5	-5 to +5
SPARK	Spark Advance	-90° to 90°	11-20	15-35	20-39
SS1	Shift Solenoid 1	ON / OFF	ON	ON	ON
SS2	Shift Solenoid 2	ON / OFF	OFF	ON	ON
TCC	TCC Solenoid	0-100%	0	0-45	90-100
TCIL	TCIL (lamp) Control	ON / OFF	OFF	OFF	OFF
VREF	Vehicle Reference	0-5.1v	4.9-5.1	4.9-5.1	4.9-5.1
WAC	A/C WOT Relay	ON / OFF	ON (if on)	OFF	OFF

1990-95 E-SERIES 4.9L I6 MFI VIN Y (A/T, M/T, E4OD)

PCM PID Acronym	Parameter Identification	PID Range	PID Value at Hot Idle	PID Value at 30	PID Value at 55
ACCS	A/C Switch Signal	ON / OFF	ON (if on)	OFF	OFF
CANP	EVAP Purge Valve	ON / OFF	OFF	ON	ON
DPFE	DPFE Sensor (Cal.)	0-5.1v	0.4-0.6	0.4-0.9	0.4-0.9
ECT (°F)	ECT Sensor	-40-304°F	160-200	160-200	160-200
ECTV	ECT Sensor	0-5.1v	0.6	0.6	0.6
EPC	EPC Solenoid (E4OD)	0-300 psi	5	5	15
EVP	EGR Valve Position	0-5.1v	0.3	1.2-2.0	2.5-3.5
EVR	EGR VR Solenoid	0-100%	0	0-40	0-40
FuelPW1	INJ Pulsewidth - Bank 1	0-999 ms	6.8-7.0	9.5-10	12-13
GEAR	Gear Position	P-R-N-D	P-R-N-D	DRIVE 3	DRIVE 4
HO2S-11	HO2S-11 (Bank 1)	0.1-1.1v	0.1-1.1	0.1-1.1	0.1-1.1
IAC	Idle Air Control	0-100%	35	44-50	59-65
IAT (°F)	IAT Sensor	-40-304°F	50-120	50-120	50-120
IATV	IAT Sensor	0-5.1v	1.5-3.5	1.5-3.5	1.5-3.5
LOOP	Loop Status	CL or OL	CLOSED	CLOSED	CLOSED
MAP	MAP Sensor	0-999 Hz	107	114-120	120-130
MLPV	MLP Switch (E4OD)	0-5.1v	0	5	5
PNP	PNP Switch Signal	Neutral/DR	NEUTRAL	DRIVE	DRIVE
RPM	Engine Speed	0-10K rpm	600-700	1050-1150	1840-1940
SHTFT1	Short Term F/T - Bank 1	-10 to 10%	-5 to +5	-5 to +5	-5 to +5
SPARK	Spark Advance	-90° to 90°	17-20	24-28	24-30
TP	TP Sensor	0-5.1v	1.0	1.2-1.3	1.5-1.6
TCC	TCC Sol. (E4OD)	ON / OFF	OFF	ON	ON
TOTV	TOT sensor (E4OD)	0-5.1v	2.10-2.40	2.10-2.40	2.10-2.40
TP	TP Sensor	0-5.1v	1.0	1.2-1.3	1.5-1.6
TP Mode	TP Sensor Mode	C/T or P/T	C/T	P/T	P/T
VPWR	Vehicle Power	0-25.5v	12-14	12-14	12-14
VREF	Vehicle Reference	0-5.1v	4.9-5.1	4.9-5.1	4.9-5.1
VSS	Vehicle Speed	0-159 mph	0	30	55

1996 E-SERIES 4.9L I6 MFI VIN Y (ALL) INPUTS

PCM PID Acronym	Parameter Identification	PID Range	PID Value at Hot Idle	PID Value at 30	PID Value at 55
ACCS	A/C Switch Signal	ON / OFF	ON (if on)	OFF	OFF
CPP	Clutch Pedal Switch	ON / OFF	ON (if in)	ON	ON
DPFE	DPFE Sensor	0-5.1v	0.4-0.6	0.4-0.9	0.6-1.0
ECT	ECT Sensor	-40-304°F	160-200	160-200	160-200
FEPS	Flash EEPROM	0-5.1v	0.1	0.1	0.1
FPM	Fuel Pump Monitor	ON / OFF	ON	ON	ON
HO2S-11	HO2S-11 (Bank 1)	0.1-1.1v	0.1-1.1	0.1-1.1	0.1-1.1
HO2S-12	HO2S-12 (Bank 1)	0.1-1.1v	0.1-1.1	0.1-1.1	0.1-1.1
HO2S-21	HO2S-21 (Bank 2)	0.1-1.1v	0.1-1.1	0.1-1.1	0.1-1.1
IAT	IAT Sensor	-40-304°F	50-120	50-120	50-120
IDM	IDM Signal	0-9999 Hz	32-38	59-65	88-95
KS1	Knock Sensor 1	0-5.1v	0	0	0
LOAD	Engine Load	0-100%	12-14	16-25	30-40
MAFV	MAF Sensor	0-5.1v	0.5-0.7	1.1-1.5	1.7-2.2
MISF	Misfire Detection	ON / OFF	OFF	OFF	OFF
OCTADJ	Octane Adjustment	CL/OPEN	CLOSED	CLOSED	CLOSED
PIP	PIP Sensor	0-9999 Hz	32-38	59-65	88-95
PNP	PNP Switch Signal	Neutral/DR	NEUTRAL	DRIVE	DRIVE
PTO	Power Takeoff Switch	ON / OFF	ON (if on)	OFF	OFF
RPM	Engine Speed	0-10K rpm	680-730	1200-1300	1810-2000
TPV	TP Sensor	0-5.1v	0.53-1.27	0.8-1.1	0.9-1.2
Vacuum	Engine Vacuum	0-30" Hg	19-20	18-20	8-15
VPWR	Vehicle Power	0-25.5v	12-14	12-14	12-14
VSS	Vehicle Speed (Hertz or MPH)	0-999 Hz (0-159 mph)	0 Hz = 0 mph	65 Hz = 30 mph	125 Hz = 55 mph
WACA	A/C WOT Relay Monitor	ON / OFF	OFF (with A/C on)	OFF	OFF

1996 E-SERIES 4.9L I6 MFI VIN Y (ALL) OUTPUTS

PCM PID Acronym	Parameter Identification	PID Range	PID Value at Hot Idle	PID Value at 30	PID Value at 55
AIRB	AIR System Bypass	0v / VBAT	0.1	0.1	0.1
AIRD	AIR System Divert	0v / VBAT	VBAT	VBAT	VBAT
EGR VR	EGR VR Solenoid	0-100%	0	0-40	0-40
FP	Fuel Pump Control	ON / OFF	ON	ON	ON
FuelPW1	INJ Pulsewidth - Bank 1	0-999 ms	4.9-5.1	5.3-10.0	9.4-13.0
FuelPW2	INJ Pulsewidth - Bank 2	0-999 ms	4.9-5.1	5.3-10.0	9.4-13.0
HTR-11	HO2S-11 Heater	ON / OFF	ON	ON	ON
HTR-12	HO2S-12 Heater	ON / OFF	ON	ON	ON
HTR-21	HO2S-21 Heater	ON / OFF	ON	ON	ON
IAC	Idle Air Control	0-100%	35	44-50	59-65
INJ1	INJ 1 Pulsewidth	0-999 ms	4.9-5.1	5.3-10.0	9.4-13.0
INJ2	INJ 2 Pulsewidth	0-999 ms	4.9-5.1	5.3-10.0	9.4-13.0
INJ3	INJ 3 Pulsewidth	0-999 ms	4.9-5.1	5.3-10.0	9.4-13.0
INJ4	INJ 4 Pulsewidth	0-999 ms	4.9-5.1	5.3-10.0	9.4-13.0
INJ5	INJ 5 Pulsewidth	0-999 ms	4.9-5.1	5.3-10.0	9.4-13.0
INJ6	INJ 6 Pulsewidth	0-999 ms	4.9-5.1	5.3-10.0	9.4-13.0
LongFT1	Long Term FT - Bank 1	-20 to 20%	-5 to +5	-5 to +5	-5 to +5
LongFT2	Long Term FT - Bank 2	-20 to 20%	-5 to +5	-5 to +5	-5 to +5
MIL	MIL (lamp) Control	ON / OFF	OFF	OFF	OFF
SHTFT1	Short Term F/T - Bank 1	-10 to 10%	-5 to +5	-5 to +5	-5 to +5
SHTFT2	Short Term FT - Bank 2	-10 to 10%	-5 to +5	-5 to +5	-5 to +5
SPARK	Spark Advance	-90º to 90º	13-16	17-20	17-22
SPOUT	Spark Output Signal	0-9999 Hz	32-38	59-65	88-95
VMV	EVAP Solenoid	0-10	0-10	0-10	0-10
VREF	Vehicle Reference	0-5.1v	4.9-5.1	4.9-5.1	4.9-5.1
WAC	A/C WOT Relay Control	ON / OFF	ON (AC on at WOT)	OFF	OFF

1996 E-SERIES 4.9L I6 MFI VIN Y (E4OD) INPUTS

PCM PID Acronym	Parameter Identification	PID Range	PID Value at Hot Idle	PID Value at 30	PID Value at 55
4x4L	4x4 Switch Signal	ON / OFF	ON (if on)	OFF	OFF
ACCS	A/C Switch Signal	ON / OFF	ON (if on)	OFF	OFF
BPP	Brake Position	ON / OFF	ON (if on)	OFF	OFF
DPFE	DPFE Sensor	0-5.1v	0.4-0.6	0.4-0.9	0.4-0.9
ECT	ECT Sensor	-40-304°F	160-200	160-200	160-200
FEPS	Flash EEPROM	0-5.1v	0.1	0.1	0.1
FPM	Fuel Pump Monitor	ON / OFF	ON	ON	ON
GEAR	Gear Position	1-2-3-4	1	3	4
HO2S-11	HO2S-11 (Bank 1)	0.1-1.1v	0.1-1.1	0.1-1.1	0.1-1.1
HO2S-12	HO2S-12 (Bank 1)	0.1-1.1v	0.1-1.1	0.1-1.1	0.1-1.1
HO2S-21	HO2S-21 (Bank 2)	0.1-1.1v	0.1-1.1	0.1-1.1	0.1-1.1
IAT	IAT Sensor	-40-304°F	50-120	50-120	50-120
IDM	IDM Signal	0-9999 Hz	35-42	58-69	71-82
KS1	Knock Sensor 1	0-5.1v	0	0	0
LOAD	Engine Load	0-100%	14-16	21-32	33-42
MAFV	MAF Sensor	0-5.1v	0.7-0.9	1.2-1.8	1.7-2.0
MISF	Misfire Detection	ON / OFF	OFF	OFF	OFF
OCTADJ	Octane Adjustment	CL/OPEN	CLOSED	CLOSED	CLOSED
PIP	PIP Sensor	0-9999 Hz	35-42	58-69	71-82
PNP	PNP Switch Signal	Neutral/DR	NEUTRAL	DRIVE	DRIVE
PTO	Power Takeoff Switch	ON / OFF	ON (if on)	OFF	OFF
RPM	Engine Speed	0-10K rpm	760-830	1200-1270	1510-1570
TCS	TCS Switch	ON / OFF	OFF	OFF	OFF
TFT	TFT Sensor Signal	-40-304°F	110-210	110-210	110-210
TPV	TP Sensor	0-5.1v	0.53-1.27	1.0-1.3	1.2-1.7
TRV/TR	TR Sensor	0v / 1.7v	0v	1.7	1.7
Vacuum	Engine Vacuum	0-30" Hg	19-20	18-20	8-15
VPWR	Vehicle Power	0-25.5v	12-14	12-14	12-14
VSS	Vehicle Speed (Hertz or MPH)	0-999 Hz (0-159 mph)	0 Hz = 0 mph	65 Hz = 30 mph	125 Hz = 55 mph
WACA	A/C WOT Relay Monitor	ON / OFF	ON (AC on at WOT)	OFF	OFF

1996 E-SERIES 4.9L I6 MFI VIN Y (E4OD) OUTPUTS

PCM PID Acronym	Parameter Identification	PID Range	PID Value at Hot Idle	PID Value at 30	PID Value at 55
AIRB	AIR System Bypass	0v / VBAT	0.1	0.1	0.1
AIRD	AIR System Divert	0v / VBAT	VBAT	VBAT	VBAT
CCS	Coast Clutch Solenoid Control	ON / OFF	OFF	OFF	OFF
EGR VR	EGR VR Solenoid	0-100%	0	0-40	0-40
EPC	EPC Solenoid	0-300 psi	5	6	7
FP	Fuel Pump Control	ON / OFF	ON	ON	ON
FuelPW1	INJ Pulsewidth - Bank 1	0-999 ms	4.9-5.3	5.5-10.0	9.5-13.0
FuelPW2	INJ Pulsewidth - Bank 2	0-999 ms	4.9-5.3	5.5-10.0	9.5-13.0
HTR-11	HO2S-11 Heater	ON / OFF	ON	ON	ON
HTR-12	HO2S-12 Heater	ON / OFF	ON	ON	ON
HTR-21	HO2S-21 Heater	ON / OFF	ON	ON	ON
IAC	Idle Air Control	0-100%	30-34	30-55	58-61
INJ1	INJ 1 Pulsewidth	0-999 ms	4.9-5.3	5.5-10.0	9.5-13.0
INJ2	INJ 2 Pulsewidth	0-999 ms	4.9-5.3	5.5-10.0	9.5-13.0
INJ3	INJ 3 Pulsewidth	0-999 ms	4.9-5.3	5.5-10.0	9.5-13.0
INJ4	INJ 4 Pulsewidth	0-999 ms	4.9-5.3	5.5-10.0	9.5-13.0
INJ5	INJ 5 Pulsewidth	0-999 ms	4.9-5.3	5.5-10.0	9.5-13.0
INJ6	INJ 6 Pulsewidth	0-999 ms	4.9-5.3	5.5-10.0	9.5-13.0
LongFT1	Long Term FT - Bank 1	-20 to 20%	-5 to +5	-5 to +5	-5 to +5
LongFT2	Long Term FT - Bank 2	-20 to 20%	-5 to +5	-5 to +5	-5 to +5
MIL	MIL (lamp) Control	ON / OFF	OFF	OFF	OFF
SHTFT1	Short Term F/T - Bank 1	-10 to 10%	-5 to +5	-5 to +5	-5 to +5
SHTFT2	Short Term FT - Bank 2	-10 to 10%	-5 to +5	-5 to +5	-5 to +5
SPARK	Spark Advance	-90° to 90°	13-16	17-20	17-22
SPOUT	Spark Output Signal	0-9999 Hz	47-52	82-88	105-110
SS1	Shift Solenoid 1	ON / OFF	ON	OFF	OFF
SS2	Shift Solenoid 2	ON / OFF	OFF	ON	OFF
TCC	TCC Solenoid	ON / OFF	OFF	ON	ON
TCIL	TCIL (lamp) Control	ON / OFF	OFF	OFF	OFF
VMV	EVAP Solenoid	0-10	0-10	0-10	0-10
VREF	Vehicle Reference	0-5.1v	4.9-5.1	4.9-5.1	4.9-5.1
WAC	A/C WOT Relay Control	ON / OFF	ON (AC on at WOT)	OFF	OFF

1990-95 E-SERIES 5.0L V8 MFI VIN N (A/T, M/T, E4OD)

PCM PID Acronym	Parameter Identification	PID Range	PID Value at Hot Idle	PID Value at 30	PID Value at 55
ACCS	A/C Switch Signal	ON / OFF	ON (if on)	OFF	OFF
CANP	EVAP Purge Valve	ON / OFF	OFF	ON	ON
CPP	Clutch Pedal Switch	ON / OFF	ON (if in)	ON	ON
DPFE	DPFE Sensor ('95)	0-5.1v	0.4-0.6	0.4-0.9	0.4-0.9
ECT (°F)	ECT Sensor	-40-304°F	160-200	160-200	160-200
ECTV	ECT Sensor	0-5.1v	0.6	0.6	0.6
EPC	EPC Solenoid (E4OD)	0-300 psi	4	4	14
EVP	EGR Valve Position	0-5.1v	0.4	0.4-0	3.5-4.5
EVR	EGR VR Solenoid	0-100%	0	0-40	0-40
FuelPW1	INJ Pulsewidth - Bank 1	0-999 ms	4.4-5.0	6.4-7.8	9.8-12.0
GEAR	Gear Position	P-R-N-D	P-R-N-D	DRIVE 3	DRIVE 4
IAC	Idle Air Control	0-100%	30-34	43-48	58-61
IAT (°F)	IAT Sensor	-40-304°F	50-120	50-120	50-120
IATV	IAT Sensor	0-5.1v	1.5-3.5	1.5-3.5	1.5-3.5
LOOP	Loop Status	CL or OL	CLOSED	CLOSED	CLOSED
MLPV	MLP Switch (E4OD)	0-5.1v	0	5	5
MAP	MAP Sensor	0-999 Hz	103-105	112-120	122-140
HO2S-11	HO2S-11 (Bank 1)	0.1-1.1v	0.1-1.1	0.1-1.1	0.1-1.1
PNP	PNP Switch Signal	Neutral/DR	NEUTRAL	DRIVE	DRIVE
RPM	Engine Speed	0-10K rpm	680-780	1240-1340	1650-1750
SHTFT1	Short Term F/T - Bank 1	-10 to 10%	-5 to +5	-5 to +5	-5 to +5
SPARK	Spark Advance	-90° to 90°	14-20	28-36	30-40
TP	TP Sensor	0-5.1v	1.0	1.2-1.3	1.5-1.6
TCC	TCC Solenoid	ON / OFF	OFF	ON	ON
TOTV	TOT sensor (E4OD)	0-5.1v	2.10-2.40	2.10-2.40	2.10-2.40
TP	TP Sensor	0-5.1v	1.0	1.2-1.3	1.5-1.6
TP Mode	TP Sensor Mode	C/T or P/T	C/T	P/T	P/T
VPWR	Vehicle Power	0-25.5v	12-14	12-14	12-14
VREF	Vehicle Reference	0-5.1v	4.9-5.1	4.9-5.1	4.9-5.1
VSS	Vehicle Speed	0-159 mph	0	30	55

1996 E-SERIES 5.0L V8 MFI VIN N (ALL) INPUTS

PCM PID Acronym	Parameter Identification	PID Range	PID Value at Hot Idle	PID Value at 30	PID Value at 55
4x4L	4x4 Switch Signal	ON / OFF	ON (if on)	OFF	OFF
ACCS	A/C Switch Signal	ON / OFF	ON (if on)	OFF	OFF
BPP	Brake Position	ON / OFF	ON (if on)	OFF	OFF
CPP	Clutch Pedal Switch	ON / OFF	ON (if in)	ON	ON
DPFE	DPFE Sensor	0-5.1v	0.4-0.6	0.4-0.9	0.4-0.9
ECT	ECT Sensor	-40-304°F	160-200	160-200	160-200
FEPS	Flash EEPROM	0-5.1v	0.1	0.1	0.1
FPM	Fuel Pump Monitor	ON / OFF	ON	ON	ON
IAT	IAT Sensor	-40-304°F	50-120	50-120	50-120
GEAR	Gear Position	1-2-3-4	1	3	4
HO2S-11	HO2S-11 (Bank 1)	0.1-1.1v	0.1-1.1	0.1-1.1	0.1-1.1
HO2S-12	HO2S-12 (Bank 1)	0.1-1.1v	0.1-1.1	0.1-1.1	0.1-1.1
HO2S-21	HO2S-21 (Bank 2)	0.1-1.1v	0.1-1.1	0.1-1.1	0.1-1.1
IDM	IDM Signal	0-9999 Hz	47-52	82-88	105-110
KS1	Knock Sensor 1	0-5.1v	0	0	0
LOAD	Engine Load	0-100%	14-16	19-25	26-35
MAFV	MAF Sensor	0-5.1v	0.7-0.9	1.1-1.6	1.7-2.4
MISF	Misfire Detection	ON / OFF	OFF	OFF	OFF
OCTADJ	Octane Adjustment	CL/OPEN	CLOSED	CLOSED	CLOSED
OSS	OSS Sensor	0-9999 Hz	0	122-133	235-250
PIP	PIP Sensor	0-9999 Hz	47-52	82-88	105-110
PNP	PNP Switch Signal	Neutral/DR	NEUTRAL	DRIVE	DRIVE
PTO	Power Takeoff Switch	ON / OFF	ON (if on)	OFF	OFF
RPM	Engine Speed	0-10K rpm	760-830	1200-1270	1590-1675
TCS	TCS Switch	ON / OFF	OFF	OFF	OFF
TFT	TFT Sensor Signal	-40-304°F	110-210	110-210	110-210
TPV	TP Sensor	0-5.1v	0.53-1.27	0.8-1.0	1.0-1.3
TRV/TR	TR Sensor	0v / 1.7v	0	1.7	1.7
Vacuum	Engine Vacuum	0-30" Hg	19-20	18-20	8-15
VPWR	Vehicle Power	0-25.5v	12-14	12-14	12-14
VSS	Vehicle Speed (Hertz or MPH)	0-999 Hz (0-159 mph)	0 Hz = 0 mph	65 Hz = 30 mph	125 Hz = 55 mph

1996 E-SERIES 5.0L V8 MFI VIN N (ALL) OUTPUTS

PCM PID Acronym	Parameter Identification	PID Range	PID Value at Hot Idle	PID Value at 30	PID Value at 55
AIRB	AIR System Bypass	0v / VBAT	0.1	0.1	0.1
AIRD	AIR System Divert	0v / VBAT	VBAT	VBAT	VBAT
EGR VR	EGR VR Solenoid	0-100%	0%	0-40	35-40
EPC	EPC Solenoid	0-300 psi	4	5	5
FP	Fuel Pump Control	ON / OFF	ON	ON	ON
FuelPW1	INJ Pulsewidth - Bank 1	0-999 ms	3.2-3.8	4.1-6.9	6.5-12
FuelPW2	INJ Pulsewidth - Bank 2	0-999 ms	3.2-3.8	4.1-6.9	6.5-12
HTR-11	HO2S-11 Heater	ON / OFF	ON	ON	ON
HTR-12	HO2S-12 Heater	ON / OFF	ON	ON	ON
HTR-21	HO2S-21 Heater	ON / OFF	ON	ON	ON
IAC	Idle Air Control	0-100%	30-34	43-48	58-61
INJ1	INJ 1 Pulsewidth	0-999 ms	3.2-3.8	4.1-6.9	6.5-12
INJ2	INJ 2 Pulsewidth	0-999 ms	3.2-3.8	4.1-6.9	6.5-12
INJ3	INJ 3 Pulsewidth	0-999 ms	3.2-3.8	4.1-6.9	6.5-12
INJ4	INJ 4 Pulsewidth	0-999 ms	3.2-3.8	4.1-6.9	6.5-12
INJ5	INJ 5 Pulsewidth	0-999 ms	3.2-3.8	4.1-6.9	6.5-12
INJ6	INJ 6 Pulsewidth	0-999 ms	3.2-3.8	4.1-6.9	6.5-12
INJ7	INJ 7 Pulsewidth	0-999 ms	3.2-3.8	4.1-6.9	6.5-12
INJ8	INJ 8 Pulsewidth	0-999 ms	3.2-3.8	4.1-6.9	6.5-12
LongFT1	Long Term FT - Bank 1	-20 to 20%	-5 to +5	-5 to +5	-5 to +5
LongFT2	Long Term FT - Bank 2	-20 to 20%	-5 to +5	-5 to +5	-5 to +5
MIL	MIL (lamp) Control	ON / OFF	OFF	OFF	OFF
SHTFT1	Short Term F/T - Bank 1	-10 to 10%	-5 to +5	-5 to +5	-5 to +5
SHTFT2	Short Term FT - Bank 2	-10 to 10%	-5 to +5	-5 to +5	-5 to +5
SPARK	Spark Advance	-90º to 90º	12-17	35-40	28-37
SPOUT	Spark Output Signal	0-9999 Hz	47-52	82-88	105-110
SS1	Shift Solenoid 1	ON / OFF	ON	OFF	OFF
SS2	Shift Solenoid 2	ON / OFF	OFF	ON	OFF
TCC	TCC Solenoid	0-100%	0	0-40	90-100
TCIL	TCIL (lamp) Control	ON / OFF	OFF	OFF	OFF
VMV	EVAP Solenoid	0-10	0-10	0-10	0-10
VREF	Voltage Reference	0-5.1v	4.9-5.1	4.9-5.1	4.9-5.1

1996 E-SERIES 5.0L V8 MFI VIN N (E4OD) INPUTS

PCM PID Acronym	Parameter Identification	PID Range	PID Value at Hot Idle	PID Value at 30	PID Value at 55
4x4L	4x4 Switch Signal	ON / OFF	ON (if on)	OFF	OFF
ACCS	A/C Switch Signal	ON / OFF	ON (if on)	OFF	OFF
BPP	Brake Position	ON / OFF	ON (if on)	OFF	OFF
DPFE	DPFE Sensor	0-5.1v	0.4-0.6	0.4-0.9	0.4-0.9
ECT	ECT Sensor	-40-304°F	160-200	160-200	160-200
FEPS	Flash EEPROM	0-5.1v	0.1	0.1	0.1
FPM	Fuel Pump Monitor	ON / OFF	ON	ON	ON
IAT	IAT Sensor	-40-304°F	50-120	50-120	50-120
GEAR	Gear Position	1-2-3-4	1	3	4
HO2S-11	HO2S-11 (Bank 1)	0.1-1.1v	0.1-1.1	0.1-1.1	0.1-1.1
HO2S-12	HO2S-12 (Bank 1)	0.1-1.1v	0.1-1.1	0.1-1.1	0.1-1.1
HO2S-21	HO2S-21 (Bank 2)	0.1-1.1v	0.1-1.1	0.1-1.1	0.1-1.1
HO2S-22	HO2S-22 (Bank 2)	0.1-1.1v	0.1-1.1	0.1-1.1	0.1-1.1
IDM	IDM Signal	0-9999 Hz	47-52	82-88	105-110
KS1	Knock Sensor 1	0-5.1v	0	0	0
LOAD	Engine Load	0-100%	14-16	19-25	26-35
MAFV	MAF Sensor	0-5.1v	0.7-0.9	1.1-1.6	1.7-2.4
MISF	Misfire Detection	ON / OFF	OFF	OFF	OFF
OCTADJ	Octane Adjustment	CL/OPEN	CLOSED	CLOSED	CLOSED
PIP	PIP Sensor	0-9999 Hz	47-52	82-88	105-110
PNP	PNP Switch Signal	Neutral/DR	NEUTRAL	DRIVE	DRIVE
PTO	Power Takeoff Switch	ON / OFF	ON (if on)	OFF	OFF
RPM	Engine Speed	0-10K rpm	760-830	1200-1270	1600-1650
TCS	TCS Switch	ON / OFF	OFF	OFF	OFF
TFT	TFT Sensor Signal	-40-304°F	110-210	110-210	110-210
TPV	TP Sensor	0-5.1v	0.53-1.27	0.8-1.0	1.0-1.3
TRV/TR	TR Sensor	0v / 1.7v	0v	1.7	1.7
Vacuum	Engine Vacuum	0-30" Hg	19-20	18-20	8-15
VPWR	Vehicle Power	0-25.5v	12-14	12-14	12-14
VSS	Vehicle Speed (Hertz or MPH)	0-999 Hz (0-159 mph)	0 Hz = 0 mph	65 Hz = 30 mph	125 Hz = 55 mph

1996 E-SERIES 5.0L V8 MFI VIN N (E4OD) OUTPUTS

PCM PID Acronym	Parameter Identification	PID Range	PID Value at Hot Idle	PID Value at 30	PID Value at 55
AIRB	AIR System Bypass	0v / VBAT	0.1	0.1	0.1
AIRD	AIR System Divert	0v / VBAT	VBAT	VBAT	VBAT
CCS	Coast Clutch Solenoid Control	0v / VBAT	VBAT	VBAT	VBAT
EGR VR	EGR VR Solenoid	0-100%	0	0-40	35-40
EVAPDC	EVAP Solenoid	0-100%	0-100	0-100	0-100
EPC	EPC Solenoid	0-300 psi	4	5	5
FP	Fuel Pump Control	ON / OFF	ON	ON	ON
FuelPW1	INJ Pulsewidth - Bank 1	0-999 ms	3.2-3.8	4.1-6.9	6.5-12
FuelPW2	INJ Pulsewidth - Bank 2	0-999 ms	3.2-3.8	4.1-6.9	6.5-12
HTR-11	HO2S-11 Heater	ON / OFF	ON	ON	ON
HTR-12	HO2S-12 Heater	ON / OFF	ON	ON	ON
HTR-21	HO2S-21 Heater	ON / OFF	ON	ON	ON
IAC	Idle Air Control	0-100%	30-34	43-48	58-61
INJ1	INJ 1 Pulsewidth	0-999 ms	3.2-3.8	4.1-6.9	6.5-12
INJ2	INJ 2 Pulsewidth	0-999 ms	3.2-3.8	4.1-6.9	6.5-12
INJ3	INJ 3 Pulsewidth	0-999 ms	3.2-3.8	4.1-6.9	6.5-12
INJ4	INJ 4 Pulsewidth	0-999 ms	3.2-3.8	4.1-6.9	6.5-12
INJ5	INJ 5 Pulsewidth	0-999 ms	3.2-3.8	4.1-6.9	6.5-12
INJ6	INJ 6 Pulsewidth	0-999 ms	3.2-3.8	4.1-6.9	6.5-12
INJ7	INJ 7 Pulsewidth	0-999 ms	3.2-3.8	4.1-6.9	6.5-12
INJ8	INJ 8 Pulsewidth	0-999 ms	3.2-3.8	4.1-6.9	6.5-12
LongFT1	Long Term FT - Bank 1	-20 to 20%	-5 to +5	-5 to +5	-5 to +5
LongFT2	Long Term FT - Bank 2	-20 to 20%	-5 to +5	-5 to +5	-5 to +5
MIL	MIL (lamp) Control	ON / OFF	OFF	OFF	OFF
SHTFT1	Short Term F/T - Bank 1	-10 to 10%	-5 to +5	-5 to +5	-5 to +5
SHTFT2	Short Term FT - Bank 2	-10 to 10%	-5 to +5	-5 to +5	-5 to +5
SPARK	Spark Advance	-90° to 90°	12-17	35-40	28-37
SPOUT	Spark Output Signal	0-9999 Hz	47-52	82-88	105-110
SS1	Shift Solenoid 1	ON / OFF	ON	OFF	OFF
SS2	Shift Solenoid 2	ON / OFF	OFF	ON	OFF
TCC	TCC Solenoid	ON / OFF	OFF	ON	ON
TCIL	TCIL (lamp) Control	ON / OFF	OFF	OFF	OFF
VREF	Vehicle Reference	0-5.1v	4.9-5.1	4.9-5.1	4.9-5.1

1997-2003 E-SERIES 5.4L V8 VIN L (E4OD) INPUTS

PCM PID Acronym	Parameter Identification	PID Range	PID Value at Hot Idle	PID Value at 30	PID Value at 55
AIR-M	Air System Monitor	ON / OFF	OFF	OFF	OFF
ACCS	A/C Switch Signal	OFF	ON	OFF	OFF
BPP	Brake Position	ON / OFF	ON (if on)	OFF	OFF
CHT	CHT Sensor	-40-500ºF	194	194	194
CID	CMP Sensor	0-999 Hz	6	11	11
CKP	CKP Sensor	0-9999 Hz	411	800-840	1000
DPFE	DPFE Sensor	0-5.1v	0.2-1.3	0.2-4.5	0.2-4.5
ECT	ECT Sensor	-40-304ºF	160-200	160-200	160-200
FEPS	Flash EEPROM	0-5.1v	0.1	0.1	0.1
FLI	Fuel Level Input	1.7 / 50%	1.7 / 50%	1.7 / 50%	1.7 / 50%
FPM	Fuel Pump Monitor	ON / OFF	ON	ON	ON
FTP	Fuel Tank Pressure	2.6v / 0 psi	2.6v / 0 psi	2.6v / 0 psi	2.6v / 0 psi
GEAR	Gear Position	1-2-3-4	1	3	4
HO2S-11	HO2S-11 (Bank 1)	0.1-1.1v	0.1-1.1	0.1-1.1	0.1-1.1
HO2S-12	HO2S-12 (Bank 1)	0.1-1.1v	0.1-1.1	0.1-1.1	0.1-1.1
HO2S-21	HO2S-21 (Bank 2)	0.1-1.1v	0.1-1.1	0.1-1.1	0.1-1.1
HO2S-22	HO2S-22 (Bank 2)	0.1-1.1v	0.1-1.1	0.1-1.1	0.1-1.1
IAT	IAT Sensor	-40-304ºF	50-120	50-120	50-120
KS1	Knock Sensor 1	0-5.1v	0	0	0
LOAD	Engine Load	0-100%	14-16	19-25	26-35
MAFV	MAF Sensor	0-5.1v	0.7-0.9	1.1-1.6	1.7-2.4
MISF	Misfire Detection	ON / OFF	OFF	OFF	OFF
OCTADJ	Octane Adjustment	CL/OPEN	CLOSED	CLOSED	CLOSED
PNP	PNP Switch Signal	Neutral/DR	NEUTRAL	DRIVE	DRIVE
RPM	Engine Speed	0-10K rpm	760-830	1200-1270	1600-1650
TCS	TCS Switch	ON / OFF	OFF	OFF	OFF
TFT	TFT Sensor Signal	-40-304ºF	110-210	110-210	110-210
TPV	TP Sensor	0-5.1v	0.53-1.27	0.8-1.0	1.0-1.3
TR1	TR1 Sensor	0v / 11.5v	0v	11.5	11.5
TR2	TR2 Sensor	0v / 11.5v	0v	11.5	11.5
TR3/TR	TR3A Sensor	0v / 1.7v	0v	1.7	1.7
TR4	TR4 Sensor	0v / 11.5v	0v	11.5	11.5
Vacuum	Engine Vacuum	0-30" Hg	19-20	18-20	8-15
VPWR	Vehicle Power	0-25.5v	12-14	12-14	12-14
VSS	Vehicle Speed (Hertz or MPH)	0-999 Hz (0-159 mph)	0 Hz = 0 mph	65 Hz = 30 mph	125 Hz = 55 mph

1997-2003 E-SERIES 5.4L V8 VIN L (E4OD) OUTPUTS

PCM PID Acronym	Parameter Identification	PID Range	PID Value at Hot Idle	PID Value at 30	PID Value at 55
AIRD	AIR System Divert	0v / VBAT	VBAT	VBAT	VBAT
AIR	AIR System Monitor	ON / OFF	OFF	OFF	OFF
CCS	Coast Clutch Solenoid Control	0v / VBAT	VBAT	VBAT	VBAT
CD1-CD8	COP Driver 1-8	0-45	5	6	8
CHTIL	CHT (lamp) Control	ON / OFF	OFF	OFF	OFF
CTO	Clean Tachometer signal	0-9999 Hz	46	90	115
EGR VR	EGR VR Solenoid	0-100%	0	0-40	35-40
EVAP CP	EVAP Purge Valve	0-100%	0-100	0-100	0-100
EVAP CV	EVAP CV Valve	0-100%	0-100	0-100	0-100
EPC	EPC Solenoid	0-300 psi	4	4	5
FP	Fuel Pump Control	ON / OFF	ON	ON	ON
FuelPW1	INJ Pulsewidth - Bank 1	0-999 ms	3.2-3.8	4.1-6.9	6.5-12
FuelPW2	INJ Pulsewidth - Bank 2	0-999 ms	3.2-3.8	4.1-6.9	6.5-12
HTR-11	HO2S-11 Heater	ON / OFF	ON	ON	ON
HTR-12	HO2S-12 Heater	ON / OFF	ON	ON	ON
HTR-21	HO2S-21 Heater	ON / OFF	ON	ON	ON
HTR-22	HO2S-22 Heater	ON / OFF	ON	ON	ON
IAC	Idle Air Control	0-100%	30-34	43-48	58-61
INJ1	INJ 1 Pulsewidth	0-999 ms	3.2-3.8	4.1-6.9	6.5-12
INJ2	INJ 2 Pulsewidth	0-999 ms	3.2-3.8	4.1-6.9	6.5-12
INJ3	INJ 3 Pulsewidth	0-999 ms	3.2-3.8	4.1-6.9	6.5-12
INJ4	INJ 4 Pulsewidth	0-999 ms	3.2-3.8	4.1-6.9	6.5-12
INJ5	INJ 5 Pulsewidth	0-999 ms	3.2-3.8	4.1-6.9	6.5-12
INJ6	INJ 6 Pulsewidth	0-999 ms	3.2-3.8	4.1-6.9	6.5-12
INJ7	INJ 7 Pulsewidth	0-999 ms	3.2-3.8	4.1-6.9	6.5-12
INJ8	INJ 8 Pulsewidth	0-999 ms	3.2-3.8	4.1-6.9	6.5-12
LongFT1	Long Term FT (B1)	-20 to 20%	-5 to +5	-5 to +5	-5 to +5
LongFT2	Long Term FT (B2)	-20 to 20%	-5 to +5	-5 to +5	-5 to +5
MIL	MIL (lamp) Control	ON / OFF	OFF	OFF	OFF
SHTFT1	Short Term F/T - Bank 1	-10 to 10%	-5 to +5	-5 to +5	-5 to +5
SHTFT2	Short Term FT - Bank 2	-10 to 10%	-5 to +5	-5 to +5	-5 to +5
SS1	Shift Solenoid 1	ON / OFF	ON	OFF	OFF
SS2	Shift Solenoid 2	ON / OFF	OFF	ON	OFF
SPARK	Spark Advance	-90° to 90°	12-17	35-40	28-37
TCC	TCC Solenoid	ON / OFF	OFF	ON	ON
VREF	Vehicle Reference	0-5.1v	4.9-5.1	4.9-5.1	4.9-5.1

1997-2003 E-SERIES 5.4L V8 CNG VIN M (E4OD) INPUTS

PCM PID Acronym	Parameter Identification	PID Range	PID Value at Hot Idle	PID Value at 30	PID Value at 55
ACCS	A/C Switch Signal	ON / OFF	ON (if on)	OFF	OFF
BPP	Brake Position	ON / OFF	ON (if on)	OFF	OFF
CHT	CHT Sensor	-40-500°F	194	194	194
CID	CMP Sensor	0-999 Hz	6-8	9-12	15-17
CKP	CKP Sensor	0-9999 Hz	440-490	810-870	1088-1089
EFTA	EFT Sensor	-40-275F	50-120	50-120	50-120
FEPS	Flash EEPROM	0-5.1v	0.1	0.1	0.1
FPM	Fuel Pump Monitor	ON / OFF	ON	ON	ON
FSV-M	Fuel Solenoid Valve	ON / OFF	ON	ON	ON
IAT	IAT Sensor	-40-304°F	50-120	50-120	50-120
FRP	FRP Sensor	0-300 psi	90-100	90-100	90-100
GEAR	Gear Position	1-2-3-4	1	3	4
HO2S-11	HO2S-11 (Bank 1)	0.1-1.1v	0.1-1.1	0.1-1.1	0.1-1.1
HO2S-21	HO2S-21 (Bank 2)	0.1-1.1v	0.1-1.1	0.1-1.1	0.1-1.1
LOAD	Engine Load	0-100%	13-19	26-33	38-46
MAFV	MAF Sensor	0-5.1v	0.7-0.9	0.9-1.7	1.2-2.4
MISF	Misfire Detection	ON / OFF	OFF	OFF	OFF
PNP	PNP Switch Signal	Neutral/DR	NEUTRAL	DRIVE	DRIVE
PTO	Power Takeoff Switch	OFF	ON (if on)	OFF	OFF
RPM	Engine Speed	0-10K rpm	900-930	1470-1490	1840-1860
TCS	TCS Switch	ON / OFF	OFF	OFF	OFF
TPV	TP Sensor	0-5.1v	0.53-1.27	0.8-1.2	0.9-1.6v
TR1	TR1 Sensor	0v / 11.5v	0v	11.5	11.5
TR2	TR2 Sensor	0v / 11.5v	0v	11.5	11.5
TR3/TR	TR3A Sensor	0v / 1.7v	0v	1.7	1.7
TR4	TR4 Sensor	0v / 11.5v	0v	11.5	11.5
Vacuum	Engine Vacuum	0-30" Hg	19-20	18-20	8-15
VPWR	Vehicle Power	0-25.5v	12-14	12-14	12-14
VSS	Vehicle Speed (Hertz or MPH)	0-999 Hz (0-159 mph)	0 Hz = 0 mph	65 Hz = 30 mph	125 Hz = 55 mph

<u>1997-2003 E-SERIES 5.4L V8 CNG VIN M (E4OD) OUTPUTS</u>

PCM PID Acronym	Parameter Identification	PID Range	PID Value at Hot Idle	PID Value at 30	PID Value at 55
CCS	Coast Clutch Solenoid Control	ON / OFF	OFF	OFF	OFF
CD1-CD8	COP Driver 1-8	0-45° dwell	5	6	8
CHTIL	CHT (lamp) Control	ON / OFF	OFF	OFF	OFF
CTO	Clean Tachometer signal	0-9999 Hz	50-57	90-100	115-125
EPC	EPC Solenoid	0-300 psi	5	6	11
FP	Fuel Pump Control	ON / OFF	ON	ON	ON
FSV	Fuel Solenoid Valve	ON / OFF	ON	ON	ON
FuelPW1	INJ Pulsewidth - Bank 1	0-999 ms	3.9-5.5	4.7-12.0	4.7-12.0
FuelPW2	INJ Pulsewidth - Bank 2	0-999 ms	3.9-5.5	4.7-12.0	4.7-12.0
HTR-11	HO2S-11 Heater	ON / OFF	ON	ON	ON
HTR-21	HO2S-21 Heater	ON / OFF	ON	ON	ON
IAC	Idle Air Control	0-100%	30-34	43-48	58-61
IMRC	Intake Manifold Runner Control	0-100%	0	0	0
INJ1	INJ 1 Pulsewidth	0-999 ms	3.9-4.6	4.7-12.0	4.7-12.0
INJ2	INJ 2 Pulsewidth	0-999 ms	3.9-4.6	4.7-12.0	4.7-12.0
INJ3	INJ 3 Pulsewidth	0-999 ms	3.9-4.6	4.7-12.0	4.7-12.0
INJ4	INJ 4 Pulsewidth	0-999 ms	3.9-4.6	4.7-12.0	4.7-12.0
INJ5	INJ 5 Pulsewidth	0-999 ms	3.9-4.6	4.7-12.0	4.7-12.0
INJ6	INJ 6 Pulsewidth	0-999 ms	3.9-4.6	4.7-12.0	4.7-12.0
INJ7	INJ 7 Pulsewidth	0-999 ms	3.9-4.6	4.7-12.0	4.7-12.0
INJ8	INJ 8 Pulsewidth	0-999 ms	3.9-4.6	4.7-12.0	4.7-12.0
LongFT1	Long Term FT - Bank 1	-20 to 20%	-5 to +5	-5 to +5	-5 to +5
LongFT2	Long Term FT - Bank 2	-20 to 20%	-5 to +5	-5 to +5	-5 to +5
MIL	MIL (lamp) Control	ON / OFF	OFF	OFF	OFF
SHTFT1	Short Term F/T - Bank 1	-10 to 10%	-5 to +5	-5 to +5	-5 to +5
SHTFT2	Short Term FT - Bank 2	-10 to 10%	-5 to +5	-5 to +5	-5 to +5
SPARK	Spark Advance	-90° to 90°	14-20	20-35	20-30
SS1	Shift Solenoid 1	ON / OFF	ON	OFF	OFF
SS2	Shift Solenoid 2	ON / OFF	OFF	ON	OFF
TCC	TCC Solenoid	0-100%	0	90-100	90-100
TCIL	TCIL (lamp) Control	ON / OFF	OFF	OFF	OFF
VREF	Vehicle Reference	5v	5v	5v	5v

1998-99 E-SERIES 5.4L NGV VIN Z (A/T, M/T, E4OD) INPUTS

PCM PID Acronym	Parameter Identification	PID Range	PID Value at Hot Idle	PID Value at 30	PID Value at 55
ACCS	A/C Switch Signal	ON / OFF	ON (if on)	OFF	OFF
BPP	Brake Position	ON / OFF	ON (if on)	OFF	OFF
CHT	CHT Sensor	-40-500°F	194	194	194
CID	CMP Sensor	0-999 Hz	6-7	9-12	15-17.5
CKP	CKP Sensor	0-9999 Hz	440-490	810-855	1088-1089
CPP	Clutch Pedal Switch	ON / OFF	ON (if in)	ON	ON
EFTA	EFT Sensor	-40-275F	50-120	50-120	50-120
FEPS	Flash EEPROM	0-5.1v	0.1	0.1	0.1
FPM	Fuel Pump Monitor	ON / OFF	ON	ON	ON
FRP (psi)	FRP Sensor	0-300 psi	90-100	90-100	90-100
FRP (V)	FRP Sensor	0-5.1v	2.7-3.7	2.7-3.7	2.7-3.7
FSV-M	Fuel Solenoid Valve	ON / OFF	ON	ON	ON
GEAR	Gear Position	1-2-3-4	1	3	4
HO2S-11	HO2S-11 (Bank 1)	0.1-1.1v	0.1-1.1v	0.1-1.1v	0.1-1.1v
IAT	IAT Sensor	-40-304°F	50-120	50-120	50-120
LOAD	Engine Load	0-100%	13-19	26-33	38-46
MAFV	MAF Sensor	0-5.1v	0.7-0.9	0.9-1.7	1.2-2.4
MISF	Misfire Detection	ON / OFF	OFF	OFF	OFF
PNP	PNP Switch Signal	Neutral/DR	NEUTRAL	DRIVE	DRIVE
PTO	Power Takeoff Switch	OFF	ON (if on)	OFF	OFF
RPM	Engine Speed	0-10K rpm	900-930	1470-1490	1840-1860
TCS	TCS Switch	ON / OFF	OFF	OFF	OFF
TFT	TFT Sensor Signal	-40-304°F	110-210	110-210	110-210
TPV	TP Sensor	0-5.1v	0.53-1.27	0.8-1.0	0.9-1.2
TR1	TR1 Sensor	0v / 11.5v	0	11.5	11.5
TR2	TR2 Sensor	0v / 11.5v	0	11.5	11.5
TR4	TR4 Sensor	0v / 11.5v	0	11.5	11.5
TRV/TR	TR Sensor	0v / 1.7v	0	1.7	1.7
Vacuum	Engine Vacuum	0-30" Hg	19-20	18-20	8-15
VPWR	Vehicle Power	0-25.5v	12-14	12-14	12-14
VSS	Vehicle Speed (Hertz or MPH)	0-999 Hz (0-159 mph)	0 Hz = 0 mph	65 Hz = 30 mph	125 Hz = 55 mph

1998-99 E-SERIES 5.4L NGV VIN Z (A/T, M/T, E4OD) OUTPUTS

PCM PID Acronym	Parameter Identification	PID Range	PID Value at Hot Idle	PID Value at 30	PID Value at 55
CCS	Coast Clutch Solenoid Control	ON / OFF	OFF	OFF	OFF
CD1-CD8	COP Driver 1-8	0-45° dwell	5	6	8
CHTIL	CHT (lamp) Control	ON / OFF	OFF	OFF	OFF
CTO	Clean Tachometer signal	0-9999 Hz	50-57	90-100	115-125
EPC	EPC Solenoid (E4OD)	0-300 psi	5	6	11
FP	Fuel Pump Control	ON / OFF	ON	ON	ON
FSV	Fuel Solenoid Valve	ON / OFF	ON	ON	ON
FuelPW1	INJ Pulsewidth - Bank 1	0-999 ms	3.9-5.5	4.7-12.0	4.7-12.0
FuelPW2	INJ Pulsewidth - Bank 2	0-999 ms	3.9-5.5	4.7-12.0	4.7-12.0
HTR-11	HO2S-11 Heater	ON / OFF	ON	ON	ON
HTR-21	HO2S-21 Heater	ON / OFF	ON	ON	ON
IAC	Idle Air Control	0-100%	30-34	43-48	58-61
IMTV	Intake Manifold Tuning Valve	0-100%	0	0	0
INJ1	INJ 1 Pulsewidth	0-999 ms	3.9-4.6	4.7-12.0	4.7-12.0
INJ2	INJ 2 Pulsewidth	0-999 ms	3.9-4.6	4.7-12.0	4.7-12.0
INJ3	INJ 3 Pulsewidth	0-999 ms	3.9-4.6	4.7-12.0	4.7-12.0
INJ4	INJ 4 Pulsewidth	0-999 ms	3.9-4.6	4.7-12.0	4.7-12.0
INJ5	INJ 5 Pulsewidth	0-999 ms	3.9-4.6	4.7-12.0	4.7-12.0
INJ6	INJ 6 Pulsewidth	0-999 ms	3.9-4.6	4.7-12.0	4.7-12.0
INJ7	INJ 7 Pulsewidth	0-999 ms	3.9-4.6	4.7-12.0	4.7-12.0
INJ8	INJ 8 Pulsewidth	0-999 ms	3.9-4.6	4.7-12.0	4.7-12.0
LongFT1	Long Term FT - Bank 1	-20 to 20%	-5 to +5	-5 to +5	-5 to +5
LongFT2	Long Term FT - Bank 2	-20 to 20%	-5 to +5	-5 to +5	-5 to +5
MIL	MIL (lamp) Control	ON / OFF	OFF	OFF	OFF
SHTFT1	Short Term F/T - Bank 1	-10 to 10%	-5 to +5	-5 to +5	-5 to +5
SHTFT2	Short Term FT - Bank 2	-10 to 10%	-5 to +5	-5 to +5	-5 to +5
SPARK	Spark Advance	-90° to 90°	15-25	20-35	20-30
SS1	Shift Solenoid 1	ON / OFF	ON	OFF	ON
SS2	Shift Solenoid 2	ON / OFF	OFF	ON	OFF
TCC	TCC Solenoid	0-100%	0	50-100	90-100
TCIL	TCIL (lamp) Control	ON / OFF	OFF	OFF	OFF
VPWR	Vehicle Power	0-25.5v	12-14	12-14	12-14
VREF	Vehicle Reference	5v	4.9-5.1	4.9-5.1	4.9-5.1

1990-95 E-SERIES 5.8L V8 MFI VIN H, R (A/T, M/T, E4OD)

PCM PID Acronym	Parameter Identification	PID Range	PID Value at Hot Idle	PID Value at 30	PID Value at 55
ACCS	A/C Switch Signal	ON / OFF	ON (if on)	OFF	OFF
CANP	EVAP Purge Valve	ON / OFF	OFF	ON	ON
CPP	Clutch Pedal Switch	ON / OFF	ON (if in)	OFF	OFF
DPFE	DPFE Sensor (Cal.)	0-5.1v	0.4-0.6	0.4-0.9	0.8-1.1
ECT (°F)	ECT Sensor	-40-304°F	160-200	160-200	160-200
ECTV	ECT Sensor	0-5.1v	0.6	0.6	0.6
EPC	EPC Solenoid (E4OD)	0-300 psi	4	4	14
EVP	EGR Valve Position	0-5.1v	0.3	0.4	0.6-3.0
EVR	EGR VR Solenoid	0-100%	0	0-40	0-40
FuelPW1	INJ Pulsewidth - Bank 1	0-999 ms	5.0-5.8	6.0-6.8	6.4-7.0
GEAR	Gear Position	P-R-N-D	P-R-N-D	DRIVE 3	DRIVE 4
HO2S-11	HO2S-11 (Bank 1)	0.1-1.1v	0.1-1.1	0.1-1.1	0.1-1.1
IAC	Idle Air Control	0-100%	30-34	28-32	63-70
IAT (°F)	IAT Sensor	-40-304°F	50-120	50-120	50-120
IATV	IAT Sensor	0-5.1v	1.5-3.5	1.5-3.5	1.5-3.5
LOOP	Loop Status	CL or OL	CLOSED	CLOSED	CLOSED
MLPV	MLP Switch (E4OD)	0-5.1v	0	5	5
MAP	MAP Sensor	0-999 Hz	108-112	110-120	130-140
PNP	PNP Switch Signal	Neutral/DR	NEUTRAL	DRIVE	DRIVE
RPM	Engine Speed	0-10K rpm	750-850	1250-1350	1600-1700
SHTFT1	Short Term F/T - Bank 1	-10 to 10%	-5 to +5	-5 to +5	-5 to +5
SPARK	Spark Advance	-90° to 90°	14-18	30-36	38-44
TP	TP Sensor	0-5.1v	0.9	1.0-1.1	1.1-1.2
TCC	TCC Solenoid	ON / OFF	OFF	ON	ON
TOTV	TOT sensor (E4OD)	0-5.1v	2.10-2.40	2.10-2.40	2.10-2.40
TP	TP Sensor	0-5.1v	1.0	1.2-1.3	1.5-1.6
TP Mode	TP Sensor Mode	C/T or P/T	C/T	P/T	P/T
VPWR	Vehicle Power	0-25.5v	12-14	12-14	12-14
VREF	Vehicle Reference	0-5.1v	4.9-5.1	4.9-5.1	4.9-5.1
VSS	Vehicle Speed	0-159 mph	0	30	55

1996 E-SERIES 5.8L V8 MFI VIN H (E4OD) INPUTS

PCM PID Acronym	Parameter Identification	PID Range	PID Value at Hot Idle	PID Value at 30	PID Value at 55
4x4L	4x4 Switch Signal	ON / OFF	ON (if on)	OFF	OFF
ACCS	A/C Switch Signal	ON / OFF	ON (if on)	OFF	OFF
BPP	Brake Position	ON / OFF	ON (if on)	OFF	OFF
DPFE	DPFE Sensor	0-5.1v	0.4-0.6	0.4-0.9	0.8-1.1
ECT	ECT Sensor	-40-304°F	160-200	160-200	160-200
FEPS	Flash EEPROM	0-5.1v	0.1	0.1	0.1
FLI	Fuel Level Input	0-100%	1.7 / 50%	1.7 / 50%	1.7 / 50%
FPM	Fuel Pump Monitor	ON / OFF	ON	ON	ON
GEAR	Gear Position	1-2-3-4	1	3	4
HO2S-11	HO2S-11 (Bank 1)	0.1-1.1v	0.1-1.1	0.1-1.1	0.1-1.1
HO2S-12	HO2S-12 (Bank 1)	0.1-1.1v	0.1-1.1	0.1-1.1	0.1-1.1
HO2S-21	HO2S-21 (Bank 1)	0.1-1.1v	0.1-1.1	0.1-1.1	0.1-1.1
IAT	IAT Sensor	-40-304°F	50-120	50-120	50-120
IDM	IDM Signal	0-9999 Hz	42-50	82-88	105-120
LOAD	Engine Load	0-100%	14-16	20-25	24-35
MAFV	MAF Sensor	0-5.1v	0.7-0.9	1.2-1.7	1.6-2.4
MISF	Misfire Detection	ON / OFF	OFF	OFF	OFF
OCTADJ	Octane Adjustment	CL/OPEN	CLOSED	CLOSED	CLOSED
PIP	PIP Sensor	0-9999 Hz	42-50	82-88	105-120
PNP	PNP Switch Signal	ON / OFF	ON	OFF	OFF
PTO	Power Takeoff Switch	OFF	ON (if on)	OFF	OFF
RPM	Engine Speed	0-10K rpm	675-715	1250-1390	1590-1700
TCS	TCS Switch	ON / OFF	OFF	OFF	OFF
TFT	TFT Sensor Signal	-40-304°F	110-210	110-210	110-210
TPV	TP Sensor	0-5.1v	0.53-1.27	0.8-1.0	0.9-1.2
TRV/TR	TR Sensor	0v / 1.7v	0v	1.7	1.7
Vacuum	Engine Vacuum	0-30" Hg	19-20	18-20	8-15
VPWR	Vehicle Power	0-25.5v	12-14	12-14	12-14
VSS	Vehicle Speed (Hertz or MPH)	0-999 Hz (0-159 mph)	0 Hz = 0 mph	65 Hz = 30 mph	125 Hz = 55 mph

1996 E-SERIES 5.8L V8 MFI VIN H (E4OD) OUTPUTS

PCM PID Acronym	Parameter Identification	PID Range	PID Value at Hot Idle	PID Value at 30	PID Value at 55
CCS	Coast Clutch Solenoid Control	ON / OFF	OFF	OFF	OFF
EGR VR	EGR VR Solenoid	0-100%	0	0-40	35-45
EVAP CP	EVAP Purge Valve	0-100%	0-100	0-100	0-100
EPC	EPC Solenoid	0-300 psi	5	5	15
FP	Fuel Pump Control	ON / OFF	ON	ON	ON
FuelPW1	INJ Pulsewidth - Bank 1	0-999 ms	4.2-4.7	4.4-7.0	7.4-12.0
FuelPW2	INJ Pulsewidth - Bank 2	0-999 ms	4.2-4.7	4.4-7.0	7.4-12.0
HTR-11	HO2S-11 Heater	ON / OFF	ON	ON	ON
HTR-12	HO2S-12 Heater	ON / OFF	ON	ON	ON
HTR-21	HO2S-21 Heater	ON / OFF	ON	ON	ON
HTR-22	HO2S-22 Heater	ON / OFF	ON	ON	ON
IAC	Idle Air Control	0-100%	30-34	28-32	63-70
INJ1	INJ 1 Pulsewidth	0-999 ms	4.2-4.7	4.4-7.0	7.4-12.0
INJ2	INJ 2 Pulsewidth	0-999 ms	4.2-4.7	4.4-7.0	7.4-12.0
INJ3	INJ 3 Pulsewidth	0-999 ms	4.2-4.7	4.4-7.0	7.4-12.0
INJ4	INJ 4 Pulsewidth	0-999 ms	4.2-4.7	4.4-7.0	7.4-12.0
INJ5	INJ 5 Pulsewidth	0-999 ms	4.2-4.7	4.4-7.0	7.4-12.0
INJ6	INJ 6 Pulsewidth	0-999 ms	4.2-4.7	4.4-7.0	7.4-12.0
INJ7	INJ 7 Pulsewidth	0-999 ms	4.2-4.7	4.4-7.0	7.4-12.0
INJ8	INJ 8 Pulsewidth	0-999 ms	4.2-4.7	4.4-7.0	7.4-12.0
LongFT1	Long Term FT - Bank 1	-20 to 20%	-5 to +5	-5 to +5	-5 to +5
LongFT2	Long Term FT - Bank 2	-20 to 20%	-5 to +5	-5 to +5	-5 to +5
MIL	MIL (lamp) Control	ON / OFF	OFF	OFF	OFF
SHTFT1	Short Term F/T - Bank 1	-10 to 10%	-5 to +5	-5 to +5	-5 to +5
SHTFT2	Short Term FT - Bank 2	-10 to 10%	-5 to +5	-5 to +5	-5 to +5
SPARK	Spark Advance	-90° to 90°	15-18	23-29	26-32
SPOUT	Spark Output Signal	0-9999 Hz	42-50	82-88	105-120
SS1	Shift Solenoid 1	ON / OFF	ON	OFF	OFF
SS2	Shift Solenoid 2	ON / OFF	OFF	ON	OFF
TCC	TCC Solenoid	0-100%	0	90-100	90-100
TCIL	TCIL (lamp) Control	ON / OFF	OFF	OFF	OFF
VREF	Vehicle Reference	0-5.1v	4.9-5.1	4.9-5.1	4.9-5.1

1997-2001 E-SERIES 6.8L V10 VIN S (E4OD) INPUTS

PCM PID Acronym	Parameter Identification	PID Range	PID Value at Hot Idle	PID Value at 30	PID Value at 55
4x4L	4x4 Switch Signal	ON / OFF	ON (if on)	OFF	OFF
ACCS	A/C Switch Signal	ON / OFF	ON (if on)	OFF	OFF
BPP	Brake Position	ON / OFF	ON (if on)	OFF	OFF
CHT	CHT Sensor	-40-500ºF	194	194	194
CID	CMP Sensor	0-999 Hz	7-10	10-13	15-17
CKP	CKP Sensor	0-9999 Hz	500-525	750-940	1195-1400
DPFE	DPFE Sensor	0-5.1v	0.2-1.3	0.2-4.5	0.2-4.5
FEPS	Flash EEPROM	0-5.1v	0.1	0.1	0.1
FLI	Fuel Level Input	1.7 / 50%	1.7 / 50%	1.7 / 50%	1.7 / 50%
FPM	Fuel Pump Monitor	ON / OFF	ON	ON	ON
FTP	Fuel Tank Pressure	2.6v / 0 psi	2.6v / 0 psi	2.6v / 0 psi	2.6v / 0 psi
GEAR	Gear Position	1-2-3-4	1	4	4
IAT	IAT Sensor	-40-304ºF	50-120	50-120	50-120
HO2S-11	HO2S-11 (Bank 1)	0.1-1.1v	0.1-1.1	0.1-1.1	0.1-1.1
HO2S-12	HO2S-12 (Bank 1)	0.1-1.1v	0.1-1.1	0.1-1.1	0.1-1.1
HO2S-21	HO2S-21 (Bank 1)	0.1-1.1v	0.1-1.1	0.1-1.1	0.1-1.1
KS1	Knock Sensor 1	0-5.1v	0	0	0
LOAD	Engine Load	0-100%	14-16	20-25	24-35
MAFV	MAF Sensor	0-5.1v	0.7-0.9	1.2-1.7	1.6-2.4
MISF	Misfire Detection	ON / OFF	OFF	OFF	OFF
OCTADJ	Octane Adjustment	CL/OPEN	CLOSED	CLOSED	CLOSED
PNP	PNP Switch Signal	ON / OFF	ON	OFF	OFF
PTO	Power Takeoff Switch	OFF	ON (if on)	OFF	OFF
RPM	Engine Speed	0-10K rpm	780-810	1380-1450	1790-1840
TCS	TCS Switch	ON / OFF	OFF	OFF	OFF
TFT	TFT Sensor Signal	-40-304ºF	110-210	110-210	110-210
TPV	TP Sensor	0-5.1v	0.53-1.27	0.8-1.1	0.9-1.2
TR1	TR1 Sensor	0v / 11.5v	0	11.5	11.5
TR2	TR2 Sensor	0v / 11.5v	0	11.5	11.5
TR3/TR	TR3A Sensor	0v / 1.7v	0	1.7	1.7
TR4	TR4 Sensor	0v / 11.5v	0	11.5	11.5
Vacuum	Engine Vacuum	0-30" Hg	19-20	18-20	8-15
VPWR	Vehicle Power	0-25.5v	12-14	12-14	12-14
VSS	Vehicle Speed (Hertz or MPH)	0-999 Hz (0-159 mph)	0 Hz = 0 mph	65 Hz = 30 mph	125 Hz = 55 mph

1997-2001 E-SERIES 6.8L V10 VIN S (E4OD) OUTPUTS

PCM PID Acronym	Parameter Identification	PID Range	PID Value at Hot Idle	PID Value at 30	PID Value at 55
CD1-10	Coil 1-10 Driver	0-90° dwell	6	8	10
CCS	Coast Clutch Solenoid Control	ON / OFF	OFF	OFF	OFF
CHTIL	CHT (lamp) Control	ON / OFF	OFF	OFF	OFF
CTO	Clean Tachometer signal	0-9999 Hz	60-70	90-110	150-185
EGR VR	EGR VR Solenoid	0-100%	0	0-40	35-55
EVAP CP	EVAP Purge Valve	0-100%	0-100	0-100	0-100
EVAP CV	EVAP CV Valve	0-100%	0-100	0-100	0-100
EPC	EPC Solenoid	0-300 psi	5	7	27
FP	Fuel Pump Control	ON / OFF	ON	ON	ON
FuelPW1	INJ Pulsewidth - Bank 1	0-999 ms	3.8-4.6	5.2-6.5	6.6-11.0
FuelPW2	INJ Pulsewidth - Bank 2	0-999 ms	3.8-4.6	5.2-6.5	6.6-11.0
HTR-11	HO2S-11 Heater	ON / OFF	ON	ON	ON
HTR-12	HO2S-12 Heater	ON / OFF	ON	ON	ON
HTR-21	HO2S-21 Heater	ON / OFF	ON	ON	ON
HTR-22	HO2S-22 Heater	ON / OFF	ON	ON	ON
IAC	Idle Air Control	0-100%	25-33	30-40	50-65
INJ1	INJ 1 Pulsewidth	0-999 ms	3.8-4.6	5.2-6.5	6.6-11.0
INJ2	INJ 2 Pulsewidth	0-999 ms	3.8-4.6	5.2-6.5	6.6-11.0
INJ3	INJ 3 Pulsewidth	0-999 ms	3.8-4.6	5.2-6.5	6.6-11.0
INJ4	INJ 4 Pulsewidth	0-999 ms	3.8-4.6	5.2-6.5	6.6-11.0
INJ5	INJ 5 Pulsewidth	0-999 ms	3.8-4.6	5.2-6.5	6.6-11.0
INJ6	INJ 6 Pulsewidth	0-999 ms	3.8-4.6	5.2-6.5	6.6-11.0
INJ7	INJ 7 Pulsewidth	0-999 ms	3.8-4.6	5.2-6.5	6.6-11.0
INJ8	INJ 8 Pulsewidth	0-999 ms	3.8-4.6	5.2-6.5	6.6-11.0
INJ9	INJ 9 Pulsewidth	0-999 ms	3.8-4.6	5.2-6.5	6.6-11.0
INJ10	INJ 10 Pulsewidth	0-999 ms	3.8-4.6	5.2-6.5	6.6-11.0
LongFT1	Long Term FT - Bank 1	-20 to 20%	-5 to +5	-5 to +5	-5 to +5
LongFT2	Long Term FT - Bank 2	-20 to 20%	-5 to +5	-5 to +5	-5 to +5
MIL	MIL (lamp) Control	ON / OFF	OFF	OFF	OFF
SHTFT1	Short Term F/T - Bank 1	-10 to 10%	-5 to +5	-5 to +5	-5 to +5
SHTFT2	Short Term F/T - Bank 2	-10 to 10%	-5 to +5	-5 to +5	-5 to +5
SPARK	Spark Advance	-90° to 90°	15-20	23-34	26-34
SS1	Shift Solenoid 1	ON / OFF	ON	OFF	OFF
SS2	Shift Solenoid 2	ON / OFF	OFF	ON	OFF
TCC	TCC Solenoid	0-100%	0	90-100	90-100
TCIL	TCIL (lamp) Control	ON / OFF	OFF	OFF	OFF
VREF	Vehicle Reference	0-5.1v	4.9-5.1	4.9-5.1	4.9-5.1

<u>1996-2003 E-SERIES 7.3L V8 DIESEL VIN F (E4OD)</u>

PCM PID Acronym	Parameter Identification	PID Range	PID Value at low idle	PID Value at high idle
4x4L	4x4 Low Switch Input	ON / OFF	ON	OFF
ACCS	A/C Switch Signal	ON / OFF	ON	OFF
AP	Accelerator Pedal Position Sensor	0-5.1v	0.5-0.9	3.8-4.2v
ARPMDES	Ancillary Engine Speed Desired	---	---	---
BARO	Barometric Pressure	0-450 kPa	---	---
BAROV	Barometric Pressure	0-5.1v	4.75 (Sea level)	4.75 (Sea level)
BPP	Brake Pedal Position	ON / OFF	ON	OFF
BPA	Brake Pressure Applied	ON / OFF	ON	OFF
CCS	Coast Clutch Solenoid Control	ON / OFF	OFF	OFF
CPP	CPP Switch Signal	0v / 5v	0v (clutch "in")	5v
CPP/TCS	TCS & Cancel Switch	0v / VBAT	Switches On: 12	0v
CRUISE	Cruise Control Module	---	---	---
DTC CNT	DTC Count	0-156	0	0
EBP	Exhaust Back Pressure	---	---	---
EBP V	Exhaust Back Pressure Actual	0-5.1v	0.8-0.95	1.25-1.75
EOT	Engine Oil Temperature	0-5.1v	0.35-4.5	0.35-4.5
EPC	Electronic Pressure Control	0-300 psi	7	15
EPC V	Electronic Pressure Control Actual	0-25.5v	7.5	12
EPR	Exhaust Pressure Regulator	0-25.5v	6-8	0-10
FuelPW1	Fuel Pulsewidth	0-999 ms	---	---
GEAR	Gear Position	1-2-3-4	1	4
GPC	Glow Plug Control Duty Cycle	0-100%	---	---
GPC TM	Glow Plug Control Time	---	---	---
GPL TM	Glow Plug Lamp Time	---	---	---
IAT (°F)	IAT Sensor	-40-304°F	50-120	50-120
IAT V	IAT Sensor Actual	0-5.1v	1.5-3.5	1.5-3.5
ICP	Injector Control Pressure Sensor	Min: 0.83v at startup	0.25-0.40	0.25-0.40
ICP	Injector Control Pressure Actual	---	---	---
IPR	Injector Control Pressure Regulator	0-100%	35	40-100
IVS	Idle Validation Switch	0v / VBAT	0v / Closed	12 / CL or OL
MAP	MAP Sensor	0-999 Hz	---	---
MAP H	MAP Sensor Actual	0-999 Hz	110-190	110-190

1996-2003 E-SERIES 7.3L V8 DIESEL VIN F (E4OD)

PCM PID Acronym	Parameter Identification	PID Range	PID value at low idle	PID value at high idle
MFDES	Mass Fuel Desired	---	---	---
MGP	Manifold Gauge Pressure	---	---	---
PBA	Parking Brake Applied	0v / 12v	Parking Brake Off: 12v	0v
RPM	Engine Speed	0-10K rpm	---	---
SCCS	Speed Control Command Switch	0-25.5v	S/C On: 12v	S/C On: 12v
SCCS-M	Speed Control Command Switch Mode	---	---	---
TCC	Transmission Converter Clutch Solenoid Control	ON / OFF	OFF	OFF
SS1	Shift Solenoid 1	ON / OFF	ON	OFF
SS2	Shift Solenoid 2	ON / OFF	OFF	ON
TCIL	TCIL (lamp) Control	ON / OFF	OFF	OFF
TCS	TCS Switch	0v / 12	0	0
TFT V	TFT Sensor	0-5.1v	2.10-2.40	2.10-2.40
TORQUE	Engine Torque	---	---	---
TPREL	Low Idle TP Sensor	---	---	---
TR	Transmission Range Sensor	0-5.1v	4.45 in 'P'	2.87
TR V	Transmission Range Sensor Actual (Volts)	0-5.1v	4.45 in 'P'	2.87 in O/D
VFDES	Volume Flow Desired	---	---	---
VPWR	Vehicle Power Supply	0-25.5v	12-14	12-14
VREF	Vehicle Reference	0-5.1v	4.9-5.1	4.9-5.1
VS SET	Vehicle Speed Setting	---	---	---
VSS	Vehicle Speed (Hertz or MPH)	0-999 Hz (0-159 mph)	0	125 Hz = 55 mph

1992-95 E-SERIES 7.5L V8 MFI VIN G (A/T, M/T, E4OD)

PCM PID Acronym	Parameter Identification	PID Range	PID Value at Hot Idle	PID Value at 30	PID Value at 55
ACCS	A/C Switch Signal	ON / OFF	ON (if on)	OFF	OFF
CANP	EVAP Purge Valve	ON / OFF	OFF	ON	ON
ECT (°F)	ECT Sensor	-40-304°F	160-200	160-200	160-200
ECTV	ECT Sensor	0-5.1v	0.6	0.6	0.6
EPC	EPC Solenoid (E4OD)	0-300 psi	4	4	14
EVP	EGR Valve Position	0-5.1v	0.3	0.4	0.4-4.0v
EVR (%)	EGR VR Solenoid	0-100%	0	0-40	0-40
FuelPW1	INJ Pulsewidth - Bank 1	0-999 ms	5.6-6.6	6.6-8.6	9.0-11.0
FuelPW2	INJ Pulsewidth - Bank 2	0-999 ms	5.6-6.6	6.6-8.6	9.0-11.0
GEAR	Gear Position	P-R-N-D	P-R-N-D	DRIVE 3	DRIVE 4
HO2S-11	HO2S-11 (B1 S1)	0.1-1.1v	0-1v	0-1v	0-1v
IAC	Idle Air Control	0-100%	30-34	28-32	63-70
IAT (°F)	IAT Sensor	-40-304°F	50-120	50-120	50-120
IATV	IAT Sensor	0-5.1v	1.5-3.5	1.5-3.5	1.5-3.5
LongFT1	Long Term FT - Bank 1	-20 to 20%	-5 to +5	-5 to +5	-5 to +5
LongFT2	Long Term FT - Bank 2	-20 to 20%	-5 to +5	-5 to +5	-5 to +5
LOOP	Loop Status	CL or OL	CLOSED	CLOSED	CLOSED
MLPV	MLP Switch (E4OD)	0-5.1v	0	5	5
MAP	MAP Sensor	0-999 Hz	108-112	110-120	130-140
PNP	PNP Switch Signal	Neutral/DR	NEUTRAL	DRIVE	DRIVE
RPM	Engine Speed	0-10K rpm	650-750	1300-1400	1730-1830
SHTFT1	Short Term F/T - Bank 1	-10 to 10%	-5 to +5	-5 to +5	-5 to +5
SHTFT2	Short Term FT - Bank 2	-10 to 10%	-5 to +5	-5 to +5	-5 to +5
SPARK	Spark Advance	-90° to 90°	18-28	36-42	38-46
TCC	TCC Solenoid	ON / OFF	OFF	ON	ON
TOTV	TOT sensor (E4OD)	0-5.1v	2.10-2.40	2.10-2.40	2.10-2.40
TP	TP Sensor	0-5.1v	1.0	1.2-1.3	1.5-1.6
TP Mode	TP Sensor Mode	C/T or P/T	C/T	P/T	P/T
VPWR	Vehicle Power	0-25.5v	12-14	12-14	12-14
VREF	Vehicle Reference	0-5.1v	4.9-5.1	4.9-5.1	4.9-5.1
VSS	Vehicle Speed	0-159 mph	0	30	55

1996 E-SERIES 7.5L V8 MFI VIN G (E4OD) INPUTS

PCM PID Acronym	Parameter Identification	PID Range	PID Value at Hot Idle	PID Value at 30	PID Value at 55
4x4L	4x4 Switch Signal	ON / OFF	ON (if on)	OFF	OFF
ACCS	A/C Switch Signal	ON / OFF	ON (if on)	OFF	OFF
BPP	Brake Switch Signal	ON / OFF	ON (if on)	OFF	OFF
DPFE	DPFE Sensor	0-5.1v	0.4-0.6	0.4-0.7	0.8-1.1
ECT	ECT Sensor	-40-304°F	160-200	160-200	160-200
FEPS	Flash EEPROM	0-5.1v	0.1	0.1	0.1
FPM	Fuel Pump Monitor	ON / OFF	ON	ON	ON
GEAR	Gear Position	1-2-3-4	1	3	4
HO2S-11	HO2S-11 (Bank 1)	0.1-1.1v	0.1-1.1	0.1-1.1	0.1-1.1
HO2S-12	HO2S-12 (Bank 1)	0.1-1.1v	0.1-1.1	0.1-1.1	0.1-1.1
HO2S-21	HO2S-21 (Bank 1)	0.1-1.1v	0.1-1.1	0.1-1.1	0.1-1.1
HO2S-22	HO2S-22 (Bank 1)	0.1-1.1v	0.1-1.1	0.1-1.1	0.1-1.1
IAT	IAT Sensor	-40-304°F	50-120	50-120	50-120
IDM	IDM Signal	0-9999 Hz	45-55	88-120	110-220
LOAD	Engine Load	0-100%	14-16	20-25	32-40
MAFV	MAF Sensor	0-5.1v	0.9-1.1	1.6-1.8	2.0-2.6
MISF	Misfire Detection	ON / OFF	OFF	OFF	OFF
OCTADJ	Octane Adjust	CL or OP	CLOSED	CLOSED	CLOSED
PIP	PIP Sensor	0-9999 Hz	45-55	88-120	110-220
PNP	PNP Switch Signal	ON / OFF	ON	OFF	OFF
PTO	Power Takeoff	ON / OFF	ON (if on)	OFF	OFF
RPM	Engine Speed	0-10K rpm	780-810	1420-1510	1700-1790
TCS	TCS Switch	ON / OFF	OFF	OFF	OFF
TFT	TFT Sensor Signal	-40-304°F	110-210	110-210	110-210
TPV	TP Sensor	0-5.1v	0.53-1.27	1.1-1.3v	1.3-1.7
TRV/TR	TR Sensor	0v / 1.7v	0	1.7	1.7
Vacuum	Engine Vacuum	0-30" Hg	19-20	18-20	8-15
VPWR	Vehicle Power	0-25.5v	12-14	12-14	12-14
VSS	Vehicle Speed (Hertz or MPH)	0-999 Hz (0-159 mph)	0 Hz = 0 mph	65 Hz = 30 mph	125 Hz = 55 mph

1996 E-SERIES 7.5L V8 MFI VIN G (E4OD) OUTPUTS

PCM PID Acronym	Parameter Identification	PID Range	PID Value at Hot Idle	PID Value at 30	PID Value at 55
AIRB	AIR System Bypass	0v / VBAT	0.1	0.1	0.1
AIRD	AIR System Divert	0v / VBAT	VBAT	VBAT	VBAT
EGR VR	EGR VR Solenoid	0-100%	0	40-45	50-60
EVAP CP	EVAP Purge Valve	0-100%	0-100	0-100	0-100
EPC	EPC Solenoid	0-300 psi	5	5	10-15
FP	Fuel Pump Control	ON / OFF	ON	ON	ON
FuelPW1	INJ Pulsewidth - Bank 1	0-999 ms	4.3-4.6	5.2-6.5	6.6-9.0
FuelPW2	INJ Pulsewidth - Bank 2	0-999 ms	4.3-4.6	5.2-6.5	6.6-9.0
HTR-11	HO2S-11 Heater	ON / OFF	ON	ON	ON
HTR-12	HO2S-12 Heater	ON / OFF	ON	ON	ON
HTR-21	HO2S-21 Heater	ON / OFF	ON	ON	ON
HTR-22	HO2S-22 Heater	ON / OFF	ON	ON	ON
IAC	Idle Air Control	0-100%	30-34	60-65	63-70
INJ1	INJ 1 Pulsewidth	0-999 ms	4.3-4.6	5.2-6.5	6.6-9.0
INJ2	INJ 2 Pulsewidth	0-999 ms	4.3-4.6	5.2-6.5	6.6-9.0
INJ3	INJ 3 Pulsewidth	0-999 ms	4.3-4.6	5.2-6.5	6.6-9.0
INJ4	INJ 4 Pulsewidth	0-999 ms	4.3-4.6	5.2-6.5	6.6-9.0
INJ5	INJ 5 Pulsewidth	0-999 ms	4.3-4.6	5.2-6.5	6.6-9.0
INJ6	INJ 6 Pulsewidth	0-999 ms	4.3-4.6	5.2-6.5	6.6-9.0
INJ7	INJ 7 Pulsewidth	0-999 ms	4.3-4.6	5.2-6.5	6.6-9.0
INJ8	INJ 8 Pulsewidth	0-999 ms	4.3-4.6	5.2-6.5	6.6-9.0
LongFT1	Long Term FT - Bank 1	-20 to 20%	-5 to +5	-5 to +5	-5 to +5
LongFT2	Long Term FT - Bank 2	-20 to 20%	-5 to +5	-5 to +5	-5 to +5
MIL	MIL (lamp) Control	ON / OFF	OFF	OFF	OFF
SHTFT1	Short Term F/T - Bank 1	-10 to 10%	-5 to +5	-5 to +5	-5 to +5
SHTFT2	Short Term FT - Bank 2	-10 to 10%	-5 to +5	-5 to +5	-5 to +5
SPARK	Spark Advance	-90° to 90°	22-26	25-28	27-32°
SPOUT	Spark Output Signal	0-9999 Hz	45-55	88-120	110-220
SS1	Shift Solenoid 1	ON / OFF	ON	OFF	OFF
SS2	Shift Solenoid 2	ON / OFF	OFF	ON	OFF
TCC	TCC Solenoid	0-100%	0%	0-100%	90-95
TCIL	TCIL (lamp) Control	ON / OFF	OFF	OFF	OFF
VREF	Vehicle Reference	0-5.1v	4.9-5.1	4.9-5.1	4.9-5.1

Villager PID Data

1996-98 VILLAGER 3.0L V6 SOHC MFI VIN W, 1 (A/T)

PCM PID Acronym	Parameter Identification	PID Range	PID Value at Hot Idle	PID Value at 30	PID Value at 55
ACR	A/C Relay Contro (with A/C & Blower On or Off)	ON / OFF	ON (if on)	OFF	OFF
ACS	AC Cycling Switch (with AC On & blower at 3-4)	ON / OFF	ON	OFF	OFF
A/F COR	Air Fuel Correction	-20 to 20%	-5 to +5	-5 to +5	-5 to +5
ECT	ECT Sensor	-40-304°F	176-220	176-220	176-220
EGRC	EGR Solenoid	ON / OFF	ON	OFF	OFF
EGRT	EGR Temp. Sensor	0-5.1v	0.3	0.2-1.0	0.2-1.0
FUEL	Fuel System Status	LEAN or RICH	LEAN / RICH	LEAN / RICH	LEAN / RICH
FuelPW1	INJ Pulsewidth - Bank 1	0-999 ms	3.38	4.4-5.5	6.6-7.3
HFAN	High Speed Cooling Fan (on with ECT more than 221°F)	ON / OFF	OFF	OFF	OFF
IAC	Idle Speed Control	0-100%	15-28	71	71
IDL SW	Throttle Idle Switch	ON / OFF	ON	OFF	OFF
LFAN	Low Speed Cooling Fan (ON with ECT at 203-210°F - AC Off)	ON / OFF	OFF	OFF	OFF
HO2S-11	HO2S-11 (Bank 1)	0.1-1.1v	0l.1-1.1	0.1-1.1	0.1-1.1
MAFV	MAF Sensor	0-5.1v	0.7	2.4	2.8
PNP SW	PNP Switch Signal	ON / OFF	ON	OFF	OFF
PSP SW	PSP Switch Signal	ON / OFF	ON: turned	OFF	OFF
RPM	Engine Speed	0-10K rpm	750	1200-1650	1850
SHTFT1	Short Term F/T - Bank 1	-10 to 10%	-5 to +5	-5 to +5	-5 to +5
IGN	Ignition Timing	-90° to 90°	15	36	32
TPV	TP Sensor	0-5.1v	0.46	0.78	1.12
VPWR	Vehicle Power	0-25.5v	12-14	12-14	12-14
VSS	Vehicle Speed	0-159 mph	0	30	55
VST	Vehicle Start Signal (on during cranking)	ON / OFF	OFF	OFF	OFF

<u>1999-2003 VILLAGER 3.3L V6 SOHC MFI VIN T (A/T)</u>

PCM PID Acronym	Parameter Identification	PID Range	PID Value at Hot Idle	PID Value at 30	PID Value at 55
ACR	A/C Relay Control	ON / OFF	ON (A/C & BLMT on)	OFF	OFF
ACS	AC Cycling Switch (AC & Blower/3-4)	ON / OFF	ON	OFF	OFF
A/F COR	Air Fuel Correction	-20 to 20%	-5 to +5	-5 to +5	-5 to +5
ECT	ECT Sensor	-40-304°F	176-220	176-220	176-220
EGRC	EGR Solenoid	ON / OFF	ON	OFF	OFF
EGRT	EGR Temp. Sensor	0-5.1v	0.3	0.2-1.0	0.2-1.0
FUEL	Fuel System Status	LEAN or RICH	LEAN / RICH	LEAN / RICH	LEAN / RICH
FuelPW1	INJ Fuel Pulsewidth Bank 1	0-999 ms	4.0-4.3	4.4-5.5	6.6-7.3
HFAN	High Speed Cooling Fan (on with ECT more than 221°F)	ON / OFF	OFF	OFF	OFF
IAC	Idle Speed Control	0-100%	15-28	71	71
IDL SW	Throttle Idle Switch	ON / OFF	ON	OFF	OFF
LFAN	Low Speed Cooling Fan (on with ECT at 203-210°F - AC Off)	ON / OFF	OFF	OFF	OFF
HO2S-11	HO2S-11 (Bank 1)	0.1-1.1v	0.1-1.1	0.1-1.1	0.1-1.1
MAFV	MAF Sensor	0-5.1v	0.7	2.4	2.8
PNP SW	PNP Switch Signal	ON / OFF	ON	OFF	OFF
PSP SW	PSP Switch Signal	ON / OFF	ON: turned	OFF	OFF
RPM	Engine Speed	0-10K rpm	750	1200-1650	1850
SHTFT1	Short Term F/T - Bank 1	-10 to 10%	-5 to +5	-5 to +5	-5 to +5
IGN	Ignition Timing	-90° to 90°	15 BTDC	36 BTDC	32° BTDC
TPV	TP Sensor	0-5.1v	0.46	0.78	1.12
VPWR	Vehicle Power	0-25.5v	12-14	12-14	12-14
VSS	Vehicle Speed	0-159 mph	0	30	55
VST	Vehicle Start Signal (on during cranking)	ON / OFF	OFF	OFF	OFF

Windstar PID Data

1996-2000 WINDSTAR 3.0L V6 VIN U (A/T) INPUTS

PCM PID Acronym	Parameter Identification	PID Range	PID Value at Hot Idle	PID Value at 30	PID Value at 55
ACCS	A/C Switch Signal	OFF	ON	OFF	OFF
ACP	AC Pressure Switch	0-5.1v	1.5-1.9	0.6-0.8	0.6-0.8
AFS	Airflow Sensor	0-999 Hz	130	130	130
BPP	Brake Position	ON / OFF	ON (if on)	OFF	OFF
CID	CMP Sensor	0-999	5-7	12-15	15-20
CKP	CKP Sensor	0-9999 Hz	518-540	975-1020	1200-1330
DPFE	DPFE Sensor	0-5.1v	0.2-1.3	0.2-4.5	0.2-4.5
ECT	ECT Sensor	-40-304°F	160-200	160-200	160-200
FEPS	Flash EEPROM	0-5.1v	0.1	0.1	0.1
FLI	Fuel Level Indicator	0-100%	50% (1.7)	50% (1.7)	50% (1.7)
FPM	Fuel Pump Monitor	ON / OFF	ON	ON	ON
GEAR	Gear Position	1-2-3-4	1	3	4
HO2S-11	HO2S-11 (Bank 1)	0.1-1.1v	0.1-1.1	0.1-1.1	0.1-1.1
HO2S-12	HO2S-12 (Bank 1)	0.1-1.1v	0.1-1.1	0.1-1.1	0.1-1.1
HO2S-21	HO2S-21 (Bank 2)	0.1-1.1v	0.1-1.1	0.1-1.1	0.1-1.1
HO2S-22	HO2S-22 (Bank 2)	0.1-1.1v	0.1-1.1	0.1-1.1	0.1-1.1
IAT	IAT Sensor	-40-304°F	50-120	50-120	50-120
LOAD	Engine Load	0-100%	15-20	20-27	35-45
MAFV	MAF Sensor	0-5.1v	0.6-0.9	0.8-1.9	1.1-2.3
MISF	Misfire Detection	ON / OFF	OFF	OFF	OFF
OCTADJ	Octane Adjustment	CL/OPEN	CLOSED	CLOSED	CLOSED
PNP	PNP Switch Signal	ON / OFF	ON	OFF	OFF
RPM	Engine Speed	0-10K rpm	680-800	1550-1700	1500-2100
TCS	TCS Switch	ON / OFF	OFF	OFF	OFF
TFT	TFT Sensor Signal	-40-304°F	110-210	110-210	110-210
TPV	TP Sensor	0-5.1v	0.53-1.27	0.9-1.2	0.9-1.9
TR1	TR1 Sensor	0v / 11.5v	0	11.5	11.5
TR2	TR2 Sensor	0v / 11.5v	0	11.5	11.5
TR4	TR4 Sensor	0v / 11.5v	0	11.5	11.5
TRV/TR	TR Sensor	0v / 1.7v	0	1.7	1.7
TSS	TSS Sensor	0-9999 Hz	35-40	47-73	100-125
Vacuum	Engine Vacuum	0-30" Hg	19-20	18-20	8-15
VPWR	Vehicle Power	0-25.5v	12-14	12-14	12-14
VSS	Vehicle Speed (Hertz or MPH)	0-999 Hz (0-159 mph)	0 Hz = 0 mph	65 Hz = 30 mph	125 Hz = 55 mph
WACA	AC WOT Monitor	ON	OFF (with A/C on)	ON	ON

1996-2000 WINDSTAR 3.0L V6 VIN U (A/T) OUTPUTS

PCM PID Acronym	Parameter Identification	PID Range	PID Value at Hot Idle	PID Value at 30	PID Value at 55
ALDFDC	Alternator Field (Hz)	0-999 Hz	0-130	0	0
CD1-3	Coil 1 Driver	0-60° dwell	6	8	12
EGR VR	EGR VR Solenoid	0-100%	0	0-40	35-50
EVAP CP	EVAP Purge Valve	0-100%	0-100	0-100	0-100
EVAP CV	EVAP CV Valve	0-100%	0-100	0-100	0-100
EPC	EPC Solenoid	0-300 psi	13-17	15-21	38-47
FP	Fuel Pump Control	ON / OFF	ON	ON	ON
FTP	Fuel Tank Pressure	0-5.1v at 0-10" H2O	2.6v at 0" H2O (gas cap "off")	2.6v at 0" H2O (gas cap "off")	2.6v at 0" H2O (gas cap "off")
FuelPW1	INJ Pulsewidth - Bank 1	0-999 ms	5.0-5.2	5.9-9.0	6.0-11.0
FuelPW2	INJ Pulsewidth - Bank 2	0-999 ms	5.0-5.2	5.9-9.0	6.0-11.0
HFC	High Speed Fan	ON / OFF	ON (if on)	OFF	OFF
HTR-11	HO2S-11 Heater	ON / OFF	ON	ON	ON
HTR-12	HO2S-12 Heater	ON / OFF	ON	ON	ON
HTR-21	HO2S-21 Heater	ON / OFF	ON	ON	ON
HTR-22	HO2S-22 Heater	ON / OFF	ON	ON	ON
IAC	Idle Air Control	0-100%	17-29	38-53	40-55
INJ1	INJ 1 Pulsewidth	0-999 ms	5.0-5.2	5.9-9.0	6.0-11.0
INJ2	INJ 2 Pulsewidth	0-999 ms	5.0-5.2	5.9-9.0	6.0-11.0
INJ3	INJ 3 Pulsewidth	0-999 ms	5.0-5.2	5.9-9.0	6.0-11.0
INJ4	INJ 4 Pulsewidth	0-999 ms	5.0-5.2	5.9-9.0	6.0-11.0
INJ5	INJ 5 Pulsewidth	0-999 ms	5.0-5.2	5.9-9.0	6.0-11.0
INJ6	INJ 6 Pulsewidth	0-999 ms	5.0-5.2	5.9-9.0	6.0-11.0
LFC	Low Speed Fan	ON / OFF	ON (if on)	OFF	OFF
LongFT1	Long Term FT (B1)	-20 to 20%	-5 to +5	-5 to +5	-5 to +5
LongFT2	Long Term FT (B2)	-20 to 20%	-5 to +5	-5 to +5	-5 to +5
MIL	MIL (lamp) Control	ON / OFF	OFF	OFF	OFF
SHTFT1	Short Term F/T - Bank 1	-10 to 10%	-5 to +5	-5 to +5	-5 to +5
SHTFT2	Short Term FT - Bank 2	-10 to 10%	-5 to +5	-5 to +5	-5 to +5
SPARK	Spark Advance	-90° to 90°	15-20	20-35	20-41
SPOUT	Spark Output Signal	0 Hz	33-37	80-87	95-105
SS1	Shift Solenoid 1	ON / OFF	ON	OFF	OFF
SS2	Shift Solenoid 2	ON / OFF	OFF	ON	ON
SS3	Shift Solenoid 3	ON / OFF	OFF	ON	ON
TCC	TCC Solenoid	0-100%	0-100	59-66	72-81
TCIL	TCIL (lamp) Control	ON / OFF	OFF	OFF	OFF
VREF	Vehicle Reference	0-5.1v	4.9-5.1	4.9-5.1	4.9-5.1
WAC	A/C WOT Relay Control	ON / OFF	ON (AC on at WOT)	OFF	OFF

1996-2003 WINDSTAR 3.8L V6 VIN R (A/T) INPUTS

PCM PID Acronym	Parameter Identification	PID Range	PID Value at Hot Idle	PID Value at 30	PID Value at 55
ACCS	A/C Switch Signal	ON / OFF	ON (if on)	OFF	OFF
ACP	AC Pressure Switch	0-5.1v	1.5-1.9	0.6-0.8	0.6-0.8
BPP	Brake Position	ON / OFF	ON (if on)	OFF	OFF
CID	CMP Sensor	0-999 Hz	5-7	10-12	13-15
CKP	CKP Sensor	0-9999 Hz	390-450	700-740	950-1050
DPFE	DPFE Sensor	0-5.1v	0.2-1.3	0.2-4.5	0.2-4.5
ECT	ECT Sensor	-40-304°F	160-200	160-200	160-200
FEPS	Flash EEPROM	0-5.1v	0.1	0.1	0.1
FTP	Fuel Tank Pressure	0-5.1v at 0-10" H2O	2.6v at 0" H2O (gas cap "off")	2.6v at 0" H2O (gas cap "off")	2.6v at 0" H2O (gas cap "off")
FPM	Fuel Pump Monitor	ON / OFF	ON	ON	ON
IAT	IAT Sensor	-40-304°F	50-120	50-120	50-120
GEAR	Gear Position	1-2-3-4	1	3	4
HO2S-11	HO2S-11 (Bank 1)	0.1-1.1v	0.1-1.1	0.1-1.1	0.1-1.1
HO2S-12	HO2S-12 (Bank 1)	0.1-1.1v	0.1-1.1	0.1-1.1	0.1-1.1
HO2S-21	HO2S-21 (Bank 2)	0.1-1.1v	0.1-1.1	0.1-1.1	0.1-1.1
HO2S-22	HO2S-22 (Bank 2)	0.1-1.1v	0.1-1.1	0.1-1.1	0.1-1.1
IMRC-M	IMRC Monitor	5v / 0v	5	5	5
LOAD	Engine Load	0-100%	15-20	19-27	31-35
MAFV	MAF Sensor	0-5.1v	0.6-0.9	0.9-1.4	1.3-2.9
MISF	Misfire Detection	ON / OFF	OFF	OFF	OFF
OCTADJ	Octane Adjustment	CL/OPEN	CLOSED	CLOSED	CLOSED
PNP	PNP Switch Signal	Neutral/DR	NEUTRAL	DRIVE	DRIVE
RPM	Engine Speed	0-10K rpm	700-730	1250-1400	1700-1870
TCS	TCS Switch	ON / OFF	OFF	OFF	OFF
TFT	TFT Sensor Signal	-40-304°F	110-210	110-210	110-210
TPV	TP Sensor	0-5.1v	0.53-1.27	0.8-1.1	0.8-1.2
TRV/TR	TR Sensor	0v / 1.7v	0v	1.7	1.7
TR1	TR1 Sensor	0v / 11.5v	0v	11.5	11.5
TR2	TR2 Sensor	0v / 11.5v	0v	11.5	11.5
TR4	TR4 Sensor	0v / 11.5v	0v	11.5	11.5
TSS	TSS Sensor	0-9999 Hz	43	88-93	99-113
Vacuum	Engine Vacuum	0-30" Hg	19-20	18-20	8-15
VPWR	Vehicle Power	0-25.5v	12-14	12-14	12-14
VSS	Vehicle Speed (Hertz or MPH)	0-999 Hz (0-159 mph)	0 Hz = 0 mph	65 Hz = 30 mph	125 Hz = 55 mph
WACA	A/C WOT Relay Monitor	ON / OFF	ON (AC on at WOT)	OFF	OFF

1996-2003 WINDSTAR 3.8L V6 VIN R (A/T) OUTPUTS

PCM PID Acronym	Parameter Identification	PID Range	PID Value at Hot Idle	PID Value at 30	PID Value at 55
ALDFDC	Alternator Field (Hz)	0-999 Hz	0-130	00-130	00-130
CD1-3	Coil 1 Driver-3	0-60° dwell	6	8	12
EGR VR	EGR VR Solenoid	0-100%	0	0-40	35-50
EVAP CP	EVAP Purge Valve	0-100%	0-100	0-100	0-100
EVAP CV	EVAP CV Valve	0-100%	0-100	0-100	0-100
EPC	EPC Solenoid	0-300 psi	15	16-25	Q42
FP	Fuel Pump Control	ON / OFF	ON	ON	ON
FuelPW1	INJ Pulsewidth - Bank 1	0-999 ms	3.5-3.8	4.9-7.9	6.1-11.0
FuelPW2	INJ Pulsewidth - Bank 2	0-999 ms	3.5-3.8	4.9-7.9	6.1-11.0
HFC	High Fan Control	ON / OFF	ON	OFF	OFF
HTR-11	HO2S-11 Heater	ON / OFF	ON	ON	ON
HTR-12	HO2S-12 Heater	ON / OFF	ON	ON	ON
HTR-21	HO2S-21 Heater	ON / OFF	ON	ON	ON
HTR-22	HO2S-22 Heater	ON / OFF	ON	ON	ON
IAC	Idle Air Control	0-100%	25-35	30-55	50-59
IMRC	IMRC Solenoid	ON / OFF	OFF	OFF	OFF
INJ1	INJ 1 Pulsewidth	0-999 ms	3.5-3.8	4.9-7.9	6.1-11.0
INJ2	INJ 2 Pulsewidth	0-999 ms	3.5-3.8	4.9-7.9	6.1-11.0
INJ3	INJ 3 Pulsewidth	0-999 ms	3.5-3.8	4.9-7.9	6.1-11.0
INJ4	INJ 4 Pulsewidth	0-999 ms	3.5-3.8	4.9-7.9	6.1-11.0
INJ5	INJ 5 Pulsewidth	0-999 ms	3.5-3.8	4.9-7.9	6.1-11.0
INJ6	INJ 6 Pulsewidth	0-999 ms	3.5-3.8	4.9-7.9	6.1-11.0
LFC	Low Fan Control	OFF	ON	OFF	OFF
LongFT1	Long Term FT - Bank 1	-20 to 20%	-5 to +5	-5 to +5	-5 to +5
LongFT2	Long Term FT - Bank 2	-20 to 20%	-5 to +5	-5 to +5	-5 to +5
MIL	MIL (lamp) Control	ON / OFF	OFF	OFF	OFF
SHTFT1	Short Term F/T - Bank 1	-10 to 10%	-5 to +5	-5 to +5	-5 to +5
SHTFT2	Short Term FT - Bank 2	-10 to 10%	-5 to +5	-5 to +5	-5 to +5
SPARK	Spark Advance	-90° to 90°	15-20	25-35	27-36
SS1	Shift Solenoid 1	ON / OFF	ON	OFF	OFF
SS2	Shift Solenoid 2	ON / OFF	OFF	ON	ON
SS3	Shift Solenoid 3	ON / OFF	OFF	ON	ON
TCC	TCC Solenoid	0-100%	0%	0-40	90-95
TCIL	TCIL (lamp) Control	ON / OFF	OFF	OFF	OFF
VREF	Vehicle Reference	0-5.1v	4.9-5.1	4.9-5.1	4.9-5.1
WAC	A/C WOT Relay Control	ON / OFF	ON (AC on at WOT)	OFF	OFF

Expedition PID Data

1997-2003 EXPEDITION 4.6L V8 SOHC 2V MFI VIN W, 6 (A/T)

PCM PID Acronym	Parameter Identification	PID Range	PID Value at Hot Idle	PID Value at 30	PID Value at 55
4x4L	4x4 Switch Signal	ON / OFF	ON (if on)	OFF	OFF
ACCS	A/C Switch Signal	ON / OFF	ON (if on)	OFF	OFF
BPP	Brake Position	ON / OFF	ON (if on)	OFF	OFF
DPFE	DPFE Sensor	0-5.1v	0.2-1.3	0.2-4.5	0.2-4.5
CHT	CHT Sensor	-40-500°F	194	194	194
FEPS	Flash EEPROM	0-5.1v	0.1	0.1	0.1
FLI	Fuel Level Input	1.7 / 50%	1.7 / 50%	1.7 / 50%	1.7 / 50%
FPM	Fuel Pump Monitor	ON / OFF	ON	ON	ON
FTP	Fuel Tank Pressure	0-5.1v at 0-10" H2O	2.6v at 0" H2O (gas cap "off")	2.6v at 0" H2O (gas cap "off")	2.6v at 0" H2O (gas cap "off")
GEAR	Gear Position	1-2-3-4	1	3	4
HO2S-11	HO2S-11 (Bank 1)	0.1-1.1v	0.1-1.1	0.1-1.1	0.1-1.1
HO2S-12	HO2S-12 (Bank 1)	0.1-1.1v	0.1-1.1	0.1-1.1	0.1-1.1
HO2S-21	HO2S-21 (Bank 2)	0.1-1.1v	0.1-1.1	0.1-1.1	0.1-1.1
HO2S-22	HO2S-22 (Bank 2)	0.1-1.1v	0.1-1.1	0.1-1.1	0.1-1.1
IAT	IAT Sensor	-40-304°F	50-120	50-120	50-120
LOAD	Engine Load	0-100%	14-16	20-25	24-35
MAFV	MAF Sensor	0-5.1v	0.7-0.9	1.2-1.7	1.6-2.4
MISF	Misfire Detection	ON / OFF	OFF	OFF	OFF
OCTADJ	Octane Adjustment	CL/OPEN	CLOSED	CLOSED	CLOSED
OSS	OSS Sensor	0-9999 Hz	0	125-131	245-255
PNP	PNP Switch Signal	ON / OFF	ON	OFF	OFF
PTO	Power Takeoff Switch	ON / OFF	ON (if on)	OFF	OFF
RPM	Engine Speed	0-10K rpm	680-830	1200-1500	1600-1800
TCS	TCS Switch	ON / OFF	OFF	OFF	OFF
TFT	TFT Sensor Signal	-40-304°F	110-210	110-210	110-210
TPV	TP Sensor	0-5.1v	0.53-1.27	0.8-1.0	0.9-1.2
TRV/TR	TR Sensor	0v / 1.7v	0v	1.7	1.7
TR1	TR1 Sensor	0v / 11.5v	0v	11.5	11.5
TR2	TR2 Sensor	0v / 11.5v	0v	11.5	11.5
TR4	TR4 Sensor	0v / 11.5v	0v	11.5	11.5
Vacuum	Engine Vacuum	0-30" Hg	19-20	18-20	8-15
VPWR	Vehicle Power	0-25.5v	12-14	12-14	12-14
VSS	Vehicle Speed (Hertz or MPH)	0-999 Hz (0-159 mph)	0 Hz = 0 mph	65 Hz = 30 mph	125 Hz = 55 mph

1997-2003 EXPEDITION 4.6L V8 SOHC MFI 2V VIN W, 6 (A/T)

PCM PID Acronym	Parameter Identification	PID Range	PID Value at Hot Idle	PID Value at 30	PID Value at 55
CCS	Coast Clutch Solenoid Control	ON / OFF	OFF	OFF	OFF
EGR VR	EGR VR Solenoid	0-100%	0	0-40	35-45
EVAP CP	EVAP Purge Valve	0-100%	0-100	0-100	0-100
EVAP CV	EVAP CV Valve	0-100%	0-100	0-100	0-100
EPC	EPC Solenoid	0-300 psi	6	40	40
FP	Fuel Pump Control	ON / OFF	ON	ON	ON
FuelPW1	INJ Pulsewidth - Bank 1	0-999 ms	2.7-4.1	4.5-8.0	5.5-9
FuelPW2	INJ Pulsewidth - Bank 2	0-999 ms	2.7-4.1	4.5-8.0	5.5-9
HTR-11	HO2S-11 Heater	ON / OFF	ON	ON	ON
HTR-12	HO2S-12 Heater	ON / OFF	ON	ON	ON
HTR-21	HO2S-21 Heater	ON / OFF	ON	ON	ON
HTR-22	HO2S-22 Heater	ON / OFF	ON	ON	ON
IAC	Idle Air Control	0-100%	30-34	28-32	63-70
INJ1	INJ 1 Pulsewidth	0-999 ms	2.7-4.1	4.5-8.0	5.5-9.0
INJ2	INJ 2 Pulsewidth	0-999 ms	2.7-4.1	4.5-8.0	5.5-9.0
INJ3	INJ 3 Pulsewidth	0-999 ms	2.7-4.1	4.5-8.0	5.5-9.0
INJ4	INJ 4 Pulsewidth	0-999 ms	2.7-4.1	4.5-8.0	5.5-9.0
INJ5	INJ 5 Pulsewidth	0-999 ms	2.7-4.1	4.5-8.0	5.5-9.0
INJ6	INJ 6 Pulsewidth	0-999 ms	2.7-4.1	4.5-8.0	5.5-9.0
INJ7	INJ 7 Pulsewidth	0-999 ms	2.7-4.1	4.5-8.0	5.5-9.0
INJ8	INJ 8 Pulsewidth	0-999 ms	2.7-4.1	4.5-8.0	5.5-9.0
LongFT1	Long Term FT - Bank 1	-20 to 20%	-5 to +5	-5 to +5	-5 to +5
LongFT2	Long Term FT - Bank 2	-20 to 20%	-5 to +5	-5 to +5	-5 to +5
MIL	MIL (lamp) Control	ON / OFF	OFF	OFF	OFF
SHTFT1	Short Term F/T - Bank 1	-10 to 10%	-5 to +5	-5 to +5	-5 to +5
SHTFT2	Short Term FT - Bank 2	-10 to 10%	-5 to +5	-5 to +5	-5 to +5
SPARK	Spark Advance	-90º to 90º	15-18	23-29	26-32
SS1	Shift Solenoid 1	ON / OFF	ON	OFF	OFF
SS2	Shift Solenoid 2	ON / OFF	OFF	ON	OFF
TCC	TCC Solenoid	0-100%	0	90-100	90-100
TCIL	TCIL (lamp) Control	ON / OFF	OFF	OFF	OFF
VREF	Vehicle Reference	0-5.1v	4.9-5.1	4.9-5.1	4.9-5.1

Blackwood PID Data

2002-03 BLACKWOOD 5.4L V8 DOHC 4V MFI VIN A (A/T) INPUTS

PCM PID Acronym	Parameter Identification	PID Range	PID Value at Hot Idle	PID Value at 30	PID Value at 55
4x4L	4x4 Switch Signal	ON / OFF	ON (if on)	OFF	OFF
ACCS	A/C Switch Signal	ON / OFF	ON (if on)	OFF	OFF
BPP	Brake Position	ON / OFF	ON (if on)	OFF	OFF
CHT	CHT Sensor	-40-500°F	194	194	194
CID	CMP Sensor	0-999 Hz	6-8	10-12	14-17
CKP	CKP Sensor	0-9999 Hz	410	800-850	900-1125
DPFE	DPFE Sensor	0-5.1v	0.2-1.3	0.2-4.5	0.2-4.5
ECT	ECT Sensor	-40-304°F	160-200	160-200	160-200
FEPS	Flash EEPROM	0-5.1v	0.1	0.1	0.1
FLI	Fuel Level Input	1.7 / 50%	1.7 / 50%	1.7 / 50%	1.7 / 50%
FPM	Fuel Pump Monitor	ON / OFF	ON	ON	ON
FTP	Fuel Tank Pressure	0-5.1v at 0-10" H2O	2.6v at 0" H2O (gas cap "off")	2.6v at 0" H2O (gas cap "off")	2.6v at 0" H2O (gas cap "off")
GEAR	Gear Position	1-2-3-4	1	3	4
HO2S-11	HO2S-11 (Bank 1)	0.1-1.1v	0.1-1.1	0.1-1.1	0.1-1.1
HO2S-12	HO2S-12 (Bank 1)	0.1-1.1v	0.1-1.1	0.1-1.1	0.1-1.1
HO2S-21	HO2S-21 (Bank 2)	0.1-1.1v	0.1-1.1	0.1-1.1	0.1-1.1
HO2S-22	HO2S-22 (Bank 2)	0.1-1.1v	0.1-1.1	0.1-1.1	0.1-1.1
IAT	IAT Sensor	-40-304°F	50-120	50-120	50-120
KS1	Knock Sensor 1	0-5.1v	0	0	0
LOAD	Engine Load	0-100%	14-16	19-25	26-35
MAFV	MAF Sensor	0-5.1v	0.7-0.9	1.1-1.6	1.7-2.4
MISF	Misfire Detection	ON / OFF	OFF	OFF	OFF
OCTADJ	Octane Adjustment	CL/OPEN	CLOSED	CLOSED	CLOSED
OSS	OSS Sensor	0-9999 Hz	0	125	350
PNP	PNP Switch Signal	ON / OFF	ON	OFF	OFF
PTO	Power Takeoff Switch	ON / OFF	ON (if on)	OFF	OFF
RPM	Engine Speed	0-10K rpm	790-710	1200-1270	1590-1675
TCS	TCS Switch	ON / OFF	OFF	OFF	OFF
TFT	TFT Sensor Signal	-40-304°F	110-210	110-210	110-210
TPO	TPO TR Sensor	0-5.1v	0.1	0.8	1.0
TPV	TP Sensor	0-5.1v	0.53-1.27	0.8-1.0	1.0-1.3
TR1	TR1 Sensor	0v / 11.5v	0	11.5	11.5
TR2	TR2 Sensor	0v / 11.5v	0	11.5	11.5
TR3/TR	TR3A Sensor	0v / 1.7v	0	1.7	1.7
TR4	TR4 Sensor	0v / 11.5v	0	11.5	11.5
Vacuum	Engine Vacuum	0-30" Hg	19-20	18-20	8-15
VPWR	Vehicle Power	0-25.5v	12-14	12-14	12-14
VSS	Vehicle Speed (Hertz or MPH)	0-999 Hz (0-159 mph)	0 Hz = 0 mph	65 Hz = 30 mph	125 Hz = 55 mph

2002-03 BLACKWOOD 5.4L V8 DOHC 4V MFI VIN A (A/T) OUTPUTS

PCM PID Acronym	Parameter Identification	PID Range	PID Value at Hot Idle	PID Value at 30	PID Value at 55
ARC	Automatic Ride Sig.	ON / OFF	OFF	OFF	OFF
CD1-CD8	COP Driver 1-8	0-45	5	6	8
CHTIL	CHT (lamp) Control	ON / OFF	OFF	OFF	OFF
EGR VR	EGR VR Solenoid	0-100%	0	0-40	35-45
EVAP CP	EVAP Purge Valve	0-100%	0-100	0-100	0-100
EVAP CV	EVAP CV Valve	0-100%	0-100	0-100	0-100
EPC	EPC Sol. Control	7.4v / 4	4	5	5
FP	Fuel Pump Control	OFF	ON	ON	ON
FuelPW1	INJ Pulsewidth - Bank 1	0-999 ms	3.2-3.8	4.0-6.9	6.5-12
FuelPW2	INJ Pulsewidth - Bank 2	0-999 ms	3.2-3.8	4.0-6.9	6.5-12
HTR-11	HO2S-11 Heater	ON / OFF	ON	ON	ON
HTR-12	HO2S-12 Heater	ON / OFF	ON	ON	ON
HTR-21	HO2S-21 Heater	ON / OFF	ON	ON	ON
HTR-22	HO2S-22 Heater	ON / OFF	ON	ON	ON
IAC	Idle Air Control	0-100%	30-34	43-48	58-61
INJ1	INJ 1 Pulsewidth	0-999 ms	3.2-3.8	4.0-6.9	6.5-12
INJ2	INJ 2 Pulsewidth	0-999 ms	3.2-3.8	4.0-6.9	6.5-12
INJ3	INJ 3 Pulsewidth	0-999 ms	3.2-3.8	4.0-6.9	6.5-12
INJ4	INJ 4 Pulsewidth	0-999 ms	3.2-3.8	4.0-6.9	6.5-12
INJ5	INJ 5 Pulsewidth	0-999 ms	3.2-3.8	4.0-6.9	6.5-12
INJ6	INJ 6 Pulsewidth	0-999 ms	3.2-3.8	4.0-6.9	6.5-12
INJ7	INJ 7 Pulsewidth	0-999 ms	3.2-3.8	4.0-6.9	6.5-12
INJ8	INJ 8 Pulsewidth	0-999 ms	3.2-3.8	4.0-6.9	6.5-12
IMTV	Intake Manifold Tune Valve	0-100%	0	0	0
LongFT1	Long Term FT - Bank 1	-20 to 20%	-5 to +5	-5 to +5	-5 to +5
LongFT2	Long Term FT - Bank 2	-20 to 20%	-5 to +5	-5 to +5	-5 to +5
MIL	MIL (lamp) Control	ON / OFF	OFF	OFF	OFF
SHTFT1	Short Term F/T - Bank 1	-10 to 10%	-5 to +5	-5 to +5	-5 to +5
SHTFT2	Short Term FT - Bank 2	-10 to 10%	-5 to +5	-5 to +5	-5 to +5
SPARK	Spark Advance	-90° to 90°	12-17	35-40	28-37
SS1	Shift Solenoid 1	ON / OFF	ON	OFF	OFF
SS2	Shift Solenoid 2	ON / OFF	OFF	ON	OFF
TCC	TCC Solenoid	0-100%	0	0-40	90-95
TCIL	TCIL (lamp) Control	ON / OFF	OFF	OFF	OFF
TPO	TPO TR Sensor	0-5.1v	0.1	0.8	1.0
VREF	Vehicle Reference	0-5.1v	4.9-5.1	4.9-5.1	4.9-5.1

Expedition & Navigator PID Data

1997-2001 EXPEDITION/NAVIGATOR 5.4L V8 VIN L (E4OD)

PCM PID Acronym	Parameter Identification	PID Range	PID Value at Hot Idle	PID Value at 30	PID Value at 55
4x4L	4x4 Switch Signal	ON / OFF	ON (if on)	OFF	OFF
ACCS	A/C Switch Signal	ON / OFF	ON (if on)	OFF	OFF
BPP	Brake Position	ON / OFF	ON (if on)	OFF	OFF
CHT	CHT Sensor	-40-500°F	194	194	194
CID	CMP Sensor	0-999 Hz	6-8	10-12	14-17
CKP	CKP Sensor	0-9999 Hz	411	800-840	1000
DPFE	DPFE Sensor	0-5.1v	0.2-1.3	0.2-4.5	0.2-4.5
ECT	ECT Sensor	-40-304°F	160-200	160-200	160-200
FEPS	Flash EEPROM	0-5.1v	0.1	0.1	0.1
FLI	Fuel Level Input	1.7 / 50%	1.7 / 50%	1.7 / 50%	1.7 / 50%
FPM	Fuel Pump Monitor	ON / OFF	ON	ON	ON
FTP	Fuel Tank Pressure	0-5.1v at 0-10" H2O	2.6v at 0" H2O (gas cap "off")	2.6v at 0" H2O (gas cap "off")	2.6v at 0" H2O (gas cap "off")
GEAR	Gear Position	1-2-3-4	1	3	4
HO2S-11	HO2S-11 (Bank 1)	0.1-1.1v	0.1-1.1	0.1-1.1	0.1-1.1
HO2S-12	HO2S-12 (Bank 1)	0.1-1.1v	0.1-1.1	0.1-1.1	0.1-1.1
HO2S-21	HO2S-21 (Bank 1)	0.1-1.1v	0.1-1.1	0.1-1.1	0.1-1.1
HO2S-22	HO2S-22 (Bank 1)	0.1-1.1v	0.1-1.1	0.1-1.1	0.1-1.1
IAT	IAT Sensor	-40-304°F	50-120	50-120	50-120
KS1	Knock Sensor 1	0-5.1v	0	0	0
LOAD	Engine Load	0-100%	14-16	19-25	26-35
MAFV	MAF Sensor	0-5.1v	0.7-0.9	1.1-1.6	1.7-2.4
MISF	Misfire Detection	ON / OFF	OFF	OFF	OFF
OCTADJ	Octane Adjustment	CL/OPEN	CLOSED	CLOSED	CLOSED
OSS	OSS Sensor	0-9999 Hz	0	400	700
PNP	PNP Switch Signal	ON / OFF	ON	OFF	OFF
PTO	Power Takeoff Switch	ON / OFF	ON (if on)	OFF	OFF
RPM	Engine Speed	0-10K rpm	760-830	1200-1270	1590-1675
TCS	TCS Switch	ON / OFF	OFF	OFF	OFF
TFT	TFT Sensor Signal	-40-304°F	110-210	110-210	110-210
TPO	TPO TR Sensor	0-5.1v	0.1	0.8	1.0
TPV	TP Sensor	0-5.1v	0.53-1.27	0.8-1.0	1.0-1.3
TR1	TR1 Sensor	0v / 11.5v	0	11.5	11.5
TR2	TR2 Sensor	0v / 11.5v	0	11.5	11.5
TR3/TR	TR3A Sensor	0v / 1.7v	0	1.7	1.7
TR4	TR4 Sensor	0v / 11.5v	0	11.5	11.5
Vacuum	Engine Vacuum	0-30" Hg	19-20	18-20	8-15
VPWR	Vehicle Power	0-25.5v	12-14	12-14	12-14
VSS	Vehicle Speed (Hertz or MPH)	0-999 Hz (0-159 mph)	0 Hz = 0 mph	65 Hz = 30 mph	125 Hz = 55 mph

1997-2001 EXPEDITION/NAVIGATOR 5.4L V8 VIN L (E4OD)

PCM PID Acronym	Parameter Identification	PID Range	PID Value at Hot Idle	PID Value at 30	PID Value at 55
ARC	Auto Ride Control	ON / OFF	OFF	OFF	OFF
CD1-CD8	COP Driver 1-8	0-45	5	6	8
CHTIL	CHT (lamp) Control	ON / OFF	OFF	OFF	OFF
EGR VR	EGR VR Solenoid	0-100%	0	0-40	35-45
EVAP CP	EVAP Purge Valve	0-100%	0-100	0-100	0-100
EVAP CV	EVAP CV Valve	0-100%	0-100	0-100	0-100
EPC	EPC Solenoid	7.4v / 4	4	5	5
FP	Fuel Pump Control	OFF	ON	ON	ON
FuelPW1	INJ Pulsewidth - Bank 1	0-999 ms	3.2-4.0	4.1-6.9	6.5-12
FuelPW2	INJ Pulsewidth - Bank 2	0-999 ms	3.2-4.0	4.1-6.9	6.5-12
HTR-11	HO2S-11 Heater	ON / OFF	ON	ON	ON
HTR-12	HO2S-12 Heater	ON / OFF	ON	ON	ON
HTR-21	HO2S-21 Heater	ON / OFF	ON	ON	ON
HTR-22	HO2S-22 Heater	ON / OFF	ON	ON	ON
IAC	Idle Air Control	0-100%	30-34	43-48	58-61
INJ1	INJ 1 Pulsewidth	0-999 ms	3.2-4.0	4.1-6.9	6.5-12
INJ2	INJ 2 Pulsewidth	0-999 ms	3.2-4.0	4.1-6.9	6.5-12
INJ3	INJ 3 Pulsewidth	0-999 ms	3.2-4.0	4.1-6.9	6.5-12
INJ4	INJ 4 Pulsewidth	0-999 ms	3.2-4.0	4.1-6.9	6.5-12
INJ5	INJ 5 Pulsewidth	0-999 ms	3.2-4.0	4.1-6.9	6.5-12
INJ6	INJ 6 Pulsewidth	0-999 ms	3.2-4.0	4.1-6.9	6.5-12
INJ7	INJ 7 Pulsewidth	0-999 ms	3.2-4.0	4.1-6.9	6.5-12
INJ8	INJ 8 Pulsewidth	0-999 ms	3.2-4.0	4.1-6.9	6.5-12
LongFT1	Long Term FT - Bank 1	-20 to 20%	-5 to +5	-5 to +5	-5 to +5
LongFT2	Long Term FT - Bank 2	-20 to 20%	-5 to +5	-5 to +5	-5 to +5
MIL	MIL (lamp) Control	ON / OFF	OFF	OFF	OFF
SHTFT1	Short Term F/T - Bank 1	-10 to 10%	-5 to +5	-5 to +5	-5 to +5
SHTFT2	Short Term FT - Bank 2	-10 to 10%	-5 to +5	-5 to +5	-5 to +5
SPARK	Spark Advance	-90° to 90°	12-17	35-40	28-37
SS1	Shift Solenoid 1	ON / OFF	ON	OFF	OFF
SS2	Shift Solenoid 2	ON / OFF	OFF	ON	OFF
TCC	TCC Solenoid	0-100%	0	0-40	90-95
TCIL	TCIL (lamp) Control	ON / OFF	OFF	OFF	OFF
TPO	TPO TR Sensor	0-5.1v	0.1	0.8	1.0
VREF	Vehicle Reference	0-5.1v	4.9-5.1	4.9-5.1	4.9-5.1

Navigator PID Data

1999-2001 NAVIGATOR 5.4L V8 DOHC VIN A (A/T) INPUTS

PCM PID Acronym	Parameter Identification	PID Range	PID Value at Hot Idle	PID Value at 30	PID Value at 55
4x4L	4x4 Switch Signal	ON / OFF	ON (if on)	OFF	OFF
ACCS	A/C Switch Signal	ON / OFF	ON (if on)	OFF	OFF
BPP	Brake Position	ON / OFF	ON (if on)	OFF	OFF
CHT	CHT Sensor	-40-500°F	194	194	194
CID	CMP Sensor	0-999 Hz	6-8	10-12	14-17
CKP	CKP Sensor	0-9999 Hz	410	800-850	900-1125
DPFE	DPFE Sensor	0-5.1v	0.2-1.3	0.2-4.5	0.2-4.5
ECT	ECT Sensor	-40-304°F	160-200	160-200	160-200
FEPS	Flash EEPROM	0-5.1v	0.1	0.1	0.1
FLI	Fuel Level Input	1.7 / 50%	1.7 / 50%	1.7 / 50%	1.7 / 50%
FPM	Fuel Pump Monitor	ON / OFF	ON	ON	ON
FTP	Fuel Tank Pressure	0-5.1v at 0-10" H2O	2.6v at 0" H2O (gas cap "off")	2.6v at 0" H2O (gas cap "off")	2.6v at 0" H2O (gas cap "off")
GEAR	Gear Position	1-2-3-4	1	3	4
HO2S-11	HO2S-11 (Bank 1)	0.1-1.1v	0.1-1.1	0.1-1.1	0.1-1.1
HO2S-12	HO2S-12 (Bank 1)	0.1-1.1v	0.1-1.1	0.1-1.1	0.1-1.1
HO2S-21	HO2S-21 (Bank 2)	0.1-1.1v	0.1-1.1	0.1-1.1	0.1-1.1
HO2S-22	HO2S-22 (Bank 2)	0.1-1.1v	0.1-1.1	0.1-1.1	0.1-1.1
IAT	IAT Sensor	-40-304°F	50-120	50-120	50-120
KS1	Knock Sensor 1	0-5.1v	0	0	0
LOAD	Engine Load	0-100%	14-16	19-25	26-35
MAFV	MAF Sensor	0-5.1v	0.7-0.9	1.1-1.6	1.7-2.4
MISF	Misfire Detection	ON / OFF	OFF	OFF	OFF
OCTADJ	Octane Adjustment	CL/OPEN	CLOSED	CLOSED	CLOSED
OSS	OSS Sensor	0-9999 Hz	0	125	350
PNP	PNP Switch Signal	ON / OFF	ON	OFF	OFF
PTO	Power Takeoff Switch	ON / OFF	ON (if on)	OFF	OFF
RPM	Engine Speed	0-10K rpm	790-710	1200-1270	1590-1675
TCS	TCS Switch	ON / OFF	OFF	OFF	OFF
TFT	TFT Sensor Signal	-40-304°F	110-210	110-210	110-210
TPO	TPO TR Sensor	0-5.1v	0.1	0.8	1.0
TPV	TP Sensor	0-5.1v	0.53-1.27	0.8-1.0	1.0-1.3
TR1	TR1 Sensor	0v / 11.5v	0	11.5	11.5
TR2	TR2 Sensor	0v / 11.5v	0	11.5	11.5
TR3/TR	TR3A Sensor	0v / 1.7v	0	1.7	1.7
TR4	TR4 Sensor	0v / 11.5v	0	11.5	11.5
Vacuum	Engine Vacuum	0-30" Hg	19-20	18-20	8-15
VPWR	Vehicle Power	0-25.5v	12-14	12-14	12-14
VSS	Vehicle Speed (Hertz or MPH)	0-999 Hz (0-159 mph)	0 Hz = 0 mph	65 Hz = 30 mph	125 Hz = 55 mph

1999-2001 NAVIGATOR 5.4L V8 DOHC VIN A (A/T) OUTPUTS

PCM PID Acronym	Parameter Identification	PID Range	PID Value at Hot Idle	PID Value at 30	PID Value at 55
ARC	Automatic Ride Sig.	ON / OFF	OFF	OFF	OFF
CD1-CD8	COP Driver 1-8	0-45	5	6	8
CHTIL	CHT (lamp) Control	ON / OFF	OFF	OFF	OFF
EGR VR	EGR VR Solenoid	0-100%	0	0-40	35-45
EVAP CP	EVAP Purge Valve	0-100%	0-100	0-100	0-100
EVAP CV	EVAP CV Valve	0-100%	0-100	0-100	0-100
EPC	EPC Sol. Control	7.4v / 4	4	5	5
FP	Fuel Pump Control	OFF	ON	ON	ON
FuelPW1	INJ Pulsewidth - Bank 1	0-999 ms	3.2-3.8	4.0-6.9	6.5-12
FuelPW2	INJ Pulsewidth - Bank 2	0-999 ms	3.2-3.8	4.0-6.9	6.5-12
HTR-11	HO2S-11 Heater	ON / OFF	ON	ON	ON
HTR-12	HO2S-12 Heater	ON / OFF	ON	ON	ON
HTR-21	HO2S-21 Heater	ON / OFF	ON	ON	ON
HTR-22	HO2S-22 Heater	ON / OFF	ON	ON	ON
IAC	Idle Air Control	0-100%	30-34	43-48	58-61
INJ1	INJ 1 Pulsewidth	0-999 ms	3.2-3.8	4.0-6.9	6.5-12
INJ2	INJ 2 Pulsewidth	0-999 ms	3.2-3.8	4.0-6.9	6.5-12
INJ3	INJ 3 Pulsewidth	0-999 ms	3.2-3.8	4.0-6.9	6.5-12
INJ4	INJ 4 Pulsewidth	0-999 ms	3.2-3.8	4.0-6.9	6.5-12
INJ5	INJ 5 Pulsewidth	0-999 ms	3.2-3.8	4.0-6.9	6.5-12
INJ6	INJ 6 Pulsewidth	0-999 ms	3.2-3.8	4.0-6.9	6.5-12
INJ7	INJ 7 Pulsewidth	0-999 ms	3.2-3.8	4.0-6.9	6.5-12
INJ8	INJ 8 Pulsewidth	0-999 ms	3.2-3.8	4.0-6.9	6.5-12
IMTV	Intake Manifold Tune Valve	0-100%	0	0	0
LongFT1	Long Term FT - Bank 1	-20 to 20%	-5 to +5	-5 to +5	-5 to +5
LongFT2	Long Term FT - Bank 2	-20 to 20%	-5 to +5	-5 to +5	-5 to +5
MIL	MIL (lamp) Control	ON / OFF	OFF	OFF	OFF
SHTFT1	Short Term F/T - Bank 1	-10 to 10%	-5 to +5	-5 to +5	-5 to +5
SHTFT2	Short Term FT - Bank 2	-10 to 10%	-5 to +5	-5 to +5	-5 to +5
SPARK	Spark Advance	-90° to 90°	12-17	35-40	28-37
SS1	Shift Solenoid 1	ON / OFF	ON	OFF	OFF
SS2	Shift Solenoid 2	ON / OFF	OFF	ON	OFF
TCC	TCC Solenoid	0-100%	0	0-40	90-95
TCIL	TCIL (lamp) Control	ON / OFF	OFF	OFF	OFF
TPO	TPO TR Sensor	0-5.1v	0.1	0.8	1.0
VREF	Vehicle Reference	0-5.1v	4.9-5.1	4.9-5.1	4.9-5.1

2001-03 NAVIGATOR 5.4L V8 4V DOHC VIN R (A/T) INPUTS

PCM PID Acronym	Parameter Identification	PID Range	PID Value at Hot Idle	PID Value at 30	PID Value at 55
4x4L	4x4 Switch Signal	ON / OFF	ON (if on)	OFF	OFF
ACCS	A/C Switch Signal	ON / OFF	ON (if on)	OFF	OFF
BPP	Brake Position	ON / OFF	ON (if on)	OFF	OFF
CHT	CHT Sensor	-40-500°F	194	194	194
CID	CMP Sensor	0-999 Hz	6-8	10-12	14-17
CKP	CKP Sensor	0-9999 Hz	410	800-850	900-1125
DPFE	DPFE Sensor	0-5.1v	0.2-1.3	0.2-4.5	0.2-4.5
ECT	ECT Sensor	-40-304°F	160-200	160-200	160-200
FEPS	Flash EEPROM	0-5.1v	0.1	0.1	0.1
FLI	Fuel Level Input	1.7 / 50%	1.7 / 50%	1.7 / 50%	1.7 / 50%
FPM	Fuel Pump Monitor	ON / OFF	ON	ON	ON
FTP	Fuel Tank Pressure	0-5.1v at 0-10" H2O	2.6v at 0" H2O (gas cap "off")	2.6v at 0" H2O (gas cap "off")	2.6v at 0" H2O (gas cap "off")
GEAR	Gear Position	1-2-3-4	1	3	4
HO2S-11	HO2S-11 (Bank 1)	0.1-1.1v	0.1-1.1	0.1-1.1	0.1-1.1
HO2S-12	HO2S-12 (Bank 1)	0.1-1.1v	0.1-1.1	0.1-1.1	0.1-1.1
HO2S-21	HO2S-21 (Bank 2)	0.1-1.1v	0.1-1.1	0.1-1.1	0.1-1.1
HO2S-22	HO2S-22 (Bank 2)	0.1-1.1v	0.1-1.1	0.1-1.1	0.1-1.1
IAT	IAT Sensor	-40-304°F	50-120	50-120	50-120
KS1	Knock Sensor 1	0-5.1v	0	0	0
LOAD	Engine Load	0-100%	14-16	19-25	26-35
MAFV	MAF Sensor	0-5.1v	0.7-0.9	1.1-1.6	1.7-2.4
MISF	Misfire Detection	ON / OFF	OFF	OFF	OFF
OCTADJ	Octane Adjustment	CL/OPEN	CLOSED	CLOSED	CLOSED
OSS	OSS Sensor	0-9999 Hz	0	125	350
PNP	PNP Switch Signal	ON / OFF	ON	OFF	OFF
PTO	Power Takeoff Switch	ON / OFF	ON (if on)	OFF	OFF
RPM	Engine Speed	0-10K rpm	790-710	1200-1270	1590-1675
TCS	TCS Switch	ON / OFF	OFF	OFF	OFF
TFT	TFT Sensor Signal	-40-304°F	110-210	110-210	110-210
TPO	TPO TR Sensor	0-5.1v	0.1	0.8	1.0
TPV	TP Sensor	0-5.1v	0.53-1.27	0.8-1.0	1.0-1.3
TR1	TR1 Sensor	0v / 11.5v	0	11.5	11.5
TR2	TR2 Sensor	0v / 11.5v	0	11.5	11.5
TR3/TR	TR3A Sensor	0v / 1.7v	0	1.7	1.7
TR4	TR4 Sensor	0v / 11.5v	0	11.5	11.5
Vacuum	Engine Vacuum	0-30" Hg	19-20	18-20	8-15
VPWR	Vehicle Power	0-25.5v	12-14	12-14	12-14
VSS	Vehicle Speed (Hertz or MPH)	0-999 Hz (0-159 mph)	0 Hz = 0 mph	65 Hz = 30 mph	125 Hz = 55 mph

<u>2001-03 NAVIGATOR 5.4L V8 4V DOHC VIN R (A/T) OUTPUTS</u>

PCM PID Acronym	Parameter Identification	PID Range	PID Value at Hot Idle	PID Value at 30	PID Value at 55
ARC	Automatic Ride Sig.	ON / OFF	OFF	OFF	OFF
CD1-CD8	COP Driver 1-8	0-45	5	6	8
CHTIL	CHT (lamp) Control	ON / OFF	OFF	OFF	OFF
EGR VR	EGR VR Solenoid	0-100%	0	0-40	35-45
EVAP CP	EVAP Purge Valve	0-100%	0-100	0-100	0-100
EVAP CV	EVAP CV Valve	0-100%	0-100	0-100	0-100
EPC	EPC Sol. Control	7.4v / 4	4	5	5
FP	Fuel Pump Control	OFF	ON	ON	ON
FuelPW1	INJ Pulsewidth - Bank 1	0-999 ms	3.2-3.8	4.0-6.9	6.5-12
FuelPW2	INJ Pulsewidth - Bank 2	0-999 ms	3.2-3.8	4.0-6.9	6.5-12
HTR-11	HO2S-11 Heater	ON / OFF	ON	ON	ON
HTR-12	HO2S-12 Heater	ON / OFF	ON	ON	ON
HTR-21	HO2S-21 Heater	ON / OFF	ON	ON	ON
HTR-22	HO2S-22 Heater	ON / OFF	ON	ON	ON
IAC	Idle Air Control	0-100%	30-34	43-48	58-61
INJ1	INJ 1 Pulsewidth	0-999 ms	3.2-3.8	4.0-6.9	6.5-12
INJ2	INJ 2 Pulsewidth	0-999 ms	3.2-3.8	4.0-6.9	6.5-12
INJ3	INJ 3 Pulsewidth	0-999 ms	3.2-3.8	4.0-6.9	6.5-12
INJ4	INJ 4 Pulsewidth	0-999 ms	3.2-3.8	4.0-6.9	6.5-12
INJ5	INJ 5 Pulsewidth	0-999 ms	3.2-3.8	4.0-6.9	6.5-12
INJ6	INJ 6 Pulsewidth	0-999 ms	3.2-3.8	4.0-6.9	6.5-12
INJ7	INJ 7 Pulsewidth	0-999 ms	3.2-3.8	4.0-6.9	6.5-12
INJ8	INJ 8 Pulsewidth	0-999 ms	3.2-3.8	4.0-6.9	6.5-12
IMTV	Intake Manifold Tune Valve	0-100%	0	0	0
LongFT1	Long Term FT - Bank 1	-20 to 20%	-5 to +5	-5 to +5	-5 to +5
LongFT2	Long Term FT - Bank 2	-20 to 20%	-5 to +5	-5 to +5	-5 to +5
MIL	MIL (lamp) Control	ON / OFF	OFF	OFF	OFF
SHTFT1	Short Term F/T - Bank 1	-10 to 10%	-5 to +5	-5 to +5	-5 to +5
SHTFT2	Short Term FT - Bank 2	-10 to 10%	-5 to +5	-5 to +5	-5 to +5
SPARK	Spark Advance	-90° to 90°	12-17	35-40	28-37
SS1	Shift Solenoid 1	ON / OFF	ON	OFF	OFF
SS2	Shift Solenoid 2	ON / OFF	OFF	ON	OFF
TCC	TCC Solenoid	0-100%	0	0-40	90-95
TCIL	TCIL (lamp) Control	ON / OFF	OFF	OFF	OFF
TPO	TPO TR Sensor	0-5.1v	0.1	0.8	1.0
VREF	Vehicle Reference	0-5.1v	4.9-5.1	4.9-5.1	4.9-5.1

Explorer PID Data

1991-95 EXPLORER 4.0L V6 VIN X (ALL) INPUTS & OUTPUTS

PCM PID Acronym	Parameter Identification	PID Range	PID Value at Hot Idle	PID Value at 30	PID Value at 55
ACCS	A/C Switch Signal	ON / OFF	ON (if on)	OFF	OFF
BARO	BARO Sensor	0-999 Hz	159	159	159
BOO	Brake Switch Signal	ON / OFF	ON (if on)	OFF	OFF
CANP	EVAP Purge Valve	0-100%	0-100	0-100	0-100
DPFE	DPFE Sensor	0-5.1v	0.4-0.6	0.5-1.0	0.7-1.1
ECT (°F)	ECT Sensor	-40-304°F	160-200	160-200	160-200
ECTV	ECT Sensor	0-5.1v	0.7	0.7	0.7
EVR	EGR VR Solenoid	0-100%	0%	0-40	0-40
FPM	Fuel Pump Monitor	ON / OFF	ON	ON	ON
FuelPW1	INJ Pulsewidth - Bank 1	0-999 ms	3.3-3.5	4.0-4.6	5.0-6.0
FuelPW2	INJ Pulsewidth - Bank 2	0-999 ms	3.3-3.5	4.0-4.6	5.0-6.0
IAC	Idle Air Control	0-100%	20-40	34-40	45-55
IAT (°F)	IAT Sensor	-40-304°F	50-120	50-120	50-120
IATV	IAT Sensor	0-5.1v	1.5-3.5	1.5-3.5	1.5-3.5
LongFT1	Long Term F/T - Bank 1	-20 to 20%	-5 to +5	-5 to +5	-5 to +5
LongFT2	Long Term F/T - Bank 2	-20 to 20%	-5 to +5	-5 to +5	-5 to +5
LOOP	Loop Status	CL or OL	CLOSED	CLOSED	CLOSED
MAF	MAF Sensor	0-5.1v	0.7	1.3-1.4	1.7-2.0v
HO2S-11	HO2S-11 (B1 S1)	0.1-1.1v	0.1-1.1	0.1-1.1	0.1-1.1
HO2S-21	HO2S-21 (B1 S1)	0.1-1.1v	0.1-1.1	0.1-1.1	0.1-1.1
PNP	PNP Switch Signal	Neutral/DR	NEUTRAL	DRIVE	DRIVE
RPM	Engine Speed	0-10K rpm	750-830	1500-1650	1800-2100
SHTFT1	Short Term F/T - Bank 1	-10 to 10%	-5 to +5	-5 to +5	-5 to +5
SHTFT2	Short Term FT - Bank 2	-10 to 10%	-5 to +5	-5 to +5	-5 to +5
SPARK	Spark Advance	-90° to 90°	11-20	26-31	20-32
TP	TP Sensor	0-5.1v	0.9	1.2-1.3	1.4-1.6
TPCT	TP sensor Minimum	0-5.1v	0.1	1.2-1.3	1.4-1.6
TP Mode	TP Sensor	C/T or P/T	C/T	P/T	P/T
VPWR	Vehicle Power	0-25.5v	12-14	12-14	12-14
VREF	Vehicle Reference	0-5.1v	4.9-5.1	4.9-5.1	4.9-5.1
VSS	Vehicle Speed	0-159 mph	0	30	55
WAC	A/C WOT Relay Control	ON / OFF	ON (AC on at WOT)	OFF	OFF

<u>1996-2001 EXPLORER/MOUNTAINEER 4.0L V6 MFI VIN X INPUTS</u>

PCM PID Acronym	Parameter Identification	PID Range	PID Value at Hot Idle	PID Value at 30	PID Value at 55
4x4L	4x4 Low Signal	ON / OFF	ON (if on)	OFF	OFF
ACCS	A/C Switch Signal	ON / OFF	ON (if on)	OFF	OFF
ARC	Auto Ride Control	ON / OFF	OFF	OFF	OFF
BPP	Brake Position	ON / OFF	ON (if on)	OFF	OFF
CID	CMP Sensor	0-999 Hz	6-8	13-15	16-19
CKP	CKP Sensor	0-9999	400-475	900-1100	1140-1220
CPP	CPP Switch Signal	ON / OFF	ON (if in)	OFF	OFF
DPFE	DPFE Sensor	0-5.1v	0.4-0.6	0.4-1.0	0.6-1.1
ECT	ECT Sensor	-40-304°F	160-200	160-200	160-200
FEPS	Flash EEPROM	0-5.1v	0.1	0.1	0.1
FLI	Fuel Level Input	1.7 / 50%	1.7 / 50%	1.7 / 50%	1.7 / 50%
FPM	Fuel Pump Monitor	ON / OFF	ON	ON	ON
FTP	Fuel Tank Pressure	0-5.1v at 0-10" H2O	2.6v at 0" H2O (gas cap "off")	2.6v at 0" H2O (gas cap "off")	2.6v at 0" H2O (gas cap "off")
GEAR	Gear Position	1-2-3-4	1	3	4
HO2S-11	HO2S-11 (Bank 1)	0.1-1.1v	0.1-1.1	0.1-1.1	0.1-1.1
HO2S-12	HO2S-12 (Bank 1)	0.1-1.1v	0.1-1.1	0.1-1.1	0.1-1.1
HO2S-21	HO2S-21 (Bank 2)	0.1-1.1v	0.1-1.1	0.1-1.1	0.1-1.1
IAT	IAT Sensor	-40-304°F	50-120	50-120	50-120
LOAD	Engine Load	0-100%	14-20	21-27	30-34
MAFV	MAF Sensor	0-5.1v	0.6-0.9	1.3-1.7	1.5-2.3
MISF	Misfire Detection	ON / OFF	OFF	OFF	OFF
OCTADJ	Octane Adjustment	CL/OPEN	CLOSED	CLOSED	CLOSED
PNP	PNP Switch Signal	ON / OFF	ON	OFF	OFF
PSP	PSP Switch Signal	LOW	HI: turning	LOW	LOW
RPM	Engine Speed	0-10K rpm	750-830	1500-1650	1800-2100
TCS	TCS Switch	ON / OFF	OFF	OFF	OFF
TFT	TFT Sensor Signal	-40-304°F	110-210	110-210	110-210
TPV	TP Sensor	0-5.1v	0.53-1.27	0.8-1.7	1.2-1.7
TRV/TR	TR Sensor	0v / 1.7v	0	1.7	1.7
TR1	TR1 Sensor	0v / 11.5v	0	11.5	11.5
TR2	TR2 Sensor	0v / 11.5v	0	11.5	11.5
TR4	TR4 Sensor	0v / 11.5v	0	11.5	11.5
TSS	TSS Sensor	0-9999 Hz	85-100	185-205	260-280
Vacuum	Engine Vacuum	0-30" Hg	19-20	18-20	8-15
VPWR	Vehicle Power	0-25.5v	12-14	12-14	12-14
VSS	Vehicle Speed (Hertz or MPH)	0-999 Hz (0-159 mph)	0 Hz = 0 mph	65 Hz = 30 mph	125 Hz = 55 mph

1996-2001 EXPLORER/MOUNTAINEER 4.0L V6 MFI VIN X OUTPUTS

PCM PID Acronym	Parameter Identification	PID Range	PID Value at Hot Idle	PID Value at 30	PID Value at 55
CD1	Coil 1 Driver	0-60°	5	6	8
CD2	Coil 2 Driver	0-60°	5	6	8
CD3	Coil 3 Driver	0-60°	5	6	8
CTO	Clean Tachometer signal	0 -9999	35-49	65-90	90-120
CCS	Coast Clutch Solenoid Control	VBAT	VBAT	VBAT	VBAT
EGR VR	EGR VR Solenoid	0-100%	0	0-40	0-40
VMV ('96)	EVAP Solenoid	0-100%	0-100	0-100	0-100
EVAP CP	EVAP Purge Valve	0-100%	0-100	0-100	0-100
EVAP CV	EVAP CV Valve	0-100%	0-100	0-100	0-100
EPC	EPC Solenoid	0-300 psi	27	36	35
FP	Fuel Pump Control	ON / OFF	ON	ON	ON
FuelPW1	INJ Pulsewidth - Bank 1	0-999 ms	3.4-3.8	3.6-7.5	6.0-9.8
FuelPW2	INJ Pulsewidth - Bank 2	0-999 ms	3.4-3.8	3.6-7.5	6.0-9.8
HTR-11	HO2S-11 Heater	ON / OFF	ON	ON	ON
HTR-12	HO2S-12 Heater	ON / OFF	ON	ON	ON
HTR-21	HO2S-21 Heater	ON / OFF	ON	ON	ON
IAC	Idle Air Control	0-100%	33	35-41	57-68
INJ1	INJ 1 Pulsewidth	0-999 ms	3.4-3.8	3.6-7.5	6.0-9.8
INJ2	INJ 2 Pulsewidth	0-999 ms	3.4-3.8	3.6-7.5	6.0-9.8
INJ3	INJ 3 Pulsewidth	0-999 ms	3.4-3.8	3.6-7.5	6.0-9.8
INJ4	INJ 4 Pulsewidth	0-999 ms	3.4-3.8	3.6-7.5	6.0-9.8
INJ5	INJ 5 Pulsewidth	0-999 ms	3.4-3.8	3.6-7.5	6.0-9.8
INJ6	INJ 6 Pulsewidth	0-999 ms	3.4-3.8	3.6-7.5	6.0-9.8
LongFT1	Long Term FT - Bank 1	-20 to 20%	-5 to +5	-5 to +5	-5 to +5
LongFT2	Long Term FT - Bank 2	-20 to 20%	-5 to +5	-5 to +5	-5 to +5
MIL	MIL (lamp) Control	ON / OFF	OFF	OFF	OFF
SHTFT1	Short Term F/T - Bank 1	-10 to 10%	-5 to +5	-5 to +5	-5 to +5
SHTFT2	Short Term FT - Bank 2	-10 to 10%	-5 to +5	-5 to +5	-5 to +5
SPARK	Spark Advance	-90° to 90°	11-20	26-31	20-32
SS1	Shift Solenoid 1	ON / OFF	ON	OFF	OFF
SS2	Shift Solenoid 2	ON / OFF	OFF	OFF	OFF
SS3	Shift Solenoid 3	ON / OFF	OFF	OFF	ON
TCC	TCC Solenoid	0-100%	0	0	90-95
TCIL	TCIL (lamp) Control	ON / OFF	OFF	OFF	OFF
VREF	Vehicle Reference	0-5.1v	4.9-5.1	4.9-5.1	4.9-5.1
WAC	A/C WOT Relay Control	ON / OFF	ON (AC on at WOT)	ON / OFF	ON / OFF

1997-2003 EXPLORER/MOUNTAINEER 4.0L V6 VIN E, K INPUTS

PCM PID Acronym	Parameter Identification	PID Range	PID Value at Hot Idle	PID Value at 30	PID Value at 55
4x4L	4x4 Low Signal	ON / OFF	ON (if on)	OFF	OFF
ACCS	A/C Switch Signal	ON / OFF	ON (if on)	OFF	OFF
ACP	A/C Pressure	OPEN/CL	OPEN	OPEN	OPEN
ARC	Auto Ride Control	ON / OFF	OFF	OFF	OFF
BPP	Brake Position	ON / OFF	ON (if on)	OFF	OFF
CID	CMP Sensor	0-999 Hz	6-8	11-15	16-19
CKP	CKP Sensor	0-9999	400-475	800-1100	1140-1220
DPFE	DPFE Sensor	0-5.1v	0.4-0.6	0.4-1.0	0.6-1.1
ECT	ECT Sensor	-40-304°F	160-200	160-200	160-200
FEPS	Flash EEPROM	0-5.1v	0.1	0.1	0.1
FLI	Fuel Level Input	1.7 / 50%	1.7 / 50%	1.7 / 50%	1.7 / 50%
FPM	Fuel Pump Monitor	ON / OFF	ON	ON	ON
FTP	Fuel Tank Pressure	0-5.1v at 0-10" H2O	2.6v at 0" H2O (gas cap "off")	2.6v at 0" H2O (gas cap "off")	2.6v at 0" H2O (gas cap "off")
IAT	IAT Sensor	-40-304°F	50-120	50-120	50-120
GEAR	Gear Position	1-2-3-4-5	1	4	5
HO2S-11	HO2S-11 (Bank 1)	0.1-1.1v	0.1-1.1	0.1-1.1	0.1-1.1
HO2S-12	HO2S-12 (Bank 1)	0.1-1.1v	0.1-1.1	0.1-1.1	0.1-1.1
HO2S-21	HO2S-21 (Bank 2)	0.1-1.1v	0.1-1.1	0.1-1.1	0.1-1.1
LOAD	Engine Load	0-100%	14-20	21-27	30-45
MAFV	MAF Sensor	0-5.1v	0.6-0.9	1.3-1.7	1.5-2.3
MISF	Misfire Detection	ON / OFF	OFF	OFF	OFF
OCTADJ	Octane Adjustment	CL/OPEN	CLOSED	CLOSED	CLOSED
PNP	PNP Switch Signal	ON / OFF	ON	OFF	OFF
PSP	PSP Switch Signal	LOW	HI: turning	LOW	LOW
RPM	Engine Speed	0-10K rpm	670-750	1400-1600	1800-2100
TCS	TCS Switch	ON / OFF	OFF	OFF	OFF
TFT	TFT Sensor Signal	-40-304°F	110-210	110-210	110-210
TPV	TP Sensor	0-5.1v	0.53-1.27	0.8-1.7	1.2-1.7
TR1	TR1 Sensor	0v / 11.5v	0v	11.5	11.5
TR2	TR2 Sensor	0v / 11.5v	0v	11.5	11.5
TR4	TR4 Sensor	0v / 11.5v	0v	11.5	11.5
TRV/TR	TR Sensor	0v / 1.7v	0v	1.7	1.7
TSS	TSS Sensor	0-9999 Hz	100-125	185-205	260-280
Vacuum	Engine Vacuum	0-30" Hg	19-20	18-20	8-15
VPWR	Vehicle Power	0-25.5v	12-14	12-14	12-14
VSS	Vehicle Speed (Hertz or MPH)	0-999 Hz (0-159 mph)	0 Hz = 0 mph	65 Hz = 30 mph	125 Hz = 55 mph

1997-2003 EXPLORER/MOUNTAINEER 4.0L V6 VIN E, K OUTPUTS

PCM PID Acronym	Parameter Identification	PID Range	PID Value at Hot Idle	PID Value at 30	PID Value at 55
CCS	Coast Clutch Solenoid Control	VBAT	VBAT	VBAT	VBAT
CD1	Coil 1 Driver	0-60°	5	6	8
CD2	Coil 2 Driver	0-60°	5	6	8
CD3	Coil 3 Driver	0-60°	5	6	8
CTO	Clean Tachometer signal	0 -9999	35-49	65-90	90-120
EGR VR	EGR VR Solenoid	0-100%	0	0-40	0-40
VMV ('96)	EVAP Solenoid	0-100%	0-100	0-100	0-100
EVAP CP	EVAP Purge Valve	0-100%	0-100	0-100	0-100
EVAP CV	EVAP CV Valve	0-100%	0-100	0-100	0-100
EPC	EPC Solenoid	0-300 psi	26	23-38	34-38
FP	Fuel Pump Control	ON / OFF	ON	ON	ON
FuelPW1	INJ Pulsewidth - Bank 1	0-999 ms	3.4-3.8	3.6-7.5	6.0-9.8
FuelPW2	INJ Pulsewidth - Bank 2	0-999 ms	3.4-3.8	3.6-7.5	6.0-9.8
HTR-11	HO2S-11 Heater	ON / OFF	ON	ON	ON
HTR-12	HO2S-12 Heater	ON / OFF	ON	ON	ON
HTR-21	HO2S-21 Heater	ON / OFF	ON	ON	ON
IAC	Idle Air Control	0-100%	25-32	35-49	30-68
INJ1	INJ 1 Pulsewidth	0-999 ms	3.4-3.8	3.6-7.5	6.0-9.8
INJ2	INJ 2 Pulsewidth	0-999 ms	3.4-3.8	3.6-7.5	6.0-9.8
INJ3	INJ 3 Pulsewidth	0-999 ms	3.4-3.8	3.6-7.5	6.0-9.8
INJ4	INJ 4 Pulsewidth	0-999 ms	3.4-3.8	3.6-7.5	6.0-9.8
INJ5	INJ 5 Pulsewidth	0-999 ms	3.4-3.8	3.6-7.5	6.0-9.8
INJ6	INJ 6 Pulsewidth	0-999 ms	3.4-3.8	3.6-7.5	6.0-9.8
LongFT1	Long Term FT - Bank 1	-20 to 20%	-5 to +5	-5 to +5	-5 to +5
LongFT2	Long Term FT - Bank 2	-20 to 20%	-5 to +5	-5 to +5	-5 to +5
MIL	MIL (lamp) Control	ON / OFF	OFF	OFF	OFF
SHTFT1	Short Term F/T - Bank 1	-10 to 10%	-5 to +5	-5 to +5	-5 to +5
SHTFT2	Short Term FT - Bank 2	-10 to 10%	-5 to +5	-5 to +5	-5 to +5
SPARK	Spark Advance	-90° to 90°	11-20	31-36	32-40
SS1	Shift Solenoid 1	ON / OFF	ON	OFF	OFF
SS2	Shift Solenoid 2	ON / OFF	OFF	OFF	OFF
SS3	Shift Solenoid 3	ON / OFF	OFF	OFF	ON
SS4	Shift Solenoid 4	ON / OFF	OFF	OFF	ON
TCC	TCC Solenoid	0-100%	0	0	90-95
TCIL	TCIL (lamp) Control	ON / OFF	OFF	OFF	OFF
VREF	Vehicle Reference	0-5.1v	4.9-5.1	4.9-5.1	4.9-5.1
WAC	A/C WOT Relay Control	ON / OFF	ON (AC on at WOT)	ON / OFF	ON / OFF

2002-03 EXPLORER/MOUNTAINEER 4.6L V8 MFI VIN W (A/T) INPUTS

PCM PID Acronym	Parameter Identification	PID Range	PID Value at Hot Idle	PID Value at 30	PID Value at 55
4x4L	4x4 Switch Signal	ON / OFF	ON (if on)	OFF	OFF
ACCS	A/C Switch Signal	ON / OFF	ON (if on)	OFF	OFF
BPP	Brake Position	ON / OFF	ON (if on)	OFF	OFF
DPFE	DPFE Sensor	0-5.1v	0.2-1.3	0.2-4.5	0.2-4.5
CHT	CHT Sensor	-40-500°F	194	194	194
FEPS	Flash EEPROM	0-5.1v	0.1	0.1	0.1
FLI	Fuel Level Input	1.7 / 50%	1.7 / 50%	1.7 / 50%	1.7 / 50%
FPM	Fuel Pump Monitor	ON / OFF	ON	ON	ON
FTP	Fuel Tank Pressure	0-5.1v at 0-10" H2O	2.6v at 0" H2O (gas cap "off")	2.6v at 0" H2O (gas cap "off")	2.6v at 0" H2O (gas cap "off")
GEAR	Gear Position	1-2-3-4	1	3	4
HO2S-11	HO2S-11 (Bank 1)	0.1-1.1v	0.1-1.1	0.1-1.1	0.1-1.1
HO2S-12	HO2S-12 (Bank 1)	0.1-1.1v	0.1-1.1	0.1-1.1	0.1-1.1
HO2S-21	HO2S-21 (Bank 2)	0.1-1.1v	0.1-1.1	0.1-1.1	0.1-1.1
HO2S-22	HO2S-22 (Bank 2)	0.1-1.1v	0.1-1.1	0.1-1.1	0.1-1.1
IAT	IAT Sensor	-40-304°F	50-120	50-120	50-120
LOAD	Engine Load	0-100%	14-16	20-25	24-35
MAFV	MAF Sensor	0-5.1v	0.7-0.9	1.2-1.7	1.6-2.4
MISF	Misfire Detection	ON / OFF	OFF	OFF	OFF
OCTADJ	Octane Adjustment	CL/OPEN	CLOSED	CLOSED	CLOSED
OSS	OSS Sensor	0-9999 Hz	0	125-131	245-255
PNP	PNP Switch Signal	ON / OFF	ON	OFF	OFF
PTO	Power Takeoff Switch	ON / OFF	ON (if on)	OFF	OFF
RPM	Engine Speed	0-10K rpm	680-830	1200-1500	1600-1800
TCS	TCS Switch	ON / OFF	OFF	OFF	OFF
TFT	TFT Sensor Signal	-40-304°F	110-210	110-210	110-210
TPV	TP Sensor	0-5.1v	0.53-1.27	0.8-1.0	0.9-1.2
TRV/TR	TR Sensor	0v / 1.7v	0	1.7	1.7
TR1	TR1 Sensor	0v / 11.5v	0	11.5	11.5
TR2	TR2 Sensor	0v / 11.5v	0	11.5	11.5
TR4	TR4 Sensor	0v / 11.5v	0	11.5	11.5
Vacuum	Engine Vacuum	0-30" Hg	19-20	18-20	8-15
VPWR	Vehicle Power	0-25.5v	12-14	12-14	12-14
VSS	Vehicle Speed (Hertz or MPH)	0-999 Hz (0-159 mph)	0 Hz = 0 mph	65 Hz = 30 mph	125 Hz = 55 mph

2002-03 EXPLORER/MOUNTAINEER 4.6L V8 MFI VIN W A/T OUTPUTS

PCM PID Acronym	Parameter Identification	PID Range	PID Value at Hot Idle	PID Value at 30	PID Value at 55
CCS	Coast Clutch Solenoid Control	ON / OFF	OFF	OFF	OFF
EGR VR	EGR VR Solenoid	0-100%	0	0-40	35-45
EVAP CP	EVAP Purge Valve	0-100%	0-100	0-100	0-100
EVAP CV	EVAP CV Valve	0-100%	0-100	0-100	0-100
EPC	EPC Solenoid	0-300 psi	6	40	40
FP	Fuel Pump Control	ON / OFF	ON	ON	ON
FuelPW1	INJ Pulsewidth - Bank 1	0-999 ms	2.7-4.1	4.5-8.0	5.5-9
FuelPW2	INJ Pulsewidth - Bank 2	0-999 ms	2.7-4.1	4.5-8.0	5.5-9
HTR-11	HO2S-11 Heater	ON / OFF	ON	ON	ON
HTR-12	HO2S-12 Heater	ON / OFF	ON	ON	ON
HTR-21	HO2S-21 Heater	ON / OFF	ON	ON	ON
HTR-22	HO2S-22 Heater	ON / OFF	ON	ON	ON
IAC	Idle Air Control	0-100%	30-34	28-32	63-70
INJ1	INJ 1 Pulsewidth	0-999 ms	2.7-4.1	4.5-8.0	5.5-9
INJ2	INJ 2 Pulsewidth	0-999 ms	2.7-4.1	4.5-8.0	5.5-9
INJ3	INJ 3 Pulsewidth	0-999 ms	2.7-4.1	4.5-8.0	5.5-9
INJ4	INJ 4 Pulsewidth	0-999 ms	2.7-4.1	4.5-8.0	5.5-9
INJ5	INJ 5 Pulsewidth	0-999 ms	2.7-4.1	4.5-8.0	5.5-9
INJ6	INJ 6 Pulsewidth	0-999 ms	2.7-4.1	4.5-8.0	5.5-9
INJ7	INJ 7 Pulsewidth	0-999 ms	2.7-4.1	4.5-8.0	5.5-9
INJ8	INJ 8 Pulsewidth	0-999 ms	2.7-4.1	4.5-8.0	5.5-9
LongFT1	Long Term FT - Bank 1	-20 to 20%	-5 to +5	-5 to +5	-5 to +5
LongFT2	Long Term FT - Bank 2	-20 to 20%	-5 to +5	-5 to +5	-5 to +5
MIL	MIL (lamp) Control	ON / OFF	OFF	OFF	OFF
SHTFT1	Short Term F/T - Bank 1	-10 to 10%	-5 to +5	-5 to +5	-5 to +5
SHTFT2	Short Term FT - Bank 2	-10 to 10%	-5 to +5	-5 to +5	-5 to +5
SPARK	Spark Advance	-90° to 90°	15-18	23-29	26-32
SS1	Shift Solenoid 1	ON / OFF	ON	OFF	OFF
SS2	Shift Solenoid 2	ON / OFF	OFF	ON	OFF
TCC	TCC Solenoid	0-100%	0	90-100	90-100
TCIL	TCIL (lamp) Control	ON / OFF	OFF	OFF	OFF
VREF	Vehicle Reference	0-5.1v	4.9-5.1	4.9-5.1	4.9-5.1

1996-2001 EXPLORER/MOUNTAINEER 5.0L V8 VIN P (A/T) INPUTS

PCM PID Acronym	Parameter Identification	PID Range	PID Value at Hot Idle	PID Value at 30	PID Value at 55
ACCS	A/C Switch Signal	ON / OFF	ON (if on)	OFF	OFF
ARC	Auto Ride Control	ON / OFF	OFF	OFF	OFF
BPP	Brake Position	ON / OFF	ON (if on)	OFF	OFF
CID	CMP Sensor	0-999	5-10	10-15	11-17
CKP	CKP Sensor	0-9999 Hz	460-500	650-720	950-1050
ECT	ECT Sensor	-40-304°F	160-200	160-200	160-200
EGR VP (1996-97)	EGR Valve Position	0-5.1v	0.1	0.5-1.5	0.5-1.8
DPFE ('98-'00)	DPFE Sensor Input	0-5.1v	0.20-1.30	0.20-2.75	0.20-2.75
DPFE ('01-'02)	DPFE Sensor Input	0-5.1v	0.95-1.05	0.96-4.65	0.95-4.65
FEPS	Flash EEPROM	0-5.1v	0.1	0.1	0.1
FLI	Fuel Level Input	1.7 / 50%	1.7 / 50%	1.7 / 50%	1.7 / 50%
FPM	Fuel Pump Monitor	ON / OFF	ON	ON	ON
FTP	Fuel Tank Pressure	0-5.1v at 0-10" H2O	2.6v at 0" H2O (gas cap "off")	2.6v at 0" H2O (gas cap "off")	2.6v at 0" H2O (gas cap "off")
IAT	IAT Sensor	-40-304°F	50-120	50-120	50-120
GEAR	Gear Position	1-2-3-4	1	3	4
LOAD	Engine Load	0-100%	10-20	19-33	22-36
MAFV	MAF Sensor	0-5.1v	0.7-0.9	0.9-1.7	1.5-0
MISF	Misfire Detection	ON / OFF	OFF	OFF	OFF
OCTADJ	Octane Adjustment	CL/OPEN	CLOSED	CLOSED	CLOSED
HO2S-11	HO2S-11 (Bank 1)	0.1-1.1v	0.1-1.1	0.1-1.1	0.1-1.1
HO2S-12	HO2S-12 (Bank 1)	0.1-1.1v	0.1-1.1	0.1-1.1	0.1-1.1
HO2S-21	HO2S-21 (Bank 2)	0.1-1.1v	0.1-1.1	0.1-1.1	0.1-1.1
HO2S-22	HO2S-22 (Bank 2)	0.1-1.1v	0.1-1.1	0.1-1.1	0.1-1.1
PNP	PNP Switch Signal	ON / OFF	ON	OFF	OFF
RPM	Engine Speed	0-10K rpm	760-830	1180-1290	1570-1695
TCS	TCS Switch	ON / OFF	OFF	OFF	OFF
TFT	TFT Sensor Signal	-40-304°F	110-210	110-210	110-210
TPV	TP Sensor	0-5.1v	0.53-1.27	0.8-1.1	1.0-1.4
TRV/TR	TR Sensor	0v / 1.7v	0	1.7	1.7
TR1	TR1 Sensor	0v / 11.5v	0	11.5	11.5
TR2	TR2 Sensor	0v / 1.7v	0	11.5	11.5
TR4	TR4 Sensor	0v / 1.7v	0	11.5	11.5
OSS	Output Shaft Speed	0-9999 Hz	0	115-145	230-280
Vacuum	Engine Vacuum	0-30" Hg	19-20	18-20	8-15
VPWR	Vehicle Power	0-25.5v	12-14	12-14	12-14
VSS	Vehicle Speed (Hertz or MPH)	0-999 Hz (0-159 mph)	0 Hz = 0 mph	65 Hz = 30 mph	125 Hz = 55 mph

1996-2001 EXPLORER/MOUNTAINEER 5.0L V8 VIN P A/T OUTPUTS

PCM PID Acronym	Parameter Identification	PID Range	PID Value at Hot Idle	PID Value at 30	PID Value at 55
CD1	Coil 1 Driver	0-90°	5	6	8
CD2	Coil 2 Driver	0-90°	5	6	8
CD3	Coil 3 Driver	0-90°	5	6	8
CD4	Coil 4 Driver	0-90°	5	6	8
CTO	Clean Tachometer signal	0-9999 Hz	40-55	80-95	100-125
EGR VR	EGR VR Solenoid	0-100%	0	0-40	0-40
EVAP CP	EVAP Purge Valve	0-100%	0-100	0-100	0-100
EVAP VMV	EVAP Purge Valve (1996-97)	0-100%	0-100	0-100	0-100
EVAP CV	EVAP CV Valve	0-100%	0	0	0
EPC	EPC Solenoid	0-300 psi	10	23-38	34-38
FP	Fuel Pump Control	ON / OFF	ON	ON	ON
FuelPW1	INJ Pulsewidth - Bank 1	0-999 ms	3.2-4.5	4.1-8.0	5.5-12.0
FuelPW2	INJ Pulsewidth - Bank 2	0-999 ms	3.2-4.5	4.1-8.0	5.5-12.0
HTR-11	HO2S-11 Heater	ON / OFF	ON	ON	ON
HTR-12	HO2S-12 Heater	ON / OFF	ON	ON	ON
HTR-21	HO2S-21 Heater	ON / OFF	ON	ON	ON
IAC	Idle Air Control	0-100%	33	35-41	57-68
INJ1	INJ 1 Pulsewidth	0-999 ms	3.2-4.5	4.1-8.0	5.5-12.0
INJ2	INJ 2 Pulsewidth	0-999 ms	3.2-4.5	4.1-8.0	5.5-12.0
INJ3	INJ 3 Pulsewidth	0-999 ms	3.2-4.5	4.1-8.0	5.5-12.0
INJ4	INJ 4 Pulsewidth	0-999 ms	3.2-4.5	4.1-8.0	5.5-12.0
INJ5	INJ 5 Pulsewidth	0-999 ms	3.2-4.5	4.1-8.0	5.5-12.0
INJ6	INJ 6 Pulsewidth	0-999 ms	3.2-4.5	4.1-8.0	5.5-12.0
INJ7	INJ 7 Pulsewidth	0-999 ms	3.2-4.5	4.1-8.0	5.5-12.0
INJ8	INJ 8 Pulsewidth	0-999 ms	3.2-4.5	4.1-8.0	5.5-12.0
LongFT1	Long Term FT - Bank 1	-20 to 20%	-5 to +5	-5 to +5	-5 to +5
LongFT2	Long Term FT - Bank 2	-20 to 20%	-5 to +5	-5 to +5	-5 to +5
MIL	MIL (lamp) Control	ON / OFF	OFF	OFF	OFF
SHTFT1	Short Term F/T - Bank 1	-10 to 10%	-5 to +5	-5 to +5	-5 to +5
SHTFT2	Short Term F/T - Bank 2	-10 to 10%	-5 to +5	-5 to +5	-5 to +5
SPARK	Spark Advance	-90° to 90°	12-17	32-40°	25-37°
SS1	Shift Solenoid 1	ON / OFF	ON	OFF	ON
SS2	Shift Solenoid 2	ON / OFF	OFF	ON	ON
TCC	TCC Solenoid	0-100%	0	0-40	90-95
TCIL	TCIL (lamp) Control	ON / OFF	OFF	OFF	OFF
VREF	Vehicle Reference	0-5.1v	4.9-5.1	4.9-5.1	4.9-5.1
WAC	A/C WOT Relay Control	ON / OFF	ON (AC on at WOT)	ON / OFF	ON / OFF

Escape PID Data

2001-03 ESCAPE 2.0L I4 DOHC 4V MFI VIN B (ALL) INPUTS

PCM PID Acronym	Parameter Identification	PID Range	PID Value at Hot Idle	PID Value at 30	PID Value at 55
ACCS	A/C Clutch Switch	ON / OFF	ON (if on)	OFF	OFF
ACP	A/C Press. Switch	CL/OPEN	Open	OPEN	OPEN
BPP	Brake Pedal Switch	ON / OFF	ON (if on)	OFF	OFF
CID	CID Sensor Hertz	0-1 KHz	5-7	13-16	20-23
CKP	CKP Sensor Hertz	0-10 KHz	400-450	985-1100	1450-1550
CHT (°F)	CHT Sensor	-40-304°F	160-200	160-200	160-200
CPP	CPP Switch	ON / OFF	ON (if in)	OFF	OFF
DPFE	DPFE Sensor Input	0-5.1v	0.95-1.05	0.96-4.65	0.95-4.65
FEPS	Flash EEPROM	0-5.1v	0.1	0.1	0.1
FLI	Fuel Level Indicator	0-100%	50% (1.7)	50% (1.7)	50% (1.7)
FPM	Fuel Pump Monitor	ON / OFF	ON	ON	ON
FTP	Fuel Tank Pressure	0-5.1v or 0-10" H2O	2.6v / 0" H2O (with cap off)	2.6v / 0" H2O (with cap off)	2.6v / 0" H2O (with cap off)
GEAR	Gear Position	1-2-3-4	1	3	4
GFDC	Generator Field DC	0-100%	0	0	0
HO2S-11	HO2S-11 (Bank 1)	0.1-1.1v	0.1-1.1	0.1-1.1	0.1-1.1
HO2S-12	HO2S-12 (Bank 1)	0.1-1.1v	0.1-1.1	0.1-1.1	0.1-1.1
IAT (°F)	IAT Sensor	-40-304°F	50-120	50-120	50-120
KS1	Knock Sensor 1	0-5.1v	0	0	0
LOAD	Engine Load	0-100%	10-20	19-30	35-44
MAFV	MAF Sensor Volts	0-5.1v	0.6-0.9	0.8-1.5	1.2-2.5
MISF	Misfire Detection	ON / OFF	OFF	OFF	OFF
OCTADJ	Octane Adjustment	Retard/NO	NO RET	NO RET	NO RET
OSS	OSS Sensor	0-9999 Hz	0	400	700-740
PATSIN	PATS Input	0-5.1v	0	0	0
PNP	PNP Switch	ON / OFF	ON (in 'P')	OFF	OFF
PSP	PSP Switch	HI / LOW	HI: turning	LOW	LOW
RPM	Engine Speed	0-10K rpm	700-800	1750-1800	2500-2660
TCS	TCS Switch	ON / OFF	OFF	ON	ON
TFT (°F)	TFT Sensor (°F)	-40-304°F	110-210	110-210	110-210
TP (V)	TP Sensor Volts	0-5.1v	0.53-1.27	0.8-1.2	1.0-1.5
Vacuum	Engine Vacuum	0-30" Hg	19-20	18-20	8-15
VPWR	Vehicle Power	0-25.5v	12-14	12-14	12-14
VSS	Vehicle Speed (MPH & Hertz Rate)	0-999 Hz (0-159 mph)	0 = 0 Hz	30 = 65 Hz	55 = 125 Hz

2001-03 ESCAPE 2.0L I4 DOHC 4V MFI VIN B (ALL) OUTPUTS

PCM PID Acronym	Parameter Identification	PID Range	PID Value at Hot Idle	PID Value at 30	PID Value at 55
CD1	Coil 1 Driver	0-90° dwell	7	10	12
CD2	Coil 2 Driver	0-90° dwell	7	10	12
EGRVR	VR Solenoid	0-100%	0	0-40	0-60
EPC	EPC Solenoid	0-500 psi	15-18	23-28	30-35
EVAP DC	EVAP Purge	0-100%	0-100	0-100	0-100
EVAPCV	EVAP CV	0-100%	0-100	0-100	0-100
FP	Fuel Pump Relay	ON / OFF	ON	ON	ON
FUELPW	INJ 1 Pulsewidth	0-99.9 ms	2.5-3.0	3.3-5.2	4.6-6.5
HFC	High Fan Control	ON / OFF	ON (if on)	OFF	OFF
HTR11	HO2S-11 Heater	ON / OFF	ON	ON	ON
HTR12	HO2S-12 Heater	ON / OFF	ON	ON	ON
IAC	Idle Air Control	0-100%	33-35	40-48	40-55
INJ1	INJ 1 Pulsewidth	0-999 ms	2.5-3.0	3.3-5.2	4.6-6.5
INJ2	INJ 2 Pulsewidth	0-999 ms	2.5-3.0	3.3-5.2	4.6-6.5
INJ3	INJ 3 Pulsewidth	0-999 ms	2.5-3.0	3.3-5.2	4.6-6.5
INJ4	INJ 4 Pulsewidth	0-999 ms	2.5-3.0	3.3-5.2	4.6-6.5
LFC	Low Fan Control	ON / OFF	ON (if on)	OFF	OFF
LONGFT	LONGFT Bank 1	-20 to 20%	-5 to +5	-5 to +5	-5 to +5
MAF	MAF Sensor	0-5.1v	0.6-0.9	1.0-1.7	1.2-2.5
MFC	Medium Fan control	ON / OFF	ON (if on)	OFF	OFF
MIL	MIL Control	ON / OFF	ON (if on)	OFF	OFF
SS1	Shift Solenoid 1	ON / OFF	OFF	ON	ON
SS2	Shift Solenoid 2	ON / OFF	ON	OFF	OFF
SHRTFT	SHRTFT Bank 1	-20 to 20%	-5 to +5	-5 to +5	-5 to +5
SPARK	Spark Advance	-90° to 90°	23-35	30-45	25-36
TCC	TCC Solenoid	0-100%	0	0	95-100
VSO	Vehicle Speed	0-1 KHz	0	65	125
VREF	Voltage Reference	0-5.1v	4.9-5.1	4.9-5.1	4.9-5.1
WAC	A/C WOT Cutout Relay	ON / OFF	ON (if on)	OFF	OFF

2001-03 ESCAPE 3.0L V6 DOHC 4V MFI VIN 1 (ALL) INPUTS

PCM PID Acronym	Parameter Identification	PID Range	PID Value at Hot Idle	PID Value at 30	PID Value at 55
ACCS	A/C Clutch Switch	ON / OFF	ON (if on)	OFF	OFF
ACP	A/C Press. Switch	CL/OPEN	Open	OPEN	OPEN
BPP	Brake Pedal Switch	ON / OFF	ON (if on)	OFF	OFF
CID	CID Sensor Hertz	0-1 KHz	5-7	12-15	14-16
CKP	CKP Sensor Hertz	0-10 KHz	400-450	850-1050	1050-1150
DPFE	DPFE Sensor Input	0-5.1v	0.95-1.05	0.96-4.65	0.95-4.65
ECT (°F)	ECT Sensor	-40-304°F	160-200	160-200	160-200
FEPS	Flash EEPROM	0-5.1v	0.1	0.1	0.1
FLI	Fuel Level Indicator	0-100%	50% (1.7)	50% (1.7)	50% (1.7)
FPM	Fuel Pump Monitor	ON / OFF	ON	ON	ON
FTP	Fuel Tank Pressure (1998-2000)	0-5.1v or 0-10" H2O	2.6v / 0" H2O (with cap off)	2.6v / 0" H2O (with cap off)	2.6v / 0" H2O (with cap off)
GEAR	Gear Position	1-2-3-4	1	3	4
HO2S-11	HO2S-11 (Bank 1)	0.1-1.1v	0.1-1.1	0.1-1.1	0.1-1.1
HO2S-12	HO2S-12 (Bank 1)	0.1-1.1v	0.1-1.1	0.1-1.1	0.1-1.1
HO2S-22	HO2S-22 (Bank 2)	0.1-1.1v	0.1-1.1	0.1-1.1	0.1-1.1
IAT (°F)	IAT Sensor	-40-304°F	50-120	50-120	50-120
KS1	Knock Sensor 1	0	0	0	0
LOAD	Engine Load	0-100%	17-21	21-30	33-40
MAFV	MAF Sensor Volts	0-5.1v	0.8-1.1	1.4-1.7	2.0-2.5
MISF	Misfire Detection	ON / OFF	OFF	OFF	OFF
OCTADJ	Octane Adjustment	CL/OPEN	CLOSED	CLOSED	CLOSED
OSS	OSS Sensor	0-9999 Hz	0	400	700-740
PATSIN	PATS Input	0-5.1v	0	0	0
PNP	PNP Switch	ON / OFF	ON (in 'P')	OFF	OFF
PSP	PSP Switch	HI / LOW	HI: turning	LOW	LOW
RPM	Engine Speed	0-10K rpm	730-750	1550-1700	1800-2000
TCS	TCS Switch	ON / OFF	OFF	ON	ON
TFT (°F)	TFT Sensor (°F)	-40-304°F	110-210	110-210	110-210
TP (V)	TP Sensor Volts	0-5.1v	0.53-1.27	0.8-1.2	1.0-1.5
TR (V)	TR Sensor Volts	0-5.1v	4.4	2.1	2.1
TSS	TSS Sensor	0-9999 Hz	45-50	90-100	110-120
Vacuum	Engine Vacuum	0-30" Hg	19-20	18-20	8-15
VPWR	Vehicle Power	0-25.5v	12-14	12-14	12-14
VSS	Vehicle Speed (MPH & Hertz Rate)	0-159 mph	0 = 0 Hz	30 = 65 Hz	55 = 125 Hz

2001-03 ESCAPE 3.0L V6 DOHC 4V MFI VIN 1 (ALL) OUTPUTS

PCM PID Acronym	Parameter Identification	PID Range	PID Value at Hot Idle	PID Value at 30	PID Value at 55
COP1	COP 1 Driver	0-90° dwell	7	10	12
COP2	COP 2 Driver	0-90° dwell	7	10	12
COP3	COP 3 Driver	0-90° dwell	7	10	12
COP4	COP 4 Driver	0-90° dwell	7	10	12
COP5	COP 5 Driver	0-90° dwell	7	10	12
COP6	COP 6 Driver	0-90° dwell	7	10	12
EGRVR	VR Solenoid	0-100%	0	0-40	0-60
EPC	EPC Solenoid	0-500 psi	25-37	42-51	50-79
EVAP DC	EVAP Purge	0-100%	0-100	80	90
EVAPCV	EVAP CV	0-100%	0-100	0-100	0-100
FP	Fuel Pump Relay	0-100%	100	100	100
FUELPW	INJ 1 Pulsewidth	0-99.9 ms	2.6-3.2	2.5-5.5	3.5-8.5
HFC	High Fan Control	ON / OFF	ON (if on)	OFF	OFF
HTR11	HO2S-11 Heater	ON / OFF	ON	ON	ON
HTR12	HO2S-12 Heater	ON / OFF	ON	ON	ON
HTR22	HO2S-22 Heater	ON / OFF	ON	ON	ON
IAC	Idle Air Control	0-100%	25-35	30-55	50-79
INJ1	INJ 1 Pulsewidth	0-999 ms	2.6-3.2	2.5-5.5	3.5-8.5
INJ2	INJ 2 Pulsewidth	0-999 ms	2.6-3.2	2.5-5.5	3.5-8.5
INJ3	INJ 3 Pulsewidth	0-999 ms	2.6-3.2	2.5-5.5	3.5-8.5
INJ4	INJ 4 Pulsewidth	0-999 ms	2.6-3.2	2.5-5.5	3.5-8.5
INJ5	INJ 5 Pulsewidth	0-999 ms	2.6-3.2	2.5-5.5	3.5-8.5
INJ6	INJ 6 Pulsewidth	0-999 ms	2.6-3.2	2.5-5.5	3.5-8.5
LFC	Low Fan Control	ON / OFF	ON (if on)	OFF	OFF
LONGFT	LONGFT Bank 1	-20 to 20%	-5 to +5	-5 to +5	-5 to +5
MIL	MIL Control	ON / OFF	ON (if on)	OFF	OFF
PATSOUT	PATS Output	0-5.1v	0.8	0.8	0.8
PATSTRT	PATS Output	0-5.1v	0	0	0
SHRTFT	SHRTFT Bank 1	-20 to 20%	-5 to +5	-5 to +5	-5 to +5
SPARK	Spark Advance	-90° to 90°	16-18	20-36	25-35
SS1	Shift Solenoid 1	ON / OFF	OFF	ON	OFF
SS2	Shift Solenoid 2	ON / OFF	ON	OFF	OFF
SS3	Shift Solenoid 3	ON / OFF	OFF	OFF	ON
TCC	TCC Solenoid	0-100%	0	50-90	95-100
VSO	Vehicle Speed	0-10 KHz	0	65	125
VREF	Voltage reference	0-5.1v	4.9-5.1	4.9-5.1	4.9-5.1
WAC	A/C WOT Cutout Relay	ON / OFF	ON (if on)	OFF	OFF

Excursion PID Data

2000-03 EXCURSION 5.4L V8 VIN L (E4OD) INPUTS

PCM PID Acronym	Parameter Identification	PID Range	PID Value at Hot Idle	PID Value at 30	PID Value at 55
4x4L	4x4 Switch Signal	ON / OFF	ON (if on)	OFF	OFF
ACCS	A/C Switch Signal	ON / OFF	ON (if on)	OFF	OFF
BPP	Brake Position	ON / OFF	ON (if on)	OFF	OFF
CHT	CHT Sensor	-40-500°F	194	194	194
CID	CMP Sensor	0-999 Hz	6-8	10-12	14-17
CKP	CKP Sensor	0-9999 Hz	411	800-840	1000
DPFE	DPFE Sensor	0-5.1v	0.2-1.3	0.2-4.5	0.2-4.5
ECT	ECT Sensor	-40-304°F	160-200	160-200	160-200
FEPS	Flash EEPROM	0-5.1v	0.1	0.1	0.1
FLI	Fuel Level Input	1.7 / 50%	1.7 / 50%	1.7 / 50%	1.7 / 50%
FPM	Fuel Pump Monitor	ON / OFF	ON	ON	ON
FTP	Fuel Tank Pressure	0-5.1v at 0-10" H2O	2.6v at 0" H2O (gas cap "off")	2.6v at 0" H2O (gas cap "off")	2.6v at 0" H2O (gas cap "off")
GEAR	Gear Position	1-2-3-4	1	3	4
HO2S-11	HO2S-11 (Bank 1)	0.1-1.1v	0.1-1.1	0.1-1.1	0.1-1.1
HO2S-12	HO2S-12 (Bank 1)	0.1-1.1v	0.1-1.1	0.1-1.1	0.1-1.1
HO2S-21	HO2S-21 (Bank 1)	0.1-1.1v	0.1-1.1	0.1-1.1	0.1-1.1
HO2S-22	HO2S-22 (Bank 1)	0.1-1.1v	0.1-1.1	0.1-1.1	0.1-1.1
IAT	IAT Sensor	-40-304°F	50-120	50-120	50-120
KS1	Knock Sensor 1	0-5.1v	0	0	0
LOAD	Engine Load	0-100%	14-16	19-25	26-35
MAFV	MAF Sensor	0-5.1v	0.7-0.9	1.1-1.6	1.7-2.4
MISF	Misfire Detection	ON / OFF	OFF	OFF	OFF
OCTADJ	Octane Adjustment	CL/OPEN	CLOSED	CLOSED	CLOSED
OSS	OSS Sensor	0-9999 Hz	0	400	700
PNP	PNP Switch Signal	ON / OFF	ON	OFF	OFF
PTO	Power Takeoff Switch	ON / OFF	ON (if on)	OFF	OFF
RPM	Engine Speed	0-10K rpm	760-830	1200-1270	1590-1675
TCS	TCS Switch	ON / OFF	OFF	OFF	OFF
TFT	TFT Sensor Signal	-40-304°F	110-210	110-210	110-210
TPO	TPO TR Sensor	0-5.1v	0.1	0.8	1.0
TPV	TP Sensor	0-5.1v	0.53-1.27	0.8-1.0	1.0-1.3
TR1	TR1 Sensor	0v / 11.5v	0	11.5	11.5
TR2	TR2 Sensor	0v / 11.5v	0	11.5	11.5
TR3/TR	TR3A Sensor	0v / 1.7v	0	1.7	1.7
TR4	TR4 Sensor	0v / 11.5v	0	11.5	11.5
Vacuum	Engine Vacuum	0-30" Hg	19-20	18-20	8-15
VPWR	Vehicle Power	0-25.5v	12-14	12-14	12-14
VSS	Vehicle Speed (Hertz or MPH)	0-999 Hz (0-159 mph)	0 Hz = 0 mph	65 Hz = 30 mph	125 Hz = 55 mph

2000-03 EXCURSION 5.4L V8 VIN L (E4OD) OUTPUTS

PCM PID Acronym	Parameter Identification	PID Range	PID Value at Hot Idle	PID Value at 30	PID Value at 55
ARC	Auto Ride Control	ON / OFF	OFF	OFF	OFF
CD1-CD8	COP Driver 1-8	0-45	5	6	8
CHTIL	CHT (lamp) Control	ON / OFF	OFF	OFF	OFF
EGR VR	EGR VR Solenoid	0-100%	0	0-40	0-60
EVAP CP	EVAP Purge Valve	0-100%	0-100	0-100	0-100
EVAP CV	EVAP CV Valve	0-100%	0-100	0-100	0-100
EPC	EPC Solenoid	7.4v / 4	4	5	5
FP	Fuel Pump Control	OFF	ON	ON	ON
FuelPW1	INJ Pulsewidth - Bank 1	0-999 ms	3.2-4.0	4.1-6.9	6.5-12
FuelPW2	INJ Pulsewidth - Bank 2	0-999 ms	3.2-4.0	4.1-6.9	6.5-12
HTR-11	HO2S-11 Heater	ON / OFF	ON	ON	ON
HTR-12	HO2S-12 Heater	ON / OFF	ON	ON	ON
HTR-21	HO2S-21 Heater	ON / OFF	ON	ON	ON
HTR-22	HO2S-22 Heater	ON / OFF	ON	ON	ON
IAC	Idle Air Control	0-100%	30-34	43-48	58-61
INJ1	INJ 1 Pulsewidth	0-999 ms	3.2-4.0	4.1-6.9	6.5-12
INJ2	INJ 2 Pulsewidth	0-999 ms	3.2-4.0	4.1-6.9	6.5-12
INJ3	INJ 3 Pulsewidth	0-999 ms	3.2-4.0	4.1-6.9	6.5-12
INJ4	INJ 4 Pulsewidth	0-999 ms	3.2-4.0	4.1-6.9	6.5-12
INJ5	INJ 5 Pulsewidth	0-999 ms	3.2-4.0	4.1-6.9	6.5-12
INJ6	INJ 6 Pulsewidth	0-999 ms	3.2-4.0	4.1-6.9	6.5-12
INJ7	INJ 7 Pulsewidth	0-999 ms	3.2-4.0	4.1-6.9	6.5-12
INJ8	INJ 8 Pulsewidth	0-999 ms	3.2-4.0	4.1-6.9	6.5-12
LongFT1	Long Term FT - Bank 1	-20 to 20%	-5 to +5	-5 to +5	-5 to +5
LongFT2	Long Term FT - Bank 2	-20 to 20%	-5 to +5	-5 to +5	-5 to +5
MIL	MIL (lamp) Control	ON / OFF	OFF	OFF	OFF
SHTFT1	Short Term F/T - Bank 1	-10 to 10%	-5 to +5	-5 to +5	-5 to +5
SHTFT2	Short Term FT - Bank 2	-10 to 10%	-5 to +5	-5 to +5	-5 to +5
SPARK	Spark Advance	-90° to 90°	12-17	35-40	28-37
SS1	Shift Solenoid 1	ON / OFF	ON	OFF	OFF
SS2	Shift Solenoid 2	ON / OFF	OFF	ON	OFF
TCC	TCC Solenoid	0-100%	0	0-40	90-95
TCIL	TCIL (lamp) Control	ON / OFF	OFF	OFF	OFF
TPO	TPO TR Sensor	0-5.1v	0.1	0.8	1.0
VREF	Vehicle Reference	0-5.1v	4.9-5.1	4.9-5.1	4.9-5.1

2000-03 EXCURSION 6.8L V10 VIN S (E4OD) INPUTS

PCM PID Acronym	Parameter Identification	PID Range	PID Value at Hot Idle	PID Value at 30	PID Value at 55
ACCS	A/C Switch Signal	ON / OFF	ON (if on)	OFF	OFF
BPP	Brake Position	ON / OFF	ON (if on)	OFF	OFF
CHT	CHT Sensor	-40-500°F	194	194	194
CID	CMP Sensor	0-999 Hz	7-10	10-13	15-17
CKP	CKP Sensor	0-9999 Hz	500-525	750-940	1195-1400
DPFE	DPFE Sensor	0-5.1v	0.2-1.3	0.2-4.5	0.2-4.5
FEPS	Flash EEPROM	0-5.1v	0.1	0.1	0.1
FLI	Fuel Level Input	1.7 / 50%	1.7 / 50%	1.7 / 50%	1.7 / 50%
FPM	Fuel Pump Monitor	ON / OFF	ON	ON	ON
FTP	Fuel Tank Pressure	2.6v / 0 psi	2.6v / 0 psi	2.6v / 0 psi	2.6v / 0 psi
GEAR	Gear Position	1-2-3-4	1	4	4
HO2S-11	HO2S-11 (Bank 1)	0.1-1.1v	0.1-1.1	0.1-1.1	0.1-1.1
HO2S-12	HO2S-12 (Bank 1)	0.1-1.1v	0.1-1.1	0.1-1.1	0.1-1.1
HO2S-21	HO2S-21 (Bank 1)	0.1-1.1v	0.1-1.1	0.1-1.1	0.1-1.1
IAT	IAT Sensor	-40-304°F	50-120	50-120	50-120
KS1	Knock Sensor 1	0-5.1v	0	0	0
LOAD	Engine Load	0-100%	14-16	20-25	24-35
MAFV	MAF Sensor	0-5.1v	0.7-0.9	1.2-1.7	1.6-2.4
MISF	Misfire Detection	ON / OFF	OFF	OFF	OFF
OCTADJ	Octane Adjustment	CL/OPEN	CLOSED	CLOSED	CLOSED
PNP	PNP Switch Signal	ON / OFF	ON	OFF	OFF
PTO	Power Takeoff Switch	OFF	ON (if on)	OFF	OFF
RPM	Engine Speed	0-10K rpm	780-810	1380-1450	1790-1840
TCS	TCS Switch	ON / OFF	OFF	OFF	OFF
TFT	TFT Sensor Signal	-40-304°F	110-210	110-210	110-210
TPV	TP Sensor	0-5.1v	0.53-1.27	0.8-1.1	0.9-1.2
TR1	TR1 Sensor	0v / 11.5v	0	11.5	11.5
TR2	TR2 Sensor	0v / 11.5v	0	11.5	11.5
TR3/TR	TR3A Sensor	0v / 1.7v	0	1.7	1.7
TR4	TR4 Sensor	0v / 11.5v	0	11.5	11.5
Vacuum	Engine Vacuum	0-30" Hg	19-20	18-20	8-15
VPWR	Vehicle Power	0-25.5v	12-14	12-14	12-14
VSS	Vehicle Speed (Hertz or MPH)	0-999 Hz (0-159 mph)	0 Hz = 0 mph	65 Hz = 30 mph	125 Hz = 55 mph
4x4L	4x4 Switch Signal	ON / OFF	ON (if on)	OFF	OFF

<u>2000-03 EXCURSION 6.8L V10 VIN S (E4OD) OUTPUTS</u>

PCM PID Acronym	Parameter Identification	PID Range	PID Value at Hot Idle	PID Value at 30	PID Value at 55
CD1-10	Coil 1-10 Driver	0-36	6	8	10°
CCS	Coast Clutch Solenoid Control	ON / OFF	OFF	OFF	OFF
CHTIL	CHT (lamp) Control	ON / OFF	OFF	OFF	OFF
CTO	Clean Tachometer signal	0-9999 Hz	60-70	90-110	150-185
EGR VR	EGR VR Solenoid	0-100%	0	0-40	0-60
EVAP CP	EVAP Purge Valve	0-100%	0-100	0-100	0-100
EVAP CV	EVAP CV Valve	0-100%	0-100	0-100	0-100
EPC	EPC Solenoid	0-300 psi	5	7	27
FP	Fuel Pump Control	ON / OFF	ON	ON	ON
FuelPW1	INJ Pulsewidth - Bank 1	0-999 ms	3.8-4.6	5.2-6.5	6.6-11.0
FuelPW2	INJ Pulsewidth - Bank 2	0-999 ms	3.8-4.6	5.2-6.5	6.6-11.0
HTR-11	HO2S-11 Heater	ON / OFF	ON	ON	ON
HTR-12	HO2S-12 Heater	ON / OFF	ON	ON	ON
HTR-21	HO2S-21 Heater	ON / OFF	ON	ON	ON
HTR-22	HO2S-22 Heater	ON / OFF	ON	ON	ON
IAC	Idle Air Control	0-100%	25-33	30-40	50-65
INJ1	INJ 1 Pulsewidth	0-999 ms	3.8-4.6	5.2-6.5	6.6-11.0
INJ2	INJ 2 Pulsewidth	0-999 ms	3.8-4.6	5.2-6.5	6.6-11.0
INJ3	INJ 3 Pulsewidth	0-999 ms	3.8-4.6	5.2-6.5	6.6-11.0
INJ4	INJ 4 Pulsewidth	0-999 ms	3.8-4.6	5.2-6.5	6.6-11.0
INJ5	INJ 5 Pulsewidth	0-999 ms	3.8-4.6	5.2-6.5	6.6-11.0
INJ6	INJ 6 Pulsewidth	0-999 ms	3.8-4.6	5.2-6.5	6.6-11.0
INJ7	INJ 7 Pulsewidth	0-999 ms	3.8-4.6	5.2-6.5	6.6-11.0
INJ8	INJ 8 Pulsewidth	0-999 ms	3.8-4.6	5.2-6.5	6.6-11.0
INJ9	INJ 9 Pulsewidth	0-999 ms	3.8-4.6	5.2-6.5	6.6-11.0
INJ10	INJ 10 Pulsewidth	0-999 ms	3.8-4.6	5.2-6.5	6.6-11.0
LongFT1	Long Term FT - Bank 1	-20 to 20%	-5 to +5	-5 to +5	-5 to +5
LongFT2	Long Term FT - Bank 2	-20 to 20%	-5 to +5	-5 to +5	-5 to +5
MIL	MIL (lamp) Control	ON / OFF	OFF	OFF	OFF
SHTFT1	Short Term F/T - Bank 1	-10 to 10%	-5 to +5	-5 to +5	-5 to +5
SHTFT2	Short Term F/T - Bank 2	-10 to 10%	-5 to +5	-5 to +5	-5 to +5
SPARK	Spark Advance	-90° to 90°	15-20	23-34	26-34
SS1	Shift Solenoid 1	ON / OFF	ON	OFF	OFF
SS2	Shift Solenoid 2	ON / OFF	OFF	ON	OFF
TCC	TCC Solenoid	0-100%	0	90-100	90-100
TCIL	TCIL (lamp) Control	ON / OFF	OFF	OFF	OFF
VREF	Vehicle Reference	0-5.1v	4.9-5.1	4.9-5.1	4.9-5.1

<u>2000-03 EXCURSION 7.3L V8 TURBO DIESEL VIN F (E4OD)</u>

PCM PID Acronym	Parameter Identification	PID Range	PID Value at low idle	PID Value at high idle
4x4L	4x4 Low Switch Input	ON / OFF	ON	OFF
ACCS	A/C Switch Signal	ON / OFF	ON	OFF
AP	Accelerator Pedal Position Sensor	0-5.1v	0.5-0.9	3.8-4.2v
ARPMDES	Ancillary Engine Speed Desired	---	---	---
BARO	Barometric Pressure	0-450 kPa	---	---
BAROV	Barometric Pressure	0-5.1v	4.75 (Sea level)	4.75 (Sea level)
BPP	Brake Pedal Position	ON / OFF	ON	OFF
BPA	Brake Pressure Applied	ON / OFF	ON	OFF
CCS	Coast Clutch Solenoid Control	ON / OFF	OFF	OFF
CPP	CPP Switch Signal	0v / 5v	0v (clutch "in")	5v
CPP/TCS	TCS & Cancel Switch	0v / VBAT	Switches On: 12	0v
CRUISE	Cruise Control Module	---	---	---
DTC CNT	DTC Count	0-156	0	0
EBP	Exhaust Back Pressure	---	---	---
EBP V	Exhaust Back Pressure Actual	0-5.1v	0.8-0.95	1.25-1.75
EOT	Engine Oil Temperature	0-5.1v	0.35-4.5	0.35-4.5
EPC	Electronic Pressure Control	0-300 psi	7	15
EPC V	Electronic Pressure Control Actual	0-25.5v	7.5	12
EPR	Exhaust Pressure Regulator	0-25.5v	6-8	0-10
FuelPW1	Fuel Pulsewidth	0-999 ms	---	---
GEAR	Gear Position	1-2-3-4	1	4
GPC	Glow Plug Control Duty Cycle	0-100%	---	---
GPC TM	Glow Plug Control Time	---	---	---
GPL TM	Glow Plug Lamp Time	---	---	---
IAT (°F)	IAT Sensor	-40-304°F	50-120	50-120
IAT V	IAT Sensor Actual	0-5.1v	1.5-3.5	1.5-3.5
ICP	Injector Control Pressure Sensor	Min: 0.83v at startup	0.25-0.40	0.25-0.40
ICP	Injector Control Pressure Actual	---	---	---
IPR	Injector Control Pressure Regulator	0-100%	35	40-100
IVS	Idle Validation Switch	0v / VBAT	0v / Closed	12 / CL or OL
MAP	MAP Sensor	0-999 Hz	---	---
MAP H	MAP Sensor Actual	0-999 Hz	110-190	110-190

2000-03 EXCURSION 7.3L V8 TURBO DIESEL VIN F (E4OD)

PCM PID Acronym	Parameter Identification	PID Range	PID value at low idle	PID value at high idle
MFDES	Mass Fuel Desired	---	---	---
MGP	Manifold Gauge Pressure	---	---	---
PBA	Parking Brake Applied	0v / 12	Parking Brake Off: 12	0v
RPM	Engine Speed	0-10K rpm	---	---
SCCS	Speed Control Command Switch	0-25.5v	S/C On: 12	S/C On: 12
SCCS-M	Speed Control Command Switch Mode	---	---	---
TCC	Transmission Converter Clutch Solenoid Control	ON / OFF	OFF	OFF
SS1	Shift Solenoid 1	ON / OFF	ON	OFF
SS2	Shift Solenoid 2	ON / OFF	OFF	ON
TCIL	TCIL (lamp) Control	ON / OFF	OFF	OFF
TCS	TCS Switch	0v / 12	0	0
TFT V	TFT Sensor	0-5.1v	2.10-2.40	2.10-2.40
TORQUE	Engine Torque	---	---	---
TPREL	Low Idle TP Sensor	---	---	---
TR	Transmission Range Sensor	0-5.1v	4.45 in P	2.87
TR V	Transmission Range Sensor Actual (Volts)	0-5.1v	4.45 in P	2.87 in O/D
VFDES	Volume Flow Desired	---	---	---
VPWR	Vehicle Power Supply	0-25.5v	12-14	12-14
VREF	Vehicle Reference	0-5.1v	4.9-5.1	4.9-5.1
VS SET	Vehicle Speed Setting	---	---	---
VSS	Vehicle Speed (Hertz or MPH)	0-999 Hz (0-159 mph)	0	125 Hz = 55 mph

FORD CAR CONTENTS

About This Section
Introduction .... Page 9-4
How to Use This Section .... Page 9-4

ASPIRE PIN TABLES
1.3L I4 MFI VIN H (All) 48 Pin **(1994-95)** .... Page 9-5
1.3L I4 MFI VIN H (All) 64 Pin **(1996-97)** .... Page 9-7

CONTOUR PIN TABLES
2.0L I4 MFI VIN 3 (All) 60 Pin **(1995)** .... Page 9-9
2.0L I4 MFI VIN 3 (All) 104 Pin **(1996-97)** .... Page 9-11
2.0L I4 MFI VIN 3 (All) 104 Pin **(1998-99)** .... Page 9-13
2.0L I4 MFI VIN 3 (All) 104 Pin **(2000)** .... Page 9-15
2.5L V6 MFI VIN G (M/T) 104 Pin **(1998-99)** .... Page 9-17
2.5L V6 MFI VIN L (All) 60 Pin **(1995)** .... Page 9-19
2.5L V6 MFI VIN L (All) 104 Pin **(1996-97)** .... Page 9-21
2.5L V6 MFI VIN L (All) 104 Pin **(1998-99)** .... Page 9-23
2.5L V6 MFI VIN L (All) 104 Pin **(2000)** .... Page 9-25

CROWN VICTORIA PIN TABLES
4.6L V8 2v MFI VIN W (A/T) 60 Pin **(1992-95)** .... Page 9-27
4.6L V8 2v MFI VIN W (A/T) 104 Pin **(1996-97)** .... Page 9-29
4.6L V8 2v MFI VIN W (A/T) 104 Pin **(1998-99)** .... Page 9-31
4.6L V8 2v MFI VIN W (A/T) 104 Pin **(2000-02)** .... Page 9-33
4.6L V8 2v MFI VIN W (A/T) 104 Pin **(2003)** .... Page 9-35
4.6L V8 CNG VIN 9 (A/T) 104 Pin **(1996-97)** .... Page 9-37
4.6L V8 CNG VIN 9 (A/T) 104 Pin **(1998-02)** .... Page 9-39
4.6L V8 CNG VIN 9 Module 60 Pin **(1998-02)** .... Page 9-41
4.6L V8 CNG VIN 9 Module Wiring **(1998-02)** .... Page 5-42
4.6L V8 CNG VIN 9 (A/T) 104 Pin **(2003)** .... Page 9-44
5.0L V8 MFI VIN F (A/T) 60 Pin **(1990-91)** .... Page 9-46
5.0L V8 MFI VIN F California (A/T) 60 Pin **(1990-91)** .... Page 9-47

ESCORT & ZX2 PIN TABLES
1.8L I4 MFI VIN 8 (All) 48 Pin **(1991-92)** .... Page 9-49
1.8L I4 MFI VIN 8 (A/T) 64 Pin **(1993)** .... Page 9-51
1.8L I4 MFI VIN 8 (M/T) 48 Pin **(1993)** .... Page 9-53
1.8L I4 MFI VIN 8 (A/T) 64 Pin **(1994-95)** .... Page 9-55
1.8L I4 MFI VIN 8 (M/T) 48 Pin **(1994-95)** .... Page 9-57
1.8L I4 MFI VIN 8 (A/T) 76 Pin **(1996)** .... Page 9-59
1.8L I4 MFI VIN 8 (M/T) 76 Pin **(1996)** .... Page 9-61
1.9L I4 CFI VIN 9 (All) 60 Pin **(1990)** .... Page 9-63
1.9L I4 MFI VIN J (All) 60 Pin **(1990)** .... Page 9-64
1.9L I4 MFI VIN J (All) 60 Pin **(1991-92)** .... Page 9-65
1.9L I4 MFI VIN J (All) 60 Pin **(1993-95)** .... Page 9-67
1.9L I4 MFI VIN J (All) 104 Pin **(1996)** .... Page 9-69
2.0L I4 MFI VIN P (All) 104 Pin **(1997)** .... Page 9-71
2.0L I4 MFI VIN P (All) 104 Pin **(1998-99)** .... Page 9-73
2.0L I4 MFI VIN P (All) 104 Pin **(2000-03)** .... Page 9-75
2.0L I4 MFI VIN 3 (All) 104 Pin **(1998-99)** .... Page 9-77
2.0L I4 MFI VIN 3 (All) 104 Pin **(2000-03)** .... Page 9-79

FOCUS PIN TABLES

2.0L I4 DURATEC VIN P (All) 104 Pin **(2000-2002)** .... Page 9-81
2.0L I4 DURATEC VIN P (All) 104 Pin **(2003)** .... Page 9-83
2.0L I4 ZETEC-E VIN 3 (All) 104 Pin **(2000-2002)** .... Page 9-85
2.0L I4 ZETEC-E VIN 3 (All) 104 Pin **(2003)** .... Page 9-87
2.3L I4 ZETEC-E VIN Z (All) 104 Pin **(2003)** .... Page 9-89

MUSTANG PIN TABLES

2.3L I4 MFI VIN A (All) 60 Pin **1990** .... Page 9-91
2.3L I4 MFI VIN M (All) 60 Pin **(1991-93)** .... Page 9-92
3.8L V6 MFI VIN 4 (All) 104 Pin **(1994-97)** .... Page 9-93
3.8L V6 MFI VIN 4 (All) 104 Pin **(1998-99)** .... Page 9-95
3.8L V6 MFI VIN 4 (All) 104 Pin **(2000)** .... Page 9-97
3.8L V6 MFI VIN 4 (All) 104 Pin **(2001)** .... Page 9-99
3.8L V6 MFI VIN 4 (All) 104 Pin **(2002-03)** .... Page 9-101
4.6L V8 MFI VIN V (All) 104 Pin **(1996-97)** .... Page 9-103
4.6L V8 MFI VIN V (M/T) 104 Pin **(1998-99)** .... Page 9-105
4.6L V8 MFI VIN V (M/T) 104 Pin **(2000-02)** .... Page 9-107
4.6L V8 MFI VIN R (M/T) 104 Pin **(2003)** .... Page 9-109
4.6L V8 MFI VIN W (All) 104 Pin **(1996-97)** .... Page 9-111
4.6L V8 MFI VIN W (All) 104 Pin **(1998-99)** .... Page 9-113
4.6L V8 MFI VIN X (All) 104 Pin **(1996-97)** .... Page 9-115
4.6L V8 MFI VIN X (All) 104 Pin **(1998-99)** .... Page 9-117
4.6L V8 MFI VIN X (All) 104 Pin **(2000-02)** .... Page 9-119
4.6L V8 MFI VIN Y (M/T) 104 Pin **(2003)** .... Page 9-121
5.0L V8 HO MFI VIN E (All) 60 Pin **(1990-92)** .... Page 9-123
5.0L V8 HO MFI VIN E (All) 60 Pin **(1993)** .... Page 9-124
5.0L V8 MFI VIN D, VIN T (All) 60 Pin **(1994-95)** .... Page 9-126

PROBE PIN TABLES

2.0L I4 MFI VIN A (A/T) 64 Pin **(1993-95)** .... Page 9-128
2.0L I4 MFI VIN A (M/T) 60 Pin **(1993-95)** .... Page 9-130
2.0L I4 MFI VIN A (A/T) 104 Pin **(1996-97)** .... Page 9-132
2.2L I4 MFI VIN C (A/T) 64 Pin **(1990-92)** .... Page 9-134
2.2L I4 MFI VIN C (M/T) 64 Pin **(1990-92)** .... Page 9-136
2.2L I4 MFI TC VIN L (All) 64 Pin **(1990-92)** .... Page 9-138
2.5L V6 MFI VIN B (All) 64 Pin **(1993)** .... Page 9-140
2.5L V6 MFI VIN B (All) 64 Pin **(1994-95)** .... Page 9-142
2.5L V6 MFI VIN B (4EAT) 100-Pin **(1996-97)** .... Page 9-144
2.5L V6 MFI VIN B (M/T) 100-Pin **(1996-97)** .... Page 9-146

TAURUS PIN TABLES
 2.5L I4 CFI VIN D (All) 60 Pin **(1990)** ....Page 9-148
 2.5L I4 CFI VIN D (All) 60 Pin **(1991)** ....Page 9-149
 3.0L V6 MFI VIN S (A/T) 104 Pin **(1996-97)** ....Page 9-151
 3.0L V6 MFI VIN S (A/T) 104 Pin **(1998-99)** ....Page 9-153
 3.0L V6 MFI VIN S (A/T) 104 Pin **(2000)** ....Page 9-155
 3.0L V6 MFI VIN S (A/T) 104 Pin **(2001)** ....Page 9-157
 3.0L V6 MFI VIN S (A/T) 104 Pin **(2002-03)** ....Page 9-159
 3.0L V6 MFI VIN U (All) 60 Pin **(1990)** ....Page 9-161
 3.0L V6 MFI VIN U (All) 60 Pin **(1991-95)** ....Page 9-162
 3.0L V6 MFI VIN U (A/T) 104 Pin **(1996-97)** ....Page 9-164
 3.0L V6 MFI VIN U (A/T) 104 Pin **(1998-99)** ....Page 9-166
 3.0L V6 MFI VIN U (A/T) 104 Pin **(2000)** ....Page 9-168
 3.0L V6 MFI VIN U (A/T) 104 Pin **(2001-03)** ....Page 9-170
 3.0L V6 MFI SHO VIN Y (M/T) 60 Pin **(1990-93)** ....Page 9-172
 3.0L V6 MFI SHO VIN Y (M/T) 60 Pin **(1994-95)** ....Page 9-174
 3.0L V6 FFV VIN 1 (A/T) 60 Pin **(1993-95)** ....Page 9-176
 3.0L V6 FFV VIN 1, 2 (A/T) 104 Pin **(1996-97)** ....Page 9-178
 3.0L V6 FFV VIN 2 (A/T) 104 Pin **(1998-99)** ....Page 9-180
 3.0L V6 FFV VIN 2 (A/T) 104 Pin **(2000)** ....Page 9-182
 3.0L V6 FFV VIN 2 (A/T) 104 Pin **(2001-03)** ....Page 9-184
 3.2L V6 MFI SHO VIN P (A/T) 60 Pin **(1993-95)** ....Page 9-186
 3.4L V6 MFI SHO VIN N (A/T) 104 Pin **(1996-97)** ....Page 9-188
 3.4L V6 MFI SHO VIN N (A/T) 104 Pin **(1998-99)** ....Page 9-190
 3.8L V6 MFI VIN 4 (A/T) 60 Pin **(1990)** ....Page 9-192
 3.8L V6 MFI VIN 4 (A/T) 60 Pin **(1991-95)** ....Page 9-194
TEMPO PIN TABLES
 2.3L I4 MFI VIN S, X (All) 60 Pin **(1990)** ....Page 9-196
 2.3L I4 MFI VIN S, X (All) 60 Pin **(1991)** ....Page 9-197
 2.3L I4 MFI VIN X (All) 60 Pin **(1992-94)** ....Page 9-198
 3.0L V6 MFI VIN U (All) 60 Pin **(1992-94)** ....Page 9-200
THUNDERBIRD PIN TABLES
 3.8L V6 MFI VIN 4 (A/T) 60 Pin **(1990-93)** ....Page 9-202
 3.8L V6 MFI VIN 4 (A/T) 60 Pin **(1991-93)** ....Page 9-204
 3.8L V6 MFI VIN 4 (A/T) 60 Pin **(1994-95)** ....Page 9-206
 3.8L V6 MFI VIN 4 (A/T) 104 Pin **(1996-97)** ....Page 9-208
 3.8L V6 MFI SC VIN C (A/T) 60 Pin **(1990-92)** ....Page 9-210
 3.8L V6 MFI SC VIN R (A/T) 60 Pin **(1990-93)** ....Page 9-212
 3.8L V6 MFI SC VIN R (A/T) 60 Pin **(1994-95)** ....Page 9-214
 3.9L V8 MFI VIN A (A/T) 150-Pin **(2002-03)** ....Page 9-216
 4.6L V8 MFI VIN W (A/T) 104 Pin **(1994-97)** ....Page 9-219
 5.0L V8 MFI VIN T (A/T) 60 Pin **(1991-93)** ....Page 9-221

About This Section

Introduction

This section contains Pin Voltage Tables for Ford cars from 1990-2003 that can be used to assist in the repair of Trouble Code and No Code faults related to the PCM.

VEHICLE COVERAGE

- 1995-97 Aspire Applications
- 1996-2000 Contour Applications
- 1990-2003 Crown Victoria Applications
- 1990-2003 Escort & ZX2 Applications
- 2000-03 Focus Applications
- 1990-2003 Mustang Applications
- 1990-97 Probe Applications
- 1990-2003 Taurus Applications
- 1990-94 Tempo Applications
- 1990-97, 2000-03 Thunderbird Applications

How to Use This Section

This section can be used to look up the location of a particular pin, a Wire Color or a "known good" value of a PCM circuit. To locate the PCM information for a particular vehicle, find the model, correct engine size (with VIN Code) and finally the year of the vehicle.

For example, to look up the PCM terminals for a 1999 Mustang 4.6L VIN W, go to Contents Page 3 and find the text string shown below:

4.6L V8 2v MFI VIN W (All) 104 Pin **(1998-99)** ....Page 9-113

Then turn to Page 9-98 to find the following PCM related information.

1998-99 Mustang 4.6L 2v V8 MFI VIN W (All) 104 Pin Connector

PCM Pin #	Wire Color	Circuit Description (104 Pin)	Value at Hot Idle
1	PK/OR	Shift Solenoid 2 Control	12v, 55 mph: 12v
2	PK/LG	MIL (lamp) Control	MIL On: 1v, Off: 12v
3	YL/BK	Digital TR1 Sensor	0v, 55 mph: 11v
12	YL/WT	Fuel Level Indicator Signal	1.7v (1/2 full)
21	DB	CKP Sensor (+) Signal	390-450 Hz

In this example, the Fuel Level Indicator circuit is connected to Pin 12 of the 104 Pin Connector by a Yellow/White wire. The value at Hot Idle shown here is the normal value for the fuel level with the tank 1/2 full.

The "All" that appears in the Title of the table indicates the information is for both automatic and manual transmission vehicle applications.

Aspire Pin Tables

1994-95 Aspire 1.3L I4 MFI VIN H (All) 'C200' 26 Pin Connector

PCM Pin #	Wire Color	Circuit Description (26 Pin)	Value at Hot Idle
C200-2	BK/OR	Power Ground	<0.1v
C200-2	BK/LG	MAF Ground	<0.050v
C200-3	GN/BK	CID Sensor Signal	5-7 Hz
C200-4	GN/RD	CKP Sensor Signal	400-425 Hz
C200-5	PK	High Pressure Switch Signal	A/C On: 0v, Off: 12v
C200-6	LG/RD	Reference Voltage	4.9-5.1v
C200-7	LG/WT	TP Sensor Signal	Idle: 0.4-1.2v
C200-8	GN/BK	MAF Sensor Signal	2.0v
C200-9	---	Not Used	---
C200-10	RD/YL	EGR Control Solenoid	12v, 55 mph: 1v
C200-11	GN/YL	Injector 1 Control	4.5 ms
C200-12	RD/WT	IAC Motor Control	2.2 ms
C200-13	GN/RD	Injector 3 Control	4.5 ms
C200-14	BK/OR	Power Ground	<0.1v
C200-15	YL/GN	Analog Signal Return	<0.050v
C200-16	BK/RD	Spark Output Signal	0.2v, 55 mph: 0.3v1
C200-17	RD/BL	ECT Sensor Signal	0.5-0.6v
C200-18	YL	EGR EVP Sensor Signal	0.67v, 55 mph: 2.2v
C200-19	GN/RD	IAT Sensor Signal	1.5-2.5v
C200-20	WT	O2S-11 (B1 S1) Signal	0.1-1.1v
C200-21	LG	Condenser Fan Control	On: 0v, Off: 12v
C200-22	BL	EGR Vent Solenoid Control	12v, 55 mph: 1v
C200-23	---	Not Used	---
C200-24	GN/BK	Injector 2 Control	4.5 ms
C200-25	RD/BL	EVAP Purge Solenoid	12v, 55 mph: 1v
C200-26	GN/BL	Injector 4 Control	4.5 ms

Pin Connector Graphic

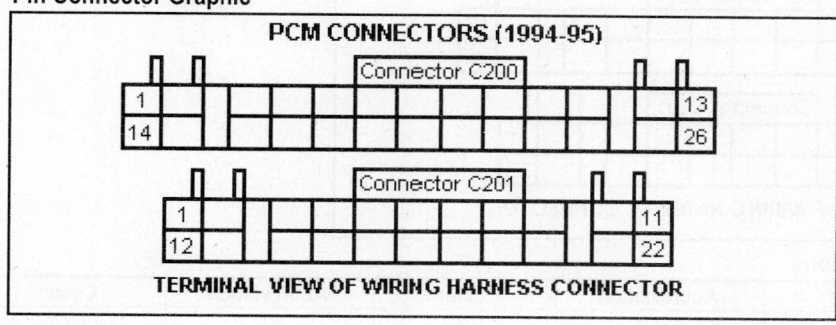

1994-95 Aspire 1.3L I4 MFI VIN H (All) 'C201' 22 Pin Connector

PCM Pin #	Wire Color	Circuit Description (22 Pin)	Value at Hot Idle
C201-1	BL/RD	Keep Alive Power	12-14v
C201-2	BK/WT	Vehicle Start Signal	KOEC: 9-11v
C201-3	BL	MIL (lamp) Control	MIL On: 1v, Off: 12v
C201-4	BK	Power Ground (M/T only)	<0.1v
C201-5	LG	Daytime Running Lamps	DRL On: 0v, Off: 12v
C201-6	BL	Self-Test Input Signal	STI On: 0v, Off: 5v
C201-7	GN/RD	VSS in Instrument Cluster	Vehicle moving: 0-5v
C201-8	GN	Brake Pedal Position Switch	Brake Off: 0v, On: 12v
C201-9	GN/WT	A/C Cycling Clutch Switch	A/C Off: 0v, On: 12v
C201-10	BR	Cooling Fan Control	On: 1v, Off: 12v
C201-11	RD/GN	Headlamp Relay Control	HDLP On: 12v, Off: 0v
C201-12	YL/WT	Vehicle Power (Main Relay)	12-14v
C201-13	BL/BK	Data Link Connector	Digital Signals
C201-14	WT/BK	Self-Test Output Signal	STO On: 0v, Off: 12v
C201-15	WT/YL	Fuel Pump Control	On: 1v, Off: 12v
C201-16	BL/OR	A/C Relay Control	ACR On: 1v, Off: 12v
C201-17	BK/RD	Rear Window Defroster	DEF On: 12v, Off: 0v
C201-18	RD	Idle Throttle Switch Signal	Closed: 0v, Open: 12v
C201-19	BL/YL	PSP Switch (ATX only)	Straight: 12v, Turned: 0v
C201-20	OR/BL	Blower Motor Switch Signal	Motor On: 12v, Off: 0v
C201-21	BL/WT	M/T: Shift Indicator Light	SIL On: 1v, Off: 12v
C201-22	GN/BK	M/T: Clutch Pedal Position	Clutch In: 0v, out: 5v
C201-22	WT	A/T: Park Neutral Position	In 'P': 0v, Others: 5v

Pin Connector Graphic

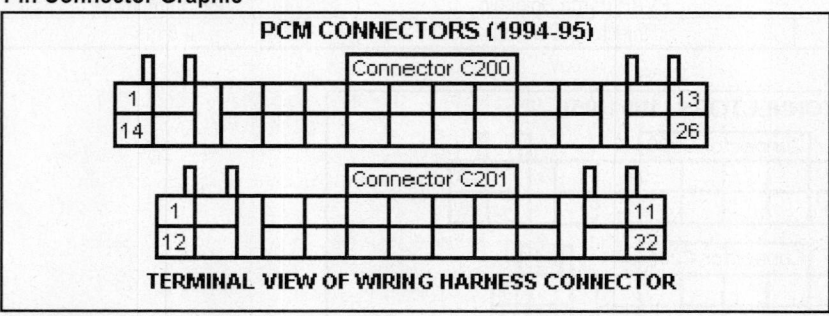

Standard Colors and Abbreviations

Abbreviation	Color	Abbreviation	Color	Abbreviation	Color
BK	Black	GY	Gray	PK	Purple
BL	Blue	GN	Green	RD	Red
BR	Brown	LG	LT Green	TN	Tan
DB	Dark Blue	OR	Orange	WT	White
DG	DK Green	PK	Pink	YL	Yellow

1996-97 1.3L I4 MFI Engine VIN H (All) 22 Pin Connector

PCM Pin #	Wire Color	Circuit Description (22 Pin)	Value at Hot Idle
C201-1	LG	Condenser Fan Control	On: 1v, Off: 12v
C201-2	BK/WT	Vehicle Start Signal	KOEC: 9-11v
C201-3	BL	MIL (lamp) Control	MIL On: 1v, Off: 12v
C201-4	BL/OR	Air Conditioning Relay Control	A/C On: 0v, Off: 12v
C201-5	BL	Self-Test Input Signal	STI On: 0v, Off: 12v
C201-6	GN/WT	A/C Cycling Clutch Signal	A/C & BLMT on: 12v
C201-7	GN/RD	Vehicle Speed Sensor	Varies: 0-5v
C201-8	---	Not Used	---
C201-9	GN	Brake Pedal Position Switch	Brake Off: 0v, On: 12v
C201-10	BK	M/T: Power Ground	<0.1v
C201-11	WT/YL	Fuel Pump Control	On: 1v, Off: 12v
C201-12	BK	Fan Control Relay	On: 1v, Off: 12v
C201-13	PK	ISO K-Line for Scan Tool	No Scan Tool: 7v
C201-14	---	Not Used	---
C201-15	RD/GN	Headlamp Switch Signal	Lamps On: 12v, Off: 0v
C201-16	BK/RD	Rear Window Defrost Switch	DEF On: 12v, Off: 0v
C201-17	WT	Clutch or P/N Position Switch	In 'P': 0v, Others: 5v
C201-18	LB	DRL (lamps) Canada only	On: 1v, Off: 12v
C201-19	OR/BL	Blower Motor Control Switch	Motor On: 12v, Off: 0v
C201-20	PK	A/C High Pressure Switch	Open: 12v, closed: 0v
C201-21, 22	---	Not Used	---

1996-97 1.3L I4 MFI Engine VIN H (All) 'C257' 16 Pin Connector

PCM Pin #	Wire Color	Circuit Description (16 Pin)	Value at Hot Idle
C257-1	---	Not Used	---
C257-2	WT	HO2S-11 (B1 S1) Signal	0.1-1.1v
C257-3	---	Not Used	---
C257-4	RD/BL	ECT Sensor Signal	0.5-0.6v
C257-5	LG/RD	Reference Voltage	4.9-5.1v
C257-6	GN/RD	IAT Sensor Signal	1.5-2.5v
C257-7	---	Not Used	---
C257-8	YL/GN	Analog Signal Return	<0.050v
C257-9	GN/BK	MAF Sensor Signal	2.0v
C257-10	OR/WT	HO2S-12 (B1 S2) Signal	0.1-1.1v
C257-11	LG/WT	TP Sensor Signal	0.5-1.2v
C257-12	PK/BK	Atmospheric Pressure Sensor	4.3v (28" Hg)
C257-13	YL	EGR EVP Sensor	0.7v
C257-14	RD	Idle Throttle Switch Signal	Closed: 0v, Open: 12v
C257-15	BR	EGR Boost Sensor	3.9v
C257-16	BL/YL	PSP Switch Signal	Straight: 12v, Turned: 0v

Standard Colors and Abbreviations

Abbreviation	Color	Abbreviation	Color	Abbreviation	Color
BK	Black	GY	Gray	PK	Purple
BL	Blue	GN	Green	RD	Red
BR	Brown	LG	LT Green	TN	Tan
DB	Dark Blue	OR	Orange	WT	White
DG	DK Green	PK	Pink	YL	Yellow

1996-97 1.3L I4 MFI Engine VIN H (All) C200 26 Pin Connector

PCM Pin #	Wire Color	Circuit Description (26 Pin)	Value at Hot Idle
C200-1	BK/LG	Power Ground	<0.1v
C200-2	BK/OR	Power Ground	<0.1v
C200-3	BL/OR	CKP Sensor (-) Signal	400-425 Hz
C200-4	GN/RD	CMP 1 Sensor Signal	5-7 Hz
C200-5	BL/RD	Keep Alive Power	12-14v
C200-6	---	Not Used	---
C200-7	---	Not Used	---
C200-8	BL	EGR Vent Solenoid Control	12v, 55 mph: 1v
C200-9	RD/WT	IAC Motor Control	2.2 ms
C200-10	---	Not Used	---
C200-11	GN/YL	Injector 1 Control	4.5 ms
C200-12	GN/RD	Injector 3 Control	4.5 ms
C200-13	---	Not Used	---
C200-14	YL/WT	Vehicle Power	12-14v
C200-15	BK/OR	Power Ground	<0.1v
C200-16	GN/BK	CMP 2 Sensor Signal	20-28 Hz
C200-17	LG	CKP Sensor (+) Signal	400-425 Hz
C200-18	---	Not Used	---
C200-19	---	Not Used	---
C200-20	BK/RD	Ignition Control Module Signal	0.2v, 55 mph: 0.3v
C200-21	RD/YL	EGR Control Solenoid	12v, 55 mph: 1v
C200-22	---	Not Used	---
C200-23	BL/RD	EVAP Purge Solenoid	12v, Cruise: 9.1v
C200-24	GN/BK	Injector 2 Control	4.5 ms
C200-25	GN/BL	Injector 4 Control	4.5 ms
C200-26	---	Not Used	---

Pin Connector Graphic

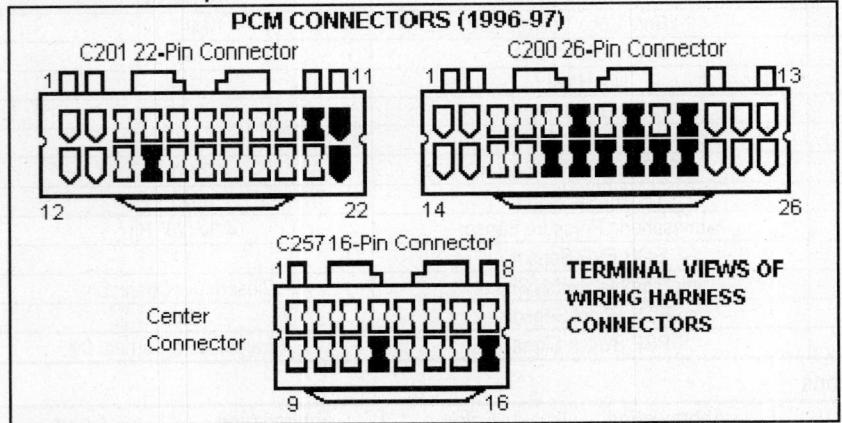

Contour Pin Tables

1995 Contour 2.0L I4 MFI VIN 3 (All) 60 Pin Connector

PCM Pin #	Wire Color	Circuit Description (60 Pin)	Value at Hot Idle
1	OR/YL	Keep Alive Power	12- 14v
2	OR	Brake On/Off Signal	Brake Off: 0v, On: 12v
3	WT/PK	VSS (+) Signal	0 Hz, 55 mph: 125 Hz
4	WT/GN	Ignition Diagnostic Monitor	20-31 Hz
5	WT/PK	A/T: TSS Sensor (+) Signal	42-50 Hz (680-720 rpm)
6	---	Not Used	---
7	WT/GN	ECT Sensor Signal	0.5-0.6v
8	PK/BK	Fuel Pump Monitor	Pump On: 12v, Off; 0v
9	BR/BL	MAF Sensor Return	<0.050v
10	PK/BL	A/C Cycling Clutch Switch	A/C On: 12v, Off: 0v
11	BK/OR	EVAP Purge Solenoid	12v, 55 mph: 1v
12	---	Not Used	---
13	BK/BL	Low Speed Fan Control	On: 1v, Off: 12v
14	---	Not Used	---
15	---	Not Used	---
16	BK/BL	Ignition Ground	<0.050v
17	BK/OR	STI Output, MIL Control	MIL On: 1v, Off: 12v
18	WT/BK	Data Bus (+) Signal	Digital Signals
19	BR/BK	Data Bus (-) Signal	Digital Signals
20	BK/RD	Case Ground	<0.050v
21	BK/OR	IAC Solenoid Control	Varies: 8-11v
22	BK/BL	Fuel Pump Control	On: 1v, Off: 12v
23	WT/BK	Knock Sensor Signal	0v
24	WT/PK	Camshaft Position Sensor	5-7 Hz
25	WT/PK	IAT Sensor Signal	1-5-2.5v
26	YL	Reference Voltage	4.9-5.1v
27	WT/BL	DPFE Sensor Signal	Hot Idle: 0.1v
28	WT/PK	PSP Switch Signal	Straight: 12v, Turned: 0v
29	WT/BK	Octane Adjust Switch Signal	Closed: 0v, Open: 9.3v
30	WT	A/T: PNP Sensor Signal	In 'P': 0v, 55 mph: 1.7v
30	WT	M/T: Clutch Pedal Position	Clutch In: 0v, Out: 5v
31	BK/WT	High Speed Fan Control	On: 1v, Off: 12v
32	---	Not Used	---
33	GN/WT	EGR VR Solenoid	12v, 30 mph: 10-12v
34	---	Not Used	---
35	BK/OR	Injector 4 Control	2.3-2.9 ms
36	WT/PK	Spark Output Signal	50% duty cycle

Pin Connector Graphic

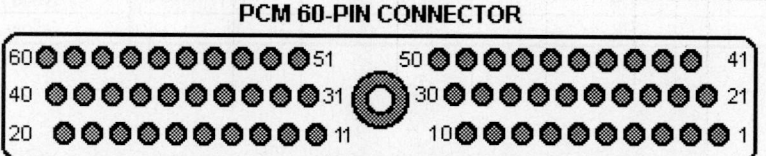

PCM 60-PIN CONNECTOR

Terminal View of 60-Pin PCM Harness Connector

1995 Contour 2.0L I4 MFI VIN 3 (All) 60 Pin Connector

PCM Pin #	Wire Color	Circuit Description (60 Pin)	Value at Hot Idle
37	GN/YL	Vehicle Power	12-14v
38	BK/RD	A/T: EPC Solenoid Control	7-8v (8-9 psi)
39	BK/BL	Injector 3 Control	2.3-2.9 ms
40	BK/YL	Power Ground	<0.1v
41	WT/BK	A/T: TCS (switch) Signal	TCS & O/D On: 12v
42	---	Not Used	---
43	---	Not Used	---
44	WT	HO2S-11 (B1 S1) Signal	0.1-1.1v
45	---	Not Used	---
46	BR	Analog Signal Return	<0.050v
47	WT	TP Sensor Signal	0.5-1.1v
48	WT/BK	Self-Test Input	STI On: 0.1v, Off: 5v
49	WT/RD	A/T: TFT Sensor Signal	2.10-2.40v
50	WT/BL	MAF Sensor Signal	1.9v
51	BK/YL	A/T: Shift Solenoid 1 Control	12v, 55 mph: 1v
52	BK/BL	A/T: Shift Solenoid 2 Control	1v, 55 mph: 12v
53	BK/WT	A/T: TCC Solenoid Control	TCC On: 1v, Off: 12v
54	BK/YL	A/C WOT Relay Control	On: 1v, Off: 12v
55	BK/OR	A/T: Shift Solenoid 3 Control	12v, 55 mph: 1v
56	WT/BK	PIP Sensor Signal	50% duty cycle
57	GN/YL	Vehicle Power	12-14v
58	BK/WT	Injector 1 Control	2.3-2.9 ms
59	BK/YL	Injector 2 Control	2.3-2.9 ms
60	BK/YL	Power Ground	<0.1v

Pin Connector Graphic

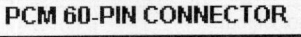
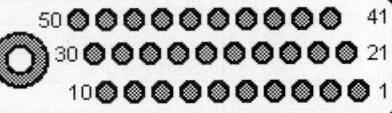

PCM 60-PIN CONNECTOR

Terminal View of 60-Pin PCM Harness Connector

Standard Colors and Abbreviations

Abbreviation	Color	Abbreviation	Color	Abbreviation	Color
BK	Black	GY	Gray	PK	Purple
BL	Blue	GN	Green	RD	Red
BR	Brown	LG	LT Green	TN	Tan
DB	Dark Blue	OR	Orange	WT	White
DG	DK Green	PK	Pink	YL	Yellow

1996-97 Contour 2.0L I4 MFI VIN 3 (All) 104 Pin Connector

PCM Pin #	Wire Color	Circuit Description (104 Pin)	Value at Hot Idle
1	BK/BL	Shift Solenoid 2 Control	12v, 55 mph: 1v
2	BK/OR	MIL (lamp) Control	MIL On: 1v, Off: 12v
3-12	---	Not Used	---
13	WT/BL	Flash EEPROM Power	0.1v
14	---	Not Used	---
15	BR/BL	Data Bus (-) Signal	Digital Signals
16	WT/BL	Data Bus (+) Signal	Digital Signals
17-20	---	Not Used	---
21	WT/RD	CKP Sensor (+) Signal	450-480 Hz
22	BR/WT	CKP Sensor (-) Signal	450-480 Hz
23	---	Not Used	---
24	BK/YL	Power Ground	<0.1v
25	BK/RD	Power Ground	<0.1v
26	BR/BL	Coil Driver 1 Control	5° dwell
27	BK/YL	Shift Solenoid 1 Control	1v, 55 mph: 12v
28	---	Not Used	---
29	PK/BK	TCS (switch) Signal	TCS & O/D On: 12v
30	WT/BK	Octane Adjust Switch	Closed: 0v, Open: 9.3v
31 ('96)	WT/PK	PSP Sensor Signal	Straight: 0v, Turning: 12v
31 ('97)	WT	PSP Sensor Signal	Straight: 0v, Turning: 12v
32-34	---	Not Used	---
35	WT/BL	HO2S-12 (B1 S2) Signal	0.1-1.1v
36	BR/BL	MAF Sensor Return	<0.050v
37	WT/RD	TFT Sensor Signal	2.10-2.40v
38	WT/GN	ECT Sensor Signal	0.5-0.6v
39	WT/PK	IAT Sensor Signal	1.5-2.5v
40	PK/BK	Fuel Pump Monitor	On: 12v, Off: 0v
41	PK/BL	A/C Cycling Clutch Switch	Closed: 12v, Open: 0v
42-44	---	Not Used	---
45	BK/BL	Low Speed Fan Control	On: 1v, Off: 12v
46	BK/WT	High Speed Fan Control	On: 1v, Off: 12v
47	BK/GN	EGR VR Solenoid	12v, 55 mph: 10-12v
48	WT/BK	Clean Tachometer Output	20-31 Hz
49-50	---	Not Used	---
51	BK/YL	Power Ground	<0.1v
52	BR/GN	Coil Driver 2 Control	5° dwell
53	---	Not Used	---
54	BK/WT	TCC Solenoid Control	0%, 55 mph: 50-90%
55	OR/YL	Keep Alive Power	12-14v
56	BK/OR	EVAP VMV Solenoid	0-10 Hz (0-100%)
57	WT/BK	Knock Sensor 1 Signal	0v
58	WT/PK	VSS (+) Signal	0 Hz, 55 mph: 125 Hz
59	---	Not Used	---
60	WT	HO2S-11 (B1 S1) Signal	0.1-1.1v
61-63	---	Not Used	---
64	WT/GN	A/T: TR Sensor	In 'P': 0v, 55 mph: 1.7v
64	WT	M/T: Clutch Pedal Position	Clutch In: 1v, Out: 5v

1996-97 Contour 2.0L I4 MFI VIN 3 (All) 104 Pin Connector

PCM Pin #	Wire Color	Circuit Description (104 Pin)	Value at Hot Idle
65	WT/BL	DPFE Sensor Signal	0.20-1.30v
66-68	---	Not Used	---
69	BK/YL	A/C WOT Relay Control	On: 1v, Off: 12v
70	BK/YL	ZETECT VCT Control	VCT Off: 12v, On: 1v
71	GN/YL	Vehicle Power	12-14v
72-73	---	Not Used	---
74	BK/BL	Injector 3 Control	2.3-2.9 ms
75	BK/WT	Injector 1 Control	2.3-2.9 ms
76	BK/YL	CMP and TSS Ground	<0.050v
77	BK/YL	Power Ground	<0.1v
78	---	Not Used	---
79	WT/BL	TCIL (lamp) Control	On: 1v, Off: 12v
80	BK/BL	Fuel Pump Control	On: 1v, Off: 12v
81	BK/RD	EPC Solenoid Control	17 psi
82	---	Not Used	---
83	BK/YL	IAC Motor Control	9v (33% duty cycle)
84	WT/PK	TSS Sensor Signal	40-56 Hz (680-710 rpm)
85	WT/PK	CID Sensor Signal	5-7 Hz
86	BK/WT	A/C Pressure Switch	A/C On: 12v (Open)
87	---	Not Used	---
88	WT/BL	MAF Sensor Signal	0.6-0.9v
89	WT	TP Sensor Signal	0.53-1.27v
90	YL	Reference Voltage	4.9-5.1v
91	BR	Analog Signal Return	<0.050v
92	OR	Brake Pedal Position Switch	Brake Off: 0v, On: 12v
93	BK/YL	HO2S-11 (B1 S1) Heater	On: 1v, Off: 12v
94	---	Not Used	---
95	BK/OR	HO2S-12 (B1 S2) Heater	On: 1v, Off: 12v
96	---	Not Used	---
97	GN/YL	Vehicle Power	12-14v
98-99	---	Not Used	---
100	BK/OR	Injector 4 Control	2.3-2.9 ms
101	BK/YL	Injector 2 Control	2.3-2.9 ms
102	BK/OR	Shift Solenoid 3 Control	8v, 55 mph: 8-9v
103	BK/YL	Power Ground	<0.1v
104	---	Not Used	---

PCM 104-PIN CONNECTOR

Terminal View of 104-Pin PCM Wiring Harness Connector

Pin Connector Graphic

Standard Colors and Abbreviations

Abbreviation	Color	Abbreviation	Color	Abbreviation	Color
BK	Black	GY	Gray	PK	Purple
BL	Blue	GN	Green	RD	Red
BR	Brown	LG	LT Green	TN	Tan
DB	Dark Blue	OR	Orange	WT	White
DG	DK Green	PK	Pink	YL	Yellow

1998-99 Contour 2.0L I4 MFI VIN 3 (All) 104 Pin Connector

PCM Pin #	Wire Color	Circuit Description (104 Pin)	Value at Hot Idle
1	BK/BL	Shift Solenoid 2 Control	12v, 55 mph: 1v
2	BK/OR	MIL (lamp) Control	MIL On: 1v, Off: 12v
3-11	---	Not Used	---
12	WT	Fuel Level Indicator Signal	1.7v (1/2 full)
13	WT/BL	Flash EEPROM Power	0.1v
15	BL	Data Bus (-) Signal	Digital Signals
16	GN	Data Bus (+) Signal	Digital Signals
17-20	---	Not Used	---
21	WT/RD	CKP Sensor (+) Signal	450-480 Hz
22	BR/WT	CKP Sensor (-) Signal	450-480 Hz
23	---	Not Used	---
24	BK/YL	Power Ground	<0.1v
25	BK/RD	Power Ground	<0.1v
26	BR/BL	Coil Driver 1 Control	5° dwell
27	BK/YL	Shift Solenoid 1 Control	1v, 55 mph: 12v
28	---	Not Used	---
29	PK/BK	TCS (switch) Signal	TCS & O/D On: 12v
30	WT/BK	Octane Adjust Switch	Closed: 0v, Open: 9.3v
31	WT/PK	PSP Sensor Signal	Straight: 0v, Turning: 12v
32-34	---	Not Used	---
35	WT/BL	HO2S-12 (B1 S2) Signal	0.1-1.1v
36	BR/BL	MAF Sensor Return	<0.050v
37	WT/RD	TFT Sensor Signal	2.10-2.40v
38	WT/GN	ECT Sensor Signal	0.5-0.6v
39	WT/PK	IAT Sensor Signal	1.5-2.5v
40	PK/BK	Fuel Pump Monitor	On: 12v, Off: 0v
41	PK/BL	A/C Cycling Clutch Switch	Closed: 12v, Open: 0v
42-43	---	Not Used	---
44	BK/RD	VCT Actuator	VCT Off: 12v, On: 1v
45	BK/BL	Low Speed Fan Control	On: 1v, Off: 12v
46	BK/WT	High Speed Fan Control	On: 1v, Off: 12v
47	BK/GN	EGR VR Solenoid	12v, 55 mph: 10-12v
48	WT/BK	Clean Tachometer Output	20-31 Hz
49-51	---	Not Used	---
51	BK/YL	Power Ground	<0.1v
52	BR/GN	Coil Driver 2 Control	5° dwell
53	---	Not Used	---
54	BK/WT	TCC Solenoid Control	0%, 55 mph: 50-90%
55	OR/YL	Keep Alive Power	12-14v
56	BK/OR	EVAP Purge Solenoid	0-10 Hz (0-100%)
57	WT/BK	Knock Sensor 1 (+) Signal	0v
58	WT/PK	VSS (+) Signal	0 Hz, 55 mph: 125 Hz
60	WT	HO2S-11 (B1 S1) Signal	0.1-1.1v
61	---	Not Used	---
62	WT/PK	FTP Sensor Signal	2.6v (cap off)
63 ('99)	WT/PK	FRP Sensor Signal	2.8v (39 psi)
64	WT/GN	A/T: TR Sensor Signal	In 'P': 0v, 55 mph: 1.7v
64	WT	M/T: Clutch or Neutral Switch	Clutch In: 1v, Out: 5v

1998-99 Contour 2.0L I4 MFI VIN 3 (All) 104 Pin Connector

PCM Pin #	Wire Color	Circuit Description (104 Pin)	Value at Hot Idle
65	WT/BL	DPFE Sensor Signal	0.20-1.30v
67	BK/OR	EVAP CV Solenoid	0-10 Hz (0-100%)
69	BK/YL	A/C WOT Relay Control	On: 1v, Off: 12v
70	BK/YL	VCT Control (ZETECT)	VCT Off: 12v, On: 1v
71	GN/YL	Vehicle Power	12-14v
72-73	---	Not Used	---
74	BK/BL	Injector 3 Control	2.3-2.9 ms
75	BK/WT	Injector 1 Control	2.3-2.9 ms
76	BR	CMP and TSS Ground	<0.050v
77	BK/YL	Power Ground	<0.1v
79	WT/BL	TCIL (lamp) Control	On: 1v, Off: 12v
80	BK/BL	Fuel Pump Control	On: 1v, Off: 12v
81	BK/RD	EPC Solenoid Control	17 psi
83	BK/YL	IAC Motor Control	9v (33% duty cycle)
84	WT/PK	TSS Sensor Signal	40-56 Hz (680-710 rpm)
85	WT/PK	CMP Sensor (+) Signal	5-7 Hz
86	BK/WT	A/C Pressure Switch	A/C On: 12v (Open)
87	---	Not Used	---
88	WT/BL	MAF Sensor Signal	0.6-0.9v
89	WT	TP Sensor Signal	0.53-1.27v
90	YL	Reference Voltage	4.9-5.1v
91	BR	Analog Signal Return	<0.050v
92	OR	Brake Pedal Position Switch	Brake Off: 0v, On: 12v
93	BK/YL	HO2S-11 (B1 S1) Heater	On: 1v, Off: 12v
95	BK/OR	HO2S-12 (B1 S2) Heater	On: 1v, Off: 12v
97	GN/YL	Vehicle Power	12-14v
100	BK/OR	Injector 4 Control	2.3-2.9 ms
101	BK/YL	Injector 2 Control	2.3-2.9 ms
102	BK/OR	Shift Solenoid 3 Control	8v, 55 mph: 8-9v
103	BK/YL	Power Ground	<0.1v

Pin Connector Graphic

PCM 104-PIN CONNECTOR

Terminal View of 104-Pin PCM Wiring Harness Connector

Standard Colors and Abbreviations

Abbreviation	Color	Abbreviation	Color	Abbreviation	Color
BK	Black	GY	Gray	PK	Purple
BL	Blue	GN	Green	RD	Red
BR	Brown	LG	LT Green	TN	Tan
DB	Dark Blue	OR	Orange	WT	White
DG	DK Green	PK	Pink	YL	Yellow

2000 Contour 2.0L I4 MFI VIN 3 (All) 104 Pin Connector

PCM Pin #	Wire Color	Circuit Description (104 Pin)	Value at Hot Idle
1	BK/BL	Shift Solenoid 2 Control	12v, 55 mph: 1v
2	BK/OR	MIL (lamp) Control	MIL On: 1v, Off: 12v
12	WT	Fuel Level Indicator Signal	1.7v (1/2 full)
13	WT/BL	Flash EEPROM Power	0.1v
15	BL	Data Bus (-) Signal	Digital Signals
16	GN	Data Bus (+) Signal	Digital Signals
17-20	---	Not Used	---
20	WT/BK	Gearshift Indicator Signal	N/A
21	WT/RD	CKP Sensor (+) Signal	450-480 Hz
22	BR/RD	CKP Sensor (-) Signal	450-480 Hz
24	BK/YL	Power Ground	<0.1v
25	BK/RD	Power Ground	<0.1v
26	BR/BL	Coil Driver 1 Control	5° dwell
27	BK/YL	Shift Solenoid 1 Control	1v, 55 mph: 12v
28	WT/BL	Speedometer Indicator Signal	Digital Signals
29	GN/BK	TCS (switch) Signal	TCS & O/D On: 12v
30	WT/BK	Octane Adjust Switch	Closed: 0v, Open: 9.3v
31	WT	PSP Sensor Signal	Straight: 0v, Turning: 12v
33	WT/PK	VSS (-) Signal	0 Hz, 55 mph: 125 Hz
34	WT/PK	A/T: TSS Sensor Signal	40-56 Hz (680-710 rpm)
35	WT/BL	HO2S-12 (B1 S2) Signal	0.1-1.1v
36	BR/BL	MAF Sensor Return	<0.050v
37	WT/RD	TFT Sensor Signal	2.10-2.40v
38	WT/GN	ECT Sensor Signal	0.5-0.6v
39	WT/PK	IAT Sensor Signal	1.5-2.5v
40	GN/BK	Fuel Pump Monitor	On: 12v, Off: 0v
41	GN/BL	A/C Cycling Clutch Switch	Closed: 12v, Open: 0v
44	BK/RD	VCT Actuator	VCT Off: 12v, On: 1v
45	BK/BL	Low Speed Fan Control	On: 1v, Off: 12v
46	BK/WT	High Speed Fan Control	On: 1v, Off: 12v
47	BK/GN	EGR VR Solenoid	12v, 55 mph: 10-12v
48	WT/BK	Clean Tachometer Output	20-31 Hz
51	BK/YL	Power Ground	<0.1v
52	BR/GN	Coil Driver 2 Control	5° dwell
54	BK/WT	TCC Solenoid Control	0%, 55 mph: 50-90%
55	OR/YL	Keep Alive Power	12-14v
56	BK/OR	EVAP Purge Solenoid	0-10 Hz (0-100%)
57	WT/BK	Knock Sensor 1 (+) Signal	0v
58	WT/PK	VSS (+) Signal	0 Hz, 55 mph: 125 Hz
59	WT/GN	Generator Load Indicator	1.5-10.v (40-250 Hz)
60	WT	HO2S-11 (B1 S1) Signal	0.1-1.1v
62	WT/PK	FTP Sensor Signal	2.6v (cap off)
63	WT/GN	FRP Sensor Signal	2.8v (39 psi)
64	WT/GN	A/T: TR Sensor Signal	In 'P': 0v, 55 mph: 1.7v
64	WT	M/T: Clutch or Neutral Switch	Clutch In: 1v, Out: 5v

1998-2000 Contour 2.0L I4 MFI VIN 3 (All) 104 Pin Connector

PCM Pin #	Wire Color	Circuit Description (104 Pin)	Value at Hot Idle
65	WT/BL	DPFE Sensor Signal	0.20-1.30v
67	BK/OR	EVAP CV Solenoid	0-10 Hz (0-100%)
69	BK/YL	A/C WOT Relay Control	On: 1v, Off: 12v
70	BK/YL	VCT Control (ZETECT)	VCT Off: 12v, On: 1v
71	GN/YL	Vehicle Power	12-14v
72	---	Not Used	---
73	BK/YL	Shift Solenoid 1 Control	12v, 55 mph: 1v
74	BK/BL	Injector 3 Control	2.3-2.9 ms
75	BK/WT	Injector 1 Control	2.3-2.9 ms
76	BR	CMP Sensor (-) Signal	5-7 Hz
77	BK/YL	Power Ground	<0.1v
79	WT/BL	TCIL (lamp) Control	On: 1v, Off: 12v
80	BK/RD	Fuel Pump Control	On: 1v, Off: 12v
81	BK/RD	EPC Solenoid Control	17 psi
82	---	Not Used	---
83	BK/YL	IAC Motor Control	9v (33% duty cycle)
84	WT/PK	TSS Sensor Signal	40-56 Hz (680-710 rpm)
85	WT/PK	CMP Sensor (+) Signal	5-7 Hz
86	BK/WT	A/C Pressure Switch	A/C On: 12v (Open)
87	BR/YL	Knock Sensor 1 (-) Signal	0v
88	WT/BL	MAF Sensor Signal	0.6-0.9v
89	WT	TP Sensor Signal	0.53-1.27v
90	YL	Reference Voltage	4.9-5.1v
91	BR	Analog Signal Return	<0.050v
92	GN/RD	Brake Pedal Position Switch	Brake Off: 0v, On: 12v
93	BK/YL	HO2S-11 (B1 S1) Heater	On: 1v, Off: 12v
95	BK/OR	HO2S-12 (B1 S2) Heater	On: 1v, Off: 12v
96	---	Not Used	---
97	GN/YL	Vehicle Power	12-14v
98	---	Not Used	---
99	BK/WT	Modulated Lockup Solenoid	N/A
100	BK/OR	Injector 4 Control	2.3-2.9 ms
101	BK/YL	Injector 2 Control	2.3-2.9 ms
102	BK/OR	Shift Solenoid 3 Control	8v, 55 mph: 8-9v
103	BK/YL	Power Ground	<0.1v
104	---	Not Used	---

Pin Connector Graphic

PCM 104-PIN CONNECTOR

Terminal View of 104-Pin PCM Wiring Harness Connector

Standard Colors and Abbreviations

Abbreviation	Color	Abbreviation	Color	Abbreviation	Color
BK	Black	GY	Gray	PK	Purple
BL	Blue	GN	Green	RD	Red
BR	Brown	LG	LT Green	TN	Tan
DB	Dark Blue	OR	Orange	WT	White
DG	DK Green	PK	Pink	YL	Yellow

1998-99 Contour SVT 2.5L 24v V6 VIN G (M/T) 104 Pin Connector

PCM Pin #	Wire Color	Circuit Description (104 Pin)	Value at Hot Idle
1	---	Not Used	---
2	BK/OR	MIL (lamp) Control	MIL On: 1v, Off: 12v
3-7	---	Not Used	---
8	BK/BL	IMRC Solenoid Control	5v
9-11	---	Not Used	---
12	WT	Fuel Level Indicator Signal	1.7v (1/2 full)
13	WT/BL	Flash EEPROM Power	0.1v
14	---	Not Used	---
15	BL	Data Bus (-) Signal	Digital Signals
16	GY	Data Bus (+) Signal	Digital Signals
17	GY/OR	Passive Antitheft System	Digital Signals
18-19	---	Not Used	---
20	BK/BL	Injector 3 Control	2.1-2.5 ms
21	WT/RD	CKP Sensor (+) Signal	410-440 Hz
22	BR/RD	CKP Sensor (-) Signal	410-440 Hz
23	---	Not Used	---
24	BK/YL	Power Ground	<0.1v
25	BK/RD	Power Ground	<0.1v
26	BR/BL	Coil Driver 1 Control	5° dwell
27-28	---	Not Used	---
29	PK/BK	Instrument Interface Module	Digital Signals
30	WT/BK	Octane Adjust Switch	Closed: 0v, Open: 9.3v
31	WT	PSP Sensor Signal	Straight: 0v, Turning: 12v
32-34	---	Not Used	---
34	WT/GN	Passive Antitheft System	Digital Signals
35	WT/BL	HO2S-12 (B1 S2) Signal	0.1-1.1v
36	BR/BL	MAF Sensor Return	<0.050v
37	---	Not Used	---
38	WT/GN	ECT Sensor Signal	0.5-0.6v
39	WT/PK	IAT Sensor Signal	1.5-2.5v
40	PK/BK	Fuel Pump Monitor	On: 12v, Off: 0v
41	PK/BL	A/C Cycling Clutch Switch	A/C On: 12v, Off: 0v
42	BK/RD	IMRC Solenoid Control	12v, 55 mph: 12v
43-44	---	Not Used	---
45	BK/BL	Low Speed Fan Control	On: 1v, Off: 12v
46	BK/WT	High Speed Fan Control	On: 1v, Off: 12v
47	BK/GN	EGR VR Solenoid	0%, 55 mph: 40%
48	WT/BK	Clean Tachometer Output	35-42 Hz
49-50	---	Not Used	---
51	BK/YL	Power Ground	<0.1v
52	BR/GN	Coil Driver 2 Control	5° dwell
53	---	Not Used	---
54	BK/BL	Fuel Pump Control	On: 1v, Off: 12v
55	OR/YL	Keep Alive Power	12-14v
56	BK/OR	EVAP Purge Solenoid	0-10 Hz (0-100%)
57	WT/BK	Knock Sensor 1 Signal	0v
58	WT/PK	VSS (+) Signal	0 Hz, 55 mph: 125 Hz
59	---	Not Used	---
60	WT	HO2S-11 (B1 S1) Signal	0.1-1.1v
61	WT/GN	HO2S-22 (B2 S2) Signal	0.1-1.1v
62	WT/PK	FTP Sensor Signal	2.6v (cap off)
63 ('99)	WT/PK	FRP Sensor Signal	2.8v (39 psi)
64	WT/GN	M/T: Clutch Position Switch	Clutch In: 0.1v, Out: 5v

1998-99 Contour SVT 2.5L 24v V6 VIN G (M/T) 104 Pin Connector

PCM Pin #	Wire Color	Circuit Description (104 Pin)	Value at Hot Idle
65	WT/BL	DPFE Sensor Signal	0.20-1.30v
66	---	Not Used	---
67	BK/OR	EVAP CV Solenoid	0-10 Hz (0-100%)
68	---	Not Used	---
69	BK/YL	A/C WOT Relay Control	On: 1v, Off: 12v
70	BK/WT	Injector 1 Control	2.1-2.5 ms
71	GN/YL	Vehicle Power	12-14v
72	---	Not Used	---
73	BK/YL	HO2S-11 (B1 S1) Heater	On: 1v, Off: 12v
74	BK/BL	Injector 3 Control	2.1-2.5 ms
75	---	Not Used	---
76	BK/YL	CMP Sensor Ground	<0.050v
77	BK/YL	Power Ground	<0.1v
78	BR/YL	Coil Driver 3 Control	5° dwell
79	WT/BL	Instrument Interface Module	Digital Signals
80-82	---	Not Used	---
83	BK/YL	IAC Motor Control	9v (33% duty cycle)
84	---	Not Used	---
85	WT/PK	CMP Sensor Signal	5-7 Hz
86	BK/WT	A/C Pressure Switch	A/C On: 12v (Open)
87	WT/RD	HO2S-21 (B2 S1) Signal	0.1-1.1v
88	WT/BL	MAF Sensor Signal	0.6-0.9v
89	WT	TP Sensor Signal	0.53-1.27v
90	YL	Reference Voltage	4.9-5.1v
91	BR	Analog Signal Return	<0.050v
92	OR	Brake Pedal Position Switch	Brake Off: 0v, On: 12v
93	BK/GN	Injector 5 Control	2.1-2.5 ms
94	BK/RD	Injector 6 Control	2.1-2.5 ms
95	BK/OR	Injector 4 Control	2.1-2.5 ms
96	BK/GN	Injector 2 Control	2.1-2.5 ms
97	GN/YL	Vehicle Power	12-14v
98	---	Not Used	---
99	BK/BL	HO2S-21 (B2 S1) Heater	On: 1v, Off: 12v
100	BK/OR	HO2S-21 (B2 S1) Heater	On: 1v, Off: 12v
101	BK/GN	HO2S-22 (B2 S2) Heater	On: 1v, Off: 12v
102	---	Not Used	---
103	BK/YL	Power Ground	<0.1v
104	---	Not Used	---

Pin Connector Graphic

PCM 104-PIN CONNECTOR

Terminal View of 104-Pin PCM Wiring Harness Connector

Standard Colors and Abbreviations

Abbreviation	Color	Abbreviation	Color	Abbreviation	Color
BK	Black	GY	Gray	PK	Purple
BL	Blue	GN	Green	RD	Red
BR	Brown	LG	LT Green	TN	Tan
DB	Dark Blue	OR	Orange	WT	White
DG	DK Green	PK	Pink	YL	Yellow

1995 Contour 2.5L V6 MFI VIN L (All) 22 Pin Connector

PCM Pin #	Wire Color	Circuit Description (22 Pin)	Value at Hot Idle
22-A	BL/RD	Keep Alive Power	12- 14v
22-B	RD/BK	Vehicle Power	12-14v
22-C	BK/RD	Vehicle Start Signal	KOEC: 9-11v
22-D	WT/RD	Switch Monitor Lamp Control	Lamp Off: 12v, On: 1v
22-E	BL	STO & MIL (lamp) Control	STO On: 5v, Off: 12v
22-F	---	Not Used	---
22-G	BL/OR	Ignition Control Module	Varies
22-H	WT	Headlamp Switch Signal	Switch On: 12v, Off: 1v
22-I	RD/WT	Self-Test Input	STI On: 1v, STI Off: 12v
22-J	PK	Rear Window Defroster	Switch On: 1v, Off: 12v
22-K	WT/BK	Torque Reduce & ECT Signal	Digital Signals
22-L	GN/BK	A/C Relay Control	On: 1v, Off: 12v
22-M	GN/RD	Vehicle Speed Sensor	0 Hz, 55 mph: 125 Hz
22-N	BL/YL	PSP Switch Signal	Straight: 12v, Turned: 0v
22-O	PK/BK	A/C Cycling Pressure Switch	Switch On: 12v, Off: 1v
22-P	OR/BK	Blower Motor Control Switch	Off or 1st: 12v, 2nd on: 1v
22-Q	WT/GN	Brake On/ Off Switch Signal	Brake Off: 0v, On: 12v
22-R	LG/BK	A/T: PNP Switch Signal	In 'P', 0v, Others: 12v
22-R	LG/BK	M/T: CPP Switch Signal	Clutch In: 0v, Out: 12v
22-S	GN	Torque Reduce Signal #1	Digital Signals
22-T	BR	Idle Switch Signal	Closed: 0v, Open: 12v
22-U	---	Not Used	---
22-V	LG/WT	Torque Reduce Signal #2	Digital Signals

1995 Contour 2.5L V6 MFI VIN L (All) 16 Pin Connector

PCM Pin #	Wire Color	Circuit Description (16 Pin)	Value at Hot Idle
16-A	GN/OR	Barometric Pressure Sensor	3.9v
16-B	RD	Measuring Core VAF Sensor	Idle: 3.06v
16-C	BK/YL	HO2S-11 (B1 S1) Signal	0.1-1.1v
16-D	BL/WT	HO2S-12 (B1 S2) Signal	0.1-1.1v
16-E	RD/GN	ECT Sensor Signal	0.5-0.6v
16-F	YL	Throttle Position Sensor	0.4-1.0v
16-G	-	Not Used	---
16-H	PK/YL	A/C High Pressure Switch	Closed: 1v, open: 12v
16-I	PK	Reference Voltage	4.9-5.1v
16-J	RD/BK	EGR Valve Position Sensor	0.4v
16-K	BK/RD	IAT Sensor Signal	1.5-2.5v
16-L	GN	DRL Signal - Canada	DRL On: 2v, Off: 12v
16-M	WT	Knock Sensor Signal	0v
16-N	-	Not Used	---
16-O	BL/BK	EVAP Purge Solenoid	12v, 55 mph: 1v
16-P	BL/GN	High Speed Fan Control	On: 1v, Off: 12v

Standard Colors and Abbreviations

Abbreviation	Color	Abbreviation	Color	Abbreviation	Color
BK	Black	GY	Gray	PK	Purple
BL	Blue	GN	Green	RD	Red
BR	Brown	LG	LT Green	TN	Tan
DB	Dark Blue	OR	Orange	WT	White
DG	DK Green	PK	Pink	YL	Yellow

1995 Contour 2.5L V6 MFI VIN L (All) 26 Pin Connector

PCM Pin #	Wire Color	Circuit Description (26 Pin)	Value at Hot Idle
26-A	BK	Power Ground	<0.1v
26-B	BK	Power Ground	<0.1v
26-C	BK/RD	Power Ground	<0.1v
26-D	BK/RD	Power Ground	<0.1v
26-E	LG/OR	CKP Sensor 1 Signal	2.5v
26-F	BL	CKP Sensor Ground	0.1v
26-G	BL/PK	CMP Sensor 1 Signal	2.5v
26-H	GN	CKP Sensor 2 Signal	2.5v
26-I	WT/GN	Variable Resonance Induction	VRIS1 On: 1v, Off: 12v
26-J	BL/RD	Variable Resonance Induction	VRIS2 On: 1v, Off: 12v
26-K	---	Not Used	---
26-L	RD/WT	Low Speed Fan Control	On: 1v, Off: 12v
26-M	GN/BK	Fuel Pressure Regulator	Startup: 3v, others: 12v
26-N	BL/OR	Condenser Fan Control	On: 1v, Off: 12v
26-O	WT/BL	EGR Vent Solenoid	12v, off-idle: 1v
26-P	GN/WT	EGR Control Solenoid	12v, off-idle: 1v
26-Q	LG/BK	Idle Air Control Valve	8-10v
26-R	---	Not Used	---
26-S	---	Not Used	---
26-T	LG	Fuel Pump Control	On: 1v, Off: 12v
26-U	RD/LG	Injector 1 Control	3.6 ms
26-V	BL/WT	Injector 2 Control	3.6 ms
26-W	BR	Injector 3 Control	3.6 ms
26-X	RD/YL	Injector 4 Control	3.6 ms
26-Y	WT	Injector 5 Control	3.6 ms
26-Z	WT/BK	Injector 6 Control	3.6 ms

Pin Connector Graphic

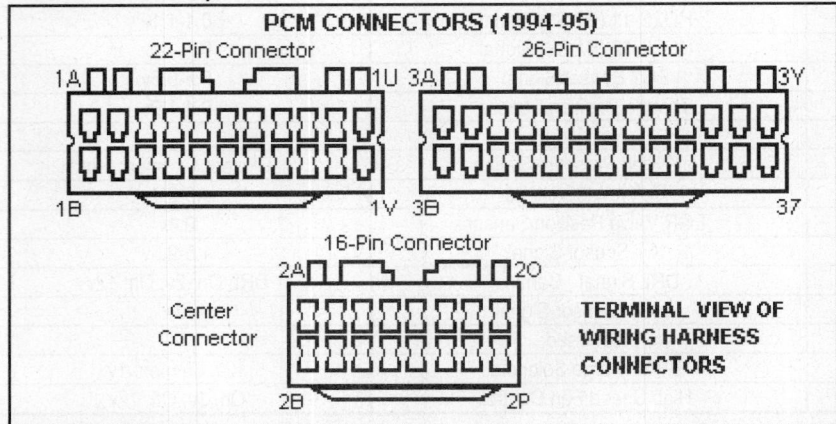

Standard Colors and Abbreviations

Abbreviation	Color	Abbreviation	Color	Abbreviation	Color
BK	Black	GY	Gray	PK	Purple
BL	Blue	GN	Green	RD	Red
BR	Brown	LG	LT Green	TN	Tan
DB	Dark Blue	OR	Orange	WT	White
DG	DK Green	PK	Pink	YL	Yellow

1996-97 Contour 2.5L V6 MFI VIN L (All) 104 Pin Connector

PCM Pin #	Wire Color	Circuit Description (104 Pin)	Value at Hot Idle
1	BK/BL	Shift Solenoid 2 Control	1v, 55 mph: 12v
2	BK/OR	MIL (lamp) Control	MIL On: 1v, Off: 12v
3-7	---	Not Used	---
8	BK/BL	IMRC Solenoid Control	5v (Off)
9-12	---	Not Used	---
13	WT/BL	Flash EEPROM Power	0.1v
14	---	Not Used	---
15	BR/BL	Data Bus (-) Signal	Digital Signals
16	WT/BL	Data Bus (+) Signal	Digital Signals
17	GY/OR	Passive Antitheft System	Digital Signals
18-20	---	Not Used	---
21	WT/RD	CKP Sensor (+) Signal	410-440 Hz
22	BR/RD	CKP Sensor (-) Signal	410-440 Hz
23	---	Not Used	---
24	BK/YL	Power Ground	<0.1v
25	BK/RD	Power Ground	<0.1v
26	BR/BL	Coil Driver 1 Control	5° dwell
27	BK/YL	Shift Solenoid 1 Control	12v, 55 mph: 1v
28	---	Not Used	---
29	PK/BK	TCS (switch) Signal	TCS & O/D On: 12v
30	WT/BK	Octane Adjust Switch	Closed: 0v, Open: 9.3v
31	WT	PSP Sensor Signal	Straight: 0v, Turning: 12v
32-34	---	Not Used	---
34	WT/GN	Passive Antitheft System	Digital Signals
35	WT/BL	HO2S-12 (B1 S2) Signal	0.1-1.1v
36	BR/BL	MAF Sensor Return	<0.050v
37	WT/RD	TFT Sensor Signal	2.10-2.40v
38	WT/GN	ECT Sensor Signal	0.5-0.6v
39	WT/PK	IAT Sensor Signal	1.5-2.5v
40	PK/BK	Fuel Pump Monitor	On: 12v, Off: 0v
41	PK/BL	A/C Cycling Clutch Switch	A/C On: 12v, Off: 0v
42	BK/RD	IMRC Solenoid Control	12v, 55 mph: 12v
43-44	---	Not Used	---
45	BK/BL	Low Speed Fan Control	On: 1v, Off: 12v
46	BK/WT	High Speed Fan Control	On: 1v, Off: 12v
47	BK/GN	EGR VR Solenoid	0%, 55 mph: 40%
48	WT/BK	Clean Tachometer Output	35-42 Hz
49-50	---	Not Used	---
51	BK/YL	Power Ground	<0.1v
52	BR/GN	Coil Driver 2 Control	5° dwell
53	---	Not Used	---
54	BK/WT	TCC Solenoid Control	0%, 55 mph: 95%
55	OR/YL	Keep Alive Power	12-14v
56	BK/OR	EVAP VMV Solenoid	0-10 Hz (0-100%)
57	WT/BK	Knock Sensor Signal	0v
58	WT/PK	VSS (+) Signal	0 Hz, 55 mph: 125 Hz
59	---	Not Used	---
60	WT	HO2S-11 (B1 S1) Signal	0.1-1.1v
61	WT/GN	HO2S-22 (B2 S2) Signal	0.1-1.1v
62-63	---	Not Used	---
64	WT/GN	TR Sensor Signal	In 'P': 0v, 55 mph: 1.7v
64	WT	M/T: Clutch Position Switch	Clutch In: 0.1v, Out: 5v

1996-97 Contour 2.5L V6 MFI VIN L (All) 104 Pin Connector

PCM Pin #	Wire Color	Circuit Description (104 Pin)	Value at Hot Idle
65	WT/BL	DPFE Sensor Signal	0.20-1.30v
66	---	Not Used	---
67	BK/OR	EVAP CV Solenoid	0-10 Hz (0-100%)
68	---	Not Used	---
69	BK/YL	A/C WOT Relay Control	On: 1v, Off: 12v
70	---	Not Used	---
71	GN/YL	Vehicle Power	12-14v
72	---	Not Used	---
73	BK/GN	Injector 5 Control	2.1-2.5 ms
74	BK/BL	Injector 3 Control	2.1-2.5 ms
75	BK/WT	Injector 1 Control	2.1-2.5 ms
76	BK/YL	CMP & TSS Ground	<0.050v
77	BK/YL	Power Ground	<0.1v
78	BR/YL	Coil Driver 3 Control	5° dwell
79	WT/BL	TCIL (lamp) Control	On: 1v, Off: 12v
80	BK/BL	Fuel Pump Control	On: 1v, Off: 12v
81	BK/RD	EPC Solenoid Control	9v (17 psi)
82	---	Not Used	---
83	BK/YL	IAC Motor Control	9v (33% duty cycle)
84	WT/PK	TSS Sensor Signal	46-50 Hz (700-730 rpm)
85	WT/PK	CMP Sensor Signal	5-7 Hz
86	BK/WT	A/C Pressure Switch	A/C On: 12v (Open)
87	WT/RD	HO2S-21 (B2 S1) Signal	0.1-1.1v
88	WT/BL	MAF Sensor Signal	0.6-0.9v
89	WT	TP Sensor Signal	0.53-1.27v
90	YL	Reference Voltage	4.9-5.1v
91	BR	Analog Signal Return	<0.050v
92	OR	Brake Pedal Position Switch	Brake Off: 0v, On: 12v
93	BK/YL	HO2S-11 (B1 S1) Heater	On: 1v, Off: 12v
94	BK/BL	HO2S-21 (B2 S1) Heater	On: 1v, Off: 12v
95	BK/OR	HO2S-12 (B1 S2) Heater	On: 1v, Off: 12v
96	BK/GN	HO2S-22 (B2 S2) Heater	On: 1v, Off: 12v
97	GN/YL	Vehicle Power	12-14v
98	---	Not Used	---
99	BK/RD	Injector 6 Control	2.1-2.5 ms
100	BK/OR	Injector 4 Control	2.1-2.5 ms
101	BK/YL	Injector 2 Control	2.1-2.5 ms
102	BK/OR	Shift Solenoid 3 Control	7-9v, 55 mph: 8-9v
103	BK/YL	Power Ground	<0.1v
104	---	Not Used	---

Pin Connector Graphic

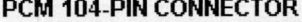

PCM 104-PIN CONNECTOR

Terminal View of 104-Pin PCM Wiring Harness Connector

1998-1999 Contour 2.5L V6 MFI VIN L (All) 104 Pin Connector

PCM Pin #	Wire Color	Circuit Description (104 Pin)	Value at Hot Idle
1	BK/BL	Shift Solenoid 2 Control	1v, 55 mph: 12v
2	BK/OR	MIL (lamp) Control	MIL On: 1v, Off: 12v
3-7	---	Not Used	---
8	BK/BL	IMRC Solenoid Control	5v (Off)
9-11	---	Not Used	---
12	WT	Fuel Level Indicator Signal	1.7v (1/2 full)
13	WT/BL	Flash EEPROM Power	0.1v
14	---	Not Used	---
15	BL	Data Bus (-) Signal	Digital Signals
16	GY	Data Bus (+) Signal	Digital Signals
17	GY/OR	Passive Antitheft System	Digital Signals
18-19	---	Not Used	---
20	BK/BL	Injector 3 Control	2.1-2.5 ms
21	WT/RD	CKP Sensor (+) Signal	410-440 Hz
22	BR/RD	CKP Sensor (-) Signal	410-440 Hz
23	---	Not Used	---
24	BK/YL	Power Ground	<0.1v
25	BK/RD	Power Ground	<0.1v
26	BR/BL	Coil Driver 1 Control	5° dwell
27	BK/YL	Shift Solenoid 1 Control	12v, 55 mph: 1v
28	---	Not Used	---
29	PK/BK	CD4E TCS (switch) Signal	TCS & O/D On: 12v
30	WT/BK	Octane Adjust Switch	Closed: 0v, Open: 9.3v
31	WT	PSP Sensor Signal	Straight: 0v, Turning: 12v
32-34	---	Not Used	---
34	WT/GN	Passive Antitheft System	Digital Signals
35	WT/BL	HO2S-12 (B1 S2) Signal	0.1-1.1v
36	BR/BL	MAF Sensor Return	<0.050v
37	WT/RD	TFT Sensor Signal	2.10-2.40v
38	WT/GN	ECT Sensor Signal	0.5-0.6v
39	WT/PK	IAT Sensor Signal	1.5-2.5v
40	PK/BK	Fuel Pump Monitor	On: 12v, Off: 0v
41	PK/BL	A/C Cycling Clutch Switch	A/C On: 12v, Off: 0v
42	BK/RD	IMRC Solenoid Motor Control	12v, 55 mph: 12v
43-44	---	Not Used	---
45	BK/BL	Low Speed Fan Control	On: 1v, Off: 12v
46	BK/WT	High Speed Fan Control	On: 1v, Off: 12v
47	BK/GN	EGR VR Solenoid	0%, 55 mph: 40%
48	WT/BK	Clean Tachometer Output	35-42 Hz
49-50	---	Not Used	---
51	BK/YL	Power Ground	<0.1v
52	BR/GN	Coil Driver 2 Control	5° dwell
53, 59	---	Not Used	---
54	BK/BL	Fuel Pump Control	On: 1v, Off: 12v
55	OR/YL	Keep Alive Power	12-14v
56	BK/OR	EVAP Purge Solenoid	0-10 Hz (0-100%)
57	WT/BK	Knock Sensor Signal	0v
58	WT/PK	VSS (+) Signal	0 Hz, 55 mph: 125 Hz
60	WT	HO2S-11 (B1 S1) Signal	0.1-1.1v
61	WT/GN	HO2S-22 (B2 S2) Signal	0.1-1.1v
62	WT/PK	FTP Sensor Signal	2.6v (cap off)
63 ('99)	WT/PK	FRP Sensor Signal	2.8v (39 psi)
64	WT/GN	TR Sensor Signal	In 'P': 0v, 55 mph: 1.7v
64	WT	M/T: Clutch Position Switch	Clutch In: 0.1v, Out: 5v

1998-1999 Contour 2.5L V6 MFI VIN L (All) 104 Pin Connector

PCM Pin #	Wire Color	Circuit Description (104 Pin)	Value at Hot Idle
65	WT/BL	DPFE Sensor Signal	0.20-1.30v
66	---	Not Used	---
67	BK/OR	EVAP CV Solenoid	0-10 Hz (0-100%)
68	---	Not Used	---
69	BK/YL	A/C WOT Relay Control	On: 1v, Off: 12v
70	BK/YL	ZETECT VCT Control	VCT Off: 12v, On: 1v
71	GN/YL	Vehicle Power	12-14v
72	---	Not Used	---
73	BK/YL	HO2S-11 (B1 S1) Heater	On: 1v, Off: 12v
74	BK/BL	Injector 3 Control	2.1-2.5 ms
75	BK/WT	Injector 1 Control	2.1-2.5 ms
76	BK/YL	CMP & TSS Ground	<0.050v
77	BK/YL	Power Ground	<0.1v
78	BR/YL	Coil Driver 3 Control	5° dwell
79	WT/BK	TCIL (lamp) Control	On: 1v, Off: 12v
80	BK/WT	TCC Solenoid Control	0%, 55 mph: 95%
81	BK/RD	EPC Solenoid Control	17 psi
82	---	Not Used	---
83	BK/YL	IAC Motor Control	9v (33% duty cycle)
84	WT/PK	TSS Sensor Signal	46-50 Hz (700-730 rpm)
85	WT/PK	CMP Sensor (+) Signal	5-7 Hz
86	BK/WT	A/C Pressure Switch	A/C On: 12v (Open)
87	WT/RD	HO2S-21 (B2 S1) Signal	0.1-1.1v
88	WT/BL	MAF Sensor Signal	0.6-0.9v
89	WT	TP Sensor Signal	0.53-1.27v
90	YL	Reference Voltage	4.9-5.1v
91	BR	Analog Signal Return	<0.050v
92	OR	Brake Pedal Position Switch	Brake Off: 0v, On: 12v
93	BK/YL	Injector 5 Control	2.1-2.5 ms
94	BK/BL	Injector 6 Control	2.1-2.5 ms
95	BK/OR	Injector 4 Control	2.1-2.5 ms
96	BK/GN	Injector 2 Control	2.1-2.5 ms
97	GN/YL	Vehicle Power	12-14v
98	---	Not Used	---
99	BK/BL	HO2S-21 (B2 S1) Heater	On: 1v, Off: 12v
100	BK/OR	HO2S-21 (B2 S1) Heater	On: 1v, Off: 12v
101	BK/GN	HO2S-22 (B2 S2) Heater	On: 1v, Off: 12v
102	BK/OR	Shift Solenoid 3 Control	7-9v, 55 mph: 8-9v
103	BK/YL	Power Ground	<0.1v
104	---	Not Used	---

Pin Connector Graphic

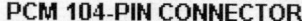

PCM 104-PIN CONNECTOR

Terminal View of 104-Pin PCM Wiring Harness Connector

2000 Contour 2.5L V6 MFI VIN L (All) 104 Pin Connector

PCM Pin #	Wire Color	Circuit Description (104 Pin)	Value at Hot Idle
1	---	Not Used	---
2	BK/OR	MIL (lamp) Control	MIL On: 1v, Off: 12v
3-5	---	Not Used	---
6	BK/YL	Shift Solenoid 1 Control	1v, 55 mph: 12v
7, 10, 14	---	Not Used	---
8	BK/BL	IMRC Solenoid Control	5v (Off)
9	WT	Fuel Level Indicator Signal	1.7v (1/2 full)
11	BK/BL	Shift Solenoid 2 Control	1v, 55 mph: 12v
12	WT/BK	Gearshift Indicator Signal	N/A
13	WT/BL	Flash EEPROM Power	0.1v
15	BL	Data Bus (-) Signal	Digital Signals
16	GY	Data Bus (+) Signal	Digital Signals
17	BK/WT	Hi Speed Fan Relay Control	On: 1v, Off: 12v
18-19	---	Not Used	---
20	BK/BL	Injector 3 Control	2.1-2.5 ms
21	WT/RD	CKP Sensor (+) Signal	410-440 Hz
22	BR/RD	CKP Sensor (-) Signal	410-440 Hz
23	---	Not Used	---
24	BK/YL	Power Ground	<0.1v
25	BK/RD	Power Ground	<0.1v
26	BR/BL	Coil Driver 1 Control	5° dwell
27	---	Not Used	---
28	WT/BL	Speedometer Indicator Signal	N/A
29	GN/BK	TCS (switch) Signal	TCS & O/D On: 12v
30	---	Not Used	---
31	WT	PSP Sensor Signal	Straight: 0v, Turning: 12v
32	BK/YL	Knock Sensor 1 (-) Signal	0v
33	BR/BL	VSS (-) Signal	0 Hz, 55 mph: 125 Hz
34, 43-46	---	Not Used	---
35	WT/BL	HO2S-12 (B1 S2) Signal	0.1-1.1v
36	BR/BL	MAF Sensor Return	<0.050v
37	WT/RD	TFT Sensor Signal	2.10-2.40v
38	WT/GN	ECT Sensor Signal	0.5-0.6v
39	WT/PK	ACT Sensor Signal	1.5-2.5v
40	GN/BK	Fuel Pump Monitor	On: 12v, Off: 0v
41	GN/BL	A/C Cycling Clutch Switch	A/C On: 12v, Off: 0v
42	BK/RD	IMRC Solenoid Motor Control	12v, 55 mph: 12v
47	GN/BK	EGR VR Solenoid	0%, 55 mph: 40%
48	WT/BK	Clean Tachometer Output	35-42 Hz
49-50, 53	---	Not Used	---
51	BK/YL	Power Ground	<0.1v
52	BR/GN	Coil Driver 2 Control	5° dwell
54	BK/RD	Fuel Pump Control	On: 1v, Off: 12v
55	OR/YL	Keep Alive Power	12-14v
56	BK/OR	EVAP Purge Solenoid	0-10 Hz (0-100%)
57	WT/BK	Knock Sensor 1 (+) Signal	0v
58	WT/BL	VSS (+) Signal	0 Hz, 55 mph: 125 Hz
59	---	Not Used	---
60	WT	HO2S-11 (B1 S1) Signal	0.1-1.1v
61	WT/GN	HO2S-22 (B2 S2) Signal	0.1-1.1v
62	WT/PK	FTP Sensor Signal	2.6v (cap off)
63	WT/GN	FRP Sensor Signal	2.8v (39 psi)
64	WT/GN	A/T: TR Sensor Signal	In 'P': 0v, 55 mph: 1.7v
64	WT	M/T: Clutch Position Switch	Clutch In: 0.1v, Out: 5v

2000 Contour 2.5L V6 MFI VIN L (All) 104 Pin Connector

PCM Pin #	Wire Color	Circuit Description (104 Pin)	Value at Hot Idle
65	WT/BL	EPT Sensor Signal	N/A
66	---	Not Used	---
67	BK/OR	EVAP CV Solenoid	0-10 Hz (0-100%)
68	BK/BL	Cooling Fan Relay Control	On: 1v, Off: 12v
69	BK/YL	A/C WOT Relay Control	On: 1v, Off: 12v
70	BK/WT	Injector 1 Control	2.1-2.5 ms
71	GN/YL	Vehicle Power	12-14v
72	---	Not Used	---
73	BK/YL	HO2S-11 (B1 S1) Heater	On: 1v, Off: 12v
74	BK/BL	Injector 3 Control	2.1-2.5 ms
75	---	Not Used	---
76	BK/YL	CMP Sensor (-) Signal	5-7 Hz
77	BK/YL	Power Ground	<0.1v
78	BR/YL	Coil Driver 3 Control	5° dwell
79	---	Not Used	---
80	BK/WT	Modulated Lockup Solenoid	N/A
81	BK/RD	EPC Solenoid Control	17 psi
82	---	Not Used	---
83	BK/YL	IAC Motor Control	9v (33% duty cycle)
84	WT/PK	TSS Sensor Signal	46-50 Hz (700-730 rpm)
85	WT/PK	CMP Sensor (+) Signal	5-7 Hz
86	BK/BL	A/C Pressure Switch	A/C On: 12v (Open)
87	WT/RD	HO2S-21 (B2 S1) Signal	0.1-1.1v
88	WT/BL	MAF Sensor Signal	0.6-0.9v
89	WT	TP Sensor Signal	0.53-1.27v
90	YL	Reference Voltage	4.9-5.1v
91	BR	Analog Signal Return	<0.050v
92	OR	Brake Pedal Position Switch	Brake Off: 0v, On: 12v
93	BK/GN	Injector 5 Control	2.1-2.5 ms
94	BK/RD	Injector 6 Control	2.1-2.5 ms
95	BK/OR	Injector 4 Control	2.1-2.5 ms
96	BK/YL	Injector 2 Control	2.1-2.5 ms
97	GN/YL	Vehicle Power	12-14v
98	---	Not Used	---
99	BK/BL	HO2S-21 (B2 S1) Heater	On: 1v, Off: 12v
100	BK/OR	HO2S-21 (B2 S1) Heater	On: 1v, Off: 12v
101	BK/GN	HO2S-22 (B2 S2) Heater	On: 1v, Off: 12v
102	BK/DG	Shift Solenoid 3 Control	7-9v, 55 mph: 8-9v
103	BK/YL	Power Ground	<0.1v
104	---	Not Used	---

Pin Connector Graphic

PCM 104-PIN CONNECTOR

Terminal View of 104-Pin PCM Wiring Harness Connector

Crown Victoria Pin Tables

1992-95 Crown Victoria 4.6L V8 MFI VIN W (A/T) 60 Pin Connector

PCM Pin #	Wire Color	Circuit Description (60 Pin)	Value at Hot Idle
1	YL	Keep Alive Power	12-14v
2 ('93-'95)	LG	Brake Pedal Position Switch	Brake Off: 0v, On: 12v
3	GY/BK	VSS (+) Signal	0 Hz, 55 mph: 125 Hz
4	TN/YL	Ignition Diagnostic Monitor	20-31 Hz
5	PK/BL	TSS Sensor (+) Signal	55 mph: 126-136 Hz
6	PK/OR	VSS (-) Signal	0 Hz, 55 mph: 125 Hz
7	LG/RD	ECT Sensor Signal	0.5-0.6v
8	DG/YL	Fuel Pump Monitor	On: 12v, Off: 0v
9	TN/BL	MAF Sensor Return	<0.050v
10	DG/OR	A/C Cycling Clutch Switch	A/C C/C on: 12v, off: 0v
11	GY/YL	Air Management 2 Solenoid	AM2 On: 1v, Off: 12v
12	LG/OR	Injector 6 Control	4.0-4.4 ms
13	TN/RD	Injector 7 Control	4.0-4.4 ms
14	BL	Injector 8 Control	4.0-4.4 ms
15	TN/BK	Injector 5 Control	4.0-4.4 ms
16	OR/RD	Ignition System Ground	<0.050v
17	TN/RD	STI Output, MIL Control	MIL On: 1v, Off: 12v
18	TN/OR	Data Bus (+) Signal	Digital Signals
19	PK/BL	Data Bus (-) Signal	Digital Signals
20	BK	PCM Case Ground	<0.050v
21	WT/BL	IAC Motor Control	8.3-11.5v
22	BL/OR	Fuel Pump Control	On: 1v, Off: 12v
23	---	Not Used	---
24	DG	CID Sensor Signal	6-7 Hz
25	GY	ACT Sensor Signal	1.5-2.5v
25	LG/PK	ACT Sensor Signal	1.5-2.5v
26	BR/WT	Reference Voltage	4.9-5.1v
27	BR/LG	DPFE EGR Sensor Signal	0.4v
28	---	Not Used	---
29	WT/RD	Octane Adjust Switch	Closed: 0v, Open: 9.1v
30 ('93-'95)	WT/PK	Neutral Drive or MLP Switch	In 'P': 0v, Others: 5v
30 ('93-'95)	BL/YL	Neutral Drive or MLP Switch	In 'P': 0v, Others: 5v
31-32	---	Not Used	---
33	BR/PK	EGR VR Solenoid	0%, 55 mph: 45%
34	DB/LG	Data Output Link	Digital Signals
35	BR/BL	Injector 4 Control	4.0-4.4 ms
36	PK	Spark Angle Word Signal	50% duty cycle
37, 57	RD	Vehicle Power	12-14v
38 ('93-'95)	WT/YL	EPC Solenoid Control	9.5v (20 psi)
39	BR/YL	Injector 4 Control	4.0-4.4 ms
40	BK/WT	Power Ground	<0.1v

Pin Connector Graphic

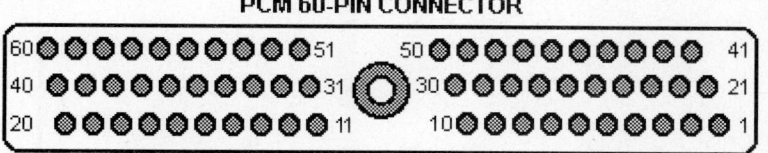

PCM 60-PIN CONNECTOR

Terminal View of 60-Pin PCM Harness Connector

1992-95 Crown Victoria 4.6L V8 MFI VIN W (A/T) 60 Pin Connector

PCM Pin #	Wire Color	Circuit Description (60 Pin)	Value at Hot Idle
41 ('93-'95)	TN/WT	TCS (switch) Signal	TCS & O/D On: 12v
42	---	Not Used	---
43	RD/BK	HO2S-11 (B1 S1) Signal	0.1-1.1v
44	GY/BL	HO2S-21 (B2 S1) Signal	0.1-1.1v
45	---	Not Used	---
46	GY/RD	Analog Signal Return	<0.050v
47	GY/WT	TP Sensor Signal	0.51-1.27v
48	WT/PK	Self-Test Indicator Signal	STI Open: 5v, Closed: 1v
49 ('93-'95)	OR/BK	TOT Sensor Signal	2.10-2.40v
50	BL/RD	MAF Sensor Signal	0.6-0.9v
51 ('93-'94)	OR/YL	Shift Solenoid 1 Control	1v, 55 mph: 12v
52 ('93-'94)	PK/OR	Shift Solenoid 2 Control	12v, 55 mph: 1v
53	BR/OR	TCC Solenoid Signal	12v, 55 mph: 9-10v
53 ('93-'95)	PK/YL	TCC Solenoid Signal	12v, 55 mph: 9-10v
54	OR/BL	A/C WOT Relay Control	On: 1v, Off: 12v
55	DB/WT	TCIL (lamp) Control	On: 1v, Off: 12v
55 ('93-'95)	WT/LG	TCIL (lamp) Control	On: 1v, Off: 12v
56	GY/OR	PIP Sensor Signal	50% dwell
58	TN	Injector 1 Control	4.0-4.4 ms
59	WT	Injector 2 Control	4.0-4.4 ms
60	BK/WT	Power Ground	<0.1v

Standard Colors and Abbreviations

Abbreviation	Color	Abbreviation	Color	Abbreviation	Color
BK	Black	GY	Gray	PK	Purple
BL	Blue	GN	Green	RD	Red
BR	Brown	LG	LT Green	TN	Tan
DB	Dark Blue	OR	Orange	WT	White
DG	DK Green	PK	Pink	YL	Yellow

1996-97 Crown Victoria 4.6L V8 VIN W (A/T) 104 Pin Connector

PCM Pin #	Wire Color	Circuit Description (104 Pin)	Value at Hot Idle
1	PK/OR	Shift Solenoid 2 Control	12v, 55 mph: 1v
2	PK/LG	MIL (lamp) Control	MIL On: 1v, Off: 12v
3-12	---	Not Used	---
13	LG/YL	Flash EEPROM Power	0.1v
14	---	Not Used	---
15	PK/LB	Data Bus (-) Signal	Digital Signals
16	TN/OR	Data Bus (+) Signal	Digital Signals
17-20	---	Not Used	---
21	DB	CKP Sensor (+) Signal	850-1120 Hz
22	GY	CKP Sensor (-) Signal	850-1120 Hz
23	---	Not Used	---
24	BK/WT	Power Ground	<0.1v
25	BK	PCM Case Ground	<0.050v
26	TN/WT	Coil Driver 1 Control	5° dwell
27	OR/YL	Shift Solenoid 1 Control	1v, 55 mph: 12v
28	---	Not Used	---
29	TN/WT	TCS (switch) Signal	TCS & O/D On: 12v
30	WT/RD	Octane Adjust Switch	Closed: 0v, Open: 9.3v
31-32	---	Not Used	---
33	PK/OR	VSS (-) Signal	0 Hz, 55 mph: 125 Hz
34	---	Not Used	---
35	RD/LG	HO2S-11 (B1 S1) Signal	0.1-1.1v
36	TN/LB	MAF Sensor Return	<0.050v
37	OR/BK	TFT Sensor Signal	2.10-2.40v
38	LG/RD	ECT Sensor Signal	0.5-0.6v
39	GY	IAT Sensor Signal	1.5-2.5v
40	DG/YL	Fuel Pump Monitor	On: 12v, Off: 0v
41	DG/OR	A/C Cycling Clutch Switch	A/C On: 12v, Off: 0v
42-44	---	Not Used	---
45	RD/OR	Low Speed Fan Control	On: 1v, Off: 12v
46	---	Not Used	---
47	BR/PK	EGR VR Solenoid	0%, 55 mph: 45%
48-50	---	Not Used	---
51	BK/WT	Power Ground	<0.1v
52	TN/OR	Coil Driver 2 Control	5° dwell
53	---	Not Used	---
54	RD/LB	TCC Solenoid Control	TCC Off Idle: 0%
55	YL/BK	Keep Alive Power	12-14v
56	LG/BK	EVAP VMV Solenoid	0-10 Hz (0-100%)
57	---	Not Used	---
58	GY/BK	VSS (+) Signal	0 Hz, 55 mph: 125 Hz
59	---	Not Used	---
60	GY/LB	HO2S-11 (B1 S1) Signal	0.1-1.1v
61	PK/LG	HO2S-22 (B2 S2) Signal	0.1-1.1v
62	RD/PK	FTP Sensor Signal	2.6v (0" H2O - cap off)
63	---	Not Used	---
64	LB/YL	TR Sensor Signal	In 'P': 0v, 55 mph: 1.7v
65	BR/LG	DPFE Sensor Signal	0.20-1.30v
66	---	Not Used	---
67	PK/WT	EVAP Purge Solenoid	0-10 Hz (0-100%)
68	---	Not Used	---
69	OR/LB	A/C WOT Relay Control	On: 1v, Off: 12v
70	---	Not Used	---
71	RD	Vehicle Power	12-14v

1996-97 Crown Victoria 4.6L V8 VIN W (A/T) 104 Pin Connector

PCM Pin #	Wire Color	Circuit Description (104 Pin)	Value at Hot Idle
72	TN/RD	Injector 7 Control	3.4-3.7 ms
73	TN/BK	Injector 5 Control	3.4-3.7 ms
74	BR/YL	Injector 3 Control	3.4-3.7 ms
75	TN	Injector 1 Control	3.4-3.7 ms
76-77	BK/WT	Power Ground	<0.1v
78	TN/LG	Coil Driver 3 Control	5° dwell
79	WT/LG	TCIL (lamp) Control	On: 1v, Off: 12v
80	LB/OR	Fuel Pump Control	On: 1v, Off: 12v
81	WT/YL	EPC Solenoid Control	9.5v (20 psi)
82	---	Not Used	---
83	WT/LB	IAC Motor Control	34% duty cycle
84	PK/LB	OSS Sensor Signal	0 Hz, 55 mph: 131 Hz
85	DG	CMP Sensor Signal	5-7 Hz
86	---	Not Used	---
87	RD/BK	HO2S-21 (B2 S1) Signal	0.1-1.1v
88	LB/RD	MAF Sensor Signal	0.6v
89	GY/WT	TP Sensor Signal	0.53-1.27v
90	BR/WT	Reference Voltage	4.9-5.1v
91	GY/RD	Analog Signal Return	<0.050v
92	LG	Brake Pedal Position Switch	Brake Off: 0v, On: 12v
93	RD/WT	HO2S-11 (B1 S1) Heater	On: 1v, Off: 12v
94	YL/LB	HO2S-21 (B2 S1) Heater	On: 1v, Off: 12v
95	WT/BK	HO2S-12 (B1 S2) Heater	On: 1v, Off: 12v
96	TN/YL	HO2S-22 (B2 S2) Heater	On: 1v, Off: 12v
97	RD	Vehicle Power	12-14v
98	LB	Injector 8 Control	3.4-3.7 ms
99	LG/OR	Injector 6 Control	3.4-3.7 ms
100	BR/LB	Injector 4 Control	3.4-3.7 ms
101	WT	Injector 2 Control	3.4-3.7 ms
102	---	Not Used	---
103	BK/WT	Power Ground	<0.1v
104	RD/YL	Coil Driver 4 Control	5° dwell

Pin Connector Graphic

PCM 104-PIN CONNECTOR

Terminal View of 104-Pin PCM Wiring Harness Connector

Standard Colors and Abbreviations

Abbreviation	Color	Abbreviation	Color	Abbreviation	Color
BK	Black	GY	Gray	PK	Purple
BL	Blue	GN	Green	RD	Red
BR	Brown	LG	LT Green	TN	Tan
DB	Dark Blue	OR	Orange	WT	White
DG	DK Green	PK	Pink	YL	Yellow

1998-99 Crown Victoria 4.6L V8 VIN W (A/T) 104 Pin Connector

PCM Pin #	Wire Color	Circuit Description (104 Pin)	Value at Hot Idle
1	OR/YL	COP 6 Driver Control	5° dwell
2	PK/LG	MIL (lamp) Control	MIL On: 1v, Off: 12v
3	BK/WT	Power Ground	<0.1v
4-5	---	Not Used	---
6	OR/YL	Shift Solenoid 'A' Control	1v, 55 mph: 12v
7-8	---	Not Used	---
9	YL/WT	Fuel Level Indicator Signal	1.7v (1/2 full)
10	---	Not Used	---
11	PK/OR	Shift Solenoid 'B' Control	12v, Cruise: 1v
12	WT/LG	TCIL (lamp) Control	Lamp On: 1v, Off: 5v
13	PK	Flash EEPROM Power	0.1v
15	PK/LB	Data Bus (-) Signal	Digital Signals
16	TN/OR	Data Bus (+) Signal	Digital Signals
17-20	---	Not Used	---
21	BK/PK	CKP Sensor (+) Signal	440-490 Hz
22	GY/YL	CKP Sensor (-) Signal	440-490 Hz
23	---	Not Used	---
24	BK/WT	Power Ground	<0.1v
25	BK	PCM Case Ground	<0.050v
26	LG/WT	COP 1 Driver Control	5° dwell
27	LG/YL	COP 5 Driver Control	1v, 55 mph: 12v
28	RD/OR	Low Speed Fan Control	On: 1v, Off: 12v
29	TN/WT	TCS (switch) Signal	TCS & O/D On: 12v
30	WT/RD	Octane Adjust Switch	Closed: 0v, Open: 9.3v
31-32	---	Not Used	---
33	PK/OR	VSS (-) Signal	0 Hz, 55 mph: 125 Hz
34	YL/BK	Digital TR1 Sensor	In 'P': 0v, 55 mph: 11v
35	RD/LG	HO2S-12 (B1 S2) Signal	0.1-1.1v
36	TN/LB	MAF Sensor Return	<0.050v
37	OR/BK	TFT Sensor Signal	2.10-2.40v
38	---	Not Used	---
39	GY	IAT Sensor Signal	1.5-2.5v
40	DG/YL	Fuel Pump Monitor	On: 12v, Off: 0v
41	DG/OR	A/C Cycling Clutch Switch	AC On: 12v, Off: 0v
42	RD/WT	ECT Sensor Signal	0.5-0.6v
43	DB/LG	Fuel Flow Rate Signal	Digital Signals
45	OR/RD	CHTIL (lamp) Control	Lamp On: 1v, Off: 5v
46	LG/PK	High Speed Fan Control	On: 1v, Off: 12v
47	BR/PK	EGR VR Solenoid	0%, 55 mph: 45%
48	---	Not Used	---
49	RD/LB	Digital TR2 Sensor	In 'P': 0v, 55 mph: 11v
50	WT/BK	Digital TR4 Sensor	In 'P': 0v, 55 mph: 11v
51	BK/WT	Power Ground	<0.1v
52	WT/PK	COP 3 Driver Control	5° dwell
53	DG/PK	COP 4 Driver Control	5° dwell
54	PK/YL	TCC Solenoid Control	0%, 55 mph: 95%
55	YL/BK	Keep Alive Power	12-14v
56	LG/BK	EVAP Purge Solenoid	0-10 Hz (0-100%)
57	---	Not Used	---
58	GY/BK	VSS + Sensor Signal	0 Hz, 55 mph: 125 Hz
59	---	Not Used	---
60	GY/LB	HO2S-11 (B1 S1) Signal	0.1-1.1v
61	PK/LG	HO2S-22 (B2 S2) Signal	0.1-1.1v

1998-99 Crown Victoria 4.6L V8 VIN W (A/T) 104 Pin Connector

PCM Pin #	Wire Color	Circuit Description (104 Pin)	Value at Hot Idle
62	RD/PK	FTP Sensor Signal	2.6v (0" H2O - cap off)
63	---	Not Used	---
64	LB/YL	Digital TR3 Sensor	In 'P': 0v, in O/D: 1.7v
65	BR/LG	DPFE Sensor Signal	0.20-1.30v
66	YL/LG	CHT Sensor Signal	0.7v (194°F)
67	PK/WT	EVAP CV Solenoid	0-10 Hz (0-100%)
68	---	Not Used	---
69	PK/YL	A/C WOT Relay Control	On: 1v, Off: 12v
70	---	Not Used	---
71	RD	Vehicle Power	12-14v
72	TN/RD	Injector 7 Control	3.4-3.7 ms
73	TN/BK	Injector 5 Control	3.4-3.7 ms
74	BR/YL	Injector 3 Control	3.4-3.7 ms
75	TN	Injector 1 Control	3.4-3.7 ms
77	BK/WT	Power Ground	<0.1v
78	PK/LB	COP 7 Driver Control	5° dwell
79	WT/RD	COP 8 Driver Control	5° dwell
80	LB/OR	Fuel Pump Control	On: 1v, Off: 12v
81	WT/YL	EPC Solenoid Control	9.5v (20 psi)
82	---	Not Used	---
83	WT/LB	IAC Motor Control	34% duty cycle
84	DB/YL	OSS Sensor Signal	0 Hz
85	DB/OR	CMP Sensor Signal	6-7 Hz
86	WT/BK	A/C High Pressure Switch	Open: 12v, Closed: 0v
87	RD/BK	HO2S-21 (B2 S1) Signal	0.1-1.1v
88	LB/RD	MAF Sensor Signal	0.6v
89	GY/WT	TP Sensor Signal	0.53-1.27v
90	BR/WT	Reference Voltage	4.9-5.1v
91	GY/RD	Analog Signal Return	<0.050v
92	LG	Brake Pedal Position Switch	Brake Off: 0v, On: 12v
93	RD/WT	HO2S-11 (B1 S1) Heater	On: 1v, Off: 12v
94	YL/LB	HO2S-21 (B2 S1) Heater	On: 1v, Off: 12v
95	WT/BK	HO2S-12 (B1 S2) Heater	On: 1v, Off: 12v
96	TN/YL	HO2S-22 (B2 S2) Heater	On: 1v, Off: 12v
97 ('98 only)	RD	Vehicle Power	12-14v
98	LB	Injector 8 Control	3.4-3.7 ms
99	LG/OR	Injector 6 Control	3.4-3.7 ms
100	BR/LB	Injector 4 Control	3.4-3.7 ms
101	WT	Injector 2 Control	3.4-3.7 ms
102	---	Not Used	---
103	BK/WT	Power Ground	<0.1v
104	PK/WT	COP 2 Driver Control	5° dwell

Pin Connector Graphic

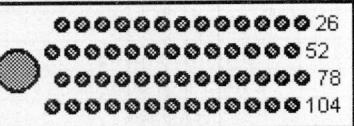

PCM 104-PIN CONNECTOR

Terminal View of 104-Pin PCM Wiring Harness Connector

2000-02 Crown Victoria 4.6L V8 VIN W (A/T) 104 Pin Connector

PCM Pin #	Wire Color	Circuit Description (104 Pin)	Value at Hot Idle
1	OR/YL	COP 6 Driver Control	5° dwell
2	PK/LG	MIL (lamp) Control	MIL On: 1v, Off: 12v
3	BK/WT	Power Ground	<0.1v
4-5	---	Not Used	---
6	OR/YL	Shift Solenoid 'A' Control	1v, 55 mph: 12v
7-8	---	Not Used	---
9	YL/WT	Fuel Level Indicator Signal	1.7v (1/2 full)
10	LB/RD	Not Used	---
11	PK/OR	Shift Solenoid 'B' Control	12v, Cruise: 1v
12	WT/LG	TCIL (lamp) Control	Lamp On: 1v, Off: 5v
13	PK	Flash EEPROM Power	0.1v
15	PK/LB	Data Bus (-) Signal	Digital Signals
16	TN/OR	Data Bus (+) Signal	Digital Signals
17-20	---	Not Used	---
21	BK/PK	CKP Sensor (+) Signal	440-490 Hz
22	GY/YL	CKP Sensor (-) Signal	440-490 Hz
23-24	---	Not Used	---
25	BK	PCM Case Ground	<0.050v
26	LG/WT	COP 1 Driver Control	5° dwell
27	LG/YL	COP 5 Driver Control	1v, 55 mph: 12v
28	RD/OR	Low Speed Fan Control	On: 1v, Off: 12v
29	TN/WT	TCS (switch) Signal	TCS & O/D On: 12v
30-32	---	Not Used	---
33	PK/OR	VSS (-) Signal	0 Hz, 55 mph: 125 Hz
34	YL/BK	Digital TR1 Sensor	In 'P': 0v, 55 mph: 11v
35	RD/LG	HO2S-12 (B1 S2) Signal	0.1-1.1v
36	TN/LB	MAF Sensor Return	<0.050v
37	OR/BK	TFT Sensor Signal	2.10-2.40v
38	---	Not Used	---
39	GY	IAT Sensor Signal	1.5-2.5v
40	DG/YL	Fuel Pump Monitor	On: 12v, Off: 0v
41	BK/YL	A/C Pressure Cutout switch	AC On: 12v, Off: 0v
42	RD/WT	ECT Sensor Signal	0.5-0.6v
43	DB/LG	Fuel Flow Rate Signal	Digital Signals
45	OR/RD	CHTIL (lamp) Control	Lamp On: 1v, Off: 5v
46	LG/PK	High Speed Fan Control	On: 1v, Off: 12v
47	BR/PK	EGR VR Solenoid	0%, 55 mph: 45%
48	---	Not Used	---
49	LB/BK	Digital TR2 Sensor	In 'P': 0v, 55 mph: 11v
50	WT/BK	Digital TR4 Sensor	In 'P': 0v, 55 mph: 11v
51	BK/WT	Power Ground	<0.1v
52	WT/PK	COP 3 Driver Control	5° dwell
53	DG/PK	COP 4 Driver Control	5° dwell
54	VT/YL	TCC Solenoid Control	0%, 55 mph: 95%
55	YL/BK	Keep Alive Power	12-14v
56	LG/BK	EVAP Purge Solenoid	0-10 Hz (0-100%)
57	---	Not Used	---
58	GY/BK	VSS + Sensor Signal	0 Hz, 55 mph: 125 Hz
59	---	Not Used	---
60	GY/BK	HO2S-11 (B1 S1) Signal	0.1-1.1v
61	PK/LG	HO2S-22 (B2 S2) Signal	0.1-1.1v
62	RD/PK	FTP Sensor Signal	2.6v (0" H2O - cap off)

2000-02 Crown Victoria 4.6L V8 VIN W (A/T) 104 Pin Connector

PCM Pin #	Wire Color	Circuit Description (104 Pin)	Value at Hot Idle
63	---	Not Used	---
64	LB/YL	Digital TR3 Sensor	In 'P': 0v, in O/D: 1.7v
65 ('00)	BR/LG	DPFE Sensor Signal	0.20-1.30v
65 ('01-'02)	BR/LG	DPFE Sensor Signal	0.95-1.05v
66	YL/LG	CHT Sensor Signal	0.7v (194°F)
67	VT/WT	EVAP CV Solenoid	0-10 Hz (0-100%)
68	---	Not Used	---
69	PK/YL	A/C WOT Relay Control	On: 1v, Off: 12v
70	---	Not Used	---
71	RD	Vehicle Power	12-14v
72	TN/RD	Injector 7 Control	3.4-3.7 ms
73	TN/BK	Injector 5 Control	3.4-3.7 ms
74	BR/YL	Injector 3 Control	3.4-3.7 ms
75	TN	Injector 1 Control	3.4-3.7 ms
77	BK/WT	Power Ground	<0.1v
78	PK/LB	COP 7 Driver Control	5° dwell
79	WT/RD	COP 8 Driver Control	5° dwell
80	LB/OR	Fuel Pump Control	On: 1v, Off: 12v
81	WT/YL	EPC Solenoid Control	9.5v (20 psi)
82	---	Not Used	---
83	WT/LB	IAC Motor Control	34% duty cycle
84	DB/YL	OSS Sensor Signal	0 Hz
85	DB/OR	CMP Sensor Signal	6-7 Hz
86	WT/BK	A/C High Pressure Switch	Open: 12v, Closed: 0v
87	RD/BK	HO2S-21 (B2 S1) Signal	0.1-1.1v
88	LB/RD	MAF Sensor Signal	0.6v
89	GY/WT	TP Sensor Signal	0.53-1.27v
90	BR/WT	Reference Voltage	4.9-5.1v
91	GY/RD	Analog Signal Return	<0.050v
92	LG	Brake Pedal Position Switch	Brake Off: 0v, On: 12v
93	RD/WT	HO2S-11 (B1 S1) Heater	On: 1v, Off: 12v
94	YL/LB	HO2S-21 (B2 S1) Heater	On: 1v, Off: 12v
95	WT/BK	HO2S-12 (B1 S2) Heater	On: 1v, Off: 12v
96	TN/YL	HO2S-22 (B2 S2) Heater	On: 1v, Off: 12v
98	LB	Injector 8 Control	3.4-3.7 ms
99	LG/OR	Injector 6 Control	3.4-3.7 ms
100	BR/LB	Injector 4 Control	3.4-3.7 ms
101	WT	Injector 2 Control	3.4-3.7 ms
102	---	Not Used	---
103	BK/WT	Power Ground	<0.1v
104	PK/WT	COP 2 Driver Control	5° dwell

Pin Connector Graphic

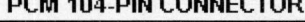

PCM 104-PIN CONNECTOR

Terminal View of 104-Pin PCM Wiring Harness Connector

2003 Crown Victoria 4.6L V8 VIN W (A/T) 104 Pin Connector

PCM Pin #	Wire Color	Circuit Description (104 Pin)	Value at Hot Idle
1	OR/YL	COP 6 Driver Control	5° dwell
2	PK/LG	MIL (lamp) Control	MIL On: 1v, Off: 12v
3	BK/WT	Power Ground	<0.1v
4-5, 14, 23	---	Not Used	---
6	OR/YL	Shift Solenoid 'A' Control	1v, 55 mph: 12v
7	YL/LG	Generator Regulator 'S' Terminal	0-130 Hz
8	RD/PK	Fuel Rail Temperature Sensor	1.7-3.5v (50-120°F)
9	YL/WT	Fuel Level Indicator Signal	1.7v (1/2 full)
10	LB/RD	MAP Sensor Signal	107 Hz
11	VT/OR	Shift Solenoid 'B' Control	12v, Cruise: 1v
12	WT/LG	TCIL (lamp) Control	Lamp On: 1v, Off: 5v
13	VT	Flash EEPROM Power	0.1v
15	PK/LB	SCP Bus (-) Signal	Digital Signals
16	TN/OR	SCP Bus (+) Signal	Digital Signals
17	GY/OR	RX Signal	Digital Signals
18	WT/LG	TX Signal	Digital Signals
19	OR/RD	Cylinder Head Temperature Lamp Control	Lamp Off: 12v, On: 1v
20	WT/LG	Fuel Door Release Solenoid Indicator	Solenoid Off: 12v, On: 1v
21	BK/PK	CKP Sensor (+) Signal	440-490 Hz
22	GY/YL	CKP Sensor (-) Signal	440-490 Hz
24, 51	BK/WT	Power Ground	<0.1v
25	BK	PCM Case Ground	<0.050v
26	LG/WT	COP 1 Driver Control	5° dwell
27	LG/YL	COP 5 Driver Control	1v, 55 mph: 12v
28	RD/OR	Low Speed Fan Control	On: 1v, Off: 12v
29	TN/WT	TCS (switch) Signal	TCS & O/D On: 12v
30	DG/RD	Light Emitting Diode Signal Ground	<0.050v
31	YL/LG	Power Steering Pressure Sensor	Straight: 0v, Turned: 12v
32	DG/VT	Knock Sensor Ground	<0.050v
33	---	Not Used	---
34	YL/BK	Digital TR1 Sensor	In 'P': 0v, 55 mph: 11v
35	RD/LG	HO2S-12 (B1 S2) Signal	0.1-1.1v
36	TN/LB	MAF Sensor Return	<0.050v
37	OR/BK	TFT Sensor Signal	2.10-2.40v
38	BK/WT	A/C Pressure Sensor Discharge Temp.	A/C On: 1.5-1.9v
39	GY	IAT Sensor Signal	1.5-2.5v
40	DG/YL	Fuel Pump Monitor	On: 12v, Off: 0v
41	PK/LB	A/C Pressure Cutout Switch	AC On: 12v, Off: 0v
42	RD/WT	ECT Sensor Signal	0.5-0.6v
43	DB/LG	Low Fuel Indicator	Digital Signals
44	GY/RD	Starter Relay Control	Relay Off: 12v, On: 9v
45	LG/RD	Generator Common	Digital Signals
46	---	Not Used	---
47	BR/PK	EGR VR Solenoid	0%, 55 mph: 45%
48	WT	Tachometer Output	DC pulse signals
49	LB/BK	Digital TR2 Sensor	In 'P': 0v, 55 mph: 11v
50	WT/BK	Digital TR4 Sensor	In 'P': 0v, 55 mph: 11v
52	WT/PK	COP 3 Driver Control	5° dwell
53	DG/VT	COP 4 Driver Control	5° dwell
54	VT/YL	TCC Solenoid Control	0%, 55 mph: 95%
55	YL/BK	Keep Alive Power	12-14v
56	LG/BK	EVAP Purge Solenoid Control	0-10 Hz (0-100%)
57	YL/RD	Knock Sensor Signal	0v
58-59	---	Not Used	---

2003 Crown Victoria 4.6L V8 VIN W (A/T) 104 Pin Connector

PCM Pin #	Wire Color	Circuit Description (104 Pin)	Value at Hot Idle
60	GY/BK	HO2S-11 (B1 S1) Signal	0.1-1.1v
61	VT/LG	HO2S-22 (B2 S2) Signal	0.1-1.1v
62	RD/PK	FTP Sensor Signal	2.6v (0" H2O - cap off)
63	OR/LG	Injection Pressure Sensor	2.8v (39 psi)
64	LB/YL	Digital TR3 Sensor	In 'P': 0v, in O/D: 1.7v
65	BR/LG	DPFE Sensor Signal	0.95-1.05v
66	YL/LG	Cylinder Head Temperature Sensor	0.7v (194°F)
67	VT/WT	EVAP Canister Vent Solenoid	0-10 Hz (0-100%)
68	GY/BK	Vehicle Speed Sensor (+) Signal	0 Hz, 55 mph: 125 Hz
69	PK/YL	A/C WOT Relay Control	On: 1v, Off: 12v
70	YL	Generator Battery Indicator Control	Lamp Off: 12v, On: 1v
71	RD	Vehicle Power (Start-Run)	12-14v
72	TN/RD	Injector 7 Control	3.4-3.7 ms
73	TN/BK	Injector 5 Control	3.4-3.7 ms
74	BR/YL	Injector 3 Control	3.4-3.7 ms
75	TN	Injector 1 Control	3.4-3.7 ms
76	BK/WT	Power Ground	<0.1v
77	---	Not Used	---
78	PK/LB	COP 7 Driver Control	5° dwell
79	WT/RD	COP 8 Driver Control	5° dwell
80	LB/OR	Fuel Pump Control	On: 1v, Off: 12v
81	WT/YL	EPC Solenoid Control	9.5v (20 psi)
82	---	Not Used	---
83	WT/LB	IAC Motor Control	34% duty cycle
84	DB/YL	OSS Sensor Signal	0 Hz
85	DB/OR	CMP Sensor Signal	6-7 Hz
86	WT/BK	A/C High Pressure Switch	Open: 12v, Closed: 0v
87	RD/BK	HO2S-21 (B2 S1) Signal	0.1-1.1v
88	LB/RD	MAF Sensor Signal	0.6v
89	GY/WT	TP Sensor Signal	0.53-1.27v
90	BR/WT	Reference Voltage	4.9-5.1v
91	GY/RD	Analog Signal Return	<0.050v
92	DG	Brake Pedal Position Switch	Brake Off: 0v, On: 12v
93	RD/WT	HO2S-11 (B1 S1) Heater	On: 1v, Off: 12v
94	YL/LB	HO2S-21 (B2 S1) Heater	On: 1v, Off: 12v
95	WT/BK	HO2S-12 (B1 S2) Heater	On: 1v, Off: 12v
96	TN/YL	HO2S-22 (B2 S2) Heater	On: 1v, Off: 12v
98	LB	Injector 8 Control	3.4-3.7 ms
99	LG/OR	Injector 6 Control	3.4-3.7 ms
100	BR/LB	Injector 4 Control	3.4-3.7 ms
101	WT	Injector 2 Control	3.4-3.7 ms
102	---	Not Used	---
103	BK/WT	Power Ground	<0.1v
104	PK/WT	COP 2 Driver Control	5° dwell

Pin Connector Graphic

PCM 104-PIN CONNECTOR

Terminal View of 104-Pin PCM Wiring Harness Connector

1996-97 Crown Victoria 4.6L V8 CNG VIN 9 104 Pin Connector

PCM Pin #	Wire Color	Circuit Description (104 Pin)	Value at Hot Idle
1	PK/OR	Shift Solenoid 2 Control	12v, 55 mph: 1v
2	PK/LG	MIL (lamp) Control	MIL On: 1v, Off: 12v
3-9	---	Not Used	---
10	LB/W	EFT Sensor 'B' Signal	1.75-3.50v (50-120ºF)
13	LG/YL	Flash EEPROM Power	0.1v
14	---	Not Used	---
15	PK/LB	Data Bus (-) Signal	Digital Signals
16	TN/OR	Data Bus (+) Signal	Digital Signals
17-20	---	Not Used	---
21	DB	CKP Sensor (+) Signal	440-490 Hz
22	GY	CKP Sensor (-) Signal	440-490 Hz
23	---	Not Used	---
24	BK/WT	Power Ground	<0.1v
25	BK	PCM Case Ground	<0.050v
26	TN/WT	Coil Driver 1 Control	5º dwell
27	OR/YL	Shift Solenoid 1 Control	1v, 55 mph: 12v
28	---	Not Used	---
29	TN/WT	TCS (switch) Signal	TCS & O/D On: 12v
30	WT/RD	Octane Adjust Switch	Closed: 0v, Open: 9.3v
31-32	---	Not Used	---
33	PK/OR	VSS (-) Signal	0 Hz, 55 mph: 125 Hz
34	---	Not Used	---
35	RD/LG	HO2S-12 (B1 S2) Signal	0.1-1.1v
36	TN/LB	MAF Sensor Return	<0.050v
37	OR/BK	TFT Sensor Signal	2.10-2.40v
38	LG/RD	ECT Sensor Signal	0.5-0.6v
39	GY	IAT Sensor Signal	1.5-2.5v
40	DG/YL	Fuel Solenoid Valve Control	On: 12v, Off: 0v
41	DG/OR	A/C Cycling Clutch Switch	A/C On: 12v, Off: 0v
42-44	---	Not Used	---
45	RD/OR	Low Speed Fan Control	On: 1v, Off: 12v
46	---	Not Used	---
47	BR/BK	EGR VR Solenoid	0%, 55 mph: 45%
48-50	---	Not Used	---
51	BK/WT	Power Ground	<0.1v
52	TN/OR	Coil Driver 2 Control	5º dwell
53	---	Not Used	---
54	RD/LB	TCC Solenoid Control	0%, 55 mph: 50%
55	YL/BK	Keep Alive Power	12-14v
56-57	---	Not Used	---
58	GY/BK	VSS (+) Signal	0 Hz, 55 mph: 125 Hz
59	---	Not Used	---
60	GY/LB	HO2S-11 (B1 S1) Signal	0.1-1.1v
61	PK/LG	HO2S-22 (B2 S2) Signal	0.1-1.1v
62	BK/YL	EFT Sensor 'A' Signal	1.75-3.50v (50-120ºF)
63	RD/PK	Injector Pressure Sensor	2.7-3.7v (105-130 psi)
64	LB/YL	TR Sensor Signal	In 'P': 0v, 55 mph: 2.1v
65	BR/LG	DPFE Sensor Signal	0.20-1.30v
66-68	---	Not Used	---
69	OR/LB	A/C WOT Relay Control	On: 1v, Off: 12v
70	---	Not Used	---
71	RD	Vehicle Power	12-14v
72	TN/RD	Injector 7 Control	3.9-4.6 ms

1996-97 Crown Victoria 4.6L V8 CNG VIN 9 104 Pin Connector

PCM Pin #	Wire Color	Circuit Description (104 Pin)	Value at Hot Idle
73	TN/BK	Injector 5 Control	3.9-4.6 ms
74	BR/YL	Injector 3 Control	3.9-4.6 ms
75	TN	Injector 1 Control	3.9-4.6 ms
76-77	BK/WT	Power Ground	<0.1v
78	TN/LG	Coil Driver 3 Control	5° dwell
79	WT/LG	TCIL (lamp) Control	On: 1v, Off: 12v
80	LB/OR	Fuel Shutoff Valve Control	Idle: 0.1v (On)
81	WT/YL	EPC Solenoid Control	9.5v (20 psi)
82	---	Not Used	---
83	WT/LB	IAC Motor Control	9.3v (34%)
84	PK/LB	OSS Sensor Signal	55 mph: 1260-1330 rpm
85	DG	CMP Sensor Signal	6-7 Hz
87	RD/BK	HO2S-21 (B2 S1) Signal	0.1-1.1v
88	LB/RD	MAF Sensor Signal	0.6-0.9v
89	GY/WT	TP Sensor Signal	0.53-1.27v
90	BR/WT	Reference Voltage	4.9-5.1v
91	GY/RD	Analog Signal Return	<0.050v
92	LG	Brake Pedal Position Switch	Brake Off: 0v, On: 12v
93	RD/WT	HO2S-11 (B1 S1) Heater	On: 1v, Off: 12v
94	YL/LB	HO2S-21 (B2 S1) Heater	On: 1v, Off: 12v
95	WT/BK	HO2S-12 (B1 S2) Heater	On: 1v, Off: 12v
96	TN/YL	HO2S-22 (B2 S2) Heater	On: 1v, Off: 12v
97	RD	Vehicle Power	12-14v
98	LB	Injector 8 Control	3.9-4.6 ms
99	LG/OR	Injector 6 Control	3.9-4.6 ms
100	BR/LB	Injector 4 Control	3.9-4.6 ms
101	WT	Injector 2 Control	3.9-4.6 ms
102	---	Not Used	---
103	BK/WT	Power Ground	<0.1v
104	TN/LB	Coil Driver 4 Control	5° dwell

Pin Connector Graphic

PCM 104-PIN CONNECTOR

Terminal View of 104-Pin PCM Wiring Harness Connector

Standard Colors and Abbreviations

Abbreviation	Color	Abbreviation	Color	Abbreviation	Color
BK	Black	GY	Gray	PK	Purple
BL	Blue	GN	Green	RD	Red
BR	Brown	LG	LT Green	TN	Tan
DB	Dark Blue	OR	Orange	WT	White
DG	DK Green	PK	Pink	YL	Yellow

1998-2002 Crown Victoria 4.6L V8 CNG VIN 9 104 Pin Connector

PCM Pin #	Wire Color	Circuit Description (104 Pin)	Value at Hot Idle
1	OR/YL	COP 6 Driver Control	5° dwell
2	PK/LG	MIL (lamp) Control	MIL On: 1v, Off: 12v
3	BK/WT	Power Ground	<0.1v
4-5	---	Not Used	---
6	OR/YL	Shift Solenoid 'A' Control	1v, 55 mph: 12v
7-10	---	Not Used	---
11	VT/OR	Shift Solenoid 'B' Control	12v, Cruise: 1v
12	WT/LG	TCIL (lamp) Control	Lamp On: 1v, Off: 5v
13	VT	Flash EEPROM Power	0.1v
14	---	Not Used	---
15	PK/LB	SCP Data Bus (-) Signal	Digital Signals
16	TN/OR	SCP Data Bus (+) Signal	Digital Signals
17-20	---	Not Used	---
21	BK/VT	CKP Sensor (+) Signal	440-490 Hz
22	GY/YL	CKP Sensor (-) Signal	440-490 Hz
23	---	Not Used	---
24	BK/WT	Power Ground	<0.1v
25	BK	PCM Case Ground	<0.050v
26	LG/WT	COP 1 Driver Control	5° dwell
27	LG/YL	COP 5 Driver Control	5° dwell
28	RD/OR	Low Speed Fan Control	On: 1v, Off: 12v
29	TN/WT	TCS (switch) Signal	TCS & O/D On: 12v
30-32	---	Not Used	---
33	PK/OR	VSS (-) Signal	0 Hz, 55 mph: 125 Hz
34	YL/BK	Digital TR1 Sensor	In 'P': 0v, 55 mph: 11v
35	RD/LG	HO2S-12 (B1 S2) Signal	0.1-1.1v
36	TN/LB	MAF Sensor Return	<0.050v
37	OR/BK	TFT Sensor Signal	2.10-2.40v
38	---	Not Used	---
39	GY	IAT Sensor Signal	1.5-2.5v
40	DG/YL	Fuel Solenoid Valve Control	On: 12v, Off: 0v
41	BK/YL	A/C Cycling Clutch Switch	AC On: 12v, Off: 0v
42	RD/WT	ECT Sensor Signal	0.5-0.6v
43-44	---	Not Used	---
45	OR/RD	CHTIL (lamp) Control	Lamp On: 1v, Off: 3.5v
46	LG/VT	High Speed Fan Control	On: 1v, Off: 12v
47	BR/VT	EGR VR Solenoid	0%, 55 mph: 45%
48	---	Not Used	---
49	LB/BK	Digital TR2 Sensor	In 'P': 0v, 55 mph: 11v
50	WT/BK	Digital TR4 Sensor	In 'P': 0v, 55 mph: 11v
51	BK/WT	Power Ground	<0.1v
52	WT/PK	COP 3 Driver Control	5° dwell
53	DG/VT	COP 4 Driver Control	5° dwell
54	VT/YL	TCC Solenoid Control	0%, 55 mph: 95%
55	YL/BK	Keep Alive Power	12-14v
56-57	---	Not Used	---
58	GY/BK	VSS (-) Signal	0 Hz, 55 mph: 125 Hz
59	---	Not Used	---
60	GY/LB	HO2S-11 (B1 S1) Signal	0.1-1.1v
61	VT/LG	HO2S-22 (B2 S2) Signal	0.1-1.1v
62	BK/YL	EFT Sensor 'A' Signal	1.7-3.5v (50-120°F)
63	R/VT	Injection Pressure Sensor	2.7-3.7v (105-130 psi)
64	LB/YL	Digital TR Sensor (TR3)	In 'P': 0v, in O/D: 1.7v

1998-2002 Crown Victoria 4.6L V8 CNG VIN 9 104 Pin Connector

PCM Pin #	Wire Color	Circuit Description (104 Pin)	Value at Hot Idle
65 ('98-'00)	BR/LG	DPFE Sensor Signal	0.20-1.30v
65 ('01-'02)	BR/LG	DPFE Sensor Signal	0.95-1.05v
66	YL/LG	CHT Sensor Signal	0.7v (194°F)
67-68	---	Not Used	---
69	PK/YL	A/C WOT Relay Control	On: 1v, Off: 12v
70	---	Not Used	---
71	RD	Vehicle Power	12-14v
72	TN/RD	Injector 7 Control	3.9-5.2 ms
73	TN/BK	Injector 5 Control	3.9-5.2 ms
74	BR/YL	Injector 3 Control	3.9-5.2 ms
75	TN	Injector 1 Control	3.9-5.2 ms
76	---	Not Used	---
77	BK/WT	Power Ground	<0.1v
78	PK/LB	COP 7 Driver Control	5° dwell
79	WT/RD	COP 8 Driver Control	5° dwell
80	LB/OR	Fuel Pump Control	On: 1v, Off: 12v
81	WT/YL	EPC Solenoid Control	9.5v (20 psi)
82	---	Not Used	---
83	WT/LB	IAC Motor Control	34% duty cycle
84	DB/YL	OSS Sensor Signal	0 Hz
85	DB/OR	CMP Sensor Signal	6-7 Hz
86	WT/BK	A/C High Pressure Switch	Open: 12v, Closed: 0v
87	RD/BK	HO2S-21 (B2 S1) Signal	0.1-1.1v
88	LB/RD	MAF Sensor Signal	0.6v
89	GY/WT	TP Sensor Signal	0.53-1.27v
90	BR/WT	Reference Voltage	4.9-5.1v
91	GY/RD	Analog Signal Return	<0.050v
92	LG	Brake Pedal Position Switch	Brake Off: 0v, On: 12v
93	RD/WT	HO2S-11 (B1 S1) Heater	On: 1v, Off: 12v
94	YL/LB	HO2S-21 (B2 S1) Heater	On: 1v, Off: 12v
95	WT/BK	HO2S-12 (B1 S2) Heater	On: 1v, Off: 12v
96	TN/YL	HO2S-22 (B2 S2) Heater	On: 1v, Off: 12v
97	RD	Vehicle Power	12-14v
98	LB	Injector 8 Control	3.9-5.2 ms
99	LG/OR	Injector 6 Control	3.9-5.2 ms
100	BR/LB	Injector 4 Control	3.9-5.2 ms
101	WT	Injector 2 Control	3.9-5.2 ms
102	---	Not Used	---
103	BK/WT	Power Ground	<0.1v
104	PK/WT	COP 2 Driver Control	5° dwell

Pin Connector Graphic

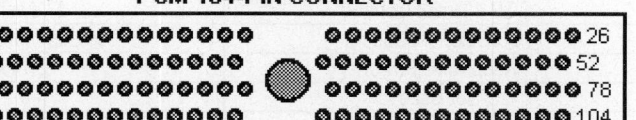

PCM 104-PIN CONNECTOR

Terminal View of 104-Pin PCM Wiring Harness Connector

1998-2002 Crown Victoria 4.6L V8 CNG Module 60 Pin Connector

PCM Pin #	Wire Color	Circuit Description (60 Pin)	Value at Hot Idle
1	YL/BK	Keep Alive Power	12-14v
2	---	Not Used	---
3	TN	Injector 1 Signal from PCM	3.9-5.2 ms
4	WT	Injector 2 Signal from PCM	3.9-5.2 ms
5	BR/YL	Injector 3 Signal from PCM	3.9-5.2 ms
6	---	Not Used	---
7	RD/PK	Fuel Tank Pressure Signal	2.6v (0" HG - cap off)
8-11	---	Not Used	---
12	BK/YL	Instrument Cluster Power	12-14v
13-17	---	Not Used	---
18	PK/LB	SCP Data Bus (-)	Digital Signals
19	TN/OR	SCP Data Bus (+)	Digital Signals
20-22	---	Not Used	---
23	BR/LB	Injector 4 Signal from PCM	3.9-5.2 ms
24	TN/BK	Injector 5 Signal from PCM	3.9-5.2 ms
25	LG/OR	Injector 6 Signal from PCM	3.9-5.2 ms
26	WT/RD	Reference Voltage	4.9-5.1v
27	---	Not Used	---
28	LB/VT	Fuel Tank Temp. Sensor #1	0.1-4.9v
29-30	---	Not Used	---
31	BK/WT	Instrument Cluster Ground	<0.1v
32	---	Not Used	---
33	TN/BK	Injector 5 Control	3.9-5.2 ms
34	---	Not Used	---
35	BR/LB	Injector 4 Control	3.9-5.2 ms
36	---	Not Used	---
37	RD	Ignition Power	12-14v
38	YL/WT	Fuel Display Output	Varies
39	BR/YL	Injector 3 Control	3.9-6.5 ms
40	BK/WT	Power Ground	<0.1v
41	---	Not Used	---
42	LG/OR	Injector 6 Control	3.9-5.2 ms
43	TN/RD	Injector 7 Signal from PCM	3.9-5.2 ms
44	LB	Injector 8 Signal from PCM	3.9-5.2 ms
45	---	Not Used	---
46	BK/LB	Sensor Signal Return	<0.050v
47	WT/YL	Fuel Tank Temp. Sensor #1	0.1-4.9v
48-52	---	Not Used	---
53	TN/RD	Injector 7 Control	3.9-5.2 ms
54	LB	Injector 8 Control	3.9-5.2 ms
55-56	---	Not Used	---
57	RD	Ignition Power	12-14v
58	TN	Injector 1 Control	3.9-5.2 ms
59	WT	Injector 2 Control	3.9-5.2 ms
60	BK/WT	Power Ground	<01v

Pin Connector Graphic

Terminal View of 60-Pin PCM Harness Connector

1998-2002 Crown Victoria 4.6L V8 CNG Module Wiring Diagram

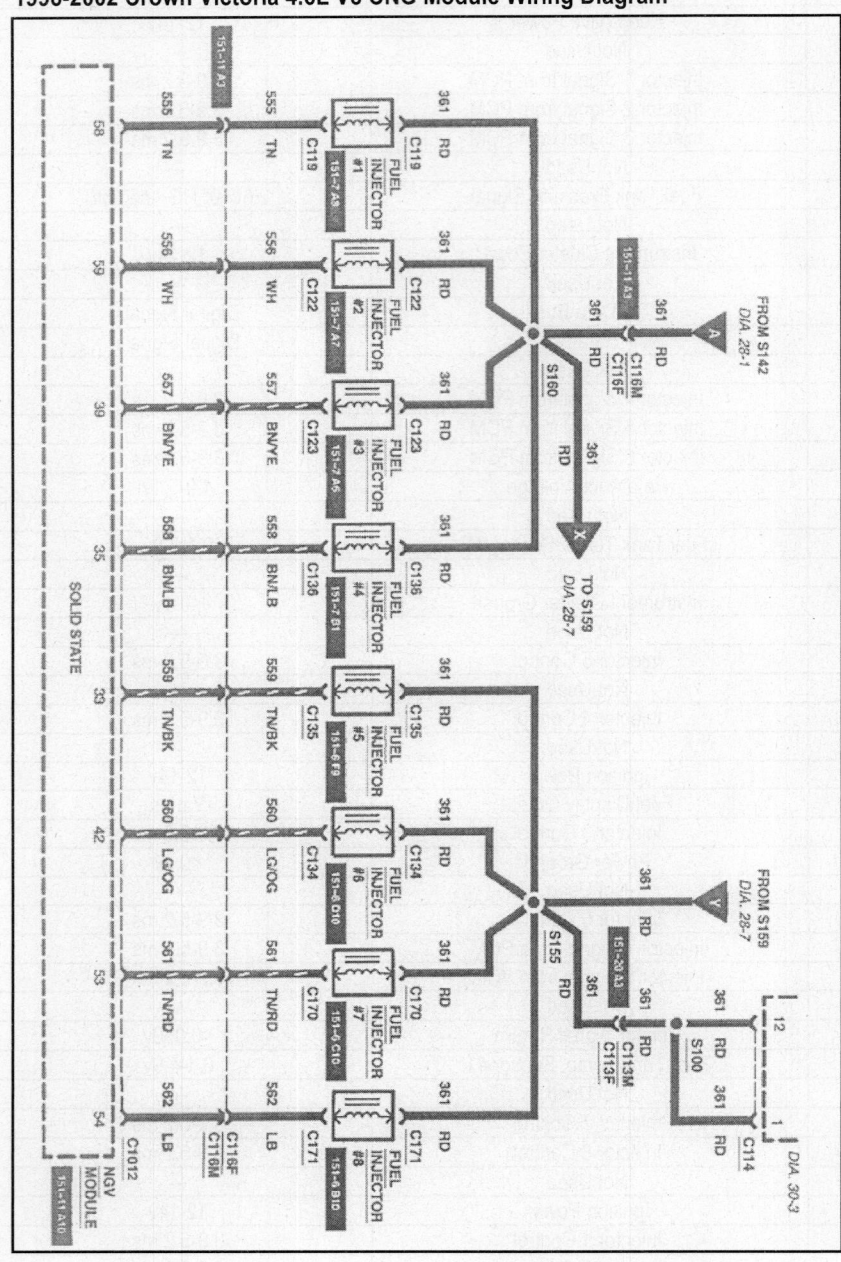

2003 Crown Victoria 4.6L V8 CNG VIN 9 (A/T) 104 Pin Connector

PCM Pin #	Wire Color	Circuit Description (104 Pin)	Value at Hot Idle
1	OR/YL	COP 6 Driver Control	5° dwell
2	PK/LG	MIL (lamp) Control	MIL On: 1v, Off: 12v
3	BK/WT	Power Ground	<0.1v
4-5, 14, 23	---	Not Used	---
6	OR/YL	Shift Solenoid 'A' Control	1v, 55 mph: 12v
7	YL/LG	Generator Regulator 'S' Terminal	0-130 Hz
8	RD/PK	Fuel Rail Temperature Sensor	1.7-3.5v (50-120°F)
9	YL/WT	Fuel Level Indicator Signal	1.7v (1/2 full)
10	LB/RD	MAP Sensor Signal	107 Hz
11	VT/OR	Shift Solenoid 'B' Control	12v, Cruise: 1v
12	WT/LG	TCIL (lamp) Control	Lamp On: 1v, Off: 5v
13	VT	Flash EEPROM Power	0.1v
15	PK/LB	SCP Bus (-) Signal	Digital Signals
16	TN/OR	SCP Bus (+) Signal	Digital Signals
17	GY/OR	RX Signal	Digital Signals
18	WT/LG	TX Signal	Digital Signals
19	OR/RD	Cylinder Head Temperature Lamp Control	Lamp Off: 12v, On: 1v
20	WT/LG	Fuel Door Release Solenoid Indicator	Solenoid Off: 12v, On: 1v
21	BK/PK	CKP Sensor (+) Signal	440-490 Hz
22	GY/YL	CKP Sensor (-) Signal	440-490 Hz
24, 51	BK/WT	Power Ground	<0.1v
25	BK	PCM Case Ground	<0.050v
26	LG/WT	COP 1 Driver Control	5° dwell
27	LG/YL	COP 5 Driver Control	1v, 55 mph: 12v
28	RD/OR	Low Speed Fan Control	On: 1v, Off: 12v
29	TN/WT	TCS (switch) Signal	TCS & O/D On: 12v
30	DG/RD	Light Emitting Diode Signal Ground	<0.050v
31	YL/LG	Power Steering Pressure Sensor	Straight: 0v, Turned: 12v
32	DG/VT	Knock Sensor Ground	<0.050v
33	---	Not Used	---
34	YL/BK	Digital TR1 Sensor	In 'P': 0v, 55 mph: 11v
35	RD/LG	HO2S-12 (B1 S2) Signal	0.1-1.1v
36	TN/LB	MAF Sensor Return	<0.050v
37	OR/BK	TFT Sensor Signal	2.10-2.40v
38	BK/WT	A/C Pressure Sensor Discharge Temp.	A/C On: 1.5-1.9v
39	GY	IAT Sensor Signal	1.5-2.5v
40	DG/YL	Natural Gas Vehicle Tank Power	On: 12v, Off: 0v
41	PK/LB	A/C Pressure Cutout Switch	AC On: 12v, Off: 0v
42	RD/WT	ECT Sensor Signal	0.5-0.6v
43	DB/LG	Low Fuel Indicator	Digital Signals
44	GY/RD	Starter Relay Control	Relay Off: 12v, On: 9v
45	LG/RD	Generator Common	Digital Signals
46	---	Not Used	---
47	BR/PK	EGR VR Solenoid	0%, 55 mph: 45%
48	WT	Tachometer Output	DC pulse signals
49	LB/BK	Digital TR2 Sensor	In 'P': 0v, 55 mph: 11v
50	WT/BK	Digital TR4 Sensor	In 'P': 0v, 55 mph: 11v
52	WT/PK	COP 3 Driver Control	5° dwell
53	DG/VT	COP 4 Driver Control	5° dwell
54	VT/YL	TCC Solenoid Control	0%, 55 mph: 95%
55	YL/BK	Keep Alive Power	12-14v
56	LG/BK	EVAP Purge Solenoid Control	0-10 Hz (0-100%)
57	YL/RD	Knock Sensor Signal	0v
58-59	---	Not Used	---

2003 Crown Victoria 4.6L V8 CNG VIN W (A/T) 104 Pin Connector

PCM Pin #	Wire Color	Circuit Description (104 Pin)	Value at Hot Idle
60	GY/BK	HO2S-11 (B1 S1) Signal	0.1-1.1v
61	VT/LG	HO2S-22 (B2 S2) Signal	0.1-1.1v
62	RD/PK	FTP Sensor Signal	2.6v (0" H2O - cap off)
63	OR/LG	Injection Pressure Sensor	2.8v (39 psi)
64	LB/YL	Digital TR3 Sensor	In 'P': 0v, in O/D: 1.7v
65	BR/LG	DPFE Sensor Signal	0.95-1.05v
66	YL/LG	Cylinder Head Temperature Sensor	0.7v (194ºF)
67	VT/WT	EVAP Canister Vent Solenoid	0-10 Hz (0-100%)
68	GY/BK	Vehicle Speed Sensor (+) Signal	0 Hz, 55 mph: 125 Hz
69	PK/YL	A/C WOT Relay Control	On: 1v, Off: 12v
70	YL	Generator Battery Indicator Control	Lamp Off: 12v, On: 1v
71	RD	Vehicle Power (Start-Run)	12-14v
72	TN/RD	Injector 7 Control	3.4-3.7 ms
73	TN/BK	Injector 5 Control	3.4-3.7 ms
74	BR/YL	Injector 3 Control	3.4-3.7 ms
75	TN	Injector 1 Control	3.4-3.7 ms
76	BK/WT	Power Ground	<0.1v
77	---	Not Used	---
78	PK/LB	COP 7 Driver Control	5º dwell
79	WT/RD	COP 8 Driver Control	5º dwell
80	LB/OR	Fuel Pump Control	On: 1v, Off: 12v
81	WT/YL	EPC Solenoid Control	9.5v (20 psi)
82	---	Not Used	---
83	WT/LB	IAC Motor Control	34% duty cycle
84	DB/YL	OSS Sensor Signal	0 Hz
85	DB/OR	CMP Sensor Signal	6-7 Hz
86	WT/BK	A/C High Pressure Switch	Open: 12v, Closed: 0v
87	RD/BK	HO2S-21 (B2 S1) Signal	0.1-1.1v
88	LB/RD	MAF Sensor Signal	0.6v
89	GY/WT	TP Sensor Signal	0.53-1.27v
90	BR/WT	Reference Voltage	4.9-5.1v
91	GY/RD	Analog Signal Return	<0.050v
92	DG	Brake Pedal Position Switch	Brake Off: 0v, On: 12v
93	RD/WT	HO2S-11 (B1 S1) Heater	On: 1v, Off: 12v
94	YL/LB	HO2S-21 (B2 S1) Heater	On: 1v, Off: 12v
95	WT/BK	HO2S-12 (B1 S2) Heater	On: 1v, Off: 12v
96	TN/YL	HO2S-22 (B2 S2) Heater	On: 1v, Off: 12v
98	LB	Injector 8 Control	3.4-3.7 ms
99	LG/OR	Injector 6 Control	3.4-3.7 ms
100	BR/LB	Injector 4 Control	3.4-3.7 ms
101	WT	Injector 2 Control	3.4-3.7 ms
102	---	Not Used	---
103	BK/WT	Power Ground	<0.1v
104	PK/WT	COP 2 Driver Control	5º dwell

Pin Connector Graphic

PCM 104-PIN CONNECTOR

Terminal View of 104-Pin PCM Wiring Harness Connector

1990-91 Crown Victoria 5.0L V8 MFI VIN F (A/T) 60 Pin Connector

PCM Pin #	Wire Color	Circuit Description (60 Pin)	Value at Hot Idle
1 ('90)	BK/OR	Keep Alive Power	12-14v
1 ('91)	YL	Keep Alive Power	12-14v
2	LG	Brake Pedal Position Switch	Brake On: 12, Off: 0v
3	DG/WT	VSS (+) Signal	0 Hz, 55 mph: 125 Hz
4	DG/YL	Ignition Diagnostic Monitor	20-31 Hz
4	BL/PK	Ignition Diagnostic Monitor	20-31 Hz
4	TN/YL	Ignition Diagnostic Monitor	20-31 Hz
5	---	Not Used	---
6 ('90)	BK/WT	VSS (-) Signal	0 Hz, 55 mph: 125 Hz
6 ('91)	GY/RD	VSS (-) Signal	0 Hz, 55 mph: 125 Hz
7 ('90)	GY/YL	ECT Sensor Signal	0.5-0.6v
7 ('91)	LG/RD	ECT Sensor Signal	0.5-0.6v
8-9	---	Not Used	---
10	LG/PK	A/C Cycling Clutch Switch	A/C On: 12v, Off: 0v
10	PK/BL	A/C Cycling Clutch Switch	A/C On: 12v, Off: 0v
11	LG/BK	Air Management 2 Solenoid	AM2 On: 1v, Off: 12v
12	TN/YL	Injector 3 Control	5.7-6.2 ms
12	BR/YL	Injector 3 Control	5.7-6.2 ms
13	TN/BK	Injector 4 Control	5.7-6.2 ms
13	BR/BL	Injector 4 Control	5.7-6.2 ms
14 ('90)	TN/BL	Injector 5 Control	5.7-6.2 ms
14 ('91)	TN	Injector 5 Control	5.7-6.2 ms
15	TN/LG	Injector 6 Control	5.7-6.2 ms
15	LG	Injector 6 Control	5.7-6.2 ms
15	LG/OR	Injector 6 Control	5.7-6.2 ms
16 ('90)	BK/OR	Ignition System Ground	<0.1v
16 ('91)	OR	Ignition System Ground	<0.1v
17	YL/BK	Self-Test Indicator & MIL	MIL On: 1v, Off: 12v
17	TN/RD	Self-Test Indicator & MIL	MIL On: 1v, Off: 12v
18-19	---	Not Used	---
20	BK	PCM Case Ground	<0.050v
21	WT/BL	IAC Motor Control	9.0-11.5v
22	TN/LG	Fuel Pump Control	On: 1v, Off: 12v
23-24, 32	---	Not Used	---
25	LG/PK	ACT Sensor Signal	1.5-2.5v
26	OR/WT	Reference Voltage	4.9-5.1v
27	BR/LG	EGR EVP Sensor	0.4v
28	---	Not Used	---
29	DG/PK	HO2S-11 (B1 S1) Signal	0.1-1.1v
29	RD/BK	HO2S-11 (B1 S1) Signal	0.1-1.1v
30	WT/PK	Neutral Drive Switch Signal	In 'P': 0v, Others: 5v
30	BL/YL	Neutral Drive Switch Signal	In 'P': 0v, Others: 5v
31	GY/YL	EVAP Purge Solenoid	12v, 55 mph: 1v
33	DG	EGR VR Solenoid	0%, 55 mph: 45%
33	BR/PK	EGR VR Solenoid	0%, 55 mph: 45%
34	BL/PK	Data Output Link	Digital Signals
35 ('90)	WT/PK	Speed Control Vent Solenoid	0%, 55 mph: 98%
35 ('91)	GY/BK	Speed Control Vent Solenoid	0%, 55 mph: 98%
36	YL/LG	Spark Output Signal	50% duty cycle
37 ('90)	RD	Vehicle Power	12-14v
37 ('91)	BK/YL	Vehicle Power	12-14v
38	GY/BK	S/C Vacuum Solenoid	0%, 55 mph: 45%
39	GY/BK	Speed Control Switch Ground	<0.050v
40	BK/LG	Power Ground	<0.1v

1990-91 Crown Victoria 5.0L V8 MFI VIN F (A/T) 60 Pin Connector

PCM Pin #	Wire Color	Circuit Description (60 Pin)	Value at Hot Idle
41	---	Not Used	---
42	TN/OR	Injector 7 Control	5.7-6.2 ms
42	TN/RD	Injector 7 Control	5.7-6.2 ms
43	DB/LG	HO2S-21 (B2 S1) Signal	0.1-1.1v
45	DB/LG	MAP Sensor	107 Hz
45	LG/BK	MAP Sensor	107 Hz
46	BK/WT	Analog Signal Return	<0.050v
46	GY/RD	Analog Signal Return	<0.050v
47	DG/LG	TP Sensor Signal	0.5-1.2v
47	GY/WT	TP Sensor Signal	0.5-1.2v
48	WT/RD	Self-Test Indicator Signal	STI Open: 5v, Closed: 1v
48	WT/BK	Self-Test Indicator Signal	STI Open: 5v, Closed: 1v
48	BR	Self-Test Indicator Signal	STI Open: 5v, Closed: 1v
49	OR	HO2S-11 (Bank 1) Ground	<0.050v
50	OR/RD	S/C Command Switch	6.7v
50	BL/BK	S/C Command Switch	6.7v
51	WT/RD	Air Management 1 Solenoid	AM1 On: 1v, Off: 12v
52 ('90)	BL	Injector 8 Control	5.7-6.2 ms
52 ('91)	TN/RD	Injector 8 Control	5.7-6.2 ms
54	OR/BL	A/C WOT Relay Control	On: 1v, Off: 12v
55	---	Not Used	---
56	DB	PIP Sensor Signal	50% dwell
57 ('90)	RD	Vehicle Power	12-14v
57 ('91)	BK/YL	Vehicle Power	12-14v
58	TN	Injector 1 Control	5.7-6.2 ms
59 ('90)	TN/WT	Injector 2 Control	5.7-6.2 ms
59 ('91)	WT	Injector 2 Control	5.7-6.2 ms
60	BK/LG	Power Ground	<0.1v

Pin Connector Graphic

Terminal View of 60-Pin PCM Harness Connector

Standard Colors and Abbreviations

Abbreviation	Color	Abbreviation	Color	Abbreviation	Color
BK	Black	GY	Gray	PK	Purple
BL	Blue	GN	Green	RD	Red
BR	Brown	LG	LT Green	TN	Tan
DB	Dark Blue	OR	Orange	WT	White
DG	DK Green	PK	Pink	YL	Yellow

1990-91 Crown Victoria 5.0L V8 VIN F (California) 60 Pin Connector

PCM Pin #	Wire Color	Circuit Description (60 Pin)	Value at Hot Idle
1 ('90)	BK/OR	Keep Alive Power	12-14v
1 ('91)	YL	Keep Alive Power	12-14v
2	---	Not Used	---
3	DG/WT	VSS (+) Signal	0 Hz, 55 mph: 125 Hz
4	DG/YL	Ignition Diagnostic Monitor	20-31 Hz
5	LG	Brake Pedal Position Switch	Brake Off: 0v, On: 12v
6	BK/WT	VSS (-) Signal	0 Hz, 55 mph: 125 Hz
7	LG/YL	ECT Sensor Signal	0.5-0.6v
8	BK/OR	Data Bus (-) Signal	Digital Signals
9	TN/BL	MAF Sensor Return	<0.050v
10	PK/BL	A/C Cycling Clutch Switch	A/C On: 12v, Off: 0v
11	---	Not Used	---
12	BR/YL	Injector 3 Control	5.7-6.2 ms
13	BR/BL	Injector 4 Control	5.7-6.2 ms
14	TN/BL	Injector 5 Control	5.7-6.2 ms
15	LG	Injector 6 Control	5.7-6.2 ms
16	BK/OR	Ignition System Ground	<0.050v
17	TN/RD	STI Output, MIL Control	MIL On: 1v, Off: 12v
18	---	Not Used	---
19	OR	Fuel Pump Monitor	On: 12v, Off: 0v
20	BK	PCM Case Ground	<0.050v
21	WT/BL	IAC Motor Control	9.0-11.5v
22	TN/LG	Fuel Pump Control	On: 1v, Off: 12v
23-24	---	Not Used	---
25	LG/PK	ACT Sensor Signal	1.5-2.5v
26	OR/WT	Reference Voltage	4.9-5.1v
27	BR/LG	EGR EVP Sensor Signal	0.4v
28	BL/BK	S/C Command Switch	6.7v
29	DG/PK	HO2S-11 (B1 S1) Signal	0.1-1.1v
30	WT/PK	Neutral Drive Switch Signal	In 'P': 0v, Others: 5v
31	GY/YL	EVAP Purge Solenoid	12v, at 55 mph: 1v
32	BL	Air Management 2 Solenoid	AM2 On: 1v, Off: 12v
33	DG	EGR VR Solenoid	0%, 55 mph: 45%
34	---	Not Used	---
35	WT/PK	Speed Control Vent Solenoid	0%, 55 mph: 98%
36	YL/LG	Spark Output Signal	50% duty cycle
37	RD	Vehicle Power	12-14v
38	YL	Air Management 1 Solenoid	AM1 On: 1v, Off: 12v
39	GY/BK	Speed Control Switch Ground	<0.050v
40	BK/LG	Power Ground	<0.1v

Pin Connector Graphic

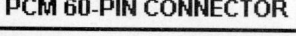

PCM 60-PIN CONNECTOR

Terminal View of 60-Pin PCM Harness Connector

1990-91 Crown Victoria 5.0L V8 VIN F (Cal) 60 Pin Connector

PCM Pin #	Wire Color	Circuit Description (60 Pin)	Value at Hot Idle
41	---	Not Used	---
42	TN/OR	Injector 7 Control	5.7-6.2 ms
43	DB/LG	HO2S-21 (B2 S1) Signal	0.1-1.1v
44	OR/BK	Data Bus (+) Signal	Digital Signals
45	LG/BK	MAP Sensor	107 Hz
46	BK/WT	Analog Signal Return	<0.050v
47	DG/LG	TP Sensor Signal	0.5-1.2v
48	WT/BK	Self-Test Indicator Signal	STI Open: 5v, Closed: 1v
49	OR	HO2S-21 (Bank 2) Ground	<0.050v
50	DB/OR	S/C Command Switch	6.7v
51	---	Not Used	---
52	BL	Injector 8 Control	5.7-6.2 ms
53	---	Not Used	---
54	OR/BL	A/C WOT Relay Control	On: 1v, Off: 12v
55	---	Not Used	---
56	DB	PIP Sensor Signal	50% dwell
57	RD	Vehicle Power	12-14v
58	TN	Injector 1 Control	5.7-6.2 ms
59	WT	Injector 2 Control	5.7-6.2 ms
60	BK/LG	Power Ground	<0.1v

Pin Connector Graphic

Terminal View of 60-Pin PCM Harness Connector

Standard Colors and Abbreviations

Abbreviation	Color	Abbreviation	Color	Abbreviation	Color
BK	Black	GY	Gray	PK	Purple
BL	Blue	GN	Green	RD	Red
BR	Brown	LG	LT Green	TN	Tan
DB	Dark Blue	OR	Orange	WT	White
DG	DK Green	PK	Pink	YL	Yellow

Escort Pin Tables

1991-92 Escort 1.8L I4 MFI VIN 8 (All) 22 Pin Connector

PCM Pin #	Wire Color	Circuit Description (22 Pin)	Value at Hot Idle
22-A	BL/RD	Keep Alive Power	12-14v
22-B	WT/RD	Vehicle Power	12-14v
22-C	PK	Start Signal	KOEC: 9-11v
22-D	WT/YL	Switch Monitor Lamp Signal	SML On: 5v, Off: 12v
22-E	YL/BK	MIL (lamp) Control	MIL On: 2.5v, Off: 12v
22-F	WT/BK	Self-Test Output Signal	STO On: 5v, Off: 12v
22-G	GN/WT	Spark Output Signal	50% duty cycle
22-H	---	Not Used	---
22-I	---	Not Used	---
22-J	BL/BK	A/C WOT Relay Control	A/C On: 12v, Off: 2.5v
22-K	LG/YL	Self-Test Input Signal	STI On: 0v, Off: 5v
22-L	BR/WT	DRL Signal	DRL Off: 12.5v, On: 2.5v
22-M	---	Not Used	---
22-N	RD/WT	Idle Throttle Switch Signal	Closed: 0v, Open: 12v
22-O	GN	M/T: BPP Switch Signal	Brake Off: 0v, On: 12v
22-O	GN/RD	A/T: Down Shift DSS to 4EAT	DSS Off: 1v, On: 12v
22-P	BL/YL	PSP Switch Signal	Straight: 0v, Turned: 12v
22-Q	GN/BK	Air Conditioning Switch Signal	A/C On: 2.5v, Off: 12v
22-R	BK/GN	Low Speed Fan Control	On: 1v, Off: 12v
22-S	OR/BL	Blower Motor Switch Signal	Off or 1st: 12v, 2nd>: 1v
22-T	BK/BL	Rear Window Defrost Switch	Switch Off: 1v, On: 12v
22-U	RD/BK	Headlamp Switch Signal	HDL On: 12v, Off: 1v
22-V	BR/YL	M/T: Neutral Drive Switch	In N: 0v, all others: 11v
22-V	BK/BL	A/T: MLP Sensor Signal	In 'N': 0v, Others: 12v

Pin Connector Graphic

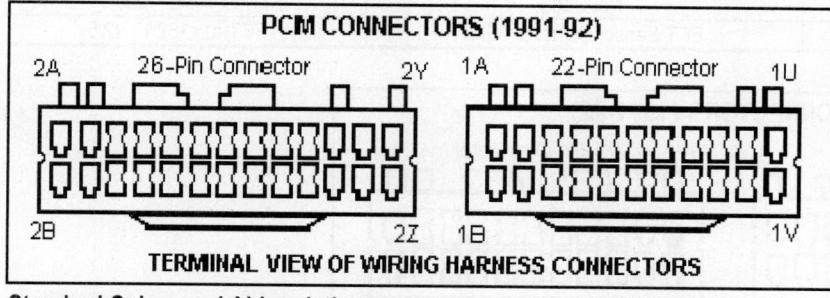

Standard Colors and Abbreviations

Abbreviation	Color	Abbreviation	Color	Abbreviation	Color
BK	Black	GY	Gray	PK	Purple
BL	Blue	GN	Green	RD	Red
BR	Brown	LG	LT Green	TN	Tan
DB	Dark Blue	OR	Orange	WT	White
DG	DK Green	PK	Pink	YL	Yellow

1991-92 Escort 1.8L I4 MFI VIN 8 (All) 26 Pin Connector

PCM Pin #	Wire Color	Circuit Description (26 Pin)	Value at Hot Idle
26-A	BK/OR	Power Ground	<0.1v
26-B	BK/OR	Power Ground	<0.1v
26-C	BK/LG	Power Ground	<0.1v
26-D	BK/WT	Analog Signal Return	<0.050v
26-E	WT	CKP Sensor Signal	400-425 Hz
26-F	---	Not Used	---
26-G	YL/BL	CID Sensor Signal	5-7 Hz
26-H	BK	Power Ground (California)	1v
26-H	BK/YL	Vehicle Power (Canada)	12-14v
26-I	---	Not Used	---
26-J	WT/RD	A/T: Vehicle Power	12-14v
26-K	LG/RD	Reference Voltage	4.9-5.1v
26-L	LG/WT	M/T: WOT Switch	At WOT: 1v
26-M	LG/WT	TP Sensor Signal to 4EAT	0.5-2.1v
26-N	RD/BL	HO2S-11 (B1 S1) Signal	0.1-1.1v
26-O	RD	Vane Airflow Meter Sensor	3.3v
26-P	RD/BK	VAT Sensor Signal	2.5v at 68°F
26-Q	BL/WT	ECT Sensor Signal	2.5v at 68°F
26-R	---	Not Used	---
26-S	BK/RD	High Speed Inlet Air Control	Below 5000 rpm: 1.5v
26-T	GN/OR	FPRC Solenoid Control	Hot engine: startup: <1.5v
26-U	YL	Fuel Injectors 1 & 3 - Bank 1	3.6 ms
26-V	YL/BK	Fuel Injectors 2 & 4 - Bank 2	3.6 ms
26-W	BL/OR	IAC Motor Control	8-10v
26-X	WT/BL	EVAP Purge Solenoid	At off-idle speeds: 1v
26-Y	---	Not Used	---
26-Z	BL/GN	ECT Sensor Signal to 4EAT	ECT at <162°F: <2.5v

Pin Connector Graphic

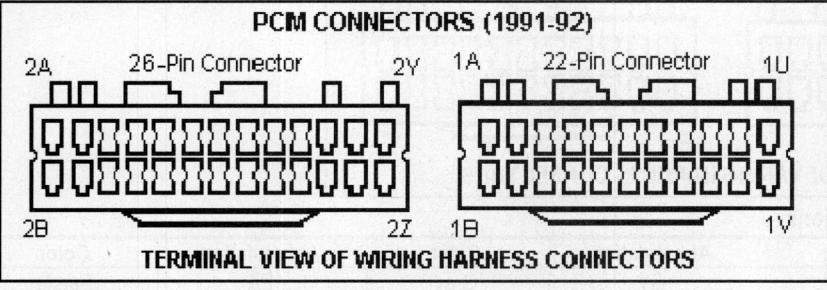

Standard Colors and Abbreviations

Abbreviation	Color	Abbreviation	Color	Abbreviation	Color
BK	Black	GY	Gray	PK	Purple
BL	Blue	GN	Green	RD	Red
BR	Brown	LG	LT Green	TN	Tan
DB	Dark Blue	OR	Orange	WT	White
DG	DK Green	PK	Pink	YL	Yellow

1993 Escort 1.8L I4 Engine VIN 8 (A/T) 22 Pin Connector

PCM Pin #	Wire Color	Circuit Description (22 Pin)	Value at Hot Idle
22-A	DB/RD	Keep Alive Power	12- 14v
22-B	WT/RD	Vehicle Power	12-14v
22-C	PK	Vehicle Start Signal	Cranking: 9-11v
22-D	WT/YL	System Monitor Lamp	Lamp On: 5v, Off: 12v
22-E	YL/BK	MIL (lamp) Control	MIL On: 1v, Off: 12v
22-F	WT/BK	Self-Test Output Signal	STO On: 5v, Off: 12v
22-G	DG/WT	Ignition Control Module Signal	0.1-0.3v
22-H	RD/BK	Headlamp Switch	Switch On: 12v, Off: 1v
22-I	LG/YL	Self-Test Input Signal	STI On: 0v, Off: 5v
22-J	BK/DB	Rear Defroster Control	Switch Off: 1v, On: 12v
22-K	BK	Power Ground	<0.1v
22-L	DB/BK	A/C Heater Relay	On: 1v, Off: 12v
22-M	DG	VSS Signal	0 Hz, 55 mph: 125 Hz
22-N	DB/YL	PSP Switch Signal	Straight: 12v, Turned: 0v
22-O	DG/DB	A/C Switch Signal	Switch On: 2.5v, Off: 12v
22-P	OR/DB	Blower Motor Switch	In 1st: 12v, Others: 1v
22-Q	DG	Brake Pedal Position Switch	Brake Off: 0v, On: 12v
22-R	BK/DB	MLP Sensor Switch	In 'P': 0v, Others: 12v
22-S	---	Not Used	---
22-T	RD/WT	Idle Switch Signal	Closed: 0v, Open: 12v
22-U	---	Not Used	---
22-V	---	Not Used	---

1993 Escort 1.8L I4 Engine VIN 8 (A/T) 16 Pin Connector

PCM Pin #	Wire Color	Circuit Description (16 Pin)	Value at Hot Idle
16-A	WT	CKP Sensor	2.5v (at Key On: 0 or 5v)
16-B	RD	Vane Airflow Meter	0.2v
16-C	RD/DB	O2S-11 (B1 S1) Signal	0.1-1.1v
16-D	BK/DG	Cooling Fan Control	On: 1v, Off: 12v
16-E	DB/WT	ECT Sensor Signal	0.5-0.6v
16-F	LG/WT	TP Sensor Signal	0.6v, 55 mph: 1.8v
16-G	WT/BK	TOT Sensor Signal	2.10-2.40v
16-H	---	Not Used	---
16-I	LG/RD	Reference Voltage	4.9-5.1v
16-J	YL/DB	CID Sensor	2.5v (at Key On: 0 or 5v)
16-K	RD/BK	IAT Sensor Signal	1.5-2.5v
16-L	DB	VSS (-) Signal	0 Hz, 55 mph: 125 Hz
16-M	WT/DB	TSS Sensor (+)	340-380 Hz
16-N	YL/DB	TSS Sensor (-)	340-380 Hz
16-O	WT/DB	EVAP Purge Solenoid	12v, 55 mph: 1v
16-P	---	Not Used	---

Standard Colors and Abbreviations

Abbreviation	Color	Abbreviation	Color	Abbreviation	Color
BK	Black	GY	Gray	PK	Purple
BL	Blue	GN	Green	RD	Red
BR	Brown	LG	LT Green	TN	Tan
DB	Dark Blue	OR	Orange	WT	White
DG	DK Green	PK	Pink	YL	Yellow

1993 Escort 1.8L I4 Engine VIN 8 (A/T) 26 Pin Connector

PCM Pin #	Wire Color	Circuit Description (26 Pin)	Value at Hot Idle
26-A	BK/OR	Power Ground	<0.1v
26-B	BK/OR	Power Ground	<0.1v
26-C	BK/LG	Power Ground	<0.1v
26-D	BK/WT	Analog Signal Return	<0.050v
26-E	YL	MLP Overdrive Switch Signal	OD Switch on: 1v, off: 12v
26-F	BR/WT	Daylight Running Lights	DRL On: 2v, Off: 12v
26-G	YL/WT	MLP Low Switch Signal	Low: 12v, all others: 1v
26-H	YL/RD	MLP Drive Switch Signal	Drive: 12v, all others: 0v
26-I	BK/RD	High Speed Inlet Air Control	Over 4900 rpm: 12v
26-K	---	Not Used	---
26-L	---	Not Used	---
26-M	DG/OR	Fuel Pressure Regulator	Startup: <3v, others: 12v
26-N	---	Not Used	---
26-O	---	Not Used	---
26-P	---	Not Used	---
26-Q	DB/OR	Idle Air Control Valve	8-10v
26-R	---	Not Used	---
26-S	---	Not Used	---
26-T	---	Not Used	---
26-U	YL	Fuel Injectors 1 & 3 (Bank 1)	3.6 ms
26-V	YL/BK	Fuel Injectors 2 & 4 (Bank 2)	3.6 ms
26-W	DB/OR	Shift Solenoid 1 Control	Off: 12v, On: 1v
26-X	DB/YL	Shift Solenoid 2 Control	Off: 12v, On: 1v
26-Y	OR	Shift Solenoid 3 Control	Off: 12v, On: 1v
26-Z	DB	TCC Solenoid Control	Shifting: 12v, Others: 1v

Pin Connector Graphic

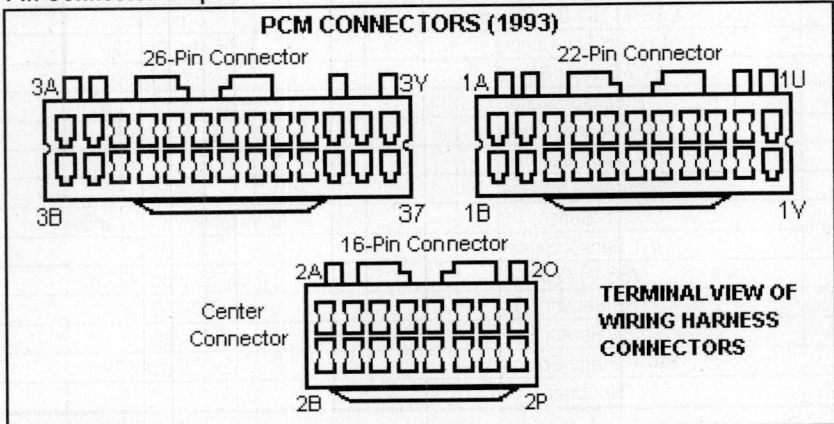

Standard Colors and Abbreviations

Abbreviation	Color	Abbreviation	Color	Abbreviation	Color
BK	Black	GY	Gray	PK	Purple
BL	Blue	GN	Green	RD	Red
BR	Brown	LG	LT Green	TN	Tan
DB	Dark Blue	OR	Orange	WT	White
DG	DK Green	PK	Pink	YL	Yellow

1993 Escort 1.8L I4 MFI VIN 8 (M/T) 22 Pin Connector

PCM Pin #	Wire Color	Circuit Description (22 Pin)	Value at Hot Idle
22-1	DB/RD	Keep Alive Power	12- 14v
22-2	WT/RD	Vehicle Power (Main Relay)	12-14v
22-3	PK	Vehicle Start Signal	Cranking: 9-11v
22-4	WT/YL	System Monitor Lamp	Lamp On: 5v, Off: 12v
22-5	YL/BK	MIL (lamp) Control	MIL On: 1v, Off: 12v
22-6	WT/BK	Self-Test Output Signal	Buzzer On: 5v, Off: 12v
22-7	DG/WT	Spark Output Signal	50% duty cycle
22-8, 22-9	---	Not Used	---
22-10	DB/BK	A/C WOT Relay Control	A/C On: 2.5v, Off: 12v
22-11	LG/YL	Self-Test Input Signal	STI On: 0v, STI Off: 5v
22-12	BR/WT	Daytime Running Lamps	DRL Off: 12.5v, On: 2.5v
22-13	---	Not Used	---
22-14	RD/WT	Idle Switch (IDL) Input	Closed: 0v, Open: 12v
22-15	DG	Brake Pedal Position Switch	Brake Off: 0v, On: 12v
22-16	DB/YL	PSP Switch Signal	Straight: 12v, Turned: 0v
22-17	DB/DB	Air Conditioning Switch Signal	A/C On: 2.5v, Off: 12v
22-18	BK/DG	Cooling Fan Control	On: 1v, Off: 12v
22-19	OR/DB	Blower Motor Switch Signal	1st: 12v, 2nd on up: 1v
22-20	BK/DB	Rear Defroster Signal	Switch On: 12v, Off: 0.1v
22-21	RD/BK	Headlamp Switch Signal	Switch On: 12v, Off: 1v
22-22	BR/YL	A/T: Neutral Drive Switch	In 'N': 0v, Others: 12v
22-22	BR/YL	M/T: Clutch Pedal Switch	Clutch In: 0v, Out: 12v

Pin Connector Graphic

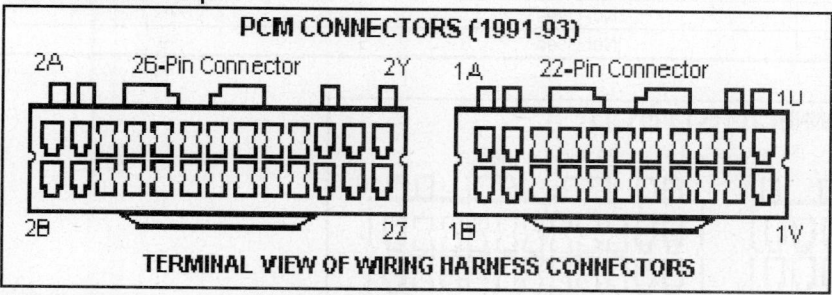

Standard Colors and Abbreviations

Abbreviation	Color	Abbreviation	Color	Abbreviation	Color
BK	Black	GY	Gray	PK	Purple
BL	Blue	GN	Green	RD	Red
BR	Brown	LG	LT Green	TN	Tan
DB	Dark Blue	OR	Orange	WT	White
DG	DK Green	PK	Pink	YL	Yellow

1993 Escort 1.8L I4 MFI VIN 8 (M/T) 26 Pin Connector

PCM Pin #	Wire Color	Circuit Description (26 Pin)	Value at Hot Idle
26-1	BK/OR	Power Ground	<0.1v
26-2	BK/OR	Power Ground	<0.1v
26-3	BK/LG	Power Ground	<0.1v
26-4	BK/WT	Analog Signal Return	<0.050v
26-5	WT	CKP Sensor Signal	2.5v (at Key On: 0 or 5v)
26-6	---	Not Used	---
26-7	YL/DB	CID Sensor Signal	2.5v (at Key On: 0 or 5v)
26-8	BK	Ground (GND)	<0.050v
26-9, 26-10	---	Not Used	---
26-11	LG/RD	Reference Voltage	4.9-5.1v
26-12	LG/WT	A/C WOT or Heater Relay	A/C On: 12v, Off: 2.5v
26-13	---	Not Used	---
26-14	RD/DB	HO2S-11 (B1 S1) Signal	0.1-1.1v
26-15	RD	Vane Airflow Meter Signal	0.1-0.3v
26-16	RD/BK	IAT Sensor Signal	1.5-2.5v
26-17	DB/WT	ECT Sensor Signal	0.5-0.6v
26-18	---	Not Used	---
26-19	BK/RD	High Speed Inlet Air Control	Over 5000 rpm: 12v
26-20	DG/OR	Fuel Pressure Regulator	Startup: <3v, others: 12v
26-21	YL	Fuel Injectors 1 & 3 - Bank 1	3.6 ms
26-22	YL/BK	Fuel Injectors 2 & 4 - Bank 2	3.6 ms
26-23	DB/OR	Idle Air Control Valve	8-10v
26-24	WT/DB	EVAP Purge Solenoid	Idle: 12-14v, off-idle: 1v
26-25	---	Not Used	---
26-26	---	Not Used	---

Pin Connector Graphic

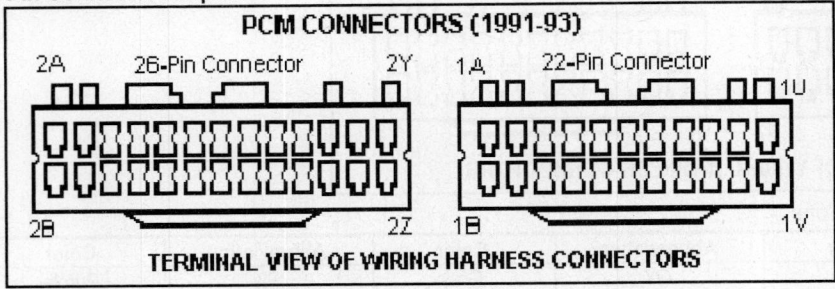

PCM CONNECTORS (1991-93)

2A   26-Pin Connector   2Y   1A   22-Pin Connector   1U

2B   2Z   1B   1V

TERMINAL VIEW OF WIRING HARNESS CONNECTORS

Standard Colors and Abbreviations

Abbreviation	Color	Abbreviation	Color	Abbreviation	Color
BK	Black	GY	Gray	PK	Purple
BL	Blue	GN	Green	RD	Red
BR	Brown	LG	LT Green	TN	Tan
DB	Dark Blue	OR	Orange	WT	White
DG	DK Green	PK	Pink	YL	Yellow

1994-95 Escort 1.8L I4 MFI VIN 8 (A/T) 22 Pin Connector

PCM Pin #	Wire Color	Circuit Description (22 Pin)	Value at Hot Idle
22-1	DB/RD	Keep Alive Power	12- 14v
22-2	PK	Vehicle Start Signal	Cranking: 9-11v
22-3	YL/BK	MIL (lamp) Control	MIL On: 1v, Off: 12v
22-4	DG/WT	CKP Sensor Signal	2.5v (at Key On: 0v or 5v)
22-5	LG/YL	Data Link Connector	Digital Signals
22-6	LG/YL	Data Link Connector	Digital Signals
22-6	BK/YL	Ignition On Power (Canada)	12-14V
22-7	DG	VSS (+) Signal	Vehicle moving: 0-5v
22-8	DGN/BK	A/C Switch Signal	Switch On: 2.5v, Off: 12v
22-9	DG	Brake Pedal Position Switch	Brake Off: 0v, On: 12v
22-10	---	Not Used	---
22-11	---	Not Used	---
22-12	WT/RD	Vehicle Power (PCM Relay)	12-14v
22-13	WT/YL	Data Link Connector	Digital Signals
22-14	WT/BK	Data Link Connector	Digital Signals
22-15	RD/BK	Park Lamp Signal	Switch On: 12v, Off: 1v
22-16	BK/DB	Rear Defroster Signal	Switch On: 12v, Off: 0.1v
22-17	DB/BK	AC/Heater WOT Relay	On: 1v, Off: 12v
22-18	DB/YL	PSP Switch Signal	Wheel Turning: 0v
22-19	OR/DB	Cooling Fan Control	On: 1v, Off: 12v
22-20	BK/DB	Ignition (Hot in Start)	KOEC: 9-11v
22-21	RD/WT	TP Idle Switch Signal	Closed: 0v, Open: 12v
22-22	---	Not Used	---

1994-95 Escort 1.8L I4 MFI VIN 8 (A/T) 16 Pin Connector

PCM Pin #	Wire Color	Circuit Description (16 Pin)	Value at Hot Idle
16-1	WT	CKP Sensor Signal	2.5v (at Key On: 0v or 5v)
16-2	RD/DB	HO2S-11 (B1 S1) Signal	0.1-1.1v
16-3	DB/WT	ECT Sensor Signal	0.5-0.6v
16-4	WT/BK	TOT Sensor Signal	3.5v at 68°F
16-5	LG/RD	Reference Voltage	4.9-5.1v
16-6	RD/BK	IAT Sensor Signal	1.5-2.5v
16-7	WT/DB	TSS Sensor (+)	340-380 rpm
16-8	WT/DB	EVAP Purge Solenoid	12v, 55 mph: 1v
16-9	RD	Vane Airflow Meter Signal	0.2v
16-10	BK/DG	Cooling Fan Signal	On: 1v, Off: 12v
16-11	LG/WT	TP Sensor Signal	0.4-1.0v
16-12	---	Not Used	---
16-13	YL/DB	CID Sensor	2.5v (at Key On: 0v or 5v)
16-14	DB	VSS (-) Signal	Vehicle moving: 0-5v
16-15	YL/DB	TSS Sensor (-)	340-380 rpm
16-16	---	Not Used	---

Standard Colors and Abbreviations

Abbreviation	Color	Abbreviation	Color	Abbreviation	Color
BK	Black	GY	Gray	PK	Purple
BL	Blue	GN	Green	RD	Red
BR	Brown	LG	LT Green	TN	Tan
DB	Dark Blue	OR	Orange	WT	White
DG	DK Green	PK	Pink	YL	Yellow

1994-95 Escort 1.8L I4 MFI VIN 8 (A/T) 26 Pin Connector

PCM Pin #	Wire Color	Circuit Description (26 Pin)	Value at Hot Idle
26-1	BK/OR	Power Ground	<0.1v
26-2	BK/LG	Power Ground	<0.1v
26-3	YL	TR Overdrive Signal	O/D On: 1v, Off: 12v
26-4	YL/WT	TR Sensor Low	In Low: 12v, Others: 1v
26-5	BK/RD	High Speed Inlet Air Control	Over 4900 rpm: 12v
26-6	---	Not Used	---
26-7	DG/OR	Fuel Pressure Regulator	Startup: <3v, others: 12v
26-8	---	Not Used	---
26-9	DB/OR	Idle Air Control Valve	8-10v
26-10	---	Not Used	---
26-11	YL	Fuel Injectors 1 & 3 (Bank 1)	3.6 ms
26-12	DB/OR	Shift Solenoid 1 Control	SS1 On: 1v, Off: 12v
26-13	OR	Shift Solenoid 3 Control	SS3 On: 1v, Off: 12v
26-14	BK/OR	Power Ground	<0.1v
26-15	BK/WT	Analog Signal Return	<0.050v
26-16	BR/WT	Daylight Running Lamps	DRL On: 1v, DRL Off: 12v
26-17	YL/RD	TR Sensor Drive	In 'D': 12v, Others: 1v
26-18	---	Not Used	---
26-19	---	Not Used	---
26-20	---	Not Used	---
26-21	---	Not Used	---
26-22	---	Not Used	---
26-23	---	Not Used	---
26-24	YL/BK	Fuel Injectors 2 & 4 (Bank 2)	3.6 ms
26-25	DB/YL	Shift Solenoid 2 Control	SS2 On: 1v, Off: 12v
26-26	DB	TCC Solenoid	Shifting 12v, Others: 1v

Pin Connector Graphic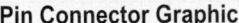

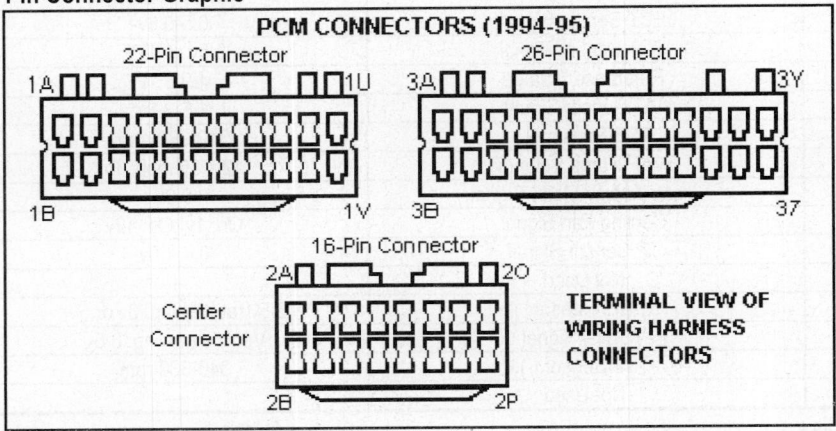

Standard Colors and Abbreviations

Abbreviation	Color	Abbreviation	Color	Abbreviation	Color
BK	Black	GY	Gray	PK	Purple
BL	Blue	GN	Green	RD	Red
BR	Brown	LG	LT Green	TN	Tan
DB	Dark Blue	OR	Orange	WT	White
DG	DK Green	PK	Pink	YL	Yellow

1994-95 Escort 1.8L I4 MFI VIN 8 (M/T) 22 Pin Connector

PCM Pin #	Wire Color	Circuit Description (22 Pin)	Value at Hot Idle
22-A	DB/RD	Keep Alive Power	12- 14v
22-B	PK	Vehicle Start Signal	Cranking: 9-11v
22-C	YL/BK	MIL (lamp) Control	MIL On: 1v, Off: 12v
22-D	DG/WT	Spark Output Signal	50% duty cycle
22-E	LG/YL	Self-Test Input Signal	STI On: 0v, STI Off: 5v
22-F	---	Not Used	---
22-G	---	Not Used	---
22-H	DG	Brake Pedal Position Switch	Brake Off: 0v, On: 12v
22-I	DGN/BK	A/C Switch Signal	Switch On: 12v, Off: 1.5v
22-J	OR/DB	Blower Motor Switch Signal	1st: 12v, others: 1v
22-K	RD/BK	Headlamp Switch Signal	Switch On: 12v, Off: 1v
22-L	WT/RD	Vehicle Power (PCM Relay)	12-14v
22-M	WT/YL	Data Link Connector	Digital Signals
22-N	WT/BK	Data Link Connector	Digital Signals
22-O	---	Not Used	---
22-P	DB/BK	A/C & Heater WOT Relay	On: 1v, Off: 12v
22-Q	BR/WT	Daytime Running Lamp	DRL Off: 12v, On: 2.5v
22-R	RD/WT	Idle Switch Signal	Closed: 0V, Open: 12v
22-S	DB/YL	PSP Switch Signal	Straight: 12v, Turned: 0v
22-T	BK/DG	Cooling Fan Control	On: 1v, Off: 12v
22-U	BK/DB	Rear Defroster Switch Signal	AC On: 12v, Off: 0v
22-V	BR/YL	A/T: Neutral Drive Switch	In 'N': 0v, Others: 12v
22-V	BR/YL	M/T: Clutch Engage Switch	Clutch In: 0v, Out: 12v

Pin Connector Graphic

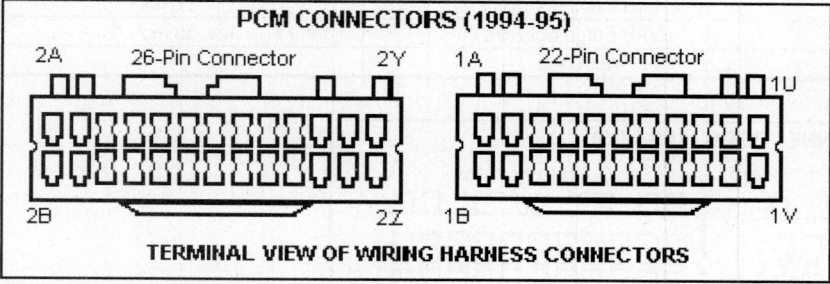

Standard Colors and Abbreviations

Abbreviation	Color	Abbreviation	Color	Abbreviation	Color
BK	Black	GY	Gray	PK	Purple
BL	Blue	GN	Green	RD	Red
BR	Brown	LG	LT Green	TN	Tan
DB	Dark Blue	OR	Orange	WT	White
DG	DK Green	PK	Pink	YL	Yellow

1994-95 Escort 1.8L I4 MFI VIN 8 (M/T) 26 Pin Connector

PCM Pin #	Wire Color	Circuit Description (26 Pin)	Value at Hot Idle
26-A	BK/OR	Power Ground	<0.1v
26-B	BK/LG	Power Ground	<0.1v
26-C	WT	CKP Sensor Signal	2.5v (at Key On: 0-5v)
26-D	YL/DB	CID Sensor Signal	2.5v (at Key On: 0-5v)
26-E	---	Not Used	---
26-F	LG/RD	Reference Voltage	4.9-5.1v
26-G	---	Not Used	---
26-H	RD	Vane Airflow Meter Signal	0.1-0.3v
26-I	DB/WT	ECT Sensor Signal	0.5-0.6v
26-J	BK/RD	High Speed Inlet Air Control	Over 5000 rpm: 12v
26-K	YL	Fuel Injectors 1 & 3 (Bank 1)	3.6 ms
26-L	DB/OR	Idle Air Control Valve	8-10v
26-M	---	Not Used	---
26-N	BK/OR	Power Ground	<0.1v
26-O	BK/WT	Analog Signal Return	<0.050v
26-P	---	Not Used	---
26-Q ('94)	BK	Power Ground	<0.1v
26-Q ('95)	BK/YL	Ignition Power	12-14v
26-R	---	Not Used	---
26-S	LG/WT	TP Sensor Signal	0.4-1.0v
26-T	RD/DB	HO2S-11 (B1 S1) Signal	0.1-1.1v
26-U	RD/BK	IAT Sensor Signal	1.5-2.5v
26-V	---	Not Used	---
26-W	DG/OR	Fuel Pressure Regulator	Startup: <3v, others: 12v
26-X	YL/BK	Fuel Injectors 2 & 4 (Bank 2)	3.6 ms
26-Y	WT/DB	EVAP Purge Solenoid	12v, 55 mph: 1v
26-Z	---	Not Used	---

Pin Connector Graphic

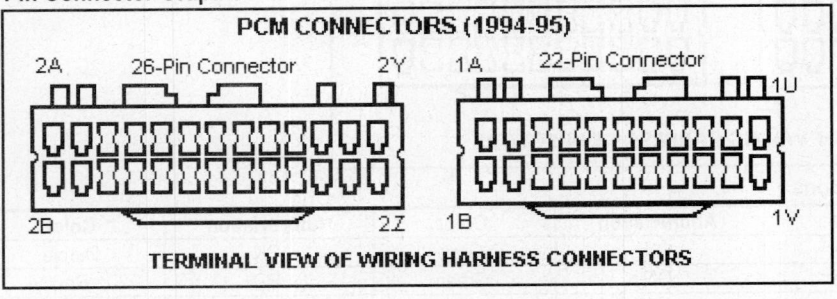

1996 Escort 1.8L I4 MFI VIN 8 (M/T) 22 Pin Connector

PCM Pin #	Wire Color	Circuit Description (22 Pin)	Value at Hot Idle
C1-1	---	Not Used	---
C1-2	PK	Vehicle Start	KOEC: 9-11v
C1-3	YL/BK	MIL (lamp) Control	MIL On: 0v, Off: 12v
C1-4	BL/BK	A/C WOT Relay Control	On: 1v, Off: 12v
C1-5	LG/YL	Self-Test Input Signal	STI On: 0v, Off: 5v
C1-6	GN/BK	A/C High Pressure Switch	Switch On: 2.5v, Off: 12v
C1-7	GN	VSS (+) Signal	0 Hz, 55 mph: 125 Hz
C1-8	BL	VSS (-) Signal	0 Hz, 55 mph: 125 Hz
C1-9	GN	Brake Pedal Position Switch	Brake Off: 0v, On: 12v
C1-10	---	Not Used	---
C1-11	LG	Fuel Pump Control	On: 1v, Off: 12v
C1-12	---	Not Used	---
C1-13	BL/WT	ISO K-Line for OBD II	Digital Signals
C1-14	---	Not Used	---
C1-15	RD/BK	Headlamp Switch Input	Switch On: 12v, Off: 1v
C1-16	GY/RD	Rear Defroster Switch Signal	DEF On: 12v, Off: 1v
C1-17	BR/YL	A/T: Neutral Drive Switch	In 'N': 0v, Others: 5v
C1-17	BR/YL	M/T: Clutch Pedal Switch	Clutch In: 0v, Out: 5v
C1-18	BR/WT	Daytime Running Lamp Relay	Relay On: 0v, Off: 12v
C1-19	OR/BL	Blower Motor Control Switch	1st: 12v, others: 1v
C1-20	BK/GN	Cooling Fan Control	On: 1v, Off: 12v
C1-21	---	Not Used	---
C1-22	---	Not Used	---

1996 Escort 1.8L I4 MFI VIN 8 (M/T) 16 Pin Connector

PCM Pin #	Wire Color	Circuit Description (16 Pin)	Value at Hot Idle
C3-1	---	Not Used	---
C3-2	RD/BL	HO2S-11 (B1 S1) Signal	0.1-1.1v
C3-3	---	Not Used	---
C3-4	BL/WT	ECT Sensor Signal	0.5-0.6v
C3-5	LG/RD	Reference Voltage	4.9-5.1v
C3-6	RD/BK	IAT Sensor Signal	1.5-2.5v
C3-7	GN/WT	HO2S-11 (B1 S1) Heater	On: 1v, Off: 12v
C3-8	BK/BL	Analog Return Signal	<0.1v
C3-9	RD	MAF Sensor Signal	0.8, 55 mph: 1.9v
C3-10	RD/YL	HO2S-12 (B1 S2) Signal	0.1-1.1v
C3-11	LG/WT	TP Sensor Signal	0.1v, 55 mph: 2.1v
C3-12	---	Not Used	---
C3-13	---	Not Used	---
C3-14	RD/WT	Idle Switch Signal	Closed: 0v, Open: 12v
C3-15	---	Not Used	---
C3-16	BL/YL	PSP Switch Signal	Straight: 12v, Turned: 0v

Standard Colors and Abbreviations

Abbreviation	Color	Abbreviation	Color	Abbreviation	Color
BK	Black	GY	Gray	PK	Purple
BL	Blue	GN	Green	RD	Red
BR	Brown	LG	LT Green	TN	Tan
DB	Dark Blue	OR	Orange	WT	White
DG	DK Green	PK	Pink	YL	Yellow

1996 Escort 1.8L I4 MFI VIN 8 (M/T) 26 Pin Connector

PCM Pin #	Wire Color	Circuit Description (26 Pin)	Value at Hot Idle
C4-1	BK/LG	Power Ground	<0.1v
C4-2	BK/OR	Power Ground	<0.1v
C4-3	OR	CKP Sensor (-) Signal	435-475 Hz
C4-4	YL/BL	CMP 1 Signal	5-7 Hz
C4-5	BL/RD	Keep Alive Power	12-14v
C4-6	BK/RD	High Speed Inlet Air Solenoid	Below 5000 rpm: 1.5v
C4-7	---	Not Used	---
C4-8	---	Not Used	---
C4-9	BL/OR	Idle Air Control Solenoid	Varies: 8-10v
C4-10	---	Not Used	---
C4-11	YL	Injector 1 Control	4.0-4.5 ms
C4-12	GN/RD	Injector 3 Control	4.0-4.5 ms
C4-13	---	Not Used	---
C4-14	WT/RD	Vehicle Power	12-14v
C4-15	BK/OR	Power Ground	<0.1v
C4-16	WT	CMP 2 Signal	5-7 Hz
C4-17	LG/WT	CKP Sensor (+) Signal	435-475 Hz
C4-18	GN/OR	Fuel Pressure Regulator	Startup: <3v, others: 12v
C4-19	---	Not Used	---
C4-20	GN/WT	Spark Output Signal	50% duty cycle
C4-21	---	Not Used	---
C4-22	---	Not Used	---
C4-23	WT/BL	EVAP Purge Solenoid	12v, 55 mph: 1v
C4-24	YL/BK	Injector 2 Control	4.0-4.5 ms
C4-25	GN/YL	Injector 4 Control	4.0-4.5 ms
C4-26	---	Not Used	---

Pin Connector Graphic

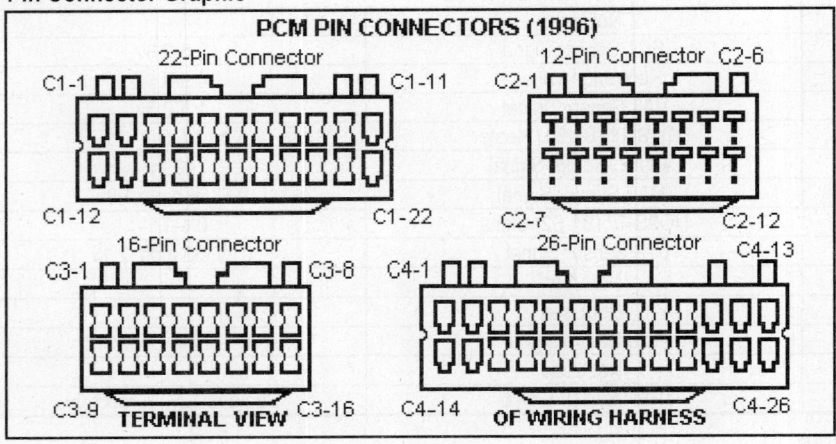

1990 Escort 1.9L I4 CFI VIN 9 (All) 60 Pin Connector

PCM Pin #	Wire Color	Circuit Description (60 Pin)	Value at Hot Idle
1	YL	Keep Alive Power	12-14v
2-3	---	Not Used	---
4	DG/YL	Ignition Diagnostic Monitor	20-31 Hz
5	RD/LG	Keep Alive Power	12-14v
6, 9	---	Not Used	---
7	LG/YL	ECT Sensor Signal	0.5-0.6v
8	PK/BK	Fuel Pump Monitor	On: 1v, Off: 12v
10	BK/YL	A/C Cycling Clutch Switch	A/C On: 12v, Off: 0v
11-15	---	Not Used	---
16	BK/OR	Ignition System Ground	<0.050v
17	TN/RD	STI Output, MIL Control	MIL On: 1v, Off: 12v
18-19	---	Not Used	---
20	BK	PCM Case Ground	<0.050v
21	BR/WT	ISC Motor (+) Control	0.3v
22	TN/LG	Fuel Pump Control	On: 1v, Off: 12v
23-24	---	Not Used	---
25	LG/PK	ACT Sensor Signal	1.5-2.5v
26	OR/WT	Reference Voltage	4.9-5.1v
27	BR/LG	PFE EGR Sensor	3.2v, 55 mph: 2.8v
28	LG/WT	Idle Tracking Switch	10.1v, 55 mph: 0v
29	DG/PK	HO2S-11 (B1 S1) Signal	0.1-1.1v
30	GY/OR	Neutral Drive Switch Signal	In 'N': 0v, Others: 5v
31-34	---	Not Used	---
35	GY/YL	EVAP Purge Solenoid	12v, 55 mph: 1v
36	YL/LG	Spark Output Signal	50% duty cycle
37	RD	Vehicle Power	12-14v
38-39	---	Not Used	---
40	BK/LG	Power Ground	<0.1v
41	WT/BL	ISC Motor (-) Control	0.3v
42-44	---	Not Used	---
45	LG/BK	MAP Sensor	107 Hz
46	BK/WT	Analog Signal Return	<0.050v
47	DG/LG	TP Sensor Signal	0.5-1.2v
48	WT/RD	Self-Test Indicator Signal	STI On: <0.1v, Off: 5v
49	OR	HO2S-11 (Bank 1) Ground	<0.050v
50	---	Not Used	---
51	TN/BL	M/T: Shift Indicator Light	SIL On: 1v, Off: 12v
52	YL	EGR VR Solenoid	0%, 55 mph: 45%
53	---	Not Used	---
54	BK/YL	A/C WOT Relay Control	On: 1v, Off: 12v
55, 59	---	Not Used	---
56	DB	PIP Sensor Signal	50% duty cycle
57	RD	Vehicle Power	12-14v
58	TN/RD	CFI Fuel Injector	1.5 ms
60	BK/LG	Power Ground	<0.1v

Pin Connector Graphic

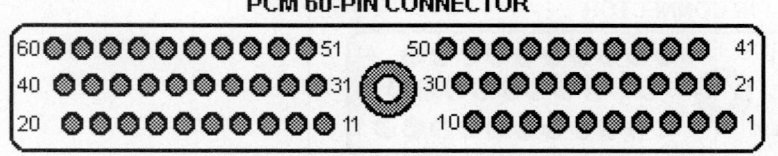

PCM 60-PIN CONNECTOR

Terminal View of 60-Pin PCM Harness Connector

1990 Escort 1.9L I4 MFI VIN J (All) 60 Pin Connector

PCM Pin #	Wire Color	Circuit Description (60 Pin)	Value at Hot Idle
1	YL	Keep Alive Power	12-14v
2-3	---	Not Used	---
4	DG/YL	Ignition Diagnostic Monitor	20-31 Hz
5-6	---	Not Used	---
7	LG/YL	ECT Sensor Signal	0.5-0.6v
8	PK/BK	Fuel Pump Monitor	On: 12v, Off: 0v
9, 11-15	---	Not Used	---
10	BK/YL	A/C Cycling Clutch Switch	A/C On: 12v, Off: 0v
16	BK/OR	Ignition System Ground	<0.050v
17	TN/BL	M/T: Shift Indicator Light	SIL On: 1v, Off: 12v
18-19	---	Not Used	---
20	BK	PCM Case Ground	<0.050v
21	OR/BK	IAC Motor Control	Idle: 9.5-10.3v
22	TN/LG	Fuel Pump Control	On: 1v, Off: 12v
23-24	---	Not Used	---
25	LG/PK	VAT Sensor	1.5-2.5v
26	OR/WT	Reference Voltage	4.9-5.1v
27-28	---	Not Used	---
29	DG/PK	HO2S-11 (B1 S1) Signal	0.1-1.1v
30	GY/OR	A/T: Neutral Drive Switch	In 'N': 0v, Others: 5v
30	GY/OR	M/T: Clutch Engage Switch	Clutch In: 0v, Out: 5v
31	---	Not Used	---
32	GY/YL	EVAP Purge Solenoid	12v, 55 mph: 1v
33-34	---	Not Used	---
35	YL	EGR Shutoff Solenoid Control	0%, 35 mph: 38%
36	YL/LG	Spark Output Signal	50% duty cycle
37	RD	Vehicle Power	12-14v
38-39	---	Not Used	---
40	BK/LG	Power Ground	<0.1v
41-42	---	Not Used	---
43	WT/BK	Vane Airflow Meter	0.7v
44	---	Not Used	---
45	LG/BK	BARO Sensor Signal	159 Hz (sea level)
46	BK/WT	Analog Signal Return	<0.050v
47	DG/LG	TP Sensor Signal	0.5-1.2v
48	WT/RD	Self-Test Indicator Signal	STI Open: 5v, Closed: 1v
49	OR	HO2S-11 (Bank 1) Ground	<0.050v
50-52, 55	---	Not Used	---
53 ('89-'90)	TN/RD	MIL (lamp) Control	MIL On: 1v, Off: 12v
54	OR/BL	A/C WOT Relay Control	On: 1v, Off: 12v
56	DB	PIP Sensor Signal	50% duty cycle
57	RD	Vehicle Power	12-14v
58	TN/RD	Injector Bank 1 (INJ 1 & 2)	4.8-5.0 ms
59	TN/OR	Injector Bank 2 (INJ 3 & 4)	4.8-5.0 ms
60	BK/LG	Power Ground	<0.1v

Pin Connector Graphic

PCM 60-PIN CONNECTOR

Terminal View of 60-Pin PCM Harness Connector

1991-92 Escort 1.9L I4 MFI VIN J (All) 60 Pin Connector

PCM Pin #	Wire Color	Circuit Description (60 Pin)	Value at Hot Idle
1	BL/RD	Keep Alive Power	12-14v
3	WT/BK	VSS (+) Signal	0 Hz, 55 mph: 125 Hz
4	RD	Ignition Diagnostic Monitor	20-31 Hz
2, 5	---	Not Used	---
6 ('91)	BL	VSS (-) Signal	0 Hz, 55 mph: 125 Hz
6 ('92)	DB	VSS (-) Signal	0 Hz, 55 mph: 125 Hz
7	BL/WT	ECT Sensor Signal	0.5-0.6v
8	YL/GN	Data Bus (-) Signal	Digital Signals
9	GN/YL	Mass Airflow Return	<0.050v
10	GN/BK	A/C Cycling Clutch Switch	A/C On: 12v, Off: 0v
11	---	Not Used	---
12	YL/OR	Injector 3 Control	4.8-5.0 ms
13	GN/OR	Injector 4 Control	4.8-5.0 ms
14	---	Not Used	---
15 ('92)	OR/WT	EVAP Purge Solenoid	12v, 55 mph: 1v
16	RD/BL	Ignition System Ground	<0.050v
17	YL/BK	STI Output, MIL Control	MIL On: 1v, Off: 12v
18, 20	---	Not Used	---
19	BK/PK	Fuel Pump Monitor	On: 12v, Off: 0v
21	BL/OR	IAC Motor Control	9.7-12.0v
22	LG	Fuel Pump Control	On: 1v, Off: 12v
23-24	---	Not Used	---
25	WT/GN	ACT Sensor Signal	1.5-2.5v
26	LG/WT	Reference Voltage	4.9-5.1v
27	BL/YL	PFE EGR Sensor	3.2v, 55 mph: 2.8v
28	YL/BL	Data Bus (+) Signal	Digital Signals
29	GN/BL	HO2S-11 (B1 S1) Signal	0.1-1.1v
30	BR/YL	A/T: Neutral Drive Switch	In 'P': 0v, Others 5v
30	BR/YL	M/T: Clutch Engage Switch	Clutch In: 0v, Out: 12v
31	RD/BK	High Speed Fan Control	On: 1v, Off: 12v
32	---	Not Used	---
33	WT/BL	EGR VR Solenoid	0%, 55 mph: 45%
34	LG/BK	Octane Adjust Switch 2	Closed: 0v, Open: 9.1v
35	YL/WT	Low Speed Fan Control	On: 1v, Off: 12v
36	LG/WT	Spark Angle Word Signal	50% duty cycle
37	WT/RD	Vehicle Power	12-14v
38-39	---	Not Used	---
40	BK/GN	Power Ground	<0.1v

Pin Connector Graphic

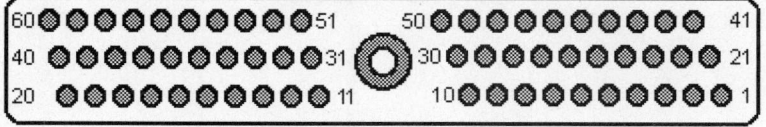

Standard Colors and Abbreviations

Abbreviation	Color	Abbreviation	Color	Abbreviation	Color
BK	Black	GY	Gray	PK	Purple
BL	Blue	GN	Green	RD	Red
BR	Brown	LG	LT Green	TN	Tan
DB	Dark Blue	OR	Orange	WT	White
DG	DK Green	PK	Pink	YL	Yellow

1991-92 Escort 1.9L I4 MFI VIN J (All) 60 Pin Connector

PCM Pin #	Wire Color	Circuit Description (60 Pin)	Value at Hot Idle
41	OR	CID+ Sensor	400-425 Hz
42	BL/GN	CID- Sensor	400-425 Hz
43	GN/WT	Octane Adjust Switch 1	Closed: 0v, Open: 9.1v
44-45	---	Not Used	---
46	LG/BK	Analog Signal Return	<0.050v
47	RD/WT	TP Sensor Signal	Hot 0.5-1.2v
48	LG/YL	Self-Test Indicator Signal	STI Open: 5v, Closed: 1v
49	BK	HO2S-11 (Bank 1) Ground	<0.050v
49	OR/BK	HO2S-11 (Bank 1) Ground	<0.050v
50	BR/BK	MAF Sensor Signal	0.6-0.9v
51-52	---	Not Used	---
53	LG/RD	Shift Indicator Control	SIL On: 1v, Off 12v
54	BL/BK	A/C WOT Relay Control	On: 1v, Off: 12v
55	---	Not Used	---
56	GN/WT	PIP Sensor Signal	50% dwell
57	WT/RD	Vehicle Power	12-14v
58	YL	Injector Bank 1 (INJ 1 & 2)	4.8-5.0 ms
59	GN/RD	Injector Bank 2 (INJ 3 & 4)	4.8-5.0 ms
60	BK/GN	Power Ground	<0.1v

Pin Connector Graphic

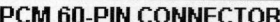
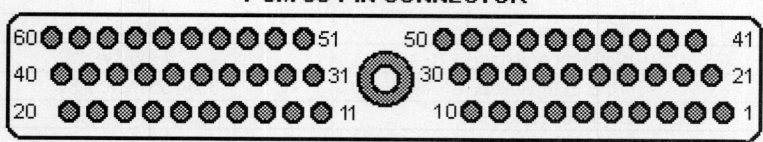

PCM 60-PIN CONNECTOR

Terminal View of 60-Pin PCM Harness Connector

Standard Colors and Abbreviations

Abbreviation	Color	Abbreviation	Color	Abbreviation	Color
BK	Black	GY	Gray	PK	Purple
BL	Blue	GN	Green	RD	Red
BR	Brown	LG	LT Green	TN	Tan
DB	Dark Blue	OR	Orange	WT	White
DG	DK Green	PK	Pink	YL	Yellow

1993-95 Escort 1.9L I4 MFI VIN J (All) 60 Pin Connector

PCM Pin #	Wire Color	Circuit Description (60 Pin)	Value at Hot Idle
1	BL/RD	Keep Alive Power	12-14v
2	WT	TOT Sensor Signal	2.10-2.40v
3	WT/BK	VSS (+) Signal	0 Hz, 55 mph: 125 Hz
4	RD	Ignition Diagnostic Monitor	20-31 Hz
5	GN	Brake Pedal Position Switch	Brake Off: 0v, On: 12v
6	BL	VSS (-) Signal	0 Hz, 55 mph: 125 Hz
6 ('94-'95)	DB	VSS (-) Signal	0 Hz, 55 mph: 125 Hz
7	BL/WT	ECT Sensor Signal	0.5-0.6v
8	YL/GN	Data Bus (-) Signal	Digital Signals
9	GN/YL	Mass Airflow Return	<0.050v
10	GN/BK	A/C Cycling Clutch Switch	A/C On: 12v, Off: 0v
11	LG/BL	Shift Solenoid 1 Control	12v, 55 mph: 1v
12	YL/OR	Injector 3 Control	4.8-5.1 ms
13	GN/OR	Injector 4 Control	4.8-5.1 ms
14	---	Not Used	---
15	OR/WT	EVAP Purge Solenoid	12v, 55 mph: 9v
16	RD/BL	Ignition System Ground	<0.1v
17	YL/BK	STI Output, MIL Control	MIL On: 1v, Off: 12v
18	BR/RD	TR Sensor Drive	12v, 55 mph: 12v
19	BK/PK	Fuel Pump Monitor	On: 12v, Off: 0v
20	BK	PCM Case Ground	<0.050v
21	OR/BK	IAC Motor Control	9.7-11.1v
22	LG	Fuel Pump Control	On: 1v, Off: 12v
23	RD/GN	TR Sensor Reverse	12v, 55 mph: 0v
24	WT/BL	A/T: TSS (+) Sensor	42-50 Hz (680-720 rpm)
25	WT/GN	IAT Sensor Signal	1.5-2.5v
26	LG/WT	Reference Voltage	4.9-5.1v
27	BL/YL	PFE EGR Sensor	3.2v, 55 mph: 3.4v
28	YL/BL	Data Bus (+) Signal	Digital Signals
29	GN/BL	HO2S-11 (B1 S1) Signal	0.1-1.1v
30	BK/BL	A/T: Neutral Drive Switch	In 'P': 0v, Others: 5v
30	BR/YL	A/T: Neutral Drive Switch	In 'P': 0v, Others: 5v
30	BK/BL	M/T: Clutch Engage Switch	Clutch In: 0v, Out 5v
30	BR/YL	M/T: Clutch Engage Switch	Clutch In: 0v, Out 5v
31	RD/BK	High Speed Fan Control	On: 1v, Off: 12v
32-34	---	Not Used	---
33	WT/BL	EGR VR Solenoid	0%, 55 mph: 45%
35	YL/WT	Low Speed Fan Control	On: 1v, Off: 12v
36	LG/WT	Spark Output Signal	50% duty cycle
37	WT/RD	Vehicle Power	12-14v
38	BR/BL	TR Sensor Overdrive	12v, 55 mph: 0v
39	---	Not Used	---
40	BK/GN	Power Ground	<0.1v

Pin Connector Graphic

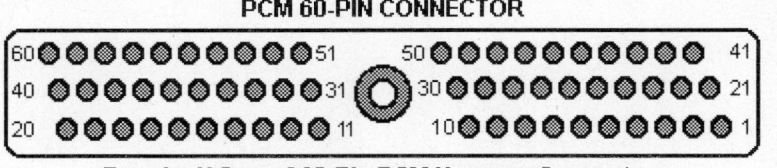

PCM 60-PIN CONNECTOR

Terminal View of 60-Pin PCM Harness Connector

1993-95 Escort 1.9L I4 MFI VIN J (All) 60 Pin Connector

PCM Pin #	Wire Color	Circuit Description (60 Pin)	Value at Hot Idle
41	OR	CID+ Sensor	5-7 Hz
42	BL/GN	CID- Sensor	<0.050v
43	GN/WT	Octane Adjust Switch 1	Closed: 0v, Open: 9.1v
44	YL/BL	A/T: TSS Sensor (-)	44-56 Hz (680-710 rpm)
45	BR/GN	TR Sensor Low	12v, 55 mph: 0v
46	LG/BK	Analog Signal Return	<0.050v
47	RD/WT	TP Sensor Signal	0.5-1.2v
48	LG/YL	Self-Test Indicator Signal	STI Open: 5v, Closed: 1v
49	BK	HO2S-11 (Bank 1) Ground	<0.050v
49	OR/BK	HO2S-11 (Bank 1) Ground	<0.050v
50	BR/BK	MAF Sensor Signal	0.7v
51	PK/BK	Shift Solenoid 2 Control	1v, 55 mph: 1v
52	OR/GN	Shift Solenoid 3 Control	12v, 55 mph: 12v
53	LG/RD	M/T: Shift Indicator Light	SIL On: 1v, Off: 12v
54	BL/BK	A/C WOT Relay Control	On: 1v, Off: 12v
55	PK/WT	TCC Solenoid Control	12v, 55 mph: 9v
56	GN/WT	PIP Sensor Signal	50% dwell
57	WT/RD	Vehicle Power	12-14v
58	YL	Injector 1 Control	4.8-5.1 ms
59	GN/RD	Injector 2 Control	4.8-5.1 ms
60	BK/GN	Power Ground	<0.1v
60	BK/GN	Power Ground	<0.1v

Pin Connector Graphic

PCM 60-PIN CONNECTOR

Terminal View of 60-Pin PCM Harness Connector

Standard Colors and Abbreviations

Abbreviation	Color	Abbreviation	Color	Abbreviation	Color
BK	Black	GY	Gray	PK	Purple
BL	Blue	GN	Green	RD	Red
BR	Brown	LG	LT Green	TN	Tan
DB	Dark Blue	OR	Orange	WT	White
DG	DK Green	PK	Pink	YL	Yellow

1996 Escort 1.9L I4 MFI VIN J (All) 104 Pin Connector

PCM Pin #	Wire Color	Circuit Description (104 Pin)	Value at Hot Idle
1	PK/BK	Shift Solenoid 2 Control	12v, 55 mph: 1v
2	YL/BK	MIL (lamp) Control	MIL On: 1v, Off: 12v
3-5	---	Not Used	---
6	BR/BK	TR Sensor Overdrive	In 'P': 0v, in O/D: 12v
7	BR/DG	TR Sensor Low	In 'P': 0v, in 1st: 12v
9	BR/YL	TR Sensor Drive	In 'P': 0v, in 2nd: 12v
8,10	---	Not Used	---
11	BL/PK	Purge Flow Sensor	1.55v, at 30 mph: 2.55v
13	PK/DG	Flash EEPROM Power	0.1v
12, 14	---	Not Used	---
15	BK/BL	Data Bus (-) Signal	Digital Signals
16	WT/BL	Data Bus (+) Signal	Digital Signals
17	RD/BK	High Speed Fan Control	On: 1v, Off: 12v
18-20	---	Not Used	---
21	BL/DG	CKP Sensor (+) Signal	435-475 Hz
22	OR	CKP Sensor (-) Signal	435-475 Hz
23	YL/BL	TSS Sensor (-)	340-380 Hz
24, 51	BK/YL	Power Ground	<0.1v
25	RD/BL	Power Ground	<0.1v
26	GY/YL	Coil Driver 1 Control	5° dwell
27	LG/BL	Shift Solenoid 1 Control	1v, 55 mph: 12v
28-29	---	Not Used	---
30	DG/WT	Octane Adjust Switch	Closed: 0v, Open: 9.3v
31, 34	---	Not Used	---
32	RD/GN	TR Sensor Reverse	In 'R': 12v, Others: 0v
33	BL	VSS (-) Signal	0 Hz, 55 mph: 125 Hz
35	GY	HO2S-12 (B1 S2) Signal	0.1-1.1v
36	OR/BL	MAF Sensor Return	<0.050v
37	WT	TFT Sensor Signal	2.10-2.40v
38	BL/WT	ECT Sensor Signal	0.5-0.6v
39	WT/GN	IAT Sensor Signal	1.5-2.5v
40	BK/PK	Fuel Pump Monitor	On: 12v, Off: 0v
41	GN/RD	A/C Cycling Clutch Switch	A/C On: 12v, Off: 0v
42-44, 46	---	Not Used	---
45	PK/WT	Low Speed Fan Control	On: 1v, Off: 12v
47	PK/RD	EGR VR Solenoid	0%, 55 mph: 45%
48	LG/RD	Clean Tachometer Output	25-38 Hz
49-50	---	Not Used	---
52	BR/RD	Coil Driver 2 Control	5° dwell
53	OR/GN	Shift Solenoid 3 Control	12v, 55 mph: 1v
54	GY/WT	TCC Solenoid Control	0%, 55 mph: 50%
55	BL/RD	Keep Alive Power	12-14v
56-57	---	Not Used	---
58	WT/BK	VSS (+) Signal	0 Hz, 55 mph: 125 Hz
59	---	Not Used	---
60	GY/DG	HO2S-11 (B1 S1) Signal	0.1-1.1v
61-63	---	Not Used	---
64	RD	A/T: Neutral Drive Switch	In 'N': 0v, in Gear: 12v
64	BR/YL	M/T: Clutch Pedal Position	Clutch In: 0v, Out: 5v

1996 Escort 1.9L I4 MFI VIN J (All) 104 Pin Connector

PCM Pin #	Wire Color	Circuit Description (104 Pin)	Value at Hot Idle
65	BL/YL	DPFE Sensor Signal	0.20-1.30v
66	---	Not Used	---
67	OR/WT	EVAP Purge Solenoid	0-100%
68	---	Not Used	---
69	BL/BK	A/C WOT Relay Control	On: 1v, Off: 12v
70	---	Not Used	---
71	WT/RD	Vehicle Power	12-14v
72-73	---	Not Used	---
74	BR/WT	Injector 3 Control	4.0-4.5 ms
75	YL	Injector 1 Control	4.0-4.5 ms
76	BL	CID- Sensor	5-7 Hz
77	BK/YL	Power Ground	<0.1v
78-79	---	Not Used	---
80	LG	Fuel Pump Control	On: 1v, Off: 12v
81-82	---	Not Used	---
83	BL/OR	IAC Motor Control	10v (30% on time)
84	WT/BL	TSS Sensor (+) Signal	340-380 Hz
85	DG	CID+ Sensor	5-7 Hz
86	DGN/BK	A/C Pressure Sensor Signal	A/C On: 1.5-1.9v
87	---	Not Used	---
88	BR/BK	MAF Sensor Signal	0.6-0.9v
89	RD/WT	TP Sensor Signal	0.53-1.27v
90	LG/WT	Reference Voltage	4.9-5.1v
91	LG/BK	Analog Signal Return	<0.050v
92	DG	Brake Pedal Position Switch	Brake Off: 0v, On: 12v
93	OR/BK	HO2S-11 (B1 S1) Heater	On: 1v, Off: 12v
94, 96	---	Not Used	---
95	PK	HO2S-12 (B1 S2) Heater	On: 1v, Off: 12v
97	WT/RD	Vehicle Power	12-14v
98-99	---	Not Used	---
100	DG/OR	Injector 4 Control	4.0-4.5 ms
101	PK/YL	Injector 2 Control	4.0-4.5 ms
102	---	Not Used	---
103	BK/YL	Power Ground	<0.1v
104	---	Not Used	---

Pin Connector Graphic

PCM 104-PIN CONNECTOR

Terminal View of 104-Pin PCM Wiring Harness Connector

Standard Colors and Abbreviations

Abbreviation	Color	Abbreviation	Color	Abbreviation	Color
BK	Black	GY	Gray	PK	Purple
BL	Blue	GN	Green	RD	Red
BR	Brown	LG	LT Green	TN	Tan
DB	Dark Blue	OR	Orange	WT	White
DG	DK Green	PK	Pink	YL	Yellow

1997 Escort 2.0L I4 MFI VIN P (All) 104 Pin Connector

PCM Pin #	Wire Color	Circuit Description (104 Pin)	Value at Hot Idle
1	PK/BK	Shift Solenoid 2 Control	12v, 55 mph: 1v
2	YL/BK	MIL (lamp) Control	MIL On: 1v, Off: 12v
3-5	---	Not Used	---
6	BR/BK	TR Sensor Overdrive	In 'P': 0v, in O/D: 11v
7	BR/GN	TR Sensor Low	In Manual 1st: 12v
8	PK/BL	IMRC Monitor Signal	5v
9	BR/YL	TR Sensor Drive	In Drive: 0v
10	---	Not Used	---
11	BL/PK	EVAP Purge Flow Sensor	0.8v, 55 mph: 3.0v
12, 14	---	Not Used	---
13	PK/GN	Flash EEPROM Power	0.1v
15	BK/BL	Data Bus (-) Signal	Digital Signals
16	WT/BL	Data Bus (+) Signal	Digital Signals
17	RD/BK	High Speed Fan Control	On: 1v, Off: 12v
18-20		Not Used	
21	BK	CKP Sensor (+) Signal	400-425 Hz
22	WT	CKP Sensor (-) Signal	400-425 Hz
23	RD	TSS Sensor (-)	340-380 Hz
24, 51	BK/YL	Power Ground	<0.1v
25	RD/BL	Shield Ground	<0.050v
26	GN/YL	Coil Driver 1 Control	5° dwell
27	LG/BL	Shift Solenoid 1 Control	1v, 55 mph: 12v
28-29	---	Not Used	---
30	GN/WT	Octane Adjust Switch	Closed: 0v, Open: 9.3v
31	YL/RD	PSP Switch Signal	Straight: 0.6v, turned: 2v
32	RD/GN	TR Sensor Reverse	In Reverse: 12v
33	BL	VSS (-) Signal	0 Hz, 55 mph: 125 Hz
34	---	Not Used	---
35	GY	HO2S-12 (B1 S2) Signal	0.1-1.1v
36	OR/BL	MAF Sensor Return	<0.050v
37	OR	TFT Sensor Signal	2.10-2.40v
38	BL/WT	ECT Sensor Signal	0.5-0.6v
39	WT/GN	IAT Sensor Signal	1.5-2.5v
40	BK/PK	Fuel Pump Monitor	On: 12v, Off: 0v
41	GN/RD	A/C Cycling Clutch Switch	A/C On: 12v, Off: 0v
42	RD/YL	IMRC Solenoid Control	12v (off)
43-44	---	Not Used	---
45	PK/WT	Low Speed Fan Control	On: 1v, Off: 12v
46	---	Not Used	---
47	PK/RD	EGR VR Solenoid	0%, 55 mph: 45%
48	LG/RD	Clean Tachometer Output	25-38 Hz
49-50	---	Not Used	---
52	BR/RD	Coil Driver 2 Control	5° dwell
53	GY/YL	Shift Solenoid 3 Control	6.7v, 55 mph: 6.7v
54	GY/WT	TCC Solenoid Control	0%, 55 mph: 95%
55	BL/RD	Keep Alive Power	12-14v
56-57	---	Not Used	---
58	WT/BK	VSS (+) Signal	0 Hz, 55 mph: 125 Hz
60	GY/DG	HO2S-11 (B1 S1) Signal	0.1-1.1v
61-63	---	Not Used	---
64	RD	TR Sensor Signal	In 'P': 0v, 55 mph: 1.7v
64	BR/YL	Clutch Pedal Position Switch	Clutch In: 0v, Out: 5v

1997 Escort 2.0L I4 MFI VIN P (All) 104 Pin Connector

PCM Pin #	Wire Color	Circuit Description (104 Pin)	Value at Hot Idle
65	BL/YL	DPFE Sensor Signal	0.20-1.30v
66	---	Not Used	---
67	OR/WT	EVAP Purge Solenoid	0%, 55 mph: 50%
68	---	Not Used	---
69	BL/BK	A/C WOT Relay Control	On: 1v, Off: 12v
70	---	Not Used	---
71	WT/RD	Vehicle Power	12-14v
72-73	---	Not Used	---
74	BR/WT	Injector 3 Control	3.3-3.7 ms
75	YL	Injector 1 Control	3.3-3.7 ms
76	BL	CID- Sensor	5-7 Hz
77	BK/YL	Power Ground	<0.1v
78-79	---	Not Used	---
80	LG	Fuel Pump Control	On: 1v, Off: 12v
81-82	---	Not Used	---
83	BL/OR	IAC Motor Control	10v (30% on time)
84	WT	TSS Sensor (+) Signal	340-380 Hz
85	GN	CID+ Sensor	5-9 Hz
86	GN/BK	A/C Pressure Sensor Signal	A/C On: 1.2-1.5v
87	---	Not Used	---
88	BR/BK	MAF Sensor Signal	0.6-0.9v
89	RD/WT	TP Sensor Signal	0.53-1.27v
90	LG/WT	Reference Voltage	4.9-5.1v
91	LG/BK	Analog Signal Return	<0.050v
92	DG	Brake Pedal Position Switch	Brake Off: 0v, On: 12v
93	OR/BK	HO2S-11 (B1 S1) Heater	On: 1v, Off: 12v
94	---	Not Used	---
95	PK	HO2S-12 (B1 S2) Heater	On: 1v, Off: 12v
96	---	Not Used	---
97	WT/RD	Vehicle Power	12-14v
98-99	---	Not Used	---
100	GN/OR	Injector 4 Control	3.3-3.7 ms
101	PK/YL	Injector 2 Control	3.3-3.7 ms
102	---	Not Used	---
103	BK/YL	Power Ground	<0.1v
104	---	Not Used	---

Pin Connector Graphic

```
        PCM 104-PIN CONNECTOR
   1 ●●●●●●●●●●●●●    ●●●●●●●●●●●●● 26
  27 ●●●●●●●●●●●●●    ●●●●●●●●●●●●● 52
  53 ●●●●●●●●●●●●●  ⬤  ●●●●●●●●●●●●● 78
  79 ●●●●●●●●●●●●●    ●●●●●●●●●●●●● 104
   Terminal View of 104-Pin PCM Wiring Harness Connector
```

Standard Colors and Abbreviations

Abbreviation	Color	Abbreviation	Color	Abbreviation	Color
BK	Black	GY	Gray	PK	Purple
BL	Blue	GN	Green	RD	Red
BR	Brown	LG	LT Green	TN	Tan
DB	Dark Blue	OR	Orange	WT	White
DG	DK Green	PK	Pink	YL	Yellow

1998-99 Escort 2.0L I4 MFI VIN P (All) 104 Pin Connector

PCM Pin #	Wire Color	Circuit Description (104 Pin)	Value at Hot Idle
1	PK/BK	Shift Solenoid 'B' Control	12v, 55 mph: 1v
2	YL/BK	MIL (lamp) Control	MIL On: 1v, Off: 12v
3-5	---	Not Used	---
6	BR/BL	TR Overdrive Switch	In O/D: 12v
7	BR/GN	TR Low Position Switch	In Manual 1st: 12v
8	PK/LB	IMRC Monitor Signal	5v
9	LG/BL	TR Drive Position Switch	In Drive: 0v
9-10	---	Not Used	---
11	BL/PK	EVAP Purge Flow Sensor	0.8v, 55 mph: 3.0v
12	WT	Fuel Level Indicator Signal	1.7v (1/2 full)
13	PK/GN	Flash EEPROM Power	0.1v
14	---	Not Used	---
15	BK/BL	Data Bus (-) Signal	Digital Signals
16	WT/BL	Data Bus (+) Signal	Digital Signals
17	RD/BK	High Speed Fan Control	On: 1v, Off: 12v
18-20	---	Not Used	---
21	BK	CKP Sensor (+) Signal	400-425 Hz
22	WT	CKP Sensor (-) Signal	400-425 Hz
23	RD	TSS Sensor (-)	340-380 Hz
24	BK/YL	Power Ground	<0.1v
25	RD/BL	Shield Ground	<0.050v
26	GN/YL	Coil 'A' Driver Control	5° dwell
27	LG/BL	Shift Solenoid 'A' Control	1v, 55 mph: 12v
28-29, 34	---	Not Used	---
30	GN/WT	Octane Adjust Switch	Closed: 0v, Open: 9.3v
31	YL/RD	PSP Switch Signal	Straight: 0.6v, turned: 2v
32	RD/GN	TR Reverse Position Switch	In Reverse: 12v
33	BL	VSS (-) Signal	0 Hz, 55 mph: 125 Hz
35	GY	HO2S-12 (B1 S2) Signal	0.1-1.1v
36	OR/BL	MAF Sensor Return	<0.050v
37	OR	TFT Sensor Signal	2.10-2.40v
38	BL/WT	ECT Sensor Signal	0.5-0.6v
39	WT/GN	IAT Sensor Signal	1.5-2.5v
40	BK/RD	Fuel Pump Monitor	On: 12v, Off: 0v
41	WT	A/C Cycling Clutch Switch	A/C On: 12v, Off: 0v
42	RD/YL	IMRC Solenoid Control	12v (off)
43	---	Not Used	---
45	PK/WT	Low Speed Fan Control	On: 1v, Off: 12v
46	---	Not Used	---
47	PK/RD	EGR VR Solenoid	0%, 55 mph: 45%
48	LG/RD	Clean Tachometer Output	25-38 Hz
49-50	---	Not Used	---
51	BK/YL	Power Ground	<0.1v
52	BR/RD	Coil 'B' Driver Control	5° dwell
53	OR/GN	Shift Solenoid 'C' Control	6.7v, 55 mph: 6.7v
54	GY/WT	TCC Solenoid Control	0%, 55 mph: 95%
55	BL/RD	Keep Alive Power	12-14v
56	OR/WT	EVAP Purge Solenoid	0-10 Hz (0-100%)
57	YL/RD	Knock Sensor 1 Signal	0v
58	WT/BK	VSS (+) Signal	0 Hz, 55 mph: 125 Hz
60	GY/GN	HO2S-11 (B1 S1) Signal	0.1-1.1v
61	---	Not Used	---
62	BK/WT	FTP Sensor Signal	2.6v (0" H2O - cap off)
63	WT/PK	Fuel Rail Pressure Sensor	2.8v (39 psi)

1998-99 Escort 2.0L I4 MFI VIN P (All) 104 Pin Connector

PCM Pin #	Wire Color	Circuit Description (104 Pin)	Value at Hot Idle
64	RD	TR Sensor Signal	In 'P': 0v, 55 mph: 1.7v
64	BR/YL	M/T: Clutch Pedal Switch	Clutch In: 0v, Out: 5v
65	BL/YL	DPFE Sensor Signal	0.20-1.30v
66	---	Not Used	---
67	BL/PK	EVAP CV Solenoid	0-10 Hz (0-100%)
68	---	Not Used	---
69	BL/BK	A/C WOT Relay Control	On: 1v, Off: 12v
71	WT/RD	Vehicle Power	12-14v
72	PK	Shift Indicator Control	SIL On: 1v, Off: 12v
73	---	Not Used	---
74	BR/WT	Injector 3 Control	3.3-3.7 ms
75	YL	Injector 1 Control	3.3-3.7 ms
76	BL	CMP Sensor Ground	5-7 Hz
77	BK/YL	Power Ground	<0.1v
78, 81, 87	---	Not Used	---
79	YL/GN	Fuel Cap Off Indicator	Digital Signal
80	WT/OR	Fuel Pump Control	On: 1v, Off: 12v
82	YL/PK	TOT Sensor Signal	2.10-2.40v
83	BL/OR	IAC Motor Control	10v (30% on time)
84	WT	TSS Sensor (+) Signal	340-380 Hz
85	GN	CMP Sensor Signal	5-9 Hz
86	GN/BK	A/C High Pressure Switch	A/C On: 1.2-1.5v
88	BR/BK	MAF Sensor Signal	0.6-0.9v
89	RD/WT	TP Sensor Signal	0.53-1.27v
90	LG/WT	Reference Voltage	4.9-5.1v
91	LG/BK	Analog Signal Return	<0.050v
92	DG	Brake Pedal Position Switch	Brake Off: 0v, On: 12v
93	OR/BK	HO2S-11 (B1 S1) Heater	On: 1v, Off: 12v
94, 96	---	Not Used	---
95	PK	HO2S-12 (B1 S2) Heater	On: 1v, Off: 12v
97	WT/RD	Vehicle Power	12-14v
98-99	---	Not Used	---
100	GN/OR	Injector 4 Control	3.3-3.7 ms
101	PK/YL	Injector 2 Control	3.3-3.7 ms
102	---	Not Used	---
103	BK/YL	Power Ground	<0.1v
104	---	Not Used	---

Pin Connector Graphic

```
            PCM 104-PIN CONNECTOR
  1 ●●●●●●●●●●●●      ●●●●●●●●●●●●● 26
 27 ●●●●●●●●●●●●      ●●●●●●●●●●●●● 52
 53 ●●●●●●●●●●●●  ⬤  ●●●●●●●●●●●●● 78
 79 ●●●●●●●●●●●●      ●●●●●●●●●●●●● 104
```
Terminal View of 104-Pin PCM Wiring Harness Connector

Standard Colors and Abbreviations

Abbreviation	Color	Abbreviation	Color	Abbreviation	Color
BK	Black	GY	Gray	PK	Purple
BL	Blue	GN	Green	RD	Red
BR	Brown	LG	LT Green	TN	Tan
DB	Dark Blue	OR	Orange	WT	White
DG	DK Green	PK	Pink	YL	Yellow

2000-02 Escort 2.0L I4 MFI VIN P (All) 104 Pin Connector

PCM Pin #	Wire Color	Circuit Description (104 Pin)	Value at Hot Idle
1	PK/BK	Shift Solenoid 'B' Control	12v, 55 mph: 1v
2	YL/BK	MIL (lamp) Control	MIL On: 1v, Off: 12v
3-5	---	Not Used	---
6	BR/BL	TR Overdrive Switch	In O/D: 12v
7	BR/GN	TR Low Position Switch	In Manual 1st: 12v
8	PK/LB	IMRC Monitor Signal	5v
9	BR/YL	TR Drive Position Switch	In Drive: 0v
10-11	---	Not Used	---
12	YL/BR	Fuel Level Indicator Signal	1.7v (1/2 full)
13	PK/GN	Flash EEPROM Power	0.1v
14	---	Not Used	---
15	BK/BL	Data Bus (-) Signal	Digital Signals
16	WT/BL	Data Bus (+) Signal	Digital Signals
17	RD/BK	High Speed Fan Control	On: 1v, Off: 12v
18-20	---	Not Used	---
21	BK	CKP Sensor (+) Signal	400-425 Hz
22	WT	CKP Sensor (-) Signal	400-425 Hz
23	RD	TSS Sensor (-)	340-380 Hz
24	BK/YL	Power Ground	<0.1v
25	RD/BL	Shield Ground	<0.050v
26	GN/YL	Coil 'A' Driver Control	5° dwell
27	LG/BL	Shift Solenoid 'A' Control	1v, 55 mph: 12v
28-30, 34	---	Not Used	---
31	YL/RD	PSP Switch Signal	Straight: 0.6v, Turned: 2v
32	RD/GN	TR Reverse Position Switch	In Reverse: 12v
33	BL	VSS (-) Signal	0 Hz, 55 mph: 125 Hz
35	GY	HO2S-12 (B1 S2) Signal	0.1-1.1v
36	OR/BL	MAF Sensor Return	<0.050v
37	OR	TFT Sensor Signal	2.10-2.40v
38	BL/WT	ECT Sensor Signal	0.5-0.6v
39	WT/GN	IAT Sensor Signal	1.5-2.5v
40	BK/RD	Fuel Pump Monitor	On: 12v, Off: 0v
41	WT	A/C Cycling Clutch Switch	A/C On: 12v, Off: 0v
42	RD/YL	IMRC Solenoid Control	12v (off)
43, 46	---	Not Used	---
45	PK/WT	Low Speed Fan Control	On: 1v, Off: 12v
47	PK/RD	EGR VR Solenoid	0%, 55 mph: 45%
48	LG/RD	Clean Tachometer Output	25-38 Hz
49-50	---	Not Used	---
51	BK/YL	Power Ground	<0.1v
52	BR/RD	Coil 'B' Driver Control	5° dwell
53	OR/GN	Shift Solenoid 'C' Control	6.7v, 55 mph: 6.7v
54	GY/WT	TCC Solenoid Control	0%, 55 mph: 95%
55	BL/RD	Keep Alive Power	12-14v
56	OR/WT	EVAP Purge Solenoid	0-10 Hz (0-100%)
57	RD	Knock Sensor 1 Signal	0v
58	WT/BK	VSS (+) Signal	0 Hz, 55 mph: 125 Hz
60	GY/GN	HO2S-11 (B1 S1) Signal	0.1-1.1v
61	---	Not Used	---
62	BK/WT	FTP Sensor Signal	2.6v (0" H2O - cap off)
63	WT/PK	Injection Pressure Switch	2.8v (39 psi)
64	RD	TR Sensor Signal	In 'P': 0v, 55 mph: 1.7v
64	BR/YL	M/T: Clutch Pedal Switch	Clutch In: 0v, Out: 5v

2000-02 Escort 2.0L I4 MFI VIN P (All) 104 Pin Connector

PCM Pin #	Wire Color	Circuit Description (104 Pin)	Value at Hot Idle
65 ('00)	BL/YL	DPFE Sensor Signal	0.20-1.30v
65 ('01-'02)	BL/YL	DPFE Sensor Signal	0.95-1.05v
67	BL/PK	EVAP CV Solenoid	0-10 Hz (0-100%)
68, 73	---	Not Used	---
69	BL/BK	A/C WOT Relay Control	On: 1v, Off: 12v
71	WT/RD	Vehicle Power	12-14v
72	PK	Shift Indicator Control	On: 1v, Off: 12v
74	BR/WT	Injector 3 Control	3.3-3.7 ms
75	YL	Injector 1 Control	3.3-3.7 ms
76	BL	CMP Sensor Ground	5-7 Hz
77	BK/YL	Power Ground	<0.1v
78, 81, 87	---	Not Used	---
79	YL/GN	Fuel Cap Off Indicator	On: 1v, Off: 12v
80	WT/OR	Fuel Pump Control	On: 1v, Off: 12v
82	YL/PK	TOT Sensor Signal	2.10-2.40v
83	BL/OR	IAC Motor Control	10v (30% on time)
84	WT	TSS Sensor (+) Signal	340-380 Hz
85	GN	CMP Sensor Signal	5-9 Hz
86	GN/BK	A/C High Pressure Switch	A/C On: 1.2-1.5v
88	BR/BK	MAF Sensor Signal	0.6-0.9v
89	RD/WT	TP Sensor Signal	0.53-1.27v
90	LG/WT	Reference Voltage	4.9-5.1v
91	LG/BK	Analog Signal Return	<0.050v
92	DG	Brake Pedal Position Switch	Brake Off: 0v, On: 12v
93	OR/BK	HO2S-11 (B1 S1) Heater	On: 1v, Off: 12v
94, 96	---	Not Used	---
95	PK	HO2S-12 (B1 S2) Heater	On: 1v, Off: 12v
97	WT/RD	Vehicle Power	12-14v
98-99	---	Not Used	---
100	GN/OR	Injector 4 Control	3.3-3.7 ms
101	PK/YL	Injector 2 Control	3.3-3.7 ms
103	BK/YL	Power Ground	<0.1v
102	---	Not Used	---
104	---	Not Used	---

Pin Connector Graphic

PCM 104-PIN CONNECTOR

Terminal View of 104-Pin PCM Wiring Harness Connector

Standard Colors and Abbreviations

Abbreviation	Color	Abbreviation	Color	Abbreviation	Color
BK	Black	GY	Gray	PK	Purple
BL	Blue	GN	Green	RD	Red
BR	Brown	LG	LT Green	TN	Tan
DB	Dark Blue	OR	Orange	WT	White
DG	DK Green	PK	Pink	YL	Yellow

1998-99 Escort 2.0L I4 MFI VIN 3 (All) 104 Pin Connector

PCM Pin #	Wire Color	Circuit Description (104 Pin)	Value at Hot Idle
1	PK/BK	Shift Solenoid 2 or 'B' Control	1v, Others: 1v
2	YL/BL	MIL (lamp) Control	MIL On: 1v, Off: 12v
3-5, 8	---	Not Used	---
6	BR/BL	TR O/D Position Signal	In 'P': 1v, In O/D: 12v
7	BR/GN	TR Low Position Signal	In 'L': 12v, Others: 1v
9	BR/YL	TR Drive Position Signal	In 'D': 12v, Others: 1v
10-11, 14	---	Not Used	---
12	YL/BL	Fuel Level Indicator Signal	1.7v (1/2 full)
13	PK/GN	Flash EEPROM Power	0.1v
15	BK/BL	SCP Data Bus (-) Signal	Digital Signals
16	WT/BL	SCP Data Bus (+) Signal	Digital Signals
17	RD/GN	High Speed Fan Control	On: 1v, Off: 12v
18-20	---	Not Used	---
21	BK	CKP Sensor (+) Signal	450-480 Hz
22	WT	CKP Sensor (-) Signal	450-480 Hz
23	RD	TSS Sensor (-)	340-380 Hz (680-710 rpm)
24, 51	BK/YL	Power Ground	<0.1v
25	RD/BL	Shield Ground	<0.050v
26	GN/YL	Coil 'A' Driver Control	5° dwell
27	LG/BL	Shift Solenoid 'A' Control	1v, Others: 12v
28, 34	---	Not Used	---
29	PK/BK	TCS (switch) Signal	TCS & O/D On: 12v
30	WT/BK	Octane Adjust Switch	Closed: 0v, Open: 9.3v
31	YL/RD	PSP Switch Signal	Straight: 1v, turned: 12v
32	RD/GN	TR Reverse Position Signal	In 'R': 12v, Others: 1v
33	BL	VSS (-) Signal	0 Hz, 55 mph: 125 Hz
35	GY	HO2S-12 (B1 S2) Signal	0.1-1.1v
36	OR/BL	MAF Sensor Return	<0.1v
37	OR/GN	TFT Sensor Signal	2.10-2.40v
38	BL/WT	ECT Sensor Signal	0.5-0.6v
39	WT/GN	IAT Sensor Signal	1.5-2.5v
40	BK/RD	Fuel Pump Monitor	On: 12v, Off: 0v
41	WT	A/C Cycling Clutch Switch	Closed: 12v, Open: 0v
44	PK/GN	Variable Valve Timing Control	12v (Off)
45	PK/WT	Low Speed Fan Control	On: 1v, Off: 12v
46, 49, 51	---	Not Used	---
47	BK/GN	EGR VR Solenoid	12v, 55 mph: 10-12v
48	LG/RD	Clean Tachometer Output	20-31 Hz
52	BR/RD	Coil 'B' Driver Control	5° dwell
53	OR/GN	Shift Solenoid 'C' Control	12v, Others: 12v
54	GY/WT	TCC Solenoid Control	12v (0%)
55	BL/RD	Keep Alive Power	12-14v
56	OR/WT	EVAP Purge Solenoid	0-10 Hz (0-100%)
57	RD	Knock Sensor 1 Signal	0v
58	WT/BK	VSS (+) Signal	0 Hz, 55 mph: 125 Hz
59	---	Not Used	---
60	GY/GN	HO2S-11 (B1 S1) Signal	0.1-1.1v
61	---	Not Used	---
62	BK/WT	FTP Sensor Signal	2.6v (cap off)
63	WT/PK	FRP Sensor Signal	2.8v (39 psi)
64	RD	A/T: TR Sensor Signal	In 'P': 0v, 55 mph: 1.7v
64	BR/YL	M/T: Clutch Pedal Position	Clutch In: 1v, Out: 5v

1998-99 Escort 2.0L I4 MFI VIN 3 (All) 104 Pin Connector

PCM Pin #	Wire Color	Circuit Description (104 Pin)	Value at Hot Idle
65	WT/BL	DPFE Sensor Signal	0.20-1.30v
66	---	Not Used	---
67	BL/PK	EVAP CV Solenoid	0-10 Hz (0-100%)
69	BL/BK	A/C WOT Relay Control	On: 1v, Off: 12v
70	BK/YL	VCT Control (ZETECT)	VCT Off: 12v, On: 1v
71	WT/RD	Vehicle Power	12-14v
72	PK	Shift Indicator Control	On: 1v, Off: 12v
73, 78	---	Not Used	---
74	BK/WT	Injector 3 Control	2.3-2.9 ms
75	YL	Injector 1 Control	2.3-2.9 ms
76	LB	CMP Sensor Ground	<0.050v
77	BK/YL	Power Ground	<0.1v
79	WT/BL	TCIL (lamp) Control	On: 1v, Off: 12v
80	WT/OR	Fuel Pump Control	On: 1v, Off: 12v
81, 87	---	Not Used	---
82	YL/PK	EPC Solenoid Control	2.6v (72%)
83	WT/LB	IAC Motor Control	9v (33% duty cycle)
84	WT	TSS Sensor (+) Signal	340-380 Hz (680-710 rpm)
85	GN	CMP Sensor Signal	5-7 Hz
86	GN/BK	A/C Pressure Switch	A/C On: 12v (Open)
88	BR/BK	MAF Sensor Signal	0.6-0.9v
89	RD/WT	TP Sensor Signal	0.53-1.27v
90	LG/WT	Reference Voltage	4.9-5.1v
91	LG/BK	Analog Signal Return	<0.050v
92	GN	Brake Pedal Position Switch	Brake Off: 0v, On: 12v
93	OR/BK	HO2S-11 (B1 S1) Heater	On: 1v, Off: 12v
95	PK	HO2S-12 (B1 S2) Heater	On: 1v, Off: 12v
96	---	Not Used	---
97	WT/RD	Vehicle Power	12-14v
98-99	---	Not Used	---
100	GN/OR	Injector 4 Control	2.3-2.9 ms
101	PK/YL	Injector 2 Control	2.3-2.9 ms
102	BK/OR	Shift Solenoid 'D' Control	8v, 55 mph: 8-9v
103	BK/YL	Power Ground	<0.1v
104	---	Not Used	---

Pin Connector Graphic

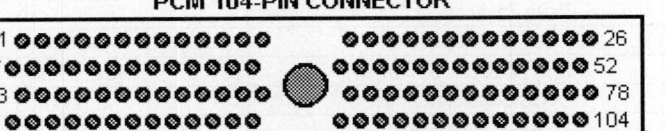

PCM 104-PIN CONNECTOR

Terminal View of 104-Pin PCM Wiring Harness Connector

Standard Colors and Abbreviations

Abbreviation	Color	Abbreviation	Color	Abbreviation	Color
BK	Black	GY	Gray	PK	Purple
BL	Blue	GN	Green	RD	Red
BR	Brown	LG	LT Green	TN	Tan
DB	Dark Blue	OR	Orange	WT	White
DG	DK Green	PK	Pink	YL	Yellow

2000-02 Escort 2.0L I4 ZETEC-E VIN 3 (All) 104 Pin Connector

PCM Pin #	Wire Color	Circuit Description (104 Pin)	Value at Hot Idle
1	GN/BL	Shift Solenoid 'B' Control	1v, Others: 1v
2	---	Not Used	---
3	WT/VT	IMRC Solenoid Control	IMRC Off: 5v, On: 2.5v
4	GN/BL	TR Reverse Position Signal	In 'R': 12v, Others: 0v
5-6	---	Not Used	---
7	GN/WT	TR First Position Signal	In '1st': 12v, Others: 1v
8	GN/OR	TR Second Position Signal	In 2nd: 12v, Others: 1v
9-10	---	Not Used	---
11	GN/BK	TR OD Position Signal	In 'P': 1v, In O/D: 12v
12	BR/WT	Sensor Ground	<0.050v
13	OG/BK	Flash EEPROM Power	0.1v
14, 18	---	Not Used	---
15	BL/BK	SCP Bus (-) Signal	Digital Signals
16	GY/OR	SCP Bus (+) Signal	Digital Signals
17	BK/WT	High Speed Fan Control	On: 1v, Off: 12v
19	GY/OR	TX Signal	Digital Signals
21	BK	CKP Sensor (+) Signal	450-480 Hz
22	WT	CKP Sensor (-) Signal	450-480 Hz
23	RD	TSS Sensor (-)	340-380 Hz (680-710 rpm)
24	BK/YL	Power Ground	<0.1v
25	RD/BL	Power Ground	<0.1v
26	GN/YL	Coil 'A' Driver Control	5° dwell
27	LG/BL	Shift Solenoid 'A' Control	1v, Others: 12v
28-30	---	Not Used	---
31	YL/RD	PSP Switch Signal	Straight: 1v, Turned: 12v
32	RD/GN	TR Reverse Position Signal	In 'R': 12v, Others: 1v
33	BL	VSS (-) Signal	0 Hz, 55 mph: 125 Hz
34	---	Not Used	---
35	GY	HO2S-12 (B1 S2) Signal	0.1-1.1v
36	OR/BL	MAF Sensor Return	<0.1v
37	OR	TFT Sensor Signal	2.10-2.40v
38	BL/WT	ECT Sensor Signal	0.5-0.6v
39	WT/GN	IAT Sensor Signal	1.5-2.5v
40	BK/RD	Fuel Pump Monitor	On: 12v, Off: 0v
41	WT	A/C Cycling Clutch Switch	Closed: 12v, Open: 0v
42-43	---	Not Used	---
44	PK/GN	Variable Valve Timing Control	12v (Off)
45	PK/WT	Low Speed Fan Control	On: 1v, Off: 12v
46-47	---	Not Used	---
48	LG/RD	Clean Tachometer Output	20-31 Hz
49-50	---	Not Used	---
51	BK/YL	Power Ground	<0.1v
52	BR/RD	Coil 'B' Driver Control	5° dwell
53	OR/GN	Shift Solenoid 'C' Control	12v, Others: 12v
54	GY/WT	TCC Solenoid Control	12v (0%)
55	BK/RD	Keep Alive Power	12-14v
56	OR/WT	EVAP Purge Solenoid	0-10 Hz (0-100%)
57	RD	Knock Sensor 1 Signal	0v
58	WT/BK	VSS (+) Signal	0 Hz, 55 mph: 125 Hz
59	---	Not Used	---
60	GY/GN	HO2S-11 (B1 S1) Signal	0.1-1.1v
61	---	Not Used	---
62	BK/WT	FTP Sensor Signal	2.6v (cap off)
63	WT/VT	Injection Pressure Sensor	2.8v (39 psi)

2000-02 Escort 2.0L I4 MFI VIN 3 (All) 104 Pin Connector

PCM Pin #	Wire Color	Circuit Description (104 Pin)	Value at Hot Idle
64	RD	A/T: TR Sensor Signal	In 'P': 0v, 55 mph: 1.7v
64	BR/YL	M/T: Clutch Pedal Position	Clutch In: 1v, Out: 5v
65-66	---	Not Used	---
67	BL/PK	EVAP CV Solenoid	0-10 Hz (0-100%)
68	---	Not Used	---
69	BL/BK	A/C WOT Relay Control	On: 1v, Off: 12v
71	WT/RD	Vehicle Power	12-14v
70	---	Not Used	---
72	PK	Shift Indicator Control	On: 1v, Off: 12v
73	---	Not Used	---
74	BR/WT	Injector 3 Control	2.3-2.9 ms
75	YL	Injector 1 Control	2.3-2.9 ms
76	BL	CMP Sensor Ground	<0.050v
77	BK/YL	Power Ground	<0.1v
78	---	Not Used	---
79	YL/GN	Fuel Cap Off Indicator	Off: 12v, On: 1v
80	WT/OR	Fuel Pump Control	Off: 12v, On: 1v
81, 87	---	Not Used	---
82	YL/VT	EPC Solenoid Control	2.6v (72%)
83	BL/OR	IAC Motor Control	9v (33% duty cycle)
84	WT	TSS Sensor (+) Signal	340-380 Hz (680-710 rpm)
85	GN	CMP Sensor Signal	5-7 Hz
86	GN/BK	A/C Pressure Switch	A/C On: 12v (Open)
88	BR/BK	MAF Sensor Signal	0.6-0.9v
89	RD/WT	TP Sensor Signal	0.53-1.27v
90	LG/WT	Reference Voltage	4.9-5.1v
91	LG/BK	Analog Signal Return	<0.050v
92	GN	Brake Pedal Position Switch	Brake Off: 0v, On: 12v
93	OR/BK	HO2S-11 (B1 S1) Heater	On: 1v, Off: 12v
95	VT	HO2S-12 (B1 S2) Heater	On: 1v, Off: 12v
96	---	Not Used	---
97	WT/RD	Vehicle Power	12-14v
98-99	---	Not Used	---
100	GN/OR	Injector 4 Control	2.3-2.9 ms
101	VT/YL	Injector 2 Control	2.3-2.9 ms
102	---	Not Used	---
103	BK/YL	Power Ground	<0.1v
104	---	Not Used	---

Pin Connector Graphic

PCM 104-PIN CONNECTOR

Terminal View of 104-Pin PCM Wiring Harness Connector

Standard Colors and Abbreviations

Abbreviation	Color	Abbreviation	Color	Abbreviation	Color
BK	Black	GY	Gray	PK	Purple
BL	Blue	GN	Green	RD	Red
BR	Brown	LG	LT Green	TN	Tan
DB	Dark Blue	OR	Orange	WT	White
DG	DK Green	PK	Pink	YL	Yellow

Focus Pin Tables

2000-2002 Focus 2.0L I4 DURATEC VIN P (All) 104 Pin Connector

PCM Pin #	Wire Color	Circuit Description (104 Pin)	Value at Hot Idle
1	GN/BL	Shift Solenoid 'B' Control	1v, at Cruise: 1v
2, 5-6	---	Not Used	---
3	WT/RD	IMRC Solenoid Control	IMRC Off: 5v, On: 2.5v
4	GN/BL	TR Reverse Position Signal	In 'R': 12v, Others: 0v
7	GN/WT	TR 1st Position Signal	Manual 1: 12v, Others: 1v
8	GN/OR	TR Drive Position Signal	In 'D': 12v, Others: 1v
9-10, 12	---	Not Used	---
11	GN/BK	TR Overdrive Position Signal	In O/D: 12v, Others: 1v
13	OR/BK	Flash EEPROM Power	0.1v
14, 18, 23	---	Not Used	---
15	BL/BK	Data Bus (-) Signal	Digital Signals
16	GY/OR	Data Bus (+) Signal	Digital Signals
17	BK/WT	High Speed Fan Control	On: 1v, Off: 12v
19	GY/OR	Passive Antitheft System	Digital Signals
20	BK/BL	Injector 3 Control	3.3-3.7 ms
21	WT/RD	CKP Sensor (+) Signal	400-425 Hz
22	BR/RD	CKP Sensor (-) Signal	400-425 Hz
24	BK/YL	Power Ground	<0.1v
25	BK	Shield Ground	<0.050v
26	BK/OR	Coil 'A' Driver Control	5° dwell
27	BK/RD	Vehicle Start Signal	KOEC: 9-11v
28	WT/VT	M/T: VSS (+) Signal	0 Hz, 55 mph: 125 Hz
29	GN/BK	TR Overdrive OFF Switch	O/D Off: 0v
30, 32	---	Not Used	---
31	BK/YL	PSP Switch Signal	Straight: 0.6v, turned: 2v
33	BR/BL	A/T: OSS (-) Sensor Signal	0 Hz, 55 mph: 120 Hz
34	WT/VT	A/T: TSS Sensor (+) Signal	340-380 Hz
35	WT/RD	HO2S-22 (B2 S2) Signal	0.1-1.1v
36	BR/BL	MAF Sensor Return	<0.050v
37	WT/GN	TFT Sensor Signal	2.10-2.40v
38	WT	ECT Sensor Signal	0.5-0.6v
39	WT/VT	IAT Sensor Signal	1.5-2.5v
40	GN/BK	Fuel Pump Driver Module	3.5v (100% on time)
41	GN/YL	A/C Cycling Clutch Switch	Closed: 12v, Open: 0v
42	BK/OR	Passive Antitheft System	Digital Signals
43, 45-46	---	Not Used	---
44	GN/OR	EPC Switch Signal	12v
47	BK/GN	EGR VR Solenoid	0%, 55 mph: 45%
48-50	---	Not Used	---
51	BK/YL	Power Ground	<0.1v
52	BK/GN	Coil 'B' Driver Control	5° dwell
53	WT/GN	Passive Antitheft System	Digital Signals
54	BK/WT	Fuel Pump Driver Module	1.5v (33% on time)
55	RD	Keep Alive Power	12-14v
56	BK/BL	EVAP Purge Solenoid	0-10 Hz (0-100%)
57	WT/BK	Knock Sensor 1 (-) Signal	0v
58	WT/BL	A/T: OSS (+) Sensor Signal	0 Hz, 55 mph: 120 Hz
59	GY	Generator Field Sense	130 Hz (30% on time)
60	WT	HO2S-21 (B2 S1) Signal	0.1-1.1v
61	---	Not Used	---
62	WT/VT	FTP Sensor Signal	2.6v (0" H2O - cap off)
63	WT/GN	FRP Sensor Signal	2.8v (39 psi)
64	GN/YL	A/T: Neutral Start Position	In 'N': 0v, Others: 5v
64	WT	M/T: Clutch Pedal Switch	Clutch In: 0v, Out: 5v

Focus Pin Tables

2000-2002 Focus 2.0L I4 DURATEC VIN P (All) 104 Pin Connector

PCM Pin #	Wire Color	Circuit Description (104 Pin)	Value at Hot Idle
65 ('00)	WT/BL	DPFE Sensor Signal	0.20-1.30v
65 ('01-'02)	WT/BL	DPFE Sensor Signal	0.95-1.05v
66	---	Not Used	---
67	BK/OR	EVAP CV Solenoid	0-10 Hz (0-100%)
68	BK/BL	Low Speed Fan Control	On: 1v, Off: 12v
69	BK/YL	A/C WOT Relay Control	On: 1v, Off: 12v
70	BK/WT	Injector 1 Control	3.3-3.7 ms
71, 97	GN/YL	Vehicle Power	12-14v
72	BL	Generator Field Control	0 Hz (0% on time)
73	GN/YL	Shift Solenoid 'A' Control	0v, 55 mph: 12v
74-75	---	Not Used	---
76	BR/WT	M/T: CMP Sensor Ground	5-7 Hz
76	BR/WT	A/T: TSS Sensor (-) Signal	340-380 Hz
77	BK/YL	Power Ground	<1v
78-79	---	Not Used	---
80	BK/WT	IMRC Solenoid Control	12v (off)
81	BK/RD	EPC Solenoid Control	2.5v (73% on time)
82	GN/BK	Shift Solenoid 'C' Control	0v, 55 mph: 4.10v
83	BK/YL	IAC Motor Control	10v (30% on time)
84, 98, 94	---	Not Used	---
85	WT/VT	CMP Sensor Signal	5-9 Hz
86	WT/RD	A/C Pressure Switch	A/C On: 12v (Open)
87	BR/YL	Knock Sensor 1 Signal	0v
88	WT/BL	MAF Sensor Signal	0.6-0.9v
89	WT	TP Sensor Signal	0.53-1.27v
90	YL	Reference Voltage	4.9-5.1v
91	BR	Analog Signal Return	<0.050v
92	GN/RD	Brake Pedal Position Switch	Brake Off: 0v, On: 12v
93	BK/YL	HO2S-21 (B2 S1) Heater	On: 1v, Off: 12v
95	BK/OR	Injector 4 Control	3.3-3.7 ms
96	BK/YL	Injector 2 Control	3.3-3.7 ms
99	GN/OR	Shift Solenoid 'D' Control	0v, 55 mph: 0v
100	BK/BL	HO2S-22 (B2 S2) Heater	On: 1v, Off: 12v
101	---	Not Used	---
102	GN/WT	Shift Solenoid 'E' Control	0v, 55 mph: 0v
103	BK/YL	Power Ground	<1v
104	---	Not Used	---

Pin Connector Graphic

PCM 104-PIN CONNECTOR

Terminal View of 104-Pin PCM Wiring Harness Connector

Standard Colors and Abbreviations

Abbreviation	Color	Abbreviation	Color	Abbreviation	Color
BK	Black	GY	Gray	PK	Purple
BL	Blue	GN	Green	RD	Red
BR	Brown	LG	LT Green	TN	Tan
DB	Dark Blue	OR	Orange	WT	White
DG	DK Green	PK	Pink	YL	Yellow

2003 Focus 2.0L I4 DURATEC VIN P (All) 104 Pin Connector

PCM Pin #	Wire Color	Circuit Description (104 Pin)	Value at Hot Idle
1	GN/BL	Shift Solenoid 'B' Control	In 2nd: 1v, Others: 1v
2	---	Not Used	---
3	WT/VT	IMRC Solenoid Monitor	5v
4	GN/BL	TR Reverse Position Signal	In 'R': 12v, Others: 0v
5-6	---	Not Used	---
7	GN/WT	TR First Position Signal	In 1st: 12v, Others: 1v
8	GN/OR	TR Second Position Signal	In 2nd: 12v, Others: 1v
9	WT/BK	Barometric Pressure Sensor	159 Hz (sea level)
10	---	Not Used	---
11	GN/BK	TR OD Position Signal	In 'P': 1v, In O/D: 12v
12	BR/WT	Sensor Ground	<0.050v
13	OG/BK	Flash EEPROM Power	0.1v
14, 18, 30	---	Not Used	---
15	BL/BK	SCP Bus (-) Signal	Digital Signals
16	GY/OR	SCP Bus (+) Signal	Digital Signals
17	BK/WT	High Speed Fan Control	On: 1v, Off: 12v
19	GY/OR	Passive Antitheft TX Signal	Digital Signals
20	BK/BL	Fuel Injector 3 Control	2.3-2.9 ms
21	WT/RD	CKP Sensor (+) Signal	450-480 Hz
22	BR/RD	CKP Sensor (-) Signal	450-480 Hz
23	BK/YL	Power Ground	<0.1v
24	BK/YL	Power Ground	<0.1v
25	BK	Power Ground	<0.1v
26	BK/OR	Coil 'A' Driver Control	5° dwell
27	BK/RD	Start Inhibit Relay Control	Relay Off: 12v, On: 1v
28	WT/VT	VSS (-) Signal	0 Hz, 55 mph: 125 Hz
29	GN/BK	Overdrive Cancel Switch	TCS & O/D On: 12v
31	WT/VT	PSP Switch Signal	Straight: 1v, Turned: 5v
32	BN/YL	Sensor Ground	<0.050v
33	BR/BL	Output Speed Sensor (-) Signal	Moving: AC pulse signals
34	WT/VT	Turbine Speed Sensor (+) Signal	340-380 Hz
35	WT/RD	HO2S-12 (B1 S2) Signal	0.1-1.1v
36	BR/BL	MAF Sensor Return	<0.1v
37	WT/GN	TFT Sensor Signal	2.10-2.40v
38	WT/VT	Cylinder Head Temperature Sensor	0.5-0.6v
39	WT/VT	MAF Sensor Signal	0.6-0.9v
40	GN/BK	Fuel Pump Monitor	On: 12v, Off: 0v
41	GN/YL	A/C Cycling Clutch Pressure Switch	Closed: 12v, Open: 0v
42	BK/OR	PATS Indicator Control	Indicator Off: 12v, On: 1v
43, 48	---	Not Used	---
44	GN/OR	Pressure Control Solenoid 'A'	Off: 12v, On: 1v
45	BK/RD	Camshaft Timing Adjuster Control	Off: 12v, On: 1v
46	BK/WT	Low Fuel Indicator	Digital Signals
47	BK/GN	Electric Vacuum Regulator Control	Off: 12v, On: 1v
49	BL/RD	Controller Area Network Bus 'L'	0-7-0v
50	GY/RD	Controller Area Network Bus 'H'	0-7-0v
51	BK/YL	Power Ground	<0.1v
52	BK/GN	Coil 'B' Driver Control	5° dwell
53	WT/GN	Passive Antitheft RX Signal	Digital Signals
54	BK/WT	Fuel Pump Control	Pump Off: 12v, On: 1v
55	RD	Keep Alive Power	12-14v
56	BK/BL	EVAP Purge Solenoid	0-10 Hz (0-100%)
57	WT/BK	Knock Sensor Signal	0v
58	WT/BL	VSS (+) Signal	Moving: AC pulse signals

2003 Focus 2.0L I4 DURATEC VIN P (All) 104 Pin Connector

PCM Pin #	Wire Color	Circuit Description (104 Pin)	Value at Hot Idle
59	GY	Generator Monitor Signal	130 Hz (45%)
60	WT	HO2S-11 (B1 S1) Signal	0.1-1.1v
61, 66	---	Not Used	---
62	WT/VT	FTP Sensor Signal	2.6v (cap off)
63	WT/GN	Fuel Rail Pressure Sensor	2.8v (39 psi)
64	GN/YL	A/T: TR Sensor Signal	In 'P': 0v, 55 mph: 1.7v
64	WT	M/T: Clutch Pedal Position	Clutch In: 1v, Out: 5v
65	WT/BL	EGR Pressure Transducer Signal	N/A
67	BK/OR	EVAP CV Solenoid	0-10 Hz (0-100%)
68	BK/YL	Ground Switched	0v
69	BK/WT	A/C WOT Relay Control	On: 1v, Off: 12v
70	BK/WT	Fuel Injector 1 Control	2.3-2.9 ms
71	GN/YL	Vehicle Power	12-14v
72	BL	Generator Communicator Control	1.5-10.5v (40-250 Hz)
73	GN/YL	Shift Solenoid A' Control	1v, Others: 1v
74	---	Not Used	---
75	BK/WT	Power Ground	<0.1v
76	BN	Turbine Speed Sensor (-) Signal	340-380 Hz
77	BK/YL	Power Ground	<0.1v
78, 84	---	Not Used	---
79	YL/GN	Fuel Cap Off Indicator	Off: 12v, On: 1v
80	BK/WT	IMRC Signal	IMRC Off: 5v, On: 2.5v
81	BK/RD	Pressure Control Solenoid 'A' (-)	<0.1v
82	GN/BK	Shift Solenoid 'C' Control	In Drive: 1v, Others: 12v
83	GN/YL	IAC Motor Control	9v (33% duty cycle)
85	WT/VT	CMP Sensor Signal	5-7 Hz
86	WT/RD	A/C Pressure Switch	A/C On: 12v (Open)
87	BR/YL	Knock Sensor Signal (-)	<0.050v
88	WT/BL	MAF Sensor Signal	0.6-0.9v
89	WT	TP Sensor Signal	0.53-1.27v
90	YL	Reference Voltage	4.9-5.1v
91	BR/GN	Analog Signal Return	<0.050v
92	GN/RD	Brake Pedal Position Switch	Brake Off: 0v, On: 12v
93	BK/YL	HO2S-11 (B1 S1) Heater	On: 1v, Off: 12v
94, 98	---	Not Used	---
95	BK/OR	Injector 4 Control	2.3-2.9 ms
96	BK/YL	Injector 3 Control	2.3-2.9 ms
97	GN/YL	Vehicle Power	12-14v
99	GN/OR	Shift Solenoid 'D' Control	In Drive: 1v, Others: 12v
100	BK/BL	HO2S-12 (B1 S2) Heater	On: 1v, Off: 12v
101	---	Not Used	---
102	GN/WT	Shift Solenoid 'E' Control	In O/D: 1v, Others: 12v
103	BK/YL	Power Ground	<0.1v
104	---	Not Used	---

Pin Connector Graphic

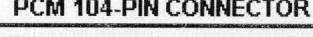

Terminal View of 104-Pin PCM Wiring Harness Connector

2000-2002 Focus 2.0L I4 ZETEC-E VIN 3 (All) 104 Pin Connector

PCM Pin #	Wire Color	Circuit Description (104 Pin)	Value at Hot Idle
1	GN/BL	Shift Solenoid 'B' Control	1v, at Cruise: 1v
2, 5-6	---	Not Used	---
3	WT/RD	IMRC Solenoid Control	IMRC Off: 5v, On: 2.5v
4	GN/BL	TR Reverse Position Signal	In 'R': 12v, Others: 0v
7	GN/WT	TR 1st Position Signal	Manual 1: 12v, Others: 1v
8	GN/OR	TR Drive Position Signal	In 'D': 12v, Others: 1v
9-10, 12	---	Not Used	---
11	GN/BK	TR Overdrive Position Signal	In O/D: 12v, Others: 1v
13	OR/BK	Flash EEPROM Power	0.1v
14, 18, 23	---	Not Used	---
15	BL/BK	Data Bus (-) Signal	Digital Signals
16	GY/OR	Data Bus (+) Signal	Digital Signals
17	BK/WT	High Speed Fan Control	On: 1v, Off: 12v
19	GY/OR	Passive Antitheft System	Digital Signals
20	BK/BL	Injector 3 Control	3.3-3.7 ms
21	WT/RD	CKP Sensor (+) Signal	400-425 Hz
22	BR/RD	CKP Sensor (-) Signal	400-425 Hz
24	BK/YL	Power Ground	<0.1v
25	BK	Shield Ground	<0.050v
26	BK/OR	Coil 'A' Driver Control	5° dwell
27	BK/RD	Vehicle Start Signal	KOEC: 9-11v
28	WT/VT	M/T: VSS (+) Signal	0 Hz, 55 mph: 125 Hz
29	GN/BK	TR Overdrive OFF Switch	O/D Off: 0v
30, 32	---	Not Used	---
31	BK/YL	PSP Switch Signal	Straight: 0.6v, turned: 2v
33	BR/BL	A/T: OSS (-) Sensor Signal	0 Hz, 55 mph: 120 Hz
34	WT/VT	A/T: TSS Sensor (+) Signal	340-380 Hz
35	WT/RD	HO2S-22 (B2 S2) Signal	0.1-1.1v
36	BR/BL	MAF Sensor Return	<0.050v
37	WT/GN	TFT Sensor Signal	2.10-2.40v
38	WT	ECT Sensor Signal	0.5-0.6v
39	WT/VT	IAT Sensor Signal	1.5-2.5v
40	GN/BK	Fuel Pump Driver Module	3.5v (100% on time)
41	GN/YL	A/C Cycling Clutch Switch	Closed: 12v, Open: 0v
42	BK/OR	Passive Antitheft System	Digital Signals
43, 45-46	---	Not Used	---
44	GN/OR	EPC Switch Signal	12v
47	BK/GN	EGR VR Solenoid	0%, 55 mph: 45%
48-50	---	Not Used	---
51	BK/YL	Power Ground	<0.1v
52	BK/GN	Coil 'B' Driver Control	5° dwell
53	WT/GN	Passive Antitheft System	Digital Signals
54	BK/WT	Fuel Pump Driver Module	1.5v (33% on time)
55	RD	Keep Alive Power	12-14v
56	BK/BL	EVAP Purge Solenoid	0-10 Hz (0-100%)
57	WT/BK	Knock Sensor 1 (-) Signal	0v
58	WT/BL	A/T: OSS (+) Sensor Signal	0 Hz, 55 mph: 120 Hz
59	GY	Generator Field Sense	130 Hz (30% on time)
60	WT	HO2S-21 (B2 S1) Signal	0.1-1.1v
61	---	Not Used	---
62	WT/VT	FTP Sensor Signal	2.6v (0" H2O - cap off)
63	WT/GN	FRP Sensor Signal	2.8v (39 psi)
64	GN/YL	A/T: Neutral Start Position	In 'N': 0v, Others: 5v
64	WT	M/T: Clutch Pedal Switch	Clutch In: 0v, Out: 5v

2000-2002 Focus 2.0L I4 ZETEC-E VIN 3 (All) 104 Pin Connector

PCM Pin #	Wire Color	Circuit Description (104 Pin)	Value at Hot Idle
65 ('00)	WT/BL	DPFE Sensor Signal	0.20-1.30v
65 ('01-'02)	WT/BL	DPFE Sensor Signal	0.95-1.05v
66	---	Not Used	---
67	BK/OR	EVAP CV Solenoid	0-10 Hz (0-100%)
68	BK/BL	Low Speed Fan Control	On: 1v, Off: 12v
69	BK/YL	A/C WOT Relay Control	On: 1v, Off: 12v
70	BK/WT	Injector 1 Control	3.3-3.7 ms
71, 97	GN/YL	Vehicle Power	12-14v
72	BL	Generator Field Control	0 Hz (0% on time)
73	GN/YL	Shift Solenoid 'A' Control	0v, 55 mph: 12v
74-75	---	Not Used	---
76	BR/WT	M/T: CMP Sensor Ground	5-7 Hz
76	BR/WT	A/T: TSS Sensor (-) Signal	340-380 Hz
77	BK/YL	Power Ground	<1v
78-79	---	Not Used	---
80	BK/WT	IMRC Solenoid Control	12v (off)
81	BK/RD	EPC Solenoid Control	2.5v (73% on time)
82	GN/BK	Shift Solenoid 'C' Control	0v, 55 mph: 4.10v
83	BK/YL	IAC Motor Control	10v (30% on time)
84, 98, 94	---	Not Used	---
85	WT/VT	CMP Sensor Signal	5-9 Hz
86	WT/RD	A/C Pressure Switch	A/C On: 12v (Open)
87	BR/YL	Knock Sensor 1 Signal	0v
88	WT/BL	MAF Sensor Signal	0.6-0.9v
89	WT	TP Sensor Signal	0.53-1.27v
90	YL	Reference Voltage	4.9-5.1v
91	BR	Analog Signal Return	<0.050v
92	GN/RD	Brake Pedal Position Switch	Brake Off: 0v, On: 12v
93	BK/YL	HO2S-21 (B2 S1) Heater	On: 1v, Off: 12v
95	BK/OR	Injector 4 Control	3.3-3.7 ms
96	BK/YL	Injector 2 Control	3.3-3.7 ms
99	GN/OR	Shift Solenoid 'D' Control	0v, 55 mph: 0v
100	BK/BL	HO2S-22 (B2 S2) Heater	On: 1v, Off: 12v
101	---	Not Used	---
102	GN/WT	Shift Solenoid 'E' Control	0v, 55 mph: 0v
103	BK/YL	Power Ground	<1v
104	---	Not Used	---

Pin Connector Graphic

PCM 104-PIN CONNECTOR

Terminal View of 104-Pin PCM Wiring Harness Connector

Standard Colors and Abbreviations

Abbreviation	Color	Abbreviation	Color	Abbreviation	Color
BK	Black	GY	Gray	PK	Purple
BL	Blue	GN	Green	RD	Red
BR	Brown	LG	LT Green	TN	Tan
DB	Dark Blue	OR	Orange	WT	White
DG	DK Green	PK	Pink	YL	Yellow

2003 Focus 2.0L I4 ZETEC-E VIN 3 (All) 104 Pin Connector

PCM Pin #	Wire Color	Circuit Description (104 Pin)	Value at Hot Idle
1	GN/BL	Shift Solenoid 'B' Control	In 2nd: 1v, Others: 1v
2	---	Not Used	---
3	WT/VT	IMRC Solenoid Monitor	5v
4	GN/BL	TR Reverse Position Signal	In 'R': 12v, Others: 0v
5-6	---	Not Used	---
7	GN/WT	TR First Position Signal	In 1st: 12v, Others: 1v
8	GN/OR	TR Second Position Signal	In 2nd: 12v, Others: 1v
9	WT/BK	Barometric Pressure Sensor	159 Hz (sea level)
10	---	Not Used	---
11	GN/BK	TR OD Position Signal	In 'P': 1v, In O/D: 12v
12	BR/WT	Sensor Ground	<0.050v
13	OG/BK	Flash EEPROM Power	0.1v
14, 18, 30	---	Not Used	---
15	BL/BK	SCP Bus (-) Signal	Digital Signals
16	GY/OR	SCP Bus (+) Signal	Digital Signals
17	BK/WT	High Speed Fan Control	On: 1v, Off: 12v
19	GY/OR	Passive Antitheft TX Signal	Digital Signals
20	BK/BL	Fuel Injector 3 Control	2.3-2.9 ms
21	WT/RD	CKP Sensor (+) Signal	450-480 Hz
22	BR/RD	CKP Sensor (-) Signal	450-480 Hz
23	BK/YL	Power Ground	<0.1v
24	BK/YL	Power Ground	<0.1v
25	BK	Power Ground	<0.1v
26	BK/OR	Coil 'A' Driver Control	5° dwell
27	BK/RD	Start Inhibit Relay Control	Relay Off: 12v, On: 1v
28	WT/VT	VSS (-) Signal	0 Hz, 55 mph: 125 Hz
29	GN/BK	Overdrive Cancel Switch	TCS & O/D On: 12v
31	WT/VT	PSP Switch Signal	Straight: 1v, Turned: 5v
32	BN/YL	Sensor Ground	<0.050v
33	BR/BL	Output Speed Sensor (-) Signal	0 Hz, 55 mph: 120 Hz
34	WT/VT	Turbine Speed Sensor (+) Signal	340-380 Hz
35	WT/RD	HO2S-12 (B1 S2) Signal	0.1-1.1v
36	BR/BL	MAF Sensor Return	<0.1v
37	WT/GN	TFT Sensor Signal	2.10-2.40v
38	WT/VT	Cylinder Head Temperature Sensor	0.5-0.6v
39	WT/VT	MAF Sensor Signal	0.6-0.9v
40	GN/BK	Fuel Pump Monitor	On: 12v, Off: 0v
41	GN/YL	A/C Cycling Clutch Pressure Switch	Closed: 12v, Open: 0v
42	BK/OR	PATS Indicator Control	Indicator Off: 12v, On: 1v
43	WT/BL	Sensor Signal	N/A
44	GN/OR	Pressure Control Solenoid 'A'	N/A
45	BK/RD	Camshaft Timing Adjuster Control	Off: 12v, On: 1v
46	BK/WT	Low Fuel Indicator	Digital Signals
47	BK/GN	Electric Vacuum Regulator Control	Off: 12v, On: 1v
48	---	Not Used	---
49	BL/RD	Controller Area Network Bus 'L'	0-7-0v
50	GY/RD	Controller Area Network Bus 'H'	0-7-0v
51	BK/YL	Power Ground	<0.1v
52	BK/GN	Coil 'B' Driver Control	5° dwell
53	WT/GN	Passive Antitheft RX Signal	Digital Signals
54	BK/WT	Fuel Pump Control	Pump Off: 12v, On: 1v
55	RD	Keep Alive Power	12-14v
56	BK/BL	EVAP Purge Solenoid	0-10 Hz (0-100%)
57	WT/BK	Knock Sensor Signal	0v

2003 Focus ZX2 2.0L I4 MFI VIN 3 (All) 104 Pin Connector

PCM Pin #	Wire Color	Circuit Description (104 Pin)	Value at Hot Idle
58	WT/BL	VSS (+) Signal	0 Hz, 55 mph: 125 Hz
59	GY	Generator Monitor Signal	130 Hz (45%)
60	WT	HO2S-11 (B1 S1) Signal	0.1-1.1v
61, 66	---	Not Used	---
62	WT/VT	FTP Sensor Signal	2.6v (cap off)
63	WT/GN	Fuel Rail Pressure Sensor	2.8v (39 psi)
64	GN/YL	A/T: TR Sensor Signal	In 'P': 0v, 55 mph: 1.7v
64	WT	M/T: Clutch Pedal Position	Clutch In: 1v, Out: 5v
65	WT/BL	EGR Pressure Transducer Signal	N/A
67	BK/OR	EVAP CV Solenoid	0-10 Hz (0-100%)
68	BK/YL	Ground Switched	0v
69	BK/WT	A/C WOT Relay Control	On: 1v, Off: 12v
70	BK/WT	Fuel Injector 1 Control	2.3-2.9 ms
71	GN/YL	Vehicle Power	12-14v
72	BL	Generator Communicator Control	1.5-10.5v (40-250 Hz)
73	GN/YL	Shift Solenoid A' Control	1v, Others: 1v
74	---	Not Used	---
75	BK/WT	Power Ground	<0.1v
76	BN	Turbine Speed Sensor (-) Signal	340-380 Hz
77	BK/YL	Power Ground	<0.1v
78, 84	---	Not Used	---
79	YL/GN	Fuel Cap Off Indicator	Off: 12v, On: 1v
80	BK/WT	IMRC Signal	IMRC Off: 5v, On: 2.5v
81	BK/RD	Pressure Control Solenoid 'A' (-)	DC pulse signals
82	GN/BK	Shift Solenoid 'C' Control	In Drive: 1v, Others: 12v
83	GN/YL	IAC Motor Control	9v (33% duty cycle)
85	WT/VT	CMP Sensor Signal	5-7 Hz
86	WT/RD	A/C Pressure Switch	A/C On: 12v (Open)
87	BR/YL	Knock Sensor Signal (-)	<0.050v
88	WT/BL	MAF Sensor Signal	0.6-0.9v
89	WT	TP Sensor Signal	0.53-1.27v
90	YL	Reference Voltage	4.9-5.1v
91	BR	Analog Signal Return	<0.050v
92	GN/RD	Brake Pedal Position Switch	Brake Off: 0v, On: 12v
93	BK/YL	HO2S-11 (B1 S1) Heater	On: 1v, Off: 12v
94, 98	---	Not Used	---
95	BK/OR	Injector 4 Control	2.3-2.9 ms
96	BK/YL	Injector 3 Control	2.3-2.9 ms
97	GN/YL	Vehicle Power	12-14v
99	GN/OR	Shift Solenoid 'D' Control	In Drive: 1v, Others: 12v
100	BK/BL	HO2S-12 (B1 S2) Heater	On: 1v, Off: 12v
101	---	Not Used	---
102	GN/WT	Shift Solenoid 'E' Control	In O/D: 1v, Others: 12v
103	BK/YL	Power Ground	<0.1v
104	---	Not Used	---

Pin Connector Graphic

PCM 104-PIN CONNECTOR

Terminal View of 104-Pin PCM Wiring Harness Connector

2003 Focus 2.3L I4 DOHC MFI VIN Z (All) 104 Pin Connector

PCM Pin #	Wire Color	Circuit Description (104 Pin)	Value at Hot Idle
1	GN/BL	Shift Solenoid 'B' Control	In 2nd: 1v, Others: 1v
2	---	Not Used	---
3	WT/VT	IMRC Solenoid Monitor	5v
4	GN/BL	TR Reverse Position Signal	In 'R': 12v, Others: 0v
5-6	---	Not Used	---
7	GN/WT	TR First Position Signal	In 1st: 12v, Others: 1v
8	GN/OR	TR Second Position Signal	In 2nd: 12v, Others: 1v
9	WT/BK	Barometric Pressure Sensor	159 Hz (sea level)
10	---	Not Used	---
11	GN/BK	TR OD Position Signal	In 'P': 1v, In O/D: 12v
12	BR/WT	Sensor Ground	<0.050v
13	OG/BK	Flash EEPROM Power	0.1v
14, 18, 30	---	Not Used	---
15	BL/BK	SCP Bus (-) Signal	Digital Signals
16	GY/OR	SCP Bus (+) Signal	Digital Signals
17	BK/WT	High Speed Fan Control	On: 1v, Off: 12v
19	GY/OR	Passive Antitheft TX Signal	Digital Signals
20	BK/BL	Fuel Injector 3 Control	2.3-2.9 ms
21	WT/RD	CKP Sensor (+) Signal	450-480 Hz
22	BR/RD	CKP Sensor (-) Signal	450-480 Hz
23	BK/YL	Power Ground	<0.1v
24	BK/YL	Power Ground	<0.1v
25	BK	Power Ground	<0.1v
26	BK/OR	Coil 'A' Driver Control	5° dwell
27	BK/RD	Start Inhibit Relay Control	Relay Off: 12v, On: 1v
28	WT/VT	VSS (-) Signal	0 Hz, 55 mph: 125 Hz
29	GN/BK	Overdrive Cancel Switch	TCS & O/D On: 12v
31	WT/VT	PSP Switch Signal	Straight: 1v, Turned: 5v
32	BN/YL	Sensor Ground	<0.050v
33	BR/BL	Output Speed Sensor (-) Signal	0 Hz, 55 mph: 120 Hz
34	WT/VT	Turbine Speed Sensor (+) Signal	340-380 Hz
35	WT/RD	HO2S-12 (B1 S2) Signal	0.1-1.1v
36	BR/BL	MAF Sensor Return	<0.1v
37	WT/GN	TFT Sensor Signal	2.10-2.40v
38	WT	Engine Coolant Temperature Sensor	0.5-0.6v
39	WT/VT	MAF Sensor Signal	0.6-0.9v
40	GN/BK	Fuel Pump Monitor	On: 12v, Off: 0v
41	GN/YL	A/C Cycling Clutch Pressure Switch	Closed: 12v, Open: 0v
42	BK/OR	PATS Indicator Control	Indicator Off: 12v, On: 1v
43	WT/BL	Sensor Signal	N/A
44	GN/OR	Pressure Control Solenoid 'A'	N/A
45	BK/RD	Camshaft Timing Adjuster Control	Off: 12v, On: 1v
46	BK/WT	Low Fuel Indicator	Indicator Off: 12v, On: 1v
47	BK/GN	Electric Vacuum Regulator Control	Off: 12v, On: 1v
48	---	Not Used	---
49	BL/RD	Controller Area Network Bus 'L'	0-7-0v
50	GY/RD	Controller Area Network Bus 'H'	0-7-0v
51	BK/YL	Power Ground	<0.1v
52	BK/GN	Coil 'B' Driver Control	5° dwell
53	WT/GN	Passive Antitheft RX Signal	Digital Signals
54	BK/WT	Fuel Pump Control	Pump Off: 12v, On: 1v
55	RD	Keep Alive Power	12-14v
56	BK/BL	EVAP Purge Solenoid	0-10 Hz (0-100%)
57	WT/BK	Knock Sensor Signal	0v

2003 Focus 2.3L I4 DOHC MFI VIN Z (All) 104 Pin Connector

PCM Pin #	Wire Color	Circuit Description (104 Pin)	Value at Hot Idle
58	WT/BL	VSS (+) Signal	0 Hz, 55 mph: 125 Hz
59	GY	Generator Monitor Signal	130 Hz (45%)
60	WT	HO2S-11 (B1 S1) Signal	0.1-1.1v
61, 66	---	Not Used	---
62	WT/VT	FTP Sensor Signal	2.6v (cap off)
63	WT/GN	Fuel Rail Pressure Sensor	2.8v (39 psi)
64	GN/YL	A/T: TR Sensor Signal	In 'P': 0v, 55 mph: 1.7v
64	WT	M/T: Clutch Pedal Position	Clutch In: 1v, Out: 5v
65	WT/BL	EGR Pressure Transducer Signal	N/A
67	BK/OR	EVAP CV Solenoid	0-10 Hz (0-100%)
68	BK/YL	Ground Switched	0v
69	BK/WT	A/C WOT Relay Control	On: 1v, Off: 12v
70	BK/WT	Fuel Injector 1 Control	2.3-2.9 ms
71	GN/YL	Vehicle Power	12-14v
72	BL	Generator Communicator Control	1.5-10.5v (40-250 Hz)
73	GN/YL	Shift Solenoid A' Control	1v, Others: 1v
74	---	Not Used	---
75	BK/WT	Power Ground	<0.1v
76	BN	Turbine Speed Sensor (-) Signal	340-380 Hz
77	BK/YL	Power Ground	<0.1v
78, 84	---	Not Used	---
79	YL/GN	Fuel Cap Off Indicator	Off: 12v, On: 1v
80	BK/WT	IMRC Signal	IMRC Off: 5v, On: 2.5v
81	BK/RD	Pressure Control Solenoid 'A' (-)	DC pulse signals
82	GN/BK	Shift Solenoid 'C' Control	In Drive: 1v, Others: 12v
83	GN/YL	IAC Motor Control	9v (33% duty cycle)
85	WT/VT	CMP Sensor Signal	5-7 Hz
86	WT/RD	A/C Pressure Switch	A/C On: 12v (Open)
87	BR/YL	Knock Sensor Signal (-)	<0.050v
88	WT/BL	MAF Sensor Signal	0.6-0.9v
89	WT	TP Sensor Signal	0.53-1.27v
90	YL	Reference Voltage	4.9-5.1v
91	BR	Analog Signal Return	<0.050v
92	GN/RD	Brake Pedal Position Switch	Brake Off: 0v, On: 12v
93	BK/YL	HO2S-11 (B1 S1) Heater	On: 1v, Off: 12v
94, 98	---	Not Used	---
95	BK/OR	Injector 4 Control	2.3-2.9 ms
96	BK/YL	Injector 3 Control	2.3-2.9 ms
97	GN/YL	Vehicle Power	12-14v
99	GN/OR	Shift Solenoid 'D' Control	In Drive: 1v, Others: 12v
100	BK/BL	HO2S-12 (B1 S2) Heater	On: 1v, Off: 12v
101	---	Not Used	---
102	GN/WT	Shift Solenoid 'E' Control	In O/D: 1v, Others: 12v
103	BK/YL	Power Ground	<0.1v
104	---	Not Used	---

Pin Connector Graphic

PCM 104-PIN CONNECTOR

Terminal View of 104-Pin PCM Wiring Harness Connector

Mustang Pin Tables

1990 Mustang 2.3L I4 MFI VIN A (All) 60 Pin Connector

PCM Pin #	Wire Color	Circuit Description (60 Pin)	Value at Hot Idle
1	BK/OR	Keep Alive Power	12-14v
2	RD/LG	Brake Pedal Position Switch	Brake Off: 0v, On: 12v
3	YL/LG	PSP Switch Signal	Straight: 0v, Turned: 10v
4	DG/YL	Ignition Diagnostic Monitor	20-31 Hz
5-6, 8-9	---	Not Used	---
7	LG/YL	ECT Sensor Signal	0.5-0.6v
10	BK/YL	A/C Cycling Clutch Switch	A/C On: 12v, Off: 0v
11-15, 18-19	---	Not Used	---
16	BK/OR	Ignition System Ground	<0.050v
17	TN	STI Output, MIL Control	MIL On: 1v, Off: 12v
20	BK/LG	PCM Case Ground	<0.050v
21	WT/BL	IAC Motor Control	8-9v
22	TN/LG	Fuel Pump Control	On: 1v, Off: 12v
23	YL/RD	Knock Sensor Signal	0v
24	LG/PK	A/C Discharge Switch	A/C On: 12v, Off: 0v
25	LG/RD	ACT Sensor Signal	1.5-2.5v
25	LG/PK	ACT Sensor Signal	1.5-2.5v
26	OR/WT	Reference Voltage	4.9-5.1v
27	BR/LG	EGR EVP Sensor Signal	0.4v
28, 31-32	---	Not Used	---
29	DG/PKK	HO2S-11 (B1 S1) Signal	0.1-1.1v
29	DG/PK	HO2S-11 (B1 S1) Signal	0.1-1.1v
30	BK/WT	Neutral Drive Switch Signal	In 'N': 0v, Others: 5v
30	YL/RD	Neutral Drive Switch Signal	In 'N': 0v, Others: 5v
30	BK/WT	Clutch Engagement Switch	Clutch In: 0v, Out: 5v
30	YL/RD	Clutch Engagement Switch	Clutch In: 0v, Out: 5v
33	DG	EGR Vent Solenoid Control	12v, 55 mph: 1v
34-35	---	Not Used	---
36	YL/LG	Spark Output Signal	50% duty cycle
37, 57	RD	Vehicle Power	12-14v
38-39, 41-44	---	Not Used	---
40	BK/LG	Power Ground	<0.1v
45	LG/BK	MAP Sensor Signal	107 Hz
46	BK/WT	Analog Signal Return	<0.050v
47	DG/LG	TP Sensor Signal	0.5-1.2v
48	WT/RD	Self-Test Indicator Signal	STI Open: 5v, Closed: 1v
49	OR	HO2S-11 (Bank 1) Ground	<0.050v
50-51, 55	---	Not Used	---
52	YL	EGR Control Solenoid	Off-idle: 12v
53	OR/YL	Converter Clutch Override	Off: 12v, On: 1v
54	OR/BL	A/C WOT Relay Control	On: 1v, Off: 12v
56	DB	PIP Sensor Signal	50% duty cycle
58	TN/OR	Injector Bank 1 (INJ 1 & 4)	3.4-4.4 ms
59	TN/RD	Injector Bank 2 (INJ 2 & 3)	3.4-4.4 ms
60	BK	Power Ground	<0.1v

Pin Connector Graphic

PCM 60-PIN CONNECTOR

Terminal View of 60-Pin PCM Harness Connector

1991-93 Mustang 2.3L I4 MFI VIN M (All) 60 Pin Connector

PCM Pin #	Wire Color	Circuit Description (60 Pin)	Value at Hot Idle
1	YL	Keep Alive Power	12-14v
2	LG	Brake Pedal Position Switch	Brake Off: 0v, On: 12v
3	GY/BK	VSS (+) Signal	0 Hz, 55 mph: 125 Hz
4	TN/YL	Ignition Diagnostic Monitor	20-31 Hz
5	DB/OR	CID Sensor Signal	5-7 Hz
6	PK/OR	VSS (-) Signal	0 Hz, 55 mph: 125 Hz
7	LG/RD	ECT Sensor Signal	0.5-0.6v
8	DG/YL	Fuel Pump Monitor	On: 12v, Off: 0v
9, 28	PK, TN/OR	Data Bus (-), (+) Signals	Digital Signals
10	BK/YL	A/C Cycling Clutch Switch	A/C On: 12v, Off: 0v
11-13, 18-19, 23	---	Not Used	---
14	BL/RD	MAF Sensor Signal	0.6v
15	TN/BL	MAF Sensor Return	<0.050v
16	OR/RD	Ignition System Ground	<0.050v
17	PK/LG	STI Output, MIL Control	MIL On: 1v, Off: 12v
20	BK	PCM Case Ground	<0.050v
21	WT/BL	IAC Motor Control	8-9v
22	BL/OR	Fuel Pump Control	On: 1v, Off: 12v
24	YL/LG	PSP Switch Signal	Straight: 0v, Turned: 10v
25	GY	ACT Sensor Signal	1.5-2.5v
26	BR/WT	Reference Voltage	4.9-5.1v
27	BR/LG	EGR EVP Sensor	0.3v
29	GY/BL	HO2S-11 (B1 S1) Signal	0.1-1.1v
30	BL/YL	Neutral Drive Switch Signal	In 'N': 0v, Others: 5v
31	GY/YL	EVAP Purge Solenoid	0%, 35 mph: 40%
32	DB/YL	Dual Plug Inhibit Signal	DPI On: 12v, Off: 0v
33	BR/PK	EGR VR Solenoid	0%, 55 mph: 45%
34-35, 38-39	---	Not Used	---
36	PK	Spark Output Signal	50% dwell
37, 57	RD	Vehicle Power	12-14v
40, 60	BK	Power Ground	<0.1v
41-43, 50, 55	---	Not Used	---
44	DG	Octane Adjust Switch	Closed: 0v, Open: 9.1v
45	LG/BK	BARO Sensor Signal	159 Hz (sea level)
46	GY/RD	Analog Signal Return	<0.050v
47	GY/WT	TP Sensor Signal	0.5-1.2v
48	WT/PK	Self-Test Indicator Signal	STI Open: 5v, Closed: 1v
49	OR	HO2S-11 (Bank 1) Ground	<0.050v
51	LG/PK	Low Speed Fan Control	On: 1v, Off: 12v
52	OR/YL	Shift Solenoid 3-4 Control	On: 1v, Off: 12v
53	PK/YL	Converter Clutch Override	0%, 55 mph: 95%
54	PK/YL	A/C WOT Relay Control	On: 1v, Off: 12v
56	GY/OR	PIP Sensor Signal	50% duty cycle
58	TN	Injector Bank 1 (INJ 1 & 4)	3.0-3.8 ms
59	WT	Injector Bank 1 (INJ 2 & 3)	3.0-3.8 ms

Pin Connector Graphic

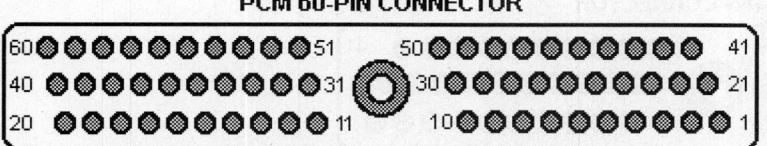

Terminal View of 60-Pin PCM Harness Connector

1994-97 Mustang 3.8L V6 MFI VIN 4 (All) 104 Pin Connector

PCM Pin #	Wire Color	Circuit Description (104 Pin)	Value at Hot Idle
1	PK/OR	Shift Solenoid 2 Control	12v, 55 mph: 1v
2	PK/LG	MIL (lamp) Control	MIL On: 1v, Off: 12v
3-10, 12, 14	---	Not Used	---
11	PK/WT	Purge Flow Sensor	0.8v, 55 mph: 3v
13	P	Flash EEPROM Power	0.1v
15	PK/LB	Data Bus (-) Signal	Digital Signals
16	TN/OR	Data Bus (+) Signal	Digital Signals
17-20, 23	---	Not Used	---
21	DB	CKP Sensor (+) Signal	850-1120 Hz
22	GY	CKP Sensor (-) Signal	850-1120 Hz
24	BK/WT	Power Ground	<0.1v
25	BK	PCM Case Ground	<0.050v
26	DB/LG	Coil Driver 1 Control	5° dwell
27	OR/YL	Shift Solenoid 1 Control	1v, 55 mph: 12v
28, 31, 32	---	Not Used	---
29	TN/WT	TCS (switch) Signal	TCS & O/D On: 12v
30	DG	Octane Adjust Switch	Closed: 0v, Open: 9.3v
33	PK/OR	VSS (-) Signal	0 Hz, 55 mph: 125 Hz
34	---	Not Used	---
35	RD/LG	HO2S-12 (B1 S2) Signal	0.1-1.1v
36	TN/LB	MAF Sensor Return	<0.050v
37	OR/BK	TFT Sensor Signal	2.10-2.40v
38	LG/RD	ECT Sensor Signal	0.5-0.6v
39	GY	IAT Sensor Signal	1.5-2.5v
40	DG/YL	Fuel Pump Monitor	On: 12v, Off: 0v
41	DG/OR	A/C Cycling Clutch Switch	A/C On: 12v, Off: 0v
42-44, 46	---	Not Used	---
45	DB	Low Speed Fan Control	On: 1v, Off: 12v
47	BR/PK	EGR VR Solenoid	0%, 55 mph: 45%
48	OR/WT	Clean Tachometer Output	33-37 Hz
49-50, 53	---	Not Used	---
51	BK/WT	Power Ground	<0.1v
52	RD/LB	Coil Driver 2 Control	5° dwell
54	BR/OR	TCC Solenoid Control	TCC Off Idle: 0%
55	YL	Keep Alive Power	12-14v
56, 57, 59	---	Not Used	---
58	GY/BK	VSS (+) Signal	0 Hz, 55 mph: 125 Hz
60	GY/LB	HO2S-11 (B1 S1) Signal	0.1-1.1v
61	PK/LG	HO2S-22 (B2 S2) Signal	0.1-1.1v
62-63, 66	---	Not Used	---
64	LB/YL	TR Sensor Signal	In 'P': 0v, 55 mph: 1.7v
64	LB/YL	Neutral Position Switch	In 'N': 0v, Others: 5v
64	LB/YL	Clutch Pedal Position Switch	Clutch In: 0v, Out: 5v

1994-97 Mustang 3.8L V6 MFI VIN 4 (All) 104 Pin Connector

PCM Pin #	Wire Color	Circuit Description (104 Pin)	Value at Hot Idle
65	BR/LG	DPFE Sensor Signal	0.20-1.30v
67	GY/YL	EVAP Purge Solenoid	0-10 Hz (0-100%)
68	---	Not Used	---
69	PK/YL	A/C WOT Relay Control	On: 1v, Off: 12v
70, 72	---	Not Used	---
71	RD	Vehicle Power	12-14v
73	TN/BK	Injector 5 Control	4.9-5.1 ms
74	BR/YL	Injector 3 Control	4.9-5.1 ms
75	TN	Injector 1 Control	4.9-5.1 ms
76	BK/WT	Power Ground	<0.1v
77	BK/WT	Power Ground	<0.1v
78	PK/WT	Coil Driver 3 Control	5° dwell
79	WT/LG	TCIL (lamp) Control	On: 1v, Off: 12v
80	LB/OR	Fuel Pump Control	On: 1v, Off: 12v
81	WT/YL	EPC Solenoid Control	9.0v (15 psi)
82	PK/YL	TCC Solenoid	12v (0%)
83	WT/LB	IAC Motor Control	30% duty cycle
84	DG/WT	TSS Sensor Signal	43 Hz (650 rpm)
85	DB/OR	CID Sensor Signal	5-7 Hz
86	TN/LG	A/C Pressure Switch	A/C On: 12v (Open)
87	RD/BK	HO2S-21 (B2 S1) Signal	0.1-1.1v
88	LB/RD	MAF Sensor Signal	0.6v
89	GY/WT	TP Sensor Signal	0.53-1.27v
90	BR/WT	Reference Voltage	4.9-5.1v
91	GY/RD	Analog Signal Return	<0.050v
92	LG	Brake Pedal Position Switch	Brake Off: 0v, On: 12v
93	RD/WT	HO2S-11 (B1 S1) Heater	On: 1v, Off: 12v
94	YL/LB	HO2S-21 (B2 S1) Heater	On: 1v, Off: 12v
95	WT/BK	HO2S-12 (B1 S2) Heater	On: 1v, Off: 12v
96	TN/YL	HO2S-22 (B2 S2) Heater	On: 1v, Off: 12v
97	RD	Vehicle Power	12-14v
98	---	Not Used	---
99	LG/OR	Injector 6 Control	4.9-5.1 ms
100	BR/LB	Injector 4 Control	4.9-5.1 ms
101	WT	Injector 2 Control	4.9-5.1 ms
102	---	Not Used	---
103	BK/WT	Power Ground	<0.1v
104	---	Not Used	---

Pin Connector Graphic

PCM 104-PIN CONNECTOR

Terminal View of 104-Pin PCM Wiring Harness Connector

Standard Colors and Abbreviations

Abbreviation	Color	Abbreviation	Color	Abbreviation	Color
BK	Black	GY	Gray	PK	Purple
BL	Blue	GN	Green	RD	Red
BR	Brown	LG	LT Green	TN	Tan
DB	Dark Blue	OR	Orange	WT	White
DG	DK Green	PK	Pink	YL	Yellow

1998-99 Mustang 3.8L V6 MFI VIN 4 (All) 104 Pin Connector

PCM Pin #	Wire Color	Circuit Description (104 Pin)	Value at Hot Idle
1	PK/OR	Shift Solenoid 2 Control	12v, Others: 1v
2	PK/LG	MIL (lamp) Control	MIL On: 1v, Off: 12v
3	YL/BK	Digital TR1 Sensor	In 'P': 0v, 55 mph: 11v
4	---	Not Used	
5	DG/YL	AIR Monitor Signal	Off: 0.1v
6-10	---	Not Used	
11	PK/WT	Purge Flow Sensor	0.8v, 55 mph: 3v
12	---	Not Used	
13	PK	Flash EEPROM Power	0.1v
14	---	Not Used	---
15	PK/LB	Data Bus (-) Signal	Digital Signals
16	TN/OR	Data Bus (+) Signal	Digital Signals
17-20	---	Not Used	---
21	DB	CKP Sensor (+) Signal	390-450 Hz
22	GY	CKP Sensor (-) Signal	390-450 Hz
23	---	Not Used	---
24	BK/WT	Power Ground	<0.1v
25	BK	PCM Case Ground	<0.050v
26	DB/LG	Coil Driver 1 Control	5° dwell
27	OR/YL	Shift Solenoid 1 Control	1v, Others: 1v
28	---	Not Used	---
29	TN/WT	TCS (switch) Signal	TCS & O/D On: 12v
30	DG	Octane Adjust Switch	Closed: 0v, Open: 9.3v
31-32	---	Not Used	---
33	PK/OR	VSS (-) Signal	0 Hz, 55 mph: 125 Hz
34	---	Not Used	---
35	RD/LG	HO2S-12 (B1 S2) Signal	0.1-1.1v
36	TN/LB	MAF Sensor Return	<0.050v
37	OR/BK	TFT Sensor Signal	2.10-2.40v
38	LG/RD	ECT Sensor Signal	0.5-0.6v
39	GY	IAT Sensor Signal	1.5-2.5v
40	DG/YL	Fuel Pump Monitor	On: 12v, Off: 0v
41	DG/OR	A/C High Pressure Switch	A/C On: 12v, Off: 1v
42-44	---	Not Used	---
45	DB	Low Speed Fan Control	On: 1v, Off: 12v
46	---	Not Used	---
47	BR/PK	EGR VR Solenoid	0%, 55 mph: 45%
48	OR/WT	Clean Tachometer Output	33-37 Hz
49	LB/BK	Digital TR2 Sensor	In 'P': 0v, 55 mph: 11v
50	WT/BK	Digital TR4 Sensor	In 'P': 0v, 55 mph: 11v
51	BK/WT	Power Ground	<0.1v
52	RD/LB	Coil Driver 2 Control	5° dwell
54	PK/YL	TCC Solenoid Control	0%, 55 mph: 95%
55	YL	Keep Alive Power	12-14v
56	LG/BK	EVAP CV Solenoid	0-10 Hz (0-100%)
57	---	Not Used	---
58	GY/BK	VSS (+) Signal	0 Hz, 55 mph: 125 Hz
59	---	Not Used	---
60	GY/LB	HO2S-11 (B1 S1) Signal	0.1-1.1v
61	PK/LG	HO2S-22 (B2 S2) Signal	0.1-1.1v
62-63	---	Not Used	---
64	LB/YL	TR Sensor	In 'P': 0v, in O/D: 1.7v
64	LB/YL	Neutral Position Switch	In 'N': 0v, Others: 5v
64	LB/YL	Clutch Pedal Position Switch	Clutch In: 0v, Out: 5v

1998-99 Mustang 3.8L V6 MFI VIN 4 (All) 104 Pin Connector

PCM Pin #	Wire Color	Circuit Description (104 Pin)	Value at Hot Idle
65	BR/LG	DPFE Sensor Signal	0.20-1.30v
66, 68	---	Not Used	---
67	GY/YL	EVAP Purge Solenoid	0-10 Hz (0-100%)
69	PK/YL	A/C WOT Relay Control	On: 1v, Off: 12v
70	WT/OR	AIR Bypass Solenoid	AIRB On: 1v, Off: 12v
71	RD	Vehicle Power	12-14v
72	---	Not Used	---
73	TN/BK	Injector 5 Control	4.9-5.1 ms
74	BR/YL	Injector 3 Control	4.9-5.1 ms
75	TN	Injector 1 Control	4.9-5.1 ms
76-77	BK/WT	Power Ground	<0.1v
78	PK/WT	Coil Driver 3 Control	6° dwell
79	WT/LG	TCIL (lamp) Control	On: 1v, Off: 12v
80	LB/OR	Fuel Pump Control	On: 1v, Off: 12v
81	WT/YL	EPC Solenoid Control	8.3v (8 psi)
82	PK/YL	TCC Solenoid Control	0%, 55 mph: 95%
83	WT/LB	IAC Motor Control	30% duty cycle
84	DG/WT	OSS Sensor Signal	0, 55 mph: 2500 rpm
85	DB/OR	CMP Sensor Signal	5-7 Hz
86	TN/LG	A/C Pressure Switch	A/C On: 12v (Open)
87	RD/BK	HO2S-21 (B2 S1) Signal	0.1-1.1v
88	LB/RD	MAF Sensor Signal	0.6v
89	GY/WT	TP Sensor Signal	0.53-1.27v
90	BR/WT	Reference Voltage	4.9-5.1v
91	GY/RD	Analog Signal Return	<0.050v
92	LG	Brake Pedal Position Switch	Brake Off: 0v, On: 12v
93	RD/WT	HO2S-11 (B1 S1) Heater	On: 1v, Off: 12v
94	YL/LB	HO2S-21 (B2 S1) Heater	On: 1v, Off: 12v
95	WT/BK	HO2S-12 (B1 S2) Heater	On: 1v, Off: 12v
96	TN/YL	HO2S-22 (B2 S2) Heater	On: 1v, Off: 12v
97	RD	Vehicle Power	12-14v
98	---	Not Used	---
99	LG/OR	Injector 6 Control	4.9-5.1 ms
100	BR/LB	Injector 4 Control	4.9-5.1 ms
101	W	Injector 2 Control	4.9-5.1 ms
102	---	Not Used	---
103	BK/WT	Power Ground	<0.1v
104	---	Not Used	---

Pin Connector Graphic

PCM 104-PIN CONNECTOR

Terminal View of 104-Pin PCM Wiring Harness Connector

Standard Colors and Abbreviations

Abbreviation	Color	Abbreviation	Color	Abbreviation	Color
BK	Black	GY	Gray	PK	Purple
BL	Blue	GN	Green	RD	Red
BR	Brown	LG	LT Green	TN	Tan
DB	Dark Blue	OR	Orange	WT	White
DG	DK Green	PK	Pink	YL	Yellow

2000 Mustang 3.8L V6 MFI VIN 4 (All) 104 Pin Connector

PCM Pin #	Wire Color	Circuit Description (104 Pin)	Value at Hot Idle
1	PK/OR	Shift Solenoid 'B' Control	12v, Others: 1v
3	YL/BK	Digital TR1 Sensor	In 'P': 0v, 55 mph: 11v
4	---	Not Used	---
5	WT	Air Injection Pump Monitor	1v, 55 mph: 1v
6-12	---	Not Used	---
13	PK	Flash EEPROM Power	0.1v
14	---	Not Used	---
15	PK/LB	Data Bus (-) Signal	Digital Signals
16	TN/OR	Data Bus (+) Signal	Digital Signals
17-20	---	Not Used	---
21	DB	CKP Sensor (+) Signal	390-450 Hz
22	GY	CKP Sensor (-) Signal	390-450 Hz
23-24	---	Not Used	---
25	BK	PCM Case Ground	<0.050v
26	DB/LG	Coil Driver 1 Control	5° dwell
27	OR/YL	Shift Solenoid 'A' Control	1v, Others: 1v
28	---	Not Used	---
29	TN/WT	TCS (switch) Signal	TCS & O/D On: 12v
30-34	---	Not Used	---
35	RD/LG	HO2S-12 (B1 S2) Signal	0.1-1.1v
36	TN/LB	MAF Sensor Return	<0.050v
37	OR/BK	TFT Sensor Signal	2.10-2.40v
38	LG/RD	ECT Sensor Signal	0.5-0.6v
39	GY	IAT Sensor Signal	1.5-2.5v
40	LB/OR	Fuel Pump Monitor	On: 12v, Off: 0v
41	DG/OR	A/C Cycling Switch	A/C On: 12v, Off: 1v
42	---	Not Used	---
45	DB	Low Speed Fan Control	On: 1v, Off: 12v
46	---	Not Used	---
47	BR/PK	EGR VR Solenoid	0%, 55 mph: 45%
48	---	Not Used	---
49	LB/BK	Digital TR2 Sensor	In 'P': 0v, 55 mph: 11v
50	WT/BK	Digital TR4 Sensor	In 'P': 0v, 55 mph: 11v
51	BK/WT	Power Ground	<0.1v
52	RD/LB	Coil Driver 2 Control	5° dwell
54	BR/OR	TCC Solenoid Control	0%, 55 mph: 95%
55	RD/WT	Keep Alive Power	12-14v
56	LG/BK	EVAP CV Solenoid	0-10 Hz (0-100%)
57-59	---	Not Used	---
60	GY/LB	HO2S-11 (B1 S1) Signal	0.1-1.1v
61	PK/LG	HO2S-22 (B2 S2) Signal	0.1-1.1v
62	RD/PK	FTP Sensor Signal	2.6v (0" H2O - cap off)
63	RD/PK	FRP Sensor Signal	2.8v (39 psi)
64	LB/YL	Digital TR3A Sensor	In 'P': 0v, in O/D: 1.7v

2000 Mustang 3.8L V6 MFI VIN 4 (All) 104 Pin Connector

PCM Pin #	Wire Color	Circuit Description (104 Pin)	Value at Hot Idle
65	BR/LG	DPFE Sensor Signal	0.20-1.30v
66	---	Not Used	---
67	GY/YL	EVAP Purge Solenoid	0-10 Hz (0-100%)
68	WT/OR	VSS (+) Signal	0 Hz, 55 mph: 125 Hz
69	PK/YL	A/C WOT Relay Control	On: 1v, Off: 12v
70	WT/OR	Electronic Air Management	On: 1v, Off: 12v
71	RD	Vehicle Power	12-14v
72	---	Not Used	---
73	TN/BK	Injector 5 Control	4.9-5.1 ms
74	BR/YL	Injector 3 Control	4.9-5.1 ms
75	T	Injector 1 Control	4.9-5.1 ms
76-77	BK/WT	Power Ground	<0.1v
78	PK/WT	Coil Driver 3 Control	6° dwell
79	---	Not Used	---
80	WT/RD	Fuel Pump Control	On: 1v, Off: 12v
81	WT/YL	EPC Solenoid Control	8.3v (8 psi)
82, 86	---	Not Used	---
83	WT/LB	IAC Motor Control	30% duty cycle
84	DG/WT	OSS Sensor Signal	0, 55 mph: 2500 rpm
85	DB/OR	CMP Sensor Signal	5-7 Hz
87	RD/BK	HO2S-21 (B2 S1) Signal	0.1-1.1v
88	LB/RD	MAF Sensor Signal	0.6v
89	GY/WT	TP Sensor Signal	0.53-1.27v
90	BR/WT	Reference Voltage	4.9-5.1v
91	GY/RD	Analog Signal Return	<0.050v
92	LG	Brake Pedal Position Switch	Brake Off: 0v, On: 12v
93	RD/WT	HO2S-11 (B1 S1) Heater	On: 1v, Off: 12v
94	YL/LB	HO2S-21 (B2 S1) Heater	On: 1v, Off: 12v
95	WT/BK	HO2S-12 (B1 S2) Heater	On: 1v, Off: 12v
96	TN/YL	HO2S-22 (B2 S2) Heater	On: 1v, Off: 12v
97	RD	Vehicle Power	12-14v
98	---	Not Used	---
99	LG/OR	Injector 6 Control	4.9-5.1 ms
100	BR/LB	Injector 4 Control	4.9-5.1 ms
101	W	Injector 2 Control	4.9-5.1 ms
102	---	Not Used	---
103	BK/WT	Power Ground	<0.1v
104	---	Not Used	---

Pin Connector Graphic

PCM 104-PIN CONNECTOR

Terminal View of 104-Pin PCM Wiring Harness Connector

Standard Colors and Abbreviations

Abbreviation	Color	Abbreviation	Color	Abbreviation	Color
BK	Black	GY	Gray	PK	Purple
BL	Blue	GN	Green	RD	Red
BR	Brown	LG	LT Green	TN	Tan
DB	Dark Blue	OR	Orange	WT	White
DG	DK Green	PK	Pink	YL	Yellow

2001 Mustang 3.8L V6 MFI VIN 4 (All) 104 Pin Connector

PCM Pin #	Wire Color	Circuit Description (104 Pin)	Value at Hot Idle
1	VT/OR	Shift Solenoid 2 Control	12v, Others: 1v
2	---	Not Used	---
3	YL/BK	Digital TR1 Sensor	In 'P': 0v, 55 mph: 11v
4-7	---	Not Used	---
8	DB/YL	IMRC Solenoid Control	5v (Off)
9-12	---	Not Used	---
13	PK	Flash EEPROM Power	0.1v
14	---	Not Used	---
15	PK/LB	Data Bus (-) Signal	Digital Signals
16	TN/OR	Data Bus (+) Signal	Digital Signals
17-20	---	Not Used	---
21	DB	CKP Sensor (+) Signal	390-450 Hz
22	GY	CKP Sensor (-) Signal	390-450 Hz
23-24	---	Not Used	---
25	BK	PCM Case Ground	<0.050v
26	DB/LG	Coil Driver 1 Control	5° dwell
27	OR/YL	Shift Solenoid 'A' Control	1v, Others: 1v
28	---	Not Used	---
29	TN/WT	TCS (switch) Signal	TCS & O/D On: 12v
30-34	---	Not Used	---
35	RD/LG	HO2S-12 (B1 S2) Signal	0.1-1.1v
36	TN/LB	MAF Sensor Return	<0.050v
37	OR/BK	TFT Sensor Signal	2.10-2.40v
38	LG/RD	ECT Sensor Signal	0.5-0.6v
39	GY	IAT Sensor Signal	1.5-2.5v
40	LB/OR	Fuel Pump Monitor	Pump On: 3.5v, Off: 0v
41	DG/OR	A/C Cycling Switch	A/C On: 12v, Off: 1v
42	BR	IMRC Solenoid Control	12v (Off)
43-44	---	Not Used	---
45	DB	Low Speed Fan Control	On: 1v, Off: 12v
46	---	Not Used	---
47	BR/PK	EGR VR Solenoid	0%, 55 mph: 45%
48	---	Not Used	---
49	LB/BK	Digital TR2 Sensor	In 'P': 0v, 55 mph: 11v
50	WT/BK	Digital TR4 Sensor	In 'P': 0v, 55 mph: 11v
51	BK/WT	Power Ground	<0.1v
52	RD/LB	Coil Driver 2 Control	5° dwell
54	BR/OR	TCC Solenoid Control	0%, 55 mph: 95%
55	RD/WT	Keep Alive Power	12-14v
56	LG/BK	EVAP CV Solenoid	0 Hz (0%)
57-59	---	Not Used	---
60	GY/LB	HO2S-11 (B1 S1) Signal	0.1-1.1v
61	PK/LG	HO2S-22 (B2 S2) Signal	0.1-1.1v
62	RD/PK	FTP Sensor Signal	2.6v (0" H2O - cap off)
63	RD/PK	FRP Sensor Signal	2.8v (39 psi)
64	LB/YL	Digital TR3A Sensor	In 'P': 0v, in O/D: 1.7v

2001 Mustang 3.8L V6 MFI VIN 4 (All) 104 Pin Connector

PCM Pin #	Wire Color	Circuit Description (104 Pin)	Value at Hot Idle
65	BR/LG	DPFE Sensor Signal	0.95-1.05v
66	YL/LG	CHT Sensor Signal	0.7v (194°F)
67	GY/YL	EVAP Purge Solenoid	0-10 Hz (0-100%)
68	WT/OR	VSS (+) Signal	0 Hz, 55 mph: 125 Hz
69	PK/YL	A/C WOT Relay Control	On: 1v, Off: 12v
71	RD	Vehicle Power	12-14v
72	---	Not Used	---
73	TN/BK	Injector 5 Control	3.8-4.9 ms
74	BR/YL	Injector 3 Control	3.8-4.9 ms
75	T	Injector 1 Control	3.8-4.9 ms
76-77	BK/WT	Power Ground	<0.1v
78	PK/WT	Coil Driver 3 Control	6° dwell
79	---	Not Used	---
80	WT/RD	Fuel Pump Control	On: 1v, Off: 12v
81	WT/YL	EPC Solenoid Control	8.3v (8 psi)
82	---	Not Used	---
83	WT/LB	IAC Motor Control	30% duty cycle
84	DG/WT	OSS Sensor Signal	0 Hz, 55 mph: 240 Hz
85	DB/OR	CMP Sensor Signal	5-7 Hz
86	---	Not Used	---
87	RD/BK	HO2S-21 (B2 S1) Signal	0.1-1.1v
88	LB/RD	MAF Sensor Signal	0.6-0.9v
89	GY/WT	TP Sensor Signal	0.53-1.27v
90	BR/WT	Reference Voltage	4.9-5.1v
91	GY/RD	Analog Signal Return	<0.050v
92	LG	Brake Pedal Position Switch	Brake Off: 0v, On: 12v
93	RD/WT	HO2S-11 (B1 S1) Heater	On: 1v, Off: 12v
94	YL/LB	HO2S-21 (B2 S1) Heater	On: 1v, Off: 12v
95	WT/BK	HO2S-12 (B1 S2) Heater	On: 1v, Off: 12v
96	TN/YL	HO2S-22 (B2 S2) Heater	On: 1v, Off: 12v
97	RD	Vehicle Power	12-14v
99	LG/OR	Injector 6 Control	3.8-4.9 ms
100	BR/LB	Injector 4 Control	3.8-4.9 ms
101	WT	Injector 2 Control	3.8-4.9 ms
102	---	Not Used	---
103	BK/WT	Power Ground	<0.1v
104	---	Not Used	---

Pin Connector Graphic

PCM 104-PIN CONNECTOR

1 ●●●●●●●●●●●● ●●●●●●●●●●●● 26
27 ●●●●●●●●●●●● ●●●●●●●●●●●● 52
53 ●●●●●●●●●●●● ●●●●●●●●●●●● 78
79 ●●●●●●●●●●●● ●●●●●●●●●●●● 104

Terminal View of 104-Pin PCM Wiring Harness Connector

Standard Colors and Abbreviations

Abbreviation	Color	Abbreviation	Color	Abbreviation	Color
BK	Black	GY	Gray	PK	Purple
BL	Blue	GN	Green	RD	Red
BR	Brown	LG	LT Green	TN	Tan
DB	Dark Blue	OR	Orange	WT	White
DG	DK Green	PK	Pink	YL	Yellow

2002-03 Mustang 3.8L V6 MFI VIN 4 (All) 104 Pin Connector

PCM Pin #	Wire Color	Circuit Description (104 Pin)	Value at Hot Idle
1	VT/OR	Shift Solenoid 'B' Control	12v, Others: 1v
2	---	Not Used	---
3	YL/BK	Digital TR1 Sensor	In 'P': 0v, 55 mph: 11v
4-6	---	Not Used	---
7	---	Not Used	---
8	DB/YL	IMRC Solenoid Control	5v (Off)
9	---	Not Used	---
10	DB/LG	Barometric Pressure Sensor Signal	159 Hz (sea level)
11-12	---	Not Used	---
13	VT	Flash EEPROM Power	0.1v
14	---	Not Used	---
15	PK/LB	SCP Bus (-) Signal	Digital Signals
16	TN/OR	SCP Bus (+) Signal	Digital Signals
17-20	---	Not Used	---
21	DB	CKP Sensor (+) Signal	390-450 Hz
22	GY	CKP Sensor (-) Signal	390-450 Hz
23-24	---	Not Used	---
25	BK	PCM Case Ground	<0.050v
26	DB/LG	Coil Driver 1 Control	5° dwell
27	OR/YL	Shift Solenoid 'A' Control	1v, Others: 1v
28	---	Not Used	---
29	TN/WT	TCS (switch) Signal	TCS & O/D On: 12v
30-34	---	Not Used	---
35	RD/LG	HO2S-12 (B1 S2) Signal	0.1-1.1v
36	TN/LB	MAF Sensor Return	<0.050v
37	OR/BK	TFT Sensor Signal	2.10-2.40v
38	---	Not Used	---
39	GY	IAT Sensor Signal	1.5-2.5v
40	LB/OR	Fuel Pump Monitor	Pump On: 3.5v, Off: 0v
41	DG/OR	A/C Cycling Switch	A/C On: 12v, Off: 1v
42	BR	IMRC Solenoid Control	12v (Off)
43-44	---	Not Used	---
45	DB	Low Speed Fan Control	On: 1v, Off: 12v
46	DB	Low Speed Fan Relay Control	Relay Off: 12v, On: 1v
47	BR/PK	EGR VR Solenoid	0%, 55 mph: 45%
48	---	Not Used	---
49	LB/BK	Digital TR2 Sensor	In 'P': 0v, 55 mph: 11v
50	WT/BK	Digital TR4 Sensor	In 'P': 0v, 55 mph: 11v
51	BK/WT	Power Ground	<0.1v
52	RD/LB	Coil Driver 2 Control	5° dwell
53	---	Not Used	---
54	BR/OR	TCC Solenoid Control	0%, 55 mph: 95%
55	RD/WT	Keep Alive Power	12-14v
56	LG/BK	EVAP CV Solenoid	0 Hz (0%)
57-59	---	Not Used	---
60	GY/LB	HO2S-11 (B1 S1) Signal	0.1-1.1v
61	VT/LG	HO2S-22 (B2 S2) Signal	0.1-1.1v
62	RD/PK	FTP Sensor Signal	2.6v (0" H2O - cap off)
63	RD/PK	FRP Sensor Signal	2.8v (39 psi)
64	LB/YL	Digital TR3A Sensor	In 'P': 0v, in O/D: 1.7v

2002-03 Mustang 3.8L V6 MFI VIN 4 (All) 104 Pin Connector

PCM Pin #	Wire Color	Circuit Description (104 Pin)	Value at Hot Idle
65	BR/LG	DPFE Sensor Signal	0.95-1.05v
66	YL/LG	CHT Sensor Signal	0.7v (194°F)
67	GY/YL	EVAP Purge Solenoid	0-10 Hz (0-100%)
68	WT/OR	VSS (+) Signal	0 Hz, 55 mph: 125 Hz
69	PK/YL	A/C WOT Relay Control	On: 1v, Off: 12v
70	---	Not Used	---
71	RD	Vehicle Power (Start-Run)	12-14v
72	---	Not Used	---
73	TN/BK	Injector 5 Control	3.8-4.9 ms
74	BR/YL	Injector 3 Control	3.8-4.9 ms
75	TN	Injector 1 Control	3.8-4.9 ms
76, 77	BK/WT	Power Ground	<0.1v
78	PK/WT	Coil Driver 3 Control	6° dwell
79	---	Not Used	---
80	WT/RD	Fuel Pump Control	Off: 12v, On: 1v
81	WT/YL	EPC Solenoid Control	8.3v (8 psi)
82	---	Not Used	---
83	WT/LB	IAC Motor Control	30% duty cycle
84	DG/WT	OSS Sensor Signal	0 Hz, 55 mph: 240 Hz
85	DB/OR	CMP Sensor Signal	5-7 Hz
86, 98	---	Not Used	---
87	RD/BK	HO2S-21 (B2 S1) Signal	0.1-1.1v
88	LB/RD	MAF Sensor Signal	0.6-0.9v
89	GY/WT	TP Sensor Signal	0.53-1.27v
90	BR/WT	Reference Voltage	4.9-5.1v
91	GY/RD	Sensor Return	<0.050v
92	LG	Brake Pedal Position Switch	Brake Off: 0v, On: 12v
93	RD/WT	HO2S-11 (B1 S1) Heater	On: 1v, Off: 12v
94	YL/LB	HO2S-21 (B2 S1) Heater	On: 1v, Off: 12v
95	WT/BK	HO2S-12 (B1 S2) Heater	On: 1v, Off: 12v
96	TN/YL	HO2S-22 (B2 S2) Heater	On: 1v, Off: 12v
97	RD	Vehicle Power	12-14v
99	LG/OR	Injector 6 Control	3.8-4.9 ms
100	BR/LB	Injector 4 Control	3.8-4.9 ms
101	WT	Injector 2 Control	3.8-4.9 ms
102	---	Not Used	---
103	BK/WT	Power Ground	<0.1v
104	---	Not Used	---

Pin Connector Graphic

PCM 104-PIN CONNECTOR

Terminal View of 104-Pin PCM Wiring Harness Connector

Standard Colors and Abbreviations

Abbreviation	Color	Abbreviation	Color	Abbreviation	Color
BK	Black	GY	Gray	PK	Purple
BL	Blue	GN	Green	RD	Red
BR	Brown	LG	LT Green	TN	Tan
DB	Dark Blue	OR	Orange	WT	White
DG	DK Green	PK	Pink	YL	Yellow

1996-97 Mustang 4.6L V8 MFI VIN V (All) 104 Pin Connector

PCM Pin #	Wire Color	Circuit Description (104 Pin)	Value at Hot Idle
1	---	Not Used	---
2	PK/LG	MIL (lamp) Control	MIL On: 1v, Off: 12v
3-4	---	Not Used	---
5	WT	EAM Monitor Signal	1v, 55 mph: 1v
6-7	---	Not Used	---
8	WT/OR	IMRC Monitor Signal	5v, 55 mph: 5v
9-10	---	Not Used	---
11	PK/WT	Purge Flow Sensor	0.8v, 55 mph: 3v
12	---	Not Used	---
13	PK	Flash EEPROM Power	0.1v
14	---	Not Used	---
15	PK/LB	Data Bus (-) Signal	Digital Signals
16	TN/OR	Data Bus (+) Signal	Digital Signals
17-20	---	Not Used	---
21	DB	CKP Sensor (+) Signal	850-1120 Hz
22	GY	CKP Sensor (-) Signal	850-1120 Hz
23	---	Not Used	---
24	BK/WT	Power Ground	<0.1v
25	BK	PCM Case Ground	<0.050v
26	DB/LG	Coil Driver 1 Control	5° dwell
27-29	---	Not Used	---
30	DG	Octane Adjust Switch	Closed: 0v, Open: 9.3v
31	---	Not Used	---
32	DG/PK	Knock Sensor 2 Signal	0v
33	PK/OR	VSS (-) Signal	0 Hz, 55 mph: 125 Hz
34	---	Not Used	---
35	RD/LG	HO2S-12 (B1 S2) Signal	0.1-1.1v
36	TN/LB	MAF Sensor Return	<0.050v
37	---	Not Used	---
38	LG/RD	ECT Sensor Signal	0.5-0.6v
39	GY	IAT Sensor Signal	1.5-2.5v
40	DG/YL	Fuel Pump Monitor	On: 12v, Off: 0v
41	DG/OR	A/C Cycling Clutch Switch	A/C On: 12v, Off: 0v
42	BR	IMRC Solenoid Control	12v, 55 mph: 12v
43-44	---	Not Used	---
45	DB	Low Speed Fan Control	On: 1v, Off: 12v
46	LG/PK	High Speed Fan Control	On: 1v, Off: 12v
47	BR/PK	EGR VR Solenoid	0%, 55 mph: 45%
48	OR/WT	Clean Tachometer Output	37-44 Hz
49-50	---	Not Used	---
51	BK/WT	Power Ground	<0.1v
52	RD/LB	Coil Driver 2 Control	5° dwell
53-54	---	Not Used	---
55	YL	Keep Alive Power	12-14v
56	---	Not Used	---
57	YL/RD	Knock Sensor 1 Signal	0v
58	GY/BK	VSS (+) Signal	0 Hz, 55 mph: 125 Hz
59	---	Not Used	---
60	GY/LB	HO2S-11 (B1 S1) Signal	0.1-1.1v
61	PK/LG	HO2S-22 (B2 S2) Signal	0.1-1.1v
62-64	---	Not Used	---

1996-97 Mustang 4.6L V8 MFI VIN V (All) 104 Pin Connector

PCM Pin #	Wire Color	Circuit Description (104 Pin)	Value at Hot Idle
65	BR/LG	DPFE Sensor Signal	0.20-1.30v
66	---	Not Used	---
67	GY/YL	EVAP Purge Solenoid	0-10 Hz (0-100%)
68	---	Not Used	---
69	PK/YL	A/C WOT Relay Control	On: 1v, Off: 12v
70	WT/OR	EAM System Control	12v, 55 mph: 12v
71	RD	Vehicle Power	12-14v
72	TN/RD	Injector 7 Control	2.4-2.8 ms
73	TN/BK	Injector 5 Control	2.4-2.8 ms
74	BR/YL	Injector 3 Control	2.4-2.8 ms
75	TN	Injector 1 Control	2.4-2.8 ms
76, 77	BK/WT	Power Ground	<0.1v
78	PK/WT	Coil Driver 3 Control	5° dwell
79	WT/RD	Low Fuel Pump Control	On: 1v, Off: 12v
80	LB/OR	High Fuel Pump Control	On: 1v, Off: 12v
81-82	---	Not Used	---
83	WT/LB	IAC Motor Control	10v (30% on time)
84	---	Not Used	---
85	DB/OR	CMP Sensor Signal	5-7 Hz
86	TN/LG	A/C Pressure Switch	A/C On: 12v (Open)
87	RD/BK	HO2S-21 (B2 S1) Signal	0.1-1.1v
88	LB/RD	MAF Sensor Signal	0.6v
89	GY/WT	TP Sensor Signal	0.53-1.27v
90	BR/WT	Reference Voltage	4.9-5.1v
91	GY/RD	Analog Signal Return	<0.050v
92	---	Not Used	---
93	RD/WT	HO2S-11 (B1 S1) Heater	On: 1v, Off: 12v
94	YL/LB	HO2S-21 (B2 S1) Heater	On: 1v, Off: 12v
95	WT/BK	HO2S-12 (B1 S2) Heater	On: 1v, Off: 12v
96	TN/YL	HO2S-22 (B2 S2) Heater	On: 1v, Off: 12v
97	RD	Vehicle Power	12-14v
98	LB	Injector 8 Control	2.4-2.8 ms
99	LG/OR	Injector 6 Control	2.4-2.8 ms
100	BR/LB	Injector 4 Control	2.4-2.8 ms
101	WT	Injector 2 Control	2.4-2.8 ms
102	---	Not Used	---
103	BK/WT	Power Ground	<0.1v
104	RD/YL	Coil Driver 4 Control	5° dwell

Pin Connector Graphic

Terminal View of 104-Pin PCM Wiring Harness Connector

Standard Colors and Abbreviations

Abbreviation	Color	Abbreviation	Color	Abbreviation	Color
BK	Black	GY	Gray	PK	Purple
BL	Blue	GN	Green	RD	Red
BR	Brown	LG	LT Green	TN	Tan
DB	Dark Blue	OR	Orange	WT	White
DG	DK Green	PK	Pink	YL	Yellow

1998-99 Mustang 4.6L 4v V8 MFI VIN V (M/T) 104 Pin Connector

PCM Pin #	Wire Color	Circuit Description (104 Pin)	Value at Hot Idle
1	---	Not Used	---
2	PK/LG	MIL (lamp) Control	MIL On: 1v, Off: 12v
3-4	---	Not Used	---
5	WT	Electronic AIR Monitor	1v, 55 mph: 1v
6-7	---	Not Used	---
8	WT/OR	IMRC Monitor Signal	5v, 55 mph: 5v
9-10	---	Not Used	---
11 ('98)	PK/WT	PF Sensor Signal	1.1-1.6v
12	YL/WT	Fuel Level Indicator Signal	1.7v (1/2 full)
13	PK	Flash EEPROM Power	0.1v
14	---	Not Used	---
15	PK/LB	Data Bus (-) Signal	Digital Signals
16	TN/OR	Data Bus (+) Signal	Digital Signals
17-20	---	Not Used	---
21	DB	CKP Sensor (+) Signal	320-420 Hz
22	GY	CKP Sensor (-) Signal	320-420 Hz
23	---	Not Used	---
24	BK/WT	Power Ground	<0.1v
25	BK	PCM Case Ground	<0.050v
26	DB/LG	Coil Driver 1 Control	5° dwell
27-29	---	Not Used	---
30	DG	Octane Adjust Switch	Closed: 0v, Open: 9.3v
31	---	Not Used	---
32	DG/PK	Knock Sensor 2 Signal	0v
33	PK/OR	VSS (-) Signal	0 Hz, 55 mph: 125 Hz
34, 37	---	Not Used	---
35	RD/LG	HO2S-12 (B1 S2) Signal	0.1-1.1v
36	TN/LB	MAF Sensor Return	<0.050v
38	LG/RD	ECT Sensor Signal	0.5-0.6v
39	GY	IAT Sensor Signal	1.5-2.5v
40	DG/YL	Fuel Pump Monitor	On: 12v, Off: 0v
41	DG/OR	A/C High Pressure Cutout	A/C On: 12v, Off: 1v
42	BR	IMRC Solenoid Control	12v, 55 mph: 12v
43-44	---	Not Used	---
45	DB	Low Speed Fan Control	On: 1v, Off: 12v
46	LG/PK	High Speed Fan Control	On: 1v, Off: 12v
47	BR/PK	EGR VR Solenoid	0%, 55 mph: 45%
48	OR/WT	Clean Tachometer Output	37-44 Hz
49-50	---	Not Used	---
51	BK/WT	Power Ground	<0.1v
52	RD/LB	Coil Driver 2 Control	5° dwell
53-54	---	Not Used	---
55	YL	Keep Alive Power	12-14v
56	LG/BK	EVAP Purge Solenoid	0-10 Hz (0-100%)
57	YL/RD	Knock Sensor 1 Signal	0v
58	GY/BK	VSS (+) Signal	0 Hz, 55 mph: 125 Hz
59	---	Not Used	---
60	GY/LB	HO2S-11 (B1 S1) Signal	0.1-1.1v
61	PK/LG	HO2S-22 (B2 S2) Signal	0.1-1.1v
62-64	---	Not Used	---

1998-99 Mustang 4.6L 4v V8 MFI VIN V (M/T) 104 Pin Connector

PCM Pin #	Wire Color	Circuit Description (104 Pin)	Value at Hot Idle
65	BR/LG	DPFE Sensor Signal	0.20-1.30v
66, 68	---	Not Used	---
67	GY/YL	EVAP Purge Solenoid	0-10 Hz (0-100%)
69	PK/YL	A/C WOT Relay Control	A/C On: 12v, Off: 1v
70	WT/OR	Electronic AIRB Solenoid	12v, Others: 12v
71	RD	Vehicle Power	12-14v
72	TN/RD	Injector 7 Control	2.4-2.8 ms
73	TN/BK	Injector 5 Control	2.4-2.8 ms
74	BR/YL	Injector 3 Control	2.4-2.8 ms
75	TN	Injector 1 Control	2.4-2.8 ms
76, 77	BK/WT	Power Ground	<0.1v
78	PK/WT	Coil Driver 3 Control	5° dwell
79	WT/RD	Low Fuel Pump Control	On: 1v, Off: 12v
80	LB/OR	Fuel Pump Control	On: 1v, Off: 12v
81-82	---	Not Used	---
82	RD/PK	FTP Sensor Signal	2.6v (0" H2O - cap off)
83	WT/LB	IAC Motor Control	9v (32% duty cycle)
84	---	Not Used	---
85	DB/OR	CMP Sensor Signal	5-7 Hz
86	TN/LG	A/C High Pressure Switch	A/C On: 12v, Off: 1v
87	RD/BK	HO2S-21 (B2 S1) Signal	0.1-1.1v
88	LB/RD	MAF Sensor Signal	0.6v
89	GY/WT	TP Sensor Signal	0.53-1.27v
90	BR/WT	Reference Voltage	4.9-5.1v
91	GY/RD	Analog Signal Return	<0.050v
92	---	Not Used	---
93	RD/WT	HO2S-11 (B1 S1) Heater	On: 1v, Off: 12v
94	YL/LB	HO2S-21 (B2 S1) Heater	On: 1v, Off: 12v
95	WT/BK	HO2S-12 (B1 S2) Heater	On: 1v, Off: 12v
96	TN/YL	HO2S-22 (B2 S2) Heater	On: 1v, Off: 12v
97	RD	Vehicle Power	12-14v
98	LB	Injector 8 Control	2.4-2.8 ms
99	LG/OR	Injector 6 Control	2.4-2.8 ms
100	BR/LB	Injector 4 Control	2.4-2.8 ms
101	WT	Injector 2 Control	2.4-2.8 ms
102	---	Not Used	---
103	BK/WT	Power Ground	<0.1v
104	RD/YL	Coil Driver 4 Control	5° dwell

Pin Connector Graphic

Standard Colors and Abbreviations

Abbreviation	Color	Abbreviation	Color	Abbreviation	Color
BK	Black	GY	Gray	PK	Purple
BL	Blue	GN	Green	RD	Red
BR	Brown	LG	LT Green	TN	Tan
DB	Dark Blue	OR	Orange	WT	White
DG	DK Green	PK	Pink	YL	Yellow

2000-02 Mustang 4.6L V8 MFI VIN V (M/T) 104 Pin Connector

PCM Pin #	Wire Color	Circuit Description (104 Pin)	Value at Hot Idle
1	OR/YL	COP 6 (Integrated) Dwell	6° dwell
2	---	Not Used	---
3	BK/WT	Power Ground	<0.1v
4-12	---	Not Used	---
13	PK	Flash EEPROM Power	0.1v
14	---	Not Used	---
15	PK/LB	Data Bus (-) Signal	Digital Signals
16	TN/OR	Data Bus (+) Signal	Digital Signals
17-18	---	Not Used	---
19	DB	Low Speed Fan Control	On: 1v, Off: 12v
20	---	Not Used	---
21	DB	CKP Sensor (+) Signal	320-420 Hz
22	GY	CKP Sensor (-) Signal	320-420 Hz
23	DG/WT	Knock Sensor 2 (-) Signal	0v
24	---	Not Used	---
25	BK	PCM Case Ground	<0.050v
26	LG/WT	COP 1 (Integrated) Dwell	6° dwell
27	LG/YL	COP 5 (Integrated) Dwell	6° dwell
28-31	---	Not Used	---
32	YL	Knock Sensor 1 (-) Signal	0v
32-34	---	Not Used	---
35	RD/LG	HO2S-12 (B1 S2) Signal	0.1-1.1v
36	TN/LB	MAF Sensor Return	<0.050v
37	---	Not Used	---
38	LG/RD	ECT Sensor Signal	0.5-0.6v
39	GY	IAT Sensor Signal	1.5-2.5v
40	LB/OR	Fuel Pump Monitor	2.5-7.5v
41	DG/OR	A/C Cycling Switch Signal	A/C On: 12v, Off: 1v
42-45	---	Not Used	---
46	LG/PK	High Speed Fan Control	On: 1v, Off: 12v
47	BR/PK	EGR VR Solenoid	0%, 55 mph: 45%
48-50	---	Not Used	---
51	BK/WT	Power Ground	<0.1v
52	WT/PK	COP 3 (Integrated) Dwell	6° dwell
53	DG/PK	COP 4 (Integrated) Dwell	6° dwell
54	---	Not Used	---
55	RD/WT	Keep Alive Power	12-14v
56	LG/BK	EVAP Purge Solenoid	0-10 Hz (0-100%)
57	YL/RD	Knock Sensor 1 (+) Signal	0v
58-59	---	Not Used	---
60	GY/LB	HO2S-11 (B1 S1) Signal	0.1-1.1v
61	PK/LG	HO2S-22 (B2 S2) Signal	0.1-1.1v
62	RD/PK	FTP Sensor Signal	2.6v (0" H2O - cap off)
63	RD/PK	FRP Sensor Signal	2.8v (39 psi)
64	---	Not Used	---
65 ('00)	BR/LG	DPFE Sensor Signal	0.20-1.30v
65 ('01-'02)	BR/LG	DPFE Sensor Signal	0.95-1.05v
66	---	Not Used	---
67	GY/YL	EVAP CV Solenoid	0-10 Hz (0-100%)
68	WT/OR	VSS (+) Signal	0 Hz, 55 mph: 125 Hz
69	PK/YL	A/C WOT Relay Control	A/C On: 12v, Off: 1v
70	---	Not Used	---

2000-02 Mustang 4.6L V8 MFI VIN V (M/T) 104 Pin Connector

PCM Pin #	Wire Color	Circuit Description (104 Pin)	Value at Hot Idle
71	RD	Vehicle Power	12-14v
72	TN/RD	Injector 7 Control	2.4-2.8 ms
73	TN/BK	Injector 5 Control	2.4-2.8 ms
74	BR/YL	Injector 3 Control	2.4-2.8 ms
75	TN	Injector 1 Control	2.4-2.8 ms
76	---	Not Used	---
77	BK/WT	Power Ground	<0.1v
78	PK/LB	COP 7 (Integrated) Dwell	6° dwell
79	WT/RD	COP 8 (Integrated) Dwell	6° dwell
80	WT/RD	Fuel Pump Control	On: 1v, Off: 12v
81-82	---	Not Used	---
83	WT/LB	IAC Motor Control	9v (32% duty cycle)
84	DG/WT	OSS Sensor (+) Signal	0 Hz, 55 mph: 470 Hz
85	DB/OR	CMP Sensor Signal	5-7 Hz
86	TN/LG	A/C High Pressure Switch	A/C On: 12v, Off: 1v
87	RD/BK	HO2S-21 (B2 S1) Signal	0.1-1.1v
88	LB/RD	MAF Sensor Signal	0.6v
89	GY/WT	TP Sensor Signal	0.53-1.27v
90	BR/WT	Reference Voltage	4.9-5.1v
91	GY/RD	Analog Signal Return	<0.050v
92	LG	Brake Pedal Position Switch	Brake Off: 0v, On: 12v
93	RD/WT	HO2S-11 (B1 S1) Heater	On: 1v, Off: 12v
94	YL/LB	HO2S-21 (B2 S1) Heater	On: 1v, Off: 12v
95	WT/BK	HO2S-12 (B1 S2) Heater	On: 1v, Off: 12v
96	TN/YL	HO2S-22 (B2 S2) Heater	On: 1v, Off: 12v
97	RD	Vehicle Power	12-14v
98	LB	Injector 8 Control	2.4-2.8 ms
99	LG/OR	Injector 6 Control	2.4-2.8 ms
100	BR/LB	Injector 4 Control	2.4-2.8 ms
101	WT	Injector 2 Control	2.4-2.8 ms
102	DG/PK	Knock Sensor 2 (+) Signal	0v
103	BK/WT	Power Ground	<0.1v
104	PK/WT	COP 2 (Integrated) Dwell	6° dwell

Pin Connector Graphic

PCM 104-PIN CONNECTOR

Terminal View of 104-Pin PCM Wiring Harness Connector

Standard Colors and Abbreviations

Abbreviation	Color	Abbreviation	Color	Abbreviation	Color
BK	Black	GY	Gray	PK	Purple
BL	Blue	GN	Green	RD	Red
BR	Brown	LG	LT Green	TN	Tan
DB	Dark Blue	OR	Orange	WT	White
DG	DK Green	PK	Pink	YL	Yellow

2003 Mustang Mach 1 4.6L V8 DOHC VIN R (All) 104 Pin Connector

PCM Pin #	Wire Color	Circuit Description (104 Pin)	Value at Hot Idle
1	VT/YL	COP 6 (Integrated) Dwell	6° dwell
2, 4-5	---	Not Used	---
3	BK/WT	Power Ground	<0.1v
6	OR/YL	Shift Solenoid 'A' Control	1v, Others: 1v
7, 9	---	Not Used	---
8	WT/OR	IMRC Monitor Signal	5v, 55 mph: 5v
10	DB/LG	Barometric Pressure Sensor Signal	159 Hz
11	VT/OR	Shift Solenoid 'B' Control	12v, Others: 1v
12, 14	---	Not Used	---
13	VT	Flash EEPROM Power	0.1v
15	PK/LB	SCP Bus (-) Signal	Digital Signals
16	TN/OR	SCP Bus (+) Signal	Digital Signals
17-20	---	Not Used	---
21	DB	CKP Sensor (+) Signal	320-420 Hz
22	GY	CKP Sensor (-) Signal	320-420 Hz
23	DG/WT	Knock Sensor 1 (-) Signal	<0.050v
24	---	Not Used	---
25	BK	PCM Case Ground	<0.050v
26	LG/WT	COP 1 (Integrated) Dwell	6° dwell
27	LG/YL	COP 5 (Integrated) Dwell	6° dwell
28	DB	Low Speed Fan Relay Control	Relay Off: 12v, On: 1v
29	TN/WT	TCS (switch) Signal	TCS & O/D On: 12v
30-31, 33	---	Not Used	---
32	YL	Knock Sensor 2 (-) Signal	<0.050v
34	YL/BK	Digital TR1 Sensor	In 'P': 0v, 55 mph: 11v
35	RD/LG	HO2S-12 (B1 S2) Signal	0.1-1.1v
36	TN/LB	MAF Sensor Return	<0.050v
37	RD/YL	Intake Air Temperature Sensor 2 Signal	1.5-2.5v
38	LG/RD	Reference Voltage	4.9-5.1v
39	GY	Intake Air Temperature Sensor 1 Signal	1.5-2.5v
40	LB/OR	Fuel Pump Monitor	2.5-7.5v
41	DG/OR	A/C Cycling Switch Signal	A/C On: 12v, Off: 1v
42	BR	IMRC Solenoid Control	12v (Off)
43-44	---	Not Used	---
45	LG/VT	Supercharger Bypass Solenoid Control	Solenoid Off: 12v, On: 1v
46	LG/VT	High Speed Fan Relay Control	Relay Off: 12v, On: 1v
47	BR/PK	EGR VR Solenoid	0%, 55 mph: 45%
49	LB/BK	Digital TR2 Sensor	In 'P': 0v, 55 mph: 11v
50	WT/BK	Digital TR4 Sensor	In 'P': 0v, 55 mph: 11v
51	BK/WT	Power Ground	<0.1v
52	WT/PK	COP 3 (Integrated) Dwell	6° dwell
53	DG/PK	COP 4 (Integrated) Dwell	6° dwell
54	---	Not Used	---
55	RD/WT	Keep Alive Power	12-14v
56	LG/BK	EVAP Purge Solenoid	0-10 Hz (0-100%)
57	YE/RD	Knock Sensor 1 (+) Signal	0v
58	---	Not Used	---
60	GY/LB	HO2S-11 (B1 S1) Signal	0.1-1.1v
61	VT/LG	HO2S-22 (B2 S2) Signal	0.1-1.1v
62	RD/PK	FTP Sensor Signal	2.6v (0" H2O - cap off)
63	RD/PK	FRP Sensor Signal	2.8v (39 psi)
64	LB/YL	Digital TR3A Sensor	In 'P': 0v, in O/D: 1.7v
65	BR/LG	DPFE Sensor Signal	0.95-1.05v
66	YE/LG	Cylinder Head Temperature Sensor	0.5-0.6v

2003 Mustang Mach 1 4.6L V8 DOHC VIN R (All) 104 Pin Connector

PCM Pin #	Wire Color	Circuit Description (104 Pin)	Value at Hot Idle
67	GY/YL	EVAP CV Solenoid	0-10 Hz (0-100%)
68	WT/OR	VSS (+) Signal	Moving: AC pulse signals
69	PK/YL	A/C WOT Relay Control	A/C On: 12v, Off: 1v
70	---	Not Used	---
71	RD	Vehicle Power (Start-Run)	12-14v
72	TN/RD	Injector 7 Control	2.4-2.8 ms
73	TN/BK	Injector 5 Control	2.4-2.8 ms
74	BR/YL	Injector 3 Control	2.4-2.8 ms
75	TN	Injector 1 Control	2.4-2.8 ms
76, 77	BK/WT	Power Ground	<0.1v
78	PK/LB	COP 7 (Integrated) Dwell	6° dwell
79	WT/RD	COP 8 (Integrated) Dwell	6° dwell
80	WT/RD	Fuel Pump Control	Off: 12v, On: 1v
81	WT/YL	EPC Solenoid Control	8.3v (8 psi)
82	WT/OR	Charge Air Cooler Pump Relay Control	Relay Off: 12v, On: 1v
83	WT/LB	IAC Motor Control	9v (32% duty cycle)
84	DG/WT	OSS Sensor (+) Signal	0 Hz, 55 mph: 470 Hz
85	DB/OR	CMP Sensor Signal	5-7 Hz
86	TN/LG	A/C High Pressure Switch	A/C On: 12v, Off: 1v
87	RD/BK	HO2S-21 (B2 S1) Signal	0.1-1.1v
88	LB/RD	MAF Sensor Signal	0.6v
89	GY/WT	TP Sensor Signal	0.53-1.27v
90	BR/WT	Reference Voltage	4.9-5.1v
91	GY/RD	Sensor Ground	<0.050v
92	LG	Brake Pedal Position Switch	Brake Off: 0v, On: 12v
93	RD/WT	HO2S-11 (B1 S1) Heater	On: 1v, Off: 12v
94	YL/LB	HO2S-21 (B2 S1) Heater	On: 1v, Off: 12v
95	WT/BK	HO2S-12 (B1 S2) Heater	On: 1v, Off: 12v
96	TN/YL	HO2S-22 (B2 S2) Heater	On: 1v, Off: 12v
97	RD	Vehicle Power (Start-Run)	12-14v
98	LB	Injector 8 Control	2.4-2.8 ms
99	LG/OR	Injector 6 Control	2.4-2.8 ms
100	BR/LB	Injector 4 Control	2.4-2.8 ms
101	WT	Injector 2 Control	2.4-2.8 ms
102	DG/PK	Knock Sensor 2 (+) Signal	0v
103	BK/WT	Power Ground	<0.1v
104	PK/WT	COP 2 (Integrated) Dwell	6° dwell

Pin Connector Graphic

PCM 104-PIN CONNECTOR

Terminal View of 104-Pin PCM Wiring Harness Connector

Standard Colors and Abbreviations

Abbreviation	Color	Abbreviation	Color	Abbreviation	Color
BK	Black	GY	Gray	PK	Purple
BL	Blue	GN	Green	RD	Red
BR	Brown	LG	LT Green	TN	Tan
DB	Dark Blue	OR	Orange	WT	White
DG	DK Green	PK	Pink	YL	Yellow

1996-97 Mustang 4.6L 2v V8 MFI VIN W (All) 104 Pin Connector

PCM Pin #	Wire Color	Circuit Description (104 Pin)	Value at Hot Idle
1	PK/OR	Shift Solenoid 2 Control	12v, 55 mph: 1v
2	PK/LG	MIL (lamp) Control	MIL On: 1v, Off: 12v
3-10	---	Not Used	---
11	PK/WT	Purge Flow Sensor	0.8v, 55 mph: 3v
12	---	Not Used	---
13	PK	Flash EEPROM Power	0.1v
14	---	Not Used	---
15	PK/LB	Data Bus (-) Signal	Digital Signals
16	TN/OR	Data Bus (+) Signal	Digital Signals
17-20	---	Not Used	---
21	DB	CKP Sensor (+) Signal	850-1120 Hz
22	GY	CKP Sensor (-) Signal	850-1120 Hz
23	---	Not Used	---
24	BK/WT	Power Ground	<0.1v
25	BK	PCM Case Ground	<0.1v
26	DB/LG	Coil Driver 1 Control	5° dwell
27	OR/YL	Shift Solenoid 1 Control	1v, 55 mph: 12v
28	---	Not Used	---
29	TN/WT	TCS (switch) Signal	TCS & O/D On: 12v
30	DG	Octane Adjust Switch	Closed: 0v, Open: 9.3v
31-32	---	Not Used	---
33	PK/OR	VSS (-) Signal	0 Hz, 55 mph: 125 Hz
34	---	Not Used	---
35	RD/LG	HO2S-12 (B1 S2) Signal	0.1-1.1v
36	TN/LB	MAF Sensor Return	<0.1v
37	OR/BK	TFT Sensor Signal	2.10-2.40v
38	LG/RD	ECT Sensor Signal	0.5-0.6v
39	GY	IAT Sensor Signal	1.5-2.5v
40	DG/YL	Fuel Pump Monitor	On: 12v, Off: 0v
41	DG/OR	A/C Cycling Clutch Switch	A/C On: 12v, Off: 0v
42-44	---	Not Used	---
45	DB	Low Speed Fan Control	On: 1v, Off: 12v
46	LG/PK	High Speed Fan Control	On: 1v, Off: 12v
47	BR/PK	EGR VR Solenoid	0%, 55 mph: 45%
48	OR/WT	Clean Tachometer Output	33-37 Hz
49-50	---	Not Used	---
51	BK/WT	Power Ground	<0.1v
52	RD/LB	Coil Driver 2 Control	5° dwell
53	---	Not Used	---
54	BR/OR	TCC Solenoid Control	0%, 55 mph: 95%
55	YL	Keep Alive Power	12-14v
56-57	---	Not Used	---
58	GY/BK	VSS (+) Signal	0 Hz, 55 mph: 125 Hz
59	---	Not Used	---
60	GY/LB	HO2S-11 (B1 S1) Signal	0.1-1.1v
61	PK/LG	HO2S-22 (B2 S2) Signal	0.1-1.1v
62-63	---	Not Used	---
64	LB/YL	TR Sensor Signal	In 'P': 0v, 55 mph: 1.7v
65	BR/LG	DPFE Sensor Signal	0.20-1.30v
66	---	Not Used	---

1996-97 Mustang 4.6L 2v V8 MFI VIN W (All) 104 Pin Connector

PCM Pin #	Wire Color	Circuit Description (104 Pin)	Value at Hot Idle
67	GY/YL	EVAP Purge Solenoid	0-10 Hz (0-100%)
68	---	Not Used	---
69	PK/YL	A/C WOT Relay Control	On: 1v, Off: 12v
70	---	Not Used	---
71	RD	Vehicle Power	12-14v
72	TN/RD	Injector 7 Control	3.5-3-7 ms
73	TN/BK	Injector 5 Control	3.5-3-7 ms
74	BR/YL	Injector 3 Control	3.5-3-7 ms
75	TN	Injector 1 Control	3.5-3-7 ms
76-77	BK/WT	Power Ground	<0.1v
78	PK/WT	Coil Driver 3 Control	5° dwell
79	WT/LG	TCIL (lamp) Control	On: 1v, Off: 12v
80	LB/OR	Fuel Pump Control	On: 1v, Off: 12v
81	WT/YL	EPC Solenoid Control	9v (15 psi)
82	---	Not Used	---
83	WT/LB	IAC Motor Control	30% duty cycle
84	DG/WT	TSS Sensor Signal	43 Hz (650 rpm)
85	DB/OR	CID Sensor Signal	5-7 Hz
86	TN/LG	A/C Pressure Switch	A/C On: 12v (Open)
87	RD/BK	HO2S-21 (B2 S1) Signal	0.1-1.1v
88	LB/RD	MAF Sensor Signal	0.6v
89	GY/WT	TP Sensor Signal	0.53-1.27v
90	BR/WT	Reference Voltage	4.9-5.1v
91	GY/RD	Analog Signal Return	<0.050v
92	LG	Brake Pedal Position Switch	Brake Off: 0v, On: 12v
93	RD/WT	HO2S-11 (B1 S1) Heater	On: 1v, Off: 12v
94	YL/LB	HO2S-21 (B2 S1) Heater	On: 1v, Off: 12v
95	WT/BK	HO2S-12 (B1 S2) Heater	On: 1v, Off: 12v
96	TN/YL	HO2S-22 (B2 S2) Heater	On: 1v, Off: 12v
97	RD	Vehicle Power	12-14v
98	LB	Injector 8 Control	3.5-3-7 ms
99	LG/OR	Injector 6 Control	3.5-3-7 ms
100	BR/LB	Injector 4 Control	3.5-3-7 ms
101	WT	Injector 2 Control	3.5-3-7 ms
102	---	Not Used	---
103	BK/WT	Power Ground	<0.1v
104	RD/YL	Coil Driver 4 Control	5° dwell

Pin Connector Graphic

PCM 104-PIN CONNECTOR

Terminal View of 104-Pin PCM Wiring Harness Connector

Standard Colors and Abbreviations

Abbreviation	Color	Abbreviation	Color	Abbreviation	Color
BK	Black	GY	Gray	PK	Purple
BL	Blue	GN	Green	RD	Red
BR	Brown	LG	LT Green	TN	Tan
DB	Dark Blue	OR	Orange	WT	White
DG	DK Green	PK	Pink	YL	Yellow

1998-99 Mustang 4.6L 2v V8 MFI VIN W (All) 104 Pin Connector

PCM Pin #	Wire Color	Circuit Description (104 Pin)	Value at Hot Idle
1	PK/OR	Shift Solenoid 2 Control	12v, 55 mph: 12v
2	PK/LG	MIL (lamp) Control	MIL On: 1v, Off: 12v
3	YL/BK	Digital TR1 Sensor	0v, 55 mph: 11v
4-10	---	Not Used	---
11 ('98)	PK/WT	PF Sensor Signal	1.1-1.6v
12	YL/WT	Fuel Level Indicator Signal	1.7v (1/2 full)
13	PK	Flash EEPROM Power	0.1v
14	---	Not Used	---
15	PK/LB	Data Bus (-) Signal	Digital Signals
16	TN/OR	Data Bus (+) Signal	Digital Signals
17-20	---	Not Used	---
21	DB	CKP Sensor (+) Signal	390-450 Hz
22	GY	CKP Sensor (-) Signal	390-450 Hz
23	---	Not Used	---
24	BK/WT	Power Ground	<0.1v
25	BK	PCM Case Ground	<0.050v
26	DB/LG	Coil Driver 1 Control	5° dwell
27	OR/YL	Shift Solenoid 1 Control	1v, 55 mph: 1v
28	---	Not Used	---
29	TN/WT	TCS (switch) Signal	TCS & O/D On: 12v
30	DG	Octane Adjust Switch	Closed: 0v, Open: 9.3v
31-32	---	Not Used	---
33	PK/OR	VSS (-) Signal	0 Hz, 55 mph: 125 Hz
34	---	Not Used	---
35	RD/LG	HO2S-12 (B1 S2) Signal	0.1-1.1v
36	TN/LB	MAF Sensor Return	<0.050v
37	OR/BK	TFT Sensor Signal	2.10-2.40v
38	LG/RD	ECT Sensor Signal	0.5-0.6v
39	GY	IAT Sensor Signal	1.5-2.5v
40	DG/YL	Fuel Pump Monitor	On: 12v, Off: 0v
41	DG/OR	A/C High Pressure Switch	AC On: 12v, Off: 0v
42, 44	---	Not Used	---
43	WT/RD	Fuel Pump Control	On: 1v, Off: 12v
45	DB	Low Speed Fan Control	On: 1v, Off: 12v
46	LG/PK	High Speed Fan Control	On: 1v, Off: 12v
47	BR/PK	EGR VR Solenoid	0%, 55 mph: 45%
48	OR/WT	Clean Tachometer Output	39-45 Hz
49	LB/BK	Digital TR2 Sensor	0v, 55 mph: 11v
50	WT/BK	Digital TR4 Sensor	0v, 55 mph: 11v
51	WT/YL	Power Ground	<0.1v
52	RD/LB	Coil Driver 2 Control	5° dwell
53	---	Not Used	---
54	PK/YL	TCC Solenoid Control	0%, 55 mph: 95%
55	Y	Keep Alive Power	12-14v
56	LG/BK	EVAP VMV Solenoid	0-10 Hz (0-100%)
57, 59	---	Not Used	---
58	GY/BK	VSS (+) Signal	0 Hz, 55 mph: 125 Hz
60	GY/LB	HO2S-11 (B1 S1) Signal	0.1-1.1v
61	PK/LG	HO2S-22 (B2 S2) Signal	0.1-1.1v
62	RD/PK	FTP Sensor Signal	2.6v (0" H2O - cap off)
63	---	Not Used	---
64	LB/YL	TR Sensor Signal	In 'P': 0v, in O/D: 1.7v
65	BR/LG	DPFE Sensor Signal	0.20-1.30v
66	---	Not Used	---

1998-99 Mustang 4.6L 2v V8 MFI VIN W (All) 104 Pin Connector

PCM Pin #	Wire Color	Circuit Description (104 Pin)	Value at Hot Idle
67	GY/YL	EVAP Purge Solenoid	0-10 Hz (0-100%)
68	---	Not Used	---
69	PK/YL	A/C WOT Relay Control	On: 1v, Off: 12v
70	---	Not Used	---
71	RD	Vehicle Power	12-14v
72	TN/RD	Injector 7 Control	3.5-3-7 ms
73	TN/BK	Injector 5 Control	3.5-3-7 ms
74	BR/YL	Injector 3 Control	3.5-3-7 ms
75	TN	Injector 1 Control	3.5-3-7 ms
76-77	BK/WT	Power Ground	<0.1v
78	PK/WT	Coil Driver 3 Control	5° dwell
79	WT/LG	TCIL (lamp) Control	On: 1v, Off: 12v
80	LB/OR	Fuel Pump Control	On: 1v, Off: 12v
81	WT/YL	EPC Solenoid Control	9v (15 psi)
82	---	Not Used	---
83	WT/LB	IAC Motor Control	20% duty cycle
84	DG/WT	TSS Sensor Signal	43 Hz (650 rpm)
85	DB/OR	CMP Sensor Signal	5-7 Hz
86	TN/LG	A/C Pressure Switch	On: 1v, Off: 12v
87	RD/BK	HO2S-21 (B2 S1) Signal	0.1-1.1v
88	LB/RD	MAF Sensor Signal	0.6v
89	GY/WT	TP Sensor Signal	0.53-1.27v
90	BR/WT	Reference Voltage	4.9-5.1v
91	GY/RD	Analog Signal Return	<0.050v
92	LG	Brake Pedal Position Switch	Brake Off: 0v, On: 12v
93	RD/WT	HO2S-11 (B1 S1) Heater	On: 1v, Off: 12v
94	YL/LB	HO2S-21 (B2 S1) Heater	On: 1v, Off: 12v
95	WT/BK	HO2S-12 (B1 S2) Heater	On: 1v, Off: 12v
96	TN/YL	HO2S-22 (B2 S2) Heater	On: 1v, Off: 12v
97	RD	Vehicle Power	12-14v
98	LB	Injector 8 Control	3.5-3-7 ms
99	LG/OR	Injector 6 Control	3.5-3-7 ms
100	LB	Injector 4 Control	3.5-3-7 ms
101	WT	Injector 2 Control	3.5-3-7 ms
102	---	Not Used	---
103	BK/WT	Power Ground	<0.1v
104	RD/YL	Coil Driver 4 Control	5° dwell

Pin Connector Graphic

PCM 104-PIN CONNECTOR

Terminal View of 104-Pin PCM Wiring Harness Connector

Standard Colors and Abbreviations

Abbreviation	Color	Abbreviation	Color	Abbreviation	Color
BK	Black	GY	Gray	PK	Purple
BL	Blue	GN	Green	RD	Red
BR	Brown	LG	LT Green	TN	Tan
DB	Dark Blue	OR	Orange	WT	White
DG	DK Green	PK	Pink	YL	Yellow

1996-97 Mustang GT 4.6L 2v V8 VIN X (All) 104 Pin Connector

PCM Pin #	Wire Color	Circuit Description (104 Pin)	Value at Hot Idle
1	PK/OR	Shift Solenoid 2 Control	12v, 55 mph: 1v
2	PK/LG	MIL (lamp) Control	MIL On: 1v, Off: 12v
3-10	---	Not Used	---
11	PK/WT	Purge Flow Sensor	0.8v, 55 mph: 3v
12	---	Not Used	---
13	PK	Flash EEPROM Power	0.1v
14	---	Not Used	---
15	PK/LB	Data Bus (-) Signal	Digital Signals
16	TN/OR	Data Bus (+) Signal	Digital Signals
17-20	---	Not Used	---
21	DB	CKP Sensor (+) Signal	850-1120 Hz
22	GY	CKP Sensor (-) Signal	850-1120 Hz
23	---	Not Used	---
24	BK/WT	Power Ground	<0.1v
25	BK	PCM Case Ground	<0.050v
26	DB/LG	Coil Driver 1 Control	5° dwell
27	OR/YL	Shift Solenoid 1 Control	1v, 55 mph: 12v
28	---	Not Used	---
29	TN/WT	TCS (switch) Signal	TCS & O/D On: 12v
30	DG	Octane Adjust Switch	Closed: 0v, Open: 9.3v
31-32	---	Not Used	---
33	PK/OR	VSS (-) Signal	0 Hz, 55 mph: 125 Hz
34	---	Not Used	---
35	RD/LG	HO2S-12 (B1 S2) Signal	0.1-1.1v
36	TN/LB	MAF Sensor Return	<0.050v
37	OR/BK	TFT Sensor Signal	2.10-2.40v
38	LG/RD	ECT Sensor Signal	0.5-0.6v
39	GY	IAT Sensor Signal	1.5-2.5v
40	DG/YL	Fuel Pump Monitor	On: 12v, Off: 0v
41	DG/OR	A/C High Pressure Switch	AC On: 12v, Off: 0v
42-44	---	Not Used	---
45	DB	Low Speed Fan Control	On: 1v, Off: 12v
46	LG/PK	High Speed Fan Control	On: 1v, Off: 12v
47	BR/PK	EGR VR Solenoid	0%, 55 mph: 45%
48	OR/WT	Clean Tachometer Output	33-37 Hz
49-50	---	Not Used	---
51	BK/WT	Power Ground	<0.1v
52	RD/LB	Coil Driver 2 Control	5° dwell
53	---	Not Used	---
54	BR/OR	TCC Solenoid Control	0%, 55 mph: 95%
55	YL	Keep Alive Power	12-14v
56-57	---	Not Used	---
58	GY/BK	VSS (+) Signal	0 Hz, 55 mph: 125 Hz
59	---	Not Used	---
60	GY/LB	HO2S-11 (B1 S1) Signal	0.1-1.1v
61	PK/LG	HO2S-22 (B2 S2) Signal	0.1-1.1v
62-63	---	Not Used	---
64	LB/YL	A/T: TR Sensor	In 'P': 0v, 55 mph: 1.7v
64	LB/YL	M/T: Clutch Pedal Position	Clutch In: 0v, Out: 5v
65	BR/LG	DPFE Sensor Signal	0.20-1.30v
66	---	Not Used	---

1996-97 Mustang GT 4.6L 2v V8 VIN X (All) 104 Pin Connector

PCM Pin #	Wire Color	Circuit Description (104 Pin)	Value at Hot Idle
67	GY/YL	EVAP Purge Solenoid	0-10 Hz (0-100%)
68	---	Not Used	---
69	PK/YL	A/C WOT Relay Control	On: 1v, Off: 12v
70	---	Not Used	---
71	RD	Vehicle Power	12-14v
72	TN/RD	Injector 7 Control	3.5-3-7 ms
73	TN/BK	Injector 5 Control	3.5-3-7 ms
74	TN/RD	Injector 3 Control	3.5-3-7 ms
75	TN	Injector 1 Control	3.5-3-7 ms
76-77	BK/WT	Power Ground	<0.1v
78	PK/WT	Coil Driver 3 Control	5° dwell
79	WT/LG	TCIL (lamp) Control	On: 1v, Off: 12v
80	LB/OR	Fuel Pump Control	On: 1v, Off: 12v
81	WT/YL	EPC Solenoid Control	9v (15 psi)
82	---	Not Used	---
83	WT/LB	IAC Motor Control	30% duty cycle
84	DG/WT	TSS Sensor Signal	43 Hz (650 rpm)
85	DB/OR	CID Sensor Signal	5-7 Hz
86	TN/LG	A/C Pressure Switch	A/C On: 12v (Open)
87	RD/BK	HO2S-21 (B2 S1) Signal	0.1-1.1v
88	LB/RD	MAF Sensor Signal	0.6v
89	GY/WT	TP Sensor Signal	0.53-1.27v
90	BR/WT	Reference Voltage	4.9-5.1v
91	GY/RD	Analog Signal Return	<0.050v
92	LG	Brake Pedal Position Switch	Brake Off: 0v, On: 12v
93	RD/WT	HO2S-11 (B1 S1) Heater	On: 1v, Off: 12v
94	YL/LB	HO2S-21 (B2 S1) Heater	On: 1v, Off: 12v
95	WT/BK	HO2S-12 (B1 S2) Heater	On: 1v, Off: 12v
96	TN/YL	HO2S-22 (B2 S2) Heater	On: 1v, Off: 12v
97	RD	Vehicle Power	12-14v
98	LB	Injector 8 Control	3.5-3-7 ms
99	LG/OR	Injector 6 Control	3.5-3-7 ms
100	LB	Injector 4 Control	3.5-3-7 ms
101	WT	Injector 2 Control	3.5-3-7 ms
102	---	Not Used	---
103	BK/WT	Power Ground	<0.1v
104	RD/YL	Coil Driver 4 Control	5° dwell

Pin Connector Graphic

PCM 104-PIN CONNECTOR

Terminal View of 104-Pin PCM Wiring Harness Connector

Standard Colors and Abbreviations

Abbreviation	Color	Abbreviation	Color	Abbreviation	Color
BK	Black	GY	Gray	PK	Purple
BL	Blue	GN	Green	RD	Red
BR	Brown	LG	LT Green	TN	Tan
DB	Dark Blue	OR	Orange	WT	White
DG	DK Green	PK	Pink	YL	Yellow

1998-99 Mustang GT 4.6L 2v V8 VIN X (All) 104 Pin Connector

PCM Pin #	Wire Color	Circuit Description (104 Pin)	Value at Hot Idle
1	PK/OR	Shift Solenoid 2 Control	12v, 55 mph: 0.1v
2	PK/LG	MIL (lamp) Control	MIL On: 1v, Off: 12v
3	YL/BK	Digital TR1 Sensor	0v, 55 mph: 11v
4-10	---	Not Used	---
11 ('88)	PK/WT	PF Sensor Signal	1.1-1.6v
12	YL/WT	Fuel Level Indicator Signal	1.7v (1/2 full)
13	PK	Flash EEPROM Power	0.1v
14	---	Not Used	
15	PK/LB	Data Bus (-) Signal	Digital Signals
16	TN/OR	Data Bus (+) Signal	Digital Signals
17-20	---	Not Used	
21	DB	CKP Sensor (+) Signal	390-450 Hz
22	GY	CKP Sensor (-) Signal	390-450 Hz
23	---	Not Used	---
24	BK/WT	Power Ground	<0.1v
25	BK	PCM Case Ground	<0.050v
26	DB/LG	Coil Driver 1 Control	5° dwell
27	OR/YL	Shift Solenoid 1 Control	1v, 55 mph: 1v
28	---	Not Used	---
29	TN/WT	TCS (switch) Signal	TCS & O/D On: 12v
30	DG	Octane Adjust Switch	Closed: 0v, Open: 9.3v
31-32	---	Not Used	---
33	PK/OR	VSS (-) Signal	0 Hz, 55 mph: 125 Hz
34	---	Not Used	---
35	RD/LG	HO2S-12 (B1 S2) Signal	0.1-1.1v
36	TN/LB	MAF Sensor Return	<0.050v
37	OR/BK	TFT Sensor Signal	2.10-2.40v
38	LG/RD	ECT Sensor Signal	0.5-0.6v
39	GY	IAT Sensor Signal	1.5-2.5v
40	DG/YL	Fuel Pump Monitor	On: 12v, Off: 0v
41	DG/OR	A/C High Pressure Switch	AC On: 12v, Off: 0v
42	---	Not Used	---
43	WT/RD	Fuel Pump Control	On: 1v, Off: 12v
44	---	Not Used	---
45	DB	Low Speed Fan Control	On: 1v, Off: 12v
46	LG/PK	High Speed Fan Control	On: 1v, Off: 12v
47	BR/PK	EGR VR Solenoid	0%, 55 mph: 45%
48	OR/WT	Clean Tachometer Output	39-45 Hz
49	LB/BK	Digital TR2 Sensor	0v, 55 mph: 11v
50	WT/BK	Digital TR4 Sensor	0v, 55 mph: 11v
51	WT/YL	Power Ground	<0.1v
52	RD/LB	Coil Driver 2 Control	5° dwell
53	---	Not Used	---
54	PK/YL	TCC Solenoid Control	0%, 55 mph: 95%
55	YL	Keep Alive Power	12-14v
56	LG/BK	EVAP VMV Solenoid	0-10 Hz (0-100%)
57, 59	---	Not Used	---
58	GY/BK	VSS (+) Signal	0 Hz, 55 mph: 125 Hz
60	GY/LB	HO2S-11 (B1 S1) Signal	0.1-1.1v
61	PK/LG	HO2S-22 (B2 S2) Signal	0.1-1.1v
62	RD/PK	FTP Sensor Signal	2.6v (0" H2O - cap off)
63	---	Not Used	---
64	DG/WT	TR Sensor Signal	In 'P': 0v, in O/D: 1.7v
65	BR/LG	DPFE Sensor Signal	0.20-1.30v

1998-99 Mustang GT 4.6L 2v V8 VIN X (All) 104 Pin Connector

PCM Pin #	Wire Color	Circuit Description (104 Pin)	Value at Hot Idle
66	---	Not Used	---
67	GY/YL	EVAP Purge Solenoid	0-10 Hz (0-100%)
68	---	Not Used	---
69	PK/YL	A/C WOT Relay Control	On: 1v, Off: 12v
70	---	Not Used	---
71	RD	Vehicle Power	12-14v
72	TN/RD	Injector 7 Signal (INJ 7)	3.5-3-7 ms
73	TN/BK	Injector 5 Control	3.5-3-7 ms
74	TN/RD	Injector 3 Control	3.5-3-7 ms
75	TN	Injector 1 Control	3.5-3-7 ms
76-77	BK/WT	Power Ground	<0.1v
78	PK/WT	Coil Driver 3 Control	5° dwell
79	WT/LG	TCIL (lamp) Control	On: 1v, Off: 12v
80	LB/OR	Fuel Pump Control	On: 1v, Off: 12v
81	WT/YL	EPC Solenoid Control	9v (15 psi)
82	---	Not Used	---
83	WT/LB	IAC Motor Control	20% duty cycle
84	DG/WT	TSS Sensor Signal	43 Hz (650 rpm)
85	DB/OR	CID Sensor Signal	5-7 Hz
86	TN/LG	A/C Pressure Switch	On: 1v, Off: 12v
87	RD/BK	HO2S-21 (B2 S1) Signal	0.1-1.1v
88	LB/RD	MAF Sensor Signal	0.6v
89	GY/WT	TP Sensor Signal	0.53-1.27v
90	BR/WT	Reference Voltage	4.9-5.1v
91	GY/RD	Analog Signal Return	<0.050v
92	LG	Brake Pedal Position Switch	Brake Off: 0v, On: 12v
93	RD/WT	HO2S-11 (B1 S1) Heater	On: 1v, Off: 12v
94	YL/LB	HO2S-21 (B2 S1) Heater	On: 1v, Off: 12v
95	WT/BK	HO2S-12 (B1 S2) Heater	On: 1v, Off: 12v
96	TN/YL	HO2S-22 (B2 S2) Heater	On: 1v, Off: 12v
97	RD	Vehicle Power	12-14v
98	LB	Injector 8 Signal (INJ 8)	3.5-3-7 ms
99	LG/OR	Injector 6 Control	3.5-3-7 ms
100	LB	Injector 4 Control	3.5-3-7 ms
101	WT	Injector 2 Control	3.5-3-7 ms
102	---	Not Used	---
103	BK/WT	Power Ground	<0.1v
104	RD/YL	Coil Driver 4 Control	5° dwell

Pin Connector Graphic

PCM 104-PIN CONNECTOR

Terminal View of 104-Pin PCM Wiring Harness Connector

Standard Colors and Abbreviations

Abbreviation	Color	Abbreviation	Color	Abbreviation	Color
BK	Black	GY	Gray	PK	Purple
BL	Blue	GN	Green	RD	Red
BR	Brown	LG	LT Green	TN	Tan
DB	Dark Blue	OR	Orange	WT	White
DG	DK Green	PK	Pink	YL	Yellow

2000-03 Mustang GT, GT Bullitt 4.6L 2v V8 VIN X (All) 104 Pin Connector

PCM Pin #	Wire Color	Circuit Description (104 Pin)	Value at Hot Idle
1	OR/YL	COP 6 (Integrated) Dwell	6° dwell
2	---	Not Used	---
3	BK/WT	Power Ground	<0.1v
4-5	---	Not Used	---
6	OR/YL	Shift Solenoid 'A' Control	1v, Others: 1v
7, 9	---	Not Used	---
8 ('02-'03)	WT/OR	IMRC Monitor Signal	5v, 55 mph: 5v
11	VT/OR	Shift Solenoid 'B' Control	12v, Others: 1v
12, 14	---	Not Used	---
13	VT	Flash EEPROM Power	0.1v
15	PK/LB	SCP Bus (-) Signal	Digital Signals
16	TN/OR	SCP Bus (+) Signal	Digital Signals
17-18, 20	---	Not Used	---
19	DB	Low Speed Fan Control	On: 1v, Off: 12v
21	DB	CKP Sensor (+) Signal	390-450 Hz
22	GY	CKP Sensor (-) Signal	390-450 Hz
23-24	---	Not Used	---
25	BK	PCM Case Ground	<0.050v
26	LG/WT	COP 1 (Integrated) Dwell	6° dwell
27	LG/YL	COP 5 (Integrated) Dwell	6° dwell
28	DB	Low Speed Fan Relay Control	Relay Off: 12v, On: 1v
29	TN/WT	TCS (switch) Signal	TCS & O/D On: 12v
30-33	---	Not Used	---
34	YL/BK	Digital TR1 Sensor	0v, 55 mph: 11v
35	RD/LG	HO2S-12 (B1 S2) Signal	0.1-1.1v
36	TN/LB	MAF Sensor Return	<0.050v
37	OR/BK	TFT Sensor Signal	2.10-2.40v
38	LG/RD	ECT Sensor Signal	0.5-0.6v
39	GY	IAT Sensor Signal	1.5-2.5v
40	LB/OR	Fuel Pump Monitor	3.5v (100%)
41	DG/OR	A/C Cycling Switch Signal	A/C On: 12v, Off: 1v
42	BR	IMRC Solenoid Control	12v (Off)
43-44	---	Not Used	---
45	LG/VT	Supercharger Bypass Solenoid Control	Solenoid Off: 12v, On: 1v
46	LG/PK	High Speed Fan Control	On: 1v, Off: 12v
47	BR/PK	EGR VR Solenoid	0%, 55 mph: 45%
49	LB/BK	Digital TR2 Sensor	0v, 55 mph: 11v
50	WT/BK	Digital TR4 Sensor	0v, 55 mph: 11v
51	BK/WT	Power Ground	<0.1v
52	WT/PK	COP 3 (Integrated) Dwell	6° dwell
53	DG/PK	COP 4 (Integrated) Dwell	6° dwell
54	BR/OR	TCC Solenoid Control	0%, 55 mph: 95%
55	RD/WT	Keep Alive Power	12-14v
56	LG/BK	EVAP Purge Solenoid	0-10 Hz (0-100%)
57-59	---	Not Used	---
60	GY/LB	HO2S-11 (B1 S1) Signal	0.1-1.1v
61	PK/LG	HO2S-22 (B2 S2) Signal	0.1-1.1v
62	RD/PK	FTP Sensor Signal	2.6v (0" H2O - cap off)
63	RD/PK	FRP Sensor Signal	3.0v (43 psi)
64	LB/YL	Digital TR3A Sensor	0v, 55 mph: 11v
65 ('00)	BR/LG	DPFE Sensor Signal	0.20-1.30v
65 ('01-'03)	BR/LG	DPFE Sensor Signal	0.95-1.05v
67	GY/YL	EVAP CV Solenoid	0-10 Hz (0-100%)
68	WT/OR	VSS (+) Signal	0 Hz, 55 mph: 125 Hz

2000-03 Mustang GT, GT Bullitt 4.6L 2v V8 VIN X (All) 104 Pin Connector

PCM Pin #	Wire Color	Circuit Description (104 Pin)	Value at Hot Idle
69	PK/YL	A/C WOT Relay Control	A/C On: 12v, Off: 1v
70	---	Not Used	---
71	RD	Vehicle Power	12-14v
72	TN/RD	Injector 7 Control	3.5-3.7 ms
73	TN/BK	Injector 5 Control	3.5-3.7 ms
74	BR/YL	Injector 3 Control	3.5-3.7 ms
75	TN	Injector 1 Control	3.5-3.7 ms
76, 77	BK/WT	Power Ground	<0.1v
78	PK/LB	COP 7 (Integrated) Dwell	6° dwell
79	WT/RD	COP 8 (Integrated) Dwell	6° dwell
80	WT/RD	Fuel Pump Control	1.3v (28%)
81	WT/YL	EPC Solenoid Control	9.5v (15-20 psi)
82	WT/OR	Charge Air Cooler Pump Relay Control	Relay Off: 12v, On: 1v
83	WT/LB	IAC Motor Control	10v (30% duty cycle)
84	DG/WT	OSS Sensor (+) Signal	0 Hz, 55 mph: 250 Hz
85	DB/OR	CMP Sensor Signal	5-7 Hz
86	TN/LG	A/C High Pressure Switch	A/C On: 12v, Off: 1v
87	RD/BK	HO2S-21 (B2 S1) Signal	0.1-1.1v
88	LB/RD	MAF Sensor Signal	0.6-0.9v
89	GY/WT	TP Sensor Signal	0.53-1.27v
90	BR/WT	Reference Voltage	4.9-5.1v
91	GY/RD	Analog Signal Return	<0.050v
92	LG	Brake Pedal Position Switch	Brake Off: 0v, On: 12v
93	RD/WT	HO2S-11 (B1 S1) Heater	On: 1v, Off: 12v
94	YL/LB	HO2S-21 (B2 S1) Heater	On: 1v, Off: 12v
95	WT/BK	HO2S-12 (B1 S2) Heater	On: 1v, Off: 12v
96	TN/YL	HO2S-22 (B2 S2) Heater	On: 1v, Off: 12v
97	RD	Vehicle Power	12-14v
98	LB	Injector 8 Control	3.5-3.7 ms
99	LG/OR	Injector 6 Control	3.5-3.7 ms
100	BR/LB	Injector 4 Control	3.5-3.7 ms
101	WT	Injector 2 Control	3.5-3.7 ms
102	---	Not Used	---
103	BK/WT	Power Ground	<0.1v
104	PK/WT	COP 2 (Integrated) Dwell	6° dwell

Pin Connector Graphic

PCM 104-PIN CONNECTOR

```
 1 ●●●●●●●●●●●●●    ●●●●●●●●●●●●● 26
27 ●●●●●●●●●●●●●    ●●●●●●●●●●●●● 52
53 ●●●●●●●●●●●●●  ⬤ ●●●●●●●●●●●●● 78
79 ●●●●●●●●●●●●●    ●●●●●●●●●●●●● 104
```

Terminal View of 104-Pin PCM Wiring Harness Connector

Standard Colors and Abbreviations

Abbreviation	Color	Abbreviation	Color	Abbreviation	Color
BK	Black	GY	Gray	PK	Purple
BL	Blue	GN	Green	RD	Red
BR	Brown	LG	LT Green	TN	Tan
DB	Dark Blue	OR	Orange	WT	White
DG	DK Green	PK	Pink	YL	Yellow

2003 Mustang Cobra 4.6L V8 DOHC VIN Y (M/T) 104 Pin Connector

PCM Pin #	Wire Color	Circuit Description (104 Pin)	Value at Hot Idle
1	VT/YL	COP 6 (Integrated) Dwell	6° dwell
2	---	Not Used	---
3	BK/WT	Power Ground	<0.1v
4-7	---	Not Used	---
8	WT/OR	IMRC Monitor Signal	5v, 55 mph: 5v
9	---	Not Used	---
10	DB/LG	Barometric Pressure Sensor Signal	159 Hz
11-12	---	Not Used	---
13	VT	Flash EEPROM Power	0.1v
14	---	Not Used	---
15	PK/LB	SCP Bus (-) Signal	Digital Signals
16	TN/OR	SCP Bus (+) Signal	Digital Signals
17-19	---	Not Used	---
20	TN/RD	Reverse Lockout Solenoid Control	Solenoid Off: 12v, On: 1v
21	DB	CKP Sensor (+) Signal	320-420 Hz
22	GY	CKP Sensor (-) Signal	320-420 Hz
23-24	---	Not Used	---
25	BK	PCM Case Ground	<0.050v
26	LG/WT	COP 1 (Integrated) Dwell	6° dwell
27	LG/YL	COP 5 (Integrated) Dwell	6° dwell
28-34	---	Not Used	---
35	RD/LG	HO2S-12 (B1 S2) Signal	0.1-1.1v
36	TN/LB	MAF Sensor Return	<0.050v
37	RD/YL	Intake Air Temperature Sensor 2 Signal	1.5-2.5v
38	---	Not Used	---
39	GY	Intake Air Temperature Sensor 1 Signal	1.5-2.5v
40	LB/OR	Fuel Pump Monitor	2.5-7.5v
41	DG/OR	A/C Cycling Switch Signal	A/C On: 12v, Off: 1v
42	BR	IMRC Solenoid Control	12v (Off)
43-44	---	Not Used	---
45	LG/VT	Supercharger Bypass Solenoid Control	Solenoid Off: 12v, On: 1v
46	---	Not Used	---
47	BR/PK	EGR VR Solenoid	0%, 55 mph: 45%
48-50	---	Not Used	---
51	BK/WT	Power Ground	<0.1v
52	WT/PK	COP 3 (Integrated) Dwell	6° dwell
53	DG/PK	COP 4 (Integrated) Dwell	6° dwell
54	---	Not Used	---
55	RD/WT	Keep Alive Power	12-14v
56	LG/BK	EVAP Purge Solenoid	0-10 Hz (0-100%)
57-59	---	Not Used	---
60	GY/LB	HO2S-11 (B1 S1) Signal	0.1-1.1v
61	VT/LG	HO2S-22 (B2 S2) Signal	0.1-1.1v
62	RD/PK	FTP Sensor Signal	2.6v (0" H2O - cap off)
63	RD/PK	FRP Sensor Signal	2.8v (39 psi)
64	---	Not Used	---
65	BR/LG	DPFE Sensor Signal	0.95-1.05v
66	YE/LG	Cylinder Head Temperature Sensor	0.5-0.6v
67	GY/YL	EVAP CV Solenoid	0-10 Hz (0-100%)
68	WT/OR	VSS (+) Signal	Moving: AC pulse signals
69	PK/YL	A/C WOT Relay Control	A/C On: 12v, Off: 1v
70	---	Not Used	---

2003 Mustang Cobra 4.6L V8 DOHC VIN Y (M/T) 104 Pin Connector

PCM Pin #	Wire Color	Circuit Description (104 Pin)	Value at Hot Idle
71	RD	Vehicle Power (Start-Run)	12-14v
72	TN/RD	Injector 7 Control	2.4-2.8 ms
73	TN/BK	Injector 5 Control	2.4-2.8 ms
74	BR/YL	Injector 3 Control	2.4-2.8 ms
75	TN	Injector 1 Control	2.4-2.8 ms
76	BK/WT	Power Ground	<0.1v
77	BK/WT	Power Ground	<0.1v
78	PK/LB	COP 7 (Integrated) Dwell	6° dwell
79	WT/RD	COP 8 (Integrated) Dwell	6° dwell
80	WT/RD	Fuel Pump Control	Off: 12v, On: 1v
81	---	Not Used	---
82	WT/DG	Charge Air Cooler Pump Relay Control	Relay Off: 12v, On: 1v
83	WT/LB	IAC Motor Control	9v (32% duty cycle)
84	DG/WT	OSS Sensor (+) Signal	0 Hz, 55 mph: 470 Hz
85	DB/OR	CMP Sensor Signal	5-7 Hz
86	TN/LG	A/C High Pressure Switch	A/C On: 12v, Off: 1v
87	RD/BK	HO2S-21 (B2 S1) Signal	0.1-1.1v
88	LB/RD	MAF Sensor Signal	0.6v
89	GY/WT	TP Sensor Signal	0.53-1.27v
90	BR/WT	Reference Voltage	4.9-5.1v
91	GY/RD	Sensor Ground	<0.050v
92	LG	Brake Pedal Position Switch	Brake Off: 0v, On: 12v
93	RD/WT	HO2S-11 (B1 S1) Heater	On: 1v, Off: 12v
94	YL/LB	HO2S-21 (B2 S1) Heater	On: 1v, Off: 12v
95	WT/BK	HO2S-12 (B1 S2) Heater	On: 1v, Off: 12v
96	TN/YL	HO2S-22 (B2 S2) Heater	On: 1v, Off: 12v
97	RD	Vehicle Power (Start-Run)	12-14v
98	LB	Injector 8 Control	2.4-2.8 ms
99	LG/OR	Injector 6 Control	2.4-2.8 ms
100	BR/LB	Injector 4 Control	2.4-2.8 ms
101	WT	Injector 2 Control	2.4-2.8 ms
102	DG/PK	Knock Sensor 2 (+) Signal	0v
103	BK/WT	Power Ground	<0.1v
104	PK/WT	COP 2 (Integrated) Dwell	6° dwell

Pin Connector Graphic

PCM 104-PIN CONNECTOR

```
1  ●●●●●●●●●●●●●    ●●●●●●●●●●●●● 26
27 ●●●●●●●●●●●●●    ●●●●●●●●●●●●● 52
53 ●●●●●●●●●●●●●  ⬡ ●●●●●●●●●●●●● 78
79 ●●●●●●●●●●●●●    ●●●●●●●●●●●●● 104
```

Terminal View of 104-Pin PCM Wiring Harness Connector

Standard Colors and Abbreviations

Abbreviation	Color	Abbreviation	Color	Abbreviation	Color
BK	Black	GY	Gray	PK	Purple
BL	Blue	GN	Green	RD	Red
BR	Brown	LG	LT Green	TN	Tan
DB	Dark Blue	OR	Orange	WT	White
DG	DK Green	PK	Pink	YL	Yellow

1990-92 Mustang 5.0L V8 HO VIN E (All) 60 Pin Connector

PCM Pin #	Wire Color	Circuit Description (60 Pin)	Value at Hot Idle
1	YL	Keep Alive Power	12-14v
2, 5, 8, 11, 18, 28	---	Not Used	---
3	GY, PK	VSS (+), (-) Signal	0 Hz, 55 mph: 125 Hz
4	TN/YL	Ignition Diagnostic Monitor	20-31 Hz
7	LG/RD	ECT Sensor Signal	0.5-0.6v
9	TN/BL	MAF Sensor Return	<0.050v
10	BK/YL	A/C Cycling Clutch Switch	A/C On: 12v, Off: 0v
12	BR/YL	Injector 3 Control	4.6-4.8 ms
13	BR/BL	Injector 4 Control	4.6-4.8 ms
14	TN/BK	Injector 5 Control	4.6-4.8 ms
15	LG/OR	Injector 6 Control	4.6-4.8 ms
16	OR/RD	Ignition System Ground	<0.050v
17	PK/LG	STI Output, MIL Control	MIL On: 1v, Off: 12v
19	DG/YL	Fuel Pump Monitor	On: 12v, Off: 0v
20	BK	PCM Case Ground	<0.050v
21	WT/BL	IAC Motor Control	8-10v
22	BL/OR	Fuel Pump Control	On: 1v, Off: 12v
23-24, 34-35	---	Not Used	---
25	GY	ACT Sensor Signal	1.5-2.5v
26	BR/WT	Reference Voltage	4.9-5.1v
27	BR/LG	EGR EVP Sensor	Idle: 0.4 v
29	GY/BL	HO2S-21 (B2 S1) Signal	0.1-1.1v
30	BL/YL	Neutral Drive (A/T), Clutch (M/T) Switch	In 'P': 0v or with Clutch In: 0v
31	GY/YL	EVAP Purge Solenoid	0%, 55 mph: 40%
32	WT/LG	Thermactor Air Divert SOL	On: 1v, Off: 12v
33	BR/PK	EGR VR Solenoid	0%, 55 mph: 45%
36	PK	Spark Output Signal	50% duty cycle
37, 57	RD	Vehicle Power	12-14v
38	WT/YL	Thermactor Air Bypass SOL	TAB On: 1v, Off: 12v
39, 41, 51, 53, 55	---	Not Used	---
40, 60	BK/WT	Power Ground	<0.1v
42	TN/RD	Injector 7 Control	4.6-4.8 ms
43	RD/BK	HO2S-11 (B1 S1) Signal	0.1-1.1v
44	GY/RD	HO2S-11 (Bank 1) Ground	<0.050v
45	LG/BK	BARO Sensor Signal	159 Hz (sea level)
46	GY/RD	Analog Signal Return	<0.050v
47	GY/WT	TP Sensor Signal	0.5-1.2v
48	WT/PK	Self-Test Indicator Signal	STI Open: 5v, Closed: 1v
49	OR	HO2S-21 (Bank 2) Ground	<0.050v
50	LB/RD	MAF Sensor Signal	0.20-1.30v
52	LB	Injector 8 Control	4.6-4.8 ms
54	PK/YL	A/C WOT Cutoff Relay	On: 1v, Off: 12v
56	GY/OR	PIP Sensor Signal	50% duty cycle
58	TN	Injector 1 Control	4.6-4.8 ms
59	WT	Injector 2 Control	4.6-4.8 ms

Pin Connector Graphic

PCM 60-PIN CONNECTOR

Terminal View of 60-Pin PCM Harness Connector

1993 Mustang 5.0L V8 HO MFI VIN E (All) 60 Pin Connector

PCM Pin #	Wire Color	Circuit Description (60 Pin)	Value at Hot Idle
1	YL	Keep Alive Power	12-14v
2	---	Not Used	---
3	GY/BK	VSS (+) Signal	0 Hz, 55 mph: 125 Hz
4	TN/YL	Ignition Diagnostic Monitor	20-31 Hz
5	---	Not Used	---
6	PK/OR	VSS (-) Signal	0 Hz, 55 mph: 125 Hz
7	LG/RD	ECT Sensor Signal	0.5-0.6v
8	---	Not Used	---
9	TN/BL	MAF Sensor Return	<0.050v
10	BK/YL	A/C Cycling Clutch Switch	A/C On: 12v, Off: 0v
11	---	Not Used	---
12	BR/YL	Injector 3 Control	4.6-5.2 ms
13	BR/BL	Injector 4 Control	4.6-5.2 ms
14	TN/BK	Injector 5 Control	4.6-5.2 ms
15	LG/OR	Injector 6 Control	4.6-5.2 ms
16	OR/RD	Ignition System Ground	<0.050v
17	PK/LG	STI Output, MIL Control	MIL On: 1v, Off: 12v
18	---	Not Used	---
19	DG/YL	Fuel Pump Monitor	On: 12v, Off: 0v
20	BK	PCM Case Ground	<0.1v
21	WT/BL	IAC Motor Control	8-9v
22	BL/OR	Fuel Pump Control	On: 1v, Off: 12v
23-24	---	Not Used	---
25	GY	IAT Sensor Signal	1.5-2.5v
26	BR/WT	Reference Voltage	4.9-5.1v
27	BR/LG	EGR EVP Sensor	0.4v
28	---	Not Used	---
29	GY/BL	HO2S-21 (B2 S1) Signal	0.1-1.1v
30	BL/YL	A/T: Park/Neutral Position	In 'P': 0v, Others: 5v
30	BL/YL	M/T: Clutch Pedal Position	Clutch In: 0v, Out: 5v
31	GY/YL	EVAP Purge Solenoid	0%, 55 mph: 50%
32	WT/LG	Air Injection Diverter Solenoid	AIRD On: 1v, Off: 12v
33	BR/PK	EGR VR Solenoid	0%, 55 mph: 45%
34-35	---	Not Used	---
36	PK	Spark Output Signal	50% duty cycle
37	RD	Vehicle Power	12-14v
38	WT/YL	Air Injection Bypass Solenoid	AIRB On: 1v, Off: 12v
39	---	Not Used	---
40	BK/WT	Power Ground	<0.1v

1993 Mustang 5.0L V8 HO MFI VIN E (All) 60 Pin Connector

PCM Pin #	Wire Color	Circuit Description (60 Pin)	Value at Hot Idle
41	---	Not Used	---
42	TN/RD	Injector 7 Control	4.6-5.2 ms
43	RD/BK	HO2S-11 (B1 S1) Signal	0.1-1.1v
44	GY/RD	HO2S-11 (Bank 1) Ground	<0.050v
45	LG/BK	BARO Sensor Signal	159 Hz (sea level)
46	GY/RD	Analog Signal Return	<0.050v
47	GY/WT	TP Sensor Signal	0.5-1.2v
48	WT/PK	Self-Test Indicator Signal	STI Open: 5v, Closed: 1v
49	OR	HO2S-21 (Bank 2) Ground	<0.050v
50	LB/RD	MAF Sensor Signal	0.6-0.9v
51	---	Not Used	---
52	LB	Injector 8 Control	4.6-5.2 ms
53	---	Not Used	---
54	PK/YL	A/C WOT Cutoff Relay	On: 1v, Off: 12v
55	---	Not Used	---
56	GY/OR	PIP Sensor Signal	50% duty cycle
57	RD	Vehicle Power	12-14v
58	TN	Injector 1 Control	4.6-5.2 ms
59	WT	Injector 2 Control	4.6-5.2 ms
60	BK/WT	Power Ground	<0.1v

Pin Connector Graphic

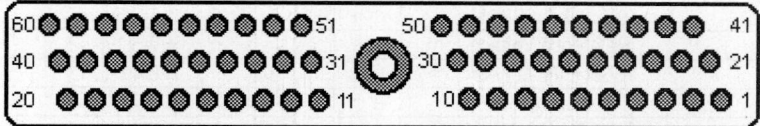

PCM 60-PIN CONNECTOR

Terminal View of 60-Pin PCM Harness Connector

Standard Colors and Abbreviations

Abbreviation	Color	Abbreviation	Color	Abbreviation	Color
BK	Black	GY	Gray	PK	Purple
BL	Blue	GN	Green	RD	Red
BR	Brown	LG	LT Green	TN	Tan
DB	Dark Blue	OR	Orange	WT	White
DG	DK Green	PK	Pink	YL	Yellow

1994-95 Mustang 5.0L V8 MFI VIN D, VIN T (All) 60 Pin Connector

PCM Pin #	Wire Color	Circuit Description (60 Pin)	Value at Hot Idle
1	YL	Keep Alive Power	12-14v
2	LG	Brake Pedal Position Switch	Brake Off: 0v, On: 12v
3	GY/BK	VSS (+) Signal	0 Hz, 55 mph: 125 Hz
4	WT/PK	Ignition Diagnostic Monitor	20-31 Hz
5	DG/WT	TSS Sensor Signal	0 Hz, 35 mph: 115 Hz
6	PK/OR	VSS (-) Signal	0 Hz, 55 mph: 125 Hz
7	LG/RD	ECT Sensor Signal	0.5-0.6v
8	DG/YL	Fuel Pump Monitor	On: 12v, Off: 0v
9	TN/BL	MAF Sensor Return	<0.050v
10	DG/OR	A/C Cycling Clutch Signal	A/C On: 12v, Off: 0v
11	GY/YL	EVAP Purge Solenoid	0%, 55 mph: 60%
12	LG/OR	Injector 6 Control	5.0-5.2 ms
13	TN/RD	Injector 7 Control	5.0-5.2 ms
14	BL	Injector 8 Control	5.0-5.2 ms
15	TN/BK	Injector 5 Control	5.0-5.2 ms
16	OR/RD	Ignition System Ground	<0.050v
17	PK/LG	STI Output, MIL Control	MIL On: 1v, Off: 12v
18	TN/OR	Data Bus (-) Signal	Digital Signals
19	PK/BL	Data Bus (+) Signal	Digital Signals
20	BK	PCM Case Ground	<0.050v
21	WT/BL	IAC Motor Control	8-9v
22	BL/OR	Fuel Pump Control	On: 1v, Off: 12v
23-24	---	Not Used	---
25	GY	IAT Sensor Signal	1.5-2.5v
26	BR/WT	Reference Voltage	4.9-5.1v
27	BR/LG	EGR EVP Sensor	0.4v
28-29	---	Not Used	---
30	BL/YL	MLP Sensor Signal	In 'P': 0v, 55 mph: 1.7v
31	WT/OR	Air Injection Bypass Solenoid	AIRB On: 1v, Off: 12v
32	LG/PK	High Speed Fan Control	On: 1v, Off: 12v
33	BR/PK	EGR VR Solenoid	0%, 55 mph: 45%
34	BR	Air Injection Diverter Solenoid	AIRD On: 1v, Off: 12v
35	BR/BL	Injector 4 Control	5.0-5.2 ms
36	PK	Spark Output Signal	50% duty cycle
37	RD	Vehicle Power	12-14v
38	WT/YL	EPC Solenoid Control	Hot Idle: 9.1v (4 psi)
39	BR/YL	Injector 3 Control	5.0-5.2 ms
40	BK/WT	Power Ground	<0.1v

1994-95 Mustang 5.0L V8 MFI VIN D, VIN T (All) 60 Pin Connector

PCM Pin #	Wire Color	Circuit Description (60 Pin)	Value at Hot Idle
41	TN/WT	TCS (switch) Signal	TCS/OD On: 1v, Off: 12v
42	PK/BL	A/C Hi Pressure Cutoff Switch	On: 1v, Off: 12v
43	RD/BK	HO2S-11 (B1 S1) Signal	0.1-1.1v
44	GY/BL	HO2S-21 (B2 S1) Signal	0.1-1.1v
45	---	Not Used	---
46	GY/RD	Analog Signal Return	<0.050v
47	GY/WT	TP Sensor Signal	0.5-1.2v
48	WT/PK	Self-Test Indicator Signal	STI Open: 5v, Closed: 1v
49	OR/BK	TOT Sensor Signal	2.10-2.40v
50	LB/RD	MAF Sensor Signal	0.6-0.9v
51	OR/YL	Shift Solenoid 1 Control	1v, 55 mph: 12v
52	PK/OR	Shift Solenoid (SS2)	12v, 55 mph: 1v
53	PK/YL	Torque Converter Clutch	0%, 55 mph: 40%
53	BR/OR	Torque Converter Clutch	0%, 55 mph: 40%
54	PK/YL	A/C WOT Relay Control	On: 1v, Off: 12v
55	DB	Low Speed Fan Control	On: 1v, Off: 12v
56	GY/OR	PIP Sensor Signal	50% duty cycle
57	RD	Vehicle Power	12-14v
58	TN	Injector 1 Control	5.0-5.2 ms
59	WT	Injector 2 Control	5.0-5.2 ms
60	BK/WT	Power Ground	<0.1v

Pin Connector Graphic

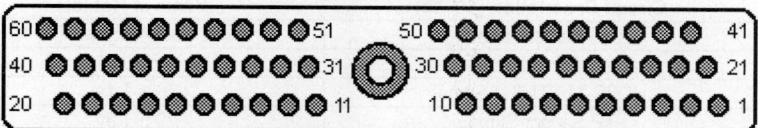

PCM 60-PIN CONNECTOR

Terminal View of 60-Pin PCM Harness Connector

Standard Colors and Abbreviations

Abbreviation	Color	Abbreviation	Color	Abbreviation	Color
BK	Black	GY	Gray	PK	Purple
BL	Blue	GN	Green	RD	Red
BR	Brown	LG	LT Green	TN	Tan
DB	Dark Blue	OR	Orange	WT	White
DG	DK Green	PK	Pink	YL	Yellow

Probe PIN Tables

1993-95 Probe 2.0L I4 MFI VIN A (A/T) 22 Pin Connector

PCM Pin #	Wire Color	Circuit Description (22 Pin)	Value at Hot Idle
22-A	BL/RD	Keep Alive Power	12- 14v
22-B	RD/BK	Vehicle Power (Power Relay)	12-14v
22-C	WT/RD	System Monitor Lamp	Lamp On: 5v, Off: 12v
22-D	BK/RD	Vehicle Start Signal	Cranking: 9-11v
22-E	WT/BL	MIL (lamp) Control	MIL On: 1v, MIL Off: 12v
22-F	WT/GN	Brake Pedal Position Switch	Brake Off: 0v, On: 12v
22-G	LG/RD	Self-Test Output Signal	STO On: 5v, Off: 12v
22-H	BR/BK	Overdrive Off Switch Signal	OD Switch On: 0.1v
22-I	BK/OR	Ignition Control Module Signal	0.2v, 55 mph: 0.3vv
22-J	WT	Headlamp Switch Signal	Lamps On: 12v, Off: 0.1v
22-K	GN	Daylight Running Lamp (DRL)	DRL On: 2.5v, Off: 12v
22-L	RD/WT	Self-Test Input	STI On: 0.1v, Off: 5v
22-M	GN/RD	VSS Signal to Cluster	Digital Signals
22-N	BR/YL	PSP Switch Signal	Straight: 12v, Turned: 0v
22-O	PK	Rear Defrost (DEF) Signal	Switch Off: 0.1v, On: 12v
22-P	OR/BK	Blower Motor Switch Signal	Off or 1st: 12v, 2nd on: 1v
22-Q	PK/GN	Overdrive Lamp (Cluster)	OD On: 1v, Off: 12v
22-R	PK/BL	A/C Cycling Clutch Switch	A/C On: 2v, Off: 12v
22-S	YL/WT	AC Relay Control	On: 1v, Off: 12v
22-T	BR	Idle Position Switch Signal	IDL closed: 0v, open: 12v
22-U	RD/BK	Vehicle Power	12-14v
22-V	GN	Manual Lever Position Signal	In 'P': 0v, Others: 12v

1993-95 Probe 2.0L I4 MFI VIN A (A/T) 16 Pin Connector

PCM Pin #	Wire Color	Circuit Description (16 Pin)	Value at Hot Idle
16-A	GN/OR	CKP Sensor Signal	390-450 Hz
16-B	LG/WT	Camshaft Position Sensor	5-7 Hz
16-C	BK/YL	HO2S-11 (B1 S1) Signal	0.1-1.1v
16-D	YL	TP Sensor Signal	0.5-1.3v
16-E	YL/BK	ECT Sensor Signal	0.5-0.6v
16-F	RD/WT	IAT Sensor Signal	1-5-2.5v
16-G	BL	Barometric Pressure Sensor	4.3v (28" Hg)
16-H	BL/YL	TOT Sensor Signal	2.10-2.40v
16-I	PK	Reference Voltage	4.9-5.1v
16-J	RD	EGR Temperature Sensor	2.25-2.55v (68-120°F)
16-K	BL/WT	Cooling Fan Temp. Sensor	0.8v
16-L	BK/RD	MAF Sensor Signal	1.9v
16-M	RD	MLP Sensor - Reverse	In 'R': 0v, Others: 0v
16-N	BL/OR	MLP Sensor - First	In 1st: 0v, Others: 0v
16-O	GN/WT	MLP Sensor - Second	In 2nd: 0v, Others: 0v
16-P	RD/BL	MLP Sensor - Drive	In 'P': 0v, Others: 12v

Standard Colors and Abbreviations

Abbreviation	Color	Abbreviation	Color	Abbreviation	Color
BK	Black	GY	Gray	PK	Purple
BL	Blue	GN	Green	RD	Red
BR	Brown	LG	LT Green	TN	Tan
DB	Dark Blue	OR	Orange	WT	White
DG	DK Green	PK	Pink	YL	Yellow

1993-95 Probe 2.0L I4 MFI VIN A (A/T) 26 Pin Connector

PCM Pin #	Wire Color	Circuit Description (26 Pin)	Value at Hot Idle
26-A	BK	Power Ground	<0.1v
26-B	BK	Power Ground	<0.1v
26-C	BK/RD	Power Ground	<0.1v
26-D	BK/BL	Analog Signal Return	<0.1v
26-E	---	Not Used	---
26-F	WT	TSS Sensor (+) Signal	680-720 rpm
26-G	BK/GN	Low Speed Fan Control	On: 1v, Off: 12v
26-H	RD	TSS Sensor (-)	680-720 rpm
26-I	BL/GN	High Speed Fan Control	On: 1v, Off: 12v
26-J	---	Not Used	---
26-K	WT/BL	EGR VR Solenoid	0%, 55 mph: 45%
26-L	LG	Fuel Pump Control	On: 1v, Off: 12v
26-M	GN	Fuel Pressure Regulator	Startup: <3v, other: 12v
26-N	WT/BK	EVAP Purge Solenoid	12v, 55 mph: 1v
26-O	LG/BK	Idle Air Control Valve	8-10v
26-P	BL	Shift Solenoid 1 Control	12v, 55 mph: 1v
26-Q	BL/BK	Shift Solenoid 2 Control	1v, 55 mph: 12v
26-R	GN/BK	Shift Solenoid 3 Control	12v, 55 mph: 1v
26-S	BL/WT	TCC Solenoid Control	12v, 55 mph: 1v
26-T	RD/WT	Downshift Timing Solenoid	DSS On: 1v, Off: 12v
26-U	RD/BK	TCC Solenoid Control	0v, 55 mph: 12v
26-V	RD/GN	Line Pressure Solenoid	9v, 55 mph: 8.4v
26-W	YL/BK	Injector 1 Control	3.5-3.7 ms
26-X	YL/WT	Injector 2 Control	3.5-3.7 ms
26-Y	YL/RD	Injector 3 Control	3.5-3.7 ms
26-Z	YL/GN	Injector 4 Control	3.5-3.7 ms

Pin Connector Graphic

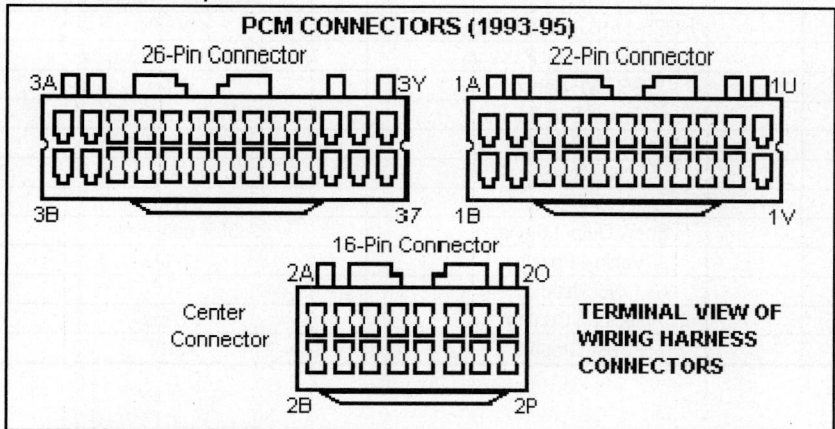

Standard Colors and Abbreviations

Abbreviation	Color	Abbreviation	Color	Abbreviation	Color
BK	Black	GY	Gray	PK	Purple
BL	Blue	GN	Green	RD	Red
BR	Brown	LG	LT Green	TN	Tan
DB	Dark Blue	OR	Orange	WT	White
DG	DK Green	PK	Pink	YL	Yellow

1993-95 Probe 2.0L I4 MFI VIN A (M/T) 60 Pin Connector

PCM Pin #	Wire Color	Circuit Description (60 Pin)	Value at Hot Idle
1	BL/RD	Keep Alive Power	12- 14v
2	---	Not Used	---
3	BL/WT	VSS (+) Signal	0 Hz, 55 mph: 125 Hz
4	BL/YL	Ignition Diagnostic Monitor	20-31 Hz
5	---	Not Used	---
6	BR/GN	VSS (-) Signal	0 Hz, 55 mph: 125 Hz
7	OR/GN	ECT Sensor Signal	0.5-0.6v
8	WT/YL	Fuel Pump Monitor	Pump On: 12v, Off; 0v
9	BR	MAF Sensor Return	<0.050v
10	---	Not Used	---
11	WT/BK	EVAP Purge Solenoid	12v, 55 mph: 1v
12	---	Not Used	---
13	OR/WT	Fuel Pressure Regulator	Startup: <3v, Others: 12v
14	OR/BK	Blower Motor Signal	Off or 1st: 12v, 2nd on: 1v
15	PK	Rear Defroster Signal	Switch Off: 0.1v, On: 12v
16	BK	Ignition Ground	<0.050v
17	BL	STI Output, MIL Control	MIL On: 1v, Off: 12v
18	---	Not Used	---
19	---	Not Used	---
20	---	Not Used	---
21	LG/BK	ISC Motor Control	8-10.1v
22	LG	Fuel Pump Control	On: 1v, Off: 12v
23	PK/BK	A/C Cycling Clutch Switch	On: 1v, Off: 12v
24	GN/WT	Camshaft Position Sensor	5-7 Hz
25	WT/LG	IAT Sensor Signal	1-5-2.5v
26	LG/PK	Reference Voltage	4.9-5.1v
27	---	Not Used	---
28	BR/YL	PSP Switch Signal	Straight: 12v, Turned: 0v
29	GN/BL	HO2S-11 (B1 S1) Ground	<0.050v
30	LG/BK	M/T: Clutch Pedal Position	Clutch In: 0v, Out: 5v
31	WT/BL	EGR EVP Sensor	0.4v
32	BL/OR	Low Speed Fan Control	On: 1v, Off: 12v
33	BL/GN	High Speed Fan Control	On: 1v, Off: 12v
34	---	Not Used	---
35	YL/GN	Injector 4 Control	3.8 ms
36	RD/BL	Spark Output Signal	50% duty cycle
37	WT/RD	Vehicle Power	12-14v
38	---	Not Used	---
39	YL/RD	Injector 3 Control	3.8 ms
40	BK/GN	Power Ground	<0.1v

1993-95 Probe 2.0L I4 MFI VIN A (M/T) 60 Pin Connector

PCM Pin #	Wire Color	Circuit Description (60 Pin)	Value at Hot Idle
41	---	Not Used	---
42	WT	Daytime Running Lamp	DRL On: 2v, Off: 12v
43	RD/GN	EGR Temperature Sensor	2.25-2.55v (68-120°F)
44	BK/YL	HO2S-11 (B1 S1) Signal	0.1-1.1v
45	W	Headlamp Switch Signal	Off or 1st: 12v, 2nd on: 1v
46	BK/BL	Analog Signal Return	<0.050v
47	LG/WT	TP Sensor Signal	0.5-1.1v
48	BL/PK	Self-Test Input	STI On: 0.1v, Off: 5v
49	---	Not Used	---
50	RD	MAF Sensor Signal	1.9v
51-53	---	Not Used	---
54	GN/BK	A/C Relay Control	On: 1v, Off: 12v
55	---	Not Used	---
56	RD/YL	PIP Sensor Signal	50% duty cycle
57	WT/RD	Vehicle Power	12-14v
58	YL/BK	Injector 1 Control	3.8 ms
59	YL/WT	Injector 2 Control	3.8 ms
60	BK/GN	Power Ground	<0.1v

Pin Connector Graphic

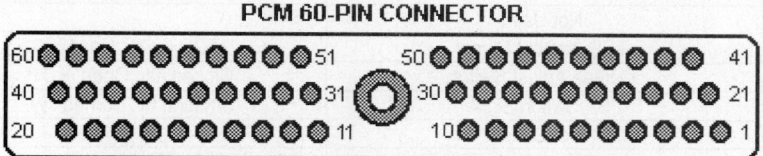

PCM 60-PIN CONNECTOR

Terminal View of 60-Pin PCM Harness Connector

Standard Colors and Abbreviations

Abbreviation	Color	Abbreviation	Color	Abbreviation	Color
BK	Black	GY	Gray	PK	Purple
BL	Blue	GN	Green	RD	Red
BR	Brown	LG	LT Green	TN	Tan
DB	Dark Blue	OR	Orange	WT	White
DG	DK Green	PK	Pink	YL	Yellow

1996-97 Probe 2.0L I4 MFI VIN A (All) 104 Pin Connector

PCM Pin #	Wire Color	Circuit Description (104 Pin)	Value at Hot Idle
1	BL/BK	Shift Solenoid 2 Control	1v, 55 mph: 12v
2	BL	MIL (lamp) Control	MIL On: 1v, Off: 12v
3-9	---	Not Used	---
10	YL/BK	Blower Motor Switch Signal	Motor On: 0.7v, Off: 12v
11-12	---	Not Used	---
13	GN/YL	Flash EEPROM Power	0.1v
14	GN	Daytime Running Lamps	Lamps Off: 12v, On: 1v
15	WT/RD	Data Bus (-) Signal	Digital Signals
16	OR/BK	Data Bus (+) Signal	Digital Signals
17	BL	High Speed Fan Control	On: 1v, Off: 12v
18	---	Not Used	---
19	GN/WT	Fuel Pressure Regulator	12v, 55 mph: 12v
20	---	Not Used	---
21	GN	CKP Sensor (+) Signal	390-450 Hz
22	BL	CKP Sensor (-) Signal	390-450 Hz
23	BK/GN	Power Ground	<0.1v
24	BK/GN	Power Ground	<0.1v
25	---	Not Used	---
26	GN	Coil Driver 1 Control	5° dwell
27	BL	Shift Solenoid 1 Control	12v, 55 mph: 1v
28	---	Not Used	---
29	BR/BK	TCS (switch) Signal	TCS & O/D On: 12v
30	OR/BL	Octane Adjust Switch	Closed: 0v, Open: 9.3v
31	BR/YL	PSP Switch Signal	Straight: 0v, Turned: 5v
32	---	Not Used	---
33	OR/BK	VSS (-) Signal	0 Hz, 55 mph: 125 Hz
34	LG	EGR BARO Pressure Sensor	Typical Range: 4.1-4.8v
35	BK/YL	HO2S-12 (B1 S2) Signal	0.1-1.1v
36	BR	MAF Sensor Return	<0.050v
37	PK/WT	TFT Sensor Signal	2.10-2.40v
38	OR/GN	ECT Sensor Signal	0.5-0.6v
39	WT/GN	IAT Sensor Signal	1.5-2.5v
40	WT/YL	Fuel Pump Monitor	On: 12v, Off: 0v
41	PK/BK	A/C Cycling Clutch Switch	A/C On: 12v, Off: 0v
42-44	---	Not Used	---
45	BL/OR	Low Speed Fan Control	On: 1v, Off: 12v
46-48	---	Not Used	---
49	WT	Headlight Switch Signal	Lights On: 12v, Off: 0.1v
46-48	---	Not Used	---
51	BK/GN	Power Ground	1v
52-53	---	Not Used	---
54	BL/LG	Modulated Converter Clutch	0%, 55 mph: 50%
55	BL/RD	Keep Alive Power	12-14v
56-57	---	Not Used	---
58	BL/WT	VSS (+) Signal	0 Hz, 55 mph: 125 Hz
59	---	Not Used	---
60	BL/WT	HO2S-11 (B1 S1) Signal	0.1-1.1v
61-63	---	Not Used	---
64	RD/BK	A/T CD4E: TR Sensor	In 'P': 0v, 55 mph: 1.7v
64	LG/BK	M/T: Clutch Pedal Position	Clutch In: 0v, Out: 5v

1996-97 Probe 2.0L I4 MFI VIN A (All) 104 Pin Connector

PCM Pin #	Wire Color	Circuit Description (104 Pin)	Value at Hot Idle
65	RD/GN	EGR Valve Position Sensor	0.20-1.30v
66	PK	Rear Defroster Switch Signal	Defrost On: 0.9v, Off: 12v
67	WT/PK	EVAP Purge Solenoid	12v, 55 mph: 1v
68	WT/BL	EGR Atmospheric Pressure	12v, 55 mph: 1v
69	GN/BK	A/C WOT Relay Control	On: 1v, Off: 12v
70	---	Not Used	---
71	WT/RD	Vehicle Power	12-14v
72	GN/BL	EGR VR Solenoid	12v, 55 mph: 1v
73	---	Not Used	---
74	YL/RD	Injector 3 Control	1.7-2.3 ms
75	YL/BK	Injector 1 Control	1.7-2.3 ms
76	BK/GN	Power Ground	<0.1v
77	BK/GN	Power Ground	<0.1v
78	---	Not Used	---
79	BR/YL	TCIL (lamp) Control	On: 1v, Off: 12v
80	LG	Fuel Pump Control	On: 1v, Off: 12v
81	BL/BR	EPC Solenoid Control	7-8v (8-9 psi)
82	---	Not Used	---
83	LG/BK	IAC Motor Control	9v (41% duty cycle)
84	BL/GN	TSS Sensor (+) Signal	42-50 Hz (680-720 rpm)
85	GY/WT	CID Sensor Signal	5-7 Hz
86	PK/YL	A/C Pressure Switch	A/C On: 12v (Open)
87	---	Not Used	---
88	RD	MAF Sensor Signal	0.6v
89	LG/WT	TP Sensor Signal	0.53-1.27v
90	LG/V	Reference Voltage	4.9-5.1v
91	BK/BL	Analog Signal Return	<0.050v
92	WT/BK	Brake Pedal Position Switch	Brake Off: 0v, On: 12v
93	YL	HO2S-11 (B1 S1) Heater	On: 1v, Off: 12v
94-96	---	Not Used	---
97	WT/RD	Vehicle Power	12-14v
98	OR/WT	EGR Check Solenoid	12v, 55 mph: 9-12v
99	---	Not Used	---
100	YL/GN	Injector 4 Control	1.7-2.3 ms
101	YL/WT	Injector 2 Control	1.7-2.3 ms
102	BL/V	Shift Solenoid 3 Control	12v, 55 mph: 1v
103	BK/GN	Power Ground	<0.1v
104	---	Not Used	---

Pin Connector Graphic

PCM 104-PIN CONNECTOR

Terminal View of 104-Pin PCM Wiring Harness Connector

Standard Colors and Abbreviations

Abbreviation	Color	Abbreviation	Color	Abbreviation	Color
BK	Black	GY	Gray	PK	Purple
BL	Blue	GN	Green	RD	Red
BR	Brown	LG	LT Green	TN	Tan
DB	Dark Blue	OR	Orange	WT	White
DG	DK Green	PK	Pink	YL	Yellow

1990-92 Probe 2.2L I4 MFI VIN C (A/T) 22 Pin Connector

PCM Pin #	Wire Color	Circuit Description (22 Pin)	Value at Hot Idle
22-A	BL/RD	Keep Alive Power	12- 14v
22-B	RD/BK	Vehicle Power (Power Relay)	12-14v
22-C	WT/YL	System Monitor Lamp	Lamp On: 5v, Off: 12v
22-D	BK/PK	Vehicle Start Signal	Cranking: 9-11v
22-E	WT/BL	MIL (lamp) Control	MIL On: 1v, MIL Off: 12v
22-F	WT/BK	Self-Test Output Signal	STO On: 5v, Off: 12v
22-G	---	Not Used	---
22-K	---	Not Used	---
22-H	WT/BK	Headlamp Switch Signal	Switch On: 12v, Off: 0.1v
22-I	RD/WT	Self-Test Input	STI On: 0.1v, Off: 5v
22-J	BK/BL	Rear Defroster Control	Switch On: 12v, Off: 0.1v
22-L	BL/BK	Condenser Fan Relay	On: 1v, Off: 12v
22-M	GN/RD	VSS Signal to Cluster	Digital Signals
22-N	BR/RD	PSP Switch Signal	Straight: 12v, Turned: 0v
22-O	BL/WT	A/C Cycling Clutch Switch	On: 1v, Off: 12v
22-P	BL/BK	Blower Motor Switch Signal	Off or 1st: 12v, 2nd on: 1v
22-Q	WT/GN	Brake Pedal Position Switch	Brake Off: 0v, On: 12v
22-R	BK/YL	A/T: PNP Position Switch	In 'P': 0v, Others: 12v
22-S	YL/WT	Speed Limiter Signal (Cluster)	Digital Signals
22-T	LG/WT	Idle Position Switch Signal	IDL closed: 0v, open: 12v
22-U	BK/YL	Vehicle Power	12-14v
22-V	YL/BK	Ignition Diagnostic Monitor	20-31 Hz

1990-92 Probe 2.2L I4 MFI VIN C (A/T) 16 Pin Connector

PCM Pin #	Wire Color	Circuit Description (16 Pin)	Value at Hot Idle
16-A	RD/WT	Vane Airflow Meter Reference	4.9-5.1v
16-B	RD/BK	Vane Airflow Meter Signal	3.3v
16-C	BK	O2S-11 (B1 S1) Signal	0.1-1.1v
16-D	BK/GN	Cooling Fan 1 Control	On: 1v, Off: 12v
16-E	YL/BK	ECT Sensor Signal	0.5-0.6v
16-F	LG/BK	TP Sensor Signal	0.5-1.3v
16-G	BL/RD	Radiator Temperature Switch	Switch closed: 0.1v
16-H	BL/WT	Manual Mode Switch	Closed: 0.1v, open: 5v
16-I	LG/RD	Reference Voltage	4.9-5.1v
16-J	YL/BL	EGR EVP Sensor	0.4v
16-K	RD	VAT Sensor Signal	2.23v (at 104°F)
16-L	---	Not Used	---
16-M	YL/GN	VSS (+) Signal	0 Hz, 55 mph: 125 Hz
16-N	YL/BL	VSS (-) Signal	0 Hz, 55 mph: 125 Hz
16-O	WT/BK	EVAP Purge Solenoid	12v, 55 mph: 1v
16-P	BR/YL	Neutral Shift Signal to Cluster	Digital Signals

Standard Colors and Abbreviations

Abbreviation	Color	Abbreviation	Color	Abbreviation	Color
BK	Black	GY	Gray	PK	Purple
BL	Blue	GN	Green	RD	Red
BR	Brown	LG	LT Green	TN	Tan
DB	Dark Blue	OR	Orange	WT	White
DG	DK Green	PK	Pink	YL	Yellow

1990-92 Probe 2.2L I4 MFI VIN C (A/T) 26 Pin Connector

PCM Pin #	Wire Color	Circuit Description (26 Pin)	Value at Hot Idle
26-A	BK	Power Ground	<0.1v
26-B	BK	Power Ground	<0.1v
26-C	BK/LG	Power Ground	<0.1v
26-D	LG/YL	Analog Signal Return	<0.050v
26-E	YL	Manual Lever Position O/D	In O/D: 12v
26-F	---	Not Used	---
26-G	YL/RD	Manual Lever Position Low	In Low: 12v
26-H	YL/BK	Manual Lever Position Drive	In Drive: 12v
26-I	---	Not Used	---
26-J	---	Not Used	---
26-K	---	Not Used	---
26-L	---	Not Used	---
26-M	WT/RD	Fuel Pressure Regulator	Startup: <3v, Others: 12v
26-N	BL/LG	TOT Sensor Signal	2.10-2.40v
26-O	WT/BL	EGR VR Solenoid	0%, 55 mph: 45%
26-P	---	Not Used	---
26-Q	WT	Idle Air Control Valve	8-10v
26-R	---	Not Used	---
26-S	---	Not Used	---
26-T	---	Not Used	---
26-U	YL	Fuel Injector 1 Bank 1	3.6 ms
26-V	YL/BK	Fuel Injector 2 Bank 2	3.6 ms
26-W	BL	Shift Solenoid 1 Control	12v, 55 mph: 1v
26-X	BL/YL	Shift Solenoid 2 Control	1v, 55 mph: 12v
26-Y	BL/RD	Shift Solenoid 3 Control	12v, 55 mph: 1v
26-Z	BL/WT	TCC Solenoid Control	12v, 55 mph: 1v

Pin Connector Graphic

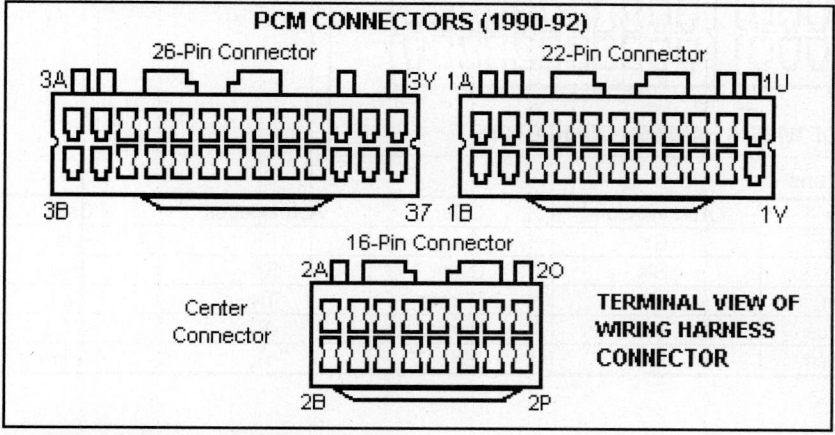

Standard Colors and Abbreviations

Abbreviation	Color	Abbreviation	Color	Abbreviation	Color
BK	Black	GY	Gray	PK	Purple
BL	Blue	GN	Green	RD	Red
BR	Brown	LG	LT Green	TN	Tan
DB	Dark Blue	OR	Orange	WT	White
DG	DK Green	PK	Pink	YL	Yellow

1990-92 Probe 2.2L I4 MFI VIN C (M/T) 22 Pin Connector

PCM Pin #	Wire Color	Circuit Description (22 Pin)	Value at Hot Idle
22-A	BL/RD	Keep Alive Power	12- 14v
22-B	RD/BK	Vehicle Power (Power Relay)	12-14v
22-C	WT/YL	System Monitor Lamp	Lamp On: 5v, Off: 12v
22-D	BK/PK	Vehicle Start Signal	Cranking: 9-11v
22-E	WT/BL	MIL (lamp) Control	MIL On: 1v, MIL Off: 12v
22-F	WT/BK	Self-Test Output Signal	STO On: 5v, Off: 12v
22-G	---	Not Used	---
22-H	---	Not Used	---
22-I	---	Not Used	---
22-J	BK/BL	Rear Defroster Control	Switch On: 12v, Off: 0.1v
22-K	---	Not Used	---
22-L	---	Not Used	---
22-M	---	Not Used	---
22-N	LG/WT	Idle Position Switch	IDL closed: 0v, Open: 12v
22-O	WT/GN	Brake Pedal Position Switch	Brake Off: 0v, On: 12v
22-P	BR/RD	PSP Switch Signal	Straight: 12v, Turned: 0v
22-Q	BL/WT	A/C Cycling Clutch Switch	On: 1v, Off: 12v
22-R	BK/GN	Cooling Fan 1 Control	On: 1v, Off: 12v
22-S	BL/BK	Blower Motor Switch	Off or 1st: 12v, 2nd on: 1v
22-T	BK/BL	Rear Defroster Switch Signal	AC On: 12v, Off: 0v
22-U	RD/BL	Clutch Engage Switch	Clutch In: 0v, Out: 11v
22-V	---	Not Used	---

Pin Connector Graphic

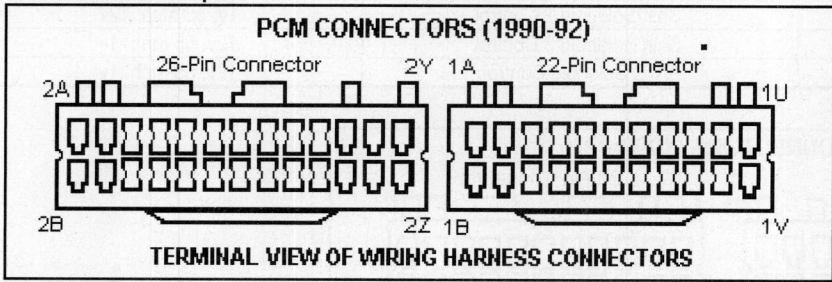

Standard Colors and Abbreviations

Abbreviation	Color	Abbreviation	Color	Abbreviation	Color
BK	Black	GY	Gray	PK	Purple
BL	Blue	GN	Green	RD	Red
BR	Brown	LG	LT Green	TN	Tan
DB	Dark Blue	OR	Orange	WT	White
DG	DK Green	PK	Pink	YL	Yellow

1990-92 Probe 2.2L I4 MFI VIN C (M/T) 26 Pin Connector

PCM Pin #	Wire Color	Circuit Description (26 Pin)	Value at Hot Idle
26-A	BK	Power Ground	<0.1v
26-B	BK	Power Ground	<0.1v
26-C	BK/LG	Power Ground	<0.1v
26-D	LG/YL	Analog Signal Return	<0.050v
26-E	---	Not Used	---
26-F	---	Not Used	---
26-G	YL/WT	Speed Limiter Signal	Digital Signals
26-H	---	Not Used	
26-I	YL/BL	Ignition Diagnostic Monitor	20-31 Hz
26-J	RD/WT	Vane Airflow Meter Reference	4.9-5.1v
26-K	LG/RD	Vehicle Reference	4.9-5.1v
26-L	YL/BL	EGR EVP Sensor	0.4v
26-M	LG/BK	TP Sensor Signal	0.5-1.1v
26-N	BK	O2S-11 (B1 S1) Signal	0.1-1.1v
26-O	RD/BK	Vane Airflow Meter Signal	3.3v
26-P	RD	VAT Sensor Signal	2.23v (at 104°F)
26-Q	YL/BK	ECT Sensor Signal	0.5-0.6v
26-R	---	Not Used	---
26-S	---	Not Used	---
26-T	WT/RD	Fuel Pressure Regulator	Startup: <3v, Others: 12v
26-U	YL	Fuel Injector 1 Bank 1	3.6 ms
26-V	YL/BK	Fuel Injector 2 Bank 2	3.6 ms
26-W	WT	Idle Air Control Valve	8-10v
26-X	WT/BL	EVAP Purge Solenoid	12v, 55 mph: 1v
26-Y	WT/BL	EGR VR Solenoid	0%, 55 mph: 45%
26-Z	---	Not Used	---

Pin Connector Graphic

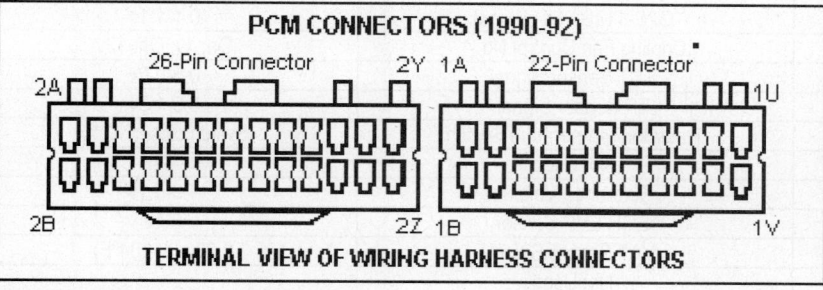

Standard Colors and Abbreviations

Abbreviation	Color	Abbreviation	Color	Abbreviation	Color
BK	Black	GY	Gray	PK	Purple
BL	Blue	GN	Green	RD	Red
BR	Brown	LG	LT Green	TN	Tan
DB	Dark Blue	OR	Orange	WT	White
DG	DK Green	PK	Pink	YL	Yellow

1990-92 Probe 2.2L I4 MFI TC VIN L (All) 22 Pin Connector

PCM Pin #	Wire Color	Circuit Description (22 Pin)	Value at Hot Idle
22-A	BL/RD	Keep Alive Power	12- 14v
22-B	RD/BK	Vehicle Power (Power Relay)	12-14v
22-C	WT/YL	System Monitor Lamp	Lamp On: 5v, Off: 12v
22-D	BK/PK	Vehicle Start Signal	Cranking: 9-11v
22-E	WT/BL	MIL (lamp) Control	MIL On: 1v, MIL Off: 12v
22-F	WT/BK	Self-Test Output Signal	STO On: 5v, Off: 12v
22-G	YL/BK	Spark Output Signal	50% duty cycle
22-H	WT/BK	Headlamp Switch Signal	Switch On: 12v, Off: 1v
22-I	RD/WT	Self-Test Input Signal	STI On: 1v, STI Off: 12v
22-J	BK/BL	Rear Defroster Control	Switch On: 12v, Off: 0.1v
22-K	---	Not Used	---
22-L	BL/BK	Condenser Fan Relay	On: 1v, Off: 12v
22-M	YL/WT	VSS Signal to Cluster	Digital Signals
22-N	BR/RD	PSP Switch Signal	Straight: 12v, Turned: 0v
22-O	BL/WT	A/C Cycling Clutch Switch	On: 1v, Off: 12v
22-P	BL/BK	Blower Motor Switch Signal	Off or 1st: 12v, 2nd on: 1v
22-Q	WT/GN	Brake Pedal Position Switch	Brake Off: 0v, On: 12v
22-R	RD/BK	A/T: Neutral Drive Switch	In 'P', 0v, Others: 12v
22-R	RD/BK	M/T: Clutch Engage Switch	Clutch In: 0v, Out: 12v
22-S	---	Not Used	---
22-T	LG/WT	Idle Switch Signal	IDL Closed: 0v, open: 12v
22-U	BK/YL	Vehicle Power	12-14v
22-V	BL/RD	Ignition Diagnostic Monitor	20-31 Hz

1990-92 Probe 2.2L I4 MFI TC VIN L (All) 16 Pin Connector

PCM Pin #	Wire Color	Circuit Description (16 Pin)	Value at Hot Idle
16-A	RD/WT	Vane Airflow Meter Reference	4.9-5.1v
16-B	RD/BK	Vane Airflow Meter Signal	3.3v
16-C	BK	O2S-11 (B1 S1) Signal	0.1-1.1v
16-D	BK/GN	Cooling Fan Control No. 1	On: 1v, Off: 12v
16-E	YL/BK	ECT Sensor Signal	0.5-0.6v
16-F	LG/BK	TP Sensor Signal	0.5-1.3v
16-G	BL/RD	Radiator Temperature Switch	Switch closed: 0.1v
16-H	---	Not Used	---
16-I	LG/RD	Reference Voltage	4.9-5.1v
16-J	YL/BL	EGR EVP Sensor	0.4v
16-K	RD	VAT Sensor Signal	2.23v (at 104ºF)
16-L	---	Not Used	---
16-M	RD/YL	Knock Sensor Signal	0v
16-N	---	Not Used	---
16-O	---	Not Used	---
16-P	---	Not Used	---

Standard Colors and Abbreviations

Abbreviation	Color	Abbreviation	Color	Abbreviation	Color
BK	Black	GY	Gray	PK	Purple
BL	Blue	GN	Green	RD	Red
BR	Brown	LG	LT Green	TN	Tan
DB	Dark Blue	OR	Orange	WT	White
DG	DK Green	PK	Pink	YL	Yellow

1990-92 Probe 2.2L I4 MFI TC VIN L (All) 26 Pin Connector

PCM Pin #	Wire Color	Circuit Description (26 Pin)	Value at Hot Idle
26-A	BK	Power Ground	<0.1v
26-B	BK	Power Ground	<0.1v
26-C	BK/LG	Power Ground	<0.1v
26-D	LG/YL	Analog Signal Return	<0.050v
26-E	BL	CKP Sensor Signal	390-450 Hz
26-F	WT	CMP Sensor VREF	4.9-5.1v
26-G	GN	CMP Sensor 1 Signal	5-7 Hz
26-H	RD	CMP Sensor 2 Signal	22-28 Hz
26-I	---	Not Used	---
26-J	---	Not Used	---
26-K	---	Not Used	---
26-L	YL/RD	MIL (lamp) Control	MIL On: 1v, Off: 12v
26-M	WT/RD	Fuel Pressure Regulator	Startup: <3v, Others: 12v
26-N	---	Not Used	---
26-O	BL/WT	EGR Vent Solenoid Control	12v, at off-idle: 1v
26-P	EGRC	EGR Control Solenoid	12v, off-idle: 1v
26-Q	WT	Idle Air Control Valve	8-10v
26-R	BR/YL	Boost Control Solenoid	On: 1v, Off: 12v
26-S	---	Not Used	---
26-T	LG	Fuel Pump Control	On: 1v, Off: 12v
26-U	YL	Fuel Injector 1 Bank 1	3.6 ms
26-V	YL/BK	Fuel Injector 2 Bank 2	3.6 ms
26-W	---	Not Used	---
26-X	---	Not Used	---
26-Y	---	Not Used	---
26-Z	---	Not Used	---

Pin Connector Graphic

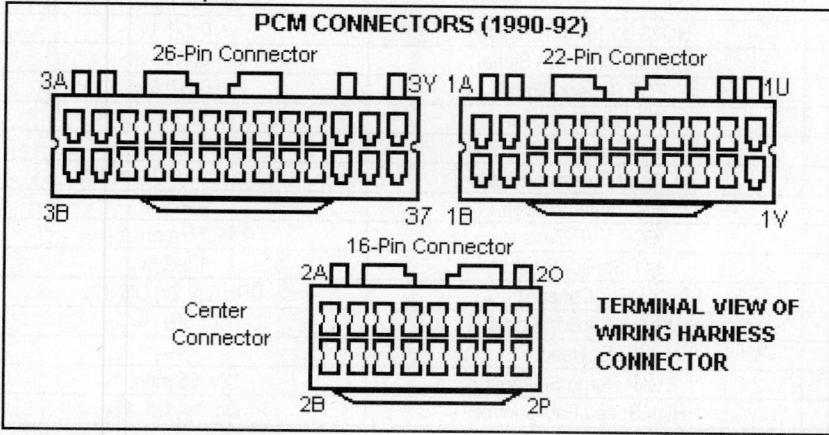

Standard Colors and Abbreviations

Abbreviation	Color	Abbreviation	Color	Abbreviation	Color
BK	Black	GY	Gray	PK	Purple
BL	Blue	GN	Green	RD	Red
BR	Brown	LG	LT Green	TN	Tan
DB	Dark Blue	OR	Orange	WT	White
DG	DK Green	PK	Pink	YL	Yellow

1993 Probe 2.5L V6 MFI VIN B (All) 22 Pin Connector

PCM Pin #	Wire Color	Circuit Description (22 Pin)	Value at Hot Idle
22-A	BL/RD	Keep Alive Power	12- 14v
22-B	RD/BK	Vehicle Power (Power Relay)	12-14v
22-C	BK/RD	Vehicle Start Signal	Cranking: 9-11v
22-D	WT/RD	System Monitor Lamp	Lamp On: 5v, Off: 12v
22-E	BL	MIL (lamp) Control	MIL On: 1v, MIL Off: 12v
22-F	LG/RD	Self-Test Output Signal	STO On: 5v, Off: 12v
22-G	BL/OR	Ignition Control Module Signal	0.1-0.3v
22-H	WT	Headlamp Switch Signal	Switch On: 12v, Off: 1v
22-I	RD/WT	Self-Test Input Signal	STI On: 1v, STI Off: 12v
22-J	PK	Rear Defroster Control	Switch On: 12v, Off: 0.1v
22-K	WT/BK	TR Switch Signal	In 'P': 0v, Others: 12v
22-L	GN/BK	A/C Heater Relay Control	On: 1v, Off: 12v
22-M	GN/RD	Vehicle Speed Sensor	0 Hz, 55 mph: 125 Hz
22-N	BK/YL	PSP Switch Signal	Straight: 12v, Turned: 0v
22-O	PK/BK	A/C Clutch Pressure Switch	Switch On: 0v, Off: 12v
22-P	OR/BK	Blower Motor Switch	Off: 12v, On: 1v
22-Q	WT/GN	Brake Pedal Position Switch	Brake Off: 0v, On: 12v
22-R	LG/BK	A/T: PNP Switch Signal	In 'P', 0v, Others: 12v
22-R	LG/BK	M/T: CPP Switch Signal	Clutch In: 0v, Out: 12v
22-S	GN	TCM RST1 Signal	Digital Signals
22-T	BR	Idle Switch Signal	Closed: 0V, Open: 12v
22-U	BK	Power Ground	<0.1v
22-V	LG/WT	TCM RST2 Signal	Digital Signals

1993 Probe 2.5L V6 MFI VIN B (All) 16 Pin Connector

PCM Pin #	Wire Color	Circuit Description (16 Pin)	Value at Hot Idle
16-A	GN/OR	Barometric Pressure Sensor	Hot Idle: 3.9v
16-B	RD	Measuring-Core VAF Sensor	3.06v, 55 mph: 1.7v
16-C	BK/YL	HO2S-11 (B1 S1) Signal	0.1-1.1v
16-D	BL/WT	HO2S-12 (B1 S2) Signal	0.1-1.1v
16-E	RD/GN	ECT Sensor Signal	0.5-0.6v
16-F	YL	TP Sensor Signal	0.7v, 55 mph: 2.1v
16-G	RD/WT	Engine Coolant Fan Signal	0.5-0.6v
16-H	PK/YL	A/C High Pressure Switch	Closed: 1v, Open: 12v
16-I	PK	Reference Voltage	4.9-5.1v
16-J	RD/BK	EGR EVP Sensor	0.4v, 55 mph: 1.8v
16-K	BK/RD	IAT Sensor Signal	1.5-2.5v
16-L	GN	DRL (Canada)	DRL On: 2v, Off: 12v
16-M	WT	Knock Sensor Signal	0v
16-N	---	Not Used	---
16-O	BL/BK	EVAP Purge Solenoid	12v, 55 mph: 1v
16-P	BL/GN	High Speed Fan Control	On: 1v, Off: 12v

Standard Colors and Abbreviations

Abbreviation	Color	Abbreviation	Color	Abbreviation	Color
BK	Black	GY	Gray	PK	Purple
BL	Blue	GN	Green	RD	Red
BR	Brown	LG	LT Green	TN	Tan
DB	Dark Blue	OR	Orange	WT	White
DG	DK Green	PK	Pink	YL	Yellow

1993 Probe 2.5L V6 MFI VIN B (All) 26 Pin Connector

PCM Pin #	Wire Color	Circuit Description (26 Pin)	Value at Hot Idle
26-A	BK	Power Ground	<0.1v
26-B	BK	Power Ground	<0.1v
26-C	BK/RD	Power Ground	<0.1v
26-D	BK/RD	Power Ground	<0.1v
26-E	LG/OR	CKP 1 Sensor	2.5v
26-F	BL	CKP Sensor Ground	<0.050v
26-G	BL/PK	CMP Sensor Signal	2.5v
26-H	GN	CKP 2 Sensor Signal	2.5v
26-I	WT/GN	Variable Resonance Induction	VRIS1 On: 1v, Off: 12v
26-J	BL/RD	Variable Resonance Induction	VRIS2 On: 1v, Off: 12v
26-K	---	Not Used	---
26-L	RD/WT	Low Speed Fan Control	On: 1v, Off: 12v
26-M	GN/BK	Fuel Pressure Regulator	Startup: <3v, other: 12v
26-N	BL/OR	Cooling Fan Control	On: 1v, Off: 12v
26-O	WT/BL	EGR Vent Solenoid	EGRV Off: 12v, On: 1v
26-P	GN/WT	EGR Control Solenoid	EGRC Off: 12v, On: 1v
26-Q	LG/BK	Idle Air Control Valve	8-10v
26-R	---	Not Used	---
26-S	---	Not Used	---
26-T	LG	Fuel Pump Control	On: 1v, Off: 12v
26-U	RD/LG	Injector 1 Control	3.6 ms
26-V	BL/WT	Injector 2 Control	3.6 ms
26-W	BR	Injector 3 Control	3.6 ms
26-X	RD/YL	Injector 4 Control	3.6 ms
26-Y	WT	Injector 5 Control	3.6 ms
26-Z	WT/BK	Injector 6 Control	3.6 ms

Pin Connector Graphic

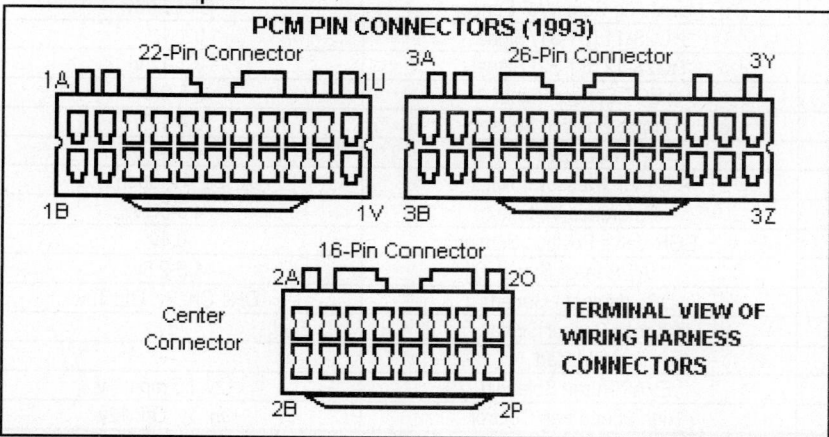

Standard Colors and Abbreviations

Abbreviation	Color	Abbreviation	Color	Abbreviation	Color
BK	Black	GY	Gray	PK	Purple
BL	Blue	GN	Green	RD	Red
BR	Brown	LG	LT Green	TN	Tan
DB	Dark Blue	OR	Orange	WT	White
DG	DK Green	PK	Pink	YL	Yellow

1994-95 Probe 2.5L V6 MFI VIN B (All) 22 Pin Connector

PCM Pin #	Wire Color	Circuit Description (22 Pin)	Value at Hot Idle
22-A	BL/RD	Keep Alive Power	12- 14v
22-B	RD/BK	Vehicle Power	12-14v
22-C	BK/RD	Vehicle Start Signal	KOEC: 9-11v
22-D	WT/RD	Switch Monitor Lamp Control	Lamp Off: 12v, On: 1v
22-E	BL	STO & MIL (lamp) Control	STO On: 5v, Off: 12v
22-F	-	Not Used	---
22-G	BL/OR	Ignition Control Module	Varies
22-H	WT	Headlamp Switch Signal	Switch On: 12v, Off: 1v
22-I	RD/WT	Self-Test Input	STI On: 1v, STI Off: 12v
22-J	PK	Rear Window Defroster	Switch On: 1v, Off: 12v
22-K	WT/BK	Torque Reduce & ECT Signal	Digital Signals
22-L	GN/BK	A/C Relay Control	On: 1v, Off: 12v
22-M	GN/RD	Vehicle Speed Sensor	0 Hz, 55 mph: 125 Hz
22-N	BK/YL	PSP Switch Signal	Straight: 12v, Turned: 0v
22-O	PK/BK	A/C Cycling Pressure Switch	Switch On: 12v, Off: 1v
22-P	OR/BK	Blower Motor Control Switch	Off or 1st: 12v, 2nd on: 1v
22-Q	WT/GN	Brake On/ Off Switch Signal	Brake Off: 0v, On: 12v
22-R	LG/BK	A/T: PNP Switch Signal	In 'P', 0v, Others: 12v
22-R	LG/BK	M/T: CPP Switch Signal	Clutch In: 0v, Out: 12v
22-S	GN	Torque Reduce Signal #1	Digital Signals
22-T	BR	Idle Switch Signal	Closed: 0v, Open: 12v
22-U	-	Not Used	---
22-V	LG/WT	Torque Reduce Signal #2	Digital Signals

1994-95 Probe 2.5L V6 MFI VIN B (All) 16 Pin Connector

PCM Pin #	Wire Color	Circuit Description (16 Pin)	Value at Hot Idle
16-A	GN/OR	Barometric Pressure Sensor	3.9v
16-B	RD	Measuring Core VAF Sensor	Idle: 3.06v
16-C	BK/YL	HO2S-11 (B1 S1) Signal	0.1-1.1v
16-D	BL/WT	HO2S-12 (B1 S2) Signal	0.1-1.1v
16-E	RD/GN	ECT Sensor Signal	0.5-0.6v
16-F	YL	Throttle Position Sensor	0.4-1.0v
16-G	-	Not Used	---
16-H	PK/YL	A/C High Pressure Switch	Closed: 1v, open: 12v
16-I	PK	Reference Voltage	4.9-5.1v
16-J	RD/BK	EGR Valve Position Sensor	0.4v
16-K	BK/RD	IAT Sensor Signal	1.5-2.5v
16-L	GN	DRL (lamps) - Canada	DRL On: 2v, Off: 12v
16-M	WT	Knock Sensor Signal	0v
16-N	-	Not Used	---
16-O	BL/BK	EVAP Purge Solenoid	12v, 55 mph: 1v
16-P	BL/GN	High Speed Fan Control	On: 1v, Off: 12v

Standard Colors and Abbreviations

Abbreviation	Color	Abbreviation	Color	Abbreviation	Color
BK	Black	GY	Gray	PK	Purple
BL	Blue	GN	Green	RD	Red
BR	Brown	LG	LT Green	TN	Tan
DB	Dark Blue	OR	Orange	WT	White
DG	DK Green	PK	Pink	YL	Yellow

1994-95 Probe 2.5L V6 MFI VIN B (All) 26 Pin Connector

PCM Pin #	Wire Color	Circuit Description (26 Pin)	Value at Hot Idle
26-A	BK	Power Ground	<0.1v
26-B	BK	Power Ground	<0.1v
26-C	BK/RD	Power Ground	<0.1v
26-D	BK/RD	Power Ground	<0.1v
26-E	LG/OR	CKP Sensor 1 Signal	2.5v
26-F	BL	CKP Sensor Ground	0.1v
26-G	BL/PK	CMP Sensor 1 Signal	2.5v
26-H	GN	CKP Sensor 2 Signal	2.5v
26-I	WT/GN	Variable Resonance Induction	VRIS1 On: 1v, Off: 12v
26-J	BL/RD	Variable Resonance Induction	VRIS2 On: 1v, Off: 12v
26-K	-	Not Used	---
26-L	RD/WT	Low Speed Fan Control	On: 1v, Off: 12v
26-M	GN/BK	Fuel Pressure Regulator	Startup: 3v, others: 12v
26-N	BL/OR	Condenser Fan Control	On: 1v, Off: 12v
26-O	WT/BL	EGR Vent Solenoid	12v, off-idle: 1v
26-P	GN/WT	EGR Control Solenoid	12v, off-idle: 1v
26-Q	LG/BK	Idle Air Control Valve	8-10v
26-R	-	Not Used	---
26-S	-	Not Used	---
26-T	LG	Fuel Pump Control	On: 1v, Off: 12v
26-U	RD/LG	Injector 1 Control	3.6 ms
26-V	BL/WT	Injector 2 Control	3.6 ms
26-W	BR	Injector 3 Control	3.6 ms
26-X	R/L	Injector 4 Control	3.6 ms
26-Y	W	Injector 5 Control	3.6 ms
26-Z	WT/BK	Injector 6 Control	3.6 ms

Pin Connector Graphic

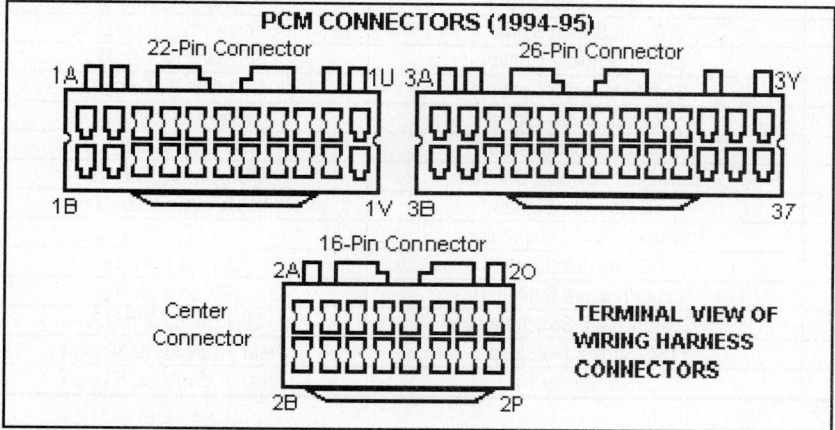

Standard Colors and Abbreviations

Abbreviation	Color	Abbreviation	Color	Abbreviation	Color
BK	Black	GY	Gray	PK	Purple
BL	Blue	GN	Green	RD	Red
BR	Brown	LG	LT Green	TN	Tan
DB	Dark Blue	OR	Orange	WT	White
DG	DK Green	PK	Pink	YL	Yellow

1996-97 Probe 2.5L V6 MFI VIN B (4EAT) 28 Pin Connector

PCM Pin #	Wire Color	Circuit Description (28 Pin)	Value at Hot Idle
C246-1	BL/GN	High Speed Fan Control	On: 1v, Off: 12v
C246-2	BR/LG	Condenser Fan Control	On: 1v, Off: 12v
C246-3	BL/OR	Low Speed Fan Control	On: 0v, Off: 12v
C246-4	GN/BK	Air Conditioning Relay	On: 1v, Off: 12v
C246-5	---	Not Used	---
C246-6	PK/YL	High Pressure Switch Signal	Closed: 1v, open: 12v
C246-7	RD/OR	ISO K-Line Signal for OBD II	Digital Signals
C246-8	---	Not Used	---
C246-9	GN/BK	FPRC Solenoid Control	ON: 1v, Off: 12v
C246-10	GN	Transmission Range Switch	AC On: 12v, Off: 0v
C246-13	BL	STI Output, MIL Control	MIL On: 1v, Off: 12v
C246-14	GN/RD	Vehicle Speed Sensor	Vehicle moving: 0-5v
C246-15	GN	DRL (lamps) - Canada	DRL On: 2v, Off: 12v
C246-16	YL/BK	Blower Motor Control Switch	Off - 1st: 12v, 2nd on: 1v
C246-17	WT/GN	Brake Pedal Position Switch	Brake Off: 0v, On: 12v
C246-20	GY	VRIS Solenoid 1 Control	VRIS1 On: 1v, Off: 12v
C246-21	BL/RD	VRIS Solenoid 2 Control	VRIS2 On: 1v, Off: 12v
C246-22	PK/BK	A/C Cycling Pressure Switch	On: 1v, Off: 12v
C246-23	BK/BR	Vehicle Start Signal	KOEC: 9-11v
C246-24	WT	Headlamp Switch Signal	Off or 1st: 12v, 2nd on: 1v
C246-25	PK	Rear Window Defrost Switch	DEF On: 1v, Off: 12v
C246-27	LG	Fuel Pump Control	On: 1v, Off: 12v

1996-97 Probe 2.5L V6 MFI VIN B (4EAT) 16 Pin Connector

PCM Pin #	Wire Color	Circuit Description (16 Pin)	Value at Hot Idle
C245-1	BL	Shift Solenoid 1 Control	SS1 On: 0v, Off: 12v
C245-2	BR/YL	Transaxle Control Indicator	On: 1v, Off: 12v
C245-3	RD/BL	TR Drive Switch Signal	In Drive: 12v, others: 0v
C245-4	GN/WT	TR 2nd Switch Signal	Second: 12v, others: 0v
C245-5	BL/BK	Shift Solenoid 2 Control	SS2 On: 1v, Off: 12v
C245-6	GN/BK	Shift Solenoid 3 Control	SS3 On: 1v, Off: 12v
C245-7	BL/OR	TR First Switch Signal	In First: 12v, others: 0v
C245-8	BL/WT	TCC Control Solenoid	TCC On: 1v, Off: 12v
C245-9	RD/WT	Downshift Solenoid Control	DSS On: 1v, Off: 12v
C245-10	RD/YL	TR Reverse Signal	In 'R': 12v, Others: 0v
C245-11	BL/YL	Transaxle Fluid Temperature	2.10-2.40v
C245-12	RD/BK	TCC Solenoid Control	12v, 55 mph: 1v
C245-13	RD/BK	Line Pressure Solenoid	12v, 55 mph: 1v
C245-14	BR/BK	Overdrive Off Switch Input	O/D On: 0v, Off: 12v
C245-15	WT	TSS Sensor (+) Signal	Near 1.25v AC at 50 mph
C245-16	RD	TSS Sensor (-)	Near 1.25v AC at 50 mph

Pin Connector Graphic

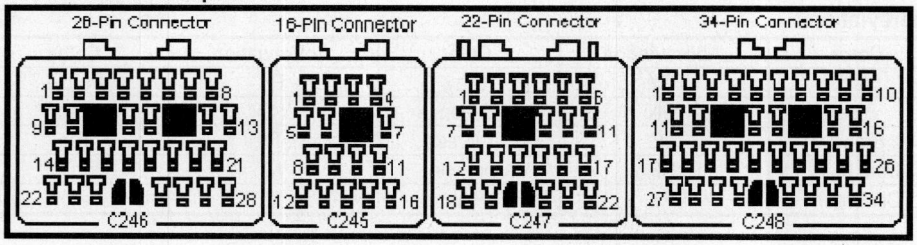

1996-97 Probe 2.5L V6 MFI VIN B (4EAT) 22 Pin Connector

PCM Pin #	Wire Color	Circuit Description (22 Pin)	Value at Hot Idle
C247-1	PK	Reference Voltage	4.9-5.1v
C247-2	RD	MC-VAF Sensor Signal	3.03v
C247-3	PK/WT	HO2S-21 (B2 S1) Signal	0.1-1.1v
C247-4	GN	HO2S-11 (B1 S1) Signal	0.1-1.1v
C247-5	RD/BL	ECT Sensor Signal	0.5-0.6v
C247-7	YL	TP Sensor Signal	0.5-1.2v
C247-8	GN	EGR Boost Sensor	4.3v (at 28" HG)
C247-9	RD/BK	EGR Valve Position Sensor	0.4v
C247-10	BK/RD	IAT Sensor Signal	1.5-2.5v
C247-11	RD/YL	HO2S-21 (B2 S1) Heater	On: 1v, Off: 12v
C247-12	RD/GN	HO2S-13 (B1 S3) Heater	On: 1v, Off: 12v
C247-16	WL	Knock Sensor Signal	0v
C247-17	BK/RD	Power Ground	<0.1v
C247-18	BK/YL	HO2S-12 (B1 S2) Signal	0.1-1.1v
C247-19	BL/WT	HO2S-13 (B1 S3) Signal	0.1-1.1v
C247-20	BL/YL	PSP Switch Signal Input	Straight: 12v, Turned: 0v
C247-21	BR	Idle Throttle Switch Signal	Closed: 0v, Open: 12v
C247-22	BK/BL	Analog Signal Return	<0.050v

1996-97 Probe 2.5L V6 MFI VIN B (4EAT) 34 Pin Connector

PCM Pin #	Wire Color	Circuit Description (34 Pin)	Value at Hot Idle
C248-1	RD/BK	Vehicle Power	12-14v
C248-2	PK/BL	HO2S-11 (B1 S1) Heater	On: 1v, Off: 12v
C248-3	PK	HO2S-21 (B2 S1) Heater	On: 1v, Off: 12v
C248-4	LG/BK	IAC Motor Control	8-10v
C248-5	RD/WT	Injector 1 Control	3.8-4.0 ms
C248-6	BL/WT	Injector 2 Control	3.8-4.0 ms
C248-7	BR	Injector 3 Control	3.8-4.0 ms
C248-8	RD/YL	Injector 4 Control	3.8-4.0 ms
C248-9	WT	Injector 5 Control	3.8-4.0 ms
C248-10	WT/BK	Injector 6 Control	3.8-4.0 ms
C248-11	BL/RD	Keep Alive Power	12-14v
C248-12	BL/PK	CMP Sensor Signal	5-7 Hz
C248-14	RD/WT	Self-Test Input Signal	STI On: 0v, Off: 12v
C248-15	GN	CKP Sensor Signal	410-445 Hz
C248-16	BL	CKP Sensor Ground	<0.050v
C248-17	RD/BK	Vehicle Power	12-14v
C248-18	PK/BK	EGR Vent Solenoid	EGRV On: 1v, Off: 12v
C248-19	GN/WT	EGR Control Solenoid	EGRC On: 1v, Off: 12v
C248-20	BL/BK	EVAP Purge Solenoid	CANP On: <1.v, Off: 12v
C248-21	BL/OR	Ignition Module Control	Varies: 0.1-0.3v
C248-27	BK/RD	Power Ground	<0.1v
C248-30	GY/GN	EGR Boost Check Solenoid	On: 1v, Off: 12v
C248-31, 32	BK	Power Ground	<0.1v
C248-33	BK	Power Ground	<0.1v
C248-34	BK	Power Ground	<0.1v

Standard Colors and Abbreviations

Abbreviation	Color	Abbreviation	Color	Abbreviation	Color
BK	Black	GY	Gray	PK	Purple
BL	Blue	GN	Green	RD	Red
BR	Brown	LG	LT Green	TN	Tan
DB	Dark Blue	OR	Orange	WT	White
DG	DK Green	PK	Pink	YL	Yellow

1996-97 Probe 2.5L V6 MFI VIN B (M/T) 28 & 22 Pin Connector

PCM Pin #	Wire Color	Circuit Description (28 Pin)	Value at Hot Idle
C246-1	BL/GN	High Speed Fan Control	On: 1v, Off: 12v
C246-2	BR/LG	Condenser Fan Control	On: 1v, Off: 12v
C246-3	BL/OR	Low Speed Fan Control	On: 1v, Off: 12v
C246-4	GN/BK	A/C Relay Control	On: 1v, Off: 12v
C246-5, 8	---	Not Used	---
C246-6	PK/YL	A/C High Pressure Switch	Closed: 1v, pen: 12v
C246-7	RD/OR	ISO K-Line for OBD II	Digital Signals
C246-9	GN/BK	FPRC Solenoid	FPRC On: 0v, Off: 12v
C246-10	GN	Neutral Position Switch	In 'N': 0v, Others: 5v
C246-10	GN	Clutch Pedal Position Switch	Clutch In: 0v, Out 5v
C246-11-12	---	Not Used	---
C246-13	BL	Self-Test Output & MIL	MIL On: 1v, Off: 12v
C246-14	GN/RD	Vehicle Speed Sensor	Vehicle moving: 0-5v
C246-15	GN	DRL (lamps) - Canada	DRL On: 2v, DRL Off: 12v
C246-16	YL/BK	Blower Motor Control Switch	Off or 1st: 12v, 2nd on: 1v
C246-17	WT/GN	Brake Pedal Position Switch	Brake Off: 0v, On: 12v
C246-18, 19	---	Not Used	---
C246-20	GY	VRIS Solenoid 1 Control	VRIS1 On: 1v, Off: 12v
C246-21	BL/RD	VRIS 2 Solenoid Control	VRIS2 On: 1v, Off: 12v
C246-22	PK/BK	A/C Cycling Pressure Switch	Closed: 1v, open: 12v
C246-23	BK/BR	Vehicle Start Signal	KOEC: 9-11v
C246-24	WT	Headlamp Switch Signal	Off or 1st: 12v, 2nd on: 1v
C246-25	PK	Rear Window Defrost Switch	Switch On: 1v, Off: 12v
C246-26, 28	---	Not Used	---
C246-27	LG	Fuel Pump Control	On: 1v, Off: 12v
PCM Pin #	Wire Color	Circuit Description (22 Pin)	Value at Hot Idle
C247-1	PK	Reference Voltage	4.9-5.1v
C247-2	RD	MCV Airflow Sensor Signal	3.03v
C247-3	PK/WT	HO2S-21 (B2 S1) Signal	0.1-1.1v
C247-4	GN	HO2S-11 (B1 S1) Signal	0.1-1.1v
C247-5	RD/BL	ECT Sensor Signal	0.5-0.6v
C247-6	---	Not Used	---
C247-7	YL	TP Sensor Signal	0.5-1.2v
C247-8	GN	EGR Boost Sensor	4.3v (at 28" HG)
C247-9	RD/BK	EGR EVP Sensor	0.4v
C247-10	BK/RD	IAT Sensor Signal	1.5-2.5v
C247-11	RD/YL	HO2S-12 (B1 S2) Heater	On: 1v, Off: 12v
C247-12	RD/GN	HO2S-13 (B1 S3) Heater	On: 1v, Off: 12v
C247-13-15	---	Not Used	---
C247-16	WT	Knock Sensor Signal	0v
C247-17	BK/RD	Power Ground	<0.1v
C247-18	BK/YL	HO2S-12 (B1 S2) Signal	0.1-1.1v
C247-19	BL/WT	HO2S-13 (B1 S3) Signal	0.1-1.1v
C247-20	BL/YL	PSP Switch Signal	Straight: 12v, Turned: 0v
C247-21	BR	Idle Throttle Switch Signal	IDL closed: 0v, Open: 12v
C247-22	BK/BL	Analog Signal Return	<0.050v

Standard Colors and Abbreviations

Abbreviation	Color	Abbreviation	Color	Abbreviation	Color
BK	Black	GY	Gray	PK	Purple
BL	Blue	GN	Green	RD	Red
BR	Brown	LG	LT Green	TN	Tan
DB	Dark Blue	OR	Orange	WT	White
DG	DK Green	PK	Pink	YL	Yellow

1996-97 Probe 2.5L V6 MFI VIN B (M/T) 'C248' 34 Pin Connector

PCM Pin #	Wire Color	Circuit Description (34 Pin)	Value at Hot Idle
C248-1	RD/BK	Vehicle Power	12-14v
C248-2	PK/BL	HO2S-11 (B1 S1) Heater	On: 1v, Off: 12v
C248-3	PK	HO2S-21 (B2 S1) Heater	On: 1v, Off: 12v
C248-4	LG/BK	IAC Motor Control	8-10v
C248-5	RD/WT	Injector 1 Control	3.8-4.0 ms
C248-6	BL/WT	Injector 2 Control	3.8-4.0 ms
C248-7	BR	Injector 3 Control	3.8-4.0 ms
C248-8	RD/YL	Injector 4 Control	3.8-4.0 ms
C248-9	WT	Injector 5 Control	3.8-4.0 ms
C248-10	WT/BK	Injector 6 Control	3.8-4.0 ms
C248-11	BL/RD	Keep Alive Power	12-14v
C248-12	BL/PK	Camshaft Position Sensor	5-7 Hz
C248-13	---	Not Used	---
C248-14	RD/WT	Self-Test Input Signal	STI On: 0v, Off: 12v
C248-15	GN	CKP Sensor Signal	410-440 Hz
C248-16	BL	CMP Sensor Ground	<0.050v
C248-17	---	Not Used	12-14v
C248-18	PK/BK	EGR Vent Solenoid Control	12v, 55 mph: 1v
C248-19	GN/WT	EGR Control Solenoid	12v, 55 mph: 1v
C248-20	BL/BK	EVAP Purge Solenoid	CANP On: <1.v, Off: 12v
C248-21	BL/OR	Ignition Control Module Signal	0.1-0.3v
C248-22	---	Not Used	---
C248-23	---	Not Used	---
C248-24	---	Not Used	---
C248-25	---	Not Used	---
C248-26	---	Not Used	---
C248-27	BK/RD	Power Ground	<0.1v
C248-28	---	Not Used	---
C248-29	---	Not Used	---
C248-30	GY/GN	EGR Boost Check Solenoid	On: 1v, Off: 12v
C248-31	BK	Power Ground	<0.1v
C248-32	BK	Power Ground	<0.1v
C248-33	BK	Power Ground	<0.1v
C248-34	BK	Power Ground	<0.1v

PCM Connector Graphic

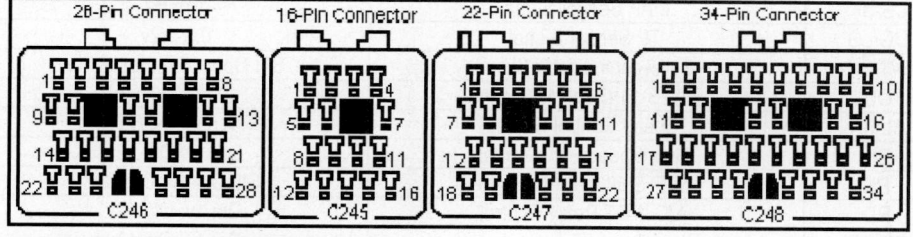

Standard Colors and Abbreviations

Abbreviation	Color	Abbreviation	Color	Abbreviation	Color
BK	Black	GY	Gray	PK	Purple
BL	Blue	GN	Green	RD	Red
BR	Brown	LG	LT Green	TN	Tan
DB	Dark Blue	OR	Orange	WT	White
DG	DK Green	PK	Pink	YL	Yellow

Taurus Pin Tables

1990 Taurus 2.5L I4 CFI VIN D (All) 60 Pin Connector

PCM Pin #	Wire Color	Circuit Description (60 Pin)	Value at Hot Idle
1	YL	Keep Alive Power	12-14v
2	RD/LG	Brake Pedal Position Switch	Brake Off: 0v, On: 12v
3	DG, OR	VSS (+), (-) Signal	0 Hz, 55 mph: 125 Hz
4	DG/YL	Ignition Diagnostic Monitor	20-31 Hz
5	RD/LG	Ignition Signal (Cranking)	KOEC: 9-11v
7	LG/YL	ECT Sensor Signal	0.5-0.6v
8	PK/BK	Fuel Pump Monitor	On: 12v, Off: 0v
9	BK, OR	Data Bus (-), (+) Signal	Digital Signals
10	PK	A/C Cycling Clutch Switch	A/C On: 12v, Off: 0v
11-12, 14-15	---	Not Used	---
13	OR/YL	Cruise Control Solenoid	Solenoid On: 1v
16	BK/OR	Ignition System Ground	<0.050v
17	TN/RD	STO & MIL Control	MIL On: 1v, Off: 12v
18-19, 32, 38	---	Not Used	---
20, 46	BK	PCM Case, Analog Signal Ground	<0.050v
21	YL, WT	ISC Motor (+), (-) Control	0.3v
22	TN/LG	Fuel Pump Control	On: 1v, Off: 12v
23	YL/LG	PSP Switch Signal	Straight: 0v, Turned: 5v
24	WT/RD	Idle Tracking Switch	11v, 55 mph: 0v
25	LG/PK	ACT Sensor Signal	1.5-2.5v
26	OR/WT	Reference Voltage	4.9-5.1v
27	GN or BR	EGR EVP Sensor	0.4v, 55 mph: 2.6v
29	GN/P	HO2S-11 (B1 S1) Signal	0.1-1.1v
29	DG/PK	HO2S-11 (B1 S1) Signal	0.1-1.1v
30	BL/YL	Neutral Drive (A/T), Clutch (M/T) Switch	In 'N' or Clutch "In": 0v
31	GY/YL	EVAP Purge Solenoid	12v, 55 mph: 1v
33	DG	EGR VR Solenoid	0%, 55 mph: 45%
34	BL/PK	Data Output Link	Digital Signals
35	WT/PK	Speed Control Vent Solenoid	Off: 0% duty cycle
36	YL/LG	Spark Output Signal	50% duty cycle
37, 57	RD	Vehicle Power	12-14v
39	OR	Speed Control Switch Ground	<0.050v
40, 60	BK/LG	Power Ground	<0.1v
42	GY/BK	S/C Vacuum Solenoid Control	Off: 0% duty cycle
43-44, 51, 59	---	Not Used	---
45	LG/BK	MAP Sensor Signal	107 Hz
47	DG/LG	TP Sensor Signal	0.5-1.2v
48	WT/BK	Self-Test Indicator Signal	STI Open: 5v, Closed: 1v
49	OR	HO2S-11 (Bank 1) Ground	<0.050v
50	BL/BK	Speed Control Switch	All speeds: 6.7v
52	PK	High Speed Fan Control	On: 1v, Off: 12v
53	PK	M/T: Shift Indicator Light	SIL On: 1v, Off: 12v
54	RD	A/C WOT Relay Control	On: 1v, Off: 12v
55	TN/OR	Low Speed Fan Control	On: 1v, Off: 12v
56	DB	PIP Sensor Signal	50% duty cycle
58	TN/RD	CFI Fuel Injector	1.6-1.7 ms

Pin Connector Graphic

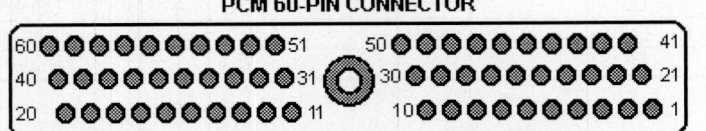

Terminal View of 60-Pin PCM Harness Connector

1991 Taurus 2.5L I4 MFI VIN D (A/T) 60 Pin Connector

PCM Pin #	Wire Color	Circuit Description (60 Pin)	Value at Hot Idle
1	YL	Keep Alive Power	12-14v
2	RD/LG	Brake Pedal Position Switch	Brake Off: 0v, On: 12v
3	DG/WT	VSS (+) Signal	0 Hz, 55 mph: 125 Hz
4	WT/PK	Ignition Diagnostic Monitor	20-31 Hz
5	GY/BK	TSS Sensor Signal	100 Hz (700 rpm)
6	OR/YL	VSS (+) Signal	0 Hz, 55 mph: 125 Hz
7	LG/RD	ECT Sensor Signal	0.5-0.6v
8	PK/BK	Fuel Pump Monitor	On: 12v, Off: 0v
9	TN/BL	MAF Sensor Return	<0.050v
10	PK/BL	A/C Cycling Clutch Switch	A/C On: 12v, Off: 0v
11	GN/YL	EVAP Purge Solenoid	12v, 55 mph: 1v
12	---	Not Used	---
13	TN/OR	Low Speed Fan Control	On: 1v, Off: 12v
14	LG/PK	High Speed Fan Control	On: 1v, Off: 12v
15	---	Not Used	---
16	OR/RD	Ignition System Ground	<0.050v
17	TN/RD	STI Output, MIL Control	MIL On: 1v, Off: 12v
18	OR/BK	Data Bus (+) Signal	Digital Signals
19	BK/OR	Data Bus (-) Signal	Digital Signals
20	BK	PCM Case Ground	<0.050v
21	OR/BK	IAC Motor Control	9.5-10.1v
22	TN/LG	Fuel Pump Control	On: 1v, Off: 12v
23	---	Not Used	---
24	DG	CID Sensor Signal	5-7 Hz
25	LG/PK	ACT Sensor Signal	1.5-2.5v
26	BR/WT	Reference Voltage	4.9-5.1v
27	BR/LG	PFE EGR Sensor	3.3v, 55 mph: 2.8v
28	YL/LG	PSP Sensor Signal	Straight: 0v, Turned: 11v
29	---	Not Used	---
30	BL/YL	MLP Sensor Signal	In 'P': 0v, 55 mph: 1.7v
31	---	Not Used	---
32	---	Not Used	---
33	BR/PK	EGR VR Solenoid	0%, 55 mph: 45%
34	---	Not Used	---
35	BR/BL	Injector 4 Control	4.8-5.0 ms
36	YL/LG	Spark Output Signal	50% duty cycle
37	RD	Vehicle Power	12-14v
38	WT/YL	EPC Solenoid Control	8-9v (9 psi)
39	BR/YL	Injector 3 Control	4.8-5.0 ms
40	BK/LG	Power Ground	<0.1v

Standard Colors and Abbreviations

Abbreviation	Color	Abbreviation	Color	Abbreviation	Color
BK	Black	GY	Gray	PK	Purple
BL	Blue	GN	Green	RD	Red
BR	Brown	LG	LT Green	TN	Tan
DB	Dark Blue	OR	Orange	WT	White
DG	DK Green	PK	Pink	YL	Yellow

1991 Taurus 2.5L I4 MFI VIN D (A/T) 60 Pin Connector

PCM Pin #	Wire Color	Circuit Description (60 Pin)	Value at Hot Idle
41-43	---	Not Used	---
44	RD/BK	HO2S-11 (B1 S1) Signal	0.1-1.1v
45	---	Not Used	---
46	GY/RD	Analog Signal Return	<0.050v
47	GN/WT	TP Sensor Signal	0.5-1.2v
48	BR	Self-Test Indicator Signal	STI Open: 5v, Closed: 1v
49	O/B	TOT Sensor Signal	2.10-2.40v
50	BL/RD	MAF Sensor Signal	0.6-0.9v
51	OR/YL	Shift Solenoid 1 Control	12v, 55 mph: 1v
52	PK/OR	Shift Solenoid 2 Control	1v, 55 mph: 12v
53	TN/WT	Lock-Up Solenoid	0%, 55 mph: 45%
54	PK/YL	A/C WOT Relay Control	On: 1v, Off: 12v
55	WT/RD	Shift Solenoid 3 Control	7-9v, 55 mph: 8-9v
56	DB	PIP Sensor Signal	50% duty cycle
57	RD	Vehicle Power	12-14v
58	TN	Injector 1 Control	4.8-5.0 ms
59	WT	Injector 2 Control	4.8-5.0 ms
60	BK/LG	Power Ground	<0.1v

Pin Connector Graphic

PCM 60-PIN CONNECTOR

Terminal View of 60-Pin PCM Harness Connector

Standard Colors and Abbreviations

Abbreviation	Color	Abbreviation	Color	Abbreviation	Color
BK	Black	GY	Gray	PK	Purple
BL	Blue	GN	Green	RD	Red
BR	Brown	LG	LT Green	TN	Tan
DB	Dark Blue	OR	Orange	WT	White
DG	DK Green	PK	Pink	YL	Yellow

1996-97 Taurus 3.0L V6 MFI VIN S (A/T) 104 Pin Connector

PCM Pin #	Wire Color	Circuit Description (104 Pin)	Value at Hot Idle
1	PK/OR	Shift Solenoid 2 Control	1v, 55 mph: 12v
2	PK/LG	MIL (lamp) Control	MIL On: 1v, Off: 12v
3-4, 6-7	---	Not Used	---
5	WT/OR	Electronic AIRB Monitor	5v
8	DB/YL	IMRC Monitor Signal	0.6v
9-12, 14	---	Not Used	---
13	PK	Flash EEPROM Power	0.1v
15	PK/LB	Data Bus (-) Signal	Digital Signals
16	TN/OR	Data Bus (+) Signal	Digital Signals
17-20, 23	---	Not Used	---
21	GY	CKP Sensor (+) Signal	850-1120 Hz
22	DB	CKP Sensor (-) Signal	850-1120 Hz
24, 51	BK/WT	Power Ground	<0.1v
25	BK	Power Ground	<0.1v
26	YL/BK	Coil Driver 1 Control	6° dwell
27	OR/YL	Shift Solenoid 1 Control	12v, 55 mph: 1v
28	TN/OR	Low Speed Fan Control	On: 1v, Off: 12v
29	TN/WT	Overdrive Cancel Switch	TCS & O/D On: 12v
30	DG	Octane Adjust Switch	Closed: 0v, Open: 9.3v
31	YL/LG	PSP Switch Signal	Straight: 0v, Turned: 12v
32, 34, 43	---	Not Used	---
33	PK/OR	VSS (-) Signal	0 Hz, 55 mph: 125 Hz
35	RD/BK	HO2S-12 (B1 S2) Signal	0.1-1.1v
36	TN/LB	MAF Sensor Return	<0.050v
37	OR/BK	TFT Sensor Signal	2.10-2.40v
38	LG/RD	ECT Sensor Signal	0.5-0.6v
39	GY	IAT Sensor Signal	1.5-2.5v
40	DG/YL	Fuel Pump Monitor	On: 12v, Off: 0v
41	PK/LB	A/C Cycling Clutch Switch	A/C On: 12v, Off: 0v
42	BR	IMRC Solenoid Control	12v (Off)
44-45, 49	---	Not Used	---
46	LG/PK	High Speed Fan Control	On: 1v, Off: 12v
47	BR/PK	EGR VR Solenoid	0%, 55 mph: 45%
48	TN/YL	Clean Tachometer Output	33-37 Hz
50, 54, 59	---	Not Used	---
52	YL/RD	Coil Driver 2 Control	6° dwell
53	PK/BK	Shift Solenoid 3 Control	12v, 55 mph: 1v
55	DG/OR	Keep Alive Power	12-14v
56	GY/YL	EVAP VMV Solenoid	0-10 Hz (0-100%)
57	YL/RD	Knock Sensor Signal	0v
58	GY/BK	VSS (+) Signal	0 Hz, 55 mph: 125 Hz
60	GY/LB	HO2S-11 (B1 S1) Signal	0.1-1.1v
61	PK/LG	HO2S-22 (B2 S2) Signal	0.1-1.1v
62	RD/PK	FTP Sensor Signal	2.6v (0" H2O - cap off)
63	---	Not Used	---
64	LB/YL	TR Sensor Signal	In 'P': 0v, 55 mph: 1.7v
65	BR/LG	DPFE Sensor Signal	0.20-1.30v
66	---	Not Used	---
67	PK/WT	EVAP Canister Vent Valve	0-10 Hz (0-100%)
68	---	Not Used	---
69	PK/YL	A/C WOT Relay Control	On: 1v, Off: 12v
70	BR	Electronic AIRB Solenoid	12v, 55 mph: 12v

1996-97 Taurus 3.0L V6 MFI VIN S (A/T) 104 Pin Connector

PCM Pin #	Wire Color	Circuit Description (104 Pin)	Value at Hot Idle
71	RD	Vehicle Power	12-14v
73	TN/BK	Injector 5 Control	2.3-5.5 ms
74	BR/YL	Injector 3 Control	2.3-5.5 ms
75	TN	Injector 1 Control	2.3-5.5 ms
76 & 77	BK/WT	Power Ground	<0.1v
78	YL/WT	Coil Driver 3 Control	6° dwell
79	WT/LG	TCIL (lamp) Control	On: 1v, Off: 12v
80	LB/OR	Fuel Pump Control	On: 1v, Off: 12v
81	WT/YL	EPC Solenoid Control	9.0v (15 psi)
82	PK/YL	TCC Solenoid Control	12v (0 psi)
83	WT/LB	IAC Motor Control	10v (55% duty cycle)
84	DG/WT	TSS Sensor Signal	50-65 Hz
85	DB/OR	CID Sensor Signal	5-7 Hz
86	BK/YL	A/C Pressure Switch	A/C On: 12v (Open)
87	RD/LG	HO2S-12 (B1 S2) Signal	0.1-1.1v
88	LB/RD	MAF Sensor Signal	0.6v
89	GY/WT	TP Sensor Signal	0.53-1.27v
90	BR/WT	Reference Voltage	4.9-5.1v
91	GY/RD	Analog Signal Return	<0.050v
92	RD/LG	Brake Pedal Position Switch	Brake Off: 0v, On: 12v
93	RD/WT	HO2S-11 (B1 S1) Heater	On: 1v, Off: 12v
94	WT/BK	HO2S-21 (B2 S1) Heater	On: 1v, Off: 12v
95	YL/LG	HO2S-12 (B1 S2) Heater	On: 1v, Off: 12v
96	TN/YL	HO2S-22 (B2 S2) Heater	On: 1v, Off: 12v
97	RD	Vehicle Power	12-14v
98	---	Not Used	---
99	LG/OR	Injector 6 Control	2.3-5.5 ms
100	BR/LB	Injector 4 Control	2.3-5.5 ms
101	WT	Injector 2 Control	2.3-5.5 ms
102	---	Not Used	---
103	BK/WT	Power Ground	<0.1v
104	---	Not Used	---

Pin Connector Graphic

Standard Colors and Abbreviations

Abbreviation	Color	Abbreviation	Color	Abbreviation	Color
BK	Black	GY	Gray	PK	Purple
BL	Blue	GN	Green	RD	Red
BR	Brown	LG	LT Green	TN	Tan
DB	Dark Blue	OR	Orange	WT	White
DG	DK Green	PK	Pink	YL	Yellow

1998-99 Taurus 3.0L 4v V6 MFI VIN S (A/T) 104 Pin Connector

PCM Pin #	Wire Color	Circuit Description (104 Pin)	Value at Hot Idle
1	PK/OR	Shift Solenoid 2 Control	1v, 55 mph: 1v
2	PK/LG	MIL (lamp) Control	MIL On: 1v, Off: 12v
3-4, 6	---	Not Used	---
5	WT/OR	Electronic AIRB Monitor	5v
7, 9-11	---	Not Used	---
8	DB/YL	IMRC Monitor Signal	5v, 55 mph: 5v
12	YL/WT	Fuel Level Indicator Signal	1.7v (1/2 full)
13	PK	Flash EEPROM Power	0.1v
15	PK/LB	Data Bus (-) Signal	Digital Signals
14, 17-20	---	Not Used	---
16	TN/OR	Data Bus (+) Signal	Digital Signals
21	GY	CKP Sensor (+) Signal	390-520 Hz
22	DB	CKP Sensor (-) Signal	390-520 Hz
24	BK/WT	Power Ground	<0.1v
25	BK	Power Ground	<0.1v
26	YL/BK	Coil Driver 1 Control	5° dwell
27	OR/YL	Shift Solenoid 1 Control	12v, 55 mph: 0.1v
28	LB	Low Speed Fan Control	On: 1v, Off: 12v
29	TN/WT	Overdrive Cancel Switch	TCS & O/D On: 12v
30	DG	Octane Adjust Switch	Closed: 0v, Open: 9.3v
31	YL/LG	PSP Switch Signal	Straight: 0v, Turned: 12v
33	PK/OR	VSS (-) Signal	0 Hz, 55 mph: 125 Hz
35	RD/BK	HO2S-12 (B1 S2) Signal	0.1-1.1v
36	TN/LB	MAF Sensor Return	<0.050v
37	OR/BK	TFT Sensor Signal	2.10-2.40v
38	LG/RD	ECT Sensor Signal	0.5-0.6v
39	GY	IAT Sensor Signal	1.5-2.5v
40	DG/YL	Fuel Pump Monitor	On: 12v, Off: 0v
41	PK/LB	A/C Cycling Clutch Switch	On: 1v, Off: 12v
42	BR	IMRC Solenoid Control	12v (Off)
45	PK/WT	EVAP CV Solenoid	0-10 Hz (0-100%)
46	LG/PK	High Speed Fan Control	On: 1v, Off: 12v
47	BR/PK	EGR VR Solenoid	0%, 55 mph: 45%
48	TN/YL	Clean Tachometer Output	33-45 Hz
49-50	---	Not Used	---
51	BK/WT	Power Ground	<0.1v
52	YL/RD	Coil Driver 2 Control	5° dwell
53	PK/BK	Shift Solenoid 3 Control	12v, 55 mph: 0.1v
54	PK/YL	TCC Solenoid Control	0%, 55 mph: 95%
56	GY/YL	EVAP Purge Solenoid	0-10 Hz (0-100%)
55	DG/OR	Keep Alive Power	12-14v
57	YL/RD	Knock Sensor Signal	0v
58	GY/BK	VSS (+) Signal	0 Hz, 55 mph: 125 Hz
59	---	Not Used	---
60	GY/LB	HO2S-11 (B1 S1) Signal	0.1-1.1v
61	PK/LG	HO2S-22 (B2 S2) Signal	0.1-1.1v
62	RD/BK	FTP Sensor Signal	2.6v (0 in. H20)
63	---	Not Used	---
64	LB/YL	TR Sensor Signal	In 'P': 0v, 55 mph: 1.7v
65	BR/LG	DPFE Sensor Signal	0.20-1.30v
66	---	Not Used	---
67	PK/WT	EVAP Purge Solenoid	0-10 Hz (0-100%)
68	---	Not Used	---

1998-99 Taurus 3.0L 4v V6 MFI VIN S (A/T) 104 Pin Connector

PCM Pin #	Wire Color	Circuit Description (104 Pin)	Value at Hot Idle
69	PK/YL	A/C WOT Relay Control	On: 1v, Off: 12v
70	BR	Electronic AIRB Solenoid	12v, 55 mph: 12v
71	RD	Vehicle Power	12-14v
72	---	Not Used	---
73	TN/BK	Injector 5 Control	2.2-2.7 ms
74	BR/YL	Injector 3 Control	2.2-2.7 ms
75	TN	Injector 1 Control	2.2-2.7 ms
76, 77	BK/WT	Power Ground	<0.1v
78	YL/WT	Coil Driver 3 Control	5° dwell
79	WT/LG	TCIL (lamp) Control	On: 1v, Off: 12v
80	LB/OR	Fuel Pump Control	On: 1v, Off: 12v
81	WT/YL	EPC Solenoid Control	9.0v (15 psi)
82	PK/YL	TCC Solenoid Control	12v (0 psi)
83	WT/LB	IAC Motor Control	10v (55% duty cycle)
84	DG/WT	TSS Sensor Signal	43 Hz (700 rpm)
85	DB/OR	CID Sensor Signal	5-7 Hz
86	BK/YL	A/C High Pressure Switch	12v (Switch Open)
87	RD/LG	HO2S-21 (B2 S1) Signal	0.1-1.1v
88	LB/RD	MAF Sensor Signal	0.6v
89	GY/WT	TP Sensor Signal	0.53-1.27v
90	BR/WT	Reference Voltage	4.9-5.1v
91	GY/RD	Analog Signal Return	<0.050v
92	RD/LG	Brake Pedal Position Switch	Brake Off: 0v, On: 12v
93	RD/WT	HO2S-11 (B1 S1) Heater	On: 1v, Off: 12v
94	WT/BK	HO2S-21 (B2 S1) Heater	On: 1v, Off: 12v
95	YL/LG	HO2S-12 (B1 S2) Heater	On: 1v, Off: 12v
96	TN/YL	HO2S-22 (B2 S2) Heater	On: 1v, Off: 12v
71	RD	Vehicle Power	12-14v
98	---	Not Used	---
99	LG/OR	Injector 6 Control	2.2-2.7 ms
100	BR/LB	Injector 4 Control	2.2-2.7 ms
101	WT	Injector 2 Control	2.2-2.7 ms
102	---	Not Used	---
103	BK/WT	Power Ground	<0.1v
104	---	Not Used	---

Pin Voltage Connector

PCM 104-PIN CONNECTOR

Terminal View of 104-Pin PCM Wiring Harness Connector

Standard Colors and Abbreviations

Abbreviation	Color	Abbreviation	Color	Abbreviation	Color
BK	Black	GY	Gray	PK	Purple
BL	Blue	GN	Green	RD	Red
BR	Brown	LG	LT Green	TN	Tan
DB	Dark Blue	OR	Orange	WT	White
DG	DK Green	PK	Pink	YL	Yellow

2000 Taurus 3.0L 4v V6 MFI VIN S (A/T) 104 Pin Connector

PCM Pin #	Wire Color	Circuit Description (104 Pin)	Value at Hot Idle
1	DG/PK	COP 4 Driver Control	6° dwell
2	PK/LG	MIL (lamp) Control	MIL On: 1v, Off: 12v
3	BK/WT	Power Ground	<0.1v
5	WT/OR	Electronic AIRB Monitor	5v
6	OR/YL	Shift Solenoid 'A' Control	12v, 55 mph: 0.1v
12	YL/WT	Fuel Level Indicator Signal	1.7v (1/2 full)
13	PK	Flash EEPROM Power	0.1v
15	PK/LB	Data Bus (-) Signal	Digital Signals
16	TN/OR	Data Bus (+) Signal	Digital Signals
17	BR/YL	SCI Receive Signal	N/A
18	GY/RD	SCI Transmit Signal	N/A
20	PK/BK	Shift Solenoid 'A' Control	12v, 55 mph: 0.1v
21	GY	CKP Sensor (+) Signal	390-520 Hz
22	DB	CKP Sensor (-) Signal	390-520 Hz
24	BK/WT	Power Ground	<0.1v
25	BK	Power Ground	<0.1v
26	LG/WT	COP 1 Driver Control	6° dwell
27	OR/YL	COP 5 Driver Control	6° dwell
28	LB	Low Speed Fan Control	On: 1v, Off: 12v
30	DB/LG	Antitheft Indicator	N/A
31	YL/LG	PSP Switch Signal	Straight: 0v, Turned: 12v
32	YL	Knock Sensor (-) Signal	0v
34	YL/BK	Digital TR1 Sensor	In 'P': 0v, 55 mph: 11v
35	RD/BK	HO2S-12 (B1 S2) Signal	0.1-1.1v
36	TN/LB	MAF Sensor Return	<0.050v
37	OR/BK	TFT Sensor Signal	2.10-2.40v
38	LG/RD	ECT Sensor Signal	0.5-0.6v
39	GY	IAT Sensor Signal	1.5-2.5v
40	DG/YL	Fuel Pump Monitor	On: 12v, Off: 0v
41	DG/WT	A/C High Pressure Switch	On: 1v, Off: 12v
44	PK	Starter Relay Control	KOEC: 9-11v
46	LG/PK	High Speed Fan Control	On: 1v, Off: 12v
47	BR/PK	EGR VR Solenoid	0%, 55 mph: 45%
48	TN/YL	Clean Tachometer Output	33-45 Hz
49	LB/BK	Digital TR2 Sensor	In 'P': 0v, 55 mph: 11v
50	WT/BK	Digital TR4 Sensor	In 'P': 0v, 55 mph: 11v
51	BK/WT	Power Ground	<0.1v
52	PK/WT	COP 2 Driver Control	6° dwell
53	OR/YL	COP 6 Driver Control	6° dwell
55	PK/LB	Keep Alive Power	12-14v
54	PK/YL	TCC Solenoid Control	0%, 55 mph: 95%
56	GY/YL	EVAP Purge Solenoid	0-10 Hz (0-100%)
55	DG/OR	Keep Alive Power	12-14v
57	YL/RD	Knock Sensor (+) Signal	0v
59	DG/WT	TSS Sensor Signal	43 Hz (700 rpm)
60	GY/LB	HO2S-11 (B1 S1) Signal	0.1-1.1v
61	PK/LG	HO2S-22 (B2 S2) Signal	0.1-1.1v
62	RD/PK	FTP Sensor Signal	2.6v (0 in. H20)
63	WT/YL	EPC Solenoid Control	9.0v (15 psi)
64	RD/BK	Digital TR3A Sensor	In 'P': 0v, 55 mph: 1.7v

2000 Taurus 3.0L 4v V6 MFI VIN S (A/T) 104 Pin Connector

PCM Pin #	Wire Color	Circuit Description (104 Pin)	Value at Hot Idle
65	BR/LG	DPFE Sensor Signal	0.20-1.30v
67	PK/WT	EVAP CV Solenoid	0-10 Hz (0-100%)
68	GY/BK	VSS (+) Signal	0 Hz, 55 mph: 125 Hz
69	PK/YL	A/C WOT Relay Control	On: 1v, Off: 12v
70	BR	Electronic AIRB Solenoid	12v, 55 mph: 12v
71	RD	Vehicle Power	12-14v
73	TN/BK	Injector 5 Control	2.2-2.7 ms
74	BR/YL	Injector 3 Control	2.2-2.7 ms
75	TN	Injector 1 Control	2.2-2.7 ms
76	BK/WT	Power Ground	<0.1v
77	BK/WT	Power Ground	<0.1v
78	YL/PKK	COP 3 Driver Control	6° dwell
79	WT/LG	TCIL (lamp) Control	On: 1v, Off: 12v
80	LB/OR	Fuel Pump Control	On: 1v, Off: 12v
81	BR/OR	EPC Solenoid Control	9.0v (15 psi)
83	WT/LB	IAC Motor Control	10v (55% duty cycle)
84	DB/YL	OSS Sensor Signal	0 Hz, 55 mph: 131 Hz
85	DB/OR	CID Sensor Signal	5-7 Hz
86	BK/YL	A/C Clutch Relay Switch	12v (Switch Open)
87	RD/LG	HO2S-21 (B2 S1) Signal	0.1-1.1v
88	LB/RD	MAF Sensor Signal	0.6v
89	GY/WT	TP Sensor Signal	0.53-1.27v
90	BR/WT	Reference Voltage	4.9-5.1v
91	GY/RD	Analog Signal Return	<0.050v
92	RD/LG	Brake Pedal Position Switch	Brake Off: 0v, On: 12v
93	RD/WT	HO2S-11 (B1 S1) Heater	On: 1v, Off: 12v
94	WT/BK	HO2S-21 (B2 S1) Heater	On: 1v, Off: 12v
95	YL/LB	HO2S-12 (B1 S2) Heater	On: 1v, Off: 12v
96	TN/YL	HO2S-22 (B2 S2) Heater	On: 1v, Off: 12v
99	LG/OR	Injector 6 Control	2.2-2.7 ms
100	BR/LB	Injector 4 Control	2.2-2.7 ms
101	WT	Injector 2 Control	2.2-2.7 ms
102	---	Not Used	---
103	BK/WT	Power Ground	<0.1v
104	---	Not Used	---

Pin Connector Graphic

Terminal View of 104-Pin PCM Wiring Harness Connector

Standard Colors and Abbreviations

Abbreviation	Color	Abbreviation	Color	Abbreviation	Color
BK	Black	GY	Gray	PK	Purple
BL	Blue	GN	Green	RD	Red
BR	Brown	LG	LT Green	TN	Tan
DB	Dark Blue	OR	Orange	WT	White
DG	DK Green	PK	Pink	YL	Yellow

2001 Taurus 3.0L 4v V6 MFI VIN S (A/T) 104 Pin Connector

PCM Pin #	Wire Color	Circuit Description (104 Pin)	Value at Hot Idle
1	PK/OR	Shift Solenoid 'B' Control	1v, 55 mph: 1v
2	PK/LG	MIL (lamp) Control	MIL On: 1v, Off: 12v
3	BK/WT	Power Ground	<0.1v
3	YL/BK	Digital TR1 Sensor	In 'P': 0v, 55 mph: 11v
4	BR/PK	Start System Voltage	Cranking: 9-11v
6	DG/WT	TSS Sensor Signal	50-65 Hz (790-820 rpm)
12	YL/WT	Fuel Level Indicator Signal	1.7v (1/2 full)
13	VT	Flash EEPROM Power	0.1v
15	PK/LB	SCP Data Bus (-) Signal	Digital Signals
16	TN/OR	SCP Data Bus (+) Signal	Digital Signals
17	BR/YL	SCI Receive (RX) Signal	Digital Signals
18	GY/RD	SCI Transmit (TX) Signal	Digital Signals
20	GY/YL	Generator Signal 2	130 Hz (45%)
21	GY	CKP Sensor (+) Signal	390-520 Hz
22	DB	CKP Sensor (-) Signal	390-520 Hz
24	BK/WT	Power Ground	<0.1v
25	BK	Chassis Ground	<0.1v
26	YL/BK	Coil 'A' Driver (Cyl 1 & 5)	6° dwell
27	OR/YL	Shift Solenoid 'A' Control	12v, 55 mph: 1v
28	LB	Low Speed Fan Control	On: 1v, Off: 12v
30	DB/LG	Antitheft Indicator	Refer to PAT System
31	YL/LG	PSP Switch Signal	Straight: 0v, Turned: 12v
32	YL	Knock Sensor (-) Signal	<0.050v
35	RD/BK	HO2S-12 (B1 S2) Signal	0.1-1.1v
36	TN/LB	MAF Sensor Return	<0.050v
37	OR/BK	TFT Sensor Signal	2.10-2.40v
38	LG/RD	ECT Sensor Signal	0.5-0.6v
39	GY	IAT Sensor Signal	1.5-2.5v
40	DG/YL	Fuel Pump Monitor	On: 12v, Off: 0v
41	DG/WT	A/C High Pressure Switch	On: 1v, Off: 12v
42	LG/YL	Generator Cutoff Relay	0.1v (relay on)
44	PK	Starter Relay Control	KOEC: 9-11v
45	LB/RD	Generator Sensor Signal	130 Hz (30%)
46	LG/VT	High Speed Fan Control	On: 1v, Off: 12v
47	BR/PK	EGR VR Solenoid	0%, 55 mph: 45%
48	TN/YL	Clean Tachometer Output	33-45 Hz
49	LB/BK	Digital TR2 Sensor	In 'P': 0v, 55 mph: 11v
50	WT/BK	Digital TR4 Sensor	In 'P': 0v, 55 mph: 11v
51	BK/WT	Power Ground	<0.1v
52	YL/RD	Coil 'B' Driver (Cyl 3 & 4)	6° dwell
53	PK/BK	Shift Solenoid 'C' Control	12v, 55 mph: 1v
54	VT/YL	TCC Solenoid Control	0%, 55 mph: 95%
55	PK/LB	Keep Alive Power	12-14v
56	GY/YL	EVAP Purge Solenoid	0-10 Hz (0-100%)
55	DG/OR	Keep Alive Power	12-14v
57	YL/RD	Knock Sensor (+) Signal	0v
60	GY/LB	HO2S-11 (B1 S1) Signal	0.1-1.1v
61	VT/LG	HO2S-22 (B2 S2) Signal	0.1-1.1v
62	RD/PK	FTP Sensor Signal	2.6v (0 in. H20)
63	WT/YL	Injector Pressure Sensor	2.74v (39 psi)
64	RD/BK	Digital TR3A Sensor	In 'P': 0v, 55 mph: 1.7v

2001 Taurus 3.0L 4v V6 MFI VIN S (A/T) 104 Pin Connector

PCM Pin #	Wire Color	Circuit Description (104 Pin)	Value at Hot Idle
65	BR/LG	DPFE Sensor Signal	0.95-1.05v
66	TN/OR	Evaporator Air Temp. Sensor	4.95-5.15v
67	VT/WT	EVAP CV Solenoid	0-10 Hz (0-100%)
68	GY/BK	VSS (+) Signal	0 Hz, 55 mph: 125 Hz
69	PK/YL	A/C WOT Relay Control	On: 1v, Off: 12v
71	RD	Vehicle Power	12-14v
73	TN/BK	Injector 5 Control	2.2-2.7 ms
74	BR/YL	Injector 3 Control	2.2-2.7 ms
75	TN	Injector 1 Control	2.2-2.7 ms
76	BK/WT	Power Ground	<0.1v
77	BK/WT	Power Ground	<0.1v
78	YL/WT	Coil 'C' Driver (Cyl 2 & 6)	6° dwell
79	WT/LG	TCIL (lamp) Control	On: 1v, Off: 12v
80	LB/OR	Fuel Pump Control	On: 1v, Off: 12v
81	BR/OR	EPC Solenoid Control	9.0v (15 psi)
82	BK/WT	Check Cap Indicator Light	On: 1v, Off: 12v
83	WT/LB	IAC Motor Control	10v (55% duty cycle)
84	DB/YL	OSS Sensor Signal	0 Hz, 55 mph: 131 Hz
85	DB/OR	CID Sensor Signal	5-7 Hz
86	BK/YL	A/C Clutch Relay Switch	12v (Switch Open)
87	RD/LG	HO2S-21 (B2 S1) Signal	0.1-1.1v
88	LB/RD	MAF Sensor Signal	0.6v
89	GY/WT	TP Sensor Signal	0.53-1.27v
90	BR/WT	Reference Voltage	4.9-5.1v
91	GY/RD	Analog Signal Return	<0.050v
92	RD/LG	Brake Pedal Position Switch	Brake Off: 0v, On: 12v
93	RD/WT	HO2S-11 (B1 S1) Heater	On: 1v, Off: 12v
94	WT/BK	HO2S-21 (B2 S1) Heater	On: 1v, Off: 12v
95	YL/LB	HO2S-12 (B1 S2) Heater	On: 1v, Off: 12v
96	TN/YL	HO2S-22 (B2 S2) Heater	On: 1v, Off: 12v
98	LG/RD	Generator Indicator Light	On: 1v, Off: 12v
99	LG/OR	Injector 6 Control	2.2-2.7 ms
100	BR/LB	Injector 4 Control	2.2-2.7 ms
101	W	Injector 2 Control	2.2-2.7 ms
103	BK/WT	Power Ground	<0.1v

Pin Voltage Connector

PCM 104-PIN CONNECTOR

```
1  ⊗⊗⊗⊗⊗⊗⊗⊗⊗⊗⊗⊗⊗      ⊗⊗⊗⊗⊗⊗⊗⊗⊗⊗⊗⊗⊗ 26
27 ⊗⊗⊗⊗⊗⊗⊗⊗⊗⊗⊗⊗⊗   ⬤  ⊗⊗⊗⊗⊗⊗⊗⊗⊗⊗⊗⊗⊗ 52
53 ⊗⊗⊗⊗⊗⊗⊗⊗⊗⊗⊗⊗⊗      ⊗⊗⊗⊗⊗⊗⊗⊗⊗⊗⊗⊗⊗ 78
79 ⊗⊗⊗⊗⊗⊗⊗⊗⊗⊗⊗⊗⊗      ⊗⊗⊗⊗⊗⊗⊗⊗⊗⊗⊗⊗⊗ 104
```

Terminal View of 104-Pin PCM Wiring Harness Connector

Standard Colors and Abbreviations

Abbreviation	Color	Abbreviation	Color	Abbreviation	Color
BK	Black	GY	Gray	PK	Purple
BL	Blue	GN	Green	RD	Red
BR	Brown	LG	LT Green	TN	Tan
DB	Dark Blue	OR	Orange	WT	White
DG	DK Green	PK	Pink	YL	Yellow

2002-03 Taurus 3.0L 4v V6 MFI VIN S (A/T) 104 Pin Connector

PCM Pin #	Wire Color	Circuit Description (104 Pin)	Value at Hot Idle
1	VT/OR	Shift Solenoid 'B' Control	1v, 55 mph: 1v
2	PK/LG	MIL (lamp) Control	MIL On: 1v, Off: 12v
3	YL/BK	Digital TR1 Sensor	In 'P': 0v, 55 mph: 11v
4	BR/PK	System Voltage (Start)	Cranking: 9-11v
5, 7-11	---	Not Used	---
6	DG/WT	TSS Sensor Signal	50-65 Hz (790-820 rpm)
12	YL/WT	Fuel Level Indicator Signal	1.7v (1/2 full)
13	VT	Flash EEPROM Power	0.1v
14, 19	---	Not Used	---
15	PK/LB	SCP Bus (-) Signal	Digital Signals
16	TN/OR	SCP Bus (+) Signal	Digital Signals
17	BR/YL	Passive Antitheft RX Signal	Digital Signals
18	GY/RD	Passive Antitheft TX Signal	Digital Signals
20	GY/YL	Generator Monitor Signal	130 Hz (45%)
21	GY	CKP Sensor (+) Signal	390-520 Hz
22	DB	CKP Sensor (-) Signal	390-520 Hz
23, 29	---	Not Used	---
24	BK/WT	Power Ground	<0.1v
25	BK	Chassis Ground	<0.1v
26	YL/BK	Coil 1 Driver (Cyl 1 & 5)	6° dwell
27	OR/YL	Shift Solenoid 'A' Control	12v, 55 mph: 1v
28	LB	Engine Cooling Fan Brake Relay Control	Relay Off: 12v, On: 1v
30	DB/LG	Antitheft Indicator	Indicator Off: 12v, On: 1v
31	YL/LG	PSP Switch Signal	Straight: 0v, Turned: 12v
32	YL	Knock Sensor (-) Signal	<0.050v
33-34, 43	---	Not Used	---
35	RD/BK	HO2S-12 (B1 S2) Signal	0.1-1.1v
36	TN/LB	MAF Sensor Return	<0.050v
37	OR/BK	TFT Sensor Signal	2.10-2.40v
38	LG/RD	ECT Sensor Signal	0.5-0.6v
39	GY	IAT Sensor Signal	1.5-2.5v
40	DG/YL	Fuel Pump Monitor	On: 12v, Off: 0v
41	DG/WT	A/C High Pressure Switch	On: 1v, Off: 12v
42	LG/YL	Engine Cooling Fan Relay Control	Relay Off: 12v, On: 1v
44	PK	Starter Relay Control	Relay Off: 12v, On: 1v
45	LB/RD	Generator Sensor Signal	130 Hz (30%)
46	LG/VT	High Speed Fan Control	On: 1v, Off: 12v
47	BR/PK	EGR VR Solenoid Control	0%, 55 mph: 45%
48	TN/YL	Clean Tachometer Output	33-45 Hz
49	LB/BK	Digital TR2 Sensor	In 'P': 0v, 55 mph: 11v
50	WT/BK	Digital TR4 Sensor	In 'P': 0v, 55 mph: 11v
51	BK/WT	Power Ground	<0.1v
52	YL/RD	Coil 2 Driver (Cyl 3 & 4)	6° dwell
53	PK/BK	Shift Solenoid 'C' Control	12v, 55 mph: 1v
54	VT/YL	TCC Solenoid Control	0%, 55 mph: 95%
55	PK/LB	Keep Alive Power	12-14v
56	GY/YL	EVAP Purge Solenoid	0-10 Hz (0-100%)
57	YL/RD	Knock Sensor (+) Signal	0v
58-59	---	Not Used	---
60	GY/LB	HO2S-11 (B1 S1) Signal	0.1-1.1v
61	VT/LG	HO2S-22 (B2 S2) Signal	0.1-1.1v
62	RD/PK	FTP Sensor Signal	2.6v (0 in. H20)
63	WT/YL	Injector Pressure Sensor	2.74v (39 psi)
64	RD/BK	Digital TR3A Sensor	In 'P': 0v, 55 mph: 1.7v

2002-03 Taurus 3.0L 4v V6 MFI VIN S (A/T) 104 Pin Connector

PCM Pin #	Wire Color	Circuit Description (104 Pin)	Value at Hot Idle
65	BR/LG	EGR DPFE Sensor Signal	0.95-1.05v
66	TN/OR	Evaporator Air Temperature Sensor	4.95-5.15v
67	VT/WT	EVAP CV Solenoid	0-10 Hz (0-100%)
68	GY/BK	VSS (+) Signal	0 Hz, 55 mph: 125 Hz
69	PK/YL	A/C WOT Relay Control	On: 1v, Off: 12v
71	RD	Vehicle Power (Start-Run)	12-14v
72	---	Not Used	---
73	TN/BK	Injector 5 Control	2.2-2.7 ms
74	BR/YL	Injector 3 Control	2.2-2.7 ms
75	TN	Injector 1 Control	2.2-2.7 ms
76, 77	BK/WT	Power Ground	<0.1v
78	YL/WT	Coil 3 Driver (Cyl 2 & 6)	6° dwell
79	WT/LG	TCIL (lamp) Control	Off: 12v, On: 1v
80	LB/OR	Fuel Pump Control	Off: 12v, On: 1v
81	BR/OR	EPC Solenoid Control	9.0v (15 psi)
82	BK/WT	Check Cap Indicator Light	Off: 12v, On: 1v
83	WT/LB	IAC Motor Control	10v (55% duty cycle)
84	DB/YL	OSS Sensor Signal	0 Hz, 55 mph: 131 Hz
85	DB/OR	CMP Sensor Signal	5-7 Hz
86	BK/YL	A/C Clutch Relay Switch	12v (Switch Open)
87	RD/LG	HO2S-21 (B2 S1) Signal	0.1-1.1v
88	LB/RD	MAF Sensor Signal	0.6v
89	GY/WT	TP Sensor Signal	0.53-1.27v
90	BR/WT	Reference Voltage	4.9-5.1v
91	GY/RD	Sensor Ground	<0.050v
92	RD/LG	Brake Pedal Position Switch	Brake Off: 0v, On: 12v
93	RD/WT	HO2S-11 (B1 S1) Heater	On: 1v, Off: 12v
94	WT/BK	HO2S-21 (B2 S1) Heater	On: 1v, Off: 12v
95	YL/LB	HO2S-12 (B1 S2) Heater	On: 1v, Off: 12v
96	TN/YL	HO2S-22 (B2 S2) Heater	On: 1v, Off: 12v
97	---	Not Used	---
98	LG/RD	Generator Indicator Light	Off: 12v, On: 1v
99	LG/OR	Injector 6 Control	2.2-2.7 ms
100	BR/LB	Injector 4 Control	2.2-2.7 ms
101	WT	Injector 2 Control	2.2-2.7 ms
102, 104	---	Not Used	---
103	BK/WT	Power Ground	<0.1v

Pin Voltage Connector

PCM 104-PIN CONNECTOR

Terminal View of 104-Pin PCM Wiring Harness Connector

Standard Colors and Abbreviations

Abbreviation	Color	Abbreviation	Color	Abbreviation	Color
BK	Black	GY	Gray	PK	Purple
BL	Blue	GN	Green	RD	Red
BR	Brown	LG	LT Green	TN	Tan
DB	Dark Blue	OR	Orange	WT	White
DG	DK Green	PK	Pink	YL	Yellow

1990 Taurus 3.0L V6 MFI VIN U (A/T) 60 Pin Connector

PCM Pin #	Wire Color	Circuit Description (60 Pin)	Value at Hot Idle
1	YL	Keep Alive Power	12-14v
2	RD/LG	Brake Pedal Position Switch	Brake Off: 0v, On: 12v
3	DG, OR/YL	VSS (+), (-) Signal	0 Hz, 55 mph: 125 Hz
4	DG/YL	Ignition Diagnostic Monitor	20-31 Hz
5, 11-12, 14-15	---	Not Used	---
7	LG/YL	ECT Sensor Signal	0.5-0.6v
8	PK/BK	Fuel Pump Monitor	On: 12v, Off: 0v
9	PK, TN	Data Bus (-), (+) Signal	Digital Signals
10	PK/BL	A/C Cycling Clutch Switch	A/C On: 12v, Off: 0v
13	OR/YL	Speed Control Solenoid (-)	Solenoid On: 1v
16	BK/OR	Ignition System Ground	<0.050v
17	TN/RD	STI Output, MIL Control	MIL On: 1v, Off: 12v
18	DG/PK	Shift Solenoid 3-4 Control	12v, 55 mph: 1v
19	OR/YL	Shift Solenoid 2-3 Control	12v, 55 mph: 1v
20	BK	PCM Case Ground	<0.050v
21	OR/BK	IAC Motor Control	7-9v
22	TN/LG	Fuel Pump Control	On: 1v, Off: 12v
23	YL/RD	Knock Sensor Signal (49 States)	0v
24	YL/LG	PSP Switch Signal	Straight: 0v, Turned: 11v
25	LG/PK	ACT Sensor Signal	1.5-2.5v
26	OR/WT	Reference Voltage	4.9-5.1v
27	BR/LG	PFE EGR Sensor	3.2v, 55 mph: 2.9v
29	DG/PK	HO2S-11 (B1 S1) Signal	0.1-1.1v
30	PK/YL	Neutral Drive Switch Signal	In 'P': 0v, Others: 12v
31	GY/YL	EVAP Purge Solenoid	12v, 55 mph: 1v
32, 41, 43, 51	---	Not Used	---
33	DG	EGR VR Solenoid	0%, 55 mph: 45%
34	BL/PK	Data Output Link	Digital Signals
35	WT/PK	Speed Control Solenoid	0%, 55 mph: 97%
36	YL/LG	Spark Output Signal	6.93v (duty cycle)
37, 57	RD	Vehicle Power	12-14v
39, 50	OR, BL/BK	S/C Switch Ground, SC Signal	<0.050v, S/C On: 6.7v
40, 60	BK/LG	Power Ground	<0.1v
42	GY/BK	S/C Vacuum Solenoid Control	S/C Off: 0% duty cycle
44	GY/WT	AXOD Transmission Solenoid	On: 1v, Off: 12v
45	LG/BK	MAP Sensor Signal	107 Hz
46	BK/WT	Analog Signal Return	<0.050v
47	DG/LG	TP Sensor Signal	0.5-1.2v
48	WT/BK	Self-Test Indicator Signal	STI Open: 5v, Closed: 1v
49	OR	HO2S-21 (B2 S1) Signal	0.1-1.1v
52	PK	High Speed Fan Control	On: 1v, Off: 12v
53	TN/BL	Lock-Up Solenoid for AXOD	12v, 55 mph: 1v
54	RD	A/C WOT Relay Control	On: 1v, Off: 12v
55	TN/OR	Low Speed Fan Control	Fan On: 1v, Off: 12v
56	DB	PIP Sensor Signal	50% duty cycle
58	TN/RD	Injector Bank 1 (INJ 1, 2 & 4)	5.8-6.1 ms
59	TN/OR	Injector Bank 2 (INJ 3, 5 & 6)	5.8-6.1 ms

PCM 60-PIN CONNECTOR

Terminal View of 60-Pin PCM Harness Connector

1991-95 Taurus 3.0L V6 MFI VIN U (A/T) 60 Pin Connector

PCM Pin #	Wire Color	Circuit Description (60 Pin)	Value at Hot Idle
1	YL	Keep Alive Power	12-14v
2	RD/LG	Brake Pedal Position Switch	Brake Off: 0v, On: 12v
3	DG/WT	VSS (+) Signal	0 Hz, 55 mph: 125 Hz
4	WT/PK	Ignition Diagnostic Monitor	20-31 Hz
4	DG/YL	Ignition Diagnostic Monitor	20-31 Hz
5	GY/BK	TSS Sensor Signal	100 Hz (700 rpm)
6	OR/YL	VSS (+) Signal	0 Hz, 55 mph: 125 Hz
7	LG/RD	ECT Sensor Signal	0.5-0.6v
8	PK/BK	Fuel Pump Monitor	On: 12v, Off: 0v
9	TN/BL	MAF Sensor Return	<0.050v
10	PK/BL	A/C Cycling Clutch Switch	A/C On: 12v, Off: 0v
11	GN/YL	EVAP Purge Solenoid	12v, 55 mph: 1v
12	LG/OR	Injector 6 Control	5.0-5.6 ms
13	TN/OR	Low Cooling Fan	On: 1v, Off: 12v
14	---	Not Used	---
15	TN/BK	Injector 5 Control	5.0-5.6 ms
16	OR/RD	Ignition System Ground	<0.050v
17	TN/RD	STI Output, MIL Control	MIL On: 1v, Off: 12v
18	TN/OR	Data Bus (+) Signal	Digital Signals
19	PK/BL	Data Bus (-) Signal	Digital Signals
20	BK	PCM Case Ground	<0.050v
21	BR/WT	IAC Motor Control	8-9v
22	TN/LG	Fuel Pump Control	On: 1v, Off: 12v
22	BL/OR	Fuel Pump Control	On: 1v, Off: 12v
23-24	---	Not Used	---
25	LG/PK	IAT Sensor Signal	1.5-2.5v
26	BR/WT	Reference Voltage	4.9-5.1v
27	BR/LG	PFE EGR Sensor	3.2v, 55 mph: 2.8v
28	YL/LG	PSP Switch Signal	Straight: 0v, Turned: 11v
29	---	Not Used	---
30	BL	MLP Sensor Signal	In 'P': 0v, 55 mph: 1.7v
30	BL/YL	MLP Sensor Signal	In 'P': 0v, 55 mph: 1.7v
31	LG/PK	High Cooling Fan	On: 1v, Off: 12v
32	---	Not Used	---
33	BR/PK	EGR VR Solenoid	0%, 55 mph: 45%
34	BL/BK	Data Output Link	Digital Signals
35	BR/BL	Injector 4 Control	5.0-5.6 ms
36	YL/LG	Spark Output Signal	50% duty cycle
37	RD	Vehicle Power	12-14v
38	WT/YL	EPC Solenoid Control	8-9v (9 psi)
39	BR/YL	Injector 3 Control	5.0-5.6 ms
40	BK/LG	Power Ground	<0.1v

1991-95 Taurus 3.0L V6 MFI VIN U (A/T) 60 Pin Connector

PCM Pin #	Wire Color	Circuit Description (60 Pin)	Value at Hot Idle
41	---	Not Used	---
42 ('94-'95)	TN/LG	A/C Pressure Sensor	A/C On: 12v (open)
43	RD/BK	HO2S-11 (B1 S1) Signal	0.1-1.1v
44	GY/BL	HO2S-21 (B2 S1) Signal	0.1-1.1v
45	---	Not Used	---
46	GY/RD	Analog Signal Return	<0.050v
47	GN/WT	TP Sensor Signal	0.5-1.2v
48	BR	Self-Test Indicator Signal	STI Open: 5v, Closed: 1v
49	OR/BK	TOT Sensor Signal	2.10-2.40v
50	BL/RD	MAF Sensor Signal	0.6-0.9v
51	OR/YL	Shift Solenoid 1 Control	12v, 55 mph: 1v
52	PK/OR	Shift Solenoid 2 Control	1v, 55 mph: 12v
53	TN/WT	TCC Solenoid Control	0%, 55 mph: 45%
54	PK/YL	A/C WOT Relay Control	On: 1v, Off: 12v
55	WT/RD	Shift Solenoid 3 Control	7-9v, 55 mph: 8-9v
56	DB	PIP Sensor Signal	50% duty cycle
56	GY/OR	PIP Sensor Signal	50% duty cycle
57	RD	Vehicle Power	12-14v
58	TN	Injector 1 Control	5.0-5.6 ms
59	WT	Injector 2 Control	5.0-5.6 ms
60	BK/LG	Power Ground	<0.1v

Pin Connector Graphic

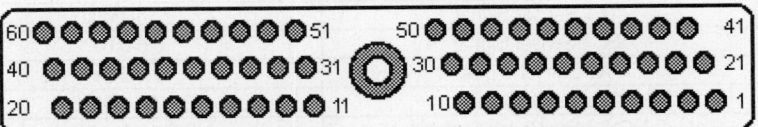

PCM 60-PIN CONNECTOR

Terminal View of 60-Pin PCM Harness Connector

Standard Colors and Abbreviations

Abbreviation	Color	Abbreviation	Color	Abbreviation	Color
BK	Black	GY	Gray	PK	Purple
BL	Blue	GN	Green	RD	Red
BR	Brown	LG	LT Green	TN	Tan
DB	Dark Blue	OR	Orange	WT	White
DG	DK Green	PK	Pink	YL	Yellow

1996-97 Taurus 3.0L V6 MFI VIN U (A/T) 104 Pin Connector

PCM Pin #	Wire Color	Circuit Description (104 Pin)	Value at Hot Idle
1	PK/OR	Shift Solenoid 2 Control	1v, 55 mph: 12v
2	PK/LG	MIL (lamp) Control	MIL On: 1v, Off: 12v
3-12	---	Not Used	---
13	PK	Flash EEPROM Power	0.1v
14	---	Not Used	---
15	PK/LB	Data Bus (-) Signal	Digital Signals
16	TN/OR	Data Bus (+) Signal	Digital Signals
17-20	---	Not Used	---
21	GY	CKP Sensor (+) Signal	510-540 Hz
22	DB	CKP Sensor (-) Signal	510-540 Hz
23	---	Not Used	---
24	BK/WT	Power Ground	<0.1v
25	BK	Power Ground	<0.1v
26	WT/BK	Coil Driver 1 Control	5° dwell
27	OR/YL	Shift Solenoid 1 Control	12v, 55 mph: 1v
28	TN/OR	Low Speed Fan Control	On: 1v, Off: 12v
29	---	Not Used	---
30	DG	Octane Adjust Switch	Closed: 0v, Open: 9.3v
31	YL/LG	PSP Switch Signal	Straight: 0v, Turned: 12v
32	---	Not Used	---
33	PK/OR	VSS (-) Signal	0 Hz, 55 mph: 125 Hz
34	---	Not Used	---
35	RD/BK	HO2S-12 (B1 S2) Signal	0.1-1.1v
36	TN/LB	MAF Sensor Return	<0.050v
37	OR/BK	TFT Sensor Signal	2.10-2.40v
38	LG/RD	ECT Sensor Signal	0.5-0.6v
39	GY	IAT Sensor Signal	1.5-2.5v
40	DG/YL	Fuel Pump Monitor	On: 12v, Off: 0v
41	PK/LB	A/C Cycling Clutch Switch	A/C On: 12v, Off: 0v
42-45	---	Not Used	---
46	LG/PK	High Speed Fan Control	On: 1v, Off: 12v
47	BR/PK	EGR VR Solenoid	0%, 55 mph: 45%
48	TN/YL	Clean Tachometer Output	42-50 Hz
49-50	---	Not Used	---
51	BK/WT	Power Ground	<0.1v
52	YL/RD	Coil Driver 2 Control	5° dwell
53	PK/BK	Shift Solenoid 3 Control	12v, 55 mph: 1v
54	---	Not Used	---
55	PK	Keep Alive Power	12-14v
56	GY/YL	EVAP VMV Solenoid	0-10 Hz (0-100%)
57	---	Not Used	---
58	GY/BK	VSS (+) Signal	0 Hz, 55 mph: 125 Hz
59	---	Not Used	---
60	GY/LB	HO2S-11 (B1 S1) Signal	0.1-1.1v
61	PK/LG	HO2S-22 (B2 S2) Signal	0.1-1.1v
62	RD/PK	FTP Sensor Signal	2.6v (0" H2O - cap off)
63	---	Not Used	---
64	LB/YL	TR Sensor Signal	In 'P': 0v, 55 mph: 1.7v

1996-97 Taurus 3.0L V6 MFI VIN U (A/T) 104 Pin Connector

PCM Pin #	Wire Color	Circuit Description (104 Pin)	Value at Hot Idle
65	BR/LG	DPFE Sensor Signal	0.20-1.30v
66, 68	---	Not Used	---
67	PK/WT	EVAP CV Solenoid	0-10 Hz (0-100%)
69	PK/YL	A/C WOT Relay Control	On: 1v, Off: 12v
70, 72	---	Not Used	---
71	RD	Vehicle Power	12-14v
73	TN/BK	Injector 5 Control	2.3-2.8 ms
74	BR/YL	Idle Air Control Valve	10v (32% duty cycle)
75	TN	Injector 1 Control	2.3-2.8 ms
76	BK/WT	Power Ground	<0.1v
77	BK/WT	Power Ground	<0.1v
78	YL/WT	Coil Driver 3 Control	5° dwell
79	---	Not Used	---
80	LB/OR	Fuel Pump Control	On: 1v, Off: 12v
81	WT/YL	EPC Solenoid Control	10.6v (40 psi)
82	PK/YL	TCC Solenoid Control	12v (0%)
83	WT/LB	Injector 3 Control	2.3-2.8 ms
84	DG/WT	TSS Sensor Signal	50-65 Hz (790-820 rpm)
85	DB/OR	CID Sensor Signal	6-8 Hz
86	BK/YL	A/C Pressure Switch	A/C On: 12v (Open)
87	RD/LG	HO2S-21 (B2 S1) Signal	0.1-1.1v
88	LB/RD	MAF Sensor Signal	0.9v
89	GY/WT	TP Sensor Signal	0.53-1.27v
90	BR/WT	Reference Voltage	4.9-5.1v
91	GY/RD	Analog Signal Return	<0.050v
92	RD/LG	Brake Pedal Position Switch	Brake Off: 0v, On: 12v
93	RD/WT	HO2S-11 (B1 S1) Heater	On: 1v, Off: 12v
94	WT/BK	HO2S-21 (B2 S1) Heater	On: 1v, Off: 12v
95	YL/LB	HO2S-12 (B1 S2) Heater	On: 1v, Off: 12v
96	TN/YL	HO2S-22 (B2 S2) Heater	On: 1v, Off: 12v
97	RD	Vehicle Power	12-14v
98	---	Not Used	---
99	LG/OR	Injector 6 Control	2.3-2.8 ms
100	BR/LB	Injector 4 Control	2.3-2.8 ms
101	WT	Injector 2 Control	2.3-2.8 ms
102	---	Not Used	---
103	BK/WT	Power Ground	<0.1v
104	---	Not Used	---

Pin Connector Graphic

PCM 104-PIN CONNECTOR

Terminal View of 104-Pin PCM Wiring Harness Connector

Standard Colors and Abbreviations

Abbreviation	Color	Abbreviation	Color	Abbreviation	Color
BK	Black	GY	Gray	PK	Purple
BL	Blue	GN	Green	RD	Red
BR	Brown	LG	LT Green	TN	Tan
DB	Dark Blue	OR	Orange	WT	White
DG	DK Green	PK	Pink	YL	Yellow

1998-99 Taurus 3.0L V6 MFI VIN U (A/T) 104 Pin Connector

PCM Pin #	Wire Color	Circuit Description (104 Pin)	Value at Hot Idle
1	VT/OR	Shift Solenoid 'B' Control	1v, 55 mph: 12v
2	PK/LG	MIL (lamp) Control	MIL On: 1v, Off: 12v
3	YL/BK	Digital TR1 Sensor	In 'P': 0v, 55 mph: 11v
4-11	---	Not Used	---
12	YL/WT	Fuel Level Indicator Signal	1.7v (1/2 full)
13	PK	Flash EEPROM Power	0.1v
15	PK/LB	Data Bus (-) Signal	Digital Signals
16	TN/OR	Data Bus (+) Signal	Digital Signals
17-18	---	Not Used	---
21	GY	CKP Sensor (+) Signal	510-540 Hz
22	DB	CKP Sensor (-) Signal	510-540 Hz
24	BK/WT	Power Ground	<0.1v
25	BK	Chassis Ground	<0.1v
26	YL/BK	Coil Driver 1 Control	6° dwell
27	OR/YL	Shift Solenoid 'A' Control	12v, 55 mph: 1v
28	LB	Low Speed Fan Control	On: 1v, Off: 12v
30	---	Not Used	---
31	YL/LG	PSP Switch Signal	Straight: 0v, Turned: 12v
32	---	Not Used	---
33	PK/OR	VSS (-) Signal	0 Hz, 55 mph: 125 Hz
35	RD/BK	HO2S-12 (B1 S2) Signal	0.1-1.1v
36	TN/LB	MAF Sensor Return	<0.050v
37	OR/BK	TFT Sensor Signal	2.10-2.40v
38	LG/RD	ECT Sensor Signal	0.5-0.6v
39	GY	IAT Sensor Signal	1.5-2.5v
40	DG/YL	Fuel Pump Monitor	On: 12v, Off: 0v
41	DG/WT	A/C High Pressure Switch	AC On: 12v, Off: 0v
42-44	---	Not Used	---
45	PK/WT	EVAP CV Solenoid	0-10 Hz (0-100%)
46	LG/PK	High Speed Fan Control	On: 1v, Off: 12v
47	BR/PK	EGR VR Solenoid	0%, 55 mph: 45%
48	TN/YL	Clean Tachometer Output	42-50 Hz
49	LB/BK	Digital TR2 Sensor	In 'P': 0v, 55 mph: 11v
50	WT/BK	Digital TR4 Sensor	In 'P': 0v, 55 mph: 11v
51	BK/WT	Power Ground	<0.1v
52	YL/RD	Coil Driver 2 Control	6° dwell
53	PK/BK	Shift Solenoid 'C' Control	12v, 55 mph: 1v
54	DG/WT	TSS Sensor Signal	50-65 Hz (820-900 rpm)
55	DG/OR	Keep Alive Power	12-14v
56	GY/YL	EVAP Purge Solenoid	0-10 Hz (0-100%)
57, 59	---	Not Used	---
58	GY/BK	VSS (+) Signal	0 Hz, 55 mph: 125 Hz
60	GY/LB	HO2S-11 (B1 S1) Signal	0.1-1.1v
61	PK/LG	HO2S-22 (B2 S2) Signal	0.1-1.1v
62	RD/PK	FTP Sensor Signal	2.6v (0" H2O - cap off)
63	---	Not Used	---
64	LB/YL	Digital TR3A Sensor	In 'P': 0v, in O/D: 1.7v

1998-99 Taurus 3.0L V6 MFI VIN U (A/T) 104 Pin Connector

PCM Pin #	Wire Color	Circuit Description (104 Pin)	Value at Hot Idle
65	BR/LG	DPFE Sensor Signal	0.20-1.30v
66-68	---	Not Used	---
69	PK/YL	A/C WOT Relay Control	On: 1v, Off: 12v
70	---	Not Used	---
71	RD	Vehicle Power	12-14v
73	TN/BK	Injector 5 Control	2.3-2.8 ms
74	BR/YL	Injector 3 Control	2.3-2.8 ms
75	TN	Injector 1 Control	2.3-2.8 ms
76-77	BK/WT	Power Ground	<0.1v
78	YL/WT	Coil Driver 3 Control	6° dwell
80	LB/OR	Fuel Pump Control	On: 1v, Off: 12v
81	WT/YL	EPC Solenoid Control	10.6v (40 psi)
81-82	---	Not Used	---
83	WT/LB	IAC Motor Control	10v (32% duty cycle)
84	DG/WT	TSS Sensor Signal	50-65 Hz (790-820 rpm)
85	DB/OR	CID Sensor Signal	6-8 Hz
86	BK/YL	A/C Pressure Switch	A/C On: 12v (Open)
87	RD/LG	HO2S-21 (B2 S1) Signal	0.1-1.1v
88	LB/RD	MAF Sensor Signal	0.9v
89	GY/WT	TP Sensor Signal	0.53-1.27v
90	BR/WT	Reference Voltage	4.9-5.1v
91	GY/RD	Analog Signal Return	<0.050v
92	RD/LG	Brake Pedal Position Switch	Brake Off: 0v, On: 12v
93	RD/WT	HO2S-11 (B1 S1) Heater	On: 1v, Off: 12v
94	WT/BK	HO2S-21 (B2 S1) Heater	On: 1v, Off: 12v
95	YL/LB	HO2S-12 (B1 S2) Heater	On: 1v, Off: 12v
96	TN/YL	HO2S-22 (B2 S2) Heater	On: 1v, Off: 12v
97 ('98)	RD	Vehicle Power	12-14v
99	LG/OR	Injector 6 Control	2.3-2.8 ms
100	BR/LB	Injector 4 Control	2.3-2.8 ms
101	WT	Injector 2 Control	2.3-2.8 ms
102	---	Not Used	---
103	BK/WT	Power Ground	<0.1v
104	---	Not Used	---

Pin Connector Graphic

PCM 104-PIN CONNECTOR

Terminal View of 104-Pin PCM Wiring Harness Connector

Standard Colors and Abbreviations

Abbreviation	Color	Abbreviation	Color	Abbreviation	Color
BK	Black	GY	Gray	PK	Purple
BL	Blue	GN	Green	RD	Red
BR	Brown	LG	LT Green	TN	Tan
DB	Dark Blue	OR	Orange	WT	White
DG	DK Green	PK	Pink	YL	Yellow

2000 Taurus 3.0L V6 MFI VIN U (A/T) 104 Pin Connector

PCM Pin #	Wire Color	Circuit Description (104 Pin)	Value at Hot Idle
1	VT/OR	Shift Solenoid 'B' Control	1v, 55 mph: 12v
2	PK/LG	MIL (lamp) Control	MIL On: 1v, Off: 12v
3	YL/BK	Digital TR1 Sensor	In 'P': 0v, 55 mph: 11v
4	BR/PK	System Voltage (Start-Run)	Cranking: 9-11v
5	WT/OR	Secondary Air Injection Relay	On: 1v, Off: 12v
6	DG/WT	TSS Sensor Signal	50-65 Hz (790-820 rpm)
7-10	---	Not Used	---
12	YL/WT	Fuel Level Indicator Signal	1.7v (1/2 full)
13	VT	Flash EEPROM Power	0.1v
14	---	Not Used	---
15	PK/LB	SCP Data Bus (-) Signal	Digital Signals
16	TN/OR	SCP Data Bus (+) Signal	Digital Signals
17	BR/YL	SCI Receive (RX) Signal	Digital Signals
18	GY/RD	SCI Transmit (TX) Signal	Digital Signals
20	GY/YL	Generator Field Control	0 Hz
21	GY	CKP Sensor (+) Signal	510-540 Hz
22	DB	CKP Sensor (-) Signal	510-540 Hz
24	BK/WT	Power Ground	<0.1v
25	BK	Case Ground	<0.1v
26	YL/BK	Coil 1 Driver (Cyl 1& 5)	6° dwell
27	OR/YL	Shift Solenoid 'A' Control	12v, 55 mph: 1v
28	LB	Low Speed Fan Control	On: 1v, Off: 12v
29	---	Not Used	---
30	DB/LG	Passive Antitheft Indicator	Lamp On: 1v, Off: 12v
31	YL/LG	PSP Switch Signal	Straight: 0v, Turned: 12v
32	YL	Knock Sensor 1 (-) Signal	0v
35	RD/BK	HO2S-12 (B1 S2) Signal	0.1-1.1v
36	TN/LB	MAF Sensor Return	<0.050v
37	OR/BK	TFT Sensor Signal	2.10-2.40v
38	LG/RD	ECT Sensor Signal	0.5-0.6v
39	GY	IAT Sensor Signal	1.5-2.5v
40	DG, PK	Fuel Pump Monitor	On: 12v, Off: 0v
41	DG/WT	A/C High Pressure Switch	AC On: 12v, Off: 0v
42	LG/YL	Generator Cutoff Relay	0.1v (with fan on)
44	PK	Starter Relay Control	KOEC: 9-11v
45	LB/RD	Generator Sensor Signal	TBD
46	LG/VT	High Speed Fan Control	On: 1v, Off: 12v
47	BR/PK	EGR VR Solenoid	0%, 55 mph: 45%
48	TN/YL	Clean Tachometer Output	42-50 Hz
49	LB/BK	Digital TR2 Sensor	In 'P': 0v, 55 mph: 11v
50	WT/BK	Digital TR4 Sensor	In 'P': 0v, 55 mph: 11v
51	BK/WT	Power Ground	<0.1v
52	YL/RD	Coil Driver 2 (Cyl 3 & 4)	6° dwell
53	PK/BK	Shift Solenoid 'C' Control	12v, 55 mph: 1v
54	VT/YL	TCC Solenoid Control	12v (0%)
55	PK/LB	Keep Alive Power	12-14v
56	GY/YL	EVAP Purge Solenoid	0-10 Hz (0-100%)
57	YL/RD	Knock Sensor 1 (+) Signal	0v
60	GY/LB	HO2S-11 (B1 S1) Signal	0.1-1.1v
61	VT/LG	HO2S-22 (B2 S2) Signal	0.1-1.1v
62	RD/PK	FTP Sensor Signal	2.6v (0" H2O - cap off)
63	WT/YL	Injector Pressure Sensor	TBD
64	LB/YL	Digital TR3A Sensor	In 'P': 0v, in O/D: 1.7v

2000 Taurus 3.0L V6 MFI VIN U (A/T) 104 Pin Connector

PCM Pin #	Wire Color	Circuit Description (104 Pin)	Value at Hot Idle
65	BR/LG	DPFE Sensor Signal	0.20-1.30v
66	TN/OR	Evaporator Air Temp. Sensor	4.95-5.15v
67	VT/WT	EVAP Purge Solenoid	0-10 Hz (0-100%)
68	GY/BK	VSS (+) Signal	0 Hz, 55 mph: 125 Hz
69	PK/YL	A/C WOT Relay Control	On: 1v, Off: 12v
70	BR	Secondary Air Injection Relay	On: 1v, Off: 12v
71	RD	Vehicle Power	12-14v
73	TN/BK	Injector 5 Control	2.3-2.8 ms
74	BR/YL	Injector 3 Control	2.3-2.8 ms
75	TN	Injector 1 Control	2.3-2.8 ms
76, 77	BK/WT	Power Ground	<0.1v
78	YL/WT	Coil 3 Driver (Cyl 2 & 6)	6° dwell
79	WT/LG	Check Transaxle Light	On: 1v, Off: 12v
80	LB/OR	Fuel Pump Control	On: 1v, Off: 12v
81	WT or BR	EPC Solenoid Control	10.6v (40 psi)
82	BK/WT	Check Cap Indicator Light	On: 1v, Off: 12v
83	WT/LB	IAC Motor Control	10v (32% duty cycle)
84	DB/YL	OSS Sensor Signal	0 Hz, 55 mph: 131 Hz
85	DB/OR	CID Sensor Signal	6-8 Hz
86	BK/YL	A/C Pressure Clutch Relay	A/C On: 12v, Off: 0v
87	RD/LG	HO2S-21 (B2 S1) Signal	0.1-1.1v
88	LB/RD	MAF Sensor Signal	0.9v
89	GY/WT	TP Sensor Signal	0.53-1.27v
90	BR/WT	Reference Voltage	4.9-5.1v
91	GY/RD	Analog Signal Return	<0.050v
92	RD/LG	Brake Pedal Position Switch	Brake Off: 0v, On: 12v
93	RD/WT	HO2S-11 (B1 S1) Heater	On: 1v, Off: 12v
94	WT/BK	HO2S-21 (B2 S1) Heater	On: 1v, Off: 12v
95	YL/LB	HO2S-12 (B1 S2) Heater	On: 1v, Off: 12v
96	TN/YL	HO2S-22 (B2 S2) Heater	On: 1v, Off: 12v
98	LG/RD	Generator Battery Indicator	Digital Signals
99	LG/OR	Injector 6 Control	2.3-2.8 ms
100	BR/LB	Injector 4 Control	2.3-2.8 ms
101	WT	Injector 2 Control	2.3-2.8 ms
102	---	Not Used	---
103	BK/WT	Power Ground	<0.1v
104	---	Not Used	---

Pin Connector Graphic

PCM 104-PIN CONNECTOR

Terminal View of 104-Pin PCM Wiring Harness Connector

Standard Colors and Abbreviations

Abbreviation	Color	Abbreviation	Color	Abbreviation	Color
BK	Black	GY	Gray	PK	Purple
BL	Blue	GN	Green	RD	Red
BR	Brown	LG	LT Green	TN	Tan
DB	Dark Blue	OR	Orange	WT	White
DG	DK Green	PK	Pink	YL	Yellow

2001-03 Taurus 3.0L V6 MFI VIN U (A/T) 104 Pin Connector

PCM Pin #	Wire Color	Circuit Description (104 Pin)	Value at Hot Idle
1	VT/OR	Shift Solenoid 'B' Control	1v, 55 mph: 12v
2	PK/LG	MIL (lamp) Control	MIL On: 1v, Off: 12v
3	YL/BK	Digital TR1 Sensor	In 'P': 0v, 55 mph: 11v
4	TN/RD	System Voltage (Start-Run)	Cranking: 9-11v
5	---	Not Used	---
6	DG/WT	TSS Sensor Signal	50-65 Hz (790-820 rpm)
7-10	---	Not Used	---
12	YL/WT	Fuel Level Indicator Signal	1.7v (1/2 full)
13	VT	Flash EEPROM Power	0.1v
14, 19	---	Not Used	---
15	PK/LB	SCP Bus (-) Signal	Digital Signals
16	TN/OR	SCP Bus (+) Signal	Digital Signals
17	BR/YL	Passive Antitheft RX Signal	Digital Signals
18	GY/RD	Passive Antitheft TX Signal	Digital Signals
20	GY/YL	Generator Field Control	130 Hz (45%)
21	GY	CKP Sensor (+) Signal	510-540 Hz
22	DB	CKP Sensor (-) Signal	510-540 Hz
23	---	Not Used	---
24	BK/WT	Power Ground	<0.1v
25	BK	Case Ground	<0.1v
26	YL/BK	Coil 1 Driver (Cyl 1& 5)	6° dwell
27	OR/YL	Shift Solenoid 'A' Control	12v, 55 mph: 1v
28	LB	Engine Cooling Fan Brake Relay Control	Off: 12v, On: 1v
29, 33-34	---	Not Used	---
30	DB/LG	Passive Antitheft Indicator	Lamp On: 1v, Off: 12v
31	YL/LG	PSP Switch Signal	Straight: 0v, Turned: 12v
32	YL	Knock Sensor 1 (-) Signal	<0.050v
35	RD/BK	HO2S-12 (B1 S2) Signal	0.1-1.1v
36	TN/LB	MAF Sensor Return	<0.050v
37	OR/BK	TFT Sensor Signal	2.10-2.40v
38	LG/RD	ECT Sensor Signal	0.5-0.6v
39	GY	IAT Sensor Signal	1.5-2.5v
40	DG/YL	Fuel Pump Monitor	On: 12v, Off: 0v
41	DG/WT	A/C High Pressure Switch	AC On: 12v, Off: 0v
42	LG/YL	Generator Cutoff Relay	0.1v (with fan on)
44	PK	Starter Relay Control	KOEC: 9-11v
45	LB/RD	Generator Sensor Signal	130 Hz (30%)
46	LG/VT	High Cooling Speed Fan Control	Off: 12v, On: 1v
47	BR/PK	EGR VR Solenoid	0%, 55 mph: 45%
48	TN/YL	Clean Tachometer Output	42-50 Hz
49	LB/BK	Digital TR2 Sensor	In 'P': 0v, 55 mph: 11v
50	WT/BK	Digital TR4 Sensor	In 'P': 0v, 55 mph: 11v
51	BK/WT	Power Ground	<0.1v
52	YL/RD	Coil Driver 2 (Cyl 3 & 4)	6° dwell
53	PK/BK	Shift Solenoid 'C' Control	12v, 55 mph: 1v
54	VT/YL	TCC Solenoid Control	12v (0%)
55	PK/LB	Keep Alive Power	12-14v
56	GY/YL	EVAP Purge Solenoid	0-10 Hz (0-100%)
57	YL/RD	Knock Sensor (+) Signal	0v
60	GY/LB	HO2S-11 (B1 S1) Signal	0.1-1.1v
61	VT/LG	HO2S-22 (B2 S2) Signal	0.1-1.1v
62	RD/PK	FTP Sensor Signal	2.6v (0" H2O - cap off)
63	WT/YL	Injector Pressure Sensor	TBD
64	RD/BK	Digital TR3A Sensor	In 'P': 0v, in O/D: 1.7v

2001-03 Taurus 3.0L V6 MFI VIN U (A/T) 104 Pin Connector

PCM Pin #	Wire Color	Circuit Description (104 Pin)	Value at Hot Idle
65	BR/LG	DPFE Sensor Signal	0.95-1.05v
66	TN/OR	Evaporator Air Temperature Sensor	4.95-5.15v
67	VT/WT	EVAP Purge Solenoid	0-10 Hz (0-100%)
68	GY/BK	VSS (+) Signal	Moving: AC pulse signals
69	PK/YL	A/C WOT Relay Control	Off: 12v, On: 1v
70	---	Not Used	---
71	RD	Vehicle Power (Start-Run)	12-14v
73	TN/BK	Injector 5 Control	2.3-2.8 ms
74	BR/YL	Injector 3 Control	2.3-2.8 ms
75	TN	Injector 1 Control	2.3-2.8 ms
76, 77	BK/WT	Power Ground	<0.1v
78	YL/WT	Coil 3 Driver (Cyl 2 & 6)	6° dwell
79	WT/LG	Check Transaxle Light	Indicator Off: 12v, On: 1v
80	LB/OR	Fuel Pump Control	Off: 12v, On: 1v
81	BR/OR	EPC Solenoid Control	10.6v (40 psi)
82	BK/WT	Check Cap Indicator Light	Indicator Off: 12v, On: 1v
83	WT/LB	IAC Motor Control	10v (32% duty cycle)
84	DB/YL	OSS Sensor Signal	0 Hz, 55 mph: 131 Hz
85	DB/OR	CID Sensor Signal	6-8 Hz
86	BK/YL	A/C Pressure Clutch Relay	Relay Off: 12v, On: 1v
87	RD/LG	HO2S-21 (B2 S1) Signal	0.1-1.1v
88	LB/RD	MAF Sensor Signal	0.9v
89	GY/WT	TP Sensor Signal	0.53-1.27v
90	BR/WT	Reference Voltage	4.9-5.1v
91	GY/RD	Sensor Return	<0.050v
92	RD/LG	Brake Pedal Position Switch	Brake Off: 0v, On: 12v
93	RD/WT	HO2S-11 (B1 S1) Heater	On: 1v, Off: 12v
94	WT/BK	HO2S-21 (B2 S1) Heater	On: 1v, Off: 12v
95	YL/LB	HO2S-12 (B1 S2) Heater	On: 1v, Off: 12v
96	TN/YL	HO2S-22 (B2 S2) Heater	On: 1v, Off: 12v
97	RD	Vehicle Power (Start-Run)	12-14v
98	LG/RD	Generator Battery Indicator	Digital Signals
99	LG/OR	Injector 6 Control	2.3-2.8 ms
100	BR/LB	Injector 4 Control	2.3-2.8 ms
101	WT	Injector 2 Control	2.3-2.8 ms
102	---	Not Used	---
103	BK/WT	Power Ground	<0.1v
104	---	Not Used	---

Pin Connector Graphic

Standard Colors and Abbreviations

Abbreviation	Color	Abbreviation	Color	Abbreviation	Color
BK	Black	GY	Gray	PK	Purple
BL	Blue	GN	Green	RD	Red
BR	Brown	LG	LT Green	TN	Tan
DB	Dark Blue	OR	Orange	WT	White
DG	DK Green	PK	Pink	YL	Yellow

1990-93 Taurus 3.0L V6 MFI SHO VIN Y (M/T) 60 Pin Connector

PCM Pin #	Wire Color	Circuit Description (60 Pin)	Value at Hot Idle
1	YL	Keep Alive Power	12-14v
2	YL/LG	PSP Switch Signal	Straight: 0v, Turned: 11v
3	DG/WT	VSS (+) Signal	0 Hz, 55 mph: 125 Hz
4	GY/OR	Ignition Diagnostic Monitor	20-31 Hz
5	RD/LG	Brake Pedal Position Switch	Brake Off: 0v, On: 12v
6	OR/YL	VSS (-) Signal	0 Hz, 55 mph: 125 Hz
7	LG/YL	ECT Sensor Signal	0.5-0.6v
7	LG/RD	ECT Sensor Signal	0.5-0.6v
8	---	Not Used	---
9	TN/BL	MAF Sensor Return	<0.050v
10	PK/BL	A/C Cycling Clutch Switch	A/C On: 12v, Off: 0v
11	OR/YL	Speed Control Solenoid (+)	S/C On: 12-14v
12	BR/YL	Injector 3 Control	3.6-4.1 ms
13	BR/BL	Injector 4 Control	3.6-4.1 ms
14	TN/BL	Injector 5 Control	3.6-4.1 ms
14	TN/BK	Injector 5 Control	3.6-4.1 ms
15	LG	Injector 6 Control	3.6-4.1 ms
15	LG/OR	Injector 6 Control	3.6-4.1 ms
16	BK/OR	Ignition System Ground	<0.050v
16	OR/RD	Ignition System Ground	<0.050v
17	TN/RD	STI Output, MIL Control	MIL On: 1v, Off: 12v
18	PK	Octane Adjust Switch	Closed: 0v, Open: 9.1v
19	PK/BK	Fuel Pump Monitor	On: 12v, Off: 0v
20	BK	PCM Case Ground	<0.050v
21	OR/BK	IAC Motor Control	8-9v
22	TN/LG	Low Fuel Pump Control	On: 1v, Off: 12v
23	Y, YL/RD	Knock Sensor Signal	0v
24	DG	CID Sensor Signal	5-7 Hz
25	GY	ACT Sensor Signal	1.5-2.5v
25	LG/PK	ACT Sensor Signal	1.5-2.5v
26	BR/WT	Reference Voltage	4.9-5.1v
26	OR/WT	Reference Voltage	4.9-5.1v
27	BR/LG	PFE EGR Sensor	3.2v, 55 mph: 2.8v
28	BL/BK	S/C Command Switch Signal	S/C On: 6.7v
29	RD/BK	HO2S-21 (B2 S1) Signal	0.1-1.1v
29	DG/PK	HO2S-21 (B2 S1) Signal	0.1-1.1v
30	PK/YL	M/T: Clutch Engage Switch	Clutch In: 0v, Out: 5v
31	GY/YL	EVAP Purge Solenoid	12v, 55 mph: 1v
32	LG/PK	IAC Motor Control	8-10v
33	DG	EGR VR Solenoid	0%, 55 mph: 45%
33	BR/PK	EGR VR Solenoid	0%, 55 mph: 45%
34	---	Not Used	---
35	WT/PK	Speed Control Vent Solenoid	0%, 55 mph: 45%
36	YL/LG	Spark Output Signal	50% duty cycle
37	RD	Vehicle Power	12-14v
38	---	Not Used	---
39	OR	Speed Control Switch Ground	<0.050v
40	BK/LG	Power Ground	<0.1v

1990-93 Taurus 3.0L V6 MFI SHO VIN Y (M/T) 60 Pin Connector

PCM Pin #	Wire Color	Circuit Description (60 Pin)	Value at Hot Idle
41	BL/OR	High Fuel Pump Control	On: 1v, Off: 12v
42	---	Not Used	---
43	DB/LG	HO2S-11 (B1 S1) Signal	0.1-1.1v
44	---	Not Used	---
45	LG/BK	BARO Sensor Signal	159 Hz (sea level)
46	BK/WT	Analog Signal Return	<0.050v
46	GY/RD	Analog Signal Return	<0.050v
47	DG/LG	TP Sensor Signal	0.5-1.2v
47	GY/WT	TP Sensor Signal	0.5-1.2v
48	BR	STI Signal	STI Open: 5v, Closed: 1v
48	WT/BK	STI Signal	STI Open: 5v, Closed: 1v
49	OR	HO2S Ground	<0.050v
50 ('90-'91)	BL/RD	MAF Sensor Signal	0.8v, 55 mph: 1.3v
50 ('92-'93)	DB/OR	MAF Sensor Signal	0.8v, 55 mph: 1.3v
51	GY/BK	S/C Vacuum Solenoid Control	0%, 55 mph: 45%
52-53	---	Not Used	---
54	PK/YL	A/C WOT Relay Control	On: 1v, Off: 12v
54	RD	A/C WOT Relay Control	On: 1v, Off: 12v
55	TN/OR	Low Speed Fan Control	On: 1v, Off: 12v
56	DB	PIP Sensor Signal	50% duty cycle
57	RD	Vehicle Power	12-14v
58	TN	Injector 1 Control	3.6-4.1 ms
59	WT	Injector 2 Control	3.6-4.1 ms
60	BK/LG	Power Ground	<0.1v

Pin Connector Graphic

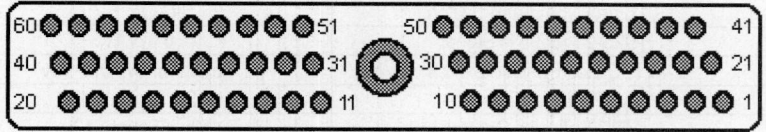

PCM 60-PIN CONNECTOR

Terminal View of 60-Pin PCM Harness Connector

Standard Colors and Abbreviations

Abbreviation	Color	Abbreviation	Color	Abbreviation	Color
BK	Black	GY	Gray	PK	Purple
BL	Blue	GN	Green	RD	Red
BR	Brown	LG	LT Green	TN	Tan
DB	Dark Blue	OR	Orange	WT	White
DG	DK Green	PK	Pink	YL	Yellow

TAURUS PIN TABLES

1994-95 Taurus 3.0L V6 MFI SHO VIN Y (M/T) 60 Pin Connector

PCM Pin #	Wire Color	Circuit Description (60 Pin)	Value at Hot Idle
1	YL	Keep Alive Power	12-14v
2	YL/LG	PSP Switch Signal	Straight: 0v, Turned: 11v
3	DG/WT	VSS (+) Signal	0 Hz, 55 mph: 125 Hz
4	TN/YL	Ignition Diagnostic Monitor	20-31 Hz
5	RD/LG	Brake Pedal Position Switch	Brake Off: 0v, On: 12v
6	OR/YL	VSS (-) Signal	0 Hz, 55 mph: 125 Hz
7	LG/RD	ECT Sensor Signal	0.5-0.6v
8	---	Not Used	---
9	TN/LB	MAF Sensor Return	<0.050v
10	PK/BL	A/C Cycling Clutch Switch	A/C On: 12v, Off: 0v
11	OR/YL	Speed Control Solenoid (+)	S/C On: 12-14v
12	BR/YL	Injector 3 Control	3.6-4.1 ms
13	BR/BL	Injector 4 Control	3.6-4.1 ms
14	TN/BK	Injector 5 Control	3.6-4.1 ms
15	LG/OR	Injector 6 Control	3.6-4.1 ms
16	OR/RD	Ignition System Ground	<0.050v
17	TN/RD	STI Output, MIL Control	MIL On: 1v, Off: 12v
18	PK	Octane Adjust Switch	Closed: 0v, Open: 9.1v
19	PK/BK	Fuel Pump Monitor	On: 12v, Off: 0v
20	BK	PCM Case Ground	<0.050v
21	OR/BK	IAC Motor Control	8-10v
22	TN/LG	Low Fuel Pump	On: 1v, Off: 12v
23	YL/RD	Knock Sensor Signal	0v
24	DG	CID Sensor Signal	5-7 Hz
25	GY	IAT Sensor Signal	1.5-2.5v
26	BR/WT	Reference Voltage	4.9-5.1v
27	BR/LG	PFE EGR Sensor	3.3v, 55 mph: 2.8v
28	BL/BK	S/Control Command Switch	S/C On: 6.7v
29	RD/BK	HO2S-21 (B2 S1) Signal	0.1-1.1v
30	PK/YL	M/T: Clutch Position Switch	Clutch In: 0v, Out: 11v
31	GY/YL	EVAP Purge Solenoid	12v, 55 mph: 1v
32	LG/PK	IMRC Solenoid Control	12v, 55 mph: 12v
33	BR/PK	EGR VR Solenoid	0%, 55 mph: 45%
34	---	Not Used	---
35	WT/PK	Speed Control Vent Solenoid	0%, 55 mph: 45%
36	YL/LG	Spark Output Signal	50% duty cycle
37	RD	Vehicle Power	12-14v
38	---	Not Used	---
39	OR	Speed Control Switch Ground	<0.050v
40	BK/LG	Power Ground	<0.1v

1994-95 Taurus 3.0L V6 MFI SHO VIN Y (M/T) 60 Pin Connector

PCM Pin #	Wire Color	Circuit Description (60 Pin)	Value at Hot Idle
41	BL/OR	High Fuel Pump Control	On: 1v, Off: 12v
41	TN/LG	High Fuel Pump Control	On: 1v, Off: 12v
42	---	Not Used	---
43	DB/LG	HO2S-11 (B1 S1) Signal	0.1-1.1v
44	---	Not Used	---
45	LG/BK	BARO Sensor Signal	159 Hz (sea level)
46	GY/RD	Analog Signal Return	<0.050v
47	GY/WT	TP Sensor Signal	0.5-2.1v
48	BR	Self-Test Indicator Signal	STI Open: 5v, Closed: 1v
49	OR	HO2S Ground	<0.050v
50	BL/RD	MAF Sensor Signal	0.6-0.9v
51	GY/BK	S/C Vacuum Solenoid Control	0%, 55 mph: 45%
52-53	---	Not Used	---
54	PK/YL	A/C WOT Relay Control	On: 1v, Off: 12v
55	TN/OR	Low Speed Fan Control	On: 1v, Off: 12v
56	DB	PIP Sensor Signal	50% duty cycle
57	RD	Vehicle Power	12-14v
58	TN	Injector 1 Control	3.6-4.1 ms
59	WT	Injector 2 Control	3.6-4.1 ms
60	BK/LG	Power Ground	<0.1v

Pin Connector Graphic

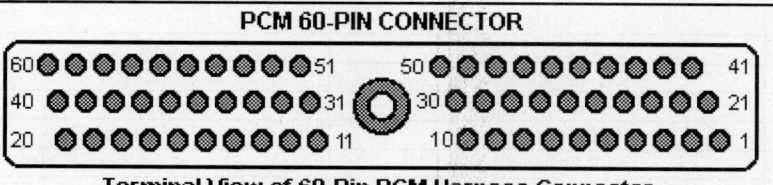

PCM 60-PIN CONNECTOR

Terminal View of 60-Pin PCM Harness Connector

Standard Colors and Abbreviations

Abbreviation	Color	Abbreviation	Color	Abbreviation	Color
BK	Black	GY	Gray	PK	Purple
BL	Blue	GN	Green	RD	Red
BR	Brown	LG	LT Green	TN	Tan
DB	Dark Blue	OR	Orange	WT	White
DG	DK Green	PK	Pink	YL	Yellow

1993-95 Taurus 3.0L V6 FFV VIN 1 (A/T) 60 Pin Connector

PCM Pin #	Wire Color	Circuit Description (60 Pin)	Value at Hot Idle
1	YL	Keep Alive Power	12-14v
2	RD/LG	Brake Pedal Position Switch	Brake Off: 0v, On: 12v
3	DG/WT	VSS (+) Signal	0 Hz, 55 mph: 125 Hz
4	WT/PK	Ignition Diagnostic Monitor	20-31 Hz
5	GY/BK	TSS Sensor Signal	100 Hz (700 rpm)
6	OR/YL	VSS (-) Signal	0 Hz, 55 mph: 125 Hz
7	LG/RD	ECT Sensor Signal	0.5-0.6v
8	PK/BK	Fuel Pump Monitor	On: 12v, Off: 0v
9	TN/BL	MAF Sensor Return	<0.050v
10	PK/BL	A/C Cycling Clutch Switch	A/C On: 12v, Off: 0v
11	GN/YL	EVAP Purge Solenoid	12v, 55 mph: 1v
12	LG/OR	Injector 6 Control	3.8-6.1 ms
13	TN/OR	Low Cooling Fan	On: 1v, Off: 12v
14	BK/LG	Cold Start Injector (CSI)	KOEC only: 5 ms
15	TN/BK	Injector 5 Control	3.8-6.1 ms
16	OR/RD	Ignition System Ground	<0.050v
17	TN/RD	STI Output, MIL Control	MIL On: 1v, Off: 12v
18	TN/OR	Data Bus (+) Signal	Digital Signals
19	PK/BL	Data Bus (-) Signal	Digital Signals
20	BK	PCM Case Ground	<0.050v
21	BR/WT	IAC Motor Control	7-9v
22	BL/OR	Fuel Pump Control	On: 1v, Off: 12v
23	---	Not Used	---
24	DB/OR	CID Sensor Signal	5-7 Hz
25	GY	IAT Sensor Signal	1.5-2.5v
26	BR/WT	Reference Voltage	4.9-5.1v
27	BR/LG	PFE EGR Sensor	3.3v, 55 mph: 2.9v
28	YL/LG	PSP Switch Signal	Straight: 0v, Turned: 11v
29	---	Not Used	---
30	BL/YL	MLP Sensor Signal	In 'P': 0v, 55 mph: 1.7v
31	LG/PK	High Cooling Fan	On: 1v, Off: 12v
32	OR/LG	Low Fuel Pump Control	On: 1v, Off: 12v
33	BR/PK	EGR VR Solenoid	0%, 55 mph: 45%
34	BL/BK	Data Output Solenoid (DOL)	Digital Signals
34	BL/PK	Data Output Solenoid (DOL)	Digital Signals
35	BR/BL	Injector 4 Control	3.8-6.1 ms
36	PK	Spark Output Signal	6.93v (duty cycle)
37	RD	Vehicle Power	12-14v
38	WT/YL	EPC Solenoid Control	8-9v (9 psi)
39	BR/YL	Injector 3 Control	3.8-6.1 ms
40	BK/LG	Power Ground	<0.1v

1993-95 Taurus 3.0L V6 FFV VIN 1 (A/T) 60 Pin Connector

PCM Pin #	Wire Color	Circuit Description (60 Pin)	Value at Hot Idle
41-42	---	Not Used	---
43	RD/BK	HO2S-11 (B1 S1) Signal	0.1-1.1v
44	GY/BL	HO2S-21 (B2 S1) Signal	0.1-1.1v
45	DG/LG	Flexible Fuel Sensor Signal	30-70 Hz (100% gasoline)
46	GY/RD	Analog Signal Return	<0.050v
47	GN/WT	TP Sensor Signal	0.5-1.2v
48	BR	Self-Test Indicator Signal	STI Open: 5v, Closed: 1v
49	OR/BK	TOT Sensor Signal	2.10-2.40v
50	BL/RD	MAF Sensor Signal	0.6-0.9v
51	OR/YL	Shift Solenoid 1 Control	12v, 55 mph: 1v
52	PK/OR	Shift Solenoid 2 Control	1v, 55 mph: 12v
53	TN/WT	TCC Solenoid Control	0%, 55 mph: 45%
54	PK/YL	A/C WOT Relay Control	On: 1v, Off: 12v
55	WT/RD	Shift Solenoid 3 Control	7-9v, 55 mph: 8-9v
56	GY/OR	PIP Sensor Signal	50% duty cycle
57	RD	Vehicle Power	12-14v
58	TN	Injector 1 Control	3.8-6.1 ms
59	WT	Injector 2 Control	3.8-6.1 ms
60	BK/LG	Power Ground	<0.1v

Pin Connector Graphic

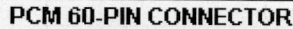
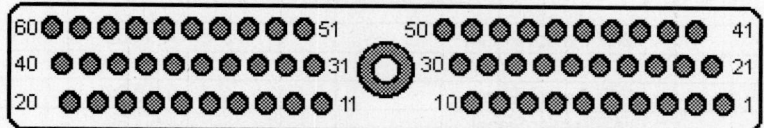

PCM 60-PIN CONNECTOR

60 ... 51   50 ... 41
40 ... 31   30 ... 21
20 ... 11   10 ... 1

Terminal View of 60-Pin PCM Harness Connector

Standard Colors and Abbreviations

Abbreviation	Color	Abbreviation	Color	Abbreviation	Color
BK	Black	GY	Gray	PK	Purple
BL	Blue	GN	Green	RD	Red
BR	Brown	LG	LT Green	TN	Tan
DB	Dark Blue	OR	Orange	WT	White
DG	DK Green	PK	Pink	YL	Yellow

1996-97 Taurus 3.0L V6 FFV VIN 1, VIN 2 (A/T) 104 Pin Connector

PCM Pin #	Wire Color	Circuit Description (104 Pin)	Value at Hot Idle
1	PK/OR	Shift Solenoid 2 Control	1v, 55 mph: 12v
2	PK/LG	MIL (lamp) Control	MIL On: 1v, Off: 12v
3-12	---	Not Used	---
13	WT/OR	Flash EEPROM Power	0.1v
14	---	Not Used	---
15	PK	Data Bus (-) Signal	Digital Signals
16	PK/LB	Data Bus (+) Signal	Digital Signals
17-20	---	Not Used	---
21	TN/OR	CKP Sensor (+) Signal	510-540 Hz
22	GY	CKP Sensor (-) Signal	510-540 Hz
23	---	Not Used	---
24	DB	Power Ground	<0.1v
25	BK/WT	Power Ground	<0.1v
26	BK	Coil Driver 1 Control	6° dwell
27	YL/BK	Shift Solenoid 1 Control	12v, 55 mph: 1v
28	OR/YL	Low Speed Fan Control	On: 1v, Off: 12v
29	---	Not Used	---
30	TN/OR	Octane Adjust Switch	Closed: 0v, Open: 9.3v
31	DG	PSP Switch Signal	Straight: 0v, Turned: 12v
32	---	Not Used	---
33	YL/LG	VSS (-) Signal	0 Hz, 55 mph: 125 Hz
34	PK/OR	Flexible Fuel Sensor Signal	30-70 Hz (100% gasoline)
35	---	Not Used	---
36	RD/BK	MAF Sensor Return	<0.050v
37	TN/LB	TFT Sensor Signal	2.10-2.40v
38	OR/BK	ECT Sensor Signal	0.5-0.6v
39	LG/RD	IAT Sensor Signal	1.5-2.5v
40	GY	Fuel Pump Monitor	On: 12v, Off: 0v
41	DG/YL	A/C High Pressure Switch	AC On: 12v, Off: 0v
42	---	Not Used	---
43	OR/LG	Data Output Link	Digital Signals
44-45	---	Not Used	---
46	LG/PK	High Speed Fan Control	On: 1v, Off: 12v
47	BR/PK	EGR VR Solenoid	0%, 55 mph: 45%
48	TN/YL	Clean Tachometer Output	42-50 Hz
49-50	---	Not Used	---
51	BK/WT	Power Ground	<0.1v
52	YL/RD	Coil Driver 2 Control	6° dwell
53	PK/BK	Shift Solenoid 3 Control	12v, 55 mph: 1v
54	---	Not Used	---
55	PK	Keep Alive Power	12-14v
56	GY/YL	EVAP VMV Solenoid	0-10 Hz (0-100%)
57	---	Not Used	---
58	GY/BK	VSS (+) Signal	0 Hz, 55 mph: 125 Hz
59	---	Not Used	---
60	GY/LB	HO2S-11 (B1 S1) Signal	0.1-1.1v
61	---	Not Used	---
62	RD/PK	FTP Sensor Signal	2.6v (0" H2O - cap off)
63	---	Not Used	---
64	LB/YL	Digital TR Sensor	In 'P': 0v, in O/D: 1.7v

1996-97 Taurus 3.0L V6 FFV VIN 1, VIN 2 (A/T) 104 Pin Connector

PCM Pin #	Wire Color	Circuit Description (104 Pin)	Value at Hot Idle
65	BR/LG	DPFE Sensor Signal	0.20-1.30v
66	---	Not Used	---
67	PK/WT	EVAP Purge Solenoid	0-10 Hz (0-100%)
68	---	Not Used	---
69	PK/YL	A/C WOT Relay Control	On: 1v, Off: 12v
70, 72	---	Not Used	---
71	RD	Vehicle Power	12-14v
73	TN/BK	Injector 5 Control	2.3-2.8 ms
74	BR/YL	Injector 3 Control	2.3-2.8 ms
75	TN	Injector 1 Control	2.3-2.8 ms
76	BK/WT	Power Ground	<0.1v
77	BK/WT	Power Ground	<0.1v
78	YL/WT	Coil Driver 3 Control	6° dwell
79	---	Not Used	---
80	LB/OR	Fuel Pump Control	On: 1v, Off: 12v
81	WT/YL	EPC Solenoid Control	10.6v (40 psi)
82	PK/YL	TCC Solenoid Control	0%, 55 mph: 95%
83	WT/LB	IAC Motor Control	10v (32% duty cycle)
84	DG/WT	TSS Sensor Signal	50-65 Hz (790-820 rpm)
85	DB/OR	CID Sensor Signal	6-8 Hz
86	BK/YL	A/C Pressure Switch	A/C On: 12v (Open)
87	RD/LG	HO2S-21 (B2 S1) Signal	0.1-1.1v
88	LB/RD	MAF Sensor Signal	0.9v
89	GY/WT	TP Sensor Signal	0.53-1.27v
90	BR/WT	Reference Voltage	4.9-5.1v
91	GY/RD	Analog Signal Return	<0.050v
92	RD/LG	Brake Pedal Position Switch	Brake Off: 0v, On: 12v
93	RD/WT	HO2S-11 (B1 S1) Heater	On: 1v, Off: 12v
94	WT/BK	HO2S-21 (B2 S1) Heater	On: 1v, Off: 12v
95, 96, 98	---	Not Used	---
97	RD	Vehicle Power	12-14v
99	LG/OR	Injector 6 Control	2.3-2.8 ms
100	BR/LB	Injector 4 Control	2.3-2.8 ms
101	WT	Injector 2 Control	2.3-2.8 ms
102	---	Not Used	---
103	BK/WT	Power Ground	<0.1v
104	---	Not Used	---

Pin Connector Graphic

PCM 104-PIN CONNECTOR

Terminal View of 104-Pin PCM Wiring Harness Connector

Standard Colors and Abbreviations

Abbreviation	Color	Abbreviation	Color	Abbreviation	Color
BK	Black	GY	Gray	PK	Purple
BL	Blue	GN	Green	RD	Red
BR	Brown	LG	LT Green	TN	Tan
DB	Dark Blue	OR	Orange	WT	White
DG	DK Green	PK	Pink	YL	Yellow

1998-99 Taurus 3.0L V6 FFV VIN 1, VIN 2 (A/T) 104 Pin Connector

PCM Pin #	Wire Color	Circuit Description (104 Pin)	Value at Hot Idle
1	PK/OR	Shift Solenoid 'B' Control	1v, 55 mph: 12v
2	PK/LG	MIL (lamp) Control	MIL On: 1v, Off: 12v
3	YL/BK	Digital TR1 Sensor	In 'P': 0v, 55 mph: 11v
5-12	---	Not Used	---
13	PK	Flash EEPROM Power	0.1v
15	PK/LB	Data Bus (-) Signal	Digital Signals
16	TN/OR	Data Bus (+) Signal	Digital Signals
17-18	---	Not Used	---
21	GY	CKP Sensor (+) Signal	510-540 Hz
22	DB	CKP Sensor (-) Signal	510-540 Hz
23	---	Not Used	---
24	BK/WT	Power Ground	<0.1v
25	BK	Power Ground	<0.1v
26	YL/BK	Coil Driver 1 Control	5° dwell
27	OR/YL	Shift Solenoid 'A' Control	12v, 55 mph: 0.1v
28	LB	Low Speed Fan Control	On: 1v, Off: 12v
29	---	Not Used	---
30	TN/OR	Octane Adjust Switch	Closed: 0v, Open: 9.3v
31	YL/LG	PSP Switch Signal	Straight: 0v, Turned: 12v
32	---	Not Used	---
33	PK/OR	VSS (-) Signal	0 Hz, 55 mph: 125 Hz
34	DG/LG	Flexible Fuel Sensor Signal	40-60 Hz (100% gasoline)
35	---	Not Used	---
36	TN/LB	MAF Sensor Return	<0.050v
37	OR/BK	TFT Sensor Signal	2.10-2.40v
38	LG/RD	ECT Sensor Signal	0.5-0.6v
39	GY	IAT Sensor Signal	1.5-2.5v
40	DG/YL	Fuel Pump Monitor	On: 12v, Off: 0v
41	DG/WT	A/C High Pressure Switch	AC On: 12v, Off: 0v
42	---	Not Used	---
43	OR/LG	Data Output Link	Digital Signals
44-45	---	Not Used	---
46	LG/PK	High Speed Fan Control	On: 1v, Off: 12v
47	BR/PK	EGR VR Solenoid	0%, 55 mph: 45%
48	TN/YL	Clean Tachometer Output	42-50 Hz
49	LB/BK	Digital TR2 Sensor	In 'P': 0v, 55 mph: 11v
50	WT/BK	Digital TR4 Sensor	In 'P': 0v, 55 mph: 11v
51	BK/WT	Power Ground	<0.1v
52	YL/RD	Coil Driver 2 Control	5° dwell
53	PK/BK	Shift Solenoid 'C' Control	12v, 55 mph: 0.1v
54	PK/YL	TCC Solenoid Control	0%, 55 mph: 95%
55	DG/OR	Keep Alive Power	12-14v
56	GY/YL	EVAP Purge Solenoid	0-10 Hz (0-100%)
57	---	Not Used	---
58	GY/BK	VSS (+) Signal	0 Hz, 55 mph: 125 Hz
59	---	Not Used	---
60	GY/LB	HO2S-11 (B1 S1) Signal	0.1-1.1v
61	---	Not Used	---
62	RD/PK	FTP Sensor Signal	2.6v (0" H2O - cap off)
63	---	Not Used	---
64	LB/YL	Digital TR Sensor	In 'P': 0v, in O/D: 1.7v

1998-99 Taurus 3.0L V6 FFV VIN 1, VIN 2 (A/T) 104 Pin Connector

PCM Pin #	Wire Color	Circuit Description (104 Pin)	Value at Hot Idle
65	BR/LG	DPFE Sensor Signal	0.20-1.30v
66	---	Not Used	---
67	PK/WT	EVAP Purge Solenoid	0-10 Hz (0-100%)
68	---	Not Used	---
69	PK/YL	A/C WOT Relay Control	On: 1v, Off: 12v
70	---	Not Used	
71	RD	Vehicle Power	12-14v
73	TN/BK	Injector 5 Control	2.3-2.8 ms
74	BR/YL	Injector 3 Control	2.3-2.8 ms
75	TN	Injector 1 Control	2.3-2.8 ms
76	BK/WT	Power Ground	<0.1v
77	BK/WT	Power Ground	<0.1v
78	YL/WT	Coil Driver 3 Control	5° dwell
80	LB/OR	Fuel Pump Control	On: 1v, Off: 12v
81	WT/YL	EPC Solenoid Control	11v (38-42 psi)
82	PK/YL	TCC Solenoid Control	12v (0%)
83	WT/LB	IAC Motor Control	10v (32% duty cycle)
84	DG/WT	TSS Sensor Signal	50-65 Hz (790-850 rpm)
85	DB/OR	CID Sensor Signal	5-7 Hz
86	BK/YL	A/C Pressure Switch	A/C On: 12v (Open)
87	RD/LG	HO2S-21 (B2 S1) Signal	0.1-1.1v
88	LB/RD	MAF Sensor Signal	0.9v
89	GY/WT	TP Sensor Signal	0.53-1.27v
90	BR/WT	Reference Voltage	4.9-5.1v
91	GY/RD	Analog Signal Return	<0.050v
92	RD/LG	Brake Pedal Position Switch	Brake Off: 0v, On: 12v
93	RD/WT	HO2S-11 (B1 S1) Heater	On: 1v, Off: 12v
94	WT/BK	HO2S-21 (B2 S1) Heater	On: 1v, Off: 12v
95-96	YL/LB	HO2S-12 (B1 S2) Heater	On: 1v, Off: 12v
97	RD	Vehicle Power	12-14v
99	LG/OR	Injector 6 Control	2.3-2.8 ms
100	BR/LB	Injector 4 Control	2.3-2.8 ms
101	WT	Injector 2 Control	2.3-2.8 ms
102	---	Not Used	---
103	BK/WT	Power Ground	<0.1v
104	---	Not Used	---

Pin Connector Graphic

Standard Colors and Abbreviations

Abbreviation	Color	Abbreviation	Color	Abbreviation	Color
BK	Black	GY	Gray	PK	Purple
BL	Blue	GN	Green	RD	Red
BR	Brown	LG	LT Green	TN	Tan
DB	Dark Blue	OR	Orange	WT	White
DG	DK Green	PK	Pink	YL	Yellow

2000 Taurus 3.0L V6 Flexible Fuel Vehicle VIN 2 (A/T) 104 Pin Connector

PCM Pin #	Wire Color	Circuit Description (104 Pin)	Value at Hot Idle
1	VT/OR	Shift Solenoid 'B' Control	1v, 55 mph: 12v
2	PK/LG	MIL (lamp) Control	MIL On: 1v, Off: 12v
3	YL/BK	Digital TR1 Sensor	In 'P': 0v, 55 mph: 11v
4	BR/PK	Start System Voltage	Cranking: 9-11v
5	---	Not Used	---
6	DG/WT	TSS Sensor Signal	50-65 Hz (790-820 rpm)
7-11	---	Not Used	---
12	YL/WT	Fuel Level Indicator Signal	1.7v (1/2 full)
13	VT	Flash EEPROM Power	0.1v
14	---	Not Used	---
15	PK/LB	SCP Bus (-) Signal	Digital Signals
16	TN/OR	SCP Bus (+) Signal	Digital Signals
17	BR/YL	Passive Antitheft RX Signal	Digital Signals
18	GY/RD	Passive Antitheft TX Signal	Digital Signals
19-20	---	Not Used	---
21	GY	CKP Sensor (+) Signal	510-540 Hz
22	DB	CKP Sensor (-) Signal	510-540 Hz
23	---	Not Used	---
24	BK/WT	Power Ground	<0.1v
25	BK	Chassis Ground	<0.1v
26	YL/BK	Coil 'A' Driver (Cyl 1 & 5)	5° dwell
27	OR/YL	Shift Solenoid 'A' Control	12v, 55 mph: 0.1v
28	LB	Low Speed Fan Control	On: 1v, Off: 12v
29, 33-34	---	Not Used	---
30	DB/LG	Antitheft Indicator Control	Indicator Off: 12v, On: 1v
31	YL/LG	PSP Switch Signal	Straight: 0v, Turned: 12v
32	YT	Knock Sensor 1 Ground	<0.050v
35	RD/BK	HO2S-12 (B1 S2) Signal	0.1-1.1v
36	TN/LB	MAF Sensor Return	<0.050v
37	OR/BK	TFT Sensor Signal	2.10-2.40v
38	LG/RD	ECT Sensor Signal	0.5-0.6v
39	GY	IAT Sensor Signal	1.5-2.5v
40	PK/BK	Fuel Pump Monitor	On: 12v, Off: 0v
41	DG/WT	A/C High Pressure Switch	AC On: 12v, Off: 0v
42-43, 45	---	Not Used	---
44	PK	Starter Relay Control	KOEC: 9-11v
46	LG/VT	High Speed Fan Control	On: 1v, Off: 12v
47	BR/PK	EGR VR Solenoid	0%, 55 mph: 45%
48	TN/YL	Clean Tachometer Output	42-50 Hz
49	LB/BK	Digital TR2 Sensor	In 'P': 0v, 55 mph: 11v
50	WT/BK	Digital TR4 Sensor	In 'P': 0v, 55 mph: 11v
51	BK/WT	Power Ground	<0.1v
52	YL/RD	Coil 'B' Driver (Cyl 3 & 4)	5° dwell
53	PK/BK	Shift Solenoid 'C' Control	12v, 55 mph: 0.1v
54	VT/YL	TCC Solenoid Control	0%, 55 mph: 95%
55	PK/LB	Keep Alive Power	12-14v
56	GY/YL	EVAP Purge Solenoid	0-10 Hz (0-100%)
57	YL/RD	Knock Sensor 1 (+) Signal	0v
58-59	---	Not Used	---
60	GY/LB	HO2S-11 (B1 S1) Signal	0.1-1.1v
61	VT/LG	HO2S-22 (B2 S2) Signal	0.1-1.1v
62	RD/PK	FTP Sensor Signal	2.6v (0" H2O - cap off)
63	WT/YL	Injector Pressure Sensor	2.7-3.7v (105-130 psi)
64	LB/YL	Digital TR3A Sensor	In 'P': 0v, in O/D: 1.7v

2000 Taurus 3.0L V6 Flexible Fuel Vehicle VIN 1, VIN 2 (A/T) 104 Pin Connector

PCM Pin #	Wire Color	Circuit Description (104 Pin)	Value at Hot Idle
65	BR/LG	DPFE Sensor Signal	0.20-1.30v
66	---	Not Used	---
67	PK/WT	EVAP CV Solenoid	0-10 Hz (0-100%)
68	GY/BK	VSS (+) Signal	0 Hz, 55 mph: 125 Hz
69	PK/YL	A/C WOT Relay Control	On: 1v, Off: 12v
71	RD	Vehicle Power	12-14v
72	---	Not Used	---
73	TN/BK	Injector 5 Control	2.3-2.8 ms
74	BR/YL	Injector 3 Control	2.3-2.8 ms
75	TN	Injector 1 Control	2.3-2.8 ms
76	BK/WT	Power Ground	<0.1v
77	BK/WT	Power Ground	<0.1v
78	YL/WT	Coil Driver 3 (Cyl 3 & 6)	5° dwell
80	LB/OR	Fuel Pump Control	On: 1v, Off: 12v
81	BR/OR	EPC Solenoid Control	11v (38-42 psi)
82	BK/WT	Check Cap Indicator Control	On: 1v, Off: 12v
83	WT/LB	IAC Motor Control	10v (32% duty cycle)
84	DB/YL	OSS Sensor Signal	0 Hz, 55 mph: 131 Hz
85	DB/OR	CID Sensor Signal	5-7 Hz
86	BK/YL	A/C Pressure Switch	A/C On: 12v (Open)
87	RD/LG	HO2S-21 (B2 S1) Signal	0.1-1.1v
88	LB/RD	MAF Sensor Signal	0.9v
89	GY/WT	TP Sensor Signal	0.53-1.27v
90	BR/WT	Reference Voltage	4.9-5.1v
91	GY/RD	Analog Signal Return	<0.050v
92	RD/LG	Brake Pedal Position Switch	Brake Off: 0v, On: 12v
93	RD/WT	HO2S-11 (B1 S1) Heater	On: 1v, Off: 12v
94	WT/BK	HO2S-21 (B2 S1) Heater	On: 1v, Off: 12v
95	YL/LB	HO2S-12 (B1 S2) Heater	On: 1v, Off: 12v
96	TN/YL	HO2S-22 (B2 S2) Heater	On: 1v, Off: 12v
97-98	---	Not Used	---
99	LG/OR	Injector 6 Control	2.3-2.8 ms
100	BR/LB	Injector 4 Control	2.3-2.8 ms
101	WT	Injector 2 Control	2.3-2.8 ms
102	---	Not Used	---
103	BK/WT	Power Ground	<0.1v
104	---	Not Used	---

Pin Connector Graphic

PCM 104-PIN CONNECTOR

Terminal View of 104-Pin PCM Wiring Harness Connector

Standard Colors and Abbreviations

Abbreviation	Color	Abbreviation	Color	Abbreviation	Color
BK	Black	GY	Gray	PK	Purple
BL	Blue	GN	Green	RD	Red
BR	Brown	LG	LT Green	TN	Tan
DB	Dark Blue	OR	Orange	WT	White
DG	DK Green	PK	Pink	YL	Yellow

2001-03 Taurus 3.0L V6 Flexible Fuel Vehicle VIN 2 (A/T) 104 Pin Connector

PCM Pin #	Wire Color	Circuit Description (104 Pin)	Value at Hot Idle
1	VT/OR	Shift Solenoid 'B' Control	1v, 55 mph: 12v
2	PK/LG	MIL (lamp) Control	MIL On: 1v, Off: 12v
3	YL/BK	Digital TR1 Sensor	In 'P': 0v, 55 mph: 11v
4	TN/RD	System Voltage (Start-Run)	Cranking: 9-11v
5	---	Not Used	---
6	DG/WT	TSS Sensor Signal	50-65 Hz (790-820 rpm)
7-10	---	Not Used	---
12	YL/WT	Fuel Level Indicator Signal	1.7v (1/2 full)
13	VT	Flash EEPROM Power	0.1v
14, 19, 23	---	Not Used	---
15	PK/LB	SCP Bus (-) Signal	Digital Signals
16	TN/OR	SCP Bus (+) Signal	Digital Signals
17	BR/YL	Passive Antitheft RX Signal	Digital Signals
18	GY/RD	Passive Antitheft TX Signal	Digital Signals
20	GY/YL	Generator Field Control	130 Hz (45%)
21	GY	CKP Sensor (+) Signal	510-540 Hz
22	DB	CKP Sensor (-) Signal	510-540 Hz
24	BK/WT	Power Ground	<0.1v
25	BK	Case Ground	<0.1v
26	YL/BK	Coil 1 Driver (Cyl 1& 5)	6° dwell
27	OR/YL	Shift Solenoid 'A' Control	12v, 55 mph: 1v
28	LB	Engine Cooling Fan Brake Relay Control	Off: 12v, On: 1v
29, 33-34	---	Not Used	---
30	DB/LG	Passive Antitheft Indicator	Lamp On: 1v, Off: 12v
31	YL/LG	PSP Switch Signal	Straight: 0v, Turned: 12v
32	YL	Knock Sensor 1 (-) Signal	<0.050v
35	RD/BK	HO2S-12 (B1 S2) Signal	0.1-1.1v
36	TN/LB	MAF Sensor Return	<0.050v
37	OR/BK	TFT Sensor Signal	2.10-2.40v
38	LG/RD	ECT Sensor Signal	0.5-0.6v
39	GY	IAT Sensor Signal	1.5-2.5v
40	DG/YL	Fuel Pump Monitor	On: 12v, Off: 0v
41	DG/WT	A/C High Pressure Switch	AC On: 12v, Off: 0v
42	LG/YL	Generator Cutoff Relay	0.1v (with fan on)
44	PK	Starter Relay Control	KOEC: 9-11v
45	LB/RD	Generator Sensor Signal	130 Hz (30%)
46	LG/VT	High Cooling Speed Fan Control	Off: 12v, On: 1v
47	BR/PK	EGR VR Solenoid Control	0%, 55 mph: 45%
48	TN/YL	Clean Tachometer Output	42-50 Hz
49	LB/BK	Digital TR2 Sensor	In 'P': 0v, 55 mph: 11v
50	WT/BK	Digital TR4 Sensor	In 'P': 0v, 55 mph: 11v
51	BK/WT	Power Ground	<0.1v
52	YL/RD	Coil Driver 2 (Cyl 3 & 4)	6° dwell
53	PK/BK	Shift Solenoid 'C' Control	12v, 55 mph: 1v
54	VT/YL	TCC Solenoid Control	12v (0%)
55	PK/LB	Keep Alive Power	12-14v
56	GY/YL	EVAP Purge Solenoid	0-10 Hz (0-100%)
57	YL/RD	Knock Sensor (+) Signal	0v
58-59	---	Not Used	---
60	GY/LB	HO2S-11 (B1 S1) Signal	0.1-1.1v
61	VT/LG	HO2S-22 (B2 S2) Signal	0.1-1.1v
62	RD/PK	FTP Sensor Signal	2.6v (0" H2O - cap off)
63	WT/YL	Injector Pressure Sensor	2.7-3.7v (105-130 psi)
64	RD/BK	Digital TR3A Sensor	In 'P': 0v, in O/D: 1.7v

2001-03 Taurus 3.0L V6 Flexible Fuel Vehicle VIN 2 (A/T) 104 Pin Connector

PCM Pin #	Wire Color	Circuit Description (104 Pin)	Value at Hot Idle
65	BR/LG	DPFE Sensor Signal	0.95-1.05v
66	TN/OR	Evaporator Air Temperature Sensor	4.95-5.15v
67	VT/WT	EVAP Purge Solenoid	0-10 Hz (0-100%)
68	GY/BK	VSS (+) Signal	Moving: AC pulse signals
69	PK/YL	A/C WOT Relay Control	Off: 12v, On: 1v
70	---	Not Used	---
71	RD	Vehicle Power (Start-Run)	12-14v
73	TN/BK	Injector 5 Control	2.3-2.8 ms
74	BR/YL	Injector 3 Control	2.3-2.8 ms
75	TN	Injector 1 Control	2.3-2.8 ms
76, 77	BK/WT	Power Ground	<0.1v
78	YL/WT	Coil 3 Driver (Cyl 2 & 6)	6° dwell
79	WT/LG	Check Transaxle Light	Indicator Off: 12v, On: 1v
80	LB/OR	Fuel Pump Control	Off: 12v, On: 1v
81	BR/OR	EPC Solenoid Control	10.6v (40 psi)
82	BK/WT	Check Cap Indicator Light	Indicator Off: 12v, On: 1v
83	WT/LB	IAC Motor Control	10v (32% duty cycle)
84	DB/YL	OSS Sensor Signal	0 Hz, 55 mph: 131 Hz
85	DB/OR	CID Sensor Signal	6-8 Hz
86	BK/YL	A/C Pressure Clutch Relay	Relay Off: 12v, On: 1v
87	RD/LG	HO2S-21 (B2 S1) Signal	0.1-1.1v
88	LB/RD	MAF Sensor Signal	0.9v
89	GY/WT	TP Sensor Signal	0.53-1.27v
90	BR/WT	Reference Voltage	4.9-5.1v
91	GY/RD	Sensor Return	<0.050v
92	RD/LG	Brake Pedal Position Switch	Brake Off: 0v, On: 12v
93	RD/WT	HO2S-11 (B1 S1) Heater	On: 1v, Off: 12v
94	WT/BK	HO2S-21 (B2 S1) Heater	On: 1v, Off: 12v
95	YL/LB	HO2S-12 (B1 S2) Heater	On: 1v, Off: 12v
96	TN/YL	HO2S-22 (B2 S2) Heater	On: 1v, Off: 12v
97	RD	Vehicle Power (Start-Run)	12-14v
98	LG/RD	Generator Battery Indicator	Digital Signals
99	LG/OR	Injector 6 Control	2.3-2.8 ms
100	BR/LB	Injector 4 Control	2.3-2.8 ms
101	WT	Injector 2 Control	2.3-2.8 ms
102	---	Not Used	---
103	BK/WT	Power Ground	<0.1v
104	---	Not Used	---

Pin Connector Graphic

PCM 104-PIN CONNECTOR

Terminal View of 104-Pin PCM Wiring Harness Connector

Standard Colors and Abbreviations

Abbreviation	Color	Abbreviation	Color	Abbreviation	Color
BK	Black	GY	Gray	PK	Purple
BL	Blue	GN	Green	RD	Red
BR	Brown	LG	LT Green	TN	Tan
DB	Dark Blue	OR	Orange	WT	White
DG	DK Green	PK	Pink	YL	Yellow

1993-95 Taurus 3.2L V6 MFI SHO VIN P (A/T) 60 Pin Connector

PCM Pin #	Wire Color	Circuit Description (60 Pin)	Value at Hot Idle
1	Y	Keep Alive Power	12-14v
2	RD/LG	Brake Pedal Position Switch	Brake Off: 0v, On: 12v
3	DG/WT	VSS (+) Signal	65 Hz at 55 mph
4	GY/OR	Ignition Diagnostic Monitor	20-31 Hz
5	GY/BK	TSS Sensor Signal	100 Hz (700 mph)
6	OR/YL	VSS (+) Signal	0 Hz, 55 mph: 125 Hz
7	LG/RD	ECT Sensor Signal	0.5-0.6v
8	PK/BK	Fuel Pump Monitor	On: 12v, Off: 0v
9	TN/BL	MAF Sensor Return	<0.050v
10	PK/BL	A/C Cycling Clutch Switch	A/C On: 12v, Off: 0v
11	GN/YL	EVAP Purge Solenoid	12v, 55 mph: 1v
12	LG/OR	Injector 6 Control	3.6-3.8 ms
13	TN/OR	Low Cooling Fan	On: 1v, Off: 12v
14	OR/YL	TCIL (lamp) Control	On: 1v, Off: 12v
14	WT/LG	TCIL (lamp) Control	On: 1v, Off: 12v
15	TN/BK	Injector 5 Control	3.6-3.8 ms
16	OR/RD	Ignition System Ground	<0.050v
17	TN/RD	STI Output, MIL Control	MIL On: 1v, Off: 12v
18	OR/BK	Data Bus (+) Signal	Digital Signals
19	BK/OR	Data Bus (-) Signal	Digital Signals
20	BK	PCM Case Ground	<0.050v
21	OR/BK	IAC Motor Control	8-9v
22	TN/LG	Fuel Pump Control	On: 1v, Off: 12v
23	YL/RD	Knock Sensor Signal	0v
24	DG	CID Sensor Signal	5-7 Hz
25	GY	IAT Sensor Signal	1.5-2.5v
26	BR/WT	Reference Voltage	4.9-5.1v
27	BR/LG	PFE EGR Sensor	0.6v
28	YL/LG	PSP Switch Signal	Straight: 0v, Turned: 11v
29	PK	Octane Adjust Switch	Closed: 0v, Open: 9.1v
30	BL/YL	MLP Sensor Signal	In 'P': 4.4v, in 'D': 2.0v
31	LG/PK	High Cooling Fan	On: 1v, Off: 12v
32	---	Not Used	---
33	BR/PK	EGR VR Solenoid	0%, 55 mph: 45%
34	LG/PK	IMRC Solenoid	12v, 55 mph: 12v
35	BR/BL	Injector 4 Control	3.6-3.8 ms
36	YL/LG	Spark Output Signal	50% duty cycle
37	RD	Vehicle Power	12-14v
38	WT/YL	EPC Solenoid Control	8.8v (9 psi)
39	BR/YL	Injector 3 Control	3.6-3.8 ms
40	BK/LG	Power Ground	<0.1v

1993-95 Taurus 3.2L V6 MFI SHO VIN P (A/T) 60 Pin Connector

PCM Pin #	Wire Color	Circuit Description (60 Pin)	Value at Hot Idle
41	WT/LG	TCS (switch) Signal	TCS & O/D On: 12v
41	TN/WT	TCS (switch) Signal	TCS & O/D On: 12v
42 ('94-'95)	DG/OR	Refrigerant Contain Switch	5v
43	DB/LG	HO2S-11 (B1 S1) Signal	0.1-1.1v
44	RD/BK	HO2S-21 (B2 S1) Signal	0.1-1.1v
45	---	Not Used	---
46	GY/RD	Analog Signal Return	<0.050v
47	GN/WT	TP Sensor Signal	0.5-2.1v
48	BR	Self-Test Indicator Signal	STI Open: 5v, Closed: 1v
49	OR/BK	TOT Sensor Signal	2.10-2.40v
50	BL/RD	MAF Sensor Signal	0.7v
51	OR/YL	Shift Solenoid 1 Control	12v, 55 mph: 1v
52	PK/OR	Shift Solenoid 2 Control	1v, 55 mph: 12v
53	TN/WT	TCC Solenoid Control	0%, 55 mph: 40%
53	PK/YL	TCC Solenoid Control	0%, 55 mph: 40%
54	PK/YL	A/C WOT Relay Control	On: 1v, Off: 12v
55	PK/BK	Shift Solenoid 3 Control	12v, 55 mph: 1v
55	WT/RD	Shift Solenoid 3 Control	12v, 55 mph: 1v
56	DB	PIP Sensor Signal	50% duty cycle
57	RD	Vehicle Power	12-14v
58	TN	Injector 1 Control	3.6-3.8 ms
59	WT	Injector 2 Control	3.6-3.8 ms
60	BK/LG	Power Ground	<0.1v

Pin Connector Graphic

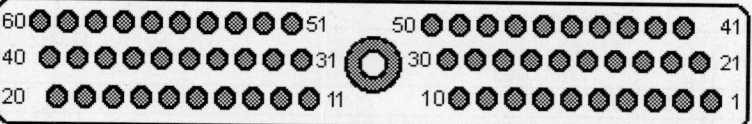

PCM 60-PIN CONNECTOR

Terminal View of 60-Pin PCM Harness Connector

Standard Colors and Abbreviations

Abbreviation	Color	Abbreviation	Color	Abbreviation	Color
BK	Black	GY	Gray	PK	Purple
BL	Blue	GN	Green	RD	Red
BR	Brown	LG	LT Green	TN	Tan
DB	Dark Blue	OR	Orange	WT	White
DG	DK Green	PK	Pink	YL	Yellow

1996-97 Taurus SHO 3.4L 4v V6 VIN N (A/T) 104 Pin Connector

PCM Pin #	Wire Color	Circuit Description (104 Pin)	Value at Hot Idle
1	OR/YL	COP 5 Driver Control	6° dwell
2	PK/LG	MIL (lamp) Control	MIL On: 1v, Off: 12v
3-4	---	Not Used	---
5	WT/OR	Electronic AIR Control	12v (Off)
6	OR/YL	Shift Solenoid 1 Control	12v, 55 mph: 1v
7	---	Not Used	---
8	DB/YL	IMRC Monitor	5v
9-10	---	Not Used	---
11	PK/OR	Shift Solenoid 2 Control	1v, 55 mph: 12v
12	WT/LG	TCIL (lamp) Control	On: 1v, Off: 12v
13	PK	Flash EEPROM Power	0.1v
14	---	Not Used	---
15	PK/LG	Data Bus (-) Signal	Digital Signals
16	TN/OR	Data Bus (+) Signal	Digital Signals
17-19	---	Not Used	---
20	PK/BK	Shift Solenoid 3 Control	1v, 55 mph: 12v
21	GY	CKP Sensor (+) Signal	380-420 Hz
22	DB	CKP Sensor (-) Signal	380-420 Hz
23, 24	BK/WT	Power Ground	<0.1v
25	BK	Power Ground	<0.1v
26	LG/WT	COP Driver 1	6° dwell
27	WT/PK	COP Driver 6	6° dwell
28	TN/OR	Low Speed Fan Control	On: 1v, Off: 12v
29	TN/WT	Overdrive Cancel Switch	OCS & OD On: 0.1v
30	DG	Octane Adjust Switch	Closed: 0v, Open: 9.3v
31	YL/LG	PSP Switch Signal	Straight: 0v, Turned: 12v
32	DG/PK	Knock Sensor 2 Signal	0v
33	PK/OR	VSS (-) Signal	0 Hz, 55 mph: 125 Hz
34, 43-45	---	Not Used	---
35	RD/BK	HO2S-12 (B1 S2) Signal	0.1-1.1v
36	TN/LB	MAF Sensor Return	<0.1v
37	OR/BK	TFT Sensor Signal	2.10-2.40v
38	LG/RD	ECT Sensor Signal	0.5-0.6v
39	GY	IAT Sensor Signal	1.5-2.5v
40	DG/YL	Fuel Pump Monitor	On: 12v, Off: 0v
41	PK/LB	A/C Cycling Clutch Switch	A/C On: 12v, Off: 0v
42	BR	IMRC Solenoid Control	12v (Off)
46	LG/PK	High Speed Fan Control	On: 1v, Off: 12v
47	BR/PK	EGR VR Solenoid	0%, 55 mph: 45%
48	TN/YL	Clean Tachometer Output	33-37 Hz
49-50, 54	---	Not Used	---
51	BK/WT	Power Ground	<0.1v
52	LG/YL	COP Driver 2	6° dwell
53	PK/LB	COP Driver 7	6° dwell
55	DG/OR	Keep Alive Power	12-14v
56	GY/YL	EVAP VMV Solenoid	0-10 Hz (0-100%)
57	YL/RD	Knock Sensor 1 Signal	0v
58	GY/BK	VSS (+) Signal	0 Hz, 55 mph: 125 Hz
60	GY/LB	HO2S-11 (B1 S1) Signal	0.1-1.1v
61	PK/LG	HO2S-22 (B2 S2) Signal	0.1-1.1v
62	RD/PK	FTP Sensor Signal	2.6v (0 in. H20)
63	---	Not Used	---
64	LB/YL	TR Sensor	In 'P': 0v, 55 mph: 1.7v

1996-97 Taurus SHO 3.4L 4v V6 VIN N (A/T) 104 Pin Connector

PCM Pin #	Wire Color	Circuit Description (104 Pin)	Value at Hot Idle
65	BR/LG	DPFE Sensor Signal	0.20-1.30v
66, 68	---	Not Used	---
67	PK/WT	EVAP Purge Solenoid	0-10 Hz (0-100%)
69	PK/YL	A/C WOT Relay Control	On: 1v, Off: 12v
70	BR	Electronic AIRB Solenoid	12v, 55 mph: 12v
71	RD	Vehicle Power	12-14v
72	TN/RD	Injector 7 Control	2.3-2-8 ms
73	TN/BK	Injector 5 Control	2.3-2-8 ms
74	BR/YL	Injector 3 Control	2.3-2-8 ms
75	TN	Injector 1 Control	2.3-2-8 ms
76, 77	BK/WT	Power Ground	<0.1v
78	DG/PK	COP Driver 3	6° dwell
79	WT/RD	COP Driver 8	6° dwell
80	LB/OR	Fuel Pump Control	On: 1v, Off: 12v
81	WT/YL	EPC Solenoid Control	9.0v (15 psi)
82	PK/YL	TCC Solenoid Control	12v (0 psi)
83	WT/LB	IAC Motor Control	10v (30% on time)
84	DG/WT	TSS Sensor Signal	43 Hz (650 rpm)
85	DB/OR	CID Sensor Signal	5-7 Hz
86	BK/YL	A/C Pressure Switch	A/C On: 12v (Open)
87	RD/LG	HO2S-21 (B2 S1) Signal	0.1-1.1v
88	LB/RD	MAF Sensor Signal	0.6v
89	GY/WT	TP Sensor Signal	0.53-1.27v
90	BR/WT	Reference Voltage	4.9-5.1v
91	GY/RD	Analog Signal Return	<0.050v
92	RD/LG	Brake Pedal Switch Signal	Brake Off: 0v, On: 12v
93	RD/WT	HO2S-11 (B1 S1) Heater	On: 1v, Off: 12v
94	WT/BK	HO2S-21 (B2 S1) Heater	On: 1v, Off: 12v
95	YL/LB	HO2S-12 (B1 S2) Heater	On: 1v, Off: 12v
96	TN/YL	HO2S-22 (B2 S2) Heater	On: 1v, Off: 12v
97	RD	Vehicle Power	12-14v
98	LB	Injector 8 Control	2.3-2-8 ms
99	LG/OR	Injector 6 Control	2.3-2-8 ms
100	BR/LB	Injector 4 Control	2.3-2-8 ms
101	WT	Injector 2 Control	2.3-2-8 ms
102	---	Not Used	---
103	BK/WT	Power Ground	<0.1v
104	PK/WT	Coil Driver 4 Control	6° dwell

Pin Connector Graphic

PCM 104-PIN CONNECTOR

Terminal View of 104-Pin PCM Wiring Harness Connector

Standard Colors and Abbreviations

Abbreviation	Color	Abbreviation	Color	Abbreviation	Color
BK	Black	GY	Gray	PK	Purple
BL	Blue	GN	Green	RD	Red
BR	Brown	LG	LT Green	TN	Tan
DB	Dark Blue	OR	Orange	WT	White
DG	DK Green	PK	Pink	YL	Yellow

1998-99 Taurus SHO 3.4L 4v V6 VIN N (A/T) 104 Pin Connector

PCM Pin #	Wire Color	Circuit Description (104 Pin)	Value at Hot Idle
1	OR/YL	COP 5 Driver Control	6° dwell
2	PK/LG	MIL (lamp) Control	MIL On: 1v, Off: 12v
3-4, 7	---	Not Used	---
5	WT/OR	Electronic AIR Monitor	1v, 55 mph: 1v
6	OR/YL	Shift Solenoid 1 Control	12v, 55 mph: 0.1v
8	DB/YL	IMRC Monitor	5v
9	YL/WT	Fuel Level Indicator Signal	1.7v (1/2 full)
10	---	Not Used	---
11	PK/OR	Shift Solenoid 2 Control	1v, 55 mph: 1v
12	WT/LG	TCIL (lamp) Control	On: 1v, Off: 12v
13	PK	Flash EEPROM Power	0.1v
14	---	Not Used	---
15	PK	Data Bus (-) Signal	Digital Signals
16	TN/OR	Data Bus (+) Signal	Digital Signals
17-19	---	Not Used	---
20	PK/BK	Shift Solenoid 3 Control	12v, 55 mph: 0.1v
21	GY	CKP Sensor (+) Signal	380-420 Hz
22	DB	CKP Sensor (-) Signal	380-420 Hz
23, 24, 51	BK/WT	Power Ground	<0.1v
25	BK	Power Ground	<0.1v
26	LG/WT	COP 1 Driver Control	6° dwell
27	WT/PK	COP 6 Driver Control	6° dwell
28	TN/OR	Low Speed Fan Control	On: 1v, Off: 12v
29	TN/WT	Overdrive Cancel Switch	OCS & O/D On: 0.1v
30	DG	Octane Adjust Switch	Closed: 0v, Open: 9.3v
31	YL/LG	PSP Switch Signal	Straight: 0v, Turned: 12v
32	DG/PK	Knock Sensor 2 Signal	0v
33	PK/OR	VSS (-) Signal	0 Hz, 55 mph: 125 Hz
35	RD/BK	HO2S-12 (B1 S2) Signal	0.1-1.1v
36	TN/LB	MAF Sensor Return	<0.050v
37	OR/BK	TFT Sensor Signal	2.10-2.40v
38	LG/RD	ECT Sensor Signal	0.5-0.6v
39	GY	IAT Sensor Signal	1.5-2.5v
40	DG/YL	Fuel Pump Monitor	On: 12v, Off: 0v
41	PK/LB	A/C Cycling Clutch Switch	AC On: 12v, Off: 0v
42	BR	IMRC Solenoid Control	12v (Off)
43, 49-50	---	Not Used	---
45	PK/WT	EVAP CV Solenoid	0-10 Hz (0-100 Hz)
46	LG/PK	High Speed Fan Control	On: 1v, Off: 12v
47	BR/PK	EGR VR Solenoid	0%, 55 mph: 45%
48	TN/YL	Clean Tachometer Output	33-37 Hz
52	LG/YL	COP 2 Driver Control	6° dwell
53	PK/LB	COP 7 Driver Control	6° dwell
54	PK/YL	TCC Solenoid Control	0%, 55 mph: 95%
55	DG/OR	Keep Alive Power	12-14v
56	GY/YL	EVAP Purge Solenoid	0-10 Hz (0-100%)
57	YL/RD	Knock Sensor Signal	0v
58	GY/BK	VSS (+) Signal	0 Hz, 55 mph: 125 Hz
59	DG/WT	TSS Sensor Signal	30-32 Hz (550-600 rpm)
60	GY/LB	HO2S-11 (B1 S1) Signal	0.1-1.1v
61	PK/LG	HO2S-22 (B2 S2) Signal	0.1-1.1v
62	RD/PK	FTP Sensor Signal	2.6v (0 in. H20)
63	---	Not Used	---
64	LB/YL	TR Sensor Signal	In 'P': 0v, 55 mph: 1.7v

1998-99 Taurus SHO 3.4L 4v V6 VIN N (A/T) 104 Pin Connector

PCM Pin #	Wire Color	Circuit Description (104 Pin)	Value at Hot Idle
65	BR/LG	DPFE Sensor Signal	0.20-1.30v
66, 68	---	Not Used	---
67	PK/WT	EVAP Purge Solenoid	0-10 Hz (0-100%)
69	PK/YL	A/C WOT Relay Control	On: 1v, Off: 12v
70	BR	Electronic AIRB Solenoid	12v, 55 mph: 12v
71	RD	Vehicle Power	12-14v
72	TN/RD	Injector 7 Control	2.3-3.2 ms
73	TN/BK	Injector 5 Control	2.3-3.2 ms
74	BR/YL	Injector 3 Control	2.3-3.2 ms
75	TN	Injector 1 Control	2.3-3.2 ms
77	BK/WT	Power Ground	<0.1v
78	DG/PK	COP 3 Driver Control	6° dwell
79	WT/RD	COP 8 Driver Control	6° dwell
80	LG/OR	Fuel Pump Control	On: 1v, Off: 12v
81	WT/YL	EPC Solenoid Control	9.3v (15 psi)
82	PK/YL	TCC Solenoid Control	12v (0 psi)
83	WT/LB	IAC Motor Control	10v (30% on time)
84	DG/WT	TSS Sensor Signal	43 Hz (650 rpm)
85	DB/OR	CID Sensor Signal	5-7 Hz
86	BK/YL	A/C High Pressure Switch	12v (Switch Open)
87	RD/LG	HO2S-21 (B2 S1) Signal	0.1-1.1v
88	LB/RD	MAF Sensor Signal	0.6v
89	GY/WT	TP Sensor Signal	0.53-1.27v
90	BR/WT	Reference Voltage	4.9-5.1v
91	GY/RD	Analog Signal Return	<0.050v
92	RD/LG	Brake Pedal Switch Signal	Brake Off: 0v, On: 12v
93	RD/WT	HO2S-11 (B1 S1) Heater	On: 1v, Off: 12v
94	WT/BK	HO2S-21 (B2 S1) Heater	On: 1v, Off: 12v
95	YL/LB	HO2S-12 (B1 S2) Heater	On: 1v, Off: 12v
96	TN/YL	HO2S-22 (B2 S2) Heater	On: 1v, Off: 12v
97	RD	Vehicle Power	12-14v
98	LB	Injector 8 Control	2.3-3.2 ms
99	LG/OR	Injector 6 Control	2.3-3.2 ms
100	BR/LB	Injector 4 Control	2.3-3.2 ms
101	WT	Injector 2 Control	2.3-3.2 ms
102	---	Not Used	---
103	BK/WT	Power Ground	<0.1v
104	PK/WT	COP 4 Driver Control	6° dwell

Pin Connector Graphic

PCM 104-PIN CONNECTOR

Terminal View of 104-Pin PCM Wiring Harness Connector

Standard Colors and Abbreviations

Abbreviation	Color	Abbreviation	Color	Abbreviation	Color
BK	Black	GY	Gray	PK	Purple
BL	Blue	GN	Green	RD	Red
BR	Brown	LG	LT Green	TN	Tan
DB	Dark Blue	OR	Orange	WT	White
DG	DK Green	PK	Pink	YL	Yellow

1990 Taurus 3.8L V6 MFI VIN 4 (A/T) 60 Pin Connector

PCM Pin #	Wire Color	Circuit Description (60 Pin)	Value at Hot Idle
1	YL	Keep Alive Power	12-14v
2	RD/LG	PSP Switch Signal	Straight: 0v, Turned: 11v
3	DG/WT	VSS (+) Signal	0 Hz, 55 mph: 125 Hz
4	DG/YL	Ignition Diagnostic Monitor	20-31 Hz
5	---	Not Used	---
6	OR/YL	VSS (-) Signal	0 Hz, 55 mph: 125 Hz
7	LG/YL	ECT Sensor Signal	0.5-0.6v
8	BK/BL	Data Bus (-) Signal	Digital Signals
8	BK/OR	Data Bus (-) Signal	Digital Signals
9	GY/WT	AXOD Transmission Solenoid	On: 1v, Off: 12v
10	PK/BL	A/C Cycling Clutch Switch	A/C On: 12v, Off: 0v
10	PK	A/C Cycling Clutch Switch	A/C On: 12v, Off: 0v
11	OR/YL	Speed Control Solenoid (+)	S/C On: 12-14v
12	BR/YL	Injector 3 Control	6.0-6.8 ms
13	BR/BL	Injector 4 Control	6.0-6.8 ms
14	TN/BL	Injector 5 Control	6.0-6.8 ms
15	LG	Injector 6 Control	6.0-6.8 ms
16	GY	Ignition System Ground	<0.050v
17	TN/RD	STI Output, MIL Control	MIL On: 1v, Off: 12v
18	DG/PK	Transmission 3-4 Switch	3-4 Switch Closed: 0.1v
19	OR/YL	Transmission 2-3 Switch	2-3 Switch Closed: 0.1v
20	---	Not Used	---
21	OR/BK	IAC Motor Control	9-10v
22	TN/LG	Fuel Pump Control	On: 1v, Off: 12v
23-24	---	Not Used	---
25	LG/PK	ACT Sensor Signal	1.5-2.5v
26	OR/WT	Reference Voltage	4.9-5.1v
27	BR/LG	PFE EGR Sensor	3.2v, 55 mph: 2.8v
28	BL/BK	Speed Control Switch	Switch On: 6.7v
29	DB/LG	HO2S-21 (B2 S1) Signal	0.1-1.1v
30	PK/YL	Neutral Position Switch	In 'N': 0v, Others: 4.6v
31	GY/YL	EVAP Purge Solenoid	12v, 55 mph: 1v
32	---	Not Used	---
33	DG	EGR VR Solenoid	0%, 55 mph: 45%
34	BL/PK	Data Output Link	Digital Signals
35	WT/PK	Speed Control Vent Solenoid	98% duty cycle (55 mph)
36	YL/LG	Spark Output Signal	6.93v (duty cycle)
37	RD	Vehicle Power	12-14v
38	---	Not Used	---
39	OR	Speed Control Switch Ground	<0.050v
40	BK/LG	Power Ground	<0.1v

1990 Taurus 3.8L V6 MFI VIN 4 (A/T) 60 Pin Connector

PCM Pin #	Wire Color	Circuit Description (60 Pin)	Value at Hot Idle
41	PK	High Speed Fan Control	On: 1v, Off: 12v
42	---	Not Used	---
43	DG/PK	HO2S-11 (B1 S1) Signal	0.1-1.1v
44	TN/OR	Data Bus (+) Signal	Digital Signals
44	OR/BK	Data Bus (+) Signal	Digital Signals
45	LG/BK	MAP Sensor Signal	107 Hz
46	BK/WT	Analog Signal Return	<0.050v
47	DG/LG	TP Sensor Signal	0.5-1.2v
48	WT/BK	Self-Test Indicator Signal	STI Open: 5v, Closed: 1v
49	OR	HO2S-11 (Bank 1) Ground	<0.050v
50	PK/BK	Fuel Pump Monitor	On: 1v, Off: 12v
51	GY/BK	S/C Vacuum Solenoid	0%, 55 mph: 45%
52	---	Not Used	---
53	TN/BL	Lockup Solenoid for AXOD	12v, 55 mph: 1v
54	RD	A/C WOT Relay Control	On: 1v, Off: 12v
55	TN/OR	Low Speed Fan Control	On: 1v, Off: 12v
56	DB	PIP Sensor Signal	50% duty cycle
57	RD	Vehicle Power	12-14v
58	TN	Injector 1 Control	6.0-6.8 ms
59	WT	Injector 2 Control	6.0-6.8 ms
60	BK/LG	Power Ground	<0.1v

Pin Connector Graphic

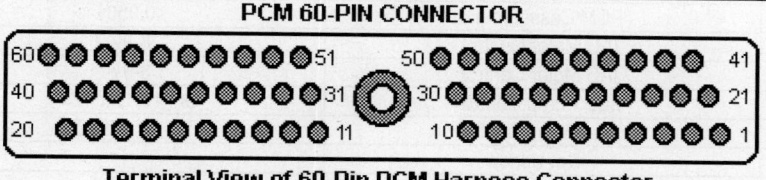

PCM 60-PIN CONNECTOR

Terminal View of 60-Pin PCM Harness Connector

Standard Colors and Abbreviations

Abbreviation	Color	Abbreviation	Color	Abbreviation	Color
BK	Black	GY	Gray	PK	Purple
BL	Blue	GN	Green	RD	Red
BR	Brown	LG	LT Green	TN	Tan
DB	Dark Blue	OR	Orange	WT	White
DG	DK Green	PK	Pink	YL	Yellow

1991-95 Taurus 3.8L V6 MFI VIN 4 (A/T) 60 Pin Connector

PCM Pin #	Wire Color	Circuit Description (60 Pin)	Value at Hot Idle
1	YL	Keep Alive Power	12-14v
2	RD/LG	Brake Pedal Position Switch	Brake Off: 0v, On: 12v
3	DG/WT	VSS (-) Signal	0 Hz, 55 mph: 125 Hz
4	WT/PK	Ignition Diagnostic Monitor	20-31 Hz
4	DG/YL	Ignition Diagnostic Monitor	20-31 Hz
5	GY/BK	TSS Sensor Signal	100 Hz (700 rpm)
6	OR/YL	VSS (+) Signal	0 Hz, 55 mph: 125 Hz
7	LG/RD	ECT Sensor Signal	0.5-0.6v
8	PK/BK	Fuel Pump Monitor	On: 12v, Off: 0v
9	TN/BL	MAF Sensor Return	<0.050v
10	PK	A/C Cycling Clutch Signal	A/C On: 12v, Off: 0v
10	PK/BL	A/C Cycling Clutch Signal	A/C On: 12v, Off: 0v
11	GN/YL	EVAP Purge Solenoid	12v, 55 mph: 1v
12	LG/OR	Injector 6 Control	5.6-6.0 ms
13	TN/OR	Low Cooling Fan	On: 1v, Off: 12v
14	---	Not Used	---
15	TN/BK	Injector 5 Control	5.6-6.0 ms
16	GY	Ignition System Ground	<0.050v
17	TN/RD	STI Output, MIL Control	MIL On: 1v, Off: 12v
18	TN/OR	Data Bus (+) Signal	Digital Signals
18	OR/BK	Data Bus (+) Signal	Digital Signals
19	PK/BL	Data Bus (-) Signal	Digital Signals
19	BK/OR	Data Bus (-) Signal	Digital Signals
20	BK	PCM Case Ground	<0.050v
21	BL/OR	IAC Motor Control	8-9v
21	BR/WT	IAC Motor Control	8-9v
22	TN/LG	Fuel Pump Control	On: 1v, Off: 12v
22	BL/OR	Fuel Pump Control	On: 1v, Off: 12v
23-24	---	Not Used	---
25	GY	IAT Sensor Signal	1.5-2.5v
26	BR/WT	Reference Voltage	4.9-5.1v
27	BR/LG	PFE EGR Sensor	3.2v, 55 mph: 2.8v
28	YL/LG	PSP Sensor Signal	Straight: 0v, Turned: 11v
29	---	Not Used	---
30	BL	MLP Sensor Signal	In 'P': 0v, 55 mph: 1.7v
30	BL/YL	MLP Sensor Signal	In 'P': 0v, 55 mph: 1.7v
31	LG/PK	High Cooling Fan	On: 1v, Off: 12v
32	---	Not Used	---
33	BR/PK	EGR VR Solenoid	0%, 55 mph: 45%
34	BL/PK	Data Output Link	Digital Signals
35	BR/BL	Injector 4 Control	5.6-6.0 ms
36	YL/LG	Spark Output Signal	50% duty cycle
37	RD	Vehicle Power	12-14v
38	WT/YL	EPC Solenoid Control	7.8v (8-9 psi)
39	BR/YL	Injector 3 Control	5.6-6.0 ms
40	BK/LG	Power Ground	<0.1v

1991-95 Taurus 3.8L V6 MFI VIN 4 (A/T) 60 Pin Connector

PCM Pin #	Wire Color	Circuit Description (60 Pin)	Value at Hot Idle
41	---	Not Used	---
42 ('94-'95)	TN/LG	Refrigerant Contain Switch	5v
43	RD/BK	HO2S-11 (B1 S1) Signal	0.1-1.1v
44	DB/LG	HO2S-21 (B2 S1) Signal	0.1-1.1v
44	GY/BL	HO2S-21 (B2 S1) Signal	0.1-1.1v
45	---	Not Used	---
46	GY/RD	Analog Signal Return	<0.050v
47	GN/WT	TP Sensor Signal	0.5-1.2v
48	BR	Self-Test Indicator Signal	STI Open: 5v, Closed: 1v
49	OR/BK	TOT Sensor Signal	2.10-2.40v
50	BL/RD	MAF Sensor Signal	0.7v
51	OR/YL	Shift Solenoid 1 Control	12v, 55 mph: 1v
52	PK/OR	Shift Solenoid 2 Control	1v, 55 mph: 12v
53	TN/WT	TCC Solenoid Control	12v, 55 mph: 1v
54 ('91-'93)	PK/YL	A/C WOT Relay Control	On: 1v, Off: 12v
54 ('94-'95)	RD	A/C WOT Relay Control	On: 1v, Off: 12v
55	WT/RD	Shift Solenoid 3 Control	12v, 55 mph: 1v
56	DB	PIP Sensor Signal	50% duty cycle
56	GY/OR	PIP Sensor Signal	50% duty cycle
57	RD	Vehicle Power	12-14v
58	TN	Injector 1 Control	5.6-6.0 ms
59	WT	Injector 2 Control	5.6-6.0 ms
60	BK/LG	Power Ground	<0.1v

Pin Connector Graphic

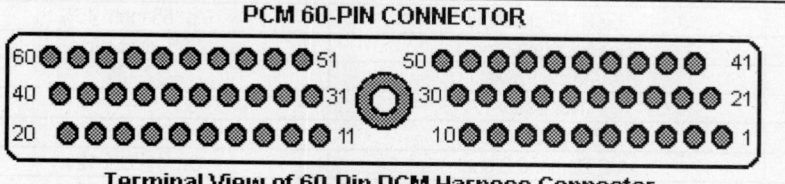

Standard Colors and Abbreviations

Abbreviation	Color	Abbreviation	Color	Abbreviation	Color
BK	Black	GY	Gray	PK	Purple
BL	Blue	GN	Green	RD	Red
BR	Brown	LG	LT Green	TN	Tan
DB	Dark Blue	OR	Orange	WT	White
DG	DK Green	PK	Pink	YL	Yellow

Tempo Pin Tables

1990 Tempo 2.3L I4 MFI VIN S, X (All) 60 Pin Connector

PCM Pin #	Wire Color	Circuit Description (60 Pin)	Value at Hot Idle
1	YL	Keep Alive Power	12-14v
2, 9	---	Not Used	---
3	DG/YL	VSS (+) Signal	0 Hz, 55 mph: 125 Hz
4	TN/YL	Ignition Diagnostic Monitor	20-31 Hz
5	RD/LG	Vehicle Power	12-14v
6	OR/YL	VSS (-) Signal	0 Hz, 55 mph: 125 Hz
7	LG/YL	ECT Sensor Signal	0.5-0.6v
8	PK/BK	Fuel Pump Monitor	On: 12v, Off: 0v
10	BK/YL	A/C Cycling Clutch Switch	A/C On: 12v, Off: 0v
11-12, 14-15	---	Not Used	---
13	TN/OR	Low Speed Fan Control	On: 1v, Off: 12v
16	BK	Ignition System Ground	<0.050v
17	TN/RD	STI Output, MIL Control	MIL On: 1v, Off: 12v
20	BK	PCM Case Ground	<0.050v
21	BR/WT	IAC Motor Control	8-10v
22	DG	Fuel Pump Control	On: 1v, Off: 12v
24	YL/LG	PSP Switch Signal	Straight 0v, Turned 11v
25	LG/PK	ACT Sensor Signal	1.5-2.5v
26	BR/WT	Reference Voltage	4.9-5.1v
26	OR/WT	Reference Voltage	4.9-5.1v
27	BR/LG	PFE EGR Sensor	3.2v, 55 mph: 2.8v
28, 32, 34-35, 41	---	Not Used	---
29	DG/PK	HO2S-11 (B1 S1) Signal	0.1-1.1v
30	WT/PK	Neutral Drive Switch Signal	In 'N': 0v, Others: 12v
31	GY/YL	EVAP Purge Solenoid	12v, at 55 mph: 1v
33	YL	EGR VR Solenoid	0%, 55 mph: 45%
36	YL/LG	Spark Output Signal	50% duty cycle
37, 57	RD	Vehicle Power	12-14v
38-39, 41-42	---	Not Used	---
40	BK/LG	Power Ground	<0.1v
43	PK	A/C Demand Signal	On: 1v, Off: 12v
43	LG/PK	A/C Demand Signal	On: 1v, Off: 12v
45	LG/BK	MAP Sensor Signal	107 Hz
46	BK/GY	Analog Signal Return	<0.050v
47	DG/LG	TP Sensor Signal	0.5-1.2v
48	WT/RD	Self-Test Indicator Signal	STI Open: 5v, Closed: 1v
49	OR	HO2S-11 (Bank 1) Ground	<0.050v
51	WT/RD	Air Management Solenoid	AM1 On: 1v, Off: 12v
52	YL	EGR VR Solenoid	0%, 55 mph: 45%
53	TN/WT	M/T: Shift Indicator Light	SIL On: 1v, Off: 12v
54	OR/BL	A/C WOT Relay Control	On: 1v, Off: 12v
56	DB	PIP Sensor Signal	50% duty cycle
58	TN/RD	Injector Bank 1 (INJ 1 & 4)	3.6-5.0 ms
59	TN/OR	Injector Bank 2 (INJ 2 & 3)	3.6-5.0 ms
60	BK/LG	Power Ground	<0.1v

Pin Connector Graphic

PCM 60-PIN CONNECTOR

Terminal View of 60-Pin PCM Harness Connector

1991 Tempo 2.3L I4 MFI VIN S, X (All) 60 Pin Connector

PCM Pin #	Wire Color	Circuit Description (60 Pin)	Value at Hot Idle
1	YL	Keep Alive Power	12-14v
2-5, 9	---	Not Used	---
3	DG/WT	VSS+ Signal	0 Hz, 55 mph: 125 Hz
4	TN/YL	Ignition Diagnostic Monitor	20-31 Hz
6	OR/YL	VSS- Signal	0 Hz, 55 mph: 125 Hz
7	LG/RD	ECT Sensor Signal	0.5-0.6v
8	PK/BK	Fuel Pump Monitor	On: 12v, Off: 0v
10	DG	A/C Cycling Clutch Switch	A/C On: 12v, Off: 0v
10	BK/YL	A/C Cycling Clutch Switch	A/C On: 12v, Off: 0v
11-15, 18-19	---	Not Used	---
16	OR/RD	Ignition System Ground	<0.1v
17	TN/RD	STI & MIL Control	MIL On: 1v, Off: 12v
20	BK	PCM Case Ground	<0.1v
21	BR/WT	IAC Motor Control	8-10v
22	DG/YL	Fuel Pump Control	On: 1v, Off: 12v
23, 28	---	Not Used	---
24	YL/LG	PSP Switch Signal	Straight 0v, Turned 11v
25	LG/PK	ACT Sensor Signal	1.5-2.5v
26	BR/WT	Reference Voltage	4.9-5.1v
27	BR/LG	PFE EGR Sensor Signal	3.2v, 55 mph: 2.8v
29	RD/BK	HO2S-11 (B1 S1) Signal	0.1-1.1v
30	WT/PK	A/T: Neutral Drive Switch	In 'N': 0v, Others: 12v
30	WT/PK	M/T: Clutch Pedal Position	Clutch In: 0v, Out: 12v
31	GY/YL	EVAP Purge Solenoid	12v, at 55 mph: 1v
32, 34-35	---	Not Used	---
33	YL	EGR VR Solenoid	0%, 55 mph: 45%
36	YL/LG	Spark Output Signal	50% duty cycle
37	RD	Vehicle Power	12-14v
38-39, 41-42	---	Not Used	---
40	BK/LG	Ground	<0.1v
43	PK	A/C Demand Signal	A/C Switch on: 12, Off: 0v
44, 50, 52, 55	---	Not Used	---
45	LG/BK	MAP Sensor Signal	107 Hz
46	GY/RD	Analog Signal Return	<0.1v
47	GN/WT	TP Sensor Signal	0.5-1.2v
48	WT/PK	Self-Test Indicator Signal	STI Open: 5v, Closed: 1v
49	OR	HO2S-11 (Bank 1) Ground	<0.1v
51	WT/RD	Air Management Solenoid	AM1 On: 1v, Off: 12v
53	TN/WT	M/T: Shift Indicator Light	SIL On: 1v, Off: 12v
54	OR/BL	A/C WOT Relay Control	On: 1v, Off: 12v
56	DB	PIP Signal	50% duty cycle
57	RD	Vehicle Power	12-14v
58	TN/RD	Injector Bank 1 (INJ 1 & 4)	3.6-5.0 ms
59	TN/OR	Injector Bank 2 (INJ 2 & 3)	3.6-5.0 ms
60	BK/LG	Power Ground	<0.1v

Pin Connector Graphic

PCM 60-PIN CONNECTOR

Terminal View of 60-Pin PCM Harness Connector

1992-94 Tempo 2.3L I4 MFI VIN X (All) Pin Connector

PCM Pin #	Wire Color	Circuit Description (60 Pin)	Value at Hot Idle
1	YL	Keep Alive Power	12-14v
2	---	Not Used	---
3	DG/WT	VSS+ Signal	0 Hz, 55 mph: 125 Hz
4	WT/PK	Ignition Diagnostic Monitor	20-31 Hz
5		Not Used	---
6	OR/YL	VSS- Signal	0 Hz, 55 mph: 125 Hz
7	LG/RD	ECT Sensor Signal	0.5-0.6v
8	PK/BK	Fuel Pump Monitor	On: 12v, Off: 0v
9	TN/BL	MAF Sensor Return	<0.1v
10	PK/BL	A/C Cycling Clutch Switch	A/C On: 12v, Off: 0v
11	GN/YL	EVAP Purge Solenoid	12v, at 55 mph: 1v
12	---	Not Used	
13	TN/OR	Low Speed Fan Control	On: 1v, Off: 12v
14	---	Not Used	---
15	---	Not Used	---
16	OR/RD	Ignition System Ground	<0.1v
17	TN/RD	STI & MIL Control	MIL On: 1v, Off: 12v
18	TN/OR	Data Bus (+) Signal	5v
19	PK/BL	Data Bus (-) Signal	Digital Signals
20	BK	PCM Case Ground	<0.1v
21	W	IAC Motor Control	8-10v
21	WT/BL	IAC Motor Control	8-10v
22	DG/YL	Fuel Pump Control	On: 1v, Off: 12v
23	---	Not Used	---
24	DG	CID Sensor Signal	5-7 Hz
25	GY	IAT Sensor Signal	1.5-2.5v
26	BR/WT	Reference Voltage	4.9-5.1v
27	BR/LG	EGR PFE Sensor Signal	3.2v, 55 mph: 2.8v
28	YL/LG	PSP Switch Signal	Straight: 0v, Turned: 11v
29	---	Not Used	---
30	PK/YL	A/T: Neutral Drive Switch	In 'N': 0v, Others: 5v
30	PK/YL	M/T: Clutch Engage Switch	Clutch In: 0v, Out: 5v
31	WT/RD	Air Management (1992)	12v, 55 mph: 1v
31	WT/RD	Air Bypass Solenoid (1993-94)	12v, 55 mph: 1v
32	---	Not Used	---
33	YL	EGR VR Solenoid	0%, 55 mph: 45%
34	---	Not Used	---
35	BR/BL	Injector 4 Control	5.1-5.3 ms
36	PK	Spark Output Signal	50% duty cycle
37	RD	Vehicle Power	12-14v
38	---	Not Used	---
39	BR/YL	Injector 3 Control	5.1-5.3 ms
40	BK/LG	Power Ground	<0.1v

1992-94 Tempo 2.3L I4 MFI VIN X (All) Pin Connector

PCM Pin #	Wire Color	Circuit Description (60 Pin)	Value at Hot Idle
41-43	---	Not Used	---
44	RD/BK	HO2S-11 (B1 S1) Signal	0.1-1.1v
45	---	Not Used	---
46	GY/RD	Analog Signal Return	<0.1v
47	GN/WT	TP Sensor (Volts)	0.5-1.2v
48	WT/PK	Self-Test Indicator Signal	STI Open: 5v, Closed: 1v
49	---	Not Used	---
50	LB/RD	MAF Sensor Signal	0.8v, 55 mph: 2.1v
50	BL/BK	MAF Sensor Signal	0.8v, 55 mph: 2.1v
51-52	---	Not Used	---
53	TN/WT	M/T: Shift Indicator Light	SIL On: 1v, Off: 12v
54	PK/YL	A/C WOT Relay Control	On: 1v, Off: 12v
55	---	Not Used	---
56	DB	PIP Signal	50% duty cycle
57	RD	Vehicle Power	12-14v
58	TN	Injector 1 Control	5.1-5.3 ms
59	WT	Injector 2 Signal	5.1-5.3 ms
60	BK/LG	Power Ground	<0.1v

Pin Connector Graphic

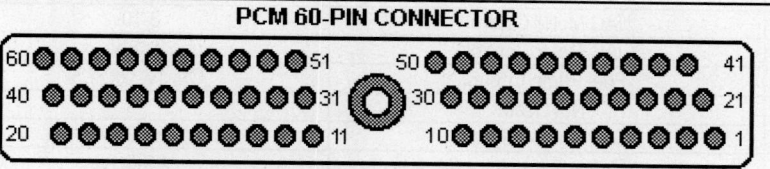

Standard Colors and Abbreviations

Abbreviation	Color	Abbreviation	Color	Abbreviation	Color
BK	Black	GY	Gray	PK	Purple
BL	Blue	GN	Green	RD	Red
BR	Brown	LG	LT Green	TN	Tan
DB	Dark Blue	OR	Orange	WT	White
DG	DK Green	PK	Pink	YL	Yellow

1992-94 Tempo 3.0L V6 MFI VIN U (All) 60 Pin Connector

PCM Pin #	Wire Color	Circuit Description (60 Pin)	Value at Hot Idle
1	YL	Keep Alive Power	12-14v
2	---	Not Used	---
3	DG/WT	VSS (+) Signal	0 Hz, 55 mph: 125 Hz
4	WT/PK	Ignition Diagnostic Monitor	20-31 Hz
5	---	Not Used	---
6	OR/YL	VSS (-) Signal	0 Hz, 55 mph: 125 Hz
7	LG/RD	ECT Sensor Signal	0.5-0.6v
8	PK/BK	Fuel Pump Monitor	On: 12v, Off: 0v
9	TN/BL	MAF Sensor Return	<0.050v
10	PK/BL	A/C Cycling Clutch Switch	A/C On: 12v, Off: 0v
11	GN/YL	EVAP Purge Solenoid	12v, at 55 mph: 1v
12	LG/OR	Injector 6 Control	4.6-5.3 ms
13	TN/OR	Low Speed Fan Control	On: 1v, Off: 12v
14	---	Not Used	---
15	TN/BK	Injector 5 Control	4.6-5.3 ms
16	OR/RD	Ignition System Ground	<0.050v
17	TN/RD	STI Output, MIL Control	MIL On: 1v, Off: 12v
18	TN/OR	Data Bus (+) Signal	Digital Signals
19	PK/BL	Data Bus (-) Signal	Digital Signals
20	BK	PCM Case Ground	<0.050v
21	WT	IAC Motor Control	8-10v
21	WT/BL	IAC Motor Control	8-10v
22	DG/YL	Fuel Pump Control	On: 1v, Off: 12v
23-24	---	Not Used	---
25	GY	IAT Sensor Signal	1.5-2.5v
26	BR/WT	Reference Voltage	4.9-5.1v
27	BR/LG	PFE EGR Sensor	3.2v, 55 mph: 2.9v
28	YL/LG	PSP Switch Signal	Straight: 0v, Turned: 11v
29	---	Not Used	---
30	PK/YL	A/T: Neutral Drive Switch	In 'N': 0v, Others: 5v
30	PK/YL	M/T: Clutch Position Switch	Clutch In: 0v, Out: 5v
31	BL	High Speed Fan Control	On: 1v, Off: 12v
32	---	Not Used	---
33	YL	EGR VR Solenoid	0%, 55 mph: 45%
34	---	Not Used	---
35	BR/BL	Injector 4 Control	4.6-5.3 ms
36	PK	Spark Output Signal	50% duty cycle
37	RD	Vehicle Power	12-14v
38	---	Not Used	---
39	BR/YL	Injector 3 Control	4.6-5.3 ms
40	BK/LG	Power Ground	<0.1v

1992-94 Tempo 3.0L V6 MFI VIN U (All) 60 Pin Connector

PCM Pin #	Wire Color	Circuit Description (60 Pin)	Value at Hot Idle
41	---	Not Used	
42 ('94)	TN/LG	A/C Cycling Clutch Switch	A/C On: 12v, Off: 0v
43	RD/BK	HO2S-11 (B1 S1) Signal	0.1-1.1v
44	GY/BL	HO2S-21 (B2 S1) Signal	0.1-1.1v
45	---	Not Used	---
46	GY/RD	Analog Signal Return	<0.050v
47	GN/WT	TP Sensor Signal	0.7v, 55 mph: 1.3v
48	WT/PK	Self-Test Indicator Signal	STI Open: 5v, Closed: 1v
49	---	Not Used	---
50	BL/RD	MAF Sensor Signal	0.7v, 55 mph: 2.1v
50	BL/BK	MAF Sensor Signal	0.7v, 55 mph: 2.1v
51-53	---	Not Used	---
54	PK/YL	A/C WOT Relay Control	On: 1v, Off: 12v
55	---	Not Used	---
56	DB	PIP Sensor Signal	50% duty cycle
57	RD	Vehicle Power	12-14v
58	TN	Injector 1 Control	4.6-5.3 ms
59	WT	Injector 2 Control	4.6-5.3 ms
60	BK/LG	Power Ground	<0.1v

Pin Connector Graphic

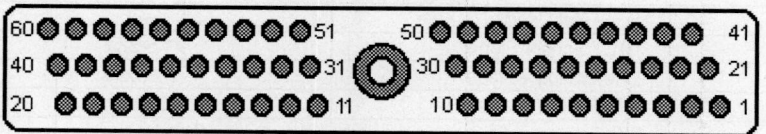

PCM 60-PIN CONNECTOR

Terminal View of 60-Pin PCM Harness Connector

Standard Colors and Abbreviations

Abbreviation	Color	Abbreviation	Color	Abbreviation	Color
BK	Black	GY	Gray	PK	Purple
BL	Blue	GN	Green	RD	Red
BR	Brown	LG	LT Green	TN	Tan
DB	Dark Blue	OR	Orange	WT	White
DG	DK Green	PK	Pink	YL	Yellow

Thunderbird Tables

1990 Thunderbird 3.8L V6 MFI VIN 4 (A/T) 60 Pin Connector

PCM Pin #	Wire Color	Circuit Description (60 Pin)	Value at Hot Idle
1	YL	Keep Alive Power	12-14v
2	---	Not Used	---
3	DG/WT	VSS (+) Signal	0 Hz, 55 mph: 125 Hz
4	DG/YL	Ignition Diagnostic Monitor	20-31 Hz
5	LG	Brake Pedal Position Switch	Brake Off: 0v, On: 12v
6	BK/WT	VSS (-) Signal	0 Hz, 55 mph: 125 Hz
7	LG/YL	ECT Sensor Signal	0.5-0.6v
8	OR/BK	Data Bus (-) Signal	Digital Signals
8	BK/OR	Data Bus (-) Signal	Digital Signals
9	---	Not Used	---
10	PK/BL	A/C Cycling Clutch Switch	A/C On: 12v, Off: 0v
11	OR/YL	Speed Control Solenoid	Solenoid On: 12-14v
12	BR/YL	Injector 3 Control	6.0-6.2 ms
13	BR/BL	Injector 4 Control	6.0-6.2 ms
14	TN/BL	Injector 5 Control	6.0-6.2 ms
15	LG	Injector 6 Control	6.0-6.2 ms
16	DB, GY	Ignition System Ground	<0.050v
17	YL/BK	STI Output, MIL Control	MIL On: 1v, Off: 12v
18-19	--	Not Used	---
20	BK	PCM Case Ground	<0.050v
21	RD/LG	IAC Motor Control	20-40%
22	TN/LG	Fuel Pump Control	On: 1v, Off: 12v
23-24	---	Not Used	---
25	LG/PK	ACT Sensor Signal	1.5-2.5v
26	OR/WT	Reference Voltage	4.9-5.1v
27	BR/LG	PFE EGR Sensor	3.2v, 55 mph: 2.8v
28	BL/BK	S/C Command Switch	Switch On: 6.7v
29	TN/OR	HO2S-21 (B2 S1) Signal	0.1-1.1v
30	RD/BL	Neutral Drive Switch Signal	In 'N': 0v, Others: 4-5v
31	GY/YL	EVAP Purge Solenoid	12v, at 55 mph: 1v
32	---	Not Used	---
33	DG	EGR VR Solenoid	0%, 55 mph: 45%
34	BL/PK	Data Output Link	Digital Signals
35	WT/PK	Speed Control Vent Solenoid	98% (on at 55 mph)
36	YL/LG	Spark Output Signal	50% duty cycle
37	RD	Vehicle Power	12-14v
38	---	Not Used	---
39	WT/BL	Speed Control Switch Ground	<0.050v
40	BK/LG	Power Ground	<0.1v

1990 Thunderbird 3.8L V6 MFI VIN 4 (A/T) 60 Pin Connector

PCM Pin #	Wire Color	Circuit Description (60 Pin)	Value at Hot Idle
41-42	---	Not Used	---
43	TN/RD	HO2S-11 (B1 S1) Signal	0.1-1.1v
44	BK/OR	Data Bus (+) Signal	Digital Signals
44	OR/BK	Data Bus (+) Signal	Digital Signals
45	DB/LG	MAP Sensor Signal	107 Hz
46	BK/WT	Analog Signal Return	<0.050v
47	DG/LG	TP Sensor Signal	0.8v, 55 mph: 1.1v
48	WT/RD	Self-Test Indicator Signal	STI Open: 5v, Closed: 1v
49	OR	HO2S-11 (Bank 1) Ground	<0.050v
50	PK/BK	Fuel Pump Monitor	On: 12v, Off: 0v
51	GY/BK	S/C Vacuum Solenoid	50-90% (On at 55 mph)
52-53	---	Not Used	---
54	OR/BL	A/C WOT Relay Control	On: 1v, Off: 12v
55	---	Not Used	---
56	DB	PIP Sensor Signal	50% duty cycle
57	RD	Vehicle Power	12-14v
58	TN	Injector 1 Control	6.0-6.2 ms
59	WT	Injector 2 Control	6.0-6.2 ms
60	BK/LG	Power Ground	<0.1v

Pin Connector Graphic

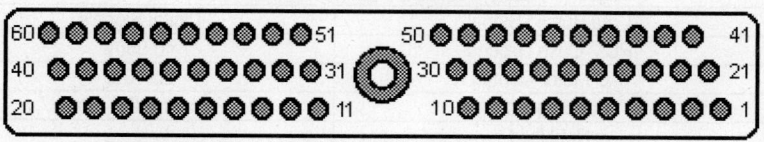

Standard Colors and Abbreviations

Abbreviation	Color	Abbreviation	Color	Abbreviation	Color
BK	Black	GY	Gray	PK	Purple
BL	Blue	GN	Green	RD	Red
BR	Brown	LG	LT Green	TN	Tan
DB	Dark Blue	OR	Orange	WT	White
DG	DK Green	PK	Pink	YL	Yellow

1991-93 Thunderbird 3.8L V6 MFI VIN 4 (A/T) 60 Pin Connector

PCM Pin #	Wire Color	Circuit Description (60 Pin)	Value at Hot Idle
1	YL	Keep Alive Power	12-14v
2	---	Not Used	---
3	DG/WT	VSS (+) Signal	0 Hz, 55 mph: 125 Hz
4	RD/BL	Ignition Diagnostic Monitor	20-31 Hz
4	WT/PK	Ignition Diagnostic Monitor	20-31 Hz
5	---	Not Used	---
6	BK/WT	VSS (-) Signal	0 Hz, 55 mph: 125 Hz
6	GY/RD	VSS (-) Signal	0 Hz, 55 mph: 125 Hz
7	LG/YL	ECT Sensor Signal	0.5-0.6v
7	LG/RD	ECT Sensor Signal	0.5-0.6v
8	PK/BL	Fuel Pump Monitor	On: 12v, Off: 0v
9	TN/BL	MAF Sensor Return	<0.050v
10	PK/BK	A/C Cycling Clutch Switch	A/C On: 12v, Off: 0v
11	GY/YL	EVAP Purge Solenoid	12v, 55 mph: 1v
12	LG	Injector 6 Control	5.6-6.3 ms
12	LG/OR	Injector 6 Control	5.6-6.3 ms
13-14	---	Not Used	---
15	TN/BL	Injector 5 Control	5.6-6.3 ms
15	TN/BK	Injector 5 Control	5.6-6.3 ms
16	GY	Ignition System Ground	<0.050v
17	YL/BK	STI Output, MIL Control	MIL On: 1v, Off: 12v
18	TN/OR	Data Bus (+) Signal	Digital Signals
19	PK/BL	Data Bus (-) Signal	Digital Signals
20	BK	PCM Case Ground	<0.050v
21	RD/LG	IAC Motor Control	20-40%
22	TN/LG	Fuel Pump Control	On: 1v, Off: 12v
23-24	---	Not Used	---
25	LG/PK	IAT Sensor Signal	1.5-2.5v
25	GY	IAT Sensor Signal	1.5-2.5v
26	OR/WT	Reference Voltage	4.9-5.1v
26	BR/WT	Reference Voltage	4.9-5.1v
27	BR/LG	PFE EGR Sensor	3.2v, 55 mph: 2.9v
28-29	---	Not Used	---
30	RD/BL	Neutral Drive Switch Signal	In 'N': 0v, Others: 5v
31-32	---	Not Used	---
33	DG	EGR VR Solenoid	0%, 55 mph: 45%
33	BR/PK	EGR VR Solenoid	0%, 55 mph: 45%
34	BL/PK	Data Output Link	Digital Signals
35	BR/BL	Injector 4 Control	5.6-6.3 ms
36	YL/LG	Spark Output Signal	6.93v (50% duty cycle)
37	RD	Vehicle Power	12-14v
38	---	Not Used	---
39	BR/YL	Injector 3 Control	5.6-6.3 ms
40	BK/LG	Power Ground	<0.1v

1991-93 Thunderbird 3.8L V6 MFI VIN 4 (A/T) 60 Pin Connector

PCM Pin #	Wire Color	Circuit Description (60 Pin)	Value at Hot Idle
41-42	---	Not Used	---
43	TN/RD	HO2S-21 (B2 S1) Signal	0.1-1.1v
43	TN/WT	HO2S-21 (B2 S1) Signal	0.1-1.1v
44	TN/OR	HO2S-11 (B1 S1) Signal	0.1-1.1v
45	---	Not Used	---
46	BK/WT	Analog Signal Return	<0.050v
46	GY/RD	Analog Signal Return	<0.050v
47	DG/LG	TP Sensor Signal	0.5-1.2v
47	GY/WT	TP Sensor Signal	0.5-1.2v
48	WT/RD	Self-Test Indicator Signal	STI Open: 5v, Closed: 1v
48	WT/PK	Self-Test Indicator Signal	STI Open: 5v, Closed: 1v
49	---	Not Used	---
50	DB or BL	MAF Sensor Signal	0.7v, 55 mph: 1.3v
51-53	---	Not Used	---
54	OR/BL	A/C WOT Relay Control	On: 1v, Off: 12v
55	---	Not Used	---
56	DB	PIP Sensor Signal	50% duty cycle
57	RD	Vehicle Power	12-14v
58	TN	Injector 1 Control	5.6-6.3 ms
59	WT	Injector 2 Control	5.6-6.3 ms
60	BK/LG	Power Ground	<0.1v

Pin Connector Graphic

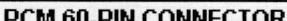
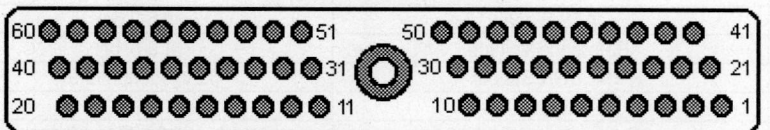

PCM 60-PIN CONNECTOR

Terminal View of 60-Pin PCM Harness Connector

Standard Colors and Abbreviations

Abbreviation	Color	Abbreviation	Color	Abbreviation	Color
BK	Black	GY	Gray	PK	Purple
BL	Blue	GN	Green	RD	Red
BR	Brown	LG	LT Green	TN	Tan
DB	Dark Blue	OR	Orange	WT	White
DG	DK Green	PK	Pink	YL	Yellow

1994-95 Thunderbird 3.8L V6 MFI VIN 4 (A/T) 60 Pin Connector

PCM Pin #	Wire Color	Circuit Description (60 Pin)	Value at Hot Idle
1	YL	Keep Alive Power	12-14v
2	LG	Brake Pedal Position Switch	Brake Off: 0v, On: 12v
3	GY/BK	VSS (+) Signal	0 Hz, 55 mph: 125 Hz
4	OR/WT	Ignition Diagnostic Monitor	20-31 Hz
5	DG/WT	TSS Sensor Signal	At 55 mph: 205-220 Hz
6	PK/OR	VSS (-) Signal	0 Hz, 55 mph: 125 Hz
7	LG/RD	ECT Sensor Signal	0.5-0.6v
8	PK/BK	Fuel Pump Monitor	On: 12v, Off: 0v
9	TN/BL	MAF Sensor Return	<0.050v
10	PK/BL	A/C Cycling Clutch Switch	A/C On: 12v, Off: 0v
11	GY/YL	EVAP Purge Solenoid	12v, 55 mph: 1v
12	LG/OR	Injector 6 Control	5.6-6.3 ms
13	TN/OR	Low Speed Fan Control	On: 1v, Off: 12v
14	LG/PK	High Speed Fan Control	On: 1v, Off: 12v
15	TN/BK	Injector 5 Control	5.6-6.3 ms
16	OR/RD	Ignition System Ground	<0.050v
17	PK/LG	STO & MIL Control	MIL On: 1v, Off: 12v
18	TN/OR	Data Bus (+) Signal	Digital Signals
19	PK/BL	Data Bus (-) Signal	Digital Signals
20	BK	PCM Case Ground	<0.050v
21	WT/BL	IAC Motor Control	9.5-11v
22	BL/OR	Fuel Pump Control	On: 1v, Off: 12v
23	---	Not Used	---
24	DB/OR	CID Sensor Signal	5-7 Hz
25	GY	ATS Signal	1.5-2.5v
26	BR/WT	Reference Voltage	4.9-5.1v
27	BR/LG	PFE EGR Sensor	3.2v, 55 mph: 2.9v
28	---	Not Used	---
29	DG	Octane Adjust Switch	Closed: 0v, Open: 9.1v
30	BL/YL	MLP Sensor Signal	In P/N: 0v, in O/D: 2.1v
31	---	Not Used	---
32	---	Not Used	---
33	BR/PK	EGR VR Solenoid	0%, 55 mph: 45%
34	---	Not Used	---
35	BR/BL	Injector 4 Control	5.6-6.3 ms
36	PK	Spark Output Signal	50% duty cycle
37	RD	Vehicle Power	12-14v
38	YL/WT	EPC Solenoid Control	8.8v (20 psi)
39	BR/YL	Injector 3 Control	5.6-6.3 ms
40	BK/WT	Power Ground	<0.1v

1994-95 Thunderbird 3.8L V6 MFI VIN 4 (A/T) 60 Pin Connector

PCM Pin #	Wire Color	Circuit Description (60 Pin)	Value at Hot Idle
41	TN/WT	TCS (switch) Signal	TCS & O/D On: 12v
42	TN/LG	A/C High Pressure Switch	AC On: 12v, Off: 0v
43	RD/BK	HO2S-21 (B2 S1) Signal	0.1-1.1v
44	GY/BL	HO2S-11 (B1 S1) Signal	0.1-1.1v
45	---	Not Used	---
46	GY/RD	Analog Signal Return	<0.050v
47	GY/WT	TP Sensor Signal	0.5-1.2v
48	WT/PK	Self-Test Indicator Signal	STI Open: 5v, Closed: 1v
49	OR/BK	TOT Sensor Signal	2.10-2.40v
50	BL/RD	MAF Sensor Signal	0.6v, 55 mph: 1.3v
51	OR/YL	Shift Solenoid 1 Control	1v, 55 mph: 12v
52	PK/OR	Shift Solenoid 2 Control	12v, 55 mph: 1v
53	BK/OR	TCC Solenoid Control	0%, 55 mph: 45%
54	PK/YL	A/C WOT Relay Control	On: 1v, Off: 12v
55	WT/LG	TCIL (lamp) Control	On: 1v, Off: 12v
56	GY/OR	PIP Sensor Signal	50% duty cycle
56	PK	PIP Sensor Signal	50% duty cycle
57	RD	Vehicle Power	12-14v
58	TN	Injector 1 Control	5.6-6.3 ms
59	WT	Injector 2 Control	5.6-6.3 ms
60	BK/LG	Power Ground	<0.1v

Pin Connector Graphic

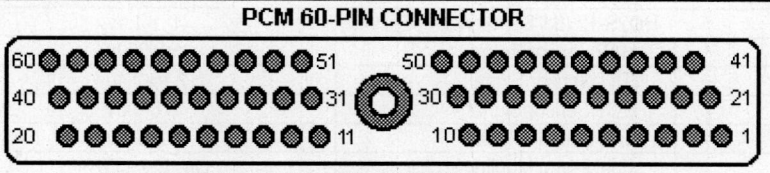

Standard Colors and Abbreviations

Abbreviation	Color	Abbreviation	Color	Abbreviation	Color
BK	Black	GY	Gray	PK	Purple
BL	Blue	GN	Green	RD	Red
BR	Brown	LG	LT Green	TN	Tan
DB	Dark Blue	OR	Orange	WT	White
DG	DK Green	PK	Pink	YL	Yellow

1996-97 Thunderbird 3.8L V6 MFI VIN 4 (A/T) 104 Pin Connector

PCM Pin #	Wire Color	Circuit Description (104 Pin)	Value at Hot Idle
1	PK/OR	Shift Solenoid 2 Control	12v, 55 mph: 1v
2	PK/LG	MIL (lamp) Control	MIL On: 1v, Off: 12v
3-10	---	Not Used	---
11	PK/WT	Purge Flow Sensor	0.8v, 55 mph: 3v
12	---	Not Used	---
13	PK	Flash EEPROM Power	0.1v
14	---	Not Used	---
15	PK/LB	Data Bus (-) Signal	Digital Signals
16	TN/OR	Data Bus (+) Signal	Digital Signals
17-20	---	Not Used	---
21	DB	CKP Sensor (+) Signal	390-450 Hz
22	GY	CKP Sensor (-) Signal	390-450 Hz
23	---	Not Used	---
24	BK/WT	Power Ground	<0.1v
25	BK	PCM Case Ground	<0.050v
26	DB/LG	Coil Driver 1 Control	8°, 55 mph: 12° dwell
27	OR/YL	Shift Solenoid 1 Control	1v, 55 mph: 12v
28	---	Not Used	---
29	TN/WT	TCS (switch) Signal	TCS & O/D On: 12v
30	DG	Octane Adjust Switch	Closed: 0v, Open: 9.3v
31-32	---	Not Used	---
33	PK/OR	VSS (-) Signal	0 Hz, 55 mph: 125 Hz
34	---	Not Used	---
35	RD/LG	HO2S-12 (B1 S2) Signal	0.1-1.1v
36	TN/LB	MAF Sensor Return	<0.050v
37	OR/BK	TFT Sensor Signal	2.10-2.40v
38	LG/RD	ECT Sensor Signal	0.5-0.6v
39	GY	IAT Sensor Signal	1.5-2.5v
40	PK/BK	Fuel Pump Monitor	On: 12v, Off: 0v
41	PK/LB	A/C Cycling Clutch Switch	AC On: 12v, Off: 0v
42-44	---	Not Used	---
45	TN/OR	Low Speed Fan Control	On: 1v, Off: 12v
46	LG/PK	High Speed Fan Control	On: 1v, Off: 12v
47	BR/PK	EGR VR Solenoid	0%, 55 mph: 45%
48-50	---	Not Used	---
51	BK/WT	Power Ground	<0.1v
52	RD/LB	Coil Driver 2 Control	8°, 55 mph: 12° dwell
53	---	Not Used	---
54	BR/OR	TCC Solenoid Control	0%, 55 mph: 95%
55	YL	Keep Alive Power	12-14v
56-57	---	Not Used	---
58	GY/BK	VSS (+) Signal	0 Hz, 55 mph: 125 Hz
59	---	Not Used	---
60	GY/LB	HO2S-11 (B1 S1) Signal	0.1-1.1v
61	PK/LG	HO2S-22 (B2 S2) Signal	0.1-1.1v
62	---	Not Used	---
63	---	Not Used	---
64	LB/YL	TR Sensor Signal	In 'P': 0v, 55 mph: 1.7v

1996-97 Thunderbird 3.8L V6 MFI VIN 4 (A/T) 104 Pin Connector

PCM Pin #	Wire Color	Circuit Description (104 Pin)	Value at Hot Idle
65	BR/LG	DPFE Sensor Signal	0.20-1.30v
66, 68	---	Not Used	---
67	GY/YL	EVAP Purge Solenoid	0-10 Hz (0-100%)
69	PK/YL	A/C WOT Relay Control	On: 1v, Off: 12v
70, 72	---	Not Used	---
71	RD	Vehicle Power	12-14v
73	TN/BK	Injector 5 Control	4.5-4.8 ms
74	BR/YL	Injector 3 Control	4.5-4.8 ms
75	TN	Injector 1 Control	4.5-4.8 ms
76	BK/WT	Power Ground	<0.1v
77	BK/WT	Power Ground	<0.1v
78	PK/WT	Coil Driver 3 Control	8°, 55 mph: 12° dwell
79	WT/LG	TCIL (lamp) Control	On: 1v, Off: 12v
80	LB/OR	Fuel Pump Control	On: 1v, Off: 12v
81	WT/YL	EPC Solenoid Control	8.8v (20 psi)
82	---	Not Used	---
83	WT/LB	IAC Motor Control	10v (30% on time)
84	DG/WT	TSS Sensor Signal	43 Hz (650 rpm)
85	DG	CID Sensor Signal	5-7 Hz
86	TN/LG	A/C Pressure Switch	On: 1v, Off: 12v
87	RD/BK	HO2S-21 (B2 S1) Signal	0.1-1.1v
88	LB/RD	MAF Sensor Signal	0.6v, 55 mph: 1.9v
89	GY/WT	TP Sensor Signal	0.8v, 55 mph: 1.1v
90	BR/WT	Reference Voltage	4.9-5.1v
91	GY/RD	Analog Signal Return	<0.050v
92	LG	Brake Pedal Position Switch	Brake Off: 0v, On: 12v
93	RD/WT	HO2S-11 (B1 S1) Heater	On: 1v, Off: 12v
94	YL/LB	HO2S-21 (B2 S1) Heater	On: 1v, Off: 12v
95	WT/BK	HO2S-12 (B1 S2) Heater	On: 1v, Off: 12v
96	TN/YL	HO2S-22 (B2 S2) Heater	On: 1v, Off: 12v
97	RD	Vehicle Power	12-14v
98	---	Not Used	---
99	LG/OR	Injector 6 Control	4.5-4.8 ms
100	BR/LB	Injector 4 Control	4.5-4.8 ms
101	WT	Injector 2 Control	4.5-4.8 ms
102	---	Not Used	---
103	BK/WT	Power Ground	<0.1v
104	---	Not Used	---

Pin Connector Graphic

PCM 104-PIN CONNECTOR

Terminal View of 104-Pin PCM Wiring Harness Connector

Standard Colors and Abbreviations

Abbreviation	Color	Abbreviation	Color	Abbreviation	Color
BK	Black	GY	Gray	PK	Purple
BL	Blue	GN	Green	RD	Red
BR	Brown	LG	LT Green	TN	Tan
DB	Dark Blue	OR	Orange	WT	White
DG	DK Green	PK	Pink	YL	Yellow

1990-92 Thunderbird 3.8L V6 SC VIN C (A/T) 60 Pin Connector

PCM Pin #	Wire Color	Circuit Description (60 Pin)	Value at Hot Idle
1	YL	Keep Alive Power	12-14v
2	---	Not Used	---
3	DG/WT	VSS (+) Signal	0 Hz, 55 mph: 125 Hz
4	DG/YL	Ignition Diagnostic Monitor	20-31 Hz
5	LG	Brake Pedal Position Switch	Brake Off: 0v, On: 12v
6	BK/WT	VSS (-) Signal	0 Hz, 55 mph: 125 Hz
7	LG/YL	ECT Sensor Signal	0.5-0.6v
8	---	Not Used	---
9	TN/BL	MAF Sensor Return	<0.050v
10	PK/BL	A/C Cycling Clutch Switch	A/C On: 12v, Off: 0v
11	OR/YL	Speed Control Solenoid (+)	S/C On: 12-14v
12	LG	Injector 3 Control	4.0-4.2 ms
12	LG/OR	Injector 3 Control	4.0-4.2 ms
13	BR/BL	Injector 4 Control	4.0-4.2 ms
14	TN/BL	Injector 5 Control	4.0-4.2 ms
15	LG	Injector 6 Control	4.0-4.2 ms
16	BL	Ignition System Ground	<0.050v
17	YL/BK	STI Output, MIL Control	MIL On: 1v, Off: 12v
18	BK/WT	Octane Adjust Switch	Closed: 0v, Open: 9.1v
19	PK/BK	Fuel Pump Monitor	On: 12v, Off: 0v
20	BK	PCM Case Ground	<0.050v
21	RD/LG	IAC Motor Control	9.9v (40%)
22	TN/LG	Fuel Pump Control	On: 1v, Off: 12v
23	YL/RD	Knock Sensor Signal	0v
24	DG	CID Sensor Signal	5-7 Hz
25	LG/PK	ACT Sensor Signal	1.5-2.5v
26	OR/WT	Reference Voltage	4.9-5.1v
27	BR/LG	PFE EGR Sensor	3.2v, 55 mph: 2.8v
28	BL/BK	S/C Command Switch	Switch On: 6.7v
29	TN/OR	HO2S-21 (B2 S1) Signal	0.1-1.1v
30	RD/BL	Neutral Drive Switch Signal	In 'N': 0v, Others: 5v
31	GY/YL	EVAP Purge Solenoid	12v, 55 mph: 1v
32	OR/WT	Automatic Shock Control	4.3v
33	YL	EGR VR Solenoid	0%, 55 mph: 45%
34	---	Not Used	---
35	WT/PK	Speed Control Vent Solenoid	0%, 55 mph: 98%
36	YL/LG	Spark Output Signal	9.93v (50% duty cycle)
37	RD	Vehicle Power	12-14v
38 ('89)	LG/PK	Supercharger Bypass Solenoid	0%, 55 mph: 45%
39	WT/BL	Speed Control Switch Ground	<0.050v
40	BK/LG	Power Ground	<0.1v

1990-92 Thunderbird 3.8L V6 SC VIN C (A/T) 60 Pin Connector

PCM Pin #	Wire Color	Circuit Description (60 Pin)	Value at Hot Idle
41	PK	High Speed Fan Control	On: 1v, Off: 12v
42	---	Not Used	---
43	TN/RD	HO2S-11 (B1 S1) Signal	0.1-1.1v
44	---	Not Used	---
45	DB/LG	MAP Sensor Signal	107 Hz
46	BK/WT	Analog Signal Return	<0.050v
47	DG/LG	TP Sensor Signal	0.5-1.2v
48	WT/RD	Self-Test Indicator Signal	STI Open: 5v, Closed: 1v
49	OR	HO2S-11 (Bank 1) Ground	<0.050v
50	DB/OR	MAF Sensor Signal	0.7v
51	GY/BK	S/C Vacuum Solenoid	0%, 55 mph: 45%
52	---	Not Used	---
53	PK	Shift Indicator Control	SIL On: 1v, Off: 12v
54	RD	A/C WOT Relay Control	On: 1v, Off: 12v
55	TN/OR	Low Speed Fan Control	On: 1v, Off: 12v
56	DB	PIP Sensor Signal	50% duty cycle
57	RD	Vehicle Power	12-14v
58	TN	Injector 1 Control	4.0-4.2 ms
59	WT	Injector 2 Control	4.0-4.2 ms
60	BK/LG	Power Ground	<0.1v

Pin Connector Graphic

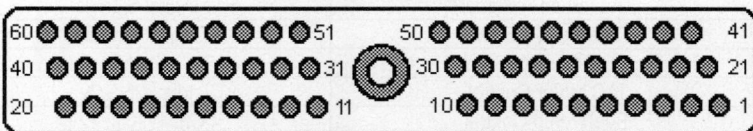

Standard Colors and Abbreviations

Abbreviation	Color	Abbreviation	Color	Abbreviation	Color
BK	Black	GY	Gray	PK	Purple
BL	Blue	GN	Green	RD	Red
BR	Brown	LG	LT Green	TN	Tan
DB	Dark Blue	OR	Orange	WT	White
DG	DK Green	PK	Pink	YL	Yellow

1990-93 Thunderbird 3.8L V6 SC VIN R (A/T) 60 Pin Connector

PCM Pin #	Wire Color	Circuit Description (60 Pin)	Value at Hot Idle
1	YL	Keep Alive Power	12-14v
2	---	Not Used	---
3	DG/WT	VSS (+) Signal	0 Hz, 55 mph: 125 Hz
4	DG/YL	Ignition Diagnostic Monitor	20-31 Hz
4	TN/YL	Ignition Diagnostic Monitor	20-31 Hz
5	---	Not Used	---
6	BK/WT	VSS (-) Signal	0 Hz, 55 mph: 125 Hz
6	GY/RD	VSS (-) Signal	0 Hz, 55 mph: 125 Hz
7	LG/YL	ECT Sensor Signal	0.5-0.6v
7	LG/RD	ECT Sensor Signal	0.5-0.6v
8 ('91-'93)	PK/BK	Fuel Pump Monitor	On: 12v, Off: 0v
9	TN/BL	MAF Sensor Return	<0.050v
10	PK/BL	A/C Cycling Clutch Switch	A/C On: 12v, Off: 0v
11	GY/YL	EVAP Purge Solenoid	12v, at 55 mph: 1v
12	TN/OR	Injector 6 Control	3.6-4.0 ms
12	LG/OR	Injector 6 Control	3.6-4.0 ms
13	TN/OR	Low Speed Fan Control	Fan On: 12v, Off: 0v
14	---	Not Used	---
15	TN/BL	Injector 5 Control	3.6-4.0 ms
15	TN/BK	Injector 5 Control	3.6-4.0 ms
16	BL, GY	Ignition System Ground	<0.050v
17	YL/BK	STI Output, MIL Control	MIL On: 1v, Off: 12v
18	TN/OR	Data Bus (+) Signal	Digital Signals
19	PK/BL	Data Bus (-) Signal	Digital Signals
20	BK	PCM Case Ground	<0.050v
21	RD/LG	IAC Motor Control	Idle: 7.6-9.2v
22	TN/LG	Fuel Pump Control	On: 1v, Off: 12v
23	YL/RD	Knock Sensor Signal	0v
24	DG/LG	CID Sensor Signal	5-7 Hz
25	LG/PK	IAT Sensor Signal	1.5-2.5v
25	GY	IAT Sensor Signal	1.5-2.5v
26	OR/WT	Reference Voltage	4.9-5.1v
26	BR/WT	Reference Voltage	4.9-5.1v
27-28	---	Not Used	---
29	BK/WT	Octane Adjust Switch	Closed: 0v, Open: 9.1v
29	GY/RD	Octane Adjust Switch	Closed: 0v, Open: 9.1v
30	RD/BL	Neutral Drive Switch Signal	In 'P': 0v, Others: 5v
31	PK	High Speed Fan Control	On: 1v, Off: 12v
31	LG/PK	High Speed Fan Control	On: 1v, Off: 12v
32	OR/WT	Automatic Ride Control	4.3v
33-34	---	Not Used	---
35	BR/BL	Injector 4 Control	3.6-4.0 ms
36	YL/LG	Spark Output Signal	50% duty cycle
37	RD	Vehicle Power	12-14v
38	---	Not Used	---
39	BR/YL	Injector 3 Control	3.6-4.0 ms
40	BK/LG	Power Ground	<0.1v

1990-93 Thunderbird 3.8L V6 SC VIN R (A/T) 60 Pin Connector

PCM Pin #	Wire Color	Circuit Description (60 Pin)	Value at Hot Idle
41-42	---	Not Used	---
43	TN/RD	HO2S-21 (B2 S1) Signal	0.1-1.1v
43	TN/WT	HO2S-21 (B2 S1) Signal	0.1-1.1v
44	TN/OR	HO2S-11 (B1 S1) Signal	0.1-1.1v
45	DB/LG	BARO Pressure Sensor	159 Hz (sea level)
46	BK/WT	Analog Signal Return	<0.050v
46	GY/RD	Analog Signal Return	<0.050v
47	DG/LG	TP Sensor Signal	Hot 0.5-1.2v
47	GY/WT	TP Sensor Signal	Hot 0.5-1.2v
48	WT/RD	Self-Test Indicator Signal	STI Open: 5v, Closed: 1v
48	WT/PK	Self-Test Indicator Signal	STI Open: 5v, Closed: 1v
49	---	Not Used	---
50	DB/OR	MAF Sensor Signal	Hot 0.7v
50	BL/RD	MAF Sensor Signal	Hot 0.7v
51-52	---	Not Used	---
52	---	Not Used	---
53	PK	Shift Indicator Lamp	SIL On: 1v, Off: 12v
54	RD or PK	A/C WOT Relay Control	On: 1v, Off: 12v
55	---	Not Used	---
56	DB	PIP Sensor Signal	50% duty cycle
57	RD	Vehicle Power	12-14v
58	TN	Injector 1 Control	3.6-4.0 ms
59	WT	Injector 2 Control	3.6-4.0 ms
60	BK/LG	Power Ground	<0.1v

Pin Connector Graphic

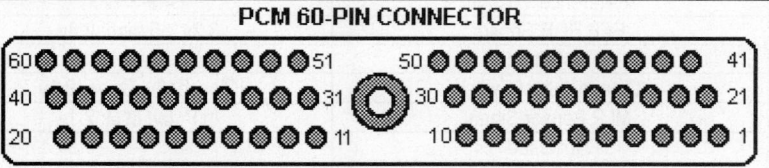

PCM 60-PIN CONNECTOR

Terminal View of 60-Pin PCM Harness Connector

Standard Colors and Abbreviations

Abbreviation	Color	Abbreviation	Color	Abbreviation	Color
BK	Black	GY	Gray	PK	Purple
BL	Blue	GN	Green	RD	Red
BR	Brown	LG	LT Green	TN	Tan
DB	Dark Blue	OR	Orange	WT	White
DG	DK Green	PK	Pink	YL	Yellow

1994-95 Thunderbird 3.8L V6 SC VIN R (A/T) 60 Pin Connector

PCM Pin #	Wire Color	Circuit Description (60 Pin)	Value at Hot Idle
1	YL	Keep Alive Power	12-14v
2	LG	Brake Pedal Position Switch	Brake Off: 0v, On: 12v
3	GY/BK	VSS (+) Signal	0 Hz, 55 mph: 125 Hz
4	OR/WT	Ignition Diagnostic Monitor	20-31 Hz
5	DG/WT	TSS Sensor Signal	215-230 Hz (55 mph)
6	PK/OR	VSS (-) Signal	0 Hz, 55 mph: 125 Hz
7	LG/RD	ECT Sensor Signal	0.5-0.6v
8	PK/BK	Fuel Pump Monitor	On: 12v, Off: 0v
9	TN/BL	MAF Sensor Return	<0.050v
10	PK/BL	A/C Cycling Clutch Switch	A/C On: 12v, Off: 0v
11	GY/YL	EVAP Purge Solenoid	12v, at 55 mph: 1v
12	LG/OR	Injector 6 Control	3.6-3.8 ms
13	TN/OR	Low Speed Fan Control	On: 1v, Off: 12v
14	LG/PK	High Speed Fan Control	On: 1v, Off: 12v
15	TN/BK	Injector 5 Control	3.6-3.8 ms
16	OR/RD	Ignition System Ground	<0.050v
17	PK/LG	STI Output, MIL Control	MIL On: 1v, Off: 12v
18	TN/OR	Data Bus (+) Signal	Digital Signals
19	PK/BL	Data Bus (-) Signal	Digital Signals
20	BK	PCM Case Ground	<0.050v
21	WT/BL	IAC Motor Control	9.2v (40%)
22	BL/OR	Fuel Pump Control	On: 1v, Off: 12v
23	YL/RD	Knock Sensor Signal	0v
24	DB/OR	CMP Sensor Signal	5-7 Hz
25	GY	IAT Sensor Signal	1.5-2.5v
26	BR/WT	Reference Voltage	4.9-5.1v
27	BR/LG	PFE EGR Sensor	3.2v, 55 mph: 2.8v
28	---	Not Used	---
29	DG	Octane Adjust Switch	Closed: 0v, Open: 9.1v
30	BL/YL	MLP Sensor Signal	In P: 0v, In D: 2.1v
31	---	Not Used	---
32	OR/WT	Automatic Shock Control	Hot 4.3v
33	BR/PK	EGR VR Solenoid	0%, 55 mph: 45%
34	OR	Cooling Fan System	Fan On: 1v, Off: 12v
35	BR/BL	Injector 4 Control	3.6-3.8 ms
36	PK	Spark Output Signal	50% duty cycle
37	RD	Vehicle Power	12-14v
38	WT/YL	EPC Solenoid Control	8.8v (20 psi)
39	BR/YL	Injector 3 Control	3.6-3.8 ms
40	BK/WT	Power Ground	<0.1v

1994-95 Thunderbird 3.8L V6 MFI SC VIN R (All) 60 Pin Connector

PCM Pin #	Wire Color	Circuit Description (60 Pin)	Value at Hot Idle
41	TN/WT	TCS (switch) Signal	TCS & O/D On: 12v
42	TN/LG	A/C Pressure Switch	A/C On: 12v, Off: 0v
43	RD/BK	HO2S-21 (B2 S1) Signal	0.1-1.1v
44	GY/BL	HO2S-11 (B1 S1) Signal	0.1-1.1v
45	DB/LG	BARO Sensor	159 Hz (sea level)
46	GY/RD	Analog Signal Return	<0.050v
47	GY/WT	TP Sensor Signal	0.5-1.2v
48	WT/PK	Self-Test Indicator Signal	STI Open: 5v, Closed: 1v
49	OR/BK	TOT Sensor Signal	2.10-2.40v
50	BL/RD	MAF Sensor Signal	0.7v, 55 mph: 2.1v
51	OR/YL	Shift Solenoid 1 Control	1v, 55 mph: 12v
52	PK/OR	Shift Solenoid 2 Control	12v, 55 mph: 1v
53	BK/OR	TCC Solenoid Control	0%, 55 mph: 95%
54	PK/YL	A/C WOT Relay Control	On: 1v, Off: 12v
55	WT/LG	TCIL (lamp) Control	On: 1v, Off: 12v
56	GY/OR	PIP Sensor Signal	50% duty cycle
57	RD	Vehicle Power	12-14v
58	TN	Injector 1 Control	3.6-3.8 ms
59	WT	Injector 2 Control	3.6-3.8 ms
60	BK/WT	Power Ground	<0.1v

Pin Connector Graphic

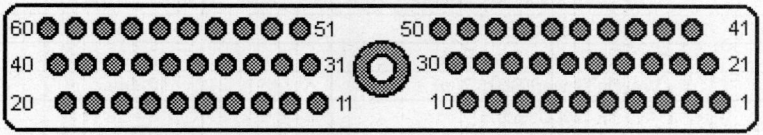

Standard Colors and Abbreviations

Abbreviation	Color	Abbreviation	Color	Abbreviation	Color
BK	Black	GY	Gray	PK	Purple
BL	Blue	GN	Green	RD	Red
BR	Brown	LG	LT Green	TN	Tan
DB	Dark Blue	OR	Orange	WT	White
DG	DK Green	PK	Pink	YL	Yellow

2002-03 Thunderbird 3.9L V8 VIN A (A/T) C175-B 58-Pin Connector

PCM Pin #	Wire Color	Circuit Description (58-Pin)	Value at Hot Idle
1	WT/RD	Electronic Throttle Control Module	Digital Signals
2	---	Not Used	---
3	GY	Data Bus (+) Signal	Digital Signals
4	BL	Data Bus (-) Signal	Digital Signals
5	BR	Signal Return	<0.050v
6-8	---	Not Used	---
9	BR/BL	A/C WOT Relay Control	On: 1v, Off: 12v
10-11	---	Not Used	---
12	BR/RD	EVAP Purge Solenoid Control	0-10 Hz (0-100%)
13	YL/GN	Flash EEPROM Power	0.1v
14	WT/VT	Manual Transmission Switch	N/A
15	WT	Electronic Throttle Control Module	Digital Signals
16	WT/VT	Electronic Throttle Control Module	Digital Signals
17	BR/WT	Electronic Throttle Control Module	Digital Signals
18-19	---	Not Used	---
20	YL/VT	ETC Reference Voltage	12-14v
21	---	Not Used	---
22	WT/BK	Manual Transmission Switch Signal	Switch Open: 12v, Closed: 1v
23	YL/RD	ETC Reference Voltage	12-14v
24-27	BK	Power Ground	<0.1v
28	OG/BL	Battery Power (Hot at all times)	12-14v
29	---	Not Used	---
30	BR/BL	EVAP Vent Solenoid Control	0-10 Hz (0-100%)
31	WT/BL	MAF Sensor Signal	0.6v, 55 mph: 3.3v
32	GN/YL	Vehicle Power (Start-Run)	12-14v
33	GN/OR	Vehicle Power (Start-Run)	12-14v
34-37	---	Not Used	---
38	BR/BL	MAF Sensor Return	<0.050v
39	---	Not Used	---
40	OR	Traction Control Disable Switch	Brake Off: 0v, On: 12v
41	WT/GN	Transmission Control Switch	TCS & O/D On: 12v
42	WT/VT	A/C Pressure Sensor Signal	0.9v (50 psi)
43	BN/YL	Power Ground	<0.1v
44	OR/YL	Keep Alive Power (CJB fuse)	12-14v
45-46	---	Not Used	---
47	WT/BK	Engine Cooling Fan Motor	Off: 12v, On: 1v
48	WT/BL	Restraint Control Module	Digital Signals
49	WT/VT	Rear Electronic Module	Digital Signals
50	---	Not Used	---
51	WT/GN	Intake Air Temperature Sensor	1.5-2.5v
52	WT/VT	Fuel Tank Pressure Sensor	2.6v (0" H2O - cap off)
53	GN/RD	Controller Area Network (+) Signal	0-7-0v
54	GN/RD	Controller Area Network (-) Signal	0-7-0v
55	YL	Reference Voltage	4.9-5.1v
56	BK/OR	Ground Switch Signal	<0.050v
57	YL/BL	Speed Control Reference Voltage	4.9-5.1v
58	---	Not Used	---

2002-03 Thunderbird 3.9L V8 VIN A (A/T) C175-T 32-Pin Connector

PCM Pin #	Wire Color	Circuit Description (32-Pin)	Value at Hot Idle
1	BR	Shift Solenoid 'A' Control	0.1, 55 mph: 12v
2	BR/RD	Shift Solenoid 'B' Control	12v, 55 mph: 12v
3-6	---	Not Used	---
7	BR/GN	Pressure Control 'A' Solenoid	8.8v
8	BR/WT	Shift Solenoid 'C' Control	12v, 55 mph: 12v
9	WT	DTR Sensor 3A Signal	In 'P': 0v, Others: 12v
10	WT/RD	DTR Sensor 4 Signal	In 'P': 0v, Others: 12v
11	---	Not Used	---
12	BR/WT	Pressure Control 'C' Solenoid	On: 1v, Off: 12v
13	BR/YL	Pressure Control 'B' Solenoid	On: 1v, Off: 12v
14	BR/RD	Sensor Signal Return	<0.050v
15	BR/BL	HO2S-12 (B1 S2) Heater	On: <1v, Off: 12v
16	BR/GN	HO2S-22 (B2 S2) Heater	On: <1v, Off: 12v
17	BR/YL	Shift Solenoid 'D' Control	0.1, 55 mph: 12v
18	WT/BL	DTR Sensor 2 Signal	In 'P': 0v, Others: 12v
19	---	Not Used	---
20	BR/WT	TCC Solenoid Control	12v, 55 mph: 0.1v
21	WT/BK	Intermediate Shaft Speed Sensor	0 Hz, 55 mph: 1320
22	WT/GN	DTR Sensor 1 Signal	In 'P': 0v, Others: 12v
23	WT/RD	TFT Sensor Signal	2.10-2.40v
24-25	---	Not Used	---
26	WT/RD	Output Shaft Speed Sensor	0 Hz, 55 mph: 1070 Hz
27	WT/VT	Turbine Shaft Speed Sensor	340 Hz, 55 mph: 1075
28	WT/BL	HO2S-12 (B1 S2) Signal	0.1-1.1v
29	WT/GN	HO2S-22 (B2 S2) Signal	0.1-1.1v
30-32	---	Not Used	---

PCM Pin Connectors Graphic

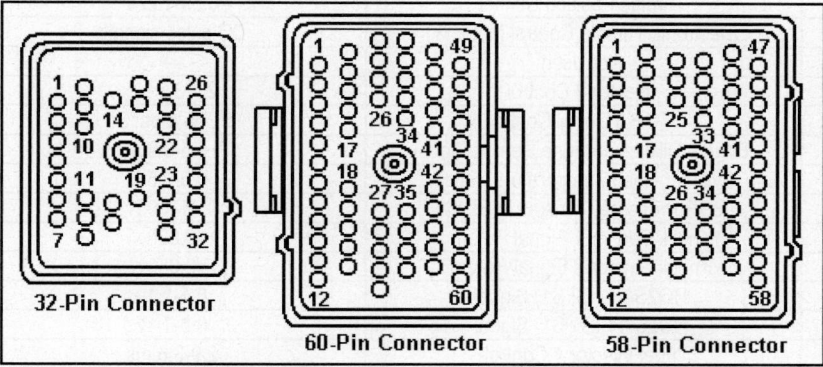

32-Pin Connector

60-Pin Connector   58-Pin Connector

Standard Colors and Abbreviations

Abbreviation	Color	Abbreviation	Color	Abbreviation	Color
BK	Black	GY	Gray	PK	Purple
BL	Blue	GN	Green	RD	Red
BR	Brown	LG	LT Green	TN	Tan
DB	Dark Blue	OR	Orange	WT	White
DG	DK Green	PK	Pink	YL	Yellow

2002-03 Thunderbird 3.9L V8 VIN A (A/T) C175-E 60 Pin Connector

PCM Pin #	Wire Color	Circuit Description (60 Pin)	Value at Hot Idle
1	WT/BK	Variable Valve Control Solenoid 1	DC pulse signals
2-3	---	Not Used	---
4	WT/BL	Fuel Rail Temperature Sensor	1.7-3.5v (50-120ºF)
5	WT	Power Steering Pressure Switch	Straight: 0v, Turned: 5v
6	---	Not Used	---
7	BR	HO2S-11 (B1 S1) Heater	On: 1v, Off: 12v
8	BR/RD	HO2S-21 (B2 S1) Signal	0.1-1.1v
9-10, 18	---	Not Used	---
11	BR/GN	Fuel Injector 5 Control	2.9-3.6 ms
12	BR/YL	COP 5 Driver Control	5º, 55 mph: 8º
13	WT/VT	Variable Valve Timing Solenoid 2	DC pulse signals
14	YL/RD	Reference Voltage	4.9-5.1v
15	BR	Throttle Position Sensor Return	0.050v
16	BR/GN	EGR Pressure Transducer Return	<0.050v
17	BR/GN	Sensor Signal Return	<0.050v
19	GY/RD	Generator Common Signal	0-130 Hz (varies)
20	BR	Fuel Injector 2 Control	2.9-3.6 ms
21	BR/YL	Fuel Injector 6 Control	2.9-3.6 ms
22	BR/WT	COP 6 Driver Control	5º, 55 mph: 8º
23	WT/GN	EGR Pressure Transducer Signal	0-3.1v
24	YL	Battery Power (at all times)	12-14v
25-26	---	Not Used	---
27	YL/BL	Electronic Throttle Control Motor (-)	DC pulse signals
28	BR/RD	Fuel Injector 5 Control	2.9-3.6 ms
29	BR/WT	Fuel Injector 7 Control	2.9-3.6 ms
30	BR	COP 7 Driver Control	5º, 55 mph: 8º
31, 33-34	---	Not Used	---
32	WT	Throttle Position Sensor	0.53-1.27v
35	WT/BL	Electronic Throttle Control Motor (+)	DC pulse signals
36	---	Not Used	---
37	BR	Injector 8 Control	2.9-3.6 ms
38	BR/RD	COP 8 Driver Control	5º, 55 mph: 8º
39	WT/GN	Engine Oil Temperature Sensor Signal	0.5-0.6v
40	WT/VT	Cylinder Head Temperature Sensor	0.5-0.6v
41	WT/BL	EGR Pressure Transducer Signal	0.95-1.05v
42	GY/BK	Knock Sensor 1 Signal Return	<0.050v
43	GY/RD	Knock Sensor 2 Signal Return	<0.050v
44	WT/RD	HO2S-21 (B2 S1) Signal	0.1-1.1v
45	WT	HO2S-11 (B1 S1) Signal	0.1-1.1v
46	BR/BL	Fuel Injector 4 Control	2.9-3.6 ms
47	BR/WT	Fuel Injector 1 Control	2.9-3.6 ms
48	BR/GN	COP 4 Driver Control	5º, 55 mph: 8º
49	WT/GN	FRP Sensor Signal	2.8v (39 psi)
50	WT/GN	Generator Monitor Signal	130 Hz (45%)
51	WT/BK	Knock Sensor 1 (+) Signal	0v
52	WT/RD	Knock Sensor 2 (+) Signal	0v
53	WT/VT	CMP Sensor 1 Signal	6 Hz
54	WT/BK	CMP Sensor 2 Signal	6 Hz
55	GY/RD	CKP Sensor (-) Signal	390 Hz
56	WT/RD	CKP Sensor (+) Signal	390 Hz
57	WT/RD	Throttle Position Sensor	0.53-1.27v
58	BR	COP 1 Driver Control	5º, 55 mph: 8º
59	BR/RD	COP 2 Driver Control	5º, 55 mph: 8º
60	BR/BL	COP 3 Driver Control	5º, 55 mph: 8º

1994-97 Thunderbird 4.6L V8 MFI VIN W (A/T) 104 Pin Connector

PCM Pin #	Wire Color	Circuit Description (104 Pin)	Value at Hot Idle
1	PK/OR	Shift Solenoid 2 Control	12v, 55 mph: 1v
2	PK/LG	MIL (lamp) Control	MIL On: 1v, Off: 12v
3-10	---	Not Used	---
11	PK/WT	Purge Flow Sensor	0.8v, 55 mph: 3v
12	---	Not Used	---
13	PK	Flash EEPROM Power	0.1v
14	---	Not Used	---
15	PK/LB	Data Bus (-) Signal	Digital Signals
16	TN/OR	Data Bus (+) Signal	Digital Signals
17-20	---	Not Used	---
21	DB	CKP Sensor (+) Signal	450-480 Hz
22	GY	CKP Sensor (-) Signal	450-480 Hz
23	---	Not Used	---
24	BK/WT	Power Ground	<0.1v
25	BK	PCM Case Ground	<0.050v
26	DB/LG	Coil Driver 1 Control	6° dwell
27	OR/YL	Shift Solenoid 1 Control	1v, 55 mph: 12v
28	---	Not Used	---
29	TN/WT	TCS (switch) Signal	TCS & O/D On: 12v
30	DG	Octane Adjust Switch	Closed: 0v, Open: 9.3v
31-32	---	Not Used	---
33	PK/OR	VSS (-) Signal	0 Hz, 55 mph: 125 Hz
34	---	Not Used	---
35	RD/LG	HO2S-12 (B1 S2) Signal	0.1-1.1v
36	TN/LB	MAF Sensor Return	<0.050v
37	OR/BK	TFT Sensor Signal	2.10-2.40v
38	LG/RD	ECT Sensor Signal	0.5-0.6v
39	GY	IAT Sensor Signal	1.5-2.5v
40	PK/BK	Fuel Pump Monitor	On: 12v, Off: 0v
41	PK/LB	A/C Cycling Clutch Switch	AC On: 12v, Off: 0v
42-44	---	Not Used	---
45	TN/OR	Low Speed Fan Control	On: 1v, Off: 12v
46	LG/PK	High Speed Fan Control	On: 1v, Off: 12v
47	BR/PK	EGR VR Solenoid	0%, 55 mph: 45%
48-50	---	Not Used	---
51	BK/WT	Power Ground	<0.1v
52	RD/LB	Coil Driver 2 Control	6° dwell
53	---	Not Used	---
54	BR/OR	TCC Solenoid Control	0%, 55 mph: 95%
55	YL	Keep Alive Power	12-14v
56-57	---	Not Used	---
58	GY/BK	VSS (+) Signal	0 Hz, 55 mph: 125 Hz
59	---	Not Used	---
60	GY/LB	HO2S-11 (B1 S1) Signal	0.1-1.1v
61	PK/LG	HO2S-22 (B2 S2) Signal	0.1-1.1v
62-63	---	Not Used	---
64	LB/YL	TR Sensor Signal	In 'P': 0v, 55 mph: 1.7v

1994-97 Thunderbird 4.6L V8 MFI VIN W (A/T) 104 Pin Connector

PCM Pin #	Wire Color	Circuit Description (104 Pin)	Value at Hot Idle
65	BR/LG	DPFE Sensor Signal	0.20-1.30v
66, 68	---	Not Used	---
67	GY/YL	EVAP Purge Solenoid	0-10 Hz (0-100%)
69	PK/YL	A/C WOT Relay Control	On: 1v, Off: 12v
70	---	Not Used	---
71	RD	Vehicle Power	12-14v
72	TN/RD	Injector 7 Control	3.1-3.5 ms
73	TN/BK	Injector 5 Control	3.1-3.5 ms
74	BR/YL	Injector 3 Control	3.1-3.5 ms
75	TN	Injector 1 Control	3.1-3.5 ms
76-77	BK/WT	Power Ground	<0.1v
78	PK/WT	Coil Driver 3 Control	6° dwell
79	WT/LG	TCIL (lamp) Control	On: 1v, Off: 12v
80	LB/OR	Fuel Pump Control	On: 1v, Off: 12v
81	WT/YL	EPC Solenoid Control	9.0v (15 psi)
82	---	Not Used	---
83	WT/LB	IAC Motor Control	10v (30% on time)
84	DG/WT	TSS Sensor Signal	43 Hz (650 rpm)
85	DG	CID Sensor Signal	5-7 Hz
86	TN/LG	A/C Pressure Switch	A/C On: 12v (Open)
87	RD/BK	HO2S-21 (B2 S1) Signal	0.1-1.1v
88	LB/RD	MAF Sensor Signal	0.6v
89	GY/WT	TP Sensor Signal	0.53-1.27v
90	BR/WT	Reference Voltage	4.9-5.1v
91	GY/RD	Analog Signal Return	<0.050v
92	LG	Brake Pedal Position Switch	Brake Off: 0v, On: 12v
93	RD/WT	HO2S-11 (B1 S1) Heater	On: 1v, Off: 12v
94	YL/LB	HO2S-21 (B2 S1) Heater	On: 1v, Off: 12v
95	WT/BK	HO2S-12 (B1 S2) Heater	On: 1v, Off: 12v
96	TN/YL	HO2S-22 (B2 S2) Heater	On: 1v, Off: 12v
97	RD	Vehicle Power	12-14v
98	LB	Injector 8 Control	3.1-3.5 ms
99	LG/OR	Injector 6 Control	3.1-3.5 ms
100	BR/LB	Injector 4 Control	3.1-3.5 ms
101	WT	Injector 2 Control	3.1-3.5 ms
102	---	Not Used	---
103	BK/WT	Power Ground	<0.1v
104	RD/YL	Coil Driver 4 Control	6° dwell

Pin Connector Graphic

PCM 104-PIN CONNECTOR

Terminal View of 104-Pin PCM Wiring Harness Connector

Standard Colors and Abbreviations

Abbreviation	Color	Abbreviation	Color	Abbreviation	Color
BK	Black	GY	Gray	PK	Purple
BL	Blue	GN	Green	RD	Red
BR	Brown	LG	LT Green	TN	Tan
DB	Dark Blue	OR	Orange	WT	White
DG	DK Green	PK	Pink	YL	Yellow

1991-93 Thunderbird 5.0L V8 MFI VIN T (A/T) 60 Pin Connector

PCM Pin #	Wire Color	Circuit Description (60 Pin)	Value at Hot Idle
1	YL	Keep Alive Power	12-14v
2	---	Not Used	---
3	DG/WT	VSS (+) Signal	0 Hz, 55 mph: 125 Hz
4	RD/BL	Ignition Diagnostic Monitor	20-31 Hz
4	WT/PK	Ignition Diagnostic Monitor	20-31 Hz
5	---	Not Used	---
6	BK/WT	VSS (-) Signal	0 Hz, 55 mph: 125 Hz
6	GY/RD	VSS (-) Signal	0 Hz, 55 mph: 125 Hz
7	LG/YL	ECT Sensor Signal	0.5-0.6v
7	LG/RD	ECT Sensor Signal	0.5-0.6v
8	PK/BK	Fuel Pump Monitor	On: 12v, Off: 0v
9	TN/BL	MAF Sensor Return	<0.050v
10	PK/BL	A/C Cycling Clutch Switch	A/C On: 12v, Off: 0v
11	GY/YL	EVAP Purge Solenoid	12v, at 55 mph: 1v
12	LG	Injector 6 Control	4.9-5.2 ms
12	LG/OR	Injector 6 Control	4.9-5.2 ms
13	TN/OR	Injector 7 Control	4.9-5.2 ms
13	TN/RD	Injector 7 Control	4.9-5.2 ms
14	BL	Injector 8 Control	4.9-5.2 ms
15	TN/BL	Injector 5 Control	4.9-5.2 ms
15	TN/BK	Injector 5 Control	4.9-5.2 ms
16	GY	Ignition System Ground	<0.050v
17	YL/BK	STI Output, MIL Control	MIL On: 1v, Off: 12v
18	TN/OR	Data Bus (+) Signal	Digital Signals
19	PK/BL	Data Bus (-) Signal	Digital Signals
20	BK	PCM Case Ground	<0.050v
21	RD/LG	IAC Motor Control	9.3v (40%)
22	TN/LG	Fuel Pump Control	On: 1v, Off: 12v
23-24	---	Not Used	---
25	GY	IAT Sensor Signal	1.5-2.5v
25	LG/PK	IAT Sensor Signal	1.5-2.5v
26	OR/WT	Reference Voltage	4.9-5.1v
26	BR/WT	Reference Voltage	4.9-5.1v
27	BR/LG	EGR EVP Sensor Signal	0.4v
28-29	---	Not Used	---
30	RD/BL	Neutral Drive Switch Signal	In P/N: 0v, all others: 4-5v
31	WT/RD	Air Management 1 Solenoid	AM1 On: 1v, Off: 12v
32	OR/WT	Automatic Ride Control	Hot 4.3v
33	DG	EGR VR Solenoid	0%, 55 mph: 45%
33	BR/PK	EGR VR Solenoid	0%, 55 mph: 45%
34	BL/PK	Data Output Link	Digital Signals
35	BR/BL	Injector 4 Control	4.9-5.2 ms
36	YL/LG	Spark Output Signal	50% duty cycle
37	RD	Vehicle Power	12-14v
38	---	Not Used	---
39	BR/YL	Injector 3 Control	4.9-5.2 ms
40	BK/LG	Power Ground	<0.1v

1991-93 Thunderbird 5.0L V8 MFI VIN T (A/T) 60 Pin Connector

PCM Pin #	Wire Color	Circuit Description (60 Pin)	Value at Hot Idle
41-42	---	Not Used	---
43	TN/RD	HO2S-21 (B2 S1) Signal	0.1-1.1v
43	TN/WT	HO2S-21 (B2 S1) Signal	0.1-1.1v
44	TN/OR	HO2S-11 (B1 S1) Signal	0.1-1.1v
45	---	Not Used	---
46	BK/WT	Analog Signal Return	<0.050v
46	GY/RD	Analog Signal Return	<0.050v
47	DG/LG	TP Sensor Signal	1.0v, 55 mph: 1.2v
47	GY/WT	TP Sensor Signal	1.0v, 55 mph: 1.2v
48	WT/RD	Self-Test Indicator Signal	STI Open: 5v, Closed: 1v
48	WT/PK	Self-Test Indicator Signal	STI Open: 5v, Closed: 1v
49	---	Not Used	---
50	DB/OR	MAF Sensor Signal	0.7v, 55 mph: 2.1v
50	BL/RD	MAF Sensor Signal	0.7v, 55 mph: 2.1v
51-53	---	Not Used	---
54	OR/BL	A/C WOT Relay Control	On: 1v, Off: 12v
55	---	Not Used	---
56	DB	PIP Sensor Signal	50% duty cycle
57	RD	Vehicle Power	12-14v
58	TN	Injector 1 Control	4.9-5.2 ms
59	WT	Injector 2 Control	4.9-5.2 ms
60	BK/LG	Power Ground	<0.1v

Pin Connector Graphic

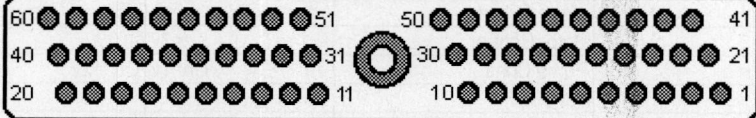

PCM 60-PIN CONNECTOR

Terminal View of 60-Pin PCM Harness Connector

Standard Colors and Abbreviations

Abbreviation	Color	Abbreviation	Color	Abbreviation	Color
BK	Black	GY	Gray	PK	Purple
BL	Blue	GN	Green	RD	Red
BR	Brown	LG	LT Green	TN	Tan
DB	Dark Blue	OR	Orange	WT	White
DG	DK Green	PK	Pink	YL	Yellow

LINCOLN CAR CONTENTS

About This Section
Introduction ....Page 10-2
How to Use This Section ....Page 10-2

CONTINENTAL PIN TABLES
3.8L V6 MFI VIN 4 (A/T) - 60 Pin **(1990)** ....Page 10-3
3.8L V6 MFI VIN 4 (A/T) - 60 Pin **(1991-94)** ....Page 10-5
4.6L V8 MFI VIN V (A/T) - 60 Pin **(1995)** ....Page 10-7
4.6L V8 MFI VIN V (A/T) - 104 Pin **(1996-97)** ....Page 10-9
4.6L V8 MFI VIN V (A/T) - 104 Pin **(1998-99)** ....Page 10-11
4.6L V8 MFI VIN V (A/T) - 104 Pin **(2000-02)** ....Page 10-13

LS 3.0L PIN TABLES
3.0L V6 MFI VIN S (A/T) - 150 Pin **(2000-01)** ....Page 10-15
3.0L V6 MFI VIN S (A/T) - 150 Pin **(2002-03)** ....Page 10-20

LS 3.9L PIN TABLES
3.9L V8 MFI VIN A (A/T) - 150 Pin **(2000-02)** ....Page 10-21
3.9L V8 MFI VIN A (A/T) - 150 Pin **(2003)** ....Page 10-24

MARK VII PIN TABLES
5.0L V8 HO MFI VIN E (A/T) - 60 Pin **(1990-92)** ....Page 10-27

MARK VIII PIN TABLES
4.6L V8 MFI VIN V (A/T) - 60 Pin **(1993-95)** ....Page 10-29
4.6L V8 MFI VIN V (A/T) - 104 Pin **(1996-97)** ....Page 10-31
4.6L V8 4v MFI VIN V (A/T) - 104 Pin **(1998)** ....Page 10-33

TOWN CAR PIN TABLES
4.6L V8 MFI VIN W (A/T) - 60 Pin **(1991-95)** ....Page 10-35
4.6L V8 MFI VIN W (A/T) - 104 Pin **(1996-97)** ....Page 10-37
4.6L V8 2v MFI VIN W (A/T) - 104 Pin **(1998-99)** ....Page 10-39
4.6L V8 2v MFI VIN W (A/T) - 104 Pin **(2000-01)** ....Page 10-41
4.6L V8 2v MFI VIN W (A/T) - 104 Pin **(2002-03)** ....Page 10-43
5.0L V8 MFI VIN F (A/T) - 60 Pin **(1990)** ....Page 10-45

About This Section

Introduction

This section contains Pin Tables for Lincoln applications from 1990-2003. It can be used to assist in the repair of Code and No Code faults related to the PCM.

VEHICLE COVERAGE
- Continental Applications (1990-2002)
- Lincoln LS Applications (2000-2003)
- Mark VII Applications (1990-92)
- Mark VIII Applications (1993-98)
- Town Car Applications (1990-2003)

How to Use This Section

This section can be used to look up the location of a particular pin, a Wire Color or a "known good" value of a PCM circuit. To locate the PCM information for a particular vehicle, find the model, correct engine size (with VIN Code) and finally the year of the vehicle.

For example, to look up the PCM terminals for a 1999 Town Car 4.6L VIN W, go to Contents Page 1 and find the text string shown below:

4.6L V8 2v MFI VIN W (A/T) - 104 Pin **(1998-99)** ....Page 10-40

Then turn to Page 10-40 to find the following PCM related information.

1998-99 Town Car 4.6L 2v V8 MFI VIN W (A/T) 104 Pin Connector

PCM Pin #	Wire Color	Circuit Description (104 Pin)	Value at Hot Idle
1	OR/YL	COP 6 Driver Control	5°, 55 mph: 8° dwell
9	YL/WT	Fuel Level Indicator Signal	1.7v (1/2 full)
11	PK/OR	Shift Solenoid 2 Control	12v, 55 mph: 1v
66	YL/LG	CHT Sensor Signal	Hot Idle: 0.6-3.7v (194°F)
85	DG	CID Sensor Signal	5-7 Hz

In this example, the Fuel Level Indicator connects to Pin 9 of the 104 Pin connector with a Yellow/White wire. The value at Hot Idle shown here is a nominal value for the fuel level with the tank 1/2 full.

The cylinder head temperature (CHT) sensor is used to indicate the engine coolant temperature on this vehicle. This device is connected to a circuit with a pull-up resistor. Due to this unique design, the CHT sensor reading at near 194°F can read in the Cold End Line (3.71v) or in the Hot End Line (0.6v). This variance should be understood when using the CHT sensor Pin Table during cooling system testing.

The "A/T" acronym that appears at the end of the Title over the table indicates the values are for vehicles with an automatic transmission.

Continental Pin Tables

1990 Continental 3.8L V6 MFI VIN 4 (A/T) 60 Pin Connector

PCM Pin #	Wire Color	Circuit Description (60 Pin)	Value at Hot Idle
1	YL	Keep Alive Power	12-14v
2	YL/LG	PSP Switch Signal	Straight: 0v, Turned: 12v
3	DG/WT	VSS (+) Signal	0 Hz, 55 mph: 125 Hz
4	DG/YL	Ignition Diagnostic Monitor	20-31 Hz
5	RD/LG	Brake Switch Signal	Brake Off: 0v, On: 12v
6	OR/YL	VSS (-) Signal	0 Hz, 55 mph: 125 Hz
7	LG/YL	ECT Sensor Signal	0.5-0.6v
8	BK/OR	Data Bus (-) Signal	Digital Signals
9	GY/WT	TOT Sensor Signal	2.10-2.40v
10	LG/PK	A/C Cycling Clutch Switch	Switch On: 12v, Off: 0v
11	OR/YL	Speed Control Solenoid (+)	Solenoid On: 12v
12	TN/YL	Injector 3 Control	6.0-6.8 ms
13	TN/BK	Injector 4 Control	6.0-6.8 ms
14	TN/BL	Injector 5 Control	6.0-6.8 ms
15	TN/LG	Injector 6 Control	6.0-6.8 ms
16	GY	Ignition System Ground	<0.050v
17	TN/RD	Self-Test Output, MIL Control	MIL On: 1v, Off: 12v
18	DG/PK	Transmission 3-4 Switch	3-4 Switch Closed: 0.1v
19	OR/YL	Transmission 2-3 Switch	2-3 Switch Closed: 0.1v
20	BK	PCM Case Ground	<0.050v
21	OR/BK	IAC Motor Control	9-10v
22	TN/LG	Fuel Pump Control	On: 1v, Off: 12v
23-24	---	Not Used	---
25	LG/PK	ACT Sensor Signal	1.5-2.5v
26	OR/WT	Reference Voltage	4.9-5.1v
27	BR/LG	PFE EGR Sensor Signal	3.2v, 55 mph: 2.9v
28	BL/BK	S/C Command Switch Signal	All speeds: 6.7v
29	DB/LG	HO2S-11 (B1 S1) Signal	0.1-1.1v
30	PK/YL	Neutral Drive Switch Signal	In 'P': 0v, Others: 5v
31	GY/YL	EVAP Purge Solenoid	12v, 55 mph: 1v
32	LG	Air Suspension Module	4.3v
33	DG	VR Solenoid Control	0%, 55 mph: 45%
34	---	Not Used	---
35	WT/PK	S/C Vent Solenoid Control	Vacuum Increasing: 1v
36	YL/LG	Spark Output Signal	6.93v (50% dwell)
37	RD	Vehicle Power	12-14v
38	---	Not Used	---
39	OR	Speed Control Switch Ground	<0.050v
40	BK/LG	Power Ground	<0.1v

1990 Continental 3.8L V6 MFI VIN 4 (A/T) 60 Pin Connector

PCM Pin #	Wire Color	Circuit Description (60 Pin)	Value at Hot Idle
41	PK	High Speed Fan Control	Fan On: 0v, Off: 12v
42	---	Not Used	---
43	DG/PK	HO2S-21 (B2 S1) Signal	0.1-1.1v
44	OR/BK	Data Bus (+) Signal	Digital Signals
45	LG/BK	MAP Sensor Signal	107 Hz (sea level)
46	BK/WT	Analog Signal Return	<0.050v
47	DG/LG	TP Sensor Signal	0.8v, 55 mph: 1.1v
48	WT/BK	Self-Test Input Signal	STI On: 0v, Off: 5v
49	OR	HO2S-11 (B1 S1) Ground	<0.050v
50	PK/BK	Fuel Pump Monitor	12v, Off: 0v
51	GY/BK	S/C Vacuum Solenoid Control	Vacuum Increasing: 1v
52	---	Not Used	---
53	TN/BL	LUS Solenoid Control	12v, 55 mph: 1v
54	RD	A/C WOT Relay Control	On: 1v, Off: 12v
55	TN/OR	Low Speed Fan Control	Fan On: 1v, Off: 12v
56	DB	PIP Sensor Signal	6.93v (50% d/cycle)
57	RD	Vehicle Power	12-14v
58	TN	Injector 1 Control	6.0-6.8 ms
59	TN/WT	Injector 2 Control	6.0-6.8 ms
60	BK/LG	Power Ground	<0.1v

Pin Connector Graphic

Terminal View of 60-Pin PCM Harness Connector

Standard Colors and Abbreviations

Abbreviation	Color	Abbreviation	Color	Abbreviation	Color
BK	Black	GY	Gray	PK	Purple
BL	Blue	GN	Green	RD	Red
BR	Brown	LG	LT Green	TN	Tan
DB	Dark Blue	OR	Orange	WT	White
DG	DK Green	PK	Pink	YL	Yellow

1991-94 Continental 3.8L V6 MFI VIN 4 (A/T) 60 Pin Connector

PCM Pin #	Wire Color	Circuit Description (60 Pin)	Value at Hot Idle
1	YL	Keep Alive Power	12-14v
2	LG	Brake Switch Signal	Brake Off: 0v, On: 12v
3	GY/BK	VSS (+) Signal	0 Hz, 55 mph: 125 Hz
4	WT/PK	Ignition Diagnostic Monitor	20-31 Hz
5	DG/WT	TSS Sensor Signal	30 mph: 126-136 Hz
6	PK/OR	VSS (-) Signal	0 Hz, 55 mph: 125 Hz
7	LG/RD	ECT Sensor Signal	0.5-0.6v
8	PK/BK	Fuel Pump Monitor	12v, Off: 0v
8	DG/YL	Fuel Pump Monitor	12v, Off: 0v
9	TN/BL	MAF Sensor Return	<0.050v
10	PK	A/C Cycling Clutch Switch	Switch On: 12v, Off: 0v
10	LG/PK	A/C Cycling Clutch Switch	Switch On: 12v, Off: 0v
11	GY/YL	EVAP Purge Solenoid	12v, 55 mph: 1v
12	LG/OR	Injector 6 Control	6.0-6.8 ms
13	DB	Low Speed Fan Control	Fan On: 1v, Off: 12v
13	TN/OR	Low Speed Fan Control	Fan On: 1v, Off: 12v
14	---	Not Used	---
15	TN/BK	Injector 5 Control	6.0-6.8 ms
16	OR/RD	Ignition System Ground	<0.050v
17	PK/LG	Self-Test Output, MIL Control	MIL On: 1v, Off: 12v
18	TN/OR	Data Bus (+) Signal	Digital Signals
19	PK/BL	Data Bus (-) Signal	Digital Signals
20	BK	PCM Case Ground	<0.050v
21	WT	IAC Motor Control	Pulse Signals: 8-14v
21	WT/BL	IAC Motor Control	Pulse Signals: 8-14v
22	BL/OR	Fuel Pump Control	On: 1v, Off: 12v
22	TN/LG	Fuel Pump Control	On: 1v, Off: 12v
23	---	Not Used	---
24	---	Not Used	---
25	GY	IAT Sensor Signal	1.5-2.5v
26	BR/WT	Reference Voltage	4.9-5.1v
27	BR/LG	PFE EGR Sensor Signal	3.2v
28	YL/LG	PSP Switch Signal	Straight: 0v, Turned: 12v
29	---	Not Used	---
30	BL/YL	MLP Sensor Signal	In 'P': 4.4v, in 'D': 2.09v
31	LG/PK	High Speed Fan Control	Fan On: 0v, Off: 12v
32	DB/WT	Auto Suspension Control	5.5-9.0v
33	BR/PK	VR Solenoid Control	0%, 55 mph: 45%
34	---	Not Used	---
35	BR/BL	Injector 4 Control	6.0-6.8 ms
36	PK	Spark Output Signal	6.93v (50% dwell)
37	RD	Vehicle Power	12-14v
38	WT/YL	EPC Solenoid Control	8.8-9.9v
39	BR/YL	Injector 3 Control	6.0-6.8 ms
40	BK	Power Ground	<0.1v
40	BK/WT	Power Ground	<0.1v

1991-94 Continental 3.8L V6 MFI VIN 4 (A/T) 60 Pin Connector

PCM Pin #	Wire Color	Circuit Description (60 Pin)	Value at Hot Idle
41	---	Not Used	---
42 ('94)	TN/LG	AC High Pressure Switch	A/C Switch On: 0.8-1.9v
43	RD/BK	HO2S-21 (B2 S1) Signal	0.1-1.1v
44 ('91-'92)	GYL/BL	HO2S-11 (B1 S1) Signal	0.1-1.1v
45 ('93-'94)	GY/RD	HO2S-11 (B1 S1) Signal	0.1-1.1v
46	GY/RD	Analog Signal Return	<0.050v
47	GY/WT	TP Sensor Signal	0.8v, 55 mph: 1.1v
48	WT/PK	Self-Test Input Signal	STI On: 0v, Off: 5v
48	WT/PK	Self-Test Input Signal	STI On: 0v, Off: 5v
49	OR/BK	TOT Sensor Signal	2.10-2.40v
50	BL/RD	MAF Sensor Signal	0.7v, 55 mph: 2.1v
51	OR/YL	Shift Solenoid 1 Control	12v, 55 mph: 1v
52	PK/OR	Shift Solenoid 2 Control	1v, 55 mph: 12v
53	TN/WT	TCC Solenoid Control	30 mph: 11-12v (90%)
54	PK/YL	A/C WOT Relay Control	On: 1v, Off: 12v
55	PK/BK	Shift Solenoid 3 Control	12v, 55 mph: 1v
56	GY/OR	PIP Sensor Signal	6.93v (50% d/cycle)
57	RD	Vehicle Power	12-14v
58	TN	Injector 1 Control	6.0-6.8 ms
59	WT	Injector 2 Control	6.0-6.8 ms
60	BK/LG	Power Ground	<0.1v

Pin Connector Graphic

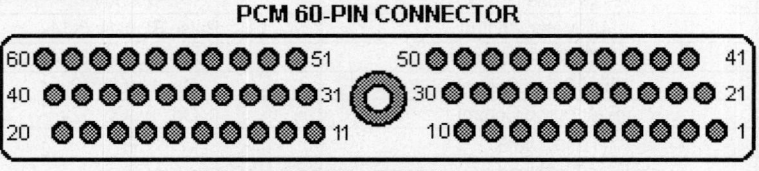

Standard Colors and Abbreviations

Abbreviation	Color	Abbreviation	Color	Abbreviation	Color
BK	Black	GY	Gray	PK	Purple
BL	Blue	GN	Green	RD	Red
BR	Brown	LG	LT Green	TN	Tan
DB	Dark Blue	OR	Orange	WT	White
DG	DK Green	PK	Pink	YL	Yellow

1995 Continental 4.6L V8 MFI VIN V (A/T) 60 Pin Connector

PCM Pin #	Wire Color	Circuit Description (60 Pin)	Value at Hot Idle
1	YL	Keep Alive Power	12-14v
2 ('93-'95)	LG	Brake Switch Signal	Brake Off: 0v, On: 12v
3	DG/WT	VSS (+) Signal	0 Hz, 55 mph: 125 Hz
4	TN/YL	Ignition Diagnostic Monitor	20-31 Hz
5	DG/WT	TSS+ Sensor	Idle: 790-820 Hz
6	PK/OR	VSS (-) Signal	0 Hz, 55 mph: 125 Hz
7	LG/RD	ECT Sensor Signal	0.5-0.6v
8	PK/BK	Fuel Pump Monitor	12v, Off: 0v
9	TN/LB	MAF Sensor Return	<0.050v
10	BK/YL	A/C Cycling Clutch Switch	Switch On: 12v, Off: 0v
11	GY/YL	EVAP Purge Solenoid	12v, 55 mph: 9v
12	LG/OR	Injector 6 Control	4.0-4.4 ms
13	TN/RD	Injector 7 Control	4.0-4.4 ms
14	LB	Injector 8 Control	4.0-4.4 ms
15	TN/BK	Injector 5 Control	4.0-4.4 ms
16	OR/RD	Ignition System Ground	<0.050v
17	PK/LG	Self-Test Output & MIL	MIL On: 1v, Off: 12v
18	TN/OR	Data Bus (+) Signal	Digital Signals
19	PK/LB	Data Bus (-) Signal	Digital Signals
20	BK	PCM Case Ground	<0.050v
21	WT/LB	IAC Motor Control	8.3-11.5v
22	LB/OR	Fuel Pump Control	On: 1v, Off: 12v
23	YL/RD	Knock Sensor Signal	0v
24	DB/OR	CID Sensor Signal	6-7 Hz
25	GY	ACT Sensor Signal	1.5-2.5v
26	BR/WT	Reference Voltage	4.9-5.1v
27	BR/LG	DPFE EGR Sensor Signal	0.4v, 55 mph: 2.9v
28	---	Not Used	---
29	DG	Octane Adjust Switch	Closed: 0v, Open: 9.1v
30	LB/YL	MLP Sensor Signal	In 'P': 0v, in O/D: 1.7v
31	---	Not Used	---
32	LG/BK	IMRC Solenoid	12v, 55 mph: 12v
33	BR/PK	VR Solenoid Control	0%, 55 mph: 45%
34	LB/PK	Data Output Link	Digital Signals
35	BR/LB	Injector 4 Control	4.0-4.4 ms
36	PK	Spark Angle Word Signal	6.93v (50% d/cycle)
37	RD	Vehicle Power	12-14v
38 ('93-'95)	WT/YL	EPC Solenoid Control	9.5v (20 psi)
39	BR/YL	Injector 3 Control	4.0-4.4 ms
40	BK/WT	Power Ground	<0.1v

1995 Continental 4.6L V8 MFI VIN V (A/T) 60 Pin Connector

PCM Pin #	Wire Color	Circuit Description (60 Pin)	Value at Hot Idle
41	TN/WT	TCS (switch) Signal	TCS & O/D On: 12v
42	---	Not Used	---
43	RD/BK	HO2S-21 (B2 S1) Signal	0.1-1.1v
44	GY/LB	HO2S-11 (B1 S1) Signal	0.1-1.1v
45	---	Not Used	---
46	GY/RD	Analog Signal Return	<0.050v
47	GY/WT	TP Sensor Signal	0.8v, 55 mph: 1.1v
48	WT/PK	Self-Test Input Signal	STI On: 0v, Off: 5v
49	OR/BK	TOT Sensor Signal	2.10-2.40v
50	LB/RD	MAF Sensor Signal	0.8v, 55 mph: 1.8v
51	OR/YL	Shift Solenoid 1 Control	1v, 55 mph: 12v
52	PK/OR	Shift Solenoid 2 Control	12v, 55 mph: 1v
53	PK/YL	TCC Solenoid Control	55 mph: 9-10v
54	PK/YL	A/C WOT Relay Control	On: 1v, Off: 12v
55	WT/LG	TCIL (lamp) Control	Lamp On: 1v, Off: 12v
56	GY/OR	PIP Sensor Signal	6.93v (50% dwell)
57	RD	Vehicle Power	12-14v
58	TN	Injector 1 Control	4.0-4.4 ms
59	WT	Injector 2 Control	4.0-4.4 ms
60	BK/WT	Power Ground	<0.1v

Pin Connector Graphic

PCM 60-PIN CONNECTOR

Terminal View of 60-Pin PCM Harness Connector

Standard Colors and Abbreviations

Abbreviation	Color	Abbreviation	Color	Abbreviation	Color
BK	Black	GY	Gray	PK	Purple
BL	Blue	GN	Green	RD	Red
BR	Brown	LG	LT Green	TN	Tan
DB	Dark Blue	OR	Orange	WT	White
DG	DK Green	PK	Pink	YL	Yellow

1996-97 Continental 4.6L V8 MFI VIN V (A/T) 104 Pin Connector

PCM Pin #	Wire Color	Circuit Description (104 Pin)	Value at Hot Idle
1	PK/OR	Shift Solenoid 2 Control	1v, 55 mph: 12v
2	PK/LG	MIL (lamp) Control	MIL On: 1v, Off: 12v
3-4	---	Not Used	---
5	BK/OR	Elect. Air Manage. System	1v, 55 mph: 1v
6-7	---	Not Used	---
8	LG/BK	IMRC Solenoid Monitor	All speeds: 5v
9-12, 14	---	Not Used	---
13	PK	Flash EEPROM Power	0.1v
15	PK/LB	Data Bus (-) Signal	Digital Signals
16	TN/OR	Data Bus (+) Signal	Digital Signals
17	---	Not Used	---
18	WT/PK	Traction Assist (PWM) Signal	Digital Signals
19-20	---	Not Used	---
21	BK/PK	CKP Sensor (+) Signal	310-330 Hz
22	GY/YL	CKP Sensor (-) Signal	310-330 Hz
23	---	Not Used	---
24	BK/WT	Power Ground	<0.1v
25	BK	PCM Case Ground	<0.050v
26	DB/LG	Coil Driver 1 Control	5°, 55 mph: 8° dwell
27	OR/YL	Shift Solenoid 1 Control	12v, 55 mph: 1v
28-29	---	Not Used	---
30	DG	Octane Adjust Switch	9.3v (Closed: 0v)
32	YL/LG	Knock Sensor 2 Signal	0v
33-34	---	Not Used	---
35	RD/LG	HO2S-12 (B1 S2) Signal	0.1-1.1v
36	TN/LB	MAF Sensor Return	<0.050v
37	OR/BK	TFT Sensor Signal	2.10-2.40v
38	LG/RD	ECT Sensor Signal	0.5-0.6v
39	GY	IAT Sensor Signal	1.5-2.5v
40	DG/YL	Fuel Pump Monitor	12v, Off: 0v
41	PK	A/C Cycling Clutch Switch	Switch On: 12v, Off: 0v
42-43	---	Not Used	---
44	BR	IMRC Solenoid Control	12v, 55 mph: 12v
46	LG/PK	High Speed Fan Control	Fan On: 1v, Off: 12v
47	BR/PK	VR Solenoid Control	0%, 55 mph: 45%
48-50	---	Not Used	---
51	BK/WT	Power Ground	<0.1v
52	RD/LB	Coil Driver 2 Control	5°, 55 mph: 8° dwell
53	PK/BK	Shift Solenoid 3 Control	12v, 55 mph: 1v
54	---	Not Used	---
55	YL	Keep Alive Power	12-14v
56	LG/BK	EVAP VMV (valve) Control	0-10 Hz (0-100%)
57	YL/RD	Knock Sensor 1 Signal	0v
58-59	---	Not Used	---
60	GY/LB	HO2S-11 (B1 S1) Signal	0.1-1.1v
61	PK/LG	HO2S-22 (B2 S2) Signal	0.1-1.1v
63	YL/WT	TP Sensor Signal 'B'	0.5-0.7v
64	LB/YL	TR Sensor Signal	In 'P': 0v, in O/D: 1.7v

1996-97 Continental 4.6L V8 MFI VIN V (A/T) 104 Pin Connector

PCM Pin #	Wire Color	Circuit Description (104 Pin)	Value at Hot Idle
65	BR/LG	DPFE EGR Sensor Signal	0.20-1.30v
68	DB	Low Speed Fan Control	Fan On: 1v, Off: 12v
69	PK/YL	A/C WOT Relay Control	On: 1v, Off: 12v
70	WT/OR	Air Mgmt. Solenoid Control	12v, 55 mph: 12v
71	RD	Vehicle Power	12-14v
72	TN/RD	Injector 7 Control	2.8-2-9 ms
73	TN/BK	Injector 5 Control	2.8-2-9 ms
74	BR/YL	Injector 3 Control	2.8-2-9 ms
75	TN	Injector 1 Control	2.8-2-9 ms
76, 77, 103	BK/WT	Power Ground	<0.1v
78	PK/WT	Coil Driver 3 Control	5°, 55 mph: 8° dwell
80	LB/OR	Fuel Pump Control	On: 1v, Off: 12v
81	WT/YL	EPC Solenoid Control	9.5v (17 psi)
82	DB/WT	TCC Solenoid Control	55 mph: 90-100%
83	WT/LB	IAC Motor Control	30%, 55 mph: 45%
84	DG/WT	TSS Sensor Signal	9.5v (17 psi)
85	DB/OR	CMP Sensor	4-6 Hz
86	TN/LG	A/C Pressure Switch Signal	Open: 12v, Closed: 0v
87	RD/BK	HO2S-21 (B2 S1) Signal	0.1-1.1v
88	LB/RD	MAF Sensor Signal	0.6v, 55 mph: 1.7v
89	GY/WT	TP Sensor Signal	0.53-1.27v
90	BR/WT	Reference Voltage	4.9-5.1v
91	GY/RD	Analog Signal Return	<0.050v
92	LG	Brake Switch Signal	Brake Off: 0v, On: 12v
93	RD/WT	HO2S-11 (B1 S1) Heater	1v, Off: 12v
94	YL/LB	HO2S-21 (B2 S1) Heater	1v, Off: 12v
95	WT/BK	HO2S-12 (B1 S2) Heater	1v, Off: 12v
96	TN/YL	HO2S-22 (B2 S2) Heater	1v, Off: 12v
97	RD	Vehicle Power	12-14v
98	LB	Injector 8 Control	2.8-2-9 ms
99	LG/OR	Injector 6 Control	2.8-2-9 ms
100	BR/LB	Injector 4 Control	2.8-2-9 ms
101	WT	Injector 2 Control	2.8-2-9 ms
102	---	Not Used	---
104	RD/YL	Coil Driver 4 Control	5°, 55 mph: 8° dwell

Pin Connector Graphic

PCM 104-PIN CONNECTOR

Terminal View of 104-Pin PCM Wiring Harness Connector

Standard Colors and Abbreviations

Abbreviation	Color	Abbreviation	Color	Abbreviation	Color
BK	Black	GY	Gray	PK	Purple
BL	Blue	GN	Green	RD	Red
BR	Brown	LG	LT Green	TN	Tan
DB	Dark Blue	OR	Orange	WT	White
DG	DK Green	PK	Pink	YL	Yellow

1998-99 Continental 4.6L V8 MFI VIN V (A/T) 104 Pin Connector

PCM Pin #	Wire Color	Circuit Description (104 Pin)	Value at Hot Idle
1	WT/PK	COP 5 Driver Control	5°, 55 mph: 8° dwell
2	PK/LG	MIL (lamp) Control	MIL On: 1v, Off: 12v
3	BK/WT	Power Ground	<0.1v
6	OR/YL	Shift Solenoid 1 Control	12v, 55 mph: 1v
8	LG/BK	IMRC Monitor	5v, 55 mph: 5v
9	YL/WT	Fuel Level Indicator Signal	1.7v (1/2 full)
11	PK/OR	Shift Solenoid 2 Control	1v, 55 mph: 1v
12	YL/WT	Fuel Level Indicator Signal	1.7v (1/2 full)
13	PK	Flash EEPROM Power	0.5v
14	LB/WT	EFTA Sensor	1.7-3.5v (50-120°F)
15	PK	Data Bus (-) Signal	<0.050v
16	TN/OR	Data Bus (+) Signal	Digital Signals
18	RD/PK	Fuel Rail Pressure Sensor	2.8v (39 psi)
20	PK/BK	Shift Solenoid 3 Control	12v, 55 mph: 1v
21	BK	CKP (+) Sensor Signal	390-450 Hz
22	GY	CKP (-) Sensor Signal	390-450 Hz
25	BK	PCM Case Ground	<0.050v
26	LG/WT	COP 1 Driver Control	5°, 55 mph: 8° dwell
27	OR/YL	COP 6 Driver Control	5°, 55 mph: 8° dwell
28	DB	Low Cooling Fan Control	Fan On: 1v, Off: 12v
29	TN/WT	TCS (switch) Signal	TCS & O/D On: 12v
30	DG	Octane Adjust Switch	9.3v (Closed: 0v)
31	YL/LG	PSP Switch Signal	Straight: 5v, Turning: 0v
32	DG/PK	Knock Sensor 2 Signal	0v
33	PK/OR	VSS (-) Signal	0 Hz, 55 mph: 125 Hz
34	YL/BK	Digital TR1 Sensor Signal	0v, 55 mph: 11v
35	RD/LG	HO2S-12 (B1 S2) Signal	0.1-1.1v
36	TN/LB	MAF Sensor Return	<0.050v
37	OR/BK	TFT Sensor Signal	2.10-2.40v
38	LG/RD	ECT Sensor Signal	0.5-0.6v
39	GY	IAT Sensor Signal	1.5-2.5v
40	DG/YL	Fuel Pump Monitor	12v, Off: 0v
41	PK	A/C High Pressure Switch	Switch On: 12v, Off: 0v
42	BR	IMRC Solenoid Control	12v, 55 mph: 12v
43	WT/RD	Fuel Pump Control	On: 1v, Off: 12v
46	LG/PK	High Speed Fan Control	Fan On: 1v, Off: 12v
47	BR/PK	VR Solenoid Control	0%, 55 mph: 45%
48	OR/WT	Clean Tachometer Output	39-45 Hz
49	LB/BK	Digital TR2 Sensor Signal	0v, 55 mph: 11v
50	WT/BK	Digital TR4 Sensor Signal	0v, 55 mph: 11v
51	BK/WT	Power Ground	<0.1v
52	PK/LB	COP 2 Driver Control	5°, 55 mph: 8° dwell
53	PK/WT	COP 7 Driver Control	5°, 55 mph: 8° dwell
54	PK/YL	TCC Solenoid Control	12v, 55 mph: 1v
55	LG/RD	Keep Alive Power	12-14v
56	LG/BK	EVAP VMV (valve) Control	0-10 Hz (0-100%)
57	YL/RD	Knock Sensor 1 Signal	0v
58	GY/BK	VSS (+) Signal	0 Hz, 55 mph: 125 Hz
59	DG/WT	TSS Sensor Signal	40-45 Hz (645-650 rpm)
60	GY/LB	HO2S-11 (B1 S1) Signal	0.1-1.1v
61	PK/LG	HO2S-22 (B2 S2) Signal	0.1-1.1v
62	YL/WT	FTP Sensor Signal	2.6v (0" H2O - cap off)
64	LB/YL	Digital TR Sensor Signal	In 'P': 0v, in O/D: 1.7v

1998-99 Continental 4.6L V8 MFI VIN V (A/T) 104 Pin Connector

PCM Pin #	Wire Color	Circuit Description (104 Pin)	Value at Hot Idle
65	BR/LG	DPFE Sensor Signal	0.20-1.30v
67	PK/WT	EVAP Vent Valve	0-10 Hz (0-100%)
69	PK/YL	A/C WOT Relay Control	Relay On: 0v, Off: 12v
71, 97	RD	Vehicle Power	12-14v
72	TN/RD	Injector 7 Control	2.8-2-9 ms
73	TN/BK	Injector 5 Control	2.8-2-9 ms
74	BR/YL	Injector 3 Control	2.8-2-9 ms
75	TN	Injector 1 Control	2.8-2-9 ms
77, 103	BK/WT	Power Ground	<0.1v
78	DG/PK	COP 3 Driver Control	5°, 55 mph: 8° dwell
79	WT/RD	COP 8 Driver Control	5°, 55 mph: 8° dwell
80	LB/OR	Fuel Pump Control	On: 1v, Off: 12v
81	WT/YL	EPC Solenoid Control	9v (17 psi)
83	WT/LB	IAC Motor Control	35% duty cycle
84	DG/WT	TSS Sensor Signal	43 Hz (650 rpm)
85	DB/OR	CMP Sensor Signal	4-6 Hz
86	TN/LG	A/C Pressure Transducer	A/C On: 0.8-1.9v
87	RD/BK	HO2S-21 (B2 S1) Signal	0.1-1.1v
88	LB/RD	MAF Sensor Signal	0.6v, 55 mph: 1.7v
89	GY/WT	TP Sensor Signal	0.53-1.27v
90	BR/WT	Reference Voltage	4.9-5.1v
91	GY/RD	Analog Signal Return	<0.050v
92	LG	BPP Switch Signal	Brake On: 12v, Off: 0v
93	RD/WT	HO2S-11 (B1 S1) Heater	1v, Off: 12v
94	YL/LB	HO2S-21 (B2 S1) Heater	1v, Off: 12v
95	WT/BK	HO2S-12 (B1 S2) Heater	1v, Off: 12v
96	TN/YL	HO2S-22 (B2 S2) Heater	1v, Off: 12v
98	LB	Injector 8 Control	2.8-2-9 ms
99	LG/OR	Injector 6 Control	2.8-2-9 ms
100	BR/LB	Injector 4 Control	2.8-2-9 ms
101	WT	Injector 2 Control	2.8-2-9 ms
104	WT/PK	Coil Driver 4 Control	5°, 55 mph: 8° dwell

Pin Connector Graphic

PCM 104-PIN CONNECTOR

```
 1 ●●●●●●●●●●●●●    ●●●●●●●●●●●●● 26
27 ●●●●●●●●●●●●●  ⬤ ●●●●●●●●●●●●● 52
53 ●●●●●●●●●●●●●    ●●●●●●●●●●●●● 78
79 ●●●●●●●●●●●●●    ●●●●●●●●●●●●● 104
```

Terminal View of 104-Pin PCM Wiring Harness Connector

Standard Colors and Abbreviations

Abbreviation	Color	Abbreviation	Color	Abbreviation	Color
BK	Black	GY	Gray	PK	Purple
BL	Blue	GN	Green	RD	Red
BR	Brown	LG	LT Green	TN	Tan
DB	Dark Blue	OR	Orange	WT	White
DG	DK Green	PK	Pink	YL	Yellow

2000-02 Continental 4.6L V8 MFI VIN V (A/T) 104 Pin Connector

PCM Pin #	Wire Color	Circuit Description (104 Pin)	Value at Hot Idle
1	LG/YL	COP 6 Driver Control	5°, 55 mph: 8° dwell
2, 4-5, 7-8	---	Not Used	---
3	BK/WT	Power Ground	<0.1v
6	OG/YL	Shift Solenoid 1 Control	12v, 55 mph: 1v
9	YL/WT	Fuel Level Input	1.7v (1/2 full)
10, 17-18	---	Not Used	---
11	VT/OR	Shift Solenoid 2 Control	1v, 55 mph: 1v
13	VT	Flash EEPROM Power	0.1v
15	PK/LB	Data Bus (-) Signal	Digital Signals
16	TN/OG	Data Bus (+) Signal	Digital Signals
20	PK/BK	Shift Solenoid 3 Control	12v, 55 mph: 1v
21	BK/PK	CKP Sensor (+) Signal	330-420 Hz
22	GY/YL	CKP Sensor (-) Signal	330-420 Hz
23	DG/WT	Knock Sensor (-) 1 Signal	0v
25	BK	Case Ground	<0.050v
26	LG/WT	COP 1 Driver Control	5°, 55 mph: 8° dwell
27	OG/YL	COP 5 Driver Control	5°, 55 mph: 8° dwell
28	DB	Low Cooling Fan Control	Fan On: 1v, Off: 12v
29-30, 33	---	Not Used	---
31	YL/LG	PSP Switch Signal	Straight: 5v, Turning: 0v
32	DG/VT	Knock Sensor 1 (+) Signal	0v
34	OG/BK	Digital TR1 Sensor Signal	0v, 55 mph: 11v
35	RD/LG	HO2S-12 (B1 S2) Signal	0.1-1.1v
36	TN/LB	MAF Sensor Return	<0.050v
37	OG/BK	TFT Sensor Signal	2.10-2.40v
38	LG/RD	ECT Sensor Signal	0.5-0.6v
39	GY	IAT Sensor Signal	1.5-2.5v
40	DG/YL	Fuel Pump Monitor	0-7v, Off: 0v
41	VT	A/C Clutch Switch Signal	A/C On: 12v, Off: 0v
42-45, 48	---	Not Used	---
46	LG/VT	High Speed Fan Control	Fan On: 1v, Off: 12v
47	BR/PK	VR Solenoid Control	0%, 55 mph: 45%
49	BK/WT	Digital TR2 Sensor Signal	0v, 55 mph: 11v
50	DG/OG	Digital TR4 Sensor Signal	0v, 55 mph: 11v
51	BK/WT	Power Ground	<0.1v
52	PK/LB	COP 3 Driver Control	5°, 55 mph: 8° dwell
53	PK/WT	COP 4 Driver Control	5°, 55 mph: 8° dwell
54	DB/WT	TCC Solenoid Control	12v, 55 mph: 1v
55	LG/RD	Keep Alive Power	12-14v
56	LG/BK	EVAP Purge Solenoid	0-10 Hz (0-100%)
57	YL/RD	Knock Sensor 1 (+) Signal	0v
58	---	Not Used	---
59	DG/WT	TSS Sensor Signal	40-45 Hz (645-650 rpm)
60	GY/LB	HO2S-11 (B1 S1) Signal	0.1-1.1v
61	VT/LG	HO2S-22 (B2 S2) Signal	0.1-1.1v
62	RD/PK	FTP Sensor Signal	2.6v (0" H2O - cap off)
63	RD/PK	Fuel Rail Pressure Sensor	2.8v (39 psi)
64	LB/YL	Digital TR3 Signal	In 'P': 0v, in O/D: 1.7v
65 ('00)	BR/LG	DPFE Sensor Signal	0.20-1.30v
65 ('01-'02)	BR/LG	DPFE Sensor Signal	0.95-1.05v
66	---	Not Used	---

2000-02 Continental 4.6L V8 MFI VIN V (A/T) 104 Pin Connector

PCM Pin #	Wire Color	Circuit Description (104 Pin)	Value at Hot Idle
67	VT/WT	EVAP CV Solenoid	0-10 Hz (0-100%)
69	PK/YL	A/C WOT Relay Control	12v (off)
70	---	Not Used	---
71	RD	Vehicle Power	12-14v
72	TN/RD	Injector 7 Control	2.8-2-9 ms
73	TN/BK	Injector 5 Control	2.8-2-9 ms
74	BR/YL	Injector 3 Control	2.8-2-9 ms
75	TN	Injector 1 Control	2.8-2-9 ms
77	BK/WT	Comm. Network Ground	<0.050v
78	DG/VT	COP 7 Driver Control	5º, 55 mph: 8º dwell
79	WT/RD	COP 8 Driver Control	5º, 55 mph: 8º dwell
80	LB/OR	Fuel Pump Control	1.5-1.7v (28-33%)
81	WT/YL	EPC Solenoid Control	8-9v (15-18 psi)
82	---	Not Used	---
83	WT/LB	IAC Motor Control	28-30%
84	---	Not Used	---
85	DB/OR	CMP Sensor Signal	4-6 Hz
86	TN/LG	A/C Pressure Sensor	A/C On: 0.6v (30 psi)
87	RD/BK	HO2S-21 (B2 S1) Signal	0.1-1.1v
88	LB/RD	MAF Sensor Signal	0.6v, 55 mph: 1.7v
89	GY/WT	TP Sensor Signal	0.53-1.27v
90	BR/WT	Reference Voltage	4.9-5.1v
91	GY/RD	Analog Signal Return	<0.050v
92	---	Not Used	---
93	RD/WT	HO2S-11 (B1 S1) Heater	1v, Off: 12v
94	YL/LB	HO2S-21 (B2 S1) Heater	1v, Off: 12v
95	WT/BK	HO2S-12 (B1 S2) Heater	1v, Off: 12v
96	TN/YL	HO2S-22 (B2 S2) Heater	1v, Off: 12v
97	RD	Vehicle Power	12-14v
98	LB	Injector 8 Control	2.8-2-9 ms
99	LG/OR	Injector 6 Control	2.8-2-9 ms
100	BR/LB	Injector 4 Control	2.8-2-9 ms
101	WT	Injector 2 Control	2.8-2-9 ms
102	YL	Knock Sensor 2 (-) Signal	0v
103	BK/WT	Communication Network GND	<0.050v
104	WT/PK	COP 2 Driver Control	5º, 55 mph: 8º dwell

Pin Connector Graphic

PCM 104-PIN CONNECTOR

Terminal View of 104-Pin PCM Wiring Harness Connector

Standard Colors and Abbreviations

Abbreviation	Color	Abbreviation	Color	Abbreviation	Color
BK	Black	GY	Gray	PK	Purple
BL	Blue	GN	Green	RD	Red
BR	Brown	LG	LT Green	TN	Tan
DB	Dark Blue	OR	Orange	WT	White
DG	DK Green	PK	Pink	YL	Yellow

LS PIN Tables

2000-01 LS 3.0L V6 VIN S (A/T) C175A 58-Pin Connector

PCM Pin #	Wire Color	Circuit Description (58-Pin)	Value at Hot Idle
1-2	---	Not Used	---
3	GY	SCP Data Bus (+)	Digital Signals
4	BL	SCP Data Bus (-)	<0.050v
6	BR/BL	EVAP CV Solenoid	0-10 Hz (0-100%)
7-8	---	Not Used	---
9	BR/BL	Ground, Switch	N/A
10-11	---	Not Used	---
12	BR/RD	EVAP Purge Solenoid	0-10 Hz (0-100%)
13	YL/GN	EEPROM Power	0.1v
14-16, 18	---	Not Used	---
17	BR	Analog Signal Return	<0.050v
19	BR/GN	Secondary AIR System Relay	On: 1v, Off: 12v
20	YL	Reference Voltage	4.9-5.1v
21-23	---	Not Used	---
24	BK/YL	Power Ground	<0.1v
25	BK/RD	Power Ground	<0.1v
26	BK/RD	Power Ground	<0.1v
27	BK/RD	Power Ground	<0.1v
28	OG/GN	Generator Field Signal	0.5-10.5v
29	GY/WT	Signal Return Normal	<0.050v
30, 35	---	Not Used	---
31	WT/BL	MAF Sensor Signal	0.6v, 55 mph: 1.7v
32	GN/YL	Vehicle Power (from relay)	12-14v
33	GN/OG	Vehicle Power (from relay)	12-14v
34	RD/BL	Secondary AIR Monitor	Pump On: 12v, Off: 0v
36	WT/GN	Normal	N/A
37	WT	PSP Switch Signal	Straight: 5v, Turning: 0v
38	BR/BL	MAF Sensor Return	<0.050v
39	---	Not Used	---
40	OG	BPP Signal from ABS	Brake Off: 0v, On: 12v
41	WT/GN	Sensor Signal	N/A
42	WT/VT	A/C Pressure Switch Signal	0.9v (50 psi)
43	BR/YL	Power Ground	<0.1v
44	OG/YL	Keep Alive Power (CJB fuse)	12-14v
45	WT/VT	Sensor Signal	N/A
46	GY/OR	Signal 2	N/A
47-48	---	Not Used	---
49	WT/VT	Sensor Signal	N/A
50	WT/BK	Generator Monitor Signal	130 Hz (45%)
51	WT/VT	IAT Sensor Signal	1.5-2.5v
52	WT/VT	FTP Sensor Signal	2.6v (0" H2O - cap off)
53-55	---	Not Used	---
56	BK/OR	Ground, Switched	N/A
57	YL/BL	Keep Alive Power	12-14v
58	WT/VT	Sensor Signal	N/A

2000-01 LS 3.0L V6 VIN S (A/T) C175B 32-Pin Connector

PCM Pin #	Wire Color	Circuit Description (32-Pin)	Value at Hot Idle
1	BR	Shift Solenoid 'A' Control	12v, 55 mph: 1v
2	BR/RD	Shift Solenoid 'B' Control	12v, 55 mph: 1v
3	BR/WT	Shift Solenoid 'C' Control	12v, 55 mph: 1v
4	BR/YL	Shift Solenoid 'D' Control	12v, 55 mph: 1v
5	BR/WT	TCC Solenoid Control	12v, 55 mph: 1v
6, 8	---	Not Used	---
7	BR/GN	Pressure Control A Solenoid	On: 1v, Off: 12v
9	WT	Digital TR Sensor (TR3A)	0v, 55 mph: 1.7v
10	WT/RD	Digital TR Sensor (TR4)	0v, 55 mph: 11v
11, 14	---	Not Used	---
12	BR/WT	Pressure Control C Solenoid	On: 1v, Off: 12v
13	BR/YL	Pressure Control B Solenoid	On: 1v, Off: 12v
15	BR/BL	HO2S-12 (B1 S2) Heater	1v, Off: 12v
16	BR/GN	HO2S-22 (B2 S2) Heater	1v, Off: 12v
17	BR	Sensor Signal Return	<0.050v
18	WT/BL	Digital TR Sensor (TR2)	0v, 55 mph: 11v
19	---	Not Used	---
21	WT/BK	Intermediate Speed Shaft	0 Hz, 55 mph: 1320 Hz
22	WT/GN	Digital TR Sensor (TR1)	0v, 55 mph: 11v
23	WT/RD	Trans. Fluid Temp. Sensor	2.10-2.40v
24-25	---	Not Used	---
26	WT/RD	Output Shaft Speed Sensor	0 Hz, 55 mph: 975 Hz
27	WT/VT	Turbine Shaft Speed Sensor	340 Hz, 55 mph: 1025 Hz
28	WT/BL	HO2S-12 (B1 S2) Signal	0.1-1.1v
29	WT/GN	HO2S-22 (B2 S2) Signal	0.1-1.1v
30	WT/BL	A/T Pressure Switch Signal	Open: 12v, Closed: 0v
31-32	---	Not Used	---

Pin Connectors Graphic

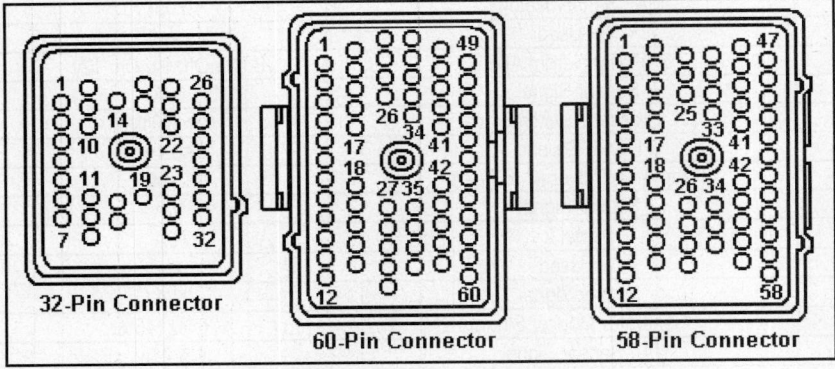

32-Pin Connector   60-Pin Connector   58-Pin Connector

Standard Colors and Abbreviations

Abbreviation	Color	Abbreviation	Color	Abbreviation	Color
BK	Black	GY	Gray	PK	Purple
BL	Blue	GN	Green	RD	Red
BR	Brown	LG	LT Green	TN	Tan
DB	Dark Blue	OR	Orange	WT	White
DG	DK Green	PK	Pink	YL	Yellow

2000-01 LS 3.0L V6 MFI VIN S (A/T) C175C 60 Pin Connector

PCM Pin #	Wire Color	Circuit Description (60 Pin)	Value at Hot Idle
1, 3-6	---	Not Used	---
2	BR/WT	Injector 1 Control	2.8-3.2 ms
7	BR	HO2S-11 (B1 S1) Heater	1v, Off: 12v
8	BR/RD	HO2S-21 (B2 S1) Heater	1v, Off: 12v
9	BR	IAC Solenoid Control	35% duty cycle
10, 15	---	Not Used	---
11	BR/GN	Injector 5 Control	2.8-3.2 ms
12	BR/GN	COP 4 Driver Control	5°, 55 mph: 8° dwell
13	BR/BL	COP 3 Driver Control	5°, 55 mph: 8° dwell
14	BR	Injector 2 Control	2.8-3.2 ms
16	BR/GN	VR Solenoid Control	0%, 55 mph: 45%
17	BR	Sensor Signal Return	<0.050v
18-19	---	Not Used	---
20	YL/BK	Sensor Voltage Reference	4.9-5.1v
21	BR/YL	Injector 6 Control	2.8-3.2 ms
22	BR/YL	COP 5 Driver Control	5°, 55 mph: 8° dwell
23	BR/RD	COP 2 Driver Control	5°, 55 mph: 8° dwell
24	BR/RD	Injector 3 Control	2.8-3.2 ms
25	---	Not Used	---
28	GY/RD	Generator Field Control	0-130 Hz (varies)
29	BR/WT	IMRC Monitor	5v, 55 mph: 5v
30	BR/WT	COP 6 Driver Control	5°, 55 mph: 8° dwell
31	BR	COP 1 Driver Control	5°, 55 mph: 8° dwell
32	BR/BL	Injector 4 Control	2.8-3.2 ms
33-35	---	Not Used	---
36	BR	Hydraulic Fan Solenoid	On: 1v, Off: 12v
37	BR	Power Steering Press. Switch	Straight: 5v, Turning: 0v
38	BR/RD	To Be Done	---
39	GY/BK	To Be Done	---
40	WT/VT	CHT Sensor Signal	0.7v (194°F)
41 ('00)	WT/BL	DPFE Sensor Signal	0.20-1.30v
41 ('01)	WT/BL	DPFE Sensor Signal	0.95-1.05v
42	GY/BK	Knock Sensor 1 Signal	0v
43	GY/RD	Knock Sensor 2 Signal	0v
44	WT/RD	HO2S-21 (B2 S1) Signal	0.1-1.1v
45	WT	HO2S-11 (B1 S1) Signal	0.1-1.1v
46-48	---	Not Used	---
49	WT/GN	Fuel Rail Pressure Sensor	2.8v (39 psi)
50	WT/RD	Generator Field Signal	130 Hz (40%)
51	WT/BK	Knock Sensor 1 Ground	0v
52	WT/RD	Knock Sensor 2 Ground	0v
53	WT/VT	CMP Sensor Signal	4-6 Hz
54	---	Not Used	---
55	GY/RD	CKP Sensor (+) Signal	390-450 Hz
56	WT/RD	CKP Sensor (-) Signal	390-450 Hz
57	WT	TP Sensor Signal	0.53-1.27v
58-60	---	Not Used	---

2002-03 LS 3.0L V6 VIN S (A/T) C175A 58-Pin Connector

PCM Pin #	Wire Color	Circuit Description (58-Pin)	Value at Hot Idle
1 ('03)	WT/RD	Accelerator Pedal Position Sensor	0.5-4.9v
2	---	Not Used	---
3	GY	SCP Data Bus (+)	Digital Signals
4	BL	SCP Data Bus (-)	<0.050v
5	BR	Signal Return	<0.050v
6-8	---	Not Used	---
9	BR/BL	A/C Clutch Relay Control	Relay Off: 12v, On: 1v
10-11	---	Not Used	---
12	BR/RD	EVAP Purge Solenoid Control	0-10 Hz (0-100%)
13	YL/GN	EEPROM Power	0.1v
14 ('03)	WT/VT	Manual Transmission Switch (+) Signal	Switch Open: 12v, Closed: 1v
15 ('03)	WT	Electronic Throttle Control Module	Digital Signals
16 ('03)	WT/BL	Electronic Throttle Control Module	Digital Signals
17 ('03)	BR	Electronic Throttle Control Module	Digital Signals
18-19	---	Not Used	---
20 ('03)	YL	ETC Reference Voltage	4.9-5.1v
21	---	Not Used	---
22 ('03)	WT/BK	Manual Transmission Switch (-) Signal	<0.050v
23 ('03)	YL/RD	ETC Reference Voltage	4.9-5.1v
24	BK/RD	Power Ground	<0.1v
25	BK/RD	Power Ground	<0.1v
26	BK/RD	Power Ground	<0.1v
27	BK/RD	Power Ground	<0.1v
28	GN/OR	Brake Pressure Switch	Brake Off: 0v, On: 12v
29	---	Not Used	---
30	BR/BL	EVAP Vent Solenoid Control	0-10 Hz (0-100%)
31	WT/BL	MAF Sensor Signal	0.6v, 55 mph: 1.7v
32	GN/YL	Vehicle Power (Start-Run)	12-14v
33	GN/OR	Vehicle Power (Start-Run)	12-14v
34-37	---	Not Used	---
38	BR/BL	MAF Sensor Signal Return	<0.050v
39	---	Not Used	---
40	OG	Keep Alive Power	12-14v
41	WT/GN	Traction Control Disable Switch	Brake Off: 0v, On: 12v
42	WT/VT	A/C Pressure Switch Signal	0.9v (50 psi)
43	BR/YL	Power Ground	<0.1v
44	OG/YL	Keep Alive Power (CJB fuse)	12-14v
45-46	---	Not Used	---
47	WT/BK	Engine Cooling Fan Motor Control	Off: 12v, On: 1v
48	WT/BL	Restraint Control Module	Digital Signals
49	WT/VT	Rear Electronic Module	Digital Signals
50	---	Not Used	---
51	WT/VT	IAT Sensor Signal	1.5-2.5v
52	WT/VT	FTP Sensor Signal	2.6v (0" H2O - cap off)
53 ('03)	GN/RD	Controller Area Network (+) Signal	0-7-0v
54 ('03)	BL/RD	Controller Area Network (-) Signal	0-7-0v
55	YL	Reference Voltage	4.9-5.1v
56	BK/OR	Sensor Return	<0.050v
57	YL/BL	Speed Control Reference Voltage	4.9-5.1v
58	---	Not Used	---

2002-03 LS 3.0L V6 VIN S (A/T) C175B 32-Pin Connector

PCM Pin #	Wire Color	Circuit Description (32-Pin)	Value at Hot Idle
1	BR	Shift Solenoid 'A' Control	12v, 55 mph: 1v
2	BR/RD	Shift Solenoid 'B' Control	12v, 55 mph: 1v
3-6	---	Not Used	---
7	BR/GN	Pressure Control Solenoid 'A' Control	8.8v
8	BR/WT	Shift Solenoid 'C' Control	12v, 55 mph: 12v
9	WT	Digital TR Sensor (TR3A)	0v, 55 mph: 1.7v
10	WT/RD	Digital TR Sensor (TR4)	0v, 55 mph: 11v
11	---	Not Used	---
12	BR/WT	Pressure Control 'C' Solenoid Control	Off: 12v, On: 1v
13	BR/YL	Pressure Control 'B' Solenoid Control	Off: 12v, On: 1v
14	BR/RD	Signal Return	<0.050v
15	BR/BL	HO2S-12 (B1 S2) Heater Control	Off: 12v, On: 1v
16	BK/GN	HO2S-22 (B2 S2) Heater Control	Off: 12v, On: 1v
17	BR/YL	Shift Solenoid 'D' Control	12v, 55 mph: 1v
18	WT/BL	Digital TR Sensor (TR2)	0v, 55 mph: 11v
19	---	Not Used	---
20	BR/WT	Torque Converter Clutch Control	12v, 55 mph: 0.1v
21	WT/BK	Intermediate Speed Shaft Signal	0 Hz, 55 mph: 1386 Hz
22	WT/GN	Digital TR Sensor (TR1)	0v, 55 mph: 11v
23	WT/RD	Transmission Fluid Temperature Sensor	2.10-2.40v
24-25	---	Not Used	---
26	WT/RD	Output Shaft Speed Sensor Signal	0 Hz, 975 Hz
27	WT/VT	Turbine Shaft Speed Sensor Signal	340 Hz, 1025 Hz
28	WT/BL	HO2S-12 (B1 S2) Signal	0.1-1.1v
29	WT/GN	HO2S-22 (B2 S2) Signal	0.1-1.1v
30-32	---	Not Used	---

Wire Harness Connectors Graphic

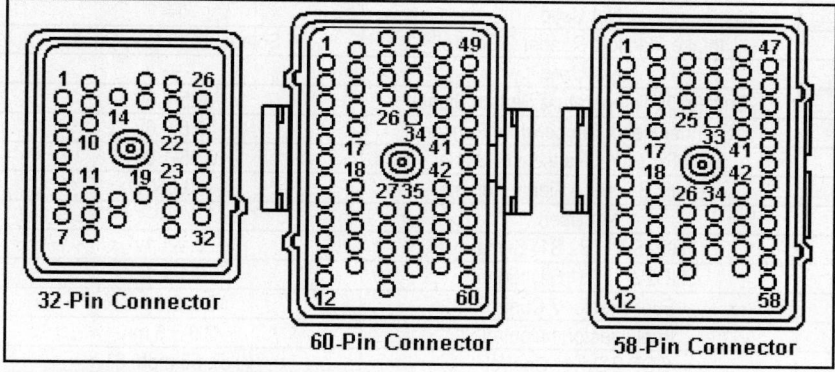

32-Pin Connector

60-Pin Connector

58-Pin Connector

Standard Colors and Abbreviations

Abbreviation	Color	Abbreviation	Color	Abbreviation	Color
BK	Black	GY	Gray	PK	Purple
BL	Blue	GN	Green	RD	Red
BR	Brown	LG	LT Green	TN	Tan
DB	Dark Blue	OR	Orange	WT	White
DG	DK Green	PK	Pink	YL	Yellow

2002-03 LS 3.0L V6 MFI VIN S (A/T) C175C 60 Pin Connector

PCM Pin #	Wire Color	Circuit Description (60 Pin)	Value at Hot Idle
1	WT/BK	Variable Valve Control Solenoid 1	DC pulse signals
2-3, 6	---	Not Used	---
4	WT/BL	Fuel Rail Temperature Sensor	1.7-3.5v (50-120°F)
5	WT	Power Steering Pressure Switch	Straight: 0v, Turned: 5v
7	BR	HO2S-11 (B1 S1) Heater	On: 1v, Off: 12v
8	BR/RD	HO2S-21 (B2 S1) Signal	0.1-1.1v
9-10	---	Not Used	---
11	BR/BL	Fuel Injector 4 Control	2.9-3.6 ms
12	BR/GN	COP 4 Driver Control	5°, 55 mph: 8°
13	WT/VT	Variable Valve Timing Solenoid 2	DC pulse signals
14	YL/BK	Reference Voltage	4.9-5.1v
15	BR	Throttle Position Sensor Return	0.050v
16	BR/GN	EGR Pressure Transducer Return	<0.050v
17	BR/GN	Sensor Signal Return	<0.050v
18	---	Not Used	---
19	GY/RD	Generator Common Signal	0-130 Hz (varies)
20	BR/GN	Fuel Injector 5 Control	2.9-3.6 ms
21	BR/RD	Fuel Injector 3 Control	2.9-3.6 ms
22	BR/YL	COP 5 Driver Control	5°, 55 mph: 8°
23	WT/GN	EGR Pressure Transducer Signal	0-3.1v
24	YL	Battery Power (at all times)	12-14v
25-26	---	Not Used	---
27 ('03)	YL/BL	Electronic Throttle Control Motor (-)	DC pulse signals
28	BR/YL	Fuel Injector 6 Control	2.9-3.6 ms
29	BR/WT	Intake Manifold Runner 1 Control	Solenoid Off: 12v, On: 1v
30	BR/WT	COP 6 Driver Control	5°, 55 mph: 8°
31, 33-34	---	Not Used	---
32	WT	Throttle Position Sensor	0.53-1.27v
35 ('03)	WT/BL	Electronic Throttle Control Motor (+)	DC pulse signals
36	---	Not Used	---
37	BR/GN	Intake Manifold Runner 2 Control	Solenoid Off: 12v, On: 1v
38	---	Not Used	---
39	WT/GN	Engine Oil Temperature Sensor Signal	0.5-0.6v
40	WT/VT	Cylinder Head Temperature Sensor	0.5-0.6v
41	WT/BL	EGR Pressure Transducer Signal	0.95-1.05v
42	GY/BK	Knock Sensor 1 Signal Return	<0.050v
43	---	Not Used	---
44	WT/RD	HO2S-21 (B2 S1) Signal	0.1-1.1v
45	WT	HO2S-11 (B1 S1) Signal	0.1-1.1v
46	BR	Fuel Injector 2 Control	2.9-3.6 ms
47	BR/WT	Fuel Injector 1 Control	2.9-3.6 ms
48	BR/RD	COP 2 Driver Control	5°, 55 mph: 8°
49	WT/GN	Fuel Rail Pressure Sensor Signal	2.8v (39 psi)
50	WT/RD	Generator Monitor Signal	130 Hz (45%)
51	WT/BK	Knock Sensor 1 (+) Signal	0v
52	---	Not Used	---
53	WT/VT	CMP Sensor 1 Signal	6 Hz
54	WT/BK	CMP Sensor 2 Signal	6 Hz
55	GY/RD	CKP Sensor (-) Signal	390 Hz
56	WT/RD	CKP Sensor (+) Signal	390 Hz
57	WT/RD	Throttle Position Sensor	0.53-1.27v
58	BR	COP 1 Driver Control	5°, 55 mph: 8°
59	---	Not Used	---
60	BR/BL	COP 3 Driver Control	5°, 55 mph: 8°

2000-02 LS 3.9L V8 VIN A (A/T) C175A 58-Pin Connector

PCM Pin #	Wire Color	Circuit Description (58-Pin)	Value at Hot Idle
1-2	---	Not Used	---
3	GY	SCP Data Bus (+)	Digital Signals
4	BL	SCP Data Bus (-)	<0.050v
5	---	Not Used	---
6	BR/BL	EVAP CV Solenoid	0-10 Hz (0-100%)
7-8	---	Not Used	---
9	BR/BL	Ground, Switch	N/A
10-11	---	Not Used	---
12	BR/RD	EVAP Purge Solenoid	0-10 Hz (0-100%)
13	YL/GN	EEPROM Power	0.1v
14-16	---	Not Used	---
17	BR	Analog Signal Return	<0.050v
19	BR/GN	Secondary AIR System Relay	On: 1v, Off: 12v
20	YL	Reference Voltage	4.9-5.1v
21-23	---	Not Used	---
24	BK/YL	Power Ground	<0.1v
25	BK/RD	Power Ground	<0.1v
26	BK/RD	Power Ground	<0.1v
27	BK/RD	Power Ground	<0.1v
28	OG/GN	Generator Field Signal	0.5-10.5v
29	GY/WT	Signal Return Normal	<0.050v
30	---	Not Used	---
31	WT/BL	MAF Sensor Signal	0.6v, 55 mph: 1.7v
32	GN/YL	Vehicle Power (switched)	12-14v
33	GN/OG	Vehicle Power (switched)	12-14v
34	RD/BL	Secondary AIR Monitor	Pump On: 12v, Off: 0v
35	---	Not Used	---
36	WT/GN	Normal	N/A
37	WT	PSP Switch Signal	Straight: 5v, Turning: 0v
38	BR/BL	MAF Sensor Return	<0.050v
39	---	Not Used	---
40	OG	BPP Signal from ABS	Brake Off: 0v, On: 12v
41	WT/GN	Sensor Signal	N/A
42	WT/VT	A/C Pressure Switch Signal	0.9v (50 psi)
43	BR/YL	Power Ground	<0.1v
44	OG/YL	Keep Alive Power	12-14v
45	WT/VT	Sensor Signal	N/A
46	GY/OR	Signal 2	N/A
47-48	---	Not Used	---
49	WT/VT	Sensor Signal	N/A
50	WT/BK	Generator Monitor Signal	130 Hz (45%)
51	WT/VT	IAT Sensor Signal	1.5-2.5v
52	WT/VT	FTP Sensor Signal	2.6v (0" H2O - cap off)
53-55	---	Not Used	---
56	BK/OR	Ground, Switched	N/A
57	YL/BL	Keep Alive Power	12-14v
58	WT/VT	Sensor Signal	N/A

2000-02 LS 3.9L V8 VIN A (A/T) C175B 32-Pin Connector

PCM Pin #	Wire Color	Circuit Description (32-Pin)	Value at Hot Idle
1	BR	Shift Solenoid 'A' Control	12v, 55 mph: 1v
2	BR/RD	Shift Solenoid 'B' Control	12v, 55 mph: 1v
3	BR/WT	Shift Solenoid 'C' Control	12v, 55 mph: 1v
4	BR/YL	Shift Solenoid 'D' Control	12v, 55 mph: 1v
5	BR/WT	TCC Solenoid Control	12v, 55 mph: 1v
6	---	Not Used	---
7	BR/GN	Pressure Control A Solenoid	On: 1v, Off: 12v
9	WT	Digital TR Sensor (TR3A)	0v, 55 mph: 1.7v
10	WT/RD	Digital TR Sensor (TR4)	0v, 55 mph: 11v
11	---	Not Used	---
12	BR/WT	Pressure Control C Solenoid	On: 1v, Off: 12v
13	BR/YL	Pressure Control B Solenoid	On: 1v, Off: 12v
14, 19	---	Not Used	---
15	BR/BL	HO2S-12 (B1 S2) Heater	1v, Off: 12v
16	BR/GN	HO2S-22 (B2 S2) Heater	1v, Off: 12v
17	BR	Sensor Signal Return	<0.050v
18	WT/BL	Digital TR Sensor (TR2)	0v, 55 mph: 11v
21	WT/BK	Intermediate Speed Shaft	0 Hz, 55 mph: 1386 Hz
22	WT/GN	Digital TR Sensor (TR1)	0v, 55 mph: 11v
23	WT/RD	Trans. Fluid Temp. Sensor	2.10-2.40v
25	---	Not Used	---
26	WT/RD	Output Shaft Speed Sensor	0 Hz, 975 Hz
27	WT/VT	Turbine Shaft Speed Sensor	340 Hz, 1025 Hz
28	WT/BL	HO2S-12 (B1 S2) Signal	0.1-1.1v
29	WT/GN	HO2S-22 (B2 S2) Signal	0.1-1.1v
30	WT/BL	A/T Pressure Switch	Open: 12v, Closed: 0v
31-32	---	Not Used	---

Pin Connectors Graphic

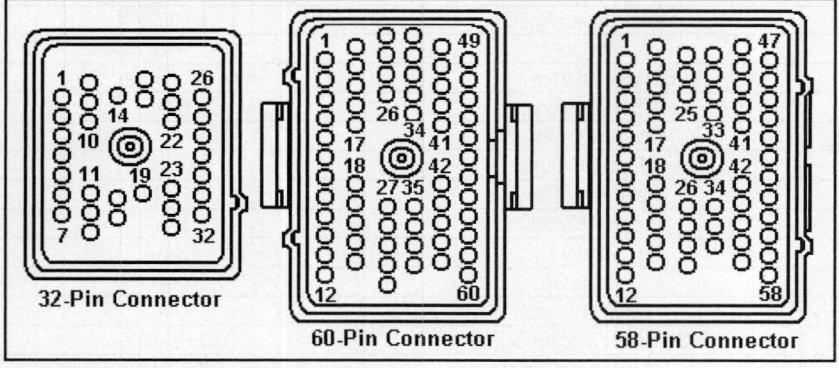

32-Pin Connector   60-Pin Connector   58-Pin Connector

Standard Colors and Abbreviations

Abbreviation	Color	Abbreviation	Color	Abbreviation	Color
BK	Black	GY	Gray	PK	Purple
BL	Blue	GN	Green	RD	Red
BR	Brown	LG	LT Green	TN	Tan
DB	Dark Blue	OR	Orange	WT	White
DG	DK Green	PK	Pink	YL	Yellow

2000-02 LS 3.9L V8 MFI VIN A (A/T) C175B 60 Pin Connector

PCM Pin #	Wire Color	Circuit Description (60 Pin)	Value at Hot Idle
1	BR/RD	COP 2 Driver Control	5°, 55 mph: 8° dwell
2	BR/WT	Injector 1 Control	2.8-3.6 ms
3-6	---	Not Used	---
7	BR	HO2S-11 (B1 S1) Heater	1v, Off: 12v
8	BR/RD	HO2S-21 (B2 S1) Heater	1v, Off: 12v
9	BR	IAC Solenoid Control	35% duty cycle
10, 15	---	Not Used	---
11	BR/GN	Injector 5 Control	2.8-3.6 ms
12	BR/WT	COP 6 Driver Control	5°, 55 mph: 8° dwell
13	BR/GN	COP 4 Driver Control	5°, 55 mph: 8° dwell
14	BR	Injector 2 Control	2.8-3.6 ms
16	BR/GN	VR Solenoid Control	0%, 55 mph: 45%
17	BR	Sensor Signal Return	<0.050v
18-19	---	Not Used	---
20	YL/BK	Sensor Voltage Reference	4.9-5.1v
21	BR/YL	Injector 6 Control	2.8-3.6 ms
22	BR/BL	COP 3 Driver Control	5°, 55 mph: 8° dwell
23	BR/YL	COP 5 Driver Control	5°, 55 mph: 8° dwell
24	BR/RD	Injector 3 Control	2.8-3.6 ms
25-27	---	Not Used	---
28	GY/RD	Generator Control Signal	0-130 Hz (varies)
29	BR/WT	Injector 7 Control	2.8-3.6 ms
30	BR	COP 7 Driver Control	5°, 55 mph: 8° dwell
31	BR	COP 1 Driver Control	5°, 55 mph: 8° dwell
32	BR/BL	Injector 4 Control	2.8-3.6 ms
33-35	---	Not Used	---
36	BR	Hydraulic Fan Solenoid	On: 1v, Off: 12v
37	BR	Injector 8 Control	2.8-3.6 ms
38	BR/RD	COP 8 Driver Control	5°, 55 mph: 8° dwell
39	---	Not Used	---
40	WT/VT	CHT Sensor Signal	0.7v (194°F)
41	WT/BL	DPFE Sensor Signal	0.95-1.05v
42	GY/BK	Knock Sensor 1 Signal	0v
43	GY/RD	Knock Sensor 2 Signal	0v
44	WT/RD	HO2S-21 (B2 S1) Signal	0.1-1.1v
45	WT	HO2S-11 (B1 S1) Signal	0.1-1.1v
46-48	---	Not Used	---
49	WT/GN	Fuel Rail Pressure Sensor	2.8v (39 psi)
50	WT/RD	Generator Field Signal	130 Hz (40%)
51	WT/BK	Knock Sensor 1 Ground	0v
52	WT/RD	Knock Sensor 2 Ground	0v
53	WT/VT	CMP Sensor Signal	6 Hz
54	---	Not Used	---
55	GY/RD	CKP Sensor (+) Signal	380 Hz
56	WT/RD	CKP Sensor (-) Signal	380 Hz
57	WT	TP Sensor Signal	0.53-1.27v
58-60	---	Not Used	---

2003 LS 3.9L V8 VIN A (A/T) C175A 58-Pin Connector

PCM Pin #	Wire Color	Circuit Description (58-Pin)	Value at Hot Idle
1	WT/RD	Accelerator Pedal Position Sensor	0.5-4.9v
2	---	Not Used	---
3	GY	SCP Data Bus (+)	Digital Signals
4	BL	SCP Data Bus (-)	<0.050v
5	BR	Signal Return	<0.050v
6-8	---	Not Used	---
9	BR/BL	A/C Clutch Relay Control	Relay Off: 12v, On: 1v
10-11	---	Not Used	---
12	BR/RD	EVAP Purge Solenoid Control	0-10 Hz (0-100%)
13	YL/GN	EEPROM Power	0.1v
14	WT/VT	Manual Transmission Switch (+) Signal	Switch Open: 12v, Closed: 1v
15	WT	Electronic Throttle Control Module	Digital Signals
16	WT/BL	Electronic Throttle Control Module	Digital Signals
17	BR	Electronic Throttle Control Module	Digital Signals
18-19	---	Not Used	---
20	YL	ETC Reference Voltage	4.9-5.1v
21	---	Not Used	---
22	WT/BK	Manual Transmission Switch (-) Signal	<0.050v
23	YL/RD	ETC Reference Voltage	4.9-5.1v
24	BK/RD	Power Ground	<0.1v
25	BK/RD	Power Ground	<0.1v
26	BK/RD	Power Ground	<0.1v
27	BK/RD	Power Ground	<0.1v
28	GN/OR	Brake Pressure Switch	Brake Off: 0v, On: 12v
29	---	Not Used	---
30	BR/BL	EVAP Vent Solenoid Control	0-10 Hz (0-100%)
31	WT/BL	MAF Sensor Signal	0.6v, 55 mph: 1.7v
32	GN/YL	Vehicle Power (Start-Run)	12-14v
33	GN/OR	Vehicle Power (Start-Run)	12-14v
34-37	---	Not Used	---
38	BR/BL	MAF Sensor Signal Return	<0.050v
39	---	Not Used	---
40	OG	Keep Alive Power	12-14v
41	WT/GN	Traction Control Disable Switch	Brake Off: 0v, On: 12v
42	WT/VT	A/C Pressure Switch Signal	0.9v (50 psi)
43	BR/YL	Power Ground	<0.1v
44	OG/YL	Keep Alive Power (CJB fuse)	12-14v
45-46	---	Not Used	---
47	WT/BK	Engine Cooling Fan Motor Control	Off: 12v, On: 1v
48	WT/BL	Restraint Control Module	Digital Signals
49	WT/VT	Rear Electronic Module	Digital Signals
50	---	Not Used	---
51	WT/VT	IAT Sensor Signal	1.5-2.5v
52	WT/VT	FTP Sensor Signal	2.6v (0" H2O - cap off)
53	GN/RD	Controller Area Network (+) Signal	0-7-0v
54	BL/RD	Controller Area Network (-) Signal	0-7-0v
55	YL	Reference Voltage	4.9-5.1v
56	BK/OR	Sensor Return	<0.050v
57	YL/BL	Speed Control Reference Voltage	4.9-5.1v
58	---	Not Used	---

2003 LS 3.9L V8 VIN A (A/T) C175B 32-Pin Connector

PCM Pin #	Wire Color	Circuit Description (32-Pin)	Value at Hot Idle
1	BR	Shift Solenoid 'A' Control	12v, 55 mph: 1v
2	BR/RD	Shift Solenoid 'B' Control	12v, 55 mph: 1v
3-6	---	Not Used	---
7	BR/GN	Pressure Control Solenoid 'A' Control	8.8v
8	BR/WT	Shift Solenoid 'C' Control	12v, 55 mph: 12v
9	WT	Digital TR Sensor (TR3A)	0v, 55 mph: 1.7v
10	WT/RD	Digital TR Sensor (TR4)	0v, 55 mph: 11v
11	---	Not Used	---
12	BR/WT	Pressure Control 'C' Solenoid Control	Off: 12v, On: 1v
13	BR/YL	Pressure Control 'B' Solenoid Control	Off: 12v, On: 1v
14	BR/RD	Signal Return	<0.050v
15	BR/BL	HO2S-12 (B1 S2) Heater Control	Off: 12v, On: 1v
16	BK/GN	HO2S-22 (B2 S2) Heater Control	Off: 12v, On: 1v
17	BR/YL	Shift Solenoid 'D' Control	12v, 55 mph: 1v
18	WT/BL	Digital TR Sensor (TR2)	0v, 55 mph: 11v
19	---	Not Used	---
20	BR/WT	Torque Converter Clutch Control	12v, 55 mph: 0.1v
21	WT/BK	Intermediate Speed Shaft Signal	0 Hz, 55 mph: 1386 Hz
22	WT/GN	Digital TR Sensor (TR1)	0v, 55 mph: 11v
23	WT/RD	Transmission Fluid Temperature Sensor	2.10-2.40v
24-25	---	Not Used	---
26	WT/RD	Output Shaft Speed Sensor Signal	0 Hz, 975 Hz
27	WT/VT	Turbine Shaft Speed Sensor Signal	340 Hz, 1025 Hz
28	WT/BL	HO2S-12 (B1 S2) Signal	0.1-1.1v
29	WT/GN	HO2S-22 (B2 S2) Signal	0.1-1.1v
30-32	---	Not Used	---

Wire Harness Connectors Graphic

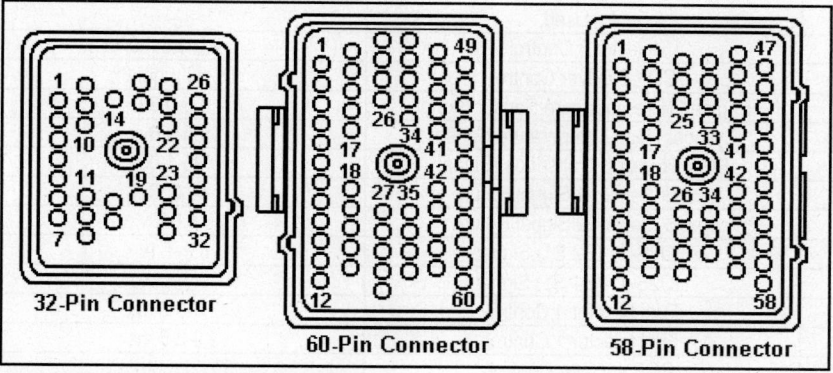

32-Pin Connector   60-Pin Connector   58-Pin Connector

Standard Colors and Abbreviations

Abbreviation	Color	Abbreviation	Color	Abbreviation	Color
BK	Black	GY	Gray	PK	Purple
BL	Blue	GN	Green	RD	Red
BR	Brown	LG	LT Green	TN	Tan
DB	Dark Blue	OR	Orange	WT	White
DG	DK Green	PK	Pink	YL	Yellow

2003 LS 3.9L V8 MFI VIN A (A/T) C175C 60 Pin Connector

PCM Pin #	Wire Color	Circuit Description (60 Pin)	Value at Hot Idle
1	WT/BK	Variable Valve Control Solenoid 1	DC pulse signals
2-3, 6	---	Not Used	---
4	WT/BL	Fuel Rail Temperature Sensor	1.7-3.5v (50-120ºF)
5	WT	Power Steering Pressure Switch	Straight: 0v, Turned: 5v
7	BR	HO2S-11 (B1 S1) Heater	On: 1v, Off: 12v
8	BR/RD	HO2S-21 (B2 S1) Signal	0.1-1.1v
9-10	---	Not Used	---
11	BR/GN	Fuel Injector 5 Control	2.9-3.6 ms
12	BR/YL	COP 5 Driver Control	5º, 55 mph: 8º
13	WT/VT	Variable Valve Timing Solenoid 2	DC pulse signals
14	YL/RD	Reference Voltage	4.9-5.1v
15	BR	Throttle Position Sensor Return	0.050v
16	BR/GN	EGR Pressure Transducer Return	<0.050v
17	BR/GN	Sensor Signal Return	<0.050v
18	---	Not Used	---
19	GY/RD	Generator Common Signal	0-130 Hz (varies)
20	BR	Fuel Injector 2 Control	2.9-3.6 ms
21	BR/YL	Fuel Injector 6 Control	2.9-3.6 ms
22	BR/WT	COP 6 Driver Control	5º, 55 mph: 8º
23	WT/GN	EGR Pressure Transducer Signal	0-3.1v
24	YL	Battery Power (at all times)	12-14v
25-26	---	Not Used	---
27	YL/BL	Electronic Throttle Control Motor (-)	DC pulse signals
28	BR/RD	Fuel Injector 5 Control	2.9-3.6 ms
29	BR/WT	Fuel Injector 7 Control	2.9-3.6 ms
30	BR	COP 7 Driver Control	5º, 55 mph: 8º
31, 33-34	---	Not Used	---
32	WT	Throttle Position Sensor	0.53-1.27v
35	WT/BL	Electronic Throttle Control Motor (+)	DC pulse signals
36	---	Not Used	---
37	BR	Injector 8 Control	2.9-3.6 ms
38	BR/RD	COP 8 Driver Control	5º, 55 mph: 8º
39	WT/GN	Engine Oil Temperature Sensor Signal	0.5-0.6v
40	WT/VT	Cylinder Head Temperature Sensor	0.5-0.6v
41	WT/BL	EGR Pressure Transducer Signal	0.95-1.05v
42	GY/BK	Knock Sensor 1 Signal Return	<0.050v
43	GY/RD	Knock Sensor 2 Signal Return	<0.050v
44	WT/RD	HO2S-21 (B2 S1) Signal	0.1-1.1v
45	WT	HO2S-11 (B1 S1) Signal	0.1-1.1v
46	BR/BL	Fuel Injector 4 Control	2.9-3.6 ms
47	BR/WT	Fuel Injector 1 Control	2.9-3.6 ms
48	BR/GN	COP 4 Driver Control	5º, 55 mph: 8º
49	WT/GN	Fuel Rail Pressure Sensor Signal	2.8v (39 psi)
50	WT/GN	Generator Monitor Signal	130 Hz (45%)
51	WT/BK	Knock Sensor 1 (+) Signal	0v
52	WT/RD	Knock Sensor 2 (+) Signal	0v
53	WT/VT	CMP Sensor 1 Signal	6 Hz
54	WT/BK	CMP Sensor 2 Signal	6 Hz
55	GY/RD	CKP Sensor (-) Signal	390 Hz
56	WT/RD	CKP Sensor (+) Signal	390 Hz
57	WT/RD	Throttle Position Sensor	0.53-1.27v
58	BR	COP 1 Driver Control	5º, 55 mph: 8º
59	BR/RD	COP 2 Driver Control	5º, 55 mph: 8º
60	BR/BL	COP 3 Driver Control	5º, 55 mph: 8º

MARK VII PIN TABLES

1990-92 Mark VII 5.0L V8 HO MFI VIN E (A/T) 60 Pin Connector

PCM Pin #	Wire Color	Circuit Description (60 Pin)	Value at Hot Idle
1	BK/OR	Keep Alive Power	12-14v
2	LG	Brake Switch Signal	Brake Off: 0v, On: 12v
3	DG/WT	VSS (+) Signal	0 Hz, 55 mph: 125 Hz
4	DG/YL	Ignition Diagnostic Monitor	20-31 Hz
4	BL/PK	Ignition Diagnostic Monitor	20-31 Hz
5	---	Not Used	---
6	PK/BL	VSS (-) Signal	0 Hz, 55 mph: 125 Hz
7	LG/YL	ECT Sensor Signal	0.5-0.6v
8	---	Not Used	---
9	TN/BL	MAF Sensor Return	<0.050v
10	LG/PK	A/C Cycling Clutch Switch	A/C On: 12v, Off: 0v
10	PK/BL	A/C Cycling Clutch Switch	A/C On: 12v, Off: 0v
11	LG/BK	Air Management 2 Solenoid	AM2 On: 1v, Off: 12v
12	TN/RD	Injector 3 Control	5.0-6.0 ms
13	TN/BK	Injector 4 Control	5.0-6.0 ms
14	TN/BK	Injector 5 Control	5.0-6.0 ms
15	TN/LG	Injector 6 Control	5.0-6.0 ms
16	BK/OR	Ignition System Ground	<0.050v
17	YL/BK	Self-Test Output, MIL Control	MIL On: 1v, Off: 12v
17	T/P	Self-Test Output, MIL Control	MIL On: 1v, Off: 12v
18-19	---	Not Used	---
20	BK	PCM Case Ground	<0.050v
21	WT/BL	IAC Motor Control	8-10v
22	BK/OR	Fuel Pump Control	On: 1v, Off: 12v
22	TN/LG	Fuel Pump Control	On: 1v, Off: 12v
23-24	---	Not Used	---
25	LG/PK	ACT Sensor Signal	1.5-2.5v
26	OR/WT	Reference Voltage	4.9-5.1v
27	BR/LG	EGR EVP Sensor Signal	0.4v
28	---	Not Used	---
29	DG/PK	HO2S-21 (B2 S1) Signal	0.1-1.1v
30	WT/PK	Neutral Drive Switch Signal	In 'N': 0v, Others: 5v
30	BL/YL	Neutral Drive Switch Signal	In 'N': 0v, Others: 5v
31	GY/YL	EVAP Purge Solenoid	12v, 55 mph: 1v
32	---	Not Used	---
33	DG	VR Solenoid Control	0%, 55 mph: 45%
34	BL/P	Data Output Link	Digital Signals
35	WT/PK	S/C Vent Solenoid Control	Vacuum Decreasing: 1v
36	YL/LG	Spark Output Signal	6.93v (50% d/cycle)
37	R, BK/YL	Vehicle Power	12-14v
38	GY/BK	S/C Vacuum Solenoid	Vacuum Increasing: 1v
39	BL/YL	Speed Control Switch Ground	<0.050v
40	BK/LG	Power Ground	<0.1v

1990-92 Mark VII 5.0L V8 HO MFI VIN E (A/T) 60 Pin Connector

PCM Pin #	Wire Color	Circuit Description (60 Pin)	Value at Hot Idle
41	OR/YL	Speed Control Solenoid (+)	S/C On at 55 mph: 12v
42	TN/OR	Injector 7 Control	5.0-6.0 ms
43	DB/LG	HO2S-11 (B1 S1) Signal	0.1-1.1v
43	BL/PK	HO2S-11 (B1 S1) Signal	0.1-1.1v
44	GY/RD	HO2S-11 (B1 S1) Ground	<0.050v
45	DB/LG	MAP Sensor Signal	107 Hz (sea level)
45	LG/BK	MAP Sensor Signal	107 Hz (sea level)
46	BK/WT	Analog Signal Return	<0.050v
47	GY/WT	TP Sensor Signal	0.8v, 55 mph: 1.1v
47	GY	TP Sensor Signal	0.8v, 55 mph: 1.1v
47	DG/BL	TP Sensor Signal	0.8v, 55 mph: 1.1v
48	WT/RD	Self-Test Input Signal	STI On: 0v, Off: 5v
48	WT/BK	Self-Test Input Signal	STI On: 0v, Off: 5v
48	TN/RD	Self-Test Input Signal	STI On: 0v, Off: 5v
49	OR	HO2S-21 (B2 S1) Ground	<0.050v
50	OR/RD	MAF Sensor Signal	0.8v, 55 mph: 1.8v
50	BL/BK	MAF Sensor Signal	0.8v, 55 mph: 1.8v
51	WT/RD	Air Management 1 Solenoid	AM1 On: 1v, Off: 12v
52	YL	Injector 8 Control	5.0-6.0 ms
53, 55	---	Not Used	---
54	OR/BL	A/C WOT Relay Control	On: 1v, Off: 12v
56	DB	PIP Sensor Signal	6.93v (50% d/cycle)
57	RD	Vehicle Power	12-14v
57	BK/YL	Vehicle Power	12-14v
58	TN	Injector 1 Control	5.0-6.0 ms
59	TN/RD	Injector 2 Control	5.0-6.0 ms
60	BK/LG	Power Ground	<0.1v

Pin Connector Graphic

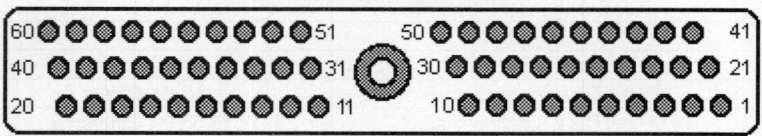

Standard Colors and Abbreviations

Abbreviation	Color	Abbreviation	Color	Abbreviation	Color
BK	Black	GY	Gray	PK	Purple
BL	Blue	GN	Green	RD	Red
BR	Brown	LG	LT Green	TN	Tan
DB	Dark Blue	OR	Orange	WT	White
DG	DK Green	PK	Pink	YL	Yellow

Mark VIII Pin Tables

1993-95 Mark VIII 4.6L V8 MFI VIN V (A/T) 60 Pin Connector

PCM Pin #	Wire Color	Circuit Description (60 Pin)	Value at Hot Idle
1	YL	Keep Alive Power	12-14v
2	LG	Brake Switch Signal	Brake Off: 0v, On: 12v
3	DG/WT	VSS (+) Signal	0 Hz, 55 mph: 125 Hz
4	TN/YL	Ignition Diagnostic Monitor	20-31 Hz
5	DG/WT	TSS Sensor Signal	At 30 mph: 126-136 Hz
6	PK/OR	VSS (-) Signal	0 Hz, 55 mph: 125 Hz
7	LG/RD	ECT Sensor Signal	0.5-0.6v
8	PK/BK	Fuel Pump Monitor	12v, Off: 0v
9	TN/LB	MAF Sensor Return	<0.050v
10	BK/YL	A/C Cycling Clutch Switch	A/C On: 12v, Off: 0v
11	GY/YL	EVAP Purge Solenoid	12v, 55 mph: 1v
12	LG/OR	Injector 6 Control	4.0-4.2 ms
13	TN/RD	Injector 7 Control	4.0-4.4 ms
14	LB	Injector 8 Control	4.0-4.4 ms
15	TN/BK	Injector 5 Control	4.0-4.4 ms
16	OR/RD	Ignition System Ground	<0.050v
17	PK/LG	Self-Test Output & MIL	MIL On: 1v, Off: 12v
18	TN/OR	Data Bus (+) Signal	Digital Signals
19	PK/LB	Data Bus (-) Signal	Digital Signals
20	BK	PCM Case Ground	<0.050v
21	WT/LB	IAC Motor Control	8.3-11.5v
22	LB/OR	Fuel Pump Control	On: 1v, Off: 12v
23	YL/RD	Knock Sensor Signal	0v
24	DB/OR	CID Sensor Signal	6-7 Hz
25	GY	ACT Sensor Signal	1.5-2.5v
26	BR/WT	Reference Voltage	4.9-5.1v
27	BR/LG	DPFE EGR Sensor Signal	0.4v, 55 mph: 2.9v
28	---	Not Used	---
29	DG	Octane Adjust Switch	Closed: 0v, Open: 9.1v
30	LB/YL	MLP Sensor Signal	In 'P': 0v, in O/D: 1.7v
31	---	Not used	---
32	LG/BK	IMRC Solenoid Control	12v, 55 mph: 12v
33	BR/PK	VR Solenoid Control	0%, 55 mph: 45%
34	LB/PK	Data Output Link	Digital Signals
35	BR/LB	Injector 4 Control	4.0-4.4 ms
36	PK	Spark Angle Word Signal	6.93v (50% d/cycle)
37	RD	Vehicle Power	12-14v
38	WT/YL	EPC Solenoid Control	9.5v (20 psi)
39	BR/YL	Injector 3 Control	4.0-4.4 ms
40	BK/WT	Power Ground	<0.1v

1993-95 Mark VIII 4.6L V8 MFI VIN V (A/T) 60 Pin Connector

PCM Pin #	Wire Color	Circuit Description (60 Pin)	Value at Hot Idle
41	TN/WT	TCS (switch) Signal	TCS & O/D On: 12v
42	---	Not Used	---
43	RD/BK	HO2S-21 (B2 S1) Signal	0.1-1.1v
44	GY/LB	HO2S-11 (B1 S1) Signal	0.1-1.1v
45	---	Not Used	---
46	GY/RD	Analog Signal Return	<0.050v
47	GY/WT	TP Sensor Signal	0.8v, 55 mph: 1.1v
48	WT/PK	Self-Test Input Signal	STI On: 0v, Off: 5v
49	OR/BK	TOT Sensor Signal	2.10-2.40v
50	LB/RD	MAF Sensor Signal	0.8v, 55 mph: 1.8v
51	OR/YL	Shift Solenoid 1 Control	1v, 55 mph: 12v
52	PK/OR	Shift Solenoid 2 Control	12v, 55 mph: 1v
53	PK/YL	TCC Solenoid Control	12v, 55 mph: 9v
54	PK/YL	A/C WOT Relay Control	On: 1v, Off: 12v
55	WT/LG	TCIL (lamp) Control	Lamp On: 1v, Off: 12v
56	GY/OR	PIP Sensor Signal	6.93v (50% dwell)
57	RD	Vehicle Power	12-14v
58	TN	Injector 1 Control	4.0-4.4 ms
59	WT	Injector 2 Control	4.0-4.4 ms
60	BK/WT	Power Ground	<0.1v

Pin Connector Graphic

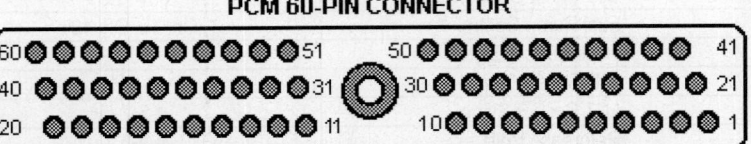

PCM 60-PIN CONNECTOR

Terminal View of 60-Pin PCM Harness Connector

Standard Colors and Abbreviations

Abbreviation	Color	Abbreviation	Color	Abbreviation	Color
BK	Black	GY	Gray	PK	Purple
BL	Blue	GN	Green	RD	Red
BR	Brown	LG	LT Green	TN	Tan
DB	Dark Blue	OR	Orange	WT	White
DG	DK Green	PK	Pink	YL	Yellow

1996-97 Mark VIII 4.6L V8 MFI VIN V (A/T) 104 Pin Connector

PCM Pin #	Wire Color	Circuit Description (104 Pin)	Value at Hot Idle
1	PK/OR	Shift Solenoid 2	1v, 55 mph: 12v
2	PK/LG	MIL (lamp) Control	MIL On: 1v, Off: 12v
3-4, 6-7	---	Not Used	---
5	W	EAM System Monitor	1v, 55 mph: 1v
8	DG/LG	IMRC Solenoid Monitor	5v, 55 mph: 5v
9-12	---	Not Used	---
13	YL/BK	Flash EEPROM Power	0.1v
15	PK/LB	Data Bus (-) Signal	Digital Signals
16	TN/OR	Data Bus (+) Signal	Digital Signals
14, 17-20	---	Not Used	---
21	DB	CKP Sensor (+) Signal	365-395 Hz
22	GY	CKP Sensor (-) Signal	365-395 Hz
23	---	Not Used	---
24	BK/WT	Power Ground	<0.1v
25	BK	PCM Case Ground	<0.050v
26	TN/WT	Coil Driver 1 Control	5º, 55 mph: 8º dwell
27	OR/YL	Shift Solenoid 1 Control	12v, 55 mph: 1v
28, 31, 34	---	Not Used	---
29	TN/WT	TCS (switch) Signal	TCS & O/D On: 12v
30	DG	Octane Adjust Switch	9.3v (Closed: 0v)
32	DG/PK	Knock Sensor 2 Signal	0v
33	PK/OR	VSS (-) Signal	0 Hz, 55 mph: 125 Hz
35	RD/LG	HO2S-12 (B1 S2) Signal	0.1-1.1v
36	TN/LB	MAF Sensor Return	<0.050v
37	OR/BK	TFT Sensor Signal	2.10-2.40v
38	LG/RD	ECT Sensor Signal	0.5-0.6v
39	GY	IAT Sensor Signal	1.5-2.5v
40	PK/BK	Fuel Pump Monitor	12v, Off: 0v
41	BK/YL	A/C Cycling Clutch Switch	A/C On: 12v
42	LG/BK	IMRC Solenoid Control	12v, 55 mph: 12v
43	LB/PK	Data Output Line	Digital Signals
44-46	---	Not Used	---
47	BR/PK	VR Solenoid Control	0%, 55 mph: 45%
48	TN/YL	Clean Tachometer Output	40-46 Hz
49-50	---	Not Used	---
51	BK/WT	Power Ground	<0.1v
52	TN/OR	Coil Driver 2 Control	5º, 55 mph: 8º dwell
53, 59	---	Not Used	---
54	PK/YL	TCC Solenoid Control	0%
55	Y	Keep Alive Power	12-14v
56	LG/BK	EVAP VMV (valve) Control	0-10 Hz (0-100%)
57	YL/RD	Knock Sensor 1 Signal	0v
58	DG/WT	VSS (+) Signal	0 Hz, 55 mph: 125 Hz
60	GY/LB	HO2S-12 (B1 S2) Signal	0.1-1.1v
61	PK/LG	HO2S-22 (B2 S2) Signal	0.1-1.1v
62-63	---	Not Used	---
64	LB/YL	TR Sensor Signal	In 'P': 0v, in O/D: 1.7v

1996-97 Mark VIII 4.6L V8 MFI VIN V (A/T) 104 Pin Connector

PCM Pin #	Wire Color	Circuit Description (104 Pin)	Value at Hot Idle
65	BR/LG	DPFE EGR Sensor Signal	0.20-1.30v
66-69	---	Not Used	---
70	OR/YL	EAM Solenoid Control	12v, 55 mph: 12v
71	RD	Vehicle Power	12-14v
72	TN/RD	Injector 7 Control	2.5-3.0 ms
73	TN/BK	Injector 5 Control	2.5-3.0 ms
74	BR/YL	Injector 3 Control	2.5-3.0 ms
75	TN	Injector 1 Control	2.5-3.0 ms
76, 77, 103	BK/WT	Power Ground	<0.1v
78	TN/LG	Coil Driver 3 Control	5°, 55 mph: 8° dwell
79	WT/LG	TCIL (lamp) Control	Lamp On: 1v, Off: 12v
80	LB/OR	Fuel Pump Control	On: 1v, Off: 12v
81	WT/YL	EPC Solenoid Control	8.3v (15 psi)
83	WT/LB	IAC Motor Control	30%, 55 mph: 45%
84	DG/WT	TSS Sensor Signal	43 Hz (647 rpm)
85	DB/OR	CMP Sensor	4-6 Hz
86	DG/WT	A/C Pressure Switch Signal	Open: 12v, Closed: 0v
87	RD/BK	HO2S-21 (B2 S1) Signal	0.1-1.1v
88	LB/RD	MAF Sensor Signal	0.6v, 55 mph: 1.7v
89	GY/WT	TP Sensor Signal	0.53-1.27v
90	BR/WT	Reference Voltage	4.9-5.1v
91	GY/RD	Analog Signal Return	<0.050v
92	LG	BPP Switch Signal	Brake Off: 0v, On: 12v
93	RD/WT	HO2S-11 (B1 S1) Heater	1v, Off: 12v
94	YL/LB	HO2S-21 (B2 S1) Heater	1v, Off: 12v
95	WT/BK	HO2S-12 (B1 S2) Heater	1v, Off: 12v
96	TN/YL	HO2S-22 (B2 S2) Heater	1v, Off: 12v
97	RD	Vehicle Power	12-14v
98	LB	Injector 8 Control	2.5-3.0 ms
99	LG/OR	Injector 6 Control	2.5-3.0 ms
100	BR/LB	Injector 4 Control	2.5-3.0 ms
101	WT	Injector 2 Control	2.5-3.0 ms
102	---	Not Used	---
104	TN/LB	Coil Driver 4 Control	5°, 55 mph: 8° dwell

Pin Connector Graphic

PCM 104-PIN CONNECTOR

Terminal View of 104-Pin PCM Wiring Harness Connector

Standard Colors and Abbreviations

Abbreviation	Color	Abbreviation	Color	Abbreviation	Color
BK	Black	GY	Gray	PK	Purple
BL	Blue	GN	Green	RD	Red
BR	Brown	LG	LT Green	TN	Tan
DB	Dark Blue	OR	Orange	WT	White
DG	DK Green	PK	Pink	YL	Yellow

1998 Mark VIII 4.6L 4v V8 MFI VIN V (A/T) 104 Pin Connector

PCM Pin #	Wire Color	Circuit Description (104 Pin)	Value at Hot Idle
1	OR/YL	COP 5 Driver Control	5°, 55 mph: 8° dwell
2	PK/LG	MIL (lamp) Control	MIL On: 1v, Off: 12v
3, 24	BK/WT	Power Ground	<0.1v
5	BK/OR	Air Injection System Monitor	1v, 55 mph: 1v
6	OR/YL	Shift Solenoid 1 Control	1v, 55 mph: 1v
8	LG/BR	IMRC Monitor	5v, 55 mph: 5v
9	YL/WT	Fuel Level Indicator Signal	1.7v (1/2 full)
11	PK/OR	Shift Solenoid 2 Control	12v, 55 mph: 1v
13	PK	Flash EEPROM Power	0.5v
15	PK	Data Bus (-) Signal	Digital Signals
16	TN/OR	Data Bus (+) Signal	Digital Signals
20	PK/BK	Shift Solenoid 3 Control	12v, 55 mph: 1v
21	BK/PK	CKP Sensor (+) Signal	390-450 Hz
22	GY/YL	CKP Sensor (-) Signal	390-450 Hz
25	BK	PCM Case Ground	<0.050v
26	LG/WT	COP 1 Driver Control	5°, 55 mph: 8° dwell
27	LG/YL	COP 6 Driver Control	5°, 55 mph: 8° dwell
28	DB	Low Cooling Fan Control	Fan On: 1v, Off: 12v
29	TN/WT	TCS (switch) Signal	TCS & O/D On: 12v
30	DG	Octane Adjust Switch	9.3v (Closed: 0v)
31	YL/LG	PSP Switch Signal	Straight: 5v, Turning: 0v
32	DG/PK	Knock Sensor 2 Signal	0v
33	PK/OR	VSS (-) Signal	0 Hz, 55 mph: 125 Hz
34	YL/BK	Digital TR1 Sensor Signal	0v, 55 mph: 11v
35	RD/LG	HO2S-12 (B1 S2) Signal	0.1-1.1v
36	TN/LB	MAF Sensor Return	<0.050v
37	OR/BK	TFT Sensor Signal	2.10-2.40v
38	LG/RD	ECT Sensor Signal	0.5-0.6v
39	GY	IAT Sensor Signal	1.5-2.5v
40	DG/YL	Fuel Pump Monitor	12v, Off: 0v
41	PK	A/C High Pressure Switch	Open: 12v, Closed: 0v
42	BR	IMRC Solenoid Control	12v, 55 mph: 12v
43	WT/RD	Fuel Pump Control	On: 1v, Off: 12v
46	LG/PK	High Speed Fan Control	Fan On: 1v, Off: 12v
47	BR/PK	VR Solenoid Control	0%, 55 mph: 45%
48	OR/WT	Clean Tachometer Output	39-45 Hz
49	LB/BK	Digital TR2 Sensor Signal	0v, 55 mph: 11v
50	WT/BK	Digital TR4 Sensor Signal	0v, 55 mph: 11v
51, 77, 103	BK/WT	Power Ground	<0.1v
52	WT/PK	COP 2 Driver Control	5°, 55 mph: 8° dwell
53	DG/PK	COP 7 Driver Control	5°, 55 mph: 8° dwell
54	PK/YL	TCC Solenoid Control	12v, 55 mph: 1v
55	BR/PK	Keep Alive Power	12-14v
56	LG/BK	EVAP VMV (valve) Control	0-10 Hz (0-100%)
57	YL/RD	Knock Sensor 1 Signal	0v
58	GY/BK	VSS (+) Signal	0 Hz, 55 mph: 125 Hz
59	DG/WT	TSS Sensor Signal	40-45 Hz (645-650 rpm)
60	GY/LB	HO2S-11 (B1 S1) Signal	0.1-1.1v
61	PK/LG	HO2S-22 (B2 S2) Signal	0.1-1.1v
62	RD/PK	Fuel Tank Pressure Sensor	2.6v (0" H2O - cap off)
64	RD/LB	Digital TR Sensor Signal	In 'P': 0v, in O/D: 1.7v

1998 Mark VIII 4.6L 4v V8 MFI VIN V (A/T) 104 Pin Connector

PCM Pin #	Wire Color	Circuit Description (104 Pin)	Value at Hot Idle
65	BR/LG	DPFE Sensor Signal	0.5v, 55 mph: 3.1v
70	WT/OR	Air Mgmt. Solenoid Control	12v, 55 mph: 12v
67	PK/WT	EVAP Vent Valve	0-10 Hz (0-100%)
69	PK/YL	A/C WOT Relay Control	Relay On: 0v, Off: 12v
71, 97	RD	Vehicle Power	12-14v
72	TN/RD	Injector 7 Control	2.5-3-0 ms
73	TN/BK	Injector 5 Control	2.5-3-0 ms
74	BR/YL	Injector 3 Control	2.5-3-0 ms
75	TN	Injector 1 Control	2.5-3-0 ms
78	PK/LB	COP 3 Driver Control	5°, 55 mph: 8° dwell
79	WT/RD	COP 8 Driver Control	5°, 55 mph: 8° dwell
80	LB/OR	Fuel Pump Control	On: 1v, Off: 12v
81	WT/YL	EPC Solenoid Control	9v (17 psi)
83	WT/LB	IAC Motor Control	32% duty cycle
84	DG/YL	OSS Sensor Signal	132-136 (1355 rpm)
85	DB/OR	CMP Sensor Signal	4-6 Hz
86	TN/LG	A/C Pressure Transducer	0.8-1.4v
87	RD/BK	HO2S-21 (B2 S1) Signal	0.1-1.1v
88	LB/RD	MAF Sensor Signal	0.6v, 55 mph: 1.7v
89	GY/WT	TP Sensor Signal	0.53-1.27v
90	BR/WT	Reference Voltage	4.9-5.1v
91	GY/RD	Analog Signal Return	<0.050v
92	LG	Brake Switch Signal	Brake On: 12v, Off: 0v
93	RD/WT	HO2S-11 (B1 S1) Heater	1v, Off: 12v
94	YL/LB	HO2S-21 (B2 S1) Heater	1v, Off: 12v
95	WT/BK	HO2S-12 (B1 S2) Heater	1v, Off: 12v
96	TN/YL	HO2S-22 (B2 S2) Heater	1v, Off: 12v
98	LB	Injector 8 Control	2.5-3-0 ms
99	LG/OR	Injector 6 Control	2.5-3-0 ms
100	BR/LB	Injector 4 Control	2.5-3-0 ms
101	WT	Injector 2 Control	2.5-3-0 ms
104	PK/WT	Coil Driver 4 Control	5°, 55 mph: 8° dwell

Pin Connector Graphic

```
                PCM 104-PIN CONNECTOR
  1 ◉◉◉◉◉◉◉◉◉◉◉◉◉      ◉◉◉◉◉◉◉◉◉◉◉◉◉ 26
 27 ◉◉◉◉◉◉◉◉◉◉◉◉◉   ⬤  ◉◉◉◉◉◉◉◉◉◉◉◉◉ 52
 53 ◉◉◉◉◉◉◉◉◉◉◉◉◉      ◉◉◉◉◉◉◉◉◉◉◉◉◉ 78
 79 ◉◉◉◉◉◉◉◉◉◉◉◉◉      ◉◉◉◉◉◉◉◉◉◉◉◉◉ 104
```

Terminal View of 104-Pin PCM Wiring Harness Connector

Standard Colors and Abbreviations

Abbreviation	Color	Abbreviation	Color	Abbreviation	Color
BK	Black	GY	Gray	PK	Purple
BL	Blue	GN	Green	RD	Red
BR	Brown	LG	LT Green	TN	Tan
DB	Dark Blue	OR	Orange	WT	White
DG	DK Green	PK	Pink	YL	Yellow

Town Car Pin Tables

1991-95 Town Car 4.6L V8 MFI VIN W (A/T) 60 Pin Connector

PCM Pin #	Wire Color	Circuit Description (60 Pin)	Value at Hot Idle
1	YL	Keep Alive Power	12-14v
1	YL/BK	Keep Alive Power	12-14v
2 ('92-'95)	LG	Brake Switch Signal	Brake Off: 0v, On: 12v
3	YL/BL	VSS (+) Signal	0 Hz, 55 mph: 125 Hz
3	G/BK	VSS (+) Signal	0 Hz, 55 mph: 125 Hz
4	TN/YL	Ignition Diagnostic Monitor	20-31 Hz
4	WT/BL	Ignition Diagnostic Monitor	20-31 Hz
5	DG/WT	OSS Sensor (RPM)	30 mph: 260-1330 rpm
6	BK/BL	VSS (-) Signal	0 Hz, 55 mph: 125 Hz
6	PK/OR	VSS (-) Signal	0 Hz, 55 mph: 125 Hz
7	LG/RD	ECT Sensor Signal	0.5-0.6v
8	DG/YL	Fuel Pump Monitor	12v, Off: 0v
8	PK/BK	Fuel Pump Monitor	12v, Off: 0v
9	TN/BL	MAF Sensor Return	<0.050v
10	PK	A/C Cycling Clutch Switch	A/C On: 12v, Off: 0v
10	DG/OR	A/C Cycling Clutch Switch	A/C On: 12v, Off: 0v
10	PK/YL	A/C Cycling Clutch Switch	A/C On: 12v, Off: 0v
11	GY/YL	EVAP Purge Solenoid	On: 1v, Off: 12v
12	LG/OR	Injector 6 Control	4.0-4.4 ms
13	TN/RD	Injector 7 Control	4.0-4.4 ms
14	BL	Injector 8 Control	4.0-4.4 ms
15	TN/BK	Injector 5 Control	4.0-4.4 ms
15	TN	Injector 5 Control	4.0-4.4 ms
16	OR	Ignition System Ground	<01v
16	OR/RD	Ignition System Ground	<01v
17	PK/LG	Self-Test Output, MIL Control	MIL On: 1v, Off: 12v
17	TN/RD	Self-Test Output, MIL Control	MIL On: 1v, Off: 12v
18	TN/OR	Data Bus (+) Signal	Digital Signals
19	PK/BL	Data Bus (-) Signal	Digital Signals
20	BK	PCM Case Ground	<0.050v
21	WT	IAC Motor Control	9.8-11.8v
21	WT/BL	IAC Motor Control	9.8-11.8v
22	BL/OR	Fuel Pump Control	On: 1v, Off: 12v
23	---	Not Used	---
24	DG	CID Sensor Signal	5-9 Hz
24	DB/OR	CID Sensor Signal	5-9 Hz
25	LG/PK	ACT Sensor Signal	1.5-2.5v
25	GY	ACT Sensor Signal	1.5-2.5v
26	BR/WT	Reference Voltage	4.9-5.1v
27	BR/LG	DPFE EGR Sensor Signal	0.4v, 55 mph: 2.9v
28 ('91)	YL/LG	PSP Switch Signal	Straight: 0v, turned: 12v
29	WT/RD	Octane Adjust Switch	Closed: 0v, Open: 9.1v
29	DG	Octane Adjust Switch	Closed: 0v, Open: 9.1v
30 ('91)	WT/PK	NDS Signal	In 'N': 0v, Others: 5v
30 ('92-'95)	BL/YL	MLP Sensor Signal	In 'P': 0v, in O/D: 2.1v
31-32	---	Not Used	---
33	BR/PK	VR Solenoid Control	0%, 55 mph: 45%
34	DB/LG	Data Output Link	Digital Signals
35	BR/BL	Injector 4 Control	4.0-4.4 ms
36	PK	Spark Angle Word Signal	6.93v (50% d/cycle)
37, 57	RD	Vehicle Power	12-14v
38 ('92-'95)	WT/YL	EPC Solenoid Signal	9.5v (16-18 psi)

1991-95 Town Car 4.6L V8 MFI VIN W (A/T) 60 Pin Connector

PCM Pin #	Wire Color	Circuit Description (60 Pin)	Value at Hot Idle
39	BR/YL	Injector 3 Control	4.0-4.4 ms
40	BK/WT	Power Ground	<0.1v
41 ('93-'94)	TN/WT	TCS (switch) Signal	TCS & O/D On: 12v
42	---	Not Used	---
43	RD/BK	HO2S-11 (B1 S1) Signal	0.1-1.1v
44	GYL/BL	HO2S-21 (B2 S1) Signal	0.1-1.1v
45	---	Not Used	---
46	GY/RD	Analog Signal Return	<0.050v
47	GY/WT	TP Sensor Signal	0.8v, 55 mph: 1.1v
48	WT/PK	Self-Test Input Signal	STI On: 0v, Off: 5v
49	OR/BK	TOT Sensor Signal	2.10-2.40v
50	BL/RD	MAF Sensor Signal	0.7v, 55 mph: 2.1v
51 ('92-'95)	OR/YL	Shift Solenoid 1 Control	12v, 55 mph: 1v
52 ('92-'95)	PK/OR	Shift Solenoid 2 Control	1v, 55 mph: 12v
53 ('92-'95)	BR/OR	TCC Solenoid Control	12v, 30 mph: 11-12v
54	PK/YL	A/C WOT Relay Control	On: 1v, Off: 12v
54	OR/BL	A/C WOT Relay Control	On: 1v, Off: 12v
55 ('93-'95)	WT/LG	TCIL (lamp) Control	Lamp On: 1v, Off: 12v
56	GY/OR	PIP Sensor Signal	6.93v (50% dwell)
58	TN	Injector 1 Control	4.0-4.4 ms
59	WT	Injector 2 Control	4.0-4.4 ms
60	BK/WT	Power Ground	<0.1v

Pin Connector Graphic

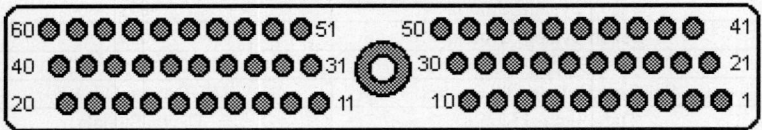

PCM 60-PIN CONNECTOR

Terminal View of 60-Pin PCM Harness Connector

Standard Colors and Abbreviations

Abbreviation	Color	Abbreviation	Color	Abbreviation	Color
BK	Black	GY	Gray	PK	Purple
BL	Blue	GN	Green	RD	Red
BR	Brown	LG	LT Green	TN	Tan
DB	Dark Blue	OR	Orange	WT	White
DG	DK Green	PK	Pink	YL	Yellow

1996-97 Town Car 4.6L V8 MFI VIN W (A/T) 104 Pin Connector

PCM Pin #	Wire Color	Circuit Description (104 Pin)	Value at Hot Idle
1	PK/OR	Shift Solenoid 2 Control	12v, 55 mph: 1v
2	PK/LG	MIL (lamp) Control	MIL On: 1v, Off: 12v
3-12, 14	---	Not Used	---
13	PK	Flash EEPROM Power	0.1v
15	PK/LB	Data Bus (-) Signal	Digital Signals
16	TN/OR	Data Bus (+) Signal	Digital Signals
17-20, 23	---	Not Used	---
21	BK/PK	CKP Sensor (+) Signal	440-490 Hz
22	GY/YL	CKP Sensor (-) Signal	440-490 Hz
24	BK/WT	Power Ground	<0.1v
25	BK	PCM Case Ground	<0.050v
26	TN/WT	Coil Driver 1 Control	5°, 55 mph: 8° dwell
27	OR/YL	Shift Solenoid 1 Control	1v, 55 mph: 12v
28, 31, 32	---	Not Used	---
29	TN/WT	TCS (switch) Signal	TCS & O/D On: 12v
30	DG	Octane Adjust Switch	9.3v (Closed: 0v)
33	PK/OR	VSS (-) Signal	0 Hz, 55 mph: 125 Hz
34	---	Not Used	---
35	RD/LG	HO2S-12 (B1 S2) Signal	0.1-1.1v
36	TN/LB	MAF Sensor Return	<0.050v
37	OR/BK	TFT Sensor Signal	2.10-2.40v
38	LG/RD	ECT Sensor Signal	0.5-0.6v
39	GY	IAT Sensor Signal	1.5-2.5v
40	DG/YL	Fuel Pump Monitor	12v, Off: 0v
41	BK/YL	A/C Cycling Clutch Switch	A/C On: 12v
42	---	Not Used	---
43	DB/LG	Data Output Link	Digital Signals
44	---	Not Used	---
45	DB	Low Speed Fan Control	Fan On: 1v, Off: 12v
46	---	Not Used	---
47	BR/PK	VR Solenoid Control	0%, 55 mph: 45%
48-50	---	Not Used	---
51	BK/WT	Power Ground	<0.1v
52	TN/OR	Coil Driver 2 Control	5°, 55 mph: 8° dwell
53	---	Not Used	---
54	PK/YL	TCC Solenoid Control	0%
55	YL/BK	Keep Alive Power	12-14v
56	LG/BK	EVAP VMV (valve) Control	0-10 Hz (0-100%)
57	---	Not Used	---
58	GY/BK	VSS (+) Signal	0 Hz, 55 mph: 125 Hz
59	---	Not Used	---
60	GY/LB	HO2S-11 (B1 S1) Signal	0.1-1.1v
61	PK/LG	HO2S-22 (B2 S2) Signal	0.1-1.1v
62	RD/PK	FTP Sensor Signal	2.6v (0" H2O - cap off)
63	---	Not Used	---
64	LB/YL	TR Sensor Signal	In 'P': 0v, in O/D: 1.7v

1996-97 Town Car 4.6L V8 MFI VIN W (A/T) 104 Pin Connector

PCM Pin #	Wire Color	Circuit Description (104 Pin)	Value at Hot Idle
65	BR/LG	DPFE EGR Sensor Signal	0.20-1.30v
66	---	Not Used	---
67	PK/WT	EVAP Vent Solenoid Control	0-10 Hz (0-100%)
68, 70	---	Not Used	---
69	PK/YL	A/C WOT Relay Control	On: 0.1v, Off: 12v
71	RD	Vehicle Power	12-14v
72	TN/RD	Injector 7 Control	3.3-3-5 ms
73	TN/BK	Injector 5 Control	3.3-3-5 ms
74	BR/YL	Injector 3 Control	3.3-3-5 ms
75	TN	Injector 1 Control	3.3-3-5 ms
76, 77	BK/WT	Power Ground	<0.1v
78	TN/LG	Coil Driver 3 Control	5°, 55 mph: 8° dwell
79	WT/LG	TCIL (lamp) Control	Lamp On: 1v, Off: 12v
80	LB/OR	Fuel Pump Control	On: 1v, Off: 12v
81	WT/YL	EPC Solenoid Control	9.5v (20 psi)
82, 86	---	Not Used	---
83	WT/LB	IAC Motor Control	30%, 55 mph: 45%
84	DG/WT	TSS Sensor Signal	43 Hz (650 rpm)
85	DG	CID Sensor Signal	5-7 Hz
87	RD/BK	HO2S-21 (B2 S1) Signal	0.1-1.1v
88	LB/RD	MAF Sensor Signal	0.6v, 55 mph: 1.7v
89	GY/WT	TP Sensor Signal	0.53-1.27v
90	BR/WT	Reference Voltage	4.9-5.1v
91	GY/RD	Analog Signal Return	<0.050v
92	LG	BPP Switch Signal	Brake Off: 0v, On: 12v
93	RD/WT	HO2S-11 (B1 S1) Heater	1v, Off: 12v
94	YL/LB	HO2S-21 (B2 S1) Heater	1v, Off: 12v
95	WT/BK	HO2S-12 (B1 S2) Heater	1v, Off: 12v
96	TN/YL	HO2S-22 (B2 S2) Heater	1v, Off: 12v
97	RD	Vehicle Power	12-14v
98	LB	Injector 8 Control	3.3-3-5 ms
99	LG/OR	Injector 6 Control	3.3-3-5 ms
100	BR/LB	Injector 4 Control	3.3-3-5 ms
101	WT	Injector 2 Control	3.3-3-5 ms
102	---	Not Used	---
103	BK/WT	Power Ground	<0.1v
104	TN/LB	Coil Driver 4 Control	5°, 55 mph: 8° dwell

Pin Connector Graphic

PCM 104-PIN CONNECTOR

Terminal View of 104-Pin PCM Wiring Harness Connector

Standard Colors and Abbreviations

Abbreviation	Color	Abbreviation	Color	Abbreviation	Color
BK	Black	GY	Gray	PK	Purple
BL	Blue	GN	Green	RD	Red
BR	Brown	LG	LT Green	TN	Tan
DB	Dark Blue	OR	Orange	WT	White
DG	DK Green	PK	Pink	YL	Yellow

1998-99 Town Car 4.6L 2v V8 MFI VIN W (A/T) 104 Pin Connector

PCM Pin #	Wire Color	Circuit Description (104 Pin)	Value at Hot Idle
1	OR/YL	COP 6 Driver Control	5°, 55 mph: 8° dwell
2	PK/LG	MIL (lamp) Control	MIL On: 1v, Off: 12v
3	BK/WT	Power Ground	<0.1v
6	OR/YL	Shift Solenoid 1 Control	1v, 55 mph: 1v
9	YL/WT	Fuel Level Indicator Signal	1.7v (1/2 full)
11	PK/OR	Shift Solenoid 2 Control	12v, 55 mph: 1v
12	WT/LG	TCIL (lamp) Control	Lamp On: 1v, Off: 12v
13	PK	Flash EEPROM Power	0.5v
15	PK/LB	Data Bus (-) Signal	Digital Signals
16	TN/OR	Data Bus (+) Signal	Digital Signals
21	BK/PK	CKP Sensor (+) Signal	440-490 Hz
22	GY/YL	CKP Sensor (-) Signal	440-490 Hz
24	BK/WT	Power Ground	<0.1v
25	BK	PCM Case Ground	<0.050v
26	LG/WT	COP 1 Driver Control	5°, 55 mph: 8° dwell
27	LG/YL	COP 5 Driver Control	1v, 30 mph: 12v
28	RD/OR	Low Speed Fan Control	Fan On: 1v, Off: 12v
29	TN/WT	TCS (switch) Signal	TCS & O/D On: 12v
30	WT/RD	Octane Adjust Switch	9.3v (Closed: 0v)
33	PK/OR	VSS (-) Signal	0 Hz, 55 mph: 125 Hz
34	YL/BK	Digital TR1 Sensor Signal	In 'P': 0v, 55 mph: 11v
35	RD/LG	HO2S-11 (B1 S1) Signal	0.1-1.1v
36	TN/LB	MAF Sensor Return	<0.050v
37	OR/BK	TFT Sensor Signal	2.10-2.40v
39	GY	IAT Sensor Signal	1.5-2.5v
40	DG/YL	Fuel Pump Monitor	12v, Off: 0v
41	BK/YL	A/C Cycling Clutch Switch	Switch On: 12v, Off: 0v
45	OR/RD	CHTIL (lamp) Control	Lamp On: 1v, Off: 12v
46	LG/RD	High Speed Fan Control	Fan On: 1v, Off: 12v
47	BR/PK	VR Solenoid Control	0%, 55 mph: 45%
49	LB/BK	Digital TR2 Sensor Signal	In 'P': 0v, 55 mph: 11v
50	WT/BK	Digital TR4 Sensor Signal	In 'P': 0v, 55 mph: 11v
51	BK/WT	Power Ground	<0.1v
52	WT/PK	COP 3 Driver Control	5°, 55 mph: 8° dwell
53	OR/PK	COP 4 Driver Control	5°, 55 mph: 8° dwell
54	PK/YL	TCC Solenoid Control	0%, 55 mph: 100%
55	RD/WT	Keep Alive Power	12-14v
56	LG/BK	EVAP Purge Valve Control	0-10 Hz (0-100%)
57	YL/RD	Knock Sensor 1 Signal	0v
58	GY/BK	VSS (+) Signal	0 Hz, 55 mph: 125 Hz
60	GY/LB	HO2S-11 (B1 S1) Signal	0.1-1.1v
61	PK/LG	HO2S-22 (B2 S2) Signal	0.1-1.1v
62	RD/PK	FTP Sensor Signal	2.6v (0" H2O - cap off)
64	DB/YL	Digital TR Sensor Signal	In 'P': 0v, in O/D: 1.7v

1998-99 Town Car 4.6L 2v V8 MFI VIN W (A/T) 104 Pin Connector

PCM Pin #	Wire Color	Circuit Description (104 Pin)	Value at Hot Idle
65	BR/LG	DPFE Sensor Signal	0.20-1.30v
66	YL/LG	CHT Sensor Signal	0.7v (194°F)
67	PK/WT	EVAP Vent Solenoid Control	0-10 Hz (0-100%)
69	OR/LB	A/C WOT Cutoff Relay	On: 1v, Off: 12v
71	RD	Vehicle Power	12-14v
72	TN/RD	Injector 7 Control	3.4-3-7 ms
73	TN/BK	Injector 5 Control	3.4-3-7 ms
74	BR/YL	Injector 3 Control	3.4-3-7 ms
75	TN	Injector 1 Control	3.4-3-7 ms
78	PK/LB	COP 7 Driver Control	5°, 55 mph: 8° dwell
79	WT/RD	COP 8 Driver Control	5°, 55 mph: 8° dwell
80	LB/OR	Fuel Pump Control	On: 1v, Off: 12v
81	WT/YL	EPC Solenoid Control	9.5v (20 psi)
83	WT/LB	IAC Motor Control	34% duty cycle
84	DB/YL	OSS Sensor Signal	0 Hz, 55 mph: 216 Hz
85	DB/OR	CMP Sensor Signal	6-7 Hz
86	WT/BK	A/C High Pressure Switch	Open: 12v, Closed: 0v
87	RD/BK	HO2S-21 (B2 S1) Signal	0.1-1.1v
88	LB/RD	MAF Sensor Signal	0.6v, 55 mph: 1.8v
89	GY/WT	TP Sensor Signal	0.53-1.27v
90	BR/WT	Reference Voltage	4.9-5.1v
91	GY/RD	Analog Signal Return	<0.050v
92	LG	BPP Switch Signal	Brake On: 12v, Off: 0v
93	RD/WT	HO2S-11 (B1 S1) Heater	1v, Off: 12v
94	YL/LB	HO2S-21 (B2 S1) Heater	1v, Off: 12v
95	WT/BK	HO2S-12 (B1 S2) Heater	1v, Off: 12v
96	TN/YL	HO2S-22 (B2 S2) Heater	1v, Off: 12v
97	RD	Vehicle Power	12-14v
98	LB	Injector 8 Control	3.4-3-7 ms
99	LG/OR	Injector 6 Control	3.4-3-7 ms
100	BR/LB	Injector 4 Control	3.4-3-7 ms
101	WT	Injector 2 Control	3.4-3-7 ms
103	BK/WT	Power Ground	<0.1v
104	PK/WT	COP 2 Driver Control	5°, 55 mph: 8° dwell

Pin Connector Graphic

PCM 104-PIN CONNECTOR

Terminal View of 104-Pin PCM Wiring Harness Connector

Standard Colors and Abbreviations

Abbreviation	Color	Abbreviation	Color	Abbreviation	Color
BK	Black	GY	Gray	PK	Purple
BL	Blue	GN	Green	RD	Red
BR	Brown	LG	LT Green	TN	Tan
DB	Dark Blue	OR	Orange	WT	White
DG	DK Green	PK	Pink	YL	Yellow

2000-01 Town Car 4.6L 2v V8 VIN W (A/T) 104 Pin Connector

PCM Pin #	Wire Color	Circuit Description (104 Pin)	Value at Hot Idle
1	OG/YL	COP 6 Driver Control	5°, 55 mph: 8° dwell
2	---	Not Used	---
3	BK/WT	Power Ground	<0.1v
4-5	---	Not Used	---
6	OG/YL	Shift Solenoid 'A' Control	1v, 55 mph: 1v
7-10	---	Not Used	---
11	VT/OR	Shift Solenoid 'B' Control	12v, 55 mph: 1v
12, 14	---	Not Used	---
13	VT	Flash EEPROM Power	0.1v
15	PK/LB	Data Bus (-) Signal	Digital Signals
16	TN/OG	Data Bus (+) Signal	Digital Signals
17-20	---	Not Used	---
21	BK/PK	CKP Sensor (+) Signal	440-490 Hz
22	GY/YL	CKP Sensor (-) Signal	440-490 Hz
23-24	---	Not Used	---
25	BK	PCM Case Ground	<0.050v
26	LG/WT	COP 1 Driver Control	5°, 55 mph: 8° dwell
27	LG/YL	COP 5 Driver Control	1v, 30 mph: 12v
28	RD/OG	Low Speed Fan Control	Fan On: 1v, Off: 12v
29	TN/WT	TCS (switch) Signal	TCS & O/D On: 12v
30-33	---	Not Used	---
34	YL/BK	Digital TR1 Sensor Signal	In 'P': 0v, 55 mph: 11v
35	RD/LG	HO2S-12 (B1 S2) Signal	0.1-1.1v
36	TN/LB	MAF Sensor Return	<0.050v
37	OG/BK	TFT Sensor Signal	2.10-2.40v
38	---	Not Used	---
39	GY	IAT Sensor Signal	1.5-2.5v
40	DG/YL	Fuel Pump Monitor	12v, Off: 0v
41	BK/YL	A/C Cycling Clutch Switch	Switch On: 12v, Off: 0v
42-45	---	Not Used	---
46	LG/VT	High Speed Fan Control	Fan On: 1v, Off: 12v
47	BR/PK	VR Solenoid Control	0%, 55 mph: 45%
48	---	Not Used	---
49	LB/BK	Digital TR2 Sensor Signal	In 'P': 0v, 55 mph: 11v
50	WT/BK	Digital TR4 Sensor Signal	In 'P': 0v, 55 mph: 11v
51	BK/WT	Power Ground	<0.1v
52	WT/PK	COP 3 Driver Control	5°, 55 mph: 8° dwell
53	DG/VT	COP 4 Driver Control	5°, 55 mph: 8° dwell
54	VT/YL	TCC Solenoid Control	0%, 55 mph: 100%
55	RD/WT	Keep Alive Power	12-14v
56	LG/BK	EVAP Purge Valve Control	0-10 Hz (0-100%)
57-59	---	Not Used	---
60	GY/LB	HO2S-11 (B1 S1) Signal	0.1-1.1v
61	VT/LG	HO2S-22 (B2 S2) Signal	0.1-1.1v
62	RD/BK	FTP Sensor Signal	2.6v (0" H2O - cap off)
63	---	Not Used	---
64	LB/YL	Digital TR Sensor Signal	In 'P': 0v, in O/D: 1.7v
65 ('00)	BR/LG	DPFE Sensor Signal	0.20-1.30v
65 ('01)	BR/LG	DPFE Sensor Signal	0.95-1.05v
66	YL/LG	CHT Sensor Signal	0.7v (194°F)
67	VT/WT	EVAP CV Solenoid	0-10 Hz (0-100%)
68	---	Not Used	---

2000-01 Town Car 4.6L 2v V8 VIN W (A/T) 104 Pin Connector

PCM Pin #	Wire Color	Circuit Description (104 Pin)	Value at Hot Idle
69	OG/LB	A/C WOT Cutoff Relay	On: 1v, Off: 12v
70	---	Not Used	---
71	RD	Vehicle Power	12-14v
72	TN/RD	Injector 7 Control	3.4-3-7 ms
73	TN/BK	Injector 5 Control	3.4-3-7 ms
74	BR/YL	Injector 3 Control	3.4-3-7 ms
75	TN	Injector 1 Control	3.4-3-7 ms
76	---	Not Used	---
77	BK/WT	Power Ground	<0.1v
78	PK/LB	COP 7 Driver Control	5°, 55 mph: 8° dwell
79	WT/RD	COP 8 Driver Control	5°, 55 mph: 8° dwell
80	LB/OR	Fuel Pump Control	On: 1v, Off: 12v
81	WT/YL	EPC Solenoid Control	9.5v (20 psi)
82	---	Not Used	---
83	WT/LB	IAC Motor Control	34% duty cycle
84	DB/YL	OSS Sensor Signal	0 Hz, 55 mph: 216 Hz
85	DB/OR	CMP Sensor Signal	6-7 Hz
86	WT/BK	A/C High Pressure Switch	Open: 12v, Closed: 0v
87	RD/BK	HO2S-21 (B2 S1) Signal	0.1-1.1v
88	LB/RD	MAF Sensor Signal	0.6v, 55 mph: 1.8v
89	GY/WT	TP Sensor Signal	0.53-1.27v
90	BR/WT	Reference Voltage	4.9-5.1v
91	GY/RD	Analog Signal Return	<0.050v
92	---	Not Used	---
93	RD/WT	HO2S-11 (B1 S1) Heater	1v, Off: 12v
94	YL/LB	HO2S-21 (B2 S1) Heater	1v, Off: 12v
95	WT/BK	HO2S-12 (B1 S2) Heater	1v, Off: 12v
96	TN/YL	HO2S-22 (B2 S2) Heater	1v, Off: 12v
97	---	Not Used	---
98	LB	Injector 8 Control	3.4-3-7 ms
99	LG/OR	Injector 6 Control	3.4-3-7 ms
100	BR/LB	Injector 4 Control	3.4-3-7 ms
101	WT	Injector 2 Control	3.4-3-7 ms
102	---	Not Used	---
103	BK/WT	Power Ground	<0.1v
104	PK/WT	COP 2 Driver Control	5°, 55 mph: 8° dwell

Pin Connector Graphic

PCM 104-PIN CONNECTOR

Terminal View of 104-Pin PCM Wiring Harness Connector

Standard Colors and Abbreviations

Abbreviation	Color	Abbreviation	Color	Abbreviation	Color
BK	Black	GY	Gray	PK	Purple
BL	Blue	GN	Green	RD	Red
BR	Brown	LG	LT Green	TN	Tan
DB	Dark Blue	OR	Orange	WT	White
DG	DK Green	PK	Pink	YL	Yellow

2002-03 Town Car 4.6L 2v V8 VIN W (A/T) 104 Pin Connector

PCM Pin #	Wire Color	Circuit Description (104 Pin)	Value at Hot Idle
1	OG/YL	COP 6 Driver Control	5º, 55 mph: 8º dwell
2, 4-5	---	Not Used	---
3	BK/WT	Power Ground	<0.1v
6	OG/YL	Shift Solenoid 'A' Control	1v, 55 mph: 1v
7 ('03)	YL/LG	Generator Regulator 'S' Terminal	0-130 Hz
8-9	---	Not Used	---
10	LG/RD	EGR System Module Signal	0.95-1.05v
11	VT/DG	Shift Solenoid 'B' Control	12v, 55 mph: 1v
12, 14	---	Not Used	---
13	VT	Flash EEPROM Power	0.1v
15	PK/LB	SCP Bus (-) Signal	Digital Signals
16	TN/OG	SCP Bus (+) Signal	Digital Signals
17	GY/OG	Passive Antitheft RX Signal	Digital Signals
18	WT/LG	Passive Antitheft TX Signal	Digital Signals
19-20	---	Not Used	---
21	BK/PK	CKP Sensor (+) Signal	440-490 Hz
22	GY/YL	CKP Sensor (-) Signal	440-490 Hz
23	---	Not Used	---
24	BK/WT	Power Ground	<0.1v
25	BK	PCM Case Ground	<0.050v
26	LG/WT	COP 1 Driver Control	5º, 55 mph: 8º dwell
27	LG/YL	COP 5 Driver Control	1v, 30 mph: 12v
28	RD/OG	Engine Cooling Fan Motor	Fan Off: 12v, On: 1v
29	TN/WT	TCS (switch) Signal	TCS & O/D On: 12v
30 ('03)	OG/RD	Antitheft Indicator Control	Indicator Off: 12v, On: 1v
31 ('03)	YL/LG	PSP Switch Signal	Straight: 0v, Turned: 12v
32 ('03)	DG/VT	Knock Sensor (+) Signal	0v
33	---	Not Used	---
34	YL/BK	Digital TR1 Sensor Signal	In 'P': 0v, 55 mph: 11v
35	RD/LG	HO2S-12 (B1 S2) Signal	0.1-1.1v
36	TN/LB	MAF Sensor Return	<0.050v
37	OG/BK	TFT Sensor Signal	2.10-2.40v
38	BK/WT	Engine Air Temperature Sensor	1.5-2.5v
39	GY	Intake Air Temperature Sensor	1.5-2.5v
40	DG/YL	Fuel Pump Monitor	12v, Off: 0v
41	PK/LB	A/C Cycling Clutch Switch	Switch On: 12v, Off: 0v
42-43	---	Not Used	---
44	GY/RD	Starter Relay Control	Relay Off: 12v, On: 1v
45	LG/RD	Generator/Battery Indicator Control	Indicator Off: 12v, On: 1v
47	BR/PK	EGR System Module Signal	0.95-1.05v
46, 48	---	Not Used	---
49	LB/BK	Digital TR2 Sensor Signal	In 'P': 0v, 55 mph: 11v
50	WT/BK	Digital TR4 Sensor Signal	In 'P': 0v, 55 mph: 11v
51	BK/WT	Power Ground	<0.1v
52	WT/PK	COP 3 Driver Control	5º, 55 mph: 8º dwell
53	DG/VT	COP 4 Driver Control	5º, 55 mph: 8º dwell
54	VT/YL	TCC Solenoid Control	0%, 55 mph: 100%
55	RD/WT	Keep Alive Power	12-14v
56	LG/BK	EVAP Purge Valve Control	0-10 Hz (0-100%)
57	YL/RD	Knock Sensor (-) Signal	<0.050v
58-59	---	Not Used	---
60	GY/LB	HO2S-11 (B1 S1) Signal	0.1-1.1v

2002-03 Town Car 4.6L 2v V8 VIN W (A/T) 104 Pin Connector

PCM Pin #	Wire Color	Circuit Description (104 Pin)	Value at Hot Idle
61	VT/LG	HO2S-22 (B2 S2) Signal	0.1-1.1v
62	RD/BK	FTP Sensor Signal	2.6v (0" H2O - cap off)
63	OG/LG	Injector Pressure Sensor	2.7-3.7v (105-130 psi)
64	LB/YL	Digital TR Sensor Signal	In 'P': 0v, in O/D: 1.7v
65	BR/LG	EGR System Module Signal	0.95-1.05v
66	YL/LG	Cylinder Head Temperature Sensor Signal	0.7v (194°F)
67	VT/WT	EVAP CV Solenoid	0-10 Hz (0-100%)
68	GY/BK	Vehicle Speed Sensor (+) Signal	Moving: AC pulse signals
69	OG/LB	A/C WOT Cutoff Relay	Off: 12v, On: 1v
70 ('03)	YL	Generator Battery Indicator	Indicator Off: 12v, On: 1v
71	RD	Vehicle Power (Start-Run)	12-14v
72	TN/RD	Injector 7 Control	3.4-3-7 ms
73	TN/BK	Injector 5 Control	3.4-3-7 ms
74	BR/YL	Injector 3 Control	3.4-3-7 ms
75	TN	Injector 1 Control	3.4-3-7 ms
76	BK/WT	Power Ground	<0.1v
77	BK/WT	Power Ground	<0.1v
78	PK/LB	COP 7 Driver Control	5°, 55 mph: 8° dwell
79	WT/RD	COP 8 Driver Control	5°, 55 mph: 8° dwell
80	LB/OR	Fuel Pump Control	On: 1v, Off: 12v
81	WT/YL	EPC Solenoid Control	9.5v (20 psi)
82	---	Not Used	---
83	WT/LB	IAC Motor Control	34% duty cycle
84	DB/YL	OSS Sensor Signal	0 Hz, 55 mph: 216 Hz
85	DB/OR	CMP Sensor Signal	6-7 Hz
86	WT/BK	A/C High Pressure Switch	Open: 12v, Closed: 0v
87	RD/BK	HO2S-21 (B2 S1) Signal	0.1-1.1v
88	LB/RD	MAF Sensor Signal	0.6v, 55 mph: 1.8v
89	GY/WT	TP Sensor Signal	0.53-1.27v
90	BR/WT	Reference Voltage	4.9-5.1v
91	GY/RD	Sensor Ground	<0.050v
92	---	Not Used	---
93	RD/WT	HO2S-11 (B1 S1) Heater	1v, Off: 12v
94	YL/LB	HO2S-21 (B2 S1) Heater	1v, Off: 12v
95	WT/BK	HO2S-12 (B1 S2) Heater	1v, Off: 12v
96	TN/YL	HO2S-22 (B2 S2) Heater	1v, Off: 12v
97	RD	System Power (Start-Run)	12-14v
98	LB	Injector 8 Control	3.4-3-7 ms
99	LG/OR	Injector 6 Control	3.4-3-7 ms
100	BR/LB	Injector 4 Control	3.4-3-7 ms
101	WT	Injector 2 Control	3.4-3-7 ms
102	---	Not Used	---
103	BK/WT	Power Ground	<0.1v
104	PK/WT	COP 2 Driver Control	5°, 55 mph: 8° dwell

Pin Connector Graphic

PCM 104-PIN CONNECTOR

Terminal View of 104-Pin PCM Wiring Harness Connector

1990 Town Car 5.0L V8 MFI VIN F (A/T) 60 Pin Connector

PCM Pin #	Wire Color	Circuit Description (60 Pin)	Value at Hot Idle
1	YL/BK	Keep Alive Power	12-14v
2	---	Not Used	---
3	YL/BL	VSS (+) Signal	0 Hz, 55 mph: 125 Hz
4	WT/BL	Ignition Diagnostic Monitor	20-31 Hz
5	DG/WT	Anti-Lock Brake Indicator	Digital Signals
6	BK/BL	VSS (-) Signal	0 Hz, 55 mph: 125 Hz
7	LG/YL	ECT Sensor Signal	0.5-0.6v
8	PK/BL	Data Bus (-) Signal	Digital Signals
9	TN/BL	MAF Sensor Return	<0.050v
10	PK	A/C Cycling Clutch Switch	A/C On: 12v, Off: 0v
11	OR/YL	Air Mgmt. 2 Solenoid Control	AM2 On: 1v, Off: 12v
12	BR/YL	Injector 3 Control	5.0-6.2 ms
13	BR/BL	Injector 4 Control	5.0-6.2 ms
14	TN/BL	Injector 5 Control	5.0-6.2 ms
15	LG	Injector 6 Control	5.0-6.2 ms
16	BK/OR	Ignition System Ground	<0.050v
17	TN/RD	Self-Test Output, MIL Control	MIL On: 1v, Off: 12v
18	---	Not Used	---
19	PK/BK	Fuel Pump Monitor	12v, Off: 0v
20	BK	PCM Case Ground	<0.050v
21	BL/OR	IAC Motor Control	5.0-11.5v
21	LG/WT	IAC Motor Control	5.0-11.5v
22	BL/OR	Fuel Pump Control	On: 1v, Off: 12v
22	TN/LG	Fuel Pump Control	On: 1v, Off: 12v
23	---	Not Used	---
24	---	Not Used	---
25	LG/PK	ACT Sensor Signal	1.5-2.5v
26	OR/WT	Reference Voltage	4.9-5.1v
27	BR/LG	EGR EVP Sensor Signal	0.4v, 55 mph: 2.9v
28	BL/BK	S/C Command Switch Signal	All speeds: 6.7v
29	DG/PK	HO2S-11 (B1 S1) Signal	0.1-1.1v
30	BL/YL	Neutral Drive Switch Signal	In 'N': 0v, Others: 5v
30	WT/PK	Neutral Drive Switch Signal	In 'N': 0v, Others: 5v
31	GY/YL	EVAP Purge Solenoid	12v, 55 mph: 1v
32	OR	Thermactor Air Divert Sol.	TAD On: 1v, Off: 12v
33	DG	VR Solenoid Control	0%, 55 mph: 45%
34	DB/LG	Data Output Link	Digital Signal
35	WT/BK	S/C Vent Solenoid Control	Vacuum Decreasing: 1v
35	WT/PK	S/C Vent Solenoid Control	Vacuum Decreasing: 1v
36	GY/OR	Spark Output Signal	6.93v (50% d/cycle)
37	RD	Vehicle Power	12-14v
38	YL	Thermactor Air Bypass SOL	TAB On: 1v, Off: 12v
39	GY/BK	Speed Control Switch Ground	<0.050v
40	BK/LG	Power Ground	<0.1v

1990 Town Car 5.0L V8 MFI VIN F (A/T) 60 Pin Connector

PCM Pin #	Wire Color	Circuit Description (60 Pin)	Value at Hot Idle
41	---	Not Used	---
42	TN/OR	Injector 7 Control	5.0-6.2 ms
43	DB/YL	HO2S-22 (B2 S2) Signal	0.1-1.1v
43	DG/YL	HO2S-22 (B2 S2) Signal	0.1-1.1v
44	TN/LG	Data Bus (+) Signal	Digital Signals
45	LG/BK	Barometric Pressure Sensor	± 4 159 Hz (Sea Level)
46	BK/WT	Analog Signal Return	<0.050v
47	DG/LG	TP Sensor Signal	0.8v, 55 mph: 1.1v
48	WT/RD	Self-Test Input Signal	STI On: 0v, Off: 5v
49	GYL/BL	HO2S-11 (B1 S1) Ground	<0.050v
50	DB/OR	MAF Sensor Signal	0.8v, 55 mph: 1.8v
51	GY/BK	Air Mgmt. 1 Solenoid Control	AM1 On: 1v, Off: 12v
52	BL	Injector 8 Control	5.0-6.2 ms
53	---	Not Used	---
54	OR/BL	A/C WOT Relay Control	On: 1v, Off: 12v
55	---	Not Used	---
56	DB	PIP Sensor Signal	6.93v (50% d/cycle)
57	RD	Vehicle Power	12-14v
58	TN	Injector 1 Control	5.0-6.2 ms
59	WT	Injector 2 Control	5.0-6.2 ms
60	BK/LG	Power Ground	<0.1v

Pin Connector Graphic

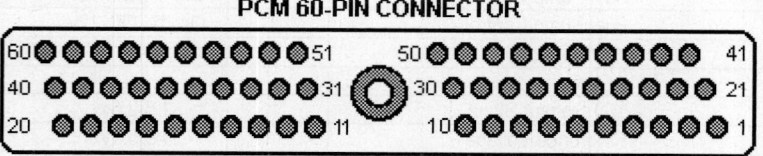

PCM 60-PIN CONNECTOR

Terminal View of 60-Pin PCM Harness Connector

Standard Colors and Abbreviations

Abbreviation	Color	Abbreviation	Color	Abbreviation	Color
BK	Black	GY	Gray	PK	Purple
BL	Blue	GN	Green	RD	Red
BR	Brown	LG	LT Green	TN	Tan
DB	Dark Blue	OR	Orange	WT	White
DG	DK Green	PK	Pink	YL	Yellow

MERCURY CAR CONTENTS

About This Section
Introduction ....Page 11-3
How to Use This Section ....Page 11-3

COLONY PARK PIN TABLES
5.0L V8 MFI VIN F (A/T) 60 Pin **(1990-91)** ....Page 11-4

COUGAR PIN TABLES
2.0L I4 MFI VIN 3 (All) 104 Pin **(1999)** ....Page 11-6
2.0L I4 MFI VIN 3 (All) 104 Pin **(2000-2002)** ....Page 11-8
2.5L V6 4v MFI VIN L (All) 104 Pin **(1999)** ....Page 11-10
2.5L V6 4v MFI VIN L (All) 104 Pin **(2000-2002)** ....Page 11-12
3.8L V6 MFI VIN 4 (A/T) 60 Pin **(1990)** ....Page 11-14
3.8L V6 MFI VIN 4 (A/T) 60 Pin **(1991-95)** ....Page 11-16
3.8L V6 MFI VIN 4 (A/T) 104 Pin **(1996-97)** ....Page 11-18
3.8L V6 SC MFI VIN C (All) 60 Pin **(1990-92)** ....Page 11-20
3.8L V6 SC MFI VIN R (All) 60 Pin **(1990-94)** ....Page 11-22
4.6L V8 MFI VIN W (A/T) 104 Pin **(1994-97)** ....Page 11-24
5.0L V8 MFI VIN T (A/T) 60 Pin **(1991-93)** ....Page 11-26

GRAND MARQUIS PIN TABLES
4.6L V8 MFI VIN W (A/T) 60 Pin **(1992-95)** ....Page 11-28
4.6L V8 MFI VIN W (A/T) 104 Pin **(1996-97)** ....Page 11-30
4.6L V8 MFI VIN W (A/T) 104 Pin **(1998-99)** ....Page 11-32
4.6L V8 MFI VIN W (A/T) 104 Pin **(2000-02)** ....Page 11-34
4.6L V8 MFI VIN W (A/T) 104 Pin **(2003)** ....Page 11-36
5.0L V8 MFI VIN F (A/T) 60 Pin **(1990-91)** ....Page 11-38
5.0L V8 MFI VIN F (A/T) California 60 Pin **(1990-91)** ....Page 11-40

MARAUDER PIN TABLES
4.6L V8 MFI VIN V (A/T) 104 Pin **(2003)** ....Page 11-42

MYSTIQUE PIN TABLES
2.0L I4 MFI VIN 3 (All) 60 Pin **(1995)** ....Page 11-44
2.0L I4 MFI VIN 3 (All) 104 Pin **(1996-97)** ....Page 11-46
2.0L I4 MFI VIN 3 (All) 104 Pin **(1998-2000)** ....Page 11-48
2.5L V6 MFI VIN L (All) 60 Pin **(1995)** ....Page 11-50
2.5L V6 MFI VIN L (All) 104 Pin **(1996-97)** ....Page 11-52
2.5L V6 MFI VIN L (All) 104 Pin **(1998-99)** ....Page 11-54
2.5L V6 MFI VIN L (All) 104 Pin **(2000)** ....Page 11-56

SABLE PIN TABLES

3.0L V6 4v MFI VIN S (A/T) 104 Pin **(1996-97)** .... Page 11-58
3.0L V6 4v MFI VIN S (A/T) 104 Pin **(1998-99)** .... Page 11-60
3.0L V6 4v MFI VIN S (A/T) 104 Pin **(2000)** .... Page 11-62
3.0L V6 4v MFI VIN S (A/T) 104 Pin **(2001)** .... Page 11-64
3.0L V6 4v MFI VIN S (A/T) 104 Pin **(2002-03)** .... Page 11-66
3.0L V6 2v MFI VIN U (A/T) 60 Pin **(1991-95)** .... Page 11-68
3.0L V6 2v MFI VIN U (A/T) 104 Pin **(1996-97)** .... Page 11-70
3.0L V6 2v MFI VIN U (A/T) 104 Pin **(1998-99)** .... Page 11-72
3.0L V6 2v MFI VIN U (A/T) 104 Pin **(2000)** .... Page 11-74
3.0L V6 2v MFI VIN U (A/T) 104 Pin **(2001-2003)** .... Page 11-76
3.8L V6 2v MFI VIN 4 (A/T) 60 Pin **(1990)** .... Page 11-78
3.8L V6 2v MFI VIN 4 (A/T) 60 Pin **(1991-95)** .... Page 11-80

TOPAZ PIN TABLES

2.3L I4 MFI VIN S & X (All) 60 Pin **(1990-91)** .... Page 11-82
2.3L I4 MFI VIN S & X (All) 60 Pin **(1990-91)** .... Page 11-84
2.3L I4 MFI VIN X (All) 60 Pin **(1992-94)** .... Page 11-86
3.0L V6 MFI VIN U (All) 60 Pin **(1992-94)** .... Page 11-88

TRACER PIN TABLES

1.8L I4 MFI VIN 8 (All) 48 Pin **(1991-92)** .... Page 11-90
1.8L I4 MFI VIN 8 (A/T) 64 Pin **(1993)** .... Page 11-92
1.8L I4 MFI VIN 8 (M/T) 48 Pin **(1993)** .... Page 11-94
1.8L I4 MFI VIN 8 (A/T) 64 Pin **(1994-95)** .... Page 11-96
1.8L I4 MFI VIN 8 (M/T) 48 Pin **(1994-95)** .... Page 11-98
1.8L I4 MFI VIN 8 (A/T) 76 Pin **(1996)** .... Page 11-100
1.8L I4 MFI VIN 8 (M/T) 76 Pin **(1996)** .... Page 11-102
1.9L I4 MFI VIN J (All) 60 Pin **(1991-92)** .... Page 11-104
1.9L I4 MFI VIN J (All) 60 Pin **(1993-95)** .... Page 11-105
1.9L I4 MFI VIN J (All) 104 Pin **(1996)** .... Page 11-107
2.0L I4 MFI VIN P (All) 104 Pin **(1997)** .... Page 11-109
2.0L I4 MFI VIN P (All) 104 Pin **(1998-99)** .... Page 11-111

About This Section

Introduction

This section of the Ford Diagnostic Manual contains Pin Tables for Mercury vehicles from 1990-2003 that can be used to assist in the repair of Code and No Code faults related to the PCM.

VEHICLE COVERAGE

- Cougar & Sable Applications (1990-2003)
- Grand Marquis Applications (1990-2003)
- Topaz Applications (1990-99)
- Mystique & Tracer Applications (1991-2000)

How to Use This Section

This Section of the manual can be used to look up the location of a particular pin, a Wire Color or a "known good" value of a PCM circuit. To locate the PCM information for a particular vehicle, find the model, correct engine size (with VIN Code) and finally the year of the vehicle.

For example, to look up the PCM terminals for a 1999 Cougar 2.5L VIN L, go to the Contents Page and find the text string shown below:

1999 Cougar 2.5L V6 4v MFI VIN L (All)... Page 11-10

Then turn to Page 11-10 to find the following PCM related information.

1999 Cougar 2.5L V6 4v MFI VIN L (All) 104 Pin Connector

PCM Pin #	Wire Color	Circuit Description (104 Pin)	Value at Hot Idle
13	WT/BL	Flash EEPROM Power	0.5v
62	WT/PK	FTP Sensor Signal	2.6v (0" H2O - cap off)
85	WT/PK	CMP Sensor Signal	5-7 Hz
99	BK/BL	HO2S-21 (B2 S1) Heater	1v, Off: 12v

In this example, the Fuel Tank Pressure Sensor circuit connects to Pin 62 of the 104 Pin connector with a White/Pink wire. The value at Hot Idle here is a nominal value for the fuel tank pressure with the cap off.

The CMP Sensor circuit is connected to Pin 85 and the reading shown here is the indicated Hertz reading (in this case, change the DVOM to the Hertz to make this measurement).

The (All) that appears in the Title line in the table indicates the values in this table are for vehicles with a manual or automatic transmission.

Colony Park Pin Tables

1990-91 Colony Park 5.0L V8 MFI VIN F (A/T) 60 Pin Connector

PCM Pin #	Wire Color	Circuit Description (60 Pin)	Value at Hot Idle
1	BK/OR	Keep Alive Power	12-14v
2	LG	Brake Switch Signal	Brake Off: 0v, On: 12v
3	DG/WT	VSS (+) Signal	0 Hz, 55 mph: 125 Hz
4	DG/YT	Ignition Diagnostic Monitor	20-31 Hz
5	---	Not Used	---
6	PK/LB	VSS (-) Signal	0 Hz, 55 mph: 125 Hz
7	LG/YL	ECT Sensor Signal	0.5-0.6v
8-9	---	Not Used	---
10	LG/PK	A/C Cycling Clutch Signal	A/C Off: 0v, On: 12v
11	LG/BK	Air Management 2 Solenoid	On: 1v, Off: 12v
12	BR/YL	Injector 3 Control	5.7-6.2 ms
13	BR/LB	Injector 4 Control	5.7-6.2 ms
14	TL/BL	Injector 5 Control	5.7-6.2 ms
15	LG	Injector 6 Control	5.7-6.2 ms
16	BK/OR	Ignition System Ground	<0.050v
17	YL/BK	Self-Test Output, MIL Control	MIL Off: 12v, On: 1v
18-19	---	Not Used	---
20	BK	Case Ground	<0.050v
21	WT/BL	IAC Motor Control	9.0-11.5v
22	TN/LG	Fuel Pump Control	Relay Off: 12v, On: 1v
23-24	---	Not Used	---
25	LG/PK	ACT Sensor Signal	1.5-2.5v
26	OR/WT	Reference Voltage	4.9-5.1v
27	BR/LG	EVP Sensor Signal	0.4v, 55 mph: 1.5v
28	---	Not Used	---
29	DG/PK	HO2S-21 (B1 S1) Signal	0.1-1.1v
30	WT/PK	Neutral Drive Switch	In 'N': 0v, Others: 5v
31	GY/YL	Canister Purge Solenoid	12v, 55 mph: 1v
32	---	Not Used	---
33	DG	VR Solenoid Control	0%, 55 mph: 0-45%
34	LB/PK	Data Output Link	Digital Signals
35	WT/PK	S/C Vent Solenoid Control	Vacuum Decreasing: 1v
36	YL/LG	Spark Output Signal	6.93v (50% d/cycle)
37	RD	Vehicle Power	12-14v
38	GY/BK	S/C Vacuum Solenoid Control	Vacuum Increasing: 1v
39	GY/BK	Speed Control Switch Ground	<0.050v
40	BK	Power Ground	<0.1v

1990-91 Colony Park 5.0L V8 MFI VIN F (A/T) 60 Pin Connector

PCM Pin #	Wire Color	Circuit Description (60 Pin)	Value at Hot Idle
41	OR/YL	Speed Control Servo (+) SIG	Solenoid On: 12v
42	TN/OR	Injector 7 Control	5.7-6.2 ms
42	TN/RD	Injector 7 Control	5.7-6.2 ms
43	DB/LG	HO2S-21 (B1 S1) Signal	0.1-1.1v
45	DB/LG	MAP Sensor Signal	107 Hz (sea level)
45	LG/BK	MAP Sensor Signal	107 Hz (sea level)
46	BK/WT	Analog Signal Return	<0.050v
47	DG/LG	TP Sensor Signal	0.8v, 55 mph: 1.1v
48	WT/RD	Self-Test Input Signal	STI On: 1v, Off: 5v
49	OR	HO2S-11 (Bank 1) Ground	<0.050v
50	OR/RD	Speed Command Control Sw.	6.7v
51	WT/RD	Air Management 1 Solenoid	AM1 On: 1v, Off: 12v
52	BL	Injector 8 Control	5.7-6.2 ms
52	TN/RD	Injector 8 Control	5.7-6.2 ms
53	---	Not Used	---
54	OR/BL	A/C WOT Relay Control	Relay Off: 12v, On: 1v
55	---	Not Used	---
56	DB	PIP Sensor Signal	6.93v (50% d/cycle)
57	RD	Vehicle Power	12-14v
58	TN	Injector 1 Control	5.7-6.2 ms
59	TN/WT	Injector 2 Control	5.7-6.2 ms
59	WT	Injector 2 Control	5.7-6.2 ms
60	BK/LG	Power Ground	<0.1v

Pin Connector Graphic

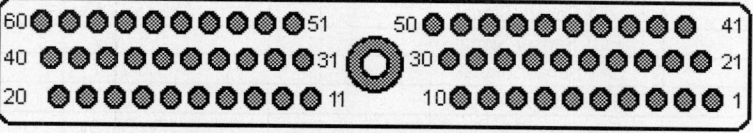

PCM 60-PIN CONNECTOR

Terminal View of 60-Pin PCM Harness Connector

Standard Colors and Abbreviations

Abbreviation	Color	Abbreviation	Color	Abbreviation	Color
BK	Black	GY	Gray	PK	Purple
BL	Blue	GN	Green	RD	Red
BR	Brown	LG	LT Green	TN	Tan
DB	Dark Blue	OR	Orange	WT	White
DG	DK Green	PK	Pink	YL	Yellow

Cougar Pin Tables

1999 Cougar 2.0L I4 4v MFI VIN 3 (All) 104 Pin Connector

PCM Pin #	Wire Color	Circuit Description (104 Pin)	Value at Hot Idle
1	BK/BL	Shift Solenoid 'B' Control	12v, 55 mph: 1v
2	BK/OR	MIL (lamp) Control	MIL On: 1v, Off: 12v
3-11	---	Not Used	---
12	WT	Fuel Level Indicator Signal	1.7v (1/2 full)
13	WT/BL	Flash EEPROM Power	0.5v
14	---	Not Used	---
15	BL	Data Bus (-) Signal	Digital Signals
16	GY	Data Bus (+) Signal	Digital Signals
17	GY/OR	Passive Anti-Theft System	Digital Signals
18	---	Not Used	---
19	BK/WT	High Speed Fan Control	Fan On: 1v, Off: 12v
20	BK/BL	Injector 3 Control	2.3-2.9 ms
21	WT/RD	CKP Sensor (+) Signal	450-480 Hz
22	BR/RD	CKP Sensor (-) Signal	450-480 Hz
24	BK/YL	Power Ground	<0.1v
25	BK/RD	Chassis Ground	<0.050v
26	BR/BL	Coil Driver 1 Control	5° dwell
27	BK/YL	Shift Solenoid 1 Control	1v, 55 mph: 12v
29	PK/BK	TCS (switch) Signal	TCS & O/D On: 12v
29	GN/BK	TCS (switch) Signal	TCS & O/D On: 12v
30	---	Not Used	---
31	WT	PSP Switch Signal	Straight: 1v, turned: 2-4v
32-33	---	Not Used	---
34	WT/GN	Passive Anti-Theft System	Digital Signals
35	WT/BL	HO2S-12 (B1 S2) Signal	0.1-1.1v
36	BR/BL	MAF Sensor Return	<0.050v
37	WT/RD	TFT Sensor Signal	2.10 to 2.40v
38	WT/GN	ECT Sensor Signal	0-5-0.6v
39	WT/PK	IAT Sensor Signal	1.5-2.5v
40	PK/BK	Fuel Pump Monitor	12v, Off: 0v
40	GN/BK	Fuel Pump Monitor	12v, Off: 0v
41	PK/BL	A/C High Pressure Cut-Off	Closed: 12v, Open: 0v
42	---	Not Used	---
43	WT	Fuel Flow Sensor	Digital Signal
44	BK/RD	VCT Actuator Control	VCT Off: 12v, On: 1v
45	BK/BL	Low Speed Fan Control	Fan On: 1v, Off: 12v
46-47	---	Not Used	---
48	WT/BK	Clean Tachometer Output	22-31 Hz
49-50	---	Not Used	---
51	BK/YL	Power Ground	<0.1v
52	BR/GN	Coil Driver 2 Control	5° dwell
53	---	Not Used	---
54	BK/BL	Fuel Pump Control	Relay Off: 12v, On: 1v
55	OR/YL	Keep Alive Power	12-14v
56	BK/OR	EVAP Purge Valve	0-10 Hz (0-100%)
57	WT/BK	Knock Sensor 1 Signal	0v
58	WT/BL	VSS (+) Signal	0 Hz, 55 mph: 125 Hz
59	---	Not Used	---
60	WT	HO2S-11 (B1 S1) Signal	0.1-1.1v
61	---	Not Used	---
62	WT/PK	FTP Sensor Signal	2.6v (0" H2O - cap off)
63	WT/GN	Fuel Pressure Transducer	2.8v (39 psi)
64	WT	Clutch Pedal Position Switch	Clutch In: 0v, Out: 5v

1999 Cougar 2.0L I4 4v MFI VIN 3 (All) 104 Pin Connector

PCM Pin #	Wire Color	Circuit Description (104 Pin)	Value at Hot Idle
65-66	---	Not Used	---
67	BK/OR	EVAP CV Solenoid	0-10 Hz (0-100%)
68	---	Not Used	
69	BK/YL	A/C WOT Relay Control	Relay Off: 12v, On: 1v
70	BK/WT	Injector 1 Control	2.3-2.9 ms
71	GN/YL	Vehicle Power	12-14v
72, 75	---	Not Used	---
73	BK/YL	HO2S-11 (B1 S1) Signal	0.1-1.1v
74	BK/BL	Injector 3 Control	2.3-2.9 ms
76	BR/WT	CMP and TSS Ground	<0.050v
77	BK/YL	Power Ground	<0.1v
78	---	Not Used	---
79	WT/BK	TCIL (lamp) Control	Lamp On: 1v, Off: 12v
80	BK/WT	TCC Solenoid Control	12v, 55 mph: 1v
81	BK/RD	EPC Solenoid Control	8.9v
82	BK/BL	PATS Sensor Signal	Digital Signal
83	BK/YL	IAC Motor Control	9v (33% d/cycle)
84	WT/PK	TSS Sensor Signal	40-56 Hz
85	WT/PK	CMP Sensor Signal	5-7 Hz
86	BK/WT	A/C Pressure Switch	Switch On: 12v (Open)
87	BR/YL	Knock Sensor 1 (+) Signal	0v
88	WT/BL	MAF Sensor Signal	0.8v
89	WT	TP Sensor Signal	0.53-1.27v
90	YL	Reference Voltage	4.9-5.1v
91	BR	Analog Signal Return	<0.050v
92	PK	Brake Pedal Position Switch	Brake Off: 0v, On: 12v
93	BK/YL	HO2S-11 (B1 S1) Heater	1v, Off: 12v
94	---	Not Used	---
95	BK/OR	Injector 4 Control	2.3-2.9 ms
96	BK/YL	Injector 2 Control	2.3-2.9 ms
97	GN/YL	Vehicle Power	12-14v
98-99	---	Not Used	---
100	BK/OR	HO2S-12 (B1 S2) Heater	1v, Off: 12v
101	---	Not Used	---
102	BK/OR	Shift Solenoid 3 Control	8v, 55 mph: 8-9v
103	BK/YL	Power Ground	<0.1v
104	---	Not Used	---

Pin Connector Graphic

Standard Colors and Abbreviations

Abbreviation	Color	Abbreviation	Color	Abbreviation	Color
BK	Black	GY	Gray	PK	Purple
BL	Blue	GN	Green	RD	Red
BR	Brown	LG	LT Green	TN	Tan
DB	Dark Blue	OR	Orange	WT	White
DG	DK Green	PK	Pink	YL	Yellow

2000-02 Cougar 2.0L I4 4v MFI VIN 3 (All) 104 Pin Connector

PCM Pin #	Wire Color	Circuit Description (104 Pin)	Value at Hot Idle
1	---	Not Used	---
2	BK/OR	MIL (lamp) Control	MIL On: 1v, Off: 12v
3-10	---	Not Used	---
6	BK/YL	Shift Solenoid 'B' Control	12v, 55 mph: 1v
8	BK/BL	IMRC Solenoid Control	0v
11	BK/BL	Shift Solenoid 'A' Control	1v, 55 mph: 12v
12	WT	Fuel Level Indicator Signal	1.7v (1/2 full)
13	WT/BL	Flash EEPROM Power	0.1v
14	---	Not Used	---
15	BL	Data Bus (-) Signal	Digital Signals
16	GY	Data Bus (+) Signal	Digital Signals
17	BK/WT	High Speed Cooling Fan	Relay Off: 12v, On: 1v
18	---	Not Used	---
19	GY/OR	PAT System	Digital Signals
20	BK/BL	Injector 3 Control	2.3-2.9 ms
21	WT/RD	CKP Sensor (+) Signal	450-480 Hz
22	BR/RD	CKP Sensor (-) Signal	450-480 Hz
24, 51	BK/YL	Power Ground	<0.1v
25	BK/RD	Chassis Ground	<0.050v
26	BR/BL	Coil Driver 1 Control	5° dwell
27	BK/YL	VTEC Solenoid	1v
28	WT/BK	VSS (-) Signal	0 Hz
29	GN/BK	TCS (switch) Signal	TCS & O/D On: 12v
30, 32-33	---	Not Used	---
31	WT	PSP Switch Signal	Straight: 1v, turned: 2-4v
34	WT/GN	Passive Anti-Theft System	Digital Signals
35	WT/BL	HO2S-12 (B1 S2) Signal	0.1-1.1v
36	BR/BL	MAF Sensor Return	<0.050v
37	WT/RD	TFT Sensor Signal	2.10 to 2.40v
38	WT/GN	ECT Sensor Signal	0-5-0.6v
39	WT/PK	IAT Sensor Signal	1.5-2.5v
40	PK/BK	Fuel Pump Monitor	12v, Off: 0v
40	GN/BK	Fuel Pump Monitor	12v, Off: 0v
41	PK/BL	A/C High Pressure Cut-Off	Closed: 12v, Open: 0v
42	---	Not Used	---
43	WT	Fuel Flow Sensor	Digital Signals
44	BK/RD	VCT Actuator Control	VCT Off: 12v, On: 1v
45	BK/BL	Low Speed Fan Control	Fan On: 1v, Off: 12v
46-47	---	Not Used	---
48	WT/BK	Clean Tachometer Output	22-31 Hz
49-50, 53	---	Not Used	---
52	BR/GN	Coil Driver 2 Control	5° dwell
54	BK/BL	Fuel Pump Control	Relay Off: 12v, On: 1v
55	OR/YL	Keep Alive Power	12-14v
56	BK/OR	EVAP Purge Valve	0-10 Hz (0-100%)
57	WT/BK	Knock Sensor 1 Signal	0v
58	WT/BL	VSS (+) Signal	0 Hz, 55 mph: 125 Hz
59, 61	---	Not Used	---
60	WT	HO2S-11 (B1 S1) Signal	0.1-1.1v
62	WT/PK	FTP Sensor Signal	2.6v (0" H2O - cap off)
63	WT/GN	Fuel Pressure Transducer	2.8v (39 psi)
64	WT	Clutch Pedal Position Switch	Clutch In: 0v, Out: 5v

2000-02 Cougar 2.0L I4 4v MFI VIN 3 (All) 104 Pin Connector

PCM Pin #	Wire Color	Circuit Description (104 Pin)	Value at Hot Idle
65 ('00)	WT/BL	DPFE Sensor Signal	0.20-1.30v
65 ('01-'02)	WT/BL	DPFE Sensor Signal	0.95-1.05v
65-66	---	Not Used	---
67	BK/OR	EVAP CV Solenoid	0-10 Hz (0-100%)
68	BK/BL	Cooling Fan Relay Control	Fan On: 1v, Off: 12v
69	BK/YL	A/C WOT Relay Control	Relay Off: 12v, On: 1v
70	BK/WT	Injector 1 Control	2.3-2.9 ms
71	GN/TN	Vehicle Power	12-14v
72, 74-75	---	Not Used	---
73	BK/YL	HO2S-11 (B1 S1) Signal	0.1-1.1v
76	BN/WT	CMP and TSS Ground	<0.050v
77	BK/YL	Power Ground	<0.1v
78	BN/YL	Ignition Transistor Ground	<0.050v
79	---	Not Used	---
80	BK/WT	TCC Solenoid Control	12v, 55 mph: 1v
81	BK/RD	EPC Solenoid Control	8.9v
82	BK/BL	PATS Sensor Signal	Digital Signals
83	BK/YL	IAC Motor Control	9v (33% d/cycle)
84	WT/VT	TSS Sensor Signal	40-56 Hz
85	WT/VT	CMP Sensor Signal	5-7 Hz
86	BK/BL	A/C Pressure Switch	Switch On: 12v (Open)
87	WT/RD	Knock Sensor 1 (+) Signal	0.1-1.1v
88	WT/BL	MAF Sensor Signal	0.8v
89	WT	TP Sensor Signal	0.53-1.27v
90	YL	Reference Voltage	4.9-5.1v
91	BN	Analog Signal Return	<0.050v
92	GN/RD	Brake Pedal Position Switch	Brake Off: 0v, On: 12v
93	BK/GN	HO2S-11 (B1 S1) Heater	1v, Off: 12v
94, 98-99	---	Not Used	---
95	BK/OR	Injector 4 Control	2.3-2.9 ms
96	BK/YL	Injector 2 Control	2.3-2.9 ms
97	GN/TN	Vehicle Power	12-14v
100	BK/OR	HO2S-12 (B1 S2) Heater	1v, Off: 12v
101	---	Not Used	---
102	BK/OR	Shift Solenoid 'C' Control	8v, 55 mph: 8-9v
103	BK/YL	Power Ground	<0.1v
104	---	Not Used	---

Pin Connector Graphic

PCM 104-PIN CONNECTOR

Terminal View of 104-Pin PCM Wiring Harness Connector

Standard Colors and Abbreviations

Abbreviation	Color	Abbreviation	Color	Abbreviation	Color
BK	Black	GY	Gray	PK	Purple
BL	Blue	GN	Green	RD	Red
BR	Brown	LG	LT Green	TN	Tan
DB	Dark Blue	OR	Orange	WT	White
DG	DK Green	PK	Pink	YL	Yellow

1999 Cougar 2.5L V6 4v MFI VIN L (All) 104 Pin Connector

PCM Pin #	Wire Color	Circuit Description (104 Pin)	Value at Hot Idle
1	BK/BL	Shift Solenoid 2 Control	1v, 55 mph: 12v
2	BK/OR	MIL (lamp) Control	MIL On: 1v, Off: 12v
4-7	---	Not Used	---
8	BK/BL	IMRC Solenoid Control	5v, 55 mph: 5v
9-11	---	Not Used	---
12	WT	Fuel Level Indicator Signal	1.7v (1/2 full)
13	WT/BL	Flash EEPROM Power	0.5v
14, 18	---	Not Used	---
15	BL	Data Bus (-) Signal	Digital Signals
16	GY	Data Bus (+) Signal	Digital Signals
17	GY/OR	Passive Anti-Theft System	Digital Signals
19	BK/WT	High Speed Fan Control	Fan On: 1v, Off: 12v
20	BK/BL	Injector 3 Control	2.1-2.5 ms
21	WT/RD	CKP Sensor (+) Signal	410-440 Hz
22	BR/RD	CKP Sensor (-) Signal	410-440 Hz
23, 28, 30	---	Not Used	---
24	BK/YL	Power Ground	<0.1v
25	BK/RD	Chassis Ground	<0.050v
26	BR/BL	Coil Driver 1 Control	5° dwell
27 ('99)	BK/YL	Shift Solenoid 1 Control	12v, 55 mph: 1v
29	PK/BK	TCS (switch) Signal	TCS & O/D On: 12v
31	WT	PSP Switch Signal	Straight: 0.1v, turned: 12v
32	BR/YL	Knock Sensor 1 Signal	0v
33	BR/BL	VSS (-) Signal	0 Hz, 55 mph: 125 Hz
34	WT/GN	Passive Anti-Theft System	Digital Signal
35	WT/BL	HO2S-12 (B1 S2) Signal	0.1-1.1v
36	BR/BL	MAF Sensor Return	<0.050v
37	WT/RD	TFT Sensor Signal	2.10 to 2.40v
38	WT/GN	ECT Sensor Signal	0-5-0.6v
39	WT/PK	IAT/ACT Sensor Signal	1.5-2.5v
40	PK/BK	Fuel Pump Monitor	12v, Off: 0v
40	GN/BK	Fuel Pump Monitor	12v, Off: 0v
41	PK/BL	A/C High Pressure Cut-Out	A/C Switch On: 12v
42	BK/RD	IMRC Solenoid Signal	12v, 55 mph: 12v
43	WT	Fuel Flow Sensor	Varies: 0-5.1v
45	BK/BL	Low Speed Fan Control	Fan On: 1v, Off: 12v
46, 50, 59	---	Not Used	---
47	BK/GN	VR Solenoid Control	0%, 55 mph: 40%
48	WT/BK	Clean Tachometer Output	35-42 Hz
51	BK/YL	Power Ground	<0.1v
52	BR/GN	Coil Driver 2 Control	5° dwell
54	BK/BL	Fuel Pump Control	Relay Off: 12v, On: 1v
55	OR/YL	Keep Alive Power	12-14v
56	BK/OR	EVAP Purge Solenoid	0-10 Hz (0-100%)
56	BK/BL	EVAP Purge Solenoid	0-10 Hz (0-100%)
57	WT/BK	Knock Sensor 1 Signal	0v
58	WT/PK	VSS (+) Signal	0 Hz, 55 mph: 125 Hz
60	WT	HO2S-11 (B1 S1) Signal	0.1-1.1v
61	WT/GN	HO2S-22 (B2 S2) Signal	0.1-1.1v
62	WT/PK	FTP Sensor Signal	2.6v (0" H2O - cap off)
63	WT/GN	Fuel Pressure Transducer	2.8v (39 psi)
64	WT/GN	A/T: TR Sensor Signal	In 'P': 0v, O/D: 1.7v
64	WT	M/T: Clutch Position Switch	Clutch In: 0v, Out: 5v

1999 Cougar 2.5L V6 4v MFI VIN L (All) 104 Pin Connector

PCM Pin #	Wire Color	Circuit Description (104 Pin)	Value at Hot Idle
65	WT/BL	DPFE Sensor Signal	0.20-1.30v
66, 68	---	Not Used	---
67	BK/OR	EVAP CV Solenoid	0-10 Hz (0-100%)
69	BK/YL	A/C WOT Relay Control	Relay Off: 12v, On: 1v
70	BK/WT	Injector 1 Control	2.1-2.5 ms
71	GN/YL	Vehicle Power	12-14v
72, 74-75	---	Not Used	---
73	BK/YL	HO2S-11 (B1 S1) Heater	1v, Off: 12v
76	BR/WT	CMP & TSS Ground	<0.050v
77	BK/YL	Power Ground	<0.1v
78	BR/YL	Coil Driver 3 Control	5° dwell
79	WT/BK	TCIL (lamp) Control	Lamp On: 1v, Off: 12v
80	BK/WT	TCC Solenoid Control	0%, 55 mph: 90-95%
81	BK/RD	EPC Solenoid Control	8.9v
82	BK/BL	PATS Sensor Signal	Digital Signal
83	BK/YL	IAC Motor Control	9v (33% d/cycle)
84	WT/PK	TSS Sensor Signal	46-50 Hz
85	WT/PK	CMP Sensor Signal	5-7 Hz
86	BK/WT	A/C Pressure Switch	Switch On: 12v (Open)
87	WT/RD	HO2S-21 (B1 S1) Signal	0.1-1.1v
88	WT/BL	MAF Sensor Signal	0.8v
89	WT	TP Sensor Signal	0.53-1.27v
90	YL	Reference Voltage	4.9-5.1v
91	BR	Analog Signal Return	<0.050v
92	OR	Brake Pedal Position Switch	Brake Off: 0v, On: 12v
92	GN/RD	Brake Pedal Position Switch	Brake Off: 0v, On: 12v
93	BK/GN	Injector 5 Control	2.1-2.5 ms
94	BK/RD	Injector 6 Control	2.1-2.5 ms
95	BK/OR	Injector 4 Control	2.1-2.5 ms
96	BK/YL	Injector 2 Control	2.1-2.5 ms
97	GN/YL	Vehicle Power	12-14v
98	---	Not Used	---
99	BK/BL	HO2S-21 (B2 S1) Heater	1v, Off: 12v
100	BK/OR	HO2S-12 (B1 S2) Heater	1v, Off: 12v
101	BK/GN	HO2S-22 (B2 S2) Heater	1v, Off: 12v
102	BK/OR	Shift Solenoid 3 Control	7-9v, 30 mph: 8-9v
103	BK/YL	Power Ground	<0.1v
104	---	Not Used	---

Pin Connector Graphic

PCM 104-PIN CONNECTOR

Terminal View of 104-Pin PCM Wiring Harness Connector

Standard Colors and Abbreviations

Abbreviation	Color	Abbreviation	Color	Abbreviation	Color
BK	Black	GY	Gray	PK	Purple
BL	Blue	GN	Green	RD	Red
BR	Brown	LG	LT Green	TN	Tan
DB	Dark Blue	OR	Orange	WT	White
DG	DK Green	PK	Pink	YL	Yellow

2000-02 Cougar 2.5L V6 4v MFI VIN L (All) 104 Pin Connector

PCM Pin #	Wire Color	Circuit Description (104 Pin)	Value at Hot Idle
1	---	Not Used	---
2	BK/OR	MIL (lamp) Control	MIL On: 1v, Off: 12v
3-5	---	Not Used	---
6	BK/YL	Shift Solenoid 'A' Control	12v, 55 mph: 1v
8	BK/BL	IMRC Solenoid Control	5v, 55 mph: 5v
9-10	---	Not Used	---
11	BK/BL	Shift Solenoid 'B' Control	1v, 55 mph: 12v
12	WT	Fuel Level Indicator Signal	1.7v (1/2 full)
13	WT/BL	Flash EEPROM Power	0.5v
15	BL	Data Bus (-) Signal	Digital Signals
16	GY	Data Bus (+) Signal	Digital Signals
17	GY/OR	Passive Anti-Theft System	Digital Signals
19	BK/WT	High Speed Fan Control	Fan On: 1v, Off: 12v
20	BK/BL	Injector 3 Control	2.1-2.5 ms
21	WT/RD	CKP Sensor (+) Signal	410-440 Hz
22	BR/RD	CKP Sensor (-) Signal	410-440 Hz
24	BK/YL	Power Ground	<0.1v
25	BK/RD	Chassis Ground	<0.050v
26	BR/BL	Coil Driver 1 Control	5° dwell
29	PK/BK	TCS (switch) Signal	TCS & O/D On: 12v
31	WT	PSP Switch Signal	Straight: 0.1v, turned: 12v
32	BR/YL	Knock Sensor 1 Signal	0v
33	BR/BL	VSS (-) Signal	0 Hz, 55 mph: 125 Hz
34	WT/GN	Passive Anti-Theft System	Digital Signal
35	WT/BL	HO2S-12 (B1 S2) Signal	0.1-1.1v
36	BR/BL	MAF Sensor Return	<0.050v
37	WT/RD	TFT Sensor Signal	2.10 to 2.40v
38	WT/GN	ECT Sensor Signal	0-5-0.6v
39	WT/PK	IAT/ACT Sensor Signal	1.5-2.5v
40	PK/BK	Fuel Pump Monitor	12v, Off: 0v
40	GN/BK	Fuel Pump Monitor	12v, Off: 0v
41	PK/BL	A/C High Pressure Cut-Out	A/C Switch On: 12v
42	BK/RD	IMRC Solenoid Signal	12v, 55 mph: 12v
43	WT	Fuel Flow Sensor	Varies: 0-5.1v
45	BK/BL	Low Speed Fan Control	Fan On: 1v, Off: 12v
47	BK/GN	VR Solenoid Control	0%, 55 mph: 40%
48	WT/BK	Clean Tachometer Output	35-42 Hz
51	BK/YL	Power Ground	<0.1v
52	BR/GN	Coil Driver 2 Control	5° dwell
54	BK/BL	Fuel Pump Control	Relay Off: 12v, On: 1v
55	OR/YL	Keep Alive Power	12-14v
56	BK/OR	EVAP Purge Solenoid	0-10 Hz (0-100%)
56	BK/BL	EVAP Purge Solenoid	0-10 Hz (0-100%)
57	WT/BK	Knock Sensor 1 Signal	0v
58	WT/PK	VSS (+) Signal	0 Hz, 55 mph: 125 Hz
60	WT	HO2S-11 (B1 S1) Signal	0.1-1.1v
61	WT/GN	HO2S-22 (B2 S2) Signal	0.1-1.1v
62	WT/PK	FTP Sensor Signal	2.6v (0" H2O - cap off)
63	WT/GN	Fuel Pressure Transducer	2.8v (39 psi)
64	WT/GN	A/T: TR Sensor Signal	In 'P': 0v, O/D: 1.7v
64	WT	M/T: Clutch Position Switch	Clutch In: 0v, Out: 5v

2000-02 Cougar 2.5L V6 4v MFI VIN L (All) 104 Pin Connector

PCM Pin #	Wire Color	Circuit Description (104 Pin)	Value at Hot Idle
65	WT/BL	DPFE Sensor Signal	0.95-1.05v
66	---	Not Used	---
67	BK/OR	EVAP CV Solenoid	0-10 Hz (0-100%)
69	BK/YL	A/C WOT Relay Control	Relay Off: 12v, On: 1v
68	---	Not Used	---
70	BK/WT	Injector 1 Control	2.1-2.5 ms
71	GN/YL	Vehicle Power	12-14v
72, 74-75	---	Not Used	---
73	BK/YL	HO2S-11 (B1 S1) Heater	1v, Off: 12v
76	BR/WT	CMP & TSS Ground	<0.050v
77	BK/YL	Power Ground	<0.1v
78	BR/YL	Coil Driver 3 Control	5° dwell
79	WT/BK	TCIL (lamp) Control	Lamp On: 1v, Off: 12v
80	BK/WT	TCC Solenoid Control	0%, 55 mph: 90-95%
81	BK/RD	EPC Solenoid Control	8.9v
82	BK/BL	PATS Sensor Signal	Digital Signal
83	BK/YL	IAC Motor Control	9v (33% d/cycle)
84	WT/PK	TSS Sensor Signal	46-50 Hz
85	WT/PK	CMP Sensor Signal	5-7 Hz
86	BK/WT	A/C Pressure Switch	Switch On: 12v (Open)
87	WT/RD	HO2S-21 (B1 S1) Signal	0.1-1.1v
88	WT/BL	MAF Sensor Signal	0.8v
89	WT	TP Sensor Signal	0.53-1.27v
90	YL	Reference Voltage	4.9-5.1v
91	BR	Analog Signal Return	<0.050v
92	OR	Brake Pedal Position Switch	Brake Off: 0v, On: 12v
92	GN/RD	Brake Pedal Position Switch	Brake Off: 0v, On: 12v
93	BK/GN	Injector 5 Control	2.1-2.5 ms
94	BK/RD	Injector 6 Control	2.1-2.5 ms
95	BK/OR	Injector 4 Control	2.1-2.5 ms
96	BK/YL	Injector 2 Control	2.1-2.5 ms
97	GN/YL	Vehicle Power	12-14v
98	---	Not Used	---
99	BK/BL	HO2S-21 (B2 S1) Heater	1v, Off: 12v
100	BK/OR	HO2S-12 (B1 S2) Heater	1v, Off: 12v
101	BK/GN	HO2S-22 (B2 S2) Heater	1v, Off: 12v
102	BK/OR	Shift Solenoid 'C' Control	7-9v, 30 mph: 8-9v
103	BK/YL	Power Ground	<0.1v
104	---	Not Used	---

Pin Connector Graphic

Terminal View of 104-Pin PCM Wiring Harness Connector

1990 Cougar 3.8L V6 MFI VIN 4 (A/T) 60 Pin Connector

PCM Pin #	Wire Color	Circuit Description (60 Pin)	Value at Hot Idle
1	YL	Keep Alive Power	12-14v
2	---	Not Used	---
3	DG/WT	VSS (+) Signal	0 Hz, 55 mph: 125 Hz
4	DG/YT	Ignition Diagnostic Monitor	20-31 Hz
5	LG	Brake Switch Signal	Brake Off: 0v, On: 12v
6	BK/WT	VSS (-) Signal	0 Hz, 55 mph: 125 Hz
7	LG/YL	ECT Sensor Signal	0.5-0.6v
8	OR/BK	Data Bus (-) Signal	Digital Signals
9	---	Not Used	---
10	PK/BL	A/C Cycling Clutch Signal	A/C Off: 0v, On: 12v
11	OR/YL	Speed Control Solenoid (+)	Solenoid On: 12-14V
12	BR/YL	Injector 3 Control	Hot Idle: 6.0-6.2 ms
13	BR/BL	Injector 4 Control	Hot Idle: 6.0-6.2 ms
14	TN/BL	Injector 5 Control	Hot Idle: 6.0-6.2 ms
15	LG	Injector 6 Control	Hot Idle: 6.0-6.2 ms
16	DB	Ignition System Ground	<0.050v
16	GY	Ignition System Ground	<0.050v
17	YL/BK	Self-Test Output, MIL Control	MIL Off: 12v, On: 1v
18	--	Not Used	---
19	---	Not Used	---
20	BK	Case Ground	<0.050v
21	RD/LG	IAC Motor Control	9-10v
22	TN/LG	Fuel Pump Control	Relay Off: 12v, On: 1v
23	---	Not Used	---
24	---	Not Used	---
25	LG/PK	ACT Sensor Signal	1.5-2.5v
26	OR/WT	Reference Voltage	4.9-5.1v
27	BR/LG	PFE Sensor Signal	3.2v, 55 mph: 2.8v
28	BL/BK	Speed Control Switch Signal	6.7v
29	TN/OR	HO2S-21 (B1 S1) Signal	0.1-1.1v
30	RD/BL	Neutral Drive Switch	In 'N': 0v, Others: 4-5v
31	GY/YL	Canister Purge Solenoid	12v, 55 mph: 1v
32	---	Not Used	---
33	DG	VR Solenoid Control	0%, 55 mph: 45%
34	BL/PK	Data Output Link	Digital Signals
35	WT/PK	S/C Vent Solenoid Control	Vacuum Decreasing: 1v
36	YL/LG	Spark Output Signal	6.93v (50% d/cycle)
37	RD	Vehicle Power	12-14v
38	---	Not Used	---
39	WT/BL	Speed Control Switch Ground	<0.050v
40	BK/LG	Power Ground	<0.1v

1990 Cougar 3.8L V6 MFI VIN 4 (A/T) 60 Pin Connector

PCM Pin #	Wire Color	Circuit Description (60 Pin)	Value at Hot Idle
41	---	Not Used	---
42	---	Not Used	---
43	TN/RD	HO2S-11 (B1 S1) Signal	0.1-1.1v
44	BK/OR	Data Bus (+) Signal	Digital Signals
45	DB/LG	MAP Sensor Signal	107 Hz (sea level)
46	BK/WT	Analog Signal Return	<0.050v
47	DG/LG	TP Sensor Signal	0.8v, 55 mph: 1.1v
48	WT/RD	Self-Test Input Signal	STI On: 1v, Off: 5v
49	OR	HO2S-11 (Bank 1) Ground	<0.050v
50	PK/BK	Fuel Pump Monitor	12v, Off: 0v
51	GY/BK	S/C Vacuum Solenoid Control	Vacuum Increasing: 1v
52	---	Not Used	---
53	---	Not Used	---
54	OR/BL	A/C WOT Relay Control	Relay Off: 12v, On: 1v
55	---	Not Used	---
56	GY	PIP Sensor Signal	6.93v (50% d/cycle)
56	DB	PIP Sensor Signal	6.93v (50% d/cycle)
57	RD	Vehicle Power	12-14v
58	TN	Injector 1 Control	6-6.2 ms
59	WT	Injector 2 Control	6-6.2 ms
60	BK/LG	Power Ground	<0.1v

Pin Connector Graphic

Terminal View of 60-Pin PCM Harness Connector

Standard Colors and Abbreviations

Abbreviation	Color	Abbreviation	Color	Abbreviation	Color
BK	Black	GY	Gray	PK	Purple
BL	Blue	GN	Green	RD	Red
BR	Brown	LG	LT Green	TN	Tan
DB	Dark Blue	OR	Orange	WT	White
DG	DK Green	PK	Pink	YL	Yellow

1991-95 Cougar 3.8L V6 MFI VIN 4 (A/T) 60 Pin Connector

PCM Pin #	Wire Color	Circuit Description (60 Pin)	Value at Hot Idle
1	YL	Keep Alive Power	12-14v
2	---	Not Used	---
3	GY/BK	VSS (+) Signal	0 Hz, 55 mph: 125 Hz
4	RD/BL	Ignition Diagnostic Monitor	20-31 Hz
4	WT/PK	Ignition Diagnostic Monitor	20-31 Hz
5	---	Not Used	---
6	BK/WT	VSS (-) Signal	0 Hz, 55 mph: 125 Hz
6	GY/RD	VSS (-) Signal	0 Hz, 55 mph: 125 Hz
7	LG/YL	ECT Sensor Signal	0.5-0.6v
7	LG/RD	ECT Sensor Signal	0.5-0.6v
8	PK/BL	Fuel Pump Monitor	12v, Off: 0v
9	TN/BL	MAF Sensor Return	<0.050v
10	PK/BK	A/C Cycling Clutch Signal	A/C Off: 0v, On: 12v
11	GY/YL	Canister Purge SOL Control	12v, 55 mph: 9-11v
12	LG	Injector 6 Control	5.6-6.3 ms
12	LG/OR	Injector 6 Control	5.6-6.3 ms
13	---	Not Used	---
14	---	Not Used	---
15	TN/BL	Injector 5 Control	5.6-6.3 ms
15	TN/BK	Injector 5 Control	5.6-6.3 ms
16	GY	Ignition System Ground	<0.050v
17	YL/BK	Self-Test Output, MIL Control	MIL Off: 12v, On: 1v
18	TN/OR	Data Bus (+) Signal	Digital Signals
19	PK/BK	Data Bus (-) Signal	Digital Signals
20	BK	Case Ground	<0.050v
21	RD/LG	IAC Motor Control	9.5-11.0v
22	TN/LG	Fuel Pump Control	Relay Off: 12v, On: 1v
23	---	Not Used	---
24	---	Not Used	---
25	LG/PK	IAT Sensor Signal	1.5-2.5v
25	GY	IAT Sensor Signal	1.5-2.5v
26	OR/WT	Reference Voltage	4.9-5.1v
26	BR/WT	Reference Voltage	4.9-5.1v
27	BR/LG	PFE Sensor Signal	3.2v, 55 mph: 2.9v
28	---	Not Used	---
29	---	Not Used	---
30	RD/BL	Neutral Drive Switch	In 'N': 0v, Others: 5v
31	---	Not Used	---
32	---	Not Used	---
33	DG	VR Solenoid Control	0%, 55 mph: 0-45%
33	BR/PK	VR Solenoid Control	0%, 55 mph: 0-45%
34	BL/PK	Data Output Link	Digital Signals
35	BR/BL	Injector 4 Control	5.6-6.3 ms
36	YL/LG	Spark Output Signal	6.93v (50% d/cycle)
37	RD	Vehicle Power	12-14v
38	---	Not Used	---
39	BR/YL	Injector 3 Control	5.6-6.3 ms
40	BK/LG	Power Ground	<0.1v

1991-95 Cougar 3.8L V6 MFI VIN 4 (A/T) 60 Pin Connector

PCM Pin #	Wire Color	Circuit Description (60 Pin)	Value at Hot Idle
41-42	---	Not Used	---
43	TN/RD	HO2S-21 (B1 S1) Signal	0.1-1.1v
43	TN/WT	HO2S-21 (B1 S1) Signal	0.1-1.1v
44	TN/OR	HO2S-11 (B1 S1) Signal	0.1-1.1v
45	---	Not Used	---
46	BK/WT	Analog Signal Return	<0.050v
46	GY/RD	Analog Signal Return	<0.050v
47	DG/LG	TP Sensor Signal	0.8v, 55 mph: 1.1v
47	GY/WT	TP Sensor Signal	0.8v, 55 mph: 1.1v
48	WT/PK	Self-Test Input Signal	STI On: 1v, Off: 5v
48	WT/RD	Self-Test Input Signal	STI On: 1v, Off: 5v
49	---	Not Used	---
50	DB/OR	MAF Sensor Signal	0.7v, 55 mph: 2.0v
50	BL/RD	MAF Sensor Signal	0.7v, 55 mph: 2.0v
51-53	---	Not Used	---
54	OR/BL	A/C WOT Relay Control	Relay Off: 12v, On: 1v
55	---	Not Used	---
56	DB	PIP Sensor Signal	6.93v (50% d/cycle)
57	RD	Vehicle Power	12-14v
58	TN	Injector 1 Control	5.6-6.3 ms
59	WT	Injector 2 Control	5.6-6.3 ms
60	BK/LG	Power Ground	<0.1v

Pin Connector Graphic

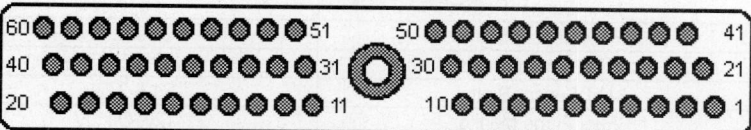

PCM 60-PIN CONNECTOR

Terminal View of 60-Pin PCM Harness Connector

Standard Colors and Abbreviations

Abbreviation	Color	Abbreviation	Color	Abbreviation	Color
BK	Black	GY	Gray	PK	Purple
BL	Blue	GN	Green	RD	Red
BR	Brown	LG	LT Green	TN	Tan
DB	Dark Blue	OR	Orange	WT	White
DG	DK Green	PK	Pink	YL	Yellow

1996-97 Cougar 3.8L V6 MFI VIN 4 (A/T) 104 Pin Connector

PCM Pin #	Wire Color	Circuit Description (104 Pin)	Value at Hot Idle
1	PK/OR	Shift Solenoid 2 Control	12v, 55 mph: 1v
2	PK/LG	MIL (lamp) Control	MIL On: 1v, Off: 12v
3-10	---	Not Used	---
11	PK/WT	Purge Flow Sensor Signal	0.5-1.5v, 30 mph: 3v
12	---	Not Used	---
13	PK	Flash EEPROM Power	0.1v
14	---	Not Used	---
15	PK/LB	Data Bus (-) Signal	Digital Signals
16	TN/OR	Data Bus (+) Signal	Digital Signals
17-20	---	Not Used	---
21	DB	CKP Sensor (+) Signal	850-1120 Hz
22	GY	CKP Sensor (-) Signal	850-1120 Hz
23	---	Not Used	---
24	BK/WT	Power Ground	<0.1v
25	BK	Case Ground	<0.050v
26	DB/LG	Coil Driver 1 Control	5° dwell
27	OR/YL	Shift Solenoid 1 Control	1v, 55 mph: 12v
28	---	Not Used	---
29	TN/WT	TCS (switch) Signal	TCS & O/D On: 12v
30	DG	Octane Adjust Switch	9.3v (Closed: 0v)
31-32	---	Not Used	---
33	PK/OR	VSS (-) Signal	0 Hz, 55 mph: 125 Hz
34	---	Not Used	---
35	RD/LG	HO2S-12 (B1 S2) Signal	0.1-1.1v
36	TN/LB	MAF Sensor Return	<0.050v
37	OR/BK	TFT Sensor Signal	2.40-2.10v
38	LG/RD	ECT Sensor Signal	0.5-0.6v
39	GY	IAT Sensor Signal	1.5-2.5v
40	PK/BK	Fuel Pump Monitor	12v, Off: 0v
41	PK/LB	A/C Cycling Clutch Switch	AC On: 12v, Off: 0v
42-44	---	Not Used	---
45	TN/OR	Low Speed Fan Control	Fan On: 1v, Off: 12v
46	LG/PK	High Speed Fan Control	Fan On: 1v, Off: 12v
47	BR/PK	VR Solenoid Control	0%, 55 mph: 45%
48-50	---	Not Used	---
51	BK/WT	Power Ground	<0.1v
52	RD/LB	Coil Driver 2 Control	5° dwell
53	---	Not Used	---
54	BR/OR	TCC Solenoid Control	0%, 55 mph: 90%
55	YL	Keep Alive Power	12-14v
56-57	---	Not Used	---
58	GY/BK	VSS (+) Signal	0 Hz, 55 mph: 125 Hz
59	---	Not Used	---
60	GY/LB	HO2S-11 (B1 S1) Signal	0.1-1.1v
61	PK/LG	HO2S-22 (B2 S2) Signal	0.1-1.1v
62-63	---	Not Used	---
64	LBYL	TR Sensor Signal	In 'P': 0v, O/D: 1.7v

1996-97 Cougar 3.8L V6 MFI VIN 4 (A/T) 104 Pin Connector

PCM Pin #	Wire Color	Circuit Description (104 Pin)	Value at Hot Idle
65	BR/LG	DPFE Sensor Signal	0.20-1.30v
66, 68	---	Not Used	---
67	GY/YL	EVAP Purge Solenoid	0-10 Hz (0-100%)
69	PK/YL	A/C WOT Relay Control	Relay Off: 12v, On: 1v
70, 72	---	Not Used	---
71	RD	Vehicle Power	12-14v
73	TN/BK	Injector 5 Control	4.5-4-8 ms
74	BR/YL	Injector 3 Control	4.5-4-8 ms
75	TN	Injector 1 Control	4.5-4-8 ms
76	BK/WT	Power Ground	<0.1v
77	BK/WT	Power Ground	<0.1v
78	PK/WT	Coil Driver 3 Control	5° dwell
79	WT/LG	TCIL (lamp) Control	Lamp On: 1v, Off: 12v
80	LB/OR	Fuel Pump Control	Relay Off: 12v, On: 1v
81	WT/YL	EPC Solenoid Control	8.8v (20 psi)
82	---	Not Used	---
83	WT/LB	IAC Motor Control	10v (30% d/cycle)
84	DG/WT	TSS Sensor Signal	43 Hz (650 rpm)
85	DG	CMP Sensor Signal	5-7 Hz
86	TN/LG	A/C Pressure Switch	Switch On: 12v (Open)
87	RD/BK	HO2S-21 (B1 S1) Signal	0.1-1.1v
88	LB/RD	MAF Sensor Signal	0.6v
89	GY/WT	TP Sensor Signal	0.53-1.27v
90	BR/WT	Reference Voltage	4.9-5.1v
91	GY/RD	Analog Signal Return	<0.050v
92	LG	Brake Pedal Position Switch	Brake Off: 0v, On: 12v
93	RD/WT	HO2S-11 (B1 S1) Heater	1v, Off: 12v
94	YL/LB	HO2S-21 (B2 S1) Heater	1v, Off: 12v
95	WT/BK	HO2S-12 (B1 S2) Heater	1v, Off: 12v
96	TN/YL	HO2S-22 (B2 S2) Heater	1v, Off: 12v
97	RD	Vehicle Power	12-14v
98	---	Not Used	---
99	LG/OR	Injector 6 Control	4.5-4-8 ms
100	BR/LB	Injector 4 Control	4.5-4-8 ms
101	WT	Injector 2 Control	4.5-4-8 ms
102	---	Not Used	---
103	BK/WT	Power Ground	<0.1v
104	---	Not Used	---

Pin Connector Graphic

PCM 104-PIN CONNECTOR

Terminal View of 104-Pin PCM Wiring Harness Connector

Standard Colors and Abbreviations

Abbreviation	Color	Abbreviation	Color	Abbreviation	Color
BK	Black	GY	Gray	PK	Purple
BL	Blue	GN	Green	RD	Red
BR	Brown	LG	LT Green	TN	Tan
DB	Dark Blue	OR	Orange	WT	White
DG	DK Green	PK	Pink	YL	Yellow

1990-92 Cougar 3.8L V6 Supercharged MFI VIN C (All) 60 Pin Connector

PCM Pin #	Wire Color	Circuit Description (60 Pin)	Value at Hot Idle
1	YL	Keep Alive Power	12-14v
2	---	Not Used	---
3	DG/WT	VSS (+) Signal	0 Hz, 55 mph: 125 Hz
4	DG/YT	Ignition Diagnostic Monitor	20-31 Hz
5	LG	Brake Switch Signal	Brake Off: 0v, On: 12v
6	BK/WT	VSS (-) Signal	0 Hz, 55 mph: 125 Hz
7	LG/YL	ECT Sensor Signal	0.5-0.6v
8	---	Not Used	---
9	TN/BL	MAF Sensor Return	<0.050v
10	PK/BL	A/C Cycling Clutch Signal	A/C Off: 0v, On: 12v
11	OR/YL	Speed Control Solenoid (+)	Solenoid On: 12-14v
12	BR/YL	Injector 3 Control	4.0-4.2 ms
13	BR/BL	Injector 4 Control	4.0-4.2 ms
14	TN/BL	Injector 5 Control	4.0-4.2 ms
15	LG	Injector 6 Control	4.0-4.2 ms
16	BL	Ignition System Ground	<0.050v
17	YL/BK	Self-Test Output, MIL Control	MIL Off: 12v, On: 1v
18	BK/WT	Octane Adjust Switch	Closed: 0v, Open: 9.1v
19	PK/BK	Fuel Pump Control	12v, Off: 0v
20	BK	Case Ground	<0.050v
21	RD/LG	IAC Motor Control	20-40%
22	TN/LG	Fuel Pump Control	Relay Off: 12v, On: 1v
23	YL/RD	Knock Sensor Signal	0v
24	DG	CID Sensor Signal	5-9 Hz
25	LG/PK	ACT Sensor Signal	1.5-2.5v
26	OR/WT	Reference Voltage	4.9-5.1v
27	BR/LG	PFE Sensor Signal	3.2v, 55 mph: 2.9v
28	BL/BK	Speed Control Switch Signal	6.7v
29	TN/OR	HO2S-21 (B1 S1) Signal	0.1-1.1v
30	RD/BL	Neutral Drive Switch	In 'N': 0v, Others: 5v
31	GY/YL	Canister Purge Solenoid	12v, 55 mph: 1v
32	OR/WT	Automatic Ride Control	4.3v
33	YL	VR Solenoid Control	0%, 55 mph: 45%
34	---	Not Used	---
35	WT/PK	S/C Vent Solenoid Control	0%, 55 mph: 98%
36	YL/LG	Spark Output Signal	6.93v (50% d/cycle)
37	RD	Vehicle Power	12-14v
38 ('89)	LG/PK	Supercharger Bypass Sol.	0% duty cycle
39	WT/BL	Speed Control Switch Ground	<0.050v
40	BK/LG	Power Ground	<0.1v

1990-92 Cougar 3.8L V6 SC MFI VIN C (All) 60 Pin Connector

PCM Pin #	Wire Color	Circuit Description (60 Pin)	Value at Hot Idle
41	PK	High Speed Fan Control	Fan On: 1v, Off: 12v
42	---	Not Used	---
43	TN/RD	HO2S-11 (B1 S1) Signal	0.1-1.1v
44	---	Not Used	---
45	DB/LG	MAP Sensor Signal	107 Hz (sea level)
46	BK/WT	Analog Signal Return	<0.050v
47	DG/LG	TP Sensor Signal	0.8v, 55 mph: 1.1v
48	WT/RD	Self-Test Input Signal	STI On: 1v, Off: 5v
49	OR	HO2S-11 (Bank 1) Ground	<0.050v
50	DB/OR	MAF Sensor Signal	0.7v, 55 mph: 2.0v
51	GY/BK	S/C Vacuum Solenoid Control	Vacuum Increasing: 1v
52	---	Not Used	---
53	PK	Shift Indicator Light Control	SIL On: 1v, Off: 12v
54	RD	A/C WOT Relay Control	Relay Off: 12v, On: 1v
55	TN/OR	Low Speed Fan Control	Fan On: 1v, Off: 12v
56	DB	PIP Sensor Signal	6.93v (50% d/cycle)
57	RD	Vehicle Power	12-14v
58	TN	Injector 1 Control	4.0-4.2 ms
59	WT	Injector 2 Control	4.0-4.2 ms
60	LG/BK	Power Ground	<0.1v

Pin Connector Graphic

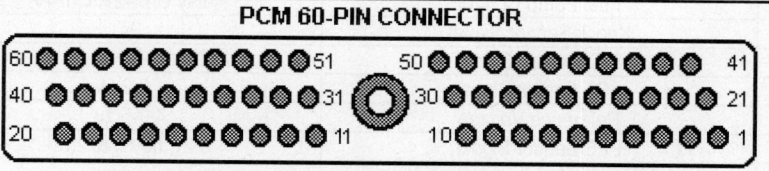

Standard Colors and Abbreviations

Abbreviation	Color	Abbreviation	Color	Abbreviation	Color
BK	Black	GY	Gray	PK	Purple
BL	Blue	GN	Green	RD	Red
BR	Brown	LG	LT Green	TN	Tan
DB	Dark Blue	OR	Orange	WT	White
DG	DK Green	PK	Pink	YL	Yellow

1990-94 Cougar 3.8L V6 SC MFI VIN R (All) 60 Pin Connector

PCM Pin #	Wire Color	Circuit Description (60 Pin)	Value at Hot Idle
1	YL	Keep Alive Power	12-14v
2	---	Not Used	---
3	DG/WT	VSS (+) Signal	0 Hz, 55 mph: 125 Hz
4	DG/YT	Ignition Diagnostic Monitor	20-31 Hz
5	LG	Brake Switch Signal	Brake Off: 0v, On: 12v
6	BK/WT	VSS (-) Signal	0 Hz, 55 mph: 125 Hz
7	LG/YL	ECT Sensor Signal	0.5-0.6v
8	---	Not Used	---
9	TN/BL	MAF Sensor Return	<0.050v
10	PK/BK	A/C Cycling Clutch Signal	A/C Off: 0v, On: 12v
11	OR/YL	Speed Control Solenoid (+)	Solenoid On: 12-14v
12	BR/YL	Injector 3 Control	4.0-4.2 ms
13	BR/BL	Injector 4 Control	4.0-4.2 ms
14	TN/BL	Injector 5 Control	4.0-4.2 ms
15	LG	Injector 6 Control	4.0-4.2 ms
16	BL	Ignition System Ground	<0.050v
17	YL/BK	Self-Test Output, MIL Control	MIL Off: 12v, On: 1v
18	BK/WT	Octane Adjust Switch	Closed: 0v, Open: 9.1v
19	PK/BK	Fuel Pump Control	12v, Off: 0v
20	BK	Case Ground	<0.050v
21	RD/LG	IAC Motor Control	Hot Idle: 9-9.4v
22	TN/LG	Fuel Pump Control	Relay Off: 12v, On: 1v
23	YL/RD	Knock Sensor Signal	0v
24	DG	CID Sensor Signal	5-9 Hz
25	LG/PK	ACT Sensor Signal	1.5-2.5v
26	OR/WT	Reference Voltage	4.9-5.1v
27	BR/LG	PFE Sensor Signal	3.2v, 55 mph: 2.9v
28	BL/BK	Speed Control Switch Signal	6.7v
29	TN/OR	HO2S-21 (B1 S1) Signal	0.1-1.1v
30	RD/BL	Neutral Drive Switch	In 'N': 0v, Others: 5v
31	GY/YL	Canister Purge Solenoid	12v, 55 mph: 1v
32	OR/WT	Automatic Ride Control	4.3v
33	YL	VR Solenoid Control	0%, 55 mph: 45%
34	---	Not Used	---
35	WT/PK	S/C Vent Solenoid Control	0%, 55 mph: 98%
36	YL/LG	Spark Output Signal	6.93v (50% d/cycle)
37	RD	Vehicle Power	12-14v
38	LG/PK	Supercharger Bypass Sol.	0% duty cycle
39	WT/BL	Speed Control Switch Ground	<0.050v
40	BK/LG	Power Ground	<0.1v

1990-94 Cougar 3.8L V6 SC MFI VIN R (All) 60 Pin Connector

PCM Pin #	Wire Color	Circuit Description (60 Pin)	Value at Hot Idle
41	PK	High Speed Fan Control	Fan On: 1v, Off: 12v
42	---	Not Used	---
43	TN/RD	HO2S-11 (B1 S1) Signal	0.1-1.1v
44	---	Not Used	---
45	DB/LG	MAP Sensor Signal	107 Hz (sea level)
46	BK/WT	Analog Signal Return	<0.050v
47	DG/LG	TP Sensor Signal	0.8v, 55 mph: 1.1v
48	WT/RD	Self-Test Input Signal	STI On: 1v, Off: 5v
49	OR	HO2S-11 (Bank 1) Ground	<0.050v
50	DB/OR	MAF Sensor Signal	0.7v, 55 mph: 2.0v
51	GY/BK	S/C Vacuum Solenoid Control	Vacuum Increasing: 1v
52	---	Not Used	---
53	PK	Shift Indicator Light Control	SIL On: 1v, Off: 12v
54	RD	A/C WOT Relay Control	Relay Off: 12v, On: 1v
55	TN/OR	Low Speed Fan Control	Fan On: 1v, Off: 12v
56	DB	PIP Sensor Signal	6.93v (50% d/cycle)
57	RD	Vehicle Power	12-14v
58	TN	Injector 1 Control	4.0-4.2 ms
59	WT	Injector 2 Control	4.0-4.2 ms
60	LG/BK	Power Ground	<0.1v

Pin Connector Graphic

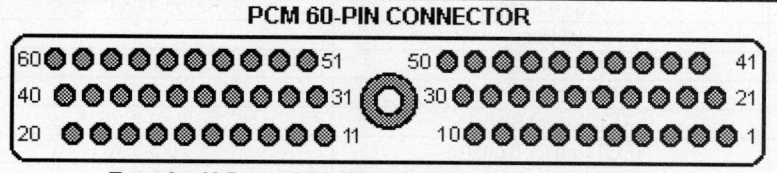

Standard Colors and Abbreviations

Abbreviation	Color	Abbreviation	Color	Abbreviation	Color
BK	Black	GY	Gray	PK	Purple
BL	Blue	GN	Green	RD	Red
BR	Brown	LG	LT Green	TN	Tan
DB	Dark Blue	OR	Orange	WT	White
DG	DK Green	PK	Pink	YL	Yellow

1994-97 Cougar 4.6L V8 2v MFI VIN W (A/T) 104 Pin Connector

PCM Pin #	Wire Color	Circuit Description (104 Pin)	Value at Hot Idle
1	PK/OR	Shift Solenoid 2 Control	12v, 55 mph: 1v
2	PK/LG	MIL (lamp) Control	MIL On: 1v, Off: 12v
3-10	---	Not Used	---
11	PK/WT	Purge Flow Sensor Signal	0-10 Hz (0-100%)
12	---	Not Used	---
13	PK	Flash EEPROM Power	0.1v
14	---	Not Used	---
15	PK/LB	Data Bus (-) Signal	Digital Signals
16	TN/OR	Data Bus (+) Signal	Digital Signals
17-20	---	Not Used	---
21	DB	CKP Sensor (+) Signal	850-1120 Hz
22	GY	CKP Sensor (-) Signal	850-1120 Hz
23	---	Not Used	---
24	BK/WT	Power Ground	<0.1v
25	BK	Case Ground	<0.050v
26	DB/LG	Coil Driver 1 Control	5° dwell
27	OR/YL	Shift Solenoid 1 Control	1v, 55 mph: 12v
28	---	Not Used	---
29	TN/WT	TCS (switch) Signal	TCS & O/D On: 12v
30	DG	Octane Adjust Switch	9.3v (Closed: 0v)
31, 32	---	Not Used	---
33	PK/OR	VSS (-) Signal	0 Hz, 55 mph: 125 Hz
34	---	Not Used	---
35	RD/LG	HO2S-12 (B1 S2) Signal	0.1-1.1v
36	TN/LB	MAF Sensor Return	<0.050v
37	OR/BK	TFT Sensor Signal	2.40-2.10v
38	LG/RD	ECT Sensor Signal	0.5-0.6v
39	GY	IAT Sensor Signal	1.5-2.5v
40	PK/BK	Fuel Pump Monitor	12v, Off: 0v
41	PK/LB	A/C Cycling Clutch Switch	AC On: 12v, Off: 0v
42-44	---	Not Used	---
45	TN/OR	Low Speed Fan Control	Fan On: 1v, Off: 12v
46	LG/PK	High Speed Fan Control	Fan On: 1v, Off: 12v
47	BR/PK	VR Solenoid Control	0%, 55 mph: 45%
48-50	---	Not Used	---
51	BK/WT	Power Ground	<0.1v
52	RD/LB	Coil Driver 2 Control	5° dwell
53	---	Not Used	---
54	BR/OR	TCC Solenoid Control	0%, 55 mph: 95%
55	YL	Keep Alive Power	12-14v
56-57	---	Not Used	---
58	GY/BK	VSS (+) Signal	0 Hz, 55 mph: 125 Hz
59	---	Not Used	---
60	GY/LB	HO2S-11 (B1 S1) Signal	0.1-1.1v
61	PK/LG	HO2S-22 (B2 S2) Signal	0.1-1.1v
62-63	---	Not Used	---
64	LBYL	TR Sensor Signal	In 'P': 0v, O/D: 1.7v

1994-97 Cougar 4.6L V8 2v MFI VIN W (A/T) 104 Pin Connector

PCM Pin #	Wire Color	Circuit Description (104 Pin)	Value at Hot Idle
65	BR/LG	DPFE Sensor Signal	0.20-1.30v
66	---	Not Used	---
67	GY/YL	EVAP Purge Solenoid	0-10 Hz (0-100%)
68, 70, 82	---	Not Used	---
69	PK/YL	A/C WOT Relay Control	Relay Off: 12v, On: 1v
71	RD	Vehicle Power	12-14v
72	TN/RD	Injector 7 Control	3.1-3-5 ms
73	TN/BK	Injector 5 Control	3.1-3-5 ms
74	BR/YL	Injector 3 Control	3.1-3-5 ms
75	TN	Injector 1 Control	3.1-3-5 ms
76, 77	BK/WT	Power Ground	<0.1v
78	PK/WT	Coil Driver 3 Control	5° dwell
79	WT/LG	TCIL (lamp) Control	Lamp On: 1v, Off: 12v
80	LB/OR	Fuel Pump Control	Relay Off: 12v, On: 1v
81	WT/YL	EPC Solenoid Control	9.0v (15 psi)
83	WT/LB	IAC Motor Control	10v (30% d/cycle)
84	DG/WT	TSS Sensor Signal	43 Hz (650 rpm)
85	DG	CMP Sensor Signal	5-7 Hz
86	TN/LG	A/C Pressure Switch	Switch On: 12v (Open)
87	RD/BK	HO2S-21 (B1 S1) Signal	0.1-1.1v
88	LB/RD	MAF Sensor Signal	0.6v
89	GY/WT	TP Sensor Signal	0.53-1.27v
90	BR/WT	Reference Voltage	4.9-5.1v
91	GY/RD	Analog Signal Return	<0.050v
92	LG	Brake Pedal Position Switch	Brake Off: 0v, On: 12v
93	RD/WT	HO2S-11 (B1 S1) Heater	1v, Off: 12v
94	YL/LB	HO2S-21 (B2 S1) Heater	1v, Off: 12v
95	WT/BK	HO2S-12 (B1 S2) Heater	1v, Off: 12v
96	TN/YL	HO2S-22 (B2 S2) Heater	1v, Off: 12v
97	RD	Vehicle Power	12-14v
98	LB	Injector 8 Control	3.1-3-5 ms
99	LG/OR	Injector 6 Control	3.1-3-5 ms
100	BR/LB	Injector 4 Control	3.1-3-5 ms
101	WT	Injector 2 Control	3.1-3-5 ms
102	---	Not Used	---
103	BK/WT	Power Ground	<0.1v
104	RD/YL	Coil Driver 4 Control	5° dwell

Pin Connector Graphic

PCM 104-PIN CONNECTOR

Terminal View of 104-Pin PCM Wiring Harness Connector

Standard Colors and Abbreviations

Abbreviation	Color	Abbreviation	Color	Abbreviation	Color
BK	Black	GY	Gray	PK	Purple
BL	Blue	GN	Green	RD	Red
BR	Brown	LG	LT Green	TN	Tan
DB	Dark Blue	OR	Orange	WT	White
DG	DK Green	PK	Pink	YL	Yellow

1991-93 Cougar 5.0L V8 MFI VIN T (A/T) 60 Pin Connector

PCM Pin #	Wire Color	Circuit Description (60 Pin)	Value at Hot Idle
1	YL	Keep Alive Power	12-14v
2	---	Not Used	---
3	DG/WT	VSS (+) Signal	0 Hz, 55 mph: 125 Hz
4	RD/BL	Ignition Diagnostic Monitor	20-31 Hz
4	WT/PK	Ignition Diagnostic Monitor	20-31 Hz
5	---	Not Used	---
6	BK/WT	VSS (-) Signal	0 Hz, 55 mph: 125 Hz
6	GY/RD	VSS (-) Signal	0 Hz, 55 mph: 125 Hz
7	LG/YL	ECT Sensor Signal	0.5-0.6v
7	LG/RD	ECT Sensor Signal	0.5-0.6v
8	PK/BK	Fuel Pump Monitor	12v, Off: 0v
9	TN/BL	MAF Sensor Return	<0.050v
10	PK/BL	A/C Cycling Clutch Signal	A/C Off: 0v, On: 12v
11	GY/YL	Canister Purge Solenoid	12v, 55 mph: 1v
12	LG	Injector 6 Control	4.9-5.2 ms
12	LG/OR	Injector 6 Control	4.9-5.2 ms
13	TN/OR	Injector 7 Control	4.9-5.2 ms
13	TN/RD	Injector 7 Control	4.9-5.2 ms
14	BL	Injector 8 Control	4.9-5.2 ms
15	TN/BL	Injector 5 Control	4.9-5.2 ms
15	TN/BK	Injector 5 Control	4.9-5.2 ms
16	GY	Ignition System Ground	<0.050v
17	YL/BK	Self-Test Output, MIL Control	MIL Off: 12v, On: 1v
18	TN/OR	Data Bus (+) Signal	Digital Signals
19	PK/BL	Data Bus (-) Signal	Digital Signals
20	BK	Case Ground	<0.050v
21	RD/LG	IAC Motor Control	9.3-10.3v
22	TN/LG	Fuel Pump Control	Relay Off: 12v, On: 1v
23-24	---	Not Used	---
25	GY	IAT Sensor Signal	1.5-2.5v
25	LG/PK	IAT Sensor Signal	1.5-2.5v
26	OR/WT	Reference Voltage	4.9-5.1v
26	BR/WT	Reference Voltage	4.9-5.1v
27	BR/LG	EVP Sensor Signal	0.4v, 55 mph: 1.5v
28	---	Not Used	---
29	---	Not Used	---
30	RD/BL	Neutral Drive Switch	In 'P': 0v, Others: 4-5v
31	WT/RD	Air Management 1 Solenoid	AM1 On: 1v, Off: 12v
32	OR/WT	Automatic Shock Control	4.3v
33	DG	VR Solenoid Control	0%, 55 mph: 0-45%
33	BR/PK	VR Solenoid Control	0%, 55 mph: 0-45%
34	BL/PK	Data Output Link	Digital Signals
35	BR/BL	Injector 4 Control	4.9-5.2 ms
36	YL/LG	Spark Output Signal	6.93v (50% d/cycle)
37	RD	Vehicle Power	12-14v
38	---	Not Used	---
39	BR/YL	Injector 3 Control	4.9-5.2 ms
40	BK/LG	Power Ground	<0.1v

1991-93 Cougar 5.0L V8 MFI VIN T (A/T) 60 Pin Connector

PCM Pin #	Wire Color	Circuit Description (60 Pin)	Value at Hot Idle
41-42	---	Not Used	---
43	TN/RD	HO2S-21 (B1 S1) Signal	0.1-1.1v
43	TN/WT	HO2S-21 (B1 S1) Signal	0.1-1.1v
44	TN/OR	HO2S-11 (B1 S1) Signal	0.1-1.1v
45	---	Not Used	---
46	BK/WT	Analog Signal Return	<0.050v
46	GY/RD	Analog Signal Return	<0.050v
47	DG/LG	TP Sensor Signal	0.8v, 55 mph: 1.1v
47	GY/WT	TP Sensor Signal	0.8v, 55 mph: 1.1v
48	WT/RD	Self-Test Input Signal	STI On: 1v, Off: 5v
48	WT/PK	Self-Test Input Signal	STI On: 1v, Off: 5v
49	---	Not Used	---
50	DB/OR	MAF Sensor Signal	1.6v
50	BL/RD	MAF Sensor Signal	1.6v
51-53	---	Not Used	---
54	OR/BL	A/C WOT Relay Control	Relay Off: 12v, On: 1v
55	---	Not Used	---
56	DB	PIP Sensor Signal	6.93v (50% d/cycle)
57	RD	Vehicle Power	12-14v
58	TN	Injector 1 Control	4.9-5.2 ms
59	WT	Injector 2 Control	4.9-5.2 ms
60	BK/LG	Power Ground	<0.1v

Pin Connector Graphic

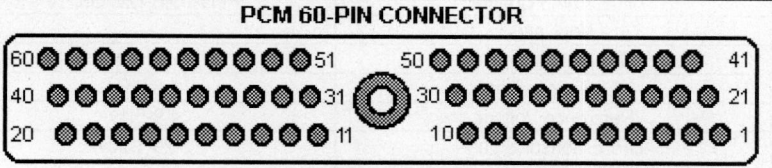

PCM 60-PIN CONNECTOR

Terminal View of 60-Pin PCM Harness Connector

Standard Colors and Abbreviations

Abbreviation	Color	Abbreviation	Color	Abbreviation	Color
BK	Black	GY	Gray	PK	Purple
BL	Blue	GN	Green	RD	Red
BR	Brown	LG	LT Green	TN	Tan
DB	Dark Blue	OR	Orange	WT	White
DG	DK Green	PK	Pink	YL	Yellow

Grand Marquis Pin Tables

1992-95 Grand Marquis 4.6L V8 VIN W (A/T) 60 Pin Connector

PCM Pin #	Wire Color	Circuit Description (60 Pin)	Value at Hot Idle
1	YL	Keep Alive Power	12-14v
2 ('93-'95)	LG	Brake Switch Signal	Brake Off: 0v, On: 12v
3	GY/BK	VSS (+) Signal	0 Hz, 55 mph: 125 Hz
4	TN/YL	Ignition Diagnostic Monitor	20-31 Hz
5	PK/BL	TSS Sensor (+) Signal	0 Hz, 30 mph: 130 Hz
6	PK/OR	VSS (-) Signal	0 Hz, 55 mph: 125 Hz
7	LG/RD	ECT Sensor Signal	0.5-0.6v
8	DG/YT	Fuel Pump Monitor	12v, Off: 0v
9	TN/BL	MAF Sensor Return	<0.050v
10	DG/OR	A/C Cycling Clutch Signal	A/C Off: 0v, On: 12v
11	GY/YL	Air Management 2 Solenoid	On: 1v, Off: 12v
12	LG/OR	Injector 6 Control	4.0-4.4 ms
13	TN/RD	Injector 7 Control	4.0-4.4 ms
14	BL	Injector 8 Control	4.0-4.4 ms
15	TN/BK	Injector 5 Control	4.0-4.4 ms
16	OR/RD	Ignition System Ground	<0.050v
16	OR	Ignition System Ground	<0.050v
17	TN/RD	Self-Test Output, MIL Control	MIL Off: 12v, On: 1v
18	TN/OR	Data Bus (+) Signal	Digital Signals
19	PK/BL	Data Bus (-) Signal	Digital Signals
20	BK	Case Ground	<0.050v
21	W	IAC Motor Control	8.3-11.5v
21	WT/BL	IAC Motor Control	8.3-11.5v
22	BL/OR	Fuel Pump Control	Relay Off: 12v, On: 1v
23	---	Not Used	---
24	DG	CID Sensor Signal	6-7 Hz
25	LG/PK	ACT Sensor Signal	1.5-2.5v
26	BR/WT	Reference Voltage	4.9-5.1v
27	BR/LG	DPFE Sensor Signal	0.20-1.30v
28	---	Not Used	---
29	WT/RD	Octane Adjust Switch	Closed: 0v, Open: 9.1v
30 ('93-'95)	WT/PK	Neutral Drive Switch	In 'P': 0v, Others: 5v
30 ('93-'95)	BL/YL	Neutral Drive Switch	In 'P': 0v, Others: 5v
30 ('93-'95)	WT/PK	MLP Sensor Signal	0v, O/D: 2.1v
30 ('93-'95)	BL/YL	MLP Sensor Signal	0v, O/D: 2.1v
31-32	---	Not Used	---
33	BR/PK	VR Solenoid Control	0%, 55 mph: 0-45%
34	BL/LG	Data Output Link	Digital Signals
35	BR/BL	Injector 4 Control	4.0-4.4 ms
36	PK	Spark Angle Word Signal	6.93v (50% d/cycle)
37	RD	Vehicle Power	12-14v
38 ('93-'95)	WT/YL	EPC Solenoid Control	9.5v (20 psi)
39	BR/YL	Injector 4 Control	4.0-4.4 ms
40	BK	Power Ground	<0.1v
40	BK/WT	Power Ground	<0.1v

1992-95 Grand Marquis 4.6L V8 VIN W (A/T) 60 Pin Connector

PCM Pin #	Wire Color	Circuit Description (60 Pin)	Value at Hot Idle
41 ('93-'95)	TN/WT	TCS Sensor Signal	TCS & O/D On: 12v
42	---	Not Used	---
43	RD/BK	HO2S-11 (B1 S1) Signal	0.1-1.1v
44	GY/BL	HO2S-21 (B1 S1) Signal	0.1-1.1v
45	---	Not Used	---
46	GY/RD	Analog Signal Return	<0.050v
47	GY/WT	TP Sensor Signal	0.8v, 55 mph: 1.1v
48	WT/PK	Self-Test Input Signal	STI On: 1v, Off: 5v
49 ('93-'95)	OR/BK	TOT Sensor Signal	2.40-2.10v
50	BL/RD	MAF Sensor Signal	0.8v
51	OR/YL	Shift Solenoid 1 Control	1v, 55 mph: 12v
52	PK/OR	Shift Solenoid 2 Control	12v, 55 mph: 1v
53	BR/OR	MCCC or TCC Solenoid	12v, 55 mph: 9-10v
54	OR/BL	A/C WOT Relay Control	Relay Off: 12v, On: 1v
55	DB/W	TCIL (lamp) Control	Lamp On: 1v, Off: 12v
55	WT/LG	TCIL (lamp) Control	Lamp On: 1v, Off: 12v
56	GY/OR	PIP Sensor Signal	6.93v (50% dwell)
57	RD	Vehicle Power	12-14v
58	TN	Injector 1 Control	4.0-4.4 ms
59	WT	Injector 2 Control	4.0-4.4 ms
60	BK	Power Ground	<0.1v

Pin Connector Graphic

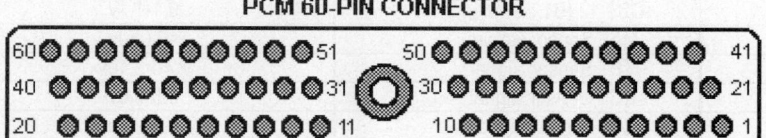

Standard Colors and Abbreviations

Abbreviation	Color	Abbreviation	Color	Abbreviation	Color
BK	Black	GY	Gray	PK	Purple
BL	Blue	GN	Green	RD	Red
BR	Brown	LG	LT Green	TN	Tan
DB	Dark Blue	OR	Orange	WT	White
DG	DK Green	PK	Pink	YL	Yellow

1996-97 Grand Marquis 4.6L V8 2v VIN W (A/T) 104 Pin Connector

PCM Pin #	Wire Color	Circuit Description (104 Pin)	Value at Hot Idle
1	PK/OR	Shift Solenoid 2 Control	12v, 55 mph: 1v
2	PK/LG	MIL (lamp) Control	MIL On: 1v, Off: 12v
3-12	---	Not Used	---
13	LG/YL	Flash EEPROM Power	0.1v
14	---	Not Used	---
15	PK/LB	Data Bus (-) Signal	Digital Signals
16	TN/OR	Data Bus (+) Signal	Digital Signals
17-20	---	Not Used	---
21	DB	CKP Sensor (+) Signal	850-1120 Hz
22	GY	CKP Sensor (-) Signal	850-1120 Hz
23	---	Not Used	---
24	BK/WT	Power Ground	<0.1v
25	BK	Case Ground	<0.050v
26	TN/WT	Coil Driver 1 Control	5° dwell
27	OR/YL	Shift Solenoid 1 Control	1v, 55 mph: 12v
28	---	Not Used	---
29	TN/WT	TCS (switch) Signal	TCS & O/D On: 12v
30	WT/RD	Octane Adjust Switch	9.3v (Closed: 0v)
31-32	---	Not Used	---
33	PK/OR	VSS (-) Signal	0 Hz, 55 mph: 125 Hz
34	---	Not Used	---
35	RD/LG	HO2S-12 (B1 S2) Signal	0.1-1.1v
36	TN/LB	MAF Sensor Return	<0.050v
37	OR/BK	TFT Sensor Signal	2.40-2.10v
38	LG/RD	ECT Sensor Signal	0.5-0.6v
39	GY	IAT Sensor Signal	1.5-2.5v
40	DG/YT	Fuel Pump Monitor	12v, Off: 0v
41	DG/OR	A/C Cycling Clutch Switch	A/C Switch On: 12v
42-44	---	Not Used	---
45	RD/OR	Low Speed Fan Control	Fan On: 1v, Off: 12v
46	---	Not Used	---
47	BR/PK	VR Solenoid Control	0%, 55 mph: 45%
48-50	---	Not Used	---
51	BK/WT	Power Ground	<0.1v
52	TN/OR	Coil Driver 2 Control	5° dwell
53	---	Not Used	---
54	RD/LB	TCC Solenoid Control	0%, 55 mph: 90%
55	YL/BK	Keep Alive Power	12-14v
56	LG/BK	EVAP VMV Solenoid	0-10 Hz (0-100%)
57	---	Not Used	---
58	GY/BK	VSS (+) Signal	0 Hz, 55 mph: 125 Hz
59	---	Not Used	---
60	GY/LB	HO2S-11 (B1 S1) Signal	0.1-1.1v
61	PK/LG	HO2S-22 (B2 S2) Signal	0.1-1.1v
62	RD/PK	FTP Sensor Signal	2.6v (0" H2O)
63	---	Not Used	---
64	LBYL	TR Sensor Signal	In 'P': 0v, O/D: 1.7v

1996-97 Grand Marquis 4.6L V8 2v VIN W (A/T) 104 Pin Connector

PCM Pin #	Wire Color	Circuit Description (104 Pin)	Value at Hot Idle
65	BR/LG	DPFE Sensor Signal	0.20-1.30v
66	---	Not Used	---
67	PK/WT	EVAP CV Solenoid	0-10 Hz (0-100%)
68	---	Not Used	---
69	OR/LB	A/C WOT Relay Control	Relay Off: 12v, On: 1v
70	---	Not Used	---
71	RD	Vehicle Power	12-14v
72	TN/RD	Injector 7 Control	3.4-3-7 ms
73	TN/BK	Injector 5 Control	3.4-3-7 ms
74	BR/YL	Injector 3 Control	3.4-3-7 ms
75	TN	Injector 1 Control	3.4-3-7 ms
76, 77	BK/WT	Power Ground	<0.1v
78	TN/LG	Coil Driver 3 Control	5° dwell
79	WT/LG	TCIL (lamp) Control	Lamp On: 1v, Off: 12v
80	LB/OR	Fuel Pump Control	Relay Off: 12v, On: 1v
81	WT/YL	EPC Solenoid Control	9.0v (15 psi)
82, 86	---	Not Used	---
83	WT/LB	IAC Motor Control	9.3v (34% duty cycle)
84	PK/LB	OSS Sensor Signal	0 Hz, 30 mph: 131 Hz
85	DG	CMP Sensor Signal	5-7 Hz
87	RD/BK	HO2S-21 (B1 S1) Signal	0.1-1.1v
88	LB/RD	MAF Sensor Signal	0.6v
89	GY/WT	TP Sensor Signal	0.53-1.27v
90	BR/WT	Reference Voltage	4.9-5.1v
91	GY/RD	Analog Signal Return	<0.050v
92	LG	Brake Pedal Position Switch	Brake Off: 0v, On: 12v
93	RD/WT	HO2S-11 (B1 S1) Heater	1v, Off: 12v
94	YL/LB	HO2S-21 (B2 S1) Heater	1v, Off: 12v
95	WT/BK	HO2S-12 (B1 S2) Heater	1v, Off: 12v
96	TN/YL	HO2S-22 (B2 S2) Heater	1v, Off: 12v
97	RD	Vehicle Power	12-14v
98	LB	Injector 8 Control	3.4-3-7 ms
99	LG/OR	Injector 6 Control	3.4-3-7 ms
100	BR/LB	Injector 4 Control	3.4-3-7 ms
101	WT	Injector 2 Control	3.4-3-7 ms
102	---	Not Used	---
103	BK/WT	Power Ground	<0.1v
104	RD/YL	Coil Driver 4 Control	5° dwell

Pin Connector Graphic

Terminal View of 104-Pin PCM Wiring Harness Connector

Standard Colors and Abbreviations

Abbreviation	Color	Abbreviation	Color	Abbreviation	Color
BK	Black	GY	Gray	PK	Purple
BL	Blue	GN	Green	RD	Red
BR	Brown	LG	LT Green	TN	Tan
DB	Dark Blue	OR	Orange	WT	White
DG	DK Green	PK	Pink	YL	Yellow

1998-1999 Grand Marquis 4.6L V8 VIN W (A/T) 104 Pin Connector

PCM Pin #	Wire Color	Circuit Description (104 Pin)	Value at Hot Idle
1	OR/YL	COP 6 Driver Control	5° dwell
2	PK/LG	MIL (lamp) Control	MIL On: 1v, Off: 12v
3	BK/WT	Power Ground	<0.1v
4-5	---	Not Used	---
6	OR/YL	Shift Solenoid 'A' Control	1v, 55 mph: 12v
9	YL/WT	Fuel Level Indicator Signal	1.7v (1/2 full)
7-8, 10	---	Not Used	---
11	PK/OR	Shift Solenoid 'B' Control	12v, Cruise: 1v
12	WT/LG	TCIL (lamp) Control	Lamp On: 1v, Off: 5v
13	PK	Flash EEPROM Power	0.1v
15	PK/LB	Data Bus (-) Signal	Digital Signals
16	TN/OR	Data Bus (+) Signal	Digital Signals
17-20	---	Not Used	---
21	BK/PKK	CKP Sensor (+) Signal	440-490 Hz
22	GY/YL	CKP Sensor (-) Signal	440-490 Hz
23	---	Not Used	---
24	BK/WT	Power Ground	<0.1v
25	BK	PCM Case Ground	<0.050v
26	LG/WT	COP 1 Driver Control	5° dwell
27	LG/YL	COP 5 Driver Control	1v, 55 mph: 12v
28	RD/OR	Low Speed Fan Control	On: 1v, Off: 12v
29	TN/WT	TCS (switch) Signal	TCS & O/D On: 12v
30	WT/RD	Octane Adjust Switch	Closed: 0v, Open: 9.3v
31-32, 38	---	Not Used	---
33	PK/OR	VSS (-) Signal	0 Hz, 55 mph: 125 Hz
34	YL/BK	Digital TR1 Sensor	In 'P': 0v, 55 mph: 11v
35	RD/LG	HO2S-12 (B1 S2) Signal	0.1-1.1v
36	TN/LB	MAF Sensor Return	<0.050v
37	OR/BK	TFT Sensor Signal	2.10-2.40v
39	GY	IAT Sensor Signal	1.5-2.5v
40	DG/YT	Fuel Pump Monitor	On: 12v, Off: 0v
41	DG/OR	A/C Cycling Clutch Switch	AC On: 12v, Off: 0v
42	RD/WT	ECT Sensor Signal	0.5-0.6v
43	DB/LG	Fuel Flow Rate Signal	Digital Signals
45	OR/RD	CHTIL (lamp) Control	Lamp On: 1v, Off: 5v
46	LG/PK	High Speed Fan Control	On: 1v, Off: 12v
47	BR/PK	EGR VR Solenoid	0%, 55 mph: 45%
48	---	Not Used	---
49	RD/LB	Digital TR2 Sensor	In 'P': 0v, 55 mph: 11v
50	WT/BK	Digital TR4 Sensor	In 'P': 0v, 55 mph: 11v
51	BK/WT	Power Ground	<0.1v
52	WT/PK	COP 3 Driver Control	5° dwell
53	DG/PK	COP 4 Driver Control	5° dwell
54	PK/YL	TCC Solenoid Control	0%, 55 mph: 95%
55	YL/BK	Keep Alive Power	12-14v
56	LG/BK	EVAP Purge Solenoid	0-10 Hz (0-100%)
57, 59	---	Not Used	---
58	GY/BK	VSS + Sensor Signal	0 Hz, 55 mph: 125 Hz
60	GY/LB	HO2S-11 (B1 S1) Signal	0.1-1.1v
61	PK/LG	HO2S-22 (B2 S2) Signal	0.1-1.1v
62	RD/PK	FTP Sensor Signal	2.6v (0" H2O - cap off)
63	---	Not Used	---
64	LBYL	Digital TR3 Sensor	In 'P': 0v, in O/D: 1.7v

1998-1999 Grand Marquis 4.6L V8 VIN W (A/T) 104 Pin Connector

PCM Pin #	Wire Color	Circuit Description (104 Pin)	Value at Hot Idle
65	BR/LG	DPFE Sensor Signal	0.20-1.30v
66	YL/LG	CHT Sensor Signal	0.7v (194°F)
67	PK/WT	EVAP CV Solenoid	0-10 Hz (0-100%)
68, 70	---	Not Used	---
69	PK/YL	A/C WOT Relay Control	On: 1v, Off: 12v
82	---	Not Used	---
71	RD	Vehicle Power	12-14v
72	TN/RD	Injector 7 Control	3.4-3.7 ms
73	TN/BK	Injector 5 Control	3.4-3.7 ms
74	BR/YL	Injector 3 Control	3.4-3.7 ms
75	TN	Injector 1 Control	3.4-3.7 ms
77	BK/WT	Power Ground	<0.1v
78	PK/LB	COP 7 Driver Control	5° dwell
79	WT/RD	COP 8 Driver Control	5° dwell
80	LB/OR	Fuel Pump Control	On: 1v, Off: 12v
81	WT/YL	EPC Solenoid Control	9.5v (20 psi)
83	WT/LB	IAC Motor Control	34% duty cycle
84	DB/YL	OSS Sensor Signal	0 Hz
85	DB/OR	CMP Sensor Signal	6-7 Hz
86	WT/BK	A/C High Pressure Switch	Open: 12v, Closed: 0v
87	RD/BK	HO2S-21 (B2 S1) Signal	0.1-1.1v
88	LB/RD	MAF Sensor Signal	0.6v
89	GY/WT	TP Sensor Signal	0.53-1.27v
90	BR/WT	Reference Voltage	4.9-5.1v
91	GY/RD	Analog Signal Return	<0.050v
92	LG	Brake Pedal Position Switch	Brake Off: 0v, On: 12v
93	RD/WT	HO2S-11 (B1 S1) Heater	On: 1v, Off: 12v
94	YL/LB	HO2S-21 (B2 S1) Heater	On: 1v, Off: 12v
95	WT/BK	HO2S-12 (B1 S2) Heater	On: 1v, Off: 12v
96	TN/YL	HO2S-22 (B2 S2) Heater	On: 1v, Off: 12v
97 ('98 only)	RD	Vehicle Power	12-14v
98	LB	Injector 8 Control	3.4-3.7 ms
99	LG/OR	Injector 6 Control	3.4-3.7 ms
100	BR/LB	Injector 4 Control	3.4-3.7 ms
101	WT	Injector 2 Control	3.4-3.7 ms
102	---	Not Used	---
103	BK/WT	Power Ground	<0.1v
104	PK/WT	COP 2 Driver Control	5° dwell

Pin Connector Graphic

```
          PCM 104-PIN CONNECTOR
   1 ◉◉◉◉◉◉◉◉◉◉◉◉◉      ◉◉◉◉◉◉◉◉◉◉◉◉◉ 26
  27 ◉◉◉◉◉◉◉◉◉◉◉◉◉   ⬤  ◉◉◉◉◉◉◉◉◉◉◉◉◉ 52
  53 ◉◉◉◉◉◉◉◉◉◉◉◉◉      ◉◉◉◉◉◉◉◉◉◉◉◉◉ 78
  79 ◉◉◉◉◉◉◉◉◉◉◉◉◉      ◉◉◉◉◉◉◉◉◉◉◉◉◉ 104
```
Terminal View of 104-Pin PCM Wiring Harness Connector

Standard Colors and Abbreviations

Abbreviation	Color	Abbreviation	Color	Abbreviation	Color
BK	Black	GY	Gray	PK	Purple
BL	Blue	GN	Green	RD	Red
BR	Brown	LG	LT Green	TN	Tan
DB	Dark Blue	OR	Orange	WT	White
DG	DK Green	PK	Pink	YL	Yellow

2000-02 Grand Marquis 4.6L V8 VIN W (A/T) 104 Pin Connector

PCM Pin #	Wire Color	Circuit Description (104 Pin)	Value at Hot Idle
1	OR/YL	COP 6 Driver Control	5° dwell
2	PK/LG	MIL (lamp) Control	MIL On: 1v, Off: 12v
3	BK/WT	Power Ground	<0.1v
4-5	---	Not Used	---
6	OR/YL	Shift Solenoid 'A' Control	1v, 55 mph: 12v
7-8	---	Not Used	---
9	YL/WT	Fuel Level Indicator Signal	1.7v (1/2 full)
10	---	Not Used	
11	VT/OR	Shift Solenoid 'B' Control	12v, Cruise: 1v
12	WT/LG	TCIL (lamp) Control	Lamp On: 1v, Off: 5v
13	VT	Flash EEPROM Power	0.1v
15	PK/LB	Data Bus (-) Signal	Digital Signals
16	TN/OG	Data Bus (+) Signal	Digital Signals
17-20	---	Not Used	---
21	BK/PKK	CKP Sensor (+) Signal	440-490 Hz
22	GY/YL	CKP Sensor (-) Signal	440-490 Hz
23-24	---	Not Used	---
25	BK	PCM Case Ground	<0.050v
26	LG/WTT	COP 1 Driver Control	5° dwell
27	LG/YL	COP 5 Driver Control	1v, 55 mph: 12v
28	RD/OG	Low Speed Fan Control	On: 1v, Off: 12v
29	TN/WT	TCS (switch) Signal	TCS & O/D On: 12v
30-32	---	Not Used	---
33	PK/OR	VSS (-) Signal	0 Hz, 55 mph: 125 Hz
34	YL/BK	Digital TR1 Sensor	In 'P': 0v, 55 mph: 11v
35	RD/LG	HO2S-12 (B1 S2) Signal	0.1-1.1v
36	TN/LB	MAF Sensor Return	<0.050v
37	OG/BK	TFT Sensor Signal	2.10-2.40v
38	---	Not Used	---
39	GY	IAT Sensor Signal	1.5-2.5v
40	DG/YL	Fuel Pump Monitor	On: 12v, Off: 0v
41	BK/YL	A/C Pressure Cutout switch	AC On: 12v, Off: 0v
42	RD/WT	ECT Sensor Signal	0.5-0.6v
43	DB/LG	Fuel Flow Rate Signal	Digital Signals
45	OG/RD	CHTIL (lamp) Control	Lamp On: 1v, Off: 5v
46	LG/VT	High Speed Fan Control	On: 1v, Off: 12v
47	BR/VT	EGR VR Solenoid	0%, 55 mph: 45%
48	---	Not Used	---
49	LB/BK	Digital TR2 Sensor	In 'P': 0v, 55 mph: 11v
50	WT/BK	Digital TR4 Sensor	In 'P': 0v, 55 mph: 11v
51	BK/WT	Power Ground	<0.1v
52	WT/PK	COP 3 Driver Control	5° dwell
53	DG/VT	COP 4 Driver Control	5° dwell
54	VT/YL	TCC Solenoid Control	0%, 55 mph: 95%
55	YL/BK	Keep Alive Power	12-14v
56	LG/BK	EVAP Purge Solenoid	0-10 Hz (0-100%)
57	---	Not Used	---
58	GY/BK	VSS + Sensor Signal	0 Hz, 55 mph: 125 Hz
59	---	Not Used	---
60	GY/BK	HO2S-11 (B1 S1) Signal	0.1-1.1v
61	VT/LG	HO2S-22 (B2 S2) Signal	0.1-1.1v
62	RD/VT	FTP Sensor Signal	2.6v (0" H2O - cap off)
63	---	Not Used	---
64	LB/YL	Digital TR3 Sensor	In 'P': 0v, in O/D: 1.7v

2000-02 Grand Marquis 4.6L V8 VIN W (A/T) 104 Pin Connector

PCM Pin #	Wire Color	Circuit Description (104 Pin)	Value at Hot Idle
65 ('00)	BN/LG	DPFE Sensor Signal	0.20-1.30v
65 ('01-'02)	BN/LG	DPFE Sensor Signal	0.95-1.05v
66	YL/LB	CHT Sensor Signal	0.7v (194°F)
67	VT/WT	EVAP CV Solenoid	0-10 Hz (0-100%)
68	---	Not Used	---
69	PK/YL	A/C WOT Relay Control	On: 1v, Off: 12v
70, 82	---	Not Used	---
71	RD	Vehicle Power	12-14v
72	TN/RD	Injector 7 Control	3.4-3.7 ms
73	TN/BK	Injector 5 Control	3.4-3.7 ms
74	BN/YL	Injector 3 Control	3.4-3.7 ms
75	TN	Injector 1 Control	3.4-3.7 ms
77	BK/WT	Power Ground	<0.1v
78	PK/LB	COP 7 Driver Control	5° dwell
79	WT/RD	COP 8 Driver Control	5° dwell
80	LB/OR	Fuel Pump Control	On: 1v, Off: 12v
81	WT/YL	EPC Solenoid Control	9.5v (20 psi)
83	WT/LB	IAC Motor Control	34% duty cycle
84	DB/YL	OSS Sensor Signal	0 Hz
85	DB/OR	CMP Sensor Signal	6-7 Hz
86	WT/BK	A/C High Pressure Switch	Open: 12v, Closed: 0v
87	RD/BK	HO2S-21 (B2 S1) Signal	0.1-1.1v
88	LB/RD	MAF Sensor Signal	0.6v
89	GY/WT	TP Sensor Signal	0.53-1.27v
90	BN/WT	Reference Voltage	4.9-5.1v
91	GY/RD	Sensor Signal Return	<0.050v
92	LG	Brake Pedal Position Switch	Brake Off: 0v, On: 12v
93	RD/WT	HO2S-11 (B1 S1) Heater	On: 1v, Off: 12v
94	YL/LB	HO2S-21 (B2 S1) Heater	On: 1v, Off: 12v
95	WT/BK	HO2S-12 (B1 S2) Heater	On: 1v, Off: 12v
96	TN/YL	HO2S-22 (B2 S2) Heater	On: 1v, Off: 12v
98	LB	Injector 8 Control	3.4-3.7 ms
99	LG/OR	Injector 6 Control	3.4-3.7 ms
100	BN/LB	Injector 4 Control	3.4-3.7 ms
101	WT	Injector 2 Control	3.4-3.7 ms
102	---	Not Used	---
103	BK/WT	Power Ground	<0.1v
104	PK/WT	COP 2 Driver Control	5° dwell

Pin Connector Graphic

PCM 104-PIN CONNECTOR

Terminal View of 104-Pin PCM Wiring Harness Connector

Standard Colors and Abbreviations

Abbreviation	Color	Abbreviation	Color	Abbreviation	Color
BK	Black	GY	Gray	PK	Purple
BL	Blue	GN	Green	RD	Red
BR	Brown	LG	LT Green	TN	Tan
DB	Dark Blue	OR	Orange	WT	White
DG	DK Green	PK	Pink	YL	Yellow

2003 Grand Marquis 4.6L V8 VIN W (A/T) 104 Pin Connector

PCM Pin #	Wire Color	Circuit Description (104 Pin)	Value at Hot Idle
1	OR/YL	COP 6 Driver Control	5° dwell
2	PK/LG	MIL (lamp) Control	MIL On: 1v, Off: 12v
3	BK/WT	Power Ground	<0.1v
4-5, 14, 23	---	Not Used	---
6	OR/YL	Shift Solenoid 'A' Control	1v, 55 mph: 12v
7	YL/LG	Generator Regulator 'S' Terminal	0-130 Hz
8	RD/PK	Fuel Rail Temperature Sensor	1.7-3.5v (50-120°F)
9	YL/WT	Fuel Level Indicator Signal	1.7v (1/2 full)
10	LB/RD	MAP Sensor Signal	107 Hz
11	VT/OR	Shift Solenoid 'B' Control	12v, Cruise: 1v
12	WT/LG	TCIL (lamp) Control	Lamp On: 1v, Off: 5v
13	VT	Flash EEPROM Power	0.1v
15	PK/LB	SCP Bus (-) Signal	Digital Signals
16	TN/OR	SCP Bus (+) Signal	Digital Signals
17	GY/OR	RX Signal	Digital Signals
18	WT/LG	TX Signal	Digital Signals
19	OR/RD	Cylinder Head Temperature Lamp Control	Lamp Off: 12v, On: 1v
20	WT/LG	Fuel Door Release Solenoid Indicator	Solenoid Off: 12v, On: 1v
21	BK/PK	CKP Sensor (+) Signal	440-490 Hz
22	GY/YL	CKP Sensor (-) Signal	440-490 Hz
24, 51	BK/WT	Power Ground	<0.1v
25	BK	PCM Case Ground	<0.050v
26	LG/WT	COP 1 Driver Control	5° dwell
27	LG/YL	COP 5 Driver Control	1v, 55 mph: 12v
28	RD/OR	Low Speed Fan Control	On: 1v, Off: 12v
29	TN/WT	TCS (switch) Signal	TCS & O/D On: 12v
30	OG/RD	Light Emitting Diode Signal Ground	<0.050v
31	YL/LG	Power Steering Pressure Sensor	Straight: 0v, Turned: 12v
32	DG/VT	Knock Sensor Ground	<0.050v
33	---	Not Used	---
34	YL/BK	Digital TR1 Sensor	In 'P': 0v, 55 mph: 11v
35	RD/LG	HO2S-12 (B1 S2) Signal	0.1-1.1v
36	TN/LB	MAF Sensor Return	<0.050v
37	OR/BK	TFT Sensor Signal	2.10-2.40v
38	BK/WT	A/C Pressure Sensor Discharge Temp.	A/C On: 1.5-1.9v
39	GY	IAT Sensor Signal	1.5-2.5v
40	DG/YL	Fuel Pump Monitor	On: 12v, Off: 0v
41	PK/LB	A/C Pressure Cutout Switch	AC On: 12v, Off: 0v
42	RD/WT	ECT Sensor Signal	0.5-0.6v
43	DB/LG	Low Fuel Indicator	Digital Signals
44	GY/RD	Starter Relay Control	Relay Off: 12v, On: 9v
45	LG/RD	Generator Common	Digital Signals
46	---	Not Used	---
47	BR/PK	EGR VR Solenoid	0%, 55 mph: 45%
48	WT	Tachometer Output	DC pulse signals
49	LB/BK	Digital TR2 Sensor	In 'P': 0v, 55 mph: 11v
50	WT/BK	Digital TR4 Sensor	In 'P': 0v, 55 mph: 11v
52	WT/PK	COP 3 Driver Control	5° dwell
53	DG/VT	COP 4 Driver Control	5° dwell
54	VT/YL	TCC Solenoid Control	0%, 55 mph: 95%
55	YL/BK	Keep Alive Power	12-14v
56	LG/BK	EVAP Purge Solenoid Control	0-10 Hz (0-100%)
57	YL/RD	Knock Sensor Signal	0v
58-59	---	Not Used	---

2003 Grand Marquis 4.6L V8 VIN W (A/T) 104 Pin Connector

PCM Pin #	Wire Color	Circuit Description (104 Pin)	Value at Hot Idle
60	GY/BK	HO2S-11 (B1 S1) Signal	0.1-1.1v
61	VT/LG	HO2S-22 (B2 S2) Signal	0.1-1.1v
62	RD/PK	FTP Sensor Signal	2.6v (0" H2O - cap off)
63	OR/LG	Injection Pressure Sensor	2.8v (39 psi)
64	LB/YL	Digital TR3 Sensor	In 'P': 0v, in O/D: 1.7v
65	BR/LG	DPFE Sensor Signal	0.95-1.05v
66	YL/LG	Cylinder Head Temperature Sensor	0.7v (194°F)
67	VT/WT	EVAP Canister Vent Solenoid	0-10 Hz (0-100%)
68	GY/BK	Vehicle Speed Sensor (+) Signal	0 Hz, 55 mph: 125 Hz
69	PK/YL	A/C WOT Relay Control	On: 1v, Off: 12v
70	YL	Generator Battery Indicator Control	Lamp Off: 12v, On: 1v
71	RD	Vehicle Power (Start-Run)	12-14v
72	TN/RD	Injector 7 Control	3.4-3.7 ms
73	TN/BK	Injector 5 Control	3.4-3.7 ms
74	BR/YL	Injector 3 Control	3.4-3.7 ms
75	TN	Injector 1 Control	3.4-3.7 ms
76	BK/WT	Power Ground	<0.1v
77	---	Not Used	---
78	PK/LB	COP 7 Driver Control	5° dwell
79	WT/RD	COP 8 Driver Control	5° dwell
80	LB/OR	Fuel Pump Control	On: 1v, Off: 12v
81	WT/YL	EPC Solenoid Control	9.5v (20 psi)
82	---	Not Used	---
83	WT/LB	IAC Motor Control	34% duty cycle
84	DB/YL	OSS Sensor Signal	0 Hz
85	DB/OR	CMP Sensor Signal	6-7 Hz
86	WT/BK	A/C High Pressure Switch	Open: 12v, Closed: 0v
87	RD/BK	HO2S-21 (B2 S1) Signal	0.1-1.1v
88	LB/RD	MAF Sensor Signal	0.6v
89	GY/WT	TP Sensor Signal	0.53-1.27v
90	BR/WT	Reference Voltage	4.9-5.1v
91	GY/RD	Analog Signal Return	<0.050v
92	DG	Brake Pedal Position Switch	Brake Off: 0v, On: 12v
93	RD/WT	HO2S-11 (B1 S1) Heater	On: 1v, Off: 12v
94	YL/LB	HO2S-21 (B2 S1) Heater	On: 1v, Off: 12v
95	WT/BK	HO2S-12 (B1 S2) Heater	On: 1v, Off: 12v
96	TN/YL	HO2S-22 (B2 S2) Heater	On: 1v, Off: 12v
97	RD	Vehicle Power (Start-Run)	12-14v
98	LB	Injector 8 Control	3.4-3.7 ms
99	LG/OR	Injector 6 Control	3.4-3.7 ms
100	BR/LB	Injector 4 Control	3.4-3.7 ms
101	WT	Injector 2 Control	3.4-3.7 ms
102	---	Not Used	---
103	BK/WT	Power Ground	<0.1v
104	PK/WT	COP 2 Driver Control	5° dwell

Pin Connector Graphic

Terminal View of 104-Pin PCM Wiring Harness Connector

1990-91 Grand Marquis 5.0L V8 MFI VIN F (A/T) 60 Pin Connector

PCM Pin #	Wire Color	Circuit Description (60 Pin)	Value at Hot Idle
1	BK/OR	Keep Alive Power	12-14v
1	YL	Keep Alive Power	12-14v
2 ('90-'91)	LG	Brake Switch Signal	Brake Off: 0v, On: 12v
3 ('90-'91)	DG/WT	VSS (+) Signal	0 Hz, 55 mph: 125 Hz
4	DG/YT	Ignition Diagnostic Monitor	20-31 Hz
4	BL/PK	Ignition Diagnostic Monitor	20-31 Hz
4	TN/YL	Ignition Diagnostic Monitor	20-31 Hz
5	---	Not Used	---
6 ('90-'91)	BK/WT	VSS (-) Signal	0 Hz, 55 mph: 125 Hz
6 ('90-'91)	GY/RD	VSS (-) Signal	0 Hz, 55 mph: 125 Hz
7	LG/YL	ECT Sensor Signal	0.5-0.6v
7	LG/RD	ECT Sensor Signal	0.5-0.6v
8-9	---	Not Used	---
10	LG/PK	A/C Cycling Clutch Signal	A/C Off: 0v, On: 12v
10	PK/BL	A/C Cycling Clutch Signal	A/C Off: 0v, On: 12v
11	LG/BK	Air Management 2 Solenoid	On: 1v, Off: 12v
12	TN/YL	Injector 3 Control	5.7-6.2 ms
12	BR/YL	Injector 3 Control	5.7-6.2 ms
13	TN/BK	Injector 4 Control	5.7-6.2 ms
13	BR/BL	Injector 4 Control	5.7-6.2 ms
14	TN/BL	Injector 5 Control	5.7-6.2 ms
14	TN	Injector 5 Control	5.7-6.2 ms
15	TN/LG	Injector 6 Control	5.7-6.2 ms
15	LG/OR	Injector 6 Control	5.7-6.2 ms
16	BK/OR	Ignition System Ground	<0.050v
16	OR	Ignition System Ground	<0.050v
17	YL	Self-Test Output, MIL Control	MIL Off: 12v, On: 1v
17	TN/RD	Self-Test Output, MIL Control	MIL Off: 12v, On: 1v
18-19	---	Not Used	---
20	BK	Case Ground	<0.050v
21	WT/BL	IAC Motor Control	9.0-11.5v
22	TN/LG	Fuel Pump Control	Relay Off: 12v, On: 1v
23-24	---	Not Used	---
25	LG/PK	ACT Sensor Signal	1.5-2.5v
26	OR/WT	Reference Voltage	4.9-5.1v
27	BR/LG	EVP Sensor Signal	0.4v, 55 mph: 1.5v
28, 32	---	Not Used	---
29	DG/PK	HO2S-11 (B1 S1) Signal	0.1-1.1v
29	RD/BK	HO2S-11 (B1 S1) Signal	0.1-1.1v
30	WT/PK	Neutral Drive Switch	In 'N': 0v, Others: 5v
30	BL/YL	Neutral Drive Switch	In 'N': 0v, Others: 5v
31	GY/YL	Canister Purge Solenoid	12v, 55 mph: 1v
33	DG	VR Solenoid Control	0%, 55 mph: 0-45%
33	BR/PK	VR Solenoid Control	0%, 55 mph: 0-45%
34	BL/P	Data Output Link	Digital Signals
34	BL/PK	Data Output Link	Digital Signals
35 ('90-'91)	WT/PK	S/C Vent Solenoid	Vacuum Decreasing: 1v
35 ('90-'91)	GY/BK	S/C Vent Solenoid	Vacuum Decreasing: 1v
36	YL/LG	Spark Output Signal	6.93v (50% d/cycle)
37, 57	RD	Vehicle Power	12-14v
38	GY/BK	S/C Vacuum Solenoid Control	Vacuum Increasing: 1v
38	LG	S/C Vacuum Solenoid Control	Vacuum Increasing: 1v
39 ('90-'91)	GY/BK	S/C Switch Ground	<0.050v
40	BK/LG	Power Ground	<0.1v

1990-91 Grand Marquis 5.0L V8 MFI VIN F (A/T) 60 Pin Connector

PCM Pin #	Wire Color	Circuit Description (60 Pin)	Value at Hot Idle
41 ('90-'91)	OR/YL	S/C Servo (+) Signal	Solenoid On: 12v
42	TN/OR	Injector 7 Control	5.7-6.2 ms
42	TN/RD	Injector 7 Control	5.7-6.2 ms
43	DB/LG	HO2S-21 (B1 S1) Signal	0.1-1.1v
44	---	Not Used	---
45	DB/LG	MAP Sensor Signal	107 Hz (sea level)
45	LG/BK	MAP Sensor Signal	107 Hz (sea level)
46	BK/WT	Analog Signal Return	<0.050v
46	GY/RD	Analog Signal Return	<0.050v
47	DG/LG	TP Sensor Signal	0.8v, 55 mph: 1.1v
47	GY/WT	TP Sensor Signal	0.8v, 55 mph: 1.1v
48	WT/RD	Self-Test Input Signal	STI On: 1v, Off: 5v
48	WT/BK	Self-Test Input Signal	STI On: 1v, Off: 5v
48	BR	Self-Test Input Signal	STI On: 1v, Off: 5v
49	OR	HO2S-11 (Bank 1) Ground	<0.050v
50 ('90-'91)	OR/RD	Speed Control Switch	6.7v
50 ('90-'91)	BL/BK	Speed Control Switch	6.7v
51	WT/RD	Air Management 1 Solenoid	AM1 On: 1v, Off: 12v
52	BL	Injector 8 Control	5.7-6.2 ms
52	TN/RD	Injector 8 Control	5.7-6.2 ms
53	---	Not Used	---
54	OR/BL	A/C WOT Relay Control	Relay Off: 12v, On: 1v
55	---	Not Used	---
56	DB	PIP Sensor Signal	6.93v (50% d/cycle)
57	BK/YL	Vehicle Power	12-14v
58	TN	Injector 1 Control	5.7-6.2 ms
59	TN/WT	Injector 2 Control	5.7-6.2 ms
59	WT	Injector 2 Control	5.7-6.2 ms
60	BK/LG	Power Ground	<0.1v

Pin Connector Graphic

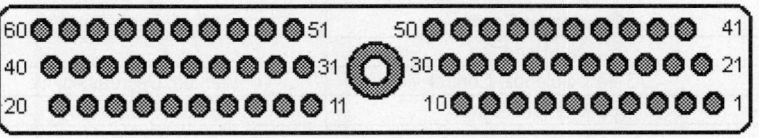

PCM 60-PIN CONNECTOR

Terminal View of 60-Pin PCM Harness Connector

Standard Colors and Abbreviations

Abbreviation	Color	Abbreviation	Color	Abbreviation	Color
BK	Black	GY	Gray	PK	Purple
BL	Blue	GN	Green	RD	Red
BR	Brown	LG	LT Green	TN	Tan
DB	Dark Blue	OR	Orange	WT	White
DG	DK Green	PK	Pink	YL	Yellow

1990-91 Grand Marquis 5.0L V8 VIN F California 60 Pin Connector

PCM Pin #	Wire Color	Circuit Description (60 Pin)	Value at Hot Idle
1	BK/OR	Keep Alive Power	12-14v
1	YL	Keep Alive Power	12-14v
2	---	Not Used	---
3	DG/WT	VSS (+) Signal	0 Hz, 55 mph: 125 Hz
4	DG/YT	Ignition Diagnostic Monitor	20-31 Hz
5	LG	Brake On/Off Signal	Brake Off: 0v, On: 12v
6	BK/WT	VSS (-) Signal	0 Hz, 55 mph: 125 Hz
7	LG/YL	ECT Sensor Signal	0.5-0.6v
8	BK/OR	Data Bus (-) Signal	Digital Signals
9	---	Not Used	---
10	PK/BL	A/C Cycling Clutch Signal	A/C Off: 0v, On: 12v
11	OR/YL	Air Management 2 Solenoid	On: 1v, Off: 12v
12	BR/YL	Injector 3 Control	5.7-6.2 ms
13	BR/BL	Injector 4 Control	5.7-6.2 ms
14	TN/BL	Injector 5 Control	5.7-6.2 ms
15	LG	Injector 6 Control	5.7-6.2 ms
16	BK/OR	Ignition System Ground	<0.050v
17	TN/RD	Self-Test Output, MIL Control	MIL Off: 12v, On: 1v
18	---	Not Used	---
19	OR	Fuel Pump Monitor	12v, Off: 0v
20	BK	Case Ground	<0.050v
21	WT/BL	IAC Motor Control	9.0-11.5v
22	TN/LG	Fuel Pump Control	Relay Off: 12v, On: 1v
23-24	---	Not Used	---
25	LG/PK	ACT Sensor Signal	1.5-2.5v
26	OR/WT	Reference Voltage	4.9-5.1v
27	BR/LG	EVP Sensor Signal	0.4v, 55 mph: 1.5v
28	BL/BK	Speed Control Switch Signal	6.7v
29	DG/PK	HO2S-11 (B1 S1) Signal	0.1-1.1v
30	WT/PK	Neutral Drive Switch	In 'P': 0v, Others: 5v
31	GY/YL	Canister Purge Solenoid	12v, 55 mph: 1v
32	LG/BK	Thermactor Air Diverter SOL	On: 1v, Off: 12v
33	DG	VR Solenoid Control	0%, 55 mph: 0-45%
34	---	Not Used	---
35	WT/PK	S/C Vent Solenoid Control	Vacuum Decreasing: 1v
36	YL/LG	Spark Output Signal	6.93v (50% d/cycle)
37	RD	Vehicle Power	12-14v
38	WT/RD	Thermactor Air Bypass SOL	On: 1v, Off: 12v
39	GY/BK	Speed Control Switch Ground	<0.050v
40	BK/LG	Power Ground	<0.1v

1990-91 Grand Marquis 5.0L V8 VIN F California 60 Pin Connector

PCM Pin #	Wire Color	Circuit Description (60 Pin)	Value at Hot Idle
41	---	Not Used	---
42	TN/OR	Injector 7 Control	5.7-6.2 ms
43	DB/LG	HO2S-21 (B1 S1) Signal	0.1-1.1v
44	OR/BK	Data Bus (+) Signal	Digital Signals
45	LG/BK	MAP Sensor Signal	107 Hz (sea level)
46	BK/WT	Analog Signal Return	<0.050v
47	DG/LG	TP Sensor Signal	0.8v, 55 mph: 1.1v
48	WT/BK	Self-Test Input Signal	STI On: 1v, Off: 5v
49	OR	HO2S-11 (Bank 1) Ground	<0.050v
50	---	Not Used	---
51	GY/BK	Air Management 1 Solenoid	AM1 On: 1v, Off: 12v
52	BL	Injector 8 Control	5.7-6.2 ms
53	---	Not Used	---
54	OR/BL	A/C WOT Relay Control	Relay Off: 12v, On: 1v
55	---	Not Used	---
56	DB	PIP Sensor Signal	6.93v (50% dwell)
57	RD	Vehicle Power	12-14v
58	TN	Injector 1 Control	5.7-6.2 ms
59	WT	Injector 2 Control	5.7-6.2 ms
60	BK/LG	Power Ground	<0.1v

Pin Connector Graphic

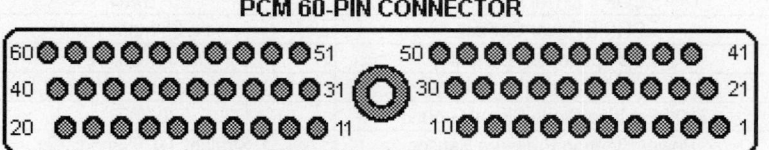

Standard Colors and Abbreviations

Abbreviation	Color	Abbreviation	Color	Abbreviation	Color
BK	Black	GY	Gray	PK	Purple
BL	Blue	GN	Green	RD	Red
BR	Brown	LG	LT Green	TN	Tan
DB	Dark Blue	OR	Orange	WT	White
DG	DK Green	PK	Pink	YL	Yellow

Marauder Pin Tables

2003 Marauder 4.6L V8 MPI VIN V 104 Pin Connector

PCM Pin #	Wire Color	Circuit Description (104 Pin)	Value at Hot Idle
1	OR/YL	COP 6 Driver Control	5° dwell
2	PK/LG	MIL (lamp) Control	MIL On: 1v, Off: 12v
3	BK/WT	Power Ground	<0.1v
4-5, 14, 23	---	Not Used	---
6	OR/YL	Shift Solenoid 'A' Control	1v, 55 mph: 12v
7	YL/LG	Generator Regulator 'S' Terminal	0-130 Hz
8	RD/PK	Fuel Rail Temperature Sensor	1.7-3.5v (50-120°F)
9	YL/WT	Fuel Level Indicator Signal	1.7v (1/2 full)
10	LB/RD	MAP Sensor Signal	107 Hz
11	VT/OR	Shift Solenoid 'B' Control	12v, Cruise: 1v
12	WT/LG	TCIL (lamp) Control	Lamp On: 1v, Off: 5v
13	VT	Flash EEPROM Power	0.1v
15	PK/LB	SCP Bus (-) Signal	Digital Signals
16	TN/OR	SCP Bus (+) Signal	Digital Signals
17	GY/OR	RX Signal	Digital Signals
18	WT/LG	TX Signal	Digital Signals
19	OR/RD	Cylinder Head Temperature Lamp Control	Lamp Off: 12v, On: 1v
20	WT/LG	Fuel Door Release Solenoid Indicator	Solenoid Off: 12v, On: 1v
21	BK/PK	CKP Sensor (+) Signal	440-490 Hz
22	GY/YL	CKP Sensor (-) Signal	440-490 Hz
24	BK/WT	Power Ground	<0.1v
25	BK	PCM Case Ground	<0.050v
26	LG/WT	COP 1 Driver Control	5° dwell
27	LG/YL	COP 5 Driver Control	1v, 55 mph: 12v
28	RD/OR	Low Speed Fan Control	On: 1v, Off: 12v
29	TN/WT	TCS (switch) Signal	TCS & O/D On: 12v
30	OR/RD	Light Emitting Diode Signal Ground	<0.050v
31	YL/LG	Power Steering Pressure Sensor	Straight: 0v, Turned: 12v
32	DG/VT	Knock Sensor Ground	<0.050v
33	---	Not Used	---
34	YL/BK	Digital TR1 Sensor	In 'P': 0v, 55 mph: 11v
35	RD/LG	HO2S-12 (B1 S2) Signal	0.1-1.1v
36	TN/LB	MAF Sensor Return	<0.050v
37	OR/BK	TFT Sensor Signal	2.10-2.40v
38	BK/WT	A/C Pressure Sensor Discharge Temp.	A/C On: 1.5-1.9v
39	GY	IAT Sensor Signal	1.5-2.5v
40	DG/YL	Fuel Pump Monitor	On: 12v, Off: 0v
41	PK/LB	A/C Pressure Cutout Switch	AC On: 12v, Off: 0v
42	RD/WT	ECT Sensor Signal	0.5-0.6v
43	DB/LG	Low Fuel Indicator	Digital Signals
44	GY/RD	Starter Relay Control	Relay Off: 12v, On: 9v
45	LG/RD	Generator Common	Digital Signals
46	---	Not Used	---
47	BR/PK	EGR VR Solenoid	0%, 55 mph: 45%
48	LG/WT	Tachometer Output	DC pulse signals
49	LB/BK	Digital TR2 Sensor	In 'P': 0v, 55 mph: 11v
50	WT/BK	Digital TR4 Sensor	In 'P': 0v, 55 mph: 11v
51	BK/WT	Power Ground	<0.1v
52	WT/PK	COP 3 Driver Control	5° dwell
53	DG/VT	COP 4 Driver Control	5° dwell
54	VT/YL	TCC Solenoid Control	0%, 55 mph: 95%
55	YL/BK	Keep Alive Power	12-14v
56	LG/BK	EVAP Purge Solenoid Control	0-10 Hz (0-100%)
57	YL/RD	Knock Sensor Signal	0v

2003 Marauder 4.6L V8 MPI VIN V 104 Pin Connector

PCM Pin #	Wire Color	Circuit Description (104 Pin)	Value at Hot Idle
58-59	---	Not Used	---
60	GY/BK	HO2S-11 (B1 S1) Signal	0.1-1.1v
61	VT/LG	HO2S-22 (B2 S2) Signal	0.1-1.1v
62	RD/PK	FTP Sensor Signal	2.6v (0" H2O - cap off)
63	OR/LG	Injection Pressure Sensor	2.8v (39 psi)
64	LB/YL	Digital TR3 Sensor	In 'P': 0v, in O/D: 1.7v
65	BR/LG	DPFE Sensor Signal	0.95-1.05v
66	LG/RD	Engine Coolant Temperature Sensor	0.7v (194ºF)
67	VT/WT	EVAP Canister Vent Solenoid	0-10 Hz (0-100%)
68	GY/BK	Vehicle Speed Sensor (+) Signal	0 Hz, 55 mph: 125 Hz
69	PK/YL	A/C WOT Relay Control	On: 1v, Off: 12v
70	YL	Generator Battery Indicator Control	Lamp Off: 12v, On: 1v
71	RD	Vehicle Power (Start-Run)	12-14v
72	TN/RD	Injector 7 Control	3.4-3.7 ms
73	TN/BK	Injector 5 Control	3.4-3.7 ms
74	BR/YL	Injector 3 Control	3.4-3.7 ms
75	TN	Injector 1 Control	3.4-3.7 ms
76	BK/WT	Power Ground	<0.1v
77	---	Not Used	---
78	PK/LB	COP 7 Driver Control	5º dwell
79	WT/RD	COP 8 Driver Control	5º dwell
80	LB/OR	Fuel Pump Control	On: 1v, Off: 12v
81	WT/YL	EPC Solenoid Control	9.5v (20 psi)
82	---	Not Used	---
83	WT/LB	IAC Motor Control	34% duty cycle
84	DB/YL	OSS Sensor Signal	0 Hz
85	DB/OR	CMP Sensor Signal	6-7 Hz
86	WT/BK	A/C High Pressure Switch	Open: 12v, Closed: 0v
87	RD/BK	HO2S-21 (B2 S1) Signal	0.1-1.1v
88	LB/RD	MAF Sensor Signal	0.6v
89	GY/WT	TP Sensor Signal	0.53-1.27v
90	BR/WT	Reference Voltage	4.9-5.1v
91	GY/RD	Analog Signal Return	<0.050v
92	DG	Brake Pedal Position Switch	Brake Off: 0v, On: 12v
93	RD/WT	HO2S-11 (B1 S1) Heater	On: 1v, Off: 12v
94	YL/LB	HO2S-21 (B2 S1) Heater	On: 1v, Off: 12v
95	WT/BK	HO2S-12 (B1 S2) Heater	On: 1v, Off: 12v
96	TN/YL	HO2S-22 (B2 S2) Heater	On: 1v, Off: 12v
97	RD	Vehicle Power (Start-Run)	12-14v
98	LB	Injector 8 Control	3.4-3.7 ms
99	LG/OR	Injector 6 Control	3.4-3.7 ms
100	BR/LB	Injector 4 Control	3.4-3.7 ms
101	WT	Injector 2 Control	3.4-3.7 ms
102	---	Not Used	---
103	BK/WT	Power Ground	<0.1v
104	PK/WT	COP 2 Driver Control	5º dwell

Pin Connector Graphic

PCM 104-PIN CONNECTOR

Terminal View of 104-Pin PCM Wiring Harness Connector

Mystique Pin Tables

1995 Mystique 2.0L I4 MFI VIN 3 (All) 60 Pin Connector

PCM Pin #	Wire Color	Circuit Description (60 Pin)	Value at Hot Idle
1	OR/YL	Keep Alive Power	12- 14v
2	OR	Brake On/Off Signal	Brake Off: 0v, On: 12v
3	WT/PK	VSS (+) Signal	0 Hz, 55 mph: 125 Hz
4	WT/GN	Ignition Diagnostic Monitor	20-31 Hz
5	WT/PK	A/T: TSS Sensor (+) Signal	42-50 Hz (680-720 rpm)
6	---	Not Used	---
7	WT/GN	ECT Sensor Signal	0.5-0.6v
8	PK/BK	Fuel Pump Monitor	Pump On: 12v, Off; 0v
9	BR/BL	MAF Sensor Return	<0.050v
10	PK/BL	A/C Cycling Clutch Switch	A/C Switch On: 12v
11	BK/OR	EVAP Canister Purge Control	12v, 55 mph: 1v
12	---	Not Used	---
13	BK/BL	Low Speed Fan Control	Fan On: 1v, Off: 12v
14-15	---	Not Used	---
16	BK/BL	Ignition Ground	<0.050v
17	BK/OR	STI Output, MIL Control	MIL On: 1v, Off: 12v
18	WT/BK	Data Bus (+) Signal	Digital Signals
19	BN/BK	Data Bus (-) Signal	Digital Signals
20	BK/RD	Case Ground	<0.050v
21	BK/OR	IAC Solenoid Control	Varies: 8-11v
22	BK/BL	Fuel Pump Control	Relay Off: 12v, On: 1v
23	WT/BK	Knock Sensor Signal	No Knock: 2.5v
24	WT/PK	Camshaft Position Sensor	5-7 Hz
25	WT/PK	IAT Sensor Signal	1.5-2.5v
26	YL	Reference Voltage	4.9-5.1v
27	WT/BL	DPFE Sensor Signal	0.20-1.30v
28	WT/PK	PSP Switch Signal	Straight: 12v, Turned: 0v
29	WT/BK	Octane Adjust Switch Signal	9.3v (Closed: 0v)
30	WT	A/T: PNP Sensor Signal	In P/N: 0v, 55 mph: 1.7v
30	WT	M/T: Clutch Pedal Position	Clutch In: 0v, Out: 5v
31	BK/WT	High Speed Fan Control	Fan On: 1v, Off: 12v
32	---	Not Used	---
33	GN/WT	VR Solenoid Control	12v, 30 mph: 10-12v
34	---	Not Used	---
35	BK/OR	Injector 4 Control	2.3-2.9 ms
36	WT/PK	Spark Output Signal	50% duty cycle
37	GN/YL	Vehicle Power	12-14v
38	BK/RD	A/T: EPC Solenoid Control	7-8v (8-9 psi)
39	BK/BL	Injector 3 Control	2.3-2.9 ms
40	BK/YL	Power Ground	<0.1v

1995 Mystique 2.0L I4 MFI VIN 3 (All) 60 Pin Connector

PCM Pin #	Wire Color	Circuit Description (60 Pin)	Value at Hot Idle
41	WT/BK	A/T: TCS (switch) Signal	TCS & O/D On: 12v
42-43	---	Not Used	---
44	WT	HO2S-11 (B1 S1) Signal	0.1-1.1v
45	---	Not Used	---
46	BN	Analog Signal Return	<0.050v
47	WT	TP Sensor Signal	0.5-1.1v
48	WT/BK	Self-Test Input	STI On: 0.1v, Off: 5v
49	WT/RD	A/T: TFT Sensor Signal	2.40-2.10v
50	WT/BL	MAF Sensor Signal	1.9v
51	BK/YL	A/T: Shift Solenoid 1 Control	12v, 55 mph: 1v
52	BK/BL	A/T: Shift Solenoid 2 Control	1v, 55 mph: 12v
53	BK/WT	A/T: TCC Solenoid Control	TCC On: 1v, Off: 12v
54	BK/YL	A/C WOT Relay Control	Relay Off: 12v, On: 1v
55	BK/OR	A/T: Shift Solenoid 3 Control	12v, 55 mph: 1v
56	WT/BK	PIP Sensor Signal	50% duty cycle
57	GN/YL	Vehicle Power	12-14v
58	BK/WT	Injector 1 Control	2.3-2.9 ms
59	BK/YL	Injector 2 Control	2.3-2.9 ms
60	BK/YL	Power Ground	<0.1v

Pin Connector Graphic

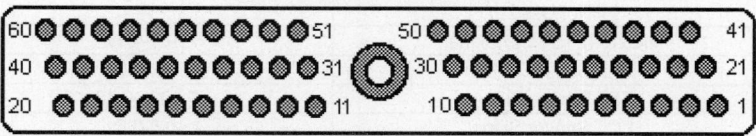

PCM 60-PIN CONNECTOR

Terminal View of 60-Pin PCM Harness Connector

Standard Colors and Abbreviations

Abbreviation	Color	Abbreviation	Color	Abbreviation	Color
BK	Black	GY	Gray	PK	Purple
BL	Blue	GN	Green	RD	Red
BR	Brown	LG	LT Green	TN	Tan
DB	Dark Blue	OR	Orange	WT	White
DG	DK Green	PK	Pink	YL	Yellow

1996-97 Mystique 2.0L I4 MFI VIN 3 (All) 104 Pin Connector

PCM Pin #	Wire Color	Circuit Description (104 Pin)	Value at Hot Idle
1	BK/BL	Shift Solenoid 2 Control	12v, 55 mph: 1v
2	BK/OR	MIL (lamp) Control	MIL On: 1v, Off: 12v
3-12	---	Not Used	---
13	WT/BL	Flash EEPROM Power	0.5v
14	---	Not Used	---
15	BR/BL	Data Bus (-) Signal	Digital Signals
16	WT/BL	Data Bus (+) Signal	Digital Signals
17-20	---	Not Used	---
21	WT/RD	CKP Sensor (+) Signal	450-480 Hz
22	BR/WT	CKP Sensor (-) Signal	450-480 Hz
23	---	Not Used	---
24	BK/YL	Power Ground	<0.1v
25	BK/RD	Chassis Ground	<0.050v
26	BR/BL	Coil Driver 1 Control	5° dwell
27	BK/YL	Shift Solenoid 1 Control	1v, 55 mph: 12v
28	---	Not Used	---
29	PK/BK	TCS (switch) Signal	TCS & O/D On: 12v
30	WT/BK	Octane Adjust Switch Signal	9.3v (Closed: 0v)
31	WT/PK	PSP Sensor Signal	Straight: 1v, Turning: 12v
31	WT	PSP Sensor Signal	Straight: 1v, Turning: 12v
32-34	---	Not Used	---
35	WT/BL	HO2S-12 (B1 S2) Signal	0.1-1.1v
36	BR/BL	MAF Sensor Return	<0.050v
37	WT/RD	TFT Sensor Signal	2.10 to 0.40v
38	WT/GN	ECT Sensor Signal	0.5-0.6v
39	WT/PK	IAT Sensor Signal	1.5-2.5v
40	PK/BK	Fuel Pump Monitor	12v, Off: 0v
41	PK/BL	A/C Cycling Clutch Switch	Closed: 12v, Open: 0v
42-44	---	Not Used	---
45	BK/BL	Low Speed Fan Control	Fan On: 1v, Off: 12v
46	BK/WT	High Fan Control	Fan On: 1v, Off: 12v
47	BK/GN	VR Solenoid Control	0 Hz, 30 mph: 40 Hz
48	WT/BK	Clean Tachometer Output	20-31 Hz
49-50	---	Not Used	---
51	BK/YL	Power Ground	<0.1v
52	BR/GN	Coil Driver 2 Control	5° dwell
53	---	Not Used	---
54	BK/WT	TCC Solenoid Control	0%, 30 mph: 80%
55	OR/YL	Keep Alive Power	12-14v
56	BK/OR	EVAP VMV Solenoid	0-10 Hz (0-100%)
57	WT/BK	Knock Sensor 1 Signal	0v
58	WT/PK	VSS (+) Signal	0 Hz, 55 mph: 125 Hz
59	---	Not Used	---
60	WT	HO2S-11 (B1 S1) Signal	0.1-1.1v
61-63	---	Not Used	---
64	WT/GN	A/T: TR Sensor Signal	In 'P': 0v, O/D: 1.7v
64	WT	M/T: Clutch Pedal Position	Clutch In: 0v, Out: 5v

1996-97 Mystique 2.0L I4 MFI VIN 3 (All) 104 Pin Connector

PCM Pin #	Wire Color	Circuit Description (104 Pin)	Value at Hot Idle
65	WT/BL	DPFE Sensor Signal	0.20-1.30v
66-68	---	Not Used	---
69	BK/YL	A/C WOT Relay Control	Relay Off: 12v, On: 1v
70	BK/YL	ZETECT VCT Control	VCT Off: 12v, On: 1v
71	GN/YL	Vehicle Power	12-14v
72-73	---	Not Used	---
74	BK/BL	Injector 3 Control	2.3-2.9 ms
75	BK/WT	Injector 1 Control	2.3-2.9 ms
76	BK/YL	CMP and TSS Sensor Ground	<0.050v
77	BK/YL	Power Ground	<0.1v
78, 82	---	Not Used	---
79	WT/BL	TCIL (lamp) Control	Lamp On: 1v, Off: 12v
80	BK/BL	Fuel Pump Control	Pump On: 1v, Off: 12v
81	BK/RD	EPC Solenoid Control	8.9v
83	BK/YL	IAC Motor Control	9v (33% d/cycle)
84	WT/PK	TSS Sensor Signal	40-56 Hz
85	WT/PK	CID Sensor Signal	5-7 Hz
86	BK/WT	A/C Pressure Switch	A/C On: 12v (Open)
87	---	Not Used	---
88	WT/BL	MAF Sensor Signal	0.8v
89	WT	TP Sensor Signal	0.53-1.27v
90	YL	Reference Voltage	4.9-5.1v
91	BR	Analog Signal Return	<0.050v
92	O	Brake Pedal Position Switch	Brake Off: 0v, On: 12v
93	BK/YL	HO2S-11 (B1 S1) Heater	1v, Off: 12v
94	---	Not Used	---
95	BK/OR	HO2S-12 (B1 S2) Heater	1v, Off: 12v
96	---	Not Used	---
97	GN/YL	Vehicle Power	12-14v
98-99	---	Not Used	---
100	BK/OR	Injector 4 Control	2.3-2.9 ms
101	BK/YL	Injector 2 Control	2.3-2.9 ms
102	BK/OR	Shift Solenoid 3 Control	8v, 55 mph: 8-9v
103	BK/YL	Power Ground	<0.1v
104	---	Not Used	---

Pin Connector Graphic

Standard Colors and Abbreviations

Abbreviation	Color	Abbreviation	Color	Abbreviation	Color
BK	Black	GY	Gray	PK	Purple
BL	Blue	GN	Green	RD	Red
BR	Brown	LG	LT Green	TN	Tan
DB	Dark Blue	OR	Orange	WT	White
DG	DK Green	PK	Pink	YL	Yellow

1998-2000 Mystique 2.0L I4 MFI VIN 3 (All) 104 Pin Connector

PCM Pin #	Wire Color	Circuit Description (104 Pin)	Value at Hot Idle
1	BK/BL	Shift Solenoid 2	12v, 55 mph: 1v
2	BK/OR	MIL (lamp) Control	MIL On: 1v, Off: 12v
3-11, 14	---	Not Used	---
12	WT	Fuel Level Indicator Signal	1.7v (1/2 full)
13	WT/BL	Flash EEPROM Power	0.5v
15	BL	Data Bus (-) Signal	Digital Signals
16	GY	Data Bus (+) Signal	Digital Signals
17-19, 23	---	Not Used	---
20	WT/BK	Transmission Indicator Signal	In 'P': 0v, O/D: 1.7v
21	WT/RD	CKP Sensor (+) Signal	450-480 Hz
22	BR/WT	CKP Sensor (-) Signal	450-480 Hz
22	BR/RD	CKP Sensor (-) Signal	450-480 Hz
24	BK/YL	Power Ground	<0.1v
25	BK/RD	Chassis Ground	<0.050v
26	BR/BL	Coil Driver 1 Control	5° dwell
27	BK/YL	Shift Solenoid 1 Control	1v, 55 mph: 12v
28	WT/BL	Speedometer Indicator Signal	Digital Signals
29	PK/BK	TCS (switch) Signal	TCS & O/D On: 12v
29	GN/BK	TCS (switch) Signal	TCS & O/D On: 12v
30	WT/BK	Octane Adjust Switch	9.3v (Closed: 0v)
31	WT/PK	PSP Sensor Signal	Straight: 1v, Turning: 12v
31	WT	PSP Sensor Signal	Straight: 1v, Turning: 12v
33	WT/PK	VSS (-) Signal	0 Hz, 55 mph: 125 Hz
34	WT/PK	TSS Sensor Signal (2000)	40-56 Hz
35	WT/BL	HO2S-12 (B1 S2) Signal	0.1-1.1v
36	BR/BL	MAF Sensor Return	<0.050v
37	WT/RD	TFT Sensor Signal	2.10 to 0.40v
38	WT/GN	ECT Sensor Signal	0.5-0.6v
39	WT/PK	IAT Sensor Signal	1.5-2.5v
40	PK/BK	Fuel Pump Monitor	12v, Off: 0v
40	GN/BK	Fuel Pump Monitor	12v, Off: 0v
41	PK/BL	A/C Cycling Clutch Switch	Closed: 12v, Open: 0v
41	GN/BL	A/C Cycling Clutch Switch	Closed: 12v, Open: 0v
44	BK/RD	VCT Actuator	VCT Off: 12v, On: 1v
45	BK/BL	Low Speed Fan Control	Fan On: 1v, Off: 12v
46	BK/WT	High Fan Control	Fan On: 1v, Off: 12v
47	BK/GN	VR Solenoid Control	12v, 30 mph: 10-12v
48	WT/BK	Clean Tachometer Output	20-31 Hz
50	---	Not Used	---
51	BK/YL	Power Ground	<0.1v
52	BR/GN	Coil Driver 2 Control	5° dwell
54	BK/WT	TCC Solenoid Control	0%, 30 mph: 80%
55	OR/YL	Keep Alive Power	12-14v
55	RD	Keep Alive Power	12-14v
56	BK/OR	EVAP Purge Valve	0-10 Hz (0-100%)
57	WT/BK	Knock Sensor (+) Signal	0v
58	WT/BK	VSS (+) Signal	0 Hz, 55 mph: 125 Hz
58	WT/PK	VSS (+) Signal	0 Hz, 55 mph: 125 Hz
59 ('20)	WT/GN	Generator Load Indicator	1.5-10.5v (40-250 Hz)
60	WT	HO2S-11 (B1 S1) Signal	0.1-1.1v
62	WT/PK	FTP Sensor Signal	2.6v (0" H2O - cap off)
63	WT/PK	FRP Sensor Signal	2.8v (39 psi)
63	WT/GN	FRP Sensor Signal	2.8v (39 psi)

1998-2000 Mystique 2.0L I4 MFI VIN 3 (All) 104 Pin Connector

PCM Pin #	Wire Color	Circuit Description (104 Pin)	Value at Hot Idle
64	WT/GN	A/T: TR Sensor Signal	In 'P': 0v, O/D: 1.7v
64	WT	M/T: Clutch or Neutral Switch	Clutch In: 0v, Out: 5v
65	WT/BL	DPFE Sensor Signal	0.20-1.30v
66, 68	---	Not Used	---
67	BK/OR	EVAP CV Solenoid	0-10 Hz (0-100%)
69	BK/YL	A/C WOT Relay Control	Relay Off: 12v, On: 1v
70	BK/YL	VCT Control (ZETECT)	VCT Off: 12v, On: 1v
71, 97	GN/YL	Vehicle Power	12-14v
73 ('20)	BK/YL	Shift Solenoid 1 Control	1v, 55 mph: 12v
74	BK/BL	Injector 3 Control	2.3-2.9 ms
75	BK/WT	Injector 1 Control	2.3-2.9 ms
76	BR	CMP and TSS Sensor Ground	<0.050v
76	BR/WT	CMP and TSS Sensor Ground	<0.050v
77	BK/YL	Power Ground	<0.1v
79	WT/BL	TCIL (lamp) Control	Lamp On: 1v, Off: 12v
80	BK/BL	Fuel Pump Control	Pump On: 1v, Off: 12v
80	BK/RD	Fuel Pump Control	Pump On: 1v, Off: 12v
81	BK/RD	EPC Solenoid Control	8.9v
83	BK/YL	IAC Motor Control	9v (33% d/cycle)
84	WT/PK	TSS Sensor Signal	40-56 Hz
85	WT/PK	CMP Sensor Signal	5-7 Hz
86	BK/WT	A/C Pressure Switch	A/C On: 12v (Open)
87	BR/YL	Knock Sensor (-) Signal	0v
88	WT/BL	MAF Sensor Signal	0.8v
89	WT	TP Sensor Signal	0.53-1.27v
90	YL	Reference Voltage	4.9-5.1v
91	BR	Analog Signal Return	<0.050v
92	OR	Brake Pedal Position Switch	Brake Off: 0v, On: 12v
92	GN/RD	Brake Pedal Position Switch	Brake Off: 0v, On: 12v
93	BK/YL	HO2S-11 (B1 S1) Heater	1v, Off: 12v
95	BK/OR	HO2S-12 (B1 S2) Heater	1v, Off: 12v
99	BK/WT	Modulated Lockup Solenoid	On: 1v, Off: 12v
100	BK/OR	Injector 4 Control	2.3-2.9 ms
101	BK/YL	Injector 2 Control	2.3-2.9 ms
102	BK/OR	Shift Solenoid 3 Control	8v, 55 mph: 8-9v
103	BK/YL	Power Ground	<0.1v
104	---	Not Used	---

Pin Connector Graphic

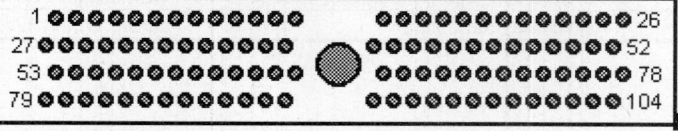

PCM 104-PIN CONNECTOR

```
 1 ●●●●●●●●●●●●●    ●●●●●●●●●●●●● 26
27 ●●●●●●●●●●●●●    ●●●●●●●●●●●●● 52
53 ●●●●●●●●●●●●●  ⬤ ●●●●●●●●●●●●● 78
79 ●●●●●●●●●●●●●    ●●●●●●●●●●●●● 104
```

Terminal View of 104-Pin PCM Wiring Harness Connector

Standard Colors and Abbreviations

Abbreviation	Color	Abbreviation	Color	Abbreviation	Color
BK	Black	GY	Gray	PK	Purple
BL	Blue	GN	Green	RD	Red
BR	Brown	LG	LT Green	TN	Tan
DB	Dark Blue	OR	Orange	WT	White
DG	DK Green	PK	Pink	YL	Yellow

1995 Mystique 2.5L V6 MFI VIN L (All) 22 Pin Connector

PCM Pin #	Wire Color	Circuit Description (22 Pin)	Value at Hot Idle
22-A	BL/RD	Keep Alive Power	12- 14v
22-B	RD/BK	Vehicle Power	12-14v
22-C	BK/RD	Vehicle Start Signal	KOEC: 9-11v
22-D	WT/RD	Switch Monitor Lamp Control	Lamp Off: 12v, On: 1v
22-E	BL	STO & MIL (lamp) Control	STO On: 5v, Off: 12v
22-F	---	Not Used	---
22-G	BL/OR	Ignition Control Module	Varies
22-H	WT	Headlamp Switch Signal	Switch On: 12v, Off: 1v
22-I	RD/WT	Self-Test Input	STI On: 1v, STI Off: 12v
22-J	PK	Rear Window Defroster	Switch On: 1v, Off: 12v
22-K	WT/BK	Torque Reduce & ECT Signal	Digital Signal
22-L	GN/BK	A/C Relay Control	Relay Off: 12v, On: 1v
22-M	GN/RD	Vehicle Speed Sensor	0 Hz, 55 mph: 125 Hz
22-N	BL/YL	PSP Switch Signal	Straight: 12v, Turned: 0v
22-O	PK/BK	A/C Cycling Pressure Switch	Switch On: 12v, Off: 1v
22-P	OR/BK	Blower Motor Control Switch	Off or 1st: 12v, 2nd on: 1v
22-Q	WT/GN	Brake On/ Off Switch Signal	Brake Off: 0v, On: 12v
22-R	LG/BK	A/T: PNP Switch Signal	In 'P', 0v, Others: 12v
22-R	LG/BK	M/T: CPP Switch Signal	Clutch In: 0v, Out: 12v
22-S	GN	Torque Reduce Signal #1	Digital Signal
22-T	BR	Idle Switch Signal	Closed: 0v, Open: 12v
22-U	---	Not Used	---
22-V	LG/WT	Torque Reduce Signal #2	Digital Signal

1995 Mystique 2.5L V6 MFI VIN L (All) 16 Pin Connector

PCM Pin #	Wire Color	Circuit Description (16 Pin)	Value at Hot Idle
16-A	GN/OR	Barometric Pressure Sensor	3.9v
16-B	RD	Measuring Core VAF Sensor	Idle: 3.06v
16-C	BK/YL	HO2S-11 (B1 S1) Signal	0.1-1.1v
16-D	BL/WT	HO2S-12 (B1 S2) Signal	0.1-1.1v
16-E	R/GN	ECT Sensor Signal	0.5-0.6v
16-F	YL	Throttle Position Sensor	0.4-1.0v
16-G	-	Not Used	---
16-N	-	Not Used	---
16-H	PK/YL	A/C High Pressure Switch	Closed: 1v, open: 12v
16-I	PK	Reference Voltage	4.9-5.1v
16-J	RD/BK	EGR Valve Position Sensor	Idle: 0.4v
16-K	BK/RD	IAT Sensor Signal	1.5-3.5v
16-L	GN	DRL Signal - Canada	DRL On: 2v, Off: 12v
16-M	WT	Knock Sensor Signal	0v
16-O	BL/BK	EVAP Canister Purge Control	12v, 55 mph: 1v
16-P	BL/GN	High Speed Fan Control	Fan On: 1v, Off: 12v

1995 Mystique 2.5L V6 MFI VIN L (All) 26 Pin Connector

PCM Pin #	Wire Color	Circuit Description (26 Pin)	Value at Hot Idle
26-A	BK	Power Ground	<0.1v
26-B	BK	Power Ground	<0.1v
26-C	BK/RD	Power Ground	<0.1v
26-D	BK/RD	Power Ground	<0.1v
26-E	LG/OR	CKP Sensor 1 Signal	2.5v
26-F	BL	CKP Sensor Ground	0.1v
26-G	BL/PK	CMP Sensor 1 Signal	2.5v
26-H	GN	CKP Sensor 2 Signal	2.5v
26-I	WT/GN	Variable Resonance Induction	VRIS1 On: 1v, Off: 12v
26-J	BL/RD	Variable Resonance Induction	VRIS2 On: 1v, Off: 12v
26-K	---	Not Used	
26-L	RD/WT	Low Speed Fan Control	Relay Off: 12v, On: 1v
26-M	GN/BK	Fuel Press. Regulator Control	Startup: 3v, others: 12v
26-N	BL/OR	Condenser Fan Control	Fan On: 1v, Off: 12v
26-O	WT/BL	EGR Vent Solenoid	12v, Cruise: 1v
26-P	GN/WT	EGR Control Solenoid	12v, Cruise: 1v
26-Q	LG/BK	Idle Air Control Valve	8-10v
26-R	---	Not Used	---
26-S	---	Not Used	---
26-T	LG	Fuel Pump Control	Relay Off: 12v, On: 1v
26-U	RD/LG	Injector 1 Control	3.6 ms
26-V	BL/WT	Injector 2 Control	3.6 ms
26-W	BR	Injector 3 Control	3.6 ms
26-X	RD/YL	Injector 4 Control	3.6 ms
26-Y	WT	Injector 5 Control	3.6 ms
26-Z	WT/BK	Injector 6 Control	3.6 ms

Pin Connector Graphic

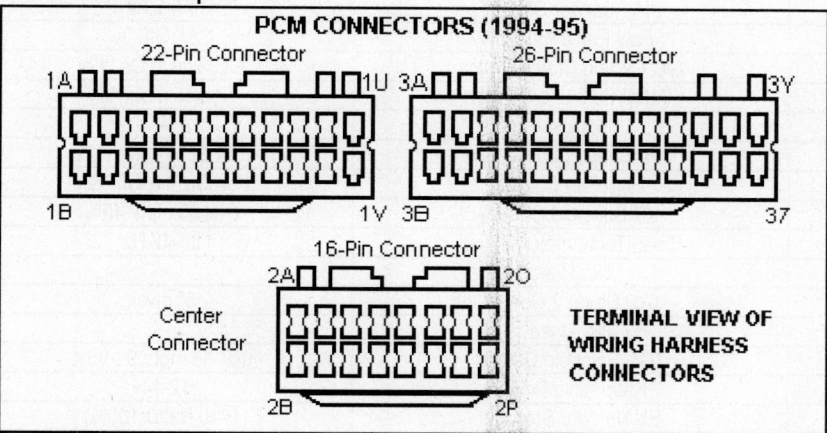

Standard Colors and Abbreviations

Abbreviation	Color	Abbreviation	Color	Abbreviation	Color
BK	Black	GY	Gray	PK	Purple
BL	Blue	GN	Green	RD	Red
BR	Brown	LG	LT Green	TN	Tan
DB	Dark Blue	OR	Orange	WT	White
DG	DK Green	PK	Pink	YL	Yellow

1996-97 Mystique 2.5L V6 MFI VIN L (All) 104 Pin Connector

PCM Pin #	Wire Color	Circuit Description (104 Pin)	Value at Hot Idle
1	BK/BL	Shift Solenoid 2	1v, 55 mph: 12v
2	BK/OR	MIL (lamp) Control	MIL On: 1v, Off: 12v
3-7	---	Not Used	---
8	BK/BL	IMRC Solenoid Control	5v, 55 mph: 5v
9-12, 14	---	Not Used	---
13	WT/BL	Flash EEPROM Power	0.5v
15	BR/BL	Data Bus (-) Signal	Digital Signals
16	WT/BL	Data Bus (+) Signal	Digital Signals
17	GY/OR	Passive Anti-Theft System	Digital Signal
18-20	---	Not Used	---
21	WT/RD	CKP Sensor (+) Signal	410-440 Hz
22	BR/RD	CKP Sensor (-) Signal	410-440 Hz
23	---	Not Used	---
24, 51	BK/YL	Power Ground	<0.1v
25	BK/RD	Chassis Ground	<0.050v
26	BR/BL	Coil Driver 1 Control	5° dwell
27	BK/YL	CD4E A/T: Shift Solenoid 1	12v, 55 mph: 1v
28	---	Not Used	
29	PK/BK	CD4E A/T: TCS (switch)	TCS & O/D On: 12v
30	WT/BK	Octane Adjust Switch	9.3v (Closed: 0v)
31	WT	PSP Sensor Signal	Straight: 1v, Turning: 12v
32-34	---	Not Used	---
34	WT/GN	Passive Anti-Theft System	Digital Signal
35	WT/BL	HO2S-12 (B1 S2) Signal	0.1-1.1v
36	BR/BL	MAF Sensor Return	<0.050v
37	WT/RD	TFT Sensor Signal	2.10 to 0.40v
38	WT/GN	ECT Sensor Signal	0.5-0.6v
39	WT/PK	IAT Sensor Signal	1.5-2.5v
40	PK/BK	Fuel Pump Monitor	12v, Off: 0v
41	PK/BL	A/C Cycling Clutch Switch	A/C Switch On: 12v
42	BK/RD	IMRC Solenoid Control	12v, 55 mph: 12v
43-44	---	Not Used	---
45	BK/BL	Low Speed Fan Control	Fan On: 1v, Off: 12v
46	BK/WT	High Fan Control	Fan On: 1v, Off: 12v
47	BK/GN	VR Solenoid Control	0%, 55 mph: 40%
48	WT/BK	Clean Tachometer Output	35-42 Hz
49-50	---	Not Used	---
52	BR/GN	Coil Driver 2 Control	5° dwell
53	---	Not Used	---
54	BK/WT	TCC Solenoid Control	0%, 55 mph: 90-95%
55	OR/YL	Keep Alive Power	12-14v
56	BK/OR	EVAP VMV Solenoid	0-10 Hz (0-100%)
57	WT/BK	Knock Sensor Signal	0v
58	WT/PK	VSS (+) Signal	0 Hz, 55 mph: 125 Hz
59	---	Not Used	---
60	WT	HO2S-11 (B1 S1) Signal	0.1-1.1v
61	WT/GN	HO2S-22 (B2 S2) Signal	0.1-1.1v
62-63	---	Not Used	---
64	WT/GN	CD4E A/T: TR Sensor Signal	In 'P': 0v, O/D: 1.7v
64	WT	M/T: Clutch Position Switch	Clutch In: 0v, Out: 5v

1996-97 Mystique 2.5L V6 MFI VIN L (All) 104 Pin Connector

PCM Pin #	Wire Color	Circuit Description (104 Pin)	Value at Hot Idle
65	WT/BL	DPFE Sensor Signal	0.20-1.30v
66, 68	---	Not Used	---
67	BK/OR	EVAP CV Solenoid	0-10 Hz (0-100%)
69	BK/YL	A/C WOT Relay Control	Relay Off: 12v, On: 1v
70	---	Not Used	---
71	GN/YL	Vehicle Power	12-14v
72, 82	---	Not Used	---
73	BK/GN	Injector 5 Control	2.1-2.5 ms
74	BK/BL	Injector 3 Control	2.1-2.5 ms
75	BK/WT	Injector 1 Control	2.1-2.5 ms
76	BK/YL	CMP & TSS Ground	<0.050v
77	BK/YL	Power Ground	<0.1v
78	BR/YL	Coil Driver 3 Control	5° dwell
79	WT/BL	TCIL (lamp) Control	Lamp On: 1v, Off: 12v
80	BK/BL	Fuel Pump Control	Pump On: 1v, Off: 12v
81	BK/RD	EPC Solenoid Control	8.9v
83	BK/YL	IAC Motor Control	9v (33% d/cycle)
84	WT/PK	TSS Sensor Signal	46-50 Hz
85	WT/PK	CMP Sensor Signal	5-7 Hz
86	BK/WT	A/C Pressure Switch	A/C On: 12v (Open)
87	WT/RD	HO2S-21 (B1 S1) Signal	0.1-1.1v
88	WT/BL	MAF Sensor Signal	0.8v
89	WT	TP Sensor Signal	0.53-1.27v
90	YL	Reference Voltage	4.9-5.1v
91	BR	Analog Signal Return	<0.050v
92	OR	Brake Pedal Position Switch	Brake Off: 0v, On: 12v
93	BK/YL	HO2S-11 (B1 S1) Heater	1v, Off: 12v
94	BK/BL	HO2S-21 (B2 S1) Heater	1v, Off: 12v
95	BK/OR	HO2S-12 (B1 S2) Heater	1v, Off: 12v
96	BK/GN	HO2S-22 (B2 S2) Heater	1v, Off: 12v
97	GN/YL	Vehicle Power	12-14v
98	---	Not Used	---
99	BK/RD	Injector 6 Control	2.1-2.5 ms
100	BK/OR	Injector 4 Control	2.1-2.5 ms
101	BK/YL	Injector 2 Control	2.1-2.5 ms
102	BK/OR	Shift Solenoid 3	7-9v, 30 mph: 8-9v
103	BK/YL	Power Ground	<0.1v
98, 104	---	Not Used	---

Pin Connector Graphic

Terminal View of 104-Pin PCM Wiring Harness Connector

Standard Colors and Abbreviations

Abbreviation	Color	Abbreviation	Color	Abbreviation	Color
BK	Black	GY	Gray	PK	Purple
BL	Blue	GN	Green	RD	Red
BR	Brown	LG	LT Green	TN	Tan
DB	Dark Blue	OR	Orange	WT	White
DG	DK Green	PK	Pink	YL	Yellow

1998-99 Mystique 2.5L V6 MFI VIN L (All) 104 Pin Connector

PCM Pin #	Wire Color	Circuit Description (104 Pin)	Value at Hot Idle
1	BK/BL	Shift Solenoid 2 Control	1v, 55 mph: 12v
2	BK/OR	MIL (lamp) Control	MIL On: 1v, Off: 12v
3-7	---	Not Used	---
8	BK/BL	IMRC Solenoid Control	5v, 55 mph: 5v
9-11, 14	---	Not Used	---
12	WT	Fuel Level Indicator Signal	1.7v (1/2 full)
12	WT/BK	Fuel Level Indicator Signal	1.7v (1/2 full)
13	WT/BL	Flash EEPROM Power	0.5v
15	BL	Data Bus (+) Signal	Digital Signals
16	GY	Data Bus (-) Signal	Digital Signals
17	GY/OR	Passive Anti-Theft System	Digital Signals
18-19	---	Not Used	---
20	BK/BL	Injector 3 Control	2.1-2.5 ms
21	WT/RD	CKP Sensor (+) Signal	410-440 Hz
22	BR/RD	CKP Sensor (-) Signal	410-440 Hz
23	---	Not Used	---
24	BK/YL	Power Ground	<0.1v
25	BK/RD	Chassis Ground	<0.050v
26	BR/BL	Coil Driver 1 Control	5° dwell
27	BK/YL	Shift Solenoid 1 Control	12v, 55 mph: 1v
28	---	Not Used	---
29	PK/BK	TCS (switch) Signal	TCS & O/D On: 12v
30	WT/BK	Octane Adjust Switch	9.3v (Closed: 0v)
31	WT	PSP Switch Signal	Straight: 1v, Turning: 12v
32-33	---	Not Used	---
34	WT/GN	Passive Anti-Theft System	Digital Signal
35	WT/BL	HO2S-12 (B1 S2) Signal	0.1-1.1v
36	BR/BL	MAF Sensor Return	<0.050v
37	WT/RD	TFT Sensor Signal	2.10 to 0.40v
38	WT/GN	ECT Sensor Signal	0.5-0.6v
39	WT/PK	IAT Sensor Signal	1.5-2.5v
40	PK/BK	Fuel Pump Monitor	12v, Off: 0v
41	PK/BL	A/C Cycling Clutch Switch	A/C Switch On: 12v
42	BK/RD	IMRC Solenoid Signal	12v, 55 mph: 12v
43-44	---	Not Used	---
45	BK/BL	Low Speed Fan Control	Fan On: 1v, Off: 12v
46	BK/WT	High Fan Control	Fan On: 1v, Off: 12v
47	BK/GN	VR Solenoid Control	0%, 55 mph: 40%
48	WT/BK	Clean Tachometer Output	35-42 Hz
49-50	---	Not Used	---
51	BK/YL	Power Ground	<0.1v
52	BR/GN	Coil Driver 2 Control	5° dwell
53, 59	---	Not Used	---
54	BK/BL	Fuel Pump Control	Pump On: 1v, Off: 12v
55	OR/YL	Keep Alive Power	12-14v
56	BK/OR	EVAP Purge Valve	0-10 Hz (0-100%)
57	WT/BK	Knock Sensor 1 Signal	0v
58	WT/PK	VSS (+) Signal	0 Hz, 55 mph: 125 Hz
60	WT	HO2S-11 (B1 S1) Signal	0.1-1.1v
61	WT/GN	HO2S-22 (B2 S2) Signal	0.1-1.1v
62	WT/PK	FTP Sensor Signal	2.6v (cap off)
63 ('99)	WT/PK	FRP Sensor Signal	2.8v (39 psi)
64	WT/GN	A/T: TR Sensor Signal	In 'P': 0v, O/D: 1.7v
64	WT	M/T: Clutch Position Switch	Clutch In: 0v, Out: 5v

1998-99 Mystique 2.5L V6 MFI VIN L (All) 104 Pin Connector

PCM Pin #	Wire Color	Circuit Description (104 Pin)	Value at Hot Idle
65	WT/BL	DPFE Sensor Signal	0.20-1.30v
66, 68, 72	---	Not Used	---
67	BK/OR	EVAP CV Solenoid	0-10 Hz (0-100%)
69	BK/YL	A/C WOT Relay Control	Relay Off: 12v, On: 1v
70	BK/YL	ZETECT VCT Control	VCT Off: 12v, On: 1v
71	GN/YL	Vehicle Power	12-14v
73	BK/YL	HO2S-11 (B1 S1) Heater	1v, Off: 12v
74	BK/BL	Injector 3 Control	2.1-2.5 ms
75	BK/WT	Injector 1 Control	2.1-2.5 ms
76	BK/YL	CMP & TSS Sensor Ground	<0.050v
77	BK/YL	Power Ground	<0.1v
78	BR/YL	Coil Driver 3 Control	5° dwell
79	WT/BK	TCIL (lamp) Control	Lamp On: 1v, Off: 12v
80	BK/WT	TCC Solenoid Control	0%, 55 mph: 90-95%
81	BK/RD	EPC Solenoid Control	8.9v
82	---	Not Used	---
83	BK/YL	IAC Motor Control	9v (33% d/cycle)
84	WT/PK	TSS Sensor Signal	46-50 Hz
85	WT/PK	CMP Sensor Signal	5-7 Hz
86	BK/WT	A/C Pressure Switch	A/C On: 12v (Open)
87	WT/RD	HO2S-21 (B1 S1) Signal	0.1-1.1v
88	WT/BL	MAF Sensor Signal	0.8v
89	WT	TP Sensor Signal	0.53-1.27v
90	YL	Reference Voltage	4.9-5.1v
91	BR	Analog Signal Return	<0.050v
92	OR	Brake Pedal Position Switch	Brake Off: 0v, On: 12v
93	BK/YL	Injector 5 Control	2.1-2.5 ms
94	BK/BL	Injector 6 Control	2.1-2.5 ms
95	BK/OR	Injector 4 Control	2.1-2.5 ms
96	BK/GN	Injector 2 Control	2.1-2.5 ms
97	GN/YL	Vehicle Power	12-14v
98	---	Not Used	---
99	BK/BL	HO2S-21 (B2 S1) Heater	1v, Off: 12v
100	BK/OR	HO2S-12 (B1 S2) Heater	1v, Off: 12v
101	BK/GN	HO2S-22 (B2 S2) Heater	1v, Off: 12v
102	BK/OR	Shift Solenoid 3 Control	7-9v, 30 mph: 8-9v
103	BK/YL	Power Ground	<0.1v
104	---	Not Used	---

Pin Connector Graphic

```
              PCM 104-PIN CONNECTOR
   1 ●●●●●●●●●●●●●      ●●●●●●●●●●●●● 26
  27 ●●●●●●●●●●●●●      ●●●●●●●●●●●●● 52
  53 ●●●●●●●●●●●●●   ⬣  ●●●●●●●●●●●●● 78
  79 ●●●●●●●●●●●●●      ●●●●●●●●●●●●● 104
   Terminal View of 104-Pin PCM Wiring Harness Connector
```

Standard Colors and Abbreviations

Abbreviation	Color	Abbreviation	Color	Abbreviation	Color
BK	Black	GY	Gray	PK	Purple
BL	Blue	GN	Green	RD	Red
BR	Brown	LG	LT Green	TN	Tan
DB	Dark Blue	OR	Orange	WT	White
DG	DK Green	PK	Pink	YL	Yellow

2000 Mystique 2.5L V6 MFI VIN L (All) 104 Pin Connector

PCM Pin #	Wire Color	Circuit Description (104 Pin)	Value at Hot Idle
1	---	Not Used	---
2	BK/OR	MIL (lamp) Control	MIL On: 1v, Off: 12v
3-5, 7	---	Not Used	---
6	BK/YL	Shift Solenoid 1 Control	12v, 55 mph: 1v
8	BK/BL	IMRC Solenoid Control	5v, 55 mph: 5v
9	WT	Fuel Level Indicator Signal	1.7v (1/2 full)
7, 10	---	Not Used	---
11	BK/BL	Shift Solenoid 2 Control	12v, 55 mph: 1v
12	WT/BK	Gear Indicator Signal	Digital Signal
13	WT/BL	Flash EEPROM Power	0.5v
14, 17-19	---	Not Used	---
15	BL	Data Bus (-) Signal	Digital Signals
16	GY	Data Bus (+) Signal	Digital Signals
20	BK/BL	Injector 3 Control	2.1-2.5 ms
21	WT/RD	CKP Sensor (+) Signal	410-440 Hz
22	BR/RD	CKP Sensor (-) Signal	410-440 Hz
23, 30, 34	---	Not Used	---
24	BK/YL	Power Ground	<0.1v
25	BK/RD	Chassis Ground	<0.050v
26	BR/BL	Coil Driver 1 Control	5° dwell
27	BK/YL	Shift Solenoid 1 Control	12v, 55 mph: 1v
28	WT/BL	Speedometer Indicator Signal	Digital Signals
29	GN/BK	TCS (switch) Signal	TCS & O/D On: 12v
31	WT	PSP Switch Signal	Straight: 1v, Turning: 12v
32	BR/YL	Knock Sensor (-) Signal	0v
33	BR/BL	VSS (-) Signal	0 Hz, 55 mph: 125 Hz
35	WT/BL	HO2S-12 (B1 S2) Signal	0.1-1.1v
36	BR/BL	MAF Sensor Return	<0.050v
37	WT/RD	TFT Sensor Signal	2.10 to 0.40v
38	WT/GN	ECT Sensor Signal	0.5-0.6v
39	WT/PK	ACT Sensor Signal	1.5-2.5v
40	GN/BK	Fuel Pump Monitor	12v, Off: 0v
41	GN/BL	A/C Cycling Clutch Switch	A/C Switch On: 12v
42	BK/RD	IMRC Motor Control Signal	12v, 55 mph: 12v
43-44, 49-50	---	Not Used	---
45	BK/BL	Low Speed Fan Control	Fan On: 1v, Off: 12v
46	BK/WT	High Fan Control	Fan On: 1v, Off: 12v
47	BK/GN	VR Solenoid Control	0%, 55 mph: 40%
48	WT/BK	Clean Tachometer Output	35-42 Hz
51	BK/YL	Power Ground	<0.1v
52	BR/GN	Coil Driver 2 Control	5° dwell
53, 59	---	Not Used	---
54	BK/RD	Fuel Pump Control	Pump On: 1v, Off: 12v
55	RD	Keep Alive Power	12-14v
56	BK/OR	EVAP Purge Valve	0-10 Hz (0-100%)
57	WT/BK	Knock Sensor + Signal	0v
58	WT/BL	VSS (+) Signal	0 Hz, 55 mph: 125 Hz
60	WT	HO2S-11 (B1 S1) Signal	0.1-1.1v
61	WT/GN	HO2S-22 (B2 S2) Signal	0.1-1.1v
62	WT/PK	FTP Sensor Signal	2.6v (cap off)
63	WT/GN	FRP Sensor Signal	2.8v (39 psi)
64	WT/GN	A/T: TR Sensor Signal	In 'P': 0v, O/D: 1.7v
64	WT	M/T: Clutch Position Switch	Clutch In: 0v, Out: 5v

2000 Mystique 2.5L V6 MFI VIN L (All) 104 Pin Connector

PCM Pin #	Wire Color	Circuit Description (104 Pin)	Value at Hot Idle
65	WT/BL	DPFE Sensor Signal	0.20-1.30v
66, 68	---	Not Used	---
67	BK/OR	EVAP CV Solenoid	0-10 Hz (0-100%)
69	BK/YL	A/C WOT Relay Control	Relay Off: 12v, On: 1v
70	BK/WT	Injector 1 Control	2.1-2.5 ms
71	GN/YL	Vehicle Power	12-14v
72, 75	---	Not Used	---
73	BK/YL	HO2S-11 (B1 S1) Heater	1v, Off: 12v
74	BK/BL	Injector 3 Control	2.1-2.5 ms
76	BR/WT	CMP & TSS Ground	<0.050v
77	BK/YL	Power Ground	<0.1v
78	BR/YL	Coil Driver 3 Control	5° dwell
79	WT/BK	TCIL (lamp) Control	Lamp On: 1v, Off: 12v
80	BK/WT	TCC Solenoid Control	0%, 55 mph: 90-95%
81	BK/RD	EPC Solenoid Control	8.9v
82	---	Not Used	---
83	BK/YL	IAC Motor Control	9v (33% d/cycle)
84	WT/PK	TSS Sensor Signal	46-50 Hz
85	WT/PK	CMP Sensor Signal	5-7 Hz
86	BK/WT	A/C Pressure Switch	A/C On: 12v (Open)
87	WT/RD	HO2S-21 (B1 S1) Signal	0.1-1.1v
88	WT/BL	MAF Sensor Signal	0.8v
89	WT	TP Sensor Signal	0.53-1.27v
90	YL	Reference Voltage	4.9-5.1v
91	BR	Analog Signal Return	<0.050v
92	GN/RD	Brake Pedal Position Switch	Brake Off: 0v, On: 12v
93	BK/GN	Injector 5 Control	2.1-2.5 ms
94	BK/RD	Injector 6 Control	2.1-2.5 ms
95	BK/OR	Injector 4 Control	2.1-2.5 ms
96	BK/YL	Injector 2 Control	2.1-2.5 ms
97	GN/YL	Vehicle Power	12-14v
98	---	Not Used	---
99	BK/BL	HO2S-21 (B2 S1) Heater	1v, Off: 12v
100	BK/OR	HO2S-12 (B1 S2) Heater	1v, Off: 12v
101	BK/GN	HO2S-22 (B2 S2) Heater	1v, Off: 12v
102	BK/OR	Shift Solenoid 3 Control	7-9v, 30 mph: 8-9v
103	BK/YL	Power Ground	<0.1v
104	---	Not Used	---

Pin Connector Graphic

Terminal View of 104-Pin PCM Wiring Harness Connector

Standard Colors and Abbreviations

Abbreviation	Color	Abbreviation	Color	Abbreviation	Color
BK	Black	GY	Gray	PK	Purple
BL	Blue	GN	Green	RD	Red
BR	Brown	LG	LT Green	TN	Tan
DB	Dark Blue	OR	Orange	WT	White
DG	DK Green	PK	Pink	YL	Yellow

Sable Pin Tables

1996-97 Sable 3.0L V6 4v MFI VIN S (A/T) 104 Pin Connector

PCM Pin #	Wire Color	Circuit Description (104 Pin)	Value at Hot Idle
1	PK/OR	Shift Solenoid 2 Control	1v, 55 mph: 12v
2	PK/LG	MIL (lamp) Control	MIL On: 1v, Off: 12v
3-4	---	Not Used	---
5	WT/OR	Electronic AIR System	0.1v (Off)
6-7	---	Not Used	---
8	DB/Y	IMRC Monitor Signal	5v, 55 mph: 5v
9-12, 14	---	Not Used	---
13	PK	Flash EEPROM Power	0.1v
15	PK/LB	Data Bus (-) Signal	Digital Signals
16	TN/OR	Data Bus (+) Signal	Digital Signals
17-20, 23	---	Not Used	---
21	GY	CKP Sensor (+) Signal	850-1120 Hz
22	DB	CKP Sensor (-) Signal	850-1120 Hz
24	BK/WT	Power Ground	<0.1v
25	BK	Chassis Ground	<0.050v
26	YL/BK	Coil Driver 1 Control	5° dwell
27	OR/YL	Shift Solenoid 1 Control	12v, 55 mph: 1v
28	TN/OR	Low Speed Fan Control	Fan On: 1v, Off: 12v
29	TN/WT	Overdrive Cancel Switch	TCS & O/D On: 12v
30	DG	Octane Adjust Switch	9.3v (Closed: 0v)
31	YL/LG	PSP Switch Signal	Straight: 0v, turned: 12v
32, 34	---	Not Used	---
33	PK/OR	VSS (-) Signal	0 Hz, 55 mph: 125 Hz
35	RD/BK	HO2S-12 (B1 S2) Signal	0.1-1.1v
36	TN/LB	MAF Sensor Return	<0.050v
37	OR/BK	TFT Sensor Signal	2.40-2.10v
38	LG/RD	ECT Sensor Signal	0.5-0.6v
39	GY	IAT Sensor Signal	1.5-2.5v
40	DG/YT	Fuel Pump Monitor	12v, Off: 0v
41	PK/LB	A/C Cycling Clutch Switch	A/C Switch On: 12v
42	BR	IMRC Solenoid Control	12v, 55 mph: 12v
43-45	---	Not Used	---
46	LG/PK	High Speed Fan Control	Fan On: 1v, Off: 12v
47	BR/PK	VR Solenoid Control	0%, 55 mph: 45%
48	TN/YL	Clean Tachometer Output	33-37 Hz
49-50	---	Not Used	---
51	BK/WT	Power Ground	<0.1v
52	YL/RD	Coil Driver 2 Control	5° dwell
53	PK/BK	Shift Solenoid 3 Control	12v, 55 mph: 1v
54	---	Not Used	---
55	DG/OR	Keep Alive Power	12-14v
56	GY/YL	EVAP VMV Solenoid	0-10 Hz (0-100%)
57	YL/RD	Knock Sensor Signal	0v
58	GY/BK	VSS (+) Signal	0 Hz, 55 mph: 125 Hz
59	---	Not Used	---
60	GY/LB	HO2S-11 (B1 S1) Signal	0.1-1.1v
61	PK/LG	HO2S-22 (B2 S2) Signal	0.1-1.1v
62	RD/PK	FTP Sensor Signal	2.6v (0" H20)
63	---	Not Used	---
64	LBYL	TR Sensor Signal	In 'P': 0v, O/D: 1.7v

1996-97 Sable 3.0L V6 4v MFI VIN S (A/T) 104 Pin Connector

PCM Pin #	Wire Color	Circuit Description (104 Pin)	Value at Hot Idle
65	BR/LG	DPFE Sensor Signal	0.20-1.30v
66, 68, 72	---	Not Used	---
64	LBYL	TR Sensor Signal	In 'P': 0v, O/D: 1.7v
67	PK/WT	EVAP CV Solenoid	0-10 Hz (0-100%)
69	PK/YL	A/C WOT Relay Control	Relay Off: 12v, On: 1v
70	BR	Elect/Secondary Air Injection	12v, 55 mph: 12v
71	RD	Vehicle Power	12-14v
73	TN/BK	Injector 5 Control	2.3-5.5 ms
74	BR/YL	Injector 3 Control	2.3-5.5 ms
75	TN	Injector 1 Control	2.3-5.5 ms
76, 77	BK/WT	Power Ground	<0.1v
78	YL/WT	Coil Driver 3 Control	5° dwell
79	WT/LG	TCIL (lamp) Control	Lamp On: 1v, Off: 12v
80	LB/OR	Fuel Pump Control	Relay Off: 12v, On: 1v
81	WT/YL	EPC Solenoid Control	9.0v (15 psi)
82	PK/YL	TCC Solenoid Control	12v (0 psi)
83	WT/LB	IAC Motor Control	10v (30% d/cycle)
84	DG/WT	TSS Sensor Signal	43 Hz (650 rpm)
85	DB/OR	CID Sensor Signal	5-7 Hz
86	BK/YL	A/C Pressure Switch	Switch On: 12v (Open)
87	RD/LG	HO2S-21 (B1 S1) Signal	0.1-1.1v
88	LB/RD	MAF Sensor Signal	0.6v
89	GY/WT	TP Sensor Signal	0.53-1.27v
90	BR/WT	Reference Voltage	4.9-5.1v
91	GY/RD	Analog Signal Return	<0.050v
92	RD/LG	Brake Switch Signal	Brake Off: 0v, On: 12v
93	RD/WT	HO2S-11 (B1 S1) Heater	1v, Off: 12v
94	WT/BK	HO2S-21 (B2 S1) Heater	1v, Off: 12v
95	YL/LG	HO2S-12 (B1 S2) Heater	1v, Off: 12v
96	TN/YL	HO2S-22 (B2 S2) Heater	1v, Off: 12v
97	RD	Vehicle Power	12-14v
98	---	Not Used	---
99	LG/OR	Injector 6 Control	2.3-5.5 ms
100	BR/LB	Injector 4 Control	2.3-5.5 ms
101	WT	Injector 2 Control	2.3-5.5 ms
102	---	Not Used	---
103	BK/WT	Power Ground	<0.1v
104	---	Not Used	---

Pin Connector Graphic

PCM 104-PIN CONNECTOR

Terminal View of 104-Pin PCM Wiring Harness Connector

Standard Colors and Abbreviations

Abbreviation	Color	Abbreviation	Color	Abbreviation	Color
BK	Black	GY	Gray	PK	Purple
BL	Blue	GN	Green	RD	Red
BR	Brown	LG	LT Green	TN	Tan
DB	Dark Blue	OR	Orange	WT	White
DG	DK Green	PK	Pink	YL	Yellow

1998-99 Sable 3.0L 4v V6 MFI VIN S (A/T) 104 Pin Connector

PCM Pin #	Wire Color	Circuit Description (104 Pin)	Value at Hot Idle
1	PK/OR	Shift Solenoid 2 Control	1v, 55 mph: 1v
2	PK/LG	MIL (lamp) Control	MIL On: 1v, Off: 12v
3-4	---	Not Used	---
5	WT/OR	Electronic AIRB Monitor	5v
8	DB/Y	IMRC Monitor Signal	5v, 55 mph: 5v
9-11	---	Not Used	---
12	YL/WT	Fuel Level Indicator Signal	1.7v (1/2 full)
13	PK	Flash EEPROM Power	0.1v
14	---	Not Used	---
15	PK/LB	Data Bus (-) Signal	Digital Signals
16	TN/OR	Data Bus (+) Signal	Digital Signals
17-20	---	Not Used	---
21	GY	CKP Sensor (+) Signal	390-520 Hz
22	DB	CKP Sensor (-) Signal	390-520 Hz
24	BK/WT	Power Ground	<0.1v
25	BK	Power Ground	<0.1v
26	YL/BK	Coil Driver 1 Control	5° dwell
27	OR/YL	Shift Solenoid 1 Control	12v, 55 mph: 0.1v
28	LB	Low Speed Fan Control	On: 1v, Off: 12v
29	TN/WT	Overdrive Cancel Switch	TCS & O/D On: 12v
30	DG	Octane Adjust Switch	Closed: 0v, Open: 9.3v
31	YL/LG	PSP Switch Signal	Straight: 0v, Turned: 12v
32	---	Not Used	---
33	PK/OR	VSS (-) Signal	0 Hz, 55 mph: 125 Hz
35	RD/BK	HO2S-12 (B1 S2) Signal	0.1-1.1v
36	TN/LB	MAF Sensor Return	<0.050v
37	OR/BK	TFT Sensor Signal	2.10-2.40v
38	LG/RD	ECT Sensor Signal	0.5-0.6v
39	GY	IAT Sensor Signal	1.5-2.5v
40	DG/YT	Fuel Pump Monitor	On: 12v, Off: 0v
41	PK/LB	A/C Cycling Clutch Switch	On: 1v, Off: 12v
42	BR	IMRC Solenoid Control	12v (Off)
43	---	Not Used	---
45	PK/WT	EVAP CV Solenoid	0-10 Hz (0-100%)
46	LG/PK	High Speed Fan Control	On: 1v, Off: 12v
47	BR/PK	EGR VR Solenoid	0%, 55 mph: 45%
48	TN/YL	Clean Tachometer Output	33-45 Hz
49-50	---	Not Used	---
51	BK/WT	Power Ground	<0.1v
52	YL/RD	Coil Driver 2 Control	5° dwell
53	PK/BK	Shift Solenoid 3 Control	12v, 55 mph: 0.1v
54	PK/YL	TCC Solenoid Control	0%, 55 mph: 95%
56	GY/YL	EVAP Purge Solenoid	0-10 Hz (0-100%)
55	DG/OR	Keep Alive Power	12-14v
57	YL/RD	Knock Sensor Signal	0v
58	GY/BK	VSS (+) Signal	0 Hz, 55 mph: 125 Hz
59	---	Not Used	---
60	GY/LB	HO2S-11 (B1 S1) Signal	0.1-1.1v
61	PK/LG	HO2S-22 (B2 S2) Signal	0.1-1.1v
62	RD/BK	FTP Sensor Signal	2.6v (0 in. H20)
63	---	Not Used	---
64	LBYL	TR Sensor Signal	In 'P': 0v, 55 mph: 1.7v

1998-99 Sable 3.0L 4v V6 MFI VIN S (A/T) 104 Pin Connector

PCM Pin #	Wire Color	Circuit Description (104 Pin)	Value at Hot Idle
65	BR/LG	DPFE Sensor Signal	0.20-1.30v
66, 68, 72	---	Not Used	---
64	LBYL	TR Sensor Signal	In 'P': 0v, 55 mph: 1.7v
67	PK/WT	EVAP Purge Solenoid	0-10 Hz (0-100%)
69	PK/YL	A/C WOT Relay Control	On: 1v, Off: 12v
70	BR	Electronic AIRB Solenoid	12v, 55 mph: 12v
71	RD	Vehicle Power	12-14v
73	TN/BK	Injector 5 Control	2.2-2.7 ms
74	BR/YL	Injector 3 Control	2.2-2.7 ms
75	TN	Injector 1 Control	2.2-2.7 ms
76, 77	BK/WT	Power Ground	<0.1v
78	YL/WT	Coil Driver 3 Control	5° dwell
79	WT/LG	TCIL (lamp) Control	On: 1v, Off: 12v
80	LB/OR	Fuel Pump Control	On: 1v, Off: 12v
81	WT/YL	EPC Solenoid Control	9.0v (15 psi)
82	VT/YL	TCC Solenoid Control	12v (0 psi)
83	WT/LB	IAC Motor Control	10v (55% duty cycle)
84	DG/WT	TSS Sensor Signal	43 Hz (700 rpm)
85	DB/OR	CID Sensor Signal	5-7 Hz
86	BK/YL	A/C High Pressure Switch	12v (Switch Open)
87	RD/LG	HO2S-21 (B2 S1) Signal	0.1-1.1v
88	LB/RD	MAF Sensor Signal	0.6v
89	GY/WT	TP Sensor Signal	0.53-1.27v
90	BR/WT	Reference Voltage	4.9-5.1v
91	GY/RD	Analog Signal Return	<0.050v
92	RD/LG	Brake Pedal Position Switch	Brake Off: 0v, On: 12v
93	RD/WT	HO2S-11 (B1 S1) Heater	On: 1v, Off: 12v
94	WT/BK	HO2S-21 (B2 S1) Heater	On: 1v, Off: 12v
95	YL/LG	HO2S-12 (B1 S2) Heater	On: 1v, Off: 12v
96	TN/YL	HO2S-22 (B2 S2) Heater	On: 1v, Off: 12v
97	RD	Vehicle Power	12-14v
98	---	Not Used	---
99	LG/OR	Injector 6 Control	2.2-2.7 ms
100	BR/LB	Injector 4 Control	2.2-2.7 ms
101	WT	Injector 2 Control	2.2-2.7 ms
98, 102	---	Not Used	---
103	BK/WT	Power Ground	<0.1v
104	---	Not Used	---

Pin Connector Graphic

```
           PCM 104-PIN CONNECTOR
     1 ●●●●●●●●●●●●●    ●●●●●●●●●●●●● 26
    27 ●●●●●●●●●●●●●         ●●●●●●●●●●●●● 52
    53 ●●●●●●●●●●●●●    ⬤    ●●●●●●●●●●●●● 78
    79 ●●●●●●●●●●●●●    ●●●●●●●●●●●●● 104
     Terminal View of 104-Pin PCM Wiring Harness Connector
```

Standard Colors and Abbreviations

Abbreviation	Color	Abbreviation	Color	Abbreviation	Color
BK	Black	GY	Gray	PK	Purple
BL	Blue	GN	Green	RD	Red
BR	Brown	LG	LT Green	TN	Tan
DB	Dark Blue	OR	Orange	WT	White
DG	DK Green	PK	Pink	YL	Yellow

2000 Sable 3.0L 4v V6 MFI VIN S (A/T) 104 Pin Connector

PCM Pin #	Wire Color	Circuit Description (104 Pin)	Value at Hot Idle
1	DG/PK	COP 4 Driver Control	6° dwell
2	PK/LG	MIL (lamp) Control	MIL On: 1v, Off: 12v
3	BK/WT	Power Ground	<0.1v
5	WT/OR	Electronic AIRB Monitor	5v
6	OR/YL	Shift Solenoid 'A' Control	12v, 55 mph: 0.1v
4, 7-11	---	Not Used	---
12	YL/WT	Fuel Level Indicator Signal	1.7v (1/2 full)
13	PK	Flash EEPROM Power	0.1v
14, 19	---	Not Used	---
15	PK/LB	Data Bus (-) Signal	Digital Signals
16	TN/OR	Data Bus (+) Signal	Digital Signals
17	BR/YL	SCI Receive Signal	N/A
18	GY/RD	SCI Transmit Signal	N/A
20	PK/BK	Shift Solenoid 'A' Control	12v, 55 mph: 0.1v
21	GY	CKP Sensor (+) Signal	390-520 Hz
22	DB	CKP Sensor (-) Signal	390-520 Hz
24	BK/WT	Power Ground	<0.1v
25	BK	Power Ground	<0.1v
26	LG/WT	COP 1 Driver Control	6° dwell
27	OR/YL	COP 5 Driver Control	6° dwell
28	LB	Low Speed Fan Control	On: 1v, Off: 12v
29	---	Not Used	---
30	DB/LG	Anti-theft Indicator	N/A
31	YL/LG	PSP Switch Signal	Straight: 0v, Turned: 12v
32	YL	Knock Sensor (-) Signal	0v
34	YL/BK	Digital TR1 Sensor	In 'P': 0v, 55 mph: 11v
35	RD/BK	HO2S-12 (B1 S2) Signal	0.1-1.1v
36	TN/LB	MAF Sensor Return	<0.050v
37	OR/BK	TFT Sensor Signal	2.10-2.40v
38	LG/RD	ECT Sensor Signal	0.5-0.6v
39	GY	IAT Sensor Signal	1.5-2.5v
40	DG/YT	Fuel Pump Monitor	On: 12v, Off: 0v
41	DG/WT	A/C High Pressure Switch	On: 1v, Off: 12v
42-43	---	Not Used	---
44	PK	Starter Relay Control	KOEC: 9-11v
46	LG/PK	High Speed Fan Control	On: 1v, Off: 12v
47	BR/PK	EGR VR Solenoid	0%, 55 mph: 45%
48	TN/YL	Clean Tachometer Output	33-45 Hz
49	LB/BK	Digital TR2 Sensor	In 'P': 0v, 55 mph: 11v
50	WT/BK	Digital TR4 Sensor	In 'P': 0v, 55 mph: 11v
51	BK/WT	Power Ground	<0.1v
52	PK/WT	COP 2 Driver Control	6° dwell
53	OR/YL	COP 6 Driver Control	6° dwell
55	PK/LB	Keep Alive Power	12-14v
54	PK/YL	TCC Solenoid Control	0%, 55 mph: 95%
56	GY/YL	EVAP Purge Solenoid	0-10 Hz (0-100%)
55	DG/OR	Keep Alive Power	12-14v
57	YL/RD	Knock Sensor (+) Signal	0v
59	DG/WT	TSS Sensor Signal	43 Hz (700 rpm)
60	GY/LB	HO2S-11 (B1 S1) Signal	0.1-1.1v
61	PK/LG	HO2S-22 (B2 S2) Signal	0.1-1.1v
62	RD/PK	FTP Sensor Signal	2.6v (0 in. H20)
63	WT/YL	EPC Solenoid Control	9.0v (15 psi)
64	RD/BK	Digital TR3A Sensor	In 'P': 0v, 55 mph: 1.7v

2000 Sable 3.0L 4v V6 MFI VIN S (A/T) 104 Pin Connector

PCM Pin #	Wire Color	Circuit Description (104 Pin)	Value at Hot Idle
65	BR/LG	DPFE Sensor Signal	0.20-1.30v
66, 72	---	Not Used	---
67	PK/WT	EVAP CV Solenoid	0-10 Hz (0-100%)
68	GY/BK	VSS (+) Signal	0 Hz, 55 mph: 125 Hz
69	PK/YL	A/C WOT Relay Control	On: 1v, Off: 12v
70	BR	Electronic AIRB Solenoid	12v, 55 mph: 12v
71	RD	Vehicle Power	12-14v
73	TN/BK	Injector 5 Control	2.2-2.7 ms
74	BR/YL	Injector 3 Control	2.2-2.7 ms
75	TN	Injector 1 Control	2.2-2.7 ms
76, 77	BK/WT	Power Ground	<0.1v
78	YLPK	COP 3 Driver Control	6° dwell
79	WT/LG	TCIL (lamp) Control	On: 1v, Off: 12v
80	LB/OR	Fuel Pump Control	On: 1v, Off: 12v
81	BR/OR	EPC Solenoid Control	9.0v (15 psi)
82	VT/YL	TCC Solenoid Control	12v (0 psi)
83	WT/LB	IAC Motor Control	10v (55% duty cycle)
84	DB/Y	OSS Sensor Signal	0 Hz, 55 mph: 131 Hz
85	DB/OR	CID Sensor Signal	5-7 Hz
86	BK/YL	A/C Clutch Relay Switch	12v (Switch Open)
87	RD/LG	HO2S-21 (B2 S1) Signal	0.1-1.1v
88	LB/RD	MAF Sensor Signal	0.6v
89	GY/WT	TP Sensor Signal	0.53-1.27v
90	BR/WT	Reference Voltage	4.9-5.1v
91	GY/RD	Analog Signal Return	<0.050v
92	RD/LG	Brake Pedal Position Switch	Brake Off: 0v, On: 12v
93	RD/WT	HO2S-11 (B1 S1) Heater	On: 1v, Off: 12v
94	WT/BK	HO2S-21 (B2 S1) Heater	On: 1v, Off: 12v
95	YL/LB	HO2S-12 (B1 S2) Heater	On: 1v, Off: 12v
96	TN/YL	HO2S-22 (B2 S2) Heater	On: 1v, Off: 12v
97-98	---	Not Used	---
99	LG/OR	Injector 6 Control	2.2-2.7 ms
100	BR/LB	Injector 4 Control	2.2-2.7 ms
101	WT	Injector 2 Control	2.2-2.7 ms
102	---	Not Used	---
103	BK/WT	Power Ground	<0.1v
104	---	Not Used	---

Pin Connector Graphic

PCM 104-PIN CONNECTOR

Terminal View of 104-Pin PCM Wiring Harness Connector

Standard Colors and Abbreviations

Abbreviation	Color	Abbreviation	Color	Abbreviation	Color
BK	Black	GY	Gray	PK	Purple
BL	Blue	GN	Green	RD	Red
BR	Brown	LG	LT Green	TN	Tan
DB	Dark Blue	OR	Orange	WT	White
DG	DK Green	PK	Pink	YL	Yellow

2001 Sable 3.0L 4v V6 MFI VIN S (A/T) 104 Pin Connector

PCM Pin #	Wire Color	Circuit Description (104 Pin)	Value at Hot Idle
1	PK/OR	Shift Solenoid 'B' Control	1v, 55 mph: 1v
2	PK/LG	MIL (lamp) Control	MIL On: 1v, Off: 12v
3	YL/BK	Digital TR1 Sensor	In 'P': 0v, 55 mph: 11v
4	BR/PK	Start System Voltage	Cranking: 9-11v
5	---	Not Used	---
6	DG/WT	TSS Sensor Signal	50-65 Hz (790-820 rpm)
7-11	---	Not Used	---
12	YL/WT	Fuel Level Indicator Signal	1.7v (1/2 full)
13	VT	Flash EEPROM Power	0.1v
15	PK/LB	SCP Data Bus (-) Signal	Digital Signals
16	TN/OR	SCP Data Bus (+) Signal	Digital Signals
17	BR/YL	SCI Receive (RX) Signal	Digital Signals
18	GY/RD	SCI Transmit (TX) Signal	Digital Signals
20	GY/YL	Generator Signal 2	TBD
21	GY	CKP Sensor (+) Signal	390-520 Hz
22	DB	CKP Sensor (-) Signal	390-520 Hz
24	BK/WT	Power Ground	<0.1v
25	BK	Chassis Ground	<0.1v
26	YL/BK	Coil 'A' Driver (Cyl 1 & 5)	6° dwell
27	OR/YLEL	Shift Solenoid 'A' Control	12v, 55 mph: 1v
28	LB	Low Speed Fan Control	On: 1v, Off: 12v
30	DB/LG	Anti-theft Indicator	Refer to PAT System
31	YL/LG	PSP Switch Signal	Straight: 0v, Turned: 12v
32	YT	Knock Sensor (-) Signal	<0.050v
35	RD/BK	HO2S-12 (B1 S2) Signal	0.1-1.1v
36	TN/LB	MAF Sensor Return	<0.050v
37	OR/BK	TFT Sensor Signal	2.10-2.40v
38	LG/RD	ECT Sensor Signal	0.5-0.6v
39	GY	IAT Sensor Signal	1.5-2.5v
40	DG/YT	Fuel Pump Monitor	On: 12v, Off: 0v
41	DG/WT	A/C High Pressure Switch	On: 1v, Off: 12v
42	LG/YL	Generator Cutoff Relay	0.1v (relay on)
44	PK	Starter Relay Control	KOEC: 9-11v
45	LB/RD	Generator Sensor Signal	130 Hz (30%)
46	LG/VT	High Speed Fan Control	On: 1v, Off: 12v
47	BR/PK	EGR VR Solenoid	0%, 55 mph: 45%
48	TN/YL	Clean Tachometer Output	33-45 Hz
49	LB/BK	Digital TR2 Sensor	In 'P': 0v, 55 mph: 11v
50	WT/BK	Digital TR4 Sensor	In 'P': 0v, 55 mph: 11v
51	BK/WT	Power Ground	<0.1v
52	YL/RD	Coil 'B' Driver (Cyl 3 & 4)	6° dwell
53	PK/BK	Shift Solenoid 'C' Control	12v, 55 mph: 1v
54	VT/YL	TCC Solenoid Control	0%, 55 mph: 95%
55	PK/LB	Keep Alive Power	12-14v
56	GY/YL	EVAP Purge Solenoid	0-10 Hz (0-100%)
55	DG/OR	Keep Alive Power	12-14v
57	YL/RD	Knock Sensor (+) Signal	0v
60	GY/LB	HO2S-11 (B1 S1) Signal	0.1-1.1v
61	VT/LG	HO2S-22 (B2 S2) Signal	0.1-1.1v
62	RD/PK	FTP Sensor Signal	2.6v (0 in. H20)
63	WT/YL	Injector Pressure Sensor	2.74v (39 psi)
64	RD/BK	Digital TR3A Sensor	In 'P': 0v, 55 mph: 1.7v

2001 Sable 3.0L 4v V6 MFI VIN S (A/T) 104 Pin Connector

PCM Pin #	Wire Color	Circuit Description (104 Pin)	Value at Hot Idle
65	BR/LG	DPFE Sensor Signal	0.95-1.05v
66	TN/OR	Evaporator Air Temp. Sensor	4.95-5.15v
67	VT/WT	EVAP CV Solenoid	0-10 Hz (0-100%)
68	GY/BK	VSS (+) Signal	0 Hz, 55 mph: 125 Hz
69	PK/YL	A/C WOT Relay Control	On: 1v, Off: 12v
70, 72	---	Not Used	---
71	RD	Vehicle Power	12-14v
73	TN/BK	Injector 5 Control	2.2-2.7 ms
74	BR/YL	Injector 3 Control	2.2-2.7 ms
75	TN	Injector 1 Control	2.2-2.7 ms
76, 77	BK/WT	Power Ground	<0.1v
78	YL/WT	Coil 'C' Driver (Cyl 2 & 6)	6° dwell
79	WT/LG	TCIL (lamp) Control	On: 1v, Off: 12v
80	LB/OR	Fuel Pump Control	On: 1v, Off: 12v
81	BR/OR	EPC Solenoid Control	9.0v (15 psi)
82	BK/WT	Check Cap Indicator Light	On: 1v, Off: 12v
83	WT/LB	IAC Motor Control	10v (55% duty cycle)
84	DB/YL	OSS Sensor Signal	0 Hz, 55 mph: 131 Hz
85	DB/OR	CID Sensor Signal	5-7 Hz
86	BK/YL	A/C Clutch Relay Switch	12v (Switch Open)
87	RD/LG	HO2S-21 (B2 S1) Signal	0.1-1.1v
88	LB/RD	MAF Sensor Signal	0.6v
89	GY/WT	TP Sensor Signal	0.53-1.27v
90	BR/WT	Reference Voltage	4.9-5.1v
91	GY/RD	Analog Signal Return	<0.050v
92	RD/LG	Brake Pedal Position Switch	Brake Off: 0v, On: 12v
93	RD/WT	HO2S-11 (B1 S1) Heater	On: 1v, Off: 12v
94	WT/BK	HO2S-21 (B2 S1) Heater	On: 1v, Off: 12v
95	YL/LB	HO2S-12 (B1 S2) Heater	On: 1v, Off: 12v
96	TN/YL	HO2S-22 (B2 S2) Heater	On: 1v, Off: 12v
97	---	Not Used	---
98	LG/RD	Generator Indicator Light	On: 1v, Off: 12v
99	LG/OR	Injector 6 Control	2.2-2.7 ms
100	BR/LB	Injector 4 Control	2.2-2.7 ms
101	WT	Injector 2 Control	2.2-2.7 ms
102	---	Not Used	---
103	BK/WT	Power Ground	<0.1v
104	---	Not Used	---

Pin Connector Graphic

PCM 104-PIN CONNECTOR

Terminal View of 104-Pin PCM Wiring Harness Connector

Standard Colors and Abbreviations

Abbreviation	Color	Abbreviation	Color	Abbreviation	Color
BK	Black	GY	Gray	PK	Purple
BL	Blue	GN	Green	RD	Red
BR	Brown	LG	LT Green	TN	Tan
DB	Dark Blue	OR	Orange	WT	White
DG	DK Green	PK	Pink	YL	Yellow

2002-03 Sable 3.0L 4v V6 MFI VIN S (A/T) 104 Pin Connector

PCM Pin #	Wire Color	Circuit Description (104 Pin)	Value at Hot Idle
1	VT/OR	Shift Solenoid 'B' Control	1v, 55 mph: 1v
2	PK/LG	MIL (lamp) Control	MIL On: 1v, Off: 12v
3	YL/BK	Digital TR1 Sensor	In 'P': 0v, 55 mph: 11v
4	TN/RD	System Voltage (Start)	Cranking: 9-11v
5, 7-11	---	Not Used	---
6	DG/WT	TSS Sensor Signal	50-65 Hz (790-820 rpm)
12	YL/WT	Fuel Level Indicator Signal	1.7v (1/2 full)
13	VT	Flash EEPROM Power	0.1v
14, 19	---	Not Used	---
15	PK/LB	SCP Bus (-) Signal	Digital Signals
16	TN/OR	SCP Bus (+) Signal	Digital Signals
17	BR/YL	Passive Antitheft RX Signal	Digital Signals
18	GY/RD	Passive Antitheft TX Signal	Digital Signals
20	GY/YL	Generator Monitor Signal	130 Hz (45%)
21	GY	CKP Sensor (+) Signal	390-520 Hz
22	DB	CKP Sensor (-) Signal	390-520 Hz
23, 29	---	Not Used	---
24	BK/WT	Power Ground	<0.1v
25	BK	Chassis Ground	<0.1v
26	YL/BK	Coil 1 Driver (Cyl 1 & 5)	6° dwell
27	OR/YL	Shift Solenoid 'A' Control	12v, 55 mph: 1v
28	LB	Engine Cooling Fan Brake Relay Control	Relay Off: 12v, On: 1v
30	DB/LG	Antitheft Indicator	Indicator Off: 12v, On: 1v
31	YL/LG	PSP Switch Signal	Straight: 0v, Turned: 12v
32	YL	Knock Sensor (-) Signal	<0.050v
33-34, 43	---	Not Used	---
35	RD/BK	HO2S-12 (B1 S2) Signal	0.1-1.1v
36	TN/LB	MAF Sensor Return	<0.050v
37	OR/BK	TFT Sensor Signal	2.10-2.40v
38	LG/RD	ECT Sensor Signal	0.5-0.6v
39	GY	IAT Sensor Signal	1.5-2.5v
40	DG/YL	Fuel Pump Monitor	On: 12v, Off: 0v
41	DG/WT	A/C High Pressure Switch	On: 1v, Off: 12v
42	LG/YL	Engine Cooling Fan Relay Control	Relay Off: 12v, On: 1v
44	PK	Starter Relay Control	Relay Off: 12v, On: 1v
45	LB/RD	Generator Sensor Signal	130 Hz (30%)
46	LG/VT	High Speed Fan Control	On: 1v, Off: 12v
47	BR/PK	EGR VR Solenoid Control	0%, 55 mph: 45%
48	TN/YL	Clean Tachometer Output	33-45 Hz
49	LB/BK	Digital TR2 Sensor	In 'P': 0v, 55 mph: 11v
50	WT/BK	Digital TR4 Sensor	In 'P': 0v, 55 mph: 11v
51	BK/WT	Power Ground	<0.1v
52	YL/RD	Coil 2 Driver (Cyl 3 & 4)	6° dwell
53	PK/BK	Shift Solenoid 'C' Control	12v, 55 mph: 1v
54	VT/YL	TCC Solenoid Control	0%, 55 mph: 95%
55	PK/LB	Keep Alive Power	12-14v
56	GY/YL	EVAP Purge Solenoid	0-10 Hz (0-100%)
57	YL/RD	Knock Sensor (+) Signal	0v
58-59	---	Not Used	---
60	GY/LB	HO2S-11 (B1 S1) Signal	0.1-1.1v
61	VT/LG	HO2S-22 (B2 S2) Signal	0.1-1.1v
62	RD/PK	FTP Sensor Signal	2.6v (0 in. H20)
63	WT/YL	Injector Pressure Sensor	2.74v (39 psi)
64	RD/BK	Digital TR3A Sensor	In 'P': 0v, 55 mph: 1.7v

2002-03 Sable 3.0L 4v V6 MFI VIN S (A/T) 104 Pin Connector

PCM Pin #	Wire Color	Circuit Description (104 Pin)	Value at Hot Idle
65	BR/LG	EGR DPFE Sensor Signal	0.95-1.05v
66	TN/OR	Evaporator Air Temperature Sensor	4.95-5.15v
67	VT/WT	EVAP CV Solenoid	0-10 Hz (0-100%)
68	GY/BK	VSS (+) Signal	0 Hz, 55 mph: 125 Hz
69	PK/YL	A/C WOT Relay Control	On: 1v, Off: 12v
71	RD	Vehicle Power (Start-Run)	12-14v
72	---	Not Used	---
73	TN/BK	Injector 5 Control	2.2-2.7 ms
74	BR/YL	Injector 3 Control	2.2-2.7 ms
75	TN	Injector 1 Control	2.2-2.7 ms
76, 77	BK/WT	Power Ground	<0.1v
78	YL/WT	Coil 3 Driver (Cyl 2 & 6)	6° dwell
79	WT/LG	TCIL (lamp) Control	Off: 12v, On: 1v
80	LB/OR	Fuel Pump Control	Off: 12v, On: 1v
81	BR/OR	EPC Solenoid Control	9.0v (15 psi)
82	BK/WT	Check Cap Indicator Light	Off: 12v, On: 1v
83	WT/LB	IAC Motor Control	10v (55% duty cycle)
84	DB/YL	OSS Sensor Signal	0 Hz, 55 mph: 131 Hz
85	DB/OR	CMP Sensor Signal	5-7 Hz
86	BK/YL	A/C Clutch Relay Switch	12v (Switch Open)
87	RD/LG	HO2S-21 (B2 S1) Signal	0.1-1.1v
88	LB/RD	MAF Sensor Signal	0.6v
89	GY/WT	TP Sensor Signal	0.53-1.27v
90	BR/WT	Reference Voltage	4.9-5.1v
91	GY/RD	Sensor Ground	<0.050v
92	RD/LG	Brake Pedal Position Switch	Brake Off: 0v, On: 12v
93	RD/WT	HO2S-11 (B1 S1) Heater	On: 1v, Off: 12v
94	WT/BK	HO2S-21 (B2 S1) Heater	On: 1v, Off: 12v
95	YL/LB	HO2S-12 (B1 S2) Heater	On: 1v, Off: 12v
96	TN/YL	HO2S-22 (B2 S2) Heater	On: 1v, Off: 12v
97	---	Not Used	---
98	LG/RD	Generator Indicator Light	Off: 12v, On: 1v
99	LG/OR	Injector 6 Control	2.2-2.7 ms
100	BR/LB	Injector 4 Control	2.2-2.7 ms
101	WT	Injector 2 Control	2.2-2.7 ms
102, 104	---	Not Used	---
103	BK/WT	Power Ground	<0.1v

Pin Voltage Connector

```
        PCM 104-PIN CONNECTOR

  1 ⊗⊗⊗⊗⊗⊗⊗⊗⊗⊗⊗⊗⊗     ⊗⊗⊗⊗⊗⊗⊗⊗⊗⊗⊗⊗⊗ 26
 27 ⊗⊗⊗⊗⊗⊗⊗⊗⊗⊗⊗⊗⊗     ⊗⊗⊗⊗⊗⊗⊗⊗⊗⊗⊗⊗⊗ 52
 53 ⊗⊗⊗⊗⊗⊗⊗⊗⊗⊗⊗⊗⊗  ⬤  ⊗⊗⊗⊗⊗⊗⊗⊗⊗⊗⊗⊗⊗ 78
 79 ⊗⊗⊗⊗⊗⊗⊗⊗⊗⊗⊗⊗⊗     ⊗⊗⊗⊗⊗⊗⊗⊗⊗⊗⊗⊗⊗ 104
```

Terminal View of 104-Pin PCM Wiring Harness Connector

Standard Colors and Abbreviations

Abbreviation	Color	Abbreviation	Color	Abbreviation	Color
BK	Black	GY	Gray	PK	Purple
BL	Blue	GN	Green	RD	Red
BR	Brown	LG	LT Green	TN	Tan
DB	Dark Blue	OR	Orange	WT	White
DG	DK Green	PK	Pink	YL	Yellow

1991-95 Sable 3.0L V6 2v MFI VIN U (A/T) 60 Pin Connector

PCM Pin #	Wire Color	Circuit Description (60 Pin)	Value at Hot Idle
1	YL	Keep Alive Power	12-14v
2	RD/LG	Brake Switch Signal	Brake Off: 0v, On: 12v
3	DG/WT	VSS (+) Signal	0 Hz, 55 mph: 125 Hz
4	WT/PK	Ignition Diagnostic Monitor	20-31 Hz
4	DG/YT	Ignition Diagnostic Monitor	20-31 Hz
5	GY/BK	TSS Sensor Signal	100 Hz (700 rpm)
6	OR/YL	VSS (+) Signal	0 Hz, 55 mph: 125 Hz
7	LG/RD	ECT Sensor Signal	0.5-0.6v
8	PK/BK	Fuel Pump Monitor	12v, Off: 0v
9	TN/BL	MAF Sensor Return	<0.050v
10	PK/BL	A/C Cycling Clutch Signal	A/C Off: 0v, On: 12v
10	PK	A/C Cycling Clutch Signal	A/C Off: 0v, On: 12v
11	GN/YL	Canister Purge Solenoid	12v, Cruise: 1v
12	LG/OR	Injector 6 Control	5.0-5.6 ms
13	TN/OR	Low Speed Cooling Fan	Fan On: 1v, Off: 12v
14	---	Not Used	---
15	TN/BK	Injector 5 Control	5.0-5.6 ms
16	OR/RD	Ignition System Ground	<0.050v
17	TN/RD	Self-Test Output, MIL Control	MIL Off: 12v, On: 1v
18	TN/OR	Data Bus (+) Signal	Digital Signals
19	PK/BL	Data Bus (-) Signal	Digital Signals
20	BK	Case Ground	<0.050v
21	BR/WT	IAC Motor Control	8-9v
22	BL/OR	Fuel Pump Control	Relay Off: 12v, On: 1v
22	TN/LG	Fuel Pump Control	Relay Off: 12v, On: 1v
23-24	---	Not Used	---
25	LG/PK	IAT Sensor Signal	1.5-2.5v
26	BR/WT	Reference Voltage	4.9-5.1v
27	BR/LG	PFE Sensor Signal	3.2v, 55 mph: 2.9v
28	YL/LG	PSP Switch Signal	Straight: 0v, Turned: 11v
29	---	Not Used	---
30	BL	A/T: MLP Sensor Signal	In 'P': 0v, O/D: 1.7v
30	BL/YL	A/T: MLP Sensor Signal	In 'P': 0v, O/D: 1.7v
31	LG/PK	High Speed Cooling Fan	Fan On: 1v, Off: 12v
32	---	Not Used	---
33	BR/PK	VR Solenoid Control	0%, 55 mph: 45%
34	BL/BK	Data Output Link	Digital Signal
34	BL/PK	Data Output Link	Digital Signal
35	BR/BL	Injector 4 Control	5.0-5.6 ms
36	YL/LG	Spark Output Signal	6.93v (50% d/cycle)
37	RD	Vehicle Power	12-14v
38	WT/YL	EPC Solenoid Control	8-9v (9 psi)
39	BR/YL	Injector 3 Control	5.0-5.6 ms
40	BK/LG	Power Ground	<0.1v

1991-95 Sable 3.0L V6 2v MFI VIN U (A/T) 60 Pin Connector

PCM Pin #	Wire Color	Circuit Description (60 Pin)	Value at Hot Idle
41	---	Not Used	---
42 ('94-'95)	TN/LG	A/C Pressure Sensor	A/C On: 12v, Off: 0v
43	RD/BK	HO2S-11 (B1 S1) Signal	0.1-1.1v
44	GY/BL	HO2S-21 (B1 S1) Signal	0.1-1.1v
45	---	Not Used	---
46	GY/RD	Analog Signal Return	<0.050v
47	GN/WT	TP Sensor Signal	0.8v, 55 mph: 1.1v
48	BR	Self-Test Input Signal	STI On: 1v, Off: 5v
49	OR/BK	TOT Sensor Signal	2.40-2.10v
50	BL/RD	MAF Sensor Signal	0.8v
51	OR/YL	Shift Solenoid 1 Control	12v, 55 mph: 1v
52	PK/OR	Shift Solenoid 2 Control	1v, 55 mph: 12v
53	TN/WT	TCC Solenoid Control	Idle: 0%, 30 mph: 0-45%
54	PK/YL	A/C WOT Relay Control	Relay Off: 12v, On: 1v
55	WT/RD	Shift Solenoid 3 Control	7-9v, 30 mph: 8-9v
56	DB	PIP Sensor Signal	6.93v (50% d/cycle)
56	GY/OR	PIP Sensor Signal	6.93v (50% d/cycle)
57	RD	Vehicle Power	12-14v
58	TN	Injector 1 Control	5.0-5.6 ms
59	WT	Injector 2 Control	5.0-5.6 ms
60	BK/LG	Power Ground	<0.1v

Pin Connector Graphic

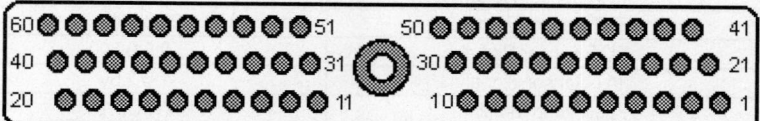

PCM 60-PIN CONNECTOR

Terminal View of 60-Pin PCM Harness Connector

Standard Colors and Abbreviations

Abbreviation	Color	Abbreviation	Color	Abbreviation	Color
BK	Black	GY	Gray	PK	Purple
BL	Blue	GN	Green	RD	Red
BR	Brown	LG	LT Green	TN	Tan
DB	Dark Blue	OR	Orange	WT	White
DG	DK Green	PK	Pink	YL	Yellow

1996-97 Sable 3.0L V6 2v MFI VIN U (A/T) 104 Pin Connector

PCM Pin #	Wire Color	Circuit Description (104 Pin)	Value at Hot Idle
1	PK/OR	Shift Solenoid 2 Control	1v, 55 mph: 12v
2	PK/LG	MIL (lamp) Control	MIL On: 1v, Off: 12v
3-12	---	Not Used	---
13	PK	Flash EEPROM Power	0.1v
14	---	Not Used	---
15	PK/LB	Data Bus (-) Signal	Digital Signals
16	TN/OR	Data Bus (+) Signal	Digital Signals
17-20	---	Not Used	---
21	GY	CKP Sensor (+) Signal	510-540 Hz
22	DB	CKP Sensor (-) Signal	510-540 Hz
23	---	Not Used	---
24	BK/WT	Power Ground	<0.1v
25	BK	Chassis Ground	<0.050v
26	WT/BK	Coil Driver 1 Control	5° dwell
27	OR/YL	Shift Solenoid 1 Control	12v, 55 mph: 1v
28	TN/OR	Low Speed Fan Control	Fan On: 1v, Off: 12v
29	---	Not Used	---
30	DG	Octane Adjust Switch	9.3v (Closed: 0v)
31	YL/LG	PSP Switch Signal	Straight: 0v, Turned: 12v
32	---	Not Used	---
33	PK/OR	VSS (-) Signal	0 Hz, 55 mph: 125 Hz
34	---	Not Used	---
35	RD/BK	HO2S-12 (B1 S2) Signal	0.1-1.1v
36	TN/LB	MAF Sensor Return	<0.050v
37	OR/BK	TFT Sensor Signal	2.40-2.10v
38	LG/RD	ECT Sensor Signal	0.5-0.6v
39	GY	IAT Sensor Signal	1.5-2.5v
40	DG/YT	Fuel Pump Monitor	12v, Off: 0v
41	PK/LB	A/C Cycling Clutch Switch	A/C Switch On: 12v
42-45	---	Not Used	---
46	LG/PK	High Speed Fan Control	Fan On: 1v, Off: 12v
47	BR/PK	VR Solenoid Control	0%, 55 mph: 45%
48	TN/YL	Clean Tachometer Output	42-50 Hz
49-50	---	Not Used	---
51	BK/WT	Power Ground	1v
52	YL/RD	Coil Driver 2 Control	5° dwell
53	PK/BK	Shift Solenoid 3 Control	12v, 55 mph: 1v
54	---	Not Used	---
55	PK	Keep Alive Power	12-14v
56	GY/YL	EVAP VMV Solenoid	0-10 Hz (0-100%)
57	---	Not Used	---
58	GY/BK	VSS (+) Signal	0 Hz, 55 mph: 125 Hz
59	---	Not Used	---
60	GY/LB	HO2S-11 (B1 S1) Signal	0.1-1.1v
61	PK/LG	HO2S-22 (B2 S2) Signal	0.1-1.1v
62	RD/PK	FTP Sensor Signal	2.6v (0" H2O - cap off)
63	---	Not Used	---
64	LBYL	TR Sensor Signal	In 'P': 0v, O/D: 1.7v

1996-97 Sable 3.0L V6 2v MFI VIN U (A/T) 104 Pin Connector

PCM Pin #	Wire Color	Circuit Description (104 Pin)	Value at Hot Idle
65	BR/LG	DPFE Sensor Signal	0.20-1.30v
66, 68	---	Not Used	---
67	PK/WT	EVAP CV Solenoid	0-10 Hz (0-100%)
69	PK/YL	A/C WOT Relay Control	Relay Off: 12v, On: 1v
70	---	Not Used	---
71	RD	Vehicle Power	12-14v
72	---	Not Used	---
73	TN/BK	Injector 5 Control	3.8-4.7 ms
74	BR/YL	Injector 3 Control	3.8-4.7 ms
75	TN	Injector 1 Control	3.8-4.7 ms
76-77	BK/WT	Power Ground	<0.1v
78	YL/WT	Coil Driver 3 Control	5° dwell
79	---	Not Used	---
80	LB/OR	Fuel Pump Control	Relay Off: 12v, On: 1v
81	WT/YL	EPC Solenoid Control	10.6v (40 psi)
82	PK/YL	TCC Solenoid Control	12v (0%)
83	WT/LB	IAC Motor Control	10v (45% d/cycle)
84	DG/WT	TSS Sensor Signal	50-65 Hz (790-820 rpm)
85	DB/OR	CID Sensor Signal	6-8 Hz
86	BK/YL	A/C Pressure Switch	Switch On: 12v (Open)
87	RD/LG	HO2S-21 (B1 S1) Signal	0.1-1.1v
88	LB/RD	MAF Sensor Signal	0.9v
89	GY/WT	TP Sensor Signal	0.53-1.27v
90	BR/WT	Reference Voltage	4.9-5.1v
91	GY/RD	Analog Signal Return	<0.050v
92	RD/LG	Brake Pedal Position Switch	Brake Off: 0v, On: 12v
93	RD/WT	HO2S-11 (B1 S1) Heater	1v, Off: 12v
94	WT/BK	HO2S-21 (B2 S1) Heater	1v, Off: 12v
95	YL/LB	HO2S-12 (B1 S2) Heater	1v, Off: 12v
96	TN/YL	HO2S-22 (B2 S2) Heater	1v, Off: 12v
97	RD	Vehicle Power	12-14v
98	---	Not Used	---
99	LG/OR	Injector 6 Control	3.8-4.7 ms
100	BR/LB	Injector 4 Control	3.8-4.7 ms
101	WT	Injector 2 Control	3.8-4.7 ms
102	---	Not Used	---
103	BK/WT	Power Ground	<0.1v
104	---	Not Used	---

Pin Connector Graphic

PCM 104-PIN CONNECTOR

Terminal View of 104-Pin PCM Wiring Harness Connector

Standard Colors and Abbreviations

Abbreviation	Color	Abbreviation	Color	Abbreviation	Color
BK	Black	GY	Gray	PK	Purple
BL	Blue	GN	Green	RD	Red
BR	Brown	LG	LT Green	TN	Tan
DB	Dark Blue	OR	Orange	WT	White
DG	DK Green	PK	Pink	YL	Yellow

1998-99 Sable 3.0L V6 MFI VIN U (A/T) 104 Pin Connector

PCM Pin #	Wire Color	Circuit Description (104 Pin)	Value at Hot Idle
1	VTN/OR	Shift Solenoid 'B' Control	1v, 55 mph: 12v
2	PK/LG	MIL (lamp) Control	MIL On: 1v, Off: 12v
3	YL/BK	Digital TR1 Sensor	In 'P': 0v, 55 mph: 11v
4-11	---	Not Used	---
12	YL/WT	Fuel Level Indicator Signal	1.7v (1/2 full)
13	PK	Flash EEPROM Power	0.1v
15	PK/LB	Data Bus (-) Signal	Digital Signals
16	TN/OR	Data Bus (+) Signal	Digital Signals
17-18	---	Not Used	---
21	GY	CKP Sensor (+) Signal	510-540 Hz
22	DB	CKP Sensor (-) Signal	510-540 Hz
24	BK/WT	Power Ground	<0.1v
25	BK	Chassis Ground	<0.1v
26	YL/BK	Coil Driver 1 Control	6° dwell
27	OR/YL	Shift Solenoid 'A' Control	12v, 55 mph: 1v
28	LB	Low Speed Fan Control	On: 1v, Off: 12v
30	---	Not Used	---
31	YL/LG	PSP Switch Signal	Straight: 0v, Turned: 12v
32	---	Not Used	---
33	PK/OR	VSS (-) Signal	0 Hz, 55 mph: 125 Hz
35	RD/BK	HO2S-12 (B1 S2) Signal	0.1-1.1v
36	TN/LB	MAF Sensor Return	<0.050v
37	OR/BK	TFT Sensor Signal	2.10-2.40v
38	LG/RD	ECT Sensor Signal	0.5-0.6v
39	GY	IAT Sensor Signal	1.5-2.5v
40	DG/YT	Fuel Pump Monitor	On: 12v, Off: 0v
41	DG/WT	A/C High Pressure Switch	AC On: 12v, Off: 0v
42-44	---	Not Used	---
45	PK/WT	EVAP CV Solenoid	0-10 Hz (0-100%)
46	LG/PK	High Speed Fan Control	On: 1v, Off: 12v
47	BR/PK	EGR VR Solenoid	0%, 55 mph: 45%
48	TN/YL	Clean Tachometer Output	42-50 Hz
49	LB/BK	Digital TR2 Sensor	In 'P': 0v, 55 mph: 11v
50	WT/BK	Digital TR4 Sensor	In 'P': 0v, 55 mph: 11v
51	BK/WT	Power Ground	<0.1v
52	YL/RD	Coil Driver 2 Control	6° dwell
53	PK/BK	Shift Solenoid 'C' Control	12v, 55 mph: 1v
54	DG/WT	TSS Sensor Signal	50-65 Hz (820-900 rpm)
55	DG/OR	Keep Alive Power	12-14v
56	GY/YL	EVAP Purge Solenoid	0-10 Hz (0-100%)
57	---	Not Used	---
58	GY/BK	VSS (+) Signal	0 Hz, 55 mph: 125 Hz
59	---	Not Used	---
60	GY/LB	HO2S-11 (B1 S1) Signal	0.1-1.1v
61	PK/LG	HO2S-22 (B2 S2) Signal	0.1-1.1v
62	RD/PK	FTP Sensor Signal	2.6v (0" H2O - cap off)
63	---	Not Used	---
64	LBYL	Digital TR3A Sensor	In 'P': 0v, in O/D: 1.7v

1998-99 Sable 3.0L V6 MFI VIN U (A/T) 104 Pin Connector

PCM Pin #	Wire Color	Circuit Description (104 Pin)	Value at Hot Idle
65	BR/LG	DPFE Sensor Signal	0.20-1.03v
67-68	---	Not Used	---
69	PK/YL	A/C WOT Relay Control	On: 1v, Off: 12v
70	---	Not Used	---
71	RD	Vehicle Power	12-14v
72	---	Not Used	---
73	TN/BK	Injector 5 Control	2.3-2.8 ms
74	BR/YL	Injector 3 Control	2.3-2.8 ms
75	TN	Injector 1 Control	2.3-2.8 ms
76, 77	BK/WT	Power Ground	<0.1v
78	YL/WT	Coil Driver 3 Control	6° dwell
80	LB/OR	Fuel Pump Control	On: 1v, Off: 12v
81	WT/YL	EPC Solenoid Control	10.6v (40 psi)
81-82	---	Not Used	---
83	WT/LB	IAC Motor Control	10v (32% duty cycle)
84	DG/WT	TSS Sensor Signal	50-65 Hz (790-820 rpm)
85	DB/OR	CID Sensor Signal	6-8 Hz
86	BK/YL	A/C Pressure Switch	A/C On: 12v (Open)
87	RD/LG	HO2S-21 (B2 S1) Signal	0.1-1.1v
88	LB/RD	MAF Sensor Signal	0.9v
89	GY/WT	TP Sensor Signal	0.53-1.27v
90	BR/WT	Reference Voltage	4.9-5.1v
91	GY/RD	Analog Signal Return	<0.050v
92	RD/LG	Brake Pedal Position Switch	Brake Off: 0v, On: 12v
93	RD/WT	HO2S-11 (B1 S1) Heater	On: 1v, Off: 12v
94	WT/BK	HO2S-21 (B2 S1) Heater	On: 1v, Off: 12v
95	YL/LB	HO2S-12 (B1 S2) Heater	On: 1v, Off: 12v
96	TN/YL	HO2S-22 (B2 S2) Heater	On: 1v, Off: 12v
97 ('98)	RD	Vehicle Power	12-14v
99	LG/OR	Injector 6 Control	2.3-2.8 ms
100	BR/LB	Injector 4 Control	2.3-2.8 ms
101	WT	Injector 2 Control	2.3-2.8 ms
102	---	Not Used	---
103	BK/WT	Power Ground	<0.1v
104	---	Not Used	---

Pin Connector Graphic

PCM 104-PIN CONNECTOR

Terminal View of 104-Pin PCM Wiring Harness Connector

Standard Colors and Abbreviations

Abbreviation	Color	Abbreviation	Color	Abbreviation	Color
BK	Black	GY	Gray	PK	Purple
BL	Blue	GN	Green	RD	Red
BR	Brown	LG	LT Green	TN	Tan
DB	Dark Blue	OR	Orange	WT	White
DG	DK Green	PK	Pink	YL	Yellow

2000 Sable 3.0L V6 MFI VIN U (A/T) 104 Pin Connector

PCM Pin #	Wire Color	Circuit Description (104 Pin)	Value at Hot Idle
1	VTN/OR	Shift Solenoid 'B' Control	1v, 55 mph: 12v
2	PK/LG	MIL (lamp) Control	MIL On: 1v, Off: 12v
3	YL/BK	Digital TR1 Sensor	In 'P': 0v, 55 mph: 11v
4	BN/PK	Start System Voltage	Cranking: 9-11v
5	WT/OR	Secondary Air Injection Relay	On: 1v, Off: 12v
6	DG/WT	TSS Sensor Signal	50-65 Hz (790-820 rpm)
7-11, 14	---	Not Used	---
12	YL/WT	Fuel Level Indicator Signal	1.7v (1/2 full)
13	VT	Flash EEPROM Power	0.1v
15	PK/LB	SCP Data Bus (-) Signal	Digital Signals
16	TN/OR	SCP Data Bus (+) Signal	Digital Signals
17	BN/Y	SCI Receive (RX) Signal	Digital Signals
18	GY/RD	SCI Transmit (TX) Signal	Digital Signals
19	---	Not Used	---
20	GY/YL	Generator Field Control	0 Hz
21	GY	CKP Sensor (+) Signal	510-540 Hz
22	DB	CKP Sensor (-) Signal	510-540 Hz
24	BK/WT	Power Ground	<0.1v
25	BK	Case Ground	<0.1v
26	YL/BK	Coil 1 Driver (Cyl 1 & 5)	6° dwell
27	OR/YL	Shift Solenoid 'A' Control	12v, 55 mph: 1v
28	LB	Low Speed Fan Control	On: 1v, Off: 12v
29	---	Not Used	---
30	DB/LG	Passive Anti-theft Indicator	Lamp On: 1v, Off: 12v
31	YL/LG	PSP Switch Signal	Straight: 0v, Turned: 12v
32	YL	Knock Sensor 1 (-) Signal	0v
33-34, 43	---	Not Used	---
35	RD/BK	HO2S-12 (B1 S2) Signal	0.1-1.1v
36	TN/LB	MAF Sensor Return	<0.050v
37	OR/BK	TFT Sensor Signal	2.10-2.40v
38	LG/RD	ECT Sensor Signal	0.5-0.6v
39	GY	IAT Sensor Signal	1.5-2.5v
40	DG, PK	Fuel Pump Monitor	On: 12v, Off: 0v
41	DG/WT	A/C High Pressure Switch	AC On: 12v, Off: 0v
42	LG/YL	Generator Cutoff Relay	0.1v (with fan on)
44	PK	Starter Relay Control	KOEC: 9-11v
45	LB/RD	Generator Sensor Signal	TBD
46	LG/VT	High Speed Fan Control	On: 1v, Off: 12v
47	BR/PK	EGR VR Solenoid	0%, 55 mph: 45%
48	TN/YL	Clean Tachometer Output	42-50 Hz
49	LB/BK	Digital TR2 Sensor	In 'P': 0v, 55 mph: 11v
50	WT/BK	Digital TR4 Sensor	In 'P': 0v, 55 mph: 11v
51	BK/WT	Power Ground	<0.1v
52	YL/RD	Coil Driver 2 (Cyl 3 & 4)	6° dwell
53	PK/BK	Shift Solenoid 'C' Control	12v, 55 mph: 1v
54	VT/YL	TCC Solenoid Control	12v (0%)
55	PK/LB	Keep Alive Power	12-14v
56	GY/YL	EVAP Purge Solenoid	0-10 Hz (0-100%)
57	YL/RD	Knock Sensor 1 (+) Signal	0v
60	GY/LB	HO2S-11 (B1 S1) Signal	0.1-1.1v
61	VT/LG	HO2S-22 (B2 S2) Signal	0.1-1.1v
62	RD/PK	FTP Sensor Signal	2.6v (0" H2O - cap off)
63	WT/YL	Injector Pressure Sensor	2.74v (39 psi)
64	LBYL	Digital TR3A Sensor	In 'P': 0v, in O/D: 1.7v

2000 Sable 3.0L V6 MFI VIN U (A/T) 104 Pin Connector

PCM Pin #	Wire Color	Circuit Description (104 Pin)	Value at Hot Idle
65	BN/LG	DPFE Sensor Signal	0.20-1.30v
66	TN/OR	Evaporator Air Temp. Sensor	4.95-5.15v
67	VT/WT	EVAP Purge Solenoid	0-10 Hz (0-100%)
68	GY/BK	VSS (+) Signal	0 Hz, 55 mph: 125 Hz
69	PK/YL	A/C WOT Relay Control	On: 1v, Off: 12v
70	BN	Secondary Air Injection Relay	On: 1v, Off: 12v
71	RD	Vehicle Power	12-14v
73	TN/BK	Injector 5 Control	2.3-2.8 ms
74	BR/YL	Injector 3 Control	2.3-2.8 ms
75	TN	Injector 1 Control	2.3-2.8 ms
76, 77	BK/WT	Power Ground	<0.1v
78	YL/WT	Coil 3 Driver (Cyl 2 & 6)	6° dwell
79	WT/LG	Check Transaxle Light	On: 1v, Off: 12v
80	LB/OR	Fuel Pump Control	On: 1v, Off: 12v
81	BR/OR	EPC Solenoid Control	10.6v (40 psi)
82	BK/WT	Check Cap Indicator Light	On: 1v, Off: 12v
83	WT/LB	IAC Motor Control	10v (32% duty cycle)
84	DB/YL	OSS Sensor Signal	0 Hz, 55 mph: 131 Hz
85	DB/OR	CID Sensor Signal	6-8 Hz
86	BK/YL	A/C Pressure Clutch Relay	A/C On: 12v, Off: 0v
87	RD/LG	HO2S-21 (B2 S1) Signal	0.1-1.1v
88	LB/RD	MAF Sensor Signal	0.9v
89	GY/WT	TP Sensor Signal	0.53-1.27v
90	BR/WT	Reference Voltage	4.9-5.1v
91	GY/RD	Analog Signal Return	<0.050v
92	RD/LG	Brake Pedal Position Switch	Brake Off: 0v, On: 12v
93	RD/WT	HO2S-11 (B1 S1) Heater	On: 1v, Off: 12v
94	WT/BK	HO2S-21 (B2 S1) Heater	On: 1v, Off: 12v
95	YL/LB	HO2S-12 (B1 S2) Heater	On: 1v, Off: 12v
96	TN/YL	HO2S-22 (B2 S2) Heater	On: 1v, Off: 12v
98	LG/RD	Generator Battery Indicator	Digital Signals
99	LG/OR	Injector 6 Control	2.3-2.8 ms
100	BR/LB	Injector 4 Control	2.3-2.8 ms
101	WT	Injector 2 Control	2.3-2.8 ms
102	---	Not Used	---
103	BK/WT	Power Ground	<0.1v
104	---	Not Used	---

Pin Connector Graphic

PCM 104-PIN CONNECTOR

Terminal View of 104-Pin PCM Wiring Harness Connector

Standard Colors and Abbreviations

Abbreviation	Color	Abbreviation	Color	Abbreviation	Color
BK	Black	GY	Gray	PK	Purple
BL	Blue	GN	Green	RD	Red
BR	Brown	LG	LT Green	TN	Tan
DB	Dark Blue	OR	Orange	WT	White
DG	DK Green	PK	Pink	YL	Yellow

2001-03 Sable 3.0L V6 MFI VIN U (A/T) 104 Pin Connector

PCM Pin #	Wire Color	Circuit Description (104 Pin)	Value at Hot Idle
1	VT/OR	Shift Solenoid 'B' Control	1v, 55 mph: 12v
2	PK/LG	MIL (lamp) Control	MIL On: 1v, Off: 12v
3	YL/BK	Digital TR1 Sensor	In 'P': 0v, 55 mph: 11v
4	TN/RD	System Voltage (Start-Run)	Cranking: 9-11v
5	---	Not Used	---
6	DG/WT	TSS Sensor Signal	50-65 Hz (790-820 rpm)
7-10	---	Not Used	---
12	YL/WT	Fuel Level Indicator Signal	1.7v (1/2 full)
13	VT	Flash EEPROM Power	0.1v
14, 19	---	Not Used	---
15	PK/LB	SCP Bus (-) Signal	Digital Signals
16	TN/OR	SCP Bus (+) Signal	Digital Signals
17	BR/YL	Passive Antitheft RX Signal	Digital Signals
18	GY/RD	Passive Antitheft TX Signal	Digital Signals
20	GY/YL	Generator Field Control	130 Hz (45%)
21	GY	CKP Sensor (+) Signal	510-540 Hz
22	DB	CKP Sensor (-) Signal	510-540 Hz
23	---	Not Used	---
24	BK/WT	Power Ground	<0.1v
25	BK	Case Ground	<0.1v
26	YL/BK	Coil 1 Driver (Cyl 1& 5)	6° dwell
27	OR/YL	Shift Solenoid 'A' Control	12v, 55 mph: 1v
28	LB	Engine Cooling Fan Brake Relay Control	Off: 12v, On: 1v
29, 33-34	---	Not Used	---
30	DB/LG	Passive Antitheft Indicator	Lamp On: 1v, Off: 12v
31	YL/LG	PSP Switch Signal	Straight: 0v, Turned: 12v
32	YL	Knock Sensor 1 (-) Signal	<0.050v
35	RD/BK	HO2S-12 (B1 S2) Signal	0.1-1.1v
36	TN/LB	MAF Sensor Return	<0.050v
37	OR/BK	TFT Sensor Signal	2.10-2.40v
38	LG/RD	ECT Sensor Signal	0.5-0.6v
39	GY	IAT Sensor Signal	1.5-2.5v
40	DG/YL	Fuel Pump Monitor	On: 12v, Off: 0v
41	DG/WT	A/C High Pressure Switch	AC On: 12v, Off: 0v
42	LG/YL	Generator Cutoff Relay	0.1v (with fan on)
44	PK	Starter Relay Control	KOEC: 9-11v
45	LB/RD	Generator Sensor Signal	130 Hz (30%)
46	LG/VT	High Cooling Speed Fan Control	Off: 12v, On: 1v
47	BR/PK	EGR VR Solenoid	0%, 55 mph: 45%
48	TN/YL	Clean Tachometer Output	42-50 Hz
49	LB/BK	Digital TR2 Sensor	In 'P': 0v, 55 mph: 11v
50	WT/BK	Digital TR4 Sensor	In 'P': 0v, 55 mph: 11v
51	BK/WT	Power Ground	<0.1v
52	YL/RD	Coil Driver 2 (Cyl 3 & 4)	6° dwell
53	PK/BK	Shift Solenoid 'C' Control	12v, 55 mph: 1v
54	VT/YL	TCC Solenoid Control	12v (0%)
55	PK/LB	Keep Alive Power	12-14v
56	GY/YL	EVAP Purge Solenoid	0-10 Hz (0-100%)
57	YL/RD	Knock Sensor (+) Signal	0v
60	GY/LB	HO2S-11 (B1 S1) Signal	0.1-1.1v
61	VT/LG	HO2S-22 (B2 S2) Signal	0.1-1.1v
62	RD/PK	FTP Sensor Signal	2.6v (0" H2O - cap off)
63	WT/YL	Injector Pressure Sensor	TBD
64	RD/BK	Digital TR3A Sensor	In 'P': 0v, in O/D: 1.7v

2001-03 Sable 3.0L V6 MFI VIN U (A/T) 104 Pin Connector

PCM Pin #	Wire Color	Circuit Description (104 Pin)	Value at Hot Idle
65	BR/LG	DPFE Sensor Signal	0.95-1.05v
66	TN/OR	Evaporator Air Temperature Sensor	4.95-5.15v
67	VT/WT	EVAP Purge Solenoid	0-10 Hz (0-100%)
68	GY/BK	VSS (+) Signal	Moving: AC pulse signals
69	PK/YL	A/C WOT Relay Control	Off: 12v, On: 1v
70	---	Not Used	---
71	RD	Vehicle Power (Start-Run)	12-14v
73	TN/BK	Injector 5 Control	2.3-2.8 ms
74	BR/YL	Injector 3 Control	2.3-2.8 ms
75	TN	Injector 1 Control	2.3-2.8 ms
76, 77	BK/WT	Power Ground	<0.1v
78	YL/WT	Coil 3 Driver (Cyl 2 & 6)	6° dwell
79	WT/LG	Check Transaxle Light	Indicator Off: 12v, On: 1v
80	LB/OR	Fuel Pump Control	Off: 12v, On: 1v
81	BR/OR	EPC Solenoid Control	10.6v (40 psi)
82	BK/WT	Check Cap Indicator Light	Indicator Off: 12v, On: 1v
83	WT/LB	IAC Motor Control	10v (32% duty cycle)
84	DB/YL	OSS Sensor Signal	0 Hz, 55 mph: 131 Hz
85	DB/OR	CID Sensor Signal	6-8 Hz
86	BK/YL	A/C Pressure Clutch Relay	Relay Off: 12v, On: 1v
87	RD/LG	HO2S-21 (B2 S1) Signal	0.1-1.1v
88	LB/RD	MAF Sensor Signal	0.9v
89	GY/WT	TP Sensor Signal	0.53-1.27v
90	BR/WT	Reference Voltage	4.9-5.1v
91	GY/RD	Sensor Return	<0.050v
92	RD/LG	Brake Pedal Position Switch	Brake Off: 0v, On: 12v
93	RD/WT	HO2S-11 (B1 S1) Heater	On: 1v, Off: 12v
94	WT/BK	HO2S-21 (B2 S1) Heater	On: 1v, Off: 12v
95	YL/LB	HO2S-12 (B1 S2) Heater	On: 1v, Off: 12v
96	TN/YL	HO2S-22 (B2 S2) Heater	On: 1v, Off: 12v
97	RD	Vehicle Power (Start-Run)	12-14v
98	LG/RD	Generator Battery Indicator	Digital Signals
99	LG/OR	Injector 6 Control	2.3-2.8 ms
100	BR/LB	Injector 4 Control	2.3-2.8 ms
101	WT	Injector 2 Control	2.3-2.8 ms
102	---	Not Used	---
103	BK/WT	Power Ground	<0.1v
104	---	Not Used	---

Pin Connector Graphic

```
              PCM 104-PIN CONNECTOR
      1 ◉◉◉◉◉◉◉◉◉◉◉◉   ◉◉◉◉◉◉◉◉◉◉◉◉◉ 26
     27 ◉◉◉◉◉◉◉◉◉◉◉◉   ◉◉◉◉◉◉◉◉◉◉◉◉◉ 52
     53 ◉◉◉◉◉◉◉◉◉◉◉◉  ●  ◉◉◉◉◉◉◉◉◉◉◉◉◉ 78
     79 ◉◉◉◉◉◉◉◉◉◉◉◉   ◉◉◉◉◉◉◉◉◉◉◉◉◉ 104
```

Terminal View of 104-Pin PCM Wiring Harness Connector

Standard Colors and Abbreviations

Abbreviation	Color	Abbreviation	Color	Abbreviation	Color
BK	Black	GY	Gray	PK	Purple
BL	Blue	GN	Green	RD	Red
BR	Brown	LG	LT Green	TN	Tan
DB	Dark Blue	OR	Orange	WT	White
DG	DK Green	PK	Pink	YL	Yellow

1990 Sable 3.8L V6 2v MFI VIN 4 (A/T) 60 Pin Connector

PCM Pin #	Wire Color	Circuit Description (60 Pin)	Value at Hot Idle
1	YL	Keep Alive Power	12-14v
2	RD/LG	PSP Switch Signal	Straight: 0v, Turned: 11v
3	DG/WT	VSS (+) Signal	0 Hz, 55 mph: 125 Hz
4	DG/YT	Ignition Diagnostic Monitor	20-31 Hz
5	---	Not Used	---
6	OR/YL	VSS (-) Signal	0 Hz, 55 mph: 125 Hz
7	LG/YL	ECT Sensor Signal	0.5-0.6v
8	BK/OR	Data Bus (-) Signal	Digital Signals
9	GY/WT	AXOD Solenoid	On: 1v, Off: 12v
10	PK/BL	A/C Cycling Clutch Signal	A/C Off: 0v, On: 12v
10	PK	A/C Cycling Clutch Signal	A/C Off: 0v, On: 12v
11	OR/YL	Speed Control Solenoid (+)	S/C On: 12-14v
12	BR/YL	Injector 3 Control	6.0-6.8 ms
13	BR/BL	Injector 4 Control	6.0-6.8 ms
14	TN/BL	Injector 5 Control	6.0-6.8 ms
15	LG	Injector 6 Control	6.0-6.8 ms
16	GY	Ignition System Ground	<0.050v
17	TN/RD	Self-Test Output, MIL Control	MIL Off: 12v, On: 1v
18	DG/PK	Transmission 3-4 Switch	3-4 Switch Closed: 0.1v
19	OR/YL	Transmission 2-3 Switch	2-3 Switch Closed: 0.1v
20 ('89)	BK	Case Ground	<0.050v
21	OR/BK	IAC Motor Control	9-10v
22	TN/LG	Fuel Pump Control	Relay Off: 12v, On: 1v
23-24	---	Not Used	---
25	LG/PK	ACT Sensor Signal	1.5-2.5v
26	OR/WT	Reference Voltage	4.9-5.1v
27	BR/LG	PFE Sensor Signal	3.2v, 55 mph: 2.9v
28	BL/BK	Speed Control Switch Signal	S/C On: 6.7v
29	DB/LG	HO2S-21 (B1 S1) Signal	0.1-1.1v
30	PK/YL	Neutral Drive Switch	In 'N': 0v, Others: 4.6v
31	GY/YL	Canister Purge Solenoid	12v, Cruise: 1v
32	---	Not Used	---
33	DG	VR Solenoid Control	0%, 55 mph: 45%
34	BL/PK	Data Output Link	Digital Signals
35	WT/PK	S/C Vent Solenoid Control	Vacuum Decreasing: 1v
36	YL/LG	Spark Output Signal	6.93v (duty cycle)
37	RD	Vehicle Power	12-14v
38	---	Not Used	---
39	OR	Speed Control Switch Ground	<0.050v
40	BK/LG	Power Ground	<0.1v

1990 Sable 3.8L V6 2v MFI VIN 4 (A/T) 60 Pin Connector

PCM Pin #	Wire Color	Circuit Description (60 Pin)	Value at Hot Idle
41	PK	High Speed Fan Control	Fan On: 1v, Off: 12v
42	---	Not Used	---
43	DG/PK	HO2S-11 (B1 S1) Signal	0.1-1.1v
44	OR/BK	Data Bus (+) Signal	Digital Signals
45	LG/BK	MAP Sensor Signal	107 Hz (sea level)
46	BK/WT	Analog Signal Return	<0.050v
47	DG/LG	TP Sensor Signal	0.8v, 55 mph: 1.1v
48	WT/BK	Self-Test Input Signal	STI On: 1v, Off: 5v
49	OR	HO2S-11 (Bank 1) Ground	<0.050v
50	PK/BK	Fuel Pump Monitor	Relay Off: 12v, On: 1v
51	GY/BK	S/C Vacuum Solenoid Control	Vacuum Increasing: 1v
52	---	Not Used	---
53	TN/BL	AXOD Lockup Solenoid	12v, 55 mph: 1v
54	RD	A/C WOT Relay Control	Relay Off: 12v, On: 1v
55	TN/OR	Low Speed Fan Control	Fan On: 1v, Off: 12v
56	DB	PIP Sensor Signal	6.93v (50% d/cycle)
57	RD	Vehicle Power	12-14v
58	TN	Injector 1 Control	6.0-6.8 ms
59	WT	Injector 2 Control	6.0-6.8 ms
60	BK/LG	Power Ground	<0.1v

Pin Connector Graphic

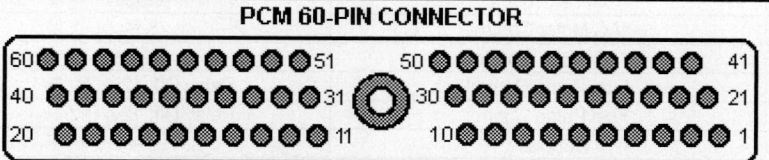

Terminal View of 60-Pin PCM Harness Connector

Standard Colors and Abbreviations

Abbreviation	Color	Abbreviation	Color	Abbreviation	Color
BK	Black	GY	Gray	PK	Purple
BL	Blue	GN	Green	RD	Red
BR	Brown	LG	LT Green	TN	Tan
DB	Dark Blue	OR	Orange	WT	White
DG	DK Green	PK	Pink	YL	Yellow

1991-95 Sable 3.8L V6 2v MFI VIN 4 (A/T) 60 Pin Connector

PCM Pin #	Wire Color	Circuit Description (60 Pin)	Value at Hot Idle
1	YL	Keep Alive Power	12-14v
2	RD/LG	Brake Switch Signal	Brake Off: 0v, On: 12v
3	DG/WT	VSS (-) Signal	0 Hz, 55 mph: 125 Hz
4	WT/PK	Ignition Diagnostic Monitor	20-31 Hz
4	DG/YT	Ignition Diagnostic Monitor	20-31 Hz
5	GY/BK	TSS Sensor Signal	100 Hz (700 rpm)
6	OR/YL	VSS (+) Signal	0 Hz, 55 mph: 125 Hz
7	LG/RD	ECT Sensor Signal	0.5-0.6v
8	PK/BK	Fuel Pump Monitor	12v, Off: 0v
9	TN/BL	MAF Sensor Return	<0.050v
10	PK	A/C Cycling Clutch Signal	A/C Off: 0v, On: 12v
10	PK/BL	A/C Cycling Clutch Signal	A/C Off: 0v, On: 12v
11	GN/YL	Canister Purge Solenoid	12v, Cruise: 1v
12	LG/OR	Injector 6 Control	5.6-6.0 ms
13	TN/OR	Low Speed Fan Control	Fan On: 1v, Off: 12v
14	---	Not Used	---
15	TN/BK	Injector 5 Control	5.6-6.0 ms
16	GY	Ignition System Ground	<0.050v
17	TN/RD	Self-Test Output, MIL Control	MIL Off: 12v, On: 1v
18	OR/BK	Data Bus (+) Signal	Digital Signals
19	PK/BL	Data Bus (-) Signal	Digital Signals
20	BK	Case Ground	<0.050v
21	BL/OR	IAC Motor Control	8-9v
21	BR/WT	IAC Motor Control	8-9v
22	TN/LG	Fuel Pump Control	Relay Off: 12v, On: 1v
22	BL/OR	Fuel Pump Control	Relay Off: 12v, On: 1v
23-24	---	Not Used	---
25	GY	IAT Sensor Signal	1.5-2.5v
26	BR/WT	Reference Voltage	4.9-5.1v
27	BR/LG	PFE Sensor Signal	3.2v, 55 mph: 2.9v
28	YL/LG	PSP Switch Signal	Straight: 0v, Turned: 11v
29	---	Not Used	---
30	BL	MLP Sensor Signal	In 'P': 0v, O/D: 1.7v
30	BL/YL	MLP Sensor Signal	In 'P': 0v, O/D: 1.7v
31	LG/PK	High Speed Fan Control	Fan On: 1v, Off: 12v
32	---	Not Used	---
33	BR/PK	VR Solenoid Control	0%, 55 mph: 45%
34	BL/PK	Data Output Link	Digital Signals
35	BR/BL	Injector 4 Control	5.6-6.0 ms
36	YL/LG	Spark Output Signal	6.93v (50% d/cycle)
37	RD	Vehicle Power	12-14v
38	WT/YL	EPC Solenoid Control	7.8v (8-9 psi)
39	BR/YL	Injector 3 Control	5.6-6.0 ms
40	BK/LG	Power Ground	<0.1v

1991-95 Sable 3.8L V6 2v MFI VIN 4 (A/T) 60 Pin Connector

PCM Pin #	Wire Color	Circuit Description (60 Pin)	Value at Hot Idle
41	---	Not Used	---
42 ('94-'95)	TN/LG	Refrigerant Containment Sw.	5v
43	RD/BK	HO2S-11 (B1 S1) Signal	0.1-1.1v
44	DB/LG	HO2S-21 (B1 S1) Signal	0.1-1.1v
44	GY/BL	HO2S-21 (B1 S1) Signal	0.1-1.1v
45	---	Not Used	---
46	GY/RD	Analog Signal Return	<0.050v
47	GN/WT	TP Sensor Signal	0.8v, 55 mph: 1.1v
48	BR	Self-Test Input Signal	STI On: 1v, Off: 5v
49	OR/BK	TOT Sensor Signal	2.40-2.10v
50	BL/RD	MAF Sensor Signal	0.7v, 55 mph: 2.0v
51	OR/YL	Shift Solenoid 1 Control	12v, 55 mph: 1v
52	PK/OR	Shift Solenoid 2 Control	1v, 55 mph: 12v
53	TN/WT	TCC Solenoid Control	12v, 55 mph: 1v
54	PK/, R	A/C WOT Relay Control	Relay Off: 12v, On: 1v
55	WT/RD	Shift Solenoid 3 Control	12v, 55 mph: 1v
56	DB	PIP Sensor Signal	6.93v (50% d/cycle)
56	GY/OR	PIP Sensor Signal	6.93v (50% d/cycle)
57	RD	Vehicle Power	12-14v
58	TN	Injector 1 Control	5.6-6.0 ms
59	WT	Injector 2 Control	5.6-6.0 ms
60	BK/LG	Power Ground	<0.1v

Pin Connector Graphic

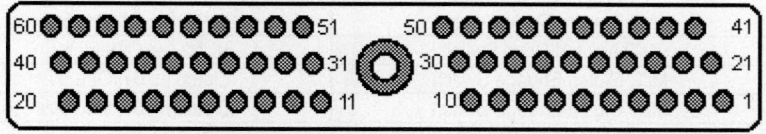

PCM 60-PIN CONNECTOR

Terminal View of 60-Pin PCM Harness Connector

Standard Colors and Abbreviations

Abbreviation	Color	Abbreviation	Color	Abbreviation	Color
BK	Black	GY	Gray	PK	Purple
BL	Blue	GN	Green	RD	Red
BR	Brown	LG	LT Green	TN	Tan
DB	Dark Blue	OR	Orange	WT	White
DG	DK Green	PK	Pink	YL	Yellow

Topaz Pin Tables

1990 Topaz 2.3L I4 MFI VIN S, VIN X (All) 60 Pin Connector

PCM Pin #	Wire Color	Circuit Description (60 Pin)	Value at Hot Idle
1	YL	Keep Alive Power	12-14v
2	---	Not Used	---
3	DG/WT	VSS (+) Signal	0 Hz, 55 mph: 125 Hz
4	DG/YT	Ignition Diagnostic Monitor	20-31 Hz
5	RD/LG	Key Power	12-14v
6	OR/YL	VSS (-) Signal	0 Hz, 55 mph: 125 Hz
7	LG/YL	ECT Sensor Signal	0.5-0.6v
8	PK/BK	Fuel Pump Monitor	12v, Off: 0v
9	---	Not Used	---
10	TN/YL	A/C Cycling Clutch Signal	A/C Off: 0v, On: 12v
11-15	---	Not Used	---
16	BK/OR	Ignition System Ground	<0.050v
16	BK/WT	Ignition System Ground	<0.050v
17	TN/RD	Self-Test Output, MIL Control	MIL Off: 12v, On: 1v
18-19	---	Not Used	---
20	BK	Case Ground	<0.050v
21	BR/WT	IAC Motor Control	8-10v
22	OR/BL	Fuel Pump Control	Relay Off: 12v, On: 1v
23	---	Not Used	---
24	YL/LG	PSP Switch Signal	Straight: 0v, Turned: 5v
25	LG/PK	ACT Sensor Signal	1.5-2.5v
26	OR/WT	Reference Voltage	4.9-5.1v
27	BR/LG	PFE Sensor Signal	3.2v, 55 mph: 2.8v
28	---	Not Used	---
29	DG/PK	HO2S-11 (B1 S1) Signal	0.1-1.1v
30	GY/OR	A/T: Neutral Drive Switch	In 'N': 0v, Others: 5v
30	WT/PK	A/T: Neutral Drive Switch	In 'N': 0v, Others: 5v
30	GY/OR	M/T: Clutch Engage Switch	Clutch In: 0v, Out: 5v
30	WT/PK	M/T: Clutch Engage Switch	Clutch In: 0v, Out: 5v
31	GY/YL	Canister Purge Solenoid	12v, 55 mph: 1v
32	---	Not Used	---
33	YL	VR Solenoid Control	0%, 55 mph: 0-45%
34-35	---	Not Used	---
36	YL/LG	Spark Output Signal	6.93v (50% d/cycle)
37	RD	Vehicle Power	12-14v
38-39	---	Not Used	---
40	BK/LG	Power Ground	<0.1v

1990 Topaz 2.3L I4 MFI VIN S, VIN X (All) 60 Pin Connector

PCM Pin #	Wire Color	Circuit Description (60 Pin)	Value at Hot Idle
41-42	---	Not Used	
43	LG/PK	A/C Demand Switch Signal	A/C On: 12v, Off: 0v
44	---	Not Used	---
45	LG/BK	MAP Sensor Signal	107 Hz (sea level)
46	BK/WT	Analog Signal Return	<0.050v
47	DG/LG	TP Sensor Signal	0.8v, 55 mph: 1.1v
48	WT/RD	Self-Test Input Signal	STI On: 1v, Off: 5v
49	OR	HO2S-11 (Bank 1) Ground	<0.050v
50	---	Not Used	---
51	WT/RD	Air Management 1 Solenoid	AM1 On: 1v, Off: 12v,
52	---	Not Used	---
53	TN/BL	M/T: Shift Indicator Light	SIL On: 1v, Off: 12v
54	OR/LB	A/C WOT Relay Control	Relay Off: 12v, On: 1v
55	---	Not Used	---
56	DB	PIP Sensor Signal	6.93v (50% d/cycle)
57	RD	Vehicle Power	12-14v
58	TN/RD	Injector Bank 1	3.8-3.9 ms
59	TN/OR	Injector Bank 2	3.8-3.9 ms
60	BK/LG	Power Ground	<0.1v

Pin Connector Graphic

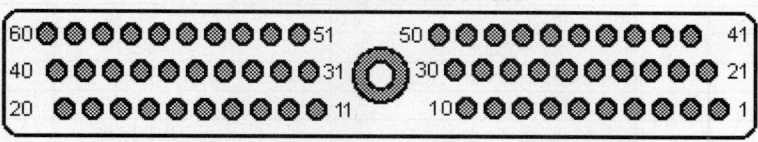

PCM 60-PIN CONNECTOR

Terminal View of 60-Pin PCM Harness Connector

Standard Colors and Abbreviations

Abbreviation	Color	Abbreviation	Color	Abbreviation	Color
BK	Black	GY	Gray	PK	Purple
BL	Blue	GN	Green	RD	Red
BR	Brown	LG	LT Green	TN	Tan
DB	Dark Blue	OR	Orange	WT	White
DG	DK Green	PK	Pink	YL	Yellow

1991 Topaz 2.3L I4 MFI VIN S, VIN X (All) 60 Pin Connector

PCM Pin #	Wire Color	Circuit Description (60 Pin)	Value at Hot Idle
1	YL	Keep Alive Power	12-14v
2	---	Not Used	---
3	DG/WT	VSS (+) Signal	0 Hz, 55 mph: 125 Hz
4	TN/YL	Ignition Diagnostic Monitor	20-31 Hz
5	---	Not Used	---
6	OR/YL	VSS (-) Signal	0 Hz, 55 mph: 125 Hz
7	LG/RD	ECT Sensor Signal	0.5-0.6v
8	PK/BK	Fuel Pump Monitor	12v, Off: 0v
9	---	Not Used	---
10	DG	A/C Cycling Clutch Signal	A/C Off: 0v, On: 12v
10	BK/YL	A/C Cycling Clutch Signal	A/C Off: 0v, On: 12v
11-15	---	Not Used	---
16	OR/RD	Ignition System Ground	<0.050v
17	TN/RD	Self-Test Output, MIL Control	MIL Off: 12v, On: 1v
18-19	---	Not Used	---
20	BK	Case Ground	<0.050v
21	BR/WT	IAC Motor Control	8-10v
22	DG/YT	Fuel Pump Control	Relay Off: 12v, On: 1v
23	---	Not Used	---
24	YL/LG	PSP Switch Signal	Straight 0v, Turned 11v
25	LG/PK	ACT Sensor Signal	1.5-2.5v
26	BR/WT	Reference Voltage	4.9-5.1v
27	BR/LG	PFE Sensor Signal	3.2v, 55 mph: 2.9v
28	---	Not Used	---
29	RD/BK	HO2S-11 (B1 S1) Signal	0.1-1.1v
30	WT/PK	Neutral Drive Switch	In 'N': 0v, Others: 12v
31	GY/YL	Canister Purge Solenoid	12v, 55 mph: 1v
32	---	Not Used	---
33	YL	VR Solenoid Control	0%, 55 mph: 0-45%
34-35	---	Not Used	---
36	YL/LG	Spark Output Signal	6.93v (50% d/cycle)
37	RD	Vehicle Power	12-14v
38-39	---	Not Used	---
40	BK/LG	Power Ground	<0.1v

1991 Topaz 2.3L I4 MFI VIN S, VIN X (All) 60 Pin Connector

PCM Pin #	Wire Color	Circuit Description (60 Pin)	Value at Hot Idle
41-42	---	Not Used	---
43	PK	A/C Demand Switch Signal	A/C On: 12v, AC Off: 0v
44	---	Not Used	---
45	LG/BK	MAP Sensor Signal	107 Hz (sea level)
46	GY/RD	Analog Signal Return	<0.050v
47	GN/WT	TP Sensor Signal	0.8v, 55 mph: 1.1v
48	WT/PK	Self-Test Input Signal	STI On: 1v, Off: 5v
49	OR	HO2S-11 (Bank 1) Ground	<0.050v
50	---	Not Used	---
51	WT/RD	Air Management 1 Solenoid	AM1 On: 1v, Off: 12v
52	---	Not Used	---
53	TN/WT	M/T: Shift Indicator Light	SIL On: 1v, Off: 12v
54	OR/BL	A/C WOT Relay Control	Relay Off: 12v, On: 1v
55	---	Not Used	---
56	DB	PIP Sensor Signal	6.93v (50% d/cycle)
57	RD	Vehicle Power	12-14v
58	TN/RD	Injector Bank 1	3.6-5.0 ms
59	TN/OR	Injector Bank 2	3.6-5.0 ms
60	BK/LG	Power Ground	<0.1v

Pin Connector Graphic

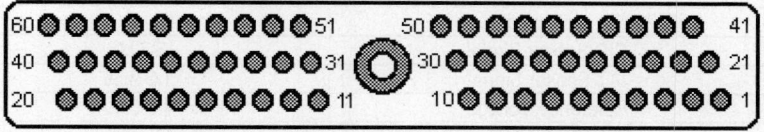

Standard Colors and Abbreviations

Abbreviation	Color	Abbreviation	Color	Abbreviation	Color
BK	Black	GY	Gray	PK	Purple
BL	Blue	GN	Green	RD	Red
BR	Brown	LG	LT Green	TN	Tan
DB	Dark Blue	OR	Orange	WT	White
DG	DK Green	PK	Pink	YL	Yellow

1992-94 Topaz 2.3L I4 MFI VIN X (All) 60 Pin Connector

PCM Pin #	Wire Color	Circuit Description (60 Pin)	Value at Hot Idle
1	YL	Keep Alive Power	12-14v
2	---	Not Used	---
3	DG/WT	VSS (+) Signal	0 Hz, 55 mph: 125 Hz
4	WT/PK	Ignition Diagnostic Monitor	20-31 Hz
5	---	Not Used	
6	OR/YL	VSS (-) Signal	0 Hz, 55 mph: 125 Hz
7	LG/RD	ECT Sensor Signal	0.5-0.6v
8	PK/BK	Fuel Pump Monitor	12v, Off: 0v
9	TN/BL	MAF Sensor Return	<0.050v
10	PK/BL	A/C Cycling Clutch Signal	A/C Off: 0v, On: 12v
11	GN/YL	Canister Purge Solenoid	12v, 55 mph: 1v
12	---	Not Used	---
13	TN/OR	Low Speed Fan Control	Fan On: 1v, Off: 12v
14-15	---	Not Used	---
16	OR/RD	Ignition System Ground	<0.050v
17	TN/RD	Self-Test Output, MIL Control	MIL Off: 12v, On: 1v
18	TN/OR	Data Bus (+) Signal	Digital Signals
19	PK/BL	Data Bus (-) Signal	Digital Signals
20	BK	Case Ground	<0.050v
21	WT	IAC Motor Control	8-10v
21	WT/BL	IAC Motor Control	8-10v
22	DG/YT	Fuel Pump Control	Relay Off: 12v, On: 1v
23	---	Not Used	---
24	DG	CID Sensor Signal	5-7 Hz
25	GY	IAT Sensor Signal	1.5-2.5v
26	BR/WT	Reference Voltage	4.9-5.1v
27	BR/LG	PFE Sensor Signal	3.2v, 55 mph: 2.9v
28	YL/LG	PSP Switch Signal	Straight: 0v, Turned: 11v
29	---	Not Used	---
30	PK/YL	A/T: Neutral Drive Switch	In 'N': 0v, Others: 5v
30	PK/YL	M/T: Clutch Engage Switch	Clutch In: 0v, Out: 5v
31	WT/RD	Air Mgmt. Bypass Solenoid	0v, at 55 mph: 12v
32	---	Not Used	---
33	YL	VR Solenoid Control	0%, 55 mph: 45%
34	---	Not Used	---
35	BR/BL	Injector 4 Control	5.1-5.3 ms
36	PK	Spark Output Signal	6.93v (50% d/cycle)
37	RD	Vehicle Power	12-14v
38	---	Not Used	---
39	BR/YL	Injector 3 Control	5.1-5.3 ms
40	BK/LG	Power Ground	<0.1v

1992-94 Topaz 2.3L I4 MFI VIN X (All) 60 Pin Connector

PCM Pin #	Wire Color	Circuit Description (60 Pin)	Value at Hot Idle
41-43	---	Not Used	---
44	RD/BK	HO2S-11 (B1 S1) Signal	0.1-1.1v
45	---	Not Used	---
46	GY/RD	Analog Signal Return	<0.050v
47	GN/WT	TP Sensor Signal	0.8v, 55 mph: 1.1v
48	WT/PK	Self-Test Input Signal	STI On: 1v, Off: 5v
49	---	Not Used	---
50	LB/RD	MAF Sensor Signal	0.8v
50	BL/BK	MAF Sensor Signal	0.8v
51-52	---	Not Used	---
53	TN/WT	M/T: Shift Indicator Light	SIL On: 1v, Off: 12v
54	PK/YL	A/C WOT Relay Control	Relay Off: 12v, On: 1v
55	---	Not Used	---
56	DB	PIP Sensor Signal	6.93v (50% d/cycle)
57	RD	Vehicle Power	12-14v
58	TN	Injector 1 Control	5.1-5.3 ms
59	WT	Injector 2 Control	5.1-5.3 ms
60	BK/LG	Power Ground	<0.1v

Pin Connector Graphic

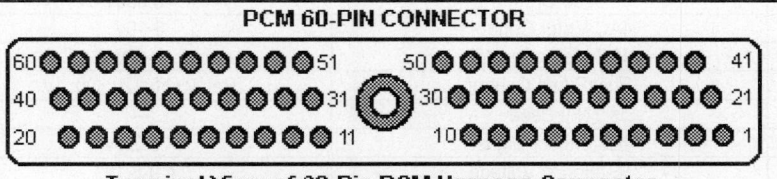

PCM 60-PIN CONNECTOR

Terminal View of 60-Pin PCM Harness Connector

Standard Colors and Abbreviations

Abbreviation	Color	Abbreviation	Color	Abbreviation	Color
BK	Black	GY	Gray	PK	Purple
BL	Blue	GN	Green	RD	Red
BR	Brown	LG	LT Green	TN	Tan
DB	Dark Blue	OR	Orange	WT	White
DG	DK Green	PK	Pink	YL	Yellow

1992-94 Topaz 3.0L V6 MFI VIN U (All) 60 Pin Connector

PCM Pin #	Wire Color	Circuit Description (60 Pin)	Value at Hot Idle
1	YL	Keep Alive Power	12-14v
2	---	Not Used	---
3	DG/WT	VSS (+) Signal	0 Hz, 55 mph: 125 Hz
4	WT/PK	Ignition Diagnostic Monitor	20-31 Hz
5	---	Not Used	---
6	OR/YL	VSS (-) Signal	0 Hz, 55 mph: 125 Hz
7	LG/RD	ECT Sensor Signal	0.5-0.6v
8	PK/BK	Fuel Pump Monitor	12v, Off: 0v
9	TN/BL	MAF Sensor Return	<0.050v
10	PK/BL	A/C Cycling Clutch Signal	A/C Off: 0v, On: 12v
11	GN/YL	Canister Purge Solenoid	12v, 55 mph: 1v
12	LG/OR	Injector 6 Control	4.6-5.3 ms
13	TN/OR	Low Speed Fan Control	Fan On: 1v, Off: 12v
14	---	Not Used	---
15	TN/BK	Injector 5 Control	4.6-5.3 ms
16	OR/RD	Ignition System Ground	<0.050v
17	TN/RD	Self-Test Output, MIL Control	MIL Off: 12v, On: 1v
18	TN/OR	Data Bus (+) Signal	Digital Signals
19	PK/BL	Data Bus (-) Signal	Digital Signals
20	BK	Case Ground	<0.050v
21	WT	IAC Motor Control	8-10v
21	WT/BK	IAC Motor Control	8-10v
22	DG/YT	Fuel Pump Control	Relay Off: 12v, On: 1v
23-24	---	Not Used	---
25	GY	IAT Sensor Signal	1.5-2.5v
26	BR/WT	Reference Voltage	4.9-5.1v
27	BR/LG	PFE Sensor Signal	3.2v, 55 mph: 2.9v
28	YL/LG	PSP Switch Signal	Straight: 0v, Turned: 11v
29	---	Not Used	---
30	PK/YL	A/T: Neutral Drive Switch	In 'N': 0v, Others: 5v
30	PK/YL	M/T: Clutch Position Switch	Clutch In: 0v, Out: 5v
31	BL	High Speed Fan Control	Fan On: 1v, Off: 12v
32	---	Not Used	---
33	YL	VR Solenoid Control	0%, 55 mph: 45%
34	---	Not Used	---
35	BR/BL	Injector 4 Control	4.6-5.3 ms
36	PK	Spark Output Signal	6.93v (50% d/cycle)
37	RD	Vehicle Power	12-14v
38	---	Not Used	---
39	BR/YL	Injector 3 Control	4.6-5.3 ms
40	BK/LG	Power Ground	<0.1v

1992-94 Topaz 3.0L V6 MFI VIN U (All) 60 Pin Connector

PCM Pin #	Wire Color	Circuit Description (60 Pin)	Value at Hot Idle
41	---	Not Used	---
42 ('94)	TN/LG	A/C Cycling Clutch Signal	A/C Off: 0v, On: 12v
43	RD/BK	HO2S-11 (B1 S1) Signal	0.1-1.1v
44	GY/BL	HO2S-21 (B1 S1) Signal	0.1-1.1v
45	---	Not Used	---
46	GY/RD	Analog Signal Return	<0.050v
47	GN/WT	TP Sensor Signal	0.5v-1.2v
48	WT/PK	Self-Test Input Signal	STI On: 1v, Off: 5v
49	---	Not Used	---
50	BL/RD	MAF Sensor Signal	0.8v
50	BL/BK	MAF Sensor Signal	0.8v
51-53	---	Not Used	---
54	PK/YL	A/C WOT Relay Control	Relay Off: 12v, On: 1v
55	---	Not Used	---
56	DB	PIP Sensor Signal	6.93v (50% d/cycle)
57	RD	Vehicle Power	12-14v
58	TN	Injector 1 Control	4.6-5.3 ms
59	WT	Injector 2 Control	4.6-5.3 ms
60	BK/LG	Power Ground	<0.1v

Pin Connector Graphic

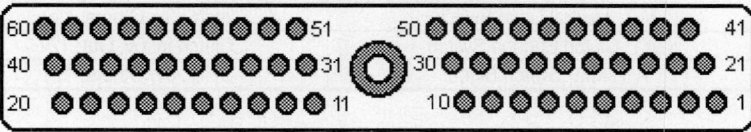

Standard Colors and Abbreviations

Abbreviation	Color	Abbreviation	Color	Abbreviation	Color
BK	Black	GY	Gray	PK	Purple
BL	Blue	GN	Green	RD	Red
BR	Brown	LG	LT Green	TN	Tan
DB	Dark Blue	OR	Orange	WT	White
DG	DK Green	PK	Pink	YL	Yellow

Tracer Pin Tables

1991-92 Tracer 1.8L I4 MFI VIN 8 (All) 22 Pin Connector

PCM Pin #	Wire Color	Circuit Description (22 Pin)	Value at Hot Idle
22-A	BL/RD	Keep Alive Power	12-14v
22-B	WT/RD	Vehicle Power	12-14v
22-C	PK	Start Signal	KOEC: 9-11v
22-D	WT/YL	Switch Monitor Lamp Signal	SML On: 5v, Off: 12v
22-E	YL/BK	MIL (lamp) Control	MIL On: 2.5v, Off: 12v
22-F	WT/BK	Self-Test Output, MIL Control	STO On: 5v, Off: 12v
22-G	GN/WT	Spark Output Signal	6.93v (50% d/cycle)
22-H	---	Not Used	---
22-I	---	Not Used	---
22-J	BL/BK	A/C WOT Relay Control	A/C On: 12v, Off: 2.5v
22-K	LG/YL	Self-Test Input Signal	STI On: 0v, Off: 5v
22-L	BR/WT	DRL (Canada)	DRL Off: 12.5v, On: 2.5v
22-M	---	Not Used	---
22-N	RD/WT	IDL Switch Signal	Closed: 0v, Open: 12v
22-O	GN	M/T: Brake Switch Signal	Brake Off: 0v, On: 12v
22-O	GN/RD	4EAT A/T: Down Shift Signal	DSS Off: 1v, On: 12v
22-P	BL/YL	PSP Switch Signal	Straight: 0v, Turned: 12v
22-Q	GN/BK	A/C Switch Signal	A/C On: 2.5v, Off: 12v
22-R	BK/GN	Cooling Fan Control	Fan On: 1v, Off: 12v
22-S	OR/BL	Blower Motor Switch Signal	Off or 1st: 12v, 2nd>: 1v
22-T	BK/BL	Rear Defroster Switch Signal	Switch Off: 1v, On: 12v
22-U	RD/BK	Headlamp Switch Signal	HDL On: 12v, Off: 1v
22-V	BR/YL	M/T: Neutral Drive Switch	In 'N': 0v, Others: 11v
22-V	BR/YL	M/T: Clutch Engage Switch	Clutch In: 0v, Out: 12v
22-V	BK/BL	A/T: MLP Switch Signal	In 'N': 0v, Others: 12v

Pin Connector Graphic

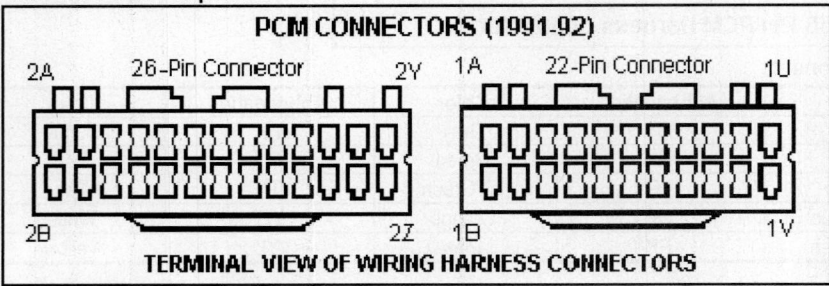

Standard Colors and Abbreviations

Abbreviation	Color	Abbreviation	Color	Abbreviation	Color
BK	Black	GY	Gray	PK	Purple
BL	Blue	GN	Green	RD	Red
BR	Brown	LG	LT Green	TN	Tan
DB	Dark Blue	OR	Orange	WT	White
DG	DK Green	PK	Pink	YL	Yellow

1991-92 Tracer 1.8L I4 MFI VIN 8 (All) 26 Pin Connector

PCM Pin #	Wire Color	Circuit Description (26 Pin)	Value at Hot Idle
26-A	BK/OR	Power Ground	<0.1v
26-B	BK/OR	Power Ground	<0.1v
26-C	BK/LG	Power Ground	<0.1v
26-D	BK/WT	Analog Signal Return	<0.050v
26-E	WT	CKP Sensor Signal	400-425 Hz
26-F	---	Not Used	---
26-G	YL/BL	CID Sensor Signal	5-7 Hz
26-H	BK	Power Ground (California)	<0.1v
26-H	BK/YL	Vehicle Power (Canada)	12-14v
26-I	---	Not Used	---
26-J	WT/RD	A/T: Vehicle Power Relay	12-14v
26-K	LG/RD	Reference Voltage	4.9-5.1v
26-L	LG/WT	M/T: WOT Switch Signal	12v, 55 mph: 12v
26-M	LG/WT	4EAT A/T: TP Sensor Signal	0.5-2.1v
26-N	RD/BL	O2S-11 (Bank 1 Sensor 1)	0.1-1.1v
26-O	RD	VAF Sensor Signal	3.3v
26-P	RD/BK	VAT Sensor Signal	1.5-2.5v
26-Q	BL/WT	ECT Sensor Signal	0.5-0.6v
26-R	---	Not Used	---
26-S	BK/RD	High Speed Inlet Air Solenoid	Below 5000 rpm: 1.5v
26-T	GN/OR	Fuel Press. Regulator Control	149 sec after startup: <3v
26-U	YL	Injector Bank 1	3.6 ms
26-V	YL/BK	Injector Bank 2	3.6 ms
26-W	BL/OR	IAC Motor Control	8-10v
26-X	WT/BL	Canister Purge Solenoid	12v, 55 mph: 1v
26-Y	---	Not Used	---
26-Z	BL/GN	ECT Signal to 4EAT Module	ECT at 162°F: <2.5v

Pin Connector Graphic

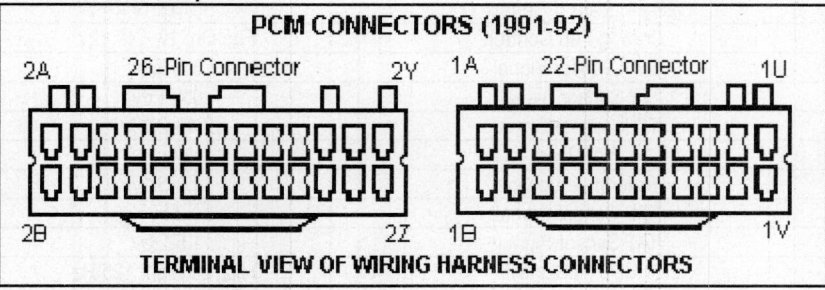

Standard Colors and Abbreviations

Abbreviation	Color	Abbreviation	Color	Abbreviation	Color
BK	Black	GY	Gray	PK	Purple
BL	Blue	GN	Green	RD	Red
BR	Brown	LG	LT Green	TN	Tan
DB	Dark Blue	OR	Orange	WT	White
DG	DK Green	PK	Pink	YL	Yellow

1993 Tracer 1.8L I4 4v MFI VIN 8 (A/T) 22 Pin Connector

PCM Pin #	Wire Color	Circuit Description (22 Pin)	Value at Hot Idle
22-A	DB/RD	Keep Alive Power	12- 14v
22-B	WT/RD	Vehicle Power	12-14v
22-C	PK	Vehicle Start Signal	Cranking: 9-11v
22-D	WT/YL	System Monitor Lamp	Lamp On: 5v, Off: 12v
22-E	YL/BK	MIL (lamp) Control	MIL On: 2.5v, Off: 12v
22-F	WT/BK	Self-Test Output, MIL Control	STO On: 5v, Off: 12v
22-G	DG/WT	Ignition Control Module Signal	0.1-0.3v
22-H	RD/BK	Headlamp Switch Signal	Switch On: 12v, Off: 1v
22-I	LG/YL	Self-Test Input Signal	STI On: 0v, Off: 5v
22-J	BK/DB	Rear Defroster Control	Switch Off: 1v, On: 12v
22-K	BK	Power Ground	<0.1v
22-L	DB/BK	A/C & Heater Relay Control	A/C On: 2.5v, AC Off: 12v
22-M	DG	VSS (+) Signal	Not moving: 0v
22-N	DB/YL	PSP Switch Signal	Straight: 12v, Turned: 0v
22-O	DG/DB	A/C Switch Signal	A/C On: 2.5v, AC Off: 12v
22-P	OR/DB	Blower Motor Switch	Off or 1st: 12v, 2nd on: 1v
22-Q	DG	Brake Switch Signal	Brake Off: 0v, On: 12v
22-R	BK/DB	MLP Switch Signal	In 'P': 0v, Others: 12v
22-S	---	Not Used	---
22-T	RD/WT	Idle Switch Signal	Closed: 0V, Open: 12v
22-U	---	Not Used	---
22-V	---	Not Used	---

1993 Tracer 1.8L I4 4v MFI VIN 8 (A/T) 16 Pin Connector

PCM Pin #	Wire Color	Circuit Description (16 Pin)	Value at Hot Idle
16-A	WT	CKP Sensor Signal	2.5v
16-B	RD	VAF Sensor Signal	0.2v
16-C	RD/DB	O2S-11 (Bank 1 Sensor 1)	0.1-1.1v
16-D	BK/DG	Cooling Fan Control	Fan On: 1v, Off: 12v
16-E	DB/W	ECT Sensor Signal	0.5-0.6v
16-F	LG/WT	TP Sensor Signal	0.4-1.0v
16-G	WT/BK	TOT Sensor Signal	2.40-2.10v
16-H	---	Not Used	---
16-I	LG/RD	Reference Voltage	4.9-5.1v
16-J	YL/DB	CID Sensor Signal	2.5v
16-K	RD/BK	IAT Sensor Signal	1.5-2.5v
16-L	DB	VSS (-) Signal	0 Hz, 55 mph: 125 Hz
16-M	WT/DB	TSS Sensor (+) Signal	340-380 Hz
16-N	YL/DB	TSS Sensor (-) Signal	340-380 Hz
16-O	WT/DB	Canister Purge Solenoid	12v, Cruise: 1v
16-P	---	Not Used	---

1993 Tracer 1.8L I4 4v MFI VIN 8 (A/T) 26 Pin Connector

PCM Pin #	Wire Color	Circuit Description (26 Pin)	Value at Hot Idle
26-A	BK/OR	Power Ground	<0.1v
26-B	BK/OR	Power Ground	<0.1v
26-C	BK/LG	Power Ground	<0.1v
26-D	BK/WT	Analog Signal Return	<0.050v
26-E	YL	MLP Overdrive Switch Signal	Switch On: 1v, Off: 12v
26-F	BR/WT	DRL (Canada)	DRL On: 2v, Off: 12v
26-G	YL/WT	MLP Sensor Low Signal	In 'L': 12v, Others: 0v
26-H	YL/RD	MLP Sensor Drive Signal	Drive: 12v, Others: 0v
26-I	BK/RD	High Speed Inlet Air Solenoid	Less than 4900 rpm: 1v
26-J	---	Not Used	---
26-K	---	Not Used	---
26-L	---	Not Used	---
26-M	DG/OR	Fuel Press. Regulator Control.	149 sec after startup: <3v
26-N	---	Not Used	---
26-O	---	Not Used	---
26-P	---	Not Used	---
26-Q	DB/OR	IAC Valve Control	8-10v
26-R	---	Not Used	---
26-S	---	Not Used	---
26-T	---	Not Used	---
26-U	YL	Injector Bank 1 (INJ 1 & 3)	3.6 ms
26-V	YL/BK	Injector Bank 2 (INJ 2 & 4)	3.6 ms
26-W	DB/OR	Shift Solenoid 1 Control	Shifting: 12v, other: 1v
26-X	DB/YL	Shift Solenoid 2 Control	Shifting: 12v, other: 1v
26-Y	OR	Shift Solenoid 3 Control	Shifting: 12v, other: 1v
26-Z	DB	TCC Solenoid Control	0v, Cruise: 12v

Pin Connector Graphic

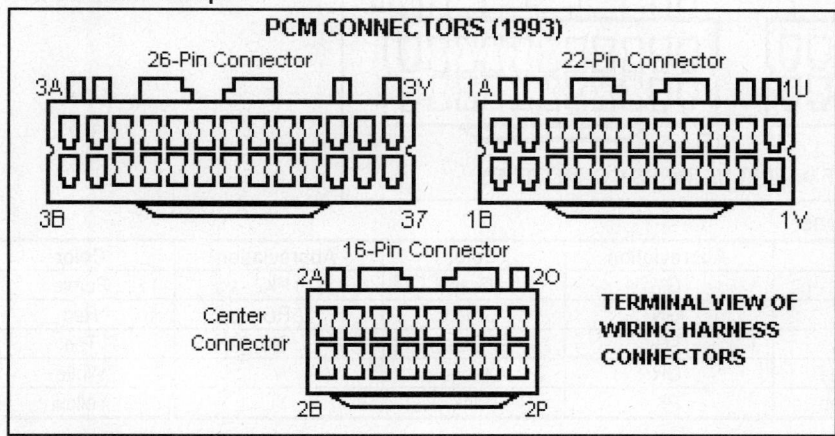

Standard Colors and Abbreviations

Abbreviation	Color	Abbreviation	Color	Abbreviation	Color
BK	Black	GY	Gray	PK	Purple
BL	Blue	GN	Green	RD	Red
BR	Brown	LG	LT Green	TN	Tan
DB	Dark Blue	OR	Orange	WT	White
DG	DK Green	PK	Pink	YL	Yellow

1993 Tracer 1.8L I4 4v MFI VIN 8 (M/T) 22 Pin Connector

PCM Pin #	Wire Color	Circuit Description (22 Pin)	Value at Hot Idle
22-1	DB/RD	Keep Alive Power	12- 14v
22-2	WT/RD	Vehicle Power (Main Relay)	12-14v
22-3	PK	Vehicle Start Signal	Cranking: 9-11v
22-4	WT/YL	System Monitor Lamp	Lamp On: 5v, Off: 12v
22-5	YL/BK	MIL (lamp) Control	MIL On: 1v, Off: 12v
22-6	WT/BK	Self-Test Output, MIL Control	Buzzer On: 5v, Off: 12v
22-7	DG/WT	Spark Output Signal	6.93v (50% d/cycle)
22-8	---	Not Used	---
22-9	---	Not Used	---
22-10	DB/BK	AC WOT Relay Control	Relay Off: 12v, On: 1v
22-11	LG/YL	Self-Test Input Signal	STI On: 0v, STI Off: 5v
22-12	BR/WT	DRL (Canada)	DRL Off: 12v, On: 2.5v
22-13	---	Not Used	---
22-14	RD/WT	Idle Switch Input	Closed: 0V, Open: 12v
22-15	DG	Brake Switch Signal	Brake Off: 0v, On: 12v
22-16	DB/YL	PSP Switch Signal	Straight: 12v, Turned: 0v
22-17	DB/DB	A/C Switch Signal	A/C On: 2.5v, Off: 12v
22-18	BK/DG	Fan Motor Control	Fan On: 1v, Off: 12v
22-19	OR/DB	Blower Motor Switch Signal	Off or 1st: 12v, 2nd on: 1v
22-20	BK/DB	Rear Defroster Signal	Switch On: 12v, Off: 0.1v
22-21	RD/BK	Headlamp Switch Signal	Switch On: 12v, Off: 1v
22-22	BR/YL	M/T: Neutral Drive Switch	In 'N': 0v, Others: 12v
22-22	BR/YL	M/T: Clutch Pedal Switch	Clutch In: 0v, Out: 12v

Pin Connector Graphic

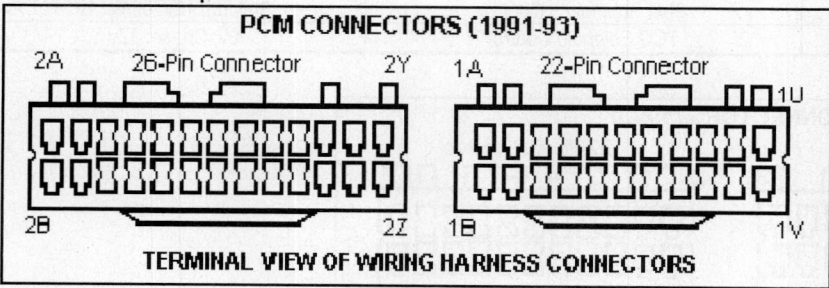

Standard Colors and Abbreviations

Abbreviation	Color	Abbreviation	Color	Abbreviation	Color
BK	Black	GY	Gray	PK	Purple
BL	Blue	GN	Green	RD	Red
BR	Brown	LG	LT Green	TN	Tan
DB	Dark Blue	OR	Orange	WT	White
DG	DK Green	PK	Pink	YL	Yellow

1993 Tracer 1.8L I4 4v MFI VIN 8 (M/T) 26 Pin Connector

PCM Pin #	Wire Color	Circuit Description (26 Pin)	Value at Hot Idle
26-1	BK/OR	Power Ground	<0.1v
26-2	BK/OR	Power Ground	<0.1v
26-3	BK/LG	Power Ground	<0.1v
26-4	BK/WT	Analog Signal Return	<0.050v
26-5	WT	CKP Sensor Signal	2.5v
26-6	---	Not Used	---
26-7	YL/DB	CID Sensor Signal	2.5v
26-8	BK	Power Ground	<0.1v
26-9	---	Not Used	---
26-10	---	Not Used	---
26-11	LG/RD	Reference Voltage	4.9-5.1v
26-12	LG/WT	A/C WOT Relay Control	Relay On: 0.1v, 12v
26-13	---	Not Used	---
26-14	RD/DB	O2S-11 (Bank 1 Sensor 1)	0.1-1.1v
26-15	RD	VAF Meter Signal	0.3v
26-16	RD/BK	IAT Sensor Signal	1.5-2.5v
26-17	DB/W	ECT Sensor Signal	0.5-0.6v
26-18	---	Not Used	---
26-19	BK/RD	High Speed Inlet Air Solenoid	More than 5000 rpm: 12v
26-20	DG/OR	Fuel Pressure Regulator	149 sec after startup: <3v
26-21	YL	Injector Bank 1 (INJ 1 & 3)	3.6 ms
26-22	YL/BK	Injector Bank 2 (INJ 2 & 4)	3.6 ms
26-23	DB/OR	IAC Valve Control	8-10v
26-24	WT/DB	Canister Purge Solenoid	12v, Cruise: 1v
26-25	---	Not Used	---
26-26	---	Not Used	---

Pin Connector Graphic

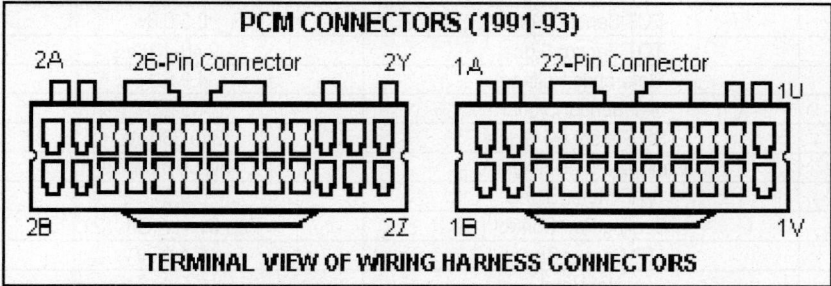

Standard Colors and Abbreviations

Abbreviation	Color	Abbreviation	Color	Abbreviation	Color
BK	Black	GY	Gray	PK	Purple
BL	Blue	GN	Green	RD	Red
BR	Brown	LG	LT Green	TN	Tan
DB	Dark Blue	OR	Orange	WT	White
DG	DK Green	PK	Pink	YL	Yellow

1994-95 Tracer 1.8L I4 4v MFI VIN 8 (A/T) 22 Pin Connector

PCM Pin #	Wire Color	Circuit Description (22 Pin)	Value at Hot Idle
22-1	DB/RD	Keep Alive Power	12- 14v
22-2	PK	Vehicle Start Signal	Cranking: 9-11v
22-3	YL/BK	MIL (lamp) Control	MIL On: 1v, Off: 12v
22-4	DG/WT	CKP Sensor Signal	2.5v
22-5	LG/YL	Data Link Connector	Digital Signals
22-6	LG/YL	Data Link Connector	Digital Signals
22-6	BK/YL	Vehicle Power (Canada)	12-14v
22-7	DG	VSS (+) Signal	0 Hz, 55 mph: 125 Hz
22-8	DG/BK	A/C Switch Signal	A/C On: 2.5v, Off: 12v
22-9	DG	Brake Switch Signal	Brake Off: 0v, On: 12v
22-10	---	Not Used	---
22-11	---	Not Used	---
22-12	WT/RD	Vehicle Power (Main Relay)	12-14v
22-13	WT/YL	Data Link Connector	Digital Signals
22-14	WT/BK	Data Link Connector	Digital Signals
22-15	RD/BK	Park Lamp Signal	Switch On: 12v, Off: 1v
22-16	BK/DB	Rear Defroster Switch Signal	Switch On: 12v, Off: 1v
22-17	DB/BK	A/C WOT Relay Control	Relay Off: 12v, On: 1v
22-18	DB/YL	PSP Switch Signal	Straight: 12v, Turned: 0v
22-19	OR/DB	Cooling Fan Control	Fan On: 1v, Off: 12v
22-20	BK/DB	Ignition Start Signal	KOEC: 9-11v
22-21	RD/WT	Throttle Position Idle Switch	Closed: 0V, Open: 12v
22-22	---	Not Used	---

1994-95 Tracer 1.8L I4 4v MFI VIN 8 (A/T) 16 Pin Connector

PCM Pin #	Wire Color	Circuit Description (16 Pin)	Value at Hot Idle
16-1	WT	CKP Sensor Signal	2.5v
16-2	RD/DB	O2S-11 (Bank 1 Sensor 1)	0.1-1.1v
16-3	DB/W	ECT Sensor Signal	0.5-0.6v
16-4	WT/BK	TOT Sensor Signal	2.40-2.10v
16-5	LG/RD	Reference Voltage	4.9-5.1v
16-6	RD/BK	IAT Sensor Signal	1.5-2.5v
16-7	WT/DB	TSS Sensor (+) Signal	340-380 Hz
16-8	WT/DB	Canister Purge Solenoid	12v, 55 mph: 1v
16-9	RD	VAF Meter Signal	0.3v
16-10	BK/DG	Cooling Fan Control	Fan On: 1v, Off: 12v
16-11	LG/WT	TP Sensor Signal	0.4-1.0v
16-12	---	Not Used	---
16-13	YL/DB	CID Sensor Signal	2.5v
16-14	DB	VSS (-) Signal	0 Hz, 55 mph: 125 Hz
16-15	YL/DB	TSS Sensor (-) Signal	340-380 Hz
16-16	---	Not Used	---

1994-95 Tracer 1.8L I4 4v MFI VIN 8 (A/T) 26 Pin Connector

PCM Pin #	Wire Color	Circuit Description (26 Pin)	Value at Hot Idle
26-1	BK/OR	Power Ground	<0.1v
26-2	BK/LG	Power Ground	<0.1v
26-3	YL	TRS Overdrive Signal	Switch On: 1v, Off: 12v
26-4	YL/WT	TRS Low Signal	In 'L': 12v, Others: 1v
26-5	BK/RD	High Speed Inlet Air Solenoid	Less than 4900 rpm: 1v
26-6	---	Not Used	---
26-7	DG/OR	Fuel Pressure Regulator	149 sec after startup: <3v
26-8	---	Not Used	---
26-9	DB/OR	IAC Valve Control	8-10v
26-10	---	Not Used	---
26-11	YL	Injector Bank 1 (INJ 1 & 3)	3.6 ms
26-12	DB/OR	Shift Solenoid 1 Control	SS1 On: 1v, Off: 12v
26-13	OR	Shift Solenoid 3 Control	SS3 On: 1v, Off: 12v
26-14	BK/OR	Power Ground	<0.1v
26-15	BK/WT	Analog Signal Return	<0.050v
26-16	BR/WT	DRL (Canada)	DRL On: 1v, Off: 12v
26-17	YL/RD	TR Sensor Drive Signal	In 'D': Others: 1v
26-18	---	Not Used	---
26-19	---	Not Used	---
26-20	---	Not Used	---
26-21	---	Not Used	---
26-22	---	Not Used	---
26-23	---	Not Used	---
26-24	YL/BK	Injector Bank 2 (INJ 2 & 4)	3.6 ms
26-25	DB/YL	Shift Solenoid 2 Control	SS2 On: 1v, Off: 12v
26-26	DB	TCC Solenoid Control	During Shifting: 12v

Pin Connector Graphic

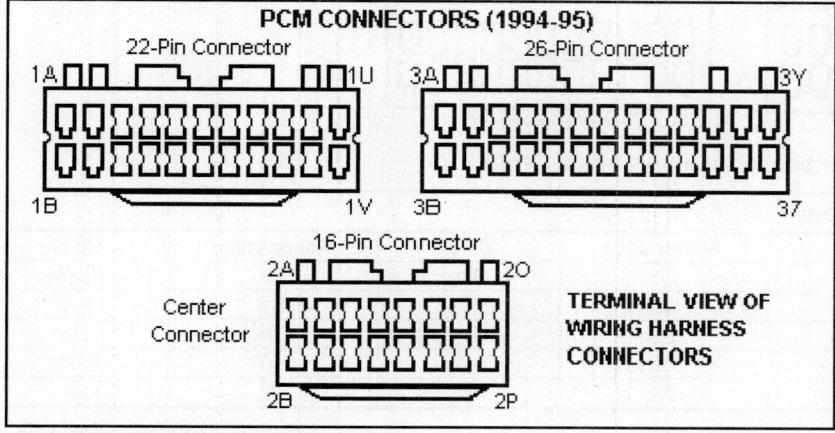

Standard Colors and Abbreviations

Abbreviation	Color	Abbreviation	Color	Abbreviation	Color
BK	Black	GY	Gray	PK	Purple
BL	Blue	GN	Green	RD	Red
BR	Brown	LG	LT Green	TN	Tan
DB	Dark Blue	OR	Orange	WT	White
DG	DK Green	PK	Pink	YL	Yellow

1994-95 Tracer 1.8L I4 4v MFI VIN 8 (M/T) 22 Pin Connector

PCM Pin #	Wire Color	Circuit Description (22 Pin)	Value at Hot Idle
22-A	DB/RD	Keep Alive Power	12- 14v
22-B	PK	Vehicle Start Signal	Cranking: 9-11v
22-C	YL/BK	MIL (lamp) Control	MIL On: 1v, Off: 12v
22-D	DG/WT	Spark Output Signal	6.93v (50% d/cycle)
22-E	LG/YL	Self-Test Input Signal	STI On: 0v, STI Off: 5v
22-F	---	Not Used	---
22-G	---	Not Used	---
22-H	DG	Brake Switch Signal	Brake Off: 0v, On: 12v
22-I	DG/BK	A/C Switch Signal	A/C On: 2.5v, Off: 12v
22-J	OR/DB	Blower Motor Switch Signal	Off or 1st: 12v, 2nd on: 1v
22-K	RD/BK	Headlamp Switch Signal	Switch On: 12v, Off: 1v
22-L	WT/RD	Vehicle Power (Main Relay)	12-14v
22-M	WT/YL	Data Link Connector	Digital Signals
22-N	WT/BK	Data Link Connector	Digital Signals
22-O	---	Not Used	---
22-P	DB/BK	A/C WOT Relay Control	Relay Off: 12v, On: 1v
22-Q	BR/WT	DRL (Canada)	DRL Off: 12v, On: 2.5v
22-R	RD/WT	Idle Position Switch Signal	Closed: 0V, Open: 12v
22-S	DB/YL	PSP Switch Signal	Straight: 12v, Turned: 0v
22-T	BK/DG	Cooling Fan Control	Fan On: 1v, Off: 12v
22-U	BK/DB	Rear Defroster Switch Signal	AC On: 12v, Off: 0v
22-V	BR/YL	M/T: Neutral Drive Switch	In 'N': 0v, Others: 12v
22-V	BR/YL	M/T: Clutch Engage Switch	Clutch In: 0v, Out: 12v

Pin Connector Graphic

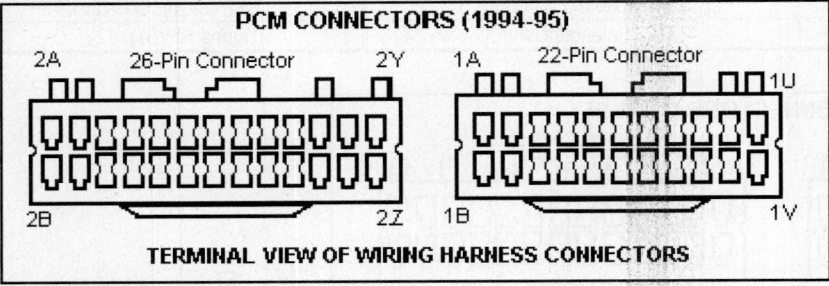

Standard Colors and Abbreviations

Abbreviation	Color	Abbreviation	Color	Abbreviation	Color
BK	Black	GY	Gray	PK	Purple
BL	Blue	GN	Green	RD	Red
BR	Brown	LG	LT Green	TN	Tan
DB	Dark Blue	OR	Orange	WT	White
DG	DK Green	PK	Pink	YL	Yellow

1994-95 Tracer 1.8L I4 4v MFI VIN 8 (M/T) 26 Pin Connector

PCM Pin #	Wire Color	Circuit Description (26 Pin)	Value at Hot Idle
26-A	BK/OR	Power Ground	<0.1v
26-B	BK/LG	Power Ground	<0.1v
26-C	W	CKP Sensor Signal	2.5v
26-D	YL/DB	CID Sensor Signal	2.5v
26-E	---	Not Used	---
26-F	LG/RD	Reference Voltage	4.9-5.1v
26-G	---	Not Used	---
26-H	RD	VAF Meter Signal	0.3v
26-I	DB/W	ECT Sensor Signal	0.5-0.6v
26-J	BK/RD	High Speed Inlet Air Solenoid	More than 5000 rpm: 12v
26-K	YL	Injector Bank 1 (INJ 1 & 3)	3.6 ms
26-L	DB/OR	IAC Valve Control	8-10v
26-M	---	Not Used	---
26-N	BK/OR	Power Ground	<0.1v
26-O	BK/WT	Analog Signal Return	<0.050v
26-P	---	Not Used	---
26-Q ('94)	BK	Power Ground	<0.1v
26-Q ('95)	BK/YL	Ignition Power	12-14v
26-R	---	Not Used	---
26-S	LG/WT	TP Sensor Signal	0.4-1.0v
26-T	RD/DB	O2S-11 (Bank 1 Sensor 1)	0.1-1.1v
26-U	RD/BK	IAT Sensor Signal	1.5-2.5v
26-V	---	Not Used	---
26-W	DG/OR	Fuel Pressure Regulator	149 sec after startup: <3v
26-X	YL/BK	Injector Bank 2 (INJ 2 & 4)	3.6 ms
26-Y	WT/DB	Canister Purge Solenoid	12v, 55 mph: 1v
26-Z	---	Not Used	---

Pin Connector Graphic

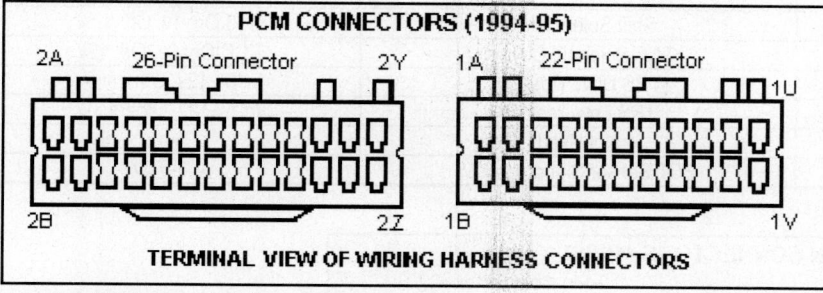

Standard Colors and Abbreviations

Abbreviation	Color	Abbreviation	Color	Abbreviation	Color
BK	Black	GY	Gray	PK	Purple
BL	Blue	GN	Green	RD	Red
BR	Brown	LG	LT Green	TN	Tan
DB	Dark Blue	OR	Orange	WT	White
DG	DK Green	PK	Pink	YL	Yellow

1996 Tracer 1.8L I4 4v MFI VIN 8 (A/T) 22 Pin Connector

PCM Pin #	Wire Color	Circuit Description (22 Pin)	Value at Hot Idle
C1-2	PK	Vehicle Start	KOEC: 9-11v
C1-3	YL/BK	MIL (lamp) Control	MIL On: 0v, Off: 12v
C1-4	BL/BK	A/C WOT Relay Control	A/C On: 2.5v, Off: 12v
C1-5	LG/YL	Self-Test Input Signal	STI On: 0v, Off: 5v
C1-6	GN/BK	A/C High Pressure Switch	A/C On: 2.5v, Off: 12v
C1-7	GN	VSS (+) Signal	0 Hz, 55 mph: 125 Hz
C1-8	BL	VSS (-) Signal	0 Hz, 55 mph: 125 Hz
C1-9	GN	Brake Switch Signal	Brake On: 12v. Off: 0v
C1-10	---	Not Used	---
C1-11	LG	Fuel Pump Control	Relay On, 0v, Off: 12v
C1-12	---	Not Used	---
C1-13	BL/WT	ISO K-Line (OBD II)	Digital Signals: 0-7-0-7v
C1-14	---	Not Used	---
C1-15	RD/BK	Headlamp Switch Signal	Switch On: 12v, Off: 1v
C1-16	GY/RD	Rear Defroster Switch Signal	DEF On: 12v, Off: 1v
C1-17	---	Not Used	---
C1-18	BR/WT	DRL Switch (Canada)	Switch On: 0v, Off: 12v
C1-19	OR/BL	Blower Motor Control Switch	Off/1st: 12v, 2nd on: 1v
C1-20	BK/GN	Cooling Fan Relay	Relay On: 0v, Off: 12v
C1-21-22	---	Not Used	---

1996 Tracer 1.8L I4 4v MFI VIN 8 (A/T) 12 Pin Connector

PCM Pin #	Wire Color	Circuit Description (12 Pin)	Value at Hot Idle
C2-1	BL/OR	Shift Solenoid 1-2	SS1 On: 1v, Off: 12 v
C2-2	OR	Shift Solenoid 3-4	SS3 On: 1v, Off: 12 v
C2-3	---	Not Used	---
C2-4	YL/RD	TRS 2nd Position	In 2nd: 12v, Others: 0v
C2-5	---	Not Used	---
C2-6	WT/BL	TSS Sensor (+) Signal	340-380 Hz
C2-7	BL/YL	Shift Solenoid 3-4	SS3 On: 1v, Off: 12 v
C2-8	BL	TCC Solenoid Control	TCC On: 0v, Off: 12v
C2-9	YL	TRS Drive Position	In 'D': 12v, Others: 0v
C2-10	BR/WT	TRS Low Position	In 'L': 12v, Others: 0v
C2-11	---	Not Used	---
C2-12	YL/BL	ECT Sensor Signal to TCM	0.5-0.6v

Pin Connector Graphic

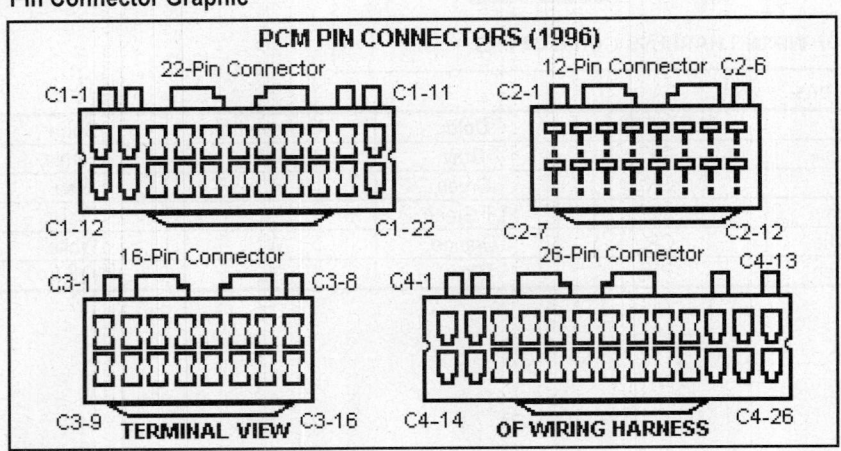

PCM PIN CONNECTORS (1996)

1996 Tracer 1.8L I4 4v MFI VIN 8 (A/T) 16 Pin Connector

PCM Pin #	Wire Color	Circuit Description (16 Pin)	Value at Hot Idle
C3-1	---	Not Used	---
C3-2	RD/BL	HO2S-11 (B1 S1) Signal	0.1-1.1v
C3-3	WT/BK	TFT Sensor Signal	2.40-2.10v
C3-4	BL/WT	ECT Sensor Signal	0.6v
C3-5	LG/RD	Reference Voltage	4.9-5.1v
C3-6	RD/BK	IAT Sensor Signal	1.5-2.5v
C3-7	GN/WT	HO2S-11 (B1 S1) Heater	1v, Off: 12v
C3-8	BK/BL	Analog Return Signal	<0.050v
C3-9	RD	MAF Sensor Signal	0.6v
C3-10	RD/YL	HO2S-12 (B1 S2) Signal	0.1-1.1v
C3-11	LG/WT	TP Sensor Signal	0.4-1.0v
C3-12	---	Not Used	---
C3-13	---	Not Used	---
C3-14	RD/WT	Idle Position Switch Signal	Closed: 0V, Open: 12v
C3-15	---	Not Used	---
C3-16	BL/YL	PSP Switch Signal	Straight: 12v, Turned: 0v

1996 Tracer 1.8L I4 4v MFI VIN 8 (A/T) 26 Pin Connector

PCM Pin #	Wire Color	Circuit Description (26 Pin)	Value at Hot Idle
C4-1	BK/LG	Power Ground	<0.1v
C4-2	BK/OR	Power Ground	<0.1v
C4-3	OR	CKP Sensor (-) Signal	435-475 Hz
C4-4	YL/BL	CMP 1 Sensor Signal	5-7 Hz
C4-5	BL/RD	Keep Alive Power	12-14v
C4-6	BK/RD	High Speed Inlet Air Solenoid	More than 5000 rpm: 12v
C4-7	---	Not Used	---
C4-8	---	Not Used	---
C4-9	BL/OR	IAC Motor Control	10.7v (30%)
C4-10	---	Not Used	---
C4-11	YL	Injector 1 Control	4-4.5 ms
C4-12	GN/RD	Injector 3 Control	4-4.5 ms
C4-13	---	Not Used	---
C4-14	WT/RD	Vehicle Power	12-14v
C4-15	BK/OR	Power Ground	<0.1v
C4-16	W	CMP 2 Sensor Signal	5-7 Hz
C4-17	LG/WT	CKP Sensor (+) Signal	435-475 Hz
C4-18	GN/OR	Fuel Pressure Regulator	149 sec after start: <3v
C4-19	---	Not Used	---
C4-20	BK/WT	Ignition Spark Output Signal	Varies: 0.1-0.3v
C4-21	---	Not Used	---
C4-22	---	Not Used	---
C4-23	WT/BL	Canister Purge Solenoid	Purge On: 0v, Off: 12v
C4-24	YL/BK	Injector 2 Control	4-4.5 ms
C4-25	GN/YL	Injector 4 Control	4-4.5 ms
C4-26	---	Not Used	---

1996 Tracer 1.8L I4 4v MFI VIN 8 (M/T) 22 Pin Connector

PCM Pin #	Wire Color	Circuit Description (22 Pin)	Value at Hot Idle
C1-1	---	Not Used	---
C1-2	PK	Vehicle Start	KOEC: 9-11v
C1-3	YL/BK	MIL (lamp) Control	MIL On: 0v, Off: 12v
C1-4	BL/BK	A/C WOT Relay Control	A/C On: 2.5v, Off: 12v
C1-5	LG/YL	Self-Test Input Signal	STI On: 0v, Off: 5v
C1-6	GN/BK	A/C High Pressure Switch	A/C On: 2.5v, Off: 12v
C1-7	GN	VSS (+) Signal	0 Hz, 55 mph: 125 Hz
C1-8	BL	VSS (-) Signal	0 Hz, 55 mph: 125 Hz
C1-9	GN	Brake Switch Signal	Brake On: 12v. Off: 0v
C1-10	---	Not Used	---
C1-11	LG	Fuel Pump Control	Relay On, 0v, Off: 12v
C1-12	---	Not Used	---
C1-13	BL/WT	ISO K-Line (OBD II)	Digital Signals: 0-7-0-7v
C1-14	---	Not Used	---
C1-15	RD/BK	Headlamp Switch Signal	Switch On: 12v, Off: 1v
C1-16	GY/RD	Rear Defroster Switch Signal	Switch On: 12v, Off: 1v
C1-17	BR/YL	M/T: Neutral Drive Switch	In 'N': 0v, Others: 5v
C1-17	BR/YL	M/T: Clutch Pedal Switch	Clutch In: 0v, Out: 5v
C1-18	BR/WT	DRL (Canada)	Relay On: 0v, Off: 12v
C1-19	OR/BL	Blower Motor Control Switch	Off or 1st: 12v, 2nd on: 1v
C1-20	BK/GN	Cooling Fan Control	Relay On: 0v, Off: 12v
C1-21	---	Not Used	---
C1-22	---	Not Used	---

1996 Tracer 1.8L I4 4v MFI VIN 8 (M/T) 16 Pin Connector

PCM Pin #	Wire Color	Circuit Description (16 Pin)	Value at Hot Idle
C3-1	---	Not Used	---
C3-2	RD/BL	HO2S-11 (B1 S1) Signal	0.1-1.1v
C3-3	---	Not Used	---
C3-4	BL/WT	ECT Sensor Signal	0.5-0.6v
C3-5	LG/RD	Reference Voltage	4.9-5.1v
C3-6	RD/BK	IAT Sensor Signal	1.5-2.5v
C3-7	GN/WT	HO2S-11 (B1 S1) Heater	1v, Off: 12v
C3-8	BK/BL	Analog Return Signal	<0.050v
C3-9	RD	MAF Sensor Signal	0.6v
C3-10	RD/YL	HO2S-12 (B1 S2) Signal	0.1-1.1v
C3-11	LG/WT	TP Sensor Signal	0.5v
C3-12, 13	---	Not Used	---
C3-14	RD/WT	Idle Position Switch Signal	Closed: 0v, Open: 12v
C3-15	---	Not Used	---
C3-16	BL/YL	PSP Switch Signal	Straight: 12v, Turned: 0v

1996 Tracer 1.8L I4 4v MFI VIN 8 (M/T) 26 Pin Connector

PCM Pin #	Wire Color	Circuit Description (26 Pin)	Value at Hot Idle
C4-1	BK/LG	Power Ground	<0.1v
C4-2	BK/OR	Power Ground	<0.1v
C4-3	OR	CKP Sensor (-) Signal	435-475 Hz
C4-4	YL/BL	CMP 1 Sensor Signal	5-7 Hz
C4-5	BL/RD	Keep Alive Power	12-14v
C4-6	BK/RD	High Speed Inlet Air Solenoid	More than 5000 rpm: 12v
C4-7	---	Not Used	---
C4-8	---	Not Used	---
C4-9	BL/OR	IAC Motor Control	Varies: 8-10v
C4-10	---	Not Used	---
C4-11	YL	Injector 1 Control	4.0-4.5 ms
C4-12	GN/RD	Injector 3 Control	4.0-4.5 ms
C4-13	---	Not Used	---
C4-14	WT/RD	Vehicle Power	12-14v
C4-15	BK/OR	Power Ground	<0.1v
C4-16	WT	CMP 2 Sensor Signal	5-7 Hz
C4-17	LG/WT	CKP Sensor (+) Signal	435-475 Hz
C4-18	GN/OR	Fuel Pressure Regulator	149 sec after startup: <3v
C4-19	---	Not Used	---
C4-20	GN/WT	Spark Output Signal	6.93v (50% d/cycle)
C4-21	---	Not Used	---
C4-22	---	Not Used	---
C4-23	WT/BL	Canister Purge Solenoid	12v, 55 mph: 1v
C4-24	YL/BK	Injector 2 Control	4.0-4.5 ms
C4-25	GN/YL	Injector 4 Control	4.0-4.5 ms
C4-26	---	Not Used	---

Pin Connector Graphic

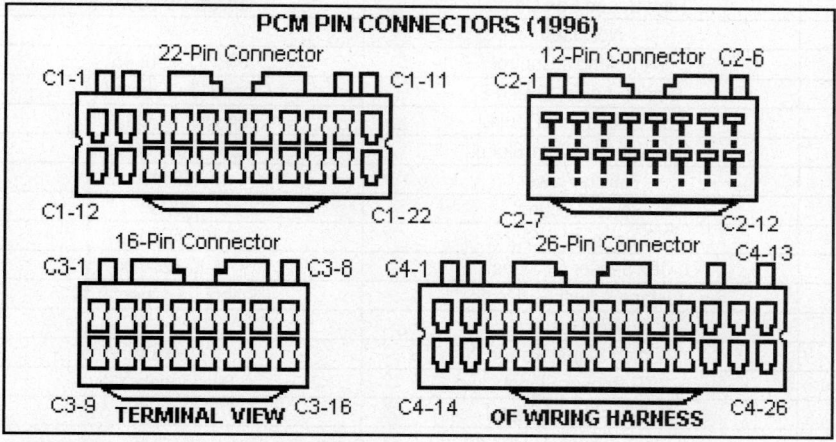

Standard Colors and Abbreviations

Abbreviation	Color	Abbreviation	Color	Abbreviation	Color
BK	Black	GY	Gray	PK	Purple
BL	Blue	GN	Green	RD	Red
BR	Brown	LG	LT Green	TN	Tan
DB	Dark Blue	OR	Orange	WT	White
DG	DK Green	PK	Pink	YL	Yellow

1991-92 Tracer 1.9L I4 MFI VIN J (All) 60 Pin Connector

PCM Pin #	Wire Color	Circuit Description (60 Pin)	Value at Hot Idle
1	BL/RD	Keep Alive Power	12-14v
2, 5, 11, 14	---	Not Used	---
3	WT/BK	VSS (+) Signal	0 Hz, 55 mph: 125 Hz
4	RD	Ignition Diagnostic Monitor	20-31 Hz
6	BL	VSS (-) Signal	0 Hz, 55 mph: 125 Hz
6	DB	VSS (-) Signal	0 Hz, 55 mph: 125 Hz
7	BL/WT	ECT Sensor Signal	0.5-0.6v
8	Y/GN	Data Bus (-) Signal	Digital Signals
9	GN/YL	MAF Sensor Return	<0.050v
10	GN/BK	A/C Cycling Clutch Signal	A/C Off: 0v, On: 12v
12	YL/OR	Injector 3 Control	4.8-5.0 ms
13	GN/OR	Injector 4 Control	4.8-5.0 ms
15 ('92)	OR/WT	Canister Purge Solenoid	12v, 55 mph: 9-11v
16	RD/BL	Ignition System Ground	<0.050v
17	YL/BK	Self-Test Output, MIL Control	MIL On: 1v, Off: 12v
18, 23-24	---	Not Used	---
19	BK/PKK	Fuel Pump Monitor	12v, Off: 0v
20	---	Not Used	---
21	BL/OR	IAC Motor Control	9.7-12.0v
22	LG	Fuel Pump Control	Relay Off: 12v, On: 1v
25	WT/GN	ACT Sensor Signal	1.5-2.5v
26	LG/WT	Reference Voltage	4.9-5.1v
27	BL/YL	PFE Sensor Signal	3.2v, 55 mph: 2.8v
28	YL/BL	Data Bus (+) Signal	Digital Signals
29	GN/BL	HO2S-11 (B1 S1) Signal	0.1-1.1v
30	BR/YL	A/T: Neutral Drive Switch	In 'P': 0v, Others 5v
30	BR/YL	M/T: Clutch Engage Switch	Clutch In: 0v, Out: 12v
31	RD/BK	High Speed Fan Control	Fan On: 1v, Off: 12v
32, 38-39	---	Not Used	---
33	WT/BL	VR Solenoid Control	0%, 55 mph: 45%
34	LG/BK	Octane Adjust Switch 2	Closed: 0v, Open: 9.1v
35	YL/WT	Low Speed Fan Control	Fan On: 1v, Off: 12v
36	LG/WT	Spark Angle Word Signal	6.93v (50% d/cycle)
37	WT/RD	Vehicle Power	12-14v
40	BK/GN	Power Ground	<0.1v
41	OR	CID Sensor (+) Signal	400-425 Hz
42	BL/GN	CID Sensor (-) Signal	400-425 Hz
43	GN/WT	Octane Adjust Switch 1	Closed: 0v, Open: 9.1v
44-45	---	Not Used	---
46	LG/BK	Analog Signal Return	<0.050v
47	RD/WT	TP Sensor Signal	0.8v, 55 mph: 1.1v
48	LG/YL	Self-Test Input Signal	STI On: 1v, Off: 5v
49	BK	HO2S-11 (Bank 1) Ground	<0.050v
49	OR/BK	HO2S-11 (Bank 1) Ground	<0.050v
50	BR/BK	MAF Sensor Signal	0.8v
51-52, 55	---	Not Used	---
53	LG/RD	Shift Indicator Light Control	SIL On: 1v, Off 12v
54	BL/BK	A/C WOT Relay Control	Relay Off: 12v, On: 1v
56	GN/WT	PIP Sensor Signal	6.93v (50% dwell)
57	WT/RD	Vehicle Power	12-14v
58	YL	Injector Bank 1 (INJ 1 & 2)	4.8-5.0 ms
59	GN/RD	Injector Bank 2 (INJ 3 & 4)	4.8-5.0 ms
60	BK/GN	Power Ground	<0.1v

1993-95 Tracer 1.9L I4 MFI VIN J (All) 60 Pin Connector

PCM Pin #	Wire Color	Circuit Description (60 Pin)	Value at Hot Idle
1	BL/RD	Keep Alive Power	12-14v
2	WT	TOT Sensor Signal	2.40-2.10v
3	WT/BK	VSS (+) Signal	0 Hz, 55 mph: 125 Hz
4	RD	Ignition Diagnostic Monitor	22-31 Hz
5	GN	Brake Switch Signal	Brake Off: 0v, On: 12v
6	BL	VSS (-) Signal	0 Hz, 55 mph: 125 Hz
6	DB	VSS (-) Signal	0 Hz, 55 mph: 125 Hz
7	BL/WT	ECT Sensor Signal	0.5-0.6v
8	Y/GN	Data Bus (-) Signal	Digital Signals
9	GN/YL	MAF Sensor Return	<0.050v
10	GN/BK	A/C Cycling Clutch Signal	A/C Off: 0v, On: 12v
11	LG/BL	Shift Solenoid 1 Control	12v, 55 mph: 1v
12	YL/OR	Injector 3 Control	4.8-5.1 ms
13	GN/OR	Injector 4 Control	4.8-5.1 ms
14	---	Not Used	---
15	OR/WT	Canister Purge Solenoid	12v, 55 mph: 9v
16	RD/BL	Ignition System Ground	<0.050v
17	YL/BK	Self-Test Output, MIL Control	MIL On: 1v, Off: 12v
18	BR/RD	A/T: TR Sensor Drive	12v, 55 mph: 12v
19	BK/PKK	Fuel Pump Monitor	12v, Off: 0v
20	BK	Case Ground	<0.050v
21	OR/BK	IAC Motor Control	9.7-11.1v
22	LG	Fuel Pump Control	Relay Off: 12v, On: 1v
23	R/GN	TR Sensor Reverse	12v, 55 mph: 10v
24	WT/BL	A/T: TSS Sensor Signal	42-50 Hz (680-720 rpm)
25	WT/GN	IAT Sensor Signal	1.5-2.5v
26	LG/WT	Reference Voltage	4.9-5.1v
27	BL/YL	PFE Sensor Signal	3.2v, 55 mph: 3.4v
28	YL/BL	Data Bus (+) Signal	Digital Signals
29	GN/BL	HO2S-11 (B1 S1) Signal	0.1-1.1v
30	BK/BL	A/T: Neutral Drive Switch	In P: 0v, all others: 5v
30	BR/YL	A/T: Neutral Drive Switch	In P: 0v, all others: 5v
30	BK/BL	M/T: Clutch Engage Switch	Clutch In: 0v, Out 5v
30	BR/YL	M/T: Clutch Engage Switch	Clutch In: 0v, Out 5v
31	RD/BK	High Speed Fan Control	Fan On: 1v, Off: 12v
32	---	Not Used	---
33	WT/BL	VR Solenoid Control	0%, 55 mph: 45%
34	---	Not Used	---
35	YL/WT	Low Speed Fan Control	Fan On: 1v, Off: 12v
36	LG/WT	Spark Output Signal	6.93v (50% d/cycle)
37	WT/RD	Vehicle Power	12-14v
38	BR/BL	TRS Overdrive Position	O/D Off: 12v, O/D On: 1v
39	---	Not Used	---
40	BK/GN	Power Ground	<0.1v

1993-95 Tracer 1.9L I4 MFI VIN J (All) 60 Pin Connector

PCM Pin #	Wire Color	Circuit Description (60 Pin)	Value at Hot Idle
41	OR	CID Sensor (+) Signal	5-7 Hz
42	BL/GN	CID Sensor (-) Signal	5-7 Hz
43	GN/WT	Octane Adjust Switch 1	Closed: 0v, Open: 9.1v
44	YL/BL	A/T: TSS Sensor Signal	44-56 Hz (680-710 rpm)
45	BR/GN	A/T: Transmission Low Signal	12v, 55 mph: 0v
46	LG/BK	Analog Signal Return	<0.050v
47	RD/WT	TP Sensor Signal	0.8v, 55 mph: 1.1v
48	LG/YL	Self-Test Input Signal	STI On: 1v, Off: 5v
49	BK	HO2S-11 (Bank 1) Ground	<0.050v
49	OR/BK	HO2S-11 (Bank 1) Ground	<0.050v
50	BR/BK	MAF Sensor Signal	0.7v, 55 mph: 2.0v
51	PK/BK	A/T: Shift Solenoid 2 Control	1v, 35 mph: 1v
52	OR/GN	A/T: Shift Solenoid 3 Control	12v, 35 mph: 12v
53	LG/RD	M/T: Shift Indicator Light	SIL On: 1v, Off: 12v
54	BL/BK	A/C WOT Relay Control	Relay Off: 12v, On: 1v
55	PK/WT	TCC Solenoid Control	12v, 55 mph: 9v
56	GN/WT	PIP Sensor Signal	6.93v (50% dwell)
57	WT/RD	Vehicle Power	12-14v
58	YL	Injector 1 Control	4.8-5.1 ms
59	GN/RD	Injector 2 Control	4.8-5.1 ms
60	BK/GN	Power Ground	<0.1v

Pin Connector Graphic

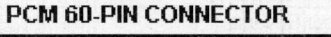
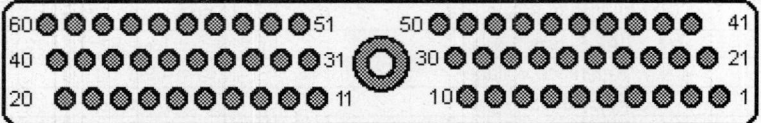

PCM 60-PIN CONNECTOR

Terminal View of 60-Pin PCM Harness Connector

Standard Colors and Abbreviations

Abbreviation	Color	Abbreviation	Color	Abbreviation	Color
BK	Black	GY	Gray	PK	Purple
BL	Blue	GN	Green	RD	Red
BR	Brown	LG	LT Green	TN	Tan
DB	Dark Blue	OR	Orange	WT	White
DG	DK Green	PK	Pink	YL	Yellow

1996 Tracer 1.9L I4 MFI VIN J (All) 104 Pin Connector

PCM Pin #	Wire Color	Circuit Description (104 Pin)	Value at Hot Idle
1	PK/BK	Shift Solenoid 2 Control	12v, 55 mph: 1v
2	YL/BK	MIL (lamp) Control	MIL On: 1v, Off: 12v
3-5	---	Not Used	---
6	BR/BK	TR Sensor Overdrive	In 'P': 0v, in OD: 12v
7	BR/DG	TR Sensor Low	In 'P': 0v, in 1st: 12v
8	---	Not Used	
9	BR/YL	TR Sensor Drive	In 'P': 0v, in 2nd: 12v
10, 12	---	Not Used	---
11	BL/PK	Purge Flow Sensor	0.5-1.5v, 30 mph: 3v
13	PK/DG	Flash EEPROM Power	0.1v
14	---	Not Used	---
15	BK/BL	Data Bus (-) Signal	Digital Signals
16	WT/BL	Data Bus (+) Signal	Digital Signals
17	RD/BK	High Speed Fan Control	Fan On: 1v, Off: 12v
18, 20	---	Not Used	---
21	BL/DG	CKP Sensor (+) Signal	435-475 Hz
22	OR	CKP Sensor (-) Signal	435-475 Hz
23	YL/BL	TSS Sensor (-) Signal	340-380 Hz
24	BK/YL	Power Ground	<0.1v
25	RD/BL	Chassis Ground	<0.050v
26	GY/YL	Coil Driver 1 Control	5° dwell
27	LG/BL	Shift Solenoid 1 Control	1v, 55 mph: 12v
28-29	---	Not Used	---
30	DG/WT	Octane Adjust Switch	9.3v (Closed: 0v)
31, 34	---	Not Used	---
32	R/GN	TR Sensor Reverse	Reverse: 12v, Others: 0v
33	BL	VSS (-) Signal	0 Hz, 55 mph: 125 Hz
35	GY	HO2S-12 (B1 S2) Signal	0.1-1.1v
36	OR/BL	MAF Sensor Return	<0.050v
37	WT	TFT Sensor Signal	2.40-2.10v
38	BL/WT	ECT Sensor Signal	0.5-0.6v
39	WT/GN	IAT Sensor Signal	1.5-2.5v
40	BK/PKK	Fuel Pump Monitor	12v, Off: 0v
41	GN/RD	A/C Cycling Clutch Switch	A/C Switch On: 12v
42-44	---	Not Used	---
45	PK/WT	Low Speed Fan Control	Fan On: 1v, Off: 12v
46	---	Not Used	---
47	PK/RD	VR Solenoid Control	0%, 55 mph: 45%
48	LG/RD	Clean Tachometer Output	25-38 Hz
49-50	---	Not Used	---
51	BK/YL	Power Ground	<0.1v
52	BR/RD	Coil Driver 2 Control	5° dwell
53	OR/GN	Shift Solenoid 3 Control	12v, 55 mph: 1v
54	GY/WT	TCC Solenoid Control	25-38 Hz
55	BL/RD	Keep Alive Power	12-14v
56-57	---	Not Used	---
58	WT/BK	VSS (+) Signal	0 Hz, 55 mph: 125 Hz
60	GY/DG	HO2S-11 (B1 S1) Signal	0.1-1.1v
61-63	---	Not Used	---
64	RD	A/T: Neutral Drive Switch	In 'N': 0v, in Gear: 12v
64	BR/YL	M/T: Clutch Position Switch	Clutch In: 0v, Out: 5v

1996 Tracer 1.9L I4 MFI VIN J (All) 104 Pin Connector

PCM Pin #	Wire Color	Circuit Description (104 Pin)	Value at Hot Idle
65	BL/YL	DPFE Sensor Signal	0.20-1.30v
66	---	Not Used	---
67	OR/WT	Canister Purge Solenoid	0-10 Hz (0-100%)
68	---	Not Used	---
69	BL/BK	A/C WOT Relay Control	Relay Off: 12v, On: 1v
70	---	Not Used	---
71	WT/RD	Vehicle Power	12-14v
72-73	---	Not Used	---
74	BR/WT	Injector 3 Control	4.0-4.5 ms
75	Y	Injector 1 Control	4.0-4.5 ms
76	BL	CMP Sensor (-) Signal	5-7 Hz
77	BK/YL	Power Ground	<0.1v
78-79	---	Not Used	---
80	LG	Fuel Pump Control	Relay Off: 12v, On: 1v
81-82	---	Not Used	---
83	BL/OR	IAC Motor Control	10v (30% d/cycle)
84	WT/BL	TSS Sensor (+) Signal	340-380 Hz
85	DG	CMP Sensor (+) Signal	5-7 Hz
86	DG/BK	A/C Pressure Sensor	A/C On: 1.5-1.9v
87	---	Not Used	---
88	BR/BK	MAF Sensor Signal	0.8v
89	RD/WT	TP Sensor Signal	0.53-1.27v
90	LG/WT	Reference Voltage	4.9-5.1v
91	LG/BK	Analog Signal Return	<0.050v
92	DG	Brake Switch Signal	Brake Off: 0v, On: 12v
93	OR/BK	HO2S-11 (B1 S1) Heater	1v, Off: 12v
94	---	Not Used	---
95	PK	HO2S-12 (B1 S2) Heater	1v, Off: 12v
96	---	Not Used	---
97	WT/RD	Vehicle Power	12-14v
98-99	---	Not Used	---
100	DG/OR	Injector 4 Control	4.0-4.5 ms
101	PK/YL	Injector 2 Control	4.0-4.5 ms
102	---	Not Used	---
103	BK/YL	Power Ground	<0.1v
104	---	Not Used	---

Pin Connector Graphic

```
        PCM 104-PIN CONNECTOR
   1  ⊗⊗⊗⊗⊗⊗⊗⊗⊗⊗⊗⊗⊗    ⊗⊗⊗⊗⊗⊗⊗⊗⊗⊗⊗⊗⊗ 26
  27  ⊗⊗⊗⊗⊗⊗⊗⊗⊗⊗⊗⊗⊗    ⊗⊗⊗⊗⊗⊗⊗⊗⊗⊗⊗⊗⊗ 52
  53  ⊗⊗⊗⊗⊗⊗⊗⊗⊗⊗⊗⊗⊗  ⊙ ⊗⊗⊗⊗⊗⊗⊗⊗⊗⊗⊗⊗⊗ 78
  79  ⊗⊗⊗⊗⊗⊗⊗⊗⊗⊗⊗⊗⊗    ⊗⊗⊗⊗⊗⊗⊗⊗⊗⊗⊗⊗⊗ 104
```

Terminal View of 104-Pin PCM Wiring Harness Connector

Standard Colors and Abbreviations

Abbreviation	Color	Abbreviation	Color	Abbreviation	Color
BK	Black	GY	Gray	PK	Purple
BL	Blue	GN	Green	RD	Red
BR	Brown	LG	LT Green	TN	Tan
DB	Dark Blue	OR	Orange	WT	White
DG	DK Green	PK	Pink	YL	Yellow

1997 Tracer 2.0L I4 2v MFI VIN P (All) 104 Pin Connector

PCM Pin #	Wire Color	Circuit Description (104 Pin)	Value at Hot Idle
1	PK/BK	Shift Solenoid 2 Control	12v, 55 mph: 1v
2	YL/BK	MIL (lamp) Control	MIL On: 1v, Off: 12v
3-5	---	Not Used	---
6	BR/BK	TR Sensor Overdrive	In O/D: 12v, Others: 0v
7	BR/GN	TR Sensor Low	In 1st: 12v, Others: 0v
8	PK/BL	IMRC Solenoid Monitor	All speeds: 5v
9	BR/YL	TR Sensor Drive	In 'D': 0v, Others: 12v
10	---	Not Used	---
11	BL/PK	EVAP Purge Flow Sensor	0.8v, 30 mph: 3.0v
12	---	Not Used	---
14	---	Not Used	---
13	PK/GN	Flash EEPROM Power	0.1v
15	BK/BL	Data Bus (-) Signal	Digital Signals
16	WT/BL	Data Bus (+) Signal	Digital Signals
17	RD/BK	High Speed Fan Control	Fan On: 1v, Off: 12v
18-20	---	Not Used	---
21	BK	CKP Sensor (+) Signal	400-425 Hz
22	WT	CKP Sensor (-) Signal	400-425 Hz
23	RD	TSS Sensor (-) Signal	340-380 Hz
24	BK/YL	Power Ground	<0.1v
25	RD/BL	Case Ground	<0.050v
26	GN/YL	Coil Driver 1 Control	5° dwell
27	LG/BL	Shift Solenoid 1 Control	1v, 55 mph: 12v
28-29, 34	---	Not Used	---
30	GN/WT	Octane Adjust Switch	9.3v (Closed: 0v)
31	YL/RD	PSP Switch Signal	Straight: 1v, Turning: 12v
32	R/GN	TR Sensor Reverse	Reverse: 12v, others: 0v
33	BL	VSS (-) Signal	0 Hz, 55 mph: 125 Hz
35	GY	HO2S-12 (B1 S2) Signal	0.1-1.1v
36	OR/BL	MAF Sensor Return	<0.050v
37	OR	TFT Sensor Signal	2.40-2.10v
38	BL/WT	ECT Sensor Signal	0.5-0.6v
39	WT/GN	IAT Sensor Signal	1.5-2.5v
40	BK/PKK	Fuel Pump Monitor	12v, Off: 0v
41	GN/RD	A/C Cycling Clutch Switch	A/C Switch On: 12v
42	RD/YL	IMRC Solenoid Control	12v, 55 mph: 12v
43-44, 46	---	Not Used	---
45	PK/WT	Low Speed Fan Control	Fan On: 1v, Off: 12v
47	PK/RD	VR Solenoid Control	0%, 55 mph: 45%
48	LG/RD	Clean Tachometer Output	25-38 Hz
49-50	---	Not Used	---
51	BK/YL	Power Ground	<0.1v
52	BR/RD	Coil Driver 2 Control	5° dwell
53	GY/YL	Shift Solenoid 3 Control	6.7v, 55 mph: 6.7v
54	GY/WT	TCC Solenoid Control	0%, 55 mph: 95%
55	BL/RD	Keep Alive Power	12-14v
56-57	---	Not Used	---
58	WT/BK	VSS (+) Signal	0 Hz, 55 mph: 125 Hz
60	GY/DG	HO2S-12 (B1 S2) Signal	0.1-1.1v
61-63	---	Not Used	---
64	RD	A/T: TR Position Sensor	In 'P': 0v, O/D: 1.7v
64	BR/YL	M/T: Clutch Pedal Position	Clutch In: 0v, Out: 5v

1997 Tracer 2.0L I4 2v MFI VIN P (All) 104 Pin Connector

PCM Pin #	Wire Color	Circuit Description (104 Pin)	Value at Hot Idle
65	BL/YL	DPFE Sensor Signal	0.20-1.30v
66	---	Not Used	---
67	OR/WT	EVAP Purge Solenoid	0-10 Hz (0-100%)
68	---	Not Used	---
69	BL/BK	A/C WOT Relay Control	Relay Off: 12v, On: 1v
70	---	Not Used	---
71	WT/RD	Vehicle Power	12-14v
72-73	---	Not Used	---
74	BR/WT	Injector 3 Control	3.3-3.7 ms
75	YL	Injector 1 Control	3.3-3.7 ms
76	BL	CMP Sensor (-) Signal	5-7 Hz
77	BK/YL	Power Ground	<0.1v
78-79	---	Not Used	---
80	LG	Fuel Pump Control	Relay Off: 12v, On: 1v
81-82	---	Not Used	---
83	BL/OR	IAC Motor Control	10v (30% d/cycle)
84	WT	TSS Sensor (+) Signal	340-380 Hz
85	GN	CMP Sensor (+) Signal	5-9 Hz
86	GN/BK	A/C Pressure Sensor	A/C On: 1.2-1.5v
87	---	Not Used	---
88	BR/BK	MAF Sensor Signal	0.8v
89	RD/WT	TP Sensor Signal	0.53-1.27v
90	LG/WT	Reference Voltage	4.9-5.1v
91	LG/BK	Analog Signal Return	<0.050 v
92	DG	Brake Switch Signal	Brake Off: 0v, On: 12v
93	OR/BK	HO2S-11 (B1 S1) Heater	1v, Off: 12v
94	---	Not Used	---
95	PK	HO2S-12 (B1 S2) Heater	1v, Off: 12v
96	---	Not Used	---
97	WT/RD	Vehicle Power	12-14v
98-99	---	Not Used	---
100	GN/OR	Injector 4 Control	3.3-3.7 ms
101	PK/YL	Injector 2 Control	3.3-3.7 ms
102	---	Not Used	---
103	BK/YL	Power Ground	<0.1v
104	---	Not Used	---

Pin Connector Graphic

```
          PCM 104-PIN CONNECTOR
    1 ◉◉◉◉◉◉◉◉◉◉◉◉◉   ◉◉◉◉◉◉◉◉◉◉◉◉◉ 26
   27 ◉◉◉◉◉◉◉◉◉◉◉◉◉  ⬤  ◉◉◉◉◉◉◉◉◉◉◉◉◉ 52
   53 ◉◉◉◉◉◉◉◉◉◉◉◉◉   ◉◉◉◉◉◉◉◉◉◉◉◉◉ 78
   79 ◉◉◉◉◉◉◉◉◉◉◉◉◉   ◉◉◉◉◉◉◉◉◉◉◉◉◉ 104
```

Terminal View of 104-Pin PCM Wiring Harness Connector

Standard Colors and Abbreviations

Abbreviation	Color	Abbreviation	Color	Abbreviation	Color
BK	Black	GY	Gray	PK	Purple
BL	Blue	GN	Green	RD	Red
BR	Brown	LG	LT Green	TN	Tan
DB	Dark Blue	OR	Orange	WT	White
DG	DK Green	PK	Pink	YL	Yellow

1998-99 Tracer 2.0L I4 MFI VIN P (All) 104 Pin Connector

PCM Pin #	Wire Color	Circuit Description (104 Pin)	Value at Hot Idle
1	PK/BK	Shift Solenoid 2 Control	0.1v, 55 mph: 0.1v
2	YL/BK	MIL (lamp) Control	MIL On: 1v, Off: 12v
3-5	---	Not Used	---
6	BR/BL	TR Sensor Overdrive	In O/D: 12v, others: 0v
7	BR/GN	TR Sensor Low	In 1st: 12v, others: 0v
8	PK/LB	IMRC Monitor Signal	All speeds: 5v
9	LG/BL	TR Drive Position	In 'D': 0v, 55 mph: 11.5v
10	---	Not Used	---
11	BL/PK	EVAP Purge Flow Sensor	0.8v, 30 mph: 3.0v
12	Y/BR	Fuel Level Indicator Signal	11.7v (1/2 full)
13	PK/GN	Flash EEPROM Power	0.1v
14	---	Not Used	---
15	BK/BL	Data Bus (-) Signal	Digital Signals
16	WT/BL	Data Bus (+) Signal	Digital Signals
17	RD/BK	High Speed Fan Control	Fan On: 1v, Off: 12v
18-20	---	Not Used	---
21	BK	CKP Sensor (+) Signal	400-425 Hz
22	WT	CKP Sensor (-) Signal	400-425 Hz
23	RD	TSS Sensor (-) Signal	340-380 Hz
24, 51	BK/YL	Power Ground	<0.1v
25	RD/BL	Shield Ground	<0.050v
26	GN/YL	Coil Driver 1 Control	5° dwell
27	LG/BL	Shift Solenoid 1 Control	0.1v, 55 mph: 12v
28-29, 34	---	Not Used	---
30	GN/WT	Octane Adjust Switch	9.3v (Closed: 0v)
31	YL/RD	PSP Sensor Signal	Straight: 0.5v, turned: 2v
32	R/GN	TR Sensor Reverse	Reverse: 12v, others: 0v
33	BL	VSS (-) Signal	0 Hz, 55 mph: 125 Hz
35	GY	HO2S-12 (B1 S2) Signal	0.1-1.1v
36	OR/BL	MAF Sensor Return	<0.050v
37	OR	TFT Sensor Signal	2.10 to 2.40v
38	BL/WT	ECT Sensor Signal	0.5-0.6v
39	WT/GN	IAT Sensor Signal	1.5-2.5v
40	BK/RD	Fuel Pump Monitor	12v, Off: 0v
41	W	A/C Cycling Clutch Switch	A/C Switch On: 12v
42	RD/YL	IMRC Solenoid Control	12v, 55 mph: 12v
43-44, 46	---	Not Used	---
45	PK/WT	Low Speed Cooling Fan	Fan On: 1v, Off: 12v
47	PK/RD	VR Solenoid Control	0%, 55 mph: 45%
48	LG/RD	Clean Tachometer Output	25-38 Hz
49-50	---	Not Used	---
52	BR/RD	Coil Driver 2 Control	5° dwell
53	OR/GN	Shift Solenoid 3 Control	12v, 55 mph: 12v
54	GY/WT	TCC Solenoid Control	0%, 55 mph: 95%
55	BL/RD	Keep Alive Power	12-14v
56	OR/WT	EVAP Purge Solenoid	0-10 Hz (0-100%)
57	RD	Knock Sensor 1 Signal	0v
58	WT/BK	VSS (+) Signal	0 Hz, 55 mph: 125 Hz
59, 61	---	Not Used	---
60	GY/GN	HO2S-11 (B1 S1) Signal	0.1-1.1v
62	BK/WT	FTP Sensor Signal	2.6v (0" H2O - cap off)
63	WT/PK	Fuel Rail Pressure Sensor	2.8v (39 psi)
64	RD	A/T: TR Sensor Signal	In 'P': 0.1v, O/D: 5v
64	BR/YL	M/T: CPP Switch Signal	Clutch In: 0v, Out: 5v

1998-99 Tracer 2.0L I4 MFI VIN P (All) 104 Pin Connector

PCM Pin #	Wire Color	Circuit Description (104 Pin)	Value at Hot Idle
65	BL/YL	DPFE Sensor Signal	0.20-1.30v
66, 68	---	Not Used	---
67	BL/PK	EVAP CV Solenoid	0-10 Hz (0-100%)
69	BL/BK	A/C WOT Relay Control	Relay Off: 12v, On: 1v
70	---	Not Used	---
71	WT/RD	Vehicle Power	12-14v
72	PK	M/T: Shift Indicator Lamp	SIL On: 1v, Off: 12v
73	---	Not Used	---
74	BR/WT	Injector 3 Control	3.3-3.7 ms
75	YL	Injector 1 Control	3.3-3.7 ms
76	BL	CMP Sensor (-) Signal	5-7 Hz
77	BK/YL	Power Ground	<0.1v
78-79	---	Not Used	---
80	WT/OR	Fuel Pump Control	Relay Off: 12v, On: 1v
81	---	Not Used	---
82	YL/PK	EPC Solenoid Control	2.5v (73%)
83	BL/OR	IAC Motor Control	10v (30% d/cycle)
84	WT	TSS Sensor (+) Signal	340-380 Hz
85	GN	CMP Sensor (+) Signal	5-7 Hz
86	GN/BK	A/C High Pressure Switch	AC On: 1.2-1.5v
87	---	Not Used	---
88	BR/BK	MAF Sensor Signal	0.8v
89	RD/WT	TP Sensor Signal	0.53-1.27v
90	LG/WT	Reference Voltage	4.9-5.1v
91	LG/BK	Analog Signal Return	<0.050v
92	GN	Brake Pedal Position Switch	Brake Off: 0v, On: 12v
93	OR/BK	HO2S-11 (B1 S1) Heater	1v, Off: 12v
94	---	Not Used	---
95	PK	HO2S-12 (B1 S2) Heater	1v, Off: 12v
96	---	Not Used	---
97	WT/RD	Vehicle Power	12-14v
98-99	---	Not Used	---
100	GN/OR	Injector 4 Control	3.3-3.7 ms
101	PK/YL	Injector 2 Control	3.3-3.7 ms
102	---	Not Used	---
103	BK/YL	Power Ground	<0.1v
104	---	Not Used	---

Pin Connector Graphic

```
        PCM 104-PIN CONNECTOR
  1 ●●●●●●●●●●●●●    ●●●●●●●●●●●●● 26
 27 ●●●●●●●●●●●●●    ●●●●●●●●●●●●● 52
 53 ●●●●●●●●●●●●●  ⬤ ●●●●●●●●●●●●● 78
 79 ●●●●●●●●●●●●●    ●●●●●●●●●●●●● 104
  Terminal View of 104-Pin PCM Wiring Harness Connector
```

Standard Colors and Abbreviations

Abbreviation	Color	Abbreviation	Color	Abbreviation	Color
BK	Black	GY	Gray	PK	Purple
BL	Blue	GN	Green	RD	Red
BR	Brown	LG	LT Green	TN	Tan
DB	Dark Blue	OR	Orange	WT	White
DG	DK Green	PK	Pink	YL	Yellow

FORD & LINCOLN TRUCK CONTENTS

About This Section
   Introduction ....Page 12-3
   How to use this Section ....Page 12-3
F-SERIES TRUCKS
   4.2L V6 MFI VIN 2 (All) 104 Pin **(1997-99)**...Page 12-4
   4.2L V6 MFI VIN 2 (All) 104 Pin **(2000-02)**...Page 12-6
   4.2L V6 MFI VIN 2 (All) 104 Pin **(2003)**...Page 12-8
   4.6L V8 MFI VIN W, 6 (All) 104 Pin **(1997-99)**.......................................Page 12-10
   4.6L V8 MFI VIN W, 6 (All) 104 Pin **(2000-02)**.....................................Page 12-12
   4.6L V8 MFI VIN W, 6 (All) 104 Pin **(2003)**...Page 12-14
   4.9L I6 MFI VIN Y (All) 60 Pin **(1990)**...Page 12-16
   4.9L I6 MFI VIN Y (E4OD) 60 Pin **(1990)**...Page 12-17
   4.9L I6 MFI VIN Y (All) 60 Pin **(1991-94)**...Page 12-18
   4.9L I6 MFI VIN Y (E4OD) 60 Pin **(1991-94)**.......................................Page 12-19
   4.9L I6 MFI VIN Y (All) California 60 Pin **(1995)**...................................Page 12-21
   4.9L I6 MFI VIN Y (All) 60 Pin **(1995)**...Page 12-23
   4.9L I6 MFI VIN Y (E4OD) California 60 Pin **(1995)**.............................Page 12-24
   4.9L I6 MFI VIN Y (E4OD) 60 Pin **(1995)**...Page 12-26
   4.9L I6 MFI VIN Y (All) 104 Pin **(1996)**...Page 12-28
   4.9L I6 MFI VIN Y (E4OD) 104 Pin **(1996)**...Page 12-30
   5.0L V8 MFI VIN N (All) 60 Pin **(1990)**...Page 12-32
   5.0L V8 MFI VIN N (All) 60 Pin **(1991-93)**...Page 12-33
   5.0L V8 MFI VIN N (All) 60 Pin **(1994-95)**...Page 12-34
   5.0L V8 MFI VIN N (E4OD) 60 Pin **(1994-95)**.....................................Page 12-36
   5.0L V8 MFI VIN N (All) 104 Pin **(1996)**...Page 12-38
   5.4L V8 MFI VIN L (E4OD) 104 Pin **(1997)**...Page 12-40
   5.4L V8 MFI VIN L (A/T) 104 Pin **(1998-99)**.......................................Page 12-42
   5.4L V8 MFI VIN L (All) 104 Pin **(2000-02)**...Page 12-44
   5.4L V8 MFI VIN L (All) 104 Pin **(2003)**...Page 12-46
   5.4L V8 CNG VIN M (E4OD) 104 Pin **(1997)**.....................................Page 12-48
   5.4L V8 CNG VIN M, VIN Z (E4OD) 104 Pin **(1998-99)**.......................Page 12-50
   5.4L V8 CNG VIN M, VIN Z (E4OD) 104 Pin **(2000-02)**.....................Page 12-52
   5.4L V8 CNG VIN M, VIN Z (E4OD) 104 Pin **(2003)**.........................Page 12-54
   5.8L V8 MFI VIN H (All) 60 Pin **(1990)**...Page 12-56
   5.8L V8 MFI VIN H (E4OD) 60 Pin **(1990)**...Page 12-57
   5.8L V8 MFI VIN H (All) 60 Pin **(1991-93)**...Page 12-58
   5.8L V8 MFI VIN H (E4OD) 60 Pin **(1991-93)**...................................Page 12-59
   5.8L V8 MFI VIN H (All) 60 Pin **(1994)**...Page 12-61
   5.8L V8 MFI VIN H (E4OD) 60 Pin **(1994)**...Page 12-63
   5.8L V8 MFI VIN H (E4OD) California 60 Pin **(1995)**.........................Page 12-65
   5.8L V8 MFI VIN H (E4OD) 60 Pin **(1995)**...Page 12-67
   5.8L V8 MFI VIN H (E4OD) 104 Pin **(1996-97)**.................................Page 12-69
   6.0L V8 Diesel VIN P (All) 104 Pin **(2003)**...Page 12-71
   6.8L V10 MFI SOHC VIN S (All) 104 Pin **(1999-2002)**.......................Page 12-74
   6.8L V10 MFI SOHC VIN S (All) 104 Pin **(2003)**...............................Page 12-76

F-SERIES TRUCKS (CONTINUED)

7.3L V8 Diesel VIN F (All) 104 Pin **(1996-97)** .... Page 12-78
7.3L V8 Diesel VIN F (All) 104 Pin **(1998-99)** .... Page 12-80
7.3L V8 Diesel VIN F (All) 104 Pin **(2000-02)** .... Page 12-82
7.3L V8 Diesel VIN F (All) 104 Pin **(2003)** .... Page 12-84
7.5L V8 MFI VIN G (All) 60 Pin **(1990)** .... Page 12-86
7.5L V8 MFI VIN G (E4OD) 60 Pin **(1990)** .... Page 12-87
7.5L V8 MFI VIN G (All) 60 Pin **(1991)** .... Page 12-89
7.5L V8 MFI VIN G (E4OD) 60 Pin **(1991)** .... Page 12-90
7.5L V8 MFI VIN G (All) 60 Pin **(1992-95)** .... Page 12-91
7.5L V8 MFI VIN G (E4OD) 60 Pin **(1992-95)** .... Page 12-92
7.5L V8 MFI VIN G (All) 104 Pin **(1996-97)** .... Page 12-93
7.5L V8 MFI VIN G (E4OD) 104 Pin **(1996-97)** .... Page 12-95

LIGHTNING PICKUPS

5.4L V8 MFI VIN 3 (E4OD) 104 Pin **(1999-2003)** .... Page 12-97

RANGER PICKUPS

2.3L I4 MFI VIN A (All) 60 Pin **(1990)** .... Page 12-99
2.3L I4 MFI VIN A (All) 60 Pin **(1991-94)** .... Page 12-100
2.3L I4 MFI VIN A California (All) 60 Pin **(1994)** .... Page 12-101
2.3L I4 MFI VIN A (All) 104 Pin **(1995-97)** .... Page 12-102
2.3L I4 MFI VIN D (All) 104 Pin **(2001-03)** .... Page 12-104
2.5L I4 MFI VIN C (All) 104 Pin **(1998-2001)** .... Page 12-106
2.9L V6 MFI VIN T (All) 60 Pin **(1990)** .... Page 12-108
2.9L V6 MFI VIN T (All) 60 Pin **(1991-92)** .... Page 12-109
3.0L V6 MFI VIN U (All) 60 Pin **(1991)** .... Page 12-110
3.0L V6 MFI VIN U (All) 60 Pin **(1992-94)** .... Page 12-111
3.0L V6 MFI VIN U (All) 104 Pin **(1995-97)** .... Page 12-112
3.0L V6 MFI VIN U (All) 104 Pin **(1998-2002)** .... Page 12-114
3.0L V6 MFI VIN U (All) 104 Pin **(2003)** .... Page 12-116
3.0L V6 FFV VIN V (All) 104 Pin **(1999-2002)** .... Page 12-118
3.0L V6 FFV VIN V (All) 104 Pin **(2003)** .... Page 12-120
4.0L V6 MFI VIN X (All) 60 Pin **(1990)** .... Page 12-122
4.0L V6 MFI VIN X (All) 60 Pin **(1991-94)** .... Page 12-123
4.0L V6 MFI VIN X (All) California 60 Pin **(1993-94)** .... Page 12-124
4.0L V6 MFI VIN X (All) 104 Pin **(1995-97)** .... Page 12-125
4.0L V6 MFI VIN X (All) 104 Pin **(1998)** .... Page 12-127
4.0L V6 MFI VIN X (All) 104 Pin **(1999-2002)** .... Page 12-129
4.0L V6 MFI VIN X (All) 104 Pin **(2003)** .... Page 12-131

BLACKWOOD PICKUP

5.4L V8 MFI DOHC 4v VIN A (All) 104 Pin **(2002-03)** .... Page 12-133

About This Section

Introduction

This section of the Diagnostic Manual contains Pin Tables for Trucks from 1990-2003. It can be used to assist in the repair of Code and No Code faults related to the PCM.

VEHICLE COVERAGE

- Blackwood Truck Applications (2002-03)
- F-Series Truck Applications (1990-2003)
- Lightning Pickup Applications (1999-2003)
- Ranger Pickup Applications (1990-2003)

How to Use This Section

This section can be used to look up the location of a particular pin, a Wire Color or a "known good" value of a PCM circuit. To locate the PCM information for a particular vehicle, find the model, correct engine size (with VIN Code) and finally the year of the vehicle.

For example, to look up the PCM terminals for a 1999 F-350 6.8L VIN S, go to Contents Page 2 and find the text string shown below:

6.8L V10 SOHC MFI VIN S (A/T) 104 Pin **(1999-2002)**...Page 12-64

Then turn to Page 12-64 to find the following PCM related information.

1999-2003 F-Series 6.8L V10 VIN S 104 Pin Connector

PCM Pin #	Wire Color	Circuit Description (104 Pin)	Value at Hot Idle
1	OR/YL	COP 6 Driver (dwell)	5°, 55 mph: 9°
9	YL/WT	Fuel Level Indicator Signal	1.7v (1/2 full)
11	PK/OR	Shift Solenoid 2 Control	12v, 55 mph: 1v
56	LG/BK	EVAP Purge Solenoid	0-10 Hz (0-100%)
65	BR/LG	EGR DPFE Sensor Signal	0.20-1.30v
66	YL/LG	CHT Sensor Signal	0.6 (194°F)

In this example, the Fuel Level Indicator circuit connects to Pin 9 of the 104 Pin connector with a Yellow/White wire. The value at Hot Idle shown here (1.7v) is a normal value with the fuel tank 1/2 full.

The cylinder head temperature (CHT) sensor is used to indicate the engine coolant temperature on this vehicle. This device is connected to a circuit with a pull-up resistor. Due to this unique design, the CHT sensor reading at near 194°F can read in the Cold End Line (3.71v) or in the Hot End Line (0.6v). This variance should be understood when using the CHT sensor Pin Table during cooling system testing.

The (A/T) in the Title over the table indicates that this coverage is only for vehicles equipped with an Automatic Transmission.

F-Series Truck PIN Tables

1997-99 F-Series 4.2L V6 MFI VIN 2 (All) 104 Pin Connector

PCM Pin #	Wire Color	Circuit Description (104 Pin)	Value at Hot Idle
1	PK/OR	Shift Solenoid 2 Control	12v, 55 mph: 1v
2	PK/LG	MIL (lamp) Control	MIL Off: 12v, On: 1v
3	YL/BK	Digital TR1 Sensor	0v, 55 mph: 11.5v
4-7	---	Not Used	---
8	WT/OR	Intake Manifold Runner No. 1	55 mph: 5v or 2.5v
9	BR	Intake Manifold Runner No. 2	55 mph: 5v or 2.5v
11	---	Not Used	---
12	YL/WT	Fuel Level Indicator Signal	1.7v (1/2 full)
13	PK	Flash EEPROM Power	0.1v
14	LB/BK	4x4 Low Indicator Lamp	7.7v (Switch On: 0v)
15	PK/LB	SCP Data Bus (-) Signal	<0.050v
16	TN/OR	SCP Data Bus (+) Signal	Digital Signals
17-20	---	Not Used	---
21	DB	CKP (-) Sensor Signal	430-475 Hz
22	GY	CKP (+) Sensor Signal	430-475 Hz
23	---	Not Used	---
24	BK/WT	Power Ground	<0.1v
25	LG/YL	Case Ground	<0.050v
26	DB/LG	Coil 1 Driver (dwell)	6°, 55 mph: 9°
27	OR/YL	Shift Solenoid 1 Control	1v, 55 mph: 1v
29	TN/WT	TCS (switch) Signal	TCS & O/D On: 12v
30	DG	Octane Adjustment	Closed: 1v, Open: 9.3v
31-32	---	Not Used	---
33	PK/OR	VSS (-) Signal	0 Hz, 55 mph: 125 Hz
35	RD/LG	HO2S-12 (B1 S2) Signal	0.1-1.1v
36	TN/LB	MAF Sensor Return	<0.050v
37	OR/BK	TFT Sensor Signal	2.10-2.40v
38	LG/RD	ECT Sensor Signal	0.5-0.6v
39	GY	IAT Sensor Signal	1.5-2.5v
40	DG/YL	Fuel Pump Monitor	On: 12v, Off: 0v
41	BK/YL	A/C Switch Signal	A/C On: 12v, Off: 0v
42	BR	Intake Manifold Runner	12v, 55 mph: 12v
43-46	---	Not Used	---
47	BR/PK	VR Solenoid Control	0%, 55 mph: 45%
48	WT/PK	Clean Tachometer Output	65 Hz, 55 mph: 175 Hz
49	LB/BK	Digital TR2 Sensor	0v, 55 mph: 11.5v
50	WT/BK	Digital TR4 Sensor	0v, 55 mph: 11.5v
51	BK/WT	Power Ground	<0.1v
52	RD/LB	Coil 3 Driver (dwell)	6°, 55 mph: 9°
53	---	Not Used	---
54	PK/YL	TCC Solenoid Control	0%, 55 mph: 95%
55	BK/LG	Keep Alive Power	12-14v
56	LG/BK	EVAP Purge Solenoid	0-10 Hz (0-100%)
57	YL/RD	Knock Sensor 1 Signal	0v
58	GY/BK	VSS (+) Signal	0 Hz, 55 mph: 125 Hz
59	---	Not Used	---
60	GY/LB	HO2S-11 (B1 S1) Signal	0.1-1.1v
61	PK/LG	HO2S-22 (B2 S2) Signal	0.1-1.1v
62	RD/PK	FTP Sensor Signal	2.6v (0" H2O - cap off)
73	---	Not Used	---
64	LB/YL	A/T: Digital TR3 Sensor	0v, 55 mph: 11.5v
64	LB/YL	M/T: CPP Switch Signal	5v (clutch "in": 0v)

1997-99 F-Series 4.2L V6 MFI VIN 2 (All) 104 Pin Connector

PCM Pin #	Wire Color	Circuit Description (104 Pin)	Value at Hot Idle
65	BR/LG	DPFE Sensor Signal	0.20-1.30v
66	YL/LG	CHT Sensor Signal	0.6v (194°F)
67	PK/WT	EVAP CV Solenoid	0-10 Hz (0-100%)
69	BK/LB	A/C WOT Relay Control	Off: 12v, On: 1v
70, 72	---	Not Used	---
71	RD	Vehicle Power	12-14v
73	TN/BK	Injector 5 Control	2.7-4.1 ms
74	BR/YL	Injector 3 Control	2.7-4.1 ms
75	TN	Injector 1 Control	2.7-4.1 ms
76, 77	BK/WT	Power Ground	<0.1v
78	PK/WT	Coil 2 Driver (dwell)	6°, 55 mph: 9°
79	WT/LG	TCIL (lamp) Control	7.7v (Switch On: 0v)
80	LB/OR	Fuel Pump Control	Off: 12v, On: 1v
81	WT/YL	EPC Solenoid Control	10v (15-20 psi)
83	WT/LB	IAC Solenoid Control	10.7v (33%)
84	DB/YL	OSS Sensor Signal	0 Hz, 55 mph: 250 Hz
85	DG	CMP Sensor Signal	6 Hz, 55 mph: 15 Hz
86	---	Not Used	---
87	RD/BK	HO2S-21 (B2 S1) Signal	0.1-1.1v
88	LB/RD	MAF Sensor Signal	0.6v, 55 mph: 2.2v
89	GY/WT	TP Sensor Signal	0.53-1.27v
90	BR/WT	Reference Voltage	4.9-5.1v
91	GY/RD	Analog Signal Return	<0.050v
92	LG	Brake Pedal Switch	0v (Brake On: 12v)
93	RD/WT	HO2S-11 (B1 S1) Heater	1v (Heater Off: 12v)
94	YL/LB	HO2S-21 (B2 S1) Heater	1v (Heater Off: 12v)
95	WT/BK	HO2S-12 (B1 S2) Heater	1v (Heater Off: 12v)
96	TN/YL	HO2S-22 (B2 S2) Heater	1v (Heater Off: 12v)
97	RD	Vehicle Power	12-14v
98	---	Not Used	---
99	LG/OR	Injector 6 Control	2.7-4.1 ms
100	BR/LB	Injector 4 Control	2.7-4.1 ms
101	WT	Injector 2 Control	2.7-4.1 ms
102	---	Not Used	---
103	BK/WT	Power Ground	<0.1v
104	RD/YL	Coil 4 Driver (dwell)	6°, 55 mph: 9°

Pin Connector Graphic

```
PCM 104-PIN CONNECTOR
 1 ●●●●●●●●●●●●●    ●●●●●●●●●●●●● 26
27 ●●●●●●●●●●●●●    ●●●●●●●●●●●●● 52
53 ●●●●●●●●●●●●●  ⬤ ●●●●●●●●●●●●● 78
79 ●●●●●●●●●●●●●    ●●●●●●●●●●●●● 104
```
Terminal View of 104-Pin PCM Wiring Harness Connector

Standard Colors and Abbreviations

Abbreviation	Color	Abbreviation	Color	Abbreviation	Color
BK	Black	GY	Gray	PK	Purple
BL	Blue	GN	Green	RD	Red
BR	Brown	LG	LT Green	TN	Tan
DB	Dark Blue	OR	Orange	WT	White
DG	DK Green	PK	Pink	YL	Yellow

2000-02 F-Series 4.2L V6 MFI VIN 2 (All) 104 Pin Connector

PCM Pin #	Wire Color	Circuit Description (104 Pin)	Value at Hot Idle
1	VT/OR	Shift Solenoid 'B' Control	12v, 55 mph: 1v
2	---	Not Used	---
3	YL/BK	Digital TR1 Sensor	0v, 55 mph: 11.5v
4-7	---	Not Used	---
8	WT/OR	Intake Manifold Runner No. 1	55 mph: 5v or 2.5v
13	PK	Flash EEPROM Power	0.1v
14	LB/BK	4x4 Low Indicator Switch	12v (Switch On: 0v)
15	PK/LB	SCP Data Bus (-) Signal	<0.050v
16	TN/OR	SCP Data Bus (+) Signal	Digital Signals
17-18	---	Not Used	---
21	DB	CKP (-) Sensor Signal	430-475 Hz
22	GY	CKP (+) Sensor Signal	430-475 Hz
23	---	Not Used	---
24	BK/WT	Power Ground	<0.1v
25	LB/YL	Case Ground	<0.050v
26	DB/LG	Coil 1 Driver (dwell)	6°, 55 mph: 9°
27	OR/YL	Shift Solenoid 'A' Control	1v, 55 mph: 1v
29	TN/WT	TCS (switch) Signal	TCS & O/D On: 12v
30-34	---	Not Used	---
35	RD/LG	HO2S-12 (B1 S2) Signal	0.1-1.1v
36	TN/LB	MAF Sensor Return	<0.050v
37	OR/BK	TFT Sensor Signal	2.10-2.40v
38	---	Not Used	---
39	GY	IAT Sensor Signal	1.5-2.5v
40	DG/YL	Fuel Pump Monitor	On: 12v, Off: 0v
41	BK/YL	A/C Switch Signal	Switch on: 12v, off: 1v
42	BR	Intake Manifold Runner	12v, 55 mph: 12v
43-46	---	Not Used	---
47	BR/PK	VR Solenoid Control	0%, 55 mph: 45%
48	---	Not Used	---
49	LB/BK	Digital TR2 Sensor	0v, 55 mph: 11.5v
50	WT/BK	Digital TR4 Sensor	0v, 55 mph: 11.5v
51	BK/WT	Power Ground	<0.1v
52	RD/LB	Coil 3 Driver (dwell)	6°, 55 mph: 9°
53	---	Not Used	---
54	VT/YL	TCC Solenoid Control	0%, 55 mph: 95%
55	RD/WT	Keep Alive Power	12-14v
56	LG/BK	EVAP Purge Solenoid	0-10 Hz (0-100%)
57	YL/RD	Knock Sensor 1 Signal	0v
58	PK	Transfer Case Speed Sensor	0 Hz, 55 mph: 471 Hz
60	GY/LB	HO2S-11 (B1 S1) Signal	0.1-1.1v
61	VT/LG	HO2S-22 (B2 S2) Signal	0.1-1.1v
62	RD/PK	FTP Sensor Signal	2.6v (0" H2O - cap off)
64	LB/YL	A/T: Digital TR3 Sensor	0v, 55 mph: 11.5v
64	LB/YL	M/T: CPP Switch Signal	5v (clutch "in": 0v)

2000-02 F-Series 4.2L V6 MFI VIN 2 (All) 104 Pin Connector

PCM Pin #	Wire Color	Circuit Description (104 Pin)	Value at Hot Idle
65	BR/LG	DPFE Sensor Signal	0.20-1.30v
66	YL/LG	CHT Sensor Signal	0.6v (194°F)
67	VT/WT	EVAP CV Solenoid	0-10 Hz (0-100%)
68	GY/BK	VSS (+) Signal	0 Hz, 55 mph: 125 Hz
69	PK/YL	A/C WOT Relay Control	Off: 12v, On: 1v
71	RD	Vehicle Power	12-14v
72	---	Not Used	---
73	TN/BK	Injector 5 Control	2.7-4.1 ms
74	BR/YL	Injector 3 Control	2.7-4.1 ms
75	TN	Injector 1 Control	2.7-4.1 ms
76	---	Not Used	---
77	BK/WT	Power Ground	<0.1v
78	PK/WT	Coil 2 Driver (dwell)	6°, 55 mph: 9°
79	WT/LG	TCIL (lamp) Control	7.7v (Switch On: 0v)
80	LB/OR	Fuel Pump Control	Off: 12v, On: 1v
81	WT/YL	EPC Solenoid Control	10v (15-20 psi)
82	---	Not Used	---
83	WT/LB	IAC Solenoid Control	10.7v (33%)
84	DB/YL	OSS Sensor Signal	0 Hz, 55 mph: 250 Hz
85	DG	CMP Sensor Signal	6 Hz, 55 mph: 15 Hz
86	---	Not Used	---
87	RD/BK	HO2S-21 (B2 S1) Signal	0.1-1.1v
88	LB/RD	MAF Sensor Signal	0.6v, 55 mph: 2.2v
89	GY/WT	TP Sensor Signal	0.53-1.27v
90	BR/WT	Reference Voltage	4.9-5.1v
91	GY/RD	Analog Signal Return	<0.050v
92	RD/LG	Brake Pedal Switch	0v (Brake On: 12v)
93	RD/WT	HO2S-11 (B1 S1) Heater	1v (Heater Off: 12v)
94	YL/LB	HO2S-21 (B2 S1) Heater	1v (Heater Off: 12v)
95	WT/BK	HO2S-12 (B1 S2) Heater	1v (Heater Off: 12v)
96	TN/YL	HO2S-22 (B2 S2) Heater	1v (Heater Off: 12v)
97	RD	Vehicle Power	12-14v
99	LG/OR	Injector 6 Control	2.7-4.1 ms
100	BR/LB	Injector 4 Control	2.7-4.1 ms
101	WT	Injector 2 Control	2.7-4.1 ms
102	---	Not Used	---
103	BK/WT	Power Ground	<0.1v
104	---	Not Used	---

Pin Connector Graphic

PCM 104-PIN CONNECTOR

```
 1 ●●●●●●●●●●●●●   ●●●●●●●●●●●●● 26
27 ●●●●●●●●●●●●●   ●●●●●●●●●●●●● 52
53 ●●●●●●●●●●●●● ◯ ●●●●●●●●●●●●● 78
79 ●●●●●●●●●●●●●   ●●●●●●●●●●●●● 104
```

Terminal View of 104-Pin PCM Wiring Harness Connector

Standard Colors and Abbreviations

Abbreviation	Color	Abbreviation	Color	Abbreviation	Color
BK	Black	GY	Gray	PK	Purple
BL	Blue	GN	Green	RD	Red
BR	Brown	LG	LT Green	TN	Tan
DB	Dark Blue	OR	Orange	WT	White
DG	DK Green	PK	Pink	YL	Yellow

2003 F-Series 4.2L V6 MFI VIN 2 (All) 104 Pin Connector

PCM Pin #	Wire Color	Circuit Description (104 Pin)	Value at Hot Idle
1	VT/OR	Shift Solenoid 'B' Control	12v, 55 mph: 1v
2	---	Not Used	---
3	YL/BK	Digital TR1 Sensor	0v, 55 mph: 11.5v
4-7	---	Not Used	---
8	WT/OR	Intake Manifold Runner No. 1	55 mph: 5v or 2.5v
13	VT	Flash EEPROM Power	0.1v
14	LB/BK	4x4 Low Indicator Switch	12v (Switch On: 0v)
15	PK/LB	Data Bus (-) Signal	Digital Signals
16	TN/OR	Data Bus (+) Signal	Digital Signals
17-18	---	Not Used	---
19	OG/LG	Fuel Pump High/Low Relay Control	Relay Off: 12v, On: 1v
20	BR/OR	Coast Clutch Solenoid Control (MSOF)	12v, 55 mph: 12v
21	DB	CKP (-) Sensor Signal	430-475 Hz
22	GY	CKP (+) Sensor Signal	430-475 Hz
23	---	Not Used	---
24	BK/WT	Power Ground	<0.1v
25	LB/YL	Case Ground	<0.050v
26	DB/LG	Coil 1 Driver (dwell)	6°, 55 mph: 9°
27	OR/YL	Shift Solenoid 'A' Control	1v, 55 mph: 1v
28	---	Not Used	---
29	TN/WT	TCS (switch) Signal	TCS & O/D On: 12v
30-34	---	Not Used	---
35	RD/LG	HO2S-12 (B1 S2) Signal	0.1-1.1v
36	TN/LB	MAF Sensor Return	<0.050v
37	OR/BK	TFT Sensor Signal	2.10-2.40v
38	---	Not Used	---
39	GY	IAT Sensor 2 Signal	1.5-2.5v
40	DG/YL	Fuel Pump Monitor	Off: 12v, On: 1v
41	BK/YL	A/C Pressure Switch Signal	Switch on: 12v, off: 1v
42	BR	Intake Manifold Runner Monitor	12v, 55 mph: 12v
43-46	---	Not Used	---
47	BR/PK	VR Solenoid Control	0%, 55 mph: 45%
48	---	Not Used	---
49	LB/BK	Digital TR2 Sensor	0v, 55 mph: 11.5v
50	WT/BK	Digital TR4 Sensor	0v, 55 mph: 11.5v
51	BK/WT	Power Ground	<0.1v
52	RD/LB	Coil 3 Driver (dwell)	6°, 55 mph: 9°
53	---	Not Used	---
54	VT/YL	TCC Solenoid Control	0%, 55 mph: 95%
55	RD/WT	Keep Alive Power	12-14v
56	LG/BK	EVAP Purge Solenoid	0-10 Hz (0-100%)
57	YL/RD	Knock Sensor 1 Signal	0v
58	---	Not Used	---
59	PK	Transfer Case Speed Sensor (MSOF)	0 Hz, 55 mph: 471 Hz
60	GY/LB	HO2S-11 (B1 S1) Signal	0.1-1.1v
61	VT/LG	HO2S-22 (B2 S2) Signal	0.1-1.1v
62	RD/PK	FTP Sensor Signal	2.6v (0" H2O - cap off)
63	---	Not Used	---
64	LB/YL	A/T: Digital TR3 Sensor	0v, 55 mph: 11.5v
64	LB/YL	M/T: CPP Switch Signal	5v (clutch "in": 0v)

2003 F-Series 4.2L V6 MFI VIN 2 (All) 104 Pin Connector

PCM Pin #	Wire Color	Circuit Description (104 Pin)	Value at Hot Idle
65	BR/LG	DPFE Sensor Signal	0.20-1.30v
66	YL/LG	Cylinder Head Temperature Sensor	0.6v (194ºF)
67	VT/WT	EVAP CV Solenoid	0-10 Hz (0-100%)
68	GY/BK	VSS (+) Signal	0 Hz, 55 mph: 125 Hz
69	PK/YL	A/C WOT Relay Control	Off: 12v, On: 1v
70	BK/WT	Check Fuel Cap Indicator Control	Indicator Off: 12v, On: 1v
71	RD	Vehicle Power	12-14v
72	---	Not Used	---
73	TN/BK	Injector 5 Control	2.7-4.1 ms
74	BR/YL	Injector 3 Control	2.7-4.1 ms
75	TN	Injector 1 Control	2.7-4.1 ms
76	---	Not Used	---
77	BK/WT	Power Ground	<0.1v
78	PK/WT	Coil 2 Driver (dwell)	6º, 55 mph: 9º
79	WT/LG	TCIL (lamp) Control	7.7v (Switch On: 0v)
80	LB/OR	Fuel Pump Control	Off: 12v, On: 1v
81	WT/YL	EPC Solenoid Control	10v (15-20 psi)
82	---	Not Used	---
83	WT/LB	IAC Solenoid Control	10.7v (33%)
84	DB/YL	OSS Sensor Signal	0 Hz, 55 mph: 250 Hz
85	DG	CMP Sensor Signal	6 Hz, 55 mph: 15 Hz
86	---	Not Used	---
87	RD/BK	HO2S-21 (B2 S1) Signal	0.1-1.1v
88	LB/RD	MAF Sensor Signal	0.6v, 55 mph: 2.2v
89	GY/WT	TP Sensor Signal	0.53-1.27v
90	BR/WT	Reference Voltage	4.9-5.1v
91	GY/RD	Sensor Return	<0.050v
92	RD/LG	Brake Pedal Switch	0v (Brake On: 12v)
93	RD/WT	HO2S-11 (B1 S1) Heater	1v (Heater Off: 12v)
94	YL/LB	HO2S-21 (B2 S1) Heater	1v (Heater Off: 12v)
95	WT/BK	HO2S-12 (B1 S2) Heater	1v (Heater Off: 12v)
96	TN/YL	HO2S-22 (B2 S2) Heater	1v (Heater Off: 12v)
97	RD	Vehicle Power	12-14v
98	---	Not Used	---
99	LG/OR	Injector 6 Control	2.7-4.1 ms
100	BR/LB	Injector 4 Control	2.7-4.1 ms
101	WT	Injector 2 Control	2.7-4.1 ms
102	---	Not Used	---
103	BK/WT	Power Ground	<0.1v
104	---	Not Used	---

Pin Connector Graphic

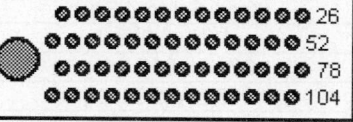

PCM 104-PIN CONNECTOR

Terminal View of 104-Pin PCM Wiring Harness Connector

Standard Colors and Abbreviations

Abbreviation	Color	Abbreviation	Color	Abbreviation	Color
BK	Black	GY	Gray	PK	Purple
BL	Blue	GN	Green	RD	Red
BR	Brown	LG	LT Green	TN	Tan
DB	Dark Blue	OR	Orange	WT	White
DG	DK Green	PK	Pink	YL	Yellow

1997-99 F-Series 4.6L V8 VIN W, VIN 6 (All) 104 Pin Connector

PCM Pin #	Wire Color	Circuit Description (104 Pin)	Value at Hot Idle
1	PK/OR	Shift Solenoid 2 Control	12v, 55 mph: 1v
2 ('97-'98)	PK/LG	MIL (lamp) Control	MIL Off: 12v, On: 1v
3	YL/BK	Digital TR1 Sensor	0v, 55 mph: 11.5v
4-11, 17-20	---	Not Used	---
12 ('97-'98)	YL/WT	Fuel Level Indicator Signal	1.7v (1/2 full)
13	VT	Flash EEPROM Power	0.1v
14	LB/BK	4x4 Indicator Switch	7.7v (Switch On: 0v)
15	PK/LB	Data Bus (-) Signal	Digital Signals
16	TN/OG	Data Bus (+) Signal	Digital Signals
21	DB	CKP (+) Sensor Signal	430-475 Hz
22	GY	CKP (-) Sensor Signal	430-475 Hz
23, 28	---	Not Used	---
24	BK/WT	Power Ground	<0.1v
25	BK/LB	Case Ground	<0.050v
26	DB/LG	Coil 1 Driver (dwell)	6°, 55 mph: 9°
27	OG/YL	Shift Solenoid 1 Control	1v, 55 mph: 12v
29	TN/WT	TCS (switch) Signal	TCS & O/D On: 12v
30 ('97-'98)	DG	Octane Adjustment	Closed: 1v, Open: 9.3v
31-32, 34	---	Not Used	---
33 ('97-'98)	PK/OR	VSS (-) Signal	0 Hz, 55 mph: 125 Hz
35	RD/LG	HO2S-12 (B1 S2) Signal	0.1-1.1v
36	TN/LB	MAF Sensor Return	<0.050v
37	OG/BK	TFT Sensor Signal	2.10-2.40v
38 ('97-'98)	LG/RD	ECT Sensor Signal	0.5-0.6v
39	GY	IAT Sensor Signal	1.5-2.5v
40	DG/YL	Fuel Pump Monitor	On: 12v, Off: 0v
41	BK/YL	A/C Switch Signal	A/C On: 12v, Off: 0v
42	BR	Inlet Air Control Valve	12v, 55 mph: 12v
43-44	---	Not Used	---
45 ('97-'98)	RD/WT	CHTIL (lamp) Control	7.7v (Switch On: 0v)
46	BR	Intake Manifold Tuning Valve	12v, 55 mph: 12v
47	BR/PK	VR Solenoid Control	0%, 55 mph: 45%
48 ('97-'98)	WT/PK	Clean Tachometer Output	39-49 Hz
49	LB/BK	Digital TR2 Sensor	0v, 55 mph: 11.5v
50	WT/BK	Digital TR4 Sensor	0v, 55 mph: 11.5v
51	BK/WT	Power Ground	<0.1v
52	RD/LB	Coil 2 Driver (dwell)	6°, 55 mph: 9°
53, 59	---	Not Used	---
54	VT/YL	TCC Solenoid Control	0%, 55 mph: 95%
55	RD/WT	Keep Alive Power	12-14v
56	LG/BK	EVAP Purge Solenoid	0-10 Hz (0-100%)
57	YL/RD	Knock Sensor 1 Signal	0v
58	PK	VSS (+) Signal	0 Hz, 55 mph: 125 Hz
60	GY/LB	HO2S-11 (B1 S1) Signal	0.1-1.1v
61	VT/LG	HO2S-22 (B2 S2) Signal	0.1-1.1v
62	RD/PK	FTP Sensor Signal	2.6v (0" H2O - cap off)
63	---	Not Used	---
64	LB/YL	A/T: Digital TR Sensor 3	0v, 55 mph: 11.5v
64	LB/YL	M/T: CPP Switch Signal	5v (clutch "in": 0v)

1997-99 F-Series 4.6L V8 VIN W, VIN 6 (All) 104 Pin Connector

PCM Pin #	Wire Color	Circuit Description (104 Pin)	Value at Hot Idle
65	BR/LG	DPFE Sensor Signal	0.20-1.30v
66	YL/LG	CHT Sensor Signal	0.6v (194°F)
67	VT/WT	EVAP CV Solenoid	0-10 Hz (0-100%)
68	GY/BK	Vehicle Speed Sensor Output	0 Hz, 55 mph: 125 Hz
69	PK/YL	A/C WOT Relay Control	12v (Relay On: 1v)
70	---	Not Used	---
71, 97	RD	Vehicle Power	12-14v
72	TN/RD	Injector 7 Control	2.7-4.1 ms
73	TN/BK	Injector 5 Control	2.7-4.1 ms
74	BR/YL	Injector 3 Control	2.7-4.1 ms
75	TN	Injector 1 Control	2.7-4.1 ms
76, 77, 103	BK/WT	Power Ground	<0.1v
78	PK/WT	Coil 3 Driver (dwell)	6°, 55 mph: 9°
79	WT/LG	TCIL (lamp) Control	12v (Lamp On: 1v)
80	LB/OR	Fuel Pump Control	Off: 12v, On: 1v
81	WT/YL	EPC Solenoid Control	9.1v (6 psi)
82, 86, 102	---	Not Used	---
83	WT/LB	IAC Solenoid Control	10.7v (33%)
84	DB/YL	OSS Sensor Signal	0 Hz, 55 mph: 250 Hz
85	DG	CMP Sensor Signal	6 Hz, 55 mph: 15 Hz
87	RD/BK	HO2S-21 (B2 S1) Signal	0.1-1.1v
88	LB/RD	MAF Sensor Signal	0.6v, 55 mph: 2.1v
89	GY/WT	TP Sensor Signal	0.53-1.27v
90	BR/WT	Reference Voltage	4.9-5.1v
91	GY/RD	Analog Signal Return	<0.050v
92	RD/LG	Brake Pedal Switch	0v (Brake On: 12v)
93	RD/WT	HO2S-11 (B1 S1) Heater	1v (Heater Off: 12v)
94	YL/LB	HO2S-21 (B2 S1) Heater	1v (Heater Off: 12v)
95	WT/BK	HO2S-12 (B1 S2) Heater	1v (Heater Off: 12v)
96	TN/YL	HO2S-22 (B2 S2) Heater	1v (Heater Off: 12v)
98	LB	Injector 8 Control	2.7-4.1 ms
99	LG/OR	Injector 6 Control	2.7-4.1 ms
100	BR/LB	Injector 4 Control	2.7-4.1 ms
101	WT	Injector 2 Control	2.7-4.1 ms
103	BK/WT	Power Ground	<0.1v
104	RD/YL	Coil 4 Driver (dwell)	6°, 55 mph: 9°

Pin Connector Graphic

PCM 104-PIN CONNECTOR

Terminal View of 104-Pin PCM Wiring Harness Connector

Standard Colors and Abbreviations

Abbreviation	Color	Abbreviation	Color	Abbreviation	Color
BK	Black	GY	Gray	PK	Purple
BL	Blue	GN	Green	RD	Red
BR	Brown	LG	LT Green	TN	Tan
DB	Dark Blue	OR	Orange	WT	White
DG	DK Green	PK	Pink	YL	Yellow

2000-02 F-Series 4.6L V8 VIN 6, VIN W (All) 104 Pin Connector

PCM Pin #	Wire Color	Circuit Description (104 Pin)	Value at Hot Idle
1	OG/YL	COP 6 Driver (dwell)	5°, 55 mph: 8°
2	---	Not Used	---
3	BK/WT	Power Ground	<0.1v
4	PK	Transfer Case Speed Sensor	0 Hz, 55 mph:
6	OG/YL	Shift Solenoid 'A' Control	1v, 55 mph: 1v
7	PK	Transfer Case VSS Signal	0 Hz, 55 mph: 471 Hz
11	VT/OR	Shift Solenoid 'B' Control	12v, 55 mph: 1v
12	WT/LG	TCIL (lamp) Control	12v (Lamp On: 1v)
13	VT	Flash EEPROM Power	0.1v
14	LB/BK	4x4 Indicator Switch	7.7v (Switch On: 0v)
15	PK/LB	SCP Data Bus (-) Signal	<0.050v
16	TN/OG	SCP Data Bus (+) Signal	Digital Signals
21	DB	CKP (-) Sensor Signal	430-475 Hz
22	GY	CKP (+) Sensor Signal	430-475 Hz
25	LB/YL	Case Ground	<0.050v
26	LG/WT	COP 1 Driver (dwell)	5°, 55 mph: 8°
27	LG/YL	COP 5 Driver (dwell)	5°, 55 mph: 8°
29	TN/WT	TCS (switch) Signal	TCS & O/D On: 12v
32	DG/VT	Knock Sensor 1 Return	<0.050v
34	YL/BK	TR1 Sensor Signal	In 'P': 0v, in O/D: 1.7v
35	RD/LG	HO2S-12 (B1 S2) Signal	0.1-1.1v
36	TN/LB	MAF Sensor Return	<0.050v
37	OG/BK	TFT Sensor Signal	2.10-2.40v
39	GY	IAT Sensor Signal	1.5-2.5v
40	DB/YL	Fuel Pump Monitor	On: 12v, Off: 0v
41	BK/YL	A/C Switch Signal	Switch on: 12v, off: 1v
46	BR	Intake Manifold Tuning Valve	12v, 55 mph: 12v
47	BR/PK	VR Solenoid Control	0%, 55 mph: 45%
49	LB/BK	Digital TR2 Sensor	0v, 55 mph: 11.5v
50	WT/BK	Digital TR4 Sensor	0v, 55 mph: 11.5v
51	BK/WT	Power Ground	<0.1v
52	WT/PK	COP 3 Driver (dwell)	5°, 55 mph: 8°
53	DG/VT	COP 4 Driver (dwell)	5°, 55 mph: 8°
54	VT/YL	TCC Solenoid Control	0%, 55 mph: 95%
55	RD/WT	Keep Alive Power	12-14v
56	LG/BK	EVAP Purge Solenoid	0-10 Hz (0-100%)
57	YL/RD	Knock Sensor 1 Signal	0v
60	GY/LB	HO2S-11 (B1 S1) Signal	0.1-1.1v
61	VT/LG	HO2S-22 (B2 S2) Signal	0.1-1.1v
62	RD/PK	FTP Sensor Signal	2.6v (0" H2O - cap off)
64	LB/YL	A/T: Digital TR3 Sensor	0v, 55 mph: 11.5v
64	LB/YL	M/T: CPP Switch Signal	5v (clutch "in": 0v)

2000-02 F-Series 4.6L V8 VIN 6, VIN W (All) 104 Pin Connector

PCM Pin #	Wire Color	Circuit Description (104 Pin)	Value at Hot Idle
65	BR/LG	EGR DPFE Sensor Signal	0.20-1.30v
66	YL/LG	CHT Sensor Signal	0.6v (194°F)
67	VT/WT	EVAP CV Solenoid	0-10 Hz (0-100%)
68	GY/BK	Vehicle Speed Control	0v
69	PK/YL	A/C WOT Relay Control	Off: 12v, On: 1v
71, 97	RD	Vehicle Power	12-14v
72	TN/RD	Injector 7 Control	2.7-4.1 ms
73	TN/BK	Injector 5 Control	2.7-4.1 ms
74	BR/YL	Injector 3 Control	2.7-4.1 ms
75	TN	Injector 1 Control	2.7-4.1 ms
77	BK/WT	Power Ground	<0.1v
78	PK/LB	COP 7 Driver (dwell)	5°, 55 mph: 8°
79	WT/RD	COP 8 Driver (dwell)	5°, 55 mph: 8°
80	LB/OR	Fuel Pump Relay Control	Off: 12v, On: 1v
81	WT/YL	EPC Solenoid Control	9.1v (6 psi)
83	WT/LB	IAC Solenoid Control	10.7v (33%)
84	DB/YL	OSS Sensor Signal	0 Hz, 55 mph: 250 Hz
85	DG	CMP Sensor Signal	6 Hz, 55 mph: 15 Hz
87	RD/BK	HO2S-21 (B2 S1) Signal	0.1-1.1v
88	LB/RD	MAF Sensor Signal	0.6v, 55 mph: 2.1v
89	GY/WT	TP Sensor Signal	0.53-1.27v
90	BR/WT	Reference Voltage	4.9-5.1v
91	GY/RD	Analog Signal Return	<0.050v
92	RD/LG	Brake Pedal Switch	0v (Brake On: 12v)
93	RD/WT	HO2S-11 (B1 S1) Heater	1v (Heater Off: 12v)
94	YL/LB	HO2S-21 (B2 S1) Heater	1v (Heater Off: 12v)
95	WT/BK	HO2S-12 (B1 S2) Heater	1v (Heater Off: 12v)
96	TN/YL	HO2S-22 (B2 S2) Heater	1v (Heater Off: 12v)
98	LB	Injector 8 Control	2.7-4.1 ms
99	LG/OR	Injector 6 Control	2.7-4.1 ms
100	BR/LB	Injector 4 Control	2.7-4.1 ms
101	WT	Injector 2 Control	2.7-4.1 ms
103	BK/WT	Power Ground	<0.1v
104	PK/WT	COP 2 Driver (dwell)	5°, 55 mph: 8°

Pin Connector Graphic

PCM 104-PIN CONNECTOR

```
 1 ●●●●●●●●●●●●●    ●●●●●●●●●●●●● 26
27 ●●●●●●●●●●●●●         ●●●●●●●●●●●● 52
53 ●●●●●●●●●●●●●    ⊗    ●●●●●●●●●●●● 78
79 ●●●●●●●●●●●●●    ●●●●●●●●●●●●●●104
```

Terminal View of 104-Pin PCM Wiring Harness Connector

Standard Colors and Abbreviations

Abbreviation	Color	Abbreviation	Color	Abbreviation	Color
BK	Black	GY	Gray	PK	Purple
BL	Blue	GN	Green	RD	Red
BR	Brown	LG	LT Green	TN	Tan
DB	Dark Blue	OR	Orange	WT	White
DG	DK Green	PK	Pink	YL	Yellow

2003 F-Series 4.6L V8 VIN 6, VIN W (All) 104 Pin Connector

PCM Pin #	Wire Color	Circuit Description (104 Pin)	Value at Hot Idle
1	OG/YL	COP 6 Driver (dwell)	5°, 55 mph: 8°
2	---	Not Used	---
3	BK/WT	Power Ground	<0.1v
4	PK	Transfer Case Speed Sensor	0 Hz, 55 mph:
5	---	Not Used	---
6	OG/YL	Shift Solenoid 'A' Control	1v, 55 mph: 1v
7-10	---	Not Used	---
11	VT/OR	Shift Solenoid 'B' Control	12v, 55 mph: 1v
12	WT/LG	TCIL (lamp) Control	12v (Lamp On: 1v)
13	VT	Flash EEPROM Power	0.1v
14	LB/BK	4x4 Indicator Switch	7.7v (Switch On: 0v)
15	PK/LB	SCP Data Bus (-) Signal	<0.050v
16	TN/OG	SCP Data Bus (+) Signal	Digital Signals
17-18	---	Not Used	---
19	OG/LG	Fuel Pump High/Low Relay Control	Relay Off: 12v, On: 1v
20	BR/OR	Coast Clutch Solenoid Control (MSOF)	12v, 55 mph: 12v
21	DB	CKP (-) Sensor Signal	430-475 Hz
22	GY	CKP (+) Sensor Signal	430-475 Hz
23-24	---	Not Used	---
25	LB/YL	Case Ground	<0.050v
26	LG/WT	COP 1 Driver (dwell)	5°, 55 mph: 8°
27	LG/YL	COP 5 Driver (dwell)	5°, 55 mph: 8°
28	---	Not Used	---
29	TN/WT	TCS (switch) Signal	TCS & O/D On: 12v
30-31	---	Not Used	---
32	DG/VT	Knock Sensor 1 Return	<0.050v
33	---	Not Used	---
34	YL/BK	TR1 Sensor Signal	In 'P': 0v, in O/D: 1.7v
35	RD/LG	HO2S-12 (B1 S2) Signal	0.1-1.1v
36	TN/LB	MAF Sensor Return	<0.050v
37	OG/BK	TFT Sensor Signal	2.10-2.40v
38	---	Not Used	---
39	GY	IAT Sensor 2 Signal	1.5-2.5v
40	DB/YL	Fuel Pump Monitor	On: 12v, Off: 0v
41	BK/YL	A/C Pressure Switch Signal	Switch on: 12v, off: 1v
42-45	---	Not Used	---
46	BR	Intake Manifold Tuning Valve	12v, 55 mph: 12v
47	BR/PK	VR Solenoid Control	0%, 55 mph: 45%
48	---	Not Used	---
49	LB/BK	Digital TR2 Sensor	0v, 55 mph: 11.5v
50	WT/BK	Digital TR4 Sensor	0v, 55 mph: 11.5v
51	BK/WT	Power Ground	<0.1v
52	WT/PK	COP 3 Driver (dwell)	5°, 55 mph: 8°
53	DG/VT	COP 4 Driver (dwell)	5°, 55 mph: 8°
54	VT/YL	TCC Solenoid Control	0%, 55 mph: 95%
55	RD/WT	Keep Alive Power	12-14v
56	LG/BK	EVAP Purge Solenoid	0-10 Hz (0-100%)
57	YL/RD	Knock Sensor 1 Signal	0v
58-59	---	Not Used	---
60	GY/LB	HO2S-11 (B1 S1) Signal	0.1-1.1v
61	VT/LG	HO2S-22 (B2 S2) Signal	0.1-1.1v
62	RD/PK	FTP Sensor Signal	2.6v (0" H2O - cap off)
63	---	Not Used	---
64	LB/YL	A/T: Digital TR3 Sensor	0v, 55 mph: 11.5v
64	LB/YL	M/T: CPP Switch Signal	5v (clutch "in": 0v)

2003 F-Series 4.6L V8 VIN 6, VIN W (All) 104 Pin Connector

PCM Pin #	Wire Color	Circuit Description (104 Pin)	Value at Hot Idle
65	BR/LG	EGR DPFE Sensor Signal	0.20-1.30v
66	YL/LG	CHT Sensor Signal	0.6v (194°F)
67	VT/WT	EVAP CV Solenoid	0-10 Hz (0-100%)
68	GY/BK	Vehicle Speed Control	0v
69	PK/YL	A/C WOT Relay Control	Off: 12v, On: 1v
70	BK/WT	Check Fuel Cap Indicator Control	Indicator Off: 12v, On: 1v
71	RD	Vehicle Power	12-14v
72	TN/RD	Injector 7 Control	2.7-4.1 ms
73	TN/BK	Injector 5 Control	2.7-4.1 ms
74	BR/YL	Injector 3 Control	2.7-4.1 ms
75	TN	Injector 1 Control	2.7-4.1 ms
76	---	Not Used	---
77	BK/WT	Power Ground	<0.1v
78	PK/LB	COP 7 Driver (dwell)	5°, 55 mph: 8°
79	WT/RD	COP 8 Driver (dwell)	5°, 55 mph: 8°
80	LB/OR	Fuel Pump Relay Control	Off: 12v, On: 1v
81	WT/YL	EPC Solenoid Control	9.1v (6 psi)
82, 86	---	Not Used	---
83	WT/LB	IAC Solenoid Control	10.7v (33%)
84	DB/YL	OSS Sensor Signal	0 Hz, 55 mph: 250 Hz
85	DG	CMP Sensor Signal	6 Hz, 55 mph: 15 Hz
87	RD/BK	HO2S-21 (B2 S1) Signal	0.1-1.1v
88	LB/RD	MAF Sensor Signal	0.6v, 55 mph: 2.1v
89	GY/WT	TP Sensor Signal	0.53-1.27v
90	BR/WT	Reference Voltage	4.9-5.1v
91	GY/RD	Sensor Return	<0.050v
92	RD/LG	Brake Pedal Switch	0v (Brake On: 12v)
93	RD/WT	HO2S-11 (B1 S1) Heater	1v (Heater Off: 12v)
94	YL/LB	HO2S-21 (B2 S1) Heater	1v (Heater Off: 12v)
95	WT/BK	HO2S-12 (B1 S2) Heater	1v (Heater Off: 12v)
96	TN/YL	HO2S-22 (B2 S2) Heater	1v (Heater Off: 12v)
97	RD	Vehicle Power	12-14v
98	LB	Injector 8 Control	2.7-4.1 ms
99	LG/OR	Injector 6 Control	2.7-4.1 ms
100	BR/LB	Injector 4 Control	2.7-4.1 ms
101	WT	Injector 2 Control	2.7-4.1 ms
102	---	Not Used	---
103	BK/WT	Power Ground	<0.1v
104	PK/WT	COP 2 Driver (dwell)	5°, 55 mph: 8°

Pin Connector Graphic

PCM 104-PIN CONNECTOR

Terminal View of 104-Pin PCM Wiring Harness Connector

Standard Colors and Abbreviations

Abbreviation	Color	Abbreviation	Color	Abbreviation	Color
BK	Black	GY	Gray	PK	Purple
BL	Blue	GN	Green	RD	Red
BR	Brown	LG	LT Green	TN	Tan
DB	Dark Blue	OR	Orange	WT	White
DG	DK Green	PK	Pink	YL	Yellow

1990 F-Series 4.9L I6 MFI VIN Y (All) 60 Pin Connector

PCM Pin #	Wire Color	Circuit Description (60 Pin)	Value at Hot Idle
1	YL	Keep Alive Power	12-14v
2, 5	---	Not Used	---
3	DG/WT	VSS (+) Signal	0 Hz, 55 mph: 125 Hz
3	GY/BK	VSS (+) Signal	0 Hz, 55 mph: 125 Hz
4	DG/YL	IDM Sensor Signal	20-31 Hz
6	BK	VSS (-) Signal	0 Hz, 55 mph: 125 Hz
7	LG/YL	ECT Sensor Signal	0.5-0.6v
8	BR	Fuel Pump Monitor	On: 12v, Off: 0v
9	BK/OR	Data Bus (-) Signal	Digital Signals
10	BK/YL	A/C Switch Signal	A/C On: 12v, Off: 0v
11	WT/BK	Air Management 2 Solenoid	AM2 Off: 12v, On: 1v
16	BK/OR	Ignition System Ground	<0.050v
17	TN/RD	Self Test Output & MIL	MIL Off: 12v, On: 1v
17	PK/LG	Self Test Output & MIL	MIL Off: 12v, On: 1v
18	TN/LG	Inferred Mileage Sensor	Digital Signals
20	BK	PCM Case Ground	<0.050v
21	GY/WT	IAC Solenoid Control	8.8-10.0v
22	TN/LG	Fuel Pump Control	Off: 12v, On: 1v
23	LG/BK	Knock Sensor Signal	0v
24	YL/LG	PSP Switch Signal	0v (turning: 12v)
25	YL/RD	ACT Sensor Signal	1.5-2.5v
26	OR/WT	Reference Voltage	4.9-5.1v
27	BR/LG	EVP Sensor Signal	0.4v, 55 mph: 2.6v
28	OR/BK	Data Bus (+) Signal	Digital Signals
29	DG/PK	HO2S-11 (B1 S1) Signal	0.1-1.1v
29	GY/BL	HO2S-11 (B1 S1) Signal	0.1-1.1v
30	GY/YL	A/T: Neutral Drive Switch	In 'N': Others: 5v
30	GY/YL	M/T: CPP Switch Signal	5v (clutch "in": 0v)
31	GY/YL	Canister Purge Solenoid	12v, 55 mph: 1v
33	DG	VR Solenoid Control	0%, 55 mph: 45%
36	YL/LG	Spark Output Signal	6.93v (50%)
37	RD	Vehicle Power	12-14v
40	BK/LG	Power Ground	<0.1v
43	LG/PK	A/C Demand Switch	A/C On: 12v, Off: 0v
45	DB/LG	MAP Sensor Signal	107 Hz (sea level)
45	LG/BK	MAP Sensor Signal	107 Hz (sea level)
46	BK/WT	Analog Signal Return	<0.050v
47	DG/LG	TP Sensor Signal	1v, 55 mph: 1.4v
48	WT/RD	Self Test Indicator Signal	STI On: 1v, Off: 5v
49	OR	HO2S-11 (B1 S1) Ground	<0.050v
51	WT/RD	Air Management 1 Solenoid	AM1 Off: 12v, On: 1v
56	DB	PIP Sensor Signal	6.93v (50%)
57	RD	Vehicle Power	12-14v
58	TN/OR	Injector Bank 1 (INJ 1, 3 & 5)	6.4-6.8 ms
59	TN/RD	Injector Bank 2 (INJ 2, 4 & 6)	6.4-6.8 ms
60	BK/LG	Power Ground	<0.1v

Pin Connector Graphic

PCM 60-PIN CONNECTOR

Terminal View of 60-Pin PCM Harness Connector

1990 F-Series 4.9L I6 MFI VIN Y (E4OD) 60 Pin Connector

PCM Pin #	Wire Color	Circuit Description (60 Pin)	Value at Hot Idle
1	YL	Keep Alive Power	12-14v
2	LG	Brake On/Off Switch	0v (Brake On: 12v)
3	GY/BK	VSS (+), (-) Signals	0 Hz, 55 mph: 125 Hz
4	DG/YL	IDM Sensor Signal	20-31 Hz
5, 13-15	---	Not Used	---
7	LG/YL	ECT Sensor Signal	0.5-0.6v
8	BR	Fuel Pump Monitor	On: 12v, Off: 0v
9, 28	BK, OR/BK	Data Bus (-), (+) Signals	Digital Signal
10	BK/YL	A/C Switch Signal	A/C On: 12v, Off: 0v
11	WT/BK	Air Management 2 Solenoid	AM2 Off: 12v, On: 1v
12	BL/BK	4x4 Switch Signal	12v (Switch On: 0v)
16	BK/OR	Ignition System Ground	<0.050v
17	PK/LG	Self Test Output & MIL	MIL Off: 12v, On: 1v
18	TN/LG	Inferred Mileage Sensor	Digital Signals
19	PK/OR	Shift Solenoid 2 Control	1v, 55 mph: 12v
20	BK	PCM Case Ground	<0.050v
21	GY/WT	IAC Solenoid Control	9.8-10.6v
22	TN/LG	Fuel Pump Control	Off: 12v, On: 1v
23	LG/BK	Knock Sensor Signal	0v
24	YL/LG	PSP Switch Signal	0v (turning: 12v)
25	YL/RD	ACT Sensor Signal	1.5-2.5v
26	OR/WT	Reference Voltage	4.9-5.1v
27	BR/LG	EVP Sensor Signal	0.4v, 55 mph: 2.6v
29	GY/BL	HO2S-11 (B1 S1) Signal	0.1-1.1v
30	BL/YL	MLP Sensor Signal	In 'P': 0v, in O/D: 5v
31	GY/YL	Canister Purge Solenoid	12v, 55 mph: 1v
32	LG/WT	OCIL (lamp) Control	7.7v (Switch On: 0v)
33	DG	VR Solenoid Control	0%, 55 mph: 45%
34-35, 39	---	Not Used	---
36	YL/LG	Spark Output Signal	6.93v (50%)
37, 57	RD	Vehicle Power	12-14v
38	WT/YL	EPC Solenoid Control	9.5v (5 psi)
40, 60	BK/LG	Power Ground	<0.1v
41	TN/WT	OCS (switch) Signal	OCS & O/D On: 12
42	OR/BK	TOT Sensor Signal	2.10-2.40v
43	LG/PK	A/C Demand Switch	A/C On: 12v, Off: 0v
44, 50, 54	---	Not Used	---
45	LG/BK	MAP Sensor Signal	107 Hz (sea level)
46	BK/WT	Analog Signal Return	<0.050v
47	DG/LG	TP Sensor Signal	1v, 55 mph: 1.4v
48	WT/RD	Self Test Indicator Signal	STI On: 1v, Off: 5v
49	OR	HO2S-11 (B1 S1) Ground	<0.050v
51	WT/RD	Air Management 1 Solenoid	AM1 Off: 12v, On: 1v
52	OR/YL	Shift Solenoid 1 Control	12v, 55 mph: 1v
53	PK/YL	TCC Solenoid Control	0v, On 55 mph: 12v
55	BR	Coast Clutch Switch	12v, 55 mph: 12v
56	DB	PIP Sensor Signal	6.93v (50%)
58	TN/OR	Injector Bank 1 (INJ 1, 3 & 5)	6.4-6.8 ms
59	TN/RD	Injector Bank 2 (INJ 2, 4 & 6)	6.4-6.8 ms

Pin Connector Graphic

PCM 60-PIN CONNECTOR

Terminal View of 60-Pin PCM Harness Connector

1991-94 F-Series 4.9L I6 VIN Y (All) 60 Pin Connector

PCM Pin #	Wire Color	Circuit Description (60 Pin)	Value at Hot Idle
1	YL	Keep Alive Power	12-14v
3	GY/BK	VSS (+) Signal	0 Hz, 55 mph: 125 Hz
4	TN/YL	IDM Sensor Signal	20-31 Hz
4	YL/BK	IDM Sensor Signal	20-31 Hz
6	PK/OR	VSS (-) Signal	0 Hz, 55 mph: 125 Hz
7	LG/RD	ECT Sensor Signal	0.5-0.6v
8	DG/YL	Fuel Pump Monitor	On: 12v, Off: 0v
9	PK/BL	Data Bus (-) Signal	Digital Signals
10	BK/YL	A/C Switch Signal	A/C On: 12v, Off: 0v
10	PK/BL	A/C Switch Signal	A/C On: 12v, Off: 0v
11	BR	Air Management 2 Solenoid	AM2 Off: 12v, On: 1v
16	OR/RD	Ignition System Ground	<0.050v
17	PK/LG	Self Test Output & MIL	MIL Off: 12v, On: 1v
20	BK	PCM Case Ground	<0.050v
21	WT/BL	IAC Solenoid Control	10.7v (33%)
22	BL/OR	Fuel Pump Control	Off: 12v, On: 1v
23	YL/RD	Knock Sensor Signal	0v
25	GY	IAT Sensor Signal	1.5-2.5v
26	BR/WT	Reference Voltage	4.9-5.1v
27	BR/LG	EVP Sensor Signal	0.4v, 55 mph: 2.6v
28	TN/OR	Data Bus (+) Signal	Digital Signals
29	GY/BL	HO2S-11 (B1 S1) Signal	0.1-1.1v
30	BL/YL	A/T: Neutral Drive Switch	In 'N': Others: 5v
30	BL/YL	M/T: CPP Switch Signal	5v (clutch "in": 0v)
31	GY/YL	Canister Purge Solenoid	12v, 55 mph: 1v
32	WT/LG	OCIL (lamp) Control	7.7v (Switch On: 0v)
33	BR/PK	VR Solenoid Control	0%, 55 mph: 45%
36	PK	Spark Output Signal	6.93v (50%)
37	RD	Vehicle Power	12-14v
38	WT/YL	EPC Solenoid Control	9.5v (5 psi)
40	BK/WT	Power Ground	<0.1v
43	PK	A/C Demand Switch	A/C On: 12v, Off: 0v
45	LG/BK	MAP Sensor Signal	107 Hz (sea level)
46	GY/RD	Analog Signal Return	<0.050v
47	GY/WT	TP Sensor Signal	1.5-1.2v
48	WT/PK	Self Test Indicator Signal	STI On: 1v, Off: 5v
49	OR	HO2S-11 (B1 S1) Ground	<0.050v
51	WT/OR	Air Management 1 Solenoid	AM1 Off: 12v, On: 1v
56	GY/OR	PIP Sensor Signal	6.93v (50%)
57	RD	Vehicle Power	12-14v
58	TN	Injector Bank 1 (INJ 1, 3 & 5)	5.6-6.4 ms
59	WT	Injector Bank 2 (INJ 2, 4 & 6)	5.6-6.4 ms
60	BK/WT	Power Ground	<0.1v

Pin Connector Graphic

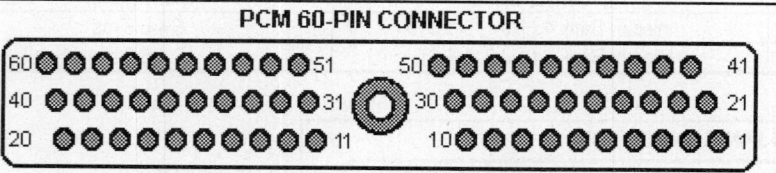

1991-94 F-Series 4.9L I6 VIN Y (E4OD) 60 Pin Connector

PCM Pin #	Wire Color	Circuit Description (60 Pin)	Value at Hot Idle
1	YL	Keep Alive Power	12-14v
2	LG	Brake Position Switch	0v (Brake On: 12v)
3	GY/BK	VSS (+) Signal	0 Hz, 55 mph: 125 Hz
4	TN/YL	IDM Sensor Signal	20-31 Hz
4	YL/BK	IDM Sensor Signal	20-31 Hz
5	---	Not Used	---
6	PK/OR	VSS (-) Signal	0 Hz, 55 mph: 125 Hz
7	LG/RD	ECT Sensor Signal	0.5-0.6v
8	DG/YL	Fuel Pump Monitor	On: 12v, Off: 0v
9	PK/BL	Data Bus (-) Signal	Digital Signals
10	BK/YL	A/C Switch Signal	A/C On: 12v, Off: 0v
10	PK/BL	A/C Switch Signal	A/C On: 12v, Off: 0v
11	BR	Air Management 2 Solenoid	AM2 Off: 12v, On: 1v
12	BL/BK	4x4 Switch Signal	12v (switch closed: 0v)
12	PK/BL	4x4 Switch Signal	12v (switch closed: 0v)
13-15	---	Not Used	---
16	OR/RD	Ignition System Ground	<0.050v
17	PK/LG	Self Test Output & MIL	MIL Off: 12v, On: 1v
18	---	Not Used	---
19	PK/OR	Shift Solenoid 2 Control	1v, 55 mph: 12v
20	BK	PCM Case Ground	<0.050v
21	WT/BL	IAC Solenoid Control	10.7v (33%)
22	BL/OR	Fuel Pump Control	Off: 12v, On: 1v
23	YL/RD	Knock Sensor Signal	0v
24	---	Not Used	---
25	GN/YL	ACT Sensor Signal	1.5-2.5v
26	BR/WT	Reference Voltage	4.9-5.1v
27	BR/LG	EVP Sensor Signal	0.4v, 55 mph: 2.6v
28	TN/OR	Data Bus (+) Signal	Digital Signals
29	GY/BL	HO2S-11 (B1 S1) Signal	0.1-1.1v
30	BL/YL	MLP Sensor Signal	In 'P': 0v, in O/D: 5v
31	GY/YL	Canister Purge Solenoid	12v, 55 mph: 1v
32	WT/LG	OCIL (lamp) Control	7.7v (Switch On: 0v)
33	BR/PK	VR Solenoid Control	0%, 55 mph: 45%
34	---	Not Used	---
35	---	Not Used	---
36	PK	Spark Output Signal	6.93v (50%)
37	RD	Vehicle Power	12-14v
38	WT/YL	EPC Solenoid Control	9.5v (5 psi)
39	---	Not Used	---
40	BK/WT	Power Ground	<0.1v

1991-94 F-Series 4.9L I6 VIN Y (E4OD) 60 Pin Connector

PCM Pin #	Wire Color	Circuit Description (60 Pin)	Value at Hot Idle
41	TN/WT	OCS (switch) Signal	12v (switch closed: 0v)
42	OR/BK	TOT Sensor Signal	2.10-2.40v
43	PK	A/C Demand Switch	A/C On: 12v, Off: 0v
44	---	Not Used	---
45	LG/BK	MAP Sensor Signal	107 Hz (sea level)
46	GY/RD	Analog Signal Return	<0.050v
47	GY/WT	TP Sensor Signal	1v, 55 mph: 1.4v
48	WT/PK	Self Test Indicator Signal	STI On: 1v, Off: 5v
49	OR	HO2S-11 (B1 S1) Ground	<0.050v
50	---	Not Used	---
51	WT/OR	Air Management 1 Solenoid	AM1 Off: 12v, On: 1v
52	OR/YL	Shift Solenoid 1 Control	12v, 55 mph: 1v
53	PK/YL	TCC Solenoid Control	0v, On 55 mph: 12v
54	---	Not Used	---
55	BR/OR	Coast Clutch Switch	12v, 55 mph: 12v
56	GY/OR	PIP Sensor Signal	6.93v (50%)
57	RD	Vehicle Power	12-14v
58	TN	Injector Bank 1 (INJ 1, 3 & 5)	5.6-6.0 ms
59	WT	Injector Bank 2 (INJ 2, 4 & 6)	5.6-6.0 ms
60	BK/WT	Power Ground	<0.1v

Pin Connector Graphic

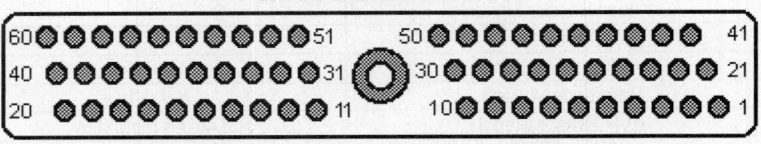

PCM 60-PIN CONNECTOR

Terminal View of 60-Pin PCM Harness Connector

Standard Colors and Abbreviations

Abbreviation	Color	Abbreviation	Color	Abbreviation	Color
BK	Black	GY	Gray	PK	Purple
BL	Blue	GN	Green	RD	Red
BR	Brown	LG	LT Green	TN	Tan
DB	Dark Blue	OR	Orange	WT	White
DG	DK Green	PK	Pink	YL	Yellow

1995 F-Series 4.9L VIN Y (All) California 60 Pin Connector

PCM Pin #	Wire Color	Circuit Description (60 Pin)	Value at Hot Idle
1	---	Not Used	---
2	LG	Brake Pedal Switch	0v (Brake On: 12v)
3	GY/BK	VSS (+) Signal	0 Hz, 55 mph: 125 Hz
4	YL/BK	IDM Sensor Signal	20-31 Hz
5	---	Not Used	---
6	PK/OR	VSS (-) Signal	0 Hz, 55 mph: 125 Hz
7	LG/RD	ECT Sensor Signal	0.5-0.6v
8	DG/YL	Fuel Pump Monitor	On: 12v, Off: 0v
9	TN/BL	MAF Sensor Return	<0.050v
10	DG/OR	A/C Switch Signal	A/C On: 12v, Off: 0v
11	GY/YL	Canister Purge Solenoid	12v, 55 mph: 1v
12	LG/OR	Injector 6 Control	5.5-7.1 ms
13	---	Not Used	---
14	---	Not Used	---
15	TN/BK	Injector 5 Control	5.5-7.1 ms
16	TN/YL	Ignition System Ground	<0.050v
17	PK/LG	Self Test Output & MIL	MIL Off: 12v, On: 1v
18	TN/OR	Data Bus (-) Signal	Digital Signals
19	PK/BK	Data Bus (+) Signal	Digital Signals
20	BK	PCM Case Ground	<0.050v
21	WT/BL	IAC Solenoid Control	10.7v (33%)
22	BL/OR	Fuel Pump Control	Off: 12v, On: 1v
23	YL/RD	Knock Sensor Signal	0v
24	---	Not Used	---
25	GY	IAT Sensor Signal	1.5-2.5v
26	BR/WT	Reference Voltage	4.9-5.1v
27	BR/LG	EVP Sensor Signal	0.4v, 55 mph: 2.6v
28-29	---	Not Used	---
30	BL/YL	A/T: Neutral Drive Switch	In 'P': 0v, Others: 5v
30	GY/YL	A/T: Neutral Drive Switch	In 'P': 0v, Others: 5v
30	BL/YL	M/T: CPP Switch Signal	5v (clutch "in": 0v)
30	GY/YL	M/T: CPP Switch Signal	5v (clutch "in": 0v)
31	WT/OR	Thermactor Air Bypass	TAB Off: 12v, On: 1v
32	WT/LG	TCIL (lamp) Control	12v (TCIL On: 0v)
33	BR/PK	VR Solenoid Control	0%, 55 mph: 45%
34	BR	Thermactor Air Diverter	TAD Off: 12v, On: 1v
35	BR/BL	Injector 4 Control	5.5-7.1 ms
36	PK	Spark Output Signal	6.93v (50%)
37	RD	Vehicle Power	12-14v
38	---	Not Used	---
39	BR/YL	Injector 3 Control	5.5-7.1 ms
40	BK/WT	Power Ground	<0.1v

1995 F-Series 4.9L VIN Y (All) California 60 Pin Connector

PCM Pin #	Wire Color	Circuit Description (60 Pin)	Value at Hot Idle
41	TN/WT	TCS (switch) Signal	TCS & O/D On: 12v
42	BL/YL	4x4 Switch Signal	12v (switch closed: 0v)
43	RD/BK	HO2S-21 (B2 S1) Signal	0.1-1.1v
44	GY/BL	HO2S-11 (B1 S1) Signal	0.1-1.1v
45	---	Not Used	---
46	GY/RD	Analog Signal Return	<0.1v
47	GY/WT	TP Sensor Signal	1v, 55 mph: 1.4v
48	WT/PK	Self Test Indicator Signal	STI On: 1v, Off: 5v
49	---	Not Used	---
50	BL/RD	MAF Sensor Signal	0.8v, 55 mph: 1.6v
51	---	Not Used	---
52	---	Not Used	---
53	---	Not Used	---
54	PK/YL	A/C WOT Cutout Control	Off: 12v, On: 1v
55	---	Not Used	---
56	GY/OR	PIP Sensor Signal	6.93v (50%)
57	RD	Vehicle Power	12-14v
58	TN	Injector 1 Control	5.5-7.1 ms
59	WT	Injector 2 Control	5.5-7.1 ms
60	BK/WT	Power Ground	<0.1v

Pin Connector Graphic

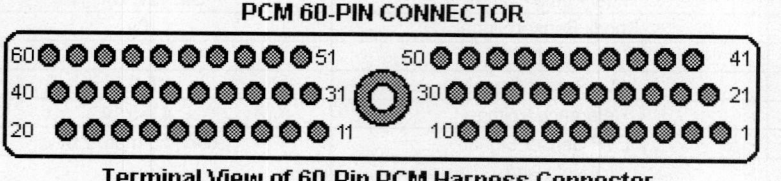

Standard Colors and Abbreviations

Abbreviation	Color	Abbreviation	Color	Abbreviation	Color
BK	Black	GY	Gray	PK	Purple
BL	Blue	GN	Green	RD	Red
BR	Brown	LG	LT Green	TN	Tan
DB	Dark Blue	OR	Orange	WT	White
DG	DK Green	PK	Pink	YL	Yellow

1995 F-Series 4.9L I6 VIN Y (All) 60 Pin Connector

PCM Pin #	Wire Color	Circuit Description (60 Pin)	Value at Hot Idle
1	YL	Keep Alive Power	12-14v
2	LG	Brake Pedal Switch	0v (Brake On: 12v)
3	GY/BK	VSS (+) Signal	0 Hz, 55 mph: 125 Hz
4	YL/BK	IDM Sensor Signal	20-31 Hz
6	PK/OR	VSS (-) Signal	0 Hz, 55 mph: 125 Hz
7	LG/RD	ECT Sensor Signal	0.5-0.6v
8	DG/YL	Fuel Pump Monitor	On: 12v, Off: 0v
9	PK/BL	Data Bus (-) Signal	Digital Signals
10	PK/BL	A/C Switch Signal	A/C On: 12v, Off: 0v
11	BR	Air Management 2 Solenoid	AM2 Off: 12v, On: 1v
12	BL/BK	4x4 Indicator Lamp	12v (switch closed: 0v)
16	OR/RD	Ignition System Ground	<0.050v
17	PK/LG	Self Test Output & MIL	MIL Off: 12v, On: 1v
20	BK	PCM Case Ground	<0.050v
21	WT/BL	IAC Solenoid Control	10.7v (33%)
22	BL/OR	Fuel Pump Control	Off: 12v, On: 1v
23	YL/RD	Knock Sensor Signal	0v
25	GY	IAT Sensor Signal	1.5-2.5v
26	BR/WT	Reference Voltage	4.9-5.1v
27	BR/LG	EVP Sensor Signal	0.4v, 55 mph: 2.6v
28	TN/OR	Data Bus (+) Signal	Digital Signals
29	GY/BL	HO2S-11 (B1 S1) Signal	0.1-1.1v
30	GY/YL	M/T: CPP Switch Signal	5v (clutch "in": 0v)
30	BL/YL	M/T: CPP Switch Signal	5v (clutch "in": 0v)
31	GY/YL	Canister Purge Solenoid	12v, 55 mph: 1v
32	WT/LG	TCIL (lamp) Control	7.7v (Switch On: 0v)
33	BR/PK	VR Solenoid Control	0%, 55 mph: 45%
36	PK	Spark Output Signal	6.93v (50%)
37	RD	Vehicle Power	12-14v
40	BK/WT	Power Ground	<0.1v
41	TN/WT	TCS (switch) Signal	0.1-0.2v
43	PK	A/C Demand Switch	A/C On: 12v, Off: 0v
45	LG/BK	MAP Sensor Signal	107 Hz (sea level)
46	GY/RD	Analog Signal Return	<0.050v
47	GY/WT	TP Sensor Signal	0.9-1.0v
48	WT/PK	Self Test Indicator Signal	STI on: 1v, Off: 5v
49	OR	HO2S-11 (B1 S1) Ground	<0.050v
51	WT/OR	Air Management 1 Solenoid	AM1 Off: 12v, On: 1v
56	GY/OR	Profile Ignition Truck Signal	6.93v (50%)
57	RD	Vehicle Power	12-14v
58	TN	Injector Bank 1 (INJ 1, 3 & 5)	6.2-7.4 ms
59	WT	Injector Bank 2 (INJ 2, 4 & 6)	6.2-7.4 ms
60	BK/WT	Power Ground	<0.1v

Pin Connector Graphic

PCM 60-PIN CONNECTOR

Terminal View of 60-Pin PCM Harness Connector

1995 F-Series 4.9L I6 VIN Y (E4OD) Calif. 60 Pin Connector

PCM Pin #	Wire Color	Circuit Description (60 Pin)	Value at Hot Idle
1	YL	Keep Alive Power	12-14v
2	LG	Brake Pedal Switch	0v (Brake On: 12v)
3	GY/BK	VSS (+) Signal	0 Hz, 55 mph: 125 Hz
4	YL/BK	IDM Sensor Signal	20-31 Hz
5	---	Not Used	---
6	PK/OR	VSS (-) Signal	0 Hz, 55 mph: 125 Hz
7	LG/RD	ECT Sensor Signal	0.5-0.6v
8	DG/YL	Fuel Pump Monitor	On: 12v, Off: 0v
9	TN/BL	MAF Sensor Return	<0.050v
10	DG/OR	A/C Switch Signal	A/C On: 12v, Off: 0v
11	GY/YL	Canister Purge Solenoid	12v, 55 mph: 1v
12	LG/OR	Injector 6 Control	5.5-7.1 ms
13-14	---	Not Used	---
15	TN/BK	Injector 5 Control	5.5-7.1 ms
16	OR/RD	Ignition System Ground	<0.050v
17	PK/LG	Self Test Output & MIL	MIL Off: 12v, On: 1v
18	TN/OR	Data Bus (-) Signal	Digital Signals
19	PK/BL	Data Bus (+) Signal	Digital Signals
20	BK	PCM Case Ground	<0.050v
21	WT/BL	IAC Solenoid Control	10.7v (33%)
22	BL/OR	Fuel Pump Control	Off: 12v, On: 1v
23	YL/RD	Knock Sensor Signal	0v
24	---	Not Used	---
25	GY	Idle Air Temperature Sensor	0.5-0.6v
26	BR/WT	Reference Voltage	4.9-5.1v
27	BR/LG	EVP Sensor Signal	0.4v, 55 mph: 2.6v
28-29	---	Not Used	---
30	BL/YL	TR Sensor Signal	In 'P': 0v, in O/D: 5v
30	GY/YL	TR Sensor Signal	In 'P': 0v, in O/D: 5v
31	WT/OR	Thermactor Air Bypass	TAB Off: 12v, On: 1v
32	WT/LG	TCIL (lamp) Control	7.7v (Switch On: 0v)
33	BR/PK	VR Solenoid Control	0%, 55 mph: 45%
34	BR	Thermactor Air Diverter	TAD Off: 12v, On: 1v
35	BR/BL	Injector 4 Control	5.5-7.1 ms
36	PK	Spark Output Signal	6.93v (50%)
37	RD	Vehicle Power	12-14v
38	WT/YL	EPC Solenoid Control	9.5v (5 psi)
39	BR/YL	Injector 3 Control	5.5-7.1 ms
40	BK/WT	Power Ground	<0.1v

1995 F-Series 4.9L I6 VIN Y (E4OD) Calif. 60 Pin Connector

PCM Pin #	Wire Color	Circuit Description (60 Pin)	Value at Hot Idle
41	TN/WT	TCS (switch) Signal	TCS & O/D On: 12v
42	BL/YL	4x4 Switch Signal	12v (switch closed: 0v)
43	RD/BK	HO2S-21 (B2 S1) Signal	0.1-1.1v
44	GY/BL	HO2S-11 (B1 S1) Signal	0.1-1.1v
45	---	Not Used	---
46	GY/RD	Analog Signal Return	<0.050v
47	GY/WT	TP Sensor Signal	1v, 55 mph: 1.4v
48	WT/PK	Self Test Indicator Signal	STI On: 1v, Off: 5v
49	OR/BK	TOT Sensor Signal	2.10-2.40v
50	BL/RD	MAF Sensor Signal	0.8v, 55 mph: 1.6v
51	OR/YL	Shift Solenoid 1 Control	12v, 55 mph: 1v
52	PK/OR	Shift Solenoid 2 Control	1v, 55 mph: 12v
53	PK/YL	TCC Solenoid Control	0v, On 55 mph: 12v
54	PK/YL	A/C WOT Relay Control	Off: 12v, On: 1v
55	BR/OR	Coast Clutch Solenoid	12v, 55 mph: 12v
56	GY/OR	PIP Sensor Signal	6.93v (50%)
57	RD	Vehicle Power	12-14v
58	TN	Injector 1 Control	5.5-7.1 ms
59	WT	Injector 2 Control	5.5-7.1 ms
60	BK/WT	Power Ground	<0.1v

Pin Connector Graphic

PCM 60-PIN CONNECTOR

60 ●●●●●●●●●● 51   50 ●●●●●●●●●● 41
40 ●●●●●●●●●● 31   ⊙ 30 ●●●●●●●●●● 21
20 ●●●●●●●●● 11   10 ●●●●●●●●●● 1

Terminal View of 60-Pin PCM Harness Connector

Standard Colors and Abbreviations

Abbreviation	Color	Abbreviation	Color	Abbreviation	Color
BK	Black	GY	Gray	PK	Purple
BL	Blue	GN	Green	RD	Red
BR	Brown	LG	LT Green	TN	Tan
DB	Dark Blue	OR	Orange	WT	White
DG	DK Green	PK	Pink	YL	Yellow

1995 F-Series 4.9L I6 VIN Y (E4OD) Federal 60 Pin Connector

PCM Pin #	Wire Color	Circuit Description (60 Pin)	Value at Hot Idle
1	YL	Keep Alive Power	12-14v
2	LG	Brake Pedal Switch	0v (Brake On: 12v)
3	GY/BK	VSS (+) Signal	0 Hz, 55 mph: 125 Hz
4	YL/BK	IDM Sensor Signal	20-31 Hz
5	---	Not Used	
6	PK/OR	VSS (-) Signal	0 Hz, 55 mph: 125 Hz
7	LG/RD	ECT Sensor Signal	0.5-0.6v
8	DG/YL	Fuel Pump Monitor	On: 12v, Off: 0v
9	PK/BL	Data Bus (-) Signal	Digital Signals
10	PK/BL	A/C Switch Signal	A/C On: 12v, Off: 0v
11	BR	Air Management 2 Solenoid	AM2 Off: 12v, On: 1v
12	BL/BK	4x4 Indicator Lamp	12v (switch closed: 0v)
13	---	Not Used	---
14	---	Not Used	---
15	---	Not Used	---
16	OR/RD	Ignition System Ground	<0.050v
17	PK/LG	Self Test Output & MIL	MIL Off: 12v, On: 1v
18	---	Not Used	---
19	PK/OR	Shift Solenoid 2 Control	12v, 55 mph: 1v
20	BK	PCM Case Ground	<0.050v
21	WT/BL	IAC Solenoid Control	10.7v (33%)
22	BL/OR	Fuel Pump Control	Off: 12v, On: 1v
23	YL/RD	Knock Sensor Signal	0v
24	---	Not Used	---
25	GY	IAT Sensor Signal	1.5-2.5v
26	BR/WT	Reference Voltage	4.9-5.1v
27	BR/LG	EVP Sensor Signal	0.4v, 55 mph: 2.6v
28	TN/OR	Data Bus (+) Signal	Digital Signals
29	GY/BL	HO2S-11 (B1 S1) Signal	0.1-1.1v
30	GY/YL	TR Sensor Signal	In 'P': 0v, in O/D: 5v
30	BL/YL	TR Sensor Signal	In 'P': 0v, in O/D: 5v
31	GY/YL	Canister Purge Solenoid	12v, 55 mph: 1v
32	WT/LG	TCIL (lamp) Control	7.7v (Switch On: 0v)
33	BR/PK	VR Solenoid Control	0%, 55 mph: 45%
34	---	Not Used	---
35	---	Not Used	---
36	PK	Spark Output Signal	6.93v (50%)
37	RD	Vehicle Power	12-14v
38	WT/YL	EPC Solenoid Control	9.5v (5 psi)
39	---	Not Used	---
40	BK/WT	Power Ground	<0.1v

1995 F-Series 4.9L I6 VIN Y (E4OD) Federal 60-P Connector

PCM Pin #	Wire Color	Circuit Description (60 Pin)	Value at Hot Idle
41	TN/WT	TCS (switch) Signal	TCS & O/D On: 12v
42	OR/BK	TOT Sensor Signal	2.10-2.40v
43	PK	A/C Demand Switch	A/C On: 12v, Off: 0v
44	---	Not Used	---
45	LG/BK	MAP Sensor Signal	107 Hz (sea level)
46	GY/RD	Analog Signal Return	<0.050v
47	GY/WT	TP Sensor Signal	1v, 55 mph: 1.4v
48	WT/PK	Self Test Indicator Signal	STI On: 1v, Off: 5v
49	OR	HO2S-11 (B1 S1) Ground	<0.050v
50	---	Not Used	---
51	WT/OR	Air Management 1 Solenoid	AM1 Off: 12v, On: 1v
52	OR/YL	Shift Solenoid 1 Control	1v, 55 mph: 12v
53	PK/YL	TCC Solenoid Control	0v, On 55 mph: 12v
54	---	Not Used	---
55	BR/OR	Coast Clutch Solenoid	12v, 55 mph: 12v
56	GY/OR	PIP Sensor Signal	6.93v (50%)
57	RD	Vehicle Power	12-14v
58	TN	Injector Bank 1 (INJ 1, 3 & 5)	6.2-7.4 ms
59	WT	Injector Bank 2 (INJ 2, 4 & 6)	6.2-7.4 ms
60	BK/WT	Power Ground	<0.1v

Pin Connector Graphic

PCM 60-PIN CONNECTOR

Terminal View of 60-Pin PCM Harness Connector

Standard Colors and Abbreviations

Abbreviation	Color	Abbreviation	Color	Abbreviation	Color
BK	Black	GY	Gray	PK	Purple
BL	Blue	GN	Green	RD	Red
BR	Brown	LG	LT Green	TN	Tan
DB	Dark Blue	OR	Orange	WT	White
DG	DK Green	PK	Pink	YL	Yellow

1996 Truck 4.9L I6 MFI VIN Y (All) 104 Pin Connector

PCM Pin #	Wire Color	Circuit Description (104 Pin)	Value at Hot Idle
1	---	Not Used	---
2	PK/LG	MIL (lamp) Control	MIL Off: 12v, On: 1v
3-12	---	Not Used	---
13	PK	Flash EPROM Power	0.1v
14	LB/BK	4x4 Low Switch Signal	12v (switch closed: 0v)
15	PK/LB	Data Bus (-) Signal	Digital Signals
16	TN/OR	Data Bus (+) Signal	Digital Signals
17-22	---	Not Used	---
23	OR/RD	Ignition Ground	<0.050v
24	BK/WT	Power Ground	<0.1v
25	BK	Case Ground	<0.050v
26-28	---	Not Used	---
29	TN/WT	TCS (switch) Signal	TCS & O/D On: 12v
30-32, 34	---	Not Used	---
33	PK/OR	PSOM (-) Signal	<0.050v
35	RD/LG	HO2S-12 (B1 S2) Signal	0.1-1.1v
36	TN/LB	MAF Sensor Return	<0.050v
37	OR/BK	TFT Sensor Signal	2.10-2.40v
38	LG/RD	ECT Sensor Signal	0.5-0.6v
39	GY	IAT Sensor Signal	1.5-2.5v
40	DG/YL	Fuel Pump Monitor	On: 12v, Off: 0v
41	DG/OR	A/C Switch Signal	A/C On: 12v, Off: 0v
42-43, 46	---	Not Used	---
44	BR	Secondary AIR Diverter	AIRD Off: 12v, On: 1v
47	BK/PK	VR Solenoid Control	0%, 55 mph: 45%
48	YL/BK	Clean Tachometer Output	39-45 Hz
49	GY/OR	PIP Sensor Signal	6.93v (50%)
50	PK	Spark Output Signal	6.93v (50%)
51	BK/WT	Power Ground	<0.1v
52-54	---	Not Used	---
55	YL	Keep Alive Power	12-14v
56	LG/BK	EVAP Purge Solenoid	0-10 Hz (0-100%)
57	YL/RD	Knock Sensor Signal	0v
58	GY/BK	PSOM (+) Signal	0 Hz, 55 mph: 125 Hz
59	DG/LG	Misfire Detection Sensor	45-55 Hz
60	GY/LB	HO2S-11 (B1 S1) Signal	0.1-1.1v
61-63	---	Not Used	---
64	LB/YL	A/T: PNP Switch Signal	In 'P': 0v, Others: 5v
64	LB/YL	M/T: CPP Switch Signal	5v (clutch "in": 0v)

1996 Truck 4.9L I6 MFI VIN Y (All) 104 Pin Connector

PCM Pin #	Wire Color	Circuit Description (104 Pin)	Value at Hot Idle
65	BR/LG	DPFE Sensor Signal	0.95-1.05v
66-68	---	Not Used	---
69	PK/YL	A/C WOT Relay Control	At idle with A/C on: 12v
70	WT/OR	Secondary AIR Bypass	AIRB Off: 12v, On: 1v
71	RD	Vehicle Power	12-14v
72	---	Not Used	---
73	TN/BK	Injector 5 Control	3.2-4.5 ms
74	BR/YL	Injector 3 Control	3.2-4.5 ms
75	TN	Injector 1 Control	3.2-4.5 ms
76	BK/WT	Power Ground	<0.1v
77	BK/WT	Power Ground	<0.1v
78	---	Not Used	---
79	WT/LG	TCIL (lamp) Control	7.7v (Switch On: 0v)
80	LB/OR	Fuel Pump Control	Off: 12v, On: 1v
81-82	---	Not Used	---
83	WT/LB	IAC Solenoid Control	10.7v (33%)
84-86	---	Not Used	---
87	RD/BK	HO2S-21 (B2 S1) Signal	0.1-1.1v
88	LB/RD	MAF Sensor Signal	0.8v, 55 mph: 1.8v
89	GY/WT	TP Sensor Signal	0.53-1.27v
90	BR/WT	Reference Voltage	4.9-5.1v
91	GY/RD	Analog Signal Return	<0.050v
92	LG	Brake Pedal Switch	0v (Brake On: 12v)
93	RD/WT	HO2S-11 (B1 S1) Heater	1v (Heater Off: 12v)
94	YL/LB	HO2S-21 (B2 S1) Heater	1v (Heater Off: 12v)
95	WT/BK	HO2S-12 (B1 S2) Heater	1v (Heater Off: 12v)
96, 98	---	Not Used	---
97	RD	Vehicle Power	12-14v
99	LG/OR	Injector 6 Control	3.2-4.5 ms
100	BR/LB	Injector 4 Control	3.2-4.5 ms
101	W	Injector 2 Control	3.2-4.5 ms
102	---	Not Used	---
103	BK/WT	Power Ground	<0.1v
104	---	Not Used	---

Pin Connector Graphic

PCM 104-PIN CONNECTOR

Terminal View of 104-Pin PCM Wiring Harness Connector

Standard Colors and Abbreviations

Abbreviation	Color	Abbreviation	Color	Abbreviation	Color
BK	Black	GY	Gray	PK	Purple
BL	Blue	GN	Green	RD	Red
BR	Brown	LG	LT Green	TN	Tan
DB	Dark Blue	OR	Orange	WT	White
DG	DK Green	PK	Pink	YL	Yellow

1996 Truck 4.9L I6 MFI VIN Y (E4OD) 104 Pin Connector

PCM Pin #	Wire Color	Circuit Description (104 Pin)	Value at Hot Idle
1	PK/OR	Shift Solenoid 2 Control	1v, 55 mph: 12v
2	PK/LG	MIL (lamp) Control	MIL Off: 12v, On: 1v
3-12	---	Not Used	---
13	PK	Flash EPROM Power	0.1v
14	LB/BK	4x4 Low Switch	12v (switch closed: 0v)
15	PK/LB	Data Bus (-) Signal	Digital Signals
16	TN/OR	Data Bus (+) Signal	Digital Signals
17-22	---	Not Used	---
23	OR/RD	Ignition Ground	<0.050v
24, 51	BK/WT	Power Ground	<0.1v
25	BK	Case Ground	<0.050v
26, 28	---	Not Used	---
27	OR/YL	Shift Solenoid 1 Control	1v, 55 mph: 12v
29	TN/WT	TCS (switch) Signal	TCS & O/D On: 12v
30-32, 34	---	Not Used	---
33	PK/OR	PSOM (-) Signal	<0.050v
35	RD/LG	HO2S-12 (B1 S2) Signal	0.1-1.1v
36	TN/LB	MAF Sensor Return	050v
37	OR/BK	TFT Sensor Signal	2.10-2.40v
38	LG/RD	ECT Sensor Signal	0.5-0.6v
39	GY	IAT Sensor Signal	1.5-2.5v
40	DG/YL	Fuel Pump Monitor	On: 12v, Off: 0v
41	DG/OR	A/C Switch Signal	A/C On: 12v, Off: 0v
42-43	---	Not Used	---
44	BR	Secondary AIR Diverter	AIRD Off: 12v, On: 1v
45-46	---	Not Used	
47	BK/PK	VR Solenoid Control	0%, 55 mph: 45%
48	YL/BK	Clean Tachometer Output	39-45 Hz
49	GY/OR	PIP Sensor Signal	6.93v (50%)
50	PK	Spark Output Signal	6.93v (50%)
52	---	Not Used	---
53	BR/OR	Coast Clutch Solenoid	12v, 55 mph: 12v
54	PK/YL	TCC Solenoid Control	0%, 55 mph: 95%
55	YL	Keep Alive Power	12-14v
56	LG/BK	EVAP Purge Solenoid	0-10 Hz (0-100%)
57	YL/RD	Knock Sensor Signal	0v
58	GY/BK	PSOM (+) Signal	0 Hz, 55 mph: 125 Hz
59	DG/LG	Misfire Detection Sensor	45-55 Hz
60	GY/LB	HO2S-11 (B1 S1) Signal	0.1-1.1v
61-63	---	Not Used	---
64	LB/YL	TR Sensor Signal	In 'P': 0v, in O/D: 5v

1996 Truck 4.9L I6 MFI VIN Y (E4OD) 104 Pin Connector

PCM Pin #	Wire Color	Circuit Description (104 Pin)	Value at Hot Idle
65	BR/LG	DPFE Sensor Signal	0.95-1.05v
66-68	---	Not Used	---
69	PK/YL	A/C WOT Relay Control	Off: 12v, On: 1v
70	WT/OR	Secondary AIR Bypass	AIRB Off: 12v, On: 1v
71	RD	Vehicle Power	12-14v
72	---	Not Used	---
73	TN/BK	Injector 5 Control	3.2-4.5 ms
74	BR/YL	Injector 3 Control	3.2-4.5 ms
75	TN	Injector 1 Control	3.2-4.5 ms
76	BK/WT	Power Ground	<0.1v
77	BK/WT	Power Ground	<0.1v
78, 82	---	Not Used	---
79	WT/LG	TCIL (lamp) Control	7.7v (Switch On: 0v)
80	LB/OR	Fuel Pump Control	Off: 12v, On: 1v
81	WT/YL	EPC Solenoid Control	10v (26 psi)
83	WT/LB	IAC Solenoid Control	10.7v (33%)
84-86	---	Not Used	---
87	RD/BK	HO2S-21 (B2 S1) Signal	0.1-1.1v
88	LB/RD	MAF Sensor Signal	0.8v, 55 mph: 1.8v
89	GY/WT	TP Sensor Signal	0.53-1.27v
90	BR/WT	Reference Voltage	4.9-5.1v
91	GY/RD	Analog Signal Return	<0.050v
92	LG	Brake Pedal Switch	0v (Brake On: 12v)
93	RD/WT	HO2S-11 (B1 S1) Heater	1v (Heater Off: 12v)
94	YL/LB	HO2S-21 (B2 S1) Heater	1v (Heater Off: 12v)
95	WT/BK	HO2S-12 (B1 S2) Heater	1v (Heater Off: 12v)
96, 98	---	Not Used	---
97	RD	Vehicle Power	12-14v
99	LG/OR	Injector 6 Control	3.2-4.5 ms
100	BR/LB	Injector 4 Control	3.2-4.5 ms
101	WT	Injector 2 Control	3.2-4.5 ms
102	---	Not Used	---
103	BK/WT	Power Ground	<0.1v
104	---	Not Used	---

Pin Connector Graphic

```
            PCM 104-PIN CONNECTOR
   1 ●●●●●●●●●●●●●      ●●●●●●●●●●●●● 26
  27 ●●●●●●●●●●●●●   ⬤  ●●●●●●●●●●●●● 52
  53 ●●●●●●●●●●●●●      ●●●●●●●●●●●●● 78
  79 ●●●●●●●●●●●●●      ●●●●●●●●●●●●● 104
```

Terminal View of 104-Pin PCM Wiring Harness Connector

Standard Colors and Abbreviations

Abbreviation	Color	Abbreviation	Color	Abbreviation	Color
BK	Black	GY	Gray	PK	Purple
BL	Blue	GN	Green	RD	Red
BR	Brown	LG	LT Green	TN	Tan
DB	Dark Blue	OR	Orange	WT	White
DG	DK Green	PK	Pink	YL	Yellow

1990 F-Series 5.0L V8 VIN N (All) 60 Pin Connector

PCM Pin #	Wire Color	Circuit Description (60 Pin)	Value at Hot Idle
1	Y, BK/R	Keep Alive Power	12-14v
3	GY	VSS (+) Signal	0 Hz, 55 mph: 125 Hz
3	DG	VSS (+) Signal	0 Hz, 55 mph: 125 Hz
4	DG, BK	IDM Sensor Signal	20-31 Hz
5	RD	Inferred Mileage Sensor	Digital Signals
6	BK	VSS (-) Signal	0 Hz, 55 mph: 125 Hz
7	LG/YL	ECT Sensor Signal	0.5-0.6v
8 ('89-'90)	BR	Fuel Pump Monitor	On: 12v, Off: 0v
9, 12, 15	---	Not Used	---
10	BK, LG	A/C Switch Signal	A/C On: 12v, Off: 0v
11	WT, BL	Air Management 2 Solenoid	AM2 Off: 12v, On: 1v
16	BK	Ignition System Ground	<0.050v
16	BK/OR	Ignition System Ground	<0.050v
17	PK, TN	Self Test Output & MIL	MIL Off: 12v, On: 1v
17	BL/YL	Self Test Output & MIL	MIL Off: 12v, On: 1v
18 ('88-'89)	TN/LG	Inferred Mileage Sensor	Digital Signals
20	BK	PCM Case Ground	<0.050v
20	BK/WT	PCM Case Ground	<0.050v
21	GY, GN	IAC Solenoid Control	9.2-10.3v
22	TN/LG, OR	Fuel Pump Control	Off: 12v, On: 1v
23	LG/BK	Knock Sensor Signal	0v
23	BK	Knock Sensor Signal	0v
24	YL, TN	PSP Switch Signal	0v (turning: 12v)
24	OR/BR	PSP Switch Signal	0v (turning: 12v)
25	YL, RD	ACT Sensor Signal	1.5-2.5v
26	OR/WT	Reference Voltage	4.9-5.1v
27	BR/LG	EVP Sensor Signal	0.4v, 55 mph: 2.6v
29	GY, DG	HO2S-11 (B1 S1) Signal	0.1-1.1v
30	DG, GY	A/T: Neutral Drive Switch	In 'P': 0v, Others: 5v
30	DG/WT	M/T: Clutch Engage Switch	5v (clutch "in": 0v)
30	GY/YL	M/T: Clutch Engage Switch	5v (clutch "in": 0v)
31 ('88-'90)	GY/YL	Canister Purge Solenoid	12v, 55 mph: 1v
33	DG, WT	VR Solenoid Control	0%, 55 mph: 45%
36	YL/LG	Spark Output Signal	6.93v (50%)
37	RD	Vehicle Power	12-14v
40	BK/WT	Power Ground	<0.1v
45	DG/LG	MAP Sensor Signal	107 Hz (sea level)
45	BL/LG	MAP Sensor Signal	107 Hz (sea level)
46	BK/WT	Analog Signal Return	<0.050v
47	DG, GN	TP Sensor Signal	1v, 55 mph: 1.4v
48	WT, BK	Self Test Indicator Signal	STI On: 1v, Off: 5v
49	OR	HO2S-11 (B1 S1) Ground	<0.050v
51	WT/RD, YL	Air Management 1 Solenoid	AM1 Off: 12v, On: 1v
56	DB, BR	PIP Sensor Signal	6.93v (50%)
57	RD	Vehicle Power	12-14v
58	TN/OR, WT	Injector Bank 1 (INJ 1, 4, 5, 8)	5.0-5.7 ms
59	TN/RD, WT	Injector Bank 2 (INJ 2, 3, 6, 7)	5.0-5.7 ms
60	BK/WT	Power Ground	<0.1v

Pin Connector Graphic

PCM 60-PIN CONNECTOR

Terminal View of 60-Pin PCM Harness Connector

1991-93 F-Series 5.0L V8 VIN N (All) 60 Pin Connector

PCM Pin #	Wire Color	Circuit Description (60 Pin)	Value at Hot Idle
1	YL	Keep Alive Power	12-14v
3	GY/BK	VSS (+) Signal	0 Hz, 55 mph: 125 Hz
4	TN/YL	IDM Sensor Signal	20-31 Hz
6	PK/OR	VSS (-) Signal	0 Hz, 55 mph: 125 Hz
7	LG/RD	ECT Sensor Signal	0.5-0.6v
8	DG/YL	Fuel Pump Monitor	On: 12v, Off: 0v
9	BK, PK	Data Bus (-) Signal	Digital Signals
10	BK/YL	A/C Switch Signal	A/C On: 12v, Off: 0v
10	PK/BL	A/C Switch Signal	A/C On: 12v, Off: 0v
11	BR	Air Management 2 Solenoid	AM2 Off: 12v, On: 1v
16	OR/RD	Ignition System Ground	<0.050v
17	PK/LG	Self Test Output & MIL	MIL Off: 12v, On: 1v
20	BK	PCM Case Ground	<0.050v
21	WT/BL	IAC Solenoid Control	10.7v (33%)
22	BL/OR	Fuel Pump Control	Off: 12v, On: 1v
23	YL/RD	Knock Sensor Signal	0v
24	YL/LG	PSP Switch Signal	0v (turning: 12v)
25	GY	ACT Sensor Signal	1.5-2.5v
26	BR/WT	Reference Voltage	4.9-5.1v
27	BR/LG	EVP Sensor Signal	0.4v, 55 mph: 2.6v
28	TN/OR	Data Bus (+) Signal	Digital Signals
29	GY/BL	HO2S-11 (B1 S1) Signal	0.1-1.1v
31 ('92-'93)	GY/YL	Canister Purge Solenoid	12v, 55 mph: 1v
33	BR/PK	VR Solenoid Control	0%, 55 mph: 45%
36	PK	Spark Output Signal	6.93v (50%)
37	RD	Vehicle Power	12-14v
40	BK/WT	Power Ground	<0.1v
45	LG/BK	MAP Sensor Signal	107 Hz (sea level)
46	GY/RD	Analog Signal Return	<0.050v
47	GY/WT	TP Sensor Signal	1v, 55 mph: 1.4v
48	WT/PK	Self Test Indicator Signal	STI On: 1v, Off: 5v
49	OR	HO2S-11 (B1 S1) Ground	<0.050v
51	WT/OR	Air Management 1 Solenoid	AM1 Off: 12v, On: 1v
56	GY/OR	PIP Sensor Signal	6.93v (50%)
57	RD	Vehicle Power	12-14v
58	TN	Injector Bank 1 (INJ 1, 4, 5, 8)	4.4-5.6 ms
59	WT	Injector Bank 2 (INJ 2, 3, 6, 7)	4.4-5.6 ms
60	BK/WT	Power Ground	<0.1v

Pin Connector Graphic

PCM 60-PIN CONNECTOR

Terminal View of 60-Pin PCM Harness Connector

Standard Colors and Abbreviations

Abbreviation	Color	Abbreviation	Color	Abbreviation	Color
BK	Black	GY	Gray	PK	Purple
BL	Blue	GN	Green	RD	Red
BR	Brown	LG	LT Green	TN	Tan
DB	Dark Blue	OR	Orange	WT	White
DG	DK Green	PK	Pink	YL	Yellow

1994-95 F-Series 5.0L V8 VIN N (All) 60 Pin Connector

PCM Pin #	Wire Color	Circuit Description (60 Pin)	Value at Hot Idle
1	YL	Keep Alive Power	12-14v
2	LG	Brake Position Switch	0v (Brake On: 12v)
3	GY/BK	VSS (+) Signal	0 Hz, 55 mph: 125 Hz
4	YL/BK	IDM Sensor Signal	20-31 Hz
5	DG/WT	TSS (+) Sensor Signal	At 55 mph: 126-136 Hz
6	PK/OR	VSS (-) Signal	0 Hz, 55 mph: 125 Hz
7	LG/RD	ECT Sensor Signal	0.5-0.6v
8	DG/YL	Fuel Pump Monitor	On: 12v, Off: 0v
9	TN/BL	MAF Sensor Return	<0.050v
10	DG/OR	A/C Switch Signal	A/C On: 12v, Off: 0v
11	GY/YL	Canister Purge Solenoid	12v, 55 mph: 1v
12	LG/OR	Injector 6 Control	5.0-5.5 ms
13	TN/RD	Injector 7 Control	5.0-5.5 ms
14	BL	Injector 8 Control	5.0-5.5 ms
15	TN/BK	Injector 5 Control	5.0-5.5 ms
16	OR/RD	Ignition System Ground	<0.050v
17	PK/LG	Self Test Output & MIL	MIL Off: 12v, On: 1v
18	TN/OR	Data Bus (+) Signal	Digital Signals
19	PK/BL	Data Bus (-) Signal	Digital Signals
20	BK	PCM Case Ground	Digital Signals
21	WT/BL	IAC Solenoid Control	10.7v (33%)
22	BL/OR	Fuel Pump Control	Off: 12v, On: 1v
23	YL/RD	Knock Sensor Signal	0v
24	---	Not Used	---
25	GY	IAT Sensor Signal	1.5-2.5v
26	BR/WT	Reference Voltage	4.9-5.1v
27	BR/LG	EVP Sensor Signal	0.4v, 55 mph: 2.6v
28	---	Not Used	---
29	BK /LG	TSS (-) Sensor Signal	<0.050v
30	BL/YL	A/T: MLP Sensor Signal	In 'P': 0v, in O/D: 5v
30	GY/YL	M/T: CPP Switch Signal	5v (clutch "in": 0v)
31	WT/OR	Air Bypass Solenoid Control	AIRB On 1v, Off: 12v
32	WT/LG	TCIL (lamp) Control	7.7v (Switch On: 0v)
33	BR/PK	VR Solenoid Control	0%, 55 mph: 45%
34	BR	Air Diverter Solenoid Control	AIRD On 1v, Off: 12v
35	BR/BL	Injector 4 Control	5.0-5.5 ms
36	PK	Spark Output Signal	6.93v (50%)
37	RD	Vehicle Power	12-14v
38	WT/YL	EPC Solenoid Control	9.5v (5 psi)
39	BR/YL	Injector 3 Control	5.0-5.5 ms
40	BK/WT	Power Ground	<0.1v

1994-95 F-Series 5.0L V8 VIN N (All) 60 Pin Connector

PCM Pin #	Wire Color	Circuit Description (60 Pin)	Value at Hot Idle
41	TN/WT	TCS (switch) Signal	TCS closed: 12, open: 0v
42	BL/BK	4x4 Indicator Light	12v (switch closed: 0v)
43	---	Not Used	---
44	GY/BL	HO2S-11 (B1 S1) Signal	0.1-1.1v
45	---	Not Used	---
46	GY/RD	Analog Signal Return	<0.050v
47	GY/WT	TP Sensor Signal	1v, 55 mph: 1.4v
48	WT/PK	Self Test Indicator Signal	STI On: 1v, Off: 5v
49	OR/BK	TOT Sensor Signal	2.10-2.40v
50	BL/RD	MAF Sensor Signal	0.7v, 55 mph: 1.9v
51	OR/YL	Shift Solenoid 1 Control	12v, 55 mph: 1v
52	PK/OR	Shift Solenoid 2 Control	1v, 55 mph: 12v
53	PK/YL	TCC Solenoid Control	0v, On 55 mph: 12v
54	---	Not Used	---
55	---	Not Used	---
56	GY/OR	PIP Sensor Signal	6.93v (50%)
57	RD	Vehicle Power	12-14v
58	TN	Injector 1 Control	5.0-5.5 ms
59	WT	Injector 2 Control	5.0-5.5 ms
60	BK/WT	Power Ground	<0.1v

Pin Connector Graphic

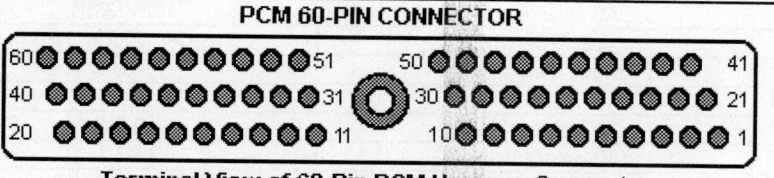

Standard Colors and Abbreviations

Abbreviation	Color	Abbreviation	Color	Abbreviation	Color
BK	Black	GY	Gray	PK	Purple
BL	Blue	GN	Green	RD	Red
BR	Brown	LG	LT Green	TN	Tan
DB	Dark Blue	OR	Orange	WT	White
DG	DK Green	PK	Pink	YL	Yellow

1994-95 F-Series 5.0L V8 VIN N (E4OD) 60 Pin Connector

PCM Pin #	Wire Color	Circuit Description (60 Pin)	Value at Hot Idle
1	YL	Keep Alive Power	12-14v
2	LG	Brake Position Switch	0v (Brake On: 12v)
3	GY/BK	VSS (+) Signal	0 Hz, 55 mph: 125 Hz
4	WT/PK	IDM Sensor Signal	20-31 Hz
4	YL/BK	IDM Sensor Signal	20-31 Hz
5	---	Not Used	---
6	PK/OR	VSS (-) Signal	0 Hz, 55 mph: 125 Hz
7	LG/RD	ECT Sensor Signal	0.5-0.6v
8	DG/YL	Fuel Pump Monitor	On: 12v, Off: 0v
9	TN/BL	MAF Sensor Return	0.050v
10	BK/YL	A/C Switch Signal	A/C On: 12v, Off: 0v
10	DG/OR	A/C Switch Signal	A/C On: 12v, Off: 0v
11	GY/YL	Canister Purge Solenoid	12v, 55 mph: 1v
12	LG/OR	Injector 6 Control	4.7-5.1 ms
13	TN/RD	Injector 7 Control	4.7-5.1 ms
14	BL	Injector 8 Control	4.7-5.1 ms
15	TN/BK	Injector 5 Control	4.7-5.1 ms
16	OR/RD	Ignition System Ground	<0.050v
17	PK/LG	Self Test Output & MIL	MIL Off: 12v, On: 1v
18	TN/OR	Data Bus (+) Signal	Digital Signals
19	PK/BL	Data Bus (-) Signal	Digital Signals
20	BK	PCM Case Ground	<0.050v
21	WT/BL	IAC Solenoid Control	10.7v (33%)
22	BL/OR	Fuel Pump Control	Off: 12v, On: 1v
23	YL/RD	Knock Sensor Signal	0v
24	---	Not Used	---
25	GY	IAT Sensor Signal	1.5-2.5v
26	BR/WT	Reference Voltage	4.9-5.1v
27	BR/LG	EVP Sensor Signal	0.4v, 55 mph: 2.6v
28	---	Not Used	---
29	---	Not Used	---
30	BL/YL	MLP Sensor Signal	In 'P': 0v, in O/D: 5v
31	WT/OR	Air Bypass Solenoid Control	AIRB Off: 12v, On: 1v
32	WT/LG	TCIL (lamp) Control	7.7v (Switch On: 0v)
33	BR/PK	VR Solenoid Control	0%, 55 mph: 45%
34	BR	Air Diverter Solenoid Control	AIRD Off: 12v, On: 1v
35	BR/BL	Injector 4 Control	4.7-5.1 ms
36	PK	Spark Output Signal	6.93v (50%)
37	RD	Vehicle Power	12-14v
38	WT/YL	EPC Solenoid Control	9.5v (5 psi)
39	BR/YL	Injector 3 Control	4.7-5.1 ms
40	BK/WT	Power Ground	<0.1v

1994-95 F-Series 5.0L V8 VIN N (E4OD) 60 Pin Connector

PCM Pin #	Wire Color	Circuit Description (60 Pin)	Value at Hot Idle
41	TN/WT	TCS (switch) Signal	TCS closed: 12, open: 0v
42	BL/BK	4x4 Indicator Light	4x4 Switch On: 0.1v
43	---	Not Used	---
44	GY/BL	HO2S-11 (B1 S1) Signal	0.1-1.1v
45	---	Not Used	---
46	GY/RD	Analog Signal Return	<0.050v
47	GY/WT	TP Sensor Signal	1v, 55 mph: 1.4v
48	WT/PK	Self Test Indicator Signal	STI On: 1v, Off: 5v
49	OR/BK	HO2S-11 (B1 S1) Ground	<0.050v
50	BL/RD	MAF Sensor Signal	0.8v, 55 mph: 1.6v
51	OR/YL	Shift Solenoid 1 Control	12v, 55 mph: 1v
52	PK/OR	Shift Solenoid 2 Control	1v, 55 mph: 12v
53	PK/YL	TCC Solenoid Control	0v, On 55 mph: 12v
54	---	Not Used	---
55	BR/OR	Coast Clutch Solenoid	12v, 55 mph: 12v
56	GY/OR	PIP Sensor Signal	6.93v (50%)
57	RD	Vehicle Power	12-14v
58	TN	Injector 1 Control	4.7-5.1 ms
59	WT	Injector 2 Control	4.7-5.1 ms
60	BK/WT	Power Ground	<0.1v

Pin Connector Graphic

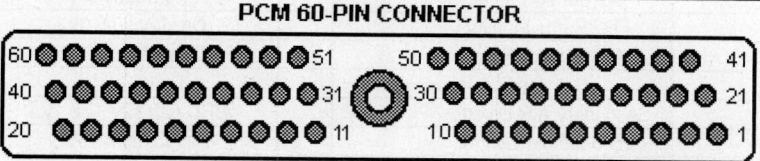

PCM 60-PIN CONNECTOR

Terminal View of 60-Pin PCM Harness Connector

Standard Colors and Abbreviations

Abbreviation	Color	Abbreviation	Color	Abbreviation	Color
BK	Black	GY	Gray	PK	Purple
BL	Blue	GN	Green	RD	Red
BR	Brown	LG	LT Green	TN	Tan
DB	Dark Blue	OR	Orange	WT	White
DG	DK Green	PK	Pink	YL	Yellow

1996 F-Series 5.0L V8 VIN N (All) 104 Pin Connector

PCM Pin #	Wire Color	Circuit Description (104 Pin)	Value at Hot Idle
1 (E40D)	PK/OR	Shift Solenoid 2 Control	1v, 55 mph: 12v
2	PK/LG	MIL (lamp) Control	MIL Off: 12v, On: 1v
3, 5	---	Not Used	---
4	LG/RD	Power Take-Off (if equipped)	0v (Off)
6 (E40D)	BK/YL	OSS (-) Sensor Signal	<0.050 v
7-12, 17-22	---	Not Used	---
13	PK	Flash EPROM Power	0.1v
14	LG/BK	4x4 Low Switch	12v (switch closed: 0v)
15	PK/LB	Data Bus (-) Signal	Digital Signals
16	TN/OR	Data Bus (+) Signal	Digital Signals
23	OR/RD	Ignition Ground	<0.050v
24, 51	BK/WT	Power Ground	<0.1v
25	BK/LB	Case Ground	<0.050v
26, 28	---	Not Used	---
27 (E40D)	OR/YL	Shift Solenoid 1 Control	1v, 55 mph: 12v
29	TN/WT	TCS (switch) Signal	TCS & O/D On: 12v
30-32, 34	---	Not Used	---
33	PK/OR	PSOM (-) Signal	<0.050v
35	RD/LG	HO2S-12 (B1 S2) Signal	0.1-1.1v
36	TN/LB	MAF Sensor Return	<0.050v
37	OR/BK	TFT Sensor Signal	2.10-2.40v
38	LG/RD	ECT Sensor Signal	0.5-0.6v
39	GY	IAT Sensor Signal	1.5-2.5v
40	DG/YL	Fuel Pump Monitor	On: 12v, Off: 0v
41	DG/OR	A/C Switch Signal	A/C On: 12v, Off: 0v
42-43, 45-46	---	Not Used	---
44	BR	Secondary AIR Diverter	AIRD Off: 12v, On: 1v
47	BK/PK	VR Solenoid Control	0%, 55 mph: 45%
48	YL/BK	Clean Tachometer Output	39-45 Hz
49	GY/OR	PIP Sensor Signal	6.93v (50%)
50	PK	Spark Output Signal	6.93v (50%)
52, 57	---	Not Used	---
53 (E40D)	BR/OR	CCS Solenoid Control	12v, 55 mph: 12v
54 (E40D)	PK/YL	TCC Solenoid Control	0%, 55 mph: 95%
55	YL	Keep Alive Power	12-14v
56	LG/BK	EVAP Purge Solenoid	0-10 Hz (0-100%)
58	GY/BK	PSOM (+) Signal	0 Hz, 55 mph: 125 Hz
59	DG/LG	Misfire Detection Sensor	45-55 Hz
60	GY/LB	HO2S-11 (B1 S1) Signal	0.1-1.1v
61-63	---	Not Used	---
64 (E40D)	LB/YL	TR Sensor Signal	In 'P': 0v, in O/D: 5v
64 (4R70W)	LB/YL	PNP Switch Signal	In 'P': 0v, Others: 5v
64	LB/YL	M/T: CPP Switch Signal	5v (clutch "in": 0v)

1996 F-Series 5.0L V8 VIN N (All) 104 Pin Connector

PCM Pin #	Wire Color	Circuit Description (104 Pin)	Value at Hot Idle
65	BR/LG	DPFE Sensor Signal	0.95-1.05v
66-69	---	Not Used	---
70	WT/OR	Secondary AIR Bypass	AIRB Off: 12v, On: 1v
71	RD	Vehicle Power	12-14v
72	TN/RD	Injector 7 Control	3.2-4.5 ms
73	TN/BK	Injector 5 Control	3.2-4.5 ms
74	BR/YL	Injector 3 Control	3.2-4.5 ms
75	TN	Injector 1 Control	3.2-4.5 ms
76	BK/WT	Power Ground	<0.1v
77	BK/WT	Power Ground	<0.1v
78, 82	---	Not Used	---
79	WT/LG	TCIL (lamp) Control	7.7v (Switch On: 0v)
80	LB/OR	Fuel Pump Control	Off: 12v, On: 1v
81 (E40D)	WT/YL	EPC Solenoid	10v (26 psi)
83	WT/LB	IAC Solenoid Control	10.7v (33%)
84 (E40D)	DB/YL	OSS (+) Sensor Signal	0 Hz, 55 mph: 128 Hz
85-86	---	Not Used	---
87	RD/BK	HO2S-21 (B2 S1) Signal	0.1-1.1v
88	LB/RD	MAF Sensor Signal	0.8v, 55 mph: 1.8v
89	GY/WT	TP Sensor Signal	0.53-1.27v
90	BR/WT	Reference Voltage	4.9-5.1v
91	GY/RD	Analog Signal Return	<0.050 v
92	LG	Brake Pedal Switch	0v (Brake On: 12v)
93	RD/WT	HO2S-11 (B1 S1) Heater	1v (Heater Off: 12v)
94	YL/LB	HO2S-21 (B2 S1) Heater	1v (Heater Off: 12v)
95	WT/BK	HO2S-12 (B1 S2) Heater	1v (Heater Off: 12v)
96	---	Not Used	---
97	RD	Vehicle Power	12-14v
98	LG	Injector 8 Control	3.2-4.5 ms
99	LG/OR	Injector 6 Control	3.2-4.5 ms
100	BR/LB	Injector 4 Control	3.2-4.5 ms
101	WT	Injector 2 Control	3.2-4.5 ms
102	---	Not Used	---
103	BK/WT	Power Ground	<0.1v
104	---	Not Used	---

Pin Connector Graphic

PCM 104-PIN CONNECTOR

Terminal View of 104-Pin PCM Wiring Harness Connector

Standard Colors and Abbreviations

Abbreviation	Color	Abbreviation	Color	Abbreviation	Color
BK	Black	GY	Gray	PK	Purple
BL	Blue	GN	Green	RD	Red
BR	Brown	LG	LT Green	TN	Tan
DB	Dark Blue	OR	Orange	WT	White
DG	DK Green	PK	Pink	YL	Yellow

1997 F-Series 5.4L V8 VIN L (E4OD) 104 Pin Connector

PCM Pin #	Wire Color	Circuit Description (104 Pin)	Value at Hot Idle
1	PK/LB	COP 6 Driver (dwell)	5°, 55 mph: 8°
2	PK/LG	MIL (lamp) Control	MIL Off: 12v, On: 1v
3	BK/WT	Power Ground	<0.1v
4-5, 7-8, 10	---	Not Used	---
6	OR/YL	Shift Solenoid 1 Control	1v, 55 mph: 12v
9	YL/WT	Fuel Level Indictor Signal	1.7v (1/2 full)
11	PK/OR	Shift Solenoid 2 Control	12v, 55 mph: 1v
12	WT/LG	TCIL (lamp) Control	7.7v (Switch On: 0v)
13	PK	Flash EEPROM Power	0.1v
14, 17-19	---	Not Used	---
15	PK/LB	Data Bus (-) Signal	Digital Signals
16	TN/OR	Data Bus (+) Signal	Digital Signals
20	BR/OR	Coast Clutch Solenoid	12v, 55 mph: 12v
21	DB	CKP (+) Sensor Signal	400-420 Hz
22	GY	CKP (-) Sensor Signal	400-420 Hz
23, 28	---	Not Used	---
24, 51	BK/WT	Power Ground	<0.1v
25	BK/LB	Case Ground	<0.050v
26	DB/YL	COP 1 Driver (dwell)	5°, 55 mph: 8°
27	PK/WT	COP 5 Driver (dwell)	5°, 55 mph: 8°
31-32, 38	---	Not Used	---
29	TN/WT	TCS (switch) Signal	TCS & O/D On: 12v
30	DG	Octane Adjustment	9.3v (shorted: 0v)
33	PK/OR	VSS (-) Signal	At 55 mph; 125 Hz
34	OR/BK	Digital TR1 Sensor	0v, 55 mph: 11.5v
35	RD/LG	HO2S-12 (B1 S2) Signal	0.1-1.1v
36	TN/LB	MAF Sensor Return	<0.050v
37	OR/BK	TFT Sensor Signal	2.10-2.40v
39	GY	IAT Sensor Signal	1.5-2.5v
40	DG/YL	Fuel Pump Monitor	On: 12v, Off: 0v
41	BK/YL	A/C Switch Signal	Switch On: 12v, Off: 0v
42-44, 48	---	Not Used	---
45	YL/LB	CHIL (lamp) Control	CHT Off: 12v, On: 1v
46	BR	Intake Manifold Tuning Valve	12v, 55 mph: 12v
47	BR/PK	VR Solenoid Control	0%, 55 mph: 45%
49	LB/BK	Digital TR2 Sensor	0v, 55 mph: 11.5v
50	GY/BK	Digital TR4 Sensor	0v, 55 mph: 11.5v
52	LG/WT	COP 3 Driver (dwell)	5°, 55 mph: 8°
53	OR/YL	COP 4 Driver (dwell)	5°, 55 mph: 8°
54	PK/YL	TCC Solenoid Control	0%, 55 mph: 95%
55	Y	Keep Alive Power	12-14v
56	LG/BK	EVAP Purge Solenoid	0-10 Hz (0-100%)
57	YL/RD	Knock Sensor Signal	0v
58	GY/BK	VSS (+) Signal	0 Hz, 55 mph: 125 Hz
59	---	Not Used	---
60	GY/LB	HO2S-11 (B1 S1) Signal	0.1-1.1v
61	---	Not Used	---
62	RD/PK	FTP Sensor Signal	2.6v (0" H2O - cap off)
63	---	Not Used	---
64	LB/YL	Digital TR3 Sensor	In 'P': 0v, in O/D: 1.7v

1997 F-Series 5.4L V8 VIN L (E4OD) 104 Pin Connector

PCM Pin #	Wire Color	Circuit Description (104 Pin)	Value at Hot Idle
65	BR/LG	DPFE Sensor Signal	0.95-1.05v
66	YL/LG	CHT Sensor Signal	0.7v (194°F)
67	PK/WT	EVAP CV Solenoid	0-10 Hz (0-100%)
68-70, 76	---	Not Used	---
71, 97	RD	Vehicle Power	12-14v
72	TN/RD	Injector 7 Control	2.7-3.5 ms
73	TN/BK	Injector 5 Control	2.7-3.5 ms
74	BR/YL	Injector 3 Control	2.7-3.5 ms
75	TN	Injector 1 Control	2.7-3.5 ms
82, 84, 86	---	Not Used	---
77	BK/WT	Power Ground	<0.1v
78	WT/PK	COP 7 Driver (dwell)	5°, 55 mph: 8°
79	WT/RD	COP 8 Driver (dwell)	5°, 55 mph: 8°
80	LB/OR	Fuel Pump Control	Off: 12v, On: 1v
81	WT/YL	EPC Solenoid Control	9.1v (4 psi)
83	WT/LB	IAC Solenoid Control	10.7v (33%)
85	DB/OR	CMP Sensor Signal	7 Hz, 55 mph: 15 Hz
87	RD/BK	HO2S-21 (B2 S1) Signal	0.1-1.1v
88	LB/RD	MAF Sensor Signal	0.8v, 55 mph: 1.6v
89	GY/WT	TP Sensor Signal	0.53-1.27v
90	BR/WT	Reference Voltage	4.9-5.1v
91	GY/RD	Analog Signal Return	<0.050v
92	LG	Brake Pedal Switch	0v (Brake On: 12v)
93	RD/WT	HO2S-11 (B1 S1) Heater	1v (Heater Off: 12v)
94	YL/LB	HO2S-21 (B2 S1) Heater	1v (Heater Off: 12v)
95	WT/BK	HO2S-12 (B1 S2) Heater	1v (Heater Off: 12v)
96	---	Not Used	---
98	LB	Injector 8 Control	2.7-3.5 ms
99	LG/OR	Injector 6 Control	2.7-3.5 ms
100	BR/LG	Injector 4 Control	2.7-3.5 ms
101	WT	Injector 2 Control	2.7-3.5 ms
102	---	Not Used	---
103	BK/WT	Power Ground	<0.1v
104	DB/LG	COP 2 Driver (dwell)	5°, 55 mph: 8°

Pin Connector Graphic

PCM 104-PIN CONNECTOR

Terminal View of 104-Pin PCM Wiring Harness Connector

Standard Colors and Abbreviations

Abbreviation	Color	Abbreviation	Color	Abbreviation	Color
BK	Black	GY	Gray	PK	Purple
BL	Blue	GN	Green	RD	Red
BR	Brown	LG	LT Green	TN	Tan
DB	Dark Blue	OR	Orange	WT	White
DG	DK Green	PK	Pink	YL	Yellow

1998-99 F-Series 5.4L V8 VIN L (A/T) 104 Pin Connector

PCM Pin #	Wire Color	Circuit Description (104 Pin)	Value at Hot Idle
1	LG/YL	COP 6 Driver (dwell)	5°, 55 mph: 8°
2 ('98)	PK/LG	MIL (lamp) Control	MIL Off: 12v, On: 1v
3, 24, 51	BK/WT	Power Ground	<0.1v
6	OR/YL	Shift Solenoid 1 Control	1v, 55 mph: 12v
7 ('99)	PK	Transfer Case Speed Sensor	0 Hz, 55 mph: 471 Hz
9 ('98)	YL/WT	Fuel Level Indicator Signal	1.7v (1/2 full)
11	PK/OR	Shift Solenoid 2 Control	12v, 55 mph: 1v
12	WT/LG	TCIL (lamp) Control	7.7v (Switch On: 0v)
13	PK	Flash EEPROM Power	0.1v
14	LB/BK	4x4 Low Indicator Switch	4x4 Switch On: 0.1v
15	PK	Data Bus (-) Signal	Digital Signals
16	TN/OR	Data Bus (+) Signal	Digital Signals
20	BR/OR	Coast Clutch Solenoid	12v, 55 mph: 12v
21	DB	CKP (+) Sensor Signal	411 Hz
22	GY	CKP (-) Sensor Signal	411 Hz
25	BK/LB	Case Ground	<0.050v
26	LG/WT	COP 1 Driver (dwell)	5°, 55 mph: 8°
27	LG/YL	COP 5 Driver (dwell)	5°, 55 mph: 8°
29	TN/WT	TCS (switch) Signal	TCS & O/D On: 12v
30 ('98)	DG	Octane Adjustment	9.3v (shorted: 0v)
32	DG/PK	Knock Sensor 1 Signal	0v
33 ('98)	PK/OR	VSS (-) Signal	0 Hz, 55 mph: 125 Hz
34	YL/BK	Digital TR1 Sensor	0v, 55 mph: 11.5v
35	RD/LG	HO2S-12 (B1 S2) Signal	0.1-1.1v
36	TN/LB	MAF Sensor Return	<0.050v
37	OR/BK	TFT Sensor Signal	2.10-2.40v
39	GY	IAT Sensor Signal	1.5-2.5v
40	DG/YL	Fuel Pump Monitor	On: 12v, Off: 0v
41	BK/YL	A/C Switch Signal	Switch Closed: 12v
45 ('98)	RD/WT	CHIL (lamp) Control	7.7v (Switch On: 0v)
46	BR	Intake Manifold Tuning Valve	12v, 55 mph: 12v
47	BR/PK	VR Solenoid Control	0%, 55 mph: 45%
48 ('98)	WT/PK	Clean Tachometer Output	55 Hz, 55 mph: 120 Hz
49	LB/BK	Digital TR2 Sensor	0v, 55 mph: 11.5v
50	WT/BK	Digital TR4 Sensor	0v, 55 mph: 11.5v
52	LG/WT	COP 3 Driver (dwell)	5°, 55 mph: 8°
53	PK/WT	COP 4 Driver (dwell)	5°, 55 mph: 8°
54	PK/YL	TCC Solenoid Control	0%, 55 mph: 95%
55	RD/WT	Keep Alive Power	12-14v
56	LG/BK	EVAP Purge Solenoid	0-10 Hz (0-100%)
57	YL/RD	Knock Sensor 1 Signal	0v
58 ('99)	GY/BK	VSS (+) Signal	0 Hz, 55 mph: 125 Hz
60	GY/LB	HO2S-11 (B1 S1) Signal	0.1-1.1v
61	PK/LG	HO2S-22 (B2 S2) Signal	0.1-1.1v
62	RD/PK	FTP Sensor Signal	2.6v (0" H2O - cap off)
64	LB/YL	Digital TR3 Sensor	In 'P': 0v, in O/D: 1.7v

1998-99 F-Series 5.4L V8 VIN L (A/T) 104 Pin Connector

PCM Pin #	Wire Color	Circuit Description (104 Pin)	Value at Hot Idle
65	BR/LG	DPFE Sensor Signal	0.95-1.05v
66	YL/LG	CHT Sensor Signal	0.7v (194°F)
67	PK/WT	EVAP CV Solenoid	0-10 Hz (0-100%)
68 ('99)	GY/BK	Transfer Case Speed Sensor	0 Hz, 55 mph: 471 Hz
69 ('99)	PK/YL	A/C WOT Relay Control	Off: 12v, On: 1v
71, 97	RD	Vehicle Power	12-14v
72	TN/RD	Injector 7 Control	2.7-3.5 ms
73	TN/BK	Injector 5 Control	2.7-3.5 ms
74	BR/YL	Injector 3 Control	2.7-3.5 ms
75	TN	Injector 1 Control	2.7-3.5 ms
77, 103	BK/WT	Power Ground	<0.1v
78	WT/PK	COP 7 Driver (dwell)	5°, 55 mph: 8°
79	WT/RD	COP 8 Driver (dwell)	5°, 55 mph: 8°
80	LB/OR	Fuel Pump Control	Off: 12v, On: 1v
81	WT/YL	EPC Solenoid Control	9.1v (4 psi)
83	WT/LB	IAC Solenoid Control	10.7v (33%)
84	DB/YL	OSS Sensor Signal	0 Hz, 55 mph: 700 Hz
85	DG	CMP Sensor Signal	7 Hz, 55 mph: 15 Hz
87	RD/BK	HO2S-21 (B2 S1) Signal	0.1-1.1v
88	LB/RD	MAF Sensor Signal	0.8v, 55 mph: 1.6v
89	GY/WT	TP Sensor Signal	0.53-1.27v
90	BR/WT	Reference Voltage	4.9-5.1v
91	GY/RD	Analog Signal Return	<0.050v
92	RD/LG	Brake Position Switch	0v (Brake On: 12v)
93	RD/WT	HO2S-11 (B1 S1) Heater	1v (Heater Off: 12v)
94	YL/LB	HO2S-21 (B2 S1) Heater	1v (Heater Off: 12v)
95	WT/BK	HO2S-12 (B1 S2) Heater	1v (Heater Off: 12v)
96	TN/YL	HO2S-22 (B2 S2) Heater	1v (Heater Off: 12v)
98	LB	Injector 8 Control	2.7-3.5 ms
99	LG/OR	Injector 6 Control	2.7-3.5 ms
100	BR/LB	Injector 4 Control	2.7-3.5 ms
101	WT	Injector 2 Control	2.7-3.5 ms
104	PK/WT	COP 2 Driver (dwell)	5°, 55 mph: 8°

Pin Connector Graphic

PCM 104-PIN CONNECTOR

Terminal View of 104-Pin PCM Wiring Harness Connector

Standard Colors and Abbreviations

Abbreviation	Color	Abbreviation	Color	Abbreviation	Color
BK	Black	GY	Gray	PK	Purple
BL	Blue	GN	Green	RD	Red
BR	Brown	LG	LT Green	TN	Tan
DB	Dark Blue	OR	Orange	WT	White
DG	DK Green	PK	Pink	YL	Yellow

2000-02 F-Series 5.4L V8 VIN L (All) 104 Pin Connector

PCM Pin #	Wire Color	Circuit Description (104 Pin)	Value at Hot Idle
1	OR/YL	COP 6 Driver (dwell)	5º, 55 mph: 8º
3	BK/WT	Power Ground	<0.1v
6	OR/YL	Shift Solenoid 'A' Control	1v, 55 mph: 12v
7	PK	Transfer Case Speed Sensor	0 Hz, 55 mph: 471 Hz
11	PK/OR	Shift Solenoid 'B' Control	12v, 55 mph: 1v
12	WT/LG	TCIL (lamp) Control	7.7v (Switch On: 0v)
13	PK	Flash EEPROM Power	0.1v
14	LB/BK	4x4 Low Indicator Switch	12v (Switch On: 0v)
15	PK/LB	Data Bus (-) Signal	Digital Signals
16	TN/OR	Data Bus (+) Signal	Digital Signals
20	BR/OR	Coast Clutch Solenoid	12v, 55 mph: 12v
21	DB	CKP (-) Sensor Signal	411 Hz
22	GY	CKP (+) Sensor Signal	411 Hz
25	LB/YL	Chassis Ground	<0.050v
26	LG/WT	COP 1 Driver (dwell)	5º, 55 mph: 8º
27	LG/YL	COP 5 Driver (dwell)	5º, 55 mph: 8º
29	TN/WT	TCS (switch) Signal	TCS & O/D On: 12v
32	DG/PK	Knock Sensor (-) Signal	0v
34	YL/BK	Digital TR1 Sensor	0v, 55 mph: 11.5v
35	RD/LG	HO2S-12 (B1 S2) Signal	0.1-1.1v
36	TN/LB	MAF Sensor Return	<0.050v
37	OR/BK	TFT Sensor Signal	2.10-2.40v
39	GY	IAT Sensor Signal	1.5-2.5v
40	DG/YL	Fuel Pump Monitor	On: 12v, Off: 0v
41	BK/YL	A/C Switch Signal	Switch Closed: 12v
46	BR	Intake Manifold Tuning Valve	12v, 55 mph: 12v
47	BR/PK	VR Solenoid Control	0%, 55 mph: 45%
49	LB/BK	Digital TR2 Sensor	0v, 55 mph: 11.5v
50	WT/BK	Digital TR4 Sensor	0v, 55 mph: 11.5v
51	BK/WT	Power Ground	<0.1v
52	WT/PK	COP 3 Driver (dwell)	5º, 55 mph: 8º
53	DG/PK	COP 4 Driver (dwell)	5º, 55 mph: 8º
54	PK/YL	TCC Solenoid Control	0%, 55 mph: 95%
55	RD/WT	Keep Alive Power	12-14v
56	LG/BK	EVAP Purge Solenoid	0-10 Hz (0-100%)
57	YL/RD	Knock Sensor 1 Signal	0v
59	DG/WT	TSS Sensor Signal	300 Hz, 55 mph: 980
60	GY/LB	HO2S-11 (B1 S1) Signal	0.1-1.1v
61	PK/LG	HO2S-22 (B2 S2) Signal	0.1-1.1v
62	RD/PK	FTP Sensor Signal	2.6v (0" H2O - cap off)
64	LB/YL	Digital TR3 Sensor	In 'P': 0v, in O/D: 1.7v

2000-02 F-Series 5.4L V8 VIN L (All) 104 Pin Connector

PCM Pin #	Wire Color	Circuit Description (104 Pin)	Value at Hot Idle
65 ('00)	BR/LG	DPFE Sensor Signal	0.20-1.30v
65 ('01-'02)	BR/LG	DPFE Sensor Signal	0.95-1.05v
66	YL/LG	CHT Sensor Signal	0.7v (194°F)
67	PK/WT	EVAP CV Solenoid	0-10 Hz (0-100%)
68	GY/BK	VSS (+) Signal	0 Hz, 55 mph: 125 Hz
69	PK/YL	A/C WOT Relay Control	Off: 12v, On: 1v
71	RD	Vehicle Power	12-14v
72	TN/RD	Injector 7 Control	2.7-3.5 ms
73	TN/BK	Injector 5 Control	2.7-3.5 ms
74	BR/YL	Injector 3 Control	2.7-3.5 ms
75	TN	Injector 1 Control	2.7-3.5 ms
76	---	Not Used	---
77	BK/WT	Power Ground	<0.1v
78	PK/LB	COP 7 Driver (dwell)	5°, 55 mph: 8°
79	WT/RD	COP 8 Driver (dwell)	5°, 55 mph: 8°
80	LB/OR	Fuel Pump Relay Control	Off: 12v, On: 1v
81	WT/YL	EPC Solenoid Control	9.1v (4 psi)
82,86	---	Not Used	---
83	WT/LB	IAC Valve Control	10.7v (33%)
84	DB/YL	OSS Sensor Signal	0 Hz, 55 mph: 228 Hz
85	DG	CMP Sensor Signal	7 Hz, 55 mph: 15 Hz
87	RD/BK	HO2S-21 (B2 S1) Signal	0.1-1.1v
88	LB/RD	MAF Sensor Signal	0.8v, 55 mph: 1.6v
89	GY/WT	TP Sensor Signal	0.53-1.27v
90	BR/WT	Reference Voltage	4.9-5.1v
91	GY/RD	Sensor Return	<0.050v
92	RD/LG	Brake Pedal Switch	0v (Brake On: 12v)
93	RD/WT	HO2S-11 (B1 S1) Heater	1v (Heater Off: 12v)
94	YL/LB	HO2S-21 (B2 S1) Heater	1v (Heater Off: 12v)
95	WT/BK	HO2S-12 (B1 S2) Heater	1v (Heater Off: 12v)
96	TN/YL	HO2S-22 (B2 S2) Heater	1v (Heater Off: 12v)
97	RD	Vehicle Power	12-14v
98	YL	Injector 8 Control	2.7-3.5 ms
99	YL/LB	Injector 6 Control	2.7-3.5 ms
100	YL/BK	Injector 4 Control	2.7-3.5 ms
101	YL/RD	Injector 2 Control	2.7-3.5 ms
102	---	Not Used	---
103	BK/WT	Power Ground	<0.1v
104	PK/WT	COP 2 Driver (dwell)	5°, 55 mph: 8°

Pin Connector Graphic

PCM 104-PIN CONNECTOR

```
 1 ●●●●●●●●●●●●●        ●●●●●●●●●●●●● 26
27 ●●●●●●●●●●●●●        ●●●●●●●●●●●●● 52
53 ●●●●●●●●●●●●●   ●    ●●●●●●●●●●●●● 78
79 ●●●●●●●●●●●●●        ●●●●●●●●●●●●● 104
```

Terminal View of 104-Pin PCM Wiring Harness Connector

Standard Colors and Abbreviations

Abbreviation	Color	Abbreviation	Color	Abbreviation	Color
BK	Black	GY	Gray	PK	Purple
BL	Blue	GN	Green	RD	Red
BR	Brown	LG	LT Green	TN	Tan
DB	Dark Blue	OR	Orange	WT	White
DG	DK Green	PK	Pink	YL	Yellow

2003 F-Series 5.4L V8 VIN L (All) 104 Pin Connector

PCM Pin #	Wire Color	Circuit Description (104 Pin)	Value at Hot Idle
1	OR/YL	COP 6 Driver (dwell)	5°, 55 mph: 8°
2	---	Not Used	---
3	BK/WT	Power Ground	<0.1v
4	PK	Transfer Case Speed Sensor (MSOF)	12v, 55 mph: 12v
5	---	Not Used	---
6	OR/YL	Shift Solenoid 'A' Control	1v, 55 mph: 12v
7-10	---	Not Used	---
11	VT/OR	Shift Solenoid 'B' Control	12v, 55 mph: 1v
12	WT/LG	TCIL (lamp) Control	7.7v (Switch On: 0v)
13	VT	Flash EEPROM Power	0.1v
14	LB/BK	4x4 Low Indicator Switch	12v (Switch On: 0v)
15	PK/LB	SCP Data Bus (-) Signal	<0.050v
16	TN/OR	SCP Data Bus (+) Signal	Digital Signals
17-19	---	Not Used	---
20	BR/OR	Coast Clutch Solenoid	12v, 55 mph: 12v
21	DB	CKP (-) Sensor Signal	411 Hz
22	GY	CKP (+) Sensor Signal	411 Hz
23-24	---	Not Used	---
25	LB/YL	Chassis Ground	<0.050v
26	LG/WT	COP 1 Driver (dwell)	5°, 55 mph: 8°
27	LG/YL	COP 5 Driver (dwell)	5°, 55 mph: 8°
28	---	Not Used	---
29	TN/WT	TCS (switch) Signal	TCS & O/D On: 12v
30-31	---	Not Used	---
32	DG/VT	Knock Sensor (-) Signal	<0.050v
33	---	Not Used	---
34	YL/BK	Digital TR1 Sensor	0v, 55 mph: 11.5v
35	RD/LG	HO2S-12 (B1 S2) Signal	0.1-1.1v
36	TN/LB	MAF Sensor Return	<0.050v
37	OR/BK	TFT Sensor Signal	2.10-2.40v
38	---	Not Used	---
39	GY	IAT Sensor Signal	1.5-2.5v
40	DG/YL	Fuel Pump Monitor	On: 12v, Off: 0v
41	BK/YL	A/C Pressure Switch Signal	Switch Closed: 12v
42-45	---	Not Used	---
46	BR	Intake Manifold Tuning Valve Control	12v, 55 mph: 12v
47	BR/PK	VR Solenoid Control	0%, 55 mph: 45%
48	---	Not Used	---
49	LB/BK	Digital TR2 Sensor	0v, 55 mph: 11.5v
50	WT/BK	Digital TR4 Sensor	0v, 55 mph: 11.5v
51	BK/WT	Power Ground	<0.1v
52	WT/PK	COP 3 Driver (dwell)	5°, 55 mph: 8°
53	DG/VT	COP 4 Driver (dwell)	5°, 55 mph: 8°
54	VT/YL	TCC Solenoid Control	0%, 55 mph: 95%
55	RD/WT	Keep Alive Power	12-14v
56	LG/BK	EVAP Purge Solenoid	0-10 Hz (0-100%)
57	YL/RD	Knock Sensor 1 Signal	0v
58	---	Not Used	---
59	DG/WT	TSS Sensor Signal	300 Hz, 55 mph: 980
60	GY/LB	HO2S-11 (B1 S1) Signal	0.1-1.1v
61	VT/LG	HO2S-22 (B2 S2) Signal	0.1-1.1v
62	RD/PK	FTP Sensor Signal	2.6v (0" H2O - cap off)
63	---	Not Used	---
64	LB/YL	Digital TR3A Sensor Signal	In 'P': 0v, in O/D: 1.7v

2003 F-Series 5.4L V8 VIN L (A/T) 104 Pin Connector

PCM Pin #	Wire Color	Circuit Description (104 Pin)	Value at Hot Idle
65	BR/LG	EGR DPFE Sensor Signal	0.95-1.05v
66	YL/LG	Cylinder Head Temperature Sensor	0.7v (194°F)
67	VT/WT	EVAP Canister Vent Solenoid	0-10 Hz (0-100%)
68	GY/BK	VSS (+) Signal	0 Hz, 55 mph: 125 Hz
69	PK/YL	A/C WOT Relay Control	Off: 12v, On: 1v
70	BK/WT	Check Fuel Cap Indicator Control	Indicator Off: 12v, On: 1v
71	RD	Vehicle Power	12-14v
72	TN/RD	Injector 7 Control	2.7-3.5 ms
73	TN/BK	Injector 5 Control	2.7-3.5 ms
74	BR/YL	Injector 3 Control	2.7-3.5 ms
75	TN	Injector 1 Control	2.7-3.5 ms
76	---	Not Used	---
77	BK/WT	Power Ground	<0.1v
78	PK/LB	COP 7 Driver (dwell)	5°, 55 mph: 8°
79	WT/RD	COP 8 Driver (dwell)	5°, 55 mph: 8°
80	LB/OR	Fuel Pump Relay Control	Off: 12v, On: 1v
81	WT/YL	EPC Solenoid Control	9.1v (4 psi)
82	---	Not Used	---
83	WT/LB	IAC Valve Control	10.7v (33%)
84	DB/YL	OSS Sensor Signal	0 Hz, 55 mph: 228 Hz
85	DG	CMP Sensor Signal	7 Hz, 55 mph: 15 Hz
86	---	Not Used	---
87	RD/BK	HO2S-21 (B2 S1) Signal	0.1-1.1v
88	LB/RD	MAF Sensor Signal	0.8v, 55 mph: 1.6v
89	GY/WT	TP Sensor Signal	0.53-1.27v
90	BR/WT	Reference Voltage	4.9-5.1v
91	GY/RD	Sensor Return	<0.050v
92	RD/LG	Brake Pedal Switch	0v (Brake On: 12v)
93	RD/WT	HO2S-11 (B1 S1) Heater	1v (Heater Off: 12v)
94	YL/LB	HO2S-21 (B2 S1) Heater	1v (Heater Off: 12v)
95	WT/BK	HO2S-12 (B1 S2) Heater	1v (Heater Off: 12v)
96	TN/YL	HO2S-22 (B2 S2) Heater	1v (Heater Off: 12v)
97	RD	Vehicle Power	12-14v
98	YL	Injector 8 Control	2.7-3.5 ms
99	YL/LB	Injector 6 Control	2.7-3.5 ms
100	BR/LB	Injector 4 Control	2.7-3.5 ms
101	YL/RD	Injector 2 Control	2.7-3.5 ms
102	---	Not Used	---
103	BK/WT	Power Ground	<0.1v
104	PK/WT	COP 2 Driver (dwell)	5°, 55 mph: 8°

Pin Connector Graphic

PCM 104-PIN CONNECTOR

Terminal View of 104-Pin PCM Wiring Harness Connector

Standard Colors and Abbreviations

Abbreviation	Color	Abbreviation	Color	Abbreviation	Color
BK	Black	GY	Gray	PK	Purple
BL	Blue	GN	Green	RD	Red
BR	Brown	LG	LT Green	TN	Tan
DB	Dark Blue	OR	Orange	WT	White
DG	DK Green	PK	Pink	YL	Yellow

1997 F-Series 5.4L CNG VIN M (E4OD) 104 Pin Connector

PCM Pin #	Wire Color	Circuit Description (104 Pin)	Value at Hot Idle
1	PK/LB	COP 6 Driver (dwell)	5º, 55 mph: 8º
2	PK/LG	MIL (lamp) Control	MIL Off: 12v, On: 1v
3	BK/WT	Power Ground	<0.1v
4-5, 7-8, 10	---	Not Used	
6	OR/YL	Shift Solenoid 1 Control	1v, 55 mph: 12v
9	YL/WT	Fuel Level Indicator Signal	1.7v (1/2 full)
11	PK/OR	Shift Solenoid 2 Control	12v, 55 mph: 1v
12	WT/LG	TCIL (lamp) Control	7.7v (Switch On: 0v)
13	PK	Flash EEPROM Power	0.1v
14, 17-19	---	Not Used	---
15	PK/LB	Data Bus (-) Signal	Digital Signals
16	TN/OR	Data Bus (+) Signal	Digital Signals
20	BR/OR	Coast Clutch Solenoid	12v, 55 mph: 12v
21	DB	CKP (+) Sensor Signal	460-500 Hz
22	GY	CKP (-) Sensor Signal	460-500 Hz
23	---	Not Used	---
24, 51	BK/WT	Power Ground	<0.1v
25	BK/LB	Case Ground	<0.050v
26	DB/YL	COP 1 Driver (dwell)	5º, 55 mph: 8º
27	PK/WT	COP 5 Driver (dwell)	5º, 55 mph: 8º
28, 31-32	---	Not Used	---
29	TN/WT	TCS (switch) Signal	TCS & O/D On: 12v
30	DG	Octane Adjustment	9.3v (shorted: 0v)
33	PK/OR	VSS (-) Signal	At 55 mph; 125 Hz
34	OR/BK	Digital TR1 Sensor	0v, 55 mph: 11.5v
36	TN/LB	MAF Sensor Return	<0.050v
37	OR/BK	TFT Sensor Signal	2.10-2.40v
38	---	Not Used	---
39	GY	IAT Sensor Signal	1.5-2.5v
40	DG/YL	Fuel Pump Monitor	On: 12v, Off: 0v
41	BK/YL	A/C Hi Pressure Cutoff Switch	AC & Switch On: 12v
42-44, 48	---	Not Used	---
45	YL/LB	CHIL (lamp) Control	CHT Off: 12v, On: 1v
46	BR	Intake Manifold Tuning Valve	12v, 55 mph: 12v
47	BR/PK	VR Solenoid Control	0%, 55 mph: 45%
49	LB/BK	Digital TR2 Sensor	0v, 55 mph: 11.5v
50	GY/BK	Digital TR4 Sensor	0v, 55 mph: 11.5v
52	LG/WT	COP 3 Driver (dwell)	5º, 55 mph: 8º
53	OR/YL	COP 4 Driver (dwell)	5º, 55 mph: 8º
54	PK/YL	TCC Solenoid Control	0%, 55 mph: 95%
55	YL	Keep Alive Power	12-14v
56	LG/BK	EVAP Purge Solenoid	0-10 Hz (0-100%)
57	YL/RD	Knock Sensor Signal	0v
58	GY/BK	VSS (+) Signal	0 Hz, 55 mph: 125 Hz
59	---	Not Used	---
60	GY/LB	HO2S-11 (B1 S1) Signal	0.1-1.1v
61	---	Not Used	---
62	RD/PK	FTP Sensor Signal	2.6v (0" H2O - cap off)
63	---	FRP Sensor Signal	2-3.7v (90-100 psi)
64	LB/YL	Digital TR3 Sensor	In 'P': 0v, in O/D: 1.7v

1997 F-Series 5.4L CNG VIN M (E4OD) 104 Pin Connector

PCM Pin #	Wire Color	Circuit Description (104 Pin)	Value at Hot Idle
65	BR/LG	DPFE Sensor Signal	0.95-1.05v
66	YL/LG	CHT Sensor Signal	0.6v (194°F)
67	PK/WT	EVAP CV Solenoid	0-10 Hz (0-100%)
68, 70	---	Not Used	---
71, 97	RD	Vehicle Power	12-14v
72	TN/RD	Injector 7 Control	2.7-3.5 ms
73	TN/BK	Injector 5 Control	2.7-3.5 ms
74	BR/YL	Injector 3 Control	2.7-3.5 ms
75	TN	Injector 1 Control	2.7-3.5 ms
76, 82, 84	---	Not Used	---
77	BK/WT	Power Ground	<0.1v
78	WT/PK	COP 7 Driver (dwell)	5°, 55 mph: 8°
79	WT/RD	COP 8 Driver (dwell)	5°, 55 mph: 8°
80	LB/OR	Fuel Pump Control	Off: 12v, On: 1v
81	WT/YL	EPC Solenoid Control	9.1v (4 psi)
83	WT/LB	IAC Solenoid Control	10.7v (33%)
85	DB/OR	CMP Sensor Signal	7 Hz, 55 mph: 15 Hz
87	RD/BK	HO2S-21 (B2 S1) Signal	0.1-1.1v
88	LB/RD	MAF Sensor Signal	0.8v, 55 mph: 1.6v
89	GY/WT	TP Sensor Signal	0.53-1.27v
90	BR/WT	Reference Voltage	4.9-5.1v
91	GY/RD	Analog Signal Return	<0.050v
92	LG	Brake Pedal Switch	0v (Brake On: 12v)
93	RD/WT	HO2S-11 (B1 S1) Heater	1v (Heater Off: 12v)
94	YL/LB	HO2S-21 (B2 S1) Heater	1v (Heater Off: 12v)
86, 96	---	Not Used	---
98	LB	Injector 8 Control	2.7-3.5 ms
99	LG/OR	Injector 6 Control	2.7-3.5 ms
100	BR/LG	Injector 4 Control	2.7-3.5 ms
101	WT	Injector 2 Control	2.7-3.5 ms
102	---	Not Used	---
103	BK/WT	Power Ground	<0.1v
104	DB/LG	COP 2 Driver (dwell)	5°, 55 mph: 8°

Pin Connector Graphic

PCM 104-PIN CONNECTOR

Terminal View of 104-Pin PCM Wiring Harness Connector

Standard Colors and Abbreviations

Abbreviation	Color	Abbreviation	Color	Abbreviation	Color
BK	Black	GY	Gray	PK	Purple
BL	Blue	GN	Green	RD	Red
BR	Brown	LG	LT Green	TN	Tan
DB	Dark Blue	OR	Orange	WT	White
DG	DK Green	PK	Pink	YL	Yellow

1998-99 F-Series 5.4L CNG VIN M, VIN Z (E4OD) 104-P Connector

PCM Pin #	Wire Color	Circuit Description (104 Pin)	Value at Hot Idle
1	LG/YL	COP 6 Driver (dwell)	5°, 55 mph: 8°
2 ('98)	PK/LG	MIL (lamp) Control	MIL Off: 12v, On: 1v
3, 24	BK/WT	Power Ground	<0.1v
4	WT/PK	Power Takeoff Signal	PTO Off: 0v, On: 12v
6	OR/YL	Shift Solenoid 1 Control	1v, 55 mph: 12v
9	YL/WT	Fuel Level Indicator Signal	1.7v (1/2 full)
11	PK/OR	Shift Solenoid 2 Control	12v, 55 mph: 1v
12	WT/LG	TCIL (lamp) Control	7.7v (Switch On: 0v)
13	PK	Flash EEPROM Power	0.1v
15	PK/LB	Data Bus (-)	<0.050v
16	TN/OR	Data Bus (+)	Digital Signals
20	BR/OR	Coast Clutch Solenoid	12v, 55 mph: 12v
21	DB	CKP (-) Sensor Signal	440-490 Hz
22	GY	CKP (+) Sensor Signal	440-490 Hz
25	BK/LB	Case Ground	<0.050v
26	LG/WT	COP 1 Driver (dwell)	5°, 55 mph: 8°
27	LG/OR	COP 5 Driver (dwell)	5°, 55 mph: 8°
29	TN/WT	TCS (switch) Signal	TCS & O/D On: 12v
30	DG	Octane Adjust Shorting Bar	9.3v (shorted: 0v)
33 ('98)	PK/OR	VSS (-) Signal	0 Hz, 55 mph: 125 Hz
34	YL/BK	Digital TR1 Sensor	0v, 55 mph: 11.5v
36	TN/LB	MAF Sensor Return	<0.050v
37	OR/BK	TFT Sensor Signal	2.10-2.40v
39	GY	IAT Sensor Signal	1.5-2.5v
40	DG/YL	Fuel Pump Monitor	On: 12v, Off: 0v
41	BK/YL	A/C Switch Signal	Switch Closed: 12v
45 ('98)	YL/LB	CHIL (lamp) Control	7.7v (Switch On: 0v)
46	BR	Intake Manifold Tuning Valve	12v, 55 mph: 12v
48 ('98)	WT/PK	Clean Tachometer Output	55 Hz, 55 mph: 120 Hz
49	LB/BK	Digital TR2 Sensor	0v, 55 mph: 11.5v
50	LG/RD	Digital TR4 Sensor	0v, 55 mph: 11.5v
51	BK/WT	Power Ground	<0.1v
52	WT/PK	COP 3 Driver (dwell)	5°, 55 mph: 8°
53	OR/YL	COP 4 Driver (dwell)	5°, 55 mph: 8°
54	PK/YL	TCC Solenoid Control	0%, 55 mph: 95%
55	RD/WT	Keep Alive Power	12-14v
56	LG/BK	EVAP Purge Solenoid	0-10 Hz (0-100%)
58 ('99)	GY/BK	VSS (+) Signal	0 Hz, 55 mph: 125 Hz
60	GY/LB	HO2S-11 (B1 S1) Signal	0.1-1.1v
62	LB	EFT Sensor Signal	1.7-3.5v (50-120°F)
63	RD/PK	Fuel Rail Pressure Sensor	2-3.7v (90-100 psi)
64	LB/YL	Digital TR 3A Sensor	In 'P': 0v, in O/D: 1.7v

1998-99 F-Series 5.4L CNG VIN M, VIN Z (E4OD) 104-P Connector

PCM Pin #	Wire Color	Circuit Description (104 Pin)	Value at Hot Idle
66	YL/LG	CHT Sensor Signal	0.6 (194°F)
68	GY/BK	VSS (-) Signal	0 Hz, 55 mph: 125 Hz
69	PK/YL	A/C WOT Relay Control	Off: 12v, On: 1v
71, 97	RD	Vehicle Power	12-14v
72	WT	Injector 7 Signal NGV Module	3.9-4.6 ms
73	WT/LB	Injector 5 Signal NGV Module	3.9-4.6 ms
74	WT/BK	Injector 3 Signal NGV Module	3.9-4.6 ms
75	WT/RD	Injector 1 Signal NGV Module	3.9-4.6 ms
77, 103	BK/WT	Power Ground	<0.1v
78	PK/LB	COP 7 Driver (dwell)	5°, 55 mph: 8°
79	WT/RD	COP 8 Driver (dwell)	5°, 55 mph: 8°
80	LB/OR	Fuel Shutoff Valve Control	Off: 12v, On: 1v
81	WT/YL	EPC Solenoid Control	9.1v (4 psi)
83	WT/LB	IAC Solenoid Control	10.7v (33%)
84	DB/YL	OSS Sensor Signal	0 Hz, 55 mph: 720 Hz
85	DG	CMP Sensor Signal	6 Hz, 55 mph: 16 Hz
87	RD/BK	HO2S-21 (B2 S1) Signal	0.1-1.1v
88	LB/RD	MAF Sensor Signal	0.8v, 55 mph: 1.6v
89	GY/WT	TP Sensor Signal	0.53-1.27v
90	BR/WT	Reference Voltage	4.9-5.1v
91	GY/RD	Analog Signal Return	<0.050v
92	RD/LG	Brake Pedal Switch	0v (Brake On: 12v)
93	RD/WT	HO2S-11 (B1 S1) Heater	1v (Heater Off: 12v)
94	YL/LB	HO2S-21 (B2 S1) Heater	1v (Heater Off: 12v)
98	YL	Injector 8 Signal NGV Module	3.9-4.6 ms
99	YL/LB	Injector 6 Signal NGV Module	3.9-4.6 ms
100	YL/BK	Injector 4 Signal NGV Module	3.9-4.6 ms
101	YL/RD	Injector 2 Signal NGV Module	3.9-4.6 ms
104	PK/WT	COP 2 Driver (dwell)	5°, 55 mph: 8°

Pin Connector Graphic

PCM 104-PIN CONNECTOR

Terminal View of 104-Pin PCM Wiring Harness Connector

Standard Colors and Abbreviations

Abbreviation	Color	Abbreviation	Color	Abbreviation	Color
BK	Black	GY	Gray	PK	Purple
BL	Blue	GN	Green	RD	Red
BR	Brown	LG	LT Green	TN	Tan
DB	Dark Blue	OR	Orange	WT	White
DG	DK Green	PK	Pink	YL	Yellow

2000-02 F-Series 5.4L V8 CNG VIN M, VIN Z (E4OD) 104-P Connector

PCM Pin #	Wire Color	Circuit Description (104 Pin)	Value at Hot Idle
1	OR/YL	COP 6 Driver (dwell)	5°, 55 mph: 8°
3	BK/WT	Power Ground	<0.1v
6	OR/YL	Shift Solenoid 'A' Control	1v, 55 mph: 12v
11	PK/OR	Shift Solenoid 'B' Control	12v, 55 mph: 1v
12	WT/LG	TCIL (lamp) Control	7.7v (Switch On: 0v)
13	VT	Flash EEPROM Power	0.1v
15	PK/LB	Data Bus (-)	<0.050v
16	TN/OR	Data Bus (+)	Digital Signals
20	BR/OR	Coast Clutch Solenoid	12v, 55 mph: 12v
21	DB	CKP (-) Sensor Signal	440-490 Hz
22	GY	CKP (+) Sensor Signal	440-490 Hz
25	LB/YL	Chassis Ground	<0.050v
26	LG/WT	COP 1 Driver (dwell)	5°, 55 mph: 8°
27	LG/YL	COP 5 Driver (dwell)	5°, 55 mph: 8°
29	TN/WT	TCS (switch) Signal	TCS & O/D On: 12v
34	YL/BK	Digital TR1 Sensor	0v, 55 mph: 11.5v
36	TN/LB	MAF Sensor Return	<0.050v
37	OR/BK	TFT Sensor Signal	2.10-2.40v
39	GY	IAT Sensor Signal	1.5-2.5v
40	DG/YL	Fuel Pump Monitor	On: 12v, Off: 0v
41	BK/YL	A/C Switch Signal	Switch Closed: 12v
49	LB/BK	Digital TR2 Sensor	0v, 55 mph: 11.5v
50	WT/BK	Digital TR4 Sensor	0v, 55 mph: 11.5v
51	BK/WT	Power Ground	<0.1v
52	WT/PK	COP 3 Driver (dwell)	5°, 55 mph: 8°
53	DG/PK	COP 4 Driver (dwell)	5°, 55 mph: 8°
54	PK/YL	TCC Solenoid Control	0%, 55 mph: 95%
55	RD/WT	Keep Alive Power	12-14v
59	DG/WT	TSS Sensor Signal	300 Hz, 55 mph: 980
60	GY/LB	HO2S-11 (B1 S1) Signal	0.1-1.1v
62	LB	EFT Sensor Signal	1.7-3.5v (50-120°F)
63	RD/PK	Fuel Rail Pressure Sensor	2-3.7v (90-100 psi)
64	LB/YL	Digital TR 3 Sensor	In 'P': 0v, in O/D: 1.7v

2000-02 F-Series 5.4L V8 CNG VIN M, VIN Z (E4OD) 104-P Connector

PCM Pin #	Wire Color	Circuit Description (104 Pin)	Value at Hot Idle
66	YL/LG	CHT Sensor Signal	0.6 (194ºF)
67	PK/WT	EVAP CV Solenoid	0-10 Hz (0-100%)
68	GY/BK	VSS (-) Signal	0 Hz, 55 mph: 125 Hz
69	PK/YL	A/C WOT Relay Control	Off: 12v, On: 1v
71, 97	RD	Vehicle Power	12-14v
72	WT	Injector 7 Signal NGV Module	3.9-4.6 ms
73	WT/LB	Injector 5 Signal NGV Module	3.9-4.6 ms
74	WT/BK	Injector 3 Signal NGV Module	3.9-4.6 ms
75	WT/RD	Injector 1 Signal NGV Module	3.9-4.6 ms
77, 103	BK/WT	Power Ground	<0.1v
78	PK/LB	COP 7 Driver (dwell)	5º, 55 mph: 8º
79	WT/RD	COP 8 Driver (dwell)	5º, 55 mph: 8º
80	LB/OR	Fuel Pump Control	Off: 12v, On: 1v
81	WT/YL	EPC Solenoid Control	9.1v (4 psi)
83	WT/LB	IAC Solenoid Control	10.7v (33%)
84	DB/YL	OSS Sensor Signal	0 Hz, 55 mph: 720 Hz
85	DG	CMP Sensor Signal	6 Hz, 55 mph: 16 Hz
87	RD/BK	HO2S-21 (B2 S1) Signal	0.1-1.1v
88	LB/RD	MAF Sensor Signal	0.8v, 55 mph: 1.6v
89	GY/WT	TP Sensor Signal	0.53-1.27v
90	BR/WT	Reference Voltage	4.9-5.1v
91	GY/RD	Analog Signal Return	<0.050v
92	RD/LG	Brake Pedal Switch	0v (Brake On: 12v)
93	RD/WT	HO2S-11 (B1 S1) Heater	1v (Heater Off: 12v)
94	YL/LB	HO2S-21 (B2 S1) Heater	1v (Heater Off: 12v)
98	YL	Injector 8 Signal NGV Module	3.9-4.6 ms
99	YL/LB	Injector 6 Signal NGV Module	3.9-4.6 ms
100	YL/BK	Injector 4 Signal NGV Module	3.9-4.6 ms
101	YL/RD	Injector 2 Signal NGV Module	3.9-4.6 ms
104	PK/WT	COP 2 Driver (dwell)	5º, 55 mph: 8º

Pin Connector Graphic

PCM 104-PIN CONNECTOR

Terminal View of 104-Pin PCM Wiring Harness Connector

Standard Colors and Abbreviations

Abbreviation	Color	Abbreviation	Color	Abbreviation	Color
BK	Black	GY	Gray	PK	Purple
BL	Blue	GN	Green	RD	Red
BR	Brown	LG	LT Green	TN	Tan
DB	Dark Blue	OR	Orange	WT	White
DG	DK Green	PK	Pink	YL	Yellow

2003 F-Series 5.4L V8 CNG VIN M, VIN Z (E4OD) 104-P Connector

PCM Pin #	Wire Color	Circuit Description (104 Pin)	Value at Hot Idle
1	OR/YL	COP 6 Driver (dwell)	5°, 55 mph: 8°
2	---	Not Used	---
3	BK/WT	Power Ground	<0.1v
4	PK	Transfer Case Speed Sensor (MSOF)	0 Hz, 55 mph: 471 Hz
4-5	---	Not Used	---
6	OR/YL	Shift Solenoid 'A' Control	1v, 55 mph: 12v
7-10	---	Not Used	---
11	PK/OR	Shift Solenoid 'B' Control	12v, 55 mph: 1v
12	WT/LG	TCIL (lamp) Control	7.7v (Switch On: 0v)
13	VT	Flash EEPROM Power	0.1v
14	---	Not Used	---
15	PK/LB	SCP Data Bus (-)	<0.050v
16	TN/OR	SCP Data Bus (+)	Digital Signals
17-19	---	Not Used	---
20	BR/OR	Coast Clutch Solenoid Control (MSOF)	12v, 55 mph: 12v
21	DB	CKP (-) Sensor Signal	440-490 Hz
22	GY	CKP (+) Sensor Signal	440-490 Hz
23-24	---	Not Used	---
25	LB/YL	Chassis Ground	<0.050v
26	LG/WT	COP 1 Driver (dwell)	5°, 55 mph: 8°
27	LG/YL	COP 5 Driver (dwell)	5°, 55 mph: 8°
28	---	Not Used	---
29	TN/WT	TCS (switch) Signal	TCS & O/D On: 12v
34	YL/BK	Digital TR1 Sensor	0v, 55 mph: 11.5v
31-35	---	Not Used	---
36	TN/LB	MAF Sensor Return	<0.050v
37	OR/BK	TFT Sensor Signal	2.10-2.40v
38	---	Not Used	---
39	GY	IAT Sensor Signal	1.5-2.5v
40	DG/YL	Fuel Pump Monitor	On: 12v, Off: 0v
41	BK/YL	A/C Switch Signal	Switch Closed: 12v
42-48	---	Not Used	---
49	LB/BK	Digital TR2 Sensor	0v, 55 mph: 11.5v
50	WT/BK	Digital TR4 Sensor	0v, 55 mph: 11.5v
51	BK/WT	Power Ground	<0.1v
52	WT/PK	COP 3 Driver (dwell)	5°, 55 mph: 8°
53	DG/PK	COP 4 Driver (dwell)	5°, 55 mph: 8°
54	PK/YL	TCC Solenoid Control	0%, 55 mph: 95%
55	RD/WT	Keep Alive Power	12-14v
59	DG/WT	TSS Sensor Signal	300 Hz, 55 mph: 980
60	GY/LB	HO2S-11 (B1 S1) Signal	0.1-1.1v
61	---	Not Used	---
62	LB	Fuel Rail Temperature Sensor	1.7-3.5v (50-120°F)
63	RD/PK	Fuel Rail Pressure Sensor	2-3.7v (90-100 psi)
64	LB/YL	Digital TR 3 Sensor	In 'P': 0v, in O/D: 1.7v

2003 F-Series 5.4L V8 CNG VIN M, VIN Z (E4OD) 104-P Connector

PCM Pin #	Wire Color	Circuit Description (104 Pin)	Value at Hot Idle
65	---	Not Used	---
66	YL/LG	CHT Sensor Signal	0.6 (194°F)
67	PK/WT	EVAP CV Solenoid	0-10 Hz (0-100%)
68	GY/BK	VSS (-) Signal	0 Hz, 55 mph: 125 Hz
69	PK/YL	A/C WOT Relay Control	Off: 12v, On: 1v
70	BK/WT	Check Fuel Cap Indicator Control	Indicator Off: 12v, On: 1v
71	RD	Vehicle Power	12-14v
72	WT	Injector 7 Signal (NGV Module)	3.9-4.6 ms
73	WT/LB	Injector 5 Signal (NGV Module)	3.9-4.6 ms
74	WT/BK	Injector 3 Signal (NGV Module)	3.9-4.6 ms
75	WT/RD	Injector 1 Signal (NGV Module)	3.9-4.6 ms
76	---	Not Used	---
77	BK/WT	Power Ground	<0.1v
78	PK/LB	COP 7 Driver (dwell)	5°, 55 mph: 8°
79	WT/RD	COP 8 Driver (dwell)	5°, 55 mph: 8°
80	LB/OR	Fuel Pump Control	Off: 12v, On: 1v
81	WT/YL	EPC Solenoid Control	9.1v (4 psi)
82, 86	---	Not Used	---
83	WT/LB	IAC Solenoid Control	10.7v (33%)
84	DB/YL	OSS Sensor Signal	0 Hz, 55 mph: 720 Hz
85	DG	CMP Sensor Signal	6 Hz, 55 mph: 16 Hz
87	RD/BK	HO2S-21 (B2 S1) Signal	0.1-1.1v
88	LB/RD	MAF Sensor Signal	0.8v, 55 mph: 1.6v
89	GY/WT	TP Sensor Signal	0.53-1.27v
90	BR/WT	Reference Voltage	4.9-5.1v
91	GY/RD	Sensor Return	<0.050v
92	RD/LG	Brake Pedal Switch	0v (Brake On: 12v)
93	RD/WT	HO2S-11 (B1 S1) Heater	1v (Heater Off: 12v)
94	YL/LB	HO2S-21 (B2 S1) Heater	1v (Heater Off: 12v)
95-96	---	Not Used	---
97	RD	Vehicle Power	12-14v
98	YL	Injector 8 Signal (NGV Module)	3.9-4.6 ms
99	YL/LB	Injector 6 Signal (NGV Module)	3.9-4.6 ms
100	YL/BK	Injector 4 Signal (NGV Module)	3.9-4.6 ms
101	YL/RD	Injector 2 Signal (NGV Module)	3.9-4.6 ms
102	---	Not Used	---
103	BK/WT	Power Ground	<0.1v
104	PK/WT	COP 2 Driver (dwell)	5°, 55 mph: 8°

Pin Connector Graphic

PCM 104-PIN CONNECTOR

Terminal View of 104-Pin PCM Wiring Harness Connector

Standard Colors and Abbreviations

Abbreviation	Color	Abbreviation	Color	Abbreviation	Color
BK	Black	GY	Gray	PK	Purple
BL	Blue	GN	Green	RD	Red
BR	Brown	LG	LT Green	TN	Tan
DB	Dark Blue	OR	Orange	WT	White
DG	DK Green	PK	Pink	YL	Yellow

1990 F-Series 5.8L V8 VIN H (All) 60 Pin Connector

PCM Pin #	Wire Color	Circuit Description (60 Pin)	Value at Hot Idle
1	YL	Keep Alive Power	12-14v
3	GY/BK	VSS (+) Signal	0 Hz, 55 mph: 125 Hz
3	DG/WT	VSS (+) Signal	0 Hz, 55 mph: 125 Hz
4	DG/YL	IDM Sensor Signal	20-31 Hz
6	BK	VSS (-) Signal	0 Hz, 55 mph: 125 Hz
7	LG/YL	ECT Sensor Signal	0.5-0.6v
8 ('89-'90)	BR	Fuel Pump Monitor	On: 12v, Off: 0v
10	BK/YL	A/C Switch Signal	A/C On: 12v, Off: 0v
11	WT/BK	Air Management 2 Solenoid	AM2 Off: 12v, On: 1v
16	BK/OR	Ignition System Ground	<0.050v
17	PK/LG	Self Test Output & MIL	MIL Off: 12v, On: 1v
18	TN/LG	Inferred Mileage Sensor	Digital Signals
20	BK	PCM Case Ground	0.050v
21	GY/WT	IAC Solenoid Control	8.3-9.1v
22	TN/LG	Fuel Pump Control	Off: 12v, On: 1v
23	LG/BK	Knock Sensor Signal	0v
24	YL/LG	PSP Switch Signal	0v (turning: 12v)
25	YL/RD	ACT Sensor Signal	1.5-2.5v
26	OR/WT	Reference Voltage	4.9-5.1v
27	BR/LG	EVP Sensor Signal	0.4v, 55 mph: 2.6v
29	GY/BL	HO2S-11 (B1 S1) Signal	0.1-1.1v
29	DG/PK	HO2S-11 (B1 S1) Signal	0.1-1.1v
30	GY/YL	A/T: Neutral Drive Switch	In 'P': 0v, Others: 5v
30	GY/YL	M/T: Clutch Engage Switch	5v (clutch "in": 0v)
31	GY/YL	Canister Purge Solenoid	12v, 55 mph: 1v
33	DG	VR Solenoid Control	0%, 55 mph: 45%
36	YL/LG	Spark Output Signal	6.93v (50%)
37	RD	Vehicle Power	12-14v
40	BK/LG	Power Ground	<0.1v
45	DG/LG	MAP Sensor Signal	107 Hz (sea level)
45	DB/LG	MAP Sensor Signal	107 Hz (sea level)
46	BK/WT	Analog Signal Return	<0.050v
47	DG/LG	TP Sensor Signal	1v, 55 mph: 1.4v
48	WT/RD	Self Test Indicator Signal	STI On: 1v, Off: 5v
49	OR	HO2S-11 (B1 S1) Ground	<0.050v
51	WT/RD	Air Management 1 Solenoid	AM1 Off: 12v, On: 1v
56	DB	PIP Sensor Signal	6.93v (50%)
57	RD	Vehicle Power	12-14v
58	TN/OR	Injector Bank 1 (INJ 1, 4, 5, 8)	5.8-6.4 ms
59	TN/RD	Injector Bank 2 (INJ 2, 3, 6, 7)	5.8-6.4 ms
60	BK/LG	Power Ground	<0.1v

Pin Connector Graphic

PCM 60-PIN CONNECTOR

Terminal View of 60-Pin PCM Harness Connector

1990 F-Series 5.8L V8 VIN H (E4OD) 60 Pin Connector

PCM Pin #	Wire Color	Circuit Description (60 Pin)	Value at Hot Idle
1	YL	Keep Alive Power	12-14v
2	LG	Brake Position Switch	0v (Brake On: 12v)
3	DG, GY	VSS (+) Signal	0 Hz, 55 mph: 125 Hz
4	DG/YL	IDM Sensor Signal	20-31 Hz
6	BK	VSS (-) Signal	0 Hz, 55 mph: 125 Hz
7	LG/YL	ECT Sensor Signal	0.5-0.6v
8	BR	Fuel Pump Monitor	On: 12v, Off: 0v
10	BK/YL	A/C Switch Signal	A/C On: 12v, Off: 0v
11	WT/BK	Air Management 2 Solenoid	AM2 Off: 12v, On: 1v
12	BL/BK	4x4 Indicator Light	Light Off: 12v, On: 1v
16	BK/OR	Ignition System Ground	<0.050v
17	PK/LG	Self Test Output & MIL	MIL Off: 12v, On: 1v
18	TN/LG	Inferred Mileage Sensor	Digital Signals
19	DG/PK	Shift Solenoid 2 Control	1v, 55 mph: 1v
20	BK	PCM Case Ground	<0.050v
21	GY/WT	IAC Solenoid Control	8.3-9.1v
22	TN/LG	Fuel Pump Control	Off: 12v, On: 1v
23	LG/BK	Knock Sensor Signal	0v
24	YL/LG	PSP Switch Signal	0v (turning: 12v)
25	YL/RD	ACT Sensor Signal	1.5-2.5v
26	OR/WT	Reference Voltage	4.9-5.1v
27	BR/LG	EVP Sensor Signal	0.4v, 55 mph: 2.6v
29	DG, GY	HO2S-11 (B1 S1) Signal	0.1-1.1v
30	BL/WT	MLP Sensor Signal	In 'P': 0v, in O/D: 5v
31	GY/YL	Canister Purge Solenoid	12v, 55 mph: 1v
32	LG/WT	OCIL (lamp) Control	7.7v (Switch On: 0v)
33	DG	VR Solenoid Control	0%, 55 mph: 45%
36	YL/LG	Spark Output Signal	6.93v (50%)
37, 57	RD	Vehicle Power	12-14v
38	BL/YL	EPC Solenoid Control	9.2v (5 psi)
40	BK/LG	Power Ground	<0.1v
41	TN/WT	OCS (switch) Signal	12v (switch closed: 0v)
42	OR/BK	TOT Sensor Signal	2.10-2.40v
43-44	---	Not Used	---
45	DB/LG	MAP Sensor Signal	107 Hz (sea level)
46	BK/WT	Analog Signal Return	<0.050v
47	DG/LG	TP Sensor Signal	1v, 55 mph: 1.4v
48	WT/RD	Self Test Indicator Signal	STI On: 1v, Off: 5v
49	OR	HO2S-11 (B1 S1) Ground	<0.050v
51	WT/RD	Air Management 1 Solenoid	AM1 Off: 12v, On: 1v
52	OR/YL	Shift Solenoid 1 Control	1v, 55 mph: 12v
53	PK/YL	TCC Solenoid Control	0v, On 55 mph: 12v
55	BR	Coast Clutch Solenoid	12v, 55 mph: 12v
56	DB	PIP Sensor Signal	6.93v (50%)
58	TN/OR	Injector Bank 1 (INJ 1, 4, 5, 8)	5.8-6.4 ms
59	TN/RD	Injector Bank 2 (INJ 2, 3, 6, 7)	5.8-6.4 ms
60	BK/LG	Power Ground	<0.1v

Pin Connector Graphic

Terminal View of 60-Pin PCM Harness Connector

1991-93 F-Series 5.8L V8 VIN H (All) 60 Pin Connector

PCM Pin #	Wire Color	Circuit Description (60 Pin)	Value at Hot Idle
1	YL	Keep Alive Power	12-14v
3	GY/BK	VSS (+) Signal	0 Hz, 55 mph: 125 Hz
4	YL/BK	IDM Sensor Signal	20-31 Hz
4	TN/YL	IDM Sensor Signal	20-31 Hz
6	PK/OR	VSS (-) Signal	0 Hz, 55 mph: 125 Hz
7	LG/RD	ECT Sensor Signal	0.5-0.6v
8	DG/YL	Fuel Pump Monitor	On: 12v, Off: 0v
9	PK, BK	Data Bus (-) Signal	Digital Signals
10	BK, PK	A/C Switch Signal	A/C On: 12v, Off: 0v
11	BR	Air Management 2 Solenoid	AM2 Off: 12v, On: 1v
16	OR/RD	Ignition System Ground	<0.050v
17	PK/LG	Self Test Output & MIL	MIL Off: 12v, On: 1v
20	BK	PCM Case Ground	<0.050v
21	WT/BL	IAC Solenoid Control	7.2-9.2v
22	BL/OR	Fuel Pump Control	Off: 12v, On: 1v
23 ('92-'93)	YL/RD	Knock Sensor Signal	0v
24 ('92-'93)	YL/LG	PSP Switch Signal	0v (turning: 12v)
25	GY	ACT Sensor Signal	1.5-2.5v
26	BR/WT	Reference Voltage	4.9-5.1v
27	BR/LG	EVP Sensor Signal	0.4v, 55 mph: 2.6v
28	TN/OR	Data Bus (+) Signal	Digital Signals
29	GY/BL	HO2S-11 (B1 S1) Signal	0.1-1.1v
30	BL/YL	A/T: Neutral Drive Switch	In 'P': 0v, Others: 5v
30	BL/YL	M/T: Clutch Engage Switch	5v (clutch "in": 0v)
33	BR/PK	VR Solenoid Control	0%, 55 mph: 45%
36	PK	Spark Output Signal	6.93v (50%)
37	RD	Vehicle Power	12-14v
40	BK/WT	Power Ground	<0.1v
45	LG/BK	MAP Sensor Signal	107 Hz (sea level)
46	GY/RD	Analog Signal Return	<0.050v
47	GY/WT	TP Sensor Signal	1v, 55 mph: 1.4v
48	WT/PK	Self Test Indicator Signal	STI On: 1v, Off: 5v
49	OR	HO2S-11 (B1 S1) Ground	<0.050v
51	WT/OR	Air Management 1 Solenoid	AM1 Off: 12v, On: 1v
56	GY/OR	PIP Sensor Signal	6.93v (50%)
57	RD	Vehicle Power	12-14v
58	TN	Injector Bank 1 (INJ 1, 4, 5, 8)	3.5-5.0 ms
59	WT	Injector Bank 2 (INJ 2, 3, 6, 7)	3.5-5.0 ms
60	BK/WT	Power Ground	<0.1v

Pin Connector Graphic

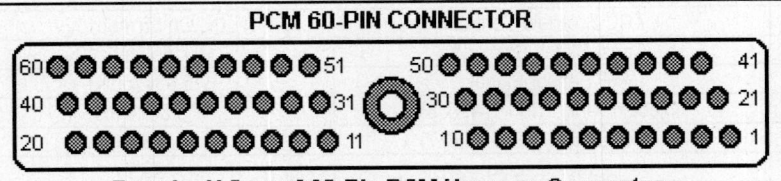

PCM 60-PIN CONNECTOR

Terminal View of 60-Pin PCM Harness Connector

1991-93 F-Series 5.8L V8 VIN H (E4OD) 60 Pin Connector

PCM Pin #	Wire Color	Circuit Description (60 Pin)	Value at Hot Idle
1	YL	Keep Alive Power	12-14v
2	LG	Brake Position Switch	0v (Brake On: 12v)
3	GY/BK	VSS (+) Signal	0 Hz, 55 mph: 125 Hz
4	YL/BK	IDM Sensor Signal	20-31 Hz
4	TN/YL	IDM Sensor Signal	20-31 Hz
5	---	Not Used	---
6	PK/OR	VSS (-) Signal	0 Hz, 55 mph: 125 Hz
7	LG/RD	ECT Sensor Signal	0.5-0.6v
8	DG/YL	Fuel Pump Monitor	On: 12v, Off: 0v
9	PK/BL	Data Bus (-) Signal	<0.050v
9	BK/OR	Data Bus (+) Signal	Digital Signals
10	BK/YL	A/C Switch Signal	A/C On: 12v, Off: 0v
10	PK/BL	A/C Switch Signal	A/C On: 12v, Off: 0v
11	BR	Air Management 2 Solenoid	AM2 Off: 12v, On: 1v
12	BL/BK	4x4 Indicator Light	Light On: 0.1v, Off: 12v
13-15	---	Not Used	---
16	OR/RD	Ignition System Ground	<0.050v
17	PK/LG	Self Test Output & MIL	MIL Off: 12v, On: 1v
18	---	Not Used	---
19	PK/OR	Shift Solenoid 2 Control	12v, 55 mph: 1v
20	BK	PCM Case Ground	<0.050v
21	WT/BL	IAC Solenoid Control	10.7v (33%)
22	BL/OR	Fuel Pump Control	Off: 12v, On: 1v
23 ('92-'93)	YL/RD	Knock Sensor Signal	0v
24 ('92-'93)	YL/LG	PSP Switch Signal	0v (turning: 12v)
25	GY	ACT Sensor Signal	1.5-2.5v
26	BR/WT	Reference Voltage	4.9-5.1v
27	BR/LG	EVP Sensor Signal	0.4v, 55 mph: 2.6v
28	TN/OR	Data Bus (+) Signal	Digital Signals
29	GY/BL	HO2S-11 (B1 S1) Signal	0.1-1.1v
29	GY/YL	HO2S-11 (B1 S1) Signal	0.1-1.1v
30	BL/YL	MLP Sensor Signal	In 'P': 0v, in O/D: 5v
31	GY/YL	Canister Purge Solenoid	12v, 55 mph: 1v
32	WT/LG	OCIL (lamp) Control	7.7v (Switch On: 0v)
33	BR/PK	VR Solenoid Control	0%, 55 mph: 45%
34	---	Not Used	---
35	---	Not Used	---
36	PK	Spark Output Signal	6.93v (50%)
37	RD	Vehicle Power	12-14v
38	WT/YL	EPC Solenoid Control	9.5v (5 psi)
39	---	Not Used	---
40	BK/WT	Power Ground	<0.1v

1991-93 F-Series 5.8L V8 VIN H (E4OD) 60 Pin Connector

PCM Pin #	Wire Color	Circuit Description (60 Pin)	Value at Hot Idle
41	TN/WT	OCS (switch) Signal	12v (switch closed: 0v)
42	OR/BK	TOT Sensor Signal	2.10-2.40v
43 ('91)	PK	A/C Demand Switch	A/C On: 12v, Off: 0v
44	---	Not Used	---
45	LG/BK	MAP Sensor Signal	107 Hz (sea level)
46	GY/RD	Analog Signal Return	<0.050v
47	GY/WT	TP Sensor Signal	1v, 55 mph: 1.4v
48	WT/PK	Self Test Indicator Signal	STI On: 1v, Off: 5v
49	OR	HO2S-11 (B1 S1) Ground	<0.050v
50	---	Not Used	---
51	WT/OR	Air Management 1 Solenoid	Solenoid Off: 12v, On: 1v
52	OR/YL	Shift Solenoid 1 Control	0.3-0.4v
53	PK/YL	TCC Solenoid Control	0v, On 55 mph: 12v
54	---	Not Used	---
55	BR/OR	Coast Clutch Solenoid	12v, 55 mph: 12v
56	GY/OR	PIP Sensor Signal	6.93v (50%)
57	RD	Vehicle Power	12-14v
58	TN	Injector Bank 1 (INJ 1, 4, 5, 8)	3.5-4.0 ms
59	WT	Injector Bank 2 (INJ 2, 3, 6, 7)	3.5-4.0 ms
60	BK/WT	Power Ground	<0.1v

Pin Connector Graphic

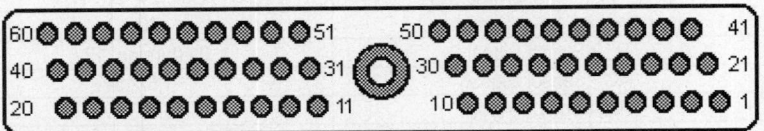

Standard Colors and Abbreviations

Abbreviation	Color	Abbreviation	Color	Abbreviation	Color
BK	Black	GY	Gray	PK	Purple
BL	Blue	GN	Green	RD	Red
BR	Brown	LG	LT Green	TN	Tan
DB	Dark Blue	OR	Orange	WT	White
DG	DK Green	PK	Pink	YL	Yellow

1994 F-Series 5.8L V8 VIN H (All) 60 Pin Connector

PCM Pin #	Wire Color	Circuit Description (60 Pin)	Value at Hot Idle
1	YL	Keep Alive Power	12-14v
2	LG	Brake Position Switch	0v (Brake On: 12v)
3	GY/BK	VSS (+) Signal	0 Hz, 55 mph: 125 Hz
4	YL/BK	IDM Sensor Signal	20-31 Hz
5	---	Not Used	---
6	PK/OR	VSS (-) Signal	0 Hz, 55 mph: 125 Hz
7	LG/RD	ECT Sensor Signal	0.5-0.6v
8	DG/YL	Fuel Pump Monitor	On: 12v, Off: 0v
9	PK/BL	Data Bus (-) Signal	<0.050v
9	BK/OR	Data Bus (+) Signal	Digital Signals
10	BK/YL	A/C Switch Signal	A/C On: 12v, Off: 0v
10	PK/BL	A/C Switch Signal	A/C On: 12v, Off: 0v
11	BR	Air Management 2 Solenoid	AM2 Off: 12v, On: 1v
12	PK/BL	4x4 Indicator Light	12v (switch closed: 0v)
13-15	---	Not Used	---
16	OR/RD	Ignition System Ground	<0.050v
17	PK/LG	Self Test Output & MIL	MIL Off: 12v, On: 1v
18-19	---	Not Used	---
20	BK	PCM Case Ground	<0.050v
21	WT/BL	IAC Solenoid Control	10.7v (33%)
22	BL/OR	Fuel Pump Control	Off: 12v, On: 1v
23	---	Not Used	---
24	---	Not Used	---
25	GY	ACT Sensor Signal	1.5-2.5v
26	BR/WT	Reference Voltage	4.9-5.1v
27	BR/LG	EVP Sensor Signal	0.4v, 55 mph: 2.6v
28	TN/OR	Data Bus (+) Signal	Digital Signals
29	GY/BL	HO2S-11 (B1 S1) Signal	0.1-1.1v
29	GY/YL	HO2S-11 (B1 S1) Signal	0.1-1.1v
30	BL/YL	A/T: Neutral Drive Switch	In 'N': Others: 5v
30	BL/YL	M/T: Clutch Engage Switch	5v (clutch "in": 0v)
31	GY/YL	Canister Purge Solenoid	12v, 55 mph: 1v
32	WT/LG	OCIL (lamp) Control	7.7v (Switch On: 0v)
33	BR/PK	VR Solenoid Control	0%, 55 mph: 45%
34	---	Not Used	---
35	---	Not Used	---
36	PK	Spark Output Signal	6.93v (50%)
37	RD	Vehicle Power	12-14v
38	---	Not Used	---
39	---	Not Used	---
40	BK/WT	Power Ground	<0.1v

1994 F-Series 5.8L V8 VIN H (All) 60 Pin Connector

PCM Pin #	Wire Color	Circuit Description (60 Pin)	Value at Hot Idle
41-42	---	Not Used	---
43	PK	A/C Demand Switch	A/C On: 12v, Off: 0v
44	---	Not Used	---
45	LG/BK	MAP Sensor Signal	107 Hz (sea level)
46	GY/RD	Analog Signal Return	<0.050v
47	GY/WT	TP Sensor Signal	1v, 55 mph: 1.4v
48	WT/PK	Self Test Indicator Signal	STI On: 1v, Off: 5v
49	OR	HO2S-11 (B1 S1) Ground	<0.050v
50	---	Not Used	---
51	WT/OR	Air Management 1 Solenoid	Solenoid Off: 12v, On: 1v
52-55	---	Not Used	---
56	GY/OR	PIP Sensor Signal	6.93v (50%)
57	RD	Vehicle Power	12-14v
58	TN	Injector Bank 1 (INJ 1, 4, 5, 8)	3.5-5.0 ms
59	WT	Injector Bank 2 (INJ 2, 3, 6, 7)	3.5-5.0 ms
60	BK/WT	Power Ground	<0.1v

Pin Connector Graphic

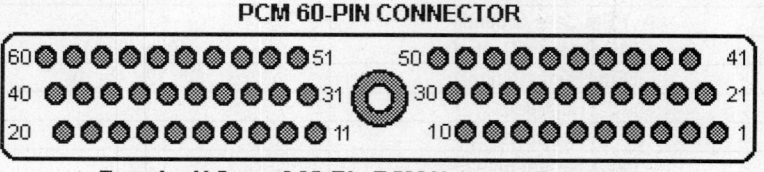

PCM 60-PIN CONNECTOR

Terminal View of 60-Pin PCM Harness Connector

Standard Colors and Abbreviations

Abbreviation	Color	Abbreviation	Color	Abbreviation	Color
BK	Black	GY	Gray	PK	Purple
BL	Blue	GN	Green	RD	Red
BR	Brown	LG	LT Green	TN	Tan
DB	Dark Blue	OR	Orange	WT	White
DG	DK Green	PK	Pink	YL	Yellow

1994 F-Series 5.8L V8 MFI VIN H (E4OD) 60 Pin Connector

PCM Pin #	Wire Color	Circuit Description (60 Pin)	Value at Hot Idle
1	YL	Keep Alive Power	12-14v
2	LG	Brake Position Switch	0v (Brake On: 12v)
3	GY/BK	VSS (+) Signal	0 Hz, 55 mph: 125 Hz
4	YL/BK	IDM Sensor Signal	20-31 Hz
5	---	Not Used	---
6	PK/OR	VSS (-) Signal	0 Hz, 55 mph: 125 Hz
7	LG/RD	ECT Sensor Signal	0.5-0.6v
8	DG/YL	Fuel Pump Monitor	On: 12v, Off: 0v
9	PK/BL	Data Bus (-) Signal	<0.050v
9	BK/OR	Data Bus (+) Signal	Digital Signals
10	BK/YL	A/C Switch Signal	A/C On: 12v, Off: 0v
10	PK/BL	A/C Switch Signal	A/C On: 12v, Off: 0v
11	BR	Air Management 2 Solenoid	AM2 Off: 12v, On: 1v
12	PK/BL	4x4 Indicator Light	12v (switch closed: 0v)
13	---	Not Used	---
14	---	Not Used	---
15	---	Not Used	---
16	OR/RD	Ignition System Ground	<0.050v
17	PK/LG	Self Test Output & MIL	MIL Off: 12v, On: 1v
18	---	Not Used	---
19	PK/OR	Shift Solenoid 2 Control	12v, 55 mph: 1v
20	BK	PCM Case Ground	<0.050v
21	WT/BL	IAC Solenoid Control	10.7v (33%)
22	BL/OR	Fuel Pump Control	Off: 12v, On: 1v
23	---	Not Used	---
24	---	Not Used	---
25	GY	ACT Sensor Signal	1.5-2.5v
26	BR/WT	Reference Voltage	4.9-5.1v
27	BR/LG	EVP Sensor Signal	0.4v, 55 mph: 2.6v
28	TN/OR	Data Bus (+) Signal	Digital Signals
29	GY/BL	HO2S-11 (B1 S1) Signal	0.1-1.1v
29	GY/YL	HO2S-11 (B1 S1) Signal	0.1-1.1v
30	BL/YL	MLP Sensor Signal	In 'P': 0v, in O/D: 5v
31	GY/YL	Canister Purge Solenoid	12v, 55 mph: 1v
32	WT/LG	OCIL (lamp) Control	7.7v (Switch On: 0v)
33	BR/PK	VR Solenoid Control	0%, 55 mph: 45%
34	---	Not Used	---
35	---	Not Used	---
36	PK	Spark Output Signal	6.93v (50%)
37	RD	Vehicle Power	12-14v
38	WT/YL	EPC Solenoid Control	9.5v (5 psi)
39	---	Not Used	---
40	BK/WT	Power Ground	<0.1v

1994 F-Series 5.8L V8 MFI VIN H (E4OD) 60 Pin Connector

PCM Pin #	Wire Color	Circuit Description (60 Pin)	Value at Hot Idle
41	TN/WT	TCS (switch) Signal	TCS closed: 12, open: 0v
42	OR/BK	TOT Sensor Signal	2.10-2.40v
43	PK	A/C Demand Switch	A/C On: 12v, Off: 0v
44	---	Not Used	---
45	LG/BK	MAP Sensor Signal	107 Hz (sea level)
46	GY/RD	Analog Signal Return	<0.050v
47	GY/WT	TP Sensor Signal	1v, 55 mph: 1.4v
48	WT/PK	Self Test Indicator Signal	STI On: 1v, Off: 5v
49	OR	HO2S-11 (B1 S1) Ground	<0.050v
50	---	Not Used	---
51	WT/OR	Air Management 1 Solenoid	Solenoid Off: 12v, On: 1v
52	OR/YL	Shift Solenoid 1 Control	1v, 55 mph: 12v
53	PK/YL	TCC Solenoid Control	0v, On 55 mph: 12v
54	---	Not Used	---
55	BR/OR	Coast Clutch Switch	12v, 55 mph: 12v
56	GY/OR	PIP Sensor Signal	6.93v (50%)
57	RD	Vehicle Power	12-14v
58	TN	Injector Bank 1 (INJ 1, 4, 5, 8)	3.5-5.0 ms
59	WT	Injector Bank 2 (INJ 2, 3, 6, 7)	3.5-5.0 ms
60	BK/WT	Power Ground	<0.1v

Pin Connector Graphic

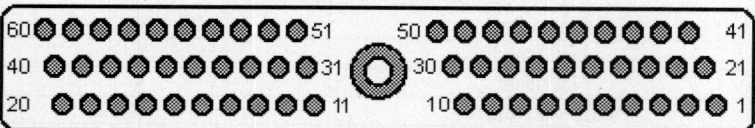

PCM 60-PIN CONNECTOR

Terminal View of 60-Pin PCM Harness Connector

Standard Colors and Abbreviations

Abbreviation	Color	Abbreviation	Color	Abbreviation	Color
BK	Black	GY	Gray	PK	Purple
BL	Blue	GN	Green	RD	Red
BR	Brown	LG	LT Green	TN	Tan
DB	Dark Blue	OR	Orange	WT	White
DG	DK Green	PK	Pink	YL	Yellow

1995 F-Series 5.8L V8 VIN H (E4OD) California 60-P Connector

PCM Pin #	Wire Color	Circuit Description (60 Pin)	Value at Hot Idle
1	YL	Keep Alive Power	12-14v
2	LG	Brake Pedal Switch	0v (Brake On: 12v)
3	GY/BK	VSS (+) Signal	At 55 mph; 125 Hz
4	YL/BK	IDM Sensor Signal	20-31 Hz
5	---	Not Used	---
6	PK/OR	VSS (-) Signal	0 Hz, 55 mph: 125 Hz
7	LG/RD	ECT Sensor Signal	0.5-0.6v
8	DG/YL	Fuel Pump Monitor	On: 12v, Off: 0v
9	TN/BL	MAF Sensor Return	<0.050v
10	DG/OR	A/C Switch Signal	A/C On: 12v, Off: 0v
11	LB/BK	EVAP Purge Solenoid	12v, 55 mph: 1v
12	LG/OR	Injector 6 Control	5.5-6.1 ms
13	TN/RD	Injector 7 Control	5.5-6.1 ms
14	LB	Injector 8 Control	5.5-6.1 ms
15	TN/BK	Injector 5 Control	5.5-6.1 ms
16	OR/RD	Ignition System Ground	<0.050v
17	PK/LG	Self Test Output & MIL	MIL Off: 12v, On: 1v
18	TN/OR	Data Bus (+) Signal	Digital Signals
19	PK/BL	Data Bus (-) Signal	Digital Signals
20	BK	PCM Case Ground	Digital Signals
21	WT/BL	IAC Solenoid Control	10.6-10.7v
22	BL/OR	Fuel Pump Control	Off: 12v, On: 1v
23	---	Not Used	---
24	---	Not Used	---
25	GY	IAT Sensor Signal	1.5-2.5v
26	BR/WT	Reference Voltage	4.9-5.1v
27	BR/LG	EVP Sensor Signal	0.4v, 55 mph: 2.6v
28	---	Not Used	---
29	BK/LG	Not Used	---
30	LB/YL	MLP Sensor Signal	In 'P': 0v, in O/D: 5v
31	---	Not Used	---
32	WT/LG	TCIL (lamp) Control	7.7v (Switch On: 0v)
33	BR/PK	VR Solenoid Control	0%, 55 mph: 45%
34	BR	Air Diverter Solenoid Control	12v
35	BR/BL	Injector 4 Control	5.5-6.1 ms
36	PK	Spark Output Signal	6.93v (50%)
37	RD	Vehicle Power	12-14v
38	WT/YL	EPC Solenoid Control	9.5v (5 psi)
39	BR/YL	Injector 3 Control	5.5-6.1 ms
40	BK/WT	Power Ground	<0.1v

1995 F-Series 5.8L V8 VIN H (E4OD) California 60-P Connector

PCM Pin #	Wire Color	Circuit Description (60 Pin)	Value at Hot Idle
41	TN/WT	TCS (switch) Signal	TCS closed: 12, open: 0v
42	BL/PK	4x4 Indicator Light	12v (switch closed: 0v)
43	RD/BK	HO2S-21 (B2 S1) Signal	0.1-1.1v
44	GY/BL	HO2S-11 (B1 S1) Signal	0.1-1.1v
45	---	Not Used	---
46	GY/RD	Analog Signal Return	<0.050v
47	GY/WT	TP Sensor Signal	1v, 55 mph: 1.4v
48	WT/PK	Self Test Indicator Signal	STI On: 1v, Off: 5v
49	OR/BK	TOT Sensor Signal	2.10-2.40v
50	LB/RD	MAF Sensor Signal	0.8v, 55 mph: 1.6v
51	OR/YL	Shift Solenoid 1 Control	1v, 55 mph: 12v
52	PK/OR	Shift Solenoid 2 Control	12v, 55 mph: 1v
53	PK/YL	TCC Solenoid Control	0v, On 55 mph: 12v
54	---	Not Used	---
55	BR/OR	Coast Clutch Solenoid	1v, 55 mph: 12v
56	GY/OR	PIP Sensor Signal	6.93v (50%)
57	RD	Vehicle Power	12-14v
58	TN	Injector 1 Control	5.5-6.1 ms
59	WT	Injector 2 Control	5.5-6.1 ms
60	BK/WT	Power Ground	<0.1v

Pin Connector Graphic

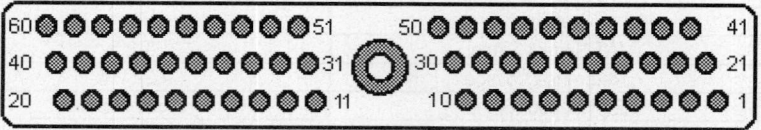

PCM 60-PIN CONNECTOR

Terminal View of 60-Pin PCM Harness Connector

Standard Colors and Abbreviations

Abbreviation	Color	Abbreviation	Color	Abbreviation	Color
BK	Black	GY	Gray	PK	Purple
BL	Blue	GN	Green	RD	Red
BR	Brown	LG	LT Green	TN	Tan
DB	Dark Blue	OR	Orange	WT	White
DG	DK Green	PK	Pink	YL	Yellow

1995 F-Series 5.8L V8 VIN H (All) 60 Pin Connector

PCM Pin #	Wire Color	Circuit Description (60 Pin)	Value at Hot Idle
1	YL	Keep Alive Power	12-14v
2	LG	Brake Pedal Switch	0v (Brake On: 12v)
3	GY/BK	VSS (+) Signal	0 Hz, 55 mph: 125 Hz
4	YL/BK	IDM Sensor Signal	20-31 Hz
5	---	Not Used	---
6	PK/OR	VSS (-) Signal	0 Hz, 55 mph: 125 Hz
7	LG/RD	ECT Sensor Signal	0.5-0.6v
8	DG/YL	Fuel Pump Monitor	On: 12v, Off: 0v
9	PK/BL	Data Bus (-) Signal	Digital Signals
10	PK/BL	A/C Switch Signal	A/C On: 12v, Off: 0v
11	BR	Air Management 2 Solenoid	AM2 Off: 12v, On: 1v
12	PK/BL	4x4 Indicator Light	12v (switch closed: 0v)
13-15	---	Not Used	---
16	OR/RD	Ignition System Ground	<0.050v
17	PK/LG	Self Test Output & MIL	MIL Off: 12v, On: 1v
18	---	Not Used	---
19	PK/OR	Shift Solenoid 2 Control	12v, 55 mph: 1v
20	BK	PCM Case Ground	<0.050v
21	WT/BL	IAC Solenoid Control	10.7v (33%)
22	BL/OR	Fuel Pump Control	Off: 12v, On: 1v
23-24	---	Not Used	---
25	GY	ACT Sensor Signal	1.5-2.5v
26	BR/WT	Reference Voltage	4.9-5.1v
27	BR/LG	EVP Sensor Signal	0.4v, 55 mph: 2.6v
28	TN/OR	Data Bus (+) Signal	Digital Signals
29	GY/BL	HO2S-11 (B1 S1) Signal	0.1-1.1v
30	BL/YL	MLP Sensor Signal	In 'P': 0v, in O/D: 5v
31	GY/YL	Canister Purge Solenoid	12v, 55 mph: 1v
32	WT/LG	OCIL (lamp) Control	7.7v (Switch On: 0v)
33	BR/PK	VR Solenoid Control	0%, 55 mph: 45%
34-35	---	Not Used	---
36	PK	Spark Output Signal	6.93v (50%)
37	R	Vehicle Power	12-14v
38	WT/YL	EPC Solenoid Control	9.5v (5 psi)
39	---	Not Used	---
40	BK/WT	Power Ground	<0.1v

1995 F-Series 5.8L V8 VIN H (All) 60 Pin Connector

PCM Pin #	Wire Color	Circuit Description (60 Pin)	Value at Hot Idle
41	TN/WT	TCS (switch) Signal	TCS closed: 12, open: 0v
42	OR/BK	TOT Sensor Signal	2.10-2.40v
43	PK	A/C Demand Switch	A/C On: 12v, Off: 0v
44	---	Not Used	---
45	LG/BK	MAP Sensor Signal	107 Hz (sea level)
46	GY/RD	Analog Signal Return	<0.050v
47	GY/WT	TP Sensor Signal	1v, 55 mph: 1.4v
48	WT/PK	Self Test Indicator Signal	STI On: 1v, Off: 5v
49	OR	HO2S-11 (B1 S1) Ground	<0.050v
50	---	Not Used	---
51	WT/OR	Air Management 1 Solenoid	AM1 Off: 12v, On: 1v
52	OR/YL	Shift Solenoid 1 Control	1v, 55 mph: 12v
53	PK/YL	TCC Solenoid Control	0v, On 55 mph: 12v
54	---	Not Used	---
55	BR/OR	Coast Clutch Solenoid	12v, 55 mph: 12v
56	GY/OR	PIP Sensor Signal	6.93v (50%)
57	RD	Vehicle Power	12-14v
58	TN	Injector Bank 1 (INJ 1, 4, 5, 8)	3.5-5.0 ms
59	WT	Injector Bank 2 (INJ 2, 3, 6, 7)	3.5-5.0 ms
60	BK/WT	Power Ground	<0.1v

Pin Connector Graphic

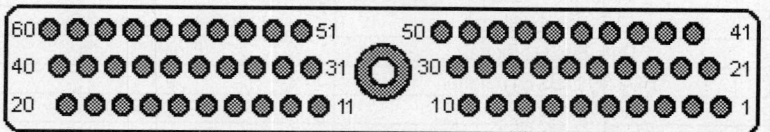

PCM 60-PIN CONNECTOR

Terminal View of 60-Pin PCM Harness Connector

Standard Colors and Abbreviations

Abbreviation	Color	Abbreviation	Color	Abbreviation	Color
BK	Black	GY	Gray	PK	Purple
BL	Blue	GN	Green	RD	Red
BR	Brown	LG	LT Green	TN	Tan
DB	Dark Blue	OR	Orange	WT	White
DG	DK Green	PK	Pink	YL	Yellow

1996-97 F-Series 5.8L V8 VIN H (E4OD) 104 Pin Connector

PCM Pin #	Wire Color	Circuit Description (104 Pin)	Value at Hot Idle
1	PK/OR	Shift Solenoid 2 Control	1v, 55 mph: 12v
2	PK/LG	MIL (lamp) Control	MIL Off: 12v, On: 1v
4	LG/RD	Power Take-Off (if equipped)	0v (Off)
5-12	---	Not Used	---
13	PK	Flash EPROM Power	0.1v
14	LG/BK	4x4 Low Indicator Switch	12v (switch closed: 0v)
15	PK/LB	Data Bus (-) Signal	Digital Signals
16	TN/OR	Data Bus (+) Signal	Digital Signals
17-22	---	Not Used	---
23	OR/RD	Ignition Ground	<0.050v
24, 51	BK/WT	Power Ground	<0.1v
25	BK/LB	Case Ground	<0.050v
26, 28	---	Not Used	---
27	OR/YL	Shift Solenoid 1 Control	1v, 55 mph: 12v
29	TN/WT	TCS (switch) Signal	TCS & O/D On: 12v
30-32, 34	---	Not Used	---
33	PK/OR	PSOM (-) Signal	<0.050v
35	RD/LG	HO2S-12 (B1 S2) Signal	0.1-1.1v
36	TN/LB	MAF Sensor Return	<0.050v
37	OR/BK	TFT Sensor Signal	2.10-2.40v
38	LG/RD	ECT Sensor Signal	0.5-0.6v
39	GY	IAT Sensor Signal	1.5-2.5v
40	DG/YL	Fuel Pump Monitor	On: 12v, Off: 0v
41	BK/YL	A/C Switch Signal	A/C On: 12v, Off: 0v
42-43, 46	---	Not Used	---
44	BR	Secondary AIR Diverter	AIRD Off: 12v, On: 1v
47	BK/PK	VR Solenoid Control	0%, 55 mph: 45%
48	YL/BK	Clean Tachometer Output	39-45 Hz
49	GY/OR	PIP Sensor Signal	6.93v (50%)
50	PK	Spark Output Signal	6.93v (50%)
52, 57	---	Not Used	---
53	BR/OR	Coast Clutch Solenoid	12v, 55 mph: 12v
54	PK/YL	TCC Solenoid Control	0%, 55 mph: 95%
55	YL	Keep Alive Power	12-14v
56	LG/BK	EVAP Purge Solenoid	0-10 Hz (0-100%)
58	GY/BK	PSOM (+) Signal	0 Hz, 55 mph: 125 Hz
59	DG/LG	Misfire Detection Sensor	45-55 Hz
60	GY/LB	HO2S-11 (B1 S1) Signal	0.1-1.1v
61-63	---	Not Used	---
64	LB/YL	TR Sensor Signal	In 'P': 0v, in O/D: 5v

1996-97 Truck 5.8L V8 VIN H (E4OD) 104 Pin Connector

PCM Pin #	Wire Color	Circuit Description (104 Pin)	Value at Hot Idle
65	BR/LG	DPFE Sensor Signal	0.95-1.05v
66-69	---	Not Used	---
70	WT/OR	Secondary AIR Bypass	AIRB Off: 12v, On: 1v
71	RD	Vehicle Power	12-14v
72	TN/RD	Injector 7 Control	4.2-4.6 ms
73	TN/BK	Injector 5 Control	4.2-4.6 ms
74	BR/YL	Injector 3 Control	4.2-4.6 ms
75	TN	Injector 1 Control	4.2-4.6 ms
76	BK/WT	Power Ground	<0.1v
77	BK/WT	Power Ground	<0.1v
78, 82	---	Not Used	---
79	WT/LG	TCIL (lamp) Control	7.7v (Switch On: 0v)
80	LB/OR	Fuel Pump Control	Off: 12v, On: 1v
81	WT/YL	EPC Solenoid Control	9.2v (5 psi)
83	WT/LB	IAC Solenoid Control	10.7v (33%)
84-86	---	Not Used	---
87	RD/BK	HO2S-21 (B2 S1) Signal	0.1-1.1v
88	LB/RD	MAF Sensor Signal	0.8v, 55 mph: 1.8v
89	GY/WT	TP Sensor Signal	0.53-1.27v
90	BR/WT	Reference Voltage	4.9-5.1v
91	GY/RD	Analog Signal Return	<0.050v
92	LG	Brake Pedal Switch	0v (Brake On: 12v)
93	RD/WT	HO2S-11 (B1 S1) Heater	1v (Heater Off: 12v)
94	YL/LB	HO2S-21 (B2 S1) Heater	1v (Heater Off: 12v)
95	WT/BK	HO2S-12 (B1 S2) Heater	1v (Heater Off: 12v)
96	---	Not Used	---
97	RD	Vehicle Power	12-14v
98	LG	Injector 8 Control	4.2-4.6 ms
99	LG/OR	Injector 6 Control	4.2-4.6 ms
100	BR/LB	Injector 4 Control	4.2-4.6 ms
101	WT	Injector 2 Control	4.2-4.6 ms
102	---	Not Used	---
103	BK/WT	Power Ground	<0.1v
104	---	Not Used	---

Pin Connector Graphic

PCM 104-PIN CONNECTOR

Terminal View of 104-Pin PCM Wiring Harness Connector

Standard Colors and Abbreviations

Abbreviation	Color	Abbreviation	Color	Abbreviation	Color
BK	Black	GY	Gray	PK	Purple
BL	Blue	GN	Green	RD	Red
BR	Brown	LG	LT Green	TN	Tan
DB	Dark Blue	OR	Orange	WT	White
DG	DK Green	PK	Pink	YL	Yellow

2003 F-Series 6.0L V8 Diesel VIN P (All) C138a Pin Connector

PCM Pin #	Wire Color	Circuit Description (46 Pin)	Value at Hot Idle
1	---	Not Used	---
2	PK/YL	A/C Relay Control	Relay Off: 12v, On: 1v
3	OG/LG	Data Output Link	Digital Signals
4	BR/PK	Starter Relay Circuit Control	Relay Off: 12v, On: 1v
5	LB/OR	Fuel Pump Relay Control	Relay Off: 12v, On: 1v
6-7	---	Not Used	---
8	TN/LG	A/C Pressure Switch Signal	Switch on: 12v, off: 1v
9	OG/LB	Speed Control Indicator Control	Indicator Off: 12v, On: 1v
10	LB/YL	Power Ground	<0.1v
11	BK/WT	Power Ground	<0.1v
12	---	Not Used	---
13	PK/YL	CAN Bus 1H Signal	0-7-0v
14	RD/WT	CAN Bus 1L Signal	0-7-0v
15	RD	Water In Fuel Indicator Control	Indicator Off: 12v, On: 1v
16	TN/LG	A/C Head Pressure Switch	0v or 12v
17	LG/RD	Parking Brake Applied Switch	Brake Off: 12v, Applied: 0v
18	RD/LG	Brake Pedal Switch	0v (Brake On: 12v)
19	PK/BK	Fuel Pump Monitor	Pump Off: 12v, On: 1v
20	LB/RD	APP Sensor 2 Ground	<0.050v
21	TN/LB	MAF Sensor Ground	<0.050v
22	---	Not Used	---
23	BK/WT	Power Ground	<0.1v
24	BK	Speed Control Ground	<0.050v
25	YL/WT	APP Sensor 3 Signal	0.8-3.5v
26	LB/BK	APP Sensor 1 Signal	0.7-4.2v
27	---	Not Used	---
28	BK/YL	Brake Pedal Switch	0v (Brake On: 12v)
29	WT/LB	APP Sensor 2 Reference Voltage	4.9-5.1v
30	VT	Generator/Battery Indicator Control	Indicator Off: 12v, On: 1v
31	LB/BK	Speed Control On/Off Switch	Switch On: 12v, Off: 0v
32	TN/OR	SCP Data Bus (+) Signal	Digital Signals
33	GY/RD	Ground	<0.050v
34	RD	Vehicle Power (Start-Run)	12-14v
35	GY/BK	VSS (+) Signal	0 Hz, 55 mph: 125 Hz
36	TN/WT	Tow Haul Switch	Off: 0v, On: 12v
37	YL	APP Sensor 2 Signal	1.4-4.1v
38	DB/LG	Barometric Absolute Pressure Sensor	159 Hz
39	VT	Flash EEPROM Power	0.1v
40	RD/WT	Battery Direct	12-14v
41	LG/BK	Manifold Absolute Pressure Sensor	0.5-4.5v
42	LB/RD	MAF Sensor Signal	0.1-4.7v
43	GY	Intake Air Temperature Sensor	1.5-2.5v
44	PK/LB	SCP Data Bus (-) Signal	<0.050v
45	BR/WT	Reference Voltage	4.9-5.1v
46	RD	Vehicle Power (Start-Run)	12-14v

2003 F-Series 6.0L V8 Diesel VIN P (All) C138b Pin Connector

PCM Pin #	Wire Color	Circuit Description (30 Pin)	Value at Hot Idle
1	PK/WT	Reference Voltage	4.9-5.1v
2	VT/YL	Pressure Control Solenoid 'A'	Solenoid Off: 12v, On: 1v
3	YL/LG	Reverse Lamps Relay Control	Relay Off: 12v, On: 1v
4	RD/WT	Transfer Case Neutral Signal	N/A
5	WT/LG	TCIL (lamp) Control	7.7v (Switch On: 0v)
6	---	Not Used	---
7	YL/WT	EPC Solenoid Control	7.5v (8 psi)
8	---	Not Used	---
9	OG/YL	Shift Solenoid Pressure Control 'A'	Solenoid Off: 12v, On: 1v
10	VT/OG	Shift Solenoid Pressure Control 'B'	Solenoid Off: 12v, On: 1v
11	PK/BK	Shift Solenoid Pressure Control 'C'	Solenoid Off: 12v, On: 1v
12	BK/LG	Shift Solenoid Pressure Control 'D'	Solenoid Off: 12v, On: 1v
13	DB/WT	Shift Solenoid Pressure Control 'E'	Solenoid Off: 12v, On: 1v
14	BR/OG	TCC Solenoid Control	TCC Off: 12v, On: 1v
15-16	---	Not Used	---
17	OG/YL	Shift Solenoid Pressure Control 'A'	Solenoid Off: 12v, On: 1v
18	LB/PK	Shift Solenoid Pressure Control 'B'	Solenoid Off: 12v, On: 1v
19	LB/RD	Shift Solenoid Pressure Control 'C'	Solenoid Off: 12v, On: 1v
20	WT/RD	Shift Solenoid Pressure Control 'D'	Solenoid Off: 12v, On: 1v
21	PK/LB	Shift Solenoid Pressure Control 'E'	Solenoid Off: 12v, On: 1v
22	BK/WT	TR Sensor TR-P Ground	<0.050v
23-24	---	Not Used	---
25	LB/YL	TR Sensor TR-P Signal	0-12-0v
26	OG/BK	TFT Sensor Signal	2.10-2.40v
27	GY/OG	ISS Sensor Signal	AC pulse signals
28	DB/YL	OSS Sensor Signal	AC pulse signals
29	DG/WT	TSS Sensor Signal	AC pulse signals
30	OG/WT	Sensor Ground	<0.050v

Pin Connector Graphic

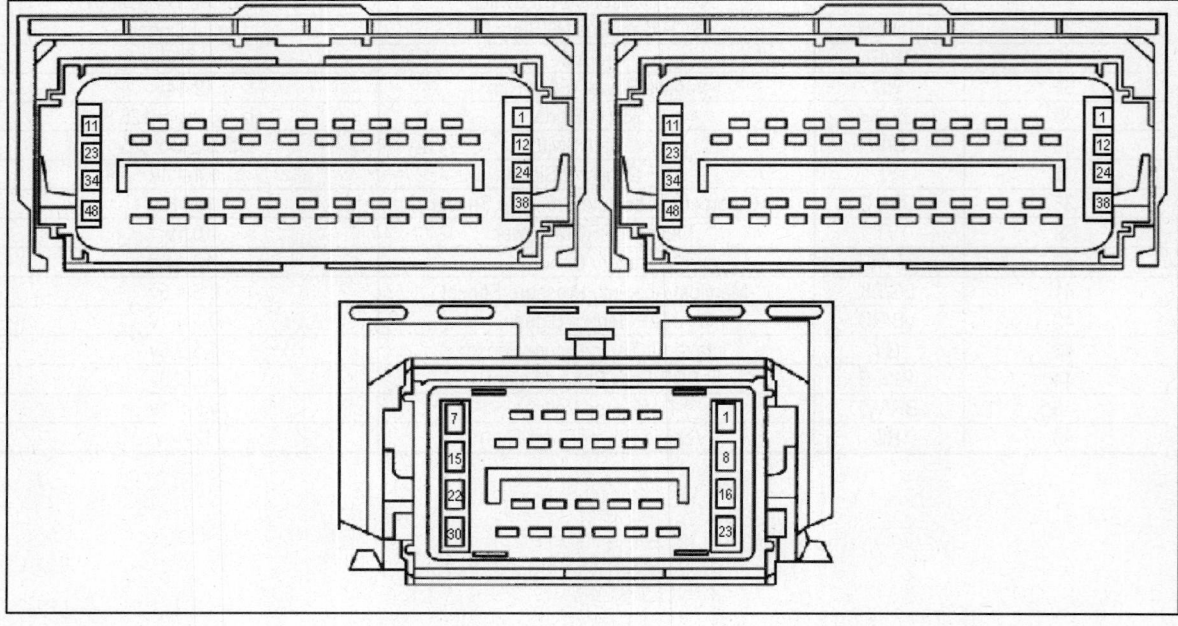

2003 F-Series 6.0L V8 Diesel VIN P C1381c Pin Connector

PCM Pin #	Wire Color	Circuit Description (46 Pin)	Value at Hot Idle
1	WT/YL	Charge Indicator Monitor	12v (Fault Detected: 0v)
2	YL/RD	Injector Pressure Regulator	12v
3	VT/OR	Glow Plug Enable	Relay Off: 12v, On: 1v
4	YL	Generator 2 Monitor	12v (Fault Detected: 0v)
5	---	Not Used	---
6	RD/OR	Electric Fan Speed Signal	0.1-4.9v
7-9	---	Not Used	---
10	BK/LB	Variable Geometry Actuator Signal	0-12-0v
11	DB	Variable Geometry Actuator Power	12-14v
12	GY/WT	Electronic Throttle Control (-) Signal	<0.050v
13	---	Not Used	---
14	DB	Electric Fan Speed Control	0-12-0v
15-16	---	Not Used	---
17	WT/LG	Glow Plug Module Signal	Off: 12v, On: 1v
18	---	Not Used	---
19	LG/YL	Injector Driver Module Command	0-6000 Hz
20	DB/OG	Injector Driver Module CID Signal	0.5-50 Hz
21	---	Not Used	---
22	BR/PK	Engine Cooling Fan Ground	<0.1v
29	TN/LB	M/T: CCP Switch Signal	Clutch In: 0v, Out: 12v
30	VT/LB	Exhaust Back Pressure Sensor	0.9v, off-idle: 2.5v
31	BK/YL	Brake Pressure Switch	Brake Off: 12v, On: 1v
32	YL/WT	Coolant Temperature Sensor	0.6 (194°F)
33	LB/OR	EGR Valve Actuator Position Sense	0.6-3.5v
34-35	---	Not Used	---
36	BR/WT	Reference Voltage	4.9-5.1v
37	RD/LG	CAN Bus 2H Signal	0-7-0v
38	OG/LB	EGR Throttle Position Sensor	0.1-5.1v
39-40	---	Not Used	---
41	GY	Crankshaft Position Sensor (-) Signal	0-6000 Hz
42	BK	Ground (Drain Wire)	<0.050v
43	OR	Camshaft Position Sensor (-) Signal	0.5-50 Hz
44	LG/RD	Engine Oil Temperature Sensor	40°F: 4.7v, 230°F: 0.358v
45	RD/WT	Manifold Air Temperature Sensor	0.2-4.7v
46	BR/LG	Electric Fan Clutch Reference Voltage	12-14v

1999-2002 F-Series 6.8L V10 VIN S (All) 104 Pin Connector

PCM Pin #	Wire Color	Circuit Description (104 Pin)	Value at Hot Idle
1	OR/YL	COP 6 Driver (dwell)	5°, 55 mph: 9°
2	PK/LG	MIL (lamp) Control	MIL Off: 12v, On: 1v
3	BK/WT	Power Ground	<0.1v
4	LB/YL	Power Take-Off (if equipped)	0v (Off)
6	OR/YL	Shift Solenoid 1 Control	1v, 55 mph: 12v
9	YL/WT	Fuel Level Indicator Signal	1.7v (1/2 full)
11	PK/OR	Shift Solenoid 2 Control	12v, 55 mph: 12v
12	WT/LG	TCIL (lamp) Control	TCIL On: 1v, Off: 2.2v
13	VT	Flash EEPROM Power	0.1v
14	LB/BK	4x4 Low Indicator Switch	12v (switch closed: 0v)
15	PK/LB	Data Bus (-)	Digital Signals
16	TN/OR	Data Bus (+)	Digital Signals
19	OR/LG	Fuel Pump High Speed Relay	Off: 12v, On: 1v
20	BR/OR	Coast Clutch Solenoid	12v, 55 mph: 12v
21	DB	CKP (-) Sensor Signal	400-420 Hz
22	GY	CKP (+) Sensor Signal	400-420 Hz
23-24	BK/WT	Power Ground	<0.1v
25	LB/YL	Case Ground	<0.050v
26	LG/WT	COP 1 Driver (dwell)	5°, 55 mph: 9°
27	LG/YL	COP 10 Driver (dwell)	5°, 55 mph: 9°
29	TN/WT	TCS (switch) Signal	TCS & O/D On: 12v
32 ('00-'02)	DG/PK	Knock Sensor 1 (-) Signal	<0.050v
33	PK/OR	VSS (-) Signal	0 Hz, 55 mph: 125 Hz
34	YL/BK	Digital TR1 Sensor	0v, 55 mph: 11.5v
35	RD/LG	HO2S-12 (B1 S2) Signal	0.1-1.1v
36	TN/LB	MAF Sensor Return	<0.050v
37	OR/BK	TFT Sensor Signal	2.10-2.40v
39	GY	IAT Sensor Signal	1.5-2.5v
40	DG/YL	Fuel Pump Monitor	On: 12v, Off: 0v
41	TN/LG	A/C Switch Signal	A/C On: 12v, Off: 0v
42	GY/RD	Injector 10 Control	4.2-4.6 ms
43	OR/LG	Data Output Link	Digital Signals
45	RD/WT	CHIL (lamp) Control	CHT Off: 12v, On: 1v
46	GY/BK	VSS (+) Signal	0 Hz, 55 mph: 125 Hz
47	BR/PK	VR Solenoid Control	0%, 55 mph: 45%
48	WT/PK	Clean Tachometer Output	60-70 Hz
49	LB/BK	Digital TR2 Sensor	0v, 55 mph: 11.5v
50	WT/BK	Digital TR4 Sensor	0v, 55 mph: 11.5v
51	BK/WT	Power Ground	<0.1v
52	WT/PK	COP 5 Driver (dwell)	5°, 55 mph: 9°
53	DGVT	COP 7 Driver (dwell)	5°, 55 mph: 9°
54	PK/YL	TCC Solenoid Control	0%, 55 mph: 95%
55	RD/WT	Keep Alive Power	12-14v
56	LG/BK	EVAP Purge Solenoid	0-10 Hz (0-100%)
57	YL/RD	Knock Sensor 1 (+) Signal	0v
58	GY/BK	VSS (+) Signal	0 Hz, 55 mph: 125 Hz
59	DG/WT	TSS Sensor Signal	300 Hz, 55 mph: 980
60	GY/LB	HO2S-11 (B1 S1) Signal	0.1-1.1v
61	---	Not Used	---
62	RD/PK	FTP Sensor Signal	2.6v at 0" H20 (cap off)
63	---	Not Used	---
64	LB/YL	A/T: Digital TR3 Sensor	In 'P': 0v, in O/D: 1.7v
64	LB/YL	M/T: CPP Switch Signal	5v (clutch "in": 0v)

1999-2002 F-Series 6.8L V10 VIN S (All) 104 Pin Connector

PCM Pin #	Wire Color	Circuit Description (104 Pin)	Value at Hot Idle
65 ('99-'00)	BR/LG	DPFE Sensor Signal	0.20-1.30v
65 ('01-'02)	BR/LG	DPFE Sensor Signal	0.95-1.05v
66	YL/LG	CHT Sensor Signal	0.6 (194ºF)
67	VT/WT	EVAP Canister Vent Solenoid Control	0-10 Hz (0-100%)
68	GY/BK	Injector 9 Control	4.2-4.6 ms
69 ('00-'02)	PK/YL	A/C Clutch Relay Control	A/C On: 12v, Off: 0v
71, 97	RD	Vehicle Power	12-14v
72	TN/RD	Injector 7 Control	4.2-4.6 ms
73	TN/BK	Injector 5 Control	4.2-4.6 ms
74	BR/YL	Injector 3 Control	4.2-4.6 ms
75	TN	Injector 1 Control	4.2-4.6 ms
76, 77, 103	BK/WT	Power Ground	<0.1v
78	PK/LB	COP 2 Driver (dwell)	5º, 55 mph: 9º
79	WT/RD	COP 8 Driver (dwell)	5º, 55 mph: 9º
80	LB/OR	Fuel Pump Relay Control	Off: 12v, On: 1v
81	WT/YL	EPC Solenoid Control	9.2v (5 psi)
82	DG/PK	COP 9 Driver (dwell)	5º, 55 mph: 9º
82 ('00-'02)	WT/RD	COP 9 Driver (dwell)	5º, 55 mph: 9º
83	WT/LB	IAC Solenoid Control	10.7v (33%)
84	DB/YL	Output Shaft Speed Sensor	0 Hz, 55 mph: 470 Hz
85	DG	CMP Sensor Signal	10 Hz, 55 mph: 16 Hz
86	---	Not Used	---
87	RD/BK	HO2S-21 (B2 S1) Signal	0.1-1.1v
88	LB/RD	MAF Sensor Signal	0.8v, 55 mph: 1.6v
89	GY/WT	TP Sensor Signal	0.53-1.27v
90	BR/WT	Reference Voltage	4.9-5.1v
91	GY/RD	Analog Signal Return	<0.050v
92	RD/LG	Brake Pedal Switch	0v (Brake On: 12v)
93	RD/WT	HO2S-11 (B1 S1) Heater	1v (Heater Off: 12v)
94	YL/LB	HO2S-21 (B2 S1) Heater	1v (Heater Off: 12v)
95	WT/BK	HO2S-12 (B1 S2) Heater	1v (Heater Off: 12v)
98	LB	Injector 8 Control	4.2-4.6 ms
99	LG/OR	Injector 6 Control	4.2-4.6 ms
100	BR/LB	Injector 4 Control	4.2-4.6 ms
101	WT	Injector 2 Control	4.2-4.6 ms
102	YL/BK	COP 4 Driver (dwell)	5º, 55 mph: 9º
104	PK/WT	COP 3 Driver (dwell)	5º, 55 mph: 9º

Pin Connector Graphic

Terminal View of 104-Pin PCM Wiring Harness Connector

Standard Colors and Abbreviations

Abbreviation	Color	Abbreviation	Color	Abbreviation	Color
BK	Black	GY	Gray	PK	Purple
BL	Blue	GN	Green	RD	Red
BR	Brown	LG	LT Green	TN	Tan
DB	Dark Blue	OR	Orange	WT	White
DG	DK Green	PK	Pink	YL	Yellow

2003 F-Series 6.8L V10 VIN S (All) 104 Pin Connector

PCM Pin #	Wire Color	Circuit Description (104 Pin)	Value at Hot Idle
1	OR/YL	COP 6 Driver (dwell)	5°, 55 mph: 9°
2	---	Not Used	---
3	BK/WT	Power Ground	<0.1v
4	LB/YL	Power Take-Off (if equipped)	0v (Off)
5	---	Not Used	---
6	OR/YL	Shift Solenoid 1 Control	1v, 55 mph: 12v
7-10	---	Not Used	---
11	PK/OR	Shift Solenoid 2 Control	12v, 55 mph: 12v
12	WT/LG	TCIL (lamp) Control	TCIL On: 1v, Off: 2.2v
13	VT	Flash EEPROM Power	0.1v
14	LB/BK	4x4 Low Indicator Switch	12v (switch closed: 0v)
15	PK/LB	SCP Data Bus (-) Signal	<0.050v
16	TN/OR	SCP Data Bus (+) Signal	Digital Signals
17-19	---	Not Used	---
20	BR/OR	Coast Clutch Solenoid Control	12v, 55 mph: 12v
21	DB	CKP (-) Sensor Signal	400-420 Hz
22	GY	CKP (+) Sensor Signal	400-420 Hz
23-24	BK/WT	Power Ground	<0.1v
25	LB/YL	Case Ground	<0.050v
26	LG/WT	COP 1 Driver (dwell)	5°, 55 mph: 9°
27	LG/YL	COP 10 Driver (dwell)	5°, 55 mph: 9°
28	---	Not Used	---
29	TN/WT	Overdrive Cancel Switch) Signal	Switch Off: 12v, On: 1v
30-31	---	Not Used	---
32	DG/VT	Knock Sensor 1 (-) Signal	<0.050v
33	PK/OR	Ground	<0.050v
34	YL/BK	Digital TR1 Sensor	0v, 55 mph: 11.5v
35	RD/LG	HO2S-12 (B1 S2) Signal	0.1-1.1v
36	TN/LB	MAF Sensor Return	<0.050v
37	OR/BK	TFT Sensor Signal	2.10-2.40v
38	---	Not Used	---
39	GY	IAT Sensor Signal	1.5-2.5v
40	DG/YL	Fuel Pump Monitor	Off: 12v, On: 1v
41	TN/LG	A/C Switch Signal	A/C On: 12v, Off: 0v
42	GY/RD	Injector 10 Control	4.2-4.6 ms
43	OR/LG	Data Output Link	Digital Signals
44-46	---	Not Used	---
47	BR/PK	VR Solenoid Control	0%, 55 mph: 45%
48	LG/WT	Clean Tachometer Output	60-70 Hz
49	LB/BK	Digital TR2 Sensor	0v, 55 mph: 11.5v
50	WT/BK	Digital TR4 Sensor	0v, 55 mph: 11.5v
51	BK/WT	Power Ground	<0.1v
52	WT/PK	COP 5 Driver (dwell)	5°, 55 mph: 9°
53	DGVT	COP 7 Driver (dwell)	5°, 55 mph: 9°
54	VT/YL	TCC Solenoid Control	0%, 55 mph: 95%
55	RD/WT	Keep Alive Power	12-14v
56	LG/BK	EVAP Purge Solenoid	0-10 Hz (0-100%)
57	YL/RD	Knock Sensor 1 (+) Signal	0v
58	GY/BK	VSS (+) Signal	0 Hz, 55 mph: 125 Hz
59	DG/WT	TSS Sensor Signal	300 Hz, 55 mph: 980
60	GY/LB	HO2S-11 (B1 S1) Signal	0.1-1.1v
61	---	Not Used	---
62	RD/PK	FTP Sensor Signal	2.6v at 0" H20 (cap off)
63	---	Not Used	---
64	LB/YL	A/T: Digital TR3 Sensor	In 'P': 0v, in O/D: 1.7v
64	LB/YL	M/T: CPP Switch Signal	5v (clutch "in": 0v)

2003 F-Series 6.8L V10 VIN S (All) 104 Pin Connector

PCM Pin #	Wire Color	Circuit Description (104 Pin)	Value at Hot Idle
65	BR/LG	DPFE Sensor Signal (California)	0.95-1.05v
66	YL/LG	Cylinder Head Temperature Sensor	0.6 (194ºF)
67	VT/WT	EVAP Canister Vent Solenoid Control	0-10 Hz (0-100%)
68	GY/BK	Injector 9 Control	4.2-4.6 ms
69	PK/YL	A/C Clutch Relay Control	A/C On: 12v, Off: 0v
70	---	Not Used	---
71	RD	Vehicle Power	12-14v
72	TN/RD	Injector 7 Control	4.2-4.6 ms
73	TN/BK	Injector 5 Control	4.2-4.6 ms
74	BR/YL	Injector 3 Control	4.2-4.6 ms
75	TN	Injector 1 Control	4.2-4.6 ms
76, 77	BK/WT	Power Ground	<0.1v
78	PK/LB	COP 2 Driver (dwell)	5º, 55 mph: 9º
79	WT/RD	COP 8 Driver (dwell)	5º, 55 mph: 9º
80	LB/OR	Fuel Pump Relay Control	Off: 12v, On: 1v
81	WT/YL	EPC Solenoid Control	9.2v (5 psi)
82	WT/RD	COP 9 Driver (dwell)	5º, 55 mph: 9º
83	WT/LB	IAC Solenoid Control	10.7v (33%)
84	DB/YL	Output Shaft Speed Sensor	0 Hz, 55 mph: 470 Hz
85	DG	CMP Sensor Signal	10 Hz, 55 mph: 16 Hz
86	---	Not Used	---
87	RD/BK	HO2S-21 (B2 S1) Signal	0.1-1.1v
88	LB/RD	MAF Sensor Signal	0.8v, 55 mph: 1.6v
89	GY/WT	TP Sensor Signal	0.53-1.27v
90	BR/WT	Reference Voltage	4.9-5.1v
91	GY/RD	Sensor Return	<0.050v
92	RD/LG	Brake Pedal Switch	0v (Brake On: 12v)
93	RD/WT	HO2S-11 (B1 S1) Heater	1v (Heater Off: 12v)
94	YL/LB	HO2S-21 (B2 S1) Heater	1v (Heater Off: 12v)
95	WT/BK	HO2S-12 (B1 S2) Heater	1v (Heater Off: 12v)
96	---	Not Used	---
97	RD	Vehicle Power	12-14v
98	LB	Injector 8 Control	4.2-4.6 ms
99	LG/OR	Injector 6 Control	4.2-4.6 ms
100	BR/LB	Injector 4 Control	4.2-4.6 ms
101	W	Injector 2 Control	4.2-4.6 ms
102	YL/BK	COP 4 Driver (dwell)	5º, 55 mph: 9º
103	BK/WT	Power Ground	<0.1v
104	PK/WT	COP 3 Driver (dwell)	5º, 55 mph: 9º

Pin Connector Graphic

PCM 104-PIN CONNECTOR

```
 1 ●●●●●●●●●●●●●   ●●●●●●●●●●●●● 26
27 ●●●●●●●●●●●●●   ●●●●●●●●●●●●● 52
53 ●●●●●●●●●●●●●  ⬤ ●●●●●●●●●●●● 78
79 ●●●●●●●●●●●●●   ●●●●●●●●●●●●● 104
```

Terminal View of 104-Pin PCM Wiring Harness Connector

Standard Colors and Abbreviations

Abbreviation	Color	Abbreviation	Color	Abbreviation	Color
BK	Black	GY	Gray	PK	Purple
BL	Blue	GN	Green	RD	Red
BR	Brown	LG	LT Green	TN	Tan
DB	Dark Blue	OR	Orange	WT	White
DG	DK Green	PK	Pink	YL	Yellow

1996-97 F-Series 7.3L V8 Diesel VIN F 104 Pin Connector

PCM Pin #	Wire Color	Circuit Description (104 Pin)	Value at Hot Idle
1	PK/OR	Shift Solenoid 2 Control	1v, 55 mph: 12v
2	PK/LG	MIL (lamp) Control	MIL Off: 12v, On: 1v
4	PK/WT	Parking Brake Applied Switch	Parking Brake Applied: 0v
5	RD/OR	Idle Validation Switch Signal	Switch up: 0v, down: 12v
5	LG/RD	Parking Brake Switch (Cal)	Parking Brake Applied: 0v
6	OR/YL	Shift Solenoid 1 Control (Cal)	SS1 Off: 12v, On: 1v
8	GY	Glow Plug Monitor High (Cal)	Plugs Off: 0v, On: 12v
9	OR	Glow Plug Monitor R/S (Cal)	Plugs Off: 0v, On: 12v
10	RD/OR	Idle Validation Switch (Cal)	0v, off-12v
11	PK/OR	Shift Solenoid 2 Control (Cal)	SS2 Off: 12v, On: 1v
12	WT/LG	TCIL (lamp) Control (Cal)	7.7v (Switch On: 0v)
13	PK	Flash EPROM Power	0.1v
14	LG/BK	4x4 Low Indicator Switch	Switch On: 0v, Off: 12v
15	PK/LB	Data Bus (-) Signal	Digital Signals
16	TN/OR	Data Bus (+) Signal	Digital Signals
17	OR/BK	TR Sensor Signal 1 (Cal)	Manual 1 or 2: 10.7v
19	WT/PK	Tachometer Signal Cal)	6.5v / 130 Hz
21	DG	CMP Sensor Signal (Cal)	6 Hz, 55 mph: 15 Hz
24	YL/BK	Accelerator Pedal (-) Sensor	<0.050v
25	BK	Case Ground	<0.050v
27	OR/YL	Shift Solenoid 1 Control	0.1v, 35 mph: 12v
28	PK/YL	TCC Solenoid Control	0v, On 55 mph: 12v
29	TN/LB	Clutch Pedal Position Switch	Clutch In: 0v, Out: 12v
29	TN/WT	TCS (switch) Signal	TCS & O/D On: 12v
30	PK/LB	Exhaust Back Press. Sensor	0.80-0.95v
31	BK/YL	Brake Pedal Applied Switch	Brake Off: 12, On: 1v
33	PK/OR	VSS (-) Signal	0 Hz, 55 mph: 125 Hz
34	LG/BK	MAP Sensor Signal	159 Hz (sea level)
37	OR/BK	TFT Sensor Signal	2.10-2.40v
38	LG/RD	EOT Sensor Signal	0.36v at 230°F
39	GY	IAT Sensor Signal	1.5-2.5v
40	DG/OR	Speed Control Ground	<0.050v
41	DG/OR	A/C Clutch Signal	A/C On: 12v, Off: 0v
42	GY/RD	Exhaust Back Pressure	6-8v (when enabled)
48	GY/WT	Electronic Feedback Line	Digital Signals
49	DG	CMP (+) Sensor Signal	0.7v
50	WT/PK	Tachometer Signal from CMP	6.5v / 130 Hz
51	BK/WT	Power Ground	<0.1v
53	BR/OR	Coast Clutch Solenoid	CCS Off: 12v, On: 1v
55	YL	Keep Alive Power	12-14v
58	GY/BK	VSS (+) Signal	0 Hz, 55 mph: 125 Hz
61	LB/BK	Speed Control Cruise Signal	Varies 0-12v
64	LB/YL	A/T: TR Sensor Signal	In 'P': 0v, in O/D: 5v

1996-97 F-Series 7.3L V8 Diesel VIN F 104 Pin Connector

PCM Pin #	Wire Color	Circuit Description (104 Pin)	Value at Hot Idle
65	LB	CMP (-) Sensor Signal	<0.050v
70	WT/BK	IDM Relay Signal	IDM Off: 12v, On: 0v
71	RD	Vehicle Power	12-14v
76	BK/WT	Power Ground	<0.1v
77	BK/WT	Power Ground	<0.1v
79	WT/LG	TCIL (lamp) Control	7.7v (Switch On: 0v)
80	BK/PK	Glow Plug Lamp Control	7.7v (Switch On: 0v)
81	WT/YL	EPC Solenoid Control	7.5v (8 psi)
83	YL/RD	Injector Pressure Regulator	12v duty cycle signal
84	DB/LG	BARO Sensor Signal	At Sea Level: 4.64v
87	DB/LG	Injection Control Pressure	0.75v (Min: 0.83v startup)
89	GY/WT	APP Sensor Signal	0.5-0.9v
90	BR/WT	Reference Voltage	4.9-5.1v
91	GY/RD	Analog Signal Return	<0.050v
92	LG	Brake Pedal Switch	0v (Brake On: 12v)
95	BR/OR	Fuel Demand Command	49 Hz
96	YL/LB	CID Sensor Signal	5 Hz
97	RD	Vehicle Power	12-14v
101	PK/OR	Glow Plug Relay Control	Off: 12v, On: 1v
103	BK/WT	Power Ground	<0.1v

Pin Connector Graphic

Standard Colors and Abbreviations

Abbreviation	Color	Abbreviation	Color	Abbreviation	Color
BK	Black	GY	Gray	PK	Purple
BL	Blue	GN	Green	RD	Red
BR	Brown	LG	LT Green	TN	Tan
DB	Dark Blue	OR	Orange	WT	White
DG	DK Green	PK	Pink	YL	Yellow

1998-99 F-Series 7.3L V8 Diesel VIN F 104 Pin Connector

PCM Pin #	Wire Color	Circuit Description (104 Pin)	Value at Hot Idle
1, 3, 7, 9	---	Not Used	---
2	PK/LG	MIL (lamp) Control	MIL Off: 12v, On: 1v
4	LB/YL	Neutral Gear Switch	Switch On: 0v, Off: 12v
5	LG/RD	Brake Warning Indicator	Parking Brake Applied: 0v
6	OR/YL	Shift Solenoid 1 Control	SS1 Off: 12v, On: 1v
8	WT/LG	Glow Plug Monitor	Plugs Off: 0v, On: 12v
10	RD/LG	Idle Validation Switch	0v, off-12v
11	PK/OR	Shift Solenoid 2 Control	SS2 Off: 12v, On: 1v
12	WT/LG	TCIL (lamp) Control	7.7v (Switch On: 0v)
13	PK	Flash EPROM Power	0.1v
14	LG/BK	4x4 Low Indicator Switch	Switch On: 0v, Off: 12v
15	PK/LB	Data Bus (-) Signal	Digital Signals
16	TN/OR	Data Bus (+) Signal	Digital Signals
17	WT/YL	Digital TR1 Sensor	0v, 55 mph: 11.5v
18, 22-23	---	Not Used	---
19	WT/PK	Tachometer Reflected Signal	6.5v / 130 Hz
20	BR/OR	Coast Clutch Solenoid	12v, 55 mph: 12v
21	DG	CMP Sensor Signal	5-7 Hz
24	YL/WT	TP Sensor Signal	0.53-1.27v
25	LB/YL	Tachometer VSS Ground	<0.050v
26-27	---	Not Used	---
28	RD	Water In Fuel Indicator Lamp	7.7v (Switch On: 0v)
29	TN/WT	A/T: TCS (switch) Signal	TCS & O/D On: 12v
29	TN/LB	M/T: CPP (switch) Signal	Clutch In: 0v, Out: 12v
30	PK/LB	Exhaust Back Pressure	0.9v, off-idle: 2.5v
31	BK/YL	Brake Pedal Applied Switch	Brake Off: 12, On: 1v
32-34	---	Not Used	---
35	WT/YL	Generator Power Switch	12-14v
36	GY/RD	Fuel Heater Signal	Heater On: 12v
37	OR/BK	TFT Sensor Signal	2.10-2.40v
38	LG/RD	Engine Oil Temperature	-40ºF: 4.7v, 230ºF: 0.358v
39	GY	IAT Sensor Signal	1.5-2.5v
40	OR/YL	Fuel Pump Relay	Off: 12v, On: 1v
41	TN/LG	A/C Switch Signal	Switch On: 0v, Off: 12v
42	GY/RD	Exhaust Pressure Regulator	Digital: 0-12-0-12v
43	OR/LG	DLC Signal	Digital Signals
44-46, 52-53	---	Not Used	---
47	WT/RD	Wastegate Control Solenoid	Solenoid Off: 12v, On: 1v
48	GY/WT	Electronic Feedback Line	0.4-2.2v -12v digital signal
49	DB/WT	Digital TR2 Sensor	0v, 55 mph: 11.5v
50	DG/YL	Digital TR4 Sensor	0v, 55 mph: 11.5v
51, 77, 103	BK/WT	Power Ground	<0.1v
54	PK/YL	TCC Solenoid Control	Clutch Off: 12v, On: 1v
55	RD/WT	Keep Alive Power	12-14v
56-57, 68-69	---	Not Used	---
58	GY/BK	VSS (+) Signal	0 Hz, 55 mph: 125 Hz
59	DB/YL	OSS Sensor Signal	0 Hz, 55 mph: 470 Hz
60	YL	Alternator No. 2 (bottom) Motor	6-10v
61	LB/BK	Speed Control Cruise Signal	Signal varies: 0-12v
62	RD/YL	ACT Sensor Signal	1.5-2.5v
63	DB/LG	BARO Sensor Signal	4.71v (sea level)
64	LB/YL	Digital TR3 Sensor	0v, 55 mph: 11.5v

1998-99 F-Series 7.3L V8 Diesel VIN F 104 Pin Connector

PCM Pin #	Wire Color	Circuit Description (104 Pin)	Value at Hot Idle
65	LB	CMP (-) Sensor Signal	<0.050v
66	LB/YL	Take Off Power Input	N/A
67	LG/RD	Generator Indicator	Indicator Off: 12v, On: 1v
70	WT/BK	Glow Plug Lamp Control	7.7v (Switch On: 0v)
71, 97	RD	Vehicle Power	12-14v
72-76, 78	---	Not Used	---
79	LG/BK	MAP Sensor Signal	0.9v
80	WT/BK	IDM Enable	Off: 12v, On: 1v
81	WT/YL	EPC Solenoid Control	7.5v (8 psi)
82, 85-86	---	Not Used	---
83	YL/RD	Injector Pressure Regulator	12v (Duty Cycle)
84	DG/WT	TSS Sensor Signal	300 Hz, 55 mph: 980
87	DB/LG	Injection Control Pressure	0.75v (Min: 0.83v startup)
88, 93	---	Not Used	---
89	GY/WT	APP Sensor Signal	0.5-0.9v
90	BR/WT	Reference Voltage	4.9-5.1v
91	GY/RD	Analog Signal Return	<0.050v
92	RD/LG	Brake Pedal Switch	0v (Brake On: 12v)
94	LB/OR	Fuel Pump Relay	
95	BR/OR	Fuel Demand Command Signal	49 Hz
96	YL/LB	CMP (+) Sensor Signal	5 Hz
98-99	---	Not Used	---
100	YL/WT	Fuel Level Indicator Signal	1.7v (1/2 full)
102, 104	---	Not Used	---
101	PK/OR	Glow Plug Relay Control	Off: 12v, On: 1v

Pin Connector Graphic

PCM 104-PIN CONNECTOR

Terminal View of 104-Pin PCM Wiring Harness Connector

Standard Colors and Abbreviations

Abbreviation	Color	Abbreviation	Color	Abbreviation	Color
BK	Black	GY	Gray	PK	Purple
BL	Blue	GN	Green	RD	Red
BR	Brown	LG	LT Green	TN	Tan
DB	Dark Blue	OR	Orange	WT	White
DG	DK Green	PK	Pink	YL	Yellow

2000-02 F-Series 7.3L V8 Diesel VIN F 104 Pin Connector

PCM Pin #	Wire Color	Circuit Description (104 Pin)	Value at Hot Idle
1	---	Not Used	---
2	PK/LG	MIL (lamp) Control	MIL Off: 12v, On: 1v
3	RD/YL	Low Current Sensor Return	<0.050v
4	LB/YL	Neutral Gear Switch	Switch On: 0v, Off: 12v
5	LG/RD	Parking Brake Applied Switch	Parking Brake Applied: 0v
6	OR/YL	Shift Solenoid 1 Control	SS1 Off: 12v, On: 1v
7, 9	---	Not Used	---
8	WT/LG	Glow Plug Monitor Left Bank	Plugs Off: 0v, On: 12v
10	RD/LG	Idle Validation Switch	12v, off-idle: 0v
11	PK/OR	Shift Solenoid 2 Control	SS2 Off: 12v, On: 1v
12	WT/LG	TCIL (lamp) Control	7.7v (Switch On: 0v)
13	PK	Generic Scan Tool Input	0.1v
14	LB/BK	4x4 Low Switch Input	Switch On: 0v, Off: 12v
15	PK/LB	Data Bus (-)	<0.050v
16	TN/OR	Data Bus (+)	Digital Signals
17	WT/YL	Digital TR1 Sensor	In 'P': 0v, in Drive: 10.7v
18	---	Not Used	---
19	WT/PK	Tachometer Reflected Signal	6.5v / 130 Hz
20	BR/OR	Coast Clutch Solenoid	CCS On: 0v, Off: 12v
21	DG	CMP Sensor Signal	Digital Signal: 0-12-0v
22-23	---	Not Used	---
24	YL/WT	APP Sensor Ground	<0.050v
25	LB/YL	Speedometer Ground	<0.050v
26-27	---	Not Used	---
28	RD	Water In Fuel Indicator Lamp	7.7v (Switch On: 0v)
29	TN/WT	A/T: TCS (switch) Signal	TCS & O/D On: 12v
29	TN/LB	M/T: CCP Switch Signal	Clutch In: 0v, Out: 12v
30	PK/LB	Exhaust Back Pressure	0.9v, off-idle: 2.5v
31	BK/YL	Brake Pressure Switch	Brake Off: 12v, On: 1v
32-34	---	Not Used	---
35	WT/YL	Generator Power Switch	12-14v
36	GY/RD	Fuel Heater / Water in Fuel	Heater On: 12v
37	OR/BK	TFT Sensor Signal	2.10-2.40v
38	LG/RD	Engine Oil Temperature	0.3-4.7v
39	GY	IAT Sensor Signal	1.5-2.5v
40	PK/BK	Fuel Pump Power	12-14v
41	TN/LG	A/C Head Pressure Switch	0v or 12v
42	GY/RD	Exhaust Backpressure Signal	Digital: 0-12-0-12v
43	OR/LG	Data Output Link	Digital Signals
44	---	Not Used	---
45	OR/LB	Speed Control Indicator	0v
46	---	Not Used	---
47	WT/RD	Wastegate Control Solenoid	Solenoid Off: 12v, On: 1v
48	GY/WT	Electronic Feedback Line	Digital Signals
49	DB/WT	Digital TR2 Sensor	In 'P': 0v, in Drive: 10.7v
50	DG/YL	Digital TR4 Sensor	In 'P': 0v, in Drive: 10.7v
51	BK/WT	Power Ground	<0.1v
52-53	---	Not Used	---
54	PK/YL	TCC Solenoid Control	TCC Off: 12v, On: 1v
55	RD/WT	Keep Alive Power	12-14v
56-57	---	Not Used	---
58	GY/BK	VSS (+) Signal	0 Hz, 55 mph: 125 Hz
59	DB/YL	OSS Sensor Signal	Varying Signal
60	YL	Generator Power	12-14v
61	LB/BK	Speed Control On/Off Switch	Switch On: 12v, Off: 0v
62	RD/YL	MAT Sensor Signal	1.5-2.5v

2000-02 F-Series 7.3L V8 Diesel VIN F 104 Pin Connector

PCM Pin #	Wire Color	Circuit Description (104 Pin)	Value at Hot Idle
63	---	Not Used	---
64	LB/YL	Digital TR3 Sensor	In P: 4.5v, in Drive: 2.2v
65	LB	CMP Sensor Ground	<0.050v
66	LB/YL	Power Take-Off Signal	0v (Off)
67	LG/RD	Battery Direct	12-14v
68-69	---	Not Used	---
70	BK/PK	Glow Plug Lamp Control	7.7v (Switch On: 0v)
71	RD	Vehicle Power	12-14v
72-76	---	Not Used	---
77	BK/WT	Power Ground	<0.1v
78	---	Not Used	---
79	LG/BK	MAP Sensor Signal	1-3v
80	WT/BK	IDM Enable Relay Control	Off: 12v, On: 1v
81	WT/YL	EPC Solenoid Control	7.5v (8 psi)
82	---	Not Used	---
83	YL/RD	Injector Pressure Regulator	12v
84	DG/WT	TSS Sensor Signal	30 mph: 130 Hz
85-86	---	Not Used	---
87	DB/LG	Injection Control Pressure	0.75v (Min: 0.83v startup)
88	---	Not Used	---
89	GY/WT	APP Sensor Signal	0.5-1.6v
90	BR/WT	Reference Voltage	4.9-5.1v
91	GY/RD	Sensor Return Signal	<0.050v
92	RD/LG	Brake Pedal Switch	0v (Brake On: 12v)
93	---	Not Used	---
94	LB/OR	Fuel Pump Relay Output	12-14v
95	BR/OR	Fuel Delivery Control Signal	49 Hz
96	YL/LB	CID Sensor Signal	5 Hz
97	RD	Vehicle Power	12-14v
98	PK	PCM Signal to Relay	Digital Signals
99	---	Not Used	---
100	YL/WT	Fuel Level Indicator Signal	1.7v (1/2 full)
101	PK/OR	Glow Plug Relay Control	Off: 12v, On: 1v
102	---	Not Used	---
103	BK/WT	Power Ground	<0.1v
104	---	Not Used	---

Pin Connector Graphic

PCM 104-PIN CONNECTOR

Terminal View of 104-Pin PCM Wiring Harness Connector

Standard Colors and Abbreviations

Abbreviation	Color	Abbreviation	Color	Abbreviation	Color
BK	Black	GY	Gray	PK	Purple
BL	Blue	GN	Green	RD	Red
BR	Brown	LG	LT Green	TN	Tan
DB	Dark Blue	OR	Orange	WT	White
DG	DK Green	PK	Pink	YL	Yellow

2003 F-Series 7.3L V8 Diesel VIN F (All) 104 Pin Connector

PCM Pin #	Wire Color	Circuit Description (104 Pin)	Value at Hot Idle
1	---	Not Used	---
2	PK/LG	MIL (lamp) Control	MIL Off: 12v, On: 1v
3	RD/YL	Manifold Intake Air Heater Monitor	Off: 12v, On: 1v
4	---	Not Used	---
5	LG/RD	Parking Brake Applied Switch	Brake Off: 12v, Applied: 0v
6	OR/YL	Shift Solenoid 1 Control	SS1 Off: 12v, On: 1v
7	---	Not Used	---
8	WT/LG	Glow Plug Monitor Left Bank	Plugs Off: 0v, On: 12v
9	---	Not Used	---
10	RD/LG	Idle Validation Switch	12v, off-idle: 0v
11	VT/OG	Shift Solenoid 2 Control	SS2 Off: 12v, On: 1v
12	WT/LG	TCIL (lamp) Control	7.7v (Switch On: 0v)
13	VT	Flash EEPROM Power Supply	0.1v
14	---	Not Used	---
15	PK/LB	SCP Data Bus (-) Signal	<0.050v
16	TN/OR	SCP Data Bus (+) Signal	Digital Signals
17	OR/BK	Digital TR1 Sensor	In 'P': 0v, in Drive: 10.7v
18	PK/YL	A/C Relay Control	Relay Off: 12v, On: 1v
19	WT/PK	Tachometer Reflected Signal	6.5v / 130 Hz
20	BR/OR	Coast Clutch Solenoid Control	CCS On: 0v, Off: 12v
21	DG	CMP Sensor Signal	Digital Signal: 0-12-0v
22-23	---	Not Used	---
24	YL/WT	APP Sensor Ground	<0.050v
25	BK/LB	Case Ground	<0.050v
26-27	---	Not Used	---
28	RD	Water In Fuel Indicator Control	Indicator Off: 12v, On: 1v
29	TN/WT	A/T: TCS (switch) Signal	TCS & O/D On: 12v
29	TN/LB	M/T: CCP Switch Signal	Clutch In: 0v, Out: 12v
30	VT/LB	Exhaust Back Pressure Sensor	0.9v, off-idle: 2.5v
31	BK/YL	Brake Pressure Switch	Brake Off: 12v, On: 1v
32	---	Not Used	---
33	PK/OG	VSS (-) Signal	<0.050v
34	---	Not Used	---
35	WT/YL	Alternator No. 1 (Top) Monitor	6-10v
36	GY/RD	Fuel Heater / Water in Fuel	Heater On: 12v
37	OR/BK	TFT Sensor Signal	2.10-2.40v
38	LG/RD	Engine Oil Temperature Sensor	0.3-4.7v
39	GY	Intake Air Temperature Sensor	1.5-2.5v
40	PK/BK	Fuel Pump Monitor	Pump Off: 12v, On: 1v
41	TN/LG	A/C Head Pressure Switch	0v or 12v
42	GY/RD	Exhaust Backpressure Signal	Digital: 0-12-0-12v
43	OR/LG	Overhead Console Fuel Consumption	0-5-0v
44-46	---	Not Used	---
47	WT/RD	Wastegate Control Solenoid Control	Solenoid Off: 12v, On: 1v
48	GY/WT	Electronic Feedback Line	0.9-3.0v
49	WT/PK	Digital TR2 Sensor	In 'P': 0v, in Drive: 10.7v
50	GY/BK	Digital TR4 Sensor	In 'P': 0v, in Drive: 10.7v
51	BK/WT	Power Ground	<0.1v
52-53	---	Not Used	---
54	VT/YL	TCC Solenoid Control	TCC Off: 12v, On: 1v
55	RD/WT	Keep Alive Power	12-14v
56-57, 63	---	Not Used	---
58	GY/BK	VSS (+) Signal	0 Hz, 55 mph: 125 Hz
59	DB/YL	OSS Sensor Signal	AC pulse signals
60	YL	Alternator No. 2 (Bottom) Control	6-10v
61	LB/BK	Speed Control On/Off Switch	Switch On: 12v, Off: 0v
62	RD/YL	MAT Sensor Signal	1.5-2.5v

2003 F-Series 7.3L V8 Diesel VIN F 104 Pin Connector

PCM Pin #	Wire Color	Circuit Description (104 Pin)	Value at Hot Idle
63	---	Not Used	---
64	LB/YL	Digital TR3 Sensor	In P: 4.5v, in Drive: 2.2v
65	LB	CMP Sensor Ground	<0.050v
66	LB/YL	Power Take-Off Signal	0v (Off)
67	LG/RD	Battery Direct	12-14v
68-69	---	Not Used	---
70	BK/PK	Glow Plug Lamp Control	7.7v (Switch On: 0v)
71	RD	Vehicle Power	12-14v
72-75	---	Not Used	---
76	BK/WT	Power Ground	<0.1v
77	BK/WT	Power Ground	<0.1v
78	---	Not Used	---
79	LG/BK	MAP Sensor Signal	1-3v
80	WT/BK	IDM Enable Relay Control	Off: 12v, On: 1v
81	WT/YL	EPC Solenoid Control	7.5v (8 psi)
82	---	Not Used	---
83	YL/RD	Injector Pressure Regulator	12v
84	DG/WT	TSS Sensor Signal	30 mph: 130 Hz
85-86	---	Not Used	---
87	DB/LG	Injection Control Pressure	0.1-3.0v
88	---	Not Used	---
89	GY/WT	APP Sensor Signal	0.5-1.6v
90	BR/WT	Reference Voltage	4.9-5.1v
91	GY/RD	Sensor Ground	<0.050v
92	RD/LG	Brake Pedal Switch	0v (Brake On: 12v)
93	---	Not Used	---
94	LB/OR	Fuel Pump Relay Control	Relay Off: 12v, On: 1v
95	BR/OR	Fuel Delivery Control	40-240 Hz
96	YL/LB	CID Sensor Signal	5-30 Hz
97	RD	Vehicle Power	12-14v
98	VT	PCM Signal to Relay	12v
99	---	Not Used	---
100	YL/WT	Fuel Level Indicator Signal	1.7v (1/2 full)
101	VT/OR	Glow Plug Relay Control	Off: 12v, On: 1v
102	---	Not Used	---
103	BK/WT	Power Ground	<0.1v
104	---	Not Used	---

Pin Connector Graphic

```
         PCM 104-PIN CONNECTOR
  1 ●●●●●●●●●●●●●   ●●●●●●●●●●●●● 26
 27 ●●●●●●●●●●●●●   ●●●●●●●●●●●●● 52
 53 ●●●●●●●●●●●●●  ●  ●●●●●●●●●●●●● 78
 79 ●●●●●●●●●●●●●   ●●●●●●●●●●●●● 104
```

Terminal View of 104-Pin PCM Wiring Harness Connector

Standard Colors and Abbreviations

Abbreviation	Color	Abbreviation	Color	Abbreviation	Color
BK	Black	GY	Gray	PK	Purple
BL	Blue	GN	Green	RD	Red
BR	Brown	LG	LT Green	TN	Tan
DB	Dark Blue	OR	Orange	WT	White
DG	DK Green	PK	Pink	YL	Yellow

1990 F-Series 7.5L V8 VIN G (All) 60 Pin Connector

PCM Pin #	Wire Color	Circuit Description (60 Pin)	Value at Hot Idle
1	YL	Keep Alive Power	12-14v
3	GY/BK	VSS (+) Signal	0 Hz, 55 mph: 125 Hz
3	DG/WT	VSS (+) Signal	0 Hz, 55 mph: 125 Hz
4	DG/YL	IDM Sensor Signal	20-31 Hz
6	BK	VSS (-) Signal	0 Hz, 55 mph: 125 Hz
7	LG/YL	ECT Sensor Signal	0.5-0.6v
8	BR	Fuel Pump Monitor	On: 12v, Off: 0v
10	BK/YL	A/C Switch Signal	A/C On: 12v, Off: 0v
11	WT/BK	Air Management 2 Solenoid	AM2 Off: 12v, On: 1v
16	BK/OR	Ignition System Ground	<0.050v
17	PK/LG	Self Test Output & MIL	MIL Off: 12v, On: 1v
18	TN/LG	Inferred Mileage Sensor	Digital Signals
20	BK	PCM Case Ground	<0.050v
21	GY/WT	IAC Solenoid Control	8.0-10.0v
22	TN/LG	Fuel Pump Control	Off: 12v, On: 1v
23 ('89-'90)	LG/BK	Knock Sensor Signal	0v
24 ('89-'90)	YL/LG	PSP Switch Signal	0v (turning: 12v)
25	YL/RD	ACT Sensor Signal	1.5-2.5v
26	OR/WT	Reference Voltage	4.9-5.1v
27	BR/LG	EVP Sensor Signal	0.4v, 55 mph: 2.6v
29	DG/PK	HO2S-11 (B1 S1) Signal	0.1-1.1v
29	GY/BL	HO2S-11 (B1 S1) Signal	0.1-1.1v
30	GY/YL	A/T: Neutral Drive Switch	In 'P': 0v, Others: 5v
30	GY/YL	M/T: Clutch Engage Switch	5v (clutch "in": 0v)
31	GY/YL	Canister Purge Solenoid	12v, 55 mph: 1v
33	DG	VR Solenoid Control	0%, 55 mph: 45%
36	YL/LG	Spark Output Signal	6.93v (50%)
37	RD	Vehicle Power	12-14v
40	BK/LG	Power Ground	<0.1v
45	DG/LG	MAP Sensor Signal	107 Hz (sea level)
45	DB/LG	MAP Sensor Signal	107 Hz (sea level)
46	BK/WT	Analog Signal Return	<0.050v
47	DG/LG	TP Sensor Signal	1v, 55 mph: 1.4v
48	WT/RD	Self Test Indicator Signal	STI On: 1v, Off: 5v
49	OR	HO2S-11 (B1 S1) Ground	<0.050v
51	WT/RD	Air Management 1 Solenoid	AM1 Off: 12v, On: 1v
56	DB	PIP Sensor Signal	6.93v (50%)
57	RD	Vehicle Power	12-14v
58	TN/OR	Injector Bank 1 (INJ 1, 4, 5, 8)	5.7-7.0 ms
59	TN/RD	Injector Bank 2 (INJ 2, 3, 6, 7)	5.7-7.0 ms
60	BK/LG	Power Ground	<0.1v

Pin Connector Graphic

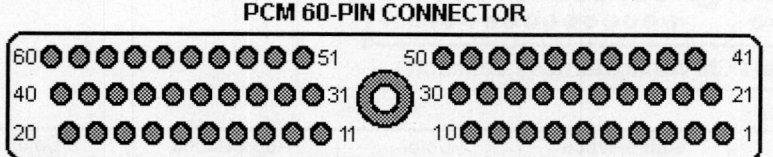

1990 F-Series 7.5L V8 VIN G (E4OD) 60 Pin Connector

PCM Pin #	Wire Color	Circuit Description (60 Pin)	Value at Hot Idle
1	YL	Keep Alive Power	12-14v
2	LG	Brake Position Switch	0v (Brake On: 12v)
3	GY, DG	VSS (+) Signal	0 Hz, 55 mph: 125 Hz
4	DG/YL	IDM Sensor Signal	20-31 Hz
6	BK	VSS (-) Signal	0 Hz, 55 mph: 125 Hz
7	LG/YL	ECT Sensor Signal	0.5-0.6v
8	BR	Fuel Pump Monitor	On: 12v, Off: 0v
10	BK/YL	A/C Switch Signal	A/C On: 12v, Off: 0v
11 ('89)	WT/BK	Air Management 2 Solenoid	AM2 Off: 12v, On: 1v
12	BL/BK	4x4 Indicator Light	4x4 Switch Closed: 1v
16	BK/OR	Ignition System Ground	<0.050v
17	PK/LG	Self Test Output & MIL	MIL Off: 12v, On: 1v
18 ('89)	TN/LG	Inferred Mileage Sensor	Digital Signals
19	DG/PK	Shift Solenoid Control	1v, 55 mph: 1v
20	BK	PCM Case Ground	<0.050v
21	GY/WT	IAC Solenoid Control	8.0-10.0v
22	TN/LG	Fuel Pump Control	Off: 12v, On: 1v
23 ('89)	LG/BK	Knock Sensor Signal	0v
24 ('89)	YL/LG	PSP Switch Signal	0v (turning: 12v)
25	YL/RD	ACT Sensor Signal	1.5-2.5v
26	OR/WT	Reference Voltage	4.9-5.1v
27	BR/LG	EVP Sensor Signal	0.4v, 55 mph: 2.6v
29	DG/PK	HO2S-11 (B1 S1) Signal	0.1-1.1v
29	GY/BL	HO2S-11 (B1 S1) Signal	0.1-1.1v
30	BL/WT	MLP Sensor Signal	In 'P': 0v, in O/D: 5v
30	BL/YL	MLP Sensor Signal	In 'P': 0v, in O/D: 5v
31	GY/YL	Canister Purge Solenoid	12v, 55 mph: 1v
32	LG/WT	OCIL (lamp) Control	7.7v (Switch On: 0v)
33	DG	VR Solenoid Control	0%, 55 mph: 45%
36	YL/LG	Spark Output Signal	6.93v (50%)
37, 57	RD	Vehicle Power	12-14v
38	BL/YL	EPC Solenoid Control	8.8v
40	BK/LG	Power Ground	<0.1v

1990 F-Series 7.5L V8 VIN G (E4OD) 60 Pin Connector

PCM Pin #	Wire Color	Circuit Description (60 Pin)	Value at Hot Idle
41	TN/WT	OCS (switch) Signal	12v (switch closed: 0v)
42	OR/BK	TOT Sensor Signal	2.10-2.40v
45	DB/LG	MAP Sensor Signal	107 Hz (sea level)
46	BK/WT	Analog Signal Return	<0.050v
47	DG/LG	TP Sensor Signal	1v, 55 mph: 1.4v
48	WT/RD	Self Test Indicator Signal	STI On: 1v, Off: 5v
49	OR	HO2S-11 (B1 S1) Ground	<0.050v
51	WT/RD	Air Management 1 Solenoid	Solenoid Off: 12v, On: 1v
52	OR/YL	Shift Solenoid 1 Control	1v, 55 mph: 12v
53	PK/YL	TCC Solenoid Control	0v, On 55 mph: 12v
55	BR	Coast Clutch Solenoid	12v, 55 mph: 12v
56	DB	PIP Sensor Signal	6.93v (50%)
58	TN/OR	Injector Bank 1 (INJ 1, 4, 5, 8)	5.7-7.0 ms
59	TN/RD	Injector Bank 2 (INJ 2, 3, 6, 7)	5.7-7.0 ms
40, 60	BK/LG	Power Ground	<0.1v

Pin Connector Graphic

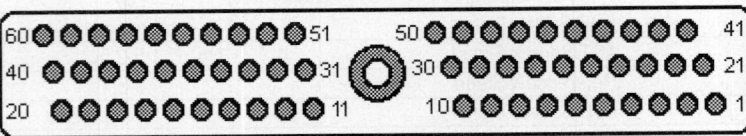

PCM 60-PIN CONNECTOR

Terminal View of 60-Pin PCM Harness Connector

Standard Colors and Abbreviations

Abbreviation	Color	Abbreviation	Color	Abbreviation	Color
BK	Black	GY	Gray	PK	Purple
BL	Blue	GN	Green	RD	Red
BR	Brown	LG	LT Green	TN	Tan
DB	Dark Blue	OR	Orange	WT	White
DG	DK Green	PK	Pink	YL	Yellow

1991 F-Series 7.5L V8 VIN G (All) 60 Pin Connector

PCM Pin #	Wire Color	Circuit Description (60 Pin)	Value at Hot Idle
1	YL	Keep Alive Power	12-14v
3	GY/BK	VSS (+) Signal	0 Hz, 55 mph: 125 Hz
4	TN/YL	IDM Sensor Signal	20-31 Hz
6	PK/OR	VSS (-) Signal	0 Hz, 55 mph: 125 Hz
7	LG/RD	ECT Sensor Signal	0.5-0.6v
8	DG/YL	Fuel Pump Monitor	On: 12v, Off: 0v
10	BK/YL	A/C Switch Signal	A/C On: 12v, Off: 0v
16	BK/OR	Ignition System Ground	<0.050v
17	PK/LG	Self Test Output & MIL	MIL Off: 12v, On: 1v
20	BK	PCM Case Ground	<0.050v
21	WT/BL	IAC Solenoid Control	8.0-10.5v
22	BL/OR	Fuel Pump Control	Off: 12v, On: 1v
25	GY	ACT Sensor Signal	1.5-2.5v
26	BR/WT	Reference Voltage	4.9-5.1v
27	BR/LG	EVP Sensor Signal	0.3v
29	GY/BL	HO2S-11 (B1 S1) Signal	0.1-1.1v
30	BL/YL	A/T: Neutral Drive Switch	In 'P': 0v, Others: 5v
30	BL/YL	M/T: Clutch Position Switch	Clutch In 0v, Out: 5v
31	GY/YL	Canister Purge Solenoid	12v, 55 mph: 1v
33	BR/PK	VR Solenoid Control	0%, 55 mph: 45%
36	PK	Spark Output Signal	6.93v (50%)
37	RD	Vehicle Power	12-14v
40	BK/WT	Power Ground	<0.1v
45	LG/BK	MAP Sensor Signal	107 Hz (sea level)
46	GY/RD	Analog Signal Return	<0.050v
47	GY/WT	TP Sensor Signal	1v, 55 mph: 1.4v
48	WT/PK	Self Test Indicator Signal	STI On: 1v, Off: 5v
49	OR	HO2S-11 (B1 S1) Ground	<0.050v
51	WT/RD	Air Management 1 Solenoid	AM1 Off: 12v, On: 1v
56	GY/OR	PIP Sensor Signal	6.93v (50%)
57	RD	Vehicle Power	12-14v
58	TN	Injector Bank 1 (INJ 1, 4, 5, 8)	5.6-6.2 ms
59	WT	Injector Bank 2 (INJ 2, 3, 6, 7)	5.6-6.2 ms
60	BK/LG	Power Ground	<0.1v

Pin Connector Graphic

PCM 60-PIN CONNECTOR

Terminal View of 60-Pin PCM Harness Connector

1991 F-Series 7.5L V8 MFI VIN G (E4OD) 60 Pin Connector

PCM Pin #	Wire Color	Circuit Description (60 Pin)	Value at Hot Idle
1	YL	Keep Alive Power	12-14v
2	LG	Brake Position Switch	0v (Brake On: 12v)
3	GY/BK	VSS (+) Signal	0 Hz, 55 mph: 125 Hz
4	TN/YL	IDM Sensor Signal	20-31 Hz
6	PK/OR	VSS (-) Signal	0 Hz, 55 mph: 125 Hz
7	LG/RD	ECT Sensor Signal	0.5-0.6v
8	DG/YL	Fuel Pump Monitor	On: 12v, Off: 0v
10	BK/YL	A/C Switch Signal	A/C On: 12v, Off: 0v
12	BL/BK	4x4 Indicator Light Signal	4x4 Switch Closed: 1v
16	BK/OR	Ignition System Ground	<0.050v
17	PK/LG	Self Test Output & MIL	MIL Off: 12v, On: 1v
19	PK/OR	Shift Solenoid 2 Control	1v, 55 mph: 1v
20	BK	PCM Case Ground	<0.050v
21	WT/BL	IAC Solenoid Control	8.0-10.5v
22	BL/OR	Fuel Pump Control	Off: 12v, On: 1v
25	GY	ACT Sensor Signal	1.5-2.5v
26	BR/WT	Reference Voltage	4.9-5.1v
27	BR/LG	EVP Sensor Signal	0.3v
29	GY/BL	HO2S-11 (B1 S1) Signal	0.1-1.1v
30	BL/YL	MLP Sensor Signal	In 'P': 4.4v, in OD: 2.1v
31	GY/YL	Canister Purge Solenoid	12v, 55 mph: 1v
32	WT/LG	OCIL (lamp) Control	7.7v (Switch On: 0v)
33	BR/PK	VR Solenoid Control	0%, 55 mph: 45%
36	PK	Spark Output Signal	6.93v (50%)
37	RD	Vehicle Power	12-14v
38	WT/YL	EPC Solenoid Control	9.5v (5 psi)
40	BK/WT	Power Ground	<0.1v
41	TN/WT	OCS (switch) Signal	12v (switch closed: 0v)
42	OR/BK	TOT Sensor Signal	2.10-2.40v
45	LG/BK	MAP Sensor Signal	107 Hz (sea level)
46	GY/RD	Analog Signal Return	<0.050v
47	GY/WT	TP Sensor Signal	1v, 55 mph: 1.4v
48	WT/PK	Self Test Indicator Signal	STI On: 1v, Off: 5v
49	OR	HO2S-11 (B1 S1) Ground	<0.050v
51	WT/RD	Air Management 1 Solenoid	AM1 Off: 12v, On: 1v
52	OR/YL	Shift Solenoid 1 Control	1v, 55 mph: 12v
53	PK/YL	TCC Solenoid Control	0v, On 55 mph: 12v
54	---	Not Used	---
55	BR/OR	Coast Clutch Solenoid	12v, 55 mph: 12v
56	GY/OR	PIP Sensor Signal	6.93v (50%)
57	RD	Vehicle Power	12-14v
58	TN	Injector Bank 1 (INJ 1, 4, 5, 8)	6.0-6.6 ms
59	WT	Injector Bank 2 (INJ 2, 3, 6, 7)	6.0-6.6 ms
60	BK/LG	Power Ground	<0.1v

Pin Connector Graphic

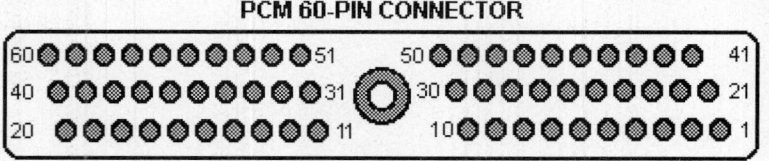

1992-95 F-Series 7.5L V8 VIN G (All) 60 Pin Connector

PCM Pin #	Wire Color	Circuit Description (60 Pin)	Value at Hot Idle
1	YL	Keep Alive Power	12-14v
2	LG	Brake Position Switch	0v (Brake On: 12v)
3	GY/BK	VSS (+) Signal	0 Hz, 55 mph: 125 Hz
4	YL/BK	IDM Sensor Signal	20-31 Hz
4	WT/PK	IDM Sensor Signal	20-31 Hz
4	TN/YL	IDM Sensor Signal	20-31 Hz
6	PK/OR	VSS (-) Signal	0 Hz, 55 mph: 125 Hz
7	LG/RD	ECT Sensor Signal	0.5-0.6v
8	DG/YL	Fuel Pump Monitor	On: 12v, Off: 0v
9	PK/BL	Data Bus (-) Signal	Digital Signals
10	PK/BL	A/C Switch Signal	A/C On: 12v, Off: 0v
11 ('93-'95)	BR	Air Management 2 Solenoid	AM2 Off: 12v, On: 1v
16	OR/RD	Ignition System Ground	<0.050v
17	PK/LG	Self Test Output & MIL	MIL Off: 12v, On: 1v
20	BK	PCM Case Ground	<0.050v
21	WT/BL	IAC Solenoid Control	8.0-10.0v
22 ('92-'93)	BL/OR	Fuel Pump Control	Off: 12v, On: 1v
23 ('92-'93)	YL/RD	Knock Sensor Signal	0v
25	GY	IAT Sensor Signal	1.5-2.5v
26	BR/WT	Reference Voltage	4.9-5.1v
27	BK/YL	EVP Sensor Signal	0.4v, 55 mph: 2.6v
27	BR/LG	EVP Sensor Signal	0.4v, 55 mph: 2.6v
28	TN/OR	Data Bus (+) Signal	Digital Signals
29	GY/BL	HO2S-11 (B1 S1) Signal	0.1-1.1v
30	GY/YL	A/T: Neutral Drive Switch	In 'P': 0v, Others: 5v
30	GY/YL	M/T: Clutch Engage Switch	5v (clutch "in": 0v)
31	GY/YL	Canister Purge Solenoid	1v, 55 mph: 12v
33	BR/PK	VR Solenoid Control	0%, 55 mph: 45%
36	PK	Spark Output Signal	6.93v (50%)
37	RD	Vehicle Power	12-14v
40	BK/WT	Power Ground	<0.1v
45	LG/BK	MAP Sensor Signal	107 Hz (sea level)
46	GY/RD	Analog Signal Return	<0.050v
47	GY/WT	TP Sensor Signal	1v, 55 mph: 1.4v
48	WT/PK	Self Test Indicator Signal	STI On: 1v, Off: 5v
49	OR	HO2S-11 (B1 S1) Ground	<0.050v
51	WT/OR	Air Management 1 Solenoid	AM1 Off: 12v, On: 1v
56	GY/OR	PIP Sensor Signal	6.93v (50%)
57	RD	Vehicle Power	12-14v
58	TN	Injector Bank 1 (INJ 1, 4, 5, 8)	5.0-5.8 ms
59	WT	Injector Bank 2 (INJ 2, 3, 6, 7)	5.0-5.8 ms
60	BK/WT	Power Ground	<0.1v

Pin Connector Graphic

PCM 60-PIN CONNECTOR

Terminal View of 60-Pin PCM Harness Connector

1992-95 F-Series 7.5L V8 VIN G (E4OD) 60 Pin Connector

PCM Pin #	Wire Color	Circuit Description (60 Pin)	Value at Hot Idle
1	YL	Keep Alive Power	12-14v
2	LG	Brake Position Switch	0v (Brake On: 12v)
3	GY/BK	VSS (+) Signal	0 Hz, 55 mph: 125 Hz
4	YL, WT	IDM Sensor Signal	20-31 Hz
6	PK/OR	VSS (-) Signal	0 Hz, 55 mph: 125 Hz
7	LG/RD	ECT Sensor Signal	0.5-0.6v
8	DG/YL	Fuel Pump Monitor	On: 12v, Off: 0v
9	PK/BL	Data Bus (-) Signal	Digital Signals
10	PK/BL	A/C Switch Signal	A/C On: 12v, Off: 0v
11 ('93-'95)	BR	Air Management 2 Solenoid	AM2 Off: 12v, On: 1v
12	BL/BK	4x4 Indicator Light Signal	12v (switch closed: 0v)
16	OR/RD	Ignition System Ground	<0.050v
17	PK/LG	Self Test Output & MIL	MIL Off: 12v, On: 1v
19	PK/OR	Shift Solenoid 2 Control	1v, 55 mph: 12v
20	BK	PCM Case Ground	<0.050v
21	WT/BL	IAC Solenoid Control	8.0-10.0v
22	BL/OR	Fuel Pump Control	Off: 12v, On: 1v
25	GY	IAT Sensor Signal	1.5-2.5v
26	BR/WT	Reference Voltage	4.9-5.1v
27	BK/YL	EVP Sensor Signal	0.4v, 55 mph: 2.6v
27	BR/LG	EVP Sensor Signal	0.4v, 55 mph: 2.6v
28	TN/OR	Data Bus (+) Signal	Digital Signals
29	GY/BL	HO2S-11 (B1 S1) Signal	0.1-1.1v
30	BL/YL	MLP Sensor Signal	In 'P': 0v, in O/D: 5v
31	GY/YL	Canister Purge Solenoid	12v, 55 mph: 1v
32	WT/LG	OCIL or TCIL (lamp) Control	7.7v (Switch On: 0v)
33	BR/PK	VR Solenoid Control	0%, 55 mph: 45%
36	PK	Spark Output Signal	6.93v (50%)
37, 57	RD	Vehicle Power	12-14v
38	WT/YL	EPC Solenoid Control	9.5v (5 psi)
40	BK/WT	Power Ground	<0.1v
41	TN/WT	OCS or TCS (switch) Signal	12v (switch closed: 0v)
42	OR/BK	TOT Sensor Signal	2.10-2.40v
45	LG/BK	MAP Sensor Signal	107 Hz (sea level)
46	GY/RD	Analog Signal Return	<0.050v
47	GY/WT	TP Sensor Signal	1v, 55 mph: 1.4v
48	WT/PK	Self Test Indicator Signal	STI On: 1v, Off: 5v
49	OR	HO2S-11 (B1 S1) Ground	<0.050v
51	WT/OR	Air Management 1 Solenoid	AM1 Off: 12v, On: 1v
52	OR/YL	Shift Solenoid 1 Control	12v, 55 mph: 1v
53	PK/YL	TCC Solenoid Control	0v, On 55 mph: 12v
55	BR/OR	Coast Clutch Solenoid	12v, 55 mph: 12v
56	GY/OR	PIP Sensor Signal	6.93v (50%)
58	TN	Injector Bank 1 (INJ 1, 4, 5, 8)	6.0-6.8 ms
59	WT	Injector Bank 2 (INJ 2, 3, 6, 7)	6.0-6.8 ms
60	BK/WT	Power Ground	<0.1v

Pin Connector Graphic

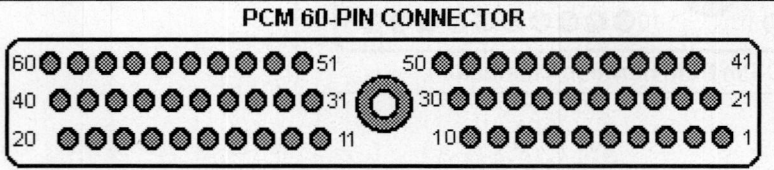

Terminal View of 60-Pin PCM Harness Connector

1996-97 F-Series 7.5L V8 MFI VIN G (All) 104 Pin

PCM Pin #	Wire Color	Circuit Description (104 Pin)	Value at Hot Idle
1	---	Not Used	---
2	PK/LG	MIL (lamp) Control	MIL Off: 12v, On: 1v
4	LG/RD	Power Take-Off (if equipped)	0v (Off)
5-12	---	Not Used	---
13	PK	Flash EPROM Power	0.1v
14	LG/BK	4x4 Low Switch	12v (switch closed: 0v)
15	PK/LB	Data Bus (-) Signal	Digital Signals
16	TN/OR	Data Bus (+) Signal	Digital Signals
17-22	---	Not Used	---
23	OR/RD	Ignition Ground	<0.050v
24	BK/WT	Power Ground	<0.1v
25	BK/LB	Case Ground	<0.050v
26-28	---	Not Used	---
29	TN/WT	TCS (switch) Signal	TCS & O/D On: 12v
30-32	---	Not Used	---
33	PK/OR	PSOM (-) Signal	<0.050v
34, 37	---	Not Used	---
35	RD/LG	HO2S-12 (B1 S2) Signal	0.1-1.1v
36	TN/LB	MAF Sensor Return	<0.050v
38	LG/RD	ECT Sensor Signal	0.5-0.6v
39	GY	IAT Sensor Signal	1.5-2.5v
40	DG/YL	Fuel Pump Monitor	On: 12v, Off: 0v
41	DG/OR	A/C Switch Signal	A/C On: 12v, Off: 0v
42-43, 46	---	Not Used	---
44	BR	Secondary AIR Diverter	AIRD Off: 12v, On: 1v
47	BK/PK	VR Solenoid Control	0%, 55 mph: 45%
48	YL/BK	Clean Tachometer Output	39-45 Hz
49	GY/OR	PIP Sensor Signal	6.93v (50%)
50	PK	Spark Output Signal	6.93v (50%)
51	BK/WT	Power Ground	<0.1v
52-54, 57	---	Not Used	---
55	YL	Keep Alive Power	12-14v
56	LG/BK	EVAP Purge Solenoid	0-10 Hz (0-100%)
58	GY/BK	PSOM (+) Signal	0 Hz, 55 mph: 125 Hz
59	DG/LG	Misfire Detection Sensor	45-55 Hz
60	GY/LB	HO2S-11 (B1 S1) Signal	0.1-1.1v
61-63	---	Not Used	---
64	LB/YL	A/T: PNP Switch Signal	In 'P': 0v, Others: 5v
64	LB/YL	M/T: CPP Switch Signal	5v (clutch "in": 0v)

1996-97 F-Series 7.5L V8 MFI VIN G (All) 104 Pin

PCM Pin #	Wire Color	Circuit Description (104 Pin)	Value at Hot Idle
65	BR/LG	DPFE Sensor Signal	0.95-1.05v
66-69	---	Not Used	---
70	WT/OR	Secondary AIR Bypass	AIRB Off: 12v, On: 1v
71	RD	Vehicle Power	12-14v
72	TN/RD	Injector 7 Control	4.3-4.6 ms
73	TN/BK	Injector 5 Control	4.3-4.6 ms
74	BR/YL	Injector 3 Control	4.3-4.6 ms
75	TN	Injector 1 Control	4.3-4.6 ms
76	BK/WT	Power Ground	<0.1v
77	BK/WT	Power Ground	<0.1v
78, 81-82	---	Not Used	---
79	WT/LG	TCIL (lamp) Control	7.7v (Switch On: 0v)
80	LB/OR	Fuel Pump Control	Off: 12v, On: 1v
83	WT/LB	IAC Solenoid Control	10.7v (33%)
84-86	---	Not Used	---
87	RD/BK	HO2S-21 (B2 S1) Signal	0.1-1.1v
88	LB/RD	MAF Sensor Signal	0.8v, 55 mph: 1.8v
89	GY/WT	TP Sensor Signal	1v, 55 mph: 1.6v
90	BR/WT	Reference Voltage	4.9-5.1v
91	GY/RD	Analog Signal Return	<0.050v
92	LG	Brake Pedal Switch	0v (Brake On: 12v)
93	RD/WT	HO2S-11 (B1 S1) Heater	1v (Heater Off: 12v)
94	YL/LB	HO2S-21 (B2 S1) Heater	1v (Heater Off: 12v)
95	WT/BK	HO2S-12 (B1 S2) Heater	1v (Heater Off: 12v)
96	---	Not Used	---
97	RD	Vehicle Power	12-14v
98	LG	Injector 8 Control	4.3-4.6 ms
99	LG/OR	Injector 6 Control	4.3-4.6 ms
100	BR/LB	Injector 4 Control	4.3-4.6 ms
101	W	Injector 2 Control	4.3-4.6 ms
102	---	Not Used	---
103	BK/WT	Power Ground	<0.1v
104	---	Not Used	---

Pin Connector Graphic

PCM 104-PIN CONNECTOR

Terminal View of 104-Pin PCM Wiring Harness Connector

Standard Colors and Abbreviations

Abbreviation	Color	Abbreviation	Color	Abbreviation	Color
BK	Black	GY	Gray	PK	Purple
BL	Blue	GN	Green	RD	Red
BR	Brown	LG	LT Green	TN	Tan
DB	Dark Blue	OR	Orange	WT	White
DG	DK Green	PK	Pink	YL	Yellow

1996-97 F-Series 7.5L V8 VIN G (E4OD) 104 Pin Connector

PCM Pin #	Wire Color	Circuit Description (104 Pin)	Value at Hot Idle
1	PK/OR	Shift Solenoid 2 Control	1v, 55 mph: 12v
2	PK/LG	MIL (lamp) Control	MIL Off: 12v, On: 1v
4	LG/RD	Power Take-Off (if equipped)	0v (Off)
5-12	---	Not Used	---
13	PK	Flash EPROM Power	0.1v
14	LG/BK	4x4 Low Switch Signal	12v (switch closed: 0v)
15	PK/LB	Data Bus (-) Signal	Digital Signals
16	TN/OR	Data Bus (+) Signal	Digital Signals
17-22	---	Not Used	---
23	OR/RD	Ignition Ground	<0.050v
24, 51	BK/WT	Power Ground	<0.1v
25	BK/LB	Case Ground	<0.050v
26, 28	---	Not Used	---
27	OR/YL	Shift Solenoid 1 Control	1v, 55 mph: 12v
29	TN/WT	TCS (switch) Signal	TCS & O/D On: 12v
30-32, 34	---	Not Used	---
33	PK/OR	PSOM (-) Signal	<0.050v
35	RD/LG	HO2S-12 (B1 S2) Signal	0.1-1.1v
36	TN/LB	MAF Sensor Return	<0.050v
37	OR/BK	TFT Sensor Signal	2.10-2.40v
38	LG/RD	ECT Sensor Signal	0.5-0.6v
39	GY	IAT Sensor Signal	1.5-2.5v
40	DG/YL	Fuel Pump Monitor	On: 12v, Off: 0v
41	DG/OR	A/C Switch Signal	A/C On: 12v, Off: 0v
42-43, 46	---	Not Used	---
44	BR	Secondary AIR Diverter	AIRD Off: 12v, On: 1v
47	BK/PK	VR Solenoid Control	0%, 55 mph: 45%
48	YL/BK	Clean Tachometer Output	39-45 Hz
49	GY/OR	PIP Sensor Signal	6.93v (50%)
50	PK	Spark Output Signal	6.93v (50%)
52, 57	---	Not Used	---
53	YL/BK	Coast Clutch Solenoid	12v, 55 mph: 12v
54	PK/YL	TCC Solenoid Control	0%, 55 mph: 95%
55	Y	Keep Alive Power	12-14v
56	LG/BK	EVAP Purge Solenoid	0-10 Hz (0-100%)
58	GY/BK	PSOM (+) Signal	0 Hz, 55 mph: 125 Hz
59	DG/LG	Misfire Detection Sensor	45-55 Hz
60	GY/LB	HO2S-11 (B1 S1) Signal	0.1-1.1v
61-63	---	Not Used	---
64	LB/YL	TR Sensor Signal	In 'P': 0v, in O/D: 5v

1996-97 F-Series 7.5L V8 VIN G (E4OD) 104 Pin Connector

PCM Pin #	Wire Color	Circuit Description (104 Pin)	Value at Hot Idle
65	BR/LG	DPFE Sensor Signal	0.95-1.05v
66-69	---	Not Used	---
70	WT/OR	Secondary AIR Bypass	AIRB Off: 12v, On: 1v
71	RD	Vehicle Power	12-14v
72	TN/RD	Injector 7 Control	4.3-4.6 ms
73	TN/BK	Injector 5 Control	4.3-4.6 ms
74	BR/YL	Injector 3 Control	4.3-4.6 ms
75	TN	Injector 1 Control	4.3-4.6 ms
76	BK/WT	Power Ground	<0.1v
77	BK/WT	Power Ground	<0.1v
78, 82	---	Not Used	---
79	WT/LG	TCIL (lamp) Control	7.7v (Switch On: 0v)
80	LB/OR	Fuel Pump Control	Off: 12v, On: 1v
81	WT/YL	EPC Solenoid Control	9.2v (5 psi)
83	WT/LB	IAC Solenoid Control	10.7v (33%)
84-86	---	Not Used	---
87	RD/BK	HO2S-21 (B2 S1) Signal	0.1-1.1v
88	LB/RD	MAF Sensor Signal	0.8v, 55 mph: 1.8v
89	GY/WT	TP Sensor Signal	0.53-1.27v
90	BR/WT	Reference Voltage	4.9-5.1v
91	GY/RD	Analog Signal Return	<0.050v
92	LG	Brake Pedal Switch	0v (Brake On: 12v)
93	RD/WT	HO2S-11 (B1 S1) Heater	1v (Heater Off: 12v)
94	YL/LB	HO2S-21 (B2 S1) Heater	1v (Heater Off: 12v)
95	WT/BK	HO2S-12 (B1 S2) Heater	1v (Heater Off: 12v)
96	---	Not Used	---
97	RD	Vehicle Power	12-14v
98	LG	Injector 8 Control	4.3-4.6 ms
99	LG/OR	Injector 6 Control	4.3-4.6 ms
100	BR/LB	Injector 4 Control	4.3-4.6 ms
101	WT	Injector 2 Control	4.3-4.6 ms
102	---	Not Used	---
103	BK/WT	Power Ground	<0.1v
104	---	Not Used	---

Pin Connector Graphic

```
         PCM 104-PIN CONNECTOR
 1 ●●●●●●●●●●●●    ●●●●●●●●●●●●● 26
27 ●●●●●●●●●●●●    ●●●●●●●●●●●●● 52
53 ●●●●●●●●●●●●   ⬢ ●●●●●●●●●●●● 78
79 ●●●●●●●●●●●●    ●●●●●●●●●●●● 104
   Terminal View of 104-Pin PCM Wiring Harness Connector
```

Standard Colors and Abbreviations

Abbreviation	Color	Abbreviation	Color	Abbreviation	Color
BK	Black	GY	Gray	PK	Purple
BL	Blue	GN	Green	RD	Red
BR	Brown	LG	LT Green	TN	Tan
DB	Dark Blue	OR	Orange	WT	White
DG	DK Green	PK	Pink	YL	Yellow

Lightning PIN Tables

1999-2003 Pickup 5.4L V8 VIN 3 (E4OD) 104 Pin Connector

PCM Pin #	Wire Color	Circuit Description (104 Pin)	Value at Hot Idle
1	OR/YL	COP 6 Driver (dwell)	5°, 55 mph: 8°
2	---	Not Used	---
3	BK/WT	Power Ground	<0.1v
4	PK	Transfer Case Speed Sensor (MSOF)	12v, 55 mph: 12v
5	---	Not Used	---
6	OR/YL	Shift Solenoid 1 Control	1v, 55 mph: 12v
7-10	---	Not Used	---
11	VT/OR	Shift Solenoid 2 Control	12v, 55 mph: 1v
12	WT/LG	TCIL (lamp) Control	7.7v (Switch On: 0v)
13	VT	Flash EEPROM Power	Digital Signals
14	LB/BK	4x4 Low Indicator Switch	12v (Switch On: 0v)
15	PK/LB	SCP Data Bus (-) Signal	<0.050v
16	TN/OR	SCP Data Bus (+) Signal	Digital Signals
17-18	---	Not Used	---
19	OR/LG	Fuel Pump High/Low Control	Off: 12v, On: 1v
20	BR/OR	Coast Clutch Solenoid	12v, 55 mph: 12v
21	DB	CKP (+) Sensor Signal	411 Hz
22	GY	CKP (-) Sensor Signal	411 Hz
23-24	---	Not Used	---
25	LB/YL	Case Ground	<0.050v
26	LG/WT	COP 1 Driver (dwell)	5°, 55 mph: 8°
27	LG/YL	COP 5 Driver (dwell)	5°, 55 mph: 8°
28	---	Not Used	---
29	TN/WT	TCS (switch) Signal	TCS & O/D On: 12v
30-31	---	Not Used	---
32	DG/PK	Knock Sensor (-) Signal	0v
33	---	Not Used	---
34	YL/BK	Digital TR1 Sensor	0v, 55 mph: 11.5v
35	RD/LG	HO2S-12 (B1 S2) Signal	0.1-1.1v
36	TN/LB	MAF Sensor Return	<0.050v
37	OR/BK	TFT Sensor Signal	2.10-2.40v
38	RD/YL	Intake Air Temperature 2 Sensor	1.5-2.5v
39	GY	Intake Air Temperature 1 Sensor	1.5-2.5v
40	RD	Fuel Pump Monitor	On: 12v, Off: 0v
41	BK/YL	A/C Switch Signal	Switch Closed: 12v
42	LG/VT	Supercharger Bypass Solenoid	Solenoid Off: 12v, On: 1v
43-45	---	Not Used	---
47	BR/PK	VR Solenoid Control	0%, 55 mph: 45%
48	---	Not Used	---
49	LB/BK	Digital TR2 Sensor	0v, 55 mph: 11.5v
50	WT/BK	Digital TR4 Sensor	0v, 55 mph: 11.5v
51	BK/WT	Power Ground	<0.1v
52	WT/PK	COP 3 Driver (dwell)	5°, 55 mph: 8°
53	DG/PK	COP 4 Driver Dwell	5°, 55 mph: 8°
54	PK/YL	TCC Solenoid Control	0%, 55 mph: 95%
55	RD/WT	Keep Alive Power	12-14v
56	LG/BK	EVAP Purge Solenoid	0-10 Hz (0-100%)
57	YL/RD	Knock Sensor (+) Signal	0v
58	---	Not Used	---
59	DG/WT	TSS (+) Sensor Signal	300 Hz, 55 mph: 980
60	GY/LB	HO2S-11 (B1 S1) Signal	0.1-1.1v
61	PK/LG	HO2S-22 (B2 S2) Signal	0.1-1.1v
62	RD/PK	FTP Sensor Signal	2.6v (0" H2O - cap off)
63	---	Not Used	---
64	LB/YL	Digital TR3 Sensor	In 'P': 0v, in O/D: 1.7v

1999-2003 Pickup 5.4L V8 VIN 3 (E4OD) 104 Pin Connector

PCM Pin #	Wire Color	Circuit Description (104 Pin)	Value at Hot Idle
65	BR/LG	EGR DPFE Sensor Signal	0.95-1.05v
66	YL/LG	Cylinder Head Temperature Sensor	0.6v (194°F)
67	PK/WT	EVAP CV Solenoid	0-10 Hz (0-100%)
68	GY/BK	VSS (+) Signal	0 Hz, 55 mph: 125 Hz
69	PK/YL	A/C WOT Relay Control	Off: 12v, On: 1v
70 ('03)	BK/WT	Check Fuel Cap Indicator Control	Indicator Off: 12v, On: 1v
71	RD	Vehicle Power	12-14v
72	WT	Injector 7 Control	2.7-3.5 ms
73	WT/LB	Injector 5 Control	2.7-3.5 ms
74	WT/BK	Injector 3 Control	2.7-3.5 ms
75	WT/RD	Injector 1 Control	2.7-3.5 ms
70	---	Not Used	---
77	BK/WT	Power Ground	<0.1v
78	PK/LB	COP 7 Driver Dwell	5°, 55 mph: 8°
79	WT/RD	COP 8 Driver Dwell	5°, 55 mph: 8°
80	LB/OR	Fuel Pump Relay Control	Off: 12v, On: 1v
81	WT/YL	EPC Solenoid Control	9.1v (4 psi)
82, 86	---	Not Used	---
83	WT/LB	IAC Solenoid Control	10.7v (33%)
84	DB/YL	OSS Sensor Signal	0 Hz, 55 mph: 700 Hz
85	DG	CMP Sensor Signal	7 Hz, 55 mph: 15 Hz
87	RD/BK	HO2S-21 (B2 S1) Signal	0.1-1.1v
88	LB/RD	MAF Sensor Signal	0.8v, 55 mph: 1.6v
89	GY/WT	TP Sensor Signal	0.53-1.27v
90	BR/WT	Reference Voltage	4.9-5.1v
91	GY/RD	Analog Signal Return	<0.050v
92	RD/LG	Brake Pedal Switch	0v (Brake On: 12v)
93	RD/WT	HO2S-11 (B1 S1) Heater	1v (Heater Off: 12v)
94	YL/LB	HO2S-21 (B2 S1) Heater	1v (Heater Off: 12v)
95	WT/BK	HO2S-12 (B1 S2) Heater	1v (Heater Off: 12v)
96	TN/YL	HO2S-22 (B2 S2) Heater	1v (Heater Off: 12v)
97	RD	Vehicle Power	12-14v
98	YL	Injector 8 Control	2.7-3.5 ms
99	YL/LB	Injector 6 Control	2.7-3.5 ms
100	YL/BK	Injector 4 Control	2.7-3.5 ms
101	YL/RD	Injector 2 Control	2.7-3.5 ms
102	---	Not Used	---
103	BK/WT	Power Ground	<0.1v
104	PK/WT	COP 2 Driver (dwell)	5°, 55 mph: 8°

Pin Connector Graphic

Terminal View of 104-Pin PCM Wiring Harness Connector

Standard Colors and Abbreviations

Abbreviation	Color	Abbreviation	Color	Abbreviation	Color
BK	Black	GY	Gray	PK	Purple
BL	Blue	GN	Green	RD	Red
BR	Brown	LG	LT Green	TN	Tan
DB	Dark Blue	OR	Orange	WT	White
DG	DK Green	PK	Pink	YL	Yellow

Ranger PIN Tables

1990 Pickup 2.3L I4 VIN A (All) 60 Pin Connector

PCM Pin #	Wire Color	Circuit Description (60 Pin)	Value at Hot Idle
1	YL/BK	Keep Alive Power	12-14v
2	LG	Brake Position Switch	0v (Brake On: 12v)
3	DG/WT	VSS (+) Signal	0 Hz, 55 mph: 125 Hz
4	BK/YL	IDM Sensor Signal	20-31 Hz
6	OR/YL	VSS (-) Signal	0 Hz, 55 mph: 125 Hz
7	LG/YL	ECT Sensor Signal	0.5-0.6v
8	OR/BL	Fuel Pump Monitor	On: 12v, Off: 0v
9	TN/OR	Data Bus (-) Signal (California)	Digital Signals
10	TN/YL	A/C Switch Signal	A/C On: 12v, Off: 0v
14	BK/LG	Mass Airflow Sensor (California)	0.6v
15	TN/BL	Mass Airflow Return (California)	<0.050v
16	BK/OR	Ignition System Ground	<0.050v
17	TN/RD	Self Test Output & MIL	MIL Off: 12v, On: 1v
18	WT/RD	Octane Adjust Sensor Signal	9.3v (shorted: 0v)
20	BK	PCM Case Ground	<0.050v
21	GY/WT	IAC Solenoid Control	10.6-11.0v
22	TN/LG	Fuel Pump Control	Off: 12v, On: 1v
24	YL/LG	PSP Switch Signal	0v (turning: 12v)
25	YL/RD	ACT Sensor Signal	1.5-2.5v
26	OR/WT	Reference Voltage	4.9-5.1v
27	BR/LG	EVP Sensor Signal	0.4v
28 (Cal)	TN/RD	Data Bus (+) Signal	Digital Signals
29	DG/PK	HO2S-11 (B1 S1) Signal	0.1-1.1v
30	WT/BK	A/T: Neutral Drive Switch	In 'P': 0v, Others: 5v
30	WT/BK	M/T: Clutch Engage Switch	5v (clutch "in": 0v)
32	GY/OR	Dual Plug Inhibit	0.1v
33	DG	VR Solenoid Control	0%, 55 mph: 45%
36	YL/LG	Spark Output Signal	6.93v (50% dwell)
37	RD	Vehicle Power	12-14v
40	BK/LG	Power Ground	<0.1v
43	LG/PK	A/C Demand Signal	A/C On: 12v
45	DG/BL	MAP Sensor Signal	107 Hz (sea level)
46	BK/WT	Analog Signal Return	<0.050v
47	DG/LG	TP Sensor Signal	1v, 55 mph: 1.4v
48	WT/RD	Self Test Input Signal	STI On: 1v, Off: 5v
49	OR	HO2S-11 (B1 S1) Ground	<0.050v
52	TN/BL	Shift Solenoid 3-4	Solenoid Off: 12v, On: 1v
53	WT	TCC Solenoid Control	0v, On 55 mph: 12v
54	PK	A/C WOT Cutout Control	Off: 12v, On: 1v
56	DB	PIP Sensor Signal	6.93v (50%)
57	RD	Vehicle Power	12-14v
58	TN	Injector Bank 1 (INJ 1 & 4)	3.7-4.4 ms
59	WT	Injector Bank 2 (INJ 2 & 3)	3.7-4.4 ms
60	BK/LG	Power Ground	<0.1v

Pin Connector Graphic

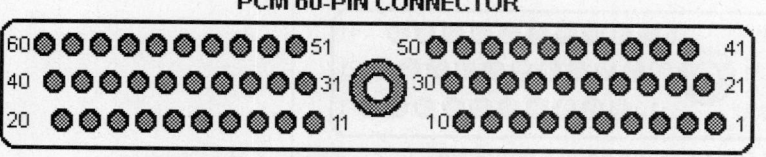

Terminal View of 60-Pin PCM Harness Connector

1991-94 Pickup 2.3L I4 VIN A (All) 60 Pin Connector

PCM Pin #	Wire Color	Circuit Description (60 Pin)	Value at Hot Idle
1	YL	Keep Alive Power	12-14v
2	LG	Brake Position Switch	0v (Brake On: 12v)
3	GY/BK	VSS (+) Signal	0 Hz, 55 mph: 125 Hz
4	TN/YL	IDM Sensor Signal	20-31 Hz
5 ('93-'94)	GY	CMP Sensor Signal	7 Hz, 55 mph: 15 Hz
6	PK/OR	VSS (-) Signal	0 Hz, 55 mph: 125 Hz
7	LG/RD	ECT Sensor Signal	0.5-0.6v
8	DG/YL	Fuel Pump Monitor	On: 12v, Off: 0v
9 ('92-'94)	BK/BL	Data Bus (-) Signal	Digital Signals
10	DG/OR	A/C Switch Signal	A/C On: 12v, Off: 0v
14	BL/RD	MAF Sensor Signal	0.7v, 55 mph: 1.9v
15	TN/BL	MAF Sensor Return	<0.050v
16	OR/RD	Ignition System Ground	<0.050v
17	PK/LG	Self Test Output & MIL	MIL Off: 12v, On: 1v
20	BK	PCM Case Ground	<0.050v
20	BK/LG	PCM Case Ground	<0.050v
21	WT/BL	IAC Solenoid Control	10.7v (33%)
22	BL/OR	Fuel Pump Control	Off: 12v, On: 1v
24	YL/LG	PSP Switch Signal	0v (turning: 12v)
25	GY	ACT Sensor Signal	1.5-2.5v
26	BR/WT	Reference Voltage	4.9-5.1v
27	BR/LG	EVP Sensor Signal	0.5v, 55 mph: 0.8v
28 ('92-'94)	TN/OR	Data Bus (+) Signal	Digital Signals
29	GY/BL	HO2S-11 (B1 S1) Signal	0.1-1.1v
30	BL/YL	A/T: Neutral Drive Switch	In 'P': 0v, Others: 5v
30	BL/YL	M/T: Clutch Engage Switch	5v (clutch "in": 0v)
32	DB/YL	Dual Plug Inhibit	0.1v
33	BR/PK	VR Solenoid Control	0%, 55 mph: 45%
36	PK	Spark Output Signal	6.93v (50% dwell)
37	RD	Vehicle Power	12-14v
40	BK/WT	Power Ground	<0.1v
43	PK	A/C Demand Signal	A/C On: 12v
44	DG	Octane Adjustment	9.3v (shorted: 0v)
46	GY/RD	Analog Signal Return	<0.050v
47	GY/WT	TP Sensor Signal	0.9v, 55 mph: 1.5v
48	WT/PK	Self Test Input Signal	STI On: 1v, Off: 5v
49	O, GY/RD	HO2S-11 (B1 S1) Ground	<0.050v
52	OR/YL	Shift Solenoid 3-4	Solenoid Off: 12v, On: 1v
53	PK/YL	TCC Solenoid Control	0v, On 55 mph: 12v
54	PK/YL	A/C WOT Cutout Control	Off: 12v, On: 1v
56	GY/OR	PIP Sensor Signal	6.93v (50%)
57	RD	Vehicle Power	12-14v
58	TN	Injector Bank 1 (INJ 1 & 4)	3.3-3.7 ms
59	WT	Injector Bank 2 (INJ 2 & 3)	3.3-3.7 ms
60	BK/WT	Power Ground	<0.1v

Pin Connector Graphic

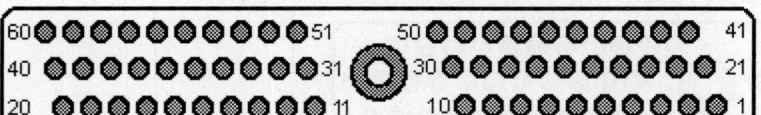

PCM 60-PIN CONNECTOR

Terminal View of 60-Pin PCM Harness Connector

1994 Pickup 2.3L VIN A California (All) 60 Pin Connector

PCM Pin #	Wire Color	Circuit Description (60 Pin)	Value at Hot Idle
1	YL	Keep Alive Power	12-14v
2	LG	Brake Position Switch	0v (Brake On: 12v)
3	GY/BK	VSS (+) Signal	0 Hz, 55 mph: 125 Hz
4	TN/YL	IDM Sensor Signal	20-31 Hz
6	PK/OR	VSS (-) Signal	0 Hz, 55 mph: 125 Hz
7	LG/RD	ECT Sensor Signal	0.5-0.6v
8	DG/YL	Fuel Pump Monitor	On: 12v, Off: 0v
9	TN/BL	MAF Sensor Return	<0.050v
10	DG/OR	A/C Switch Signal	A/C On: 12v, Off: 0v
16	OR/RD	Ignition System Ground	<0.050v
17	PK/LG	Self-Test Output & MIL	MIL Off: 12v, On: 1v
18	TN/OR	Data Bus (+) Signal	Digital Signals
19	PK/BL	Data Bus (-) Signal	Digital Signals
20	BK/LG	PCM Case Ground	<0.050v
21	WT/BL	IAC Solenoid Control	9.5-11.5v
22	BL/OR	Fuel Pump Control	Off: 12v, On: 1v
24	DB/OR	CMP Sensor Signal	7 Hz, 55 mph: 15 Hz
25	GY	IAT Sensor Signal	1.5-2.5v
26	BR/WT	Reference Voltage	4.9-5.1v
27	BR/LG	EVP Sensor Signal	0.4v, 55 mph: 1.3v
28	YL/LG	PSP Switch Signal	0v (turning: 12v)
29	DG	HO2S-11 (B1 S1) Signal	0.1-1.1v
30	BL/YL	A/T: Park Neutral Switch	In 'P': 0v, Others: 5v
30	BL/YL	M/T: Clutch Engage Switch	5v (clutch "in": 0v)
31	DB/YL	Dual Plug Inhibit Switch	0.1v
33	BR/PK	VR Solenoid Control	0%, 55 mph: 45%
35	BR/BK	Injector 4 Control	3.3-3.5 ms
36	PK	Spark Output Signal	6.93v (50% dwell)
37	RD	Vehicle Power	12-14v
39	BR/YL	Injector 3 Control	3.3-3.5 ms
40	BK/WT	Power Ground	<0.1v
44	GY/BK	Octane Adjustment	9.3v (shorted: 0v)
46	GY/RD	Analog Signal Return	<0.050v
47	GY/WT	TP Sensor Signal	1v, 55 mph: 1.4v
48	WT/PK	Self Test Input Signal	STI On: 1v, Off: 5v
50	BL/RD	MAF Sensor Signal	0.7v, 55 mph: 1.9v
52	OR/YL	Shift Solenoid 3-4	Solenoid Off: 12v, On: 1v
53	PK/YL	TCC Solenoid Control	0v, On 55 mph: 12v
54	PK/YL	A/C WOT Cutout Control	Off: 12v, On: 1v
56	GY/OR	PIP Sensor Signal	6.93v (50%)
57	RD	Vehicle Power	12-14v
58	TN	Injector 1 Control	3.3-3.5 ms
59	WT	Injector 2 Control	3.3-3.5 ms
60	BK/WT	Power Ground	<0.1v

Pin Connector Graphic

Terminal View of 60-Pin PCM Harness Connector

1995-97 Pickup 2.3L I4 VIN A (All) 104 Pin Connector

PCM Pin #	Wire Color	Circuit Description (104 Pin)	Value at Hot Idle
1	BK/WT	Shift Solenoid 2 Control	12v, 55 mph: 1v
2	PK/LG	MIL (lamp) Control	MIL Off: 12v, On: 1v
3-10, 12	---	Not Used	---
11	BL/LG	EVAP Purge Flow Sensor	0.8v, at 55 mph: 3v
13	PK	Flash EEPROM Power	0.1v
14	GY/BK	4x4 Low Switch Signal	12v (switch closed: 0v)
15	PK/LB	Data Bus (-) Signal	Digital Signals
16	TN/OR	Data Bus (+) Signal	Digital Signals
17-20, 23	---	Not Used	---
21	DB	CKP (+) Sensor Signal	390-450 Hz
22	GY	CKP (-) Sensor Signal	390-450 Hz
24, 51	BK/WT	Power Ground	<0.1v
25	BK	Case Ground	<0.050v
26	TN/WT	Coil 1 Driver (dwell)	6°, 55 mph: 9°
27	OR/YL	Shift Solenoid 1 Control	1v, 55 mph: 12v
28	BR/OR	Coast Clutch Solenoid	12v, 55 mph: 12v
29	TN/WT	TCS (switch) Signal	TCS & O/D On: 12v
30	DG	Octane Adjustment	9.3v (shorted: 0v)
31	YL/LG	PSP Switch Signal	0v (turning: 12v)
32, 34	---	Not Used	---
33	PK/OR	VSS (-) Signal	0 Hz, 55 mph: 125 Hz
35	RD/LG	HO2S-12 (B1 S2) Signal	0.1-1.1v
36	TN/LB	MAF Sensor Return	<0.050v
37	OR/BK	TFT Sensor Signal	2.10-2.40v
38	LG/RD	ECT Sensor Signal	0.5-0.6v
39	GY	IAT Sensor Signal	1.5-2.5v
40	DG/YL	Fuel Pump Monitor	On: 12v, Off: 0v
41	TN/YL	A/C Switch Signal	A/C On: 12v, Off: 0v
42-46	---	Not Used	---
47	BR/PK	VR Solenoid Control	0%, 55 mph: 45%
48	TN/YL	Clean Tachometer Output	25-38 Hz
49-50	---	Not Used	---
52	TN/OR	Coil 2 Driver (dwell)	5°, 55 mph: 8°
53	PK/BK	Shift Solenoid 3 Control	1v, 55 mph: 12v
54	PK/YL	TCC Solenoid Control	0%, 55 mph: 95%
55	YL	Keep Alive Power	12-14v
56-57	---	Not Used	---
58	GY/BK	VSS (+) Signal	0 Hz, 55 mph: 125 Hz
59	---	Not Used	---
60	GY/LB	HO2S-11 (B1 S1) Signal	0.1-1.1v
61-63	---	Not Used	---
64	LB/YL	A/T: TR Sensor Signal	In 'P': 0v, in O/D: 5v
64	LB/YL	M/T: CPP Switch Signal	5v (clutch "in": 0v)
65	BR/LG	DPFE Sensor Signal	0.20-1.30v
66	---	Not Used	---
67	GY/YL	EVAP Purge Solenoid	0-10 Hz (0-100%)

1995-97 Pickup 2.3L I4 VIN A (All) 104 Pin Connector

PCM Pin #	Wire Color	Circuit Description (104 Pin)	Value at Hot Idle
68	---	Not Used	---
69	PK/YL	A/C WOT Cutout Control	Off: 12v, On: 1v
70	---	Not Used	
71	RD	Vehicle Power	12-14v
72-73	---	Not Used	---
74	BR/YL	Injector 3 Control	4.0-4.5 ms
75	TN	Injector 1 Control	4.0-4.5 ms
76	BK/WT	Power Ground	<0.1v
77	BK/WT	Power Ground	<0.1v
78	TN/LB	Coil 3 Driver (dwell)	5°, 55 mph: 8°
79	WT/LG	TCIL (lamp) Control	7.7v (Switch On: 0v)
80	LB/OR	Fuel Pump Control	Off: 12v, On: 1v
81	WT/YL	EPC Solenoid Control	11v (24 psi)
82	---	Not Used	---
83	WT/LB	IAC Solenoid Control	10.7v (33%)
84	DG/WT	TSS Sensor Signal	105 Hz (775 rpm)
85	DB/OR	CMP Sensor Signal	7 Hz, 55 mph: 15 Hz
86-87	---	Not Used	---
88	LB/RD	MAF Sensor Signal	0.8v, 55 mph: 1.6v
89	GY/WT	TP Sensor Signal	1v, 55 mph: 1.7v
90	BR/WT	Reference Voltage	4.9-5.1v
91	GY/RD	Analog Signal Return	<0.050v
92	LG	Brake Pedal Switch	0v (Brake On: 12v)
93	RD/WT	HO2S-11 (B1 S1) Heater	1v (Heater Off: 12v)
94	---	---	---
95	WT/BK	HO2S-12 (B1 S2) Heater	1v (Heater Off: 12v)
96	---	---	---
97	RD	Vehicle Power	12-14v
98-99	---	Not Used	---
100	BR/LB	Injector 4 Control	4.0-4.5 ms
101	WT	Injector 2 Control	4.0-4.5 ms
102	---	Not Used	---
103	BK/WT	Power Ground	<0.1v
104	TN/LG	Coil 3 Driver (dwell)	5°, 55 mph: 8°

Pin Connector Graphic

PCM 104-PIN CONNECTOR

Terminal View of 104-Pin PCM Wiring Harness Connector

Standard Colors and Abbreviations

Abbreviation	Color	Abbreviation	Color	Abbreviation	Color
BK	Black	GY	Gray	PK	Purple
BL	Blue	GN	Green	RD	Red
BR	Brown	LG	LT Green	TN	Tan
DB	Dark Blue	OR	Orange	WT	White
DG	DK Green	PK	Pink	YL	Yellow

2001-03 Pickup 2.3L VIN D (All) 104 Pin Connector

PCM Pin #	Wire Color	Circuit Description (104 Pin)	Value at Hot Idle
1	VT/OR	Shift Solenoid 'B' Control	1v, 55 mph: 12v
2	PK/LG	MIL (lamp) Control	MIL Off: 12v, On: 1v
3	YL/BK	Digital TR1 Sensor	0v, 55 mph: 11.5v
4-5, 7	---	Not Used	---
6	DG/WT	TSS Sensor Signal	0 Hz, 55 mph: 385 Hz
8	WT	Swirl Control Motor Signal	Digital Signals
9-11	---	Not Used	---
12	YL/WT	Fuel Level Indicator Signal	1.7v (1/2 full)
13	VT	Flash EEPROM Power	0.1v
14	LB/BK	4WD Indicator Low Signal	12v (switch closed: 0v)
15	PK/LB	SCP Data Bus (-) Signal	<0.050v
16	TN/OR	SCP Data Bus (+) Signal	Digital Signals
17	GY/OR	Passive Antitheft RX Signal	Digital Signals
18	WT/LG	Passive Antitheft TX Signal	Digital Signals
19	BK/PK	ECT Sensor Signal	0.5-0.6v
20, 23	---	Not Used	---
21	DB	CKP (+) Sensor Signal	440-490 Hz
22	GY	CKP (-) Sensor Signal	440-490 Hz
24	BK/WT	Power Ground	<0.1v
25	BK	Chassis Ground	<0.050v
26	TN/WT	Coil 1 Driver (dwell)	5°, 55 mph: 8°
27	OR/YL	Shift Solenoid 'A' Control	1v, 55 mph: 12v
28	BR/OR	Coast Clutch Solenoid Control	1v, 55mph: 12v
29	TN/WT	TCS (switch) Signal	TCS & O/D On: 12v
30	DB/LG	Anti-Theft Indicator Signal	Digital Signals
31	YL/LG	Power Steering Pressure Switch	0v (turning: 12v)
32	YL	Knock Sensor Signal	0v
33-34	---	Not Used	---
35	RD/LG	HO2S-12 (B1 S2) Signal	0.1-1.1v
36	TN/LB	MAF Sensor Return	<0.050v
37	OR/BK	TFT Sensor Signal	2.10-2.40v
38	---	Not Used	
39	GY	IAT Sensor Signal	1.5-2.5v
40	DG/YL	Fuel Pump Monitor	On: 12v, Off: 0v
41	VT	A/C Demand Signal	A/C On: 12v, Off: 0v
42	WT/OR	Swirl Control Motor Signal	Digital Signals
43	---	Not Used	---
44	DB/OR	Starter Relay Control Circuit	Relay Off: 12v, On: 1v
45	OR/WT	Engine Cooling Fan Control	Fan Off: 12v, On: 1v
46	DB	Electric Thermostat to PCM	0.1-4.9v
47	---	Not Used	---
48	TN/YL	Clean Tachometer Output	25-38 Hz
49	LB/BK	Digital TR2 Sensor	0v, 55 mph: 11.5v
50	WT/BK	Digital TR4 Sensor	0v, 55 mph: 11.5v
51	BK/WT	Power Ground	<0.1v
52	TN/OR	Coil 2 Driver (dwell)	5°, 55 mph: 8°
53	PK/BK	Shift Solenoid 'C' Control	1v, 55 mph: 12v
54	VT/YL	TCC Solenoid Control	0%, 55 mph: 95%
55	YL	Battery Power	12-14v
56	LG/BK	EVAP Purge Control Valve	0-10 Hz (0-100%)
57	YL/RD	Knock Sensor Signal	0V
58	---	Not Used	---
59	GY/OR	ISS Sensor Signal	0 Hz, 55 mph: 1150 Hz
60	GY/LB	HO2S-11 (B1 S1) Signal	0.1-1.1v
61	---	Not Used	---
62	RD/PK	FTP Sensor Signal	2.6v (1/2 full)
63	LG/BK	MAP Sensor Signal	1-2v

2001-02 Pickup 2.3L VIN D (All) 104 Pin Connector

PCM Pin #	Wire Color	Circuit Description (104 Pin)	Value at Hot Idle
64	LB/YL	A/T: TR Sensor Signal	In 'P': 0v, in O/D: 5v
64	LB/YL	M/T: CPP Switch Signal	5v (clutch "in": 0v)
65	---	Not Used	---
66	YL/LG	CHT Sensor Signal	0.6v (194°F)
67	VT/WT	EVAP Canister Vent Solenoid	0-10 Hz (0-100%)
68	GY/BK	VSS (+) Signal	0 Hz, 55 mph: 125 Hz
69	PK/YL	A/C Clutch Relay Control	Off: 12v, On: 1v
70	---	Not Used	---
71	RD	Vehicle Power	12-14v
72	GY/RD	EGR Stepper Motor (B1) Control	Digital Signals
73	VT/OR	EGR Stepper Motor (A1) Control	Digital Signals
74	BR/YL	Injector 3 Control	2.7-3.5 ms
75	TN	Injector 1 Control	2.7-3.5 ms
76, 77	BK/WT	Power Ground	<0.1v
78	---	Not Used	---
79	WT/LG	TCIL (lamp) Control	7.7v (Switch On: 0v)
80	LB/OR	Fuel Pump Relay Control	Off: 12v, On: 1v
81	WT/YL	EPC Solenoid Control	9.1v (4 psi)
82	BK/WT	Check Gas Cap Indicator Control	Indicator Off: 12v, On: 1v
83	WT/LB	IAC Solenoid Control	10.7v (33%)
84	DB/YL	OSS Sensor Signal	0 Hz, 55 mph: 385 Hz
85	DB/OR	CMP Sensor Signal	7 Hz, 55 mph: 15 Hz
86	BK/YL	A/C Switch Signal	A/C On: 12v, Off: 0v
87	---	Not Used	---
88	LB/RD	MAF Sensor Signal	0.8v, 55 mph: 1.6v
89	GY/WT	TP Sensor Signal	0.53-1.27v
90	BR/WT	Reference Voltage	4.9-5.1v
91	GY/RD	Sensor Return	<0.050v
92	RD/LG	Brake Pedal Switch	0v (Brake On: 12v)
93	RD/WT	HO2S-11 (B1 S1) Heater	1v (Heater Off: 12v)
94	---	Not Used	---
95	WT/BK	HO2S-12 (B1 S2) Heater	1v (Heater Off: 12v)
96	---	Not Used	---
97	RD	Vehicle Power	12-14v
98	TN/RD	EGR Stepper Motor (B2) Control	Digital Signals
99	DG	EGR Stepper Motor (A2) Control	Digital Signals
100	BR/LB	Injector 4 Control	2.7-3.5 ms
101	WT	Injector 2 Control	2.7-3.5 ms
102	---	Not Used	---
103	BK/WT	Power Ground	<0.1v
104	---	Not Used	---

Pin Connector Graphic

PCM 104-PIN CONNECTOR

Terminal View of 104-Pin PCM Wiring Harness Connector

Standard Colors and Abbreviations

Abbreviation	Color	Abbreviation	Color	Abbreviation	Color
BK	Black	GY	Gray	PK	Purple
BL	Blue	GN	Green	RD	Red
BR	Brown	LG	LT Green	TN	Tan
DB	Dark Blue	OR	Orange	WT	White
DG	DK Green	PK	Pink	YL	Yellow

1998-2001 Pickup 2.5L VIN C (All) 104 Pin Connector

PCM Pin #	Wire Color	Circuit Description (104 Pin)	Value at Hot Idle
1	PK/OR	Shift Solenoid 2/B Control	1v, 55 mph: 12v
2	PK/LG	MIL (lamp) Control	MIL Off: 12v, On: 1v
3	YL/BK	Digital TR1 Sensor	0v, 55 mph: 11.5v
12	YL/WT	Fuel Level Indicator Signal	1.7v (1/2 full)
13	PK	Flash EEPROM Power	0.1v
15	PK/LB	Data Bus (-) Signal	Digital Signals
16	TN/OR	Data Bus (+) Signal	Digital Signals
21	DB	CKP (+) Sensor Signal	440-490 Hz
22	GY	CKP (-) Sensor Signal	440-490 Hz
24	BK/WT	Power Ground	<0.1v
25	BK	Case Ground	<0.050v
26	TN/WT	Coil 1 Driver (dwell)	5°, 55 mph: 8°
27	OR/YL	Shift Solenoid 1/A Control	1v, 55 mph: 12v
28	BR/OR	Coast Clutch Solenoid	1v, 55mph: 12v
29	TN/WT	TCS (switch) Signal	TCS & O/D On: 12v
31	YL/LG	PSP Switch Signal	0v (turning: 12v)
35	RD/LG	HO2S-12 (B1 S2) Signal	0.1-1.1v
36	TN/LB	MAF Sensor Return	<0.050v
37	OR/BK	TFT Sensor Signal	2.10-2.40v
38	LG/RD	ECT Sensor Signal	0.5-0.6v
39	GY	IAT Sensor Signal	1.5-2.5v
40	DG/YL	Fuel Pump Monitor	On: 12v, Off: 0v
41	PK	A/C Demand Signal	A/C On: 12v, Off: 0v
47	BR/PK	VR Solenoid Control	0%, 55 mph: 45%
48	TN/YL	Clean Tachometer Output	25-38 Hz
49	LB/BK	Digital TR2 Sensor	0v, 55 mph: 11.5v
50	WT/BK	Digital TR4 Sensor	0v, 55 mph: 11.5v
51	BK/WT	Power Ground	1v
52	TN/OR	Coil 2 Driver (dwell)	5°, 55 mph: 8°
53	PK/BK	Shift Solenoid 'C' Control	1v, 55 mph: 12v
54	PK/YL	TCC Solenoid Control	0%, 55 mph: 95%
55	YL	Keep Alive Power	12-14v
56	LG/BK	EVAP Purge Solenoid	0-10 Hz (0-100%)
58	GY/BK	VSS (+) Signal	0 Hz, 55 mph: 125 Hz
60	GY/LB	HO2S-11 (B1 S1) Signal	0.1-1.1v
62	RD/PK	FTP Sensor Signal	2.6v (1/2 full)
64	LB/YL	A/T: TR Sensor Signal	In 'P': 0v, in O/D: 5v
64	LB/YL	M/T: CPP Switch Signal	5v (clutch "in": 0v)
65	BR/LG	DPFE Sensor Signal	0.95-1.05v
67	PK/WT	EVAP CV Solenoid	0-10 Hz (0-100%)

1998-2001 Pickup 2.5L VIN C (All) 104 Pin Connector

PCM Pin #	Wire Color	Circuit Description (104 Pin)	Value at Hot Idle
69	PK/YL	A/C WOT Relay Control	Off: 12v, On: 1v
71	RD	Vehicle Power	12-14v
74	BR/YL	Injector 3 Control	2.7-3.5 ms
75	TN	Injector 1 Control	2.7-3.5 ms
76	BK/WT	Power Ground	<0.1v
77	BK/WT	Power Ground	<0.1v
78	TN/LG	Coil 3 Driver (dwell)	5°, 55 mph: 8°
79	WT/LG	TCIL (lamp) Control	7.7v (Switch On: 0v)
80	LB/OR	Fuel Pump Control	Off: 12v, On: 1v
81	WT/YL	EPC Solenoid Control	9.1v (4 psi)
83	WT/LB	IAC Solenoid Control	10.7v (33%)
84	DG/WT	TSS Sensor Signal	0 Hz, 55 mph: 385 Hz
85	DB/OR	CMP Sensor Signal	7 Hz, 55 mph: 15 Hz
86	BK/YL	A/C Switch Signal	A/C On: 12v, Off: 0v
88	LB/RD	MAF Sensor Signal	0.8v, 55 mph: 1.6v
89	GY/WT	TP Sensor Signal	0.53-1.27v
90	BR/WT	Reference Voltage	4.9-5.1v
91	GY/RD	Analog Signal Return	<0.050v
92	RD/LG	Brake Pedal Switch	0v (Brake On: 12v)
93	RD/WT	HO2S-11 (B1 S1) Heater	1v (Heater Off: 12v)
95	WT/BK	HO2S-12 (B1 S2) Heater	1v (Heater Off: 12v)
97	RD	Vehicle Power	12-14v
100	BR/LB	Injector 4 Control	2.7-3.5 ms
101	WT	Injector 2 Control	2.7-3.5 ms
103	BK/WT	Power Ground	<0.1v
104	TN/LB	Coil 4 Driver (dwell)	5°, 55 mph: 8°

Pin Connector Graphic

```
      PCM 104-PIN CONNECTOR
 1 ⊚⊚⊚⊚⊚⊚⊚⊚⊚⊚⊚⊚⊚      ⊚⊚⊚⊚⊚⊚⊚⊚⊚⊚⊚⊚⊚ 26
27 ⊚⊚⊚⊚⊚⊚⊚⊚⊚⊚⊚⊚⊚      ⊚⊚⊚⊚⊚⊚⊚⊚⊚⊚⊚⊚⊚ 52
53 ⊚⊚⊚⊚⊚⊚⊚⊚⊚⊚⊚⊚⊚  ⬤  ⊚⊚⊚⊚⊚⊚⊚⊚⊚⊚⊚⊚⊚ 78
79 ⊚⊚⊚⊚⊚⊚⊚⊚⊚⊚⊚⊚⊚      ⊚⊚⊚⊚⊚⊚⊚⊚⊚⊚⊚⊚⊚ 104
```

Terminal View of 104-Pin PCM Wiring Harness Connector

Standard Colors and Abbreviations

Abbreviation	Color	Abbreviation	Color	Abbreviation	Color
BK	Black	GY	Gray	PK	Purple
BL	Blue	GN	Green	RD	Red
BR	Brown	LG	LT Green	TN	Tan
DB	Dark Blue	OR	Orange	WT	White
DG	DK Green	PK	Pink	YL	Yellow

1990 Pickup 2.9L V6 VIN T (All) 60 Pin Connector

PCM Pin #	Wire Color	Circuit Description (60 Pin)	Value at Hot Idle
1	YL/BK	Keep Alive Power	12-14v
2	LG	Brake Position Switch	0v (Brake On: 12v)
3	DG/WT	VSS (+) Signal	0 Hz, 55 mph: 125 Hz
4	DG/YL	IDM Sensor Signal	20-31 Hz
5, 9	---	Not Used	---
6	OR/YL	VSS (-) Signal	0 Hz, 55 mph: 125 Hz
7	LG/YL	ECT Sensor Signal	0.5-0.6v
8	OR/BL	Fuel Pump Monitor	On: 12v, Off: 0v
10	TN/YL	A/C Switch Signal	A/C On: 12v, Off: 0v
11-13	---	Not Used	---
14	DB/OR	MAF Sensor Signal (California)	0.8v, 55 mph: 1.6v
15	TN/BL	MAF Sensor Return	<0.050v
16	BK/OR	Ignition System Ground	<0.050v
17	TN/RD	Self Test Output & MIL	MIL Off: 12v, On: 1v
18-19	---	Not Used	---
20	BK	PCM Case Ground	<0.050v
21	GY/WT	IAC Solenoid Control	9.3-11.0v
22	TN/LG	Fuel Pump Control	Off: 12v, On: 1v
25	LG/PK	ACT Sensor Signal	1.5-2.5v
26	OR/WT	Reference Voltage	4.9-5.1v
29	DG/PK	HO2S-11 (B1 S1) Signal	0.1-1.1v
30	WT/BK	A/T: Neutral Drive Switch	In 'P': 0v, Others: 5v
30	WT/BK	M/T: Clutch Engage Switch	5v (clutch "in": 0v)
36	YL/LG	Spark Output Signal	6.93v (50% dwell)
37	RD	Vehicle Power	12-14v
40	BK/LG	Power Ground	<0.1v
45	DB/LG	MAP Sensor Signal	107 Hz (sea level)
46	BK/WT	Analog Signal Return	<0.050v
47	DG/LG	TP Sensor Signal	1v, 55 mph: 1.4v
48	WT/RD	Self-Test Input Signal	STI On: 1v, Off: 5v
49	OR	HO2S-11 (B1 S1) Ground	<0.050v
52	TN/BL	Shift Solenoid 3-4 Control	Solenoid Off: 12v, On: 1v
53	WT	TCC Solenoid Control	0v, On 55 mph: 12v
54	PK	A/C WOT Relay Control	Off: 12v, On: 1v
56	DB	PIP Sensor Signal	6.93v (50%)
57	RD	Vehicle Power	12-14v
58	LG/WT	Injector Bank 1 (INJ 1, 2 & 4)	3.3-4.5 ms
58	TN	Injector Bank 1 (INJ 1, 2 & 4)	3.3-4.5 ms
59	TN/RD, WT	Injector Bank 2 (INJ 3, 5 & 6)	3.3-4.5 ms
60	BK/LG	Power Ground	<0.1v

Pin Connector Graphic

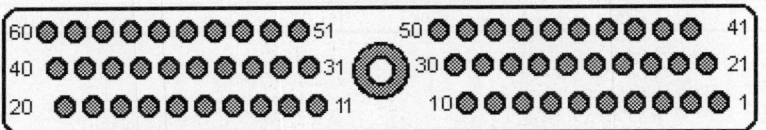

PCM 60-PIN CONNECTOR

Terminal View of 60-Pin PCM Harness Connector

Standard Colors and Abbreviations

Abbreviation	Color	Abbreviation	Color	Abbreviation	Color
BK	Black	GY	Gray	PK	Purple
BL	Blue	GN	Green	RD	Red
BR	Brown	LG	LT Green	TN	Tan
DB	Dark Blue	OR	Orange	WT	White
DG	DK Green	PK	Pink	YL	Yellow

1991-92 Pickup 2.9L V6 VIN T (All) 60 Pin Connector

PCM Pin #	Wire Color	Circuit Description (60 Pin)	Value at Hot Idle
1	YL	Keep Alive Power	12-14v
2	LG	Brake Position Switch	0v (Brake On: 12v)
3	GY/BK	VSS (+) Signal	0 Hz, 55 mph: 125 Hz
4	TN/YL	IDM Sensor Signal	20-31 Hz
6	PK/OR	VSS (-) Signal	0 Hz, 55 mph: 125 Hz
7	LG/RD	ECT Sensor Signal	0.5-0.6v
8	DG/YL	Fuel Pump Monitor	On: 12v, Off: 0v
9 ('92)	PK/BL	Data Bus (-) Signal	Digital Signals
10	DG/OR	A/C Switch Signal	A/C On: 12v, Off: 0v
11-15, 18-19	---	Not Used	---
16	OR/RD	Ignition System Ground	<0.050v
17	PK/LG	Self Test Output & MIL	MIL Off: 12v, On: 1v
20	BK	PCM Case Ground	<0.050v
21	WT/BL	IAC Solenoid Control	10.7v (33%)
22	BL/OR	Fuel Pump Control	Off: 12v, On: 1v
25	GY	ACT Sensor Signal	1.5-2.5v
26	BR/WT	Reference Voltage	4.9-5.1v
28 ('92)	TN/OR	Data Bus (+) Signal	Digital Signals
29	GY/BL	HO2S-11 (B1 S1) Signal	0.1-1.1v
30	BL/YL	A/T: Neutral Drive Switch	In 'P': 0v, Others: 5v
30	BL/YL	M/T: Clutch Engage Switch	5v (clutch "in": 0v)
36	PK	Spark Output Signal	6.93v (50% dwell)
37	RD	Vehicle Power	12-14v
40	BK/WT	Power Ground	<0.1v
45	LG/BK	MAP Sensor Signal	107 Hz (sea level)
46	GY/RD	Analog Signal Return	<0.050v
47	GY/WT	TP Sensor Signal	1v, 55 mph: 1.4v
48	WT/PK	Self-Test Input Signal	STI On: 1v, Off: 5v
49	OR	HO2S-11 (B1 S1) Ground	<0.050v
52	OR/YL	Shift Solenoid 3-4 Control	Solenoid Off: 12v, On: 1v
53	PK/YL	TCC Solenoid Control	0v, On 55 mph: 12v
54	PK/YL	A/C WOT Relay Control	Off: 12v, On: 1v
56	GY/OR	PIP Sensor Signal	6.93v (50%)
57	RD	Vehicle Power	12-14v
58	TN	Injector Bank 1 (INJ 1, 2 & 4)	3.3-3.5 ms
59	WT	Injector Bank 2 (INJ 3, 5 & 6)	3.3-3.5 ms
60	BK/WT	Power Ground	<0.1v

Pin Connector Graphic

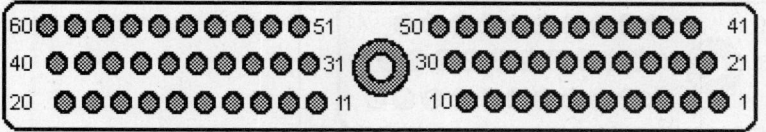

PCM 60-PIN CONNECTOR

Terminal View of 60-Pin PCM Harness Connector

Standard Colors and Abbreviations

Abbreviation	Color	Abbreviation	Color	Abbreviation	Color
BK	Black	GY	Gray	PK	Purple
BL	Blue	GN	Green	RD	Red
BR	Brown	LG	LT Green	TN	Tan
DB	Dark Blue	OR	Orange	WT	White
DG	DK Green	PK	Pink	YL	Yellow

1991 Pickup 3.0L V6 VIN U (All) 60 Pin Connector

PCM Pin #	Wire Color	Circuit Description (60 Pin)	Value at Hot Idle
1	YL	Keep Alive Power	12-14v
2	LG	Brake Position Switch	0v (Brake On: 12v)
3	GY/BK	VSS (+) Signal	0 Hz, 55 mph: 125 Hz
4	WT/PK	IDM Sensor Signal	20-31 Hz
6	PK/OR	VSS (-) Signal	0 Hz, 55 mph: 125 Hz
7	LG/RD	ECT Sensor Signal	0.5-0.6v
8	DG/YL	Fuel Pump Monitor	On: 12v, Off: 0v
10	DG/OR	A/C Switch Signal	A/C On: 12v, Off: 0v
14	BL/RD	MAF Sensor Signal	0.8v, 55 mph: 1.6v
15	TN/BL	MAF Sensor Return	<0.050v
16	OR/RD	Ignition System Ground	<0.050v
17	PK/LG	Self Test Output & MIL	MIL Off: 12v, On: 1v
20	BK	PCM Case Ground	<0.050v
21	WT/BL	IAC Solenoid Control	9.3-11.0v
22	BL/OR	Fuel Pump Control	Off: 12v, On: 1v
25	GY	ACT Sensor Signal	1.5-2.5v
26	BR/WT	Reference Voltage	4.9-5.1v
29	GY/BL	HO2S-11 (B1 S1) Signal	0.1-1.1v
30	BL/YL	A/T: Neutral Drive Switch	In 'P': 0v, Others: 5v
30	BL/YL	M/T: Clutch Engage Switch	5v (clutch "in": 0v)
31	GY/YL	Canister Purge Solenoid	CANP On at 55 mph: 1v
36	PK	Spark Output Signal	6.93v (50% dwell)
37	RD	Vehicle Power	12-14v
40	BK/WT	Power Ground	<0.1v
45	LG/BK	BARO Sensor Signal	159 Hz (sea level)
46	GY/RD	Analog Signal Return	<0.050v
47	GY/WT	TP Sensor Signal	0.7v, 55 mph: 1.1v
48	WT/PK	Self-Test Input Signal	STI On: 1v, Off: 5v
49	OR	HO2S-11 (B1 S1) Ground	<0.050v
53	PK/YL	Converter Clutch Override	0v, On 55 mph: 12v
53	PK/YL	TCC Solenoid Control	0v, On 55 mph: 12v
54	PK/YL	A/C WOT Relay Control	Off: 12v, On: 1v
56	GY/OR	PIP Sensor Signal	6.93v (50%)
57	RD	Vehicle Power	12-14v
58	TN	Injector Bank 1 (INJ 1, 2 & 4)	3.7-3.9 ms
59	WT	Injector Bank 2 (INJ 3, 5 & 6)	3.7-3.9 ms
60	BK/LG	Power Ground	<0.1v

Pin Connector Graphic

PCM 60-PIN CONNECTOR

```
60 ◉◉◉◉◉◉◉◉◉◉ 51    50 ◉◉◉◉◉◉◉◉◉◉ 41
40 ◉◉◉◉◉◉◉◉◉◉ 31    30 ◉◉◉◉◉◉◉◉◉◉ 21
20 ◉◉◉◉◉◉◉◉◉◉ 11    10 ◉◉◉◉◉◉◉◉◉◉ 1
```

Terminal View of 60-Pin PCM Harness Connector

Standard Colors and Abbreviations

Abbreviation	Color	Abbreviation	Color	Abbreviation	Color
BK	Black	GY	Gray	PK	Purple
BL	Blue	GN	Green	RD	Red
BR	Brown	LG	LT Green	TN	Tan
DB	Dark Blue	OR	Orange	WT	White
DG	DK Green	PK	Pink	YL	Yellow

1992-94 Pickup 3.0L V6 VIN U (All) 60 Pin Connector

PCM Pin #	Wire Color	Circuit Description (60 Pin)	Value at Hot Idle
1	YL	Keep Alive Power	12-14v
2	LG	Brake Position Switch	0v (Brake On: 12v)
3	GY/BK	VSS (+) Signal	0 Hz, 55 mph: 125 Hz
4	WT/PK	IDM Sensor Signal	20-31 Hz
6	PK/OR	VSS (-) Signal	0 Hz, 55 mph: 125 Hz
7	LG/RD	ECT Sensor Signal	0.5-0.6v
8	DG/YL	Fuel Pump Monitor	On: 12v, Off: 0v
9	TN/BL	MAF Sensor Return	<0.050v
10	DG/OR	A/C Switch Signal	A/C On: 12v, Off: 0v
11	GY/YL	Canister Purge Solenoid	12v, 55 mph: 1v
12	LG/OR	Injector 6 Control	4.0-4.3 ms
14 ('92-'93)	PK/YL	TCC Solenoid Control	0v, On 55 mph: 12v
15	TN/BK	Injector 6 Control	4.0-4.3 ms
16	OR/RD	Ignition System Ground	<0.050v
17	PK/LG	Self Test Output & MIL	MIL Off: 12v, On: 1v
18	TN/OR	Data Bus (+) Signal	Digital Signals
19	PK/BL	Data Bus (-) Signal	Digital Signals
20	BK	PCM Case Ground	<0.050v
20	BK/LG	PCM Case Ground	<0.050v
21	WT/BL	IAC Solenoid Control	10.8-12.6v
22	BL/OR	Fuel Pump Control	Off: 12v, On: 1v
25	GY	ACT Sensor Signal	1.5-2.5v
26	BR/WT	Reference Voltage	4.9-5.1v
27 ('93-'94)	BR/LG	DPFE Sensor Signal	0.5v, 55 mph: 0.8v
30	BL/YL	A/T: Park Neutral Switch	In 'P': 0v, Others: 5v
30	BL/YL	M/T: Clutch Engage Switch	5v (clutch "in": 0v)
33 ('93-'94)	BR/PK	VR Solenoid Control	0%, 55 mph: 45%
35	BR/BL	Injector 4 Control	4.0-4.3 ms
36	PK	Spark Output Signal	6.93v (50%)
37, 57	RD	Vehicle Power	12-14v
39	BR/YL	Injector 3 Control	4.0-4.3 ms
40	BK/WT	Power Ground	<0.1v
43	RD/BK	HO2S-21 (B2 S1) Signal	0.1-1.1v
44	GY/BL	HO2S-11 (B1 S1) Signal	0.1-1.1v
46	GY/RD	Analog Signal Return	<0.050v
47	GY/WT	TP Sensor Signal	1v, 55 mph: 1.4v
48	WT/PK	Self-Test Input Signal	STI On: 1v, Off: 5v
50	BL/RD	MAF Sensor Signal	0.8v, 55 mph: 1.8v
51	OR/YL	Shift Solenoid 3-4 Control	Solenoid Off: 12v, On: 1v
53 ('94)	PK/YL	TCC Solenoid Control	0v, On 55 mph: 12v
54	PK/YL	A/C WOT Relay Control	Off: 12v, On: 1v
55	---	Not Used	---
56	GY/OR	PIP Sensor Signal	6.93v (50%)
58	TN	Injector 1 Control	4.0-4.3 ms
59	WT	Injector 2 Control	4.0-4.3 ms
60	BK/WT	Power Ground	<0.1v

Pin Connector Graphic

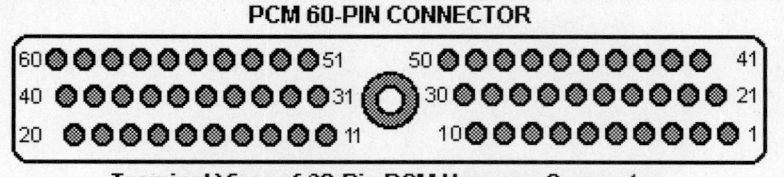

PCM 60-PIN CONNECTOR

Terminal View of 60-Pin PCM Harness Connector

1995-97 Pickup 3.0L V6 VIN U (All) 104 Pin Connector

PCM Pin #	Wire Color	Circuit Description (104 Pin)	Value at Hot Idle
1	BK/WT	Shift Solenoid 2 Control	1v, 55 mph: 12v
2	PK/LG	MIL (lamp) Control	MIL Off: 12v, On: 1v
3-10, 12	---	Not Used	---
11	BL/LG	EVAP Purge Flow Sensor	0.8v, 55 mph: 3.0v
13	PK	Flash EEPROM Power	0.1v
14	GY/BK	4x4 Low Switch Signal	12v (switch closed: 0v)
15	PK/LB	Data Bus (-) Signal	Digital Signals
16	TN/OR	Data Bus (+) Signal	Digital Signals
17-20, 23	---	Not Used	---
21	DB	CKP (+) Sensor Signal	518-540 Hz
22	GY	CKP (-) Sensor Signal	518-540 Hz
24	BK/WT	Power Ground	<0.1v
25	BK	Chassis Ground	<0.050v
26	TN/WT	Coil 1 Driver (dwell)	6°, 55 mph: 9°
27	OR/YL	Shift Solenoid 1 Control	12v, 55 mph: 1v
28	BR/OR	Coast Clutch Solenoid	1v, 55 mph: 12v
29	TN/WT	TCS (switch) Signal	TCS & O/D On: 12v
30	DG	Octane Adjustment	9.3v (shorted: 0v)
31-32	---	Not Used	---
33	PK/OR	VSS (-) Signal	0 Hz, 55 mph: 125 Hz
35	RD/LG	HO2S-12 (B1 S2) Signal	0.1-1.1v
36	TN/LB	MAF Sensor Return	<0.050v
37	OR/BK	TFT Sensor Signal	2.10-2.40v
38	LG/RD	ECT Sensor Signal	0.5-0.6v
39	GY	IAT Sensor Signal	1.5-2.5v
40	DG/YL	Fuel Pump Monitor	On: 12v, Off: 0v
41	TN/YL	A/C Switch Signal	A/C On: 12v, Off: 0v
42-46	---	Not Used	---
47	BR/PK	VR Solenoid Control	0%, 55 mph: 45%
48	TN/YL	Clean Tachometer Output	42-50 Hz
49-50	---	Not Used	---
51	BK/WT	Power Ground	1v
52	TN/OR	Coil 2 Driver (dwell)	6°, 55 mph: 9°
53	PK/BK	Shift Solenoid 3 Control	12v, 55 mph: 1v
54	PK/YL	TCC Solenoid Control	0% (TCC Off)
55	YL	Keep Alive Power	12-14v
56-57, 59	---	Not Used	---
58	GY/BK	VSS (+) Signal	0 Hz, 55 mph: 125 Hz
60	GY/LB	HO2S-11 (B1 S1) Signal	0.1-1.1v
61-63	---	Not Used	---
64	LB/YL	A/T: TR Sensor Signal	In 'P': 0v, in O/D: 5v
64	LB/YL	M/T: Clutch Pedal Switch	5v (clutch "in": 0v)

1995-97 Pickup 3.0L V6 VIN U (All) 104 Pin Connector

PCM Pin #	Wire Color	Circuit Description (104 Pin)	Value at Hot Idle
65	BR/LG	DPFE Sensor Signal	0.20-1.30v
66	---	Not Used	---
67	GY/YL	Canister Purge Solenoid	12v, 55 mph: 1v
68	---	Not Used	---
69	PK/YL	A/C WOT Cutoff Relay	Off: 12v, On: 1v
70	---	Not Used	---
71	RD	Vehicle Power	12-14v
72	---	Not Used	---
73	TN/BK	Injector 5 Control	4.5-4.8 ms
74	BR/YL	Injector 3 Control	4.5-4.8 ms
75	TN	Injector 1 Control	4.5-4.8 ms
76-77	BK/WT	Power Ground	<0.1v
78	TN/LB	Coil 3 Driver (dwell)	6º, 55 mph: 9º
79	WT/LG	TCIL (lamp) Control	7.7v (Switch On: 0v)
80	LB/OR	Fuel Pump Control	Off: 12v, On: 1v
81	WT/YL	EPC Solenoid Control	10.6v (40 psi)
82	---	Not Used	---
83	WT/LB	IAC Solenoid Control	10.7v (33%)
84	DG/WT	TSS Sensor Signal	50-65 Hz
85	DB/OR	CMP Sensor Signal	7 Hz, 55 mph: 15 Hz
86	---	Not Used	---
87	RD/BK	HO2S-11 (B1 S1) Signal	0.1-1.1v
88	LB/RD	MAF Sensor Signal	0.9v
89	GY/WT	TP Sensor Signal	0.53-1.27v
90	BR/WT	Reference Voltage	4.9-5.1v
91	GY/RD	Analog Signal Return	<0.050v
92	LG	Brake Pedal Switch	0v (Brake On: 12v)
93	RD/WT	HO2S-11 (B1 S1) Heater	1v (Heater Off: 12v)
94	YL/LB	HO2S-21 (B2 S1) Heater	1v (Heater Off: 12v)
95	WT/BK	HO2S-12 (B1 S2) Heater	1v (Heater Off: 12v)
96	---	Not Used	---
97	RD	Vehicle Power	12-14v
98	---	Not Used	---
99	LG/OR	Injector 6 Control	4.5-4.8 ms
100	BR/LB	Injector 4 Control	4.5-4.8 ms
101	WT	Injector 2 Control	4.5-4.8 ms
102	---	Not Used	---
103	BK/WT	Power Ground	<0.1v
104	---	Not Used	---

Pin Connector Graphic

PCM 104-PIN CONNECTOR

Terminal View of 104-Pin PCM Wiring Harness Connector

Standard Colors and Abbreviations

Abbreviation	Color	Abbreviation	Color	Abbreviation	Color
BK	Black	GY	Gray	PK	Purple
BL	Blue	GN	Green	RD	Red
BR	Brown	LG	LT Green	TN	Tan
DB	Dark Blue	OR	Orange	WT	White
DG	DK Green	PK	Pink	YL	Yellow

1998-2002 Pickup 3.0L V6 MFI VIN U (All) 104 Pin Connector

PCM Pin #	Wire Color	Circuit Description (104 Pin)	Value at Hot Idle
1	PK/OR	Shift Solenoid 2/B Control	1v, 55 mph: 12v
2	PK/LG	MIL (lamp) Control	MIL Off: 12v, On: 1v
3	YL/BK	Digital TR1 Sensor	0v, 55 mph: 11.5v
4-11	---	Not Used	---
12	YL/WT	Fuel Level Indicator Signal	1.7v (1/2 full)
13	PK	Flash EEPROM Power	0.1v
14	LB/BK	4x4 Low Indicator Switch	12v (switch closed: 0v)
15	PK/LB	Data Bus (-) Signal	Digital Signals
16	TN/OR	Data Bus (+) Signal	Digital Signals
17 ('02)	GY/OR	Passive Antitheft RX Signal	Digital Signals
18 ('02)	WT/LG	Passive Antitheft TX Signal	Digital Signals
19-20	---	Not Used	---
21	DB	CKP (+) Sensor Signal	440-490 Hz
22	GY	CKP (-) Sensor Signal	440-490 Hz
23	---	Not Used	---
24	BK/WT	Power Ground	<0.1v
25	BK	Case Ground	<0.050v
26	TN/WT	Coil 1 Driver (dwell)	6°, 55 mph: 9°
27	OR/YL	Shift Solenoid 'A' Control	1v, 55 mph: 12v
28	BR/OR	Shift Solenoid 'D' Control	1v, 55mph: 12v
29	TN/WT	TCS (switch) Signal	TCS & O/D On: 12v
30 ('02)	DB/LG	Antitheft Indicator Control	Indicator Off: 12v, On: 1v
31-34	---	Not Used	---
35	RD/LG	HO2S-12 (B1 S2) Signal	0.1-1.1v
36	TN/LB	MAF Sensor Return	<0.050v
37	OR/BK	TFT Sensor Signal	2.10-2.40v
38	LG/RD	ECT Sensor Signal	0.5-0.6v
39	GY	IAT Sensor Signal	1.5-2.5v
40	DG/YL	Fuel Pump Monitor	On: 12v, Off: 0v
41	PK	A/C Demand Signal	A/C On: 12v, Off: 0v
42-47	---	Not Used	---
48	TN/YL	Clean Tachometer Output	25-38 Hz
49	LB/BK	Digital TR2 Sensor	0v, 55 mph: 11.5v
50	WT/BK	Digital TR4 Sensor	0v, 55 mph: 11.5v
51	BK/WT	Power Ground	1v
52	TN/OR	Coil 3 Driver (dwell)	6°, 55 mph: 9°
53	PK/BK	Shift Solenoid 3/C Control	1v, 55 mph: 12v
54	PK/YL	TCC Solenoid Control	0%, 55 mph: 95%
55	YL	Keep Alive Power	12-14v
56	LG/BK	EVAP Purge Solenoid	0-10 Hz (0-100%)
57	---	Not Used	---
58	GY/BK	VSS (+) Signal	0 Hz, 55 mph: 125 Hz
59	---	Not Used	---
60	GY/LB	HO2S-11 (B1 S1) Signal	0.1-1.1v
61, 63, 66	---	Not Used	---
62	RD/PK	FTP Sensor Signal	2.6v (1/2 full)
64	LB/YL	A/T: TR Sensor Signal	In 'P': 0v, in O/D: 5v
64	LB/YL	M/T: CPP Switch Signal	5v (clutch "in": 0v)

1998-2002 Pickup 3.0L V6 MFI VIN U (All) 104 Pin Connector

PCM Pin #	Wire Color	Circuit Description (104 Pin)	Value at Hot Idle
65 ('98-'00)	BR/LG	DPFE Sensor Signal	0.20-1.30v
65 ('01-'02)	BR/LG	DPFE Sensor Signal	0.95-1.05v
67	PK/WT	EVAP CV Solenoid	0-10 Hz (0-100%)
68	---	Not Used	---
69	PK/YL	A/C WOT Relay Control	Off: 12v, On: 1v
70	---	Not Used	---
71	RD	Vehicle Power	12-14v
72	---	Not Used	---
73	TN/BK	Injector 5 Control	2.7-3.5 ms
74	BR/YL	Injector 3 Control	2.7-3.5 ms
75	TN	Injector 1 Control	2.7-3.5 ms
76-77	BK/WT	Power Ground	<0.1v
78	TN/LG	Coil 2 Driver (dwell)	6°, 55 mph: 9°
79	WT/LG	TCIL (lamp) Control	7.7v (Switch On: 0v)
80	LB/OR	Fuel Pump Control	Off: 12v, On: 1v
81	WT/YL	EPC Solenoid Control	9.1v (4 psi)
82	---	Not Used	---
83	WT/LB	IAC Solenoid Control	10.7v (33%)
84	DG/WT	TSS Sensor Signal	120 Hz, 55 mph: 260 Hz
85	DB/OR	CMP Sensor Signal	7 Hz, 55 mph: 15 Hz
86	BK/YL	A/C Switch Signal	A/C On: 12v, Off: 0v
87	RD/BK	HO2S-12 (B1 S2) Signal	0.1-1.1v
88	LB/RD	MAF Sensor Signal	0.8v, 55 mph: 1.6v
89	GY/WT	TP Sensor Signal	0.53-1.27v
90	BR/WT	Reference Voltage	4.9-5.1v
91	GY/RD	Analog Signal Return	<0.050v
92	RD/LG	Brake Pedal Switch	0v (Brake On: 12v)
93	RD/WT	HO2S-11 (B1 S1) Heater	1v (Heater Off: 12v)
94	YL/LB	HO2S-12 (B1 S2) Heater	1v (Heater Off: 12v)
95	WT/BK	HO2S-13 (B1 S3) Heater	1v (Heater Off: 12v)
96	---	Not Used	---
97	RD	Vehicle Power	12-14v
98	---	Not Used	---
99	LG/OR	Injector 6 Control	2.7-3.5 ms
100	BR/LB	Injector 4 Control	2.7-3.5 ms
101	WT	Injector 2 Control	2.7-3.5 ms
102	---	Not Used	---
103	BK/WT	Power Ground	<0.1v
104	---	Not Used	---

Pin Connector Graphic

PCM 104-PIN CONNECTOR

Terminal View of 104-Pin PCM Wiring Harness Connector

Standard Colors and Abbreviations

Abbreviation	Color	Abbreviation	Color	Abbreviation	Color
BK	Black	GY	Gray	PK	Purple
BL	Blue	GN	Green	RD	Red
BR	Brown	LG	LT Green	TN	Tan
DB	Dark Blue	OR	Orange	WT	White
DG	DK Green	PK	Pink	YL	Yellow

2003 Pickup 3.0L V6 MFI VIN U (All) 104 Pin Connector

PCM Pin #	Wire Color	Circuit Description (104 Pin)	Value at Hot Idle
1	VT/OR	Shift Solenoid 'B' Control	1v, 55 mph: 12v
2	PK/LG	MIL (lamp) Control	MIL Off: 12v, On: 1v
3	YL/BK	Digital TR1 Sensor	0v, 55 mph: 11.5v
4-5	---	Not Used	---
6	DG/WT	TSS Sensor Signal	120 Hz, 55 mph: 260 Hz
7-11	---	Not Used	---
12	YL/WT	Fuel Level Indicator Signal	1.7v (1/2 full)
13	VT	Flash EEPROM Power	0.1v
14	LB/BK	4x4 Low Indicator Switch	12v (switch closed: 0v)
15	PK/LB	SCP Data Bus (-) Signal	<0.050v
16	TN/OR	SCP Data Bus (+) Signal	Digital Signals
17	GY/OR	Passive Antitheft RX Signal	Digital Signals
18	WT/LG	Passive Antitheft TX Signal	Digital Signals
19-20	---	Not Used	---
21	DB	CKP (+) Sensor Signal	440-490 Hz
22	GY	CKP (-) Sensor Signal	440-490 Hz
23	---	Not Used	---
24	BK/WT	Power Ground	<0.1v
25	BK	Case Ground	<0.050v
26	TN/WT	Coil 1 Driver (dwell)	6°, 55 mph: 9°
27	OR/YL	Shift Solenoid 'A' Control	1v, 55 mph: 12v
28	BR/OR	Coast Clutch Solenoid	1v, 55mph: 12v
29	TN/WT	TCS (switch) Signal	TCS & O/D On: 12v
30	DB/LG	Antitheft Indicator Control	Indicator Off: 12v, On: 1v
31-34	---	Not Used	---
35	RD/LG	HO2S-12 (B1 S2) Signal	0.1-1.1v
36	TN/LB	MAF Sensor Return	<0.050v
37	OR/BK	TFT Sensor Signal	2.10-2.40v
38	LG/RD	ECT Sensor Signal	0.5-0.6v
39	GY	IAT Sensor Signal	1.5-2.5v
40	DG/YL	Fuel Pump Monitor	Off: 12v, On: 1v
41	VT	A/C Demand Signal	A/C On: 12v, Off: 0v
42-43	---	Not Used	---
44	DB/OR	Starter Relay Control Circuit	Relay Off: 12v, On: 1v
45-47	---	Not Used	---
48	TN/YL	Clean Tachometer Output	25-38 Hz
49	LB/BK	Digital TR2 Sensor	0v, 55 mph: 11.5v
50	WT/BK	Digital TR4 Sensor	0v, 55 mph: 11.5v
51	BK/WT	Power Ground	1v
52	TN/OR	Coil 3 Driver (dwell)	6°, 55 mph: 9°
53	PK/BK	Shift Solenoid 'C' Control	1v, 55 mph: 12v
54	VT/YL	TCC Solenoid Control	0%, 55 mph: 95%
55	YL	Keep Alive Power	12-14v
56	LG/BK	EVAP Purge Solenoid Control	0-10 Hz (0-100%)
57	YL/RD	Knock Sensor Signal	0v
58	---	Not Used	---
59	GY/OR	Intermediate Speed Shaft Sensor	0 Hz, 55 mph: 1150 Hz
60	GY/LB	HO2S-11 (B1 S1) Signal	0.1-1.1v
61	---	Not Used	---
62	RD/PK	FTP Sensor Signal	2.6v (1/2 full)
63	---	Not Used	---
64	LB/YL	A/T: TR Sensor Signal	In 'P': 0v, in O/D: 5v
64	LB/YL	M/T: CPP Switch Signal	5v (clutch "in": 0v)

2003 Pickup 3.0L V6 MFI VIN U (All) 104 Pin Connector

PCM Pin #	Wire Color	Circuit Description (104 Pin)	Value at Hot Idle
65-66	---	Not Used	---
67	VT/WT	EVAP Canister Purge Solenoid	0-10 Hz (0-100%)
68	GY/BK	Vehicle Speed Sensor Signal	0 Hz, 55 mph: 125 Hz
69	PK/YL	A/C WOT Relay Control	Off: 12v, On: 1v
70	---	Not Used	---
71	RD	Vehicle Power (Start-Run)	12-14v
72	---	Not Used	---
73	TN/BK	Injector 5 Control	2.7-3.5 ms
74	BR/YL	Injector 3 Control	2.7-3.5 ms
75	TN	Injector 1 Control	2.7-3.5 ms
76, 77	BK/WT	Power Ground	<0.1v
78	TN/LG	Coil 2 Driver (dwell)	6°, 55 mph: 9°
79	WT/LG	TCIL (lamp) Control	7.7v (Switch On: 0v)
80	LB/OR	Fuel Pump Control	Off: 12v, On: 1v
81	WT/YL	EPC Solenoid Control	9.1v (4 psi)
82	BK/WT	Check Fuel Cap Indicator Control	Indicator Off: 12v, On: 1v
83	WT/LB	IAC Solenoid Control	10.7v (33%)
84	DG/YL	OSS Sensor Signal	120 Hz, 55 mph: 260 Hz
85	DB/OR	CMP Sensor Signal	7 Hz, 55 mph: 15 Hz
86	BK/YL	A/C Switch Signal	A/C On: 12v, Off: 0v
87	RD/BK	HO2S-12 (B1 S2) Signal	0.1-1.1v
88	LB/RD	MAF Sensor Signal	0.8v, 55 mph: 1.6v
89	GY/WT	TP Sensor Signal	0.53-1.27v
90	BR/WT	Reference Voltage	4.9-5.1v
91	GY/RD	Analog Signal Return	<0.050v
92	RD/LG	Brake Pedal Switch	0v (Brake On: 12v)
93	RD/WT	HO2S-11 (B1 S1) Heater	1v (Heater Off: 12v)
94	YL/LB	HO2S-12 (B1 S2) Heater	1v (Heater Off: 12v)
95	WT/BK	HO2S-13 (B1 S3) Heater	1v (Heater Off: 12v)
96	---	Not Used	---
97	RD	Vehicle Power (Start-Run)	12-14v
98	---	Not Used	---
99	LG/OR	Injector 6 Control	2.7-3.5 ms
100	BR/LB	Injector 4 Control	2.7-3.5 ms
101	WT	Injector 2 Control	2.7-3.5 ms
102	---	Not Used	---
103	BK/WT	Power Ground	<0.1v
104	---	Not Used	---

Pin Connector Graphic

PCM 104-PIN CONNECTOR

Terminal View of 104-Pin PCM Wiring Harness Connector

Standard Colors and Abbreviations

Abbreviation	Color	Abbreviation	Color	Abbreviation	Color
BK	Black	GY	Gray	PK	Purple
BL	Blue	GN	Green	RD	Red
BR	Brown	LG	LT Green	TN	Tan
DB	Dark Blue	OR	Orange	WT	White
DG	DK Green	PK	Pink	YL	Yellow

1999-2002 Pickup 3.0L V6 Flexible Fuel Vehicle VIN V (All) 104 Pin Connector

PCM Pin #	Wire Color	Circuit Description (104 Pin)	Value at Hot Idle
1	VT/OR	Shift Solenoid 'B' Control	1v, 55 mph: 12v
2	PK/LG	MIL (lamp) Control	MIL Off: 12v, On: 1v
3	YL/BK	Digital TR1 Sensor	0v, 55 mph: 11.5v
4-11	---	Not Used	---
12	YL/WT	Fuel Level Indicator Signal	1.7v (1/2 full)
13	PK	Flash EEPROM Power	0.1v
14 ('99)	LB/BK	4x4 Low Indicator Switch	12v (switch closed: 0v)
15	PK/LB	Data Bus (-) Signal	<0.050v
16	TN/OR	Data Bus (+) Signal	Digital Signals
17-20	---	Not Used	---
21	DB	CKP (+) Sensor Signal	440-490 Hz
22	GY	CKP (-) Sensor Signal	440-490 Hz
23	---	Not Used	---
24	BK/WT	Power Ground	<0.1v
25	BK	Case Ground	<0.050v
26	TN/WT	Coil 1 Driver (dwell)	6°, 55 mph: 9°
27	OR/YL	Shift Solenoid 1/A Control	1v, 55 mph: 12v
28	BR/OR	Coast Clutch Solenoid	1v, 55mph: 12v
29	TN/WT	TCS (switch) Signal	TCS & O/D On: 12v
30	---	Not Used	---
31 ('99)	YL/LG	PSP Switch Signal	0v (turning: 12v)
32-33	---	Not Used	---
34	DG/LG	Fuel Composition Sensor	40-60 Hz
35	RD/LG	HO2S-13 (B1 S3) Signal	0.1-1.1v
36	TN/LB	MAF Sensor Return	<0.050v
37	OR/BK	TFT Sensor Signal	2.10-2.40v
38	LG/RD	ECT Sensor Signal	0.5-0.6v
39	GY	IAT Sensor Signal	1.5-2.5v
40	DG/YL	Fuel Pump Monitor	On: 12v, Off: 0v
41	PK	A/C Demand Signal	A/C On: 12v, Off: 0v
42-46	---	Not Used	---
47	BR/PK	VR Solenoid Control	0%, 55 mph: 45%
48	TN/YL	Clean Tachometer Output	25-38 Hz
49	LB/BK	Digital TR2 Sensor	0v, 55 mph: 11.5v
50	WT/BK	Digital TR4 Sensor	0v, 55 mph: 11.5v
51	BK/WT	Power Ground	<0.1v
52	TN/OR	Coil 3 Driver (dwell)	6°, 55 mph: 9°
53	PK/BK	Shift Solenoid 3 Control	1v, 55 mph: 12v
54	PK/YL	TCC Solenoid Control	0%, 55 mph: 95%
55	Y	Keep Alive Power	12-14v
56	LG/BK	EVAP Purge Solenoid	0-10 Hz (0-100%)
58	GY/BK	VSS (+) Signal	0 Hz, 55 mph: 125 Hz
60	GY/LB	HO2S-11 (B1 S1) Signal	0.1-1.1v
62	RD/PK	FTP Sensor Signal	2.6v (1/2 full)
62	RD/PK	FTP Sensor Signal	2.6v (1/2 full)
64	LB/YL	A/T: TR3A Sensor Signal	In 'P': 0v, in O/D: 5v
64	LB/YL	M/T: CPP Switch Signal	5v (clutch "in": 0v)

1999-2002 Pickup 3.0L V6 Flexible Fuel Vehicle VIN V (All) 104 Pin Connector

PCM Pin #	Wire Color	Circuit Description (104 Pin)	Value at Hot Idle
65 ('99-'00)	BR/LG	DPFE Sensor Signal	0.20-1.30v
65 ('01-'02)	BR/LG	DPFE Sensor Signal	0.95-1.05v
67	PK/WT	EVAP CV Solenoid	0-10 Hz (0-100%)
69	PK/YL	A/C WOT Relay Control	Off: 12v, On: 1v
69 ('00-'02)	PK/YL	A/C Clutch Relay Control	Off: 12v, On: 1v
71	RD	Vehicle Power	12-14v
73	TN/BK	Injector 5 Control	2.7-3.5 ms
74	BR/YL	Injector 3 Control	2.7-3.5 ms
75	TN	Injector 1 Control	2.7-3.5 ms
76-77	BK/WT	Power Ground	<0.1v
78	TN/LG	Coil 2 Driver (dwell)	6°, 55 mph: 9°
79	WT/LG	TCIL (lamp) Control	7.7v (Switch On: 0v)
80	LB/OR	Fuel Pump Control	Off: 12v, On: 1v
81	WT/YL	EPC Solenoid Control	9.1v (4 psi)
83	WT/LB	IAC Solenoid Control	10.7v (33%)
84	DG/WT	OSS Sensor Signal	0 Hz, 530 Hz
85	DB/OR	CMP Sensor Signal	7 Hz, 55 mph: 15 Hz
86	BK/YL	A/C Switch Signal	A/C On: 12v, Off: 0v
87	RD/BK	HO2S-12 (B1 S2) Signal	0.1-1.1v
88	LB/RD	MAF Sensor Signal	0.8v, 55 mph: 1.6v
89	GY/WT	TP Sensor Signal	0.53-1.27v
90	BR/WT	Reference Voltage	4.9-5.1v
91	GY/RD	Analog Signal Return	<0.050v
92	RD/LG	Brake Pedal Switch	0v (Brake On: 12v)
93	RD/WT	HO2S-11 (B1 S1) Heater	1v (Heater Off: 12v)
94	YL/LB	HO2S-12 (B1 S2) Heater	1v (Heater Off: 12v)
95	WT/BK	HO2S-13 (B1 S3) Heater	1v (Heater Off: 12v)
97	RD	Vehicle Power	12-14v
99	LG/OR	Injector 6 Control	2.7-3.5 ms
100	BR/LB	Injector 4 Control	2.7-3.5 ms
101	WT	Injector 2 Control	2.7-3.5 ms
103	BK/WT	Power Ground	<0.1v

Pin Connector Graphic

```
          PCM 104-PIN CONNECTOR
  1 ◉◉◉◉◉◉◉◉◉◉◉◉◉    ◉◉◉◉◉◉◉◉◉◉◉◉◉ 26
 27 ◉◉◉◉◉◉◉◉◉◉◉◉◉    ◉◉◉◉◉◉◉◉◉◉◉◉◉ 52
 53 ◉◉◉◉◉◉◉◉◉◉◉◉◉ ⬤ ◉◉◉◉◉◉◉◉◉◉◉◉◉ 78
 79 ◉◉◉◉◉◉◉◉◉◉◉◉◉    ◉◉◉◉◉◉◉◉◉◉◉◉◉ 104
```

Terminal View of 104-Pin PCM Wiring Harness Connector

Standard Colors and Abbreviations

Abbreviation	Color	Abbreviation	Color	Abbreviation	Color
BK	Black	GY	Gray	PK	Purple
BL	Blue	GN	Green	RD	Red
BR	Brown	LG	LT Green	TN	Tan
DB	Dark Blue	OR	Orange	WT	White
DG	DK Green	PK	Pink	YL	Yellow

2003 Pickup 3.0L V6 Flexible Fuel Vehicle VIN V (All) 104 Pin Connector

PCM Pin #	Wire Color	Circuit Description (104 Pin)	Value at Hot Idle
1	VT/OR	Shift Solenoid 'B' Control	1v, 55 mph: 12v
2	PK/LG	MIL (lamp) Control	MIL Off: 12v, On: 1v
3	YL/BK	Digital TR1 Sensor	0v, 55 mph: 11.5v
4-5	---	Not Used	---
6	DG/WT	TSS Sensor Signal	120 Hz, 55 mph: 260 Hz
7-11	---	Not Used	---
12	YL/WT	Fuel Level Indicator Signal	1.7v (1/2 full)
13	VT	Flash EEPROM Power	0.1v
14	LB/BK	4x4 Low Indicator Switch	12v (switch closed: 0v)
15	PK/LB	SCP Data Bus (-) Signal	<0.050v
16	TN/OR	SCP Data Bus (+) Signal	Digital Signals
17	GY/OR	Passive Antitheft RX Signal	Digital Signals
18	WT/LG	Passive Antitheft TX Signal	Digital Signals
19	---	Not Used	---
20	YL/WT	Fuel Composition Sensor	40-60 Hz
21	DB	CKP (+) Sensor Signal	440-490 Hz
22	GY	CKP (-) Sensor Signal	440-490 Hz
23	---	Not Used	---
24	BK/WT	Power Ground	<0.1v
25	BK	Case Ground	<0.050v
26	TN/WT	Coil 1 Driver (dwell)	6°, 55 mph: 9°
27	OR/YL	Shift Solenoid 'A' Control	1v, 55 mph: 12v
28	BR/OR	Coast Clutch Solenoid	1v, 55mph: 12v
29	TN/WT	TCS (switch) Signal	TCS & O/D On: 12v
30	DB/LG	Antitheft Indicator Control	Indicator Off: 12v, On: 1v
31-34	---	Not Used	---
35	RD/LG	HO2S-12 (B1 S2) Signal	0.1-1.1v
36	TN/LB	MAF Sensor Return	<0.050v
37	OR/BK	TFT Sensor Signal	2.10-2.40v
38	LG/RD	ECT Sensor Signal	0.5-0.6v
39	GY	IAT Sensor Signal	1.5-2.5v
40	DG/YL	Fuel Pump Monitor	Off: 12v, On: 1v
41	VT	A/C Demand Signal	A/C On: 12v, Off: 0v
42-43	---	Not Used	---
44	DB/OR	Starter Relay Control Circuit	Relay Off: 12v, On: 1v
45-47	---	Not Used	---
48	TN/YL	Clean Tachometer Output	25-38 Hz
49	LB/BK	Digital TR2 Sensor	0v, 55 mph: 11.5v
50	WT/BK	Digital TR4 Sensor	0v, 55 mph: 11.5v
51	BK/WT	Power Ground	1v
52	TN/OR	Coil 3 Driver (dwell)	6°, 55 mph: 9°
53	PK/BK	Shift Solenoid 'C' Control	1v, 55 mph: 12v
54	VT/YL	TCC Solenoid Control	0%, 55 mph: 95%
55	YL	Keep Alive Power	12-14v
56	LG/BK	EVAP Purge Solenoid Control	0-10 Hz (0-100%)
57	YL/RD	Knock Sensor Signal	0v
58	---	Not Used	---
59	GY/OR	Intermediate Speed Shaft Sensor	0 Hz, 55 mph: 1150 Hz
60	GY/LB	HO2S-11 (B1 S1) Signal	0.1-1.1v
61	---	Not Used	---
62	RD/PK	FTP Sensor Signal	2.6v (1/2 full)
63	---	Not Used	---
64	LB/YL	A/T: TR Sensor Signal	In 'P': 0v, in O/D: 5v
64	LB/YL	M/T: CPP Switch Signal	5v (clutch "in": 0v)

2003 Pickup 3.0L V6 Flexible Fuel Vehicle VIN V (All) 104 Pin Connector

PCM Pin #	Wire Color	Circuit Description (104 Pin)	Value at Hot Idle
65-66	---	Not Used	---
67	VT/WT	EVAP Canister Purge Solenoid	0-10 Hz (0-100%)
68	GY/BK	Vehicle Speed Sensor Signal	0 Hz, 55 mph: 125 Hz
69	PK/YL	A/C WOT Relay Control	Off: 12v, On: 1v
70	---	Not Used	---
71	RD	Vehicle Power (Start-Run)	12-14v
72	---	Not Used	---
73	TN/BK	Injector 5 Control	2.7-3.5 ms
74	BR/YL	Injector 3 Control	2.7-3.5 ms
75	TN	Injector 1 Control	2.7-3.5 ms
76, 77	BK/WT	Power Ground	<0.1v
78	TN/LG	Coil 2 Driver (dwell)	6°, 55 mph: 9°
79	WT/LG	TCIL (lamp) Control	7.7v (Switch On: 0v)
80	LB/OR	Fuel Pump Control	Off: 12v, On: 1v
81	WT/YL	EPC Solenoid Control	9.1v (4 psi)
82	BK/WT	Check Fuel Cap Indicator Control	Indicator Off: 12v, On: 1v
83	WT/LB	IAC Solenoid Control	10.7v (33%)
84	DG/YL	OSS Sensor Signal	120 Hz, 55 mph: 260 Hz
85	DB/OR	CMP Sensor Signal	7 Hz, 55 mph: 15 Hz
86	BK/YL	A/C Switch Signal	A/C On: 12v, Off: 0v
87	RD/BK	HO2S-12 (B1 S2) Signal	0.1-1.1v
88	LB/RD	MAF Sensor Signal	0.8v, 55 mph: 1.6v
89	GY/WT	TP Sensor Signal	0.53-1.27v
90	BR/WT	Reference Voltage	4.9-5.1v
91	GY/RD	Analog Signal Return	<0.050v
92	RD/LG	Brake Pedal Switch	0v (Brake On: 12v)
93	RD/WT	HO2S-11 (B1 S1) Heater	1v (Heater Off: 12v)
94	YL/LB	HO2S-21 (B2 S1) Heater	1v (Heater Off: 12v)
95	WT/BK	HO2S-12 (B1 S2) Heater	1v (Heater Off: 12v)
96	---	Not Used	---
97	RD	Vehicle Power (Start-Run)	12-14v
98	---	Not Used	---
99	LG/OR	Injector 6 Control	2.7-3.5 ms
100	BR/LB	Injector 4 Control	2.7-3.5 ms
101	WT	Injector 2 Control	2.7-3.5 ms
102	---	Not Used	---
103	BK/WT	Power Ground	<0.1v
104	---	Not Used	---

Pin Connector Graphic

PCM 104-PIN CONNECTOR

Terminal View of 104-Pin PCM Wiring Harness Connector

Standard Colors and Abbreviations

Abbreviation	Color	Abbreviation	Color	Abbreviation	Color
BK	Black	GY	Gray	PK	Purple
BL	Blue	GN	Green	RD	Red
BR	Brown	LG	LT Green	TN	Tan
DB	Dark Blue	OR	Orange	WT	White
DG	DK Green	PK	Pink	YL	Yellow

1990 Pickup 4.0L V6 MFI VIN X (All) 60 Pin Connector

PCM Pin #	Wire Color	Circuit Description (60 Pin)	Value at Hot Idle
1	YL/BK	Keep Alive Power	12-14v
2	LG	Brake Position Switch	0v (Brake On: 12v)
3	DG/WT	VSS (+) Signal	0 Hz, 55 mph: 125 Hz
4	DG/YL	IDM Sensor Signal	20-31 Hz
6	OR/YL	VSS (-) Signal	0 Hz, 55 mph: 125 Hz
7	LG/YL	ECT Sensor Signal	0.5-0.6v
8	OR/BL	Fuel Pump Monitor	On: 12v, Off: 0v
10	TN/YL	A/C Switch Signal	A/C On: 12v, Off: 0v
14	DB/OR	MAF Sensor Signal	0.8v
15	TN/BL	MAF Sensor Return	<0.050v
16	BK/OR	Ignition System Ground	<0.050v
17	TN/RD	Self Test Output & MIL	MIL Off: 12v, On: 1v
20	BK	PCM Case Ground	<0.050v
21	GY/WT	IAC Solenoid Control	8.0-11.5v
22	TN/LG	Fuel Pump Control	Off: 12v, On: 1v
25	LG/PK	ACT Sensor Signal	1.5-2.5v
26	OR/WT	Reference Voltage	4.9-5.1v
29	DG/PK	HO2S-11 (B1 S1) Signal	0.1-1.1v
30	WT/BK	A/T: Neutral Drive Switch	In 'P': 0v, Others: 5v
30	WT/BK	M/T: Clutch Engage Switch	5v (clutch "in": 0v)
36	YL/LG	Spark Output Signal	6.93v (50%)
37	RD	Vehicle Power	12-14v
40	BK/LG	Power Ground	<0.1v
45	DB/LG	BARO Sensor Signal	159 Hz (sea level)
46	BK/WT	Analog Signal Return	<0.050v
47	DG/LG	TP Sensor Signal	1v, 55 mph: 1.4v
48	WT/RD	Self-Test Input Signal	STI On: 1v, Off: 5v
49	OR	HO2S-11 (B1 S1) Ground	<0.050v
52	TN/BL	Shift Solenoid 3-4 Control	Solenoid Off: 12v, On: 1v
53	WT	TCC Solenoid Control	0v, On 55 mph: 12v
54	PK	A/C WOT Relay Control	Off: 12v, On: 1v
56	BK/BL	PIP Sensor Signal	6.93v (50%)
57	RD	Vehicle Power	12-14v
58	TN	Injector Bank 1 (INJ 1, 2 & 4)	3.0-3.2 ms
59	WT	Injector Bank 2 (INJ 3, 5 & 6)	3.0-3.2 ms
60	BK/LG	Power Ground	<0.1v

Pin Connector Graphic

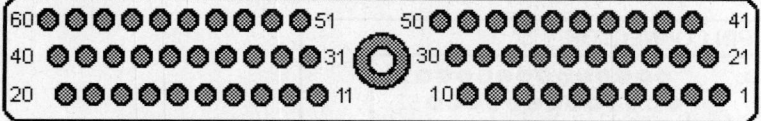

PCM 60-PIN CONNECTOR

Terminal View of 60-Pin PCM Harness Connector

Standard Colors and Abbreviations

Abbreviation	Color	Abbreviation	Color	Abbreviation	Color
BK	Black	GY	Gray	PK	Purple
BL	Blue	GN	Green	RD	Red
BR	Brown	LG	LT Green	TN	Tan
DB	Dark Blue	OR	Orange	WT	White
DG	DK Green	PK	Pink	YL	Yellow

1991-94 Pickup 4.0L V6 VIN X (All) 60 Pin Connector

PCM Pin #	Wire Color	Circuit Description (60 Pin)	Value at Hot Idle
1	YL	Keep Alive Power	12-14v
2	LG	Brake Position Switch	0v (Brake On: 12v)
3	GY/BK	VSS (+) Signal	0 Hz, 55 mph: 125 Hz
4	TN/YL	IDM Sensor Signal	20-31 Hz
6	PK/OR	VSS (-) Signal	0 Hz, 55 mph: 125 Hz
7	LG/RD	ECT Sensor Signal	0.5-0.6v
8	DG/YL	Fuel Pump Monitor	On: 12v, Off: 0v
9 ('92-'94)	PK/BL	Data Bus (-) Signal	Digital Signals
10	DG/OR	A/C Switch Signal	A/C On: 12v, Off: 0v
14	BL/RD	MAF Sensor Signal	0.8v
15	TN/BL	MAF Sensor Return	<0.050v
16	OR/RD	Ignition System Ground	<0.050v
17	PK/LG	Self Test Output & MIL	MIL Off: 12v, On: 1v
20	BK	PCM Case Ground	<0.050v
20	BK/LG	PCM Case Ground	<0.050v
21	WT/BL	IAC Solenoid Control	8.0-11.5v
22	BL/OR	Fuel Pump Control	Off: 12v, On: 1v
25	GY	IAT Sensor Signal	1.5-2.5v
26	BR/WT	Reference Voltage	4.9-5.1v
28 ('92-'94)	TN/OR	Data Bus (+) Signal	Digital Signals
29	GY/BL	HO2S-11 (B1 S1) Signal	0.1-1.1v
30	BL/YL	A/T: Neutral Drive Switch	In 'P': 0v, Others: 5v
30	BL/YL	M/T: Clutch Engage Switch	5v (clutch "in": 0v)
31	GY/DB	Canister Purge Solenoid	CANP On at 55 mph: 1v
31	GY/YL	Canister Purge Solenoid	CANP On at 55 mph: 1v
36	PK	Spark Output Signal	6.93v (50%)
37	RD	Vehicle Power	12-14v
39 ('93-'94)	RD/BK	HO2S-21 (B2 S1) Signal	0.1-1.1v
40	BK/WT	Power Ground	<0.1v
44	DG	Octane Adjustment	9.3v (shorted: 0v)
45 ('91)	LG/BK	BARO Sensor Signal	159 Hz (sea level)
46	GY/RD	Analog Signal Return	<0.050v
47	GY/WT	TP Sensor Signal	1v, 55 mph: 1.4v
48	WT/PK	Self-Test Input Signal	STI On: 1v, Off: 5v
49	OR	HO2S-11 (B1 S1) Ground	<0.050v
49	GY/RD	HO2S-11 (B1 S1) Ground	<0.050v
52	OR/YL	Shift Solenoid 3-4 Control	Solenoid Off: 12v, On: 1v
53	PK/YL	TCC Solenoid Control	0v, On 55 mph: 12v
54	PK/YL	A/C WOT Relay Control	Off: 12v, On: 1v
56	GY/OR	PIP Sensor Signal	6.93v (50%)
57	RD	Vehicle Power	12-14v
58	TN	Injector Bank 1 (INJ 1, 2 & 4)	3.3-3.5 ms
59	WT	Injector Bank 2 (INJ 3, 5 & 6)	3.3-3.5 ms
60	BK/LG	Power Ground	<0.1v
60	BK/WT	Power Ground	<0.1v

Pin Connector Graphic

PCM 60-PIN CONNECTOR

Terminal View of 60-Pin PCM Harness Connector

1993-94 Pickup 4.0L V6 VIN X MFI (All) California 60 Pin

PCM Pin #	Wire Color	Circuit Description (60 Pin)	Value at Hot Idle
1	YL	Keep Alive Power	12-14v
2	LG	Brake Position Switch	0v (Brake On: 12v)
3	GY/BK	VSS (+) Signal	0 Hz, 55 mph: 125 Hz
4	TN/YL	IDM Sensor Signal	20-31 Hz
5, 13-14	---	Not Used	---
6	PK/OR	VSS (-) Signal	0 Hz, 55 mph: 125 Hz
7	LG/RD	ECT Sensor Signal	0.5-0.6v
8	DG/YL	Fuel Pump Monitor	On: 12v, Off: 0v
9	TN/BL	MAF Sensor Return	<0.050v
10	DG/OR	A/C Switch Signal	A/C On: 12v, Off: 0v
11	GY/YL	Canister Purge Solenoid	CANP On at 55 mph: 1v
12	BR/BL	Injector 6 Control	3.3-5.7 ms
12	LG/OR	Injector 6 Control	3.3-5.7 ms
15	TN/BK	Injector 5 Control	3.3-5.7 ms
16	OR/RD	Ignition System Ground	<0.050v
17	PK/LG	Self-Test Output & MIL	MIL Off: 12v, On: 1v
18	TN, PK	Data Bus (+), (-) Signals	Digital Signals, <0.050v
20	BK/LG	PCM Case Ground	<0.050v
21	WT/BL	IAC Solenoid Control	10.7v (33%)
22	BL/OR	Fuel Pump Control	Off: 12v, On: 1v
23, 28, 31-32	---	Not Used	---
24	DB/OR	CMP Sensor Signal	7 Hz, 55 mph: 15 Hz
25	GY	IAT Sensor Signal	1.5-2.5v
26	BR/WT	Reference Voltage	4.9-5.1v
27	BR/LG	DPFE Sensor Signal	0.20-1.30v
29	DG	Octane Adjustment	9.3v (shorted: 0v)
30	BL/YL	A/T: Neutral Drive Switch	In 'P': 0v, Others: 5v
30	BL/YL	M/T: Clutch Engage Switch	5v (clutch "in": 0v)
33	BR/PK	VR Solenoid Control	0%, 55 mph: 45%
34, 38, 41-42	---	Not Used	---
35	BR/BL	Injector 4 Control	3.3-5.7 ms
36	PK	Spark Output Signal	6.93v (50%)
37, 57	RD	Vehicle Power (Start-Run)	12-14v
39	BR/YL	Injector 3 Control	3.3-5.7 ms
40	BK/WT	Power Ground	<0.1v
43	RD/BK	HO2S-21 (B2 S1) Signal	0.1-1.1v
44	GY/BL	HO2S-11 (B1 S1) Signal	0.1-1.1v
45, 49, 52, 55	---	Not Used	---
46	GY/RD	Analog Signal Return	<0.050v
47	GY/WT	TP Sensor Signal	1v, 55 mph: 1.4v
48	WT/PK	Self-Test Input Signal	STI On: 1v, Off: 5v
50	BL/RD	MAF Sensor Signal	0.8v
51	OR/YL	Shift Solenoid 3-4 Control	Solenoid Off: 12v, On: 1v
53	PK/YL	TCC Solenoid Control	0v, On 55 mph: 12v
54	PK/YL	A/C WOT Relay Control	Off: 12v, On: 1v
56	GY/OR	PIP Sensor Signal	6.93v (50%)
58	TN	Injector 1 Control	3.3-5.7 ms
59	WT	Injector 2 Control	3.3-5.7 ms
60	BK/WT	Power Ground	<0.1v

Pin Connector Graphic

PCM 60-PIN CONNECTOR

Terminal View of 60-Pin PCM Harness Connector

1999-2000 Pickup 4.0L V6 MFI VIN X (All) 104 Pin Connector

PCM Pin #	Wire Color	Circuit Description (104 Pin)	Value at Hot Idle
1	PK/OR	Shift Solenoid 'B' Control	1v, 55 mph: 12v
2	PK/LG	MIL (lamp) Control	MIL Off: 12v, On: 1v
3	YL/BK	Digital TR1 Sensor	0v, 55 mph: 11.5v
4-5	---	Not Used	---
6	DB/YL	OSS Sensor Signal	115 Hz, 55 mph: 260 Hz
7-11	---	Not Used	---
12	YL/WT	Fuel Level Indicator Signal	1.7v (1/2 full)
13	PK	Flash EEPROM Power	0.1v
14	LB/BK	4x4 Low Indicator Switch	12v (switch closed: 0v)
15	PK/LB	Data Bus (-) Signal	Digital Signals
16	TN/OR	Data Bus (+) Signal	Digital Signals
17-20	---	Not Used	---
21	DB	CKP (+) Sensor Signal	440-490 Hz
22	GY	CKP (-) Sensor Signal	440-490 Hz
23	---	Not Used	---
24	BK/WT	Power Ground	<0.1v
25	BK	Case Ground	<0.050v
26	TN/WT	Coil 1 Driver (dwell)	6°, 55 mph: 9°
27	OR/YL	Shift Solenoid 'A' Control	1v, 55 mph: 12v
28	BR/OR	Shift Solenoid D Control	1v, 55 mph: 12v
29	TN/WT	TCS (switch) Signal	TCS & O/D On: 12v
30-34	---	Not Used	---
35	RD/LG	HO2S-13 (B1 S3) Signal	0.1-1.1v
36	TN/LB	MAF Sensor Return	<0.050v
37	OR/BK	TFT Sensor Signal	2.10-2.40v
38	LG/RD	ECT Sensor Signal	0.5-0.6v
39	GY	IAT Sensor Signal	1.5-2.5v
40	DG/YL	Fuel Pump Monitor	On: 12v, Off: 0v
41	PK	A/C Demand Signal	A/C On: 12v, Off: 0v
42-47	---	Not Used	---
48	TN/YL	Clean Tachometer Output	25-38 Hz
49	LB/BK	Digital TR2 Sensor	0v, 55 mph: 11.5v
50	WT/BK	Digital TR4 Sensor	0v, 55 mph: 11.5v
51	BK/WT	Power Ground	<0.1v
52	TN/OR	Coil 2 Driver (dwell)	6°, 55 mph: 9°
53	PK/BK	Shift Solenoid 'C' Control	1v, 55 mph: 12v
54	PK/YL	TCC Solenoid Control	0%, 55 mph: 95%
55	Y	Keep Alive Power	12-14v
56	LG/BK	EVAP Purge Solenoid	0-10 Hz (0-100%)
57	---	Not Used	---
58	GY/BK	VSS (+) Signal	0 Hz, 55 mph: 125 Hz
59	---	Not Used	---
60	GY/LB	HO2S-11 (B1 S1) Signal	0.1-1.1v
61	---	Not Used	---
62	RD/PK	FTP Sensor Signal	2.6v (1/2 full)
63	---	Not Used	---
64	LB/YL	A/T: TR3A Sensor Signal	In 'P': 0v, in O/D: 5v
64	LB/YL	M/T: CPP Switch Signal	5v (clutch "in": 0v)

1999-2000 Pickup 4.0L V6 MFI VIN X (All) 104 Pin Connector

PCM Pin #	Wire Color	Circuit Description (104 Pin)	Value at Hot Idle
65	BR/LG	DPFE Sensor Signal	0.20-1.30v
66	---	Not Used	---
67	PK/WT	EVAP CV Solenoid	0-10 Hz (0-100%)
68	---	Not Used	---
69	PK/YL	A/C Clutch Relay Control	Off: 12v, On: 1v
70	---	Not Used	---
71	RD	Vehicle Power	12-14v
72	---	Not Used	---
73	TN/BK	Injector 5 Control	2.7-3.5 ms
74	BR/YL	Injector 3 Control	2.7-3.5 ms
75	TN	Injector 1 Control	2.7-3.5 ms
76-77	BK/WT	Power Ground	<0.1v
78	TN/LG	Coil 3 Driver (dwell)	6°, 55 mph: 9°
79	WT/LG	TCIL (lamp) Control	7.7v (Switch On: 0v)
80	LB/OR	Fuel Pump Relay Control	Off: 12v, On: 1v
81	WT/YL	EPC Solenoid Control	9.1v (4 psi)
82	---	Not Used	---
83	WT/LB	IAC Solenoid Control	10.7v (33%)
84	DG/WT	TSS Sensor Signal	115 Hz, 55 mph: 260 Hz
85	DB/OR	CMP Sensor Signal	7 Hz, 55 mph: 15 Hz
86	BK/YL	A/C Switch Signal	A/C On: 12v, Off: 0v
87	RD/BK	HO2S-12 (B1 S2) Signal	0.1-1.1v
88	LB/RD	MAF Sensor Signal	0.8v, 55 mph: 1.6v
89	GY/WT	TP Sensor Signal	0.53-1.27v
90	BR/WT	Reference Voltage	4.9-5.1v
91	GY/RD	Analog Signal Return	<0.050v
92	RD/LG	Brake Pedal Switch	0v (Brake On: 12v)
93	RD/WT	HO2S-11 (B1 S1) Heater	1v (Heater Off: 12v)
94	YL/LB	HO2S-12 (B1 S2) Heater	1v (Heater Off: 12v)
95	WT/BK	HO2S-13 (B1 S3) Heater	1v (Heater Off: 12v)
96	---	Not Used	---
97	RD	Vehicle Power	12-14v
98	---	Not Used	---
99	LG/OR	Injector 6 Control	2.7-3.5 ms
100	BR/LB	Injector 4 Control	2.7-3.5 ms
101	WT	Injector 2 Control	2.7-3.5 ms
102	---	Not Used	---
103	BK/WT	Power Ground	<0.1v
104	---	Not Used	---

Pin Connector Graphic

```
              PCM 104-PIN CONNECTOR
    1 ●●●●●●●●●●●●●    ●●●●●●●●●●●●● 26
   27 ●●●●●●●●●●●●●    ●●●●●●●●●●●●● 52
   53 ●●●●●●●●●●●●●  ●  ●●●●●●●●●●●●● 78
   79 ●●●●●●●●●●●●●    ●●●●●●●●●●●●● 104
    Terminal View of 104-Pin PCM Wiring Harness Connector
```

Standard Colors and Abbreviations

Abbreviation	Color	Abbreviation	Color	Abbreviation	Color
BK	Black	GY	Gray	PK	Purple
BL	Blue	GN	Green	RD	Red
BR	Brown	LG	LT Green	TN	Tan
DB	Dark Blue	OR	Orange	WT	White
DG	DK Green	PK	Pink	YL	Yellow

2002-03 Pickup 4.0L V6 MFI VIN E (All) 104 Pin Connector

PCM Pin #	Wire Color	Circuit Description (104 Pin)	Value at Hot Idle
1	VT/OR	Shift Solenoid 'B' Control	1v, 55 mph: 12v
2	PK/LG	MIL (lamp) Control	MIL Off: 12v, On: 1v
3	YL/BK	Digital TR1 Sensor	0v, 55 mph: 11.5v
4-5	---	Not Used	---
6	DG/WT	TSS Sensor Signal	115 Hz, 55 mph: 260 Hz
7-11	---	Not Used	---
12	YL/WT	Fuel Level Indicator Signal	1.7v (1/2 full)
13	VT	Flash EEPROM Power	0.1v
14	LB/BK	4x4 Low Indicator Switch	12v (switch closed: 0v)
15	PK/LB	SCP Data Bus (-) Signal	<0.050v
16	TN/OR	SCP Data Bus (+) Signal	Digital Signals
17	GY/OR	Passive Antitheft RX Signal	Digital Signals
18	WT/LG	Passive Antitheft TX Signal	Digital Signals
19-20	---	Not Used	---
21	DB	CKP (+) Sensor Signal	440-490 Hz
22	GY	CKP (-) Sensor Signal	440-490 Hz
23	---	Not Used	---
24	BK/WT	Power Ground	<0.1v
25	BK	Case Ground	<0.050v
26	TN/WT	Coil 1 Driver (dwell)	6°, 55 mph: 9°
27	OR/YL	Shift Solenoid 'A' Control	1v, 55 mph: 12v
28	BR/OR	Shift Solenoid 'D' Control	1v, 55 mph: 12v
29	TN/WT	TCS (switch) Signal	TCS & O/D On: 12v
30	DB/LG	Antitheft Indicator Control	Indicator Off: 12v, On: 1v
31	---	Not Used	---
32	DG/VT	Knock Sensor Signal	0v
33-34	---	Not Used	---
35	RD/LG	HO2S-13 (B1 S3) Signal	0.1-1.1v
36	TN/LB	MAF Sensor Return	<0.050v
37	OR/BK	TFT Sensor Signal	2.10-2.40v
38	LG/RD	ECT Sensor Signal	0.5-0.6v
39	GY	IAT Sensor Signal	1.5-2.5v
40	DG/YL	Fuel Pump Monitor	On: 12v, Off: 0v
41	VT	A/C Demand Signal	A/C On: 12v, Off: 0v
42-43	---	Not Used	---
44	DB/OR	Starter Relay Control Circuit	Relay Off: 12v, On: 1v
45-46	---	Not Used	---
47	BN/PK	EGR Vacuum Regulator Control	0%, 55 mph: 45%
48	TN/YL	Clean Tachometer Output	25-38 Hz
49	LB/BK	Digital TR2 Sensor	0v, 55 mph: 11.5v
50	WT/BK	Digital TR4 Sensor	0v, 55 mph: 11.5v
51	BK/WT	Power Ground	<0.1v
52	TN/OR	Coil 2 Driver (dwell)	6°, 55 mph: 9°
53	PK/BK	Shift Solenoid 'C' Control	1v, 55 mph: 12v
54	VT/YL	TCC Solenoid Control	0%, 55 mph: 95%
55	BK/LG	Keep Alive Power	12-14v
56	LG/BK	EVAP Purge Solenoid Control	0-10 Hz (0-100%)
57	YL/RD	Knock Sensor Signal	0v
58	---	Not Used	---
59	GY/OR	Intermediate Speed Sensor Signal	0 Hz, 55 mph: 125 Hz
60	GY/LB	HO2S-11 (B1 S1) Signal	0.1-1.1v
61	---	Not Used	---
62	RD/PK	FTP Sensor Signal	2.6v (1/2 full)
63	---	Not Used	---
64	LB/YL	A/T: TR3A Sensor Signal	In 'P': 0v, in O/D: 5v
64	LB/YL	M/T: CPP Switch Signal	5v (clutch "in": 0v)

2002-03 Pickup 4.0L V6 MFI VIN X (All) 104 Pin Connector

PCM Pin #	Wire Color	Circuit Description (104 Pin)	Value at Hot Idle
65	BR/LG	EGR DPFE Sensor Signal	0.95-1.05v
66	---	Not Used	---
67	VT/WT	EVAP Canister Vent Solenoid Control	0-10 Hz (0-100%)
68	GY/BK	Vehicle Speed Sensor Signal	0 Hz, 55 mph: 125 Hz
69	PK/YL	A/C Clutch Relay Control	Off: 12v, On: 1v
70	---	Not Used	---
71	RD	Vehicle Power (Start-Run)	12-14v
72	---	Not Used	---
73	TN/BK	Injector 5 Control	2.7-3.5 ms
74	BR/YL	Injector 3 Control	2.7-3.5 ms
75	TN	Injector 1 Control	2.7-3.5 ms
76-77	BK/WT	Power Ground	<0.1v
78	TN/LG	Coil 3 Driver (dwell)	6°, 55 mph: 9°
79	WT/LG	TCIL (lamp) Control	7.7v (Switch On: 0v)
80	LB/OR	Fuel Pump Relay Control	Off: 12v, On: 1v
81	WT/YL	EPC Solenoid Control	9.1v (4 psi)
82	BK/WT	Check Fuel Cap Indicator Control	Indicator Off: 12v, On: 1v
83	WT/LB	IAC Solenoid Control	10.7v (33%)
84	DB/YL	OSS Sensor Signal	115 Hz, 55 mph: 260 Hz
85	DB/OR	CMP Sensor Signal	7 Hz, 55 mph: 15 Hz
86	BK/YL	A/C Switch Signal	A/C On: 12v, Off: 0v
87	RD/BK	HO2S-12 (B1 S2) Signal	0.1-1.1v
88	LB/RD	MAF Sensor Signal	0.8v, 55 mph: 1.6v
89	GY/WT	TP Sensor Signal	0.53-1.27v
90	BR/WT	Reference Voltage	4.9-5.1v
91	GY/RD	Analog Signal Return	<0.050v
92	RD/LG	Brake Pedal Switch	0v (Brake On: 12v)
93	RD/WT	HO2S-11 (B1 S1) Heater	1v (Heater Off: 12v)
94	YL/LB	HO2S-21 (B2 S1) Heater	1v (Heater Off: 12v)
95	WT/BK	HO2S-12 (B1 S2) Heater	1v (Heater Off: 12v)
96	---	Not Used	---
97	RD	Vehicle Power	12-14v
98	---	Not Used	---
99	LG/OR	Injector 6 Control	2.7-3.5 ms
100	BR/LB	Injector 4 Control	2.7-3.5 ms
101	WT	Injector 2 Control	2.7-3.5 ms
102	---	Not Used	---
103	BK/WT	Power Ground	<0.1v
104	---	Not Used	---

Pin Connector Graphic

PCM 104-PIN CONNECTOR

Terminal View of 104-Pin PCM Wiring Harness Connector

Standard Colors and Abbreviations

Abbreviation	Color	Abbreviation	Color	Abbreviation	Color
BK	Black	GY	Gray	PK	Purple
BL	Blue	GN	Green	RD	Red
BR	Brown	LG	LT Green	TN	Tan
DB	Dark Blue	OR	Orange	WT	White
DG	DK Green	PK	Pink	YL	Yellow

Blackwood Pin Tables

2002-03 Pickup 5.4L V8 4v VIN A (A/T) 104 Pin Connector

PCM Pin #	Wire Color	Circuit Description (104 Pin)	Value at Hot Idle
1	OR/YL	COP 6 Driver (dwell)	5°, 55 mph: 8°
2	---	Not Used	---
3	BK/WT	Power Ground	<0.1v
4-5	---	Not Used	---
7-10	OR/YL	Shift Solenoid 1 Control	1v, 55 mph: 12v
4-5	---	Not Used	---
11	VT/OR	Shift Solenoid 2 Control	12v, 55 mph: 12v
12	WT/LG	TCIL (lamp) Control	TCIL Off: 12v, On: 1v
13	VT	Flash EEPROM Power	0.1v
14	LB/BK	4x4 Low Indicator Switch	Switch On: 0v, Off: 12v
15	PK/LB	Data Bus (-) Signal	Digital Signals
16	TN/OR	Data Bus (+) Signal	Digital Signals
17-18	---	Not Used	---
19	DB/WT	Air Suspension Control	Digital Signals
20	BR/OR	Coast Clutch Solenoid	12v, 55 mph: 12v
21	DB	CKP (+) Sensor Signal	400-420 Hz
22	GY	CKP (-) Sensor Signal	400-420 Hz
23	---	Not Used	---
24	BK/WT	Power Ground	<0.1v
25	LB/YL	Case Ground	<0.050v
26	LG/WT	COP 1 Driver (dwell)	5°, 55 mph: 8°
27	OR/YL	COP 5 Driver (dwell)	5°, 55 mph: 8°
28	---	Not Used	---
29	TN/WT	TCS (switch) Signal	TCS & O/D On: 12v
32	DG/VT	Knock Sensor 1 Signal	0v
33	---	Not Used	---
34	YL/BK	Digital TR1 Sensor	0v, 55 mph: 11.5v
35	RD/LG	HO2S-12 (B1 S2) Signal	0.1-1.1v
36	TN/LB	MAF Sensor Return	<0.050v
37	OR/BK	TFT Sensor Signal	2.10-2.40v
38	LG/RD	ECT Sensor Signal	0.5-0.6v
39	GY	IAT Sensor Signal	1.5-2.5v
40	DG/YL	Fuel Pump Monitor	On: 12v, Off: 0v
41	BK/YL	A/C High Pressure Switch	Switch On: 12v (open)
42	PK/OR	TP Sensor Signal	0.53-1.27v
43	OR/LG	Overhead Console Module	Digital Signals
43	OR/LG	Data Output Link	Digital Signals
46	BR	IMRC (valve) Control	12v
47	BR/PK	VR Solenoid Control	0%, 55 mph: 45%
48	---	Not Used	---
49	LB/BK	Digital TR2 Sensor	0v, 55 mph: 11.5v
50	WT/BK	Digital TR4 Sensor	0v, 55 mph: 11.5v
51	BK/WT	Power Ground	<0.1v
52	WT/PK	COP 3 Driver (dwell)	5°, 55 mph: 8°
53	DG/VT	COP 4 Driver (dwell)	5°, 55 mph: 8°
54	VT/YL	TCC Solenoid Control	0%, 55 mph: 95%
55	RD/WT	Keep Alive Power	12-14v
56	LG/BK	EVAP Purge Solenoid	0-10 Hz (0-100%)
57	YL/RD	Knock Sensor 1 Signal	0v
58	---	Not Used	---
59	DG/WT	TSS Sensor Signal	370 Hz (700 rpm)
60	GY/LB	HO2S-11 (B1 S1) Signal	0.1-1.1v
61	VT/LG	HO2S-22 (B2 S2) Signal	0.1-1.1v
62	RD/PK	FTP Sensor Signal	2.6v (0" H2O - cap off)
63	---	Not Used	---
64	LB/YL	Digital TR3A Sensor	In 'P': 0v, in O/D: 1.7v

2002-03 Pickup 5.4L V8 4v VIN A (A/T) 104 Pin Connector

PCM Pin #	Wire Color	Circuit Description (104 Pin)	Value at Hot Idle
65	BR/LG	DPFE Sensor Signal	0.95-1.05v
66	YL/LG	CHT Sensor Signal	0.6 to 1.7v (194°F)
67	VT/WT	EVAP CV Solenoid	0-10 Hz (0-100%)
68	GY/BK	Vehicle Speed Out Signal	0 Hz, 55 mph: 125 Hz
69	PK/YL	A/C WOT Relay Control	Off: 12v, On: 1v
70	BK/WT	Check Fuel Cap Indicator Control	Indicator Off: 12v, On: 1v
71	RD	Vehicle Power (Start-Run)	12-14v
72	TN/RD	Injector 7 Control	3.2-3.8 ms
73	TN/BK	Injector 5 Control	3.2-3.8 ms
74	BR/YL	Injector 3 Control	3.2-3.8 ms
75	TN	Injector 1 Control	3.2-3.8 ms
77	BK/WT	Power Ground	<0.1v
78	PK/LB	COP 7 Driver (dwell)	5°, 55 mph: 8°
79	WT/RD	COP 8 Driver (dwell)	5°, 55 mph: 8°
80	LB/OR	Fuel Pump Relay Control	Off: 12v, On: 1v
81	WT/YL	EPC Solenoid Control	9.1v (5 psi)
82, 86	---	Not Used	---
83	WT/LB	IAC Solenoid Control	10.7v (33%)
84	DB/YL	OSS Sensor Signal	0 Hz, 55 mph: 125 Hz
85	DG	CMP Sensor Signal	6 Hz, 55 mph: 14-17 Hz
87	RD/BK	HO2S-21 (B2 S1) Signal	0.1-1.1v
88	LB/RD	MAF Sensor Signal	0.8v, 55 mph: 2.3v
89	GY/WT	TP Sensor Signal	0.53-1.27v
90	BR/WT	Reference Voltage	4.9-5.1v
91	GY/RD	Analog Signal Return	<0.050v
92	RD/LG	Brake Pedal Switch Signal	0v (brake on: 12v)
93	RD/WT	HO2S-11 (B1 S1) Heater	Off: 12v, On: 1v
94	YL/LB	HO2S-21 (B2 S1) Heater	Off: 12v, On: 1v
95	WT/BK	HO2S-12 (B1 S2) Heater	Off: 12v, On: 1v
96	TN/YL	HO2S-22 (B2 S2) Heater	Off: 12v, On: 1v
97	RD	Vehicle Power (Start-Run)	12-14v
98	LB	Injector 8 Control	3.2-3.8 ms
99	LG/OR	Injector 6 Control	3.2-3.8 ms
100	BR/LB	Injector 4 Control	3.2-3.8 ms
101	WT	Injector 2 Control	3.2-3.8 ms
102	---	Not Used	---
103	BK/WT	Power Ground	<0.1v
104	PK/WT	COP 2 Driver (dwell)	5°, 55 mph: 8°

Pin Connector Graphic

```
           PCM 104-PIN CONNECTOR
 1 ●●●●●●●●●●●●●      ●●●●●●●●●●●●● 26
27 ●●●●●●●●●●●●●      ●●●●●●●●●●●●● 52
53 ●●●●●●●●●●●●●    ⬤ ●●●●●●●●●●●●● 78
79 ●●●●●●●●●●●●●      ●●●●●●●●●●●●● 104
```

Terminal View of 104-Pin PCM Wiring Harness Connector

Standard Colors and Abbreviations

Abbreviation	Color	Abbreviation	Color	Abbreviation	Color
BK	Black	GY	Gray	PK	Purple
BL	Blue	GN	Green	RD	Red
BR	Brown	LG	LT Green	TN	Tan
DB	Dark Blue	OR	Orange	WT	White
DG	DK Green	PK	Pink	YL	Yellow

FORD SUV CONTENTS

About This Section
Introduction ....Page 13-3
How to use this Section ....Page 13-3

AVIATOR PIN TABLES
4.6L V8 MFI VIN H (A/T) 104 Pin **(2003)** ....Page 13-4

BRONCO PIN TABLES
4.9L I6 MFI VIN Y (All) 60 Pin **(1990)** ....Page 13-7
4.9L I6 MFI VIN Y (E4OD) 60 Pin **(1990)** ....Page 13-8
4.9L I6 MFI VIN Y (All) 60 Pin **(1991-94)** ....Page 13-9
4.9L I6 MFI VIN Y (E4OD) 60 Pin **(1991-94)** ....Page 13-10
4.9L I6 MFI VIN Y (All) California 60 Pin **(1995)** ....Page 13-11
4.9L I6 MFI VIN Y (All) 60 Pin **(1995)** ....Page 13-12
4.9L I6 MFI VIN Y (E4OD) California 60 Pin **(1995)** ....Page 13-13
4.9L I6 MFI VIN Y (E4OD) 60 Pin **(1995)** ....Page 13-15
5.0L V8 MFI VIN N (All) 60 Pin **(1990)** ....Page 13-17
5.0L V8 MFI VIN N (A/T) 60 Pin **(1991-93)** ....Page 13-18
5.0L V8 MFI VIN N [4R70W] 60 Pin **(1994-95)** ....Page 13-19
5.0L V8 MFI VIN N (All) 60 Pin **(1994-95)** ....Page 13-21
5.0L V8 MFI VIN N (E4OD) 60 Pin **(1994-95)** ....Page 13-22
5.0L V8 MFI VIN N (All) 104 Pin **(1996)** ....Page 13-24
5.0L V8 MFI VIN N (E4OD) - **(1996)** ....Page 13-26
5.8L V8 MFI VIN H (All) 60 Pin **(1990)** ....Page 13-28
5.8L V8 MFI VIN H (E4OD) 60 Pin **(1990)** ....Page 13-29
5.8L V8 MFI VIN H (All) 60 Pin **(1991-93)** ....Page 13-30
5.8L V8 MFI VIN H (E4OD) 60 Pin **(1991-93)** ....Page 13-31
5.8L V8 MFI VIN H (E4OD) 60 Pin **(1994)** ....Page 13-32
5.8L V8 MFI VIN H (E4OD) California 60 Pin **(1995)** ....Page 13-33
5.8L V8 MFI VIN H (E4OD) 60 Pin **(1995)** ....Page 13-35
5.8L V8 MFI VIN H (E4OD) 104 Pin **(1996)** ....Page 13-36

ESCAPE PIN TABLES
2.0L I4 4v SFI VIN B (All) 104 Pin **(2001-03)** ....Page 13-38
3.0L V6 4v SFI VIN 1 (A/T) 104 Pin **(2001-03)** ....Page 13-40

EXCURSION PIN TABLES
5.4L V8 SOHC MFI VIN L (A/T) 104 Pin **(2000-02)** ....Page 13-42
5.4L V8 SOHC MFI VIN L (A/T) 104 Pin **(2003)** ....Page 13-44
6.8L V10 SOHC VIN S (A/T) 104 Pin **(2000-02)** ....Page 13-46
6.8L V10 SOHC VIN S (A/T) 104 Pin **(2003)** ....Page 13-48
6.8L V10 SOHC Bi-Fuel VIN Z (A/T) 104 Pin **(2000-01)** ....Page 13-50
6.8L V10 SOHC Bi-Fuel VIN Z (A/T) 104 Pin **(2002-03)** ....Page 13-52
7.3L V8 OHV Diesel VIN F (A/T) 104 Pin **(2000-01)** ....Page 13-54
7.3L V8 OHV Diesel VIN F (A/T) 104 Pin **(2002-03)** ....Page 13-56

FORD SUV CONTENTS

EXPEDITION PIN TABLES
4.6L V8 MFI VIN 6, VIN W (A/T) 104 Pin **(1997)** .... Page 13-58
4.6L V8 MFI VIN 6, VIN W (A/T) 104 Pin **(1998-99)** .... Page 13-60
4.6L V8 SOHC VIN 6 (A/T) 104 Pin **(2000-02)** .... Page 13-62
4.6L V8 SOHC VIN W (A/T) 104 Pin **(2003)** .... Page 13-64
5.4L V8 SOHC MFI VIN L (E4OD) 104 Pin **(1997)** .... Page 13-67
5.4L V8 SOHC MFI VIN L (E4OD) 104 Pin **(1998-99)** .... Page 13-69
5.4L V8 SOHC MFI VIN L (A/T) 104 Pin **(2000-02)** .... Page 13-71
5.4L V8 SOHC MFI VIN L (A/T) 122 Pin **(2003)** .... Page 13-73

EXPLORER PIN TABLES
4.0L V6 SOHC MFI VIN E (A/T) 104 Pin **(1997)** .... Page 13-76
4.0L V6 SOHC MFI VIN E (A/T) 104 Pin **(1998-99)** .... Page 13-78
4.0L V6 SOHC MFI VIN E (A/T) 150-Pin **(2000-02)** .... Page 13-80
4.0L V6 SOHC MFI VIN K (A/T) 104 Pin **(2002-03)** .... Page 13-82
4.0L V6 MFI VIN X (All) 60 Pin **(1991-94)** .... Page 13-85
4.0L V6 MFI VIN X (All) California 60 Pin **(1993-94)** .... Page 13-86
4.0L V6 MFI VIN X (All) 60 Pin **(1995)** .... Page 13-88
4.0L V6 MFI VIN X (All) 104 Pin **(1996-97)** .... Page 13-90
4.0L V6 MFI VIN X (All) 104 Pin **(1998-2002)** .... Page 13-92
4.6L V8 MFI VIN W (All) 150-Pin **(2002-03)** .... Page 13-94
5.0L V8 MFI VIN P (A/T) 104 Pin **(1996-97)** .... Page 13-97
5.0L V8 MFI VIN P (A/T) 104 Pin **(1998-2001)** .... Page 13-99

MOUNTAINEER PIN TABLES
4.0L V6 MFI VIN E (A/T) 104 Pin **(1998-99)** .... Page 13-101
4.0L V6 SOHC MFI VIN E (A/T) 150-Pin **(2000-02)** .... Page 13-103
4.0L V6 SOHC MFI VIN K (A/T) 104 Pin **(2002-03)** .... Page 13-105
4.6L V8 MFI VIN W (All) 150-Pin **(2002-03)** .... Page 13-108
5.0L V8 MFI VIN P (A/T) 104 Pin **(1997)** .... Page 13-111
5.0L V8 MFI VIN P (A/T) 104 Pin **(1998-2001)** .... Page 13-113

NAVIGATOR PIN TABLES
5.4L V8 MFI VIN L (4R100) 104 Pin **(1998-99)** .... Page 13-115
5.4L V8 MFI DOHC 4v VIN A 104 Pin **(1999-2001)** .... Page 13-117
5.4L V8 MFI DOHC 4v VIN R 104 Pin **(2001-02)** .... Page 13-119
5.4L V8 MFI DOHC 4v VIN R 122 Pin **(2003)** .... Page 13-121

About This Section

Introduction

This section of the Chilton Diagnostic Manual contains Pin Tables for Sports Utility Vehicles from 1990-2003. It can be used to assist in the repair of Code and No Code faults related to the PCM.

VEHICLE COVERAGE

- Aviator Applications (2003)
- Bronco Applications (1990-96)
- Bronco II Applications (1990)
- Escape Applications (2001-03)
- Excursion Applications (2000-2003)
- Expedition Applications (1997-2003)
- Explorer Applications (1991-2003)
- Mountaineer Applications (1996-2003)
- Navigator Applications (1997-2003)

How to Use This Section

This Section of the Handbook can be used to look up the location of a particular pin, a Wire Color or a "known good" value of a PCM circuit. To locate the PCM information for a particular vehicle, find the model, correct engine size (with VIN Code) and finally the year of the vehicle.

For example, to look up the PCM terminals for a 1999 Expedition 4.6L VIN W, go to Contents Page 1 and find the text string shown below:

4.6L V8 2v MFI VIN 6, VIN W (A/T) 104 Pin **(1998-99)** ....Page 13-60

Then turn to Page 13-60 to find the following PCM related information.

1998-99 Expedition 4.6L 2v V8 MFI VIN W (A/T)

PCM Pin #	Wire Color	Circuit Description (104 Pin)	Value at Hot Idle
1	OR/YL	COP 6 Driver (dwell)	5°, 55 mph: 8°
9	YL/WT	Fuel Level Indicator Signal	1.7v (1/2 full)
11	PK/OR	Shift Solenoid 2 Control	12v, 55 mph: 1v
66	YL/LG	CHT Sensor Signal	0.6 (194°F)

In this example, the Fuel Level Indicator circuit connects to Pin 9 of the 104 Pin connector with a Yellow/White wire. The value at Hot Idle value shown here (1.7v) is a normal value with the tank 1/2 full.

The cylinder head temperature (CHT) sensor is used to indicate the engine coolant temperature on this vehicle. This device is connected to a circuit with a pull-up resistor. Due to this unique design, the CHT sensor reading at near 194°F can read in the Cold End Line (3.71v) or in the Hot End Line (0.6v). This variance should be understood when using the CHT sensor Pin Table during cooling system testing.

The (A/T) in the Title over the table indicates the coverage is A/T.

Aviator Pin Tables

2003 Aviator 4.6L V8 DOHC 4v VIN H C175B 58 Pin Connector

PCM Pin #	Wire Color	Circuit Description (58 Pin)	Value at Hot Idle
1, 5	---	Not Used	---
2	DG/YL	Fuel Pump Power	Pump On: 12v, Off: 0v
3	TN/OR	SCP Data Bus (+) Signal	Digital Signals
4	PK/LB	SCP Data Bus (-) Signal	<0.050v
6	VT/WT	EVAP CV Solenoid	0-10 Hz (0-100%)
7	GY/BK	Vehicle Speed Sensor (+)	0 Hz, 55 mph: 125 Hz
8	LG/RD	Generator Monitor Signal	130 Hz (45%)
9	PK/YL	A/C Clutch Relay	12v (relay on: 1v)
10	YL/WT	Throttle Position Sensor 2 Signal	0.5-4.1v
11	WT/LG	Passive Antitheft TX Signal	Digital Signals
12	LG/BK	EVAP Canister Purge Solenoid Control	0-10 Hz (0-100%)
13	WT/VT	Module Programming Signal	0.1v
14	GY/OR	Passive Antitheft RX Signal	Digital Signals
15-16, 18	---	Not Used	---
17	GY/RD	Sensor Return	<0.050v
19	WT/BK	Auxiliary Condenser Cooling Fan Relay	Relay Off: 12v, On: 1v
20	BR/WT	Reference Voltage	4.9-5.1v
21	---	Not Used	---
22	DB/LG	Passive Antitheft Indicator Control	Indicator Off: 12v, On: 1v
23-24	LB/BK	4WD Indicator Low Signal	12v (switch on: 1v)
25-26	BK/WT	Communication Network Ground	<0.050v
27	BK/WT	Communication Network Ground	<0.050v
28	BK/YL	Brake Pressure Switch	12v (Brake On: 0v)
29	OR/LB	Speed Control Motor 'A' Control	DC pulse signals
30	BK/YL	A/C Cyclic Pressure Switch	A/C Off: 0v, On: 12v
31	LB/RD	MAF Sensor Signal	0.7v, 55 mph: 1.8v
32	VT	Vehicle Power (Start-Run)	12-14v
33	VT	Vehicle Power (Start-Run)	12-14v
34-35	---	Not Used	---
36	LG/WT	Speed Control Motor 'B' Control	DC pulse signals
37	YL/LG	Power Steering Pressure Switch	Straight: 0v, Turned: 12v
38	TN/LB	MAF Sensor Return	<0.050v
39	OR	Starter Motor Relay Circuit	Relay Off: 0v, On: 12v
40	LG	Brake Position Switch	Brake Off: 12v, On: 1v
41	TN/WT	Transmission Overdrive Cancel Switch	Switch Off: 0v, On: 12v
42, 48-50	---	Not Used	---
43	BK	Power Ground	<0.1v
44	RD/WT	Battery Power	12-14v
45	BK	Speed Control Switch Ground	<0.050v
46	BR/WT	Speed Control Motor 'B' Control	DC pulse signals
47	RD/YL	A/C Head Pressure Switch	Switch Open: 12v, Closed: 0v
51	GY	Air Charge Temperature Sensor	1.5-2.5v
52	RD/PK	Fuel Tank Pressure Sensor	2.6v (0" H20 - cap off)
53-55	---	Not Used	---
56	DG/OR	Speed Control Switch	Off: 0v, On: 6.7v
57	LB/BK	Speed Control On/Off to Amplifier	0v (switch on: 6.7v)
58	LB/OR	Fuel Pump Relay Control	On: 1v, Off: 12v

Pin Connector Graphic

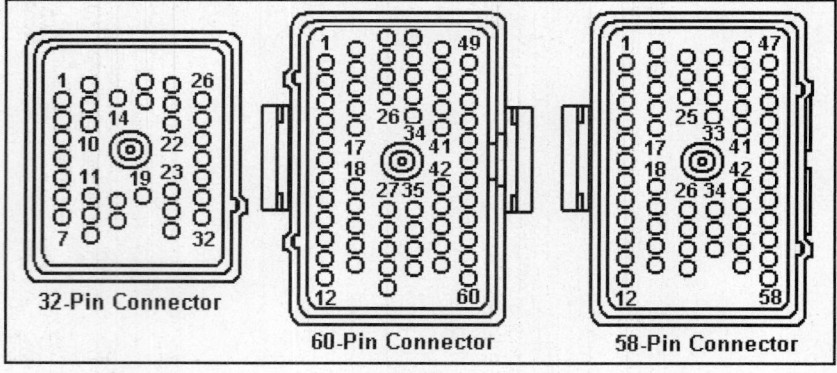

32-Pin Connector   60-Pin Connector   58-Pin Connector

2003 Aviator 4.6L V8 DOHC 4v VIN H C175T 32 Pin Connector

PCM Pin #	Wire Color	Circuit Description (32 Pin)	Value at Hot Idle
1	OR/YL	Shift Solenoid 'A' Control	12v, 55 mph: 1v
2	VT/OR	Shift Solenoid 'B' Control	12v, 55 mph: 1v
3	PK/BK	Shift Solenoid 'C' Control	12v, 55 mph: 1v
4	BR/OR	Shift Solenoid 'D' Control	12v, 55 mph: 1v
5	VT/YL	TCC Solenoid Control	12v, 55 mph: 1v
6	---	Not Used	---
7	WT/YL	Electronic Pressure Solenoid Control	On: 1v, Off: 12v
8	---	Not Used	---
9	LB/YL	Digital TR Sensor (TR3A)	0v, 55 mph: 1.7v
10	WT/BK	Digital TR Sensor (TR4)	0v, 55 mph: 11v
11	---	Not Used	---
12	WT	Pressure Control Solenoid 'C'	On: 1v, Off: 12v
13	LB/PK	Pressure Control Solenoid 'B'	On: 1v, Off: 12v
14	---	Not Used	---
15	WT/BK	HO2S-12 (B1 S2) Heater Control	Heater Off: 12v, On: 1v
16	TN/YL	HO2S-22 (B2 S2) Heater Control	Heater Off: 12v, On: 1v
17	GY/RD	Sensor Return	<0.050v
18	LB/BK	Digital TR Sensor (TR2)	0v, 55 mph: 11v
19-20	---	Not Used	---
21	YL/LG	Overdrive Drum Speed Sensor	200 Hz, 55 mph: 1185 Hz
22	YL/BK	Digital TR Sensor (TR1)	0v, 55 mph: 11v
23	OR/BK	Transmission Fluid Temperature Sensor	2.10-2.40v
24-25	---	Not Used	---
26	DB/YL	Output Shaft Speed Sensor	0 Hz, 985 Hz
27	DG/WT	Turbine Shaft Speed Sensor	360 Hz, 890 Hz
28	RD/LG	HO2S-12 (B1 S2) Signal	0.1-1.1v
29	VT/LG	HO2S-22 (B2 S2) Signal	0.1-1.1v
30-32	---	Not Used	---

Pin Connector Graphic

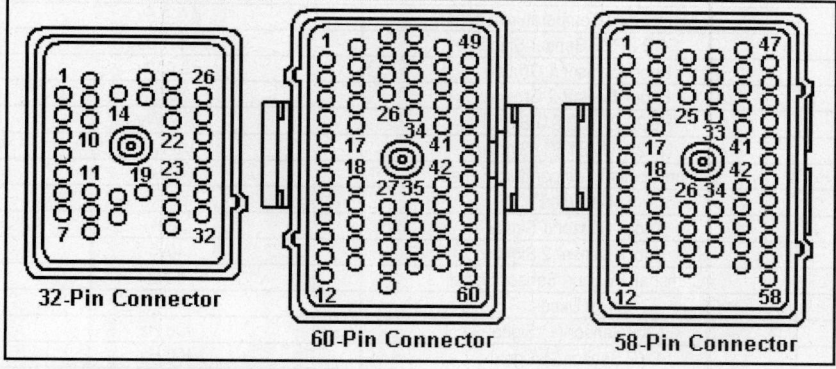

32-Pin Connector

60-Pin Connector

58-Pin Connector

Standard Colors and Abbreviations

Abbreviation	Color	Abbreviation	Color	Abbreviation	Color
BK	Black	GY	Gray	PK	Purple
BL	Blue	GN	Green	RD	Red
BR	Brown	LG	LT Green	TN	Tan
DB	Dark Blue	OR	Orange	WT	White
DG	DK Green	PK	Pink	YL	Yellow

2003 Aviator 4.6L V8 DOHC 4v VIN H C175E 60 Pin Connector

PCM Pin #	Wire Color	Circuit Description (60 Pin)	Value at Hot Idle
1	PK/WT	Coil On Plug (COP) 2 Driver (dwell)	5°, 55 mph: 7°
2	TN	Fuel Injector 1 Control	3.3-3.8 ms
3	DB/LG	Intake Manifold Communication Control	0%, over 3500 rpm: 100%
4-6	---	Not Used	
7	RD/WT	HO2S-11 (B1 S1) Heater Control	Heater Off: 12v, On: 1v
8	YL/LB	HO2S-21 (B2 S1) Heater Control	Heater Off: 12v, On: 1v
9	WT/LB	IAC Solenoid Control	10.7v (33%)
10	---	Not Used	
11	TN/BK	Fuel Injector 5 Control	3.3-3.8 ms
12	OR/YL	Coil On Plug (COP) 6 Driver (dwell)	5°, 55 mph: 7°
13	PK/LB	Coil On Plug (COP) 7 Driver (dwell)	5°, 55 mph: 7°
13	PK/WT	Coil 1 Driver (dwell)	5°, 55 mph: 7°
14	WT	Fuel Injector 2 Control	3.3-3.8 ms
15	---	Not Used	---
16	BR/PK	Electric VR Solenoid Control	0%, 55 mph: 45%
17	GY/RD	Sensor Return	<0.050v
18-19	---	Not Used	
20	BR/WT	Sensor Voltage Reference	4.9-5.1v
21	LG/OR	Fuel Injector 6 Control	3.3-3.8 ms
22	LG/YL	Coil On Plug (COP) 5 Driver (dwell)	5°, 55 mph: 7°
23	WT/PK	Coil On Plug (COP) 3 Driver (dwell)	5°, 55 mph: 7°
24	BR/YL	Fuel Injector 3 Control	3.3-3.8 ms
25-28	---	Not Used	---
29	TN/RD	Fuel Injector 3 Control	3.3-3.8 ms
30	DG/VT	Coil On Plug (COP) 4 Driver (dwell)	5°, 55 mph: 7°
31	LG/WT	Coil On Plug (COP) 1 Driver (dwell)	5°, 55 mph: 7°
32	BR/LB	Fuel Injector 4 Control	3.3-3.8 ms
33-37	---	Not Used	---
38	WT/RD	Fuel Injector 8 Control	3.3-3.8 ms
39	---	Not Used	
40	YL/LG	Cylinder Head Temperature Sensor	0.5-0.6v
41	BR/LG	EGR DPFE Sensor Signal	0.95-1.05v
42	YL	Knock Sensor 1 Ground	<0.050v
43	DG/WT	Knock Sensor 2 Ground	0v
44	RD/BK	HO2S-21 (B2 S1) Signal	0.1-1.1v
45	GY/LB	HO2S-11 (B1 S1) Signal	0.1-1.1v
46	LG/RD	Engine Coolant Temperature Sensor	0.5-0.6v
47-48	---	Not Used	---
51	YL/RD	Knock Sensor 1 Signal	0v
52	DG/VT	Knock Sensor 2 Signal	0v
53	DB/OR	Camshaft Position Sensor Signal	6 Hz
54	---	Not Used	---
55	DB	CKP Sensor (+) Signal	400 Hz
56	GY	CKP Sensor (-) Signal	400 Hz
57	GY/WT	TP Sensor Signal	0.53-1.27v
58	LG/WT	Intake Manifold Communication Sensor	0.5-4.5v
59	LG/BK	Manifold Absolute Pressure Sensor	4v, Over 3500 rpm: 1v
60	---	Not Used	---

Pin Connector Graphic

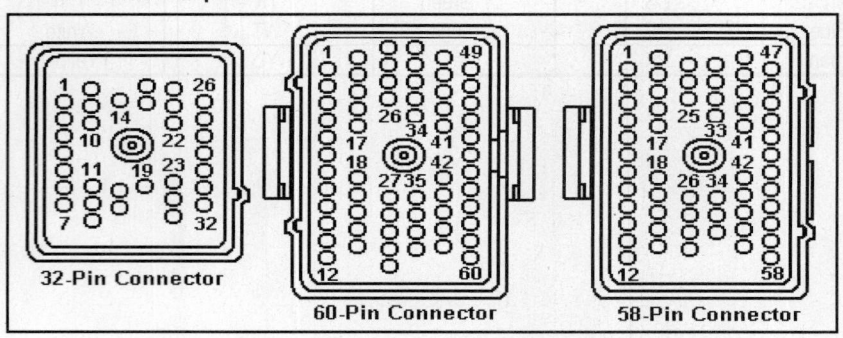

32-Pin Connector

60-Pin Connector

58-Pin Connector

Bronco Pin Tables

1990 Bronco 4.9L I6 MFI VIN Y 60 Pin Connector

PCM Pin #	Wire Color	Circuit Description (60 Pin)	Value at Hot Idle
1	YL	Keep Alive Power	12-14v
3	GY/BK	VSS (+) Signal	0 Hz, 55 mph: 125 Hz
4	DG/YL	Ignition Diagnostic Monitor	20-31 Hz
6	BK	VSS (-) Signal	0 Hz, 55 mph: 125 Hz
7	LG/YL	ECT Sensor Signal	0.5-0.6v
8	BR	Fuel Pump Monitor	On: 12v, Off: 0v
9	BK/OR	Data Bus (-) Signal	Digital Signals
10	BK/YL	A/C Switch Signal	A/C On: 12v, Off: 0v
11	WT/BK	Air Management 2 Solenoid	AM2 On: 1v, Off: 12v
16	BK/OR	Ignition System Ground	<0.050v
17	PK/LG	Self-Test Output & MIL	MIL On: 1v, Off: 12v
18	TN/LG	Inferred Mileage Sensor	Digital Signals
20	BK	PCM Case Ground	<0.050v
21	GY/WT	IAC Solenoid Control	8.8-10.0v
22	TN/LG	Fuel Pump Control	On: 1v, Off: 12v
23	LG/BK	Knock Sensor Signal	0v
24	YL/LG	PSP Switch Signal	Straight: 0v, Turned: 12v
25	YL/RD	ACT Sensor Signal	1.5-2.5v
26	OR/WT	Reference Voltage	4.9-5.1v
27	BR/LG	EVP Sensor Signal	0.4v
28	OR/BK	Data Bus (+) Signal	Digital Signals
29	GY/BL	HO2S-11 (B1 S1) Signal	0.1-1.1v
30	GY/YL	A/T: Neutral Drive Switch	In 'N': 0v, Others: 5v
30	GY/YL	M/T: CPP Switch Signal	Clutch In: 0v, Out: 5v
31	GY/YL	EVAP Purge Solenoid	12v, 55 mph: 1v
33	DG	VR Solenoid Control	0%, 55 mph: 45%
36	YL/LG	Spark Output Signal	6.93v (50%)
37	RD	Vehicle Power	12-14v
40	BK/LG	Power Ground	<0.1v
41-42	---	Not Used	---
43	LG/PK	A/C Demand Switch	A/C On: 12v, Off: 0v
45	LG/BK	MAP Sensor Signal	107 Hz (sea level)
46	BK/WT	Analog Signal Return	<0.050v
47	DG/LG	TP Sensor Signal	0.5-1.2v
48	WT/RD	Self-Test Indicator Signal	STI On: 1v, Off: 5v
49	OR	HO2S-11 (B1 S1) Ground	<0.050v
50	---	Not Used	---
51	WT/RD	Air Management 1 Solenoid	1v, 55 mph: 12v
52-55	---	Not Used	---
56	DB	PIP Sensor Signal	6.93v (50%)
57	RD	Vehicle Power	12-14v
58	TN/OR	Injector Bank 1 (INJ 1, 3 & 5)	6.4-6.8 ms
59	TN/RD	Injector Bank 2 (INJ 2, 4 & 6)	6.4-6.8 ms
60	BK/LG	Power Ground	<0.1v

Pin Connector Graphic

PCM 60-PIN CONNECTOR

Terminal View of 60-Pin PCM Harness Connector

Standard Colors and Abbreviations

Abbreviation	Color	Abbreviation	Color	Abbreviation	Color
BK	Black	GY	Gray	PK	Purple
BL	Blue	GN	Green	RD	Red
BR	Brown	LG	LT Green	TN	Tan
DB	Dark Blue	OR	Orange	WT	White
DG	DK Green	PK	Pink	YL	Yellow

1990 Bronco 4.9L I6 MFI VIN Y (E4OD) 60 Pin Connector

PCM Pin #	Wire Color	Circuit Description (60 Pin)	Value at Hot Idle
1	YL	Keep Alive Power	12-14v
2	LG	Brake Pedal Switch	Brake Off: 12v, On: 1v
3	GY/BK	VSS (+) Signal	0 Hz, 55 mph: 125 Hz
4	DG/YL	Ignition Diagnostic Monitor	20-31 Hz
6	BK	VSS (-) Signal	0 Hz, 55 mph: 125 Hz
7	LG/YL	ECT Sensor Signal	0.5-0.6v
8	BR	Fuel Pump Monitor	On: 12v, Off: 0v
9	BK/OR	Data Bus (-) Signal	Digital Signals
10	BK/YL	A/C Switch Signal	A/C On: 12v, Off: 0v
11	WT/BK	Air Management 2 Solenoid	AM2 On: 1v, Off: 12v
12	BL/BK	4x4 Switch Signal	12v (switch closed: 0v)
16	BK/OR	Ignition System Ground	<0.050v
17	PK/LG	Self-Test Output & MIL	MIL On: 1v, Off: 12v
18	TN/LG	Inferred Mileage Sensor	Digital Signals
19	PK/OR	Shift Solenoid 2 Control	1v, 55 mph: 12v
20	BK/LG	Case Ground	<0.1v
21	GY/WT	IAC Solenoid Control	9.8-10.6v
22	TN/LG	Fuel Pump Control	On: 1v, Off: 12v
23	LG/BK	Knock Sensor Signal	0v
24	YL/LG	PSP Switch Signal	Straight: 0v, Turned: 12v
25	YL/RD	ACT Sensor Signal	1.5-2.5v
26	OR/WT	Reference Voltage	4.9-5.1v
27	BR/LG	EVP Sensor Signal	0.4v
28	OR/BK	Data Bus (+) Signal	Digital Signals
29	GY/BL	HO2S-11 (B1 S1) Signal	0.1-1.1v
30	BL/YL	MLP Sensor Signal	In 'P': 0v, in O/D: 5v
31	GY/YL	EVAP Purge Solenoid	12v, 55 mph: 1v
32	LG/WT	Overdrive Cancel Indicator	OCIL On: 1v, Off: 12v
33	DG	VR Solenoid Control	0%, 55 mph: 45%
36	YL/LG	Spark Output Signal	6.93v (50%)
37	RD	Vehicle Power	12-14v
38	WT/YL	EPC Solenoid Control	9.5v (5 psi)
40	BK/LG	Power Ground	<0.1v
41	WT	Overdrive Cancel Switch	OCS & O/D On: 0.1v
42	OR/BK	TOT Sensor Signal	2.10-2.40v
43	LG/PK	A/C Demand Switch	A/C On: 12v, Off: 0v
45	LG/BK	MAP Sensor Signal	107 Hz (sea level)
46	BK/WT	Analog Signal Return	<0.050v
47	DG/LG	TP Sensor Signal	0.5-1.2v
48	WT/RD	Self-Test Indicator Signal	STI On: 1v, Off: 5v
49	OR	HO2S-11 (B1 S1) Ground	<0.050v
51	WT/RD	Air Management 1 Solenoid	1v, 55 mph: 12v
52	OR/YL	Shift Solenoid 1 Control	12v, 55 mph: 1v
53	PK/YL	TCC Solenoid Control	12v, 55 mph: 1v
55	BR	Coast Clutch Switch Signal	12v, 55 mph: 12v
56	DB	PIP Sensor Signal	6.93v (50%)
57	RD	Vehicle Power	12-14v
58	TN/OR	Injector Bank 1 (INJ 1, 3 & 5)	6.4-6.8 ms
59	TN/RD	Injector Bank 2 (INJ 2, 4 & 6)	6.4-6.8 ms
60	BK/LG	Power Ground	<0.1v

Pin Connector Graphic

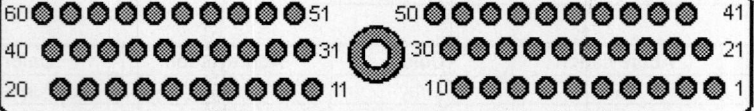

PCM 60-PIN CONNECTOR

Terminal View of 60-Pin PCM Harness Connector

1991-94 Bronco 4.9L I6 MFI VIN Y 60 Pin Connector

PCM Pin #	Wire Color	Circuit Description (60 Pin)	Value at Hot Idle
1	YL	Keep Alive Power	12-14v
3	GY/BK	VSS (+) Signal	0 Hz, 55 mph: 125 Hz
4	TN/YL	Ignition Diagnostic Monitor	20-31 Hz
4	YL/BK	Ignition Diagnostic Monitor	20-31 Hz
6	PK/OR	VSS (-) Signal	0 Hz, 55 mph: 125 Hz
7	LG/RD	ECT Sensor Signal	0.5-0.6v
8	DG/YL	Fuel Pump Monitor	On: 12v, Off: 0v
9	PK/BL	Data Bus (-) Signal	Digital Signals
10	BK/YL	A/C Switch Signal	A/C On: 12v, Off: 0v
10	PK/BL	A/C Switch Signal	A/C On: 12v, Off: 0v
11	BR	Air Management 2 Solenoid	AM2 On: 1v, Off: 12v
16	OR/RD	Ignition System Ground	<0.050v
17	PK/LG	Self-Test Output & MIL	MIL On: 1v, Off: 12v
20	BK	PCM Case Ground	<0.050v
21	WT/BL	IAC Solenoid Control	10.7v (33%)
21	BL/WT	IAC Solenoid Control	10.7v (33%)
22	BL/OR	Fuel Pump Control	On: 1v, Off: 12v
23	YL/RD	Knock Sensor Signal	0v
25	GN/YL	IAT Sensor Signal	1.5-2.5v
25	GY	IAT Sensor Signal	1.5-2.5v
26	BR/WT	Reference Voltage	4.9-5.1v
27	BR/LG	EVP Sensor Signal	0.4v
28	TN/OR	Data Bus (+) Signal	Digital Signals
29	GY/BL	HO2S-11 (B1 S1) Signal	0.1-1.1v
30	BL/YL	A/T: Neutral Drive Switch	In 'N': 0v, Others: 5v
30	BL/YL	M/T: CPP Switch Signal	Clutch In: 0v, Out: 5v
31	GY/YL	EVAP Purge Solenoid	12v, 55 mph: 1v
32	WT/LG	Overdrive Cancel Indicator	OCIL On: 1v, Off: 12v
33	BR/PK	VR Solenoid Control	0%, 55 mph: 45%
36	PK	Spark Output Signal	6.93v (50%)
37	RD	Vehicle Power	12-14v
40	BK/WT	Power Ground	<0.1v
43	PK	A/C Demand Switch	A/C On: 12v, Off: 0v
45	LG/BK	MAP Sensor Signal	107 Hz (sea level)
46	GY/RD	Analog Signal Return	<0.050v
47	GY/WT	TP Sensor Signal	1.5-1.2v
48	WT/PK	Self-Test Indicator Signal	STI On: 1v, Off: 5v
49	OR	HO2S-11 (B1 S1) Ground	<0.050v
51	WT/OR	Air Management 1 Solenoid	1v, 55 mph: 12v
56	GY/OR	PIP Sensor Signal	6.93v (50%)
57	RD	Vehicle Power	12-14v
58	TN	Injector Bank 1 (INJ 1, 3 & 5)	5.6-6.4 ms
59	WT	Injector Bank 2 (INJ 2, 4 & 6)	5.6-6.4 ms
60	BK/WT	Power Ground	<0.1v

Pin Connector Graphic

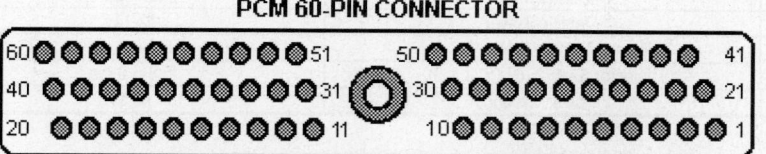

1991-94 Bronco 4.9L I6 MFI VIN Y (E4OD) 60 Pin Connector

PCM Pin #	Wire Color	Circuit Description (60 Pin)	Value at Hot Idle
1	YL	Keep Alive Power	12-14v
2	LG	Brake Pedal Switch	Brake Off: 12v, On: 1v
3	GY/BK	VSS (+) Signal	0 Hz, 55 mph: 125 Hz
4	TN or YL/BK	Ignition Diagnostic Monitor	20-31 Hz
5, 13-15	---	Not Used	---
6	PK/OR	VSS (-) Signal	0 Hz, 55 mph: 125 Hz
7	LG/RD	ECT Sensor Signal	0.5-0.6v
8	DG/YL	Fuel Pump Monitor	On: 12v, Off: 0v
9	PK/BL	Data Bus (-) Signal	Digital Signals
10	BK or PK/BL	A/C Switch Signal	A/C On: 12v, Off: 0v
11	BR	Air Management 2 Solenoid	AM2 On: 1v, Off: 12v
12	BL or PK	4x4 Switch Signal	12v (switch closed: 0v)
16	OR/RD	Ignition System Ground	<0.050v
17	PK/LG	Self-Test Output & MIL	MIL On: 1v, Off: 12v
18, 24	---	Not Used	---
19	PK/OR	Shift Solenoid 2 Control	1v, Off: 12v
20	BK	PCM Case Ground	<0.050v
21	WT or BK	IAC Solenoid Control	10.7v (33%)
22	BL/OR	Fuel Pump Control	On: 1v, Off: 12v
23	YL/RD	Knock Sensor Signal	0v
25	GN/YL or GY	ACT Sensor Signal	1.5-2.5v
26	BR/WT	Reference Voltage	4.9-5.1v
27	BR/LG	EVP Sensor Signal	0.4v
28	TN/OR	Data Bus (+) Signal	Digital Signals
29	GY/BL	HO2S-11 (B1 S1) Signal	0.1-1.1v
30	BL/YL	MLP Sensor Signal	In 'P': 0v, in O/D: 5v
31	GY/YL	EVAP Purge Solenoid	12v, 55 mph: 1v
32	WT/LG	Overdrive Cancel Indicator	OCIL On: 1v, Off: 12v
33	BR/PK	VR Solenoid Control	0%, 55 mph: 45%
34-35, 39, 44	---	Not Used	---
36	PK	Spark Output Signal	6.93v (50%)
37	RD	Vehicle Power	12-14v
38	WT/YL	EPC Solenoid Control	9.5v (5 psi)
40	BK/WT	Power Ground	<0.1v
41	TN/WT	Overdrive Cancel Switch	OCS Closed: 0.1v
42	OR/BK	TOT Sensor Signal	2.10-2.40v
43	PK	A/C Demand Switch	A/C On: 12v, Off: 0v
45	LG/BK	MAP Sensor Signal	107 Hz (sea level)
46	GY/RD	Analog Signal Return	<0.050v
47	GY/WT	TP Sensor Signal	0.5-1.2v
48	WT/PK	Self-Test Indicator Signal	STI On: 1v, Off: 5v
49	OR	HO2S-11 (B1 S1) Ground	<0.050v
50, 54	---	Not Used	---
51	WT/OR	Air Management 1 Solenoid	1v, 55 mph: 12v
52	OR/YL	Shift Solenoid 1 Control	12v, 55 mph: 1v
53	PK/YL	TCC Solenoid Control	12v, 55 mph: 9v
55	BR/OR	Coast Clutch Switch Signal	12v, 55 mph: 12v
56	GY/OR	PIP Sensor Signal	6.93v (50%)
57	RD	Vehicle Power	12-14v
58	TN	Injector Bank 1 (INJ 1, 3 & 5)	5.6-6.0 ms
59	W	Injector Bank 2 (INJ 2, 4 & 6)	5.6-6.0 ms
60	BK/WT	Power Ground	<0.1v

Pin Connector Graphic

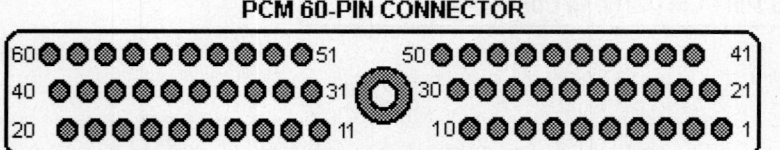

Terminal View of 60-Pin PCM Harness Connector

1995 Bronco 4.9L I6 MFI VIN Y California 60 Pin Connector

PCM Pin #	Wire Color	Circuit Description (60 Pin)	Value at Hot Idle
2	LG	Brake Pedal Switch	Brake Off: 12v, On: 1v
3	GY/BK	VSS (+) Signal	0 Hz, 55 mph: 125 Hz
4	YL/BK	Ignition Diagnostic Monitor	20-31 Hz
6	PK/OR	VSS (-) Signal	0 Hz, 55 mph: 125 Hz
7	LG/RD	ECT Sensor Signal	0.5-0.6v
8	DG/YL	Fuel Pump Monitor	On: 12v, Off: 0v
9	TN/BL	MAF Sensor Return	<0.050v
10	DG/OR	A/C Switch Signal	A/C On: 12v, Off: 0v
11	GY/YL	EVAP Purge Solenoid	12v, 55 mph: 1v
12	LG/OR	Injector 6 Control	5.5-7.1 ms
15	TN/BK	Injector 5 Control	5.5-7.1 ms
16	TN/YL	Ignition System Ground	<0.050v
17	PK/LG	Self-Test Output & MIL	MIL On: 1v, Off: 12v
18	TN/OR	Data Bus (-) Signal	Digital Signals
19	PK/BK	Data Bus (+) Signal	Digital Signals
20	BK	PCM Case Ground	<0.050v
21	WT/BL	IAC Solenoid Control	10.7v (33%)
22	BL/OR	Fuel Pump Control	On: 1v, Off: 12v
23	YL/RD	Knock Sensor Signal	0v
25	GY	IAT Sensor Signal	1.5-2.5v
26	BR/WT	Reference Voltage	4.9-5.1v
27	BR/LG	EVP Sensor Signal	0.4v
30	BL/YL	A/T: Neutral Drive Switch	In 'N': 0v, Others: 5v
30	GY/YL	M/T: CPP Switch Signal	Clutch In: 0v, Out: 5v
31	WT/OR	Thermactor Air Bypass	TAB On: 1v, Off: 12v
32	WT/LG	TCIL (lamp) Control	TCIL On: 1v, Off: 12v
33	BR/PK	VR Solenoid Control	0%, 55 mph: 45%
34	BR	Thermactor Air Diverter	TAD On: 1v, Off: 12v
35	BR/BL	Injector 4 Control	5.5-7.1 ms
36	PK	Spark Output Signal	6.93v (50%)
37, 57	RD	Vehicle Power	12-14v
39	BR/YL	Injector 3 Control	5.5-7.1 ms
40	BK/WT	Power Ground	<0.1v
41	TN/WT	TCS (switch) Signal	TCS & O/D On: 12v
42	BL/YL	4x4 Switch Signal	12v (switch closed: 0v)
43	RD/BK	HO2S-21 (B2 S1) Signal	0.1-1.1v
44	GY/BL	HO2S-11 (B1 S1) Signal	0.1-1.1v
46	GY/RD	Analog Signal Return	<0.050v
47	GY/WT	TP Sensor Signal	0.5-1.2v
48	WT/PK	Self-Test Indicator Signal	STI On: 1v, Off: 5v
50	BL/RD	MAF Sensor Signal	0.8v
54	PK/YL	A/C WOT Relay Control	On: 1v, Off: 12v
56	GY/OR	PIP Sensor Signal	6.93v (50%)
58	TN	Injector 1 Control	5.5-7.1 ms
59	WT	Injector 2 Control	5.5-7.1 ms
60	BK/WT	Power Ground	<0.1v

Pin Connector Graphic

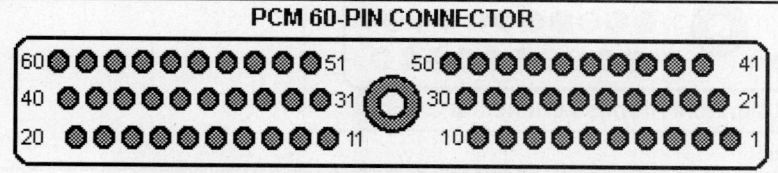

PCM 60-PIN CONNECTOR

Terminal View of 60-Pin PCM Harness Connector

1995 Bronco 4.9L I6 VIN Y (All) 60 Pin Connector

PCM Pin #	Wire Color	Circuit Description (60 Pin)	Value at Hot Idle
1	YL	Keep Alive Power	12-14v
2	LG	Brake Switch Signal	Brake Off: 12v, On: 1v
3	GY/BK	VSS (+) Signal	0 Hz, 55 mph: 125 Hz
4	YL/BK	Ignition Diagnostic Monitor	20-31 Hz
6	PK/OR	VSS (-) Signal	0 Hz, 55 mph: 125 Hz
7	LG/RD	ECT Sensor Signal	0.5-0.6v
8	DG/YL	Fuel Pump Monitor	On: 12v, Off: 0v
9	PK/BL	Data Bus (-) Signal	Digital Signals
10	PK/BL	A/C Switch Signal	A/C On: 12v, Off: 0v
11	BR	Air Management 2 Solenoid	AM2 On: 1v, Off: 12v
12	BL/BK	4x4 Indicator Lamp	12v (switch closed: 0v)
16	OR/RD	Ignition System Ground	<0.050v
17	PK/LG	Self-Test Output & MIL	MIL On: 1v, Off: 12v
20	BK	PCM Case Ground	<0.050v
21	WT/BL	IAC Solenoid Control	10.7v (33%)
22	BL/OR	Fuel Pump Control	On: 1v, Off: 12v
23	YL/RD	Knock Sensor Signal	0v
25	GY	IAT Sensor Signal	1.5-2.5v
26	BR/WT	Reference Voltage	4.9-5.1v
27	BR/LG	EVP Sensor Signal	0.4v
28	TN/OR	Data Bus (+) Signal	Digital Signals
29	GY/BL	HO2S-11 (B1 S1) Signal	0.1-1.1v
30	GY/YL	A/T: Neutral Drive Switch	In 'N': 0v, Others: 5v
30	BL/YL	M/T: CPP Switch Signal	Clutch In: 0v, Out: 5v
31	GY/YL	EVAP Purge Solenoid	12v, 55 mph: 1v
32	WT/LG	TCIL (lamp) Control	TCIL On: 1v, Off: 12v
33	BR/PK	VR Solenoid Control	0%, 55 mph: 45%
36	PK	Spark Output Signal	6.93v (50%)
37	RD	Vehicle Power	12-14v
40	BK/WT	Power Ground	<0.1v
41	TN/WT	TCS (switch) Signal	0.1-0.2v
43	PK	A/C Demand Switch	A/C On: 12v, Off: 0v
45	LG/BK	MAP Sensor Signal	107 Hz (sea level)
46	GY/RD	Analog Signal Return	<0.050v
47	GY/WT	TP Sensor Signal	0.53-1.27v
48	WT/PK	Self-Test Indicator Signal	STI On: 1v, Off: 5v
49	OR	HO2S-11 (B1 S1) Ground	<0.050v
51	WT/OR	Air Management 1 Solenoid	AM1 On: 1v, 55 mph: 12v
56	GY/OR	PIP Sensor Signal	6.93v (50%)
57	RD	Vehicle Power	12-14v
58	TN	Injector Bank 1 (INJ 1, 3 & 5)	6.2-7.4 ms
59	WT	Injector Bank 2 (INJ 2, 4 & 6)	6.2-7.4 ms
60	BK/WT	Power Ground	<0.1v

Pin Connector Graphic

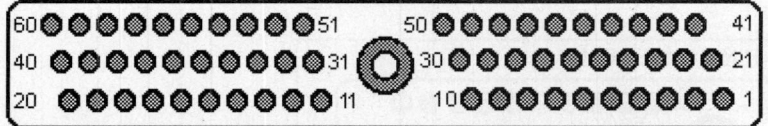

PCM 60-PIN CONNECTOR

Terminal View of 60-Pin PCM Harness Connector

Standard Colors and Abbreviations

Abbreviation	Color	Abbreviation	Color	Abbreviation	Color
BK	Black	GY	Gray	PK	Purple
BL	Blue	GN	Green	RD	Red
BR	Brown	LG	LT Green	TN	Tan
DB	Dark Blue	OR	Orange	WT	White
DG	DK Green	PK	Pink	YL	Yellow

1995 Bronco 4.9L I6 VIN Y (E4OD) California 60 Pin Connector

PCM Pin #	Wire Color	Circuit Description (60 Pin)	Value at Hot Idle
1	YL	Keep Alive Power	12-14v
2	LG	Brake Pedal Switch	Brake Off: 12v, On: 1v
3	GY/BK	VSS (+) Signal	0 Hz, 55 mph: 125 Hz
4	YL/BK	Ignition Diagnostic Monitor	20-31 Hz
5	---	Not Used	---
6	PK/OR	VSS (-) Signal	0 Hz, 55 mph: 125 Hz
7	LG/RD	ECT Sensor Signal	0.5-0.6v
8	DG/YL	Fuel Pump Monitor	On: 12v, Off: 0v
9	TN/BL	MAF Sensor Return	<0.050v
10	DG/OR	A/C Switch Signal	A/C On: 12v, Off: 0v
11	GY/YL	EVAP Purge Solenoid	12v, 55 mph: 1v
12	LG/OR	Injector 6 Control	5.5-7.1 ms
13-14	---	Not Used	---
15	TN/BK	Injector 5 Control	5.5-7.1 ms
16	TN/YL	Ignition System Ground	<0.050v
17	PK/LG	Self-Test Output & MIL	MIL On: 1v, Off: 12v
18	TN/OR	Data Bus (-) Signal	Digital Signals
19	PK/BK	Data Bus (+) Signal	Digital Signals
20	BK	PCM Case Ground	<0.050v
21	WT/BL	IAC Solenoid Control	10.7v (33%)
22	BL/OR	Fuel Pump Control	On: 1v, Off: 12v
23	YL/RD	Knock Sensor Signal	0v
24	---	Not Used	---
25	GY	IAT Sensor Signal	1.5-2.5v
26	BR/WT	Reference Voltage	4.9-5.1v
27	BR/LG	EVP Sensor Signal	0.4v
28-29	---	Not Used	---
30	BL/YL	TR Sensor Signal	In 'P': 0v, in O/D: 5v
30	GY/YL	TR Sensor Signal	In 'P': 0v, in O/D: 5v
31	WT/OR	Thermactor Air Bypass	TAB On: 1v, Off: 12v
32	WT/LG	TCIL (lamp) Control	TCIL On: 1v, Off: 12v
33	BR/PK	VR Solenoid Control	0%, 55 mph: 45%
34	BR	Thermactor Air Diverter	TAD On: 1v, Off: 12v
35	BR/BL	Injector 4 Control	5.5-7.1 ms
36	PK	Spark Output Signal	6.93v (50%)
37	RD	Vehicle Power	12-14v
38	WT/YL	EPC Solenoid Control	9.5v (5 psi)
39	BR/YL	Injector 3 Control	5.5-7.1 ms
40	BK/WT	Power Ground	<0.1v

1995 Bronco 4.9L I6 VIN Y (E4OD) California 60 Pin Connector

PCM Pin #	Wire Color	Circuit Description (60 Pin)	Value at Hot Idle
41	TN/WT	TCS (switch) Signal	TCS & O/D On: 12v
42	BL/YL	4x4 Switch Signal	12v (switch closed: 0v)
43	RD/BK	HO2S-21 (B2 S1) Signal	0.1-1.1v
44	GY/BL	HO2S-11 (B1 S1) Signal	0.1-1.1v
45	---	Not Used	---
46	GY/RD	Analog Signal Return	<0.050v
47	GY/WT	TP Sensor Signal	0.5-1.2v
48	WT/PK	Self-Test Indicator Signal	STI On: 1v, Off: 5v
49	OR/BK	TOT Sensor Signal	2.10-2.40v
50	BL/RD	MAF Sensor Signal	0.8v
51	OR/YL	Shift Solenoid 1 Control	12v, 55 mph: 1v
52	PK/OR	Shift Solenoid 2 Control	1v, 30 mph: 12v
53	PK/YL	TCC Solenoid Control	12v, 55 mph: 9v
54	PK/YL	A/C WOT Relay Control	On: 1v, Off: 12v
55	BR/OR	Coast Clutch Switch Signal	12v, 55 mph: 12v
56	GY/OR	PIP Sensor Signal	6.93v (50%)
57	RD	Vehicle Power	12-14v
58	TN	Injector 1 Control	5.5-7.1 ms
59	WT	Injector 2 Control	5.5-7.1 ms
60	BK/WT	Power Ground	<0.1v

Pin Connector Graphic

PCM 60-PIN CONNECTOR

Terminal View of 60-Pin PCM Harness Connector

Standard Colors and Abbreviations

Abbreviation	Color	Abbreviation	Color	Abbreviation	Color
BK	Black	GY	Gray	PK	Purple
BL	Blue	GN	Green	RD	Red
BR	Brown	LG	LT Green	TN	Tan
DB	Dark Blue	OR	Orange	WT	White
DG	DK Green	PK	Pink	YL	Yellow

1995 Bronco 4.9L I6 VIN Y (E4OD) 60 Pin Connector

PCM Pin #	Wire Color	Circuit Description (60 Pin)	Value at Hot Idle
1	YL	Keep Alive Power	12-14v
2	LG	Brake Switch Signal	Brake Off: 12v, On: 1v
3	GY/BK	VSS (+) Signal	0 Hz, 55 mph: 125 Hz
4	YL/BK	Ignition Diagnostic Monitor	20-31 Hz
5	---	Not Used	---
6	PK/OR	VSS (-) Signal	0 Hz, 55 mph: 125 Hz
7	LG/RD	ECT Sensor Signal	0.5-0.6v
8	DG/YL	Fuel Pump Monitor	On: 12v, Off: 0v
9	PK/BL	Data Bus (-) Signal	Digital Signals
10	PK/BL	A/C Switch Signal	A/C On: 12v, Off: 0v
11	BR	Air Management 2 Solenoid	AM2 On: 1v, Off: 12v
12	BL/BK	4x4 Indicator Lamp	12v (switch closed: 0v)
13-15	---	Not Used	---
16	OR/RD	Ignition System Ground	<0.050v
17	PK/LG	Self-Test Output & MIL	MIL On: 1v, Off: 12v
18	---	Not Used	---
19	PK/OR	Shift Solenoid 2 Control	12v, 55 mph: 1v
20	BK	PCM Case Ground	<0.050v
21	WT/BL	IAC Solenoid Control	10.7v (33%)
22	BL/OR	Fuel Pump Control	On: 1v, Off: 12v
23	YL/RD	Knock Sensor Signal	0v
24	---	Not Used	---
25	GY	IAT Sensor Signal	1.5-2.5v
26	BR/WT	Reference Voltage	4.9-5.1v
27	BR/LG	EVP Sensor Signal	0.4v
28	TN/OR	Data Bus (+) Signal	Digital Signals
29	GY/BL	HO2S-11 (B1 S1) Signal	0.1-1.1v
30	GY/YL	TR Sensor Signal	In 'P': 0v, in O/D: 5v
30	BL/YL	TR Sensor Signal	In 'P': 0v, in O/D: 5v
31	GY/YL	EVAP Purge Solenoid	12v, 55 mph: 1v
32	WT/LG	TCIL (lamp) Control	TCIL On: 1v, Off: 12v
33	BR/PK	VR Solenoid Control	0%, 55 mph: 45%
34-35	---	Not Used	---
36	PK	Spark Output Signal	6.93v (50%)
37	RD	Vehicle Power	12-14v
38	WT/YL	EPC Solenoid Control	9.5v (5 psi)
39	---	Not Used	---
40	BK/WT	Power Ground	<0.1v

1995 Bronco 4.9L I6 VIN Y (E4OD) 60 Pin Connector

PCM Pin #	Wire Color	Circuit Description (60 Pin)	Value at Hot Idle
41	TN/WT	TCS (switch) Signal	TCS & O/D On: 12v
42	OR/BK	TOT Sensor Signal	2.10-2.40v
43	PK	A/C Demand Switch	A/C On: 12v, Off: 0v
44	---	Not Used	---
45	LG/BK	MAP Sensor Signal	107 Hz (sea level)
46	GY/RD	Analog Signal Return	<0.050v
47	GY/WT	TP Sensor Signal	0.5-1.2v
48	WT/PK	Self-Test Indicator Signal	STI On: 1v, Off: 5v
49	OR	HO2S-11 (B1 S1) Ground	<0.050v
50	---	Not Used	---
51	WT/OR	Air Management 1 Solenoid	1v, 55 mph: 12v
52	OR/YL	Shift Solenoid 1 Control	1v, 30 mph: 12v
53	PK/YL	TCC Solenoid Control	12v, 55 mph: 9v
54	---	Not Used	---
55	BR/OR	Coast Clutch Solenoid	12v, 55 mph: 12v
56	GY/OR	PIP Sensor Signal	6.93v (50%)
57	RD	Vehicle Power	12-14v
58	TN	Injector Bank 1 (INJ 1, 3 & 5)	6.2-7.4 ms
59	WT	Injector Bank 2 (INJ 2, 4 & 6)	6.2-7.4 ms
60	BK/WT	Power Ground	<0.1v

Pin Connector Graphic

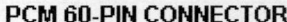

PCM 60-PIN CONNECTOR

60 ⊙⊙⊙⊙⊙⊙⊙⊙⊙⊙ 51   50 ⊙⊙⊙⊙⊙⊙⊙⊙⊙⊙ 41
40 ⊙⊙⊙⊙⊙⊙⊙⊙⊙⊙ 31 ⊙ 30 ⊙⊙⊙⊙⊙⊙⊙⊙⊙⊙ 21
20 ⊙⊙⊙⊙⊙⊙⊙⊙⊙⊙ 11   10 ⊙⊙⊙⊙⊙⊙⊙⊙⊙⊙ 1

Terminal View of 60-Pin PCM Harness Connector

Standard Colors and Abbreviations

Abbreviation	Color	Abbreviation	Color	Abbreviation	Color
BK	Black	GY	Gray	PK	Purple
BL	Blue	GN	Green	RD	Red
BR	Brown	LG	LT Green	TN	Tan
DB	Dark Blue	OR	Orange	WT	White
DG	DK Green	PK	Pink	YL	Yellow

1990 Bronco 5.0L V8 MFI VIN N (All) 60 Pin Connector

PCM Pin #	Wire Color	Circuit Description (60 Pin)	Value at Hot Idle
1	YL	Keep Alive Power	12-14v
3	GY/BK	VSS (+) Signal	0 Hz, 55 mph: 125 Hz
3	DG/WT	VSS (+) Signal	0 Hz, 55 mph: 125 Hz
4	DG/YL	Ignition Diagnostic Monitor	20-31 Hz
6	BK	VSS (-) Signal	0 Hz, 55 mph: 125 Hz
7	LG/YL	ECT Sensor Signal	0.5-0.6v
8	BR	Fuel Pump Monitor	On: 12v, Off: 0v
10	BK/YL	A/C Switch Signal	A/C On: 12v, Off: 0v
11	WT/BK	Air Management 2 Solenoid	AM2 On: 1v, Off: 12v
16	BK/OR	Ignition System Ground	<0.050v
17	PK/LG	Self-Test Output & MIL	MIL On: 1v, Off: 12v
18	TN/LG	Inferred Mileage Sensor	Digital Signals
20	BK	PCM Case Ground	<0.050v
21	GY/WT	IAC Solenoid Control	9.0v
22	TN/LG	Fuel Pump Control	On: 1v, Off: 12v
23	LG/BK	Knock Sensor Signal	0v
24	YL/LG	PSP Switch Signal	Straight: 0v, Turned: 12v
25	YL/RD	ACT Sensor Signal	1.5-2.5v
26	OR/WT	Reference Voltage	4.9-5.1v
27	BR/LG	EVP Sensor Signal	0.4v
29	GY/BK	HO2S-11 (B1 S1) Signal	0.1-1.1v
29	DG/PK	HO2S-11 (B1 S1) Signal	0.1-1.1v
30	GY/YL	A/T: Neutral Drive Switch	In 'P': 0v, Others: 5v
30	BL/YL	M/T: CPP Switch Signal	Clutch In: 0v, Out: 5v
31	GY/YL	EVAP Purge Solenoid	12v, 55 mph: 1v
33	DG	VR Solenoid Control	0%, 55 mph: 45%
36	YL/LG	Spark Output Signal	6.93v (50%)
37	RD	Vehicle Power	12-14v
40	BK/LG	Power Ground	<0.1v
45	DG/LG	MAP Sensor Signal	107 Hz (sea level)
45	DB/LG	MAP Sensor Signal	107 Hz (sea level)
46	BK/WT	Analog Signal Return	<0.050v
47	DG/LG	TP Sensor Signal	0.5-1.2v
48	WT/RD	Self-Test Indicator Signal	STI On: 1v, Off: 5v
49	OR	HO2S-11 (B1 S1) Ground	<0.050v
51	WT/RD	Air Management 1 Solenoid	1v, 55 mph: 12v
56	DB	PIP Sensor Signal	6.93v (50%)
57	RD	Vehicle Power	12-14v
58	TN/OR	Injector Bank 1 (INJ 1, 4, 5, 8)	5.0-5.7 ms
59	TN/RD	Injector Bank 2 (INJ 2, 3, 6, 7)	5.0-5.7 ms
60	BK/LG	Power Ground	<0.1v

Pin Connector Graphic

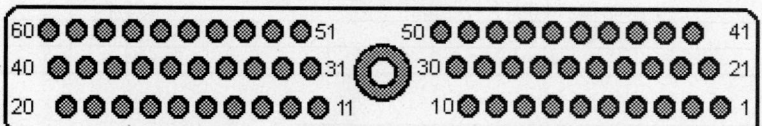

PCM 60-PIN CONNECTOR

Terminal View of 60-Pin PCM Harness Connector

Standard Colors and Abbreviations

Abbreviation	Color	Abbreviation	Color	Abbreviation	Color
BK	Black	GY	Gray	PK	Purple
BL	Blue	GN	Green	RD	Red
BR	Brown	LG	LT Green	TN	Tan
DB	Dark Blue	OR	Orange	WT	White
DG	DK Green	PK	Pink	YL	Yellow

1991-93 Bronco 5.0L V8 MFI VIN N (A/T) 60 Pin Connector

PCM Pin #	Wire Color	Circuit Description (60 Pin)	Value at Hot Idle
1	YL	Keep Alive Power	12-14v
2, 5	---	Not Used	---
3	GY/BK	VSS (+) Signal	0 Hz, 55 mph: 125 Hz
4	TN/YL	Ignition Diagnostic Monitor	20-31 Hz
6	PK/OR	VSS (-) Signal	0 Hz, 55 mph: 125 Hz
7	LG/RD	ECT Sensor Signal	0.5-0.6v
8	DG/YL	Fuel Pump Monitor	On: 12v, Off: 0v
9	BK/OR	Data Bus (-) Signal	Digital Signals
9	PK/BL	Data Bus (-) Signal	Digital Signals
10	BK/YL	A/C Switch Signal	A/C On: 12v, Off: 0v
10	PK/BL	A/C Switch Signal	A/C On: 12v, Off: 0v
11	BR	Air Management 2 Solenoid	AM2 On: 1v, Off: 12v
12-15, 18-19	---	Not Used	---
16	OR/RD	Ignition System Ground	<0.050v
17	PK/LG	Self-Test Output & MIL	MIL On: 1v, Off: 12v
20	BK	PCM Case Ground	<0.050v
21	WT/BL	IAC Solenoid Control	10.7v (33%)
22	BL/OR	Fuel Pump Control	On: 1v, Off: 12v
23	YL/RD	Knock Sensor Signal	0v
24	YL/LG	PSP Switch Signal	Straight: 0v, Turned: 12v
25	GY	ACT Sensor Signal	1.5-2.5v
26	BR/WT	Reference Voltage	4.9-5.1v
27	BR/LG	EVP Sensor Signal	0.4v
28	TN/OR	Data Bus (+) Signal	Digital Signals
29	GY/BL	HO2S-11 (B1 S1) Signal	0.1-1.1v
30	GY/YL	A/T: Neutral Drive Switch	In 'P': 0v, Others: 5v
30	BL/YL	M/T: CPP Switch Signal	Clutch In: 0v, Out: 5v
31 ('92-'93)	GY/YL	EVAP Purge Solenoid	12v, 55 mph: 1v
32, 41-44	---	Not Used	---
33	BR/PK	VR Solenoid Control	0%, 55 mph: 45%
34-35, 38-39	---	Not Used	---
36	PK	Spark Output Signal	6.93v (50%)
37	RD	Vehicle Power	12-14v
40	BK/WT	Power Ground	<0.1v
45	LG/BK	MAP Sensor Signal	107 Hz (sea level)
46	GY/RD	Analog Signal Return	<0.050v
47	GY/WT	TP Sensor Signal	0.5-1.2v
48	WT/PK	Self-Test Indicator Signal	STI On: 1v, Off: 5v
49	OR	HO2S-11 (B1 S1) Ground	<0.050v
50	---	Not Used	---
51	WT/OR	Air Management 1 Solenoid	1v, 55 mph: 12v
52-55	---	Not Used	---
56	GY/OR	PIP Sensor Signal	6.93v (50%)
57	RD	Vehicle Power	12-14v
58	TN	Injector Bank 1 (INJ 1, 4, 5, 8)	4.4-5.6 ms
59	WT	Injector Bank 2 (INJ 2, 3, 6, 7)	4.4-5.6 ms
60	BK/WT	Power Ground	<0.1v

Pin Connector Graphic

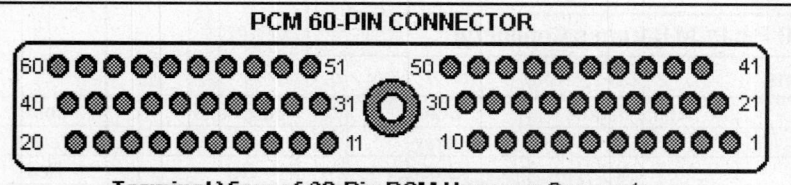

PCM 60-PIN CONNECTOR

Terminal View of 60-Pin PCM Harness Connector

1994-95 Bronco 5.0L V8 VIN N [4R70W] 60 Pin Connector

PCM Pin #	Wire Color	Circuit Description (60 Pin)	Value at Hot Idle
1	YL	Keep Alive Power	12-14v
2	LG	Brake Pedal Switch	Brake Off: 12v, On: 1v
3	GY/BK	VSS (+) Signal	0 Hz, 55 mph: 125 Hz
4	YL/BK	Ignition Diagnostic Monitor	20-31 Hz
5	DG/WT	TSS (+) Sensor Signal	30 mph: 126-136 Hz
6	PK/OR	VSS (-) Signal	0 Hz, 55 mph: 125 Hz
7	LG/RD	ECT Sensor Signal	0.5-0.6v
8	DG/YL	Fuel Pump Monitor	On: 12v, Off: 0v
9	TN/BL	MAF Sensor Return	<0.050v
10	DG/OR	A/C Switch Signal	A/C On: 12v, Off: 0v
11	GY/YL	EVAP Purge Solenoid	12v, 55 mph: 1v
12	LG/OR	Injector 6 Control	5.0-5.5 ms
13	TN/RD	Injector 7 Control	5.0-5.5 ms
14	BL	Injector 8 Control	5.0-5.5 ms
15	TN/BK	Injector 5 Control	5.0-5.5 ms
16	OR/RD	Ignition System Ground	<0.050v
17	PK/LG	Self-Test Output & MIL	MIL On: 1v, Off: 12v
18	TN/OR	Data Bus (+) Signal	Digital Signals
19	PK/BL	Data Bus (-) Signal	Digital Signals
20	BK	PCM Case Ground	<0.050v
21	WT/BL	IAC Solenoid Control	10.7v (33%)
22	BL/OR	Fuel Pump Control	On: 1v, Off: 12v
23	YL/RD	Knock Sensor Signal	0v
24	---	Not Used	---
25	GY	IAT Sensor Signal	1.5-2.5v
26	BR/WT	Reference Voltage	4.9-5.1v
27	BR/LG	EVP Sensor Signal	0.4v
28	---	Not Used	---
29	BK /LG	TSS (-) Sensor Signal	30 mph: 126-136 Hz
30	BL/YL	MLP Sensor Signal	In 'P': 0v, in O/D: 5v
31	WT/OR	Air Bypass Solenoid Control	AIRB On 1v, Off: 12v
32	WT/LG	TCIL (lamp) Control	TCIL On: 1v, Off: 12v
33	BR/PK	VR Solenoid Control	0%, 55 mph: 45%
34	BR	Air Diverter Solenoid Control	AIRD On 1v, Off: 12v
35	BR/BL	Injector 4 Control	5.0-5.5 ms
36	PK	Spark Output Signal	6.93v (50%)
37	RD	Vehicle Power	12-14v
38	WT/YL	EPC Solenoid Control	9.5v (5 psi)
39	BR/YL	Injector 3 Control	5.0-5.5 ms
40	BK/WT	Power Ground	<0.1v

1994-95 Bronco 5.0L V8 VIN N [4R70W] 60 Pin Connector

PCM Pin #	Wire Color	Circuit Description (60 Pin)	Value at Hot Idle
41	TN/WT	TCS (switch) Signal	TCS Closed: 0.1v
42	BL/BK	4x4 Indicator Light	12v (switch closed: 0v)
43	---	Not Used	---
44	GY/BL	HO2S-11 (B1 S1) Signal	0.1-1.1v
45	---	Not Used	---
46	GY/RD	Analog Signal Return	<0.050v
47	GY/WT	TP Sensor Signal	0.5-1.2v
48	WT/PK	Self-Test Indicator Signal	STI On: 1v, Off: 5v
49	OR/BK	TOT Sensor Signal	2.10-2.40v
50	BL/RD	MAF Sensor Signal	0.7v
51	OR/YL	Shift Solenoid 1 Control	12v, 55 mph: 1v
52	PK/OR	Shift Solenoid 2 Control	1v, 30 mph: 12v
53	PK/YL	TCC Solenoid Control	12v, 55 mph: 1v
54-55	---	Not Used	---
56	GY/OR	PIP Sensor Signal	6.93v (50%)
57	RD	Vehicle Power	12-14v
58	TN	Injector 1 Control	5.0-5.5 ms
59	WT	Injector 2 Control	5.0-5.5 ms
60	BK/WT	Power Ground	<0.1v

Pin Connector Graphic

PCM 60-PIN CONNECTOR

Terminal View of 60-Pin PCM Harness Connector

Standard Colors and Abbreviations

Abbreviation	Color	Abbreviation	Color	Abbreviation	Color
BK	Black	GY	Gray	PK	Purple
BL	Blue	GN	Green	RD	Red
BR	Brown	LG	LT Green	TN	Tan
DB	Dark Blue	OR	Orange	WT	White
DG	DK Green	PK	Pink	YL	Yellow

1994-95 Bronco 5.0L MFI VIN N (All) 60 Pin Connector

PCM Pin #	Wire Color	Circuit Description (60 Pin)	Value at Hot Idle
1	YL	Keep Alive Power	12-14v
2	---	Not Used	---
3	GY/BK	VSS (+) Signal	0 Hz, 55 mph: 125 Hz
4	WT/PK	Ignition Diagnostic Monitor	20-31 Hz
5	---	Not Used	---
6	PK/OR	VSS (-) Signal	0 Hz, 55 mph: 125 Hz
7	LG/RD	ECT Sensor Signal	0.5-0.6v
8	DG/YL	Fuel Pump Monitor	On: 12v, Off: 0v
9	PK/BL	Data Bus (-) Signal	Digital Signals
10	PK/BL	A/C Switch Signal	A/C On: 12v, Off: 0v
11	BR	Air Management 2 Solenoid	AM2 On: 1v, Off: 12v
12-15	---	Not Used	---
16	OR/RD	Ignition System Ground	<0.050v
17	PK/LG	Self-Test Output & MIL	MIL On: 1v, Off: 12v
18-19	---	Not Used	---
20	BK	PCM Case Ground	<0.050v
21	WT/BL	IAC Solenoid Control	10.7v (33%)
22	BL/OR	Fuel Pump Control	On: 1v, Off: 12v
23	YL/RD	Knock Sensor Signal	0v
24	YL/LG	PSP Switch Signal	Straight: 0v, Turned: 12v
25	GY	IAT Sensor Signal	1.5-2.5v
26	BR/WT	Reference Voltage	4.9-5.1v
27	BR/LG	EVP Sensor Signal	0.4v
28	TN/OR	Data Bus (+) Signal	Digital Signals
29	GY/YL	HO2S-11 (B1 S1) Signal	0.1-1.1v
30	BL/YL	MLP Sensor Signal	In 'P': 0v, in O/D: 5v
30	BL/YL	Clutch Pedal Position Switch	Clutch In: 0v, Out: 5v
31	GY/YL	EVAP Purge Solenoid	12v, 55 mph: 1v
32	---	Not Used	---
33	BR/PK	VR Solenoid Control	0%, 55 mph: 45%
36	PK	Spark Output Signal	6.93v (50%)
37	RD	Vehicle Power	12-14v
40	BK/WT	Power Ground	<0.1v
41	TN/WT	TCS (switch) Signal	TCS Closed: 0.1v
42	OR/BK	TOT Sensor Signal	2.10-2.40v
43	PK	A/C Demand Switch	A/C On: 12v, Off: 0v
44	---	Not Used	---
45	LG/BK	MAP Sensor Signal	107 Hz (sea level)
46	GY/RD	Analog Signal Return	<0.050v
47	GY/WT	TP Sensor Signal	0.5-1.2v
48	WT/PK	Self-Test Indicator Signal	STI On: 1v, Off: 5v
49	OR	HO2S-11 (B1 S1) Ground	<0.050v
50	---	Not Used	---
51	WT/OR	Air Management 1 Solenoid	1v, 55 mph: 12v
52-55	---	Not Used	---
56	GY/OR	PIP Sensor Signal	6.93v (50%)
57	RD	Vehicle Power	12-14v
58	TN	Injector Bank 1 (INJ 1, 4, 5, 8)	4.4-5.4 ms
59	WT	Injector Bank 2 (INJ 2, 3, 6, 7)	4.4-5.4 ms
60	BK/WT	Power Ground	<0.1v

Pin Connector Graphic

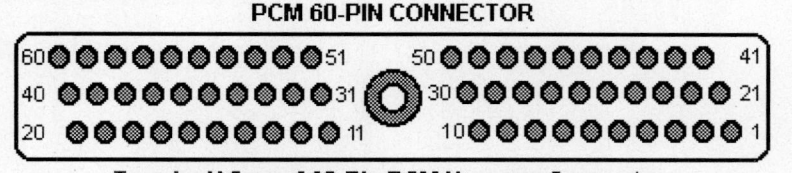

1994-95 Bronco 5.0L V8 VIN N (E4OD) 60 Pin Connector

PCM Pin #	Wire Color	Circuit Description (60 Pin)	Value at Hot Idle
1	YL	Keep Alive Power	12-14v
2	LG	Brake On/Off Switch	Brake Off: 12v, On: 1v
3	GY/BK	VSS (+) Signal	0 Hz, 55 mph: 125 Hz
4	WT/PK	Ignition Diagnostic Monitor	20-31 Hz
5	---	Not Used	---
6	PK/OR	VSS (-) Signal	0 Hz, 55 mph: 125 Hz
7	LG/RD	ECT Sensor Signal	0.5-0.6v
8	DG/YL	Fuel Pump Monitor	On: 12v, Off: 0v
9	TN/BL	MAF Sensor Return	<0.050v
10	BK/YL	A/C Switch Signal	A/C On: 12v, Off: 0v
10	DG/OR	A/C Switch Signal	A/C On: 12v, Off: 0v
11	GY/YL	EVAP Purge Solenoid	12v, 55 mph: 1v
12	LG/OR	Injector 6 Control	4.7-5.1 ms
13	TN/RD	Injector 7 Control	4.7-5.1 ms
14	BL	Injector 8 Control	4.7-5.1 ms
15	TN/BK	Injector 5 Control	4.7-5.1 ms
16	OR/RD	Ignition System Ground	<0.050v
17	PK/LG	Self-Test Output & MIL	MIL On: 1v, Off: 12v
18	TN/OR	Data Bus (+) Signal	Digital Signals
19	PK/BL	Data Bus (-) Signal	Digital Signals
20	BK	PCM Case Ground	<0.050v
21	WT/BL	IAC Solenoid Control	10.7v (33%)
22	BL/OR	Fuel Pump Control	On: 1v, Off: 12v
23	YL/RD	Knock Sensor Signal	0v
24	---	Not Used	---
25	GY	IAT Sensor Signal	1.5-2.5v
26	BR/WT	Reference Voltage	4.9-5.1v
27	BR/LG	EVP Sensor Signal	0.4v
28-29	---	Not Used	---
30	BL/YL	MLP Sensor Signal	In 'P': 0v, in O/D: 5v
31	WT/OR	Air Bypass Solenoid Control	AIRB On: 1v, Off: 12v
32	WT/LG	TCIL (lamp) Control	TCIL On: 1v, Off: 12v
33	BR/PK	VR Solenoid Control	0%, 55 mph: 45%
34	BR	Air Diverter Solenoid	AIRD On: 1v, Off: 12v
35	BR/BL	Injector 4 Control	4.7-5.1 ms
36	PK	Spark Output Signal	6.93v (50%)
37	RD	Vehicle Power	12-14v
38	WT/YL	EPC Solenoid Control	9.5v (5 psi)
39	BR/YL	Injector 3 Control	4.7-5.1 ms
40	BK/WT	Power Ground	<0.1v

1994-95 Bronco 5.0L V8 VIN N (E4OD) 60 Pin Connector

PCM Pin #	Wire Color	Circuit Description (60 Pin)	Value at Hot Idle
41	TN/WT	TCS (switch) Signal	TCS Closed: 0.1v
42	BL/BK	4x4 Indicator Light	4x4 Switch On: 0.1v
43	---	Not Used	---
44	GY/BL	HO2S-11 (B1 S1) Signal	0.1-1.1v
45	---	Not Used	---
46	GY/RD	Analog Signal Return	<0.050v
47	GY/WT	TP Sensor Signal	0.5-1.2v
48	WT/PK	Self-Test Indicator Signal	STI On: 1v, Off: 5v
49	OR/BK	HO2S-11 (B1 S1) Ground	<0.050v
50	BL/RD	MAF Sensor Signal	0.8v
51	OR/YL	Shift Solenoid 1 Control	12v, 55 mph: 1v
52	PK/OR	Shift Solenoid 2 Control	1v, 30 mph: 12v
53	PK/YL	TCC Solenoid Control	12v, 55 mph: 9v
54	---	Not Used	---
55	BR/OR	Coast Clutch Solenoid	12v, 55 mph: 12v
56	GY/OR	PIP Sensor Signal	6.93v (50%)
57	RD	Vehicle Power	12-14v
58	TN	Injector 1 Control	4.7-5.1 ms
59	WT	Injector 2 Control	4.7-5.1 ms
60	BK/WT	Power Ground	<0.1v

Pin Connector Graphic

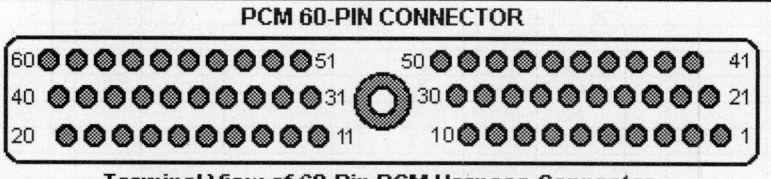

PCM 60-PIN CONNECTOR

Terminal View of 60-Pin PCM Harness Connector

Standard Colors and Abbreviations

Abbreviation	Color	Abbreviation	Color	Abbreviation	Color
BK	Black	GY	Gray	PK	Purple
BL	Blue	GN	Green	RD	Red
BR	Brown	LG	LT Green	TN	Tan
DB	Dark Blue	OR	Orange	WT	White
DG	DK Green	PK	Pink	YL	Yellow

1996 Bronco 5.0L MFI VIN N (All) 104 Pin Connector

PCM Pin #	Wire Color	Circuit Description (104 Pin)	Value at Hot Idle
1	---	Not Used	---
2	PK/LG	MIL (lamp) Control	MIL On: 1v, Off: 12v
4	LG/RD	Power Take-Off Signal	0.1v (off)
5-12	---	Not Used	---
13	PK	Flash EPROM Power	0.1v
14	LG/BK	4x4 Low Switch Signal	12v (Closed: 0v)
15	PK/LB	Data Bus (-) Signal	Digital Signals
16	TN/OR	Data Bus (+) Signal	Digital Signals
18-22	---	Not Used	---
23	OR/RD	Ignition Ground	<0.050v
24	BK/WT	Power Ground	<0.1v
25	BK/LB	Case Ground	<0.050v
26-28	---	Not Used	---
29	TN/WT	TCS (switch) Signal	TCS & O/D On: 12v
30-32	---	Not Used	---
33	PK/OR	PSOM (-) Signal	<0.050v
34	---	Not Used	---
35	RD/LG	HO2S-12 (B1 S2) Signal	0.1-1.1v
36	TN/LB	MAF Sensor Return	<0.050v
37	OR/BK	TFT Sensor Signal	2.10-2.40v
38	LG/RD	ECT Sensor Signal	0.5-0.6v
39	GY	IAT Sensor Signal	1.5-2.5v
40	DG/YL	Fuel Pump Monitor	On: 12v, Off: 0v
41	DG/OR	A/C Switch Signal	A/C On: 12v, Off: 0v
42-43	---	Not Used	---
44	BR	Secondary AIR Diverter	AIRD On: 1v, Off: 12v
45-46	---	Not Used	---
47	BK/PK	VR Solenoid Control	0%, 55 mph: 45%
48	YL/BK	Clean Tachometer Output	39-45 Hz
50	---	Not Used	---
51	BK/WT	Power Ground	<0.1v
52-54	---	Not Used	---
55	YL	Keep Alive Power	12-14v
56	LG/BK	EVAP Purge Solenoid	0-10 Hz (0-100%)
57	---	Not Used	---
58	GY/BK	PSOM (+) Signal	0 Hz, 55 mph: 125 Hz
59	DG/LG	Misfire Detection Sensor	45-55 Hz
60	GY/LB	HO2S-11 (B1 S1) Signal	0.1-1.1v
61-63	---	Not Used	---
64	LB/YL	A/T: Neutral Position Switch	In 'P': 0v, Others: 5v
64	LB/YL	M/T: Clutch Engage Switch	Clutch In: 0v, Out: 5v
65	BR/LG	DPFE Sensor Signal	0.20-1.30v
66-69	---	Not Used	---

1996 Bronco 5.0L MFI VIN N (All) 104 Pin Connector

PCM Pin #	Wire Color	Circuit Description (104 Pin)	Value at Hot Idle
70	WT/OR	Secondary AIR Bypass	AIRB On: 1v, Off: 12v
71	RD	Vehicle Power	12-14v
72	TN/RD	Injector 7 Control	3.2-4.5 ms
73	TN/BK	Injector 5 Control	3.2-4.5 ms
74	BR/YL	Injector 3 Control	3.2-4.5 ms
75	TN	Injector 1 Control	3.2-4.5 ms
76	BK/WT	Power Ground	<0.1v
77	BK/WT	Power Ground	<0.1v
78	---	Not Used	---
79	WT/LG	TCIL (lamp) Control	TCIL On: 1v, Off: 12v
80	LB/OR	Fuel Pump Control	On: 1v, Off: 12v
81-82	---	Not Used	---
83	WT/LB	IAC Solenoid Control	10.8v (33%)
84	DB/YL	OSS (+) Sensor Signal	0 Hz, 30 mph: 130 Hz
85	---	Not Used	---
87	RD/BK	HO2S-21 (B2 S1) Signal	0.1-1.1v
88	LB/RD	MAF Sensor Signal	0.8v, 55 mph: 2.1v
89	GY/WT	TP Sensor Signal	0.53-1.27v
90	BR/WT	Reference Voltage	4.9-5.1v
91	GY/RD	Analog Signal Return	<0.050v
92	LG	Brake Position Switch	Brake Off: 12v, On: 1v
93	RD/WT	HO2S-11 (B1 S1) Heater	On: 1v, Off: 12v
94	YL/LB	HO2S-21 (B2 S1) Heater	On: 1v, Off: 12v
95	WT/BK	HO2S-12 (B1 S2) Heater	On: 1v, Off: 12v
96	---	Not Used	---
97	RD	Vehicle Power	12-14v
98	LG	Injector 8 Control	3.2-4.5 ms
99	LG/OR	Injector 6 Control	3.2-4.5 ms
100	BR/LB	Injector 4 Control	3.2-4.5 ms
101	W	Injector 2 Control	3.2-4.5 ms
102	---	Not Used	---
103	BK/WT	Power Ground	<0.1v
104	---	Not Used	---

Pin Connector Graphic

```
              PCM 104-PIN CONNECTOR
     1 ⊙⊙⊙⊙⊙⊙⊙⊙⊙⊙⊙⊙⊙   ⊙⊙⊙⊙⊙⊙⊙⊙⊙⊙⊙⊙⊙ 26
    27 ⊙⊙⊙⊙⊙⊙⊙⊙⊙⊙⊙⊙⊙   ⊙⊙⊙⊙⊙⊙⊙⊙⊙⊙⊙⊙⊙ 52
    53 ⊙⊙⊙⊙⊙⊙⊙⊙⊙⊙⊙⊙⊙  ●  ⊙⊙⊙⊙⊙⊙⊙⊙⊙⊙⊙⊙⊙ 78
    79 ⊙⊙⊙⊙⊙⊙⊙⊙⊙⊙⊙⊙⊙   ⊙⊙⊙⊙⊙⊙⊙⊙⊙⊙⊙⊙⊙ 104
```

Terminal View of 104-Pin PCM Wiring Harness Connector

Standard Colors and Abbreviations

Abbreviation	Color	Abbreviation	Color	Abbreviation	Color
BK	Black	GY	Gray	PK	Purple
BL	Blue	GN	Green	RD	Red
BR	Brown	LG	LT Green	TN	Tan
DB	Dark Blue	OR	Orange	WT	White
DG	DK Green	PK	Pink	YL	Yellow

1996 Bronco 5.0L VIN N (E4OD) 104 Pin Connector

PCM Pin #	Wire Color	Circuit Description (104 Pin)	Value at Hot Idle
1	PK/OR	Shift Solenoid 2 Control	12v, 55 mph: 12v
2	PK/LG	MIL (lamp) Control	MIL On: 1v, Off: 12v
3, 5, 7, 12	---	Not Used	---
4	LG/RD	Power Take-Off Signal	0.1v (off)
6	BK/YL	OSS (-) Sensor Signal	At 30 mph: 125-131 Hz
13	PK	Flash EPROM Power	0.1v
14	LG/BK	4x4 Low Switch Signal	12v (Closed: 0v)
15	PK/LB	Data Bus (-) Signal	Digital Signals
16	TN/OR	Data Bus (+) Signal	Digital Signals
17-22, 26	---	Not Used	---
23	OR/RD	Ignition Ground	<0.050v
24	BK/WT	Power Ground	<0.1v
25	BK/LB	Case Ground	<0.050v
27	OR/YL	Shift Solenoid 1 Control	1v, 55 mph: 12v
28	---	Not Used	---
29	TN/WT	TCS (switch) Signal	TCS & O/D On: 12v
30-32	---	Not Used	---
33	PK/OR	PSOM (-) Signal	<0.050v
34	---	Not Used	---
35	RD/LG	HO2S-12 (B1 S2) Signal	0.1-1.1v
36	TN/LB	MAF Sensor Return	<0.050v
37	OR/BK	TFT Sensor Signal	2.10-2.40v
38	LG/RD	ECT Sensor Signal	0.5-0.6v
39	GY	IAT Sensor Signal	1.5-2.5v
40	DG/YL	Fuel Pump Monitor	On: 12v, Off: 0v
41	DG/OR	A/C Switch Signal	A/C On: 12v, Off: 0v
42-43	---	Not Used	---
44	BR	Secondary AIR Diverter	AIRD On: 1v, Off: 12v
45-46	---	Not Used	---
47	BK/PK	VR Solenoid Control	0%, 55 mph: 45%
48	YL/BK	Clean Tachometer Output	39-45 Hz
49	GY/OR	PIP Sensor Signal	6.93v (50%)
50	PK	Spark Output Signal	6.93v (50%)
51	BK/WT	Power Ground	<0.1v
52, 57	---	Not Used	---
53	BR/OR	Coast Clutch Solenoid	12v, 55 mph: 12v
54	PK/YL	TCC Solenoid Control	0%, 55 mph: 95%
55	YL	Keep Alive Power	12-14v
56	LG/BK	EVAP Purge Solenoid	0-10 Hz (0-100%)
58	GY/BK	PSOM (+) Signal	0 Hz, 55 mph: 125 Hz
59	DG/LG	Misfire Detection Sensor	45-55 Hz
60	GY/LB	HO2S-11 (B1 S1) Signal	0.1-1.1v
61-63	---	Not Used	---
64	LB/YL	TR Sensor Signal	In 'P': 0v, in O/D: 5v
65	BR/LG	DPFE Sensor Signal	0.20-1.30v
66-69	---	Not Used	---

1996 Bronco 5.0L VIN N (E4OD) 104 Pin Connector

PCM Pin #	Wire Color	Circuit Description (104 Pin)	Value at Hot Idle
70	WT/OR	Secondary AIR Bypass	AIRB On: 1v, Off: 12v
71	RD	Vehicle Power	12-14v
72	TN/RD	Injector 7 Control	3.2-4.5 ms
73	TN/BK	Injector 5 Control	3.2-4.5 ms
74	BR/YL	Injector 3 Control	3.2-4.5 ms
75	TN	Injector 1 Control	3.2-4.5 ms
76	BK/WT	Power Ground	<0.1v
77	BK/WT	Power Ground	<0.1v
78	---	Not Used	---
79	WT/LG	TCIL (lamp) Control	TCIL On: 1v, Off: 12v
80	LB/OR	Fuel Pump Control	On: 1v, Off: 12v
81	WT/YL	EPC Solenoid Control	10v (26 psi)
82	---	Not Used	---
83	WT/LB	IAC Solenoid Control	10.8v (33%)
84-86	---	Not Used	---
87	RD/BK	HO2S-21 (B2 S1) Signal	0.1-1.1v
88	LB/RD	MAF Sensor Signal	0.8v, 55 mph: 2.1v
89	GY/WT	TP Sensor Signal	0.53-1.27v
90	BR/WT	Reference Voltage	4.9-5.1v
91	GY/RD	Analog Signal Return	<0.050v
92	LG	Brake Position Switch	Brake Off: 12v, On: 1v
93	RD/WT	HO2S-11 (B1 S1) Heater	On: 1v, Off: 12v
94	YL/LB	HO2S-21 (B2 S1) Heater	On: 1v, Off: 12v
95	WT/BK	HO2S-12 (B1 S2) Heater	On: 1v, Off: 12v
96	---	Not Used	---
97	RD	Vehicle Power	12-14v
98	LG	Injector 8 Control	3.2-4.5 ms
99	LG/OR	Injector 6 Control	3.2-4.5 ms
100	BR/LB	Injector 4 Control	3.2-4.5 ms
101	WT	Injector 2 Control	3.2-4.5 ms
102	---	Not Used	---
103	BK/WT	Power Ground	<0.1v
104	---	Not Used	---

Pin Connector Graphic

PCM 104-PIN CONNECTOR

Terminal View of 104-Pin PCM Wiring Harness Connector

Standard Colors and Abbreviations

Abbreviation	Color	Abbreviation	Color	Abbreviation	Color
BK	Black	GY	Gray	PK	Purple
BL	Blue	GN	Green	RD	Red
BR	Brown	LG	LT Green	TN	Tan
DB	Dark Blue	OR	Orange	WT	White
DG	DK Green	PK	Pink	YL	Yellow

1990 Bronco 5.8L MFI VIN H (All) 60 Pin Connector

PCM Pin #	Wire Color	Circuit Description (60 Pin)	Value at Hot Idle
1	YL	Keep Alive Power	12-14v
2, 5, 9	---	Not Used	---
3	DG/WT	VSS (+) Signal	0 Hz, 55 mph: 125 Hz
4	DG/YL	Ignition Diagnostic Monitor	20-31 Hz
6	BK	VSS (-) Signal	0 Hz, 55 mph: 125 Hz
7	LG/YL	ECT Sensor Signal	0.5-0.6v
8	BR	Fuel Pump Monitor	On: 12v, Off: 0v
10	BK/YL	A/C Switch Signal	A/C On: 12v, Off: 0v
11	WT/BK	Air Management 2 Solenoid	AM2 On: 1v, Off: 12v
12-15, 19	---	Not Used	---
16	BK/OR	Ignition System Ground	<0.050v
17	PK/LG	Self-Test Output & MIL	MIL On: 1v, Off: 12v
18	TN/LG	Inferred Mileage Sensor	Digital Signals
20	BK	PCM Case Ground	0.050v
21	GY/WT	IAC Solenoid Control	8.0-9.1v
22	TN/LG	Fuel Pump Control	On: 1v, Off: 12v
23	LG/BK	Knock Sensor Signal	0v
24	YL/LG	PSP Switch Signal	Straight: 0v, Turned: 12v
25	YL/RD	ACT Sensor Signal	1.5-2.5v
26	OR/WT	Reference Voltage	4.9-5.1v
27	BR/LG	EVP Sensor Signal	0.4v
28, 32	---	Not Used	---
29	GY/BL	HO2S-11 (B1 S1) Signal	0.1-1.1v
29	DG/PK	HO2S-11 (B1 S1) Signal	0.1-1.1v
30	GY/YL	A/T: Neutral Drive Switch	In 'P': 0v, Others: 5v
30	GY/YL	M/T: Clutch Engage Switch	Clutch In: 0v, Out: 5v
31	GY/YL	EVAP Purge Solenoid	12v, 55 mph: 1v
33	DG	VR Solenoid Control	0%, 55 mph: 45%
34-35, 38-39	---	Not Used	---
36	YL/LG	Spark Output Signal	6.93v (50%)
37	R	Vehicle Power	12-14v
40	BK/LG	Power Ground	<0.1v
42-45, 50	BK/LG	Power Ground	<0.1v
45	DG/LG	MAP Sensor Signal	107 Hz (sea level)
45	DB/LG	MAP Sensor Signal	107 Hz (sea level)
46	BK/WT	Analog Signal Return	<0.050v
47	DG/LG	TP Sensor Signal	0.5-1.2v
48	WT/RD	Self-Test Indicator Signal	STI On: 1v, Off: 5v
49	OR	HO2S-11 (B1 S1) Ground	<0.050v
51	WT/RD	Air Management 1 Solenoid	1v, 55 mph: 12v
52-56	---	Not Used	---
57	RD	Vehicle Power	12-14v
56	DB	PIP Sensor Signal	6.93v (50%)
58	TN/OR	Injector Bank 1 (INJ 1, 4, 5, 8)	5.8-6.4 ms
59	TN/RD	Injector Bank 2 (INJ 2, 3, 6, 7)	5.8-6.4 ms
60	BK/LG	Power Ground	<0.1v

Pin Connector Graphic

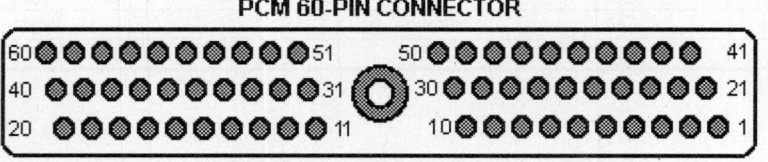

PCM 60-PIN CONNECTOR

Terminal View of 60-Pin PCM Harness Connector

1990 Bronco 5.8L V8 VIN H (E4OD) 60 Pin Connector

PCM Pin #	Wire Color	Circuit Description (60 Pin)	Value at Hot Idle
1	YL	Keep Alive Power	12-14v
2	LG	Brake Pedal Switch	Brake Off: 12v, On: 1v
3	DG/WT	VSS (+) Signal	0 Hz, 55 mph: 125 Hz
4	DG/YL	Ignition Diagnostic Monitor	20-31 Hz
5	---	Not Used	---
6	BK	VSS (-) Signal	0 Hz, 55 mph: 125 Hz
7	LG/YL	ECT Sensor Signal	0.5-0.6v
8	BR	Fuel Pump Monitor	On: 12v, Off: 0v
9	---	Not Used	---
10	BK/YL	A/C Switch Signal	A/C On: 12v, Off: 0v
11	WT/BK	Air Management 2 Solenoid	AM2 On: 1v, Off: 12v
12	BL/BK	4x4 Indicator Light	12v (Closed: 0v)
16	BK/OR	Ignition System Ground	<0.050v
17	PK/LG	Self-Test Output & MIL	MIL On: 1v, Off: 12v
18	TN/LG	Inferred Mileage Sensor	Digital Signals
19	DG/PK	Shift Solenoid 2 Control	1v, 55 mph: 1v
20	BK/LG	Case, Power Ground	<0.1v
21	GY/WT	IAC Solenoid Control	8.0-9.1v
22	TN/LG	Fuel Pump Control	On: 1v, Off: 12v
23	LG/BK	Knock Sensor Signal	0v
24	YL/LG	PSP Switch Signal	Straight: 0v, Turned: 12v
25	YL/RD	ACT Sensor Signal	1.5-2.5v
26	OR/WT	Reference Voltage	4.9-5.1v
27	BR/LG	EVP Sensor Signal	0.4v
28	---	Not Used	---
29	DG/PK	HO2S-11 (B1 S1) Signal	0.1-1.1v
30	BL/WT	MLP Sensor Signal	In 'P': 0v, in O/D: 5v
31	GY/YL	EVAP Purge Solenoid	12v, 55 mph: 1v
32	LG/WT	Overdrive Cancel Indicator	OCIL On: 1v, Off: 12v
33	DG	VR Solenoid Control	0%, 55 mph: 45%
36	YL/LG	Spark Output Signal	6.93v (50%)
37	RD	Vehicle Power	12-14v
38	BL/YL	EPC Solenoid Control	9.2v (5 psi)
40	BK/LG	Case, Power Ground	<0.1v
41	TN/WT	Overdrive Cancel Switch	OCS Closed: 0.1v
42	OR/BK	TOT Sensor Signal	2.10-2.40v
45	DB/LG	MAP Sensor Signal	107 Hz (sea level)
46	BK/WT	Analog Signal Return	<0.050v
47	DG/LG	TP Sensor Signal	0.5-1.2v
48	WT/RD	Self-Test Indicator Signal	STI On: 1v, Off: 5v
49	OR	HO2S-11 (B1 S1) Ground	<0.050v
51	WT/RD	Air Management 1 Solenoid	1v, 55 mph: 12v
52	OR/YL	Shift Solenoid 1 Control	1v, 55 mph: 12v
53	PK/YL	TCC Solenoid Control	12v, 55 mph: 1v
55	BR	Coast Clutch Solenoid	12v, 55 mph: 12v
56	DB	PIP Sensor Signal	6.93v (50%)
57	RD	Vehicle Power	12-14v
58	TN/OR	Injector Bank 1 (INJ 1, 4, 5, 8)	5.8-6.4 ms
59	TN/RD	Injector Bank 2 (INJ 2, 3, 6, 7)	5.8-6.4 ms
60	BK/LG	Case, Power Ground	<0.1v

Pin Connector Graphic

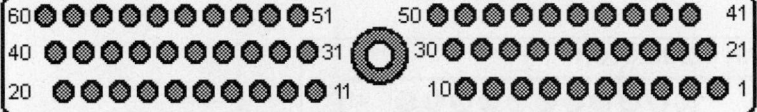

PCM 60-PIN CONNECTOR

Terminal View of 60-Pin PCM Harness Connector

1991-93 Bronco 5.8L MFI VIN H (All) 60 Pin Connector

PCM Pin #	Wire Color	Circuit Description (60 Pin)	Value at Hot Idle
1	YL	Keep Alive Power	12-14v
2	---	Not Used	---
3	GY/BK	VSS (+) Signal	0 Hz, 55 mph: 125 Hz
4	YL/BK	Ignition Diagnostic Monitor	20-31 Hz
4	TN/YL	Ignition Diagnostic Monitor	20-31 Hz
5	---	Not Used	---
6	PK/OR	VSS (-) Signal	0 Hz, 55 mph: 125 Hz
7	LG/RD	ECT Sensor Signal	0.5-0.6v
8	DG/YL	Fuel Pump Monitor	On: 12v, Off: 0v
9	PK/BL	Data Bus (-) Signal	Digital Signals
9	BK/OR	Data Bus (-) Signal	Digital Signals
10	BK/YL	A/C Switch Signal	A/C On: 12v, Off: 0v
10	PK/BL	A/C Switch Signal	A/C On: 12v, Off: 0v
11	BR	Air Management 2 Solenoid	AM2 On: 1v, Off: 12v
12-15	---	Not Used	---
16	OR/RD	Ignition System Ground	<0.050v
17	PK/LG	Self-Test Output & MIL	MIL On: 1v, Off: 12v
18-19	---	Not Used	---
20	BK	PCM Case Ground	<0.050v
21	WT/BL	IAC Solenoid Control	7.2-9.2v
22	BL/OR	Fuel Pump Control	On: 1v, Off: 12v
23 ('92-'93)	YL/RD	Knock Sensor Signal	0v
24 ('92-'93)	YL/LG	PSP Switch Signal	Straight: 0v, Turned: 12v
25	GY	ACT Sensor Signal	1.5-2.5v
26	BR/WT	Reference Voltage	4.9-5.1v
27	BR/LG	EVP Sensor Signal	0.4v
28	TN/OR	Data Bus (+) Signal	Digital Signals
29	GY/BL	HO2S-11 (B1 S1) Signal	0.1-1.1v
29	GY/YL	HO2S-11 (B1 S1) Signal	0.1-1.1v
30	BL/YL	A/T: Neutral Drive Switch	In 'P': 0v, Others: 5v
30	BL/YL	M/T: Clutch Engage Switch	Clutch In: 0v, Out: 5v
31-32	---	Not Used	---
33	BR/PK	VR Solenoid Control	0%, 55 mph: 45%
34-35	---	Not Used	---
36	PK	Spark Output Signal	6.93v (50%)
37	RD	Vehicle Power	12-14v
38-39	---	Not Used	---
40	BK/WT	Power Ground	<0.1v
41-44	---	Not Used	---
45	LG/BK	MAP Sensor Signal	107 Hz (sea level)
46	GY/RD	Analog Signal Return	<0.050v
47	GY/WT	TP Sensor Signal	0.5-1.2v
48	WT/PK	Self-Test Indicator Signal	STI On: 1v, Off: 5v
49	Or	HO2S-11 (B1 S1) Ground	<0.050v
50	---	---	---
51	WT/OR	Air Management 1 Solenoid	AM1 On: 1v, 55 mph: 12v
52-55	---	Not Used	---
56	GY/OR	PIP Sensor Signal	6.93v (50%)
57	RD	Vehicle Power	12-14v
58	TN	Injector Bank 1 (INJ 1, 4, 5, 8)	3.5-5.0 ms
59	WT	Injector Bank 2 (INJ 2, 3, 6, 7)	3.5-5.0 ms
60	BK/WT	Power Ground	<0.1v

Pin Connector Graphic

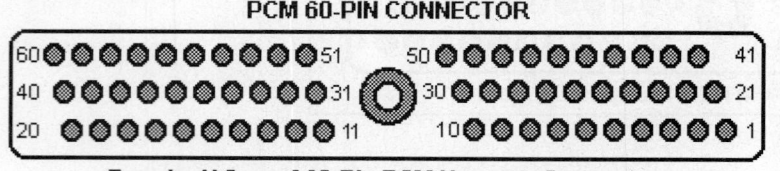

Terminal View of 60-Pin PCM Harness Connector

1991-93 Bronco 5.8L V8 VIN H (E4OD) 60 Pin Connector

PCM Pin #	Wire Color	Circuit Description (60 Pin)	Value at Hot Idle
1	YL	Keep Alive Power	12-14v
2	LG	Brake Pedal Switch	Brake Off: 12v, On: 1v
3	GY/BK	VSS (+) Signal	0 Hz, 55 mph: 125 Hz
4	YL/BK	Ignition Diagnostic Monitor	20-31 Hz
4	TN/YL	Ignition Diagnostic Monitor	20-31 Hz
5, 13-15, 18	---	Not Used	---
6	PK/OR	VSS (-) Signal	0 Hz, 55 mph: 125 Hz
7	LG/RD	ECT Sensor Signal	0.5-0.6v
8	DG/YL	Fuel Pump Monitor	On: 12v, Off: 0v
9	PK/BL	Data Bus (+), (+) Signals	Digital Signals, <0.050v
10	BK/YL	A/C Switch Signal	A/C On: 12v, Off: 0v
10	PK/BL	A/C Switch Signal	A/C On: 12v, Off: 0v
11	BR	Air Management 2 Solenoid	AM2 On: 1v, Off: 12v
12	PK/BL	4x4 Indicator Light	12v (switch closed: 0v)
16	OR/RD	Ignition System Ground	<0.050v
17	PK/LG	Self-Test Output & MIL	MIL On: 1v, Off: 12v
19	PK/OR	Shift Solenoid 2 Control	12v, 55 mph: 1v
20	BK	PCM Case Ground	<0.050v
21	WT/BL	IAC Solenoid Control	7.0-10.1v
22	BL/OR	Fuel Pump Control	On: 1v, Off: 12v
23 ('92-'93)	YL/RD	Knock Sensor Signal	0v
24 ('92-'93)	YL/LG	PSP Switch Signal	Straight: 0v, Turned: 12v
25	GY	ACT Sensor Signal	1.5-2.5v
26	BR/WT	Reference Voltage	4.9-5.1v
27	BR/LG	EVP Sensor Signal	0.4v
28	TN/OR	Data Bus (+) Signal	Digital Signals
29	GY/BL	HO2S-11 (B1 S1) Signal	0.1-1.1v
29	GY/YL	HO2S-11 (B1 S1) Signal	0.1-1.1v
30	BL/YL	MLP Sensor Signal	In 'P': 0v, in O/D: 5v
31	GY/YL	EVAP Purge Solenoid	12v, 55 mph: 1v
32	WT/LG	Overdrive Cancel Indicator	OCIL On: 1v, Off: 12v
33	BR/PK	VR Solenoid Control	0%, 55 mph: 45%
34-35, 39, 44	---	Not Used	---
36	PK	Spark Output Signal	6.93v (50%)
37, 57	RD	Vehicle Power	12-14v
38	WT/YL	EPC Solenoid Control	9.5v (5 psi)
40	BK/WT	Power Ground	<0.1v
41	TN/WT	Overdrive Cancel Switch	OCS Closed: 0.1v
42	OR/BK	TOT Sensor Signal	2.10-2.40v
43 ('91)	PK	A/C Demand Switch	A/C On: 12v, Off: 0v
45	LG/BK	MAP Sensor Signal	107 Hz (sea level)
46	GY/RD	Analog Signal Return	<0.050v
47	GY/WT	TP Sensor Signal	0.5-1.2v
48	WT/PK	Self-Test Indicator Signal	STI On: 1v, Off: 5v
49	OR	HO2S-11 (B1 S1) Ground	<0.050v
50, 54	---	Not Used	---
51	WT/OR	Air Management 1 Solenoid	1v, 55 mph: 12v
52	OR/YL	Shift Solenoid 1 Control	1v, 55 mph: 12v
53	PK/YL	TCC Solenoid Control	1v, 55 mph: 12v
55	BR/OR	Coast Clutch Switch Signal	12v, 55 mph: 12v
56	GY/OR	PIP Sensor Signal	6.93v (50%)
58	TN	Injector Bank 1 (INJ 1, 4, 5, 8)	3.5-4.0 ms
59	WT	Injector Bank 2 (INJ 2, 3, 6, 7)	3.5-4.0 ms
60	BK/WT	Power Ground	<0.1v

Pin Connector Graphic

PCM 60-PIN CONNECTOR

60 ... 51   50 ... 41
40 ... 31   30 ... 21
20 ... 11   10 ... 1

Terminal View of 60-Pin PCM Harness Connector

1994 Bronco 5.8L V8 MFI VIN H (E4OD) 60 Pin Connector

PCM Pin #	Wire Color	Circuit Description (60 Pin)	Value at Hot Idle
1	YL	Keep Alive Power	12-14v
2	LG	Brake Pedal Switch	Brake Off: 12v, On: 1v
3	GY/BK	VSS (+) Signal	0 Hz, 55 mph: 125 Hz
4	YL/BK	Ignition Diagnostic Monitor	20-31 Hz
5, 13-15	---	Not Used	---
6	PK/OR	VSS (-) Signal	0 Hz, 55 mph: 125 Hz
7	LG/RD	ECT Sensor Signal	0.5-0.6v
8	DG/YL	Fuel Pump Monitor	On: 12v, Off: 0v
9	PK/BL	Data Bus (-) Signal	Digital Signals
9	BK/OR	Data Bus (-) Signal	Digital Signals
10	BK/YL	A/C Switch Signal	A/C On: 12v, Off: 0v
10	PK/BL	A/C Switch Signal	A/C On: 12v, Off: 0v
11	BR	Air Management 2 Solenoid	AM2 On: 1v, Off: 12v
12	PK/BL	4x4 Indicator Light	12v (switch closed: 0v)
16	OR/RD	Ignition System Ground	<0.050v
17	PK/LG	Self-Test Output & MIL	MIL On: 1v, Off: 12v
18, 23-24	---	Not Used	---
19	PK/OR	Shift Solenoid 2 Control	12v, 55 mph: 1v
20	BK	PCM Case Ground	<0.050v
21	WT/BL	IAC Solenoid Control	7.0-10.1v
22	BL/OR	Fuel Pump Control	On: 1v, Off: 12v
25	GY	ACT Sensor Signal	1.5-2.5v
26	BR/WT	Reference Voltage	4.9-5.1v
27	BR/LG	EVP Sensor Signal	0.4v
28	TN/OR	Data Bus (+) Signal	Digital Signals
29	GY/BL	HO2S-11 (B1 S1) Signal	0.1-1.1v
29	GY/YL	HO2S-11 (B1 S1) Signal	0.1-1.1v
30	BL/YL	MLP Sensor Signal	In 'P': 0v, in O/D: 5v
31	GY/YL	EVAP Purge Solenoid	12v, 55 mph: 1v
32	WT/LG	Overdrive Cancel Indicator	OCIL On: 1v, Off: 12v
33	BR/PK	VR Solenoid Control	0%, 55 mph: 45%
34-35, 39, 44	---	Not Used	---
36	PK	Spark Output Signal	6.93v (50%)
37	RD	Vehicle Power	12-14v
38	WT/YL	EPC Solenoid Control	9.5v (5 psi)
40	BK/WT	Power Ground	<0.1v
41	TN/WT	TCS (switch) Signal	TCS Closed: 0.1v
42	OR/BK	TOT Sensor Signal	2.10-2.40v
43	PK	A/C Demand Switch	A/C On: 12v, Off: 0v
45	LG/BK	MAP Sensor Signal	107 Hz (sea level)
46	GY/RD	Analog Signal Return	<0.050v
47	GY/WT	TP Sensor Signal	0.5-1.2v
48	WT/PK	Self-Test Indicator Signal	STI On: 1v, Off: 5v
49	OR	HO2S-11 (B1 S1) Ground	<0.050v
50, 54	---	Not Used	---
51	WT/OR	Air Management 1 Solenoid	1v, 55 mph: 12v
52	OR/YL	Shift Solenoid 1 Control	1v, 30 mph: 12v
53	PK/YL	TCC Solenoid Control	12v, 55 mph: 9v
55	BR/OR	Coast Clutch Switch Signal	12v, 55 mph: 12v
56	GY/OR	PIP Sensor Signal	6.93v (50%)
57	RD	Vehicle Power	12-14v
58	TN	Injector Bank 1 (INJ 1, 4, 5, 8)	3.5-5.0 ms
59	WT	Injector Bank 2 (INJ 2, 3, 6, 7)	3.5-5.0 ms
60	BK/WT	Power Ground	<0.1v

Pin Connector Graphic

PCM 60-PIN CONNECTOR

60 ●●●●●●●●●● 51   50 ●●●●●●●●●● 41
40 ●●●●●●●●●● 31   ⊙   30 ●●●●●●●●●● 21
20 ●●●●●●●●●● 11   10 ●●●●●●●●●● 1

Terminal View of 60-Pin PCM Harness Connector

1995 Bronco 5.8L V8 VIN H (E4OD) California 60 Pin Connector

PCM Pin #	Wire Color	Circuit Description (60 Pin)	Value at Hot Idle
1	YL	Keep Alive Power	12-14v
2	LG	Brake Pedal Switch	Brake Off: 12v, On: 1v
3	GY/BK	VSS (+) Signal	0 Hz, 55 mph: 125 Hz
4	YL/BK	Ignition Diagnostic Monitor	20-31 Hz
5	---	Not Used	---
6	PK/OR	VSS (-) Signal	0 Hz, 55 mph: 125 Hz
7	LG/RD	ECT Sensor Signal	0.5-0.6v
8	DG/YL	Fuel Pump Monitor	On: 12v, Off: 0v
9	TN/BL	MAF Sensor Return	<0.050v
10	DG/OR	A/C Switch Signal	A/C On: 12v, Off: 0v
11	GY/YL	EVAP Purge Solenoid	12v, 55 mph: 1v
12	LG/OR	Injector 6 Control	5.5-6.1 ms
13	TN/RD	Injector 7 Control	5.5-6.1 ms
14	BL	Injector 8 Control	5.5-6.1 ms
15	TN/BK	Injector 5 Control	5.5-6.1 ms
16	OR/RD	Ignition System Ground	<0.050v
17	PK/LG	Self-Test Output & MIL	MIL On: 1v, Off: 12v
18	TN/OR	Data Bus (+) Signal	Digital Signals
19	PK/BL	Data Bus (-) Signal	Digital Signals
20	BK	PCM Case Ground	<0.050v
21	WT/BL	IAC Solenoid Control	10.7v (33%)
22	BL/OR	Fuel Pump Control	On: 1v, Off: 12v
23-24	---	Not Used	---
25	GY	IAT Sensor Signal	1.5-2.5v
26	BR/WT	Reference Voltage	4.9-5.1v
27	BR/LG	EVP Sensor Signal	0.4v
28-29	---	Not Used	---
30	BL/YL	MLP Sensor Signal	In 'P': 0v, in O/D: 5v
31	---	Not Used	---
32	WT/LG	TCIL (lamp) Control	TCIL On: 1v, Off: 12v
33	BR/PK	VR Solenoid Control	0%, 55 mph: 45%
34	BR	Air Diverter Solenoid Control	12v, 55 mph: 12v
35	BR/BL	Injector 4 Control	5.5-6.1 ms
36	PK	Spark Output Signal	6.93v (50%)
37	RD	Vehicle Power	12-14v
38	WT/YL	EPC Solenoid Control	9.5v (5 psi)
39	BR/YL	Injector 3 Control	5.5-6.1 ms
40	BK/WT	Power Ground	<0.1v

1995 Bronco 5.8L V8 VIN H (E4OD) California 60 Pin Connector

PCM Pin #	Wire Color	Circuit Description (60 Pin)	Value at Hot Idle
41	TN/WT	TCS (switch) Signal	TCS Closed: 0.1v
42	BL/PK	4x4 Indicator Light	12v (switch closed: 0v)
43	RD/BK	HO2S-21 (B2 S1) Signal	0.1-1.1v
44	GY/BL	HO2S-11 (B1 S1) Signal	0.1-1.1v
45	---	Not Used	---
46	GY/RD	Analog Signal Return	<0.050v
47	GY/WT	TP Sensor Signal	0.5-1.2v
48	WT/PK	Self-Test Indicator Signal	STI On: 1v, Off: 5v
49	OR/BK	TOT Sensor Signal	2.10-2.40v
50	BL/RD	MAF Sensor Signal	0.8v
51	OR/YL	Shift Solenoid 1 Control	1v, 30 mph: 12v
52	PK/OR	Shift Solenoid 2 Control	12v, 55 mph: 1v
53	PK/YL	TCC Solenoid Control	12v, 55 mph: 9v
54	---	Not Used	---
55	BR/OR	Coast Clutch Solenoid	12v, 55 mph: 12v
56	GY/OR	PIP Sensor Signal	6.93v (50%)
57	RD	Vehicle Power	12-14v
58	TN	Injector 1 Control	5.5-6.1 ms
59	WT	Injector 2 Control	5.5-6.1 ms
60	BK/WT	Power Ground	<0.1v

Pin Connector Graphic

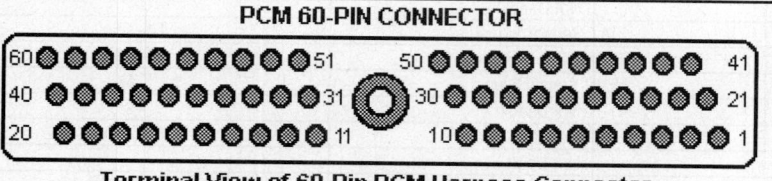

PCM 60-PIN CONNECTOR

Terminal View of 60-Pin PCM Harness Connector

Standard Colors and Abbreviations

Abbreviation	Color	Abbreviation	Color	Abbreviation	Color
BK	Black	GY	Gray	PK	Purple
BL	Blue	GN	Green	RD	Red
BR	Brown	LG	LT Green	TN	Tan
DB	Dark Blue	OR	Orange	WT	White
DG	DK Green	PK	Pink	YL	Yellow

1995 Bronco 5.8L V8 VIN H (E4OD) 60 Pin Connector

PCM Pin #	Wire Color	Circuit Description (60 Pin)	Value at Hot Idle
1	YL	Keep Alive Power	12-14v
2	LG	Brake Pedal Switch	Brake Off: 12v, On: 1v
3	GY/BK	VSS (+) Signal	0 Hz, 55 mph: 125 Hz
4	YL/BK	Ignition Diagnostic Monitor	20-31 Hz
5, 13-15, 18	---	Not Used	---
6	PK/OR	VSS (-) Signal	0 Hz, 55 mph: 125 Hz
7	LG/RD	ECT Sensor Signal	0.5-0.6v
8	DG/YL	Fuel Pump Monitor	On: 12v, Off: 0v
9	PK/BL	Data Bus (-) Signal	Digital Signals
9	BK/OR	Data Bus (-) Signal	Digital Signals
10	BK/YL	A/C Switch Signal	A/C On: 12v, Off: 0v
10	PK/BL	A/C Switch Signal	A/C On: 12v, Off: 0v
11	BR	Air Management 2 Solenoid	AM2 On: 1v, Off: 12v
12	PK/BL	4x4 Indicator Light	12v (switch closed: 0v)
16	OR/RD	Ignition System Ground	<0.050v
17	PK/LG	Self-Test Output & MIL	MIL On: 1v, Off: 12v
19	PK/OR	Shift Solenoid 2 Control	12v, 55 mph: 1v
20	BK	PCM Case Ground	<0.050v
21	WT/BL	IAC Solenoid Control	7.0-10.1v
22	BL/OR	Fuel Pump Control	On: 1v, Off: 12v
23-24, 34-35	---	Not Used	---
25	GY	ACT Sensor Signal	1.5-2.5v
26	BR/WT	Reference Voltage	4.9-5.1v
27	BR/LG	EVP Sensor Signal	0.4v
28	TN/OR	Data Bus (+) Signal	Digital Signals
29	GY/BL	HO2S-11 (B1 S1) Signal	0.1-1.1v
29	GY/YL	HO2S-11 (B1 S1) Signal	0.1-1.1v
30	BL/YL	MLP Sensor Signal	In 'P': 0v, in O/D: 5v
31	GY/YL	EVAP Purge Solenoid	12v, 55 mph: 1v
32	WT/LG	Overdrive Cancel Indicator	OCIL On: 1v, Off: 12v
33	BR/PK	VR Solenoid Control	0%, 55 mph: 45%
36	PK	Spark Output Signal	6.93v (50%)
37	RD	Vehicle Power	12-14v
38	WT/YL	EPC Solenoid Control	9.5v (5 psi)
39, 44, 50, 54	---	Not Used	---
40	BK/WT	Power Ground	<0.1v
41	TN/WT	TCS (switch) Signal	TCS Closed: 0.1v
42	OR/BK	TOT Sensor Signal	2.10-2.40v
43	PK	A/C Demand Switch	A/C On: 12v, Off: 0v
45	LG/BK	MAP Sensor Signal	107 Hz (sea level)
46	GY/RD	Analog Signal Return	<0.050v
47	GY/WT	TP Sensor Signal	0.5-1.2v
48	WT/PK	Self-Test Indicator Signal	STI On: 1v, Off: 5v
49	OR	HO2S-11 (B1 S1) Ground	<0.050v
51	WT/OR	Air Management 1 Solenoid	1v, 55 mph: 12v
52	OR/YL	Shift Solenoid 1 Control	1v, 30 mph: 12v
53	PK/YL	TCC Solenoid Control	12v, 55 mph: 9v
55	BR/OR	Coast Clutch Switch Signal	12v, 55 mph: 12v
56	GY/OR	PIP Sensor Signal	6.93v (50%)
57	RD	Vehicle Power	12-14v
58	TN	Injector Bank 1 (INJ 1, 4, 5, 8)	3.5-5.0 ms
59	WT	Injector Bank 2 (INJ 2, 3, 6, 7)	3.5-5.0 ms
60	BK/WT	Power Ground	<0.1v

Pin Connector Graphic

PCM 60-PIN CONNECTOR

Terminal View of 60-Pin PCM Harness Connector

1996 Bronco 5.8L V8 MFI VIN H (E4OD) 104 Pin Connector

PCM Pin #	Wire Color	Circuit Description (104 Pin)	Value at Hot Idle
1	PK/OR	Shift Solenoid 2 Control	12v, 55 mph: 12v
2	PK/LG	MIL (lamp) Control	MIL On: 1v, Off: 12v
4	LG/RD	Power Take-Off Signal	0.1v (off)
5-12, 17, 22	---	Not Used	---
13	PK	Flash EPROM Power	0.1v
14	LG/BK	4x4 Low Switch Signal	12v (Closed: 0v)
15	PK/LB	Data Bus (-) Signal	Digital Signals
16	TN/OR	Data Bus (+) Signal	Digital Signals
23	OR/RD	Ignition Ground	<0.050v
24	BK/WT	Power Ground	<0.1v
25	BK/LB	Case Ground	<0.050v
26, 28	---	Not Used	---
27	OR/YL	Shift Solenoid 1 Control	1v, 55 mph: 12v
29	TN/WT	TCS (switch) Signal	TCS & O/D On: 12v
30-32, 34	---	Not Used	---
33	PK/OR	PSOM (+) Signal	<0.050v
35	RD/LG	HO2S-12 (B1 S2) Signal	0.1-1.1v
36	TN/LB	MAF Sensor Return	<0.050v
37	OR/BK	TFT Sensor Signal	2.10-2.40v
38	LG/RD	ECT Sensor Signal	0.5-0.6v
39	GY	IAT Sensor Signal	1.5-2.5v
40	DG/YL	Fuel Pump Monitor	On: 12v, Off: 0v
41	BK/YL	A/C Switch Signal	A/C On: 12v, Off: 0v
42-43, 46	---	Not Used	---
44	BR	Secondary AIR Diverter	AIRD On: 1v, Off: 12v
47	BK/PK	VR Solenoid Control	0%, 55 mph: 45%
48	YL/BK	Clean Tachometer Output	39-45 Hz
49	GY/OR	PIP Sensor Signal	6.93v (50%)
50	PK	Spark Output Signal	6.93v (50%)
51	BK/WT	Power Ground	<0.1v
52, 57	---	Not Used	---
53	BR/OR	Coast Clutch Solenoid	12v, 55 mph: 12v
54	PK/YL	TCC Solenoid Control	0%, 55 mph: 95%
55	Y	Keep Alive Power	12-14v
56	LG/BK	EVAP Purge Solenoid	0-10 Hz (0-100%)
58	GY/BK	PSOM (-) Signal	0 Hz, 55 mph: 125 Hz
59	DG/LG	Misfire Detection Sensor	45-55 Hz
60	GY/LB	HO2S-11 (B1 S1) Signal	0.1-1.1v
61-63	---	Not Used	---
64	LB/YL	TR Sensor Signal	In 'P': 0v, in O/D: 5v

1996 Bronco 5.8L V8 MFI VIN H (E4OD) 104 Pin Connector

PCM Pin #	Wire Color	Circuit Description (104 Pin)	Value at Hot Idle
65	BR/LG	DPFE Sensor Signal	0.20-1.30v
66-69	---	Not Used	---
70	WT/OR	Secondary AIR Bypass	AIRB On: 1v, Off: 12v
71	RD	Vehicle Power	12-14v
72	TN/RD	Injector 7 Control	4.2-4.6 ms
73	TN/BK	Injector 5 Control	4.2-4.6 ms
74	BR/YL	Injector 3 Control	4.2-4.6 ms
75	TN	Injector 1 Control	4.2-4.6 ms
76	BK/WT	Power Ground	<0.1v
77	BK/WT	Power Ground	<0.1v
78, 82	---	Not Used	---
79	WT/LG	TCIL (lamp) Control	TCIL On: 1v, Off: 12v
80	LB/OR	Fuel Pump Control	On: 1v, Off: 12v
81	WT/YL	EPC Solenoid Control	9.2v (5 psi)
83	WT/LB	IAC Solenoid Control	10.7v (33%)
84-86	---	Not Used	---
87	RD/BK	HO2S-21 (B2 S1) Signal	0.1-1.1v
88	LB/RD	MAF Sensor Signal	0.8v, 55 mph: 2.1v
89	GY/WT	TP Sensor Signal	0.53-1.27v
90	BR/WT	Reference Voltage	4.9-5.1v
91	GY/RD	Analog Signal Return	<0.050v
92	LG	Brake Position Switch	Brake Off: 12v, On: 1v
93	RD/WT	HO2S-11 (B1 S1) Heater	On: 1v, Off: 12v
94	YL/LB	HO2S-21 (B2 S1) Heater	On: 1v, Off: 12v
95	WT/BK	HO2S-12 (B1 S2) Heater	On: 1v, Off: 12v
96	---	Not Used	---
97	RD	Vehicle Power	12-14v
98	LG	Injector 8 Control	4.2-4.6 ms
99	LG/OR	Injector 6 Control	4.2-4.6 ms
100	BR/LB	Injector 4 Control	4.2-4.6 ms
101	WT	Injector 2 Control	4.2-4.6 ms
102	---	Not Used	---
103	BK/WT	Power Ground	<0.1v
104	---	Not Used	---

Pin Connector Graphic

PCM 104-PIN CONNECTOR

Terminal View of 104-Pin PCM Wiring Harness Connector

Standard Colors and Abbreviations

Abbreviation	Color	Abbreviation	Color	Abbreviation	Color
BK	Black	GY	Gray	PK	Purple
BL	Blue	GN	Green	RD	Red
BR	Brown	LG	LT Green	TN	Tan
DB	Dark Blue	OR	Orange	WT	White
DG	DK Green	PK	Pink	YL	Yellow

Escape Pin Tables

2001-03 Escape 2.0L I4 4v ZETEC VIN B (All) 104 Pin Connector

PCM Pin #	Wire Color	Circuit Description (104 Pin)	Value at Hot Idle
1	VT/OR	Shift Solenoid 'B' Control	12, 55 mph: 1v
2-12	---	Not Used	---
13	VT	Flash EEPROM Power	0.1v
14	---	Not Used	---
15	PK/LB	Data Bus (-) Signal	<0.050v
16	TN/OR	Data Bus (+) Signal	Digital Signals
17	DB	High Speed Fan Control	Fan Off: 12v, On: 1v
18	---	Not Used	---
19	BR/OR	Passive Antitheft TX Signal	Digital Signals
20	BR/YL	Fuel Injector 3 Control	2.5-3.0 ms
21	BK/PK	CKP Sensor (+) Signal	400-500 Hz
22	GY/YL	CKP Sensor (-) Signal	400-500 Hz
23	BR/LG	Power Ground	<0.1v
24	BK	Power Ground	<0.1v
25	BR/WT	Power Ground	<0.1v
26	DG/VT	Coil 1 Driver (dwell)	5°, 55 mph: 7°
27	DB/OR	Starter Relay Circuit	Relay Off: 12v, On: 1v
28	GY/BK	VSS (+) Signal	0 Hz, 55 mph: 125 Hz
29	OR/YL	Overdrive Cancel Switch	OCS Off: 0v, On: 12v
30	---	Not Used	---
31	YL/LG	Power Steering Pressure Switch	Straight: 0v, Turning: 12v
32	---	Not Used	---
33	BR/LG	M/T: Power Ground	<0.1v
34	WT/LB	Turbine Shaft Speed Sensor	340 Hz, 55 mph: 1090 Hz
35	RD/LG	HO2S-12 (B1 S2) Signal	0.1-1.1v
36	TN/BK	MAF Sensor Return	<0.050v
37	OR/BK	Transmission Fluid Temperature Sensor	2.10-2.40v
38	LG/RD	Cylinder Head Temperature Sensor	0.6 (194°F)
39	PK/BK	Sensor Return	<0.050v
40	DG/YL	Fuel Pump Relay Control	Off: 12v, On: 1v
41	DG/OR	A/C Dual Pressure Switch	A/C Off: 0v, On: 12v
43	BK	Medium Speed Fan Control Relay	Relay Off: 12v, On: 1v
44-50	---	Not Used	---
51	BK	Power Ground	<0.1v
52	LG/WT	Coil 2 Driver (dwell)	5°, 55 mph: 7°
53	RD/LG	Passive Antitheft RX Signal	Digital Signals
54	LB/OR	Fuel Pump Relay Control	Relay Off: 12v, On: 1v
55	RD/LG	Battery Power	12-14v
56	LG/BK	EVAP Vapor Management Valve	0-10 Hz (0-100%)
57	YL/RD	Knock Sensor 1 (+) Signal	0v
58	DB/YL	A/T: Output Shaft Speed Sensor	0 Hz, 55 mph: 720 Hz
58	DB/YL	M/T: VSS (-) Signal	0 Hz, 55 mph: 125 Hz
59	GY/OR	Generator Load Indicator	1.5-10.v (40-250 Hz)
60	GY/LG	HO2S-11 (B1 S1) Signal	0.1-1.1v
61	---	Not Used	---
62	RD/PK	Fuel Tank Pressure Sensor	2.6v (0" HG - cap off)
63	---	Not Used	---
64	LB/YL	Digital TR Sensor	0v, in O/D: 1.7v

2001-03 Escape 2.0L I4 4v ZETEC VIN B (All) 104 Pin Connector

PCM Pin #	Wire Color	Circuit Description (104 Pin)	Value at Hot Idle
65	BR/LG	EGR DPFE Sensor Signal	0.95-1.05v
66	---	Not Used	---
67	VT/WT	EVAP Canister Vent Solenoid Control	0-10 Hz (0-100%)
68	LB	Low Speed Fan Relay Control	Fan Off: 12v, On: 1v
69	PK/YL	A/C WOT Relay Control	Relay Off: 12v, On: 1v
70	TN	Fuel Injector 1 Control	2.5-3.0 ms
71	WT/RD	Vehicle Power (Start-Run)	12-14v
72	BR/WT	Generator Control Signal	0-130 Hz
73	OR/YL	Shift Solenoid 'A' Control	1v, 55 mph: 1v
74-75	---	Not Used	---
76	BK	Power Ground	<0.1v
77	BK	Power Ground	<0.1v
78	---	Not Used	---
79	VT	Electric VR Solenoid Control	0%, 55 mph: 45%
80	---	Not Used	---
81	WT/YL	EPC Solenoid Control	9.5v
82	---	Not Used	---
83	WT/LB	IAC Solenoid Control	10.7v (33%)
84	---	Output Shaft Speed (+) Signal	0 Hz, 55 mph: 720 Hz
85	DB/OR	CMP Sensor (+) Signal	5-7 Hz
86	BK/YL	A/C Dual Pressure Switch	A/C On: 12v (Open)
87	DG/VT	Knock Sensor 1 (-) Signal	<0.050v
88	LB/RD	MAF Sensor Signal	0.6-0.9v
89	GY/WT	TP Sensor Signal	0.53-1.27v
90	BR/WT	Reference Voltage	4.9-5.1v
91	OR	Signal Return	<0.050v
92	LG	Brake Position Switch	Brake Off: 12v, On: 1v
93	RD/WT	HO2S-11 (B1 S1) Heater Control	Off: 12v, On: 1v
94	---	Not Used	---
95	BR/LB	Injector 4 Control	2.5-3.0 ms
96	WT	Injector 2 Control	2.5-3.0 ms
97	WT/RD	Vehicle Power	12-14v
98	---	Not Used	---
99	VT/YL	Torque Converter Clutch	0%, 55 mph: 95%
100	WT/BK	HO2S-12 (B1 S2) Heater	On: 1v, Off: 12v
101	---	Not Used	---
102	PK/BK	Shift Solenoid 3 (3-2T/TCCS)	12v, 55 mph: 8.8v
103	BK	Power Ground	<0.1v
104	---	Not Used	---

Pin Connector Graphic

PCM 104-PIN CONNECTOR

```
 1 ●●●●●●●●●●●●●   ●●●●●●●●●●●●● 26
27 ●●●●●●●●●●●●●   ●●●●●●●●●●●●● 52
53 ●●●●●●●●●●●●●  ⬤  ●●●●●●●●●●●●● 78
79 ●●●●●●●●●●●●●   ●●●●●●●●●●●●● 104
```

Terminal View of 104-Pin PCM Wiring Harness Connector

Standard Colors and Abbreviations

Abbreviation	Color	Abbreviation	Color	Abbreviation	Color
BK	Black	GY	Gray	PK	Purple
BL	Blue	GN	Green	RD	Red
BR	Brown	LG	LT Green	TN	Tan
DB	Dark Blue	OR	Orange	WT	White
DG	DK Green	PK	Pink	YL	Yellow

2001-03 Escape 3.0L V6 4v DURATEC VIN 1 (A/T) 104 Pin Connector

PCM Pin #	Wire Color	Circuit Description (104 Pin)	Value at Hot Idle
1	WT/PK	Coil On Plug (COP) 4 Driver (dwell)	6°, 55 mph: 8°
2	---	Not Used	---
3	BR/LG	Power Ground	<0.1v
4-5	---	Not Used	---
6	OR/YL	Shift Solenoid 'A' Control	12v, 55 mph: 12v
7-10	---	Not Used	---
11	VT/OR	Shift Solenoid 'B' Control	1v, 55 mph: 12v
13	VT	Flash EEPROM Power	0.1v
14	---	Not Used	---
15	PK/LB	Data Bus (-) Signal	<0.050v
16	TN/OR	Data Bus (+) Signal	Digital Signals
17	RD/BK	Passive Antitheft RX Signal	Digital Signals
18	BR/OR	Passive Antitheft Output	0.8v
19	---	Not Used	---
20	PK/BK	Shift Solenoid 3 (3-2T/TCCS)	12v, 55 mph: 8.8v
21	BK/PK	CKP Sensor (+) Signal	400-450 Hz
22	GY/YL	CKP Sensor (-) Signal	400-450 Hz
23	---	Not Used	---
24	BK	Power Ground	<0.1v
25	BK/WT	Power Ground	<0.1v
26	LG/WT	Coil On Plug (COP) 1 Driver (dwell)	6°, 55 mph: 8°
27	PK/WT	Coil On Plug (COP) 5 Driver (dwell)	6°, 55 mph: 8°
28	DB	Low Speed Fan Relay Control	Relay Off: 12v, On: 1v
29	OR/YL	Overdrive Cancel Switch	OCS Off: 0v, On: 12v
30	---	Not Used	---
31	YL/LG	Power Steering Pressure Switch	Straight: 0v, Turning: 12v
32	DG/VT	Knock Sensor (-) Signal	<0.050v
33-34	---	Not Used	---
35	RD/LG	HO2S-12 (B1 S2) Signal	0.1-1.1v
36	TN/BK	MAF Sensor Return	<0.050v
37	OR/BK	Transmission Fluid Temperature Sensor	2.10-2.40v
38	LG/RD	Engine Coolant Temperature Sensor	0.5-0.6v
39	PK/BK	Sensor Ground	<0.050v
40	DG/YL	Fuel Pump Relay	Relay Off: 12v, On: 1v
41	DG/OR	A/C Dual Pressure Switch Signal	A/C Off: 0v, On: 12v
42	BK	Medium Speed Fan Relay Control	Relay Off: 12v, On: 1v
43	---	Not Used	---
44	DB/OR	Starter Relay Control Circuit	Relay Off: 12v, On: 1v
45	---	Not Used	---
46	LB	High Speed Fan Relay Control	Relay Off: 12v, On: 1v
47	VT	Electric VR Solenoid Control	0%, 55 mph: 45%
48-50	---	Not Used	---
51	BK	Power Ground	<0.1v
52	DG/VT	Coil On Plug (COP) 2 Driver (dwell)	6°, 55 mph: 8°
53	LG/YL	Coil On Plug (COP) 6 Driver (dwell)	6°, 55 mph: 8°
54	VT/YL	Torque Converter Clutch Control	0%, 55 mph: 95%
55	RD/LG	Battery Power	12-14v
56	LG/BK	EVAP Vapor Management Valve	0-10 Hz (0-100%)
57	YL/RD	Knock Sensor 1 (+) Signal	0v
58	DB/YL	A/T: Output Shaft Speed Sensor (+)	0 Hz, 55 mph: 720 Hz
58	DB/YL	M/T: VSS (+) Signal	0 Hz, 55 mph: 125 Hz
59	WT/LB	Turbine Shaft Speed Sensor	50 Hz, 55 mph: 120 Hz
60	GY/LG	HO2S-11 (B1 S1) Signal	0.1-1.1v
61	VT/LG	HO2S-22 (B2 S2) Signal	0.1-1.1v
62	RD/PK	Fuel Tank Pressure Sensor	2.6v (0" HG - cap off)
63	---	Not Used	---
64	LB/YL	Digital TR Sensor	0v, in O/D: 1.7v

2001-02 Escape 3.0L V6 4v DURATEC VIN 1 (A/T) 104 Pin Connector

PCM Pin #	Wire Color	Circuit Description (104 Pin)	Value at Hot Idle
65	BR/LG	EGR DPFE Sensor Signal	0.95-1.05v
66	---	Not Used	---
67	VT/WT	EVAP Canister Vent Solenoid Control	0-10 Hz (0-100%)
68	GY/BK	Vehicle Speed Sensor (+) Signal	0 Hz, 55 mph: 125 Hz
69	PK/YL	A/C WOT Relay Control	Off: 12v, On: 1v
70	---	Not Used	---
71	WT/RD	Vehicle Power (Start-Run)	12-14v
72	---	Not Used	---
73	TN/BK	Fuel Injector 5 Control	2.6-3.2 ms
74	BR/YL	Fuel Injector 3 Control	2.5-3.2 ms
75	TN	Fuel Injector 1 Control	2.5-3.2 ms
76	BK	Power Ground	<0.1v
77	---	Not Used	---
78	OR/YL	Coil On Plug (COP) 3 Driver (dwell)	6°, 55 mph: 8°
79	---	Not Used	---
80	LB/OR	Fuel Pump Relay	Relay Off: 12v, On: 1v
81	WT/YL	EPC Solenoid Control	9.5v
82	---	Not Used	---
83	WT/LB	IAC Solenoid Control	10.7v (33%)
84	DB/YL	Output Shaft Speed (-) Signal	0 Hz, 55 mph: 720 Hz
84	DB/YL	M/T: VSS (-) Signal	0 Hz, 55 mph: 125 Hz
85	DB/OR	Camshaft Position Sensor	5-7 Hz
86	BK/YL	A/C Dual Pressure Switch	A/C On: 12v (Open)
87	RD/BK	HO2S-21 (B2 S1) Signal	0.1-1.1v
88	LB/RD	MAF Sensor Signal	0.6-0.9v
89	GY/WT	TP Sensor Signal	0.53-1.27v
90	BR/WT	Reference Voltage	4.9-5.1v
91	OR	Signal Return	<0.050v
92	LG	Brake Position Switch	Brake Off: 12v, On: 1v
93	RD/WT	HO2S-11 (B1 S1) Heater	On: 1v, Off: 12v
94	YL/LB	HO2S-21 (B2 S1) Heater	On: 1v, Off: 12v
95	WT/BK	HO2S-12 (B1 S2) Heater	On: 1v, Off: 12v
96	TN/YL	HO2S-22 (B2 S2) Heater	On: 1v, Off: 12v
97	WT/RD	Vehicle Power	12-14v
98	---	Not Used	---
99	LG/OR	Injector 6 Control	2.5-3.2 ms
100	BR/LB	Injector 4 Control	2.5-3.2 ms
101	WT	Injector 2 Control	2.5-3.2 ms
102, 104	---	Not Used	---
103	BK	Power Ground	<0.1v

Pin Connector Graphic

PCM 104-PIN CONNECTOR

```
 1 ●●●●●●●●●●●●●    ●●●●●●●●●●●●● 26
27 ●●●●●●●●●●●●●  ⬤ ●●●●●●●●●●●●● 52
53 ●●●●●●●●●●●●●    ●●●●●●●●●●●●● 78
79 ●●●●●●●●●●●●●    ●●●●●●●●●●●●● 104
```

Terminal View of 104-Pin PCM Wiring Harness Connector

Standard Colors and Abbreviations

Abbreviation	Color	Abbreviation	Color	Abbreviation	Color
BK	Black	GY	Gray	PK	Purple
BL	Blue	GN	Green	RD	Red
BR	Brown	LG	LT Green	TN	Tan
DB	Dark Blue	OR	Orange	WT	White
DG	DK Green	PK	Pink	YL	Yellow

Excursion PIN Tables
2000-02 Excursion 5.4L V8 VIN L (A/T) 104 Pin Connector

PCM Pin #	Wire Color	Circuit Description (104 Pin)	Value at Hot Idle
1	OR/YL	COP 6 Driver (dwell)	5°, 55 mph: 8°
2	PK/LG	MIL (lamp) Control	MIL On: 1v, Off: 12v
3	BK/WT	Power Ground	<0.1v
4-5	---	Not Used	---
6	OR/YL	Shift Solenoid 1 Control	1v, 55 mph: 12v
7-8	---	Not Used	---
9	YL/WT	Fuel Level Indicator Signal	1.7v (1/2 full)
10	---	Not Used	---
11	PK/ORG	Shift Solenoid 2 Control	12v, 55 mph: 12v
12	WT/LG	TCIL (lamp) Control	TCIL On: 1v, Off: 12v
13	VT	Flash EEPROM Power	0.1v
14	LB/BK	4x4 Low Indicator Switch	Switch On: 0v, Off: 12v
15	PK/LB	Data Bus (-) Signal	Digital Signals
16	TN/OR	Data Bus (+) Signal	Digital Signals
17-19	---	Not Used	---
20	BR/O	Coast Clutch Solenoid	12v, 55 mph: 12v
21	DB	CKP (+) Sensor Signal	380-410 Hz
22	GY	CKP (-) Sensor Signal	380-410 Hz
23-24	BK/WT	Power Ground	<0.1v
25	LB/YL	Case Ground	<0.050v
26	LG/WT	COP 1 Driver (dwell)	5°, 55 mph: 8°
27	LG/YL	COP 5 Driver (dwell)	5°, 55 mph: 8°
28	---	Not Used	---
29	TN/WT	TCS (switch) Signal	TCS & O/D On: 12v
30-31	---	Not Used	---
32	DG/PK	Knock Sensor (-) Signal	<0.050v
33	PK/OR	VSS (-) Signal	0 Hz, 55 mph: 125 Hz
34	WT/YL	Digital TR1 Sensor	0v, 55 mph: 11.5v
35	RD/LG	HO2S-12 (B1 S2) Signal	0.1-1.1v
36	TN/LB	MAF Sensor Return	<0.050v
37	OR/BK	TFT Sensor Signal	2.10-2.40v
38	---	Not Used	---
39	GY	IAT Sensor Signal	1.5-2.5v
40	DG/YL	Fuel Pump Monitor	On: 12v, Off: 0v
41	TN/LG	A/C Head Pressure Switch	A/C On: 12v, Off: 0v
42	---	Not Used	---
43	OR/LG	Data Output Link	Digital Signals
44	---	Not Used	---
45	RD/WT	ECT Signal to Dash	Digital Signals
46	---	Not Used	---
47	BR/PK	VR Solenoid Control	0%, 55 mph: 45%
48	WT/PK	Clean Tachometer Output	65 Hz, 55 mph: 175 Hz
49	DB/WT	Digital TR2 Sensor	0v, 55 mph: 11.5v
50	DG/YL	Digital TR4 Sensor	0v, 55 mph: 11.5v
51	BK/WT	Power Ground	<0.1v
52	WT/PK	COP 3 Driver (dwell)	5°, 55 mph: 8°
53	DG/PK	COP 4 Driver (dwell)	5°, 55 mph: 8°
54	PK/YL	TCC Solenoid Control	0%, 55 mph: 95%
55	RD/WT	Keep Alive Power	12-14v
56	LG/BK	EVAP Purge Solenoid	0-10 Hz (0-100%)
57	YL/RD	Knock Sensor (+) Signal	0v
58	GY/BK	VSS (+) Signal	0 Hz, 55 mph: 125 Hz
59	DG/WT	TSS Sensor Signal	0 Hz, 700 Hz
60	GY/LB	HO2S-11 (B1 S1) Signal	0.1-1.1v
61	---	Not Used	---
62	RD/PK	FTP Sensor Signal	2.6v at 0" H20 (cap off)
63	---	Not Used	---
64	LB/YL	Digital TR3A Sensor	In 'P': 0v, in O/D: 1.7v

2000-02 Excursion 5.4L V8 VIN L (A/T) 104 Pin Connector

PCM Pin #	Wire Color	Circuit Description (104 Pin)	Value at Hot Idle
68	---	Not Used	---
65 ('00)	BR/LG	DPFE Sensor Signal	0.20-1.30v
65 ('01-'02)	BR/LG	DPFE Sensor Signal	0.95-1.05v
66	YL/LG	CHT Sensor Signal	0.6 (194°F)
67	PK/WT	EVAP CV Solenoid	0-10 Hz (0-100%)
69	PK/YL	AC Switch Signal	A/C On: 12v, Off: 0v
70	---	Not Used	---
71	RD	Vehicle Power	12-14v
72	TN/RD	Injector 7 Control	3.8-4.6 ms
73	TN/BK	Injector 5 Control	3.8-4.6 ms
74	BR/YL	Injector 3 Control	3.8-4.6 ms
75	TN	Injector 1 Control	3.8-4.6 ms
76, 77	BK/WT	Power Ground	<0.1v
78	PK/LB	COP 7 Driver (dwell)	5°, 55 mph: 8°
79	WT/RD	COP 8 Driver (dwell)	5°, 55 mph: 8°
80	LB/OR	Fuel Pump Relay Control	On: 1v, Off: 12v
81	WT/YL	EPC Solenoid Control	9.2v (5 psi)
82	---	Not Used	---
83	WT/LB	IAC Solenoid Control	10.7v (33%)
84	DB/YL	OSS Sensor (+) Signal	0 Hz, 55 mph: 2050 Hz
85	DG	CMP Sensor Signal	9 Hz, 55 mph: 16 Hz
86	---	Not Used	---
87	RD/BK	HO2S-21 (B2 S1) Signal	0.1-1.1v
88	LB/RD	MAF Sensor Signal	0.8v, 55 mph: 1.6v
89	GY/WT	TP Sensor Signal	0.9v, 55 mph: 1.3v
90	BR/WT	Reference Voltage	4.9-5.1v
91	GY/RD	Sensor Ground	<0.050v
92	RD/LG	Brake Position Switch	Brake Off: 12v, On: 1v
93	RD/WT	HO2S-11 (B1 S1) Heater	On: 1v, Off: 12v
94	YL/LB	HO2S-21 (B2 S1) Heater	On: 1v, Off: 12v
95	WT/BK	HO2S-12 (B1 S2) Heater	On: 1v, Off: 12v
97	RD	Vehicle Power	12-14v
98	LB	Injector 8 Control	3.8-4.6 ms
99	LG/OR	Injector 6 Control	3.8-4.6 ms
100	WT/LB	Injector 4 Control	3.8-4.6 ms
101	WT	Injector 2 Control	3.8-4.6 ms
102	---	Not Used	---
103	BK/WT	Power Ground	<0.1v
104	PK/WT	COP 2 Driver (dwell)	5°, 55 mph: 8°

Pin Connector Graphic

```
           PCM 104-PIN CONNECTOR
    1 ●●●●●●●●●●●●●      ●●●●●●●●●●●●● 26
   27 ●●●●●●●●●●●●●      ●●●●●●●●●●●●● 52
   53 ●●●●●●●●●●●●●  ▓   ●●●●●●●●●●●●● 78
   79 ●●●●●●●●●●●●●      ●●●●●●●●●●●●● 104
    Terminal View of 104-Pin PCM Wiring Harness Connector
```

Standard Colors and Abbreviations

Abbreviation	Color	Abbreviation	Color	Abbreviation	Color
BK	Black	GY	Gray	PK	Purple
BL	Blue	GN	Green	RD	Red
BR	Brown	LG	LT Green	TN	Tan
DB	Dark Blue	OR	Orange	WT	White
DG	DK Green	PK	Pink	YL	Yellow

2003 Excursion 5.4L V8 VIN L (A/T) 104 Pin Connector

PCM Pin #	Wire Color	Circuit Description (104 Pin)	Value at Hot Idle
1	OR/YL	COP 6 Driver (dwell)	5°, 55 mph: 8°
2	---	Not Used	---
3	BK/WT	Power Ground	<0.1v
4	LB/YL	Customer Access Signal	Digital Signals
5	---	Not Used	---
6	OR/YL	Shift Solenoid 'A' Control	1v, 55 mph: 12v
7-10	---	Not Used	---
11	VT/OR	Shift Solenoid 'B' Control	12v, 55 mph: 1v
12	WT/LG	TCIL (lamp) Control	7.7v (Switch On: 0v)
13	VT	Flash EEPROM Power	0.1v
14	---	Not Used	---
15	PK/LB	SCP Data Bus (-) Signal	<0.050v
16	TN/OR	SCP Data Bus (+) Signal	Digital Signals
17-19	---	Not Used	---
20	BR/OR	Coast Clutch Solenoid	12v, 55 mph: 12v
21	DB	CKP (-) Sensor Signal	411 Hz
22	GY	CKP (+) Sensor Signal	411 Hz
23	BK/WT	Power Ground	<0.1v
24	BK/WT	Power Ground	<0.1v
25	LB/YL	Chassis Ground	<0.050v
26	LG/WT	COP 1 Driver (dwell)	5°, 55 mph: 8°
27	LG/YL	COP 5 Driver (dwell)	5°, 55 mph: 8°
28	---	Not Used	---
29	TN/WT	TCS (switch) Signal	TCS & O/D On: 12v
30-31	---	Not Used	---
32	DG/VT	Knock Sensor (-) Signal	<0.050v
33	PK/OR	Power Ground	<0.1v
34	YL/BK	Digital TR1 Sensor	0v, 55 mph: 11.5v
35	RD/LG	HO2S-12 (B1 S2) Signal	0.1-1.1v
36	TN/LB	MAF Sensor Return	<0.050v
37	OR/BK	Transmission Fluid Temperature Sensor	2.10-2.40v
38	---	Not Used	---
39	GY	Intake Air Temperature Sensor	1.5-2.5v
40	DG/YL	Fuel Pump Monitor	Relay Off: 12v, On: 1v
41	TN/LG	A/C Pressure Switch Signal	Switch Closed: 12v
42	---	Not Used	---
43	OR/LG	Data Output Link	5v
44-46	---	Not Used	---
47	BR/PK	Electric VR Solenoid Control	0%, 55 mph: 45%
48	LG/WT	Customer Access (Tachometer)	DC signals
49	LB/BK	Digital TR2 Sensor	0v, 55 mph: 11.5v
50	WT/BK	Digital TR4 Sensor	0v, 55 mph: 11.5v
51	BK/WT	Power Ground	<0.1v
52	WT/PK	COP 3 Driver (dwell)	5°, 55 mph: 8°
53	DG/VT	COP 4 Driver (dwell)	5°, 55 mph: 8°
54	VT/YL	TCC Solenoid Control	0%, 55 mph: 95%
55	RD/WT	Keep Alive Power	12-14v
56	LG/BK	EVAP Vapor Management Valve	0-10 Hz (0-100%)
57	YL/RD	Knock Sensor 1 Signal	0v
58	GY/BK	Vehicle Speed Sensor Signal	Moving: DC signals
59	DG/WT	TSS Sensor Signal	300 Hz, 55 mph: 980
60	GY/LB	HO2S-11 (B1 S1) Signal	0.1-1.1v
61	---	Not Used	---
62	RD/PK	Fuel Tank Pressure Sensor	2.6v (0" H2O - cap off)
63	---	Not Used	---
64	LB/YL	Digital TR3A Sensor Signal	In 'P': 0v, in O/D: 1.7v

2003 Excursion 5.4L V8 VIN L (A/T) 104 Pin Connector

PCM Pin #	Wire Color	Circuit Description (104 Pin)	Value at Hot Idle
65	BR/LG	EGR DPFE Sensor Signal	0.95-1.05v
66	YL/LG	Cylinder Head Temperature Sensor	0.6 (194°F)
67	VT/WT	EVAP Canister Vent Solenoid Control	0-10 Hz (0-100%)
68	---	Not Used	---
69	PK/YL	AC Switch Signal	A/C On: 12v, Off: 0v
70	---	Not Used	---
71	RD	Vehicle Power (Start-Run)	12-14v
72	TN/RD	Fuel Injector 7 Control	3.8-4.6 ms
73	TN/BK	Fuel Injector 5 Control	3.8-4.6 ms
74	BR/YL	Fuel Injector 3 Control	3.8-4.6 ms
75	TN	Fuel Injector 1 Control	3.8-4.6 ms
76	BK/WT	Power Ground	<0.1v
77	BK/WT	Power Ground	<0.1v
78	PK/LB	COP 7 Driver (dwell)	5°, 55 mph: 8°
79	WT/RD	COP 8 Driver (dwell)	5°, 55 mph: 8°
80	LB/OR	Fuel Pump Relay Control	Relay Off: 12v, On: 1v
81	WT/YL	EPC Solenoid Control	9.2v (5 psi)
82	---	Not Used	---
83	WT/LB	IAC Solenoid Control	10.7v (33%)
84	DB/YL	OSS Sensor (+) Signal	0 Hz, 55 mph: 2050 Hz
85	DG	CMP Sensor Signal	9 Hz, 55 mph: 16 Hz
86	---	Not Used	---
87	RD/BK	HO2S-21 (B2 S1) Signal	0.1-1.1v
88	LB/RD	MAF Sensor Signal	0.8v, 55 mph: 1.6v
89	GY/WT	TP Sensor Signal	0.9v, 55 mph: 1.3v
90	BR/WT	Reference Voltage	4.9-5.1v
91	GY/RD	Sensor Return	<0.050v
92	RD/LG	Brake Position Switch	Brake Off: 12v, On: 1v
93	RD/WT	HO2S-11 (B1 S1) Heater Control	On: 1v, Off: 12v
94	YL/LB	HO2S-21 (B2 S1) Heater Control	On: 1v, Off: 12v
95	WT/BK	HO2S-12 (B1 S2) Heater Control	On: 1v, Off: 12v
97	RD	Vehicle Power (Start-Run)	12-14v
98	LB	Fuel Injector 8 Control	3.8-4.6 ms
99	LG/OR	Fuel Injector 6 Control	3.8-4.6 ms
100	BR/LB	Fuel Injector 4 Control	3.8-4.6 ms
101	WT	Fuel Injector 2 Control	3.8-4.6 ms
102	---	Not Used	---
103	BK/WT	Power Ground	<0.1v
104	PK/WT	COP 2 Driver (dwell)	5°, 55 mph: 8°

Pin Connector Graphic

```
            PCM 104-PIN CONNECTOR
   1 ⊗⊗⊗⊗⊗⊗⊗⊗⊗⊗⊗⊗⊗    ⊗⊗⊗⊗⊗⊗⊗⊗⊗⊗⊗⊗⊗ 26
  27 ⊗⊗⊗⊗⊗⊗⊗⊗⊗⊗⊗⊗⊗   ⬤  ⊗⊗⊗⊗⊗⊗⊗⊗⊗⊗⊗⊗⊗ 52
  53 ⊗⊗⊗⊗⊗⊗⊗⊗⊗⊗⊗⊗⊗       ⊗⊗⊗⊗⊗⊗⊗⊗⊗⊗⊗⊗⊗ 78
  79 ⊗⊗⊗⊗⊗⊗⊗⊗⊗⊗⊗⊗⊗    ⊗⊗⊗⊗⊗⊗⊗⊗⊗⊗⊗⊗⊗ 104
```

Terminal View of 104-Pin PCM Wiring Harness Connector

Standard Colors and Abbreviations

Abbreviation	Color	Abbreviation	Color	Abbreviation	Color
BK	Black	GY	Gray	PK	Purple
BL	Blue	GN	Green	RD	Red
BR	Brown	LG	LT Green	TN	Tan
DB	Dark Blue	OR	Orange	WT	White
DG	DK Green	PK	Pink	YL	Yellow

2000-01 Excursion 6.8L V10 VIN S (A/T) 104 Pin Connector

PCM Pin #	Wire Color	Circuit Description (104 Pin)	Value at Hot Idle
1	OR/YL	COP 6 Driver (dwell)	5°, 55 mph: 8°
2	PK/LG	MIL (lamp) Control	MIL On: 1v, Off: 12v
3	BK/WT	Power Ground	<0.1v
4	LB/YL	Power Takeoff Signal	PTO Off: 0v, On: 12v
5	---	Not Used	---
6	OR/YL	Shift Solenoid 1 Control	1v, 55 mph: 12v
7-10	---	Not Used	---
9	YL/WT	Fuel Level Indicator Signal	1.7v (1/2 full)
11	PK/OR	Shift Solenoid 2 Control	12v, 55 mph: 12v
12	WT/LG	TCIL (lamp) Control	TCIL On: 1v, Off: 12v
13	PK	Flash EEPROM Power	0.1v
14	LB/BK	4x4 Low Switch Signal	Switch On: 0v, Off: 12v
15	PK/LB	Data Bus (-) Signal	Digital Signals
16	TN/OR	Data Bus (+) Signal	Digital Signals
17-19	---	Not Used	---
20	BR/OR	Coast Clutch Solenoid	12v, 55 mph: 12v
21	DB	CKP (-) Sensor Signal	500-525 Hz
22	GY	CKP (+) Sensor Signal	500-525 Hz
23-24	BK/WT	Power Ground	<0.1v
25	LB/YL	Case Ground	<0.050v
26	LG/WT	COP 1 Driver (dwell)	5°, 55 mph: 8°
27	LG/YL	COP 10 Driver (dwell)	5°, 55 mph: 8°
28	---	Not Used	---
29	TN/WT	TCS (switch) Signal	TCS & O/D On: 12v
30-31	---	Not Used	---
32	DG/PK	Knock Sensor 1 Return	<0.050v
33	PK/OR	VSS (-) Signal	0 Hz, 55 mph: 125 Hz
34	WT/YL	Digital TR1 Sensor	0v, 55 mph: 11.5v
35	RD/LG	HO2S-12 (B1 S2) Signal	0.1-1.1v
36	TN/LB	MAF Sensor Return	<0.050v
37	OR/BK	TFT Sensor Signal	2.10-2.40v
38	---	Not Used	---
39	GY	IAT Sensor Signal	1.5-2.5v
40	DG/YL	Fuel Pump Monitor	On: 12v, Off: 0v
41	TN/LG	A/C Head Pressure Switch	A/C On: 12v, Off: 0v
42	GY/RD	Injector 10 Control	3.8-4.6 ms
43	OR/LG	Data Output Link	Digital Signals
45	RD/WT	ECT Sensor Signal to Dash	0.6v
46	---	Not Used	---
47	BR/PK	VR Solenoid Control	0%, 55 mph: 45%
48	WT/PK	Clean Tachometer Output	65 Hz, 55 mph: 175 Hz
49	DB/WT	Digital TR2 Sensor	0v, 55 mph: 11.5v
50	DG/YL	Digital TR4 Sensor	0v, 55 mph: 11.5v
51	BK/WT	Power Ground	<0.1v
52	WT/PK	COP 5 Driver (dwell)	5°, 55 mph: 8°
53	DG/PK	COP 7 Driver (dwell)	5°, 55 mph: 8°
54	PK/YL	TCC Solenoid Control	0%, 55 mph: 95%
55	RD/WT	Keep Alive Power	12-14v
56	LG/BK	EVAP Purge Solenoid	0-10 Hz (0-100%)
57	YL/RD	Knock Sensor (+) Signal	0v
58	GY/BK	VSS (+) Signal	0 Hz, 55 mph: 125 Hz
59	DG/WT	TSS Sensor Signal	100 Hz, 55 mph: 270 Hz
60	GY/LB	HO2S-11 (B1 S1) Signal	0.1-1.1v
61	---	Not Used	---
62	RD/PK	Fuel Tank Pressure Sensor	2.6v (0" H2O - cap off)
63	---	Not Used	---
64	LB/YL	Digital TR3A Sensor	In 'P': 0v, in O/D: 1.7v

2000-01 Excursion 6.8L V10 VIN S (A/T) 104 Pin Connector

PCM Pin #	Wire Color	Circuit Description (104 Pin)	Value at Hot Idle
65	BR/LG	DPFE Sensor Signal	0.9v, 55 mph: 3.1v
66	YL/LG	CHT Sensor Signal	0.6 (194°F)
68	GY/BK	Injector 9 Control	3.8-4.6 ms
69	PK/YL	AC Switch Signal	A/C On: 12v, Off: 0v
70	---	Not Used	---
71	RD	Vehicle Power (Start-Run)	12-14v
72	TN/RD	Injector 7 Control	3.8-4.6 ms
73	TN/BK	Injector 5 Control	3.8-4.6 ms
74	BR/YL	Injector 3 Control	3.8-4.6 ms
75	TN	Injector 1 Control	3.8-4.6 ms
76	BK/WT	Power Ground	<0.1v
77	BK/WT	Power Ground	<0.1v
78	PK/LB	COP 2 Driver (dwell)	5°, 55 mph: 8°
79	WT/RD	COP 8 Driver (dwell)	5°, 55 mph: 8°
80	LB/OR	Fuel Pump Control	On: 1v, Off: 12v
81	WT/YL	EPC Solenoid Control	9.2v (5 psi)
82	WT/RD	COP 9 Driver (dwell)	5°, 55 mph: 8°
83	WT/LB	IAC Solenoid Control	10.7v (33%)
84	DB/YL	OSS Sensor Signal	0 Hz, 55 mph: 2050 Hz
85	DG	CMP Sensor Signal	9 Hz, 55 mph: 16 Hz
86	---	Not Used	---
87	RD/BK	HO2S-21 (B2 S1) Signal	0.1-1.1v
88	LB/RD	MAF Sensor Signal	0.8v, 55 mph: 1.6v
89	GY/WT	TP Sensor Signal	0.9v, 55 mph: 1.3v
90	BR/WT	Reference Voltage	4.9-5.1v
91	GY/RD	Sensor Return	<0.050v
92	RD/LG	Brake Position Switch	Brake Off: 12v, On: 1v
93	RD/WT	HO2S-11 (B1 S1) Heater	On: 1v, Off: 12v
94	YL/LB	HO2S-21 (B2 S1) Heater	On: 1v, Off: 12v
97	RD	Vehicle Power	12-14v
98	LB	Injector 8 Control	3.8-4.6 ms
99	LG/OR	Injector 6 Control	3.8-4.6 ms
100	BR/LB	Injector 4 Control	3.8-4.6 ms
101	WT	Injector 2 Control	3.8-4.6 ms
102	YL/BK	COP 4 Driver (dwell)	5°, 55 mph: 8°
103	BK/WT	Power Ground	<0.1v
104	PK/WT	COP 3 Driver (dwell)	5°, 55 mph: 8°

Pin Connector Graphic

```
          PCM 104-PIN CONNECTOR
  1 ●●●●●●●●●●●●●    ●●●●●●●●●●●●● 26
 27 ●●●●●●●●●●●●●         ●●●●●●●●●●●●● 52
 53 ●●●●●●●●●●●●●    ⬤    ●●●●●●●●●●●●● 78
 79 ●●●●●●●●●●●●●    ●●●●●●●●●●●●● 104
```

Terminal View of 104-Pin PCM Wiring Harness Connector

Standard Colors and Abbreviations

Abbreviation	Color	Abbreviation	Color	Abbreviation	Color
BK	Black	GY	Gray	PK	Purple
BL	Blue	GN	Green	RD	Red
BR	Brown	LG	LT Green	TN	Tan
DB	Dark Blue	OR	Orange	WT	White
DG	DK Green	PK	Pink	YL	Yellow

2002-03 Excursion 6.8L V10 VIN S (A/T) 104 Pin Connector

PCM Pin #	Wire Color	Circuit Description (104 Pin)	Value at Hot Idle
1	OR/YL	COP 6 Driver (dwell)	5º, 55 mph: 8º
2	---	Not Used	---
3	BK/WT	Power Ground	<0.1v
4	LB/YL	Power Takeoff Signal	PTO Off: 0v, On: 12v
5	---	Not Used	---
6	OR/YL	Shift Solenoid 1 Control	1v, 55 mph: 12v
7-8	---	Not Used	---
9	YL/WT	Fuel Level Indicator Signal	1.7v (1/2 full)
10	---	Not Used	---
11	PK/OR	Shift Solenoid 2 Control	12v, 55 mph: 12v
12	WT/LG	TCIL (lamp) Control	TCIL On: 1v, Off: 12v
13	PK	Flash EEPROM Power	0.1v
14	LB/BK	4x4 Low Switch Signal	Switch On: 0v, Off: 12v
15	PK/LB	Data Bus (-) Signal	Digital Signals
16	TN/OR	Data Bus (+) Signal	Digital Signals
17-19	---	Not Used	---
20	BR/OR	Coast Clutch Solenoid	12v, 55 mph: 12v
21	DB	CKP (-) Sensor Signal	500-525 Hz
22	GY	CKP (+) Sensor Signal	500-525 Hz
23	BK/WT	Power Ground	<0.1v
24	BK/WT	Power Ground	<0.1v
25	LB/YL	Case Ground	<0.050v
26	LG/WT	COP 1 Driver (dwell)	5º, 55 mph: 8º
27	LG/YL	COP 10 Driver (dwell)	5º, 55 mph: 8º
28	---	Not Used	---
29	TN/WT	TCS (switch) Signal	TCS & O/D On: 12v
30-31	---	Not Used	---
32	DG/VT	Knock Sensor 1 Return	<0.050v
33	PK/OR	VSS (-) Signal	0 Hz, 55 mph: 125 Hz
34	WT/YL	Digital TR1 Sensor	0v, 55 mph: 11.5v
35	RD/LG	HO2S-12 (B1 S2) Signal	0.1-1.1v
36	TN/LB	MAF Sensor Return	<0.050v
37	OR/BK	TFT Sensor Signal	2.10-2.40v
38	---	Not Used	---
39	GY	IAT Sensor Signal	1.5-2.5v
40	DG/YL	Fuel Pump Monitor	On: 12v, Off: 0v
41	TN/LG	A/C Head Pressure Switch	A/C On: 12v, Off: 0v
42	GY/RD	Injector 10 Control	3.8-4.6 ms
43	OR/LG	Data Output Link	Digital Signals
44-46	---	Not Used	---
47	BR/PK	VR Solenoid Control	0%, 55 mph: 45%
48	LG/WT	Clean Tachometer Output	65 Hz, 55 mph: 175 Hz
49	LB/BK	Digital TR2 Sensor	0v, 55 mph: 11.5v
50	WT/BK	Digital TR4 Sensor	0v, 55 mph: 11.5v
51	BK/WT	Power Ground	<0.1v
52	WT/PK	COP 5 Driver (dwell)	5º, 55 mph: 8º
53	DG/VT	COP 7 Driver (dwell)	5º, 55 mph: 8º
54	VT/YL	TCC Solenoid Control	0%, 55 mph: 95%
55	RD/WT	Keep Alive Power	12-14v
56	LG/BK	EVAP Vapor Management Valve	0-10 Hz (0-100%)
57	YL/RD	Knock Sensor (+) Signal	0v
58	YL/BK	VSS (+) Signal	0 Hz, 55 mph: 125 Hz
59	DG/WT	TSS Sensor Signal	100 Hz, 55 mph: 270 Hz
60	GY/LB	HO2S-11 (B1 S1) Signal	0.1-1.1v
61	---	Not Used	---
62	RD/PK	Fuel Tank Pressure Sensor	2.6v (0" H2O - cap off)
63	---	Not Used	---
64	LB/YL	Digital TR3A Sensor	In 'P': 0v, in O/D: 1.7v

2002-03 Excursion 6.8L V10 VIN S (A/T) 104 Pin Connector

PCM Pin #	Wire Color	Circuit Description (104 Pin)	Value at Hot Idle
65	BR/LG	EGR DPFE Sensor Signal	0.9v, 55 mph: 3.1v
66	YL/LG	Cylinder Head Temperature Sensor	0.6 (194°F)
68	GY/BK	Injector 9 Control	3.8-4.6 ms
69	PK/YL	AC Switch Signal	A/C On: 12v, Off: 0v
70	---	Not Used	---
71	RD	Vehicle Power (Start-Run)	12-14v
72	TN/RD	Injector 7 Control	3.8-4.6 ms
73	TN/BK	Injector 5 Control	3.8-4.6 ms
74	BR/YL	Injector 3 Control	3.8-4.6 ms
75	TN	Injector 1 Control	3.8-4.6 ms
76	BK/WT	Power Ground	<0.1v
77	BK/WT	Power Ground	<0.1v
78	PK/LB	COP 2 Driver (dwell)	5°, 55 mph: 8°
79	WT/RD	COP 8 Driver (dwell)	5°, 55 mph: 8°
80	LB/OR	Fuel Pump Control	On: 1v, Off: 12v
81	WT/YL	EPC Solenoid Control	9.2v (5 psi)
82	WT/RD	COP 9 Driver (dwell)	5°, 55 mph: 8°
83	WT/LB	IAC Solenoid Control	10.7v (33%)
84	DB/YL	OSS Sensor Signal	0 Hz, 55 mph: 2050 Hz
85	DG	CMP Sensor Signal	9 Hz, 55 mph: 16 Hz
86	---	Not Used	---
87	RD/BK	HO2S-21 (B2 S1) Signal	0.1-1.1v
88	LB/RD	MAF Sensor Signal	0.8v, 55 mph: 1.6v
89	GY/WT	TP Sensor Signal	0.9v, 55 mph: 1.3v
90	BR/WT	Reference Voltage	4.9-5.1v
91	GY/RD	Sensor Return	<0.050v
92	RD/LG	Brake Position Switch	Brake Off: 12v, On: 1v
93	RD/WT	HO2S-11 (B1 S1) Heater	Heater Off: 12v, On: 1v
94	YL/LB	HO2S-21 (B2 S1) Heater	Heater Off: 12v, On: 1v
95	WT/BK	HO2S-12 (B1 S2) Heater	Heater Off: 12v, On: 1v
96	---	Not Used	---
97	RD	Vehicle Power	12-14v
98	LB	Injector 8 Control	3.8-4.6 ms
99	LG/OR	Injector 6 Control	3.8-4.6 ms
100	BR/LB	Injector 4 Control	3.8-4.6 ms
101	WT	Injector 2 Control	3.8-4.6 ms
102	YL/BK	COP 4 Driver (dwell)	5°, 55 mph: 8°
103	BK/WT	Power Ground	<0.1v
104	PK/WT	COP 3 Driver (dwell)	5°, 55 mph: 8°

Pin Connector Graphic

PCM 104-PIN CONNECTOR

Terminal View of 104-Pin PCM Wiring Harness Connector

Standard Colors and Abbreviations

Abbreviation	Color	Abbreviation	Color	Abbreviation	Color
BK	Black	GY	Gray	PK	Purple
BL	Blue	GN	Green	RD	Red
BR	Brown	LG	LT Green	TN	Tan
DB	Dark Blue	OR	Orange	WT	White
DG	DK Green	PK	Pink	YL	Yellow

2000-01 Excursion 6.8L V10 Bi-Fuel VIN S (A/T) 104 Pin Connector

PCM Pin #	Wire Color	Circuit Description (104 Pin)	Value at Hot Idle
1	OR/YL	COP 6 Driver (dwell)	5°, 55 mph: 8°
2	PK/LG	MIL (lamp) Control	MIL On: 1v, Off: 12v
3	BK/WT	Power Ground	<0.1v
4	LB/YL	Power Takeoff Signal	PTO Off: 0v, On: 12v
5	---	Not Used	---
6	OR/YL	Shift Solenoid 1 Control	1v, 55 mph: 12v
7-8	---	Not Used	---
9	YL/WT	Fuel Level Indicator Signal	1.7v (1/2 full)
10	---	Not Used	---
11	PK/OR	Shift Solenoid 2 Control	12v, 55 mph: 12v
12	WT/LG	TCIL (lamp) Control	TCIL On: 1v, Off: 12v
13	PK	Flash EEPROM Power	0.1v
14	LB/BK	4x4 Low Switch Signal	Switch On: 0v, Off: 12v
15	PK/LB	Data Bus (-) Signal	Digital Signals
16	TN/OR	Data Bus (+) Signal	Digital Signals
17-18	---	Not Used	---
19	OR/LG	Fuel Pump Speed Relay	Relay On: 12v, Off: 0v
20	BR/OR	Coast Clutch Solenoid	12v, 55 mph: 12v
21	DB	CKP (-) Sensor Signal	500-525 Hz
22	GY	CKP (+) Sensor Signal	500-525 Hz
23-24	BK/WT	Power Ground	<0.1v
25	LB/YL	Case Ground	<0.050v
26	LG/WT	COP 1 Driver (dwell)	5°, 55 mph: 8°
27	LG/YL	COP 10 Driver (dwell)	5°, 55 mph: 8°
28	---	Not Used	---
29	TN/WT	TCS (switch) Signal	TCS & O/D On: 12v
30-31	---	Not Used	---
32	DG/PK	Knock Sensor 1 Return	<0.050v
33	PK/OR	VSS (-) Signal	0 Hz, 55 mph: 125 Hz
34	WT/YL	Digital TR1 Sensor	0v, 55 mph: 11.5v
35	RD/LG	HO2S-12 (B1 S2) Signal	0.1-1.1v
36	TN/LB	MAF Sensor Return	<0.050v
37	OR/BK	TFT Sensor Signal	2.10-2.40v
38	---	Not Used	---
39	GY	IAT Sensor Signal	1.5-2.5v
40	DG/YL	Fuel Pump Monitor	On: 12v, Off: 0v
41	TN/LG	A/C Head Pressure Switch	A/C On: 12v, Off: 0v
42	GY/RD	Injector 10 Control	3.8-4.6 ms
43	OR/LG	Data Output Link	Digital Signals
45	RD/WT	ECT Sensor Signal to Dash	0.6v
46	GY/BK	Vehicle Speed Sensor	0 Hz, 55 mph: 125 Hz
47	BR/PK	VR Solenoid Control	0%, 55 mph: 45%
48	WT/PK	Clean Tachometer Output	65 Hz, 55 mph: 175 Hz
49	DB/WT	Digital TR2 Sensor	0v, 55 mph: 11.5v
50	DG/YL	Digital TR4 Sensor	0v, 55 mph: 11.5v
51	BK/WT	Power Ground	<0.1v
52	WT/PK	COP 5 Driver (dwell)	5°, 55 mph: 8°
53	DG/PK	COP 7 Driver (dwell)	5°, 55 mph: 8°
54	PK/YL	TCC Solenoid Control	0%, 55 mph: 95%
55	RD/WT	Keep Alive Power	12-14v
56	LG/BK	EVAP Purge Solenoid	0-10 Hz (0-100%)
57	YL/RD	Knock Sensor (+) Signal	0v
58	GY/BK	VSS (+) Signal	0 Hz, 55 mph: 125 Hz
59	DG/WT	TSS Sensor Signal	100 Hz, 55 mph: 270 Hz
60	GY/LB	HO2S-11 (B1 S1) Signal	0.1-1.1v
61	---	Not Used	---
62	RD/PK	Fuel Tank Pressure Sensor	2.6v (0" H2O - cap off)
63	---	Not Used	---
64	LB/YL	Digital TR3A Sensor	In 'P': 0v, in O/D: 1.7v

2000-01 Excursion 6.8L V10 Bi-Fuel VIN S (A/T) 104 Pin Connector

PCM Pin #	Wire Color	Circuit Description (104 Pin)	Value at Hot Idle
65	BR/LG	EGR DPFE Sensor Signal	0.9v, 55 mph: 3.1v
66	YL/LG	Cylinder Head Temperature Sensor	0.6 (194°F)
67	VT/WT	EVAP Canister Vent Solenoid Control	0-10 Hz (0-100%)
68	GY/BK	Injector 9 Control	3.8-4.6 ms
69	PK/YL	AC Switch Signal	A/C On: 12v, Off: 0v
70	---	Not Used	---
71	RD	Vehicle Power	12-14v
72	TN/RD	Injector 7 Control	3.8-4.6 ms
73	TN/BK	Injector 5 Control	3.8-4.6 ms
74	BR/YL	Injector 3 Control	3.8-4.6 ms
75	TN	Injector 1 Control	3.8-4.6 ms
76	BK/WT	Power Ground	<0.1v
77	BK/WT	Power Ground	<0.1v
78	PK/LB	COP 2 Driver (dwell)	5°, 55 mph: 8°
79	WT/RD	COP 8 Driver (dwell)	5°, 55 mph: 8°
80	LB/OR	Fuel Pump Control	On: 1v, Off: 12v
81	WT/YL	EPC Solenoid Control	9.2v (5 psi)
82	WT/RD	COP 9 Driver (dwell)	5°, 55 mph: 8°
83	WT/LB	IAC Solenoid Control	10.7v (33%)
84	DB/YL	OSS Sensor Signal	0 Hz, 55 mph: 2050 Hz
85	DG	CMP Sensor Signal	9 Hz, 55 mph: 16 Hz
86	---	Not Used	---
87	RD/BK	HO2S-21 (B2 S1) Signal	0.1-1.1v
88	LB/RD	MAF Sensor Signal	0.8v, 55 mph: 1.6v
89	GY/WT	TP Sensor Signal	0.9v, 55 mph: 1.3v
90	BR/WT	Reference Voltage	4.9-5.1v
91	GY/RD	Sensor Ground	<0.050v
92	RD/LG	Brake Position Switch	Brake Off: 12v, On: 1v
93	RD/WT	HO2S-11 (B1 S1) Heater	Heater Off: 12v, On: 1v
94	YL/LB	HO2S-21 (B2 S1) Heater	Heater Off: 12v, On: 1v
95	WT/BK	HO2S-12 (B1 S2) Heater	Heater Off: 12v, On: 1v
96	---	Not Used	---
97	RD	Vehicle Power (Start-Run)	12-14v
98	LB	Injector 8 Control	3.8-4.6 ms
99	LG/OR	Injector 6 Control	3.8-4.6 ms
100	BR/LB	Injector 4 Control	3.8-4.6 ms
101	WT	Injector 2 Control	3.8-4.6 ms
102	YL/BK	COP 4 Driver (dwell)	5°, 55 mph: 8°
103	BK/WT	Power Ground	<0.1v
104	PK/WT	COP 3 Driver (dwell)	5°, 55 mph: 8°

Pin Connector Graphic

```
          PCM 104-PIN CONNECTOR
   1 ●●●●●●●●●●●●●    ●●●●●●●●●●●●● 26
  27 ●●●●●●●●●●●●●    ●●●●●●●●●●●●● 52
  53 ●●●●●●●●●●●●●   ⬤   ●●●●●●●●●●●● 78
  79 ●●●●●●●●●●●●●    ●●●●●●●●●●●●● 104
```

Terminal View of 104-Pin PCM Wiring Harness Connector

Standard Colors and Abbreviations

Abbreviation	Color	Abbreviation	Color	Abbreviation	Color
BK	Black	GY	Gray	PK	Purple
BL	Blue	GN	Green	RD	Red
BR	Brown	LG	LT Green	TN	Tan
DB	Dark Blue	OR	Orange	WT	White
DG	DK Green	PK	Pink	YL	Yellow

2002-03 Excursion 6.8L V10 Bi-Fuel VIN S (A/T) 104 Pin Connector

PCM Pin #	Wire Color	Circuit Description (104 Pin)	Value at Hot Idle
1	OR/YL	COP 6 Driver (dwell)	5°, 55 mph: 8°
2	---	Not Used	---
3	BK/WT	Power Ground	<0.1v
4	LB/YL	Power Takeoff Signal	PTO Off: 0v, On: 12v
5	---	Not Used	---
6	OR/YL	Shift Solenoid 1 Control	1v, 55 mph: 12v
7-8	---	Not Used	---
9	YL/WT	Fuel Level Indicator Signal	1.7v (1/2 full)
10	---	Not Used	---
11	PK/OR	Shift Solenoid 2 Control	12v, 55 mph: 12v
12	WT/LG	TCIL (lamp) Control	TCIL On: 1v, Off: 12v
13	PK	Flash EEPROM Power	0.1v
14	LB/BK	4x4 Low Switch Signal	Switch On: 0v, Off: 12v
15	PK/LB	Data Bus (-) Signal	Digital Signals
16	TN/OR	Data Bus (+) Signal	Digital Signals
17-18	---	Not Used	---
19	OR/LG	Fuel Pump Speed Relay	Relay On: 12v, Off: 0v
20	BR/OR	Coast Clutch Solenoid	12v, 55 mph: 12v
21	DB	CKP (-) Sensor Signal	500-525 Hz
22	GY	CKP (+) Sensor Signal	500-525 Hz
23-24	BK/WT	Power Ground	<0.1v
25	LB/YL	Case Ground	<0.050v
26	LG/WT	COP 1 Driver (dwell)	5°, 55 mph: 8°
27	LG/YL	COP 10 Driver (dwell)	5°, 55 mph: 8°
28	---	Not Used	---
29	TN/WT	TCS (switch) Signal	TCS & O/D On: 12v
30-31	---	Not Used	---
32	DG/PK	Knock Sensor 1 Return	<0.050v
33	PK/OR	VSS (-) Signal	0 Hz, 55 mph: 125 Hz
34	WT/YL	Digital TR1 Sensor	0v, 55 mph: 11.5v
35	RD/LG	HO2S-12 (B1 S2) Signal	0.1-1.1v
36	TN/LB	MAF Sensor Return	<0.050v
37	OR/BK	TFT Sensor Signal	2.10-2.40v
39	GY	IAT Sensor Signal	1.5-2.5v
40	DG/YL	Fuel Pump Monitor	On: 12v, Off: 0v
41	TN/LG	A/C Head Pressure Switch	A/C On: 12v, Off: 0v
42	GY/RD	Injector 10 Control	3.8-4.6 ms
43	OR/LG	Data Output Link	Digital Signals
45	RD/WT	ECT Sensor Signal to Dash	0.6v
46	GY/BK	Vehicle Speed Sensor	0 Hz, 55 mph: 125 Hz
47	BR/PK	VR Solenoid Control	0%, 55 mph: 45%
48	WT/PK	Clean Tachometer Output	65 Hz, 55 mph: 175 Hz
49	DB/WT	Digital TR2 Sensor	0v, 55 mph: 11.5v
50	DG/YL	Digital TR4 Sensor	0v, 55 mph: 11.5v
51	BK/WT	Power Ground	<0.1v
52	WT/PK	COP 5 Driver (dwell)	5°, 55 mph: 8°
53	DG/PK	COP 7 Driver (dwell)	5°, 55 mph: 8°
54	PK/YL	TCC Solenoid Control	0%, 55 mph: 95%
55	RD/WT	Keep Alive Power	12-14v
56	LG/BK	EVAP Purge Solenoid	0-10 Hz (0-100%)
57	YL/RD	Knock Sensor (+) Signal	0v
58	GY/BK	VSS (+) Signal	0 Hz, 55 mph: 125 Hz
59	DG/WT	TSS Sensor Signal	100 Hz, 55 mph: 270 Hz
60	GY/LB	HO2S-11 (B1 S1) Signal	0.1-1.1v
61	---	Not Used	---
62	RD/PK	Fuel Tank Pressure Sensor	2.6v (0" H2O - cap off)
63	---	Not Used	---
64	LB/YL	Digital TR3A Sensor	In 'P': 0v, in O/D: 1.7v

2002-03 Excursion 6.8L V10 Bi-Fuel VIN S (A/T) 104 Pin Connector

PCM Pin #	Wire Color	Circuit Description (104 Pin)	Value at Hot Idle
65	BR/LG	DPFE Sensor Signal	0.9v, 55 mph: 3.1v
66	YL/LG	CHT Sensor Signal	0.6 (194°F)
67	VT/WT	EVAP Canister Vent Solenoid Control	0-10 Hz (0-100%)
68	GY/BK	Injector 9 Control	3.8-4.6 ms
69	PK/YL	AC Switch Signal	A/C On: 12v, Off: 0v
70	---	Not Used	---
71	RD	Vehicle Power (Start-Run)	12-14v
72	TN/RD	Injector 7 Control	3.8-4.6 ms
73	TN/BK	Injector 5 Control	3.8-4.6 ms
74	BR/YL	Injector 3 Control	3.8-4.6 ms
75	TN	Injector 1 Control	3.8-4.6 ms
76	BK/WT	Power Ground	<0.1v
77	BK/WT	Power Ground	<0.1v
78	PK/LB	COP 2 Driver (dwell)	5°, 55 mph: 8°
79	WT/RD	COP 8 Driver (dwell)	5°, 55 mph: 8°
80	LB/OR	Fuel Pump Control	On: 1v, Off: 12v
81	WT/YL	EPC Solenoid Control	9.2v (5 psi)
82	WT/RD	COP 9 Driver (dwell)	5°, 55 mph: 8°
83	WT/LB	IAC Solenoid Control	10.7v (33%)
84	DB/YL	OSS Sensor Signal	0 Hz, 55 mph: 2050 Hz
85	DG	CMP Sensor Signal	9 Hz, 55 mph: 16 Hz
86	DB/YL	OSS Sensor Signal	0 Hz, 55 mph: 250 Hz
87	RD/BK	HO2S-21 (B2 S1) Signal	0.1-1.1v
88	LB/RD	MAF Sensor Signal	0.8v, 55 mph: 1.6v
89	GY/WT	TP Sensor Signal	0.9v, 55 mph: 1.3v
90	BR/WT	Reference Voltage	4.9-5.1v
91	GY/RD	Sensor Ground	<0.050v
92	RD/LG	Brake Position Switch	Brake Off: 12v, On: 1v
93	RD/WT	HO2S-11 (B1 S1) Heater	Heater Off: 12v, On: 1v
94	YL/LB	HO2S-21 (B2 S1) Heater	Heater Off: 12v, On: 1v
95	WT/BK	HO2S-12 (B1 S2) Heater	Heater Off: 12v, On: 1v
96	---	Not Used	---
97	RD	Vehicle Power (Start-Run)	12-14v
98	LB	Injector 8 Control	3.8-4.6 ms
99	LG/OR	Injector 6 Control	3.8-4.6 ms
100	BR/LB	Injector 4 Control	3.8-4.6 ms
101	WT	Injector 2 Control	3.8-4.6 ms
102	YL/BK	COP 4 Driver (dwell)	5°, 55 mph: 8°
103	BK/WT	Power Ground	<0.1v
104	PK/WT	COP 3 Driver (dwell)	5°, 55 mph: 8°

Pin Connector Graphic

PCM 104-PIN CONNECTOR

Terminal View of 104-Pin PCM Wiring Harness Connector

Standard Colors and Abbreviations

Abbreviation	Color	Abbreviation	Color	Abbreviation	Color
BK	Black	GY	Gray	PK	Purple
BL	Blue	GN	Green	RD	Red
BR	Brown	LG	LT Green	TN	Tan
DB	Dark Blue	OR	Orange	WT	White
DG	DK Green	PK	Pink	YL	Yellow

2000-01 Excursion 7.3L V8 Diesel VIN F 104 Pin Connector

PCM Pin #	Wire Color	Circuit Description (104 Pin)	Value at Hot Idle
1	---	Not Used	---
2	PK/LG	MIL (lamp) Control	MIL On: 1v, Off: 12v
3	RD/YL	Low Current Sensor Return	<0.050v
4	LB/YL	Neutral Gear Switch	Switch On: 0v, Off: 12v
5	LG/RD	Parking Brake Applied Switch	Park Brake Applied: 0v
6	OR/YL	Shift Solenoid 1 Control	SS1 On: 1v, Off: 12v
7	---	Not Used	---
8	WT/LG	Glow Plug Monitor Left Bank	Digital Signal: 0-12-0v
9	WT/PK	Glow Plug Monitor Right Bank	Digital Signal: 0-12-0v
10	RD/OR	Idle Validation Switch	12v, off-idle: 0v
11	RD/GN	Shift Solenoid 'B' Control	SS2 On: 1v, Off: 12v
12	WT/LG	TCIL (lamp) Control	Lamp On: 1v, Off: 12v
13	PK	Generic Scan Tool Input	0.1v
14	OR/LB	4x4 Low Switch Input	Switch On: 0v, Off: 12v
15	PK/LB	Data Bus (-)	<0.050v
16	TN/OR	Data Bus (+)	Digital Signals
17	WT/YL	Digital TR1 Sensor	In 'P': 0v, in 'D': 10.7v
18	---	Not Used	---
19	WT/PK	Tachometer Reflected Signal	6.5v / 130 Hz
20	BR/OR	Coast Clutch Solenoid	CCS On: 0v, Off: 12v
21	DG	CMP Sensor Signal	Digital Signal: 0-12-0v
22-23	---	Not Used	---
24	YL/WT	APP Sensor Ground	<0.050v
25	LB/YL	Speedometer Ground	<0.050v
26-27	---	Not Used	---
28	RD	Water In Fuel Indicator Lamp	Lamp on: 1v, Off: 12v
29	TN/WT	TCS (switch) Signal	TCS & O/D On: 12v
30	PK/LB	Exhaust Back Pressure	0.9v, off-idle: 2.5v
31	RD/LG	Brake Pressure Switch	Brake On: 0v, Off: 12v
32-34	---	Not Used	---
35	WT/YL	Generator Power Switch	12-14v
36	GY/RD	Fuel Heater/Water in Fuel	Heater On: 12v
37	OR/BK	TFT Sensor Signal	0.3-4.5v
38	LG/RD	Engine Oil Temperature	0.3-4.7v
39	GY	IAT Sensor Signal	0.2-4.5v
40	PK/BK	Fuel Pump Power	12-14v
41	TN/LG	A/C Head Pressure Switch	0v or 12v
42	GY/RD	Exhaust Backpressure Signal	Digital Signal: 0-12-0v
43-46	---	Not Used	---
47	WT/RD	Wastegate Control Solenoid	Solenoid Off: 12v, On: 1v
48	GY/WT	Electronic Feedback Line	Digital Signals
49	DB/WT	Digital TR2 Sensor	In 'P': 0v, in 'D': 10.7v
50	DG/YL	Digital TR4 Sensor	In 'P': 0v, in 'D': 10.7v
51	BK/WT	Power Ground	<0.1v
52-53	---	Not Used	---
54	PK/YL	TCC Solenoid Control	TCC On: 1v, Off: 12v
55	RD/WT	Keep Alive Power	12-14v
56-57	---	Not Used	---
58	GY/BK	VSS (+) Signal	0 Hz, 55 mph: 125 Hz
59	DB/YL	OSS Sensor Signal	Varying Signal
60	---	Not Used	---
61	LB/BK	Speed Control On/Off Switch	Switch On: 12v, Off: 0v
62	RD/YL	ACT Sensor Signal	1.5-2.5v
63	---	Not Used	---
64	LB/YL	Digital TR3A Sensor3	In Park: 4.5v, in Drive: 2.2v

2000-01 Excursion 7.3L V8 Diesel VIN F 104 Pin Connector

PCM Pin #	Wire Color	Circuit Description (104 Pin)	Value at Hot Idle
65	LB	CMP Sensor Ground	<0.050v
66-69	---	Not Used	---
70	BK/PK	Wait To Start Indicator Control	Indicator Off: 12v, On: 1v
71	RD	Vehicle Power	12-14v
72-76	---	Not Used	---
77	BK/WT	Power Ground	<0.1v
78	---	Not Used	---
79	LG/BK	MAP Sensor Signal	1-3v
80	WT/BK	IDM Enable Relay Control	On: 1v, Off: 12v
81	WT/YL	EPC Solenoid Control	7.5v (8 psi)
82	---	Not Used	---
83	YL/RD	Injector Pressure Regulator Control	Duty Cycle: 0-12-0v
84	DG/WT	TSS Sensor Signal	At 30 mph: 126-136 Hz
85-86	---	Not Used	---
87	DB/LG	Injection Control Pressure	0.75v (Min: 0.83v startup)
88	---	Not Used	---
89	GY/WT	APP Sensor Signal	0.5-1.6v
90	BR/WT	Reference Voltage	4.9-5.1v
91	GY/RD	Sensor Return Signal	<0.050v
92	RD/LG	Brake Position Switch	Brake Off: 12v, On: 1v
93	---	Not Used	---
94	LB/OR	Fuel Pump Relay Output	12-14v
95	BR/OR	Fuel Delivery Control Signal	49 Hz
96	YL/LB	CMP Sensor Signal	6v (5 Hz)
97	RD	Vehicle Power (Start-Run)	12-14v
98	PK	Manifold Intake Air Heater Relay Control	Relay Off: 12v, On: 1v
99	---	Not Used	---
100	YL/WT	Fuel Level Indicator Signal	1.7v (1/2 full)
101	PK/OR	Glow Plug Relay Control	Relay Off: 12v, On: 1v
102	---	Not Used	---
103	BK/WT	Power Ground	<0.1v
104	---	Not Used	---

Pin Connector Graphic

Terminal View of 104-Pin PCM Wiring Harness Connector

Standard Colors and Abbreviations

Abbreviation	Color	Abbreviation	Color	Abbreviation	Color
BK	Black	GY	Gray	PK	Purple
BL	Blue	GN	Green	RD	Red
BR	Brown	LG	LT Green	TN	Tan
DB	Dark Blue	OR	Orange	WT	White
DG	DK Green	PK	Pink	YL	Yellow

2002-03 Excursion 7.3L V8 Diesel VIN F (A/T) 104 Pin Connector

PCM Pin #	Wire Color	Circuit Description (104 Pin)	Value at Hot Idle
1-2	---	Not Used	---
3	RD/YL	Manifold Intake Air Heater Monitor	Off: 12v, On: 1v
4	---	Not Used	---
5	LG/RD	Parking Brake Applied Switch	Brake Off: 12v, Applied: 0v
6	OR/YL	Shift Solenoid 1 Control	SS1 Off: 12v, On: 1v
7	---	Not Used	---
8	WT/LG	Glow Plug Control Module Communication	Digital Signal: 0-12-0v
9	---	Not Used	---
10	RD/LG	Idle Validation Switch	12v, off-idle: 0v
11	VT/OG	Shift Solenoid 2 Control	SS2 Off: 12v, On: 1v
12	WT/LG	TCIL (lamp) Control	7.7v (Switch On: 0v)
13	VT	Flash EEPROM Power Supply	0.1v
14	---	Not Used	---
15	PK/LB	SCP Data Bus (-) Signal	<0.050v
16	TN/OR	SCP Data Bus (+) Signal	Digital Signals
17	YL/BK	Digital TR1 Sensor	In 'P': 0v, in Drive: 10.7v
18	PK/YL	A/C Relay Control	Relay Off: 12v, On: 1v
19	LG/WT	Tachometer Reflected Signal	6.5v / 130 Hz
20	BR/OR	Coast Clutch Solenoid Control	CCS On: 0v, Off: 12v
21	DG	CMP Sensor Signal	Digital Signal: 0-12-0v
22-23	---	Not Used	---
24	YL/WT	APP Sensor Ground	<0.050v
25	LB/YL	Case Ground	<0.050v
26-28	---	Not Used	---
29	TN/WT	TCS (switch) Signal	TCS & O/D On: 12v
30	VT/LB	Exhaust Back Pressure Sensor	0.9v, off-idle: 2.5v
31	BK/YL	Brake Pressure Switch	Brake Off: 12v, On: 1v
32-34	---	Not Used	---
35	WT/YL	Alternator No. 1 (Top) Monitor	6-10v
36	GY/RD	Fuel Heater / Water in Fuel	Heater On: 12v, Off: 0v
37	OR/BK	Transmission Fluid Temperature Sensor	2.10-2.40v
38	LG/RD	Engine Oil Temperature Sensor	0.3-4.7v
39	GY	Intake Air Temperature Sensor	1.5-2.5v
40	PK/BK	Fuel Pump Monitor	Pump Off: 12v, On: 1v
41	TN/LG	A/C Head Pressure Switch	0v or 12v
42	GY/RD	Exhaust Backpressure Regulator Control	Duty Cycle (0-12-0v)
43	OR/LG	Overhead Console Fuel Consumption	0-5-0v
44	---	Not Used	---
45	OG/LB	Cruise Control Set Indicator	Indicator Off: 12v, On: 1v
47	WT/RD	Wastegate Control Solenoid Control	Solenoid Off: 12v, On: 1v
48	GY/WT	Injector Driver Module	Digital Signal: 0.9-3.0v
49	LB/PK	Digital TR2 Sensor	In 'P': 0v, in Drive: 10.7v
50	WT/BK	Digital TR4 Sensor	In 'P': 0v, in Drive: 10.7v
51	BK/WT	Power Ground	<0.1v
52-53	---	Not Used	---
54	VT/YL	TCC Solenoid Control	TCC Off: 12v, On: 1v
55	RD/WT	Direct Battery	12-14v
56-57	---	Not Used	---
58	GY/BK	VSS (+) Signal	0 Hz, 55 mph: 125 Hz
59	DB/YL	OSS Sensor Signal	AC pulse signals
60	YL	Alternator No. 2 (Bottom) Control	6-10v
61	LB/BK	Speed Control On/Off Switch	Switch On: 12v, Off: 0v
62	RD/YL	MAT Sensor Signal	1.5-2.5v
63	---	Not Used	---
64	LB/YL	Digital TR3 Sensor	In Park: 4.5v, in Drive: 2.2v

2002-03 Excursion 7.3L V8 Diesel VIN F (A/T) 104 Pin Connector

PCM Pin #	Wire Color	Circuit Description (104 Pin)	Value at Hot Idle
65	LB	CMP Sensor Ground	<0.050v
66	LB/YL	Power Take-Off Enable	0v (Off)
67	LG/RD	Generator/Battery Indicator Control	Indicator Off: 12v, On: 1v
68-70	---	Not Used	---
71	RD	Vehicle Power (Start-Run)	12-14v
72-76	---	Not Used	---
77	BK/WT	Power Ground	<0.1v
78	---	Not Used	---
79	LG/BK	Manifold Absolute Pressure Sensor	1-3v
80	WT/BK	Injector Drive Module Relay Control	Relay Off: 12v, On: 1v
81	WT/YL	EPC Solenoid Control	7.5v (8 psi)
82	---	Not Used	---
83	YL/RD	Injector Pressure Regulator Control	Duty Cycle: 0-12-0v
84	DG/WT	Turbine Shaft Speed Sensor	30 mph: 130 Hz
85-86	---	Not Used	---
87	DB/LG	Injection Control Pressure Sensor	0.1-3.0v
88	---	Not Used	---
89	GY/WT	APP Sensor Signal	0.5-1.6v
90	BR/WT	Reference Voltage	4.9-5.1v
91	GY/RD	Sensor Ground	<0.050v
92	RD/LG	Brake Pedal Switch	Brake Off: 0v, On: 12v
93	---	Not Used	---
94	LB/OR	Fuel Pump Relay Control	Relay Off: 12v, On: 1v
95	BR/OR	Fuel Delivery Control	40-240 Hz
96	YL/LB	CID Sensor Signal	5-30 Hz
97	RD	Vehicle Power (Start-Run)	12-14v
98	VT	Manifold Intake Air Heater Relay Control	Relay Off: 12v, On: 1v
99-100	---	Not Used	---
101	VT/OR	Glow Plug Relay Control	Relay Off: 12v, On: 1v
102	---	Not Used	---
103	BK/WT	Power Ground	<0.1v
104	---	Not Used	---

Pin Connector Graphic

PCM 104-PIN CONNECTOR

Terminal View of 104-Pin PCM Wiring Harness Connector

Standard Colors and Abbreviations

Abbreviation	Color	Abbreviation	Color	Abbreviation	Color
BK	Black	GY	Gray	PK	Purple
BL	Blue	GN	Green	RD	Red
BR	Brown	LG	LT Green	TN	Tan
DB	Dark Blue	OR	Orange	WT	White
DG	DK Green	PK	Pink	YL	Yellow

Expedition PIN Tables

1997 Expedition 4.6L V8 VIN W, VIN 6 (A/T) 104 Pin Connector

PCM Pin #	Wire Color	Circuit Description (104 Pin)	Value at Hot Idle
1	PK/OR	Shift Solenoid 2 Control	12v, 55 mph: 1v
2	PK/LG	MIL (lamp) Control	MIL On: <1v, Off: 12v
3	YL/BK	Digital TR1 Sensor	0v, 55 mph: 11.5v
12	YL/WT	Fuel Level Indicator Signal	1.7v (1/2 full)
13	PK	Flash EEPROM Power	0.1v
14	LB/BK	4x4 Indicator Signal	4x4 On: 1v, Off: 12v
15	PK	Data Bus (-) Signal	Digital Signals
16	TN/OR	Data Bus (+) Signal	Digital Signals
19	DB/WT	Automatic Ride Control	Digital Signals
21	DB	CKP (-) Sensor Signal	430-475 Hz
22	GY	CKP (+) Sensor Signal	430-475 Hz
24	BK/WT	Power Ground	<0.1v
25	LG/YL	Case Ground	<0.050v
26	DB/LG	Coil 1 Driver (dwell)	5°, 55 mph: 8°
27	OR/YL	Shift Solenoid 1 Control	1v, 55 mph: 1v
29	TN/WT	TCS (switch) Signal	TCS & O/D On: 12v
30	DG	Octane Adjustment	9.3v (switch shorted: 0v)
33	PK/OR	VSS (-) Signal	0 Hz, 55 mph: 125 Hz
35	RD/LG	HO2S-12 (B1 S2) Signal	0.1-1.1v
36	TN/LB	MAF Sensor Return	<0.050v
37	OR/BK	TFT Sensor Signal	2.10-2.40v
38	LG/RD	ECT Sensor Signal	0.5-0.6v
39	GY	IAT Sensor Signal	1.5-2.5v
40	DG/YL	Fuel Pump Monitor	On: 12v, Off: 0v
41	BK/YL	A/C Switch Signal	A/C On: 12v, Off: 0v
43	OR/LG	Overhead Trip Computer	Digital Signals
45	RD/WT	CHTIL (lamp) Control	Lamp On: 1v, Off: 12v
46	BR	Intake Manifold Tuning Valve	12v, 55 mph: 12v
47	BR/PK	VR Solenoid Control	0%, 55 mph: 45%
48	WT/PK	Clean Tachometer Output Signal	39-49 Hz
49	LB/BK	Digital TR2 Sensor	0v, 55 mph: 11.5v
50	WT/BK	Digital TR4 Sensor	0v, 55 mph: 11.5v
51	BK/WT	Power Ground	<0.1v
52	RD/LB	Coil 2 Driver (dwell)	5°, 55 mph: 8°
54	PK/YL	TCC Solenoid Control	0%, 55 mph: 95%
55	BK/LG	Keep Alive Power	12-14v
56	LG/BK	EVAP Purge Solenoid	0-10 Hz (0-100%)
57	YL/RD	Knock Sensor 1 Signal	0v
58	GY/BK	VSS (+) Signal	0 Hz, 55 mph: 125 Hz
60	GY/LB	HO2S-11 (B1 S1) Signal	0.1-1.1v
61	PK/LG	HO2S-22 (B2 S2) Signal	0.1-1.1v
62	RD/PK	FTP Sensor Signal	2.6v (0" H2O - cap off)
64	LB/YL	TR Sensor Signal	In 'P': 0v, in O/D: 1.7v

1997 Expedition 4.6L V8 VIN W, VIN 6 (A/T) 104 Pin Connector

PCM Pin #	Wire Color	Circuit Description (104 Pin)	Value at Hot Idle
65	BR/LG	DPFE Sensor Signal	0.20-1.30v
66	YL/LG	CHT Sensor Signal	0.6 to 1.7v (194°F)
67	PK/WT	EVAP CV Solenoid	0-10 Hz (0-100%)
69	PK/YL	A/C WOT Relay Control	Relay On: 0.1v, Off: 12v
71	RD	Vehicle Power	12-14v
72	TN/RD	Injector 7 Control	2.7-4.1 ms
73	TN/BK	Injector 5 Control	2.7-4.1 ms
74	BR/YL	Injector 3 Control	2.7-4.1 ms
75	TN	Injector 1 Control	2.7-4.1 ms
76, 77	BK/WT	Power Ground	<0.1v
78	PK/WT	Coil 3 Driver (dwell)	5°, 55 mph: 8°
79	WT/LG	TCIL (lamp) Control	TCIL On: 1v, Off: 12v
80	LB/OR	Fuel Pump Control	On: 1v, Off: 12v
81	WT/YL	EPC Solenoid Control	9.1v (6 psi)
83	WT/LB	IAC Solenoid Control	10.7v (33%)
84	DB/YL	OSS Sensor Signal	0 Hz, 55 mph: 250 Hz
85	DG	CID Sensor Signal	6 Hz, 55 mph: 15 Hz
87	RD/BK	HO2S-21 (B2 S1) Signal	0.1-1.1v
88	LB/RD	MAF Sensor Signal	0.6v, 55 mph: 2.1v
89	GY/WT	TP Sensor Signal	0.53-1.27v
90	BR/WT	Reference Voltage	4.9-5.1v
91	GY/RD	Analog Signal Return	<0.050v
92	LG	Brake Switch Signal	Brake Off: 12v, On: 1v
93	RD/WT	HO2S-11 (B1 S1) Heater	On: 1v, Off: 12v
94	YL/LB	HO2S-21 (B2 S1) Heater	On: 1v, Off: 12v
95	WT/BK	HO2S-12 (B1 S2) Heater	On: 1v, Off: 12v
96	TN/YL	HO2S-22 (B2 S2) Heater	On: 1v, Off: 12v
97	RD	Vehicle Power	12-14v
98	LB	Injector 8 Control	2.7-4.1 ms
99	LG/OR	Injector 6 Control	2.7-4.1 ms
100	BR/LB	Injector 4 Control	2.7-4.1 ms
101	WT	Injector 2 Control	2.7-4.1 ms
103	BK/WT	Power Ground	<0.1v
104	RD/YL	Coil 4 Driver (dwell)	5°, 55 mph: 8°

Pin Connector Graphic

```
PCM 104-PIN CONNECTOR
1  ●●●●●●●●●●●●    ●●●●●●●●●●●●● 26
27 ●●●●●●●●●●●●    ●●●●●●●●●●●●● 52
53 ●●●●●●●●●●●●  ● ●●●●●●●●●●●●● 78
79 ●●●●●●●●●●●●    ●●●●●●●●●●●●● 104
```

Terminal View of 104-Pin PCM Wiring Harness Connector

Standard Colors and Abbreviations

Abbreviation	Color	Abbreviation	Color	Abbreviation	Color
BK	Black	GY	Gray	PK	Purple
BL	Blue	GN	Green	RD	Red
BR	Brown	LG	LT Green	TN	Tan
DB	Dark Blue	OR	Orange	WT	White
DG	DK Green	PK	Pink	YL	Yellow

1998-99 Expedition 4.6L V8 VIN 6, VIN W (A/T) 104 Pin Connector

PCM Pin #	Wire Color	Circuit Description (104 Pin)	Value at Hot Idle
1	PK/OR	Shift Solenoid 2 Control	12v, 55 mph: 1v
2	PK/LG	MIL (lamp) Control	MIL On: <1v, Off: 12v
3	YL/BK	Digital TR1 Sensor	0v, 55 mph: 11.5v
4-11	---	Not Used	---
12	YL/WT	Fuel Level Indicator Signal	1.7v (1/2 full)
13	PK	Flash EEPROM Power	0.1v
14	LB/BK	4x4 Indicator Signal	4x4 On: 1v, Off: 12v
15	PK	Data Bus (-) Signal	Digital Signals
16	TN/OR	Data Bus (+) Signal	Digital Signals
17	---	Not Used	---
19	DB/WT	Automatic Ride Control	Digital Signals
20	---	Not Used	---
21	DB	CKP (-) Sensor Signal	430-475 Hz
22	GY	CKP (+) Sensor Signal	430-475 Hz
24	BK/WT	Power Ground	<0.1v
25	LG/YL	Case Ground	<0.050v
26	DB/LG	Coil 1 Driver (dwell)	5°, 55 mph: 8°
27	OR/YL	Shift Solenoid 1 Control	1v, 55 mph: 1v
28	---	Not Used	---
29	TN/WT	TCS (switch) Signal	TCS & O/D On: 12v
30	DG	Octane Adjustment	9.3v (switch shorted: 0v)
31-32	---	Not Used	---
33	PK/OR	VSS (-) Signal	0 Hz, 55 mph: 125 Hz
35	RD/LG	HO2S-12 (B1 S2) Signal	0.1-1.1v
36	TN/LB	MAF Sensor Return	<0.050v
37	OR/BK	TFT Sensor Signal	2.10-2.40v
38	LG/RD	ECT Sensor Signal	0.5-0.6v
39	GY	IAT Sensor Signal	1.5-2.5v
40	DG/YL	Fuel Pump Monitor	On: 12v, Off: 0v
41	BK/YL	A/C Switch Signal	A/C On: 12v, Off: 0v
42	---	Not Used	---
43	OR/LG	Overhead Trip Computer	Digital Signals
44	---	Not Used	---
45	RD/WT	CHTIL (lamp) Control	Lamp On: 1v, Off: 12v
46	BR	Intake Manifold Tuning Valve	12v, 55 mph: 12v
47	BR/PK	VR Solenoid Control	0%, 55 mph: 45%
48	WT/PK	Clean Tachometer Output Signal	39-49 Hz
49	LB/BK	Digital TR2 Sensor	0v, 55 mph: 11.5v
50	WT/BK	Digital TR4 Sensor	0v, 55 mph: 11.5v
51	BK/WT	Power Ground	<0.1v
52	RD/LB	Coil 2 Driver (dwell)	5°, 55 mph: 8°
54	PK/YL	TCC Solenoid Control	0%, 55 mph: 95%
55	BK/LG	Keep Alive Power	12-14v
56	LG/BK	EVAP Purge Solenoid	0-10 Hz (0-100%)
57	YL/RD	Knock Sensor 1 Signal	0v
58	GY/BK	VSS (+) Signal	0 Hz, 55 mph: 125 Hz
59	---	Not Used	---
60	GY/LB	HO2S-11 (B1 S1) Signal	0.1-1.1v
61	PK/LG	HO2S-22 (B2 S2) Signal	0.1-1.1v
62	RD/PK	FTP Sensor Signal	2.6v (0" H2O - cap off)
63	---	Not Used	---
64	LB/YL	TR Sensor Signal	In 'P': 0v, in O/D: 1.7v

1998-99 Expedition 4.6L V8 VIN 6, VIN W (A/T) 104 Pin Connector

PCM Pin #	Wire Color	Circuit Description (104 Pin)	Value at Hot Idle
65	BR/LG	DPFE Sensor Signal	0.20-1.30v
66	YL/LG	CHT Sensor Signal	0.6 to 1.7v (194°F)
67	PK/WT	EVAP CV Solenoid	0-10 Hz (0-100%)
69	PK/YL	A/C WOT Relay Control	Relay On: 0.1v, Off: 12v
71	RD	Vehicle Power	12-14v
72	TN/RD	Injector 7 Control	2.7-3.5 ms
73	TN/BK	Injector 5 Control	2.7-3.5 ms
74	BR/YL	Injector 3 Control	2.7-3.5 ms
75	TN	Injector 1 Control	2.7-3.5 ms
76, 77	BK/WT	Power Ground	<0.1v
78	PK/WT	Coil 3 Driver (dwell)	5°, 55 mph: 8°
79	WT/LG	TCIL (lamp) Control	TCIL On: 1v, Off: 12v
80	LB/OR	Fuel Pump Control	On: 1v, Off: 12v
81	WT/YL	EPC Solenoid Control	9.1v (6 psi)
83	WT/LB	IAC Solenoid Control	10.7v (33%)
84	DB/YL	OSS Sensor Signal	0 Hz, 55 mph: 250 Hz
85	DG	CID Sensor Signal	6 Hz, 55 mph: 15 Hz
87	RD/BK	HO2S-21 (B2 S1) Signal	0.1-1.1v
88	LB/RD	MAF Sensor Signal	0.6v, 55 mph: 2.1v
89	GY/WT	TP Sensor Signal	0.53-1.27v
90	BR/WT	Reference Voltage	4.9-5.1v
91	GY/RD	Analog Signal Return	<0.050v
92	LG	Brake Switch Signal	Brake Off: 12v, On: 1v
93	RD/WT	HO2S-11 (B1 S1) Heater	On: 1v, Off: 12v
94	YL/LB	HO2S-21 (B2 S1) Heater	On: 1v, Off: 12v
95	WT/BK	HO2S-12 (B1 S2) Heater	On: 1v, Off: 12v
96	TN/YL	HO2S-22 (B2 S2) Heater	On: 1v, Off: 12v
97	RD	Vehicle Power	12-14v
98	LB	Injector 8 Control	2.7-3.5 ms
99	LG/OR	Injector 6 Control	2.7-3.5 ms
100	BR/LB	Injector 4 Control	2.7-3.5 ms
101	WT	Injector 2 Control	2.7-3.5 ms
103	BK/WT	Power Ground	<0.1v
104	RD/YL	Coil 4 Driver (dwell)	5°, 55 mph: 8°

Pin Connector Graphic

PCM 104-PIN CONNECTOR

Terminal View of 104-Pin PCM Wiring Harness Connector

Standard Colors and Abbreviations

Abbreviation	Color	Abbreviation	Color	Abbreviation	Color
BK	Black	GY	Gray	PK	Purple
BL	Blue	GN	Green	RD	Red
BR	Brown	LG	LT Green	TN	Tan
DB	Dark Blue	OR	Orange	WT	White
DG	DK Green	PK	Pink	YL	Yellow

2000-02 Expedition 4.6L V8 VIN 6, VIN W (A/T) 104 Pin Connector

PCM Pin #	Wire Color	Circuit Description (104 Pin)	Value at Hot Idle
1	OR/YL	COP 6 Driver (dwell)	5°, 55 mph: 8°
2	---	Not Used	---
3	BK/WT	Power Ground	<0.1v
4-5	---	Not Used	---
6	OR/YL	Shift Solenoid 'A' Control	1v, 55 mph: 1v
7-10	---	Not Used	---
11	PK/OR	Shift Solenoid 'B' Control	12v, 55 mph: 1v
12	WT/LG	TCIL (lamp) Control	Lamp On: 1v, Off: 12v
13	PK	Flash EEPROM Power	0.1v
14	LB/BK	4x4 Low Switch	Switch On: 0v, Off: 12v
15	PK/LB	Data Bus (-) Signal	Digital Signals
16	TN/OR	Data Bus (+) Signal	Digital Signals
17-18	---	Not Used	---
19	DB/WT	S/C Module to Acceleration	Digital Signals
20	---	Not Used	---
21	DB	CKP (-) Sensor Signal	430-475 Hz
22	GY	CKP (+) Sensor Signal	430-475 Hz
23	DG/WT	Dual Knock Sensor (-) Signal	<0.050v
24	---	Not Used	---
25	LB/YL	Case Ground	<0.050v
26	LG/WT	COP 1 Driver (dwell)	5°, 55 mph: 8°
27	LG/YL	COP 5 Driver (dwell)	5°, 55 mph: 8°
28	---	Not Used	---
29	TN/WT	TCS (switch) Signal	TCS & O/D On: 12v
30-31	---	Not Used	---
32	DG/PK	Dual Knock Sensor (-) Signal	<0.050v
33	---	Not Used	---
34	YL/BK	TR Sensor Signal	In 'P': 0v, in O/D: 1.7v
35	RD/LG	HO2S-12 (B1 S2) Signal	0.1-1.1v
36	TN/LB	MAF Sensor Return	<0.050v
37	OR/BK	TFT Sensor Signal	2.10-2.40v
38	---	Not Used	---
39	GY	IAT Sensor Signal	1.5-2.5v
40	DG/YL	Fuel Pump Monitor	On: 12v, Off: 0v
41	BK/YL	A/C High Pressure Switch	A/C On: 12v, Off: 0v
42	PK/OR	TP Sensor Signal	0.9v, 55 mph: 1.3v
43	OR/LG	Data Output Link	Digital Signals
44-45	---	Not Used	---
46	BR	IMRC (valve) Control	12v
47	BR/PK	VR Solenoid Control	0%, 55 mph: 45%
48	---	Not Used	---
49	LB/BK	Digital TR2 Sensor	0v, 55 mph: 11.5v
50	WT/BK	Digital TR4 Sensor	0v, 55 mph: 11.5v
51	BK/WT	Power Ground	<0.1v
52	WT/PK	COP 3 Driver (dwell)	5°, 55 mph: 8°
53	DG/PK	COP 4 Driver (dwell)	5°, 55 mph: 8°
54	PK/YL	TCC Solenoid Control	0%, 55 mph: 95%
55	RD/WT	Keep Alive Power	12-14v
56	LG/BK	EVAP Purge Solenoid	0-10 Hz (0-100%)
57	YL/RD	Dual Knock Sensor (+) Signal	0v
58-59	---	Not Used	---
60	GY/LB	HO2S-11 (B1 S1) Signal	0.1-1.1v
61	PK/LG	HO2S-22 (B2 S2) Signal	0.1-1.1v
62	RD/PK	FTP Sensor Signal	2.6v (0" H2O - cap off)
63	---	Not Used	---
64	LB/YL	Digital TR3A Sensor	0v, 55 mph: 11.5v

2000-02 Expedition 4.6L V8 VIN 6, VIN W (A/T) 104 Pin Connector

PCM Pin #	Wire Color	Circuit Description (104 Pin)	Value at Hot Idle
65 ('00)	BR/LG	DPFE Sensor Signal	0.20-1.30v
65 ('01-'02)	BR/LG	DPFE Sensor Signal	0.95-1.05v
66	YL/LG	CHT Sensor Signal	0.6 to 1.7v (194ºF)
67	PK/WT	EVAP CV Solenoid	0-10 Hz (0-100%)
68	GY/BK	VSS Out Signal	Digital Signals
69	PK/YL	A/C WOT Relay Control	On: 1v, Off: 12v
70	---	Not Used	---
71	RD	Vehicle Power	12-14v
72	TN/RD	Injector 7 Control	2.7-4.1 ms
73	TN/BK	Injector 5 Control	2.7-4.1 ms
74	BR/YL	Injector 3 Control	2.7-4.1 ms
75	TN	Injector 1 Control	2.7-4.1 ms
76	---	Not Used	---
77	BK/WT	Power Ground	<0.1v
78	PK/LB	COP 7 Driver (dwell)	5º, 55 mph: 8º
79	WT/RD	COP 8 Driver (dwell)	5º, 55 mph: 8º
80	LB/OR	Fuel Pump Relay Control	On: 1v, Off: 12v
81	WT/YL	EPC Solenoid Control	9.1v (6 psi)
83	WT/LB	IAC Solenoid Control	10.7v (33%)
84	DB/YL	OSS Sensor Signal	0 Hz, 55 mph: 250 Hz
85	DG	CMP Sensor Signal	6 Hz, 55 mph: 15 Hz
87	RD/BK	HO2S-21 (B2 S1) Signal	0.1-1.1v
88	LB/RD	MAF Sensor Signal	0.6v, 55 mph: 2.1v
89	GY/WT	TP Sensor Signal	0.53-1.27v
90	BR/WT	Reference Voltage	4.9-5.1v
91	GY/RD	Sensor Ground	<0.050v
92	RD/LG	Brake Position Switch	Brake Off: 12v, On: 1v
93	RD/WT	HO2S-11 (B1 S1) Heater	On: 1v, Off: 12v
94	YL/LB	HO2S-21 (B2 S1) Heater	On: 1v, Off: 12v
95	WT/BK	HO2S-12 (B1 S2) Heater	On: 1v, Off: 12v
96	TN/YL	HO2S-22 (B2 S2) Heater	On: 1v, Off: 12v
97	RD	Vehicle Power	12-14v
98	LB	Injector 8 Control	2.7-4.1 ms
99	LG/OR	Injector 6 Control	2.7-4.1 ms
100	BR/LB	Injector 4 Control	2.7-4.1 ms
101	WT	Injector 2 Control	2.7-4.1 ms
102	DG/WT	Dual Knock Sensor (-) Signal	0v
103	BK/WT	Power Ground	<0.1v
104	PK/WT	COP 2 Driver (dwell)	5º, 55 mph: 8º

Pin Connector Graphic

PCM 104-PIN CONNECTOR

Terminal View of 104-Pin PCM Wiring Harness Connector

Standard Colors and Abbreviations

Abbreviation	Color	Abbreviation	Color	Abbreviation	Color
BK	Black	GY	Gray	PK	Purple
BL	Blue	GN	Green	RD	Red
BR	Brown	LG	LT Green	TN	Tan
DB	Dark Blue	OR	Orange	WT	White
DG	DK Green	PK	Pink	YL	Yellow

2003 Expedition 4.6L V8 MFI VIN W (A/T) C138b Pin Connector

PCM Pin #	Wire Color	Circuit Description (46 Pin)	Value at Hot Idle
1	BK/WT	Power Ground	<0.1v
2	PK/YL	A/C WOT Relay Control	Relay Off: 12v, On: 1v
3	PK/OR	TP Sensor Signal	0.9v, 55 mph: 1.3v
4	DB/OG	Starter Relay Control Circuit	Relay Off: 12v, On: 1v
5	---	Not Used	---
6	OG/LB	Cooling Fan Speed Signal	0.5-4.9v
7	TN/LG	Generator Communicator Command	1.5-10.5v (40-250 Hz)
8	RD/PK	Fuel Tank Pressure Sensor	2.6v (0" H2O - cap off)
9	OR/RD	Antitheft Indicator Control	Indicator Off: 12v, On: 1v
10	LB/YL	Power Ground	<0.1v
11	BK/WT	Power Ground	<0.1v
12	VT/YL	Speed Control Servo Signal 'C'	DC pulse signals
13	GY/BK	VSS (+) Signal	0 Hz, 55 mph: 125 Hz
14	BR/PK	Generator Monitor Signal	130 Hz (45%)
15	BK/YL	Brake Pedal Position Switch	Brake Off: 0v, On: 12v
16	---	Not Used	---
17	DG/OR	Steering Wheel Control Ground	<0.050v
18	YL/LG	Power Steering Pressure Switch	Straight: 0v, Turned: 12v
19	GY	IAT Sensor Signal	1.5-2.5v
20	DG/YL	Fuel Pump Relay Output	Off: 0v, On: 12v
21	---	Not Used	---
22	TN/WT	TCS (switch) Signal	TCS & O/D On: 12v
23	BK/WT	Power Ground	<0.1v
24	PK/BK	Speed Control Servo Signal 'B'	DC pulse signals
25	LG/VT	Cooling Fan Control	Digital Signal: 0-12-0v
26	TN	Speed Control Servo Common	<0.050v
27	LB/OR	Fuel Pump Relay Control	Relay Off: 12v, On: 1v
28-29	---	Not Used	---
30	TN/LB	MAF Sensor Return	<0.050v
31	LB/RD	MAF Sensor Signal	0.6v, 55 mph: 2.1v
32	TN/OR	SCP Data Bus (+) Signal	Digital Signals
33	GY/RD	Sensor Ground	<0.050v
34	RD	Vehicle Power (Start-Run)	12-14v
35	DG/WT	Speed Control Servo Signal 'A'	DC pulse signals
36	VT/WT	EVAP Canister Vent Solenoid Control	0-10 Hz (0-100%)
37	LB/BK	Speed Control Switch	0v (switch on: 6.7v)
38	LG/BK	EVAP Purge Solenoid Control	0-10 Hz (0-100%)
39	VT	Module Programming Signal	0.1v
40	RD/WT	Direct Battery	12-14v
41	OR/LB	A/C Pressure Switch Signal	0v (switch on: 12v)
42	GY/OR	Passive Antitheft RX Signal	Digital Signals
43	WT/LG	Passive Antitheft TX Signal	Digital Signals
44	PK/LB	Data Bus (-) Signal	<0.050v
45	BR/WT	Reference Voltage	4.9-5.1v
46	RD	Vehicle Power (Start-Run)	12-14v

2003 Expedition 4.6L V8 MFI VIN W (A/T) C138t Pin Connector

PCM Pin #	Wire Color	Circuit Description (30 Pin)	Value at Hot Idle
1	---	Not Used	---
2	VT/LG	HO2S-22 (B2 S2) Signal	0.1-1.1v
3	RD/LG	HO2S-12 (B1 S2) Signal	0.1-1.1v
4-10	---	Not Used	
11	VT/YL	TCC Solenoid Control	0%, 55 mph: 95%
12	OR/YL	Shift Solenoid 'A' Control	1v, 55 mph: 1v
13	VT/OR	Shift Solenoid 'B' Control	12v, 55 mph: 1v
14	VT/OR	Coast Clutch Solenoid Control	12v, 55 mph: 12v
15-16	---	Not Used	---
17	LB/YL	Digital TR3A Sensor	0v, 55 mph: 11.5v
18	LB/BK	Digital TR2 Sensor	0v, 55 mph: 11.5v
19	WT/BK	Digital TR4 Sensor	0v, 55 mph: 11.5v
20	YL/BK	Digital TR1 Sensor	In 'P': 0v, in O/D: 1.7v
21	WT/BK	HO2S-12 (B1 S2) Heater	On: 1v, Off: 12v
22	---	Not Used	---
23	WT/YL	EPC Solenoid Control	9.1v (6 psi)
24	---	Not Used	
25	DB/YL	OSS Sensor Signal	0 Hz, 55 mph: 250 Hz
26	DG/WT	Turbine Shaft Speed Sensor	360 Hz, 890 Hz
27	GY/RD	Sensor Ground	<0.050v
28	OR/BK	Transmission Fluid Temperature Sensor	2.10-2.40v
29	TN/YL	HO2S-22 (B2 S2) Heater	On: 1v, Off: 12v
30	---	Not Used	---

Pin Connector Graphic

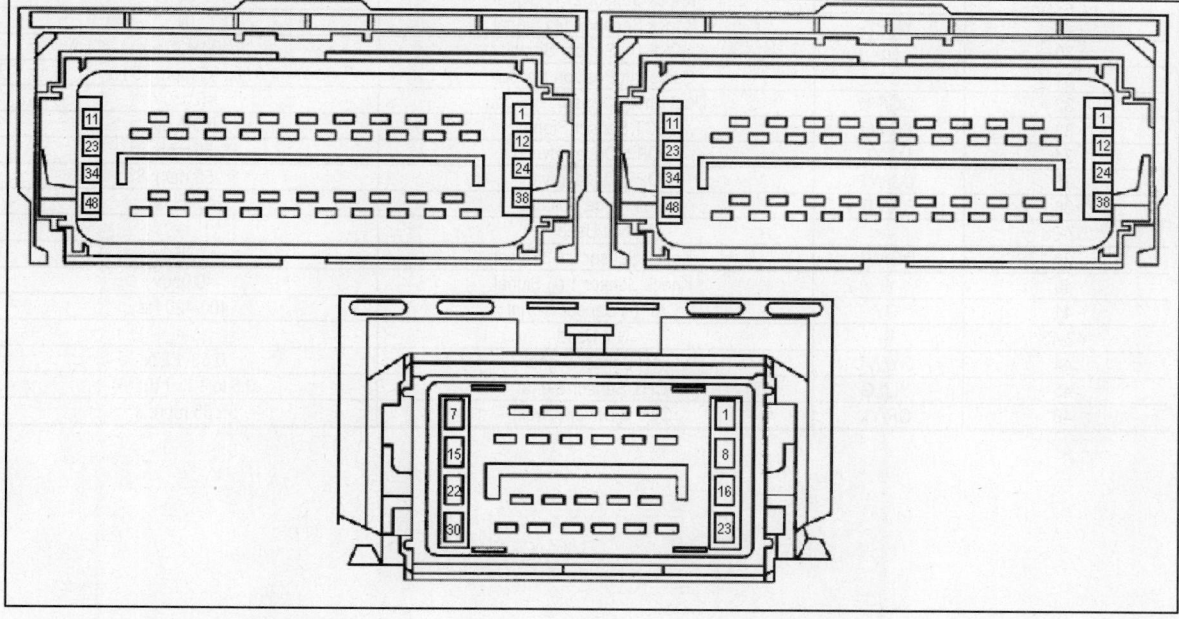

2003 Expedition 4.6L V8 MFI VIN W (A/T) C1381e Pin Connector

PCM Pin #	Wire Color	Circuit Description (46 Pin)	Value at Hot Idle
1	PK/WT	COP 2 Driver (dwell)	5°, 55 mph: 8°
2	WT/LB	IAC Solenoid Control	10.7v (33%)
3	TN/BK	Injector 5 Control	2.7-4.1 ms
4	LG/OR	Injector 6 Control	2.7-4.1 ms
5	TN/RD	Injector 7 Control	2.7-4.1 ms
6	LB	Injector 8 Control	2.7-4.1 ms
7	BR/YL	Injector 3 Control	2.7-4.1 ms
8	BR/LB	Injector 4 Control	2.7-4.1 ms
9	GY/YL	Vehicle Power (Start-Run)	12-14v
10	---	Not Used	---
11	WT/RD	COP 8 Driver (dwell)	5°, 55 mph: 8°
12	PK/LB	COP 7 Driver (dwell)	5°, 55 mph: 8°
13	BR	Intake Manifold Runner Control Module	12v
14	TN	Injector 1 Control	2.7-4.1 ms
15	WT	Injector 2 Control	2.7-4.1 ms
16-19	---	Not Used	---
20	YL/LB	HO2S-21 (B2 S1) Heater	On: 1v, Off: 12v
21	RD/WT	HO2S-11 (B1 S1) Heater	On: 1v, Off: 12v
22	BR/PK	EGR Vacuum Regulator Control	0%, 55 mph: 45%
23	DG/VT	COP 4 Driver (dwell)	5°, 55 mph: 8°
24	WT/PK	COP 3 Driver (dwell)	5°, 55 mph: 8°
25	GY/RD	Sensor Ground	<0.050v
26	GY/LB	HO2S-11 (B1 S1) Signal	0.1-1.1v
27	RD/BK	HO2S-21 (B2 S1) Signal	0.1-1.1v
28	DG/VT	Knock Sensor 2 (+) Signal	0v
29	YL/RD	Knock Sensor 1 (+) Signal	0v
30	DB	CKP (+) Sensor Signal	400-420 Hz
31	DG	Camshaft Position Sensor	6 Hz, 55 mph: 15 Hz
32	DG/PK	Dual Knock Sensor (-) Signal	<0.050v
33	BR/LG	ECT Sensor Signal	0.5-0.6v
34	LG/YL	COP 5 Driver (dwell)	5°, 55 mph: 8°
35	LG/WT	COP 1 Driver (dwell)	5°, 55 mph: 8°
36	BR/WT	Reference Voltage	4.9-5.1v
37-38	---	Not Used	---
39	DG/WT	Knock Sensor 2 (-) Signal	<0.050v
40	YL	Knock Sensor 1 (-) Signal	<0.050v
41	GY	CKP (-) Sensor Signal	400-420 Hz
42-43	---	Not Used	---
44	GY/WT	TP Sensor Signal	0.53-1.27v
45	YL/LG	CHT Sensor Signal	0.6 to 1.7v (194°F)
46	OR/YL	COP 6 Driver (dwell)	5°, 55 mph: 8°

1997 Expedition 5.4L V8 VIN L (E4OD) 104 Pin Connector

PCM Pin #	Wire Color	Circuit Description (104 Pin)	Value at Hot Idle
1	LG/YL	COP 6 Driver (dwell)	5°, 55 mph: 8°
2	PK/LG	MIL (lamp) Control	MIL On: 1v, Off: 12v
3	BK/WT	Power Ground	<0.1v
6	OR/YL	Shift Solenoid 1 Control	1v, 30 mph: 12v
9	YL/WT	Fuel Level Indicator Signal	1.7v at 50%
11	PK/OR	Shift Solenoid 2 Control	12v, 55 mph: 1v
12	WT/LG	TCIL (lamp) Control	TCIL On: 1v, Off: 12v
13	PK	Flash EEPROM Power	0.1v
14	LB/BK	4x4 Indicator Signal	4x4 On: 1v, Off: 12v
15	PK	Data Bus (-) Signal	Digital Signals
16	TN/OR	Data Bus (+) Signal	Digital Signals
19	DB/WT	Suspension Control Module	Digital Signals
20	BR/OR	Coast Clutch Solenoid	12v, 55 mph: 12v
21	DB	CKP (+) Sensor Signal	400-420 Hz
22	GY	CKP (-) Sensor Signal	400-420 Hz
24	BK/WT	Power Ground	<0.1v
25	BK	Case Ground	<0.050v
26	LG/WT	COP 1 Driver (dwell)	5°, 55 mph: 8°
27	OR/YL	COP 5 Driver (dwell)	5°, 55 mph: 8°
29	TN/WT	TCS (switch) Signal	TCS & O/D On: 12v
30	DG	Octane Adjustment Signal	9.3v (Closed: 0v)
33	PK/OR	VSS (-) Signal	0 Hz, 55 mph: 125 Hz
34	YL/BK	Digital TR1 Sensor	0v, 55 mph: 11.5v
35	RD/LG	HO2S-12 (B1 S2) Signal	0.1-1.1v
36	TN/LB	MAF Sensor Return	<0.050v
37	OR/BK	TFT Sensor Signal	2.10-2.40v
39	GY	IAT Sensor Signal	1.5-2.5v
40	DG/YL	Fuel Pump Monitor	On: 12v, Off: 0v
41	BK/YL	A/C High Pressure Switch	A/C On: 12v, Off: 0v
43	OR/LG	Overhead Trip Computer	Digital Signals
45	RD/WT	CHIL (lamp) Control	CHT On: 1v, Off: 12v
46	BR	IMTV (tuning valve) Control	1v, 55 mph: 1v
47	BR/PK	VR Solenoid Control	0%, 55 mph: 45%
48	WT/PK	Clean Tachometer Output	46 Hz
49	LB/BK	Digital TR2 Sensor	0v, 55 mph: 11.5v
50	WT/BK	Digital TR4 Sensor	0v, 55 mph: 11.5v
51	BK/WT	Power Ground	<0.1v
52	PK/WT	COP 3 Driver (dwell)	5°, 55 mph: 8°
53	PK/B	COP 4 Driver (dwell)	5°, 55 mph: 8°
54	PK/YL	TCC Solenoid Control	0%, 55 mph: 95%
55	BK/LG	Keep Alive Power	12-14v
56	LG/BK	EVAP Purge Solenoid	0-10 Hz (0-100%)
57	YL/RD	Knock Sensor Signal	0v
58	GY/BK	VSS (+) Signal	0 Hz, 55 mph: 125 Hz
60	GY/LB	HO2S-11 (B1 S1) Signal	0.1-1.1v
61	PK/LG	HO2S-22 (B2 S2) Signal	0.1-1.1v
62	RD/PK	FTP Sensor Signal	2.6v (0" H2O - cap off)
64	LB/YL	Digital TR3A Sensor	In 'P': 0v, in O/D: 1.7v

1997 Expedition 5.4L V8 VIN L (E4OD) 104 Pin Connector

PCM Pin #	Wire Color	Circuit Description (104 Pin)	Value at Hot Idle
65	BR/LG	DPFE Sensor Signal	0.20-1.30v
66	LG	CHT Sensor Signal	4.52v (68°F)
67	PK/WT	EVAP CV Solenoid	0-10 Hz (0-100%)
71	RD	Vehicle Power	12-14v
72	TN/RD	Injector 7 Control	3.2-4.0 ms
73	TN/BK	Injector 5 Control	3.2-4.0 ms
74	BR/YL	Injector 3 Control	3.2-4.0 ms
75	TN	Injector 1 Control	3.2-4.0 ms
77	BK/WT	Power Ground	<0.1v
78	WT/PK	COP 7 Driver (dwell)	5°, 55 mph: 8°
79	WT/RD	COP 8 Driver (dwell)	5°, 55 mph: 8°
80	LB/OR	Fuel Pump Control	On: 1v, Off: 12v
81	WT/YL	EPC Solenoid Control	9.1v (4 psi)
83	WT/LB	IAC Solenoid Control	10.7v (33%)
85	DG	CMP Sensor Signal	5-7 Hz
87	RD/BK	HO2S-21 (B2 S1) Signal	0.1-1.1v
88	LB/RD	MAF Sensor Signal	0.8v
89	GY/WT	TP Sensor Signal	0.53-1.27v
90	BR/WT	Reference Voltage	4.9-5.1v
91	GY/RD	Analog Signal Return	<0.050v
92	LG	Brake Pedal Switch Signal	Brake Off: 12v, On: 1v
93	RD/WT	HO2S-11 (B1 S1) Heater	On: 1v, Off: 12v
94	YL/LB	HO2S-21 (B2 S1) Heater	On: 1v, Off: 12v
95	WT/BK	HO2S-12 (B1 S2) Heater	On: 1v, Off: 12v
96	TN/YL	HO2S-22 (B2 S2) Heater	On: 1v, Off: 12v
97	RD	Vehicle Power	12-14v
98	LB	Injector 8 Control	3.2-4.0 ms
99	LG/OR	Injector 6 Control	3.2-4.0 ms
100	BR/LB	Injector 4 Control	3.2-4.0 ms
101	WT	Injector 2 Control	3.2-4.0 ms
103	BK/WT	Power Ground	<0.1v
104	DG/PK	COP 2 Driver (dwell)	5°, 55 mph: 8°

Pin Connector Graphic

PCM 104-PIN CONNECTOR

Terminal View of 104-Pin PCM Wiring Harness Connector

Standard Colors and Abbreviations

Abbreviation	Color	Abbreviation	Color	Abbreviation	Color
BK	Black	GY	Gray	PK	Purple
BL	Blue	GN	Green	RD	Red
BR	Brown	LG	LT Green	TN	Tan
DB	Dark Blue	OR	Orange	WT	White
DG	DK Green	PK	Pink	YL	Yellow

1998-99 Expedition 5.4L VIN L (E4OD) 104 Pin Connector

PCM Pin #	Wire Color	Circuit Description (104 Pin)	Value at Hot Idle
1	LG/YL	COP 6 Driver (dwell)	5°, 55 mph: 8°
2	PK/LG	MIL (lamp) Control	MIL On: 1v, Off: 12v
3, 24	BK/WT	Power Ground	<0.1v
6	OR/YL	Shift Solenoid 1 Control	1v, 30 mph: 12v
9	YL/WT	Fuel Level Indicator Signal	1.7v (1/2 full)
11	PK/OR	Shift Solenoid 2 Control	12v, 55 mph: 1v
12	WT/LG	TCIL (lamp) Control	TCIL On: 1v, Off: 12v
13	PK	Flash EEPROM Power	0.1v
14	LB/BK	4x4 Indicator Signal	4x4 Switch Off: 0.1v
15	PK	Data Bus (-) Signal	Digital Signals
16	TN/OR	Data Bus (+) Signal	Digital Signals
19	DB/WT	Suspension Control Module	Digital Signals
20	BR/OR	Coast Clutch Solenoid	12v, 55 mph: 12v
21	DB	CKP (+) Sensor Signal	430-475 Hz
22	GY	CKP (-) Sensor Signal	430-475 Hz
25	BK	Case Ground	<0.050v
26	LG/WT	COP 1 Driver (dwell)	5°, 55 mph: 8°
27	OR/YL	COP 5 Driver (dwell)	5°, 55 mph: 8°
29	TN/WT	TCS (switch) Signal	TCS & O/D On: 12v
30	DG	Octane Adjustment Signal	9.3v (Closed: 0v)
33	PK/OR	VSS (-) Signal	0 Hz, 55 mph: 125 Hz
34	YL/BK	Digital TR1 Sensor	0v, 55 mph: 11.5v
35	RD/LG	HO2S-12 (B1 S2) Signal	0.1-1.1v
36	TN/LB	MAF Sensor Return	<0.050v
37	OR/BK	TFT Sensor Signal	2.10-2.40v
39	GY	IAT Sensor Signal	1.5-2.5v
40	DG/YL	Fuel Pump Monitor	On: 12v, Off: 0v
41	BK/YL	A/C Request Signal	A/C On: 12v, Off: 0v
43	OR/LG	Overhead Trip Computer	Digital Signals
45	RD/WT	CHIL (lamp) Control	CHT On: 1v, Off: 12v
46	BR	IMTV (tuning valve) Control	12v, 55 mph: 12v
47	BR/PK	VR Solenoid Control	0%, 55 mph: 45%
48	WT/PK	Clean Tachometer Output	46 Hz
49	LB/BK	Digital TR2 Sensor	0v, 55 mph: 11.5v
50	WT/BK	Digital TR4 Sensor	0v, 55 mph: 11.5v
51	BK/WT	Power Ground	<0.1v
52	PK/WT	COP 3 Driver (dwell)	5°, 55 mph: 8°
53	PK/LB	COP 4 Driver (dwell)	5°, 55 mph: 8°
54	PK/YL	TCC Solenoid Control	0%, 55 mph: 95%
55	BK/LG	Keep Alive Power	12-14v
56	LG/BK	EVAP Purge Solenoid	0-10 Hz (0-100%)
57	YL/RD	Knock Sensor 1 Signal	0v
58	GY/BK	VSS (+) Signal	0 Hz, 55 mph: 125 Hz
60	GY/LB	HO2S-11 (B1 S1) Signal	0.1-1.1v
61	PK/LG	HO2S-22 (B2 S2) Signal	0.1-1.1v
62	RD/PK	FTP Sensor Signal	2.6v (0" H2O - cap off)
64	LB/YL	Digital TR3A Sensor	In 'P': 0v, in O/D: 1.7v

1998-99 Expedition 5.4L VIN L (E4OD) 104 Pin Connector

PCM Pin #	Wire Color	Circuit Description (104 Pin)	Value at Hot Idle
65	BR/LG	DPFE Sensor Signal	0.2-1.30v
66	YL/LG	CHT Sensor Signal	0.6v (194°F)
67	PK/WT	EVAP CV Solenoid	0-10 Hz (0-100%)
71	RD	Vehicle Power	12-14v
72	TN/RD	Injector 7 Control	2.7-3.5 ms
73	TN/BK	Injector 5 Control	2.7-3.5 ms
74	BR/YL	Injector 3 Control	2.7-3.5 ms
75	TN	Injector 1 Control	2.7-3.5 ms
77	BK/WT	Power Ground	<0.1v
78	WT/PK	COP 7 Driver (dwell)	5°, 55 mph: 8°
79	WT/RD	COP 8 Driver (dwell)	5°, 55 mph: 8°
80	LB/OR	Fuel Pump Control	On: 1v, Off: 12v
81	WT/YL	EPC Solenoid Control	9.1v (4 psi)
83	WT/LB	IAC Solenoid Control	10.7v (33%)
85	DG	CMP Sensor Signal	6 Hz, 55 mph: 13-17 Hz
87	RD/BK	HO2S-21 (B2 S1) Signal	0.1-1.1v
88	LB/RD	MAF Sensor Signal	0.8v, 55 mph: 1.9v
89	GY/WT	TP Sensor Signal	0.53-1.27v
90	BR/WT	Reference Voltage	4.9-5.1v
91	GY/RD	Analog Signal Return	<0.050v
92	LG	Brake Pedal Switch Signal	Brake Off: 12v, On: 1v
93	RD/WT	HO2S-11 (B1 S1) Heater	On: 1v, Off: 12v
94	YL/LB	HO2S-21 (B2 S1) Heater	On: 1v, Off: 12v
95	WT/BK	HO2S-12 (B1 S2) Heater	On: 1v, Off: 12v
96	TN/YL	HO2S-22 (B2 S2) Heater	On: 1v, Off: 12v
97	RD	Vehicle Power	12-14v
98	LB	Injector 8 Control	2.7-3.5 ms
99	LG/OR	Injector 6 Control	2.7-3.5 ms
100	BR/LB	Injector 4 Control	2.7-3.5 ms
101	WT	Injector 2 Control	2.7-3.5 ms
103	BK/WT	Power Ground	<0.1v
104	DG/PK	COP 2 Driver (dwell)	5°, 55 mph: 8°

Pin Connector Graphic

Standard Colors and Abbreviations

Abbreviation	Color	Abbreviation	Color	Abbreviation	Color
BK	Black	GY	Gray	PK	Purple
BL	Blue	GN	Green	RD	Red
BR	Brown	LG	LT Green	TN	Tan
DB	Dark Blue	OR	Orange	WT	White
DG	DK Green	PK	Pink	YL	Yellow

2000-02 Expedition 5.4L V8 VIN L (E4OD) 104 Pin Connector

PCM Pin #	Wire Color	Circuit Description (104 Pin)	Value at Hot Idle
1	OR/YL	COP 6 Driver (dwell)	5°, 55 mph: 8°
3	BK/WT	Power Ground	<0.1v
6	OR/YL	Shift Solenoid 1 Control	1v, 30 mph: 12v
11	PK/OR	Shift Solenoid 2 Control	12v, 55 mph: 1v
12	WT/LG	TCIL (lamp) Control	TCIL On: 1v, Off: 12v
13	PK	Flash EEPROM Power	0.1v
14	LB/BK	4x4 Low Indicator Switch	4x4 Switch Off: 0.1v
15	PK/LB	Data Bus (-) Signal	Digital Signals
16	TN/OR	Data Bus (+) Signal	Digital Signals
19	DB/WT	Suspension Control Module	Digital Signals
20	BR/OR	Coast Clutch Solenoid	12v, 55 mph: 12v
21	DB	CKP (+) Sensor Signal	430-475 Hz
22	GY	CKP (-) Sensor Signal	430-475 Hz
25	LB/YL	Case Ground	<0.050v
26	LG/WT	COP 1 Driver (dwell)	5°, 55 mph: 8°
27	LG/YL	COP 5 Driver (dwell)	5°, 55 mph: 8°
29	TN/WT	TCS (switch) Signal	TCS & O/D On: 12v
32	DG/PK	Knock Sensor 1 Return	<0.050v
34	YL/BK	Digital TR1 Sensor	0v, 55 mph: 11.5v
35	RD/LG	HO2S-12 (B1 S2) Signal	0.1-1.1v
36	TN/LB	MAF Sensor Return	<0.050v
37	OR/BK	TFT Sensor Signal	2.10-2.40v
38	LG/RD	ECT Sensor Signal	0.5-0.6v
39	GY	IAT Sensor Signal	1.5-2.5v
40	DG/YL	Fuel Pump Monitor	On: 12v, Off: 0v
41	BK/YL	A/C Request Signal	A/C On: 12v, Off: 0v
42	PK/OR	TP Sensor Signal	0.9v, 55 mph: 1.3v
43	OR/LG	Data Output Link	5v
46	BR	IMRC (valve) Control	12v
47	BR/PK	VR Solenoid Control	0%, 55 mph: 45%
49	LB/BK	Digital TR2 Sensor	0v, 55 mph: 11.5v
50	WT/BK	Digital TR4 Sensor	0v, 55 mph: 11.5v
51	BK/WT	Power Ground	<0.1v
52	WT/PK	COP 3 Driver (dwell)	5°, 55 mph: 8°
53	DG/PK	COP 4 Driver (dwell)	5°, 55 mph: 8°
54	PK/YL	TCC Solenoid Control	0%, 55 mph: 95%
55	RD/WT	Keep Alive Power	12-14v
56	LG/BK	EVAP Purge Solenoid	0-10 Hz (0-100%)
57	YL/RD	Knock Sensor 1 Signal	0v
59	DG/WT	TSS Sensor Signal	100 Hz, 30 mph: 700 Hz
60	GY/LB	HO2S-11 (B1 S1) Signal	0.1-1.1v
61	PK/LG	HO2S-22 (B2 S2) Signal	0.1-1.1v
62	RD/PK	FTP Sensor Signal	2.6v (0" H2O - cap off)
64	LB/YL	Digital TR3A Sensor	In 'P': 0v, in O/D: 1.7v

2000-02 Expedition 5.4L V8 VIN L (E4OD) 104 Pin Connector

PCM Pin #	Wire Color	Circuit Description (104 Pin)	Value at Hot Idle
65 ('00)	BR/LG	DPFE Sensor Signal	0.2-1.30v
65 ('01-'02)	BR/LG	DPFE Sensor Signal	0.95-1.05v
66	YL/LG	CHT Sensor Signal	0.6v (194°F)
67	PK/WT	EVAP CV Solenoid	0-10 Hz (0-100%)
68	GY/BK	Vehicle Speed Sensor	0 Hz, 55 mph: 125 Hz
71	RD	Vehicle Power	12-14v
72	TN/RD	Injector 7 Control	2.7-3.5 ms
73	TN/BK	Injector 5 Control	2.7-3.5 ms
74	BR/YL	Injector 3 Control	2.7-3.5 ms
75	TN	Injector 1 Control	2.7-3.5 ms
77	BK/WT	Power Ground	<0.1v
78	PK/LB	COP 7 Driver (dwell)	5°, 55 mph: 8°
79	WT/RD	COP 8 Driver (dwell)	5°, 55 mph: 8°
80	LB/OR	Fuel Pump Control	On: 1v, Off: 12v
81	WT/YL	EPC Solenoid Control	9.1v (4 psi)
83	WT/LB	IAC Solenoid Control	10.7v (33%)
84	DB/YL	OSS Sensor Signal	0 Hz, 55 mph: 350 Hz
85	DG	CMP Sensor Signal	6 Hz, 55 mph: 13-17 Hz
87	RD/BK	HO2S-21 (B2 S1) Signal	0.1-1.1v
88	LB/RD	MAF Sensor Signal	0.8v, 55 mph: 1.9v
89	GY/WT	TP Sensor Signal	0.53-1.27v
90	BR/WT	Reference Voltage	4.9-5.1v
91	GY/RD	Analog Signal Return	<0.050v
92	RD/LG	Brake Pedal Switch Signal	Brake Off: 12v, On: 1v
93	RD/WT	HO2S-11 (B1 S1) Heater	On: 1v, Off: 12v
94	YL/LB	HO2S-21 (B2 S1) Heater	On: 1v, Off: 12v
95	WT/BK	HO2S-12 (B1 S2) Heater	On: 1v, Off: 12v
96	TN/YL	HO2S-22 (B2 S2) Heater	On: 1v, Off: 12v
97	RD	Vehicle Power	12-14v
98	LB	Injector 8 Control	2.7-3.5 ms
99	LG/OR	Injector 6 Control	2.7-3.5 ms
100	BR/LB	Injector 4 Control	2.7-3.5 ms
101	W	Injector 2 Control	2.7-3.5 ms
103	BK/WT	Power Ground	<0.1v
104	PK/WT	COP 2 Driver (dwell)	5°, 55 mph: 8°

Pin Connector Graphic

PCM 104-PIN CONNECTOR

Terminal View of 104-Pin PCM Wiring Harness Connector

Standard Colors and Abbreviations

Abbreviation	Color	Abbreviation	Color	Abbreviation	Color
BK	Black	GY	Gray	PK	Purple
BL	Blue	GN	Green	RD	Red
BR	Brown	LG	LT Green	TN	Tan
DB	Dark Blue	OR	Orange	WT	White
DG	DK Green	PK	Pink	YL	Yellow

2003 Expedition 5.4L V8 MFI VIN L (A/T) C175a Pin Connector

PCM Pin #	Wire Color	Circuit Description (46 Pin)	Value at Hot Idle
1	BK/WT	Power Ground	<0.1v
2	PK/YL	A/C WOT Relay Control	Relay Off: 12v, On: 1v
3	PK/OR	TP Sensor Signal	0.9v, 55 mph: 1.3v
4	DB/OG	Starter Relay Control Circuit	Relay Off: 12v, On: 1v
5	---	Not Used	---
6	OG/LB	Cooling Fan Speed Signal	0.5-4.9v
7	TN/LG	Generator Communicator Command	1.5-10.5v (40-250 Hz)
8	RD/PK	Fuel Tank Pressure Sensor	2.6v (0" H2O - cap off)
9	OR/RD	Antitheft Indicator Control	Indicator Off: 12v, On: 1v
10	LB/YL	Power Ground	<0.1v
11	BK/WT	Power Ground	<0.1v
12	VT/YL	Speed Control Servo Signal 'C'	DC pulse signals
13	GY/BK	VSS (+) Signal	0 Hz, 55 mph: 125 Hz
14	BR/PK	Generator Monitor Signal	130 Hz (45%)
15	BK/YL	Brake Pedal Position Switch	Brake Off: 0v, On: 12v
16	---	Not Used	---
17	DG/OR	Steering Wheel Control Ground	<0.050v
18	YL/LG	Power Steering Pressure Switch	Straight: 0v, Turned: 12v
19	GY	IAT Sensor Signal	1.5-2.5v
20	DG/YL	Fuel Pump Relay Output	Off: 0v, On: 12v
21	---	Not Used	---
22	TN/WT	TCS (switch) Signal	TCS & O/D On: 12v
23	BK/WT	Power Ground	<0.1v
24	PK/BK	Speed Control Servo Signal 'B'	DC pulse signals
25	LG/VT	Cooling Fan Control	Digital Signal: 0-12-0v
26	TN	Speed Control Servo Common	<0.050v
27	LB/OR	Fuel Pump Relay Control	Relay Off: 12v, On: 1v
28-29	---	Not Used	---
30	TN/LB	MAF Sensor Return	<0.050v
31	LB/RD	MAF Sensor Signal	0.6v, 55 mph: 2.1v
32	TN/OR	SCP Data Bus (+) Signal	Digital Signals
33	GY/RD	Sensor Ground	<0.050v
34	RD	Vehicle Power (Start-Run)	12-14v
35	DG/WT	Speed Control Servo Signal 'A'	DC pulse signals
36	VT/WT	EVAP Canister Vent Solenoid Control	0-10 Hz (0-100%)
37	LB/BK	Speed Control Switch	N/A
38	LG/BK	EVAP Purge Solenoid Control	0-10 Hz (0-100%)
39	VT	Module Programming Signal	0.1v
40	RD/WT	Direct Battery	12-14v
41	OR/LB	A/C Pressure Switch Signal	0v (switch on: 12v)
42	GY/OR	Passive Antitheft RX Signal	Digital Signals
43	WT/LG	Passive Antitheft TX Signal	Digital Signals
44	PK/LB	Data Bus (-) Signal	<0.050v
45	BR/WT	Reference Voltage	4.9-5.1v
46	RD	Vehicle Power (Start-Run)	12-14v

2003 Expedition 5.4L V8 MFI VIN L (A/T) C138t Pin Connector

PCM Pin #	Wire Color	Circuit Description (30 Pin)	Value at Hot Idle
1	---	Not Used	---
2	VT/LG	HO2S-22 (B2 S2) Signal	0.1-1.1v
3	RD/LG	HO2S-12 (B1 S2) Signal	0.1-1.1v
4-10	---	Not Used	---
11	VT/YL	TCC Solenoid Control	0%, 55 mph: 95%
12	OR/YL	Shift Solenoid 'A' Control	1v, 55 mph: 1v
13	VT/OR	Shift Solenoid 'B' Control	12v, 55 mph: 1v
14	BR/OR	Coast Clutch Solenoid Control	12v, 55 mph: 12v
15-16	---	Not Used	---
17	LB/YL	Digital TR3A Sensor	0v, 55 mph: 11.5v
18	LB/BK	Digital TR2 Sensor	0v, 55 mph: 11.5v
19	WT/BK	Digital TR4 Sensor	0v, 55 mph: 11.5v
20	YL/BK	Digital TR1 Sensor	In 'P': 0v, in O/D: 1.7v
21	WT/BK	HO2S-12 (B1 S2) Heater	On: 1v, Off: 12v
22	---	Not Used	---
23	WT/YL	EPC Solenoid Control	9.1v (6 psi)
24	---	Not Used	---
25	DB/YL	OSS Sensor Signal	0 Hz, 55 mph: 250 Hz
26	DG/WT	Turbine Shaft Speed Sensor	360 Hz, 890 Hz
27	GY/RD	Sensor Ground	<0.050v
28	OR/BK	Transmission Fluid Temperature Sensor	2.10-2.40v
29	TN/YL	HO2S-22 (B2 S2) Heater	On: 1v, Off: 12v
30	---	Not Used	---

Pin Connector Graphic

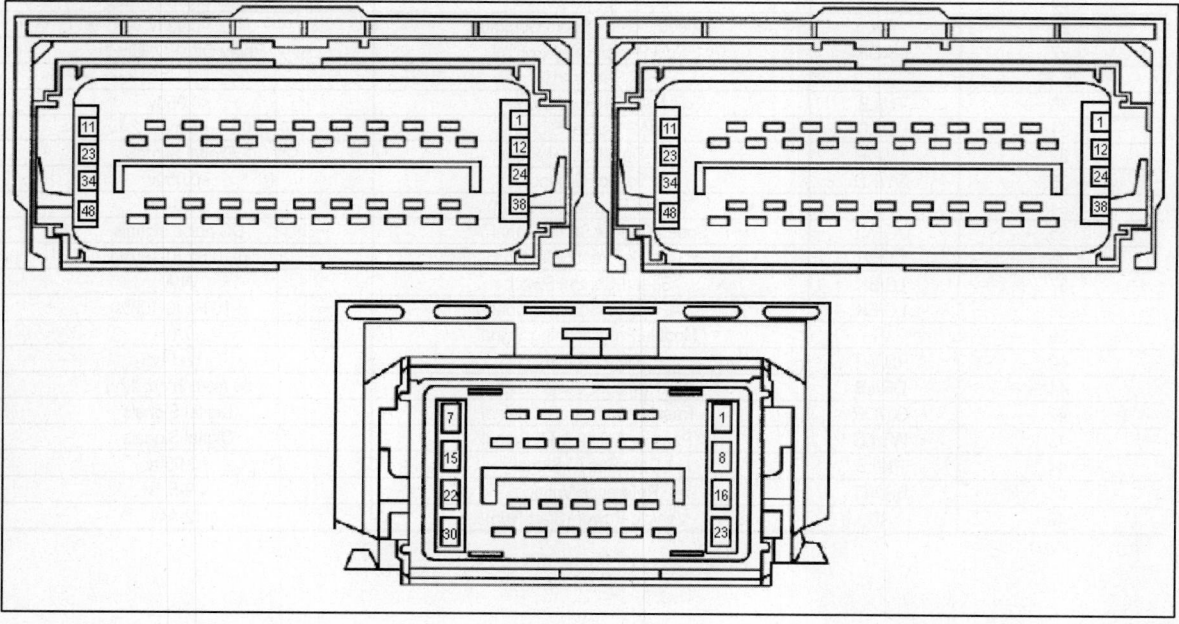

2003 Expedition 5.4L V8 MFI VIN L (A/T) C1381e Pin Connector

PCM Pin #	Wire Color	Circuit Description (46 Pin)	Value at Hot Idle
1	PK/WT	COP 2 Driver (dwell)	5°, 55 mph: 8°
2	WT/LB	IAC Solenoid Control	10.7v (33%)
3	TN/BK	Injector 5 Control	2.7-4.1 ms
4	LG/OR	Injector 6 Control	2.7-4.1 ms
5	TN/RD	Injector 7 Control	2.7-4.1 ms
6	LB	Injector 8 Control	2.7-4.1 ms
7	BR/YL	Injector 3 Control	2.7-4.1 ms
8	BR/LB	Injector 4 Control	2.7-4.1 ms
9	GY/YL	Vehicle Power (Start-Run)	12-14v
10	---	Not Used	---
11	WT/RD	COP 8 Driver (dwell)	5°, 55 mph: 8°
12	PK/LB	COP 7 Driver (dwell)	5°, 55 mph: 8°
13	BR	Intake Manifold Runner Control Module	12v
14	TN	Injector 1 Control	2.7-4.1 ms
15	WT	Injector 2 Control	2.7-4.1 ms
16-19	---	Not Used	---
20	YL/LB	HO2S-21 (B2 S1) Heater	On: 1v, Off: 12v
21	RD/WT	HO2S-11 (B1 S1) Heater	On: 1v, Off: 12v
22	BR/PK	EGR Vacuum Regulator Control	0%, 55 mph: 45%
23	DG/VT	COP 4 Driver (dwell)	5°, 55 mph: 8°
24	WT/PK	COP 3 Driver (dwell)	5°, 55 mph: 8°
25	GY/RD	Sensor Ground	<0.050v
26	GY/LB	HO2S-11 (B1 S1) Signal	0.1-1.1v
27	RD/BK	HO2S-21 (B2 S1) Signal	0.1-1.1v
28	DG/VT	Knock Sensor 2 (+) Signal	0v
29	YL/RD	Knock Sensor 1 (+) Signal	0v
30	DB	CKP (+) Sensor Signal	400-420 Hz
31	DG	Camshaft Position Sensor	6 Hz, 55 mph: 15 Hz
32	DG/PK	Dual Knock Sensor (-) Signal	<0.050v
33	BR/LG	ECT Sensor Signal	0.5-0.6v
34	LG/YL	COP 5 Driver (dwell)	5°, 55 mph: 8°
35	LG/WT	COP 1 Driver (dwell)	5°, 55 mph: 8°
36	BR/WT	Reference Voltage	4.9-5.1v
37-38	---	Not Used	---
39	DG/WT	Knock Sensor 2 (-) Signal	<0.050v
40	YL	Knock Sensor 1 (-) Signal	<0.050v
41	GY	CKP (-) Sensor Signal	400-420 Hz
42-43	---	Not Used	---
44	GY/WT	TP Sensor Signal	0.53-1.27v
45	YL/LG	CHT Sensor Signal	0.6 to 1.7v (194°F)
46	OR/YL	COP 6 Driver (dwell)	5°, 55 mph: 8°

Explorer PIN Tables

1997 Explorer 4.0L V6 MFI VIN E (A/T) 104 Pin Connector

PCM Pin #	Wire Color	Circuit Description (104 Pin)	Value at Hot Idle
1	PK/OR	Shift Solenoid 2 Control	12v, 55 mph: 12v
2	PK/LG	MIL (lamp) Control	MIL On: 1v, Off: 12v
3	YL/BK	Digital TR1 Sensor	In 'P': 0v, 55 mph: 11.5v
6	DB/YL	OSS Sensor Signal	0 Hz, 55 mph: 350 Hz
12	YL/WT	Fuel Level Indicator Signal	1.7v (1/2 full)
13	PK	Flash EPROM Power	0.1v
14	LB/BK	4x4 Low Switch Signal	12v (Switch On: 0v)
15	PK	Data Bus (-) Signal	Digital Signals
16	TN/OR	Data Bus (+) Signal	Digital Signals
19	OR/BK	Accelerator Ride Control	0.1v (Off)
21	DB	CKP (+) Sensor Signal	430-460 Hz
22	GY	CKP (-) Sensor Signal	430-460 Hz
24	BK/WT	Power Ground	<0.1v
25	BK	Case Ground	<0.050v
26	YL/BK	Coil 1 Driver (dwell)	5°, 55 mph: 8°
27	OR/YL	Shift Solenoid 1 Control	1v, 55 mph: 12v
28	BR/OR	Coast Clutch Solenoid	12v, 55 mph: 12v
29	TN/WT	TCS (switch) Signal	TCS & O/D On: 12v
30	DG	Octane Adjustment	9.3v (Closed: 0v)
33	PK/OR	VSS (-) Signal	0 Hz, 55 mph: 125 Hz
35	RD/LG	HO2S-12 (B1 S2) Signal	0.1-1.1v
36	TN/LB	MAF Sensor Return	<0.050v
37	OR/BK	TFT Sensor Signal	2.10-2.40v
38	LG/RD	ECT Sensor Signal	0.5-0.6v
39	DG	IAT Sensor Signal	1.5-2.5v
40	DG/YL	Fuel Pump Monitor	On: 12v, Off: 0v
41	DG/OR	A/C Switch Signal	A/C On: 12v, Off: 0v
43	LB/PK	Neutral Tow Connector Signal	N/A
46	BR	Inlet Air Control Valve	0.9v (15%)
47	BK/PK	VR Solenoid Control	0%, 55 mph: 45%
48	TN/YL	Clean Tachometer Output	39-45 Hz
49	LB/BK	Digital TR2 Sensor	0v, 55 mph: 11.5v
50	YL/LG	Digital TR4 Sensor	0v, 55 mph: 11.5v
51	BK/WT	Power Ground	<0.1v
52	YL/RD	Coil 2 Driver (dwell)	5°, 55 mph: 8°
53	PK/BK	Shift Solenoid 3 Control	12v, 55 mph: 1v
54	PK/YL	TCC Solenoid Control	0%, 55 mph: 95%
55	Y	Keep Alive Power	12-14v
56	LG/BK	EVAP Purge Solenoid	0-10 Hz (0-100%)
57	YL/RD	Knock Sensor 1 Signal	0v
58	GY/BK	VSS (+) Signal	0 Hz, 55 mph: 125 Hz
60	GY/LB	HO2S-11 (B1 S1) Signal	0.1-1.1v
61	PK/LG	HO2S-22 (B2 S2) Signal	0.1-1.1v
62	RD/PK	FTP Sensor Signal	2.6v (0" H2O - cap off)
64	LB/YL	Digital TR3A Sensor	In 'P': 0v, 30 mph: 1.7v

1997 Explorer 4.0L V6 MFI VIN E (A/T) 104 Pin Connector

PCM Pin #	Wire Color	Circuit Description (104 Pin)	Value at Hot Idle
65	BR/LG	DPFE Sensor Signal	0.20-1.30v
67	PK/WT	EVAP CV Solenoid	0-10 Hz (0-100%)
69	PK/YL	A/C WOT Relay Control	On: 1v, Off: 12v
71	RD	Vehicle Power	12-14v
73	TN/BK	Injector 5 Control	3.6-7.5 ms
74	BR/YL	Injector 3 Control	3.6-7.5 ms
75	TN	Injector 1 Control	3.6-7.5 ms
76, 77	BK/WT	Power Ground	<0.1v
78	YL/WT	Coil 3 Driver (dwell)	5°, 55 mph: 8°
79	WT/LG	TCIL (lamp) Control	TCIL On: 1v, Off: 12v
80	LB/OR	Fuel Pump Control	On: 1v, Off: 12v
81	WT/YL	EPC Solenoid Control	10v (26 psi)
83	WT/LB	IAC Solenoid Control	10.8v (33%)
84	DG/WT	TSS (+) Sensor Signal	100-125 Hz
85	DB/OR	CMP Sensor Signal	7 Hz, 55 mph: 18 Hz
87	RD/BK	HO2S-21 (B2 S1) Signal	0.1-1.1v
88	LB/RD	MAF Sensor Signal	0.8v, 55 mph: 2.1v
89	GY/WT	TP Sensor Signal	0.53-1.27v
90	BR/WT	Reference Voltage	4.9-5.1v
91	GY/RD	Analog Signal Return	<0.050v
92	LG	Brake Position Switch	Brake Off: 12v, On: 1v
93	RD/WT	HO2S-11 (B1 S1) Heater	On: 1v, Off: 12v
94	YL/LB	HO2S-21 (B2 S1) Heater	On: 1v, Off: 12v
95	WT/BK	HO2S-12 (B1 S2) Heater	On: 1v, Off: 12v
96	TN/YL	HO2S-22 (B2 S2) Heater	On: 1v, Off: 12v
97	RD	Vehicle Power	12-14v
99	LG/OR	Injector 6 Control	3.6-7.5 ms
100	BR/LB	Injector 4 Control	3.6-7.5 ms
101	WT	Injector 2 Control	3.6-7.5 ms
103	BK/WT	Power Ground	<0.1v

Pin Connector Graphic

PCM 104-PIN CONNECTOR

1 ● ● ● ● ● ● ● ● ● ● ● ● ●   ● ● ● ● ● ● ● ● ● ● ● ● ● 26
27 ● ● ● ● ● ● ● ● ● ● ● ● ●   ● ● ● ● ● ● ● ● ● ● ● ● ● 52
53 ● ● ● ● ● ● ● ● ● ● ● ● ●   ● ● ● ● ● ● ● ● ● ● ● ● ● 78
79 ● ● ● ● ● ● ● ● ● ● ● ● ●   ● ● ● ● ● ● ● ● ● ● ● ● ● 104

Terminal View of 104-Pin PCM Wiring Harness Connector

Standard Colors and Abbreviations

Abbreviation	Color	Abbreviation	Color	Abbreviation	Color
BK	Black	GY	Gray	PK	Purple
BL	Blue	GN	Green	RD	Red
BR	Brown	LG	LT Green	TN	Tan
DB	Dark Blue	OR	Orange	WT	White
DG	DK Green	PK	Pink	YL	Yellow

1998-99 Explorer 4.0L V6 VIN E 104 Pin Connector

PCM Pin #	Wire Color	Circuit Description (104 Pin)	Value at Hot Idle
1	PK/OR	Shift Solenoid 'B' Control	12v, 55 mph: 12v
2	PK/LG	MIL (lamp) Control	MIL On: 1v, Off: 12v
3	YL/BK	Digital TR1 Sensor	In 'P': 0v, 55 mph: 11.5v
4-5, 7-11	---	Not Used	---
6	DB/YL	OSS Sensor Signal	0 Hz, 55 mph: 350 Hz
12	YL/WT	Fuel Level Indicator Signal	1.7v (1/2 full)
13	VT	Flash EPROM Power	0.1v
14	LB/BK	4x4 Indicator Signal	4x4 On: 1v, Off: 12v
15	PK/LB	Data Bus (-) Signal	Digital Signals
16	TN/OR	Data Bus (+) Signal	Digital Signals
17-20	---	Not Used	---
19	OR/BK	Accelerator Ride Control	0.1v (Off)
21	DB	CKP (+) Sensor Signal	430-460 Hz
22	GY	CKP (-) Sensor Signal	<0.050v
24, 51	BK/WT	Power Ground	<0.1v
25	BK	Case Ground	<0.050v
26	TN/WT	Coil 1 Driver (dwell)	5°, 55 mph: 8°
27	OR/YL	Shift Solenoid 'A' Control	1v, 55 mph: 12v
28	BR/OR	Coast Clutch Solenoid	12v, 55 mph: 12v
28 (2000-01)	BR/OR	Shift Solenoid 'D' Control	12v, 55 mph: 12v
29	TN/WT	TCS (switch) Signal	TCS & O/D On: 12v
30 ('98-'99)	DG	Octane Adjustment	9.3v (Closed: 0v)
31-32, 34	---	Not Used	---
32 ('98-'99)	DG/VT	Knock Sensor 1 Ground	<0.050v
33 ('98-'99)	PK/OR	VSS (-) Signal	0 Hz, 55 mph: 125 Hz
35	RD/LG	HO2S-12 (B1 S2) Signal	0.1-1.1v
36	TN/LB	MAF Sensor Return	<0.050v
37	OR/BK	TFT Sensor Signal	2.10-2.40v
38	LG/RD	ECT Sensor Signal	0.5-0.6v
39	GY	IAT Sensor Signal	1.5-2.5v
40	DG/YL	Fuel Pump Monitor	On: 12v, Off: 0v
41	PK	A/C Switch Signal	A/C On: 12v, Off: 0v
42	---	Not Used	---
43 ('98-'99)	LB/PK	Neutral Tow Connector Signal	N/A
44	---	Not Used	---
45	YL/WT	PCM Signal to the GEM	Digital Signals
46 ('98-'99)	BR	Inlet Air Control Valve	0.9v (15%)
47	BRD/PK	VR Solenoid Control	0%, 55 mph: 45%
48	TN/YL	Clean Tachometer Output	39-45 Hz
49	LB/BK	Digital TR2 Sensor	0v, 55 mph: 11.5v
50	WT/BK	Digital TR4 Sensor	0v, 55 mph: 11.5v
52	TN/OR	Coil 2 Driver (Control)	5°, 55 mph: 8°
53	P/BK	Shift Solenoid 'C' Control	12v, 55 mph: 1v
54	PK/YL	TCC Solenoid Control	0%, 55 mph: 95%
55	YL	Keep Alive Power	12-14v
56	LG/BK	EVAP Purge Solenoid	0-10 Hz (0-100%)
57	---	Not Used	---
58	GY/BK	VSS (+) Signal	0 Hz, 55 mph: 125 Hz
59	---	Not Used	---
60	GY/LB	HO2S-11 (B1 S1) Signal	0.1-1.1v
61	---	Not Used	---
62	RD/PK	FTP Sensor Signal	2.6v (0" H2O - cap off)
64	LB/YL	Digital TR3 Sensor	In 'P': 0v, 30 mph: 1.7v

1998-99 Explorer 4.0L V6 VIN E 104 Pin Connector

PCM Pin #	Wire Color	Circuit Description (104 Pin)	Value at Hot Idle
65	BR/LG	DPFE Sensor Signal	0.20-1.30v
66	---	Not Used	---
67	PK/WT	EVAP CV Solenoid	0-10 Hz (0-100%)
68	---	Not Used	---
69	PK/YL	A/C WOT Relay Control	On: 1v, Off: 12v
71	RD	Vehicle Power	12-14v
70, 72	---	Not Used	---
73	TN/BK	Injector 5 Control	3.6-7.5 ms
74	BR/YL	Injector 3 Control	3.6-7.5 ms
75	TN	Injector 1 Control	3.6-7.5 ms
76, 77	BK/WT	Power Ground	<0.1v
78	TN/LG	Coil 3 Driver (dwell)	5º, 55 mph: 8º
79	WT/LG	TCIL (lamp) Control	TCIL On: 1v, Off: 12v
80	LB/OR	Fuel Pump Control	On: 1v, Off: 12v
81	WT/YL	EPC Solenoid Control	10v (26 psi)
82	---	Not Used	---
83	WT/LB	IAC Solenoid Control	10.8v (33%)
84	DG/WT	TSS (+) Sensor Signal	100-125 Hz
85	DB/OR	CMP Sensor Signal	7 Hz, 55 mph: 18 Hz
86	BK/YL	A/C Pressure Switch Signal	Switch Open: 12v
87	RD/BK	HO2S-21 (B2 S1) Signal	0.1-1.1v
88	LB/RD	MAF Sensor Signal	0.8v, 55 mph: 2.1v
89	GY/WT	TP Sensor Signal	0.53-1.27v
90	BR/WT	Reference Voltage	4.9-5.1v
91	GY/RD	Sensor Ground	<0.050v
92	RD/LG	Brake Position Switch	Brake Off: 12v, On: 1v
93	RD/WT	HO2S-11 (B1 S1) Heater	On: 1v, Off: 12v
94	YL/LB	HO2S-21 (B2 S1) Heater	On: 1v, Off: 12v
95	WT/BK	HO2S-12 (B1 S2) Heater	On: 1v, Off: 12v
96, 98	---	Not Used	---
97	RD	Vehicle Power	12-14v
99	LG/OR	Injector 6 Control	3.6-7.5 ms
100	BR/LB	Injector 4 Control	3.6-7.5 ms
101	WT	Injector 2 Control	3.6-7.5 ms
102	---	Not Used	---
103	BK/WT	Power Ground	<0.1v
104	---	Not Used	---

Pin Connector Graphic

Terminal View of 104-Pin PCM Wiring Harness Connector

Standard Colors and Abbreviations

Abbreviation	Color	Abbreviation	Color	Abbreviation	Color
BK	Black	GY	Gray	PK	Purple
BL	Blue	GN	Green	RD	Red
BR	Brown	LG	LT Green	TN	Tan
DB	Dark Blue	OR	Orange	WT	White
DG	DK Green	PK	Pink	YL	Yellow

2000-02 Explorer 4.0L V6 SOHC MFI VIN E 104 Pin Connector

PCM Pin #	Wire Color	Circuit Description (104 Pin)	Value at Hot Idle
1	PK/OR	Shift Solenoid 'B' Control	12v, 55 mph: 12v
2	PK/LG	MIL (lamp) Control	MIL On: 1v, Off: 12v
3	YL/BK	Digital TR1 Sensor	In 'P': 0v, 55 mph: 11.5v
4-5	---	Not Used	---
6	DB/YL	OSS Sensor Signal	0 Hz, 55 mph: 350 Hz
7-11	---	Not Used	---
12	YL/WT	Fuel Level Indicator Signal	1.7v (1/2 full)
13	PK	Flash EPROM Power	0.1v
14	LB/BK	4x4 Low Indicator Switch	Switch On: 1v, Off: 12v
15	PK/LB	Data Bus (-) Signal	Digital Signals
16	TN/OR	Data Bus (+) Signal	Digital Signals
17-18	---	Not Used	---
19	OR/BK	Load Leveling Acceleration	Digital Signals
20	---	Not Used	---
21	DB	CKP (+) Sensor Signal	430-460 Hz
22	GY	CKP (-) Sensor Signal	<0.050v
23	---	Not Used	---
24	BK/WT	Power Ground	<0.1v
25	BK	Chassis Ground	<0.050v
26	TN/WT	Coil 1 Driver (dwell)	5°, 55 mph: 8°
27	OR/YL	Shift Solenoid 'A' Control	1v, 55 mph: 12v
28	BR/OR	Shift Solenoid 'D' Control	12v, 55 mph: 12v
29	TN/WT	TCS (switch) Signal	TCS & O/D On: 12v
30	---	Not Used	---
32	DG/VT	Knock Sensor (-) Return	<0.050v
33-34	---	Not Used	---
35	RD/LG	HO2S-12 (B1 S2) Signal	0.1-1.1v
36	TN/LB	MAF Sensor Return	<0.050v
37	OR/BK	TFT Sensor Signal	2.10-2.40v
38	LG/RD	ECT Sensor Signal	0.5-0.6v
39	GY	IAT Sensor Signal	1.5-2.5v
40	DG/YL	Fuel Pump Monitor	On: 12v, Off: 0v
41	PK	A/C High Pressure Switch	Switch On: 12v (open)
42	---	Not Used	---
43	LB/PK	Message Center Fuel Flow	Digital Signals
44	---	Not Used	---
45	YL/WT	PCM Signal to GEM	Digital Signals
46	BK/WT	Fuel Cap Indicator (Cluster)	Digital Signals
47	BR/PK	EGR VR Regulator Solenoid	0%, 55 mph: 45%
48	TN/YL	Clean Tachometer Output	39-45 Hz
49	LB/BK	Digital TR2 Sensor	0v, 55 mph: 11.5v
50	WT/BK	Digital TR4 Sensor	0v, 55 mph: 11.5v
51	BK/WT	Power Ground	<0.1v
52	TN/OR	Coil 2 Driver (Control)	5°, 55 mph: 8°
53	PK/BK	Shift Solenoid 'C' Control	12v, 55 mph: 1v
54	PK/YL	TCC Solenoid Control	0%, 55 mph: 95%
55	YL	Keep Alive Power	12-14v
56	LG/BK	EVAP Purge Solenoid	0-10 Hz (0-100%)
57	YL/RD	Knock Sensor (+) Signal	0v
58	GY/BK	VSS (+) Signal	0 Hz, 55 mph: 125 Hz
59	---	Not Used	---
60	GY/LB	HO2S-11 (B1 S1) Signal	0.1-1.1v
61	PK/LG	HO2S-22 (B2 S2) Signal	0.1-1.1v
63	---	Not Used	---
62	RD/PK	FTP Sensor Signal	2.6v (0" H2O - cap off)
64	LB/YL	Digital TR3 Sensor	In 'P': 0v, 30 mph: 1.7v

2000-02 Explorer 4.0L V6 SOHC MFI VIN E 104 Pin Connector

PCM Pin #	Wire Color	Circuit Description (104 Pin)	Value at Hot Idle
65 ('00)	BR/LG	DPFE Sensor Signal	0.20-1.30v
65	BR/LG	DPFE Sensor Signal	0.95-1.05v
66	---	Not Used	
67	VT/WT	EVAP CV Solenoid	0-10 Hz (0-100%)
69	PK/YL	A/C WOT Relay Control	On: 1v, Off: 12v
70	---	Not Used	
71	RD	Vehicle Power	12-14v
73	TN/BK	Injector 5 Control	3.6-7.5 ms
74	BR/YL	Injector 3 Control	3.6-7.5 ms
75	TN	Injector 1 Control	3.6-7.5 ms
76, 77	BK/WT	Power Ground	<0.1v
78	TN/LG	Coil 3 Driver (dwell)	5°, 55 mph: 8°
79	WT/LG	TCIL (lamp) Control	TCIL On: 1v, Off: 12v
80	LB/OR	Fuel Pump Relay Control	On: 1v, Off: 12v
81	WT/YL	EPC Solenoid Control	10v (26 psi)
82	---	Not Used	---
83	WT/LB	IAC Solenoid Control	10.8v (33%)
84	DG/WT	TSS Sensor Signal	100-125 Hz
85	DB/OR	CMP Sensor Signal	7 Hz, 55 mph: 18 Hz
86	BK/YL	A/C Pressure Sensor	Switch open: 12v
87	RD/BK	HO2S-21 (B2 S1) Signal	0.1-1.1v
88	LB/RD	MAF Sensor Signal	0.8v, 55 mph: 2.1v
89	GY/WT	TP Sensor Signal	0.53-1.27v
90	BR/WT	Reference Voltage	4.9-5.1v
91	GY/RD	Sensor Ground	<0.050v
92	RD/LG	Brake Position Switch	Brake Off: 12v, On: 1v
93	RD/WT	HO2S-11 (B1 S1) Heater	On: 1v, Off: 12v
94	YL/LB	HO2S-21 (B2 S1) Heater	On: 1v, Off: 12v
95	WT/BK	HO2S-12 (B1 S2) Heater	On: 1v, Off: 12v
96	TN/YL	HO2S-22 (B2 S2) Heater	On: 1v, Off: 12v
97	RD	Vehicle Power	12-14v
98	---	Not Used	---
99	LG/OR	Injector 6 Control	3.6-7.5 ms
100	BR/LB	Injector 4 Control	3.6-7.5 ms
101	WT	Injector 2 Control	3.6-7.5 ms
102	---	Not Used	---
103	BK/WT	Power Ground	<0.1v
104	---	Not Used	---

Pin Connector Graphic

Terminal View of 104-Pin PCM Wiring Harness Connector

Standard Colors and Abbreviations

Abbreviation	Color	Abbreviation	Color	Abbreviation	Color
BK	Black	GY	Gray	PK	Purple
BL	Blue	GN	Green	RD	Red
BR	Brown	LG	LT Green	TN	Tan
DB	Dark Blue	OR	Orange	WT	White
DG	DK Green	PK	Pink	YL	Yellow

2002-03 Explorer 4.0L V6 MFI VIN K 58 Pin Connector

PCM Pin #	Wire Color	Circuit Description (58 Pin)	Value at Hot Idle
1, 5, 15-16	---	Not Used	---
2	PK/BK	Fuel Pump Power	Pump On: 12v, Off: 0v
3	TN/OR	SCP Data Bus (+)	Digital Signals
4	PK/LB	SCP Data Bus (-)	Digital Signals
6	VT/WT	EVAP Canister Vent Valve Control	0-10 Hz (0-100%)
7	GY/BK	Vehicle Speed Sensor (+) Signal	0 Hz, 55 mph: 125 Hz
8	LG/RD	Generator Monitor Signal	130 Hz (45%)
9	PK/YL	A/C Clutch Relay Control	12v (relay on: 1v)
10	YL/WT	4WD Indicator Control	Indicator Off: 12v, On: 1v
11	WT/LG	Passive Antitheft TX Signal	Digital Signals
12	LG/BK	EVAP Purge Solenoid	0-10 Hz (0-100%)
13	VT	Module Programming Signal	0.1v
14	GY/OR	Passive Antitheft RX Signal	Digital Signals
18-19, 21	---	Not Used	---
17	GY/RD	Sensor Ground	<0.050v
20	BR/WT	Reference Voltage	4.9-5.1v
22	DB/LG	Passive Antitheft Indicator Control	Indicator Off: 12v, On: 1v
23	LB/BK	4WD Indicator Low Signal	12v (switch on: 1v)
24	BK/WT	Power Ground	<0.1v
25	BK/WT	Power Ground	<0.1v
26	BK/WT	Power Ground	<0.1v
27	BK/WT	Power Ground	<0.1v
28	BK/YL	Brake Pressure Switch	12v (Brake On: 0v)
29	OR/LB	Cruise Set Indicator Control	Indicator Off: 12v, On: 1v
30	BK/YL	A/C Clutch Relay (switched)	12v (relay on: 1v)
31	LB/RD	MAF Sensor Signal	0.7v, 55 mph: 1.8v
32	VT	Vehicle Power (Start-Run)	12-14v
33	VT	Vehicle Power (Start-Run)	12-14v
34-35	---	Not Used	---
36	OR/LB	Cruise Set Indicator Control	12v (switch set: 0v)
37	---	Not Used	---
38	TN/LB	MAF Sensor Return	<0.050v
39	OR	Starter Motor Relay Circuit	Relay Off: 12v, On: 1v
40	RD/LG	Brake Position Switch	Brake Off: 12v, On: 1v
41	TN/WT	Overdrive Cancel Switch	0v (switch on: 12v)
42, 48-49	---	Not Used	---
43	BK	Power Ground	<0.1v
44	RD/WT	Keep Alive Power (CJB fuse)	12-14v
45	BK	Speed Control Switch Ground	<0.050v
46	OR/LB	Cruise Switch Indicator Control	Indicator Off: 12v, On: 1v
47	RD/YL	A/C Pressure Switch Signal	0v (switch on: 12v)
50	VT	A/C Demand Switch	0v (A/C On: 12v)
51	GY	Intake Air Temperature Sensor	1.5-2.5v
52	RD/PK	Fuel Tank Pressure Sensor	2.6v (0" H20 - cap off)
53-55	---	Not Used	---
56	DG/OR	Speed Control Switch Ground	<0.050v
57	LB/BK	Speed Control Switch Input	0v (switch on: 6.7v)
58	LB/OR	Fuel Pump Relay Control	Relay Off: 12v, On: 1v

Pin Connector Graphic

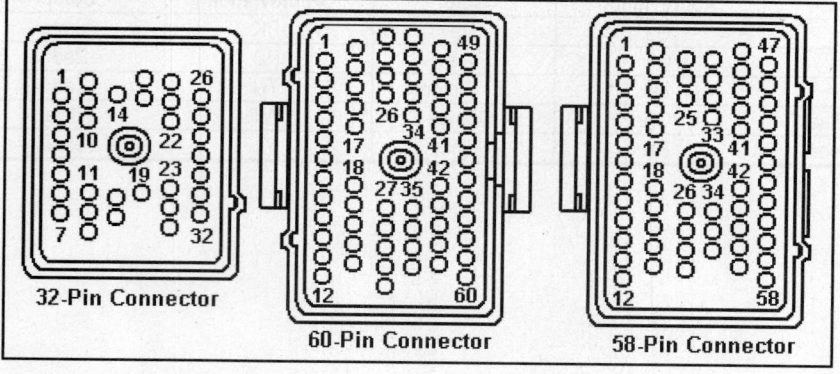

32-Pin Connector

60-Pin Connector

58-Pin Connector

2002-03 Explorer 4.0L V6 MFI VIN K 32 Pin Connector

PCM Pin #	Wire Color	Circuit Description (32 Pin)	Value at Hot Idle
1	OR/YL	Shift Solenoid 'A' Control	12v, 55 mph: 1v
2	VT/OR	Shift Solenoid 'B' Control	12v, 55 mph: 1v
3	PK/BK	Shift Solenoid 'C' Control	12v, 55 mph: 1v
4	BR/OR	Shift Solenoid 'D' Control	12v, 55 mph: 1v
5	VT/YL	TCC Solenoid Control	12v, 55 mph: 1v
6	---	Not Used	---
7	WT/YL	Electronic Pressure Control	On: 1v, Off: 12v
8	---	Not Used	---
9	LB/YL	Digital TR Sensor (TR3A)	0v, 55 mph: 1.7v
10	WT/BK	Digital TR Sensor (TR4)	0v, 55 mph: 11v
11	---	Not Used	---
12	WT	Motor Position #4	On: 1v, Off: 12v
13	LB/PK	Pressure Control 'B' Solenoid	On: 1v, Off: 12v
14	---	Not Used	---
15	WT/BK	HO2S-12 (B1 S2) Heater	1v, Off: 12v
16	TN/YL	HO2S-22 (B2 S2) Heater	1v, Off: 12v
17	GY/RD	Sensor Reference Ground	<0.050v
18	LB/BK	Digital TR Sensor (TR2)	0v, 55 mph: 11v
19-20	---	Not Used	---
21	YL/LG	Overdrive Drum Speed Input	200 Hz, 55 mph: 1185 Hz
22	YL/BK	Digital TR Sensor (TR1)	0v, 55 mph: 11v
23	OR/BK	Transmission Fluid Temperature Sensor	2.10-2.40v
24-25	---	Not Used	---
26	DB/YL	Output Shaft Speed Sensor	0 Hz, 985 Hz
27	DG/WT	Turbine Shaft Speed Sensor	360 Hz, 890 Hz
28	RD/LG	HO2S-12 (B1 S2) Signal	0.1-1.1v
29	VT/LG	HO2S-22 (B2 S2) Signal	0.1-1.1v
30-32	---	Not Used	---

Pin Connector Graphic

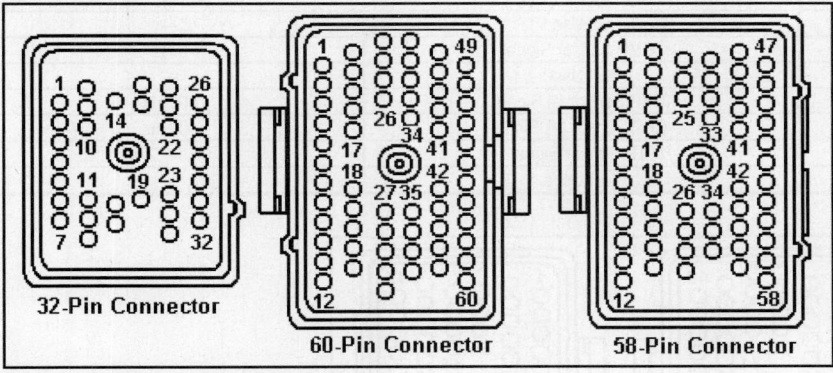

32-Pin Connector

60-Pin Connector

58-Pin Connector

Standard Colors and Abbreviations

Abbreviation	Color	Abbreviation	Color	Abbreviation	Color
BK	Black	GY	Gray	PK	Purple
BL	Blue	GN	Green	RD	Red
BR	Brown	LG	LT Green	TN	Tan
DB	Dark Blue	OR	Orange	WT	White
DG	DK Green	PK	Pink	YL	Yellow

2002-03 Explorer 4.0L V6 MFI VIN K 60 Pin Connector

PCM Pin #	Wire Color	Circuit Description (60 Pin)	Value at Hot Idle
1	---	Not Used	---
2	TN	Injector 1 Control	3.3-3.8 ms
3-6	---	Not Used	---
7	RD/WT	HO2S-11 (B1 S1) Heater	1v, Off: 12v
8	YL/LB	HO2S-21 (B2 S1) Heater	1v, Off: 12v
9	WT/LB	IAC Solenoid Control	10.7v (33%)
10	---	Not Used	---
11	TN/BK	Injector 5 Control	3.3-3.8 ms
12	---	Not Used	---
13	PK/WT	Coil 1 Driver (dwell)	5°, 55 mph: 7°
14	WT	Injector 2 Control	3.3-3.8 ms
15	---	Not Used	---
16	BR/PK	VR Solenoid Control	0%, 55 mph: 45%
17	GY/RD	Sensor Reference Ground	<0.050v
18-19	---	Not Used	---
20	BR/WT	Sensor Voltage Reference	4.9-5.1v
21	LG/OR	Injector 6 Control	3.3-3.8 ms
22	---	Not Used	---
23	RD/LB	Coil 2 Driver (dwell)	5°, 55 mph: 7°
24	BR/YL	Injector 3 Control	3.3-3.8 ms
25-30	---	Not Used	---
31	DB/LG	Coil 3 Driver (dwell)	5°, 55 mph: 7°
32	BR/LB	Injector 4 Control	3.3-3.8 ms
33-40	---	Not Used	---
41	BR/LG	DPFE Sensor Signal	0.95-1.05v
42	YL	Knock Sensor 1 Signal	0v
43	DG/WT	Knock Sensor 2 Signal	0v
44	RD/BK	HO2S-21 (B2 S1) Signal	0.1-1.1v
45	GY/LB	HO2S-11 (B1 S1) Signal	0.1-1.1v
46	LG/RD	ECT Sensor Signal	0.5-0.6v
47-50	---	Not Used	---
51	YL/RD	Knock Sensor 1 Ground	0v
52	---	Not Used	---
53	DG	CMP Sensor Signal	6 Hz
54	---	Not Used	---
55	DB	CKP Sensor (-) Signal	400 Hz
56	GY	CKP Sensor (+) Signal	400 Hz
57	GY/WT	TP Sensor Signal	0.53-1.27v
58-60	---	Not Used	---

Pin Connector Graphic

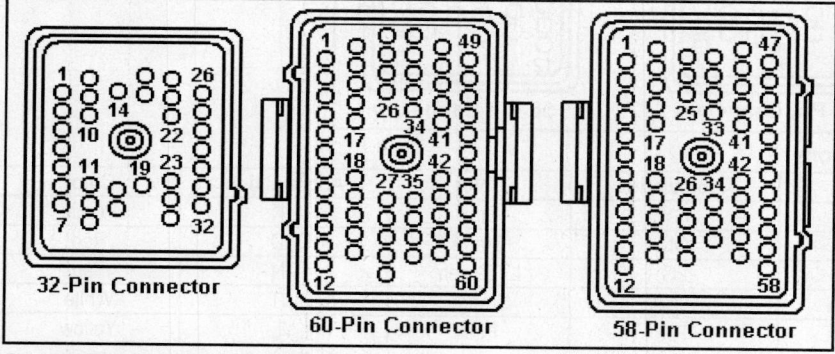

32-Pin Connector

60-Pin Connector

58-Pin Connector

Standard Colors and Abbreviations

Abbreviation	Color	Abbreviation	Color	Abbreviation	Color
BK	Black	GY	Gray	PK	Purple
BL	Blue	GN	Green	RD	Red
BR	Brown	LG	LT Green	TN	Tan
DB	Dark Blue	OR	Orange	WT	White
DG	DK Green	PK	Pink	YL	Yellow

1991-94 Explorer 4.0L V6 MFI VIN X 60 Pin Connector

PCM Pin #	Wire Color	Circuit Description (60 Pin)	Value at Hot Idle
1	YL	Keep Alive Power	12-14v
2	LG	Brake Pedal Switch	Brake Off: 12v, On: 1v
3	GY/BK	VSS (+) Signal	0 Hz, 55 mph: 125 Hz
4	TN/YL	Ignition Diagnostic Monitor	20-31 Hz
6	PK/OR	VSS (-) Signal	0 Hz, 55 mph: 125 Hz
7	LG/RD	ECT Sensor Signal	0.5-0.6v
8	DG/YL	Fuel Pump Monitor	On: 12v, Off: 0v
9 ('92-'94)	PK/BL	Data Bus (-) Signal	Digital Signals
10	DG/OR	A/C Switch Signal	A/C On: 12v, Off: 0v
14	BL/RD	MAF Sensor Signal	0.7v
15	TN/BL	MAF Sensor Return	<0.050v
16	OR/RD	Ignition System Ground	<0.050v
17	PK/LG	Self-Test Output & MIL	MIL On: 1v, Off: 12v
20	BK	PCM Case Ground	<0.050v
21	WT/BL	IAC Solenoid Control	10.7v (33%)
22	BL/OR	Fuel Pump Control	On: 1v, Off: 12v
25	GY	IAT Sensor Signal	1.5-2.5v
26	BR/WT	Reference Voltage	4.9-5.1v
28 ('92-'94)	TN/OR	Data Bus (+) Signal	Digital Signals
29	GY/BL	HO2S-11 (B1 S1) Signal	0.1-1.1v
30	BL/YL	A/T: Neutral Drive Switch	In 'P': 0v, Others: 5v
30	BL/YL	M/T: Clutch Engage Switch	Clutch In: 0v, Out: 5v
31	GY/DB	EVAP Purge Solenoid	12v, 55 mph: 1v
31	GY/YL	EVAP Purge Solenoid	12v, 55 mph: 1v
36	PK	Spark Output Signal	6.93v (50%)
37	RD	Vehicle Power	12-14v
39 ('93-'94)	RD/BK	HO2S-12 (B1 S2) Signal	0.1-1.1v
40	BK/WT	Power Ground	<0.1v
44	DG	Octane Adjustment	9.3v (Closed: 0v)
45 ('91)	LG/BK	BARO Sensor Signal	159 Hz (sea level)
46	GY/RD	Analog Signal Return	<0.050v
47	GY/WT	TP Sensor Signal	0.5-1.2v
48	WT/PK	Self-Test Input Signal	STI On: 1v, Off: 5v
49	OR	HO2S-11 (B1 S1) Ground	<0.050v
49	GY/RD	HO2S-11 (B1 S1) Ground	<0.050v
52	OR/YL	Shift Solenoid 3/4	1v, 55 mph: 12v
53	PK/YL	TCC Solenoid Control	12v, 55 mph: 1v
54	PK/YL	A/C WOT Relay Control	On: 1v, Off: 12v
55	---	Not Used	---
56	GY/OR	PIP Sensor Signal	6.93v (50%)
57	RD	Vehicle Power	12-14v
58	TN	Injector Bank 1 (INJ 1, 2, 4)	3.3-5.7 ms
59	WT	Injector Bank 2 (INJ 3, 5, 6)	3.3-5.7 ms
60	BK/LG	Power Ground	<0.1v
60	BK/WT	Power Ground	<0.1v

Pin Connector Graphic

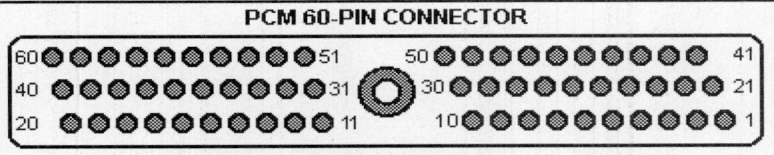

PCM 60-PIN CONNECTOR

Terminal View of 60-Pin PCM Harness Connector

Standard Colors and Abbreviations

Abbreviation	Color	Abbreviation	Color	Abbreviation	Color
BK	Black	GY	Gray	PK	Purple
BL	Blue	GN	Green	RD	Red
BR	Brown	LG	LT Green	TN	Tan
DB	Dark Blue	OR	Orange	WT	White
DG	DK Green	PK	Pink	YL	Yellow

1993-94 Explorer 4.0L V6 MFI VIN X California 60 Pin Connector

PCM Pin #	Wire Color	Circuit Description (60 Pin)	Value at Hot Idle
1	YL	Keep Alive Power	12-14v
2	LG	Brake Pedal Switch	Brake Off: 12v, On: 1v
3	GY/BK	VSS (+) Signal	0 Hz, 55 mph: 125 Hz
4	TN/YL	Ignition Diagnostic Monitor	20-31 Hz
5	---	Not Used	---
6	PK/OR	VSS (-) Signal	0 Hz, 55 mph: 125 Hz
7	LG/RD	ECT Sensor Signal	0.5-0.6v
8	DG/YL	Fuel Pump Monitor	On: 12v, Off: 0v
9	TN/BL	MAF Sensor Return	<0.050v
10	DG/OR	A/C Switch Signal	A/C On: 12v, Off: 0v
11	GY/YL	EVAP Purge Solenoid	12v, 55 mph: 1v
12	LG/OR	Injector 6 Control	3.3-5.7 ms
13-14	---	Not Used	---
15	TN/BK	Injector 5 Control	3.3-5.7 ms
16	OR/RD	Ignition System Ground	<0.050v
17	PK/LG	Self-Test Output & MIL	MIL On: 1v, Off: 12v
18	TN/OR	Data Bus (+) Signal	Digital Signals
19	PK/BL	Data Bus (-) Signal	Digital Signals
20	BK/LG	PCM Case Ground	<0.050v
21	WT/BL	IAC Solenoid Control	10.7v (33%)
22	BL/OR	Fuel Pump Control	On: 1v, Off: 12v
23	---	Not Used	---
24	DB/OR	CMP Sensor Signal	5-7 Hz
25	GY	IAT Sensor Signal	1.5-2.5v
26	BR/WT	Reference Voltage	4.9-5.1v
27	BR/LG	DPFE Sensor Signal	0.20-1.20v
28	---	Not Used	---
29	DG	Octane Adjustment	9.3v (Closed: 0v)
30	BL/YL	A/T: Neutral Drive Switch	In 'P': 0v, Others: 5v
30	BL/YL	M/T: Clutch Engage Switch	Clutch In: 0v, Out: 5v
31-32	---	Not Used	---
33	BR/PK	VR Solenoid Control	0%, 55 mph: 45%
34	---	Not Used	---
35	BR/BL	Injector 4 Control	3.3-5.7 ms
36	PK	Spark Output Signal	6.93v (50%)
37	RD	Vehicle Power	12-14v
38	---	Not Used	---
39	BR/YL	Injector 3 Control	3.3-5.7 ms
40	BK/WT	Power Ground	<0.1v

1993-94 Explorer 4.0L V6 MFI VIN X California 60 Pin Connector

PCM Pin #	Wire Color	Circuit Description (60 Pin)	Value at Hot Idle
41-42	---	Not Used	---
43	RD/BK	HO2S-21 (B2 S1) Signal	0.1-1.1v
44	GY/BL	HO2S-11 (B1 S1) Signal	0.1-1.1v
45	---	Not Used	---
46	GY/RD	Analog Signal Return	<0.050v
47	GY/WT	TP Sensor Signal	0.5-1.2v
48	WT/PK	Self-Test Input Signal	STI On: 1v, Off: 5v
49	---	Not Used	---
50	BL/RD	MAF Sensor Signal	0.7v
51	OR/YL	Shift Solenoid 3-4 Control	1v, 55 mph: 12v
52	---	Not Used	---
53	PK/YL	TCC Solenoid Control	12v, 55 mph: 1v
54	PK/YL	A/C WOT Relay Control	On: 1v, Off: 12v
55	---	Not Used	---
56	GY/OR	PIP Sensor Signal	6.93v (50%)
57	RD	Vehicle Power	12-14v
58	TN	Injector 1 Control	3.3-5.7 ms
59	WT	Injector 2 Control	3.3-5.7 ms
60	BK/WT	Power Ground	<0.1v

Pin Connector Graphic

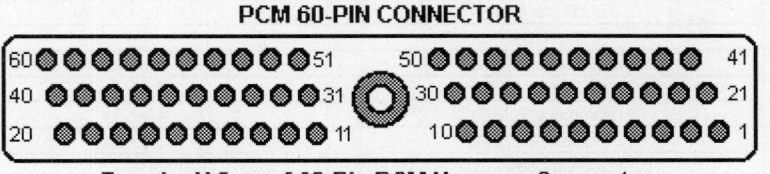

PCM 60-PIN CONNECTOR

60 ●●●●●●●●●● 51   50 ●●●●●●●●●● 41
40 ●●●●●●●●●● 31 ◎ 30 ●●●●●●●●●● 21
20 ●●●●●●●●●● 11   10 ●●●●●●●●●● 1

Terminal View of 60-Pin PCM Harness Connector

Standard Colors and Abbreviations

Abbreviation	Color	Abbreviation	Color	Abbreviation	Color
BK	Black	GY	Gray	PK	Purple
BL	Blue	GN	Green	RD	Red
BR	Brown	LG	LT Green	TN	Tan
DB	Dark Blue	OR	Orange	WT	White
DG	DK Green	PK	Pink	YL	Yellow

1995 Explorer 4.0L V6 MFI VIN X 60 Pin Connector

PCM Pin #	Wire Color	Circuit Description (60 Pin)	Value at Hot Idle
1	YL	Keep Alive Power	12-14v
2	LG	Brake Pedal Switch	Brake Off: 12v, On: 1v
3	GY/BK	VSS (+) Signal	0 Hz, 55 mph: 125 Hz
4	TN/YL	Ignition Diagnostic Monitor	20-31 Hz
5	DG/WT	TSS (+) Sensor Signal	100 Hz, 55 mph: 270 Hz
6	PK/OR	VSS (-) Signal	0 Hz, 55 mph: 125 Hz
7	LG/RD	ECT Sensor Signal	0.5-0.6v
8	DG/YL	Fuel Pump Monitor	On: 12v, Off: 0v
9	TN/BL	MAF Sensor Return	<0.050v
10	DG/OR	A/C Switch Signal	A/C On: 12v, Off: 0v
11	GY/YL	EVAP Purge Solenoid	12v, 55 mph: 1v
12	LG/OR	Injector 6 Control	3.3-5.7 ms
13	WT/LG	TCIL (lamp) Control	TCIL On: 1v, Off: 12v
14	---	Not Used	---
15	TN/BK	Injector 5 Control	3.3-5.7 ms
16	OR/RD	Ignition System Ground	<0.050v
17	PK/LG	Self-Test Output & MIL	MIL On: 1v, Off: 12v
18	TN/OR	Data Bus (+) Signal	Digital Signals
19	PK/BL	Data Bus (-) Signal	Digital Signals
20	BK	PCM Case Ground	<0.050v
21	WT/BL	IAC Solenoid Control	10.7v (33%)
22	BL/OR	Fuel Pump Control	On: 1v, Off: 12v
23	---	Not Used	---
24	DB/OR	CMP Sensor Signal	5-7 Hz
25	GY	IAT Sensor Signal	1.5-2.5v
26	BR/WT	Reference Voltage	4.9-5.1v
27	BR/LG	DPFE Sensor Signal	0.4v
28	---	Not Used	---
29	DG	Octane Adjustment	9.3v (Closed: 0v)
30	BL/YL	A/T: Neutral Drive Switch	In 'P': 0v, Others: 5v
30	BL/YL	M/T: Clutch Engage Switch	Clutch In: 0v, Out: 5v
31	BL/PK	Fuel Flow Rate Signal	Digital Signal
32	BR/OR	Coasting Clutch Solenoid	12v, 55 mph: 12v
33	BR/PK	VR Solenoid Control	0%, 55 mph: 45%
34	OR/BK	Automatic Ride Control	4.3v
35	BR/BL	Injector 4 Control	3.3-5.7 ms
36	PK	Spark Output Signal	6.93v (50%)
37	RD	Vehicle Power	12-14v
38	WT/YL	EPC Solenoid Control	10.9v (26 psi)
39	BR/YL	Injector 3 Control	3.3-5.7 ms
40	BK/WT	Power Ground	<0.1v

1995 Explorer 4.0L V6 MFI VIN X 60 Pin Connector

PCM Pin #	Wire Color	Circuit Description (60 Pin)	Value at Hot Idle
41	TN/WT	TCS (switch) Signal	TCS & O/D On: 12v
42	BL/BK	Low Range Indicator Signal	N/A
43	RD/BK	HO2S-22 (B2 S2) Signal	0.1-1.1v
44	GY/BL	HO2S-11 (B1 S1) Signal	0.1-1.1v
45	---	Not Used	---
46	GY/RD	Analog Signal Return	<0.050v
47	GY/WT	TP Sensor Signal	0.5-1.2v
48	WT/PK	Self-Test Input Signal	STI On: 1v, Off: 5v
49	OR/BK	TFT Sensor Signal	2.10-2.40v
50	BL/RD	MAF Sensor Signal	0.7v
51	OR/YL	Shift Solenoid 1 Control	1v, 30 mph: 12v
52	PK/OR	Shift Solenoid 2 Control	12v, 55 mph: 1v
53	PK/BK	TCC Solenoid Control	12v, 55 mph: 9v
54	PK/YL	A/C WOT Relay Control	On: 1v, Off: 12v
55	PK/OR	Shift Solenoid 3 Control	12v, 55 mph: 1v
56	GY/OR	PIP Sensor Signal	6.93v (50%)
57	RD	Vehicle Power	12-14v
58	TN	Injector 1 Control	3.3-5.7 ms
59	WT	Injector 2 Control	3.3-5.7 ms
60	BK/WT	Power Ground	<0.1v

Pin Connector Graphic

PCM 60-PIN CONNECTOR

Terminal View of 60-Pin PCM Harness Connector

Standard Colors and Abbreviations

Abbreviation	Color	Abbreviation	Color	Abbreviation	Color
BK	Black	GY	Gray	PK	Purple
BL	Blue	GN	Green	RD	Red
BR	Brown	LG	LT Green	TN	Tan
DB	Dark Blue	OR	Orange	WT	White
DG	DK Green	PK	Pink	YL	Yellow

1996-97 Explorer 4.0L V6 MFI VIN X 104 Pin Connector

PCM Pin #	Wire Color	Circuit Description (104 Pin)	Value at Hot Idle
1	PK/OR	Shift Solenoid 2 Control	12v, 55 mph: 12v
2	PK/LG	MIL (lamp) Control	MIL On: 1v, Off: 12v
13	VT	Flash EPROM Power	0.1v
14	LB/BK	4x4 Low Switch Signal	12v (switch closed: 0v)
15	PK/LB	Data Bus (-) Signal	Digital Signals
16	TN/OR	Data Bus (+) Signal	Digital Signals
19	OR/BK	Accelerator Ride Control	0.1v (Off)
21	DB	CKP (+) Sensor Signal	430-460 Hz
22	GY	CKP (-) Sensor Signal	430-460 Hz
24	BK/WT	Power Ground	<0.1v
25	BK	Case Ground	<0.050v
26	YL/BK	Coil 1 Driver (dwell)	5°, 55 mph: 8°
27	OR/YL	Shift Solenoid 1 Control	1v, 55 mph: 12v
28	BR/OR	Coast Clutch Solenoid	12v, 55 mph: 12v
29	TN/WT	TCS (switch) Signal	TCS & O/D On: 12v
30	DG	Octane Adjustment	9.3v (Closed: 0v)
33	PK/OR	VSS (-) Signal	0 Hz, 55 mph: 125 Hz
35	RD/LG	HO2S-12 (B1 S2) Signal	0.1-1.1v
36	TN/LB	MAF Sensor Return	<0.050v
37	OR/BK	TFT Sensor Signal	2.10-2.40v
38	LG/RD	ECT Sensor Signal	0.5-0.6v
39	GY	IAT Sensor Signal	1.5-2.5v
40	DG/YL	Fuel Pump Monitor	On: 12v, Off: 0v
41	DG/OR	A/C Switch Signal	A/C On: 12v, Off: 0v
43	LB/PK	Data Output Link	5v
47	BK/PK	VR Solenoid Control	0%, 55 mph: 45%
48	TN/YL	Clean Tachometer Output	39-45 Hz
51	BK/WT	Power Ground	<0.1v
52	WT/RD	Coil 2 Driver (dwell)	5°, 55 mph: 8°
53	PK/BK	Shift Solenoid 3 Control	12v, 55 mph: 1v
54	PK/YL	TCC Solenoid Control	0%, 55 mph: 95%
55	YL	Keep Alive Power	12-14v
56	LG/BK	EVAP Purge Solenoid	0-10 Hz (0-100%)
58	GY/BK	VSS (+) Signal	0 Hz, 55 mph: 125 Hz
60	GY/LB	HO2S-11 (B1 S1) Signal	0.1-1.1v
62	RD/PK	FTP Sensor Signal	2.6v (0" H2O - cap off)
64	LB/YL	A/T: TR Sensor Signal	In 'P': 0v, 55 mph: 1.7v
64	LB/YL	M/T: Clutch Pedal Switch	Clutch In: 0v, Out: 12v

1997 Explorer 4.0L V6 MFI VIN E (A/T) 104 Pin Connector

PCM Pin #	Wire Color	Circuit Description (104 Pin)	Value at Hot Idle
65	BR/LG	DPFE Sensor Signal	0.20-1.30v
67	PK/WT	EVAP CV Solenoid	0-10 Hz (0-100%)
69	PK/YL	A/C WOT Relay Control	On: 1v, Off: 12v
71	RD	Vehicle Power	12-14v
73	TN/BK	Injector 5 Control	3.6-7.5 ms
74	BR/YL	Injector 3 Control	3.6-7.5 ms
75	TN	Injector 1 Control	3.6-7.5 ms
76, 77	BK/WT	Power Ground	<0.1v
78	YL/WT	Coil 3 Driver (dwell)	5°, 55 mph: 8°
79	WT/LG	TCIL (lamp) Control	TCIL On: 1v, Off: 12v
80	LB/OR	Fuel Pump Control	On: 1v, Off: 12v
81	WT/YL	EPC Solenoid Control	10v (26 psi)
83	WT/LB	IAC Solenoid Control	10.8v (33%)
84	DG/WT	TSS (+) Sensor Signal	100-125 Hz
85	DB/OR	CMP Sensor Signal	7 Hz, 55 mph: 18 Hz
87	RD/BK	HO2S-21 (B2 S1) Signal	0.1-1.1v
88	LB/RD	MAF Sensor Signal	0.8v, 55 mph: 2.1v
89	GY/WT	TP Sensor Signal	0.53-1.27v
90	BR/WT	Reference Voltage	4.9-5.1v
91	GY/RD	Analog Signal Return	<0.050v
92	LG	Brake Position Switch	Brake Off: 12v, On: 1v
93	RD/WT	HO2S-11 (B1 S1) Heater	On: 1v, Off: 12v
94	YL/LB	HO2S-21 (B2 S1) Heater	On: 1v, Off: 12v
95	WT/BK	HO2S-12 (B1 S2) Heater	On: 1v, Off: 12v
96	TN/YL	HO2S-22 (B2 S2) Heater	On: 1v, Off: 12v
97	RD	Vehicle Power	12-14v
99	LG/OR	Injector 6 Control	3.6-7.5 ms
100	BR/LB	Injector 4 Control	3.6-7.5 ms
101	WT	Injector 2 Control	3.6-7.5 ms
103	BK/WT	Power Ground	<0.1v

Pin Connector Graphic

PCM 104-PIN CONNECTOR

Terminal View of 104-Pin PCM Wiring Harness Connector

Standard Colors and Abbreviations

Abbreviation	Color	Abbreviation	Color	Abbreviation	Color
BK	Black	GY	Gray	PK	Purple
BL	Blue	GN	Green	RD	Red
BR	Brown	LG	LT Green	TN	Tan
DB	Dark Blue	OR	Orange	WT	White
DG	DK Green	PK	Pink	YL	Yellow

1998-99 Explorer 4.0L V6 VIN E 104 Pin Connector

PCM Pin #	Wire Color	Circuit Description (104 Pin)	Value at Hot Idle
1	PK/OR	Shift Solenoid 'B' Control	12v, 55 mph: 12v
2	PK/LG	MIL (lamp) Control	MIL On: 1v, Off: 12v
3	YL/BK	Digital TR1 Sensor	In 'P': 0v, 55 mph: 11.5v
4-5, 7-11	---	Not Used	---
6	DB/YL	OSS Sensor Signal	0 Hz, 55 mph: 350 Hz
12	YL/WT	Fuel Level Indicator Signal	1.7v (1/2 full)
13	VT	Flash EPROM Power	0.1v
14	LB/BK	4x4 Indicator Signal	4x4 On: 1v, Off: 12v
15	PK/LB	Data Bus (-) Signal	Digital Signals
16	TN/OR	Data Bus (+) Signal	Digital Signals
17-20	---	Not Used	---
19	OR/BK	Accelerator Ride Control	0.1v (Off)
21	DB	CKP (+) Sensor Signal	430-460 Hz
22	GY	CKP (-) Sensor Signal	<0.050v
24, 51	BK/WT	Power Ground	<0.1v
25	BK	Case Ground	<0.050v
26	TN/WT	Coil 1 Driver (dwell)	5°, 55 mph: 8°
27	OR/YL	Shift Solenoid 'A' Control	1v, 55 mph: 12v
28	BR/OR	Coast Clutch Solenoid	12v, 55 mph: 12v
28 (2000-01)	BR/OR	Shift Solenoid 'D' Control	12v, 55 mph: 12v
29	TN/WT	TCS (switch) Signal	TCS & O/D On: 12v
30 ('98-'99)	DG	Octane Adjustment	9.3v (Closed: 0v)
31-32, 34	---	Not Used	---
32 ('98-'99)	DG/VT	Knock Sensor 1 Ground	<0.050v
33 ('98-'99)	PK/OR	VSS (-) Signal	0 Hz, 55 mph: 125 Hz
35	RD/LG	HO2S-12 (B1 S2) Signal	0.1-1.1v
36	TN/LB	MAF Sensor Return	<0.050v
37	OR/BK	TFT Sensor Signal	2.10-2.40v
38	LG/RD	ECT Sensor Signal	0.5-0.6v
39	GY	IAT Sensor Signal	1.5-2.5v
40	DG/YL	Fuel Pump Monitor	On: 12v, Off: 0v
41	PK	A/C Switch Signal	A/C On: 12v, Off: 0v
42	---	Not Used	---
43 ('98-'99)	LB/PK	Neutral Tow Connector Signal	N/A
44	---	Not Used	---
45	YL/WT	PCM Signal to the GEM	Digital Signals
46 ('98-'99)	BR	Inlet Air Control Valve	0.9v (15%)
47	BRD/PK	VR Solenoid Control	0%, 55 mph: 45%
48	TN/YL	Clean Tachometer Output	39-45 Hz
49	LB/BK	Digital TR2 Sensor	0v, 55 mph: 11.5v
50	WT/BK	Digital TR4 Sensor	0v, 55 mph: 11.5v
52	TN/OR	Coil 2 Driver (Control)	5°, 55 mph: 8°
53	P/BK	Shift Solenoid 'C' Control	12v, 55 mph: 1v
54	PK/YL	TCC Solenoid Control	0%, 55 mph: 95%
55	YL	Keep Alive Power	12-14v
56	LG/BK	EVAP Purge Solenoid	0-10 Hz (0-100%)
57	---	Not Used	---
58	GY/BK	VSS (+) Signal	0 Hz, 55 mph: 125 Hz
59	---	Not Used	---
60	GY/LB	HO2S-11 (B1 S1) Signal	0.1-1.1v
61	---	Not Used	---
62	RD/PK	FTP Sensor Signal	2.6v (0" H2O - cap off)
64	LB/YL	Digital TR3 Sensor	In 'P': 0v, 30 mph: 1.7v

1998-2002 Explorer 4.0L V6 MFI VIN X 104 Pin Connector

PCM Pin #	Wire Color	Circuit Description (104 Pin)	Value at Hot Idle
65	BR/LG	DPFE Sensor Signal	0.20-1.30v
66	---	Not Used	---
67	PK/WT	EVAP CV Solenoid	0-10 Hz (0-100%)
68	---	Not Used	---
69	PK/YL	A/C WOT Relay Control	On: 1v, Off: 12v
70	---	Not Used	---
71	RD	Vehicle Power	12-14v
72	---	Not Used	---
73	TN/BK	Injector 5 Control	3.4-3.8 ms
74	BR/YL	Injector 3 Control	3.4-3.8 ms
75	TN	Injector 1 Control	3.4-3.8 ms
76	BK/WT	Power Ground	<0.1v
77	BK/WT	Power Ground	<0.1v
78	TN/LG	Coil 3 Driver (dwell)	5°, 55 mph: 8°
79	WT/LG	TCIL (lamp) Control	TCIL On: 1v, Off: 12v
80	LB/OR	Fuel Pump Control	On: 1v, Off: 12v
81	WT/YL	EPC Solenoid Control	10v (26 psi)
82	---	Not Used	---
83	WT/LB	IAC Solenoid Control	10.7v (33%)
84	DG/WT	TSS Sensor Signal	100-125 Hz
85	DB/OR	CMP Sensor Signal	7 Hz, 55 mph: 18 Hz
86	BK/YL	A/C Pressure Sensor	Switch open: 12v
87	RD/BK	HO2S-21 (B2 S1) Signal	0.1-1.1v
88	LB/RD	MAF Sensor Signal	0.8v, 55 mph: 2.1v
89	GY/WT	TP Sensor Signal	0.53-1.27v
90	BR/WT	Reference Voltage	4.9-5.1v
91	GY/RD	Sensor Ground	<0.050v
92	RD/LG	Brake Position Switch	Brake Off: 12v, On: 1v
93	RD/WT	HO2S-11 (B1 S1) Heater	On: 1v, Off: 12v
94	YL/LB	HO2S-21 (B2 S1) Heater	On: 1v, Off: 12v
95	WT/BK	HO2S-12 (B1 S2) Heater	On: 1v, Off: 12v
97	RD	Vehicle Power	12-14v
98	---	Not Used	---
99	LG/OR	Injector 6 Control	3.4-3.8 ms
100	BR/LB	Injector 4 Control	3.4-3.8 ms
101	WT	Injector 2 Control	3.4-3.8 ms
102	---	Not Used	---
103	BK/WT	Power Ground	<0.1v
104	---	Not Used	---

Pin Connector Graphic

PCM 104-PIN CONNECTOR

```
 1 ●●●●●●●●●●●●●     ●●●●●●●●●●●●● 26
27 ●●●●●●●●●●●●●   ● ●●●●●●●●●●●●● 52
53 ●●●●●●●●●●●●●     ●●●●●●●●●●●●● 78
79 ●●●●●●●●●●●●●     ●●●●●●●●●●●●● 104
```

Terminal View of 104-Pin PCM Wiring Harness Connector

Standard Colors and Abbreviations

Abbreviation	Color	Abbreviation	Color	Abbreviation	Color
BK	Black	GY	Gray	PK	Purple
BL	Blue	GN	Green	RD	Red
BR	Brown	LG	LT Green	TN	Tan
DB	Dark Blue	OR	Orange	WT	White
DG	DK Green	PK	Pink	YL	Yellow

2002-03 Explorer 4.6L V8 VIN W 58 Pin Connector

PCM Pin #	Wire Color	Circuit Description (58 Pin)	Value at Hot Idle
1, 5	---	Not Used	---
2	PK/BK	Fuel Pump Power	Pump On: 12v, Off: 0v
3	TN/OR	SCP Data Bus (+)	Digital Signals
4	PK/LB	SCP Data Bus (-)	Digital Signals
6	VT/WT	EVAP Canister Vent Solenoid Control	0-10 Hz (0-100%)
7	GY/BK	Vehicle Speed Sensor (+) Signal	0 Hz, 55 mph: 125 Hz
8	LG/RD	Generator Monitor Signal	130 Hz (45%)
9	PK/YL	A/C Clutch Relay	12v (relay on: 1v)
11	WT/LG	TX Signal	Digital Signals
12	LG/BK	EVAP Purge Solenoid	0-10 Hz (0-100%)
13	VT	Module Programming Signal	0.1v
14	GY/OR	RX Signal	Digital Signals
15-16, 21	---	Not Used	---
17	GY/RD	Sensor Ground	<0.050v
18-19	---	Not Used	---
20	BR/WT	Reference Voltage	4.9-5.1v
22	DB/LG	Passive Antitheft Indicator Control	Indicator Off: 12v, On: 1v
23	LB/BK	4WD Indicator Low Signal	12v (switch on: 1v)
24-27	BK/WT	Power Ground	<0.1v
28	BK/YL	Brake Pressure Switch	12v (Brake On: 0v)
29	OR/LB	Cruise Set Indicator Control	N/A
30	BK/YL	A/C Clutch Relay (switched)	12v (relay on: 1v)
31	LB/RD	MAF Sensor Signal	0.7v, 55 mph: 1.8v
32	VT	Vehicle Power (Start-Run)	12-14v
33	VT	Vehicle Power (Start-Run)	12-14v
34-35	---	Not Used	---
36	OR/LB	Cruise Set Indicator Control	12v (switch set: 0v)
37	---	Not Used	---
38	TN/LB	MAF Sensor Return	<0.050v
39	OR	Starter Motor Relay Signal	0v
40	RD/LG	Brake Position Switch	Brake Off: 12v, On: 1v
41	TN/WT	Overdrive Cancel Switch	0v (switch on: 12v)
42	---	Not Used	---
43	BK	Power Ground	<0.1v
44	RD/WT	Keep Alive Power (CJB fuse)	12-14v
45	BK	Speed Control Switch Ground	<0.050v
46	OR/LB	Cruise Switch Indicator	12v
47	RD/YL	A/C Pressure Switch Signal	0v (switch on: 12v)
48-49	---	Not Used	---
50	VT	A/C Demand Switch	0v (A/C On: 12v)
51	GY	IAT Sensor Signal	1.5-2.5v
52	RD/PK	FTP Sensor Signal	2.6v (0" H20 - cap off)
53-55	---	Not Used	---
56	DG/OR	Speed Control Switch Ground	<0.050v
57	LB/BK	Speed Control Switch Input	0v (switch on: 6.7v)
58	LB/OR	Fuel Pump Relay Control	On: 1v, Off: 12v

Pin Connector Graphic

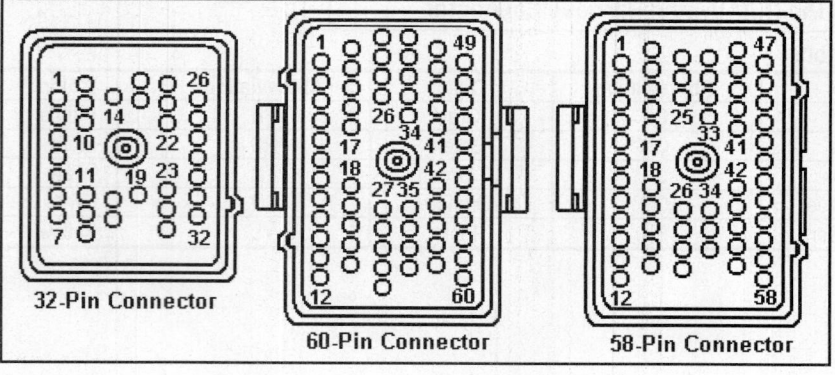

32-Pin Connector

60-Pin Connector

58-Pin Connector

2002-03 Explorer 4.6L V8 VIN W 32 Pin Connector

PCM Pin #	Wire Color	Circuit Description (32 Pin)	Value at Hot Idle
1	OR/YL	Shift Solenoid 'A' Control	12v, 55 mph: 1v
2	VT/OR	Shift Solenoid 'B' Control	12v, 55 mph: 1v
3	PK/BK	Shift Solenoid 'C' Control	12v, 55 mph: 1v
4	BR/OR	Shift Solenoid 'D' Control	12v, 55 mph: 1v
5	VT/YL	TCC Solenoid Control	12v, 55 mph: 1v
6	---	Not Used	---
7	WT/YL	Electronic Pressure Control	On: 1v, Off: 12v
8	---	Not Used	---
9	LB/YL	Digital TR Sensor (TR3A)	0v, 55 mph: 1.7v
10	WT/BK	Digital TR Sensor (TR4)	0v, 55 mph: 11v
11	---	Not Used	---
12	WT	Motor Position #4	On: 1v, Off: 12v
13	LB/PK	Pressure Control 'B' Solenoid	On: 1v, Off: 12v
14	---	Not Used	---
15	WT/BK	HO2S-12 (B1 S2) Heater	1v, Off: 12v
16	TN/YL	HO2S-22 (B2 S2) Heater	1v, Off: 12v
17	GY/RD	Sensor Reference Ground	<0.050v
18	LB/BK	Digital TR Sensor (TR2)	0v, 55 mph: 11v
19-20	---	Not Used	---
21	YL/LG	Overdrive Drum Speed Input	200 Hz, 55 mph: 1185 Hz
22	YL/BK	Digital TR Sensor (TR1)	0v, 55 mph: 11v
23	OR/BK	Transmission Fluid Temperature Sensor	2.10-2.40v
24-25	---	Not Used	---
26	DB/YL	Output Shaft Speed Sensor	0 Hz, 985 Hz
27	DG/WT	Turbine Shaft Speed Sensor	360 Hz, 890 Hz
28	RD/LG	HO2S-12 (B1 S2) Signal	0.1-1.1v
29	VT/LG	HO2S-22 (B2 S2) Signal	0.1-1.1v
30-32	---	Not Used	---

Pin Connector Graphic

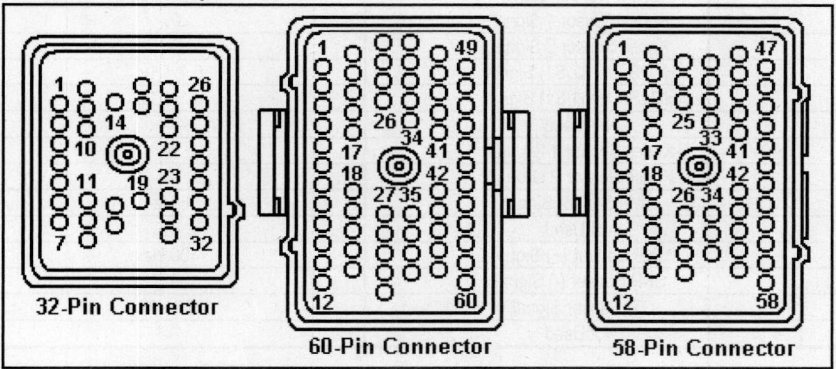

32-Pin Connector

60-Pin Connector

58-Pin Connector

Standard Colors and Abbreviations

Abbreviation	Color	Abbreviation	Color	Abbreviation	Color
BK	Black	GY	Gray	PK	Purple
BL	Blue	GN	Green	RD	Red
BR	Brown	LG	LT Green	TN	Tan
DB	Dark Blue	OR	Orange	WT	White
DG	DK Green	PK	Pink	YL	Yellow

2002-03 Explorer 4.6L V8 VIN W 60 Pin Connector

PCM Pin #	Wire Color	Circuit Description (60 Pin)	Value at Hot Idle
1	PK/WT	COP 2 Driver (dwell)	5°, 55 mph: 7°
2	TN	Injector 1 Control	3.3-3.8 ms
3-6	---	Not Used	---
7	RD/WT	HO2S-11 (B1 S1) Heater	1v, Off: 12v
8	YL/LB	HO2S-21 (B2 S1) Heater	1v, Off: 12v
9	WT/LB	IAC Solenoid Control	10.7v (33%)
10, 15	---	Not Used	---
11	TN/BK	Injector 5 Control	3.3-3.8 ms
12	OR/YL	COP 6 Driver Control	5°, 55 mph: 7°
13	PK/LB	COP 7 Driver Control	5°, 55 mph: 7°
14	WT	Injector 2 Control	3.3-3.8 ms
16	BR/PK	VR Solenoid Control	0%, 55 mph: 45%
17	GY/RD	Sensor Reference Ground	<0.050v
18-19	---	Not Used	---
20	BR/WT	Sensor Voltage Reference	4.9-5.1v
21	LG/OR	Injector 6 Control	3.3-3.8 ms
22	LG/YL	COP 5 Driver Control	5°, 55 mph: 7°
23	BR/YL	COP 5 Driver Control	5°, 55 mph: 7°
24	BR/YL	Injector 3 Control	3.3-3.8 ms
25-28	---	Not Used	---
29	TN/RD	Injector 7 Control	3.3-3.8 ms
30	DG/VT	COP 4 Driver Control	5°, 55 mph: 7°
31	LG/WT	COP 1 Driver (dwell)	5°, 55 mph: 7°
32	BR/LB	Injector 4 Control	3.3-3.8 ms
33-36	---	Not Used	---
37	LB	Injector 8 Control	3.3-3.8 ms
38	WT/RD	COP 8 Driver Control	5°, 55 mph: 7°
39	---	Not Used	---
40	YL/LB	CHT Sensor Signal	0.7v (194°F)
41	BR/LG	DPFE Sensor Signal	0.95-1.05v
42	YL	Knock Sensor 1 Signal	0v
43	DG/WT	Knock Sensor 2 Signal	0v
44	RD/BK	HO2S-21 (B2 S1) Signal	0.1-1.1v
45	GY/LB	HO2S-11 (B1 S1) Signal	0.1-1.1v
46-50	---	Not Used	---
51	YL/RD	Knock Sensor 1 Ground	0v
52	DG/VT	Knock Sensor 2 Ground	0v
53	DG	CMP Sensor Signal	6 Hz
54	---	Not Used	---
55	DB	CKP Sensor (+) Signal	400 Hz
56	GY	CKP Sensor (-) Signal	400 Hz
57	GY/WT	TP Sensor Signal	0.53-1.27v
58-60	---	Not Used	---

Pin Connector Graphic

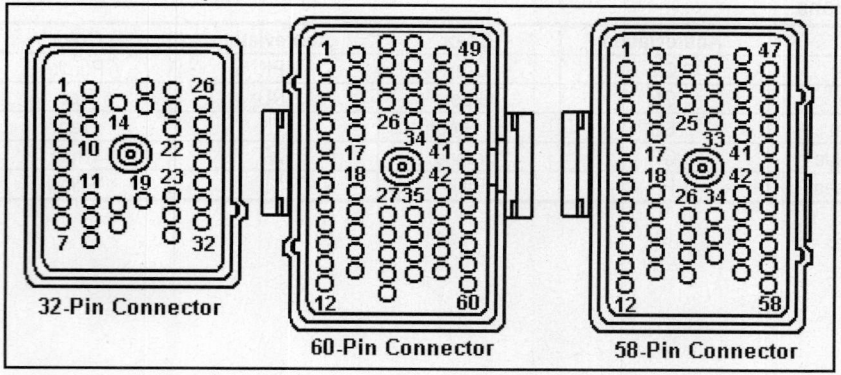

32-Pin Connector

60-Pin Connector

58-Pin Connector

1996-97 Explorer 5.0L V8 VIN P 104 Pin Connector

PCM Pin #	Wire Color	Circuit Description (104 Pin)	Value at Hot Idle
1	PK/OR	Shift Solenoid 2 Control	12v, 55 mph: 12v
2	PK/LG	MIL (lamp) Control	MIL On: 1v, Off: 12v
3-12	---	Not Used	---
13	VT	Flash EPROM Power	0.1v
14	---	Not Used	---
15	PK/LB	Data Bus (-) Signal	Digital Signals
16	TN/OR	Data Bus (+) Signal	Digital Signals
17-18	---	Not Used	---
19	OR/BK	Accelerator Ride Control	0.1v (Off)
20	---	Not Used	---
21	BK/PK	CKP (+) Sensor Signal	430-460 Hz
22	GY/YL	CKP (-) Sensor Signal	430-460 Hz
23	---	Not Used	---
24	BK/WT	Power Ground	<0.1v
25	BK	Case Ground	<0.050v
26	DB/LG	Coil 1 Driver (dwell)	5°, 55 mph: 8°
27	OR/YL	Shift Solenoid 1 Control	1v, 55 mph: 12v
28	---	Not Used	---
29	TN/WT	TCS (switch) Signal	TCS & O/D On: 12v
30	DG	Octane Adjustment	9.3v (Closed: 0v)
31-32	---	Not Used	---
33	PK/OR	VSS (-) Signal	0 Hz, 55 mph: 125 Hz
34	---	Not Used	---
35	RD/LG	HO2S-12 (B1 S2) Signal	0.1-1.1v
36	TN/LB	MAF Sensor Return	<0.050v
37	OR/BK	TFT Sensor Signal	2.10-2.40v
38	LG/RD	ECT Sensor Signal	0.5-0.6v
39	GY	IAT Sensor Signal	1.5-2.5v
40	DG/YL	Fuel Pump Monitor	On: 12v, Off: 0v
41	DG/OR	A/C Switch Signal	A/C On: 12v, Off: 0v
42	---	Not Used	---
43	LB/PK	Data Output Link	5v
44-46	---	Not Used	---
47	BK/PK	VR Solenoid Control	0%, 55 mph: 45%
48	TN/YL	Clean Tachometer Output	39-45 Hz
49-50	---	Not Used	---
51	BK/WT	Power Ground	<0.1v
52	RD/LB	Coil 2 Driver (dwell)	5°, 55 mph: 8°
53	---	Not Used	---
54	DB/WT	TCC Solenoid Control	0%, 55 mph: 95%
55	Y	Keep Alive Power	12-14v
56	LG/BK	EVAP Purge Solenoid	0-10 Hz (0-100%)
57	---	Not Used	---
58	GY/BK	VSS (+) Signal	0 Hz, 55 mph: 125 Hz
59	---	Not Used	---
60	GY/LB	HO2S-11 (B1 S1) Signal	0.1-1.1v
61	PK/LG	HO2S-22 (B2 S2) Signal	0.1-1.1v
62	RD/PK	FTP Sensor Signal	2.6v (0" H2O - cap off)
63	OR/YL	EVP Sensor Signal	0.20-1.30v
64	LB/YL	A/T: TR Sensor Signal	In 'P': 0v, 55 mph: 1.7v

1996-97 Explorer 5.0L V8 VIN P 104 Pin Connector

PCM Pin #	Wire Color	Circuit Description (104 Pin)	Value at Hot Idle
65-66	---	Not Used	---
67	PK/WT	EVAP CV Solenoid	0-10 Hz (0-100%)
68	---	Not Used	---
69	PK/YL	A/C WOT Relay Control	On: 1v, Off: 12v
70	---	Not Used	---
71	RD	Vehicle Power	12-14v
72	TN/RD	Injector 7 Control	3.2-4.5 ms
73	TN/BK	Injector 5 Control	3.2-4.5 ms
74	BR/YL	Injector 3 Control	3.2-4.5 ms
75	TN	Injector 1 Control	3.2-4.5 ms
76	BK/WT	Power Ground	<0.1v
77	BK/WT	Power Ground	<0.1v
78	PK/WT	Coil 4 Driver (dwell)	5°, 55 mph: 8°
79	WT/LG	TCIL (lamp) Control	TCIL On: 1v, Off: 12v
80	LB/OR	Fuel Pump Control	On: 1v, Off: 12v
81	WT/YL	EPC Solenoid Control	10v (26 psi)
83	WT/LB	IAC Solenoid Control	10.8v (33%)
84	DB/YL	OSS Sensor Signal	0 Hz, 55 mph: 230 Hz
85	DB/OR	CMP Sensor Signal	5-10 Hz
86	---	Not Used	---
87	RD/BK	HO2S-21 (B2 S1) Signal	0.1-1.1v
88	LB/RD	MAF Sensor Signal	0.8v, 55 mph: 2.1v
89	GY/WT	TP Sensor Signal	0.53-1.27v
90	BR/WT	Reference Voltage	4.9-5.1v
91	GY/RD	Analog Signal Return	<0.050v
92	LG	Brake Position Switch	Brake Off: 12v, On: 1v
93	RD/WT	HO2S-11 (B1 S1) Heater	On: 1v, Off: 12v
94	YL/LB	HO2S-21 (B2 S1) Heater	On: 1v, Off: 12v
95	WT/BK	HO2S-12 (B1 S2) Heater	On: 1v, Off: 12v
96	TN/YL	HO2S-22 (B2 S2) Heater	On: 1v, Off: 12v
97	RD	Vehicle Power	12-14v
98	LB	Injector 8 Control	3.2-4.5 ms
99	LG/OR	Injector 6 Control	3.2-4.5 ms
100	BR/LB	Injector 4 Control	3.2-4.5 ms
101	WT	Injector 2 Control	3.2-4.5 ms
102	---	Not Used	---
103	BK/WT	Power Ground	<0.1v
104	RD/YL	Coil 4 Driver (dwell)	5°, 55 mph: 8°

Pin Connector Graphic

PCM 104-PIN CONNECTOR

Terminal View of 104-Pin PCM Wiring Harness Connector

Standard Colors and Abbreviations

Abbreviation	Color	Abbreviation	Color	Abbreviation	Color
BK	Black	GY	Gray	PK	Purple
BL	Blue	GN	Green	RD	Red
BR	Brown	LG	LT Green	TN	Tan
DB	Dark Blue	OR	Orange	WT	White
DG	DK Green	PK	Pink	YL	Yellow

1998-2001 Explorer 5.0L V8 VIN P 104 Pin Connector

PCM Pin #	Wire Color	Circuit Description (104 Pin)	Value at Hot Idle
1	PK/OR	Shift Solenoid 'B' Control	12v, 55 mph: 1v
2	PK/LG	MIL (lamp) Control	MIL On: 1v, Off: 12v
3	YL/BK	Digital TR1 Sensor	In 'P': 0v, Others: 11.5v
4-11	---	Not Used	---
12	YL/WT	Fuel Level Indicator Signal	1.7v (1/2 full)
13	VT	Flash EPROM Power	0.1v
14	---	Not Used	---
15	PK/LB	Data Bus (-) Signal	Digital Signals
16	TN/OR	Data Bus (+) Signal	Digital Signals
17-18	---	Not Used	---
19	OR/BK	Air Suspension Control Signal	Digital Signals
20	---	Not Used	---
21	BK/PK	CKP (+) Sensor Signal	460-500 Hz
22	GY/YL	CKP (-) Sensor Signal	460-500 Hz
24	BK/WT	Power Ground	<0.1v
25	BK	Case Ground	<0.050v
26	DB/LG	Coil 1 Driver (dwell)	5°, 55 mph: 8°
27	OR/YL	Shift Solenoid 'A' Control	1v, 55 mph: 1v
28	---	Not Used	---
29	TN/WT	TCS or Manual Switch	TCS or M/D pressed: 12v
30 ('98-'99)	DG	Octane Adjustment	9.3v (Closed: 0v)
31-32	---	Not Used	---
33 ('98-'99)	PK/OR	VSS (-) Signal	0 Hz, 55 mph: 125 Hz
35	RD/LG	HO2S-12 (B1 S2) Signal	0.1-1.1v
36	TN/LB	MAF Sensor Return	<0.050v
37	OR/BK	TFT Sensor Signal	2.10-2.40v
38	LG/RD	ECT Sensor Signal	0.5-0.6v
39	GY	IAT Sensor Signal	1.5-2.5v
40	DG/YL	Fuel Pump Monitor	On: 12v, Off: 0v
41	VT	A/C Switch Signal	Switch On: 12v (open)
42, 45-46	---	Not Used	---
43 ('98-'99)	LB/PK	Neutral Tow Connector Signal	N/A
47	BR/PK	VR Solenoid Control	0%, 55 mph: 45%
48	TN/YL	Clean Tachometer Output	40-55 Hz
49	LB/BK	Digital TR2 Sensor	In 'P': 0v, Others: 11.5v
50	WT/BK	Digital TR4 Sensor	In 'P': 0v, Others: 11.5v
51	BK/WT	Power Ground	<0.1v
52	RD/LB	Coil 2 Driver (Control)	5°, 55 mph: 8°
53	---	Not Used	---
54	DB/WT	TCC Solenoid Control	0%, 55 mph: 95%
55	YL	Keep Alive Power	12-14v
56	LG/BK	EVAP Purge Solenoid	0-10 Hz (0-100%)
57	---	Not Used	---
58	GY/BK	VSS (+) Signal	0 Hz, 55 mph: 125 Hz
59	---	Not Used	---
60	GY/LB	HO2S-11 (B1 S1) Signal	0.1-1.1v
61	PK/LG	HO2S-22 (B2 S2) Signal	0.1-1.1v
62	RD/PK	FTP Sensor Signal	2.6v (0" H2O - cap off)
63	---	Not Used	---
64	LB/YL	Digital TR3 Sensor	In 'P': 0v, 55 mph: 1.7v

1998-2001 Explorer 5.0L V8 VIN P 104 Pin Connector

PCM Pin #	Wire Color	Circuit Description (104 Pin)	Value at Hot Idle
65 ('98-'00)	BR/LG	DPFE Sensor Signal	0.20-1.30v
65 ('01-'00)	BR/LG	DPFE Sensor Signal	0.95-1.05v
66	---	Not Used	---
67	PK/WT	EVAP CV Solenoid	0-10 Hz (0-100%)
68	---	Not Used	---
69	PK/YL	A/C WOT Relay Control	On: 1v, Off: 12v
70	---	Not Used	---
71	RD	Vehicle Power	12-14v
72	TN/RD	Injector 7 Control	3.2-4.5 ms
73	TN/BK	Injector 5 Control	3.2-4.5 ms
74	BR/YL	Injector 3 Control	3.2-4.5 ms
75	TN	Injector 1 Control	3.2-4.5 ms
76	BK/WT	Power Ground	<0.1v
77	BK/WT	Power Ground	<0.1v
78	PK/WT	Coil 3 Driver (dwell)	5°, 55 mph: 8°
79	WT/LG	TCIL (lamp) Control	TCIL On: 1v, Off: 12v
80	LB/OR	Fuel Pump Control	On: 1v, Off: 12v
81	WT/YL	EPC Solenoid Control	9.5v (10 psi)
82	---	Not Used	---
83	WT/LB	IAC Solenoid Control	10.8v (12-30%)
84	DB/YL	OSS Sensor Signal	At 55 mph: 230-280 Hz
85	DB/OR	CMP Sensor Signal	5-10 Hz
86	BK/YL	A/C Pressure Switch Signal	Open: 12v (<24 psi)
87	RD/BK	HO2S-21 (B2 S1) Signal	0.1-1.1v
88	LB/RD	MAF Sensor Signal	0.8v, 55 mph: 2.1v
89	GY/WT	TP Sensor Signal	0.53-1.27v
90	BR/WT	Reference Voltage	4.9-5.1v
91	GY/RD	Analog Signal Return	<0.050v
92	RD/LG	Brake Position Switch	Brake Off: 12v, On: 1v
93	RD/WT	HO2S-11 (B1 S1) Heater	On: 1v, Off: 12v
94	YL/LB	HO2S-21 (B2 S1) Heater	On: 1v, Off: 12v
95	WT/BK	HO2S-12 (B1 S2) Heater	On: 1v, Off: 12v
96	TN/YL	HO2S-22 (B2 S2) Heater	On: 1v, Off: 12v
97	RD	Vehicle Power	12-14v
98	LB	Injector 8 Control	3.2-4.5 ms
99	LG/OR	Injector 6 Control	3.2-4.5 ms
100	BR/LB	Injector 4 Control	3.2-4.5 ms
101	WT	Injector 2 Control	3.2-4.5 ms
102	---	Not Used	---
103	BK/WT	Power Ground	<0.1v
104	RD/YL	Coil 4 Driver (dwell)	5°, 55 mph: 8°

Pin Connector Graphic

PCM 104-PIN CONNECTOR

Terminal View of 104-Pin PCM Wiring Harness Connector

Standard Colors and Abbreviations

Abbreviation	Color	Abbreviation	Color	Abbreviation	Color
BK	Black	GY	Gray	PK	Purple
BL	Blue	GN	Green	RD	Red
BR	Brown	LG	LT Green	TN	Tan
DB	Dark Blue	OR	Orange	WT	White
DG	DK Green	PK	Pink	YL	Yellow

Mountaineer PIN Tables

1998-99 Mountaineer 4.0L V6 VIN E 104 Pin Connector

PCM Pin #	Wire Color	Circuit Description (104 Pin)	Value at Hot Idle
1	PK/OR	Shift Solenoid 'B' Control	12v, 55 mph: 12v
2	PK/LG	MIL (lamp) Control	MIL On: 1v, Off: 12v
3	YL/BK	Digital TR1 Sensor	In 'P': 0v, 55 mph: 11.5v
4-5	---	Not Used	---
6	DB/YL	OSS Sensor Signal	0 Hz, 55 mph: 350 Hz
7-11	---	Not Used	---
12	YL/WT	Fuel Level Indicator Signal	1.7v (1/2 full)
13	VT	Flash EPROM Power	0.1v
14	LB/BK	4x4 Indicator Signal	4x4 On: 1v, Off: 12v
15	PK/LB	Data Bus (-) Signal	Digital Signals
16	TN/OR	Data Bus (+) Signal	Digital Signals
17-20	---	Not Used	---
19	OR/BK	Accelerator Ride Control	0.1v (Off)
21	DB	CKP (+) Sensor Signal	430-460 Hz
22	GY	CKP (-) Sensor Signal	<0.050v
24	BK/WT	Power Ground	<0.1v
25	BK	Case Ground	<0.050v
26	TN/WT	Coil 1 Driver (dwell)	5º, 55 mph: 8º
27	OR/YL	Shift Solenoid 'A' Control	1v, 55 mph: 12v
28	BR/OR	Coast Clutch Solenoid	12v, 55 mph: 12v
29	TN/WT	TCS (switch) Signal	TCS & O/D On: 12v
30	DG	Octane Adjustment	9.3v (Closed: 0v)
31-32	---	Not Used	---
32	DG/VT	Knock Sensor 1 Ground	<0.050v
33	PK/OR	VSS (-) Signal	0 Hz, 55 mph: 125 Hz
34	---	Not Used	---
35	RD/LG	HO2S-12 (B1 S2) Signal	0.1-1.1v
36	TN/LB	MAF Sensor Return	<0.050v
37	OR/BK	TFT Sensor Signal	2.10-2.40v
38	LG/RD	ECT Sensor Signal	0.5-0.6v
39	GY	IAT Sensor Signal	1.5-2.5v
40	DG/YL	Fuel Pump Monitor	On: 12v, Off: 0v
41	PK	A/C Switch Signal	A/C On: 12v, Off: 0v
42	---	Not Used	---
43	LB/PK	Neutral Tow Connector Signal	N/A
44	---	Not Used	---
45	YL/WT	PCM Signal to the GEM	Digital Signals
46	BR	Inlet Air Control Valve	0.9v (15%)
47	BRD/PK	VR Solenoid Control	0%, 55 mph: 45%
48	TN/YL	Clean Tachometer Output	39-45 Hz
49	LB/BK	Digital TR2 Sensor	0v, 55 mph: 11.5v
50	WT/BK	Digital TR4 Sensor	0v, 55 mph: 11.5v
51	BK/WT	Power Ground	<0.1v
52	TN/OR	Coil 2 Driver (Control)	5º, 55 mph: 8º
53	P/BK	Shift Solenoid 'C' Control	12v, 55 mph: 1v
54	PK/YL	TCC Solenoid Control	0%, 55 mph: 95%
55	YL	Keep Alive Power	12-14v
56	LG/BK	EVAP Purge Solenoid	0-10 Hz (0-100%)
57	---	Not Used	---
58	GY/BK	VSS (+) Signal	0 Hz, 55 mph: 125 Hz
59	---	Not Used	---
60	GY/LB	HO2S-11 (B1 S1) Signal	0.1-1.1v
61	---	Not Used	---
62	RD/PK	FTP Sensor Signal	2.6v (0" H2O - cap off)
63	---	Not Used	---
64	LB/YL	Digital TR3 Sensor	In 'P': 0v, 30 mph: 1.7v

1998-99 Mountaineer 4.0L V6 VIN E 104 Pin Connector

PCM Pin #	Wire Color	Circuit Description (104 Pin)	Value at Hot Idle
65	BR/LG	DPFE Sensor Signal	0.20-1.30v
66	---	Not Used	---
67	PK/WT	EVAP CV Solenoid	0-10 Hz (0-100%)
68	---	Not Used	---
69	PK/YL	A/C WOT Relay Control	On: 1v, Off: 12v
71	RD	Vehicle Power	12-14v
70	---	Not Used	---
72	---	Not Used	---
73	TN/BK	Injector 5 Control	3.6-7.5 ms
74	BR/YL	Injector 3 Control	3.6-7.5 ms
75	TN	Injector 1 Control	3.6-7.5 ms
76	BK/WT	Power Ground	<0.1v
77	BK/WT	Power Ground	<0.1v
78	TN/LG	Coil 3 Driver (dwell)	5°, 55 mph: 8°
79	WT/LG	TCIL (lamp) Control	TCIL On: 1v, Off: 12v
80	LB/OR	Fuel Pump Control	On: 1v, Off: 12v
81	WT/YL	EPC Solenoid Control	10v (26 psi)
82	---	Not Used	---
83	WT/LB	IAC Solenoid Control	10.8v (33%)
84	DG/WT	TSS (+) Sensor Signal	100-125 Hz
85	DB/OR	CMP Sensor Signal	7 Hz, 55 mph: 18 Hz
86	BK/YL	A/C Pressure Switch Signal	Switch Open: 12v
87	RD/BK	HO2S-21 (B2 S1) Signal	0.1-1.1v
88	LB/RD	MAF Sensor Signal	0.8v, 55 mph: 2.1v
89	GY/WT	TP Sensor Signal	0.53-1.27v
90	BR/WT	Reference Voltage	4.9-5.1v
91	GY/RD	Sensor Ground	<0.050v
92	RD/LG	Brake Position Switch	Brake Off: 12v, On: 1v
93	RD/WT	HO2S-11 (B1 S1) Heater	On: 1v, Off: 12v
94	YL/LB	HO2S-21 (B2 S1) Heater	On: 1v, Off: 12v
95	WT/BK	HO2S-12 (B1 S2) Heater	On: 1v, Off: 12v
96	---	Not Used	---
97	RD	Vehicle Power	12-14v
98	---	Not Used	---
99	LG/OR	Injector 6 Control	3.6-7.5 ms
100	BR/LB	Injector 4 Control	3.6-7.5 ms
101	WT	Injector 2 Control	3.6-7.5 ms
102	---	Not Used	---
103	BK/WT	Power Ground	<0.1v
104	---	Not Used	---

Pin Connector Graphic

PCM 104-PIN CONNECTOR

Terminal View of 104-Pin PCM Wiring Harness Connector

Standard Colors and Abbreviations

Abbreviation	Color	Abbreviation	Color	Abbreviation	Color
BK	Black	GY	Gray	PK	Purple
BL	Blue	GN	Green	RD	Red
BR	Brown	LG	LT Green	TN	Tan
DB	Dark Blue	OR	Orange	WT	White
DG	DK Green	PK	Pink	YL	Yellow

2000-02 Mountaineer 4.0L V6 SOHC MFI VIN E 104 Pin Connector

PCM Pin #	Wire Color	Circuit Description (104 Pin)	Value at Hot Idle
1	PK/OR	Shift Solenoid 'B' Control	12v, 55 mph: 12v
2	PK/LG	MIL (lamp) Control	MIL On: 1v, Off: 12v
3	YL/BK	Digital TR1 Sensor	In 'P': 0v, 55 mph: 11.5v
4-5	---	Not Used	---
6	DB/YL	OSS Sensor Signal	0 Hz, 55 mph: 350 Hz
7-11	---	Not Used	---
12	YL/WT	Fuel Level Indicator Signal	1.7v (1/2 full)
13	PK	Flash EPROM Power	0.1v
14	LB/BK	4x4 Low Indicator Switch	Switch On: 1v, Off: 12v
15	PK/LB	Data Bus (-) Signal	Digital Signals
16	TN/OR	Data Bus (+) Signal	Digital Signals
17-18	---	Not Used	---
19	OR/BK	Load Leveling Acceleration	Digital Signals
20	---	Not Used	---
21	DB	CKP (+) Sensor Signal	430-460 Hz
22	GY	CKP (-) Sensor Signal	<0.050v
23	---	Not Used	---
24	BK/WT	Power Ground	<0.1v
25	BK	Chassis Ground	<0.050v
26	TN/WT	Coil 1 Driver (dwell)	5°, 55 mph: 8°
27	OR/YL	Shift Solenoid 'A' Control	1v, 55 mph: 12v
28	BR/OR	Shift Solenoid 'D' Control	12v, 55 mph: 12v
29	TN/WT	TCS (switch) Signal	TCS & O/D On: 12v
30	---	Not Used	---
32	DG/VT	Knock Sensor (-) Return	<0.050v
33-34	---	Not Used	---
35	RD/LG	HO2S-12 (B1 S2) Signal	0.1-1.1v
36	TN/LB	MAF Sensor Return	<0.050v
37	OR/BK	TFT Sensor Signal	2.10-2.40v
38	LG/RD	ECT Sensor Signal	0.5-0.6v
39	GY	IAT Sensor Signal	1.5-2.5v
40	DG/YL	Fuel Pump Monitor	On: 12v, Off: 0v
41	PK	A/C High Pressure Switch	Switch On: 12v (open)
42	---	Not Used	---
43	LB/PK	Message Center Fuel Flow	Digital Signals
44	---	Not Used	---
45	YL/WT	PCM Signal to GEM	Digital Signals
46	BK/WT	Fuel Cap Indicator (Cluster)	Digital Signals
47	BR/PK	EGR VR Regulator Solenoid	0%, 55 mph: 45%
48	TN/YL	Clean Tachometer Output	39-45 Hz
49	LB/BK	Digital TR2 Sensor	0v, 55 mph: 11.5v
50	WT/BK	Digital TR4 Sensor	0v, 55 mph: 11.5v
51	BK/WT	Power Ground	<0.1v
52	TN/OR	Coil 2 Driver (Control)	5°, 55 mph: 8°
53	PK/BK	Shift Solenoid 'C' Control	12v, 55 mph: 1v
54	PK/YL	TCC Solenoid Control	0%, 55 mph: 95%
55	YL	Keep Alive Power	12-14v
56	LG/BK	EVAP Purge Solenoid	0-10 Hz (0-100%)
57	YL/RD	Knock Sensor (+) Signal	0v
58	GY/BK	VSS (+) Signal	0 Hz, 55 mph: 125 Hz
59	---	Not Used	---
60	GY/LB	HO2S-11 (B1 S1) Signal	0.1-1.1v
61	PK/LG	HO2S-22 (B2 S2) Signal	0.1-1.1v
63	---	Not Used	---
62	RD/PK	FTP Sensor Signal	2.6v (0" H2O - cap off)
64	LB/YL	Digital TR3 Sensor	In 'P': 0v, 30 mph: 1.7v

2000-02 Mountaineer 4.0L V6 SOHC MFI VIN E 104 Pin Connector

PCM Pin #	Wire Color	Circuit Description (104 Pin)	Value at Hot Idle
65 ('00)	BR/LG	DPFE Sensor Signal	0.20-1.30v
65	BR/LG	DPFE Sensor Signal	0.95-1.05v
66	---	Not Used	---
67	VT/WT	EVAP CV Solenoid	0-10 Hz (0-100%)
69	PK/YL	A/C WOT Relay Control	On: 1v, Off: 12v
70	---	Not Used	---
71	RD	Vehicle Power	12-14v
73	TN/BK	Injector 5 Control	3.6-7.5 ms
74	BR/YL	Injector 3 Control	3.6-7.5 ms
75	TN	Injector 1 Control	3.6-7.5 ms
76	BK/WT	Power Ground	<0.1v
77	BK/WT	Power Ground	<0.1v
78	TN/LG	Coil 3 Driver (dwell)	5°, 55 mph: 8°
79	WT/LG	TCIL (lamp) Control	TCIL On: 1v, Off: 12v
80	LB/OR	Fuel Pump Relay Control	On: 1v, Off: 12v
81	WT/YL	EPC Solenoid Control	10v (26 psi)
82	---	Not Used	---
83	WT/LB	IAC Solenoid Control	10.8v (33%)
84	DG/WT	TSS Sensor Signal	100-125 Hz
85	DB/OR	CMP Sensor Signal	7 Hz, 55 mph: 18 Hz
86	BK/YL	A/C Pressure Sensor	Switch open: 12v
87	RD/BK	HO2S-21 (B2 S1) Signal	0.1-1.1v
88	LB/RD	MAF Sensor Signal	0.8v, 55 mph: 2.1v
89	GY/WT	TP Sensor Signal	0.53-1.27v
90	BR/WT	Reference Voltage	4.9-5.1v
91	GY/RD	Sensor Ground	<0.050v
92	RD/LG	Brake Position Switch	Brake Off: 12v, On: 1v
93	RD/WT	HO2S-11 (B1 S1) Heater	On: 1v, Off: 12v
94	YL/LB	HO2S-21 (B2 S1) Heater	On: 1v, Off: 12v
95	WT/BK	HO2S-12 (B1 S2) Heater	On: 1v, Off: 12v
96	TN/YL	HO2S-22 (B2 S2) Heater	On: 1v, Off: 12v
97	RD	Vehicle Power	12-14v
98	---	Not Used	---
99	LG/OR	Injector 6 Control	3.6-7.5 ms
100	BR/LB	Injector 4 Control	3.6-7.5 ms
101	WT	Injector 2 Control	3.6-7.5 ms
102	---	Not Used	---
103	BK/WT	Power Ground	<0.1v
104	---	Not Used	---

Pin Connector Graphic

PCM 104-PIN CONNECTOR

```
 1 ©©©©©©©©©©©©©    ©©©©©©©©©©©©© 26
27 ©©©©©©©©©©©©©    ©©©©©©©©©©©©© 52
53 ©©©©©©©©©©©©©  ⬤ ©©©©©©©©©©©©© 78
79 ©©©©©©©©©©©©©    ©©©©©©©©©©©©© 104
```

Terminal View of 104-Pin PCM Wiring Harness Connector

Standard Colors and Abbreviations

Abbreviation	Color	Abbreviation	Color	Abbreviation	Color
BK	Black	GY	Gray	PK	Purple
BL	Blue	GN	Green	RD	Red
BR	Brown	LG	LT Green	TN	Tan
DB	Dark Blue	OR	Orange	WT	White
DG	DK Green	PK	Pink	YL	Yellow

2002-03 Mountaineer 4.0L V6 MFI VIN K 58 Pin Connector

PCM Pin #	Wire Color	Circuit Description (58 Pin)	Value at Hot Idle
1, 5, 15-16	---	Not Used	---
2	PK/BK	Fuel Pump Power	Pump On: 12v, Off: 0v
3	TN/OR	SCP Data Bus (+)	Digital Signals
4	PK/LB	SCP Data Bus (-)	Digital Signals
6	VT/WT	EVAP Canister Vent Valve Control	0-10 Hz (0-100%)
7	GY/BK	Vehicle Speed Sensor (+) Signal	0 Hz, 55 mph: 125 Hz
8	LG/RD	Generator Monitor Signal	130 Hz (45%)
9	PK/YL	A/C Clutch Relay Control	12v (relay on: 1v)
10	YL/WT	4WD Indicator Control	Indicator Off: 12v, On: 1v
11	WT/LG	Passive Antitheft TX Signal	Digital Signals
12	LG/BK	EVAP Purge Solenoid	0-10 Hz (0-100%)
13	VT	Module Programming Signal	0.1v
14	GY/OR	Passive Antitheft RX Signal	Digital Signals
18-19, 21	---	Not Used	---
17	GY/RD	Sensor Ground	<0.050v
20	BR/WT	Reference Voltage	4.9-5.1v
22	DB/LG	Passive Antitheft Indicator Control	Indicator Off: 12v, On: 1v
23	LB/BK	4WD Indicator Low Signal	12v (switch on: 1v)
24	BK/WT	Power Ground	<0.1v
25	BK/WT	Power Ground	<0.1v
26	BK/WT	Power Ground	<0.1v
27	BK/WT	Power Ground	<0.1v
28	BK/YL	Brake Pressure Switch	12v (Brake On: 0v)
29	OR/LB	Cruise Set Indicator Control	Indicator Off: 12v, On: 1v
30	BK/YL	A/C Clutch Relay (switched)	12v (relay on: 1v)
31	LB/RD	MAF Sensor Signal	0.7v, 55 mph: 1.8v
32	VT	Vehicle Power (Start-Run)	12-14v
33	VT	Vehicle Power (Start-Run)	12-14v
34-35	---	Not Used	---
36	OR/LB	Cruise Set Indicator Control	12v (switch set: 0v)
37	---	Not Used	---
38	TN/LB	MAF Sensor Return	<0.050v
39	OR	Starter Motor Relay Circuit	Relay Off: 12v, On: 1v
40	RD/LG	Brake Position Switch	Brake Off: 12v, On: 1v
41	TN/WT	Overdrive Cancel Switch	0v (switch on: 12v)
42, 48-49	---	Not Used	---
43	BK	Power Ground	<0.1v
44	RD/WT	Keep Alive Power (CJB fuse)	12-14v
45	BK	Speed Control Switch Ground	<0.050v
46	OR/LB	Cruise Switch Indicator Control	Indicator Off: 12v, On: 1v
47	RD/YL	A/C Pressure Switch Signal	0v (switch on: 12v)
50	VT	A/C Demand Switch	0v (A/C On: 12v)
51	GY	Intake Air Temperature Sensor	1.5-2.5v
52	RD/PK	Fuel Tank Pressure Sensor	2.6v (0" H20 - cap off)
53-55	---	Not Used	---
56	DG/OR	Speed Control Switch Ground	<0.050v
57	LB/BK	Speed Control Switch Input	0v (switch on: 6.7v)
58	LB/OR	Fuel Pump Relay Control	Relay Off: 12v, On: 1v

Pin Connector Graphic

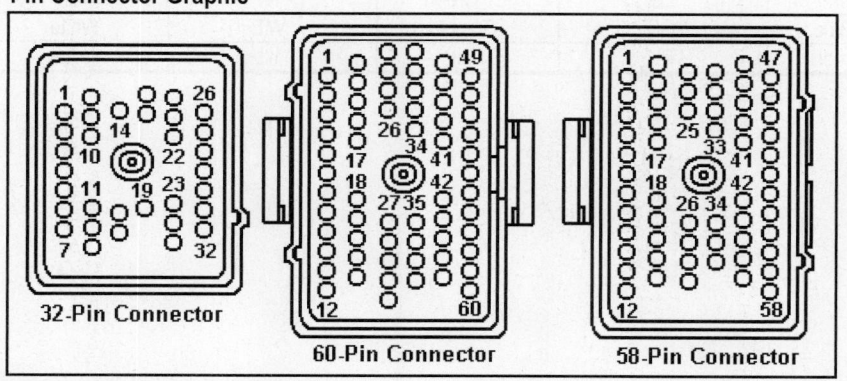

32-Pin Connector   60-Pin Connector   58-Pin Connector

2002-03 Mountaineer 4.0L V6 MFI VIN K 32 Pin Connector

PCM Pin #	Wire Color	Circuit Description (32 Pin)	Value at Hot Idle
1	OR/YL	Shift Solenoid 'A' Control	12v, 55 mph: 1v
2	VT/OR	Shift Solenoid 'B' Control	12v, 55 mph: 1v
3	PK/BK	Shift Solenoid 'C' Control	12v, 55 mph: 1v
4	BR/OR	Shift Solenoid 'D' Control	12v, 55 mph: 1v
5	VT/YL	TCC Solenoid Control	12v, 55 mph: 1v
6	---	Not Used	---
7	WT/YL	Electronic Pressure Control	On: 1v, Off: 12v
8	---	Not Used	---
9	LB/YL	Digital TR Sensor (TR3A)	0v, 55 mph: 1.7v
10	WT/BK	Digital TR Sensor (TR4)	0v, 55 mph: 11v
11	---	Not Used	---
12	WT	Motor Position #4	On: 1v, Off: 12v
13	LB/PK	Pressure Control 'B' Solenoid	On: 1v, Off: 12v
14	---	Not Used	---
15	WT/BK	HO2S-12 (B1 S2) Heater	1v, Off: 12v
16	TN/YL	HO2S-22 (B2 S2) Heater	1v, Off: 12v
17	GY/RD	Sensor Reference Ground	<0.050v
18	LB/BK	Digital TR Sensor (TR2)	0v, 55 mph: 11v
19-20	---	Not Used	---
21	YL/LG	Overdrive Drum Speed Input	200 Hz, 55 mph: 1185 Hz
22	YL/BK	Digital TR Sensor (TR1)	0v, 55 mph: 11v
23	OR/BK	Transmission Fluid Temperature Sensor	2.10-2.40v
24-25	---	Not Used	---
26	DB/YL	Output Shaft Speed Sensor	0 Hz, 985 Hz
27	DG/WT	Turbine Shaft Speed Sensor	360 Hz, 890 Hz
28	RD/LG	HO2S-12 (B1 S2) Signal	0.1-1.1v
29	VT/LG	HO2S-22 (B2 S2) Signal	0.1-1.1v
30-32	---	Not Used	---

Pin Connector Graphic

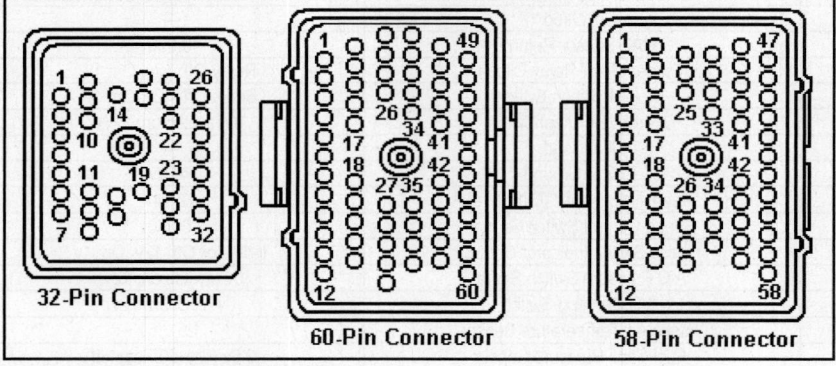

32-Pin Connector

60-Pin Connector

58-Pin Connector

Standard Colors and Abbreviations

Abbreviation	Color	Abbreviation	Color	Abbreviation	Color
BK	Black	GY	Gray	PK	Purple
BL	Blue	GN	Green	RD	Red
BR	Brown	LG	LT Green	TN	Tan
DB	Dark Blue	OR	Orange	WT	White
DG	DK Green	PK	Pink	YL	Yellow

2002-03 Mountaineer 4.0L V6 MFI VIN K 60 Pin Connector

PCM Pin #	Wire Color	Circuit Description (60 Pin)	Value at Hot Idle
1	---	Not Used	---
2	TN	Injector 1 Control	3.3-3.8 ms
3-6	---	Not Used	---
7	RD/WT	HO2S-11 (B1 S1) Heater	1v, Off: 12v
8	YL/LB	HO2S-21 (B2 S1) Heater	1v, Off: 12v
9	WT/LB	IAC Solenoid Control	10.7v (33%)
10	---	Not Used	---
11	TN/BK	Injector 5 Control	3.3-3.8 ms
12	---	Not Used	---
13	PK/WT	Coil 1 Driver (dwell)	5°, 55 mph: 7°
14	WT	Injector 2 Control	3.3-3.8 ms
15	---	Not Used	---
16	BR/PK	VR Solenoid Control	0%, 55 mph: 45%
17	GY/RD	Sensor Reference Ground	<0.050v
18-19	---	Not Used	---
20	BR/WT	Sensor Voltage Reference	4.9-5.1v
21	LG/OR	Injector 6 Control	3.3-3.8 ms
22	---	Not Used	---
23	RD/LB	Coil 2 Driver (dwell)	5°, 55 mph: 7°
24	BR/YL	Injector 3 Control	3.3-3.8 ms
25-30	---	Not Used	---
31	DB/LG	Coil 3 Driver (dwell)	5°, 55 mph: 7°
32	BR/LB	Injector 4 Control	3.3-3.8 ms
33-40	---	Not Used	---
41	BR/LG	DPFE Sensor Signal	0.95-1.05v
42	YL	Knock Sensor 1 Signal	0v
43	DG/WT	Knock Sensor 2 Signal	0v
44	RD/BK	HO2S-21 (B2 S1) Signal	0.1-1.1v
45	GY/LB	HO2S-11 (B1 S1) Signal	0.1-1.1v
46	LG/RD	ECT Sensor Signal	0.5-0.6v
47-50	---	Not Used	---
51	YL/RD	Knock Sensor 1 Ground	0v
52	---	Not Used	---
53	DG	CMP Sensor Signal	6 Hz
54	---	Not Used	---
55	DB	CKP Sensor (-) Signal	400 Hz
56	GY	CKP Sensor (+) Signal	400 Hz
57	GY/WT	TP Sensor Signal	0.53-1.27v
58-60	---	Not Used	---

Pin Connector Graphic

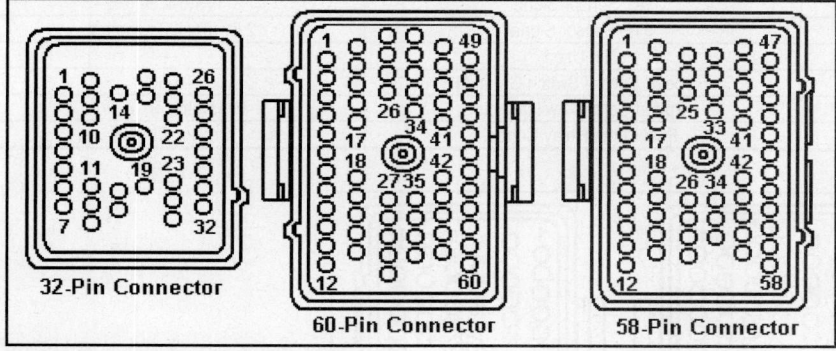

32-Pin Connector

60-Pin Connector

58-Pin Connector

Standard Colors and Abbreviations

Abbreviation	Color	Abbreviation	Color	Abbreviation	Color
BK	Black	GY	Gray	PK	Purple
BL	Blue	GN	Green	RD	Red
BR	Brown	LG	LT Green	TN	Tan
DB	Dark Blue	OR	Orange	WT	White
DG	DK Green	PK	Pink	YL	Yellow

2002-03 Mountaineer 4.6L V8 VIN W 58 Pin Connector

PCM Pin #	Wire Color	Circuit Description (58 Pin)	Value at Hot Idle
1, 5	---	Not Used	---
2	PK/BK	Fuel Pump Power	Pump On: 12v, Off: 0v
3	TN/OR	SCP Data Bus (+)	Digital Signals
4	PK/LB	SCP Data Bus (-)	Digital Signals
6	VT/WT	EVAP Canister Vent Solenoid Control	0-10 Hz (0-100%)
7	GY/BK	Vehicle Speed Sensor (+) Signal	0 Hz, 55 mph: 125 Hz
8	LG/RD	Generator Monitor Signal	130 Hz (45%)
9	PK/YL	A/C Clutch Relay	12v (relay on: 1v)
11	WT/LG	TX Signal	Digital Signals
12	LG/BK	EVAP Purge Solenoid	0-10 Hz (0-100%)
13	VT	Module Programming Signal	0.1v
14	GY/OR	RX Signal	Digital Signals
15-16, 21	---	Not Used	---
17	GY/RD	Sensor Ground	<0.050v
18-19	---	Not Used	---
20	BR/WT	Reference Voltage	4.9-5.1v
22	DB/LG	Passive Antitheft Indicator Control	Indicator Off: 12v, On: 1v
23	LB/BK	4WD Indicator Low Signal	12v (switch on: 1v)
24-27	BK/WT	Power Ground	<0.1v
28	BK/YL	Brake Pressure Switch	12v (Brake On: 0v)
29	OR/LB	Cruise Set Indicator Control	N/A
30	BK/YL	A/C Clutch Relay (switched)	12v (relay on: 1v)
31	LB/RD	MAF Sensor Signal	0.7v, 55 mph: 1.8v
32	VT	Vehicle Power (Start-Run)	12-14v
33	VT	Vehicle Power (Start-Run)	12-14v
34-35	---	Not Used	---
36	OR/LB	Cruise Set Indicator Control	12v (switch set: 0v)
37	---	Not Used	---
38	TN/LB	MAF Sensor Return	<0.050v
39	OR	Starter Motor Relay Signal	0v
40	RD/LG	Brake Position Switch	Brake Off: 12v, On: 1v
41	TN/WT	Overdrive Cancel Switch	0v (switch on: 12v)
42	---	Not Used	---
43	BK	Power Ground	<0.1v
44	RD/WT	Keep Alive Power (CJB fuse)	12-14v
45	BK	Speed Control Switch Ground	<0.050v
46	OR/LB	Cruise Switch Indicator	12v
47	RD/YL	A/C Pressure Switch Signal	0v (switch on: 12v)
48-49	---	Not Used	---
50	VT	A/C Demand Switch	0v (A/C On: 12v)
51	GY	IAT Sensor Signal	1.5-2.5v
52	RD/PK	FTP Sensor Signal	2.6v (0" H20 - cap off)
53-55	---	Not Used	---
56	DG/OR	Speed Control Switch Ground	<0.050v
57	LB/BK	Speed Control Switch Input	0v (switch on: 6.7v)
58	LB/OR	Fuel Pump Relay Control	On: 1v, Off: 12v

Pin Connector Graphic

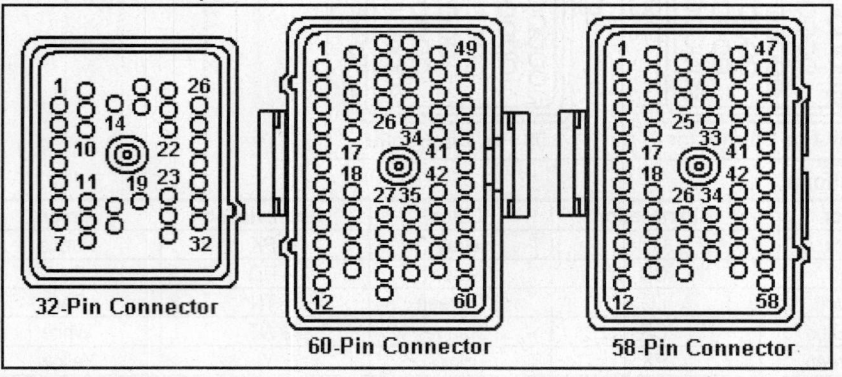

32-Pin Connector

60-Pin Connector

58-Pin Connector

2002-03 Mountaineer 4.6L V8 VIN W 32 Pin Connector

PCM Pin #	Wire Color	Circuit Description (32 Pin)	Value at Hot Idle
1	OR/YL	Shift Solenoid 'A' Control	12v, 55 mph: 1v
2	VT/OR	Shift Solenoid 'B' Control	12v, 55 mph: 1v
3	PK/BK	Shift Solenoid 'C' Control	12v, 55 mph: 1v
4	BR/OR	Shift Solenoid 'D' Control	12v, 55 mph: 1v
5	VT/YL	TCC Solenoid Control	12v, 55 mph: 1v
6	---	Not Used	---
7	WT/YL	Electronic Pressure Control	On: 1v, Off: 12v
8	---	Not Used	
9	LB/YL	Digital TR Sensor (TR3A)	0v, 55 mph: 1.7v
10	WT/BK	Digital TR Sensor (TR4)	0v, 55 mph: 11v
11	---	Not Used	---
12	WT	Motor Position #4	On: 1v, Off: 12v
13	LB/PK	Pressure Control 'B' Solenoid	On: 1v, Off: 12v
14	---	Not Used	---
15	WT/BK	HO2S-12 (B1 S2) Heater	1v, Off: 12v
16	TN/YL	HO2S-22 (B2 S2) Heater	1v, Off: 12v
17	GY/RD	Sensor Reference Ground	<0.050v
18	LB/BK	Digital TR Sensor (TR2)	0v, 55 mph: 11v
19-20	---	Not Used	---
21	YL/LG	Overdrive Drum Speed Input	200 Hz, 55 mph: 1185 Hz
22	YL/BK	Digital TR Sensor (TR1)	0v, 55 mph: 11v
23	OR/BK	Transmission Fluid Temperature Sensor	2.10-2.40v
24-25	---	Not Used	---
26	DB/YL	Output Shaft Speed Sensor	0 Hz, 985 Hz
27	DG/WT	Turbine Shaft Speed Sensor	360 Hz, 890 Hz
28	RD/LG	HO2S-12 (B1 S2) Signal	0.1-1.1v
29	VT/LG	HO2S-22 (B2 S2) Signal	0.1-1.1v
30-32	---	Not Used	---

Pin Connector Graphic

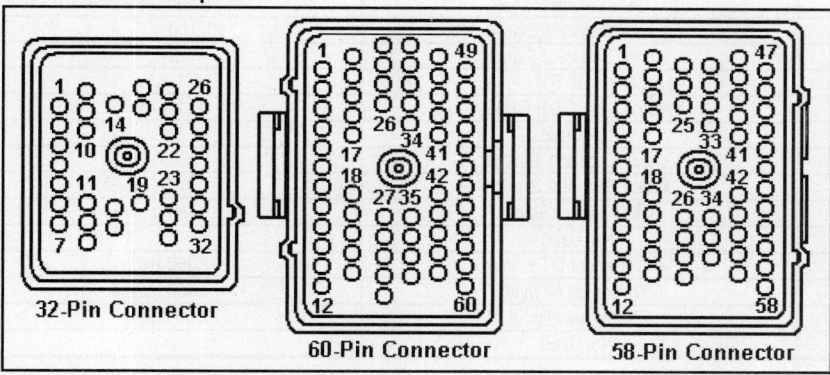

32-Pin Connector   **60-Pin Connector**   **58-Pin Connector**

Standard Colors and Abbreviations

Abbreviation	Color	Abbreviation	Color	Abbreviation	Color
BK	Black	GY	Gray	PK	Purple
BL	Blue	GN	Green	RD	Red
BR	Brown	LG	LT Green	TN	Tan
DB	Dark Blue	OR	Orange	WT	White
DG	DK Green	PK	Pink	YL	Yellow

2002-03 Mountaineer 4.6L V8 VIN W 60 Pin Connector

PCM Pin #	Wire Color	Circuit Description (60 Pin)	Value at Hot Idle
1	PK/WT	COP 2 Driver (dwell)	5°, 55 mph: 7°
2	TN	Injector 1 Control	3.3-3.8 ms
3-6	---	Not Used	---
7	RD/WT	HO2S-11 (B1 S1) Heater	1v, Off: 12v
8	YL/LB	HO2S-21 (B2 S1) Heater	1v, Off: 12v
9	WT/LB	IAC Solenoid Control	10.7v (33%)
10, 15	---	Not Used	---
11	TN/BK	Injector 5 Control	3.3-3.8 ms
12	OR/YL	COP 6 Driver Control	5°, 55 mph: 7°
13	PK/LB	COP 7 Driver Control	5°, 55 mph: 7°
14	WT	Injector 2 Control	3.3-3.8 ms
16	BR/PK	VR Solenoid Control	0%, 55 mph: 45%
17	GY/RD	Sensor Reference Ground	<0.050v
18-19	---	Not Used	---
20	BR/WT	Sensor Voltage Reference	4.9-5.1v
21	LG/OR	Injector 6 Control	3.3-3.8 ms
22	LG/YL	COP 5 Driver Control	5°, 55 mph: 7°
23	BR/YL	COP 5 Driver Control	5°, 55 mph: 7°
24	BR/YL	Injector 3 Control	3.3-3.8 ms
25-28	---	Not Used	---
29	TN/RD	Injector 7 Control	3.3-3.8 ms
30	DG/VT	COP 4 Driver Control	5°, 55 mph: 7°
31	LG/WT	COP 1 Driver (dwell)	5°, 55 mph: 7°
32	BR/LB	Injector 4 Control	3.3-3.8 ms
33-36	---	Not Used	---
37	LB	Injector 8 Control	3.3-3.8 ms
38	WT/RD	COP 8 Driver Control	5°, 55 mph: 7°
39	---	Not Used	---
40	YL/LB	CHT Sensor Signal	0.7v (194°F)
41	BR/LG	DPFE Sensor Signal	0.95-1.05v
42	YL	Knock Sensor 1 Signal	0v
43	DG/WT	Knock Sensor 2 Signal	0v
44	RD/BK	HO2S-21 (B2 S1) Signal	0.1-1.1v
45	GY/LB	HO2S-11 (B1 S1) Signal	0.1-1.1v
46-50	---	Not Used	---
51	YL/RD	Knock Sensor 1 Ground	0v
52	DG/VT	Knock Sensor 2 Ground	0v
53	DG	CMP Sensor Signal	6 Hz
54	---	Not Used	---
55	DB	CKP Sensor (+) Signal	400 Hz
56	GY	CKP Sensor (-) Signal	400 Hz
57	GY/WT	TP Sensor Signal	0.53-1.27v
58-60	---	Not Used	---

Pin Connector Graphic

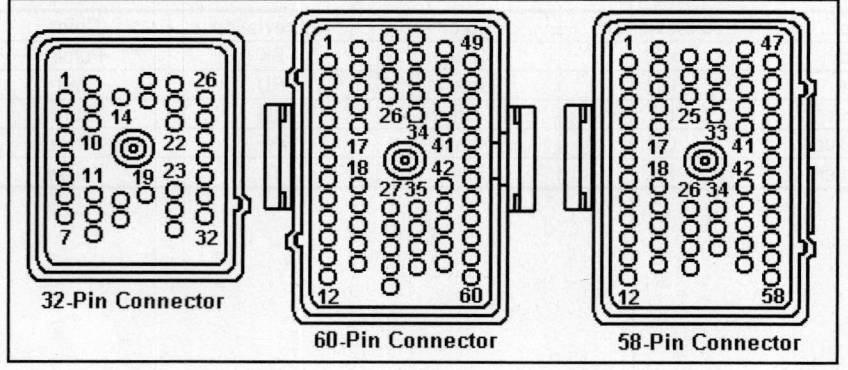

32-Pin Connector

60-Pin Connector

58-Pin Connector

1996-97 Mountaineer 5.0L V8 VIN P 104 Pin Connector

PCM Pin #	Wire Color	Circuit Description (104 Pin)	Value at Hot Idle
1	PK/OR	Shift Solenoid 2 Control	12v, 55 mph: 12v
2	PK/LG	MIL (lamp) Control	MIL On: 1v, Off: 12v
3-12, 14, 17	---	Not Used	---
13	VT	Flash EPROM Power	0.1v
15	PK/LB	Data Bus (-) Signal	Digital Signals
16	TN/OR	Data Bus (+) Signal	Digital Signals
18, 20, 23	---	Not Used	---
19	OR/BK	Accelerator Ride Control	0.1v (Off)
21	BK/PK	CKP (+) Sensor Signal	430-460 Hz
22	GY/YL	CKP (-) Sensor Signal	430-460 Hz
24	BK/WT	Power Ground	<0.1v
25	BK	Case Ground	<0.050v
26	DB/LG	Coil 1 Driver (dwell)	5º, 55 mph: 8º
27	OR/YL	Shift Solenoid 1 Control	1v, 55 mph: 12v
28	---	Not Used	---
29	TN/WT	TCS (switch) Signal	TCS & O/D On: 12v
30	DG	Octane Adjustment	9.3v (Closed: 0v)
31-32, 34	---	Not Used	---
33	PK/OR	VSS (-) Signal	0 Hz, 55 mph: 125 Hz
35	RD/LG	HO2S-12 (B1 S2) Signal	0.1-1.1v
36	TN/LB	MAF Sensor Return	<0.050v
37	OR/BK	TFT Sensor Signal	2.10-2.40v
38	LG/RD	ECT Sensor Signal	0.5-0.6v
39	GY	IAT Sensor Signal	1.5-2.5v
40	DG/YL	Fuel Pump Monitor	On: 12v, Off: 0v
41	DG/OR	A/C Switch Signal	A/C On: 12v, Off: 0v
43	LB/PK	Data Output Link	5v
42, 44-46	---	Not Used	---
47	BK/PK	VR Solenoid Control	0%, 55 mph: 45%
48	TN/YL	Clean Tachometer Output	39-45 Hz
49-50, 53	---	Not Used	---
51	BK/WT	Power Ground	<0.1v
52	RD/LB	Coil 2 Driver (dwell)	5º, 55 mph: 8º
54	DB/WT	TCC Solenoid Control	0%, 55 mph: 95%
55	Y	Keep Alive Power	12-14v
56	LG/BK	EVAP Purge Solenoid	0-10 Hz (0-100%)
57	---	Not Used	---
58	GY/BK	VSS (+) Signal	0 Hz, 55 mph: 125 Hz
59	---	Not Used	---
60	GY/LB	HO2S-11 (B1 S1) Signal	0.1-1.1v
61	PK/LG	HO2S-22 (B2 S2) Signal	0.1-1.1v
62	RD/PK	FTP Sensor Signal	2.6v (0" H2O - cap off)
63	OR/YL	EVP Sensor Signal	0.20-1.30v
64	LB/YL	A/T: TR Sensor Signal	In 'P': 0v, 55 mph: 1.7v

1996-97 Mountaineer 5.0L V8 VIN P 104 Pin Connector

PCM Pin #	Wire Color	Circuit Description (104 Pin)	Value at Hot Idle
65-66	---	Not Used	---
67	PK/WT	EVAP CV Solenoid	0-10 Hz (0-100%)
68	---	Not Used	---
69	PK/YL	A/C WOT Relay Control	On: 1v, Off: 12v
70	---	Not Used	---
71	RD	Vehicle Power	12-14v
72	TN/RD	Injector 7 Control	3.2-4.5 ms
73	TN/BK	Injector 5 Control	3.2-4.5 ms
74	BR/YL	Injector 3 Control	3.2-4.5 ms
75	TN	Injector 1 Control	3.2-4.5 ms
76	BK/WT	Power Ground	<0.1v
77	BK/WT	Power Ground	<0.1v
78	PK/WT	Coil 4 Driver (dwell)	5°, 55 mph: 8°
79	WT/LG	TCIL (lamp) Control	TCIL On: 1v, Off: 12v
80	LB/OR	Fuel Pump Control	On: 1v, Off: 12v
81	WT/YL	EPC Solenoid Control	10v (26 psi)
83	WT/LB	IAC Solenoid Control	10.8v (33%)
84	DB/YL	OSS Sensor Signal	0 Hz, 55 mph: 230 Hz
85	DB/OR	CMP Sensor Signal	5-10 Hz
86	---	Not Used	---
87	RD/BK	HO2S-21 (B2 S1) Signal	0.1-1.1v
88	LB/RD	MAF Sensor Signal	0.8v, 55 mph: 2.1v
89	GY/WT	TP Sensor Signal	0.53-1.27v
90	BR/WT	Reference Voltage	4.9-5.1v
91	GY/RD	Analog Signal Return	<0.050v
92	LG	Brake Position Switch	Brake Off: 12v, On: 1v
93	RD/WT	HO2S-11 (B1 S1) Heater	On: 1v, Off: 12v
94	YL/LB	HO2S-21 (B2 S1) Heater	On: 1v, Off: 12v
95	WT/BK	HO2S-12 (B1 S2) Heater	On: 1v, Off: 12v
96	TN/YL	HO2S-22 (B2 S2) Heater	On: 1v, Off: 12v
97	RD	Vehicle Power	12-14v
98	LB	Injector 8 Control	3.2-4.5 ms
99	LG/OR	Injector 6 Control	3.2-4.5 ms
100	BR/LB	Injector 4 Control	3.2-4.5 ms
101	WT	Injector 2 Control	3.2-4.5 ms
102	---	Not Used	---
103	BK/WT	Power Ground	<0.1v
104	RD/YL	Coil 4 Driver (dwell)	5°, 55 mph: 8°

Pin Connector Graphic

PCM 104-PIN CONNECTOR

Terminal View of 104-Pin PCM Wiring Harness Connector

Standard Colors and Abbreviations

Abbreviation	Color	Abbreviation	Color	Abbreviation	Color
BK	Black	GY	Gray	PK	Purple
BL	Blue	GN	Green	RD	Red
BR	Brown	LG	LT Green	TN	Tan
DB	Dark Blue	OR	Orange	WT	White
DG	DK Green	PK	Pink	YL	Yellow

1998-2001 Mountaineer 5.0L V8 VIN P 104 Pin Connector

PCM Pin #	Wire Color	Circuit Description (104 Pin)	Value at Hot Idle
1	PK/OR	Shift Solenoid 'B' Control	12v, 55 mph: 1v
2	PK/LG	MIL (lamp) Control	MIL On: 1v, Off: 12v
3	YL/BK	Digital TR1 Sensor	In 'P': 0v, Others: 11.5v
4-11	---	Not Used	---
12	YL/WT	Fuel Level Indicator Signal	1.7v (1/2 full)
13	VT	Flash EPROM Power	0.1v
14	---	Not Used	---
15	PK/LB	Data Bus (-) Signal	Digital Signals
16	TN/OR	Data Bus (+) Signal	Digital Signals
17-18	---	Not Used	---
19	OR/BK	Air Suspension Control Signal	Digital Signals
20	---	Not Used	---
21	BK/PK	CKP (+) Sensor Signal	460-500 Hz
22	GY/YL	CKP (-) Sensor Signal	460-500 Hz
24	BK/WT	Power Ground	<0.1v
25	BK	Case Ground	<0.050v
26	DB/LG	Coil 1 Driver (dwell)	5°, 55 mph: 8°
27	OR/YL	Shift Solenoid 'A' Control	1v, 55 mph: 1v
28	---	Not Used	---
29	TN/WT	TCS or Manual Switch	TCS or M/D pressed: 12v
30 ('98-'99)	DG	Octane Adjustment	9.3v (Closed: 0v)
31-32	---	Not Used	---
33 ('98-'99)	PK/OR	VSS (-) Signal	0 Hz, 55 mph: 125 Hz
35	RD/LG	HO2S-12 (B1 S2) Signal	0.1-1.1v
36	TN/LB	MAF Sensor Return	<0.050v
37	OR/BK	TFT Sensor Signal	2.10-2.40v
38	LG/RD	ECT Sensor Signal	0.5-0.6v
39	GY	IAT Sensor Signal	1.5-2.5v
40	DG/YL	Fuel Pump Monitor	On: 12v, Off: 0v
41	VT	A/C Switch Signal	Switch On: 12v (open)
42, 45-46	---	Not Used	---
43 ('98-'99)	LB/PK	Neutral Tow Connector Signal	N/A
47	BR/PK	VR Solenoid Control	0%, 55 mph: 45%
48	TN/YL	Clean Tachometer Output	40-55 Hz
49	LB/BK	Digital TR2 Sensor	In 'P': 0v, Others: 11.5v
50	WT/BK	Digital TR4 Sensor	In 'P': 0v, Others: 11.5v
51	BK/WT	Power Ground	<0.1v
52	RD/LB	Coil 2 Driver (Control)	5°, 55 mph: 8°
53, 57	---	Not Used	---
54	DB/WT	TCC Solenoid Control	0%, 55 mph: 95%
55	YL	Keep Alive Power	12-14v
56	LG/BK	EVAP Purge Solenoid	0-10 Hz (0-100%)
58	GY/BK	VSS (+) Signal	0 Hz, 55 mph: 125 Hz
59	---	Not Used	---
60	GY/LB	HO2S-11 (B1 S1) Signal	0.1-1.1v
61	PK/LG	HO2S-22 (B2 S2) Signal	0.1-1.1v
62	RD/PK	FTP Sensor Signal	2.6v (0" H2O - cap off)
63	---	Not Used	---
64	LB/YL	Digital TR3 Sensor	In 'P': 0v, 55 mph: 1.7v

1998-2001 Mountaineer 5.0L V8 VIN P 104 Pin Connector

PCM Pin #	Wire Color	Circuit Description (104 Pin)	Value at Hot Idle
65 ('98-'00)	BR/LG	DPFE Sensor Signal	0.20-1.30v
65 ('01-'00)	BR/LG	DPFE Sensor Signal	0.95-1.05v
66, 68, 70	---	Not Used	---
67	PK/WT	EVAP CV Solenoid	0-10 Hz (0-100%)
69	PK/YL	A/C WOT Relay Control	On: 1v, Off: 12v
71	RD	Vehicle Power	12-14v
72	TN/RD	Injector 7 Control	3.2-4.5 ms
73	TN/BK	Injector 5 Control	3.2-4.5 ms
74	BR/YL	Injector 3 Control	3.2-4.5 ms
75	TN	Injector 1 Control	3.2-4.5 ms
76, 77	BK/WT	Power Ground	<0.1v
78	PK/WT	Coil 3 Driver (dwell)	5°, 55 mph: 8°
79	WT/LG	TCIL (lamp) Control	TCIL On: 1v, Off: 12v
80	LB/OR	Fuel Pump Control	On: 1v, Off: 12v
81	WT/YL	EPC Solenoid Control	9.5v (10 psi)
82	---	Not Used	---
83	WT/LB	IAC Solenoid Control	10.8v (12-30%)
84	DB/YL	OSS Sensor Signal	At 55 mph: 230-280 Hz
85	DB/OR	CMP Sensor Signal	5-10 Hz
86	BK/YL	A/C Pressure Switch Signal	Open: 12v (<24 psi)
87	RD/BK	HO2S-21 (B2 S1) Signal	0.1-1.1v
88	LB/RD	MAF Sensor Signal	0.8v, 55 mph: 2.1v
89	GY/WT	TP Sensor Signal	0.53-1.27v
90	BR/WT	Reference Voltage	4.9-5.1v
91	GY/RD	Analog Signal Return	<0.050v
92	RD/LG	Brake Position Switch	Brake Off: 12v, On: 1v
93	RD/WT	HO2S-11 (B1 S1) Heater	On: 1v, Off: 12v
94	YL/LB	HO2S-21 (B2 S1) Heater	On: 1v, Off: 12v
95	WT/BK	HO2S-12 (B1 S2) Heater	On: 1v, Off: 12v
96	TN/YL	HO2S-22 (B2 S2) Heater	On: 1v, Off: 12v
97	RD	Vehicle Power	12-14v
98	LB	Injector 8 Control	3.2-4.5 ms
99	LG/OR	Injector 6 Control	3.2-4.5 ms
100	BR/LB	Injector 4 Control	3.2-4.5 ms
101	WT	Injector 2 Control	3.2-4.5 ms
102	---	Not Used	---
103	BK/WT	Power Ground	<0.1v
104	RD/YL	Coil 4 Driver (dwell)	5°, 55 mph: 8°

Pin Connector Graphic

PCM 104-PIN CONNECTOR

Terminal View of 104-Pin PCM Wiring Harness Connector

Standard Colors and Abbreviations

Abbreviation	Color	Abbreviation	Color	Abbreviation	Color
BK	Black	GY	Gray	PK	Purple
BL	Blue	GN	Green	RD	Red
BR	Brown	LG	LT Green	TN	Tan
DB	Dark Blue	OR	Orange	WT	White
DG	DK Green	PK	Pink	YL	Yellow

Navigator PIN Tables

1998-99 Navigator 5.4L V8 MFI VIN L (4R100) 104 Pin Connector

PCM Pin #	Wire Color	Circuit Description (104 Pin)	Value at Hot Idle
1	LG/YL	COP 5 Driver (dwell)	5°, 55 mph: 8°
2	PK/LG	MIL (lamp) Control	MIL On: 1v, Off: 12v
3	BK/WT	Power Ground	<0.1v
6	OR/YL	Shift Solenoid 1 Control	1v, 55 mph: 12v
9	YL/WT	Fuel Level Indicator Signal	1.7v (1/2 full)
11	PK/OR	Shift Solenoid 2 Control	12v, 55 mph: 12v
12	WT/LG	TCIL (lamp) Control	TCIL On: 1v, Off: 12v
13	VT	Flash EEPROM Power	0.1v
14	LB/BK	4x4 Indicator Control	4x4 lamp On: 1v, Off: 0v
15	PK	Data Bus (-) Signal	Digital Signals
16	TN/OR	Data Bus (+) Signal	Digital Signals
19	DB/WT	Air Suspension Control	Digital Signals
20	BR/OR	Coast Clutch Solenoid	12v, 55 mph: 12v
21	DB	CKP (+) Sensor Signal	400-420 Hz
22	GY	CKP (-) Sensor Signal	400-420 Hz
24	BK/WT	Power Ground	<0.1v
25	LG/YL	Case Ground	<0.050v
26	LG/WT	COP 1 Driver (dwell)	5°, 55 mph: 8°
27	OR/YL	COP 6 Driver (dwell)	5°, 55 mph: 8°
29	TN/WT	TCS (switch) Signal	TCS & O/D On: 12v
30	DG	Octane Adjust Shorting Bar	9.3v (Closed: 0v)
33	PK/OR	VSS (-) Signal	0 Hz, 55 mph: 125 Hz
34	YL/BK	Digital TR1 Sensor	0v, 55 mph: 11.5v
35	RD/LG	HO2S-12 (B1 S2) Signal	0.1-1.1v
36	TN/LB	MAF Sensor Return	<0.050v
37	OR/BK	TFT Sensor Signal	2.10-2.40v
39	GY	IAT Sensor Signal	1.5-2.5v
40	DG/YL	Fuel Pump Monitor	On: 12v, Off: 0v
41	BK/YL	A/C High Pressure Switch	Switch On: 12v (open)
43	OR/LG	Overhead Console Module	Digital Signals
45	RD/WT	CHIL (lamp) Control	CHT On: 1v, Off: 12v
46	BR	IMTV (tuning valve) Control	12v
47	BR/PK	VR Solenoid Control	0%, 55 mph: 45%
48	WT/PK	Clean Tachometer Output	46 Hz, 55 mph: 115 Hz
49	LB/BK	Digital TR2 Sensor	0v, 55 mph: 11.5v
50	WT/BK	Digital TR4 Sensor	0v, 55 mph: 11.5v
51	BK/WT	Power Ground	<0.1v
52	PK/WT	COP Driver 2 Driver	5°, 55 mph: 8°
53	PK/B	COP 7 Driver (dwell)	5°, 55 mph: 8°
54	PK/YL	TCC Solenoid Control	0%, 55 mph: 95%
55	BK/LG	Keep Alive Power	12-14v
56	LG/BK	EVAP Purge Solenoid	0-10 Hz (0-100%)
57	YL/RD	Knock Sensor 1 Signal	0v
58	GY/BK	VSS (+) Signal	0 Hz, 55 mph: 125 Hz
60	GY/LB	HO2S-11 (B1 S1) Signal	0.1-1.1v
61	PK/LG	HO2S-22 (B2 S2) Signal	0.1-1.1v
62	RD/PK	FTP Sensor Signal	2.6v (0" H2O - cap off)
64	LB/YL	Digital TR3A Sensor	In 'P': 0v, in O/D: 1.7v

1998-99 Navigator 5.4L V8 MFI VIN L (4R100) 104 Pin Connector

PCM Pin #	Wire Color	Circuit Description (104 Pin)	Value at Hot Idle
65	BR/LG	DPFE Sensor Signal	0.20-1.30v
66	YL/LG	CHT Sensor Signal	0.6 to 1.7v (194ºF)
67	PK/WT	EVAP CV Solenoid	0-10 Hz (0-100%)
71	RD	Vehicle Power	12-14v
72	TN/RD	Injector 7 Control	3.2-3.8 ms
73	TN/BK	Injector 5 Control	3.2-3.8 ms
74	BR/YL	Injector 3 Control	3.2-3.8 ms
75	TN	Injector 1 Control	3.2-3.8 ms
77	BK/WT	Power Ground	<0.1v
78	WT/PK	COP 3 Driver (dwell)	5º, 55 mph: 8º
79	WT/RD	COP 8 Driver (dwell)	5º, 55 mph: 8º
80	LB/OR	Fuel Pump Control	On: 1v, Off: 12v
81	WT/YL	EPC Solenoid Control	9.1v (5 psi)
83	WT/LB	IAC Solenoid Control	10.7v (33%)
85	DG	CMP Sensor Signal	6 Hz, 55 mph: 15-17 Hz
87	RD/BK	HO2S-21 (B2 S1) Signal	0.1-1.1v
88	LB/RD	MAF Sensor Signal	0.8v, 55 mph: 2.3v
89	GY/WT	TP Sensor Signal	0.53-1.27v
90	BR/WT	Reference Voltage	4.9-5.1v
91	GY/RD	Analog Signal Return	<0.050v
92	LG	Brake Pedal Switch Signal	Brake Off: 12v, On: 1v
93	RD/WT	HO2S-11 (B1 S1) Heater	On: 1v, Off: 12v
94	YL/LB	HO2S-21 (B2 S1) Heater	On: 1v, Off: 12v
95	WT/BK	HO2S-12 (B1 S2) Heater	On: 1v, Off: 12v
96	TN/YL	HO2S-22 (B2 S2) Heater	On: 1v, Off: 12v
97	RD	Vehicle Power	12-14v
98	LB	Injector 8 Control	3.2-3.8 ms
99	LG/OR	Injector 6 Control	3.2-3.8 ms
100	BR/LB	Injector 4 Control	3.2-3.8 ms
101	WT	Injector 2 Control	3.2-3.8 ms
103	BK/WT	Power Ground	<0.1v
104	DG/PK	COP 4 Driver (dwell)	5º, 55 mph: 8º

Pin Connector Graphic

PCM 104-PIN CONNECTOR

Terminal View of 104-Pin PCM Wiring Harness Connector

Standard Colors and Abbreviations

Abbreviation	Color	Abbreviation	Color	Abbreviation	Color
BK	Black	GY	Gray	PK	Purple
BL	Blue	GN	Green	RD	Red
BR	Brown	LG	LT Green	TN	Tan
DB	Dark Blue	OR	Orange	WT	White
DG	DK Green	PK	Pink	YL	Yellow

1999-2001 Navigator 5.4L V8 VIN A (A/T) 104 Pin Connector

PCM Pin #	Wire Color	Circuit Description (104 Pin)	Value at Hot Idle
1	OR/YL	COP 6 Driver (dwell)	5°, 55 mph: 8°
3	BK/WT	Power Ground	<0.1v
6	OR/YL	Shift Solenoid 1 Control	1v, 55 mph: 12v
11	VT/O	Shift Solenoid 2 Control	12v, 55 mph: 12v
12	WT/LG	TCIL (lamp) Control	TCIL On: 1v, Off: 12v
13	VT	Flash EEPROM Power	0.1v
14	LB/BK	4x4 Low Indicator Switch	Switch On: 0v, Off: 12v
15	PK/LB	Data Bus (-) Signal	Digital Signals
16	TN/OR	Data Bus (+) Signal	Digital Signals
19	DB/WT	Air Suspension Control	Digital Signals
20	BR/OR	Coast Clutch Solenoid	12v, 55 mph: 12v
21	DB	CKP (+) Sensor Signal	400-420 Hz
22	GY	CKP (-) Sensor Signal	400-420 Hz
24	BK/WT	Power Ground	<0.1v
25	LB/YL	Case Ground	<0.050v
26	LG/WT	COP 1 Driver (dwell)	5°, 55 mph: 8°
27	OR/YL	COP 5 Driver (dwell)	5°, 55 mph: 8°
29	TN/WT	TCS (switch) Signal	TCS & O/D On: 12v
32	DG/VT	Knock Sensor 1 Signal	0v
34	YL/BK	Digital TR1 Sensor	0v, 55 mph: 11.5v
35	RD/LG	HO2S-12 (B1 S2) Signal	0.1-1.1v
36	TN/LB	MAF Sensor Return	<0.050v
37	OR/BK	TFT Sensor Signal	2.10-2.40v
38	LG/RD	ECT Sensor Signal	0.5-0.6v
39	GY	IAT Sensor Signal	1.5-2.5v
40	DG/YL	Fuel Pump Monitor	On: 12v, Off: 0v
41	BK/YL	A/C High Pressure Switch	Switch On: 12v (open)
42	PK/OR	TP Sensor Signal	0.53-1.27v
43	OR/LG	Overhead Console Module	Digital Signals
43 ('00-'02)	OR/LG	Data Output Link	Digital Signals
46	BR	IMRC (valve) Control	12v
47	BR/PK	VR Solenoid Control	0%, 55 mph: 45%
49	LB/BK	Digital TR2 Sensor	0v, 55 mph: 11.5v
50	WT/BK	Digital TR4 Sensor	0v, 55 mph: 11.5v
51	BK/WT	Power Ground	<0.1v
52	WT/PK	COP 3 Driver (dwell)	5°, 55 mph: 8°
53	DG/VT	COP 4 Driver (dwell)	5°, 55 mph: 8°
54	VTR/YL	TCC Solenoid Control	0%, 55 mph: 95%
55	RD/WTH	Keep Alive Power	12-14v
56	LG/BK	EVAP Purge Solenoid	0-10 Hz (0-100%)
57	YL/RD	Knock Sensor 1 Signal	0v
59	DG/WT	TSS Sensor Signal	370 Hz (700 rpm)
60	GY/LB	HO2S-11 (B1 S1) Signal	0.1-1.1v
61	VTR/LG	HO2S-22 (B2 S2) Signal	0.1-1.1v
62	RD/PK	FTP Sensor Signal	2.6v (0" H2O - cap off)
64	LB/YL	Digital TR3A Sensor	In 'P': 0v, in O/D: 1.7v

1999-2001 Navigator 5.4L V8 VIN A (A/T) 104 Pin Connector

PCM Pin #	Wire Color	Circuit Description (104 Pin)	Value at Hot Idle
65 ('99-'00)	BR/LG	DPFE Sensor Signal	0.20-1.30v
65 ('01-'02)	BR/LG	DPFE Sensor Signal	0.95-1.05v
66	YL/LG	CHT Sensor Signal	0.6 to 1.7v (194°F)
67	VT/WT	EVAP CV Solenoid	0-10 Hz (0-100%)
68	GY/BK	Vehicle Speed Out Signal	0 Hz, 55 mph: 125 Hz
69	PK/YL	A/C WOT Relay Control	On: 1v, Off: 12v
71	RD	Vehicle Power	12-14v
72	TN/RD	Injector 7 Control	3.2-3.8 ms
73	TN/BK	Injector 5 Control	3.2-3.8 ms
74	BR/YL	Injector 3 Control	3.2-3.8 ms
75	TN	Injector 1 Control	3.2-3.8 ms
77	BK/WT	Power Ground	<0.1v
78	PK/LB	COP 7 Driver (dwell)	5°, 55 mph: 8°
79	WT/RD	COP 8 Driver (dwell)	5°, 55 mph: 8°
80	LB/OR	Fuel Pump Relay Control	On: 1v, Off: 12v
81	WT/YL	EPC Solenoid Control	9.1v (5 psi)
83	WT/LB	IAC Solenoid Control	10.7v (33%)
84	DB/YL	OSS Sensor Signal	0 Hz, 55 mph: 125 Hz
85	DG	CMP Sensor Signal	6 Hz, 55 mph: 14-17 Hz
87	RD/BK	HO2S-21 (B2 S1) Signal	0.1-1.1v
88	LB/RD	MAF Sensor Signal	0.8v, 55 mph: 2.3v
89	GY/WT	TP Sensor Signal	0.53-1.27v
90	BR/WT	Reference Voltage	4.9-5.1v
91	GY/RD	Analog Signal Return	<0.050v
92	RD/LG	Brake Pedal Switch Signal	Brake Off: 12v, On: 1v
93	RD/WT	HO2S-11 (B1 S1) Heater	On: 1v, Off: 12v
94	YL/LB	HO2S-21 (B2 S1) Heater	On: 1v, Off: 12v
95	WT/BK	HO2S-12 (B1 S2) Heater	On: 1v, Off: 12v
96	TN/YL	HO2S-22 (B2 S2) Heater	On: 1v, Off: 12v
97	RD	Vehicle Power	12-14v
98	LB	Injector 8 Control	3.2-3.8 ms
99	LG/OR	Injector 6 Control	3.2-3.8 ms
100	BR/LB	Injector 4 Control	3.2-3.8 ms
101	W	Injector 2 Control	3.2-3.8 ms
103	BK/WT	Power Ground	<0.1v
104	PK/WT	COP 2 Driver (dwell)	5°, 55 mph: 8°

Pin Connector Graphic

Standard Colors and Abbreviations

Abbreviation	Color	Abbreviation	Color	Abbreviation	Color
BK	Black	GY	Gray	PK	Purple
BL	Blue	GN	Green	RD	Red
BR	Brown	LG	LT Green	TN	Tan
DB	Dark Blue	OR	Orange	WT	White
DG	DK Green	PK	Pink	YL	Yellow

2001-02 Expedition 5.4L V8 VIN R (E4OD) 104 Pin Connector

PCM Pin #	Wire Color	Circuit Description (104 Pin)	Value at Hot Idle
1	OR/YL	COP 6 Driver (dwell)	5º, 55 mph: 8º
2	---	Not Used	
3	BK/WT	Power Ground	<0.1v
6	OR/YL	Shift Solenoid 1 Control	1v, 30 mph: 12v
7-10	---	Not Used	
11	VT/OR	Shift Solenoid 2 Control	12v, 55 mph: 1v
12	WT/LG	TCIL (lamp) Control	TCIL On: 1v, Off: 12v
13	VT	Flash EEPROM Power	0.1v
14	LB/BK	4x4 Low Indicator Switch	4x4 Switch Off: 0.1v
15	PK/LB	Data Bus (-) Signal	Digital Signals
16	TN/OR	Data Bus (+) Signal	Digital Signals
17-18	---	Not Used	---
19	DB/WT	Suspension Control Module	Digital Signals
20	BR/OR	Coast Clutch Solenoid	12v, 55 mph: 12v
21	DB	CKP (+) Sensor Signal	430-475 Hz
22	GY	CKP (-) Sensor Signal	430-475 Hz
25	LB/YL	Case Ground	<0.050v
26	LG/WT	COP 1 Driver (dwell)	5º, 55 mph: 8º
27	LG/YL	COP 5 Driver (dwell)	5º, 55 mph: 8º
28	---	Not Used	---
29	TN/WT	TCS (switch) Signal	TCS & O/D On: 12v
30-31	---	Not Used	---
32	DG/PK	Knock Sensor 1 Return	<0.050v
33	---	Not Used	---
34	YL/BK	Digital TR1 Sensor	0v, 55 mph: 11.5v
35	RD/LG	HO2S-12 (B1 S2) Signal	0.1-1.1v
36	TN/LB	MAF Sensor Return	<0.050v
37	OR/BK	TFT Sensor Signal	2.10-2.40v
38	---	Not Used	---
39	GY	IAT Sensor Signal	1.5-2.5v
40	DG/YL	Fuel Pump Monitor	On: 12v, Off: 0v
41	BK/YL	A/C Request Signal	A/C On: 12v, Off: 0v
42	PK/OR	TP Sensor Signal	0.9v, 55 mph: 1.3v
43	OR/LG	Data Output Link	5v
44-45	---	Not Used	---
46	BR	IMRC (valve) Control	12v
47	BR/PK	VR Solenoid Control	0%, 55 mph: 45%
48	---	Not Used	---
49	LB/BK	Digital TR2 Sensor	0v, 55 mph: 11.5v
50	WT/BK	Digital TR4 Sensor	0v, 55 mph: 11.5v
51	BK/WT	Power Ground	<0.1v
52	WT/PK	COP 3 Driver (dwell)	5º, 55 mph: 8º
53	DG/VT	COP 4 Driver (dwell)	5º, 55 mph: 8º
54	VT/YL	TCC Solenoid Control	0%, 55 mph: 95%
55	RD/WT	Keep Alive Power	12-14v
56	LG/BK	EVAP Purge Solenoid	0-10 Hz (0-100%)
57	YL/RD	Knock Sensor 1 Signal	0v
58	---	Not Used	---
59	DG/WT	TSS Sensor Signal	100 Hz, 30 mph: 700 Hz
60	GY/LB	HO2S-11 (B1 S1) Signal	0.1-1.1v
61	VT/LG	HO2S-22 (B2 S2) Signal	0.1-1.1v
62	RD/PK	FTP Sensor Signal	2.6v (0" H2O - cap off)
63	---	Not Used	---
64	LB/YL	Digital TR3A Sensor	In 'P': 0v, in O/D: 1.7v

2001-02 Navigator 5.4L V8 VIN R (E4OD) 104 Pin Connector

PCM Pin #	Wire Color	Circuit Description (104 Pin)	Value at Hot Idle
65	BR/LG	DPFE Sensor Signal	0.95-1.05v
66	YL/LG	CHT Sensor Signal	0.6v (194°F)
67	VT/WT	EVAP CV Solenoid	0-10 Hz (0-100%)
68	GY/BK	Vehicle Speed Sensor	0 Hz, 55 mph: 125 Hz
69	PK/YL	A/C WOT Relay Control	Relay Off: 12v, On: 1v
70	---	Not Used	---
71	RD	Vehicle Power	12-14v
72	TN/RD	Injector 7 Control	2.7-3.5 ms
73	TN/BK	Injector 5 Control	2.7-3.5 ms
74	BR/YL	Injector 3 Control	2.7-3.5 ms
75	TN	Injector 1 Control	2.7-3.5 ms
76	BK/WT	Power Ground	<0.1v
77	---	Not Used	---
78	PK/LB	COP 7 Driver (dwell)	5°, 55 mph: 8°
79	WT/RD	COP 8 Driver (dwell)	5°, 55 mph: 8°
80	LB/OR	Fuel Pump Control	On: 1v, Off: 12v
81	WT/YL	EPC Solenoid Control	9.1v (4 psi)
82	---	Not Used	---
83	WT/LB	IAC Solenoid Control	10.7v (33%)
84	DB/YL	OSS Sensor Signal	0 Hz, 55 mph: 350 Hz
85	DG	CMP Sensor Signal	6 Hz, 55 mph: 13-17 Hz
86	---	Not Used	---
87	RD/BK	HO2S-21 (B2 S1) Signal	0.1-1.1v
88	LB/RD	MAF Sensor Signal	0.8v, 55 mph: 1.9v
89	GY/WT	TP Sensor Signal	0.53-1.27v
90	BR/WT	Reference Voltage	4.9-5.1v
91	GY/RD	Analog Signal Return	<0.050v
92	RD/LG	Brake Pedal Switch Signal	Brake Off: 12v, On: 1v
93	RD/WT	HO2S-11 (B1 S1) Heater	On: 1v, Off: 12v
94	YL/LB	HO2S-21 (B2 S1) Heater	On: 1v, Off: 12v
95	WT/BK	HO2S-12 (B1 S2) Heater	On: 1v, Off: 12v
96	TN/YL	HO2S-22 (B2 S2) Heater	On: 1v, Off: 12v
97	RD	Vehicle Power	12-14v
98	LB	Injector 8 Control	2.7-3.5 ms
99	LG/OR	Injector 6 Control	2.7-3.5 ms
100	BR/LB	Injector 4 Control	2.7-3.5 ms
101	WT	Injector 2 Control	2.7-3.5 ms
102	---	Not Used	---
103	BK/WT	Power Ground	<0.1v
104	PK/WT	COP 2 Driver (dwell)	5°, 55 mph: 8°

Pin Connector Graphic

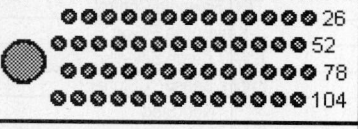

PCM 104-PIN CONNECTOR

Terminal View of 104-Pin PCM Wiring Harness Connector

Standard Colors and Abbreviations

Abbreviation	Color	Abbreviation	Color	Abbreviation	Color
BK	Black	GY	Gray	PK	Purple
BL	Blue	GN	Green	RD	Red
BR	Brown	LG	LT Green	TN	Tan
DB	Dark Blue	OR	Orange	WT	White
DG	DK Green	PK	Pink	YL	Yellow

2003 Navigator 5.4L V8 MFI VIN R (E4OD) C175a Pin Connector

PCM Pin #	Wire Color	Circuit Description (46 Pin)	Value at Hot Idle
1	BK/WT	Power Ground	<0.1v
2	PK/YL	A/C WOT Relay Control	Relay Off: 12v, On: 1v
3	PK/OR	TP Sensor Signal	0.9v, 55 mph: 1.3v
4	DB/OG	Starter Relay Control Circuit	Relay Off: 12v, On: 1v
5	---	Not Used	---
6	OG/LB	Cooling Fan Speed Signal	0.5-4.9v
7	TN/LG	Generator Communicator Command	1.5-10.5v (40-250 Hz)
8	RD/PK	Fuel Tank Pressure Sensor	2.6v (0" H2O - cap off)
9	OR/RD	Antitheft Indicator Control	Indicator Off: 12v, On: 1v
10	LB/YL	Power Ground	<0.1v
11	BK/WT	Power Ground	<0.1v
12	VT/YL	Speed Control Servo Signal 'C'	DC pulse signals
13	GY/BK	VSS (+) Signal	0 Hz, 55 mph: 125 Hz
14	BR/PK	Generator Monitor Signal	130 Hz (45%)
15	BK/YL	Brake Pedal Position Switch	Brake Off: 0v, On: 12v
16	---	Not Used	---
17	DG/OR	Steering Wheel Control Ground	<0.050v
18	YL/LG	Power Steering Pressure Switch	Straight: 0v, Turned: 12v
19	GY	IAT Sensor Signal	1.5-2.5v
20	DG/YL	Fuel Pump Relay Output	Off: 0v, On: 12v
21	---	Not Used	---
22	TN/WT	TCS (switch) Signal	TCS & O/D On: 12v
23	BK/WT	Power Ground	<0.1v
24	PK/BK	Speed Control Servo Signal 'B'	DC pulse signals
25	LG/VT	Cooling Fan Control	Digital Signal: 0-12-0v
26	TN	Speed Control Servo Common	<0.050v
27	LB/OR	Fuel Pump Relay Control	Relay Off: 12v, On: 1v
28-29	---	Not Used	---
30	TN/LB	MAF Sensor Return	<0.050v
31	LB/RD	MAF Sensor Signal	0.6v, 55 mph: 2.1v
32	TN/OR	SCP Data Bus (+) Signal	Digital Signals
33	GY/RD	Sensor Ground	<0.050v
34	RD	Vehicle Power (Start-Run)	12-14v
35	DG/WT	Speed Control Servo Signal 'A'	DC pulse signals
36	VT/WT	EVAP Canister Vent Solenoid Control	0-10 Hz (0-100%)
37	LB/BK	Speed Control Switch	N/A
38	LG/BK	EVAP Purge Solenoid Control	0-10 Hz (0-100%)
39	VT	Module Programming Signal	0.1v
40	RD/WT	Direct Battery	12-14v
41	OR/LB	A/C Pressure Switch Signal	0v (switch on: 12v)
42	GY/OR	Passive Antitheft RX Signal	Digital Signals
43	WT/LG	Passive Antitheft TX Signal	Digital Signals
44	PK/LB	Data Bus (-) Signal	<0.050v
45	BR/WT	Reference Voltage	4.9-5.1v
46	RD	Vehicle Power (Start-Run)	12-14v

2003 Navigator 5.4L V8 MFI VIN R (E4OD) C138t Pin Connector

PCM Pin #	Wire Color	Circuit Description (30 Pin)	Value at Hot Idle
1	---	Not Used	---
2	VT/LG	HO2S-22 (B2 S2) Signal	0.1-1.1v
3	RD/LG	HO2S-12 (B1 S2) Signal	0.1-1.1v
4-10	---	Not Used	---
11	VT/YL	TCC Solenoid Control	0%, 55 mph: 95%
12	OR/YL	Shift Solenoid 'A' Control	1v, 55 mph: 1v
13	VT/OR	Shift Solenoid 'B' Control	12v, 55 mph: 1v
14	BR/OR	Coast Clutch Solenoid Control	12v, 55 mph: 12v
15-16	---	Not Used	---
17	LB/YL	Digital TR3A Sensor	0v, 55 mph: 11.5v
18	LB/BK	Digital TR2 Sensor	0v, 55 mph: 11.5v
19	WT/BK	Digital TR4 Sensor	0v, 55 mph: 11.5v
20	YL/BK	Digital TR1 Sensor	In 'P': 0v, in O/D: 1.7v
21	WT/BK	HO2S-12 (B1 S2) Heater	On: 1v, Off: 12v
22	---	Not Used	---
23	WT/YL	EPC Solenoid Control	9.1v (6 psi)
24	---	Not Used	---
25	DB/YL	OSS Sensor Signal	0 Hz, 55 mph: 250 Hz
26	DG/WT	Turbine Shaft Speed Sensor	360 Hz, 890 Hz
27	GY/RD	Sensor Ground	<0.050v
28	OR/BK	Transmission Fluid Temperature Sensor	2.10-2.40v
29	TN/YL	HO2S-22 (B2 S2) Heater	On: 1v, Off: 12v
30	---	Not Used	---

Pin Connector Graphic

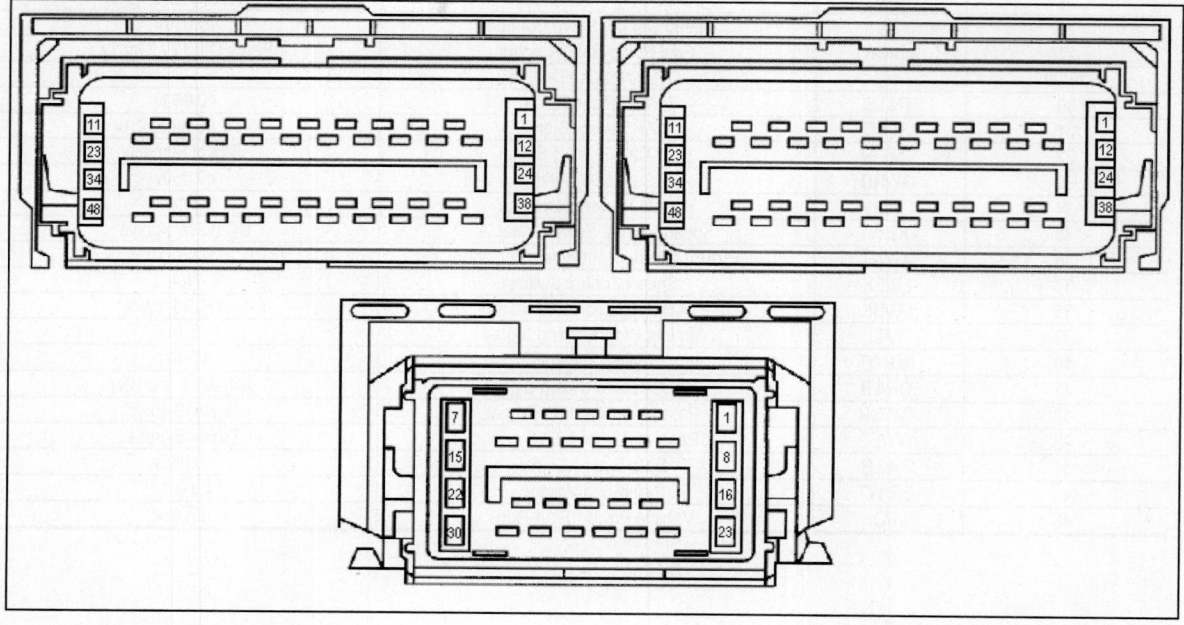

2003 Navigator 5.4L V8 MFI VIN R (E4OD) C1381e Pin Connector

PCM Pin #	Wire Color	Circuit Description (46 Pin)	Value at Hot Idle
1	PK/WT	COP 2 Driver (dwell)	5°, 55 mph: 8°
2	WT/LB	IAC Solenoid Control	10.7v (33%)
3	TN/BK	Injector 5 Control	2.7-4.1 ms
4	LG/OR	Injector 6 Control	2.7-4.1 ms
5	TN/RD	Injector 7 Control	2.7-4.1 ms
6	LB	Injector 8 Control	2.7-4.1 ms
7	BR/YL	Injector 3 Control	2.7-4.1 ms
8	BR/LB	Injector 4 Control	2.7-4.1 ms
9	GY/YL	Vehicle Power (Start-Run)	12-14v
10	---	Not Used	---
11	WT/RD	COP 8 Driver (dwell)	5°, 55 mph: 8°
12	PK/LB	COP 7 Driver (dwell)	5°, 55 mph: 8°
13	BR	Intake Manifold Runner Control Module	12v
14	TN	Injector 1 Control	2.7-4.1 ms
15	WT	Injector 2 Control	2.7-4.1 ms
16-19	---	Not Used	---
20	YL/LB	HO2S-21 (B2 S1) Heater	On: 1v, Off: 12v
21	RD/WT	HO2S-11 (B1 S1) Heater	On: 1v, Off: 12v
22	BR/PK	EGR Vacuum Regulator Control	0%, 55 mph: 45%
23	DG/VT	COP 4 Driver (dwell)	5°, 55 mph: 8°
24	WT/PK	COP 3 Driver (dwell)	5°, 55 mph: 8°
25	GY/RD	Sensor Ground	<0.050v
26	GY/LB	HO2S-11 (B1 S1) Signal	0.1-1.1v
27	RD/BK	HO2S-21 (B2 S1) Signal	0.1-1.1v
28	DG/VT	Knock Sensor 2 (+) Signal	0v
29	YL/RD	Knock Sensor 1 (+) Signal	0v
30	DB	CKP (+) Sensor Signal	400-420 Hz
31	DG	Camshaft Position Sensor	6 Hz, 55 mph: 15 Hz
32	DG/PK	Dual Knock Sensor (-) Signal	<0.050v
33	BR/LG	ECT Sensor Signal	0.5-0.6v
34	LG/YL	COP 5 Driver (dwell)	5°, 55 mph: 8°
35	LG/WT	COP 1 Driver (dwell)	5°, 55 mph: 8°
36	BR/WT	Reference Voltage	4.9-5.1v
37-38	---	Not Used	---
39	DG/WT	Knock Sensor 2 (-) Signal	<0.050v
40	YL	Knock Sensor 1 (-) Signal	<0.050v
41	GY	CKP (-) Sensor Signal	400-420 Hz
42-43	---	Not Used	---
44	GY/WT	TP Sensor Signal	0.53-1.27v
45	YL/LG	CHT Sensor Signal	0.6 to 1.7v (194°F)
46	OR/YL	COP 6 Driver (dwell)	5°, 55 mph: 8°

FORD VAN CONTENTS

About This Section
Introduction ....Page 14-3
How to use this Section ....Page 14-3

AEROSTAR PIN TABLES
3.0L V6 MFI VIN U (All) 60 Pin **(1990)** ....Page 14-4
3.0L V6 MFI VIN U (All) 60 Pin **(1991)** ....Page 14-5
3.0L V6 MFI VIN U (All) 60 Pin **(1992-95)** ....Page 14-6
3.0L V6 MFI VIN U (A/T) 104 Pin **(1996-97)** ....Page 14-8
4.0L V6 OHV MFI VIN X (A/T) 60 Pin **(1990)** ....Page 14-10
4.0L V6 OHV MFI VIN X (A/T) 60 Pin **(1991)** ....Page 14-11
4.0L V6 OHV MFI VIN X (A/T) 60 Pin **(1992-95)** ....Page 14-12
4.0L V6 OHV MFI VIN X (A/T) 104 Pin **(1996-97)** ....Page 14-14

E-SERIES VAN PIN TABLES
4.2L V6 OHV MFI VIN 2 (A/T) 104 Pin **(1997-99)** ....Page 14-16
4.2L V6 OHV MFI VIN 2 (A/T) 104 Pin **(2000-03)** ....Page 14-18
4.6L V8 SOHC MFI VIN 6 (A/T) 104 Pin **(1997-99)** ....Page 14-20
4.6L V8 SOHC VIN W (All) 104 Pin **(2000-03)** ....Page 14-22
4.9L I6 OHV MFI VIN Y (All) 60 Pin **(1990)** ....Page 14-24
4.9L I6 OHV MFI VIN Y (E4OD) 60 Pin **(1990)** ....Page 14-25
4.9L I6 OHV MFI VIN Y (All) 60 Pin **(1991)** ....Page 14-26
4.9L I6 OHV MFI VIN Y (E4OD) 60 Pin **(1991)** ....Page 14-27
4.9L I6 OHV MFI VIN Y (All) 60 Pin **(1992-93)** ....Page 14-28
4.9L I6 OHV MFI VIN Y (E4OD) 60 Pin **(1992-93)** ....Page 14-29
4.9L I6 OHV MFI VIN Y (All) 60 Pin **(1994-95)** ....Page 14-30
4.9L I6 OHV MFI VIN Y (E4OD) 60 Pin **(1994-95)** ....Page 14-31
4.9L I6 OHV MFI VIN Y (E4OD) California 60 Pin **(1995)** ....Page 14-32
4.9L I6 OHV MFI VIN Y (A/T) 104 Pin **(1996)** ....Page 14-34
4.9L I6 OHV MFI VIN Y (E4OD) 104 Pin **(1996)** ....Page 14-36
5.0L V8 OHV MFI VIN N (A/T) 60 Pin **(1990)** ....Page 14-38
5.0L V8 OHV MFI VIN N (A/T) 60 Pin **(1991)** ....Page 14-39
5.0L V8 OHV MFI VIN N (A/T) 60 Pin **(1992-93)** ....Page 14-40
5.0L V8 OHV MFI VIN N (E4OD) 60 Pin **(1992-93)** ....Page 14-41
5.0L V8 OHV MFI VIN N (A/T) 60 Pin **(1994-95)** ....Page 14-42
5.0L V8 OHV MFI VIN N (A/T) 104 Pin **(1996)** ....Page 14-44
5.0L V8 OHV MFI VIN N (E4OD) 104 Pin **(1996)** ....Page 14-46

E-SERIES VAN PIN TABLES (CONTINUED)

5.4L V8 SOHC MFI VIN L (E4OD) 104 Pin **(1997)** .... Page 14-48
5.4L V8 SOHC MFI VIN L (E4OD) 104 Pin **(1998-99)** .... Page 14-50
5.4L V8 SOHC MFI VIN L (A/T) 104 Pin **(2000-02)** .... Page 14-52
5.4L V8 SOHC MFI VIN L (A/T) 104 Pin **(2003)** .... Page 14-54
5.4L V8 SOHC CNG VIN M (A/T) 104 Pin **(1998-99)** .... Page 14-56
5.4L V8 SOHC CNG VIN M (A/T) 104 Pin **(2000-02)** .... Page 14-58
5.4L V8 SOHC CNG VIN M (A/T) 104 Pin **(2003)** .... Page 14-60
5.4L V8 SOHC NGV VIN M (A/T) **(1998-02)** .... Page 14-62
5.4L V8 NGV VIN M (A/T) Module Wiring **(1998-02)** .... Page 14-63
5.8L V8 OHV MFI VIN H (A/T) 60 Pin **(1990)** .... Page 14-64
5.8L V8 OHV MFI VIN H (E4OD) 60 Pin **(1990)** .... Page 14-65
5.8L V8 OHV MFI VIN H (A/T) 60 Pin **(1991)** .... Page 14-66
5.8L V8 OHV MFI VIN H (A/T) 60 Pin **(1992-94)** .... Page 14-67
5.8L V8 OHV MFI VIN H (E4OD) California 60 Pin **(1995)** .... Page 14-68
5.8L V8 OHV MFI VIN H (E4OD) 60 Pin **(1995)** .... Page 14-70
5.8L V8 OHV MFI VIN H (E4OD) 104 Pin **(1996)** .... Page 14-72
6.8L V10 OHC MFI VIN S (E4OD) 104 Pin **(1997)** .... Page 14-74
6.8L V10 OHC MFI VIN S (E4OD) 104 Pin **(1998-99)** .... Page 14-76
6.8L V10 OHC MFI VIN S (A/T) 104 Pin **(2000-02)** .... Page 14-78
6.8L V10 OHC MFI VIN S (A/T) 104 Pin **(2003)** .... Page 14-80
7.3L V8 Turbo Diesel VIN F (A/T) 104 Pin **(1996-99)** .... Page 14-82
7.3L V8 Turbo Diesel VIN F (A/T) 104 Pin **(2000-02)** .... Page 14-84
7.3L V8 Turbo Diesel VIN F (A/T) 104 Pin **(2003)** .... Page 14-86
7.5L V8 OHV MFI VIN G (A/T) 60 Pin **(1990)** .... Page 14-88
7.5L V8 OHV MFI VIN G (E4OD) 60 Pin **(1990)** .... Page 14-89
7.5L V8 OHV MFI VIN G (E4OD) 60 Pin **(1991)** .... Page 14-90
7.5L V8 OHV MFI VIN G (E4OD) 60 Pin **(1992-95)** .... Page 14-91
7.5L V8 OHV MFI VIN G (E4OD) 104 Pin **(1996)** .... Page 14-93

VILLAGER PIN TABLES

3.0L V6 SOHC MFI VIN W (A/T) 116-Pin **(1993-95)** .... Page 14-95
3.0L V6 SOHC MFI VIN W, 1 (A/T) 88-Pin **(1996-98)** .... Page 14-97
3.3L V6 SOHC MFI VIN T (A/T) 104 Pin **(1999-2000)** .... Page 14-99
3.3L V6 SOHC MFI VIN T (A/T) 104 Pin **(2001-02)** .... Page 14-101

WINDSTAR PIN TABLES

3.0L V6 OHV MFI VIN U (A/T) 104 Pin **(1995-97)** .... Page 14-103
3.0L V6 OHV MFI VIN U (A/T) 104 Pin **(1998-2000)** .... Page 14-105
3.8L V6 OHV MFI VIN 4 (A/T) 104 Pin **(1995)** .... Page 14-107
3.8L V6 OHV MFI VIN 4 (A/T) 104 Pin **(1996-97)** .... Page 14-109
3.8L V6 OHV MFI VIN 4 (A/T) 104 Pin **(1998-99)** .... Page 14-111
3.8L V6 OHV MFI VIN 4 (A/T) 104 Pin **(2000)** .... Page 14-113
3.8L V6 OHV MFI VIN 4 (A/T) 104 Pin **(2001-03)** .... Page 14-115

About This Section

Introduction

This section of the Domestic Ford Handbook contains Pin Tables for Vans from 1990-2003. It can be used to assist in the repair of Code and No Code faults related to the PCM.

VEHICLE COVERAGE

- Aerostar Applications (1990-97)
- E-Series Van Applications (1990-2003)
- Villager Applications (1993-2002)
- Windstar Applications (1995-2003)

How to Use This Section

This Section of the Handbook can be used to look up the location of a particular pin, a Wire Color or a "known good" value of a PCM circuit. To locate the PCM information for a particular vehicle, find the model, correct engine size (with VIN Code) and finally the year of the vehicle.

For example, to look up the PCM terminals for a 1999 Windstar 3.0L VIN U, go to Contents Page 1 and find the text string shown below:

3.0L V6 OHV MFI VIN U (A/T) - 104 Pin **(1998-2000)**..Page 14-112

Then turn to Page 14-112 to find the following PCM related information.

1999 Windstar 3.0L OHV V6 VIN U (A/T) 104 Pin Connector

PCM Pin #	Wire Color	Circuit Description (104 Pin)	Value at Hot Idle
1	BK/WT	Shift Solenoid 2/B Control	1v, 55 mph: 12v
3	OR/BK	Digital TR1 Sensor	In 'P': 0v, 55 mph: 11v
13	PK	Flash EEPROM	0.1v
39	GY	Intake Air Temperature Sensor	1.5-2.5v
52	RD/LB	Coil 2 Driver (Dwell)	5°, at 55 mph: 8°

In this example, the Digital TR1 Sensor circuit connects to the Pin 3 of the 104 Pin connector with an OR/BK wire. The value at Hot Idle shown here (0v) is the normal value for the TR Sensor 1 with the gearshift selector in position 1. The value changes to 11.5v at 55 mph.

The Coil 2 Driver (Dwell) values (5°, at 55 mph: 8° can be used to check how the Ignition Module circuits (inside the PCM) control the ignition coil dwell time. This idle dwell reading (5° dwell) should change to a higher value at 30 and 55 mph (8°) to indicate that the coil "on" time is longer at higher engine speeds.

The (A/T) in the Title over the table indicates the coverage is A/T.

Aerostar PIN Tables

1990 Aerostar 3.0L V6 MFI VIN U (A/T) 60 Pin Connector

PCM Pin #	Wire Color	Circuit Description (60 Pin)	Value at Hot Idle
1	YL	Keep Alive Power	12-14v
2	RD/LG	Brake Switch Signal	Brake Off: 12v, On: 1v
3	DG/WT	VSS (+) Signal	0 Hz, at 55 mph: 125 Hz
4	DG/YL	Ignition Diagnostic Monitor	20-31 Hz
6	BK/YL	VSS (-) Signal	0 Hz, at 55 mph: 125 Hz
6	BK/WT	VSS (-) Signal	0 Hz, at 55 mph: 125 Hz
7	LG/YL	Engine Coolant Temperature Sensor	0.5-0.6v
8	PK/BK	Fuel Pump Monitor	On: 12v, Off: 0v
8	OR/BL	Fuel Pump Monitor	On: 12v, Off: 0v
10	BK/YL	A/C Pressure Switch Signal	A/C On: 12v, off: 0v
16	BK/OR	Ignition System Ground	<0.050v
17	TN/RD	Self-Test Output & MIL	Lamp On: 1v, Off: 12v
20	BK	PCM Case Ground	<0.050v
21	OR/BK	IAC Solenoid Control	10.6-11.8v
22	TN/LG	Fuel Pump Control	On: 1v, Off: 12v
23	YL/RD	Knock Sensor Signal	0v
24	YL/LG	Power Steering Pressure Switch	Straight: 0v,Turned: 12v
25	LG/PK	ACT Sensor Signal	1.5-2.5v
26	OR/WT	Reference Voltage	4.9-5.1v
29	DG/PK	HO2S-11 (B1 S1) Signal	0.1-1.1v
30	BL/YL	A/T: Neutral Drive Switch	In 'N': 0v, Others: 5v
30	WT/BK	A/T: Neutral Drive Switch	In 'N': 0v, Others: 5v
30	WT/BK	M/T: Clutch Engage Switch	Clutch Out: 5v, In: 0v
30	BL/BK	M/T: Clutch Engage Switch	Clutch Out: 5v, In: 0v
31	GY/YL	Canister Purge Solenoid	12v, 55 mph: 1v
34	BL/PK	Data Output Link	5v
36	YL/LG	Spark Output Signal	6.93v (50%)
37	RD	Vehicle Power	12-14v
40	BK/LG	Power Ground	<0.1v
45	DB/LG	MAP Sensor Signal	107 Hz
46	BK/WT	Analog Signal Return	<0.050v
47	DG/LG	TP Sensor Signal	0.7v, 55 mph: 2.1v
48	WT/RD	Self-Test Input Signal	STI On: 1v, Off: 5v
49	OR	HO2S-11 (B1 S1) Ground	<0.050v
52	TN/BL	Shift Solenoid 3-4 Control	1v, 55 mph: 12v
53	OR/YL	TCC Solenoid Control	12v, 55 mph: 1v
53	WT	TCC Solenoid Control	12v, 55 mph: 1v
54	GY/WT	A/C WOT Relay Control	On: 1v, Off: 12v
54	RD	A/C WOT Relay Control	On: 1v, Off: 12v
56	DB	PIP Sensor Signal	6.93v (50%)
57	RD	Vehicle Power	12-14v
58	TN/OR	Injector Bank 1 (INJ 1, 2 & 4)	5.6-6.0 ms
59	TN/RD	Injector Bank 2 (INJ 3, 5 & 6)	5.6-6.0 ms
60	BK/LG	Power Ground	<0.1v

Pin Connector Graphic

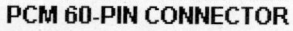

PCM 60-PIN CONNECTOR

60 ●●●●●●●●●● 51　50 ●●●●●●●●●● 41
40 ●●●●●●●●●● 31　○　30 ●●●●●●●●●● 21
20 ●●●●●●●●●● 11　10 ●●●●●●●●●● 1

Terminal View of 60-Pin PCM Harness Connector

1991 Aerostar 3.0L V6 MFI VIN U (A/T) 60 Pin Connector

PCM Pin #	Wire Color	Circuit Description (60 Pin)	Value at Hot Idle
1	YL	Keep Alive Power	12-14v
2	LG	Brake Switch Signal	Brake Off: 12v, On: 1v
3	GY/BK	VSS (+) Signal	0 Hz, at 55 mph: 125 Hz
4	TN/YL	Ignition Diagnostic Monitor	20-31 Hz
6	BK/YL	VSS (-) Signal	0 Hz, at 55 mph: 125 Hz
7	LG/YL	Engine Coolant Temperature Sensor	0.5-0.6v
8	PK/BK	Fuel Pump Monitor	On: 12v, Off: 0v
10	BK/YL	A/C Pressure Switch Signal	A/C On: 12v, off: 0v
16	OR/RD	Ignition System Ground	<0.050v
17	TN/RD	Self-Test Output & MIL	MIL On: 1v, Off: 12v
20	BK	PCM Case Ground	<0.050v
21	WT/BL	IAC Solenoid Control	10.5-11.8v
22	BL/OR	Fuel Pump Control	On: 1v, Off: 12v
24	YL/LG	Power Steering Pressure Switch	Straight: 0v, Turned: 12v
25	LG/PK	ACT Sensor Signal	1.5-2.5v
26	BR/WT	Reference Voltage	4.9-5.1v
29	RD/BK	HO2S-11 (B1 S1) Signal	0.1-1.1v
30	BL/YL	A/T: Neutral Drive Switch	In 'N': 0v, Others: 5v
30	BL/YL	M/T: Clutch Engage Switch	Clutch Out: 5v, In: 0v
31	GY/YL	Canister Purge Solenoid	12v, 55 mph: 1v
34	BL/PK	Data Output Link	5v
36	PK	Spark Output Signal	6.93v (50%)
37	RD	Vehicle Power	12-14v
40	BK/LG	Power Ground	<0.1v
45	DB/LG	MAP Sensor Signal	107 Hz
46	GY/RD	Analog Signal Return	<0.050v
47	GY/WT	TP Sensor Signal	0.7v, 55 mph: 2.1v
48	WT/PK	Self-Test Input Signal	STI On: 1v, Off: 5v
49	OR	HO2S-11 (B1 S1) Ground	<0.050v
52	OR/YL	Shift Solenoid 3-4 Control	1v, 55 mph: 12v
53	WT	TCC Solenoid Control	12v, 55 mph: 1v
54	GY/WT	A/C WOT Relay Control	On: 1v, Off: 12v
56	GY/OR	PIP Sensor Signal	6.93v (50%)
57	RD	Vehicle Power	12-14v
58	TN	Injector Bank 1 (INJ 1, 2 & 4)	5.6-6.0 ms
59	WT	Injector Bank 2 (INJ 3, 5 & 6)	5.6-6.0 ms
60	BK/LG	Power Ground	<0.1v

Pin Connector Graphic

PCM 60-PIN CONNECTOR

Terminal View of 60-Pin PCM Harness Connector

Standard Colors and Abbreviations

Abbreviation	Color	Abbreviation	Color	Abbreviation	Color
BK	Black	GY	Gray	PK	Purple
BL	Blue	GN	Green	RD	Red
BR	Brown	LG	LT Green	TN	Tan
DB	Dark Blue	OR	Orange	WT	White
DG	DK Green	PK	Pink	YL	Yellow

1992-95 Aerostar 3.0L V6 MFI VIN U (A/T) 60 Pin Connector

PCM Pin #	Wire Color	Circuit Description (60 Pin)	Value at Hot Idle
1	YL	Keep Alive Power	12-14v
2	RD/LG	Brake Switch Signal	Brake Off: 12v, On: 1v
3	GY/BK	VSS (+) Signal	0 Hz, at 55 mph: 125 Hz
4	WT/PK	Ignition Diagnostic Monitor	20-31 Hz
5	---	Not Used	---
6	PK/OR	VSS (-) Signal	0 Hz, at 55 mph: 125 Hz
7	LG/RD	Engine Coolant Temperature Sensor	0.5-0.6v
8	DG/YL	Fuel Pump Monitor	On: 12v, Off: 0v
9	TN/BL	MAF Sensor Return	<0.050v
10	DG/OR	A/C Pressure Switch Signal	A/C On: 12v, off: 0v
11	GY/YL	Canister Purge Solenoid	12v, 55 mph: 1v
12	LB/OR	Injector 6 Control	5.0-5.5 ms
13	---	Not Used	---
14 ('92)	PK/BK	TCC Solenoid Control	0%, at 55 mph: 95%
15	TN/BK	Injector 5 Control	5.0-5.5 ms
16	BK/OR	Ignition System Ground	<0.050v
16	OR/RD	Ignition System Ground	<0.050v
17	PK/LG	Self-Test Output & MIL	MIL On: 1v, Off: 12v
18	TN/OR	Data Bus (+)	Digital Signals
19	PK/BL	Data Bus (-)	Digital Signals
20	BK	PCM Case Ground	<0.050v
21	WT/BL	IAC Solenoid Control	10.8-12.6v
22	BL/OR	Fuel Pump Control	On: 1v, Off: 12v
23-24	---	Not Used	---
25	GY	ACT Sensor Signal	1.5-2.5v
26	BR/WT	Reference Voltage	4.9-5.1v
27-29	---	Not Used	---
30	BL/YL	A/T: Neutral Drive Switch	In 'N': 0v, Others: 5v
30	BL/YL	M/T: Clutch Engage Switch	Clutch Out: 5v, In: 0v
31-33	---	Not Used	---
34	BL/PK	Data Output Link	5v
35	BR/LB	Injector 4 Control	5.0-5.5 ms
36	PK	Spark Output Signal	6.93v (50%)
37	RD	Vehicle Power	12-14v
38	---	Not Used	---
39	BR/YL	Injector 3 Control	5.0-5.5 ms
40	BK/LG	Power Ground	<0.1v

1992-95 Aerostar 3.0L V6 MFI VIN U (A/T) 60 Pin Connector

PCM Pin #	Wire Color	Circuit Description (60 Pin)	Value at Hot Idle
41-43	---	Not Used	---
44	GY/BL	HO2S-11 (B1 S1) Signal	0.1-1.1v
45	---	Not Used	---
46	GY/RD	Analog Signal Return	<0.050v
47	GY/WT	TP Sensor Signal	0.7v, 55 mph: 2.1v
48	WT/PK	Self-Test Input Signal	STI On: 1v, Off: 5v
49	---	Not Used	---
50	BL/RD	MAF Sensor Signal	0.9v, 55 mph: 2.2v
51	OR/YL	Shift Solenoid 3-4 Control	1v, 55 mph: 12v
52	---	Not Used	---
53 ('93-'94)	PK/BK	TCC Solenoid Control	12v, 55 mph: 1v
53 ('93-'94)	PK/BK	TCC Solenoid Control	12v, 55 mph: 1v
54	PK/YL	A/C WOT Relay Control	On: 1v, Off: 12v
55	---	Not Used	---
56	GY/OR	PIP Sensor Signal	6.93v (50%)
57	RD	Vehicle Power	12-14v
58	TN	Injector 1 Control	5.0-5.5 ms
59	WT	Injector 2 Control	5.0-5.5 ms
60	BK/LG	Power Ground	<0.1v

Pin Connector Graphic

PCM 60-PIN CONNECTOR

Terminal View of 60-Pin PCM Harness Connector

Standard Colors and Abbreviations

Abbreviation	Color	Abbreviation	Color	Abbreviation	Color
BK	Black	GY	Gray	PK	Purple
BL	Blue	GN	Green	RD	Red
BR	Brown	LG	LT Green	TN	Tan
DB	Dark Blue	OR	Orange	WT	White
DG	DK Green	PK	Pink	YL	Yellow

1996-97 Aerostar 3.0L V6 VIN U (A/T) 104 Pin Connector

PCM Pin #	Wire Color	Circuit Description (104 Pin)	Value at Hot Idle
1	BK/WT	Shift Solenoid 2 Control	1v, 55 mph: 12v
2	PK/LG	MIL (lamp) Control	MIL On: 1v, Off: 12v
3-10, 12, 14	---	Not Used	---
11	BL/LG	Purge Flow Sensor	0.8v, 55 mph: 3.0v
13	PK	Flash EEPROM	0.1v
15	PK/LB	Data Bus (-)	Digital Signals
16	TN/OR	Data Bus (+)	Digital Signals
17-20, 23	---	Not Used	---
21	DB	CKP (+) Sensor	518-540 Hz
22	GY	CKP (-) Sensor	518-540 Hz
24	BK/WT	Power Ground	<0.050v
25	BK	Case Ground	<0.050v
26	YL/BK	Coil 1 Driver (Dwell)	5°, at 55 mph: 8°
27	OR/YL	Shift Solenoid 1 Control	12v, 55 mph: 1v
28	BR/OR	Coast Clutch Solenoid	12v, 55 mph: 12v
29	TN/WT	TCS (switch) Signal	TCS & O/D On: 12v
30	DG	Octane Adjustment	9.3v (switch shorted: 0v)
31-32	---	Not Used	---
33	PK/OR	VSS (-) Signal	0 Hz, at 55 mph: 125 Hz
35	RD/LG	HO2S-12 (B1 S2) Signal	0.1-1.1v
36	TN/LB	MAF Sensor Return	<0.050v
37	OR/BK	Transmission Fluid Temperature Sensor	2.10-2.40v
38	LG/RD	Engine Coolant Temperature Sensor	0.5-0.6v
39	GY	Intake Air Temperature Sensor	1.5-2.5v
40	DG/YL	Fuel Pump Monitor	On: 12v, Off: 0v
41	TN/YL	A/C Pressure Switch Signal	A/C On: 12v, off: 0v
42, 44-46	---	Not Used	---
43	LB/PK	Data Output Link Signal	5v
47	BR/PK	VR Solenoid Control	0%, at 55 mph: 45%
48	TN/YL	Clean Tachometer Output	42-50 Hz
49-50	---	Not Used	---
51	BK/WT	Power Ground	<0.1v
52	YL/RD	Coil 2 Driver (Dwell)	5°, at 55 mph: 8°
53	PK/BK	Shift Solenoid 3 Control	12v, 55 mph: 1v
54	PK/YL	TCC Solenoid Control	0%, at 55 mph: 95%
55	YL	Keep Alive Power	12-14v
56-57, 59	---	Not Used	---
58	GY/BK	VSS (+) Signal	0 Hz, at 55 mph: 125 Hz
60	GY/LB	HO2S-11 (B1 S1) Signal	0.1-1.1v
61-63	---	Not Used	---
64	LB/YL	Transmission Range Sensor	In 'P': 0v, in O/D: 1.7v
65	BR/LG	EGR DPFE Sensor Signal	0.20-1.30v
66, 68	---	Not Used	---
67	GY/YL	EVAP Purge Solenoid	0-10 Hz (0-100%)
69	PK/YL	A/C WOT Relay Control	On: 1v, Off: 12v
70	---	Not Used	---

1996-97 Aerostar 3.0L V6 VIN U (A/T) 104 Pin Connector

PCM Pin #	Wire Color	Circuit Description (104 Pin)	Value at Hot Idle
71	RD	Vehicle Power	12-14v
72	---	Not Used	---
73	TN/BK	Injector 5 Control	4.5-4.8 ms
74	BR/YL	Injector 3 Control	4.5-4.8 ms
75	TN	Injector 1 Control	4.5-4.8 ms
76, 77	BK/WT	Power Ground	<0.1v
78	YL/WT	Coil 3 Driver (Dwell)	5°, at 55 mph: 8°
79	WT/LG	TCIL (lamp) Control	Lamp On: 1v, Off: 12v
80	LB/OR	Fuel Pump Control	On: 1v, Off: 12v
81	WT/YL	EPC Solenoid Control	10.7v (27 psi)
82	---	Not Used	---
83	WT/LB	IAC Solenoid Control	10v (45% duty cycle)
84	DG/WT	Turbine Shaft Speed Sensor	120 Hz, 30 mph: 300 Hz
85	DB/OR	Camshaft Position Sensor	7 Hz, 55 mph: 13-17 Hz
86-87	---	Not Used	---
88	LB/RD	MAF Sensor Signal	0.9v, 55 mph: 2.2v
89	GY/WT	TP Sensor Signal	0.53-1.27v
90	BR/WT	Reference Voltage	4.9-5.1v
91	GY/RD	Analog Signal Return	<0.050v
92	LG	Brake Position Switch	Brake Off: 12v, On: 1v
93	RD/WT	HO2S-11 (B1 S1) Heater	1v, Off: 12v
94	---	Not Used	---
95	WT/BK	HO2S-12 (B1 S2) Heater	1v, Off: 12v
96	---	Not Used	---
97	RD	Vehicle Power	12-14v
98	---	Not Used	---
99	LG/OR	Injector 6 Control	4.5-4.8 ms
100	BR/LB	Injector 4 Control	4.5-4.8 ms
101	WT	Injector 2 Control	4.5-4.8 ms
102	---	Not Used	---
103	BK/WT	Power Ground	<0.1v
104	---	Not Used	---

Pin Connector Graphic

PCM 104-PIN CONNECTOR

Terminal View of 104-Pin PCM Wiring Harness Connector

Standard Colors and Abbreviations

Abbreviation	Color	Abbreviation	Color	Abbreviation	Color
BK	Black	GY	Gray	PK	Purple
BL	Blue	GN	Green	RD	Red
BR	Brown	LG	LT Green	TN	Tan
DB	Dark Blue	OR	Orange	WT	White
DG	DK Green	PK	Pink	YL	Yellow

1990 Aerostar 4.0L V6 OHV VIN X (A/T) 60 Pin Connector

PCM Pin #	Wire Color	Circuit Description (60 Pin)	Value at Hot Idle
1	YL	Keep Alive Power	12-14v
2	RD/LG	Brake Switch Signal	Brake Off: 12v, On: 1v
3	DG/WT	VSS (+) Signal	0 Hz, at 55 mph: 125 Hz
4	DG/YL	Ignition Diagnostic Monitor	20-31 Hz
6	BK/YL	VSS (-) Signal	0 Hz, at 55 mph: 125 Hz
7	LG/YL	Engine Coolant Temperature Sensor	0.5-0.6v
8	PK/BK	Fuel Pump Monitor	On: 12v, Off: 0v
9	BK/OR	Data Bus (+)	Digital Signals
10	BK/YL	A/C Pressure Switch Signal	A/C On: 12v, off: 0v
14	DB/OR	MAF Sensor Signal	0.7v, 55 mph: 2.4v
15	TN/BL	MAF Sensor Return	<0.050v
16	BK/OR	Ignition System Ground	<0.050v
17	TN/RD	Self-Test Output & MIL	MIL On: 1v, Off: 12v
20	BK	PCM Case Ground	<0.050v
21	OR/BK	IAC Solenoid Control	10.2v (39%)
22	TN/LG	Fuel Pump Control	On: 1v, Off: 12v
24	YL/LG	Power Steering Pressure Switch	Straight: 0v, Turned: 12v
25	LG/PK	ACT Sensor Signal	1.5-2.5v
26	OR/WT	Reference Voltage	4.9-5.1v
28	OR/BK	Data Bus (-)	Digital Signals
29	DG/PK	HO2S-11 (B1 S1) Signal	0.1-1.1v
30	BL/YL	A/T: Neutral Drive Switch	In 'N': 0v, Others: 5v
31	GY/YL	Canister Purge Solenoid	12v, 55 mph: 1v
34	BL/PK	Data Output Link Signal	Digital Signals
36	YL/LG	Spark Output Signal	6.93v (50%)
37	RD	Vehicle Power	12-14v
40	BK/LG	Power Ground	<0.1v
44	OR	Octane Adjustment	9.3v (switch shorted: 0v)
45	DB/LG	BARO Sensor Signal	159 Hz (Sea Level)
46	BK/WT	Analog Signal Return	<0.050v
47	DG/LG	TP Sensor Signal	0.7v, 55 mph: 2.1v
48	WT/RD	Self-Test Input Signal	STI On: 1v, Off: 5v
49	OR	HO2S-11 (B1 S1) Ground	<0.050v
52	TN/BL	Shift Solenoid 3-4 Control	1v, 55 mph: 12v
53	WT	TCC Solenoid Control	12v, 55 mph: 1v
54	GY/WT	A/C WOT Relay Control	On: 1v, Off: 12v
56	DB	PIP Sensor Signal	6.93v (50%)
57	RD	Vehicle Power	12-14v
58	TN/OR	Injector Bank 1 (INJ 1, 2 & 4)	3.0-3.2 ms
59	TN	Injector Bank 2 (INJ 3, 5 & 6)	3.0-3.2 ms
60	BK/LG	Power Ground	<0.1v

Pin Connector Graphic

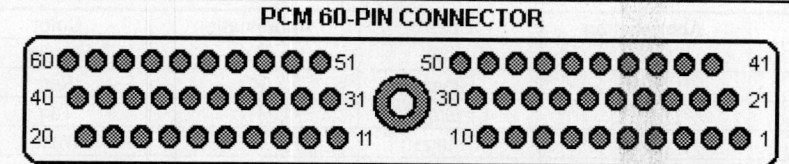

PCM 60-PIN CONNECTOR

Terminal View of 60-Pin PCM Harness Connector

1991 Aerostar 4.0L V6 OHV VIN X (A/T) 60 Pin Connector

PCM Pin #	Wire Color	Circuit Description (60 Pin)	Value at Hot Idle
1	YL	Keep Alive Power	12-14v
2	RD/LG	Brake Switch Signal	Brake Off: 12v, On: 1v
3	DG/WT	VSS (+) Signal	0 Hz, at 55 mph: 125 Hz
4	TN/YL	Ignition Diagnostic Monitor	20-31 Hz
6	BK/YL	VSS (-) Signal	0 Hz, at 55 mph: 125 Hz
7	LG/RD	Engine Coolant Temperature Sensor	0.5-0.6v
8	PK/BK	Fuel Pump Monitor	On: 12v, Off: 0v
9	BK/OR	Data Bus (-)	Digital Signals
10	BK/YL	A/C Pressure Switch Signal	A/C On: 12v, off: 0v
14	BL/RD	MAF Sensor Signal	0.7v, 55 mph: 2.4v
15	TN/BL	MAF Sensor Return	<0.050v
16	OR/RD	Ignition System Ground	<0.050v
17	TN/RD	Self-Test Output & MIL	MIL On: 1v, Off: 12v
20	BK	PCM Case Ground	<0.050v
21	OR/BK	IAC Solenoid Control	10.2v (39%)
22	TN/LG	Fuel Pump Control	On: 1v, Off: 12v
25	LG/PK	ACT Sensor Signal	1.5-2.5v
26	BR/WT	Reference Voltage	4.9-5.1v
28	OR/BK	Data Bus (+)	Digital Signals
29	RD/BK	HO2S-11 (B1 S1) Signal	0.1-1.1v
30	BL/YL	A/T: Neutral Drive Switch	In 'P': 0v, Others: 5v
31	GY/YL	Canister Purge Solenoid	12v, 55 mph: 1v
34	BL/PK	Data Output Link Signal	Digital Signals
36	YL/LG	Spark Output Signal	6.93v (50%)
37	RD	Vehicle Power	12-14v
40	BK/LG	Power Ground	<0.1v
44	OR	Octane Adjustment	9.3v (switch shorted: 0v)
45	DB/LG	BARO Sensor Signal	159 Hz (Sea Level)
46	GY/RD	Analog Signal Return	<0.050v
47	GY/WT	TP Sensor Signal	0.7v, 55 mph: 2.1v
48	WT/RD	Self-Test Input Signal	STI On: 1v, Off: 5v
49	OR	HO2S-11 (B1 S1) Ground	<0.050v
52	TN/WT	Shift Solenoid 3-4 Control	1v, 55 mph: 12v
53	WT	TCC Solenoid Control	12v, 55 mph: 1v
54	GY/WT	A/C WOT Relay Control	On: 1v, Off: 12v
56	DB	PIP Sensor Signal	6.93v (50%)
57	RD	Vehicle Power	12-14v
58	TN/OR	Injector Bank 1 (INJ 1, 2 & 4)	3.3-3.5 ms
59	TN/RD	Injector Bank 2 (INJ 3, 5 & 6)	3.3-3.5 ms
60	BK/LG	Power Ground	<0.1v

Pin Connector Graphic

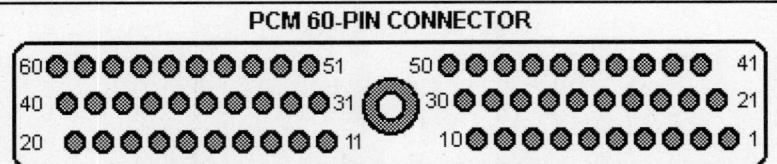

PCM 60-PIN CONNECTOR

Terminal View of 60-Pin PCM Harness Connector

1992-95 Aerostar 4.0L V6 VIN X (A/T) 60 Pin Connector

PCM Pin #	Wire Color	Circuit Description (60 Pin)	Value at Hot Idle
1	YL	Keep Alive Power	12-14v
2	RD/LG	Brake Switch Signal	Brake Off: 12v, On: 1v
3	GY/BK	VSS (+) Signal	0 Hz, at 55 mph: 125 Hz
3	DG/WT	VSS (+) Signal	0 Hz, at 55 mph: 125 Hz
4	TN/YL	Ignition Diagnostic Monitor	20-31 Hz
5	---	Not Used	---
6	PK/OR	VSS (-) Signal	0 Hz, at 55 mph: 125 Hz
7	LG/RD	Engine Coolant Temperature Sensor	0.5-0.6v
8	DG/YL	Fuel Pump Monitor	On: 12v, Off: 0v
9	PK/BL	Data Bus (-)	Digital Signals
10	DG/OR	A/C Pressure Switch Signal	A/C On: 12v, off: 0v
11-13	---	Not Used	---
14	BL/RD	MAF Sensor Signal	0.7v, 55 mph: 2.4v
15	TN/BL	MAF Sensor Return	<0.050v
16	OR/RD	Ignition System Ground	<0.050v
17	PK/LG	Self-Test Output & MIL	MIL On: 1v, Off: 12v
18-19	---	Not Used	---
20	BK	PCM Case Ground	<0.050v
21	WT/BL	IAC Solenoid Control	10.2v (39%)
22	BL/OR	Fuel Pump Control	On: 1v, Off: 12v
23-24	---	Not Used	---
25	GY	Intake Air Temperature Sensor	1.5-2.5v
26	BR/WT	Reference Voltage	4.9-5.1v
27	---	Not Used	---
28	TN/OR	Data Bus (+)	Digital Signals
29	GY/BL	HO2S-11 (B1 S1) Signal	0.1-1.1v
29	WT/BL	HO2S-11 (B1 S1) Signal	0.1-1.1v
30	BL/YL	A/T: Neutral Drive Switch	In 'N': 0v, Others: 5v
31	GY/YL	Canister Purge Solenoid	12v, 55 mph: 1v
32-33	---	Not Used	---
34	BL/PK	Data Output Link Signal	Digital Signals
35	---	Not Used	---
36 ('92-'93)	PK	Spark Angle Word	6.93v (50%)
36 ('93-'94)	PK	Spark Output Signal	6.93v (50%)
37	RD	Vehicle Power	12-14v
38-39	---	Not Used	---
40	BK/LG	Power Ground	<0.1v

1992-95 Aerostar 4.0L V6 VIN X (A/T) 60 Pin Connector

PCM Pin #	Wire Color	Circuit Description (60 Pin)	Value at Hot Idle
41-43	---	Not Used	---
44	DG	Octane Adjustment	9.3v (switch shorted: 0v)
45	---	Not Used	---
46	GY/RD	Analog Signal Return	<0.050v
47	GY/WT	TP Sensor Signal	0.7v, 55 mph: 2.1v
48	WT/PK	Self-Test Input Signal	STI On: 1v, Off: 5v
49	OR	HO2S-11 (B1 S1) Ground	<0.050v
50-51	---	Not Used	---
52	OR/YL	Shift Solenoid 3-4 Control	1v, 55 mph: 12v
53	PK/BK	TCC Solenoid Control	12v, 55 mph: 1v
54	PK/YL	A/C WOT Relay Control	On: 1v, Off: 12v
55	---	Not Used	---
56	GY/OR	PIP Sensor Signal	6.93v (50%)
57	RD	Vehicle Power	12-14v
58	TN	Injector Bank 1 (INJ 1, 2 & 4)	3.3-3.6 ms
59	WT	Injector Bank 2 (INJ 3, 5 & 6)	3.3-3.6 ms
60	BK/LG	Power Ground	<0.1v

Pin Connector Graphic

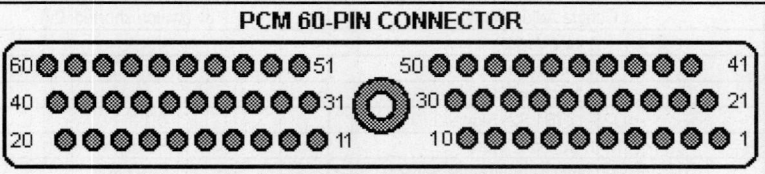

PCM 60-PIN CONNECTOR

Terminal View of 60-Pin PCM Harness Connector

Standard Colors and Abbreviations

Abbreviation	Color	Abbreviation	Color	Abbreviation	Color
BK	Black	GY	Gray	PK	Purple
BL	Blue	GN	Green	RD	Red
BR	Brown	LG	LT Green	TN	Tan
DB	Dark Blue	OR	Orange	WT	White
DG	DK Green	PK	Pink	YL	Yellow

1996-97 Aerostar 4.0L V6 VIN X (A/T) 104 Pin Connector

PCM Pin #	Wire Color	Circuit Description (104 Pin)	Value at Hot Idle
1	BK/WT	Shift Solenoid 2 Control	12v, 55 mph: 12v
2	PK/LG	MIL (lamp) Control	MIL On: 1v, Off: 12v
3-10	---	Not Used	---
11	BL/LG	Purge Flow Sensor	0.8v, 30 mph: 3v
12	---	Not Used	---
13	VT	Flash EPROM Power	0.1v
14	---	Not Used	---
15	PK/LB	Data Bus (-)	Digital Signals
16	TN/OR	Data Bus (+)	Digital Signals
17-20	---	Not Used	---
21	DB	CKP (+) Sensor	450-480 Hz
22	GY	CKP (-) Sensor	450-480 Hz
23	---	Not Used	---
24	BK/WT	Power Ground	<0.1v
25	BK	Case Ground	<0.050v
26	YL/BK	Coil 1 Driver (Dwell)	5°, at 55 mph: 8°
27	OR/YL	Shift Solenoid 1 Control	1v, 55 mph: 12v
28	BR/OR	Coast Clutch Solenoid	12v, 55 mph: 12v
29	TN/WT	TCS (switch) Signal	TCS & O/D On: 12v
30	DG	Octane Adjustment	9.3v (switch shorted: 0v)
31-32	---	Not Used	---
33	PK/OR	PSOM (-) Signals	<0.050v
34	---	Not Used	---
35	RD/LG	HO2S-12 (B1 S2) Signal	0.1-1.1v
36	TN/LB	MAF Sensor Return	<0.050v
37	OR/BK	Transmission Fluid Temperature Sensor	2.10-2.40v
38	LG/RD	Engine Coolant Temperature Sensor	0.5-0.6v
39	G/Y	Intake Air Temperature Sensor	1.5-2.5v
40	DG/YL	Fuel Pump Monitor	On: 12v, Off: 0v
41	TN/YL	A/C Pressure Switch Signal	A/C On: 12v, off: 0v
42	---	Not Used	---
43	LB/PK	Overhead Trip Computer	Digital Signals
44-46	---	Not Used	---
47	BR/PK	VR Solenoid Control	0%, at 55 mph: 45%
48	TN/YL	Clean Tachometer Output	38-42 Hz
49-50	---	Not Used	---
51	BK/WT	Power Ground	<0.1v
52	YL/RD	Coil 2 Driver (Dwell)	5°, at 55 mph: 8°
53	PK/BK	Shift Solenoid 3 Control	12v, 55 mph: 1v
54	PK/YL	TCC Solenoid Control	0%, at 55 mph: 95%
55	YL	Keep Alive Power	12-14v
56-57	---	Not Used	---
58	GY/BK	PSOM (+) Signals	Digital signals
59	---	Not Used	---
60	GY/LB	HO2S-11 (B1 S1) Signal	0.1-1.1v
61-63	---	Not Used	---
64	LB/YL	Digital Transmission Range Sensor	In 'P': 0v, in O/D: 1.7v

1996-97 Aerostar 4.0L V6 VIN X (A/T) 104 Pin Connector

PCM Pin #	Wire Color	Circuit Description (104 Pin)	Value at Hot Idle
65	BR/LG	EGR DPFE Sensor Signal	0.20-1.30v
66	---	Not Used	---
67	GY/YL	EVAP Purge Solenoid	0-10 Hz (0-100%)
68	---	Not Used	---
69	PK/YL	A/C WOT Relay Control	On: 1v, Off: 12v
70	---	Not Used	---
71	RD	Vehicle Power	12-14v
72	---	Not Used	---
73	TN/BK	Injector 5 Control	3.9-4.1 ms
74	BR/YL	Injector 3 Control	3.9-4.1 ms
75	TN	Injector 1 Control	3.9-4.1 ms
76	BK/WT	Power Ground	<0.1v
77	BK/WT	Power Ground	<0.1v
78	YL/WT	Coil 3 Driver (Dwell)	5°, at 55 mph: 8°
79	WT/LG	TCIL (lamp) Control	Lamp On: 1v, Off: 12v
80	LB/OR	Fuel Pump Control	On: 1v, Off: 12v
81	WT/YL	EPC Solenoid Control	10.2v (24 psi)
82	---	Not Used	---
83	WT/LB	IAC Solenoid Control	10.7v (33%)
84	DG/WT	Turbine Shaft Speed Sensor	120 Hz, 30 mph: 300 Hz
85	DB/OR	Camshaft Position Sensor	7 Hz, 55 mph: 18 Hz
86-87	---	Not Used	---
88	LB/RD	MAF Sensor Signal	0.8v, 55 mph: 1.6v
89	GY/WT	TP Sensor Signal	0.53-1.27v
90	BR/WT	Reference Voltage	4.9-5.1v
91	GY/RD	Analog Signal Return	<0.050v
92	LG	Brake Position Switch	Brake Off: 12v, On: 1v
93	RD/WT	HO2S-11 (B1 S1) Heater	0.1-1.1v
94	---	Not Used	---
95	WT/BK	HO2S-22 (B2 S2) Heater	0.1-1.1v
96	---	Not Used	---
97	RD	Vehicle Power	12-14v
98	---	Not Used	---
99	LG/OR	Injector 6 Control	4.5-4.8 ms
100	BR/LB	Injector 4 Control	4.5-4.8 ms
101	W	Injector 2 Control	4.5-4.8 ms
102	---	Not Used	---
103	BK/WT	Power Ground	<0.1v
104	---	Not Used	---

Pin Connector Graphic

Terminal View of 104-Pin PCM Wiring Harness Connector

Standard Colors and Abbreviations

Abbreviation	Color	Abbreviation	Color	Abbreviation	Color
BK	Black	GY	Gray	PK	Purple
BL	Blue	GN	Green	RD	Red
BR	Brown	LG	LT Green	TN	Tan
DB	Dark Blue	OR	Orange	WT	White
DG	DK Green	PK	Pink	YL	Yellow

E-Series Van PIN Tables

1997-99 E-Series Van 4.2L V6 VIN 2 (A/T) 104-P Connector

PCM Pin #	Wire Color	Circuit Description (104 Pin)	Value at Hot Idle
1	PK/OR	Shift Solenoid 2 Control	12v, 55 mph: 1v
2	PK/LG	MIL (lamp) Control	MIL On: <1v, Off: 12v
3	YL/BK	Digital TR1 Sensor Signal	0v, at 55 mph: 11.5v
8	DB/YL	Electric IMRC Monitor	5v, 55 mph: 5v
12	YL/WT	Fuel Level Indicator Signal	1.7v (1/2 full)
13	VT	Flash EEPROM	0.1v
14	LB/BK	4x4 Indicator Signal	Off: 7.7v, On: 0v
15	PK/LB	Data Bus (-) Signal	Digital Signals
16	TN/OR	Data Bus (+) Signal	Digital Signals
19	DB/WT	Automatic Ride Control	Digital Signals
21	DB	CKP (+) Sensor	430-500 Hz
22	GY	CKP (-) Sensor	430-500 Hz
24	BK/WT	Power Ground	<0.1v
25	LG/YL	Case Ground	<0.050v
25	BK/LB	Case Ground	<0.050v
26	DB/LG	Coil 1 Driver (Dwell)	5°, at 55 mph: 8°
27	OR/YL	Shift Solenoid 1 Control	1v, 55 mph: 1v
29	TN/WT	TCS (switch) Signal	TCS & O/D On: 12v
30 ('97-'99)	DG	Octane Adjustment	9.3v (switch shorted: 0v)
32	DG/PK	Knock Sensor 1 Ground	<0.050v
33	PK/OR	VSS (-) Signal	0 Hz, at 55 mph: 125 Hz
35	RD/LG	HO2S-12 (B1 S2) Signal	0.1-1.1v
36	TN/LB	MAF Sensor Return	<0.050v
37	OR/BK	Transmission Fluid Temperature Sensor	2.10-2.40v
38	LG/RD	Engine Coolant Temperature Sensor	0.5-0.6v
39	GY	Intake Air Temperature Sensor	1.5-2.5v
40	DG/YL	Fuel Pump Monitor	On: 12v, Off: 0v
41	TN/LG	A/C Cycling Clutch Switch	A/C On: 12v, off: 0v
42	DB/LG	Electric IMRC Control	12v, 55 mph: 12v
43	OR/LG	Overhead Trip Computer	Digital Signals
45	RD/WT	CHTIL (lamp) Control	Lamp On: 1v, Off: 12v
46	BR	Intake Manifold Tuning Valve	12v, 55 mph: 12v
47	BR/PK	VR Solenoid Control	0%, at 55 mph: 45%
48	WT/PK	Clean Tachometer Output Signal	35-49 Hz
49	LB/BK	Digital TR2 Sensor	0v, at 55 mph: 11.5v
50	LG/RD	Digital TR4 Sensor	0v, at 55 mph: 11.5v
51	BK/WT	Power Ground	<0.1v
52	RD/LB	Coil 2 Driver (Dwell)	5°, at 55 mph: 8°
54	PKWT	TCC Solenoid Control	0%, at 55 mph: 95%
54	PK/YL	TCC Solenoid Control	0%, at 55 mph: 95%
55	YL	Keep Alive Power	12-14v
56	LG/BK	EVAP Purge Solenoid	0-10 Hz (0-100%)
57	YL/RD	Knock Sensor 1 Signal	0v
58	GY/BK	VSS (+) Signal	0 Hz, at 55 mph: 125 Hz
60	GY/LB	HO2S-11 (B1 S1) Signal	0.1-1.1v
61	PK/LG	HO2S-22 (B2 S2) Signal	0.1-1.1v
62	RD/PK	FTP Sensor Signal	2.6v (0" H2O - cap off)
64	LB/YL	Transmission Range Sensor	In 'P': 0v, in O/D: 1.7v

1997-99 E-Series Van 4.2L V6 VIN 2 (A/T) 104-P Connector

PCM Pin #	Wire Color	Circuit Description (104 Pin)	Value at Hot Idle
65	BR/LG	EGR DPFE Sensor Signal	0.20-1.30v
66	YL/LG	Cylinder Head Temperature Sensor	0.6 to 1.7v (194ºF)
67	PKWT	EVAP CV Solenoid	0-10 Hz (0-100%)
69	BK/LB	A/C WOT Relay Control	On: 1v, Off: 12v
70-71	RD	Vehicle Power	12-14v
73	TN/BK	Injector 5 Control	2.7-4.1 ms
74	BR/YL	Injector 3 Control	2.7-4.1 ms
75	TN	Injector 1 Control	2.7-4.1 ms
76-77	BK/WT	Power Ground	<0.1v
78	PKWT	Coil 3 Driver (Dwell)	5º, at 55 mph: 8º
79	WT/LG	TCIL (lamp) Control	Lamp On: 1v, Off: 12v
80	LB/OR	Fuel Pump Control	On: 1v, Off: 12v
81	WT/OR	EPC Solenoid Control	10v (15-20 psi)
83	WT/LB	IAC Solenoid Control	10v (31% duty cycle)
84	DB/YL	Output Shaft Speed Sensor	0 Hz, 55 mph: 250 Hz
85	DG	CID Sensor Signal	6 Hz, 55 mph: 16 Hz
87	RD/BK	HO2S-21 (B2 S1) Signal	0.1-1.1v
88	LB/RD	MAF Sensor Signal	0.6v, 55 mph: 2.2v
89	GY/WT	TP Sensor Signal	0.9v, 55 mph: 1.1v
90	BR/WT	Reference Voltage	4.9-5.1v
91	GY/RD	Analog Signal Return	<0.050v
92	LG	Brake Switch Signal	Brake Off: 12v, On: 1v
93	RD/WT	HO2S-11 (B1 S1) Heater	1v, Off: 12v
94	YL/LB	HO2S-21 (B2 S1) Heater	1v, Off: 12v
95	WT/BK	HO2S-12 (B1 S2) Heater	1v, Off: 12v
96	TN/YL	HO2S-22 (B2 S2) Heater	1v, Off: 12v
97	RD	Vehicle Power	12-14v
99	LG/OR	Injector 6 Control	2.7-4.1 ms
100	BR/LB	Injector 4 Control	2.7-4.1 ms
101	WT	Injector 2 Control	2.7-4.1 ms
102	---	Not Used	---
103	BK/WT	Power Ground	<0.1v
104	---	Not Used	---

Pin Connector Graphic

PCM 104-PIN CONNECTOR

Terminal View of 104-Pin PCM Wiring Harness Connector

Standard Colors and Abbreviations

Abbreviation	Color	Abbreviation	Color	Abbreviation	Color
BK	Black	GY	Gray	PK	Purple
BL	Blue	GN	Green	RD	Red
BR	Brown	LG	LT Green	TN	Tan
DB	Dark Blue	OR	Orange	WT	White
DG	DK Green	PK	Pink	YL	Yellow

2000-03 E-Series Van 4.2L V6 VIN 2 (A/T) 104-P Connector

PCM Pin #	Wire Color	Circuit Description (104 Pin)	Value at Hot Idle
1	VT/OR	Shift Solenoid 2 Control	12v, 55 mph: 1v
2	---	Not Used	---
3	ORBK	Digital TR1 Sensor Signal	0v, at 55 mph: 11.5v
4-5	---	Not Used	---
6	DG/WT	Turbine Shaft Speed Sensor	120 Hz, 30 mph: 300 Hz
7	---	Not Used	---
8	DB/YL	Intake Manifold Runner Control	5v, 55 mph: 5v
9-12	---	Not Used	---
13	VT	Flash EEPROM	0.1v
14	---	Not Used	---
15	PK/LB	SCP Bus (-) Signal	<0.050v
16	TN/OR	SCP Bus (+) Signal	Digital Signals
17	GY/OR	Passive Antilock RX Signal	Digital Signals
18	WT/LG	Passive Antilock TX Signal	Digital Signals
19	---	Not Used	---
20	VT	Generator 'S' Terminal	Frequency Signal
21	BK/PK	CKP (+) Sensor	430-500 Hz
22	GY/YL	CKP (-) Sensor	430-500 Hz
23	---	Not Used	---
24	BK/WT	Power Ground	<0.1v
25	---	Not Used	---
26	DB/LG	Coil 1 Driver (Dwell)	5°, at 55 mph: 8°
27	OR/YL	Shift Solenoid 1 Control	1v, 55 mph: 1v
28	DB/LG	Low Speed Fan Relay Control	Relay Off: 12v, On: 1v
29	---	Not Used	---
30	DB/LG	Passive Antitheft Indicator Control	Indicator Off: 12v, On: 1v
31	YL/LG	Power Steering Pressure Switch	Straight: 0v, Turned: 12v
32-34	---	Not Used	---
35	RD/LG	HO2S-12 (B1 S2) Signal	0.1-1.1v
36	TN/LB	MAF Sensor Return	<0.050v
37	OR/BK	Transmission Fluid Temperature Sensor	2.10-2.40v
38	LG/RD	Engine Coolant Temperature Sensor	0.5-0.6v
39	GY	Intake Air Temperature Sensor	1.5-2.5v
40	DG/YL	Fuel Pump Monitor	Off: 0v, On: 12v
41	BK/YL	A/C Pressure Switch Signal	A/C On: 12v, Off: 0v
42	BR	Intake Manifold Runner Control Module	12v, 55 mph: 12v
43	---	Not Used	---
44	LG/YL	Starter Relay Circuit	Off: 12v, On: 1v
45	RD/PK	Generator 'L' Terminal	0v or 12v
46	LG/VT	High Speed Cooling Fan Relay Control	Relay Off: 12v, On: 1v
47	BR/PK	EGR VR Solenoid Control	0%, at 55 mph: 45%
48	---	Not Used	---
49	LB/BK	Digital TR2 Sensor	0v, at 55 mph: 11.5v
50	WT/BK	Digital TR4 Sensor	0v, at 55 mph: 11.5v
51	BK/WT	Power Ground	<0.1v
52	RD/LB	Coil 2 Driver (Dwell)	5°, at 55 mph: 8°
53	PK/BK	Shift Solenoid 'C' Control	12v, 55 mph: 1v
54	RD/LB	TCC Solenoid Control	0%, at 55 mph: 95%
55	RD	Direct Battery	12-14v
56	LG/BK	EVAP Canister Purge Solenoid	0-10 Hz (0-100%)
57-59	---	Not Used	---
60	GY/LB	HO2S-11 (B1 S1) Signal	0.1-1.1v
61	VT/LG	HO2S-22 (B2 S2) Signal	0.1-1.1v
62	RD/PK	Fuel Tank Pressure Sensor	2.6v (0" H2O - cap off)
63	---	Not Used	---
64	LB/YL	Digital Transmission Range Sensor	In 'P': 0v, in O/D: 1.7v

2000-02 E-Series Van 4.2L V6 VIN 2 (A/T) 104-P Connector

PCM Pin #	Wire Color	Circuit Description (104 Pin)	Value at Hot Idle
65 ('00)	BR/LG	EGR DPFE Sensor Signal	0.20-1.30v
65 ('01-'02)	BR/LG	EGR DPFE Sensor Signal	0.95-1.05v
66	---	Not Used	---
67	VT/WT	EVAP Canister Vent Solenoid Control	0-10 Hz (0-100%)
68	---	Not Used	---
69	PK/YL	A/C Relay Control	On: 1v, Off: 12v
70	---	Not Used	---
71	RD	Vehicle Power (Start-Run)	12-14v
73	TN/BK	Injector 5 Control	2.7-4.1 ms
74	BR/YL	Injector 3 Control	2.7-4.1 ms
75	TN	Injector 1 Control	2.7-4.1 ms
76	BK/WT	Power Ground	<0.1v
77	---	Not Used	---
78	PK/WT	Coil 3 Driver (Dwell)	5°, at 55 mph: 8°
79	---	Not Used	---
80	LB/OR	Fuel Pump Relay Control	Off: 12v, On: 1v
81	WT/YL	EPC Solenoid Control	10v (15-20 psi)
82	---	Not Used	---
83	WT/LB	IAC Solenoid Control	10v (31% duty cycle)
84	DB/YL	Output Shaft Speed Sensor	0 Hz, 55 mph: 250 Hz
85	DG/OR	Camshaft Position Sensor	6 Hz, 55 mph: 16 Hz
86	TN/LG	A/C Pressure Switch	Switch On: 12v (Open)
87	RD/BK	HO2S-21 (B2 S1) Signal	0.1-1.1v
88	LB/RD	MAF Sensor Signal	0.6v, 55 mph: 2.2v
89	GY/WT	TP Sensor Signal	0.9v, 55 mph: 1.1v
90	BR/WT	Reference Voltage	4.9-5.1v
91	GY/RD	Sensor Ground	<0.050v
92	---	Not Used	---
93	RD/WT	HO2S-11 (B1 S1) Heater Control	1v, Off: 12v
94	YL/LB	HO2S-21 (B2 S1) Heater Control	1v, Off: 12v
95	WT/BK	HO2S-12 (B1 S2) Heater Control	1v, Off: 12v
96	TN/YL	HO2S-22 (B2 S2) Heater Control	1v, Off: 12v
97-98	---	Not Used	---
99	LG/OR	Injector 6 Control	2.7-4.1 ms
100	BR/LB	Injector 4 Control	2.7-4.1 ms
101	WT	Injector 2 Control	2.7-4.1 ms
102	---	Not Used	---
103	BK/WT	Power Ground	<0.1v
104	---	Not Used	---

Pin Connector Graphic

PCM 104-PIN CONNECTOR

Terminal View of 104-Pin PCM Wiring Harness Connector

Standard Colors and Abbreviations

Abbreviation	Color	Abbreviation	Color	Abbreviation	Color
BK	Black	GY	Gray	PK	Purple
BL	Blue	GN	Green	RD	Red
BR	Brown	LG	LT Green	TN	Tan
DB	Dark Blue	OR	Orange	WT	White
DG	DK Green	PK	Pink	YL	Yellow

1997-99 E-Series Van 4.6L V8 VIN 6 (A/T) 104-P Connector

PCM Pin #	Wire Color	Circuit Description (104 Pin)	Value at Hot Idle
1	VTN/ORG	Shift Solenoid 2 Control	12v, 55 mph: 1v
2	PK/LG	MIL (lamp) Control	MIL On: <1v, Off: 12v
3	YL/BK	Digital TR1 Sensor Signal	0v, at 55 mph: 11.5v
12	YL/WT	Fuel Level Indicator Signal	1.7v (1/2 full)
13	VT	Flash EEPROM	0.1v
14	LB/BK	4x4 Indicator Signal	Off: 12v, On: 1v
15	PK/LB	Data Bus (-) Signal	Digital Signals
16	TN/OR	Data Bus (+) Signal	Digital Signals
19	DB/WT	Automatic Ride Control	Digital Signals
21	DB	CKP (+) Sensor	430-475 Hz
22	GY	CKP (-) Sensor	430-475 Hz
24	BK/WT	Power Ground	<0.1v
25	BK/LB	Case Ground	<0.050v
26	DB/LG	Coil 1 Driver (Dwell)	5°, at 55 mph: 8°
27	OR/YL	Shift Solenoid 1 Control	1v, 55 mph: 1v
29	TN/WT	TCS (switch) Signal	TCS & O/D On: 12v
30	DG	Octane Adjustment	9.3v (switch shorted: 0v)
33	PK/OR	VSS (-) Signal	0 Hz, at 55 mph: 125 Hz
35	RD/LG	HO2S-12 (B1 S2) Signal	0.1-1.1v
36	TN/LB	MAF Sensor Return	<0.050v
37	OR/BK	Transmission Fluid Temperature Sensor	2.10-2.40v
38	LG/RD	Engine Coolant Temperature Sensor	0.5-0.6v
39	GY	Intake Air Temperature Sensor	1.5-2.5v
40	DG/YL	Fuel Pump Monitor	On: 12v, Off: 0v
41	BK/YL	A/C Cycling Clutch Switch	A/C On: 12v, off: 0v
42	BR	Inlet Air Control Solenoid	12v, 55 mph: 12v
43	OR/LG	Overhead Trip Computer	Digital Signals
45	RD/WT	CHTIL (lamp) Control	Lamp On: 1v, Off: 12v
46	BR	Intake Manifold Tuning Valve	12v, 55 mph: 12v
47	BR/PK	VR Solenoid Control	0%, at 55 mph: 45%
48	WT/PK	Clean Tachometer Output Signal	39-49 Hz
49	LB/BK	Digital TR2 Sensor	0v, at 55 mph: 11.5v
50	LG/RD	Digital TR4 Sensor	0v, at 55 mph: 11.5v
51	BK/WT	Power Ground	<0.1v
52	RD/LB	Coil 2 Driver (Dwell)	5°, at 55 mph: 8°
54	PK/YL	TCC Solenoid Control	0%, at 55 mph: 95%
55	Y	Keep Alive Power	12-14v
56	LG/BK	EVAP Purge Solenoid	0-10 Hz (0-100%)
57	YL/RD	Knock Sensor 1 Signal	0v
58	GY/BK	VSS (+) Signal	0 Hz, at 55 mph: 125 Hz
60	GY/LB	HO2S-11 (B1 S1) Signal	0.1-1.1v
61	PK/LG	HO2S-22 (B2 S2) Signal	0.1-1.1v
62	RD/PK	Fuel Tank Pressure Sensor	2.6v (0" H2O - cap off)

1997-99 E-Series Van 4.6L V8 VIN 6 (A/T) 104-P Connector

PCM Pin #	Wire Color	Circuit Description (104 Pin)	Value at Hot Idle
64	LB/YL	Clutch Pedal Position Switch	5v (Clutch In: 0v)
64	LB/YL	Transmission Range Sensor	In 'P': 0v, in O/D: 1.7v
65	BR/LG	EGR DPFE Sensor Signal	0.20-1.30v
66	YL/LG	Cylinder Head Temperature Sensor	0.6 to 1.7v (194°F)
67	PKWT	EVAP CV Solenoid	0-10 Hz (0-100%)
69	PK/YL	A/C WOT Relay Control	On: 1v, Off: 12v
71, 97	RD	Vehicle Power	12-14v
72	TN/RD	Injector 7 Control	2.7-4.1 ms
73	TN/BK	Injector 5 Control	2.7-4.1 ms
74	BR/YL	Injector 3 Control	2.7-4.1 ms
75	TN	Injector 1 Control	2.7-4.1 ms
76, 77, 103	BK/WT	Power Ground	<0.1v
78	PK/WT	Coil 3 Driver (Dwell)	5°, at 55 mph: 8°
79	WT/LG	TCIL (lamp) Control	Lamp On: 1v, Off: 12v
80	LB/OR	Fuel Pump Control	On: 1v, Off: 12v
81	WT/OR	EPC Solenoid Control	9.1v (6 psi)
83	WT/LB	IAC Solenoid Control	10v (31% duty cycle)
84	DG/WT	Output Shaft Speed Sensor	0 Hz, 55 mph: 250 Hz
85	DG	CID Sensor Signal	6 Hz, 55 mph: 15 Hz
87	RD/BK	HO2S-21 (B2 S1) Signal	0.1-1.1v
88	LB/RD	MAF Sensor Signal	0.6v, 55 mph: 2.1v
89	GY/WT	TP Sensor Signal	0.53-1.27v
90	BR/WT	Reference Voltage	4.9-5.1v
91	GY/RD	Analog Signal Return	<0.050v
92	LG	Brake Switch Signal	Brake Off: 12v, On: 1v
93	RD/WT	HO2S-11 (B1 S1) Heater	1v, Off: 12v
94	YL/LB	HO2S-21 (B2 S1) Heater	1v, Off: 12v
95	WT/BK	HO2S-12 (B1 S2) Heater	1v, Off: 12v
96	TN/YL	HO2S-22 (B2 S2) Heater	1v, Off: 12v
98	LB	Injector 8 Control	2.7-4.1 ms
99	LG/OR	Injector 6 Control	2.7-4.1 ms
100	BR/LB	Injector 4 Control	2.7-4.1 ms
101	WT	Injector 2 Control	2.7-4.1 ms
102	---	Not Used	---
104	RD/YL	Coil 4 Driver (Dwell)	5°, at 55 mph: 8°

Pin Connector Graphic

PCM 104-PIN CONNECTOR

Terminal View of 104-Pin PCM Wiring Harness Connector

Standard Colors and Abbreviations

Abbreviation	Color	Abbreviation	Color	Abbreviation	Color
BK	Black	GY	Gray	PK	Purple
BL	Blue	GN	Green	RD	Red
BR	Brown	LG	LT Green	TN	Tan
DB	Dark Blue	OR	Orange	WT	White
DG	DK Green	PK	Pink	YL	Yellow

2000-03 E-Series 4.6L V8 VIN W (All) 104 Pin Connector

PCM Pin #	Wire Color	Circuit Description (104 Pin)	Value at Hot Idle
1	OR/YL	COP 6 Driver (Dwell)	5°, at 55 mph: 8°
2	PK/LG	MIL (lamp) Control	MIL On: <1v, Off: 12v
3	BK/WT	Power Ground	<0.1v
4-5	---	Not Used	---
6	OR/YL	Shift Solenoid 1 Control	1v, 55 mph: 1v
9	YL/WT	Fuel Pump Monitor	On: 12v, Off: 0v
10	---	Not Used	---
11	VT/OR	Shift Solenoid 2 Control	12v, 55 mph: 1v
12	WT/LG	TCIL (lamp) Control	Indicator Off: 12v, On: 1v
13	VT	Flash EEPROM	0.1v
14	OG/LB	4WD Low Signal	0v or 12v
15	PK/LB	SCP Bus (-) Signal	<0.050v
16	TN/OR	SCP Bus (+) Signal	Digital Signals
17-19	---	Not Used	---
20	BR/OR	Coast Clutch Solenoid Control	12v, 55 mph: 12v
21	DB	CKP (+) Sensor	390-430 Hz
22	GY	CKP (-) Sensor	390-430 Hz
23	---	Not Used	---
24	BK/WT	Power Ground	<0.1v
25	BK/LB	Case Ground	<0.050v
26	LG/WT	COP 1 Driver (Dwell)	5°, at 55 mph: 8°
27	LG/YL	COP 5 Driver (Dwell)	5°, at 55 mph: 8°
28	---	Not Used	---
29	TN/WT	PTO (switch) Signal	PTO On: 12v
30-31	---	Not Used	---
32	DG/VT	Knock Sensor 1 Ground	<0.050v
33	---	Not Used	---
34	YL/BK	Transmission Range Sensor	In 'P': 0v, in O/D: 1.7v
35	RD/LG	HO2S-12 (B1 S2) Signal	0.1-1.1v
36	TN/LB	MAF Sensor Return	<0.050v
37	OR/BK	Transmission Fluid Temperature Sensor	2.10-2.40v
38	LG/RD	Engine Oil Temperature Sensor	-40°F: 4.7v, 230°F: 0.36v
39	GY	Intake Air Temperature Sensor	1.5-2.5v
40	DG/YL	Fuel Pump Monitor	On: 12v, Off: 0v
41	TN/LG	A/C Pressure Switch Signal	A/C On: 12v, Off: 0v
42-44	---	Not Used	---
45	YL/LB	Engine Coolant Temperature Sensor	0.5-0.6v
46	BR	Intake Manifold Tuning Valve	12v, 55 mph: 12v
47	BR/PK	EGR VR Solenoid Control	0%, at 55 mph: 45%
48	WT/PK	Clean Tachometer Output Signal	DC pulse signals
49	LB/BK	Digital TR2 Sensor	0v, at 55 mph: 11.5v
50	LG/RD	Digital TR4 Sensor	0v, at 55 mph: 11.5v
51	BK/WT	Power Ground	<0.1v
52	WT/PK	COP 3 Driver (Dwell)	5°, at 55 mph: 8°
53	DG/VT	COP 4 Driver (Dwell)	5°, at 55 mph: 8°
54	VT/YL	TCC Solenoid Control	0%, at 55 mph: 95%
55	YL	Keep Alive Power	12-14v
56	LG/BK	EVAP Purge Solenoid	0-10 Hz (0-100%)
57	YL/RD	Knock Sensor (-) Signal	<0.050v
58	---	Not Used	---
59	DG/WT	Turbine Shaft Speed Sensor	120 Hz, 30 mph: 300 Hz
60	GY/LB	HO2S-11 (B1 S1) Signal	0.1-1.1v
61	VT/LG	HO2S-22 (B2 S2) Signal	0.1-1.1v
62	RD/PK	Fuel Tank Pressure Sensor	2.6v (0" H2O - cap off)
63	---	Not Used	---
64	LB/YL	Digital TR3A Sensor	0v, at 55 mph: 11.5v

2000-03 E-Series Van 4.6L V8 VIN W (All) 104 Pin Connector

PCM Pin #	Wire Color	Circuit Description (104 Pin)	Value at Hot Idle
65 '00)	BR/LG	EGR DPFE Sensor Signal	0.20-1.30v
65 ('01-'02)	BR/LG	EGR DPFE Sensor Signal	0.95-1.05v
66	YL/LG	Cylinder Head Temperature Sensor	0.6 to 1.7v (194°F)
67	VT/WT	EVAP CV Solenoid	0-10 Hz (0-100%)
68	GY/BK	Vehicle Speed Output	0 Hz, at 55 mph: 125 Hz
69	BK/LB	A/C Relay Control	On: 1v, Off: 12v
71	RD	Vehicle Power (Start Run)	12-14v
72	TN/RD	Injector 7 Control	2.7-4.1 ms
73	TN/BK	Injector 5 Control	2.7-4.1 ms
74	BR/YL	Injector 3 Control	2.7-4.1 ms
75	TN	Injector 1 Control	2.7-4.1 ms
76	BK/WT	Power Ground	<0.1v
77	BK/WT	Power Ground	<0.1v
78	PK/LB	COP 7 Driver (Dwell)	5°, at 55 mph: 8°
79	WT/RD	COP 8 Driver (Dwell)	5°, at 55 mph: 8°
80	LB/OR	Fuel Pump Control	On: 1v, Off: 12v
81	WT/YL	EPC Solenoid Control	9.1v (6 psi)
83	WT/LB	IAC Solenoid Control	10v (31% duty cycle)
84	DG/YL	Output Shaft Speed Sensor	0 Hz, 55 mph: 250 Hz
85	DG	Camshaft Position Sensor	6 Hz, 55 mph: 15 Hz
87	RD/BK	HO2S-21 (B2 S1) Signal	0.1-1.1v
88	LB/RD	MAF Sensor Signal	0.6v, 55 mph: 2.1v
89	GY/WT	TP Sensor Signal	0.53-1.27v
90	BR/WT	Reference Voltage	4.9-5.1v
91	GY/RD	Analog Signal Return	<0.050v
92	LG	Brake Position Switch	Brake Off: 12v, On: 1v
93	RD/WT	HO2S-11 (B1 S1) Heater	1v, Off: 12v
94	YL/LB	HO2S-21 (B2 S1) Heater	1v, Off: 12v
95	WT/BK	HO2S-12 (B1 S2) Heater	1v, Off: 12v
96	TN/YL	HO2S-22 (B2 S2) Heater	1v, Off: 12v
97	RD	Vehicle Power (Start Run)	12-14v
98	LB	Injector 8 Control	2.7-4.1 ms
99	LG/OR	Injector 6 Control	2.7-4.1 ms
100	BR/LB	Injector 4 Control	2.7-4.1 ms
101	WT	Injector 2 Control	2.7-4.1 ms
103	BK/WT	Power Ground	<0.1v
104	PK/WT	COP 2 Driver (Dwell)	5°, at 55 mph: 8°

Pin Connector Graphic

PCM 104-PIN CONNECTOR

Terminal View of 104-Pin PCM Wiring Harness Connector

Standard Colors and Abbreviations

Abbreviation	Color	Abbreviation	Color	Abbreviation	Color
BK	Black	GY	Gray	PK	Purple
BL	Blue	GN	Green	RD	Red
BR	Brown	LG	LT Green	TN	Tan
DB	Dark Blue	OR	Orange	WT	White
DG	DK Green	PK	Pink	YL	Yellow

1990 E-Series 4.9L I6 OHV MFI VIN Y (All) 60 Pin Connector

PCM Pin #	Wire Color	Circuit Description (60 Pin)	Value at Hot Idle
1	BK/OR	Keep Alive Power	12-14v
3	DG/WT	VSS (+) Signal	0 Hz, at 55 mph: 125 Hz
4	DG/YL	Ignition Diagnostic Monitor	20-31 Hz
6	OR/YL	VSS (-) Signal	0 Hz, at 55 mph: 125 Hz
7	LG/YL	Engine Coolant Temperature Sensor	0.5-0.6v
8	OR/BL	Fuel Pump Monitor	On: 12v, Off: 0v
9	BK/OR	Data Bus (-)	Digital Signals
10	BK/YL	A/C Pressure Switch Signal	A/C On: 12v, off: 0v
11	WT/BK	Air Management 2 Solenoid	On: 1v, Off: 12v
16	BK/OR	Ignition System Ground	<0.050v
17	TN/RD	Self-Test Output & MIL	MIL On: 1v, Off: 12v
20	BK	PCM Case Ground	<0.050v
21	GY/WT	IAC Solenoid Control	9.3-10.1v
22	TN/LG	Fuel Pump Control	On: 1v, Off: 12v
23	LG/BK	Knock Sensor 1 Signal	0v
24	YL/LG	Power Steering Pressure Switch	Straight: 0v,Turned: 12v
25	YL/RD	ACT Sensor Signal	1.5-2.5v
26	OR/WT	Reference Voltage	4.9-5.1v
27	BR/LG	EVP Sensor Signal	0.4v, 55 mph: 3.1v
28	OR/BK	Data Bus (+)	Digital Signals
29	DG/PK	HO2S-11 (B1 S1) Signal	0.1-1.1v
30	BL/YL	A/T: Neutral Drive Switch	In 'P': 0v, Others: 5v
30	BL/YL	M/T: Clutch Engage Switch	Clutch Out: 5v, In: 0v
31	GY/YL	Canister Purge Solenoid	12v, 55 mph: 1v
33	DG	VR Solenoid Control	0%, at 55 mph: 45%
36	YL/LG	Spark Output Signal	6.93v (50%)
37	RD	Vehicle Power	12-14v
40	BK/LG	Power Ground	<0.1v
43	LG/PK	A/C Demand Switch	A/C On: 12v, Off: 0v
45	DB/LG	MAP Sensor Signal	107 Hz
46	BK/WT	Analog Signal Return	<0.050v
47	DG/LG	TP Sensor Signal	1v, 55 mph: 2.1v
48	WT/RD	Self-Test Indicator Signal	STI On: 1v, Off: 5v
49	OR	HO2S-11 (B1 S1) Ground	<0.050v
51	WT/RD	Air Management 1 Solenoid	On: 1v, Off: 12v
56	DB	PIP Sensor Signal	6.93v (50%)
57	RD	Vehicle Power	12-14v
58	TN/OR	Injector Bank 1 (INJ 1, 3 & 5)	6.4-6.8 ms
59	TN/RD	Injector Bank 2 (INJ 2, 4 & 6)	6.4-6.8 ms
60	BK/LG	Power Ground	<0.1v

Pin Connector Graphic

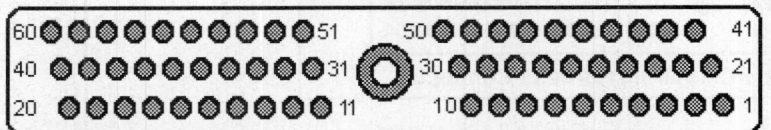

PCM 60-PIN CONNECTOR

Terminal View of 60-Pin PCM Harness Connector

Standard Colors and Abbreviations

Abbreviation	Color	Abbreviation	Color	Abbreviation	Color
BK	Black	GY	Gray	PK	Purple
BL	Blue	GN	Green	RD	Red
BR	Brown	LG	LT Green	TN	Tan
DB	Dark Blue	OR	Orange	WT	White
DG	DK Green	PK	Pink	YL	Yellow

1990 E-Series 4.9L I6 OHV VIN Y (E4OD) 60 Pin Connector

PCM Pin #	Wire Color	Circuit Description (60 Pin)	Value at Hot Idle
1	BK/OR	Keep Alive Power	12-14v
2	LG	Brake Switch Signal	Brake Off: 12v, On: 1v
3	DG/WT	VSS (+) Signal	0 Hz, at 55 mph: 125 Hz
4	DG/YL	Ignition Diagnostic Monitor	20-31 Hz
5	---	Not Used	---
6	OR/YL	VSS (-) Signal	0 Hz, at 55 mph: 125 Hz
7	LG/YL	Engine Coolant Temperature Sensor	0.5-0.6v
8	OR/BL	Fuel Pump Monitor	On: 12v, Off: 0v
9	BK/OR	Data Bus (-)	Digital Signals
10	BK/YL	A/C Pressure Switch Signal	A/C On: 12v, off: 0v
11	WT/BK	Air Management 2 Solenoid	On: 1v, Off: 12v
12-15	---	Not Used	---
16	BK/OR	Ignition System Ground	<0.050v
17	TN/RD	Self-Test Output & MIL	MIL On: 1v, Off: 12v
18	---	Not Used	---
19	DG/PK	Shift Solenoid 2 Control	12v, 55 mph: 1v
20	BK	PCM Case Ground	<0.050v
21	GY/WT	IAC Solenoid Control	9.3-10.1v
22	TN/LG	Fuel Pump Control	On: 1v, Off: 12v
23	LG/BK	Knock Sensor 1 Signal	0v
24	YL/LG	Power Steering Pressure Switch	Straight: 0v, Turned: 12v
25	YL/RD	Intake Air Temperature Sensor	1.5-2.5v
26	OR/WT	Reference Voltage	4.9-5.1v
27	BR/LG	EVP Sensor Signal	0.4v, 55 mph: 3.1v
28	OR/BK	Data Bus (+)	Digital Signals
29	DG/PK	HO2S-11 (B1 S1) Signal	0.1-1.1v
30	BL/YL	MLP Sensor Signal	In 'P': 0v, in O/D: 5v
31	GY/YL	Canister Purge Solenoid	12v, 55 mph: 1v
32	LG/WT	OCIL (lamp) Control	Lamp On: 1v, Off: 12v
33	DG	VR Solenoid Control	0%, at 55 mph: 45%
34-35	---	Not Used	---
36	YL/LG	Spark Output Signal	6.93v (50%)
37	RD	Vehicle Power	12-14v
38	BL/YL	EPC Solenoid Control	9.1v (4 psi)
39	---	Not Used	---
40	BK/LG	Power Ground	<0.1v
41	TN/BL	Overdrive Cancel Switch	OCS & O/D On: 0.1v
42	OR/BK	TOT Sensor Signal	2.10-2.40v
43	LG/PK	A/C Demand Switch	A/C On: 12v, Off: 0v
44	---	Not Used	---
45	DB/LG	MAP Sensor Signal	107 Hz
46	BK/WT	Analog Signal Return	<0.050v
47	DG/LG	TP Sensor Signal	0.7v, 55 mph: 2.1v
48	WT/RD	Self-Test Indicator Signal	STI On: 1v, Off: 5v
49	OR	HO2S-11 (B1 S1) Ground	<0.050v
50	---	Not Used	---
51	WT/RD	Air Management 1 Solenoid	On: 1v, Off: 12v
52	OR/YL	Shift Solenoid 1 Control	1v, 55 mph: 1v
53	PK/YL	TCC Solenoid Control	12v, 55 mph: 1v
54	---	Not Used	---
55	BR	Coast Clutch Solenoid	12v, 55 mph: 12v
56	DB	PIP Sensor Signal	6.93v (50%)
57	R	Vehicle Power	12-14v
58	TN/OR	Injector Bank 1 (INJ 1, 3 & 5)	6.4-6.8 ms
59	TN/RD	Injector Bank 2 (INJ 2, 4 & 6)	6.4-6.8 ms
60	BK/LG	Power Ground	<0.1v

1991 E-Series 4.9L I6 OHV VIN Y (All) 60 Pin Connector

PCM Pin #	Wire Color	Circuit Description (60 Pin)	Value at Hot Idle
1	BK/OR	Keep Alive Power	12-14v
2, 5	---	Not Used	---
3	DG/WT	VSS (+) Signal	0 Hz, at 55 mph: 125 Hz
4	TN/YL	Ignition Diagnostic Monitor	20-31 Hz
6	OR/YL	VSS (-) Signal	0 Hz, at 55 mph: 125 Hz
7	LG/PK	Engine Coolant Temperature Sensor	0.5-0.6v
8	DG/YL	Fuel Pump Monitor	On: 12v, Off: 0v
9	BK/OR	Data Bus (-) California	Digital Signals
10	BK/YL	A/C Pressure Switch Signal	A/C On: 12v, off: 0v
11	BR	Air Management 2 Solenoid	On: 1v, Off: 12v
12-15	---	Not Used	---
16	OR/RD	Ignition System Ground	<0.050v
17	TN/RD	Self-Test Output & MIL	MIL On: 1v, Off: 12v
18-19	---	Not Used	---
20	BK	PCM Case Ground	<0.050v
21	GY/WT	IAC Solenoid Control	8.5-10.2v
22	TN/LG	Fuel Pump Control	On: 1v, Off: 12v
23	LG/BK	Knock Sensor 1 Signal	0v
24	---	Not Used	---
25	YL/RD	ACT Sensor Signal	1.5-2.5v
26	BR/WT	Reference Voltage	4.9-5.1v
27	BR/LG	EVP Sensor Signal	0.4v, 55 mph: 3.1v
28	OR/BK	Data Bus (+)	Digital Signals
29	RD/BK	HO2S-11 (B1 S1) Signal	0.1-1.1v
30	WT/RD	A/T: Neutral Drive Switch	In 'P': 0v, Others: 5v
30	WT/RD	M/T: Clutch Engage Switch	Clutch Out: 5v, In: 0v
31	GY/YL	Canister Purge Solenoid	12v, 55 mph: 1v
32	---	Not Used	---
33	BR/PK	VR Solenoid Control	0%, at 55 mph: 45%
34-35	---	Not Used	---
36	YL/LG	Spark Output Signal	6.93v (50%)
37	Rd	Vehicle Power	12-14v
38-39	---	Not Used	---
40	BK/LG	Power Ground	<0.1v
41-42	---	Not Used	---
43	PK	A/C Demand Switch	A/C On: 12v, Off: 0v
44	---	Not Used	---
45	DB/LG	MAP Sensor Signal	107 Hz
46	GY/RD	Analog Signal Return	<0.050v
47	GY/WT	TP Sensor Signal	0.7v, 55 mph: 2.1v
48	WT/RD	Self-Test Indicator Signal	STI On: 1v, Off: 5v
49	OR	HO2S-11 (B1 S1) Ground	<0.050v
50	---	Not Used	---
51	OR	Air Management 1 Solenoid	On: 1v, Off: 12v
52-55	---	Not Used	---
56	DB	PIP Sensor Signal	6.93v (50%)
57	RD	Vehicle Power	12-14v
58	TN/OR	Injector Bank 1 (INJ 1, 3 & 5)	5.0-6.4 ms
59	TN/RD	Injector Bank 2 (INJ 2, 4 & 6)	5.0-6.4 ms
60	BK/LG	Power Ground	<0.1v

Pin Connector Graphic

PCM 60-PIN CONNECTOR

Terminal View of 60-Pin PCM Harness Connector

1991 E-Series 4.9L I6 OHV VIN Y (E4OD) 60 Pin Connector

PCM Pin #	Wire Color	Circuit Description (60 Pin)	Value at Hot Idle
1	BK/OR	Keep Alive Power	12-14v
2	LG	Brake Switch Signal	Brake Off: 12v, On: 1v
3	DG/WT	VSS (+) Signal	0 Hz, at 55 mph: 125 Hz
4	TN/YL	Ignition Diagnostic Monitor	20-31 Hz
5	---	Not Used	---
6	OR/YL	VSS (-) Signal	0 Hz, at 55 mph: 125 Hz
7	LG/PK	Engine Coolant Temperature Sensor	0.5-0.6v
8	DG/YL	Fuel Pump Monitor	On: 12v, Off: 0v
9	BK/OR	Data Bus (-)	Digital Signals
10	BK/YL	A/C Pressure Switch Signal	A/C On: 12v, off: 0v
11	BR	Air Management 2 Solenoid	On: 1v, Off: 12v
12-15	---	Not Used	---
16	OR/RD	Ignition System Ground	<0.050v
17	TN/RD	Self-Test Output & MIL	MIL On: 1v, Off: 12v
18	---	Not Used	---
19	PK	Shift Solenoid 2 Control	12v, 55 mph: 1v
20	BK	PCM Case Ground	<0.050v
21	GY/WT	IAC Solenoid Control	8.5-10.3v
22	TN/LG	Fuel Pump Control	On: 1v, Off: 12v
23	LG/BK	Knock Sensor 1 Signal	0v
24	---	Not Used	---
25	YL/RD	ACT Sensor Signal	1.5-2.5v
26	BR/WT	Reference Voltage	4.9-5.1v
27	BR/LG	EVP Sensor Signal	0.4v, 55 mph: 3.1v
28	OR/BK	Data Bus (+)	Digital Signals
29	RD/BK	HO2S-11 (B1 S1) Signal	0.1-1.1v
30	WT/RD	MLP Sensor Signal	In 'P': 0v, in O/D: 5v
31	GY/YL	Canister Purge Solenoid	12v, 55 mph: 1v
32	WT/LG	OCIL (lamp) Control	Lamp On: 1v, Off: 12v
33	BR/PK	VR Solenoid Control	0%, at 55 mph: 45%
34-35	---	Not Used	---
36	YL/LG	Spark Output Signal	6.93v (50%)
37	RD	Vehicle Power	12-14v
38	BL/YL	EPC Solenoid Control	7.7v (5 psi)
39	---	Not Used	---
40	BK/LG	Power Ground	<0.1v
41	TN/WT	Overdrive Cancel Switch	OCS & O/D On: 0.1v
42	OR/BK	TOT Sensor Signal	2.10-2.40v
43	PK	A/C Demand Switch	A/C On: 12v, Off: 0v
44	---	Not Used	---
45	DB/LG	MAP Sensor Signal	107 Hz
46	GY/RD	Analog Signal Return	<0.050v
47	GY/WT	TP Sensor Signal	0.7v, 55 mph: 2.1v
48	WT/RD	Self-Test Indicator Signal	STI On: 1v, Off: 5v
49	OR	HO2S-11 (B1 S1) Ground	<0.050v
50	---	Not Used	---
51	OR	Air Management 1 Solenoid	On: 1v, Off: 12v
52	OR	Shift Solenoid 1 Control	1v, 55 mph: 12v
53	PK/YL	TCC Solenoid Control	0%, at 55 mph: 95%
54	---	Not Used	---
55	BR/OR	Coast Clutch Switch Signal	12v, 55 mph: 12v
56	DB	PIP Sensor Signal	6.93v (50%)
57	RD	Vehicle Power	12-14v
58	TN/OR	Injector Bank 1 (INJ 1, 3 & 5)	5.6-6.4 ms
59	TN/RD	Injector Bank 2 (INJ 2, 4 & 6)	5.6-6.4 ms
60	BK/LG	Power Ground	<0.1v

1992-93 E-Series 4.9L I6 OHV VIN Y (All) 60 Pin Connector

PCM Pin #	Wire Color	Circuit Description (60 Pin)	Value at Hot Idle
1	YL	Keep Alive Power	12-14v
2, 5	---	Not Used	---
3	GY/BK	VSS (+) Signal	0 Hz, at 55 mph: 125 Hz
4	TN/YL	Ignition Diagnostic Monitor	20-31 Hz
6	PK/OR	VSS (-) Signal	0 Hz, at 55 mph: 125 Hz
7	LG/RD	Engine Coolant Temperature Sensor	0.5-0.6v
8	DG/YL	Fuel Pump Monitor	On: 12v, Off: 0v
9	PK/BL	Data Bus (-)	Digital Signals
10	BK/YL	A/C Pressure Switch Signal	A/C On: 12v, off: 0v
11	BR	Air Management 2 Solenoid	On: 1v, Off: 12v
12-15	---	Not Used	---
16	OR/RD	Ignition System Ground	<0.050v
17	PK/LG	Self-Test Output & MIL	MIL On: 1v, Off: 12v
18-19, 24	---	Not Used	---
20	BK	PCM Case Ground	<0.050v
21	WT/BL	IAC Solenoid Control	8.2-10.1v
22	BL/OR	Fuel Pump Control	On: 1v, Off: 12v
23	YL/RD	Knock Sensor 1 Signal	0v
25	GY	ACT Sensor Signal	1.5-2.5v
26	BR/WT	Reference Voltage	4.9-5.1v
27	BR/LG	EVP Sensor Signal	0.4v, 55 mph: 3.1v
28	TN/OR	Data Bus (+)	Digital Signals
29	GY/BL	HO2S-11 (B1 S1) Signal	0.1-1.1v
30	BL/YL	A/T: Neutral Drive Switch	In 'P': 0v, Others: 5v
30	BL/YL	M/T: Clutch Engage Switch	Clutch Out: 5v, In: 0v
31	GY/YL	Canister Purge Solenoid	12v, 55 mph: 1v
32	---	Not Used	---
33	BR/PK	VR Solenoid Control	0%, at 55 mph: 45%
34-35	---	Not Used	---
36	PK	Spark Output Signal	6.93v (50%)
37	RD	Vehicle Power	12-14v
38-39	---	Not Used	---
40	BK/WT	Power Ground	<0.1v
41-42	---	Not Used	---
43	PK	A/C Demand Switch	A/C On: 12v, Off: 0v
44	---	Not Used	---
45	LG/BK	MAP Sensor Signal	107 Hz
46	GY/RD	Analog Signal Return	<0.050v
47	GY/WT	TP Sensor Signal	0.7v, 55 mph: 2.1v
48	WT/PK	Self-Test Indicator Signal	STI On: 1v, Off: 5v
49	OR	HO2S-11 (B1 S1) Ground	<0.050v
50	---	Not Used	---
51	WT/OR	Air Management 1 Solenoid	On: 1v, Off: 12v
52-54	---	Not Used	---
55	---	Not Used	---
56	GY/OR	PIP Sensor Signal	6.93v (50%)
57	RD	Vehicle Power	12-14v
58	TN	Injector Bank 1 (INJ 1, 3 & 5)	6.8-7.0 ms
59	WT	Injector Bank 2 (INJ 2, 4 & 6)	6.8-7.0 ms
60	BK/WT	Power Ground	<0.1v

Pin Connector Graphic

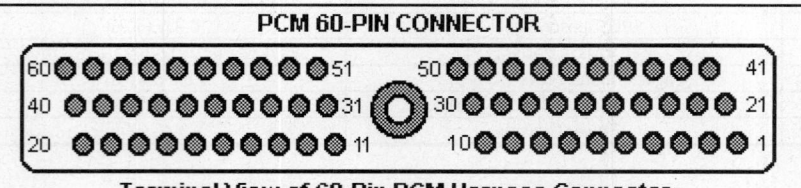

PCM 60-PIN CONNECTOR

Terminal View of 60-Pin PCM Harness Connector

1992-93 E-Series 4.9L I6 VIN Y (E4OD) 60 Pin Connector

PCM Pin #	Wire Color	Circuit Description (60 Pin)	Value at Hot Idle
1	YL	Keep Alive Power	12-14v
2	LG	Brake Switch Signal	Brake Off: 12v, On: 1v
3	GY/BK	VSS (+) Signal	0 Hz, at 55 mph: 125 Hz
4	TN/YL	Ignition Diagnostic Monitor	20-31 Hz
5	---	Not Used	---
6	PK/OR	VSS (-) Signal	0 Hz, at 55 mph: 125 Hz
7	LG/RD	Engine Coolant Temperature Sensor	0.5-0.6v
8	DG/YL	Fuel Pump Monitor	On: 12v, Off: 0v
9	PK/BL	Data Bus (-)	Digital Signals
10	BK/YL	A/C Pressure Switch Signal	A/C On: 12v, off: 0v
11	BR	Air Management 2 Solenoid	On: 1v, Off: 12v
12-15	---	Not Used	---
16	OR/RD	Ignition System Ground	<0.050v
17	PK/LG	Self-Test Output & MIL	MIL On: 1v, Off: 12v
18	---	Not Used	---
19	PK/OR	Shift Solenoid 2 Control	12v, 55 mph: 1v
20	BK	PCM Case Ground	<0.050v
21	WT/BL	IAC Solenoid Control	8.2-10.1v
22	BL/OR	Fuel Pump Control	On: 1v, Off: 12v
23	YL/RD	Knock Sensor 1 Signal	0v
24	---	Not Used	---
25	GY	ACT Sensor Signal	1.5-2.5v
26	BR/WT	Reference Voltage	4.9-5.1v
27	BR/LG	EVP Sensor Signal	0.4v, 55 mph: 3.1v
28	TN/OR	Data Bus (+)	Digital Signals
29	GY/BL	HO2S-11 (B1 S1) Signal	0.1-1.1v
30	BL/YL	MLP Sensor Signal	In 'P': 0v, in O/D: 5v
31	GY/YL	Canister Purge Solenoid	12v, 55 mph: 1v
32	WT/LG	OCIL (lamp) Control	Lamp On: 1v, Off: 12v
33	BR/PK	VR Solenoid Control	0%, at 55 mph: 45%
34-35	---	Not Used	---
36	PK	Spark Output Signal	6.93v (50%)
37	RD	Vehicle Power	12-14v
38	WT/YL	EPC Solenoid Control	7.7-8.7v
39	---	Not Used	---
40	BK/WT	Power Ground	<0.1v
41	TN/WT	Overdrive Cancel Switch	OCS & O/D On: 0.1v
42	OR/BK	TOT Sensor Signal	2.10-2.40v
43	PK	A/C Demand Switch	A/C On: 12v, Off: 0v
44	---	Not Used	---
45	LG/BK	MAP Sensor Signal	107 Hz
46	GY/RD	Analog Signal Return	<0.050v
47	GY/WT	TP Sensor Signal	0.7v, 55 mph: 2.1v
48	WT/PK	Self-Test Indicator Signal	STI On: 1v, Off: 5v
49	OR	HO2S-11 (B1 S1) Ground	<0.050v
50	---	Not Used	---
51	WT/OR	Air Management 1 Solenoid	On: 1v, Off: 12v
52	OR/YL	Shift Solenoid 1 Control	1v, 55 mph: 12v
53	PK/YL	TCC Solenoid Control	0%, at 55 mph: 95%
54	---	Not Used	---
55	BR/OR	Coast Clutch Solenoid	12v, 55 mph: 12v
56	GY/OR	PIP Sensor Signal	6.93v (50%)
57	RD	Vehicle Power	12-14v
58	TN	Injector Bank 1 (INJ 1, 3 & 5)	6.2-7.4 ms
59	WT	Injector Bank 2 (INJ 2, 4 & 6)	6.2-7.4 ms
60	BK/WT	Power Ground	<0.1v

1994-95-Series 4.9L I6 OHV VIN Y (All) 60 Pin Connector

PCM Pin #	Wire Color	Circuit Description (60 Pin)	Value at Hot Idle
1	YL	Keep Alive Power	12-14v
2	LG	Brake Switch Signal	Brake Off: 12v, On: 1v
3	GY/BK	VSS (+) Signal	0 Hz, at 55 mph: 125 Hz
4	WT/PK	Ignition Diagnostic Monitor	20-31 Hz
5	---	Not Used	---
6	PK/OR	VSS (-) Signal	0 Hz, at 55 mph: 125 Hz
7	LG/RD	Engine Coolant Temperature Sensor	0.5-0.6v
8	DG/YL	Fuel Pump Monitor	On: 12v, Off: 0v
9	TN/BL	MAF Sensor Return	<0.050v
10	DG/OR	A/C Pressure Switch Signal	A/C On: 12v, off: 0v
11	GY/YL	Canister Purge Solenoid	12v, 55 mph: 1v
12	LG/OR	Injector 6 Control	5.5-7.1 ms
13-14	---	Not Used	---
15	TN/BK	Injector 5 Control	5.5-7.1 ms
16	OR/RD	Ignition System Ground	<0.1v
17	PK/LG	Self-Test Output & MIL	MIL On: 1v, Off: 12v
18	TN/OR	Data Bus (-)	Digital Signals
19	PK/OR	Data Bus (+)	Digital Signals
20	BK	PCM Case Ground	<0.050v
21	WT/BL	IAC Solenoid Control	8.6-10.7v
22	BL/OR	Fuel Pump Control	On: 1v, Off: 12v
23	YL/RD	Knock Sensor 1 Signal	0v
24	---	Not Used	---
25	GY	Intake Air Temperature Sensor	1.5-2.5v
26	BR/WT	Reference Voltage	4.9-5.1v
27	BR/LG	EVP Sensor Signal	0.4v, 55 mph: 3.1v
28-29	---	Not Used	---
30	BL/YL	A/T: PNP Sensor Signal	In 'P': 0v, Others: 5v
30	BL/YL	M/T: Clutch Pedal Position	Clutch Out: 5v, In: 0v
31	WT/OR	Thermactor Air Bypass	TAB On: 1v, Off: 12v
32	WT/LG	TCIL (lamp) Control	Lamp On: 1v, Off: 12v
33	BR/PK	VR Solenoid Control	0%, at 55 mph: 45%
34	BR	Thermactor Air Diverter	TAD On: 1v, Off: 12v
35	BR/BL	Injector 4 Control	5.5-7.1 ms
36	PK	Spark Output Signal	6.93v (50%)
37	RD	Vehicle Power	12-14v
38	---	Not Used	---
39	BR/YL	Injector 3 Control	5.5-7.1 ms
40	BK/WT	Power Ground	<0.1v
41	TN/WT	TCS (switch) Signal	TCS & O/D On: 12v
42	---	Not Used	---
43	RD/BR	HO2S-21 (B2 S1) Signal	0.1-1.1v
44	GY/BL	HO2S-11 (B1 S1) Signal	0.1-1.1v
45	---	Not Used	---
46	GY/RD	Analog Signal Return	<0.050v
47	GY/WT	TP Sensor Signal	0.7v, 55 mph: 2.1v
48	WT/PK	Self-Test Indicator Signal	STI On: 1v, Off: 5v
49	---	Not Used	---
50	BL/RD	MAF Sensor Signal	0.8v, 55 mph: 1.6v
51-53	---	Not Used	---
54	PK/YL	A/C WOT Relay Control	On: 1v, Off: 12v
55	---	Not Used	---
56	GY/OR	PIP Sensor Signal	6.93v (50%)
57	RD	Vehicle Power	12-14v
58	TN	Injector 1 Control	5.5-7.1 ms
59	WT	Injector 2 Control	5.5-7.1 ms
60	BK/WT	Power Ground	<0.1v

1994-95 E-Series 4.9L I6 VIN Y (E4OD) 60 Pin Connector

PCM Pin #	Wire Color	Circuit Description (60 Pin)	Value at Hot Idle
1	YL	Keep Alive Power	12-14v
2	LG	Brake Switch Signal	Brake Off: 12v, On: 1v
3	GY/BK	VSS (+) Signal	0 Hz, at 55 mph: 125 Hz
4	WT/PK	Ignition Diagnostic Monitor	20-31 Hz
5	---	Not Used	---
6	PK/OR	VSS (-) Signal	0 Hz, at 55 mph: 125 Hz
7	LG/RD	Engine Coolant Temperature Sensor	0.5-0.6v
8	DG/YL	Fuel Pump Monitor	On: 12v, Off: 0v
9	PK/BL	Data Bus (-)	Digital Signals
10	BK/YL	A/C Pressure Switch Signal	A/C On: 12v, off: 0v
11	BR	Air Management 2 Solenoid	On: 1v, Off: 12v
12-15	---	Not Used	---
16	OR/RD	Ignition System Ground	<0.050v
17	PK/LG	Self-Test Output & MIL	MIL On: 1v, Off: 12v
18	---	Not Used	---
19	PK/OR	Shift Solenoid 2 Control	12v, 55 mph: 1v
20	BK	PCM Case Ground	<0.050v
21	WT/BL	IAC Solenoid Control	10.7v (32%)
22	BL/OR	Fuel Pump Control	On: 1v, Off: 12v
23	YL/RD	Knock Sensor 1 Signal	0v
24	---	Not Used	---
25	GY	Intake Air Temperature Sensor	1.5-2.5v
26	BR/WT	Reference Voltage	4.9-5.1v
27	BR/LG	EVP Sensor Signal	0.4v, 55 mph: 3.1v
28	TN/OR	Data Bus (+)	Digital Signals
29	GY/BL	HO2S-11 (B1 S1) Signal	0.1-1.1v
30	BL/YL	MLP Sensor Signal	In 'P': 0v, in Drive: 5v
31	GY/YL	Canister Purge Solenoid	12v, 55 mph: 1v
32	WT/LG	TCIL (lamp) Control	Lamp On: 1v, Off: 12v
33	BR/PK	VR Solenoid Control	0%, at 55 mph: 45%
34-35	---	Not Used	---
36	PK	Spark Output Signal	6.93v (50%)
37	RD	Vehicle Power	12-14v
38	WT/YL	EPC Solenoid Control	8.1v (4 psi)
39	---	Not Used	---
40	BK/WT	Power Ground	<0.1v
41	TN/WT	TCS (switch) Signal	TCS & O/D On: 12v
42	OR/BK	TOT Sensor Signal	2.10-2.40v
43	PK	A/C Demand Switch	A/C On: 12v, Off: 0v
44	---	Not Used	---
45	LG/BK	MAP Sensor Signal	107 Hz (sea level)
46	GY/RD	Analog Signal Return	<0.050v
47	GY/WT	TP Sensor Signal	0.7v, 55 mph: 2.1v
48	WT/PK	Self-Test Indicator Signal	STI On: 1v, Off: 5v
49	OR	HO2S-11 (B1 S1) Ground	<0.050v
50	---	Not Used	---
51	WT/OR	Air Management 1 Solenoid	On: 1v, Off: 12v
52	OR/YL	Shift Solenoid 1 Control	1v, 55 mph: 1v
53	PK/YL	TCC Solenoid Control	0%, at 55 mph: 95%
54	---	Not Used	---
55	BR/OR	Coast Clutch Solenoid	12v, 55 mph: 12v
56	GY/OR	PIP Sensor Signal	6.93v (50%)
57	RD	Vehicle Power	12-14v
58	TN	Injector Bank 1 (INJ 1, 3 & 5)	6.2-7.4 ms
59	WT	Injector Bank 2 (INJ 2, 4 & 6)	6.2-7.4 ms
60	BK/WT	Power Ground	<0.1v

1995 E-Series 4.9L I6 MFI VIN Y California 60 Pin Connector

PCM Pin #	Wire Color	Circuit Description (60 Pin)	Value at Hot Idle
1	YL	Keep Alive Power	12-14v
2	LG	Brake Switch Signal	Brake Off: 12v, On: 1v
3	GY/BK	VSS (+) Signal	0 Hz, at 55 mph: 125 Hz
4	WT/PK	Ignition Diagnostic Monitor	20-31 Hz
5	---	Not Used	---
6	PK/OR	VSS (-) Signal	0 Hz, at 55 mph: 125 Hz
7	LG/RD	Engine Coolant Temperature Sensor	0.5-0.6v
8	DG/YL	Fuel Pump Monitor	On: 12v, Off: 0v
9	PK/BL	Data Bus (-)	Digital Signals
10	BK/YL	A/C Pressure Switch Signal	A/C On: 12v, off: 0v
11	BR	Air Management 2 Solenoid	On: 1v, Off: 12v
12-15	---	Not Used	---
16	OR/RD	Ignition System Ground	<0.050v
17	PK/LG	Self-Test Output & MIL	MIL On: 1v, Off: 12v
18	---	Not Used	---
19	PK/BL	Shift Solenoid 2 Control	12v, 55 mph: 1v
20	BK	PCM Case Ground	<0.050v
21	WT/BL	IAC Solenoid Control	10.7v (33%)
22	BL/OR	Fuel Pump Control	On: 1v, Off: 12v
23	YL/RD	Knock Sensor 1 Signal	0v
24	---	Not Used	---
25	GY	Intake Air Temperature Sensor	1.5-2.5v
26	BR/WT	Reference Voltage	4.9-5.1v
27	BR/LG	EVP Sensor Signal	0.4v, 55 mph: 3.1v
28	TN/OR	Data Bus (+)	Digital Signals
29	GY/BL	HO2S-11 (B1 S1) Signal	0.1-1.1v
30	BL/YL	MLP Sensor Signal	In 'P': 0v, in Drive: 5v
31	GY/YL	Canister Purge Solenoid	12v, 55 mph: 1v
32	WT/LG	TCIL (lamp) Control	Lamp On: 1v, Off: 12v
33	BR/PK	VR Solenoid Control	0%, at 55 mph: 45%
34-35	---	Not Used	---
36	PK	Spark Output Signal	6.93v (50%)
37	RD	Vehicle Power	12-14v
38	WT/YL	EPC Solenoid Control	8.1v (4 psi)
39	---	Not Used	---
40	BK/WT	Power Ground	<0.1v

1995 E-Series 4.9L I6 MFI VIN Y (Cal) 60 Pin Connector

PCM Pin #	Wire Color	Circuit Description (60 Pin)	Value at Hot Idle
41	TN/WT	TCS (switch) Signal	TCS & O/D On: 12v
42	OR/BK	TOT Sensor Signal	2.10-2.40v
43	PK	A/C Demand Switch	A/C On: 12v, Off: 0v
44	---	Not Used	---
45	LG/BK	MAP Sensor Signal	107 Hz (sea level)
46	GY/RD	Analog Signal Return	<0.050v
47	GY/WT	TP Sensor Signal	0.7v, 55 mph: 2.1v
48	WT/PK	Self-Test Indicator Signal	STI On: 1v, Off: 5v
49	OR	HO2S-11 (B1 S1) Ground	<0.050v
50	---	Not Used	---
51	WT/OR	Air Management 1 Solenoid	On: 1v, Off: 12v
52	OR/YL	Shift Solenoid 1 Control	1v, 55 mph: 12v
53	PK/YL	TCC Solenoid Control	12v, 55 mph: 1v
54	---	Not Used	---
55	BR/OR	Coast Clutch Solenoid	12v, 55 mph: 12v
56	GY/OR	PIP Sensor Signal	6.93v (50%)
57	RD	Vehicle Power	12-14v
58	TN	Injector Bank 1 (INJ 1, 3 & 5)	6.2-7.4 ms
59	WT	Injector Bank 2 (INJ 2, 4 & 6)	6.2-7.4 ms
60	BK/WT	Power Ground	<0.1v

Pin Connector Graphic

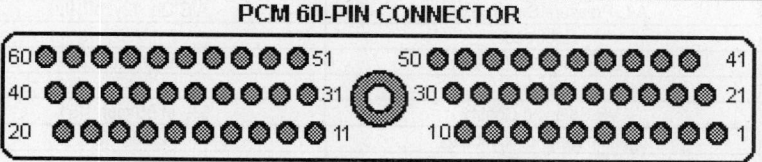

PCM 60-PIN CONNECTOR

Terminal View of 60-Pin PCM Harness Connector

Standard Colors and Abbreviations

Abbreviation	Color	Abbreviation	Color	Abbreviation	Color
BK	Black	GY	Gray	PK	Purple
BL	Blue	GN	Green	RD	Red
BR	Brown	LG	LT Green	TN	Tan
DB	Dark Blue	OR	Orange	WT	White
DG	DK Green	PK	Pink	YL	Yellow

1996 E-Series 4.9L I6 OHV VIN Y (A/T) 104 Pin Connector

PCM Pin #	Wire Color	Circuit Description (104 Pin)	Value at Hot Idle
1	---	Not Used	---
2	PK/LG	MIL (lamp) Control	MIL On: 1v, Off: 12v
3-12	---	Not Used	---
13	PK	Flash EEPROM	0.1v
14	---	Not Used	---
15	PK/LB	Data Bus (-)	Digital Signals
16	TN/OR	Data Bus (+)	Digital Signals
17-22	---	Not Used	---
23	OR/RD	Ignition System Ground	<0.050v
24	BK/WT	Power Ground	<0.1v
25	BK	Chassis Ground	<0.050v
26-28	---	Not Used	---
29	TN/WT	Transmission Cancel Switch	12v (Closed: 1v)
30-32	---	Not Used	---
33	OR/YL	PSOM (-) Signal	<0.050v
34	---	Not Used	---
35	RD/LG	HO2S-12 (B1 S2) Signal	0.1-1.1v
36	TN/LB	MAF Sensor Return	<0.050v
37	---	Not Used	---
38	LG/RD	Engine Coolant Temperature Sensor	0.5-0.6v
39	GY	Intake Air Temperature Sensor	1.5-2.5v
40	DG/YL	Fuel Pump Monitor	On: 12v, Off: 0v
41	BK/YL	A/C Pressure Switch Signal	A/C On: 12v, off: 0v
42-43	---	Not Used	---
44	BR	Secondary AIR Diverter	On: 1v, Off: 12v
45-46	---	Not Used	---
47	BR/PK	VR Solenoid Control	0%, at 55 mph: 45%
48	WT/PK	Ignition Diagnostic Monitor	20-31 Hz
49	GY/OR	PIP Sensor Signal	6.93v (50%)
50	PK	Spark Output Signal	6.93v (50%)
51	BK/WT	Power Ground	<0.1v
52-54	---	Not Used	---
55	YL	Keep Alive Power	12-14v
56	LG/BK	EVAP VMV Solenoid	0-10 Hz (0-100%)
57	YL/RD	Knock Sensor 1 Signal	0v
58	GY/BK	PSOM (+) Signal	0 Hz, at 55 mph: 125 Hz
59	DG/LG	Misfire Detection Sensor	35 Hz (150 mv AC)
60	GY/LB	HO2S-11 (B1 S1) Signal	0.1-1.1v
61-63	---	Not Used	---
64	LB/YL	PNP Switch Signal	In 'P': 0v, Others: 5v

1996 E-Series 4.9L I6 OHV VIN Y (A/T) 104 Pin Connector

PCM Pin #	Wire Color	Circuit Description (104 Pin)	Value at Hot Idle
65	BR/LG	EGR DPFE Sensor Signal	0.20-1.30v
66-68	---	Not Used	---
69	PK/YL	A/C WOT Relay Control	On: 1v, Off: 12v
70	WT/OR	Secondary Air Bypass Solenoid	AIRB On: 1v, Off: 12v
71	RD	Vehicle Power	12-14v
72	---	Not Used	---
73	TN/BK	Injector 5 Control	3.8-4.7 ms
74	BR/YL	Injector 3 Control	3.8-4.7 ms
75	TN	Injector 1 Control	3.8-4.7 ms
76	BK/WT	Power Ground	<0.1v
77	BK/WT	Power Ground	<0.1v
78	---	Not Used	---
79	WT/LG	TCIL (lamp) Control	Lamp On: 1v, Off: 12v
80	LB/OR	Fuel Pump Control	On: 1v, Off: 12v
81-82	---	Not Used	---
83	WT/LB	IAC Solenoid Control	10.7v (33%)
84-86	---	Not Used	---
87	RD/BK	HO2S-21 (B2 S1) Signal	0.1-1.1v
88	LB/RD	MAF Sensor Signal	0.8v, 55 mph: 1.6v
89	GY/WT	TP Sensor Signal	0.53-1.27v
90	BR/WT	Reference Voltage	4.9-5.1v
91	GY/RD	Analog Signal Return	<0.050v
92	LG	Brake Switch Signal	Brake Off: 12v, On: 1v
93	RD/WT	HO2S-11 (B1 S1) Heater	1v, Off: 12v
94	YL/LB	HO2S-21 (B2 S1) Heater	1v, Off: 12v
95	WT/BK	HO2S-12 (B1 S2) Heater	1v, Off: 12v
96, 98	---	Not Used	---
97	R	Vehicle Power	12-14v
99	LG/OR	Injector 6 Control	3.8-4.7 ms
100	BR/LB	Injector 4 Control	3.8-4.7 ms
101	W	Injector 2 Control	3.8-4.7 ms
102	---	Not Used	---
103	BK/WT	Power Ground	<0.1v
104	---	Not Used	---

Pin Connector Graphic

PCM 104-PIN CONNECTOR

```
 1 ●●●●●●●●●●●●●   ●●●●●●●●●●●●● 26
27 ●●●●●●●●●●●●●   ●●●●●●●●●●●●● 52
53 ●●●●●●●●●●●●● ● ●●●●●●●●●●●●● 78
79 ●●●●●●●●●●●●●   ●●●●●●●●●●●●● 104
```

Terminal View of 104-Pin PCM Wiring Harness Connector

Standard Colors and Abbreviations

Abbreviation	Color	Abbreviation	Color	Abbreviation	Color
BK	Black	GY	Gray	PK	Purple
BL	Blue	GN	Green	RD	Red
BR	Brown	LG	LT Green	TN	Tan
DB	Dark Blue	OR	Orange	WT	White
DG	DK Green	PK	Pink	YL	Yellow

1996 E-Series 4.9L I6 OHV VIN Y (E4OD) 104 Pin Connector

PCM Pin #	Wire Color	Circuit Description (104 Pin)	Value at Hot Idle
1	PK/OR	Shift Solenoid 2 Control	12v, 55 mph: 1v
2	PK/LG	MIL (lamp) Control	MIL On: 1v, Off: 12v
3-12	---	Not Used	---
13	PK	Flash EEPROM	0.1v
14	---	Not Used	---
15	PK/LB	Data Bus (-)	Digital Signals
16	TN/OR	Data Bus (+)	Digital Signals
17-22	---	Not Used	---
23	OR/RD	Ignition System Ground	<0.050v
24	BK/WT	Power Ground	<0.1v
25	BK	Chassis Ground	<0.050v
26	---	Not Used	---
27	OR/YL	Shift Solenoid 1 Control	1v, 55 mph: 12v
28	---	Not Used	---
29	TN/WT	Transmission Cancel Switch	12v (Closed: 1v)
30-32	---	Not Used	---
33	OR/YL	PSOM (-) Signal	<0.050v
34	---	Not Used	---
35	RD/LG	HO2S-12 (B1 S2) Signal	0.1-1.1v
36	TN/LB	MAF Sensor Return	<0.050v
37	OR/BK	Transmission Fluid Temperature Sensor	2.10-2.40v
38	LG/RD	Engine Coolant Temperature Sensor	0.5-0.6v
39	GY	Intake Air Temperature Sensor	1.5-2.5v
40	DG/YL	Fuel Pump Monitor	On: 12v, Off: 0v
41	BK/YL	A/C Pressure Switch Signal	A/C On: 12v, off: 0v
42-43	---	Not Used	---
44	BR	Secondary AIR Diverter	On: 1v, Off: 12v
45-46	---	Not Used	---
47	BR/PK	VR Solenoid Control	0%, at 55 mph: 45%
48	WT/PK	Ignition Diagnostic Monitor	20-31 Hz
49	GY/OR	PIP Sensor Signal	6.93v (50%)
50	PK	Spark Output Signal	6.93v (50%)
51	BK/WT	Power Ground	<0.1v
52	---	Not Used	---
53	BR/OR	Coast Clutch Solenoid	12v, 55 mph: 12v
54	DB/WT	Torque Converter Clutch	0%, at 55 mph: 95%
55	Y	Keep Alive Power	12-14v
56	LG/BK	EVAP VMV Solenoid	0-10 Hz (0-100%)
57	YL/RD	Knock Sensor 1 Signal	0v
58	GY/BK	PSOM (+) Signal	0 Hz, at 55 mph: 125 Hz
59	DG/LG	Misfire Detection Sensor	35 Hz (150 mv AC)
60	GY/LB	HO2S-11 (B1 S1) Signal	0.1-1.1v
61-63	---	Not Used	---
64	LB/YL	Transmission Range Sensor	In 'P': 0v, in O/D: 1.7v

1996 E-Series 4.9L I6 OHV VIN Y (E4OD) 104 Pin Connector

PCM Pin #	Wire Color	Circuit Description (104 Pin)	Value at Hot Idle
65	BR/LG	EGR DPFE Sensor Signal	0.20-1.30v
66-68	---	Not Used	---
69	PK/YL	A/C WOT Relay Control	On: 1v, Off: 12v
70	WT/OR	Secondary Air Bypass Solenoid	AIRB On: 1v, Off: 12v
71	RD	Vehicle Power	12-14v
72	---	Not Used	---
73	TN/BK	Injector 5 Control	3.8-4.7 ms
74	BR/YL	Injector 3 Control	3.8-4.7 ms
75	TN	Injector 1 Control	3.8-4.7 ms
76	BK/WT	Power Ground	<0.1v
77	BK/WT	Power Ground	<0.1v
78	---	Not Used	---
79	WT/LG	TCIL (lamp) Control	Lamp On: 1v, Off: 12v
80	LB/OR	Fuel Pump Control	On: 1v, Off: 12v
81	WT/YL	EPC Solenoid Control	10.6v (40 psi)
82	---	Not Used	---
83	WT/LB	IAC Solenoid Control	10.7v (33%)
84-86	---	Not Used	---
87	RD/BK	HO2S-21 (B2 S1) Signal	0.1-1.1v
88	LB/RD	MAF Sensor Signal	0.8v, 55 mph: 1.6v
89	GY/WT	TP Sensor Signal	0.53-1.27v
90	BR/WT	Reference Voltage	4.9-5.1v
91	GY/RD	Analog Signal Return	<0.050v
92	LG	Brake Switch Signal	Brake Off: 12v, On: 1v
93	RD/WT	HO2S-11 (B1 S1) Heater	1v, Off: 12v
94	YL/LB	HO2S-21 (B2 S1) Heater	1v, Off: 12v
95	WT/BK	HO2S-12 (B1 S2) Heater	1v, Off: 12v
96	---	Not Used	---
97	RD	Vehicle Power	12-14v
98	---	Not Used	---
99	LG/OR	Injector 6 Control	3.8-4.7 ms
100	BR/LB	Injector 4 Control	3.8-4.7 ms
101	WT	Injector 2 Control	3.8-4.7 ms
102	---	Not Used	---
103	BK/WT	Power Ground	<0.1v
104	---	Not Used	---

Pin Connector Graphic

PCM 104-PIN CONNECTOR

Terminal View of 104-Pin PCM Wiring Harness Connector

Standard Colors and Abbreviations

Abbreviation	Color	Abbreviation	Color	Abbreviation	Color
BK	Black	GY	Gray	PK	Purple
BL	Blue	GN	Green	RD	Red
BR	Brown	LG	LT Green	TN	Tan
DB	Dark Blue	OR	Orange	WT	White
DG	DK Green	PK	Pink	YL	Yellow

1990 E-Series 5.0L V8 VIN N (A/T) 60 Pin Connector

PCM Pin #	Wire Color	Circuit Description (60 Pin)	Value at Hot Idle
1	YL	Keep Alive Power	12-14v
2, 5	---	Not Used	---
3	GY/BK	VSS (+) Signal	0 Hz, at 55 mph: 125 Hz
3	DG/WT	VSS (+) Signal	0 Hz, at 55 mph: 125 Hz
4	DG/YL	Ignition Diagnostic Monitor	20-31 Hz
6	BK	VSS (-) Signal	0 Hz, at 55 mph: 125 Hz
7	LG/YL	Engine Coolant Temperature Sensor	0.5-0.6v
8	BR	Fuel Pump Monitor	On: 12v, Off: 0v
9	---	Not Used	---
10	BK/YL	A/C Pressure Switch Signal	A/C On: 12v, off: 0v
11	WT/BK	Air Management 2 Solenoid	On: 1v, Off: 12v
12-15	---	Not Used	---
16	BK/OR	Ignition System Ground	<0.050v
17	PK/LG	Self-Test Output & MIL	MIL On: 1v, Off: 12v
18-19	---	Not Used	---
20	BK	PCM Case Ground	<0.050v
21	GY/WT	IAC Solenoid Control	9v (31%)
22	TN/LG	Fuel Pump Control	On: 1v, Off: 12v
23	LG/BK	Knock Sensor 1 Signal	0v
24	YL/LG	Power Steering Pressure Switch	Straight: 0v, Turned: 12v
25	YL/RD	ACT Sensor Signal	1.5-2.5v
26	OR/WT	Reference Voltage	4.9-5.1v
27	BR/LG	EVP Sensor Signal	0.4v, 55 mph: 3.1v
28	---	Not Used	---
29	GY/BK	HO2S-11 (B1 S1) Signal	0.1-1.1v
29	DG/PK	HO2S-11 (B1 S1) Signal	0.1-1.1v
30	GY/YL	A/T: Neutral Drive Switch	In 'P': 0v, Others: 5v
30	BL/YL	A/T: Neutral Drive Switch	In 'P': 0v, Others: 5v
31	GY/YL	Canister Purge Solenoid	12v, 55 mph: 1v
32, 34-35	---	Not Used	---
33	DG	VR Solenoid Control	0%, at 55 mph: 45%
36	YL/LG	Spark Output Signal	6.93v (50%)
37	RD	Vehicle Power	12-14v
38-39	---	Not Used	---
40	BK/LG	Power Ground	<0.1v
41-44	---	Not Used	---
45	DG/LG	MAP Sensor Signal	107 Hz (sea level)
45	DB/LG	MAP Sensor Signal	107 Hz (sea level)
46	BK/WT	Analog Signal Return	<0.50v
47	DG/LG	TP Sensor Signal	0.7v, 55 mph: 2.1v
48	WT/RD	Self-Test Indicator Signal	STI On: 1v, Off: 5v
49	OR	HO2S-11 (B1 S1) Ground	<0.050v
50, 52-55	---	Not Used	---
51	WT/RD	Air Management 1 Solenoid	On: 1v, Off: 12v
56	DB	PIP Sensor Signal	6.93v (50%)
57	RD	Vehicle Power	12-14v
58	TN/OR	Injector Bank 1 (INJ 1, 4, 5, 8)	5.0-5.7 ms
59	TN/RD	Injector Bank 2 (INJ 2, 3, 6, 7)	5.0-5.7 ms
60	BK/LG	Power Ground	<0.1v

Pin Connector Graphic

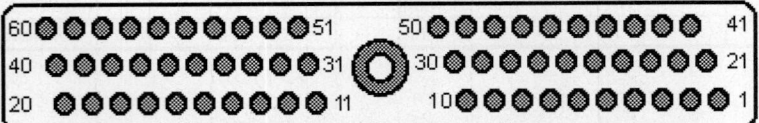

PCM 60-PIN CONNECTOR

Terminal View of 60-Pin PCM Harness Connector

1991 E-Series 5.0L V8 OHV VIN N (A/T) 60 Pin Connector

PCM Pin #	Wire Color	Circuit Description (60 Pin)	Value at Hot Idle
1	BK/OR	Keep Alive Power	12-14v
2	---	Not Used	---
3	DG/WT	VSS (+) Signal	0 Hz, at 55 mph: 125 Hz
4	TN/YL	Ignition Diagnostic Monitor	20-31 Hz
5	---	Not Used	---
6	OR/YL	VSS (-) Signal	0 Hz, at 55 mph: 125 Hz
7	LG/RD	Engine Coolant Temperature Sensor	0.5-0.6v
8	DG/YL	Fuel Pump Monitor	On: 12v, Off: 0v
9	BK/OR	Data Bus (-)	Digital Signals
10	BK/YL	A/C Pressure Switch Signal	A/C On: 12v, off: 0v
11	BR	Air Management 2 Solenoid	On: 1v, Off: 12v
12-15	---	Not Used	---
16	OR/RD	Ignition System Ground	<0.050v
17	TN/RD	Self-Test Output & MIL	MIL On: 1v, Off: 12v
18-19	---	Not Used	---
20	BK	PCM Case Ground	<0.050v
21	GY/WT	IAC Solenoid Control	10.7v (33%)
22	TN/LG	Fuel Pump Control	On: 1v, Off: 12v
23	LG/BK	Knock Sensor 1 Signal	0v
24	YL/LG	Power Steering Pressure Switch	Straight: 0v, Turned: 12v
25	YL/RD	ACT Sensor Signal	1.5-2.5v
26	BR/WT	Reference Voltage	4.9-5.1v
27	BR/LG	EVP Sensor Signal	0.4v, 55 mph: 3.1v
28	OR/BK	Data Bus (+)	Digital Signals
29	RD/BK	HO2S-11 (B1 S1) Signal	0.1-1.1v
30	---	Not Used	---
31	GY/YL	Canister Purge Solenoid	12v, 55 mph: 1v
32	---	Not Used	---
33	BR/PK	VR Solenoid Control	0%, at 55 mph: 45%
34-35	---	Not Used	---
36	YL/LG	Spark Output Signal	6.93v (50%)
37	RD	Vehicle Power	12-14v
38-39	---	Not Used	---
40	BK/LG	Power Ground	<0.1v
41-44	---	Not Used	---
45	DB/LG	MAP Sensor Signal	107 Hz (sea level)
46	GY/RD	Analog Signal Return	<0.050v
47	GY/WT	TP Sensor Signal	0.7v, 55 mph: 2.1v
48	WT/RD	Self-Test Indicator Signal	STI On: 1v, Off: 5v
49	OR	HO2S-11 (B1 S1) Ground	<0.050v
50	---	Not Used	---
51	WT/OR	Air Management 1 Solenoid	On: 1v, Off: 12v
52-55	---	Not Used	---
56	DB	PIP Sensor Signal	6.93v (50%)
57	RD	Vehicle Power	12-14v
58	TN/OR	Injector Bank 1 (INJ 1, 4, 5, 8)	4.4-5.6 ms
59	TN/RD	Injector Bank 2 (INJ 2, 3, 6, 7)	4.4-5.6 ms
60	BK/LG	Power Ground	<0.1v

Pin Connector Graphic

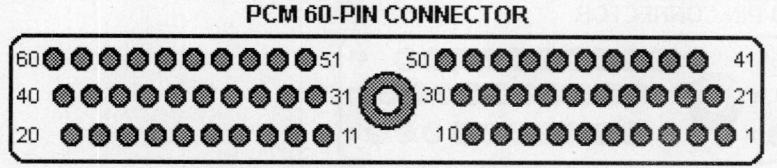

PCM 60-PIN CONNECTOR

Terminal View of 60-Pin PCM Harness Connector

1992-93 E-Series 5.0L V8 VIN N (A/T) 60 Pin Connector

PCM Pin #	Wire Color	Circuit Description (60 Pin)	Value at Hot Idle
1	YL	Keep Alive Power	12-14v
2	---	Not Used	---
3	GY/BK	PSOM (-) Signal	<0.050v
4	TN/YL	Ignition Diagnostic Monitor	20-31 Hz
5	---	Not Used	---
6	PK/OR	PSOM (+) Signal	0 Hz, at 55 mph: 125 Hz
7	LG/RD	Engine Coolant Temperature Sensor	0.5-0.6v
8	DG/YL	Fuel Pump Monitor	On: 12v, Off: 0v
9	PK/LB	Data Bus (-)	Digital Signals
10	BK/YL	A/C Pressure Switch Signal	A/C On: 12v, off: 0v
11	BR	Air Management 2 Solenoid	On: 1v, Off: 12v
12-15	---	Not Used	---
16	OR/RD	Ignition System Ground	<0.050v
17	PK/LG	Self-Test Output & MIL	MIL On: 1v, Off: 12v
18-19	---	Not Used	---
20	BK	Case Ground	<0.050v
21	WT/LB	IAC Solenoid Control	10.7v (33%)
22	LB/OR	Fuel Pump Control	On: 1v, Off: 12v
23	YL/RD	Knock Sensor 1 Signal	0v
24	---	Not Used	---
25	GY	ACT Sensor Signal	1.5-2.5v
26	BR/WT	Reference Voltage	4.9-5.1v
27	BR/LG	EVP Sensor Signal	0.4v, 55 mph: 3.1v
28	TN/OR	Data Bus (+)	Digital Signals
29	GY/LB	HO2S-11 (B1 S1) Signal	0.1-1.1v
30	LB/YL	Neutral Drive Switch	In 'P': 0v, Others: 5v
31	GY/YL	Canister Purge Solenoid	12v, 55 mph: 1v
32	---	Not Used	---
33	BR/PK	VR Solenoid Control	0%, at 55 mph: 45%
34-35	---	Not Used	---
36	PK	Spark Output Signal	6.93v (50%)
37	RD	Vehicle Power	12-14v
38-39	---	Not Used	---
40	BK/WT	Power Ground	<0.1v
41-44	---	Not Used	---
45	LG/BK	MAP Sensor Signal	107 Hz (sea level)
46	GY/RD	Analog Signal Return	<0.050v
47	GY/WT	TP Sensor Signal	0.7v, 55 mph: 2.1v
48	WT/PK	Self-Test Indicator Signal	STI On: 1v, Off: 5v
49	OR	HO2S-11 (B1 S1) Ground	<0.050v
50	---	Not Used	---
51	WT/OR	Air Management 1 Solenoid	On: 1v, Off: 12v
52-55	---	Not Used	---
56	GY/OR	PIP Sensor Signal	6.93v (50%)
57	RD	Vehicle Power	12-14v
58	TN	Injector Bank 1 (INJ 1, 4, 5, 8)	4.4-5.6 ms
59	WT	Injector Bank 2 (INJ 2, 3, 6, 7)	4.4-5.6 ms
60	BK/WT	Power Ground	<0.1v

Pin Connector Graphic

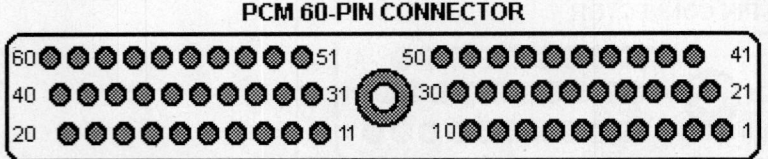

PCM 60-PIN CONNECTOR

Terminal View of 60-Pin PCM Harness Connector

1992-93 E-Series 5.0L V8 VIN N (E4OD) 60 Pin Connector

PCM Pin #	Wire Color	Circuit Description (60 Pin)	Value at Hot Idle
1	YL	Keep Alive Power	12-14v
2	LG	Brake Switch Signal	Brake Off: 12v, On: 1v
3	GY/BK	PSOM (-) Signal	<0.050v
4	TN/YL	Ignition Diagnostic Monitor	20-31 Hz
5, 12-15	---	Not Used	---
6	PK/OR	PSOM (+) Signal	0 Hz, at 55 mph: 125 Hz
7	LG/RD	Engine Coolant Temperature Sensor	0.5-0.6v
8	DG/YL	Fuel Pump Monitor	On: 12v, Off: 0v
9	PK/LB	Data Bus (-)	Digital Signals
10	BK/YL	A/C Pressure Switch Signal	A/C On: 12v, off: 0v
11	BR	Air Management 2 Solenoid	On: 1v, Off: 12v
16	OR/RD	Ignition System Ground	<0.050v
17	PK/LG	Self-Test Output & MIL	MIL On: 1v, Off: 12v
18, 24	---	Not Used	---
19	PK/OR	Shift Solenoid 1 Control	1v, 55 mph: 12v
20	BK	Case Ground	<0.050v
21	WT/LB	IAC Solenoid Control	10.7v (33%)
22	LB/OR	Fuel Pump Control	On: 1v, Off: 12v
23	YL/RD	Knock Sensor 1 Signal	0v
25	GY	ACT Sensor Signal	1.5-2.5v
26	BR/WT	Reference Voltage	4.9-5.1v
27	BR/LG	EVP Sensor Signal	0.4v, 55 mph: 3.1v
28	TN/OR	Data Bus (+)	Digital Signals
29	GY/LB	HO2S-11 (B1 S1) Signal	0.1-1.1v
30	LB/YL	Neutral Drive Switch	In 'P': 0v, Others: 5v
31	GY/YL	Canister Purge Solenoid	12v, 55 mph: 1v
32	WT/LG	TCIL (lamp) Control	Lamp On: 1v, Off: 12v
33	BR/PK	VR Solenoid Control	0%, at 55 mph: 45%
34-35, 39	---	Not Used	---
36	PK	Spark Output Signal	6.93v (50%)
37	RD	Vehicle Power	12-14v
38	WT/YL	EPC Solenoid Control	9.5v (5 psi)
40	BK/WT	Power Ground	<0.1v
41	TN/WT	Overdrive Cancel Switch	OCS Switch On: 0.1v
42	OR/BK	TOT Sensor Signal	2.10-2.40v
43-44	---	Not Used	---
45	LG/BK	MAP Sensor Signal	107 Hz (sea level)
46	GY/RD	Analog Signal Return	<0.050v
47	GY/WT	TP Sensor Signal	0.7v, 55 mph: 2.1v
48	WT/PK	Self-Test Indicator Signal	STI On: 1v, Off: 5v
49	OR	HO2S-11 (B1 S1) Ground	<0.050v
50, 54	---	Not Used	---
51	WT/OR	Air Management 1 Solenoid	On: 1v, Off: 12v
52	OR/YL	Shift Solenoid 2 Control	12v, 55 mph: 1v
53	PK/YL	TCC Solenoid Control	0%, at 55 mph: 95%
55	BR/OR	Coast Clutch Solenoid	12v, 55 mph: 12v
56	GY/OR	PIP Sensor Signal	6.93v (50%)
57	RD	Vehicle Power	12-14v
58	TN	Injector Bank 1 (INJ 1, 4, 5, 8)	4.4-5.6 ms
59	WT	Injector Bank 2 (INJ 2, 3, 6, 7)	4.4-5.6 ms
60	BK/WT	Power Ground	<0.1v

Pin Connector Graphic

PCM 60-PIN CONNECTOR

Terminal View of 60-Pin PCM Harness Connector

1994-95 E-Series 5.0L V8 VIN N (A/T) 60 Pin Connector

PCM Pin #	Wire Color	Circuit Description (60 Pin)	Value at Hot Idle
1	YL	Keep Alive Power	12-14v
2	LG	Brake Switch Signal	Brake Off: 12v, On: 1v
3	GY/BK	VSS (+) Signal	0 Hz, at 55 mph: 125 Hz
4	WT/PK	Ignition Diagnostic Monitor	20-31 Hz
5	DG/WT	Turbine Shaft Speed Sensor	120 Hz, 30 mph: 300 Hz
6	PK/OR	VSS (-) Signal	0 Hz, at 55 mph: 125 Hz
7	LG/RD	Engine Coolant Temperature Sensor	0.5-0.6v
8	DG/YL	Fuel Pump Monitor	On: 12v, Off: 0v
9	TN/BL	MAF Sensor Return	<0.050v
10	BK/YL	A/C Pressure Switch Signal	A/C On: 12v, off: 0v
11	GY/YL	Canister Purge Solenoid	12v, 55 mph: 1v
12	LG/OR	Injector 6 Control	4.7-5.5 ms
13	TN/RD	Injector 7 Control	4.7-5.5 ms
14	BL	Injector 8 Control	4.7-5.5 ms
15	TN/BK	Injector 5 Control	4.7-5.5 ms
16	OR/RD	Ignition System Ground	<0.050v
17	PK/LG	Self-Test Output & MIL	MIL On: 1v, Off: 12v
18	TN/OR	Data Bus (+)	Digital Signals
19	PK/BL	Data Bus (-)	Digital Signals
20	BK	PCM Case Ground	<0.050v
21	WT/BL	IAC Solenoid Control	10.7v (33%)
22	BL/OR	Fuel Pump Control	On: 1v, Off: 12v
23	YL/RD	Knock Sensor 1 Signal	0v
24	---	Not Used	---
25	GY	Intake Air Temperature Sensor	1.5-2.5v
26	BR/WT	Reference Voltage	4.9-5.1v
27	BR/LG	EVP Sensor Signal	0.4v, 55 mph: 3.1v
28	---	Not Used	---
29	WT/BL	Turbine Shaft Speed Sensor	120 Hz, 30 mph: 300 Hz
30	BL/YL	MLP Sensor Signal	In 'P': 0v, in O/D: 5v
31	WT/OR	Air Bypass Solenoid Control	AIRB On: 1v, Off: 12v
32	WT/LG	TCIL (lamp) Control	Lamp On: 1v, Off: 12v
33	BR/PK	VR Solenoid Control	0%, at 55 mph: 45%
34	BR	Air Diverter Solenoid Control	On: 1v, Off: 12v
35	BR/BL	Injector 4 Control	4.7-5.5 ms
36	PK	Spark Output Signal	6.93v (50%)
37	RD	Vehicle Power	12-14v
38	WT/YL	EPC Solenoid Control	9.5v (5 psi)
39	BR/YL	Injector 3 Control	4.7-5.5 ms
40	BK/WT	Power Ground	<0.1v

1994-95 E-Series 5.0L V8 VIN N (A/T) 60 Pin Connector

PCM Pin #	Wire Color	Circuit Description (60 Pin)	Value at Hot Idle
41	TN/WT	TCS (switch) Signal	TCS & O/D On: 12v
42-43	---	Not Used	---
44	GY/BL	HO2S-11 (B1 S1) Signal	0.1-1.1v
45	---	Not Used	---
46	GY/RD	Analog Signal Return	<0.050v
47	GY/WT	TP Sensor Signal	0.7v, 55 mph: 2.1v
48	WT/PK	Self-Test Indicator Signal	STI On: 1v, Off: 5v
49	OR/BK	TOT Sensor Signal	2.10-2.40v
50	BL/RD	MAF Sensor Signal	0.8v, 55 mph: 1.6v
51	OR/YL	Shift Solenoid 1 Control	1v, 55 mph: 12v
52	PK/OR	Shift Solenoid 2 Control	12v, 55 mph: 1v
53	PK/YL	TCC Solenoid Control	0%, at 55 mph: 95%
54-55	---	Not Used	---
56	GY/OR	PIP Sensor Signal	6.93v (50%)
57	RD	Vehicle Power	12-14v
58	TN	Injector 1 Control	4.7-5.5 ms
59	WT	Injector 2 Control	4.7-5.5 ms
60	BK/WT	Power Ground	<0.1v

Pin Connector Graphic

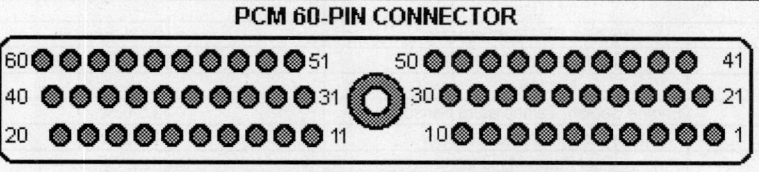

Standard Colors and Abbreviations

Abbreviation	Color	Abbreviation	Color	Abbreviation	Color
BK	Black	GY	Gray	PK	Purple
BL	Blue	GN	Green	RD	Red
BR	Brown	LG	LT Green	TN	Tan
DB	Dark Blue	OR	Orange	WT	White
DG	DK Green	PK	Pink	YL	Yellow

1996 E-Series 5.0L V8 OHV VIN N (A/T) 104 Pin Connector

PCM Pin #	Wire Color	Circuit Description (104 Pin)	Value at Hot Idle
1	---	Not Used	---
2	PK/LG	MIL (lamp) Control	MIL On: 1v, Off: 12v
3, 5	---	Not Used	---
4	LG/RD	Power Take-Off Signal	0.1v (Off)
6	GY	Output Shaft Speed Sensor	0 Hz, 55 mph: 130 Hz
7-12	---	Not Used	---
13	PK	Flash EPROM Power	0.1v
14	---	Not Used	---
15	PK/LB	Data Bus (-)	Digital Signals
16	TN/OR	Data Bus (+)	Digital Signals
17-22	---	Not Used	---
23	OR/RD	Ignition System Ground	<0.050v
24	BK/WT	Power Ground	<0.1v
25	BK	Case Ground	<0.050v
26	---	Not Used	---
27	OR/YL	Shift Solenoid 1 Control	1v, 55 mph: 12v
28	---	Not Used	---
29	TN/WT	TCS (switch) Signal	TCS & O/D On: 12v
30-32	---	Not Used	---
33	PK/OR	PSOM (-) Signal	<0.050v
34	---	Not Used	---
35	RD/LG	HO2S-12 (B1 S2) Signal	0.1-1.1v
36	TN/LB	MAF Sensor Return	<0.050v
37	---	Not Used	---
38	LG/RD	Engine Coolant Temperature Sensor	0.5-0.6v
39	GY	Intake Air Temperature Sensor	1.5-2.5v
40	DG/YL	Fuel Pump Monitor	On: 12v, Off: 0v
41	BK/YL	A/C Pressure Switch Signal	A/C On: 12v, off: 0v
42-43	---	Not Used	---
44	BR	Secondary AIR Diverter	On: 1v, Off: 12v
45-46	---	Not Used	---
47	BK/PK	VR Solenoid Control	0%, at 55 mph: 45%
48	WT/PK	Clean Tachometer Output	38-42 Hz
49	GY/OR	PIP Sensor Signal	6.93v (50%)
50	PK	Spark Output Signal	6.93v (50%)
51	BK/WT	Power Ground	<0.1v
52-54	---	Not Used	---
55	YL	Keep Alive Power	12-14v
56	LG/BK	EVAP VMV Solenoid	0-10 Hz (0-100%)
57	---	Not Used	---
58	GY/BK	PSOM (+) Signal	0 Hz, at 55 mph: 125 Hz
59	DB	Misfire Detection Sensor	45-55 Hz
60	GY/LB	HO2S-11 (B1 S1) Signal	0.1-1.1v
61-63	---	Not Used	---
64	LB/YL	PNP Switch Signal	In 'P': 0v, Others: 5v

1996 E-Series 5.0L V8 OHV VIN N (A/T) 104 Pin Connector

PCM Pin #	Wire Color	Circuit Description (104 Pin)	Value at Hot Idle
65	BR/LG	EGR DPFE Sensor Signal	0.20-1.30v
66-69	---	Not Used	---
70	WT/OR	Secondary Air Bypass Solenoid	AIRB On: 1v, Off: 12v
71	RD	Vehicle Power	12-14v
72	TN/RD	Injector 7 Control	3.2-4.5 ms
73	TN/BK	Injector 5 Control	3.2-4.5 ms
74	BR/YL	Injector 3 Control	3.2-4.5 ms
75	TN	Injector 1 Control	3.2-4.5 ms
76	BK/WT	Power Ground	<0.1v
77	BK/WT	Power Ground	<0.1v
78	---	Not Used	---
79	WT/LG	TCIL (lamp) Control	Lamp On: 1v, Off: 12v
80	LB/OR	Fuel Pump Control	On: 1v, Off: 12v
81-82	---	Not Used	---
83	WT/LB	IAC Solenoid Control	10.7v (33%)
84	DG/YL	Output Shaft Speed Sensor	0 Hz, 55 mph: 130 Hz
85-86	---	Not Used	---
87	RD/BK	HO2S-21 (B2 S1) Signal	0.1-1.1v
88	LB/RD	MAF Sensor Signal	0.8v, 55 mph: 2.3v
89	GY/WT	TP Sensor Signal	0.9v, 55 mph: 1.2v
90	BR/WT	Reference Voltage	4.9-5.1v
91	GY/RD	Analog Signal Return	<0.050v
92	LG	Brake Position Switch	Brake Off: 12v, On: 1v
93	RD/WT	HO2S-11 (B1 S1) Heater	1v, Off: 12v
94	YL/LB	HO2S-21 (B2 S1) Heater	1v, Off: 12v
95	WT/BK	HO2S-12 (B1 S2) Heater	1v, Off: 12v
96	---	Not Used	---
97	RD	Vehicle Power	12-14v
98	LG	Injector 8 Control	3.2-4.5 ms
99	LG/OR	Injector 6 Control	3.2-4.5 ms
100	BR/LB	Injector 4 Control	3.2-4.5 ms
101	WT	Injector 2 Control	3.2-4.5 ms
102	---	Not Used	---
103	BK/WT	Power Ground	<0.1v
104	---	Not Used	---

Pin Connector Graphic

PCM 104-PIN CONNECTOR

```
 1 ◉◉◉◉◉◉◉◉◉◉◉◉◉      ◉◉◉◉◉◉◉◉◉◉◉◉◉ 26
27 ◉◉◉◉◉◉◉◉◉◉◉◉◉      ◉◉◉◉◉◉◉◉◉◉◉◉◉ 52
53 ◉◉◉◉◉◉◉◉◉◉◉◉◉   ⬤  ◉◉◉◉◉◉◉◉◉◉◉◉◉ 78
79 ◉◉◉◉◉◉◉◉◉◉◉◉◉      ◉◉◉◉◉◉◉◉◉◉◉◉◉ 104
```

Terminal View of 104-Pin PCM Wiring Harness Connector

Standard Colors and Abbreviations

Abbreviation	Color	Abbreviation	Color	Abbreviation	Color
BK	Black	GY	Gray	PK	Purple
BL	Blue	GN	Green	RD	Red
BR	Brown	LG	LT Green	TN	Tan
DB	Dark Blue	OR	Orange	WT	White
DG	DK Green	PK	Pink	YL	Yellow

1996 E-Series 5.0L V8 VIN N (E4OD) 104 Pin Connector

PCM Pin #	Wire Color	Circuit Description (104 Pin)	Value at Hot Idle
1	PK/OR	Shift Solenoid 2 Control	12v, 55 mph: 12v
2	PK/LG	MIL (lamp) Control	MIL On: 1v, Off: 12v
3, 5	---	Not Used	---
4	LG/RD	Power Take-Off Signal	0.1v (Off)
6	GY	Output Shaft Speed Sensor	0 Hz, 30 mph: 130 Hz
7-12	---	Not Used	---
13	VT	Flash EPROM Power	0.1v
14	---	Not Used	---
15	PK/LB	Data Bus (-)	Digital Signals
16	TN/OR	Data Bus (+)	Digital Signals
17-22	---	Not Used	---
23	OR/RD	Ignition System Ground	<0.050v
24	BK/WT	Power Ground	<0.1v
25	BK	Case Ground	<0.050v
26	---	Not Used	---
27	OR/YL	Shift Solenoid 1 Control	1v, 55 mph: 12v
28	---	Not Used	---
29	TN/WT	TCS (switch) Signal	TCS & O/D On: 12v
30-32	---	Not Used	---
33	PK/OR	PSOM (-) Signal	<0.050v
34	---	Not Used	---
35	RD/LG	HO2S-12 (B1 S2) Signal	0.1-1.1v
36	TN/LB	MAF Sensor Return	<0.050v
37	OR/BK	Transmission Fluid Temperature Sensor	2.10-2.40v
38	LG/RD	Engine Coolant Temperature Sensor	0.5-0.6v
39	GY	Intake Air Temperature Sensor	1.5-2.5v
40	DG/YL	Fuel Pump Monitor	On: 12v, Off: 0v
41	BK/YL	A/C Pressure Switch Signal	A/C On: 12v, off: 0v
42-43	---	Not Used	---
44	BR	Secondary AIR Diverter	On: 1v, Off: 12v
45-46	---	Not Used	---
47	BK/PK	VR Solenoid Control	0%, at 55 mph: 45%
48	WT/PK	Clean Tachometer Output	38-42 Hz
49	GY/OR	PIP Sensor Signal	6.93v (50%)
50	PK	Spark Output Signal	6.93v (50%)
51	BK/WT	Power Ground	<0.1v
52	---	Not Used	---
53	BR/OR	Coast Clutch Solenoid	12v, 55 mph: 12v
54	DB/WT	TCC Solenoid Control	0%, at 55 mph: 95%
55	YL	Keep Alive Power	12-14v
56	LG/BK	EVAP VMV Solenoid	0-10 Hz (0-100%)
57	---	Not Used	---
58	GY/BK	PSOM (+) Signal	0 Hz, at 55 mph: 125 Hz
59	DB	Misfire Detection Sensor	45-55 Hz
60	GY/LB	HO2S-11 (B1 S1) Signal	0.1-1.1v
61-63	---	Not Used	---
64	LB/YL	Digital Transmission Range Sensor	In 'P': 0v, in O/D: 1.7v

1996 E-Series 5.0L V8 VIN N (E4OD) 104 Pin Connector

PCM Pin #	Wire Color	Circuit Description (104 Pin)	Value at Hot Idle
65	BR/LG	EGR DPFE Sensor Signal	0.20-1.30v
66-69	---	Not Used	---
70	WT/OR	Secondary Air Bypass Solenoid	AIRB On: 1v, Off: 12v
71	RD	Vehicle Power	12-14v
72	TN/RD	Injector 7 Control	3.2-4.5 ms
73	TN/BK	Injector 5 Control	3.2-4.5 ms
74	BR/YL	Injector 3 Control	3.2-4.5 ms
75	TN	Injector 1 Control	3.2-4.5 ms
76	BK/WT	Power Ground	<0.1v
77	BK/WT	Power Ground	<0.1v
78	---	Not Used	---
79	WT/LG	TCIL (lamp) Control	Lamp On: 1v, Off: 12v
80	LB/OR	Fuel Pump Control	On: 1v, Off: 12v
81	WT/YL	EPC Solenoid Control	10v (26 psi)
82	---	Not Used	---
83	WT/LB	IAC Solenoid Control	10.7v (33%)
84	DG/YL	Output Shaft Speed Sensor	0 Hz, 30 mph: 130 Hz
85-86	---	Not Used	---
87	RD/BK	HO2S-21 (B2 S1) Signal	0.1-1.1v
88	LB/RD	MAF Sensor Signal	0.8v, 55 mph: 2.3v
89	GY/WT	TP Sensor Signal	0.9v, 55 mph: 1.2v
90	BR/WT	Reference Voltage	4.9-5.1v
91	GY/RD	Analog Signal Return	<0.050v
92	LG	Brake Position Switch	Brake Off: 12v, On: 1v
93	RD/WT	HO2S-11 (B1 S1) Heater	1v, Off: 12v
94	YL/LB	HO2S-21 (B2 S1) Heater	1v, Off: 12v
95	WT/BK	HO2S-12 (B1 S2) Heater	1v, Off: 12v
96	---	Not Used	---
97	RD	Vehicle Power	12-14v
98	LG	Injector 8 Control	3.2-4.5 ms
99	LG/OR	Injector 6 Control	3.2-4.5 ms
100	BR/LB	Injector 4 Control	3.2-4.5 ms
101	WT	Injector 2 Control	3.2-4.5 ms
102	---	Not Used	---
103	BK/WT	Power Ground	<0.1v
104	---	Not Used	---

Pin Connector Graphic

PCM 104-PIN CONNECTOR

Terminal View of 104-Pin PCM Wiring Harness Connector

Standard Colors and Abbreviations

Abbreviation	Color	Abbreviation	Color	Abbreviation	Color
BK	Black	GY	Gray	PK	Purple
BL	Blue	GN	Green	RD	Red
BR	Brown	LG	LT Green	TN	Tan
DB	Dark Blue	OR	Orange	WT	White
DG	DK Green	PK	Pink	YL	Yellow

1997 E-Series 5.4L V8 VIN L (E4OD) 104 Pin Connector

PCM Pin #	Wire Color	Circuit Description (104 Pin)	Value at Hot Idle
1	PK/LB	COP 6 Driver (Dwell)	5°, at 55 mph: 8°
2	PK/LG	MIL (lamp) Control	MIL On: 1v, Off: 12v
3	BK/WT	Power Ground	<0.1v
4-5	---	Not Used	---
6	OR/YL	Shift Solenoid 1 Control	1v, 55 mph: 12v
7-8	---	Not Used	---
9	YL/WT	Fuel Level Indicator Signal	1.7v (1/2 full)
10	---	Not Used	---
11	PK/OR	Shift Solenoid 2 Control	12v, 55 mph: 1v
12	WT/LG	TCIL (lamp) Control	Lamp On: 1v, Off: 12v
13	PK	Flash EEPROM	0.1v
14	---	Not Used	---
15	PK/LB	Data Bus (-)	Digital Signals
16	TN/OR	Data Bus (+)	Digital Signals
17-19	---	Not Used	---
20	BR/OR	Coast Clutch Solenoid	12v, 55 mph: 12v
21	DB	CKP (+) Sensor	400-420 Hz
22	GY	CKP (-) Sensor	400-420 Hz
23	---	Not Used	---
24	BK/WT	Power Ground	<0.1v
25	BK/LB	Case Ground	<0.050v
26	DB/YL	COP 1 Driver (Dwell)	5°, at 55 mph: 8°
27	PKWT	COP 5 Driver (Dwell)	5°, at 55 mph: 8°
28	---	Not Used	---
29	TN/WT	TCS (switch) Signal	TCS & O/D On: 12v
30	DG	Octane Adjustment	9.3v (switch shorted: 0v)
31-32	---	Not Used	---
33	PK/OR	VSS (-) Signal	0 Hz, at 55 mph: 125 Hz
34	OR/BK	Transmission Range Sensor 1	0v, at 55 mph: 11.5v
35	RD/LG	HO2S-12 (B1 S2) Signal	0.1-1.1v
36	TN/LB	MAF Sensor Return	<0.050v
37	OR/BK	Transmission Fluid Temperature Sensor	2.10-2.40v
38	---	Not Used	---
39	GY	Intake Air Temperature Sensor	1.5-2.5v
40	DG/YL	Fuel Pump Monitor	On: 12v, Off: 0v
41	BK/YL	A/C High Pressure Switch	Switch Closed: 12v
42-44	---	Not Used	---
45	YL/LB	CHIL (lamp) Control	Lamp On: 1v, Off: 12v
46	---	Not Used	---
47	BR/PK	VR Solenoid Control	0%, at 55 mph: 45%
48	---	Not Used	---
49	LB/BK	Digital TR2 Sensor Signal	0v, at 55 mph: 11.5v
50	GY/BK	Digital TR4 Sensor Signal	0v, at 55 mph: 11.5v
51	BK/WT	Power Ground	<0.1v
52	LG/WT	COP 3 Driver (Dwell)	5°, at 55 mph: 8°
53	OR/YL	COP 4 Driver (Dwell)	5°, at 55 mph: 8°
54	PK/YL	TCC Solenoid Control	0%, at 55 mph: 95%
55	YL	Keep Alive Power	12-14v
56	LG/BK	EVAP VMV Solenoid	0-10 Hz (0-100%)
57	YL/RD	Knock Sensor 1 Signal	0v
58	GY/BK	VSS (+) Signal	0 Hz, at 55 mph: 125 Hz
59	---	Not Used	---
60	GY/LB	HO2S-11 (B1 S1) Signal	0.1-1.1v
61, 63	---	Not Used	---
62	RD/PK	Fuel Tank Pressure Sensor	2.6v (0" H2O - cap off)
64	LB/YL	Transmission Range Sensor 3	In 'P': 0v, in O/D: 1.7v

1997 E-Series 5.4L V8 VIN L (E4OD) 104 Pin Connector

PCM Pin #	Wire Color	Circuit Description (104 Pin)	Value at Hot Idle
65	BR/LG	EGR DPFE Sensor Signal	0.20-1.30v
66	YL/LG	Cylinder Head Temperature Sensor	0.6v (194°F)
67	PKWT	EVAP CV Solenoid	0-10 Hz (0-100%)
68-70	---	Not Used	---
71	RD	Vehicle Power	12-14v
72	TN/RD	Injector 7 Control	3.2-3.8 ms
73	TN/BK	Injector 5 Control	3.2-3.8 ms
74	BR/YL	Injector 3 Control	3.2-3.8 ms
75	TN	Injector 1 Control	3.2-3.8 ms
76	---	Not Used	---
77	BK/WT	Power Ground	<0.1v
78	WT/PK	COP 7 Driver (Dwell)	5°, at 55 mph: 8°
79	WT/RD	COP 8 Driver (Dwell)	5°, at 55 mph: 8°
80	LB/OR	Fuel Pump Control	On: 1v, Off: 12v
82	---	Not Used	---
81	WT/YL	EPC Solenoid Control	9.1v (4 psi)
83	WT/LB	IAC Solenoid Control	10.7v (33%)
84	---	Not Used	---
85	DB/OR	Camshaft Position Sensor	6 Hz, 55 mph: 14-18 Hz
86	---	Not Used	---
87	RD/BK	HO2S-21 (B2 S1) Signal	0.1-1.1v
88	LB/RD	MAF Sensor Signal	0.8v, 55 mph: 1.6v
89	GY/WT	TP Sensor Signal	0.53-1.27v
90	BR/WT	Reference Voltage	4.9-5.1v
91	GY/RD	Analog Signal Return	<0.050v
92	LG	Brake Position Switch	Brake Off: 12v, On: 1v
93	RD/WT	HO2S-11 (B1 S1) Heater	1v, Off: 12v
94	YL/LB	HO2S-21 (B2 S1) Heater	1v, Off: 12v
95	WT/BK	HO2S-12 (B1 S2) Heater	1v, Off: 12v
96	---	Not Used	---
97	RD	Vehicle Power	12-14v
98	LB	Injector 8 Control	3.2-3.8 ms
99	LG/OR	Injector 6 Control	3.2-3.8 ms
100	BR/LG	Injector 4 Control	3.2-3.8 ms
101	WT	Injector 2 Control	3.2-3.8 ms
102	---	Not Used	---
103	BK/WT	Power Ground	<0.1v
104	DB/LG	COP 2 Driver (Dwell)	5°, at 55 mph: 8°

Pin Connector Graphic

PCM 104-PIN CONNECTOR

Terminal View of 104-Pin PCM Wiring Harness Connector

Standard Colors and Abbreviations

Abbreviation	Color	Abbreviation	Color	Abbreviation	Color
BK	Black	GY	Gray	PK	Purple
BL	Blue	GN	Green	RD	Red
BR	Brown	LG	LT Green	TN	Tan
DB	Dark Blue	OR	Orange	WT	White
DG	DK Green	PK	Pink	YL	Yellow

1998-99 E-Series 5.4L V8 VIN L (E4OD) 104 Pin Connector

PCM Pin #	Wire Color	Circuit Description (104 Pin)	Value at Hot Idle
1	PK/LB	COP 6 Driver (Dwell)	5°, at 55 mph: 8°
2	PK/LG	MIL (lamp) Control	MIL On: 1v, Off: 12v
3	BK/WT	Power Ground	<0.1v
4-5	---	Not Used	---
6	OR/YL	Shift Solenoid 1 Control	1v, 55 mph: 12v
7-8	---	Not Used	---
9	YL/WT	Fuel Level Indicator Signal	1.7v (1/2 full)
10	---	Not Used	---
11	PK/OR	Shift Solenoid 2 Control	12v, 55 mph: 1v
12	WT/LG	TCIL (lamp) Control	Lamp On: 1v, Off: 12v
13	PK	Flash EEPROM	0.1v
14	---	Not Used	---
15	PK/LB	Data Bus (-)	Digital Signals
16	TN/OR	Data Bus (+)	Digital Signals
17-19	---	Not Used	---
20	BR/OR	Coast Clutch Solenoid	12v, 55 mph: 12v
21	DB	CKP (+) Sensor	410 Hz, 55 mph: 1 KHz
22	GY	CKP (-) Sensor	410 Hz, 55 mph: 1 KHz
23	---	Not Used	---
24	BK/WT	Case Ground	<0.050v
25	BK/LB	Case Ground	<0.050v
26	DB/YL	COP 1 Driver (Dwell)	5°, at 55 mph: 8°
27	PKWT	COP 5 Driver (Dwell)	5°, at 55 mph: 8°
28	---	Not Used	---
29	TN/WT	TCS (switch) Signal	TCS & O/D On: 12v
30	DG	Octane Adjustment	9.3v (switch shorted: 0v)
31-32	---	Not Used	---
33	PK/OR	VSS (-) Signal	0 Hz, at 55 mph: 125 Hz
34	YL/BK	Digital TR1 Sensor	0v, at 55 mph: 11.5v
35	RD/LG	HO2S-12 (B1 S2) Signal	0.1-1.1v
36	TN/LB	MAF Sensor Return	<0.050v
37	OR/BK	Transmission Fluid Temperature Sensor	2.10-2.40v
38	---	Not Used	---
39	GY	Intake Air Temperature Sensor	1.5-2.5v
40	DG/YL	Fuel Pump Monitor	On: 12v, Off: 0v
41	BK/YL	A/C High Pressure Switch	Switch Closed: 12v
42-44	---	Not Used	---
45	YL/LB	CHIL (lamp) Control	Lamp On: 1v, Off: 12v
46	BR	Intake Manifold Tuning Valve	0.1v
47	BR/PK	VR Solenoid Control	0%, at 55 mph: 45%
48	WT/PK	Clean Tachometer Output Signal	46 Hz, 55 mph: 115 Hz
49	LB/BK	Digital TR2 Sensor	0v, at 55 mph: 11.5v
50	LG/RD	Digital TR4 Sensor	0v, at 55 mph: 11.5v
51	BK/WT	Power Ground	<0.1v
52	LG/WT	COP 3 Driver (Dwell)	5°, at 55 mph: 8°
53	OR/YL	COP 4 Driver (Dwell)	5°, at 55 mph: 8°
54	PK/YL	TCC Solenoid Control	0%, at 55 mph: 95%
55	Y	Keep Alive Power	12-14v
56	LG/BK	EVAP Purge Solenoid	0-10 Hz (0-100%)
57	YL/RD	Knock Sensor 1 Signal	0v
58	GY/BK	VSS (+) Signal	0 Hz, at 55 mph: 125 Hz
60	GY/LB	HO2S-11 (B1 S1) Signal	0.1-1.1v
61	---	Not Used	---
62	RD/PK	Fuel Tank Pressure Sensor	2.6v (0" H2O - cap off)
63	---	Not Used	---
64	LB/YL	Digital TR Sensor	In 'P': 0v, in O/D: 1.7v

1998-99 E-Series 5.4L V8 VIN L (E4OD) 104 Pin Connector

PCM Pin #	Wire Color	Circuit Description (104 Pin)	Value at Hot Idle
65	BR/LG	EGR DPFE Sensor Signal	0.20-1.30v
66	YL/LG	Cylinder Head Temperature Sensor	0.6v (194°F)
67	PKWT	EVAP CV Solenoid	0-10 Hz (0-100%)
68-70	---	Not Used	---
71	RD	Vehicle Power	12-14v
72	TN/RD	Injector 7 Control	3.2-3.8 ms
73	TN/BK	Injector 5 Control	3.2-3.8 ms
74	BR/YL	Injector 3 Control	3.2-3.8 ms
75	TN	Injector 1 Control	3.2-3.8 ms
76	---	Not Used	---
77	BK/WT	Power Ground	<0.1v
78	WT/PK	COP 7 Driver (Dwell)	5°, at 55 mph: 8°
79	WT/RD	COP 8 Driver (Dwell)	5°, at 55 mph: 8°
80	LB/OR	Fuel Pump Control	On: 1v, Off: 12v
81	WT/YL	EPC Solenoid Control	9.1v (5 psi)
82	---	Not Used	---
83	WT/LB	IAC Solenoid Control	10.7v (33%)
84	DB/YL	Output Shaft Speed Sensor	0 Hz, 55 mph: 228 Hz
85	DG	Camshaft Position Sensor	6 Hz, 55 mph: 14-18 Hz
86	---	Not Used	---
87	RD/BK	HO2S-21 (B2 S1) Signal	0.1-1.1v
88	LB/RD	MAF Sensor Signal	0.8v, 55 mph: 1.6v
89	GY/WT	TP Sensor Signal	0.53-1.27v
90	BR/WT	Reference Voltage	4.9-5.1v
91	GY/RD	Analog Signal Return	<0.050v
92	LG	Brake Position Switch	Brake Off: 12v, On: 1v
93	RD/WT	HO2S-11 (B1 S1) Heater	1v, Off: 12v
94	YL/LB	HO2S-21 (B2 S1) Heater	1v, Off: 12v
95	WT/BK	HO2S-12 (B1 S2) Heater	1v, Off: 12v
96	---	Not Used	---
97	RD	Vehicle Power	12-14v
98	LB	Injector 8 Control	3.2-3.8 ms
99	LG/OR	Injector 6 Control	3.2-3.8 ms
100	BR/LG	Injector 4 Control	3.2-3.8 ms
101	WT	Injector 2 Control	3.2-3.8 ms
102	---	Not Used	---
103	BK/WT	Power Ground	<0.1v
104	DB/LG	COP 2 Driver (Dwell)	5°, at 55 mph: 8°

Pin Connector Graphic

PCM 104-PIN CONNECTOR

Terminal View of 104-Pin PCM Wiring Harness Connector

Standard Colors and Abbreviations

Abbreviation	Color	Abbreviation	Color	Abbreviation	Color
BK	Black	GY	Gray	PK	Purple
BL	Blue	GN	Green	RD	Red
BR	Brown	LG	LT Green	TN	Tan
DB	Dark Blue	OR	Orange	WT	White
DG	DK Green	PK	Pink	YL	Yellow

2000-02 E-Series 5.4L V8 VIN L (A/T) 104 Pin Connector

PCM Pin #	Wire Color	Circuit Description (104 Pin)	Value at Hot Idle
1	OR/YL	COP 6 Driver (Dwell)	5°, 55 mph: 8°
2	PK/LG	MIL (lamp) Control	MIL On: 1v, Off: 12v
3	BK/WT	Power Ground	<0.1v
4	PK/LB	Power Takeoff Signal	PTO Off: 0v, On: 12v
5	---	Not Used	---
6	OR/YL	Shift Solenoid 1 Control	1v, 55 mph: 1v
7-8	---	Not Used	---
9	YL/WT	Fuel Pump Monitor	On: 1v, Off: 12v
10	---	Not Used	---
11	PK/OR	Shift Solenoid 2 Control	12v, 55 mph: 1v
12	WT/LG	TCIL (lamp) Control	Lamp On: 1v, Off: 12v
13	PK	Flash EEPROM	0.1v
14	OR/LB	4x4 Low Switch	Switch On: 0v, Off: 12v
15	PK/LB	Data Bus (-)	Digital Signals
16	TN/OR	Data Bus (+)	Digital Signals
17-19	---	Not Used	---
20	BR/OR	Coast Clutch Solenoid	12v, 55 mph: 12v
21	DB	CKP (+) Sensor	410 Hz, 55 mph: 1 KHz
22	GY	CKP (-) Sensor	410 Hz, 55 mph: 1 KHz
24	BK/WT	Case Ground	<0.050v
25	BK/LB	Case Ground	<0.050v
26	DB/YL	COP 1 Driver (Dwell)	5°, at 55 mph: 8°
27	PKWT	COP 5 Driver (Dwell)	5°, at 55 mph: 8°
28	---	Not Used	---
29	TN/WT	TCS (switch) Signal	TCS & O/D On: 12v
30-31	---	Not Used	---
32	DG/PK	Knock Sensor 1 Return	<0.050v
33	---	Not Used	---
34	YL/BK	Digital TR1 Sensor	0v, at 55 mph: 11.5v
35	RD/LG	HO2S-12 (B1 S2) Signal	0.1-1.1v
36	TN/LB	MAF Sensor Return	<0.050v
37	OR/BK	Transmission Fluid Temperature Sensor	2.10-2.40v
38	---	Not Used	---
39	GY	Intake Air Temperature Sensor	1.5-2.5v
40	DG/YL	Fuel Pump Monitor	On: 12v, Off: 0v
41	BK/YL	A/C Pressure Switch Signal	A/C On: 12v, Off: 0v
42-44	---	Not Used	---
45	YL/LB	Engine Coolant Temperature Sensor	0.5-0.6v
46	---	Not Used	---
47	BR/PK	VR Solenoid Control	0%, at 55 mph: 45%
48	---	Not Used	---
49	LB/BK	Digital TR2 Sensor	0v, at 55 mph: 11.5v
50	LG/RD	Digital TR4 Sensor	0v, at 55 mph: 11.5v
51	BK/WT	Power Ground	<0.1v
52	WT/PK	COP 3 Driver (Dwell)	5°, at 55 mph: 8°
53	DG/PK	COP 4 Driver (Dwell)	5°, at 55 mph: 8°
54	PK/YL	TCC Solenoid Control	0%, at 55 mph: 95%
55	YL	Keep Alive Power	12-14v
56	LG/BK	EVAP Purge Solenoid	0-10 Hz (0-100%)
57	YL/RD	Knock Sensor 1 Signal	0v
58	---	Not Used	---
59	DG/WT	Turbine Shaft Speed Sensor	120 Hz, 30 mph: 300 Hz
60	GY/LB	HO2S-11 (B1 S1) Signal	0.1-1.1v
61	---	Not Used	---
62	RD/PK	Fuel Tank Pressure Sensor	2.6v (0" H2O - cap off)
63	---	Not Used	---
64	LB/YL	Digital TR3A Sensor	In 'P': 0v, in O/D: 1.7v

2000-02 E-Series 5.4L V8 VIN L (A/T) 104 Pin Connector

PCM Pin #	Wire Color	Circuit Description (104 Pin)	Value at Hot Idle
65 ('00)	BR/LG	EGR DPFE Sensor Signal	0.20-1.30v
65 ('01-'02)	BR/LG	EGR DPFE Sensor Signal	0.95-1.05v
66	YL/LG	Cylinder Head Temperature Sensor	0.6v (194°F)
67	PKWT	EVAP CV Solenoid	0-10 Hz (0-100%)
68	GY/BK	Vehicle Speed Signal	0 Hz, 55 mph: 124 Hz
69-70	---	Not Used	---
71	RD	Vehicle Power	12-14v
72	TN/RD	Injector 7 Control	3.2-3.8 ms
73	TN/BK	Injector 5 Control	3.2-3.8 ms
74	BR/YL	Injector 3 Control	3.2-3.8 ms
75	TN	Injector 1 Control	3.2-3.8 ms
76	BK/WT	Power Ground	<0.1v
77	BK/WT	Power Ground	<0.1v
78	PK/LB	COP 7 Driver (Dwell)	5°, at 55 mph: 8°
79	WT/RD	COP 8 Driver (Dwell)	5°, at 55 mph: 8°
80	LB/OR	Fuel Pump Control	On: 1v, Off: 12v
81	WT/YL	EPC Solenoid Control	8.0v (3.5 psi)
82	---	Not Used	---
83	WT/LB	IAC Solenoid Control	8.7v (43%)
84	DB/YL	Output Shaft Speed Sensor	0 Hz, 55 mph: 228 Hz
85	DG	Camshaft Position Sensor	7 Hz, 55 mph: 13-17 Hz
86	---	Not Used	---
87	RD/BK	HO2S-21 (B2 S1) Signal	0.1-1.1v
88	LB/RD	MAF Sensor Signal	0.8v, 55 mph: 1.9v
89	GY/WT	TP Sensor Signal	0.53-1.27v
90	BR/WT	Reference Voltage	4.9-5.1v
91	GY/RD	Analog Signal Return	<0.050v
92	LG	Brake Position Switch	Brake Off: 12v, On: 1v
93	RD/WT	HO2S-11 (B1 S1) Heater	1v, Off: 12v
94	YL/LB	HO2S-21 (B2 S1) Heater	1v, Off: 12v
95	WT/BK	HO2S-12 (B1 S2) Heater	1v, Off: 12v
96	---	Not Used	---
97	RD	Vehicle Power	12-14v
98	LB	Injector 8 Control	3.2-3.8 ms
99	LG/OR	Injector 6 Control	3.2-3.8 ms
100	BR/LG	Injector 4 Control	3.2-3.8 ms
101	WT	Injector 2 Control	3.2-3.8 ms
102	---	Not Used	---
103	BK/WT	Power Ground	<0.1v
104	PKWT	COP 2 Driver (Dwell)	5°, at 55 mph: 8°

Pin Connector Graphic

PCM 104-PIN CONNECTOR

```
 1 ●●●●●●●●●●●●●    ●●●●●●●●●●●●● 26
27 ●●●●●●●●●●●●●    ●●●●●●●●●●●●● 52
53 ●●●●●●●●●●●●●  ⬤  ●●●●●●●●●●●●● 78
79 ●●●●●●●●●●●●●    ●●●●●●●●●●●●● 104
```

Terminal View of 104-Pin PCM Wiring Harness Connector

Standard Colors and Abbreviations

Abbreviation	Color	Abbreviation	Color	Abbreviation	Color
BK	Black	GY	Gray	PK	Purple
BL	Blue	GN	Green	RD	Red
BR	Brown	LG	LT Green	TN	Tan
DB	Dark Blue	OR	Orange	WT	White
DG	DK Green	PK	Pink	YL	Yellow

2003 E-Series 5.4L V8 VIN L (A/T) 104 Pin Connector

PCM Pin #	Wire Color	Circuit Description (104 Pin)	Value at Hot Idle
1	OR/YL	COP 6 Driver (Dwell)	5°, at 55 mph: 8°
2	PK/LG	MIL (lamp) Control	MIL On: 1v, Off: 12v
3	BK/WT	Power Ground	<0.1v
4-5	---	Not Used	---
6	OR/YL	Shift Solenoid 'A' Control	1v, 55 mph: 1v
7-8	---	Not Used	---
9	YL/WT	Low Fuel Indicator	Indicator Off: 12v, On: 1v
10	---	Not Used	---
11	VT/OR	Shift Solenoid 'B' Control	12v, 55 mph: 1v
12	WT/LG	TCIL (lamp) Control	Lamp Off: 12v, On: 1v
13	VT	Flash EEPROM	0.1v
14	OR/LB	4x4 Low Switch	Switch Off: 12v, On: 1v
15	PK/LB	SCP Data Bus (-)	<0.050v
16	TN/OR	SCP Data Bus (+)	Digital Signals
17-19	---	Not Used	---
20	BR/OR	Coast Clutch Solenoid Control	12v, 55 mph: 12v
21	DB	CKP (+) Sensor	410 Hz, 55 mph: 1 KHz
22	GY	CKP (-) Sensor	410 Hz, 55 mph: 1 KHz
23	---	Not Used	---
24	BK/WT	Case Ground	<0.050v
25	BK/LB	Power Ground	<0.1v
26	LG/WT	COP 1 Driver (Dwell)	5°, at 55 mph: 8°
27	LG/YL	COP 5 Driver (Dwell)	5°, at 55 mph: 8°
28	---	Not Used	---
29	TN/WT	TCS (switch) Signal	TCS & O/D On: 12v
30-31	---	Not Used	---
32	DG/VT	Knock Sensor 1 Ground	<0.050v
33	---	Not Used	---
34	YL/BK	Digital TR1 Sensor	0v, at 55 mph: 11.5v
35	RD/LG	HO2S-12 (B1 S2) Signal	0.1-1.1v
36	TN/LB	MAF Sensor Return	<0.050v
37	OR/BK	Transmission Fluid Temperature Sensor	2.10-2.40v
38	---	Not Used	---
39	GY	Intake Air Temperature Sensor	1.5-2.5v
40	DG/YL	Fuel Pump Monitor	On: 12v, Off: 0v
41	BK/YL	A/C Pressure Switch Signal	A/C On: 12v, Off: 0v
42-44	---	Not Used	---
45	YL/LB	Engine Coolant Temperature Sensor	0.5-0.6v
46	---	Not Used	---
47	BR/PK	VR Solenoid Control	0%, at 55 mph: 45%
48	---	Not Used	---
49	LB/BK	Digital TR2 Sensor	0v, at 55 mph: 11.5v
50	LG/RD	Digital TR4 Sensor	0v, at 55 mph: 11.5v
51	BK/WT	Power Ground	<0.1v
52	WT/PK	COP 3 Driver (Dwell)	5°, at 55 mph: 8°
53	DG/VT	COP 4 Driver (Dwell)	5°, at 55 mph: 8°
54	VT/YL	TCC Solenoid Control	0%, at 55 mph: 95%
55	YL	Keep Alive Power	12-14v
56	LG/BK	EVAP Canister Purge Valve	0-10 Hz (0-100%)
57	YL/RD	Knock Sensor 1 Ground	<0.050v
58	---	Not Used	---
59	DG/WT	Turbine Shaft Speed Sensor	120 Hz, 30 mph: 300 Hz
60	GY/LB	HO2S-11 (B1 S1) Signal	0.1-1.1v
61	---	Not Used	---
62	RD/PK	Fuel Tank Pressure Sensor	2.6v (0" H2O - cap off)
63	---	Not Used	---
64	LB/YL	Digital TR3A Sensor	In 'P': 0v, in O/D: 1.7v

2003 E-Series 5.4L V8 VIN L (A/T) 104 Pin Connector

PCM Pin #	Wire Color	Circuit Description (104 Pin)	Value at Hot Idle
65	BR/LG	EGR DPFE Sensor Signal	0.95-1.05v
66	YL/LG	Cylinder Head Temperature Sensor	0.6v (194°F)
67	VT/WT	EVAP Canister Vent Valve	0-10 Hz (0-100%)
68	GY/BK	Vehicle Speed Sensor	0 Hz, 55 mph: 124 Hz
69-70	---	Not Used	---
71	RD	Vehicle Power (Start-Run)	12-14v
72	TN/RD	Injector 7 Control	3.2-3.8 ms
73	TN/BK	Injector 5 Control	3.2-3.8 ms
74	BR/YL	Injector 3 Control	3.2-3.8 ms
75	TN	Injector 1 Control	3.2-3.8 ms
76	BK/WT	Power Ground	<0.1v
77	BK/WT	Power Ground	<0.1v
78	PK/LB	COP 7 Driver (Dwell)	5°, at 55 mph: 8°
79	WT/RD	COP 8 Driver (Dwell)	5°, at 55 mph: 8°
80	LB/OR	Fuel Pump Relay Control	Off: 12v, On: 1v
81	WT/YL	EPC Solenoid Control	8.0v (3.5 psi)
82	---	Not Used	---
83	WT/LB	IAC Solenoid Control	8.7v (43%)
84	DB/YL	Output Shaft Speed Sensor	0 Hz, 55 mph: 228 Hz
85	DG	Camshaft Position Sensor	7 Hz, 55 mph: 13-17 Hz
86	---	Not Used	---
87	RD/BK	HO2S-21 (B2 S1) Signal	0.1-1.1v
88	LB/RD	Mass Airflow Sensor	0.8v, 55 mph: 1.9v
89	GY/WT	Throttle Position Sensor	0.53-1.27v
90	BR/WT	Reference Voltage	4.9-5.1v
91	GY/RD	Sensor Ground	<0.050v
92	LG	Brake Position Switch	Brake Off: 12v, On: 1v
93	RD/WT	HO2S-11 (B1 S1) Heater Control	1v, Off: 12v
94	YL/LB	HO2S-21 (B2 S1) Heater Control	1v, Off: 12v
95	WT/BK	HO2S-12 (B1 S2) Heater Control	1v, Off: 12v
96	---	Not Used	---
97	RD	Vehicle Power (Start-Run)	12-14v
98	LB	Injector 8 Control	3.2-3.8 ms
99	LG/OR	Injector 6 Control	3.2-3.8 ms
100	BR/LG	Injector 4 Control	3.2-3.8 ms
101	WT	Injector 2 Control	3.2-3.8 ms
102	---	Not Used	---
103	BK/WT	Power Ground	<0.1v
104	PKWT	COP 2 Driver (Dwell)	5°, at 55 mph: 8°

Pin Connector Graphic

PCM 104-PIN CONNECTOR

```
 1 ●●●●●●●●●●●●●    ●●●●●●●●●●●●● 26
27 ●●●●●●●●●●●●●  ◉ ●●●●●●●●●●●●● 52
53 ●●●●●●●●●●●●●    ●●●●●●●●●●●●● 78
79 ●●●●●●●●●●●●●    ●●●●●●●●●●●●● 104
```

Terminal View of 104-Pin PCM Wiring Harness Connector

Standard Colors and Abbreviations

Abbreviation	Color	Abbreviation	Color	Abbreviation	Color
BK	Black	GY	Gray	PK	Purple
BL	Blue	GN	Green	RD	Red
BR	Brown	LG	LT Green	TN	Tan
DB	Dark Blue	OR	Orange	WT	White
DG	DK Green	PK	Pink	YL	Yellow

1998-89 E-Series 5.4L V8 CNG VIN M (A/T) 104 Pin Connector

PCM Pin #	Wire Color	Circuit Description (104 Pin)	Value at Hot Idle
1	OR/YL	COP 6 Driver (Dwell)	5°, at 55 mph: 8°
2	PK/LG	MIL (lamp) Control	MIL On: 1v, Off: 12v
3	BK/WT	Power Ground	<0.1v
4	WT/PK	Power Takeoff Signal	PTO Off: 0v, On: 12v
5	---	Not Used	---
6	OG/YL	Shift Solenoid 'A' Control	1v, 55 mph: 12v
7-10	---	Not Used	---
11	VTN/ORG	Shift Solenoid 'B' Control	12v, 55 mph: 1v
12	WT/LG	TCIL (lamp) Control	Lamp On: 1v, Off: 12v
13	VT	Flash EEPROM	0.1v
14	---	Not Used	---
15	PK/LB	Data Bus (-)	Digital Signals
16	TN/OR	Data Bus (+)	Digital Signals
17-20	---	Not Used	---
20	BR/OG	Coast Clutch Solenoid	12v, 55 mph: 12v
21	DB	CKP (+) Sensor	440-490 Hz
22	GY	CKP (-) Sensor	440-490 Hz
24	BK/WT	Power Ground	<0.1v
25	BK/LB	Case Ground	<0.050v
26	LG/WTH	COP 1 Driver (Dwell)	5°, at 55 mph: 8°
27	LG/OR	COP 5 Driver (Dwell)	5°, at 55 mph: 8°
29	TN/WT	TCS (switch) Signal	TCS & O/D On: 12v
30-32	---	Not Used	---
33	PK/ORG	VSS (-) Signal Return	<0.050v
34	YL/BK	Transmission Range Sensor 1	0v, at 55 mph: 11.5v
36	TN/LB	MAF Sensor Return	<0.050v
37	OG/BK	Transmission Fluid Temperature Sensor	2.10-2.40v
38	---	Not Used	---
39	GY	Intake Air Temperature Sensor	1.5-2.5v
40	DG/YL	Fuel Pump Monitor	On: 12v, Off: 0v
41	BK/YL	A/C Pressure Switch Signal	A/C On: 12v, Off: 0v
42-44	---	Not Used	---
45	YL/LB	Engine Coolant Temperature Sensor	0.5-0.6v
48	WT/PK	Clean Tachometer Output	55 Hz, 55 mph: 120 Hz
49	LB/BK	Digital TR2 Sensor Signal	0v, at 55 mph: 11.5v
50	LG/RD	Digital TR4 Sensor Signal	0v, at 55 mph: 11.5v
51	BK/WT	Power Ground	<0.1v
52	WT/PK	COP 3 Driver (Dwell)	5°, at 55 mph: 8°
53	DG/PK	COP 4 Driver (Dwell)	5°, at 55 mph: 8°
54	VT/YL	TCC Solenoid Control	0%, at 55 mph: 95%
55	YL	Keep Alive Power	12-14v
56-57	---	Not Used	---
58	GY/BK	VSS (+) Signal	0 Hz, at 55 mph: 125 Hz
59	DG/WT	Turbine Shaft Speed Sensor	120 Hz, 30 mph: 300 Hz
60	GY/LB	HO2S-11 (B1 S1) Signal	0.1-1.1v
62	LB	Fuel Rail Temperature Sensor	1.7-3.5v (50-120°F)
63	RD/PK	Injection Pressure Sensor	2-3.7v (90-100 psi)
64	LB/YL	Digital TR3A Sensor Signal	In 'P': 0v, in O/D: 1.7v

1998-99 E-Series 5.4L V8 CNG VIN M 104 Pin Connector

PCM Pin #	Wire Color	Circuit Description (104 Pin)	Value at Hot Idle
65	---	Not Used	---
66	YL/LG	Cylinder Head Temperature Sensor	0.6v (194°F)
68-70	---	Not Used	---
71	RD	Vehicle Power	12-14v
72	TN/RD	Injector 7 (to NGV Module)	3.9-4.6 ms
73	TN/BK	Injector 5 (to NGV Module)	3.9-4.6 ms
74	BR/YL	Injector 3 (to NGV Module)	3.9-4.6 ms
75	TN	Injector 1 (to NGV Module)	3.9-4.6 ms
77	BK/WT	Power Ground	<0.1v
78	PK/LB	COP 7 Driver (Dwell)	5°, at 55 mph: 8°
79	WT/RD	COP 8 Driver (Dwell)	5°, at 55 mph: 8°
80	LB/OR	Fuel Pump Control	On: 1v, Off: 12v
81	WT/YL	EPC Solenoid Control	9.1v (4 psi)
82	---	Not Used	---
83	WT/LB	IAC Solenoid Control	10.7v (33%)
84	---	Not Used	---
85	DG	Camshaft Position Sensor	6 Hz, 55 mph: 14-18 Hz
86	---	Not Used	---
87	RD/BK	HO2S-21 (B2 S1) Signal	0.1-1.1v
88	LB/RD	MAF Sensor Signal	0.8v, 55 mph: 1.6v
89	GY/WT	TP Sensor Signal	0.53-1.27v
90	BR/WT	Reference Voltage	4.9-5.1v
91	GY/RD	Sensor Signal Return	<0.050v
92	LG	Brake Position Switch	Brake Off: 12v, On: 1v
93	RD/WT	HO2S-11 (B1 S1) Heater	1v, Off: 12v
94	YL/LB	HO2S-21 (B2 S1) Heater	1v, Off: 12v
95-96	---	Not Used	---
97	RD	Vehicle Power	12-14v
98	LB	Injector 8 (to NGV Module)	3.9-4.6 ms
99	LG/OR	Injector 6 (to NGV Module)	3.9-4.6 ms
100	BR/LB	Injector 4 (to NGV Module)	3.9-4.6 ms
101	WT	Injector 2 (to NGV Module)	3.9-4.6 ms
102	---	Not Used	---
103	BK/WT	Power Ground	<0.1v
104	PK/WT	COP 2 Driver (Dwell)	5°, at 55 mph: 8°

Pin Connector Graphic

PCM 104-PIN CONNECTOR

Terminal View of 104-Pin PCM Wiring Harness Connector

Standard Colors and Abbreviations

Abbreviation	Color	Abbreviation	Color	Abbreviation	Color
BK	Black	GY	Gray	PK	Purple
BL	Blue	GN	Green	RD	Red
BR	Brown	LG	LT Green	TN	Tan
DB	Dark Blue	OR	Orange	WT	White
DG	DK Green	PK	Pink	YL	Yellow

2000-02 E-Series 5.4L V8 CNG VIN M 104 Pin Connector

PCM Pin #	Wire Color	Circuit Description (104 Pin)	Value at Hot Idle
1	OR/YL	COP 6 Driver (Dwell)	5°, at 55 mph: 8°
2	PK/LG	MIL (lamp) Control	MIL On: 1v, Off: 12v
3	BK/WT	Power Ground	<0.1v
4	VTN/LB	Auxiliary Power Feed	PTO Off: 0v, On: 12v
5	---	Not Used	---
6	OG/YL	Shift Solenoid 'A' Control	1v, 55 mph: 12v
7-10	---	Not Used	---
11	VTN/ORG	Shift Solenoid 'B' Control	12v, 55 mph: 1v
12	WT/LG	TCIL (lamp) Control	Lamp On: 1v, Off: 12v
13	VT	Flash EEPROM	0.1v
14	---	Not Used	---
15	PK/LB	Data Bus (-)	Digital Signals
16	TN/OR	Data Bus (+)	Digital Signals
17-19	---	Not Used	---
20	BR/OG	Coast Clutch Solenoid	12v, 55 mph: 12v
21	DB	CKP (+) Sensor	440-490 Hz
22	GY	CKP (-) Sensor	440-490 Hz
23	---	Not Used	---
24	BK/WT	Power Ground	<0.1v
25	BK/LB	Case Ground	<0.050v
26	LG/WTH	COP 1 Driver (Dwell)	5°, at 55 mph: 8°
27	LG/YL	COP 5 Driver (Dwell)	5°, at 55 mph: 8°
28	LG/OR	COP 5 Driver (Dwell)	5°, at 55 mph: 8°
29	TN/WT	TCS (switch) Signal	TCS & O/D On: 12v
30-33	LG/OR	COP 5 Driver (Dwell)	5°, at 55 mph: 8°
34	YL/BK	Transmission Range Sensor 1	0v, at 55 mph: 11.5v
36	TN/LB	MAF Sensor Return	<0.050v
37	OG/BK	Transmission Fluid Temperature Sensor	2.10-2.40v
38	---	Not Used	---
39	GY	Intake Air Temperature Sensor	1.5-2.5v
40	DG/YL	Fuel Pump Monitor	On: 12v, Off: 0v
41	BK/YL	A/C Pressure Switch Signal	A/C On: 12v, Off: 0v
42-44	---	Not Used	---
45	YL/LB	Engine Coolant Temperature Sensor	0.5-0.6v
46-47	---	Not Used	---
48	WT/PK	Clean Tachometer Output	55 Hz, 55 mph: 120 Hz
49	LB/BK	Digital TR2 Sensor Signal	0v, at 55 mph: 11.5v
50	LG/RD	Digital TR4 Sensor Signal	0v, at 55 mph: 11.5v
51	BK/WT	Power Ground	<0.1v
52	WT/PK	COP 3 Driver (Dwell)	5°, at 55 mph: 8°
53	DG/PK	COP 4 Driver (Dwell)	5°, at 55 mph: 8°
54	VT/YL	TCC Solenoid Control	0%, at 55 mph: 95%
55	YL	Keep Alive Power	12-14v
56-57	---	Not Used	---
59	DG/WT	Turbine Shaft Speed Sensor	120 Hz, 30 mph: 300 Hz
60	GY/LB	HO2S-11 (B1 S1) Signal	0.1-1.1v
61	---	Not Used	---
62	LB	Fuel Rail Temperature Sensor	1.7-3.5v (50-120°F)
63	RD/PK	Injection Pressure Sensor	2-3.7v (90-100 psi)
64	LB/YL	Digital TR3A Sensor Signal	In 'P': 0v, in O/D: 1.7v
65	---	Not Used	---

2000-02 E-Series 5.4L V8 CNG VIN M 104 Pin Connector

PCM Pin #	Wire Color	Circuit Description (104 Pin)	Value at Hot Idle
66	YL/LG	Cylinder Head Temperature Sensor	0.6v (194°F)
67	---	Not Used	---
68	GY/BK	Vehicle Speed Signal	0 Hz, at 55 mph: 125 Hz
69-70	---	Not Used	---
71	RD	Vehicle Power	12-14v
72	TN/RD	Injector 7 (to NGV Module)	3.9-4.6 ms
73	TN/BK	Injector 5 (to NGV Module)	3.9-4.6 ms
74	BR/YL	Injector 3 (to NGV Module)	3.9-4.6 ms
75	TN	Injector 1 (to NGV Module)	3.9-4.6 ms
76	---	Not Used	---
77	BK/WT	Power Ground	<0.1v
78	PK/LB	COP 7 Driver (Dwell)	5°, at 55 mph: 8°
79	WT/RD	COP 8 Driver (Dwell)	5°, at 55 mph: 8°
80	LB/OR	Fuel Pump Control	On: 1v, Off: 12v
81	WT/YL	EPC Solenoid Control	9.1v (4 psi)
82	---	Not Used	---
83	WT/LB	IAC Solenoid Control	10.7v (33%)
84	DB/YL	Output Shaft Speed Sensor	0 Hz, 55 mph: 228 Hz
85	DG	Camshaft Position Sensor	6 Hz, 55 mph: 14-18 Hz
86	---	Not Used	---
87	RD/BK	HO2S-21 (B2 S1) Signal	0.1-1.1v
88	LB/RD	MAF Sensor Signal	0.8v, 55 mph: 1.6v
89	GY/WT	TP Sensor Signal	0.53-1.27v
90	BR/WT	Reference Voltage	4.9-5.1v
91	GY/RD	Sensor Signal Return	<0.050v
92	LG	Brake Position Switch	Brake Off: 12v, On: 1v
93	RD/WT	HO2S-11 (B1 S1) Heater	1v, Off: 12v
94	YL/LB	HO2S-21 (B2 S1) Heater	1v, Off: 12v
95-96	---	Not Used	---
97	RD	Vehicle Power	12-14v
98	LB	Injector 8 (to NGV Module)	3.9-4.6 ms
99	LG/OR	Injector 6 (to NGV Module)	3.9-4.6 ms
100	BR/LB	Injector 4 (to NGV Module)	3.9-4.6 ms
101	WT	Injector 2 (to NGV Module)	3.9-4.6 ms
102	---	Not Used	---
103	BK/WT	Power Ground	<0.1v
104	PK/WT	COP 2 Driver (Dwell)	5°, at 55 mph: 8°

Pin Connector Graphic

PCM 104-PIN CONNECTOR

Terminal View of 104-Pin PCM Wiring Harness Connector

Standard Colors and Abbreviations

Abbreviation	Color	Abbreviation	Color	Abbreviation	Color
BK	Black	GY	Gray	PK	Purple
BL	Blue	GN	Green	RD	Red
BR	Brown	LG	LT Green	TN	Tan
DB	Dark Blue	OR	Orange	WT	White
DG	DK Green	PK	Pink	YL	Yellow

2003 E-Series 5.4L V8 CNG VIN M (A/T) 104 Pin Connector

PCM Pin #	Wire Color	Circuit Description (104 Pin)	Value at Hot Idle
1	OR/YL	COP 6 Driver (Dwell)	5°, at 55 mph: 8°
2	PK/LG	MIL (lamp) Control	MIL On: 1v, Off: 12v
3	BK/WT	Power Ground	<0.1v
4-5	---	Not Used	---
6	OR/YL	Shift Solenoid 'A' Control	1v, 55 mph: 1v
7-8	---	Not Used	---
9	YL/WT	Low Fuel Indicator	Indicator Off: 12v, On: 1v
10	---	Not Used	---
11	VT/OR	Shift Solenoid 'B' Control	12v, 55 mph: 1v
12	WT/LG	TCIL (lamp) Control	Lamp Off: 12v, On: 1v
13	VT	Flash EEPROM	0.1v
14	OR/LB	4x4 Low Switch	Switch Off: 12v, On: 1v
15	PK/LB	SCP Data Bus (-)	<0.050v
16	TN/OR	SCP Data Bus (+)	Digital Signals
17-19	---	Not Used	---
20	BR/OR	Coast Clutch Solenoid Control	12v, 55 mph: 12v
21	DB	CKP (+) Sensor	410 Hz, 55 mph: 1 KHz
22	GY	CKP (-) Sensor	410 Hz, 55 mph: 1 KHz
23	---	Not Used	---
24	BK/WT	Case Ground	<0.050v
25	BK/LB	Power Ground	<0.1v
26	LG/WT	COP 1 Driver (Dwell)	5°, at 55 mph: 8°
27	LG/YL	COP 5 Driver (Dwell)	5°, at 55 mph: 8°
28	---	Not Used	---
29	TN/WT	TCS (switch) Signal	TCS & O/D On: 12v
30-31	---	Not Used	---
32	DG/VT	Knock Sensor 1 Ground	<0.050v
33	---	Not Used	---
34	YL/BK	Digital TR1 Sensor	0v, at 55 mph: 11.5v
35	RD/LG	HO2S-12 (B1 S2) Signal	0.1-1.1v
36	TN/LB	MAF Sensor Return	<0.050v
37	OR/BK	Transmission Fluid Temperature Sensor	2.10-2.40v
38	---	Not Used	---
39	GY	Intake Air Temperature Sensor	1.5-2.5v
40	DG/YL	Fuel Pump Monitor	On: 12v, Off: 0v
41	BK/YL	A/C Pressure Switch Signal	A/C On: 12v, Off: 0v
42-44	---	Not Used	---
45	YL/LB	Engine Coolant Temperature Sensor	0.5-0.6v
46	---	Not Used	---
47	BR/PK	VR Solenoid Control	0%, at 55 mph: 45%
48	---	Not Used	---
49	LB/BK	Digital TR2 Sensor	0v, at 55 mph: 11.5v
50	LG/RD	Digital TR4 Sensor	0v, at 55 mph: 11.5v
51	BK/WT	Power Ground	<0.1v
52	WT/PK	COP 3 Driver (Dwell)	5°, at 55 mph: 8°
53	DG/VT	COP 4 Driver (Dwell)	5°, at 55 mph: 8°
54	VT/YL	TCC Solenoid Control	0%, at 55 mph: 95%
55	YL	Keep Alive Power	12-14v
56	LG/BK	EVAP Canister Purge Valve	0-10 Hz (0-100%)
57	YL/RD	Knock Sensor 1 Ground	<0.050v
58	---	Not Used	---
59	DG/WT	Turbine Shaft Speed Sensor	120 Hz, 30 mph: 300 Hz
60	GY/LB	HO2S-11 (B1 S1) Signal	0.1-1.1v
61	---	Not Used	---
62	LB	Fuel Rail Temperature Sensor	1.7-3.5v (50-120°F)
63	RD/PK	Injection Pressure Sensor	2-3.7v (90-100 psi)
64	LB/YL	Digital TR3A Sensor	In 'P': 0v, in O/D: 1.7v

2003 E-Series 5.4L V8 CNG VIN M (A/T) 104 Pin Connector

PCM Pin #	Wire Color	Circuit Description (104 Pin)	Value at Hot Idle
65	BR/LG	EGR DPFE Sensor Signal	0.95-1.05v
66	YL/LG	Cylinder Head Temperature Sensor	0.6v (194°F)
67	VT/WT	EVAP Canister Vent Valve	0-10 Hz (0-100%)
68	GY/BK	Vehicle Speed Sensor	0 Hz, 55 mph: 124 Hz
69-70	---	Not Used	---
71	RD	Vehicle Power (Start-Run)	12-14v
72	TN/RD	Injector 7 Control	3.2-3.8 ms
73	TN/BK	Injector 5 Control	3.2-3.8 ms
74	BR/YL	Injector 3 Control	3.2-3.8 ms
75	TN	Injector 1 Control	3.2-3.8 ms
76	BK/WT	Power Ground	<0.1v
77	BK/WT	Power Ground	<0.1v
78	PK/LB	COP 7 Driver (Dwell)	5°, at 55 mph: 8°
79	WT/RD	COP 8 Driver (Dwell)	5°, at 55 mph: 8°
80	LB/OR	Fuel Pump Relay Control	Off: 12v, On: 1v
81	WT/YL	EPC Solenoid Control	8.0v (3.5 psi)
82	---	Not Used	---
83	WT/LB	IAC Solenoid Control	8.7v (43%)
84	DB/YL	Output Shaft Speed Sensor	0 Hz, 55 mph: 228 Hz
85	DG	Camshaft Position Sensor	7 Hz, 55 mph: 13-17 Hz
86	---	Not Used	---
87	RD/BK	HO2S-21 (B2 S1) Signal	0.1-1.1v
88	LB/RD	Mass Airflow Sensor	0.8v, 55 mph: 1.9v
89	GY/WT	Throttle Position Sensor	0.53-1.27v
90	BR/WT	Reference Voltage	4.9-5.1v
91	GY/RD	Sensor Ground	<0.050v
92	LG	Brake Position Switch	Brake Off: 12v, On: 1v
93	RD/WT	HO2S-11 (B1 S1) Heater Control	1v, Off: 12v
94	YL/LB	HO2S-21 (B2 S1) Heater Control	1v, Off: 12v
95	WT/BK	HO2S-12 (B1 S2) Heater Control	1v, Off: 12v
96	---	Not Used	---
97	RD	Vehicle Power (Start-Run)	12-14v
98	LB	Injector 8 Control	3.2-3.8 ms
99	LG/OR	Injector 6 Control	3.2-3.8 ms
100	BR/LG	Injector 4 Control	3.2-3.8 ms
101	WT	Injector 2 Control	3.2-3.8 ms
102	---	Not Used	---
103	BK/WT	Power Ground	<0.1v
104	PKWT	COP 2 Driver (Dwell)	5°, at 55 mph: 8°

Pin Connector Graphic

PCM 104-PIN CONNECTOR

Terminal View of 104-Pin PCM Wiring Harness Connector

Standard Colors and Abbreviations

Abbreviation	Color	Abbreviation	Color	Abbreviation	Color
BK	Black	GY	Gray	PK	Purple
BL	Blue	GN	Green	RD	Red
BR	Brown	LG	LT Green	TN	Tan
DB	Dark Blue	OR	Orange	WT	White
DG	DK Green	PK	Pink	YL	Yellow

1998-02 E-Series 5.4L V8 NGV Module 60 Pin Connector

PCM Pin #	Wire Color	Circuit Description (60 Pin)	Value at Hot Idle
1	YL	Keep Alive Power	12-14v
2	---	Not Used	---
3	TN	Injector 1 Signal from PCM	3.9-4.6 ms
4	WT	Injector 2 Signal from PCM	3.9-4.6 ms
5	BR/YL	Injector 3 Signal from PCM	3.9-4.6 ms
6	---	Not Used	---
7	RD/PK	Fuel Tank Pressure Signal	2.6v (0" HG - cap off)
8-15	---	Not Used	---
16	BK/LB	NGV Timer	N/A
17	---	Not Used	---
18	PK/LB	SCP Data Bus (-)	<0.050V
19	TN/OG	SCP Data Bus (+)	Digital Signals
20	BK/LB	Power Ground	<0.1v
21-22	---	Not Used	--
23	BR/LB	Injector 4 Signal from PCM	3.9-6.5 ms
24	TN/BK	Injector 5 Signal from PCM	3.9-4.6 ms
25	LG/OR	Injector 6 Signal from PCM	3.9-4.6 ms
26	BR/WT	Reference Voltage	4.9-5.1v
27-32	---	Not Used	---
33	TN/BK	Injector 5 Control	3.9-6.5 ms
34	---	Not Used	---
35	BR/LB	Injector 4 Control	3.9-6.5 ms
36	---	Not Used	---
37	RD	Vehicle Power Input	12-14v
38	YL/WT	Fuel Gauge Signal	Varies
39	BR/YL	Injector 3 Control	2.9-6.5 ms
40	BK	Power Ground	<0.1v
41	---	Not Used	---
42	LG/OR	Injector 6 Control	3.9-6.5 ms
43	TN/RD	Injector 7 Signal from PCM	3.9-6.5 ms
44	LB	Injector 8 Signal from PCM	3.9-6.5 ms
45	---	Not Used	---
46	GY/RD	Sensor Signal Return	<0.050v
47	OG/LG	Fuel Tank Temperature	Varies: 0.1-4.9v
48-52	---	Not Used	---
53	TN/RD	Injector 7 Control	3.9-6.5 ms
54	LB	Injector 8 Control	3.9-6.5 ms
55-56	---	Not Used	---
57	RD	Vehicle Power	12-14v
58	TN	Injector 1 Control	3.9-6.5 ms
59	WT	Injector 2 Control	3.9-6.5 ms
60	BK	Power Ground	<01v

Pin Connector Graphic

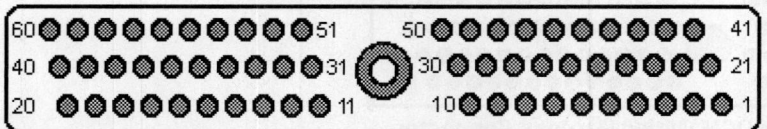

PCM 60-PIN CONNECTOR

Terminal View of 60-Pin PCM Harness Connector

Standard Colors and Abbreviations

Abbreviation	Color	Abbreviation	Color	Abbreviation	Color
BK	Black	GY	Gray	PK	Purple
BL	Blue	GN	Green	RD	Red
BR	Brown	LG	LT Green	TN	Tan
DB	Dark Blue	OR	Orange	WT	White
DG	DK Green	PK	Pink	YL	Yellow

1998-02 E-Series 5.4L V8 NGV (A/T) Module Wiring

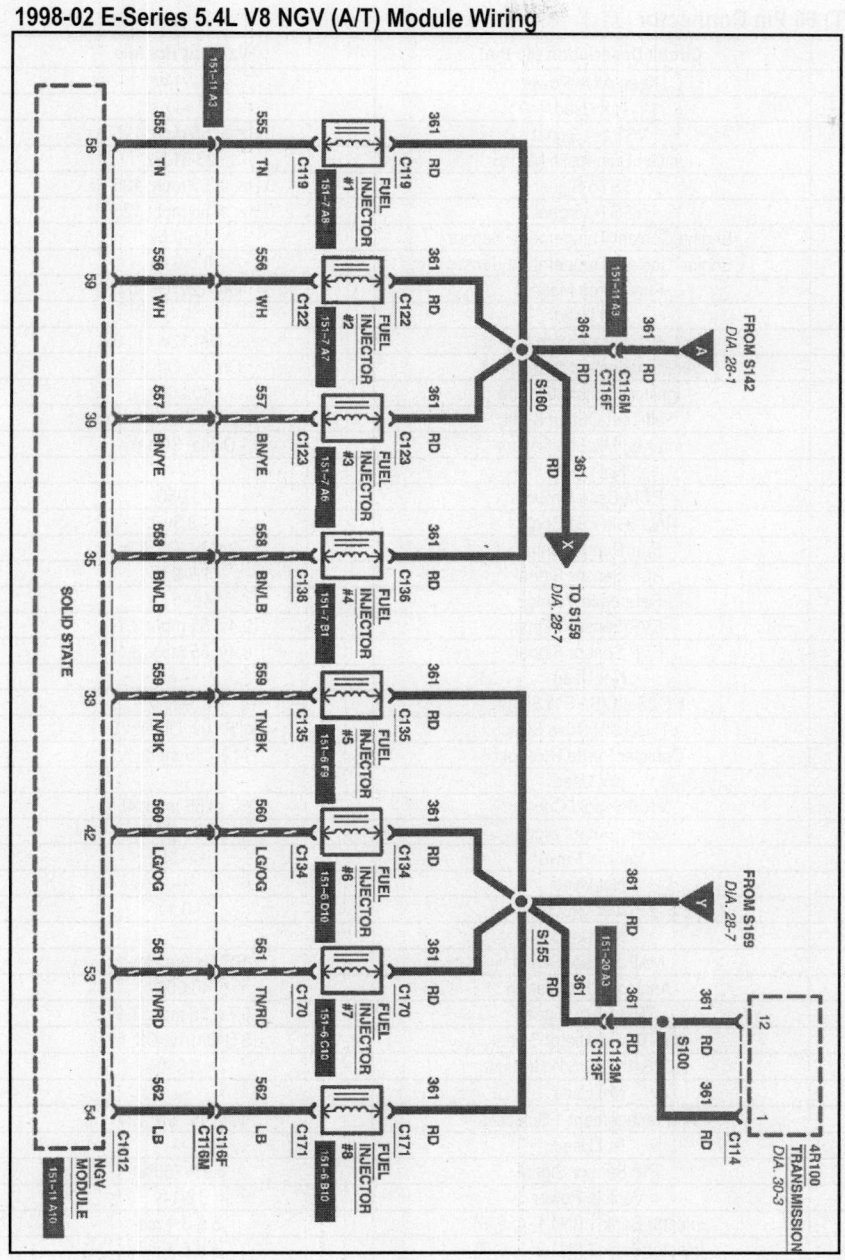

1990 E-Series 5.8L V8 VIN H (A/T) 60 Pin Connector

PCM Pin #	Wire Color	Circuit Description (60 Pin)	Value at Hot Idle
1	BK/OR	Keep Alive Power	12-14v
2, 5	---	Not Used	---
3	DG/WT	VSS (+) Signal	0 Hz, at 55 mph: 125 Hz
4	DG/YL	Ignition Diagnostic Monitor	20-31 Hz
6	OR/YL	VSS (-) Signal	0 Hz, at 55 mph: 125 Hz
6	GY/BK	VSS (-) Signal	0 Hz, at 55 mph: 125 Hz
7	LG/YL	Engine Coolant Temperature Sensor	0.5-0.6v
7	GN/YL	Engine Coolant Temperature Sensor	0.5-0.6v
8	OR/BL	Fuel Pump Monitor	On: 12v, Off: 0v
9, 12-15	---	Not Used	---
10	BK/YL	A/C Pressure Switch Signal	A/C On: 12v, off: 0v
11	WT/BK	Air Management 2 Solenoid	On: 1v, Off: 12v
16	BK/OR	Ignition System Ground	<0.050v
17	TN/RD	Self-Test Output & MIL	MIL On: 1v, Off: 12v
18	TN/LG	Inferred Mileage Sensor	Digital Signals
19, 23-24	---	Not Used	---
20	BK	PCM Case Ground	<0.050v
21	GY/WT	IAC Solenoid Control	8.3v
22	TN/LG	Fuel Pump Control	On: 1v, Off: 12v
25	YL/RD	ACT Sensor Signal	1.5-2.5v
26	OR/WT	Reference Voltage	4.9-5.1v
27	BR/LG	EVP Sensor Signal	0.4v, 55 mph: 3.1v
27	DG/LG	EVP Sensor Signal	0.4v, 55 mph: 3.1v
28	---	Not Used	---
29	DG/PK	HO2S-11 (B1 S1) Signal	0.1-1.1v
30	BL/WT	A/T: Neutral Drive Switch	In 'P': 0v, Others: 5v
31	GY/YL	Canister Purge Solenoid	12v, 55 mph: 1v
32, 34-35	---	Not Used	---
33	DG	VR Solenoid Control	0%, at 55 mph: 45%
36	YL/GN	Spark Output Signal	6.93v (50%)
37	RD	Vehicle Power	12-14v
38-39	---	Not Used	---
40	BK/LG	Power Ground	<0.1v
41-44	---	Not Used	---
45	DB/LG	MAP Sensor Signal	107 Hz (sea level)
46	BK/WT	Analog Signal Return	<0.050v
47	DG/LG	TP Sensor Signal	0.7v, 55 mph: 2.1v
48	WT/RD	Self-Test Indicator Signal	STI On: 1v, Off: 5v
49	OR	HO2S-11 (B1 S1) Ground	<0.050v
50	---	Not Used	---
51	WT/RD	Air Management 1 Solenoid	On: 1v, Off: 12v
52-55	---	Not Used	---
56	DB	PIP Sensor Signal	6.93v (50%)
57	R	Vehicle Power	12-14v
58	TN/OR	Injector Bank 1 (INJ 1, 4, 5, 8)	5.6-6.4 ms
59	TN/RD	Injector Bank 2 (INJ 2, 3, 6, 7)	5.6-6.4 ms
60	BK/LG	Power Ground	<0.1v

Pin Connector Graphic

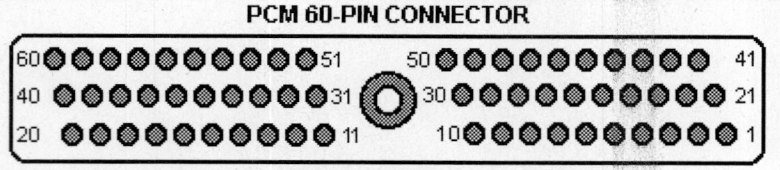

Terminal View of 60-Pin PCM Harness Connector

1990 E-Series 5.8L V8 VIN H (E4OD) 60 Pin Connector

PCM Pin #	Wire Color	Circuit Description (60 Pin)	Value at Hot Idle
1	BK/OR	Keep Alive Power	12-14v
2	LG	Brake Switch Signal	Brake Off: 12v, On: 1v
3	DG/WT	VSS (+) Signal	0 Hz, at 55 mph: 125 Hz
4	DG/YL	Ignition Diagnostic Monitor	20-31 Hz
5, 9, 12-15	---	Not Used	---
6	OR/YL	VSS (-) Signal	0 Hz, at 55 mph: 125 Hz
7	LG/YL	Engine Coolant Temperature Sensor	0.5-0.6v
8	OR/BL	Fuel Pump Monitor	On: 12v, Off: 0v
10	BK/YL	A/C Pressure Switch Signal	A/C On: 12v, off: 0v
11	WT/BK	Air Management 2 Solenoid	On: 1v, Off: 12v
16	BK/OR	Ignition System Ground	<0.050v
17	TN/RD	Self-Test Output & MIL	MIL On: 1v, Off: 12v
18	TN/LG	Inferred Mileage Sensor	Digital Signals
19	DG/PK	Shift Solenoid 2 Control	12v, 30 mph: 1v
20	BK	PCM Case Ground	<0.050v
21	GY/WT	IAC Solenoid Control	8.3v
22	TN/LG	Fuel Pump Control	On: 1v, Off: 12v
23, 28	---	Not Used	---
24	YL/LG	Power Steering Pressure Switch	Straight: 0v, Turned: 12v
25	YL/RD	ACT Sensor Signal	1.5-2.5v
26	OR/WT	Reference Voltage	4.9-5.1v
27	BR/LG	EVP Sensor Signal	0.4v, 55 mph: 3.1v
29	DG/PK	HO2S-11 (B1 S1) Signal	0.1-1.1v
30	BL/WT	MLP Sensor Signal	In 'P': 0v, in O/D: 5v
31	GY/YL	Canister Purge Solenoid	12v, 55 mph: 1v
32	LG/WT	OCIL (lamp) Control	Lamp On: 1v, Off: 12v
33	DG	VR Solenoid Control	0%, at 55 mph: 45%
34-35	---	Not Used	---
36	YL/LG	Spark Output Signal	6.93v (50%)
37	RD	Vehicle Power	12-14v
38	BL/YL	EPC Solenoid Control	9.2v (5 psi)
39, 43-44	---	Not Used	---
40	BK/LG	Power Ground	<0.1v
41	TN/BL	Overdrive Cancel Switch	OCS Closed: 0.1v
42	OR/BK	TOT Sensor Signal	2.10-2.40v
45	DB/LG	MAP Sensor Signal	107 Hz
46	BK/WT	Analog Signal Return	<0.050v
47	DG/LG	TP Sensor Signal	0.7v, 55 mph: 2.1v
48	WT/RD	Self-Test Indicator Signal	STI On: 1v, Off: 5v
49	OR	HO2S-11 (B1 S1) Ground	<0.050v
50, 54	---	Not Used	---
51	WT/RD	Air Management 1 Solenoid	On: 1v, Off: 12v
52	OR/YL	Shift Solenoid 1 Control	In 'P': 1v, 30 mph 12v
53	PK/YL	TCC Solenoid Control	12v, 55 mph: 1v
55	BR	Coast Clutch Solenoid	12v, 55 mph: 12v
56	DB	PIP Sensor Signal	6.93v (50%)
57	RD	Vehicle Power	12-14v
58	TN/OR	Injector Bank 1 (INJ 1, 4, 5, 8)	5.8-6.4 ms
59	TN/RD	Injector Bank 2 (INJ 2, 3, 6, 7)	5.8-6.4 ms
60	BK/LG	Power Ground	<0.1v

Pin Connector Graphic

PCM 60-PIN CONNECTOR

Terminal View of 60-Pin PCM Harness Connector

1991 E-Series 5.8L V8 OHV VIN H (All) 60 Pin Connector

PCM Pin #	Wire Color	Circuit Description (60 Pin)	Value at Hot Idle
1	BK/OR	Keep Alive Power	12-14v
2	LG	Brake Switch Signal	Brake Off: 12v, On: 1v
3	DG/WT	VSS (+) Signal	0 Hz, at 55 mph: 125 Hz
4	TN/YL	Ignition Diagnostic Monitor	20-31 Hz
5, 12-15	---	Not Used	---
6	OR/YL	VSS (-) Signal	0 Hz, at 55 mph: 125 Hz
7	LG/RD	Engine Coolant Temperature Sensor	0.5-0.6v
8	DG/YL	Fuel Pump Monitor	On: 12v, Off: 0v
9	BK/OR	Data Bus (-)	Digital Signals
10	BK/YL	A/C Pressure Switch Signal	A/C On: 12v, off: 0v
11	BR	Air Management 2 Solenoid	On: 1v, Off: 12v
16	OR/RD	Ignition System Ground	<0.050v
17	TN/RD	Self-Test Output & MIL	MIL On: 1v, Off: 12v
18, 24	---	Not Used	---
19	PK	Shift Solenoid 2 Control	12v, 55 mph: 1v
20	BK	PCM Case Ground	<0.050v
21	GY/WT	IAC Solenoid Control	7.0-8.5v
22	TN/LG	Fuel Pump Control	On: 1v, Off: 12v
23	LG/BK	Knock Sensor 1 Signal	0v
25	YL/RD	ACT Sensor Signal	1.5-2.5v
26	BR/WT	Reference Voltage	4.9-5.1v
27	BR/LG	EVP Sensor Signal	0.3v, 55 mph: 2.3v
28	OR/BK	Data Bus (+)	Digital Signals
29	RD/BK	HO2S-11 (B1 S1) Signal	0.1-1.1v
30	WT/RD	MLP Sensor Signal	In 'P': 0v, in O/D: 5v
31	GY/YL	Canister Purge Solenoid	12v, 55 mph: 1v
32	WT/LG	OCIL (lamp) Control	Lamp On: 1v, Off: 12v
33	BR/PK	VR Solenoid Control	0%, at 55 mph: 45%
34-35, 39	---	Not Used	---
36	YL/LG	Spark Output Signal	6.93v (50%)
37	RD	Vehicle Power	12-14v
38	BL/YL	EPC Solenoid Control	9.5v (5 psi)
40	BK/LG	Power Ground	<0.1v
41	TN/WT	Overdrive Cancel Switch	Switch Closed: 0.1v
42	OR/BK	TOT Sensor Signal	2.10-2.40v
43-44, 50, 54	---	Not Used	---
44	---	Not Used	---
45	DB/LG	MAP Sensor Signal	107 Hz
46	GY/RD	Analog Signal Return	<0.050v
47	GY/WT	TP Sensor Signal	0.7v, 55 mph: 2.1v
48	WT/RD	Self-Test Indicator Signal	STI On: 1v, Off: 5v
49	OR	HO2S-11 (B1 S1) Ground	<0.050v
51	WT/RD	Air Management 1 Solenoid	On: 1v, Off: 12v
52	OR	Shift Solenoid 1 Control	1v, 55 mph: 12v
53	PK/YL	TCC Solenoid Control	0%, at 55 mph: 95%
55	BR/OR	Coast Clutch Solenoid	12v, 55 mph: 12v
56	DB	PIP Sensor Signal	6.93v (50%)
57	RD	Vehicle Power	12-14v
58	TN/OR	Injector Bank 1 (INJ 1, 4, 5, 8)	5.0-5.8 ms
59	TN/RD	Injector Bank 2 (INJ 2, 3, 6, 7)	5.0-5.8 ms
60	BK/LG	Power Ground	<0.1v

Pin Connector Graphic

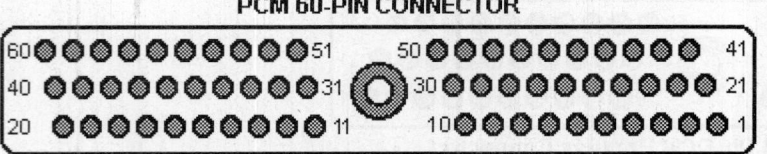

Terminal View of 60-Pin PCM Harness Connector

1992-94 E-Series 5.8L V8 VIN H (A/T) 60 Pin Connector

PCM Pin #	Wire Color	Circuit Description (60 Pin)	Value at Hot Idle
1	YL	Keep Alive Power	12-14v
2	LG	Brake Switch Signal	Brake Off: 12v, On: 1v
3	GY/BK	VSS (+) Signal	0 Hz, at 55 mph: 125 Hz
4	TN/YL	Ignition Diagnostic Monitor	20-31 Hz
4	WT/PK	Ignition Diagnostic Monitor	20-31 Hz
5	---	Not Used	---
6	PK/OR	VSS (-) Signal	0 Hz, at 55 mph: 125 Hz
7	LG/RD	Engine Coolant Temperature Sensor	0.5-0.6v
8	DG/YL	Fuel Pump Monitor	On: 12v, Off: 0v
9	PK/BL	Data Bus (-)	Digital Signals
10	BK/YL	A/C Pressure Switch Signal	A/C On: 12v, off: 0v
11	BR	Air Management 2 Solenoid	On: 1v, Off: 12v
12-15	---	Not Used	---
16	OR/RD	Ignition System Ground	<0.050v
17	PK/LG	Self-Test Output & MIL	MIL On: 1v, Off: 12v
18	---	Not Used	---
19	PK/OR	Shift Solenoid 2 Control	12v, 55 mph: 1v
20	BK	PCM Case Ground	<0.050v
21	WT/BL	IAC Solenoid Control	10.7v (33%)
22	BL/OR	Fuel Pump Control	On: 1v, Off: 12v
23	YL/RD	Knock Sensor 1 Signal	0v
24	---	Not Used	---
25	GY	ACT Sensor Signal	1.5-2.5v
26	BR/WT	Reference Voltage	4.9-5.1v
27	BR/LG	EVP Sensor Signal	0.4v, 55 mph: 3.1v
28	TN/OR	Data Bus (+)	Digital Signals
29	GY/BL	HO2S-11 (B1 S1) Signal	0.1-1.1v
30	BL/YL	MLP Sensor Signal	In 'P': 0v, in O/D: 5v
31	GY/YL	Canister Purge Solenoid	12v, 55 mph: 1v
32	WT/LG	OCIL (lamp) Control	Lamp On: 1v, Off: 12v
33	BR/PK	VR Solenoid Control	0%, at 55 mph: 45%
34-35	---	Not Used	---
36	PK	Spark Output Signal	6.93v (50%)
37	RD	Vehicle Power	12-14v
38	WT/YL	EPC Solenoid Control	9.5v (5 psi)
39	---	Not Used	---
40	BK/WT	Power Ground	<0.1v
41 ('92-'93)	TN/WT	Overdrive Cancel Switch	OCS On: 12v, Off: 0v
41 ('94)	TN/WT	TCS (switch) Signal	TCS & O/D On: 12v
42	OR/BK	TOT Sensor Signal	2.10-2.40v
43 ('94)	PK	A/C Demand Switch	A/C On: 12v, Off: 0v
44	---	Not Used	---
45	LG/BK	MAP Sensor Signal	107 Hz (sea level)
46	GY/RD	Analog Signal Return	<0.050v
47	GY/WT	TP Sensor Signal	0.7v, 55 mph: 2.1v
48	WT/PK	Self-Test Indicator Signal	STI On: 1v, Off: 5v
49	OR	HO2S-11 (B1 S1) Ground	<0.050v
50	---	Not Used	---
51	WT/OR	Air Management 1 Solenoid	AM1 On: 1v, Off: 12v
52	OR/YL	Shift Solenoid 1 Control	1v, 55 mph: 12v
53	PK/YL	TCC Solenoid Control	0%, at 55 mph: 95%
54	---	Not Used	---
55	BR/OR	Coast Clutch Solenoid	12v, 55 mph: 12v
56	GY/OR	PIP Sensor Signal	6.93v (50%)
57	RD	Vehicle Power	12-14v
58	TN	Injector Bank 1 (INJ 1, 3 & 5)	3.5-4.0 ms
59	WT	Injector Bank 2 (INJ 2, 4 & 6)	3.5-4.0 ms
60	BK/WT	Power Ground	<0.1v

1995 E-Series 5.8L V8 VIN H (A/T) California 60 Pin Connector

PCM Pin #	Wire Color	Circuit Description (60 Pin)	Value at Hot Idle
1	Y	Keep Alive Power	12-14v
2	LG	Brake Switch Signal	Brake Off: 12v, On: 1v
3	GY/BK	VSS (+) Signal	0 Hz, at 55 mph: 125 Hz
4	WT/PK	Ignition Diagnostic Monitor	20-31 Hz
5	---	Not Used	
6	PK/OR	VSS (-) Signal	0 Hz, at 55 mph: 125 Hz
7	LG/RD	Engine Coolant Temperature Sensor	0.5-0.6v
8	DG/YL	Fuel Pump Monitor	On: 12v, Off: 0v
9	TN/BL	MAF Sensor Return	<0.050v
10	BK/YL	A/C Pressure Switch Signal	A/C On: 12v, off: 0v
11	GY/YL	Canister Purge Solenoid	12v, 55 mph: 1v
12	LG/OR	Injector 6 Control	5.5-6.1 ms
13	TN/RD	Injector 7 Control	5.5-6.1 ms
14	BL	Injector 8 Control	5.5-6.1 ms
15	TN/BK	Injector 5 Control	5.5-6.1 ms
16	OR/RD	Ignition System Ground	<0.050v
17	PK/LG	Self-Test Output & MIL	MIL On: 1v, Off: 12v
18	TN/OR	Data Bus (+)	Digital Signals
19	PK/BL	Data Bus (-)	Digital Signals
20	BK	PCM Case Ground	<0.050v
21	WT/BL	IAC Solenoid Control	10.7v (33%)
22	BL/OR	Fuel Pump Control	On: 1v, Off: 12v
23	---	Not Used	---
24	---	Not Used	---
25	GY	Intake Air Temperature Sensor	1.5-2.5v
26	BR/WT	Reference Voltage	4.9-5.1v
27	BR/LG	EVP Sensor Signal	0.4v, 55 mph: 3.1v
28	---	Not Used	---
29	---	Not Used	---
30	BL/YL	MLP Sensor Signal	In 'P': 0v, in O/D: 1.7v
31	---	Not Used	---
32	WT/LG	TCIL (lamp) Control	Lamp On: 1v, Off: 12v
33	BR/PK	VR Solenoid Control	0%, at 55 mph: 45%
34	BR	Air Diverter Solenoid Control	On: 1v, Off: 12v
35	BR/BL	Injector 4 Control	5.5-6.1 ms
36	PK	Spark Output Signal	6.93v (50%)
37	R	Vehicle Power	12-14v
38	WT/YL	EPC Solenoid Control	9.5v (5 psi)
39	BR/YL	Injector 3 Control	5.5-6.1 ms
40	BK/WT	Power Ground	<0.1v

1995 E-Series 5.8L V8 VIN H (A/T) California 60 Pin Connector

PCM Pin #	Wire Color	Circuit Description (60 Pin)	Value at Hot Idle
41	TN/WT	TCS (switch) Signal	TCS & O/D On: 12v
42	---	Not Used	---
43	RD/BK	HO2S-21 (B2 S1) Signal	0.1-1.1v
44	GY/BL	HO2S-11 (B1 S1) Signal	0.1-1.1v
45	---	Not Used	---
46	GY/RD	Analog Signal Return	<0.050v
47	GY/WT	TP Sensor Signal	0.7v, 55 mph: 2.1v
48	WT/PK	Self-Test Indicator Signal	STI On: 1v, Off: 5v
49	OR/BK	TOT Sensor Signal	2.10-2.40v
50	BL/RD	MAF Sensor Signal	0.8v, 55 mph: 1.6v
51	OR/YL	Shift Solenoid 1 Control	12v, 55 mph: 1v
52	PK/OR	Shift Solenoid 2 Control	1v, 55 mph: 12v
53	PK/YL	TCC Solenoid Control	0%, at 55 mph: 95%
54	---	Not Used	---
55	BR/OR	Coasting Clutch Solenoid	12v, 55 mph: 12v
56	GY/OR	PIP Sensor Signal	6.93v (50%)
57	RD	Vehicle Power	12-14v
58	TN	Injector 1 Control	5.5-6.1 ms
59	WT	Injector 2 Control	5.5-6.1 ms
60	BK/WT	Power Ground	<0.1v

Pin Connector Graphic

PCM 60-PIN CONNECTOR

Terminal View of 60-Pin PCM Harness Connector

Standard Colors and Abbreviations

Abbreviation	Color	Abbreviation	Color	Abbreviation	Color
BK	Black	GY	Gray	PK	Purple
BL	Blue	GN	Green	RD	Red
BR	Brown	LG	LT Green	TN	Tan
DB	Dark Blue	OR	Orange	WT	White
DG	DK Green	PK	Pink	YL	Yellow

1995 E-Series 5.8L V8 VIN H (E4OD) 60 Pin Connector

PCM Pin #	Wire Color	Circuit Description (60 Pin)	Value at Hot Idle
1	YL	Keep Alive Power	12-14v
2	LG	Brake Switch Signal	Brake On: 12v: Off: 0v
3	GY/BK	VSS (+) Signal	0 Hz, at 55 mph: 125 Hz
4	WT/PK	Ignition Diagnostic Monitor	20-31 Hz
5	---	Not Used	---
6	PK/OR	VSS (-) Signal	0 Hz, at 55 mph: 125 Hz
7	LG/RD	Engine Coolant Temperature Sensor	0.5-0.6v
8	DG/YL	Fuel Pump Monitor	On: 12v, Off: 0v
9	PK/BL	Data Bus (-)	Digital Signals
10	BK/YL	A/C Pressure Switch Signal	A/C On: 12v, off: 0v
11	BR	Air Management 2 Solenoid	On: 1v, Off: 12v
12-15	---	Not Used	---
16	OR/RD	Ignition System Ground	<0.050v
17	PK/LG	Self-Test Output & MIL	MIL On: 1v, Off: 12v
18	---	Not Used	---
19	PK/OR	Shift Solenoid 2 Control	12v, 55 mph: 1v
20	BK	PCM Case Ground	<0.050v
21	WT/BL	IAC Solenoid Control	10.7v (33%)
22	BL/OR	Fuel Pump Control	On: 1v, Off: 12v
23	YL/RD	Knock Sensor 1 Signal	0v
24	---	Not Used	---
25	GY	ACT Sensor Signal	1.5-2.5v
26	BR/WT	Reference Voltage	4.9-5.1v
27	BR/LG	EVP Sensor Signal	0.4v, 55 mph: 3.1v
28	TN/OR	Data Bus (+)	Digital Signals
29	GY/BL	HO2S-11 (B1 S1) Signal	0.1-1.1v
30	BL/YL	MLP Sensor Signal	In 'P': 0v, in O/D: 5v
31	GY/YL	Canister Purge Solenoid	12v, 55 mph: 1v
32	WT/LG	OCIL (lamp) Control	Lamp On: 1v, Off: 12v
33	BR/PK	VR Solenoid Control	0%, at 55 mph: 45%
34-35	---	Not Used	---
36	PK	Spark Output Signal	6.93v (50%)
37	RD	Vehicle Power	12-14v
38	WT/YL	EPC Solenoid Control	9.5v (5 psi)
39	---	Not Used	---
40	BK/WT	Power Ground	<0.1v

1995 E-Series 5.8L V8 VIN H (E4OD) 60 Pin Connector

PCM Pin #	Wire Color	Circuit Description (60 Pin)	Value at Hot Idle
41	TN/WT	TCS (switch) Signal	TCS & O/D On: 12v
42	OR/BK	TOT Sensor Signal	2.10-2.40v
43	PK	A/C Demand Switch	A/C On: 12v, Off: 0v
44	---	Not Used	---
45	LG/BK	MAP Sensor Signal	107 Hz (sea level)
46	GY/RD	Analog Signal Return	<0.050v
47	GY/WT	TP Sensor Signal	0.7v, 55 mph: 2.1v
48	WT/PK	Self-Test Indicator Signal	STI On: 1v, Off: 5v
49	OR	HO2S-11 (B1 S1) Ground	<0.050v
50	---	Not Used	---
51	WT/OR	Air Management 1 Solenoid	On: 1v, Off: 12v
52	OR/YL	Shift Solenoid 1 Control	1v, 55 mph: 12v
53	PK/YL	TCC Solenoid Control	0%, at 55 mph: 95%
54	---	Not Used	---
55	BR/OR	Coast Clutch Solenoid	12v, 55 mph: 12v
56	GY/OR	PIP Sensor Signal	6.93v (50%)
57	RD	Vehicle Power	12-14v
58	TN	Injector Bank 1 (INJ 1, 4, 5, 8)	3.5-5.0 ms
59	WT	Injector Bank 2 (INJ 2, 4, 6, 7)	3.5-5.0 ms
60	BK/WT	Power Ground	<0.1v

Pin Connector Graphic

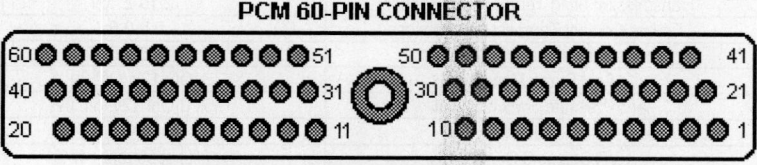

PCM 60-PIN CONNECTOR

Terminal View of 60-Pin PCM Harness Connector

Standard Colors and Abbreviations

Abbreviation	Color	Abbreviation	Color	Abbreviation	Color
BK	Black	GY	Gray	PK	Purple
BL	Blue	GN	Green	RD	Red
BR	Brown	LG	LT Green	TN	Tan
DB	Dark Blue	OR	Orange	WT	White
DG	DK Green	PK	Pink	YL	Yellow

1996 E-Series 5.8L V8 VIN H (E4OD) 104 Pin Connector

PCM Pin #	Wire Color	Circuit Description (104 Pin)	Value at Hot Idle
1	PK/OR	Shift Solenoid 2 Control	12v, 55 mph: 12v
2	PK/LG	MIL (lamp) Control	MIL On: 1v, Off: 12v
3	---	Not Used	---
4	LG/RD	Power Take-Off Signal	PTO On: 12v, Off: 0v
5-12	---	Not Used	---
13	PK	Flash EPROM Power	0.1v
14	---	Not Used	---
15	PK/LB	Data Bus (-)	Digital Signals
16	TN/OR	Data Bus (+)	Digital Signals
17-22	---	Not Used	---
23	OR/RD	Ignition System Ground	<0.050v
24	BK/WT	Power Ground	<0.1v
25	BK	Case Ground	<0.050v
26	---	Not Used	---
27	OR/YL	Shift Solenoid 1 Control	1v, 55 mph: 12v
28		Not Used	
29	TN/WT	TCS (switch) Signal	TCS & O/D On: 12v
30-32	---	Not Used	---
33	PK/OR	PSOM (-) Signal	<0.050v
34	---	Not Used	---
35	RD/LG	HO2S-12 (B1 S2) Signal	0.1-1.1v
36	TN/LB	MAF Sensor Return	<0.050v
37	OR/BK	Transmission Fluid Temperature Sensor	2.10-2.40v
38	LG/RD	Engine Coolant Temperature Sensor	0.5-0.6v
39	GY	Intake Air Temperature Sensor	1.5-2.5v
40	DG/YL	Fuel Pump Monitor	On: 12v, Off: 0v
41	BK/YL	A/C Pressure Switch Signal	A/C On: 12v, off: 0v
42-46	---	Not Used	---
47	BK/PK	VR Solenoid Control	0%, at 55 mph: 45%
48	WT/PK	Clean Tachometer Output	38-42 Hz
49	GY/OR	PIP Sensor Signal	6.93v (50%)
50	PK	Spark Output Signal	6.93v (50%)
51	BK/WT	Power Ground	<0.1v
52	---	Not Used	---
53	BR/OR	Coast Clutch Solenoid	12v, 55 mph: 12v
54	DB/WT	TCC Solenoid Control	0%, at 55 mph: 95%
55	YL	Keep Alive Power	12-14v
56	LG/BK	EVAP VMV Solenoid	0-10 Hz (0-100%)
57	---	Not Used	---
58	GY/BK	PSOM (+) Signal	0 Hz, at 55 mph: 125 Hz
59	DB	Misfire Detection Sensor	45-55 Hz
60	GY/LB	HO2S-11 (B1 S1) Signal	0.1-1.1v
61-63	---	Not Used	---
64	LB/YL	Digital Transmission Range Sensor	In 'P': 0v, in O/D: 1.7v

1996 E-Series 5.8L V8 VIN H (E4OD) 104 Pin Connector

PCM Pin #	Wire Color	Circuit Description (104 Pin)	Value at Hot Idle
65	BR/LG	EGR DPFE Sensor Signal	1v, 55 mph: 1v
66-70	---	Not Used	---
71	RD	Vehicle Power	12-14v
72	TN/RD	Injector 7 Control	4.2-4.6 ms
73	TN/BK	Injector 5 Control	4.2-4.6 ms
74	BR/YL	Injector 3 Control	4.2-4.6 ms
75	TN	Injector 1 Control	4.2-4.6 ms
76	BK/WT	Power Ground	<0.1v
77	BK/WT	Power Ground	<0.1v
78	---	Not Used	---
79	WT/LG	TCIL (lamp) Control	Lamp On: 1v, Off: 12v
80	LB/OR	Fuel Pump Control	On: 1v, Off: 12v
81	WT/YL	EPC Solenoid Control	10v (26 psi)
82	---	Not Used	
83	WT/LB	IAC Solenoid Control	10.7v (33%)
84-86	---	Not Used	---
87	RD/BK	HO2S-21 (B2 S1) Signal	0.1-1.1v
88	LB/RD	MAF Sensor Signal	0.8v, 55 mph: 2.3v
89	GY/WT	TP Sensor Signal	0.9v, 55 mph: 1.2v
90	BR/WT	Reference Voltage	4.9-5.1v
91	GY/RD	Analog Signal Return	<0.050v
92	LG	Brake Position Switch	Brake Off: 12v, On: 1v
93	RD/WT	HO2S-11 (B1 S1) Heater	1v, Off: 12v
94	YL/LB	HO2S-21 (B2 S1) Heater	1v, Off: 12v
95	WT/BK	HO2S-12 (B1 S2) Heater	1v, Off: 12v
96	---	Not Used	---
97	R	Vehicle Power	12-14v
98	LG	Injector 8 Control	4.2-4.6 ms
99	LG/OR	Injector 6 Control	4.2-4.6 ms
100	BR/LB	Injector 4 Control	4.2-4.6 ms
101	W	Injector 2 Control	4.2-4.6 ms
102	---	Not Used	---
103	BK/WT	Power Ground	<0.1v
104	---	Not Used	---

Pin Connector Graphic

PCM 104-PIN CONNECTOR

Terminal View of 104-Pin PCM Wiring Harness Connector

Standard Colors and Abbreviations

Abbreviation	Color	Abbreviation	Color	Abbreviation	Color
BK	Black	GY	Gray	PK	Purple
BL	Blue	GN	Green	RD	Red
BR	Brown	LG	LT Green	TN	Tan
DB	Dark Blue	OR	Orange	WT	White
DG	DK Green	PK	Pink	YL	Yellow

1997 E-Series 6.8L V10 VIN S (E4OD) 104 Pin Connector

PCM Pin #	Wire Color	Circuit Description (104 Pin)	Value at Hot Idle
1	PK/LB	COP 6 Driver (Dwell)	5°, at 55 mph: 8°
2	PK/LG	MIL (lamp) Control	MIL On: 1v, Off: 12v
3	BK/WT	Power Ground	<0.1v
4-5	---	Not Used	---
6	OR/YL	Shift Solenoid 1 Control	1v, 55 mph: 12v
7-8	---	Not Used	---
9	YL/WT	Fuel Level Indicator Signal	1.7v (1/2 full)
10	---	Not Used	---
11	PK/OR	Shift Solenoid 2 Control	12v, 55 mph: 12v
12	---	Not Used	---
13	VT	Flash EEPROM	0.1v
14	---	Not Used	---
15	PK/LB	Data Bus (-)	Digital Signals
16	TN/OR	Data Bus (+)	Digital Signals
17-19	---	Not Used	---
20	BR/OR	Coast Clutch Solenoid	12v, 55 mph: 12v
21	DB	CKP (+) Sensor	400-420 Hz
22	GY	CKP (-) Sensor	400-420 Hz
23-24	BK/WT	Power Ground	<0.1v
25	BK/LB	Case Ground	<0.050v
26	DB/YL	COP 1 Driver (Dwell)	5°, at 55 mph: 8°
27	PKWT	COP 10 Driver (Dwell)	5°, at 55 mph: 8°
28	WT/LG	TCIL (lamp) Control	Lamp On: 1v, Off: 12v
29	TN/WT	TCS (switch) Signal	TCS & O/D On: 12v
30	DG	Octane Adjustment	9.3v (switch shorted: 0v)
31-32	---	Not Used	---
33	PK/OR	VSS (-) Signal	0 Hz, at 55 mph: 125 Hz
34	YL/BK	Digital TR1 Sensor	0v, at 55 mph: 11.5v
35	RD/LG	HO2S-12 (B1 S2) Signal	0.1-1.1v
36	TN/LB	MAF Sensor Return	<0.1v
37	OR/BK	Transmission Fluid Temperature Sensor	2.10-2.40v
38	---	Not Used	---
39	GY	Intake Air Temperature Sensor	1.5-2.5v
40	DG/YL	Fuel Pump Monitor	On: 12v, Off: 0v
41	BK/YL	A/C High Pressure Switch	A/C On: 12v, off: 0v
42	GY/RD	Injector 10 Control	4.2-4.6 ms
43-44	---	Not Used	---
45	YL/LB	CHIL (lamp) Control	CHT Off: 12v, On: 1v
46	---	Not Used	---
47	BR/PK	VR Solenoid Control	0%, at 55 mph: 45%
48	PK/BL	Clean Tachometer Output	60-70 Hz
49	LB/BK	Digital TR2 Sensor	0v, at 55 mph: 11.5v
50	LG/RD	Digital TR4 Sensor	0v, at 55 mph: 11.5v
51	BK/WT	Power Ground	<0.1v
52	PKWT	COP 5 Driver (Dwell)	5°, at 55 mph: 8°
53	WT/PK	COP 7 Driver (Dwell)	5°, at 55 mph: 8°
54	PK/YL	TCC Solenoid Control	0%, at 55 mph: 95%
55	Y	Keep Alive Power	12-14v
56	LG/BK	EVAP VMV Solenoid	0-10 Hz (0-100%)
57	YL/RD	Knock Sensor 1 Signal	0v
58	GY/BK	VSS (+) Signal	0 Hz, at 55 mph: 125 Hz
59	---	Not Used	
60	GY/LB	HO2S-11 (B1 S1) Signal	0.1-1.1v
61	---	Not Used	---
62	RD/PK	Fuel Tank Pressure Sensor	2.6v (0" H2O - cap off)
63	---	Not Used	---
64	LB/YL	Digital Transmission Range Sensor	In 'P': 0v, in O/D: 1.7v

1997 E-Series 6.8L V10 VIN S (E4OD) 104 Pin Connector

PCM Pin #	Wire Color	Circuit Description (104 Pin)	Value at Hot Idle
65	BR/LG	EGR DPFE Sensor Signal	0.20-1.30v
66	YL/LG	Cylinder Head Temperature Sensor	0.6v (194°F)
67	PKWT	EVAP Purge Solenoid	0-10 Hz (0-100%)
68	GY/BK	Injector 9 Control	4.2-4.6 ms
69-70	---	Not Used	---
71	RD	Vehicle Power (Start-Run)	12-14v
72	TN/RD	Injector 7 Control	4.2-4.6 ms
73	TN/BK	Injector 5 Control	4.2-4.6 ms
74	BR/YL	Injector 3 Control	4.2-4.6 ms
75	TN	Injector 1 Control	4.2-4.6 ms
76	BK/WT	Power Ground	<0.1v
77	BK/WT	Power Ground	<0.1v
78	DB/LG	COP 2 Driver (Dwell)	5°, at 55 mph: 8°
79	WT/RD	COP 8 Driver (Dwell)	5°, at 55 mph: 8°
80	LB/OR	Fuel Pump Control	On: 1v, Off: 12v
81	WT/YL	EPC Solenoid Control	9.2v (5 psi)
82	DG/PK	COP 9 Driver (Dwell)	5°, at 55 mph: 8°
83	WT/LB	IAC Solenoid Control	10.7v (33%)
84	---	Not Used	---
85	DG	Camshaft Position Sensor	9 Hz, 55 mph: 16 Hz
86	---	Not Used	---
87	RD/BK	HO2S-21 (B2 S1) Signal	0.1-1.1v
88	LB/RD	MAF Sensor Signal	0.8v, 55 mph: 1.6v
89	GY/WT	TP Sensor Signal	0.9v, 55 mph: 1.3v
90	BR/WT	Reference Voltage	4.9-5.1v
91	GY/RD	Analog Signal Return	<0.050v
92	LG	Brake Position Switch	Brake Off: 12v, On: 1v
93	RD/WT	HO2S-11 (B1 S1) Heater	1v, Off: 12v
94	YL/LB	HO2S-21 (B2 S1) Heater	1v, Off: 12v
95	WT/BK	HO2S-12 (B1 S2) Heater	1v, Off: 12v
96	---	Not Used	---
97	RD	Vehicle Power (Start-Run)	12-14v
98	LB	Injector 8 Control	4.2-4.6 ms
99	LG/OR	Injector 6 Control	4.2-4.6 ms
100	BR/LB	Injector 4 Control	4.2-4.6 ms
101	W	Injector 2 Control	4.2-4.6 ms
102	OR/YL	COP 4 Driver (Dwell)	5°, at 55 mph: 8°
103	BK/WT	Power Ground	<0.1v
104	LG/WT	COP 3 Driver (Dwell)	5°, at 55 mph: 8°

Pin Connector Graphic

PCM 104-PIN CONNECTOR

Terminal View of 104-Pin PCM Wiring Harness Connector

Standard Colors and Abbreviations

Abbreviation	Color	Abbreviation	Color	Abbreviation	Color
BK	Black	GY	Gray	PK	Purple
BL	Blue	GN	Green	RD	Red
BR	Brown	LG	LT Green	TN	Tan
DB	Dark Blue	OR	Orange	WT	White
DG	DK Green	PK	Pink	YL	Yellow

1998-99 E-Series 6.8L V10 VIN S (E4OD) 104 Pin Connector

PCM Pin #	Wire Color	Circuit Description (104 Pin)	Value at Hot Idle
1	PK/LB	COP 6 Driver (Dwell)	5°, at 55 mph: 8°
2	PK/LG	MIL (lamp) Control	MIL On: 1v, Off: 12v
3	BK/WT	Power Ground	<0.1v
4	PK/LB	Power Takeoff Signal	PTO Off: 0v, On: 12v
5	---	Not Used	---
6	OR/YL	Shift Solenoid 1 Control	1v, 55 mph: 12v
7-8	---	Not Used	---
9	YL/WT	Fuel Level Indicator Signal	1.7v (1/2 full)
10	---	Not Used	---
11	PK/OR	Shift Solenoid 2 Control	12v, 55 mph: 12v
12	---	Not Used	---
13	VT	Flash EEPROM	0.1v
14	---	Not Used	---
15	PK/LB	Data Bus (-)	Digital Signals
16	TN/OR	Data Bus (+)	Digital Signals
17-19	---	Not Used	---
20	BR/OR	Coast Clutch Solenoid	12v, 55 mph: 12v
21	DB	CKP (+) Sensor	500-525 Hz
22	GY	CKP (-) Sensor	500-525 Hz
23-24	BK/WT	Power Ground	<0.1v
25	BK/LB	Case Ground	<0.050v
26	DB/YL	COP 1 Driver (Dwell)	5°, at 55 mph: 8°
27	PKWT	COP 10 Driver (Dwell)	5°, at 55 mph: 8°
28	WT/LG	TCIL (lamp) Control	Lamp On: 1v, Off: 12v
29	TN/WT	TCS (switch) Signal	TCS & O/D On: 12v
30	DG	Octane Adjustment	9.3v (switch shorted: 0v)
31-32	---	Not Used	---
33	PK/OR	VSS (-) Signal	0 Hz, at 55 mph: 125 Hz
34	YL/BK	Digital TR1 Sensor	0v, at 55 mph: 11.5v
35	RD/LG	HO2S-12 (B1 S2) Signal	0.1-1.1v
36	TN/LB	MAF Sensor Return	<0.050v
37	OR/BK	Transmission Fluid Temperature Sensor	2.10-2.40v
38	---	Not Used	---
39	GY	Intake Air Temperature Sensor	1.5-2.5v
40	DG/YL	Fuel Pump Monitor	On: 12v, Off: 0v
41	BK/YL	A/C Pressure Switch Signal	A/C On: 12v, off: 0v
42	GY/RD	Injector 10 Control	3.8-4.6 ms
43-44	---	Not Used	---
45	YL/LB	CHIL (lamp) Control	CHT Off: 12v, On: 1v
46	---	Not Used	
47	BR/PK	VR Solenoid Control	0%, at 55 mph: 45%
48	PK/BL	Clean Tachometer Output	65 Hz, 55 mph: 175 Hz
49	LB/BK	Digital TR2 Sensor	0v, at 55 mph: 11.5v
50	LG/RD	Digital TR4 Sensor	0v, at 55 mph: 11.5v
51	BK/WT	Power Ground	<0.1v
52	PKWT	COP 5 Driver (Dwell)	5°, at 55 mph: 8°
53	WT/PK	COP 7 Driver (Dwell)	5°, at 55 mph: 8°
54	PK/YL	TCC Solenoid Control	0%, at 55 mph: 95%
55	YL	Keep Alive Power	12-14v
56	LG/BK	EVAP Purge Solenoid	0-10 Hz (0-100%)
57	YL/RD	Knock Sensor 1 Signal	0v
58	GY/BK	VSS (+) Signal	0 Hz, at 55 mph: 125 Hz
59	---	Not Used	---
60	GY/LB	HO2S-11 (B1 S1) Signal	0.1-1.1v
61-63	---	Not Used	---
62	RD/PK	Fuel Tank Pressure Sensor	2.6v (0" H2O - cap off)
64	LB/YL	Digital Transmission Range Sensor	In 'P': 0v, in O/D: 1.7v

1998-99 E-Series 6.8L V10 VIN S (E4OD) 104 Pin Connector

PCM Pin #	Wire Color	Circuit Description (104 Pin)	Value at Hot Idle
65	BR/LG	EGR DPFE Sensor Signal	0.20-1.30v
66	YL/LG	Cylinder Head Temperature Sensor	0.6v (194°F)
67	PKWT	EVAP Purge Solenoid	0-10 Hz (0-100%)
68	GY/BK	Injector 9 Control	3.8-4.6 ms
69-70	---	Not Used	---
71	RD	Vehicle Power	12-14v
72	TN/RD	Injector 7 Control	3.8-4.6 ms
73	TN/BK	Injector 5 Control	3.8-4.6 ms
74	BR/YL	Injector 3 Control	3.8-4.6 ms
75	TN	Injector 1 Control	3.8-4.6 ms
76	BK/WT	Power Ground	<0.1v
77	BK/WT	Power Ground	<0.1v
78	DB/LG	COP 2 Driver (Dwell)	5°, at 55 mph: 8°
79	WT/RD	COP 8 Driver (Dwell)	5°, at 55 mph: 8°
80	LB/OR	Fuel Pump Control	On: 1v, Off: 12v
81	WT/YL	EPC Solenoid Control	9.2v (5 psi)
82	DG/PK	COP 9 Driver (Dwell)	5°, at 55 mph: 8°
83	WT/LB	IAC Solenoid Control	10.7v (33%)
84	---	Not Used	---
85	DG	Camshaft Position Sensor	9 Hz, 55 mph: 16 Hz
86	---	Not Used	---
87	RD/BK	HO2S-21 (B2 S1) Signal	0.1-1.1v
88	LB/RD	MAF Sensor Signal	0.8v, 55 mph: 1.6v
89	GY/WT	TP Sensor Signal	0.9v, 55 mph: 1.3v
90	BR/WT	Reference Voltage	4.9-5.1v
91	GY/RD	Analog Signal Return	<0.050v
92	LG	Brake Position Switch	Brake Off: 12v, On: 1v
93	RD/WT	HO2S-11 (B1 S1) Heater	1v, Off: 12v
94	YL/LB	HO2S-21 (B2 S1) Heater	1v, Off: 12v
95	WT/BK	HO2S-12 (B1 S2) Heater	1v, Off: 12v
96	---	Not Used	---
97	RD	Vehicle Power	12-14v
98	LB	Injector 8 Control	3.8-4.6 ms
99	LG/OR	Injector 6 Control	3.8-4.6 ms
100	BR/LB	Injector 4 Control	3.8-4.6 ms
101	WT	Injector 2 Control	3.8-4.6 ms
102	OR/YL	COP 4 Driver (Dwell)	5°, at 55 mph: 8°
103	BK/WT	Power Ground	<0.1v
104	LG/WT	COP 3 Driver (Dwell)	5°, at 55 mph: 8°

Pin Connector Graphic

Standard Colors and Abbreviations

Abbreviation	Color	Abbreviation	Color	Abbreviation	Color
BK	Black	GY	Gray	PK	Purple
BL	Blue	GN	Green	RD	Red
BR	Brown	LG	LT Green	TN	Tan
DB	Dark Blue	OR	Orange	WT	White
DG	DK Green	PK	Pink	YL	Yellow

2000-2002 E-Series 6.8L V10 VIN S (A/T) 104 Pin Connector

PCM Pin #	Wire Color	Circuit Description (104 Pin)	Value at Hot Idle
1	OR/YL	COP 6 Driver (Dwell)	5°, at 55 mph: 8°
2	PK/LG	MIL (lamp) Control	MIL On: 1v, Off: 12v
3	BK/WT	Power Ground	<0.1v
4	PK/LB	Power Takeoff Signal	PTO Off: 0v, On: 12v
5	---	Not Used	---
6	OR/YL	Shift Solenoid 'A' Control	1v, 55 mph: 12v
7-10	---	Not Used	---
9	YL/WT	Fuel Pump Sender	TBD
11	PK/OR	Shift Solenoid 'B' Control	12v, 55 mph: 12v
12	WT/LG	Overdrive Cancel Switch	TBD
13	VT	Flash EEPROM	0.1v
14	OR/LB	4x4 Low Switch	Switch On: 0v, Off: 12v
15	PK/LB	Data Bus (-)	Digital Signals
16	TN/OR	Data Bus (+)	Digital Signals
17-19	---	Not Used	---
20	BR/OR	Coast Clutch Solenoid	12v, 55 mph: 12v
21	DB	CKP (+) Sensor	500-525 Hz
22	GY	CKP (-) Sensor	500-525 Hz
23-24	BK/WT	Power Ground	<0.1v
25	BK/LB	Case Ground	<0.050v
26	LG/WT	COP 1 Driver (Dwell)	5°, at 55 mph: 8°
27	WT/RD	COP 10 Driver (Dwell)	5°, at 55 mph: 8°
28	---	Not Used	---
29	TN/WT	TCS (switch) Signal	TCS & O/D On: 12v
30-31	---	Not Used	---
32	DG/PK	Knock Sensor 1 Return	<0.050v
33	---	Not Used	---
34	YL/BK	Digital TR1 Sensor	0v, at 55 mph: 11.5v
35	RD/LG	HO2S-12 (B1 S2) Signal	0.1-1.1v
36	TN/LB	MAF Sensor Return	<0.050v
37	OR/BK	Transmission Fluid Temperature Sensor	2.10-2.40v
38	---	Not Used	---
39	GY	Intake Air Temperature Sensor	1.5-2.5v
40	DG/YL	Fuel Pump Monitor	On: 12v, Off: 0v
41	BK/YL	A/C Pressure Switch Signal	A/C On: 12v, Off: 0v
42	GY/RD	Injector 10 Control	3.8-4.6 ms
43-44	---	Not Used	---
45	YL/LB	Engine Coolant Temperature Sensor	0.5-0.6v
46	GY/BK	Vehicle Speed Signal	0 Hz, at 55 mph: 125 Hz
47	BR/PK	VR Solenoid Control	0%, at 55 mph: 45%
48	WT/PK	Clean Tachometer Output	65 Hz, 55 mph: 175 Hz
49	LB/BK	Digital TR2 Sensor	0v, at 55 mph: 11.5v
50	LG/RD	Digital TR4 Sensor	0v, at 55 mph: 11.5v
51	BK/WT	Power Ground	<0.1v
52	LG/YL	COP 5 Driver (Dwell)	5°, at 55 mph: 8°
53	PK/LB	COP 7 Driver (Dwell)	5°, at 55 mph: 8°
54	PK/YL	TCC Solenoid Control	0%, at 55 mph: 95%
55	Y	Keep Alive Power	12-14v
56	LG/BK	EVAP Purge Solenoid	0-10 Hz (0-100%)
57	YL/RD	Knock Sensor 1 Signal	0v
58-59	---	Not Used	---
60	GY/LB	HO2S-11 (B1 S1) Signal	0.1-1.1v
61	---	Not Used	---
62	RD/PK	Fuel Tank Pressure Sensor	2.6v (0" H2O - cap off)
63	---	Not Used	---
64	LB/YL	Digital TR3A Sensor Signal	In 'P': 0v, in O/D: 1.7v

2000-2002 E-Series 6.8L V10 VIN S (A/T) 104 Pin Connector

PCM Pin #	Wire Color	Circuit Description (104 Pin)	Value at Hot Idle
65 ('00)	BR/LG	EGR DPFE Sensor Signal	0.20-1.30v
65 ('01-'02)	BR/LG	EGR DPFE Sensor Signal	0.95-1.05v
66	YL/LG	Cylinder Head Temperature Sensor	0.6v (194°F)
67	PKWT	EVAP Purge Solenoid	0-10 Hz (0-100%)
68	GY/BK	Injector 9 Control	3.8-4.6 ms
71	RD	Vehicle Power (Start-Run)	12-14v
72	TN/RD	Injector 7 Control	3.8-4.6 ms
73	TN/BK	Injector 5 Control	3.8-4.6 ms
74	BR/YL	Injector 3 Control	3.8-4.6 ms
75	TN	Injector 1 Control	3.8-4.6 ms
76	BK/WT	Power Ground	<0.1v
77	BK/WT	Power Ground	<0.1v
78	PKWT	COP 2 Driver (Dwell)	5°, at 55 mph: 8°
79	WT/RD	COP 8 Driver (Dwell)	5°, at 55 mph: 8°
80	LB/OR	Fuel Pump Control	On: 1v, Off: 12v
81	WT/YL	EPC Solenoid Control	9.2v (5 psi)
82	YL/BK	COP 9 Driver (Dwell)	5°, at 55 mph: 8°
83	WT/LB	IAC Solenoid Control	10.7v (33%)
84	DB/YL	Output Shaft Speed Sensor	0 Hz, 55 mph: 2 KHz
85	DG	Camshaft Position Sensor	9 Hz, 55 mph: 16 Hz
87	RD/BK	HO2S-21 (B2 S1) Signal	0.1-1.1v
88	LB/RD	MAF Sensor Signal	0.8v, 55 mph: 1.6v
89	GY/WT	TP Sensor Signal	0.9v, 55 mph: 1.3v
90	BR/WT	Reference Voltage	4.9-5.1v
91	GY/RD	Analog Signal Return	<0.050v
92	LG	Brake Position Switch	Brake Off: 12v, On: 1v
93	RD/WT	HO2S-11 (B1 S1) Heater	1v, Off: 12v
94	YL/LB	HO2S-21 (B2 S1) Heater	1v, Off: 12v
95	WT/BK	HO2S-12 (B1 S2) Heater	1v, Off: 12v
96	---	Not Used	---
97	RD	Vehicle Power (Start-Run)	12-14v
98	LB	Injector 8 Control	3.8-4.6 ms
99	LG/OR	Injector 6 Control	3.8-4.6 ms
100	BR/LB	Injector 4 Control	3.8-4.6 ms
101	WT	Injector 2 Control	3.8-4.6 ms
102	DG/PK	COP 4 Driver (Dwell)	5°, at 55 mph: 8°
103	BK/WT	Power Ground	<0.1v
104	WT/PK	COP 3 Driver (Dwell)	5°, at 55 mph: 8°

Pin Connector Graphic

PCM 104-PIN CONNECTOR

1 [connector pins] 26
27 [connector pins] 52
53 [connector pins] 78
79 [connector pins] 104

Terminal View of 104-Pin PCM Wiring Harness Connector

Standard Colors and Abbreviations

Abbreviation	Color	Abbreviation	Color	Abbreviation	Color
BK	Black	GY	Gray	PK	Purple
BL	Blue	GN	Green	RD	Red
BR	Brown	LG	LT Green	TN	Tan
DB	Dark Blue	OR	Orange	WT	White
DG	DK Green	PK	Pink	YL	Yellow

2003 E-Series 6.8L V10 VIN S (All) 104 Pin Connector

PCM Pin #	Wire Color	Circuit Description (104 Pin)	Value at Hot Idle
1	OR/YL	COP 6 Driver (dwell)	5°, 55 mph: 9°
2	---	Not Used	---
3	BK/WT	Power Ground	<0.1v
4	LB/YL	Power Take-Off (if equipped)	0v (Off)
5	---	Not Used	---
6	OR/YL	Shift Solenoid 1 Control	1v, 55 mph: 12v
7-10	---	Not Used	---
11	PK/OR	Shift Solenoid 2 Control	12v, 55 mph: 12v
12	WT/LG	TCIL (lamp) Control	TCIL On: 1v, Off: 2.2v
13	VT	Flash EEPROM Power	0.1v
14	LB/BK	4x4 Low Indicator Switch	12v (switch closed: 0v)
15	PK/LB	SCP Data Bus (-) Signal	<0.050v
16	TN/OR	SCP Data Bus (+) Signal	Digital Signals
17-19	---	Not Used	---
20	BR/OR	Coast Clutch Solenoid Control	12v, 55 mph: 12v
21	DB	CKP (-) Sensor Signal	400-420 Hz
22	GY	CKP (+) Sensor Signal	400-420 Hz
23-24	BK/WT	Power Ground	<0.1v
25	LB/YL	Case Ground	<0.050v
26	LG/WT	COP 1 Driver (dwell)	5°, 55 mph: 9°
27	LG/YL	COP 10 Driver (dwell)	5°, 55 mph: 9°
28	---	Not Used	---
29	TN/WT	Overdrive Cancel Switch) Signal	Switch Off: 12v, On: 1v
30-31	---	Not Used	---
32	DG/VT	Knock Sensor 1 (-) Signal	<0.050v
33	PK/OR	Ground	<0.050v
34	YL/BK	Digital TR1 Sensor	0v, 55 mph: 11.5v
35	RD/LG	HO2S-12 (B1 S2) Signal	0.1-1.1v
36	TN/LB	MAF Sensor Return	<0.050v
37	OR/BK	TFT Sensor Signal	2.10-2.40v
38	---	Not Used	---
39	GY	IAT Sensor Signal	1.5-2.5v
40	DG/YL	Fuel Pump Monitor	Off: 12v, On: 1v
41	TN/LG	A/C Switch Signal	A/C On: 12v, Off: 0v
42	GY/RD	Injector 10 Control	4.2-4.6 ms
43	OR/LG	Data Output Link	Digital Signals
44-46	---	Not Used	---
47	BR/PK	VR Solenoid Control	0%, 55 mph: 45%
48	LG/WT	Clean Tachometer Output	60-70 Hz
49	LB/BK	Digital TR2 Sensor	0v, 55 mph: 11.5v
50	WT/BK	Digital TR4 Sensor	0v, 55 mph: 11.5v
51	BK/WT	Power Ground	<0.1v
52	WT/PK	COP 5 Driver (dwell)	5°, 55 mph: 9°
53	DGVT	COP 7 Driver (dwell)	5°, 55 mph: 9°
54	VT/YL	TCC Solenoid Control	0%, 55 mph: 95%
55	RD/WT	Keep Alive Power	12-14v
56	LG/BK	EVAP Purge Solenoid	0-10 Hz (0-100%)
57	YL/RD	Knock Sensor 1 (+) Signal	0v
58	GY/BK	VSS (+) Signal	0 Hz, 55 mph: 125 Hz
59	DG/WT	TSS Sensor Signal	300 Hz, 55 mph: 980
60	GY/LB	HO2S-11 (B1 S1) Signal	0.1-1.1v
61	---	Not Used	---
62	RD/PK	FTP Sensor Signal	2.6v at 0" H20 (cap off)
63	---	Not Used	---
64	LB/YL	A/T: Digital TR3 Sensor	In 'P': 0v, in O/D: 1.7v
64	LB/YL	M/T: CPP Switch Signal	5v (clutch "in": 0v)

2003 E-Series 6.8L V10 VIN S (All) 104 Pin Connector

PCM Pin #	Wire Color	Circuit Description (104 Pin)	Value at Hot Idle
65	BR/LG	DPFE Sensor Signal (California)	0.95-1.05v
66	YL/LG	Cylinder Head Temperature Sensor	0.6 (194°F)
67	VT/WT	EVAP Canister Vent Solenoid Control	0-10 Hz (0-100%)
68	GY/BK	Injector 9 Control	4.2-4.6 ms
69	PK/YL	A/C Clutch Relay Control	A/C On: 12v, Off: 0v
70	---	Not Used	---
71	RD	Vehicle Power	12-14v
72	TN/RD	Injector 7 Control	4.2-4.6 ms
73	TN/BK	Injector 5 Control	4.2-4.6 ms
74	BR/YL	Injector 3 Control	4.2-4.6 ms
75	TN	Injector 1 Control	4.2-4.6 ms
76	BK/WT	Power Ground	<0.1v
77	BK/WT	Power Ground	<0.1v
78	PK/LB	COP 2 Driver (dwell)	5°, 55 mph: 9°
79	WT/RD	COP 8 Driver (dwell)	5°, 55 mph: 9°
80	LB/OR	Fuel Pump Relay Control	Off: 12v, On: 1v
81	WT/YL	EPC Solenoid Control	9.2v (5 psi)
82	WT/RD	COP 9 Driver (dwell)	5°, 55 mph: 9°
83	WT/LB	IAC Solenoid Control	10.7v (33%)
84	DB/YL	Output Shaft Speed Sensor	0 Hz, 55 mph: 470 Hz
85	DG	CMP Sensor Signal	10 Hz, 55 mph: 16 Hz
86	---	Not Used	---
87	RD/BK	HO2S-21 (B2 S1) Signal	0.1-1.1v
88	LB/RD	MAF Sensor Signal	0.8v, 55 mph: 1.6v
89	GY/WT	TP Sensor Signal	0.53-1.27v
90	BR/WT	Reference Voltage	4.9-5.1v
91	GY/RD	Sensor Return	<0.050v
92	RD/LG	Brake Pedal Switch	0v (Brake On: 12v)
93	RD/WT	HO2S-11 (B1 S1) Heater	1v (Heater Off: 12v)
94	YL/LB	HO2S-21 (B2 S1) Heater	1v (Heater Off: 12v)
95	WT/BK	HO2S-12 (B1 S2) Heater	1v (Heater Off: 12v)
96	---	Not Used	---
97	RD	Vehicle Power	12-14v
98	LB	Injector 8 Control	4.2-4.6 ms
99	LG/OR	Injector 6 Control	4.2-4.6 ms
100	BR/LB	Injector 4 Control	4.2-4.6 ms
101	W	Injector 2 Control	4.2-4.6 ms
102	YL/BK	COP 4 Driver (dwell)	5°, 55 mph: 9°
103	BK/WT	Power Ground	<0.1v
104	PK/WT	COP 3 Driver (dwell)	5°, 55 mph: 9°

Pin Connector Graphic

PCM 104-PIN CONNECTOR

Terminal View of 104-Pin PCM Wiring Harness Connector

Standard Colors and Abbreviations

Abbreviation	Color	Abbreviation	Color	Abbreviation	Color
BK	Black	GY	Gray	PK	Purple
BL	Blue	GN	Green	RD	Red
BR	Brown	LG	LT Green	TN	Tan
DB	Dark Blue	OR	Orange	WT	White
DG	DK Green	PK	Pink	YL	Yellow

1996-99 E-Series 7.3L V8 Diesel VIN F (A/T) 104 Pin Connector

PCM Pin #	Wire Color	Circuit Description (104 Pin)	Value at Hot Idle
1	PK/OR	Shift Solenoid 2 Control	12v, 55 mph: 12v
2	PK/LG	MIL (lamp) Control	MIL On: 1v, Off: 12v
3	RD/YL	Manifold Intake Air Heater Relay	Heater On: 12v, Off: 0v
4	OR/GR	Parking Brake Applied Switch	Switch Up: 12, Down: 0v
5	RD/OR	Idle Validation Switch	Switch Up: 0v, Down: 12v
5 (Cal)	LG/RD	Parking Brake Applied Switch	Park Brake Applied: 0v
6 (Cal)	OR/YL	Shift Solenoid 1 Control	SS1 On: 1v, Off: 12v
7	---	Not Used	
8 (Cal)	GY	Glow Plug Monitor High Side	Plugs Off: 0v, On: 12v
9 (Cal)	OR	Glow Plug Monitor (Bank 1)	Plugs Off: 0v, On: 12v
10 (Cal)	RD/OR	Idle Validation Switch	0v, off-12v
11 (Cal)	PK/OR	Shift Solenoid 2 Control	SS2 On: 1v, Off: 12v
12 (Cal)	WT/LG	TCIL (lamp) Control	Lamp On: 1v, Off: 12v
13	VT	Flash EPROM Power	0.1v
14	LG/BK	4x4 Low Switch	Switch On: 0v, Off: 12v
15	PK/LB	Data Bus (-)	Digital Signals
16	TN/OR	Data Bus (+)	Digital Signals
17 (Cal)	OR/BK	Digital TR1 Sensor Signal	Manual 1 or 2: 10.7v
19 (Cal)	WT/PK	Tachometer Reflected Signal	6.5v (130 Hz)
20	---	Not Used	---
21 (Cal)	DG	Camshaft Position Sensor	7 Hz
22-23	---	Not Used	---
24	YL/BK	ACP Sensor Ground	<0.050v
25	BK	Case Ground	<0.050v
26	---	Not Used	---
27	OR/YL	Shift Solenoid 1 Control	1v, 55 mph: 12v
28	PK/YL	TCC Solenoid Control	12v, 55 mph: 1v
29	TN/LB	Clutch Pedal Position Switch	Clutch In: 0v, Out: 12v
29	TN/WT	TCS (switch) Signal	TCS & O/D On: 12v
30	PK/LB	Exhaust Back Pressure	0.9v, Cruise: 2.5v
31	RD/LG	Brake Pedal Applied Switch	Pedal Up: 12v, Down: 0v
32	---	Not Used	---
33	PK/OR	VSS (-) Signal	0 Hz, at 55 mph: 125 Hz
34	LG/BK	MAP Sensor Signal	107 Hz
35-36	---	Not Used	---
37	OR/BK	Transmission Fluid Temperature Sensor	2.10-2.40v
38	LG/RD	Engine Oil Temperature	-40°F: 4.7v, 230°F: 0.36v
39	GY	Intake Air Temperature Sensor	1.5-2.5v
40	DG/OR	Speed Control Ground	<0.050v
41	BK/YL	AC Clutch (ACC) Signal	AC & Clutch On: 12v
42	GY/RD	Exhaust Backpressure Signal	0-12v (duty cycle)
43-47	---	Not Used	---
48	GY/WT	Electronic Feedback Line	Digital Signals
49	DG	Camshaft Position Sensor	6.5v (130 Hz0
50	WT/PK	Tachometer Signal (CMP)	6.5v (130 Hz)
51	BK/WT	Power Ground	<0.1v
52	---	Not Used	---
53	BR/OR	Coast Clutch Solenoid	12v, 55 mph: 12v
54	---	Not Used	---
55	YL	Keep Alive Power	12-14v
56-57	---	Not Used	---
58	GY/BK	VSS (+) Signal	0 Hz, at 55 mph: 125 Hz
61	LB/BK	Speed Control Cruise Signal	0 or 12v
62	---	Not Used	---
63	DB/LG	BARO Sensor	4.6v (sea level)
64	LB/YL	Digital Transmission Range Sensor	In 'P': 0v, in O/D: 1.7v

1996-99 E-Series 7.3L V8 Diesel VIN F (A/T) 104 Pin Connector

PCM Pin #	Wire Color	Circuit Description (104 Pin)	Value at Hot Idle
65	LB	Camshaft Position Sensor Ground	<0.050v
66-69	---	Not Used	---
70	WT/BK	Glow Plug Lamp Control	Lamp On: 0v, Off: 12v
71	RD	Vehicle Power (Start-Run)	12-14v
72-75	---	Not Used	---
76	BK/WT	Power Ground	<0.1v
77	BK/WT	Power Ground	<0.1v
78	---	Not Used	---
79	WT/LG	TCIL (lamp) Control	Lamp On: 1v, Off: 12v
80	BK/PK	IDM Relay Control	On: 1v, Off: 12v
81	WT/YL	EPC Solenoid Control	7.5v (8 psi)
82	---	Not Used	---
83	YL/RD	Injector Pressure Regulator	12v, 55 mph: 12v
84	DB/LG	BARO Sensor Signal	At Sea Level: 4.64v
85-86	---	Not Used	---
87	DB/LG	Injection Control Pressure	0.75v (Min: 0.83v startup)
88	---	Not Used	---
89	GY/WT	ACP Sensor Signal	0.5-0.9v
90	BR/WT	Reference Voltage	4.9-5.1v
91	GY/RD	Analog Signal Return	<0.050v
92	LG	Brake Position Switch	Brake Off: 12v, On: 1v
93-94	---	Not Used	---
95	BR/OR	Fuel Delivery Control Signal	1v / 49 Hz
96	YL/LB	Camshaft Position Sensor	5 Hz (6v)
97	RD	Vehicle Power (Start-Run)	12-14v
98-99	---	Not Used	---
100	YL/WT	Fuel Level Indicator Signal	1.7v (1/2 full)
101	PK/OR	Glow Plug Relay Control	On: 1v, Off: 12v
103	BK/WT	Power Ground	<0.1v
104	---	Not Used	---

Pin Connector Graphic

PCM 104-PIN CONNECTOR

Terminal View of 104-Pin PCM Wiring Harness Connector

Standard Colors and Abbreviations

Abbreviation	Color	Abbreviation	Color	Abbreviation	Color
BK	Black	GY	Gray	PK	Purple
BL	Blue	GN	Green	RD	Red
BR	Brown	LG	LT Green	TN	Tan
DB	Dark Blue	OR	Orange	WT	White
DG	DK Green	PK	Pink	YL	Yellow

2000-02 E-Series 7.3L V8 Diesel VIN F (A/T) 104 Pin Connector

PCM Pin #	Wire Color	Circuit Description (104 Pin)	Value at Hot Idle
1	---	Not Used	---
2	PK/LG	MIL (lamp) Control	MIL On: 1v, Off: 12v
3	RD/YL	Manifold Intake Air Heater Monitor Signal	Heater On: 12v, Off: 0v
4	OR/GR	Driveline Disconnect Switch	Switch On: 0v, Off: 12v
5	LG/RD	Brake Warning Switch Input	Applied: 0v, Off: 12v
6	OR/YL	Shift Solenoid 1/A Control	SS1 On: 1v, Off: 12v
7	---	Not Used	---
8	WT/PK	Glow Plug Monitor High Side	Plugs Off: 0v, On: 12v
8 (Cal)	WT/LG	Glow Plug Control Module	Digital Signals
9	WT	Glow Plug Monitor Right Bank	Plugs Off: 0v, On: 12v
10	RD/OR	Idle Validation Switch	12, Cruise: 0v
11	PK/OR	Shift Solenoid 'B' Control	SS2 On: 1v, Off: 12v
12	WT/LG	TCIL (lamp) Control	Lamp On: 1v, Off: 12v
13	PK	Generic Scan Tool Input	0.1v
14	OR/LB	4x4 Low Switch	Switch On: 0v, Off: 12v
15	PK/LB	Data Bus (-)	Digital Signals
16	TN/OR	Data Bus (+)	Digital Signals
17	OR/BK	Digital TR1 Sensor	In 'P': 0v, Drive: 10.7v
18	---	Not Used	---
19	WT/PK	Clean Tachometer Output	6.5v (130 Hz)
20	BR/OR	Coast Clutch Solenoid	CCS On: 0v, Off: 12v
21	DG	Camshaft Position Sensor	7 Hz
22-23	---	Not Used	---
24	YL/BK	APP Sensor Ground	<0.050v
25	BK/LB	Case Ground	<0.050v
26-27	---	Not Used	---
28	RD	Water In Fuel Indicator Lamp	Lamp on: 1v, Off: 12v
29	TN/WT	A/T: TCS (switch) Signal	TCS & O/D On: 12v
30	PK/LB	Exhaust Back Pressure	0.9v, off-idle: 2.5v
31	RD/LG	Brake Pressure Switch	Brake On: 0v, Off: 12v
32	---	Not Used	---
33	PK/OR	VSS (-) Signal	<0.050v
34	WT/LG	Glow Plug Monitor Left Bank	Plugs Off: 0v, On: 12v
35	WT/YL	Generator Power Switch	12-14v
36	GY/RD	Alternator #1 (top) Monitor	6-10v
36	GY/RD	Fuel Line Heater	Heater On: 12v, Off: 0v
37	OR/BK	Transmission Fluid Temperature Sensor	0.3-4.5v
38	LG/RD	Engine Oil Temperature	0.3-4.7v
39	GY	Intake Air Temperature Sensor	0.2-4.5v
40	PK/BK	Speed Control Ground	<0.050v
41	BK/YL	A/C Pressure Switch Signal	A/C On: 12v, Off: 0v
42	GY/RD	Exhaust Backpressure Signal	Digital Signals
43-47	---	Not Used	---
48	GY/WT	Electronic Feedback Line	Digital Signals
49	WT/PK	Digital TR2 Sensor	In 'P': 0v, Drive: 10.7v
50	GY/BK	Digital TR4 Sensor	In 'P': 0v, Drive: 10.7v
51	BK/WT	Power Ground	<0.1v
52-53	---	Not Used	---
54	PK/YL	TCC Solenoid Control	TCC On: 1v, Off: 12v
55	YL	Keep Alive Power	12-14v
56-57	---	Not Used	---
58	GY/BK	VSS (+) Signal	0 Hz, at 55 mph: 125 Hz
59	---	Not Used	---
60	YL	Alternator #2 (bottom) Monitor	6-10v
61	LB/BK	Speed Control Cruise Signal	S/C Switch On: 12v
62-63	---	Not Used	---
64	LB/YL	Digital TR Sensor3	In 'P': 4.5v, Drive: 2.2v

2000-02 E-Series 7.3L V8 Diesel VIN F (A/T) 104 Pin Connector

PCM Pin #	Wire Color	Circuit Description (104 Pin)	Value at Hot Idle
65	LB	Camshaft Position Sensor Ground	<0.050v
66	TN/LG	A/C Pressure Switch	0-4.9v
67	GY/LB	Battery Direct	12-14v
68-69	---	Not Used	---
70	BK/PK	Glow Plug Lamp Control	Lamp On: 1v, Off: 12v
71	RD	Vehicle Power (Start-Run)	12-14v
72-75	---	Not Used	---
76	BK/WT	Power Ground	<0.1v
77	BK/WT	Power Ground	<0.1v
78	---	Not Used	---
79	LG/BK	MAP Sensor Signal	1-3v
80	WT/BK	IDM Relay Control	On: 1v, Off: 12v
81	WT/YL	EPC Solenoid Control	7.5v (8 psi)
82	---	Not Used	---
83	YL/RD	Injector Pressure Regulator	12v
84	DG/WT	Turbine Shaft Speed Sensor	30 mph: 126-136 Hz
85-86	---	Not Used	---
87	DB/LG	Injection Control Pressure	0.75v (Min: 0.83v startup)
88	---	Not Used	---
89	GY/WT	APP Sensor Signal	0.5-1.6v
90	BR/WT	Reference Voltage	4.9-5.1v
91	GY/RD	Analog Signal Return	<0.050v
92	LG	Brake Position Switch	Brake Off: 12v, On: 1v
93	---	Not Used	---
94	LB/OR	Fuel Pump Control	On: 1v, Off: 12v
95	BR/OR	Fuel Delivery Control Signal	49 Hz
96	YL/LB	Camshaft Position Sensor	6v (5 Hz)
97	RD	Vehicle Power (Start-Run)	12-14v
98	PK	Manifold Intake Air Heater Control	12-14v
99	---	Not Used	---
100	YL/WT	Fuel Level Indicator Signal	1.7v (1/2 full)
101	PK/OR	Glow Plug Relay Control	On: 1v, Off: 12v
102	---	Not Used	---
103	BK/WT	Power Ground	<0.1v
104	---	Not Used	---

Pin Connector Graphic

PCM 104-PIN CONNECTOR

Terminal View of 104-Pin PCM Wiring Harness Connector

Standard Colors and Abbreviations

Abbreviation	Color	Abbreviation	Color	Abbreviation	Color
BK	Black	GY	Gray	PK	Purple
BL	Blue	GN	Green	RD	Red
BR	Brown	LG	LT Green	TN	Tan
DB	Dark Blue	OR	Orange	WT	White
DG	DK Green	PK	Pink	YL	Yellow

2003 E-Series 7.3L V8 Diesel VIN F (All) 104 Pin Connector

PCM Pin #	Wire Color	Circuit Description (104 Pin)	Value at Hot Idle
1	---	Not Used	---
2	PK/LG	MIL (lamp) Control	MIL Off: 12v, On: 1v
3	RD/YL	Manifold Intake Air Heater Monitor	Off: 12v, On: 1v
4	---	Not Used	---
5	LG/RD	Parking Brake Applied Switch	Brake Off: 12v, Applied: 0v
6	OR/YL	Shift Solenoid 1 Control	SS1 Off: 12v, On: 1v
7	---	Not Used	---
8	WT/LG	Glow Plug Monitor Left Bank	Plugs Off: 0v, On: 12v
9	---	Not Used	---
10	RD/LG	Idle Validation Switch	12v, off-idle: 0v
11	VT/OG	Shift Solenoid 2 Control	SS2 Off: 12v, On: 1v
12	WT/LG	TCIL (lamp) Control	7.7v (Switch On: 0v)
13	VT	Flash EEPROM Power Supply	0.1v
14	---	Not Used	---
15	PK/LB	SCP Data Bus (-) Signal	<0.050v
16	TN/OR	SCP Data Bus (+) Signal	Digital Signals
17	OR/BK	Digital TR1 Sensor	In 'P': 0v, in Drive: 10.7v
18	PK/YL	A/C Relay Control	Relay Off: 12v, On: 1v
19	WT/PK	Tachometer Reflected Signal	6.5v / 130 Hz
20	BR/OR	Coast Clutch Solenoid Control	CCS On: 0v, Off: 12v
21	DG	CMP Sensor Signal	Digital Signal: 0-12-0v
22-23	---	Not Used	---
24	YL/WT	APP Sensor Ground	<0.050v
25	BK/LB	Case Ground	<0.050v
26-27	---	Not Used	---
28	RD	Water In Fuel Indicator Control	Indicator Off: 12v, On: 1v
29	TN/WT	A/T: TCS (switch) Signal	TCS & O/D On: 12v
29	TN/LB	M/T: CCP Switch Signal	Clutch In: 0v, Out: 12v
30	VT/LB	Exhaust Back Pressure Sensor	0.9v, off-idle: 2.5v
31	BK/YL	Brake Pressure Switch	Brake Off: 12v, On: 1v
32	---	Not Used	---
33	PK/OG	VSS (-) Signal	<0.050v
34	---	Not Used	---
35	WT/YL	Alternator No. 1 (Top) Monitor	6-10v
36	GY/RD	Fuel Heater / Water in Fuel	Heater On: 12v
37	OR/BK	TFT Sensor Signal	2.10-2.40v
38	LG/RD	Engine Oil Temperature Sensor	0.3-4.7v
39	GY	Intake Air Temperature Sensor	1.5-2.5v
40	PK/BK	Fuel Pump Monitor	Pump Off: 12v, On: 1v
41	TN/LG	A/C Head Pressure Switch	0v or 12v
42	GY/RD	Exhaust Backpressure Signal	Digital: 0-12-0-12v
43	OR/LG	Overhead Console Fuel Consumption	0-5-0v
44-46	---	Not Used	---
47	WT/RD	Wastegate Control Solenoid Control	Solenoid Off: 12v, On: 1v
48	GY/WT	Electronic Feedback Line	0.9-3.0v
49	WT/PK	Digital TR2 Sensor	In 'P': 0v, in Drive: 10.7v
50	GY/BK	Digital TR4 Sensor	In 'P': 0v, in Drive: 10.7v
51	BK/WT	Power Ground	<0.1v
52-53	---	Not Used	---
54	VT/YL	TCC Solenoid Control	TCC Off: 12v, On: 1v
55	RD/WT	Keep Alive Power	12-14v
56-57	---	Not Used	---
58	GY/BK	VSS (+) Signal	0 Hz, 55 mph: 125 Hz
59	DB/YL	OSS Sensor Signal	AC pulse signals
60	YL	Alternator No. 2 (Bottom) Control	6-10v
61	LB/BK	Speed Control On/Off Switch	Switch On: 12v, Off: 0v
62	RD/YL	MAT Sensor Signal	1.5-2.5v
63	---	Not Used	---
64	LB/YL	Digital TR3 Sensor	In P: 4.5v, in Drive: 2.2v

2003 E-Series 7.3L V8 Diesel VIN F 104 Pin Connector

PCM Pin #	Wire Color	Circuit Description (104 Pin)	Value at Hot Idle
65	LB	CMP Sensor Ground	<0.050v
66	LB/YL	Power Take-Off Signal	0v (Off)
67	LG/RD	Battery Direct	12-14v
68-69	---	Not Used	---
70	BK/PK	Glow Plug Lamp Control	7.7v (Switch On: 0v)
71	RD	Vehicle Power	12-14v
72-75	---	Not Used	---
76	BK/WT	Power Ground	<0.1v
77	BK/WT	Power Ground	<0.1v
78	---	Not Used	---
79	LG/BK	MAP Sensor Signal	1-3v
80	WT/BK	IDM Enable Relay Control	Off: 12v, On: 1v
81	WT/YL	EPC Solenoid Control	7.5v (8 psi)
82	---	Not Used	---
83	YL/RD	Injector Pressure Regulator	12v
84	DG/WT	TSS Sensor Signal	30 mph: 130 Hz
85-86	---	Not Used	---
87	DB/LG	Injection Control Pressure	0.1-3.0v
88	---	Not Used	---
89	GY/WT	APP Sensor Signal	0.5-1.6v
90	BR/WT	Reference Voltage	4.9-5.1v
91	GY/RD	Sensor Ground	<0.050v
92	RD/LG	Brake Pedal Switch	0v (Brake On: 12v)
93	---	Not Used	---
94	LB/OR	Fuel Pump Relay Control	Relay Off: 12v, On: 1v
95	BR/OR	Fuel Delivery Control	40-240 Hz
96	YL/LB	CID Sensor Signal	5-30 Hz
97	RD	Vehicle Power	12-14v
98	VT	PCM Signal to Relay	12v
99	---	Not Used	---
100	YL/WT	Fuel Level Indicator Signal	1.7v (1/2 full)
101	VT/OR	Glow Plug Relay Control	Off: 12v, On: 1v
102	---	Not Used	---
103	BK/WT	Power Ground	<0.1v
104	---	Not Used	---

Pin Connector Graphic

PCM 104-PIN CONNECTOR

```
 1 ⦻⦻⦻⦻⦻⦻⦻⦻⦻⦻⦻⦻⦻    ⦻⦻⦻⦻⦻⦻⦻⦻⦻⦻⦻⦻⦻ 26
27 ⦻⦻⦻⦻⦻⦻⦻⦻⦻⦻⦻⦻⦻    ⦻⦻⦻⦻⦻⦻⦻⦻⦻⦻⦻⦻⦻ 52
53 ⦻⦻⦻⦻⦻⦻⦻⦻⦻⦻⦻⦻⦻  ⬤ ⦻⦻⦻⦻⦻⦻⦻⦻⦻⦻⦻⦻⦻ 78
79 ⦻⦻⦻⦻⦻⦻⦻⦻⦻⦻⦻⦻⦻    ⦻⦻⦻⦻⦻⦻⦻⦻⦻⦻⦻⦻⦻ 104
```

Terminal View of 104-Pin PCM Wiring Harness Connector

Standard Colors and Abbreviations

Abbreviation	Color	Abbreviation	Color	Abbreviation	Color
BK	Black	GY	Gray	PK	Purple
BL	Blue	GN	Green	RD	Red
BR	Brown	LG	LT Green	TN	Tan
DB	Dark Blue	OR	Orange	WT	White
DG	DK Green	PK	Pink	YL	Yellow

1990 E-Series 7.5L V8 MFI VIN G (A/T) 60 Pin Connector

PCM Pin #	Wire Color	Circuit Description (60 Pin)	Value at Hot Idle
1	BK/OR	Keep Alive Power	12-14v
2, 5, 9	---	Not Used	---
3	DG/WT	VSS (+) Signal	0 Hz, at 55 mph: 125 Hz
4	DG/YL	Ignition Diagnostic Monitor	20-31 Hz
6	OR/YL	VSS (-) Signal	0 Hz, at 55 mph: 125 Hz
6	GY/BK	VSS (-) Signal	0 Hz, at 55 mph: 125 Hz
7	LG/YL	Engine Coolant Temperature Sensor	0.5-0.6v
7	GN/YL	Engine Coolant Temperature Sensor	0.5-0.6v
8	OR/BL	Fuel Pump Monitor	On: 12v, Off: 0v
8	RD	Fuel Pump Monitor	On: 12v, Off: 0v
10	BK/YL	A/C Pressure Switch Signal	A/C On: 12v, off: 0v
11	WT/BK	Air Management 2 Solenoid	On: 1v, Off: 12v
12-15, 19	---	Not Used	---
16	BK/OR	Ignition System Ground	<0.050v
17	PK/LG	Self-Test Output & MIL	MIL On: 1v, Off: 12v
17	TN/RD	Self-Test Output & MIL	MIL On: 1v, Off: 12v
18	TN/LG	Inferred Mileage Sensor	Digital Signals
20	BK	PCM Case Ground	<0.050v
21	GY/WT	IAC Solenoid Control	8.1-10.1v
22	TN/LG	Fuel Pump Control	On: 1v, Off: 12v
23	LG/BK	Knock Sensor 1 Signal	0v
24	YL/LG	Power Steering Pressure Switch	Straight: 0v, Turned: 12v
25	YL/RD	ACT Sensor Signal	1.5-2.5v
26	OR/WT	Reference Voltage	4.9-5.1v
27	BR/LG	EVP Sensor Signal	0.4v, 55 mph: 3.1v
27	DG/LG	EVP Sensor Signal	0.4v, 55 mph: 3.1v
28, 32, 34-35	---	Not Used	---
29	DG/PK	HO2S-11 (B1 S1) Signal	0.1-1.1v
30	BL/WT	A/T: Neutral Drive Switch	In 'P': 0v, Others: 5v
30	BR/WT	A/T: Neutral Drive Switch	In 'P': 0v, Others: 5v
31	GY/YL	Canister Purge Solenoid	12v, 55 mph: 1v
33	DG	VR Solenoid Control	0%, at 55 mph: 45%
36	YL/GN	Spark Output Signal	6.93v (50%)
37	RD	Vehicle Power	12-14v
38-39	---	Not Used	---
40	BK/LG	Power Ground	<0.1v
41-44	---	Not Used	---
45	DB/LG	MAP Sensor Signal	107 Hz
46	BK/WT	Analog Signal Return	<0.050v
47	DG/LG	TP Sensor Signal	1.0v
48	WT/RD	Self-Test Indicator Signal	STI On: 1v, Off: 5v
49	OR	HO2S-11 (B1 S1) Ground	<0.050v
50	---	Not Used	---
51	WT/RD	Air Management 1 Solenoid	On: 1v, Off: 12v
52-55	---	Not Used	---
56	DB	PIP Sensor Signal	6.93v (50%)
57	RD	Vehicle Power	12-14v
58	TN/OR	Injector Bank 1 (INJ 1, 4, 5, 8)	5.7-7.0 ms
59	TN/RD	Injector Bank 2 (INJ 2, 3, 6, 7)	5.7-7.0 ms
60	BK/LG	Power Ground	<0.1v

Pin Connector Graphic

PCM 60-PIN CONNECTOR

Terminal View of 60-Pin PCM Harness Connector

1990 E-Series 7.5L V8 VIN G (E4OD) 60 Pin Connector

PCM Pin #	Wire Color	Circuit Description (60 Pin)	Value at Hot Idle
1	BK/OR	Keep Alive Power	12-14v
2	LG	Brake Switch Signal	Brake Off: 12v, On: 1v
3	DG/WT	VSS (+) Signal	0 Hz, at 55 mph: 125 Hz
4	DG/YL	Ignition Diagnostic Monitor	20-31 Hz
5, 9, 12-15	---	Not Used	---
6	OR/YL	VSS (-) Signal	0 Hz, at 55 mph: 125 Hz
7	LG/YL	Engine Coolant Temperature Sensor	0.5-0.6v
8	OR/BL	Fuel Pump Monitor	On: 12v, Off: 0v
10	BK/YL	A/C Pressure Switch Signal	A/C On: 12v, off: 0v
11	WT/BK	Air Management 2 Solenoid	On: 1v, Off: 12v
16	BK/OR	Ignition System Ground	<0.050v
17	PK/LG	Self-Test Output & MIL	MIL On: 1v, Off: 12v
17	TN/RD	Self-Test Output & MIL	MIL On: 1v, Off: 12v
18	TN/LG	Inferred Mileage Sensor	Digital Signals
19	DG/PK	Shift Solenoid 2 Control	12v, 55 mph: 1v
20	BK	PCM Case Ground	<0.050v
21	GY/WT	IAC Solenoid Control	8.1-10.1v
22	TN/LG	Fuel Pump Control	On: 1v, Off: 12v
24	YL/LG	Power Steering Pressure Switch	Straight: 0v, Turned: 12v
25	YL/RD	ACT Sensor Signal	1.5-2.5v
26	OR/WT	Reference Voltage	4.9-5.1v
27	BR/LG	EVP Sensor Signal	0.4v, 55 mph: 3.1v
28, 34-35, 39	---	Not Used	---
29	DG/PK	HO2S-11 (B1 S1) Signal	0.1-1.1v
30	BL/WT	MLP Sensor Signal	In 'P': 0v, in O/D: 5v
31	GY/YL	Canister Purge Solenoid	12v, 55 mph: 1v
32	LG/WT	OCIL (lamp) Control	Lamp On: 1v, Off: 12v
33	DG	VR Solenoid Control	0%, at 55 mph: 45%
36	YL/LG	Spark Output Signal	6.93v (50%)
37	RD	Vehicle Power	12-14v
38	BL/YL	EPC Solenoid Control	9.5v (5 psi)
40	BK/LG	Power Ground	<0.1v
41	TN/BL	Overdrive Cancel Switch	OCS Closed: 1v
42	OR/BK	TOT Sensor Signal	2.10-2.40v
43-44	---	Not Used	---
45	DB/LG	MAP Sensor Signal	107 Hz (sea level)
46	BK/WT	Analog Signal Return	<0.050v
47	DG/LG	TP Sensor Signal	0.7v, 55 mph: 2.1v
48	WT/RD	Self-Test Indicator Signal	STI On: 1v, Off: 5v
49	OR	HO2S-11 (B1 S1) Ground	<0.050v
50, 54	---	Not Used	---
51	WT/RD	Air Management 1 Solenoid	AM1 On: 1v, Off: 12v
52	OR/YL	Shift Solenoid 1 Control	1v, 55 mph: 12v
53	PK/YL	TCC Solenoid Control	0%, at 55 mph: 95%
55	BR	Coast Clutch Solenoid	12v, 55 mph: 12v
56	DB	PIP Sensor Signal	6.93v (50%)
57	RD	Vehicle Power	12-14v
58	TN/OR	Injector Bank 1 (INJ 1, 4, 5, 8)	5.7-7.0 ms
59	TN/RD	Injector Bank 2 (INJ 2, 3, 6, 7)	5.7-7.0 ms
60	BK/LG	Power Ground	<0.1v

Pin Connector Graphic

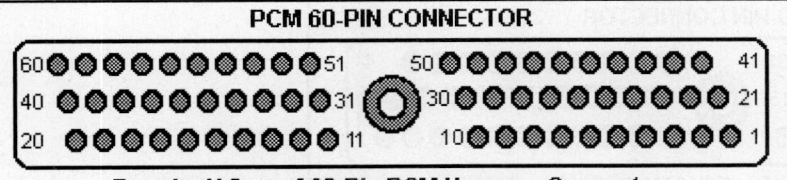

Terminal View of 60-Pin PCM Harness Connector

1991 E-Series 7.5L V8 VIN G (E4OD) 60 Pin Connector

PCM Pin #	Wire Color	Circuit Description (60 Pin)	Value at Hot Idle
1	BK/OR	Keep Alive Power	12-14v
2	LG	Brake Switch Signal	Brake Off: 12v, On: 1v
3	DG/WT	VSS (+) Signal	0 Hz, at 55 mph: 125 Hz
4	TN/YL	Ignition Diagnostic Monitor	20-31 Hz
5, 9, 12-15	---	Not Used	---
6	OR/YL	VSS (-) Signal	0 Hz, at 55 mph: 125 Hz
7	LG/RD	Engine Coolant Temperature Sensor	0.5-0.6v
8	DG/YL	Fuel Pump Monitor	On: 12v, Off: 0v
9	---	Not Used	---
10	BK/YL	A/C Pressure Switch Signal	A/C On: 12v, off: 0v
16	OR/RD	Ignition System Ground	<0.050v
17	TN/RD	Self-Test Output & MIL	MIL On: 1v, Off: 12v
18, 23-24	---	Not Used	---
19	PK	Shift Solenoid 2 Control	1v, 55 mph: 12v
20	BK	PCM Case Ground	<0.050v
21	GY/WT	IAC Solenoid Control	8.1-10.1v
22	TN/LG	Fuel Pump Control	On: 1v, Off: 12v
25	YL/RD	ACT Sensor Signal	1.5-2.5v
26	BR/WT	Reference Voltage	4.9-5.1v
27	BR/LG	EVP Sensor Signal	0.4v, 55 mph: 3.1v
28, 34-35	---	Not Used	---
29	RD/BK	HO2S-11 (B1 S1) Signal	0.1-1.1v
30	WT/RD	MLP Sensor Signal	In 'P': 0v, in O/D: 5v
31	GY/YL	Canister Purge Solenoid	12v, 55 mph: 1v
32	WT/LG	OCIL (lamp) Control	Lamp On: 1v, Off: 12v
33	BR/PK	VR Solenoid Control	0%, at 55 mph: 45%
36	YL/LG	Spark Output Signal	6.93v (50%)
37	RD	Vehicle Power	12-14v
38	BL/YL	EPC Solenoid Control	9.5v (5 psi)
39, 43-44	---	Not Used	---
40	BK/LG	Power Ground	<0.1v
41	TN/WT	Overdrive Cancel Switch	OCS pressed: 12v
42	OR/BK	TOT Sensor Signal	2.10-2.40v
45	DB/LG	MAP Sensor Signal	107 Hz (sea level)
46	GY/RD	Analog Signal Return	<0.050v
47	GY/WT	TP Sensor Signal	0.7v, 55 mph: 2.1v
48	WT/OR	Self-Test Indicator Signal	STI On: 1v, Off: 5v
49	OR	HO2S-11 (B1 S1) Ground	<0.050v
50, 54	---	Not Used	---
51	WT/RD	Air Management 1 Solenoid	On: 1v, Off: 12v
52	OR	Shift Solenoid 1 Control	1v, 55 mph: 12v
53	PK/YL	TCC Solenoid Control	0%, at 55 mph: 95%
55	BR/OR	Coast Clutch Solenoid	12v, 55 mph: 12v
56	DB	PIP Sensor Signal	6.93v (50%)
57	RD	Vehicle Power	12-14v
58	TN/OR	Injector Bank 1 (INJ 1, 4, 5, 8)	6.0-6.6 ms
59	TN/RD	Injector Bank 2 (INJ 2, 3, 6, 7)	6.0-6.6 ms
60	BK/LG	Power Ground	<0.1v

Pin Connector Graphic

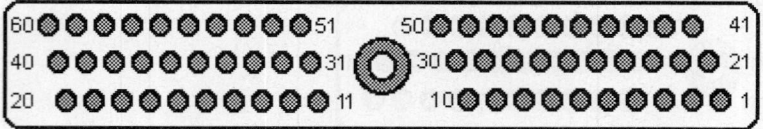

Terminal View of 60-Pin PCM Harness Connector

1992-95 E-Series 7.5L V8 VIN G (E4OD) 60 Pin Connector

PCM Pin #	Wire Color	Circuit Description (60 Pin)	Value at Hot Idle
1	YL	Keep Alive Power	12-14v
2	LG	Brake Switch Signal	Brake Off: 12v, On: 1v
3	GY/BK	VSS (+) Signal	0 Hz, at 55 mph: 125 Hz
4	TN/LG	Ignition Diagnostic Monitor	20-31 Hz
4	WT/PK	Ignition Diagnostic Monitor	20-31 Hz
5	---	Not Used	---
6	PK/OR	VSS (-) Signal	0 Hz, at 55 mph: 125 Hz
7	LG/RD	Engine Coolant Temperature Sensor	0.5-0.6v
8	DG/YL	Fuel Pump Monitor	On: 12v, Off: 0v
9	PK/BL	Data Bus (-)	Digital Signals
10	BK/YL	A/C Pressure Switch Signal	A/C On: 12v, off: 0v
11	BR	Air Management 2 Solenoid	On: 1v, Off: 12v
12-15	---	Not Used	---
16	OR/RD	Ignition System Ground	<0.050v
17	PK/LG	Self-Test Output & MIL	MIL On: 1v, Off: 12v
18	---	Not Used	---
19	PK/OR	Shift Solenoid 2 Control	1v, 55 mph: 12v
20	BK	PCM Case Ground	<0.050v
21	WT/BL	IAC Solenoid Control	8.1-10.1v
22	BL/OR	Fuel Pump Control	On: 1v, Off: 12v
23-24	---	Not Used	---
25	GY	ACT Sensor Signal	1.5-2.5v
26	BR/WT	Reference Voltage	4.9-5.1v
27	BR/LG	EVP Sensor Signal	0.4v, 55 mph: 3.1v
28	TN/OR	Data Bus (+)	Digital Signals
29	GY/BL	HO2S-11 (B1 S1) Signal	0.1-1.1v
30	BL/YL	MLP Sensor Signal	In 'P': 0v, in O/D: 5v
31	GY/YL	Canister Purge Solenoid	12v, 55 mph: 1v
32	WT/LG	OCIL (lamp) Control	Lamp On: 1v, Off: 12v
33	BR/PK	VR Solenoid Control	0%, at 55 mph: 45%
34-35	---	Not Used	---
36	PK	Spark Output Signal	6.93v (50%)
37	RD	Vehicle Power	12-14v
38	WT/YL	EPC Solenoid Control	9.5v (5 psi)
39	---	Not Used	---
40	BK/WT	Power Ground	<0.1v

1992-95 E-Series 7.5L V8 VIN G (E4OD) 60 Pin Connector

PCM Pin #	Wire Color	Circuit Description (60 Pin)	Value at Hot Idle
41	TN/WT	Overdrive Cancel Switch	OCS Closed: 0.1v
42	OR/BK	TOT Sensor Signal	2.10-2.40v
43-44	---	Not Used	---
45	LG/BK	MAP Sensor Signal	107 Hz (sea level)
46	GY/RD	Analog Signal Return	<0.050v
47	GY/WT	TP Sensor Signal	0.7v, 55 mph: 2.1v
48	WT/PK	Self-Test Indicator Signal	STI On: 1v, Off: 5v
49	OR	HO2S-11 (B1 S1) Ground	<0.050v
50	---	Not Used	---
51	WT/OR	Air Management 1 Solenoid	On: 1v, Off: 12v
52	OR/YL	Shift Solenoid 1 Control	1v, 55 mph: 12v
53	PK/YL	TCC Solenoid Control	0%, at 55 mph: 95%
54	---	Not Used	---
55	BR/OR	Coast Clutch Solenoid	12v, 55 mph: 12v
56	GY/OR	PIP Sensor Signal	6.93v (50%)
57	RD	Vehicle Power	12-14v
58	TN	Injector Bank 1 (INJ 1, 4, 5, 8)	6.0-6.6 ms
59	WT	Injector Bank 2 (INJ 2, 3, 6, 7)	6.0-6.6 ms
60	BK/WT	Power Ground	<0.1v

Pin Connector Graphic

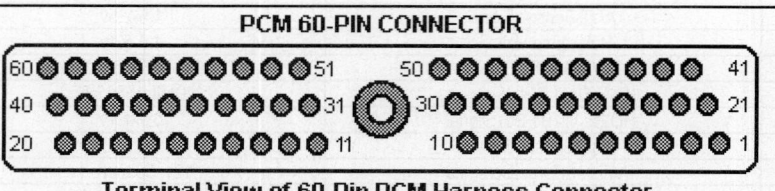

PCM 60-PIN CONNECTOR

Terminal View of 60-Pin PCM Harness Connector

Standard Colors and Abbreviations

Abbreviation	Color	Abbreviation	Color	Abbreviation	Color
BK	Black	GY	Gray	PK	Purple
BL	Blue	GN	Green	RD	Red
BR	Brown	LG	LT Green	TN	Tan
DB	Dark Blue	OR	Orange	WT	White
DG	DK Green	PK	Pink	YL	Yellow

1996 E-Series 7.5L V8 VIN G (E4OD) 104 Pin Connector

PCM Pin #	Wire Color	Circuit Description (104 Pin)	Value at Hot Idle
1	PK/OR	Shift Solenoid 2 Control	12v, 55 mph: 12v
2	PK/LG	MIL (lamp) Control	MIL On: 1v, Off: 12v
3-12	---	Not Used	---
13	VT	Flash EPROM Power	0.1v
14	---	Not Used	---
15	PK/LB	Data Bus (-)	Digital Signals
16	TN/OR	Data Bus (+)	Digital Signals
17-22	---	Not Used	---
23	OR/RD	Ignition System Ground	<0.050v
24	BK/WT	Power Ground	<0.1v
25	BK	Case Ground	<0.050v
26	---	Not Used	---
27	OR/YL	Shift Solenoid 1 Control	1v, 55 mph: 12v
28	---	Not Used	---
29	TN/WT	TCS (switch) Signal	TCS & O/D On: 12v
30-32	---	Not Used	---
33	PK/OR	PSOM (-) Signal	<0.050v
34	---	Not Used	---
35	RD/LG	HO2S-12 (B1 S2) Signal	0.1-1.1v
36	TN/LB	MAF Sensor Return	<0.050v
37	OR/BK	Transmission Fluid Temperature Sensor	2.10-2.40v
38	LG/RD	Engine Coolant Temperature Sensor	0.5-0.6v
39	GY	Intake Air Temperature Sensor	1.5-2.5v
40	DG/YL	Fuel Pump Monitor	On: 12v, Off: 0v
41	DG/OR	A/C Pressure Switch Signal	AC On: 12v, Off: 0v
42-43	---	Not Used	---
44	BR	Secondary AIR Diverter	On: 1v, Off: 12v
45-46	---	Not Used	---
47	BK/PK	VR Solenoid Control	0%, at 55 mph: 45%
48	WT/PK	Clean Tachometer Output	38-42 Hz
49	GY/OR	PIP Sensor Signal	6.93v (50%)
50	PK	Spark Output Signal	6.93v (50%)
51	BK/WT	Power Ground	<0.1v
52	---	Not Used	---
53	YL/BK	Coast Clutch Solenoid	12v, 55 mph: 12v
54	PK/YL	TCC Solenoid Control	0%, at 55 mph: 95%
55	Y	Keep Alive Power	12-14v
56	LG/BK	EVAP VMV Solenoid	0-10 Hz (0-100%)
57	---	Not Used	---
58	GY/BK	PSOM (+) Signal	0 Hz, at 55 mph: 125 Hz
59	DG/LG	Misfire Detection Sensor	45-55 Hz
60	GY/LB	HO2S-11 (B1 S1) Signal	0.1-1.1v
61-63	---	Not Used	---
64	LB/YL	Digital Transmission Range Sensor	In 'P': 0v, in O/D: 1.7v

1996 E-Series 7.5L V8 VIN G (E4OD) 104 Pin Connector

PCM Pin #	Wire Color	Circuit Description (104 Pin)	Value at Hot Idle
65	BR/LG	EGR DPFE Sensor Signal	1v, 55 mph: 1v
66-69	---	Not Used	
70	WT/OR	Secondary Air Bypass Solenoid	AIRB On: 1v, Off: 12v
71	RD	Vehicle Power (Start-Run)	12-14v
72	TN/RD	Injector 7 Control	4.3-4.6 ms
73	TN/BK	Injector 5 Control	4.3-4.6 ms
74	BR/YL	Injector 3 Control	4.3-4.6 ms
75	TN	Injector 1 Control	4.3-4.6 ms
76	BK/WT	Power Ground	<0.1v
77	BK/WT	Power Ground	<0.1v
78	---	Not Used	---
79	WT/LG	TCIL (lamp) Control	Lamp On: 1v, Off: 12v
80	LB/OR	Fuel Pump Control	On: 1v, Off: 12v
81	WT/YL	EPC Solenoid Control	10v (26 psi)
82	---	Not Used	---
83	WT/LB	IAC Solenoid Control	10.7v (33%)
84-86	---	Not Used	---
87	RD/BK	HO2S-21 (B2 S1) Signal	0.1-1.1v
88	LB/RD	MAF Sensor Signal	0.8v, 55 mph: 2.3v
89	GY/WT	TP Sensor Signal	0.9v, 55 mph: 1.2v
90	BR/WT	Reference Voltage	4.9-5.1v
91	GY/RD	Analog Signal Return	<0.050v
92	LG	Brake Position Switch	Brake Off: 12v, On: 1v
93	RD/WT	HO2S-11 (B1 S1) Heater	1v, Off: 12v
94	YL/LB	HO2S-21 (B2 S1) Heater	1v, Off: 12v
95	WT/BK	HO2S-12 (B1 S2) Heater	1v, Off: 12v
96	---	Not Used	---
97	RD	Vehicle Power	12-14v
98	LG	Injector 8 Control	4.3-4.6 ms
99	LG/OR	Injector 6 Control	4.3-4.6 ms
100	BR/LB	Injector 4 Control	4.3-4.6 ms
101	W	Injector 2 Control	4.3-4.6 ms
102	---	Not Used	---
103	BK/WT	Power Ground	<0.1v
104	---	Not Used	---

Pin Connector Graphic

```
           PCM 104-PIN CONNECTOR
    1 ●●●●●●●●●●●●●      ●●●●●●●●●●●●● 26
   27 ●●●●●●●●●●●●●      ●●●●●●●●●●●●● 52
   53 ●●●●●●●●●●●●●   ⬤  ●●●●●●●●●●●●● 78
   79 ●●●●●●●●●●●●●      ●●●●●●●●●●●●● 104
```

Terminal View of 104-Pin PCM Wiring Harness Connector

Standard Colors and Abbreviations

Abbreviation	Color	Abbreviation	Color	Abbreviation	Color
BK	Black	GY	Gray	PK	Purple
BL	Blue	GN	Green	RD	Red
BR	Brown	LG	LT Green	TN	Tan
DB	Dark Blue	OR	Orange	WT	White
DG	DK Green	PK	Pink	YL	Yellow

Villager PIN Tables

1993-95 Villager 3.0L V6 VIN W (A/T) 116-Pin Connector

PCM Pin #	Wire Color	Circuit Description (116-Pin)	Value at Hot Idle
1	BL	Ignition Control Signal	2000 rpm: 1.2-1.3v
2	GN/WT	Tachometer Output Signal	2000 rpm: 1.2-1.3v
3	RD	Ignition Check Signal	9.0-12.0v
4	WT/GN	PCM Relay Control	Key off: 12-14v for 2 sec.
5	OR/BK	Fuel Flow Signal to IPC	Digital Signals
6	BK	Power Ground	<0.1v
7	YL/RD	Data Link Connector	5v
8	WT/PK	EGR Temperature Sensor	0.4-5v, 55 mph: 0.2-1v
9	BL/OR	Low Speed Fan Control	Fan On: 1v, Off: 12v
10	BR/WT	High Speed Fan Control	Fan On: 1v, Off: 12v
11	GY/RD	A/C Relay Control	On: 1v, Off: 12v
12	PK	Power Steering Pressure Switch	Straight: 5v, Turned: 0v
13	BK	Power Ground	<0.1v
14	YL/BL	Data Link Connector Signal	5v
15	YL/BK	Data Link Connector Signal	5v
16	WT/BL	MAF Sensor Signal	1.7v, 55 mph: 2.8v
17	OR/BL	MAF Sensor Ground	<0.050v
18	LG/RD	Engine Coolant Temperature Sensor	0.5-0.6v
19	LG/BK	HO2S-11 (B1 S1) Signal	0.1-1.1v
20	RD	TP Sensor Signal	0.4-0.1v
21	WT/BK	Analog Signal Return	<0.050v
22	RD/WT	Camshaft Position Sensor	0.2-0.4v
23	BL/WT	Diagnostic Test Connector	5v
24	GY/RD	MIL (lamp) Control	Lamp On: 1v, Off: 12v
25-26	---	Not Used	---
27	WT	Knock Sensor 1 Signal	0v
28	RD/GN	TP Sensor Signal to TCM	0.4-0.1v
29	WT/BK	Analog Signal Return	<0.050v
30	RD/WT	CKP Sensor Reference	0.2-0.4v
31	RD/BL	CKP Sensor Signal	2.5-2.7v
32	GN/YL	Speedometer Signal to PCM	Moving: Digital Signals
33	BR/YL	IDL Switch (-) Signal	Throttle Closed: 8-10v
34	BL/BK	Vehicle Start Signal	KOEC: 9-11v
35	GN/BK	PNP Signal to TCM	In 'P': 0v, Others: 5v
36	BL/YL	Ignition Power	12-14v
37	BR	Reference Voltage	4.9-5.1v
38	RD/BK	Vehicle Power	12-14v
39	BK/RD	Power Ground	<0.1v
40	RD/BL	CKP Sensor Signal	2.5-2.7v

1993-95 Villager 3.0L V6 VIN W (A/T) 116-Pin Connector

PCM Pin #	Wire Color	Circuit Description (116-Pin)	Value at Hot Idle
41	WT/RD	A/C High Pressure Switch	A/C On: 2.5v, Off: 12-14v
42-43	---	Not Used	---
44	RD/YL	IDL Switch (+) Signal	Closed: 8-10v, Open: 0v
45	---	Not Used	---
46	YL	Keep Alive Power	12-14v
47	RD/BK	Vehicle Power	12-14v
48	BK/RD	Power Ground	<0.1v
101	GN/OR	Injector 1 Control	4.2-4.6 ms
102	GY	EGR Control Solenoid	0.7v, 2000 rpm: 12v
103	GN/RD	Injector 3 Control	4.2-4.6 ms
104	BL/RD	Fuel Pump Control	On: 1v, Off: 12v
105	YL/GN	Injector 5 Control	4.2-4.6 ms
106 ('94-'95)	BK	HO2S-11 (B1 S1) Signal	0.1-1.1v
107	BK	Power Ground	<0.1v
108	BK	Power Ground	<0.1v
109	RD/BK	Vehicle Power	12-14v
110	GN	Injector 2 Control	4.2-4.6 ms
111	---	Not Used	---
112	YL/PK	Injector 4 Control	4.2-4.6 ms
113	LB	Auxiliary Air Valve Control	8-11v, with load: 4-7v
114	GY/BL	Injector 6 Control	4.2-4.6 ms
115	---	Not Used	---
116	BK	Power Ground	<0.1v

Pin Connector Graphic

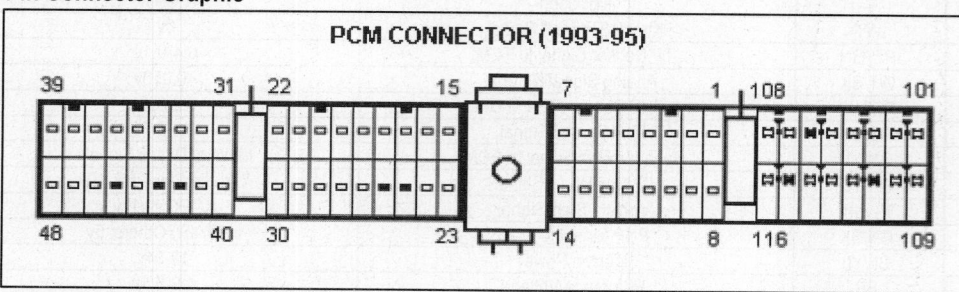

Standard Colors and Abbreviations

Abbreviation	Color	Abbreviation	Color	Abbreviation	Color
BK	Black	GY	Gray	PK	Purple
BL	Blue	GN	Green	RD	Red
BR	Brown	LG	LT Green	TN	Tan
DB	Dark Blue	OR	Orange	WT	White
DG	DK Green	PK	Pink	YL	Yellow

1996-98 Villager 3.0L V6 VIN W, 1 (A/T) 88-Pin Connector

PCM Pin #	Wire Color	Circuit Description (88-Pin)	Value at Hot Idle
1	BL	Ignition Control Signal	0.4-0.6v
2	RD	Ignition Check Signal	9.0-9.5v
3	GN/WT	Tachometer Output Signal	1.0v
4	WT/GN	Main Power Relay Control	On: 1v, Off: 12v
5-6	---	Not Used	---
7	BR/BK	TCM Communication Line	Digital Signals
8	BL/RD	Fuel Pump Control	On: 1v, Off: 12v
9	BL/BK	A/C High Pressure Switch	5v
10	BK	Power Ground	<0.1v
11	---	Not Used	---
12	BL/WT	Self-Test Input Signal	STI On: 1v, Off: 5v
13	BR/WT	High Speed Fan 1 & 2 Control	Fan On: 1v, Off: 12v
14	BL/OR	Low Speed Fan Control	Fan On: 1v, Off: 12v
15	GY/RD	A/C Relay Control	A/C & BLMT On: 1v
16-17	---	Not Used	---
18	PK	MIL (lamp) Control	MIL On: 1v, Off: 12v
19	BK	Power Ground	<0.1v
20	BL/BK	Vehicle Start Signal	KOEC: 9-11v
21	WT/RD	A/C Dual Pressure Switch	A/C & BLMT On: 2-2.5v
22	GN/BK	PNP Switch Signal	In 'P': 0v, Others: 5v
23	RD	TP Sensor Signal	0.4-1.1v
24	GN/WT	A/T Communication Line	Digital Signals
25	PK	Power Steering Pressure Switch	Straight: 5v, Turned: 0v
26	GN/YL	Speedometer Signal to PCM	Moving: Digital Signals
27	---	Not Used	---
28	YL/GN	Intake Air Temperature Sensor	1.5-2.5v
29	W	A/T Communication Line	Digital Signals
30	GN/YL	A/T Communication Line	Digital Signals
31-32	---	Not Used	---
33	WT/RD	TP Sensor Signal to TCM	0.4-1.1v
34-37	---	Not Used	---
38	BL/YL	Vehicle Power	12-14v
39	BK/RD	. Main Ground	<0.1v
40	GN/BK	Camshaft Position Sensor	0.2-0.4v
41	GN/YL	Camshaft Position Sensor	2.0-3.0v
42	---	Not Used	---
43	BK/RD	Power Ground	<0.1v
44	GN/BK	Camshaft Position Sensor	0.2-0.4v
45	GN/YL	Camshaft Position Sensor	2.0-3.0v
46	LG	HO2S-11 (B1 S1) Signal	0.1-1.1v
47	WT/BL	MAF Sensor Signal	1.6v, 55 mph: 2.8v
48	OR/BL	MAF Sensor Reference	12-14v
49	BR	Reference Voltage	4.9-5.1v
50	BK/YL	Analog Signal Return	<0.050v
51	LG/RD	Engine Coolant Temperature Sensor	0.5-0.6v
52	WT	HO2S-12 (B1 S2) Signal	0.1-1.1v
53	LG	CKP (+) Sensor	0.4v
54	WT	Knock Sensor 1 Signal	0v
55	LB	IAC or AAC Valve Control	With Load: 4-7v
56	BK/WT	Vehicle Power	12-14v
57	---	Not Used	---
58	YL/GN	OBD II DLC ISO K-Line	5v
59-60	---	Not Used	---

1996-98 Villager 3.0L V6 VIN W, 1 (A/T) 88-Pin Connector

PCM Pin #	Wire Color	Circuit Description (88-Pin)	Value at Hot Idle
61	BK/WT	Vehicle Power	12-14v
62	WT/PK	EGR Temperature Sensor	0.4-5v, 55 mph: 0.2-1v
63	---	Not Used	---
64	YL/RD	DLC Signal to Scan Tool	0v
65	YL/BK	DLC Signal to Scan Tool	0v
66-67	---	Not Used	---
68	YL/BL	DLC Signal to Scan Tool	0v
69	---	Not Used	---
70	YL	Keep Alive Power	12-14v
71	---	Not Used	---
72	GN/OR	Injector 1 Control	4.2-4.6 ms
73	GY	EGR/EVAP Solenoid Control	12v, >3200 rpm: 1v
74	GN/RD	Injector 3 Control	4.2-4.6 ms
75	---	Not Used	---
76	BK	Power Ground	<0.1v
77	GN	Injector 2 Control	4.2-4.6 ms
78	---	Not Used	---
79	YL/PK	Injector 4 Control	4.2-4.6 ms
80	---	Not Used	---
81	YL/GN	Injector 5 Control	4.2-4.6 ms
82	BK	Power Ground	<0.1v
83	BK/WT	Vehicle Power	12-14v
84	GY/BL	Injector 6 Control	4.2-4.6 ms
85	BK	HO2S-11 (B1 S1) Heater	Heater Off: 12v, On: 1v
86	YL	HO2S-12 (B1 S2) Heater	Heater Off: 12v, On: 1v
87	---	Not Used	---
88	BK	Power Ground	<0.1v

Pin Connector Graphic

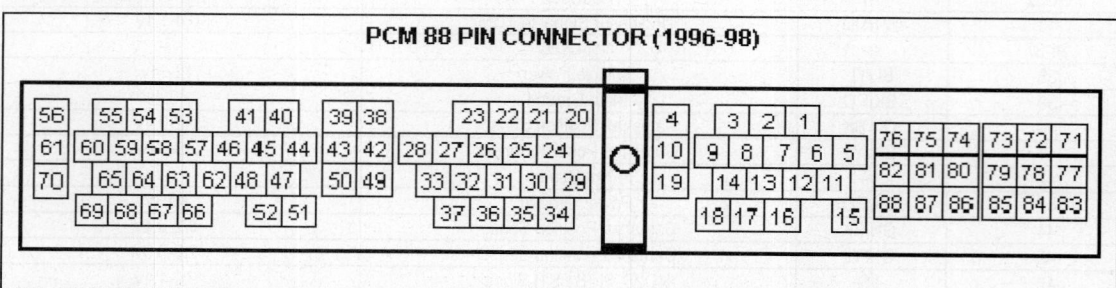

PCM 88 PIN CONNECTOR (1996-98)

Standard Colors and Abbreviations

Abbreviation	Color	Abbreviation	Color	Abbreviation	Color
BK	Black	GY	Gray	PK	Purple
BL	Blue	GN	Green	RD	Red
BR	Brown	LG	LT Green	TN	Tan
DB	Dark Blue	OR	Orange	WT	White
DG	DK Green	PK	Pink	YL	Yellow

1999-2000 Villager 3.3L V6 SOHC VIN T 104 Pin Connector

PCM Pin #	Wire Color	Circuit Description (104 Pin)	Value at Hot Idle
1	BL	Ignition Control Signal	0.4-0.6v
2	W	IDM Sensor Signal	5-7 Hz
3	GN/WT	Tachometer Output Signal	30-35 Hz
4	WT/GN	Main Power Relay Control	On: 1v, Off: 12v
5	GN/BK	EVAP Purge Solenoid	0-10 Hz (0-100%)
6	---	Not Used	---
7	GN/PK	TCM Communication Line	Digital Signals
8	OR/BK	Instrument Cluster Signal	N/A
9	BL/BK	A/C High Pressure Switch	Any speed: 5v
10	BK	Power Ground	<0.1v
11	BL/RD	Main Relay Control	On: 1v, Off: 12v
12	LG	Self-Test Input Signal	STI On: 1v, Off: 5v
13-16	---	Not Used	---
17	BL/WT	DLC (connector) Signals	Digital Signals
18	PK	MIL (lamp) Control	MIL On: 1v, Off: 12v
19	BK	Power Ground	<0.1v
20	BL/BK	Vehicle Start Signal	KOEC: 9-11v
21	WT/RD	A/C Dual Pressure Switch	A/C & BLMT On: 2-2.5v
22	GY/BK	PNP Switch Signal	In 'P': 0v, Others: 5v
23	RD	TP Sensor Signal	0.4-1.1v
24	GN/WT	A/T Communication Line	Digital Signals
25	WT	TCM Signal (TCM Pin 6)	Digital Signals
26	BL/YL	Wide Open Throttle Signal	12v, WOT: 0v
27	GN/YL	Vehicle Speed Sensor	Digital Signals
28	---	Not Used	---
29	BL/RD	A/T Communication Line	Digital Signals
30	BL/WT	A/T Communication Line	Digital Signals
31	PK/BK	TCM Signal (TCM Pin 7)	Digital Signals
32	BR/WT	High Speed Fan Control	Fan On: 1v, Off: 12v
33	RD/GN	BARO/MAP Solenoid Control	On: 1v, Off: 12v
34	BL/OR	Low Speed Fan Control	Fan On: 1v, Off: 12v
35	PK	Power Steering Pressure Switch	Straight: 0v, Turned: 12v
36-37	---	Not Used	---
38	BL/YL	Vehicle Power	12-14v
39	BK/RD	Power Ground	<0.1v
40	WT	Camshaft Position Sensor	0.2-0.4v
41	PK	MAP Sensor Signal	0.9v
42	---	Not Used	---
43	LG	CKP Sensor Signal	0.4v
44	---	Not Used	---
45	BK/RD	Power Ground	<0.1v
46	WT	Camshaft Position Sensor	0.2-0.4v
47	OR	Camshaft Position Sensor	2.0-3.0v
48	LG	HO2S-11 (B1 S1) Signal	0.1-1.1v
49-51	BR	Reference Voltage	4.9-5.1v
52	WT/BL	MAF Sensor Signal	1.6v, 55 mph: 2.8v
53	BL/OR	MAF Sensor Return	<0.050v
54	WT	HO2S-12 (B1 S2) Heater	1v, Off: 12v

1999-2000 Villager 3.3L V6 SOHC VIN T 104 Pin Connector

PCM Pin #	Wire Color	Circuit Description (104 Pin)	Value at Hot Idle
55-56	---	Not Used	---
57	BR	Vehicle Reference	4.9-5.1v
58	BK/YL	Analog Sensor Ground	<0.050v
59	LG/RD	Engine Coolant Temperature Sensor	0.5-0.6v
60	RD/YL	Fuel Pump Control	On: 1v, Off: 12v
61	YL/GN	Intake Air Temperature Sensor	1.5-2.5v
62	RD	EVAP Pressure Sensor	2.6v at 0" Hg (cap off)
63	WT	Knock Sensor Signal	0v
64-65	---	Not Used	---
66	BK/WT	Main Relay Power to PCM	12-14v
67	---	Not Used	---
68	YL/GN	DLC Signal to Scan Tool	No Scan Tool: 0v
69-70	---	Not Used	---
71	BK/WT	Vehicle Power (Main Relay)	12-14v
72-73	---	Not Used	---
74	YL/RD	DLC (connector) Signal	Digital Signals
75	YL/RD	DLC (connector) Signal	Digital Signals
76	WT/PK	EGR Temperature Sensor	0.4-5v, 55 mph: 0.2-1v
77-79	---	Not Used	---
80	YL	Direct Battery	12-14v
81	LB	ISC Solenoid Control	7-9v
82	GN/OR	Injector 1 Control	4.0-4.6 ms
83	GY	VR Solenoid Control	0%, at 55 mph: 45%
84	GN/RD	Injector 3 Control	4.0-4.6 ms
85, 87	---	Not Used	---
86	YL/GN	Injector 5 Control	4.0-4.6 ms
88	LG/BK	EVAP CV Solenoid	0-10 Hz (0-100%)
89	GN	Injector 2 Control	4.0-4.6 ms
90	---	Not Used	---
91	YL/PK	Injector 4 Control	4.0-4.6 ms
92	---	Not Used	---
93	GY/BL	Injector 6 Control	4.0-4.6 ms
94-95	---	Not Used	---
96	BK	Power Ground	<0.1v
97	BK/WT	Vehicle Power (Main Relay)	12-14v
98	PK	BARO/MAP Solenoid Control	On: 1v, Off: 12v
99	BK	HO2S-11 (B1 S1) Heater	1v, Off: 12v
100	BL/GN	EVAP Vacuum Cut Valve	0-10 Hz (0-100%)
101	---	Not Used	---
102	YL	HO2S-12 (B1 S2) Signal	0.1-1.1v
103	---	Not Used	---
104	BK	Power Ground	<0.1v

Pin Connector Graphic

Standard Colors and Abbreviations

Abbreviation	Color	Abbreviation	Color	Abbreviation	Color
BK	Black	GY	Gray	PK	Purple
BL	Blue	GN	Green	RD	Red
BR	Brown	LG	LT Green	TN	Tan
DB	Dark Blue	OR	Orange	WT	White
DG	DK Green	PK	Pink	YL	Yellow

2001-02 Villager 3.3L V6 SOHC VIN T 104 Pin Connector

PCM Pin #	Wire Color	Circuit Description (104 Pin)	Value at Hot Idle
1	BL	Ignition Control Signal	0.4-0.6v
2	WT	IDM Sensor Signal	5-7 Hz
3	GN/WT	Tachometer Output Signal	30-35 Hz
4	WT/GN	Main Power Relay Control	On: 1v, Off: 12v
5	GN/BK	EVAP Purge Solenoid	0-10 Hz (0-100%)
6	---	Not Used	---
7	GN/PK	TCM Communication Line	Digital Signals
8	OR/BK	Instrument Cluster Signal	N/A
9	BL/BK	A/C High Pressure Switch	Any speed: 5v
10	BK	Power Ground	<0.1v
11	BL/RD	Main Relay Control	On: 1v, Off: 12v
12	LG	Self-Test Input Signal	STI On: 1v, Off: 5v
13-16	---	Not Used	---
17	BL/WT	DLC (connector) Signals	Digital Signals
18	PK	MIL (lamp) Control	MIL On: 1v, Off: 12v
19	BK	Power Ground	<0.1v
20	BL/BK	Vehicle Start Signal	KOEC: 9-11v
21	WT/RD	A/C Dual Pressure Switch	A/C & BLMT On: 2-2.5v
22	GY/BK	PNP Switch Signal	In 'P': 0v, Others: 5v
23	RD	TP Sensor Signal	0.4-1.1v
24	GN/WT	A/T Communication Line	Digital Signals
25	WT	TCM Signal (TCM Pin 6)	Digital Signals
26	BL/YL	Wide Open Throttle Signal	12v, WOT: 0v
27	GN/YL	Vehicle Speed Sensor	Digital Signals
28	---	Not Used	---
29	BL/RD	A/T Communication Line	Digital Signals
30	BL/WT	A/T Communication Line	Digital Signals
31	PK/BK	TCM Signal (TCM Pin 7)	Digital Signals
32	BR/WT	High Speed Fan Control	Fan On: 1v, Off: 12v
33	RD/GN	BARO/MAP Solenoid Control	On: 1v, Off: 12v
34	BL/OR	Low Speed Fan Control	Fan On: 1v, Off: 12v
35	PK	Power Steering Pressure Switch	Straight: 0v, Turned: 12v
36-37	---	Not Used	---
38	BL/YL	Vehicle Power	12-14v
39	BK/RD	Power Ground	<0.1v
40	WT	Camshaft Position Sensor	0.2-0.4v
41	PK	MAP Sensor Signal	0.9v
42	---	Not Used	---
43	LG	CKP Sensor Signal	0.4v
44	---	Not Used	---
45	BK/RD	Power Ground	<0.1v
46	WT	Camshaft Position Sensor	0.2-0.4v
47	OR	Camshaft Position Sensor	2.0-3.0v
48	LG	HO2S-11 (B1 S1) Signal	0.1-1.1v
49-51	BR	Reference Voltage	4.9-5.1v
52	WT/BL	MAF Sensor Signal	1.6v, 55 mph: 2.8v
53	BL/OR	MAF Sensor Return	<0.050v
54	WT	HO2S-12 (B1 S2) Heater	1v, Off: 12v

2001-02 Villager 3.3L V6 SOHC VIN T 104 Pin Connector

PCM Pin #	Wire Color	Circuit Description (104 Pin)	Value at Hot Idle
55-56	---	Not Used	---
57	BR	Vehicle Reference	4.9-5.1v
58	BK/YL	Analog Sensor Ground	<0.050v
59	LG/RD	Engine Coolant Temperature Sensor	0.5-0.6v
60	RD/YL	Fuel Pump Control	On: 1v, Off: 12v
61	YL/GN	Intake Air Temperature Sensor	1.5-2.5v
62	RD	EVAP Pressure Sensor	2.6v at 0" Hg (cap off)
63	WT	Knock Sensor Signal	0v
64-65	---	Not Used	---
66	BK/WT	Main Relay Power to PCM	12-14v
67	---	Not Used	---
68	YL/GN	DLC Signal to Scan Tool	No Scan Tool: 0v
69-70	---	Not Used	---
71	BK/WT	Vehicle Power (Main Relay)	12-14v
72-73	---	Not Used	---
74	YL/RD	DLC (connector) Signal	Digital Signals
75	YL/RD	DLC (connector) Signal	Digital Signals
76	WT/PK	EGR Temperature Sensor	0.4-5v, 55 mph: 0.2-1v
77-79	---	Not Used	---
80	YL	Direct Battery	12-14v
81	LB	ISC Solenoid Control	7-9v
82	GN/OR	Injector 1 Control	4.0-4.6 ms
83	GY	VR Solenoid Control	0%, at 55 mph: 45%
84	GN/RD	Injector 3 Control	4.0-4.6 ms
85	---	Not Used	---
86	YL/GN	Injector 5 Control	4.0-4.6 ms
87	---	Not Used	---
88	LG/BK	EVAP CV Solenoid	0-10 Hz (0-100%)
89	GN	Injector 2 Control	4.0-4.6 ms
90	---	Not Used	---
91	YL/PK	Injector 4 Control	4.0-4.6 ms
92	---	Not Used	---
93	GY/BL	Injector 6 Control	4.0-4.6 ms
94-95	---	Not Used	---
96	BK	Power Ground	<0.1v
97	BK/WT	Vehicle Power (Main Relay)	12-14v
98	PK	BARO/MAP Solenoid Control	On: 1v, Off: 12v
99	BK	HO2S-11 (B1 S1) Heater	1v, Off: 12v
100	BL/GN	EVAP Vacuum Cut Valve	0-10 Hz (0-100%)
101	---	Not Used	---
102	YL	HO2S-12 (B1 S2) Signal	0.1-1.1v
103	---	Not Used	---
104	BK	Power Ground	<0.1v

Pin Connector Graphic

97	98	99	100		101	102	103	104
89	90	91	92		93	94	95	96
81	82	83	84		85	86	87	88

15	16	17	18		19
11	12	13	14		10
5	6	7	8	9	
1	2	3	4		4

34	35		36	37		57	58
29	30	31	32	33		44	45
24	25	26	27	28			
20	21	22		23		38	39

59	60	61	62	76		77	78	79		80
52	53	54	55	56	72	73	74	75		71
46	47	48	49	50	51	67	68	69	70	
40	41	42	43		63	64	65			88

Standard Colors and Abbreviations

Abbreviation	Color	Abbreviation	Color	Abbreviation	Color
BK	Black	GY	Gray	PK	Purple
BL	Blue	GN	Green	RD	Red
BR	Brown	LG	LT Green	TN	Tan
DB	Dark Blue	OR	Orange	WT	White
DG	DK Green	PK	Pink	YL	Yellow

Windstar PIN Tables

1995-97 Windstar 3.0L V6 VIN U (A/T) 104 Pin Connector

PCM Pin #	Wire Color	Circuit Description (104 Pin)	Value at Hot Idle
1	PK/OR	Shift Solenoid 2 Control	1v, 55 mph: 12v
2	PK/LG	MIL (lamp) Control	MIL On: 1v, Off: 12v
3	---	Not Used	---
4 ('95)	RD/OR	Cooling Fan Monitor	Fan On: 12v, Off: 0v
5-9	---	Not Used	---
10 ('96-'97)	YL/BK	Coil 1 Driver (Dwell) (Cyl 1, 5)	5°, at 55 mph: 8°
11 ('96-'97)	YL/WT	Coil 2 Driver (Dwell) (Cyl 3, 4)	5°, at 55 mph: 8°
12 ('96-'97)	YL/RD	Coil 3 Driver (Dwell) (Cyl 2, 6)	5°, at 55 mph: 8°
13	YL/BK	Flash EEPROM	0.1v
14	---	Not Used	---
15	PK/LB	Data Bus (-)	Digital Signals
16	TN/OR	Data Bus (+)	Digital Signals
17	LG/PK	High Speed Fan Control	Fan On: 1v, Off: 12v
18-22	---	Not Used	---
23 ('95)	OR/RD	Ignition System Ground	<0.050v
24	BK/WT	Power Ground	<0.1v
25	BK	Case Ground	<0.050v
26	---	Not Used	---
27	OR/YL	Shift Solenoid 1 Control	12v, 55 mph: 1v
28	---	Not Used	---
29	TN/WT	TCS (switch) Signal	12v (Closed: 0v)
30	DG	Octane Adjustment	9.3v (switch shorted: 0v)
31-32	---	Not Used	---
33	OR/YL	VSS (-) Signal	0 Hz, at 55 mph: 125 Hz
34	---	Not Used	---
35	RD/LG	HO2S-12 (B1 S2) Signal	0.1-1.1v
36	TN/LB	MAF Sensor Return	<0.050v
37	OR/BK	Transmission Fluid Temperature Sensor	2.10-2.40v
38	LG/RD	Engine Coolant Temperature Sensor	0.5-0.6v
39	GY	Intake Air Temperature Sensor	1.5-2.5v
40	DG/YL	Fuel Pump Monitor	On: 12v, Off: 0v
41	BK/YL	A/C Pressure Switch Signal	A/C On: 12v, off: 0v
42	---	Not Used	---
43	OR/LG	Fuel Flow Rate Signal	Digital Signals
44-46	---	Not Used	---
47	BK/PK	VR Solenoid Control	0%, at 55 mph: 45%
48	WT/PK	Clean Tachometer Output	33-37 Hz
49 ('95)	GY/OR	Profile Ignition Pickup	6.93v (50%)
50 ('95)	PK	Spark Output Signal	6.93v (50%)
51	BK/WT	Power Ground	<0.1v
52	---	Not Used	---
53	PK/BK	Shift Solenoid 3 Control	12v, 55 mph: 1v
54	---	Not Used	---
55	YL	Keep Alive Power	12-14v
56	LG/BK	EVAP VMV Solenoid	0-10 Hz (0-100%)
57	---	Not Used	---
58	DG/WT	VSS (+) Signal	0 Hz, at 55 mph: 125 Hz
59	---	Not Used	---
60	GY/LB	HO2S-11 (B1 S1) Signal	0.1-1.1v
61	PK/LG	HO2S-22 (B2 S2) Signal	0.1-1.1v
62-63	---	Not Used	---
64	LB/YL	Transmission Range Sensor	In 'P': 0v, in O/D: 1.7v

1995-97 Windstar 3.0L V6 VIN U (A/T) 104 Pin Connector

PCM Pin #	Wire Color	Circuit Description (104 Pin)	Value at Hot Idle
65	BR/LG	EGR DPFE Sensor Signal	0.20-1.30v
66-68	---	Not Used	---
69	PK/YL	A/C WOT Relay Control	On: 1v, Off: 12v
70	---	Not Used	---
71	RD	Vehicle Power	12-14v
72	---	Not Used	---
73	TN/BK	Injector 5 Control	5.0-5.2 ms
74	BR/YL	Injector 3 Control	5.0-5.2 ms
75	TN	Injector 1 Control	5.0-5.2 ms
76	BK/WT	Power Ground	<0.1v
77	BK/WT	Power Ground	<0.1v
78	---	Not Used	---
79	WT/LG	TCIL (lamp) Control	Lamp On: 1v, Off: 12v
80	LB/OR	Fuel Pump Control	On: 1v, Off: 12v
81	WT/YL	EPC Solenoid Control	9.9v (15 psi)
82	RD/LB	TCC Solenoid Control	0%, at 55 mph: 95%
83	WT/LB	IAC Solenoid Control	10.7v (33%)
84	GY/BK	Turbine Shaft Speed Sensor	120 Hz, 30 mph: 300 Hz
85	DB/OR	Camshaft Position Sensor	6 Hz, 55 mph: 14-18 Hz
86	TN/LG	A/C Pressure Switch Signal	Switch On: 12v (Open)
87	RD/BK	HO2S-21 (B2 S1) Signal	0.1-1.1v
88	LB/RD	MAF Sensor Signal	0.8v, 55 mph: 1.6v
89	GY/WT	TP Sensor Signal	0.53-1.27v
90	BR/WT	Reference Voltage	4.9-5.1v
91	GY/RD	Analog Signal Return	<0.050v
92	RD/LG	Brake Position Switch	Brake Off: 12v, On: 1v
93	RD/WT	HO2S-11 (B1 S1) Heater	1v, Off: 12v
94	YL/LB	HO2S-21 (B2 S1) Heater	1v, Off: 12v
95	WT/BK	HO2S-12 (B1 S2) Heater	1v, Off: 12v
96	TN/YL	HO2S-22 (B2 S2) Heater	1v, Off: 12v
97	RD	Vehicle Power	12-14v
98	DB	Low Speed Fan Control	Fan On: 1v, Off: 12v
99	LG/OR	Injector 6 Control	5.0-5.2 ms
100	BR/LB	Injector 4 Control	5.0-5.2 ms
101	WT	Injector 2 Control	5.0-5.2 ms
102	---	Not Used	---
103	BK/WT	Power Ground	<0.1v
104	---	Not Used	---

Pin Connector Graphic

PCM 104-PIN CONNECTOR

Terminal View of 104-Pin PCM Wiring Harness Connector

Standard Colors and Abbreviations

Abbreviation	Color	Abbreviation	Color	Abbreviation	Color
BK	Black	GY	Gray	PK	Purple
BL	Blue	GN	Green	RD	Red
BR	Brown	LG	LT Green	TN	Tan
DB	Dark Blue	OR	Orange	WT	White
DG	DK Green	PK	Pink	YL	Yellow

1998-2000 Windstar 3.0L V6 VIN U (A/T) 104 Pin Connector

PCM Pin #	Wire Color	Circuit Description (104 Pin)	Value at Hot Idle
1	PK/OR	Shift Solenoid 2/B Control	1v, 55 mph: 12v
2 ('98-'99)	PK/LG	MIL (lamp) Control	MIL On: 1v, Off: 12v
3	OR/BK	Digital TR1 Sensor	In 'P': 0v, 55 mph: 11v
4-5	---	Not Used	---
6	DB/YL	Output Shaft Speed Sensor	0 Hz, 55 mph: 2 KHz
7-12	---	Not Used	---
13	VT	Flash EEPROM	0.1v
14	---	Not Used	---
15	PK/LB	Data Bus (-)	Digital Signals
16	TN/OR	Data Bus (+)	Digital Signals
17 ('98-'99)	LG/PK	High Speed Fan Control	Fan On: 1v, Off: 12v
18-20	---	Not Used	---
21	BK/PK	CKP (+) Sensor	518-540 Hz
22	GY/YL	CKP (-) Sensor	518-540 Hz
23	---	Not Used	---
24	BK/WT	Power Ground	<0.1v
25	BK	Case Ground	<0.050v
26	DB/LG	Coil 1 Driver (Dwell)	5°, at 55 mph: 8°
27	OR/YL	Shift Solenoid 1/A Control	12v, 55 mph: 1v
28	DB	Low Speed Fan Control	Fan On: 1v, Off: 12v
29-34	---	Not Used	---
35	RD/LG	HO2S-12 (B1 S2) Signal	0.1-1.1v
36	TN/LB	MAF Sensor Return	<0.050v
37	OR/BK	Transmission Fluid Temperature Sensor	2.10-2.40v
38	LG/RD	Engine Coolant Temperature Sensor	0.5-0.6v
39	GY	Intake Air Temperature Sensor	1.5-2.5v
40	DG/YL	Fuel Pump Monitor	On: 12v, Off: 0v
41	BK/YL	A/C High Pressure Switch	A/C On: 12v, off: 0v
42	---	Not Used	---
43	OR/LG	Fuel Flow Output Signal	Digital Signals
44-45	---	Not Used	---
46	LG/PK	High Speed Fan Control	Fan On: 1v, Off: 12v
47	BR/PK	VR Solenoid Control	0%, at 55 mph: 45%
48	---	Not Used	---
49	LB/BK	Digital TR2 Sensor	In 'P': 0v, 55 mph: 11v
50	WT/BK	Digital TR4 Sensor	In 'P': 0v, 55 mph: 11v
51	BK/WT	Power Ground	<0.1v
52	RD/LB	Coil 2 Driver (Dwell)	5°, at 55 mph: 8°
53	PK/BK	Shift Solenoid 3/C Control	12v, 55 mph: 1v
54	RD/LB	TCC Solenoid Control	0%, at 55 mph: 95%
55	Y	Keep Alive Power	12-14v
56	LG/BK	EVAP Purge Solenoid	0-10 Hz (0-100%)
57-59	---	Not Used	---
60	GY/LB	HO2S-11 (B1 S1) Signal	0.1-1.1v
61	PK/LG	HO2S-22 (B2 S2) Signal	0.1-1.1v
62	RD/PK	Fuel Tank Pressure Sensor	2.6v at 0" Hg (cap off)
63	---	Not Used	---
64	LB/YL	Transmission Range Sensor	In 'P': 0v, in O/D: 1.7v

1998-2000 Windstar 3.0L V6 VIN U (A/T) 104 Pin Connector

PCM Pin #	Wire Color	Circuit Description (104 Pin)	Value at Hot Idle
65	BR/LG	EGR DPFE Sensor Signal	0.20-1.30v
66	---	Not Used	---
67	PKWT	EVAP CV Solenoid	0-10 Hz (0-100%)
68	---	Not Used	---
69	PK/YL	A/C WOT Relay Control	On: 1v, Off: 12v
70	---	Not Used	---
71	RD	Vehicle Power	12-14v
72	---	Not Used	---
73	TN/BK	Injector 5 Control	5.0-5.2 ms
74	BR/YL	Injector 3 Control	5.0-5.2 ms
75	TN	Injector 1 Control	5.0-5.2 ms
76	BK/WT	Power Ground	<0.1v
77	BK/WT	Power Ground	<0.1v
78	PKWT	Coil 3 Driver (Dwell)	5°, at 55 mph: 8°
79	---	Not Used	---
80	LB/OR	Fuel Pump Control	On: 1v, Off: 12v
81	WT/YL	EPC Solenoid Control	9.9v (15 psi)
82	---	Not Used	---
83	WT/LB	IAC Solenoid Control	10.7v (33%)
84	GY/BK	Turbine Shaft Speed Sensor	120 Hz, 30 mph: 300 Hz
85	DB/OR	Camshaft Position Sensor	6 Hz, 55 mph: 15-20 Hz
86	TN/LG	A/C Pressure Switch Signal	Switch On: 12v (Open)
87	RD/BK	HO2S-21 (B2 S1) Signal	0.1-1.1v
88	LB/RD	MAF Sensor Signal	0.8v, 55 mph: 1.9v
89	GY/WT	TP Sensor Signal	0.9v, 55 mph: 1.2v
90	BR/WT	Reference Voltage	4.9-5.1v
91	GY/RD	Analog Signal Return	<0.050v
92	---	Not Used	---
93	RD/WT	HO2S-11 (B1 S1) Heater	1v, Off: 12v
94	YL/LB	HO2S-21 (B2 S1) Heater	1v, Off: 12v
95	WT/BK	HO2S-12 (B1 S2) Heater	1v, Off: 12v
96	TN/YL	HO2S-22 (B2 S2) Heater	1v, Off: 12v
97	RD	Vehicle Power	12-14v
98	---	Not Used	---
99	LG/OR	Injector 6 Control	5.0-5.2 ms
100	BR/LB	Injector 4 Control	5.0-5.2 ms
101	WT	Injector 2 Control	5.0-5.2 ms
102	---	Not Used	---
103	BK/WT	Power Ground	<0.1v
104	---	Not Used	---

Pin Connector Graphic

```
         PCM 104-PIN CONNECTOR
    1 ●●●●●●●●●●●●●     ●●●●●●●●●●●●● 26
   27 ●●●●●●●●●●●●●     ●●●●●●●●●●●●● 52
   53 ●●●●●●●●●●●●●  ⬤  ●●●●●●●●●●●●● 78
   79 ●●●●●●●●●●●●●     ●●●●●●●●●●●●● 104

   Terminal View of 104-Pin PCM Wiring Harness Connector
```

Standard Colors and Abbreviations

Abbreviation	Color	Abbreviation	Color	Abbreviation	Color
BK	Black	GY	Gray	PK	Purple
BL	Blue	GN	Green	RD	Red
BR	Brown	LG	LT Green	TN	Tan
DB	Dark Blue	OR	Orange	WT	White
DG	DK Green	PK	Pink	YL	Yellow

1995 Windstar 3.8L V6 VIN 4 (A/T) 104 Pin Connector

PCM Pin #	Wire Color	Circuit Description (104 Pin)	Value at Hot Idle
1	PK/OR	Shift Solenoid 2 Control	1v, 55 mph: 12v
2	PK/LG	MIL (lamp) Control	MIL On: 1v, Off: 12v
3	---	Not Used	---
4	RD/OR	Low Speed Fan Control	Fan On: 12v, Off: 0v
5-12	---	Not Used	---
13	YL/BK	Flash EEPROM	0.1v
14	---	Not Used	---
15	PK/LB	Data Bus (-)	Digital Signals
16	TN/OR	Data Bus (+)	Digital Signals
17	LG/PK	High Speed Fan Control	Fan On: 1v, Off: 12v
18-22	---	Not Used	---
23	OR/RD	Ignition System Ground	<0.050v
24	BK/WT	Power Ground	<0.1v
25	BK	Case Ground	<0.050v
26	---	Not Used	---
27	OR/YL	Shift Solenoid 1 Control	12v, 55 mph: 1v
28	---	Not Used	---
29	TN/WT	TCS (switch) Signal	0v (Closed: 12v)
30	DG	Octane Adjustment	9.3v (switch shorted: 0v)
31-32	---	Not Used	---
33	OR/YL	VSS (-) Signal	0 Hz, at 55 mph: 125 Hz
34	---	Not Used	---
35	RD/LG	HO2S-12 (B1 S2) Signal	0.1-1.1v
36	TN/LB	MAF Sensor Return	<0.050v
37	OR/BK	Transmission Fluid Temperature Sensor	2.10-2.40v
38	LG/RD	Engine Coolant Temperature Sensor	0.5-0.6v
39	GY	Intake Air Temperature Sensor	1.5-2.5v
40	DG/YL	Fuel Pump Monitor	On: 12v, Off: 0v
41	BK/YL	A/C High Pressure Switch	A/C On: 12v, off: 0v
42	---	Not Used	---
43	OR/LG	Fuel Flow Rate Signal	Digital Signals
44-46	---	Not Used	---
47	BR/PK	VR Solenoid Control	0%, at 55 mph: 45%
48	WT/PK	Clean Tachometer Output	33-37 Hz
49	GY/OR	Profile Ignition Pickup	6.93v (50%)
50	PK	Spark Output Signal	6.93v (50%)
51	BK/WT	Power Ground	<0.1v
52	---	Not Used	---
53	PK/BK	Shift Solenoid 3 Control	12v, 55 mph: 1v
54	---	Not Used	---
55	YL	Keep Alive Power	12-14v
56	LG/BK	EVAP Purge Solenoid	0-10 Hz (0-100%)
57	---	Not Used	---
58	DG/WT	VSS (+) Signal	0 Hz, at 55 mph: 125 Hz
59	---	Not Used	---
60	GY/LB	HO2S-11 (B1 S1) Signal	0.1-1.1v
61	PK/LG	HO2S-12 (B1 S2) Signal	0.1-1.1v
62-63	---	Not Used	---
64	LB/YL	Digital Transmission Range Sensor	In 'P': 0v, in O/D: 1.7v

1995 Windstar 3.8L V6 VIN 4 (A/T) 104 Pin Connector

PCM Pin #	Wire Color	Circuit Description (104 Pin)	Value at Hot Idle
65	BR/LG	EGR DPFE Sensor Signal	0.20-1.30v
66-68	---	Not Used	---
69	PK/YL	A/C WOT Relay Control	On: 1v, Off: 12v
70	---	Not Used	---
71	RD	Vehicle Power	12-14v
72	---	Not Used	---
73	TN/BK	Injector 5 Control	3.5-3.8 ms
74	BR/YL	Injector 3 Control	3.5-3.8 ms
75	TN	Injector 1 Control	3.5-3.8 ms
76	BK/WT	Power Ground	<0.1v
77	BK/WT	Power Ground	<0.1v
78	---	Not Used	---
79	WT/LG	TCIL (lamp) Control	Lamp On: 1v, Off: 12v
80	LB/OR	Fuel Pump Control	On: 1v, Off: 12v
81	WT/YL	EPC Solenoid Control	9v (15 psi)
82	RD/LB	TCC Solenoid Control	0%, at 55 mph: 95%
83	WT/LB	IAC Solenoid Control	10.7v (33%)
84	GY/BK	Turbine Shaft Speed Sensor	120 Hz, 30 mph: 300 Hz
85	DB/OR	Camshaft Position Sensor	6 Hz, 55 mph: 14-18 Hz
86	TN/LG	A/C Pressure Switch	Switch On: 12v (Open)
87	RD/BK	HO2S-21 (B2 S1) Signal	0.1-1.1v
88	LB/RD	MAF Sensor Signal	0.8v, 55 mph: 1.6v
89	GY/WT	TP Sensor Signal	0.53-1.27v
90	BR/WT	Reference Voltage	4.9-5.1v
91	GY/RD	Analog Signal Return	<0.050v
92	RD/LG	Brake Position Switch	Brake Off: 12v, On: 1v
93	RD/WT	HO2S-11 (B1 S1) Heater	1v, Off: 12v
94	YL/LB	HO2S-21 (B2 S1) Heater	1v, Off: 12v
95	WT/BK	HO2S-12 (B1 S2) Heater	1v, Off: 12v
96	TN/YL	HO2S-22 (B2 S2) Heater	1v, Off: 12v
97	RD	Vehicle Power	12-14v
98	DB	Low Speed Fan Control	Fan On: 1v, Off: 12v
99	LG/OR	Injector 6 Control	3.5-3.8 ms
100	BK/LB	Injector 4 Control	3.5-3.8 ms
101	WT	Injector 2 Control	3.5-3.8 ms
102	---	Not Used	---
103	BK/WT	Power Ground	<0.1v
104	---	Not Used	---

Pin Connector Graphic

PCM 104-PIN CONNECTOR

Terminal View of 104-Pin PCM Wiring Harness Connector

Standard Colors and Abbreviations

Abbreviation	Color	Abbreviation	Color	Abbreviation	Color
BK	Black	GY	Gray	PK	Purple
BL	Blue	GN	Green	RD	Red
BR	Brown	LG	LT Green	TN	Tan
DB	Dark Blue	OR	Orange	WT	White
DG	DK Green	PK	Pink	YL	Yellow

1996-97 Windstar 3.8L V6 VIN 4 (A/T) 104 Pin Connector

PCM Pin #	Wire Color	Circuit Description (104 Pin)	Value at Hot Idle
1	PK/OR	Shift Solenoid 2 Control	1v, 55 mph: 12v
2	PK/LG	MIL (lamp) Control	MIL On: 1v, Off: 12v
3-7	---	Not Used	---
8	OR/WT	IMRC Monitor 2	6.9v, 55 mph: 6.9v
9	TN	IMRC Monitor 1	6.9v, 55 mph: 6.9v
10-12	---	Not Used	---
13	YL/BK	Flash EEPROM	0.1v
14	---	Not Used	---
15	PK/LB	Data Bus (-)	Digital Signals
16	TN/OR	Data Bus (+)	Digital Signals
17-20	---	Not Used	---
21	DB	CKP (+) Sensor	390-450 Hz
22	GY	CKP (-) Sensor	390-450 Hz
23	---	Not Used	---
24	BK/WT	Power Ground	<0.1v
25	BK	Case Ground	<0.050v
26	DB/LG	Coil 1 Driver (Dwell)	5°, at 55 mph: 8°
27	OR/YL	Shift Solenoid 1 Control	12v, 55 mph: 1v
28	DB	Low Speed Fan Control	Fan On: 1v, Off: 12v
29	TN/WT	TCS (switch) Signal	0v (Closed: 12v)
30	DG	Octane Adjustment	9.3v (switch shorted: 0v)
31-32	---	Not Used	---
33	OR/YL	VSS (-) Signal	0 Hz, at 55 mph: 125 Hz
34	---	Not Used	---
35	RD/LG	HO2S-12 (B1 S2) Signal	0.1-1.1v
36	TN/LB	MAF Sensor Return	<0.050v
37	OR/BK	Transmission Fluid Temperature Sensor	2.10-2.40v
38	LG/RD	Engine Coolant Temperature Sensor	0.5-0.6v
39	GY	Intake Air Temperature Sensor	1.5-2.5v
40	DG/YL	Fuel Pump Monitor	On: 12v, Off: 0v
41	BK/YL	A/C High Pressure Switch	A/C On: 12v, off: 0v
42	BR	IMRC Solenoid Control	12v, 55 mph: 12v
43	OR/LG	Fuel Flow Rate Signal	Digital Signals
44-45	---	Not Used	---
46	LG/PK	High Speed Fan Control	Fan On: 1v, Off: 12v
47	BR/PK	VR Solenoid Control	0%, at 55 mph: 45%
48	WT/PK	Clean Tachometer Output	33-37 Hz
49-50	---	Not Used	---
51	BK/WT	Power Ground	<0.1v
52	RD/LB	Coil 2 Driver (Dwell)	5°, at 55 mph: 8°
53	PK/BK	Shift Solenoid 3 Control	12v, 55 mph: 1v
54	---	Not Used	---
55	YL	Keep Alive Power	12-14v
56	LG/BK	EVAP Purge Solenoid	0-10 Hz (0-100%)
57	---	Not Used	---
58	DG/WT	VSS (+) Signal	0 Hz, at 55 mph: 125 Hz
59	---	Not Used	---
60	GY/LB	HO2S-11 (B1 S1) Signal	0.1-1.1v
61	PK/LG	HO2S-12 (B1 S2) Signal	0.1-1.1v
62-63	---	Not Used	---
64	LB/YL	Digital Transmission Range Sensor	In 'P': 0v, in O/D: 1.7v

1995-97 Windstar 3.8L V6 VIN 4 (A/T) 104 Pin Connector

PCM Pin #	Wire Color	Circuit Description (104 Pin)	Value at Hot Idle
65	BR/LG	EGR DPFE Sensor Signal	0.20-1.30v
66-68	---	Not Used	---
69	PK/YL	A/C WOT Relay Control	On: 1v, Off: 12v
70	---	Not Used	---
71	RD	Vehicle Power (Start-Run)	12-14v
72	---	Not Used	---
73	TN/BK	Injector 5 Control	3.5-3.8 ms
74	BR/YL	Injector 3 Control	3.5-3.8 ms
75	TN	Injector 1 Control	3.5-3.8 ms
76	BK/WT	Power Ground	<0.1v
77	BK/WT	Power Ground	<0.1v
78	RD/LB	Coil 3 Driver (Dwell)	5°, at 55 mph: 8°
79	WT/LG	TCIL (lamp) Control	Lamp On: 1v, Off: 12v
80	LB/OR	Fuel Pump Control	On: 1v, Off: 12v
81	WT/YL	EPC Solenoid Control	9v (15 psi)
82	RD/LB	TCC Solenoid Control	0%, at 55 mph: 95%
83	WT/LB	IAC Solenoid Control	10.7v (33%)
84	GY/BK	Turbine Shaft Speed Sensor	120 Hz, 30 mph: 300 Hz
85	DB/OR	Camshaft Position Sensor	6 Hz, 55 mph: 14-18 Hz
86	TN/LG	A/C Pressure Switch	Switch On: 12v (Open)
87	RD/BK	HO2S-21 (B2 S1) Signal	0.1-1.1v
88	LB/RD	MAF Sensor Signal	0.8v, 55 mph: 1.6v
89	GY/WT	TP Sensor Signal	0.53-1.27v
90	BR/WT	Reference Voltage	4.9-5.1v
91	GY/RD	Analog Signal Return	<0.050v
92	RD/LG	Brake Position Switch	Brake Off: 12v, On: 1v
93	RD/WT	HO2S-11 (B1 S1) Heater	1v, Off: 12v
94	YL/LB	HO2S-21 (B2 S1) Heater	1v, Off: 12v
95	WT/BK	HO2S-12 (B1 S2) Heater	1v, Off: 12v
96	TN/YL	HO2S-22 (B2 S2) Heater	1v, Off: 12v
97	RD	Vehicle Power (Start-Run)	12-14v
98	---	Not Used	---
99	LG/OR	Injector 6 Control	3.5-3.8 ms
100	BK/LB	Injector 4 Control	3.5-3.8 ms
101	WT	Injector 2 Control	3.5-3.8 ms
103	BK/WT	Power Ground	<0.1v

Pin Connector Graphic

PCM 104-PIN CONNECTOR

Terminal View of 104-Pin PCM Wiring Harness Connector

Standard Colors and Abbreviations

Abbreviation	Color	Abbreviation	Color	Abbreviation	Color
BK	Black	GY	Gray	PK	Purple
BL	Blue	GN	Green	RD	Red
BR	Brown	LG	LT Green	TN	Tan
DB	Dark Blue	OR	Orange	WT	White
DG	DK Green	PK	Pink	YL	Yellow

1998-99 Windstar 3.8L V6 VIN 4 (A/T) 104 Pin Connector

PCM Pin #	Wire Color	Circuit Description (104 Pin)	Value at Hot Idle
1	PK/OR	Shift Solenoid 2 Control	1v, 55 mph: 12v
2	PK/LG	MIL (lamp) Control	MIL On: 1v, Off: 12v
3	OR/BK	Digital TR1 Sensor	In 'P': 0v, 55 mph: 10.7v
4-5	---	Not Used	---
6	DG/WT	Turbine Shaft Speed Sensor	43 Hz, 55 mph: 108 Hz
7	---	Not Used	---
8	OR/WT	IMRC Monitor Signal	5v, 55 mph: 5v
9-12	---	Not Used	---
13	VT	Flash EEPROM	0.1v
14	---	Not Used	---
15	PK/LB	Data Bus (-)	Digital Signals
16	TN/OR	Data Bus (+)	Digital Signals
19	---	Not Used	---
20	PK	Generator Field Signal	130 Hz (37%)
21	BK/PK	CKP (-) Sensor	390-450 Hz
22	GY/YL	CKP (+) Sensor	390-450 Hz
23	---	Not Used	---
24	BK/WT	Power Ground	<0.1v
25	BK	Case Ground	<0.050v
26	DB/LG	Coil 1 Driver (Dwell)	5°, at 55 mph: 8°
27	OR/YL	Shift Solenoid 1 Control	12v, 55 mph: 1v
28	DB	Low Speed Fan Control	Fan On: 1v, Off: 12v
29	---	Not Used	---
30	DB/LG	PATSIL Control	Indicator On: 1v
31	YL/LG	Power Steering Pressure Switch	Straight: 0v, Turned: 12v
32-34	---	Not Used	---
35	RD/LG	HO2S-12 (B1 S2) Signal	0.1-1.1v
36	TN/LB	MAF Sensor Return	<0.050v
37	OR/BK	Transmission Fluid Temperature Sensor	2.10-2.40v
38	LG/RD	Engine Coolant Temperature Sensor	0.5-0.6v
39	GY	Intake Air Temperature Sensor	1.5-2.5v
40	DG/YL	Fuel Pump Monitor	On: 12v, Off: 0v
41	BK/YL	A/C Pressure Switch Signal	A/C On: 12v, off: 0v
42	BR	IMRC Solenoid Control	12v, 55 mph: 12v
43-45	---	Not Used	---
46	LG/PK	High Speed Fan Control	Fan On: 1v, Off: 12v
47	BR/PK	VR Solenoid Control	0%, at 55 mph: 45%
48	---	Not Used	---
49	LG/BK	Digital TR2 Sensor	In 'P': 0v, 55 mph: 10.7v
50	WT/BK	Digital TR4 Sensor	In 'P': 0v, 55 mph: 10.7v
51	BK/WT	Power Ground	<0.1v
52	RD/LB	Coil 2 Driver (Dwell)	5°, at 55 mph: 8°
53	PK/BK	Shift Solenoid 3/C Control	12v, 55 mph: 1v
54	RD/LB	TCC Solenoid Control	12v (0% duty cycle)
55	RD	Keep Alive Power	12-14v
56	LG/BK	EVAP Purge Solenoid	0-10 Hz (0-100%)
57-59	---	Not Used	---
60	GY/LB	HO2S-11 (B1 S1) Signal	0.1-1.1v
61	PK/LG	HO2S-22 (B2 S2) Signal	0.1-1.1v
62	RD/PK	Fuel Tank Pressure Sensor	2.6v at 0" Hg (cap off)
63	---	Not Used	---
64	LB/YL	Digital TR3 Sensor Signal	In 'P': 0v, in O/D: 1.7v

1998-99 Windstar 3.8L V6 VIN 4 (A/T) 104 Pin Connector

PCM Pin #	Wire Color	Circuit Description (104 Pin)	Value at Hot Idle
65	BR/LG	EGR DPFE Sensor Signal	0.20-1.30v
66	---	Not Used	
67	PKWT	EVAP CV Solenoid	0-10 Hz (0-100%)
68	---	Not Used	
69	PK/YL	A/C WOT Relay Control	On: 1v, Off: 12v
70	---	Not Used	---
71	RD	Vehicle Power	12-14v
72	---	Not Used	---
73	TN/BK	Injector 5 Control	3.0-4.0 ms
74	BR/YL	Injector 3 Control	3.0-4.0 ms
75	TN	Injector 1 Control	3.0-4.0 ms
76-77	BK/WT	Power Ground	<0.1v
78	PKWT	Coil 3 Driver (Dwell)	5°, at 55 mph: 8°
79	---	Not Used	
80	LB/OR	Fuel Pump Control	On: 1v, Off: 12v
81	WT/YL	EPC Solenoid Control	9.2v (15 psi)
82	---	Not Used	---
83	WT/LB	IAC Solenoid Control	10.7v (33%)
84	GY/BK	Output Shaft Speed Sensor	0 Hz, 55 mph: 475 Hz
85	DB/OR	Camshaft Position Sensor	6 Hz, 55 mph: 13-15 Hz
86	TN/LG	A/C Pressure Switch Signal	1v (54 psi)
87	RD/BK	HO2S-21 (B2 S1) Signal	0.1-1.1v
88	LB/RD	MAF Sensor Signal	0.8v, 55 mph: 1.8v
89	GY/WT	TP Sensor Signal	0.53-1.27v
90	BR/WT	Reference Voltage	4.9-5.1v
91	GY/RD	Analog Signal Return	<0.050v
92	---	Not Used	
93	RD/WT	HO2S-11 (B1 S1) Heater	1v, Off: 12v
94	YL/LB	HO2S-21 (B2 S1) Heater	1v, Off: 12v
95	WT/BK	HO2S-12 (B1 S2) Heater	1v, Off: 12v
96	TN/YL	HO2S-22 (B2 S2) Heater	1v, Off: 12v
97	RD	Vehicle Power	12-14v
98	---	Not Used	---
99	LG/OR	Injector 6 Control	3.0-4.0 ms
100	BR/LB	Injector 4 Control	3.0-4.0 ms
101	WT	Injector 2 Control	3.0-4.0 ms
102	---	Not Used	---
103	BK/WT	Power Ground	<0.1v
104	---	Not Used	---

Pin Connector Graphic

```
        PCM 104-PIN CONNECTOR
  1 ●●●●●●●●●●●●●    ●●●●●●●●●●●●●26
 27 ●●●●●●●●●●●●●  ⬤  ●●●●●●●●●●●●●52
 53 ●●●●●●●●●●●●●  ⬤  ●●●●●●●●●●●●●78
 79 ●●●●●●●●●●●●●    ●●●●●●●●●●●●●104
```

Terminal View of 104-Pin PCM Wiring Harness Connector

Standard Colors and Abbreviations

Abbreviation	Color	Abbreviation	Color	Abbreviation	Color
BK	Black	GY	Gray	PK	Purple
BL	Blue	GN	Green	RD	Red
BR	Brown	LG	LT Green	TN	Tan
DB	Dark Blue	OR	Orange	WT	White
DG	DK Green	PK	Pink	YL	Yellow

2000 Windstar 3.8L V6 VIN 4 (A/T) 104 Pin Connector

PCM Pin #	Wire Color	Circuit Description (104 Pin)	Value at Hot Idle
1	VT/OR	Shift Solenoid 2 Control	1v, 55 mph: 12v
2	---	Not Used	---
3	OR/BK	Digital TR1 Sensor	In 'P': 0v, 55 mph: 10.7v
4-5	---	Not Used	---
6	DG/WT	Turbine Shaft Speed Sensor	43 Hz, 55 mph: 108 Hz
7	---	Not Used	---
8	OR/WT	IMRC Monitor Signal	5v, 55 mph: 5v
9-12	---	Not Used	---
13	VT	Flash EEPROM	0.1v
14	---	Not Used	---
15	PK/LB	SCP Data Bus (-)	<0.050v
16	TN/OR	SCP Data Bus (+)	Digital Signals
17	GN/OR	PATSIN Signal	12v
18	WT/LG	PATSOUT Signal	12v
19	---	Not Used	---
20	VT	Generator Field Signal	130 Hz (37%)
21	BK/PK	CKP (-) Sensor	390-450 Hz
22	GY/YL	CKP (+) Sensor	390-450 Hz
23	---	Not Used	---
24	BK/WT	Power Ground	<0.1v
25	---	Not Used	---
26	DB/LG	Coil 1 Driver (Dwell)	5°, at 55 mph: 8°
27	OR/YL	Shift Solenoid 'A' Control	12v, 55 mph: 1v
28	DB	Low Speed Fan Control	Fan On: 1v, Off: 12v
29	---	Not Used	---
30	DB/LG	Passive Antitheft Indicator Control	Indicator Off: 12v, On: 1v
31	YL/LG	Power Steering Pressure Switch	Straight: 0v, Turned: 12v
32-34	---	Not Used	---
35	RD/LG	HO2S-12 (B1 S2) Signal	0.1-1.1v
36	TN/LB	MAF Sensor Return	<0.050v
37	OR/BK	Transmission Fluid Temperature Sensor	2.10-2.40v
38	LG/RD	Engine Coolant Temperature Sensor	0.5-0.6v
39	GY	Intake Air Temperature Sensor	1.5-2.5v
40	DG/YL	Fuel Pump Monitor	Off: 0v, On: 12v
41	BK/YL	A/C Pressure Switch Signal	A/C On: 12v, off: 0v
42	BR	IMRC Solenoid Control	12v, 55 mph: 12v
43	---	Not Used	---
44	LG/YL	PATSTRT	0v
45	RD/PK	Generator Communication	0-130 Hz
46	LG/VT	High Speed Fan Relay Control	Relay Off: 12v, On: 1v
47	BR/PK	EGR VR Solenoid Control	0%, at 55 mph: 45%
48	---	Not Used	---
49	LG/BK	Digital TR2 Sensor	In 'P': 0v, 55 mph: 10.7v
50	WT/BK	Digital TR4 Sensor	In 'P': 0v, 55 mph: 10.7v
51	BK/WT	Power Ground	<0.1v
52	RD/LB	Coil 2 Driver (Dwell)	5°, at 55 mph: 8°
53	PK/BK	Shift Solenoid 'C' Control	12v, 55 mph: 1v
54	RD/LB	TCC Solenoid Control	12v (0% duty cycle)
55	RD	Direct Battery	12-14v
56	LG/BK	EVAP Canister Purge Solenoid	0-10 Hz (0-100%)
57-59	---	Not Used	---
60	GY/LB	HO2S-11 (B1 S1) Signal	0.1-1.1v
61	PK/LG	HO2S-22 (B2 S2) Signal	0.1-1.1v
62	RD/PK	Fuel Tank Pressure Sensor	2.6v at 0" Hg (cap off)
63	---	Not Used	---
64	LB/YL	Digital TR3 Sensor Signal	In 'P': 0v, in O/D: 1.7v

2000 Windstar 3.8L V6 VIN 4 (A/T) 104 Pin Connector

PCM Pin #	Wire Color	Circuit Description (104 Pin)	Value at Hot Idle
65	BR/LG	EGR DPFE Sensor Signal	0.20-1.30v
66	---	Not Used	---
67	VT/WT	EVAP CV Solenoid	0-10 Hz (0-100%)
68	---	Not Used	---
69	PK/YL	A/C WOT Relay Control	On: 1v, Off: 12v
70	---	Not Used	---
71	RD	Vehicle Power (Start-Run)	12-14v
72	---	Not Used	---
73	TN/BK	Injector 5 Control	3.0-4.0 ms
74	BR/YL	Injector 3 Control	3.0-4.0 ms
75	TN	Injector 1 Control	3.0-4.0 ms
76	BK/WT	Power Ground	<0.1v
77	---	Not Used	---
78	PKWT	Coil 3 Driver (Dwell)	5°, at 55 mph: 8°
79	---	Not Used	---
80	LB/OR	Fuel Pump Control	On: 1v, Off: 12v
81	WT/YL	EPC Solenoid Control	9.2v (15 psi)
82	---	Not Used	---
83	WT/LB	IAC Solenoid Control	10.7v (33%)
84	DB/YL	Output Shaft Speed Sensor	0 Hz, 55 mph: 475 Hz
85	DB/OR	Camshaft Position Sensor	6 Hz, 55 mph: 13-15 Hz
86	TN/LG	A/C Pressure Switch Signal	1v (54 psi)
87	RD/BK	HO2S-21 (B2 S1) Signal	0.1-1.1v
88	LB/RD	MAF Sensor Signal	0.8v, 55 mph: 1.8v
89	GY/WT	TP Sensor Signal	0.53-1.27v
90	BR/WT	Reference Voltage	4.9-5.1v
91	GY/RD	Sensor Ground	<0.050v
92	---	Not Used	---
93	RD/WT	HO2S-11 (B1 S1) Heater	1v, Off: 12v
94	YL/LB	HO2S-21 (B2 S1) Heater	1v, Off: 12v
95	WT/BK	HO2S-12 (B1 S2) Heater	1v, Off: 12v
96	TN/YL	HO2S-22 (B2 S2) Heater	1v, Off: 12v
97-98	---	Not Used	---
99	LG/OR	Injector 6 Control	3.0-4.0 ms
100	BR/LB	Injector 4 Control	3.0-4.0 ms
101	WT	Injector 2 Control	3.0-4.0 ms
102	---	Not Used	---
103	BK/WT	Power Ground	<0.1v
104	---	Not Used	---

Pin Connector Graphic

PCM 104-PIN CONNECTOR

Terminal View of 104-Pin PCM Wiring Harness Connector

Standard Colors and Abbreviations

Abbreviation	Color	Abbreviation	Color	Abbreviation	Color
BK	Black	GY	Gray	PK	Purple
BL	Blue	GN	Green	RD	Red
BR	Brown	LG	LT Green	TN	Tan
DB	Dark Blue	OR	Orange	WT	White
DG	DK Green	PK	Pink	YL	Yellow

2001-03 Windstar 3.8L V6 VIN 4 (A/T) 104 Pin Connector

PCM Pin #	Wire Color	Circuit Description (104 Pin)	Value at Hot Idle
1	VT/OR	Shift Solenoid 2 Control	1v, 55 mph: 12v
2	---	Not Used	---
3	OR/BK	Digital TR1 Sensor	In 'P': 0v, 55 mph: 10.7v
4-5	---	Not Used	---
6	DG/WT	Turbine Shaft Speed Sensor	43 Hz, 55 mph: 108 Hz
7	---	Not Used	---
8	OR/WT	IMRC Monitor Signal	5v, 55 mph: 5v
9-12	---	Not Used	---
13	VT	Flash EEPROM	0.1v
14	---	Not Used	---
15	PK/LB	SCP Data Bus (-)	<0.050v
16	TN/OR	SCP Data Bus (+)	Digital Signals
17	GY/OR	PATSIN Signal	12v
18	WT/LG	PATSOUT Signal	12v
19	---	Not Used	---
20	VT	Generator Field Signal	130 Hz (37%)
21	BK/PK	CKP (-) Sensor	390-450 Hz
22	GY/YL	CKP (+) Sensor	390-450 Hz
23	---	Not Used	---
24	BK/WT	Power Ground	<0.1v
25	---	Not Used	---
26	DB/LG	Coil 1 Driver (Dwell)	5°, at 55 mph: 8°
27	OR/YL	Shift Solenoid 'A' Control	12v, 55 mph: 1v
28	DB	Low Speed Fan Control	Fan On: 1v, Off: 12v
29	---	Not Used	---
30	DB/LG	Passive Antitheft Indicator Control	Indicator Off: 12v, On: 1v
31	YL/LG	Power Steering Pressure Switch	Straight: 0v, Turned: 12v
32-34	---	Not Used	---
35	RD/LG	HO2S-12 (B1 S2) Signal	0.1-1.1v
36	TN/LB	MAF Sensor Return	<0.050v
37	OR/BK	Transmission Fluid Temperature Sensor	2.10-2.40v
38	LG/RD	Engine Coolant Temperature Sensor	0.5-0.6v
39	GY	Intake Air Temperature Sensor	1.5-2.5v
40	DG/YL	Fuel Pump Monitor	Off: 0v, On: 12v
41	BK/YL	A/C Pressure Switch Signal	A/C On: 12v, off: 0v
42	BR	IMRC Solenoid Control	12v, 55 mph: 12v
43	---	Not Used	---
44	LG/YL	PATSTRT	0v
45	RD/PK	Generator Communication	0-130 Hz
46	LG/VT	High Speed Fan Relay Control	Relay Off: 12v, On: 1v
47	BR/PK	EGR VR Solenoid Control	0%, at 55 mph: 45%
48	---	Not Used	---
49	LG/BK	Digital TR2 Sensor	In 'P': 0v, 55 mph: 10.7v
50	WT/BK	Digital TR4 Sensor	In 'P': 0v, 55 mph: 10.7v
51	BK/WT	Power Ground	<0.1v
52	RD/LB	Coil 2 Driver (Dwell)	5°, at 55 mph: 8°
53	PK/BK	Shift Solenoid 'C' Control	12v, 55 mph: 1v
54	RD/LB	TCC Solenoid Control	12v (0% duty cycle)
55	RD	Direct Battery	12-14v
56	LG/BK	EVAP Canister Purge Solenoid	0-10 Hz (0-100%)
57-59	---	Not Used	---
60	GY/LB	HO2S-11 (B1 S1) Signal	0.1-1.1v
61	VT/LG	HO2S-22 (B2 S2) Signal	0.1-1.1v
62	RD/PK	Fuel Tank Pressure Sensor	2.6v at 0" Hg (cap off)
63	---	Not Used	---
64	LB/YL	Digital TR3 Sensor Signal	In 'P': 0v, in O/D: 1.7v

2001-03 Windstar 3.8L V6 VIN 4 (A/T) 104 Pin Connector

PCM Pin #	Wire Color	Circuit Description (104 Pin)	Value at Hot Idle
65	BR/LG	EGR DPFE Sensor Signal	0.95-1.05v
66	---	Not Used	---
67	VT/WT	EVAP CV Solenoid	0-10 Hz (0-100%)
68	---	Not Used	---
69	PK/YL	A/C WOT Relay Control	On: 1v, Off: 12v
70	---	Not Used	---
71	RD	Vehicle Power (Start-Run)	12-14v
72	---	Not Used	---
73	TN/BK	Injector 5 Control	3.0-4.0 ms
74	BR/YL	Injector 3 Control	3.0-4.0 ms
75	TN	Injector 1 Control	3.0-4.0 ms
76	BK/WT	Power Ground	<0.1v
77	---	Not Used	---
78	PKWT	Coil 3 Driver (Dwell)	5°, at 55 mph: 8°
79	---	Not Used	---
80	LB/OR	Fuel Pump Relay Control	Off: 12v, On: 1v
81	WT/YL	EPC Solenoid Control	9.2v (15 psi)
82	---	Not Used	---
83	WT/LB	IAC Solenoid Control	10.7v (33%)
84	DB/YL	Output Shaft Speed Sensor	0 Hz, 55 mph: 475 Hz
85	DB/OR	Camshaft Position Sensor	6 Hz, 55 mph: 13-15 Hz
86	TN/LG	A/C Pressure Switch Signal	1v (54 psi)
87	RD/BK	HO2S-21 (B2 S1) Signal	0.1-1.1v
88	LB/RD	MAF Sensor Signal	0.8v, 55 mph: 1.8v
89	GY/WT	TP Sensor Signal	0.53-1.27v
90	BR/WT	Reference Voltage	4.9-5.1v
91	GY/RD	Sensor Ground	<0.050v
92	---	Not Used	---
93	RD/WT	HO2S-11 (B1 S1) Heater	1v, Off: 12v
94	YL/LB	HO2S-21 (B2 S1) Heater	1v, Off: 12v
95	WT/BK	HO2S-12 (B1 S2) Heater	1v, Off: 12v
96	TN/YL	HO2S-22 (B2 S2) Heater	1v, Off: 12v
97-98	---	Not Used	---
99	LG/OR	Injector 6 Control	3.0-4.0 ms
100	BR/LB	Injector 4 Control	3.0-4.0 ms
101	WT	Injector 2 Control	3.0-4.0 ms
102	---	Not Used	---
103	BK/WT	Power Ground	<0.1v
104	---	Not Used	---

Pin Connector Graphic

PCM 104-PIN CONNECTOR

Terminal View of 104-Pin PCM Wiring Harness Connector

Standard Colors and Abbreviations

Abbreviation	Color	Abbreviation	Color	Abbreviation	Color
BK	Black	GY	Gray	PK	Purple
BL	Blue	GN	Green	RD	Red
BR	Brown	LG	LT Green	TN	Tan
DB	Dark Blue	OR	Orange	WT	White
DG	DK Green	PK	Pink	YL	Yellow